HORMONE/BEHAVIOR RELATIONS OF CLINICAL IMPORTANCE: ENDOCRINE SYSTEMS INTERACTING WITH BRAIN AND BEHAVIOR

HORMONE/BEHAVIOR RELATIONS OF CLINICAL IMPORTANCE: ENDOCRINE SYSTEMS INTERACTING WITH BRAIN AND BEHAVIOR

Edited by

Robert T. Rubin

Department of Psychiatry and Biobehavioral Sciences, University of California, Los Angeles, Los Angeles, California

Donald W. Pfaff

The Rockefeller University, New York, New York

ELSEVIER

AMSTERDAM • BOSTON • HEIDELBERG • LONDON • NEW YORK • OXFORD
PARIS • SAN DIEGO • SAN FRANCISCO • SINGAPORE • SYDNEY • TOKYO
Academic Press is an imprint of Elsevier

ACADEMIC PRESS

Academic Press is an imprint of Elsevier
Linacre House, Jordan Hill, Oxford, OX2 8DP, UK
525 B Street, Suite 1900, San Diego, CA 92101-4495, USA

Material in the work originally appeared in *Hormones, Brain and Behavior, 2nd Edition*, edited by D. W. Pfaff, A. P. Arnold, A. M. Etgen, S. E. Fahrbach and R.T. Rubin (Elsevier Inc 2009)

British Library Cataloguing in Publication Data
A catalogue record for this book is available from the British Library

Library of Congress Catalog Number: 2009927235

ISBN: 978-0-12-374926-0

For information on all Elsevier publications
visit our website at www.elsevierdirect.com

Contents

Contributors

R.A. Allison (29, Transsexualism)
CIGNA Medical Group of Arizona, Phoenix, AZ, USA

L. Baor (28, Assisted Reproduction in Infertile Women)
Tel-Aviv, Israel

S.L. Berga (22, Premenstrual Dysphoric Disorder)
Emory University, Atlanta, GA, USA

K.J. Berkley (36, Pain: Sex/Gender Differences)
Florida State University, Tallahassee, FL, USA

M. Bettendorf (30, Disorders of Salt and Fluid Balance)
University Clinic of Heidelberg, Heidelberg, Germany

S. Bocklandt (10, The Biology of Sexual Orientation and Gender Identity)
UCLA School of Medicine, Los Angeles, CA, USA

L. Borg (35, Short-Acting Opiates vs. Long-Acting Opioids)
The Rockefeller University, New York, NY, USA

G.D. Braunstein (19, Diseases of Hypothalamic Origin)
Cedars-Sinai Medical Center, Los Angeles, CA, USA

R.S. Bridges (13, Prolactin Actions in the Brain)
Tufts University School of Veterinary Medicine, North Grafton, MA, USA

W. Byne (11, Sexual Orientation in Men and Women)
Mount Sinai School of Medicine, New York, NY, USA

L. Cahill (6, Sex Differences in Human Brain Structure and Function)
University of California, Irvine, CA, USA

J.D. Carmichael (19, Diseases of Hypothalamic Origin)
Cedars-Sinai Medical Center, Los Angeles, CA, USA

B.J. Carroll (21, Mood Disorders)
Pacific Behavioral Research Foundation, Carmel, CA, USA

M.B.C.M. Cools (26, Genetic Defects of Female Sexual Differentiation)
Ghent University, Ghent, Belgium

T.J. Creel (7, Sex Differences in CNS Neurotransmitter Influences on Behavior)
Saint Vincent College, Latrobe, PA, USA

L. Czibere (1, Genetic Transmission of Behavior and Its Neuroendocrine Correlates)
Max Planck Institute of Psychiatry, Munich, Germany

M. Daly (12, Sex Differences in Competitive Confrontation and Risk-taking)
McMaster University, Hamilton, ON, Canada

A.B. Dessens (26, Genetic Defects of Female Sexual Differentiation)
Sophia Children's Hospital/Erasmus MC, Rotterdam, The Netherlands

S.L.S. Drop (26, Genetic Defects of Female Sexual Differentiation)
Sophia Childen's Hospital/Erasmus MC, Rotterdam, The Netherlands

N.N. Ekhator (16, Brain Peptides: From Laboratory to Clinic)
University of Cincinnati, Cincinnati, OH, USA

J. Emens (17, Melatonin Actions in the Brain)
Oregon Health and Science University, Portland, OR, USA

A.J. Fiocco (25, Aging and Alzheimer's Disease)
University of California, San Francisco, CA, USA

C.A. Frye (15, Neurosteroids: From Basic Research to Clinical Perspectives)
University at Albany-State University of New York, Albany, NY, USA

S.N. Garfinkel (20, Stress and Anxiety Disorders)
University of Michigan School of Medicine, Ann Arbor, MI, USA

N.E. Gentile (2, Hypothalamic-Pituitary-Adrenal Cortical Axis)
Saint Vincent College, Latrobe, PA, USA

T.D. Geracioti, Jr. (16, Brain Peptides: From Laboratory to Clinic)
University of Cincinnati, Cincinnati, OH, USA

E.S. Ginsburg (32, Alcohol Abuse: Endocrine Concomitants)
Brigham and Women's Hospital, Boston, MA, USA

M.B. Goetz (38, Human Immunodeficiency Virus and AIDS)
VA Greater Los Angeles Healthcare System, Los Angeles, CA, USA

L.J. Gooren (11, Sexual Orientation in Men and Women)
VU University Medical Center, Amsterdam, The Netherlands

D.R. Grattan (13, Prolactin Actions in the Brain)
University of Otago, Dunedin, New Zealand

J.A. Grootegoed (26, Genetic Defects of Female Sexual Differentiation)
Erasmus MC, Rotterdam, The Netherlands

B. Hambsch (1, Genetic Transmission of Behavior and Its Neuroendocrine Correlates)
Max Planck Institute of Psychiatry, Munich, Germany

V.L. Harsh (4, Hypothalamic-Pituitary-Gonadal Axis in Women)
National Institutes of Health, Bethesda, MD, USA

M. Hines (8, Gonadal Hormones and Sexual Differentiation of Human Brain and Behavior)
University of Cambridge, Cambridge, UK

A. Holdcroft (36, Pain: Sex/Gender Differences)
Imperial College London, London, UK

J. Imperato-McGinley (27, Genetic Defects of Male Sexual Differentiation)
Weill Medical College of Cornell University, New York, NY, USA

R.T. Joffe (3, Hypothalamic-Pituitary-Thyroid Axis)
New Jersey Medical School, Maplewood, NJ, USA

B.A. Kaminski (9, Human Puberty: Physiology and Genetic Regulation)
Rainbow Babies and Children's Hospital, Cleveland, OH, USA

J. Kaskow (16, Brain Peptides: From Laboratory to Clinic)
University of Pittsburgh Medical Center, Pittsburgh, PA, USA

I. Kravets (35, Short-Acting Opiates vs. Long-Acting Opioids)
The Rockefeller University, New York, NY, USA

M.J. Kreek (35, Short-Acting Opiates vs. Long-Acting Opioids)
The Rockefeller University, New York, NY, USA

R. Landgraf (1, Genetic Transmission of Behavior and Its Neuroendocrine Correlates)
Max Planck Institute of Psychiatry, Munich, Germany

Z. Laron (14, Growth Hormone and Insulin-Like Growth Factor-1: Effects on the Brain)
Tel Aviv University, Tel Aviv, Israel

T. Lenhard (30, Disorders of Salt and Fluid Balance)
University Clinic of Heidelberg, Heidelberg, Germany

A.J. Lewy (17, Melatonin Actions in the Brain)
Oregon Health and Science University, Portland, OR, USA

I. Liberzon (20, Stress and Anxiety Disorders)
University of Michigan School of Medicine, Ann Arbor, MI, USA

J. Licinio (24, Anorexia Nervosa and Bulimia Nervosa)
University of Miami Miller School of Medicine, Miami, FL, USA

L.H.J. Looijenga (26, Genetic Defects of Female Sexual Differentiation)
Erasmus MC, Rotterdam, The Netherlands

C. Lord (25, Aging and Alzheimer's Disease)
McMaster University Women's Health Concerns Clinic, Hamilton, ON, Canada

S.J. Lupien (25, Aging and Alzheimer's Disease)
Université de Montréal, Montreal, QC, Canada

L.F. Martinez (22, Premenstrual Dysphoric Disorder)
University of California, San Diego, La Jolla, CA, USA

B.E. Masel (37, Traumatic Brain Injury)
Transitional Learning Center at Galveston, Galveston, TX, USA

J.M. McKlveen (2, Hypothalamic-Pituitary-Adrenal Cortical Axis)
Saint Vincent College, Latrobe, PA, USA

N.K. Mello (32, Alcohol Abuse: Endocrine Concomitants; 34, Cocaine, Hormones and Behavior)
McLean Hospital and Harvard Medical School, Boston, MA, USA

S. Meltzer-Brody (4, Hypothalamic-Pituitary-Gonadal Axis in Women)
University of North Carolina at Chapel Hill, Chapel Hill, NC, USA

J.H. Mendelson (32, Alcohol Abuse: Endocrine Concomitants; 34, Cocaine, Hormones and Behavior)
McLean Hospital and Harvard Medical School, Boston, MA, USA

A.H. Miller (18, Neuroendocrine-Immune Interactions: Implications for Health and Behavior)
Emory University School of Medicine, Atlanta, GA, USA

Y. Miyasaki (38, Human Immunodeficiency Virus and AIDS)
VA Greater Los Angeles Healthcare System, Los Angeles, CA, USA

A.Z. Murphy (36, Pain: Sex/Gender Differences)
Georgia State University, Atlanta, GA, USA

T.F. Newton (38, Human Immunodeficiency Virus and AIDS)
VA Greater Los Angeles Healthcare System, Los Angeles, CA, USA

A.N. Nord (7, Sex Differences in CNS Neurotransmitter Influences on Behavior)
Saint Vincent College, Latrobe, PA, USA

S. Nowakowski (22, Premenstrual Dysphoric Disorder)
University of California, San Diego, La Jolla, CA, USA

T.W.W. Pace (18, Neuroendocrine-Immune Interactions: Implications for Health and Behavior)
Emory University School of Medicine, Atlanta, GA, USA

M.R. Palmert (9, Human Puberty: Physiology and Genetic Regulation)
The Hospital for Sick Children, Toronto, ON, Canada

B.L. Parry (22, Premenstrual Dysphoric Disorder)
University of California, San Diego, La Jolla, CA, USA

G.J. Paz-Filho (24, Anorexia Nervosa and Bulimia Nervosa)
University of Miami Miller School of Medicine, Miami, FL, USA

R.E. Poland (33, Effect of Smoking on Hormones, Brain and Behavior)
The Research and Education Institute for Texas Health Resources, Arlington, TX, USA

N. Pound (12, Sex Differences in Competitive Confrontation and Risk-taking)
Brunel University, Uxbridge, UK

C.L. Raison (18, Neuroendocrine-Immune Interactions: Implications for Health and Behavior)
Emory University School of Medicine, Atlanta, GA, USA

U. Rao (33, Effect of Smoking on Hormones, Brain and Behavior)
UT Southwestern Medical Center, Dallas, TX, USA

M.E. Rhodes (2, Hypothalamic-Pituitary-Adrenal Cortical Axis; 7, Sex Differences in CNS Neurotransmitter Influences on Behavior)
Saint Vincent College, Latrobe, PA, USA

A. Richter-Unruh (26, Genetic Defects of Female Sexual Differentiation)
Endokrinologikum MC, Bochum, Germany

D.R. Ripepi (2, Hypothalamic-Pituitary-Adrenal Cortical Axis)
Saint Vincent College, Latrobe, PA, USA

J. Rough (17, Melatonin Actions in the Brain)
Oregon Health and Science University, Portland, OR, USA

R.T. Rubin (21, Mood Disorders)
VA Greater Los Angeles Healthcare System, Los Angeles, CA, USA

D.R. Rubinow (4, Hypothalamic-Pituitary-Gonadal Axis in Women)
University of North Carolina at Chapel Hill, Chapel Hill, NC, USA

C.M. Ryan (31, Diabetes Mellitus and Neurocognitive Dysfunction)
University of Pittsburgh School of Medicine, Pittsburgh, PA, USA

F.J. Sánchez (10, The Biology of Sexual Orientation and Gender Identity)
UCLA School of Medicine, Los Angeles, CA, USA

C. Sarapas (23, Post-Traumatic Stress Disorder)
James J. Peters VA Medical Center, Bronx, NY, USA

P.J. Schmidt (4, Hypothalamic-Pituitary-Gonadal Axis in Women)
National Institutes of Health, Bethesda, MD, USA

S. Schwab (30, Disorders of Salt and Fluid Balance)
University Clinic Erlangen-Nürnberg, Erlangen, Germany

T. Sidhartha (33, Effect of Smoking on Hormones, Brain and Behavior)
UT Southwestern Medical Center, Dallas, TX, USA

S. Sindi (25, Aging and Alzheimer's Disease)
McGill University, Montreal, QC, Canada

A.P. Sinha Hikim (5, Hypothalamic-Pituitary-Gonadal Axis in Men)
David Geffen School of Medicine at UCLA, Torrance, CA, USA

J. Songer (17, Melatonin Actions in the Brain)
Oregon Health and Science University, Portland, OR, USA

J.R. Strawn (16, Brain Peptides: From Laboratory to Clinic)
University of Cincinnati, Cincinnati, OH, USA

R.S. Swerdloff (5, Hypothalamic-Pituitary-Gonadal Axis in Men)
David Geffen School of Medicine at UCLA, Torrance, CA, USA

R. Temple (37, Traumatic Brain Injury)
Transitional Learning Center at Galveston, Galveston, TX, USA

C. Touma (1, Genetic Transmission of Behavior and Its Neuroendocrine Correlates)
Max Planck Institute of Psychiatry, Munich, Germany

E. Vilain (10, The Biology of Sexual Orientation and Gender Identity)
UCLA School of Medicine, Los Angeles, CA, USA

C. Wang (5, Hypothalamic-Pituitary-Gonadal Axis in Men)
David Geffen School of Medicine at UCLA, Torrance, CA, USA

C.W. Wilkinson (25, Aging and Alzheimer's Disease)
Geriatric Research Education and Clinical Center, VA Puget Sound Health Care System, Seattle, WA, USA and University of Washington, Seattle, WA, USA

M. Wilson (12, Sex Differences in Competitive Confrontation and Risk-taking)
McMaster University, Hamilton, ON, Canada

M. Wortman (16, Brain Peptides: From Laboratory to Clinic)
University of Cincinnati, Cincinnati, OH, USA

R. Yehuda (23, Post-Traumatic Stress Disorder)
James J. Peters VA Medical Center, Bronx, NY, USA

E.A. Young (20, Stress and Anxiety Disorders)
University of Michigan School of Medicine, Ann Arbor, MI, USA

Y. Zhou (35, Short-Acting Opiates vs. Long-Acting Opioids)
The Rockefeller University, New York, NY, USA

Y.-S. Zhu (27, Genetic Defects of Male Sexual Differentiation)
Weill Medical College of Cornell University, New York, NY, USA

About the Editors

ROBERT T. RUBIN, M.D., Ph.D., is Professor and Vice-Chair of Psychiatry and Biobehavioral Sciences at the David Geffen School of Medicine at UCLA. He also is Chief of the Department of Psychiatry and Mental Health at the VA Greater Los Angeles Healthcare System. Prior to these appointments, from 1992 to 2005, he was Highmark Blue Cross Blue Shield Professor of Neurosciences and Professor of Psychiatry at the Drexel University College of Medicine, Allegheny General Hospital Campus, Pittsburgh, Pennsylvania. Prior to joining the Allegheny system in 1992, he was Professor of Psychiatry and Biobehavioral Sciences in the UCLA School of Medicine. He is certified in psychiatry by the American Board of Psychiatry and Neurology, and he has a Ph.D. in physiology. For 40 years, his research has focused on the neuroendocrinology of stress and depression. Currently, he is studying the influence of acetylcholine neurotransmission in the brain on the activity of the hypothalamic-pituitary-adrenal cortical axis.

DONALD W. PFAFF (The Rockefeller University, New York, New York) heads the Laboratory of Neurobiology and Behavior at The Rockefeller University. He received his scientific training at Harvard University and MIT and is a Member of the National Academy of Science and a Fellow of the American Academy of Arts and Sciences. Pfaff's laboratory focuses on steroid hormones and brain function, interactions among transcription factors, luteinizing-hormone-releasing-hormone neurons, and genes influencing neuronal functions. He is the author or editor of over 15 books and more than 800 research publications.

Principles of Translational Neuroendocrinology

R T Rubin, University of California, Los Angeles, Los Angeles, CA, USA
D W Pfaff, Rockefeller University, New York, NY, USA

The field of neuroendocrinology was launched in the late 1930s by the pioneering experiments of Geoffrey Harris in Oxford, England. Over the years, his results proved the ability of hypothalamic secretions to influence the activity of the anterior pituitary gland. Other workers showed that the brain was a major influence on hypothalamic secretions, leading to the aphorism that "the brain is the largest gland in the body." This area of work was so difficult that for some decades it remained a boutique area of neuroscience, mostly limited to experimental laboratory models chosen for their simplicity rather than for their clinical importance. Now, enough new knowledge has been collected to firmly support the concept of a translational neuroendocrinology. To provide a contemporary overview of this concept, selected chapters from the recent electronic publication, *Hormones, Brain and Behavior*, 2nd edn., edited by Donald W. Pfaff, Arthur P. Arnold, Susan E. Fahrbach, Anne M. Etgen, and Robert T. Rubin, are presented in this volume. Several chapters at the beginning provide essential basic science background for the clinically oriented chapters that follow.

The volume is divided into two main sections. (1) *Endocrine Systems Interacting with Brain and Behavior* focuses on the normal interactions among endocrine axes, brain function, and behavioral components and is organized around the endocrine axes known to play a role in, and reciprocally to be influenced by, brain states and behaviors. (2) *Endocrinologically Important Behavioral Syndromes* focuses on aberrant hormone–brain–behavior interactions that result in identifiable pathologies or syndromes. There is no overarching causality inferred; indeed, a disruption in an endocrine system, brain function, and/or voluntary behavior, can be etiologically related to the pathologies discussed. Some chapter titles have been modified to reflect the essence of their content and a few chapters have been shortened, but content remains the same as in *Hormones, Brain and Behavior*, 2nd edn.

As an introduction, it is useful to briefly discuss several fundamental principles of hormone action that can affect both normal and abnormal brain states and behaviors. These principles, highlighted in bold type, are stated and exemplified in Pfaff et al. (2003), referenced at the end of this introduction. An understanding of these principles will allow the reader to appreciate many of the relationships and mechanisms discussed in the following sections.

Hormone Effects Are Strong and Reliable. One important principle is that **hormones can both facilitate and repress behavioral responses**. A prominent example is the regulation of eating. The feelings of hunger and satiety, which control feeding behavior, are regulated by the interplay of central nervous system (CNS), gastrointestinal, pancreatic, and adipose tissue hormones. Ghrelin, produced in the stomach, stimulates the production of neuropeptide Y (NPY) in the arcuate nucleus of the hypothalamus; these orexigenic hormones stimulate eating behavior. When the stomach becomes filled, the secretion of ghrelin decreases. Corticotropin-releasing hormone (CRH), also produced in the brain, is an anorexigenic hormone that counteracts the effect of NPY. In addition, cholecystokinin, secreted by the gastrointestinal tract, signals satiety in the CNS via the vagus nerve. Over a longer time period, as adiposity increases, leptin is secreted by adipose tissue; leptin also is an anorexigenic hormone that inhibits NPY in the brain and increases muscle sensitivity to insulin by increasing fatty acid oxidation. Adiponectin, also released by adipose tissue, similarly increases muscle sensitivity to insulin by increasing fatty acid oxidation.

While the control of a single behavior, feeding, is regulated by an interplay of several hormones, conversely **a single hormone can have many effects and can affect complex behaviors**. For example, CRH, mentioned above, produced in the arcuate nucleus of the hypothalamus, stimulates the release of the neurotransmitter, norepinephrine, from the locus ceruleus, which increases anxiety and anxiety-related behavior. In addition, CRH produced in the paraventricular nucleus of the hypothalamus stimulates the secretion of adrenocorticotropic hormone (ACTH) from the anterior pituitary gland, which in turn stimulates the secretion of adrenal cortical hormones. These adrenal hormones have multiple effects on glucose metabolism, salt and water balance, and mood states and related behaviors.

For every hormone, **there are optimal hormone concentrations**; too much or too little can be damaging. For example, inadequate circulating thyroid hormones (hypothyroidism) can result in metabolic slowing, with mental dullness, depression, and psychosis. In contrast, excessive thyroid hormones (hyperthyroidism) can lead to metabolic overactivity and severe anxiety states. Optimal hormone concentrations can vary by time of day; for example, the secretion cascade of CRH from the hypothalamus, ACTH from the pituitary, and cortisol (the major glucocorticoid in humans) from the adrenal cortex is most active in the morning just before awakening and declines throughout the day and evening to reach a low point at about 2–3 a.m. If this hormone rhythm is disturbed by administration of glucocorticoids for treatment purposes or by pituitary or adrenal tumors such that high, constant circulating glucocorticoids are maintained throughout the day and night, pathological physical changes (Cushing's syndrome) and behavioral disturbances, including insomnia, euphoria, depression, and psychosis, can result.

Hormone Effects Can Depend on Family, Gender, and Development. Many examples indicate that the sex of the individual can influence behavioral responses. Sexual dimorphism (sex differences in structure) occurs at many levels, from genes to whole organs, and correlates highly with sexual diergism (sex differences in physiological function). Many chapters in this volume highlight sexual dimorphism and diergism, ranging from CNS neurotransmitters to multiple hormone axes to complex behaviors such as confrontation and risk taking. There also are many examples supporting the principle that **hormone actions early in development can influence hormone responsiveness in the CNS during adulthood**, for example, exposure of a neonate or infant to severe stressors can sensitize the hypothalamic–pituitary–adrenal cortical axis to exhibit exaggerated stress responses later in life, often accompanied by pathological behavioral responses such as anxiety and depression. In addition, it is common knowledge that hormonal alterations during specific developmental stages such as puberty, menopause, and aging can profoundly affect behavior differentially by sex.

Temporal Parameters Influence Hormonal Effects on Behavior. The duration of hormone exposure can make a big difference. In some cases longer is better; in other cases brief pulses are optimal. For example, thyroid hormones have comparatively little variation throughout the day, the month, and the year, which leads to a relatively constant effect on basal metabolic rate and overall physiological functioning, an important requirement for survival. In contrast, in women of child-bearing age, the hypothalamic–pituitary–gonadal axis shows a prominent monthly cycle that is necessary for ovulation, conception, and implantation of the fertilized ovum in the uterus. When this cycling is lost, as at the time of menopause, fertility also is lost. On a much shorter timescale are the circadian (about 24h) and ultradian (shorter than 24h; pulsatile) secretion patterns of many hormones, such as those of the hypothalamic–pituitary–adrenal cortical axis, as discussed above. These examples give meaning to the principle, **hormonal secretions and responses are affected by biological clocks**.

Spatial Parameters Influence Hormonal Effects on Behavior. A fundamental example is the common action of thyroid hormones to regulate basal metabolic rate in all tissues of the body, as indicated above, and thus to maintain optimum physiological conditions for behavioral responses of the organism to CNS commands, such as the fight or flight response. The normally small fluctuation in thyroid hormone secretion is maintained by tightly controlled feedback of thyroid hormones to the pituitary and hypothalamus. Other hormones have different (but complementary) effects in the CNS and periphery, such as increased prolactin secretion during pregnancy acting in the CNS to promote nest-building behavior and acting in the breast to promote milk production. In addition, hormones can act at all levels of the neuraxis, and the nature of the behavioral effect can depend on the site of action. For example, estrogenic and androgenic steroid hormones act in the hypothalamus to influence mating behaviors, whereas acting in the brainstem and neocortex they can influence a wide range of emotional and cognitive processes. These examples underscore the principle that **CNS actions consonant with peripheral actions form coordinated, unified effects**.

Molecular and Biophysical Mechanisms of Hormone Action Can Give Clues to Therapeutic Strategies. Rapid hormonal effects can facilitate later genomic actions. For example, steroid hormones have both rapid, nongenomic effects and effects on the genome after they are transported to the cell nucleus. When the gonadal steroid hormone, estrogen, is administered experimentally in two pulses, the early pulse amplifies the transcription-facilitating action of the second pulse, either through yet-to-be-identified cell membrane receptors for steroid hormones, or perhaps because estrogen is highly lipophilic, concentrates in the center of the membrane lipid bilayer, and thus might facilitate the passage of the second estrogen pulse through the cell membrane. Similarly, testosterone administration to male rats can elevate neuronal activity and reduce anxiety-related behaviors within 30–40 min, too fast for most transcriptional effects.

Gene duplication and splicing products for hormone receptors in the CNS often have different behavioral effects. Estrogen receptor (ER)-β is likely a gene duplication product and, when activated, has different effects at both the molecular and the behavioral levels than activation of ER-α, the classical estrogen receptor. At the molecular level, ER-β activation inhibits the transcription-facilitating effect of ER-α. At the behavioral level, primary reproductive behaviors often depend only on the activation of ER-α, whereas social recognition and suppression of aggression, two ancillary behaviors facilitating reproduction, depend on ER-β acting in concert with ER-α. In addition, the interaction of estrogen with both receptors in the cell nucleus is modulated by an array of nuclear proteins that both facilitate and inhibit hormone-receptor binding.

Kallman's syndrome is hypogonadotrophic hypogonadism, often coupled with anosmia (inability to smell). One cause is the lack of migration of gonadotropin-releasing hormone (GnRH)-secreting cells from the olfactory pit, where they form, into the hypothalamus, thus leaving them inactive. The GnRH gene is expressed normally, but the gene coding for an extracellular matrix protein necessary for cell migration is abnormal. A second cause of Kallman's syndrome has been identified as resulting from multiple mutations distributed across the gene for the GnRH receptor, leading to reduced binding of GnRH and/or the ability of the receptor to trigger signal-transduction pathways in pituitary gonadotrophs and CNS neurons. Both causes lead to similar endocrine and behavioral consequences. These examples support the principle that **hormone receptors and other nuclear proteins influence hormone effects**.

Environmental and Evolutionary Variables Influence Hormone/Behavior Relations. It is well established that **hormone effects on behavior depend on context**, and this is particularly true for humans. For example, sex hormones generally increase aggressive behavior in teenage boys, but whether the behavior is socially appropriate, such as energetic sports, or inappropriate, such as fighting, depends on the influence of several social factors – family support versus family stressors, socioeconomic status, and school environment. Large discrepancies in socioeconomic status (thus increasing the chance of personal humiliation of an adolescent boy), large schools (thus increasing the sense of anonymity), and an absence of socially sanctioned rites of passage (thus failing to provide a positive view of a boy's role in adult society), all increase the chance of inappropriately violent behavior.

In addition, **behavioral and environmental contexts alter hormone release**. For example, a sensitive measure of social effects on the release of the reproductive pituitary hormones, luteinizing hormone (LH) and follicle-stimulating hormone (FSH), is the age at which girls enter menarche, that is, the age at which LH is released in sufficient quantity to initiate menstrual cycles. Familial factors, including the presence of a male, basic approval of the girl by her family, and the absence of conflict all appear to be influential, in that earlier maturation of LH release patterns is associated with less positive family relations.

A second example of environmental influences on age of menarche in girls could be considered an effect of family, of stress, of nutrition, or of all three. Trained ballet dancers enter menarche at a significantly later age in comparison to other girls. Beginning and persisting with ballet could result in part from family pressures, and the dancing itself could represent a type of environmental stressor. A dominating factor, however, is that dancers need to have a very low percentage of body fat, which is likely to delay reproductive maturation.

Of importance, **neuroendocrine mechanisms have been conserved throughout evolution** to provide biologically adaptive body/brain/behavior coordination, which implies that studies of hormone/brain/behavior mechanisms in laboratory animals can contribute to a medical understanding of disorders involving the same neurons and same biochemical reactions in humans. This has allowed researchers to develop the field of translational neuroendocrinology, as described in the chapters that follow.

Outlook. The now-established field of translational neuroendocrinology will get more complex before it becomes simpler, for several reasons. First, even though major efforts by many laboratories have succeeded in unraveling the relations between two large integrative systems – endocrine systems and the CNS – neuroendocrinology will need to bring into focus its interface with immune cells. In addition to the presence of microglia, immune cells of the CNS, we now know that mast cells invade the forebrain, and there is a wide distribution of dendritic (antigen-presenting) cells in the CNS. Second, new research techniques will add greatly to the depth and complexity of our knowledge. Nucleic acid technology represents an obvious application of this supposition, as do the increasing analytic power of protein chemistry and the ability of biophysical techniques to sort out events at the nerve-cell and endocrine-cell membranes. Third, for the neuroendocrine field, as for other areas of medicine, sampling the genotype of the patient will foster the development of individualized neuroendocrine medicine, our own particular example of pharmacogenomics. It also is the fervent hope of the psychoneuroendocrinologist that our understanding of human behavior can be extended from relatively simple phenotypes to a much wider field of behaviors for which effective treatments, including endocrine therapies, can be developed.

References

Pfaff DW, Phillips IM, and Rubin RT (2003) *Principles of Hormone/Behavior Relations*. San Diego, CA: Academic Press.

PART I

ENDOCRINE SYSTEMS INTERACTING WITH BRAIN & BEHAVIOR

1 Genetic Transmission of Behavior and Its Neuroendocrine Correlates

B Hambsch, R Landgraf, L Czibere, and C Touma, Max Planck Institute of Psychiatry, Munich, Germany

Chapter Outline

Glossary

anxiety An evolutionary conserved, polygenic, multifactorial trait, presumed to have a complex inheritance and to involve the interaction of multiple genes with epigenetic/environmental factors; exaggerated anxiety may evolve to anxiety disorders.

corticosteroids The principal glucocorticoids that are synthesized by the adrenal cortex and secreted in response to stressors (mainly cortisol in humans and corticosterone in murine rodents).

emotionality A psychological trait of complex etiology, which produces a quasi-continuously distributed phenotype and moderates an organism's response to stress.

endophenotype A quantitative biological trait associated with a complex genetic disorder.

hypothalamic–pituitary–adrenal (HPA) axis The HPA axis is the endocrine core of the stress system, which involves hypothalamic corticotropin-releasing hormone (CRH), pituitary adrenocorticotrophic hormone (ACTH), and adrenal corticosteroids.

knockout A genetic technique in which DNA sequences are incorporated, disrupted, or removed from an organism to disable the expression or function of a gene of interest. The technique is essentially the opposite of a gene knockin, based on comparable technical strategies.

neuropeptides The primary products of protein synthesis, consisting of amino acids, produced by and acting on neurons as neuromodulators/neurotransmitters and acting on peripheral organs as neurohormones.

nociception The sensation of pain, based on the function of the underlying physiological system. The mitigation of pain sensation is called analgesia.

opioids The chemically heterogeneous group of ligands acting on opioid receptors that exert, among others, psychotropic and analgetic effects. The known endogenous opioids are neuropeptides and the most important exogenous opioids are the alkaloids, morphin and codein.

stress-related disorder The illness whose causation, onset, or development is substantially influenced by stress and its neurobiological correlates.

1.1 Introduction

The behavior of an organism is determined by both its genetic composition and environmental influences. The components of neuroendocrine signaling play a pivotal role in transmitting genetic predispositions into complex behavioral patterns as well as appropriately modulating behavior under a variety of nongenetic environmental conditions.

In this chapter, we address the question of how neuroendocrine systems can modulate behavior, including stress reactivity and emotionality, in rodents and humans via the activation of specific receptors and how the expression of the corresponding genes is influenced by both genetic and environmental parameters. Therefore, the focus of this chapter is directed to the hypothalamic–pituitary–adrenocortical (HPA) axis as well as nonapeptide, tachykinin, and opioid systems, with particular emphasis on neuroendocrine and behavioral phenomena related to stress, anxiety, and depression-like behaviors.

Apart from the classical pharmacological intervention by receptor agonists and antagonists, we will concentrate on the powerful tools of genetic approaches and the generation of genetically manipulated mice. Since directed displacements, mutations, or insertions enable research to identify and characterize the role of single molecules in living organisms, this method, in particular, sheds light on the complex mechanisms of neuroendocrine signaling. In addition, attention is paid to selective phenotypic breeding approaches, which may be beneficial to the identification of candidate genes that are involved in complex neuroendocrine and behavioral traits. All these approaches have their own advantages and limitations and are therefore considered complementary for research. Thus, combined methodologies are the focus of this chapter.

1.2 Stress and the HPA System

The concept of stress has a very long research history (going back to the ancient Greeks), as it deals with the daily social and nonsocial stimuli that are challenging or threatening to the survival, health, and reproductive success of animals including humans (for review, see Axelrod and Reisine (1984), Munck et al. (1984), Holst (1998), Sapolsky et al. (2000), Kim and Diamond (2002), Romero (2004), de Kloet et al. (2005), and Korte et al. (2005)). Stress, in its broadest sense, is well known to significantly impact a variety of bodily functions. Its disruptive effects (e.g., on the immune system, reproduction, cognition, and behavior of vertebrates) have broadly been demonstrated (for review, see Axelrod and Reisine (1984), Munck et al. (1984), Holst (1998), Sapolsky et al. (2000), Kim and Diamond (2002), Romero (2004), de Kloet et al. (2005), and Korte et al. (2005)). Furthermore, stress hormones such as glucocorticoids and catecholamines have been implicated in a wide range of human disorders including depression, anxiety, cancer, cardiovascular diseases, diabetes, and dementia (for review, see Munck et al. (1984), Kim and Diamond (2002), de Kloet et al. (1998, 2005), McEwen and Sapolsky (1995), Holsboer (2000), Korte (2001), Engelmann et al. (2004), Landgraf and Neumann (2004), and Swaab et al. (2005)).

When confronted with a stressor (environmental, physiological, or psychological), an individual typically displays a stress response, consisting of a suite of physiological and behavioral alterations, to cope with the challenge. One of the main mediators of this response is the HPA system, which is not only responsive to stressors but also to other types of activity that are associated with emotional arousal (e.g., courtship or sexual behaviors) (Holst, 1998; Sapolsky et al., 2000; Romero, 2004; de Kloet et al., 2005; Korte et al., 2005). We will just briefly introduce those aspects relevant for understanding the genetics of neuroendocrine and behavioral functions related to stress and emotions.

The HPA axis can be dissected into the central (neuropeptide) components and the related peripheral (endocrine) organs (for review, see Sapolsky et al. (2000), de Kloet et al. (2005), Engelmann et al. (2004), Herman and Cullinan (1997), Bale (2005), and Aguilera et al. (2007)). Both basal and stress-induced release of adrenocorticotropic hormone (ACTH) from the anterior lobe of the pituitary are primarily regulated by corticotropin-releasing hormone (CRH), a 41-amino-acid peptide first isolated by Vale et al. (1981). Shortly after its discovery, it became apparent that CRH is also implicated in other components of the stress response, such as arousal and autonomic activity (for review, see Bale (2005), Steckler and Holsboer (1999), Bale and Vale (2004), and Muller and Wurst (2004)). Therefore, CRH-producing parvocellular neurons of the paraventricular nucleus (PVN) of the hypothalamus can be regarded as the focal point for modulating HPA-axis activity. Furthermore, the PVN's central role in integrating information relevant for eliciting a stress response is also evidenced by its multiple connections to other brain centers. Prominent neuronal inputs include, for example, excitatory afferents from the amygdala and inhibitory afferents from the hippocampus, as well as other brain regions implicated in the neuroanatomy of emotions and cognition, such as the limbic system and the prefrontal cortex, are tightly linked to the PVN (for review, see Engelmann et al. (2004), Herman and Cullinan (1997), Bale (2005), and Aguilera et al. (2007)). From the PVN, hypothalamic CRH neurons project to the anterior pituitary, where CRH is released into the hypophyseal–portal circulation and activates the HPA axis by triggering the release of ACTH from pituitary corticotropes through activation of CRH type 1 receptors (CRH-R1). In addition, CRH-binding sites are also found in various peripheral tissues, such as the adrenal medulla, heart, prostate, gut, liver, kidney, and testis. CRH receptors belong to the G-protein-coupled receptor superfamily and CRH binding stimulates the intracellular accumulation of cyclic adenosine monophosphate (cAMP). Two distinct CRH receptor subtypes designated CRH-R1 and CRH-R2 have been characterized, encoded by distinct genes that are also differentially expressed (for review, see Bale (2005), Steckler and Holsboer (1999), Bale and Vale (2004), and Muller and Wurst (2004)). CRH-R1 is the most abundant subtype found in the anterior pituitary and is also widely distributed in the brain. CRH-R2 is expressed mainly in the peripheral vasculature and the heart, as well as in subcortical structures in the brain.

At the pituitary level, the effects of CRH are amplified by arginine vasopressin (AVP) which, in particular during chronic activation of the HPA axis, is increasingly co-expressed and co-secreted from PVN neurons. After release into the circulation, ACTH, in turn, stimulates the production and secretion of glucocorticoids (GCs) from the *zona fasciculata* of the adrenal cortex. Which GC is predominantly produced, largely depends on the species. Cortisol, for example, is the major GC in humans and most primates, carnivores, and ungulates. In murine rodents, birds, and reptiles, however, mainly corticosterone is produced. These GCs can be regarded as final effectors of the HPA axis orchestrating the organism's response to challenges, acting on numerous organ systems including the brain, and modulating physiology and behavior (for review, see Munck et al. (1984), Holst (1998), Sapolsky et al. (2000), Romero (2004), de Kloet et al. (2005), Korte et al. (2005), Engelmann et al. (2004), and Herman and Cullinan (1997)). Controlling the concentration of circulating GCs is therefore of utmost importance and several negative feedback loops are involved in regulating HPA-axis activity. Cortisol and corticosterone, for example, feedback directly at the level of the pituitary and the PVN to control the expression of ACTH and CRH, respectively, and also largely influence the activity of the hippocampus, amygdala, and prefrontal cortex.

The mode of action of corticosteroids in the brain as well as in the periphery mainly involves two related receptor molecules, the high-affinity mineralocorticoid receptor (MR) and the glucocorticoid receptor (GR), which has about tenfold lower affinity to GCs than the MR (for review, see Munck et al. (1984), Sapolsky et al. (2000), Romero (2004), de Kloet et al. (2005), and de Kloet et al. (1998)). Both receptors are co-expressed abundantly in neurons of limbic structures, serving different tasks within the stress-response cascade. The MR is implicated in the appraisal process and the onset of the stress response, while the GR, which is only activated by higher concentrations of GCs, terminates the stress reactions, mobilizes energy resources, and thereby facilitates recovery. GR also promotes memory storage in preparation for future events. Both receptors act either as homodimers (GR–GR, MR–MR) or as heterodimers (GR–MR) to activate or suppress the activity of manifold genes including proopiomelanocortin (POMC; the precursor protein of ACTH), AVP, and CRH. Upon binding to other transcription factors, however, activated MR and GR can also indirectly influence gene activity via protein–protein interactions. Besides these relatively slow actions of GCs on gene expression levels, very rapid effects via cell-membrane-associated receptors have also been revealed (for review, see Keller-Wood and Dallman (1984), Orchinik (1998), Makara and Haller (2001), Dallman (2005), Tasker (2006), and Tasker et al. (2006)). Furthermore, MR and GR are together with chaperones and several heat-shock proteins – important components of the cytoplasmatic corticosteroid receptor complex that dissociates after ligand binding and can initiate multiple intracellular alterations modifying, for example, neuronal excitability (for review, see Munck et al. (1984), Sapolsky et al. (2000), Romero (2004), and de Kloet et al. (1998, 2005)).

1.2.1 Dysregulation of the Hypothalamic–Pituitary–Adrenal Axis in Affective Disorders

HPA-axis abnormalities have been found to be associated with many psychiatric disorders including major depression (MD), bipolar disorder, schizophrenia, and anxiety disorders. The dysfunctions observed in patients with MD, for example (reviewed in de Kloet et al. (2005), Holsboer (2000), Bale (2005), Wong and Licinio (2001), Gold and Chrousos (2002), Nestler et al. (2002), Hasler et al. (2004), Pariante et al. (2004), and Ising et al. (2005)), include hyper- or hypo-(re)activity of the HPA axis, loss of circadian rhythmicity of GC secretion, and impairments in the negative feedback suppression of the stress-hormone system. Additionally, features of MD such as anxiety, insomnia, and the intensity of mood changes are highly correlated with cortisol responses. These observations suggest that HPA-axis abnormalities are closely tied with the disease state. This is further underlined by findings that in MD patients (mainly of the melancholic subtype), the expression of CRH and AVP in the PVN is enhanced, increased CRH levels are found in the cerebrospinal fluid (CSF), the adrenals show hypertrophy, and basal corticosteroid and ACTH levels are elevated (see reviews cited above). Moreover, in response to stressors, the HPA axis is persistently activated and when challenged in different functional tests, it shows feedback resistance at the level of the PVN and pituitary. Tests investigating the negative feedback mechanisms, such as the dexamethasone (DEX) suppression test (DST) and the combined DEX suppression/CRH stimulation test (DEX/CRH), consistently indicate dysfunctions in HPA-axis autoregulation in both unipolar and bipolar depression (reviewed in de Kloet et al. (2005), Holsboer (2000), Gold and Chrousos (2002), Nestler et al. (2002), Pariante et al. (2004), and Ising et al. (2005)). Furthermore, the observation that normalization of the neuroendocrine responses correlates with successful remission of affective symptoms suggests that reinstating GR-mediated HPA-axis regulation may be at least a correlate, if not the mechanism of action of drugs used in the treatment of mood disorders (reviewed in de Kloet et al. (2005), Holsboer (2000), Gold and Chrousos (2002), Nestler et al. (2002), Pariante et al. (2004), and Ising et al. (2005)). A crucial involvement of HPA axis hyper- or hypo-activity in affective disorders is also indicated by other clinical observations. For example, patients with Cushing's disease whose adrenals produce excessive cortisol frequently suffer from depression and overall psychopathology decreases significantly after correction of hypercortisolism (Fava et al., 1987; Sonino et al., 1998). Similarly, patients with Addison's disease have adrenal glands that produce insufficient amounts of cortisol and depressive symptoms occur in this disorder as well (Fava et al., 1987; Thomsen et al., 2006). Thus, an inverted U-shaped function for the effects of GC concentrations on mood is suggested. If this is a causal relationship, however, and whether changes in hormone production cause depression or vice versa, still needs to be clarified.

Abnormalities in HPA-axis function have also been reported in schizophrenia and anxiety disorders. Similar to melancholically depressed patients, basal hyperactivity of the HPA axis and enlarged pituitary volumes are also observed in schizophrenic patients and patients with panic disorder and generalized anxiety. Interestingly, however, patients with atypical depression and post-traumatic stress disorder (PTSD) consistently show decreased GC release. Studies using the DST or the DEX/CRH test, for example, have demonstrated that this decrease in GC levels is a result of enhanced sensitivity of the GR to negative feedback from circulating cortisol (Gold and Chrousos, 2002; Yehuda et al., 2002; Rydmark et al., 2006). These observations strongly suggest that intact regulation of the HPA axis is critical for normal regulation of emotion.

Indeed, the HPA-axis abnormalities observed in psychiatric patients have been largely correlated with changes in the CRH system and GR expression in brain structures involved in behavioral emotionality and stress responsiveness (reviewed in de Kloet et al. (2005), Holsboer (2000), Bale (2005), Wong and Licinio (2001), Gold and Chrousos (2002), Nestler et al. (2002), Hasler et al. (2004), Pariante et al. (2004), and Ising et al. (2005)). Analysis of postmortem tissue from psychiatric patients, for example, demonstrated reduced GR mRNA in the hippocampus of individuals suffering from unipolar and bipolar depression, as well as schizophrenia (Webster et al., 2002; Perlman et al., 2004). In addition, GR mRNA expression was reduced in the basolateral amygdala of schizophrenic and bipolar-disorder patients as well as in the frontal cortex of patients with unipolar depression (Webster et al., 2002; Perlman et al., 2004). Studies in patients with PTSD, on the other hand, have revealed enhanced sensitivity of GR that is associated with increased negative-feedback inhibition of the HPA axis (Yehuda et al., 2002; Raison and Miller 2003). Importantly, it has been shown that both antidepressants and mood stabilizers increase GR mRNA expression in the brain, leading to enhanced HPA-axis feedback regulation and thereby lowered levels of CRH and cortisol. In addition, a number of clinical trials indicate that drugs influencing GR activation, for example, the GR antagonist mifepristone (RU486), show clinical benefits in patients with depression, bipolar disorder, or anxiety (Belanoff et al., 2001; Young et al., 2004; Schatzberg, 2003; DeBattista et al., 2006; Flores et al., 2006).

Nevertheless, it should be seriously taken into consideration that MD, like other psychiatric disorders, shows a high degree of complexity, involving a multitude of molecular, neuroendocrine, and behavioral alterations as well as an intense gene–environment interaction, making it difficult to dissociate the primary causes from secondary consequences of the disease. A better understanding of these pathways, however, is essential in order to develop more targeted treatment strategies. Recent progress in generating transgenic mouse models has advanced our knowledge in this regard, as they allow investigations of the specific role of particular neurotransmitters, receptors, or neuropeptides in the brain, for example, elucidating mechanisms regulating HPA-axis activity and emotional behaviors.

1.2.2 Animal Models Elucidating the Molecular Basis of Neuroendocrine–Behavior Interactions

1.2.2.1 Mice with targeted mutations modulating HPA-axis function

The above-mentioned clinical observations raise a number of questions that are amenable to fundamental-research approaches making use of mice with specific mutations modifying the function and regulation of the HPA system. For example, the question can be investigated as to what extent central neuropeptides, such as CRH and AVP, act as neurotransmitters coordinating behavioral adaptations to stressful situations. Similarly, the role of MR and GR signaling in precipitating behavioral and neuroendocrine symptoms of affective disorders can be studied in these mouse mutants as well as the effect of antidepressant drugs upon complex signaling cascades and pathways in the brain.

1.2.2.1(i) CRH mutant mice

As outlined above, CRH plays a prominent role in mood disorders and is one of the neuropeptides that is particularly upregulated during depressive episodes. However, in animals as well as in humans, CRH is not only involved in activating the stress response, but also increases arousal and affects emotional behaviors (Bale, 2005; Steckler and Holsboer, 1999; Bale and Vale, 2004; Muller and Wurst, 2004). In order to study the functions of CRH and its receptors from the molecular over the neuroendocrine to the behavioral level, several mutant mouse lines have been generated (for review, see Bale (2005), Steckler and Holsboer (1999), Bale and Vale (2004), Muller and Wurst (2004), Urani et al. (2005), and Muller and Holsboer (2006)), revealing distinct effects of CRH overexpression as well as CRH deficiency.

Insertion of an additional copy of the CRH gene into the genome (nonselective overexpression of rat CRH under the murine methallothionine-1 gene promoter or the Thy1-promotor) leads to a clearly elevated CRH expression in nearly all areas of the mouse brain where it is normally found, but also in some regions where it is not found in wild-type mice (for details, see Stenzel-Poore et al. (1992), Stenzel-Poore et al. (1994), Groenink et al. (2002), and Groenink et al. (2003)). These transgenic animals display a general hyperactivation of the HPA axis with increased levels of ACTH and corticosterone, but present a blunted response to stressors. Concerning their behavioral emotionality, most lines of CRH-overexpressing mice show an increase in anxiety-related behavior (Groenink et al., 2003), but either no difference (Heinrichs et al., 1996, 1997; Dirks et al., 2002) or less immobility than control mice van Gaalen et al. (2002) in the forced swim test (FST) – a test assessing depression-like behavior in terms of passive versus more active coping with an aversive situation (for review, see Cryan and Holmes (2005)). The interpretation of the behavioral findings, however, needs to be done with caution as these animals exhibit prominent endocrine and physical changes similar to those seen in patients with Cushing's disease.

In contrast to the CRH-overexpressing mutants, the phenotype of CRH-knockout mice, on the other hand, includes an attenuated corticosterone production, that is, GC deficiency, and a hyporeactivity of the HPA axis to stressors (Muglia et al., 1995; Jacobson et al., 2000). Interestingly, however, the animals show no gross abnormalities in several behavioral paradigms (Weninger et al., 1999), indicating that compensatory mechanisms might be involved.

1.2.2.1(ii) CRH receptor mutant mice

CRH-R1-knockout mice demonstrate severe disturbances in HPA-axis function, in particular, a decrease in basal GC levels and hyporeactivity to stressors, including diminished stress-induced release of ACTH and corticosterone (Smith et al., 1998; Timpl et al., 1998). Furthermore, these animals display less anxiety-related behaviors in paradigms such as the elevated plus-maze (EPM) test and the dark-light box test (Smith et al., 1998; Timpl et al., 1998; Contarino et al., 1999). However, for the interpretation of the behavioral findings, it should be taken into consideration that CRH-R1-knockout mice also showed impairments in a spatial memory task (Y-maze test; Contarino et al., 1999), which might be related to their GC deficiency or the compensatory activation of the hypothalamic AVP system (Muller et al., 2000).

In order to investigate the role of the CRH-R1 independent of its control over the HPA axis, conditional knockout mice have been generated using the Cre/loxP recombinase system (Tsien et al., 1996) and the CaMKII-alpha promoter to postnatally inactivate the CRH-R1 in the forebrain and the limbic system without affecting the expression in the hypothalamus and pituitary, that is, leaving the HPA system intact (Muller et al., 2003). As expected, these animals show normal HPA-axis activity and regulation under basal conditions as well as in response to stressors. However, the reduction of anxiety-related behavior observed in the conventional CRH-R1 knockouts is still present, indicating that regional, in particular, limbic CRH receptors modulate behavioral emotionality (Muller et al., 2003).

In contrast to the phenotype of CRH-R1 deficient mice, CRH-R2-knockout animals present an exaggerated HPA-axis response to stressors (hypersensitivity to restraint stress) and spend less time on the open arms of the EPM or in the center of an open field (OF), both indicative of increased anxiety (Bale et al., 2000; Coste et al., 2000; Kishimoto et al., 2000). Interestingly, however, gender differences as well as effects of the mother's genotype on the emotionality of the offspring have been observed in this mouse line (Bale et al., 2002), revealing a more complex interplay between genes and environment influencing the behavioral phenotype.

CRH-R1 and CRH-R2 double-knockout mice, that is, animals without a functional CRH-receptor system, have also been generated and are viable (Bale et al., 2002; Preil et al., 2001). The neuroendocrine phenotype of these double mutants is determined by the functional lack of the CRH-R1, leading to a profound decrease in HPA-axis activity and reactivity (basal GC deficiency and hyposecretion of ACTH and corticosterone in response to stressors) (Bale et al., 2002; Preil et al., 2001). Behaviorally, no conclusive results have been published for these mice. Female double mutants, for example, are reported to display less anxiety-related behaviors than wild-type mice in the EPM and the OF test, but the males did not differ in this regard (Bale et al., 2002).

Taken together, these findings on single and double CRH-receptor knockouts indicate that CRH-R1 and

CRH-R2 play opposing roles in terms of controlling HPA-axis activity and reactivity as well as with regard to mediating anxiety-modulating effects (anxiolytic or anxiogenic) in the limbic system.

1.2.2.1(iii) Glucocorticoid receptor mutant mice

Both corticosteroid receptors (MR and GR) act as ligand-dependent transcription factors and are widely expressed in the brain as well as in peripheral organs. The GR is of particular importance for the feedback regulation of GCs on the HPA axis at the level of the pituitary, PVN, and hippocampus (for review, see Sapolsky et al. (2000), Kim and Diamond (2002), de Kloet et al. (2005), de Kloet et al. (1998), Holsboer (2000), Aguilera et al. (2007), Pariante et al. (2004), and Ising et al. (2005)). As one of the most common findings in biological psychiatry is a dysregulation of the stress-hormone system and impaired GC signaling has been proposed to be a key mechanism in the pathogenesis of depression (for review, see de Kloet et al. (2005), Holsboer (2000), Pariante et al. (2004), and Ising et al. (2005)), several mouse lines with altered expression of the GR have been generated. These GR mutants include conventional knockouts, GR-antisense expressing mice, several conditional knockouts, as well as different GR-overexpressing mouse lines (for review, see Urani et al. (2005), Muller and Holsboer (2006), Gass et al. (2001), and Howell and Muglia (2006)).

In general, functional disruption of the GR is lethal (shortly after birth) due to severely disturbed development of the lungs (Tronche et al., 1999). Confirming this, impaired lung function and a very high postnatal mortality rate (more than 90%) has been observed in conventional GR-knockout mice (knockdown achieved by targeting exon 2 of the GR gene; Cole et al., 1995). Surviving individuals of this mouse line show all characteristics of GC insensitivity, that is, demonstrating an inability to elicit transcription of a functional GR despite a residual capacity to bind DEX (Cole et al., 2001). In particular, the impaired negative-feedback regulation of the HPA axis leads to strongly elevated plasma ACTH and corticosterone levels (about 15-fold and 2.5-fold increases, respectively). Related to this HPA-axis hyperactivity, an increased behavioral reactivity in the OF test and deficits in spatial learning and memory, assessed in the Morris water maze, have also been described for homozygous mutants (GR −/−), which is discussed to be largely dependent on impaired GC-receptor activation in the hippocampal formation (Oitzl et al., 1997). Heterozygous mice (GR +/−) with only a 50% reduction of GR expression, on the other hand, show no dysregulation of the HPA system under basal conditions and a similar leaning performance as wild-type mice (Oitzl et al., 1997; Ridder et al., 2005). However, in response to stressors, an increased secretion of corticosterone is observed, resulting in higher peak levels and delayed recovery (Ridder et al., 2005). A battery of tests investigating anxiety-related behaviors and depression-like behaviors without prior stress exposure found no differences between GR +/− mice and wild types (Ridder et al., 2005). However, the learned helplessness paradigm, which involves repeatedly subjecting the animals to stressful events (inescapable electric footshocks), revealed increased despair-related behaviors in the heterozygous GR mutants, that is, they showed fewer escapes, longer escape latencies, and more escape failures than wild-type mice (Ridder et al., 2005). Furthermore, an abnormally elevated response in the combined DEX/CRH test and a downregulation of brain-derived neurotrophic factor (BDNF) in the hippocampus have also been described for these mice (Ridder et al., 2005), indicating several similarities to the neuroendocrine and behavioral alterations observed in depressed patients (see Section 1.1).

Another model to study the consequences of a lifelong defect in GR function has been developed by mutant mice overexpressing an antisense GR gene, which leads to a reduction in GR expression (Pepin et al., 1992a). These GR antisense mice show a reduced CRH expression in the hypothalamus but no alterations compared to wild-type mice with regard to basal HPA-axis activity and ACTH and corticosterone concentrations at different time points across the circadian cycle (Stec et al., 1994; Barden et al., 1997; Karanth et al., 1997; Dijkstra et al., 1998). The expected dysregulation of the HPA system only becomes evident under stressful/challenging conditions, for example, resulting in enhanced release of ACTH (Pepin et al., 1992a,b) and can be reversed by antidepressant treatment (Montkowski et al., 1995; Barden, 1996). Moreover, GR antisense mice show nonsuppression in the DST, which is in agreement with observations in melancholically depressed patients (Barden et al., 1997). At the behavioral level, mice transgenic for the GR-antisense sequence present several cognitive deficits (that are reversible by antidepressant treatment with moclobemide) and alterations in emotionality (Montkowski et al., 1995; Rousse et al., 1997). Surprisingly, mutants show less anxiety-related behavior in the EPM test (Montkowski et al., 1995; Rochford et al., 1997;

Strohle et al., 1998) and also in response to psychological stressors, such as exposure to a rat, which is a natural predator for mice (Linthorst et al., 2000). Moreover, they display enhanced responses to novelty, such as locomotor hyperactivity (Beaulieu et al., 1994), and are less immobile than wild-type mice in the FST (Montkowski et al., 1995), which is in line with the arousal and activity-inducing effects of CRH released in response to stressors.

Besides the conventional GR-knockout mice described above, several nervous system-specific (conditional) GR-knockout mice have also been generated. Deletion of the GR only in the central nervous system (CNS) was achieved, for example, using the Cre/loxP recombinase system under the rat Nestin promoter (Tronche et al., 1999). These so-called GR^{NesCre} mice are viable and lack the GR in all neurons and glial cells, which leads to, as could be expected, a hyperactive HPA system and a Cushing-like phenotype, including increased plasma corticosterone levels and reduced ACTH levels under basal conditions (Tronche et al., 1999). This is likely due to the deletion of the GR in the PVN, which largely affects the negative feedback of GCs regulating HPA-axis activity. In line with this are findings that these mice respond to immobilization stress with higher ACTH and corticosterone secretion and also show alterations in behavioral emotionality (Tronche et al., 1999). Compared to wild-type animals, GR^{NesCre} mice show less anxiety-related behavior as they enter the aversive compartments of the EPM and the dark-light box more often, but on the other hand do not present increased despair-like behavior in the FST (Tronche et al., 1999). However, the interpretation of these behavioral results should be done with care, as they might be confounded by the profound alterations in HPA-system regulation, for example, at the level of the PVN.

In order to avoid these functional changes in brain sites fundamental for HPA-axis regulation, another conditional GR-knockout model has been generated, again using the Cre/loxP system. In these so-called FBGRKO mice (Howell and Muglia, 2006; Boyle et al., 2005), the GR is deleted only in the forebrain and the limbic system (applying Cre recombinase expressed under the control of the CaMKII-alpha promoter). The deletion starts around postnatal day 21 and is not complete until about 6 months of age, thereby avoiding the developmental effects of the loss of GR function. The GR is progressively knocked out in neurons throughout the hippocampus, cortex, striatum, nucleus accumbens, and both the basolateral and basomedial amygdala (Howell and Muglia, 2006; Boyle et al., 2005). Partial loss of GR was also noted in the bed nucleus of the stria terminalis, but importantly, the GR is not deleted in the PVN, the central nucleus of the amygdala, or the anterior pituitary, that is, leaving the regulation of the HPA axis intact (Boyle et al., 2005). Analysis of HPA-system function in this mouse model revealed that loss-of-forebrain GR caused a twofold increase in basal and a 50% increase in circadian release of corticosterone and ACTH as well as an increased responsiveness to stressors (Howell and Muglia, 2006; Boyle et al., 2005; Boyle et al., 2006). Furthermore, FBGRKO mice subjected to the DST showed no suppression of corticosterone release following DEX administration (Boyle et al., 2005). This finding demonstrates the importance of extra-hypothalamic GR in the feedback regulation of the HPA axis despite preservation of direct negative feedback effects of DEX in the PVN and the pituitary. The failure of FBGRKO mice to demonstrate DEX-induced suppression suggests that increased drive to PVN neurons and corticotrophs from GR-deleted forebrain sites is capable of overcoming the suppressive effects of DEX in the paradigm utilized (Howell and Muglia, 2006; Boyle et al., 2005). These changes in HPA-axis regulation were also found to be associated with increased AVP mRNA expression in the PVN (Boyle et al., 2005). No changes in the PVN, however, were found regarding the expression of CRH, contrasting the finding in GR^{NesCre} mice (see above). This suggests that while CRH synthesis is a major target of hypothalamic GR, AVP synthesis is likely a major target for increased drive to the PVN brought about by disruption of extra-hypothalamic GR (Howell and Muglia, 2006). Disruption of forebrain GR also alters emotionally relevant behavior. FBGRKO mice, for example, show increased depression-like behaviors, as indicated by the increased time spent immobile/floating in the FST and TST (Howell and Muglia, 2006; Boyle et al., 2005). Differences between wild type and mutants were also noted in the sucrose preference test used as an index of anhedonia-related behavior. When mice were presented with a choice between 1% sucrose solution and water, FBGRKO animals consumed significantly less sucrose solution compared to controls (Howell and Muglia, 2006; Boyle et al., 2005). Regarding anxiety-related behavior, more ambiguous results were reported. Although the findings in both, the EPM test and the dark-light box test, indicate less-anxious behavior in FBGRKO mice, it should be taken into account that the mutants were also significantly more active in these tests, possibly reflecting increased stress-induced locomotor activity (Howell and Muglia, 2006; Boyle et al., 2006). Importantly, when locomotor activity was measured under basal conditions, no differences were found between FBGRKO and control mice, as was also the case for the GR^{NesCre} line (see above). This suggests that the observed phenotype may be related to increased locomotor agitation in response to the aversive stimuli associated with the test situation (e.g., elevated or brightly lit areas) as opposed to reflecting decreased anxiety-like behavior (Howell and Muglia, 2006). The FBGRKO mouse model has also been investigated in terms of pharmacological treatment with antidepressants, potentially restoring the behavioral and neuroendocrine alterations. Indeed, chronic treatment with the tricyclic antidepressant imipramine reversed the despair phenotype in the FST and also changed some of the HPA-system dysfunctions (Howell and Muglia, 2006; Boyle et al., 2005). In particular, the circadian hyperactivity of the HPA axis was reversed by the imipramine treatment, but not the impairment in negative-feedback inhibition, indicating that the MR might play a more important role in mediating antidepressant-like effects than the GR (see also Muller and Holsboer (2006)).

As also outlined above, the GR regulates transcription by two major mechanisms: first, as dimer, binding to GC-response elements (GREs) in the promoter of target genes; second, as monomer, modulating the activity of other transcription factors via protein–protein interactions (for reviews, see de Kloet et al. (1998, 2005), Holsboer (2000), and Pariante et al. (2004)). Dissecting these two modes of action has been achieved in a mutant mouse line by introducing a point mutation into one of the dimerization domains of the GR using a knockin strategy to replace the endogenous GR gene with the mutated one (Reichardt et al., 1998). These so-called GR^{dim} mice express GR molecules that cannot dimerize, but can still act as monomers. In contrast to mice carrying disrupted alleles of the GR, GR^{dim} mice are viable and have been studied regarding physiological and behavioral alterations in adulthood (Reichardt et al., 1998a). Although CRH levels in the hypothalamus are normal in these mutants, ACTH and corticosterone concentrations in the plasma are markedly elevated compared to wild-type mice, indicating that the mechanism of protein–protein interactions are important for the negative feedback at the level of the hypothalamus (Reichardt et al., 1998b). Regarding behavioral emotionality, GR^{dim} mice do not differ from control animals in their locomotor and explorative activity, but display spatial memory deficits in the Morris water-maze test (Oitzl et al., 2001). Furthermore, neither anxiety-related behaviors nor immobility in the FST are affected in this mouse line (Oitzl et al., 2001).

Overexpressing a gene of interest is another strategy to study its potential role in pathophysiological mechanisms underlying complex disorders such as MD. Therefore, GR overexpression has been applied by researchers using different genetic approaches to increase GR abundance in mice, resulting in two transgenic mouse lines.

In the first model, the so-called YGR mouse line, a global GR overexpression has been achieved by insertion of two additional copies of the GR gene using a yeast artificial chromosome (YAC) (Reichardt et al., 2000). These mice overexpress GR mRNA by about 25% and the GR protein by about 50% (Reichardt et al., 2000). As expected, the feedback regulation of the HPA system is strongly enhanced in these animals, suggesting a stress-resistant neuroendocrine and behavioral phenotype (Ridder et al., 2005; Reichardt et al., 2000). Indeed, YGR mice show a reduced secretion of corticosterone in response to immobilization stress and also present a stronger suppression of plasma corticosterone concentration after administration of DEX (Ridder et al., 2005). In terms of behavioral emotionality, no alterations have been observed in YGR mice compared to wild-type animals in several test paradigms assessing anxiety-related and depression-like behaviors (Ridder et al., 2005). However, after repeated stress exposure in the learned helplessness paradigm, YGR mice perform better, that is, display reduced helplessness than controls (showing a higher number of escapes and shorter escape latencies) (Ridder et al., 2005). Furthermore, BDNF protein levels in the hippocampus of YGR mice are increased, which is in line with decreased levels of this neurotrophic factor observed in GR +/− mice (Ridder et al., 2005). Nevertheless, a limitation of this model is that the GR overexpression is not limited to the CNS.

For the second mouse model (referred to as GRov), therefore, a conditional gene expression approach was applied, achieving GR overexpression by generating GR mRNA under the control of the forebrain-specific calcium-calmodulin-dependent kinase II-alpha (CaMKII-alpha) promoter (Wei et al., 2004). This transgene leads to significant increases in GR expression in the forebrain and the limbic system, including the frontal cortex, amygdala, PVN, and hippocampus, but leaves the anterior pituitary unaffected (Howell and Muglia, 2006; Wei et al., 2004). Analysis of HPA-axis function in these mice revealed no differences in basal activity, including the circadian secretion of ACTH and corticosterone, CRH expression in the PVN, and also the response to mild stressors, which was comparable between wild type and mutants (Howell and Muglia, 2006; Wei et al., 2004). However, following exposure to a more severe stressor (restraint stress), GRov mice exhibit a blunted initial stress response followed by a delayed turnoff (Wei et al., 2007). This deficit in negative feedback is paradoxical in the face of elevated GR levels and continues to worsen with increasing age (Wei et al., 2007). On the behavioral level, mild cognitive deficits and increased anxiety-related behaviors are observed in GRov animals compared to wild-type mice (Wei et al., 2004, 2007). In the EPM test as well as the dark-light box test, GRov mice made fewer entries into and spent less time in the more aversive compartments of the apparatus (open arms of the EPM and lit area of the dark-light box) (Howell and Muglia, 2006; Wei et al., 2004). No differences were found in the number of total arm entries, suggesting equivalent locomotor activity in the EPM test across genotypes. Interestingly, chronic treatment with the antidepressants imipramine or desipramine effectively reversed this anxiety phenotype (Howell and Muglia, 2006; Wei et al., 2004).

Overexpression of the GR in the forebrain also led to increased depression-like behavior. GRov mice showed increased immobility in the FST 30 min after a mild stressor (saline injection) and acute treatment with either fluoxetine, desipramine, or imipramine all decreased floating time 30 min after injection (Howell and Muglia, 2006; Wei et al., 2004). Interestingly, however, when the experiment was repeated the next day, it was found that mutant mice showed an enhanced sensitivity to antidepressants. GRov animals treated with desipramine spent significantly less time floating compared to similarly treated controls and a trend toward increased sensitivity to imipramine was observed (Wei et al., 2004). The authors suggest that this phenomenon may reflect increased emotional instability. Further evidence indicating an enhanced emotional lability in GRov mice comes from studies examining sensitivity to cocaine (Wei et al., 2004). No differences in locomotor activation were found following acute or chronic (5 days) dosing. It was noted, however, that 9 days after this chronic dosing, GRov mice showed increased locomotor activation following administration of the same dose of cocaine compared with control mice, indicating enhanced behavioral sensitization (Wei et al., 2004). The authors conclude that GRov mice, therefore, may represent a mouse model with a behavioral phenotype relevant to bipolar disorder.

Taken together, targeting key components of the HPA system such as CRH and its receptors (CRH-R1 and CRH-R2) as well as the GC receptors (MR and GR) in mice has so far yielded valuable insights into the mechanisms underlying emotional behavior and its neuroendocrine correlates. Although it is, of course, impossible to recapitulate all aspects of a complex human disease such as MD in a mouse, mice genetically engineered to model specific key symptoms prevalent in human depression can be successfully employed to discover neurobiological endophenotypes (defined as quantitative biological traits associated with a complex genetic disorder) bridging the gap between behavioral phenotype and genotype (Hasler et al., 2004; Urani et al., 2005; Muller and Holsboer, 2006; Cryan and Holmes, 2005; Tecott, 2003). Thus, focusing on individual endpoints of the disease rather than the entire syndrome, that is, following an endophenotype-based approach, can yield valuable insights into the genetic/epigenetic and neurobiological underpinnings of psychiatric disorders (see also Kas et al. (2007)).

1.2.2.2 Nontargeted genetic approaches

As outlined above, during the last years, genetically modified mice carrying specific mutations modulating HPA-axis function and regulation have mainly been used to study the consequences of alterations in specific gene products on anxiety and depression-like behavior as well as neuroendocrine functions (for review, see Urani et al. (2005), Muller and Holsboer (2006), Cryan and Holmes (2005), and Tecott (2003)). Although these transgenic (overexpression or knockout) animal models are extremely valuable for dissecting the functional role of a given molecule, the genetic background can exert unwanted effects on the mutants' phenotype. For example, flanking genes – that is, DNA sequences on either side of a targeted mutation that derive from the embryonic stem-cell donor strain – or complex interactions between the mutation and background genes, including compensatory mechanisms and developmental alterations, can mask the causal link between molecule and behavior (Muller and Holsboer, 2006; Cryan and Holmes, 2005; Tecott, 2003). Furthermore, as the majority of psychiatric disorders are complex – that is, minor changes of many genes as well as environmental factors play a major role – emotional abnormalities in mice with single-gene mutations must be carefully interpreted and discussed critically, in particular, when attempting to create a genetic animal model of MD (de Kloet et al., 2005; Muller and Holsboer, 2006; Cryan and Holmes, 2005; Tecott, 2003; Kas et al., 2007).

As an alternative approach, selective breeding has proved to be a powerful strategy to unravel the genetic basis of psychiatric disorders, providing unique information about pleiotropy (multiple effects of a single gene), epistasis (interaction of genes residing at different loci, i.e., nonallelic interaction), and gene-by-environment interaction (Phillips et al., 2002; Swallow and Garland, 2005). The breeding program of such approaches usually begins with evaluating the trait of interest in a genetically heterogeneous population, for example, a commercially available outbred strain of mice. Individuals with responses at either extreme of the response curve are then selectively bred together for their opposing trait phenotypes over multiple generations. Thus, heritability characteristics of the trait can be evaluated and later generations of these inbred lines can be examined for underlying neurobiology and polygenetic or pleiotropic correlates of the trait of interest. Successfully established examples of this classical genetic approach include mice and rats selected for extremes in anxiety-related behavior (Kromer et al., 2005; Landgraf et al., 2007), helplessness/avoidance/depression-like behavior (El Yacoubi et al., 2003; Steimer and Driscoll, 2003; Henn and Vollmayr, 2005), aggressiveness (Lagerspetz et al., 1968; Veenema et al., 2003; Gammie et al., 2006), novelty-seeking behavior (Stead et al., 2006), and nest-building behavior (Lynch, 1980). Interestingly, most selection experiments involving rodents focus on behavioral traits. However, as evidence from human and animal studies reveals a vital link between individual stress sensitivity and the predisposition toward mood disorders, (de Kloet et al., 2005; Holsboer, 2000; Bale, 2005) applying selection for HPA-axis reactivity might be a promising approach yielding insights into the genetic and mechanistic basis of complex traits underlying MD, including the different subtypes of the disease.

Only very recently, first results of such a selective breeding experiment were reported involving laboratory mice and applying increased or decreased HPA-axis reactivity as the selected trait (Touma et al., 2008). The response to selection for high, intermediate, or low corticosterone increase in a so-called stress reactivity test (SRT) turned out to be quite strong for males and females. Already in the first generation, that is, offspring derived from breeding pairs selected from the founder population of outbred CD-1 mice, significant differences in stress reactivity were found (Touma et al., 2008). These differences could even be increased further from generation to generation by assortative breeding (Touma et al., 2008). By generation VI, the mean corticosterone increase of males and females in the high reactivity (HR) and low reactivity (LR) line was well within the range or even exceeded the responses observed in the respective founder pairs. This rather rapid and stable response to selection for extremes in stress reactivity strongly indicates a genetic basis of the respective phenotype and is in line with the findings of other studies selecting animals for GC secretion in response to stressors (Japanese quail (Satterlee and Johnson, 1988), rainbow trout (Pottinger and Carrick, 1999), and zebra finch (Evans et al., 2006)). Together with the considerable individual variation found in the parental generation (Touma et al., 2008), this suggests that HPA-axis reactivity (and regulation) is a highly heritable trait probably determined by a set of major-impact genes that presumably have been conserved during evolution (see also Overli (2007)).

Tests investigating the emotionality of these HR, intermediate reactivity (IR), and LR mice, including anxiety-related behavior, exploratory drive, locomotor activity, and depression-like behavior, point to phenotypic similarities with behavioral changes observed in depressive patients, in particular, when the two subtypes of melancholic and atypical depression are considered (Touma et al., 2008). In general, LR mice showed more passive-aggressive coping styles, while HR males and females were hyperactive in some behavioral paradigms (Touma et al., 2008), resembling signs of retardation and retreat versus restlessness and agitation often seen in atypical and melancholic depression, respectively (subtypes of depression specified in DSM-IV). Morphometric and neuroendocrine investigations addressing functional alterations of the HPA axis in the three breeding lines further support this view. For example, the ACTH challenge test proved that the differential stress reactivity between HR, IR, and LR animals is not brought about by profound disturbances in the periphery, such as differences in the capacity of the adrenal cortex to produce and secrete glucocorticoids (Touma et al., 2008). Additionally, monitoring the circadian rhythm of glucocorticoid secretion revealed clearly increased trough levels in HR mice, resulting in a flattened diurnal rhythm, again adding to the neuroendocrine similarities to patients suffering from melancholic depression (Touma et al., 2008). Although the basal expression of CRH and AVP in the PVN was not different between the three breeding lines, this suggests that distinct mechanisms influencing the function and regulation of the HPA axis are involved in the respective behavioral and neurobiological endophenotypes. Thus, the HR/IR/LR mouse lines generated by selective breeding can be a valuable model to elucidate molecular genetic, neuroendocrine, and behavioral parameters associated with altered stress reactivity, thereby improving our understanding of affective disorders, presumably including the symptomatology and pathophysiology of specific subtypes of MD (Touma et al., 2008).

1.3 The Oxytocin and Vasopressin Systems

Both oxytocin (OXT) and AVP neurons of the hypothalamus are among the most intensively studied neurons in the mammalian brain and, although it does not necessarily follow, they are probably also among the best understood. This understanding ranges from genetic polymorphisms, precursor processing, cellular trafficking, stimulus–secretion coupling, modes of release, receptor characterization up to the involvement of these neuropeptidergic systems in cognitive abilities, emotionality, social behaviors, and psychopathology (for review, see Ring (2005) and Landgraf (2006)). Secreted peripherally from the neurohypophysis/posterior pituitary upon appropriate stimulation, both AVP and OXT act as hormones to retain body water, co-stimulate ACTH and trigger the milk ejection reflex, respectively. In addition to their peripheral secretion, the nonapeptides are released somato-dendritically within their sites of origin, the hypothalamic PVN, and supraoptic nuclei, semi-independent of the electrical activity of the cell bodies (Ludwig et al., 2002).

This section concentrates on central release patterns, within distinct brain areas, resulting in receptor-mediated effects on a wide variety of brain functions. Importantly, while peripheral and central release patterns may be independently regulated, they may, nevertheless, interact at multiple levels to ensure homeostasis, reproductive success, and adequate behavior. There is no doubt that mammalian behavior has a genetic component and there is abundant evidence from the study of rodents that both AVP and OXT systems contribute to it. This heritable component of behavior includes the question as to which extent behavioral differences among species and individuals up to psychopathologies are caused by genetic polymorphisms. Recent advances in behavioral and molecular biology, genetic engineering, and large-scale DNA sequencing have provided ample evidence of behavioral endophenotypes under genetic control in diverse species, including mice, rats, and humans. We concentrate on both from-inside-to-outside approaches that modify the genome and examine phenotypic consequences as well as phenotype-based from-outside-to-inside approaches, particularly on selective and bidirectional breeding strategies that try to identify genetic polymorphisms likely to contribute to well-defined behavioral phenotypes and their natural variation. All of these approaches and tools have their own advantages and limitations; thus, combined approaches are highly recommended to be used. While many studies are more or less exclusively descriptive in nature, we will try to focus on those studies that approach causality by tracing behavior to its genetic basis and vice versa. Neuropeptides such as AVP and OXT are attractive targets in this context. As primary gene products they are prone to direct structural changes by mutations. In addition to genetic polymorphisms in the coding region, resulting in structurally changed gene products, even subtle variations in the promoter structure of genes can alter the pattern of neuropeptide release and/or receptor characteristics in the brain. While, along a continuum of central release, the optimum ensures adequate behavior and fitness, deficient or overexpression and over-release contribute to behavioral disturbances up to psychopathology. Due to the resulting remarkable number and diversity, the dynamics of their release patterns, the varying modes of intercellular signaling, and the multiplicity of receptors to which they bind, neuropeptides such as AVP and OXT are considered ideal neuromodulator candidates underlying behavioral regulation.

In both the periphery and brain, OXT actions are mediated by the OXT receptor (OXTR), which belongs to the G-protein-coupled receptor family, linked to phospholipase C. AVP acts on three receptor subtypes, V1a, V1b, and V2. In the brain, AVP effects are mediated by V1aR and V1bR, linked to phosphatidylinositol turnover and intracellular calcium. Until recently, the V1aR was thought to be the only subtype expressed widely in the mammalian brain, while the V1bR was localized primarily in the pituitary. The recent finding that V1bR is also expressed in numerous brain areas, though to a lesser extent, makes it necessary to determine the relative contribution of both receptor subtypes to AVP-mediated behaviors.

A variety of experimental approaches have been applied to elucidate the role both AVP and OXT systems play in behavioral regulation. The data from different approaches are often paradoxical and at times inconsistent, with effects varying with species, strain, gender, dose and mode of administration, testing context, etc. What appears to be necessary, for example, in light of the closely related neuropeptides AVP and OXT, is to examine if differences in behavioral outcomes may be accounted for by differing effects on AVP associated with OXT deletions and vice versa. Similarly, long-lasting deficits in a receptor subtype can be accompanied by alterations of the corresponding ligand and vice versa. Given the complex circuitries and modes of interneuronal signaling these neuropeptides are involved in (Landgraf and Neumann, 2004), it is tempting to speculate that no single component can be activated or deleted without co-affecting a multiplicity of related parameters, both up- and downstream. While these dynamic interactions may well reflect the physiological functioning of the complex system, it is at times difficult, if not impossible, to clearly distinguish between primary and secondary actions, physiologically relevant and pharmacological effects, etc.

1.3.1 The Oxytocin System

1.3.1.1 Oxytocin

OXT-knockout mice provide an example of the complex nature of behavioral differences observed in genetically engineered animals, suggesting that these differences emerging in a developmental knockout may not necessarily be related to the absence of the gene or its product at the time of testing. Pharmacological studies indicate that activation of OXT-signaling pathways may exert an anxiolytic-like effect on behavior in both infant and adult rodents (Neumann, 2002; McCarthy et al., 1996; Windle et al., 2005). More recently, Waldherr and Neumann (2007) succeeded in demonstrating that OXT released within the rat brain during mating triggers anxiolytic effects, thus converging its implication in reproduction and emotionality. However, OXT-knockout male mice were less anxious in the infant separation test, the elevated plus-maze, and the acoustic startle test; furthermore, they were shown to be more aggressive in both home-cage (Winslow et al., 2000) and seminatural-environment testing conditions (Ragnauth et al., 2005). Confirming these findings, OXT-knockout males from OXT −/−, but not +/− dams, were recently shown to have higher levels of aggression (Takayanagi et al., 2005). In a former study (Ferguson et al., 2000), OXT-knockout male mice failed to develop social memory, with olfactory detection of nonsocial stimuli being intact. Treatment with OXT, but not AVP, rescued social memory in knockout animals, and treatment with an OXT-receptor antagonist produced a social amnesic-like effect in wild-type males (Ferguson et al., 2001). Similarly, wild-type females that received an antisense oligonucleotide targeting the mRNA of the OXT receptor gene in the medial amygdala became completely impaired in social recognition (Choleris et al., 2007). Like for the α- and β-estrogen-receptor-knockout mice, even in the OXT-knockout mice impaired social recognition is reflected in impaired capability of recognizing and avoiding parasitized conspecifics. In addition, the OXT-knockout animals are impaired in utilizing other mice as a source of information in mate choices and parasite avoidance (Kavaliers et al., 2003). That social memory is distinct from normal sociability and normal preference for social novelty has recently been underlined by Crawley et al. (2007). Scoring social approach behaviors in OXT-knockout males, these authors describe no genotype differences in two independently generated lines of OXT mutants compared to wild type, clearly suggesting that more specific memory and social affiliation deficits previously described (e.g., Ferguson et al., 2000; Carter, 2003; Young and Wang, 2004; Pedersen and Boccia, 2006) are not global to social approach behaviors in general.

In contrast to males, female OXT-knockout mice exhibited more anxiety-related behavior in an elevated plus-maze test compared to female wild-type animals (Mantella et al., 2003). Interestingly, they also showed signs of a heightened corticosterone response following exposure to an anticipatory and physical stress (Mantella et al., 2004, 2005). These findings support previous data suggesting a stress-attenuating action of centrally released OXT (Landgraf and Neumann, 2004; Ebner et al., 2000; Wotjak et al., 2001; Terenzi and Ingram, 2005; Wigger and Neumann, 2002).

In contrast to Mantella et al. (2003), Choleris et al. (2003) described female OXT-knockout mice to be less anxious than their wild-type littermates and, additionally, deficient in social recognition. The remarkable parallelism to similar alterations found in estrogen receptor β- and α-knockout mice led these authors to the suggestion of a gene micronet, which links hypothalamic and limbic forebrain neurons in the estrogen control over the OXT regulation of social recognition. Indeed, consistent with their similar behavioral profiles, OXT and estrogens act in a tightly interrelated manner, with OXT and its receptor actively being regulated by estrogens.

Generally, OXT-deficient mice have normal body weight, but there are several abnormalities of ingestive behavior in these animals. Dehydration-induced anorexia, for example, is attenuated in OXT-deficient mice (Rinaman et al., 2005), while consumption of solutions that contain NaCl is enhanced (Amico et al., 2003; Vollmer et al., 2006). This is not solely attributable to any specific effect on sodium appetite, as OXT-deficient mice will also overconsume palatable sucrose solutions (Miedlar et al., 2007) and both sweet and nonsweet carbohydrate solutions (Sclafani et al., 2007).

Apart from genetic evidence, many authors have speculated about a role for OXT in autism. This is plausible, since a core feature of the disease is impaired social interaction and affiliation that are mediated by endogenous OXT. Several lines of preclinical and clinical evidence support this hypothesis: plasma OXT levels were found to be lowered in autism, and correlated with social impairment (Green et al., 2001; Carter, 2007) and autism spectrum disorder patients showed a significant reduction in repetitive behaviors and improvement in social cognition after OXT versus placebo infusion (Hollander et al., 2007). However, it is of note in this context that endogenous OXT circulating in plasma is unlikely to reflect central release pattern and to induce any effects on brain functions (Landgraf and Neumann, 2004; Landgraf, 2006).

1.3.1.2 The oxytocin receptor

Similar to its ligand, OXTR-knockout mice were viable and had no obvious deficits in fertility and reproductive behaviors. While exhibiting normal parturition, OXTR-knockout dams demonstrated defects in lactation and maternal nurturing. In response to social isolation, infant OXTR-knockout males emitted fewer ultrasonic vocalization calls than wild-type littermates, indicative of reduced anxiety. Adult knockout males also showed signs of elevated aggression and deficits in social discrimination abilities (Takayanagi et al., 2005).

Recently, to examine the specific roles of the OXTR during development and in particular brain regions, conditional OXTR-knockout mice were created by Lee et al. (2007). Due to the use of late-onset promoters to control the spatial and temporal gene inactivation, this technique has the potential to overcome some limitations of the conventional knockout technology. Preliminary data indicate that both forebrain-specific and total OXTR-knockout mice showed signs of normal olfaction, with the former being able to lactate and displaying no deficits in maternal behavior (Lee et al., 2008).

In addition to these from-inside-to-outside approaches, opposite strategies have been used to shed light on neurobiological correlates of extreme phenotypes in both rats and mice. We have therefore embarked on a selective breeding paradigm in both species to enrich for the high-anxiety-related behavior (HAB) versus low-axiety-related behavior (LAB) traits, thus proving the way for genetic analyses to identify specific genetic variants that associate with variations in anxiety-related and comorbid behaviors. Selective inbreeding to conserve genetic polymorphisms causally underlying trait anxiety and to increase genetic homozygosity and therefore to reduce genetic variance can lead to reduced viability of animals (inbreeding depression), and can result in major random alterations in the genetic composition of the selected lines due to genetic drift. This may potentially result in marked differences that are unrelated to the selected phenotype, severely limiting one's ability to identify true genetic factors that are responsible for the phenotypic differences between lines. A series of measures has been employed to minimize the risk related to genetic drift: (1) in each selectively bred line, several independent sublines are run; (2) identified genetic polymorphisms are tested in a freely segregating F2 panel; they are accepted to be causal, if they co-segregate with the phenotype; (3) in addition to the divergent lines, an intermediate control line is bred for normal anxiety-related behavior; and (4) replication is informative in terms of the impact of founder effects and genetic drift and should be run concurrently with the main breeding paradigm in the same or another species, thus also facilitating complementary interspecies genetics (Kas et al., 2007). Hence, we generated both HAB/LAB rats and mice (Landgraf et al., 2007).

Not only selective breeding but also knockout approaches have typical limitations and constraints, including compensation and genetic background issues. Thus, gene knockouts are not as clean a system for testing function as is commonly assumed. Congenic footprints in gene expression, for example, are a general phenomenon that hampers the interpretation of knockout experiments (Schalkwyk et al., 2007). A wide variety of complementary approaches, including loss- and gain-of-function techniques and selective breeding, have thus to be considered useful for studying the genetic impact on hormones and behavior.

The rapid and continued bidirectional phenotypic divergence between HAB and LAB lines demonstrates that the trait is clearly heritable. The HAB and LAB phenotypes are therefore already largely predictable based purely on parental phenotype. Accordingly, cross-fostering data demonstrated that differences in maternal behavior have a relatively minor impact on the future phenotype of the pups, and cross-mating between HAB and LAB animals confirmed that their phenotype is strongly dependent on genetic background. Remarkably, in HAB/LAB rats and mice, OXT and its gene, in contrast to AVP, do not seem to play a crucial role in rats and mice bred for either high or low anxiety-related behavior. As shown in a wide variety of techniques, including microdialysis, *in situ* hybridization, allele-specific transcription analyses, and expression profiling, centrally released OXT does not appear to contribute to the marked differences in inborn trait anxiety and comorbid depression-like behavior.

Rodent models of social behavior have been used as a tool for better understanding disorders of social behavior in humans, including autism. The involvement of the OXT system in human social behaviors and related brain activity has recently been confirmed by intranasal administration of OXT and effects on abilities to recognize emotions, willingness to trust anonymous partners in an economic task, responses to facial expressions and others (Kirsch et al., 2005; Kosfeld et al., 2005; Domes et al., 2007). Recent molecular-genetic studies of the OXTR in humans have strengthened the evidence regarding the role of the OXT system in behavioral regulation, both in health and pathology. Approximately 30 single nucleotide polymorphisms (SNPs) are known in the human OXTR gene region. Selecting subsets of these SNPs that tag the common haplotypes of a region for genotyping, Ebstein and co-workers (Israel et al., 2008) observed a significant association between single SNPs and haplotypes and symptoms of autism-spectrum disorders characterized by deficits in social interaction and communication. This result confirms similar studies in other ethnic groups (Wu et al., 2005; Jacob et al., 2007) and a gene expression study (Bittel et al., 2007), all suggesting that SNP and haplotypes in the OXTR gene confer risk for autism.

1.3.2 The Vasopressin System

1.3.2.1 Vasopressin

There is a vast literature linking the AVP system to cognitive abilities, male aggression, and emotionality. Compelling evidence for a critical involvement of centrally released AVP in anxiety-related and depression-like behaviors is provided by the HAB/LAB rat and mouse models. As mentioned before, these animals, bidirectionally and selectively bred for extreme poles of trait anxiety, can actually be harnessed as a useful tool to identify molecular-genetic and neuroendocrine correlates of the corresponding genetic predisposition. Indeed, in highly anxious HAB rats, an SNP in the promoter structure of the AVP gene has been identified that drives the gene to overexpress AVP in the PVN of the hypothalamus. This SNP has been shown to reduce the binding of the transcriptional repressor CBF-A, thus causing AVP overexpression, as confirmed by allele-specific transcription approaches (Murgatroyd et al., 2004). Correlative evidence indicated that it is indeed AVP that critically contributes to features typical of the HAB phenotype including high anxiety and depression-like behaviors. The latter could be normalized by chronic paroxetine treatment, associated with an alteration of AVP expression toward LAB rats (Keck et al., 2003). Similar behavioral effects were obtained by V1aR antagonist treatment, particularly attenuating depression-like behavior of HABs (Wigger et al., 2004).

To examine the generalizability of an AVP involvement in trait anxiety, the same breeding protocol was applied in CD1 mice, again resulting in an extreme divergence of anxiety-related behavior (Kromer et al., 2005). In this model, we focused on LAB mice and the possibility that an AVP underexpression might underlie the phenomenon of total nonanxiety. Indeed, in LAB animals, AVP expression and release within the brain were found to be reduced, due to polymorphisms in the AVP gene. First, in exon 1, an SNP was identified that causes an amino acid exchange in the signal peptide of the AVP precursor, this substitution being likely to cause an insufficient intracellular processing and trafficking of the precursor leading to a strong deficit in AVP bioavailability. It is of interest to note, in this context, that deficits in OXT processing in children with autism may be important in the development of this syndrome (Green et al., 2001). The causal involvement of the AVP deficit in the phenotype was tested in a freely segregating F2 panel. In this panel, the SNP was found to co-segregate with drinking behavior (reflecting the antidiuretic effect of endogenous AVP) and, partially, anxiety-related behavior (indicating that the AVP gene is one of multiple genes exerting a minor, albeit significant, anxiogenic effect). Importantly, locomotor activity was not linked to the signal peptide SNP (Kessler et al., 2007). Second, a deletion in the promoter structure of the AVP gene of LAB mice was identified, explaining a presumably additive reduction in AVP expression in this line. Indeed, in all allele-specific transcription and microarray approaches, confirmed by quantitative RT-PCR, LAB AVP gene expressed less AVP than HAB and normal CD1 animals (Bunck et al., 2008). Interestingly, this promoter deletion is strictly linked to the signal peptide SNP, again stressing the co-segregation with the nonanxious phenotype in the F2 panel. Both polymorphisms are thus likely to add to a dramatic deficit in AVP release within and transport from the hypothalamic PVN. This deficit in AVP availability finally causes signs of diabetes insipidus and contributes to nonanxiety. Additional analyses in the CD1 population identified the HAB-specific sequence as the most common genotype, thus explaining why CD1 controls are closer to HAB mice regarding their neuroendocrine (AVP expression) and behavioral phenotypes than HAB × LAB cross-mated and strictly intermediate F1 controls. Remarkably, supporting findings in LAB mice, Mlynarik et al. (2007) succeeded in demonstrating signs of attenuated depression-like behavior in AVP-deficient Brattleboro rats and Bruins et al. (2006) observed signs of reduced anxiety/cognitive dysfunctions in diabetes insipidus patients suffering from an AVP deficit.

Beginning in the 1970s, an association between AVP and psychiatric disorders was proposed (Gold et al., 1978). Although endogenous AVP circulating in plasma is unlikely to reflect central release pattern and to induce any effects on brain functions (Landgraf and Neumann, 2004; Landgraf, 2006), elevated plasma AVP levels have repeatedly been described to be linked to psychosis and depression (van Londen et al., 1997; de Wied and Sigling, 2002; Goekoop et al., 2006). Postmortem studies on brains of depressed patients, supposedly more relevant to behavior and psychopathology than plasma levels, revealed an increase in AVP-expressing neurons in the PVN (Purba et al., 1996; Meynen et al., 2006).

1.3.2.2 The vasopressin V1a receptor

A critical involvement of the V1aR subtype in anxiety-related behavior became likely when Landgraf et al. (1995) succeeded in showing that an antisense oligo-induced downregulation of V1aR in the rat septum induced anxiolytic-like effects and a loss of the ability to adequately remember social stimuli. Since then, this loss-of-function result has repeatedly been confirmed by gain-of-function studies, for instance, by virus vector approaches (Landgraf et al., 2003; Pitkow et al., 2001). Whereas the knockout of the V1aR gene in male mice resulted in a marked impairment of social recognition and interaction abilities, it had no influence on performance in nonsocial olfactory learning and memory tests (Bielsky et al., 2004) and on depression-like behavior (Egashira et al., 2007). When, however, the V1aR was re-expressed in the lateral septum of V1aR-knockout mice via a virus vector, a complete rescue of social recognition capabilities was achieved, together with an increase in anxiety-related behavior (Bielsky et al., 2005). In addition to impaired social interaction abilities, Egashira et al. (2007) reported reduced anxiety-related behavior in male V1a-knockout mice. This role of the V1aR in emotionality could not be confirmed in female mice, which performed normally in a number of anxiety tests. The authors hypothesize that this phenomenon is due to the sexual dimorphism in the extra-hypothalamic AVP system, with males having more AVP-containing fibers than females (Bielsky et al., 2005; de Vries et al., 2008). Evidence for a control of spatial memory by V1aR was provided by Egashira et al. (2004) who tested knockout mice on a radial maze.

The mechanisms of how genetic variation of the V1aR translates into differences in social behavior are best known for voles (Young et al., 1999; Hammock and Young, 2004; Lim et al., 2004). The regulatory region with major impact on social behavior is a few hundred base pairs upstream of exon 1. The presence or absence of highly repetitive microsatellites of several hundred nucleotides in the promoter region was found to be consistent with differences in partner preference and monogamy in several species of voles. In contrast, for the coding sequence, a 99% homology in monogamous prairie and promiscuous montane voles was reported (Young et al., 1999). Transfer of the coding region or the entire V1a gene including the microsatellites from monogamous prairie voles to other rodents resulted in modified V1aR distributions in the brain and typical behavioral changes in the target organism, including increased affiliative behavior similar to prairie voles, improved social discrimination abilities and more active social behavior, and increased partner preference formation when expressing V1aR from monogamous prairie voles (Landgraf et al., 2003; Young et al., 1999; Lim et al., 2004). Breeding experiments with prairie voles demonstrated predictable individual differences in receptor-distribution patterns and in some social behaviors among males with microsatellites differing by less than 50 base pairs in length (Hammock and Young, 2005). The observation of associations between V1aR binding patterns and differences in social behavior as well as the detection of repetitive structures upstream of the primate V1aR led to the hypothesis of a general connection between the evolution of social bonding and monogamy and the expansion of microsatellites in the V1aR promoter region (Hammock and Young, 2005). This is still a matter of debate, as recent analyses of a large number of rodents and other mammals showed that the evolution of monogamy in some vole species is generally independent of the expansion of microsatellites upstream of the V1a gene (Fink et al., 2006; Heckel and Fink, 2008). In addition to the evolution from poly- to monogamy relying on one gene, a possible co-variation of the AVP ligand remains unclear.

There is a number of studies describing an association between AVP V1aR polymorphisms and human social behavior. In an elegant series of experiments, Ebstein and co-workers (Bachner-Melman et al., 2005; Israel et al., 2008; Granota et al., 2007) provided evidence that this receptor subtype is linked to sibling social interactions. Even musical memory and dance abilities that are related to social contact, courtship, and openness to communication seem to be associated with V1aR-haplotype frequencies. Furthermore, in a so-called dictator game, the contribution of common genetic polymorphisms to a unique human trait, altruism/prosocial behavior, has been determined (Knafo et al., 2008). This approach, which has advantages over the standard self-report questionnaires, suggested that V1aR variations partially explain individual variance in altruistic behavior, thus influencing pro-self versus pro-social styles of behavior. Accordingly, also associations between autism and V1aR have been reported (Kim et al., 2002; Wassink et al., 2004; Yirmiya et al., 2006). Importantly, this kind of associations appears to be mainly mediated by the role of the V1aR gene in shaping socialization skills, confirming this role of the AVP system throughout vertebrates. The involvement of centrally or peripherally released AVP in autism susceptibility remains to be shown.

1.3.2.3 The vasopressin V1b receptor

The AVP V1bR is most highly expressed in the anterior pituitary where it is thought to play a role in co-stimulating the neuroendocrine response to stress (Volpi et al., 2004). It is only recently that V1bR mRNA as well as V1bR immunoreactive neurons have been found in the rodent brain, including the olfactory bulb, septum, and hippocampus. Importantly, particularly in the septum, a specific V1bR antagonist has been described to exert antidepressive activity and, to a lesser extent, anxiolytic activity (Griebel et al., 2005). Furthermore, this receptor subtype is likely to be important for adequate responses to both acute and chronic stressors (Serradeil-Le Gal et al., 2005).

These pharmacological data are not always consistent with behavioral alterations described in V1bR-knockout mice, including reduced aggression and mild deficits in social memory, compared to wild-type controls (Wersinger et al., 2002, 2004) or failure to affect measures of anxiety- and depression-like behaviors (Wersinger et al., 2002; Caldwell et al., 2006). According to Wersinger et al. (2007), V1b-knockout mice have longer attack latencies in a resident–intruder test, but no global deficit in all aggressive behaviors. Experience modulates their aggression, though, never to the level observed in wild-type controls. Interestingly, when given the opportunity to predate a cricket, these mice display comparable latencies to attack as wild-type littermates do, indicating that the aggression phenotype is specific to social forms of aggression. Indeed, in a competitive test, where the animals were food-deprived, V1b-knockout mice showed signs of aggression that, although being higher than during nonfasting conditions, was lower than that found in wild-type animals (Wersinger et al., 2007).

No significant difference was found in basal plasma levels of ACTH and corticosterone between mice lacking functional V1b receptors and wild-type controls. While there was no difference in the ACTH response to acute and chronic restraint in the latter, V1b-knockout animals subjected to 14 sessions of daily restraint showed decreased ACTH responses, suggesting that this receptor subtype is necessary only for ACTH responses during chronic stress (Lolait et al., 2007). In a follow-up study, V1b-knockout mice had markedly compromized plasma ACTH and corticosterone responses to acute but not extended exposure to lipopolysaccharide. The stress-hormone levels stimulated by high doses of ethanol were decreased in V1b-knockout mice compared to wild-type littermates, suggesting a significant role of the V1b receptor in the HPA axis response to acute immune stress and ethanol intoxication (Lolait et al., 2007). Caldwell et al. (2006) reported that there is no interaction between either the V1a or V1b subtypes and ethanol on motor coordination, mood, or voluntary ethanol consumption.

In a more recent study, Stewart et al. (2008) evaluated the involvement of the V1b receptor subtype in the HPA-axis activation to acute administration of selective serotonin reuptake inhibitor (SSRI), such as fluoxetine and desipramine. Both antidepressive drugs were found to attenuate plasma ACTH and corticosterone levels in male and female V1b knockout mice when compared to their wild-type counterparts, suggesting the AVP and its V1b play a major role in driving the normal HPA-axis response to acute SSRI administration. Interestingly, in this study, evidence of a sexual dichotomy in the regulation of AVP, OXT, and CRH gene expression in the hypothalamic PVN is demonstrated following antidepressant administration with male fluoxetine-treated V1b-knockout or wild-type mice not changing their expression profiles at PVN level.

To examine whether genomic variations in the human V1bR might contribute to the liability to develop affective disorders, van West et al. (2004) identified SNPs in the gene and analyzed them in association with recurrent MD. The authors came to the conclusion that a major SNP haplotype of V1bR protects against MD. In a more recent study (Dempster et al., 2007), both single-marker and haplotype analyses provided evidence indicating an association between the V1bR gene and childhood-onset mood disorders in females, essentially supporting the data obtained by van West et al. (2004).

1.4 Tachykinins

Tachykinins represent an evolutionary rather well-conserved peptide family, as tachykinin-related peptides have already been shown to exist in invertebrates like locust or even hydra (Severini et al., 2002). Amphibians and other vertebrates also display a broad variation of tachykinins. Sequence and – of course – function do not fully correspond to mammalian tachykinins, but some of them are potent agonists of mammalian tachykinin receptors as will be discussed in detail below. They all share a common hydrophobic C-terminal sequence, defined as FXGLM-NH_2, with X representing any amino acid with an apolar side chain. Numerous studies demonstrate the indispensability of this sequence for the interaction with one of the three known mammalian neurokinin receptors for tachykinins, NK_1, NK_2, and NK_3. Nevertheless it took about 70 years of intense research from the first identification of substance P (SP) by von Euler and Gaddum in 1931 to neurokinin A (NKA) and neurokinin B (NKB) and others to reach the current state of knowledge in tachykinin variety (Severini et al., 2002; Page, 2004). Among mammalian tachykinins, SP is one of the best-studied and very well-characterized neuropeptides, although recent work focusing on novel members – based on C-terminal sequence similarity – requires a more careful view on older studies, since they are exclusively based on immunoreactivity (Page, 2004; Brain and Cox, 2006).

1.4.1 Different Types of Tachykinins and Receptors

Three genes encoding tachykinins have been described so far in mammals: TAC1, TAC3, and TAC4 in humans and *Tac1*, *Tac2*, and *Tac4* in rodents (mouse and rat). *Tac2* of mice and rats displays high homology to the human TAC3 gene, both encoding NKB (Duarte et al., 2006). *Tac1* and *Tac4* are similar concerning their intron–exon structure and both express at least four different splicing variants.

The following peptides are derived from *Tac1* transcripts: SP, NKA, neuropeptide K (NPK), and neuropeptide γ (NPγ), with all these peptides representing members of the tachykinin-peptide family. In contrast, *Tac4* transcripts variants encode the peptides hemokinin 1 (HK-1) and the endokinins A, B, C and D (EKA, EKB, EKC, EKD), with EKC and EKD merely representing tachykinin-like peptides. Here, the C-terminal sequence lacks two characteristic amino acids, changing the tachykinin motif FXGLM to FQGLL, thereby weakening hydrophobicity. Another specialty of *Tac4* gene products is represented by the fact that these endokinins seem to be specific to human cells and have not been described as being expressed in rodents. Only recent studies suggest that similar endokinins might be encoded by *Tac4* in rabbits (Page, 2004).

Three different receptors are described and classified as neurokinin receptors and consequently named NK_1, NK_2, and NK_3. They are G-protein-coupled receptors and show a high degree of expression in the brain (Patak et al., 2005). All tachykinins featuring the FXGLM peptide motif are binding to any of the three NK receptors, but are differing in ligand affinity. Although mammalian tachykinins are the natural ligands of NK_1, NK_2, and NK_3, nonmammalian vertebrate tachykinins – like uperolein (from *Uperuleia marmorata*), physalaemin (from *Physalaemus biligonigerus*), kassinin (from *Kassina senegalensis*), and phyllomedusin (from *Phyllomedusa bicolor*) – are able to bind with fairly high affinity. Beneath other tachykinin-like peptides (the locustatachykinins) that differ in the common hydrophobic C-terminal pentapeptide, are some genuine tachykinins, isolated from invertebrate species like eleidosin (from *Eledone aldovrandi*), and sialokinin I and II (from *Aedes aegypti*). The latter display similar receptor-binding effects in mammals, thereby representing potent agonists to NK receptors (Severini et al., 2002).

Although there is dispute on the role and specificity of *Tac* gene products on different NK receptors, the following ranking in ligand affinity is widely established:

1. NK_1: SP>NKA>NKB (physalaemin and eleidosin have higher affinities to NK_1 than SP).
2. NK_2: NKA> NKB>SP (kassinin has higher affinity to NK_2 than NKB).
3. NK_3: NKB>NKA>SP (kassinin, physalaemin and eleidosin have higher affinity to NK_3 than NKA) (Cascieri et al., 1992).

Some studies also highlight the role of the amino-terminal domain of tachykinins, that seem to confer not receptor affinity itself, but receptor signaling and desensitization. For NK_1 it was demonstrated that – although NKA and NKB are agonists – only NPK and NPγ are capable of desensitizing the receptor and inducing intracellular IP_3 signaling (Vigna, 2003). Further receptor binding studies revealed that NPK and NPγ act as agonists of the NK_2 receptor, similar to NKA (van Giersbergen et al., 1992).

1.4.2 Function of Tachykinin Signaling

Beneath the influence of tachykinins on blood-pressure regulation and nociception, numerous studies have also proven a direct anxiogenic effect of either SP or other tachykinin gene-related peptides. The three tachykinin receptor types have also been reported to be involved in modulating anxiety-related behavior (Severini et al., 2002). Numerous studies utilizing receptor antagonists have shown anxiolytic or antidepressive-like effects in animal studies (Ebner and Singewald, 2006). Specific knockout models of either NK_1 or *Tac1* also demonstrate a decrease in anxiety-related behavior, where mice with targeted deletion of the *Tac1* gene were still displaying normal development, but showed decreased anxiety-related behavior in the elevated zero-maze and the OF test combined with decreased immobility time in the FST. In contrast to that, NK_1-deficient animals have been demonstrated to show diminished levels of depression-like behavior (Bilkei-Gorzo and Zimmer, 2005), with only one study suggesting anxiolytic effects (Bilkei-Gorzo et al., 2002). Also, a disruption of the gene encoding NK_3 has recently been described to result in deficits concerning learning and memory. Whereas wild-type and knockout mice did not differ in parameters of spontaneous locomotion, anxiety-related measures, and depression-like behavior, they displayed deficits in the acquisition of conditioned-avoidance responding and the Morris water maze. Knockout mice also had slightly elevated body weight compared to wild-type mice with the difference being significant at a few developmental stages (Siuciak et al., 2007).

Nevertheless, anxiogenic effects of NK_1 signaling via tachykinin activation resulted in an increased interest in NK_1 antagonists as potential therapeutics in treatment of anxiety disorders and depression, for detailed effects and an overview on these studies, see the review by Ebner and Singewald (2006). Although some substances already displayed good response and low rates of subsidiary effects (e.g., aprepitant) when compared to other commercial antidepressants, currently not a single study is conducted that is promising the release of novel therapeutics for the treatment of depression, based on tachykinin signaling for the near future (Czeh et al., 2006).

In addition, anxiolytic effects have also been demonstrated in gerbils by selectively antagonizing NK_2 and NK_3 receptors. Furthermore, antagonists are increasing social interaction time and decreasing immobility time in the tonic immobility test, suggesting antidepressant-like effects (Salome et al., 2006).

Especially one antagonist of NK_2 (saredutant) has been demonstrated to cause increased social interaction, reduced depression-like effects, and reduced ultrasonic vocalization in rats (Louis et al., 2008).

Altogether, the mammalian tachykinin system looks like a perfect proof of nature to mediate the same effects via a whole battery of different receptors and substances. So, parallel subsystems can compensate the possible defect in a specific trait, concerning not only essential behavioral phenomena, but also vitally important physiological processes, like blood pressure regulation and nociception.

Beneath the obvious possibility of applying NK_1 antagonists as therapeutics in tachykinin-signaling systems, there are interesting variants in cross talk with other endocrine systems. Morphines have been described to antagonize the NK_1 receptor (Kosson et al., 2005). Moreover, the idea of alternatively processed tachykinin-gene-derived peptides performing opposing behavioral effects has been supported by a number of different studies. A truncated version of SP, not containing the typical C-terminal amino acid sequence (SP 1–6 or 1–7) still remains highly active, but is reducing anxiety-related behavior and enhancing memory (Hallberg and Nyberg, 2003). These effects cannot be mediated by NK receptors, and other mechanisms mediating these behavioral effects are currently unknown. Similarly, it has been described for NPγ that post-translational processing produces a truncated peptide (1–9) and NKA (Wang et al., 1993). So, a better understanding of the molecular pathways of tachykinin signaling modulating animal and human behavior may contribute to the development of novel therapeutics in the treatment of psychopathologies.

1.5 Opioid Receptors

Morphine, named after Morpheus, a son of the Greek god of dreams, is the main active agent in opium. Intake of morphine is associated with analgesia as well as intense reward properties such as euphoria, excitation, and high and intense pleasure. Unlike that, withdrawal in addiction is associated with aversion, dysphoria, and discomfort. These symptoms are thought to be mediated mainly through the μ(mu)-opioid receptors, that are, together with δ(delta) and κ(kappa), by far, the best-investigated and best-characterized opioid receptors, although about 17 other subtypes exist. The main functional role of μ-, δ-, and κ-opioid receptors is modulating inhibitory neurotransmission in the brain. According to this, they are highly expressed in γ-aminobutyric acid (GABA)ergic neurons, for example, found in the mesolimbic–mesocortical dopaminergic system known to play a central role in reward pathways. Moreover, opioid receptors modulate nociception, responses to stress, respiration, gastrointestinal motility, endocrine and immune physiology, as well as addiction and anxiety behaviors.

Opioid receptors belong to the family of G-protein-coupled receptors. Together with the later-discovered orphanin FQ/nociceptin receptor (Darland et al., 1998), they form a four-member gene subfamily. Receptor activation with inhibitory effects, as shown for coupling to inhibitory G-protein (G_i) complexes in striatal dopamine neurons, reduces release of neurotransmitters, whereas activation in connection with a stimulatory (G_o) complex has excitatory effects (as measured by assessing agonist stimulation of membrane binding of the nonhydrolyzable analog of guanosine-5′-triphosphate (GTP), guanosine-5′-o-(3-[^{35}S]thio) triphosphate ([^{35}S]GTPγS)). Moreover, additional signaling pathways have been described, for example, coupling to Ca^{2+}- or K^+-channels.

In humans, genes coding for the μ-, δ-, and κ-opioid receptors are called MOR, DOR, and KOR or MOP, DOP, and KOP, according to *Oprm*, *Oprd1*, and *Oprk1* for the mouse nomenclature. Mice lacking all of the three opioid receptor genes have been generated by interbreeding of individually generated opioid-receptor mutants. These triple mutants are viable and healthy, indicating that the opioid system, in general, is rather involved in modulating behavior, than promoting physiological survival (Gaveriaux-Ruff and Kieffer, 2002).

The genetic organization of the sevenfold transmembrane opioid receptor proteins is highly similar in exon–intron structure. Coding regions extend over three exons, with exon 1 encoding the extracellular and transmembrane domain I, exon 2 encoding transmembrane domains II–IV, and exon 3 encoding the transmembrane domains V–VII, followed by the cytoplasmic C-terminal region. The μ-opioid receptor gene differs slightly in the 3′-end, where alternatively spliced codons are found on additional coding exons (recently, splice variants of exon 1 have been described, too (see below)). In close correlation to their genetic organization, all three receptor genes are highly similar at the level of their predicted protein sequence (Kieffer and Gaveriaux-Ruff, 2002).

Despite these similarities, the different opioid receptors react differently upon binding of the same ligand. The pharmacodynamic response of an endogenous or exogenous opioid depends on the receptor type, the affinity to that receptor, and whether the opioid is an agonist or an antagonist. Three major classes of endogeneous opioid peptides have been described as exerting their agonistic effects upon opioid-receptor binding: endorphins, enkephalins, and dynorphins. All these receptor ligands are small peptides, 5–31 amino acids in length that feature the canonical Tyr-Gly-Gly-Phe-Met/Leu N-terminal motive in their primary sequence. This motive is indispensable to activate opioid receptors. Typically for neuroendocrine messenger peptides, all these receptor ligands derive from protein precursors that contain several secreted peptides: POMC, proenkephalin (Penk), and prodynorphin (Pdyn). The multiple opioid peptides arising from these genes, respectively, bind to all three receptors with high affinity and low selectivity (ligands of the orphanin FQ/nociceptin family derive from their precursor proorphanin). In close similarity to the opioid-receptor triple-knockouts, mutants homozygous for the different opioid-peptide precursor genes did not show obvious developmental defects, were able to reproduce, and expanded at the expected Mendelian frequency (Kieffer and Gaveriaux-Ruff, 2002).

1.5.1 μ-Opioid Receptors

Due to the almost low selectivity of antagonists, research on the specific physiological mode of action of μ-opioid receptors on behavior, discreted from the effect of other opioid receptors, was hard to address. Therefore, mice, deficient for the Oprm-gene, have been generated to produce model organisms disengaged from μ-opioid-receptor expression. Knockouts were produced in five different laboratories, either by deletion of exon 1 (Tian et al., 1997; Sora et al., 1997; Schuller et al., 1999), insertion of a Neo cassette in exon 2 (Matthes et al., 1996), or deletion of both exons 2 and 3 (Loh et al., 1998). In all lines homozygous for the particular mutation, binding of μ-opioid-receptor agonists [3H] DAMGO and [3H] endomorphin-2 was abolished in quantitative autoradiographic mappings, demonstrating that μ-opioid receptor sites were completely deleted, independent from the targeting strategy. In addition, these mappings revealed subtle downregulations of δ- and κ-opioid-receptor sites in mutant brains (Kitchen et al., 1997), although these alterations in expression levels remained regionally restricted and did not change the anatomical distribution of the remaining opioid receptor sites.

μ-Receptors have been found to localize to different regions in the CNS: they are distributed in the brainstem nuclei, nucleus of the solitary tract, respiratory nuclei, as well as periaqueductal and periventricular zones of the midbrain. They have been further localized in the striatum, nucleus accumbens, anterior limbic forebrain, locus ceruleus, several layers of the cerebral cortex, cingulate, orbifrontal, medial prefrontal and insula of the cortex, and in some nuclei of the amygdala (Reyes et al., 2007).

1.5.1.1 μ-Opioid receptors in nociception, stress response, and post-traumatic stress disorder

As expected from the antinociceptive properties of morphine, Oprm-mutants showed increased sensitivity to pain. They displayed an enhanced response in the tail pressure test, in the hot-plate test (Sora et al., 1997), and in the early phase upon formalin injection into ascites, suggesting that μ-opioid receptors inhibit thermal, mechanical, and irritant chemical nociception. This is supported by the fact that *in vivo* effects of morphine analgesia in these tests are also strongly reduced at doses that produce maximal analgesia in wild-type mice (Matthes et al., 1996).

Stress exposure can induce analgesia that could be partially reversed by opioid antagonists. Thus, it has been proposed that endogenous opioid peptides released upon stress exposure are responsible for stress-induced analgesia (SIA). Indeed, μ-opioid receptor-deficient mice displayed less SIA to thermal pain, compared to wild-type mice 15 min after forced swim stress but not in the 5-min initial phase, indicating that μ-opioid receptors are implicated at a later phase of SIA.

However, in contrast to increased pain sensitivity in the SIA paradigm, ACTH and corticosterone levels (see above) were shown to be elevated in μ-opioid receptor-deficient mice under basal conditions (Kieffer and Gaveriaux-Ruff, 2002).

In addition, morphine, normally elevating plasma ACTH and corticosterone, failed to do so in mutant mice. This could be due to desensitization to the chronically elevated HPA-axis activity in mutants and, as was shown for the noradrenergic locus coeruleus, to a chronically reduced activation by stress hormones. Locus ceruleus neurons contain a high concentration of μ-opioid receptors that are prominently distributed on somatodendritic processes. Here it has to be mentioned that CRH was evidenced to serve as a neurotransmitter in this system, with intracerebroventricular-administered CRH increasing the spontaneous discharge rate of locus ceruleus neurons. Following this, immunoreactivities for CRH and μOR were observed at ultrastructural level at postsynaptic sites in dendrites in the locus ceruleus of adult male Sprague-Dawley rats, with 57% of dendrites exhibiting CRH-receptor also exhibiting μ-opioid-receptor immunoreactivity. This CRH-opioid convergence (most likely innervated by enkephalin-containing axon terminals), in turn, was shown to be involved in the neural circuitry underlying stress responses and opiate action. Chronic but not acute morphine treatment – acute morphine effects do not cause increased production of stress hormones – selectively sensitized the locus ceruleus system to CRH. So, 3ng of CRH produced a near-maximal activation of locus ceruleus neurons of rats chronically treated with morphine but were ineffective in vehicle-treated rats. Moreover, the chronic opiate-induced locus ceruleus sensitization altered the behavioral repertoire in response to swim stress (Reyes et al., 2007). Finally, it is tempting to speculate that mechanisms like the strategic co-localization of CRH- and μ-opioid receptor in locus ceruleus dendrites, leading to hypersensitivity to stress in morphine dependence, may be the reason for hyposensitivity of the locus ceruleus–norepinephrine system in μ-opioid-receptor-knockout mice. Then, due to the lack of μ-opioid receptor-mediated desensibilization, hyposensitivity to stress hormones could be the reason for elevated ACTH and corticosterone levels as a compensatory mechanism to elevate CRH signaling. This could also lead to an altered behavioral repertoire, like the observed reduction of SIA in mutant mice.

Enhanced SIA that could be reversed by naloxone has also been described in human patients suffering from post-traumatic stress disorder (PTSD). PTSD is characterized by emotional numbing and anhedonia. When exposed to traumatic reminders, PTSD patients experience analgesia or numbness and opioid blockade reduces this analgesic effect.

Therefore, recent work (Liberzon et al., 2007) has addressed the μ-opioid system in male patients with PTSD compared to non-PTSD male control groups, with and without combat exposure by functional neuroimaging. Receptor binding in the baseline state was examined by positron emission tomography (PET) using the μ-opioid-receptor-selective radiotracer $[_{11}C]$-carfentanil. In both traumatized groups a decreased binding potential was found in the rostral component of the extended amygdala system and in the nucleus accumbens, suggesting an adaptive change to general combat experience. However, only the combat group without PTSD exhibited decreased binding potential in the more caudal amygdala, suggesting that the PTSD subjects failed to downregulate μ-opioid receptors in this brain region. Because the amygdala is known to play an active role in emotion, fear processing, and SIA, these findings could provide a neuroanatomical substrate for the enhanced SIA observed in PTSD patients.

Rather, an adaptive change to general combat experience than one specific to PTSD has also been demonstrated in the cortical regions of the insula, dorsal cingulate cortex, and medial prefrontal cortex in both trauma-exposed groups by lower μ-opioid-receptor-binding potential. This observation is consistent with the role of cingulate cortex and medial prefrontal cortex, described in affective regulation and fear extinction. Indeed, PTSD patients and combat controls had significantly higher binding potential in orbifrontal cortex and subgenual cingulate cortex compared with normal controls. The combat controls exhibited even higher binding potential in the orbifrontal cortex than the PTSD group. Again, this may reflect insufficient or failed adaptation. One could speculate, if the μ-opioid receptor does play an inhibitory role in orbifrontal cortex via stimulatory G_o-complex signaling in GABAergic neurons, then insufficient upregulation leads to increased activity in this region associated with negative mood states and self-induced sadness.

PET analysis in female patients suffering from autobiographical experience has also been performed during the cued recall of this episode. Thereby a sustained state of sadness has been induced that was additionally measured by the positive and negative affect schedule (PANAS) (Watson et al., 1988). Here, the main effect of sustained sadness induction was a reduction in $[_{11}C]$-carfentanil binding to μ-opioid receptors in the rostral anterior cingulate, ventral basal ganglia, amygdala, and inferior temporal cortex. The reduction in the anterior cingulate and ventral basal ganglia was further correlated with the increase in negative affect during the challenge, whereas reduction in ventral basal ganglia and amygdala was correlated with positive affect ratings. Interestingly, the left amygdala μ-opioid neurotransmission was significantly more deactivated in response to the emotional challenge than the right site (Ribeiro et al., 2005).

1.5.1.2 μ-Opioid receptors in reward, pleasure, and anxiety

Activation of the μ-opioid receptor by an agonist such as morphine causes euphoria and an intense feeling of pleasure. With repeated use, as tolerance and dependence develop, the effect of the drug is more and more weakened. Finally, withdrawal symptoms can be associated with depression-like mood states and the acute feeling of anxiety.

How is the feeling of pleasure, initially caused by morphine, measured in a mouse model? Two behavioral tests are currently used to approach pleasure induced by certain substances. In the self-administration test, the preference for a substance is directly measured as the amount consumed in comparison to another freely accessible substance (e.g., the two-bottle choice test). In the place preference paradigm, two distinct neutral environments are experienced that are subsequently paired spatially and temporally with distinct drug states. The animal is later given the opportunity to choose to enter and explore either environment, and the time spent in either environment is considered an index of the reinforcing value of the drug.

According to this, morphine reward was measured by both, conditioned place preference and self-administration, and shown to be totally absent in μ-opioid-receptor-knockout animals. Moreover, dependence induced by withdrawal after chronic treatment with the general opioid antagonist naloxone was not developed in these animals (Matthes et al., 1996). Despite the fact that morphine selectivity is rather low for μ-opiod receptors in mouse brain, these data suggest that morphine does not require δ- and κ-opiod receptors to develop full activity. This proposition is further supported by the fact that morphine analgesia is intact in δ- and κ-opiod-receptor-knockout mice (Kieffer and Gaveriaux-Ruff, 2002).

Knockout mice also displayed reduced alcohol consumption. In the two-bottle choice test they consumed significantly less alcohol than wild-type mice and showed less alcohol reward in a test of conditioned place preference (Hall et al., 2001).

Consistent to this, histological investigations in ethanol-drinking rats revealed insights into mechanisms of behavioral alterations in the development of addiction to alcohol: acute ethanol administration produced an upregulation of μ-opiod-receptor densities, notably in the nucleus accumbens and in the basolateral amygdala. However, long-term ethanol intake downregulated μ-opiod-receptor density in the rat striatum and nucleus accumbens, a region known to play a role in reward (Ghozland et al., 2005).

Moreover, anxiolytic properties have been demonstrated in acute alcohol consumption as evidenced by the increased time spent on the open arms of the EPM, while repeated ethanol deprivation in addiction results in increased anxiety-like behavior during withdrawal symptoms.

Is there any connection between histology data and anxiolysis? Anxiety levels of μ-opiod-receptor-knockout mice do not support this notion. They displayed decreased levels of anxiety on the EPM under basal conditions and the anxiolytic properties of alcohol were not modified (LaBuda and Fuchs, 2001). Hence, due to the fact that mutants did not differ from wild-type mice, the acute anxiolytic properties of alcohol seem to be independent of μ-opiod receptors and their expression upregulated in the nucleus accumbens and the basolateral amygdala.

However, a somehow different picture is sketched in the OF test, another behavioral task, operating with the avoidence of bright, open areas in anxiety: here, ethanol produced a significant hyper-locomotion in the first 5 min following injection and increased the time spent in the lit compartment of the arena to 20 min following injection in wild-type animals. However, there was no effect of alcohol injection either on locomotion or on time spent in the lit area in mutant mice, suggesting that μ-opiod receptors are involved in locomotion and the anxiolytic properties of alcohol.

Moreover, chronic ethanol-treated mutants spent significantly less time in the lit compartment compared to control diet-treated knockouts, suggesting elevated anxiety-related behavior in an advanced state of addiction. Disruption of ethanol treatment reduced the time spent in the lit compartment even more on the second and third withdrawal episodes in knockouts, a decrease appearing not until the third withdrawal for wild types. These data suggest that μ-opiod receptors are involved in stabilizing anxiolysis in dependence and that the striatum and nucleus accumbens could be involved (Ghozland et al., 2005).

Somehow, unexpected data were obtained for the impact of morphine in the lateral septum on anxiety-related behavior. This area is mainly constituted of GABAergic neurons and expresses high densities of GABA receptors. Anxiolysis is observed when lateral septum GABAergic transmission is facilitated by means of local stereotactic injections of $GABA_A$ agonist. However, unilateral injections of morphine into the lateral septum, but not medial septum, resulted in increased anxiety-like behavior and reduced locomotor activity in the EPM, suggesting that the anxiogenic-like effects of morphine were specific to this septal subdivision. Possible lateralization of the lateral septum in the modulation of anxiety seems very unlikely since unilateral morphine injections were found to be equally anxiogenic, whichever was the targeted (left or right) hemisphere. Bilateral infusions of morphine into the lateral septum failed to reproduce the anxiogenic-like effects of unilateral injections in the EPM, but increased, at high doses, the number crossing the compartments, indicating a stimulant effect of bilateral morphine on locomotor activity. Subsequent immunohistochemical investigations on c-Fos expression were at least able to explain parts of the finding. Unilateral injections activated the mesolimbic pathway, which could account for their rewarding properties, but surprisingly failed to do so in the amygdala, a brain region known to play a critical role in anxiety and fear modulation. Thus, a reason for increased anxiety-like behavior was not provided in this study. In addition to activation of the mesolimbic pathway, morphine injected bilaterally induced a strong stimulation in the dorsal caudate putamen and the primary motor cortex. This result suggests that bilateral injections enabled the dopaminergic nigrostriatal pathway, possibly accounting for stimulation of locomotion (Le Merrer et al., 2006).

1.5.1.3 μ-Opioid receptor ligand binding in different splice variants

Up to now a total of 25 brain-specific splice variants have been isolated in the mouse *Oprm* gene, which are derived from combinations of 16 exons spread over 250 kb, in total, and emerge in region-specific and cell-specific pre-mRNA splicing and differential receptor targeting. Splicing mechanisms used are shown to include alternative 5′ and 3′ splicing, exon inclusion and skipping, mutually exclusive exons, intron retention, and alternative promoters (see Pan (2005) for review). Ten splice variants have been described in mice featuring the same exons 1, 2, and 3, but differ in alternative splicing from exon 3 to additional downstream exons. Thus, all these variants share the same protein structure predicted from exons 1–3, but have different carboxyterminal ends. Hence, alterations in the protein sequence of these isoforms are restricted to the intracellular domain modulating agonist-mediated internalization and receptor resensitization. In addition, marked differences have been described for agonist-induced G-protein activation.

However, the most relevant differences seem to be varying ligand affinities: μ-opioid receptor-specific opioids such as morphine, DAMGO, and morphine-6b-glucuronide (M6G) all competed binding with high affinity, at carboxyterminal variants, while δ- and κ-opioid-receptor-selective alkaloids were unable to compete at high concentrations. In addition, subtle but significant differences of the binding profiles have been described, in particular for the major endogeneous opioid peptides, endorphins, enkephalins, and dynorphins, and, beyond the major classes, enhanced affinity has also been described for endomorphin-1 and -2 (Zadina et al., 1997). These opioid peptides were isolated from the mammalian brain and were shown to display high selectivity to μ-opioid receptors, as revealed by their naloxone-reversible antinociception in mice after either intracerebroventricular or intrathecal administration (Mizoguchi et al., 2002). Endomorphin-1 and -2 even seem to be more potent than other endogenous opioids at these receptors, hence even leading the ranking in ligand affinity of β-endorphin, superior to enkephalins and dynorphins.

Alternative splicing at the 5′-end of the mouse *Oprm* gene has also been described: the new splice variants all featured exon 11, located 5′ proximal to exon 1 as the most upstream coding exon of the mRNA-transcript, either alternatively spliced to or in combination with exon 1. The promotor region of exon 11, in turn, has been reported as differing substantially from the exon 1 promoter that has been proposed to contain two promoters, a proximal and a distal one, 500 bp apart and differentially regulated. Western blots with specific antibodies suggested that exon 11-containing variants were expressed in the mouse brain.

As described above, knocking out exon 1 completely diminished the morphine-induced analgesic response (Sora et al., 1997). However, M6G- and heroin-induced analgesia was maintained with decreased potency. Therefore, these results strongly implied the existence of an alternative μ-opioid receptor transcript lacking exon 1, maybe involving exon 11 and encoding an M6G or heroin receptor.

1.5.2 Endorphins

Endorphins, and in particular β-endorphins, are the endogenous opiates that have received the most research attention. They are released into the circulation from the anterior-pituitary corticotropes and melanotropes of the hypothalamus and adrenal medulla. Apart from this, the synthesis of β-endorphins in the brain is limited to two cell groups: the arcuate nucleus and in a small population of neurons of the nucleus of the solitary tract in the brainstem. The relevant neurons in the arcuate nucleus project anteriorly to other parts of the hypothalamus, including the medial preoptic area, and in additon to the amygdala. Dorsally, neurons project to the PVN of the hypothalamus and then on to the brainstem to structures involved in the autonomic nervous system (Bancroft, 2005).

Despite the fact that β-endorphin has nearly equal affinity for the μ- and δ-opioid receptors, it acts mainly through μ-opioid receptors. According to expression and ligand affinity, β-endorphin has many behavioral effects, including those on sexual behavior, pleasure, appetite, and hedonic value.

1.5.2.1 Maturation of the β-endorphin-precursor proopiomelanocortin

β-Endorphin$_{31}$ is a peptide of 31 amino acids, the sequence (RYGGF MTSEK SQTPL VTLFK NALKN AFKKG E) being an integral part of the POMC protein. POMC is precursor for several neurohormones, including ACTH, β- and γ-lipotrophin, α-, β-, and γ-melanocyte-stimulating hormone (MSH), and β-endorphins. However, before products acquire biological activity, POMC has to undergo maturation by proteolytic cleavage in post-translational processing.

Post-translational processing is accomplished by highly specific cleavage enzymes, called prohormone convertases. In the neuronal endocrine tissue of mammals, prohormone convertases 1/3 (PC1/3) and 2 (PC2) have been identified to cleave protein precursors at paired basic residues to generate neuropeptides and peptide hormones. Gene expression of PC1/3 and PC2 was shown to parallel that of β-endorphin$_{31}$ in the arcuate nucleus and the lateral hypothalamus, but furthermore they are found in PVN, ventromedial nucleus, and in a small group of neurons within the dorsal medial posterior part of the arcuate nucleus (Helwig et al., 2006).

According to this, POMC is parted by PC1/3 into large intermediates, including ACTH and β-lipotrophin. Subsequent PC2 activity generates γ-MSH and can further process ACTH and β-lipotrophin. Proximate PC2-breakdown products of ACTH are α-MSH and the corticotropin-like intermediate lobe peptide, whereas β-endorhin$_{31}$ and γ-lipotropin are derivatives of β-lipotrophin. The latter could also be processed by PC1/3 and, indeed, β-endorhin$_{31}$ levels analyzed by radioimmunoassay in PC2-deficient mouse hypothalamus were shown to be either increased (despite loss of PC2) or at most diminished by two-thirds (Helwig et al., 2006) (see **Figure 1**).

PC2 is also involved in breakdown of β-endorphin$_{31}$. Initially generated β-endorphin$_{31}$ is cleaved to β-endorphin$_{27}$ by PC2, which in turn could be further processed by carboxypeptidase E, removing the terminal basic residue. Resulting β-endorphin$_{26}$, as well as β-endorphin$_{27}$, are considered opiate-receptor antagonists opposing the effects of β-endorphin$_{31}$. However, immunohistochemistry in the hypothalamus revealed that less than 30% of total immunoreactivity is contributed by β-endorphin$_{26}$ and β-endorphin$_{27}$, whereas more than 60% was constituted by β-endorphin$_{31}$.

In addition to β-endorphin$_{26,27,31}$ derivates, α-endorphin (containing β-endorphin amino acids 1–17) and γ-endorphin (containing β-endorphin amino acids 1–16) have been described.

Remarkably, a photoperiod-dependent production of β-endorphin$_{31}$ by proteolytic processing of POMC has been described for Siberian hamsters (Helwig et al., 2006). These animals exhibit the physiological drive to reduce food intake in shortened day

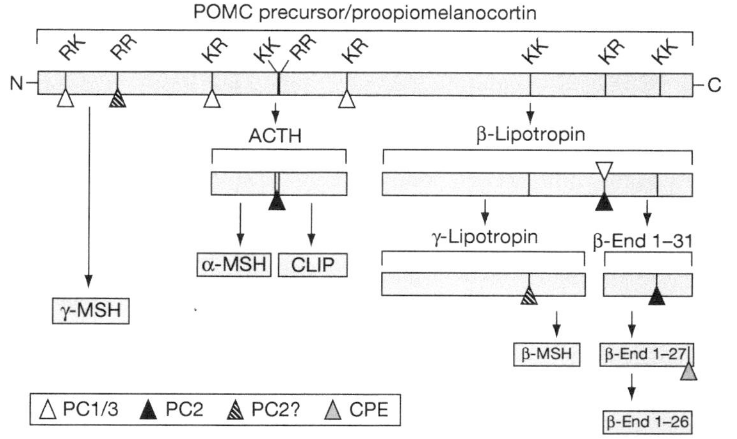

Figure 1 Proteolytic processing of POMC. Reproduced from Helwig M, Khorooshi RM, Tups A, et al. (2006) *Journal of Neuroendocrinology* 18: 413–425, with permission from Fachbereich Biologie.

periods, persistent even if enough food is provided. Resulting photoperiodic variations in body weight – caused by α- and β-MSH signaling – were paralleled by alterations of PC2 gene expression in the arcuate nucleus and dorsal medial posterior part of the arcuate nucleus. When the photoperiod was shortest, PC2 mRNA levels have been shown to be elevated. The increasing proteolytic activity on its part caused an increase of 76% of β-endorphin$_{31}$-immunoreactive neurons in short days compared to long days. So, β-endorphin$_{31}$-peptide production was shown to be elevated, despite the fact that previous studies demonstrated reduced POMC mRNA levels in shortened photoperiod. However, PC1/3 expression was not regulated by photoperiodic inputs.

1.5.2.2 β-Endorphin in motivation, reward, and hedonic value

POMC consists of three exons and encodes several biologically active peptides. To generate mice, specifically lacking β-endorphin, a stop codon was introduced in exon 3 of the murine POMC locus on chromosome 12, preceding the β-endorphin coding region. To exclude adverse effects of genomic targeting on the remaining POMC products, the peptide content of MSH and ACTH was analyzed in homozygous mutant mice. Immunoreactivities of MSH and ACTH were unchanged compared to controls while β-endorphin was totally absent (Rubinstein et al., 1996).

In contrast to the expected impact of β-endorphin on μ-opioid receptor signaling, morphine analgesia, locomotor activity, and withdrawal symptoms were unchanged in β-endorphin-knockout mice. Furthermore, no change in baseline sensitivity to thermal pain was reported in mutants.

However, modifications of rewarding behavior and a reduced hedonic value of food have been described for β-endorphin deficient mice in operant conditioning. The incentive value of rewarding stimuli was measured by quantifying the reinforcing efficacy of food pellets. Thereby, reward requires bar presses under a progressive ratio or, in other words, additional bar presses of a defined number for each subsequent reinforcer. Tests were done under *ad libitum* feeding conditions to provide independence from energy homeostasis.

Although the total number of reinforcer earned did not differ significantly between wild-type and mutant mice, β-endorphin-deficient mice had reduced breakpoints for food reinforcers (Hayward et al., 2002), defined as the highest number of lever presses completed before 15 min elapsed without the mouse receiving a reinforcer. As a prerequisite for this instrumental performance, the loss of β-endorphin was shown not to influence preference for sucrose in a two-bottle free-choice drinking paradigm, suggesting that appetitive behavior, but not consuming behavior is modulated. These data suggest that β-endorphin positively contributes to the incentive motivation in enhancing the hedonic value of food.

A more detailed effect of β-endorphin on motivation and hedonic value was described for sexual behavior in mice. High doses of β-endorphin infused into the medial amygdala inhibited the initial appetitive phase and prevented males from mounting and intromission (Bancroft, 2005). On the other hand, infusion of low peptide doses had facilitative effects, most probably by acting on the ventral tegmental area activating the mesolimbic dopaminergic system.

A likewise dose-related effect in humans was observed in women masturbating to orgasm. Low doses of naloxone enhanced pleasure during orgasm, while higher doses had the opposite effect, reducing sexual arousal as well as orgasmic pleasure. These data suggest that β-endorphin acts dose-dependently on sexual appetite, a prerequisite for pleasure (Bancroft, 2005).

1.5.2.3 β-Endorphin in stress, anxiety, and post-traumatic stress disorder

β-Endorphin abnormalities have been implicated in PTSD. Two studies reported increased immunoreactivity to β-endorphin in PTSD patients, either in response to exercises in plasma or among continued CSF sampling under basal conditions (Liberzon et al., 2007). Both suggest the overactivation of β-endorphin signaling in PTSD, supporting the idea that affected subjects failed to downregulate μ-opioid receptor-mediated signaling systems in brain regions involved (see Sections 84.4.1 and 84.4.2). An earlier report found lower plasma levels in PTSD (Hoffman et al., 1989).

Increased levels of β-endorphin have also been investigated after acute administration of ethanol in various brain regions, including the nucleus accumbens, the pituitary gland, and the hypothalamus. According to this, decreased plasma levels of β-endorphin have been detected in patients undergoing alcohol withdrawal, an effect that has been correlated with self-rated anxiety and depression. Again, this is in accordance with the finding in Oprm-deficient mutants indicating that μ-opiod receptors in nucleus accumbens and other regions are involved in stabilizing alcohol-mediated anxiolysis (see above). So, elevated levels of β-endorphin may contribute to alcohol-mediated anxiolysis via activating μ-opiod receptors, a signaling system that seems to be affected in alcohol withdrawal.

The β-endorphin/μ-opiod-receptor-signaling system also seems to be involved in anxiolysis in acute response to fearful stimuli. Cat odor, placed on the cage of male Wistar rats, induced a robust anxiogenic-like action (Areda et al., 2005). They spent less time in the vicinity of the cloth and displayed markedly reduced grooming behavior, persistent for some time after exposure. Thirty minutes after induction of predator-evoked anxiety, a significant gain of POMC and μ-opiod receptors mRNA was detected in forebrain structures. POMC mRNA was significantly increased in the amygdala, fronto-parietal cortex, and mesolimbic area. μ-Opiod receptor mRNA was significantly elevated in the fronto-parietal cortex, mesolimbic area, and striatum, but not in the amygdala. These observations are consistent with earlier findings that morphine eliminates ultrasonic vocalizations evoked by cat odor and support the theory that the β-endorphin/μ-opiod-receptor-signaling system mediates mitigation in response to anxiogenic stimuli.

1.5.3 κ-Opioid Receptors

κ-Opioid receptors are named for their prototypic ligand ketocyclazocine. There are three variants (κ1, κ2, and κ3) described, mediating modulation of mood in opposition to μ- and δ-opioid receptors. Thus, stimulation of κ-opioid receptors produces intense feelings of discomfort, fear, derealization, depersonalization, visual and auditory disturbances, and uncontrollable unpleasant thoughts. According to this, κ-opioid-receptor antagonists have been used in rats to block receptor activation and induce antidepressant-like effects. On the other hand, receptor activation by agonists also produces analgesia, sparking interest in κ-opioid receptors as attractive targets for the development of analgesic drugs with low abuse potential.

In the CNS, κ-opioid receptors are found in the cerebral cortex, substantia nigra, interpeduncular nucleus, striatum, and hippocampus. In addition, κ-opioid receptor mRNA is also found in the ventral tegmental area and nucleus accumbens of rats.

1.5.3.1 κ-Opioid receptors in reward and aversion

κ-Opioid receptor-deficient mice have been generated by targeting exon 1 of the *opkr1* gene (Simonin et al., 1998). Deletion of exon 1 totally abolished binding of the highly kappa selective agonist [3H] CI-977 in mutant knockout mice, as revealed by autoradiographic mapping. The anatomical distribution of remaining μ- and δ-opioid receptor sites was unchanged, albeit a small upregulation of delta sites has been investigated in mutants (Kieffer and Gaveriaux-Ruff, 2002).

As expected, the aversive properties of the κ-opioid receptor agonist U 50488H, measured by conditioned place aversion, has been shown to be absent in mutant mice. In the conditioned place aversion test, a negative stimulus – for example, negative mood states evoked by withdrawal symptoms or pharmaceutical substances – is paired with exposure of a distinct environment to the animal. In subsequent trials, the time period spent in this compartment is used as an indicator of preference. As a prerequisite of this test, κ-opioid receptor-deficient mice had to show normal locomotor activities.

The conditioned place preference paradigms also revealed absence of tetrahydrocannabinol (THC) reward in μ-opioid receptor deficient mice (see above) in low-dosage-condition paradigms. THC is the main psychoactive compound present in *cannabis sativa* that can be either rewarding or aversive, depending on the mode of administration. In high-dosage-condition paradigms, when THC place aversion is maximal in wild-type controls, opkr1 mutant animals did not develop aversion, indicating κ-opioid receptors mediate the negative aspect of THC activity (Gaveriaux-Ruff and Kieffer, 2002). This suggests that the mode of μ- and κ-opioid-receptor functions forms the basis for the dual euphoric/dysphoric actions of cannabinoids. This is in line with previous reports suggesting an opposing activity of the two receptors in modulating mesolimbic reward pathways.

1.5.3.2 κ-Opioid receptors in anxiety and ethanol-induced anxiolysis

Associations between increased anxiety states, enhanced ethanol consumption, and subsequent anxiolysis have been reported in both, human and animal, studies. Enhanced ethanol consumption has been reported in mice, which repeatedly experienced social defeat using the model of sensory contact. In this model, mice are placed into halves of a cage separated by a perforated transparent partition that permits the animals to see, hear, and perceive the smell of the neighbor, but prevents direct physical contact. Once a day, the partition is removed to allow agonistic interaction. Undoubted superiority of one of the partners is evident within two to three tests, with one partner demonstrating aggressive (winner) and the other one displaying defensive (loser) behavior. In subsequent observations, it is notable that winners approached the partition more frequently and spent more time near the partition than losers. Accessibility to ethanol changes the behavior of mice keeping away from partition. Here, the ethanol-consuming losers approached the partition almost as often as winners. However, activation of κ-opioid receptors by its agonist U-50488H also increased approach behavior near the partition in ethanol naive losers. In winners, stimulation of κ-opioid receptors decreased approach behaviors in the partition test, an effect that was most pronounced in winners that had consumed ethanol. Interestingly, losers that had consumed ethanol reacted to U-50488H similarly as winners did. It seemed as if this drug increases approach behavior in animals when approach behavior is low and decreases approach behavior when approach behavior is high (Kudryavtseva et al., 2006). These opposing effects of κ-opioid receptor activation and ethanol could be explained in part by their action in the mesolimbic–mesocortical dopaminergic system. Activation of the dopaminergic cell bodies in the ventral tegmental area of Sprague-Dawley rats by ethanol causes an increase in the extracellular dopamine (DA) concentration in a terminal region of this pathway, the nucleus accumbens (Lindholm et al., 2007). In reverse, stimulation of the κ-opioid receptors by U50488H in the nucleus accumbens decreases DA transmission, as measured by microdialysis in combination with high-performance liquid chromatography. Furthermore, Nor-BNI, a selective κ-receptor antagonist acting at κ-receptors located primarily on presynaptic DAergic nerve terminals in the nucleus accumbens also significantly increased extracellular DA concentrations in the nucleus accumbens of ethanol pretreated rats whereas no effect was observed in the control group. In addition, the κ-receptor mRNA levels in the ventral tegmental area and the nucleus accumbens are reduced in response to repeated ethanol and/or cocaine exposure. Thus, inducing dopamine release may contribute to the anxiolytic properties of ethanol and κ-opioid receptor blocking; an effect that may be counteracted by activated κ-opioid receptor-mediated DA retention. Therefore, increased DA release may contribute to increased approach behavior in winners and ethanol-consuming losers, whereas κ-opioid-receptor activation may decrease heightened approach behavior. However, unfortunately, these DA alterations cannot explain κ-opioid-receptor-mediated increase of approach behavior in ethanol naive losers.

1.5.4 Dynorphins

Dynorphins are peptides of 9–17 amino acids in length. Their sequences are integrated into the 28 kDa precursor protein Pdyn. Dynorphins act through κ-opioid receptors and are widely distributed in the CNS. They are found in the spinal cord, hypothalamus, in particular the arcuate nucleus, and in OXT- and AVP-positive neurons in the supraoptic nucleus. Moreover, they are detected in the striatum, amygdala, ventral tegmental area, and CA3-region of hippocampus (Yakovleva et al., 2006). In the latter, they are located in small axons and axon terminals in stratum lucidum, mainly in mossy fiber terminals, since they were large and form synapses with multiple dendritic spines, but not in neuronal cell bodies, as revealed by electron microscopy.

Pdyn forms a high-molecular-weight complex consisting of oligomers crosslinked by disulfide bonds at the very N-terminus. There is evidence that Pdyn is not processed during transport between soma and axon-terminals, as Pdyn was shown to coexist and exceed levels of some of its cleavage products in areas containing axon terminals. Moreover processing is not affected by stimulation of neuronal activity.

Maturation of prodynorphin gives rise to dynorphin A (Dyn A, YGGFLRRIRPKLKWDNQ), dynorphin B (Dyn B, YGGFLRRQFKVVTR), and α-neoendorphin (YGGFLRKYP). These peptides are endogenous ligands for κ-opioid receptors and play a role in memory acquisition, motor control, pain processing, and modulation of reward (induced by intake of addictive substances). An alternative cleavage product, big dynorphin (Big Dyn), was described to consist of Dyn A and Dyn B sequences. In contrast to Dyn B and α-neoendorphin, primarily acting through κ-opioid receptors, BigDyn activates NMDA receptors; Dyn A uses both systems (Kuzmin et al., 2006).

1.5.4.1 Prodynorphin in analgesia, reward, and aversion

The prodynorphin gene on chromosome 2 consists of 4 exons and was inactivated by deleting exon 3 and part of exon 4 in mice. This genetic mutation deleted the translation initiation codon of the prodynorphin locus to generate Pdyn-deficient mice. No dynorphin peptide was detected in the brains of homozygous mutants by radioimmunoassay.

As expected from the analgesic properties of κ-opioid-receptor activation, Pdyn-knockout mice showed increased pain response in the hot-plate and tail-flick tests upon behavioral characterization

(Kieffer and Gaveriaux-Ruff, 2002). These data evidenced that κ-opioid receptors activated by dynorphins promote antinociceptive properties.

In addition, locomotor activities were also normal in Pdyn-deficient mice and THC-evoked conditioned place aversion was abolished in high-dosage paradigms (see above). These data suggest that the κ-opioid signaling system mediates the aversive site of hedonic homeostasis in motivational circuits.

Moreover, Pdyn-knockout mice demonstrated diminished age-associated impairment in spatial learning, as tested in the Morris water-maze task (Kuzmin et al., 2006). Indeed, it has been shown that endogenous dynorphins are upregulated with age. Due to the fact that less impairment was observed in aged-mutant compared to control mice, an adverse action of endogenous dynorphins on memory formation was demonstrated. This adverse effect is most likely caused by activated κ-opioid receptors that inhibit Ca^{2+}-dependent glutamate secretion in the hippocampus, downstream to Pdyn-expressing CA3 neurons.

According to this, injections of synthetic dynorphin into the hippocampus also impair memory formation in the Morris water maze and, furthermore, in the aversive spacial learning task of passive avoidance (PA). Based on classical fear conditioning, this maze is separated into a light and a dark compartment. As soon as mice explore the dark compartment, a scrambled electrical current is delivered through the grid floor, thereby coupling spatial cues to aversive experiences. In subsequent sessions, retention latencies for entering the dark compartment are tested. Intracerebroventricular-administered Dyn A or Dyn B peptides caused facilitation of PA retention when injected 5 min before PA training, suggesting impairment in aversive spatial learning. In addition, the selective κ-opioid receptor antagonist Nor-BNI injected prior to the PA experiment did not produce any effect on PA memory retention by itself, but blocked memory facilitation produced by Dyn A or Dyn B. This suggests that impairment in aversive spatial learning produced by Dyn A or Dyn B peptides is mediated via κ-opioid-receptor activation. However, in part, the increase in PA latencies reflects enhanced sensitivity to the foot shock. Both, Dyn A and Dyn B elevated nociceptive latencies after intracerebroventricular administration in the hot-plate test.

In contrast to Dyn A and Dyn B effects, BigDyn enhanced PA retention in the PA task, with Nor-BNI having no impact in modulating this outcome. Furthermore, BigDyn enhanced locomotion, exploratory activity, and had anxiolytic-like effects, as tested in the OF and the EPM test, but did not affect pain sensitivity. The latter suggests that BigDyn is not converted to Dyn A and Dyn B or that the amounts of these products are much lower than those required to produce behavioral effects. Moreover, these observations demonstrate BigDyn action on NMDA, but not on κ-opioid receptors.

1.5.5 δ-Opioid receptors

Understanding of δ-opioid-receptor function is more limited than those of the other opioid receptors. This is mainly due to the fact that pharmacological tools specific for δ-opioid receptors have only become available recently. Nevertheless, as was also shown for the other opioid receptors, activation of the δ-opioid receptors produces analgesia.

The distribution of δ-opioid receptors within the brain is well distinct from κ-opioid receptor, but resembles patterns of μ-opioid receptor sites. Recently, mice expressing enhanced green fluorescent protein (EGFP), fused to δ-opioid-receptor C-terminus, have confirmed distribution patterns revealed by autoradiographic mappings (Scherrer et al., 2006). To produce DOR-EGFP mice, EGFP coding sequence was introduced in frame into the 3′ end of exon 3 and 5′ from the stop codon of the *Oprd1* gene. The DOR-EGFP fusion protein, resulting from targeting, did not show any changes in ligand binding, signaling, or trafficking, but the genomic modification slightly increased *Oprd1* transcription. Accordingly, receptor number and maximal activation increased in DOR-EGFP mice. Green fluorescence from DOR-EGFP fusion protein was detectable until birth in the caudate putamen, appeared in the hippocampus at day 3, and increased throughout the brain reaching maximum intensity at day 15. Expression in adult mutant mice was prominent in the olfactory bulb, caudate putamen, hippocampal GABAergic neurons, and in most striatal cholinergic neurons.

1.5.5.1 δ-Opioid receptors in depression, anxiety, and ethanol-induced anxiolysis

In addition to the insights obtained for the distribution pattern of δ-opioid receptors, DOR-EGFP mice have been proven to be a useful tool for studying *in vivo* receptor trafficking. DOR-EGFP fluorescence flow was monitored in striatal primary neurons exposed to different specific δ-opioid receptor agonists. Both, [met]-enkephalin (see below) and SNC80 triggered the clustering of receptors into bright spots along the plasma membrane of cell bodies and processes formerly exhibiting intense punctuate fluorescence. DOR-EGFP clusters then progressively internalized, producing a typical vesicular punctate pattern. Finally, after 20 min, the fluorescent spots converged into bigger vesicles of unknown nature. These data indicate that DOR-EGFP respond to agonists by rapid endocytosis of receptor clusters in a dose-dependent manner. Given that surface receptor number is reduced via endocytosis, weakening the dose response of subsequent activation, the observed internalization could reflect an effect described as desensitization.

Moreover, the agonist SNC80 was shown to increase locomotor activity in DOR-EGFP mice, an effect that does not manifest in δ-opioid-receptor-knockout mice. Although these data indicate that activation of δ-opioid receptor signaling elevates locomotion, one should keep in mind that – under basal conditions – mice deficient in δ-opioid receptors showed significant hyperlocomotion (Filliol et al., 2000). δ-opioid-receptor-knockout mice have been generated by targeting of exon 1, including the translation-initiation codon (Filliol et al., 2000), or exon 2 (Zhu et al., 1999) of the *opdr1* gene. Deletion of respective exons produced δ-opioid-receptor-knockout alleles. As revealed by autoradiographic mapping, binding of selective agonists was abolished in mice homozygous for either mutation, and the anatomical distribution of the remaining opioid receptor sites was unchanged, although a tendency to regional decrease of μ- and κ-opioid-receptor sites in mutant mice could be detected.

Most strikingly, during behavioral characterization of δ-opioid-receptor-knockout mice, increased levels of anxiety in the EPM and the dark-light box have been observed. Furthermore, mutant mice displayed a strong increase in time spent immobile in the forced swim paradigm, indicating a depressive-like phenotype (Filliol et al., 2000). These data suggest that the activity of *Oprd1*-encoded receptors may contribute to diminishing levels of anxiety and depression. This assumption is further supported by pharmacological interventions with either [met]-enkephalin or SNC80 in wild-type mice, exerting antidepressant like effects upon δ-opioid-receptor activation. Thus, taking into consideration that μ-opioid-receptor-knockout mice displayed decreased levels of anxiety (see above), signaling through δ- and μ-opioid-receptor systems may have opposing effects in modulating mood states associated with depression and anxiety.

Interestingly, the elevated anxiety-like behavior in δ-opioid-receptor-knockout mice reversed to wild-type levels after ethanol self-administration. Initially, when tested in a two-bottle choice test, mutants consumed alcohol to the same extent as wild-type mice, indicating no additional rewarding or beneficial effects of alcohol consumption upon mood states. However, when submitted to an operant paradigm, these mice self-administered more alcohol than wild-types, even maintaining a strong preference for alcohol after the procedure (Gaveriaux-Ruff and Kieffer, 2002). δ-opioid-receptor activation therefore may have an impact on altered emotional states in the abuse of alcohol.

1.5.6 Enkephalins

Preproenkephalin (Penk), a 267-amino-acid protein precursor, encoded by 4 exons on the mouse chromosome 4 contains several copies of enkephalin pentapeptides; one copy of [leu]-enkephalin, and four copies of [met]-enkephalin. [leu]-enkephalin (YGGFL) acts through δ-opioid receptors, whereas [met]-enkephalin (YGGFM) is able to act on both, μ- and δ-opioid receptors. Enkephalins are widely expressed in the CNS, with highest concentrations in the basal ganglia, brainstem, thalamus, substantia gelatinosa, and amygdala. In the central nucleus of the amygdala, [met]-enkephalin was shown to co-localize with glutamate decarboxylase (GAD), the GABA-synthesizing enzyme specific for interneurons.

1.5.6.1 Enkephalins in nociception and anxiety

Two strains of knockout mice have been described, generated by targeting the 5′-part of Penk locus exon number 3. In one mutant (König et al., 1996) a truncated exon followed by an unexpected partial duplication of exon 3 resulted from specific deletion. Both gene knockouts were proven to become operant, since no enkephalin peptide could be detected any more in homozygous mutants. In addition, a strong increase in μ- and δ-opioid receptor sites has been observed in those animals. Up to threefold increment was apparent in regions important for the processing of emotional aspects of behavior, including the central nucleus of the amygdala for μ-opioid receptor or the ventral pallidum for δ-opioid receptor sites. Since enkephalin signaling was shown to involve both opioid receptor types, this may reflect an adaptive change to lack of receptor activation.

Upon behavioral analysis, enhanced nociception has been observed in Penk mutant mice. They showed increased pain response in the hot-plate and tail-flick test, respectively (König et al., 1996). In contrast, decreased pain perception was shown in response to formalin injection into ascites, suggesting that enkephalins inhibit thermal and mechanical nociceptive stimuli via δ- and μ-opioid-receptor activation, but not irritant chemical nociception as shown for μ-opioid receptor signaling (see above).

Next, the two different strains of Penk mutant mice were shown to be more responsive to fear conditioning and displayed increased responsiveness to anxiety-evoking environments, suggesting that lack of enkephalins is beneficial to the formation of anxiety. Consistently, virus-mediated overexpression of enkephalin in the amygdala potentiated the anxiolytic effects of diazepam in rats, as determined by the EPM test (Primeaux et al., 2006). Diazepam is thought to mediate anxiolytic actions by interneurons expressing the α_2-subunit of $GABA_A$ receptors, largely localized in limbic structures, including the amygdala. Penk mRNA on its part is also shown to be expressed in various nuclei of the amygdala including the central, basolateral, and posteroventral portion of the medial nucleus. Furthermore, most of the enkephalin-containing afferents to the centromedial amygdala arise from intraamygdaloid sources or the bed nucleus of the stria terminalis (Poulin et al., 2006). This increase of the anxiolytic properties of diazepam, mediated by overexpression of enkephalin, was shown to be attenuated by naltrindole, a specific δ-opioid-receptor antagonist. These data suggest that the anxiolytic effects of diazepam involve the activation of δ-opioid-receptor signaling in the amygdala caused by enkephalins. Naltrindole administration did not alter baseline anxiety in control or enkephalin-overexpressing rats, demonstrating that this effect is depending on diazepam (Poulin et al., 2006). The latter is surprising, since increased levels of anxiety have been observed in diminished δ-opioid receptor signaling independent of diazepin targets in rodents, as demonstrated for *oprd1*-knockout mice (see above).

1.5.6.2 Enkephalins in stress-induced anhedonia and depression

The amount of enkephalin peptides has also been shown to be reduced in the nucleus accumbens of mice in the chronic mild-stress model of depression (Bilkei-Gorzo et al., 2007). In this model, mice are subjected to chronic stress procedures, such as rat exposure, restraint stress, and tail suspension for a 4-week period. This treatment causes a strong decrease in sucrose preference, a putative indicator of anhedonia in rodents. Anhedonia on its part, but not chronic stress *per se*, is associated with key analogs of depressive symptoms, such as increased floating during forced swimming and decreased exploration of novelty (Strekalova et al., 2004). So, anhedonia and depression-like behaviors may be related to lower levels of enkephalin release or responsiveness of enkephalin neurotransmission in the nucleus accumbens. In addition, it has been demonstrated that enkephalin levels are increased by the administration of imipramine. This most classical antidepressant is thought to exert its effect through inhibition of enkephalin breakdown. So, blocking enkephalin-degrading enzymes, like enkephalinase and aminopeptidase N, also reduces depression-related behaviors.

The assumption that lower levels of enkephalin may contribute to anhedonia and depression-like behaviors is further supported by pharmacological interventions with [met]-enkephalin or SNC80 (see above). Both exert antidepressant-like effects upon δ-opioid-receptor activation. Taking into consideration that *oprd1*-knockout mice also showed an increased frequency of depression-like behavior, these data suggest that the activation of δ-opioid receptors by enkephalin may contribute to diminishing levels of depression.

However, as already shown for β-endorphin (see above), modification of rewarding behavior and a reduced hedonic value of food have also been described for Penk deficient mice in operant conditioning. Penk-knockout animals also caused reduced breakpoints for food reinforcers, suggesting that appetitive behavior is modulated in these animals and that enkephalins also positively contribute to the incentive-motivation in enhancing the hedonic value of food (Hayward et al., 2002). So, lower levels of enkephalin could also contribute to anhedonia by reducing hedonic behavior.

1.6 Conclusion

There is compelling evidence that variations in genes and their products of neuropeptidergic systems contribute to variations in stress coping, social behavior, emotionality, and psychopathology. However, it should also be noted that given the estimation that several dozen genes may be critically involved in, for example, anxiety regulation and the etiology of anxiety disorders, single genes exert minor effects, which have to be confirmed by multiple approaches. Given the heterogeneity of emotional behaviors and psychiatric diseases, the challenge of mapping these traits in humans with modest-effect genes and loci will continue. While we are only at the very earliest stages of understanding the genetics of behavior, this understanding will nevertheless provide valuable clues to the regulation of human emotionality, thereby providing opportunities for preclinical investigation and therapeutic intervention.

Questions that are fundamental to a broad conceptualization of genetic underpinnings of behavior include coordinated actions of different systems (cross talk at multiple levels?), concomitant alterations in ligand and its receptor(s), mechanisms of how genetic polymorphisms translate into behavioral alterations, gene–gene interactions, and epigenetic influences. Despite enormous efforts to address and answer these questions, there has been little progress in the identification of genetic risk factors and biomarkers of psychopathology, and there is still considerable scepticism on the validity of animal models of human behavior. It should be stressed at this point that many data obtained in mice and rats are interesting on their own, even if they were valid only in rodents. Most of current research, however, is *a priori* translational in nature, taking advantage of the high conservation of neurobiological pathways and mechanisms and, thus, the homology of their genetic underpinnings between rodents and humans. While conceptionally difficult to model, many physiological and pathological behaviors, especially their endophenotypes, exist across many species at a more basic behavioral level, such as within domains related to anxiety, arousal, cognition, and social interaction. Importantly, no single-gene knockout, overexpression, or selection line is likely to represent the full genotypic and phenotypic complexity of the psychopathology of interest. Multiple animal models and complementary approaches therefore offer unique opportunities to study the effect of genetic polymorphisms, susceptibility genes, environmental factors, and their interactions on complex behaviors, including psychopathology.

References

Aguilera G, Kiss A, Liu Y, and Kamitakahara A (2007) Negative regulation of corticotropin releasing factor expression and limitation of stress response. *Stress* 10: 153–161.

Amico JA, Mantella RC, and Vollmer RR (2003) Consumption of solutions containing sodium chloride is enhanced in female oxytocin-deficient mice. *Behavioral Neuroscience* 117: 32–37.

Areda T, Koks S, Philips MA, Vasar E, Karis A, and Asser T (2005) Alterations in opioid system of the rat brain after cat odor exposure. *Neuroscience Letters* 377: 136–139.

Axelrod J and Reisine TD (1984) Stress hormones: Their interaction and regulation. *Science* 224: 452–459.

Bachner-Melman R, Dina C, Zohar AH, et al. (2005) AVPRl a and SLC6A4 gene polymorphisms are associated with creative dance performance. *PLoS Genetics* 1: e42.

Bale TL (2005) Sensitivity to stress: Dysregulation of CRF pathways and disease development. *Hormones and Behavior* 48: 1–10.

Bale TL, Contarino A, Smith GW, et al. (2000) Mice deficient for corticotropin-releasing hormone receptor-2 display anxiety-like behaviour and are hypersensitive to stress. *Nature Genetics* 24: 410–414.

Bale TL, Picetti R, Contarino A, Koob GF, Vale WW, and Lee KF (2002) Mice deficient for both corticotropin-releasing factor receptor I (CRFR1) and CRFR2 have an impaired stress response and display sexually dichotomous anxiety-like behavior. *Journal of Neuroscience* 22: 193–199.

Bale TL and Vale WW (2004) CRF and CRF receptors: Rle in stress responsivity and other behaviors. *Annual Review of Pharmacology and Toxicology* 44: 525–557.

Bancroft J (2005) The endocrinology of sexual arousal. *Journal of Endocrinology* 186: 411–427.

Barden N (1996) Modulation of glucocorticoid receptor gene expression by antidepressant drugs. *Pharmacopsychiatry* 29: 12–22.

Barden N, Stec IS, Montkowski A, Holsboer F, and Reul JM (1997) Endocrine profile and neuroendocrine challenge tests in transgenic mice expressing antisense RNA against the glucocorticoid receptor. *Neuroendocrinology* 66: 212–220.

Beaulieu S, Rousse I, Gratton A, Barden N, and Rochford J (1994) Behavioral and endocrine impact of impaired type II glucocorticoid receptor function in a transgenic mouse model. *Annals of the New York Academy of Sciences* 746: 388–391.

Belanoff JK, Flores BH, Kalezhan M, Sund B, and Schatzberg AF (2001) Rapid reversal of psychotic depression using mifepristone. *Journal of Clinical Psychopharmacology* 21: 516–521.

Bielsky IF, Hu SB, Ren X, Terwilliger EF, and Young LJ (2005) The Vla vasopressin receptor is necessary and sufficient for normal social recognition: A gene replacement study. *Neuron* 47: 503–513.

Bielsky IF, Hu SB, Szegda KL, Westphal H, and Young LJ (2004) Profound impairment in social recognition and reduction in anxiety-like behavior in vasopressin Vl a receptor knockout mice. *Neuropsychopharmacology* 29: 483–493.

Bilkei-Gorzo A, Michel K, Noble F, Roques BP, and Zimmer A (2007) Preproenkephalin knockout mice show no depression-related phenotype. *Neuropsychopharmacology* 32: 2330–2337.

Bilkei-Gorzo A, Racz I, Michel K, and Zimmer A (2002) Diminished anxiety- and depression-related behaviors in mice with selective deletion of the Tac1 gene. *Journal of Neuroscience* 22: 10046–10052.

Bilkei-Gorzo A and Zimmer AM (2005) Mutagenesis and knockout models: NK 1 and substance P. *Handbook of Experimental Pharmacology* 169: 143–162.

Bittel DC, Kibiryeva N, McNulty SG, Driscoll DJ, Butler MG, and White RA (2007) Whole genome microarray analysis of gene expression in an imprinting center deletion mouse model of Prader-Willi syndrome. *American Journal of Medical Genetics A* 143: 422–429.

Boyle MP, Brewer JA, Funatsu M, Wozniak DF, Tsien JZ, Izumi Y, and Muglia LJ (2005) Acquired deficit of forebrain glucocorticoid receptor produces depression-like changes in adrenal axis regulation and behavior. *Proceedings of the National Academy of Sciences of United States of America* 102: 473–478.

Boyle MP, Kolber BJ, Vogt SK, Wozniak DF, and Muglia LJ (2006) Forebrain glucocorticoid receptors modulate anxiety-associated locomotor activation and adrenal responsiveness. *Journal of Neuroscience* 26: 1971–1978.

Brain SD and Cox HM (2006) Neuropeptides and their receptors: Innovative science providing novel therapeutic targets. *British Journal of Pharmacology* 147 (supplement 1): S202–S211.

Bruins J, Kovacs GL, Abbes AP, et al. (2006) Minor disturbances in central nervous system function in familial neurohypophysial diabetes insipidus. *Psychoneuroendocrinology* 31: 80–91.

Caldwell HK, Stewart J, Wiedholz LM, et al. (2006) The acute intoxicating effects of ethanol are not dependent on the vasopressin 1a or 1b receptors. *Neuropeptides* 40: 325–337.

Carter CS (2003) Developmental consequences of oxytocin. *Physiology and Behavior* 79: 383–397.

Carter CS (2007) Sex differences in oxytocin and vasopressin: Implications for autism spectrum disorders? *Behavioural Brain Research* 176: 170–186.

Cascieri MA, Huang RR, Fong TM, Cheung AH, Sadowski S, Ber E, and Strader CD (1992) Determination of the amino acid residues in substance P conferring selectivity and specificity for the rat neurokinin receptors. *Molecular Pharmacology* 41: 1096–1099.

Choleris E, Gustafsson JA, Korach KS, Muglia LJ, Pfaff DW, and Ogawa S (2003) An estrogen-dependent four-gene micronet regulating social recognition: A study with oxytocin and estrogen receptor-alpha and -beta knockout mice. *Proceedings of the National Academy of Sciences of United States of America* 100: 6192–6197.

Choleris E, Little SR, Mong JA, Puram SV, Langer R, and Pfaff DW (2007) Microparticle-based delivery of oxytocin receptor antisense DNA in the medial amygdala blocks social recognition in female mice. *Proceedings of the National Academy of Sciences of United States of America* 104: 4670–4675.

Cole TJ, Blendy JA, Monaghan AP, et al. (1995) Targeted disruption of the glucocorticoid receptor gene blocks adrenergic chromaffin cell development and severely retards lung maturation. *Genes and Development* 9: 1608–1621.

Cole TJ, Myles K, Purton JF, et al. (2001) GRKO mice express an aberrant dexamethasone-binding glucocorticoid receptor, but are profoundly glucocorticoid resistant. *Molecular and Cellular Endocrinology* 173: 193–202.

Contarino A, Dellu F, Koob GF, et al. (1999) Reduced anxiety-like and cognitive performance in mice lacking the corticotropin-releasing factor receptor 1. *Brain Research* 835: 1–9.

Coste SC, Kesterson RA, Heldwein KA, et al. (2000) Abnormal adaptations to stress and impaired cardiovascular function in mice lacking corticotropin-releasing hormone receptor-2. *Nature Genetics* 24: 403–409.

Crawley JN, Chen T, Puri A, et al. (2007) Social approach behaviors in oxytocin knockout mice: Comparison of two independent lines tested in different laboratory environments. *Neuropeptides* 41: 145–163.

Cryan JF and Holmes A (2005) The ascent of mouse: advances in modelling human depression and anxiety. *Nature Reviews Drug Discovery* 4: 775–790.

Czeh B, Fuchs E, and Simon M (2006) NK1 receptor antagonists under investigation for the treatment of affective disorders. *Expert Opinion on Investigational Drugs* 15: 479–486.

Dallman MF (2005) Fast glucocorticoid actions on brain: back to the future. *Frontiers in Neuroendocrinology* 26: 103–108.

Darland T, Heinricher MM, and Grandy DK (1998) Orphanin FQ/nociceptin: A role in pain and analgesia, but so much more. *Trends in Neurosciences* 21: 215–221.

DeBattista C, Belanoff J, Glass S, et al. (2006) Mifepristone versus placebo in the treatment of psychosis in patients with psychotic major depression. *Biological Psychiatry* 60: 1343–1349.

de Kloet ER, Joels M, and Holsboer F (2005) Stress and the brain: From adaptation to disease. *Nature Reviews Neuroscience* 6: 463–475.

de Kloet ER, Vreugdenhil E, Oitzl MS, and Joels M (1998) Brain corticosteroid receptor balance in health and disease. *Endocrine Reviews* 19: 269–301.

Dempster EL, Burcescu I, Wigg K, et al. (2007) Evidence of an association between the vasopressin V1b receptor gene (AVPRIB) and childhood-onset mood disorders. *Archives of General Psychiatry* 64: 1189–1195.

de Vries GJ (2008) Sex differences in vasopressin and oxytocin innervation of the brain. *Progress in Brain Research* 170: 17–27.

de Wied D and Sigling HO (2002) Neuropeptides involved in the pathophysiology of schizophrenia and major depression. *Neurotoxicity Research* 4: 453–468.

Dijkstra I, Tilders FJ, Aguilera G, et al. (1998) Reduced activity of hypothalamic corticotropin-releasing hormone neurons in transgenic mice with impaired glucocorticoid receptor function. *Journal of Neuroscience* 18: 3909–3918.

Dirks A, Groenink L, Bouwknecht JA, et al. (2002) Overexpression of corticotropin-releasing hormone in transgenic mice and chronic stress-like autonomic and physiological alterations. *European Journal of Neuroscience* 16: 1751–1760.

Domes G, Heinrichs M, Glascher J, Buchel C, Braus DF, and Herpertz SC (2007) Oxytocin attenuates amygdala responses to emotional faces regardless of valence. *Biological Psychiatry* 62: 1187–1190.

Duarte CR, Schutz B, and Zimmer A (2006) Incongruent pattern of neurokinin B expression in rat and mouse brains. *Cell and Tissue Research* 323: 43–51.

Ebner K and Singewald N (2006) The role of substance P in stress and anxiety responses. *Amino Acids* 31: 251–272.

Ebner K, Wotjak CT, Landgraf R, and Engelmann M (2000) A single social defeat experience selectively stimulates the release of oxytocin, but not vasopressin, within the septal brain area of male rats. *Brain Research* 872: 87–92.

Egashira N, Tanoue A, Higashihara F, et al. (2004) V1a receptor knockout mice exhibit impairment of spatial memory in an eight-arm radial maze. *Neuroscience Letters* 356: 195–198.

Egashira N, Tanoue A, Matsuda T, et al. (2007) Impaired social interaction and reduced anxiety-related behavior in vasopressin V1a receptor knockout mice. *Behavioural Brain Research* 178: 123–127.

El Yacoubi M, Bouali S, Popa D, et al. (2003) Behavioral, neurochemical, and electrophysiological characterization of a genetic mouse model of depression. *Proceedings of the National Academy of Sciences of United States of America* 100: 6227–6232.

Engelmann M, Landgraf R, and Wotjak CT (2004) The hypothalamic–neurohypophysial system regulates the hypothalamic–pituitary–adrenal axis under stress: An old concept revisited. *Frontiers in Neuroendocrinology* 25: 132–149.

Evans MR, Roberts ML, Buchanan KL, and Goldsmith AR (2006) Heritability of corticosterone response and changes in life history traits during selection in the zebra finch. *Journal of Evolutionary Biology* 19: 343–352.

Fava GA, Sonino N, and Morphy MA (1987) Major depression associated with endocrine disease. *Psychiatric Developments* 5: 321–348.

Ferguson JN, Aldag JM, Insel TR, and Young LJ (2001) Oxytocin in the medial amygdala is essential for social recognition in the mouse. *Journal of Neuroscience* 21: 8278–8285.

Ferguson JN, Young LJ, Hearn EF, Matzuk MM, Insel TR, and Winslow JT (2000) Social amnesia in mice lacking the oxytocin gene. *Nature Genetics* 25: 284–288.

Filliol D, Ghozland S, Chluba J, et al. (2000) Mice deficient for delta- and mu-opioid receptors exhibit opposing alterations of emotional responses. *Nature Genetics* 25: 195–200.

Fink S, Excoffier L, and Heckel G (2006) Mammalian monogamy is not controlled by a single gene. *Proceedings of the National Academy of Sciences of United States of America* 103: 10956–10960.

Flores BH, Kenna H, Keller J, Solvason HB, and Schatzberg AF (2006) Clinical and biological effects of mifepristone treatment for psychotic depression. *Neuropsychopharmacology* 31: 628–636.

Gammie SC, Garland T, Jr., and Stevenson SA (2006) Artificial selection for increased maternal defense behavior in mice. *Behavioral Genetics* 36: 713–722.

Gass P, Reichardt HM, Strekalova T, Henn F, and Tronche F (2001) Mice with targeted mutations of glucocorticoid and mineralocorticoid receptors: models for depression and anxiety? *Physiology and Behavior* 73: 811–825.

Gaveriaux-Ruff C and Kieffer BL (2002) Opioid receptor genes inactivated in mice: The highlights. *Neuropeptides* 36: 62–71.

Ghozland S, Chu K, Kieffer BL, and Roberts AJ (2005) Lack of stimulant and anxiolytic-like effects of ethanol and accelerated development of ethanol dependence in mu-opioid receptor knockout mice. *Neuropharmacology* 49: 493–501.

Goekoop JG, de Winter RP, de Rijk R, Zwinderman KH, Frankhuijzen-Sierevogel A, and Wiegant VM (2006) Depression with above-normal plasma vasopressin: Validation by relations with family history of depression and mixed anxiety and retardation. *Psychiatry Research* 141: 201–211.

Gold PW and Chrousos GP (2002) Organization of the stress system and its dysregulation in melancholic and atypical depression: High vs low CRH/NE states. *Molecular Psychiatry* 7: 254–275.

Gold PW, Goodwin FK, and Reus VI (1978) Vasopressin in affective illness. *Lancet* 1: 1233–1236.

Granota RY, Frankel Y, Gritsenko V, et al. (2007) Provisional evidence that the arginine vasopressin la receptor gene is associated with musical memory. *Evolution and Human Behavior* 28: 313–318.

Green L, Fein D, Modahl C, Feinstein C, Waterhouse L, and Morris M (2001) Oxytocin and autistic disorder: Alterations in peptide forms. *Biological Psychiatry* 50: 609–613.

Griebel G, Stemmelin J, Gal CS, and Soubrie P (2005) Non-peptide vasopressin V1b receptor antagonists as potential drugs for the treatment of stress-related disorders. *Current Pharmaceutical Design* 11: 1549–1559.

Groenink L, Dirks A, Verdouw PM, Schipholt M, Veening JG, van der Gugten J, and Olivier B (2002) HPA axis dysregulation in mice overexpressing corticotropin releasing hormone. *Biological Psychiatry* 51: 875–881.

Groenink L, Pattij T, de Jongh R, van der Gugten J, Oosting RS, Dirks A, and Olivier B (2003) 5-HT1 A receptor knockout mice and mice overexpressing corticotropin-releasing hormone in models of anxiety. *European Journal of Pharmacology* 463: 185–197.

Hall FS, Sora I, and Uhl GR (2001) Ethanol consumption and reward are decreased in mu-opiate receptor knockout mice. *Psychopharmacology (Berl)* 154: 43–49.

Hallberg M and Nyberg F (2003) Neuropeptide conversion to bioactive fragments – an important pathway in neuromodulation. *Current Protein and Peptide Science* 4: 31–44.

Hammock EA and Young LJ (2004) Functional microsatellite polymorphism associated with divergent social structure in vole species. *Molecular Biology and Evolution* 21: 1057–1063.

Hammock EA and Young LJ (2005) Microsatellite instability generates diversity in brain and sociobehavioral traits. *Science* 308: 1630–1634.

Hasler G, Drevets WC, Manji HK, and Charney DS (2004) Discovering endophenotypes for major depression. *Neuropsychopharmacology* 29: 1765–1781.

Hayward MD, Pintar JE, and Low MJ (2002) Selective reward deficit in mice lacking beta-endorphin and enkephalin. *Journal of Neuroscience* 22: 8251–8258.

Heckel G and Fink S (2008) Evaluation of the arginine vasopressin 1a receptor and implications for mammalian social behavior. *Progress in Brain Research* 170: 321–330.

Heinrichs SC, Min H, Tamraz S, Carmouche M, Boehme SA, and Vale WW (1997) Anti-sexual and anxiogenic behavioral consequences of corticotropin-releasing factor overexpression are centrally mediated. *Psychoneuroendocrinology* 22: 215–224.

Heinrichs SC, Stenzel-Poore MP, Gold LH, et al. (1996) Learning impairment in transgenic mice with central overexpression of corticotropin-releasing factor. *Neuroscience* 74: 303–311.

Helwig M, Khorooshi RM, Tups A, et al. (2006) PC1/3 and PC2 gene expression and post-translational endoproteolytic pro-opiomelanocortin processing is regulated by photoperiod in the seasonal Siberian hamster (Phodopus sungorus). *Journal of Neuroendocrinology* 18: 413–425.

Henn FA and Vollmayr B (2005) Stress models of depression: forming genetically vulnerable strains. *Neuroscience and Biobehavioral Reviews* 29: 799–804.

Herman JP and Cullinan WE (1997) Neurocircuitry of stress: central control of the hypothalamo-pituitary–adrenocortical axis. *Trends in Neurosciences* 20: 78–84.

Hoffman L, Burges WP, Wilson G, and Montgomery J (1989) Low plasma beta-endorphin in post-traumatic stress disorder. *Australian and New Zealand Journal of Psychiatry* 23: 269–273.

Hollander E, Bartz J, Chaplin W, et al. (2007) Oxytocin increases retention of social cognition in autism. *Biological Psychiatry* 61: 498–503.

Holsboer F (2000) The corticosteroid receptor hypothesis of depression. *Neuropsychopharmacology* 23: 477–501.

Holst D von (1998) The concept of stress and its relevance for animal behavior. *Advances in the Study of Behavior* 27: 1–131.

Howell MP and Muglia LJ (2006) Effects of genetically altered brain glucocorticoid receptor action on behavior and adrenal axis regulation in mice. *Frontiers in Neuroendocrinology* 27: 275–284.

Ising M, Kunzel HE, Binder EB, Nickel T, Modell S, and Holsboer F (2005) The combined dexamethasone/CRH test as a potential surrogate marker in depression. *Progress in Neuropsychopharmacology and Biological Psychiatry* 29: 1085–1093.

Israel S, Lerer E, Shalev I, et al. (2008) Molecular genetic studies of the arginine vasopressin 1a receptor (AVPRla) and the oxytocin receptor (OXTR) in human behaviour: From autism to altruism with some notes in between. *Progress in Brain Research* 170: 435–449.

Jacob S, Brune CW, Carter CS, Leventhal BL, Lord C, and Cook EH, Jr. (2007) Association of the oxytocin receptor gene (OXTR) in Caucasian children and adolescents with autism. *Neuroscience Letters* 417: 6–9.

Jacobson L, Muglia LJ, Weninger SC, Pacak K, and Majzoub JA (2000) CRH deficiency impairs but does not block pituitary–adrenal responses to diverse stressors. *Neuroendocrinology* 71: 79–87.

Karanth S, Linthorst AC, Stalla GK, Barden N, Holsboer F, and Reul JM (1997) Hypothalamic–pituitary–adrenocortical axis changes in a transgenic mouse with impaired glucocorticoid receptor function. *Endocrinology* 138: 3476–3485.

Kas MJ, Fernandes C, Schalkwyk LC, and Collier DA (2007) Genetics of behavioural domains across the neuropsychiatric spectrum; of mice and men. *Molecular Psychiatry* 12: 324–330.

Kavaliers M, Colwell DD, Choleris E, Agmo A, Muglia LJ, Ogawa S, and Pfaff DW (2003) Impaired discrimination of and aversion to parasitized male odors by female oxytocin knockout mice. *Genes, Brain and Behavior* 2: 220–230.

Keck ME, Welt T, Muller MB, et al. (2003) Reduction of hypothalamic vasopressinergic hyperdrive contributes to clinically relevant behavioral and neuroendocrine effects of chronic paroxetine treatment in a psychopathological rat model. *Neuropsychopharmacology* 28: 235–243.

Keller-Wood ME and Dallman MF (1984) Corticosteroid inhibition of ACTH secretion. *Endocrine Reviews* 5: 1–24.

Kessler MS, Murgatroyd C, Bunck M, et al. (2007) Diabetes insipidus and, partially, low anxiety-related behaviour are linked to a SNP-associated vasopressin deficit in LAB mice. *European Journal of Neuroscience* 26: 2857–2864.

Kieffer BL and Gaveriaux-Ruff C (2002) Exploring the opioid system by gene knockout. *Progress in Neurobiology* 66: 285–306.

Kim JJ and Diamond DM (2002) The stressed hippocampus, synaptic plasticity and lost memories. *Nature Reviews Neuroscience* 3: 453–462.

Kim SJ, Young LJ, Gonen D, et al. (2002) Transmission disequilibrium testing of arginine vasopressin receptor 1A (AVPR1A) polymorphisms in autism. *Molecular Psychiatry* 7: 503–507.

Kirsch P, Esslinger C, Chen Q, et al. (2005) Oxytocin modulates neural circuitry for social cognition and fear in humans. *Journal of Neuroscience* 25: 11489–11493.

Kishimoto T, Radulovic J, Radulovic M, et al. (2000) Deletion of crhr2 reveals an anxiolytic role for corticotropin-releasing hormone receptor-2. *Nature Genetics* 24: 415–419.

Kitchen I, Slowe S, Matthes HWD, Simonin F, Befort K, and Kieffer B (1997) Quantitative autoradiographic mapping of mu-, delta- and kappa-opioid receptors in the brain of mu-receptor knockout mice. *British Journal of Pharmacology* 120: 124.

Knafo A, Israel S, Darvasi A, et al. (2008) Individual differences in allocation of funds in the dictator game associated with length of the arginine vasopressin 1a receptor RS3 promoter region and correlation between RS3 length and hippocampal mRNA. *Genes, Brain and Behavior* 7: 266–275.

Konig M, Zimmer AM, Steiner H, et al. (1996) Pain responses, anxiety and aggression in mice deficient in pre-proenkephalin. *Nature* 383: 535–538.

Korte SM (2001) Corticosteroids in relation to fear, anxiety and psychopathology. *Neuroscience and Biobehavioral Reviews* 25: 117–142.

Korte SM, Koolhaas JM, Wingfield JC, and McEwen BS (2005) The Darwinian concept of stress: Benefits of allostasis and costs of allostatic load and the trade-offs in health and disease. *Neuroscience and Biobehavioral Reviews* 29: 3–38.

Kosfeld M, Heinrichs M, Zak PJ, Fischbacher U, and Fehr E (2005) Oxytocin increases trust in humans. *Nature* 435: 673–676.

Kosson P, Bonney I, Carr DB, and Lipkowski AW (2005) Endomorphins interact with tachykinin receptors. *Peptides* 26: 1667–1669.

Kromer SA, Kessler MS, Milfay D, et al. (2005) Identification of glyoxalase-I as a protein marker in a mouse model of extremes in trait anxiety. *Journal of Neuroscience* 25: 4375–4384.

Kudryavtseva N, Gerrits MA, Avgustinovich DF, Tenditnik MV, and van Ree JM (2006) Anxiety and ethanol consumption in victorious and defeated mice; effect of kappa-opioid receptor activation. *European Neuropsychopharmacology* 16: 504–511.

Kuzmin A, Madjid N, Terenius L, Ogren SO, and Bakalkin G (2006) Big dynorphin, a prodynorphin-derived peptide produces NMDA receptor-mediated effects on memory, anxiolytic-like and locomotor behavior in mice. *Neuropsychopharmacology* 31: 1928–1937.

LaBuda CJ and Fuchs PN (2001) The anxiolytic effect of acute ethanol or diazepam exposure is unaltered in mu-opioid receptor knockout mice. *Brain Research Bulletin* 55: 755–760.

Lagerspetz KY, Tirri R, and Lagerspetz KM (1968) Neurochemical and endocrinological studies of mice selectively bred for aggressiveness. *Scandinavian Journal of Psychology* 9: 157–160.

Landgraf R (2006) The involvement of the vasopressin system in stress-related disorders. *CNS and Neurological Disorders - Drug Targets* 5: 167–179.

Landgraf R, Frank E, Aldag JM, et al. (2003) Viral vector-mediated gene transfer of the vole Vla vasopressin receptor in the rat septum: improved social discrimination and active social behaviour. *European Journal of Neuroscience* 18: 403–411.

Landgraf R, Gerstberger R, Montkowski A, Probst JC, Wotjak CT, Holsboer F, and Engelmann M (1995) Vl vasopressin receptor antisense oligodeoxynucleotide into septum reduces vasopressin binding, social discrimination abilities, and anxiety-related behavior in rats. *Journal of Neuroscience* 15: 4250–4258.

Landgraf R, Kessler MS, Bunck M, et al. (2007) Candidate genes of anxiety-related behavior in HAB/LAB rats and mice: Focus on vasopressin and glyoxalase-I. *Neuroscience and Biobehavioral Reviews* 31: 89–102.

Landgraf R and Neumann ID (2004) Vasopressin and oxytocin release within the brain: A dynamic concept of multiple and variable modes of neuropeptide communication. *Frontiers in Neuroendocrinology* 25: 150–176.

Lee HJ, Caldwell HK, Macbeth AH, and Young WS, III (2008) Behavioural studies using temporal and spatial inactivation of the oxytocin receptor. *Progress in Brain Research* 170: 73–77.

Le Merrer J, Cagniard B, and Cazala P (2006) Modulation of anxiety by mu-opioid receptors of the lateral septal region in mice. *Pharmacology Biochemistry and Behavior* 83: 465–479.

Liberzon I, Taylor SF, Phan KL, et al. (2007) Altered central mu-opioid receptor binding after psychological trauma. *Biological Psychiatry* 61: 1030–1038.

Lim MM, Wang Z, Olazabal DE, Ren X, Terwilliger EF, and Young LJ (2004) Enhanced partner preference in a promiscuous species by manipulating the expression of a single gene. *Nature* 429: 754–757.

Lindholm S, Rosin A, Dahlin I, Georgieva J, and Franck J (2007) Ethanol alters the effect of kappa receptor ligands on dopamine release in the nucleus accumbens. *Physiology and Behavior* 92: 167–171.

Linthorst AC, Flachskamm C, Barden N, Holsboer F, and Reul JM (2000) Glucocorticoid receptor impairment alters CNS responses to a psychological stressor: An *in vivo*

microdialysis study in transgenic mice. *European Journal of Neuroscience* 12: 283–291.
S'Loh HH, Liu HC, Cavalli A, Yang W, Chen YF, and Wei LN (1998) mu Opioid receptor knockout in mice: Effects on ligand-induced analgesia and morphine lethality. *Molecular Brain Research* 54: 321–326.
Lolait SJ, Stewart LQ, Roper JA, Harrison G, Jessop DS, Young WS, III, and O'Carroll AM (2007) Attenuated stress response to acute lipopolysaccharide challenge and ethanol administration in vasopressin V1b receptor knockout mice. *Journal of Neuroendocrinology* 19: 543–551.
Louis C, Stemmelin J, Boulay D, Bergis O, Cohen C, and Griebel G (2008) Additional evidence for anxiolytic- and antidepressant-like activities of saredutant (SR48968), an antagonist at the neurokinin-2 receptor in various rodent-models. *Pharmacology Biochemistry and Behavior* 89: 36–45.
Ludwig M, Sabatier N, Dayanithi G, Russell JA, and Leng G (2002) The active role of dendrites in the regulation of magnocellular neurosecretory cell behavior. *Progress in Brain Research* 139: 247–256.
Lynch CB (1980) Response to divergent selection for nesting behavior in *Mus musculus*. *Genetics* 96: 757–765.
Makara GB and Haller J (2001) Non-genomic effects of glucocorticoids in the neural system. Evidence, mechanisms and implications. *Progress in Neurobiology* 65: 367–390.
Mantella RC, Vollmer RR, and Amico JA (2005) Corticosterone release is heightened in food or water deprived oxytocin deficient male mice. *Brain Research* 1058: 56–61.
Mantella RC, Vollmer RR, Li X, and Amico JA (2003) Female oxytocin-deficient mice display enhanced anxiety-related behavior. *Endocrinology* 144: 2291–2296.
Mantella RC, Vollmer RR, Rinaman L, Li X, and Amico JA (2004) Enhanced corticosterone concentrations and attenuated Fos expression in the medial amygdala of female oxytocin knockout mice exposed to psychogenic stress. *American Journal of Physiology - Regulatory, Integrative and Comparative Physiology* 287: R1494–R1504.
Matthes HW, Maldonado R, Simonin F, et al. (1996) Loss of morphine-induced analgesia, reward effect and withdrawal symptoms in mice lacking the mu-opioid-receptor gene. *Nature* 383: 819–823.
McCarthy MM, McDonald CH, Brooks PJ, and Goldman D (1996) An anxiolytic action of oxytocin is enhanced by estrogen in the mouse. *Physiology and Behavior* 60: 1209–1215.
McEwen BS and Sapolsky RM (1995) Stress and cognitive function. *Current Opinion in Neurobiology* 5: 205–216.
Meynen G, Unmehopa UA, van Heerikhuize JJ, Hofman MA, Swaab DF, and Hoogendijk WJ (2006) Increased arginine vasopressin mRNA expression in the human hypothalamus in depression: A preliminary report. *Biological Psychiatry* 60: 892–895.
Miedlar JA, Rinaman L, Vollmer RR, and Amico JA (2007) Oxytocin gene deletion mice overconsume palatable sucrose solution but not palatable lipid emulsions. *American Journal of Physiology - Regulatory, Integrative and Comparative Physiology* 293: R1063–R1068.
Mizoguchi H, Tseng LF, Suzuki T, Sora I, and Narita M (2002) Differential mechanism of G-protein activation induced by endogenous mu-opioid peptides, endomorphin and beta-endorphin. *Japanese Journal of Pharmacology* 89: 229–234.
Mlynarik M, Zelena D, Bagdy G, Makara GB, and Jezova D (2007) Signs of attenuated depression-like behavior in vasopressin deficient Brattleboro rats. *Hormones and Behavior* 51: 395–405.
Montkowski A, Barden N, Wotjak C, Stec I, Ganster J, Meaney M, et al. (1995) Long-term antidepressant treatment reduces behavioural deficits in transgenic mice with impaired glucocorticoid receptor function. *Journal of Neuroendocrinology* 7: 841–845.
Muglia L, Jacobson L, Dikkes P, and Majzoub JA (1995) Corticotropin-releasing hormone deficiency reveals major fetal but not adult glucocorticoid need. *Nature* 373: 427–432.
Muller MB and Holsboer F (2006) Mice with mutations in the HPA-system as models for symptoms of depression. *Biological Psychiatry* 59: 1104–1115.
Muller MB, Landgraf R, Preil J, et al. (2000) Selective activation of the hypothalamic vasopressinergic system in mice deficient for the corticotropin-releasing hormone receptor 1 is dependent on glucocorticoids. *Endocrinology* 141: 4262–4269.
Muller MB and Wurst W (2004) Getting closer to affective disorders: the role of CRH receptor systems. *Trends in Molecular Medicine* 10: 409–415.
Muller MB, Zimmermann S, Sillaber I, et al. (2003) Limbic corticotropin-releasing hormone receptor 1 mediates anxiety-related behavior and hormonal adaptation to stress. *Nature Neuroscience* 6: 1100–1107.
Munck A, Guyre PM, and Holbrook NJ (1984) Physiological functions of glucocorticoids in stress and their relation to pharmacological actions. *Endocrine Reviews* 5: 25–44.
Murgatroyd C, Wigger A, Frank E, et al. (2004) Impaired repression at a vasopressin promoter polymorphism underlies overexpression of vasopressin in a rat model of trait anxiety. *Journal of Neuroscience* 24: 7762–7770.
Nestler EJ, Barrot M, DiLeone RJ, Eisch AJ, Gold SJ, and Monteggia LM (2002) Neurobiology of depression. *Neuron* 34: 13–25.
Neumann ID (2002) Involvement of the brain oxytocin system in stress coping: interactions with the hypothalamo-pituitary-adrenal axis. *Progress in Brain Research* 139: 147–162.
Oitzl MS, de Kloet ER, Joels M, Schmid W, and Cole TJ (1997) Spatial learning deficits in mice with a targeted glucocorticoid receptor gene disruption. *European Journal of Neuroscience* 9: 2284–2296.
Oitzl MS, Reichardt HM, Joels M, and de Kloet ER (2001) Point mutation in the mouse glucocorticoid receptor preventing DNA binding impairs spatial memory. *Proceedings of the National Academy of Sciences of Unites States Of America* 98: 12790–12795.
Orchinik M (1998) Glucocorticoids, stress, and behavior: Shifting the timeframe. *Hormones and Behavior* 34: 320–327.
Overli O, Sorensen C, Pulman KGT, Pottinger TG, Korzan W, Summers CH, and Nilsson GE (2007) Evolutionary background for stress-coping styles: Relationships between physiological, behavioral, and cognitive traits in non-mammalian vertebrates. *Neuroscience and Biobehavioral Reviews* 31: 396–412.
Page NM (2004) Hemokinins and endokinins. *Cellular and Molecular Life Sciences* 61: 1652–1663.
Pan YX (2005) Diversity and complexity of the mu opioid receptor gene: Alternative pre-mRNA splicing and promoters. *DNA Cell Biology* 24: 736–750.
Pariante CM, Thomas SA, Lovestone S, Makoff A, and Kerwin RW (2004) Do antidepressants regulate how cortisol affects the brain? *Psychoneuroendocrinology* 29: 423–447.
Patak E, Pinto FM, Story ME, et al. (2005) Functional and molecular characterization of tachykinins and tachykinin receptors in the mouse uterus. *Biology of Reproduction* 72: 1125–1133.
Pedersen CA and Boccia ML (2006) Vasopressin interactions with oxytocin in the control of female sexual behavior. *Neuroscience* 139: 843–851.
Pepin MC, Pothier F, and Barden N (1992a) Antidepressant drug action in a transgenic mouse model of the endocrine changes seen in depression. *Molecular Pharmacology* 42: 991–995.

Pepin MC, Pothier F, and Barden N (1992b) Impaired type II glucocorticoid-receptor function in mice bearing antisense RNA transgene. *Nature* 355: 725–728.

Perlman WR, Webster MJ, Kleinman JE, and Weickert CS (2004) Reduced glucocorticoid and estrogen receptor alpha messenger ribonucleic acid levels in the amygdala of patients with major mental illness. *Biological Psychiatry* 56: 844–852.

Phillips TJ, Belknap JK, Hitzemann RJ, Buck KJ, Cunningham CL, and Crabbe JC (2002) Harnessing the mouse to unravel the genetics of human disease. *Genes, Brain and Behavior* 1: 14–26.

Pitkow LJ, Sharer CA, Ren X, Insel TR, Terwilliger EF, and Young LJ (2001) Facilitation of affiliation and pair-bond formation by vasopressin receptor gene transfer into the ventral forebrain of a monogamous vole. *Journal of Neuroscience* 21: 7392–7396.

Pottinger TG and Carrick TR (1999) Modification of the plasma cortisol response to stress in rainbow trout by selective breeding. *General and Comparative Endocrinology* 116: 122–132.

Poulin JF, Chevalier B, Laforest S, and Drolet G (2006) Enkephalinergic afferents of the centromedial amygdala in the rat. *Journal of Comparative Neurology* 496: 859–876.

Preil J, Muller MB, Gesing A, et al. (2001) Regulation of the hypothalamic-pituitary-adrenocortical system in mice deficient for CRH receptors 1 and 2. *Endocrinology* 142: 4946–4955.

Primeaux SD, Wilson SP, McDonald AJ, Mascagni F, and Wilson MA (2006) The role of delta opioid receptors in the anxiolytic actions of benzodiazepines. *Pharmacology Biochemistry and Behavior* 85: 545–554.

Purba JS, Hoogendijk WJ, Hofman MA, and Swaab DF (1996) Increased number of vasopressin- and oxytocin-expressing neurons in the paraventricular nucleus of the hypothalamus in depression. *Archives of General Psychiatry* 53: 137–143.

Ragnauth AK, Devidze N, Moy V, et al. (2005) Female oxytocin gene-knockout mice, in a semi-natural environment, display exaggerated aggressive behavior. *Genes, Brain and Behavior* 4: 229–239.

Raison CL and Miller AH (2003) When not enough is too much: the role of insufficient glucocorticoid signaling in the pathophysiology of stress-related disorders. *American Journal of Psychiatry* 160: 1554–1565.

Reichardt HM, Kaestner KH, Tuckermann J, et al. (1998a) DNA binding of the glucocorticoid receptor is not essential for survival. *Cell* 93: 531–541.

Reichardt HM, Kaestner KH, Wessely O, Gass P, Schmid W, and Schutz G (1998b) Analysis of glucocorticoid signalling by gene targeting. *Journal of Steroid Biochemistry and Molecular Biology* 65: 111–115.

Reichardt HM, Umland T, Bauer A, Kretz O, and Schutz G (2000) Mice with an increased glucocorticoid receptor gene dosage show enhanced resistance to stress and endotoxic shock. *Molecular and Cellar Biology* 20: 9009–9017.

Reyes BAS, Glaser JD, and van Bockstaele EJ (2007) Ultrastructural evidence for co-localization of corticotropin-releasing factor receptor and mu-opioid receptor in the rat nucleus locus coeruleus. *Neuroscience Letters* 413: 216–221.

Ribeiro SC, Kennedy SE, Smith YR, Stohler CS, and Zubieta JK (2005) Interface of physical and emotional stress regulation through the endogenous opioid system and mu-opioid receptors. *Progress in Neuropsychopharmacology and Biological Psychiatry* 29: 1264–1280.

Ridder S, Chourbaji S, Hellweg R, et al. (2005) Mice with genetically altered glucocorticoid receptor expression show altered sensitivity for stress-induced depressive reactions. *Journal of Neuroscience* 25: 6243–6250.

Rinaman L, Vollmer RR, Karam J, Phillips D, Li X, and Amico JA (2005) Dehydration anorexia is attenuated in oxytocin-deficient mice. *American Journal of Physiology - Regulatory, Integrative and Comparative Physiology* 288: R1791–R1799.

Ring RH (2005) The central vasopressinergic system: examining the opportunities for psychiatric drug development. *Current Pharmaceutical Design* 11: 205–225.

Rochford J, Beaulieu S, Rousse I, Glowa JR, and Barden N (1997) Behavioral reactivity to aversive stimuli in a transgenic mouse model of impaired glucocorticoid (type II) receptor function: Effects of diazepam and FG-7142. *Psychopharmacology* 132: 145–152.

Romero LM (2004) Physiological stress in ecology: lessons from biomedical research. *Trends in Ecology and Evolution* 19: 249–255.

Rousse I, Beaulieu S, Rowe W, Meaney MJ, Barden N, and Rochford J (1997) Spatial memory in transgenic mice with impaired glucocorticoid receptor function. *Neuro Report* 8: 841–845.

Rubinstein M, Mogil JS, Japon M, Chan EC, Allen RG, and Low MJ (1996) Absence of opioid stress-induced analgesia in mice lacking beta-endorphin by site-directed mutagenesis. *Proceedings of the National Academy of Sciences of United States of America* 93: 3995–4000.

Rydmark I, Wahlberg K, Ghatan PH, et al. (2006) Neuroendocrine, cognitive and structural imaging characteristics of women on longterm sickleave with job stress-induced depression. *Biological Psychiatry* 60: 867–873.

Salome N, Stemmelin J, Cohen C, and Griebel G (2006) Selective blockade of NK2 or NK3 receptors produces anxiolytic- and antidepressant-like effects in gerbils. *Pharmacology Biochemistry and Behavior* 83: 533–539.

Sapolsky RM, Romero LM, and Munck AU (2000) How do glucocorticoids influence stress responses? Integrating permissive, suppressive, stimulatory, and preparative actions. *Endocrine Reviews* 21: 55–89.

Satterlee DG and Johnson WA (1988) Selection of Japanese quail for contrasting blood corticosterone response to immobilization. *Poultry Science* 67: 25–32.

Schalkwyk LC, Fernandes C, Nash MW, Kurrikoff K, Vasar E, and Koks S (2007) Interpretation of knockout experiments: the congenic footprint. *Genes, Brain and Behavior* 6: 299–303.

Schatzberg AF (2003) New approaches to managing psychotic depression. *Journal of Clinical Psychiatry* 64 (supplement 1): 19–23.

Scherrer G, Tryoen-Toth P, Filliol D, et al. (2006) Knockin mice expressing fluorescent delta-opioid receptors uncover G protein-coupled receptor dynamics in vivo. *Proceedings of the National Academy of Sciences of United States of America* 103: 9691–9696.

Schuller AG, King MA, Zhang J, et al. (1999) Retention of heroin and morphine-6 beta-glucuronide analgesia in a new line of mice lacking exon 1 of MOR-1. *Nature Neuroscience* 2: 151–156.

Sclafani A, Rinaman L, Vollmer RR, and Amico JA (2007) Oxytocin knockout mice demonstrate enhanced intake of sweet and nonsweet carbohydrate solutions. *American Journal of Physiology – Regulatory, Integrative and Comparative Physiology* 292: R1828–R1833.

Serradeil-Le Gal C, Wagnon J, III, Tonnerre B, Roux R, Garcia G, Griebel G, and Aulombard A (2005) An overview of SSR149415, a selective nonpeptide vasopressin V(1b) receptor antagonist for the treatment of stress-related disorders. *CNS Drug Reviews* 11: 53–68.

Severini C, Improta G, Falconieri-Erspamer G, Salvadori S, and Erspamer V (2002) The tachykinin peptide family. *Pharmacological Reviews* 54: 285–322.

Simonin F, Valverde O, Smadja C, et al. (1998) Disruption of the kappa-opioid receptor gene in mice enhances sensitivity to chemical visceral pain, impairs pharmacological actions of the selective kappa-agonist U-50, 488H and attenuates morphine withdrawal. *EMBO Journal* 17: 886–897.

Siuciak JA, McCarthy SA, Martin AN, et al. (2007) Disruption of the neurokinin-3 receptor (NK3) in mice leads to cognitive deficits. *Psychopharmacology (Berl)* 194: 185–195.

Smith GW, Aubry JM, Dellu F, et al. (1998) Corticotropin releasing factor receptor 1-deficient mice display decreased anxiety, impaired stress response, and aberrant neuroendocrine development. *Neuron* 20: 1093–1102.

Sonino N, Fava GA, Raffi AR, Boscaro M, and Fallo F (1998) Clinical correlates of major depression in Cushing's disease. *Psychopathology* 31: 302–306.

Sora I, Takahashi N, Funada M, et al. (1997) Opiate receptor knockout mice define mu receptor roles in endogenous nociceptive responses and morphine-induced analgesia. *Proceedings of the National Academy of Sciences of United States of America* 94: 1544–1549.

Stead JD, Clinton S, Neal C, et al. (2006) Selective breeding for divergence in novelty-seeking traits: heritability and enrichment in spontaneous anxiety-related behaviors. *Behavioral Genetics* 36: 697–712.

Stec I, Barden N, Reul JM, and Holsboer F (1994) Dexamethasone nonsuppression in transgenic mice expressing antisense RNA to the glucocorticoid receptor. *Journal of Psychiatric Research* 28: 1–5.

Steckler T and Holsboer F (1999) Corticotropin-releasing hormone receptor subtypes and emotion. *Biological Psychiatry* 46: 1480–1508.

Steimer T and Driscoll P (2003) Divergent stress responses and coping styles in psychogenetically selected Roman high-(RHA) and low-(RLA) avoidance rats: behavioural, neuroendocrine and developmental aspects. *Stress* 6: 87–100.

Stenzel-Poore MP, Cameron VA, Vaughan J, Sawchenko PE, and Vale W (1992) Development of Cushing's syndrome in corticotropin-releasing factor transgenic mice. *Endocrinology* 130: 3378–3386.

Stenzel-Poore MP, Heinrichs SC, Rivest S, Koob GF, and Vale WW (1994) Overproduction of corticotropin-releasing factor in transgenic mice: a genetic model of anxiogenic behavior. *Journal of Neuroscience* 14: 2579–2584.

Stewart LQ, Roper JA, Scott Young W, 3rd, O'Carroll AM, and Lolait SJ (2008) The role of the arginine vasopressin Avp 1b receptor in the acute neuroendocrine action of antidepressants. *Psychoneuroendocrinology* 33: 405–415.

Strekalova T, Spanagel R, Bartsch D, Henn FA, and Gass P (2004) Stress-induced anhedonia in mice is associated with deficits in forced swimming and exploration. *Neuropsychopharmacology* 29: 2007–2017.

Strohle A, Poettig M, Barden N, Holsboer F, and Montkowski A (1998) Age- and stimulus-dependent changes in anxiety-related behaviour of transgenic mice with GR dysfunction. *Neuro Report* 9: 2099–2102.

Swaab DF, Bao AM, and Lucassen PJ (2005) The stress system in the human brain in depression and neurodegeneration. *Ageing Research Reviews* 4: 141–194.

Swallow JG and Garland T (2005) Selection experiments as a tool in evolutionary and comparative physiology: Insights into complex traits - An introduction to the symposium. *Integrative and Comparative Biology* 45: 387–390.

Takayanagi Y, Yoshida M, Bielsky IF, et al. (2005) Pervasive social deficits, but normal parturition, in oxytocin receptor-deficient mice. *Proceedings of the National Academy of Sciences of United States of America* 102: 16096–16101.

Tasker JG (2006) Rapid glucocorticoid actions in the hypothalamus as a mechanism of homeostatic integration. *Obesity (Silver. Spring)* 14 (supplement 5): 259S–265S.

Tasker JG, Di S, and Malcher-Lopes R (2006) Minireview: rapid glucocorticoid signaling via membrane-associated receptors. *Endocrinology* 147: 5549–5556.

Tecott LH (2003) The genes and brains of mice and men. *American Journal of Psychiatry* 160: 646–656.

Terenzi MG and Ingram CD (2005) Oxytocin-induced excitation of neurones in the rat central and medial amygdaloid nuclei. *Neuroscience* 134: 345–354.

Thomsen AF, Kvist TK, Andersen PK, and Kessing LV (2006) The risk of affective disorders in patients with adrenocortical insufficiency. *Psychoneuroendocrinology* 31: 614–622.

Tian M, Broxmeyer HE, Fan Y, et al. (1997) Altered hematopoiesis, behavior, and sexual function in mu opioid receptor-deficient mice. *Journal of Experimental Medicine* 185: 1517–1522.

Timpl P, Spanagel R, Sillaber I, Kresse A, Reul JM, Stalla GK, Blanquet V, Steckler T, Holsboer F, and Wurst W (1998) Impaired stress response and reduced anxiety in mice lacking a functional corticotropin-releasing hormone receptor 1. *Nature Genetics* 19: 162–166.

Touma C, Bunck M, Glasl L, et al. (2008) Mice selected for high versus low stress reactivity: A new animal model for affective disorders. *Psychoneuroendocrinology* [Epub ahead of print; doi:10.1016/j.psyneuen.2008.03.013].

Tronche F, Kellendonk C, Kretz O, et al. (1999) Disruption of the glucocorticoid receptor gene in the nervous system results in reduced anxiety. *Nature Genetics* 23: 99–103.

Tsien JZ, Chen DF, Gerber D, et al. (1996) Subregion- and cell type-restricted gene knockout in mouse brain. *Cell* 87: 1317–1326.

Urani A, Chourbaji S, and Gass P (2005) Mutant mouse models of depression: candidate genes and current mouse lines. *Neuroscience and Biobehavioral Reviews* 29: 805–828.

Vale W, Spiess J, Rivier C, and Rivier J (1981) Characterization of a 41-residue ovine hypothalamic peptide that stimulates secretion of corticotropin and beta-endorphin. *Science* 213: 1394–1397.

van Gaalen MM, Stenzel-Poore MP, Holsboer F, and Steckler T (2002) Effects of transgenic overproduction of CRH on anxiety-like behaviour. *European Journal of Neuroscience* 15: 2007–2015.

van Giersbergen PL, Shatzer SA, Burcher E, and Buck SH (1992) Comparison of the effects of neuropeptide K and neuropeptide gamma with neurokinin A at NK2 receptors in the hamster urinary bladder. *Naunyn-Schmiedebergs Archives of Pharmacology* 345: 51–56.

van Londen L, Goekoop JG, van Kempen GM, et al. (1997) Plasma levels of arginine vasopressin elevated in patients with major depression. *Neuropsychopharmacology* 17: 284–292.

van West D, Del Favero J, Aulchenko Y, et al. (2004) A major SNP haplotype of the arginine vasopressin 1B receptor protects against recurrent major depression. *Molecular Psychiatry* 9: 287–292.

Veenema AH, Meijer OC, de Kloet ER, and Koolhaas JM (2003) Genetic selection for coping style predicts stressor susceptibility. *Journal of Neuroendocrinology* 15: 256–267.

Vigna SR (2003) The role of the amino-terminal domain of tachykinins in neurokinin-1 receptor signaling and desensitization. *Neuropeptides* 37: 30–35.

Vollmer RR, Li X, Karam JR, and Amico JA (2006) Sodium ingestion in oxytocin knockout mice. *Experimental Neurology* 202: 441–448.

Volpi S, Rabadan-Diehl C, and Aguilera G (2004) Regulation of vasopressin V1b receptors and stress adaptation. *Annals of the New York Academy of Sciences* 1018: 293–301.

Waldherr M and Neumann ID (2007) Centrally released oxytocin mediates mating-induced anxiolysis in male rats. *Proceedings of the National Academy of Sciences of United States of America* 104: 16681–16684.

Wang Y, Bockman CS, Lovas S, Abel PW, Murphy RF, and Conlon JM (1993) Neuropeptide gamma-(1-9)-peptide: a major product of the posttranslational processing of gamma-preprotachykinin in rat tissues. *Journal of Neurochemistry* 61: 1231–1235.

Wassink TH, Piven J, Vieland VJ, Pietila J, Goedken RJ, Folstein SE, and Sheffield VC (2004) Examination of AVPR1a as an autism susceptibility gene. *Molecular Psychiatry* 9: 968–972.

Watson D, Clark LA, and Tellegen A (1988) Development and validation of brief measures of positive and negative affect: the PANAS scales. *Journal of Personality and Social Psychology* 54: 1063–1070.

Webster MJ, Knable MB, O'Grady J, Orthmann J, and Weickert CS (2002) Regional specificity of brain glucocorticoid receptor mRNA alterations in subjects with schizophrenia and mood disorders. *Molecular Psychiatry* 7: 985–994, 924.

Wei Q, Hebda-Bauer EK, Pletsch A, et al. (2007) Overexpressing the glucocorticoid receptor in forebrain causes an aging-like neuroendocrine phenotype and mild cognitive dysfunction. *Journal of Neuroscience* 27: 8836–8844.

Wei Q, Lu XY, Liu L, et al. (2004) Glucocorticoid receptor overexpression in forebrain: a mouse model of increased emotional lability. *Proceedings of the National Academy of Sciences of United States of America* 101: 11851–11856.

Wersinger SC, Dunn AJ, Muglia LJ, et al. (1999) Stress-induced behaviors require the corticotropin-releasing hormone (CRH) receptor, but not CRH. *Proceedings of the National Academy of Sciences of United States of America* 96: 8283–8288.

Wersinger SR, Caldwell HK, Christiansen M, and Young WS, III (2007) Disruption of the vasopressin 1b receptor gene impairs the attack component of aggressive behavior in mice. *Genes, Brain and Behavior* 6: 653–660.

Wersinger SR, Ginns EI, O'Carroll AM, Lolait SJ, and Young WS, III (2002) Vasopressin VIb receptor knockout reduces aggressive behavior in male mice. *Molecular Psychiatry* 7: 975–984.

Wersinger SR, Kelliher KR, Zufall F, Lolait SJ, O'Carrol AM, and Young WS, III (2004) Social motivation is reduced in vasopressin 1b receptor null mice despite normal performance in an olfactory discrimination task. *Hormones and Behavior* 46: 638–645.

Wigger A and Neumann ID (2002) Endogenous opioid regulation of stress-induced oxytocin release within the hypothalamic paraventricular nucleus is reversed in late pregnancy: A microdialysis study. *Neuroscience* 112: 121–129.

Wigger A, Sanchez MM, Mathys KC, et al. (2004) Alterations in central neuropeptide expression, release, and receptor binding in rats bred for high anxiety: Critical role of vasopressin. *Neuropsychopharmacology* 29: 1–14.

Windle RJ, Kershaw YM, Shanks N, Wood SA, Lightman SL, and Ingram CD (2004) Oxytocin attenuates stress-induced c-fos mRNA expression in specific forebrain regions associated with modulation of hypothalamo-pituitary–adrenal activity. *Journal of Neuroscience* 24: 2974–2982.

Winslow JT, Hearn EF, Ferguson J, Young LJ, Matzuk MM, and Insel TR (2000) Infant vocalization, adult aggression, and fear behavior of an oxytocin null mutant mouse. *Hormones and Behavior* 37: 145–155.

Wong ML and Licinio J (2001) Research and treatment approaches to depression. *Nature Reviews Neuroscience* 2: 343–351.

Wotjak CT, Naruo T, Muraoka S, Simchen R, Landgraf R, and Engelmann M (2001) Forced swimming stimulates the expression of vasopressin and oxytocin in magnocellular neurons of the rat hypothalamic paraventricular nucleus. *European Journal of Neuroscience* 13: 2273–2281.

Wu S, Jia M, Ruan Y, et al. (2005) Positive association of the oxytocin receptor gene (OXTR) with autism in the Chinese Han population. *Biological Psychiatry* 58: 74–77.

Yakovleva T, Bazov I, Cebers G, et al. (2006) Prodynorphin storage and processing in axon terminals and dendrites. *FASEB Journal* 20: 2124–2126.

Yehuda R, Halligan SL, Grossman R, Golier JA, and Wong C (2002) The cortisol and glucocorticoid receptor response to low dose dexamethasone administration in aging combat veterans and holocaust survivors with and without posttraumatic stress disorder. *Biological Psychiatry* 52: 393–403.

Yirmiya N, Rosenberg C, Levi S, et al. (2006) Association between the arginine vasopressin la receptor (AVPR1a) gene and autism in a family-based study: mediation by socialization skills. *Molecular Psychiatry* 11: 488–494.

Young LJ and Wang Z (2004) The neurobiology of pair bonding. *Nature Neuroscience* 7: 1048–1054.

Young AH, Gallagher P, Watson S, Del Estal D, Owen BM, and Ferrier IN (2004) Improvements in neurocognitive function and mood following adjunctive treatment with mifepristone (RU-486) in bipolar disorder. *Neuropsychopharmacology* 29: 1538–1545.

Young LJ, Nilsen R, Waymire KG, MacGregor GR, and Insel TR (1999) Increased affiliative response to vasopressin in mice expressing the V1a receptor from a monogamous vole. *Nature* 400: 766–768.

Zadina JE, Hackler L, Ge LJ, and Kastin AJ (1997) A potent and selective endogenous agonist for the mu-opiate receptor. *Nature* 386: 499–502.

Zhu YX, King MA, Schuller AGP, et al. (1999) Retention of supraspinal delta-like analgesia and loss of morphine tolerance in delta opioid receptor knockout mice. *Neuron* 24: 243–252.

2 Hypothalamic–Pituitary–Adrenal Cortical Axis

M E Rhodes, J M McKlveen, D R Ripepi, and N E Gentile, Saint Vincent College, Latrobe, PA, USA

Chapter Outline

Glossary

allostasis The action of maintaining stability, or homeostasis, through change; more specifically, the action of a wide range of physiological factors, most notably hormones, which are continuously changing to maintain the stability of critical parameters (e.g., autonomic responses, glycemia, osmolarity, and core temperature) outside of the normal homeostatic range despite environmental challenges.

allostatic load The many events of daily life that elevate the activity of physiological systems that promote adaptation and homeostasis, but cause some measure of wear and tear on the organism from which pathological consequences could ensue.

amygdala Bilateral brain nuclei in the temporal lobes of the forebrain that are concerned primarily with fear processing and which stimulate the secretion of corticotropin-releasing hormone (CRH) and, in turn, the rest of the hypothalamic–pituitary–adrenal cortical (HPA) axis.

circadian rhythm (diurnal rhythm) A rhythmic activity cycle, based on 24-h intervals, which is exhibited in the physiological functions of many organisms, including hormone secretions.

glucocorticoids Hormones produced by the adrenal cortex that increase glucose production in the liver, inhibit glucose metabolism by body tissues, and promote lipid breakdown in fat tissue. The principal glucocorticoid in humans is cortisol (hydrocortisone) and, in laboratory rodents, is corticosterone. When administered in high therapeutic doses, glucocorticoids suppress immunological function, reduce inflammation, and decrease connective tissue and new bone formation.

hippocampus An area of the brain containing nerve cells that inhibit the secretion of CRH and, in turn, the rest of the HPA axis.

homeostasis It refers to the set of physiological mechanisms that maintain certain critical parameters (e.g., autonomic responses, glycemia, osmolarity, and core temperature) within a narrow range to allow an organism to survive.

mineralocorticoids The hormones produced by the adrenal cortex that reduce the excretion of sodium and enhance the excretion of potassium and hydrogen ions by the kidney. The principal mineralocorticoid in humans and laboratory rodents is aldosterone.

paraventricular nucleus (PVN) The hypothalamic nucleus is broadly divided into two distinct regions: (1) the parvocellular region which contains neurons that send axons to the median eminence where CRH and vasopressin (AVP) are released into the hypophyseal portal system and (2) the magnocellular region which contains neurons that send axons through the median eminence to the neurohypophysis (posterior pituitary) where AVP and oxytocin are released into the systemic circulation.

vasopressin (AVP; antidiuretic hormone; ADH) A hormone produced by cells in the hypothalamus that is transported down the pituitary stalk to (1) the anterior pituitary gland where, along with CRH, it synergistically stimulates the secretion of adrenocorticotropic hormone (ACTH) and (2) the posterior pituitary gland, at which point it is carried by the blood stream to the kidneys, where it reduces the excretion of water.

2.1 Introduction

Stress and related concepts can be traced as far back as written science. The stress system coordinates the generalized stress response, which takes place when a stressor of any kind exceeds a threshold. The main components of the stress system are the corticotropin-releasing hormone (CRH) system and locus ceruleus–norepinephrine (LC/NE) autonomic system and their peripheral effectors, the pituitary–adrenal axis and the limbs of the autonomic nervous system (ANS). Activation of the stress system leads to behavioral and peripheral changes that improve the ability of the organism to adjust to environmental challenges, maintain homeostasis, and increase its chances of survival.

There has been an exponential increase in knowledge regarding the interactions among the components of the stress system, and between the stress system and other brain elements involved in the regulation of emotion, cognitive function, and behavior, as well as with the endocrine axes responsible for reproduction, growth, and immunity. CRH inhibits gonadotropin-releasing-hormone (GnRH) secretion during stress, thus suppressing reproduction. Via somatostatin, CRH inhibits growth-hormone (GH), thyrotropin-releasing-hormone (TRH), and thyroid-stimulating-hormone (TSH) secretion, thus suppressing growth and thyroid function. The end hormones of the hypothalamic–pituitary–adrenal (HPA) axis, glucocorticoids, directly inhibit pituitary gonadotropin, GH, and TSH secretion, decrease target tissue responsiveness to sex steroids and growth factors, and suppress 5′-deiodinase, which converts the relatively inactive tetraiodothyronine (T4) to triiodothyronine (T3), further suppressing reproductive, growth, and thyroid functions (Gabry et al., 2002). During chronic stress, glucocorticoids also produce insulin-mediated effects on adipose tissue promoting visceral adiposity and metabolic syndrome and its sequelae, and inhibit bone osteoblastic activity promoting osteoporosis and immune function promoting susceptibility to infection and inflammatory disorders (Gabry et al., 2002).

Stress system dysfunction is characterized by sustained maladaptive hyperactivity or hypoactivity of the HPA axis and is associated with various pathological conditions. These include a range of psychiatric (anxiety, depression, addiction, and withdrawal), eating (anorexia nervosa (AN), bulimia, and obesity), metabolic (diabetes and metabolic syndrome), and inflammatory disorders (rheumatoid arthritis).

2.1.1 The Stress System

Similar to other organisms, humans strive to maintain a stable internal environment that is optimal for their functioning using a set of physiological mechanisms, a phenomenon known as homeostasis (Chrousos and Gold, 1992). External variables, whether physical, chemical, or biological, continually challenge this homeostasis. Adverse changes sensed by the subject are referred to as stressful stimuli or stressors. Body systems continually inform the nervous system of their status in relation to one another and to the external environment. The brain processes this incoming data and initiates appropriate responses to maintain homeostasis (Johnson et al., 1992). Stressful stimuli trigger the body's adaptive responses, which attempt to keep the organism functioning even though the circumstances may not be optimal. Adaptive responses to adversity are proportional to stimulus intensity and range from a simple localized reaction to a generalized and systemic state that affects the entire organism (Gabry et al., 2002). The chronic state of reestablishing or maintaining maladaptive homeostasis is called allostasis (McEwen, 2000, 2002, 2007).

The many events of daily life that elevate the activities of adaptive physiological systems that promote homeostasis but cause some measure of wear and tear to the systems of an organism are termed the allostatic load (McEwen, 2000). Stress represents the state in which the brain interprets the amount of stressors as being excessive, thus responding in a generalized way. Stimuli perceived as stressful may be psychological, physical, or biological. In the context of profound and threatening stimuli, cognitive, affective, and physiological responses work together to improve the chances of survival (Smith et al., 1989). Attention is focused solely on the perceived danger as autonomic function generates a fight-or-flight readiness. Behavioral effects of the stress response include increased awareness, improved cognition, euphoria, and enhanced analgesia (Charmandari et al., 2005; Chrousos and Gold, 1992). Physiological adaptations include activation of the HPA axis and increased cardiovascular tone, respiratory rate, and metabolism (Habib et al., 2001; Sapolsky et al., 2000). There is a generalized shutdown of functions whose execution could compromise the likelihood of surviving danger (e.g., sleep, food intake, growth, and reproduction; Johnson et al., 1992). This state was well described by Selye (1976) as the general adaptation syndrome (GAS) and has been conventionally known as the stress response.

Humans vary in their capacity to tolerate allostatic load and the intensity of their response to acute and chronic stress. Although the stress response is essential for survival, a dysregulated stress response could also produce disease (Gold and Chrousos, 2002; Gold et al., 1996; Habib et al., 2001). A state of stress proneness is commonly seen in mood and anxiety disorders (Bremner et al., 1996; Gold et al., 1988). In addition to major depression, a spectrum of other conditions may be associated with increased and prolonged stress responses reflected by increased activation of the HPA axis. These include anxiety, obsessive–compulsive disorder, chronic active alcoholism, alcohol and narcotic withdrawal, Alzheimer's disease, poorly controlled diabetes mellitus types I and II, AN and malnutrition, excessive exercising, childhood sexual abuse, and hyperthyroidism (Gilligan et al., 2000; Kiefer and Wiedemann, 2004).

2.1.2 The HPA Axis

The anatomical structures that mediate the stress response are found in both the central nervous system (CNS) and the peripheral tissues. As mentioned previously, in addition to the HPA axis, other neuronal systems play important roles in the regulation of adaptive responses to stress. These include the CRH system, the LC/NE autonomic system, and the limbs of the ANS (Armario, 2006; Habib et al., 2001; Whitnall, 1993). The principal endocrine effectors of the stress response are released from the hypothalamus, pituitary, and adrenal glands. CRH is released from parvocellular neurons of the paraventricular nucleus (PVN) of the hypothlamus into the hypophyseal–pituitary portal system and is the principal regulator of anterior-pituitary adrenocorticotropic hormone (ACTH) secretion. Arginine vasopressin (AVP) is a potent synergistic factor with CRH in the stimulation of ACTH release. The principal target for circulating ACTH is the adrenal cortex, where it stimulates the synthesis and secretion of glucocorticoids from the *zona fasciculata.* Glucocorticoids are the downstream effectors of the HPA axis and regulate physiological responses to stress via ubiquitously distributed intracellular receptors (Munck, 2005; Munck et al., 1984).

In nonstressful situations, both CRH and AVP are secreted by parvocellular neurons of the PVN into the hypophyseal portal system in a circadian and highly concordant pulsatile fashion (Chrousos and Gold, 1992). The amplitude of the CRH and AVP pulses increases in the early-morning hours, resulting eventually in increases of both the amplitude and frequency of ACTH and cortisol secretory bursts in the general circulation. These diurnal variations are disrupted by changes in lighting, feeding schedules, as well as by stress (Geracioti et al., 1992; Hiroshige and Wada-Okada, 1973).

During acute stress, the amplitude and synchronization of CRH and AVP pulsations into the hypophyseal portal system increase (Chrousos, 1998, 2000a). During excessive acute stress, especially that associated with hypotension or a decrease of blood volume, AVP of magnocellular neuron origin is secreted into both the hypophyseal portal system via collateral neuraxons and the systemic circulation (Engelmann et al., 2004). Depending on the type and duration of stress, other factors such as angiotension II, as well as various cytokines and lipid mediators of inflammation, are secreted and act on the hypothalamic, pituitary, and adrenal components of the HPA axis to mostly potentiate its activity (Gold and Chrousos, 1999).

Direct and indirect afferents from various limbic, hypothalamic, and brainstem brain regions innervate the PVN and regulate the release of CRH via numerous neurotransmitter systems. A variety of other factors are involved in regulating and modulating HPA-axis activity, including neuropeptides, glucocorticoids, and gonadal steroid hormones, as well as age, weight, and environmental influences (Suescun et al., 1997; Whitnall, 1993).

2.1.3 Corticotropin-Releasing Hormone

CRH was first described by Vale et al. (1981) as a 41-residue hypothalamic peptide that stimulates secretion of ACTH and β-endorphin. More recently, three additional members of the CRH peptide family have been identified, including urocortin (Ucn) 1, Ucn 2, and Ucn 3 (Hsu and Hsueh, 2001; Lewis et al., 2001; Reyes et al., 2001; Vaughan et al., 1995). Ucn 2 and Ucn 3 are also known as stresscopin-related peptide and stresscopin, respectively. CRH and the urocortins are widespread throughout the brain correlating well with the diverse array of physiological functions associated with this peptide family (Rivier and Vale, 1983; Smith and Vale, 2006). CRH is the key hypothalamic peptide controlling HPA-axis activity; therefore, it plays a central role in the stress response (Gold and Chrousos, 2002; Smith and Vale, 2006).

In the hypothalamus, several nuclei contain CRH cell bodies. These include the PVN, medial preoptic area (MPOA), dorsomedial nucleus, arcuate nucleus (ARC), posterior hypothalamus, and mammillary nuclei. Among these, the PVN contains the majority of CRH cell bodies that stimulate anterior-pituitary ACTH secretion (Gabry et al., 2002). CRH is present in a small group of PVN neurons that project to the lower brainstem and spinal cord. These are involved in regulating ANS function (Palkovits et al., 1985; Swanson et al., 1983).

Within seconds after exposure to stress, the synthesis of CRH is increased in peptidergic neurons of the PVN, leading to increased release of CRH into the median eminence. CRH is transported through the hypophyseal portal system to the pituitary and stimulates ACTH and β-endorphins from the anterior pituitary, ultimately leading to the secretion of glucocorticoids from the adrenal cortex (**Figure 1**; Claes, 2004a). In response to stress, the amplitude and synchronization of CRH pulsations in the hypophyseal portal system increase markedly leading to increased ACTH and glucocorticoid secretory episodes (Claes, 2004b; Tsigos and Chrousos, 1994). The extent and time course of changes of CRH in the PVN and the median eminence following stress are highly dependent on the nature of the stressor as well as the state of the organism. CRH, in addition to its role as the primary regulator of ACTH release from anterior-pituitary corticotropes, also has been implicated in the regulation of the ANS, learning and memory, feeding, and reproduction-related behaviors.

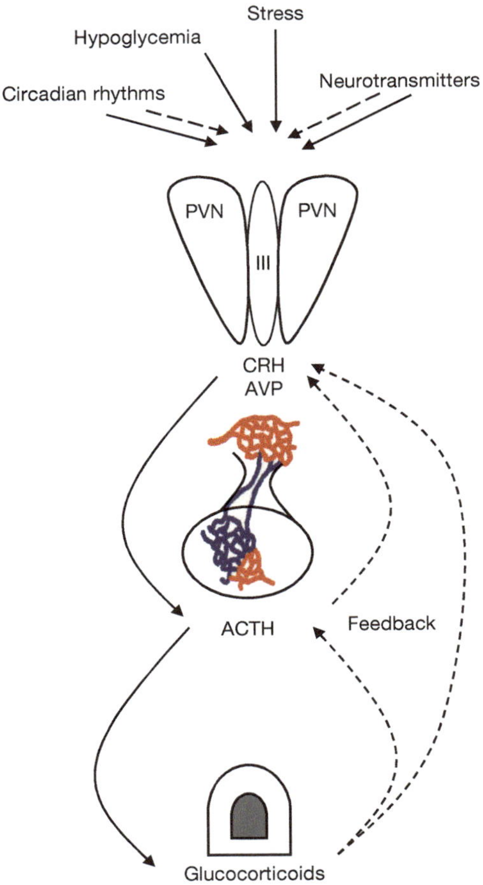

Figure 1 Factors influencing hypothalamic–pituitary–adrenal cortical (HPA)-axis activity. Glucocorticoid feedback directly modulates corticotropin-releasing hormone (CRH) and adrenocorticotropic hormone (ACTH) secretion, while factors such as stress, circadian rhythms, and neurotransmitters modulate HPA-axis activity by influencing the hypophyseal portal secretion of CRH and arginine vasopressin (AVP) from the paraventricular nucleus (PVN) of the hypothalamus which borders the third ventricle (III). Solid lines represent stimulation; dashed lines represent negative feedback or inhibition.

CRH neurons in the cerebral cortex may be important in several behavioral actions of the peptide. CRH interneurons are contained in the second and third layers of the cerebral cortex and project to layers I and IV. In addition, scattered cells, which may be pyramidal cells, are present in the deeper layers of the neocortex. Although CRH-containing neurons are found throughout the neocortex, they are found in higher densities in the prefrontal insular and cingulate areas (Swanson et al., 1983). The distribution of CRH in these areas corresponds to, and may explain, its effects on cognitive processing (Palkovits et al., 1985).

CRH cell bodies are also present in the central nucleus of the amygdala, the substantia innominata, and in the bed nucleus of the stria terminalis (BNST). CRH neurons in the central nucleus of the amygdala project to the parvocellular regions of the PVN and to the parabrachial nucleus of the brainstem. CRH neurons originating in the BNST send terminals to the parabrachial nuclei and dorsal vagal complex in the brainstem to coordinate autonomic activity. CRH neurons also interconnect the amygdala with the BNST and the hypothalamus (Owens and Nemeroff, 1991). The distribution and projections of CRH also contribute to the neuroendocrine, autonomic, and behavioral effects of CRH (Grigoriadis et al., 1993).

In the spinal cord, CRH cell bodies have been described in laminae V, VI, VII, and X, as well as in the intermediolateral column of the thoracic and lumbar regions. Spinal CRH neurons may play an important role in modulating sensory input via ascending projections to the thalamus and the brainstem reticular formation (Owens and Nemeroff, 1991). Several groups of CRH neurons occur throughout the brainstem. In the medulla, CRH neurons exist in the nucleus of solitary tract (NST), the dorsal vagal complex, the spinal trigeminal nucleus, and the reticular formation (Habib et al., 2001). Neurons of the NST relay sensory information to the PVN from cranial nerves that innervate large areas of the abdominal and thoracic viscera. The NST also receives projections from limbic structures that regulate the behavioral responses to stress. Furthermore, stress-receptive neurons in the A2/C2 region of the NST densely innervate the medial parvocellular subdivision of the PVN. In the pons, CRH cell bodies in the parabrachial nucleus project to the medial preoptic nucleus of the hypothalamus. In the midbrain, CRH cells are found in the Edinger–Westphal nucleus, the periaqueductal gray, and the dorsal raphe nucleus, and the cells of these nuclei project to various limbic structures. In addition, CRH neurons project from the dorsolateral tegmental nucleus to the medial prefrontal cortex. The locus ceruleus (LC) also contains CRH neurons that are believed to contribute to the positive feedback between CRH and the LC/NE systems.

The physiological actions of the CRH, Ucn 1, Ucn 2, and Ucn 3 peptides are mediated through two distinct mammalian receptor subtypes: CRH type 1 receptor (CRH-R1) and CRH type 2 receptor (CRH-R2) (Keck, 2006). CRH-R1 and CRH-R2 belong to the seven-transmembrane, G-protein-coupled family of receptors, and the two receptor types display markedly different tissue distribution and pharmacological profiles (Chalmers et al., 1995; Gabry et al., 2002; Steckler and Holsboer, 1999). CRH-R1 is expressed at high levels in the brain and pituitary and at low levels in peripheral tissues. The highest levels of CRH-R1 expression are found in the anterior pituitary, olfactory bulb, cerebral cortex, hippocampus, and cerebellum (Smith and Vale, 2006). CRH-R2, on the other hand, is expressed mainly in peripheral tissues, including the heart, skeletal muscle, skin, and gastrointestinal tract. CRH-R2, albeit at lower levels than CRH-R1, is also expressed in mammalian brain areas such as the lateral septum, BNST, amygdala, and hypothalamus (Chalmers et al., 1995). CRH with higher affinity binds to CRH-R1 than to CRH-R2. Furthermore, Ucn 1 has high affinity for both CRH-R1 and CRH-R2, while Ucn 2 and Ucn 3 are highly selective for CRH-R2 (Smith and Vale, 2006).

The neuroendocrine properties of CRH are mediated through CRH-R1 on anterior-pituitary corticotropes. Binding of CRH to CRH-R1 results in the stimulation of adenylate cyclase and subsequent activation of the cyclic adenosine monophosphate (cAMP) pathway, in turn, resulting in the release of ACTH from the pituitary corticotropes. Supporting this action, studies have shown that mice deficient in CRH-R1 have severely attenuated HPA-axis responses to stress and show decreased anxiety-like behaviors (Smith and Vale, 2006). The role of CRH-R2 in the stress response is less clear. Studies have shown that administration of CRH-R2 agonists and antagonists into discrete brain regions reveals both anxiolytic and anxiogenic roles for CRH-R2 (Smith and Vale, 2006).

CRH is the key hypothalamic peptide controlling HPA-axis activity; therefore, it plays a central role in the stress response. Glucocorticoids are potent inhibitors of CRH release from the PVN. The inhibition of CRH release by glucocorticoids is mediated directly at the level of the PVN, as well as indirectly through actions on receptors in the hippocampus. However, glucocorticoids exert a stimulatory role on CRH neurons in the amygdala and some neurons of the LC/NE system. The latter effects may be of importance in perpetuating the effects of severe stress by creating a positive-feedback loop (Calogero et al., 1988; Castro and Moreira, 1996; Spinedi et al., 1988). The CRH system is an integral component of the stress response, coordinating endocrine, autonomic, and behavioral functions to increase survival. CRH is the principal hormone regulating the HPA axis; however, the synergistic influence of AVP is also important.

2.1.4 Arginine Vasopressin

AVP, also known as antidiuretic hormone (ADH) and a key modulator of the HPA axis, has been linked to disorders such as anxiety, depression, and learning and memory deficits (Caldwell et al., 2008; Dinan and Scott, 2005; Scott and Dinan, 2002; Surget and Belzung, 2008). AVP is a cyclic nonapeptide, differing from oxytocin (another related hormone) by only two amino acid residues (Caldwell et al., 2008; Donaldson and Young, 2008; Surget and Belzung, 2008). AVP is released in a circadian manner, with peak release occurring in the daytime. AVP is also released during activated HPA-axis activity and potentiates CRH-mediated ACTH release (Caldwell et al., 2008).

AVP-producing neurons are spatially located throughout the brain in areas, including the BNST, medial amygdala, and suprachiasmatic nucleus (SCN) (Caldwell et al., 2008). The primary area of AVP production is in the magnocellular regions of the PVN and supraoptic nucleus (SON) of the hypothalamus (Caldwell et al., 2008). Magnocellular neurons of the PVN and SON project to the posterior pituitary and release AVP directly into the systemic circulation to regulate osmotic homeostasis (Aguilera and Rabadan-Diehl, 2000b; Antoni, 1986, 1993; Scott and Dinan, 2002; Smith and Vale, 2006; Surget and Belzung, 2008).

In addition to magnocellular neurons, parvocellular neurons of the PVN synthesize AVP and secrete it into the hypophyseal portal circulation in the external zone of the median eminence, where this peptide functions as another hypothalamic secretagog of ACTH (Smith and Vale, 2006). AVP synergizes with CRH during stress to stimulate the secretion of abundant quantities of ACTH. A subset of parvocellular neurons of the PVN synthesizes and secretes both CRH and AVP, whereas another subset secretes AVP only. Rodent studies have shown that under basal conditions, 50% of CRH-containing neurons in the PVN co-express AVP, and this ratio increases under acute and chronic stress conditions (Aguilera, 1994; Aguilera and Rabadan-Diehl, 2000b; Aguilera et al., 2008).

The axon terminals of the parvocellular neurons of the PVN project to different sites, including noradrenergic neurons of the brainstem and the hypophyseal portal system in the median eminence. PVN CRH and AVP neurons also send projections to and activate proopiomelanocortin (POMC)-containing neurons in the ARC of the hypothalamus, which in turn reciprocally project to the PVN. Other projections of PVN CRH and AVP neurons include the LC/NE system in the brainstem and pain-control neurons of the hind brain and spinal cord (Gabry et al., 2002).

The physiological actions of AVP are mediated by three receptor subtypes: the V_{1a}, V_{1b} (also known as the V_3), and V_2 (renal) receptors (Caldwell et al., 2008; Scott and Dinan, 2002). Similar to CRH receptors, each AVP receptor subtype is G-protein coupled and contains seven transmembrane domains (Surget and Belzung, 2008). In contrast to CRH receptor signaling, binding of AVP to the V_{1b} receptor activates phospholipase C via Gq proteins, stimulating protein kinase C and intracellular calcium, and ultimately potentiating ACTH release. V_{1b} receptors are located in pituitary corticotropes and in several brain regions, including the cortex, amygdala, hippocampus, and hypothalamus (Dinan and Scott, 2005; Surget and Belzung, 2008). V_{1a} receptors are also located throughout the brain, including within the PVN. V_2 receptors are located in the collecting tubules of the nephrons of the kidneys (Dinan and Scott, 2005).

AVP expression in parvocellular neurons of the PVN and V_{1b} receptor density in pituitary corticotropes is significantly increased following chronic stress (Aguilera and Rabadan-Diehl, 2000a,b; Kovacs et al., 2000; Kovacs and Sawchenko, 1996; Smith and Vale, 2006). As mentioned earlier, AVP production and storage in the PVN also appears to increase in CRH-containing neurons following chronic stress (Dinan and Scott, 2005; Scott and Dinan, 2002). Therefore, during chronic or prolonged stress, there appears to be an increase in the regulatory control of ACTH secretion by AVP and its receptors; nonetheless, current evidence indicates that the primary regulator of ACTH secretion in most stress paradigms is CRH (Aguilera et al., 2007, 2008; Surget and Belzung, 2008).

In addition to its role in HPA-axis regulation, AVP also acts as a neuromodulator and/or neurotransmitter, depending upon the mode of AVP release (Landgraf, 2006; Landgraf and Neumann, 2004). Central AVP has been implicated in processes such as learning and memory (Alescio-Lautier et al., 1993, 2000), cardiovascular function (Crofton et al., 1988; Versteeg et al., 1983), flank-marking behavior (Dubois-Dauphin et al., 1996, 1990), various social behaviors (Donaldson and Young, 2008; Heinrichs and Domes, 2008), thermoregulation, and motor behaviors (Kasting, 1988, 1989).

There are currently no knockouts of the AVP gene; however, the Brattleboro rat, a strain bred to lack AVP, has been indispensable for studying the effects of this peptide (Bohus and de Wied, 1998; Bundzikova et al., 2008; Caldwell et al., 2008; Grant, 2000). Studies with these rats suggest that reduced vasopressinergic activity is associated with attenuated depressive-like behaviors (Mlynarik et al., 2007). Also important in the study of AVP has been the development of specific antagonists for V_{1a} and V_{1b} receptors. Studies using these compounds have led to the association of the: (1) V_{1a} receptor with the anxiety-like effects of AVP and (2) V_{1b} receptor with AVP-mediated stress responses (Caldwell et al., 2008; Surget and Belzung, 2008). Because of their potential anxiolytic and antidepressant effects, these compounds show promise in the treatment of affective disorders (Surget and Belzung, 2008).

AVP is released from the posterior pituitary to maintain osmotic homeostasis. As part of the HPA axis, AVP is considered a key modulator of the stress response, particularly in chronic stress. AVP has been linked to disorders such as anxiety, depression, and learning and memory deficits. Regulation of the proportional secretion of CRH and AVP, modulation of pituitary V_{1b} receptors, and negative-feedback sensitivity play a role in allostasis and adaptation, by modulating HPA-axis activity.

2.1.5 Adrenocorticotropic Hormone

ACTH (corticotropin) is a 39-amino-acid peptide hormone produced by cells of the anterior pituitary gland and carried by the peripheral circulation to its effector organ, the adrenal cortex, where it stimulates the synthesis and secretion of glucocorticoids from the *zona fasciculata* by binding to melanocortin type 2 receptors (MC2-Rs) (Smith and Vale, 2006). Activation of MC2-R induces stimulation of cAMP-mediated events that increase steroidogenesis and the secretion of glucocorticoids, mineralocorticoids, and adrenal androgens (Cone and Mountjoy, 1993; Mountjoy et al., 1992; Smith and Vale, 2006). Advances in the measurement of ACTH and its related peptides have elucidated their extensive distribution in the body outside the pituitary gland, with differential processing in different tissues. ACTH and other peptides, including β-endorphin, a peptide known for its analgesic and euphoric effects in the brain, are produced in the pituitary from the chemical breakdown of a large precursor protein, POMC. POMC is processed in the anterior lobe to yield an N-terminal peptide, whose function is unclear, and the peptides ACTH and β-lipotropin (β-LPH). ACTH and β-LPH are secreted by the anterior pituitary. α-Melanocyte-stimulating hormone (α-MSH) and corticotropin-like intermediate lobe peptide are contained within the ACTH molecule. These peptides are found in species with developed intermediate lobes (e.g., amphibians, reptiles, and rats); however, they are not secreted as separate hormones in humans. Within the β-LPH molecule exists the amino acid sequence for γ-LPH, β-endorphin, β-MSH, and metenkephalin (Rhodes, 2007).

Anterior-pituitary ACTH is secreted by basophilic corticotropes that represent 15–20% of the total anterior-pituitary cell population. Under the electron microscope, ACTH-producing cells appear irregularly shaped and full of secretory granules. The entire ACTH molecule ($ACTH_{1-39}$) is not needed for biological activity. The first 16 amino acids, beginning with the N-terminal amino acid, are all that is required for minimal biological activity, although full biological activity is present only with a polypeptide over 22 amino acids long. $ACTH_{1-39}$ has a circulating half-life of 7–12 min and exhibits a normal circadian (24-h) rhythm: ACTH hormone levels are highest between 7 and 8 a.m., an hour or so after awakening, and lowest in early morning between 2 and 3 a.m. The development of immunoradiometric assays specific for intact human $ACTH_{1-39}$ has improved the reliability of ACTH measurement, although glucocorticoid concentrations in the blood can also be measured as an index of ACTH secretion.

The regulation of ACTH secretion primarily involves the stimulatory effect of CRH and AVP, hypothalamic hormones released directly into the portal blood supplying the anterior pituitary, and the inhibitory effect of glucocorticoids. A number of other factors, such as angiotensin II, catecholamines, circadian activity, and acute and chronic stress, stimulate ACTH secretion (**Figure 1**). ACTH secretion is also intimately linked with immune function. Interleukins are chemical messengers secreted by cells of the immune system that affect the behavior of the rest of the immune system. In addition to their effects on immune function, interleukin-1 (IL-1), IL-2, and IL-6 appear to stimulate HPA-axis activity (Arzt et al., 1999; Auernhammer et al., 1998; Bugajski, 1996; Turnbull and Rivier, 1995). IL-1 enhances

ACTH release perhaps due to enhancement of CRH release, or by modulating the actions of other ACTH secretagogs. IL-2 augments POMC gene expression in the anterior pituitary and enhances ACTH release. The potent ACTH release by IL-6 may derive from stimulation of AVP release (Besedovsky and del Rey, 2000; Dunn, 2000; Gabry et al., 2002; Gold and Chrousos, 1999).

The mechanisms regulating ACTH secretion during stress are multifactorial and include the stimulatory effect of the hypothalamus, mainly from CRH and AVP, and the inhibitory influence of the adrenal cortex from glucocorticoids (Aguilera, 1994). ACTH is regulated by negative-feedback mechanisms involving glucocorticoid binding to hormone receptors in the hippocampus, hypothalamus, and pituitary gland (**Figure 1**). ACTH itself also feeds back on the hypothalamus, exerting negative feedback. ACTH is regulated by the CNS, as well as by metabolic control of CRH secretion independent of glucocorticoids (Dallman et al., 2003; Rhodes and Rubin, 1999).

Cushing's syndrome is a clinical condition resulting from chronic elevation of circulating glucocorticoids. Clinical signs and symptoms of Cushing's syndrome include obesity of the face and trunk, weakness and atrophy of limb muscles, increased blood pressure, imbalance of glucose metabolism, and psychological changes. There are two main types of Cushing's syndrome: ACTH-dependent and ACTH-independent. ACTH-dependent Cushing's syndrome results from increased pituitary secretion of ACTH, usually from a pituitary tumor (Cushing's disease), inappropriate ACTH secretion by nonpituitary tumors, often in the lungs, and inappropriate CRH secretion by nonhypothalamic tumors, in turn stimulating excessive pituitary ACTH secretion. These conditions, all involving excess ACTH production, cause enlargement of the adrenal glands and excessive cortisol secretion from continuous stimulation (Rhodes, 2007). ACTH-independent Cushing's syndrome is caused by primary tumors or abnormalities of the adrenal cortex itself, resulting in excessive cortisol secretion and suppression of ACTH production by the pituitary. Prolonged administration of glucocorticoids for the treatment of certain illnesses may also cause ACTH-independent Cushing's syndrome (Rhodes, 2007).

ACTH is released from the anterior pituitary in response to various stimuli and its release is sustained or inhibited by CNS afferents and by intricate feedback systems, with ACTH at the heart of this dynamic homeostatic network of feedback loops (**Figure 1**). As part of the HPA axis, ACTH is considered as one of the major stress hormones. It is now apparent that AVP and CRH cooperate as the major factors involved in the control of ACTH release.

2.1.6 Glucocorticoids

The *zona fasciculata* cells of the adrenal cortex synthesize and secrete glucocorticoids in response to ACTH secreted by the anterior pituitary (Jacobson, 2005). Glucocorticoids, cortisol in humans and corticosterone in rodents, are the final effectors of the HPA axis. Stimulation of the HPA axis causes secretion of glucocorticoid hormones which act in the brain and periphery to promote adaptation to allostasis (Herman and Seroogy, 2006). These hormones exert a multitude of functions that affect virtually every cell in the body and exert their effects through their ubiquitously distributed intracellular receptors. Glucocorticoids function primarily to redistribute energy resources and are intimately involved in restoration or defense of homeostasis after challenge (Herman and Seroogy, 2006; Jacobson, 2005). Glucocorticoids play an important role in energy mobilization because they stimulate gluconeogenesis, which promotes lipolysis and an increase in protein catabolism (Mastorakos et al., 2005). Glucocorticoids also act within the brain to increase appetite as well as locomotor and food-seeking activities, which are important behaviors influencing energy expenditure (McEwen, 2000). In addition, these hormones play a key regulatory role in the regulation of HPA-axis activity and in the termination of the stress response by acting, via negative-feedback loops, on the hypothalamus as well as extrahypothalamic regulatory centers, such as the hippocampus, frontal complex, and pituitary gland (Chrousos, 2000a; Gabry et al., 2002).

Circulation of glucocorticoids occurs by low-affinity binding to albumin and by high-affinity binding to corticosteroid-binding globulin. The lipid-soluble glucocorticoid hormones easily penetrate the blood–brain barrier to affect mood, cognitive functions, and sleep in humans (Gabry et al., 2002; Jacobson, 2005). In general, glucocorticoids exert powerful suppressive effects on the immune and inflammatory systems, and synthetic glucocorticoids are often used to control symptoms stemming from many inflammatory and autoimmune diseases (Vinson et al., 2007).

In the rat (a nocturnal organism), morning plasma levels of corticosterone under resting conditions are generally in the range of 10–100 $ng ml^{-1}$ and they vary diurnally, with the highest levels just before or during the dark period (Marquez et al., 2005). In humans, the diurnal variation of plasma cortisol is lower, with the highest levels just after awakening (*c.* 200 $ng ml^{-1}$) and lower levels in the evening (*c.* 100 $ng ml^{-1}$) (Armario, 2006). HPA activity exhibits circadian rhythmicity in the absence of glucocorticoids, but the set point and amplitude of this rhythm are highly sensitive to feedback inhibition by glucocorticoids (Jacobson, 2005).

Tight control of glucocorticoid secretion is imperative to minimize the deleterious effects of glucocorticoid excess. Experimental and clinical evidence demonstrates that increased circulating glucocorticoid concentrations along with proinflammatory cytokines, associated with chronic stress and major depression, contribute to the behavioral changes associated with these conditions (Leonard, 2006). During chronic stress, glucocorticoids also produce insulin-mediated effects on adipose tissue promoting visceral adiposity and metabolic syndrome and its sequelae. Glucocorticoid excess is also observed in conditions such as Cushing's syndrome. As mentioned earlier, endogenous Cushing's syndrome results from increased secretion of cortisol by the adrenal cortex and is due to ACTH hypersecretion or autonomous hyperfunction of the adrenocortical cells. The profound catabolic activities of glucocorticoids lie behind some of the primary clinical manifestations of Cushing's syndrome: delayed growth and bone maturation, hypogonadism, decreased lean body mass, and frequent fungal or saprophytic infections (Gabry et al., 2002).

Negative-feedback inhibition by glucocorticoids on extrahypothalamic, hypothalamic, and pituitary tissues limits the duration of the total tissue exposure to glucocorticoids, thus minimizing the catabolic, antireproductive, and immunosuppressive effects of these hormones (Gabry et al., 2002). Glucocorticoid feedback inhibition of the HPA axis occurs through at least two distinct mechanisms involving several time domains, referred to as fast, delayed (or intermediate), and slow feedback (Jacobson, 2005; Keller-Wood and Dallman, 1984). Two receptors mediate the effects of glucocorticoids. This dual receptor system is composed of the glucocorticoid receptor type I (or mineralocorticoid receptor), which responds positively to low levels of glucocorticoids, and the classic glucocorticoid receptor (type II), which responds to both basal and stress levels of glucocorticoids. Glucocorticoid type II receptors are expressed widely in many peripheral tissues and brain regions, while glucocorticoid type I receptors are restricted to peripheral aldosterone targets such as the kidneys and colon, and to a limited number of brain regions (Jacobson, 2005). The glucocorticoid sensitivity of target tissues is defined not only by glucocorticoid receptors that are present, but also by other molecules that participate in the glucocorticoid signal transduction pathway, including the heat-shock proteins, several transcription coregulator molecules, and other transcription factors (Lamberts et al., 1996; Thrivikraman et al., 2000).

Glucocorticoids are the final effectors of the HPA axis and key mediators of the stress system. Because glucocorticoid receptors are widely expressed in the brain, the precise anatomical locations determining glucocorticoid negative feedback remain poorly defined; however, two regions of the brain (the PVN and the hippocampus) appear to be key sites for feedback to the HPA axis. The PVN expresses high levels of glucocorticoid type I receptors. In addition, the hippocampus expresses high levels of both type I and type II receptors, and infusion of glucocorticoids into this brain area reduces basal and stress-induced HPA-axis activity (Smith and Vale, 2006). During periods of chronic stress, elevated glucocorticoids from maladaptive responses to allostatic load may produce damage to body systems resulting in various pathologies.

2.2 Brain Regulation of Stress Responses

As mentioned previously, the direct and indirect afferents from various limbic, hypothalamic, and brainstem regions innervate the PVN and regulate the release of CRH via numerous neurotransmitter systems. Majority of afferent projections to the PVN originate from four distinct brain regions: brainstem neurons, nuclei of the lamina terminalis, extra-PVN hypothalamic nuclei, and forebrain limbic structures (reviewed by Smith and Vale (2006)). These afferent projections are summarized in **Figure 2**. Efferents of the NST innervate the PVN and relay the peripheral status of an organism via cranial nerves that innervate thoracic and abdominal areas (Rhodes and Rubin, 1999). Catecholaminergic input from the NST appears to be a major activator of the HPA axis and increases CRH expression and release through an α-1 adrenergic receptor mechanism (Plotsky, 1987; Plotsky

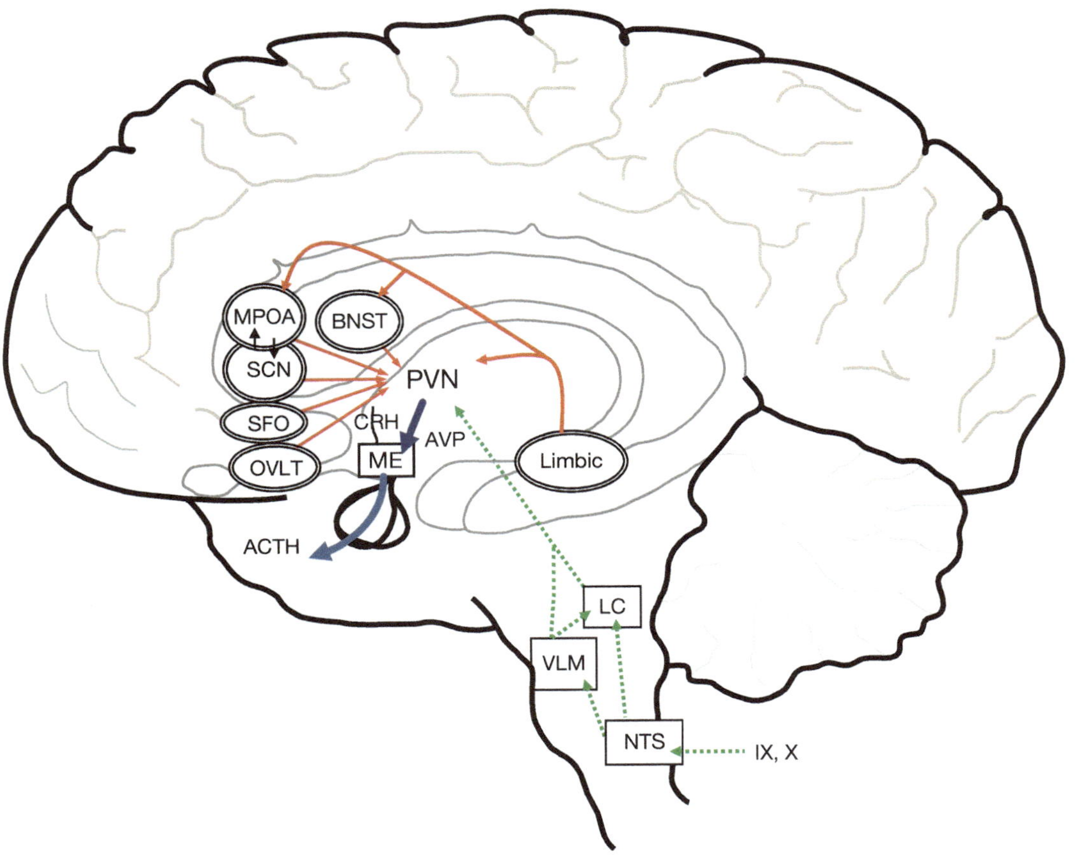

Figure 2 Schematic representation of the primary innervation of the paraventricular nucleus (PVN) of the hypothalamus, which produces CRH and AVP. CRH and AVP are secreted into the median eminence (ME) and transported by the hypophyseal portal vessels to the anterior pituitary. The PVN receives direct and/or indirect afferents from limbic structures (Limbic), including the hippocampus, prefrontal cortex, and amygdala. The PVN also receives direct and/or indirect afferents from lamina terminalis structures such as the vascular organ of the lamina terminalis (OVLT) and the subfornical organ (SFO). Indirect pathways pass primarily through the medial preoptic area (MPOA), suprachiasmatic nucleus (SCN), and bed nucleus of the stria terminalis (BNST). Additionally, major adrenergic inputs to the PVN include the locus ceruleus (LC), ventrolateral medulla (VLM), and the nucleus of the solitary tract (NTS), which receives peripheral afferents via the glosso-pharyngeal (IX) and vagus (X) nerves.

et al., 1989). In addition, the NST regulates the HPA axis via the neuropeptide somatostation, substance P, and enkephalin (Saphier et al., 1994; Sawchenko et al., 1988a,b).

The lamina terminalis, which lies outside of the blood–brain barrier, is localized on the rostral border of the third ventricle and includes the subfornical organ (SFO), the vascular organ of the lamina terminalis (OVLT), and the median preoptic nucleus. Blood osmotic composition is relayed via afferents to the magnocellular areas of the PVN and the SON via these cell groups. Thus, lamina terminalis projections regulate the release of CRH from the PVN and AVP from both hypothalamic nuclei, serving as a vital link between HPA and neurohypophysial activation (Antoni, 1993; Engelmann et al., 2004; Smith and Vale, 2006).

Gamma-aminobutyric acid (GABA)-ergic neurons in the dorsomedial hypothalamic nucleus (DMH) and MPOA of the hypothalamus innervate the PVN. The MPOA is an important linking structure between the limbic system and the PVN (**Figure 2**) and it expresses high levels of gonadal steroid receptors. The MPOA therefore may represent an important site of integration among gonadal hormones, the limbic system, and the HPA axis (Simerly et al., 1990). The SCN also directly and indirectly (by way of the MPOA) innervates the PVN and SON, and regulates the circadian rhythm in HPA-axis activity (Hofman, 1997; Madeira and Lieberman, 1995).

Another hypothalamic area, the ARC of the hypothalamus, regulates energy homeostasis. Information concerning blood glucose, insulin, and leptin composition is relayed via ARC efferents to PVN neurons (Higuchi et al., 2005; Sahu, 2004). The ARC is also important in the regulation of luteinizing hormone, GH, and prolactin secretion. GnRH neurons of the ARC are important in the reproductive axis and are inhibited by CRH (Gold and Chrousos, 2002).

Limbic forebrain areas directly and indirectly regulate the HPA axis (**Figure 2**). Behavioral aspects of the stress response, including memory formation, alertness and arousal changes, and emotional responses, such as anxiety, fear, sadness, rage, and anger, are rooted in the limbic system. Limbic forebrain areas include the hippocampus, prefrontal cortex, and amygdala. Majority of limbic system projections innervate the PVN indirectly via the BNST, hypothalamus (MPOA), and the brainstem (Smith and Vale, 2006). The LC also contributes to behavioral aspects of stress-response adaptation, including alertness and arousal, memory, and emotion. Stress-related and affective disorders have been linked to dysfunction of catecholamergic neurons in the LC (Rassnick et al., 1994; Southwick et al., 1999ab, 2005).

Numerous other neuroendocrine factors are involved in regulating and modulating HPA-axis activity, including various neurotransmitters, neuropeptides, adrenal steroid hormones, and gonadal steroid hormones, as well as non-neuroendocrine factors such as age, weight, and environmental influences (Antoni, 1993; Suescun et al., 1997; Whitnall, 1993). GABA agonists and benzodiazepines exert an inhibitory effect on CRH neurons (Gabry et al., 2002). In contrast, cholinergic and serotonergic neurons stimulate CRH release (Rhodes et al., 2001a,b; Rhodes and Rubin, 1999). NE and opioid peptides have both stimulatory and inhibitory effects on CRH release, depending on the dosage administered and the receptor type involved. Glucocorticoids are potent inhibitors of CRH release from the PVN and, therefore, HPA-axis activity. The inhibition of CRH release by glucocorticoids is mediated directly at the level of the PVN, as well as indirectly through actions on glucocorticoid receptors in the hippocampus. However, as mentioned earlier, glucocorticoids exert a stimulatory role on CRH neurons in the amygdala and some neurons of the LC/NE system which may create a positive-feedback loop that perpetuates maladaptive responses to allostatic load (Calogero et al., 1988; Castro and Moreira, 1996; Spinedi et al., 1988).

2.3 Physiological Responses to Stress

Selye's landmark paper first defined stress as a nonspecific response to demands placed upon the body (Selye, 1936). The stress response was defined as unfolding in specific stages comprising the GAS: (1) the alarm phase where the disruptive effects of the stressor alter homeostatic processes, leading to balancing effects of adrenal hormones, (2) the resistance phase where the deleterious effects of homeostatic processes begin, and (3) the exhaustion phase characterized by the classic triad of the stress response – adrenal hypertrophy, thymolymphatic dystrophy, and stress ulceration (Selye, 1936).

Stress is associated with inhibition of gastric secretion and motility, inhibition of small intestinal motility, and enhancement of large bowel transit. Other responses include mucin depletion, diminution of mucosal blood flow, mast cell degranulation, oxidative injury, and increased susceptibility to inflammation and stress ulceration (Pothoulakis et al., 1998). The gut and the brain are highly integrated and communicate in a bidirectional fashion largely through the ANS and HPA axis. Within the CNS, the locus of gut control appears to reside predominantly within the limbic system. A better understanding of the interactions of the CNS, HPA axis, and enteric nervous system will significantly improve our understanding of gastrointestinal disorders (Crowell et al., 2004; Jones et al., 2006; Wessinger et al., 2005).

A close reciprocal relationship also exists between the HPA and hypothalamic–pituitary–ovarian (HPO) axes. Basal HPA activity changes as the estrous cycle fluctuates and interference with the adrenal corticosteroid circadian rhythm results in irregular estrous cycles (Rhodes et al., 2002, 2004). The reproductive axis is inhibited at all levels by various components of the HPA axis. Thus, either directly or via the ARC-POMC neurons that secrete β-endorphin, CRH suppresses GnRH neurons of the ARC and preoptic nuclei. The presence of CRH-R1 mRNA in a subset of gonadotropes highlights the pituitary as a potentially important site of interaction between the HPA and HPO axes (Westphal et al., 2009). In addition, glucocorticoids exert inhibitory effects at the hypothalamic, pituitary, and gonadal levels. Moreover, glucocorticoids render target tissues resistant to the actions of sex steroids (Gabry et al., 2002).

The growth axis is also inhibited at many levels during stress. Prolonged activation of the HPA axis leads to the suppression of GH secretion, perhaps

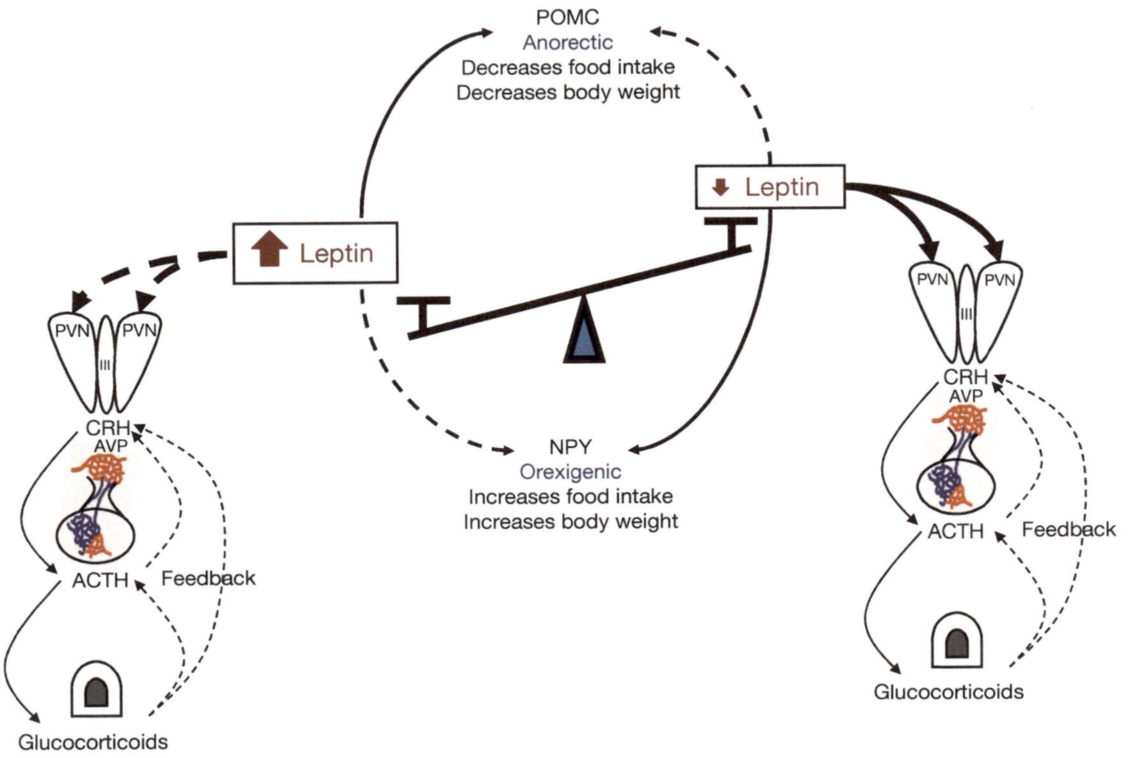

Figure 3 Connections among the hypothalamic–pituitary–adrenal cortical (HPA) axis, leptin, proopiomelanocortin (POMC), and neuropeptide Y (NPY). Increased leptin levels with increased adiposity suppress HPA-axis activity. Decreased leptin levels during fasting, starvation, and anorexia increase HPA-axis activity. Thus, leptin is central to the regulation of energy homeostasis. Solid lines represent stimulation; dashed lines represent negative feedback or inhibition.

resulting from CRH-mediated increases in somatostatin secretion, with resultant inhibition of GH secretion (Chrousos and Gold, 1992). Glucocorticoids not only have profound inhibitory effects on GH and gonadal steroid production, but also antagonize the actions of these hormones on fat-tissue catabolism, and muscle and bone anabolism. Thus, the chronic activation of the stress system is expected to decrease lipolysis and increase visceral adiposity, decrease lean body (bone and muscle) mass, and suppress osteoblastic activity.

Leptin, a hormone produced in white adipose tissue, relays information regarding the status of body energy stores to the CNS, and is thus important in appetite suppression, food-intake behaviors, and energy metabolism (Fehm et al., 2006; Kelesidis and Mantzoros, 2006; Mantzoros, 1999; Mantzoros and Moschos, 1998; Rubin et al., 2002). Leptin receptors are predominantly found not only in the ARC, but also in the brainstem and other CNS regions (Sahu, 2003, 2004, 2008). Of particular interest to leptin sensitivity are two types of neurons in the ARC, neuropeptide Y (NPY)- and POMC-containing neurons. NPY is a potent orexigenic neuropeptide, thus stimulating food intake and body weight. POMC and its related peptides are anorectic, thus reducing food intake and body weight (Sahu, 1998a,b, 2003). Increases in leptin stimulate POMC and reduce NPY signaling (**Figure 3**). Mutation in the leptin gene or the leptin receptor is associated with massive obesity in humans and rodents (Sahu, 2004).

Leptin production is most directly controlled by the amount of body mass present in the individual (i.e., weight loss results in lower circulating leptin levels, while increasing adiposity results in increased levels of leptin; **Figure 3**). Leptin release is also stimulated by the release of glucocorticoids and insulin. During starvation, decreased levels of leptin signal activation of the HPA axis, while higher levels of leptin suppress the HPA axis, signaling that glucocorticoid levels are sufficient. In other words, by suppressing the HPA axis, increases in leptin work to suppress the appetite-stimulating effects of glucocorticoids that would be appropriate during starvation (Ahima et al., 2000; Jacobson, 2005; Sahu, 2004). However, chronic elevation of glucocorticoids, due to chronic activation of the HPA axis, may attenuate the ability of leptin to inhibit HPA-axis activity, leading to increased food intake and obesity. The effects of leptin therefore represent examples of the consequences resulting from maladaptive responses to allostatic load. With the growing prevalence of type II diabetes mellitus and obesity, understanding the association among stress, the HPA axis, and leptin is important.

2.4 HPA Dysregulation: Conditions with Altered HPA-Axis Activity

2.4.1 Hyperactive Conditions

Affective disorders, such as depression, certain types of anxiety, and obsessive–compulsive disorder, are related to stress and characterized by increased HPA-axis hormone secretion (Abelson et al., 2007; Arborelius et al., 1999; Keck, 2006; Surget and Belzung, 2008). Arguably, the best example of hyperactivation of the HPA axis occurs in major depression (Arborelius et al., 1999; Claes, 2004a; Gabry et al., 2002; Tichomirowa et al., 2005). AVP appears to play a prominent role in stress-associated responses, and many studies have linked AVP to stress-related diseases such as anxiety and depression (Caldwell et al., 2008; Landgraf, 2006). Patients with major depression have higher plasma AVP compared to healthy controls, and higher ACTH and cortisol responses to AVP administration (Bao et al., 2008; Dinan et al., 2004; Rubin et al., 1999; van Londen et al., 1997). Furthermore, patients with depression have greater amounts of AVP mRNA in both the SON and the PVN (Meynen et al., 2006). In addition to affective disorders, a broad spectrum of other conditions may be associated with increased and/or prolonged activation of the HPA axis. These include Alzheimer's disease, alcoholism, alcohol and narcotic withdrawal, poorly controlled diabetes, the third trimester of pregnancy, gastrointestinal disorders, excessive exercise, and AN and malnutrition (**Figure 4**).

Drug addiction is a chronic mental illness characterized by compulsive drug use despite harmful consequences, the development of tolerance, the appearance of withdrawal symptoms upon cessation of drug use, and relapse to drug-taking behavior after periods of abstinence (Bruijnzeel and Gold, 2005). Exposure to stress and associated increases in HPA activity have been shown to increase vulnerability to addiction (Goeders, 2003). Chronic activation of the HPA axis resulting from continued drug use blunts HPA-axis responses, such that, upon cessation of drug use, homeostatic responses overshoot, leading to increased hypothalamic and extrahypothalamic CRH production and increased vulnerability to negative affective episodes (Bruijnzeel and Gold, 2005).

HPA alterations appear to be dependent upon the substance of abuse. Withdrawal from nicotine and alcohol is characterized by increased depression and anxiety, cravings, irritability, difficulty in concentrating, and restlessness (Bruijnzeel and Gold, 2005). However, nicotine withdrawal has also been associated with decreased HPA-axis activity and decreased catecholamine secretion (Bruijnzeel and Gold, 2005; Gabry et al., 2002). In contrast, alcohol use and withdrawal are characterized by increased HPA-axis activity, which normalize after approximately 2 weeks as body systems stabilize and the severity of the withdrawal syndrome wanes (Clarke et al., 2008; Kiefer et al., 2006; Kiefer and Wiedemann, 2004). Interestingly, HPA hormone levels may actively modulate alcohol-related behaviors, with low CRH levels being associated with more intense craving and increased probability of relapse after acute detoxification (Adinoff et al., 2005; Junghanns et al., 2003).

Chronic stress can have detrimental effects on memory and may represent yet another important example of damage resulting from maladaptive responses to allostatic load. Damage to the hippocampus is a characteristic of Alzheimer's disease patients. Hippocampal atrophy disrupts negative feedback to the HPA axis, resulting in hypercortisolemia (Magri et al., 2006; Pomara et al., 2003), which in turn, may cause hippocampal dysfunction and interfere with the memory performance, potentially creating a vicious cycle (Lupien et al., 2005, 1999; Pomara et al., 2003; Wolf, 2003; Wolf et al., 2001). HPA-axis hyperactivity and the related hippocampal atrophy are characteristics of Alzheimer's disease. Although HPA dysreguation is not the primary cause of the disease, elevated levels of glucocorticoids may enhance or accelerate the structural damage in the brain of Alzheimer's disease patients (Wolf, 2003). Therefore, the association between stress and memory appears important to understanding the pathophysiology and progression of memory dysfunction associated with Alzheimer's disease.

The concept of the metabolic syndrome has developed in response to increased findings that insulin resistance is linked to altered lipid profiles which are characteristic of diabetes mellitus and cardiovascular disease (Bjorntorp, 1993; Rosmond, 2002; Vicennati and Pasquali, 2000). Risk factors for metabolic

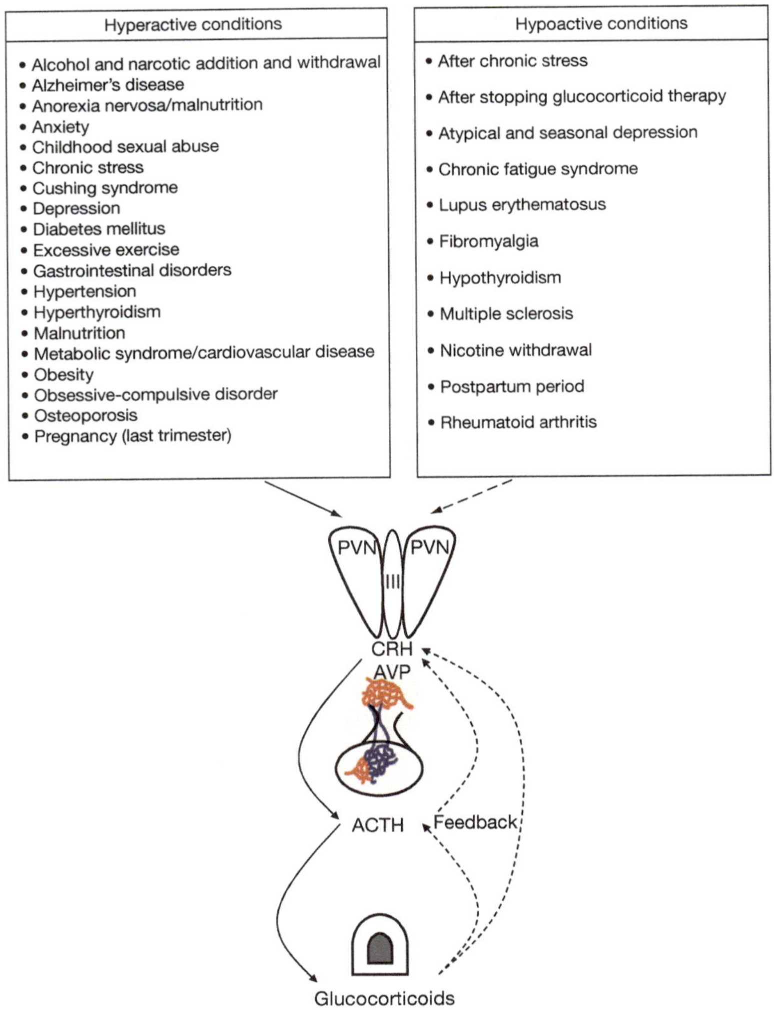

Figure 4 Hyperactive and hypoactive conditions influencing hypothalamic–pituitary–adrenal cortical (HPA)-axis activity. Solid lines represent stimulation; dashed lines represent negative feedback or inhibition.

syndrome include: elevated insulin levels, visceral obesity, high levels of triglycerides, low levels of high-density lipoprotein (HDL) cholesterol, and hypertension (Rosmond, 2002). Glucocorticoids induce insulin resistance, and HPA-axis activation contributes to the poor control of diabetic patients during periods of emotional stress or concurrently with inflammatory and other diseases (Gabry et al., 2002). Indeed, a mild chronic activation of the HPA axis has been demonstrated in diabetic patients under moderate or poor glycemic control (Chrousos, 2000b). Glucocorticoid-induced and progressively increasing visceral adiposity causes further insulin resistance and deterioration of glycemic control of patients with diabetes mellitus (Buckingham, 2006; Chrousos, 2000b). Increasing visceral adiposity, in turn, may exaggerate the chronic elevation of glucocorticoids associated with chronic stress (Vicennati et al., 2006). Thus chronic activation of the stress system in diabetes may contribute to a vicious cycle of increasing hyperglycemia, hypercholesterolemia, and insulin needs (Chrousos, 2000b).

Stress studies involving organisms do not determine plasma CRH, because it is not a valid index of hypothalamic CRH (Plotsky et al., 1990). Although CRH is not detected in the circulation under normal circumstances, very high levels have been measured in the plasma of pregnant women, reflecting CRH of placental origin (Makrigiannakis et al., 1997; Margioris et al., 1988). Pregnancy in the third trimester is the only known physiological state in humans in which CRH circulates in plasma at levels high enough to cause the activation of the HPA axis. Although circulating CRH is bound with high affinity to CRH-binding protein, it appears free plasma CRH is sufficient to produce the escalating circulating cortisol concentrations in the third trimester of pregnancy. The administration of CRH-R1 antagonists to pregnant sheep has been shown to delay the onset of parturition, suggesting that placental CRH may have a role in precipitating labor (Gabry et al., 2002).

As mentioned previously, the gut and the brain are highly integrated and communicate in a bidirectional fashion largely through the ANS and HPA axis (Jones et al., 2006). Physical and psychological stressors are widely accepted as triggers and/or modifiers of the clinical course of diverse gastrointestinal disorders (Caso et al., 2008). It is common for humans to respond to stress with nausea or diarrhea, and a strong association has been reported between stress, associated HPA-axis activation, and gastrointestinal disorders such as peptic ulcer, irritable bowel syndrome, and inflammatory bowel disease (Caso et al., 2008; Gabry et al., 2002). CRH has been implicated as an important mediator of stress-induced abnormalities in intestinal mucosal function in animal models (Wallon et al., 2008). Endogenous CRH in the brain plays a significant role in the autonomic mediation of stress-induced inhibition of upper gastrointestinal and stimulation of lower gastrointestinal motor function through activation of brain CRH receptors. The inhibition of gastric emptying by CRH may be mediated by interaction with the CRH-R2 receptor, while CRH-R1 receptors are involved in the colonic and anxiogenic responses to stress (Monnikes et al., 2001). Also, glucocorticoids influence the etiology and development of gastrointestinal disorders. Much evidence supports the idea that glucocorticoid hormones released in response to acute stress act as gastroprotective substances; however, chronic stress is associated with peptic ulceration and irritable bowel syndrome (Filaretova et al., 2001, 2007; Gabry et al., 2002).

Excessive exercise may be associated with increased and prolonged activation of the HPA axis (Gilligan et al., 2000). Excessive exercise is not only observed in the training regimen of highly trained athletes, it may also be observed in individuals exhibiting exercise addiction. Exercise addiction is generally described as an addiction of a psychological and/or physiological nature to regular physical activity, characterized by withdrawal symptoms after 2–3 days without exercise (Khatri and Blumenthal, 2007). Conditions such as overtraining have been explored because of the tendency for excessive exercisers to ineffectively allow time for muscle recovery and rest. The balance between training volumes and physiological recovery is often a very delicate one; exercise-induced stress that exceeds the capacity of neuroendocrine adaptation may occur even in the absence of overtraining (Angeli et al., 2004). The stress associated with intense training leads to an increase in plasma catecholamines and glucocorticoids. Sustained physical conditioning in highly trained athletes is associated with a decreased HPA response to exercise; however, this population exhibits a chronic mild hypercortisolism at baseline that may be an adaptive change to chronic exercise (Mastorakos and Pavlatou, 2005; Mastorakos et al., 2005). In highly trained runners, HPA activation during acute exercise is inversely proportional to the level of physical training, perhaps resulting from maladaptative responses to allostatic load. There is some suggestion of an overlap between excessive exercise and eating disorders, but there is little empirical evidence to support this notion (Khatri and Blumenthal, 2007).

AN is widely recognized as a syndrome of weight loss with features including weight loss below 85% of the individual's expected body weight and an intense fear of excessive weight gain coupled with abnormal perceptions of body-weight status (Jimerson, 2002; Jimerson and Wolfe, 2004). Hypercortisolism is a manifestation of self-starvation; restricted feeding or increased physical activity activates the HPA axis and causes a loss of body weight (Bergh and Sodersten, 1996). NPY, a neuropeptide of fundamental importance in the stress circuitry, is emerging as an important neuromodulatory agent that affects behavior, anterior-pituitary hormone secretion, autonomic control, and other neurotransmitter systems (**Figure 3**; Gabry et al., 2002). It is of interest that AN and malnutrition are characterized by increased levels of cerebrospinal fluid NPY, which could provide an explanation why the HPA axis in these subjects is activated (Kaye et al., 1990; Laue et al., 1991; Licinio et al., 1996), whereas the LC/NE system is hypoactive (Laue et al., 1991).

It has been suggested that HPA-axis activity may serve as a window to an organism's allostatic load, an index of the cumulative effects of continued maintenance of homeostasis within the body (McEwen, 2000). A high allostatic load, reflected by hyperactivity of the HPA axis, may lead to a number of unhealthy outcomes and an increased morbidity, because conditions such as depression and other affective disorders are associated with cardiovascular disease and diabetes.

2.4.2 Hypoactive Conditions

Chronically reduced secretion of CRH may result in pathological hypoarousal and characterizes another group of conditions associated with blunted activation, or overall suppression, of the HPA axis (**Figure 4**). Patients with atypical and seasonal depression (i.e., seasonal affective disorder), fibromyalgia, and chronic fatigue syndrome fall into this category. Hypothyroid patients also have decreased HPA-axis activity. As mentioned earlier, nicotine withdrawal has been associated with decreased cortisol and catecholamine secretion (Gabry et al., 2002).

The period after being cured of hypercortisolism (i.e., following glucocorticoid therapy), the postpartum period, and periods following cessation of chronic stress are also associated with atypical depression, suppressed PVN CRH secretion, and decreased HPA-axis activity (Chrousos, 1998). Patients with rheumatoid arthritis are more likely to have an increased incidence of atypical depression and decreased HPA-axis activity. Hypoactivity of the HPA axis is also associated with other autoimmune processes such as lupus erythematosus and multiple sclerosis (Gabry et al., 2002; Heesen et al., 2007; Kudielka and Kirschbaum, 2005).

2.5 Conclusion

The HPA axis is a three-gland component of the endocrine system and a key regulator of the stress response. The HPA axis is the major endocrine output of a central CRH system that coordinates the autonomic, neuroendocrine, behavioral, and immune responses following alterations in homeostasis. Afferent projections from limbic, midbrain, and brainstem nuclei, inhibitory feedback from glucocorticoids, and circadian activity regulate the magnitude and duration of HPA-axis activity at baseline and following acute and chronic stress. The HPA axis is also regulated at the level of the hypothalamus by numerous other factors, including neuropeptides, gonadal steroid hormones, age, weight, and environmental influences. A broad spectrum of conditions, including affective, metabolic, and eating disorders, may be associated with sustained maladaptive hyperactivity or hypoactivity of the HPA axis as a result of excessive or prolonged allostatic load. The development of agonists and antagonists specific to CRH and AVP receptor subtypes represents a promising avenue toward the treatment of disorders associated with HPA-axis hypoactivity or hyperactivity. Understanding the physiology of the HPA axis and its regulation should further our understanding of the neurochemical basis of behavior during stress, and may uncover new therapeutic strategies for stress-related disorders.

References

Abelson JL, Khan S, Liberzon I, and Young EA (2007) HPA axis activity in patients with panic disorder: Review and synthesis of four studies. *Depression and Anxiety* 24: 66–76.

Adinoff B, Junghanns K, Kiefer F, and Krishnan-Sarin S (2005) Suppression of the HPA axis stress-response: Implications for relapse. *Alcoholism, Clinical and Experimental Research* 29: 1351–1355.

Aguilera G (1994) Regulation of pituitary ACTH secretion during chronic stress. *Frontiers in Neuroendocrinology* 15: 321–350.

Aguilera G, Kiss A, Liu Y, and Kamitakahara A (2007) Negative regulation of corticotropin releasing factor expression and limitation of stress response. *Stress* 10: 153–161.

Aguilera G and Rabadan-Diehl C (2000a) Regulation of vasopressin V1b receptors in the anterior pituitary gland of the rat. *Experimental Physiology* 85: 19S–26S.

Aguilera G and Rabadan-Diehl C (2000b) Vasopressinergic regulation of the hypothalamic–pituitary–adrenal axis: Implications for stress adaptation. *Regulatory Peptides* 96: 23–29.

Aguilera G, Subburaju S, Young S, and Chen J (2008) The parvocellular vasopressinergic system and responsiveness of the hypothalamic pituitary adrenal axis during chronic stress. *Progress in Brain Research* 170: 29–39.

Ahima RS, Saper CB, Flier JS, and Elmquist JK (2000) Leptin regulation of neuroendocrine systems. *Frontiers in Neuroendocrinology* 21: 263–307.

Alescio-Lautier B, Metzger D, and Soumireu-Mourat B (1993) Central behavioral effects of vasopressin: Point and perspectives. *Reviews in the Neurosciences* 4: 239–266.

Alescio-Lautier B, Paban V, and Soumireu-Mourat B (2000) Neuromodulation of memory in the hippocampus by vasopressin. *European Journal of Pharmacology* 405: 63–72.

Angeli A, Minetto M, Dovio A, and Paccotti P (2004) The overtraining syndrome in athletes: A stress-related disorder. *Journal of Endocrinological Investigation* 27: 603–612.

Antoni FA (1986) Hypothalamic control of adrenocorticotropin secretion: Advances since the discovery of 41-residue corticotropin-releasing factor. *Endocrine Reviews* 7: 351–378.

Antoni FA (1993) Vasopressinergic control of pituitary adrenocorticotropin secretion comes of age. *Frontiers in Neuroendocrinology* 14: 76–122.

Arborelius L, Owens MJ, Plotsky PM, and Nemeroff CB (1999) The role of corticotropin-releasing factor in depression and anxiety disorders. *Journal of Endocrinology* 160: 1–12.

Armario A (2006) The hypothalamic–pituitary–adrenal axis: What can it tell us about stressors? *CNS Neurological Disorders – Drug Targets* 5: 485–501.

Arzt E, Pereda MP, Castro CP, Pagotto U, Renner U, and Stalla GK (1999) Pathophysiological role of the cytokine network in the anterior pituitary gland. *Frontiers in Neuroendocrinology* 20: 71–95.

Auernhammer CJ, Chesnokova V, and Melmed S (1998) Leukemia inhibitory factor modulates interleukin-1beta-induced activation of the hypothalamo-pituitary-adrenal axis. *Endocrinology* 139: 2201–2208.

Bao AM, Meynen G, and Swaab DF (2008) The stress system in depression and neurodegeneration: Focus on the human hypothalamus. *Brain Research Reviews* 57: 531–553.

Bergh C and Sodersten P (1996) Anorexia nervosa, self-starvation and the reward of stress. *Nature Medicine* 2: 21–22.

Besedovsky HO and del Rey A (2000) The cytokine-HPA axis feed-back circuit. *Zeitschrift fur Rheumatologie* 59 (supplement 2): II/26–II/30.

Bjorntorp P (1993) Visceral obesity: A "civilization syndrome." *Obesity Research* 1: 206–222.

Bohus B and de Wied D (1998) The vasopressin deficient Brattleboro rats: A natural knockout model used in the search for CNS effects of vasopressin. *Progress in Brain Research* 119: 555–573.

Bremner JD, Krystal JH, Southwick SM, and Charney DS (1996) Noradrenergic mechanisms in stress and anxiety: II. Clinical studies. *Synapse* 23: 39–51.

Bruijnzeel AW and Gold MS (2005) The role of corticotropin-releasing factor-like peptides in cannabis, nicotine, and alcohol dependence. *Brain Research. Brain Research Reviews* 49: 505–528.

Buckingham JC (2006) Glucocorticoids: Exemplars of multi-tasking. *British Journal of Pharmacology* 147(supplement 1): S258–S268.

Bugajski J (1996) Role of prostaglandins in the stimulation of the hypothalamic–pituitary–adrenal axis by adrenergic and neurohormone systems. *Journal of Physiology and Pharmacology* 47: 559–575.

Bundzikova J, Pirnik Z, Zelena D, Mikkelsen JD, and Kiss A (2008) Response of substances co-expressed in hypothalamic magnocellular neurons to osmotic challenges in normal and Brattleboro rats. *Cellular and Molecular Neurobiology* 28: 1033–1047.

Caldwell HK, Lee HJ, Macbeth AH, and Young WS, 3rd (2008) Vasopressin: Behavioral roles of an "original" neuropeptide. *Progress in Neurobiology* 84: 1–24.

Calogero AE, Bernardini R, Gold PW, and Chrousos GP (1988) Regulation of rat hypothalamic corticotropin-releasing hormone secretion *in vitro*: Potential clinical implications. *Advances in Experimental Medicine and Biology* 245: 167–181.

Caso JR, Leza JC, and Menchen L (2008) The effects of physical and psychological stress on the gastro-intestinal tract: Lessons from animal models. *Current Molecular Medicine* 8: 299–312.

Castro M and Moreira AC (1996) Regulation of corticotropin-releasing hormone secretion by ACTH at different times after adrenalectomy. *Brazilian Journal of Medical and Biological Research* 29: 1573–1578.

Chalmers DT, Lovenberg TW, and De Souza EB (1995) Localization of novel corticotropin-releasing factor receptor (CRF2) mRNA expression to specific subcortical nuclei in rat brain: Comparison with CRF1 receptor mRNA expression. *Journal of Neuroscience* 15: 6340–6350.

Charmandari E, Tsigos C, and Chrousos G (2005) Endocrinology of the stress response. *Annual Review of Physiology* 67: 259–284.

Chrousos GP (1998) Stressors, stress, and neuroendocrine integration of the adaptive response. The 1997 Hans Selye memorial lecture. *Annals of the New York Academy of Sciences* 851: 311–335.

Chrousos GP (2000a) The HPA axis and the stress response. *Endocrine Research* 26: 513–514.

Chrousos GP (2000b) The role of stress and the hypothalamic–pituitary–adrenal axis in the pathogenesis of the metabolic syndrome: Neuro-endocrine and target tissue-related causes. *International Journal of Obesity and Related Metabolic Disorders* 24(supplement 2): S50–S55.

Chrousos GP and Gold PW (1992) The concepts of stress and stress system disorders. Overview of physical and behavioral homeostasis. *Journal of the American Medical Association* 267: 1244–1252.

Claes SJ (2004a) Corticotropin-releasing hormone (CRH) in psychiatry: From stress to psychopathology. *Annals of Medicine* 36: 50–61.

Claes SJ (2004b) CRH, stress, and major depression: A psychobiological interplay. *Vitamins and Hormones* 69: 117–150.

Clarke TK, Treutlein J, Zimmermann US, et al. (2008) HPA-axis activity in alcoholism: Examples for a gene–environment interaction. *Addiction Biology* 13: 1–14.

Cone RD and Mountjoy KG (1993) Molecular genetics of the ACTH and melanocyte-stimulating hormone receptors. *Trends in Endocrinology and Metabolism* 4: 242–247.

Crofton JT, Share L, and Brooks DP (1988) Pressor responsiveness to and secretion of vasopressin during the estrous cycle. *American Journal of Physiology* 255: R1041–R1048.

Crowell MD, Jones MP, Harris LA, Dineen TN, Schettler VA, and Olden KW (2004) Antidepressants in the treatment of irritable bowel syndrome and visceral pain syndromes. *Current Opinion in Investigational Drugs* 5: 736–742.

Dallman MF, Akana SF, Laugero KD, Gomez F, Manalo S, Bell ME, and Bhatnagar S (2003) A spoonful of sugar: Feedback signals of energy stores and corticosterone regulate responses to chronic stress. *Physiology and Behavior* 79: 3–12.

Dinan TG, O'Brien S, Lavelle E, and Scott LV (2004) Further neuroendocrine evidence of enhanced vasopressin V3 receptor responses in melancholic depression. *Psychological Medicine* 34: 169–172.

Dinan TG and Scott LV (2005) Anatomy of melancholia: Focus on hypothalamic–pituitary–adrenal axis overactivity and the role of vasopressin. *Journal of Anatomy* 207: 259–264.

Donaldson ZR and Young LJ (2008) Oxytocin, vasopressin, and the neurogenetics of sociality. *Science* 322: 900–904.

Dubois-Dauphin M, Barberis C, and De Bilbao F (1996) Vasopressin receptors in the mouse (*Mus musculus*) brain: Sex-related expression in the medical preoptic area and hypothalamus. *Brain Research* 743: 32–39.

Dubois-Dauphin M, Pevet P, Tribollet E, and Dreifuss J (1990) Vasopressin in the brain of the golden hamster: The distribution of vasopressin binding sites and of immunoreactivity to the vasopressin-related glycopeptide. *Journal of Comparative Neurology* 300: 535–548.

Dunn AJ (2000) Cytokine activation of the HPA axis. *Annals of the New York Academy of Sciences* 917: 608–617.

Engelmann M, Landgraf R, and Wotjak CT (2004) The hypothalamic–neurohypophysial system regulates the hypothalamic–pituitary–adrenal axis under stress: An old concept revisited. *Frontiers in Neuroendocrinology* 25: 132–149.

Fehm HL, Kern W, and Peters A (2006) The selfish brain: Competition for energy resources. *Progress in Brain Research* 153: 129–140.

Filaretova L, Bagaeva T, Podvigina T, and Makara G (2001) Various ulcerogenic stimuli are potentiated by glucocorticoid deficiency in rats. *Journal of Physiology (Paris)* 95: 59–65.

Filaretova L, Podvigina T, Bagaeva T, Bobryshev P, and Takeuchi K (2007) Gastroprotective role of glucocorticoid hormones. *Journal of Pharmacological Sciences* 104: 195–201.

Gabry KE, Gold PW, and Chrousos GP (2002) Hypothalamic–pituitary–adrenal axis: Introduction to physiology and pathophysiology. In: Pfaff D (ed.) *Hormones, Brain, and Behavior,* vol. 4, pp. 841–866. New York: Academic Press.

Geracioti TD, Jr., Orth DN, Ekhator NN, Blumenkopf B, and Loosen PT (1992) Serial cerebrospinal fluid corticotropin-releasing hormone concentrations in healthy and depressed humans. *Journal of Clinical Endocrinology and Metabolism* 74: 1325–1330.

Gilligan PJ, Robertson DW, and Zaczek R (2000) Corticotropin releasing factor (CRF) receptor modulators: Progress and opportunities for new therapeutic agents. *Journal of Medicinal Chemistry* 43: 1641–1660.

Goeders NE (2003) The impact of stress on addiction. *European Neuropsychopharmacology* 13: 435–441.

Gold PW and Chrousos GP (1999) The endocrinology of melancholic and atypical depression: Relation to neurocircuitry and somatic consequences. *Proceedings of the Association of American Physicians* 111: 22–34.

Gold PW and Chrousos GP (2002) Organization of the stress system and its dysregulation in melancholic and atypical depression: High vs low CRH/NE states. *Molecular Psychiatry* 7: 254–275.

Gold PW, Pigott TA, Kling MA, Kalogeras K, and Chrousos GP (1988) Basic and clinical studies with corticotropin-releasing hormone. Implications for a possible role in panic disorder. *Psychiatric Clinics of North America* 11: 327–334.

Gold PW, Wong ML, Chrousos GP, and Licinio J (1996) Stress system abnormalities in melancholic and atypical depression: Molecular, pathophysiological, and therapeutic implications. *Molecular Psychiatry* 1: 257–264.

Grant FD (2000) Genetic models of vasopressin deficiency. *Experimental Physiology* 85: 203S–209S.

Grigoriadis DE, Heroux JA, and De Souza EB (1993) Characterization and regulation of corticotropin-releasing factor receptors in the central nervous, endocrine and immune systems. *Ciba Foundation Symposium* 172: 85–101; discussion 101–107.

Habib KE, Gold PW, and Chrousos GP (2001) Neuroendocrinology of stress. *Endocrinology and Metabolism Clinics of North America* 30: 695–728; vii–viii.

Heesen C, Gold SM, Huitinga I, and Reul JM (2007) Stress and hypothalamic–pituitary–adrenal axis function in experimental autoimmune encephalomyelitis and multiple sclerosis – a review. *Psychoneuroendocrinology* 32: 604–618.

Heinrichs M and Domes G (2008) Neuropeptides and social behaviour: Effects of oxytocin and vasopressin in humans. *Progress in Brain Research* 170: 337–350.

Herman JP and Seroogy K (2006) Hypothalamic–pituitary–adrenal axis, glucocorticoids, and neurologic disease. *Neurologic Clinics* 24: 461–481, vi.

Higuchi H, Hasegawa A, and Yamaguchi T (2005) Transcriptional regulation of neuronal genes and its effect on neural functions: Transcriptional regulation of neuropeptide Y gene by leptin and its effect on feeding. *Journal of Pharmacological Sciences* 98: 225–231.

Hiroshige T and Wada-Okada S (1973) Diurnal changes of hypothalamic content of corticotropin-releasing activity in female rats at various stages of the estrous cycle. *Neuroendocrinology* 12: 316–319.

Hofman MA (1997) Lifespan changes in the human hypothalamus. *Experimental Gerontology* 32: 559–575.

Hsu SY and Hsueh AJ (2001) Human stresscopin and stresscopin-related peptide are selective ligands for the type 2 corticotropin-releasing hormone receptor. *Nature Medicine* 7: 605–611.

Jacobson L (2005) Hypothalamic–pituitary–adrenocortical axis regulation. *Endocrinology and Metabolism Clinics of North America* 34: 271–292, vii.

Jimerson DC (2002) Leptin and the neurobiology of eating disorders. *Journal of Laboratory and Clinical Medicine* 139: 70–71.

Jimerson DC and Wolfe BE (2004) Neuropeptides in eating disorders. *CNS Spectrums* 9: 516–522.

Johnson EO, Kamilaris TC, Chrousos GP, and Gold PW (1992) Mechanisms of stress: A dynamic overview of hormonal and behavioral homeostasis. *Neuroscience and Biobehavioral Reviews* 16: 115–130.

Jones MP, Dilley JB, Drossman D, and Crowell MD (2006) Brain–gut connections in functional GI disorders: Anatomic and physiologic relationships. *Neurogastroenterology and Motility* 18: 91–103.

Junghanns K, Backhaus J, Tietz U, et al. (2003) Impaired serum cortisol stress response is a predictor of early relapse. *Alcohol and Alcoholism* 38: 189–193.

Kasting NW (1988) Simultaneous and independent release of vasopressin and oxytocin in the rat. *Canadian Journal of Physiology and Pharmacology* 66: 22–26.

Kasting NW (1989) Criteria for establishing a physiological role for brain peptides a case in point: The role of vasopressin in thermoregulation during fever and antipyresis. *Brain Research Reviews* 14: 143–153.

Kaye WH, Berrettini W, Gwirtsman H, and George DT (1990) Altered cerebrospinal fluid neuropeptide Y and peptide YY immunoreactivity in anorexia and bulimia nervosa. *Archives of General Psychiatry* 47: 548–556.

Keck ME (2006) Corticotropin-releasing factor, vasopressin and receptor systems in depression and anxiety. *Amino Acids* 31: 241–250.

Kelesidis T and Mantzoros CS (2006) The emerging role of leptin in humans. *Pediatric Endocrinology Reviews* 3: 239–248.

Keller-Wood M and Dallman M (1984) Corticosteroid inhibition of ACTH secretion. *Endocrine Reviews* 5: 1–24.

Khatri P and Blumenthal JA (2007) Exercise. In: Fink G (ed.) *Encyclopedia of Stress,* vol. 1, pp. 982–986. Oxford: Elsevier.

Kiefer F, Jahn H, Otte C, Naber D, and Wiedemann K (2006) Hypothalamic–pituitary–adrenocortical axis activity: A target of pharmacological anticraving treatment? *Biological Psychiatry* 60: 74–76.

Kiefer F and Wiedemann K (2004) Neuroendocrine pathways of addictive behaviour. *Addiction Biology* 9: 205–212.

Kovacs KJ, Foldes A, and Sawchenko PE (2000) Glucocorticoid negative feedback selectively targets vasopressin transcription in parvocellular neurosecretory neurons. *Journal of Neuroscience* 20: 3843–3852.

Kovacs KJ and Sawchenko PE (1996) Sequence of stress-induced alterations in indices of synaptic and transcriptional activation in parvocellular neurosecretory neurons. *Journal of Neuroscience* 16: 262–273.

Kudielka BM and Kirschbaum C (2005) Sex differences in HPA axis responses to stress: A review. *Biological Psychology* 69: 113–132.

Lamberts SW, Huizenga AT, de Lange P, de Jong FH, and Koper JW (1996) Clinical aspects of glucocorticoid sensitivity. *Steroids* 61: 157–160.

Landgraf R (2006) The involvement of the vasopressin system in stress-related disorders. *CNS Neurological Disorders – Drug Targets* 5: 167–179.

Landgraf R and Neumann ID (2004) Vasopressin and oxytocin release within the brain: A dynamic concept of multiple and variable modes of neuropeptide communication. *Frontiers in Neuroendocrinology* 25: 150–176.

Laue L, Gold PW, Richmond A, and Chrousos GP (1991) The hypothalamic–pituitary–adrenal axis in anorexia nervosa and bulimia nervosa: Pathophysiologic implications. *Advances in Pediatrics* 38: 287–316.

Leonard BE (2006) HPA and immune axes in stress: Involvement of the serotonergic system. *Neuroimmunomodulation* 13: 268–276.

Lewis K, Li C, Perrin MH, et al. (2001) Identification of urocortin III, an additional member of the corticotropin-releasing factor (CRF) family with high affinity for the CRF2 receptor. *Proceedings of the National Academy of Sciences of the United States of America* 98: 7570–7575.

Licinio J, Wong ML, and Gold PW (1996) The hypothalamic–pituitary–adrenal axis in anorexia nervosa. *Psychiatry Research* 62: 75–83.

Lupien SJ, Fiocco A, Wan N, Maheu F, Lord C, Schramek T, and Tu MT (2005) Stress hormones and human memory function across the lifespan. *Psychoneuroendocrinology* 30: 225–242.

Lupien SJ, Gillin CJ, and Hauger RL (1999) Working memory is more sensitive than declarative memory to the acute effects of corticosteroids: A dose–response study in humans. *Behavioral Neuroscience* 113: 420–430.

Madeira MD and Lieberman AR (1995) Sexual dimorphism in the mammalian limbic system. *Progress in Neurobiology* 45: 275–333.

Magri F, Cravello L, Barili L, et al. (2006) Stress and dementia: The role of the hypothalamic–pituitary–adrenal axis. *Aging Clinical and Experimental Research* 18: 167–170.

Makrigiannakis A, Psychoyos A, Zoumakis E, Margioris AN, Stournaras C, and Gravanis A (1997) Endometrial corticotropin-releasing hormone: Expression, regulation, and potential physiological implications. *Annals of the New York Academy of Sciences* 816: 116–128.

Mantzoros CS (1999) The role of leptin in human obesity and disease: A review of current evidence. *Annals of Internal Medicine* 130: 671–680.

Mantzoros CS and Moschos SJ (1998) Leptin: In search of role(s) in human physiology and pathophysiology. *Clinical Endocrinology (Oxf)* 49: 551–567.

Margioris AN, Grino M, Rabin D, and Chrousos GP (1988) Human placenta and the hypothalamic–pituitary–adrenal axis. *Advances in Experimental Medicine and Biology* 245: 389–398.

Marquez C, Nadal R, and Armario A (2005) Responsiveness of the hypothalamic–pituitary–adrenal axis to different novel environments is a consistent individual trait in adult male outbred rats. *Psychoneuroendocrinology* 30: 179–187.

Mastorakos G and Pavlatou M (2005) Exercise as a stress model and the interplay between the hypothalamus-pituitary–adrenal and the hypothalamus-pituitary–thyroid axes. *Hormones and Metabolic Research* 37: 577–584.

Mastorakos G, Pavlatou M, Diamanti-Kandarakis E, and Chrousos GP (2005) Exercise and the stress system. *Hormones (Athens)* 4: 73–89.

McEwen BS (2000) Allostasis, allostatic load, and the aging nervous system: Role of excitatory amino acids and excitotoxicity. *Neurochemical Research* 25: 1219–1231.

McEwen BS (2002) Sex, stress and the hippocampus: Allostasis, allostatic load and the aging process. *Neurobiology of Aging* 23: 921–939.

McEwen BS (2007) Physiology and neurobiology of stress and adaptation: Central role of the brain. *Physiological Reviews* 87: 873–904.

Meynen G, Unmehopa UA, van Heerikhuize JJ, Hofman MA, Swaab DF, and Hoogendijk WJ (2006) Increased arginine vasopressin mRNA expression in the human hypothalamus in depression: A preliminary report. *Biological Psychiatry* 60: 892–895.

Mlynarik M, Zelena D, Bagdy G, Makara GB, and Jezova D (2007) Signs of attenuated depression-like behavior in vasopressin deficient Brattleboro rats. *Hormones and Behavior* 51: 395–405.

Monnikes H, Tebbe JJ, Hildebrandt M, et al. (2001) Role of stress in functional gastrointestinal disorders. Evidence for stress-induced alterations in gastrointestinal motility and sensitivity. *Digestive Diseases* 19: 201–211.

Mountjoy KG, Robbins LS, Mortrud MT, and Cone RD (1992) The cloning of a family of genes that encode the melanocortin receptors. *Science* 257: 1248–1251.

Munck A (2005) Glucocorticoid receptors and physiology: A personal history. *Steroids* 70: 335–344.

Munck A, Guyre PM, and Holbrook NJ (1984) Physiological functions of glucocorticoids in stress and their relation to pharmacological actions. *Endocrine Reviews* 5: 25–44.

Owens MJ and Nemeroff CB (1991) Physiology and pharmacology of corticotropin-releasing factor. *Pharmacological Reviews* 43: 425–473.

Palkovits M, Brownstein MJ, and Vale W (1985) Distribution of corticotropin-releasing factor in rat brain. *Federation Proceedings* 44: 215–219.

Plotsky P, Otto S, Toyama T, and Sutton S (1990) Lack of correlation between immunoreactive corticotropin-releasing factor concentration profiles in hypophysial–portal and peripheral plasma. *Journal of Neuroendocrinology* 2: 65–69.

Plotsky PM (1987) Facilitation of immunoreactive corticotropin-releasing factor secretion into the hypophysial–portal circulation after activation of catecholaminergic pathways or central norepinephrine injection. *Endocrinology* 121: 924–930.

Plotsky PM, Cunningham ET, Jr., and Widmaier EP (1989) Catecholaminergic modulation of corticotropin-releasing factor and adrenocorticotropin secretion. *Endocrine Reviews* 10: 437–458.

Pomara N, Greenberg WM, Branford MD, and Doraiswamy PM (2003) Therapeutic implications of HPA axis abnormalities in Alzheimer's disease: Review and update. *Psychopharmacology Bulletin* 37: 120–134.

Pothoulakis C, Castagliuolo I, and Leeman SE (1998) Neuroimmune mechanisms of intestinal responses to stress. Role of corticotropin-releasing factor and neurotensin. *Annals of the New York Academy of Sciences* 840: 635–648.

Rassnick S, Sved AF, and Rabin BS (1994) Locus coeruleus stimulation by corticotropin-releasing hormone suppresses *in vitro* cellular immune responses. *Journal of Neuroscience* 14: 6033–6040.

Reyes TM, Lewis K, Perrin MH, et al. (2001) Urocortin II: A member of the corticotropin-releasing factor (CRF) neuropeptide family that is selectively bound by type 2 CRF receptors. *Proceedings of the National Academy of Sciences of the United States of America* 98: 2843–2848.

Rhodes ME (2007) Adrenocorticotropic hormone (ACTH). In: Fink G (ed.) *Encyclopedia of Stress,* vol. 1, pp. 69–72. Oxford: Academic Press.

Rhodes ME, Balestreire EM, Czambel RK, and Rubin RT (2002) Estrous cycle influences on sexual diergism of HPA axis responses to cholinergic stimulation in rats. *Brain Research Bulletin* 59: 217–225.

Rhodes ME, Kennell JS, Belz EE, Czambel RK, and Rubin RT (2004) Rat estrous cycle influences the sexual diergism of HPA axis stimulation by nicotine. *Brain Research Bulletin* 64: 205–213.

Rhodes ME, O'Toole SM, Czambel RK, and Rubin RT (2001a) Male–female differences in rat hypothalamic–pituitary–adrenal axis responses to nicotine stimulation. *Brain Research Bulletin* 54: 681–688.

Rhodes ME, O'Toole SM, Wright SL, Czambel RK, and Rubin RT (2001b) Sexual diergism in rat hypothalamic–pituitary–adrenal axis responses to cholinergic stimulation and antagonism. *Brain Research Bulletin* 54: 101–113.

Rhodes ME and Rubin RT (1999) Functional sex differences ('sexual diergism') of central nervous system cholinergic

systems, vasopressin, and hypothalamic–pituitary–adrenal axis activity in mammals: A selective review. *Brain Research. Brain Research Reviews* 30: 135–152.
Rivier C and Vale W (1983) Modulation of stress-induced ACTH release by corticotropin-releasing factor, catecholamines and vasopressin. *Nature* 305: 325–327.
Rosmond R (2002) The glucocorticoid receptor gene and its association to metabolic syndrome. *Obesity Research* 10: 1078–1086.
Rubin RT, O'Toole SM, Rhodes ME, Sekula LK, and Czambel RK (1999) Hypothalamo–pituitary–adrenal cortical responses to low-dose physostigmine and arginine vasopressin administration: Sex differences between major depressives and matched control subjects. *Psychiatry Research* 89: 1–20.
Rubin RT, Rhodes ME, and Czambel RK (2002) Sexual diergism of baseline plasma leptin and leptin suppression by arginine vasopressin in major depressives and matched controls. *Psychiatry Research* 113: 255–268.
Sahu A (1998a) Evidence suggesting that galanin (GAL), melanin-concentrating hormone (MCH), neurotensin (NT), proopiomelanocortin (POMC) and neuropeptide Y (NPY) are targets of leptin signaling in the hypothalamus. *Endocrinology* 139: 795–798.
Sahu A (1998b) Leptin decreases food intake induced by melanin-concentrating hormone (MCH), galanin (GAL) and neuropeptide Y (NPY) in the rat. *Endocrinology* 139: 4739–4742.
Sahu A (2003) Leptin signaling in the hypothalamus: Emphasis on energy homeostasis and leptin resistance. *Frontiers in Neuroendocrinology* 24: 225–253.
Sahu A (2004) Minireview: A hypothalamic role in energy balance with special emphasis on leptin. *Endocrinology* 145: 2613–2620.
Sahu A (2008) Effects of chronic central leptin infusion on proopiomelanocortin and neurotensin gene expression in the rat hypothalamus. *Neuroscience Letters* 440: 125–129.
Saphier D, Welch JE, Farrar GE, Nguyen NQ, Aguado F, Thaller TR, and Knight DS (1994) Interactions between serotonin, thyrotropin-releasing hormone, and substance P in the CNS regulation of adrenocortical secretion. *Psychoneuroendocrinology* 19: 779–797.
Sapolsky RM, Romero LM, and Munck AU (2000) How do glucocorticoids influence stress responses? Integrating permissive, suppressive, stimulatory, and preparative actions. *Endocrine Reviews* 21: 55–89.
Sawchenko PE, Benoit R, and Brown MR (1988a) Somatostatin 28-immunoreactive inputs to the paraventricular and supraoptic nuclei: Principal origin from non-aminergic neurons in the nucleus of the solitary tract. *Journal of Chemical Neuroanatomy* 1: 81–94.
Sawchenko PE, Plotsky PM, Pfeiffer SW, Cunningham ET, Jr., Vaughan J, Rivier J, and Vale W (1988b) Inhibin beta in central neural pathways involved in the control of oxytocin secretion. *Nature* 334: 615–617.
Scott LV and Dinan TG (2002) Vasopressin as a target for antidepressant development: An assessment of the available evidence. *Journal of Affective Disorders* 72: 113–124.
Selye H (1936) A syndrome produced by diverse nocuous agents. *Nature* 138: 32.
Selye H (1976) Forty years of stress research: Principal remaining problems and misconceptions. *Canadian Medical Association Journal* 115: 53–56.
Simerly RB, Chang C, Muramatsu M, and Swanson LW (1990) Distribution of androgen and estrogen receptor mRNA-containing cells in the rat brain: An *in situ* hybridization study. *Journal of Comparative Neurology* 294: 76–95.
Smith MA, Kling MA, Whitfield HJ, et al. (1989) Corticotropin-releasing hormone: From endocrinology to psychobiology. *Hormone Research* 31: 66–71.
Smith SM and Vale WW (2006) The role of the hypothalamic–pituitary–adrenal axis in neuroendocrine responses to stress. *Dialogues in Clinical Neuroscience* 8: 383–395.
Southwick SM, Bremner JD, Rasmusson A, Morgan CA, 3rd, Arnsten A, and Charney DS (1999a) Role of norepinephrine in the pathophysiology and treatment of posttraumatic stress disorder. *Biological Psychiatry* 46: 1192–1204.
Southwick SM, Paige S, Morgan CA, 3rd, Bremner JD, Krystal JH, and Charney DS (1999b) Neurotransmitter alterations in PTSD: Catecholamines and serotonin. *Seminars in Clinical Neuropsychiatry* 4: 242–248.
Southwick SM, Vythilingam M, and Charney DS (2005) The psychobiology of depression and resilience to stress: Implications for prevention and treatment. *Annual Review of Clinical Psychology* 1: 255–291.
Spinedi E, Johnston CA, Chisari A, and Negro-Vilar A (1988) Role of central epinephrine on the regulation of corticotropin-releasing factor and adrenocorticotropin secretion. *Endocrinology* 122: 1977–1983.
Steckler T and Holsboer F (1999) Corticotropin-releasing hormone receptor subtypes and emotion. *Biological Psychiatry* 46: 1480–1508.
Suescun MO, Chisari A, Gaillard RC, and Spinedi E (1997) Age and sex steroid environment modulate hypothalamo-pituitary-adrenal axis function in mice. *Endocrinology and Metabolism* 4: 179–184.
Surget A and Belzung C (2008) Involvement of vasopressin in affective disorders. *European Journal of Pharmacology* 583: 340–349.
Swanson LW, Sawchenko PE, Rivier J, and Vale WW (1983) Organization of ovine corticotropin-releasing factor immunoreactive cells and fibers in the rat brain: An immunohistochemical study. *Neuroendocrinology* 36: 165–186.
Thrivikraman KV, Nemeroff CB, and Plotsky PM (2000) Sensitivity to glucocorticoid-mediated fast-feedback regulation of the hypothalamic–pituitary–adrenal axis is dependent upon stressor specific neurocircuitry. *Brain Research* 870: 87–101.
Tichomirowa MA, Keck ME, Schneider HJ, Paez-Pereda M, Renner U, Holsboer F, and Stalla GK (2005) Endocrine disturbances in depression. *Journal of Endocrinological Investigation* 28: 89–99.
Tsigos C and Chrousos GP (1994) Physiology of the hypothalamic–pituitary–adrenal axis in health and dysregulation in psychiatric and autoimmune disorders. *Endocrinology and Metabolism Clinics of North America* 23: 451–466.
Turnbull AV and Rivier C (1995) Regulation of the HPA axis by cytokines. *Brain Behavior and Immunity* 9: 253–275.
Vale W, Spiess J, River C, and River J (1981) Characterization of a 41-residue ovine hypothalamic peptide that stimulates secretion of corticotropin and beta-endorphin. *Science* 213: 1394–1397.
van Londen L, Goekoop JG, van Kempen GM, et al. (1997) Plasma levels of arginine vasopressin elevated in patients with major depression. *Neuropsychopharmacology* 17: 284–292.
Vaughan J, Donaldson C, Bittencourt J, et al. (1995) Urocortin, a mammalian neuropeptide related to fish urotensin I and to corticotropin-releasing factor. *Nature* 378: 287–292.
Versteeg CA, Cransberg K, De Jong W, and Bohus B (1983) Reduction of a centrally induced pressor response by neurohypophyseal peptides: The involvement of lower brainstem mechanisms. *European Journal of Pharmacology* 94: 133–140.
Vicennati V, Ceroni L, Genghini S, Patton L, Pagotto U, and Pasquali R (2006) Sex difference in the relationship between the hypothalamic–pituitary–adrenal axis and sex hormones in obesity. *Obesity (Silver Spring)* 14: 235–243.

Vicennati V and Pasquali R (2000) Abnormalities of the hypothalamic–pituitary–adrenal axis in nondepressed women with abdominal obesity and relations with insulin resistance: Evidence for a central and a peripheral alteration. *Journal of Clinical Endocrinology and Metabolism* 85: 4093–4098.

Vinson GP, Whitehouse BJ, and Hinson JP (2007) Adrenal cortex. In: Fink G (ed.) *Encyclopedia of Stress,* vol. 1, pp. 38–46. Oxford: Elsevier.

Wallon C, Yang PC, Keita AV, et al. (2008) Corticotropin-releasing hormone (CRH) regulates macromolecular permeability via mast cells in normal human colonic biopsies *in vitro*. *Gut* 57: 50–58.

Wessinger S, Jones MP, and Crowell MD (2005) Editorial overview: Serotonergic agents in functional GI disorders: Targeting the brain–gut axis. *Current Opinion in Investigational Drugs* 6: 663–666.

Westphal NJ, Evans RT, and Seasholtz AF (2009) Novel expression of type 1 CRH receptor in multiple endocrine cell types in the murine anterior pituitary. *Endocrinology* 150: 260–267.

Whitnall M (1993) Regulation of the hypothalamic corticotropin-releasing hormone neurosecretory system. *Progress in Neurobiology* 40: 573–629.

Wolf OT (2003) HPA axis and memory. *Best Practice and Research. Clinical Endocrinology and Metabolism* 17: 287–299.

Wolf OT, Convit A, McHugh PF, et al. (2001) Cortisol differentially affects memory in young and elderly men. *Behavioral Neuroscience* 115: 1002–1011.

3 Hypothalamic–Pituitary–Thyroid Axis

R T Joffe, New Jersey Medical School, Maplewood, NJ, USA

Chapter Outline

Glossary

depression It refers to both an emotional state and the disorder, major depression.

thyroid An endocrine gland responsible for the production and secretion of thyroid hormones.

thyrotropin The pituitary hormone responsible for regulation of thyroid gland secretion.

thyroxine The main thyroid hormone secreted by the thyroid axis.

triiodothyronine The active thyroid hormone produced from thyroxine.

3.1 Introduction

The psychiatric sequelae of perturbations of various hormonal systems have been extensively documented in the literature. In particular, the thyroid gland has been linked to psychiatric symptomatology in a rich and extensive clinical and experimental literature that dates back more than 100 years. Rich clinical descriptions of prominent psychiatric symptoms and syndromes in both hyper- and hypothyroidism have appeared frequently in the literature. These observations from endocrine patients have led to the conclusion that perturbations of the thyroid axis may be implicated in the etiology of psychiatric symptomatology and perhaps even psychiatric disorders. However, despite a substantial research effort, particularly since the 1950s, the evaluation of the thyroid axis in psychiatric patients, particularly those with mood disorders, has been far less conclusive about the potential role of thyroid hormones in the biological basis of these illnesses. This chapter presents a broad and critical overview of the relationship between psychiatric symptoms and syndromes and disturbances of the hypothalamic–pituitary–thyroid axis in both endocrine and psychiatric patients. Particular attention is paid not only to the theoretical importance of the accumulated literature, but also to the potential clinical utility of the findings presented.

3.2 Hypothalamic–Pituitary–Thyroid Axis

The thyroid gland produces two main thyroid hormones, thyroxine (T4) and triiodothyronine (T3). The functional unit of the thyroid gland is the thyroid secretory follicle (Larsen et al., 2008), in which thyroid hormone is synthesized first by iodide organification and then by iodothyronine formation (Taurog, 1986) to form T4 and T3. Although other iodothyronines are formed, their biological functions appear to be limited (Chan and Singer, 1993). The major secretory product of the thyroid gland is T4. Of circulating T3, only approximately 15–20% is directly secreted by the thyroid and the remainder is derived from peripheral conversion of T4 to T3 by monodeiodination (l,2). The process of monodeiodination of T4 can lead to the production of either active T3 or inactive reverse T3 (rT3) (Berry et al., 1991; Cheron et al., 1979).

Thyroid hormones are found in all tissues of the body and exert their effect on growth and development as well as a complex series of metabolic, thermoregulatory, and specific organ functions (Chan and Singer, 1993) through the binding of T3 to a nuclear receptor that regulates and modulates specific gene transcription (Sterling, 1979). The homeostatic control of thyroid hormone function occurs through the hypothalamic–pituitary–thyroid axis. The synthesis and release of thyroid hormones is controlled through feedback mechanisms by the glycoprotein thyrotropin (thyroid-stimulating hormone, TSH) secreted by the anterior pituitary gland. Both T4 and T3 on the pituitary thyrotroph affect the release of TSH. The secretion of TSH by the pituitary is also regulated by thyrotropin-releasing hormone (TRH; Chan and Singer, 1993), which is a tripeptide released by the hypothalamus. TRH stimulates the release of TSH by the pituitary. The TSH response to TRH is highly sensitive to a variety of environmental and physical factors, including various neurotransmitters such as dopamine, norepinephrine, and serotonin, as well as to circulating levels of T4 and T3 (Chan and Singer, 1993). However, the overall effect of the TRH–TSH–thyroid hormone axis is to maintain normal thyroid hormone levels in order to produce the appropriate physiological effects.

3.3 Thyroid Disease

Psychiatric symptoms are a frequent component of the clinical presentation of both hyper- and hypothyroidism. In a minority of patients, psychiatric symptoms may be the only manifestation of thyroid disease. The psychiatric symptoms of both clinical hyper- and hypothyroidism have been well described.

3.3.1 Hyperthyroidism

Alterations of mood, behavior, and cognition have all been described in patients with hyperthyroidism. These descriptions which are quite heterogeneous extend back to the 1800s, starting with Parry's original reports (Parry, 1825). In addition to the usual physical symptoms of hyperthyroidism, mood symptoms include irritability, agitation, emotional lability, fatigue, and depression (Fava et al., 1987; Folks, 1984; Loosen, 1986; Wilson and Jefferson, 1985; Zack and Ackerman, 1988). A variety of other psychiatric syndromes can occur. For example, the broad range of anxiety disorders, including generalized anxiety, panic disorder with or without agoraphobia, and other phobic disorders, have been described in patients with hyperthyroidism (Raj and Sheehan, 1987; Stein, 1986; Turner, 1984; Weller, 1984). Some patients present with apathetic hyperthyroidism and have depression as a prominent manifestation of their illness (Young, 1984), and psychotic features may be an infrequent manifestation of thyrotoxicosis (Jefferson, 1988; Kamlana and Holms, 1986). Occasional manifestations of hyperthyroidism are variants of bipolar disorder, including rapid-cycling bipolar disorder (Corn and Checkley, 1983; Walter-Ryan and Fahs, 1987). Behavioral features of hyperthyroidism are consistent with alterations of mood and may include restlessness, agitation, and hyperactivity (Loosen, 1986). Cognitive disturbances in hyperthyroidism are also common and, as in behavioral changes, may be consistent with changes in mood, but they may also occur independent of such changes. These include alterations of attention, concentration, and memory (MacCrimmon et al., 1979; Perrild et al., 1986). In extreme cases of thyrotoxicosis, these cognitive alterations may be so extreme as to mimic a dementia or delirium (Reus, 1993). Although a heterogeneous group of psychiatric and cognitive symptoms may accompany hyperthyroidism, the exact pathophysiology of these symptoms is poorly understood. There is, however, indirect evidence that such symptoms may not be directly related to elevated thyroid hormone levels. For example, the normalization of thyroid function does often, but not always, lead to a resolution of psychiatric symptomatology (Kathol et al., 1986), and on occasion a variety of psychotropic treatments are required to obtain resolution of these psychiatric syndromes even though the patient has been euthyroid (Reus, 1993). Furthermore, there is the description of resolution of some psychiatric symptoms with popranolol treatment of Graves disease, despite a continuation of the hyperthyroid state (Trzepacz et al., 1988). In summary, it is evident that a broad range of psychiatric symptoms and psychiatric syndromes may be associated with hyperthyroidism. However, the heterogeneity of these symptoms and the evidence of no direct link to elevated thyroid hormone levels makes it very difficult to draw any definitive conclusions about the role that pertubations of the thyroid axis may have in the etiology of any specific psychiatric symptom or syndrome.

3.3.2 Hypothyroidism

In contrast to the psychiatric symptomatology of hyperthyroidism, there is a much more direct link between clinical hypothyroidism and symptoms of depression. In addition to the classical physical symptoms of clinical hypothyroidism, these patients frequently present with depressive symptoms or a complete depressive syndrome. The link between depression and clinical hypothyroidism extends back to writings in the late nineteenth century (Gull, 1874; Ord, 1878) and has been consistently documented in the literature (Loosen, 1986). However, the psychiatric manifestations of clinical hypothyroidism are not restricted to depression and may also present with features of dementia, psychosis, and, on rarer occasions, mania rather than depression (Nordgren and von Scheele, 1976; Shaw et al., 1985; Zolese and Henryk-Gutt, 1987).

The strong association between depression and clinical hypothyroidism has led to the commonly accepted notion that deficient thyroid hormones are of etiological importance in depression. There are several factors that argue against this assumption. First, other psychiatric symptoms, including mania, are also less commonly observed in patients with clinical hypothyroidism (Reus, 1993). Second, although depression is common, it is not an invariable consequence of clinical hypothyroidism. Last, the correction of the thyroid hormone deficit frequently, but not always, leads to the resolution of depressive symptoms and, in some instances, further intervention with antidepressant therapy is required (Reed and Bland, 1977; Tachman and Guthrie, 1984).

In summary, the link between depression and clinical hypothyroidism is well established. These observations, extending back to the 1800s, have promoted the extensive study of the thyroid axis in patients with psychiatric disorders, particularly depression. Although this has not necessarily clarified the etiology of depression, it has led to an understanding of its biological basis and also advanced some therapeutic options that have been of benefit to patients with depression.

3.3.3 Subclinical Hypothyroidism

Subclinical hypothyroidism is defined as elevation of basal or stimulated TSH in the absence of decreases in circulating levels of T4 and T3. Various grades of subclinical hypothyroidism have been defined (Evered et al., 1973). Grade I hypothyroidism is clinical hypothyroidism in which there are decreased levels of T4 and T3 and a corresponding increase in TSH, as well as enhanced thyrotropin response to TRH. Grade II hypothyroidism refers to a condition in which there is no clinical evidence of hypothyroidism with normal circulating T4 and T3, an elevated basal TSH level, and an augmented TSH response to TRH. Grade III hypothyroidism is similar to grade II hypothyroidism in that there is no clinical evidence of thyroid disease and circulating T4 and T3 levels are normal, but in this instance basal TSH levels are also normal but there is an elevated TSH response to TRH. The most common cause of subclinical hypothyroidism is autoimmune thyroiditis (Cooper, 1987). It is likely that approximately 5% of the general population has evidence of various grades of subclinical hypothyroidism (Cooper, 1987), although the incidence of this condition increases, particularly in women, over the age of 60, with an estimated prevalence of 10–15%. The clinical implications of subclinical hypothyroidism are still uncertain. Although the impact of subclinical hypothyroidism, particularly grade II hypothyroidism, on various physiological parameters remains to be clarified (Cooper, 1987; Bell et al., 1985), it has not been convincingly demonstrated that there is an excessive frequency of grade II hypothyroidism in any psychiatric population, particularly in those with depressive illness or bipolar disorder (Gold et al., 1981). However, studies have been limited by methodological problems, particularly representative patient samples and appropriate control groups (Gold et al., 1981). There may be a particular link between subclinical hypothyroidism and rapid-cycling bipolar disorder (discussed further in Section 3.4.2.1). There are, however, some preliminary data suggesting that depression associated with subclinical hypothyroidism is less likely to respond to tricyclic antidepressants (Haggerty et al., 1990; Joffe and Levitt, 1992). In our own study (Joffe and Levitt, 1993), we found in a cohort of 139 patients with unipolar nonpsychotic major depressive disorder that those who had grade II hypothyroidism had a response rate to antidepressants that was one-third of that of the remainder of the cohort. It therefore follows that the evaluation of baseline TSH levels may be useful in depressed patients in order to ensure that factors that may mitigate against antidepressant response can be readily identified. It should be further noted that there is only preliminary evidence that the addition of thyroid hormones enhances the response to antidepressants (Targum et al., 1984) in patients with an elevated TSH.

Another series of studies have examined the prevalence of antithyroid antibodies in depressed patients. As previously noted, autoimmune thyroiditis is the most common cause of the various grades of subclinical hypothyroidism (Haggerty et al., 1990). Several of these studies (Gold et al., 1982; Nemeroff et al., 1985; Joffe, 1987; Haggerty et al., 1990) have estimated the prevalence of antithyroid antibodies to be between 8% and 20%. However, these studies have largely lacked control groups to determine whether there is an increased prevalence compared with the normal population. In fact, many of these studies (Gold et al., 1982; Nemeroff et al., 1985; Joffe, 1987; Haggerty et al., 1990) have reported prevalence rates that are not substantially different from the prevalence of positive antithyroid antibodies of approximately 7% observed in the general population (Tunbridge et al., 1977). In summary, although there is no clear evidence for the enhanced prevalence of any of the grades of subclinical hypothyroidism or positive antithyroid antibodies in patients with mood disorders, there is evidence that such perturbations of the thyroid axis may impact the course of illness and treatment response.

3.3.4 Euthyroid Hypothyroxinemia

Transient and isolated elevations of thyroxine are reported in patients with varying medical and psychiatric diagnoses (Cohen and Swigar, 1979; Gavin et al., 1979; McConnon, 1984; Kramlinger et al., 1984; Spratt et al., 1982). This biochemical phenomenon does not appear to have any specific clinical consequences analogous to hyperthyroidism and is thought to result from impaired conversion of T4 to T3, resulting in increased circulating levels of T4 and reduced circulating levels of T3, or less uncommon to a genetic alteration in thyroid hormone-binding proteins (Premachandra et al., 1976).

Transient hypothyroxinemia has been reported in acute psychiatric patients (Cohen and Swigar, 1979; Kramlinger et al., 1984; Spratt et al., 1982) in up to one-third of patients presenting with acute psychiatric symptoms. Invariably, this normalizes within 10 days to 2 weeks; it is commonly associated with mood disorders that may occur with a broad range of psychiatric diagnoses and should not be confused with primary thyroid disease. In the following sections, a detailed review of the relationship between the hypothalamic–pituitary–thyroid axis and specific psychiatric disorders is undertaken.

3.4 Major Psychiatric Disorders

3.4.1 Depression

In view of the clearly established relationship between clinical hypothyroidism and depressive symptomatology, the thyroid axis has been extensively studied in major depressive disorder. It had been assumed that the strong linkage between reduced thyroid function and depression observed in the endocrine patient carries over to the depressed patient and that thyroid hormone deficiency would be a major factor in the etiology of this psychiatric illness. Regrettably, this has not been the case, but there is a rich and varied database on all aspects of thyroid function and mood disorder. In the following sections, the literature on basal thyroid hormone levels, the provocative tests of the thyroid axis, changes in thyroid hormone levels with antidepressant treatment, and the use of thyroid hormones to manage depressive disorders is critically evaluated.

3.4.1.1 Basal thyroid hormone levels

A large number of studies have examined circulating levels of free T4, T3, T3, and TSH in depressive disorders. A more limited number of studies have examined TRH levels and more particularly the TRH test in depressive disorder. Studies that have examined basal circulating levels of T4 or free T4 have produced inconsistent findings (Board et al., 1959; Dewhurst et al., 1968; Whybrow et al., 1972; Rybakowski and Sowinski, 1973; Takahashi et al., 1974; Yamaguchi et al., 1975; Kolakowska and Swigar, 1977; Hatotani et al., 1977; Rinieris et al., 1978; Linnoila et al., 1979; Targum et al., 1984; Joffe et al.,

1985; Orsulak et al., 1985; Wahby et al., 1989). With regard to T4, studies have in general reported normal or significantly higher levels in depressed patients compared with either healthy or psychiatric control groups. However, other studies have reported lower levels of T4 in depressed versus healthy control subjects (Rybakowski and Sowinski, 1973; Kolakowska and Swigar, 1977; Linnoila et al., 1979).

The inconsistency of findings may be related to several methodological difficulties. For example, in the early studies, protein-bound iodine was used as a measure of T4 before the development of radioimmunoassay (Board et al., 1959; Whybrow et al., 1972). In addition, in many studies, sample sizes were small and the definition of major depression or its equivalent has changed over time. With regard to T3, findings are more consistent. In general, studies report that total T3 or free T3 levels are decreased in depressed patients as compared with controls (Joffe et al., 1985; Orsulak et al., 1985; Wahby et al., 1989).

Reverse T3 derived from the monodeiodination of T4 does not have any effect on the regulation of the hypothalamic–pituitary–thyroid axis (Snyder and Utiger, 1972). However, consistent alterations do occur in depressed patients. Studies of T3 levels consistently report increases in blood and cerebrospinal fluid (Kirkegaard and Faber, 1981; Linnoila et al., 1982) in unipolar depressed patients compared with bipolar depression and healthy controls. One study by Kjellman et al. (1983) was inconclusive. Taken together, however, these studies suggest that there is abnormal metabolism of thyroid hormones with alterations in the breakdown of T4 in depressed patients. The physiological significance of this finding and its implications for understanding of the biology of depression remain uncertain.

The data on basal TSH levels have been dealt with in Section 3.3.3. However, there is another body of literature that has examined the circadian variation of serum TSH levels in depressed subjects. Under normal circumstances, circulating levels of TSH increased during the night (Patel et al., 1972). A variety of studies, including uncontrolled (Weeke and Weeke, 1978, 1980) compared to healthy volunteers (Kjellman et al., 1984; Undén et al., 1986) and measured before and after recovery in depressed subjects (Kijne et al., 1982), have consistently demonstrated an absence of circadian variation in TSH levels in depressed patients, which may be related to the severity of depression (Weeke and Weeke, 1978). These findings are consistent with disturbances in the circadian variation of various other hormones in depressive illness. Furthermore, taken together with elevations in T4 and the alteration in thyroid hormone levels with treatment as well as the blunted TRH test (described later), this may be consistent with a state of overactivity of the hypothalamic–pituitary–thyroid axis in depression. Although studies of basal thyroid hormone levels are largely inconsistent, there is substantial consistency in studies that have examined measures of thyroxine before and after antidepressant treatment. Unfortunately, most of these studies have involved tricyclic antidepressants, although data on newer antidepressants, particularly the selective serotonin reuptake inhibitors, are very limited but consistent (Gitlin et al., 2004). Nonetheless, the data consistently show that antidepressant treatment leads to substantial decrements on the order of 10–20% in measures of T4 and free T4 (Gibbons et al., 1960; Whybrow et al., 1972; Ferrari, 1973; Kirkegaard et al., 1975a,b, 1977; Kirkegaard and Faber, 1981, 1986; Undén et al., 1986; Joffe and Singer, 1987, 1990b; Baumgartner et al., 1988; Muller and Boning, 1988; Brady and Anton, 1989; Mason et al., 1989; Yamaguchi et al., 1975). The reductions in T4, although substantial, are limited so that the levels before and after antidepressant treatment are generally within the normal range reported for clinical thyroid disorders. In several of these studies (Baumgartner et al., 1988; Joffe and Singer, 1990b), responders to antidepressant treatment had substantially greater reductions in T4 and free T4 compared with nonresponders. Changes in other thyroid function tests with antidepressant treatment were not as consistent as T4. In fact, there were no significant changes in measures of either T3 or TSH.

There are limited studies on basal TRH levels in depressed subjects. Two studies (Kirkegaard et al., 1979; Banki et al., 1988) examined cerebrospinal fluid levels of TRH in depressed subjects. Kirkegaard et al. (1979) observed that mean TRH levels were higher both prior to and after recovery in the depressed patient group compared with neurological controls, whereas Banki and colleagues reported higher cerebrospinal-fluid-TRH concentrations in depressed patients compared with other psychiatric patient groups. Although these two studies have consistent findings, these should be interpreted with caution considering the relatively small sample size and the fact that the physiological significance of levels of TRH in the cerebrospinal fluid are not clearly understood, particularly as they pertain to depression. TRH is widely distributed in the central nervous system and, therefore, although the measurement of TRH in the cerebrospinal fluid remains a good surrogate measure for levels in the brain, it is unclear to what extent it reflects the levels of TRH that may be particularly relevant to the depressed patient.

The TRH test has been extensively studied in depressed patients. This test, involving the measurement of TSH levels at regular intervals following the administration of TRH, is a standard endocrine test. In psychiatric patients, particularly depressed subjects, the approach has been to administer superphysiological doses of TRH, usually 400–500 μg as a single bolus, and to evaluate the TSH response by measuring serum levels over a period of 90 min (Sternbach et al., 1982; Loosen, 1985; Loosen and Prange, 1982). The interpretation of the TRH test has been problematic, especially when attempts have been made to use this as a diagnostic test for depression. Multiple factors, including age, gender, nutritional status, various medical syndromes, stress, and a variety of drugs, may affect the outcome of the TRH test (Snyder and Utiger, 1972). In addition to these various factors, there may be technical difficulties in the administration of this test (Snyder and Utiger, 1972). In the original studies (Loosen, 1985; Loosen and Prange, 1982), a definition of blunting was suggested that would provide the opportunity to use this test for the diagnosis of depression. Specifically, it was suggested that a maximal TSH response less than 5 units would be diagnostic of depression. Unfortunately, only a minority of depressed patients, up to one-third, have a positive test, providing a relatively low sensitivity and rendering it of limited diagnostic value. Furthermore, there are several false negatives among other psychiatric diagnoses, including alcoholism (Garbutt et al., 1983; Loosen et al., 1979), borderline personality disorder even in the absence of depression (Garbutt et al., 1983), and chronic pain syndrome (Krishnan and France, 1984). The TRH test has been largely superseded by ultrasensitive measurements of TSH but the studies of the TRH test, although of limited clinical value, do provide supportive evidence for changes in the thyroid axis in depressed patients.

The exact pathophysiological basis for the abnormal TRH test in depression has not been elucidated and several theories have been promulgated (Loosen, 1985; Loosen and Prange, 1982). One of the most parsimonious explanations is that depression is a state of relative overactivity of the thyroid axis. In subjects with thyroid-axis overactivity, elevations of T4, minimal alterations in T3, blunting of the circadian variation of TSH, and a blunted TRH test are observed (Iglesia et al., 1985; Hartnell et al., 1987). The elevations of TRH in the cerebrospinal fluid are also consistent with this hypothesis, suggesting an overactivity of the hypothalamic–pituitary–thyroid axis. Other hypotheses that could explain the varied abnormalities of thyroid function tests observed in depressed patients suggest that the thyroid abnormalities may be secondary to neurotransmitter dysfunction or the alteration of other endocrine systems, suggesting that disturbances of the thyroid axis are an epiphenomenon of other biological alterations in depression rather than being primarily involved in the etiological basis of the disorder.

3.4.1.2 Use of thyroid hormones to treat depression

A variety of thyroid hormones have been used to treat depression. These include T3, T4, TSH, and TRH. Here these studies are systematically reviewed and their clinical implications critically evaluated.

3.4.1.2(i) Triiodothyronine

The most commonly used thyroid hormone for the treatment of depression is the L-isomer of T3 called liothyronine. This hormone has been used in various ways in the treatment of depression. First, T3 has been used to treat depression as a monotherapy. These studies were carried out in the 1950s and, although they showed efficacy in some patients, their findings are limited by the heterogeneous nature of the patient sample (many of the patients would not meet current criteria for depressive illness) and also by the lack of objective criteria to assess treatment outcome. There is little interest in the use of T3 alone to treat depressive disorders. Second, some earlier studies have suggested that when 25–50 μg^{-1} of T3 are added at the outset of an antidepressant trial, there is an accelerated response to the antidepressant (Prange et al., 1969; Wilson et al., 1970; Coppen et al., 1972; Wheatley, 1972). This finding has not been replicated in all studies (Feighner et al., 1972; Steiner et al., 1980). Although there is no consensus in the studies published, a recent meta-analysis concluded that addition of T3 at the outset of an antidepressant trial did produce an acceleration effect, particularly in women (Altshuler et al., 2001). This finding is of considerable interest given that there have been numerous attempts to find reliable mechanisms for reducing the lag in therapeutic onset that occurs with all classes of antidepressants. T3 may offer some opportunity to achieve this, and it is, therefore, surprising that no attempt has been made to replicate these earlier studies.

Last, T3 has also been used to enhance therapeutic response in patients who have failed to receive an adequate response to antidepressants, particularly tricyclic antidepressants (Earle, 1970; Banki, 1975, 1977; Ogura et al., 1974; Tsutsui et al., 1979; Goodwin et al., 1982; Schwarcz et al., 1984; Gitlin et al., 1987; Thase et al., 1989; Joffe and Singer, 1990a; Joffe et al., 1993). Most of the studies are open and uncontrolled and suggest that approximately 55% of patients who are nonresponsive to tricyclic antidepressants achieve a more complete therapeutic response within 2–3 weeks after the addition of 25–50 μg^{-1} T3 to their antidepressant. In contrast to T3 acceleration, in which the effect appears to be more pronounced in women, the antidepressant augmentation effect of T3 does not appear to be related to gender or to any particular clinical correlate. The controlled studies with tricyclics are few (Gitlin et al., 1987; Joffe and Singer, 1990a; Joffe et al., 1993), but are largely consistent with the open studies. The exception is the study by Gitlin et al. (1987), who reported no difference between T3 and placebo in the augmentation of tricyclic antidepressants using a crossover design in 16 treatment-resistant depressed subjects. In the latest of these studies, T3 has been found to be comparably effective to lithium, which is the augmentation strategy whose effectiveness has been most substantiated. In a meta-analysis, Aronson et al. (1996) showed that T3 was significantly more effective than placebo when all studies of T3 augmentation were examined, using various sensitivity analyses to exclude particular studies with different methodological features. There is no doubt that T3 is effective as an augmentation agent in antidepressant or more specific tricyclic nonresponders.

Lately, there have been both open-label and controlled trials of T3 augmentation. The open-label studies are all positive (Abraham et al., 2006; Agid and Lerer, 2003; Iosofescu et al., 2005). The controlled trials are different in design in that T3 was added at initiation of antidepressant treatment, and both acceleration and augmentation effects were evaluated. While the first of these studies were equivocal, Cooper-Kazaz et al. (2007) did observe a significant acceleration effect and a moderate antidepressant enhancement effect in more than 100 subjects reseiving a selective serotonin reuptake inhibitor but this was not replicated in another study using a similar design (Appelhof et al., 2004). In the recent STAR*D trial, T3 performed comparably to other augmentation strategies in achieving remission in patients who did not respond to citalopram and the dropout rate due to side effects was lower in the T3 treatment group (Nierenberg et al., 2006).

The general clinical utility of T3 still remains an issue to be clarified. The mechanism of action of T3 is uncertain. The notion that it may be correcting some subtle thyroid deficiency is not supported by the fact that all patients entered in the T3 augmentation studies to date were euthyroid, including normal basal TSH levels. Furthermore, in one study T4 has been shown to be less effective than T3, which is unlikely if this was simply a hormone-replacement effect. If, indeed, T3 is acting as a thyroid hormone replacement, it fails to explain the approximately 50–60% response rate observed, unless one postulates that there is a subgroup of patients with depression who have subtle thyroid deficiency. There is no evidence, however, to support such a contention. Because thyroid hormones, including triiodothyronine, have substantial effects on other endocrine and neurotransmitter systems, it requires further research to clarify what action may be involved in this antidepressant effect. In particular, it remains to be determined whether its mechanism of action is mediated through a direct effect on the hypothalamic–pituitary–thyroid axis or whether it can be considered to have a pharmacological effect through its action on other neurochemical or biological systems, analogous to the use of corticosteroids to treat a variety of medical disorders unrelated to perturbations of the hypothalamic–pituitary–adrenal axis.

3.4.1.2(ii) Thyroxine

The vast majority of studies of thyroid hormone treatment of unipolar depression has involved the use of T3. Although there are some case series and anecdotal reports suggesting the efficacy of T4 in the treatment of depression, there is no substantial body of rigorously controlled studies that has established the efficacy of this thyroid hormone in the treatment of depression. One controlled study (Joffe and Singer, 1990b) observed T3 to be more effective than T4 in the augmentation of therapeutic response in tricyclic nonresponders. This study requires further replication with a larger sample size and an appropriate placebo control group, so as to clarify the role of T4 in the treatment of unipolar depression. It would also be of considerable clinical and therapeutic importance to determine whether T3 indeed has greater benefit than T4 in the treatment of depression.

A recent study from Bunevicius et al. (1999) compared the effects of T4 alone with those of T4 plus T3 in 33 patients receiving replacement therapy for hypothyroidism. They observed over two 5-week treatment periods that patients on combined T3 plus T4 had greater improvement in mood and neuropsychological functioning compared with those on T4 alone. These data suggested that T3 may have differential effects compared with T4 on brain function. This study prompted firther study of this issue. Subsequent studies and a meta-analysis could not confirm the advantage of combined treatment over T4 alone (Joffe et al., 2007). However, there was evidence that despite lack of objective measures of a difference between T4 plus T3 versus T4 alone, patients had a significant preference for combined treatment when this was assessed in several of the studies (Joffe et al., 2007).

3.4.1.2(iii) Thyrotropin

There are limited data on the use of TSH to treat depression. The rationale for the use of this pituitary hormone is that it stimulates thyroid function and, therefore, has antidepressant activity. In the only study of this, Prange et al. (1969) reported that a single dose of TSH administered intravenously to depressed women the day before the initiation of a standard tricyclic antidepressant trial led to a more rapid antidepressant response compared with saline.

These data are of considerable interest, but require further replication before making definitive conclusions about the efficacy of TSH in the treatment of depression. Furthermore, even if effective, the required mode of administration, by intravenous route, seriously limits the clinical applications of this hormone.

3.4.1.2(iv) Thyrotropin-releasing hormone

Several studies have examined the efficacy of TRH in the treatment of depression. There are two primary reasons for using this peptide to treat depression. First, TRH is stimulatory to the thyroid axis and, therefore, is considered to have possible antidepressant activity. Second, this peptide has a broad range of effects on the brain, including the stimulation of local motor activity, reversal of drug-induced sedation or anesthesia, and a wide range of other somatic and central nervous system activity (Griffiths, 1985). Studies have involved administration of TRH by intravenous routes (Kastin et al., 1972; Prange et al., 1972; Coppen et al., 1974; Ehrensing et al., 1974; Hollister et al., 1974; Van Den Burgh et al., 1975, 1976; Furlong et al., 1976; Vogel et al., 1977) and other routes (Mountjoy et al., 1970; Kieley et al., 1976). Studies have been almost entirely double-blind, usually placebo controlled, and in one instance (Karlberg et al., 1978) compared to amitriptyline. At best, the antidepressant effect of TRH has been minimal and transient, and it has been impossible to discern whether this is indeed a true antidepressant effect or whether it is a nonspecific stimulatory effect consistent with its action observed in animals (Griffiths, 1985). It can, therefore, be concluded from these studies that TRH does not have a substantial antidepressant effect and that the limited effect observed probably has limited therapeutic application. In a study, Callahan et al. (1997) compared intrathecal to intravenous TRH using a double-blind design in two patients with bipolar disorder. The aim was to determine whether a more substantial antidepressant effect was observed if the drug was administered intrathecally so as to bypass concerns about poor blood–brain barrier penetration of intravenously administered TRH. Both routes of administration were observed to be comparably effective until tolerance developed to each. On the re-administration of the intrathecal, an antidepressant effect was again observed. Here, a further study is required of the different routes of administration of TRH to see whether it may have clinical utility in selected patients, particularly those who have very severe refractory illness. In summary, the various thyroid hormones have been used at various stages of depressive illness. For the most part, data are conflicting and inconclusive. T3 use to augment antidepressant response has achieved the most clinical application and, although its efficacy is well established, its clinical utility and its place relative to other putative augmentation strategies still require clarification using clinical trial methodology.

3.4.2 Bipolar Disorder

Bipolar affective disorder, previously known as manic-depressive illness, is characterized by episodes of mania with or without the presence of depression. The link between bipolar disorder and the thyroid axis has also been established, analogous to unipolar depressive disorder. The following outlines the specific issues related to the link between the thyroid axis and bipolar disorder when they differ from unipolar depression.

3.4.2.1 Thyroid hormone levels

Findings from basal and stimulated thyroid function tests in depression apply as well to bipolar disorder. In fact, in many of the earlier studies, bipolar and unipolar disorder were not differentiated. Moreover, studies that examine the effect of antidepressants on basal thyroid hormone levels outlined in Section 3.4.1.2 probably also included patients with bipolar depression. There is no reason to think that the effects of antidepressants on thyroid hormone levels differ in bipolar versus unipolar depression. There is, however, a specific issue related to the link between rapid-cycling bipolar disorder and subclinical hypothyroidism, particularly grade II hypothyroidism. There has been much controversy about whether rapid cycling is associated with a higher prevalence of subclinical hypothyroidism (Cho et al., 1979; Cowdry et al., 1983; Wehr et al., 1988; Joffe et al., 1988; Bauer et al., 1990). The problem with these studies is that, for the most part, they lack any or an adequate control group. Although the data indicate that the prevalence of grade II and other forms of subclinical hypothyroidism is not increased in patients with rapid cycling, these forms of thyroid hypofunction may in fact cause increased vulnerability to the rapid-cycling form of bipolar disorder. Therefore, in any patient presenting with bipolar illness, the rapid-cycling form in particular, an evaluation of the thyroid axis is required, especially because many of these patients have a history of lithium treatment.

3.4.2.2 Effect of mood-stabilizing treatments on thyroid hormone levels

The major mood stabilizers, particularly lithium and carbamazepine, have substantial effects on thyroid hormone levels. Lithium has been shown to affect many stages of thyroid hormone production, including the uptake of iodide by the thyroid gland, inhibition of various stages of thyroid hormone synthesis, and blocking of the secretion of thyroid hormones (Mannisto, 1973; Berens and Wolff, 1975). However, lithium has its major inhibitory effect on the release of thyroid hormones from the thyroid gland (Berens and Wolff, 1975). Extrathyroidally, lithium is a potent inhibitor of the conversion of T4 to T3 (Voss et al., 1977). There is also accumulating evidence that lithium may have inhibitory effects on both TRH and TSH, suggesting that by altering feedback mechanisms it may diminish thyroid function at several levels of the hypothalamic–pituitary–thyroid axis (Bakker, 1982). It, therefore, follows that lithium treatment in patients with bipolar disorder leads to decreases in thyroid function tests. Studies examining the short-term effects of lithium in both healthy volunteers and clinical samples suggest that there are decreases in measures of thyroid function with compensatory increases in TSH levels that usually normalize over an approximately 3-month period (Child et al., 1977; Perrild et al., 1984; Sedvall et al., 1968; Cooper and Simpson, 1969; Burrow et al., 1971; Emerson et al., 1973; Rifkin et al., 1974). In this respect, the short-term effects of lithium are comparable to the effects of other antidepressant treatments. During this longer-term lithium treatment, approximately 5–15% of patients develop evidence of clinical hypothyroidism (Amdisen and Andersen, 1982; Cowdry et al., 1983; Smigan et al., 1984; Joffe et al., 1988; Yassa et al., 1988). It is also clear that in addition to clinical hypothyroidism, which requires thyroid replacement therapy, subclinical hypothyroidism frequently accompanies long-term lithium treatment (Cowdry et al., 1983; Joffe et al., 1988). As a result, regular monitoring of thyroid function tests, particularly T4 and especially TSH, is required in detecting the development of overt hypothyroidism with the long-term lithium therapy. Of the other mood stabilizers, carbamazepine has been most studied with regard to its effects on the thyroid axis. Analogous to lithium, carbamazepine appears to have limited thyrostatic effects, in that it leads to substantially lower levels of T4 and measures of free T4 as well as T3, but no significant alteration in TSH (Rootwelt et al., 1978; Leiwendahl et al., 1978; Roy-Byrne et al., 1984). Although changes in thyroid hormone levels are consistent with carbamazepine, cases of clinical hypothyroidism on carbamazepine have been very rarely reported (Aanderud and Strandjord, 1980). The effects of other anticonvulsants on thyroid function on affectively ill patients have not been systematically determined. In particular, it would be of interest to determine whether divalproex sodium, the most commonly used anticonvulsant in mood disorders, has effects on thyroid function similar to those observed with carbamazepine and perhaps lithium in patients with primary affective disorder.

3.4.2.3 Use of thyroid hormones to treat bipolar disorder

The literature on the use of thyroid hormones to treat bipolar disorder has been largely limited to the use of T4. There are virtually no data on the use of T3 or the other thyroid peptides, TRH and TSH.

There is, however, preliminary evidence that T4 may be effective for the treatment of patients with bipolar affective disorders. There are several case studies and open trials (Stancer and Persad, 1982; Bauer and Whybrow, 1990) that suggest that high doses of T4 may decrease the frequency and severity of cycling in patients with bipolar affective disorder, particularly the rapid-cycling form. The doses of T4 used maintain patients in a state of chemical hyperthyroidism, with levels of T4 just above the upper limit of normal established for clinical thyroid disease. The data also suggest that T4 is much more effective when used as an adjunctive treatment to other mood stabilizers rather than as monotherapy (Bauer and Whybrow, 1990). Although these data are encouraging, further systematic large-scale studies are required to first clearly establish the role of T4 in the treatment of bipolar illness.

3.4.3 Other Psychiatric Disorders

3.4.3.1 Anxiety disorder

Although anxiety may be a prominent component of thyroid dysregulation in patients with clinical thyroid disease, studies of patients with a broad range of anxiety disorders have yielded little substantial evidence to support a role of thyroid dysfunction in the biological basis of anxiety disorders (Stein and Uhde, 1993). The database is considerably smaller than that for major depressive disorder and bipolar disorder, but essentially most of the studies carried out in major depression have been extended to patients with a broad range of anxiety disorder.

In summary, there is no evidence for consistent alterations in basal thyroid hormone levels, in the TRH test, or in the frequency of clinical or subclinical thyroid disorders. Moreover, it is of interest that treatment with antidepressants for anxiety disorders does not cause consistent alterations of thyroid function tests, as has been observed in patients with major depression (Stein and Uhde, 1993). Moreover, in a small study, T3 was an effective augmenter of treatment for panic disorder (Lydiard and Ballenger, 1987; Uhde and Stein, 1988).

3.4.3.2 Schizophrenia

The data on the thyroid axis in schizophrenia are extremely limited. Studies suggest that there are no consistent abnormalities of the thyroid axis reported in patients with schizophrenia. Furthermore, treatment with antipsychotics does not lead to replicable changes in thyroid function tests in schizophrenic patients (Joffe and Levitt, 1993).

3.5 Conclusion

Links between the hypothalamic–pituitary–thyroid axis and psychiatric symptomatology are well established, with a rich literature documenting this in both endocrine and psychiatric patients. A direct relationship appears to exist between psychiatric symptomatology and thyroid hormone levels in patients with clinical thyroid disease, particularly hypothyroidism. This relationship appears to be much more complex in patients with major mood disorders, particularly depression. Although there are consistent changes reported in the thyroid axis in patients with major depression, the interpretation of these findings is still subject to considerable conjecture and the definitive role of the thyroid axis in the biological basis of depressive disorders remains to be established. There are, however, some direct clinical implications of these data. First, patients with depressive and bipolar disorders should be routinely screened for the presence of thyroid disease. This applies particularly to patients on long-term lithium treatment. However, even patients on antidepressant therapy should be screened for the presence of varying degrees of clinical and subclinical hypothyroidism because this may mitigate against robust antidepressant effects. Second, regardless of the underlying abnormalities of the thyroid axis in major depression, thyroid hormones may be useful as treatments for mood disorders, particularly refractory mood disorders. There is a growing database establishing the efficacy of T3 in the treatment of refractory depression, although the exact clinical utility of this hormone in this disease entity remains to be clarified. There are also suggestions that T4 may be useful in at least some forms of rapid-cycling bipolar disorder. Further work, however, is required to elucidate the exact nature of the abnormality of the hypothalamic–pituitary–thyroid axis in major mood disorders and its relevance for understanding the biology of these disorders.

References

Aanderud S and Strandjord RE (1980) Hypothyroidism induced by anti-epileptic therapy. *Acta Neurologica Scandinavica* 61: 330–332.
Abraham G, Milev R, and Stuart Lawson J (2006) T3 augmentation of SSRI resistant depression. *Journal of Affective Disorders* 91: 211–215.
Abuzzahab FS and Dahlam HC (1975) Long-term effects of lithium carbonate on serum protein bound iodine in affective disorders. *Progress in Neuropsychopharmacology and Biological Psychiatry* 2: 269–277.

Agid O and Lerer BE (2003) Algorithm-based treatment of major depression in an outpatient clinic: Clinical correlates of response to a specific serotonin reuptake inhibitor and to triiodothyronine augmentation. *International Journal of Neuropsychopharmacology* 6: 41–49.
Altshuler LL, Bauer M, Frye MA, et al. (2001) Does thyroid supplementation accelerate tricyclic antidepressant response? A review and meta-analysis of the literature. *American Journal of Psychiatry* 158: 1617–1622.
Amdisen A and Andersen CJ (1982) Lithium treatment and thyroid function. A survey of 237 patients in long-term lithium treatment. *Pharmacopsychiatry* 15: 149–155.
Appelhof BC, Brouwer JP, van Dyck R, et al. (2004) Triiodothyronine addition to paroxetine in the treatment of major depressive disorder. *Journal of Clinical Endocrinology and Metabolism* 89: 6271–6276.
Aronson R, Gffman HJ, Joffe RT, and Naylor CD (1996) Triiodothyronine augmentation in the treatment of refractory depression. *Archives of General Psychiatry* 53: 842–848.
Bakker K (1982) The influence of lithium carbonate on the hypothalamic–pituitary–thyroid axis. *Agressologie* 23: 89–93.
Banki CM (1975) Triiodothyronine in the treatment of depression. *Orvosi Hetilap* 116: 2543–2547.
Banki CM (1977) Cerebrospinal fluid amine metabolites after combined amitriptyline–triiodothyronine treatment of depressed women. *European Journal of Clinical Pharmacology* 11: 311–315.
Banki CM, Bissette G, Arato M, and Nemeroff CB (1988) Elevation of immunoreactive CSF TRH in depressed patients. *American Journal of Psychiatry* 145: 1526–1531.
Bauer MS and Whybrow PC (1990) Rapid-cycling bipolar affective disorder. II: Treatment of refractory rapid cycling with high-dose levothyroxine: A preliminary study. *Archives of General Psychiatry* 47: 435–440.
Bauer MS, Whybrow PC, and Winokur A (1990) Rapid cycling bipolar affective disorder. 1: Association with grade I hypothyroidism. *Archives of General Psychiatry* 47: 427–432.
Baumgartner A, Graf KJ, Kurten T, and Meinhold H (1988) Repeated measurements of thyroxine, free-thyroxine, triiodothyronine and reverse triiodothyronine in patients with major depressive disorder and schizophrenia and in health subjects. *Psychiatry Research* 24: 283–305.
Bell GM, Todd WA, Forfar JC, et al. (1985) End-organ responses to thyroxine therapy in subclinical hypothyroidism. *Clinical Endocrinology (Oxford)* 22: 83–89.
Berens SC and Wolff J (1975) The endocrine effects of lithium. In: Johnson FN (ed.) *Lithium Research and Therapy*, pp. 443–472. New York: Academic Press.
Berry MJ, Banu L, and Larsen PR (1991) Type I iodothyronine deiodinase is a seleno-cysteine-containing enzyme. *Nature (London)* 349: 438–440.
Board F, Wadeson R, and Persky H (1959) Depressive affect and endocrine functions. *Archives of Neurology and Psychiatry* 78: 612–620.
Brady KT and Anton RF (1989) The thyroid axis and desipramine treatment in depression. *Biological Psychiatry* 25: 703–709.
Bunevicius R, Kazanavicius G, Zalinkevicius R, and Prange AJ, Jr. (1999) Effects of thyroxine as compared with thyroxine plus triiodothyronine in patients with hypothyroidism. *New England Journal of Medicine* 340: 424–429.
Burrow GN, Burke WR, Himmelhoch JM, Spencer RP, and Hershman JM (1971) Effect of lithium on thyroid function. *Journal of Clinical Endocrinology and Metabolism* 32: 647–652.
Callahan AM, Frye MA, Marangell LB, George MC, Ketter TA, and Post RM (1997) Comparative antidepressant effects of intravenous and intrathecal thyrotropin releasing hormone: Confounding effects of tolerance and implications for therapeutics. *Biological Psychiatry* 41: 264–272.
Chan BL and Singer W (1993) The hypothalamic–pituitary–thyroid axis: Clinical and theoretical principles. In: Joffe RT and Levitt AJ (eds.) *The Thyroid Axis and Psychiatric Illness*, pp. 147–168. Washington, DC: American Psychiatric Press.
Cheron RG, Kaplan MM, and Larsen PR (1979) Physiological and pharmacological influences on thyroxine to 3,5,3′-triiodothyronine conversion and nuclear 3,5,3′-triiodothyronine binding in rat anterior pituitary. *Journal of Clinical Investigation* 64: 1402–1414.
Child C, Nolan G, and Jubiz W (1977) Changes in serum thyroxine, triiodothyronine and thyrotropin induced by lithium in normal subjects and in rats. *Clinical Pharmacology and Therapeutics* 20: 715–719.
Cho JT, Bone S, Donner DL, Colt E, and Fieve RR (1979) The effect of lithium treatment on thyroid function in patients with primary affective disorder. *American Journal of Psychiatry* 136: 115–116.
Cohen KL and Swigar ME (1979) Thyroid function screening in psychiatric patients. *Journal of the American Medical Association* 242: 254–255.
Cooper DS (1987) Subclinical hypothyroidism. *Journal of the American Medical Association* 258: 246–247.
Cooper TB and Simpson GM (1969) Preliminary report of a longitudinal study on the effects of lithium on iodine metabolism. *Current Therapeutic Research* 11: 603–608.
Cooper-Kazaz A, Apter JT, Cohen R, et al. (2007) Combined treatment with sertraline and liothyronine in major depression: A randomized, double-blind, placebo-controlled trial. *Archives of General Psychiatry* 64: 679–688.
Coppen A, Montgomery S, Peet M, Bailey J, Marks V, and Woods P (1974) Thyrotropin releasing hormone in the treatment of depression. *Lancet* 1: 433–435.
Coppen A, Whybrow PC, Noguera R, Maggs R, Prange AJ, Jr., and Goodwin FK (1972) Comparative antidepressant value of L-tryptophan and imipramine with and without attempted potentiation by liothyronine. *Archives of General Psychiatry* 26: 234–244.
Corn TH and Checkley SA (1983) A case of recurrent mania with recurrent hyperthyroidism. *British Journal of Psychiatry* 143: 74–76.
Cowdry RW, Wehr TA, Zis AP, and Goodwin FK (1983) Thyroid abnormalities associated with rapid-cycling bipolar illness. *Archives of General Psychiatry* 40: 414–420.
Dewhurst KE, Kabir DJ, Exley D, Harris GW, and Mandelbrote BM (1968) Blood levels of thyrotropic hormone, protein-bound iodine, and cortisol in schizophrenia and affective states. *Lancet* 2: 1160–1162.
Earle BV (1970) Thyroid hormone and tricyclic antidepressants in resistant depression. *American Journal of Psychiatry* 126: 1667–1669.
Ehrensing RH, Kastin AJ, Schalch DS, Friesen HG, Vargas JR, and Schally AV (1974) Affective state and thyrotropin and prolactin responses after repeated injections of thyrotropin-releasing hormone in depressed patients. *American Journal of Psychiatry* 131: 714–718.
Emerson CH, Dyson WL, and Utiger RD (1973) Serum thyrotropin and thyroxine concentrations in patients receiving lithium carbonate. *Journal of Clinical Endocrinology and Metabolism* 36: 338–346.
Evered DC, Ormston BJ, and Smith PA (1973) Grades of hypothyroidism. *British Medical Journal* 1: 657–659.
Fava GA, Sonino N, and Morphy MA (1987) Major depression associated with endocrine disease. *Psychiatric Developments* 4: 321–348.

Feighner JP, King L, Schuckit MA, Croughan J, and Briscoe W (1972) Hormonal potentiation of imipramine and ECT in primary depression. *American Journal of Psychiatry* 128: 1230–1238.

Ferrari G (1973) On some biochemical aspects of affective disorders. *Rivista Sperimentale di Freniatria e Medicina Legale Delle Alienazioni Mentali* 93: 1167–1175.

Folks DG (1984) Organic affective disorder and underlying thyrotoxicosis. *Psychosomatics* 25: 243–245.

Furlong FW, Brown GM, and Beeching MF (1976) Thyrotropin-releasing hormone: Differential antidepressant and endocrinological effects. *American Journal of Psychiatry* 133: 1187–1190.

Garbutt JC, Loosen PT, Tipermas A, and Prange AJ, Jr. (1983) The TRH test in borderline personality disorder. *Psychiatry Research* 9: 107–113.

Gavin LA, Rosenthal M, and Cavalieri RR (1979) The diagnostic dilemma of isolated hyperthyroxinemia in acute illness. *Journal of the American Medical Association* 242: 251–253.

Gibbons JL, Gibson G, Maxwell AE, and Willcox DR (1960) An endocrine study of depressive illness. *Journal of Psychosomal Research* 5: 32–41.

Gitlin M, Altshuler LI, Frye MA, Suri R, Huynh L, Fairbanks L, Bauer M, and Korenman S (2004) Peripheral thyroid hormones and response to selective serotonin reuptake inhibitors. *Journal of Psychiatric Neuroscience* 29: 383–386.

Gitlin MJ, Weiner H, Fairbanks L, Hershman JM, and Friedfeld N (1987) Failure of T3 to potentiate tricyclic antidepressant response. *Journal of Affective Disorders* 13: 267–272.

Gold MS, Pottash ALC, and Extein I (1981) Hypothyroidism and depression: Evidence from complete thyroid function evaluation. *Journal of the American Medical Association* 242: 1919–1922.

Gold MS, Pottash ALC, and Extein I (1982) Symptomless autoimmune thyroiditis in depression. *Psychiatry Research* 6: 261–269.

Golstein J, Van Cauter RE, Linkowski P, Vanhaelst L, and Mendlewicz J (1980) Thyrotropin nyctohemeral pattern in primary depression: Differences between unipolar and bipolar women. *Life Sciences* 27: 1695–1703.

Goodwin FK, Prange AJ, Jr., Post RM, Muscettola G, and Lipton MA (1982) Potentiation of antidepressant effects by L-triiodothyronine in tricyclic nonresponders. *American Journal of Psychiatry* 139: 34–38.

Griffiths ED (1985) Thyrotropin-releasing hormone: Endocrine and central effects. *Psychoneuroendocrinology* 3: 225–235.

Gull WW (1874) On a cretinoid state supervening in adult life in women. *Transactions of the Clinical Society of London* 7: 180–185.

Haggerty J, Jr., Garbutt JC, Evans DL, et al. (1990) Subclinical hypothyroidism: A review of neuropsychiatric aspects. *International Journal of Psychiatry Medicine* 20: 193–208.

Hartnell JM, Pekary AE, and Hershman JM (1987) Comparison of the effects of pulsatile and continuous TRH infusion on TSH release in men. *Metabolism – Clinical and Experimental* 36: 878–882.

Hatotani N, Nomura J, Yamaguchi T, and Kitayama I (1977) Clinical and experimental studies on the pathogenesis of depression. *Psychoneuroendocrinology* 2: 115–130.

Hollister LE, Berger P, Ogle FL, Arnold RC, and Johnson A (1974) Protirelin (TRH) in depression. *Archives of General Psychiatry* 31: 468–470.

Iglesia SR, Llobra M, and Montoya E (1985) Sequential changes in the pituitary–thyroid axis after chronic TRH administration: Effects on euthyroid and thyroxine treated female rats. *Acta Endocrinologica (Copenhagen)* 109: 237–242.

Iosofescu DV, Nierenberg AA, Mischoulon D, et al. (2005) An open study of triiodothyronine augmentation of selective serotonin reuptake inhibitors in treatment-resistant major depressive disorders. *Journal of Clinical Psychiatry* 66: 1038–1042.

Jefferson JW (1988) Haldol decanoate and thyroid disease. *Journal of Clinical Psychiatry* 49: 457–458.

Joffe RT (1987) Antithyroid antibodies in major depression. *Acta Psychiatrica Scandinavica* 76: 598–599.

Joffe RT, Blank DW, Post RM, and Uhde TW (1985) Decreased triiodothyronines in depression: A preliminary report. *Biological Psychiatry* 20: 922–925.

Joffe RT, Brimacombe M, and Levitt AJ (2007) Treatment of clinical hypothyroidism with thyroxine and triiodothyronine: A literature review and meta-analysis. *Psychosomatics* 48: 379–384.

Joffe RT, Kutcher S, and MacDonald C (1988) Thyroid function and bipolar affective disorder. *Psychiatry Research* 25: 117–121.

Joffe RT and Levitt A (1992) Major depression and subclinical (Grade II) hypothyroidism. *Psychoneuroendocrinology* 17: 215–221.

Joffe RT and Levitt A (ed.) (1993) The thyroid and schizophrenia. In: *The Thyroid Axis and Psychiatric Illness*, pp. 317–325. Washington, DC: American Psychiatric Press.

Joffe RT and Singer W (1987) Effect of phenelzine on thyroid function in depressed patients. *Biological Psychiatry* 22: 1033–1034.

Joffe RT and Singer W (1990a) Effect of tricyclic antidepressants on thyroid hormone levels in depressed patients. *Pharmacopsychiatry* 23: 67–69.

Joffe RT and Singer W (1990b) Acomparison of triiodothyronine and thyroxine in the potentiation of tricyclic antidepressants. *Psychiatry Research* 32: 241–252.

Joffe RT, Singer W, Levitt AJ, and MacDonald C (1993) A placebo-controlled comparison of lithium and triiodothyronine augmentation of tricyclic antidepressants in unipolar refractory depression. *Archives of General Psychiatry* 50: 387–393.

Kamlana SH and Holms L (1986) Paranoid reaction and underlying thyrotoxicosis. *British Journal of Psychiatry* 149: 376–377.

Karlberg BE, Kjellman BF, and Kagedel B (1978) Treatment of endogenous depression with oral thyrotropin-releasing hormone and amitriptyline. *Acta Psychiatrica Scandinavica* 58: 389–400.

Kastin AJ, Ehrensing RH, Schalch DS, and Anderson MS (1972) Improvement in mental depression with decreased thyrotropin response after administration of thyrotropin releasing hormone. *Lancet* 2: 740–742.

Kathol RG, Turner R, and Delahunt J (1986) Depression and anxiety associated with hyperthyroidism: Response to antithyroid therapy. *Psychosomatics* 27: 501–505.

Kieley WF, Adrian AD, Lee JH, et al. (1976) Therapeutic failure of oral thyrotropin-releasing hormone in depression. *Psychosomatic Medicine* 38: 233–241.

Kijne B, Aggernaes H, Fog-Moller F, et al. (1982) Circadian variation of serum thyrotropin in endogenous depression. *Psychiatry Research* 6: 277–282.

Kirkegaard C, Bjorum N, Cohn D, Faber J, Lauridsen UB, and Nekup JR (1977) Studies on the influence of biogenic amines and psychoactive drugs on the prognostic value of the TRH stimulation test in endogenous depression. *Psychoneuroendocrinology* 2: 131–136.

Kirkegaard C and Faber J (1981) Altered serum levels of thyroxine, triiodothyronines and diiodothyronines in endogenous depression. *Acta Endocrinologica (Copenhagen)* 96: 199–207.

Kirkegaard C and Faber J (1986) Influence of free-thyroid hormone levels on the TSH response to TRH in endogenous depression. *Psychoneuroendocrinology* 11: 491–497.

Kirkegaard C, Faber J, Hummer L, and Rogowski P (1979) Increased levels of TRH in cerebrospinal fluid from patients with endogenous depression. *Psychoneuroendocrinology* 4: 227–235.

Kirkegaard C, Norlem N, Lauridsen UB, Bjorum N, and Christiansen C (1975a) Protirelin stimulation test and thyroid function during treatment of depression. *Archives of Genearl Psychiatry* 32: 1115–1118.

Kirkegaard C, Norlem N, Lauridsen UB, and Bjorum N (1975b) Prognostic value of thyrotropin releasing hormone stimulation test in endogenous depression. *Acta Psychiatrica Scandinavica* 52: 170–177.

Kjellman BF, Beck-Friss J, Ljunggren JG, and Wetterberg L (1984) 24 Hour serum levels of TSH in affective disorders. *Acta Psychiatrica Scandinavica* 69: 491–502.

Kjellman BF, Ljunggren JG, Beck-Friis J, and Wetterberg L (1983) Reverse T3 levels in affective disorders. *Psychiatry Research* 10: 1–9.

Kolakowska T and Swigar ME (1977) Thyroid function in depression and alcohol abuse: A retrospective study. *Archives of General Psychiatry* 34: 984–988.

Kramlinger KG, Gharib H, and Swanson DQ (1984) Normal serum thyroxine values in patients with acute psychiatric illness. *American Journal of Medicine* 766: 779–801.

Krishnan AR and France RD (1984) TRH stimulation test in chronic pain. *Papers, 138th Annual Meeting of the American Psychiatric Association*, Los Angeles.

Larsen PR, Davies TF, Shlumberger MJ, and Hay ID (2008) Thyroid physiology and diagnostic evaluation of patients with thyroid disorders. In: Kronrnberg HM, Melmed S, Polonsky KS, and Larsen PR (eds.) *Kronenberg Williams Textbook of Endocrinology,* 11th edn., pp. 300–311. Philadelphia, PA: Saunders.

Leiwendahl K, Majuri H, and Helenius T (1978) Thyroid function tests in patients on long-term treatment with various anticonvulsant drugs. *Clinical Endocrinology (Oxford)* 8: 185–191.

Linnoila M, Lamberg BA, Potter WZ, Gold PW, and Goodwin FK (1982) High reverse T3 levels in manic and unipolar depressed women. *Psychiatry Research* 6: L271–L276.

Loosen PT (1985) TRH-induced TSH response in psychiatric patients: A possible neuroendocrine marker. *Psychoneuroendocrinology* 10: 237–260.

Loosen PT (1986) Hormones of the hypothalamic–pituitary–thyroid axis: A psychoneuroendocrine perspective. *Pharmacopsychiatry* 19: 401–415.

Loosen PT and Prange AJ, Jr. (1982) Serum thyrotropin response to thyrotropin-releasing hormone in psychiatric patients: A review. *American Journal of Psychiatry* 139: 405–416.

Loosen PT, Prange AJ, Jr., and Wilson LC (1979) TRH (protirelin) in depressed alcoholic men: Behavioral changes and endocrine responses. *Archives of General Psychiatry* 36: 540–547.

Lydiard RB and Ballenger JC (1987) Antidepressants in panic disorder and agoraphobia. *Journal of Affective Disorders* 13: 153–168.

MacCrimmon DJ, Wallace JE, Goldberg WM, and Streiner DL (1979) Emotional disturbance and cognitive deficits in hyperthyroidism. *Psychosomatic Medicine* 41: 331–340.

Mannisto PT (1973) Thyroid iodine metabolism *in vitro*. II: Effect of lithium ion. *Annales Medicinae Experimentalis et Biologiae* 51: 42–45.

Mason JW, Kennedy JL, Kosten TR, and Giller EL, Jr. (1989) Serum thyroxine levels in schizophrenic and affective disorder diagnostic subgroups. *Journal of Nervous and Mental Disease* 177: 351–358.

McConnon JK (1984) Thirty-five cases of transient hyperthyroidism. *Canadian Medical Association* 130: 1159–1161.

Mountjoy CQ, Price JS, Weller M, et al. (1970) A double-blind cross-over sequential trial of oral thyrotropin-releasing hormone in depression. *Lancet* 2: 958–960.

Muller B and Boning GJ (1988) Changes in the pituitary–thyroid axis accompanying major affective disorders. *Acta Psychiatrica Scandinavica* 77: 143–150.

Nemeroff CB, Simon JS, Haggerty JJ, Jr., and Evans DL (1985) Antithyroid antibodies in depressed patients. *American Journal of Psychiatry* 142: 840–843.

Nierenberg AA, Fava M, Trivedi MH, et al. (2006) A comparison of lithium and T(3) augmentation following two failed medication treatments for depression: A STAR*D report. *American Journal of Psychiatry* 163: 1519–1530.

Nordgren L and von Scheele C (1976) Myxedematous madness without myxedema. *Acta Medica Scandinavica* 199: 233–236.

Ogura C, Okuma T, Uchida Y, Imai S, and Yogi H (1974) Combined thyroid (triiodothyronine)-tricyclic antidepressant treatment in depressed states. *Folia Psychiatrica et Neurologica Japonica* 28: 179–186.

Ord WM (1878) On myxedema: A term proposed to be applied to an essential condition in the cretinoid affection occasionally observed in middle-aged women. *Royal Medical Chirurgal Society Transactions* 61: 57–78.

Orsulak BJ, Crowley G, Schlesser MA, Giles D, Fairchild C, and Rush AJ (1985) Free triiothyronine (T3) and thyroxine (T4) in a group of unipolar depressed patients and normal subjects. *Biological Psychiatry* 20: 1047–1054.

Parry CH (1825) *Collections from the Unpublished Writings of the Late Caleb Hillier Parry,* vol. I. London: Underwoods.

Patel YC, Alford FP, and Berger HG (1972) The 24-hour plasma thyrotrophin profile. *Clinical Science (Oxford)* 43: 71–77.

Perrild H, Hansen JM, Arnung K, Olsen PZ, and Danielsen U (1986) Intellectual impairment after hyperthyroidism. *Acta Endocrinologica (Copenhagen)* 112: 185–19l.

Perrild H, Hegdus L, and Arnung K (1984) Sex related goiterogenic effect of lithium carbonate in healthy young subjects. *Acta Endocrinologica (Copenhagen)* 106: 203–208.

Prange AJ, Jr., Wilson IC, Lara PP, et al. (1972) Effects of thyrotropin-releasing hormone in depression. *Lancet* 2: 999–1001.

Prange AJ, Jr., Wilson IC, Raybon AM, and Lipton MA (1969) Enhancement of the imipramine antidepressant activity by thyroid hormone. *American Journal of Psychiatry* 126: 457–469.

Premachandra BN, Gossain VV, and Perlstein IB (1976) Increased free thyroxine in a euthyroid patient with thyroxinebinding globulin deficiency. *Journal of Clinical Endocrinology and Metabolism* 42: 309–318.

Raj A and Sheehan DV (1987) Medical evaluation of panic attacks. *Journal of Clinical Psychiatry* 48: 309–313.

Reed K and Bland RC (1977) Masked myxedema madness. *Acta Psychiatrica Scandinavica* 56: 421–426.

Reus VI (1993) Psychiatric aspects ofthyroid disease. In: Joffe RT and Levitt AJ (eds.) *The Thyroid Axis and Psychiatric Illness*, pp. 171–194. Washington, DC: American Psychiatric Press.

Rifkin A, Quitkin F, Blumberg AG, and Klein DF (1974) The effect of lithium on thyroid functioning: A controlled study. *Journal of Psychiatric Research* 10: 115–120.

Rinieris PM, Christodoulou CN, Souvatzoglou GA, Koutras DA, and Stefanis CN (1978) Free-thyroxine index in psychotic and neurotic depression. *Acta Psychiatrica Scandinavica* 58(1): 56–60.

Rootwelt K, Ganes T, and Johannessen SI (1978) Effects of carbamazepine, phenytoin and phenobarbitone on serum levels of thyroid hormones and thyrotropin in humans. *Scandinavian Journal of Clinical and Laboratory Investigation* 38: 731–736.

Roy-Byrne PP, Joffe RT, Uhde TW, and Post RM (1984) Carbamazepine and thyroid function in affectively ill patients. *Archives of General Psychiatry* 41: 1150–1153.

Rybakowski J and Sowinski J (1973) Free-thyroxine index and absolute free-thyroxine in affectie disorders. *Lancet* 1: 889.

Schwarcz G, Halaris A, Baxter L, Escobar J, Thompson M, and Young M (1984) Normal thyroid function in desipramine nonresponders compared to responders by the addition of L-triiodothyronine. *American Journal of Psychiatry* 141: 1614–1616.

Sedvall G, Jonsson B, Pettersson U, et al. (1968) Effects of lithium salts on plasma protein bound iodine and uptake of I in thyroid gland of man and rat. *Life Sciences* 7: 1257–1264.

Shaw E, Halper J, Yi JE, and Asch S (1985) Diagnosis of myxedema madness. *American Journal of Psychiatry* 142: 655.

Smigan L, Wahlin A, Jacobsson L, and von Kuossing L (1984) Lithium therapy and thyroid function tests: A prospective study. *Neuropsychobiology* 11: 39–43.

Snyder PJ and Utiger RD (1972) Response to thyrotropin releasing hormone (TRH) in normal man. *Journal of Clinical Endocrinology and Metabolism* 34: 380–385.

Spratt DI, Pont A, Miller MB, McDougall IR, Bayer MF, and McLaughlin WT (1982) Hyperthyroxinemia in patients with acute psychiatric disorders. *American Journal of Medicine* 78: 41–48.

Stancer HC and Persad E (1982) Treatment of intractable rapid-cycling manic-depressive disorder with levothyroxine. *Archives of General Psychiatry* 38: 311–312.

Steiner M (1980) Triiodothyronine, imipramine and depression. *American Journal of Psychiatry* 137: 383–384.

Stein MB (1986) Panic disorder and medical illness. *Psychosomatics* 27: 833–840.

Stein MB and Uhde TW (1993) Anxiety disorders. In: Joffe RT and Levitt AJ (eds.) *The Thyroid Axis and Psychiatric Illness*, pp. 255–277. Washington, DC: American Psychiatric Press.

Sterling K (1979) Thyroid homone action at the cell level. *New England Journal of Medicine* 300: 117–123.

Sternbach H, Gerner RH, and Gwirtsman HE (1982) The thyrotropin releasing hormone stimulation test: A review. *Journal of Clinical Psychiatry* 43: 4–6.

Tachman ML and Guthrie GP, Jr. (1984) Hypothyroidism: Diversity of presentation. *Endocrine Reviews* 5: 456–465.

Takahashi S, Kaondo H, Yoshimura M, and Ochi Y (1974) Thyrotropin responses to TRH in depressive illness: Relationship to clinical subtypes after duration of depressive episode. *Folia Psychiatrica et Neurologica Japonica* 28: 255–265.

Targum SD, Greenberg RD, Harmon RL, Kessler K, Salerian AJ, and Fram DH (1984) Thyroid hormone and the TRH stimulation test in refractory depression. *Journal of Clinical Psychiatry* 45: 345–347.

Taurog A (1986) Hormone synthesis: Thyroid iodine metabolism. In: Ingbar H and Braverman LE (eds.) *The Thyroid,* 5th edn., pp. 53–97. Philadelphia, PA: Lippincott.

Thase ME, Kupfer DJ, and Jarrett DB (1989) Treatment of imipramine-resistant recurrent depression, I: An open clinical trial of adjunctive I.-triiodothyronine. *Journal of Clinical Psychiatry* 50: 385–388.

Trzepacz PT, McCue M, Klein I, Greenhouse J, and Levey GS (1988) Psychiatric and neuropsychological response to propranolol in Graves' disease. *Biological Psychiatry* 23: 678–688.

Tsutsui S, Yamazaki Y, Namba T, and Tsushima M (1979) Combined therapy of T3 and antidepressants in depression. *Journal of International Medical Research* 7: 138–146.

Tunbridge WMG, Evered DC, Hall R, et al. (1977) The spectrum of thyroid disease in a community: The Wickham survey. *Clinical Endocrinology (Oxford)* 7: 481–493.

Turner TH (1984) Agoraphobia and hyperthyroidism. *British Journal of Psychiatry* 145: 215–216.

Uhde TW and Stein MB (1988) The biology and pharmacological treatment of panic disorder. In: Hand I and Wittchen HU (eds.) *Panic and Phobias II: Treatments and Variables Affecting Outcome*, pp. 18–35. Berlin: Springer.

Undén F, Ljunggren JG, Kjellman BF, Beck-Friis J, and Wetterberg L (1986) Twenty four hour serum levels of T4 and T3 in relation to decreased TSH serum levels and decreased TSH response to TRH in affective disorders. *Acta Psychiatrica Scandinavica* 73: 358–365.

Van Den Burgh W, Van Praag HM, Bos ERH, Piers DA, Van Zanten AK, and Doorenbos H (1975) Thyrotropin releasing hormone (TRH) as a possible quick acting but short lasting antidepressant. *Psychological Medicine* 5: 404–412.

Van Den Burgh W, Van Praag HM, Bos ERH, Piers DA, Van Zanten AK, and Doorenbos H (1976) TRH by slow, continuous infusion: An antidepressant? *Psychological Medicine* 6: 393–397.

Vogel HP, Benkert O, Illig R, Muller-Oerlinghansen B, and Poppenberg A (1977) Psychoendocrinological and therapeutic effects ofTRH in depression. *Acta Psychiatrica Scandinavica* 56: 223–232.

Voss C, Schober HC, and Hartman N (1977) Einfluss von lithium auf die *in vitro* dejodierung von I.-thyroxine in der Rattenleber. *Acta Biologica et Medica Germanica* 36: 1061–1065.

Wahby V, Ibrahim G, Friedenthal S, Giller E, Kosten T, and Mason J (1989) Serum concentrations of circulating thyroid hormones in a group of depressed men. *Neuropsychobiology* 22: 8–10.

Walter-Ryan WG and Fahs JJ (1987) The problem with parsimony: Mania and hyperthyroidism. *Journal of Clinical Psychiatry* 48: 289–290.

Weeke A and Weeke J (1978) Disturbed circadian variation ofserum thyrotropin in patients with endogenous depression. *Acta Psychiatrica Scandinavica* 57: 281–289.

Weeke A and Weeke J (1980) The 24-hour pattern of serum TSH in patients with endogenous depression. *Acta Psychiatrica Scandinavica* 62: 69–74.

Wehr TA, Sack DA, Rosenthal NE, and Cowdry RW (1988) Rapid cycling affective disorder: Contributing factors and treatment responses in 51 patients. *American Journal of Psychiatry* 145: 179–184.

Weller MPI (1984) Agoraphobia and hyperthyroidism. *British Journal of Psychiatry* 144: 553–554.

Wheatley D (1972) Potentiation of amitriptyline by thyroid hormone. *Archives of General Psychiatry* 26: 229–233.

Whybrow PC, Coppen A, Prange AJ, Jr., Noguera R, and Bailey JE (1972) Thyroid function and the response to liothyronine in depression. *Archives of General Psychiatry* 26: 242–245.

Wilson WH and Jefferson JW (1985) Thyroid disease, behavior, and psychopharmacology. *Psychosomatics* 26: 481–492.

Wilson IC, Prange AJ, Jr., McClane TK, Rabon AM, and Lipton MA (1970) Thyroid hormone enhancement of imipramine in non-retarded depression. *New England Journal of Medicine* 282: 1063–1067.

Yamaguchi T, Nomura A, Nishikubo M, Tsujimura T, and Hatotani N (1975) Studies on thyroid therapy and thyroid

function in depressive patients. *Folia Psychiatrica et Neurologica Japonica* 29: 221–230.
Yassa R, Saunders A, Nastas EC, and Camille Y (1988) Lithium induced thyroid disorders: A prevalence study. *Journal of Clinical Psychiatry* 49: 14–16.
Young LD (1984) Organic affective disorder associated with thyrotoxicosis. *Psychosomatics* 25: 490–492.
Zack J and Ackerman SH (1988) Thyroid function, metabolic regulation, and depression. *Psychosomatic Medicine* 50: 454–468.
Zolese G and Henryk-Gutt R (1987) Mania induced by biochemical imbalance resulting from low energy diet in a patient with undiagnosed myxodema. *British Journal of Medicine* 295: 1026–1027.

4 Hypothalamic–Pituitary–Gonadal Axis in Women

D R Rubinow, University of North Carolina at Chapel Hill, Chapel Hill, NC, USA
P J Schmidt, National Institutes of Health, Bethesda, MD, USA
S Meltzer-Brody, University of North Carolina at Chapel Hill, Chapel Hill, NC, USA
V L Harsh, National Institutes of Health, Bethesda, MD, USA

Published by Elsevier Inc.

Chapter Outline

Glossary

affective disorders A group of psychiatric disorders characterized by disturbances of mood, including mania (elevated, expansive, or irritable mood with hyperactivity, pressured speech, and inflated self-esteem) or depression (low mood with loss of interest in usual activities, sleep and/or appetite disturbance, agitation, and feelings of worthlessness, guilt, or lowered self-esteem), or combinations of both states.

dehydroepiandrosterone (DHEA) A neurosteroid or steroid prohormone produced from cholesterol by the adrenal glands, the gonads, adipose tissue, brain, and skin. DHEA is the precursor of androstenedione, which can undergo further conversion to produce testosterone, estrone, and estradiol.

neurosteroid A steroid synthesized in neuronal tissue. The actions of neurosteroids are thought to be mediated by membrane (nongenomic) receptors, including those for GABA.

saccadic eye velocity (SEV) Saccades are voluntary, quick, simultaneous movements of both eyes in the same direction during visual search actions. Initiated by several areas of the brain, including the frontal lobe (e.g., Brodmann area 8), saccades serve as a mechanism for fixation, rapid eye movement, and the fast phase of optokinetic nystagmus and bring the retinal image being viewed onto the fovea. The velocity of saccadic eye movements are measured with a specialized apparatus and are considered to be a reliable neurophysiologic measure of $GABA_A$ receptor sensitivity.

4.1 Introduction

Studies in animals have made abundantly clear the important role played by gonadal steroids in the regulation of behavior. Given the importance of reproductive behavior in the survival of the species, the potency and range (e.g., learning and memory, appetite, aggression, and affiliation) of these behavioral effects are not surprising. The role of gonadal steroids in human behavior is both more complex and more poorly delineated. In this chapter we examine the role of gonadal steroids in behavior in women by employing two strategies: first, we suggest that findings from both molecular biological and animal *in vivo* studies illustrate the exquisite context dependency of responses to gonadal steroid signals; and second, we review the role of both gonadal steroids and context in several reproductive endocrine-related mood disorders in women (menstrual cycle-related mood disorders, perimenopausal depression, and postpartum depression).

4.2 Cell as Context

The discovery of the estrogen receptor by Jensen and Jacobson (1962) ushered in a new era of investigation of the underpinnings of physiology. Steroid receptors were determined to be members of a large family of intracellular proteins that serve as transcription factors when activated by their cognate hormone. By influencing the transcription of cellular proteins, steroid hormones could potentially regulate all aspects of cellular function. This model, while clearly powerful, left unanswered at least one question: How can different tissues show a different response to the same hormonal ligand despite the presence of receptors for that ligand in both tissues? As the mechanics of transcription became elucidated, it became clear that activated steroid receptors influence transcription not as solitary agents but by forming combinations with other intracellular proteins.

These protein–protein interactions were such that an activated receptor might enhance, reduce, initiate, or terminate transcription of a particular gene solely as a function of the specific proteins with which it interacted. One group of these proteins – coregulators – altered the efficiency of activated steroid receptor transactivation by interacting with general transcription factors (GTFs) and RNA polymerase and, thereby, creating an environment at the promoter that favored or prevented transcription (McKenna et al., 1999; Hall and McDonnell, 2005). The relative expression of these coregulators could provide one explanation for observed differences in the hormone responsivity of different tissues (McKenna et al., 1999). Differential response (across tissues, environments, or individuals) to a hormone signal, then, might be accomplished in a variety of ways: presence or absence of tissue-specific coactivator or corepressor proteins (Smith et al., 1997; Jackson et al., 1997); presence of coregulators (e.g., peroxisome proliferator-activated receptor gamma coactivator-1 (PGC-1)); for which expression is environmentally (e.g., temperature) regulated, thereby enabling environmental control over coactivator-regulated transcription (Puigserver et al., 1998); squelching (transcriptional interference), the competition for rate-limiting cofactors by activated receptors (e.g., sequestering of steroid receptor coactivator-1 (SRC-1) by estrogen receptor (ER), which inhibits transactivation by activated progesterone receptor (PR)) (Meyer et al., 1989; Ankenbauer et al., 1988); hormonal constraint, in which the transactivational effects of an activated receptor are blocked by the formation of receptor heterodimers or through countervailing inputs to signal cointegrator proteins (Pettersson et al., 1997; Uht et al., 1998, 1997) (e.g., cAMP response element binding (CREB)-binding protein (CBP), p300); tissue-specific alternative promoters (D'Souza et al., 1995); differential activational state of coregulators (Hall et al., 2005); post-translational covalent modification of DNA and histone proteins, leading to different genetic expression (Callinan and Feinberg, 2006). While the response, then, to a hormone signal cannot be inferred absent an understanding of the external hormonal milieu in which the signal occurs, it is the cell that not only integrates incoming signals but also provides a specific context that determines the range of possible responses to the stimulus.

4.3 Developmental Stage as Context: Critical Periods

The response to a biological stimulus may vary dramatically as a function of the developmental context in which it occurs. This principle was articulated in a classic paper by Phoenix et al. (1959) describing the long-term consequences of prenatal androgen exposure in female guinea pigs, that is, defeminization of reproductive behavior and increased sensitivity to androgen-induced male mating behavior in adulthood. Phoenix et al. interpreted their results as demonstrating an organizational effect of prenatal steroids on structure and subsequent function of the brain, to be contrasted with the transient, activational effects. Activational effects disappeared in the absence of the steroid, while organizational effects represented permanent changes in structure or function that, once set in motion, no longer required the presence of the steroid. The restriction of organizational effects to developmental windows, outside of which the steroid would no longer be able to similarly impact the brain (e.g., regulate the size of sexually dimorphic brain structures), was demonstrated in multiple subsequent studies. Differences across species were demonstrated in the timing of these windows (e.g., prenatal for guinea pig and perinatal for rat (Gorski, 1991)) as well as in the ability to demonstrate the effect. For example, perinatal (but not adult) administration of testosterone (T) to the female zebra finch will result in the development of male song behavior (Schlinger and Arnold, 1991), while administration during adulthood of T to the female canary will produce male song behavior and morphologic changes in the vocal control nuclei characteristic of males (Nottebohm, 1980; DeVoogd and Nottebohm, 1981). While the organizational–activational dichotomy is far more fluid and brain plasticity far greater than the notion of critical periods would suggest (Sodersten, 1984), support for developmental stage-dependent biological actions continues to accumulate. For example, the response to steroids is developmental stage dependent: Toran-Allerand (1994) demonstrated that the effects of estrogen on neuronal proliferation are facilitatory early in development, inhibitory during adulthood, and facilitatory again in the face of brain injury; Garey et al. (2003) demonstrated that the effects on locomotion of ERβ knockout are seen in old, but not young, animals; Miranda et al. (1999) observed that estradiol modulates spine density in the dentate gyrus in old but not young female rats; and Adams et al. (2001) demonstrated that the effect of estradiol to increase *N*-methyl-D-aspartate (NMDA) R1 receptor density/spines appears only in older rats. Further, alterations in perinatal gonadal steroids may direct the formation of gonadal-steroid-sensitive neurocircuitry that creates the capacity for different behavioral responses upon re-exposure to steroid post puberty (Gorski et al., 1978). Changes in the internal or external milieu (Ward and Stehm, 1991) at a critical developmental stage then may permanently alter the context in which neural signals during adulthood are processed.

4.4 Environment/Experience as Context

The brain is a nonlinear transform system, in which the response to a stimulus can be altered as a function of past history or present environment. Multiple demonstrations of this process are to be found in the animal literature. Kindling is an experimental model of epilepsy in which a subictal electrical stimulus administered repeatedly over time elicits an ever increasing response, ultimately culminating in a seizure (Post et al., 1986). In addition, once kindling is established, the stimulus retains the ability to precipitate a seizure even after a long period with no stimulus administration. In such a fashion, what initially appeared as an innocuous stimulus acquires an enduring ability to elicit a very different response from that initially produced. Behavioral sensitization refers to a similar, amplified behavioral response (e.g., aggression) to repeated exposure to a pharmacologic stimulus (Post et al., 1986). Two elements of this process are of further interest. First, Antelman et al. (1991) suggested that even absent repeated administration, exposure to certain drugs may yield an amplified response upon re-administration, simply by virtue of the passage of time. There is a memory following exposure that alters the response when the stimulus is re-presented. Such an example of time-dependent sensitization was seen in the enormously amplified adrenocorticotropic hormone (ACTH) response to re-administration of interleukin II in cancer patients 4 weeks following prior interleukin II exposure (Denicoff et al., 1989). Second, Post et al. have demonstrated that expression of behavioral sensitization may be context dependent, in that the exaggerated response elicited to cocaine in the test cage will not be manifest if, after sensitization is achieved, the cocaine is administered in the home cage (Weiss et al., 1989). Both past experience and environment, then, may alter subsequent response. Learned helplessness (Seligman, 1972), an animal model of depression, provides another demonstration of the same phenomenon. The behavioral, neurochemical, and immune responses to a series of shocks differ dramatically in rats that are or are not able to control (terminate or escape) the administration of the shock (Laudenslager et al., 1983). As the shock is administered in a yoked fashion (i.e., the executive rat controls the shock administered to its cage and to the cage of its yoked, helpless counterpart), the difference in response appears to derive from the ability to control a noxious stimulus (not protection from the stimulus). Further, if the executive rat is put in the position of being unable to terminate the shock (helpless), it does not develop the changes seen in helpless rats, suggesting that prior experience with mastery or control may confer a protective advantage in the face of subsequent adversity. More recent studies suggest that it is the perception of control rather than the control itself that protects against the adverse effects of stressors (Maier et al., 2006). One of the most impressive demonstrations of developmentally related alterations in context is provided by the work of Meaney et al. (described below). Levine (1975) observed that rat pups that were separated from their mother and handled during the perinatal period developed, as adults, a different physiologic response pattern to stressors from that in nonhandled pups. These authors suggested that this differential response pattern resulted from differences in the behavior of the mothers when reunited with their pups. In an elegant series of studies, Meany and colleagues (Liu et al., 1997) confirmed this hypothesis and showed that the separation and handling elicited an increase in licking and grooming behavior from the mother that permanently determined the nature of the offspring's response to stressors. Meany et al. then went on to demonstrate in cross-fostering that it was the maternal licking and grooming behavior, not genetic factors, that influence the licking and grooming behavior (as well as the stress responsivity) of the female offspring, and that the adopted licking and grooming behavior and stress responsivity were passed down to subsequent generations (Francis et al., 1999). Subsequent studies revealed that the licking and grooming reversed epigenetic changes (methylation) that otherwise gave rise to enduring enhanced stress reactivity (Szyf et al., 2005). This series of studies then demonstrates that maternal behavior can alter the developmental context such that permanent and dramatic differences in response – from the transcriptional to the behavioral level – are programmed into the offspring.

Several studies also demonstrate the exquisite sensitivity of reproductive physiology and behavior to environmental alterations during development. Ward and Weisz (1984, 1980) demonstrated that male offspring of a rat dam stressed during gestation were demasculinized, with lower T levels (on critical gestational days) and deficient adult male mating behavior (Ward et al., 1991). Moore et al. (1992) observed that the size of the sexually dimorphic nucleus of the bulbo-cavernosus as well as adult male mating behavior were in part determined by maternal licking of the ano-genital region of the pup, which in turn appeared to be elicited by androgens in the rat pup urine. vom Saal (1989) and others have demonstrated the impact of intrauterine position on ano-genital morphology and adult nesting behavior: female fetuses surrounded by males showed deficient nesting behavior as adults. Finally, reproductive hormones interact with environmental factors during development to determine the adult behavioral repertoire. Adult aggressive behavior in mice can be attenuated by pre-pubertal castration; the attenuation, however, is blunted to the extent to which the mouse has already been exposed to aggressive encounters (Schechter and Gandelman, 1981). The examples listed above serve to demonstrate that current and past environments and experience can create a context in which the same hormonal or environmental stimulus may elicit any of a range of behavioral responses.

4.5 Reproductive Endocrine System

4.5.1 Hypothalamic–Pituitary–Ovarian Axis and Gonadal Steroids

Under the control of neural inputs, gonadotropin-releasing hormone (GnRH) neurons in the hypothalamus secrete the decapeptide GnRH into the portal hypophyseal blood to regulate the release of follicle-stimulating hormone (FSH) and luteinizing hormone (LH) by cells in the anterior pituitary (i.e., gonadotropes). FSH and LH are released into the systemic circulation to act directly on cells in the ovary and stimulate the release of hormones (e.g., estradiol and progesterone) from the ovary. GnRH secretion is in turn regulated by both pituitary and ovarian hormones. In addition, a variety of other local or peripheral neuromodulators (e.g., β-endorphin, corticotropin-releasing hormone (CRH), and neurosteroid metabolites of progesterone) regulate GnRH secretion (**Figure 1**).

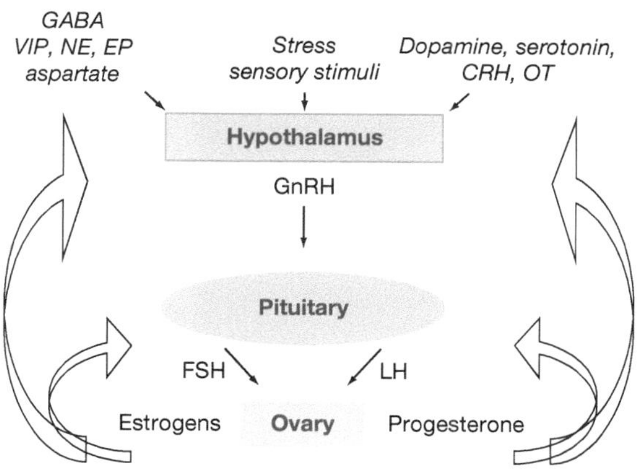

Figure 1 The hypothalamic–pituitary–ovarian axis. Secretory products of the axis are in bold type, and modulators of the axis are in italics. Solid arrows indicate stimulation, and hollow arrows indicate inhibition. The ovarian products display feedback effects at both the pituitary level and the hypothalamic level. Reproduced from Rubinow DR and Schmidt PJ (1995) Psychoneuroendocrinology. In: Kaplan I and Sadock D (eds.) *Comprehensive Textbook of Psychiatry*, 6th edn., pp. 104–112. Baltimore, MD: Lippincott Williams and Wilkins, with permission from Lippincott Williams and Wilkins.

4.5.2 Dynamics of the Menstrual Cycle, Menopause Transition, Pregnancy, and Postpartum

4.5.2.1 Menstrual cycle

The first day of menstruation is, by convention, the first day of the menstrual cycle, when estrogen and progesterone levels are low (**Figure 2**). GnRH is secreted in a pulsatile fashion from the hypothalamus and stimulates the secretion of FSH from the pituitary. FSH stimulates the secretion of estrogen from the ovarian follicles, resulting in the proliferation of the uterine lining. Estrogen and another ovarian hormone, inhibin, exert negative feedback on FSH release from the pituitary. At the end of the first menstrual cycle week, one follicle is selected and becomes the predominant follicle. That follicle undergoes maturation and secretes increasing amounts of estrogen.

The amplitude and particularly the frequency of GnRH pulses increase during the second menstrual cycle week, with the increasingly frequent GnRH pulses giving rise to a surge of LH secretion, the trigger for the expulsion of the egg from the follicle (ovulation) between 35 and 44 h after the onset of the LH surge. Before the LH surge the rising estrogen levels through undetermined mechanisms suddenly exert a positive, rather than a negative, feedback on gonadotropin secretion and are responsible for the changes in GnRH secretion that trigger the LH surge. Ovulation marks the end of the follicular phase.

After ovulation and under the influence of LH stimulation, the remains of the ovarian follicle, the corpus luteum, secrete large amounts of progesterone and, to a smaller extent, estradiol. During this phase of the menstrual cycle, the luteal phase, the amplitude of the GnRH pulses increases, and the frequency greatly decreases under the influence of brain opiates. If fertilization and implantation of the egg do not take place, the corpus luteum atrophies. Progesterone levels precipitously decline, and that decline initiates the shedding of the uterine lining, menstruation, within approximately 12–16 days of ovulation. During the last few days of the luteal phase, declining estradiol levels remove the negative feedback on FSH secretion, thereby initiating the rise in FSH levels that will give rise to the next menstrual cycle.

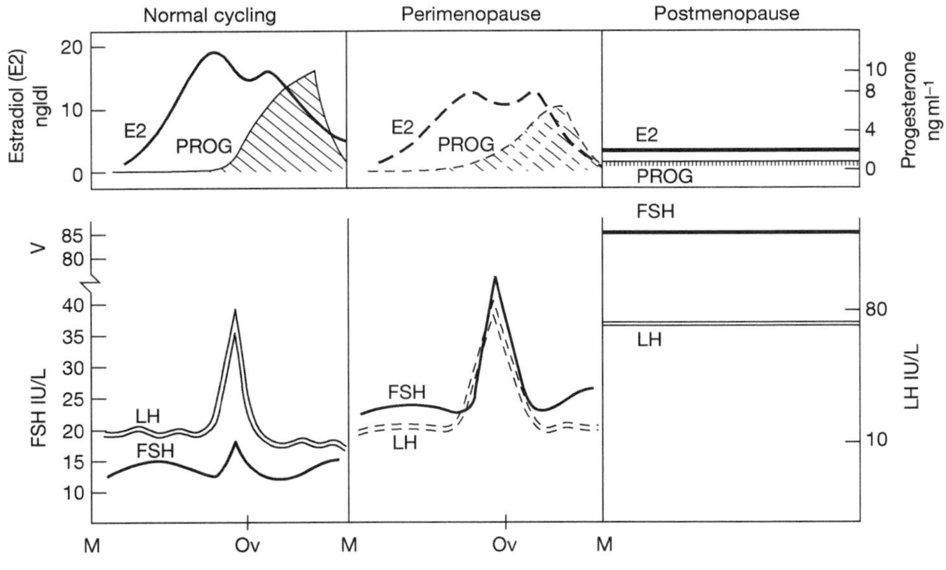

Figure 2 Levels of the ovarian steroids estradiol (E2) and progesterone (PROG) (top) and the pituitary gonadotropic hormones follicle-stimulating hormone (FSH) and luteinizing hormone (LH) at three phases of reproductive life (Ov, ovulation; M, menses). The illustrated hormonal patterns for the climacteric do not reflect intra- and interindividual variability in frequency of ovulation and length of menstrual cycle during this phase. Reprinted from Schmidt PJ and Rubinow DR (1991) Menopause-related affective disorders: A justification for further study. *American Journal of Psychiatry* 148: 844–852, with permission from the American Journal of Psychiatry, (Copyright 1991). American Psychiatric Association.

4.5.2.2 Menopause transition

The process of reproductive aging in women has several unique features that distinguish it from reproductive senescence in most animal species. The average age of the menopause in women is estimated to be 50–51 years; however, the transition from normal reproductive life to the last year after the final menstrual period (referred to as either the menopause transition or the perimenopause) lasts an average of 5 years but can range from 1 to 15 years in duration. The endocrinology of the menopause transition has yet to be fully characterized and represents a complex interplay of actions at all levels of the hypothalamic–pituitary–ovarian (HPO) axis (Santoro, 2005). Nonetheless, evidence suggests that this phase of reproductive aging occurs in stages. The early menopause transition is associated with lower ovarian inhibin secretion, which in turn reduces the restraint on both the hypothalamus and pituitary and results in elevated pituitary gonadotropin (FSH) secretion. In addition to reduced ovarian inhibin secretion, age-related increases in GnRH production could contribute to the elevated pituitary gonadotropin levels (Maffucci and Gore, 2006; Yin and Gore, 2006). Importantly, ovarian estradiol secretion is normal or at times elevated during this early stage of the menopause transition (Santoro, 2005). As ovarian aging proceeds into the late menopause transition, despite occasional episodes of normal cycling, women are exposed to periods of estrogen withdrawal, fewer ovulatory cycles, and prolonged hypogonadism, ultimately leading to the last menstrual period. Five to 10 years after the menopause, the activity of the HPO axis becomes relatively more stable and is characterized by persistent hypogonadism and relatively tonic (not phasic) elevated gonadotropin secretion (Hall and Gill, 2001). In addition to changes in ovarian estradiol and progesterone secretion, production rates of several other hormones (e.g., androgens and growth factors) decline with aging during a period overlapping with the menopause transition (Burger et al., 2000; Davison et al., 2005).

4.5.2.3 Pregnancy and the postpartum

The successful implantation of the developing embryo initiates a series of physiologic events within the lining of the uterus, including the formation of the decidua, the development of the maternal-feto-placental vascular system, and the secretion of human chorionic gonadotropin (hCG) (Yen et al., 1999). In turn, hCG stimulates the maternal ovary and corpus luteum to produce increasing quantities of sex steroids (e.g., progesterone, 17α-hydroxyprogesterone, estradiol, and estrone). The high levels of the mother's ovarian hormones maintain pregnancy until the eighth or ninth week of pregnancy when the placenta is sufficiently developed to produce these hormones independent of corpus luteal activity (Csapo et al., 1973).

Hormone production by the fetal–placental–maternal unit is responsible for normal growth and development as well as the onset of parturition. Compared to the adult ovary, however, neither the placenta nor the fetal adrenal cortex have complete sets of synthetic enzymes, and, therefore, the adrenal cortex, placenta, as well as the maternal system combine to produce the profile of steroid hormones that characterize pregnancy.

In the fetus, the principal sites of steroid hormone synthesis are the hypothalamic pituitary system, the fetal zone of the adrenal cortex, and the liver. The majority of hypothalamic and pituitary hormones (e.g., CRH) are detectable in the fetus by 14–18 weeks of pregnancy. The fetal adrenal cortex is active by 10–12 weeks of gestation and accounts for the majority of the adrenal hormones produced by the fetus. Cortisol secretion is present by the 16th week of gestation and is regulated by fetal pituitary ACTH. The steroids synthesized by the fetal adrenal zone are limited by and a reflection of the type of synthetic enzymes present in the tissue. The steroid precursor pregnenolone is shunted to the delta five pathway (see **Figure 3**), converted to dehydroepiandrosterone (DHEA), which is then promptly sulfated to form DHEAS. Since DHEAS is not a substrate for any of the fetal adrenal enzymes, DHEAS leaves the adrenal (Miller, 1998; Yen et al., 1999) and passes into the liver, where DHEAS is metabolized into 16α-hydroxy DHEAS (i.e., a second

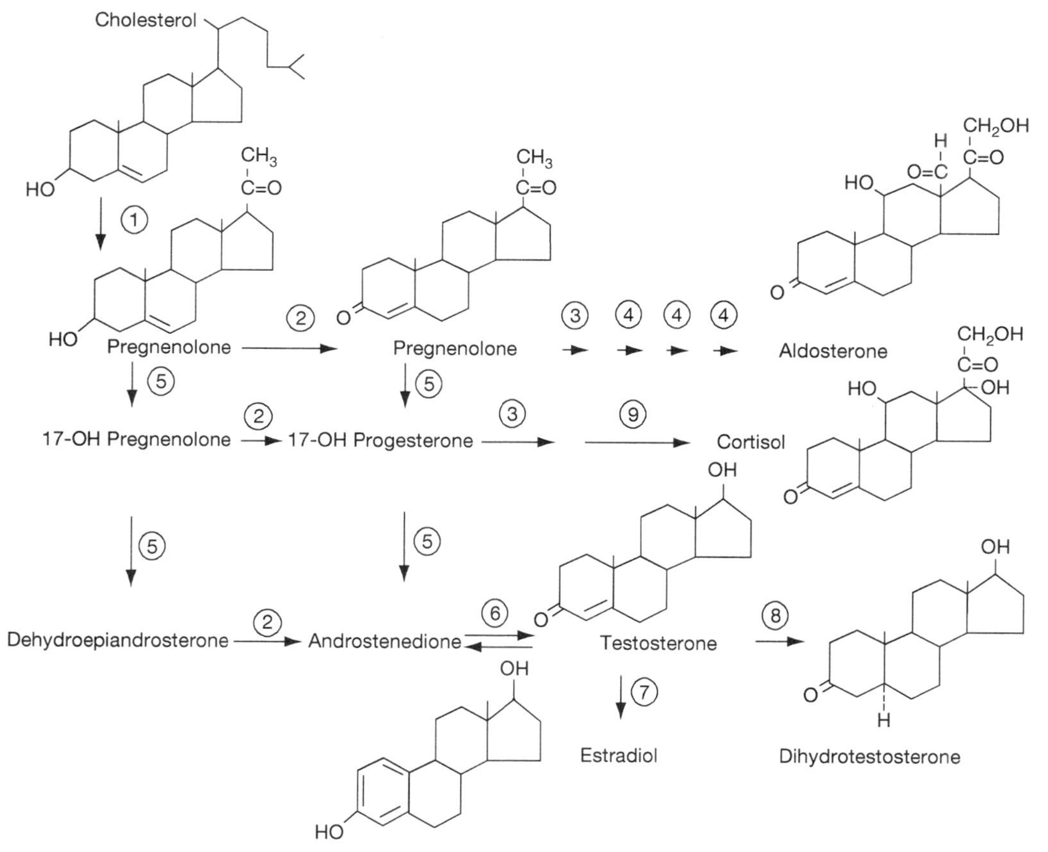

Figure 3 Synthetic pathways for steroid hormones. Circled numbers identify synthetic enzymes: 1= cytochrome P450 (CYP) 11A (cholesterol desmolase); 2 = 3β-hydroxysteroid dehydrogenase; 3 = CYP21 (21-hydroxylase); 4 = CYP11B2 (11β-hydroxylase, 18-hydroxylase, 18-oxidase); 5 = CYP17 (17α-hydroxylase, 17,20-lyase); 6 = 17β-hydroxysteroid dehydrogenase (or oxidoreductase); 7 = aromatase; 8 = 5α-reductase; 9 = CYP11B1 (11β-hydroxylase). Reprinted from Rubinow DR and Schmidt PJ (1996) Androgens, brain, and behavior. *American Journal of Psychiatry* 153: 974–984, with permission from the American Journal of Psychiatry, (Copyright 1996). American Psychiatric Association.

hydroxyl group is added). Large quantities of this latter compound pass through the fetal–placental unit where it is desulfated and aromatized to form estriol, an estrogenic steroid with a structure similar to that of estradiol. By the 12th week of gestation, estriol concentrations in the maternal circulation rapidly increase by approximately a 100-fold coinciding with the enlargement of the adrenal cortex and increased ACTH secretion by the fetal pituitary gland. The role of estrogens in pregnancy has been questioned (Miller, 1998). As described above, 90% of the estrogens produced by the fetal–placental unit are in the form of estriol. Estriol has a potency approximately a tenth that of estradiol in all functions with the exception of its ability to increase utero-placental blood flow, in which it is equipotent to estradiol. Thus, if there is a role of estrogen in pregnancy, it would be to generate the high vascular flow needed to maintain normal growth and development of the fetus.

By 12 weeks of gestation the ovarian interstitial cells have the enzymatic capacity to synthesize steroids; however, there is no evidence that fetal ovaries are active during pregnancy, in contrast to the fetal testes, which are actively producing and secreting androgens from the Leydig cells by about 8 weeks with a peak around 15–18 weeks of gestation.

In addition to metabolizing the massive amounts of fetal-source DHEAS into estriol, the placenta also serves a vital role in the production of progesterone. The placenta produces progesterone largely from the maternal source of low-density lipoprotein (LDL) cholesterol, which is converted to pregnenolone and then into progesterone. By term, progesterone is produced in quantities of approximately 250 mg per day, resulting in maternal plasma levels of progesterone of approximately 130 ng ml^{-1} (approximately 10 times plasma levels observed in the mid-luteal phase of the normal menstrual cycle). Progesterone plays a critical role in maintaining pregnancy, as reflected by miscarriages induced by the administration of PR antagonists. In addition to its actions to inhibit uterine contractility, progesterone is thought to reduce cellular immune responsivity and contribute to the longevity of pregnancy.

Pregnancy is accompanied by a sustained elevation in the secretion of several steroid and peptide hormones, followed by a sudden drop in hormone levels over the first few days after delivery. During the third trimester of pregnancy, plasma progesterone levels in the mother are approximately 130 ng ml^{-1} and estradiol reaches plasma levels of approximately 10–15 ng ml^{-1}, levels that are increased ten- and 50-fold, respectively, beyond maximum menstrual cycle levels (Tulchinsky et al., 1972). After parturition, progesterone and estradiol levels drop to early follicular phase levels within days (Speroff et al., 1983).

During the postpartum period the secretion of estradiol and progesterone as well as ovulation are sufficiently compromised to result in relative hypogonadism and the absence of follicular development. The absence of ovarian activity during the postpartum is a reflection of reduced gonadotropin secretion. Normal LH pulsatility reappears after 6–8 weeks; however, lactation may prolong the restraint on hypothalamic GnRH secretion and LH pulsatility. The exact mechanism mediating the suppression of pulsatile hypothalamic GnRH secretion during the postpartum and lactation remains to be fully clarified (McNeilly, 2002). There is substantial individual variability in the duration of postpartum hypogonadism and amenorrhea prior to the resumption of normal follicular development and cyclic ovarian steroid production.

4.6 Reproductive Endocrine Systems and the Pathophysiology of Mood Disorders

Recent advances in cell biology, pharmacology, and neuroimaging techniques have contributed greatly to hypotheses about the causes of mood disorders and potential treatments. Our knowledge of the intricacies of cellular signaling, transcriptional regulation, and the processes of cellular resilience, neuroplasticity, and apoptosis in the central nervous system (CNS) have increased the number of potential candidate pathophysiologic processes that could mediate mood disorders. In addition, both brain imaging and neuropathological studies have permitted the mapping of many brain regions involved in the regulation of affect and cognition, abnormalities of which may underlie disturbances in mood. In this section, those systems currently implicated in the pathophysiology of mood disorders are reviewed, followed by a presentation of the effects of reproductive endocrine events (or specific gonadal steroids) on the regulation of these same systems.

4.6.1 Neurotransmitters

Affective disorders have traditionally been considered to reflect an underlying dysregulation of one or more of the classic neurotransmitter systems.

Thus, preclinical studies as well as those in humans suggested that mood disorders arose from either deficiencies or excesses within the synapse of serotonin, dopamine, noradrenaline, acetylcholine, or gamma-aminobutyric acid (GABA) in brain regions that subserve the regulation of mood and behavior. Indeed, the treatments of depression were reported to influence these same systems as an integral part of their therapeutic actions. Reports of abnormal levels of these neurotransmitters and their metabolites in the cerebrospinal fluid (CSF), urine, plasma, or in peripheral cells in depression supported this concept (Potter and Manji, 1994). Moreover, pharmacologic challenge studies employing agents that targeted these neurotransmitter systems demonstrated differences between depressed subjects and controls in several outcomes, including neuroendocrine, behavioral, and temperature measures. For example, the acute depletion of either serotonin or noradrenaline/dopamine in humans induced depression in antidepressant-treated subjects (Booij et al., 2003) as well as changes in the pattern of activation in the prefrontal cortex (Neumeister et al., 2004; Bremner et al., 2003). More recently, postmortem and *in vivo* radioligand imaging studies have identified abnormalities in neurotransmitter receptor levels or functions that distinguish depressed patients from controls, including decreased serotonin 1A receptors (postmortem and *in vivo*) and α2- and β-adrenergic receptors (postmortem) (Manji et al., 2001).

4.6.2 Cell Signaling Pathways

Although the mood stabilizers – lithium, valproate, and carbamazepine – do not act like antidepressants on monoamine activity, they do influence several of the signal transduction pathways regulated by traditional neurotransmitters and antidepressants. For example, *in vitro* studies have reported that mood stabilizers (lithium, valproate) and antidepressants alter the levels of many components and targets of these systems: cAMP levels and CREB (increased), brain-derived neurotrophic factor (BDNF) (increased), ERK-MAP kinase activity (increased), bcl-2 (increased), Wnt cascade-glycogen synthesis kinase-3 beta (GSK-3 beta) (decreased), and beta catenin (increased) (Manji et al., 2001). The roles of these small molecules in cellular resilience, neurogenesis, and cell death became apparent and were integrated into hypotheses about the pathophysiology of depression. Indeed, prevention of hippocampal neurogenesis was shown to block the behavioral response to antidepressants (Santarelli et al., 2003). Moreover, interest was renewed in some neurotransmitter systems not previously considered of major importance in mood regulation (e.g., glutamatergic) because of their roles in neuroplasticity (Manji et al., 2001).

4.6.3 Brain Regional Morphological Changes

Abnormalities in brain regional activity also supported the concept that neurogenesis could be an integral part of both the pathophysiology and treatment of affective disorders. For example, brain imaging studies have identified abnormalities (both increases and decreases) in the function (e.g., blood flow) of the following brain regions: amygdala (increased); dorsomedial and dorsoanterolateral prefrontal cortex (decreased); and subgenual and pregenual areas of the cingulate gyrus (increased). Structural imaging studies have confirmed abnormalities in similar brain regions, and postmortem studies have identified both glial and neuronal cell loss in some of these same brain regions in patients with affective disorders compared to controls. Moreover, the functional abnormalities in some but not all of these brain regions reverse with successful antidepressant treatment (Drevets, 2000, 2001). Finally, human and animal brain mapping studies have shown that many of these brain regions are involved in the regulation of emotion, including the integration of the emotional, cognitive, and physiologic responses to stress, the ability to experience pleasure, the identification of internal cues and vegetative state, the response to reward, as well as decision making.

4.6.4 The Hypothalamic–Pituitary–Adrenal Axis

Stress is considered to be a central component in the pathophysiology of mood disorders. Some of the most consistent neuroendocrine abnormalities in depression have been hypercortisolemia, enhanced CRH secretion, blunted feedback inhibition, a blunted ACTH response to CRH administration, as well as several other abnormalities in the regulation of the hypothalamic–pituitary–adrenal (HPA) axis.

Under normal conditions, information about specific stressors is transmitted from higher cortical centers to the mediobasal hypothalamus, where CRH, along with several other factors, is released into the hypophysial–portal circulation. CRH acts through at least two CRH receptors on the corticotropes in the anterior pituitary gland (Plotsky et al., 1998). In addition, hormones such as arginine vasopressin (AVP) modulate the stimulatory effects of CRH on ACTH secretion. ACTH circulates in the blood and, when in contact with the adrenal cortex, stimulates production of both glucocorticoids and the adrenal androgen DHEA. Glucocorticoid production is regulated by feedback systems at the levels of the pituitary and the hypothalamus mediated through both the type 1 and type 2 corticosteroid receptors, also present in other areas of the CNS. After glucocorticoids bind to their receptor, the ligand–receptor complex undergoes a series of events and ultimately binds glucocorticoid response elements in the genome. However, as with other members of the steroid family of receptors, glucocorticoids also may act through nongenomic or membrane-related mechanisms. In addition to their roles in metabolism, the stress response, immunity, and the inflammatory response, glucocorticoids may play an important role in cellular resilience and neuroplasticity in brain regions, including the adult hippocampus (Gallagher et al., 1996).

4.6.5 Role of Gonadal Steroids in Modulating the Systems Involved in Mood Disorders

4.6.5.1 Neuroregulation

The neuroregulatory effects of gonadal steroids are myriad, well known, and widely appreciated. Indeed, wherever one finds a system believed to play a role in the etiology or treatment of depression, modulatory effects of gonadal steroids are observed as well. The neurotransmitter systems implicated in depression – serotonin, norepinephrine, dopamine, acetylcholine, GABA, and glutamate – all are regulated by estradiol (McEwen, 2002; McEwen and Alves, 1999). Estradiol's regulation of the serotonergic system, for example, is extensive, involving serotonin synthesis (tryptophan hydroxylase), receptors (transcripts, protein, and binding) (e.g., 5-HT1A, 2A), and transporter (Rubinow et al., 1998). In addition, estrogen increases sensitivity to dopamine and cocaine and is believed to contribute to the increased vulnerability to substance abuse in women (Hu and Becker, 2003; Hruska and Silbergeld, 1980).

As noted above, several nonclassical neural signaling systems have been identified as potential mediators of the therapeutic actions of antidepressants and ECT (e.g., CREB and BDNF; Nestler et al., 1989) based on observations that these systems are modulated by a range of therapies effective in depression (e.g., serotonergic and noradrenergic agents and ECT) and exhibit a pattern of change consistent with the latency to therapeutic efficacy for most antidepressants (Duman et al., 1997). For example, antidepressants increase the expression and activity of CREB in certain brain regions (e.g., hippocampus; Nibuya et al., 1996) and regulate (in a brain region-specific manner) activity of genes with a cAMP response element (Duman et al., 1997). Genes for BDNF and its receptor trkB have been proposed as potential targets for antidepressant-related changes in CREB activity (Duman et al., 1997). Similarly, estradiol has been reported to influence many of these same neuroregulatory processes. Specifically, ovariectomy has been reported to decrease, and estradiol increase, BDNF levels in the forebrain and hippocampus (Sohrabji et al., 1994b). Estrogen also increases CREB activity (Zhou et al., 1996), trkA (Sohrabji et al., 1994a), and decreases GSK-3 beta activity (Wnt pathway) (Cardona-Gomez et al., 2004) in the rat brain in a direction similar to that of mood stabilizer drugs. In contrast, an estradiol-induced decrease in BDNF has been reported to mediate estradiol's regulation of dendritic spine formation in hippocampal neurons (Murphy et al., 1998). Thus, the therapeutic potential of gonadal steroids in depression is suggested not only by their widespread actions on neurotransmitter systems but also by certain neuroregulatory actions shared by both ovarian steroids and traditional therapies for depression (i.e., antidepressants, ECT).

Modulation of neural and glial survival during aging provides yet another means by which reproductive steroids may influence the susceptibility to neuropsychiatric illness, given the putative role of neurodegeneration in depression (Ongur et al., 1998; Rajkowska et al., 1999; Rajkowska, 2000). Indeed, both reproductive steroids and mood-regulating therapies regulate cell death and survival through effects on cell survival proteins (e.g., Bcl-2, BAX), signal transduction (e.g., MAPK, Wnt, Akt), and free radical species generation (Watters et al., 1997; Garcia-Segura et al., 1998; Gouras et al., 2000; Zhang et al., 2002a,b).

4.6.5.2 Neural systems

Several studies have employed neuroimaging techniques (i.e., PET or functional magnetic resonance imaging (fMRI)) to examine the effects of ovarian steroids on regional cerebral blood flow under conditions of cognitive activation. Recent brain imaging studies in asymptomatic women confirm for the first time in humans that physiologic levels of ovarian steroids have the capacity to modulate the neurocircuitry thought to be involved in both normal and pathological affective states. First, Berman et al. (1997) performed cognition-activated O^{15}PET scans in women during conditions of GnRH agonist-induced hypogonadism and gonadal steroid replacement. They observed the elimination of Wisconsin Card Sort-activated regional cerebral blood flow (rCBF) in the dorsolateral prefrontal cortex as well as an attenuation of cortical activation in the inferior parietal lobule and posterior inferior temporal cortex (bilaterally) during GnRH agonist-induced hypogonadism (Berman et al., 1997). The characteristic pattern of cortical activation reemerged during both estradiol and progesterone addback. In addition, they observed a differential pattern of hippocampal activation with estradiol increasing and progesterone decreasing activation relative to hypogonadism. This was the first demonstration that ovarian steroids have activational effects on rCBF during cognitive stimulation in the brain regions (i.e., prefrontal cortex (PFC)) implicated in disorders of affect and cognition. In other studies, Shaywitz et al. (1999) reported in a randomized, double-blind, placebo-controlled crossover trial that postmenopausal women did not perform differently on estrogen therapy (ET) compared with placebo, but fMRI during ET showed significantly increased activation in the inferior parietal lobule and right superior frontal gyrus during verbal encoding, with significant decreases in the inferior parietal lobule during nonverbal coding. Goldstein et al. (2005) observed an increase in amygdalar activity and arousal (as measured by fMRI and skin conductance, respectively) during the late follicular phase of the menstrual cycle (higher estradiol levels) compared to the early follicular phase (characterized by relatively low estradiol levels). Protopopescu et al. (2005) employed an affective pictures task in an fMRI study and observed increased orbitofrontal cortex (OFC) activity (a region that in some studies exerts inhibitory control over amygdalar functioning) during the luteal compared with the follicular phase. Finally, Dreher et al. (2007) performed an event-related fMRI study of reward processing across the menstrual cycle in women with premenstrual dysphoria (PMD) and controls. The paradigm employed disentangles transient reward error prediction (PFC) from sustained response to reward uncertainty (ventral striatum). Data in the controls demonstrate, for the first time in humans, that ovarian steroids modulate reward system function, with increased follicular phase activation of the OFC and amygdala during reward anticipation and of the midbrain, striatum, and left ventrolateral PFC during reward delivery (Dreher et al., 2007). These findings then suggest that cognitive and affective information processes may serve as probes to identify candidate circuits for the mediation of gonadal steroid-dependent affective dysregulation.

4.6.5.3 Stress axis

Extensive studies in animals demonstrate that both gender and reproductive steroids regulate basal and stimulated HPA-axis function. In general, low-dose, short-term administration of estradiol inhibits HPA-axis responses in ovariectomized animals (Redei et al., 1994; Young et al., 2001; Dayas et al., 2000; Komesaroff et al., 1998), whereas higher doses and longer treatment regimens enhance HPA-axis reactivity to stressors (Burgess and Handa, 1992; Carey et al., 1995; Viau and Meaney, 1991). The regulatory effects of changes in reproductive steroids or menstrual cycle phase on the HPA axis in women are less well studied. Although some studies using psychological stressors identified increased stimulated cortisol in the luteal phase (Marinari et al., 1976; Kirschbaum et al., 1999), others using psychological (Collins et al., 1985; Ablanalp et al., 1977) or physiological (e.g., insulin-induced hypoglycemia, exercise) (Long et al., 2000; Galliven et al., 1997) stressors failed to find a luteal phase increase in HPA-axis activity.

Altemus et al. (2001) demonstrated that exercise-stimulated HPA responses were increased in the mid-luteal compared with the follicular phase. However, in contrast to a large animal literature documenting the ability of estradiol to increase HPA-axis secretion, Roca et al. (2003) found that progesterone, but not estradiol, significantly increased exercise-stimulated AVP, ACTH, and cortisol secretion compared with a leuprolide-induced hypogonadal condition or estradiol replacement. The mechanism by which progesterone augments stimulated HPA-axis activity is currently unknown but could include the following: modulation of cortisol feedback restraint of the axis (Keller-Wood et al., 1988; Turner, 1997; Redei et al., 1994; Patchev and Almeida, 1996; Young, 1995); neurosteroid-related downregulation of GABA receptors (Smith et al., 1998a); upregulation of AVP (consistent with luteal phase reductions in the threshold for AVP release (Spruce et al., 1985). Alternatively, Ochedalski et al. (2007) suggest that progesterone enhances oxytocin-induced CRH.

Pregnancy is also associated with marked changes in adrenocortical function. Plasma levels of cortisol, desoxycorticosterone, aldosterone, and corticosteroid-binding globulin all increase considerably during gestation. Cortisol levels, for example, rise to 3 or 4 times normal, peak during delivery, and return to normal levels quickly after delivery (Smith and Thomson, 1991), although a high rate of Dexamethasone Suppression Test (DST) nonsuppression persists in the postpartum (Wisner and Stowe, 1997). Both the fetal hypothalamus and the placenta produce identical forms of CRH, and there is a 20-fold increase in placental CRH mRNA in the last 5 weeks of pregnancy. One study in normal pregnant women showed increases in CRH from $50\,pg\,ml^{-1}$ at 28 weeks gestation to over $1400\,pg\,ml^{-1}$ at 40 weeks (Campbell et al., 1987). In contrast to hypothalamic CRH, placental CRH is regulated by a positive feedback effect of cortisol (Weiss, 2000). Thus, as cortisol increases during the last trimester, there is an increase in CRH production causing a progressive increase in both ACTH and cortisol secretion.

4.7 Sexual Dimorphisms in Psychiatric Disorders

4.7.1 Introduction

With the description by Pfaff (1966) of sexual dimorphisms in rat brain morphology and by Raisman and Field (1971) of gender-related differences in the synaptic density of the arcuate nucleus in the rat, the notion that the brain could differ functionally, as well as structurally, as a product of gender (i.e., it was sexually dimorphic) formally entered the conceptual lexicon of neuroscience. It has subsequently been demonstrated that most of the many observed sexual dimorphisms occur as a consequence of the organizing or activating effects of gonadal steroids. Nonetheless, despite the elegance of sexually dimorphic brain organization as an explanation for dimorphic behaviors, the complexity of the process underlying the development of sexual dimorphisms has assumed daunting proportions. First, it is often difficult to interpret the meaning of the dimorphisms. For example, lesions of the sexually dimorphic nucleus of the preoptic area (SDN-POA) do not compromise male copulatory behavior (Arendash and Gorski, 1983) (despite the role of the POA in reproductive behavior), and De Vries and Boyle (1998) have suggested that sex differences in some brain regions actually serve to mediate similar behaviors in males and females (e.g., the same parental behavior in male and female prairie voles is mediated by sex differences in vasopressin). Second, a number of asymmetries complicate ascription of dimorphisms to the presence or absence of a particular steroid hormone. Female zebra finches, for example, will develop song behavior (seen usually only in males) if administered androgen or estradiol perinatally, but males deprived of androgen perinatally show no disruption of song behavior as adults (Breedlove, 1992; Gurney, 1982). Third, some sexual dimorphisms appear to be organized and are independent of subsequent steroid exposure (Enriquez et al., 1991), others are activated (i.e., are dependent on subsequent steroid exposure) but not organized (i.e., they are not permanently influenced by perinatal steroid manipulation) (Goldman and Nottebohm, 1983), while still others are both organized and activated (e.g., the perinatally androgenized female zebra finch requires androgen as an adult to express song behavior) (Schlinger et al., 1991; De Vries and Boyle, 1998). Further, Reisert and Pilgrim (1991) have evidence suggesting that dimorphisms in the course of development of mesencephalic and diencephalic neurons are under genetic control (i.e., they are determined well before the appearance of any differences in gonadal steroid levels), similar to the genetically determined pouch or scrotum in marsupials (Renfree et al., 1995). More recently, Rissman and colleagues demonstrated that some sexually dimorphic behaviors are determined by genes on the sex chromosomes that are independent of Sry and subsequent gonadal function (Gatewood et al., 2006). Fourth, the activational–organizational dichotomy is far more fluid and plasticity much greater than the concept of critical periods allows. In contrast to the female zebra finch (who shows no male song behavior if androgenized during adulthood only), the female canary receiving androgen during adulthood will both develop male song behavior and show male-like morphologic changes in the vocal control nuclei, including marked dendritic branching (Nottebohm, 1980; DeVoogd et al., 1981). Not only is the timing of hormonal administration (and species of the animal) important in determining outcome, but the manner of administration as well may dictate the response. For example, Sodersten (1984) demonstrated that one could induce female-typical behavior in gonadectomized adult male rats by pulsatile, but not chronic, administration of estradiol, followed by progesterone. Complexities notwithstanding, gonadal steroids appear capable of programming gonadal-steroid-sensitive circuitry in the brain, behavioral capacities, and differential response to the same physiologic stimulus.

Given the relative lack of access to the brain in human studies compared with similar investigations in animals, the existence of gender-related differences has provided a major source of inference about the role of gonadal steroids in brain function and behavior. Reported gender dimorphisms in psychiatry include the following: prevalence, phenomenology (including characteristic symptoms, susceptibility to recurrence, stress responsivity, age of onset), and treatment response characteristics. Specific examples of such dimorphisms are listed below.

4.7.1.1 Depression

Studies consistently demonstrate a twofold increased prevalence of depression in women compared with men (Robins and Regier, 1991; Weissman and Klerman, 1988; Kessler et al., 1993). This increased prevalence has been observed in a variety of countries (Weissman et al., 1993). A two- to threefold increased prevalence of dysthymia and threefold increase in seasonal affective disorder (Leibenluft et al., 1995) in women have also been noted (Diagnostic, 1994), while bipolar illnesses are equi-prevalent in men and women (Psychiatric, 1991; Weissman and Klerman 1977, 1985) (reviewed in Leibenluft (1996)). Prepubertal depression prevalence rates are not higher in girls (Anderson et al., 1987; McGee et al., 1992), possibly reflecting ascertainment bias/reporting bias (depressed boys may be more likely to come to the attention of healthcare providers) or the possibility that prepubertal major depression is premonitory of bipolar illness (Leibenluft, 1999). With some exceptions, the age of onset (Weissman et al., 1993; Kessler et al., 1993; Frank et al., 1988; Thase et al., 1994; Burke et al., 1990; Winokur et al., 1982; and also see Kornstein et al. (1995), Fava et al. (1996), Spicer et al. (1973), and Nolen-Hoeksema (1987)) type of symptoms, severity, and likelihood of chronicity and recurrence (Weissman et al., 1993; Kessler et al., 1993; Frank et al., 1988; Kornstein et al., 1995; Kessler et al., 1994; Simpson et al., 1997; Zlotnick et al., 1996; and also see Srikant and Patel (1985), Aneshensel (1985), Ernst and Angst (1992), Keitner et al. (1991), and Winokur et al. (1993)) display few differences between men and women. Women are more likely to present with anxiety, atypical symptoms, or somatic symptoms (Frank et al., 1988; Winokur et al., 1993; Kornstein et al., 1995; Leibenluft et al., 1995; Ernst et al., 1992; Young et al., 1990; Angst and Dobler-Mikola, 1984); are more likely to report symptoms, particularly in self-ratings (Angst and Dobler-Mikola 1984; Leibenluft et al., 1995; Frank et al., 1988); are more likely to report antecedent stressful events (Bebbington et al., 1988; Karp and Frank, 1995); manifest a more robust effect of stress on the likelihood of developing depression during adolescence (Silberg et al., 1999); and display increased comorbidity of anxiety and eating disorders (Blazer et al., 1994; Regier et al., 1990; Judd, 1994; Fava et al., 1996), thyroid disease (Reus, 1989; Whybrow, 1995), and migraine headaches (Moldin et al., 1993), as well as lower lifetime prevalence of substance abuse and dependence (Kornstein et al., 1995; Fava et al., 1996; Doran et al., 1986). Reported differences in treatment response characteristics in women compared with men include poor response to tricyclics (Old Age Depression Interest Group, 1993; Raskin, 1974; Glassman et al., 1977; Coppen et al., 1972), particularly in younger women (Raskin, 1974), superior response to selective serotonin reuptake inhibitors (SSRIs) or monoamine oxidase inhibitors (MAOIs) (Davidson and Pelton, 1986; Steiner et al., 1993; Kornstein et al., 2000), and a greater likelihood of response to triiodothyronine (T3) augmentation (Prange et al., 1969; Whybrow, 1995). The extent to which these differences reflect gender-related differences in pharmacokinetics (Dawkins and Potter, 1991; Yonkers et al., 1992; Moody et al., 1967; Preskorn and Mac, 1985; Gex-Fabry et al., 1990; Greenblatt et al., 1987; Warrington, 1991) remains to be determined. Finally, while the prevalence of bipolar disorder is comparable in men and women, women are more likely to develop rapid cycling (Leibenluft, 1996) and may be more susceptible to antidepressant-induced rapid cycling (Altshuler et al., 1995).

4.7.1.2 Physiological dimorphisms

The epidemiologic observations described above are increasingly complemented by demonstrations of sexual dimorphisms in brain structure and physiology in humans. Structural and functional brain imaging studies, for example, have shown the following: (1) differences in functional organization of the brain, with brain activation response to rhyming task lateralized in men but not women (Shaywitz et al., 1995); (2) sex-specific decreases in regional brain volume (caudate in males and globus pallidus, putamen in females) during development (Giedd et al., 1999); (3) increased neuronal density in the temporal cortex in women (Witelson, 1991); (4) greater interhemispheric-coordinated activation of brain regions in women (Azari et al., 1995); (5) larger volume hypothalamic nucleus (INAH 3) in men (Allen et al., 1989); (6) differences in both resting blood flow and the activation pattern accompanying self-induced mood change (George et al., 1996); (7) decreased 5-HT2 binding in the frontal, parietal, temporal, and cingular cortices in women (Biver et al., 1996); (8) differences in whole-brain serotonin synthesis (interpreted as decreased in women but possibly increased if corrected for plasma-free tryptophan levels) (Nishizawa et al., 1997); (9) higher and more symmetric cerebral blood flow in women (Rodriguez et al., 1988; Gur et al., 1982, 1987; Shaw et al., 1979; Esposito et al., 1996); (10) greater asymmetry in the planum temporale in men (Kulynych et al., 1994); and (11) greater brain glucose metabolism (19%) in women (Baxter et al., 1987; Andreason et al., 1993). The potential relevance of gonadal steroids in some of these differences has also been demonstrated with the same technologies (see above). The contribution of these and other effects of gonadal steroids to observed gender dimorphisms must, obviously, await further determination.

Given the complexity of factors that impact on gender throughout development, it is very difficult to infer the degree to which differential exposure to gonadal steroids determines gender-related behavioral differences. A better opportunity to determine the behavioral relevance of fluctuations in gonadal steroids is provided by mood disorders that appear linked to changes in levels of reproductive steroids. In the following sections, we review the role of endocrine factors in three reproductive endocrine-related mood disorders, with the focus in each case on data supporting the importance of hormonal concentrations or context in the precipitation/pathophysiology of the disorder.

4.8 Premenstrual Dysphoria

Endocrine studies of premenstrual dysphoria (PMD) can be divided arbitrarily into those occurring before or after 1983. A comprehensive review of the early studies was performed by Reid and Yen (1981). As described in Rubinow and Roy-Byrne (1984), most of these studies suffered from methodologic flaws, the most serious being the inadequacy of diagnostic criteria. Unlike other disorders in medicine, PMD is a time-oriented, not a symptom-oriented, diagnosis and requires prospective demonstration that symptoms are confined to the luteal phase and disappear at or soon after the onset of menses. Since 1983, the use of two sets of diagnostic guidelines – *Diagnostic and Statistical Manual of Mental Disorders Fourth Edition*, 1994 (American Psychiatric Association, 1994) and NIMH Premenstrual Syndrome Workshop Guidelines, unpublished – has permitted greater homogeneity of samples across studies, a requirement for comparison and generalization of results obtained. Data subsequently generated provide little, if any, evidence for a role of hormone excess or deficiency in the etiology of PMD.

4.9 Hormonal Studies of PMD

Hormonal studies in women with PMD have employed several different strategies: (1) the measurement of basal hormone levels at selected points in the menstrual cycle; (2) evaluation of dynamic endocrine function employing endocrine challenge paradigms; and (3) the manipulation of menstrual cycle physiology in order to examine the plasticity of the linkage between the menstrual cycle and PMD symptoms. The most frequently employed strategy has been the comparison of luteal phase basal hormone levels with those from the follicular phase in women with PMD or with comparable values from a non-PMD control group.

4.9.1 Hypothalamic–Pituitary–Ovarian Axis

Given the coincidence of symptoms with the luteal phase in women with PMD, early investigators sought, as an etiology, a disturbance in reproductive endocrine function. Comparisons of basal plasma hormone levels in women with PMD and controls have revealed no consistent diagnosis-related differences. Specifically, no diagnosis-related differences in the plasma levels, areas under the curve, or patterns of hormone secretion have been observed for estradiol, progesterone, FSH, or LH (Rubinow et al., 1988; Backstrom et al., 1983; Redei and Freeman, 1995; Facchinetti et al., 1993). Results for studies of androgen levels have been inconsistent, demonstrating both normal and decreased T levels (Backstrom and Aakvaag, 1981; Eriksson et al., 1992; Bloch et al., 1998) and elevated and decreased free T levels (Eriksson et al., 1992; Bloch et al., 1998). In addition, two of four studies failed to find any diagnosis-related differences in the pattern of LH pulsatility or in the gonadotropin response to GnRH (Reame et al., 1992; Smith et al., 2004; Facchinetti et al., 1990, 1993). Finally, studies of a variety of other hormonal factors have been similarly unrevealing (Schmidt et al., 1993; Facchinetti et al., 1987; Chuong et al., 1985; Taylor et al., 1984; Ashby et al., 1988; Malmgren et al., 1987; Veeninga and Westenberg, 1992; Tulenheimo et al., 1987; Hamilton and Gallant, 1988; Bloch et al.,

1998). Several studies do, however, suggest that levels of estrogen, progesterone, or neurosteroids (e.g., pregnenolone sulfate) may be correlated with symptom severity in women with PMD (Schechter et al., 1996; Halbreich et al., 1986; Wang et al., 1996b).

Recent speculations about the etiology of PMD have focused on putative abnormal neurosteroid levels. Observations central to these speculations include the following: (1) the GABA receptor (the presumed mediator of anxiolysis) is positively modulated by the 5α- and β-reduced metabolites of progesterone (allopregnanolone and pregnanolone, respectively) (Majewska et al., 1986); (2) withdrawal of progesterone in rats produces anxiety and insensitivity to benzodiazepines due to withdrawal of allopregnanolone, with consequent induction of $GABA_A$ α4-subunit levels and inhibition of GABA currents (Smith et al., 1998a,b); (3) decreased plasma allopregnanolone levels are seen in major depressive disorder and in depression associated with alcohol withdrawal, with an increase in levels seen in plasma and CSF following successful antidepressant treatment (Ströhle et al., 1999; Romeo et al., 1996, 1998; Uzunova et al., 1998; Eser et al., 2006; Schule et al., 2007, 2006); (4) allopregnanolone displays anxiolytic effects in several animal anxiety models (Bitran et al., 1991, 1993; Wieland et al., 1991) and may be involved in the stress response (Purdy et al., 1991); (5) antidepressants may promote the reductive activity of one of the synthetic enzymes (3-α-hydroxysteroid oxidoreductase), thus favoring the formation of allopregnanolone (Uzunov et al., 1996; Griffin and Mellon, 1999); (6) cerebral cortical inhibition increases during the luteal phase, a presumed effect of increased allopregnanolone levels and a finding absent in women with PMD (Smith et al., 2002, 2003; Maguire et al., 2005); (7) PMD patients show differences from controls in pregnanolone-modulated saccadic eye velocity (SEV) and sedation in the luteal phase (Sundstrom et al., 1998a) (although the reported differences seem attributable to an SEV response to vehicle in those with PMD and a blunted sedation response in the follicular phase in controls); high severity PMD patients show blunted SEV and sedation responses to $GABA_A$ receptor agonists – pregnanolone (Sundstrom et al., 1998a) or midazolam(Sundstrom et al., 1997b) – compared with low severity PMD patients; (8) women with PMD have blunted allopregnanolone responses to stress and evidence of altered metabolism of progesterone to allopregnanolone (Girdler et al., 2001; Klatzkin et al., 2006). While several investigators observed decreased serum allopregnanolone levels in women with PMD compared to controls on menstrual cycle day 26 (Rapkin et al., 1997), during the luteal phase only (Monteleone et al., 2000), or during the follicular phase only (Bicikova et al., 1998), PMD patients in the last two studies had lower progesterone levels, which may explain the observed decreased allopregnanolone levels. This explanation is supported by the observation of Girdler et al. (2001) that women with PMD had both higher progesterone and allopregnanolone levels during the luteal phase compared with controls. Further, other studies showed no diagnosis-related differences in allopregnanolone or pregnanolone (Schmidt et al., 1994; Wang et al., 1996a) nor any difference in allopregnanolone levels in women with PMD before and after successful treatment with citalopram (Sundstrom and Backstrom, 1998b). Wang et al. (1996a) did find that if two cycles differed in the area under the curve (AUC) of a hormone by more than 10%, the cycle with the lower levels of allopregnanolone and higher levels of estradiol, pregnanolone, and pregnanolone sulfate was accompanied by higher levels of symptom severity.

In general, no differences have been observed in basal plasma cortisol levels, urinary free cortisol, the circadian pattern of plasma cortisol secretion, or basal plasma ACTH levels (Rubinow and Schmidt, 1995). Both decreased ACTH levels in PMD patients across the menstrual cycle and no differences from controls have been reported (Redei and Freeman, 1993; Rosenstein et al., 1996; Bloch et al., 1998; Rabin et al., 1990). In contrast, the cortisol responses to the $serotonin_{2C}$ ($5\text{-}HT_{2C}$) agonist/$5\text{-}HT_{2A}$ antagonist m-chlorophenylpiperazine (m-CPP) (Su et al., 1997), a psychological stressor (Girdler et al., 2001), and CRH (Rabin et al., 1990) or naloxone (Facchinetti et al., 1994) were blunted in patients with PMD during the luteal phase. Finally, in a study of CSF, Eriksson et al. (1994) observed no differences in CSF monoamine metabolites in PMD patients compared with controls, nor were there menstrual cycle-related differences in either group. Similarly, Parry et al. (1991) found no cycle-related differences (midcycle vs. premenstrual) in CSF ACTH, β-endorphin, GABA, 5-hydroxyindole acetic acid (5-HIAA), homovanillic acid (HVA), or norepinephrine; a slight but significant premenstrual increase in CSF 3-methoxy-4-hydroxyphenyl glycol (MHPG) was noted.

Roca et al. (2003) reported that women with PMD also fail to show the luteal-phase enhancement of exercise-stimulated HPA-axis activity seen in normal control women. The differential HPA-axis response to exercise stimulation in women with PMD provides strong additional evidence for the dysregulation of stress response physiology in this disorder. PMD patients failed to show the luteal-phase increase in AVP, ACTH, and cortisol seen in controls; indeed, stimulated hormone levels in women with PMD were higher (albeit nonsignificantly) in the follicular phase. Differences seen were not attributable to differences in the level of stress achieved, as similar stimulated levels of lactate were obtained in both menstrual cycle phases in patients and controls. In addition to the abnormal response to menstrual cycle phase, women with PMD showed (at a trend level) reduced adrenal response to ACTH in both cycle phases. The failure of prior studies to demonstrate these significant differences in HPA-axis function may reflect the nature of the stimulation paradigms employed: graded exercise stimulation is a more robust activator of the axis than most others used (e.g., CRH, m-CPP) and additionally permits a similar degree of stress across individuals by indexing the stimulus parameters to those required to elicit 90% of the individual's maximal aerobic capacity.

In conclusion, there are no consistently demonstrated endocrine or other biological abnormalities in PMD. Further, for the overwhelming majority of biologic factors for which diagnostic group-related differences have been suggested or demonstrated, the difference is not confined to the luteal phase but rather appears in both follicular and luteal phases (Roy-Byrne et al., 1987; Lee et al., 1990; Howard et al., 1992; Parry et al., 1990, 1989; Sherwood et al., 1986; Rosenstein et al., 1994, 1996; Bancroft et al., 1991; Su et al., 1997; Rabin et al., 1990; Eriksson et al., 1992; Bloch et al., 1998; Rabin et al., 1990). Even if these differences are confirmed, their persistence across the menstrual cycle would appear to argue against their direct role in the expression of a disorder confined to the luteal phase. Presently, then, there is no clearly demonstrated luteal-phase-specific physiologic abnormality in PMD.

4.9.2 Context (Hormones as Triggers or Treatments)

PMD does not, therefore, appear to reflect an abnormality of the reproductive endocrine axis. Indeed, we administered a PR blocker, mifepristone, with or without hCG to women with PMD during the early to mid-luteal phase and demonstrated that hormonal events and gonadal steroid levels of the mid- to late luteal phase were irrelevant to PMD, as they could be eliminated without altering subsequent symptom appearance (Schmidt et al., 1991). It, nonetheless, remained possible that the follicular-phase or early luteal-phase gonadal steroids might be critical to the appearance of PMD, a speculation supported by reports of the therapeutic efficacy in PMD of ovarian suppression through either medical (GnRH agonist; danocrine) (Muse et al., 1984; Hammarback and Backstrom, 1988; Brown et al., 1994; West and Hiller, 1994; Hussain et al., 1992; Mortola et al., 1991; Bancroft et al., 1987; Mezrow et al., 1994; Sarno et al., 1987; Halbreich et al., 1991) or surgical (oophorectomy) (Casson et al., 1990; Casper and Hearn, 1990) means. Consequently, we evaluated the effect of elimination of ovarian steroid secretion on PMD symptoms as well as the effect of ovarian steroid replacement in those whose symptoms were responsive to ovarian steroid suppression. We confirmed the therapeutic efficacy of GnRH agonist-induced ovarian suppression (Schmidt et al., 1998) and, consistent with data from Mortola and Muse (Mortola et al., 1991; Muse, 1989), demonstrated that either estrogen or progesterone could precipitate the return of typical symptoms in women with PMD (Schmidt et al., 1998). In contrast, a group of control women lacking PMD showed no perturbation of mood during GnRH agonist-induced hypogonadism nor during hormone addback with either progesterone or estradiol, despite achieving hormone levels comparable to those seen in the women with PMD. Women with PMD, therefore, are differentially sensitive to gonadal steroids such that they experience mood destabilization with levels or changes in gonadal steroids that are without effect on mood in women without a history of PMD. Gonadal steroids, then, are necessary but not sufficient for PMD: they can trigger PMD, but only in women who are otherwise vulnerable to experience mood-state destabilization. Thus, PMD could represent a disorder of mood state that is triggered by hormone-related events occurring prior to the mid- to late luteal phase of the menstrual cycle. Nonetheless, the system(s) that underlies the vulnerability to gonadal steroid-induced mood disturbances in PMD remains to be identified. One potential candidate system that may mediate the differential behavioral sensitivity is the central serotonergic system.

Several observations have suggested the importance of interactions between the serotonin system and gonadal steroids in the pathophysiology of PMD. First, in a potential animal model of menstrual cycle-related irritability (resident intruder model) (Ho et al., 2001), female rat aggression is ovarian steroid dependent and is prevented by serotonin reuptake inhibitors (as is PMD) (Dimmock et al., 2000). Second, serotonin has a role in behaviors (e.g., appetite, impulsivity, mood, sleep, and sexual interest) that vary with the menstrual cycle in PMD. Third, women with PMD have altered imipramine binding and platelet 5HT uptake compared to controls (Steege et al., 1992; Rojansky et al., 1991; Taylor et al., 1984; Rapkin et al., 1987; Ashby et al., 1988, 1990) as well as altered platelet paroxetine binding (which normalize with successful treatment with GnRH agonist) (Bixo et al., 2001). Fourth, pharmacologic challenge studies, although limited by the absence of selective agonists/antagonists of the 5HT system, suggest that 5HT regulation differs between women with and without PMD. For example, blunted endocrine responses to serotonergic agonists (e.g., L-tryptophan, m-CPP) have been described in PMD (although not confined to the luteal phase) (Dimmock et al., 2000; Bancroft et al., 1991; Su et al., 1997). In addition, the $5HT_{1A}$ system, implicated in one study as disturbed in PMD (Yatham, 1993), is involved in the regulation of GABA activity (Krezel et al., 2001; Adell et al., 2002; Sibille et al., 2000; Kishimoto et al., 2000; Koyama et al., 1999; Stutzmann and LeDoux, 1999), abnormalities of which have been described or inferred in PMD (Gulinello et al., 2001; Wang et al., 1996b; Monteleone et al., 2000; Rapkin et al., 1997; Sundstrom et al., 1998a; Halbreich et al., 1996; Smith et al., 2002, 1998a,b; Sundstrom et al., 1997a,b; Sundstrom et al., 1998a; Schmidt et al., 1994). Finally, serotonin reuptake inhibitors, but not nonserotonergic antidepressants, are efficacious in the treatment of PMD (suggesting increased SERT activity in PMD) (Freeman, 1997), and the therapeutic efficacy of serotonin agonists can be reversed by tryptophan depletion (Menkes et al., 1994) or serotonin receptor blockade (Roca et al., 2002). While alterations in serotonin function are clearly relevant to the successful treatment of PMD symptoms, it remains unclear whether alterations in serotonin function underlie the differential mood response to ovarian steroids in PMD.

4.10 Perimenopausal Depression

The majority of women do not develop depression during the menopause transition, and, therefore, reproductive aging is not uniformly associated with either depressive symptoms or the syndrome of depression. Nonetheless, several studies report an association between the menopause transition and an increased risk for depression (Schmidt, 2005). Indeed, five recent longitudinal studies all have documented an increased risk for depression during the menopause transition with odd ratios ranging from 1.8 to 2.9 compared with the premenopause (Bromberger et al., 2001; Freeman et al., 2004; Schmidt et al., 2004; Cohen et al., 2006; Freeman et al., 2006). In particular, two studies (Cohen et al., 2006; Freeman et al., 2006) observed a 2–2.5 times greater risk for the first onset of depression during the late menopause transition compared to the premenopause. These data suggest that events surrounding the final menstrual period may predispose some women to develop clinically significant depressive illness. Although several factors could precipitate depression in these women, endocrine events are suggested by the stage of the menopause transition (i.e., late) during which depressions appeared. The late transition is characterized by more prolonged hypogonadism than the early perimenopause, during which estradiol secretion may be increased. Thus, the timing of appearance of the depressions observed suggests an endocrine mechanism related to the perimenopause (estradiol withdrawal and/or recent-onset of prolonged hypogonadism) in the pathophysiology of perimenopausal depression.

4.11 Hormonal Studies of Perimenopausal Depression

There have been no consistent abnormalities of reproductive or adrenal hormones identified in women with perimenopausal depression compared to controls. Nonetheless, the relevance of changes in pituitary–ovarian function to depression during the perimenopause is suggested by evidence that mood symptoms may change concordantly with FSH levels (Daly et al., 2003) and that estradiol therapy has acute mood-enhancing effects in perimenopausal women with depression (Schmidt et al., 2000; Soares et al., 2001).

Several additional reports indirectly support a role for reproductive hormones during the perimenopause in depression: hormone replacement beneficially affects both hot flushes and mood in hypogonadal women (Steingold et al., 1985; Brincat et al., 1984; Montgomery et al., 1987; Ditkoff et al., 1991; Sherwin and Gelfand, 1985); and lower gonadotropin levels are observed in postmenopausal depressed women compared to asymptomatic comparison groups (Brambilla et al., 1990; Amsterdam et al., 1983; Altman et al., 1975; Guicheney et al., 1988). The observed improvement in depressive symptoms after hormone replacement suggests the contribution of hypoestrogenism to mood disturbances, permitting the speculation that depressed perimenopausal women are relatively more estrogen deficient than nondepressed perimenopausal women. Perimenopausal women with depressive symptoms have been reported to have lower plasma estrone (E1) levels (Ballinger, 1990) than nondepressed perimenopausal women, and an association has been described between increased plasma FSH levels and depression (Huerta et al., 1995) (contradicting studies cited above). In contrast, three studies of perimenopausal and postmenopausal women observed either no diagnosis-related differences in plasma estradiol (E_2) and FSH (Saletu et al., 1996) or no correlation between plasma levels of estrogens or androgens and severity of depressive symptoms (Barrett-Connor et al., 1999; Cawood and Bancroft, 1996).

In a study of 21 women with their first episode of depression occurring during the perimenopause and 21 asymptomatic perimenopausal controls (Schmidt et al., 2002), we were unable to confirm previous reports of lower basal plasma levels of LH (Brambilla et al., 1990; Amsterdam et al., 1983; Altman et al., 1975; Guicheney et al., 1988) or E1 (Ballinger, 1990) in perimenopausal and postmenopausal women with depression compared to matched controls. In addition, we observed no diagnosis-related differences in basal plasma levels of FSH, E_2, T, or free T. Our data are consistent with those of Barrett-Connor et al. (1999) and of Cawood and Bancroft (1996), who found no correlation between mood symptoms and plasma levels of E1, E_2, or T. Notwithstanding the limitations of basal hormonal measures, data suggest that depressed perimenopausal women are not distinguished from nondepressed perimenopausal women by being more estrogen deficient.

Age-related differences in the function of several physiologic systems have been observed in both animals and humans. Some of these differences may occur coincident with the perimenopause and, therefore, may potentially contribute to mood dysregulation at this time. Although postmenopausal women have been reported to exhibit increased stress-induced plasma norepinephrine levels compared to premenopausal women (Matthews, 1992), only one previous study (Ballinger, 1990) reported elevated urinary cortisol levels in perimenopausal women reporting depressive symptoms compared to asymptomatic controls. Unfortunately, to date, no systematic study has been performed of HPA-axis function in perimenopausal women with a depressive syndrome.

A role for the adrenal androgen DHEA and DHEAS in the regulation of mood state has been suggested by both its effects on neural physiology (Majewska et al., 1990; Compagnone and Mellon, 1998; Baulieu and Robel, 1998) and its potential synthesis within the CNS (Robel and Baulieu, 1994; Zwain and Yen, 1999). Moreover, in clinical trials, DHEA administration has been reported to improve mood in some (Morales et al., 1994; Wolkowitz et al., 1999, 1997; Bloch et al., 1999), but not all, studies (Wolf et al., 1997). Finally, abnormalities of DHEA secretion have been observed in depressive disorders, with both increased and decreased levels observed relative to nondepressed controls (Goodyer et al., 1996, 1998; Ferrari et al., 1997; Heuser et al., 1998). DHEA's potential role in the onset of depression may be particularly relevant at midlife given the declining levels of DHEA production with aging and the accelerated decrease in DHEA levels reported in women, but not men, during midlife (Laughlin and Barrett-Connor, 2000; Cumming et al., 1982). Plasma levels of DHEA and DHEAS decline progressively from the third decade at a rate of about 2–3% per year (Gray et al., 1991), reaching about 50% of peak levels during the fifth to sixth decades (Orentreich et al., 1984; Orentreich et al., 1992; Belanger et al., 1994). It is possible, therefore, that declining secretion (or abnormally low secretion) of DHEA may interact with perimenopause-related changes in ovarian function to trigger the onset of depression in some women. In fact, in perimenopausal and postmenopausal women, mood is correlated with DHEA(S) levels, with lower DHEA levels associated with more depression and higher levels associated with greater well-being (Barrett-Connor et al., 1999; Cawood and Bancroft, 1996). We measured morning plasma levels of DHEA, DHEAS, and cortisol in a separate sample of women with their first onset of depression during the perimenopause and in nondepressed women matched for age and reproductive status. Depressed perimenopausal women had significantly lower levels of both plasma DHEA and DHEAS but not cortisol compared to controls (Schmidt et al., 2002). Thus, DHEA, but not adrenal glucocorticoid secretion, differed in depressed and nondepressed perimenopausal women.

Despite the antidepressant efficacy of estradiol and the linkage of perimenopausal depression to a time of estrogen withdrawal, we still do not have direct evidence that estradiol withdrawal triggers the onset of depression in these women. Preliminary observations from a study of the effects on mood of estradiol withdrawal suggest, however, that women with a history of depression during the perimenopause, but not those lacking such a history, experience a recurrence of depressive symptoms when estradiol is withdrawn under blinded, placebo-controlled conditions (Schmidt et al., unpublished observation). Future studies will focus on the mechanisms by which declining or low estradiol levels induce changes in CNS function sufficient to trigger depression in some women.

4.12 Gonadal Steroids as Treatments of Mood Disorders

The potential role of gonadal/adrenal steroids in the treatment of depression was suggested over a 100 years ago (Easterbrook, 1900). The observed responses to the administration of these crude extracts prompted some of the first theories about the cause of these disorders, specifically, that a hormone-deficiency state was present. Although speculation still continues about the existence of a hormone-deficiency state in perimenopause-related mood disorders, our current understanding of the complexity of the mood and behavioral response to exogenous hormone replacement has demonstrated the inadequacy of the simple reproductive hormone deficiency model of mood disorders.

4.12.1 Estrogen Treatment

In one of the first placebo-controlled trials of estrogen in depression, Werner et al. (1934) reported the beneficial effects of theelin (estrone suspension, USP) injections in nine women with involutional melancholia compared to eight women with involutional melancholia receiving saline injections. Subsequent controlled studies employing synthetic forms of estrogen in the treatment of depression have yielded mixed results. Estrogen has been reported to improve mood (albeit inconsistently) (Coope, 1981; Campbell, 1976; George et al., 1973) in the following samples: (1) perimenopausal and postmenopausal women reporting depressive symptoms (Montgomery et al., 1987; Saletu et al., 1995; Sherwin, 1988), (2) postmenopausal women with depression unresponsive to traditional antidepressant therapy (Klaiber et al., 1979), and (3) nondepressed menopausal women not experiencing hot flushes (Ditkoff et al., 1991). However, several methodologic problems complicate the interpretation of these data. First, the diagnosis of depression was not confirmed by a structured psychiatric diagnostic interview in any study, leaving unclear the proportion of women with a depressive syndrome (as opposed to depressive symptoms, which have different etiologies and show a more variable therapeutic response). Second, only one study (Ditkoff et al., 1991) evaluated the effects of estrogen independent of the presence of hot flushes. Since hot flush-induced sleep disturbances may adversely impact mood, estrogen's salutary effects on hot flushes may have confounded interpretation of the effect of estrogen on mood symptoms. Finally, the antidepressant efficacy of estrogen may differ in perimenopausal and postmenopausal women (Montgomery et al., 1987) and as a function of dose (Saletu et al., 1995; Klaiber et al., 1982).

We examined the therapeutic efficacy of estradiol replacement in 34 women (approximately half of whom had no prior history of depression) with perimenopausal depression under double-blind, placebo-controlled conditions (Schmidt et al., 2000). After 3 weeks of estradiol, symptom rating scale scores (CES-D, HAM-D, and visual analog scales) were significantly decreased compared to baseline scores and significantly lower than the scores in the women receiving placebo. Women receiving placebo showed no significant improvement compared with their baseline scores. A full or partial therapeutic response was seen in 80% of subjects on estradiol and 22% of those on placebo consistent with the observed effect size in a recent meta-analysis of studies examining estrogen's effects on mood (Zweifel and O'Brien, 1997). Six of the seven women with a current diagnosis of major depressive episode were considered responders and 19 out of 24 women with minor depression responded to estrogen. Finally, no significant effect of the presence of hot flushes was observed for the majority of symptoms. The efficacy of estradiol in depression in the absence of hot flushes suggests that estrogen's effect on depression is not solely a product of its ability to reduce the distress of hot flushes.

This study demonstrated that a dose of 0.05 mg per day of estradiol administered by a skin patch is associated with a significant improvement in mood in depressed perimenopausal women with or without hot flushes. Our findings are consistent with data from Montgomery et al. (1987) and Saletu et al. (1995) suggesting the beneficial effects of estrogen on mood in perimenopausal women reporting depressive symptoms. In addition, we extended prior findings by demonstrating that estradiol has salutary effects on mood in women who meet standardized diagnostic criteria for depression and that estradiol's effect on mood occurs independent of its effect on hot flushes. Others have replicated and expanded these observations (Soares et al., 2001; Morrison et al., 2004). First, Soares et al. reported a significant and beneficial effect of ET compared to placebo in women with perimenopause-related major depression (as defined by the PRIME MD) (Spitzer et al., 1995) and, additionally, reported that baseline plasma estradiol levels did not predict response to estrogen treatment (Soares et al., 2001). Second, Morrison et al. (2004) observed that estrogen was no more effective than placebo in postmenopausal depressed women in contrast to previous results in perimenopausal women. These data emphasize that the stage of reproductive senescence may predict response to estrogen, as originally reported by Appleby et al. (1981). Thus, perimenopausal women who are undergoing changes in reproductive function may be more responsive to estrogen than postmenopausal women whose hormonal changes have long since stabilized.

4.12.2 Dehydroepiandrosterone Treatment

Midlife in both men and women is also characterized by a steady decline in the production of androgens, such as DHEA, androstenediol, and androstenedione, which are mostly of adrenal origin.

We examined the effects of DHEA on mood in men and women with midlife-onset depression in a double-blind, placebo-controlled crossover-design study. Patients were treated with 30 mg t.i.d. DHEA for 3 weeks followed by 3 weeks of 150 mg t.i.d. (total of 6 weeks). Results in both men and women suggest the antidepressant efficacy of DHEA. DHEA but not placebo significantly improved depression ratings on all mood rating scales. Symptoms that improved significantly after 6 weeks of DHEA compared with baseline or placebo were as follows: low energy, anhedonia, lack of motivation, emotional flattening (numbness), sadness, excessive worry, and inability to cope (Bloch et al., 1999; Schmidt et al., 2005).

While baseline plasma DHEA levels and mood were not correlated, the increase in plasma DHEAS levels was significantly correlated with subsequent response, and responders had significantly higher DHEAS levels (when covaried for age) during DHEA treatment compared to nonresponders. As the clearance of DHEA is very rapid, several orders of magnitude greater than that of DHEAS (Longcope, 1995), plasma levels of DHEAS are both more stable and a more integrated measure of DHEA replacement than are plasma DHEA levels. Therefore, it is of interest that while unrelated to the levels of DHEA achieved, the change in mood is related to DHEAS levels, suggesting that the ability to see therapeutic improvement is dependent on the extent of DHEA augmentation. However, the lack of correlations between therapeutic response and baseline DHEA/DHEAS levels emphasizes our inability to infer any relationship between DHEA levels and the onset of these mood disorders; that is, one cannot infer that the mood disorder in any way reflects a deficiency of DHEA.

4.13 Postpartum Psychiatric Disorders

Affective syndromes that occur during the postpartum period have traditionally been divided into three categories: (1) postpartum blues, (2) postpartum depression (PPD), and (3) puerperal psychosis. PPD is associated with more persistent symptoms and a higher rate of morbidity than the blues but is less severe (depressions of minor to moderate severity) than postpartum psychotic depressions. The 2- to 3-month prevalence rates of postpartum depression in studies using conventional diagnostic criteria (e.g., RDC, DSM-III) have been reported to be in the range of 8.2–14.9% (Cutrona, 1983; Kumar and Mordecai-Robson, 1984; O'Hara, 1986; Wisner et al., 2002). Some studies (Brockington et al., 1982; Kumar et al., 1984; Cox et al., 1993) have reported that the incidence of depression is increased significantly during the first 3 months after birth as compared to during prepregnancy, pregnancy, or the period after the first postpartum year. Others have disputed this association, arguing that the prevalence of depression during the postpartum period is no greater than that in comparably aged nonpuerperal women (O'Hara et al., 1991a; Josefsson et al., 2001; Evans et al., 2001; Yonkers et al., 2001; Halbreich, 2004). In fact, recent epidemiologic studies (Josefsson et al., 2001; Evans et al., 2001) observed that the last trimester of pregnancy also was associated with an increased prevalence of depression comparable to the postpartum. Thus, the postpartum is not associated with an increased prevalence of major or minor depression. Nonetheless, it is not the increased prevalence of depression, but the linkage of the onset of depression to a specific phase of reproductive change that distinguishes this condition.

A number of studies have attempted to determine the relationship between postpartum mood symptoms and gonadal steroid level changes by examining basal levels, or changes in levels, during pregnancy and the postpartum period. O'Hara et al. showed that women with PPD (diagnosed at 9 weeks postpartum by self-administered Beck Depression Inventories) were not distinguished from controls by basal plasma estradiol or progesterone levels (with the exception of lower plasma estradiol levels during week 36 of gestation and day 2 postpartum) (O'Hara et al., 1991a) nor by differences in the rate of change of either estradiol or progesterone during the peripartum. Similarly, Harris et al. (1996) observed no associations between salivary progesterone levels and PPD during the peripartum. In contrast, another study showed higher progesterone, but not estradiol, levels at day 7 postpartum in women who went on to develop PPD at 6–10 weeks after delivery compared with control mothers who did not develop PPD (Abou-Saleh et al., 1998). In addition to levels of estradiol and progesterone, studies have focused on measures that may predict a woman's vulnerability to develop gonadal steroid-induced depression. Examples of such measures include apomophine-induced growth hormone response and alterations in neurosteroid levels in postpartum psychiatric illness. Wieck et al. (1991) demonstrated that an increased growth hormone response to apomorphine on postpartum day 4 (before the usual onset of illness) was associated with an increased risk of a recurrent episode of depression. The authors speculated that these findings reflected increased sensitivity of central dopamine receptors, which may be triggered by the sharp fall in circulating estrogen concentrations after delivery (i.e., estradiol uncouples D_2 receptors) (Maus et al., 1989) with an acute upregulation in D_2 receptors possibly resulting in psychiatric disturbance following the sudden postpartum drop in estradiol levels). As described above, neurosteroid metabolites of gonadal steroids are known to have acute, nongenomic modulatory effects at GABA and glutamate receptors. Levels of one such potent progesterone metabolite, allopregnanolone, rise progressively during pregnancy (Luisi et al., 2000) and drop abruptly after parturition (as levels are closely correlated with plasma progesterone levels) (Schmidt et al., 1994). Preliminary data (Daly, unpublished data) suggest that women with a history of PPD show a significant correlation between decreasing levels of this anxiolytic neurosteroid and mood symptoms. Pearson Murphy et al. (2001) also have suggested a role for alterations in progesterone metabolites in PPD, with higher levels of 5α-dihydroprogesterone observed in depressed patients compared to controls during the last trimester of pregnancy.

To summarize the above data, no consistent differences in gonadal steroid levels have been demonstrated, either in pregnancy or the postpartum, between women with and without PPD, suggesting that the condition does not represent a simple gonadal steroid excess or deficiency state. Our data would, nonetheless, suggest that alterations in the levels of gonadal steroids are implicated in the development of the condition, either during the period of elevated levels or during withdrawal from such levels.

Higher cortisol levels at the end of pregnancy have been reported in association with more severe blues, and cortisol levels have been shown to correlate with postpartum mood in breastfeeding mothers during the first week postpartum (Bonnin, 1992). Most studies, however, have failed to show any association of blues or PPD with plasma or salivary cortisol or with urinary metabolites (O'Hara et al., 1991b; Kuevi et al., 1983; Harris et al., 1994; Abou-Saleh et al., 1998; Feksi et al., 1984).

Abnormalities of CRH-stimulated ACTH (but not cortisol) have been reported in mixed samples of PPD and blues (Magiakou et al., 1996b). Magiakou et al. (1996a) showed that women with the blues or PPD had a more severe and longer-lasting suppression of hypothalamic CRH secretion in the postpartum period than euthymic mothers. In addition, Bloch et al. (2005) observed greater CRH-stimulated cortisol in euthymic women with a history of PPD compared with controls during a hormone-addback-state simulating pregnancy. These dynamic abnormalities of the HPA axis suggest that adaptative response to stress may be compromised in women who experience or are susceptible to PPD.

Finally, no clear relationship between thyroid dysfunction and PPD exists, and although thyroid dysfunction may contribute to postpartum mood disorders, other factors would appear to play more defining roles in the development of the condition.

In summary, gonadal steroids appear to play a key role in the development of PPD, but the exact nature of this role has yet to be fully determined. Only a subgroup of women appears to have an underlying biological sensitivity that ultimately manifests as PPD.

4.14 Hormone Treatment Studies

4.14.1 Estrogen Treatment

A small study of 11 women at high risk for puerperal psychosis (seven with a history of puerperal psychosis and four with a history of puerperal major depression) showed that high-dose estrogen treatment, commencing immediately postpartum and administered for 4 weeks, resulted in a lower than expected 1-year relapse rate (9% compared to an expected 35–60% without prophylaxis) (Sichel et al., 1995). Estrogen was administered initially as Premarin 5 mg B.I.D. p.o. and then tapered over 4 weeks. The authors speculated that estrogen administration may attenuate the rapid puerperal drop in estradiol levels, thereby cushioning the negative impact of the usual postpartum estrogen-withdrawal state on serotonergic and dopaminergic neurotransmission.

Gregoire et al. (1996) performed a double-blind, placebo-controlled study of estradiol in 61 women, who developed major depression within 3 months of delivery. Eighty percent of the patients receiving an estrogen patch had Edinburgh Postnatal Depression Scale scores under the threshold for major depression (less than 14) after 3 months of treatment, compared with 31% of the placebo-treated group. Nearly half of the estrogen- and placebo-treated patients were also on antidepressant medication; however, the efficacy of estrogen in reducing depression was not influenced by whether the women were concurrently taking antidepressants. While preliminary, these data provide indirect evidence for a role of estradiol in the appearance of PPD.

4.14.2 Progesterone Treatment

An open study on the use of progesterone for prophylaxis against PPD was carried out by Dalton (1985), who reported a recurrence rate of postnatal depression in nine of 94 women (10%) using prophylactic progesterone, contrasted with a recurrence rate of 68% among 221 untreated women who had experienced postnatal depression previously (Dalton, 1980). Progesterone was administered as 100 mg IM daily for 7 days (starting at completion of labor), followed by progesterone suppositories for 2 months or until the onset of menstruation. However, methodologic factors (open trial without comparison group and placebo control, lack of standardized rating instruments, and selection bias (patients in the study had sought out progesterone treatment)) greatly limit interpretation of this report.

A double-blind, placebo-controlled study of 180 postpartum women, randomly treated with either the long-acting progestogen contraceptive norethisterone enanthate or placebo, showed an increased risk of developing depressive symptoms following treatment with norethisterone (Lawrie et al., 1998). Although the intensity of depressive symptoms (as measured by the Edinburgh Postnatal Depression Score and the Montgomery–Asberg Depression Rating Scale) differed significantly between the treatment and placebo groups, the rates of depressive illness were not examined. Less than one-fourth of the women approached agreed to participate in the study, which may have introduced some subject bias. The authors noted that findings with synthetic progestogens (such as norethisterone) may not be generalizable to progesterone. Despite methodologic limitations, current evidence does not support a role for progesterone in the treatment of PPD.

4.15 Gonadal Triggers in Context

The study of PPD is compromised by many methodological limitations. Confounding effects include breastfeeding, sleep deprivation, obstetrical complications, and psychosocial stressors. Rating instruments designed for use in the postpartum period may be unsuitable for assessing mood changes during pregnancy. In addition, time of day/seasonal sampling effects may occur. Of particular note, the probable heterogeneity of postpartum mood syndromes can complicate findings.

As the preceding studies have shown, postpartum depression is not consistently associated with any particular hormonal deficiency syndrome. Prompted in part by our findings that mood changes in women with PMD represented abnormal responses to normal hormonal changes (Schmidt et al., 1998), we sought to further examine the role of alterations in gonadal steroid levels in PPD by creating a scaled-down model of pregnancy and parturition and determining whether the induced hormonal changes precipitated mood symptoms in euthymic women with a past history of PPD. The simulated pregnancy was accomplished by inducing hypogonadism using the GnRH agonist leuprolide acetate to suppress ovarian function, then adding back supraphysiologic doses of estradiol and progesterone, and then withdrawing both steroids under double-blind conditions. This methodology helped replicate two crucial hormonal phases of pregnancy: elevated gonadal steroid levels during pregnancy, and the abrupt decline in levels following parturition. Five of eight women with a history of PPD, and none of the control women, developed significant mood symptoms during the addback and withdrawal period, consistent with some reports that depressive symptoms may be more common during both the final trimester and in the postpartum period. This study (Bloch et al., 2000) provides some of the first direct evidence in support of the involvement of estrogen and progesterone in the development of PPD and suggests that women with a history of PPD are differentially sensitive to the mood-destabilizing effects of gonadal steroids.

In summary, gonadal steroids appear to play a key role in the development of PPD, but the exact nature of this role has yet to be fully determined. Only a subgroup of women (8–15%) appear to have an underlying biological sensitivity that ultimately manifests as PPD. Our data would suggest that alterations in the levels of gonadal steroids are implicated in the development of the condition, either during the period of elevated levels or during withdrawal from such levels.

4.16 Context

The differential sensitivity to gonadal steroids seen in women with histories of PMD and PPD (and possibly perimenopausal depression) emphasizes that the response to a biological signal cannot be inferred absent an understanding of the context in which the signal occurs. This context includes current physiological and external environments, prior experience, past history of exposure to the stimulus, and genetic makeup. With the imminent mapping of the human genome, this last contextual determinant becomes of great practical interest as a potential explanation for differential response to steroids. Data already exist from both animal and human studies in support of this hypothesis. Spearow et al. (1999) demonstrated greater than 16-fold differences in sensitivity to estrogen (reproductive disruption) across six different mouse strains, with genotype accounting for more of the variation than the dose of E_2. Similarly, strain/genetic (and task-dependent) differences in behavioral sensitivity to allopregnanolone were observed by Finn et al. (1997). Huizenga et al. (1998a) demonstrated not only the intraperson stability of baseline cortisol and feedback sensitivity (to dexamethasone) suggesting a genetic influence (Huizenga et al., 1998b), but also a higher sensitivity to exogenously administered glucocorticoid (dexamethasone) in association with a polymorphism in exon 2 of the glucocorticoid receptor. Association studies also suggest a progressively increased rate and severity of prostate cancer as the number of CAG trinucleotide repeats in exon 1 of the androgen receptor decreases (Giovannucci et al., 1997). This observation is accompanied by the subsequent observation that androgen receptors with decreased CAG repeats demonstrate increased transcriptional efficiency (Beilin and Zajac, 1999). Steroid receptor polymorphisms, then, may alter the steroid signaling pathway in such a way as to produce or contribute to a different behavioral/phenotypic response to a hormone signal. As appealing an explanation as this is for the differential sensitivity observed in PMD and PPD, the demonstrations in animal studies that perinatal steroid manipulations alter the organization of gonadal steroid-sensitive circuitry (Phoenix et al., 1959) as well as gonadal steroid-activated gene expression (Salo et al., 1997) caution us that gene–environment interactions may yield markedly different phenotypic expressions of the same genotype. Nonetheless, possible genetic susceptibility loci for PMD and puerperal psychosis have been identified. In a gene-based haplotyping study, Huo et al. (2006) found four single nucleotide polymorphisms (SNPs) in the fourth intron of the ERα gene that were significantly associated with PMD. Coyle et al. (2000) observed that a variable number tandem repeat (VNTR) polymorphism in intron 2 of the serotonin transporter was significantly associated with bipolar affective puerperal psychoses, and in a genome-wide association study, Jones et al. (2007) observed significant linkage signals on chromosome 16p13 and 8q24.

At the very least, however, it is time to recognize the importance of context, at the behavioral level no less than at the cellular level, in determining the response to a steroid signal. By understanding the mechanisms underlying the differential sensitivity to gonadal steroids exemplified by women with PMD, PPD, and perimenopausal depression, we will be in a far better position to answer what is arguably the most important question in behavioral neuroscience: Why do different people respond differently to the same stimulus?

References

Ablanalp JM, Livingston L, Rose RM, et al. (1977) Cortisol and growth hormone responses to psychological stress during the menstrual cycle. *Psychosomatic Medicine* 39: 158–177.

Abou-Saleh MT, Ghubash R, Karim L, et al. (1998) Hormonal aspects of postpartum depression. *Psychoneuroendocrinology* 23: 465–475.

Adams MM, Shah RA, Janssen WGM, et al. (2001) Different modes of hippocampal plasticity in response to estrogen in young and aged female rats. *Proceedings of the National*

Academy of Sciences of the United States of America 98: 8071–8076.
Adell A, Celada P, Abellan MT, et al. (2002) Origin and functional role of the extracellular serotonin in the midbrain raphe nuclei. *Brain Research Reviews* 39: 154–180.
Allen L, Hines M, Shryne J, et al. (1989) Two sexually dimorphic cell groups in the human brain. *Journal of Neuroscience* 9: 497–506.
Altemus M, Roca C, Galliven E, et al. (2001) Increased vasopressin and adrenocorticotropin responses to stress in the midluteal phase of the menstrual cycle. *Journal of Clinical Endocrinology and Metabolism* 86: 2525–2530.
Altman N, Sachar EJ, Gruen PH, et al. (1975) Reduced plasma LH concentration in postmenopausal depressed women. *Psychosomatic Medicine* 37: 274–276.
Altshuler LL, Post RM, Leverich GS, et al. (1995) Antidepressant-induced mania and cycle acceleration: A controversy revisited. *American Journal of Psychiatry* 152: 1130–1138.
American Psychiatric Association (1994) *Diagnostic and Statistical Manual of Mental Disorders*, 4th edn. Washington, DC: American Psychiatric Association.
Amsterdam JD, Winokur A, Lucki I, et al. (1983) Neuroendocrine regulation in depressed postmenopausal women and healthy subjects. *Acta Psychiatrica Scandinavica* 67: 43–49.
Anderson JC, Williams S, McGee R, et al. (1987) DSM-III disorders in preadolescent children: Prevalence in a large sample from the general population. *Archives of General Psychiatry* 44: 69–77.
Andreason PJ, Zametkin AJ, Guo AC, et al. (1993) Gender-related differences in regional cerebral glucose metabolism in normal volunteers. *Psychiatry Research* 51: 175–183.
Aneshensel CS (1985) The natural history of depressive symptoms. *Research in Community and Mental Health* 5: 45–74.
Angst J and Dobler-Mikola A (1984) Do the diagnostic criteria determine the sex ratio in depression? *Journal of Affective Disorders* 7: 189–198.
Ankenbauer W, Strahle U, and Schutz G (1988) Synergistic action of glucocorticoid and estradiol responsive elements. *Proceedings of the National Academy of Sciences of the United States of America* 85: 7526–7530.
Antelman SM, Caggiula AR, Kocan D, et al. (1991) One experience with lower or higher intensity stressors, respectively enhances or diminishes responsiveness to haloperidol weeks later: Implications for understanding drug variability. *Brain Research* 566: 276–283.
Appleby L, Montgomery J, and Studd J (1981) Oestrogens and affective disorders. In: Studd J (ed.) *Progress in Obstetrics and Gynaecology*, pp. 289–302. Edinburgh: Churchill Livingstone.
Arendash GW and Gorski RA (1983) Effects of discrete lesions of the sexually dimorphic nucleus of the preoptic area or other medial preoptic regions on the sexual behavior of male rats. *Brain Research Bulletin* 10: 147–150.
Ashby CR Jr., Carr LA, Cook CL, et al. (1988) Alteration of platelet serotonergic mechanisms and monoamine oxidase activity in premenstrual syndrome. *Biological Psychiatry* 24: 225–233.
Ashby CR Jr., Carr LA, Cook CL, et al. (1990) Alteration of 5-HT uptake by plasma fractions in the premenstrual syndrome. *Journal of Neural Transmission (Gen. Sect.)* 79: 41–50.
Azari NP, Pettigrew KD, Pietrini P, et al. (1995) Sex – differences in patterns of hemispheric cerebral metabolism-a multiple-regression discriminant-analysis of positron emission tomographic data. *International Journal of Neuroscience* 81: 1–20.
Backstrom T and Aakvaag A (1981) Plasma prolactin and testosterone during the luteal phase in women with premenstrual tension syndrome. *Psychoneuroendocrinology* 6: 245–251.
Backstrom T, Sanders D, Leask R, et al. (1983) Mood, sexuality, hormones, and the menstrual cycle: II. Hormone levels and their relationship to the premenstrual syndrome. *Psychosomatic Medicine* 45: 503–507.
Ballinger S (1990) Stress as a factor in lowered estrogen-levels in the early postmenopause. *Annals of the New York Academy of Sciences* 592: 95–113.
Bancroft J, Boyle H, Warner P, et al. (1987) The use of an LHRH agonist, buserelin, in the long-term management of premenstrual syndromes. *Clinical Endocrinology* 27: 171–182.
Bancroft J, Cook A, Davidson D, et al. (1991) Blunting of neuroendocrine responses to infusion of L-tryptophan in women with perimenstrual mood change. *Psychological Medicine* 21: 305–312.
Barrett-Connor E, von Muhlen D, Laughlin GA, et al. (1999) Endogenous levels of dehydroepiandrosterone sulfate, but not other sex hormones, are associated with depressed mood in older women: The Rancho Bernardo study. *Journal of the American Geriatrics Society* 47: 685–691.
Baulieu E-E and Robel P (1998) Dehydroepiandrosterone (DHEA) and dehydroepiandrosterone sulfate (DHEAS) as neuroactive neurosteroids. *Proceedings of the National Academy of Sciences of the United States of America* 95: 4089–4091.
Baxter LR, Jr., Mazziotta JC, Phelps ME, et al. (1987) Cerebral glucose metabolic rates in normal human females versus normal males. *Psychiatry Research* 21: 237–245.
Bebbington PE, Brugha T, MacCarthy B, et al. (1988) The Camberwell Collaborative Depression Study, I: Depressed probands-adversity and the form of depression. *British Journal of Psychiatry* 152: 754–765.
Beilin J and Zajac JD (1999) Function of the human androgen receptor varies according to CAG repeat number within the normal range. *Abstract Presented at the 81st Annual Meeting of the Endocrine Society* 500.
Belanger A, Candas B, Dupont A, et al. (1994) Changes in serum concentrations of conjugated and unconjugated steroids in 40- to 80-year-old men. *Journal of Clinical Endocrinology and Metabolism* 79: 1086–1090.
Berman KF, Schmidt PJ, Rubinow DR, et al. (1997) Modulation of cognition-specific cortical activity by gonadal steroids: A positron-emission tomography study in women. *Proceedings of the National Academy of Sciences of the United States of America* 94: 8836–8841.
Bicikova M, Dibbelt L, Hill M, et al. (1998) Allopregnanolone in women with premenstrual syndrome. *Hormone and Metabolic Research* 30: 227–230.
Bitran D, Hilvers RJ, and Kellogg CK (1991) Anxiolytic effects of 3α-hydroxy-5α[β]-pregnan-20-one: Endogenous metabolites of progesterone that are active at the $GABA_A$ receptor. *Brain Research* 561: 157–161.
Bitran D, Purdy RH, and Kellogg CK (1993) Anxiolytic effect of progesterone is associated with increases in cortical allopregnanolone and $GABA_A$ receptor function. *Pharmacology, Biochemistry, and Behavior* 45: 423–428.
Biver F, Lotstra F, Monclus M, et al. (1996) Sex difference in $5HT_2$ receptor in the living human brain. *Neuroscience Letters* 204: 25–28.
Bixo M, Allard P, Backstrom T, et al. (2001) Binding of [^{3}H] paroxetine to serotonin uptake sites and of [^{3}H]lysergic acid diethylamide to 5-HT_{2A} receptors in platelets from women with premenstrual dysphoric disorder during gonadotropin releasing hormone treatment. *Psychoneuroendocrinology* 26: 551–564.
Blazer DG, Kessler RC, McGonagle KA, et al. (1994) The prevalence and distribution of major depression in a national

community sample: The National Comorbidity Survey. *American Journal of Psychiatry* 151: 979–986.
Bloch M, Rubinow DR, Schmidt PJ, et al. (2005) Cortisol response to ovine corticotropin-releasing hormone in a model of pregnancy and parturition in euthymic women with and without a history of postpartum depression. *Journal of Clinical Endocrinology and Metabolism* 90: 695–699.
Bloch M, Schmidt PJ, Danaceau M, et al. (2000) Effects of gonadal steroids in women with a history of postpartum depression. *American Journal of Psychiatry* 157: 924–930.
Bloch M, Schmidt PJ, Danaceau MA, et al. (1999) Dehydroepiandrosterone treatment of mid-life dysthymia. *Biological Psychiatry* 45: 1533–1541.
Bloch M, Schmidt PJ, Su T-P, et al. (1998) Pituitary–adrenal hormones and testosterone across the menstrual cycle in women with premenstrual syndrome and controls. *Biological Psychiatry* 43: 897–903.
Bonnin F (1992) Cortisol levels in saliva and mood changes in early puerperium. *Journal of Affective Disorders* 26: 231–240.
Booij L, Van der Does AJW, and Riedel WJ (2003) Monoamine depletion in psychiatric and healthy populations: Review. *Molecular Psychiatry* 8: 951–973.
Brambilla F, Maggioni M, Ferrari E, et al. (1990) Tonic and dynamic gonadotropin secretion in depressive and normothymic phases of affective disorders. *Psychiatry Research* 32: 229–239.
Breedlove SM (1992) Sexual dimorphism in the vertebrate nervous system. *Journal of Neuroscience* 12: 4133–4142.
Bremner JD, Vythilingam M, Ng CK, et al. (2003) Regional brain metabolic correlates of α-methylparatyrosine-induced depressive symptoms: Implications for the neural circuitry of depression. *Journal of the American Medical Association* 289: 3125–3134.
Brincat M, Studd JWW, O'Dowd T, et al. (1984) Subcutaneous hormone implants for the control of climacteric symptoms: A prospective study. *Lancet* 323: 16–18.
Brockington IF, Winokur G, and Dean C (1982) Puerperal psychosis. In: Brockington IF and Kumar R (eds.) *Motherhood and Mental Illness*, pp. 37–69. London: Academic Press.
Bromberger JT, Meyer PM, Kravitz HM, et al. (2001) Psychologic distress and natural menopause: A multiethnic community study. *American Journal of Public Health* 91: 1435–1442.
Brown CS, Ling FW, Andersen RN, et al. (1994) Efficacy of depot leuprolide in premenstrual syndrome: Effect of symptom severity and type in a controlled trial. *Obstetrics and Gynecology* 84: 779–786.
Burger HG, Dudley EC, Cui J, et al. (2000) A prospective longitudinal study of serum testosterone, dehydroepiandrosterone sulfate, and sex hormone-binding globulin levels through the menopause transition. *Journal of Clinical Endocrinology and Metabolism* 85: 2832–2838.
Burgess LH and Handa RJ (1992) Chronic estrogen-induced alterations in adrenocorticotropin and corticosterone secretion, and glucocorticoid receptor-mediated functions in female rats. *Endocrinology* 131: 1261–1269.
Burke KC, Burke JD, Regier DA, et al. (1990) Age at onset of selected mental-disorders in five community populations. *Archives of General Psychiatry* 47: 511–518.
Callinan PA and Feinberg AP (2006) The emerging science of epigenomics. *Human Molecular Genetics* 15: R95–R101.
Campbell S (1976) Double-blind psychometric studies on the effects of natural estrogens on post-menopausal women. In: Campbell S (ed.) *The Management of the Menopause and Post Menopausal Years*, pp. 149–158. Lancaster: MTP Press.
Campbell EA, Linton EA, Wolfe CD, et al. (1987) Plasma corticotropin-releasing hormone concentrations during pregnancy and parturition. *Journal of Clinical Endocrinology and Metabolism* 64: 1054–1059.
Cardona-Gomez P, Perez M, Avila J, et al. (2004) Estradiol inhibits GSK3 and regulates interaction of estrogen receptors, GSK3, and beta-catenin in the hippocampus. *Molecular and Cellular Neuroscience* 25: 363–373.
Carey MP, Deterd CH, de Koning J, et al. (1995) The influence of ovarian steroids on hypothalamic–pituitary–adrenal regulation in the female rat. *Journal of Endocrinology* 144: 311–321.
Casper RF and Hearn MT (1990) The effect of hysterectomy and bilateral oophorectomy in women with severe premenstrual syndrome. *American Journal of Obstetrics and Gynecology* 162: 105–109.
Casson P, Hahn PM, VanVugt DA, et al. (1990) Lasting response to ovariectomy in severe intractable premenstrual syndrome. *American Journal of Obstetrics and Gynecology* 162: 99–105.
Cawood EH and Bancroft J (1996) Steroid hormones, the menopause, sexuality and well-being of women. *Psychological Medicine* 26: 925–936.
Chuong CJ, Coulam CB, Kao PC, et al. (1985) Neuropeptide levels in premenstrual syndrome. *Fertility and Sterility* 44: 760–765.
Cohen LS, Soares CN, Vitonis AF, et al. (2006) Risk for new onset of depression during the menopausal transition. The Harvard study of moods and cycles. *Archives of General Psychiatry* 63: 385–390.
Collins A, Eneroth P, and Landgren B (1985) Psychoneuroendocrine stress responses and mood as related to the menstrual cycle. *Psychosomatic Medicine* 47: 512–527.
Compagnone NA and Mellon SH (1998) Dehydroepiandrosterone: A potential signalling molecule for neocortical organization during development. *Proceedings of the National Academy of Sciences of the United States of America* 95: 4678–4683.
Coope J (1981) Is oestrogen therapy effective in the treatment of menopausal depression? *Journal of the Royal College of General Practitioners* 31: 134–140.
Coppen A, Whybrow PC, Noguera R, et al. (1972) The comparative antidepressant value of L-tryptophan and imipramine with and without attempted potentiation by liothyronine. *Archives of General Psychiatry* 26: 234–241.
Cox JL, Murray D, and Chapman G (1993) A controlled study of the onset, duration and prevalence of postnatal depression. *British Journal of Psychiatry* 163: 27–31.
Coyle N, Jones I, Robertson E, et al. (2000) Variation at the serotonin transporter gene influence susceptibility to bipolar affective puerperal psychosis. *Lancet* 356: 1490–1491.
Csapo AI, Pulkkinen MO, and Wiest WG (1973) Effects of luteectomy and progesterone replacement therapy in early pregnant patients. *American Journal of Obstetrics and Gynecology* 115: 759–765.
Cumming DC, Rebar RW, Hopper BR, et al. (1982) Evidence for an influence of the ovary on circulating dehydroepiandrosterone sulfate levels. *Journal of Clinical Endocrinology and Metabolism* 54: 1069–1071.
Cutrona CE (1983) Causal attributions and perinatal depression. *Journal of Abnormal Psychology* 92: 161–172.
D'Souza VN, Man NT, Morris GE, et al. (1995) A novel dystrophin isoform is required for normal retinal electrophysiology. *Human Molecular Genetics* 4: 837–842.

Dalton K (1980) *Depression after Childbirth*. Oxford: Oxford University Press.

Dalton K (1985) Progesterone prophylaxis used successfully in postnatal depression. *Practitioner* 229: 507–508.

Daly RC, Danaceau MA, Rubinow DR, et al. (2003) Concordant restoration of ovarian function and mood in perimenopausal depression. *American Journal of Psychiatry* 160: 1842–1846.

Davidson J and Pelton S (1986) Forms of atypical depression and their response to antidepressant drugs. *Psychiatry Research* 17: 87–95.

Davison SL, Donath S, Montalto JG, et al. (2005) Androgen levels in adult females: Changes with age, menopause, and oophorectomy. *Journal of Clinical Endocrinology and Metabolism* 90: 3847–3853.

Dawkins K and Potter WZ (1991) Gender differences in pharmacokinetics and pharmacodynamics of psychotropics: Focus on women. *Psychopharmacology Bulletin* 27: 417–426.

Dayas CV, Xu Y, Buller KM, et al. (2000) Effects of chronic oestrogen replacement on stress-induced activation of hypothalamic–pituitary–adrenal axis control pathways. *Journal of Neuroendocrinology* 12: 784–794.

Denicoff KD, Durkin TM, Lotze MT, et al. (1989) The neuroendocrine effects of interleukin-2 treatment. *Journal of Clinical Endocrinology and Metabolism* 69: 402–410.

De Vries GJ and Boyle PA (1998) Double duty for sex differences in the brain. *Behavioural Brain Research* 92: 205–213.

DeVoogd T and Nottebohm F (1981) Gonadal hormones induce dendritic growth in the adult avian brain. *Science* 214: 202–204.

Dimmock PW, Wyatt KM, Jones PW, et al. (2000) Efficacy of selective serotonin-reuptake inhibitors in premenstrual syndrome: A systematic review. *Lancet* 356: 1131–1136.

Ditkoff EC, Crary WG, Cristo M, et al. (1991) Estrogen improves psychological function in asymptomatic postmenopausal women. *Obstetrics and Gynecology* 78: 991–995.

Doran AR, Rubinow DR, Roy A, et al. (1986) CSF somatostatin and abnormal response to dexamethasone administration in schizophrenic and depressed patients. *Archives of General Psychiatry* 43: 365–369.

Dreher J, Schmidt PJ, Kohn P, et al. (2007) Menstrual cycle phase modulates reward-related neural function in women. *Proceedings of the National Academy of Sciences of the United States of America* 104: 2465–2470.

Drevets WC (2000) Neuroimaging studies of mood disorders. *Biological Psychiatry* 48: 813–829.

Drevets WC (2001) Neuroimaging and neuropathological studies of depression: Implications for the cognitive-emotional features of mood disorders. *Current Opinion in Neurobiology* 11: 240–249.

Duman RS, Heninger GR, and Nestler EJ (1997) A molecular and cellular theory of depression. *Archives of General Psychiatry* 54: 597–606.

Easterbrook CC (1900) Organo-therapeutics in mental diseases. *British Medical Journal* 2: 813–823.

Enriquez P, Cales JM, Sanchez-Santed F, et al. (1991) Effects of early postnatal gonadal steroids on extinction of a continuously food-rewarded running response. *Physiology and Behavior* 49: 57–61.

Eriksson E, Alling C, Andersch B, et al. (1994) Cerebrospinal fluid levels of monoamine metabolites: A preliminary study of their relation to menstrual cycle phase, sex steroids, and pituitary hormones in healthy women and in women with premenstrual syndrome. *Neuropsychopharmacology* 11: 201–213.

Eriksson E, Sundblad C, Lisjo P, et al. (1992) Serum levels of androgens are higher in women with premenstrual irritability and dysphoria than in controls. *Psychoneuroendocrinology* 17: 195–204.

Ernst C and Angst J (1992) The Zurich Study. XII. Sex differences in depression. Evidence from longitudinal epidemiological data. *European Archives of Psychiatry and Clinical Neuroscience* 241: 222–230.

Eser D, Schule C, Baghai TC, et al. (2006) Neuroactive steroids in depression and anxiety disorders: Clinical studies. *Neuroendocrinology* 84: 244–254.

Esposito G, Van Horn JD, Weinberger DR, et al. (1996) Gender differences in cerebral blood flow as a function of cognitive state with PET. *Journal of Nuclear Medicine* 37: 559–564.

Evans J, Heron J, Francomb H, et al. (2001) Cohort study of depressed mood during pregnancy and after childbirth. *British Medical Journal* 323: 257–260.

Facchinetti F, Fioroni L, Martignoni E, et al. (1994) Changes of opioid modulation of the hypothalamo-pituitary–adrenal axis in patients with severe premenstrual syndrome. *Psychosomatic Medicine* 56: 418–422.

Facchinetti F, Genazzani AD, Martignoni E, et al. (1990) Neuroendocrine correlates of premenstrual syndrome: Changes in the pulsatile pattern of plasma LH. *Psychoneuroendocrinology* 15: 269–277.

Facchinetti F, Genazzani AD, Martignoni E, et al. (1993) Neuroendocrine changes in luteal function in patients with premenstrual syndrome. *Journal of Clinical Endocrinology and Metabolism* 76: 1123–1127.

Facchinetti F, Martignoni E, Petraglia F, et al. (1987) Premenstrual fall of plasma B-endorphin in patients with premenstrual syndrome. *Fertility and Sterility* 47: 570–573.

Fava M, Abraham M, Alpert J, et al. (1996) Gender differences in Axis I comorbidity among depressed outpatients. *Journal of Affective Disorders* 38: 129–133.

Feksi A, Harris B, Walker RF, et al. (1984) Maternity blues and hormone levels in saliva. *Journal of Affective Disorders* 6: 351–355.

Ferrari E, Locatelli M, Arcaini A, et al. (1997) Chronobiological study of some neuroendocrine features of major depression in elderly people. *Abstract Presented at the 79th Annual Meeting of The Endocrine Society*.

Finn DA, Roberts AJ, Lotrich F, et al. (1997) Genetic differences in behavioral sensitivity to a neuroactive steroid. *Journal of Pharmacology and Experimental Therapeutics* 280: 820–828.

Francis D, Diorio J, Liu D, et al. (1999) Nongenomic transmission across generations of maternal behavior and stress responses in the rat. *Science* 286: 1155–1158.

Frank E, Carpenter LL, and Kupfer DJ (1988) Sex differences in recurrent depression: Are there any that are significant? *American Journal of Psychiatry* 145: 41–45.

Freeman EW (1997) Premenstrual syndrome: Current perspectives on treatment and etiology. *Current Opinion in Obstetrics and Gynecology* 9: 147–153.

Freeman EW, Sammel MD, Liu L, et al. (2004) Hormones and menopausal status as predictors of depression in women in transition to menopause. *Archives of General Psychiatry* 61: 62–70.

Freeman EW, Sammel MD, Lin H, et al. (2006) Associations of hormones and menopausal status with depressed mood in women with no history of depression. *Archives of General Psychiatry* 63: 375–382.

Gallagher M, Landfield PW, McEwen B, et al. (1996) Hippocampal neurodegeneration in aging. *Science* 274: 484–485.

Galliven EA, Singh A, Michelson D, et al. (1997) Hormonal and metabolic responses to exercise across time of day and

menstrual cycle phase. *Journal of Applied Physiology* 83: 1822–1831.
Garcia-Segura LM, Cardona-Gomez P, Naftolin F, et al. (1998) Estradiol upregulates Bcl-2 expression in adult brain neurons. *NeuroReport* 9: 593–597.
Garey J, Goodwillie A, Frohlich J, et al. (2003) Genetic contributions to generalized arousal of brain and behavior. *Proceedings of the National Academy of Sciences of the United States of America* 100: 11019–11022.
Gatewood JD, Wills A, Shetty S, et al. (2006) Sex chromosome complement and gonadal sex influence aggressive and parental behaviors in mice. *Journal of Neuroscience* 26: 2335–2342.
George GCW, Utian WH, Beumont PJV, et al. (1973) Effect of exogenous oestrogens on minor psychiatric symptoms in postmenopausal women. *South African Medical Journal* 47: 2387–2388.
George MS, Ketter TA, Parekh PI, et al. (1996) Gender differences in regional cerebral blood flow during transient self-induced sadness or happiness. *Biological Psychiatry* 40: 859–871.
Gex-Fabry M, Balant-Gorgia AE, Balant LP, et al. (1990) Clomipramine metabolism: Model-based analysis of variability factors from drug monitoring data. *Clinical Pharmacokinetics* 19: 241–255.
Giedd JN, Blumenthal J, Jeffries NO, et al. (1999) Brain development during childhood and adolescence: A longitudinal MRI study. *Nature Neuroscience* 2: 861–863.
Giovannucci E, Stampfer MJ, Krithivas K, et al. (1997) The CAG repeat within the androgen receptor gene and its relationship to prostate cancer. *Proceedings of the National Academy of Sciences of the United States of America* 94: 3320–3323.
Girdler SS, Straneva PA, Light KC, et al. (2001) Allopregnanolone levels and reactivity to mental stress in premenstrual dysphoric disorder. *Biological Psychiatry* 49: 788–797.
Glassman AH, Perel JM, Shostak M, et al. (1977) Clinical implications of imipramine plasma levels for depressive illness. *Archives of General Psychiatry* 34: 197–204.
Goldman S and Nottebohm F (1983) Neuronal production, migration, and differentiation in a vocal control nucleus of the adult female canary brain. *Proceedings of the National Academy of Sciences of the United States of America* 80: 2390.
Goldstein JM, Jerram M, Poldrack R, et al. (2005) Hormonal cycle modulates arousal circuitry in women using functional magnetic resonance imaging. *Journal of Neuroscience* 25: 9309–9316.
Goodyer IM, Herbert J, and Altham PME (1998) Adrenal steroid secretion and major depression in 8- to 16-year-olds, III. Influence of cortisol/DHEA ratio at presentation on subsequent rates of disappointing life events and persistent major depression. *Psychological Medicine* 28: 265–273.
Goodyer IM, Herbert J, Altham PME, et al. (1996) Adrenal secretion during major depression in 8- to 16-year-olds, I. Altered diurnal rhythms in salivary cortisol and dehydroepiandrosterone (DHEA) at presentation. *Psychological Medicine* 26: 245–256.
Gorski RA (1991) Sexual differentiation of the endocrine brain and its control. In: Motta M (ed.) *Brain Endocrinology*, pp. 71–104. New York: Raven Press.
Gorski RA, Gordon JH, Shryne JE, et al. (1978) Evidence for a morphological sex difference within the medial preoptic area of the rat brain. *Brain Research* 148: 333–346.
Gouras GK, Xu H, Gross RS, et al. (2000) Testosterone reduces neuronal secretion of Alzheimer's beta-amyloid peptides. *Proceedings of the National Academy of Sciences of the United States of America* 97: 1202–1205.
Gray A, Feldman HA, McKinlay JB, et al. (1991) Age, disease and changing sex hormone levels in middle-aged men: Results of the Massachusetts male aging study. *Journal of Clinical Endocrinology and Metabolism* 73: 1016–1023.
Greenblatt DJ, Friedman H, Burstein ES, et al. (1987) Trazodone kinetics: Effect of age, gender, and obesity. *Clinical Pharmacology and Therapeutics* 42: 193–200.
Gregoire AJP, Kumar R, Everitt B, et al. (1996) Transdermal oestrogen for treatment of severe postnatal depression. *Lancet* 347: 930–933.
Griffin LD and Mellon SH (1999) Selective serotonin reuptake inhibitors directly alter activity of neurosteroidogenic enzymes. *Proceedings of the National Academy of Sciences of the United States of America* 96: 13512–13517.
Guicheney P, Léger D, Barrat J, et al. (1988) Platelet serotonin content and plasma tryptophan in peri- and postmenopausal women: Variations with plasma oestrogen levels and depressive symptoms. *European Journal of Clinical Investigation* 18: 297–304.
Gulinello M, Gong QH, Li X, et al. (2001) Short-term exposure to a neuroactive steroid increases $\alpha 4$ $GABA_A$ receptor subunit levels in association with increased anxiety in the female rat. *Brain Research* 910: 55–66.
Gur RC, Gur RE, Obrist WD, et al. (1982) Sex and handedness differences in cerebral blood flow during rest and during cognitive activity. *Science* 217: 659–661.
Gur RC, Gur RE, Obrist WD, et al. (1987) Age and regional cerebral blood flow at rest and during cognitive activity. *Archives of General Psychiatry* 44: 617–621.
Gurney ME (1982) Behavioral correlates of sexual differentiation in the zebra finch song system. *Brain Research* 231: 153–172.
Halbreich U (2004) Prevalence of mood symptoms and depressions during pregnancy: Implications for clinical practice and research. *CNS Spectrums* 9: 177–184.
Halbreich U, Endicott J, Goldstein S, et al. (1986) Premenstrual changes and changes in gonadal hormones. *Acta Psychiatrica Scandinavica* 74: 576–586.
Halbreich U, Petty F, Yonkers K, et al. (1996) Low plasma gamma-aminobutyric acid levels during the late luteal phase of women with premenstrual dysphoric disorder. *American Journal of Psychiatry* 153: 718–720.
Halbreich U, Rojansky N, and Palter S (1991) Elimination of ovulation and menstrual cyclicity (with danazol) improves dysphoric premenstrual syndromes. *Fertility and Sterility* 56: 1066–1069.
Hall JE and Gill S (2001) Neuroendocrine aspects of aging in women. *Endocrinology Metabolism Clinics of North America* 30: 631–646.
Hall JM and McDonnell DP (2005) Coregulators in nuclear estrogen receptor action: From concept to therapeutic targeting. *Molecular Interventions* 5: 343–357.
Hamilton JA and Gallant S (1988) Premenstrual symptom changes and plasma β-endorphin/β-lipotropin throughout the menstrual cycle. *Psychoneuroendocrinology* 13: 505–514.
Hammarback S and Backstrom T (1988) Induced anovulation as a treatment of premenstrual tension syndrome: A double-blind cross-over study with GnRH-agonist versus placebo. *Acta Obstetricia et Gynecologica Scandinavica* 67: 159–166.
Harris B, Lovett L, Newcombe RG, et al. (1994) Maternity blues and major endocrine changes: Cardiff puerperal mood and hormone study II. *British Medical Journal* 308: 949–953.
Harris B, Lovett L, Smith J, et al. (1996) Cardiff puerperal mood and hormone study. III. Postnatal depression at 5 to 6 weeks postpartum, and its hormonal correlates across the peripartum period. *British Journal of Psychiatry* 168: 739–744.

Heuser I, Deuschle M, Luppa P, et al. (1998) Increased diurnal plasma concentrations of dehydroepiandrosterone in depressed patients. *Journal of Clinical Endocrinology and Metabolism* 83: 3130–3133.
Ho HP, Olsson M, Westberg L, et al. (2001) The serotonin reuptake inhibitor fluoxetine reduces sex steroid-related aggression in female rats: An animal model of premenstrual irritability? *Neuropsychopharmacology* 24: 502–510.
Howard R, Mason P, Taghavi E, et al. (1992) Brainstem auditory evoked responses (BAERs) during the menstrual cycle in women with and without premenstrual syndrome. *Biological Psychiatry* 32: 682–690.
Hruska RE and Silbergeld EK (1980) Estrogen treatment enhances dopamine receptor sensitivity in the rat striatum. *European Journal of Pharmacology* 61: 397–400.
Hu M and Becker JB (2003) Effects of sex and estrogen on behavioral sensitization to cocaine in rats. *Journal of Neuroscience* 23: 693–699.
Huerta R, Mena A, Malacara JM, et al. (1995) Symptoms at perimenopausal period: Its association with attitudes toward sexuality, life-style, family function, and FSH levels. *Psychoneuroendocrinology* 20: 135–148.
Huizenga NATM, Koper JW, DeLange P, et al. (1998a) A polymorphism in the glucocorticoid receptor gene may be associated with an increased sensitivity to glucocorticoids *in vivo*. *Journal of Clinical Endocrinology and Metabolism* 83: 144–151.
Huizenga NATM, Koper JW, DeLange P, et al. (1998b) Interperson variability but intraperson stability of baseline plasma cortisol concentrations, and its relation to feedback sensitivity of the hypothalamo-pituitary–adrenal axis to a low dose of dexamethasone in elderly individuals. *Journal of Clinical Endocrinology and Metabolism* 83: 47–54.
Huo L, Straub RE, Schmidt PJ, et al. (2006) Risk for premenstrual dysphoric disorder is associated with genetic variation in ESR1, the estrogen receptor alpha gene. *Biological Psychiatry* 62: 925–933.
Hussain SY, Massil JH, Matta WH, et al. (1992) Buserelin in premenstrual syndrome. *Gynecological Endocrinology* 6: 57–64.
Jackson TA, Richer JK, Bain DL, et al. (1997) The partial agonist activity of antagonist-occupied steroid receptors is controlled by a novel hinge domain-binding co-activator L7/SPA and the corepressors N-COR or SMRT. *Molecular Endocrinology* 11: 693–705.
Jensen EV and Jacobson HI (1962) Basic guides to the mechanism of estrogen action. *Recent Progress in Hormone Research* 18: 387–414.
Jones I, Hamshere M, Nangle J, et al. (2007) Bipolar affective puerperal psychosis: Genome-wide significant evidence for linkage to chromosome 16. *American Journal of Psychiatry* 164: 1099–1104.
Josefsson A, Berg G, Nordin C, et al. (2001) Prevalence of depressive symptoms in late pregnancy and postpartum. *Acta Obstetricia et Gynecologica Scandinavica* 80: 251–255.
Judd LL (1994) When anxiety disorders are comorbid with major depression: Social and clinical burden. *Abstract, APA 147th Annual Meeting*.
Karp JF and Frank E (1995) Combination therapy and the depressed woman. *Depression* 3: 91–98.
Keitner GI, Ryan CE, Miller IW, et al. (1991) 12-Month outcome of patients with major depression and comorbid psychiatric or medical illness (compound depression). *American Journal of Psychiatry* 148: 345–350.
Keller-Wood M, Silbiger J, and Wood CE (1988) Progesterone attenuates the inhibition of adrenocorticotropin responses by cortisol in nonpregnant ewes. *Endocrinology* 123: 647–651.
Kessler RC, McGonagle KA, Nelson CB, et al. (1994) Sex and depression in the National Comorbidity Survey II: Cohort effects. *Journal of Affective Disorders* 30: 15–26.
Kessler RC, McGonagle KA, Swartz M, et al. (1993) Sex and depression in the National Comorbidity Survey I: Lifetime prevalence, chronicity and recurrence. *Journal of Affective Disorders* 29: 85–96.
Kirschbaum C, Kudielka BM, Gaab J, et al. (1999) Impact of gender, menstrual cycle phase, and oral contraceptives on the activity of the hypothalamic–pituitary–adrenal axis. *Psychosomatic Medicine* 61: 154–162.
Kishimoto K, Koyama S, and Akaike N (2000) Presynaptic modulation of synaptic γ-aminobutyric acid transmission by tandospirone in rat basolateral amygdala. *European Journal of Pharmacology* 407: 257–265.
Klaiber EL, Broverman DM, Vogel W, et al. (1979) Estrogen therapy for severe persistent depressions in women. *Archives of General Psychiatry* 36: 550–554.
Klaiber EL, Broverman DM, Vogel W, et al. (1982) Estrogens and central nervous system function: Electroencephalography, cognition, and depression. In: Friedman RC (ed.) *Behavior and the Menstrual Cycle*, pp. 267–289. New York: Dekker.
Klatzkin RR, Morrow AL, Light KC, et al. (2006) Histories of depression, allopregnanolone responses to stress, and premenstrual symptoms in women. *Biological Psychiatry* 71: 2–11.
Komesaroff PA, Esler M, Clarke IJ, et al. (1998) Effects of estrogen and estrous cycle on glucocorticoid and catecholamine responses to stress in sheep. *American Journal of Physiology* 275: E671–E678.
Kornstein SG, Schatzberg AF, Thase ME, et al. (2000) Gender differences in treatment response to sertraline versus imipramine in chronic depression. *American Journal of Psychiatry* 157: 1445–1452.
Kornstein SG, Schatzberg AF, Yonkers KA, et al. (1995) Gender differences in presentation of chronic major depression. *Psychopharmacology Bulletin* 31: 711–718.
Koyama S, Kubo C, Rhee J-S, et al. (1999) Presynaptic serotonergic inhibition of GABAergic synaptic transmission in mechanically dissociated rat basolateral amygdala neurons. *Journal of Physiology* 518(pt 2): 525–538.
Krezel W, Dupont S, Krust A, et al. (2001) Increased anxiety and synaptic plasticity in estrogen receptor beta-deficient mice. *Proceedings of the National Academy of Sciences of the United States of America* 98: 12278–12282.
Kuevi V, Causon R, Dixson AF, et al. (1983) Plasma amine and hormone changes in post-partum blues. *Clinical Endocrinology* 19: 39–46.
Kulynych JJ, Vladar K, Jones DW, et al. (1994) Gender differences in the normal lateralization of the supratemporal cortex: MRI surface-rendering morphometry of Heschl's gyrus and the planum temporale. *Cerebral Cortex* 4: 107–118.
Kumar R and Mordecai-Robson K (1984) A prospective study of emotional disorders in childbearing women. *British Journal of Psychiatry* 144: 35–47.
Laudenslager ML, Ryan SM, Drugan RC, et al. (1983) Coping and immunosuppression: Inescapable but not escapable shock suppresses lymphocyte proliferation. *Science* 221: 568–570.
Laughlin GA and Barrett-Connor E (2000) Sexual dimorphism in the influence of advanced aging on adrenal hormone levels: The Rancho Bernardo study. *Journal of Clinical Endocrinology and Metabolism* 85: 3561–3568.
Lawrie TA, Hofmeyr GJ, DeJager M, et al. (1998) A double-blind randomised placebo controlled trial of postnatal norethisterone enanthate: The effect on postnatal

depression and serum hormones. *British Journal of Obstetrics and Gynaecology* 105: 1082–1090.
Lee KA, Shaver JF, Giblin EC, et al. (1990) Sleep patterns related to menstrual cycle phase and premenstrual affective symptoms. *Sleep* 13: 403–409.
Leibenluft E (1996) Women with bipolar illness: Clinical and research issues. *American Journal of Psychiatry* 153: 163–173.
Leibenluft E (1999) Gender differences in major depressive disorder and bipolar disorder. *CNS Spectrums* 4: 25–33.
Leibenluft E, Hardin TA, and Rosenthal NE (1995) Gender differences in seasonal affective disorder. *Depression* 3: 13–19.
Levine S (1975) Infantile experience and resistance to physiological stress. *Science* 126: 405–406.
Liu D, Diorio J, Tannenbaum B, et al. (1997) Maternal care, hippocampal glucocorticoid receptors, and hypothalamic–pituitary–adrenal responses to stress. *Science* 277: 1659–1662.
Long TD, Ellingrod VL, Kathol RG, et al. (2000) Lack of menstrual cycle effects on hypothalamic–pituitary–adrenal axis response to insulin-induced hypoglycaemia. *Clinical Endocrinology (Oxf.)* 52: 781–787.
Longcope C (1995) The metabolism of dehydroepiandrosterone sulfate. *Seminars in Reproductive Medicine* 13: 270–274.
Luisi S, Petraglia F, Benedetto C, et al. (2000) Serum allopregnanolone levels in pregnant women: Changes during pregnancy, at delivery, and in hypertensive patients. *Journal of Clinical Endocrinology and Metabolism* 85: 2429–2433.
Maffucci JA and Gore AC (2006) Age-related changes in hormones and their receptors in animal models of female reproductive senescence. In: Conn PM (ed.) *Handbook of Models for Human Aging*, pp. 533–552. New York: Academic Press/Elsevier.
Magiakou MA, Mastorakos G, Rabin D, et al. (1996a) Hypothalamic cortico-releasing hormone suppression during the postpartum period: Implications for the increase in psychiatric manifestations at this time. *Journal of Clinical Endocrinology and Metabolism* 81: 1912–1917.
Magiakou MA, Mastorakos G, Rabin D, et al. (1996b) The maternal hypothalamic–pituitary–adrenal axis in the third trimester of human pregnancy. *Clinical Endocrinology* 44: 419–428.
Maguire JL, Stell BM, Rafizadeh M, et al. (2005) Ovarian cycle-linked changes in GABAA receptors mediating tonic inhibition alter seizure susceptibility and anxiety. *Nature Neuroscience* 8: 797–804.
Maier SF, Amat J, Baratta MV, et al. (2006) Behavioral control, the medial prefrontal cortex, and resilience. *Dialogues in Clinical Neurosciences* 8: 397–406.
Majewska MD, Demirgören S, Spivak CE, et al. (1990) The neurosteroid dehydroepiandrosterone sulfate is an allosteric antagonist of the $GABA_A$ receptor. *Brain Research* 526: 143–146.
Majewska MD, Harrison NL, Schwartz RD, et al. (1986) Steroid hormone metabolites are barbiturate-like modulators of the GABA receptor. *Science* 232: 1004–1007.
Malmgren R, Collins A, and Nilsson CG (1987) Platelet serotonin uptake and effects of vitamin B6-treatment in premenstrual tension. *Neuropsychobiology* 18: 83–88.
Manji HK, Drevets WC, and Charney DS (2001) The cellular neurobiology of depression. *Nature Medicine* 7: 541–547.
Marinari KT, Leschner AI, and Doyle MP (1976) Menstrual cycle status and adrenocortical reactivity to psychological stress. *Psychoneuroendocrinology* 1: 213.
Matthews KA (1992) Myths and realities of the menopause. *Psychosomatic Medicine* 54: 1–9.
Maus M, Bertrand P, Drouva S, et al. (1989) Differential modulation of D1 and D2 dopamine-sensitive adenylate cyclases by 17β-estradiol in cultured striatal neurons and anterior pituitary cells. *Journal of Neurochemistry* 52: 410–418.
McEwen BS (2002) Basic neurobiology of ovarian steroids: Clinical implications. *Dialogues in Clinical Neurosciences* 4: 163–175.
McEwen BS and Alves SE (1999) Estrogen actions in the central nervous system. *Endocrine Reviews* 20: 279–307.
McGee R, Feehan M, Williams S, et al. (1992) DSM-III disorders from age 11 to age 15 years. *Journal of the American Academy of Child and Adolescent Psychiatry* 31: 50–59.
McKenna NJ, Lanz RB, and O'Malley BW (1999) Nuclear receptor coregulators: Cellular and molecular biology. *Endocrine Reviews* 20: 321–344.
McNeilly AS (2002) Lactational endocrinology: The biology of lam. *Advances in Experimental Medicine and Biology* 503: 199–205.
Menkes DB, Coates DC, and Fawcett JP (1994) Acute tryptophan depletion aggravates premenstrual syndrome. *Journal of Affective Disorders* 32: 37–44.
Meyer M-E, Gronemeyer H, Turcott B, et al. (1989) Steroid hormone receptors compete for factors that mediate their enhancer function. *Cell* 57: 433–442.
Mezrow G, Shoupe D, Spicer D, et al. (1994) Depot leuprolide acetate with estrogen and progestin add-back for long-term treatment of premenstrual syndrome. *Fertility and Sterility* 62: 932–937.
Miller WL (1998) Steroid hormone biosynthesis and actions in the materno-feto-placental unit. *Clinical Perinatology* 25: 799–817.
Miranda P, Williams CL, and Einstein G (1999) Granule cells in aging rats are sexually dimorphic in their response to estradiol. *Journal of Neuroscience* 19: 3316–3325.
Moldin SO, Scheftner WA, Rice JP, et al. (1993) Association between major depressive disorder and physical illness. *Psychological Medicine* 23: 755–761.
Monteleone P, Luisi S, Tonetti A, et al. (2000) Allopregnanolone concentrations and premenstrual syndrome. *European Journal of Endocrinology* 142: 269–273.
Montgomery JC, Brincat M, Tapp A, et al. (1987) Effect of oestrogen and testosterone implants on psychological disorders in the climacteric. *Lancet* 329: 297–299.
Moody JP, Tait AC, and Todrick A (1967) Plasma levels of imipramine and desmethylimipramine during therapy. *British Journal of Psychiatry* 113: 183–193.
Moore CL, Dou H, and Juraska JM (1992) Maternal stimulation affects the number of motor neurons in a sexually dimorphic nucleus of the lumbar spinal cord. *Brain Research* 572: 52–56.
Morales AJ, Nolan JJ, Nelson JC, et al. (1994) Effects of replacement dose of dehydroepiandrosterone in men and women of advancing age. *Journal of Clinical Endocrinology and Metabolism* 78: 1360–1367.
Morrison MF, Kallan MJ, Ten Have T, et al. (2004) Lack of efficacy of estradiol for depression in postmenopausal women: A randomized, controlled trial. *Biological Psychiatry* 55: 406–412.
Mortola JF, Girton L, and Fischer U (1991) Successful treatment of severe premenstrual syndrome by combined use of gonadotropin-releasing hormone agonist and estrogen/progestin. *Journal of Clinical Endocrinology and Metabolism* 71: 252A–252F.
Murphy DD, Cole NB, and Segal M (1998) Brain-derived neurotrophic factor mediates estradiol-induced dendritic spine formation in hippocampal neurons. *Proceedings of the*

National Academy of Sciences of the United States of America 95: 11412–11417.
Muse K (1989) Gonadotropin-releasing hormone agonist-suppressed premenstrual syndrome (PMS): PMS symptom induction by estrogen, progestin, or both. *Abstract, Society of Gynecologic Investigation* 118.
Muse KN, Cetel NS, Futterman LA, et al. (1984) The premenstrual syndrome: Effects of medical ovariectomy *New England Journal of Medicine* 311: 1345–1349.
Nestler EJ, Terwilliger RZ, and Duman RS (1989) Chronic antidepressant administration alters the subcellular distribution of cyclic AMP-dependent protein kinase in rat frontal cortex. *Journal of Neurochemistry* 53: 1644–1647.
Neumeister A, Nugent AC, Waldeck T, et al. (2004) Neural and behavioral responses to tryptophan depletion in unmedicated patients with remitted major depressive disorder and controls. *Archives of General Psychiatry* 61: 765–773.
Nibuya M, Nestler EJ, and Duman RS (1996) Chronic antidepressant administration increases the expression of cAMP response element-binding protein (CREB) in rat hippocampus. *Journal of Neuroscience* 16: 2365–2372.
Nishizawa S, Benkelfat C, Young SN, et al. (1997) Differences between males and females in rates of serotonin synthesis in human brain. *Proceedings of the National Academy of Sciences of the United States of America* 94: 5308–5313.
Nolen-Hoeksema S (1987) Sex differences in unipolar depression: Evidence and theory. *Psychological Bulletin* 101: 259–282.
Nottebohm F (1980) Testosterone triggers growth of brain vocal control nuclei in adult female canaries. *Brain Research* 189: 429–436.
O'Hara MW (1986) Social support, life events, and depression during pregnancy and the puerperium. *Archives of General Psychiatry* 43: 569–573.
O'Hara MW, Schlechte JA, Lewis DA, et al. (1991a) Controlled prospective study of postpartum mood disorders: Psychological, environmental, and hormonal variables. *Journal of Abnormal Psychology* 100: 63–73.
O'Hara MW, Schlechte JA, Lewis DA, et al. (1991b) Prospective study of postpartum blues: Biologic and psychosocial factors. *Archives of General Psychiatry* 48: 801–806.
Ochedalski T, Subburaju S, Wynn PC, et al. (2007) Interaction between oestrogen and oxytocin on hypothalamic–pituitary–adrenal axis activity. *Journal of Neuroendocrinology* 19: 189–197.
Old Age Depression Interest Group (1993) How long should the elderly take antidepressants? A double-blind placebo-controlled study of continuation/prophylaxis therapy with dothiepin. *British Journal of Psychiatry* 162: 175–182.
Ongur D, Drevets WC, and Price JL (1998) Glial reduction in the subgenual prefrontal cortex in mood disorders. *Proceedings of the National Academy of Sciences of the United States of America* 95: 13290–13295.
Orentreich N, Brind JL, Rizer RL, et al. (1984) Age changes and sex differences in serum dehydroepiandrosterone sulfate concentrations throughout adulthood. *Journal of Clinical Endocrinology and Metabolism* 59: 551–555.
Orentreich N, Brind JL, Vogelman JH, et al. (1992) Long-term longitudinal measurements of plasma dehydroepiandrosterone sulfate in normal men. *Journal of Clinical Endocrinology and Metabolism* 75: 1002–1004.
Parry BL, Berga SL, Kripke DF, et al. (1990) Altered waveform of plasma nocturnal melatonin secretion in premenstrual syndrome. *Archives of General Psychiatry* 47: 1139–1146.
Parry BL, Gerner RH, Wilkins JN, et al. (1991) CSF and endocrine studies of premenstrual syndrome. *Neuropsychopharmacology* 5: 127–137.
Parry BL, Mendelson WB, Duncan WB, et al. (1989) Longitudinal sleep EEG, temperature, and activity measurements across the menstrual cycle in patients with premenstrual depression and in age-matched controls. *Psychiatry Research* 30: 285–303.
Patchev VK and Almeida OFX (1996) Gonadal steroids exert facilitating and buffering effects on glucocorticoid-mediated transcriptional regulation of corticotropin-releasing hormone and corticosteroid receptor genes in rat brain. *Journal of Neuroscience* 16: 7077–7084.
PearsonMurphy BE, Steinberg SI, Hu FY, et al. (2001) Neuroactive ring A-reduced metabolites of progesterone in human plasma during pregnancy: Elevated levels of 5α-dihydroprogesterone in depressed patients during the latter half of pregnancy. *Journal of Clinical Endocrinology and Metabolism* 86: 5981–5987.
Pettersson K, Grandien K, Kuiper GGJM, et al. (1997) Mouse estrogen receptor β forms estrogen-response element-binding heterodimers with estrogen receptor α. *Molecular Endocrinology* 11: 1486–1496.
Pfaff DW (1966) Morphological changes in the brains of adult male rats after neonatal castration. *Journal of Endocrinology* 36: 415–416.
Phoenix CH, Goy RW, Gerall AA, et al. (1959) Organizing action of prenatally administered testosterone propionate on the tissues mediating mating behavior in the female guinea pig. *Endocrinology* 65: 369–382.
Plotsky PM, Owens MJ, and Nemeroff CB (1998) Psychoneuroendocrinology of depression: Hypothalamic–pituitary–adrenal axis. *Psychiatric Clinics of North America* 21: 293–307.
Post RM, Rubinow DR, and Ballenger JC (1986) Conditioning and sensitisation in the longitudinal course of affective illness. *British Journal of Psychiatry* 149: 191–201.
Potter WZ and Manji HK (1994) Catecholamines in depression: An update. *Clinical Chemistry* 40: 279–287.
Prange AJ, Jr., Wilson IC, Rabin AM, et al. (1969) Enhancement of imipramine antidepressant activity by thyroid hormone. *American Journal of Psychiatry* 126: 457–469.
Preskorn SH and Mac DS (1985) Plasma levels of amitriptyline: Effects of age and sex. *Journal of Clinical Psychiatry* 46: 276–277.
Protopopescu X, Pan H, Altemus M, et al. (2005) Orbitofrontal cortex activity related to emotional processing changes across the menstrual cycle. *Proceedings of the National Academy of Sciences of the United States of America* 102: 16060–16065.
Puigserver P, Wu Z, Park CW, et al. (1998) A cold-inducible coactivator of nuclear receptors linked to adaptive thermogenesis. *Cell* 92: 829–839.
Purdy RH, Morrow AL, Moore PH, Jr., et al. (1991) Stress-induced elevations of gamma-aminobutyric acid type A receptor-active steroids in the rat brain. *Proceedings of the National Academy of Sciences of the United States of America* 88: 4553–4557.
Rabin DS, Schmidt PJ, Campbell G, et al. (1990) Hypothalamic–pituitary–adrenal function in patients with the premenstrual syndrome. *Journal of Clinical Endocrinology and Metabolism* 71: 1158–1162.
Raisman G and Field PM (1971) Sexual dimorphism in the preoptic area of the rat. *Science* 173: 731–733.
Rajkowska G (2000) Postmortem studies in mood disorder indicate altered numbers of neurons and glial cells. *Biological Psychiatry* 48: 766–777.
Rajkowska G, Miguel-Hidalgo JJ, Wei J, et al. (1999) Morphometric evidence for neuronal and glial prefrontal cell pathology in major depression. *Biological Psychiatry* 45: 1085–1098.

Rapkin AJ, Edelmuth E, Chang LC, et al. (1987) Whole-blood serotonin in premenstrual syndrome. *Obstetrics and Gynecology* 70: 533–537.
Rapkin AJ, Morgan M, Goldman L, et al. (1997) Progesterone metabolite allopregnanolone in women with premenstrual syndrome. *Obstetrics and Gynecology* 90: 709–714.
Raskin A (1974) Age-sex differences in response to antidepressant drugs. *Journal of Nervous and Mental Disease* 159: 120–130.
Reame NE, Marshall JC, and Kelch RP (1992) Pulsatile LH secretion in women with premenstrual syndrome (PMS): Evidence for normal neuroregulation of the menstrual cycle. *Psychoneuroendocrinology* 17: 205–213.
Redei E and Freeman EW (1993) Preliminary evidence for plasma adrenocorticotropin levels as biological correlates of premenstrual symptoms. *Acta Endocrinology* 128: 536–542.
Redei E and Freeman EW (1995) Daily plasma estradiol and progesterone levels over the menstrual cycle and their relation to premenstrual symptoms. *Psychoneuroendocrinology* 20: 259–267.
Redei E, Li L, Halasz I, et al. (1994) Fast glucocorticoid feedback inhibition of ACTH secretion in the ovariectomized rat: Effect of chronic estrogen and progesterone. *Neuroendocrinology* 60: 113–123.
Regier DA, Burke JD, and Burke KC (1990) Comorbidity of affective and anxiety disorders in the NIMH Epidemiologic Catchment Area Program. In: Maser JD and Cloninger CR (eds.) *Comorbidity of Mood and Anxiety Disorders*, pp. 113–122. Washington, DC: American Psychiatric Press.
Reid RL and Yen SSC (1981) Premenstrual syndrome. *American Journal of Obstetrics and Gynecology* 139: 85–104.
Reisert L and Pilgrim C (1991) Sexual differentiation of monoaminergic neurons – genetic or epigenetic? *Trends in Neuroscience* 14: 468–473.
Renfree MB, Harry JL, and Shaw G (1995) The marsupial male: A role model for sexual development. *Philosophical Transactions of the Royal Society, B* 350: 243–251.
Reus VI (1989) Behavioral aspects of thyroid disease in women. *Psychiatric Clinics of North America* 12: 153–165.
Robel P and Baulieu E-E (1994) Neurosteroids, biosynthesis and function. *Trends in Endocrinology and Metabolism* 5: 1–8.
Robins LN and Regier DA (1991) *Psychiatric Disorders in America: The Epidemiologic Catchment Area Study*. New York: The Free Press.
Roca CA, Schmidt PJ, Altemus M, et al. (2003) Differential menstrual cycle regulation of hypothalamic–pituitary–adrenal axis in women with premenstrual syndrome and controls. *Journal of Clinical Endocrinology and Metabolism* 88: 3057–3063.
Roca CA, Schmidt PJ, Smith MJ, et al. (2002) Effects of metergoline on symptoms in women with premenstrual dysphoric disorder. *American Journal of Psychiatry* 159: 1876–1881.
Rodriguez G, Warkentin S, Risberg J, et al. (1988) Sex differences in regional cerebral blood flow. *Journal of Cerebral Blood Flow and Metabolism* 8: 783–789.
Rojansky N, Halbreich U, Zander K, et al. (1991) Imipramine receptor binding and serotonin uptake in platelets of women with premenstrual changes. *Gynecologic and Obstetric Investigation* 31: 146–152.
Romeo E, Brancati A, de Lorenzo A, et al. (1996) Marked decrease of plasma neuroactive steroids during alcohol withdrawal. *Clinical Neuropharmacology* 19: 366–369.
Romeo E, Strohle A, Spalletta G, et al. (1998) Effects of antidepressant treatment on neuroactive steroids in major depression. *American Journal of Psychiatry* 155: 910–913.
Rosenstein DL, Elin RJ, Hosseini JM, et al. (1994) Magnesium measures across the menstrual cycle in premenstrual syndrome. *Biological Psychiatry* 35: 557–561.
Rosenstein DL, Kalogeras KT, Kalafut M, et al. (1996) Peripheral measures of arginine vasopressin, atrial natriuretic peptide and adrenocorticotropic hormone in premenstrual syndrome. *Psychoneuroendocrinology* 21: 347–359.
Roy-Byrne PP, Rubinow DR, Hoban MC, et al. (1987) TSH and prolactin responses to TRH in patients with premenstrual syndrome. *American Journal of Psychiatry* 144: 480–484.
Rubinow DR, Hoban MC, Grover GN, et al. (1988) Changes in plasma hormones across the menstrual cycle in patients with menstrually related mood disorder and in control subjects. *American Journal of Obstetrics and Gynecology* 158: 5–11.
Rubinow DR and Roy-Byrne PP (1984) Premenstrual syndromes: Overview from a methodologic perspective. *American Journal of Psychiatry* 141: 163–172.
Rubinow DR and Schmidt PJ (1995) The neuroendocrinology of menstrual cycle mood disorders. *Annals of the New York Academy of Sciences* 771: 648–659.
Rubinow DR and Schmidt PJ (1996) Androgens, brain, and behavior. *American Journal of Psychiatry* 153: 974–984.
Rubinow DR, Schmidt PJ, and Roca CA (1998) Estrogen–serotonin interactions: Implications for affective regulation. *Biological Psychiatry* 44: 839–850.
Saletu B, Brandstatter N, Metka M, et al. (1995) Double-blind, placebo-controlled, hormonal, syndromal and EEG mapping studies with transdermal oestradiol therapy in menopausal depression. *Psychopharmacology* 122: 321–329.
Saletu B, Brandstatter N, Metka M, et al. (1996) Hormonal, syndromal and EEG mapping studies in menopausal syndrome patients with and without depression as compared with controls. *Maturitas* 23: 91–105.
Salo LK, Makela SI, Stancel GM, et al. (1997) Neonatal exposure to diethylstilbesterol permanently alters the basal and 17 beta-estradiol induced expression of c-fos proto-oncogene in mouse urethroprostatic complex. *Molecular and Cellular Endocrinology* 126: 133–141.
Santarelli L, Saxe M, Gross C, et al. (2003) Requirement of hippocampal neurogenesis for the behavioral effects of antidepressants. *Science* 301: 805–809.
Santoro N (2005) The menopausal transition. *American Journal of Medicine* 118: 8S–13S.
Sarno AP, Miller EJ, Jr., and Lundblad EG (1987) Premenstrual syndrome: Beneficial effects of periodic, low-dose danazol. *Obstetrics and Gynecology* 70: 33–36.
Schechter D and Gandelman R (1981) Inter-male aggression in mice: Influence of gonadectomy and prior fighting experience. *Aggressive Behavior* 7: 187–193.
Schechter D, Strasser TJ, Endicott J, et al. (1996) Role of ovarian steroids in modulating mood in premenstrual syndrome. *Abstract, Society of Biological Psychiatry 51st Annual Meeting* 646.
Schlinger BA and Arnold AP (1991) Androgen effects on the development of the zebra finch song system. *Brain Research* 561: 99–105.
Schmidt PJ (2005) Mood, depression, and reproductive hormones in the menopausal transition. *American Journal of Medicine* 118: 545–585.
Schmidt PJ, Daly RE, Bloch M, et al. (2005) Dehydroepiandrosterone monotherapy in midlife-onset major and minor depression. *Archive of General Psychiatry* 62: 154–162.
Schmidt PJ, Grover GN, Roy-Byrne PP, et al. (1993) Thyroid function in women with premenstrual syndrome. *Journal of Clinical Endocrinology and Metabolism* 76: 671–674.

Schmidt PJ, Haq NA, and Rubinow DR (2004) A longitudinal evaluation of the relationship between reproductive status and mood in perimenopausal women. *American Journal of Psychiatry* 161: 2238–2244.

Schmidt PJ, Murphy JH, Haq N, et al. (2002) Basal plasma hormone levels in depressed perimenopausal women. *Psychoneuroendocrinology* 27: 907–920.

Schmidt PJ, Nieman LK, Danaceau MA, et al. (1998) Differential behavioral effects of gonadal steroids in women with and in those without premenstrual syndrome. *New England Journal of Medicine* 338: 209–216.

Schmidt PJ, Nieman L, Danaceau MA, et al. (2000) Estrogen replacement in perimenopause-related depression: A preliminary report. *American Journal of Obstetrics and Gynecology* 183: 414–420.

Schmidt PJ, Nieman LK, Grover GN, et al. (1991) Lack of effect of induced menses on symptoms in women with premenstrual syndrome. *New England Journal of Medicine* 324: 1174–1179.

Schmidt PJ, Purdy RH, Moore PH, Jr., et al. (1994) Circulating levels of anxiolytic steroids in the luteal phase in women with premenstrual syndrome and in control subjects. *Journal of Clinical Endocrinology and Metabolism* 79: 1256–1260.

Schmidt PJ and Rubinow DR (1991) Menopause-related affective disorders: A justification for further study. *American Journal of Psychiatry* 148: 844–852.

Schule C, Baghai TC, di Michele F, et al. (2007) Effects of combination treatment with mood stabilizers and mirtazapine on plasma concentrations of neuroactive steroids in depressed patients. *Psychoneuroendocrinology* 32: 669–680.

Schule C, Romeo E, Uzunov DP, et al. (2006) Influence of mirtazapine on plasma concentrations of neuroactive steroids in major depression and on 3α-hydroxysteroid dehydrogenase activity. *Molecular Psychiatry* 11: 261–272.

Seligman ME (1972) Learned helplessness. *Annual Review of Medicine* 23: 407–412.

Shaw T, Meyer JS, Mortel K, et al. (1979) Effects of normal aging, sex, and risk factors for stroke on regional cerebral blood flow (rCBF) in normal volunteers. In: Gotoh F, Hagai H, and Tazaki Y (eds.) *Cerebral Blood Flow and Metabolism*, pp. 462–463. Copenhagen: Munksgaard.

Shaywitz BA, Shaywitz SE, Pugh KR, et al. (1995) Sex differences in the functional organization of the brain for language. *Nature* 373: 607–609.

Shaywitz SE, Shaywitz BA, Pugh KR, et al. (1999) Effect of estrogen on brain activation patterns in postmenopausal women during working memory tasks. *Journal of the American Medical Association* 281: 1197–1202.

Sherwin BB (1988) Affective changes with estrogen and androgen replacement therapy in surgically menopausal women. *Journal of Affective Disorders* 14: 177–187.

Sherwin BB and Gelfand MM (1985) Differential symptom response to parenteral estrogen and/or androgen administration in the surgical menopause. *American Journal of Obstetrics and Gynecology* 151: 153–160.

Sherwood RA, Rocks BF, Stewart A, et al. (1986) Magnesium and the premenstrual syndrome. *Annals of Clinical Biochemistry* 23: 667–670.

Sibille E, Pavlides C, Benke D, et al. (2000) Genetic inactivation of the serotonin$_{1A}$ receptor in mice results in downregulation of major GABA$_A$ receptor α subunits, reduction of GABA$_A$ receptor binding and benzodiazepine-resistant anxiety. *Journal of Neuroscience* 20: 2758–2765.

Sichel DA, Cohen LS, Robertson LM, et al. (1995) Prophylactic estrogen in recurrent postpartum affective disorder. *Biological Psychiatry* 38: 814–818.

Silberg J, Pickles A, Rutter M, et al. (1999) The influence of genetic factors and life stress on depression among adolescent girls. *Archives of General Psychiatry* 56: 225–232.

Simpson HB, Nee JC, and Endicott J (1997) First-episode major depression. Few sex differences in course. *Archives of General Psychiatry* 54: 633–639.

Smith MJ, Adams LF, Schmidt PJ, et al. (2002) Effects of ovarian hormone on human cortical excitability. *Annals of Neurology* 51: 599–603.

Smith MJ, Adams LF, Schmidt PJ, et al. (2003) Abnormal luteal phase excitability of the motor cortex in women with premenstrual syndrome. *Biological Psychiatry* 54: 757–762.

Smith SS, Gong QH, Hsu F-C, et al. (1998a) GABA$_A$ receptor alpha-4 subunit suppression prevents withdrawal properties of an endogenous steroid. *Nature* 392: 926–930.

Smith SS, Gong QH, Li X, et al. (1998b) Withdrawal from 3α-OH-5α-pregnan-20-one using a pseudopregnancy model alters the kinetics of hippocampal GABA$_A$-gated current and increases the GABA$_A$ receptor α4 subunit in association with increased anxiety. *Journal of Neuroscience* 18: 5275–5284.

Smith CL, Nauaz Z, and O'Malley BW (1997) Co-activator and co-repressor regulation of the agonist/antagonist activity of the mixed antiestrogen, 4-hydroxytamoxifen. *Molecular Endocrinology* 11: 657–666.

Smith MJ, Schmidt PJ, Su T-P, et al. (2004) Gonadotropin-releasing hormone-stimulated gonadotropin levels in women with premenstrual dysphoria. *Gynecological Endocrinology 19* 333–343.

Smith R and Thomson M (1991) Neuroendocrinology of the hypothalamo-pituitary–adrenal axis in pregnancy and the puerperium. *Baillière's Clinical Endocrinology and Metabolism* 5: 167–186.

Soares CD, Almeida OP, Joffe H, et al. (2001) Efficacy of estradiol for the treatment of depressive disorders in perimenopausal women: A double-blind, randomized, placebo-controlled trial. *Archives of General Psychiatry* 58: 529–534.

Sodersten P (1984) Sexual differentiation: Do males differ from females in behavioral sensitivity to gonadal hormones? In: De Vries GJ, De Bruin JPC, Uylings HBM, et al. (eds.) *Progress in Brain Research,* vol. 61, pp. 257–270. Amsterdam: Elsevier.

Sohrabji F, Greene LA, Miranda RC, et al. (1994a) Reciprocal regulation of estrogen and NGF receptors by their ligands in PC12 cells. *Journal of Neurobiology* 25: 974–988.

Sohrabji F, Miranda RC, and Toran-Allerand CD (1994b) Estrogen differentially regulates estrogen and nerve growth factor receptor mRNAs in adult sensory neurons. *Journal of Neuroscience* 14: 459–471.

Spearow JL, Doemeny P, Sera R, et al. (1999) Genetic variation in susceptibility to endocrine disruption by estrogen in mice. *Science* 285: 1259–1261.

Speroff L, Glass RH, and Kase NG (1983) *Clinical Gynecologic Endocrinology and Infertility*. Baltimore, MD: Williams and Wilkins.

Spicer CC, Hare EH, and Slater E (1973) Neurotic and psychotic forms of depressive illness: Evidence from age-incidence in a national sample. *British Journal of Psychiatry* 123: 535–541.

Spitzer RL, Kroenke K, Linzer M, et al. (1995) Health-related quality of life in primary care patients with mental disorders: Results from the PRIME-MD 1000 study. *Journal of the American Medical Association* 274: 1511–1517.

Spruce BA, Baylis PH, Burd J, et al. (1985) Variation in osmoregulation of arginine vasopressin during the human menstrual cycle. *Clinical Endocrinology* 22: 37–42.

Srikant CB and Patel YC (1985) Somatostatin receptors. *Advances in Experimental Medicine and Biology* 188: 291–304.

Steege JF, Stout AL, Knight DL, et al. (1992) Reduced platelet tritium-labeled imipramine binding sites in women with premenstrual syndrome. *American Journal of Obstetrics and Gynecology* 167: 168–172.

Steiner M, Wheadon DE, and Kreider MS (1993) Antidepressant response to paroxetine by gender. *Abstract, APA 146th Annual Meeting*.

Steingold KA, Laufer L, Chetkowski RJ, et al. (1985) Treatment of hot flashes with transdermal estradiol administration. *Journal of Clinical Endocrinology and Metabolism* 61: 627-632.

Ströhle A, Romeo E, Hermann B, et al. (1999) Concentrations of 3α-reduced neuroactive steroids and their precursors in plasma of patients with major depression and after clinical recovery. *Biological Psychiatry* 45: 274–277.

Stutzmann GE and LeDoux JE (1999) GABAergic antagonists block the inhibitory effects of serotonin in the lateral amygdala: A mechanism for modulation of sensory inputs related to fear conditioning. *Journal of Neuroscience* 19: RC81–RC84.

Su T-P, Schmidt PJ, Danaceau M, et al. (1997) Effect of menstrual cycle phase on neuroendocrine and behavioral responses to the serotonin agonist m-chlorophenylpiperazine in women with premenstrual syndrome and controls. *Journal of Clinical Endocrinology and Metabolism* 82: 1220–1228.

Sundstrom I, Andersson A, Nyberg S, et al. (1998a) Patients with premenstrual syndrome have a different sensitivity to a neuroactive steroid during the menstrual cycle compared to control subjects. *Neuroendocrinology* 67: 126–138.

Sundstrom I, Ashbrook D, and Backstrom T (1997a) Reduced benzodiazepine sensitivity in patients with premenstrual syndrome: A pilot study. *Psychoneuroendocrinology* 22: 25–38.

Sundstrom I and Backstrom T (1998b) Citalopram increases pregnanolone sensitivity in patients with premenstrual syndrome: An open trial. *Psychoneuroendocrinology* 23: 73–88.

Sundstrom I, Nyberg S, and Backstrom T (1997b) Patients with premenstrual syndrome have reduced sensitivity to midazolam compared to control subjects. *Neuropsychopharmacology* 17: 370–381.

Szyf M, Weaver ICG, Champagne FA, et al. (2005) Maternal programming of steroid receptor expression and phenotype through DNA methylation in the rat. *Frontiers in Neuroendocrinology* 26: 139–162.

Taylor DL, Mathew RJ, Ho BT, et al. (1984) Serotonin levels and platelet uptake during premenstrual tension. *Neuropsychobiology* 12: 16–18.

Thase ME, Reynolds CF, Frank E, et al. (1994) Do depressed men and women respond similarly to cognitive-behavior therapy. *American Journal of Psychiatry* 151: 500–505.

Toran-Allerand CD (1994) Developmental interactions of estrogens with the neurotrophins and their receptors. In: Micevych P and Hammer RP (eds.) *Neurobiological Effects of Sex Steroid Hormones*, pp. 391–411. Cambridge: Cambridge University Press.

Tulchinsky D, Hobel CJ, and Yeager EM (1972) Plasma estrone, estradiol, estriol, progesterone and 17-hydroxyprogesterone in human pregnancy. *American Journal of Obstetrics and Gynecology* 112: 1095–1100.

Tulenheimo A, Laatikainen T, and Salminen K (1987) Plasma β-endorphin immunoreactivity in premenstrual tension. *British Journal of Obstetrics and Gynaecology* 94: 26–29.

Turner BB (1997) Influence of gonadal steroids on brain corticosteriod receptors: A minireview. *Neurochemical Research* 22: 1375–1385.

Uht R, Anderson CM, Webb P, et al. (1998) Steroid hormone interactions at the AP-1 site. *Abstracts of the 1998 American Neuroendocrine Society* .

Uht RM, Anderson CM, Webb P, et al. (1997) Transcriptional activities of estrogen and glucocorticoid receptors are functionally integrated at the AP-1 response element. *Endocrinology* 138: 2900–2908.

Uzunov DP, Cooper TB, Costa E, et al. (1996) Fluoxetine-elicited changes in brain neurosteroid content measured by negative ion mass fragmentography. *Proceedings of the National Academy of Sciences of the United States of America* 93: 12599–13604.

Uzunova V, Sheline Y, Davis JM, et al. (1998) Increase in the cerebrospinal fluid content of neurosteroids in patients with unipolar major depression who are receiving fluoxetine or fluvoxamine. *Proceedings of the National Academy of Sciences of the United States of America* 95: 3239–3244.

Veeninga AT and Westenberg HGM (1992) Serotonergic function and late luteal phase dysphoric disorder. *Psychopharmacology* 108: 153–158.

Viau V and Meaney MJ (1991) Variations in the hypothalamic–pituitary–adrenal response to stress during the estrous cycle in the rat. *Endocrinology* 129: 2503–2511.

vom Saal FS (1989) Sexual differentiation in litter-bearing mammals: Influence of sex of adjacent fetuses *in utero*. *Journal of Animal Sciences* 67: 1824–1840.

Wang G-J, Volkow ND, Overall J, et al. (1996a) Reproducibility of regional brain metabolic responses to lorazepam. *Journal of Nuclear Medicine* 37: 1609–1613.

Wang M, Seippel L, Purdy RH, et al. (1996b) Relationship between symptom severity and steroid variation in women with premenstrual syndrome: Study on serum pregnenolone, pregnenolone sulfate, 5α-pregnane-3,20-dione and 3α-hydroxy-5α-pregnan-20-one. *Journal of Clinical Endocrinology and Metabolism* 81: 1076–1082.

Ward IL and Stehm KE (1991) Prenatal stress feminizes juvenile play patterns in male rats. *Physiology and Behavior* 50: 601–605.

Ward IL and Weisz J (1980) Maternal stress alters plasma testosterone in fetal males. *Science* 207: 328–329.

Ward IL and Weisz J (1984) Differential effects of maternal stress on circulating levels of corticosterone, progesterone, and testosterone in male and female rat fetuses and their mothers. *Endocrinology* 114: 1635–1644.

Warrington SJ (1991) Clinical implications of the pharmacology of sertraline. *Interinational Clinical Psychopharmacology* 6: 11–21.

Watters JJ, Campbell JS, Cunningham MJ, et al. (1997) Rapid membrane effects of steroids in neuroblastoma cells: Effects of estrogen on mitogen activated protein kinase signalling cascade and c-fos immediate early gene transcription. *Endocrinology* 138: 4030–4033.

Weiss G (2000) Endocrinology of parturition. *Journal of Clinical Endocrinology and Metabolism* 85: 4421–4425.

Weiss SR, Post RM, Pert A, et al. (1989) Context-dependent cocaine sensitization: Differential effect of haloperidol on development versus expression. *Pharmacology, Biochemistry, and Behavior* 34: 655–661.

Weissman MM, Bland R, Joyce PR, et al. (1993) Sex differences in rates of depression: Cross-national perspectives. *Journal of Affective Disorders* 29: 77–84.

Weissman MM and Klerman GL (1977) Sex differences in the epidemiology of depression. *Archives of General Psychiatry* 34: 98–111.

Weissman MM and Klerman GL (1985) Gender and depression. *Trends in Neuroscience* 8: 416–420.

Werner AA, Johns GA, Hoctor EF, et al. (1934) Involutional melancholia: Probable etiology and treatment. *Journal of the American Medical Association* 103: 13–16.

West CP and Hillier H (1994) Ovarian suppression with the gonadotrophin-releasing hormone agonist goserelin (Zoladex) in management of the premenstrual tension syndrome. *Human Reproduction* 9: 1058–1063.

Whybrow PC (1995) Sex differences in thyroid axis function: Relevance to affective disorder and its treatment. *Depression* 3: 33–42.

Wieck A, Kumar R, Hirst AD, et al. (1991) Increased sensitivity of dopamine receptors and recurrence of affective psychosis after childbirth. *British Medical Journal* 303: 613–616.

Wieland S, Lan NC, Mirasedeghi S, et al. (1991) Anxiolytic activity of the progesterone metabolite 5α-pregnan-3α-ol-one. *Brain Research* 565: 263–268.

Winokur G, Coryell W, Keller M, et al. (1993) A prospective follow-up of patients with bipolar and primary unipolar-affective disorder. *Archives of General Psychiatry* 50: 457–465.

Winokur G, Tsuang MT, and Crowe RR (1982) The Iowa 500: Affective disorder in relatives of manic and depressed patients. *American Journal of Psychiatry* 139: 209–212.

Wisner KL, Parry BL, and Piontek CM (2002) Postpartum depression. *New England Journal of Medicine* 347: 194–199.

Wisner KL and Stowe ZN (1997) Psychobiology of postpartum mood disorders. *Seminars in Reproductive Endocrinology* 15: 77–89.

Witelson SA (1991) Neural sexual mosaicism: Sexual differentiation of the human temporo-parietal region for functional asymmetry. *Psychoneuroendocrinology* 16: 131–153.

Wolf OT, Neumann O, Hellhammer DH, et al. (1997) Effects of a two-week physiological dehydroepiandrosterone substitution on cognitive performance and well-being in healthy elderly women and men. *Journal of Clinical Endocrinology and Metabolism* 82: 2363–2367.

Wolkowitz OM, Reus VI, Keebler A, et al. (1999) Double-blind treatment of major depression with dehydroepiandrosterone. *American Journal of Psychiatry* 156: 646–649.

Wolkowitz OM, Reus VI, Roberts E, et al. (1997) Dehydroepiandrosterone (DHEA) treatment of depression. *Biological Psychiatry* 41: 311–318.

Yatham LN (1993) Is $5HT_{1A}$ receptor subsensitivity a trait marker for late luteal phase dysphoric disorder? A pilot study. *Canadian Journal of Psychiatry* 38: 662–664.

Yen SSC, Jaffe RB, and Barbieri RL (1999) *Reproductive Endocrinology: Physiology, Pathophysiology, and Clinical Management*. Philadelphia, PA: Saunders.

Yin W and Gore AC (2006) Neuroendocrine control of reproductive aging: Roles of GnRH neurons. *Reproduction* 131: 403–414.

Yonkers KA, Kando JC, Cole JO, et al. (1992) Gender differences in pharmacokinetics and pharmacodynamics of psychotropic medication. *American Journal of Psychiatry* 149: 587–595.

Yonkers KA, Ramin SM, Rush AJ, et al. (2001) Onset and persistence of postpartum depression in an inner-city maternal health clinic system. *American Journal of Psychiatry* 158: 1856–1863.

Young EA (1995) The role of gonadal steroids in hypothalamic–pituitary–adrenal axis regulation. *Critical Reviews in Neurobiology* 9: 371–381.

Young EA, Altemus M, Parkinson V, et al. (2001) Effects of estrogen antagonists and agonists on the ACTH response to restraint stress in female rats. *Neuropsychopharmacology* 25: 881–891.

Young MA, Fogg LF, Scheftner WA, et al. (1990) Sex differences in the lifetime prevalence of depression: Does varying the diagnostic criteria reduce the female/male ratio? *Journal of Affective Disorders* 18: 187–192.

Zhang L, Li B, Ma W, et al. (2002a) Dehydroepiandrosterone (DHEA) and its sulfated derivative (DHEAS) regulate apoptosis during neurogenesis by triggering the Akt signaling pathway in opposing ways. *Molecular Brain Research* 98: 58–66.

Zhang L, Li B, Zhao W, et al. (2002b) Sex-related differences in MAPKs activation in rat astrocytes: Effects of estrogen on cell death. *Molecular Brain Research* 103: 1–11.

Zhou Y, Watters JJ, and Dorsa DM (1996) Estrogen rapidly induces the phosphorylation of the cAMP response element binding protein in rat brain. *Endocrinology* 137: 2163–2166.

Zlotnick C, Shea MT, Pilkonis PA, et al. (1996) Gender, type of treatment, dysfunctional attitudes, social support, life events, and depressive symptoms over naturalistic follow-up. *American Journal of Psychiatry* 153: 1021–1027.

Zwain IH and Yen SSC (1999) Dehydroepiandrosterone: Biosynthesis and metabolism in the brain. *Endocrinology* 140: 880–887.

Zweifel JE and O'Brien WH (1997) A meta-analysis of the effect of hormone replacement therapy upon depressed mood. *Psychoneuroendocrinology* 22: 189–212.

5 Hypothalamic–Pituitary–Gonadal Axis in Men

R S Swerdloff, C Wang, and A P Sinha Hikim, David Geffen School of Medicine at UCLA, Torrance, CA, USA

Chapter Outline

5.1 Hypothalamic Control

5.1.1 Hypothalamic Regulation of Gonadotropin-Releasing Hormone

The hypothalamus is the principal integrative unit responsible for the normal pulsatile secretion of gonadotropin-releasing hormone (GnRH), which is delivered through the hypothalamic-hypophyseal portal blood system to the pituitary gland. Although GnRH has been identified in many areas of the central nervous system (CNS), it is most concentrated in the medial basal, arcuate, and suprachiasmatic nuclei in the hypothalamus and travels by axonomic flow to the axon terminals of the median eminence. The pulsatile release of GnRH provides the signals for the timing of the release of luteinizing hormone (LH) and follicle-stimulating hormone (FSH), which under normal circumstances occurs approximately every 60–90 min. GnRH acts by binding to the GnRH receptors on the surface of the gonadotrophs. A number of endocrine, paracrine, and autocrine factors regulate the GnRH gene expression (Norwitz et al., 1999). The secretion of GnRH is regulated in a complex fashion by neuronal input from higher cognitive and sensory centers and by the circulating levels of sex steroids and peptide hormones, such as prolactin, activin, inhibin, and leptin. The local effectors of GnRH synthesis and release include a number of neuropeptides, opioids, catecholamines, indolamines, nitric oxide and excitatory amino acids, γ-aminobutyric acid (GABA), dopamine, neuropeptide Y, vasoactive intestinal peptide (VIP), and corticotropin-releasing hormone (CRH). Recent studies have shown a critical role of kisspeptin-54 and its receptor G-protein-coupled receptor 54 (GPR54) in the regulation of hypothalamic GnRH secretion (de Roux et al., 2003; Seminara et al., 2003; Kotani et al., 2001; Muir et al., 2001; Ohtaki et al., 2001; Dhillo et al., 2007). Kisspeptin is expressed in the arcuate, periventricular, and anterioventral periventricular nuclei of the brain. The GPR54 receptors are present in the CNS. Kisspeptin stimulates GnRH secretion, and LH and FSH responses to kisspeptin can be blocked with GnRH antagonists. The inhibitor effects of testosterone (T) and estradiol (E2) on gonadotropin secretion are mediated by inhibition of kisspeptin production in the hypothalamus (Navarro et al., 2004; Smith et al., 2005). Chronic administration of kisspeptin, paradoxically, inhibits LH secretion in monkeys and causes testicular atrophy in rats (Seminara et al., 2006; Thompson et al., 2006). In addition to the changes in the kisspeptin–GPR-54 system, the catecholamines, excitatory amino acids, and nitric oxide in physiologic amounts are stimulatory, whereas opioid peptides and β-endorphin are inhibitory. T, either directly or through its metabolic products (E2 and dihydrotestosterone (DHT)), has predominantly inhibitory effects on the secretion and release of GnRH, LH, and FSH. Prolactin is a potent inhibitor of GnRH secretion, thus explaining its role in inhibiting LH and T secretion in the clinical condition of hyperprolactinemia.

5.1.2 GnRH Synthesis and Secretion

GnRH, the central initiator in the reproductive hormonal cascade, was first isolated from a million sheep and pig hypothalami as a decapeptide (pGlu-HIS-TRP-Ser-Tyr-GLY-Leu-Arg-Pro-Gly NH2; Schally et al., 1971; Matsuo et al., 1971). GnRH is generated in the neurons of the hypothalamus from a precursor polypeptide by enzymatic processing and is secreted in a pulsatile manner into the hypophyseal portal circulation to stimulate the biosynthesis and secretion of LH and FSH. Despite the structural variants of GnRH, the amino acid sequences of the 14 GnRHs from vertebrate and lower species revealed features that have been conserved for over 500 million years of evolution. Initially, GnRH was thought to function exclusively as a stimulator of gonadotropin release, but it has become apparent that the peptide has other functions in vertebrates and lower species. In vertebrates, GnRH and variants are expressed both in the hypothalamus and in extrahypothalamic regions of the CNS (e.g., the midbrain, spinal cord, and sympathic ganglia) and in non-neuronal tissue (e.g., the gonads, placenta, and breast).

5.1.3 Origin and Migration of GnRH Neurons during Development

The neurons that secrete GnRH originate in the region of the olfactory apparatus (Schwanzel-Fukuda and Pfaff, 1989). These neurons migrate along with the olfactory and vomeronasal nerves into the forebrain and then into their final location in the medial basal, preoptic, and arcuate nuclei in the hypothalamus (Schwanzel-Fukuda and Pfaff, 1989). This orderly migration of GnRH neurons requires the coordinated action of direction-finding molecules, adhesion proteins, such as the KALIG-1 gene product anosmin-1 (Soussi-Yanicostas et al., 2002), and enzymes that help the neuronal cells burrow their way through the intercellular matrix. The mutation of KAL1 and FGFR1 genes (Dodé et al., 2003) can arrest the migratory process and result in the failure of GnRH neurons to arrive at appropriate hypothalamic secretory sites. This, in turn, leads to GnRH functional deficiency. In at least a subset of patients with idiopathic hypogonadotropic hypogonadism, developmental migratory disorders are evident. Recent studies on GnRH and GnRH-receptor-deficient mice showed that GnRH presence, or action, is not required for the developmental migration of GnRH neurons into the brain or for the projection of GnRH neurosecretory afferent neurons (Gill et al., 2008).

5.2 Pituitary

5.2.1 Gonadotropin-Secreting Cells in the Pituitary

Gonadotrophs constitute approximately 10–15% of anterior pituitary cells (Kovacs and Horvath, 1985; Moriarty, 1973, 1976). They are dispersed throughout the anterior pituitary, close to the capillaries and often in close approximation to the lactotroph. Gonadotrophs from the male and female gland cannot be distinguished on morphologic grounds. Gonadotrophs are easily demonstrable in the fetal and pre-pubertal pituitary gland; however, their numbers are low before sexual maturation. Immunocytologic evidence indicates that a single cell type in the pituitary secretes both LH and FSH (Kovacs and Horvath, 1985; Childs et al., 1990; Moriarty, 1973, 1976). Despite this evidence, some gonadotrophs stain only for FSH or LH and it is unclear if these monohormonal-appearing cells represent separate cell types or identical cells in different secretory phases (Childs et al., 1983; Lloyd and Childs, 1988). It has been long recognized that castration leads to an increase in size, as well as in the number, of gonadotrophs (gonadotroph hypertrophy and hyperplasia) and that patients with idiopathic hypogonadotropic hypogonadism reveal fewer hypoplastic gonadotrophs (Kovacs and Horvath, 1985).

Pulsatile secretion of GnRH results in a one-to-one pulse of LH (Crowley et al., 1985). FSH is regulated in a more complex fashion but is also driven by GnRH.

5.2.2 Molecular Basis of Pituitary Development

The coordinated, temporal expression of a number of homeodomain transcription factors directs the embryological development of the pituitary and its differentiated cell types. Three homeobox genes are essential for early organogenesis (Parks et al., 1997, 1999; Watkins-Chow and Camper, 1998). Cell specialization and the proliferation of differentiated cell types require the expression of transcription factors PROP-1 and PIT-1. The PROP-1 genes encode a transcription factor with a single DNA-binding domain (Parks et al., 1997, 1999). PIT-1 apparently has at least two DNA-binding domains (Ingraham et al., 1988). Whereas PIT-1 mutations are associated with deficiencies of growth hormone, thyroid-stimulating hormone (TSH), and prolactin, mutations in PROP-1 are associated with deficiencies of LH and FSH in addition to the deficiencies of growth hormone, prolactin, and TSH. Much of our understanding of the molecular mechanisms of pituitary development comes from phenotypic correlations of the developmental disorders of the pituitary and the genetic analysis of mutations associated with these disorders.

5.2.3 GnRH Receptors

GnRH receptors are present in the pituitary gonadotroph cells and in extrapituitary tissues (suggesting actions of GnRH at other sites). The GnRH receptors have now been extensively analyzed for structure and function. GnRH pituitary receptors from the rat, human, sheep, cow, and pig share over 80% amino acid identity. The GnRH receptor has the characteristic features of most other GPRs, with an N-terminal domain followed by seven α-helical transmembrane domains connected by three extracellular domains and three intercellular loop domains. The unique feature of the mammalian GnRH receptor is the absence of a terminal tail that is present in other GPRs and in all of the nonmammalian GnRH receptors.

The intracellular signal transduction of GnRH and its receptor have also been extensively studied (Kaiser et al., 1997; Sealfon and Millar, 1994; Turgeon et al., 1996; Stojilkovic et al., 1994; Stojilkovic and Catt, 1995; Flanagan et al., 1997; Schertler et al., 1993; Naor et al., 1998). The consensus view is that the primary pathway of GnRH action is the activation of the calcium-dependent phospholipase C (PLC)-β through a guanine-nucleotide-binding protein. PLC hydrolyzes phosphatidylinositol (PIP2) to inositol triphosphate (IP3) and diacylglycerol (DAG). IP3 binds to receptors in the endoplasmic reticulum to transiently release Ca^{2+} from these intrasellar stores and this elicits spikes of LH release. DAG activates protein kinase C (PKC), which phosphorylates proteins involved in the more sustained synthesis and release of LH. GnRH binding to the receptor also activates L-type voltage-operated Ca^{2+} channels, which results in the influx of extracellular Ca^{2+} required for recharging intracellular stores and prolonged (second-phase) LH release. In addition, to the effects of GnRH on PLC, there appears to be an important role for phospholipase D (PLD) in the catalytic production of DAG.

The relative stimulation of gonadotropin subunits in the pituitary is influenced by GnRH pulse frequency. High pulse frequencies stimulate LH-β mRNA levels more than FSH-β, whereas the converse is true at low pulse frequencies. Thus, it is believed that the combination of alterations in pulse frequency, together with differential phasing and duration of the various signaling pathways, provides the potential for the fine regulation of gonadotropin secretion. Additional feedback by gonadal steroids and peptide hormones, such as activin and inhibin, on the gonadotrope provides additional modulatory effects. Low doses of GnRH delivered in a pulsatile fashion to mimic normal physiology stimulate LH and FSH secretion and correct clinical manifestations of GnRH deficiency (Conn and Crowley, 1991, 1994; Barbieri, 1992; Filicori, 1994a,b). In contrast, high doses of GnRH or GnRH agonist analogs cause desensitization of the GnRH receptor and result in the suppression of LH and FSH secretion. GnRH antagonists, on the other hand, competitively inhibit the binding of GnRH to its receptor. Surprisingly, GnRH analogs (antagonists and agonists) have been used more extensively in clinical practice to suppress the reproductive axis than the use of authentic GnRH or its agonist to stimulate the system (Millar et al., 1987; Conn and Crowley, 1991, 1994; Barbieri, 1992; Filicori, 1994a,b; Handelsman and Swerdloff, 1986; Rajfer et al., 1987; Bhasin et al., 1987; Handelsman et al., 1988; Salameh et al., 1991, 1994; Tom et al., 1992). GnRH agonists are the mainstay of medical suppression of T for treatment of metastatic cancer of the prostate and true isosexual precocious puberty in boys (The Leuprolide Study Group, 1984; Nathan and Palmert, 2005). GnRH antagonists are more rapid acting and inhibit FSH and LH secretion more completely than agonists. GnRH antagonists, combined with T, have also been tested in male contraceptive trials (Bagatell et al., 1993; Swerdloff et al., 1998a,b; Herbst et al., 2004).

5.2.4 Biochemistry of LH and FSH

LH and FSH are part of the family of glycopeptide hormones that also include TSH and the various chorionic gonadotropins. Each of these hormones is heterodimeric, consisting of an α- and a β-subunit. The primary structures of the α-subunits of these glycopeptides are nearly identical within a species; the biologic specificity is conferred by the dissimilar β-subunit. Formation of the heterodimer, tightly linked internally by disulfide bonds, is essential for receptor binding and consequent biologic activity.

5.2.5 LH and FSH Subunit Genes

In humans, as in rats and mice, a single gene codes for the α-subunit of the four glycopeptide hormones (Fiddes and Goodman, 1979, 1981; Boothby et al., 1981; Gharib et al., 1990; Gibson et al., 1980; Burnside et al., 1988; Godine et al., 1982; Gordone et al., 1988; Nilson et al., 1983). GnRH induces the transcription of the human gonadotroph α-subunit gene. The LH-β gene is relatively small in size (approximately 1.5 kb in length) and similar to other glycoprotein hormone β genes (Talmadge et al., 1984;

Otani et al., 1988; Albanese et al., 1996). There are great similarities between the LH-β and chorionic gonadotroph (CG)-β genes. The general organization of the FSH-β gene is similar to that of other glycoprotein hormone β genes (three exons and two introns) (Gharib et al., 1989, 1990; Kim et al., 1988, 1990; Watkins et al., 1987; Jameson et al., 1988; Maurer and Kim, 1989; Hirai et al., 1990). The last two exons contain the coding sequence of the gene. An analysis of the ovine FSH-β gene promoter has revealed two functional activating protein 1 (AP-1) enhancers that are important for its expression and regulation by GnRH (Stahl et al., 1998). It is believed that the GnRH-regulated expression of the FSH-β gene involves the activation of PKC-signaling pathways (Stahl et al., 1998; Sauders et al., 1998; Brown and McNeilly, 1999). The secretion of LH and FSH is activated by the pulsatile delivery of GnRH and modified locally by multiple factors, including prolactin; sex steroids such as T, E2, and DHT; and testicular peptide substances such as inhibin and possibly activin. Because GnRH is the predominant regulatory factor responsible for LH and FSH secretion, its pulsatile pattern is responsible for the pulsatile release of LH and FSH. LH and FSH are secreted into the systemic circulation where they act predominantly on the testis to regulate spermatogenesis and Sertoli cell function in men.

5.2.6 Synthesis and Post-Translational Processing of the Subunits

The α- and β-subunits of LH and FSH are encoded by separate genes (Chin et al., 1983; Jameson et al., 1983; Fiddes and Goodman, 1979, 1981; Boothby et al., 1981; Godine et al., 1982). The α-subunit is initially synthesized as a precursor with a molecular mass of 14 kDa, whereas LH-β is synthesized as a precursor of molecular mass of 15–17 kDa (Kefier et al., 1980; Godine et al., 1980, 1981; Landefeld and Kepa, 1979; Counis et al., 1981; Chin et al., 1981). Precursors are processed by the enzymatic removal of amino-terminal leader peptides and also by the addition of carbohydrates. Glycosylation occurs co-translationally in several steps (Weintraub et al., 1980; Ruddon et al., 1980; Wachter and Lernmaez, 1976; Kornfeld and Kornfeld, 1976). Although the functional roles of the carbohydrate side chains of the gonadotropic hormones remain somewhat uncertain (Weintraub et al., 1985; Sairam and Bhargavi, 1985; Gibson et al., 1980; Burnside et al., 1988), it appears that they may serve multiple roles in hormone assembly, secretion, action (Weintraub et al., 1985), and the metabolic clearance rate of the glycoprotein (Gibson et al., 1980).

5.2.7 LH and FSH Receptor Structure

Most of LH and FSH receptors are found in the gonads, although they are also reported to be present in other tissues, including the brain. The gonadotroph receptors are members of the GPRs superfamily; cDNAs for the receptors encode polypeptides having a large extracellular ligand binding and N-terminus, a transmembrane domain composed of seven hydrotropic α-helices and a small intracytoplasmic C-terminus. Most LH receptors have molecular weights of approximately 90 kDa and some have molecular weights of 170 kDa. These sizes suggest that they exist both as a full-length and homodimer receptor (Kusuda and Dufau, 1988; Roche and Ryan, 1989). The amino sequences of the LH and FSH transmembrane domains are similar, whereas the extracellular and cytoplasmic domains are approximately 50% identical.

Chemical cross-linking studies suggest that both the hormone subunits are near the target organ receptor interface (Ji and Ji, 1981). Considerable work has been done in order to understand the three-dimensional (3D) structure of the gonadotropins and their role in hormone-receptor interactions.

5.2.8 Clearance and Secretory Rhythms of LH and FSH

The clearance of the two gonadotrophic hormones differs, with LH having a shorter plasma half-life (<20 min) in the blood than FSH (1–4 h; Yen et al., 1968, 1970; Veldhuis and Johnson, 1988; Urban et al., 1991). There is a diurnal rhythm of both gonadotropins (in young adult men) with higher circulating levels in the early morning and lowest levels in the evening. Puberty is heralded by nighttime pulsatile serum patterns before obvious increased levels are noted in the daytime. The diurnal pattern of adolescent and younger men is markedly blunted in the elderly male (Bremner et al., 1983; Diver et al., 2003).

5.2.9 Roles of LH and FSH in the Male

LH binds to specific receptors on the Leydig cells and stimulates T production *in vitro* and *in vivo*. LH is required for maintaining the very high intratesticular levels of T essential for spermatogenesis (Sharpe, 1987; Steinberger, 1971; Heller and Clermont, 1964; Boccabella, 1963; Sharpe et al., 1988; Weinbauer and Nieschlag, 1990). Circulating T is also essential for maintaining sexual function, secondary sex characteristics, and a number of other androgen-dependent processes, such as the maintenance of normal bone mineral density and stimulation of protein anabolism necessary for the maintenance of normal muscle mass. FSH mediates its effects on the testis predominantly through its action on the Sertoli cells. FSH is known to bind to a specific GPR on these cells and stimulate the production of a large number of proteins, including the inhibin-related peptides, androgen-binding protein, transferrin, androgen receptor (AR), and γ-glutamyl transpeptidase, among others. The role of FSH in the spermatogenic process is discussed later. It is quite clear that the high intratesticular levels of T, induced by LH (mediated through actions on the Sertoli cells), play a key role in spermatogenic maturation.

5.2.10 Gonadal Feedback Regulation of LH and FSH

5.2.10.1 Gonadal steroids

The testes regulate LH and FSH secretion through feedback effects of sex-steroid hormones and other peptides. The castration of experimental animals (Gay and Midgley, 1969; Swerdloff et al., 1971) and men (Walsh et al., 1973) predictably results in elevated serum LH and FSH levels. The mRNAs for LH and FSH α- and β-subunits all rise after the castration of male rats (Badger et al., 1978; Yamamoto et al., 1970; Gay and Midgley, 1969). The primary regulator of LH secretion is the circulating level of T. T replacement in castrated male rats blocks the rise in LH-α and LH-β mRNAs and serum LH levels, but has little effect on FSH-β mRNA levels (Gharib et al., 1986). T administration to normal men and male rats also inhibits LH secretion (Swerdloff et al., 1979; Winters et al., 1979; Sherins and Lonaux, 1973). It appears that the inhibitory effects are largely accomplished at both the hypothalamic and pituitary levels. This conclusion is based on the observation that T decreases the frequency of LH pulses (GnRH entrainment) in normal eugonadal men (Sherins and Lonaux, 1973; Matsumoto and Bremner, 1984; Stewart-Bentley et al., 1974; Veldhuis et al., 1984; Santen, 1975) and decreases the amplitude of LH pulses in GnRH-treated GnRH-deficient men (Scheckter et al., 1989; Finkelstein et al., 1991b). Data from multiple sources suggest that the LH suppression induced by T is partly mediated by T directly through the AR (Kerrigan et al., 1994; Knuth et al., 1984; Veldhuis et al., 1992; Vermeulen and Deslypere, 1985) and through aromatization to E2 (Finkelstein et al., 1991a; Hayes et al., 2000; Raven et al., 2006; Rochira et al., 2006). E2-mediated suppression of LH secretion appears to occur predominantly by action at the pituitary level (Santen, 1975; Finkelstein et al., 1991b). Because the administration of finasteride (5α-reductase inhibitor) to eugonadal men does not increase LH and FSH levels (Gormley and Ridgmaster, 1991), it appears that the conversion of T to DHT is not required for the inhibitory effects of T on either LH or FSH. This is supported by the demonstration that DHT, a nonaromatizable potent androgen, appeared to be a poorer suppressor of gonadotropic production than T (Schaison et al., 1980). Nevertheless, chronic pharmacologic doses of DHT inhibit both LH and FSH despite suppressed serum levels of both T and E2 (Wang et al., 1998; Kunelius et al., 2002; Ly et al., 2001).

The role of progesterone on normal gonadotropin physiology is unknown, but it is clear that pharmacological doses of progestational steroids synergize with T in suppressing LH and FSH secretion (Nieschlag et al., 2003; Wang and Swerdloff, 2004). This observation is the basis of combined androgen–progestin male contraceptives.

5.2.10.2 Gonadal peptides (inhibin, activins, and follistatins) and feedback regulation of FSH

It has long been recognized that men with azoospermia and normal serum T levels often have selective elevations of serum FSH. A search for a testicular peptide that could selectively suppress FSH was rewarded by the identification of a family of inhibin-related peptides (Robertson et al., 1985; Ling et al., 1985; Mason et al., 1985; Forage et al., 1986; Miyamoto et al., 1985; Esch et al., 1987). These studies identified inhibin as a dimeric protein consisting of an α-subunit combined with 2β-subunits, A and B. The heterodimers inhibin A and B are equipotent in their FSH-inhibiting actions (Ying, 1988; Vale et al., 1988; Burger and Igarashi, 1988). The βA-subunits can form homodimers called activin A or heterodimerize with βB-subunits to form activin B. Activins A and B both stimulate FSH secretion *in vitro* (Ying, 1988; Vale et al., 1988). It is now known that activins and inhibin B are produced in the Sertoli cell of the spermatogenic tubules and stimulated by circulating levels of FSH. Despite these observations, the role of inhibin and activin in normal adult male physiology is unclear because serum inhibin levels do not show an inverse correlation with serum FSH levels in men with a variety of testicular disorders. Chronic FSH and spermatogenic suppression results in only a 25% increase in inhibin-B levels indicating that FSH-independent secretion of inhibin B occurs (Matthiesson et al., 2003).

Another class of FSH inhibitors are called follistatins (Ueno et al., 1987; Robertson et al., 1987). Follistatins are glycosylated single-chain polypeptides with structural homology to pancreatic-secretory-inhibiting protein (PSTI) and human epidermal growth factor (hEGF; Ying, 1988). The precise role of follistatin is also not known, but it may serve as an inhibitor in the regulation of FSH. Follistatins seem to act predominantly through their binding of activin.

Activins regulate both pituitary and intragonadal function in the male and the female (Vale et al., 1988). Activin B appears to be an autocrine or paracrine mediator in the pituitary acting to enhance FSH-β gene expression and secretion by the gonadotrophs (Bilezikjion et al., 2006). Activins have additional paracrine roles in the testis through the enhancement of steroidogenesis and spermatogonial proliferation (Mather et al., 1992). Despite these multiple putative roles for activin, it is unclear if activin is predominantly an embryonic organizing substance, a paracrine regulator, or a circulatory regulatory peptide involved in the regulation of FSH secretion. Serum levels of activin increase progressively with age in men (Loria et al., 1998). Despite the increase with age, no correlations of serum activin-A levels and FSH in men or women have been demonstrated. Even less is known about the physiological role of activin B because reliable assays are not available.

5.2.10.3 Summary

The predominant inhibitors of LH and FSH appear to be the gonadal steroids T and E2. There are many situations in which decreased spermatogenesis is associated with selective increases in serum FSH; most instances of selective elevations of FSH levels in men have been associated with low sperm counts. Several inhibitory peptides (e.g., activin and follistatin) have been postulated as selecting FSH inhibitors, but evidence to relate changing serum levels of either peptide with increased FSH has not been shown in men. Thus, our understanding of FSH regulation is incomplete.

5.3 Testes-Leydig Cell Compartment

5.3.1 Testicular Steroidogenesis

As summarized in **Figure 1**, T is synthesized in the Leydig cells of the interstitial compartment of the testis, by an enzymatic sequence of steps, from cholesterol (Swerdloff and Wang, 2000). Cholesterol is predominantly formed by *de novo* synthesis from acetate, although preformed cholesterol, either from intracellular cholesterol ester stores or from extracellular circulating low-density lipoproteins, also contribute.

The first and rate-limiting step in gonadal and adrenal steroidogenesis is the transfer of the steroidogenic substrate cholesterol from the outer mitochondrial membrane to the inner membrane mediated by the cholesterol-transport protein, steroidgenic acute regulatory protein (StAR). Transgenic XY male mice deficient in StAR have adrenal and testicular failure and female external genitalia. StAR mRNA expression is stimulated by LH and adrenocorticotropic

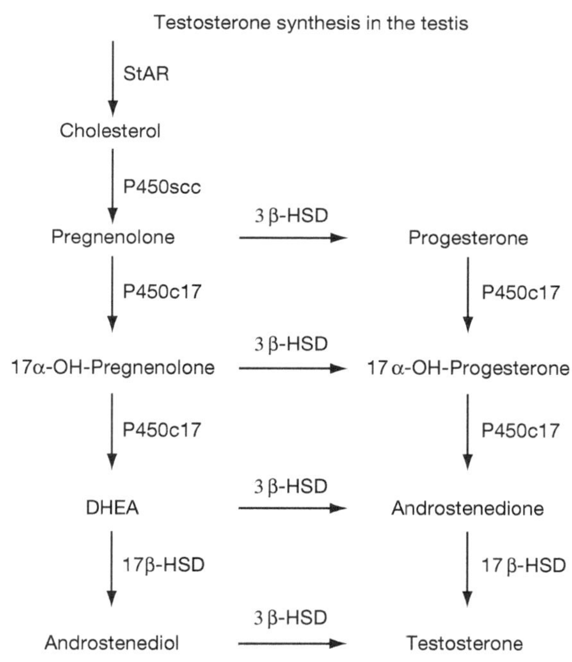

Figure 1 StAR protein mobilizes cholesterol from cellular stores to the mitochondria along the intratesticular steroidogenic pathways for synthesis of testosterone. Whereas both the delta 5 (left) and delta 4 (right) pathways exist, the delta 5 pathway predominates in the testis. Modified from Swerdloff RS and Wang C (2000) The testis and male sexual function. In: Goldman L and Bennet JC (eds.) *Cecil Textbook of Medicine*, 21st edn., pp. 1306–1317. Philadelphia, PA: W. B. Saunders, with permission from Elsevier.

hormone (ACTH) via the transmembrane-coupled cyclic adenosine monophosphate (cAMP) system. The promoter of the StAR gene contains binding sites for two transcription factors, steroidogenic factor-1 (SF-1) and estrogen receptor (ER; Sandhoff et al., 1998; Shen et al., 1994), suggesting their role in StAR transcription regulation. In contrast, another transcription factor, DAX-1, blocks steroidogenesis by transcriptional inhibition of the StAR gene (Zazopoulos et al., 1997). A dysfunctional mutation of StAR gene results in the human syndrome congenital lipoid adrenal hyperplasia. Once cholesterol is transported into the inner mitochondrial membrane, P450 side-chain-cleavage enzyme (p450scc), previously thought to be the rate-limiting step in androgen synthesis, converts cholesterol to pregnenolone via three integrated biochemical reactions – 20α-hydroxylation, 22R-hydroxylation, and scission of the cholesterol side chain at the 20–22 carbon bond (Duque et al., 1978; Takikawa et al., 1978). The 3β-hydroxysteroid dehydrogenase (3β-HSD), 17α-hydroxylase/17 20-lyase complex enzymes (p450c17), and 17β-hydroxysteroid dehydrogenase (17β-HSD) are other key catalysts of T synthesis. Dysfunctional mutations of all these genes result in androgen deficiency. Because these defects occur prior to the development of genitalia, ambiguous genitalia and impaired spermatogenesis are seen.

5.3.2 T Transport and Metabolism

T leaves the testes through the blood stream and 95% of circulating T in the adult man is derived from testicular secretion; the remainder arises from metabolic conversion of precursor steroids, predominantly secreted by the adrenal cortex (e.g., androstenedione; androstenediol; and dehydroepiandrosterone, DHEA). Hormonal production rates are calculated either by estimating the metabolic clearance rates and mean circulating T level or by estimating testicular arterioventricular (AV) differences in testicular blood-flow rate. These methods give a relatively consistent estimate for T production in normal adult men of approximately 5–10 mg day^{-1}. T is converted to metabolic products with serum conversion rates of 4% to DHT (Ishimaru et al., 1978) and 0.2% E2 (Longcope et al., 1984).

T circulates in the blood in concentrations above its aqueous solubility by binding to circulating plasma proteins, leaving only a small fraction (~1–2%) circulating in a free state. T binds avidly to sex-hormone-binding globulin (SHBG) and weakly to other proteins, such as albumin. SHBG is produced in the liver and its production is regulated in a complex fashion to amplify or reduce the availability of the biologically active forms of the hormone. Thus, SHBG production is stimulated by E2 and thyroxine and reduced by T, growth hormone, and glucocorticoids. Other important factors regulating SHBG include upregulation by acute or chronic liver disease and downregulation by obesity, protein-losing states, and genetic causes of SHBG deficiency. Under physiological conditions, 45–70% of T is bound to SHBG, 30–55% is bound to albumin, and approximately 2% remains in the nonbound form. The free fraction is the most biologically available. However, because albumin binding is weak, albumin-bound T can be considered, for all practical purposes, to be readily available; the combination of the free and the albumin-bound T is called the bioavailable T.

Circulating T levels have both distinct (pulsatile) and diurnal rhythms. Although the pulsatile LH secretion influences the pattern of T secretion, this effect is greatly buffered by the presence of transport proteins, such as SHBG, and, to a lesser degree, albumin.

5.3.3 T Secretion during Fetal Development, Childhood, Puberty, and Senescence

T is secreted in increased amounts during three independent periods in the life of a human male. These are during intrauterine life, neonatal period, and during puberty. The initial secretory surge in intrauterine life is critical for genital sexual differentiation, whereas the intrauterine and neonatal surges may play an important role in gender-associated behavior and structural and functional gender differences in cognition. Serum levels of T reach their peak in early adulthood and begin to decline as early as 30 years of age. This middle-aged decline in circulating and in total and free T becomes of clinical significance in older age when circulating levels fall below the thresholds for optimal multiple-tissue biological effects.

5.3.3.1 *Fetal Leydig cell steroidogenesis*

T production by the fetal Leydig cells is essential for the masculinization of the embryonic Wolffian system, external genitalia, and perhaps, the brain. T is first detected in the human fetus at 9 weeks. There is a brisk rise in serum and testicular T to a peak at 15–18 weeks of fetal life and then declines. It has been assumed that first trimester Leydig cell steroidogenesis and the associated sexual differentiation of the internal and external genitalia are independent of LH-HCG-LH receptor interactions because the LH-CG receptor is not detectable until 12 weeks of age (Huhtaniemi et al., 1977; Molsberry et al., 1982). Perhaps, this reflects our inability to adequately detect low levels of the receptor at this developmental time period because the human syndrome Leydig cell hypoplasia, caused by autosomal-recessive mutations of the LH-CG gene, results in female or ambiguous genitalia, low serum T, and XY testes. Less uncertain is the independence of the testicular steroidogenesis from LH control in the first 8 weeks of fetal life, as demonstrated by normal penile structural development in boys with congenital hypopituitarism and genital masculinization in a boy with LH-β dysfunctional gene mutation (Jameson, 1996). The importance of LH in late fetal development is evidenced by the reduced penile size in boys with congenital hypopituitarism. Although the early control of steroidogenesis is still in question, it is clear that Leydig cell steroidogenesis in the second and third trimester of fetal life requires the activation of the LH-CG receptor system by CG and subsequently LH (third trimester).

5.3.3.2 Neonatal T secretion

There is a secondary surge in T secretion in the neonatal period (first 3 months of life; Kaplan et al., 1976; Penny et al., 1979). It is presumed to be driven by the elevated LH levels at this time. The role of this neonatal elevation of T is uncertain, but it has been speculated to have an influence on gender-related brain structure and function.

5.3.3.3 Adrenarche and puberty

5.3.3.3(i) Adrenarche

Adrenarche occurs at approximately 7 or 8 years of age when the zona reticularis of the adrenal undergoes maturation, leading to increased section of androgen precursors, such as androstenedione, DHEA, and its metabolite DHEA sulfate (DHEA-S). Although the physiological events initiating adrenarche are incompletely understood, the process is probably under the control of ACTH and independent of the control of LH and FSH. Adrenarche usually heralds subsequent activity in the hypothalamic–pituitary–gonadal axis. Androstenedione and DHEA are technically androgenic prehormones and do not bind to the AR. In part, the prepubertal growth spurt and the early development of pubic and axillary hair are mediated by the conversion of these precursors to T and DHT at the peripheral tissues sites.

5.3.3.3(ii) Puberty

Puberty occurs when a hypothalamic clock is activated, resulting in increased GnRH and gonadotropin secretion. In the interval before the onset of puberty, LH and FSH are secreted in low amounts and are subject to feedback control by the small amounts of circulating T from the testes. The initiation of puberty is determined by an increase in the pulsatile pattern of hypothalamic-GnRH secretion. The timing of this event seems to be regulated by kisspeptin-1 activation of GPR-54. Glutamine and GABA may influence the timing of the KISS-1- and GPR-54-mediated events (Banerjee and Clayton, 2007). As puberty progresses, negative-feedback sensitivity of the hypothalamus and pituitary to circulating steroids lessens and increasing concentrations of both gonadal steroids and gonadotropin hormones ensue. The increasing concentrations of intratesticular T and circulating FSH stimulate the Sertoli cells to produce factors leading to the maturation of spermatogenesis and inhibition of germ-cell apoptosis. The phenotypic equivalents of the hormonal changes in puberty have been well documented. Pediatricians and endocrinologists routinely perform staging of the genital and pubic hair development according to Tanner's stages (**Table 1**).

Table 1 Pubertal stages in boys

Stage	*Pubic hair*	*Genital*
1	Absence of pubic hair	Childlike penis, testes, and scrotum (testes 2 ml)
2	Sparse, lightly pigmented hair, mainly at the base of the penis	Scrotum enlarged with early rugation and pigmentation; testes begin to enlarge (3–5 ml)
3	Hair becomes coarse, darker, more curled, and more extensive	Penis has grown in length and diameter; testes now 8–10 ml; scrotum more rugated
4	Hair adult in quality but distribution does not include medial aspect of thighs	Penis further enlarged with development of the glans; scrotum and testes (10–13 ml) further enlarged
5	Hair is adult and extends to thighs	Penis and scrotum fully adult; testes 15 ml and greater

Modified from Marshall WA and Tanner JM (1970) Variation in the pattern of pubertal changes in boys. *Archives of Disease in Childhood* 45: 13–23, with permission of the BMJ Publishing Group.

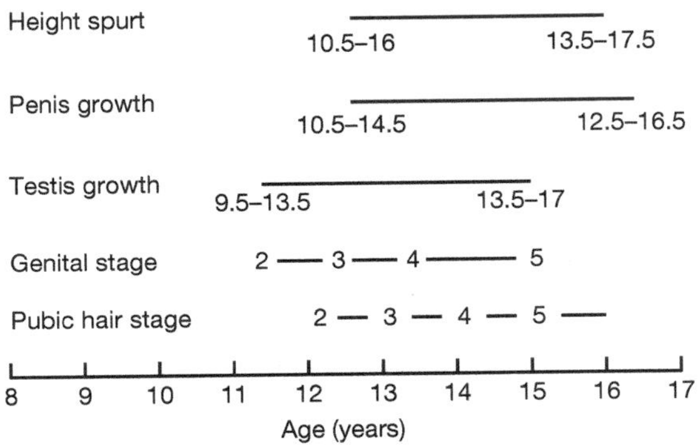

Figure 2 Diagram of the timing of the various components of puberty. The range of ages in which each parameter begins and is completed is shown for each bar. These data are from European children obtained in the late 1960s. There may be a slight trend for earlier onset of puberty since the 1970s. Data from Marshall WA and Tanner JM (1970) Variation in the pattern of pubertal changes in boys. *Archives of Disease in Childhood* 45: 13–23, with permission of the BMJ Publishing Group.

The majority of the extratesticular end organ events of puberty are secondary to the increased circulating levels of T and its metabolic products (DHT and E2). The penis and scrotum grow and become pigmented. As spermatogenesis advances, the testes increase in size from 1–2 ml at the outset to 15–35 ml in adulthood. Increase in testis size is an early sign of puberty. There is a progressive increase in facial, axillary, chest, abdominal, thigh, and pubic hair; frontal scalp hair regresses, and the voice deepens (**Figure 2**). Genital and secondary sexual hair development requires the conversion of T to DHT for its full effects. T, probably acting through E2, increases bone mass and stimulates the closure of the epiphyses. T acting directly also increases the muscle mass of pubertal boys and causes increases in hematocrit and red cell mass to adult male levels.

5.3.3.3(iii) Aberrations of timing of puberty

Delayed puberty in boys is usually defined as a temporary (physiological) form of hypothalamic-hypogonadotropic hypogonadism in which sexual development has not begun by age 13.5. Once initiated, puberty should be completed within approximately 4.5 years. Although delayed sexual maturation is an inevitable component of the prepubertal onset of hypogonadism, the majority of boys with delayed development have a constitutional delayed physiological clock and eventually attain full secondary sexual characteristics and function. Careful documentation of the changing physical findings and measurement of serum LH, FSH, and T may prove valuable clues to the beginning of puberty. Inquiring and testing for hyposmia or anosmia may indicate a common variant of congenital hypogonadotrophic hypogonadism (Kallmann's syndrome). Several genes have been implicated in the pathogenesis of idiopathic hypogonadotropic hypogonadism (IHH) (KAL1, FERRI, GnRHR, NELF, and GPR54; Raivio et al., 2007). A family history of delay in puberty may encourage patience and observation. The decision regarding how early to treat depends on the perceived degree of psychological stress associated with the maturational delay. The major concern about treatment is the early fusion of the epiphyses, which compromises optimal height. With proper dosing and monitoring of bone age, this is unusual because bone age is usually retarded in delayed puberty. In adolescent boys with delayed puberty and low levels of gonadotropins, periodic withdrawal of T treatment is used to determine if spontaneous puberty has occurred. Many adult men, diagnosed and treated for hypogonadotropic hypogonadism at ages 15–19, have proved to have normal reproductive function when taken off T therapy many years later. A recent report (Raivio et al., 2007) suggests that idiopathic hypogonadotropic hypogonadism identified after age 18 may also be spontaneously reversible with or without prior T therapy. GnRH-receptor loss-of-function mutations have been found in patients with IHH both with and without impaired olfactory function (Beranova et al., 2001).

Precocious puberty in boys is defined as the onset of pubertal (genital and secondary sexual) development beginning before age 9 (2.5 SD above the mean age of progression to stage 2). Unlike girls, sexual precocity in boys is usually associated with an underlying definable disorder. Sexual precocity can be subcategorized into true isosexual precocious puberty and incomplete isosexual precocity or pseudoprecocious puberty. The distinction is that true precocious puberty is associated with increases in GnRH-stimulated LH and FSH secretion (of hypothalamic–pituitary origin), whereas pseudoprecocious puberty is independent of GnRH stimulation of LH and FSH secretion. True precocious puberty is often associated with CNS disease (two-thirds of boys), including hypothalamic tumors, cysts, inflammatory conditions, and seizure disorders. The diagnosis is based on the finding of sexual precocity, inappropriately elevated serum LH levels, and associated elevations of serum T. Actuating mutations in the LH receptor may be a cause of precocious puberty in boys (Reiter and Norjavaara, 2005; Piersma et al., 2007). CNS visualizations by magnetic resonance imaging (MRI) can localize most lesions. Pseudoprecocious puberty is characterized by increased T with suppressed LH levels. Diagnosis includes human chorionic gonadotrophin-secretory tumors (e.g., testes, liver, hypothalamic, and pineal tumors), congenital virilizing adrenal hyperplasia (CAH), testicular testosterone-secreting neoplasms, and constitutively active LH-receptor mutations, resulting in uncontrolled T (testitoxicosis) secretion. The treatment of true precocious puberty is the removal of the CNS lesion if possible and treatment with GnRH analogs. The treatment of pseudoprecocious puberty depends on the cause, but includes glucocorticoids for CAH and ketoconazole (which suppresses steroidogenesis) with or without adrenal anti-androgens (e.g., spironolactone and flutamide).

5.3.3.4 Male senescence: Decreased T and other anabolic hormones

5.3.3.4(i) T deficiency in the elderly

Older men have significantly lower blood concentrations of T, other anabolic hormones (e.g., growth hormone), or prehormones (e.g., DHEA and DHEA-S; **Table 2**). Unlike in women, aging in men is not associated with an abrupt cessation of gonadal hormone secretion but rather a gradual decline, beginning as a young adult and progressing throughout life. Multiple cross-sectional and longitudinal studies have shown a progressive decrease in both total and free serum T levels with aging (**Figure 3**). Androgen deficiency, defined as serum T, free or bioavailable T serum levels below the reference range for young healthy adult males, occurs in an increasing rate for each decade of life after age 40. In the United States, men are projected to comprise 43% of the population ≥65 years of age by the year 2030. While the estimate differs among populations and thresholds vary in different studies, it has been estimated that 20% of healthy ambulatory men in their 60s and 30% in their 70s have lower T levels than the reference range of healthy younger men (Hijazi and Cunningham, 2005). Some, but not all, such older men with low serum T levels have symptoms or signs similar to those of hypogonadal younger men. These symptoms, such as loss of energy, depressed mood, decreased libido, erectile dysfunction (ED), decreased muscle mass and strength,

Table 2 Hormonal changes associated with aging[a]

LH/FSH/ testosterone	*ACTH–DHEA(S)*	*GHRH–GH–IGH axis*
↑LH[b], ↑FSH	No change in ACTH	↓GHRH message and receptor
↓T (↓ Leydig cells)	↓DHEA and DHEA-S	↓GH secretory pulses
↓Free T	↓DHEA and DHEA-S	↓Circulating GH
↑SHBG	Response to ACTH	↓Serum IGF-I

[a]ACTH, adreonocoticotropin; DHEA, dehydroepiandrosterone; DHEA-S, dehydroepiandrosterone sulfate; FSH, follicle-stimulating hormone; GH, growth hormone; GHRH, growth-hormone-releasing hormone; GnRH, gonadotropin-releasing hormone; IGF-I, insulin-like growth factor I; LH, luteinizing hormone; SHBG, sex-hormone-binding globulin; T, testosterone; up arrows indicate increase; down arrows indicate decrease.
[b]↓LH pulse amplitude and ↓ responsiveness to GnRH
Modified from Swerdloff RS and Wang C (2008). The testis and male sexual function. In: Gaedman L and Ausiello D (eds.) *Cecil Textbook of Medicine*, 23rd edn., pp. 1786–1791. Philadelphia, PA: W. B. Saunders, with permission from Elsevier.

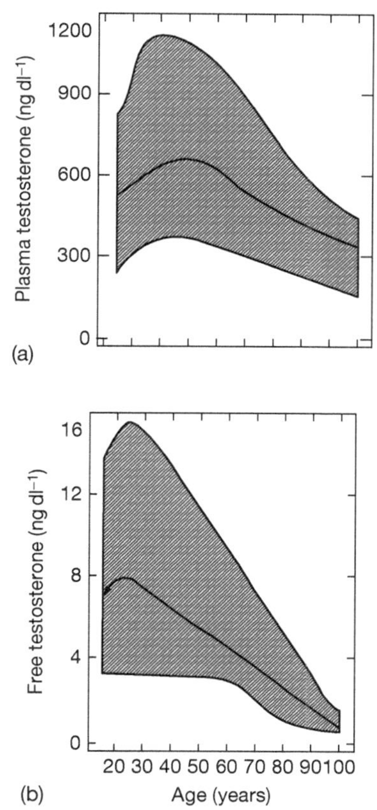

Figure 3 Relationship between (a) plasma testosterone levels and age and (b) free testosterone levels and age in normal males. Reproduced from Baker HWG, Berger HG, and de Kretser DM (1976) Changes in the pituitary–testicular system with age. *Clinical Endocrinology* 5: 349–372, with permission of Blackwell Publishing.

increased fat mass, frailty, osteopenia, and osteoporosis, are not specific for T deficiency (Kaufman and Vermeulen, 2005). Such decreases additively lead to impaired balance, mobility, and frailty. Furthermore, older men have increased body fat, particularly visceral fat, which may lead to the age-associated increase in the metabolic syndrome. Lack of specificity of symptoms and a validated instrument that can accurately quantitate hypogonadism has caused these symptoms to be attributed, by some clinicians and regulatory agencies, to normal aging. Thus, treatment of older men with hypogonadism has been a topic of considerable debate.

A number of short-term studies have demonstrated the beneficial effects of T replacement in elderly men with relatively low serum T level (Sih et al., 1997; Tenover, 1992; Snyder et al., 1999a,b). Long-term, placebo-controlled studies are in progress. T-replacement therapy, in most studies, decreases fat mass, increases lean body mass, and improves strength. Because ED in the older man is multifactorial, with impaired vasodilatory function in the penis predominating in many cases, T-replacement therapy in older men may enhance libido; however, ED is often not improved. In this situation, combined treatment with a PDE-5I and T may result in a satisfactory response (Greenstein et al., 2005; Rosenthal et al., 2006). Improved sense of well-being, mood, and energy levels are also generally observed after treatment with T (Wang et al., 1996, 2000). The Institute of Medicine in the US has reviewed the available information and argued that more data on efficacy of such treatment are desirable before determining the large-scale risk of T or other androgen-replacement therapy in elderly men (Liverman and Blazer, 2004).

5.3.3.4(ii) Adrenal deficiency of androgen precursors in older men

A marked decline in the circulating levels of adrenal androgens, especially DHEA and DHEA-S, has been recognized. Serum levels of DHEA and DHEA-S peak at approximately the third decade of life and then decline at approximately 2% per year, resulting in levels of 10–20% of baseline by age 80. This decline in DHEA and DHEA-S is not accompanied by a decrease in ACTH (Labrie et al., 1997).

DHEA is a precursor to true androgens, such as T and DHT, but does not bind to the AR itself. It is unclear whether DHEA binds to a unique nuclear receptor to initiate its action. Studies have been reported that DHEA administered to aging experimental animals and humans may improve the sense of well-being, reduce anxiety and depression, enhance memory, prevent development of cancer, decrease body fat, decrease the risk of cardiovascular disease, and provide other beneficial effects on immune function. Most studies in humans used oral doses of 1–5 mg day^{-1}. An oral dose of 50 mg kg day^{-1} increases T and DHT levels to or above the normal physiological range for women, but not men. Much higher doses of DHEA can increase T to male ranges, but at the expense of very high serum DHEA concentrations. The results of placebo-controlled trials in normal older adults revealed minimal benefit (Nair et al., 2006). In the United States, DHEA is available without prescription as a health supplement and is widely used, creating a situation in which large-scale multicenter, prospective, placebo-controlled trials are difficult to perform. There is no substantial reason to administer DHEA to older men who may have low serum DHEA levels.

5.3.4 T as a Hormone, Prehormone, and Paracrine Factor

T serves both paracrine (effects on the spermatogenic compartment and Wolffian duct) and endocrine functions. The paracrine actions are discussed in the section on control of spermatogenesis.

T exerts its effects at different end organs either through a direct action or after conversion to the active metabolite, DHT, by binding to the AR. T is aromatized to E2 which acts through the ERα or -β (**Figure 4**). Approximately, 4% of T is converted peripherally to DHT by two isoforms of an enzyme 5α-reductase. These two forms, named type I and type II, are each specified by distinctive genes. The conversion to DHT occurs in the skin, mediated by type I 5α-reductase and in the prostate stroma by type II 5α-reductase. Although only a small amount of T is converted to E2 in multiple end-organ tissues by aromatase, these low levels are adequate for biological effect because E2 is a potent hormone. Despite the relatively low conversion rate, more than 80% of the circulating E2 in adult men is the result of the aromatization of T in multiple organs. It should be noted that although the blood conversion rates of T to DHT and E2 are relatively small, the actual conversion rates in tissue compartments may be higher. Various end organs differ in their 5α-reductase and aromatase activity and in their

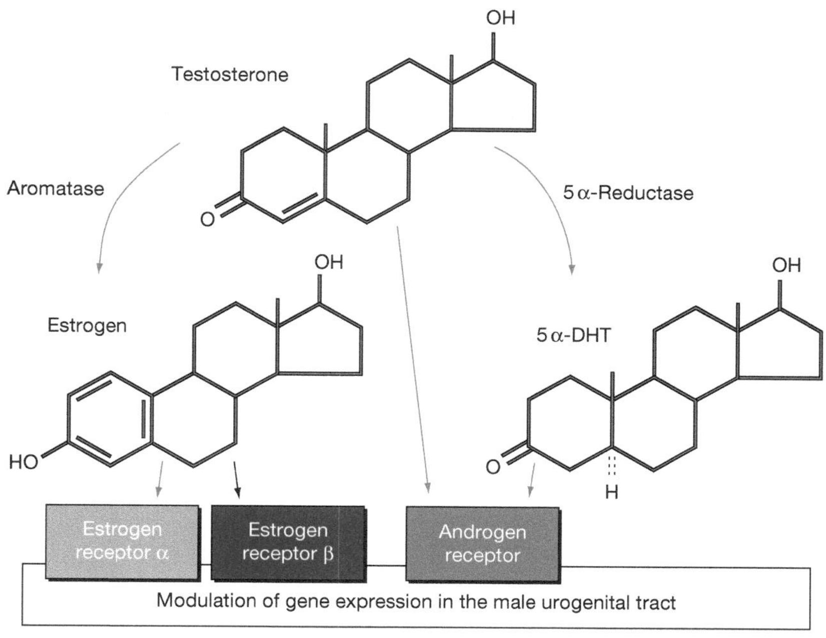

Figure 4 In the male urogenital tract, testosterone action is mediated either directly (androgen receptor), after conversion to estradiol (estrogen receptor α or β) or after conversion to DHT (androgen receptor). Reproduced from Kuiper GGJM, Carlquist M, and Gustafsson JA (1998) Estrogen is a male and female hormone. *Science and Medicine* 5: 36–45, with permission from Science and Medicine.

requirements for conversion of T to DHT for androgenic activity. DHT has greater biological potency than T due to its higher affinity for, and slower disassociation from, the AR. Congenital and acquired defects in these two enzymes as well as in the estrogen and ARs result in distinct syndromes with characteristic phenotypes.

5.3.5 Androgen Receptor

5.3.5.1 AR gene, protein structure, and regulatory proteins

The AR is a member of the nuclear receptor family, which includes the steroid, vitamin D, thyroid hormone, retinoic acid, and a number of orphan receptors. These receptors share certain structural features such as a centrally located DNA-binding domain, a C-terminal hormone-binding domain, and an N-terminal regulatory region. The central DNA-binding domain is the most highly conserved zinc-finger segment of the nuclear receptors and serves to mediate the contact between the receptor protein and its DNA targets (Umesono and Evans, 1989). The ligand hormone-binding domain is the portion of the receptor that binds the ligand (e.g., T) with high affinity. Although there is considerable homology in the family, relative specificity exists. Although the N-terminus of the AR is necessary for the full activity of the receptor, details of the critical elements of the regulatory region are still incomplete.

There are homopolymeric repeats in exon 1 of the N-terminus of the human AR gene. The length of the repeated segments of the AR varies considerably within the population, effects receptor function, and is associated with differing racial, ethnic, and familial susceptibility to disease (McPhaul et al., 1991; Edwards et al., 1992). Long glutamine (CAG) repeats are associated with lower AR activity. Examples of the clinical importance of the variations in the length of the glutamine homopolymeric repeats in the AR include a progressive neurologic disorder, spinal and bulbar muscular atrophy (expanded CAG repeats), and increased susceptibility to aggressive prostate cancer (short CAG repeats). Polymorphisms of the CAG repeats of the receptor influence the degree of modulation of androgens and the AR in hypogonadism (Zittsman and Nieschlag, 2007).

AR assays have been developed to measure the abundance, location, ligand binding, and function of the AR. Methodologies include ligand-binding assays, immunoblot and other immunobinding methods, and reporter-gene-linked assays.

The functional characteristics of the AR are dependent on its binding to ligand. The nonligand-bound AR exists in cells as a complex of associated proteins, including chaperones such as heat-shock proteins 70 and 90 (Fang et al., 1996). The binding of the ligand to the AR results in the disassociation from these proteins and the transformation of its 3D structure to allow its binding to the DNA-targeted sites. The process of receptor–ligand-induced effects on target organs is further modulated by the actions of coactivators, co-repressors, and co-integrators.

5.3.5.2 AR defects

Multiple deficits in the AR gene have been described. These include loss-of-function mutations that interrupt the open reading frame, DNA-binding domain, and hormone-binding domain. In each category, multiple, specific defects have been reported. Defects of the AR are not rare. The AR gene is located on the X chromosome, thus in the male, a mutation in only one allele is associated with phenotypic manifestations. There are surprisingly a relatively large number of patients demonstrating varying degrees of androgen resistance. The range of phenotypic abnormalities represents a wide spectrum. Complete androgen insensitivity is referred to as testicular feminization (none of the internal or external male structures develop) with feminization (breast development) occurring as a result of aromatization of T to E2. These patients appear, superficially, to be normally developed females with amenorrhea and the absence of body hair. Careful exam reveals testes located in the labia or abdominal cavity and the vagina is shortened with absence of uterus. In cases in which the AR function is less impaired, ambiguous genitalia with varying degrees of virilization are found (Reifenstein's syndrome, partial androgen insensitivity, and incomplete testicular feminization). Some are females with the internal genitalia of androgen insensitivity, but partial virilization of the external genitalia. Finally, there are more subtle defects resulting in a complete male phenotype with either gynecomastia or impaired spermatogenesis as the only described findings.

5.3.6 T Target Organs

Table 3 lists multiple target sites for T. The biological effects of androgen are mediated primarily by binding and activating the AR. The conversion of T to DHT results in a significant, local androgen-amplification process allowing effects in certain tissues that would not take place in the absence of 5α-reductase. The majority of the effects of T are believed to occur through genomic mechanisms as a result of specific binding of the dimerized AR complex to androgen-response elements resulting in gene transcription. New observations of direct androgen effects at a cell membrane by nongenomic processes suggest additional mechanisms for androgen action. There is much to learn about the mechanism of action of T through inherited defects such as 5α-reductase deficiency and aromatase deficiency. T's biological effects include the maintenance of normal mood, male sexual drive (libido) and behavior, secondary sexual characteristics (e.g., male hair distribution), muscle mass and strength, bone mineral density, lipid profile, prostate and seminal vesicle function, and a number of metabolic parameters. The role of T in cognition has not been fully explored.

Table 3 Androgen target organs

Brain
Pituitary
Reproductive organs
Prostate, seminal vesicles, epididymis
Testes
Penis
Breast
Skin
Hair follicles: sexual hair, scalp hair
Sebaceous gland
Upper respiratory septum
Larynx
Bone
Hematopoietic system
Muscle
Vascular system
Liver

5.3.7 Role of T in Normal Sexual Function and Erectile Physiology

Normal sexual function in men requires normal sexual desire (libido) and erectile, ejaculatory, and orgasmic capacity. The process is complex, involving cognitive, sensory, hormonal, autonomic neuronal, and penile vascular integrative actions for normal function. Defects occur at multiple levels. Although considerable progress has occurred in therapeutic options, an understanding of the normal physiology is essential for proper assessment and treatment of men with sexual dysfunction.

The brain is the integrative center of the sexual response system. It processes sensory input, stored fantasy information, purposeful thoughts, spontaneous nocturnal reflex activity, and hormonal signals (i.e., T) to create the hypothalamic-neuronal message that traverses the spinal cord to the thoracic, 9–12 sympathetic, and sacral, parasympathetic, outflow tracts. The nonadrenergic noncholinergic (NANC) autonomic-plexus nerves initiate the vasodilatation of the cavernosal arterial and corporal cavernosa sinusoids of the penis, through the release of local vasodilators, such as nitric oxide (NO) and VIP, and vasoconstrictors, such as endothelins, from the vascular endothelium and smooth-muscle cells of the sinusoids (**Figure 5**). A family of enzymes (NO synthases) regulates NO synthesis, which produces smooth-muscle dilatation through the activation of cyclase guanosine monophosphate (GMP) and modification of calcium flux. Cyclic GMP levels are rapidly reversible through inactivation by PDE. The neurogenic mechanisms leading to vasodilatation of the cavernosal arterioles and sinusoids lead to a rapid increase in penile blood flow and an expansion of the vascular channels; this, in turn, inhibits venous return through compression of the venous channels against the tunica albuginea and limits the drainage of the obliquely penetrating veins. Following orgasm, detumescence occurs, owing to less vasodilation (NO) and greater vasoconstrictive signals (α_2-adrenergic).

T seems to have its primary effect on erectile function by enhancing libido with secondary effects on penile NO-synthase activity. Libido is highly sensitive to T, thus explaining the preservation of erectile capacity in many men with partial androgen deficiency. In contrast, ED is common in men, including older men, despite normal serum T levels; the latter effect appears to be the result of impaired penile vasodilatory capacity. In rats, it has been shown that androgen increases NO-synthase activity in the penis (Garban et al., 1995;

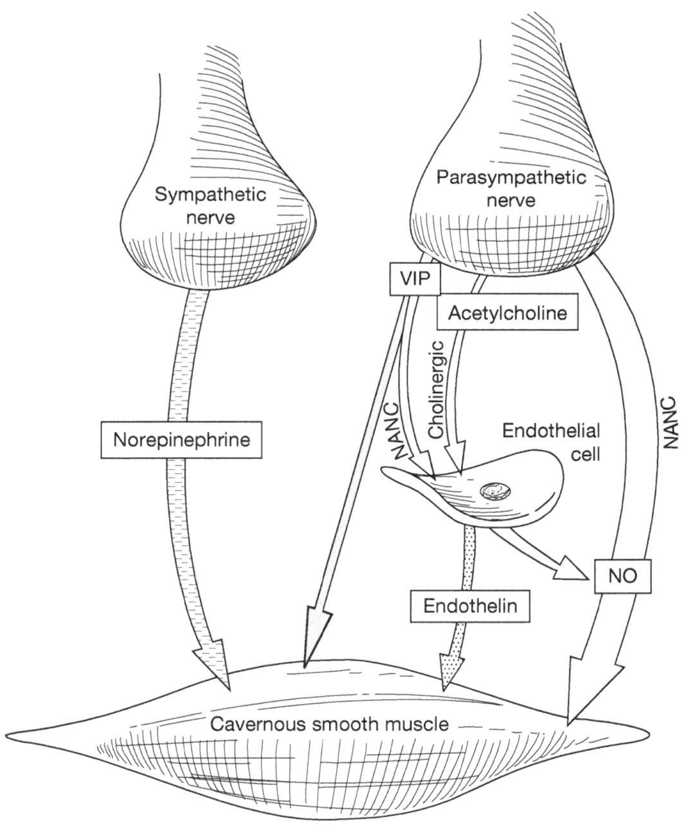

Figure 5 The interaction among cholinergic, adrenergic, and nonadrenergic, noncholinergic (NANC) neuronal pathways and their contribution to penile smooth muscle contraction (patterned arrows) and dilation (open arrows). NO, nitric oxide; VIP, vasoactive intestinal polypeptide. Reproduced from Lue TF (1998) Physiology of penile erection and pathophysiology of erectile dysfunction and priapism. In: Walsh PC, Retick A, Vaughn B, and Wein A (eds.) *Campbell's Urology*, 7th edn., p. 1164. Philadelphia, PA: W. B. Saunders, with permission from Elsevier.

Penson et al., 1996). It is not clear if a direct action of T on the erectile process in the penis is also present in humans. ED is often reversible through local (intracavernosal or transurethral) administration of potent vasodilators (prostaglandins, papaverine, and phentolamine; Giuliano and Rampin, 2000) or by oral administration of penile-specific PDE-5I inhibitors (i.e., sildenafil, tadalafil, and vardenafil; Bivalacqua et al., 2000; Meuleman et al., 2001; Brant et al., 2007). Combined androgen deficiency, with decreased libido and decreased penile responsiveness due to impaired NO-synthase activity, may be common in elderly men. With the availability of effective penile-vasodilatory medications to ensure erectile capacity, complaints of diminished libido may be effectively treated with androgen supplementation. Recent studies on T-deficient men with ED suggest that some men who fail ED treatment with either T or PDE-5Is alone may benefit from combined therapy (Greenstein et al., 2005; Rosenthal BD et al., 2006).

5.3.8 T Deficiency: Male Hypogonadism

Hypogonadism refers to patients with low circulating levels of T, although most androgen-deficient men are infertile. Primary hypogonadism indicates that the abnormality originates in the testis; secondary hypogonadism indicates a defect in the hypothalamus or pituitary, resulting in decreased gonadotropins (LH and FSH) and a secondary impairment of testicular function. Combined primary and secondary hypogonadism occurs with aging and a number of systemic diseases, such as alcoholism, liver disease, and sickle cell disease. Decreased androgen action mimicking androgen deficiency may occur in patients with AR defects (androgen resistance), postreceptor signaling abnormalities, and inability to convert T to the active metabolite DHT (5α-reductase abnormalities).

5.3.8.1 Etiologies

Many of the causes of primary and secondary hypogonadism are listed in **Tables 4** and **5**.

5.3.8.1(i) Primary testicular hypogonadism

Primary hypogonadism refers to a condition of androgen deficiency with or without infertility in which the pathologic process lies at the testis level. A list of common etiologies is included in **Table 4**.

5.3.8.1(i)(a) **Infectious and inflammatory disorders** After puberty, mumps is associated with clinical orchiditis in 25% of the cases and 60% of those affected become infertile. During acute orchitis, the testes are inflamed, painful, and swollen. After the acute inflammatory phase, the testes gradually decrease in size, although swelling can persist for months. The testes may return to normal size and function or undergo atrophy. Spermatogenic changes

Table 4 Causes of primary testicular failure and end organ resistance

Congenital disorders	Chromosomal disorders: Klinefelter's and related syndromes (i.e., XXY, XXY/XY, XYY, XX males)
	Testosterone biosynthetic enzyme defects
	Myotonia dystrophy
	Developmental disorders: prenatal diethylstilbestrol syndrome; cryptorchidism
Acquired defects	Orchitis: mumps and other viruses; granulomatous (i.e., tuberculosis, leprosy); HIV
	Infiltrative diseases (i.e., hemochromatosis, amyloidosis)
	Surgical, traumatic injuries, and torsion of testis
	Irradiation
	Toxins (i.e., alcohol, fungicides, insecticides, heavy metals, cotton seed oil, DDT, and other environmental estrogens)
	Drugs: cytotoxic agents; inhibitors of testosterone synthesis and androgen action (i.e., ketoconazole, cimetidine, flutamide, cyproterone, spironolactone); ethanol and recreational drugs
	Autoimmune testicular failure: isolated; associated with other organ specific disorders (i.e., Addison's disease, Hashimoto's thyroiditis, insulin-dependent diabetes)
	Systemic diseases (i.e., cirrhosis, chronic renal failure, sickle cell disease, AIDS, amyloidosis)
Androgen resistance syndromes	
5α-Reductase deficiency	

Modified from Swerdloff RS and Wang C (2008) The testis and male sexual function. In: Gaedman L and Ausiello D (eds.) *Cecil Textbook of Medicine*, 23rd edn., pp. 1786–1791. Philadelphia, PA: W. B. Saunders, with permission from Elsevier.

Table 5 Causes of hypogonadotropic hypogonadism

Idiopathic or congenital	GnRH deficiency	Isolated deficiency of GnRH: with anosmia (Kallmann's syndrome); with other abnormalities (Prader–Willi syndrome, Laurence–Moon–Biedl syndrome, basal encephalocele) Partial deficiency of GnRH (fertile eunuch syndrome)
	Multiple hypothalamic/pituitary hormone deficiency	
	Pituitary hypoplasia or aplasia	
Acquired	Trauma, postsurgery, postirradiation	Pituitary adenomas: prolactinomas and other functional and nonfunctional tumors
	Neoplastic	Craniopharyngiomas, germinomas, gliomas, leukemia, and lymphomas
	Pituitary infarction and carotid aneurysm	Sarcoidosis, tuberculosis, coccidiodomycosis, histoplasmosis, syphilis, abscess, histiocytosis X, and hemochromatosis
	Infiltrative and infectious diseases of hypothalamus and pituitary	
	Autoimmune hypophysitis	Anorexia nervosa, starvation, renal failure liver failure
	Malnutrition and systemic disease	
	Exogenous hormones and drugs	Antiandrogens, estrogens and antiestrogens, progestogens, glucocorticoids, cimetidine, spironolactone, digoxin, and drug-induced hyperprolactinemia (metoclopramide, tranquilizers, other dopamine antagonists, antihypertensives)

Modified from Swerdloff RS and Wang C (2008) The testis and male sexual function. In: Gaedman L and Ausiello D (eds.) *Cecil Textbook of Medicine*, 23rd edn., pp. 1786–1791. Philadelphia, PA: W. B. Saunders, with permission from Elsevier.

occur more often and earlier than Leydig cell dysfunction. Thus, patients with postorchitic infertility may have normal T and LH levels with increased serum FSH. With time, elevated LH and lowered serum T levels may appear. Leprosy may also cause orchiditis and gonadal insufficiency. HIV infection is often associated with hypogonadism, which can be either hypogonadotropic or hypergonadotropic. Hemochromatosis and amyloidosis are examples of infiltrative diseases of the testis that can result in hypogonadism.

5.3.8.1(i)(b) Trauma The exposed position of the testes in the scrotum makes it particularly susceptible to injury. Surgical injury during scrotal surgery for hernias, varicocele, and vasectomy can result in permanent testicular damage.

5.3.8.1(i)(c) Irradiation Irradiation to the testes due to accidental exposure in the treatment of an associated malignant disease produces testicular damage.

5.3.8.1(i)(d) Drugs Chemotherapy, in particular alkylating agents such as in busulfan, for malignant disorders frequently leads to irreversible germ cell damage. Toxins may also directly damage the testes. Many agents, such as fungicides and insecticides (e.g., dibromochloropropane, DBCP), heavy metals (lead, cadmium, etc.), and cottonseed oil (gossypol), produce damage to the germ cells. Leydig cells are relatively less susceptible to most chemotherapeutic drugs than the Sertoli and germ cells. Serum T levels are usually normal despite infertility in the exposed men. Some medications may interfere with T biosynthesis (e.g., ketoconazole, spironolactone, and cyproterone). Ethanol, independent of its effect in causing liver disease, inhibits T biosynthesis. Marijuana, heroin, methadone, medroxyprogesterone acetate, and estrogens all lower T, but mainly by decreasing the pituitary secretion of LH.

5.3.8.1(i)(e) Autoimmune testicular failure Antibodies against the microsomal fraction of the Leydig cell may occur either as an isolated disorder or as part of a multiglandular disorder involving, to variable degrees, the thyroid, pituitary, adrenals, pancreas, and other organs.

5.3.8.1(i)(f) Testicular defects associated with systemic diseases Abnormalities of the hypothalamic–pituitary–testicular axis occur in a number of systemic diseases. These include liver failure, renal failure, severe malnutrition, sickle cell anemia, advanced malignancies, cystic fibrosis, and amyloidosis. Approximately one-half of men undergoing chronic hemodialysis for renal failure experience decreased libido, infertility, and impotence. The effects of cirrhosis of the liver on testicular function are complex and may be either independent of, or associated with, the direct toxic effects of the continued use of alcohol. Gynecomastia, testicular atrophy, and impotence are concomitant signs of cirrhosis. Decreased spermatogenesis with peritubular fibrosis occurs in at least 50% of the patients. In contrast to the decrease in serum T levels, E2 levels are usually elevated. This results in an increased ratio of serum E2 to T with an increased proclivity for gynecomastia. Patients with sickle cell anemia often have impaired testicular function. Boys with sickle cell anemia may have impaired sexual maturation, and men with this disorder are often infertile. The defect in sickle cell anemia seems to be ischemic in origin, probably with accelerated apoptosis; it may occur either at the testicular or at the hypothalamic–pituitary level.

5.3.8.1(ii) Secondary gonadal insufficiency (hypogonadotropic hypogonadism)

Hypogonadotropic hypogonadism represents a deficiency in the secretion of gonadotropins (LH and FSH) due to an intrinsic or functional abnormality in the hypothalamus or pituitary glands. Such disorders result in secondary Leydig cell dysfunction (**Table 5**). The clinical manifestations depend on the age of onset of the disorder.

5.3.8.1(ii)(a) **Congenital hypogonadotrophic disorders** These include Kallmann's syndrome (hypogonadotropic hypogonadism and anosmia or hyposmia) and obesity-associated syndromes such as Prader–Willi syndrome and Laurence–Moon–Biedel syndrome. The molecular basis of IHH is complex with many known genetic mutations; the three most common gene mutations are in KAL1, FGFR1, and GnRHR (Bhagavath and Layman, 2007).

5.3.8.1(ii)(b) **Acquired hypogonadotropic disorders and functional disorders** Anorexia nervosa and weight loss are examples of functional defects resulting in low serum T levels. Anorexia nervosa, predominantly a disorder of adolescent girls, is characterized by excessive weight loss as a result of dietary restriction or bulimia. Occasionally, this disorder is seen in men, but in that instance this usually implies a variant of a more severe psychiatric disorder. Men and women present with manifestations of hypogonadotropic hypogonadism. Starvation from other than a psychologic basis may also reduce gonadotropic secretion, although females seem more susceptible to this disorder. Although strenuous exercise commonly produces reproductive dysfunction in female athletes (long-distance runners and dancers), it has minimal effects on testicular function in men. Severe stress and systemic illness also lower gonadotropin and T levels. Structured hypothalamic–pituitary disorders include neoplastic, granulomatous, infiltrative, and post-traumatic lesions in the region of the hypothalamus and pituitary. Prolactinomas present differently in men and women. Unlike in women in whom small tumors are detected early because of symptoms of amenorrhea and galactorrhea, in men they are usually large (greater than 1 cm in diameter, macroadenomas) by the time of their detection. It is unclear whether the large size of the adenoma at the time of presentation in men is due to the late diagnosis caused by the failure of patients and physicians to appreciate the early signs or to more rapid growth of these tumors in men. Male patients with prolactin-secreting macroadenomas usually present with hypogonadism, ED, and visual manifestations due to suprasellar extension. Large nonprolactin-secreting pituitary tumors (growth hormone, GH; ACTH; glycopeptide; and null cell) may also produce gonadotropin insufficiency from damage of the adjacent normal pituitary gland, resulting in decreased serum LH and T levels.

5.3.8.1(iii) Androgen resistance (androgen-sensitive end organ deficiency)

Certain conditions have clinical phenotypes mimicking T deficiency in the absence of lowered T levels. These are either drug-induced (anti androgens) or congenital defects in the AR, postreceptor defects, or 5α-reductase deficiency.

5.3.8.2 Clinical manifestations of hypogonadism: Clinical history and physical examination

The medical history should focus on testicular descent, pubertal development, shaving frequency, changes in body hair, and present and past systemic illnesses. A complete sexual history includes changes in libido; erectile and ejaculatory functions; and frequency of masturbation, coital activity, and fertility (including that of the present and previous partners). Information should be obtained on previous orchitis, sinopulmonary complaints, sexually transmitted diseases, human immunodeficiency virus (HIV) status, genitourinary infections, and previous surgical procedures that might have affected the reproductive tract (e.g., vasectomy, hernia repair, prostatectomy, and varicocele ligation). Social history should include tobacco, alcohol intake, and drug abuse. Medication and drug history should include any agent that could affect hormonal, spermatogenic, and erectile function. These include recreational drugs; anabolic steroids; psychiatric, antihypertensive, anti androgenic, cytotoxic, and alternative medicine therapies; environmental toxins; and exposure to heat (including saunas and Jacuzzis) and radiation.

The generalized physical examination is supplemented by height and span measurements; assessment of muscle mass and adiposity; characterization of facial, pubic, and body-hair distribution; presence of acne and facial wrinkling; breast examination for gynecomastia; measurement of penile length and urethra integrity; digital rectal prostate examination; and visual-field assessment. The scrotal examination should include an assessment of midline fusion (e.g., bifid scrotum and hypospadias); measurement of testes size (a ruler will suffice, but a Prader or Takihara orchidometer is preferred) and consistency; presence of intratesticular masses; abnormalities of the epididymis; bilateral presence of a vas deferens; and presence of varicoceles, hydroceles, or hernias. The normal testis-size range is from 3.6 to 5.5 cm in length, 2.1 to 3.2 cm in width, and 15 to 35 ml in volume in white and black men. Asian men have a slightly smaller mean testicular size. A decrease in testicular volume usually implies decreased spermatogenic cells because the tubular tissue accounts for more than 80% of testis volume.

5.3.8.3 Laboratory tests in assessment of hypogonadism

Because a strong diurnal rhythm in T secretion results in the highest serum levels in the morning hours and lowest levels in the evening, the measurement of T, LH, and FSH is routinely determined from morning blood samples. Liquid chromatography tandem mass spectroscopy (LC/MS/MS) methodology for measuring serum T is now the gold standard (Wang et al., 2004). Measurement of serum unbound (free) T (by the equilibrium dialysis method) or bioavailable (free plus albumin-bound) T may be helpful in the diagnosis of hypogonadism when the total T levels are borderline low. This is particularly true in the evaluation of older men with increased sex hormone-binding globulin levels (Swerdloff and Wang, 2008). Elevated LH and FSH distinguish primary from secondary hypogonadism (both have low serum T levels).

Serum prolactin levels should be measured in all cases of hypogonadotrophic hypogonadism, pituitary mass lesions, and galactorrhea. DHT is measured in cases of abnormal differentiation of the genitalia and when 5α-reductase deficiency is suspected. Serum E2 should be measured in cases of gynecomastia. The assessment of other T precursors and products may be required in special circumstances, including suspected congenital enzyme defects. Semen analysis is a cornerstone of the laboratory examination for infertility.

5.3.8.4 Treatment of androgen deficiency

5.3.8.4(i) Indications

The main medical indication for androgen-replacement therapy is male hypogonadism (**Table 6**). The diagnosis is based on clinical symptoms and signs and a reduced serum T level. If a morning serum T level is repeatedly less than 250 ng dl^{-1} (8.5 nmol l^{-1}), the patient is most probably hypogonadal and T replacement is indicated. If the serum T level is between 250 and 300 ng dl^{-1} with normal serum LH levels, the patient may not be hypogonadal and androgen replacement may not improve the symptoms (e.g., sexual dysfunction; Bhasin et al., 2006).

5.3.8.4(ii) Contraindications to T therapy

Absolute contraindications for androgen-replacement therapy include carcinoma of the prostate and the male breast. These cancers are androgen dependent for growth and proliferation. Androgens should be used with caution in older men with enlarged prostates and urinary symptoms.

5.3.8.4(iii) Androgen preparations

Table 7 lists commonly used androgen preparations. T esters such as T enanthate (or cypionate) are the most widely used preparations in the United States

Table 6 Indications for androgen therapy

Androgen deficiency (hypogonadism)
Microphallus (neonatal)
Delayed puberty in boys
Low testosterone levels in elderly men
Angioneurotic edema
Others possible uses or under investigation: hormonal male contraception
Wasting disease associated with cancer, HIV, or chronic infection
Postmenopausal and oophorectomized female
Aging men with low testosterone levels

Modified from Swerdloff RS and Wang C (2008) The testis and male sexual function. In: Gaedman L and Ausiello D (eds.) *Cecil Textbook of Medicine*, 23rd edn., pp. 1786–1791. Philadelphia, PA: W. B. Saunders, with permission from Elsevier.

Table 7 Androgen preparations

Route	*Preparation*	*Dose and frequency of administration*
Oral[a]	Testosterone undecanoate (available in Canada, Mexico, Europe, Asia)	40–80 mg 2–3 times per day
	Bucal testosterone tablets	30 mg tablets twice per day
Implants	Testosterone implants	200 mg pellets, three inserted once every 4–6 months
Transdermal	Scrotal patch	One patch delivers 4 or 6 mg testosterone per day
	Nonscrotal patch	
	Androderm	Two patches each deliver 2.5 mg testosterone per day or one patch delivers 5 mg testosterone per day
	Testoderm TTS	One patch delivers 5 mg testosterone per day
	Gel: AndroGel, (Testogel)	5–10 g 1% hydroalcoholic gel per day
	Testim	5–10 g 1% hydroalcoholic gel per day
	Injectables	
	Testosterone Enanthate and cypionate	75–100 mg per week
	Testosterone undecenoate (available in Europe as Nebido)	150–200 mg per 2 weeks 1000 mg i.m. every 3 months

[a]Oral modified 17α-alkylated androgens such as methyltestosterone, fluoxymesterone, oxymethalone, stanozool, and oxandrolone are not recommended for use in the treatment of androgen deficiency states because of potential hypertoxicity and adverse effects on serum lipids.

Modified from Swerdloff RS and Wang C (2008) The testis and male sexual function. In: Gaedman L and Ausiello D (eds.) *Cecil Textbook of Medicine*, 23rd edn., pp. 1786–1791. Philadelphia, PA: W. B. Saunders; and Schubert M, Minnemann T, Heber D, et al. (2004) Intramuscular testosterone undecancate pharmacokinetic aspects of a novel testosterone formulation during longterm treatment of men with hypogonadism. *Journal of Endocrinology and Metabolism* 89: 5429–5434, with permission from Elsevier.

and throughout the world. The recommended dose is 150–200 mg administered intramuscularly once every 2–3 weeks.

Modified 17α-alkylated androgens, which are available in oral preparations, are not recommended for androgen replacement. These agents may lead to abnormalities in liver function tests, marked decreases in high-density lipoprotein (HDL) cholesterol, and increases in total cholesterol levels compared to the T esters. Orally active T undecenoate is not available in the United States, but is used in Canada, Europe, and other places in the world. This ester is absorbed into the lymphatics and has variable bioavailability.

Implants are pellets of crystalline T. The serum T levels are maintained in the physiological range for 4–6 months. Implants are not popular in the United States, but are widely used in Australia and the United Kingdom.

Transdermal skin patches represent the most recent development in androgen delivery systems. The nonscrotal patch(es) deliver 5 or 6 mg of T per day, which is the physiological production rate. A T gel preparation, which is applied to the skin once daily, has been developed and approved for use in the United States. Its pharmacokinetics results in fairly consistent blood levels throughout a 24-h period. It can be administered in graded amounts to provide desired blood levels with the normal range (Swerdloff et al., 2000).

5.3.8.4(iii) Benefits versus risks of androgen therapy

Table 8 shows the benefits and potential side effects of androgen treatment. In hypogonadal men, androgen replacement leads to the development and maintenance of secondary sexual characteristics, anabolic effects on muscle and bone, and improvement of libido, sexual dysfunction, and mood. It has less effect on ED than on libido in older men.

5.4 Spermatogenesis and Sperm Transport

The spermatogenic compartment consists of the Sertoli and germ cells and is intimately interactive with the interstitial compartment (**Figure 6**). The Sertoli cells bridge the entire space between the basement membrane and the lumen of the tubules. They are the target of androgenic and FSH stimulation of spermatogenesis and also the source of a multitude of paracrine regulators of spermatogenesis and gonadotropin secretion (e.g., inhibin and activin).

Table 8 Benefits and risks of androgen therapies

Benefits	*Risks*
Development or maintenance of secondary sex characteristics	Fluid retention Gynecomastia
Improvement of libido and sexual function	Acne or oily skin
Increase in muscle mass and strength	Increase in hematocrit
Increase in bone mineral density	Decrease in HDL cholesterol
Decrease in body and visceral fat	
Improvement of mood	Sleep apnea
Effect on cognition (?)	Prostate diseases (lower urinary track syndrome): benign prostate hyperplasia; growth of invasive carcinoma of prostate
Effect on quality of life (?)	

Modified from Swerdloff RS and Wang C (2008) The testis and male sexual function. In: Gaedman L and Ausiello D (eds.) *Cecil Textbook of Medicine*, 23rd edn., pp. 1786–1791. Philadelphia, PA: W. B. Saunders, with permission from Elsevier.

Germ cell maturation is dependent on the proper hormonal (FSH) and paracrine (T) milieu for proliferation to occur. Not all germ cells reach maturity. The spontaneous death of certain germ cells is a constant feature of germ cell homeostasis. In fact, considerable data indicate that the major effects of both T and FSH are to limit the amount of germ cell death (apoptosis).

5.4.1 Hormonal Regulation of Spermatogenesis

The hormonal control of spermatogenesis has been studied since the 1920s when its dependence on pituitary gonadotropins was first described (Smith, 1927; Greep et al., 1936). This process is thought to be primarily under the control of the pituitary gonadotropins FSH and LH (via the stimulation of T) and by poorly defined paracrine and other interactions between Sertoli and germ cells (Steinberger, 1971; Sharpe, 1994; Zirkin et al., 1994). Sertoli cells possess receptors for both FSH and androgen, and it is likely that these hormones exert their stimulatory effects on Sertoli cells, which in turn results in the stimulation of intratubular factors for the survival of germ cells through a paracrine mechanism (Jegou, 1993; Parvinen, 1993; Spiteri-Grech and Nieschlag, 1993; Pescovitz et al., 1994; Sharpe, 1994). Despite the considerable attention that hormonal control of spermatogenesis has received, the specific role and relative contributions of FSH and T on the regulation of spermatogenesis is debatable.

5.4.1.1 Gonadotropins and androgen regulation of spermatogenesis

T levels are much higher in the testes than in the serum. The T produced by the Leydig cells under the stimulation of LH is not only secreted into the system circulation but diffuses the interstitial space into the tubular compartment resulting in high intratesticular levels. Previous studies, mainly from rodents, have shown that quantitatively normal spermatogenesis – assessed by measurements of homogenization-resistant advanced (steps 17–19) spermatids – can be restored by the exogenous administration of T alone in relatively high dosages in adult rats made azoospermic by treatment with polydimethylsiloxane implants of T and E2 (Awoniyi et al., 1989a), by active immunization against either GnRH or LH (Awoniyi et al., 1989b, 1990), or by treatment with GnRH antagonist (Rea et al., 1986; Bhasin et al., 1988). These results of quantitative maintenance or complete restoration of spermatogenesis by T alone in rats in the absence of both radioimmunoassayable LH and FSH suggest that FSH may not be essential for the regulation of spermatogenesis in the adult rat. However, because T supplementation increases both the serum concentrations and pituitary content of FSH in GnRH-antagonist treated rats, the observed quantitative maintenance of spermatogenesis in these rats (Rea et al., 1986; Bhasin et al., 1988) cannot be attributed with certainty to T alone.

Other data suggest a significant role for FSH even in male rodents. Studies have, further, shown that spermatogenesis is not quantitatively restored in GnRH-immunized rats (McLachlan et al., 1994), in gonadotropin-deficient (hpg) mice (Singh et al., 1995), and after hypophysectomy (Awoniyi et al., 1990) even when rats received the larger amount of T used in the earlier studies. These studies further emphasized the need for both FSH and T in the restoration of spermatogenesis (Awoniyi et al., 1990; Sun et al., 1989; Santulli et al., 1990; Kerr et al., 1992).

Sinha Hikim and Swerdloff (1995) showed that short-term recombinant human FSH replacement to GnRH-antagonist-treated rats attenuated the GnRH-antagonist-induced reduction in total germ cell numbers and pachytene spermatocytes and the number of advanced spermatids (steps 17–19), and increased the

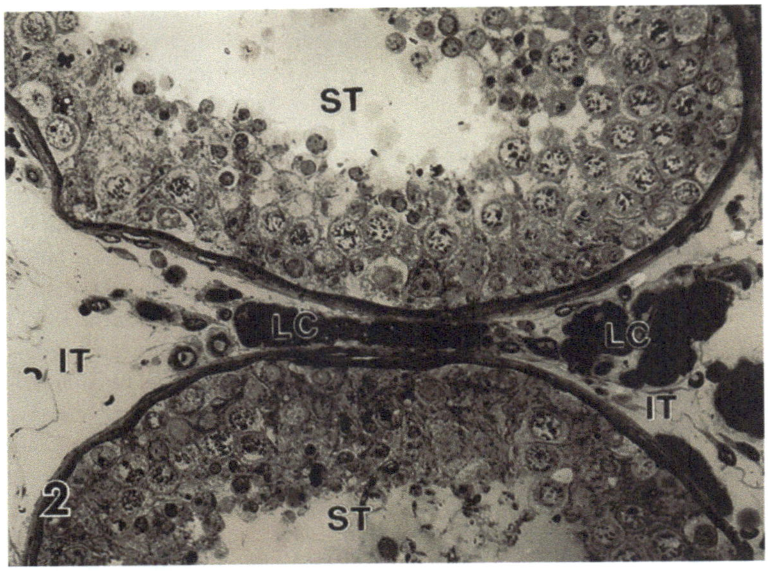

Figure 6 Light micrograph of the glutaraldehyde-fixed epoxy-embedded testicular section from a normal human male showing seminiferous tubules (ST) and interstitium (IT). The seminiferous tubules contain Sertoli cells (S) and germ cells at various phases of maturation. The interstitium consists of Leydig cells (LC), blood vessels, and lymphatic space. Reproduced from Swerdloff RS and Wang C (2000) The testis and male sexual function. In: Goldman L and Bennet JC (eds.) *Cecil Textbook of Medicine*, 21st edn., p. 1307. Philadelphia, PA: W. B. Saunders, with permission from Elsevier.

number of B spermatogonia available for entry into meiosis and to maintain the number of preleptotene spermatocytes throughout the treatment period. The observed beneficial effects of recombinant human FSH in spermatogenesis in GnRH-antagonist-treated rats are most likely not due to the stimulation of Leydig cell function (via paracrine interaction between the Sertoli and Leydig cells) because the addition of FSH to GnRH antagonist had no discernible effect on intratesticular or plasma T levels, accessory sex-organ weight, and the total volume of the Leydig cells when compared with GnRH antagonist alone.

Mice deficient in FSH-β exhibited a striking decrease in testis weight, seminiferous tubule volume, and epididymal sperm number (up to 75%) compared to their littermate controls (Kumar et al., 1997). This 75% reduction in the epididymal sperm number is similar to the reported 76% decrease in the transformation of round-to-elongated spermatids following immunoneutralization of FSH in the adult rat (Vaishnav and Moudgal, 1994). Thus, the reduction in the epididymal sperm number in selective FSH-deficient mice is most likely attributed to a decrease in the number of elongated spermatids during spermiogenesis. However, the absence of any apparent fertility defect, despite a 75% reduction in the epididymal sperm number, in these mice suggests that there are far more sperm produced in the adult mice than required to achieve fertility. The quantitative, if not absolute, importance of FSH in the regulation of spermatogenesis was, however, reinforced by Allan et al. (2004) using generically modified mouse models to selectively dissociate FSH effects from those of LH effects on spermatogenesis. These models used two distinct strategies to define FSH actions, which include a selective gain-of-function by autonomous pituitary-independent transgenic FSH (tg-FSH) or ligand-independent activated FSH-receptor expression (tg-FSHR$^+$) in gonadotropin-deficient hpg mice and an LH-receptor-deficient mouse. Stereological evaluation revealed that tg-FSH or tg-FSHR$^+$ fully restored Sertoli cell numbers in hpg mice. Selective induction of FSH activity increased total spermatogonia and spermatocyte numbers in hpg mice up to 57% and 44%, respectively, of values measured in normal testes. In addition, FSH alone was able to increase in round and elongated spermatid numbers in hpg testes to 16% and 6%, respectively, of normal values but failed to produce mature sperm. These data suggest a definite role for FSH in stimulating spermatogonial proliferation and meiotic germ cell development and a limited role in postmeiotic germ cell development. Also, consistent with a role for FSH in spermatogenesis, FSH-receptor knockout (FORKO) males had a reduced fertility and displayed small testes and marked loss of round and elongated spermatids (Krishnamurthy et al., 2000, 2001). But as expected, LH (T) is required for the successful completion of spermatogenesis and full fertility (Sinha Hikim et al., 2005; Kumar, 2005, 2007; O'Bryan and de Kretser, 2006). The latter point was subsequently underscored by the development of an LHβ-subunit KO mouse which was infertile and displayed the blockage of spermatogenesis at the round spermatid stage (Ma et al., 2004). Together, these data indicate that meiosis can be stimulated by either FSH or androgen, but the post-meiotic maturation of germ cells is strictly androgen dependent. The development of these genetically altered mice has allowed a complication-free dissection of the relative importance of FSH and LH in the regulation of spermatogenesis.

The role of FSH in the regulation of spermatogenesis is also well documented in primates. For example, administration of exogenous T via SILASTIC brand implants in adult macaque monkeys for 20 weeks induces azoospermia in some animals and variable degrees of spermatogenic suppression in others (Narula et al., 2002). Interestingly, such variability in T-induced spermatogenic suppression is not associated with differences in residual intratesticular androgens, LH, or inhibin B levels, but rather is associated with differences in the degree of FSH suppression between azoospermic and nonazoospermic animals (Narula et al., 2002). These results suggest that FSH is a key factor in supporting spermatogenesis in monkeys.

T treatment of healthy men suppresses both gonadotropins and intratesticular T also induces azoospermia in some individuals and variable degrees of suppression in others (World Health Organization Task Force on Methods for the Regulation of Male Fertility, 1990, 1996; Zhengwei et al., 1998). If these T-treated men are supplemented with LH or human chorionic gonadotropin (hCG; Matsumoto et al., 1984, 1986), their spermatogenesis recovers qualitatively; the sperm count remains suppressed (25–50 million/ml) from the pre-treatment concentrations (75–100 million/ml). FSH supplementation also stimulates spermatogenesis, though not quantitatively, in these men (Matsumoto et al., 1983). Additional data on the relative roles of FSH and LH in regulating spermatogenesis come from a recent prospective, randomized, three-arm, 6-week study using a suppress-and-repress model (Matthiesson et al., 2006). In that study, endogenous gonadotropins were suppressed using a combined androgen/progestin male contraceptive regimen together with their selective maintenance via administration of recombinant human (rh) FSH or hCG as an LH substitute. Either gonadotropin is able to quantitatively maintain spermatogenesis. FSH appears to be potentially more effective than LH (intratesticular androgen) in maintaining pachytene spermatocyte numbers. Given that round spermatid numbers were not different between FSH- and hCG-treated groups, despite the latter having reduced pachytene spermatocytes numbers, LH (intratesticular T) could be more effective in the completion of meiosis, possibly through improved conversion of pachytene spermatocytes to round spermatids. Available data of men with a mutation in the gene encoding either the FSH-R (Tapanainen et al., 1997) or the FSHβ-subunit (Phillip et al., 1998; Lindstedt et al., 1997) further provided an opportunity to evaluate the role of FSH in human spermatogenesis. For example, men homozygous for an inactivating mutation of FSH-receptor gene result in variable suppression of both spermatogenesis and fertility (Tapanainen et al., 1997). Men with markedly impaired secretion of FSH caused by a homozygous mutation in the gene for FSH β-subunit (Phillip et al., 1998; Lindstedt et al., 1997) have azoospermia and severely reduced testis size. On the basis of these two sporadic cases, it was proposed that FSH is an absolute requirement for spermatogenesis in men. However, it is more likely that some other conditions, in addition to FSH deficiency, cause the azoospermia, since one of the men was resistant to FSH therapy, and the other also had low T and high LH, indicating primary Leydig cell failure (Themmen and Huhtaniemi, 2000).

5.4.1.2 Gonadotropins and androgen regulation of programmed germ cell death

Programmed cell death (apoptosis) is an evolutionarily conserved cell-death process that plays a major role in normal development, homeostasis, and aging of multicellular organisms. In adult mammals, including humans, germ cell death is (as are cellular proliferation and differentiation) a process of fundamental importance and plays a critical role in determining the quantitative degree of sperm output (Sharpe, 1994). A growing body of evidence demonstrates that both spontaneous (during normal spermatogenesis) and accelerated germ cell death triggered by the withdrawal of gonadotropins or intratesticular T occur almost exclusively via apoptosis in rodents (Swerdloff et al., 1998b; Sinha Hikim and Swerdloff, 1999). The authors have extended the experimental paradigm from rodents to monkeys (Lue et al., 2006; Jia et al., 2007), and more recently, in humans (Wang et al., 2007) and demonstrated that germ cell apoptosis plays an important role in the organized regression of spermatogenesis after hormone deprivation and/or testicular hyperthermia. This mode of cell death serves as a quality-control system for proofreading the production of normal sperm (Brash, 1996). Therefore, any deregulation of the apoptotic process during spermatogenesis would lead to defective sperm production. While increased germ cell apoptosis could potentially lead to infertility, decreased cell death involving spermatogonia could do the same by disrupting germ cell homeostasis due to accumulation of premeiotic cells. Ablation of the Bax gene by homologous recombination results in male sterility due to accumulation of atypical premeiotic germ cells but with accelerated apoptosis of mature germ cells leading to complete cessation of sperm production (Knudson et al., 1995). Consistent with the proapoptotic role of Bax, male Bax-deficient mice showed marked accumulation of early germ cells up to the B spermatogonia/preleptotene cells due to prevention of the early wave of spermatogonial apoptosis. The consequence of this is defective spermatogenesis in the adult. Thus, the early wave of spermatogonial apoptosis may regulate the ratio of germ cells to Sertoli cells, thereby ensuring that adult Sertoli cell function is not compromised by excessive germ cells. These results also suggest that a proper balance between germ cell proliferation and death is critical for normal spermatogenesis.

The signaling events leading to apoptosis can be divided into two major pathways, involving either mitochondria (intrinsic) or death receptors (extrinsic). Recently, using a rat model of hormone deprivation by a potent GnRH-antagonist treatment, the authors have demonstrated the involvement of the intrinsic pathway signaling, characterized by perturbation of the BAX/BCL-2 rheostat in the mitochondria, cytosolic translocation of cytochrome c and DIABLO, and activation of the initiator caspase 9 and the executioner caspase 3, and poly (ADP-ribose) polymerase (PARP) cleavage in male germ cell apoptosis after hormone deprivation (Vera et al., 2006). In additional studies, using the generalized lymphoproliferation disease (*gld*) mice that harbor a loss-of-function of FAS ligand (FASL), we have shown that germ cells from wild-type (WT) and the FASL-defective *gld* mice are equally sensitive to apoptosis induced by hormone withdrawal, thus suggesting that the FAS signaling system may be dispensable for hormone deprivation-induced apoptosis (Vera et al., 2006). p38 mitogen-activated PK activation and induction of inducible NOS represent essential upstream signaling events that promote male germ cell apoptosis in rats, monkeys, and in humans after hormone deprivation (Vera et al., 2006; Jia et al., 2007).

5.4.1.3 Gonadotropins and androgens as germ cell survival factors

Although the requirement of the pituitary gonadotropins LH and FSH and testicular T for optimal regulation of spermatogenesis has been well documented, studies have also demonstrated the important role of these hormones as germ cell survival factors. Tapanainen et al. (1993) have shown the induction of germ cell apoptosis in the testes of immature rats after short-term (up to 4 days) hypophysectomy and its prevention by either gonadotropin or T. However, hCG and T were not as effective as FSH in preventing hypophysectomy-induced germ cell apoptosis. Furthermore, because FSH supplementation also increases LH-receptor content, mRNA levels, and the steroidogenic capacity of the Leydig cells in immature hypophysectomized rats (Vihko et al., 1991), the observed protective effect of FSH on germ cell apoptosis in these animals cannot be attributed with certainty to FSH alone. The involvement of apoptosis in specific germ cells undergoing degeneration has been recognized in the adult rat after gonadotropin deprivation by GnRH antagonist treatment (Sinha Hikim et al., 1995). Treatment with GnRH antagonist for 5 days resulted in selective activation of germ cell apoptosis involving preleptotene and pachytene spermatocytes, step-7 spermatids, and step-19 spermatids; such apoptotic germ cells are rarely if ever observed in a normal rat. A subsequent study showed that the rate of apoptosis induced by the GnRH antagonist was intimately associated with overall germ cell loss (Sinha Hikim et al., 1997). Additional studies have also provided evidence that gonadotropins and T are the critical germ cell survival factors. For example, similar stage-specific activation of germ cell apoptosis can also be found during the early regression of spermatogenesis after complete T withdrawal following ethane dimethane sulfonate treatment (Henriksen et al., 1995) or after selective deprivation of gonadotropins and testicular T by E2 treatment (Blanco-Rodriguez and Martinez-Garcia, 1996, 1998). It has also been found that the incubation of segments of adult human seminiferous tubules under serum-free conditions induces, within 4 h, germ cell apoptosis and that such induction of apoptosis can effectively be prevented by T (Erkkila et al.,

1997). These studies, thus, confirm and extend earlier findings showing the role of gonadotropins and androgens as germ cell survival factors.

To determine the role and relative contribution of FSH and LH in regulating specific germ cell survival, Swerdloff et al. (1998a,b) and Sinha Hikim et al. (1999) examined the extent to which recombinant human FSH or LH were able to prevent the GnRH-antagonist-induced germ cell apoptosis in the adult rat. Using gel fractionation and *in situ* analysis of germ cell apoptotic DNA fragmentation, these studies demonstrated that an increase in germ cell apoptosis after GnRH-antagonist treatment can effectively be prevented by the concomitant administration of either LH or FSH. FSH, however, was not as effective as LH in preventing the GnRH-antagonist-induced germ cell apoptosis. These findings in the adult rat are very different from that in the immature rat, where hCG (LH surrogate) or T alone was unable to prevent the hypophysectomy-induced germ cell apoptosis. Because FSH and ARs are found almost exclusively in the Sertoli cells, it is likely that these hormones exert their stimulatory effects on the Sertoli cells, which in turn results in the stimulation of intratubular factor(s) essential for the survival of germ cells through a paracrine mechanism (Sharpe, 1994; Kierszenbaum, 1994). The idea that cells require an adequate supply of survival factors elaborated by other cells in order to survive is not new and has been elegantly demonstrated in other tissues (Raff, 1992). Collectively, these results suggest that pituitary gonadotropins LH (via stimulation of T) and, to a lesser extent, FSH are the important extrinsic regulators of germ cell apoptosis and that the hormonal regulation of germ cell apoptosis is decisively different in immature versus adult rats.

5.4.1.4 Sertoli cell control of spermatogenesis

Although circulating hormones clearly play an important role in initiating and regulating the process of spermatogenesis, the Sertoli cell barrier prevents most substances from entering the seminiferous tubule compartment and directly influencing germ cell development. Therefore, the tubules must independently produce their own regulatory substances. Sertoli cells, the somatic cells of the seminiferous tubules, play an especially prominent role in the cell–cell interactions necessary for germ cell development by their physical association with and the transfer of molecules to the developing germ cells (Russell and Griswold, 1993; Jegou, 1993; Jegou et al., 1999; Kierszenbaum, 1994). Sertoli cells also establish an important physiological separation between basal (consisting of primarily spermatogonia) and adluminal (consisting of meiotic and postmeiotic) compartments of germ cells by the formation of Sertoli–Sertoli cell barrier. This compartmentalized organization of the seminiferous epithelium allows the bidirectional movement of Sertoli cell products to different populations of developing germ cells. Sertoli cells secrete many products, including activin, inhibin, Müllerian-inhibiting substance, stem cell factor (c-kit ligand), interleukin (IL)-1, IL6, transferrin, ceruloplasmin, cathepsin L, α_2-macroglobulin, transforming growth factor-α (TGFα), TGFβ, androgen-binding protein, retinol-binding protein, sulfated glycoproteins 1 and 2, and testibumin. All of these products have been proposed to have putative actions on germ cells. Evidence indicates that some of these Sertoli cell products are also regulated by specific hormones. For example, stem cell factor (SCF) gene expression (both at the transcriptional and post-transcriptional levels) in the rat seminiferous tubule was upregulated by FSH in a stage-specific manner (Yan et al., 1999). In contrast, T, E2, TGFα, TGFβ, and activin had no effect on SCF gene expression. Meng et al. (2000) have provided for a paracrine role of glial-cell-line derived neurotropic factor (GDNF), secreted by the Sertoli cells, in modulating the cell fate of undifferentiated spermatogonia. Gene-targeted mice with one GDNF-null allele show a depletion of stem cell reserves, whereas mice over-expressing GDNF show an accumulation of undifferentiated spermatogonia. Of note, FSH can upregulate GDNF (Walker and Chang, 2005). Additional genes known to be regulated by FSH include lactatedehydrogenase, aromatase, plasminogen activator, and insulin-like growth factor (Walker and Chang, 2005). Sertoli cells also have typical ARs that mediate the effects of T on spermatogenesis. The fact that T exerts its effects on somatic cells rather than germ cells was highlighted by recent germ cell-transplantation studies (Johnston et al., 2001), in which spermatogonia from AR-deficient mice developed into spermatozoa in WT recipient mice. Likewise, a Sertoli cell-selective KO of the AR causes spermatogenic arrest in meiosis and complete absence of elongated spermatids (De Gendt et al., 2004). Also, to identify the testicular genes regulated by T, microarray analyses were performed using testes of hpg mice with or without T supplementation for up to 24 h (Sadate-Ngatchou et al., 2004). Interestingly, more genes were downregulated than upregulated by about a 2:1 ratio at 4, 8, or 12 h, whereas after 24 h most of the transcripts were upregulated. This study suggests that the primary action of T in murine testis could be to repress the expression of those genes required by the cells for growth and proliferation, thereby ensuring proper maturation of Sertoli and/or peritubular myoid cells (Sadate-Ngatchou et al., 2004).

In summary, Sertoli cells thus appear to mediate the biological actions of both circulating hormones and the paracrine regulatory factors.

5.4.2 Sperm Transport

After spermatogenesis is completed, mature spermatozoa are released into the excretory system and travel through the rete testes and epididymis where they functionally mature before traversing the vas deferens. The semen gains constituents from the seminal vesicles, prostate, and bulbourethral glands before ejaculation. The transport of spermatozoa from the testes to the ejaculatory ducts takes approximately 2 weeks.

5.4.3 Environmental Agents and the Reproductive System

There is an increasing concern about environmental toxins and their potential adverse effects on the reproductive system. Environmental agents with known or potential reproductive effects include polychlorinated biphenyls (PCBs), a variety of pesticides, heavy metals, dioxins, synthetic hormones, hormone antagonists, alkylated phenols, and various natural fungal and plant products. Endocrine disrupters are exogenous compounds that alter reproductive development and adult reproductive function. Many of these compounds and their metabolites seem to act through hormone-receptor mechanisms (e.g., estrogen agonists or antiandrogens), whereas others may affect hormone metabolism or have direct toxic effects. Endocrine disrupters may act at different levels of the reproductive system from the brain (neurotoxins) to end organ steroid receptors. The implications for the male reproductive system include developmental abnormalities (e.g., cryptorchidism and ambiguous genitalia), infertility, androgen deficiency and resistance, and testicular malignancies. Recent studies suggest that the effects of endocrine disruptors may be perpetuated across generations through epigenetic mechanisms.

5.4.4 Male Infertility

Infertility is defined as the failure of a couple to achieve a pregnancy after at least 1 year of frequent unprotected intercourse. If a pregnancy has not occurred after 3 years, infertility most likely will persist without medical treatment.

5.4.4.1 Prevalence and incidence

Studies in the United States and Europe showed a 1-year prevalence of infertility in 15% of couples. The prevalence in developing countries is likely to be higher because of the higher prevalence of genital-tract infection. As shown in multicenter studies, approximately 30–35% of subfertility can be attributed to predominantly female factors, 25–30% to male factors, and 25–30% to problems in both partners; in the remaining percentage no cause can be identified.

5.4.4.2 Etiology

Hypothalamic–pituitary disorders are infrequent causes of male infertility (1–20%) and are discussed in the section on hypogonadism and androgen deficiency. Primarily testicular disorders are the most frequent identifiable cause of infertility (30–40%; see **Table 4**). Post-testicular defects (disorders of sperm transport) account for 10–20%, and nonidentifiable causes account for another 40–50%. The last category represents a large number of patients with impaired spermatogenesis for which our knowledge of the regulation of spermatogenesis is inadequate to define the pathophysiology of the defect. A number of microdeletions of the Y chromosome have been identified as a cause of infertility in approximately 15–25% of the infertile patients previously classified as idiopathic azoospermia or severe oligozoospermia (de Kretser, 1997; Reijo et al., 1995; Najmabadi et al., 1996).

5.4.4.3 Approach to the diagnosis of male infertility

The approach to the diagnosis of an infertile couple includes the management of the male and female partner. The examination of the ejaculate is the cornerstone for the investigation of an infertile man (**Table 9**). Semen samples are collected, when possible, at the physician's office or at home preferably after 2–7 days abstinence from sexual intercourse.

The generally accepted reference values for a semen analysis are given in **Table 10**. A normal sperm concentration is greater than 20 million/ml; however, men with lower sperm counts can be fertile. Over 50% of the spermatozoa should be motile and over 25% should demonstrate a rapidly progressive motility pattern.

In patients with abnormal semen analyses, the measurement of serum FSH, LH, and T is indicated.

Table 9 Male infertility: basic laboratory tests

Semen analyses	*Hormone analyses*[a]
Volume	Serum LH, FSH
pH	Serum testosterone
Microscopy: agglutination, debris	If LH and testosterone are low, serum prolactin
Sperm: concentration, motility, morphology, vitality	
Leukocytes	
Immature germ cells	
Sperm autoantibodies	
Sperm/semen biochemistry	
Sperm function tests	

[a]In patients with abnormal semen analyses.
Modified from Swerdloff RS and Wang C (2008) The testis and male sexual function. In: Gaedman L and Ausiello D (eds.) *Cecil Textbook of Medicine*, 23rd edn., pp. 1786–1791. Philadelphia, PA: W. B. Saunders, with permission from Elsevier

Table 10 Semen analysis: reference ranges

Parameter	*Reference range*
Semen volume	>2 ml
Sperm	
Concentration	>20 million/ml
Total count	>40 million/ejaculate
Motility	>50% motile
	>25% rapid progressively motile
Morphology	>15 % normal[a]
Vitality (live)	>75%
Leukocytes	<1 million/ml

[a]Value based on the strict criteria for assessment of sperm morphology in studies using *in vitro* fertilization as an end point.
Reproduced from World Health Organization (1999) *Laboratory Manual for Examination of Human Semen and Sperm Cervical Mucus Interaction*, 4th edn. Cambridge: Cambridge University Press, with permission from the WHO Press.

Elevated FSH levels usually indicate severe germinal epithelium damage and may be associated with a guarded prognosis. A decreased serum inhibin B level also reflects poor Sertoli cell function and may be a marker of spermatogenic dysfunction. Elevated serum LH and FSH levels together with a low serum T level indicate primary testicular failure involving both Leydig cell and Sertoli cell dysfunction. Low serum FSH, LH, and T levels suggest hypothalamic–pituitary dysfunction; a serum prolactin level should be measured and additional appropriate investigations (as discussed in the section on secondary hypogonadism) may be required. The presence of low sperm concentration and suppressed LH, with increased, normal, or low serum T levels (without clinical manifestations of androgen deficiency), may suggest exogenous androgen therapy. The hormonal pattern in androgen insensitivity (an uncommon cause of male infertility) is elevated LH, normal FSH, and high-normal to increased serum T levels.

5.4.4.4 Management of male infertility

Successful treatment of male infertility by medical means has been limited to the small proportion of infertile men with gonadotrophin deficiency who have been effectively treated with gonadotropin replacement and men with antisperm antibodies who have been treated with glucocorticoids. Patients with oligospermia are frequently managed by assisted reproductive technologies such as *in vitro* fertilization (IVF; Buyse and Feingold, 1979; Aynsley-Green et al., 1976) and intracytoplasmic sperm injection (ICSI; Howards, 1992; Pryor and Howards, 1987; Tekeda and Ueda, 1977; Griffin, 1992). When there are no sperm in the ejaculate but there are sperm in the testes, ICSI can be performed with spermatozoa isolated from testicular biopsies or fine-needle aspirates (Hess et al., 1997; Smith et al., 1994). Successful pregnancies have been obtained even with injection of fresh or cryopreserved immature, but differentiated, germ cells (e.g., spermatids). The ability of testicular spermatozoa to fertilize human oocytes has been reported for azoospermic men with maturation arrest (Krege et al., 1998), defective spermatogenesis (Reijo et al., 1995), deletion of DAZ (Najmabadi et al., 1996), and Klinefelter syndrome (Krausz et al., 1999). The benefits of surgical treatment of men with varicoceles by internal spermatic vein ligation have been controversial. Studies seem to support that such surgery is beneficial and should be performed when the varicocele is first diagnosed (de Kretser, 1997). In some centers, surgical ligation of the spermatic vein has been replaced by vascular catheter embolization of the internal spermatic vein.

5.5 Sexual Dysfunction

Sexual dysfunction can be divided into four main categories – loss of libido, ED, anorgasmic states, and ejaculatory insufficiency.

5.5.1 Decreased Libido

Loss of libido refers to the reduction in sexual interest, initiative, and frequency and intensity of responses to internal or external erotic stimuli. The causes of decreased or absent libido are many, ranging from physical (e.g., serotonin uptake, inhibiting antidepressants, etc.) to psychological (Kandeel et al., 2001). The most common causes of disordered desire are psychogenic and androgen deficiency.

5.5.2 Ejaculatory Failure and Impaired Orgasm

Ejaculatory insufficiency refers to the absent of or reduced seminal emission or impaired ejaculatory contraction. It is usually associated with neurologic conditions and medication therapy. An anorgasmic state is a distressing but relatively uncommon condition in men in which the normal process of erection and ejaculation occurs in the absence of the subjective sensation of pleasure initiated at the time of emission and ejaculation.

5.5.3 Erectile Dysfunction

ED can be defined as the inability of a man to obtain rigidity sufficient to permit coitus of adequate duration to satisfy himself and his partner.

5.5.3.1 Prevalence

Estimates suggest that 10–15% of all American males suffer from ED, with the incidence progressively increasing as men get older. Data from the Massachusetts Aging Study report that 52% of men ages 40–70 experience some degree of ED.

5.5.3.2 Etiology

The causes of ED are many, but can be generally categorized as psychological, endocrine, systemic illness, neurological, iatrogenic, drug-related, and age-related.

5.5.3.3 Clinical management of ED

5.5.3.3(i) Oral medications for ED

Yohimbine is an indol alquinolonic alkaloid with centrally acting effects, including α_2-adrenergic blockade and cholinergic and dopaminergic stimulation. Despite its widespread use, placebo-controlled studies have failed to show a significant effect. Trazodone possesses both serotonin and α_2-adrenergic antagonistic properties. It appears to be moderately effective in approximately one-third of patients, with the main side effect being sedation. Oral sildenafil has become the most widely used and the most financially successful new drug for this disorder. Sildenafil is a competitive and selective inhibitor of GMP PDE-5 (the primary PDE in cavernosal tissue). The inhibition of PDE-5 causes the persistence of normally stimulated GMP in the corpora cavernosa, resulting in protracted cavernosal tumescence and rigidity. Sildenafil is effective in 60–80% of men with ED. The usual starting dose of sildenafil is 50 mg 1 h before anticipated intercourse, increasing in 25-mg increments up to 100 mg when required. The most serious side effect is cardiovascular collapse, particularly in patients taking long-acting nitrate or nitroglycerin preparations. Because of its mechanism of action, sildenafil is used on demand, with administration approximately 20–60 min before intercourse. A number of new PDE inhibitors are being developed. Apomorphine is a selective dopamine-receptor agonist that stimulates the CNS, generating an arousal response that includes a penile erection. Apomorphine had not been approved by the Food and Drug Administration (FDA). The intraurethral prostaglandin E_1 suppository (alprostadil) is believed to work locally on the corpora cavernosa as a vasodilatory agent. The suppository is apparently successful in improving erectile function in one-third to two-thirds of cases.

5.5.3.3(ii) Intracavernosal injections of vasodilating drugs

Until the availability of oral sildenafil, intracavernosal injection with prostaglandin E_1 and other vasodilators (papaverine and phentolamine) was the mainstay of pharmacological therapy for ED. The medications are injected using a 27–30-gauge needle.

5.5.3.3(iii) Penile prosthesis

This treatment is the last resort for ED nonresponsive to medical and psychological treatments and is rarely used nowadays.

References

Albanese C, Colin IM, Crowley WF, Ito M, Pestell RG, Weiss J, and Jameson JL (1996) The gonadotropin genes: Evolution of distinct mechanisms for hormonal control. *Recent Progress in Hormone Research* 51: 23–58; discussion: pp. 59–61.

Allan CM, Garcia A, Spaliviero J, Zhang F-P, Jimenez M, Huhtaniemi I, and Handelsman J (2004) Complete Sertoli cell proliferation induced by follicle-stimulating hormone (FSH) independently of luteinizing hormone activity: Evidence from genetic models of isolated FSH action. *Endocrinology* 145: 1587–1593.

Awoniyi CA, Santulli R, Chandrashekar V, Schanbacher BD, and Zirkin BR (1989b) Quantitative restoration of advanced spermatogenic cells in adult male rats made azoospermic by active immunization against luteinizing hormone or gonadotropin-releasing hormone. *Endocrinology* 125: 1303–1309.

Awoniyi CA, Santulli R, Sprando RL, Ewing LL, and Zirkin BR (1989a) Restoration of advanced spermatogenic cells in the experimentally regressed rat testis: Quantitative relationship to testosterone concentration within the testis. *Endocrinology* 124: 1217–1223.

Awoniyi CA, Sprando RL, Santulli R, Chandrashekar V, Ewing LL, and Zirkin BR (1990) Restoration of spermatogenesis by exogenously administered testosterone in rats made azoospermic by hypophysectomy or withdrawal of luteinizing hormone alone. *Endocrinology* 127: 177–184.

Aynsley-Green A, Zachmann M, Illig R, Rampini S, and Prader A (1976) Congenital bilateral anorchia in childhood: A clinical, endocrine and therapeutic evaluation of twenty-one case. *Clinical Endocrinology* 5: 381–391.

Badger TM, Edcox CE, Meyer ER, Bell RD, and Cicero TJ (1978) Simultaneous changes in tissue and serum levels of LH, FSH, and LH/FSH releasing factor after castration in the male rat. *Endocrinology* 102(1): 136–141.

Bagatell CJ, Matsumoto AM, Christensen RB, Rivier JE, and Bremner WJ (1993) Comparison of a gonadotropin releasing-hormone antagonist plus testosterone (T) versus T alone as potential male contraceptive regimens. *Journal of Clinical Endocrinology and Metabolism* 77: 427–432.

Baker HWG, Berger HG, de Kretser DM, et al. (1996) Changes in the pituitary–testicular system with age. *Clinical Endocrinology* 5: 349–372.

Banerjee I and Clayton P (2007) The genetic basis for the timing of human puberty. *Journal of Neuroendocrine* 19: 831–838.

Barbieri RL (1992) Clinical applications of GnRH and its analogues. *Trends in Endocrinology and Metabolism* 3: 30–34.

Beranova M, Oliveira LM, Bédécarrats GY, et al. (2001) Prevalence, phenotypic spectrum, and modes of inheritance of gonadotropin-releasing hormone receptor mutations in idiopathic hypogonadotropic hypogonadism. *Journal of Endocrinology and Metabolism* 86(4): 1580–1588.

Bhagavath B and Layman LC (2007) The genetics of hypogonadotropic hypogonadism. *Seminars in Reproductive Medicine* 5(4): 272–286.

Bhasin S, Cunningham GR, Hayes FJ, Matsumoto AM, Snyder PJ, Swerdloff RS, and Montori VM (2006) Testosterone therapy in adult men with androgen deficiency syndromes: An endocrine society clinical practice guideline. *Journal of Endocrinology and Metabolism* 91(6): 1995–2010.

Bhasin S, Fielder T, Peacock N, Sod-Moriah A, and Swerdloff RS (1988) Dissociating antifertility effects of GnRH-antagonist from its adverse effects on mating behaviour in male rats. *American Journal of Physiology* 254: E84–E91.

Bhasin S, Yuan QX, Steiner BS, and Swerdloff RS (1987) Hormonal effects of GnRH agonist in men: Effects of long term treatment with agonist infusion and androgen. *Journal of Clinical Endocrinology and Metabolism* 65: 568–574.

Bilezikjian LM, Blount AL, Donaldson CJ, and Vale WW (2006) Pituitary actions of ligands of the TGF-beta family: Activins and inhibins. *Reproduction* 132(2): 207–215.

Bivalacqua TJ, Champion HC, Hellstrom WJ, and Kadowitz PJ (2000) Pharmacotherapy for erectile dysfunction. *Trends in Pharmacological Sciences* 1: 484–489.

Blanco-Rodriguez J and Martinez-Garcia C (1996) Induction of apoptotic cell death in the seminiferous tubule of the adult rat testis: Assessment of germ cell types that exhibit the ability to enter apoptosis after hormone suppression by estradiol treatment. *International Journal of Andrology* 19: 237–247.

Blanco-Rodriguez J and Martinez-Garcia C (1998) Apoptosis precedes detachment of germ cells from the seminiferous epithelium after hormone suppression by short-term oestradiol treatment of rats. *International Journal of Andrology* 21: 109–115.

Boccabella AV (1963) Reinitiation and restoration of spermatogenesis with testosterone propionate and other hormones after a long term posthypophysectomy regression period. *Endocrinology* 72: 787–798.

Boothby M, Ruddon RW, Anderson C, McWilliams D, and Boime I (1981) A single gonadotropin alpha-subunit gene in normal tissue and tumor-derived cell lines. *Journal of Biological Chemistry* 256: 5121–5127.

Brant WO, Bella AJ, and Lue YH (2007) Treatment options for erectile dysfunction. *Endocrinology and Metabolism Clinics of North America* 36: 465–479.

Brash DE (1996) Cellular proofreading. *Nature Medicine* 2: 525–526.

Bremner WJ, Vitiello MV, and Prinz PN (1983) Loss of circadian rhythmicity in blood testosterone levels with aging in normal men. *Journal of Clinical Endocrinology and Metabolism* 56: 1278.

Brown P and McNeilly AS (1999) Transcriptional regulation of pituitary gonadotropin subunit genes. *Reviews of Reproduction* 4: 117–124.

Burger HG and Igarashi M (1988) Inhibin: Definition and nomenclature, including related substances. *Journal of Clinical Endocrinology and Metabolism* 66: 885–886.

Burnside J, Buckland PR, and Chin WW (1988) Isolation and characterization of the genes encoding the alpha subunit of the rat pituitary glyprotein hormones. *Genes* 70: 67–74.

Buyse M and Feingold M (1979) Syndromes associated with abnormal external genitalia. In: Vallet HL and Porter IH (eds.) *Genetic Mechanisms of Sexual Development*, pp. 425–435. New York: Academic Press.

Childs GV, Ellison DG, and Unabia G (1990) Functional differentiation of gonadotropes and thyrotropes. In: Chin WW and Boime I (eds.) *Glycoprotein Hormones Serono Symposia, Norwell*, pp. 1–10. New York: Raven Press.

Childs GV, Hyde C, Naor Z, and Catt K (1983) Heterogeneous LH and FSH storage patterns in subtypes of gonadotropes separated by centrifugal elutriation. *Endocrinology* 113: 2120–2128.

Chin WW, Godine JE, Mein DR, Chang AS, Tan LK, and Habener JF (1983) Nucleotide sequence of the cDNA encoding the precursors of the beta subunit of rat lutropin. *Proceedings of the National Academy of Sciences of the United States of America* 80: 4649–4653.

Chin WW, Maloof F, and Habener JF (1981) TSH biosynthesis. Cellular processing, assembly and release of subunits. *Journal of Biological Chemistry* 256: 3059–3066.

Conn PM and Crowley WF, Jr. (1991) Gonadotropin releasing hormone and its analogues. *New England Journal of Medicine* 324: 93–103.

Conn PM and Crowley WF, Jr. (1994) Gonadotropin-releasing hormone and its analogs. *Annual Review of Medicine* 45: 391–405.

Counis R, Ribot G, Corbani M, Poissonnier M, and Jutisz M (1981) Cell-free translation of the rat pituitary messenger RNA coding for the precursors of alpha and beta subunits of lutropin. *FEBS Letters* 123: 151–155.

Crowley WF, Jr., Filicori M, Spratt DI, and Santoro NF (1985) The physiology of gonadotropin-releasing hormone (GnRH) secretion in men and women. *Recent Progress in Hormone Research* 41: 473–531.

De Gendt K, Swinner JV, and Saunders PTK (2004) A Sertoli cell-selective knockout of the androgen receptor causes spermatogenic arrest in meiosis. *Proceedings of the National Academy of Sciences of the United States of America* 101: 1327–1332.

de Kretser DM (1997) Male Infertility. *Lancet* 349: 787.

de Roux N, Genin E, Carel JC, Matsuda F, Chaussain JL, and Milgrom E (2003) Hypogonadotropic hypogonadism due to loss of function of the KiSS1 derived peptide receptor GPR54. *Proceedings of the National Academy of Sciences of the United States of America* 100: 10972–10976.

Dhillo WS, Murphy KG, and Bloom SR (2007) The neuroendocrine physiology of kisspeptin in the human. *Reviews in Endocrine and Metabolic Disorders* 8: 41–46.

Diver MJ, Imtiaz KE, Ahmad AM, Vora JP, and Fraser WD (2003) Diurnal rhythms of serum total, free and bioavailable testosterone and of SHBG in middle-aged men compared with those in young men. *Clinical Endocrinology* 58: 710–717.

Dodé C, Levilliers J, Dupont JM, et al. (2003) Loss-of-function mutations in FGFR1 cause autosomal dominant Kallmann syndrome. *Nature Genetics* 33(4): 463–465.

Duque C, Morisaki M, Ikekawa N, and Shikita M (1978) The enzyme activity of bovine adrenocortical cytochrome P-450 producing pregnenolone from cholesterol: Kinetic and electrophoretic studies of the reactivity of hydrocholesterol intermediates. *Biochemical and Biophysical Research Communications* 82: 174–178.

Edwards A, Hammond HA, Jin L, Caskey CT, and Chakraborty R (1992) Genetic variation at five trimeric and tetrameric tandem repeat loci in four human population groups. *Genomics* 12: 241–253.

Erkkila K, Henriksen K, Hirvonen V, Rannikko S, Salo J, Parvinen M, and Dunkel L (1997) Testosterone regulates apoptosis in adult human seminiferous tubules *in vitro*. *Journal of Clinical Endocrinology and Metabolism* 82: 2314–2321.

Esch ES, Shimasaki S, Cooksey K, et al. (1987) Complementary deoxyribonucleic acid (cDNA) cloning and DNA sequence analysis of rat ovarian inhibins. *Molecular Endocrinology* 1: 388–396.

Fang Y, Fliss AE, Robins DM, and Caplan AJ (1996) Hsp90 regulates androgen receptor hormone binding affinity *in vivo*. *Journal of Biological Chemistry* 271: 28697–28702.

Fiddes JC and Goodman HM (1979) Isolation, cloning and sequence analysis of the cDNA for the alpha subunit of hCG. *Nature* 281: 351–356.

Fiddes JC and Goodman HM (1981) The gene encoding the common alpha subunit of the four human glycoprotein hormones. *Journal of Molecular and Applied Genetics* 1: 3–18.

Filicori M (1994a) Gonadotrophin-releasing hormone agonists. A guide to use and selection. *Drugs* 48: 41–58.

Filicori M (1994b) Gonadotrophin-releasing hormone agonists. A guide to use and selection – erratum. *Drugs* 48: 326.

Finkelstein JS, O'Dea LS, Whitcomb RW, and Crowley WE, Jr. (1991b) Sex steroid control of gonadotropin secretion in the human male. II: Effects of estradiol administration in normal and gonadotropin releasing hormone deficient men. *Journal of Clinical Endocrinology and Metabolism* 73: 621–628.

Finkelstein JS, Whitcomb RW, O'Dea LS, Longcope C, Schoenfeld DA, and Crowley WF, Jr. (1991a) Sex steroid control of gonadotropin secretion in the human male. I. Effects to testosterone administration in normal and gonadotropin-releasing hormone deficient men. *Journal of Clinical Endocrinology and Metabolism* 73: 609–620.

Flanagan CA, Millar RP, and Illing N (1997) Advances in understanding gonadotrophin-releasing hormone receptor structure and ligand interactions. *Reviews of Reproduction* 2: 113–120.

Forage RG, Ring JM, Brown RW, et al. (1986) Cloning and sequence analysis of cDNA species coding for the two subunits of inhibin from bovine follicular fluid. *Proceedings of the National Academy of Sciences of the United States of America* 83: 3091–3095.

Garban H, Marquez D, Cai L, Rajfer J, and Gonzalez-Cadavid NF (1995) Restoration of normal adult penile erectile response in aged rats by long-term treatment with androgens. *Biology of Reproduction* 53: 1365–1372.

Gay VL and Midgley AR, Jr. (1969) Response of the adult rate to orchidectomy and ovariectomy as determined by LH radioimmunoassay. *Endocrinology* 84: 1359–1364.

Gharib SD, Bowers SM, Need LR, and Chin WW (1986) Regulation of rat LH subunit mRNAs by gonadal steroid hormones. *Journal of Clinical Investigation* 77: 582–589.

Gharib SD, Roy A, Wierman ME, and Chin WW (1989) Isolation and characterization of the gene encoding the beta-subunit of rat follicle-stimulating hormone. *DNA* 8: 339–349.

Gharib SD, Weirman ME, Shupnick MA, and Chin WW (1990) Molecular biology of the pituitary gonadotropins. *Endocrine Reviews* 11: 177–199.

Gibson R, Komfeld S, and Schlesinger S (1980) A role for oligosaccharides in glycoprotein synthesis. *Trends in Biochemical Sciences* 249: 290–295.

Gill JC, Wadas B, Chen P, et al. (2008) The GnRH neuronal population is normal in size and distribution in GnRH-deficient and GnRH receptor-mutant hypogonadal mice. *Endocrinology* 149: 4596–4604.

Giuliano F and Rampin O (2000) Central control of erection and its pharmacological modification. *Current Opinion in Urology* 10: 629–633.

Godine JE, Chin WW, and Habener JF (1980) LH and FSH hormones. Cell free translations of mRNAs encoding for subunit precursors. *Journal of Biological Chemistry* 255: 8780–8783.

Godine JE, Chin WW, and Habener JF (1981) Cell free synthesis and processing of the precursors to the subunits of LH. *Journal of Biological Chemistry* 256: 2475–2479.

Godine JE, Chin WW, and Habener JF (1982) Alpha Subunit structure of rat pituitary glycoprotein hormones. Primary structure of the precursor determined from the nucleotide sequence of cloned cDNAs. *Journal of Biological Chemistry* 257: 8368–8371.

Gordone DF, Wood WM, and Ridgway EC (1988) Organization and nucleotide sequence of the mouse alpha subunit gene of the pituitary glycoprotein hormones. *DNA* 7: 679–690.

Gormley G and Ridgmaster R (1991) Effect of finasteride, a 5-alpha reductase inhibitor, on serum gonadotropins in normal men. *Endocrinology* 73: 145A.

Greep R, Fevold H, and Hisaw F (1936) Effects of two hypophyseal gonadotropic hormones on the reproductive system of the male rat. *Anatomical Record* 65: 261–271.

Greenstein A, Mabjeesh NJ, Sofer M, Kaver I, Matzkin H, and Chen J (2005) Does Sildenafil combined with testosterone gel improve erectile dysfunction in hypogonadal men in whom testosterone supplemental therapy alone failed? *Journal of Urology* 173(2): 530–532.

Griffin JE (1992) Androgen resistance – the clinical and molecular spectrum. *New England Journal of Medicine* 326: 611.

Handelsman DJ, Peng S, Sikka SC, Rajfer J, and Swerdloff RS (1988) Effects of gonadotropin-releasing hormone analogs on *cis*-platinum-induced spermatogenic damage. *International Journal of Andrology* 11: 425–435.

Handelsman DJ and Swerdloff RS (1986) Pharmacokinetics of gonadotropin-releasing hormone (GnRH) and its analogs. *Endocrine Reviews* 7: 95–105.

Hayes FJ, Seminara SB, Decruz S, Boepple PA, and Crowley WF, Jr. (2000) Aromatase inhibition in the human male reveals a hypothalamic site of estrogen feedback. *Journal of Endocrinology and Metabolism* 85: 3027–3035.

Herbst KL, Coviello AD, Page S, Amory JK, Anawalt BD, and Bremner WJ (2004) A single dose of the potent gonadotropin-releasing hormone antagonist acyline suppresses gonadotropins and testosterone for 2 weeks in healthy young men. *Journal of Clinical Endocrinology and Metabolism* 89: 5959–5965.

Heller CG and Clermont Y (1964) Kinetics of the germinal epithelium in man. *Recent Progress in Hormone Research* 20: 545–575.

Henriksen K, Hakovirta H, and Parvinen M (1995) Testosterone inhibits and induces apoptosis in rat seminiferous tubules in a stage-specific manner: *In situ* quantification in squash preparations after administration of ethane dimethane sulfonate. *Endocrinology* 136: 3285–3291.

Hess RA, Bunick D, Lee KH, Bahr J, Taylor JA, Korach KS, and Lubahn DB (1997) A role for oestrogens in the male reproductive system. *Nature* 390: 509–512.

Hijazi RA and Cunningham GR (2005) Andropause: Is androgen replacement therapy indicated for the aging male? *Annual Review of Medicine* 56: 117–137.

Hirai T, Takikawa H, and Kato Y (1990) The gene for the beta subunit of porcine FSH: Absence of consensus oestrogenresponsive element and presence of retroposons. *Journal of Molecular Endocrinology* 5: 147–158.

Howards SS (1992) Varicocele. *Infertility and Reproductive Medicine Clinics of North America* 3: 429–441.

Huhtaniemi IT, Korebrot CC, and Jaffe RB (1977) hCG binding and stimulation of testosterone synthesis in the human fetal testis. *Journal of Clinical Endocrinology and Metabolism* 44: 963–967.

Ingraham HA, Chen RP, Mangalam HJ, et al. (1988) A tissue-specific transcription factor containing a homeodomain specifies a pituitary phenotype. *Cell* 55: 519–529.

Ishimaru T, Edmiston WA, Pages L, and Horton R (1978) Splanchnic extraction and conversion of testosterone and dihydrotestosterone in man. *Journal of Clinical Endocrinology and Metabolism* 46: 528–533.

Jameson JL (1996) Inherited disorders of gonadotropin hormones. *Molecular and Cellular Endocrinology* 125: 143–149.

Jameson JL, Becker CG, Lindell CM, and Habener JF (1988) Human FSH beta subunit gene encodes multiple mRNAs. *Molecular Endocrinology* 2: 806–815.

Jameson JL, Chin WW, Hollenberg AN, Chang AS, and Habener JF (1983) The gene encoding the beta subunit of rat *I.H. Journal of Biological Chemistry* 259: 15474–15480.

Jegou B (1993) The Sertoli-germ cell communication network in mammals. *Internatonal. Review of Cytology* 147: 25–96.

Jegou B, Pineau C, and Dupaix A (1999) Paracrine control of testis function. In: Wang C (ed.) *Male Reproductive Function*, pp. 41–64. Boston, MA: Kluwer Academic Publishers.

Ji I and Ji TH (1981) Both α and β subunits of human choriogonadotropin photoaffinity label the hormone receptor. *Proceedings of the National Academy of Sciences of the United States of America* 78: 5464–5469.

Jia Y, Sinha Hikim AP, Swerdloff RS, et al. (2007) Signaling pathways for germ cell death in adult Cynomolgus monkeys (*Macaca fascicularis*) induced by mild testicular hyperthermia and exogenous testosterone treatment. *Biology of Reproduction* 77: 83–92.

Johnston DS, Russell LD, Friel PJ, and Griswold MD (2001) Murine germ cells do not require functional androgen receptors to complete spermatogenesis following spermatogonial stem cell transplantation. *Endocrinology* 142: 2405–2408.

Kaiser UB, Conn PM, and Chin WW (1997) Studies of gonadotropin-releasing hormone (GnRH) action using GnRH receptor-expressing pituitary cell lines. *Endocrine Reviews* 18: 46–70.

Kandeel FR, Koussa VK, and Swerdloff RS (2001) Male sexual function and its disorders: Physiology, pathophysiology, clinical investigation, and treatment. *Endocrine Reviews* 22: 342–388.

Kaplan SL, Grumbach MM, and Aubert ML (1976) The ontogenesis of pituitary hormones and hypothalamic factors in the human fetus: Maturation of central nervous system regulation of anterior pituitary function. *Recent Progress in Hormone Research* 32: 161–243.

Kaufman JM and Vermeulen A (2005) The decline of androgen levels in elderly men and its clinical and therapeutic implications. *Endocrine Reviews* 6: 833–876.

Kefier D, Fetherston J, and Boime L (1980) Isolation of mRNA from bovine pituitary: The cell tree synthesis of the alpha and beta subunits of luteinizing hormone. *European Journal of Biochemistry* 108: 367–372.

Kerr JB, Maddocks S, and Sharpe RM (1992) Testosterone and FSH have independent, synergistic and stage-dependent effects upon spermatogenesis in the rat testis. *Cell and Tissue Research* 268: 179–189.

Kerrigan JR, Veldhuis JD, and Rogol AD (1994) Androgen-receptor blockade enhances pulsatile luteinizing hormone production in late pubertal males: Evidence for a hypothalamic site of physiologic androgen feedback action. *Pediatric Research* 35: 102–106.

Kierszenbaum AL (1994) Mammalian spermatogenesis *in vivo* and *in vitro*: A partnership of spermatogenic and somatic cell lineages. *Endocrine Reviews* 15: 116–134.

Kim KE, Day KH, Howard P, Salton SR, Roberts JL, and Maurer RA (1990) DNA sequences required for expression of the LH beta promoter in primary cultures of rat pituitary cells. *Molecular and Cellular Endocrinology* 74: 101–107.

Kim KE, Gordon DF, and Maurer RA (1988) Nucleotide sequence of the bovine gene for FSH beta subunit. *DNA* 7: 227–233.

Knudson CM, Tung KSK, Tourtellotte WG, Brow GAJ, and Korsmeyer SJ (1995) Bax-deficient mice lymphoid hyperplasia and male germ cell death. *Science* 270: 96–99.

Knuth UA, Hano R, and Nieschlag E (1984) Effect of flutamide or cyproterone acetate on pituitary and testicular hormones in normal men. *Journal of Endocrinology and Metabolism* 59: 963–969.

Kornfeld R and Kornfeld S (1976) Comparative aspects of glycoprotein structure. *Annual Review of Biochemistry* 45: 217–237.

Kotani M, Detheux M, Vandenbogaerde A, et al. (2001) The metastasis suppressor gene KiSS-1 encodes kisspeptins, the natural ligands of the orphan G protein-coupled receptor GPR54. *Journal of Biolcgical Chemistry* 276: 34631–34636.

Kovacs K and Horvath E (1985) Morphology of adenophyseal cells and pituitary adenomas. In: Martini L (ed.) *The Pituitary Gland*, pp. 25–56. New York: Raven Press.

Krausz C, Quintana-Murci L, Barbaux S, et al. (1999) A high frequency of Y chromosome deletions in males with nonidiopathic infertility. *Journal of Clinical Endocrinology and Metabolism* 84: 3606–3612.

Krege JH, Hodgin JB, Couse JF, et al. (1998) Generation and reproductive phenotypes of mice lacking estrogen receptor beta. *Proceedings of the National Academy of Sciences of the United States of America* 95: 15677–15682.

Krishnamurthy H, Danilovich N, Morales CR, and Siaram MR (2000) Qualitative and quantitative decline in spermatogenesis of the follicle-stimulating hormone receptor knockout (FORKO) mouse. *Biology of Reproduction* 62: 1146–1159.

Krishnamurthy H, Babu PS, Morales CR, and Siaram MR (2001) Delay in sexual maturity of the follicle-stimulating hormone receptor knockout male mouse. *Biology of Reproduction* 65: 522–531.

Kuiper GGJM, Carlquist M, and Gustafsson JA (1998) Estrogen is a male and female hormone. *Science and Medicine* 5: 36–45.

Kumar TR (2005) What have you learned about gonadotropin function from gonadotropin subunit and receptor knockout mice? *Reproduction* 130: 293–302.

Kumar TR (2007) Functional analysis of LHβ knockout mice. *Molecular and Cellular Endocrinology* 269: 81–84.

Kumar TR, Wang Y, Lu N, and Matzuk MM (1997) Follicle stimulating hormone is required for ovarian follicle maturation but not male fertility. *Nature Genetics* 15: 201–204.

Kunelius P, Lukkarinen O, Hannuksela ML, Itkonen O, and Tapanainen JS (2002) The effects of transdermal dihydrotestosterone in the aging male: A prospective, randomized double blind study. *Journal of Endocrinology and Metabolism* 87: 1467–1472.

Kusuda S and Dufau ML (1988) Characterization of ovarian gonadotropin receptor. Monomer and associated form of the receptor. *Journal of Biological Chemistry* 263: 30–46.

Labrie F, Bélanger A, Cusan L, Gomez J-L, and Candas B (1997) Marked decline in serum concentrations of adrenal C 19 sex steroid precursors and conjugated androgen metabolites during aging. *Journal of Clinical Endocrinology and Metabolism* 89: 2396–2402.

Landefeld T and Kepa J (1979) The cell free synthesis of bovine lutropin beta subunit. *Biochemical and Biophysical Research Communications* 90: 1111–1118.

Lindstedt G, Ernest I, Nystrom E, and Janson PO (1997) Fall av manlig infertilitet. *Klinsk Kemi Norden* 3: 81–87.

Ling N, Ying S-Y, Ueno N, Esch E, Denoroy L, and Guillernin R (1985) Isolation and partial characterization of a Mr 3200 protein with inhibin activity from porcine follicular fluid. *Proceedings of the National Academy of Sciences of the United States of America* 182: 7217–7221.

Liverman CT and Blazer DG (2004) *Testosterone and Aging: Clinical Research Directions*. Washington, DC: National Academies Press.

Lloyd JM and Childs CV (1988) Differential storage and release of LH and FSH from individual gonadotropes separated by centrifugal elutriation. *Endocrinology* 122: 1281–1290.

Longcope C, Sato K, McKay C, and Horton R (1984) Aromatization by splanchnic tissue in men. *Journal of Clinical Endocrinology and Metabolism* 58: 1089–1093.

Loria P, Petraglia F, Concari M, et al. (1998) Influence of age and sex on serum concentrations of total dimeric activin A. *European Journal of Endocrinology* 139: 487–492.

Lue TF (1998) Physiology of penile erection and pathophysiology of erectile dysfunction and priapism. In: Walsh PC, Retick A, Vaughn B, and Wein A (eds.) *Campbell's Urology,* 7th edn., p. 1164. Philadelphia, PA: Saunders.

Lue YH, Wang C, Liu Y-X, et al. (2006) Transient testicular warming enhances the suppressive effect of testosterone on spermatogenesis in adult cynomolgus monkeys (*Macaca fascicularis*). *Journal of Endocrinology and Metabolism* 91: 539–545.

Ly LP, Jimenez M, Zhuang TN, Cleermajer DS, Conway AJ, and Handelsman DJ (2001) A double-blind placebo-controlled, randomized clinical trial of transdermal dihydrotestosterone gel on muscular strength, mobility, and quality of life in older men with partial androgen deficiency. *Journal of Endocrinology and Metabolism* 86: 4078–4088.

Ma X, Dong Y, Matzuk MM, and Kumar TR (2004) Targeted disruption of luteinizing hormone beat-subunit leads to hypogonadism, defects in gonadal steroidogenesis, and infertility. *Proceedings of the National Academy of Sciences of the United States of America* 101: 17294–17299.

Marshall WA and Tanner JM (1970) Variation in pattern of pubertal changes in boys. *Archives of Disease in Childhood* 45: 13–23.

Mason AJ, Hayflick JS, Ling N, et al. (1985) Complementary DNA sequences of ovarian follicular fluid inhibin show precursor structure and homology with transforming growth factor-beta. *Nature* 318: 659–663.

Mather JP, Woodruff TK, and Krummen LA (1992) Paracrine regulation of reproductive function by inhibin and activin. *Proceedings of the Society for Experimental Biology and Medicine* 201: 1–15.

Matthiesson KL, McLachlan RI, O'Donnel L, Frydenberg M, Robertson DM, Stanton PG, and Meachem SJ (2006) The relative role of follicle-stimulating hormone and luteinizing hormone in maintaining spermatogonial maturation and spermiation in normal men. *Journal of Endocrinology and Metabolism* 91: 3962–3969.

Matthiesson KL, Robertson DM, Burger HG, and McLachlan RI (2003) Response of serum inhibin B and pro-alphaC levels to gonadotrophic stimulation in normal men before and after steroidal contraceptive treatment. *Human Reproduction* 18: 734–743.

Matsumoto AM and Bremner WJ (1984) Modulation of pulsatile gonadotropin secretion by testosterone in men. *Journal of Clinical Endocrinology and Metabolism* 58: 609–614.

Matsumoto AM, Karpas AE, and Bremner WJ (1986) Chronic human chorionic gonadotropin administration in normal men: Evidence that follicle-stimulating hormone is necessary for the maintenance of quantitative normal spermatogenesis in men. *Journal of Clinical Endocrinology and Metabolism* 62: 1184–1190.

Matsumoto AM, Karpas AE, Paulsen CA, and Bremner WJ (1983) Reinitiation of sperm production in gonadotropin-suppressed normal men by administration of FSH. *Journal of Clinical Investigation* 72: 1005–1015.

Matsumoto AM, Paulsen CA, and Bremner WJ (1984) Stimulation of sperm production by human LH in gonadotropin suppressed normal men. *Journal of Clinical Endocrinology and Metabolism* 59: 882–887.

Matsuo H, Baba Y, Nair RM, Arimura A, and Schally AV (1971) Structure of the porcine LH- and FSH-releasing hormone. I: The proposed amino acid sequence. *Biochemical and Biophysical Research Communications* 43: 1334–1339.

Maurer RA and Kim KE (1989) Analysis of gonadotropin gene structure and expression. In: Chin WW and Boime L (eds.) *Glycoprotein Hormones Serono Symposia, Norwell,* pp. 227–246. New York: Raven Press.

McLachlan RI, Wreford NG, Tsonis C, de Kretser D, and Robertson DM (1994) Testosterone effects on spermatogenesis in the gonadotropin-releasing hormone-immunized rat. *Biology of Reproduction* 50: 271–280.

McPhaul MJ, Marcelli M, Tilley WD, Griffin JE, Isidro-Gutierrez RF, and Wilson JD (1991) Molecular basis of androgen resistance in a family with a qualitative abnormality of the androgen receptor and responsive to high-dose androgen therapy. *Journal of Clinical Investigation* 87: 1413–1421.

Meng X, Lindahl M, Hyvonen ME, et al. (2000) Regulation of cell fate decision of undifferentiated spermatogonia by GDNF. *Science* 287: 1489–1493.

Meuleman E, Cuzin B, Opsomer RJ, et al. (2001) A dose-escalation study to assess the efficacy and safety of sildenafil citrate in men with erectile dysfunction. *BJU International* 87: 75–81.

Millar RP, King JA, Davidson JS, and Milton RC (1987) Gonadotrophin-releasing hormone – diversity of functions and clinical applications. *South African Medical Journal* 72: 748–755.

Miyamoto K, Hasegawa Y, Fukuda M, Nomura M, Igarashi M, Kangawa K, and Matsuo H (1985) Isolation of porcine follicular fluid inhibin of 32K daltons. *Biochemical and Biophysical Research Communications* 129: 396–403.

Molsberry RL, Carr BR, Mendelsohn CR, and Simpson ER (1982) Human chorionic gonadotropins binding to human fetal testes as a function of gestational age. *Journal of Clinical Endocrinology and Metabolism* 55: 791–794.

Morales AJ, Nolan JJ, Nelson JC, and Yen SSC (1994) Effects of replacement dose of dehydroepiandrosterone in men and women of advancing age. *Journal of Clinical Endocrinology and Metabolism* 78: 1360–1367.

Moriarty GC (1973) Adenohypophysis: Ultrasound cytochemistry. A review. *Journal of Histochemistry and Cytochemistry* 21: 855–892.

Moriarty GC (1976) Immunocytochemistry of the pituitary glycoprotein hormones. *Journal of Histochemistry and Cytochemistry* 24(7): 846–863.

Muir AI, Chamberlain L, Elshourbagy NA, et al. (2001) AXOR12, a novel human G protein-coupled receptor, activated by the peptid KiSS-1. *Journal of Biological Chemistry* 276: 28969–28975.

Murk DD and Cheng CY (2004) Sertoli–Sertoli and Sertoli–germ cell interactions and their significance in germ cell movement in the seminiferous epithelium and spermatogenesis. *Endocrine Reviews* 25: 747–806.

Nair KS, Rizza RA, O'Brien P, et al. (2006) DHEA in elderly women and DHEA or testosterone in elderly men. *New England Journal of Medicine* 355: 1647–1659.

Najmabadi H, Huang V, Yen P, et al. (1996) Substantial prevalence of microdeletions of the Y-chromosome in infertile men with idiopathic azoospermia and oligozoospermia detected using a sequence-tagged site-based mapping strategy. *Journal of Clinical Endocrinology and Metabolism* 81: 1347–1352.

Naor Z, Harris D, and Shacham S (1998) Mechanism of GnRH receptor signaling: Combinatorial cross-talk of Ca^{2+} and protein kinase C. *Frontiers in Neuroendocrinology* 19: 1–19.

Narula A, Gu YQ, O'Donnell L, Stanton PG, Robertson DM, McLachlan RI, and Bremner WJ (2002) Variability in sperm suppression during testosterone administration to adult monkeys is related to follicle stimulating hormone suppression and not to intratesticular androgens. *Journal of Endocrinology and Metabolism* 87: 3399–3406.

Nathan N and Palmert MR (2005) Regulation and disorders of pubertal timing. *Endocrinology and Metabolism Clinics of North America* 34: 617–641.

Navarro VM, Castellano JM, Fernandez-Fernandez R, et al. (2004) Developmental and hormonally regulated messenger ribonucleic acid expression of KiSS-1 and its putative receptor, GPR54, in rat hypothalamus and potent luteinizing hormone-releasing activity of KiSS-1 peptide. *Endocrinology* 145: 4565–4574.

Nieschlag E, Zitzmann M, and Kamischke A (2003) Use of progestins in male contraception. *Steroids* 68: 965–972.

Nilson JH, Thomason AR, Cserbak MT, Moncman CL, and Woychik RP (1983) Nucleotide sequence of a cDNA for the common alpha subunit of the bovine pituitary glycoprotein hormones. *Journal of Biological Chemistry* 258: 4679–4682.

Norwitz ER, Jeong KH, and Chin WW (1999) Molecular mechanisms of gonadotropin-releasing hormone receptor gene regulation. *Journal of the Society for Gynecologic Investigation* 6: 169–178.

O'Bryan MK and De Kretser D (2006) Mouse models for genes involved in impaired spermatogenesis. *International Journal of Andrology* 29: 76–89.

Ohtaki N, Yamaguchi A, Goi T, et al. (2001) Somatic alterations of the DPC4 Madr2 genes in colorectal cancers and relationship to metastasis. *International Journal of Oncology* 18: 265–270.

Otani T, Otani E, Krych M, Chaplin DD, and Boime I (1988) Identification of a promoter region in the CG beta gene cluster. *Journal of Biological Chemistry* 263: 7322–7329.

Parks JS, Adess ME, and Brown MR (1997) Genes regulating hypothalamic and pituitary development. *Acta Paediatrica. (Stockholm) Supplementum* 423: 28–32.

Parks JS, Brown MR, Hurley DL, Phelps CJ, and Wajnarajch MP (1999) Heritable disorders of pituitary development. *Journal of Clinical Endocrinology and Metabolism* 84: 4362–4370.

Parvinen M (1993) Cyclic function of Sertoli cells. In: Russell LD and Griswold MD (eds.) *The Sertoli Cell*, pp. 331–347. Cache Clearwater, FL: River Press.

Penny R, Parlow AF, and Frasier O (1979) Testosterone and estradiol concentrations in paired maternal and cord sera and their correlation with the concentration of chorionic gonadotropin. *Pediatrics* 64: 604–608.

Penson DF, Ng C, Cai L, Rajfer J, and Gonzalez-Cadavid NF (1996) Androgen and pituitary control of penile nitric oxide synthase and erectile function in the rat. *Biology of Reproduction* 55: 567–574.

Pescovitz OH, Srivastava CH, Breyer PR, and Monts BA (1994) Paracrine control of spermatogenesis. *TEM* 5: 126–131.

Phillip M, Arbelle JC, Seger Y, and Pavnari F (1998) Male hypogonadism due to a mutation in the gene for the β-subunit of follicle stimulating hormone. *New England Journal of Medicine* 338: 1729–1732.

Piersma D, Verhoef-Post M, Berns EM, and Themmen AP (2007) LH receptor gene mutations and polymorphisms: An overview. *Molecular and Cellular Endocrinology* 260–262: 282–286.

Pryor JL and Howards SS (1987) Varicocele. *The Urologic Clinics of North America* 14: 499–513.

Raff MC (1992) Social controls on cell survival and cell death. *Nature* 356: 397–400.

Raivio T, Falardeau J, Dwyer A, et al. (2007) Reversal of idiopathic hypogonadotropic hypogonadism. *New England Journal of Medicine* 357(9): 863–873.

Rajfer J, Sikka SC, and Swerdloff RS (1987) Lack of direct effect of GnRH agonist on human testicular steroidogenesis. *Journal of Clinical Endocrinology and Metabolism* 64: 62–67.

Raven G, de Jong FH, Kaufman JM, and de Ronde W (2006) In men, peripheral estradiol levels directly reflect the action of estrogens at the hypothalamo-pituitary level to inhibit gonadotropin secretion. *Journal of Endocrinology and Metabolism* 91: 3324–3328.

Rea MA, Marshall GR, Weinbauer GF, and Nieschlag E (1986) Testosterone maintains pituitary and serum FSH and spermatogenesis in GnRH-antagonist suppressed rats. *Journal of Endocrinology* 108: 101–107.

Reijo R, Lee TY, Salo P, et al. (1995) Diverse spermatogenic effects in humans caused by Y chromosome deletions encompassing a novel RNA-binding protein gene. *Nature Genetics* 10: 383–393.

Reiter EO and Norjavaara E (2005) Testotoxicosis: Current viewpoint. *Pediatric Endocrinology Reviews* 3: 77–86.

Robertson DM, Foulds LM, Leversha L, et al. (1985) Isolation of inhibin from bovine follicular fluid. *Biochemical and Biophysical Research Communications* 126: 220–226.

Robertson DM, Klein R, de Vos FL, et al. (1987) The isolation of polypeptides with FSH suppressing activity from bovine follicular fluid which are structurally different to inhibin. *Biochemical and Biophysical Research Communications* 149: 744–749.

Roche PC and Ryan RJ (1989) Purification, characterization, and amino-terminal sequence of rat ovarian receptor for luteinizing hormone/human choriogonadotropin. *Journal of Biological Chemistry* 264: 4636–4641.

Rochira V, Zirilli L, Genazzani AD, et al. (2006) Hypothalamic–pituitary–gonadal axis in two men with aromatase deficiency: Evidence that circulating estrogens are required at the hypothalamic level for the integrity of gonadotropin negative feedback. *European Journal of Endocrinology* 155: 513–522.

Rosenthal BD, May NR, Metro M, Harkaway RC, and Ginsberg PC (2006) Adjunctive use of AndroGel (testosterone gel) with sildenafil to treat erectile dysfunction in men with acquired androgen deficiency syndrome after failure using sildenafil alone. *Urology* 67(3): 571–574.

Ruddon RW, Hanson CA, Bryon AH, et al. (1980) Synthesis and secretion of human chorionic gonadotropin subunits by cultured human malignant cells. *Journal of Biological Chemistry* 255: 1000–1007.

Russell LD, Ettlin RA, Sinha Hikim AP, and Clegg ED (1990) *Histological and Histopathological Evaluation of the Testes*. Clearwater, FL: Cache River Press.

Russell LD and Griswold MD (eds.) (1993) *The Sertoli Cell*. Clearwater, FL: Cache River Press.

Sadate-Ngatchou PI, Pouchnik DJ, and Griswold MD (2004) Identification of testosterone-regulated genes in testes of mypogonadal mice using oligonucleotide microarray. *Molecular Endocrinology* 18: 422–433.

Sairam MR and Bhargavi GN (1985) A role for the glycosylation of the alpha-subunit in the transduction of biological signal in glycoprotein hormones. *Science* 229: 65–67.

Salameh W, Bhasin S, Steiner BS, et al. (1991) Marked suppression of gonadotropins and testosterone by an antagonist analog of gonadotropin-releasing hormone in men. *Fertility and Sterility* 55: 156–164.

Salameh W, Bhasin S, Steiner BS, et al. (1994) Comparative effects of two different delivery systems on gonadotropin-releasing hormone (GnRH) antagonist induced suppression of gonadotropins and testosterone in man. *Journal of Andrology* 15: 22–28.

Sandhoff TW, Hales DB, Hales KH, and McLean MP (1998) Transcriptional regulation of the rat steroidogenic acute regulatory protein gene by steroidogenic factor 1. *Endocrinology* 139: 4820–4831.

Santen RJ (1975) Aromatization of testosterone to estradiol required for inhibition of LH secretion in men? *Journal of Clinical Investigation* 56: 1555–1563.

Santulli R, Sprando RL, Awoniyi CA, Ewing LL, and Zirkin BR (1990) To what extent can spermatogenesis be maintained in the hypophysectomized adult rat testis with exogenously administered testosterone? *Endocrinology* 126: 95–102.

Sauders BD, Sabbagh E, Chin WW, and Kaiser UB (1998) Differential use of signal transduction pathway in the GnRH-mediated regulation of gonadotropin subunit gene expression. *Endocrinology* 139: 1835–1843.

Schaison G, Young J, Pholsena M, Nahoul K, and Couzinet B (1993) Failure of combined follicle-stimulating hormone-testosterone administration to initiate and/or maintain spermatogenesis in men with hypogonadotropic hypogonadism. *Journal of Clinical Endocrinology and Metabolism* 77: 1545–1549.

Schally AV, Arimura A, Baba Y, et al. (1971) Isolation and properties of the FSH and LH-releasing hormone. *Biochemical and Biophysical Research Communications* 43: 393–399.

Scheckter CB, Matsumoto AM, and Bremner WJ (1989) Testosterone administration inhibits gonadotropin secretion by an effect directly at the human pituitary. *Journal of Clinical Endocrinology and Metabolism* 68: 397–401.

Schertler GF, Villa C, and Henderson R (1993) Projection structure of rhodopsin. *Nature* 362: 770–772.

Schubert M, Minnemann T, Heber D, et al. (2004) Intramuscular testosterone undecanoate pharmacokinetic aspects of a novel testosterone formulation during longterm treatment of men with hypogonadism. *Journal of Endocrinology and Metabolism* 89: 5429–5434.

Schwanzel-Fukuda M and Pfaff DW (1989) Origin of luteinizing hormone releasing hormone neurons. *Nature* 338: 161–164.

Sealfon SC and Millar RP (1994) The gonadotropin-releasing hormone receptor: Structural determinants and regulatory control. *Oxford Reviews Reproductive Biology* 17: 255–283.

Seminara SB, Dipietro MJ, Ramaswamy S, Crowley WF, Jr., and Plant TM (2006) Continuous human metastin 45–54 infusion desensitizes G protein-coupled receptor 54-induced gonadotropin-releasing hormone release monitored indirectly in the juvenile male Rhesus monkey (*Macaca mulatta*): A finding with therapeutic implications. *Endocrinology* 147: 2122–2126.

Seminara SB, Messager S, Chatzidaki EE, et al. (2003) The GPR54 gene as a regulator of puberty. *New England Journal of Medicine* 349: 1614–1627.

Shaha C (2007) Modulators of spermatogenic cell survival. *Society for Reproduction and Fertility – Supplements* 63: 173–186.

Sharpe RM (1987) Testosterone and spermatogenesis. *Journal of Endocrinology* 113: 1–2.

Sharpe RM (1994) Regulation of spermatogenesis. In: Knobil E and Neil JD (eds.) *The Physiology of Reproduction*, pp. 1363–1434. New York: Raven Press.

Sharpe RM, Donachie K, and Cooper L (1988) Reevaluation of intratesticular level of testosterone for quantitative maintenance of spermatogenesis in the rat. *Journal of Endocrinology* 117: 19–26.

Shen WH, Moore CCD, Ikeda Y, Parker KL, and Ingraham HA (1994) Nuclear receptor steroidogenic facto 1 regulates the Müllerian inhibiting substance gene: A link to the sex determination cascade. *Cell* 77: 651–661.

Sherins RJ and Lonaux DL (1973) Studies on the role of sex-steroids in the feedback control of FSH concentrations in men. *Journal of Clinical Endocrinology and Metabolism* 36: 886–893.

Sih R, Morley JE, Kaiser FE, Perry HM, Patrick P, and Ross C (1997) Testosterone replacement in older hypogonadal men: A 12-month randomized controlled trial. *Journal of Clinical Endocrinology and Metabolism* 82: 1661–1667.

Singh J, O'Neil C, and Handelsman DJ (1995) Induction of spermatogenesis by androgens in gonadotropin-deficient (hpg) mice. *Endocrinology* 136: 5311–5321.

Sinha Hikim AP, Lue Y-H, Wang C, and Swerdloff RS (1999) Spermatogenesis and germ cell death. In: Wang C (ed.) *Male*

Reproductive Function, pp. 19–39. Boston, MA: Kluwer Academic Publishers.
Sinha Hikim AP, Rajavashisth TB, Sinha Hikim I, et al. (1997) Significance of apoptosis in the temporal and stage-specific loss of germ cells in the adult rat after gonadotropin deprivation. *Biology of Reproduction* 57: 1193–1201.
Sinha Hikim AP and Swerdloff RS (1995) Temporal and stage-specific effects of recombinant human follicle stimulating hormone on the maintenance of spermatogenesis in gonadotropin-releasing hormone antagonist-treated rat. *Endocrinology* 136: 253–261.
Sinha Hikim AP and Swerdloff RS (1999) Hormonal and genetic control of germ cell apoptosis in the testis. *Reviews of Reproduction* 4: 38–47.
Sinha Hikim AP, Vera Y, Elhag RI, et al. (2005) Mouse model of male germ cell apoptosis in response to a lack of hormonal stimulation. *Indian Journal of Experimental Biology* 43: 1048–1057.
Sinha Hikim AP, Wang C, Leung A, and Swerdloff RS (1995) Involvement of apoptosis in the induction of germ cell degeneration in adult rats after gonadotropin-releasing hormone antagonist treatment. *Endocrinology* 136: 2770–2775.
Smith EP, Boyd J, Frank GR, et al. (1994) Estrogen resistance caused by a mutation in the estrogen-receptor gene in man. *New England Journal of Medicine* 332: 1056–1061.
Smith JT, Cunningham MJ, Rissman EF, Clifton DK, and Steiner RA (2005) Regulation of KiSS 1 gene expression in the brain of the female mouse. *Endocrinology* 146: 3686–3692.
Smith P (1927) The disabilities caused by hypophysectomy and their repair. *Journal of the American Medical Association* 88: 158–161.
Snyder PJ, Peachey H, Hannoush P, et al. (1999a) Effect of testosterone treatment on bone mineral density in men over 65 years of age. *Journal of Clinical Endocrinology and Metabolism* 84: 1966–1972.
Snyder PJ, Peachey H, Hannoush P, et al. (1999b) Effect of testosterone treatment on body composition and muscle strength in men over 65 years of age. *Journal of Clinical Endocrinology and Metabolism* 84: 2647–2653.
Soussi-Yanicostas N, de Castro F, Julliard AK, Perfettini I, Chédotal A, and Petit C (2002) Anosmin-1, defective in the X-linked form of Kallmann syndrome, promotes axonal branch formation from olfactory bulb output neurons. *Cell* 109(2): 217–228.
Spiteri-Grech J and Nieschlag E (1993) Paracrine factors relevant to the regulation of spermatogenesis – a review. *Journal of Reproduction and Fertilility* 98: 1–14.
Stahl BD, Huang HJ, Sebastian J, Ghosh BR, and Miller WL (1998) Transcriptional activation of ovine FSH beta gene by GnRH hormone: Involvement of two activating protein-1 binding sites and protein kinase C. *Endocrinology* 139: 4455–4465.
Steinberger E (1971) Hormonal control of mammalian spermatogenesis. *Physiological Reviews* 51: 1–22.
Stewart-Bentley M, Odell W, and Horton R (1974) The feedback control of LH in normal men. *Journal of Clinical Endocrinology and Metabolism* 38: 545–553.
Stojilkovic SS and Catt KJ (1995) Expression and signal transduction pathways of gonadotropin-releasing hormone receptors. *Recent Progress in Hormone Research* 50: 161–205.
Stojilkovic SS, Reinhart J, and Catt KJ (1994) Gonadotropin-releasing hormone receptors: Structure and signal transduction pathways. *Endocrine Reviews* 15: 462–499.
Sun YT, Irby DC, Robertson DM, and de Kretser DM (1989) The effects of exogenously administered testosterone on spermatogenesis in intact and hypophysectomized rats. *Endocrinology* 125: 1000–1010.
Swerdloff RS, Bagatell CJ, Wang C, Anawalt BD, Berman N, Steiner B, and Bremner WJ (1998a) Suppression of spermatogenesis in man induced by Nal-Glu gonadotropin releasing hormone antagonist and testosterone enanthate (TE) is maintained by TE alone. *Journal of Clinical Endocrinology Metabolism* 83: 3527–3533.
Swerdloff RS, Campfield LA, Palacios A, and McClure RD (1979) Suppression of human spermatogenesis by depot androgen: Potential for male contraception. *Journal of Steroid Biochemistry* 11: 663–670.
Swerdloff RS, Lue Y, Wang C, Rajavashisth T, and Sinha Hikim A (1998b) Hormonal regulation of germ cell apoptosis. In: Zirkin BR (ed.) *Germ Cell Development, Division, Disruption and Death*, pp. 150–164. New York: Springer.
Swerdloff RS, Walsh PC, Jacobs HS, and Odell WD (1971) Serum LH and FSH during sexual maturation in the male rat: Effect of castration and cryptorchidism. *Endocrinology* 88: 120–128.
Swerdloff RS and Wang C (2000) The testis and male sexual function. In: Goldman L and Bennet JC (eds.) *Cecil Textbook of Medicine,* 21st edn., pp. 1306–1317. Philadelphia, PA: W. B. Saunders.
Swerdloff RS and Wang C (2008) The testis and male sexual function. In: Gaedman L and Ausiello D (eds.) *Cecil Textbook of Medicine,* 23rd edn., pp. 1786–1791. Philadelphia, PA: W. B. Saunders.
Swerdloff RS, Wang C, Cunningham G, et al. (2000) Long term pharmacokinetics of transdermal testosterone gel in hypogonadal men. *Journal of Clinical Endocrinology and Metabolism* 85: 4500–4510.
Swerdloff RS and Wang C (2008) Free Testosterone measurement by the analog displacement direct assay: Old concerns and new evidence. *Clinical Chemistry* 54(3): 458–460.
Takikawa O, Gomi T, Suhara K, Itagaki E, Takemori S, and Katagiri M (1978) Properties of an adrenal cytochrome P-450 (P-450SCC) for the side chain cleavage of cholesterol. *Archives of Biochemistry and Biophysics* 190: 300–306.
Talmadge K, Vamvakopoulos NC, and Fiddes JC (1984) Evolution of the genes for the beta subunits of hCG and LH. *Nature* 307: 37–40.
Tapanainen JS, Aittomaki K, Min J, Vaskivuo T, and Huhtaniemi IT (1997) Men homozygous for an inactivating mutation of the follicle-stimulating (FSH) receptor gene present variable suppression of spermatogenesis and fertility. *Nature Genetics* 15: 205–206.
Tapanainen JS, Tilly JK, Vihko KK, and Hsueh AJW (1993) Hormonal control of apoptotic cell death in the testis: Gonadotropins and androgens as testicular cell survival factors. *Molecular Endocrinology* 7: 643–650.
Tekeda R and Ueda M (1977) Pituitary–gonadal function in male patients with myotonic dystrophy: Serum luteinizing hormone, follicle stimulating hormone and testosterone levels and histological damage of the testis. *Acta Endocrinology (Copenhagen)* 84: 382.
Tenover JS (1992) Effects of testosterone supplementation in the aging male. *Journal of Clinical Endocrinology and Metabolism* 75: 1092–1098.
The Leuprolide Study Group (1984) Leuprolide versus diethylshiltesterol for metastatic prostate cancer. *New England Journal of Medicine* 311(20): 1281–1286.
Themmen APN and Huhtaniemi IT (2000) Mutations of gonadotropins and gonadotropin receptors: Elucidating the

physiology and pathophysiology of pituitary–gonadal function. *Endocrine Reviews* 21: 551–583.

Thompson EL, Murphy KG, Patterson M, et al. (2006) Chronic subcutaneous administration of kisspeptin-54 causes testicular degeneration in adult male rats. *American Journal of Physiology. Endocrinology and Metabolism* 291: E1074–E1082.

Tom L, Bhasin S, Salameh W, et al. (1992) Induction of azoospermia in normal men with combined Nal-Glu gonadotropin-releasing hormone (GnRH) antagonist and testosterone enanthate. *Journal of Clinical Endocrinology and Metabolism* 75: 476–483.

Turgeon JL, Kimura Y, Waring DW, and Mellon PL (1996) Steroid and pulsatile gonadotropin-releasing hormone (GnRH) regulation of luteinizing hormone and GnRH receptor in a novel gonadotrope cell line. *Molecular Endocrinology* 10: 439–450.

Ueno N, Ling N, Ying SY, Esch E, Shimasakis S, and Guillemin R (1987) Isolation and partial characterization of follistatin, a novel Mr 35000 monomeric protein which inhibits the release of FSH. *Proceedings of the National Academy of Sciences of the United States of America* 84: 8282–8286.

Umesono K and Evans RM (1989) Determinants of target gene specificity for steroid/thyroid hormone receptors. *Cell* 57: 1139–1146.

Urban RJ, Padmanabhan V, Beitins I, and Veldhuis JD (1991) Metabolic clearance of human follicle-stimulating hormone assessed by radioimmunoassay, immunoradiometric assay, and *in vitro* Sertoli cell bioassay. *Journal of Clinical Endocrinology and Metabolism* 73: 818–823.

Vaishnav M and Moudgal NR (1994) Role of FSH in regulating testicular germ cell transformation in the rat: A study using DNA flow cytometry. *Andrologia* 26: 111–117.

Vale W, Rivier C, Hsueh AJW, et al. (1988) Chemical and biological characterization of the inhibin family of protein hormones. *Recent Progress in Hormone Research* 44: 1–34.

Veldhuis JD and Johnson ML (1988) *In vivo* dynamics of luteinizing hormone secretion and clearance in man: Assessment by deconvolution mechanics. *Journal of Clinical Endocrinology and Metabolism* 66: 1291–1300.

Veldhuis JD, Rogol AD, Samojlik E, and Ertel NH (1984) Role of endogenous opiates in the expression of negative feedback actions of androgen and estrogen on pulsatile properties of LH secretion in man. *Journal of Clinical investigation* 74: 47–55.

Veldhuis JD, Urban RJ, and Dufau ML (1992) Evidence that androgen negative feedback regulated hypothalamic gonadotropin-releasing hormone impulse strength and the burst-like secretion of biologically active luteinizing hormone in men. *Journal of Endocrinology and Metabolism* 74: 1227–1235.

Vera Y, Erkkila K, Wang C, et al. (2006) Involvement of p38 mitogen-activated protein kinase and inducible nitric oxide synthase in apoptotic signaling of murine and human male germ cells after hormone deprivation. *Molecular Endocrinology* 20: 1597–1609.

Vermeulen A and Deslypere JP (1985) Long-term transdermal dihydrotestosterone therapy: Effects on pituitary gonadal axis and plasma lipoproteins. *Maturitas* 7: 281–287.

Vihko KK, LaPolt PS, Nishimori K, and Hsueh AJW (1991) Stimulatory effects of recombinant follicle-stimulating hormone on Leydig cell function and spermatogenesis in immature hypophysectomized rats. *Endocrinology* 129: 1926–1932.

Wachter CJ and Lernmaez WJ (1976) The role of polyprenol-lined sugars in glycoprotein synthesis. *Annual Review of Biochemistry* 45: 95–112.

Walker WH and Cheng J (2005) FSH and testosterone signalling in Sertoli cells. *Reproduction* 130: 15–28.

Walsh PC, Swerdloff RS, and Odell WD (1973) Feedback regulation of gonadotropin secretion in men. *Journal of Urology* 110: 84–89.

Wang C, Alexander G, Berman N, et al. (1996) Testosterone replacement therapy improves mood in hypogonadal men. *Journal of Clinical Endocrinology and Metabolism* 81: 3578–3583.

Wang C, Catlin DH, Demers LM, Starcevic B, and Swerdloff RS (2004) Measurement of total serum testosterone in adult men: Comparison of current laboratory methods versus liquid chromatography-tandem mass spectrometry. *Journal of Endocrinology and Metabolism* 89(2): 534–543.

Wang C, Cui YG, Wang XH, et al. (2007) Transient scrotal hyperthermia and levonorgestrel enhance testosterone induced spermatogenesis suppression in men through increased germ cell apoptosis. *Journal of Endocrinology and Metabolism* 92: 3292–3304.

Wang C, Iranmanesh A, Berman N, et al. (1998) Comparative pharmacokinetics of three doses of percutaneous dihydrotestosterone gel in healthy elderly men – a clinical research center study. *Journal of Endocrinology and Metabolism* 83: 2749–2757.

Wang C and Swerdloff RS (2004) Male hormonal contraception. *American Journal of Obstetric Gynecology* 190: S60–S68.

Wang C, Swerdloff RS, Iranmanesh A, et al. (2000) Transdermal testosterone gel improves sexual function, mood, muscle strength, and body composition parameters in hypogonadal men. *Journal of Clinical Endocrinology and Metabolism* 85: 2839–2853.

Watkins-Chow DE and Camper SA (1998) How many homeobox genes does it take to make a pituitary gland? *Trends in Genetics* 14: 284–290.

Watkins PC, Eddy R, Beck AK, et al. (1987) DNA sequence and regional assignment of the human follicle-stimulating hormone beta-subunit gene to the short arm of chromosome 11. *DNA* 6: 205–212.

Weinbauer GF and Nieschlag E (1990) The role of testosterone in spermatogenesis. In: Nieschlag E and Behere HM (eds.) *Testosterone Action, Deficiency, Substitution*, pp. 25–30. New York: Springer.

Weintraub BD, Stannard BS, Linnekin D, and Marshall M (1980) Relationship of glycosylation to *de novo* thyroid-stimulating hormone biosynthesis and secretion by mouse pituitary, tumor cells. *Journal of Biological Chemistry* 255: 5715–5723.

Weintraub BD, Stannard BS, Magner JA, et al. (1985) Glycosylation and posttranslational processing of thyroid-stimulating hormone: Clinical implications. *Recent Progress in Hormone Research* 41: 577–606.

Winters SJ, Sherins RJ, and Loriaux DL (1979) Studies on the role of sex steroids in the feedback control of gonadotropin concentration in men. *Journal of Clinical Endocrinology and Metabolism* 48: 553–558.

World Health Organization (1999) *Laboratory Manual for Examination of Human Semen and Sperm Cervical Mucus Interaction,* 4th edn. Cambridge: Cambridge University Press.

World Health Organization Task Force on Methods for the Regulation of Male Fertility (1990) Contraceptive efficacy of testosterone-induced azoospermia in normal men. *Lancet* 336: 955–959.

World Health Organization Task Force on Methods for the Regulation of Male Fertility (1996) Contraceptive efficacy of testosterone-induced azoospermia and

oligozoospermia in normal men. *Fertility and Sterility* 65: 821–829.
Yamamoto M, Diebell ND, and Bodgdanove EM (1970) Analysis of initial and delayed effects of orchidectomy and ovariectomy on pituitary and serum LH levels in adult and immature rats. *Endocrinology* 86: 1102–1111.
Yan W, Linderborg J, Suominen J, and Toppari J (1999) Stage-specific regulation of stem cell factor gene expression in the rat seminiferous epithelium. *Endocrinology* 140: 1499–1504.
Yen SSC, Llerena O, Little B, and Pearson OH (1968) Disappearance rates of endogenous luteinizing hormone and chorionic gonadotropin in man. *Journal of Clinical Endocrinology and Metabolism* 28: 1763–1767.
Yen SSC, Llerena LA, Pearson OH, and Littell AS (1970) Disappearance rates of endogenous follicle-stimulating hormone in serum following surgical hypophysectomy in man. *Journal of Clinical Endocrinology and Metabolism* 30: 325–329.
Ying SY (1988) Inhibins, activins and follistatins: Gonadal proteins modulating the secretion of follicle-stimulating hormone. *Endocrine Reviews* 9: 267–293.
Zazopoulos E, Lalli E, Stocco DM, and Sassone-Corsi P (1997) DNA binding and transcriptional repression by DAX-1 blocks steroidogenesis. *Nature* 390: 311–315.
Zhengwei Y, Wreford NG, Royce P, de Kretser DM, and McLachlan RI (1998) Stereological evaluation of human spermatogenesis after suppression by testosterone treatment: Heterogeneous pattern of spermatogenic impairment. *Journal of Endocrinology and Metabolism* 83: 1284–1291.
Zirkin BR, Awoniyi C, Griswold MD, Russell LD, and Sharpe R (1994) Is FSH required for adult spermatogenesis? *Journal of Andrology* 15: 273–276.
Zittsman M and Nieschlag E (2007) Androgen receptor gene CAG repeat length and body mass index modulate the safety of long-term intramuscular testosterone undecanoate therapy in hypogonadal men. *Journal of Endocrinology and Metabolism* 92(10): 3844–3853.

Further Reading

Gromoll J, Simoni M, and Nieschlag E (1996) An activating mutation of the follicle stimulating hormone receptor autonomously sustains spermatogenesis in a hypophysectomized man. *Journal of Clinical Endocrinology and Metabolism* 81: 1367–1370.
Swerdloff RS, Steiner B, Callegari C, and Bhasin S (1996) GnRH analogues and male contraception. In: Bhasin S, Gabelnick H, Spieler J, Swerdloff RS, and Wang C (eds.) *Biology, Pharmacology, and Clinical Applications of Androgen*, pp. 355–365. New York: Wiley-Liss.

6 Sex Differences in Human Brain Structure and Function

L Cahill, University of California, Irvine, CA, USA

Chapter Outline

6.1 Introduction

The issue of sex influences on brain function appears to be rapidly moving to center stage in neuroscience, after many years of being considered by the neuroscience mainstream as an issue primarily, or exclusively, of concern to those interested in sex behaviors. As must be abundantly clear to any reader of this volume, sex influences on brain function are ubiquitous. They are found in studies ranging from human behavior to single neurons in cell culture, and everywhere in between (Cahill, 2006). The implications of this simple fact are enormous. Conclusions in essentially every domain of neuroscience, both basic and clinically applied, are likely affected by the issue, whether investigators know it or not. Indeed, so strong is the evidence for ubiquitous sex influences on brain function that one may argue that the burden of proof has finally shifted – from those neuroscientists investigating sex influences in their work having to justify why, to those not doing so having to justify why not (Wetherington, 2007).

Although substantial evidence supporting this conclusion comes from animal research, this chapter aims to contribute a perspective from studies of the human brain. Here too, driven largely by the advent of human-brain-imaging methods, investigators are uncovering previously unsuspected sex influences. Each such discovery challenges us to explain the findings. Ignoring or dismissing sex influences, while unfortunately still an appealing option for some investigators, is not only scientifically invalid, but harmful to the advance of knowledge in our field.

It is not possible in this chapter to fully describe the extant literature regarding sex influences on human brain structure and function. Hence, this chapter aims to describe some of the more intriguing findings of recent years, and highlights the issue more directly by focusing on a particular domain in which sex influences are increasingly apparent, namely, the domain of emotional memory. Before beginning, it is helpful to address a common misconception about the issue of sex influences still held by many investigators of human brain function.

6.2 Are Sex Influences in the Human Brain Small and Unreliable?

Many investigators of human brain and behavior unfortunately appear to hold the view that sex differences are small and unreliable. These colleagues overwhelmingly cite two issues they believe support their view: (1) sex differences in the size/shape of the corpus callosum; and (2) sex differences in the functional organization or language. We address each in turn.

It is indeed the case that sex differences in the size/shape of various aspects of the corpus callosum have been much debated, with replication failures a clear issue (see **Chapter 8, Gonadal Hormones and Sexual Differentiation of Human Brain and Behavior**), although there exists consensus that some small sex differences in the corpus callosum exist, particularly in its anterior and posterior regions (Luders et al., 2003). Why, however, should investigators invariably cite this example as evidence for a general unreliability of sex differences in brain structure? Why instead should not investigators cite reliable sex differences in the size of the amygdala (men > women), or of the hippocampus (women > men), as evidence of the reliability of structural sex differences in the human brain (Goldstein et al., 2001)? Why should not investigators cite extremely large sex differences in the texture of white matter (a magnetic resonance imaging (MRI)-based assessment of the orderliness of fibers within white matter; see Kovalev and Kruggel (2007)) as evidence that sex differences in brain anatomy are quite large? In fact, there exists no evidence of which the author is aware that the average effect size in the domain of sex differences in brain anatomy/function is any smaller than that seen in other domains of neuroscience, the corpus callosum notwithstanding.

The second widely adopted, supposed example of the unreliability of sex influences on human brain function concerns language. Since a report by Shaywitz and colleagues (1995), considerable attention focused on whether language function, as determined by imaging techniques, is more left-hemisphere dependent in men than it is in women. Although several investigations replicated the essential Shaywitz findings, a few did not. From these failures, it appears, came the view that the key findings were unreliable. However, as convincingly shown by Clements et al. (2006), the apparent failures to replicate actually stem from methodological differences between studies. Furthermore, Clements et al. provide their own evidence, as well as a clear literature summary, both strongly pointing to the validity of the conclusion that language is more left-lateralized in males. Thus, the view that sex differences in the laterality of language seen in imaging studies are somehow unreliable is simply incorrect.

6.3 Sex Influences on Human Brain Function Generally Considered

Biological sex influences brain function to a far greater extent that neuroscience in the main has recognized to date (Cahill, 2006). Multiple sex differences have been reported in every brain lobe, including in cognitive brain regions such as the neocortex and hippocampus. In human subject work, these discoveries happened in large measure, thanks to the widespread advent of modern imaging techniques, which have revealed sex-related differences in brain correlates of many brain functions (Cahill, 2006). Here some of the more notable findings are highlighted.

Some sex differences in the human brain are relatively global in nature. For example, widespread areas of cortex are significantly thicker in women than in men (Luders et al., 2005). But the majority of sex differences are much more focal in nature, leading to a mosaic concept – sex differences, as a rule, of many types exist in many areas at many levels of investigations, all of which undoubtedly interact (Cahill, 2006). Consider the size of human brain regions. Goldstein et al. (2001) examined the size of cortical and subcortical regions throughout the human brain. They found significant sex differences (adjusted for total brain size) in every lobe, some favoring men and some women (see **Figure 1**). Interestingly, they also found that the magnitude of the effects related to the degree to which the regions expressed sex-hormone receptors during development (as inferred from animal research), suggesting that developmental mechanisms help account for the adult sex differences in size. Indeed, both global and more focal sex differences in brain structure exist in pre-adolescent children (Blanton et al., 2004).

Another example of relatively focal sex differences in human brain comes from a study employing a three-dimensional analytic technique on MRI scans to quantify cortical complexity. Luders et al. (2004) reported that the degree of gyrification was significantly larger parts of the frontal and parietal cortex in women than in men, many of the effects being quite large. Further, no lobar region showed significantly larger cortical complexity in men. The authors suggest that these sex differences may exist in part as a compensation for smaller overall brain volume in females. They also suggest that the differences may result in part from differential developmental trajectories in the sexes. In fact, differences in developmental trajectories are among the most striking of sex differences in the human brain (Giedd et al., 1996). Sex differences have even been reported in the effects of both prenatal and postnatal environmental factors on the subsequent size of particular brain regions in adulthood (Buss et al., 2007).

Sex differences in human brain sometimes occur not so much in the size of individual brain regions,

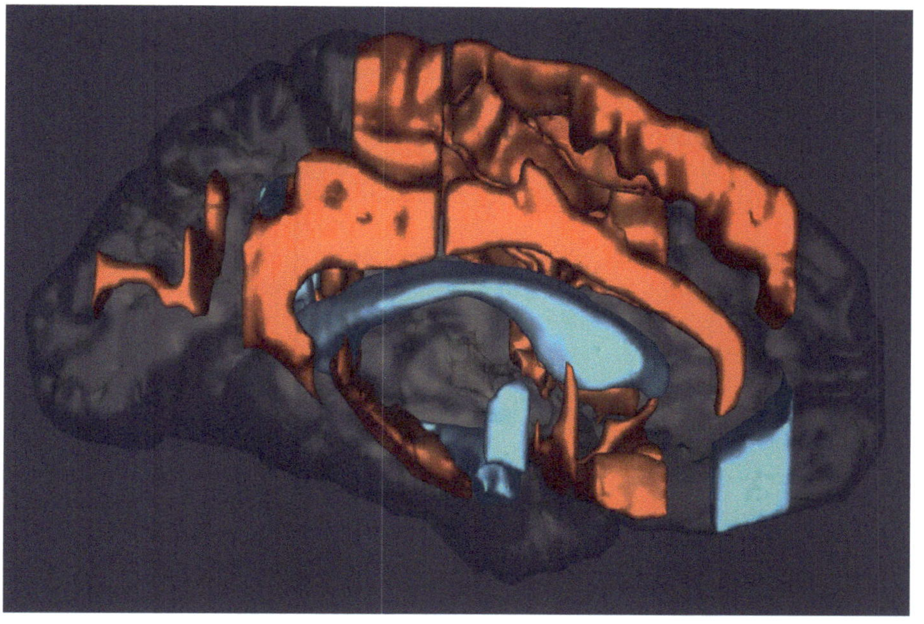

Figure 1 Sex differences in the size of brain regions (relative to whole brain) exist in every brain lobe. Reproduced from Goldstein J, Seidman L, Horton N, et al. (2001) Normal sexual dimorphism of the adult human brain assessed by *in vivo* magnetic resonance imaging. *Cerebral Cortex* 11: 490–497, with permission from Oxford University Press.

but in their size relative to other regions. For example, Gur et al. (2004) reported that the relative size of the orbitofrontal compared to amygdala regions is significantly greater in women compared to men, possibly pointing toward a differential ability to exert emotional control in the sexes. Interestingly, the ratio between the two structures is differentially altered in schizophrenia: the orbitofrontal-to-amygdala size ratio is increased in schizophrenic relative to healthy men, yet it is decreased in schizophrenic relative to healthy women. Thus, schizophrenia in men feminizes this measure, whereas schizophrenia in females masculinizes it.

A large challenge for the domain of sex influences on human brain function, as for all of behavioral neuroscience, is to attach behavioral meaning to observed sex differences. Some progress in this effort is being made. For example, Gur et al. (1999) correlated brain gray and white matter to cognitive performance in healthy adults. They confirmed previous findings that women have a higher percentage of gray matter, whereas men have a higher percentage of white matter. They found that both gray and white matter volumes correlated positively with a global index of cognitive ability in both men and women, but that the relationship was much steeper in women, leading the authors to suggest that there exists a more efficient use of white matter in the female brain. No matter the validity of this particular conclusion, the study represents an area in which far more work is needed, namely, relating sex differences in structure/function in the human brain to behavior.

More rapid progress toward this goal might be achieved if investigators of sex differences also addressed potential influences of cerebral hemisphere in their studies. Indeed, this point was made as long ago as 1964, when Lansdell discovered apparent sex differences in hemispheric asymmetries of mylenization in the human brain, commenting that "the sex of patients is a factor which should be heeded in investigations of the laterality of cerebral function." A recent illustration of this view comes from Frings et al. (2006), who examined sex-related differences in activation of the hippocampus during memory processing. Healthy men and women received functional magnetic resonance imaging (fMRI) scans while performing a spatial memory task in a virtual environment. Although the men and women performed the task equally well, the left hemisphere hippocampus was activated in women performing the task, whereas the right hemisphere amygdala was activated in men. This finding may reflect a fundamental difference in brain organization between the sexes, a difference in the use of cognitive strategies, or both. Regardless, it illustrates the need to attend to potential influences of sex and hemisphere in imaging studies of human memory.

6.4 Sex Differences in Emotional Memory

The lessons above concerning the need to attend to potential influences of sex and cerebral hemisphere in studies of human memory are perhaps most evident at present in the domain of emotional memory, to which we now turn.

Extensive evidence indicates that the amygdala modulates the storage of different forms of memory, in particular conscious (declarative) memory, and does so via extensive interactions with the endogenous stress response produced by emotionally salient events (McGaugh et al., 1996, 2004). This view fits very well with the known anatomical connectivity of the primate amygdala. A meta-analysis of cortico-cortical connectivity in the monkey by Young and Scannell (1994) revealed that the amygdala is remarkably, and uniquely, well suited to widely modulate mnemonic functions in the rest of the brain. In fact, across many species, learning tasks, and laboratories, stimulation of the amygdala (and in particular its basolateral complex) potently modulates – enhances or impairs – memory storage processes. Most often, stimulation has been given immediately after learning, allowing the conclusion that the effects of the stimulation on memory resulted from an influence on consolidation processes. Evidence also indicates that the amygdala's ability to modulate memory consolidation depends crucially on endogenous stress hormones. For example, amygdala stimulation may improve or impair memory storage, depending on adrenal gland function (McGaugh, 2004). All peripherally administered drugs and hormones require the basolateral amygdala to affect memory. Lesions or functional inactivation of the basolateral amygdala blocks the memory modulating actions of essentially all drugs and hormones tested to date. Thus, extensive research involving both animal and human subjects converges on a neurobiological mechanism by which emotional arousal sculpts the contents of memory (McGaugh, 2004).

6.5 Amygdala Activity and Emotional Memory in Humans – Emergence of Sex Effects

If the amygdala functions, at least in part, to modulate the storage of memory for emotional events, then it should be possible to detect a relationship between the degree to which the amygdala is activated in response to emotional events, and the degree to which those events are subsequently recalled. In fact, such a relationship is now well established. In the first of these studies, Cahill et al. (1996) scanned healthy male subjects with positron emission tomography (PET) for regional cerebral glucose while they viewed either a series of relatively emotionally arousing (negative) films, or a matched but much more emotionally neutral set of films, and examined memory for the films 3 weeks later. The results showed that right hemisphere amygdala activity while viewing the emotional films correlated significantly with long-term recall, but not with recall of emotionally neutral films. Several other laboratories have now confirmed this finding, providing additional support for the view that the amygdala plays a special, presumably modulatory, role in memory storage for emotional events, as predicted by animal research.

At this point, it is noted that studies reporting amygdala effects predominantly on the right side of the brain involved only male subjects, whereas studies reporting amygdala effects predominantly on the left side of the brain involved only female subjects. This finding suggested that subject sex might be one determinant of the hemispheric lateralization of amygdala function.

6.6 Sex-Related Hemispheric Lateralization of the Amygdala Relationship to Emotional Memory

To determine whether subject sex was influencing lateralization of the amygdala relationship to long-term memory for emotional material, we scanned 11 men and 11 women with PET for regional cerebral glucose (Cahill et al., 2001) twice – once while watching a series of emotionally arousing film clips, and again while watching a series of more emotionally neutral clips. Memory for the films was assessed in a surprise free recall test 3 weeks later. We found that right, but not left hemisphere amygdala activity significantly related to enhanced memory for the emotional film clips in men, whereas left, and not right, hemisphere amygdala activity related to enhanced memory for the emotional films in women. Canli et al. (2002) confirmed this sex-related lateralization in an fMRI study of amygdala function. Subjects in this study were scanned while viewing a series of emotionally arousing or neutral slides. Activity of the right, and not left amygdala in males related significantly to memory for the most emotional slides, whereas activity of the left, and not right amygdala related to memory for the emotional slides in women.

We again demonstrated a sex-related hemispheric lateralization in an fMRI study (Cahill et al., 2004), examining amygdala activity at encoding and subsequent memory for emotional images. Consistent with the previous studies, we found that activity of the right hemisphere amygdala was significantly more related to subsequent memory for the emotional images in men than in women, but activity of the left hemisphere amygdala was significantly more related to subsequent memory for the emotional images in women than in men (see **Figure 2**). Unlike the studies just mentioned, Cahill et al. (2004) also reported a significant interaction between the variables of hemisphere and sex in the amygdala relationship to memory for emotional material.

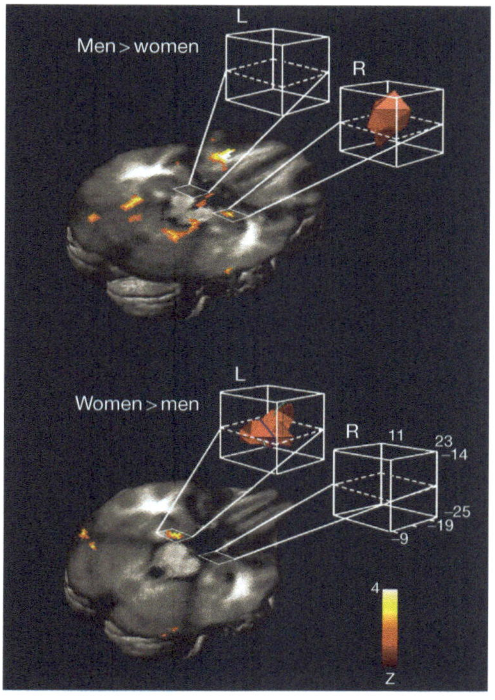

Figure 2 Sex-related hemispheric lateralization of amygdala function in long-term memory for emotionally arousing films. Activity of the right, and not left amygdala in males while viewing emotionally arousing films related significantly to memory for the films 2 weeks later. Activity of the left, and not right amygdala in women related significantly to memory for the films. Reproduced from Cahill L, Uncapher M, Kilpatrick L, Alkire M, and Turner J (2004) Sex-related hemispheric lateralization of amygdala function in emotionally-influenced memory: An fMRI investigation. *Learning and Memory* 11: 261–266, with permission from Cold Spring Harbor Laboratory Press.

Still another study directly comparing amygdala function in men and women further documents the sex-related lateralization (Mackiewicz et al., 2006). In this study, the effect was evident for more ventral amygdala aspects, which correspond largely to the basolateral nuclei which, as discussed above, are the nuclei most clearly implicated in memory modulation in animals. Collectively then, a sex-related hemispheric lateralization of amygdala function with respect to long-term memory for emotional events is evident across many studies of amygdala function from many laboratories, including four studies that have directly compared amygdala function in men and women in this context.

The sex-related hemispheric lateralization of amygdala function in emotional memory raises a simple question: What does it mean?

6.7 Sex Difference in Human Amygdala Functional Connectivity at Rest

To begin addressing the functional meaning of the amygdala lateralization, we asked a related question: Does the sex difference in amygdala function in response to emotional stimuli stem, at least in part, from a preexisting sex difference in the functional connectivity of the human amygdala at rest? To ask this question, we examined the patterns of functional covariance between the left and right hemisphere amygdalae and the rest of the brain in a large sample of men and women given blood-flow PET scans while resting with their eyes closed (Kilpatrick et al., 2006). The results of this analysis revealed that activity of the right hemisphere amygdala covaried to a much larger extent with other brain regions in men than it did in women; conversely, activity of the left hemisphere amygdala covaried with other brain regions far more in women than in men. These results are shown in **Figure 3**. Consistent with findings from several earlier investigations, no difference existed between the sexes in the overall levels of amygdala activity; rather, the sexes differed in the pattern of amygdala connectivity with the rest of the brain. These essential findings have been robustly replicated in several cohorts (Savic and Lindstrom, 2008). Thus, it appears that the sex-related hemispheric lateralization of amygdala function in emotional memory results in part from a robust, preexisting sex difference in the functional connectivity of the left and right hemisphere amygdalae at rest. The results also indicate

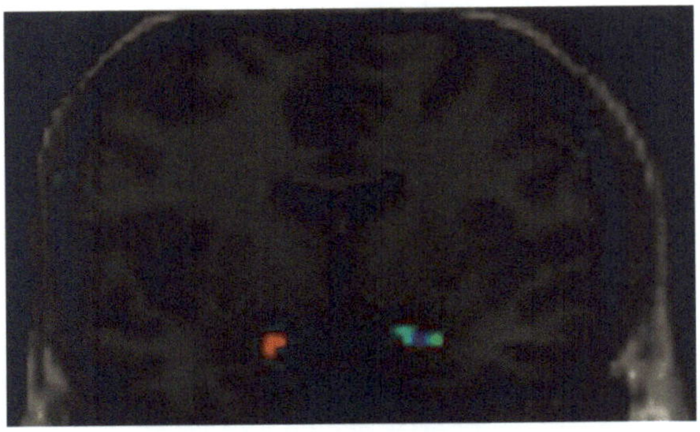

Figure 3 Amygdala seed voxels displaying significant sex-related differences in amygdala functional connectivity during resting conditions. Red areas indicate greater functional connectivity with other brain regions in women than in men. Blue areas indicate greater functional connectivity with other brain regions in men than in women. Reproduced from Kilpatrick LA, Zald DH, Pardo JV, and Cahill LF (2006) Sex-related differences in amygdala functional connectivity during resting conditions. *Neuroimage* 30: 452–461, with permission from Elsevier.

further that sex can no longer safely be ignored by any investigators of human amygdala function, since pronounced sex differences in its function presumably must exist in all experimental situations.

6.8 Relationship of the Sex-Related Amygdala Hemispheric Specialization to Hemispheric Global/Local Processing Bias

We have also sought to better understand what the sex-related lateralization of amygdala function may mean by integrating it with other knowledge about hemisphere lateralization of function. In particular, a good deal of evidence suggests that the two cerebral hemispheres differentially process more global versus local aspects of a stimulus or scene. Evidence from a variety of sources indicates that the right hemisphere is biased toward the processing of more global, holistic aspects of a stimulus or scene, while the left hemisphere is biased toward more local, finer detail processing of the same stimulus or scene (Beeman and Bowden, 2000; Fink et al., 1996, 1997).

Combining our evidence of a sex-related hemispheric laterality of amygdala function in memory for emotional material (males/right and females/left) with the view that the hemispheres differentially process global versus local information (holistic/right and detail/left) allowed us to posit that there may exist a sex-related difference in the effects of a β-adrenergic blockade on emotional memory. We know from animal research that amygdala function is impaired by β-blockers, drugs that induce blockade of β-adrenergic receptors. We also know, on anatomical grounds, that each amygdala largely modulates its own hemisphere. Hence, we reasoned that a β-blocker, by impairing amygdala function, might impair the presumed modulatory effect of the right hemisphere amygdala on the more global processing of the right hemisphere in men, thereby reducing their memory for the more global (central) aspects of an emotional story. Similarly, we reasoned that the same β-blocker might impair the presumed modulatory effect of the left hemisphere amygdala on the more local processing of the left hemisphere, thereby reducing memory for the details of the same emotional story in women. To test this hypothesis, we re-analyzed published data from two studies demonstrating an impairing effect of β-adrenergic blockade on memory for an emotionally arousing story (Cahill and van Stegeren, 2003). **Figure 4** shows the results of this analysis. Note in particular the results for story phase 2 (P2 on the *x*-axis) in which the emotional story elements were introduced (concerning severe injuries to a small boy in an accident while his mother watched), and for which the hypothesis at issue most clearly holds. The P2 results reveal a double dissociation of gender and type of to-be-remembered information (central vs. peripheral) on propranolol's impairing effect on memory: propranolol significantly impaired P2 memory of central information in men but not women, yet impaired P2 memory of peripheral detail in women but not men.

These results are consistent with the view that, under emotionally arousing conditions, activation of right amygdala/hemisphere function produces a relative enhancement of memory for central information in males, and activation of left amygdala/hemisphere function in females produces a relative enhancement of memory for peripheral details in women. Future work needs to address the generalizability of this conclusion to other emotional learning situations.

6.8.1 Other Influences of Sex on Neural and Hormonal Mechanisms of Emotional Memory

As investigators increasingly examine potential sex influences on emotion and memory, many more sex

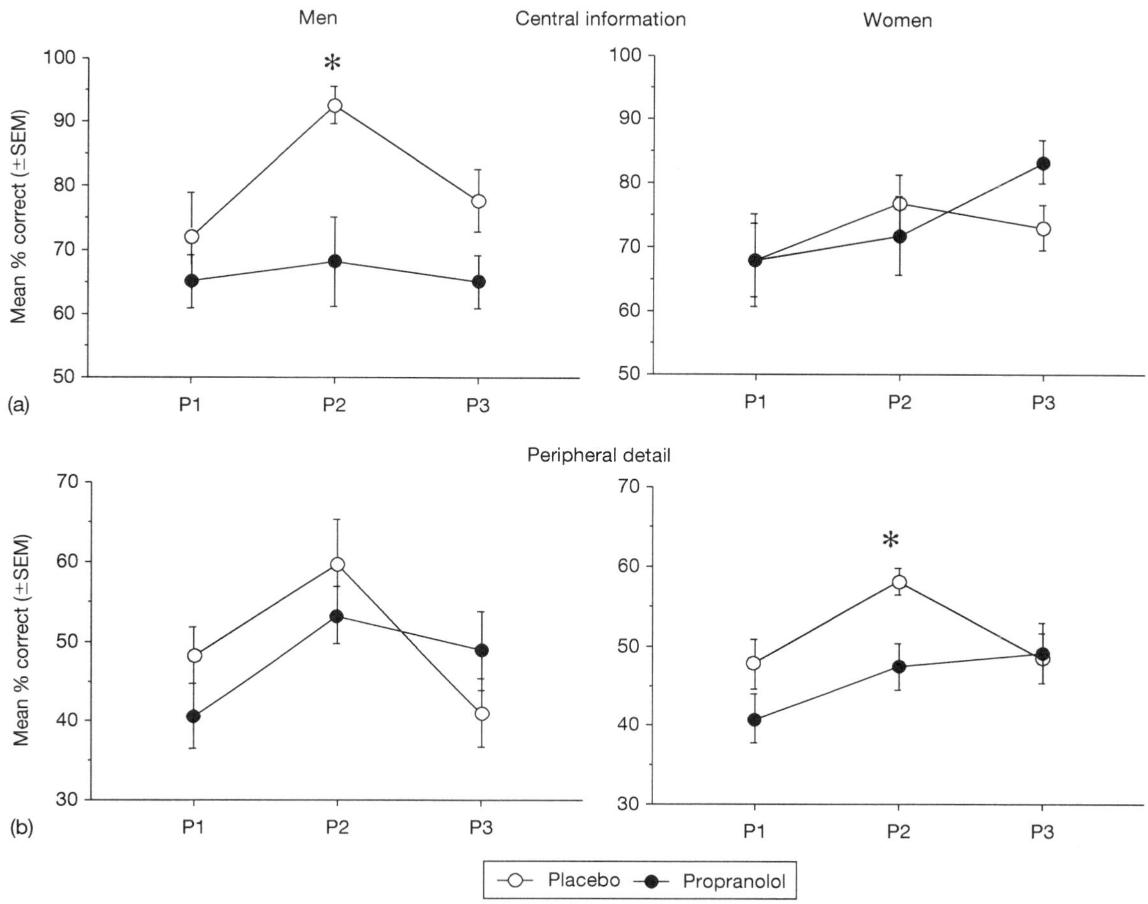

Figure 4 Recognition test scores for the three-phase emotional story phase. (a) Values for questions defined as pertaining to central information. (b) Values for questions defined as pertaining to peripheral detail. Values represent mean percent correct (±SEM) on the recognition test in each experimental group. P1, P2, P3 indicate story phases 1, 2, 3, respectively. Emotional story elements were introduced in P2. * = $p < 0.01$ placebo compared with corresponding P2 propranolol group (*post hoc*, two-tailed, unpaired *t*-test comparison). Reproduced from Cahill L and van Stegeren A (2003) Sex-related impairment of memory for emotional events with β-adrenergic blockade. *Neurobiology of Learning and Memory* 79: 81–88, with permission from Elsevier.

effects are now being uncovered. For example, Gasbarri et al. (2007) examined EEG responses to emotional and neutral stimuli in healthy men and women. The P300 response was assessed from electrodes located over the left and right hemispheres as men and women viewed emotional images. They report that for the negative (and presumably most arousing) slides, the P300 was greater when recorded over the left hemisphere in women than it was in men. Conversely, the P300 was greater when recorded over the right hemisphere in men than it was in women, a pattern (women left/men right) similar to that observed in earlier studies regarding the amygdala. They also demonstrate that sex-related differences in how the brain processes emotional events occur within 300 ms of the event onset.

Other work has examined the effects of a post-learning stressor (cold pressor stress (CPS) induced by forearm immersion in ice water) on memory consolidation. In one study (Andreano and Cahill, 2006), subjects received CPS or a control procedure immediately after hearing a short story. Memory for the story was assessed in an incidental, free recall test 1 week later. CPS produced a retrograde enhancing effect on memory in men, but not in women, despite having produced a similar cortisol response in both groups. Interestingly, the effect in men exhibited a classic inverted-U relationship between cortisol release by CPS and memory: increasing levels of cortisol associated with increased memory up to a point, after which the relationship declined. These data constitute the first demonstration of an inverted-U relationship between endogenous stress hormone release and memory in humans since Yerkes and Dodson first conceived of the inverted-U concept in 1908.

There are numerous reports of menstrual cycle influences on cognition, including learning. For example, Milad and colleagues (2007) examined sex differences exist in the acquisition and extinction of Pavlovian fear conditioning in healthy men and women. Acquisition was significantly faster in men than in women. During extinction, no overall sex difference was found, but the menstrual cycle significantly influenced the rate of extinction in females. Given such menstrual effects, we conducted a follow-up study (Andreano et al., 2008) to examine whether influences of menstrual cycle hormones may help explain the overall lack of an enhancing effect of CPS on consolidation in women in our previous study (Andreano and Cahill, 2006). Healthy, naturally cycling women were defined a being in one of three hormonally defined menstrual cycle stages at the time of learning a short story: (1) early follicular, with low levels of estrogen and progesterone; (2) later follicular, with significantly elevated levels of estrogen only and (3) mid-luteal, with significantly elevated levels of progesterone only. Subjects received CPS immediately after hearing a short story, and their memory for the story was tested 1 week later. The most intriguing finding was that the relationship between cortisol release and memory differed depending on the status of the menstrual cycle. Specifically, there was no relationship between cortisol and memory in early follicular women (low estrogen and progesterone), a strong positive relationship between cortisol and memory in mid-luteal women (high progesterone), and a nearly significant negative relationship between cortisol and memory in late-follicular women (high estrogen). This pattern of results helps explain why no overall relationship between cortisol release and memory was detected in our earlier study, which failed to account for menstrual effects. The findings also suggest that progesterone somehow potentiates the ability of cortisol to influence memory in women. In fact, other evidence suggests that progesterone increases amygdala reactivity to emotional stimuli in women, providing a potential neurobiological correlation of the progesterone effect (van Wingen et al., 2008).

6.9 Summary

The issue of sex influences on human brain function is rapidly achieving the attention it clearly deserves from cognitive neuroscientists. Sex differences in nervous system function so great that they can negate or even reverse conclusions about brain function depending on which sex is considered. When one in addition considers the abundant evidence from animal research, it becomes clear that investigators may no longer safely assume that sex influences may be ignored in any study of human brain function. This conclusion is perhaps especially clear in the domain of emotional memory, where evidence indicates that while the relevant neural mechanisms may be broadly similar in men and women, they also differ in significant ways. These developments are part of a much broader realization among neuroscientists focused on understanding human brain function that sex matters in way previously unsuspected. Investigators also increasingly realize that fully understanding disorders of emotional memory (such as post-traumatic stress disorder (PTSD) and depression) with established sex differences in their incidence and/or nature requires that we better understand sex influences on emotional memory in our basic science.

Acknowledgment

This study is supported by an NIMH RO1 57508 to L.C.

References

Andreano JM and Cahill L (2006) Glucocorticoid release and memory consolidation in men and women. *Psychological Science* 17: 466–470.

Andreano JM, Arjomandi H, and Cahill L (2008) Menstrual cycle modulation of the relationship between cortisol and long-term memory. *Psychoneuroendocrinology* 33(6): 874–882.

Beeman MJ and Bowden EM (2000) The right hemisphere maintains solution-related activation for yet-to-be-solved problems. *Memory and Cognition* 28: 1231–1241.

Blanton R, Levitt J, Peterson J, et al. (2004) Gender differences in the left inferior frontal gyrus in normal children. *Neuroimage* 22: 626–636.

Buss C, Lord C, Wadiwalla M, Hellhammer D, lupien S, Meaney M, and Pruessner J (2007) Maternal care modulates the relationship between prenatal risk and hippocampal volume in women but not in men. *Journal of Neuroscience* 27: 2592–2595.

Cahill L (2006) Why sex matters for neuroscience. *Nature Reviews Neuroscience* 7: 477–484.

Cahill L, Haier RJ, Fallon J, et al. (1996) Amygdala activity at encoding correlated with long-term, free recall of emotional information. *Proceedings of the National Academy of Sciences United States of America* 93: 8016–8021.

Cahill L, Haier RJ, White NS, et al. (2001) Sex-related difference in amygdala activity during emotionally influenced memory storage. *Neurobiology of Learning and Memory* 75: 1–9.

Cahill L, Uncapher M, Kilpatrick L, Alkire M, and Turner J (2004) Sex-related hemispheric lateralization of amygdala function in emotionally-influenced memory: An fMRI investigation. *Learning and Memory* 11: 261–266.

Cahill L and van Stegeren A (2003) Sex-related impairment of memory for emotional events with β-adrenergic blockade. *Neurobiology of Learning and Memory* 79: 81–88.

Canli T, Desmond JE, Zhao Z, and Gabrieli JD (2002) Sex differences in the neural basis of emotional memories. *Proceedings of the National Academy of Sciences of the United States of America* 99: 10789–10794.

Clements A, Rimrodt S, Abel J, et al. (2006) Sex diffrences in cerebral laterality of language and visuospatial processing. *Brain and Language* 98: 150–158.

Fink GR, Halligan PW, Marshall JC, Frith CD, Frackowiak RS, and Dolan RJ (1996) Where in the brain does visual attention select the forest and the trees? *Nature* 382: 626–628.

Fink GR, Marshall JC, Halligan PW, and Dolan RJ (1999) Hemispheric asymmetries in global/local processing are modulated by perceptual salience. *Neuropsychologia* 37: 31–40.

Frings I, Wagner k, Unterrrainer J, Spreer J, Halsband U, and Schulze-Bonhage A (2006) Gender-related differences in the lateralization of hippocampal activation and cognitive strategy. *Neuroreport* 17: 417–421.

Gasbarri A, Arnone B, Lucchese F, Pacitti F, and Cahill L (2007) Sex-related hemispheric laterality of emotional picture processing: An event-related potential study. *Brain Research* 1139: 178–186.

Giedd J, Vaituzis A, Hamburger S, et al. (1996) Quantitative MRI of the temporal lobe, amygdala, and hippocampus in normal human development: Ages 4–18 years. *Journal of Comparative Neurology* 366: 223–230.

Goldstein J, Seidman L, Horton N, et al. (2001) Normal sexual dimorphism of the adult human brain assessed by *in vivo* magnetic resonance imaging. *Cerebral Cortex* 11: 490–497.

Gur RE, Kohler C, Turetsky B, Kanes S, Bilker W, Brennan A, and Gur RC (2004) A sexually dimorphic ratio of orbitofrontal to amygdala volume is altered in schizophrenia. *Biological Psychiatry* 55: 512–517.

Gur RC, Turetsky B, Matsui M, Bilker W, Hughett P, and Gur RE (1999) Sex differences in brain gray and white matter in healthy young adults: Correlations with cognitive performance. *Journal of Neuroscience* 19: 4065–4072.

Kilpatrick LA, Zald DH, Pardo JV, and Cahill LF (2006) Sex-related differences in amygdala functional connectivity during resting conditions. *Neuroimage* 30: 452–461.

Kovalev V and Kruggel F (2007) Texture anistropy of the Brains's white matter as revealed by anatomical MRI. *IEEE Transactions on Image Processing* 26: 678–685.

Lansdell H (1964) Sex differences in hemispheric asymmetries of the human brain. *Nature* 203: 550.

Luders E, Narr KL, Thompson PM, et al. (2005) Gender effects on cortical thickness and the influence of scaling. *Human Brain Mapping* 26: 314–324.

Luders E, Narr K, Thompson P, Rex D, Jancke L, Steinmetz H, and Toga A (2004) Gender differences in cortical complexity. *Nature Neuroscience* 7: 799–800.

Luders E, Rex D, Narr R, et al. (2003) Relationships between sulcal asymmetries and corpus callosum size: Gender and handedness effects. *Cerebral Cortex* 13: 1084–1093.

Mackiewicz KL, Sarinopoulos I, Cleven KL, and Nitschke JB (2006) The effect of anticipation and the specificity of sex differences for amygdala and hippocampus function in emotional memory. *Proceedings of the National Academy of Sciences of the United States of American* 103(38): 14200–14205.

McGaugh JL (2004) The amygdala modulates the consolidation of memories of emotionally arousing experiences. *Annual Review of Neuroscience* 27: 1–28.

McGaugh JL, Cahill L, and Roozendaal B (1996) Involvement of the amygdala in memory storage: Interaction with other brain systems. *Proceedings of the National Academy of Sciences of the United States of America* 93: 13508–13514.

Milad MR, Goldstein JM, Orr SP, Wedig MM, Klibanski A, Pitman RK, and Rauch SL (2006) Fear conditioning and extinction: Influence of sex and menstrual cycle in healthy humans. *Behavioral Neuroscience* 120(6): 1196–1203.

Savic I and Lindstrom P (2008) PET and MRI show differences in cerebral asymmetry and functional connectivity between homo- and heterosexual subjects. *Proceedings of National Academy of Sciences of the United States of America* 105(27): 9403–9408.

Van Wingen G, van Broekhoven F, Verkes R, Petersson K, Backstrom T, Buitelaar J, and Fernandez G (2008) Progesterone selectively increases amygdala reactivity in women. *Molecular Psychiatry* 13: 325–333.

Wetherington CL (2007) Sex-gender differences in drug abuse: A shift in the burden of proof? *Experimental and Clinical Psychopharmacology* 15(5): 411–417.

Young MP and Scannell JW (1994) Analysis of connectivity: Neural systems in the cerebral cortex. *Review of Neuroscience* 5: 227–250.

Further Reading

Amaral DG and Price JL (1984) Amygdalo-cortical projections in the monkey (*Macaca fascicularis*). *Journal of Comparative Neurology* 230: 465–496.

Cahill L, Gorski L, and Le K (2003) Enhanced human memory consolidation with post-learning stress: Interaction with the degree of arousal at encoding. *Learning and Memory* 10: 270–274.

Canli T, Desmond JE, Zhao Z, Glover G, and Gabrieli JD (1998) Hemispheric asymmetry for emotional stimuli detected with fMRI. *Neuroreport* 9: 3233–3239.

Canli T, Zhao Z, Brewer J, Gabrieli JD, and Cahill L (2000) Event-related activation in the human amygdala associates with later memory for individual emotional experience. *Journal of Neuroscience* 20(RC99): 1–5.

Lansdell H (1964) Sex differences in hemispheric asymmetries of the human brain. *Nature* 203: 550.

Lewis PA, Critchley HD, Rotshtein P, and Dolan RJ (2007) Neural correlates of processing valence and arousal in affective words. *Cerebral Cortex* 17: 742–748.

Li H, Pin S, Zeng Z, Wang MM, Andreasson KA, and McCullough LD (2005) Sex differences in cell death. *Annals of Neurology* 58: 317–321.

Packard M, Cahill L, and McGaugh JL (1994) Amygdala modulation of hippocampal-dependent and caudate nucleus-dependent memory processes. *Proceedings of the National Academy of Sciences of the United States of America* 91: 8477–8481.

Wolf OT, Schommer NC, Hellhammer DH, McEwen BS, and Kirschbaum C (2001) The relationship between stress induced cortisol levels and memory differs between men and women. *Psychoneuroendocrinology* 26: 711–720.

7 Sex Differences in CNS Neurotransmitter Influences on Behavior

M E Rhodes, T J Creel, and A N Nord, Saint Vincent College, Latrobe, PA, USA

Chapter Outline

Glossary

catecholamines Family of amines derived from tyrosine that mediate important physiological effects as neurotransmitters and hormones. Examples of catecholamines include dopamine, norepinephrine (noradrenaline), and epinephrine (adrenaline).

circadian rhythms (diurnal rhythms) A rhythmic activity cycle, based on 24-h intervals, that is exhibited in physiological functions, including hormone secretions, of many organisms.

glucocorticoids Hormones produced by the adrenal cortex that increase glucose production in the liver, inhibit glucose metabolism by body tissues, and promote lipid breakdown in fat tissue. The principal glucocorticoid, in humans, is cortisol (hydrocortisone) and, in laboratory rodents, is corticosterone. When administered in high, therapeutic doses, glucocorticoids suppress immunological function, reduce inflammation, and decrease connective tissue and new bone formation.

lordosis Dorsiflexion of the lumbar curvature of the spine exhibited by many mammalian estrous female species upon male mounting and during copulation.

mineralocorticoids Hormones produced by the adrenal cortex that reduce the excretion of sodium and enhance the excretion of potassium and hydrogen ions by the kidney. The principal mineralocorticoid in humans and laboratory rodents is aldosterone.

sexual diergism Functional (physiological) sex differences, often deriving from anatomical differences (sexual dimorphism), that may underlie differences in behavior.

sexual dimorphism Anatomical differences between males and females.

tyrosine hydroxylase Enzyme that converts tyrosine to L-DOPA in the biochemical pathway responsible for dopamine synthesis.

vasopressin (AVP; antidiuretic hormone; ADH) A hormone, produced by cells in the hypothalamus, that is transported down the pituitary stalk to (1) the anterior pituitary gland where, along with CRH, it stimulates the secretion of ACTH and (2) the posterior pituitary gland, where it is carried by the bloodstream to the kidneys, where it reduces the excretion of water.

7.1 Introduction

A plethora of research has documented sexual dimorphism in the mammalian central nervous system (CNS). It is widely accepted that sexual dimorphism in mammals, in part, underlies sex differences in function (Arnold and Gorski, 1984; Robinson et al., 1986). A prominent example can be found in rats, where close correlations between structural dimorphisms and sex differences in sexual, social, and parental behaviors have been well established (De Vries et al., 1992; Lonstein and De Vries, 2000). Many studies have reported sexual dimorphism as differences in the number and morphology of cells in various regions of the brain. However, relatively few of these studies have supplemented the observations of anatomical dimorphisms with information on sex differences in corresponding physiology or behavior. For example, overall brain size differs between the sexes: male brains are larger in volume than female brains in both humans and laboratory animals, and females have a higher percentage of gray matter while males have a higher percentage of white matter; however, the neurochemical and behavioral ramifications of these differences are not clear (Spring et al., 2007; Cosgrove et al., 2007).

Understanding sex differences in brain–behavior relationships of complex behaviors is difficult because measures of complex behaviors are influenced by many factors, such as sensory processing, motor activity, attention, and stress response, all of which exhibit sex differences or are influenced by gonadal hormones (Kelly et al., 1999). The value of using both sexes in research is that general brain mechanisms underlying behavior may be illuminated by the differences between males and females (Kelly et al., 1999), and understanding sexual differentiation may lead to a deeper insight into the general processes involved in differentiation and specialization of neurons and their patterns of connectivity. Insight into the etiology of sex differences in the brain provides an important foundation to determine mechanisms underlying behavior and disease, and may serve as a guide in the development of sex-specific treatments for brain disorders (Cosgrove et al., 2007). With an increasing interest in the study of sex differences, the development of new strategies and methods for studying sex differences in brain and behavior will better enable scientists currently researching sex differences, or those scientists new to the field, in obtaining this valuable insight (Becker et al., 2005).

Whereas the term sexual dimorphism (Greek: *di*; two, + *morph*; shape or form) denotes differences in the shape of structures, in many instances, it is used to indicate functional sex differences as well (Loy and Sheldon, 1987). Because of the increasing importance and potential ambiguity of discussions pertaining to sex differences in physiological function versus sexually dimorphic areas of the mammalian CNS, we shall use the term sexual diergism (Greek: *di*; two, + *erg*; work or function) to characterize sex differences in physiological and biochemical functions, as well as functions underlying behavior (**Figure 1**) (Rhodes and Rubin, 1999).

In many instances, sexual dimorphism, often shown to be dependent on the presence of sex hormones during a restricted or critical period of development (Arnold, 1996; De Vries et al., 1984b, 1994a,b; De Vries and al-Shamma, 1990), underlies sexual diergism (**Figure 2**). Organizational effects of these sex hormones occur during development, while activational effects in adulthood ensure appropriate and timely sex-specific behaviors (McCarthy and Konkle, 2005). Sexually diergic behaviors are often influenced by both organizational and activational actions of sex hormones, but some behaviors are predominantly influenced by one or the other. In laboratory rodents, maze learning and locomotor activity are predominantly influenced by organizational actions, while wheel-running activity is predominantly influenced by activational actions (Beatty, 1979). The diergic consequence of some of these anatomical and cytological sex differences is clear; for example, the spinal motor nucleus of the bulbocavernosus – which contains motoneurons innervating the striated muscles of the penis – is present in male rats but absent in females (Breedlove and Arnold, 1980). In most cases, however, it is difficult to elucidate the functional significance of a particular morphological sex difference.

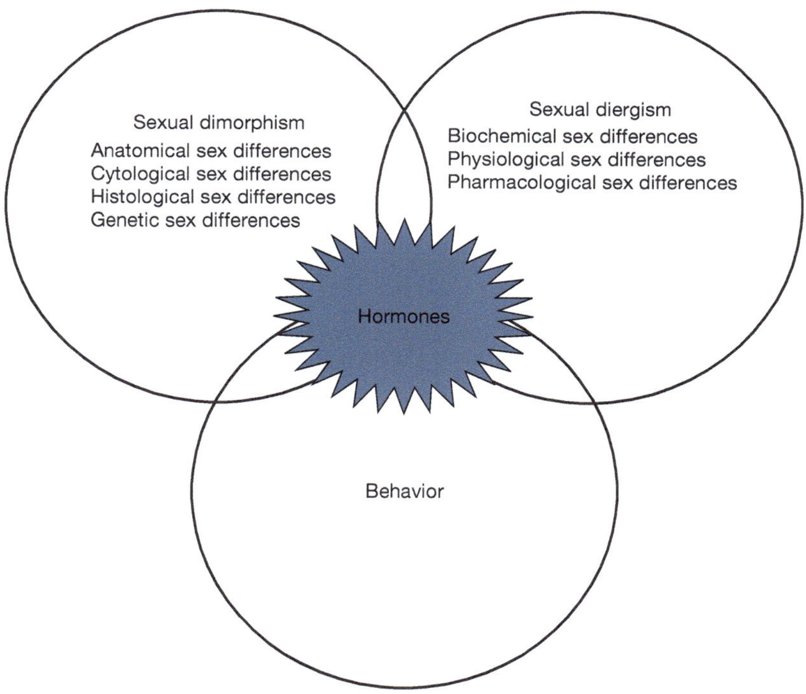

Figure 1 Defining the concepts of sexual dimorphism and sexual diergism. Sexual diergism is becoming an important variable in the study of behavior, diseases and disorders, and pharmacological and toxicological principles such as drug choice, toxicity, tolerance, dependence, and receptor regulation. Hormones, as well as neurotransmitters, are key players in the development and maintenance of sexual dimorphism, sexual diergism, and behavior. Modified from Rhodes ME and Rubin RT (1999) Functional sex differences (sexual diergism) of central nervous system cholinergic systems, vasopressin, and hypothalamic–pituitary–adrenal axis activity in mammals: A selective review. *Brain Research Reviews* 30(2): 135–152, with permission from Elsevier.

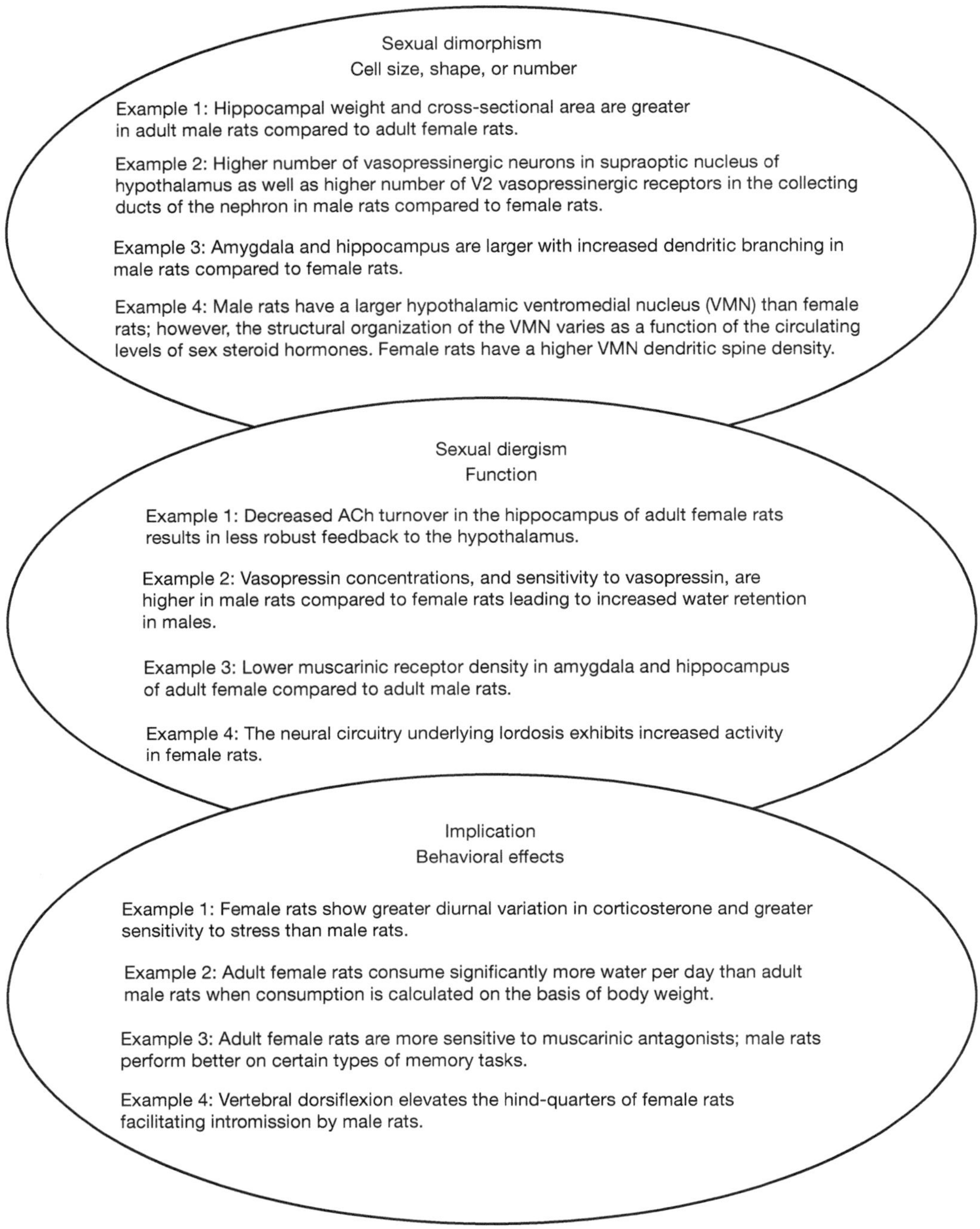

Figure 2 Sexual dimorphism, sexual diergism, and their behavioral effects. Modified from Rhodes ME and Rubin RT (1999) Functional sex differences (sexual diergism) of central nervous system cholinergic systems, vasopressin, and hypothalamic–pituitary–adrenal axis activity in mammals: A selective review. *Brain Research Reviews* 30(2): 135–152, with permission from Elsevier.

Three factors make it difficult to relate structural sex differences in the brain to sexual diergism (De Vries, 1990). First, sexually dimorphic structures are found in brain areas implicated in more than one physiological function. Second, the neuronal connections of a sexually dimorphic area are often poorly understood. Third, not many of the reported structural sex differences are dramatically influenced by sex steroids in adulthood, although many sexually diergic physiological functions are influenced by these hormones (De Vries, 1990).

The influence of sex hormones on brain morphology and function with regard to the hypothalamic–pituitary–gonadal axis has been well documented (Becu-Villalobos et al., 1997), but little is known about the sexual diergism and CNS control of the hypothalamic–pituitary–adrenal (HPA) axis (Rhodes and Rubin, 1999). This chapter focuses on the sexually diergic aspects of the mammalian CNS cholinergic systems and HPA-axis activity, as the neurotransmitter acetylcholine has been the focus of our neuroendocrine studies with male and female humans and laboratory animals over the past 10 years. Pertinent sex differences and behavioral influences of other CNS neurotransmitter systems, including dopaminergic, Gamma-aminobutyric acid (GABA)ergic, noradrenergic, serotonergic, and vasopressinergic systems, are also reviewed. Although most of the literature and data analyzed for this review relate to laboratory rodent models, other species, including humans, are discussed. Sexual dimorphism will be discussed briefly where considered essential for the understanding of a particular diergism. In many cases diergism results from dimorphism, but the intricate design of the mammalian brain can obscure evidence of a connection between a dimorphic brain region and a sexually diergic physiology or behavior. The emphasis herein is on sexual diergism and its relevance to behavior, disease, and the use of pharmacological compounds in therapeutics and research.

7.1.1 Sexual Dimorphism of the Mammalian CNS

Sexual dimorphism is often discussed in regard to the developmental sensitivity of certain neurons or areas of the brain to gonadal hormones such as estrogen, testosterone, and progesterone (PROG). In becoming sexually differentiated, a particular brain region is influenced by its hormonal environment, including hormone concentrations in the male and female circulations and hormone metabolism in the CNS itself. The magnitude of the difference in sexual behaviors in a particular species may or may not be strongly correlated with a hormone-dependent sexual dimorphism. For example, the size of the anterior hypothalamic nuclei in male and female mice is comparable, and males and females show identical levels of sexual behavior when given testosterone as adults (Cooke et al., 1998). In the male rat, however, the volume of the sexually dimorphic nucleus of the hypothalamus is larger than in the female rat, and this size difference is positively correlated with male sexual behavior (Madeira and Lieberman, 1995).

The presence of gonadal steroid receptors and the actions of steroids on such receptors should no longer be regarded as the only condition for the development of sexual dimorphism and subsequent diergism of the CNS. Sexual differentiation of the brain and behavior may be triggered by genetic mechanisms, independent of any steroid-dependent mechanism (Arnold, 1996; Pilgrim and Hutchison, 1994; Pilgrim and Reisert, 1992; Reisert and Pilgrim, 1991). For example, a human genetic susceptibility to obsessive–compulsive disorder may be present; onset of this disorder is earlier in men than in women, and this may be associated with differential activity of alleles associated with the monoamine oxidase A gene (Lochner et al., 2004). In addition, the environment has a significant impact on CNS dimorphism. For example, the structure of rat forebrain areas, such as cerebral cortex and hippocampus, is responsive to the outside environment in a dimorphic way: male rats have an increased number of cortical pyramidal neuron dendritic spines compared to female rats, and this sex difference is further increased by a stimulating housing environment (Juraska, 1991). The sex difference in the size of the cortex (dimorphism) may be due to a difference in the size of dendritic trees, ultimately leading to sex differences in physiological function (diergism) manifested by the manner in which the animal interacts with its environment (behavior). This behavior, in turn, may dynamically influence the animal's environment to sustain, increase, or decrease a respective sex difference. This example illustrates how a dimorphic brain area can underlie a diergism, with consequences for a particular behavior (**Figure 2**).

7.1.2 Sexual Diergism – Physiological Differences between the Sexes

Sexually dimorphic areas may differ in their neurotransmitter content and connectivity. Knowledge of transmitter concentrations provides little insight into the functional activity of the system being studied, but the latter can be inferred from the concentration of neurotransmitter along with receptor density and affinity and the activity of enzymes responsible for the synthesis and metabolism of the transmitter. Additional information about the sexual diergism of a sexually dimorphic region can be gained when combining transmitter information with imaging methodologies, electrophysiological studies, gonadal steroid implantation studies, and lesion studies, among others (see **Chapter 6, Sex Differences in Human Brain Structure and Function**).

7.2 Acetylcholine

7.2.1 Cholinergic Nervous System

Cholinergic systems in the brain are highly interactive with other neurotransmitter and neuromodulator systems (Karczmar, 1993), and they are anatomically diffuse, thus enhancing their capacity to mediate behavior and cognition (Woolf, 1991). Rather than being organized into discrete nuclei, cholinergic neurons are arranged into widely interconnected columns. Of the four main groupings, two are of putative relevance to HPA-axis regulation (Fibiger, 1991; Wainer et al., 1993). First is the cholinergic basal forebrain complex, which spans several discrete nuclei from the medial septal nucleus rostrally to the substantia innominata and globus pallidus caudally, and which has widespread telencephalic projections, including the cerebral cortex, olfactory tubercle, amygdala, and hippocampus. The second system is the pedunculopontine and laterodorsal tegmental cholinergic nuclei, which are distributed in a continuum in the tegmentum of the midbrain and pons. Relevant projections from this system are to the thalamus, lateral hypothalamus, lateral preoptic area, and medial prefrontal cortex.

Within these systems are eight cell groups (Ch1–Ch8) (Mesulam, 1995), with specific areas of origin (e.g., Ch1 – medial septal nucleus, Ch4 – nucleus basalis of Meynert) and projection fields (e.g., Ch1 and Ch2 – hippocampal complex, Ch4 – cerebral cortex and amygdala). These Ch1–Ch8 cell groups appear to be discrete functional units, in that they are heterogeneous with respect to co-localized receptors, neuropeptides, and enzymes (Reiner and Fibiger, 1995).

With reference to the muscarinic subtype of cholinergic receptors, which are G-protein-coupled, five subtypes have been cloned (m1–m5) and identified pharmacologically (M1–M5) (Eglen and Watson, 1996; Ehlert et al., 1995; Reiner and Fibiger, 1995; Wess, 1996, 2004; Wess et al., 1995, 2003). M1 receptors are far more numerous than M2 in the primate brain. The highest density of M1 receptors is in the limbic and association cortex, the highest density of M2 is in primary sensory and cortical motor areas, and M3 and M4 are present primarily outside the CNS. The nicotinic subtype of the cholinergic receptor is a ligand-gated ion channel related to $GABA_A$, glycine, and *N*-methyl-D-aspartate (NMDA) receptors. Activation of central nicotinic receptors presynaptically facilitates the release of a variety of neurotransmitters (Arneric et al., 1995).

7.2.2 Sexual Diergism in Choline, Choline Transport, and Acetylcholine

Sex differences have been reported for virtually all central cholinergic markers. In normal animals, basal acetylcholine (ACh) concentrations (Hortnagl et al., 1993b), high-affinity choline uptake (Miller, 1983), and choline acetyltransferase (ChAT) activity (Brown and Brooksbank, 1979) appear to be more sensitive to stimulation or blockade and less stable with age (i.e., markers tend to decrease) in females compared to males. However, ChAT activity in the mouse brain has been reported to decrease with age to a significantly higher degree in males compared to females (ChAT activity decreases were apparent by 17 months in males and 25 months in females) (Frick et al., 2002). Acetylcholinesterase (AChE) activity appears to be more sensitive to antagonism in males than in females but exhibits less stability with age in females compared to males (Loy and Sheldon, 1987; Luine and McEwen, 1983; Luine et al., 1986; Smolen et al., 1987). However, AChE-positive staining with 192 IgG-saporin is similar, regardless of brain region, in the male and female rat brain (Galani et al., 2002). Despite the overall increased sensitivity of cholinergic markers in females, cholinergic areas within the mammalian limbic system of males, such as the amygdala, hippocampus, and hypothalamus, are larger than the corresponding female areas, based on volume, sectional area, weight, and cell number (Madeira and Lieberman, 1995).

Fluctuations in circulating gonadal hormones also influence central cholinergic markers. Estrous cycling in females influences choline transport across the blood–brain barrier (BBB) (Shimon et al., 1988). Higher hippocampal levels of ACh occur in 3-month-old female rats during diestrus and proestrus compared to estrous and to age-matched male rats (Hortnagl et al., 1993b). These hippocampal differences in ACh levels at different stages of the estrous cycle parallel differences in ACh levels in the preoptic area of the hypothalamus (Egozi et al., 1986) and cycle-dependent variations in choline uptake into isolated brain microvessels (Shimon et al., 1988).

Male and female rats differ with respect to basal and stress-induced ACh concentrations in certain brain regions. The nucleus basalis magnocellularis (NBM) provides cholinergic afferents to the neocortex; therefore, it plays a role in the maintenance of cortical functions, including learning, consciousness, and control of spontaneous movements. Both male and female rats show a distinct 24-h rhythm of ACh release in the NBM, which is high during the dark phase and low during the light phase; however, female rats showed a greater ACh release and more cholinergic neurons in the NBM compared to male rats (Takase et al., 2007). This greater ACh release was supported by findings that spontaneous motor activity also showed a distinct 24-h rhythm that was greater in females compared to males (Takase et al., 2007). In contrast to ACh concentrations in the NBM, basal 24-h profiles of hippocampal ACh release were greater in male rats than in female rats, and hippocampal ACh following stress was greater in males compared to diestrous and proestrous females (Mitsushima et al., 2003; Masuda et al., 2005).

Adult female rats are more susceptible to the hippocampal neurotoxic action of the cholinotoxin AF64A than are age-matched males (Hortnagl et al., 1993a). AF64A, a cytotoxic analog of choline, interferes with the sodium-dependent high-affinity choline transport system (Pittel et al., 1987), which is localized on cholinergic nerve terminals and represents the rate-limiting step in ACh synthesis (Tucek, 1985). A more vigorous cholinergic response in female rat brains has also been demonstrated for other noxious stimuli, including heat shock and chronic ethanol administration (Papasozomenos and Su, 1991; Witt et al., 1986). The higher vulnerability of the female hippocampus to an antagonist or neurotoxin may be related to the fact that a greater number of high-affinity binding sites for glucocorticoids exists in the hippocampus of female rats than in males. This may result from a negative feedback system that is less robust in females than males, causing females to have a greater diurnal variation in plasma corticosterone (Turner, 1992; Turner and Weaver, 1985). Glucocorticoids also have decreased binding in pituitary cytosol in female rats, suggesting sexual diergism of direct and indirect feedback mechanisms (Turner, 1990). Female rats also show a greater glucocorticoid response after various forms of stress (Kant et al., 1983; Williams et al., 1985). Recent evidence that the axonal wiring of the hypothalamic infundibular area and median eminence of rats is sexually dimorphic may also contribute to sexual diergic patterns of pulsatile hormone secretions from the pituitary (Ciofi et al., 2006).

7.2.3 Sexual Diergism in Cholinergic Enzymes

Decreases in ACh synthesis and the synthesis of cholinergic enzymes have been reported in whole brain or large sections of aged rodent brain (Dwyer et al., 1980; Gibson et al., 1981), and similar changes have been found in postmortem brains of patients with Alzheimer's disease (Bartus et al., 1982; Coyle et al., 1983). Age-related declines in memory have also been shown in normal-aged rodents and monkeys (Kubanis and Zornetzer, 1981; Lippa et al., 1980), perhaps related to diminished ChAT activity of the basal forebrain (Luine et al., 1986). Luine et al. (1986) showed small but consistent decreases in the activity of ChAT in specific basal forebrain nuclei of aged (24-month-old) compared to young (4-month-old) rats. The location of the age-dependent decrease was dependent on the sex of the animal: senescent female rats showed approximately 30% lower ChAT and 40% lower AChE in the ventral globus pallidus compared to young females, whereas aged males did not show decreased ChAT activity in the ventral globus pallidus compared to young males but had 50% lower activity in the medial aspect of the horizontal nucleus of the diagonal band. ChAT activity in the mouse brain has been reported to decrease with age to a significantly higher degree in males compared to females (Frick et al., 2002). Hippocampal ChAT activity has also been shown to increase following stress (formalin test) to a greater degree in female rats compared to males (Aloisi et al., 1994; Ceccarelli et al., 2002).

Smolen et al. (1987) showed that, in male mice, brain AChE activity did not return to control levels following a single exposure to organophosphate (OP), whereas in female mice control levels were regained, albeit 20 days following exposure. These investigators suggested an increased susceptibility in males to neurotoxicity and long-term behavioral changes following OP poisoning, which possibly may have been due to sex-related pharmacokinetic differences, leading to metabolites with sustained activity in the male mice.

The flinders sensitive line (FSL) of rats was selectively bred to be more sensitive to the OP diisopropylfluorophosphate, and therefore more sensitive to irreversible blockade of AChE activity, than their control counterparts, the flinders resistant line (FRL) (Overstreet et al., 1996). Because of their behavioral features and enhanced cholinergic sensitivity, FSL rats have been suggested as an animal model of both multiple chemical sensitivity (MCS) and depression. This may have relevance for studies of biological warfare based on sex differences in the sensitivity of human AChE to agents such as pyridostigmine, OPs, and nitrogen mustard gases. For example, MCS is a clinical phenomenon in which individuals, after acute or intermittent exposure to one or more chemicals, usually irreversible cholinesterase inhibitors such as OP pesticides or nerve gases, report hypersensitivity to a wide variety of chemically unrelated compounds (Overstreet et al., 1996). MCS is a disabling syndrome strikingly similar to the illness reported among Gulf War veterans (Overstreet et al., 1996; Jamal, 1998; Weiss, 1998). There have been more female than male MCS patients reported, reaching a ratio of 4:1 in some studies (Miller, 1995). Female FSL rats are more sensitive to cholinergic agonists than their male counterparts (Netherton and Overstreet, 1983), which may have relevance for the greater incidence of MCS in women (Overstreet et al., 1988).

This is in contrast to results from our laboratory showing that young adult men and male laboratory animals are more sensitive to the effects of cholinergic stimulation than young adult women and female laboratory animals (Rhodes et al., 2001b, 1999b). This is also in contrast to the studies with mice by Smolen et al. (1987) described earlier. These findings, taken together, suggest that studies of sex differences using mice, where males are more sensitive to OPs, may not represent the best model of study for extrapolation to humans, where females may be more sensitive, based on the incidence of MCS. Since cholinergic hypersensitivity that develops chronically in depression and MCS parallels the chronic breeding process for FSL rats, these rats appear to be the more appropriate model for studies of MCS and AChE sensitivity, and perhaps depression as well.

7.2.4 Sexual Diergism in Cholinergic Receptor Activity

Sexual dimorphism has been reported for muscarinic receptors, where males have a greater number of receptors than females in certain regions of the hypothalamus and adenohypophysis of mature rats (Avissar et al., 1981a,b; Egozi et al., 1982). Therefore, it is possible that diergism exists as well. Differences in age-related brain changes in these receptors, declines of 25% and 60% in male and female rats, respectively, strongly suggest that the aging process in some cholinergic pathways of the rat brain may be characterized by sexual dimorphism.

Binding of agonists to muscarinic receptors follows a heterogeneous mode of interaction, thought to result from the existence of both high- and low-affinity binding sites (Sokolovsky, 1984). The density of muscarinic receptors changes differentially with age in various brain regions. It decreases with age in the cerebral cortex, hippocampus, striatum, and the olfactory bulb of both male and female rats. In old (27–30 months) female rats, high-affinity binding sites were preserved in all brain regions, suggesting that the decreases represent a loss of low-affinity binding sites (Gurwitz et al., 1987). In contrast, in old (27–30 months) male rats, high-affinity binding sites were decreased in the hypothalamus and increased in the brainstem. These data highlight sex differences in the aging process of central cholinergic mechanisms (Gurwitz et al., 1987).

Animal studies have found that male and female rodents have different sensitivities to the effects of nicotine. For example, female rats are less sensitive than male rats to the discriminative stimulus effects of nicotine, nicotine-induced suppression of Y-maze activity, and nicotine stimulation of active avoidance learning; whereas, female rats are more sensitive than male rats to nicotine-induced pre-pulse inhibition, antinociception, locomotor activity, social interaction, and food intake (Grunberg et al., 1987; Faraday et al., 1999; Damaj, 2001; Yilmaz et al., 1997; Cheeta et al., 2001). In contrast, female mice were found to be less sensitive to the acute antinociceptive and anxiolytic effects of nicotine (Damaj, 2001). Nicotine levels following repeated intravenous nicotine injection are higher in female rats than in male rats and these differences were attenuated by gonadectomy, suggesting that gonadal hormones influence nicotine pharmacokinetics and therefore may influence nicotine-induced sex differences in behaviors such as locomotor activity (Harrod et al., 2007). Sexual diergism in the upregulation of CNS nicotinic acetylcholine receptors has also been demonstrated: Both female control rats and female rats withdrawn from nicotine administration for 20 days exhibited higher nicotinic receptor densities than their male counterparts (Koylu et al., 1997). These findings suggest the importance of sex differences in pharmacological responses to nicotine, as well as physiological and psychological dependence and co-dependences with alcohol and other drugs associated with nicotine use, and they may have implications for understanding sex differences in smoking and smoking cessation (Parrott and Craig, 1995; Pomerleau et al., 1994, 1997; Svikis et al., 1986). Although female smokers show lower sensitivity than male smokers to the discriminative stimulus effects of nicotine, women are more sensitive to mood changes after smoking and during abstinence, resulting in shorter or less-frequent abstinence periods compared to men (Perkins et al., 1999).

7.2.5 Influence of Gonadal Steroids on Cholinergic Systems

The activity of ChAT appears to be under gonadal steroid regulation. Starting with proestrus, female rats show an increased specific activity of ChAT through metestrus, followed by a decrease in diestrus (Miller, 1983). In the hypothalamus, estradiol induces *de novo* ChAT protein synthesis (Luine et al., 1980). Therefore, the increased specific activity of ChAT during estrus and metestrus could occur from estrogen-enhanced synthesis of new ChAT or activation of existing enzyme. Estradiol administration increases ChAT activity in the nucleus of the horizontal limb of the diagonal band in ovariectomized (OVX) females, but not in castrated males, and it decreases ChAT activity in the nucleus of the vertical limb of the diagonal band in castrated males, but not in OVX females (Luine and McEwen, 1983). ChAT from Ammon's horn also has been shown to increase with estradiol administration in OVX females (Luine, 1985).

Unlike at younger ages, where hippocampal weights are similar between sexes, the hippocampus has been shown to be significantly heavier in adult male rats than in adult female rats (Loy and Sheldon, 1987). Exposure to gonadal steroids may contribute to these sex differences in hippocampal weight; however, studies addressing hippocampal weight across the estrous cycle are lacking. Biochemical sex differences have also been shown in the hippocampal formation (Madeira and Lieberman, 1995), including developmental cholinergic enzyme activities, adrenergic receptors, corticosterone receptors, GABA receptors, and cholecystokinin content (Loy and Sheldon, 1987; Madeira and Lieberman, 1995).

Nerve growth factor (NGF) is a physiological trophic factor for basal forebrain cholinergic neurons, and factors affecting NGF activity, such as gonadal sex hormones, can have a significant effect on cholinergic systems (Gibbs, 1994). Yanase et al. (1988) reported that intracerebroventricular (ICV) infusions of antibodies to NGF in neonatal rats could prevent the development of sexually diergic behaviors (Hasegawa et al., 1991, Yanase et al., 1988). Consequently, NGF may play an important role in mediating early postnatal effects of gonadal steroids on the development of specific hypothalamic projections which underlie sexually diergic behaviors in adulthood (Gibbs, 1994). Levels of NGF in the hippocampal formation of female rats decrease significantly with age, but remain relatively constant in males (Nishizuka et al., 1991), suggesting that cholinergic neurons may receive less trophic support in aged females than in aged males. These results parallel the finding that, as adult rats age, the hippocampus shrinks significantly less in male than in female rats (Loy and Sheldon, 1987).

Based on histochemical visualization of AChE-labeled cells in the ventral globus pallidus and nucleus basalis, no significant sex-dependent changes in cholinergic cell number have been apparent in aged, compared to young, mice or rats (Hornberger et al., 1985). This lack of sexual dimorphism, together with the finding of decreased and sexually diergic ChAT and AChE activity in the aged ventral globus pallidus, suggest that decreased cholinergic enzyme activity with age is likely related not only to gonadal hormone influences, but also to a slowing of neuronal metabolism, attenuated genomic activity, or deficits in post-translational processing of cellular proteins (Miller, 1983). Sexual diergism of any of these mechanisms, with subsequent decreased enzyme quantity or activity, could result in the observed greater decrease in cholinergic function with age in females than in males.

In addition to its effects on cholinergic enzyme activities, estrogen has been shown to increase basal cholinergic activity relative to other neurotransmitters, most notably dopamine. A shift toward greater cholinergic activity relative to dopaminergic activity in the striatum has been demonstrated in normal female rats and estrogen-treated males and females. A shift toward cholinergic predominance has been shown to favor the development of parkinsonian-type extrapyramidal side effects (Miller, 1983). However, clinical reports show an increased preponderance of Parkinson's disease in men (Diamond et al., 1990), indicating that gonadal steroids cannot be the only factor influencing this sex difference. Data from clinical studies have indicated that estrogen influences the development of movement disorders and tardive dyskinesias (Bedard et al., 1979; Bickerstaff, 1969), perhaps by exerting a dopamine-depleting effect upon dopaminergic pathways or by enhancing upregulation of dopamine receptors, as is observed in response to dopamine receptor-blocking antipsychotic drugs (McDermott et al., 1994). Tardive dyskinesia does occur more often in women, and its incidence increases with age in both sexes (Smith and Dunn, 1979; Smith et al., 1979). Further discussion of sex differences in dopaminergic systems is detailed later in this chapter.

7.2.6 Cholinergic Sexual Diergism in Relation to Learning, Memory, and Other Behaviors

Male rodents reportedly perform better than female rodents in a variety of mazes (Means and Dent, 1991; Mishima et al., 1986); they make fewer errors and reach criterion levels more quickly. Although this sexual diergism in performance of spatial tasks has been well documented, little is known about the underlying mechanisms. Treatments that increase cholinergic function generally improve performance of spatial tasks, whereas treatments that decrease cholinergic function impair it (Berger-Sweeney et al., 1995; Fibiger, 1991). In experiments involving spatial and cued tasks, females are more sensitive than males to the impairing effects of scopolamine, a centrally acting muscarinic antagonist (Berger-Sweeney et al., 1995). Decreased muscarinic receptor density in sexually dimorphic areas such as the amygdala and hippocampus may be one mechanism underlying diergism in memory task performance (**Figure 2**). Cholinergic lesions to the medial septum and vertical limb of the diagonal band fail to impair spatial learning in male rats, and facilitate spatial learning in female rats, suggesting that neurons in these areas support sex-specific spatial learning processes (Jonasson et al., 2004).

Considerable evidence also exists from human cognitive studies that males perform better than females on tasks requiring spatial skills (Halpern, 1986; Maccoby and Jacklin, 1974). In humans, sex differences have also been demonstrated for certain verbal abilities, including verbal fluency and speech production, where females typically demonstrate better verbal abilities than males (Hampson and Kimura, 1992). It is still not known, however, whether the sexual diergism in spatial cognition of higher mammals and the sexual diergism in verbal abilities in humans are rooted in differences in early learning experiences, emotional differences, or cultural biases. Of particular importance, it is also uncertain to what extent, if any, sex differences in spatial and verbal abilities can be attributed to hormonally organized dimorphisms in neural structures subserving cognitive functions (Williams and Meck, 1991).

Sexual dimorphism and diergism of the septohippocampal system has been extensively documented, as has its partial organization by testosterone acting via aromatization to estradiol (Williams and Meck, 1991). In addition to hormonal influences on the septohippocampal system, evidence supports the existence of environmental influences. Dendritic branching patterns in dentate gyrus cells are influenced by rearing conditions in female but not in male rats (Juraska, 1984). Male rats raised in isolation have more granule-cell dendritic material than isolation-reared females, but females have longer dendrites with more extensive branching than do males after being raised in a complex environment (Juraska, 1984). Enhanced branching of dentate gyrus cell dendrites and axons increases synaptic contact with adjacent areas, such as cholinergic neurons in the hippocampus – an area associated with memory function. These data suggest an association among sexually dimorphic cholinergic components of the hippocampus, environmental influences, and sexually diergic spatial memory performance.

Overall, from a diergic standpoint, the cholinergic nervous system appears to be more responsive in female than in male mammals, but from a dimorphic standpoint, it appears to be more prominent and less affected by aging in males than in females. As mentioned, the cholinergic nervous system is one of the most abundant and important transmitter networks. Cholinergic regulation of important regions such as the hippocampus, hypothalamus, and cerebral cortex has significant consequences as it becomes altered by the hormonal milieu, the environment, and age. Differences in age-related changes in the cholinergic receptors of both male and female rats indicate that the aging process is a major factor influencing both the basal activity of the cholinergic nervous system and its responsiveness to stimulation.

7.2.7 Acetylcholine and the HPA axis

As mentioned, the direct and indirect projections from both the basal forebrain and the mesopontine tegmental cholinergic systems innervate the hypothalamic areas, including the paraventricular nucleus (PVN) and supraoptic nucleus (SON), and regulate the release of corticotropin-releasing hormone (CRH) from the PVN and arginine vasopressin (AVP) from both nuclei (Antoni, 1993). Numerous factors are involved in regulating and modulating HPA-axis activity, including neurotransmitters, neuropeptides, adrenal steroid hormones, and gonadal steroid hormones, as well as factors such as age, weight, and environmental influences (Suescun et al., 1997; Whitnall, 1993). Whereas CRH is the primary stimulus to adrenocorticotropic hormone (ACTH) secretion, there are contributions from AVP, catecholamines, and angiotensin-II (Antoni, 1993). Considerable evidence indicates that CNS cholinergic systems stimulate both CRH and AVP secretion (Assenmacher et al., 1987; Calogero, 1995; Gregg, 1985; Michels et al., 1991; Swaab et al., 1995; Tsagarakis and Grossman, 1990; Tuomisto and Mannisto, 1985).

7.2.7.1 Sexual diergism in basal HPA-axis activity

Pronounced sex differences exist in several aspects of basal HPA-axis function in rats. One is the higher secretion and plasma concentration of corticosterone in females (Critchlow et al., 1963; Kitay, 1961; Atkinson and Waddell, 1997). Sex hormones are important in this sexual diergism. Gonadectomy has been shown to reverse the circadian variation in basal levels of corticosterone in the rat, where castration increased corticosterone secretion in male rats compared to controls and ovariectomy decreased corticosterone secretion in female rats compared to controls (Seale et al., 2004). Sex differences in the secretion of corticosterone also suggest a dimorphism and/or diergism at one or more levels of negative feedback within the axis. For example, sex differences in neurotransmitter activity (as discussed above regarding differential ACh release between male and female rats (Takase et al., 2007) in the hippocampus, an area known to regulate HPA-axis activity (Sapolsky et al., 1991; Herman et al., 1989), may be associated with sex differences in plasma–corticosterone concentrations. Female rats exhibit a greater magnitude and duration of HPA response to stressors such as handling, ether, restraint, shocks, and high-conflict situations, suggesting that negative feedback in females may be less responsive than in males (Brett et al., 1986; Critchlow et al., 1963; Griffin and Whitacre, 1991; Heinsbroek et al., 1988; Kant et al., 1983; Kitay, 1961; Weinberg et al., 1982; Williams et al., 1985).

7.2.7.2 Influence of gonadal steroids on HPA-axis activity

Testosterone can inhibit HPA-axis function, and estrogen can enhance it (Handa et al., 1994a). A mechanism by which androgens and estrogens modulate stress responses is through binding to their respective genomic receptors in the CNS. The distribution and regulation of androgen and estrogen receptors suggest areas of the brain where gonadal steroids can influence the HPA-axis stress response. In both male and female rats, estrogen administration increases basal corticosterone secretion as well as ACTH and corticosterone responses to physical and psychological stressors (Burgess and Handa, 1992; Handa et al., 1994b). Furthermore, when circulating estrogen levels are high, such as during proestrus, corticosterone levels are also high, as is the response of the HPA axis to a stressor (Viau and Meaney, 1991).

7.2.7.3 Sexual diergism of HPA-axis responses to stimulation

Inflammation represents a major stressor of the HPA axis, and sexual diergism of the HPA stress response occurs during the acute phase of inflammatory processes (Spinedi et al., 1992). Female mice exhibit more pronounced corticosterone secretion than male mice following an injection of endotoxin (Spinedi et al., 1992). The response of the HPA axis to endotoxin stimulation in female mice was similar during different stages of the estrous cycle, suggesting independence from the influence of sex steroids.

Most stress studies, however, have been done in rats. Similar to mice, there are sex differences in the response of the rat HPA axis to stress, with female rats reacting quantitatively greater than male rats (Handa et al., 1994a); for example, significantly greater increases in circulating ACTH occurred in female rats in an open field compared to males (Brett et al., 1983).

Male and female rats subjected to conditioned taste aversion exhibit similar behavioral responses, but very different HPA responses, when re-exposed to the aversive solution (Weinberg et al., 1982). Although drinking was significantly suppressed upon re-exposure in both sexes, indicating the acquisition of a conditioned aversion, female rats showed a decrease in circulating corticosterone relative to their preacquisition basal levels, whereas male rats maintained their elevated circulating levels. This effect was interpreted as opposing tendencies to consume and to avoid the solution, producing less conflict in the females than in the males. As suggested by the decrease in corticosterone, drinking produced a reduction in arousal in deprived females but not in deprived males (Brett et al., 1986). Further, the response of the HPA axis appeared to be uncompromised with age in senescent male rats but was disrupted in senescent female rats, the latter exhibiting a corticosterone elevation to deprivation or upon re-exposure and a corticosterone suppression upon drinking (Brett et al., 1986).

Females have a higher prevalence of many painful disorders (e.g., migraine, neuralgia, and arthritis) and research has consistently shown that females are more sensitive than males to nociceptive stimuli, including those that occur in internal organs (Giamberardino, 2003; Giamberardino et al., 1997; Wizemann and Pardue, 2001). The differences in pain sensitivity, however, are often small, exist for only certain forms of stimulation, and are affected by many situational variables such as the presence of disease, experimental setting, and even nutritive status (Berkley et al., 2006; Berkley, 1997). Most stressors, including restraint, induce less analgesia in female rats than in males (Bodnar et al., 1988). With reference to pain-induced stress, female rats orally self-administer greater amounts of the opioid fentanyl than male rats when exposed to a footshock stressor, and plasma corticosterone concentrations are positively correlated with fentanyl self-administration (Klein et al., 1997). Interestingly, following fentanyl self-administration, male rats exhibited greater withdrawal behaviors following naloxone challenge. These results suggested that sex differences play a role in drug-taking, and perhaps drug-quitting, behavior by rats (Klein et al., 1997).

Sex differences in opioid analgesia itself appear to be species dependent, with greater opioid analgesia observed in female than in male humans, and the opposite sex difference (or no sex difference) most often observed in rats and mice (Craft, 2003). The application of footshock to male mice induces a decrease in the response to painful stimuli in the tail-flick test, the decrease being mediated by endogenous opioids and prevented by the administration of naloxone (Menendez et al., 1993). Cortisol inhibited the analgesic response to footshock in males, suggesting that sex differences in stress-induced analgesia may result from differences in circulating glucocorticoids, since female mice have higher concentrations and stressors induce less analgesia in female than in male rats (Bodnar et al., 1988). In humans, menstrual cycle phase, dysmenorrhea, stimulus site, tissue depth, and sex all have interacting effects on pain threshold (Giamberardino et al., 1997). Both corticosteroids and sex hormones, therefore, may be involved in sex-specific responses to pain.

Other neurotransmitters also affect the HPA axis and may be involved in its sexual diergism. Histamine acts as a mediator of some neuroendocrine functions, including ACTH and corticosterone secretion following environmental stress (Bugajski and Gadek, 1983; Ghi et al., 1992; Kjaer et al., 1994). Histamine has been proposed as a possible presynaptic mediator of the release of a variety of neurotransmitters (Ferretti et al., 1998). Hypothalamic histamine release was demonstrated to be significantly lower in female rats compared to male rats, independent of age (Ferretti et al., 1998). Therefore, histamine may play an important role in the responsiveness of the female HPA axis to stress via decreased presynaptic inhibition in females compared to males.

As mentioned earlier, from a dimorphic standpoint, male mammals have increased size of some hypothalamic nuclei (e.g., preoptic area), increased size of cholinergic nuclei (e.g., lateral septum, bed nucleus of the stria terminalis (BNST)) that control secretion of hormones from the hypothalamus, and increased vasopressinergic innervation to the hypothalamus (discussed below), and therefore the potential for increased activation of the HPA axis. From a diergic standpoint, however, the HPA axis of female mammals has been reported to be more responsive to stimuli than the HPA axis of male mammals, as discussed above. However, whereas female rats often show a greater absolute increase in HPA-axis activity in response to stimulation, male rats release more corticosterone relative to baseline or control values in response to certain stimuli (Aloisi et al., 1994; Kirschbaum et al., 1992; Rivier, 1993; Spinedi et al., 1992; Weinstock et al., 1992). For example, male rats had greater corticosterone responses relative to baseline than did females in response to footshock (Rivier, 1993), restraint of various durations (Aloisi et al., 1994), and open-field testing (Weinstock et al., 1992). Orchiectomized male mice also had increased HPA-axis activity relative to baseline compared to OVX female mice in response to endotoxin (Spinedi et al., 1992). Similarly, men have both a greater absolute cortisol response and a greater cortisol response relative to baseline compared to women when faced with public speaking or performing mental arithmetic in front of an audience (Kirschbaum et al., 1992). While differences in the absolute increase in HPA-hormone concentrations may be important, sex differences in percent increases above baseline may have greater physiological relevance. Therefore, percent increases from baseline also should be considered in comparisons of HPA-axis activity between the sexes.

7.2.7.4 Sexual diergism of HPA-axis responses to cholinergic stimulation and antagonism

Our research for the past 10 years has focused on sexual diergism in CNS cholinergic systems, as reflected in HPA responses to cholinergic stimulation and blockade in female and male depressed patients and matched controls, laboratory rats, cholinergic receptor knockout (KO) mice, and *in vitro* HPA-axis tissue preparations. Our research was prompted by the unexpected finding that HPA-hormone responses (ACTH, AVP, and cortisol) following cholinergic stimulation by the acetylcholinesterase inhibitor – physostigmine – in normal men were significantly greater than those in normal women (Rubin et al., 1999b, 2006b), whereas in patients with major depression, the opposite occurred: HPA-hormone responses were significantly greater in depressed women than in depressed men (Rubin et al., 1999a).

To investigate mechanisms for this sexual diergism in humans, we have conducted numerous studies using laboratory animals, in which we have had considerably more latitude for pharmacological manipulations. Our objective has been to determine if laboratory rats and mice show the same sexual diergism that human subjects show, in order to use them as models for human studies, and to use agonists and antagonists that are relatively specific for muscarinic and nicotinic cholinergic subsystems, to determine the relative influence of these subsystems on HPA-axis function. Overall, the animal studies have yielded results similar to those in humans.

We first performed a dose–response study in 8-week-old, jugular vein-cannulated, female and male Sprague-Dawley rats (Rhodes et al., 2001b), to determine the physostigmine dose that maximally stimulated the HPA axis with minimal adverse effects, similar to that which we had determined for women and men (Rubin et al., 1999b). In the animals, this dose was 0.1 $mg\,kg^{-1}$ and has been used in our subsequent animal studies.

This first animal study involved IP saline or physostigmine administration, with and without scopolamine and mecamylamine pretreatment (Rhodes et al., 2001b). Scopolamine and mecamylamine are muscarinic and nicotinic cholinergic receptor antagonists, respectively. Similar to the human studies, physostigmine administration (0.1 $mg\,kg^{-1}$) increased ACTH responses to a significantly greater extent in the male rats compared to the females. Scopolamine pretreatment (muscarinic blockade) enhanced the ACTH responses to physostigmine in both sexes, more so in the males, whereas mecamylamine pretreatment (nicotinic blockade) abolished the ACTH responses to physostigmine in both sexes. These data suggested that presynaptic muscarinic receptors and nicotinic cholinergic mechanisms stimulate pituitary ACTH secretion.

As with ACTH, physostigmine administration increased AVP concentrations to a significantly greater extent in the male rats compared to the females. Scopolamine pretreatment resulted in a slight enhancement of the AVP responses to physostigmine in both sexes. In contrast to its effect on ACTH, mecamylamine pretreatment abolished the AVP response to physostigmine in the males but considerably enhanced it in the females. Together, the ACTH and AVP data suggested differential muscarinic versus nicotinic cholinergic regulation of the HPA axis in male versus female rats (Rhodes et al., 2001b).

In contrast to ACTH responses, physostigmine administration increased corticosterone concentrations to a significantly greater extent in the female rats compared to the males. Scopolamine pretreatment produced no discernible effects on corticosterone responses to physostigmine in either sex. Mecamylamine pretreatment, however, significantly reduced the corticosterone response to physostigmine in both sexes. The corticosterone data suggest that nicotinic cholinergic mechanisms were driving corticosterone stimulation by physostigmine.

We are currently formulating human studies to investigate the effects of scopolamine and mecamylamine pretreatment and physostigmine. Scopolamine pretreatment completely abolished the AVP and ACTH responses to physostigmine in both sexes. In contrast, mecamylamine pretreatment had no effect on the AVP response to physostigmine in the men, but it enhanced it in the women to almost that in the men. Similar to scopolamine, mecamylamine pretreatment reduced the ACTH responses to physostigmine in both sexes (unpublished results). Our preliminary human studies suggest that, as in female and male rats, muscarinic and nicotinic components of central cholinergic systems affect the central HPA axis (hypothalamic release of AVP and pituitary secretion of ACTH) differently in women and men.

Our subsequent studies in jugular vein-cannulated rats have attempted to further elucidate these sex differences. In the second study, nicotine itself was administered to female and male rats, and plasma

ACTH, AVP, and corticosterone responses were determined (Rhodes et al., 2001a). Female rats had significantly greater, dose-related ACTH and corticosterone responses to nicotine than did males, whereas the male rats had a significantly greater, dose-related AVP response to nicotine than did the females. Hormone responses following nicotine were similar to hormone responses following scopolamine pretreatment with physostigmine. These results more directly supported the concept that nicotinic receptors influence the HPA axis differentially in male and female rats.

In these first two studies using jugular vein-cannulated rats, sample sizes were not sufficient to permit analysis of female hormone responses by estrous-cycle stage. In our third study, we therefore studied larger numbers of female rats following physostigmine administration in all four defined stages of the estrous cycle (Rhodes et al., 2002). Proestrous (the cycle stage with highest plasma estrogen concentrations) and estrous females had significantly higher ACTH and corticosterone responses compared to metestrous and diestrous females. In contrast, AVP responses were not significantly different across cycle stages. Male rats had higher ACTH and AVP responses than did females in all estrous cycle stages, which further supported the concept of greater male than female HPA-axis responses to cholinergic challenge. Similar to our first and second study, corticosterone responses were significantly greater in the female rats compared to the males, regardless of estrous cycle stage.

In a parallel fourth study, we studied the influence of estrous cycle stage on HPA-hormone responses to nicotine (Rhodes et al., 2004). Proestrous and estrous females had higher ACTH responses to nicotine ($0.3\,mg\,kg^{-1}$) compared to metestrous and diestrous females and compared to males. ACTH responses to nicotine ($0.5\,mg\,kg^{-1}$) were similar, regardless of estrous cycle stage or sex. In contrast to ACTH, males had higher AVP responses to both NIC doses than did females in all estrous stages (Rhodes et al., 2004). These results suggest that the nicotinic system specifically contributes to differential HPA-axis responses to cholinergic challenge across the estrous cycle. Again similar to our first and second study, corticosterone responses were significantly greater in the female rats compared to the males, regardless of estrous cycle stage.

A fundamental issue in our animal studies has been baseline HPA-hormone concentrations, before experimental manipulations have begun. To achieve as low baseline HPA-axis activity as possible, female and male jugular vein-cannulated rats were individually housed in standard cages and in cages with environmental enrichment with Kong toys (rubber chew-toys) and Nestlets (shredding material for nest building) (Belz et al., 2003). Both sexes housed with environmental enrichment had significantly lower baseline ACTH and corticosterone concentrations than did those housed without environmental enrichment. As well, HPA-hormone responses to the mild stress of saline injection were significantly lower in rats housed with environmental enrichment, and this effect was most pronounced in the females. These results were important, because low and stable baselines are essential for accurately discerning pharmacological and other challenges to the HPA axis.

We recently have extended our studies to an additional animal experimental model, M1 and M2 muscarinic receptor-KO and wild-type (WT) mice, to investigate the influence of components of the muscarinic cholinergic subsystem on the sexual diergism of HPA-axis responses to cholinergic stimulation. We first explored the role of M2 receptors in the sexual diergism of HPA-axis responses to cholinergic stimulation by administration of the nonselective muscarinic agonist oxotremorine, physostigmine, and saline (a mild stressor and used as control) to male and female M2-receptor-KO and WT mice of the same genetic background (Rhodes et al., 2005). Because M2 receptors are primarily presynaptic autoreceptors, we hypothesized that M2-KO mice would have greater hormone responses to cholinergic stimulation than WT mice. Both male and female M2-KO mice were significantly more responsive to the mild stress of saline injection than were their WT counterparts. Oxotremorine and physostigmine increased ACTH and corticosterone in all four groups to a significantly greater degree in KO males (Rhodes et al., 2005). Consistent with our findings in rats, as noted above, the increase in ACTH also was significantly greater in WT males compared to WT females. In contrast to ACTH, yet also consistent with our findings in rats, the increase in corticosterone was significantly greater in females compared to males, independent of genotype. Following pretreatment with scopolamine, ACTH and corticosterone responses to oxotremorine and to saline in the M2-KO mice were comparable to those in the WT mice, as would be expected from blockade of M1 postsynaptic receptors. These findings suggested that the M2 receptor regulates ACTH responses to cholinergic stimulation in males but not in females.

We have recently performed similar studies on M1 muscarinic receptor KO and WT mice (Rhodes et al., 2006). Because M1 receptors are postsynaptic receptors, we hypothesized that hormone responses in M1-KO mice would be lower than responses in their WT counterparts because the absence of M1 receptors, and presence of M2 autoreceptors, would reduce cholinergic stimulation of the HPA axis. As expected, M1-KO females and males were not more responsive to the mild stress of saline injection than their WT counterparts, and they had significantly lower ACTH and corticosterone responses to oxotremorine than their WT counterparts. Consistent with our previous findings, male M1-KO and WT mice had significantly greater ACTH responses to oxotremorine than did M1-KO and WT females. These findings suggested that the M1 receptor also regulates ACTH responses to cholinergic stimulation, in both sexes.

Our studies in humans and laboratory animals do not permit pharmacologic manipulation of the separate components of the HPA axis, a three-gland endocrine system. We therefore developed a novel, *in vitro* perfusion system to enable study of pharmacological and hormonal challenges to individual tissue components of the male and female HPA axis of rats (Moidel et al., 2006). Following nicotine addition, *in vitro* CRH and ACTH responses from female tissues were greater than responses from male tissues, similar to the responses following nicotine administration *in vivo* (Rhodes et al., 2001a). The ability of the *in vitro* system to replicate *in vivo* HPA-axis responses supports its potential as a new method for pharmacological studies. As well, it especially permits determination of CRH release from hypothalamic tissue, whereas peripheral CRH in intact humans and laboratory animals does not accurately reflect CRH release into the pituitary portal system (Plotsky et al., 1990). We continue to use this *in vitro* perfusion system to investigate at which level(s) of the HPA axis sexual diergism occurs in response to cholinergic stimulation.

As indicated above, HPA-hormone responses (ACTH, AVP, and cortisol) following cholinergic stimulation by the acetylcholinesterase inhibitor, physostigmine, in normal men were significantly greater than those in normal women (Rubin et al., 1999b), whereas HPA-hormone responses following cholinergic stimulation were significantly greater in depressed women than in depressed men (Rubin et al., 1999a).

Because cholinergic deficits are a prominent aspect of aging and diseases, such as Alzheimer's disease, we examined the persistence of sexually diergic HPA-axis responses to cholinergic stimulation with aging by performing a cholinergic stimulation study in elderly women and men (Rubin et al., 2002). The elderly subjects had greater HPA responses to physostigmine than did their younger counterparts. Notably, there was a loss of the sexual diergism between male and female hormone responses to physostigmine that occurred in the younger subjects. This change with aging is likely related to the altered sex-steroid hormonal milieu in elderly subjects.

Because of the sensitivity of growth hormone (GH) secretion to cholinergic stimulation (Giustina and Veldhuis, 1998; Bertherat et al., 1995), in addition to HPA-axis hormones, we measured GH in the samples from our initial study of depressed patients and controls (Rubin et al., 2003). Baseline GH was higher in the women than in the men, but it was not significantly different between the female or the male patients and their respective matched controls. The female depressed patients had a significantly greater GH response to physostigmine than did their female controls, whereas the male depressed patients and controls had similar GH responses. These findings suggest sexual diergism in cholinergically stimulated GH, as well as in HPA-axis hormones, between patients with major depression and matched normal controls.

Because GH decreases considerably with age, we also measured GH in the plasma samples from our study of HPA-axis hormone responses to physostigmine in healthy elderly women and men (Rubin et al., 2002, 2006a). We hypothesized that (1) elderly women and men would have similar GH responses, because of relatively low circulating estrogen in the women and (2) the elderly women would have significantly lower baseline GH and GH responses to cholinergic challenge than the young women we studied previously. Physostigmine significantly increased GH compared to saline, to a similar degree in the elderly women and men. In contrast to our hypothesis, the elderly women had a significantly greater GH response to physostigmine than did the young women, whereas GH responses were similar in the elderly and young men. A likely mechanism for the greater GH response in the elderly women is increased sensitivity of central cholinergic systems that inhibit somatostatin and/or enhance growth hormone-releasing hormone release from the hypothalamus, in spite of reduced circulating estrogen.

From the evidence reviewed concerning sex differences, cholinergic systems, and HPA-axis activity, it is clear that male and female mammals respond to stimulation of the HPA axis in different ways at several points of regulation. Numerous factors regulate HPA-axis activity, including neurotransmitter systems, neuropeptides, adrenal and gonadal steroid hormones, as well as environmental influences. Correlating studies of sexual dimorphism, sexual diergism, and behavior may provide insights toward our understanding of the complex relationships among these factors and HPA-axis regulation. Given the growing interest in neuroendocrine–immunological interrelationships, further investigation into sexually diergic responses to stress, pain, and the release of corticosteroids may provide valuable insights into the influence of the CNS and the HPA axis on immunological function as well.

7.3 Dopamine

Dopamine exerts an important modulatory role in the control of motor activity, reward-related mechanisms, emotion, and cognitive processes (Heijtz et al., 2002). It has long been recognized that sexual diergism exists in dopaminergic function, particularly within the striatum. Although these differences often can be attributed to sexual dimorphism, they become more apparent when considering the unique behavioral responses of males versus females to drugs that affect dopamine release and the incidence and severity of disease states associated with the dopaminergic system.

7.3.1 Dopaminergic Age-Related Sex Differences

Aging is associated with changes in dopaminergic function and these changes may underlie certain changes in behavior and the proneness to diseases (Morgan et al., 1987). Decreased tyrosine hydroxylase activity and D_2-dopamine receptor density has been reported within the striatum of aged male and female rats. These changes were especially evident in the males, in whom decreases in receptor number were associated with increases in their affinity (Fernandez-Ruiz et al., 1992). In the limbic forebrain, tyrosine hydroxylase activity also was decreased during aging, but only in males (Fernandez-Ruiz et al., 1992). These results suggest that dopaminergic neurotransmission is more stable across age in females versus males.

It would appear that these age-related sex differences are dependent on hormonal changes before and during sexual maturation. For example, male rats exhibit a more prominent rise in D_1 and D_2 receptor density in the striatum prior to puberty compared to female rats, which may compensate for the greater decline in receptor density that occurs in males between the onset of puberty and adulthood. Despite these sex differences in overproduction and elimination of dopamine receptors near puberty, adult receptor densities overall were similar in this study (Andersen et al., 1997). In addition to striatal dopamine receptor changes during sexual maturation, sharp decreases in hypothalamic D_2 receptor concentrations have also been observed, to a greater extent in females compared to males (Herdon and Wilson, 1985). These changes may be linked with the alterations in hormone levels (e.g., prolactin) that occur during sexual maturation. These findings suggest that gonadal hormones are important in the age-related sexual diergism associated with the dopaminergic system.

7.3.2 Sexual Diergism, Gonadal Hormones, and Dopamine

The nigrostriatal dopamine system is sexually dimorphic and gonadal hormones modulate striatal dopamine in female rats but not in male rats (Becker, 1990, 1999, 2000; Castner et al., 1993). Ovariectomy decreases striatal dopamine release and turnover and estrogen replacement restores dopaminergic function to that of intact females in estrus. In contrast, castration of male rats has no effect on striatal dopaminergic function (Becker, 1990). Furthermore, ovariectomy results in attenuation of amphetamine-induced rotational behavior, electrical stimulation-induced rotational behavior, and amphetamine-induced stereotyped behaviors; whereas castration of male rats has no effect on these behaviors (Becker, 1990; Castner et al., 1993). In addition, treatment of OVX female rats with estradiol results in enhancement of amphetamine-stimulated striatal dopamine release and stereotyped head and forelimb movements (Morissette and Di Paolo, 1993; Becker, 1990).

Dopamine reuptake into nerve terminals is the primary mechanism of inactivation of this neurotransmitter at the synapse. Striatal dopamine reuptake sites are lower in male compared to intact female rats and these sites fluctuate during the estrous cycle in females (Morissette and Di Paolo, 1993). As well, dopaminergic reuptake sites are lower in OVX rats compared to intact cycling females and the density of these sites is similar in castrated male rats, intact male rats, and OVX female rats (Morissette and Di Paolo, 1993). Studies across various stages of the estrous cycle provide further support to the importance of estrogen in the regulation of dopaminergic activity. During proestrus and estrus, when circulating estrogen and PROG are increased, amphetamine-stimulated striatal dopamine release is enhanced and striatal dopamine reuptake sites are most concentrated in coincidence with peak dopamine and serotonin concentrations (Becker, 1990, 1999).

These results suggest that estrogen plays a major role in the sexual diergism of dopaminergic functioning within the striatum. It has been postulated that estrogen may offer neuroprotection to the striatal dopaminergic system, and that this neuroprotection may reflect lower incidences in Parkinson's disease, as well as drug toxicities and addiction (e.g., resulting from amphetamine or methamphetamine use), in women compared to men (Dluzen and McDermott, 2000; Dluzen and Mickley, 2005; Dluzen et al., 2003). These results also suggest that gonadal hormones could influence the activity of psychoactive drugs, particularly those with mechanisms that influence dopamine release, turnover, and reuptake sites. For example, cocaine has a high affinity for dopamine reuptake sites and inhibits dopamine reuptake. Following cocaine administration, female rats exhibit more hyperactivity and stereotyped behaviors than male rats (Schindler and Carmona, 2002; Morissette and Di Paolo, 1993; Walker et al., 2001). Female rats also show increased striatal dopamine release following administration of the typical antipsychotic haloperidol (Walker et al., 2006), and a greater vulnerability to antipsychotic-induced catalepsy compared to male rats (Miller, 1983). This vulnerability to antipsychotics has a human counterpart in that women show an increased frequency of extrapyramidal syndrome following antipsychotic administration compared to men (Seeman, 1985, 2000). These results suggest that nigrostriatal dopamine neurotransmission in females is more tightly regulated by autoreceptor and transporter mechanisms than in males.

These sex differences in dopaminergic activity are not exclusive to the striatum. Gonadal hormones modulate behavioral and neurochemical indices of dopamine activity in the nucleus accumbens, in a manner very similar to striatal dopamine activity (Becker, 1999). In addition to these brain areas, dopamine concentrations from pituitary stalk plasma are significantly greater in diestrous female rats compared to male rats. Female gonadal hormones appear to contribute to this sexual diergism considering that OVX female rats and prepubertal female rats had lower dopamine concentrations from pituitary stalk plasma than diestrous female rats (Gudelsky et al., 1984; Guillamon and Segovia, 1997). Furthermore, treatment of OVX female rats with estradiol resulted in significant elevations of dopamine concentrations in the pituitary stalk (Gudelsky and Porter, 1981). These findings support the stimulatory action of female gonadal hormones on dopamine release in other areas of the brain in addition to the striatum.

Recent evidence, using positron emission tomography (PET) scans, suggests that men have markedly greater dopamine release from the striatum, anterior putamen, and anterior and posterior caudate nuclei than women following administration of amphetamine (Munro et al., 2006). There is a high density of D_2 autoreceptors in the human striatum and a much lower density in extrastriatal regions, specifically the frontal cortex, temporal cortex, and thalamus. PET evidence indicates that women have higher D_2-like receptor-binding potentials than men in these nonstriatal regions, particularly the frontal cortex (Kaasinen et al., 2001). This difference could be representative of either a difference in receptor density or affinity or a combination of the two factors.

These dopaminergic sex differences from both animal and human studies instigate speculation on sex differences in susceptibility to neuropsychiatric and addictive disorders related to dopamine release and dopamine receptor density and affinity (Kaasinen et al., 2001). Indeed, attention deficit/hyperactivity disorder (ADHD), schizophrenia, Tourette's syndrome, and drug addiction are more common in men than in women (Leung and Chue, 2000; Andersen et al., 1997), and the striatum and nucleus acccumbens have been implicated in the pathophysicology of Tourette's syndrome, ADHD, substance abuse, and schizophrenia (Andersen et al., 1997).

7.4 Gamma-Aminobutyric acid

7.4.1 Sex Differences in GABAergic Systems

Sexual dimorphism and diergism of the GABAergic nervous system have been extensively studied in the hypothalamus. GABAergic neurons in the hypothalamus mediate pituitary hormone release, perhaps under regulation by gonadal steroids. In the medial preoptic and ventromedial nuclei of the hypothalamus, areas involved in the regulation of gonadotropin secretion and sexual behavior, GABAergic activity was found to be about twofold greater in males than in females (Grattan and Selmanoff, 1997). GABA concentrations in the preoptic area were found to be higher in males than females, regardless of estrous cycle stage; however, GABA concentrations in the ventromedial nucleus were found to be higher in proestrous females compared to males and diestrous and estrous females (Frankfurt et al., 1984). Activity of the GABAergic synthetic enzyme glutamic acid decarboxylase (GAD) in the mouse brain appears to decrease with age to a significantly higher degree in males compared to females (GAD activity decreases were apparent by 17 months in males and 25 months in females) (Frick et al., 2002). In addition, a large body of pharmacological evidence also suggests that GABAergic mechanisms in the hypothalamus are involved in differentially regulating reproductive behavior in male and female rats (McCarthy et al., 1991a,b; Agmo and Paredes, 1985; Agmo et al., 1989; Zhou et al., 2005).

Ovulation is triggered by estrogen-dependent activation of gonadotropin-releasing hormone (GnRH) neurons in the preoptic area of the hypothalamus. Evidence from rodent models indicates that this activation is indirect and mediated primarily by a specific region of the preoptic area, the anteroventral periventricular nucleus. Nearly all neurons in the anteroventral periventricular nucleus of female rats express vesicular glutamate transporter 2, a marker of hypothalamic glutamatergic neurons, as well as GAD and the vesicular GABA transporter, markers of GABAergic neurons. These dual-phenotype neurons are the main targets of estrogen in the region and are more than twice as numerous in females as in males (Ottem et al., 2004). Additionally, these dual-phenotype neurons innervate GnRH neurons, and at the time of the ovulation surge, GABA transporter-containing neurons decrease and glutamate transporter-containing vesicles increase in these terminals with GnRH neurons (Ottem et al., 2004).

GABA-immunoreactive neurons in the medial section of the BNST are higher in number in female rats than in males, a sexual dimorphism that may underlie functional differences (Stefanova et al., 1997). The BNST is larger in males than it is in females, and projections from the BNST are also sexually dimorphic, innervating the anteroventral periventricular nucleus of males far more densely than that of females (Polston et al., 2004). This sex-specific inhibitory input to the anteroventral periventricular nucleus may represent a key component of forebrain pathways in males that render them incapable of displaying female-typical physiological and behavioral responses (Polston et al., 2004).

Parental care in the rat appears to be sexually diergic; females become maternal immediately after parturition whereas males do not display pup care and may even cannabalize the infants (Segovia et al., 1996). These sex differences in behavior may be regulated by GABAergic inputs to the hypothalamus, as postnatal exposure to diazepam, a benzodiazepine that potentiates the activity of GABA, facilitates the induction of maternal behavior in adult virgin female rats (Segovia et al., 1996). Furthermore, neonatal administration of the GABA agonist diazepam to male rats facilitates the induction of maternal behavior in adults, while administration of the GABA antagonist picrotoxin to female rats appears to disrupt maternal behavior in adults (Segovia et al., 1996).

7.4.2 Influence of Gonadal Steroids on GABAergic Sex Differences

In vitro studies of the sexual differentiation of GABAergic hypothalamic neurons have shown that sexual dimorphism appears to be independent of gonadal hormones, because short- and long-term treatment with estradiol or testosterone failed to affect cell numbers or neuronal GABA uptake (Lieb et al., 1994; Segovia et al., 1996). Pharmacological sex differences in benzodiazepine binding have been reported in certain rat strains (Shephard et al., 1982), and these differences may also be hormonally independent based on the finding that neither GABA receptors labeled with ^{3}H-muscimol nor benzodiazepine appear to vary over the estrous cycle in rats (Hamon et al., 1983).

Other studies suggest that, in some cases, GABAergic sex differences may be mediated by gonadal hormones. The arcuate nucleus is a region of the rat hypothalamus important in the regulation of luteinizing hormone (LH), GH, and prolactin secretion. In the adult female rat, neurons in the arcuate nucleus show a phased synaptic remodeling during the estrous cycle that is characterized by a decrease in axosomatic synapses when circulating estrogen levels are high, such as the morning and afternoon of proestrus (Mong et al., 1996). When circulating estrogen decreases after estrous, the number of axosomatic synapses rises to a baseline condition (Olmos et al., 1989). These synaptic changes can be induced in OVX rats by a single dose of estradiol (Perez et al., 1993; Parducz et al., 1993) and immunocytochemical studies show that these axosomatic synapses are primarily GABAergic (Parducz et al., 1993). Thus, a decrease in the number of inhibitory GABAergic synapses when circulating estrogen concentrations are high may contribute to the estrogen-stimulated increase in arcuate neuronal firing, leading to the LH surge of the estrous cycle (Yeoman and Jenkins, 1989). This mechanism may have implications in the increased susceptibility of female mammals to stress and seizures at certain times during the estrous cycle.

7.4.3 Sexual Diergism in GABAergic Systems

As mentioned earlier, in the rat medial preoptic nucleus and ventromedial nucleus of the hypothalamus, areas involved in the regulation of gonadotropin secretion and sexual behavior, GABAergic neuronal activity has been reported as twofold greater in males than in females (Grattan and Selmanoff, 1997). Similar rates (twofold greater in male vs. diestrous female rats) of GABA turnover and GAD mRNA have been reported in the diagonal band of Broca, anteroventral periventricular nucleus, and median eminence of rats (Searles et al., 2000). The functional relationships between these sex differences and sexually diergic phenotypes such as gonadotropin secretion, sexual behavior, and seizure threshold remain to be determined.

From a behavioral perspective, human studies have demonstrated that women are more prone than men to many types of anxiety disorders (Altemus, 2006). Because women complain of premenstrual and postpartum exacerbations of anxiety and panic states, hormonal fluctuations have been advanced as possible contributors to understanding the differences between the sexes and anxiety disorders. Since estrogen has been shown to upregulate the $GABA_A$ receptor (Maggi and Perez, 1986), it is likely that the cyclic withdrawal of estrogen and progestins kindles neuronal systems and promotes anxiety states by mechanisms similar to those which have been implicated in the provocation of perimenstrual epilepsy (Narbone et al., 1990). The periodic emmergence and withdrawal of circulating estrogen and PROG may be responsible for women, as a group, being more sensitive than men to the anxiogenic effects of nonspecific stress (Seeman, 1997).

Sex differences have been observed in the susceptibility to seizures induced by the administration of the GABA antagonist bicuculline (Wilson, 1992). Female rats had higher seizure thresholds compared to males because of enhanced cortical $GABA_A$ receptor affinity and higher circulating levels of PROG, a positive allosteric modulator of the GABA receptor. Male rats showed greater seizure susceptibility than females as indicated by the lower dose of bicuculline required to induce seizures (Wilson, 1992). However, this effect is in contrast to the sex differences in susceptibility reported for picrotoxin-induced convulsions and maximal electroshock seizures (Woolley and Timiras, 1962a,b) and the lower seizure induction observed in males compared to intact and OVX females after picrotoxin (Schwartz-Giblin et al., 1989; Pericic et al., 1996).

Epileptic and absence seizures have also been reported to be more common in males than in females (Ravizza et al., 2003; Kyrozis et al., 2006). One of the brain areas that have been implicated in the higher susceptibility of males to seizures is the substantia nigra reticulata (SNR) (Ravizza et al., 2003; Kyrozis et al., 2006). Several studies support the existence of GABAergic differences in male versus female infantile SNR neurons (Galanopoulou, 2005). Therefore, in infantile rats, drugs or conditions that modulate the activity of $GABA_A$ receptors may have different effects on SNR differentiation during development. As a result, these conditions have the potential of causing long-term changes in the function of the SNR in the control of seizures, movement, and the susceptibility to and course of epilepsy and movement disorders (Galanopoulou, 2005). The SNR is by no means the only brain area implicated in sexually diergic seizure susceptibility. Recently, sex differences in the expression of $GABA_A$ receptor gamma2 subunits in the thalamic reticular and ventrobasal nuclei in a rat absence seizure model may suggest a mechanism for the observed sex differences in this type of seizure (Li et al., 2007).

GABAergic systems also appear to modulate HPA-axis activity in a sexually diergic manner. Diazepam decreases plasma corticosterone in female but not in male rats, and picrotoxin increases corticosterone in females approximately threefold greater than in males (Pericic et al., 1985). Changes in HPA activity resulting from acute and chronic stress also may influence GABAergic systems, because acute swim stress in female, but not male, mice results in a significant increase in GABA binding to membranes prepared from the forebrain (Akinci and Johnston, 1993). In contrast, chronic restraint stress resulted in increased hippocampal GABA concentrations in male rats, but not in female rats (Luine, 2002). Acute and chronic stress generally produce deficits on hippocampal-dependent memory processes in male rats, while enhancing these processes in females (Conrad et al., 1996, 2004; Luine, 2002; Luine et al., 1994). Therefore, changes in hippocampal GABA concentrations may underlie sexually diergic memory performance following various forms of stress.

In a unique reversal of function, GABA serves as the major source of neuronal excitation in the developing brain (McCarthy et al., 2002). During embryonic and postnatal development, neuronal excitation is mediated by both GABA and glutamate. The excitatory depolarizing action of GABA is gradually replaced by inhibitory hyperpolarizations beginning around postnatal day 7 (Davis et al., 1999). During a perinatal sensitive period, estradiol modulates many of the cellular responses that differentiate male and female brains. Perinatal treatment with the GABA agonist muscimol reduces the size of the male sexually dimorphic nucleus of the preoptic area in rats while having no significant effect on the same region in females, indicating that the influence of the GABAergic systems in development may be independent of the influence of estrogen (Bach et al., 1992). Recent evidence, however, suggests that the mechanism of gonadal hormone action perinatally involves enhancing, and extending the duration of, the developmental excitatory effects of GABA, resulting in divergence of the signal-transduction pathways activated in males versus females resulting in a pivotal point in brain development that dictates the sex differences evident in adult physiology and behavior (McCarthy et al., 2002).

The accessory olfactory (vomeronasal) system implicated in reproductive physiology and behavior in mammals is sexually dimorphic (Garcia-Falgueras et al., 2006a; Segovia et al., 2006). The rat vomeronasal system and olfactory accessory bulb has a greater volume and cell number in males than in females (Segovia et al., 1999). In the rat olfactory accessory bulb, GABAergic excitatory neurotransmission during development appears to be related to the organization of a female rather than a male brain because administration of a GABA agonist induces a feminine pattern of this structure in male rats; whereas the administration of the GABA antagonist picrotoxin caused a masculine pattern in the organization of this structure in the female rat (Segovia et al., 1999, 1996).

In addition to their classical mechanisms, steroids are involved in neuromodulatory, nongenomic actions. Steroids in this category include PROG, allopregnenolone (ALLO), the glucocorticoid derivative tetrahydrodeoxycorticosterone (THDOC), dehydroepiandrosterone (DHEA), pregnenolone (PREG), and their sulfates esters (DHEA-S and PREGS), among others. THDOC and PROG and its metabolites act primarily as positive allosteric modulators of the GABA receptor, while DHEA-S and PREGS, for example, act as negative allosteric modulators (antagonists). Although these steroids have not been implicated to the extent of estrogen and testosterone in sexual dimorphism and differentiation, their ability to modulate receptor function, particularly GABAergic function, and their unequal concentration profiles in males and females make them likely candidates to contribute to sexual diergism. For example, male rats show greater GABAergic responses following THDOC than female rats in the amygdala and preoptic area of the hypothalamus (Wilson and Biscardi, 1997). Because DHEA-S is the major steroid in human plasma, and because its concentration is sex and age dependent, it would seem that this neurosteroid may also influence the development and maintenance of sexually diergic behaviors.

Although evidence exists for both hormonally dependent and independent mechanisms driving sexual dimorphism and diergism of GABAergic systems, the findings reviewed herein suggest that sex differences in the GABAergic output, particularly from the hypothalamus and perhaps from other regions of the brain, are under the influence of circulating estrogen. Many of the reviewed articles also establish a close relationship between GABA-mediated conditions (such as seizure susceptibility and sexual behavior) and the GABA-directed development of certain brain regions.

7.5 Norepinephrine

7.5.1 Sexual Dimorphism and Diergism of Noradrenergic Systems

A limited number of studies have indicated areas of sexual dimorphism and sexual diergism in the catecholaminergic systems. Typically, males have been found to exhibit greater noradrenergic sensitivity and greater levels of norepinephrine (NE) in specific regions of the brain compared to females (van Doornen, 1986). Most notably, regions such as the preoptic, suprachiasmatic, paraventricular, periventricular, and arcuate nuclei of the hypothalamus and the median eminence have been found to contain higher concentrations of NE in male rats compared to female rats (Crowley et al., 1978). Also, in spontaneously hypertensive rats, the NE concentrations in both the posterior hypothalamic region and the pons of males were significantly higher than in females (Chen and Meng, 1991). In a behavioral turning preference study, females showed greater NE concentrations in the left caudate-putamen while males showed greater NE concentrations in both the left and right caudate-putamen (Dark et al., 1984).

At birth, the superior cervical ganglion (SCG) of male and female rats contains equal numbers of neurons. There is also no sex difference at this time in NE content or tyrosine hydroxylase activity in the SCG. During the first 2 prenatal weeks, neuronal death in the SCG results in the loss of significantly more neurons in females than in males, and in the adult, this sex difference in SCG neuron number is maintained (Beaston-Wimmer and Smolen, 1991). It is believed that high levels of neonatal testosterone are responsible for the decrease in cell death, and therefore higher levels of NE, found in the male SCG. The regulatory actions of testosterone in the neonate are believed to serve as an anticipatory mechanism for the ultimate difference in size of adult males versus females; males have both a larger body weight and neurotransmitter target mass and therefore may require more neurons to innervate adrenergic targets. However, the mechanisms by which gonadal hormones such as testosterone act to regulate this process are currently unidentified.

In contrast, other regions of the brain exhibit higher levels of NE activity in females than in males. The locus ceruleus–norepinephrine (LC-NE) system, serving as a major site of NE influence on the CNS, demonstrates sexual dimorphism where the LC in female rats exhibits a larger volume and greater number of neurons than the LC in male rats (Pinos et al., 2001). This example of dimorphism manifests during postpuberal development, during which time females exhibit an extended length of neurogenesis in the LC from postnatal days (P) 45 through 60, whereas in males this growth period ends as of P45. It is believed that the presence of estradiol in females governs neurogenesis in the LC during maturation to produce this sexual dimorphism; however, the exact mechanism by which this occurs remains unclear.

Sexual diergism in the LC was demonstrated in studies involving the activation of the LC-NE system following acute stress. While both male and female rats responded to hypotensive stress with an increase in LC activity, females experienced substantially greater LC activation than males (Curtis et al., 2006). This difference in stress sensitivity was attributed to morphological differences in the number of protein receptors in the LC for the stress-related LC-activating agent, CRH. The female LC contained greater levels of CRH receptor proteins compared to males; administration of CRH in females produced a 30% increase in LC activation compared to only a 10% increase in males. This heightened postsynaptic sensitivity to CRF in the female LC ultimately results in an elevated response to hypotensive stress. It is believed that increased sensitivity of the LC-NE system to CRH is a contributing factor to both the elevated sensitivity of the female HPA axis to stress and the greater prevalence of stress-related disorders in females.

Sexual diergism in NE has also been demonstrated in chronic stress studies. Following repeated long-term maternal deprivation, only female rats showed an increase in plasma NE levels as adults (Renard et al., 2005). This effect was even more pronounced when maternal deprivation was combined with episodes of varied chronic stress. This further supports the concept that elevated sensitivity of noradrenergic systems in females may serve as a predisposition to stress-related disorders.

Other stress studies have demonstrated sexual diergism in the adrenergic regulation of vasopressin release. Females demonstrated greater elevation in plasma vasopressin concentrations following ICV administration of NE than males. This diergic effect was maximal during proestrus, when estradiol and PROG levels are at their highest levels, and minimal during diestrus when these ovarian hormones are at their lowest levels. These results suggest that gonadal hormones act through NE pathways to regulate vaspressinergic neurons (Stone et al., 1989). Furthermore, the stimulatory effect of estrogen on both the expression and stress-induced release of vasopressin also appears to be modulated by the α_1-adrenoreceptor (Viau and Meaney, 2004), because estrogen-induced elevations in median eminence AVP content were reduced by peripheral administration of the α_1-adrenoreceptor antagonist prazosin (Viau and Meaney, 2004). These findings suggest that gonadal hormones are important in the sexual diergism associated with adrenergic regulation of vasopressin, as well as cardiovascular function.

Further studies have also indicated sexual dimorphism in rats in the regulation of oxytocin (OT) via noradrenergic pathways (Stone et al., 1989). Upon lesioning of the ascending ventral noradrenergic bundle (VNAB), males exhibited a significantly elevated stress response and consequent OT release to immobilization, while females did not (Carter and Lightman, 1987). These results provide functional evidence of a sexually dimorphic inhibitory role of the VNAB in the control of OT secretion.

Sexual dimorphism and diergism of brain noradrenergic systems therefore appear to be region specific. Males typically have greater NE activity in hypothalamic and autonomic regions as well as the caudate-putamen; whereas females typically have greater NE activity in the stress-sensitive LC region. As with many of the other neurotransmitter systems, gonadal hormones appear to mediate these sex differences in several of these brain areas.

7.6 Serotonin

7.6.1 Sexual Dimorphism and Diergism of Serotonergic Systems

Of the few studies addressing sexual dimorphism and sexual diergism of serotonergic systems, most have focused on sex differences in serotonin content in specific brain regions (Carlsson and Carlsson, 1988; Carlsson et al., 1985) and the expression of serotonin (5-HT) receptors (Zhang et al., 1999). Recent studies with serotonin receptor KO mice have examined the relationship between sex differences and serotonergic systems by examining mice bred to lack 5-HT_{1B} presynaptic autoreceptors (Jones and Lucki, 2005). Female 5-HT_{1B} receptor KO mice demonstrated significantly reduced immobility compared to male 5-HT_{1B} receptor KO mice or WT mice of both sexes on the tail-suspension test and forced swimming test. These behavioral effects were attributed to disinhibition of serotonin release in the female 5-HT_{1B} receptor KO mice, because microdialysis studies confirmed that female 5-HT_{1B} receptor KO mice had significantly higher baseline levels of hippocampal serotonin than the other mouse groups (Jones and Lucki, 2005). In contrast, in normal mice, males have been found to have higher hippocampal serotonin concentrations compared to females (Jones and Lucki, 2005).

The forced swim test has been shown to induce a decrease in serotonergic activity in the hippocampus and hypothalamus in female rats and to induce an increase in serotonergic activity in the hypothalamus in male rats (Drossopoulou et al., 2004). Moreover, hypothalamic serotonin 5-HT_{1A} mRNA levels were decreased in female rats and hippocampal 5-HT_{1A} mRNA levels were increased in male rats following the forced swim test. Stress-induced alterations in corticosteroid concentrations may regulate serotonin function (Drossopoulou et al., 2004), and serotonin actions in the hypothalamus may be involved in mediating hormonal responses to stress (Grippo et al., 2005). Several studies have also shown that estrogen can affect serotonin receptor density and serotonergic function (Sumner and Fink, 1997; Sumner et al., 1999; Fink et al., 1996). For example, serotonin release is decreased in female rats during estrous when circulating levels of estrogen and PROG are high, and male rats and diestrous female rats release more serotonin following administration of the selective serotonin-reuptake inhibitors (SSRI) paroxetine than estrous female rats (Gundlah et al., 1998). Therefore, hormonal fluctuations may contribute to the variable diergic and behavioral effects reported following the forced swim test.

As discussed earlier in other sections of this chapter, the hypothalamic medial preoptic area contains several sexually dimorphic neurotransmitter systems that influence male and female sexual behaviors. Depletion of serotonin in the medial preoptic area facilitates lordosis in male and female rats. Males have more serotonin receptors in the preoptic area than females; however, females have more serotonergic fibers in the preoptic area than males (De Vries, 1990). Gonadal hormones appear to contribute to these sex differences, because estradiol increases serotonin receptors in the preoptic area of females but lowers them in males. Therefore, sex differences in serotonin receptor density and neuron concentration within the medial preoptic area may contribute to differences in sexual behaviors.

A cluster of neurons in the medial preoptic area of the rat brain, called the sexually dimorphic nucleus of the preoptic area (SDN-POA), is several times greater in volume in males than in females (Handa et al., 1986). This nucleus is surrounded by a relatively dense distribution of serotonin immunoreactive fibers which itself is sexually dimorphic (proportionally more dense in females compared to males.) Administration of the serotonin biosynthesis inhibitor *p*-chlorophenylalanine to pregnant rats results in significant increases in the volume of the SDN-POA in female neonates, and these increases result in SDN-POA volumes in females similar to volumes found in control males (Handa et al., 1986). Therefore, in addition to circulating gonadal hormones, serotonin itself has been implicated in the sexually dimorphic development of the SDN-POA.

The dorsal raphe nuclei (DRN) and the medial raphe nuclei (MRN) are two structures that have been consistently implicated in the brain circuitry associated with fear and anxiety reactions (Dominguez et al., 2003). The MRN of women has been shown to have a larger number of cells than the MRN of men (Cordero et al., 2000). Sexual dimorphism of DRN and MRN structures may account for the finding that serotonergic activity in the DRN of female rats was consistently higher than in the DRN of males, and the finding that female rats display more aversive responses than males in the elevated plus-maze test (Dominguez et al., 2003). Studies in primates and rodents have also established that estrogen and PROG receptors are expressed in the DRN (Alves et al., 1998; Klink et al., 2002a,b), suggesting that circulating gonadal hormones underlie the sexual dimorphism and/or sexual diergism associated with these two structures.

The amygdala is activated following stressful stimuli and coordinates reactions that cope with the stress, including emotional, autonomic, and neuroendocrine responses. Serotonin and dopamine are important neurotransmitters in the regulation of emotional responses of the amygdala. Aggression and anxiety appear to be inversely proportional to central serotonergic activity and directly proportional to central dopaminergic activity. For example, increased serotonin release in the amygdala reduces freezing response and 5-HT_{1A} receptor activation in the amygdala reduces anxiety (Mitsushima et al., 2006). In contrast, blocking D_1 receptors in the amygdala impairs the retrieval of the conditioned fear response and reduces anxiety (Mitsushima et al., 2006). Mean extracellular concentrations of serotonin and dopamine in the basolateral amygdala are higher in male rats than in female rats. Following restraint stress, serotonin release in the amygdala increases in both males and females; however, the increases are greater in females compared to males (Mitsushima et al., 2006). Moreover, restraint stress significantly increased dopamine concentrations in female rats, but not in males. Behavioral responses, such as fecal pellet production and freezing behavior time, were greater in male rats compared to female rats. These data suggest that sex differences in amygdala-driven emotional responses to stress may derive from sex-specific changes in serotonin and dopamine function in the amygdala.

Men have been found to have a higher serotonin-synthesis rate and higher serotonin receptor binding in specific brain regions (e.g., amygdala and hippocampus) compared to women, which may in part explain the differing prevalence and incidence reports of depression and anxiety disorders between the sexes (women greater than men) (Jones and Lucki, 2005; Khan et al., 2005; Nishizawa et al., 1997). There is ongoing controversy about whether men and women respond equally well to antidepressant medications (Parker et al., 2003), however, some studies suggest that women respond greater to SSRIs (Khan et al., 2005; Kornstein et al., 2000). In contrast, women with generalized anxiety disorder, particularly those with a later age of onset, may have a poorer response to SSRIs than men.

Evidence from both human and animal studies indicates variable responses and behaviors associated with serotonin that are brain region specific. Therefore, overall conclusions regarding sexual diergism of serotonergic systems in the brain are difficult to formulate. The discrepancies in both human and animal studies support the need for future studies that address sexual dimorphism and diergism of serotonergic systems.

7.7 Vasopressin

7.7.1 Sexual Dimorphism of AVP

In addition to AVP's role as a neurosecretory product of the hypothalamic paraventricular nuclei (PVN) SON, vasopressinergic neurons innervate many areas of the CNS, ranging from the olfactory bulb to the spinal cord (De Vries, 1990). Sexually dimorphic AVP nuclei are found in the lateral septum, medial amygdaloid nucleus, hypothalamus, and BNST, where male mammals have a greater cell number and fiber density than female mammals (De Vries, 1990; De Vries and Boyle, 1998; Garcia-Falgueras et al., 2006b). In the BNST and amygdala, male rats have 2–3 times as many AVP neurons as do females (De Vries and Boyle, 1998). Gonadal steroids organize this sexual dimorphism during development and maintain it in adulthood (Miller et al., 1989). In humans, the BNST has been reported to be larger and contain more neurons in males than in females only in adulthood, suggesting that sexual differentiation of the human brain may extend into adulthood (Chung et al., 2002). The higher density of AVP neurons in the lateral septum has been correlated with higher levels of aggression in male rats (De Vries and Boyle, 1998). AVP pathways in the lateral septum and BNST also may participate in the regulation of hormone release from the neurohypophysis (Miller et al., 1989).

7.7.2 Sexual Diergism of AVP

Plasma AVP and OT are synthesized in the magnocellular neurons of the hypothalamic PVN, SON, and accessory nuclei (de Wied et al., 1993). Magnocellular neurons containing AVP and OT release these hormones into the circulation via the neurohypophysis. Parvicellular AVP- and OT-containing neurons of the PVN project to the external lamina of the median eminence, where the hormones are released into the hypothalamic–hypophyseal–portal vessel system to regulate anterior pituitary function. Magno- and parvicellular AVP- and OT-containing neurons also project to other areas inside the CNS (Landgraf, 1992). Neuropeptides such as AVP are sexually diergic modulators of synaptic transmission in the mammalian CNS (de Wied et al., 1993). Central AVP has been implicated in processes such as learning and memory (De Wied, 1969), cardiovascular function (Crofton et al., 1986; Versteeg et al., 1983), flank-marking behavior (Dubois-Dauphin et al., 1996, 1990), thermoregulation and motor behaviors (Kasting, 1989; Kasting et al., 1980). Competitive binding studies with nonradioactive AVP analogs, electrophysiological studies, *in situ* hybridization studies, and functional studies have suggested that AVP produces these physiological and behavioral responses by interacting with central AVP receptors (V1A), and not the antidiuretic receptor (V2).

Under basal conditions, plasma AVP concentration and 24-h urinary excretion of AVP are higher in male than in female rats, regardless of the phase of the estrous cycle (Crofton et al., 1986, 1988). This may result from increased synthetic and secretory activities in the SON of males compared to females (Madeira et al., 1993; Paula-Barbosa et al., 1993). These differences are due to a higher rate of secretion of AVP from the posterior pituitary in male rats, because there are no sex-related differences in metabolic clearance of the hormone (Crofton et al., 1986). These sex differences in AVP directly influence behavior; adult female rats consume significantly more water per day than adult male rats (McGivern and Handa, 1996; McGivern et al., 1996) (**Figure 2**). Whether basal AVP concentrations in the CNS parallel these sex differences remains to be determined.

Immobilization results in a marked stimulation of both AVP and OT plasma concentrations in females, whereas in males the OT response is only slight, and there is no significant AVP response (Williams et al., 1985). It has also been shown that intrahypothalamic release of AVP in response to the elevation of plasma osmolality is greater in female rats than in males (Ota et al., 1994). In contrast, in hemorrhaged male rats (20% loss of blood volume) the concentration of AVP in PVN dialysate increased significantly from baseline values, whereas the posthemorrhage increase in female rats failed to reach significance (Ota et al., 1994). These differences may have been caused by sexually diergic osmoreceptor and baroreceptor feedback mechanisms that were slightly attenuated in female rats but remained sensitive in male rats. It is also possible that these differences were caused by sexual dimorphism in the distribution of muscarinic and nicotinic cholinergic receptors in the hypothalamus.

Ota et al. (1992) demonstrated that activation of nicotinic receptors in the SON of the hypothalamus resulted in increased release of AVP from the posterior pituitary of male rats. Microinjection into the SON of oxotremorine, a specific muscarinic agonist approximately equal in potency to ACh but lacking nicotinic activity (Cho et al., 1962), was without effect on AVP release. Oxotremorine and nicotine microinjection into the PVN produced the opposite responses, wherein oxotremorine was highly effective in stimulating AVP release, and nicotine was essentially inactive (Shoji et al., 1989). Understanding sex differences in response to selective nicotinic and muscarinic stimulation in brain regions, such as the hypothalamus, is of considerable importance and warrants future investigation (see Section 7.2.7.4).

In a study of the plasma concentrations of AVP in patients with major depression and normal control subjects, the control men had higher mean plasma AVP and OT than did the control women, and a trend existed toward increased plasma AVP in the male depressed patients (van Londen et al., 1997). Similar results have been reported in an elderly population, where plasma AVP concentrations were twofold higher in men compared to women (Asplund and Aberg, 1991). These results are supported by our own finding of greater AVP secretion in normal men compared to normal women in response to cholinergic stimulation with low-dose physostigmine (Rubin et al., 1999b). However, we did not find differences in basal AVP between men and women. Greater ACTH and cortisol responses to physostigmine have also been reported in older women compared to older men (average age 71) (Peskind et al., 1996). This finding is consistent with reports of greater pituitary and adrenocortical responses to CRH in older women compared to older men (Greenspan et al., 1993; Heuser et al., 1994). In contrast, we found increased AVP, ACTH, and cortisol responses to low-dose physostigmine in younger men compared to younger women (average age 35 years) (Rubin et al., 1999b). Studies have shown that the total number of CRH-stained neurons in the human PVN increases with age, and that men have a significantly larger number of CRH neurons than women (Bao and Swaab, 2007). These results indicate that age may be a determining factor for sexually diergic AVP and HPA responses to cholinergic stimulation.

Plasma AVP is reported to be lowest during and just after menstruation, when estradiol and PROG concentrations are both low (Forsling et al., 1981). At ovulation, after the preovulatory rise in estradiol and PROG, AVP concentration peaks. This suggests that secretion of AVP is influenced by female sex hormones as well as age (Forsling et al., 1981).

In addition to coordinating many physiological and behavioral rhythms, the human suprachiasmatic nucleus (SCN) modulates endocrine function as well, including the activity of the PVN and SON (Hofman, 1997; Madeira and Lieberman, 1995). The SCN is sexually dimorphic, being elongated in women and more spherical in men. Shape differences could have functional consequences, altering contacts between the SCN and structures in its anatomical vicinity. However, the number and volume of AVP-containing cells in the SCN are similar in both sexes (Swaab et al., 1985; Swaab, 1995), and the number of SCN neurons expressing AVP decreases with age in both men and women. In contrast, the number of vasoactive intestinal polypeptide (VIP)-expressing neurons in the SCN of women does not change with age, whereas in men a complex pattern of age-related changes occurs. Between ages 10 and 30 years, SCN of men contains twice as many VIP neurons as that of women. A subsequent decrease in the number of VIP neurons in males occurs between 40 and 60 years of age, resulting in a smaller number of VIP neurons in men than in women. After 60 years of age no sex differences are apparent (Swaab et al., 1994). Since the SCN affects activity from the PVN and SON, changes in other neuropeptides, such as VIP, could indirectly influence AVP release.

Both α_1- and α_2-adrenergic receptors have been isolated from hypothalamic nuclei, including the PVN and SON (Leibowitz et al., 1982). As indicated previously, ICV administration of norepinephrine has been shown to stimulate AVP release in both male and female conscious rats (Stone et al., 1989). In females, the elevation of plasma AVP is much greater than in males, and the magnitude of the difference depends on the phase of the estrous cycle, implicating hormonal involvement. Much evidence indicates that stimulation of AVP release by centrally administered norepinephrine results from activation of the α_1-receptor subtype (Armstrong et al., 1986; Brooks et al., 1986; Hiwatari and Johnston, 1985; Willoughby et al., 1987). These findings suggest that females exhibit a greater α_1-adrenergic receptor density or sensitivity to the central administration of norepinephrine, and that increased AVP is influenced by the hormonal milieu, discussed next.

7.7.3 Influence of Gonadal Steroids on AVP Secretion

The ability of gonadal steroids to control neuropeptide transmission provides an efficient way to exert diverse and diergic effects on brain function. Gonadal hormones have effects on the expression of cholecystokinin (CCK), substance P, galanin, neurotensin, GnRH, and endogenous opioid peptides in the hypothalamus, BNST, and amygdala, regions implicated in steroid-sensitive functions such as the regulation of gonadotropin release and sexual behavior (De Vries et al., 1994a). As mentioned earlier, the most prominent sexual dimorphism in AVP systems occurs in the AVP projections of the BNST and the medial amygdaloid nucleus (MA). Male rodents have 2–3 times more AVP-immunoreactive (ir) cells in the BNST than female rodents, and their projections are much denser (De Vries et al., 1981; De Vries and al-Shamma, 1990). In addition to being sexually dimorphic, the AVP-ir projections of the BNST and MA exhibit a higher affinity for sex steroids in male rodents. After orchiectomy, BNST and MA cells and their projections completely lose their AVP immunoreactivity and can no longer be labeled for AVP mRNA; treatment with gonadal steroids prevents these changes (De Vries et al., 1984a; van Leeuwen et al., 1985).

Several other neuropeptide systems show dramatic fluctuations in their expression under the influence of gonadal steroids. However, a complete elimination of the expression of a particular neuropeptide by gonadectomy has been reported only for the AVP cells of the BNST and MA (De Vries et al., 1994b). As shown by *in situ* hybridization and quantitative autoradiographic measurement of propressophysin mRNA concentrations in cells of the BNST of male and female rats, the sexual diergism of AVP pathways in the BNST and lateral septum results from a difference in the biosynthetic capacity of extrahypothalamic vasopressinergic neurons (Miller et al., 1989). This suggests that gonadal steroids differentially enhance the expression of the AVP gene in these cells.

The pronounced modulation of the biosynthetic capacity of AVP by gonadal steroids suggests that its physiological functions would exhibit sexual diergism as well; indeed, AVP neurons in the BNST have been implicated in the regulation of gonadotropin secretion (Kawakami and Kimura, 1974), male copulatory behavior (Bohus, 1977), and antipyresis (Cooper et al., 1979). Glucocorticoids may also regulate the synthesis and release of AVP in both rats (Ferrini et al., 1997) and humans (Erkut et al., 1998). Computerized autoradiography studies have shown that AVP mRNA concentrations in the parvicellular cells of the PVN were increased after adrenalectomy and decreased following dexamethasone treatment in both sexes, but the reduction by dexamethasone was more pronounced in female rats (Ferrini et al., 1997).

The role of AVP in CNS function has expanded to include its activity in stimulating ACTH secretion from the anterior pituitary and its actions as a neurotransmitter in other brain areas (Antoni, 1993; De Vries et al., 1984b). AVP pathways innervate multimodal areas of the brain, indicating that, in addition to producing neurohypophysial hormones, these fibers influence numerous other functions and behavior. Studies of the sexual diergism of AVP should further elucidate the importance of this neuropeptide as both a neurotransmitter and hormone.

7.8 Implications and Relevance of Sexual Diergism

Given the data presented herein, it should not be surprising that many sexually diergic behaviors are influenced by sexually diergic neurotransmitter systems in the mammalian CNS. There is now sufficient knowledge of the biological basis of sex differences to validate the scientific study of sex differences and to allow the generation of hypotheses. The next step is to move from the descriptive to the experimental and establish conditions that must be in place to facilitate and encourage the scientific study of the mechanisms and origins of sex differences (Wizemann and Pardue, 2001). Basic genetic and physiological differences, in combination with environmental factors, result in behavioral and cognitive differences between males and females. Sex differences in the brain, sex-typed behavior and gender identity, and sex differences in cognitive ability should be studied at all points across the life span. Further research is needed on the natural variations between and within the sexes in behavior, cognition, and perception, with expanded investigation of sex and age differences in brain organization and function (Wizemann and Pardue, 2001).

Caution should be used when interpreting studies which focus on the sexual dimorphism of a single population of neurons in the brain, since behaviors are often associated with several neurotransmitter systems working in tandem. Many studies concerning sexual dimorphism and diergism have involved the use of nonhuman mammals, and extrapolation from animals to humans regarding structural, physiological, and behavioral sex differences may not be appropriate. For example, male mice and men appear to be more sensitive to some effects of AChE inhibition than their female counterparts, whereas certain breeds of female rats, such as FSL rats, are more sensitive than male rats. Both MCS, resulting from increased sensitivity of AChE, and depression, hypothesized to result from cholinergic supersensitivity, occur more frequently in women (Goodwin and Jamison, 1990), as mentioned earlier. For MCS and depression, the FSL rat appears to be a good experimental model, but the complex interactions among neurotransmitters and their influences on behavior make extrapolation from rats to humans difficult.

In many cases diergism originates from dimorphism, but because of the intricacy of the mammalian brain, the connection between the two is not always clear. Indeed, it has been proposed that sex differences within the brain may allow male and female mammals to display remarkably similar behaviors, despite major differences in their physiological and hormonal conditions (De Vries and Boyle, 1998; Kirkpatrick and Bryant, 1995). This possibility, that sexual dimorphism acts to mitigate sexual diergism, may be counterintuitive, but it is certainly intriguing (De Vries and Boyle, 1998; De Vries, 2004) and deserves further study.

The data in this chapter suggest that sexual diergism should be an important variable in the study of behavior, normal and abnormal physiology, and pharmacological principles such as tolerance, dependence, and receptor up- and downregulation. We therefore conclude this chapter with examples of the relevance of studying sexual diergisms that exist or that may be discovered in the mammalian CNS. The emphasis of these examples, although not exclusive, will be on CNS cholinergic systems and the HPA axis.

7.8.1 Behavioral Relevance of Sexual Diergism

Many behaviors are expressed differently in male and female mammals. These include sexual behavior, play and social behavior, learning and gender role behavior, posture during urination, scent-marking behavior, vocalization, regulation of food and water intake, and body weight (Becu-Villalobos et al., 1997; Dorner, 1981). The most obvious functional differences between males and females are those involved in reproductive physiology, and the best-studied animal model is the rat.

Sex differences in behavior may sometimes be based on an anatomical difference, but it is also possible that behavioral differences result from sexual diergism of neurotransmitter systems, or perhaps a combination of dimorphism and diergism. In the rat brain, the ventromedial hypothalamus is important in the regulation of reproductive behavior such as lordosis. Both estrogen and testosterone (by aromatization to estrogen) can promote lordosis behavior in OVX rats (Luine et al., 1987). The estrogen-inducible PROG receptors in the ventromedial nucleus also appear to play a role (Schumacher et al., 1992; Parsons et al., 1984). Estrogens have also been shown to induce receptors for OT in the hypothalamus, and blockade of OT receptors interferes with the expression of lordosis behavior (Pedersen and Boccia, 2002; Devidze et al., 2005). Pharmacological studies suggest that muscarinic cholinergic systems increase activity of the ventromedial hypothalamus and facilitate the lordosis response in estrogen-stimulated rats (Clemens et al., 1989; Pfaff, 2005). The enhanced ventromedial hypothalamic output subsequent to stimulation by muscarinic cholinergic afferents primes lower pathways in the circuit for lordosis behavior (Menard and Dohanich, 1990; Pfaff, 2005).

Other neurotransmitters have also been implicated in this sexually diergic behaviors. Norepinephrine was one of the first neurotransmitters to be implicated in female lordosis behavior (Meyerson, 1984). Depletion of preoptic and hypothalamic serotonin receptors facilitates lordosis in both males and females (Luine et al., 1987). Females have lower dendritic spines in the ventromedial nucleus than males, and antagonism of glutamate AMPA and kainate receptors prevents estradiol-induced dendritic branching in the hypothalamic ventromedial nucleus in females but not in males (Todd et al., 2007). β-Endorphin might also inhibit female sexual behavior, perhaps via LH-releasing-hormone neurons in the preoptic area, on which β-endorphin fibers synapse (Pfais and Gorzalka, 1987). CCK injections in female rats can facilitate or inhibit lordosis, depending on the level of sexual receptivity and the site of injection (Bloch et al., 1987; Mendelson and Gorzalka, 1988).

Both estrogen and testosterone treatments reverse the effects of castration on male sexual behavior (Chen and Tu, 1992). Dopamine both facilitates and inhibits male sexual behavior, while serotonin only has an inhibitory effect (De Vries, 1990). Factors that feminize sexual behavior and decrease the volume of the sexually dimorphic nucleus/preoptic area of the male rat brain, such as prenatal stress, have been suggested as possible contributors to homosexuality in men (Allen et al., 1989; Anderson et al., 1986). However, differences in the volume of sexually dimorphic nuclei were not found between homosexual and heterosexual men (Swaab and Hofman, 1990). Morphological and receptor analysis of brains from humans with different sexual orientations and from individuals exposed to atypical steroid hormones may lead to further insights concerning the possible influence of sex hormones on the structure and function of the human brain.

Although clearly a major influence, as indicated above, the actions of gonadal steroids should no longer be regarded as the only mechanism underlying the development of sexual dimorphism and related diergism. Sexual differentiation of the brain and behavior may be triggered by genetic mechanisms, independent of any influence of steroid hormones (Arnold, 1996). The environment also appears to have an important impact on the dimorphism and differentiation of the CNS, thus influencing behavior (Juraska, 1991).

Sexual dimorphism of the hippocampus likely underlies the diergism of hippocampus-dependent behaviors, including open-field activity, maze learning and contextual fear conditioning, in which male rats perform better than females (Aloisi, 1997; Maren et al., 1994; Pilgrim and Reisert, 1992). The structure of rat forebrain areas, such as the cerebral cortex and hippocampus, is responsive to the outside environment in a sexually dimorphic way: male rats have thicker dendritic branching in pyramidal neurons of the cortex and dentate gyral cells of the hippocampus, which was further increased in both areas by a stimulating environment in male, but not in female, rats (Juraska, 1991). Sex differences in the size of an organism's dendritic tree for a given population of neurons (dimorphism) may ultimately lead to sex differences in their physiological function (diergism), manifested by the manner in which the organism interacts with its environment. Coming full circle, the manner in which each sex interacts with the environment, shaped by dimorphism and diergism, may have been influenced by the environment itself during development (see **Chapter 8, Gonadal Hormones and Sexual Differentiation of Human Brain and Behavior**).

7.8.2 Sexual Diergism in Relationship to Disease

The higher vulnerability of the CNS, particularly the septo-hippocampal pathway, to exogenous neurotoxins in the female rat may imply a higher risk for impairment in cognitive function and may have relevance for neurological disease states in humans. Men appear to have a higher frequency of epilepsy (Kyrozis et al., 2006; Galanopoulou, 2005) while a greater number of women appear to suffer from nonepileptic attacks or pseudoseizures (van Merode et al., 1997). As mentioned earlier, the prevalence of major depression in women is about twice that in men (Coryell et al., 1992; Leon et al., 1993; Young et al., 1990). Women are somewhat more susceptible to Alzheimer's disease, even after adjustment for their longer survival (Bachman et al., 1993). Women's cognitive impairments may also be more severe than men's, perhaps as a result of an accelerated degeneration of the cholinergic nervous system in women (Henderson and Buckwalter, 1994).

Sex differences also exist in the prevalence of other neuropsychiatric disorders (Earls, 1987). Alcoholism and other drug abuse, antisocial personality, attention-deficit disorder, schizophrenia, Tourette's syndrome, and completed suicide predominate in men, whereas depression, anxiety, eating disorders, dementia, and attempted suicide are more common in women (Earls, 1987; Goodwin and Jamison, 1990; Kudielka and Kirschbaum, 2005; Palanza, 2001; Swaab et al., 2003). Females appear to be more vulnerable than males to the reinforcing effects of psychostimulants, opiates, and nicotine during many phases of the addiction process (Lynch et al., 2002). The extensive overproduction of dopamine receptors in the striatum and nucleus accumbens during prepubertal development may explain why males are more often afflicted by ADHD and Tourette's syndrome, since dopaminergic increases in these regions can produce hyperactivity and stereotypies (Andersen et al., 1997). In contrast, a shift toward greater cholinergic activity relative to dopaminergic activity in the striatum has been demonstrated in normal female rats and estrogen-treated males and females and is associated with parkinsonian-type extrapyramidal side effects (Miller, 1983). Extrapyramidal side effects, for example, to neuroleptic drugs, occur more frequently in women than in men (Bedard et al., 1979; Bickerstaff, 1969; Smith and Dunn, 1979; Smith et al., 1979).

Although there is much debate about the reasons for differing rates of psychiatric illness between the sexes, it is generally agreed that in women the incidence of many of these illnesses increases dramatically after puberty (Earls, 1987). Before puberty, boys are more susceptible to psychiatric problems than girls (Seeman, 1997). These differences again suggest the importance of hormonal influences on both normal and pathological CNS function. Sex differences in biological and behavioral responses to stressors may also impact the manifestation of psychiatric illness. For example, males may rely on the typical fight-or-flight as an adaptive response to stress, whereas females may rely on social interactions to provide physical and perhaps psychological protection (tend-and-befriend) as an adaptive response to stress (Taylor et al., 2000; Klein and Corwin, 2002). Social contact increases OT release in rats and has been associated with the beneficial effects of social support during stress exposure (Westenbroek et al., 2003). Female rats, and women, may seek social contact during stress, which could increase OT concentrations, perhaps contributing to a tend-and-befriend adaptive response.

7.8.3 Therapeutic Implications of Sexual Diergism

Sex differences in pharmacokinetics and pharmacodynamics have long been recognized in animals (Fletcher et al., 1994). In human pharmacological studies, sex differences in pharmacokinetics have been identified for numerous drugs; however, differences are generally only subtle. For a few drugs, such as verapamil, β-blockers, and SSRIs, sex differences in pharmacokinetics have been shown to result in different pharmacological responses, but their clinical relevance remains unproven (Meibohm et al., 2002; Meibohm and Derendorf, 2002; Beierle et al., 1999). The study of sexual diergism in patients affected by various diseases should help elucidate their mechanisms and refine their treatments.

Women are often underrepresented in clinical trials, particularly Phase I studies in which therapeutic doses of drugs are developed (Yonkers et al., 1992). This exclusion compromises the ability to determine the nature and importance of sex-related differences in pharmacokinetics, pharmacodynamics, drug efficacy, and drug toxicity.

Sexually diergic responses to drugs have not been extensively investigated. The implications of such studies would be to uncover drugs, both new and those on the market, which produce sex-selective therapeutic and untoward effects. For example, responses and adverse reactions to antidepressants may vary by gender (Yonkers et al., 1992); women may have an enhanced response to nontricyclic antidepressants compared to men, necessitating the need to vary dosages or drug-class selections based on the sex of the depressed patient. There is ongoing controversy about whether men and women respond equally well to antidepressant medications, and preliminary evidence suggests that SSRIs are more effective in the presence of estrogen (Gorman, 2006; Khan et al., 2005). Women have been shown to respond better to treatment with sertraline, an SSRI, than to imipramine, a tricyclic antidepressant; whereas men responded better to imipramine than to sertraline (Kornstein et al., 2000). However, analysis of response rates by hormone status showed that postmenopausal women had similar rates of response to both drugs.

Differences in metabolism may determine differences in the onset and duration of action of drugs, thus influencing dosage forms, times, and selections based on gender. For example, women metabolize methylprednisolone faster than men, but they are more sensitive to negative feedback compared to men (Fletcher et al., 1994). Plasma concentrations of sertraline have been reported as being 50–70% higher in women and elderly men compared to young men. Although SSRIs have failed to show a relationship between concentration and antidepressant response, these increases might be clinically relevant, because they may increase therapy-limiting and concentration-dependent adverse effects as well as drug interactions arising from inhibition of cytochrome P450 isoenzymes (Meibohm et al., 2002). The response to antipsychotic drugs, women requiring lower doses on a $mg\,kg^{-1}$ basis, may also be different as a result of hormone concentrations or receptor densities (Yonkers et al., 1992). Estrogen may also increase CNS cholinergic function on the basis of its antidopaminergic effect (Yonkers et al., 1992; Miller, 1983).

Microinjection of selective agonists and antagonists may be a useful technique for the elucidation of sexual diergism in a particular brain region or cell group. As discussed earlier, microinjection studies of cholinergic agonists into the PVN and SON have shown that these regions respond differently to specific cholinergic agonists, but only in male rats (Shoji et al., 1989; Ota et al., 1992). Pharmacokinetic sex differences in the permeability of the BBB, drug metabolism (Kato, 1974), or blood flow to particular regions of the brain cannot be disregarded, although these considerations are often not discussed even in regard to the results of an experiment quantifying a particular sexual dimorphism or diergism where an agent has been injected peripherally (Berger-Sweeney et al., 1995; Klein et al., 1997; Peskind et al., 1996; Shoji et al., 1989; Smolen et al., 1987). Male rats and female rats in estrus have somewhat greater permeability of the BBB compared with female rats of similar age and weight in proestrus (Saija et al., 1990), and OVX female rats have greater crossing of the BBB by a radiolabeled amino acid compared to cycling female rats (Saija et al., 1990).

Finally, prenatal exposure to drugs put the developing fetus at risk for behavioral alterations or susceptibility to disease. Research into mechanisms which cause alterations in sexually diergic behavior may increase our understanding of the role of prenatal drug exposure on sexually diergic behavior as it develops throughout the life span.

7.9 Conclusion

We have used sexual diergism to represent functional or physiological sex differences, whether or not they result from sexual dimorphism of a particular brain region. This selective review focused on sexual diergism of the mammalian CNS neurotransmitter systems and, when available, behavioral influences of sexual diergism. Examples were presented, and implications and relevance of such sex differences to behavior, diseases, and therapeutics were discussed.

The sexual dimorphism that exists between the male and female CNS has been well documented. Exclusion of some of the extensive literature on CNS sexual dimorphism that emphasized only anatomical and morphological sex differences was inevitable. We do not mean these reports were not worthy of attention; rather our intent was to construct a comprehensive review emphasizing certain specific aspects of sexual diergism of the mammalian CNS.

Most investigations of the functional significance of anatomical dimorphisms have focused on sex-specific behaviors, particularly mating and parenting. There are, however, behaviors exhibited by males and females for which the anatomical substrate differs. In these cases, sexual dimorphism, and diergism, may exist to maintain similarities, rather than divergency, in physiology and behavior (De Vries and Boyle, 1998; Kirkpatrick and Bryant, 1995). In many instances, however, dimorphism generates diergism; the dimorphism often being dependent on the presence of sex hormones during a restricted period of development. Moreover, sexual differences of the CNS and behavior may have an underlying genetic component, independent of any sex steroid-dependent mechanism, and the environment can have a significant impact on the dimorphism and sexual differentiation of the CNS. Furthermore, it has been proposed that sex differences in brain structure may also prevent sex differences in function and behavior, by compensating for sex differences in physiological conditions (e.g., hormonal status) (De Vries, 2004). Future studies of both dimorphism and diergism across the life span should further our understanding of the neurochemical basis of behavior and the mechanisms of disease, setting the stage for more rational pharmacological therapeutics.

References

Agmo A and Paredes R (1985) GABAergic drugs and sexual behaviour in the male rat. *European Journal of Pharmacology* 112: 371–378.

Agmo A, Soria P, and Paredes R (1989) GABAergic drugs and lordosis behavior in the female rat. *Hormones and Behavior* 23: 368–380.

Akinci MK and Johnston GA (1993) Sex differences in the effects of acute swim stress on binding to $GABA_A$ receptors in mouse brain. *Journal of Neurochemistry* 60: 2212–2216.

Allen LS, Hines M, Shryne JE, and Gorski RA (1989) Two sexually dimorphic cell groups in the human brain. *Journal of Neuroscience* 9: 497–506.

Aloisi AM (1997) Sex differences in pain-induced effects on the septo-hippocampal system. *Brain Research Reviews* 25: 397–406.

Aloisi AM, Steenbergen HL, van de Poll NE, and Farabollini F (1994) Sex-dependent effects of restraint on nociception and pituitary–adrenal hormones in the rat. *Physiological Behaviour* 55: 789–793.

Altemus M (2006) Sex differences in depression and anxiety disorders: Potential biological determinants. *Hormones and Behavior* 50: 534–538.

Alves SE, Weiland NG, Hayashi S, and McEwen BS (1998) Immunocytochemical localization of nuclear estrogen receptors and progestin receptors within the rat dorsal raphe nucleus. *Journal of Comparative Neurology* 391: 322–334.

Andersen SL, Rutstein M, Benzo JM, Hostetter JC, and Teicher MH (1997) Sex differences in dopamine receptor overproduction and elimination. *NeuroReport* 8: 1495–1498.

Anderson RH, Fleming DE, Rhees RW, and Kinghorn E (1986) Relationships between sexual activity, plasma testosterone, and the volume of the sexually dimorphis nucleus of the preoptic area in prenatally stressed and non-stressed rats. *Brain Research* 370: 1–10.

Antoni FA (1993) Vasopressinergic control of pituitary adrenocorticotropin secretion comes of age. *Frontiers in Neuroendocrinology* 14: 76–122.

Armstrong WE, Gallagher MJ, and Sladek CD (1986) Noradrenergic stimulation of supraoptic neuronal activity and vasopressin release *in vitro*: Mediation by an alpha 1-receptor. *Brain Research* 365: 192–197.

Arneric SP, Sullivan JP, and Williams M (1995) Neuronal nicotinic acetylcholine receptors: Novel targets for central nervous system therapeutics. In: Bloom FE and Kupfer DJ (eds.) *Psychopharmacology: The Fourth Generation of Progress*, pp. 95–110. New York: Raven Press.

Arnold AP (1996) Genetically triggered sexual differentiation of brain and behavior. *Hormones and Behavior* 30: 495–505.

Arnold AP and Gorski RA (1984) Gonadal steroid induction of structural sex differences in the central nervous system. *Annual Review of Neuroscience* 7: 413–442.

Asplund R and Aberg H (1991) Diurnal variation in the levels of antidiuretic hormone in the elderly. *Journal of Internal Medicine* 229: 131–134.

Assenmacher I, Szafarczyk A, Alonso G, Ixart G, and Barbanel G (1987) Physiology of neural pathways affecting CRH secretion. *Annals of the New York Academy of Sciences* 512: 149–161.

Atkinson HC and Waddell BJ (1997) Circadian variation in basal plasma corticosterone and adrenocorticotropin in the rat: Sexual dimorphism and changes across the estrous cycle. *Endocrinology* 138: 3842–3848.

Avissar S, Egozi Y, and Sokolovsky M (1981a) Biochemical characterization and sex dimorphism of muscarinic receptors in rat adenohypophysis. *Neuroendocrinology* 32: 303–309.

Avissar S, Egozi Y, and Sokolovsky M (1981b) Studies on muscarinic receptors in mouse and rat hypothalamus: A comparison of sex and cyclical differences. *Neuroendocrinology* 32: 295–302.

Bach F, Flugge G, and Wuttke W (1992) GABAergic influence on the development of the sexually dimorphic nucleus of male and female rats. *Brain Research* 573: 341–344.

Bachman DL, Wolf PA, Linn RT, et al. (1993) Incidence of dementia and probable Alzheimer's disease in a general population: The Framingham Study. *Neurology* 43: 515–519.

Bao AM and Swaab DF (2007) Gender difference in age-related number of corticotropin-releasing hormone-expressing neurons in the human hypothalamic paraventricular nucleus and the role of sex hormones. *Neuroendocrinology* 85: 27–36.

Bartus RT, Dean RL, III, Beer B, and Lippa AS (1982) The cholinergic hypothesis of geriatric memory dysfunction. *Science* 217: 408–414.

Beaston-Wimmer P and Smolen AJ (1991) Gender differences in neurotransmitter expression in the rat superior cervical ganglion. *Brain Research – Developmental Brain Research* 58: 123–128.

Beatty WW (1979) Gonadal hormones and sex differences in nonreproductive behaviors in rodents: Organizational and activational influences. *Hormones and Behavior* 12: 112–163.

Becker JB (1990) Direct effect of 17 beta-estradiol on striatum: Sex differences in dopamine release. *Synapse* 5: 157–164.

Becker JB (1999) Gender differences in dopaminergic function in striatum and nucleus accumbens. *Pharmacology, Biochemistry, and Behavior* 64: 803–812.

Becker JB (2000) Oestrogen effects on dopaminergic function in striatum. *Novartis Foundation Symposium* 230: 134–145 (discussion 145–154).

Becker JB, Arnold AP, Berkley KJ, et al. (2005) Strategies and methods for research on sex differences in brain and behavior. *Endocrinology* 146: 1650–1673.

Becu-Villalobos D, Gonzalez Iglesias A, Diaz-Torga G, Hockl P, and Libertun C (1997) Brain sexual differentiation and gonadotropins secretion in the rat. *Cellular and Molecular Neurobiology* 17: 699–715.

Bedard PJ, Langelier P, Dankova J, et al. (1979) Estrogens, progesterone, and the extrapyramidal system. *Advances in Neurology* 24: 411–422.

Beierle I, Meibohm B, and Derendorf H (1999) Gender differences in pharmacokinetics and pharmacodynamics. *International Journal of Clinical Pharmacology and Therapeutics* 37: 529–547.

Belz EE, Kennell JS, Czambel RK, Rubin RT, and Rhodes ME (2003) Environmental enrichment lowers stress-responsive hormones in singly housed male and female rats. *Pharmacology, Biochemistry, and Behavior* 76: 481–486.

Berger-Sweeney J, Arnold A, Gabeau D, and Mills J (1995) Sex differences in learning and memory in mice: Effects of sequence of testing and cholinergic blockade. *Behavioral Neuroscience* 109: 859–873.

Berkley KJ (1997) Sex differences in pain. *Behavioral and Brain Sciences* 20: 371–380 (discussion 435–513).

Berkley KJ, Zalcman SS, and Simon VR (2006) Sex and gender differences in pain and inflammation: A rapidly maturing field. *American Journal of Physiology – Regulatory, Integrative and Comparative Physiology* 291: R241–R244.

Bertherat J, Bluet-Pajot MT, and Epelbaum J (1995) Neuroendocrine regulation of growth hormone. *European Journal of Endocrinology* 132: 12–24.

Bickerstaff ER (1969) Involuntary movements and the pill. *British Medical Journal* 4: 556–557.

Bloch GJ, Babcock AM, Gorski RA, and Micevych PE (1987) Cholecystokinin stimulates and inhibits lordosis behavior in female rats. *Physiology and Behavior* 39: 217–224.

Bodnar RJ, Romero MT, and Kramer E (1988) Organismic variables and pain inhibition: Roles of gender and aging. *Brain Research Bulletin* 21: 947–953.

Bohus B (1977) The influence of pituitary neuropeptides on sexual behavior. *Problemes Actuels d Endocrinologie et de Nutrition* 235–246.

Breedlove SM and Arnold AP (1980) Hormone accumulation in a sexually dimorphic motor nucleus of the rat spinal cord. *Science* 210: 564–566.

Brett LP, Chong GS, Coyle S, and Levine S (1983) The pituitary–adrenal response to novel stimulation and ether stress in young adult and aged rats. *Neurobiology of Aging* 4: 133–138.

Brett LP, Levine R, and Levine S (1986) Bidirectional responsiveness of the pituitary–adrenal system in old and young male and female rats. *Neurobiology of Aging* 7: 153–159.

Brooks DP, Share L, and Crofton JT (1986) Central adrenergic control of vasopressin release. *Neuroendocrinology* 42: 416–420.

Brown R and Brooksbank BW (1979) Developmental changes in choline acetyltransferase and glutamate decarboxylase activity in various regions of the brain of the male, female, and neonatally androgenized female rat. *Neurochemical Research* 4: 127–136.

Bugajski J and Gadek A (1983) Central H1- and H2-histaminergic stimulation of pituitary–adrenocortical response under stress in rats. *Neuroendocrinology* 36: 424–430.

Burgess LH and Handa RJ (1992) Chronic estrogen-induced alterations in adrenocorticotropin and corticosterone secretion, and glucocorticoid receptor-mediated functions in female rats. *Endocrinology* 131: 1261–1269.

Calogero A (1995) Neurotransmitter regulation of the hypothalamic corticotropin-releasing hormone neuron. *Annals of the New York Academy of Sciences* 771: 31–40.

Carlsson M and Carlsson A (1988) A regional study of sex differences in rat brain serotonin. *Progress in Neuropsychopharmacology and Biological Psychiatry* 12: 53–61.

Carlsson M, Svensson K, Eriksson E, and Carlsson A (1985) Rat brain serotonin: Biochemical and functional evidence for a sex difference. *Journal of Neural Transmission* 63: 297–313.

Carter DA and Lightman SL (1987) Modulation of oxytocin secretion by ascending noradrenergic pathways: Sexual dimorphism in rats. *Brain Research* 406: 313–316.

Castner SA, Xiao L, and Becker JB (1993) Sex differences in striatal dopamine: *In vivo* microdialysis and behavioral studies. *Brain Research* 610: 127–134.

Ceccarelli I, Masi F, Fiorenzani P, and Aloisi AM (2002) Sex differences in the citrus lemon essential oil-induced increase of hippocampal acetylcholine release in rats exposed to a persistent painful stimulation. *Neuroscience Letters* 330: 25–28.

Cheeta S, Irvine EE, Tucci S, Sandhu J, and File SE (2001) In adolescence, female rats are more sensitive to the anxiolytic effect of nicotine than are male rats. *Neuropsychopharmacology* 25: 601–607.

Chen TJ and Tu WW (1992) Sex differences in estrogen and androgen receptors in hamster brain. *Life Science* 50: 1639–1647.

Chen YF and Meng QC (1991) Sexual dimorphism of blood pressure in spontaneously hypertensive rats is androgen dependent. *Life Science* 48: 85–96.

Cho A, Haslett WL, and Jenden DJ (1962) The peripheral action of oxotremorine, a metabolite of tremorine. *Journal of Pharmacology and Experimental Therapeutics* 138: 249–257.

Chung WC, De Vries GJ, and Swaab DF (2002) Sexual differentiation of the bed nucleus of the stria terminalis in humans may extend into adulthood. *Journal of Neuroscience* 22: 1027–1033.

Ciofi P, Leroy D, and Tramu G (2006) Sexual dimorphism in the organization of the rat hypothalamic infundibular area. *Neuroscience* 141: 1731–1745.

Clemens LG, Barr P, and Dohanich GP (1989) Cholinergic regulation of female sexual behavior in rats demonstrated by manipulation of endogenous acetylcholine. *Physiology and Behavior* 45: 437–442.

Conrad CD, Galea LA, Kuroda Y, and McEwen BS (1996) Chronic stress impairs rat spatial memory on the Y maze, and this effect is blocked by tianeptine pretreatment. *Behavioral Neuroscience* 110: 1321–1334.

Conrad CD, Jackson JL, Wieczorek L, Baran SE, Harman JS, Wright RL, and Korol DL (2004) Acute stress impairs spatial memory in male but not female rats: Influence of estrous cycle. *Pharmacology, Biochemistry, and Behavior* 78: 569–579.

Cooke B, Hegstrom CD, Villeneuve LS, and Breedlove SM (1998) Sexual differentiation of the vertebrate brain: Principles and mechanisms. *Frontiers in Neuroendocrinology* 19: 323–362.

Cooper KE, Kasting NW, Lederis K, and Veale WL (1979) Evidence supporting a role for endogenous vasopressin in natural suppression of fever in the sheep. *Journal of Physiology* 295: 33–45.

Cordero ME, Valenzuela CY, Torres R, and Rodriguez A (2000) Sexual dimorphism in number and proportion of neurons in the human median raphe nucleus. *Brain Research – Developmental Brain Research* 124: 43–52.

Coryell W, Endicott J, and Keller M (1992) Major depression in a nonclinical sample. Demographic and clinical risk factors for first onset. *Archives of General Psychiatry* 49: 117–125.

Cosgrove KP, Mazure CM, and Staley JK (2007) Evolving knowledge of sex differences in brain structure, function, and chemistry. *Biological Psychiatry* 62: 847–855.

Coyle JT, Price DL, and DeLong MR (1983) Alzheimer's disease: A disorder of cortical cholinergic innervation. *Science* 219: 1184–1190.

Craft RM (2003) Sex differences in drug- and non-drug-induced analgesia. *Life Science* 72: 2675–2688.

Critchlow V, Liebelt RA, Bar-Sela M, Mountcastle W, and Lipscomb HS (1963) Sex difference in resting pituitary–adrenal function in the rat. *American Journal of Physiology* 205: 807–815.

Crofton JT, Ratliff DL, Brooks DP, and Share L (1986) The metabolic clearance rate of and pressor responses to vasopressin in male and female rats. *Endocrinology* 118: 1777–1781.

Crofton JT, Share L, and Brooks DP (1988) Pressor responsiveness to and secretion of vasopressin during the estrous cycle. *American Journal of Physiology* 255: R1041–R1048.

Crowley WR, O'Donohue TL, and Jacobowitz DM (1978) Sex differences in catecholamine content in discrete brain nuclei of the rat: Effects of neonatal castration or testosterone treatment. *Acta Endocrinology (Copenh)* 89: 20–28.

Curtis AL, Bethea T, and Valentino RJ (2006) Sexually dimorphic responses of the brain norepinephrine system to stress and corticotropin-releasing factor. *Neuropsychopharmacology* 31: 544–554.

Damaj MI (2001) Influence of gender and sex hormones on nicotine acute pharmacological effects in mice. *Journal of Pharmacology and Experimental Therapeutics* 296: 132–140.

Dark KA, Ellman G, Peeke HV, Galin D, and Reus VI (1984) Sex differences and asymmetries of catecholamines: Relation to turning preferences. *Pharmacology, Biochemistry, and Behavior* 20: 327–330.

Davis AM, Ward SC, Selmanoff M, Herbison AE, and McCarthy MM (1999) Developmental sex differences in amino acid neurotransmitter levels in hypothalamic and limbic areas of rat brain. *Neuroscience* 90: 1471–1482.

De Vries GJ (1990) Sex differences in neurotransmitter systems. *Journal of Neuroendocrinology* 2: 1–13.

De Vries GJ (2004) Minireview: Sex differences in adult and developing brains: Compensation, Compensation, Compensation. *Endocrinology* 145: 1063–1068.

De Vries GJ, Al-Shamma A, and Zhou L (1994a) The sexually dimorphis vasopressin innervation of the brain as a model for steroid modulation of neuropeptide transmission. *Annals of the New York Academy of Sciences* 743: 95–120.

De Vries GJ and al-Shamma HA (1990) Sex differences in hormonal responses of vasopressin pathways in the rat brain. *Journal of Neurobiology* 21: 686–693.

De Vries GJ and Boyle PA (1998) Double duty for sex differences in the brain. *Behavioural Brain Research* 92: 205–213.

De Vries GJ, Buijs RM, and Sluiter AA (1984a) Gonadal hormone actions on the morphology of the vasopressinergic innervation of the adult rat brain. *Brain Research* 298: 141–145.

De Vries GJ, Buijs RM, and Swaab DF (1981) Ontogeny of the vasopressinergic neurons of the suprachiasmatic nucleus and their extrahypothalamic projections in the rat brain – presence of a sex difference in the lateral septum. *Brain Research* 218: 67–78.

De Vries GJ, Buijs RM, and Van Leeuwen FW (1984b) Sex differences in vasopressin and other neurotransmitter systems in the brain. *Progress in Brain Research* 61: 185–203.

De Vries GJ, Crenshaw BD, and Al-Shamma HA (1992) Oxytocin in maternal, sexual and social behaviors. Gonadal steroid modulation of vasopressin pathways. *Annals of the New York Academy of Sciences* 652: 387–396.

De Vries GJ, Wang Z, Bullock NA, and Numan S (1994b) Sex differences in the effects of testosterone and its metabolites on vasopressin messenger RNA levels in the bed nucleus of the stria terminalis of rats. *Journal of Neuroscience* 14: 1789–1794.

De Wied D (1969) Effects of peptide hormones on behavior. In: Ganong WF and Martini LE (eds.) *Frontiers in Neuroendocrinology*, pp. 97–140. Oxford: University of London Press.

de Wied D, Diamant M, and Fodor M (1993) Central nervous system effects of the neurohypophyseal hormones and related peptides. *Frontiers in Neuroendocrinology* 14: 251–302.

Devidze N, Mong JA, Jasnow AM, Kow LM, and Pfaff DW (2005) Sex and estrogenic effects on coexpression of mRNAs in single ventromedial hypothalamic neurons. *Proceedings of the National Academy of Sciences of the United States of America* 102: 14446–14451.

Diamond SG, Markham CH, Hoehn MM, McDowell FH, and Muenter MD (1990) An examination of male–female differences in progression and mortality of Parkinson's disease. *Neurology* 40: 763–766.

Dluzen DE and McDermott JL (2000) Gender differences in neurotoxicity of the nigrostriatal dopaminergic system: Implications for Parkinson's disease. *Journal of Gender Specific Medicine* 3: 36–42.

Dluzen DE and Mickley KR (2005) Gender differences in modulatory effects of tamoxifen upon the nigrostriatal dopaminergic system. *Pharmacology, Biochemistry, and Behavior* 80: 27–33.

Dluzen DE, Tweed C, Anderson LI, and Laping NJ (2003) Gender differences in methamphetamine-induced mRNA associated with neurodegeneration in the mouse nigrostriatal dopaminergic system. *Neuroendocrinology* 77: 232–238.

Dominguez R, CruzMorales SE, Carvalho MC, Xavier M, and Brandao ML (2003) Sex differences in serotonergic activity in dorsal and median raphe nucleus. *Physiology and Behavior* 80: 203–210.

Dorner G (1981) Sexual differentiation of the brain. *Vitamins and Hormones* 38: 325–381.

Drossopoulou G, Antoniou K, Kitraki E, Papathanasiou G, Papalexi E, Dalla C, and Papadopoulou-Daifoti Z (2004) Sex differences in behavioral, neurochemical and neuroendocrine effects induced by the forced swim test in rats. *Neuroscience* 126: 849–857.

Dubois-Dauphin M, Barberis C, and De Bilbao F (1996) Vasopressin receptors in the mouse *Mus musculus* brain: Sex-related expression in the medical preoptic area and hypothalamus. *Brain Research* 743: 32–39.

Dubois-Dauphin M, Pevet P, Tribollet E, and Dreifuss J (1990) Vasopressin in the brain of the golden hamster: The distribution of vasopressin binding sites and of immunoreactivity to the vasopressin-related glycopeptide. *Journal of Comparative Neurology* 300: 535–548.

Dwyer BE, Frando JL, and Wasterlain CG (1980) Rat brain protein synthesis declines during post-developmental aging. *Journal of Neurochemistry* 35: 746–749.
Earls F (1987) Sex differences in psychiatric disorders: Origins and developmental influences. *Psychiatric Developments* 5: 1–23.
Eglen RM and Watson N (1996) Selective muscarinic receptor agonists and antagonists. *Pharmacology and Toxicology* 78: 59–78.
Egozi Y, Avissar S, and Sokolovsky M (1982) Muscarinic mechanisms and sex hormone secretion in rat adenohypophysis and preoptic area. *Neuroendocrinology* 35: 93–97.
Egozi Y, Kloog Y, and Sokolovsky M (1986) Acetylcholine rhythm in the preoptic area of the rat hypothalamus is synchronized with the estrous cycle. *Brain Research* 383: 310–313.
Ehler FJ, Roeske WR, and Yamamura HI (1995) Molecular biology, pharmacology, and brain distribution of subtypes of the muscarinic receptor. In: Bloom FE and Kupfer DJE (eds.) *Psychopharmacology: The Fourth Generation of Progress*, pp. 111–124. New York: Raven Press.
Erkut Z, Pool C, and Swaab D (1998) Glucocorticoids suppress corticotropin-releasing hormone and vasopressin expression in human hypothalamic neurons. *Journal of Clinical Endocrinology and Metabolism* 83: 2066–2073.
Faraday MM, Scheufele PM, Rahman MA, and Grunberg NE (1999) Effects of chronic nicotine administration on locomotion depend on rat sex and housing condition. *Nicotine and Tobacco Research* 1: 143–151.
Fernandez-Ruiz J, De Miguel R, Hernandez ML, Cebeira M, and Ramos JA (1992) Comparisons between brain dopaminergic neurons of juvenile and aged rats: Sex-related differences. *Mechanisms of Ageing and Development* 63: 45–55.
Ferretti C, Blengio M, Ghi P, Adage T, Portaleone P, and Gamalero SR (1998) Hypothalamic histamine release in normal and stressed rats is affected by sex and aging. *Pharmacology, Biochemistry, and Behavior* 59: 255–260.
Ferrini MG, Grillo CA, Piroli G, de Kloet ER, and De Nicola AF (1997) Sex difference in glucocorticoid regulation of vasopressin mRNA in the paraventricular hypothalamic nucleus. *Cellular and Molecular Neurobiology* 17: 671–686.
Fibiger HD (1991) Cholinergic mechanisms in learning, memory and dementia: A review of recent evidence. *Trends in Neuroscience* 14: 220–280.
Fink G, Sumner BE, Rosie R, Grace O, and Quinn JP (1996) Estrogen control of central neurotransmission: Effect on mood, mental state, and memory. *Cellular and Molecular Neurobiology.* 16: 325–344.
Fletcher CV, Acosta E, and Strykowski JM (1994) Gender differences in human pharmacokinetics and pharmacodynamics. *Journal of Adolescent Health* 15: 619–629.
Forsling ML, Akerlund M, and Stromberg P (1981) Variations in plasma concentrations of vasopressin during the menstrual cycle. *Journal of Endocrinology* 89: 263–266.
Frankfurt M, Fuchs E, and Wuttke W (1984) Sex differences in gamma-aminobutyric acid and glutamate concentrations in discrete rat brain nuclei. *Neuroscience Letters* 50: 245–250.
Frick KM, Burlingame LA, Delaney SS, and Berger-Sweeney J (2002) Sex differences in neurochemical markers that correlate with behavior in aging mice. *Neurobiology of Aging* 23: 145–158.
Galani R, Jeltsch H, Lehmann O, Bertrand F, and Cassel JC (2002) Effects of 192 IgG-saporin on acetylcholinesterase histochemistry in male and female rats. *Brain Research Bulletin* 58: 179–186.
Galanopoulou AS (2005) GABA receptors as broadcasters of sexually differentiating signals in the brain. *Epilepsia* 46 (supplement 5): 107–112.
Garcia-Falgueras A, Junque C, Gimenez M, Caldu X, Segovia S, and Guillamon A (2006a) Sex differences in the human olfactory system. *Brain Research* 1116: 103–111.
Garcia-Falgueras A, Pinos H, Fernandez R, Collado P, Pasaro E, Segovia S, and Guillamon A (2006b) Sexual dimorphism in hybrids rats. *Brain Research* 1123: 42–50.
Ghi P, Blengio M, Ferretti C, and Portaleone P (1992) Stress and brain histaminergic system: Effects of weak electric foot shock. *Pharmacology, Biochemistry, and Behavior* 41: 317–320.
Giamberardino MA (2003) Referred muscle pain/hyperalgesia and central sensitization. *Journal of Rehabilitation Medicine* 41: 85–88.
Giamberardino MA, Berkley KJ, Iezzi S, de Bigontina P, and Vecchiet L (1997) Pain threshold variations in somatic wall tissues as a function of menstrual cycle, segmental site and tissue depth in non-dysmenorrheic women, dysmenorrheic women and men. *Pain* 71: 187–197.
Gibbs RB (1994) Estrogen and nerve growth factor-related systems in brain. *Annals of the New York Academy of Sciences* 743: 165–199.
Gibson G, Peterson C, and Jenden DJ (1981) Brain acetylcholine synthesis declines with senescence. *Science* 213: 674–676.
Giustina A and Veldhuis JD (1998) Pathophysiology of the neuroregulation of growth hormone secretion in experimental animals and the human. *Endocrine Reviews* 19: 717–797.
Goodwin FK and Jamison KR (1990) *Manic-Depressive Illness*, pp. 157–185. New York: Oxford University Press.
Gorman JM (2006) Gender differences in depression and response to psychotropic medication. *Gender Medicine* 3: 93–109.
Grattan DR and Selmanoff M (1997) Sex differences in the activity of gamma-aminobutyric acidergic neurons in the rat hypothalamus. *Brain Research* 775: 244–249.
Greenspan SL, Rowe JW, Maitland LA, McAloon-Dyke M, and Elahi D (1993) The pituitary–adrenal glucocorticoid response is altered by gender and disease. *Journal of Gerontology* 48: M72–M77.
Gregg CM (1985) The compartmentalized hypothalamo-neurohypophysial system: Evidence for a neurohypophysial action of acetylcholine on vasopressin release. *Neuroendocrinology* 40: 423–429.
Griffin A and Whitacre CC (1991) Sex and strain differences in the circadian rhythm fluctuation of endocrine and immune function in the rat: Implications for rodent models of autoimmune disease. *Journal of Neuroimmunology* 35: 53–64.
Grippo AJ, Sullivan NR, Damjanoska KJ, et al. (2005) Chronic mild stress induces behavioral and physiological changes, and may alter serotonin 1A receptor function, in male and cycling female rats. *Psychopharmacology (Berl)* 179: 769–780.
Grunberg NE, Winders SE, and Popp KA (1987) Sex differences in nicotine's effects on consummatory behavior and body weight in rats. *Psychopharmacology (Berl)* 91: 221–225.
Gudelsky GA and Porter JC (1981) Sex-related differences in the release of dopamine into hypophysial portal blood. *Endocrinology* 109: 1394–1398.
Gudelsky GA, Simonovic M, and Meltzer HY (1984) Dopaminergic and serotonergic control of neuroendocrine function. *Monographs in Neural Sciences* 10: 85–102.
Guillamon A and Segovia S (1997) Sex differences in the vomeronasal system. *Brain Research Bulletin* 44: 377–382.
Gundlah C, Simon LD, and Auerbach SB (1998) Differences in hypothalamic serotonin between estrous phases and

gender: An *in vivo* microdialysis study. *Brain Research* 785: 91–96.
Gurwitz D, Egozi Y, Henis YI, Kloog Y, and Sokolovsky M (1987) Agonist and antagonist binding to rat brain muscarinic receptors: Influence of aging. *Neurobiology of Aging* 8: 115–122.
Halpern DF (1986) *Sex Differences in Cognitive Abilities*, pp. 10–40. Hillsdale, NJ: Eelbaum.
Hamon M, Goetz C, Euvrard C, et al. (1983) Biochemical and functional alterations of central GABA receptors during chronic estradiol treatment. *Brain Research* 279: 141–152.
Hampson E and Kimura D (1992) Sex differences and hormonal influences on cognitive function in humans. In: Becker JB, Breedlove SM, and Crews D (eds.) *Behavioral Endocrinology*, pp. 357–398. Cambridge, MA: MIT Press.
Handa RJ, Burgess LH, Kerr JE, and O'Keefe JA (1994a) Gonadal steroid hormone receptors and sex differences in the hypothalamo-pituitary–adrenal axis. *Hormones and Behavior* 28: 464–476.
Handa RJ, Hines M, Schoonmaker JN, Shryne JE, and Gorski RA (1986) Evidence that serotonin is involved in the sexually dimorphic development of the preoptic area in the rat brain. *Brain Research* 395: 278–282.
Handa RJ, Nunley KM, Lorens SA, Louie JP, McGivern RF, and Bollnow MR (1994b) Androgen regulation of adrenocorticotropin and corticosterone secretion in the male rat following novelty and foot shock stressors. *Physiology and Behavior* 55: 117–124.
Harrod SB, Booze RM, and Mactutus CF (2007) Sex differences in nicotine levels following repeated intravenous injection in rats are attenuated by gonadectomy. *Pharmacology, Biochemistry, and Behavior* 86: 32–36.
Hasegawa N, Takeo T, and Sakuma Y (1991) Differential regulation of estrogen-dependent sexual development of rat brain by growth factors. *Neuroscience Letters* 123: 183–186.
Heijtz RD, Beraki S, Scott L, Aperia A, and Forssberg H (2002) Sex differences in the motor inhibitory and stimulatory role of dopamine D1 receptors in rats. *European Journal of Pharmacology* 445: 97–104.
Heinsbroek R, Van Haaren F, and van de Poll N (1988) Sex differences in passive avoidance behavior of rats: Sex-dependent susceptibility to shock-induced behavioral depression. *Physiology and Behavior* 43: 201–206.
Henderson VW and Buckwalter JG (1994) Cognitive deficits of men and women with Alzheimer's disease. *Neurology* 44: 90–96.
Herdon HJ and Wilson CA (1985) Changes in hypothalamic dopamine D-2 receptors during sexual maturation in male and female rats. *Brain Research* 343: 151–153.
Herman J, Schafer MK-H, Young E, Thompson R, Douglass J, Akil H, and Watson S (1989) Evidence for hippocampal regulation of neuroendocrine neurons of the hypothalamo-pituitary–adrenocortical axis. *Journal of Neuroscience* 9: 3072–3082.
Heuser IJ, Gotthardt U, Schweiger U, Schmider J, Lammers C-H, Dettling M, and Holsboer F (1994) Age-associated changes of pituitary–adrenocortical hormone regulation in humans: Importance of gender. *Neurobiology of Aging* 15: 227–231.
Hiwatari M and Johnston CI (1985) Involvement of vasopressin in the cardiovascular effects of intracerebroventricularly administered alpha 1-agonists in the conscious rat. *Journal of Hypertension* 3: 613–620.
Hofman M (1997) Lifespan changes in the human hypothalamus. *Experimental Gerontology* 32: 559–575.
Hornberger JC, Buell SJ, Flood DG, McNeill TH, and Coleman PD (1985) Stability of numbers but not size of mouse forebrain cholinergic neurons to 53 months. *Neurobiology of Aging* 6: 269–275.
Hortnagl H, Berger ML, Havelec L, and Hornykiewicz O (1993a) Role of glucocorticoids in the cholinergic degeneration in rat hippocampus induced by ethylcholine aziridinum. *Journal of Neuroscience* 13: 2939–2945.
Hortnagl H, Hansen L, Kindel G, Schneider B, el Tamer A, and Hanin I (1993b) Sex differences and estrous cycle-variations in the AF64A-induced cholinergic deficit in the rat hippocampus. *Brain Research Bulletin* 31: 129–134.
Jamal G (1998) Gulf War Syndrome – A model for the complexity of biological and environmental interaction with human health. *Adverse Drug Reactions and Toxicological Reviews* 17: 1–17.
Jonasson Z, Cahill JF, Tobey RE, and Baxter MG (2004) Sexually dimorphic effects of hippocampal cholinergic deafferentation in rats. *European Journal of Neuroscience* 20: 3041–3053.
Jones MD and Lucki I (2005) Sex differences in the regulation of serotonergic transmission and behavior in 5-HT receptor knockout mice. *Neuropsychopharmacology* 30: 1039–1047.
Juraska JD (1991) Sex differences in 'cognitive' regions of the rat brain. *Psychoneuroendocrinology* 16: 155–159.
Juraska JM (1984) Sex differences in developmental plasticity in visual cortex and hippocampal dentate gyrus. *Progress in Brain Research* 61: 205–214.
Kaasinen V, Nagren K, Hietala J, Farde L, and Rinne JO (2001) Sex differences in extrastriatal dopamine d(2)-like receptors in the human brain. *American Journal of Psychiatry* 158: 308–311.
Kant GJ, Lenox RH, and Bunnell BN (1983) Comparison of stress response in male and female rats: Pituitary cyclic AMP and plasma prolactin, growth hormone and corticosterone. *Psychoneuroendocrinology* 8: 421–428.
Karczmar A (1993) Brief presentation of the story and present status of studies of the vertebrate cholinergic system. *Neuropsychopharmacology* 9: 181–199.
Kasting NW (1989) Criteria for establishing a physiological role for brain peptides a case in point: The role of vasopressin in thermoregulation during fever and antipyresis. *Brain Research Review* 14: 143–153.
Kasting NW, Veale WL, and Cooper KE (1980) Convulsive and hypothermic effects of vasopressin in the brain of the rat. *Canadian Journal of Physiology and Pharmacology* 58: 316–319.
Kato R (1974) Sex-related differences in drug metabolism. *Drug Metabolism Reviews* 3: 1–32.
Kawakami M and Kimura R (1974) Study on the bed nucleus of stria terminalis in relation to gonadotropin control. *Endocrinology (Jpn)* 21: 125–130.
Kelly S, Ostrowksi N, and Wilson M (1999) Gender differences in brain and behavior: Hormonal and neural bases. *Pharmacology, Biochemistry, and Behavior* 64: 655–664.
Khan A, Brodhead AE, Schwartz KA, Kolts RL, and Brown WA (2005) Sex differences in antidepressant response in recent antidepressant clinical trials. *Journal of Clinical Psychopharmacology* 25: 318–324.
Kirkpatrick B and Bryant NL (1995) Sexual dimorphism in the brain: It's worse than you thought. *Biological Psychiatry* 38: 347–348.
Kirschbaum C, Wust S, and Hellhammer D (1992) Consistent sex differences in cortisol responses to psychological stress. *Psychosomatic Medicine* 54: 648–657.
Kitay JI (1961) Sex differences in adrenal coretical secretion in the rat. *Endocrinology* 68: 818–824.
Kjaer A, Larsen PJ, Knigge U, and Warberg J (1994) Histaminergic activation of the hypothalamic–pituitary–adrenal axis. *Endocrinology* 135: 1171–1177.
Klein LC and Corwin EJ (2002) Seeing the unexpected: How sex differences in stress responses may provide a new

perspective on the manifestation of psychiatric disorders. *Current Psychiatry Reports* 4: 441–448.
Klein LC, Popke EJ, and Grunberg NE (1997) Sex differences in effects of predictable and unpredictable footshock on fentanyl self-administration in rats. *Experimental and Clinical Psychopharmacology* 5: 99–106.
Klink R, Robichaud M, and Debonnel G (2002a) Gender and gonadal status modulation of dorsal raphe nucleus serotonergic neurons. Part I: Effects of gender and pregnancy. *Neuropharmacology* 43: 1119–1128.
Klink R, Robichaud M, and Debonnel G (2002b) Gender and gonadal status modulation of dorsal raphe nucleus serotonergic neurons. Part II. Regulatory mechanisms. *Neuropharmacology* 43: 1129–1138.
Kornstein S, Schatzberg A, Thase M, et al. (2000) Gender differences in treatment response to sertraline versus imipramine in chronic depression. *American Journal of Psychiatry* 157: 1445–1452.
Koylu E, Demirgoren S, London ED, and Pogun S (1997) Sex difference in up-regulation of nicotinic acetylcholine receptors in rat brain. *Life Science* 61: PL185–PL190.
Kubanis P and Zornetzer SF (1981) Age-related behavioral and neurobiological changes: A review with an emphasis on memory. *Behavioral and Neural Biology* 31: 115–172.
Kudielka BM and Kirschbaum C (2005) Sex differences in HPA axis responses to stress: A review. *Biological Psychology* 69: 113–132.
Kyrozis A, Chudomel O, Moshe SL, and Galanopoulou AS (2006) Sex-dependent maturation of $GABA_A$ receptor-mediated synaptic events in rat substantia nigra reticulata. *Neuroscience Letters* 398: 1–5.
Landgraf R (1992) Central release of vasopressin: Stimuli, dynamics, consequences. In: Ermisch A, Landgraf R, and Ruhle H-JE (eds.) *Circumventricular Organs and Brain Fluid Envioronment: Molucular and Functional Aspects*, vol. 91, pp. 29–39. Amsterdam: Elsevier.
Leibowitz SF, Jhanwar-Uniyal M, Dvorkin B, and Makman MH (1982) Distribution of alpha-adrenergic, beta-adrenergic and dopaminergic receptors in discrete hypothalamic areas of rat. *Brain Research* 233: 97–114.
Leon A, Klerman GL, and Wickramaratne P (1993) Continuing female predominance in depressive illness. *American Journal of Public Health* 83: 754–757.
Leung A and Chue P (2000) Sex differences in schizophrenia: A review of the literature. *Acta Psychiatrica Scandinavic* 401(supplement): 3–38.
Li H, Huguenard JR, and Fisher RS (2007) Gender and age differences in expression of $GABA_A$ receptor subunits in rat somatosensory thalamus and cortex in an absence epilepsy model. *Neurobiology of Disease* 25: 623–630.
Lieb K, Reisert I, and Pilgrim C (1994) Differentiation of hypothalamic GABAergic neurons *in vitro*: Absence of effects of sex and gonadal steroids. *Experimental Brain Research* 99: 435–440.
Lippa A, Pelham RW, Beer B, Critchett DJ, Dean RL, and Bartus RT (1980) Brain cholinergic dysfunction and memory in aged rats. *Neurobiology of Aging* 1: 13–19.
Lochner C, Hemmings SM, Kinnear CJ, et al. (2004) Gender in obsessive–compulsive disorder: Clinical and genetic findings. *European Neuropsychopharmacology* 14: 105–113.
Lonstein J and De Vries G (2000) Sex differences in the parental behavior of rodents. *Neuroscience and Biobehavioral Reviews* 24: 669–686.
Loy R and Sheldon RA (1987) Sexually dimorphic development of cholinergic enzymes in the rat septohippocampal system. *Brain Research* 431: 156–160.
Luine V (2002) Sex differences in chronic stress effects on memory in rats. *Stress* 5: 205–216.
Luine V, Park D, Joh T, Reis D, and McEdwen B (1980) Immunochemical demonstration of increased choline acetyltransferase concentration in preoptic area after estradiol administration. *Brain Research* 191: 273–277.
Luine V, Villegas M, Martinez C, and McEwen BS (1994) Repeated stress causes reversible impairments of spatial memory performance. *Brain Research* 639: 167–170.
Luine VN (1985) Estradiol increases choline acetyltransferase activity in specific basal forebrain nuclei and projection areas of female rats. *Experimental Neurology* 89: 484–490.
Luine VN and McEwen BS (1983) Sex differences in cholinergic enzymes of diagonal band nuclei in the rat preoptic area. *Neuroendocrinology* 36: 475–482.
Luine VN, Renner KJ, Heady S, and Jones KJ (1986) Age and sex-dependent decreases in ChAT in basal forebrain nuclei. *Neurobiology of Aging* 7: 193–198.
Luine VN, Thornton JE, Frankfurt M, and MacLusky NJ (1987) Effects of hypothalamic serotonin depletion on lordosis behavior and gonadal hormone receptors. *Brain Research* 426: 47–54.
Lynch WJ, Roth ME, and Carroll ME (2002) Biological basis of sex differences in drug abuse: Preclinical and clinical studies. *Psychopharmacology (Berl)* 164: 121–137.
Maccoby EE and Jacklin CN (1974) *The Psychology of Sex Differences*, pp. 100–165. Stanford, CA: Stanford University Press.
Madeira MD and Lieberman AR (1995) Sexual dimorphism in the mammalian limbic system. *Progress in Neurobiology* 45: 275–333.
Madeira MD, Sousa N, Cadete-Leite A, Lieberman AR, and Paula-Barbosa MM (1993) The supraoptic nucleus of the adult rat hypothalamus displays marked sexual dimorphism which is dependent on body weight. *Neuroscience* 52: 497–513.
Maggi A and Perez J (1986) Estrogen-induced up-regulation of gamma-aminobutyric acid receptors in the CNS of rodents. *Journal of Neurochemistry* 47: 1793–1797.
Maren S, De Oca B, and Franselow MS (1994) Sex differences in hippocampal long-term potentiation (LTP) and Pavlovian fear conditioning in rats: Positive correlation between LTP and contextual learning. *Brain Research* 661: 25–34.
Masuda J, Mitsushima D, Funabashi T, and Kimura F (2005) Sex and housing conditions affect the 24-h acetylcholine release profile in the hippocampus in rats. *Neuroscience* 132: 537–542.
McCarthy MM, Auger AP, and Perrot-Sinal TS (2002) Getting excited about GABA and sex differences in the brain. *Trends in Neuroscience* 25: 307–312.
McCarthy MM and Konkle AT (2005) When is a sex difference not a sex difference? *Frontiers in Neuroendocrinology* 26: 85–102.
McCarthy MM, Masters DB, Fiber JM, Lopez-Colome AM, Beyer C, Komisaruk BR, and Feder HH (1991a) GABAergic control of receptivity in the female rat. *Neuroendocrinology* 53: 473–479.
McCarthy MM, Pfaff DW, and Schwartz-Giblin S (1991b) Midbrain central gray $GABA_A$ receptor activation enhances, and blockade reduces, sexual behavior in the female rat. *Experimental Brain Research* 86: 108–116.
McDermott JL, Liu B, and Diuzen DE (1994) Sex differences and effects of estrogen on dopamine and DOPAC release from the striatum of male and female. *Experimental Neurology* 125: 306–311.
McGivern RF and Handa RJ (1996) Prenatal exposure to drugs of abuse: Methodological considerations and effects on sexual differentiation. *NIDA Research Monographs* 164: 78–124.
McGivern RF, Henschel D, Hutcheson M, and Pangburn T (1996) Sex difference in daily water consumption of

rats: Effect of housing and hormones. *Physiology and Behavior* 59: 653–658.
Means LW and Dent MF (1991) The effects of number of trials per day, retention interval, gender and time of day on acquisition of a two-choice win-stay, water-escapae working memory task in the rat. *Journal of Neuroscience Methods* 39: 77–87.
Meibohm B, Beierle I, and Derendorf H (2002) How important are gender differences in pharmacokinetics? *Clinical Pharmacokinetics* 41: 329–342.
Meibohm B and Derendorf H (2002) Pharmacokinetic/pharmacodynamic studies in drug product development. *Journal of Pharmaceutical Sciences* 91: 18–31.
Menard CS and Dohanich GP (1990) Physostigmine facilitation of lordosis in naturally cycling female rats. *Pharmacology, Biochemistry, and Behavior* 36: 853–858.
Mendelson SD and Gorzalka BB (1988) Cholecystokinin-octapeptide produces inhibition of lordosis behavior in female rats. *Pharmacology, Biochemistry, and Behavior* 21: 755–759.
Menendez L, Andres-Trelles F, Hidalgo A, and Baamonde A (1993) Opioid footshock-induced analgesia in mice acutely falls by stress prolongation. *Physiology and Behavior* 53: 1115–1119.
Mesulam M-M (1995) Structure and function of cholinergic pathways in the cerebral cortex, limbic system, basal ganglia, and thalamus of the human brain. In: Bloom FE and Kupfer DJE (eds.) *Psychopharmacology: The Fourth Generation of Progress*, pp. 135–146. New York: Raven Press.
Meyerson BJ (1984) Hormone-dependent socio-sexual behaviors and neurotransmitters. *Progress in Brain Research.* 61: 271–281.
Michels KM, Meeker RB, and Hayward JN (1991) Muscarinic cholinergic control of vasopressin secretion from the acute hypothalamoneurohypophysial explant. *Neuroendocrinology* 54: 219–226.
Miller CS (1995) Chemical sensitivity attributed to pesticide exposure versus remodeling. *Archives of Environmental Health* 50: 119–129.
Miller JC (1983) Sex differences in dopaminergic and cholinergic activity and function in the nigro-striatal system of the rat. *Psychoneuroendocrinology* 8: 225–236.
Miller MA, Vician L, Clifton DK, and Dorsa DM (1989) Sex differences in vasopressin neurons in the bed nucleus of the stria terminalis by *in situ* hybridization. *Peptides* 10: 615–659.
Mishima N, Higashitani F, Kazuhiko T, and Yoshioka R (1986) Sex differences in appetitive learning of mice. *Physiology and Behavior* 37: 263–268.
Mitsushima D, Masuda J, and Kimura F (2003) Sex differences in the stress-induced release of acetylcholine in the hippocampus and corticosterone from the adrenal cortex in rats. *Neuroendocrinology* 78: 234–240.
Mitsushima D, Yamada K, Takase K, Funabashi T, and Kimura F (2006) Sex differences in the basolateral amygdala: The extracellular levels of serotonin and dopamine, and their responses to restraint stress in rats. *European Journal of Neuroscience* 24: 3245–3254.
Moidel MA, Belz EE, Czambel RK, Rubin RT, and Rhodes ME (2006) Novel *in vitro* perfusion system for the determination of hypothalamic–pituitary–adrenal axis responses. *Journal of Pharmacological and Toxicological Methods* 53: 264–271.
Mong JA, Kurzweil RL, Davis AM, Rocca MS, and McCarthy MM (1996) Evidence for sexual differentiation of glia in rat brain. *Hormones and Behavior* 30: 553–562.
Morgan DG, May PC, and Finch CE (1987) Dopamine and serotonin systems in human and rodent brain: Effects of age and neurodegenerative disease. *Journal of the American Geriatrics Society* 35: 334–345.
Morissette M and Di Paolo T (1993) Sex and estrous cycle variations of rat striatal dopamine uptake sites. *Neuroendocrinology* 58: 16–22.
Munro CA, McCaul ME, Wong DF, et al. (2006) Sex differences in striatal dopamine release in healthy adults. *Biological Psychiatry* 59: 966–974.
Narbone MC, Ruello C, Oliva A, Baviera G, D'Amico D, Bramanti P, and Di Perri R (1990) Hormonal disregulation and catamenial epilepsy. *Functional Neurology* 5: 49–53.
Netherton RA and Overstreet DH (1983) Genetic and sex differences in the cholinergic modulation of thermoregulation. In: Lomas P and Schonbaum EE (eds.) pp. 74–77. Basel: Karger.
Nishizawa S, Benkelfat C, Young SN, et al. (1997) Differences between males and females in rates of serotonin synthesis in human brain. *Proceedings of the National Academy of Sciences of the United States of America* 94: 5308–5313.
Nishizuka M, Katoh-Sembra R, Eto K, Arai Y, Iizuka R, and Kato K (1991) Age and sex related differences in nerve growth factor distribution in the rat brain. *Brain Research Bulletin* 27: 685–688.
Olmos G, Naftolin F, Perez J, Tranque PA, and Garcia-Segura LM (1989) Synaptic remodeling in the rat arcuate nucleus during the estrous cycle. *Neuroscience* 32: 663–667.
Ota M, Crofton JT, and Share L (1994) Hemorrhage-induced vasopressin release in the paraventricular nucleus measured by *in vivo* microdialysis. *Brain Research* 658: 49–54.
Ota M, Crofton JT, Toba K, and Share L (1992) Effect on vasopressin release of microinjection of cholinergic agonists into the rat supraoptic nucleus. *Proceedings of the Society for Experimental Biology and Medicine* 201: 208–214.
Ottem EN, Godwin JG, Krishnan S, and Petersen SL (2004) Dual-phenotype GABA/glutamate neurons in adult preoptic area: Sexual dimorphism and function. *Journal of Neuroscience* 24: 8097–8105.
Overstreet DH, Miller CS, Janowsky DS, and Russell RW (1996) Potential animal model of multiple chemical sensitivity with cholinergic supersensitivity. *Toxicology* 111: 119–134.
Overstreet DH, Russell RW, Crocker AD, Gillin JC, and Janowsky DS (1988) Genetic and pharmacological models of cholinergic supersensitivity and affective disorders. *Experientia* 44: 465–472.
Palanza P (2001) Animal models of anxiety and depression: How are females different? *Neuroscience and Biobehavioral Reviews* 25: 219–233.
Papasozomenos S and Su Y (1991) Altered phosphoregulation of protein in heat-shocked rats and patients with Alzheimer's disease. *Proceedings of the National Academy of Sciences of the United States of America* 88: 4543–4547.
Parducz A, Perez J, and Garcia-Segura LM (1993) Estradiol induces plasticity of gabaergic synapses in the hypothalamus. *Neuroscience* 53: 395–401.
Parker G, Parker K, Austin MP, Mitchell P, and Brotchie H (2003) Gender differences in response to differing antidepressant drug classes: Two negative studies. *Psychological Medicine* 33: 1473–1477.
Parrott A and Craig D (1995) Psychological functions served by nicotine chewing gum. *Addictive Behaviors* 20: 271–278.
Parsons B, Rainbow TC, and McEwen BS (1984) Organizational effects of testosterone via aromatization on feminine reproductive behavior and neural progestin receptors in rat brain. *Endocrinology* 115: 1412–1417.
Paula-Barbosa MM, Sousa N, and Madeira MD (1993) Ultrastructural evidence of sexual dimorphism in supraoptic neurons: A morphometric study. *Journal of Neurocytology* 22: 697–706.

Pedersen CA and Boccia ML (2002) Oxytocin maintains as well as initiates female sexual behavior: Effects of a highly selective oxytocin antagonist. *Hormones and Behavior* 41: 170–177.

Perez J, Luquin S, Naftolin F, and Garcia-Segura LM (1993) The role of estradiol and progesterone in phased synaptic remodelling of the rat arcuate nucleus. *Brain Research* 608: 38–44.

Pericic D, Manev H, and Bujas M (1996) Gonadal hormones and picrotoxin-induced convulsions in male and female rats. *Brain Research* 736: 174–179.

Pericic D, Manev H, and Lakic N (1985) Sex differences in the response of rats to drugs affecting GABAergic transmission. *Life Science* 36: 541–547.

Perkins KA, Donny E, and Caggiula AR (1999) Sex differences in nicotine effects and self-administration: Review of human and animal evidence. *Nicotine and Tobacco Research* 1: 301–315.

Peskind ER, Raskind MA, Wingerson D, Pascualy M, Thal LJ, Dobie DJ, and Wilkinson CW (1996) Hypothalamic–pituitary–adrenocortical axis responses to physostigmine: Effects of Alzheimer's disease and gender. *Biological Psychiatry* 40: 61–68.

Pfaff D (2005) Hormone-driven mechanisms in the central nervous system facilitate the analysis of mammalian behaviours. *Journal of Endocrinology* 184: 447–453.

Pfais JG and Gorzalka BB (1987) Opioids and sexual behavior. *Neuroscience and Biobehavioral Review* 11: 1–34.

Pilgrim C and Hutchison JB (1994) Developmental regulation of sex differences in the brain: Can the role of gonadal steroids be redefined? *Neuroscience* 60: 843–855.

Pilgrim C and Reisert I (1992) Differences between male and female brains – developmental mechanisms and implications. *Hormone and Metabolic Research* 24: 353–359.

Pinos H, Collado P, Rodriguez-Zafra M, Rodriguez C, Segovia S, and Guillamon A (2001) The development of sex differences in the locus coeruleus of the rat. *Brain Research Bulletin* 56: 73–78.

Pittel Z, Fisher A, and Heldman E (1987) Reversible and irreversible inhibition of high-affinity choline transport caused by ethylcholine ariridinium ion. *Journal of Neurochemistry* 49: 468–474.

Plotsky P, Otto S, Toyama T, and Sutton S (1990) Lack of correlation between immunoreactive corticotropin-releasing factor concentration profiles in hypophysial-portal and peripheral plasma. *Journal of Neuroendocrinology* 2: 65–69.

Polston EK, Gu G, and Simerly RB (2004) Neurons in the principal nucleus of the bed nuclei of the stria terminalis provide a sexually dimorphic GABAergic input to the anteroventral periventricular nucleus of the hypothalamus. *Neuroscience* 123: 793–803.

Pomerleau C, Aubin H, and Pomerleau O (1997) Self-reported alcohol use patterns in a sample of male and female heavy smokers. *Journal of Addictive Diseases* 16: 19–24.

Pomerleau C, Tate J, Lumley M, and Pomerleau O (1994) Gender differences in prospectively versus retrospectively assessed smoking withdrawal symptoms. *Journal of Substance Abuse* 6: 433–440.

Ravizza T, Friedman LK, Moshe SL, and Veliskova J (2003) Sex differences in GABA(A)ergic system in rat substantia nigra pars reticulata. *International Journal of Developmental Neuroscience* 21: 245–254.

Reiner PB and Fibiger HC (1995) Functional heterogencity of central cholinergic systems. In: Blood FE and Kupfer DJE (eds.) *Psychopharmacology: The Fourth Generation of Progress*, pp. 147–153. New York: Raven Press.

Reisert I and Pilgrim C (1991) Sexual differentiation of monoaminergic neurons – genetic or epigenetic? *Trends in Neurosciences* 14: 468–473.

Renard GM, Suarez MM, Levin GM, and Rivarola MA (2005) Sex differences in rats: Effects of chronic stress on sympathetic system and anxiety. *Physiology and Behavior* 85: 363–369.

Rhodes ME, Balestreire EM, Kenneth Czambel R, and Rubin RT (2002) Estrous cycle influences on sexual diergism of HPA axis responses to cholinergic stimulation in rats. *Brain Research Bulletin* 59: 217–225.

Rhodes ME, Billings TE, Czambel RK, and Rubin RT (2005) Pituitary–adrenal responses to cholinergic stimulation and acute mild stress are differentially elevated in male and female M(2) muscarinic receptor knockout mice. *Journal of Neuroendocrinology* 17: 817–826.

Rhodes ME, Kennell JS, Belz EE, Czambel RK, and Rubin RT (2004) Rat estrous cycle influences the sexual diergism of HPA axis stimulation by nicotine. *Brain Research Bulletin* 64: 205–213.

Rhodes ME, Laco JE, Evans BB, Billings TE, Czambel RK, and Rubin RT (2006) HPA axis responses to oxotremorine in male and female M1 muscarinic receptor knockout and wild-type mice. *Society for Neuroscience Abstract* 563.

Rhodes ME, O'Toole SM, Czambel RK, and Rubin RT (2001a) Male-female differences in rat hypothalamic–pituitary–adrenal axis responses to nicotine stimulation. *Brain Research Bulletin* 54: 681–688.

Rhodes ME, O'Toole SM, Wright SL, Czambel RK, and Rubin RT (2001b) Sexual diergism in rat hypothalamic–pituitary–adrenal axis responses to cholinergic stimulation and antagonism. *Brain Research Bulletin* 54: 101–113.

Rhodes ME and Rubin RT (1999) Functional sex differences ('sexual diergism') of central nervous system cholinergic systems, vasopressin, and hypothalamic-pituitary-adrenal axis activity in mammals: A selective review. *Brain Research Reviews* 30: 135–152.

Rivier C (1993) Female rats release more corticosterone than males in response to alcohol: Influence of circulating sex steroids and possible consequences for blood alcohol levels. *Alcoholism: Clinical and Experimental Research* 17: 854–859.

Robinson SM, Fox TO, and Pearlstein RA (1986) Sex differences in the shape of the sexually dimorphic nucleus of the preoptic area and suprachiasmatic nucleus of the rat: 3-D computer reconstructions and morphometrics. *Brain Research* 371: 380–384.

Rubin RT, Abbasi SA, Rhodes ME, and Czambel RK (2003) Growth hormone responses to low-dose physostigmine administration: Functional sex differences (sexual diergism) between major depressives and matched controls. *Psychological Medicine* 33: 655–665.

Rubin RT, Miller TH, Rhodes ME, and Czambel RK (2006a) Growth hormone responses to low-dose physostigmine in elderly vs. young women and men. *Gerontology* 52: 76–84.

Rubin RT, O'Toole SM, Rhodes ME, Sekula LK, and Czambel RK (1999a) Hypothalamo-pituitary–adrenal cortical responses to low-dose physostigmine and arginine vasopressin administration: Sex differences between major depressives and matched control subjects. *Psychiatry Research* 89: 1–20.

Rubin RT, Rhodes ME, Miller TH, Jakab RL, and Czambel RK (2006b) Sequence of pituitary–adrenal cortical hormone responses to low-dose physostigmine administration in young adult women and men. *Life Science* 79: 2260–2268.

Rubin RT, Rhodes ME, O'Toole S, and Czambel RK (2002) Sexual diergism of hypothalamo-pituitary–adrenal cortical responses to low-dose physotigmine in elderly vs. young women and men. *Neuropsychopharmacology* 26: 672–681.

Rubin RT, Sekula LK, O'Toole S, Rhodes ME, and Czambel RK (1999b) Pituitary–adrenal cortical responses to low-dose physostigmine and arginine vasopressin administration in

normal women and men. *Neuropsychopharmacology* 20: 434–446.
Saija A, Princi P, D'Amico N, De Pasquale R, and Costa G (1990) Aging and sex influence the permeability iof the blood-brain barrier in the rat. *Life Science* 47: 2261–2267.
Sapolsky RM, Zola-Morgan S, and Squire LR (1991) Inhibition of glucocorticoid secretion by the hippocampal formation in the primate. *Journal of Neuroscience* 11: 3695–3704.
Schindler CW and Carmona GN (2002) Effects of dopamine agonists and antagonists on locomotor activity in male and female rats. *Pharmacology, Biochemistry, and Behavior* 72: 857–863.
Schumacher M, Coirini H, Flanagan LM, Frankfurt M, Pfaff DW, and McEwen BS (1992) Ovarian steroid modulation of oxytocin receptor binding in the ventromedial hypothalamus. *Annals of the New York Academy of Sciences* 652: 374–386.
Schwartz-Giblin S, Korotzer A, and Pfaff DW (1989) Steroid hormone effects on picrotoxin-induced seizures in female and male rats. *Brain Research* 476: 240–247.
Seale JV, Wood SA, Atkinson HC, et al. (2004) Gonadectomy reverses the sexually diergic patterns of circadian and stress-induced hypothalamic–pituitary–adrenal axis activity in male and female rats. *Journal of Neuroendocrinology* 16: 516–524.
Searles RV, Yoo MJ, He JR, Shen WB, and Selmanoff M (2000) Sex differences in GABA turnover and glutamic acid decarboxylase (GAD(65) and GAD(67)) mRNA in the rat hypothalamus. *Brain Research* 878: 11–19.
Seeman MV (1985) Sex and schizophrenia. *Canadian Journal of Psychiatry* 30: 313–315.
Seeman MV (1997) Psychopathology in women and men: Focus on female hormones. *American Journal of Psychiatry* 154: 1641–1647.
Seeman MV (2000) Women and schizophrenia. *Medscape Womens Health* 5: 2.
Segovia S, del Cerro MC, Ortega E, et al. (1996) Role of $GABA_A$ receptors in the organization of brain and behavioural sex differences. *NeuroReport* 7: 2553–2557.
Segovia S, Garcia-Falgueras A, Carrillo B, et al. (2006) Sexual dimorphism in the vomeronasal system of the rabbit. *Brain Research* 1102: 52–62.
Segovia S, Guillamon A, del Cerro MC, Ortega E, Perez-Laso C, Rodriguez-Zafra M, and Beyer C (1999) The development of brain sex differences: A multisignaling process. *Behavioural Brain Research* 105: 69–80.
Shephard RA, Nielsen EB, and Broadhurst PL (1982) Sex and strain differences in benzodiazepine receptor binding in Roman rat strains. *European Journal of Pharmacology* 77: 327–330.
Shimon M, Egozi Y, Kloog Y, Sokolovsky M, and Cohen S (1988) Kinetics of choline uptake into isolated rat forebrain microvessels: Evidence of endocrine modulation. *Journal of Neurochemistry* 50: 1719–1724.
Shoji M, Share L, Crofton JT, and Brooks D (1989) The effect on vasopressin release of microinjection of cholinergic agonists into the paraventricular nucleus of conscious rats. *Journal of Neuroendocrinology* 1: 401–406.
Smith JM and Dunn DD (1979) Sex differences in the prevalence of severe tardive dyskinesia. *American Journal of Psychiatry* 136: 1080–1082.
Smith JM, Kucharski LT, Oswald WT, and Waterman LJ (1979) Tardive dyskinesia: Effect of age, sex, and criterion level of symptomatology on prevalence estimates. *Psychopharmacological Bulletin* 15: 69–71.
Smolen A, Smolen TN, Han PC, and Collins A (1987) Sex difference in the recovery of brain acetylcholinesterase activity following a single exposure to DFP. *Pharmacology, Biochemistry, and Behavior* 26: 813–820.
Sokolovsky M (1984) Muscarinic receptors in the central nervous system. *International Review of Neurobiology* 25: 138–183.
Spinedi E, Suescun MO, Hadid R, Daneva T, and Gaillard RC (1992) Effects of gonadectomy and sex hormone therapy on the endotoxin-stimulated hypothalamo-pituitary–adrenal axis: Evidence for a neuroendocrine-immunological sexual dimorphism. *Endocrinology* 131: 2430–2436.
Spring S, Lerch JP, and Henkelman RM (2007) Sexual dimorphism revealed in the structure of the mouse brain using three-dimensional magnetic resonance imaging. *NeuroImage* 35: 1424–1433.
Stefanova N, Bozhilova-Pastirova A, and Ovtscharoff W (1997) Distribution of GABA-immunoreactive nerve cells in the bed nucleus of the stria terminalis in male and female rats. *European Journal of Histochemistry* 41: 23–28.
Stone JD, Crofton JT, and Share L (1989) Sex differences in central adrenergic control of vasopressin release. *American Journal of Physiology* 257: R1040–R1045.
Suescun MO, Chisari A, Gaillard RC, and Spinedi E (1997) Age and sex steroid environment modulate hypothalamo-pituitary–adrenal axis function in mice. *Endocrinology and Metabolism* 4: 179–184.
Sumner BE and Fink G (1997) The density of 5-hydoxytryptamine2A receptors in forebrain is increased at pro-oestrus in intact female rats. *Neuroscience Letters* 234: 7–10.
Sumner BE, Grant KE, Rosie R, Hegele-Hartung C, Fritzemeier KH, and Fink G (1999) Effects of tamoxifen on serotonin transporter and 5-hydroxytryptamine(2A) receptor binding sites and mRNA levels in the brain of ovariectomized rats with or without acute estradiol replacement. *Brain Research – Molecular Brain Research* 73: 119–128.
Svikis D, Hatsukami D, Hughes J, Carroll K, and Pickens R (1986) Sex differences in tabacco withdrawal syndrome. *Addictive Behaviors* 11: 459–462.
Swaab DF (1995) Ageing of the human hypothalamus. *Hormone Research* 43: 8–11.
Swaab DF, Chung WC, Kruijver FP, Hofman MA, and Hestiantoro A (2003) Sex differences in the hypothalamus in the different stages of human life. *Neurobiology of Aging* 24(supplement 1): S1–S16 (discussion S17–9).
Swaab DF, Fliers E, and Partiman TS (1985) The suprachiasmatic nucleus of the human brain in relation to sex, age, and senile dementia. *Brain Research* 342: 37–44.
Swaab DF and Hofman MA (1990) An enlarged suprachiasmatic nucleus in homosexual men. *Brain Research* 537: 141–148.
Swaab DF, Zhou JN, Ehlart T, and Hoffman MD (1995) Development of vasoactive intestinal polypeptide (VIP) neurons in the human suprachiasmatic nucleus (SCN) in relation to birth and sex. *Developmental Brain Research* 79: 249–259.
Swaab DF, Zhou JN, Ehlhart T, and Hofman MA (1994) Development of vasoactive intestinal polypeptide neurons in the human suprachiasmatic nucleus in relation to birth and sex. *Brain Research – Developmental Brain Research* 79: 249–259.
Takase K, Mitsushima D, Funabashi T, and Kimura F (2007) Sex difference in the 24-h acetylcholine release profile in the premotor/supplementary motor area of behaving rats. *Brain Research* 1154: 105–115.
Taylor SE, Klein LC, Lewis BP, Gruenewald TL, Gurung RA, and Updegraff JA (2000) Biobehavioral responses to stress in females: Tend-and-befriend, not fight-or-flight. *Psychological Review* 107: 411–429.
Todd BJ, Schwarz JM, Mong JA, and McCarthy MM (2007) Glutamate AMPA/kainate receptors, not GABA(A) receptors, mediate estradiol-induced sex differences in the hypothalamus. *Developmental Neurobiology* 67: 304–315.
Tsagarakis S and Grossman A (1990) Central neuroregulation of hypothalamic corticotropin-releasing hormone (CRH-41) secretion. *Journal of Endocrinological Investigation* 13: 765–775.

Tucek S (1985) Regulation of acetylcholine synthesis in the brain. *Journal of Neurochemistry* 44: 11–24.

Tuomisto J and Mannisto P (1985) Neurotransmitter regulation of anterior pituitary hormones. *Pharmacological Reviews* 37: 249–332.

Turner BB (1990) Sex difference in glucocorticoid binding in rat pituitary is estrogen dependent. *Life Science* 46: 1399–1406.

Turner BB (1992) Sex differences in the binding of type I and type II corticosteroid receptors in rat hippocampus. *Brain Research* 581: 229–236.

Turner BB and Weaver DA (1985) Sexual dimorphism of glucocorticoid binding in the rat brain. *Brain Research* 343: 16–23.

van Doornen LJ (1986) Sex differences in physiological reactions to real life stress and their relationship to psychological variables. *Psychophysiology* 23: 657–662.

van Leeuwen FW, Caffe AR, and De Vries GJ (1985) Vasopressin cells in the bed nucleus of the stria terminalis of the rat: Sex differences and the influence of androgens. *Brain Research* 325: 391–394.

van Londen L, Goekoop JG, van Kempen GM, et al. (1997) Plasma levels of arginine vasopressin elevated in patients with major depression. *Neuropsychopharmacology* 17: 284–292.

van Merode T, de Krom MD, and Knottnerus JA (1997) Gender-related differences in non-epileptic attacks: A study of patients' cases in the literature. *Seizure* 6: 311–316.

Versteeg CA, Cransberg K, De Jong W, and Bohus B (1983) Reduction of a centrally induced pressor response by neurohypophyseal peptides: The involvement of lower brainstem mechanisms. *European Journal of Pharmacology* 94: 133–140.

Viau V and Meaney MJ (1991) Basal and stress hypothalamic–pituitary–adrenal activity in cycling and ovariectomized-steroid treated rats. *Endocrinology* 129: 2503–2511.

Viau V and Meaney MJ (2004) Alpha1 adrenoreceptors mediate the stimulatory effects of oestrogen on stress-related hypothalamic–pituitary–adrenal activity in the female rat. *Journal of Neuroendocrinology* 16: 72–78.

Wainer BH, Steininger TL, Roback JD, Burke-Watson MA, Mufson EJ, and Kordower J (1993) Ascending cholinergic pathways: Functional organization and implications for disease models. *Progress in Brain Research* 98: 9–30.

Walker QD, Cabassa J, Kaplan KA, Li ST, Haroon J, Spohr HA, and Kuhn CM (2001) Sex differences in cocaine-stimulated motor behavior: Disparate effects of gonadectomy. *Neuropsychopharmacology* 25: 118–130.

Walker QD, Ray R, and Kuhn CM (2006) Sex differences in neurochemical effects of dopaminergic drugs in rat striatum. *Neuropsychopharmacology* 31: 1193–1202.

Weinberg J, Gunnar MR, Brett LP, Gonzalez CA, and Levine S (1982) Sex differences in biobehavioral responses to conflict in a taste aversion paradigm. *Physiology and Behavior* 29: 201–210.

Weinstock M, Matlina E, Maor GI, Rosen H, and McEwen BS (1992) Prenatal stress selectively alters the reactivity of the hypothalamic–pituitary adrenal system in the female rat. *Brain Research* 595: 195–200.

Weiss B (1998) Neurobehavioral properties of chemical sensitivity syndromes. *Neurotoxicology* 19: 259–268.

Wess J (1996) Molecular biology of muscarinic acetylcholine receptors. *Critical Reviews in Neurobiology* 10: 69–99.

Wess J (2004) Muscarinic acetylcholine receptor knockout mice: Novel phenotypes and clinical implications. *Annual Review of Pharmacology and Toxicology* 44: 423–450.

Wess J, Blin N, Mutschler E, and Bluml K (1995) Muscarinic acetylcholine receptors: Structural basis of ligand binding and G protein coupling. *Life Science* 56: 915–922.

Wess J, Duttaroy A, Zhang W, et al. (2003) M1-M5 muscarinic receptor knockout mice as novel tools to study the physiological roles of the muscarinic cholinergic system. *Receptors Channels* 9: 279–290.

Westenbroek C, Ter Horst GJ, Roos MH, Kuipers SD, Trentani A, and den Boer JA (2003) Gender-specific effects of social housing in rats after chronic mild stress exposure. *Progress in Neuropsychopharmacology and Biological Psychiatry* 27: 21–30.

Whitnall M (1993) Regulation of the hypothalamic corticotropin-releasing hormone neurosecretory system. *Progress in Neurobiology* 40: 573–629.

Williams CL and Meck WH (1991) The organizational effects of gonadal steroids on sexually dimorphic spatial ability. *Psychoneuroendocrinology* 16: 155–176.

Williams TD, Carter DA, and Lightman SL (1985) Sexual dimorphism in the posterior pituitary response to stress in the rat. *Endocrinology* 116: 738–740.

Willoughby JO, Jervois PM, Menadue MF, and Blessing WW (1987) Noradrenaline, by activation of alpha-1-adrenoceptors in the region of the supraoptic nucleus, causes secretion of vasopressin in the unanaesthetized rat. *Neuroendocrinology* 45: 219–226.

Wilson MA (1992) Influences of gender, gonadectomy, and estrous cycle on GABA/BZ receptors and benzodiazepine responses in rats. *Brain Research Bulletin* 29: 165–172.

Wilson MA and Biscardi R (1997) Influence of gender and brain region on neurosteroid modulation of GABA responses in rats. *Life Science* 60: 1679–1691.

Witt ED, Mantione CR, and Hanin I (1986) Sex differences in muscarinic receptor binding after chronic ethanol administration in the rat. *Psychopharmacology (Berl)* 90: 537–542.

Wizemann TM and Pardue M-L (2001) *Exploring the Biological Contributions to Human Health.* Washington, DC: National Academy of Sciences.

Woolf NJ (1991) Cholinergic systems in mammalian brain and spinal cord. *Progress in Neurobiology* 37: 475–524.

Woolley DE and Timiras PS (1962a) Estrous and circadian periodicity and electroshock convulsions in rats. *American Journal of Physiology* 202: 379–382.

Woolley DE and Timiras PS (1962b) The gonad-brain relationship: Effects of female sex hormones on electroshock convulsions in the rat. *Endocrinology* 70: 196–209.

Yanase M, Honmura A, Akaishi T, and Sakuma Y (1988) Nerve growth factor-mediated sexual differentiation of the rat hypothalamus. *Neuroscience Research* 6: 181–185.

Yeoman RR and Jenkins AJ (1989) Arcuate area of the female rat maintained *in vitro* exhibits increased afternoon electrical activity. *Neuroendocrinology* 49: 144–149.

Yilmaz O, Kanit L, Okur BE, and Pogun S (1997) Effects of nicotine on active avoidance learning in rats: Sex differences. *Behavioural Pharmacology* 8: 253–260.

Yonkers KA, Kando JC, Cole JO, and Blumenthal S (1992) Gender differences in pharmacokinetics and pharmacodynamics of psychotropic medication. *American Journal of Psychiatry* 149: 587–595.

Young MA, Scheftner WA, Fawcett J, and Klerman GL (1990) Gender differences in the clinical features of unipolar major depressive disorder. *Journal of Nervous and Mental Disease* 178: 200–203.

Zhang L, Ma W, Barker JL, and Rubinow DR (1999) Sex differences in expression of serotonin receptors (subtypes 1A and 2A) in rat brain: A possible role of testosterone. *Neuroscience* 94: 251–259.

Zhou J, Pfaff DW, and Chen G (2005) Sex differences in estrogenic regulation of neuronal activity in neonatal cultures of ventromedial nucleus of the hypothalamus. *Proceedings of the National Academy of Sciences of the United States of America* 102: 14907–14912.

8 Gonadal Hormones and Sexual Differentiation of Human Brain and Behavior

M Hines, University of Cambridge, Cambridge, UK

Chapter Outline

8.1 Introduction

Gonadal hormones have powerful influences on sexual differentiation of brain and behavior in a wide range of mammals. This chapter evaluates evidence regarding similar influences on sexual differentiation of human behavior, and explores neural mechanisms that could underlie these influences. The chapter begins by defining terms and outlining theoretical models, derived from empirical research in other species. The main purpose of this section is to establish basic principles on which to base hypotheses and guide interpretation of human findings. The second section of the chapter describes and evaluates approaches that have been used to study hormonal influences on sexual differentiation of human behavior. Because of ethical constraints, these approaches are largely nonexperimental, and this section of the chapter covers some of the limitations of various approaches. The third, and central, section of the chapter reviews findings obtained using these approaches, while the fourth section describes sex differences in the human brain that could underlie the behavioral outcomes of hormonal exposure. The fifth and final section reviews methodological and interpretational considerations and suggests directions for future research.

8.2 Definitions and Theoretical Models

8.2.1 Definitions

8.2.1.1 Organization and activation

In general, gonadal hormones have two types of influences on brain and behavior, termed organizational and activational. Both organizational and activational influences can involve changes in neural structure, but they are distinguished by their timing and permanence. Organizational influences typically occur early in life, during critical or sensitive periods of development, and they are permanent. The hormone must be present at a specific time to exert its effect, and, although it is present only briefly, its effect persists across the life span and is not reversed by subsequent hormone withdrawal. These early, permanent effects of hormones are thought to occur because hormones direct some aspects of neural development during early life and thus influence the underlying organization of the brain (Arnold and Gorski, 1984). This is why they are called organizational effects (Phoenix et al., 1959). Activational influences of hormones occur later in life, typically in adulthood, and are reversed by hormone withdrawal. This chapter focuses on organizational, rather than activational, influences of hormones, because neurobehavioral sexual differentiation in mammals occurs early in life and involves permanent changes in brain and behavior.

8.2.1.2 Sex differences and gender differences

Sex differences or gender differences are characteristics that differ, on the average, for males versus females. Some researchers suggest that sex differences are biological differences between males and females and gender differences are culturally based differences. However, sex differences in human behavior typically involve both biological and cultural components. Therefore, like others (e.g., Maccoby, 1988), the terms sex difference and gender difference are used interchangeably here, to refer to characteristics on which males and females differ, on average.

8.2.2 Theoretical Models

8.2.2.1 The classic model

The classic model of hormonal influences on sexual differentiation posits that the presence of testicular hormones during early life causes male-typical development, while their absence causes female-typical development. Empirical evidence generally supports this model for sexual differentiation of a wide range of brain structures and behaviors, at least in rodents and nonhuman primates. For instance, exposure of XX rodents to testosterone (T) during critical periods of prenatal or neonatal development produces adult animals who show male-typical sexual behavior (e.g., mounting) but not female-typical sexual behavior (e.g., lordosis). Similarly, castration of XY animals early in life produces adults who show reduced male-typical behavior and increased female-typical behavior (Goy and McEwen, 1980). The same hormone treatments also produce permanent changes in the brain. Perhaps the best-known example involves a subregion of the anterior hypothalamus/preoptic area (AH/POA) called the sexually dimorphic nucleus of the preoptic area (SDN-POA). This nucleus is several times larger in male, than in female, rats (Gorski et al., 1978, 1980), and treating XX animals with testicular hormones during early life enlarges the nucleus while withdrawing these hormones from developing males reduces it (Gorski et al., 1978; Jacobson et al., 1981). In contrast to the dramatic effects produced by manipulating testicular hormones, removal of the ovaries at comparable early stages of development generally has little or no impact on male-typical or female-typical behavior or on the SDN-POA (Goy and McEwen, 1980; Jacobson et al., 1981). In addition, treating XX rodents with estrogen during early development generally has similar influences to treatment with T; it promotes male-typical behavioral and brain development, including development of the SDN-POA, and impairs female-typical behavior (Goy and McEwen, 1980; MacLusky and Naftolin, 1981). These outcomes occur because T is normally converted, within the brain, to estrogen before interacting with receptors to produce male-typical development, at least in regard to many brain regions and behaviors in rodents.

In general, the classic influences of testicular hormones occur during critical periods of prenatal and neonatal development. These critical periods vary somewhat from species to species, but appear to correspond to times when T concentrations are higher in developing males than females. In the rat, such periods occur from about the 17th to the 19th day of an approximately 21-day gestation and from the first to the tenth postnatal day. Within this overall critical period, there are separate periods when specific sexually differentiated characteristics are most sensitive to hormonal influences.

The term, masculinization, has been used to refer to enhancement of characteristics that are more common (or larger) in males than in females, and the most extensively studied example of masculinization is male reproductive behavior, particularly mounting behavior directed at sexually receptive females. Similarly, the term, feminization, has been used to refer to enhancement of characteristics that are more common (or larger) in females than in males and an example is the lordosis posture (arching the back and deflecting the tail) shown by sexually receptive females. These two processes, mounting and lordosis, are influenced by hormones at slightly different times during early development (Christensen and Gorski, 1978), and by timing hormonal manipulations to hit or avoid periods during which specific behaviors differentiate, animals can be both masculinized and femininized (i.e., show both mounting and lordosis) or demasculinized and defeminized (i.e., show neither mounting nor lordosis), as well as be conventionally masculine (i.e., show mounting but not lordosis) or conventionally feminine (i.e., show lordosis but not mounting). Hence, sexual differentiation has been conceptualized as involving two separate dimensions of masculinization and feminization. Although mounting and lordosis have been studied most extensively, sexual differentiation involves many different outcomes, in addition to mounting and lordosis, and it is likely that these outcomes also differentiate at somewhat different times, or involve different specific mechanisms. Therefore, sexual differentiation is likely to involve a large number of separable dimensions that distinguish the average male from the average female.

8.2.2.2 The gradient model

Hormones influence not only behavioral differences between the sexes, but also differences within each sex. Hormone influences are graded; when more of a hormone is administered, there is a bigger effect on behavior (Collaer and Hines, 1995). In addition, naturally occurring variations in hormones relate to behavior within each sex. For instance, some female rats mount other females and some do not, and females who are exposed prenatally to blood that has contacted male littermates (because of their position relative to male siblings *in utero*) show more mounting as adults than those who are not so positioned (Meisel and Ward, 1981). Studies in other rodents, including mice and gerbils, have produced similar results (Clark and Galef, Jr., 1998). In support of T as the responsible agent, gerbil fetuses positioned between two males have higher levels of T than those positioned between two females (Clark et al., 1991). Because of the graded influences of the early hormone environment on behavior, a modification of the classic model of hormonal influences, called the gradient model, has been proposed (Collaer and Hines, 1995). In this model, not only do testicular hormones cause differences between normal male and female animals, but also small amounts of hormones produce movement along sexually differentiated gradients within each sex.

8.2.2.3 Active feminization

The classic model suggests that ovarian hormones are not needed for female-typical development, and so it has sometimes been called a passive feminization model. In contrast, some researchers argue that some sexually differentiated characteristics are feminized by ovarian hormones (Fitch and Denenberg, 1998; Toran-Allerand, 1984). For instance, the presence of ovarian hormones near the time of puberty may permanently enhance some aspects of female-typical sexual behavior in the rat (Dunlap et al., 1978; Gerall et al., 1972). Ovarian hormones have also been found to promote female-typical development of some structural characteristics in the rodent cerebral cortex, including asymmetries in cortical thickness and the size of the corpus callosum (Diamond et al., 1981; Fitch et al., 1990, 1991). Like the effects on female-typical sexual behavior, these feminizing effects of estrogen on cortical development appear to occur after the neonatal period. Thus, there may be a critical period when ovarian hormones actively feminize some characteristics, at least in rodents, and this critical period may occur later in life than the critical period for the effects of T. It would be interesting to know how this critical period corresponds to times when estrogen levels (or levels of other ovarian hormones) are higher in females than in males. Additional research is also needed to demonstrate that the effects of ovarian hormones are truly organizational in the sense of being irreversible by hormone withdrawal, and that they occur only during a particular developmental window (Hines, 1998).

8.2.2.4 Complexity and multiple models

Sexual differentiation is regulated by gonadal hormones during early development across a wide range of mammals and behavioral outcomes (Hines, 2004), but the details involved can vary from species to species and from behavior to behavior. For example, it is well established that androgen is converted to estrogen before acting through estrogen receptors to cause male-typical neural and behavioral development in rodents (McCarthy, 2008). However, there is some evidence that this is not true for all sexually differentiated rodent behaviors, male-typical rough-and-tumble play being a likely exception (Meaney and Stewart, 1981). In addition, the role of estrogen may be reduced in nonhuman primates, where sexual behavior appears to be masculinized by androgenic hormones that cannot be converted to estrogen (Goy, 1978). However, in the same species (the rhesus macaque), long-term prenatal exposure to the synthetic estrogen, diethylstilbestrol (DES), has been found to enhance male-typical rough-and-tumble play (Goy and Deputte, 1996). Thus, although the specific aspects of behavior that are influenced directly by androgen, versus estrogen derived from androgen, appear to differ for rats versus rhesus macaques, it is an oversimplification to say that androgen is converted to estrogen before acting in rodents (or even just in rats), whereas androgen acts directly in nonhuman primates (or even just in rhesus macaques). Instead, this type of statement is more valid when applied to a specific behavioral outcome in a specific species. Another detail that can vary from one endpoint to another in the rat, and could well vary across species, involves downstream effectors of hormones involved in sexual differentiation. In rats, for instance, estradiol produces male-typical sexual behavior and dendritic spine formation in the preoptic area by influencing synthesis of prostaglandin E_2 (PGE_2); removing PGE_2 from developing males has the same demasculinizing effect on these endpoints as does administering estrogen, while administering PGE_2 to developing females has the opposite effects. However, similar influences are not seen for a second estrogen-sensitive aspect of preoptic area anatomy, the SDN-POA (Amateau and McCarthy, 2004).

8.2.3 Summary

Conceptualization of sexual differentiation has proceeded from a one-dimensional (1D) continuum with masculine at one end and feminine at the other, through a 2D space, defined by separate masculine and feminine axes, to a multidimensional space with axes for each sexually differentiated characteristic. This multidimensional conceptualization allows not only for differences in critical periods for hormones to act on specific behaviors, but also for different mechanisms (e.g., different hormone metabolites and different cofactors) to be involved in differentiation of each characteristic. It also allows different sexually differentiated characteristics to conform to different models of hormone action, such as a gradient version of the classic model for some characteristics and a model involving active feminization for others. A multidimensional model is also consistent with evidence that individuals can vary in sex-related behavior from one dimension to another, being strongly sex-typical in some respects, but less so, or even sex-atypical, in others.

8.3 Hormonal Influences on Human Sexual Differentiation: Sources of Information

It generally is unethical to manipulate hormones during human development for experimental purposes. Therefore, true experiments similar to those conducted in other species are largely impossible. However, other approaches have been used to evaluate the relevance of animal models in understanding human neural and behavioral sexual differentiation. These include two general types of studies. The first examines individuals who have experienced dramatic alterations in hormones prenatally, for instance, because of genetic disorders or because their mothers were prescribed hormones during pregnancy. The second relates normal variability in the early hormone environment to normal variability in subsequent behavior.

Evidence from genetic syndromes and situations where women have been prescribed hormones during pregnancy indicate that sexual differentiation of the human internal and external genitalia conform to the classical model of hormonal influences. High levels of testicular hormones promote male-typical development, and, in the absence of these hormones, female-typical structures develop (Wilson et al., 1981). Although both internal and external genital structures are influenced by the presence or absence of testicular hormones, some details of the mechanisms for differentiation of the internal genitalia differ from those for the external genitalia.

In the case of the internal genitalia, both XX and XY fetuses begin with Müllerian ducts as well as Wolffian ducts. Information on the Y chromosome directs the gonads to differentiate into testes, and by week 8 of gestation, almost all XY fetuses have functioning testes (in rare cases, genetic anomalies, such as a missing portion of the Y chromosome, prevent this). One hormone produced by the testes, Müllerian-inhibiting factor (MIF), then causes the Müllerian ducts to regress, while a second hormone, T, causes the Wolffian ducts to develop into male internal genitalia (vas deferens, seminal vesicles, and prostate). In contrast, in XX fetuses, the gonads differentiate into ovaries instead of testes. Then, in the absence of testicular hormones, the Wolffian ducts regress and the Müllerian ducts differentiate as feminine internal genitalia (uterus, fallopian tubes, and upper vagina).

In contrast to the internal genitalia, where two sets of structures are initially present in both XX and XY fetuses and one regresses, the external genitalia begin as one set of structures, identical in both XX and XY fetuses. In the presence of testicular androgens, particularly dihydrotestosterone (DHT), these structures become penis and scrotum. In the absence of testicular hormones, the same structures become clitoris and labia. Therefore, although both the internal and external genitalia differentiate under the influence of testicular hormones, the processes differ in that, for the internal genitalia, both XX and XY fetuses begin with two sets of structures, one of which is lost, whereas, for the external genitalia, both XX and XY fetuses begin with the same single set of structures that then develop differently depending on the hormonal environment. Thus, although testicular hormones are important for both internal and external structures and operate in accord with the classic model of hormone action, the specific mechanisms involved differ.

In addition, the specific hormones influencing the two sets of structures differ, with T and MIF influencing internal structures and DHT bearing primary responsibility for external structures.

These processes of physical sexual differentiation have been established in experimental studies involving hormone manipulations in nonhuman mammals, as well as by observing the consequences of hormone abnormality for human development. Essentially identical mechanisms appear to govern sexual differentiation of the internal and external genitalia in humans as in other mammals. XX individuals exposed to high levels of androgens, for example, because of genetic problems, are born with masculinized external genitalia. However, because they were not exposed to MIF, their internal genital structures are female. Similarly, XY individuals who produce normal levels of testicular hormones, but whose cells cannot respond to androgen because of a genetic defect, are born with female-appearing external genitalia, but lack female internal genitalia because these have been inhibited by MIF from their testes. These syndromes and others are described more fully in the next section.

8.3.1 Syndromes Involving Prenatal Hormonal Abnormality

The nomenclature for medical conditions producing genital ambiguity at birth was revised in 2006 (Hughes et al., 2006). Conditions once called pseudohermaphroditism or intersex are now defined as disorders of sex development (DSDs). DSDs that produce genital ambiguity sufficient to cause difficulty assigning a newborn as a girl or a boy probably occur in about 1 in 5000 to 1 in 10 000 births in North America and Europe, although the exact frequency is not known. Some estimates of intersex conditions include less-dramatic abnormalities, for example, hypospadias (a condition where the urethral opening does not reach the tip of the penis) or gynecomastia (excessive development of the breasts in males), and, because of this broader definition, can produce numbers as high as 1 in 100 births, but our interest here is in the more dramatic DSDs that are known to involve prenatal hormone perturbation. These include (1) XX individuals exposed to high levels of androgens, because of classical congenital adrenal hyperplasia (CAH); (2) XY individuals exposed to reduced androgen, because their cells have deficient or defective androgen receptors (androgen insensitivity syndrome: AIS); and (3) XY individuals with defects in androgen biosynthesis (5-alpha reductase (5-αR) or 17-hydroxysteroid dehydrogenase (17-HSD) deficiencies).

Some other conditions that involve prenatal hormonal abnormality, usually without ambiguity of the genitalia at birth, have also been studied. These include (1) XY individuals with idiopathic hypogonadotropic hypogonadism (IHH), a syndrome involving deficiency of hypothalamic hormones that promote the production of testicular hormones; and (2) Individuals exposed to lower than normal levels of ovarian hormones prenatally, because their second X chromosome is absent or imperfect, resulting in ovarian regression (Turner syndrome (TS)).

A third set of conditions involves XY individuals assigned (or re-assigned) as female early in life, because of problems with the appearance of their external genitalia. These individuals have a prenatal hormonal environment typical of a male, in contrast with their female sex of rearing. The conditions include (1) cloacal exstrophy; (2) penile agenesis (aphallia); and (3) ablatio penis.

8.3.1.1 Congenital adrenal hyperplasia

Classical CAH is an autosomal, recessive disorder that results in overproduction of androgen, beginning prenatally. The underlying deficiency is in enzymes needed to produce adrenal steroids. In over 90% of cases, the deficient enzyme is 21-hydroxylase (21-OH), and the incidence of CAH caused by 21-OH deficiency in Europe and the United States is estimated at between 1 in 5000 and 1 in 15 000 births (New, 1998). The lack of 21-OH prevents cortisol production. The negative-feedback system detects the low levels of cortisol and additional metabolic precursors are produced. Because of the blockage in cortisol production, however, the precursors are shunted into the androgen pathway, resulting in an overproduction of adrenal androgens, as well as progesterone and 17-hydroxyprogesterone. Androgen levels in female fetuses with classical CAH are markedly elevated (Pang et al., 1980; Wudy et al., 1999) and girls with the disorder are typically born with some degree of genital virilization. In rare cases, the virilization is so severe that girls are mistaken for, and assigned and reared as, boys (Money and Daléry, 1976). Typically, however, they are diagnosed with CAH near the time of birth, and assigned and reared as girls. Then, they are treated with hormones to regulate hormones postnatally, and their genitalia are usually feminized surgically.

T levels in male fetuses with CAH appear to be generally within the normal male range, although androstenedione, a weak androgen, appears to be elevated (Pang et al., 1980; Wudy et al., 1999). Assumedly, adrenal androgens are initially elevated, with feedback mechanisms reducing testicular androgen production subsequently, resulting in the near-normal levels noted later in pregnancy. Boys with CAH are born with normal-appearing male genitalia and, in areas without universal screening at birth, CAH in boys is usually detected because of salt-losing crises caused by aldosterone deficiency. This typically occurs within a few weeks of birth, but in some cases, affected boys are not identified until the elevated adrenal androgens induce precocious puberty in early childhood. In areas without universal screening, boys are more likely than girls to die from CAH and associated salt-losing crises in infancy.

There are different forms of CAH, and the different forms are associated with differences in the degree of androgen abnormality. The most severe form is classical, salt-losing CAH, followed by classical, simple-virilizing CAH, and then late-onset CAH. Late onset CAH is thought to involve only postnatal androgen elevation, whereas classical forms involve prenatal hormonal perturbation as well.

8.3.1.2 Androgen insensitivity syndrome

Androgen insensitivity refers to a deficiency in the ability of androgen receptors to respond to androgens (Grumbach et al., 2003). The insensitivity can be complete (CAIS) or partial (PAIS). Both forms are transmitted as X-linked, recessive traits, and so occur almost exclusively in genetic males. Individuals with CAIS appear female at birth, and typically, are raised as girls with no suspicion of the underlying disorder or the Y chromosome. At puberty, estrogen derived from testicular androgen causes feminine breast development. Typically, CAIS is detected when menstruation fails to occur, because of the lack of female internal reproductive structures. Physical appearance in PAIS varies, ranging from essentially that of a CAIS individual, through various degrees of genital ambiguity, to uncomplicated hypospadias, infertility, or even gynecomastia in an otherwise healthy-appearing male. Estimates of the incidence of CAIS vary widely, although it appears to be far rarer than CAH. The incidence of PAIS is not known, perhaps in part because its milder manifestations can go undetected.

8.3.1.3 Androgen biosynthesis deficiencies (5-αR and 17-HSD deficiencies)

These deficiencies are transmitted as autosomal, recessive traits (Imperato-McGinley et al., 1974; Imperato-McGinley, 1994; Rosler and Kohn, 1983). They are rare in the general population, but can occur frequently in populations where inbreeding is common. In one area of the Dominican Republic, the incidence of 5-αR deficiency has been estimated at 1 in 90 males (Imperato-McGinley et al., 1974). The enzyme 5-αR converts T to DHT, and patients deficient in the enzyme have low levels of DHT but normal-to-high levels of T (Imperato-McGinley, 1994). Because DHT is needed for normal virilization of the external genitalia prenatally (Wilson et al., 1981), 5-αR deficiency results in female-appearing or ambiguous genitalia at birth, and individuals with the disorder are usually assigned and reared as girls. At puberty, however, T and other androgens cause virilization, including growth of the phallus and scrotum, deepening of the voice and development of male-typical musculature. The enzyme 17-HSD is needed to produce T from its immediate precursor, androstenedione Patients deficient in this enzyme have low levels of T and DHT, but elevated levels of androstenedione (Imperato-McGinley et al., 1979b; Rosler and Kohn, 1983). The natural history of 17-HSD is similar to that of 5-αR deficiency. The genital appearance at birth is feminine or ambiguous, but physical virilization occurs at puberty. In populations where these disorders are common, they sometimes have descriptive names, such as *machihembra* (first woman, then man) (Imperato-McGinley et al., 1979a) or Turnim Man (Herdt and Davidson, 1988).

8.3.1.4 Hypogonadotropic hypogonadism

Individuals with hypogonadotropic hypogonadism (HH) have low levels of pituitary gonadotropins or their hypothalamic-releasing factor. As a consequence, their gonads lack sufficient stimulation to produce normal levels of hormones (Grumbach and Styne, 2003). The disorder can occur after puberty, or congenitally (Whitcomb and Crowley, 1993). If the disorder is congenital, it is usually detected when the child does not undergo normal puberty. Males with congenital HH usually have normal-appearing genitalia at birth, perhaps because maternal gonadotropins stimulated their testes to produce hormones prenatally (Hier and Crowley, 1982). Thus, it cannot be assumed that their hormone levels are lower than normal before birth. However, beginning at birth, and perhaps to some extent before, their levels of testicular hormones would be lower than normal males.

8.3.1.5 Turner syndrome

TS results from an absent or imperfect X chromosome, and is thought to involve a random genetic error. In 50–60% of cases the second X chromosome is entirely missing. Other cases involve mosaicisms (mixtures of cell lines, some of which can be normal) or other abnormalities of the X chromosome (White, 1994; Zinn et al., 1993). TS occurs in approximately 1 in 2000 to 1 in 5000 live female births in North America and Western Europe (Lippe, 1991). The external genitalia are female, but in the majority of TS girls, the ovaries regress sometime after the 3rd month of gestation (Singh and Carr, 1966), impairing or eliminating their ability to produce hormones. The syndrome has several stigmata, including short stature, skeletal growth disturbances, cardiovascular and renal abnormalities, otitis media, primary gonadal failure, absence of secondary sexual characteristics, and infertility (Lippe, 1991). Short stature is universal in TS, and over 90% of affected females experience primary gonadal failure and infertility, but other stigmata vary dramatically from individual to individual (Lippe, 1991).

8.3.1.6 Cloacal exstrophy

Cloacal exstrophy is a severe defect of the ventral abdominal wall (Groner and Zeigler, 1996; Hurwitz and Manzoni, 1997), involving abnormalities and insufficiencies, in the urinary and bowel systems, that were almost always fatal prior to 1960. Now, with changes in medical management, children with this syndrome often survive. Cloacal exstrophy occurs in approximately 1 in 200 000 to 1 in 400 000 births (Hurwitz and Manzoni, 1997), and is more common in XY than XX individuals. In XY individuals, the testes appear histologically normal but are typically undescended, and the penis is usually either absent or represented as two separate and incomplete structures. Even when present as a single structure, the penis is usually small and poorly formed. In XX individuals there also are abnormalities of the genitalia. Because of this, many surviving XX and XY individuals with cloacal exstrophy have been surgically feminized and reared as girls (Hurwitz and Manzoni, 1997). Those who are XY would appear to have been exposed to male-typical levels of testicular hormones prenatally and neonatally, until surgical removal of the testes.

8.3.1.7 Penile agenesis (aphallia)

In this condition, an XY individual is born without a penis, despite the presence of a normal scrotum and functioning testes (Kessler and McLaughlin, 1973; Richart and Benirschke, 1960). The causes of the condition are unknown, although it is usually associated with abnormalities of the urinary and gastrointestinal tracts (Farah and Reno, 1972; Kirshbaum, 1950). Estimates of its incidence range from 1 in 50 000 (Young et al., 1971) to 1 in 10–30 million (Kessler and McLaughlin, 1973), and mortality is high. As a result, there are very few individuals with aphallia. However, those who survive are often surgically feminized and reared as girls. Like XY individuals with cloacal exstrophy, their prenatal and early neonatal hormonal milieu would seem to resemble that of healthy males.

8.3.1.8 Ablatio penis

In rare instances, accidents can cause severe damage, or even complete ablation, of the penis in an otherwise healthy infant. In some such cases, XY infants have been surgically feminized, and reassigned and reared as girls. These individuals would have been exposed to normal male levels of testicular hormones prenatally and postnatally until the time when the testes were removed (usually at the time of sex reassignment).

8.3.2 Hormone Administration during Pregnancy

With the exception of ablatio penis, the syndromes described above involve genetic abnormalities or other disorders intrinsic to the individual, and they can have continued manifestations across the life span, independent of the prenatal hormonal abnormalities associated with them. In contrast to these endogenous causes of gonadal hormonal abnormality, there are exogenous causes. In these situations, the hormonal abnormality is limited in time and is less likely to be accompanied by nonhormonal symptoms, such as are often associated with genetic disorders. Exogenous causes of hormone exposure include situations where hormones have been prescribed to pregnant women, usually for medical reasons.

The most commonly prescribed hormone was the synthetic estrogen, DES. DES was prescribed to millions of women in the United States from the late 1940s to the early 1960s (Heinonen, 1973; Herbst and Bern, 1981; Noller and Fish, 1974). It was mistakenly thought to provide protection against miscarriage and was prescribed to women with a history of miscarriage, with threatened miscarriage (e.g., because of bleeding during pregnancy) and, in some cases, as a routine precaution. Double-blind, placebo-controlled studies eventually demonstrated that DES did not protect against miscarriage, and it was removed from use in the United States when it was associated with an increased risk of vaginal and cervical adenocarcinoma in the early 1970s (Herbst et al., 1971). In nonhuman mammals, DES and other estrogens promote male-typical brain and behavioral development when administered prenatally or neonatally (Goy and McEwen, 1980; Hines et al., 1987). These masculinizing and defeminizing effects of estrogen were originally considered paradoxical. However, as noted earlier, it is now well established that T from the testes is converted to estradiol within the brain, and, in normal male animals, estradiol acting through neural estrogen receptors is responsible for many aspects of male-typical neural and behavioral development. Although most of the evidence that estrogen masculinizes and defeminizes comes from studies of rats and other rodents, there also is evidence that long-term prenatal exposure to DES can masculinize play behavior, and some aspects of mounting behavior, in rhesus macaques (Goy and Deputte, 1996). Thus, if DES or other estrogens influence human sexual differentiation, they would be hypothesized to have masculinizing or defeminizing effects on developing females.

Progestins are the second main type of hormone that has been prescribed to pregnant women, and are of two general types – progestational and androgenic. Progestational progestins interfere with the actions of androgens, whereas androgenic progestins mimic the actions of androgen. Thus, these two types of progestins would be predicted to have opposite effects – the first impairing male-typical development and the second promoting it. (See Collaer and Hines (1995), for additional discussion.)

8.3.3 Normal Variability in Hormones

Another approach to studying hormonal influences on human sexual differentiation relates normal variability in hormones to behavior. This approach includes at least five types of studies. The first three relate hormone levels during early development to subsequent behavior, hormones being obtained from umbilical-cord blood at birth, from amniotic fluid, or from maternal blood during pregnancy. A fourth approach is based on evidence, described above, that female rodents gestating near males show increased male-typical behavior, and compares the behavior of female twins gestating with male co-twins to those gestating with female co-twins. Finally, a fifth approach correlates physical characteristics that differ for males and females and are thought to result from early hormonal exposure to behavioral characteristics that show sex differences. The most commonly studied physical characteristic is the ratio of the second to the fourth digit on the right hand (2D:4D), a ratio that is higher (i.e., closer to 1.0) on average in females than in males.

Each approach to studying hormonal influences on human neural and behavioral sexual differentiation has limitations. Studies of genetic disorders, and of individuals whose mothers were treated with hormones, are typically limited to small samples. In addition, disorders usually involve problems in addition to hormonal abnormality, which could themselves cause behavioral change. Also, for individuals born with ambiguous genitalia, the genital appearance and the family's, society's, or the individual's own reaction to the ambiguity could produce behavioral change, independent of hormonal influences on the brain.

Although studies of normal variability avoid these problems, they have problems of their own. Studies involving umbilical cord, amniotic fluid, or maternal samples typically involve a single assessment of hormone levels, usually uncontrolled for time of day. Because of circadian and other fluctuations in hormones, such single measures can be unreliable. Also, a single sample provides limited information compared to a disorder causing prolonged prenatal abnormality or to sustained treatment with hormones. In addition, samples are generally taken when clinically necessary or otherwise available and so not necessarily at critical periods for sexual differentiation. This is particularly problematic for umbilical cord measures, because sex differences in T are relatively small at birth. The timing of amniotic fluid samples might be better. Evidence from studies of fetal blood samples suggest a peak in T production in male fetuses from about week 8–24 of gestation (Reyes et al., 1973), and amniotic fluid is typically sampled at about week 16. Similarly, blood samples from pregnant women are often taken for clinical purposes at about this point in gestation. It has been suggested that amniotic fluid samples show a relatively constant T elevation in males across gestation, rather than a peak at around week 16 (Sarkar et al., 2007), but other results using amniotic fluid samples show higher levels in male fetuses for weeks 15–21 than weeks 36 and 40, producing a sex difference at the earlier, but not the later, time (Carson et al., 1982). These results resemble those obtained with fetal blood samples. Finally, although T in amniotic fluid might be assumed to be closer than maternal T during pregnancy to the primary measure of interest – T in fetal blood – maternal T during pregnancy has been reported to correlate well with T in fetal blood ($r = 0.414$) (Gitau et al., 2005).

Interpretation of studies comparing same-sex and other-sex twins is difficult, because hormonal influences are generally confounded with social influences. If behavioral differences are seen, they could result from prenatal exposure to hormones from a male versus a female twin, or from social exposure to this twin and his or her behavior postnatally. In animal studies, offspring can be cross-fostered to eliminate the postnatal impact of sibling sex on behavior, but this is not possible in humans. A few studies have attempted to address this problem, for example, by comparing the behavior of twins to their male and female nontwin siblings as well as their co-twins; this chapter includes these controlled studies.

Finally, finger ratios show only small sex differences, and they vary with ethnicity more than with sex, although sex differences are seen within ethnic groups (Manning et al., 2007). Also, finger ratios may reflect hormone levels only at a particular, probably very early, stage of gestation. In addition, the ease of conducting studies of finger ratios has probably led to spurious results. Nevertheless, there is evidence that individuals with CAH have reduced (i.e., more male-typical) ratios (Brown et al., 2002b; Okten et al., 2002; but cf. Buck et al. (2003)), suggesting that finger ratios provide some evidence about androgen exposure at some point during prenatal development.

Despite specific problems associated with the various approaches to studying hormonal influences on human sexual differentiation, data from different approaches will be discussed in relation to various human behaviors that show sex differences. The existence of several different approaches to studying the role of gonadal hormones in human sexual differentiation allows for evaluation of the convergence of evidence. This convergent evidence approach increases confidence in conclusions when a range of approaches, each with specific, but different, interpretational difficulties, point to the same conclusion.

8.4 Hormonal Influences on Human Sexual Differentiation: Human Behavioral Sex Differences

Animal models suggest that behaviors that show sex differences are susceptible to influences of gonadal hormones, whereas those that do not are not (Hines, 1982, 2004). Thus, only human behaviors that show sex differences would be hypothesized to be influenced by the early gonadal hormone environment. The question of which behaviors these are has itself been debated, and the study of sex differences presents particular difficulties (Maccoby and Jacklin, 1974). One problem is that researchers' own preconceptions, or sex-related stereotypes, influence their work. Because expectations can influence perceptions of outcomes or even actual outcomes, this can distort results. A second problem is that a finding of differences between groups is easier to publish than a finding of no differences. This problem is exacerbated for research on sex differences, since sex can easily be measured and is often routinely analyzed, even when there are no specific hypotheses about sex differences. Because statistical decision rules result in a certain percentage of false-positive results (5% with alpha set at 0.05), there is a high probability that spurious results suggesting sex differences will be published. Therefore, this chapter focuses on sex differences that have been documented in numerous studies, by independent research groups, and on those that have been studied in relation to hormones during development.

The conclusion that a behavior or psychological characteristic shows a sex difference does not necessarily mean that males and females are dramatically different. Typically, it means that when groups of men and women or boys and girls are compared, the groups show average differences. The size of these average differences varies from characteristic to characteristic. Where possible, results of meta-analyses, which combine data from many studies to get reliable estimates of the sizes of group differences, will be provided. The estimate of effect size used is d, obtained by calculating the difference in means for the two groups (males minus females) and dividing by the combined standard deviation. In the behavioral sciences, positive or negative d values of 0.8 or greater are considered large, those around 0.5, moderate, and those around 0.2, small (Cohen, 1988). Effect size values smaller than 0.2 are considered negligible. To put the size of behavioral sex differences into a familiar context, the sex difference in height at age 18 and into adulthood in the United States and in Great Britain has a magnitude of two standard deviations: $d = 2.0$ (International Committee on Radiological Protection, 1975; Tanner et al., 1966).

8.4.1 Core Gender Identity

Core gender identity, or the sense of self as male or female, shows a sex difference. The vast majority of XY individuals think of themselves as boys or men and the vast majority of XX individuals think of themselves as girls or women. However, even this basic aspect of sexual identity is not always consistent with genetic sex or with a person's physical appearance as male or female. In adults, the incidence of psychological identity as the other sex (called gender-identity disorder or transsexualism) is not known with certainty, but based on data from European countries with access to total population statistics, it is estimated that approximately 1 in 20 000 to 30 000 genetic males and 1 in 50 000 to 100 000 genetic females seek sex-reassignment surgery (American Psychiatric Association, 2000). Although gender-identity disorder can also occur in children, estimates of its incidence in childhood are not available. The sex difference in core gender identity is the largest of all psychological sex differences. It is not typically represented quantitatively, but results from small data sets where quantification has been attempted suggest a very large magnitude ($d \sim 11.0$) (Hines et al., 2003a, 2004).

8.4.2 Sexual Orientation

Sexual orientation also shows a sex difference. The great majority of males are sexually attracted to, and erotically interested in, females, whereas for the great majority of females, sexual attraction and erotic interest is focused on males. Again, this is not universal. Kinsey's data suggested that about 10% of men and about 5% of women are bisexual or homosexual (Kinsey et al., 1948, 1953). More recent studies provide lower estimates, at least for males having homosexual experience, ranging from 2% to 6% in the United States, France, and Great Britain (Billy et al., 1993; Johnson et al., 1992; Spira et al., 1992), although estimates that include homosexual attractions as well as behavior are higher (16–21% of males; 17–19% of females) for the same three countries (Sell et al., 1995). In the study by Sell et al., reports of homosexual behavior in the past 5 years were also higher than in prior studies (6.2%, 4.5%, and 10.7% for males and 3.6%, 2.1%, and 3.3% for females in the United States, United Kingdom, and France, respectively), though still somewhat lower than Kinsey's estimates. The sex difference in sexual orientation in fantasy and behavior appears to be very large ($d \sim 6$–7.0) (Hines et al., 2003a, 2004; Meyer-Bahlburg et al., 2008).

8.4.3 Gender-Role Behavior

In addition to core gender identity and sexual orientation, several other human behaviors, sometimes called gender-role behaviors, show sex differences. They include (1) childhood play behaviors; (2) specific cognitive abilities; (3) emotional, temperamental, or personality characteristics; (4) psychiatric disorders; and (5) manifestations of neural asymmetry. Sex differences in these areas are substantially smaller than those in core gender identity, sexual orientation, or even height.

8.4.3.1 Childhood play

Several aspects of juvenile play, including toy, activity, and playmate preferences, show sex differences. In regard to toys, boys prefer vehicles, weapons, and building toys, and girls prefer dolls, kitchen accessories, and cosmetics and dress-up toys (Maccoby and Jacklin, 1974; Pasterski et al., 2005; Sutton-Smith et al., 1963). Some sex differences in toy preferences appear early in life, by the age of 12 months (Snow et al., 1983), and typically grow larger as childhood progresses (Golombok and Hines, 2002). The size of the sex difference also depends on the means of assessment (e.g., questionnaire vs. direct observation) and the specific toys compared. Sex differences in toy preferences can be large ($d > 0.80$) (Alexander and Hines, 1994; Berenbaum and Hines, 1992; Pasterski et al., 2005; Sutton-Smith et al., 1963). In addition to showing differences in toy preferences, boys are more physically active than girls and engage in more rough, active play, including rough-and-tumble interactions that involve playful aggression and overall body contact. Meta-analytic findings suggest that the sex difference in activity level begins prenatally and is moderate in size (Eaton and Enns, 1986). Individual studies suggest that the sex difference in rough-and-tumble play is also moderate in size (DiPietro, 1981; Hines and Kaufman, 1994; Maccoby, 1988). Finally, boys and girls differ in preferred play partners with approximately 80–90% of partners being of the same sex (Hines and Kaufman, 1994; Maccoby, 1988). Like the sex difference in toy preferences, this sex difference becomes larger as childhood progresses. At age 4.5 years, children spend about 3 times as much time with peers of the same sex and this increases to about tenfold at age 6.5 years (Maccoby and Jacklin, 1987).

8.4.3.2 Cognitive abilities

There is no sex difference in general intelligence (see, e.g., Collaer and Hines (1995) and Hines (2004)). This may seem unsurprising, given that intelligence tests are currently specifically designed to avoid sex differences. However, even before an effort was made to avoid sex differences, intelligence tests were largely gender neutral (Loehlin, 2000). Nevertheless, some measures of specific cognitive abilities show sex differences. These include aspects of spatial, mathematical, and verbal abilities, and perceptual speed and accuracy.

8.4.3.2(i) Spatial abilities

Mental rotations, or the ability to rotate 2D or 3D stimuli in the mind rapidly and accurately, shows a sex difference favoring males. The difference is present in children (Linn and Petersen, 1985) and adults (Voyer et al., 1995), and may increase with age, although it is hard to be certain because different tasks are used with different age groups. The sex difference on 3D tasks ($d = 0.92$) appears to be larger than that on 2D tasks ($d = 0.26$) (Linn and Petersen, 1985; Voyer et al., 1995), although this may be because the 2D tests are relatively easy for the students typically studied. In one investigation involving people of average intelligence, both 2D and 3D tasks showed large sex differences (Hines et al., 2003b). Tests of spatial perception also show sex differences favoring males. These tasks require accurate positioning of a stimulus (e.g., a line) within a distracting array (e.g., a tilted frame). As was the case for mental rotations tests, the sex difference in spatial perception appears larger in adults than in children ($d = 0.56$ vs. 0.38, respectively), but the tests may be too difficult for children, reducing the apparent sex difference (Voyer et al., 1995). Also, as with mental rotations tasks, the size of the sex difference may increase with task difficulty. A spatial perception task, adapted from the Benton Judgment of Line Orientation task to increase its difficulty, appears to show a sex difference as large, or larger than, the sex difference for 3D mental rotations (Collaer et al., 2007). In contrast to mental rotations and spatial perception tasks, measures of a third aspect of visuospatial ability, spatial visualization, do not show appreciable sex differences ($d < 0.20$) (Linn and Petersen, 1985; Voyer et al., 1995). These tasks require complex, sequential manipulation of spatial information and typically have more than one solution strategy. Measures include tests that require identification of simple figures within complex designs (e.g., embedded figures, hidden patterns, etc.), construction of specified shapes from 3D blocks (e.g., block design), and imagining what unfolded shapes would look like when folded to form 3D objects (e.g., paper folding, surface development, etc.). A final area of spatial performance at which males excel is targeting, for instance, throwing darts or balls at bulls' eyes. This sex difference is seen in children and adults and is large ($d > 1.0$) (Hines et al., 2003b; Jardine and Martin, 1983; Watson and Kimura, 1991). However, targeting differs from paper-and-pencil measures of spatial abilities as it involves a motor, as well as a spatial, component.

8.4.3.2(ii) Mathematical abilities

Meta-analytic results (Hyde et al., 1990) suggest that the overall sex difference in mathematical abilities is negligible ($d = -0.05$), but that measures of problem solving show small sex differences favoring males, particularly among older, highly selected samples, such as college students ($d = 0.32$). Some standardized tests, again used with highly selected samples, also favor males. This is true of the mathematics subtests of the scholastic aptitude tests (SATs: $d = 0.38$) and the graduate record exam (GRE: $d = 0.77$) which are used in the United States to select students for bachelors and doctoral degree programs, respectively. In contrast, in childhood, tests of computational skills show small sex differences favoring females ($d = -0.21$), and there are no sex differences in computational skills in adults ($d = 0.00$) or in understanding of mathematical concepts at any age ($d = -0.06$).

8.4.3.2(iii) Verbal abilities

Meta-analysis (Hyde and Linn, 1988) suggests a negligible female advantage for general verbal ability ($d = -0.11$) in children as well as adults, but some sex differences on other types of verbal tasks; males show a negligible advantage on analogies ($d = 0.16$), and females show a small advantage on speech production ($d = -0.33$). There is also evidence of a moderate advantage for females on verbal fluency (e.g., the ability to generate words that begin with specified letters ($d = -0.53$) (Kolb and Whishaw, 1985; Spreen and Strauss, 1991)). Female infants begin to talk earlier than males do, and from 16 to 24 months of age girls have a larger vocabulary than boys (Halpern, 2000). However, this vocabulary advantage is gone later in life, and most tests of verbal abilities show essentially no sex differences in adults ($d = -0.02$ for vocabulary, $d = -0.03$ for reading comprehension, and $d = 0.03$ for the verbal subtest of the SAT) (Hyde and Linn, 1988).

8.4.3.2(iv) Perceptual speed and accuracy

Among high school students, the sex difference in perceptual speed and accuracy favors females and ranges in size from $d = 0.29$ to 0.66 (mean = 0.48), at least as assessed using the clerical speed and accuracy subtest of the differential aptitudes test (DAT) (Feingold, 1988). Sex differences of similar size have been observed for similar tests from the educational testing service ($d = 0.49$) (Ekstrom et al., 1976). The magnitude of the sex difference on the DAT appears to have declined somewhat over the years, from a mean of 0.62 in 1947 to 0.34 in 1980 (Feingold, 1988).

Are sex differences in other cognitive abilities also declining? In some cases, the answer appears to be yes. Feingold (1988) looked at measures of specific cognitive abilities from the 1940s to the 1980s and found that sex differences on almost all abilities declined linearly over the decades, but he concluded that measures of algebraic problem solving, such as the SAT, were an exception and that these continued to favor males, especially at the upper end of the distribution. Nevertheless, even the sex difference in performance on the SAT-M at the upper extreme has declined substantially since 1982, when the sex ratio for children scoring over 700 at age 13 was 13 boys to 1 girl, as opposed to more recent figures of 2.8 boys to 1 girl (Halpern et al., 2007). In contrast to abilities studied by Feingold, two studies suggest that the sex difference on 3D mental rotations tasks remained stable from the 1970s to the 1990s (Sanders et al., 1982; Voyer et al., 1995). Social and educational changes could underlie the reduction in sex differences in performance on some cognitive tasks, but this does not rule out the possibility that the remaining sex differences relate, in part, to hormones.

8.4.3.3 Emotion, temperament, and personality

Across cultures and from childhood through adulthood, males are more aggressive than females (Maccoby and Jacklin, 1974). Meta-analytic results suggest that the sex difference is moderate in size ($d = 0.50$) and may be larger in young children than in adults ($d = 0.58$ vs. 0.27), although this apparent age difference could reflect the use of different measures of aggression at different points in the life span (Hyde, 1984). Some individual measures of physical aggression appear to show large sex differences ($d > 0.80$) (Pasterski et al., 2007).

In regard to nurturing interests, it is widely assumed that women have more interest in nurturing than men do. Additionally, in most cultures, and in most families within our culture, women spend more time caring for children than do men. Interest in infants also shows sex differences (Berman, 1980) and this too can be measured using questionnaires. In fact, questionnaires regarding interest in infants show bigger sex differences than are seen when behavior is observed (Berman, 1980). The sex difference in interest in infants varies over the life span and appears to be largest in young adults, perhaps because issues related to parenting are most salient at that age (Berman, 1980). Among young adults, questionnaire measures of interest in infants or of nurturing qualities show sex differences in the moderate-to-large range (Leveroni and Berenbaum, 1998; Mathews et al., in press). Some measures of empathy also show large sex differences (Feingold, 1994).

8.4.3.4 Psychopathology

There are sex differences in some psychiatric disorders (American Psychiatric Association, 2000; Rutter et al., 2003). Depression is more common in females than in males, at least from adolescence, although it may be more severe in males. Schizophrenia afflicts males and females in about equal numbers, but the age of onset is earlier in males than in females. In general, it has been suggested that males are more likely to be afflicted with early-onset disorders involving neurodevelopmental impairment and females are more likely to be afflicted with adolescent-onset emotional disorders (Rutter et al., 2003). Some disorders show dramatic sex differences in incidence. For instance, classic autism is 4 times more common, and the less-severe autistic spectrum condition (ASC), Asperger Syndrome (AS) is 9 times more common in males than in females (Rutter, 1978; Wing, 1981). The tic-related disorder, Tourette syndrome, is also several times as common in males as in females (Alexander and Peterson, 2004).

8.4.3.5 Neural asymmetries

Most people, both male and female, are right-handed. However, there are more men than women among the left-handed minority. In addition, when the degree of preference for the right hand is assessed across a range of tasks, women show stronger or more consistent right-hand preferences than men (Hines and Gorski, 1985). Sex differences have also been reported in the specialization of the two cerebral hemispheres for language. Again, most people show left hemispheric dominance, but more members of one sex than the other appear in the minority that does not. In contrast to handedness, where men are more likely to be atypical, women are more likely to show atypical cerebral dominance for language. Evidence of this sex difference comes from studies of language disruption following neural injury and from studies of normal language function. In regard to the former, both men and women are more likely to show disruption following injury to the left than the right hemisphere, but the impairment has been reported to be less severe in women (McGlone, 1980), perhaps because language and speech are less exclusively focused in the left hemisphere in females. In the intact brain, language lateralization is usually assessed by simultaneous presentation of verbal stimuli to each cerebral hemisphere separately. The number of stimuli accurately identified by each hemisphere is then calculated. Data from these procedures also suggest that women show less-dramatic language lateralization than men, although meta-analytic results suggest that the sex difference for these types of procedures, overall, may be negligible ($d = 0.10$) (Voyer, 1996).

8.5 Hormones and Sexual Differentiation of Human Behavior: Findings

8.5.1 Core Gender Identity

Gonadal hormones appear to contribute to, but do not determine, core gender identity. For example, the vast majority of women exposed to high levels of androgen during early development, because of CAH, have a female core gender identity, but they are at increased risk of gender-identity disorder or for changing sex from female to male. One study (Zucker et al., 1996) found that of 53 XX, CAH patients seen at one clinic during a defined time period, one had been diagnosed with transsexualism (or gender-identity disorder) and was now living as a man, despite assignment and rearing as a female. Gender-identity disorder was estimated to occur in one in 30 400 cases of XX individuals in the general population, resulting in odds of one in 608 that this co-incidence of gender dysphoria and CAH was a chance happening. Another study (Meyer-Bahlburg et al., 1996) reported that four XX, CAH patients in the New York area, who had been assigned and reared as females, were now living as men. The authors estimated that the probability of this occurring by chance was 1 in 420 million. Also, among women with CAH who do not wish to live as men, and who do not have gender dysphoria, identification with the female gender has been found to be reduced compared to that of unaffected female relatives (Hines et al., 2004).

Studies of children also suggest that girls with CAH express reduced satisfaction with being a girl. In one study, seven of 15 girls with CAH said they were content to be or preferred to be a girl, compared to 14 of 15 controls (Ehrhardt et al., 1968). In a second study, six of 17 girls with CAH said that, if given a choice, they might have chosen to be a boy or would have been undecided as to whether to be a boy or a girl compared to one of 17 unaffected sisters of CAH children (Ehrhardt and Baker, 1974). However, in both studies, severe gender dysphoria was reported to be rare or nonexistent. A third study (Slijper et al., 1998) found that two of 18 girls with CAH in one clinic population met the diagnostic criteria for gender-identity disorder of childhood, as did five of 29 children raised as girls who had been exposed to high levels of androgen prenatally because of other DSDs, including PAIS, cloacal exstrophy, or ovotesticular DSDs. (This last syndrome was previously called true hermaphroditism.)

XY individuals whose genitalia appear female at birth, because of CAIS, are assigned and reared as females and do not wish to change sex as adults. Reports regarding their gender identity uniformly conclude that they are content with the female sex of assignment (Hines et al., 2003a; Masica et al., 1971; Mazur, 2005; Wisniewski et al., 2000). These findings suggest that lack of stimulation by androgen, at least when combined with an unambiguously female sex of rearing, produces a female core gender identity. A second X chromosome is apparently not needed, nor are ovaries, and the presence of a Y chromosome does not prevent this outcome.

Studies of individuals with deficiencies in enzymes needed to produce androgen are also relevant to the role of androgen in gender-identity development. Imperato-McGinley et al. (1974, 1979a) reported on 18 individuals with 5 α-R deficiency who lived in an isolated community in the Dominican Republic. These XY individuals were born with undervirilized genitalia, were assigned and reared as females, and were reportedly content in the female role as children. However, following physical virilization at puberty, 17 of the 18 lived as males. This outcome was interpreted to support a role for androgen in the development of male gender identity. However, the presumed critical period for hormonal influences on gender identity is prenatal or neonatal, certainly prior to puberty. These individuals did not have sufficient androgen to virilize their external genitalia prior to puberty, raising the question of how their brains were exposed to enough androgen to do so. One possibility is that T is converted to estrogen before acting on whatever neural regions are involved in male gender identity, whereas T is converted to DHT before acting on the genitalia. However, females with CAIS are assumedly exposed to normal male levels of estrogen, derived from testicular androgens, but unlike individuals with enzymatic deficiencies, they do not have a male core gender identity, arguing against this explanation.

Other studies of individuals with deficiencies in enzymes on the androgen pathway suggest similar, though less-dramatic, findings for individuals in Papua New Guinea, Mexico, Brazil, and the Middle East. A 2005 review found that 56–63% of individuals with 5-αR deficiency and 39–64% of those with 17-HSD deficiency who were raised as girls, changed to live as men, and that the likelihood of change did not appear to relate to the degree of external genital virilization at birth (Cohen-Kettenis, 2005). Even within a single family, XY individuals with the same genetic mutation and enzymatic deficiency, have been found to choose to live in different sexes (Wilson, 2001). Explanations for postpubertal gender changes, in addition to the early hormonal environment, include possible ambiguity in the sex of rearing (Herdt and Davidson, 1988; Money, 1976), or the advantages of being a male (as opposed to a sterile female) in the societies where the syndromes have been studied (Herdt and Davidson, 1988; Wilson, 1979). In the United States and much of Europe, individuals with these enzymatic deficiencies are sometimes assigned and reared as females and have their testes removed prior to puberty to prevent virilization. Patients treated in this way tend to maintain a female gender identity (Wilson et al., 1993; Zucker, 2002), arguing against a prenatal or neonatal hormonal influence as the sole explanation for the change in sex seen in other situations. Androgen exposure at puberty, either acting directly on the brain or by producing a male body type, or the cultural advantages of being a male in certain societies, are likely to be contributory factors.

It has been suggested that XY individuals with cloacal exstrophy, penile agenesis, or aphallia who have been surgically feminized and reared as girls often experience serious gender-identity problems (Reiner et al., 1999; Reiner and Gearhart, 2004). However, other researchers have reported fewer such problems (Meyer-Bahlburg, 2005; Schober et al., 2002). Outcomes for core gender identity also vary in the two well-documented cases where gender reassignment has occurred in early life, because of ablatio penis. One widely publicized case involved a pair of male, identical twins, one of whom was sex reassigned because of a surgical accident that cauterized his penis during a phimosis repair at the age of 8 months. Reassignment to female occurred by the age of 17 months. Although the reassignment was viewed as successful during early childhood (Money and Ehrhardt, 1972), by adulthood this individual was living as a man and reported that he had been unhappy as a female for many years (Diamond and Sigmundson, 1997). Although this outcome might suggest that early exposure to testicular hormones determines male gender identity, for at least the first 8 months of life, and perhaps somewhat longer, this individual's sex of rearing was male. Also, there is little information on the rearing environment after the child was reassigned to the female sex and there is no evidence as to how well the parents were able to adapt to treating a child who had once been their son as their daughter. Another similar case produced a different outcome. This time the penis was damaged during electrocautery circumcision at the age of 2 months and the child was reassigned as female sometime before the age of 7 months. This individual has been evaluated at the age of 16 and 26 years and has a female core gender identity with no evidence of gender dysphoria (Bradley et al., 1998). The situation of a Y chromosome, a male-typical early hormonal environment, and a female sex of rearing may represent a bipotential situation, where familial factors, genetic constitution, or even medical management regimes may channel gender-identity formation in one direction or the other.

Summary. Data on clinical syndromes and on sex reassignment following surgical accidents suggest that the gonadal hormonal environment, particularly androgen exposure during early development, influences the development of core gender identity, but does not completely determine it. The risk of gender dysphoria in individuals reared as females is increased following androgen exposure during early life, but it is still rare. Most individuals assigned and reared as female are content with that identity even if they were exposed to higher than normal levels of androgens during early development. XY individuals who are unable to respond to androgen because of CAIS show female gender identity, but unambiguous rearing as females complicates interpretation of this finding. Males with enzymatic disorders that impair production of certain androgens often choose to live as men after virilizing puberty, despite having been reared as females, whereas others appear content to remain living as females, particularly if virilizing puberty is prevented. These findings argue that factors other than the prenatal environment influence the formation of gender identity. Outcomes for males reared as girls, because of cloacal exstrophy or other causes of severely undervirilized external genitalia, are variable. Additionally, outcomes differ for two well-documented cases where the complete male hormonal cascade was present both prenatally and during the early neonatal period, and sex reassignment as female occurred after a period of postnatal socialization in the male sex. In one case, the reassignment to female was successful and in the other, it was not. More information is needed before strong conclusions can be made about the likelihood of successful sex change in these difficult cases where severe penile damage occurs, or where the external genitalia are severely undervirilized at birth. However, the ability of a genetically male (XY) infant, exposed to normal levels of testicular hormones prenatally and in early infancy, and reared as a boy for at least the first 2 months of life to be reared subsequently as a girl and maintain a female gender identity into adulthood is remarkable. It suggests that human gender identity is not determined by the sex chromosomes or the early hormone environment but that it can also be dramatically influenced by socialization. In fact, the most obvious conclusion based on data from these two reassigned infants, as well as on individuals with other hormonal abnormalities, is that, given the right social environment, it is possible for an XY individual to develop a female gender identity, despite a functional Y chromosome and exposure to male-typical levels of androgens during early development.

8.5.2 Sexual Orientation

The two cases mentioned above where boys were reassigned as girls following surgical damage to the penis have been evaluated in regard to sexual orientation as well as core gender identity. The child in whom the damage occurred at the age of 8 months had erotic interest exclusively in women as an adult (Diamond and Sigmundson, 1997), whereas the child in whom the damage occurred at the age of 2 months was bisexual (Bradley et al., 1998). Although they differ in that one individual was bisexual and the other interested only in female erotic partners, both cases suggest that early exposure to male-typical levels of testicular hormones influences sexual orientation away from the primary or exclusive erotic interest in men that is typical of females.

Most studies find that women with CAH show reduced heterosexual orientation or interest (Meyer-Bahlburg et al., 2008). Although some studies find no change in sexual orientation (Kuhnle and Bullinger, 1997; Lev-Ran, 1974), these studies generally have methodological weaknesses that could explain the negative findings, such as inclusion of large numbers of patients with less severe forms of the disorder, limitations in the methodology used to assess sexual orientation, or demand characteristics that mitigate against participants reporting atypical sexual orientation (Hines, 2004). Evidence from several independent studies that homosexual or bisexual orientation is more likely in women with the more severe, salt-losing form of CAH than in the simple-virilizing form (Dittmann et al., 1992; Meyer-Bahlburg et al., 2008; Mulaikal et al., 1987), and in women born with more dramatically virilized genitalia (Gastaud et al., 2007), strengthens the likelihood that androgen is the responsible agent. In the most rigorous study conducted to date (Meyer-Bahlburg et al., 2008), 40 women with the more severe, salt-losing form of CAH were more likely to show reduced heterosexual orientation than 21 women with the less severe, simple-virilizing form, or than 24 unaffected females. In this study, 82 women with nonclassical CAH also showed reduced heterosexual orientation compared to the 24 female-relative controls, although the difference was not as large as that between the salt-losers and the controls. Because nonclassical CAH is thought to involve postnatal, but not prenatal, androgen excess, this result is surprising. Assuming it proves replicable, it suggests either that postnatal androgen excess influences sexual orientation in females, or that nonclassical CAH involves prenatal as well as postnatal overproduction of androgen. In this study, lifetime sexual orientation was not exclusively or almost exclusively heterosexual in 47% of those with salt-wasting CAH, 33% of those with simple virilizing CAH, 24% of those with nonclassical CAH, and 5% of relative controls. The figures for classical CAH, either salt-wasting or simple virilizing, are similar to those reported by others (Dittmann et al., 1992; Hines et al., 2004; Money et al., 1984; Mulaikal et al., 1987).

One difficulty in interpreting data on women with CAH is that other aspects of the syndrome might influence behavior independent of hormonal influences on the developing brain. This is particularly relevant to studies of sexual behavior, because girls with CAH are born with masculinized genitalia. Even with surgical feminization early in life, the genitalia are not identical to those of other women, and are often problematic, in some cases producing pain with intercourse (Schober, 1999). In addition, the knowledge of virilization at birth and experience with genital surgery could themselves influence behavior (Fausto-Sterling, 1992; Quadagno et al., 1977). Data on women exposed to DES, a hormone that would be hypothesized to promote male-typical development of the brain and behavior, without virilizing the external genitalia, or causing the other problems associated with CAH, could help resolve this issue. DES-exposed women have been found to show reduced heterosexual orientation in three samples studied by one group of researchers (Ehrhardt et al., 1985; Meyer-Bahlburg et al., 1995). The first sample included 30 DES-exposed women, 12 of their unexposed sisters, and 30 unexposed, control women recruited from the same source as the DES-exposed group (a doctor treating women with abnormal PAP smears, a control group selected because DES exposure is associated with abnormal smears). The incidence of a lifelong homosexual or bisexual orientation was higher in the DES-exposed women than in the controls (24% vs. 0%). Among the 12 sister pairs where one had been exposed to DES and the other had not, 42% of the DES-exposed group versus 8% of the sisters indicated a lifelong homosexual or bisexual orientation. The second sample included a new group of 30 DES-exposed women, eight of their unexposed sisters, and 30 controls matched for demographic background. The DES-exposed group was again more likely than the control group to indicate a lifelong homosexual or bisexual orientation (35% vs. 13%), and this was also the case for the eight sister pairs (36% vs. 0%). The final sample included 37 women who were identified from medical records of their mothers' pregnancies indicating treatment with at least 1000 mg of DES. They were matched to women of similar age who were identified from the same set of medical records, but whose records indicated no DES treatment. In this study, 16% of the DES-exposed group and 5% of the controls indicated a lifelong homosexual or bisexual orientation. In contrast to these studies, an independent research group found that 3946 women exposed prenatally to DES did not differ from 1740 unexposed women in response to a question as to whether their sexual partners had been only the opposite sex, mostly the opposite sex, mostly the same sex, only the same sex, or they had had no sexual contact (Titus-Ernstoff et al., 2003). The large sample size in this study is impressive, but the assessment of sexual orientation by a single question regarding sexual partners may have lacked sufficient sensitivity; sexual interest and fantasies generally are more informative, given social constraints on sexual behavior.

Thus, reduced heterosexual orientation appears to be associated with high androgen exposure, prenatally, in women, but it is not clear that this is also the case for prenatal exposure to estrogen. In addition, as was the case for core gender identity, the relationship between androgen and sexual orientation is not one-to-one. Although women with CAH are more likely to be bisexual or homosexual than are other women, most are exclusively or almost exclusively heterosexual.

Does the early hormonal environment influence sexual orientation in men? In addition to having a typically female core gender identity, XY individuals with CAIS are female-typical in their sexual orientation. They do not differ in this respect from population norms (Wisniewski et al., 2000) or from matched controls (Hines et al., 2003a) and are more likely to be exclusively interested in female sexual partners than are women with CAH (Money et al., 1984). However, this female-typical sexual orientation cannot be attributed solely to their inability to respond to androgen, since their rearing is that of a typical female. Reports on sexual orientation in genetic males with enzymatic deficiencies that reduce androgen exposure are sketchy, but those who change to live as males after puberty also appear to adopt a sexual orientation toward females (Wilson et al., 1993; Zucker, 2002). This suggests that reduction in androgen levels from those of a typical male during prenatal and neonatal development does not necessarily prevent erotic interest in females in adulthood. In cases in the United States and Western Europe, where the testes are removed before puberty in individuals with these same enzymatic deficiencies, sexual orientation, like core gender identity, usually resembles that of women, in general (Wilson et al., 1993; Zucker, 2002). Again, the conflicting outcomes in different cultural contexts, and with testicular removal before versus after virilizing puberty, suggest that sexual orientation, although influenced by the early hormonal environment, is not determined by it, and may be influenced by social and cultural factors, or by hormones and physical changes at puberty, as well.

Influences of prenatal exposure to DES (or other estrogens) or progestins on sexual orientation in males have also been studied. These investigations were based in part on the assumption that estrogen and other so-called 'female' hormones would feminize developing males. It is now known that estrogen has little if any feminizing influences on XY individuals during development and, as noted above, it may play a role in male-typical development. However, estrogen can interfere with the action of androgen, and estrogen was sometimes prescribed in combination with progestins, some of which act as antiandrogens, providing a basis for predicting an influence on sexual differentiation in males.

The sexual orientation of men exposed to DES or to progestins prenatally does not appear to differ from that of other men. One study (Kester et al., 1980) compared four groups of hormone-exposed men to matched controls. One group was exposed to DES alone ($n = 17$), one to DES plus progesterone (a hormone that can act as an antiandrogen, $n = 21$), one to progesterone alone ($n = 10$), and one to progestins alone ($n = 13$). None of the hormone-exposed groups differed in sexual orientation in fantasy or behavior from control groups made up of similar numbers of men matched for demographic background. A second study compared two groups of DES-exposed men to matched controls, with similar results (Meyer-Bahlburg et al., 1987). Interviewer assessments of sexual orientation in fantasy or behavior were similar in 31 men identified from maternal records indicating DES treatment during pregnancy and in 29 men whose records did not indicate DES treatment. Similarly, 34 DES-exposed men and 15 controls recruited from a urological practice did not differ on the same assessments of sexual fantasy and behavior. The large-scale study of DES-treated offspring also found no differences in self-report of same-sex sexual partners in 1342 DES-exposed, compared to 1342 unexposed, men (Titus-Ernstoff et al., 2003), although, as noted above, the assessment procedure in that study may have lacked sensitivity.

Many studies have related sexual orientation to finger ratios (2D:4D), particularly in men. An early study reported that homosexual men had more male-typical ratios than heterosexual men (Williams et al., 2000), a surprising finding. Overall these studies have produced varied results, some suggesting less male-typical 2D:4D ratios, some suggesting more male-typical ratios and some suggesting no differences in ratios in homosexual versus heterosexual men (McFadden et al., 2005). These inconsistent results may reflect the ease with which studies relating finger ratios to behavior can be conducted and a consequent proliferation of spurious findings. A study of sex differences that was commissioned by the British Broadcasting Corporation (BBC), and involved participants who completed procedures online, found that both right- and left-hand 2D:4D ratios were more male-typical in 102 499 heterosexual men than in 11 060 homosexual or bisexual men, consistent with the typical pattern of androgenic influences on behavior (Collaer et al., 2007). Among women, 2D:4D did not differ for 84 417 heterosexual, compared to 9153 homosexual or bisexual, women.

Summary. The early hormonal environment appears to influence sexual orientation. Women exposed to high levels of androgens prenatally, because of CAH, show reduced erotic interest in men. Also, XY individuals who are insensitive to androgen, and reared as females, show primary erotic interest in men, and genetic males exposed to the complete male hormonal cascade, but sex reassigned in infancy and reared as females, because of penile damage early in life, develop either bisexual interest or erotic interest in women. All of these findings support a prenatal role for androgen in promoting male-typical sexual orientation. Prenatal exposure to estrogen does not appear to alter sexual orientation in men, nor does exposure to progesterone or synthetic progestins. Evidence regarding prenatal exposure of females to estrogen is equivocal.

Individuals with enzymatic deficiencies that impair androgen production appear capable of developing erotic interest in women following spontaneous virilization at puberty. However, when the testes are removed prior to puberty, and virilization does not occur, erotic interest appears to be in men. Whether these different outcomes relate to cultural influences, influences of hormones at puberty, or other factors is not known. Regardless, it appears that, although the early hormonal environment contributes to sexual orientation, it is not the only factor determining it.

8.5.3 Childhood Play

The clearest evidence of hormonal influences on sexual differentiation of human behavior has come from studies of childhood play. Girls with CAH show increased interest in male-typical, and reduced interest in female-typical, toys, activities, and playmates. These findings have been reported based on interview and questionnaire data from girls with CAH and their mothers, as well as direct observation of toy choices in a playroom (Berenbaum and Hines, 1992; Nordenstrom et al., 2002; Pasterski et al., 2005; Dittmann et al., 1990; Ehrhardt et al., 1968; Ehrhardt and Baker, 1974). Similar results have been reported in studies from the United States, Canada, Germany, The Netherlands, and Sweden, and girls with CAH have been found to differ from matched controls as well as from female relatives (sisters and cousins) who do not have CAH (Hines, 2004). The amount of male-typical behavior shown by girls with CAH correlates with the severity of the disorder (Nordenstrom et al., 2002), and with the degree of genital virilization at birth (Hall et al., 2004), strengthening the case for androgen being the responsible agent. Convergent evidence suggesting that the effect is caused by androgen exposure, rather than other aspects of CAH, also comes from studies of girls whose mothers were prescribed hormones during pregnancy for medical reasons. Girls whose mothers took androgenic progestins show increased male-typical play behavior similar to that seen in girls with CAH (Ehrhardt and Money, 1967), and girls whose mothers were prescribed the anti-androgen, medroxyprogesterone acetate (MPA) during pregnancy, show less male-typical and more female-typical play (Ehrhardt et al., 1977).

Prenatal exposure of females to DES does not appear to influence juvenile play or sex-typed interests. Interviews and questionnaires have been used to assess childhood activities, retrospectively, in 60 women exposed to DES, and in a variety of control groups (Ehrhardt et al., 1989; Lish et al., 1991, 1992). No consistent differences have been seen. Thus, it seems likely that hormonal influences on the development of childhood play behavior in humans are exerted directly by androgen, rather than following conversion to estrogen. Alternatively, it might be suggested that the genital virilization, at birth, in CAH girls and in girls exposed to androgenic progestins played a role in their behavior, for example, by altering parental encouragement of sex-typical play (Fausto-Sterling, 1992). However, the impact of MPA on sex-typical play argues against this interpretation, since MPA would not produce noticeable alterations in the external genitalia of females. In addition, parents are instructed by medical personnel to treat their CAH daughters as they would any other girl, and their responses to interview and questionnaire items (Berenbaum and Hines, 1992; Ehrhardt and Baker, 1974), as well as observation of parent–child interactions in the playroom (Pasterski et al., 2005) suggest that they do so. In fact, Pasterski et al. found that parents encouraged female-typical play more, rather than less, in their daughters with CAH compared to their unaffected daughters.

One study found reduced rough-and-tumble play in boys with CAH (Hines and Kaufman, 1994), but no alterations in playmate preferences or toy choices (Berenbaum and Hines, 1992; Hines and Kaufman, 1994). In contrast, another study reported increased intensity of energy expenditure in outdoor play and sports (Ehrhardt and Baker, 1974) in boys with CAH, although, in regard to general activity level there appears to be no difference between boys with and without CAH (Pasterski et al., 2007). In regard to progestins, exposure to the antiandrogen, MPA, has been associated with reduced male-typical play (Meyer-Bahlburg et al., 1977), although a study of boys exposed to 17-alpha-hydroxyprogesterone caproate (17-αHC), a different antiandrogenic progestin, found no evidence of this (Kester, 1984). A third study of boys exposed to estrogen and progestins reported reduced athleticism at age 6, but no alterations in other aspects of sex-typical play and no reduction in athleticism at age 16 (Yalom et al., 1973). There are no prospective data on childhood play in XY girls with CAIS, although retrospective assessments suggest their childhood behavior is female-typical (Hines et al., 2003a; Wisniewski et al., 2000).

Childhood behavior has also been related to normal variability in the early hormonal environment. In a longitudinal, population sample of over 8000, 3.5-year-old children, mothers of 112 behaviorally masculine girls were found to have higher levels of T during pregnancy than mothers of 116 behaviorally feminine girls; a random sample of 106 girls, who showed normative female behavior, had mothers whose T levels were intermediate to those seen in the mothers of the masculine, and the feminine, girls (Hines et al., 2002). No relationship was seen between maternal T during pregnancy and childhood behavior in boys. Studies of T in twins and in parents and offspring suggest 40–60% heritability for T (Harris et al., 1998). In addition, T is correlated in mothers and daughters, but not in mothers and sons (Harris et al., 1998), perhaps because the primary source of T in both mothers and daughters is the adrenal gland, whereas, in sons, T comes mainly from the testes. Thus, the most likely explanation of the relationship between maternal hormones and child behavior in girls is that, because of genetic similarity, mothers with relatively high T also have daughters with relatively high T, including during the fetal period (Hines et al., 2002).

T measured in amniotic fluid has also been related to subsequent play behavior. Two studies found insignificant relationships in both boys and girls, but these negative results may reflect small samples (Knickmeyer et al., 2005) or insensitive measures (van de Beek et al., 2008). A third study of 100 girls and 112 boys found the predicted relationship between T in amniotic fluid and sex-typed play for both girls and boys (Auyeung et al., in press), using a measure that has been found to be sensitive to androgen exposure in females with CAH (Hines et al., 2004), as well as in the study of maternal T during pregnancy (Hines et al., 2002).

Studies of twins have not supported the hypothesis that girls with male co-twins are masculinized by sharing the uterus with them and are not influenced by the male twin's T (Henderson and Berenbaum, 1997; Iervolino, 2003). No differences have been found between girls with male versus female twins, even in a study of over 6000 twins (Iervolino, 2003). In contrast, the sex of older siblings influences sex-typed play in both twins and nontwins (Iervolino, 2003; Rust et al., 2000); children's behavior is more male-typical if they have older brothers and more female-typical if they have older sisters. These findings illustrate the difficulty of interpreting data from studies of same and other sex twins, unless the problem of separating social from hormonal factors is addressed, for example, by looking at the impact of siblings as well as co-twins.

Summary.There is substantial evidence that prenatal androgen exposure influences sex-typed childhood behavior. Several studies from a number of independent research groups have found that girls with CAH show more male-typical interests in toys, playmates, and activities in comparison to various female control groups. In addition, these effects are graded; girls with more severe forms of CAH, involving the most dramatic androgen elevation, show the most male-typical play, as do girls with more severe genital virilization. Girls exposed prenatally to androgenic progestins also show increased male-typical play, girls and boys exposed to antiandrogenic progestins show reduced male-typical play, and normal variability in T, prenatally, has been linked to sex-typed behavior postnatally. Thus, several approaches all point to the same conclusion – that the amount of prenatal androgen exposure influences the amount of postnatal male-typical behavior. Boys with CAH do not show increased male-typical behavior, but they also appear to experience minimal androgen elevation prenatally, making their behavior less informative than that of girls with CAH as to the effects of prenatal androgen excess.

8.5.4 Cognition

8.5.4.1 General intelligence

Early reports concluded that prenatal androgen exposure increased general intelligence. Intelligence test scores among individuals with CAH (or exposed prenatally to androgenic progestins) were found to be substantially higher than population norms (Ehrhardt and Money, 1967; Money and Lewis, 1966). Subsequently, it was claimed that prenatal exposure to natural progesterone also enhanced intelligence, based on reports of high levels of academic achievement in the offspring of mothers from a clinic that used progesterone to treat difficult pregnancies (Dalton, 1968, 1976). Because natural progesterone acts as an antiandrogen, these results would appear to contradict those reported for CAH and other causes of androgen exposure. Later, high intelligence in androgen-exposed individuals and academic excellence in progesterone-exposed offspring were related to factors other than hormones. When CAH patients were compared to their relatives or to controls matched for demographic background, there were no differences in intelligence (Baker and Ehrhardt, 1974; McGuire and Omenn, 1975; Wenzel et al., 1978). The elevation seen in comparison to population norms probably reflected selection biases; individuals with higher intelligence were more likely to enrol in the research project. Regarding prenatal progesterone exposure, inappropriate statistical analyses contributed to the apparent academic enhancement, and re-analyses of the original data as well as subsequent research found no evidence of academic enhancement (Lynch and Mychalkiw, 1978; Lynch et al., 1978, discussed in Collaer and Hines (1995)). Other studies of individuals exposed to a variety of progestins, to estrogen and progestin or to DES, have also found no differences in general intelligence in comparison to relative controls (Hines and Shipley, 1984; Hines and Sandberg, 1993; Reinisch and Karow, 1977). The lack of an influence of androgens, progestins, or estrogens on general intelligence is not surprising, since general intelligence does not show a sex difference, and so would not be expected to relate to gonadal hormones.

There are also reports of reduced, rather than enhanced, intelligence in individuals with CAH compared to matched controls (Helleday et al., 1994a; Johannsen et al., 2006). These findings may reflect difficulty matching for all the background factors associated with intellectual development, or individuals of relatively low intelligence with CAH may be more likely to enrol in studies than are potential controls of similarly low intelligence. Alternatively, aggressive corticosteroid treatment or salt-losing crises in infancy could impair intellectual development in some individuals with CAH. Examination of these possible influences on intellectual development is important for clinical management of children with CAH, although perhaps of limited relevance to theories of sexual differentiation.

8.5.4.2 Specific cognitive abilities

Researchers have also investigated whether hormones influence sexual differentiation of specific cognitive abilities that show sex differences. Among the largest sex differences in human cognition are those seen in mental rotations ability (i.e., determining what objects would look like when rotated in space). Studies of mental rotations performance in girls and women with CAH have produced variable results. One study found the predicted enhanced performance, but involved a small sample (17 girls with CAH compared to 13 female relatives) (Resnick et al., 1986). Other studies, generally using larger samples, have not found significant differences on mental rotations performance in females with and without CAH (Baker and Ehrhardt, 1974; Helleday et al., 1994a; Hines et al., 2003b; Malouf et al., 2006).

Researchers interested in cognitive effects of prenatal androgen exposure have also looked at other spatial abilities. Some of this research was conducted before the magnitude of sex differences on different types of spatial tasks had been identified, and, in some cases, types of tasks that do not normally show robust sex differences were used, perhaps accounting for negative results. For instance, McGuire et al. (1975) found no differences between 15 females or 16 males with CAH and matched controls on a spatial visualization task. Similarly, Baker and Ehrhardt (1974) found no differences between 13 girls with CAH and 11 unaffected female relatives on an embedded-figures test. In other cases, positive results were reported. In one study, eight girls with CAH performed better than eight matched female controls on the Healey Pictorial Completion task, described as a "nonverbal evaluative" task "useful in identifying particular types of learning disability" (Perlman, 1973); no differences were seen for five boys with CAH compared to five matched male controls. In another study, seven girls with CAH performed better than six unaffected female relatives on a measure requiring children to select shapes that combined with other shapes to form squares (Hampson et al., 1998), and five boys with CAH performed worse than four unaffected male relatives on the same task. Both of the nonrotational tasks used in these two studies, of very small samples, showed surprisingly large sex differences in controls in the studies, but would not have been expected to do so based on meta-analyses of similar types of tests (Linn and Petersen, 1985; Voyer et al., 1995). One possible explanation of the findings is that many clinics have ten or 20 patients whose cognitive abilities can be assessed, and many studies of spatial abilities are carried out, sometimes using tests that do not normally show appreciable sex differences, and so would not be expected to relate to hormones. By chance, a few of these studies produce positive results, and these positive findings are published, whereas negative findings on similarly small samples are viewed as unreliable and unpublishable.

A meta-analysis, in 2008, of seven studies of spatial abilities in females with CAH concluded that they have a moderate-sized advantage ($d = 0.34$–0.47) (Puts et al., 2008). However, meta-analytic results depend on the data included, and this report included all studies using any measure of spatial ability, regardless of whether the measure shows reliable sex differences. In addition, subsamples of study participants were sometimes excluded from the effect sizes calculated for studies, and results for only some of the measures used were reported. These exclusions of subjects and measures generally enhanced effect sizes. If analysis is limited to all female participants with and without CAH in the four studies that used measures of mental rotations abilities (see **Table 1**), the average difference between females with and without CAH is negligible ($d = 0.16$), and is reduced to 0.11, when weighted for sample size. The median effect size is 0.01. These results suggest a negligible association between CAH and mental rotations performance in females.

Boys with CAH generally perform similarly to relative and matched controls on measures of spatial abilities (Baker and Ehrhardt, 1974; McGuire et al., 1975; Perlman, 1973; Resnick et al., 1986). However, two studies have found reduced spatial abilities in males with CAH, one, noted above, in a very small sample of five boys with CAH and four without, and not using a mental rotations task (Hampson et al., 1998), and one using two different mental rotations tasks in a sample of 29 males with CAH and 30 unaffected, male-relative controls (Hines et al., 2003b). Although T levels appear to be largely normal prenatally in males with CAH, weak androgens may be elevated (Pang et al., 1980; Wudy et al., 1999). Also, there is some evidence that T is reduced, shortly after birth, in boys with CAH (Pang et al., 1979), perhaps caused by feedback-related reduction in testicular androgen production coupled with

Table 1 Effects sizes for differences in mental rotations performance in females with and without congenital adrenal hyperplasia (CAH)

Study	*Participants*	*Measure*	*Effect size (d)*
Resnick et al. (1986)	17 CAH; 13 Rel	Card Rotations[a]	0.79
		Vandenberg and Kuse[b]	0.92
			Mean = 0.86
Helleday et al. (1994a)	22 CAH; 22 Mat	Figure Rotation[a]	−0.13
Hines et al. (2003b)	40 CAH; 29 Rel	PMAT[a]	0.12
		Vandenberg and Kuse[b]	0.17
			Mean = 0.15
Malouf et al. (2006)	24 CAH: 12 SV; 12 SW	Cube comparisons[b]	−0.46 (SV)
			−0.01 (SW)
	18 Con: 10 Rel; 8 PCOS		Mean = −0.24

[a]Two-dimensional mental rotations task.
[b]Three-dimensional mental rotations task.
Overall mean effect size = 0.16.
Overall mean effect size weighted for sample size = 0.11.
Median effect size = 0.01.
Rel, relative controls; Mat, matched controls; SV, simple virilizing; SW, salt wasting; Con, controls; PCOS, controls with polycystic ovarian syndrome; PMAT, thurstone primary mental abilities test.

postnatal treatment with corticosteroids that reduces adrenal androgen production. Assuming that the results suggesting reduced spatial abilities in males with CAH prove replicable, one possible explanation is that spatial abilities continue to develop in the early neonatal period, when cortical development is proceeding rapidly and boys with CAH have reduced androgen (Hines et al., 2003b). It has also been suggested that unusually high levels of androgen during early development might impair spatial abilities (Puts et al., 2008), although this explanation would not be predicted from animal models of androgen effects. A third possibility is that the spatial deficits result from salt-losing crises in infancy, as these are more common in males than in females with CAH (Hines et al., 2003b).

Concerning tasks at which females excel, such as verbal fluency and perceptual speed and accuracy, prenatal exposure to androgen, such as occurs in CAH, might be expected to defeminize or impair performance. However, to date, there is little or no evidence to support this hypothesis. Resnick et al. (1986) found no alterations in perceptual speed and accuracy or verbal abilities in either males or females with CAH. Baker and Ehrhardt (1974) also looked at verbal measures, including verbal fluency, and found no differences in CAH boys or girls compared to relative controls. McGuire et al. (1975) found no differences in either sex on measures of perceptual speed or verbal abilities, and Sinforiani et al. (1994) reported no differences between a combined group of seven female and 12 male patients versus matched controls on similar tests. Helleday et al. (1994a) also found no differences on these types of measures in their study of women with CAH compared to matched female controls. A single study reported impaired performance on a measure of perceptual speed and accuracy, particularly in CAH girls (Hampson et al., 1998). However, given the small sample (seven girls with CAH and six without), and the lack of similar findings in other studies, this could be a chance finding.

Some studies have found impaired mathematical abilities, particularly computational ability, in children with CAH, although this is not universally the case (Helleday et al., 1994a). One study reported impairment in girls with CAH but not boys (Perlman, 1973), one in CAH girls and CAH boys (Baker and Ehrhardt, 1974), and one in a combined group of boys and girls with CAH (Sinforiani et al., 1994). Computational ability shows a small sex difference favoring females, but only in young children (Hyde et al., 1990). No studies have reported enhanced mathematical performance in girls or boys exposed to androgen prenatally, despite speculations that the better performance of males on standardized tests of mathematical ability, such as the SAT, is due to prenatal androgen exposure (Benbow and Stanley, 1983; Benbow, 1988). However, neither SAT performance nor performance on other tests of mathematical abilities that show large sex differences have been reported for individuals with CAH or other causes of early hormonal abnormality.

Studies of DES-exposed women have not found alterations in cognitive performance. Two studies from one research group compared the performance of DES-exposed women and their unexposed sisters on cognitive tasks that show sex differences. The first found no differences between 25 sister pairs, one in each pair exposed to DES for at least 20 weeks prenatally and one not exposed at all, on a 2D mental rotations task or on a measure of verbal fluency (Hines and Shipley, 1984). The second found no differences between a second sample of 42 DES-exposed women and their 26 unexposed sisters on 3D mental rotations, spatial perception, perceptual speed and accuracy, verbal fluency, vocabulary, or Raven's Progressive Matrices. A third study, from a different research group, investigated cognitive abilities in DES-exposed men and women who had participated in a double-blind, placebo-controlled study of the efficacy of DES for preventing miscarriage. American college testing (ACT) subtest scores were compared in 325 female offspring (175 DES-treated and 150 placebo-treated) and 347 male offspring (172 DES-treated and 175 placebo-treated). Four of the ACT subtests showed sex differences, but none of these differed in DES-exposed versus placebo-exposed females (Wilcox et al., 1992). Males exposed to DES scored higher than males exposed to placebo on the social sciences subtest, a test at which males typically outperform females, but not on other subtests, and the authors attributed this single, unpredicted finding to chance. Another study, of ten boys exposed prenatally to DES, found them to show impaired performance compared to unexposed brothers on a composite of three performance subtests from the Wechsler scales (Picture Completion, Object Assembly, and Block Design), tests that show small-to-negligible sex differences. The ten pairs of brothers did not differ on composites of verbal or sequencing tasks (Reinisch and Sanders, 1992); given the small sample, the results for the performance tasks may have been spurious.

Girls and women with TS perform normally on measures of vocabulary and general intelligence (Garron, 1977), but show deficits in several specific cognitive functions. The best-known impairment is in spatial abilities, although other abilities, including numerical ability, memory or attention, and verbal fluency, have also been found to be impaired (Ross and Zinn, 1999; Rovet, 1990; Temple and Carney, 1996; Waber, 1979). Some of the tests on which performance is reduced favor males, whereas others favor females or show no sex differences. Impairments on tasks that show sex differences favoring either males or females are larger than impairments on sex-neutral tasks (Collaer et al., 2002). The particularly marked alterations on sex-linked tasks strengthen suggestions that the cognitive deficits are related to hormonal deficiency. Impairment on tasks at which females excel may also support a role for ovarian hormones in feminization of some cognitive characteristics (Collaer et al., 2002).

Men with IHH have been found to show impaired performance on several spatial tasks, including a 2D-mental rotations task, a measure of spatial perception, and measures of spatial visualization (Buchsbaum and Henkin, 1980; Cappa et al., 1988; Hier and Crowley, 1982). Some of these tasks show sex differences, but others do not. IHH men are not impaired on verbal tasks that do not show sex differences, such as vocabulary (Cappa et al., 1988; Hier and Crowley, 1982), although impairment on a measure of verbal fluency has been reported (Cappa et al., 1988). In one study (Hier and Crowley, 1982), men who became hypogonadal after puberty did not show impaired spatial performance, although those who had been hypogonadal from early life did, and treatment with T in adulthood did not improve performance in the impaired group. This suggests that, if hormones underlie the spatial deficiency, the absence of testicular hormones during early development, rather than later in life, is the crucial factor.

One study found that ten patients with CAIS showed reduced performance, compared to both male and female relatives, on Block Design and other performance subtests from the Wechsler scales (Imperato-McGinley et al., 1991). The subtests do not show consistent sex differences, making it uncertain if the impairment relates to androgen insensitivity or to some other aspect of the syndrome.

Normal variability in hormones has also been related to cognitive development. One study related T in amniotic fluid, during the second trimester of pregnancy, to cognitive performance of 28 female and 30 male offspring at the age of 4 years (Finegan et al., 1992). For girls, T related negatively to counting and sorting, number questions, and block building and showed an inverted-U-shaped relationship to language comprehension and conceptual grouping. None of these relationships was consistent with predictions. For boys, T did not relate significantly to any of 11 abilities measured. Another study related hormones in umbilical cord blood at birth to cognitive performance at the age of 6.5 years (Jacklin et al., 1988). In 53 boys, there were no significant relationships between any hormone and cognitive performance. In 43 girls, androgens (both T and androstenedione) related negatively to spatial ability, a result opposite prediction. None of the cognitive measures used in either study show substantial sex differences. A follow up at age 7, of the children whose amniotic fluid had been sampled prenatally, found that T in girls related positively to the speed of mental rotations, but not to accuracy (Grimshaw et al., 1995).

Laboratory-based studies relating finger ratios (2D:4D) to spatial performance have produced variable results, and a meta-analysis in 2008 concluded that there is no relationship between 2D:4D and spatial abilities (Puts et al., 2008). A large-scale, online study, however, found the predicted negative relationship between the 2D:4D ratio and mental rotations performance in both men and women (Peters et al., 2007). A similar relationship was seen, again in both males and females, for a spatial perception task that also shows a large sex difference (Collaer et al., 2007). Although finger ratios accounted for negligible percentages of the variance (less than 0.01% for mental rotations; Peters et al., 2007) the results are consistent with predictions based on the masculinizing effects of T. The variable results in laboratory studies of finger ratios may reflect the small size of the sex difference in 2D:4D.

Summary. Early reports that prenatal exposure to T or progesterone enhanced intelligence have proved unreliable. Hormonal influences on specific cognitive abilities that show sex differences are theoretically more likely. Although research to date has produced inconsistent findings, inconsistencies may reflect the use of small samples which can produce spurious positive findings, as well as negative findings (e.g., due to limited experimental power). A second problem has been the use of measures that do not show large sex differences and so would be unlikely to relate to hormones. Despite these problems, evidence from studies of females with CAH suggests that prenatal exposure to very high levels of androgens may slightly improve mental rotations performance, a conclusion that is reinforced by a large-scale study relating finger ratios (thought to reflect prenatal androgen levels) to mental rotations performance, as well as judgment of line orientation, in both men and women. The size of the effect appears to be negligible, however. Prenatal exposure to the synthetic estrogen, DES, does not appear to influence cognitive development.

Hormones could influence human cognitive development neonatally, and neonatal androgen deficiency could contribute to impaired spatial abilities in males with CAH. Cortical systems underlying higher cognitive processes are still developing during the first year of postnatal life. In addition, males experience an androgen surge during the first few months of infancy (Forest et al., 1973; Winter et al., 1976), and estrogen levels show an elevation in females during the first few postnatal months (Bidlingmaier et al., 1974, 1987; Channing et al., 1984; Winter et al., 1976). Thus, the neonatal period could be important for hormonal influences on human cognition. The finding of impaired visuospatial ability in men with IHH is consistent with this possibility, since the hormonal deficit in IHH is likely to begin at birth. Similarly, data suggesting that females with TS show cognitive defeminization as well as demasculinization provide some support for the hypothesis that ovarian hormones actively feminize cognitive sexual differentiation, and this influence could occur neonatally, since infants with TS are not treated with hormones. However, additional evidence on IHH and other situations that involve neonatal hormonal abnormality or variability are needed to evaluate the hypothesis that the neonatal period is important for cognitive sexual differentiation.

8.5.5 Emotion, Temperament, and Personality

8.5.5.1 Aggression

Two studies used interviews to assess fighting in girls with and without CAH and found no differences. In one study, 15 girls with CAH did not differ from 15 matched controls (Ehrhardt et al., 1968), and in a second, 17 females with CAH did not differ from 11 unaffected sisters (Ehrhardt and Baker, 1974). A third study involved three samples of individuals with CAH and unaffected relative controls (Berenbaum and Resnick, 1997). Eighteen female adolescents and adults with CAH showed enhanced aggressive response tendencies compared to 13 female controls in the same age range, but a second group of 11 CAH females of similar age, compared to five control females, showed this effect only for one of two measures of aggression. There were also no differences between 20 younger girls with CAH and ten female controls. Males with CAH did not differ from male controls in any of the three samples. These variable or negative results could reflect small sample size and insensitivity of measurement. A 2007 report on 38 girls with CAH found increased maternal reports of fighting and physical aggression compared to 25 unaffected female relatives, but no differences between 29 boys with CAH compared to 21 unaffected male relatives (Pasterski et al., 2007). A report on 17 girls and eight boys prenatally exposed to androgenic progestins also suggested increased physical aggression in response to provocation in both the hormone-exposed boys and girls compared to their unexposed siblings of the same sex (Reinisch, 1981). In this study, possible confounding influences of genital virilization were not a problem, because the hormone-exposed children had normal-appearing genitalia at birth.

8.5.5.2 Empathy

Normal variability in T measured in amniotic fluid has been found to relate negatively to empathy as assessed by questionnaire in 99, 6–8-year-old boys, but not in 91, 6–8-year-old girls; for a subset of children in the study, empathy assessed by the ability to judge emotions from pictures of eyes related negatively to T in both 39 boys and 37 girls (Chapman et al., 2006). Another report on a subset from the same sample of children at age 4 found that T related negatively in 24 boys, but not in 14 girls, to the frequency of using intentional propositions to describe interactions between cartoon images of moving triangles whose interactions suggested social relationships and psychological motivations (Knickmeyer et al., 2006b). Females also used more mental- and affective-state terms to describe the interactions than did males, but these sex differences did not relate to T in either boys or girls. Females with CAH ($n = 40$) also have been found to show reduced scores on a standardized subscale of the 16 Personality Factor Questionnaire (16 PF) called 'tendermindedness' in comparison to 30 unaffected female relatives (Mathews et al., in press). This 16 PF subscale taps empathy and shows a sex difference favoring females (Feingold, 1994; Mathews et al., in press).

8.5.5.3 Interest in parenting

As noted above, girls with CAH show reduced interest in toys typically preferred by girls, including dolls (Berenbaum and Hines, 1992; Pasterski et al., 2005). Three studies also suggest that CAH girls are less interested in babysitting and other aspects of child-care, including plans to have children of their own (Dittmann et al., 1990; Ehrhardt et al., 1968; Ehrhardt and Baker, 1974). The reduced interest in child-care may be more apparent in CAH girls older than 16 (Dittmann et al., 1990). These studies relied on a small number of interview questions. A more extensive questionnaire assessment of 23 girls and 16 boys with CAH compared to 12 female and 22 male relatives produced equivocal results (Leveroni and Berenbaum, 1998; see Hines (2004) for discussion.) However, a subsequent study using the same questionnaire in a larger sample of 35 females with CAH, 26 males with CAH, and 26 unaffected relatives of each sex found clear results, showing greater interest toward infants in unaffected females than in any of the other three groups (Mathews et al., in press). No differences were seen in males with and without CAH. Interests related to parenting and children have also been explored in women exposed prenatally to the synthetic estrogen, DES. An initial study suggested reduced interest in parenting, but this was just one of many variables assessed, and none of the others differed for DES-exposed women and controls (Ehrhardt et al., 1989). This single, significant finding was not replicated in two subsequent studies (Lish et al., 1991, 1992).

8.5.5.4 Other personality characteristics

A broad measure of personality has also been investigated in 22 individuals with CAH compared to 22 matched female controls. Females with CAH, ages 17–34, gave more male-typical responses than matched controls on measures of Indirect Aggression and Detachment, two of eight personality scales that showed sex differences in the study (Helleday et al., 1993). Similar differences were not seen in CAH males compared to controls.

8.5.6 Psychopathology

For both tic-related disorders and ASC, it has been suggested that the disorders may represent an extreme form of masculinity, produced by exposure to unusually high levels of androgens, prenatally (Alexander and Peterson, 2004; Baron-Cohen, 2002). Several studies suggest that characteristics associated with autism (e.g., lack of empathy, as noted above) correlate with T measured in amniotic fluid of typically developing children (Chapman et al., 2006; Knickmeyer et al., 2006b, 2005; Lutchmaya et al., 2002a,b). Some of the studies find the relationship between T and behavior only when data are analyzed for males and females combined, raising the possibility that the correlation simply reflects sex differences in both behavior and amniotic fluid T, but this is not always the case. In addition, 34 females with CAH have been found to score higher than 24 unaffected female relatives on a questionnaire measure of characteristics associated with ASC (Knickmeyer et al., 2006a). However, none of the females with CAH had scores high enough to arouse suspicion of an ASC. Only two participants in the study had scores sufficiently high to do so, and these were brothers – one of whom had CAH and one of whom did not – suggesting that genetic or other family factors are more important than androgens in producing ASC. Also, only certain subtests of the questionnaire differed for CAH versus unaffected girls and on one scale on which males did not score higher than females, females with CAH scored in the less-autistic (more-masculine) direction than females without CAH. This last result suggests that when being more masculine and more autistic are in conflict, prenatal T makes an individual more masculine, rather than more autistic. The idea that ASC represent an extreme male brain has also been called into question by evidence that the pattern of perceptual and spatial performance in ASC is not consistently male-typical (Falter et al., 2008).

Research investigating the possibility that androgen plays a role in tic-related disorders has compared behaviors that show sex differences in individuals with and without the disorders (Alexander and Peterson, 2004). Results of this initial investigation suggested that individuals with tic-related disorder, either obsessive–compulsive disorder or Tourette syndrome, show more male-typical behavioral profiles than healthy controls.

Summary. Studies of aggression following prenatal exposure to high levels of androgen support an influence on physically aggressive behavior. There is also evidence linking androgen, prenatally, to empathy, postnatally, with lower androgen associated with increased empathy. Similarly, prenatal androgen exposure has been linked to reduced interest in infants and child-care. A promising research area involves identifying the causes of sex differences in psychopathology, as identification of these causes could elucidate the etiology of specific disorders (Rutter et al., 2003). Relatively little research has been devoted to this area, but it has been suggested that hormonal influences on sexual differentiation contribute to ASC and tic-related disorders. Findings, to date, suggest that the early hormonal environment may contribute to some behaviors, such as empathy, that are associated with ASC, but that other influences, particularly genetic effects, are important for ASC as well. In addition, although T has been associated with ASC-related characteristics in individuals without ASC, there is as yet no link between T and actual ASC diagnosis. Only one study has been published exploring the link between androgen exposure and tic-related disorder. This study produced promising results that await replication and extension.

8.5.7 Neural Asymmetries

8.5.7.1 Hand preferences

Three studies from three different research groups have found that women exposed to DES prenatally show reduced right-hand preferences or increased left-handedness (Schachter, 1994; Scheirs and Vingerhoets, 1995; Smith and Hines, 2000). A fourth study found increased left-handedness, for writing, in DES-exposed men, but not women (Titus-Ernstoff et al., 2003). Reduced right-hand or increased left-hand preferences are the predicted outcome, if estrogen masculinizes or defeminizes human development, in that they reflect a more male-typical pattern. The finding of reduced right-hand preferences has been noted in comparing DES-exposed women to unrelated controls as well as to their unexposed sisters. In addition, in one study where the time of DES-exposure could be confirmed from medical records, the effect occurred only with exposure prior to week 9 of gestation (Smith and Hines, 2000), suggesting a critical period for this influence early in development. Some instances of left-handedness have been termed pathological because they appear to result from early injury to the left hemisphere of the brain (Satz et al., 1985). DES was often prescribed because of threatened miscarriage, making it possible that the shift toward left-hand preferences in DES-exposed offspring relates to prenatal trauma (Scheirs and Vingerhoets, 1995). Women exposed to high levels of androgen prenatally, because of the genetic disorder, CAH, have been found to show reduced right-hand preferences in one study (Nass et al., 1987), and other studies have reported increased left-handedness in females and males with CAH combined (Kelso et al., 1999, 2000) or in males, but not females, with CAH (Mathews et al., 2004). There is also a report that 20% of a combined group of males and females with CAH were left-handed, a figure thought to differ from the general population, although no statistical comparison was reported (Tirosh et al., 1993). Some studies report no differences, however (Helleday et al., 1994b; Resnick et al., 1986).

8.5.7.2 *Language lateralization*

One study of DES-exposed women found evidence of a more male-typical pattern of language lateralization in comparison to unexposed sisters (Hines and Shipley, 1984). However, a subsequent report from the same research group on a new, larger sample of women did not replicate the finding (Smith and Hines, 2000). In regard to CAH, one study found no differences between females with and without CAH (Helleday et al., 1994b), whereas another found increased language lateralization in individuals with CAH, particularly females, compared to right-handed, but not left-handed, controls (Tirosh et al., 1993). TS is associated with reduced language lateralization (see, e.g., Hines and Gorski (1985), for a review). This could reflect further defeminization of the brain in comparison to healthy females, in general, in this syndrome which involves deficiencies in ovarian hormones. It is not clear, however, whether reduced language lateralization can be attributed to hormonal abnormality, since TS involves structural and functional problems in many organ systems.

8.6 Hormonal Influences on Neural Sexual Differentiation

8.6.1 Sex Differences in Neural Structure and Function

8.6.1.1 *Brain size*

Perhaps the most obvious neural sex difference is in brain size. The male brain is larger and heavier, on average, than the female brain. This difference might be expected, given that the male body is larger and heavier than the female body, as are its component parts. Although some methods of adjusting statistically for body size suggest that the sex difference in brain size remains (Ankney, 1992), others suggest that it does not (Ho et al., 1980). The larger male body could easily account for the larger brain, since the sex difference in the brain is only about half the size ($d = 1.05$) (Ho et al., 1980) of the sex difference in height ($d = 2.00$) (International Committee on Radiological Protection, 1975; Tanner et al., 1966). Finally, in at least some brain regions, neurons are packed more densely in females than in males (Witelson et al., 1995), raising questions as to the functional significance of overall size differences. More subtle aspects of the brain and its architecture probably provide more productive territory for unearthing the foundations of behavioral sex differences.

8.6.1.2 *Anterior hypothalamic/preoptic area*

The AH/POA has been a particular focus of research on human sex differences, probably because this region contains a dramatic, structural sex difference in several species of mammals (the SDN-POA, also called the central subdivision of the medial preoptic nucleus) (Gorski et al., 1980; Hines et al., 1985; Tobet et al., 1986). The AH/POA is rich in androgen and estrogen receptors (MacLusky et al., 1979; Stumpf et al., 1975), and, in nonhuman mammals, lesion and implant studies implicate it in several behaviors that show sex differences, including male and female sexual behavior and maternal behavior (Allen et al., 1989; Goy and McEwen, 1980).

Four subregions of the AH/POA, called the interstitial nuclei of the anterior hypothalamus, numbers one to four (INAH1–4), have been assessed in the human brain in a search for sex differences comparable to the SDN-POA of other species. An early report suggested that INAH-1 was larger in males, than females, and the authors called the nucleus the human SDN-POA (Swaab and Fliers, 1985). Three subsequent investigations did not replicate the reported sex difference in INAH-1, however (Allen et al., 1989; Byne et al., 2000; LeVay, 1991). Instead, all three found a sex difference in INAH-3, which was larger in males, than in females. One of the studies also found INAH-3 to contain more neurons in men than in women (Byne et al., 2000). Based on the location of INAH-3, and the types of neurons it contains, it also seems the most likely counterpart to the SDN-POA of other species (Allen et al., 1989).

The misnomer of human SDN-POA for INAH-1, which does not show a sex difference, has led to questionable conclusions. For example, evidence that the nucleus does not show a sex difference in children and that it does not differ in homosexual versus heterosexual men, led to the conclusions that gender-related behavior in humans is not influenced by the early hormonal environment and that sexual orientation is independent of brain differences (Swaab and Hofman, 1988). However, if the nucleus does not show a sex difference, it cannot provide a basis for conclusions about human sexual differentiation. In addition, there are many sex differences in the human brain, so the lack of a relationship between any one region and any one behavior cannot support conclusions about more general processes of gender development. Similarly, the lack of a relationship between a single brain region and a single behavior does not exclude a neural basis for any psychological outcome, including homosexuality.

Of course, all behaviors and psychological traits, including those that relate to sex, must have some neural basis, be it structural, neurochemical, or other. What positive results have been found relating sex differences in brain structure to sex differences in behavior? One finding is that the size of INAH-3, the nucleus most likely to be a human counterpart to the SDN-POA, is smaller (i.e., more female-typical) in men who are homosexual or bisexual than in those who are heterosexual (LeVay, 1991). Although most of the homosexual and bisexual men in the study had died of acquired immune deficiency syndrome (AIDS), a subgroup of heterosexual men had also died of AIDS, and the volume of INAH-3 in this subgroup was similar to that in other heterosexual men, suggesting that AIDS or its consequences were not responsible for the difference in INAH-3 in heterosexual versus homosexual men. The other three INAH nuclei did not show sex differences in this study and did not differ for heterosexual versus nonheterosexual men. This, too, suggests that the difference in INAH-3 was not caused by general neural changes related to AIDS. The link between sexual orientation in men and the size of INAH-3 has been replicated (Byne et al., 2001). However, although INAH-3 was smaller in homosexual, than in heterosexual, men in the replication study, the number of neurons in the nucleus, which like the size of the nucleus is smaller in females than in males, did not differ for the two groups of men.

8.6.1.3 The bed nucleus of the stria terminalis

The BNST, like the AH/POA, contains steroid-responsive neurons, and has been linked by lesion and implant studies in nonhuman mammals to behavioral sex differences (Hines et al., 1985, 1992a). In addition, in rodents, a subregion of the BNST, called the posterodorsal BNST (BNSTpd) is larger in males than in females (Hines et al., 1985, 1992a). In humans, a similar region has also been reported to be larger in males than in females (Allen and Gorski, 1990). A separate region of the BNST (called the central BST or BSTc) has also been reported to show a sex difference in humans and to relate to core gender identity (Zhou et al., 1995). Six male-to-female (M-to-F) transsexuals were found to have a smaller (i.e., more female-typical) BSTc than either heterosexual or homosexual males. The volume of BSTc did not relate to sexual orientation, suggesting a specific association with core gender identity. A subsequent study from the same group found that the BSTc also contained fewer neurons expressing the neurohormone somatostatin in the M to F group, and the female control group, than in the male control group. In addition, in this study, the brain of one female-to-male (F-to-M) transsexual showed a similar number of somatostatin-expressing neurons to the male controls. A third report from the same group found that the sex difference in BSTc was not present in children, only in adults (Chung et al., 2002). This is puzzling, because transsexuals typically recall feeling different in terms of their gender identification, beginning very early in life. One possible explanation is that the link between BSTc and core gender identity is the result of cross-gender identification or experiences associated with it. Alternatively, the structural difference in BSTc could be a delayed expression of early influences, such as prenatal or neonatal hormones. The findings linking the BSTc to core gender identity are intriguing, but have yet to be replicated independently. Other cautions regarding the findings include the small samples studied so far and the lack of animal models for sex differences in BSTc, as opposed to other regions of the BNST.

8.6.1.4 The anterior commissure

The anterior commissure, a fiber tract connecting the left and right cerebral hemispheres, has also been studied in relationship to sex and sexual orientation. Nonheterosexual men have been reported to have larger anterior commissures than heterosexual men (Allen and Gorski, 1992). Women in the study also had larger anterior commissures than men, and so this difference was viewed as showing female-typical brain structure in nonheterosexual men. However, a separate study reported a sex difference in the opposite direction in the anterior commissure, finding it to be larger in males than in females (Demeter et al., 1988), and a third study found no sex difference in the anterior commissure and no difference between heterosexual and homosexual men (Lasco et al., 2002).

8.6.1.5 The suprachiasmatic nucleus

The suprachiasmatic nucleus (SCN) has been reported to be larger and to contain more vasopressin neurons in homosexual, than heterosexual, men

(Swaab and Hofman, 1990). No similar sex differences have been reported in the rodent SCN. On the contrary, the SCN has been used as a control region in studies of hormonal influences on other neural sex differences in the rat, because it does not show marked binding of gonadal steroids and does not show a volumetric sex difference (Gorski et al., 1978, 1980; Hines et al., 1985, 1992a; Madeira et al., 1995). Also, in humans, although a sex difference has been reported in the SCN, it involves shape, not size (Hofman et al., 1988; Swaab et al., 1985). The nucleus has been found to be relatively elongated in women, as opposed to spherical in men, while volume, cell numbers, and cell density are similar in males and females. Hence, the report of differences in the SCN of homosexual and heterosexual men may not relate to sexual differentiation, given that it does not correspond clearly to patterns of sex differences.

8.6.1.6 The corpus callosum

An initial report on a sample of brains obtained at autopsy suggested that the splenium of the corpus callosum (defined as the posterior fifth viewed in mid-sagittal section) was larger and more bulbous in female than in male brains (de Lacoste-Utamsing and Holloway, 1982). Maximum splenial width and spenial area and total callosal area were also larger (relative to total brain weight) in females, than in males. Similar findings were reported for a second sample of adult brains (Holloway and de Lacoste, 1986), and the sex difference in maximum splenial width adjusted for brain weight was also observed in a sample of fetal brains (de Lacoste et al., 1986).

These reports were controversial. One review suggested that over 20 subsequent studies failed to replicate the sex difference in the splenium (Byne and Parsons, 1993) and a second, based on a meta-analysis of 49 studies, concluded that there was no sex difference (Bishop and Wahlsten, 1997). The validity of these reviews depends on the quality of the studies included and their use of methodology similar to that used in the original report. However, studies of the corpus callosum have used vastly different methodologies. For instance, some have defined the splenium as the posterior fourth, rather than the posterior fifth, of the callosum, have not adjusted values to reflect the overall sex difference in brain size, or have marked off fourths or fifths of the callosum along a straight line from front to back, rather than on a curved line bisecting it. (The curved-line procedure was used in the original reports of sex differences to adjust for individual differences in curvature of the callosum.) Some studies also used clinical images (such as those obtained using magnetic resonance imaging (MRI)), instead of autopsy material. Although both autopsy material and MRI images could produce reliable results, images obtained for clinical purposes are not necessarily designed to produce accurate midline images of the callosum. Because the fibers of the callosum fan out dramatically after passing the midline, an image that was not precisely positioned in the midline could add substantial error variability, making it difficult to detect group differences. (These methodological issues are discussed more extensively elsewhere (Allen et al., 1991; Elster et al., 1990; Hines and Collaer, 1993).) Studies using procedures similar to those used in the original reports have found sex differences in the callosum (Allen et al., 1991; Clarke et al., 1989; Elster et al., 1990), although they are not as dramatic as they originally appeared. In addition, there is debate over the appropriateness of adjusting for overall callosal or brain size (Jancke et al., 1997). One relatively recent approach has been template deformation morphometry (TDM). TDM registers each subject to a template callosum, avoiding the problem of using overall brain size, while still adjusting for size differences. Using TDM analysis of MR images, the splenium has been found to be larger in the female, than the male, brain (Dubb et al., 2003).

The isthmus of the callosum, lying just anterior to the splenium, and defined as the posterior third minus the posterior fifth, has also been reported to vary with sex (Witelson, 1985, 1989), although in this case, hand preferences must also be considered. The isthmus is smaller in individuals who consistently use their right-hand for motor tasks than in those who do not. In addition, in consistent right-handers, the isthmus is larger in females than in males, whereas there is no sex difference in those who are not consistent right-handers. These reports, like those of sex differences in the splenium, were controversial and there were reported failures to replicate the initial findings (Kertesz et al., 1987; Nasrallah et al., 1986). Methodological differences similar to those described for studies of the splenium are also relevant here. In addition, some studies used different procedures to classify hand-preference groups, or based classifications only on the hand used for writing, rather than on preferences across a range of motor tasks. A study using similar methodology to that of the original report produced similar findings (Habib et al., 1991). Witelson and colleagues have also linked the size of the isthmus to sexual orientation, reporting that 12 right-handed homosexual men had larger mid-sagittal isthmal areas, assessed using MRI, than ten right-handed heterosexual men (Witelson et al., 2008). Other callosal regions, including the anterior half, posterior mid-body, total callosum, and splenium also showed group differences of moderate size ($d = 0.41$–0.66), but only the difference in the isthmus ($d = 0.83$) was statistically significant.

The anterior-most portion of the callosum also appears to show a sex difference. One report found the genu of the callosum, defined as the anterior fourth as viewed in mid-sagittal section, to be larger in males, than in females (Reinarz et al., 1988), while a second, defining a smaller area of the callosum as the genu, also found the region to be larger in males (Witelson, 1989). This difference has also been found using TDM (Dubb et al., 2003).

The size and shape of the corpus callosum have been related to sex-linked cognitive functions. One study found that the mid-sagittal surface of the callosum was greater in those who showed right-hemispheric dominance for language (O'Kusky et al., 1988). Another found that the mid-sagittal surface area of posterior regions of the callosum, particularly the splenial region, related negatively to left-hemispheric language lateralization and positively to verbal fluency in women (Hines et al., 1992b), reflecting positive correlations between female-typical brain structure and female-typical cognitive function.

Sex differences in the size of callosal subregions, similar to those seen in humans, have not been reported in rodents. However, under certain rearing conditions, there are sex differences in the numbers of various types of fibers in posterior regions of the callosum (Juraska and Kopcik, 1988; Juraska, 1991). When rats are reared after weaning in complex environments (large group cages containing objects that are changed daily), females have more myelinated axons in the posterior callosum than males do. However, when reared in isolation (single cages with no objects other than those needed to provide food and water), this sex difference is not seen. Other studies have found that similar environmental manipulations influence dendritic growth in various regions of the cerebral cortex, and can produce, enhance, reduce, or even reverse sex differences in these neural characteristics (Juraska, 1991).

These findings illustrate the complexity of studying sex differences in the brain, given their apparent dependence on postnatal experience. It has sometimes been assumed that a structural sex difference in the brain indicates an inborn or irreversible state. In addition to Juraska's work on regions of the brain that show sex differences, other recent work also suggests that brain structure is more easily modified by the postnatal environment than has been generally understood. Brain changes, including dendritic growth as well as the birth of neurons, can continue into adulthood and neuronal survival can depend on adult experience (Kempermann et al., 1997, 1998). Hence, structural sex differences in the brains of males and females cannot be assumed to reflect an inborn state, or even to be the result of experience during very early, as opposed to adult, life.

8.6.1.7 The cerebral cortex

Studies of sex differences in the human brain are not limited to investigations of structure. Technologies such as functional MRI and positron emission tomography (PET) allow the functioning of the living brain to be investigated as well. These technologies were developed relatively recently and continue to be refined, and research on sex differences in human brain function has not yet reached a stage where firm, general conclusions are possible. However, it appears that to, a large extent, male and female brains function similarly (Frost et al., 1999; Mansour et al., 1996), even for male and female individuals who are completing measures of mental rotations and verbal fluency and showing sex differences in performance (Halari et al., 2005). Some sex differences have been reported, but they are limited, involving, for instance, asymmetries of function during very specific language-related tasks (Shaywitz et al., 1995). The question of sex differences in the functioning brain is far more complex than it might appear at first. Many factors influence results and could be invoked to explain inconsistencies among individual studies. They include, but are unlikely to be limited to: the age of subjects, their hand preferences, whether the participant is resting or conducting a task, the specific task being conducted, the difficulty level of the task, the skill of the individual being tested or their prior experience with the specific task, the response modality, the specific imaging technique being used, and the statistical procedures used to quantify function and evaluate male/female differences.

Finally, male and female brains may use different mechanisms to achieve the same behavioral outcome. For instance, one study using PET found that, among men, scores on the SAT math subtests correlated with activation in the temporal lobes, whereas for women matched for math ability there was no correlation between performance and activation in this brain region or any other identifiable brain region (Haier and Benbow, 1995). Similarly, the activational asymmetries noted above during language tasks involved tasks that do not show sex differences (Shaywitz et al., 1995), and performance on standard tests of intelligence that do not show sex differences has been found to correlate with the volume of different neural structures in men and women as assessed using MRI. In women, the amount of gray and white matter in frontal regions relate to performance, whereas in men parietal regions do (Haier et al., 2005).

8.6.2 Hormones and the Human Brain

There is little information on hormone-related alterations in human brain development. Two strategies might be used to study such alterations. One would look at identified sex differences in the brains of individuals who developed in atypical hormonal milieus. At present, the most reliable neural sex difference is that in INAH-3. However, current technologies cannot visualize this sex difference in the living brain; it can be visualized only in brains obtained at autopsy. This makes it difficult to study in the limited number of individuals with atypical hormonal histories, and to date, no such study has been published. A second strategy would be to use techniques such as MRI and PET to provide information on the structure and function of the living brain. As mentioned above, information on sex differences that can be visualized using these techniques is still developing. Because of this, research using this approach has been largely exploratory.

One study used MRI to compare brain structure in 15 individuals with CAH and 50 controls. Controls had been referred for imaging because of headaches or psychological disturbances and were selected to exclude those who showed evidence of a structural brain lesion (Sinforiani et al., 1994). The CAH group differed from controls only in that four of the 15 showed increased signal intensity in white matter, but these disturbances were not localized to a particular region. In addition, they did not appear to correlate with any cognitive or affective impairment in the CAH group. A second study of CAH individuals reported that both they and their siblings without CAH showed structural brain abnormalities on MRI scans (Plante et al., 1996). All five siblings and most of 11 CAH individuals showed reduced or reversed asymmetry in the perisylvian region of the brain. One individual with CAH also showed evidence of a neuromigratory disturbance in the posterior portion of the left hemisphere. Among 16 matched controls, only one showed atypical asymmetry and none showed any evidence of a neuromigratory disturbance. The CAH families in this study also showed an increase in learning disabilities in both CAH children and their unaffected siblings. The authors concluded that the findings represent an association of learning disability and altered brain asymmetries with the gene for CAH. Selection bias or chance might also explain the findings, however, especially given that some of the siblings of CAH children would not be expected to carry the gene for CAH.

At least one hypothesis-driven study of brain function in individuals with CAH has been conducted. This study used fMRI to compare adolescents with CAH to healthy controls and examined responses to negative facial emotions. The seven females with CAH, but not the seven males, showed more activation of the amygdala than their respective controls, and this response resembled that of the control males, a result interpreted to suggest masculinized functioning of the amygdala in the girls with CAH (Ernst et al., 2007).

Brain structure and function in individuals with TS has also been investigated. Reports suggest ventricular enlargement, and alterations in orbitofrontal, parietal, and occipital regions of the cerebral cortex, as well as elsewhere (Clark et al., 1990; Molko et al., 2004; Murphy et al., 1996, 1997; Reiss et al., 1993, 1995). Consistent findings include decreased tissue in the parietal lobes, and increased volume of the amygdala (Brown et al., 2002a; Good et al., 2003; Molko et al., 2004). There is also evidence that connections between posterior and anterior temporal regions are disrupted (Molko et al., 2004). It is not known if any of the observed neural alterations in TS are caused by early hormonal abnormality, as opposed to other aspects of the syndrome, such as X-chromosome abnormality.

8.7 Summary and Conclusions

It appears that gonadal hormones influence some aspects of human behavioral development in a manner similar to that documented in experimental studies of animals. This has been shown most clearly in regard to childhood play behavior, where information from clinical syndromes, from situations where hormones have been administered to pregnant women, and from studies of normal variability converge to indicate that T levels during early development promote more male-typical patterns of behavior. Though less extensive, there is also good evidence that sexual orientation, core gender identity, physical aggression, empathy, and hand preferences relate to the early hormonal environment, with higher levels of androgens, and in the case of hand preferences, and perhaps sexual orientation, estrogen, promoting male-typical development. However, in all cases, hormonal exposure increases the likelihood of more male-typical characteristics, without completely determining that they will develop. Clearly, other factors also are important in the development of behavioral sex differences. Indeed, research from other perspectives has documented the importance of social learning in the development of sex differences in children's toy preferences and physically aggressive behavior, suggesting that the early hormonal environment is one of several factors influencing sexual differentiation of human behavior.

Influences of the prenatal hormonal environment on other behaviors that show sex differences, including sex-specific cognitive abilities, and interest in infants, are less well established. In regard to interest in infants, this may reflect the small number of published reports. Results are currently equivocal for cognitive abilities, in part because studies have often relied on small samples or on measures that do not show reliable sex differences. Information regarding hormonal influences on brain structure and functional characteristics that can be visualized using techniques such as MRI and PET is still developing, and this area is likely to expand in the coming years. At present, one obstacle is a lack of convincingly demonstrated, reliable sex differences that can be visualized in the living human brain.

8.7.1 Fitting a Theoretical Model

Most of the findings supporting a role for gonadal hormones in human neurobehavioral development are consistent with a gradient version of the classic model of hormonal influences. However, relatively little information is available about hormonal influences on sexual differentiation of cognitive abilities and other characteristics that depend on the cerebral cortex. Evidence supporting active feminization in rodents appears to involve cortical structures. Thus, human characteristics dependent on the cortex might be the most likely candidates for active feminization by ovarian hormones. In addition, the cerebral cortex develops relatively late. This is also consistent with the suggested influences of ovarian hormones on cortical development in rodents, since these appear to occur later than the classic influences of testicular hormones on sexual differentiation. Therefore, although the classic model of hormonal influences, particularly in its gradient version, currently appears to be the best description of hormonal influences on human neurobehavioral development, as it is in other species, the possibility that ovarian hormones actively feminize some sexually differentiated characteristics cannot be ruled out. Investigation of this possibility might focus on cognitive abilities and other characteristics that develop late and are mediated by cortical mechanisms.

8.7.2 Mechanisms of Hormone Action

In addition to the possibility that the classic model of hormonal influences does not explain all of human sexual differentiation, research to date suggests that, as is true for sexual differentiation of the genitalia, different testicular hormones control development of different sexually differentiated neural and behavioral traits. As noted above, either androgen or estrogen may influence sexual differentiation of hand preferences. However, childhood play behavior seems to be influenced by androgen only. This suggests that testicular hormones influence hand preferences following conversion to estrogen, whereas they influence childhood play behavior directly.

In addition to different types of hormones exerting effects on different behaviors, hormones influence behaviors at different times. Determination of the separate critical periods during which hormones influence specific characteristics is also likely to be a focus of future research. Current information suggests that the critical period for hormonal influences on hand preferences occurs very early in gestation, beginning before week 10 (Smith and Hines, 2000). Specific critical periods, when other hormonal influences, are exerted are as yet unspecified. However, as noted above, much of cortical development occurs postnatally. It may be, therefore, that critical periods for hormonal influences on at least some cognitive functions that show sex differences occur relatively late, perhaps including periods that coincide with early postnatal elevations in estrogen in females or androgen in males.

The specific neural mechanisms that control hormonal influences on sexual differentiation of human behavior are also largely unknown. In rodents, hormones appear to act, in part, by enhancing neurite outgrowth, by preventing cell death in certain brain regions and by promoting it in others. In the internal genitalia, hormones promote development of some structures and inhibit development of others. In the external genitalia, they act on identical structures to cause them to develop differently. Whether similarly diverse mechanisms are involved in hormonal influences on human brain development has yet to be established. In fact, there is currently almost no information on the neural consequences of early exposure to hormones on human brain structure or visible function. Imaging techniques, such as MRI and PET, promise important advances in this area, but these have yet to be realized.

8.7.3 Clinical and Theoretical Importance

Finally, findings in this area of research have clinical implications as well as theoretical importance. The birth of a child with ambiguous genitalia can present a dilemma regarding sex assignment. Treatment of DSDs has been reconsidered in light of increasing evidence that prenatal exposure to testicular hormones has permanent influences on human psychological development. Historically, the tendency was to assign ambiguous cases as girls, in part because greater ability to surgically create female, than male, external genitalia. However, some individuals assigned as females are dissatisfied and wish to live as males. As described above, this has occurred with an XY individual whose penis was ablated in infancy and who was reassigned as female, in a small proportion of XX individuals with CAH, and in some others, either XX or XY, who were exposed prenatally to levels of androgens in excess of what would typically be the case in an XX individual. At the same time, other people were exposed to similar hormone environments prenatally, presented similarly as infants, were assigned as female and are content in this role. It is not known why some such assignments are successful, whereas others are not. Obviously, the clinical intention is to assign individuals successfully in all cases. To do so, it would be helpful to have more extensive outcome data to establish the frequency of unsuccessful outcomes for specific syndromes. More research is also needed to determine precisely what social or cultural factors influence successful outcomes. At present, it is not possible to predict the optimal direction of sex assignment with 100% certainty from chromosomes, hormone levels, the appearance of the external or internal genitalia, or diagnosis. It is possible that additional research on sex differences in brain structure and function could help in this area. For instance, if there were structural or functional sex differences that could be visualized in the living human brain at or near the time of birth, this information, along with other characteristics that are currently considered, might increase the accuracy of predictions regarding the optimal sex assignment in these difficult cases.

References

Alexander GM and Hines M (1994) Gender labels and play styles: Their relative contribution to children's selection of playmates. *Child Development* 65: 869–879.

Alexander GM and Peterson BS (2004) Testing the prenatal hormone hypothesis of tic-related disorders: Gender identity and gender role behavior. *Development and Psychopathology* 16: 407–420.

Allen LS and Gorski RA (1990) Sex difference in the bed nucleus of the stria terminalis of the human brain. *Journal of Comparative Neurology* 302: 697–706.

Allen LS and Gorski RA (1992) Sexual orientation and the size of the anterior commissure in the human brain. *Proceedings of the National Academy of Sciences of the United States of America* 89: 7199–7202.

Allen LS, Hines M, Shryne JE, and Gorski RA (1989) Two sexually dimorphic cell groups in the human brain. *Journal of Neuroscience* 9: 497–506.

Allen LS, Richey MF, Chai YM, and Gorski RA (1991) Sex differences in the corpus callosum of the living human being. *Journal of Neuroscience* 11: 933–942.

Amateau SK and McCarthy MM (2004) Induction of PGE2 by estradiol mediates developmental masculinization of sex behavior. *Nature Neuroscience* 7: 643–650.

American Psychiatric Association (2000) *Diagnostic and Statistical Manual of Mental Disorders: DSM-IV-TR, Fourth Edition Text Revision*. Washington, DC: American Psychiatric.

Ankney CD (1992) Sex differences in relative brain size: The mismeasure of women too? *Intelligence* 16: 329–336.

Arnold AP and Gorski RA (1984) Gonadal steroid induction of structural sex differences in the central nervous system. *Annual Review of Neuroscience* 7: 413–442.

Auyeung B, Baron-Cohen S, Chapman E (in press) Fetal testosterone predicts sexually differentiated childhood behavior in girls and in boys. *Psychological Science*.

Baker SW and Ehrhardt AA (1974) Prenatal androgen, intelligence and cognitive sex differences. In: Friedman RC, Richart RN, and Vande Wiele RL (eds.) *Sex Differences in Behavior*, pp. 53–76. New York: Wiley.

Baron-Cohen S (2002) The extreme male brain theory of autism. *Trends in Cognitive Science* 6: 248–254.

Benbow CP (1988) Sex differences in mathematical reasoning ability in intellectually talented preadolescents: Their nature, effects and possible causes. *Behavioral and Brain Sciences* 11: 169–232.

Benbow CP and Stanley JC (1983) Sex differences in mathematical reasoning ability: More facts. *Science* 222: 1029–1031.

Berenbaum SA and Hines M (1992) Early androgens are related to childhood sex-typed toy preferences. *Psychological Science* 3: 203–206.

Berenbaum SA and Resnick SM (1997) Early androgen effects on aggression in children and adults with congenital adrenal hyperplasia. *Psychoneuroendocrinology* 22: 505–515.

Berman PA (1980) Are women more responsive than men to the young? A review of developmental and situational variables. *Psychological Bulletin* 88: 668–695.

Bidlingmaier F, Strom TM, Dörr G, Eisenmenger W, and Knorr D (1987) Estrone and estradiol concentrations in human ovaries, testes, and adrenals during the first two years of life. *Journal of Clinical Endocrinology and Metabolism* 65: 862–867.

Bidlingmaier F, Versmold H, and Knorr D (1974) Plasma estrogens in newborns and infants. In: Forest M and Bertrand J (eds.) *Sexual Endocrinology of the Perinatal Period*, pp. 299–314. Paris: Inserm.

Billy JOG, Tanfer K, Grady WR, and Klepinger DH (1993) The sexual behavior of men in the United States. *Family Planning Perspectives* 25: 52–60.

Bishop KM and Wahlsten D (1997) Sex differences in the human corpus callosum: Myth or reality? *Neuroscience and Biobehavioral Reviews* 21: 581–601.

Bradley SJ, Oliver GD, Chernick AB, and Zucker KJ (1998) Experiment of nurture: Ablatio penis at 2 months, sex reassignment at 7 months and a psychosexual follow-up in young adulthood. *Pediatrics* 102(1): e9.

Brown WE, Kesler SR, Eliez S, et al. (2002a) Brain development in Turner syndrome: A magnetic resonance imaging study. *Psychiatry Research* 116: 187–196.

Brown WM, Hines M, Fane BA, and Breedlove SM (2002b) Masculinized finger length patterns in human males and females with congenital adrenal hyperplasia. *Hormones and Behavior* 42: 380–386.

Buchsbaum MS and Henkin RI (1980) Perceptual abnormalities in patients with chromatin negative gonadal dysgenesis and hypogonadotropic hypogonadism. *International Journal of Neurosciences* 11: 201–209.

Buck JJ, Williams RM, Hughes IA, and Acerini CL (2003) *In-utero* androgen exposure and 2nd to 4th digit length ratio – comparisons between healthy controls and females with classical congenital adrenal hyperplasia. *Human Reproduction* 18: 976–979.

Byne W, Lasco MS, Kemether E, et al. (2000) The interstitial nuclei of the human anterior hypothalamus: An investigation of sexual variation in volume and cell size, number and density. *Brain Research* 856: 254–258.

Byne W and Parsons B (1993) Human sexual orientation: The biological theories reappraised. *Archives of General Psychiatry* 50: 228–239.

Byne W, Tobet SA, Mattiace LA, et al. (2001) The interstitial nuclei of the human anterior hypothalamus: An investigation of variation with sex, sexual orientation, and HIV status. *Hormones and Behavior* 40: 86–92.

Cappa SF, Guariglia C, Papagno C, et al. (1988) Patterns of lateralization and performance levels for verbal and spatial tasks in congenital androgen deficiency. *Behavioural Brain Research* 31: 177–183.

Carson DJ, Okuno A, Lee PA, Stetten G, Didolkar SM, and Migeon CJ (1982) Amniotic fluid steroid levels: Fetuses with adrenal hyperplasia, 46,XXY fetuses, and normal fetuses. *American Journal of Diseases of Children* 136: 218–222.

Channing CP, Chacon M, Tanabe K, Gagliano P, and Tildon T (1984) Follicular fluid inhibin activity and steroid levels in ovarian tissue obtained at autopsy from human infants from 18 to 200 days of age. *Fertility and Sterility* 42: 861–869.

Chapman E, Baron-Cohen S, Auyeung B, Knickmeyer R, Taylor K, and Hackett G (2006) Fetal testosterone and empathy: Evidence from the Empathy Quotient (EQ) and the 'reading the mind in the eyes' test. *Social Neuroscience* 1: 135–148.

Christensen LW and Gorski RA (1978) Independent masculinization of neuroendocrine systems by intracerebral implants of testosterone or estradiol in the neonatal female rat. *Brain Research* 146: 325–340.

Chung WCJ, De Vries GJ, and Swaab D (2002) Sexual differentiation of the bed nucleus of the stria terminalis in humans may extend into adulthood. *Journal of Neuroscience* 22: 1027–1033.

Clark C, Klonoff H, and Hayden M (1990) Regional cerebral glucose metabolism in Turner syndrome. *Canadian Journal of Neurological Sciences* 17: 140–144.

Clark MM, Crews D, and Galef BG, Jr. (1991) Concentrations of sex steroid hormones in pregnant and fetal Mongolian gerbils. *Physiology and Behavior* 49: 239–243.

Clark MM and Galef BG, Jr. (1998) Effects of intrauterine position on the behavior and genital morphology of litter-bearing rodents. *Developmental Neuropsychology* 14: 197–211.

Clarke S, Kraftsik R, van der Loos H, and Innocenti GM (1989) Forms and measures of adult and developing human corpus callosum: Is there sexual dimorphism? *Journal of Comparative Neurology* 280: 213–220.

Cohen J (1988) *Statistical Power Analysis for the Behavioral Sciences*, 2nd edn. Hillsdale, NJ: Lawrence Erlbaum Associates.

Cohen-Kettenis PT (2005) Gender change in 46, XY persons with 5alpha-reductase-2 deficiency and 17beta-hydroxysteroid dehydrogenase-3 deficiency. *Archives of Sexual Behavior* 34: 399–410.

Collaer ML, Geffner M, Kaufman FR, Buckingham B, and Hines M (2002) Cognitive and behavioral characteristics of Turner syndrome: Exploring a role for ovarian hormones in female sexual differentiation. *Hormones and Behavior* 41: 139–155.

Collaer ML and Hines M (1995) Human behavioral sex differences: A role for gonadal hormones during early development? *Psychological Bulletin* 118: 55–107.

Collaer ML, Reimers S, and Manning J (2007) Visuospatial performance on an internet line judgment task and potential hormonal markers: Sex, sexual orientation, and 2D:4D. *Archives of Sexual Behavior* 36: 177–192.

Dalton K (1968) Antenatal progesterone and intelligence. *British Journal of Psychiatry* 114: 1377–1382.

Dalton K (1976) Prenatal progesterone and educational attainments. *British Journal of Psychiatry* 129: 438–442.

de Lacoste MC, Holloway RL, and Woodward DJ (1986) Sex differences in the fetal human corpus callosum. *Human Neurobiology* 5: 93–96.

de Lacoste-Utamsing C and Holloway RL (1982) Sexual dimorphism in the human corpus callosum. *Science* 216: 1431–1432.

Demeter S, Ringo JL, and Doty RW (1988) Morphometric analysis of the human corpus callosum and anterior commissure. *Human Neurobiology* 6: 219–226.

Diamond M and Sigmundson HK (1997) Sex reassignment at birth: Long-term review and clinical implications. *Archives of Pediatric and Adolescent Medicine* 151: 298–304.

Diamond MC, Dowling GA, and Johnson RE (1981) Morphological cerebral cortical asymmetry in male and female rats. *Experimental Neurology* 71: 261–268.

DiPietro JA (1981) Rough and tumble play: A function of gender. *Developmental Psychology* 17: 50–58.

Dittmann RW, Kappes ME, and Kappes MH (1992) Sexual behavior in adolescent and adult females with congenital adrenal hyperplasia. *Psychoneuroendocrinology* 17: 153–170.

Dittmann RW, Kappes MH, Kappes ME, et al. (1990) Congenital adrenal hyperplasia I: Gender-related behavior and attitudes in female patients and sisters. *Psychoneuroendocrinology* 15: 401–420.

Dubb A, Gur RC, Avants B, and Gee J (2003) Characterization of sexual dimorphism in the human corpus callosum. *Neuroimage* 20: 512–519.
Dunlap JL, Gerall AA, and Carlton SF (1978) Evaluation of prenatal androgen and ovarian secretions on receptivity in female and male rats. *Journal of Comparative and Physiological Psychology* 92: 280–288.
Eaton WO and Enns LR (1986) Sex differences in human motor activity level. *Psychological Bulletin* 100: 19–28.
Ehrhardt AA and Baker SW (1974) Fetal androgens, human central nervous system differentiation, and behavior sex differences. In: Friedman RC, Richart RM, and van de Wiele RL (eds.) *Sex Differences in Behavior*, pp. 33–52. New York: Wiley.
Ehrhardt AA, Epstein R, and Money J (1968) Fetal androgens and female gender identity in the early-treated adrenogenital syndrome. *Johns Hopkins Medical Journal* 122: 165–167.
Ehrhardt AA, Grisanti GC, and Meyer-Bahlburg HFL (1977) Prenatal exposure to medroxyprogesterone acetate (MPA) in girls. *Psychoneuroendocrinology* 2: 391–398.
Ehrhardt AA, Meyer-Bahlburg HFL, Rosen LR, et al. (1985) Sexual orientation after prenatal exposure to exogenous estrogen. *Archives of Sexual Behavior* 14: 57–77.
Ehrhardt AA, Meyer-Bahlburg HFL, Rosen LR, et al. (1989) The development of gender-related behavior in females following prenatal exposure to diethylstilbestrol (DES). *Hormones and Behavior* 23: 526–541.
Ehrhardt AA and Money J (1967) Progestin-induced hermaphroditism: IQ and psychosexual identity in a study of ten girls. *Journal of Sex Research* 3: 83–100.
Ekstrom RB, French JW, and Harman HH (1976) *Manual for Kit of Factor-Referenced Cognitive Tests*. Princeton, NJ: Educational Testing Service.
Elster AD, DiPersio DA, and Moody DM (1990) Sexual dimorphism of the human corpus callosum studied by magnetic resonance imaging: Fact, fallacy and statistical confidence. *Brain Development* 12: 321–325.
Ernst M, Maheu FS, Schroth E, et al. (2007) Amygdala function in adolescents with congenital adrenal hyperplasia: A model for the study of early steroid abnormalities. *Neuropsychologia* 45: 2104–2113.
Falter CM, Plaisted KC, and Davis G (2008) Visuo-spatial processing in autism – testing the predictions of extreme male brain theory. *Journal of Autism and Developmental Disorders* 38: 507–517.
Farah R and Reno G (1972) Congenital absence of the penis. *Journal of Urology* 107: 154–155.
Fausto-Sterling A (1992) *Myths of Gender*. New York: Basic Books.
Feingold A (1988) Cognitive gender differences are disappearing. *American Psychologist* 43: 95–103.
Feingold A (1994) Gender differences in personality: A meta-analysis. *Psychological Bulletin* 116: 429–256.
Finegan JK, Niccols GA, and Sitarenios G (1992) Relations between prenatal testosterone levels and cognitive abilities at 4 years. *Developmental Psychology* 28: 1075–1089.
Fitch RH, Berrebi AS, Cowell PE, Schrott LM, and Denenberg VH (1990) Corpus callosum: Effects of neonatal hormones on sexual dimorphism in the rat. *Brain Research* 515: 111–116.
Fitch RH, Cowell PE, Schrott LM, and Denenberg VH (1991) Corpus callosum: Ovarian hormones and feminization. *Brain Research* 542(2): 313–317.
Fitch RH and Denenberg VH (1998) A role for ovarian hormones in sexual differentiation of the brain. *Behavioral and Brain Sciences* 21: 311–352.
Forest MG, Cathiard AM, and Bertrand JA (1973) Evidence of testicular activity in early infancy. *Journal of Clinical Endocrinology and Metabolism* 41: 751–760.
Frost JA, Binder JR, Springer JA, et al. (1999) Language processing is strongly left lateralized in both sexes – evidence from functional MRI. *Brain* 122: 199–208.
Garron DC (1977) Intelligence among persons with Turner's syndrome. *Behavior Genetics* 7: 105–127.
Gastaud F, Bouvattier C, Duranteau L, et al. (2007) Impaired sexual and reproductive outcomes in women with classical forms of congenital adrenal hyperplasia. *Journal of Clinical Endocrinology and Metabolism* 92: 1391–1396.
Gerall AA, Dunlap JL, and Hendricks SE (1972) Effect of ovarian secretions on female behavior potentiality in the rat. *Journal of Comparative and Physiological Psychology* 82: 449–465.
Gitau R, Adams D, Fisk NM, and Glover V (2005) Fetal plasma testosterone correlates positively with cortisol. *Archives of Disease in Childhood Fetal and Neonatal Edition* 90: F166–F169.
Golombok S and Hines M (2002) Sex differences in social behavior. In: Smith PK and Hart CH (eds.) *Blackwell Handbook of Childhood Social Development*, pp. 117–136. Malden, MA: Blackwell.
Good CD, Lawrence K, Thomas NS, et al. (2003) Dosage-sensitive X-linked locus influences the development of amygdala and orbitofrontal cortex, and fear recognition in humans. *Brain* 126: 2431–2446.
Gorski RA, Gordon JH, Shryne JE, and Southam AM (1978) Evidence for a morphological sex difference within the medial preoptic area of the rat brain. *Brain Research* 148: 333–346.
Gorski RA, Harlan RE, Jacobson CD, Shryne JE, and Southam AM (1980) Evidence for the existence of a sexually dimorphic nucleus in the preoptic area of the rat. *Journal of Comparative Neurology* 193: 529–539.
Goy RW (1978) Development of play and mounting behaviour in female rhesus virilized prenatally with esters of testosterone or dihydrotestosterone. In: Chivers DJ and Herbert J (eds.) *Recent Advances in Primatology*, pp. 449–462. New York: Academic Press.
Goy RW and Deputte BL (1996) The effects of diethylstilbestrol (DES) before birth on the development of masculine behavior in juvenile female rhesus monkeys. *Hormones and Behavior* 30: 379–386.
Goy RW and McEwen BS (1980) *Sexual Differentiation of the Brain*. Cambridge, MA: MIT Press.
Grimshaw GM, Sitarenios G, and Finegan JK (1995) Mental rotation at 7 years: Relations with prenatal testosterone levels and spatial play experiences. *Brain and Cognition* 29: 85–100.
Groner JI and Zeigler MM (1996) Cloacal exstrophy. In: Puri P (ed.) *Newborn Surgery*, pp. 461–465. Oxford: Butterworth-Heinemann.
Grumbach MM, Hughes IA, and Conte FA (2003) Disorders of sex differentiation. In: Larsen PR, Kronenberg HM, Melmed S, and Polonsky KS (eds.) *Williams Textbook of Endocrinology,* 10th edn., pp. 842–1002. Philadelphia, PA: Saunders.
Grumbach MM and Styne DM (2003) Puberty: Ontogeny, neuroendocrinology, physiology and disorders. In: Larsen PR, Kronenberg HM, Melmed S, and Polonsky KS (eds.) *Williams Textbook of Endocrinology,* 10th edn., pp. 1115–1286. Philadelphia, PA: Saunders.
Habib M, Gayraud D, Oliva A, Regis J, Salamon G, and Khalil R (1991) Effects of handedness and sex on the morphology of the corpus callosum: A study with brain magnetic resonance imaging. *Brain and Cognition* 16: 41–61.
Haier RJ and Benbow CP (1995) Sex differences and lateralization in temporal lobe glucose metabolism during mathematical reasoning. *Developmental Neuropsychology* 11: 405–414.

Haier RJ, Jung RE, Yeo RA, Head K, and Alkire MT (2005) The neuroanatomy of general intelligence: Sex matters. *Neuroimage* 25: 320–327.

Halari R, Sharma T, Hines M, Andrew C, Simmons A, and Kumari V (2005) Comparable fMRI activity with differential behavioural performance on mental rotation and overt verbal fluency tasks in healthy men and women. *Experimental Brain Research* 169: 1–14.

Hall CM, Jones JA, Meyer-Bahlburg HFL, et al. (2004) Behavioral and physical masculinization are related to genotype in girls with congenital adrenal hyperplasia. *Journal of Clinical Endocrinology and Metabolism* 89: 419–424.

Halpern DF (2000) *Sex Differences in Cognitive Abilities*, 3rd edn. Mahwah, NJ: Erlbaum. Hartmann.

Halpern DF, Benbow CP, Geary DC, Gur RC, Hyde JS, and Gernsbacher MA (2007) The science of sex differences in science and mathematics. *Psychological Science in the Public Interest* 8: 1–51.

Hampson E, Rovet J, and Altmann D (1998) Spatial reasoning in children with congenital adrenal hyperplasia due to 21-hydroxylase deficiency. *Developmental Neuropsychology* 14: 299–320.

Harris JA, Vernon PA, and Boomsma DI (1998) The heritability if testosterone: A study of Dutch adolescent twins and their parents. *Behavior Genetics* 28: 165–171.

Heinonen OP (1973) Diethylstilbestrol in pregnancy: Frequency of exposure and usage patterns. *Cancer (Philadelphia)* 31: 573–577.

Helleday J, Bartfai A, Ritzen EM, and Forsman M (1994a) General intelligence and cognitive profile in women with congenital adrenal hyperplasia (CAH). *Psychoneuroendocrinology* 19: 343–356.

Helleday J, Edman G, Ritzen EM, and Siwers B (1993) Personality characteristics and platelet MAO activity in women with congenital adrenal hyperplasia (CAH). *Psychoneuroendocrinology* 18: 343–354.

Helleday J, Siwers B, Ritzen EM, and Hugdahl K (1994b) Normal lateralization for handedness and ear advantage in a verbal dichotic listening task in women with congenital adrenal hyperplasia (CAH). *Neuropsychologia* 32: 875–880.

Henderson BA and Berenbaum SA (1997) Sex-typed play in opposite-sex twins. *Developmental Psychobiology* 31: 115–123.

Herbst AL and Bern HA (1981) *Developmental Effects of Diethylstilbestrol (DES) in Pregnancy*. New York: Thieme-Stratton.

Herbst AL, Ulfeder H, and Poskanzer DC (1971) Adenocarcinoma of the vagina. Association of maternal stilbestrol therapy with tumor appearance in young women. *New England Journal of Medicine* 284: 878–881.

Herdt GH and Davidson J (1988) The Sambia 'Turnim-Man': Sociocultural and clinical aspects of gender formation in male pseudohermaphrodites with 5-alpha-reductase deficiency in Papua New Guinea. *Archives of Sexual Behavior* 17: 33–56.

Hier DB and Crowley WF (1982) Spatial ability in androgen-deficient men. *New England Journal of Medicine* 306: 1202–1205.

Hines M (1982) Prenatal gonadal hormones and sex differences in human behavior. *Psychological Bulletin* 92: 56–80.

Hines M (1998) Activation/organization, masculinization/feminization: What are they and how are they distinguished? *Behavioral and Brain Sciences* 21: 332–333.

Hines M (2004) *Brain Gender*. New York: Oxford University Press.

Hines M, Ahmed SF, and Hughes I (2003a) Psychological outcomes and gender-related development in complete androgen insensitivity syndrome. *Archives of Sexual Behavior* 32: 93–101.

Hines M, Allen LS, and Gorski RA (1992a) Sex differences in subregions of the medial nucleus of the amygdala and the bed nucleus of the stria terminalis. *Brain Research* 579: 321–326.

Hines M, Alsum P, Roy M, Gorski RA, and Goy RW (1987) Estrogenic contributions to sexual differentiation in the female guinea pig: Influences of diethylstilbestrol and tamoxifen on neural, behavioral and ovarian development. *Hormones and Behavior* 21: 402–417.

Hines M, Brook C, and Conway GS (2004) Androgen and psychosexual development: Core gender identity, sexual orientation and recalled childhood gender role behavior in women and men with congenital adrenal hyperplasia (CAH). *Journal of Sex Research* 41: 1–7.

Hines M, Chiu L, McAdams LA, Bentler PM, and Lipcamon J (1992b) Cognition and the corpus callosum: Verbal fluency, visuospatial ability and language lateralization related to midsagittal surface areas of callosal subregions. *Behavioral Neuroscience* 106: 3–14.

Hines M and Collaer ML (1993) Gonadal hormones and sexual differentiation of human behavior: Developments from research on endocrine syndromes and studies of brain structure. *Annual Review of Sex Research* 4: 1–48.

Hines M, Davis FC, Coquelin A, Goy RW, and Gorski RA (1985) Sexually dimorphic regions in the medial preoptic area and the bed nucleus of the stria terminalis of the guinea pig brain: A description and an investigation of their relationship to gonadal steroids in adulthood. *Journal of Neuroscience* 5: 40–47.

Hines M, Fane BA, Pasterski VL, Mathews GA, Conway GS, and Brook C (2003b) Spatial abilities following prenatal androgen abnormality: Targeting and mental rotations performance in individuals with congenital adrenal hyperplasia (CAH). *Psychoneuroendocrinology* 28: 1010–1026.

Hines M, Golombok S, Rust J, Johnston K, and Golding J ALSPAC Study Team (2002) Testosterone during pregnancy and childhood gender role behavior: A longitudinal population study. *Child Development* 73: 1678–1687.

Hines M and Gorski RA (1985) Hormonal influences on the development of neural asymmetries. In: Benson DF and Zaidel E (eds.) *The Dual Brain: Hemispheric Specialization in Humans*, pp. 75–96. New York: Guilford Press.

Hines M and Kaufman FR (1994) Androgen and the development of human sex-typical behavior: Rough-and-tumble play and sex of preferred playmates in children with congenital adrenal hyperplasia (CAH). *Child Development* 65: 1042–1053.

Hines M and Shipley C (1984) Prenatal exposure to diethylstilbestrol (DES) and the development of sexually dimorphic cognitive abilities and cerebral lateralization. *Developmental Psychology* 20: 81–94.

Ho KC, Roessmann U, Straumfjord JD, and Monroe G (1980) Analysis of brain weight: I. Adult brain weight in relation to sex, race and age. *Archives of Pathology and Laboratory Medicine* 104: 635–639.

Hofman MA, Fliers E, Goudsmit E, and Swaab DF (1988) Morphometric analysis of the suprachiasmatic and paraventricular nuclei in the human brain. *Journal of Anatomy* 160: 127–143.

Holloway RL and de Lacoste MC (1986) Sexual dimorphism in the human corpus callosum: An extension and replication study. *Human Neurobiology* 5: 87–91.

Hughes IA, Houk C, Ahmed SF, and Lee PA (2006) Consensus statement on management of intersex disorders. *Archives of Disease in Childhood* 91: 554–563.

Hurwitz RS and Manzoni GM (1997) Cloacal exstrophy. In: O'Donnell B and Kopp S (eds.) *Pediatric Urology,* 3rd edn., pp. 514–525. Oxford: Butterworth-Heinemann.
Hyde JS (1984) How large are gender differences in aggression? A developmental meta-analysis. *Developmental Psychology* 20: 722–736.
Hyde JS, Fennema E, and Lamon SJ (1990) Gender differences in mathematics performance: A meta-analysis. *Psychological Bulletin* 107: 139–155.
Hyde JS and Linn MC (1988) Gender differences in verbal ability: A meta-analysis. *Psychological Bulletin* 104: 53–69.
Iervolino AC (2003) *Genetic and Environmental Influences on Gender-Role Behaviour During the Preschool Years: A Study of 3- and 4-Year Old Twins*. PhD Thesis, University of London.
Imperato-McGinley J (1994) 5-Alpha-reductase deficiency: Human and animal models. *European Urology* 25(supplement 1): 20–23.
Imperato-McGinley J, Guerrero L, Gautier T, and Peterson RE (1974) Steroid 5-alpha-reductase deficiency in man: An inherited form of male pseudohermaphroditism. *Science* 186: 1213–1215.
Imperato-McGinley J, Peterson RE, Gautier T, and Sturla E (1979a) Androgens and the evolution of male-gender identity among male pseudohermaphrodites with 5 alpha reductase deficiency. *New England Journal of Medicine* 300: 1233–1237.
Imperato-McGinley J, Peterson RE, Stoller R, and Goodwin WE (1979b) Male pseudohermaphroditism secondary to 17-beta-dehydroxysteroid dehydrogenase deficiency: Gender role change with puberty. *Journal of Clinical Endocrinology and Metabolism* 49: 391–395.
Imperato-McGinley J, Pichardo M, Gautier T, Voyer D, and Bryden MP (1991) Cognitive abilities in androgen-insensitive subjects: Comparison with control males and females from the same kindred. *Clinical Endocrinology* 34: 341–347.
International Committee on Radiological Protection (1975) Report of the Task Group on Reference Man, ICRP Publication No. 23. New York: Pergamon Press.
Jacklin CN, Wilcox KT, and Maccoby EE (1988) Neonatal sex-steroid hormones and cognitive abilities at six years. *Developmental Psychobiology* 21: 567–574.
Jacobson CD, Csernus VJ, Shryne JE, and Gorski RA (1981) The influence of gonadectomy, androgen exposure, or a gonadal graft in the neonatal rat on the volume of the sexually dimorphic nucleus of the preoptic area. *Journal of Neuroscience* 1: 1142–1147.
Jancke L, Staiger JF, Schlaug G, Huang YX, and Steinmetz H (1997) The relationship between corpus callosum size and forebrain volume. *Cerebral Cortex* 7: 48–56.
Jardine R and Martin NG (1983) Spatial ability and throwing accuracy. *Behavior Genetics* 13: 331–340.
Johannsen TH, Ripa CPL, Reinisch JM, Schwartz M, Mortensen EL, and Main KM (2006) Impaired cognitive function in women with congenital adrenal hyperplasia. *Journal of Clinical Endocrinology and Metabolism* 91: 1376–1381.
Johnson AM, Wadsworth J, Wellings K, Bradshaw S, and Field J (1992) Sexual lifestyles and HIV risk. *Nature* 360: 410–412.
Juraska JM (1991) Sex differences in 'cognitive' regions of the rat brain: Environmental influences. *Psychoneuroendocrinology* 16: 105–119.
Juraska JM and Kopcik JR (1988) Sex and environmental influences on the size and ultrastructure of the rat corpus callosum. *Brain Research* 450: 1–8.
Kelso WM, Nicholls MER, and Warne GL (1999) Effects of prenatal androgen exposure on cerebral lateralization in patients with congenital adrenal hyperplasia (CAH). *Brain and Cognition* 40: 156.
Kelso WM, Nicholls MER, Warne GL, and Zacharin M (2000) Cerebral lateralization and cognitive functioning in patients with congenital adrenal hyperplasia. *Neuropsychology* 14: 370–378.
Kempermann G, Brandon EP, and Gage FH (1998) Environmental stimulation of 129/SvJ mice causes increased cell proliferatin and neurogenesis in the adult dentate gyrus. *Current Biology* 8: 939–942.
Kempermann G, Kuhn HG, and Gage FH (1997) More hippocampal neurons in adult mice living in an enriched environment. *Nature* 386: 493–495.
Kertesz A, Polk M, Howell J, and Black SE (1987) Cerebral dominance, sex, and callosal size in MRI. *Neurology* 37: 1385–1388.
Kessler WO and McLaughlin AP (1973) Agenesis of penis: Embryology and management. *Urology* 1: 226–229.
Kester P, Green R, Finch SJ, and Williams K (1980) Prenatal female hormone administration and psychosexual development in human males. *Psychoneuroendocrinology* 5: 269–285.
Kester PA (1984) Effects of prenatally administered 17 alpha hydroxyprogesterone caproate on adolescent males. *Archives of Sexual Behavior* 13: 441–455.
Kinsey A, Pomeroy W, and Martin C (1948) *Sexual Behavior in the Human Male*. Philadelphia, PA: Saunders.
Kinsey A, Pomeroy W, Martin C, and Gebhard PH (1953) *Sexual Behavior in the Human Female*. Philadelphia, PA: Saunders.
Kirshbaum JD (1950) Congenital absence of the external genitals (persistent primitive cloaca). *Journal of Pediatrics* 37: 102–105.
Knickmeyer R, Baron-Cohen S, Fane B, et al. (2006a) Androgen and autistic traits: A study of individuals with congenital adrenal hyperplasia. *Hormones and Behavior* 50: 148–153.
Knickmeyer R, Baron-Cohen S, Raggatt P, and Taylor K (2005) Foetal testosterone, social relationships, and restricted interests in children. *Journal of Child Psychology and Psychiatry* 45: 1–13.
Knickmeyer R, Baron-Cohen S, Raggatt P, Taylor K, and Hackett G (2006b) Fetal testosterone and empathy. *Hormones and Behavior* 49: 282–292.
Kolb B and Whishaw IQ (1985) *Fundamentals of Human neuropsychology,* 2nd edn. New York: W.H. Freeman.
Kuhnle U and Bullinger M (1997) Outcome of congenital adrenal hyperplasia. *Pediatric Surgery International* 12: 511–515.
Lasco MS, Jordan TJ, Edgar MA, Petito CK, and Byne W (2002) A lack of dimorphism of sex or sexual orientation in the human anterior commissure. *Brain Research* 936: 95–98.
Lev-Ran A (1974) Sexuality and educational levels of women with the late-treated adrenogenital syndrome. *Archives of Sexual Behavior* 3: 27–32.
LeVay S (1991) A difference in hypothalamic structure between heterosexual and homosexual men. *Science* 253: 1034–1037.
Leveroni CL and Berenbaum SA (1998) Early androgen effects on interest in infants: Evidence from children with congenital adrenal hyperplasia. *Developmental Neuropsychology* 14: 321–340.
Linn MC and Petersen AC (1985) Emergence and characterization of sex differences in spatial ability: A meta-analysis. *Child Development* 56: 1479–1498.
Lippe B (1991) Turner syndrome. *Endocrinology and Metabolism Clinics of North America* 20: 121–152.
Lish JD, Ehrhardt AA, Meyer-Bahlburg HL, Rosen LR, Gruen RS, and Veridiano NP (1991) Gender-related behavior development in females exposed to diethylstilbestrol (DES) *in utero*: An attempted replication. *Journal of the American Academy of Child and Adolescent Psychiatry* 30: 29–37.

Lish JD, Meyer-Bahlburg HFL, Ehrhardt AA, Travis BG, and Veridiano NP (1992) Prenatal exposure to diethylstilbestrol (DES): Childhood play behavior and adult gender-role behavior in women. *Archives of Sexual Behavior* 21: 423–441.

Loehlin JC (2000) Group differences in intelligence. In: Sternberg RJ (ed.) *Handbook of Intelligence*, pp. 176–193. Cambridge: Cambridge University Press.

Lutchmaya S, Baron-Cohen S, and Raggatt P (2002a) Foetal testosterone and eye contact in 12-month-old human infants. *Infant Behavior and Development* 25: 327–335.

Lutchmaya S, Baron-Cohen S, and Raggatt P (2002b) Foetal testosterone and vocabulary size in 18- and 24-month-old infants. *Infant Behavior and Development* 24: 418–442.

Lynch A and Mychalkiw W (1978) Prenatal progesterone II. Its role in the treatment of pre-eclamptic toxaemia and its effect on the offspring's intelligence: A reappraisal. *Early Human Development* 2: 323–339.

Lynch A, Mychalkiw W, and Hutt SJ (1978) Prenatal progesterone I. Its effect on development and on intellectual and academic achievement. *Early Human Development* 2: 305–322.

Maccoby EE (1988) Gender as a social category. *Developmental Psychology* 24: 755–765.

Maccoby EE and Jacklin CN (1974) *The Psychology of Sex Differences*. Stanford, CA: Stanford University Press.

Maccoby EE and Jacklin CN (1987) Gender segregation in children. In: Reece HW (ed.) *Advances in Child Development and Behavior*, pp. 239–287. New York: Academic Press.

MacLusky N, Lieberburg I, and McEwen BS (1979) The development of estrogen receptor systems in the rat brain: Perinatal development. *Brain Research* 178: 129–142.

MacLusky NJ and Naftolin F (1981) Sexual differentiation of the central nervous system. *Science* 211: 1294–1303.

Madeira MD, Sousa N, Santer RM, Paulabarbosa MM, and Gunderson HJG (1995) Age and sex do not affect the volume, cell numbers or cell-size of the suprachiasmatic nucleus of the rat – an unbiased stereological study. *Journal of Comparative Neurology* 361: 585–601.

Malouf MA, Migeon CJ, Carson KA, Pertrucci L, and Wisniewski AB (2006) Cognitive outcome in adult women affected by congenital adrenal hyperplasia due to 21-hydroxylase deficiency. *Hormone Research* 65: 142–150.

Manning J, Churchill AJG, and Peters M (2007) The effects of sex, ethnicity, and sexual orientation on self-measured digit ratio (2D:4D). *Archives of Sexual Behavior* 36: 223–233.

Mansour CS, Haier RJ, and Buchsbaum MS (1996) Gender comparisons of cerebral glucose metabolic rate in healthy adults during a cognitive task. *Personality and Individual Differences* 20: 183–191.

Masica DN, Money J, and Ehrhardt A (1971) Fetal feminization and female gender identity in the testicular feminizing syndrome of androgen insensitivity. *Archives of Sexual Behavior* 1: 131–142.

Mathews GA, Fane B, Pasterski VL, Conway GS, Brook C, and Hines M (2004) Androgenic influences on neural asymmetry: Handedness and language lateralization in congenital adrenal hyperplasia (CAH). *Psychoneuroendocrinology* 29: 810–822.

Mathews GA, Fane BA, Conway GS, Brook C, Hines M (in press) Prenatal androgen abnormality and personality development. *Hormones and Behavior*.

Mazur T (2005) Gender dysphoria and gender change in androgen insensitivity or micropenis. *Archives of Sexual Behavior* 34: 411–421.

McCarthy MM (2008) Estradiol and the developing brain. *Physiological Reviews* 88: 91–134.

McFadden D, Loehlin JC, Breedlove SM, Lippa RA, Manning JT, and Rahman Q (2005) A reanalysis of five studies on sexual orientation and the relative length of the 2nd and 4th fingers (the 2D:4D ratio). *Archives of Sexual Behavior* 34: 341–356.

McGlone J (1980) Sex differences in human brain asymmetry: A critical survey. *Behavioral and Brain Sciences* 3: 215–263.

McGuire LS and Omenn GS (1975) Congenital adrenal hyperplasia: I. Family studies of IQ. *Behavioral Genetics* 5: 165–173.

McGuire LS, Ryan KO, and Omenn GS (1975) Congenital adrenal hyperplasia II: Cognitive and behavioral studies. *Behavior Genetics* 5: 175–188.

Meaney MJ and Stewart J (1981) Neonatal androgens influence the social play of prepubescent rats. *Hormones and Behavior* 15: 197–213.

Meisel RL and Ward IL (1981) Fetal female rats are masculinized by littermates located caudally in the uterus. *Science* 213: 239–242.

Meyer-Bahlburg HFL (2005) Gender identity outcome in female-raised 46, XY persons with penile agenesis, cloacal exstrophy of the bladder, or penile ablation. *Archives of Sexual Behavior* 34: 423–438.

Meyer-Bahlburg HFL, Dolezal C, Baker SW, and New MI (2008) Sexual orientation in women with classical or non-classical congenital adrenal hyperplasia as a function of degree of prenatal androgen excess. *Archives of Sexual Behavior* 37: 85–99.

Meyer-Bahlburg HFL, Ehrhardt AA, Rosen LR, et al. (1995) Prenatal estrogens and the development of homosexual orientation. *Developmental Psychology* 31: 12–21.

Meyer-Bahlburg HFL, Ehrhardt AA, Whitehead ED, and Vann FH (1987) Sexuality in males with a history of prenatal exposure to diethylstilbestrol (DES). In: *Proceedings of the Workshop on Psychosexual and Reproductive Issues Affecting Patients with Cancer–1987*, pp. 79–82. New York: American Cancer Society.

Meyer-Bahlburg HFL, Grisanti GC, and Ehrhardt AA (1977) Prenatal effects of sex hormones on human male behavior: Medroxyprogesterone acetate (MPA). *Psychoneuroendocrinology* 2: 383–390.

Meyer-Bahlburg HFL, Gruen RS, New MI, et al. (1996) Gender change from female to male in classical congenital adrenal hyperplasia. *Hormones and Behavior* 30: 319–332.

Molko N, Cachia A, Riviere D, et al. (2004) Brain anatomy in Turner syndrome: Evidence for impaired social and spatial-numerical networks. *Cerebral Cortex* 14: 840–850.

Money J (1976) Gender identity and hermaphroditism. *Science* 191: 872.

Money J and Daléry J (1976) Iatrogenic homosexuality: Gender identity in seven 46, XX chromosomal females with hyperadrenocortical hermaphroditism born with a penis, three reared as boys, four reared as girls. *Journal of Homosexuality* 1: 357–371.

Money J and Ehrhardt A (1972) *Man and Woman: Boy and Girl*. Baltimore, MD: Johns Hopkins University Press.

Money J and Lewis V (1966) IQ, genetics and accelerated growth: Adrenogenital syndrome. *Johns Hopkins Hospital Bulletin* 118: 365–373.

Money J, Schwartz M, and Lewis V (1984) Adult erotosexual status and fetal hormonal masculinization and demasculinization: 46 XX congenital virilizing adrenal hyperplasia and 46 XY androgen-insensitivity syndrome compared. *Psychoneuroendocrinology* 9: 405–414.

Mulaikal RM, Migeon CJ, and Rock JA (1987) Fertility rates in female patients with congenital adrenal hyperplasia due to 21-hydroxylase deficiency. *New England Journal of Medicine* 316: 178–182.

Murphy DGM, DeCarli C, McIntosh AR, et al. (1996) Sex differences in human brain morphometry and metabolism: An *in vivo* quantitative magnetic resonance imaging and positron emission tomography study on the effect of aging. *Archives of General Psychiatry* 53: 585–594.

Murphy DGM, Mentis MJ, Pietrini P, et al. (1997) A PET study of Turner's syndrome: Effects of sex steroids and the X chromosome on brain. *Biological Psychiatry* 41: 285–298.

Nasrallah HA, Andreasen NC, Coffman JA, Olson SC, Dunn V, and Ehrhardt JC (1986) The corpus callosum is not larger in left-handers. *Society for Neuroscience Abstracts* 12: 720.

Nass R, Baker S, Speiser P, et al. (1987) Hormones and handedness: Left-hand bias in female congenital adrenal hyperplasia patients. *Neurology* 37: 711–715.

New M (1998) Diagnosis and management of congenital adrenal hyperplasia. *Annual Review of Medicine* 49: 311–328.

Noller KL and Fish CR (1974) Diethylstilbestrol usage: Its interesting past, important present and questionable future. *Medical Clinics of America* 58: 793–810.

Nordenstrom A, Servin A, Bohlin G, Larsson A, and Wedell A (2002) Sex-typed toy play behavior correlates with the degree of prenatal androgen exposure assessed by CYP21 genotype in girls with congenital adrenal hyperplasia. *Journal of Clinical Endocrinology and Metabolism* 87: 5119–5124.

O'Kusky J, Strauss E, Kosaka B, et al. (1988) The corpus callosum is larger with right-hemisphere cerebral speech dominance. *Annals of Neurology* 24: 379–383.

Okten A, Kalyonku M, and Yaris N (2002) The ratio of second- and fourth-digit lengths and congenital adrenal hyperplasia due to 21-hydroxylase deficiency. *Early Human Development* 70: 47–54.

Pang S, Levine LS, Cederqvist LL, et al. (1980) Amniotic fluid concentrations of delta 5 and delta 4 steroids in fetuses with congenital adrenal hyperplasia due to 21-hydroxylase deficiency and in anencephalic fetuses. *Journal of Clinical Endocrinology and Metabolism* 51: 223–229.

Pang S, Levine LS, Chow DM, Faiman C, and New MI (1979) Serum androgen concentrations in neonates and young infants with congenital adrenal hyperplasia due to 21-hydroxylase deficiency. *Clinical Endocrinology* 11: 575–584.

Pasterski VL, Geffner ME, Brain C, Hindmarsh P, Brook C, and Hines M (2005) Prenatal hormones and postnatal socialization by parents as determinants of male-typical toy play in girls with congenital adrenal hyperplasia. *Child Development* 76: 264–278.

Pasterski VL, Hindmarsh P, Geffner M, Brook C, Brain C, and Hines M (2007) Increased aggression and activity level in 3- to 11-year-old girls with congenital adrenal hyperplasia (CAH). *Hormones and Behavior* 52: 368–374.

Perlman SM (1973) Cognitive abilities of children with hormone abnormalities: Screening by psychoeducational tests. *Journal of Learning Disabilities* 6: 21–29.

Peters M, Manning J, and Reimers S (2007) The effects of sex, sexual orientation, and digit ratio (2D:4D) on mental rotation performance. *Archives of Sexual Behavior* 36: 251–260.

Phoenix CH, Goy RW, Gerall AA, and Young WC (1959) Organizing action of prenatally administered testosterone propionate on the tissues mediating mating behavior in the female guinea pig. *Endocrinology (Baltimore)* 65: 163–196.

Plante E, Boliek C, Binkiewicz A, and Erly WK (1996) Elevated androgen, brain development and language/learning disabilities in children with congenital adrenal hyperplasia. *Developmental Medicine and Child Neurology* 38: 423–437.

Puts DA, McDaniel MA, Jordan CL, and Breedlove SM (2008) Spatial ability and prenatal androgens: Meta-analyses of congenital adrenal hyperplasia and digit ratio (2D:4D) studies. *Archives of Sexual Behavior* 37: 100–111.

Quadagno DM, Briscoe R, and Quadagno JS (1977) Effects of perinatal gonadal hormones on selected nonsexual behavior patterns: A critical assessment of the nonhuman and human literature. *Psychological Bulletin* 84: 62–80.

Reinarz SJ, Coffman CE, Smoker WRK, and Godersky FC (1988) MR imaging of the corpus callosum: Normal and pathologic findings and correlation with CT. *American Journal of Radiology* 151: 791–798.

Reiner WG and Gearhart JP (2004) Discordant sexual identity in some genetic males with cloacal exstrophy assigned to female sex at birth. *New England Journal of Medicine* 350: 333–341.

Reiner WG, Gearhart JP, and Jeffs R (1999) Psychosexual dysfunction in males with genital anomalies: Late adolescence, Tanner stages IV to VI. *Journal of the American Academy of Child and Adolescent Psychiatry* 38: 865–872.

Reinisch JM (1981) Prenatal exposure to synthetic progestins increases potential for aggression in humans. *Science* 211: 1171–1173.

Reinisch JM and Karow WG (1977) Prenatal exposure to synthetic progestins and estrogens: Effects on human development. *Archives of Sexual Behavior* 6: 257–288.

Reinisch JM and Sanders SA (1992) Effects of prenatal exposure to diethylstilbestrol (DES) on hemispheric laterality and spatial ability in human males. *Hormones and Behavior,* 26: 62–75.

Reiss AL, Freund LS, and Plotnick LP, et al. (1993) The effects of X monosomy on brain development: Monozygotic twins discordant for Turner's syndrome. *Annals of Neurology* 34: 95–107.

Reiss AL, Mazzocco MM, Greenlaw R, Freund LS, and Ross JL (1995) Neurodevelopmental effects of X monosomy. A volumetric imaging study. *Annals of Neurology* 38: 731–738.

Resnick SM, Berenbaum SA, Gottesman II, and Bouchard T (1986) Early hormonal influences on cognitive functioning in congenital adrenal hyperplasia. *Developmental Psychology* 22: 191–198.

Reyes FI, Winter JSD, and Faiman C (1973) Studies on human sexual development. I. Fetal gonadal and adrenal sex steroids. *Journal of Clinical Endocrinology and Metabolism* 37: 74–78.

Richart RM and Benirschke K (1960) Penile agenesis: Report of case, review of world literature and discussion of pertinent embryology. *Archives of Pathology* 70: 252–260.

Rosler A and Kohn G (1983) Male pseudohermaphroditism due to 17 beta-hydroxysteroid dehydrogenase deficiency: Studies on the natural history of the defect and effect of androgens on gender role. *Journal of Steroid Biochemistry* 19: 663–674.

Ross JL and Zinn AR (1999) Turner syndrome: Potential hormonal and genetic influences on the neurocognitive profile. In: Tager-Flusberg H (ed.) *Neurodevelopmental Disorders*, pp. 251–267. Cambridge, MA: MIT Press.

Rovet JF (1990) The cognitive and neuropsychological characteristics of females with Turner syndrome. In: Berch DB and Bender BG (eds.) *Sex Chromosome Abnormalities and Human Behavior*, pp. 38–77. Boulder, CO: Westview.

Rust J, Golombok S, Hines M, Johnston K, Golding J, and ALSPAC Study Team (2000) The role of brothers and sisters in the gender development of preschool children. *Journal of Experimental Child Psychology* 77: 292–303.

Rutter M (1978) Language disorder and infantile autism. In: Rutter M and Schopler E (eds.) *Autism: A Reappraisal of Concepts and Treatment*, pp. 1–26. New York: Plenum.

Rutter M, Caspi A, and Moffitt TE (2003) Using sex differences in psychopathology to study causal mechanisms: Unifying issues and research strategies. *Journal of Child Psychology and Psychiatry* 44: 1092–1115.

Sanders B, Soares MP, and D'Aquila JM (1982) The sex difference on one test of spatial visualization: A nontrivial difference. *Child Development* 53: 1106–1110.

Sarkar P, Bergman K, Fisk NM, O'Connor TG, and Glover V (2007) Amniotic fluid testosterone: Relationship with cortisol and gestational age. *Clinical Endocrinology* 67: 743–747.

Satz P, Orsini DL, Saslow E, and Henry R (1985) Early brain injury and pathological left-handedness: Clues to a syndrome. In: Benson DF and Zaidel E (eds.) *The Dual Brain: Hemispheric Specialization in Humans*, pp. 117–125. New York: Guilford.

Schachter SC (1994) Handedness in women with intrauterine exposure to diethystilbesterol. *Neuropsychologia* 32: 619–623.

Scheirs JGM and Vingerhoets AJJM (1995) Handedness and other laterality indices in women prenatally exposed to DES. *Journal of Clinical and Experimental Neuropsychology* 17: 725–730.

Schober JM (1999) Quality-of-life studies in patients with ambiguous genitalia. *World Journal of Urology* 17: 249.

Schober JM, Carmichael PA, Hines M, and Ransley PG (2002) The ultimate challenge of cloacal exstrophy. *Journal of Urology* 167: 300–304.

Sell RL, Wells JA, and Wypij D (1995) The prevalence of homosexual behavior and attraction in the United States, the United Kingdom and France: Results of national population-based samples. *Archives of Sexual Behavior* 24: 235–248.

Shaywitz BA, Shaywitz SE, Pugh KR, et al. (1995) Sex differences in the functional organization of the brain for language. *Nature* 373: 607–609.

Sinforiani E, Livieri C, Mauri M, et al. (1994) Cognitive and neuroradiological findings in congenital adrenal hyperplasia. *Psychoneuroendocrinology* 19: 55–64.

Singh RP and Carr DH (1966) The anatomy and histology of XO human embryos and fetuses. *Anatomical Record* 155: 369–384.

Slijper FME, Drop SLS, Molenaar JC, and de Muinck Keizer-Schrama SMPF (1998) Long-term psychological evaluation of intersex children. *Archives of Sexual Behavior* 27: 125–144.

Smith LL and Hines M (2000) Language lateralization and handedness in women prenatally exposed to diethylstilbestrol (DES). *Psychoneuroendocrinology* 25: 497–512.

Snow ME, Jacklin CN, and Maccoby EE (1983) Sex of child differences in father–child interaction at one year of age. *Child Development* 54: 227–232.

Spira A, Bajos N, Bejin A, et al. (1992) AIDS and sexual behavior in France. *Nature* 360: 407–409.

Spreen O and Strauss E (1991) *A Compendium of Neuropsychological Tests*. New York: Oxford University Press.

Stumpf WE, Sar M, and Keefer DA (1975) Atlas of estrogen target cells in rat brain. In: Stumpf WE and Grant CD (eds.) *Anatomical Neuroendocrinology*, pp. 104–119. Basel: Karger.

Sutton-Smith B, Rosenberg BG, and Morgan EF, Jr. (1963) Development of sex differences in play choices during preadolescence. *Child Development* 34: 119–126.

Swaab D and Hofman M (1988) Sexual differentiation of the human hypothalamus: Ontogeny of the sexually dimorphic nucleus of the preoptic area. *Developmental Brain Research* 44: 314–318.

Swaab DF and Fliers E (1985) A sexually dimorphic nucleus in the human brain. *Science* 228: 1112–1115.

Swaab DF, Fliers E, and Partiman TS (1985) The suprachiasmatic nucleus of the human brain in relationship to sex, age and senile dementia. *Brain Research* 342: 37–44.

Swaab DF and Hofman MA (1990) An enlarged suprachiasmatic nucleus in homosexual men. *Brain Research* 537: 141–148.

Tanner JM, Whitehouse RH, and Takaishi M (1966) Standards from birth to maturity for height, weight, height velocity and weight velocity: British children, 1965. *Archives of Disease in Childhood* 41: 454–471.

Temple CM and Carney R (1996) Reading skills in children with Turner's syndrome: An analysis of hyperlexia. *Cortex* 32: 335–345.

Tirosh E, Rod R, Cohen A, and Hochberg Z (1993) Congenital adrenal hyperplasia and cerebral lateralizations. *Pediatric Neurology* 9: 198–201.

Titus-Ernstoff L, Perez K, Hatch E, et al. (2003) Psychosexual characteristics of men and women exposed prenatally to diethylstilbestrol. *Epidemiology* 14: 155.

Tobet SA, Zahniser DJ, and Baum MJ (1986) Sexual dimorphism in the preoptic/anterior hypothalamic area of ferrets: Effects of adult exposure to sex steroids. *Brain Research* 364: 249–257.

Toran-Allerand CD (1984) On the genesis of sexual differentiation of the central nervous system: Morphogenetic consequences of steroidal exposure and the possible role of alpha-fetoprotein. *Progress in Brain Research* 61: 63–98.

van de Beek C, Van Goozen SHM, Buitelaar JK, and Cohen-Kettenis PT (2008) Prenatal sex hormones (maternal and amniotic fluid) and gender-related play behavior in 13-month-old infants. *Archives of Sexual Behavior* (doi:10.1007/s10508-007-9291-z).

Voyer D (1996) On the magnitude of laterality effects and sex differences in functional lateralities. *Laterality* 1: 51–83.

Voyer D, Voyer S, and Bryden MP (1995) Magnitude of sex differences in spatial abilities: A meta-analysis and consideration of critical variables. *Psychological Bulletin* 117: 250–270.

Waber DP (1979) Neuropsychological aspects of Turner's syndrome. *Developmental Neurology* 1979: 58–70.

Watson NV and Kimura D (1991) Nontrivial sex differences in throwing and intercepting: Relation to psychometrically-defined spatial functions. *Personality and Individual Differences* 12: 375–385.

Wenzel U, Schneider M, Zachmann M, Knorr-Murset G, Weber A, and Prader A (1978) Intelligence of patients with congenital adrenal hyperplasia due to 21-hydroxylase deficiency, their parents and unaffected siblings. *Helvetica Paediatrica Acta* 33: 11–16.

Whitcomb RW and Crowley WF (1993) Male hypogonadotropic hypogonadism. *Endocrinology and Metabolism Clinics of North America* 22: 125–163.

White BJ (1994) The Turner syndrome: Origin, cytogenetic variants, and factors influencing the phenotype. In: Broman SH and Grafman J (eds.) *Atypical Cognitive Deficits in Developmental Disorders: Implications for Brain Function*, pp. 183–195. Hillsdale, NJ: Erlbaum.

Wilcox AJ, Maxey J, and Herbst AL (1992) Prenatal hormone exposure and performance on college entrance examinations. *Hormones and Behavior* 24: 433–439.

Williams TJ, Pepitone ME, Christensen BM, et al. (2000) Finger length patterns indicate an influence of fetal androgens on human sexual orientation. *Nature* 30: 455–456.

Wilson JD (1979) Sex hormones and sexual behavior. *New England Journal of Medicine* 300: 1269–1270.

Wilson JD (2001) Androgens, androgen receptors and male gender role behavior. *Hormones and Behavior* 40: 358–366.

Wilson JD, George FW, and Griffin JE (1981) The hormonal control of sexual development. *Science* 211: 1278–1284.

Wilson JD, Griffin JE, and Russell DW (1993) Steroid 5 alpha reductase 2 deficiency. *Endocrine Reviews* 14: 577–593.

Wing L (1981) Asperger syndrome: A clinical account. *Psychological Medicine* 11: 115–130.

Winter JSD, Hughes IA, Reyes FI, and Faiman C (1976) Pituitary–gonadal relations in infancy: 2. Patterns of serum gonadal steroid concentration sin man from birth to two years of age. *Journal of Clinical Endocrinology and Metabolism* 42: 679–686.

Wisniewski AB, Migeon CJ, Meyer-Bahlburg HFL, et al. (2000) Complete androgen insensitivity syndrome. Long-term medical, surgical, and psychosexual outcome. *Journal of Clinical Endocrinology and Metabolism* 85: 2664–2669.

Witelson SF (1985) The brain connection: The corpus callosum is larger in left-handers. *Science* 229: 665–668.

Witelson SF, Glezer II, and Kigar DL (1995) Women have greater density of neurons in posterior temporal cortex. *Journal of Neuroscience* 15: 3418–3428.

Witelson SF, Kigar DL, Scamvougeras A, et al. (2008) Corpus callosum anatomy in right-handed homosexual and heterosexual men. *Archives of Sexual Behavior* (doi:10.1007/s10508-007-9276-y).

Witelson WF (1989) Hand and sex differences in the isthmus and genu of the human corpus callosum: A postmortem morphological study. *Brain* 112: 799–835.

Wudy S, Dorr HG, Solleder C, Djalali M, and Homoki J (1999) Profiling steroid hormones in amniotic fluid of midpregnancy by routine stable isotope dilution/gas chromatography-mass spectrometry: Reference values and concentrations in fetuses at risk for 21-hydroxylase deficiency. *Journal of Clinical Endocrinology and Metabolism* 84: 2724–2728.

Yalom ID, Green R, and Fisk N (1973) Prenatal exposure to female hormones: Effect on psychosexual development in boys. *Archives of General Psychiatry* 28: 554–561.

Young HH, Cockett ATK, Stoller R, Ashley FL, and Goodwin WE (1971) The management of agenesis of the phallus. *Pediatrics* 47: 81–87.

Zhou J, Hofman MA, Gooren LJG, and Swaab DF (1995) A sex difference in the human brain and its relation to transsexuality. *Nature* 378: 68–70.

Zinn AR, Page DC, and Fisher EMC (1993) Turner syndrome: The case of the missing sex chromosome. *Trends in Genetics* 9: 90–93.

Zucker KJ (2002) Intersexuality and gender identity differentiation. *Journal of Pediatric Adolescent Gynecology* 15: 3–13.

Zucker KJ, Bradley SJ, Oliver G, Blake J, Fleming S, and Hood J (1996) Psychosexual development of women with congenital adrenal hyperplasia. *Hormones and Behavior* 30: 300–318.

9 Human Puberty: Physiology and Genetic Regulation

B A Kaminski, Rainbow Babies and Children's Hospital, Cleveland, OH, USA
M R Palmert, The Hospital for Sick Children, Toronto, ON, Canada

Chapter Outline

Glossary

constitutional delay in growth and puberty (CDGP) A variant of normal growth identified by prepubertal short stature and delay in pubertal onset. Growth and development are appropriately correlated with bone age, which is delayed, rather than chronologic age. Catch-up growth, pubertal onset, and pubertal growth spurt occur later than average, but result in normal adult stature and sexual development.

idiopathic hypogonadotropic hypogonadism (IHH) Absolute or functional deficiency of hypothalamic gonadotropin-releasing hormone (GnRH) leading to absence of pituitary luteinizing hormone (LH) and follicle-stimulating hormone (FSH). The gonads are not stimulated to secrete sex steroids (estrogen, testosterone) and normal pubertal development does not occur.

Kallmann syndrome IHH with anosmia/hyposmia.

9.1 Introduction

Puberty is an important developmental process and life stage that leads to sexual maturation and reproductive capability. Although the physiology and progression of puberty is relatively common among individuals, the timing of puberty varies across the normal population and is affected by both genetic and environmental influences. The alterations in human behavior related to the hormonal changes of puberty have been extensively reviewed in the previous edition of this chapter (Styne and Grumbach, 2002) with emphasis on the available data relating pubertal development to associated behaviors, as well as the limitations of those data. We will not re-review that literature in depth, but instead focus on a discussion of normal physiology and the regulation of pubertal development, with emphasis on genetics, a topic that was not stressed in the previous review. This discussion is followed by a brief summary of recent data about behavior as it relates to timing of pubertal development.

9.2 Prepubertal Development

During the prenatal, infant, and prepubertal stages of life, the hypothalamic–pituitary–gonadal (HPG) axis undergoes several periods of changing activity. It is active *in utero*, has a peak of activity during infancy, and then enters a period of relative quiescence until activity resumes during puberty.

9.2.1 Prenatal and Postnatal Development

The HPG axis begins development in the early–mid-gestational period with the hypothalamus producing gonadotropin-releasing hormone (GnRH) at around 10 weeks and the pituitary producing luteinizing hormone (LH) and follicle-stimulating hormone (FSH) by 10–15 weeks gestation (Kaplan et al., 1976). As a result, human exposure to sex steroid hormones begins *in utero* and the fetus develops in an active steroid hormone environment with relatively high levels of estrogen and testosterone.

In the immediate postnatal period, there is a brief surge of gonadotropins. During the first few months after birth, FSH and LH increase again and exhibit a pulsatile pattern, with higher FSH and estradiol levels in females, and higher LH and testosterone levels in males. In females the FSH levels peak by 3–6 months, with a decline to prepubertal levels by 12–24 months. The LH increase is less pronounced in females while in males the LH surge predominates, beginning at 1–2 weeks of life, peaking at 1–2 months, and declining by 4–6 months. In males, FSH levels peak at 3 months and decline by 9 months (Quigley, 2002).

This early exposure to gonadotropins and sex steroids, both *in utero* and during infancy (which is often described as mini-puberty), can have physical consequences. In some neonates this hormonal stimulation is evidenced by the estrogenic effects of palpable breast buds (in both males and females), and hypertrophic labia minora. More rarely, the fall in estrogen levels due to removal from the maternal sex steroid environment may result in withdrawal menstrual bleeding in some females, while others may have a prominent clitoral shaft reflecting androgenic stimulation.

9.2.2 The Juvenile Pause

The juvenile pause is the period of relative quiescence of the HPG axis between infancy and the initiation of puberty. Although it can start later, the juvenile pause typically begins by 6 months of age in males and 12 months of age in females, as the levels of gonadotropins and sex steroids decrease to the low levels that characterize mid-childhood before pubertal maturation occurs (Styne, 1994).

9.2.3 Ontogeny of Gonadotropin Secretion

Pituitary LH and FSH are released in pulsatile fashion and are evoked by the pulsatile secretion of GnRH from the hypothalamus. In the infant, high-amplitude pulses are generated (Waldhauser et al., 1981), while during the juvenile pause, the amplitude and frequency of the pulses, which occur mainly at night, are greatly reduced (Dunkel et al., 1990). Studies over the last 15 years have demonstrated that the subsequent transition from childhood quiescence to pubertal patterns of GnRH secretion is gradual, with small but progressive increases in LH and FSH until the beginning of puberty when secretion increases markedly (Palmert and Boepple, 2001). As puberty begins, the pulses continue to be mainly sleep associated, with progression to daytime pulses in late puberty and into adulthood (Boyar et al., 1972). In the monkey it has been demonstrated that the amplitude, and to a lesser extent the frequency, of the sleep-associated pulses both increase from prepuberty to puberty, although the change in pulse frequency is debated and may be method dependent (Watanabe and Terasawa, 1989; Wennink et al., 1990; Apter et al., 1993; Neely et al., 1995; Wu et al., 1996; Mitamura et al., 1999; Terasawa and Fernandez, 2001).

9.3 Physical Changes of Puberty

Puberty is defined as the period of transition between the juvenile and mature stages of development during which secondary sexual characteristics develop and reproductive capacity is achieved. The physical changes during puberty include the development of pubic and axillary hair in both males and females, the development of gender-specific secondary sexual characteristics, and the pubertal growth spurt.

During puberty, the gonadotrophs in the anterior pituitary produce increased LH and FSH pulses. These in turn stimulate the maturation of the gonads and production of sex steroids (primarily estradiol, testosterone) in a process termed gonadarche. In females, the increased levels of estrogens stimulate breast development, or thelarche. The mucosal surface of the vagina also changes in response to estrogen, becoming thicker and pinker in color. In males the increased gonadotropin levels stimulate testicular growth and production of testosterone (the predominant androgen) followed by growth of the penis and scrotum. The increased testosterone is also important in stimulating spermatogenesis.

Adrenarche refers to the maturation of the zona reticularus of the adrenal gland, which leads to increased production of adrenal androgens, including dehydroepiandrosterone (DHEA), sulfated version of DHEA (DHEAS), and androstenedione. This in turn contributes to the development of pubic hair (pubarche), axillary hair, body odor, and acne. Gonadarche and adrenarche often overlap, although they are separate processes with independent regulation (Sklar et al., 1980; Ilondo et al., 1982; Wierman et al., 1986). Adrenarche can occur 1–2 years before gonadarche and has been shown to be a gradual process beginning in early childhood, rather than an abrupt change during the immediate prepubertal period (Palmert et al., 2001). The triggers for adrenarche are unknown; however, body weight and body mass index (BMI) as well as *in utero* and neonatal physiology likely modulate the timing of adrenarche (Remer and Manz, 1999; Ong et al., 2004).

The sequence of secondary sexual development has been classified into stages by Marshall and Tanner (1969, 1970), known as the Tanner stages or sexual maturation ratings (SMRs). The five Tanner stages are used in classifying breast and pubic hair development in females, and genital and pubic hair development in males. Tanner stage 1 is prepubertal and stage 5 is adult development. Detailed descriptions of the Tanner stages can be found in **Table 1**. In females, the earliest secondary sexual characteristic is usually thelearche, although pubarche may occur first in 10–15% of females. On average, menarche occurs 2.5 years after the onset of puberty. In males the earliest sign of maturation is an increase in testicular volume, which usually occurs prior to the appearance of pubic hair and penile growth.

Table 1 Stages of development of secondary sexual characteristics

Males: Development of external genitalia
Stage 1: Prepubertal. Testicular length <2.5 cm.
Stage 2: Testes are ≥2.5 cm in longest diameter; scrotum thinning and reddening.
Stage 3: Further growth of testes. Growth of penis in width and length.
Stage 4: Testes become larger with darkening of scrotum. Penis further enlarged.
Stage 5: Adult size and shape genetalia.
Females: Breast development
Stage 1: Prepubertal. Elevation of papilla only.
Stage 2: Breast buds palpable, enlargement of areola.
Stage 3: Further enlargement of breast and areola, no separation of contours.
Stage 4: Areola and papilla form secondary mound above level of breast.
Stage 5: Adult contour of breast with projection of papilla only.
Males and females: Pubic hair development
Stage 1: Prepubertal. No pubic hair.
Stage 2: Sparse growth of slightly pigmented long, straight, or slightly curly pubic hair at base of penis or along labia.
Stage 3: Hair becomes thicker, curlier; spreads over mons pubis.
Stage 4: Adult-type hair that does not yet spread to medial surface of thighs.
Stage 5: Adult-type hair with adult distribution, spread to medial surface of thighs.

Modified from Marshall WA and Tanner JM (1969) Variations in pattern of pubertal changes in girls. *Archives of Disease in Childhood*. 44(235): 91–303; and Marshall WA and Tanner JM (1970) Variations in the pattern of pubertal changes in boys. *Archives of Disease in Childhood* 45(239): 13–23, with permission from the BMJ Publishing Group.

9.3.1 Bone Age

Bone age is a reflection of physiological status and a measure of integrated sex steroid levels. It has a better correlation with pubertal stage than with chronological age, height, or weight (Marshall, 1974). In delayed puberty, for example, bone age correlates with the onset of secondary sexual development better than chronological age does. Bone age (or skeletal maturation) is determined by comparing radiographs of the hand, knee, or elbow with standards obtained from a normal population (Greulich and Pyle, 1959; Tanner et al., 1975). There is a gender difference in skeletal maturation, with girls having more advanced skeletal maturation than boys of the same chronologic age. Bone age can not only help assess pubertal development, but it can also be used to predict final adult height, for example, using the Bayley–Pinneau tables (Bayley and Pinneau, 1952).

Of note, because the combination of skeletal age with assessment of secondary sexual development provides a complete picture of an individual's pubertal status, this combined assessment would theoretically be more accurate than chronologic age alone for studies of behaviors related to the physiologic changes of puberty. Realistically, however, the ethical ramifications of using multiple radiographs in a psychological study and the invasiveness of repeated physical examinations often result in the single use of chronologic age as a control variable. It should be acknowledged, though, that bone age may be a less reliable indicator of pubertal stage in overweight individuals since it is known, for example, that overweight boys with delayed puberty have less delayed bone ages than their non-overweight counterparts (Nathan et al., 2006b).

9.4 Timing of Pubertal Onset

The onset of puberty occurs across a wide range of ages in normal adolescents. The development 2–2.5 standard deviations before the mean age of pubertal onset is termed precocious puberty, whereas the development 2–2.5 standard deviations beyond the mean age is termed delayed puberty. Most of the variation in pubertal timing does not derive from underlying pathology, but rather stems from differences in the maturational program of GnRH secretion.

Precocious puberty has traditionally been defined as the development of secondary sexual characteristics prior to the age of 8 years in girls and 9 years in boys, and delayed puberty has been defined as the lack of secondary sexual characteristics by age of 13 years in girls and 14 years in boys. These ages were determined based on the evidence that 95% of girls experience thelarche or pubic hair development between the ages of 8.5 and 13 years, and 95% of boys experience testicular enlargement ($\geq$4 cc in volume) between 9.5 and 13.5 years (Marshall and Tanner, 1969, 1970).

Historically, there had been a secular trend toward earlier pubertal development (Wyshak and Frisch, 1982), but the timing of puberty was thought to have been fairly stable over the last several decades. However, data from the Pediatric Research in Office Settings (PROS) network have called into question whether the trend toward earlier puberty has continued in North America (Herman-Giddens et al., 1997). In this study of over 17 000 females, aged 3–12 years, the mean age for attainment of Tanner 2 breast development in white girls was 10.0 ± 1.8 years, while Tanner 2 pubic hair was attained at a mean age of 10.5 ± 1.7 years. The onset of puberty occurred earlier in African-American girls with a mean of 8.9 ± 1.9 years for Tanner 2 breast development and 8.8 ± 2.0 years for Tanner 2 pubic hair. Of note, 6.7% of white and 27.2% of African-American girls had evidence of breast and/or pubic hair development before 8 years of age (the traditional cutoff for precocious puberty). These findings prompted recommendations by some authors to change the age limit for precocious puberty to 7 years in white girls and 6 years in African-American girls (Kaplowitz and Oberfield, 1999). This recommendation was accompanied by additional recommendations for further evaluation of girls who have abnormal tempo of puberty, headaches, focal neurologic findings, underlying neurological problems, or an adverse effect of early puberty on the child or family's emotional state.

The suggestion to redefine the age limits of precocious puberty was met with substantial controversy and concern that lowering the limits of what can be considered normal development would result in failure to recognize patients with true pathologic causes for early puberty. Several retrospective cohort studies described cases of brain and pituitary tumors, McCune–Albright syndrome, and congenital adrenal hyperplasia that would have been missed using the revised age limits (Chalumeau et al., 2003; Midyett et al., 2003; Grunt et al., 2004). These data have led some to continue to recommend a full evaluation, including magnetic resonance imaging (MRI), for all girls presenting with puberty at less than 8 years of age (Ng et al., 2003; Stanhope, 2003).

In addition, acceptance of the new recommendations was lessened by concerns raised about the validity of the PROS data. These concerns included the accuracy of the study examinations (breast development was assessed by visualization only, without palpation, which might be problematic in overweight girls), the true randomness of the sample population, and the lack of a full endocrinologic evaluation to rule out underlying pathology for the girls with pubertal development at less than 8 years of age. Concern also arose for the finding that the age of breast development decreased, while the average age of menarche occurred at a similar age as other recent studies. It should be noted, however, that in a separate study of Spanish youth, girls who experienced earlier onset of breast development were also noted to have a longer interval before the onset of regular menses than girls with later breast development (Marti-Henneberg and Vizmanos, 1997). Although the authors and others have attempted to address the concerns about the PROS data (Emans and Biro, 1998; Finlay and Jones, 2000; Rosenfield et al., 2000; Herman-Giddens et al., 2001a, 2004; Kaplowitz et al., 2001), uncertainties remain about the redefinition of precocious puberty among females. Despite the controversy about specific ages, several other studies have corroborated the trend of earlier pubertal development (although none as early as in the PROS study) and differences in pubertal timing among different ethnicities in North America (Sun et al., 2002; Wu et al., 2002).

There is some, although more muted, controversy surrounding the determination of the normal limits of pubertal timing and the definition of precocious puberty in boys. Two separate analyses of the National Health and Nutrition Examination Survey (NHANES) III data revealed conflicting results. Herman-Giddens et al. (2001b) reported earlier mean Tanner 2 genital development than previously established norms, whereas Sun et al. (2002) reported mean ages at Tanner 2 genital development that were more consistent with traditional standards (although the median values were much younger). Overall, due to uncertainties about the data (Lee et al., 2001; Reiter and Lee, 2001), as well as the higher proportions of underlying pathology in boys versus girls with precocious puberty, there has been no general call for the lowering of age limits for normal pubertal development in boys.

9.5 Genetic Basis of Pubertal Timing

The timing of puberty varies greatly in the general population and, although the exact mechanisms that regulate the onset of puberty and the end of the prepubertal juvenile pause are not fully understood, the timing is known to be influenced by both environmental and genetic factors (reviewed in Palmert and Hirschhorn (2003), Parent et al. (2003), Plant and Barker-Gibb (2004), and Towne et al. (2005)). Evidence for genetic regulation derives from pubertal timing being highly correlated within racial/ethnic population groups, within families, and between monozygotic compared to dizygotic twins. The combined data suggest that 50–80% of the variation in pubertal timing is determined by genetic factors (Palmert and Hirschhorn, 2003; Parent et al., 2003; Plant and Barker-Gibb, 2004; Towne et al., 2005). There is also strong evidence for environmental and physiologic effects on the timing of puberty. Examples include the late puberty that often accompanies chronic disease and the decline in the age of menarche over the last 150 years, with the mean age at menarche (AAM) in mid-nineteenth-century Europe being 17–18 years (Ritzen, 2003). This shift toward earlier AAM is presumably due to the environmental effects of improved health and nutrition (Wyshak and Frisch, 1982; Parent et al., 2003). However, the changing environmental and secular influences on the timing of puberty do not negate the significant role that genetic background plays in regulating the variation of pubertal timing within the population at any particular point in history.

Although much progress has been made in identifying genetic causes of reproductive endocrine disorders, such as idiopathic hypogonadotropic hypogonadism (IHH), the specific genetic factors that regulate the variation in pubertal timing in the general population remain elusive. The identification of these genes is made more difficult because pubertal timing is a complex genetic trait, where a direct one-to-one relationship between genotype and phenotype does not exist (Darvasi, 1998). This complex nature is due to the broad spectrum of phenotypic variation within the general population that results from multigenic influences and the interaction between genetic variants and environmental exposures (Palmert and Boepple, 2001).

9.5.1 Approaches to Identifying Genetic Factors

Three major approaches are used to find genes underlying complex traits in humans: (1) resequencing of candidate genes, (2) genome-wide linkage analyses, and (3) association studies (reviewed in Hodges and Palmert (2007)).

Resequencing of candidate genes involves identifying and selecting candidate genes and determining if sequence variants are present that can explain the phenotypic variation. This is an expensive method and best used when there is a high likelihood that the candidate gene(s) harbors identifiable causal variants. For example, many single-gene disorders are explained by highly penetrant, rare variants with readily apparent functional significance such as missense, frameshift, or nonsense mutations. In these analyses, the candidate genes are initially chosen based on their location in the genome (positional candidates) or by their known or putative function (biological candidates). There are several biological candidates which could account for the variation in pubertal timing in the general population and these are discussed in further detail subsequently.

Linkage analysis uses an unbiased genome-wide scan to identify genes that modulate a specific trait. While this approach has proven to be useful for identifying genes responsible for single-gene disorders (since the causative mutations in a single gene cosegregate with disease in families), it has proven more difficult, though still possible, to use linkage to identify genes underlying complex traits (Altmuller et al., 2001; Jimenez-Sanchez et al., 2001; Glazier et al., 2002). This is presumably because complex traits result from the interaction of variation in multiple genes and a single genetic variant likely has only modest effects on the trait, making it more difficult to detect. In addition, environmental factors influence the phenotype of complex traits, thus making the phenotype–genotype correlations more variable. Nevertheless, there are several successful linkage studies for common diseases/complex traits, including type 2 diabetes (Horikawa et al., 2000) and schizophrenia (Stefansson et al., 2002, 2003), confirming that this is a viable strategy. No genes that modulate the timing of puberty have been identified thus far using linkage analysis, although recently Guo et al. (2006) found significant linkage to the genomic region 22q13 in a genome-wide linkage analysis investigating the genetic basis for variation in AAM; suggestive linkage peaks were also found at 22q11 and 11q23. However, other studies have not replicated this finding and it is not clear whether the lack of replication of significant linkage stems from genetic or environmental heterogeneity or statistical fluctuations. Thus, important progress may be made in investigating the genetic basis of pubertal timing using linkage analysis but replication of positive findings will be necessary.

Association studies assess the frequency of a genetic variant in an affected population versus a control population. If the frequency of a variant differs significantly between the groups, an association is said to be present. One advantage of association studies is that they have better power than traditional linkage analysis to detect common variants with modest genetic effects (Risch and Merikangas, 1996). However, several limitations must also be considered including that traditional association studies depend on candidate genes, while linkage analyses are unbiased. Multiple association studies have investigated the genetic basis of pubertal timing, specifically the genetics underlying the AAM. Some of these studies are promising; for example, several independent reports have associated variants in cytochrome P450c17α (CYP17), which is involved in estrogen biosynthesis, with early menarche (Feigelson et al., 1997; Dunning et al., 1998; Weston et al., 1998; Haiman et al., 1999; Lai et al., 2001; Ambrosone et al., 2003; Gorai et al., 2003). On the other hand, conflicting reports have been published regarding an association between the age of menarche and other genes, including catechol-O-methyltransferase (COMT) which is important in estrogen metabolism (Gorai et al., 2003; Eriksson et al., 2005), estrogen receptor α (Weel et al., 1999; Stavrou et al., 2002; Gorai et al., 2003; Boot et al., 2004; Long et al., 2005), the sex-hormone binding globulin (SHBG) gene (Xita et al., 2005), and the androgen receptor (Comings et al., 2002; Jorm et al., 2004). It is interesting to note that the genome-wide linkage study by Guo et al. (2006) reports linkage of age of menarche to the genomic region that contains COMT. These associations between genes involved in sex steroid metabolism and AAM are intriguing. However, it remains to be determined whether these particular regulators simply modulate estrogen responsiveness or whether they affect the maturation and central activation of the HPG axis, including the increased GnRH secretion that signifies the onset of central puberty.

In our work, we have used trio-based and case-control-based association studies to test for associations between variants in GnRH and the GnRH receptor (GnRHR) and late pubertal development. We found only nominally significant associations between three single-nucleotide polymorphism (SNPs) in the *GNRHR* gene and late pubertal development, indicating that genetic variation in *GNRH* and *GNRHR* does not appear to be a substantial modulator of pubertal timing in the general population (Sedlmeyer et al., 2005). This finding is important because it indicates that if major genes exist, they may encode proteins that function downstream of the GnRH receptor, proteins that function as upstream regulators of GnRH secretion, or proteins in completely independent, yet unidentified, pathways.

Recent advances in genomics have made the approaches to identifying genes involved in complex traits easier to accomplish. Sequencing of the human genome has facilitated the identification of putative candidate genes, which is important for resequencing and association studies, as well as linkage studies because it facilitates progression from a region of linkage to the identification of the gene(s) and sequence variant(s) that are responsible for phenotypic variation. In addition, the identification of over 9 million SNPs raises the possibility that an increasing number of future linkage and association studies may be done using dense panels of SNPs, rather than the more widely spaced traditional markers or relatively small candidate gene-based SNP panels. (For further discussion, the reader is referred to several excellent reviews regarding the emerging use of whole genome association (WGA) studies (Kruglyak and Nickerson, 2001; Sachidanandam et al., 2001; Carlson et al., 2004; Hirschhorn and Daly, 2005; Frayling, 2007).)

9.5.2 Insights from Single Gene Disorders

Although the genes contributing to the variation in the timing of puberty in the general population have not yet been identified, several genes have been found that have critical roles in the HPG axis and the pathogenesis of abnormal pubertal timing. The study of genes that lead to IHH and Kallmann syndrome (KS) have provided important insights into the development and regulation of the HPG axis (reviewed in Kalantaridou and Chrousos (2002), Silveira et al. (2002a), and Palmert and Hirschhorn (2003)). For example, this work has defined roles for the genes that lead to IHH (*GNRHR, GPR54, FGFR1*) (de Roux et al., 1997; Layman et al., 1998; de Roux et al., 2003; Seminara et al., 2003; Pitteloud et al., 2006); to the X-linked (*KAL1*) (Franco et al., 1991; Legouis et al., 1991) and autosomal (*FGFR1, PROK2, PROKR2*) forms of KS (Hardelin et al., 1992; Dode et al., 2003, 2006; Pitteloud et al., 2005; Pitteloud et al., 2006); to obesity and IHH (*LEP, LEPR*, and *PC1*) (Jackson et al., 1997; Montague et al., 1997; Clement et al., 1998; Strobel et al., 1998; Farooqi et al., 1999); and to abnormal HPG development (*DAX1, SF-1, HESX-1, LHX3*, and *PROP-1*) (Muscatelli et al., 1994; Zanaria et al., 1994; Silveira et al., 2002a).

9.5.2.1 *Idiopathic hypogonadotropic hypogonadism*

IHH with normal olfaction has been primarily associated with mutations in *GNRHR* as well as *GPR54*, the G-protein-coupled receptor for kisspeptins (products of *KISS1*) (de Roux et al., 1997, 2003; Bedecarrats et al., 2003a,b; Karges et al., 2003; Seminara et al., 2003; Wolczynski et al., 2003; Messager et al., 2005; Shahab et al., 2005). *GNRHR* was the first gene discovered to cause autosomal recessive IHH (de Roux et al., 1997; Layman et al., 1998). The GnRHR is a G-protein-coupled receptor expressed on the surface of pituitary gonadotropes. Mutations result in impaired GnRH binding, intracellular trafficking, or signal transduction, causing subtypes of IHH ranging from partial to complete resistance to GnRH. Up to 20 mutations in *GNRHR* have been identified (de Roux et al., 1997, 1999; Layman et al., 1998; Caron et al., 1999; Pralong et al., 1999; Kottler et al., 2000; Beranova et al., 2001; Costa et al., 2001; Soderlund et al., 2001; Silveira et al., 2002b), the majority of which occur as compound heterozygotes. Several studies have attempted to quantify the frequency of *GNRHR* mutations in IHH (with normosmia) with resulting estimates ranging from 3.5% to 10.4% (Beranova et al., 2001; Bhagavath et al., 2005). One report cited a case of constitutional delay of growth and puberty (CDGP) associated with a homozygous partial loss of function mutation in the *GNRHR* (Lin et al., 2006); however, more extensive analysis suggests that genetic variation in *GNRH* or *GNRHR* is not a common cause of late puberty in the general population (Sedlmeyer et al., 2005).

Over the last 4 years, research into the KISS-1/GPR54 system has revealed that it is a critical component of the HPG axis and is necessary for pubertal onset, as demonstrated in both animal and human studies. The KISS-1/GPR54 signaling complex was first described as an important regulator of the HPG axis in 2003 when two independent groups reported deletions and inactivating mutations of *GPR54* in patients with IHH (de Roux et al., 2003; Seminara et al., 2003). *KISS-1* produces kisspeptins, which are the ligands that bind to the G-protein-coupled receptor GPR54. Evidence suggests that kisspeptins exert direct effects on the GnRH neuron which expresses GPR54 (Irwig et al., 2004; Messager et al., 2005), and central administration of kisspeptin activates GnRH neurons and elicits GnRH release and LH secretion *in vivo* (Gottsch et al., 2004b; Irwig et al., 2004; Messager et al., 2005). The binding of kisspeptins to GPR54 appears to be required for the onset of puberty based on evidence such as: (1) *KISS-1* and *GPR54* mRNA levels increase during puberty in both rats and primates (Navarro et al., 2004a; Shahab et al., 2005); (2) administration of kisspeptins can induce early puberty and early activation of the HPG axis in female rats (Navarro et al., 2004b) as well as inducing sustained GnRH discharges in primates when administered at the end of the juvenile pause (Plant, 2006); (3) both *GPR54* and *KISS-1* knockout mice fail to enter puberty and demonstrate phenotypes consistent with hypogonadism (Funes et al., 2003; Seminara et al., 2003; Lapatto et al., 2007); (4) after administration of metastin (a kisspepetin-derived peptide and GPR54 agonist) *GPR54* knockout mice fail to increase gonadotropins, while *KISS1* knockout mice demonstrate increased gonadotropin levels (Lapatto, et al. 2007); and (5) mutations in *GPR54* have been found in patients with IHH (de Roux et al., 2003; Seminara et al., 2003; Gottsch et al., 2006). In addition, GnRH neurons develop increased sensitivity to kisspeptins during pubertal maturation in mice and rats, possibly through increased GPR54 signaling efficiency (Han et al., 2005). Taken together, these findings demonstrate that activation of GPR54 by kisspeptins plays a pivotal role in the onset of puberty; however, it is not yet known whether the KISS-1/GPR54 system is the initial trigger of puberty or whether it acts as a downstream effector of other regulatory factors (Roth et al., 2007; Seminara, 2007). It has also been suggested that, in addition to contributing to the onset of puberty, kisspeptin-GPR54 signaling may play an ongoing regulatory role in the HPG axis, including regulation of LH and GnRH secretion and feedback response to sex steroids in adult animals.

9.5.2.2 Kallmann syndrome

Investigation of KS (hypogonadotropic hypogonadism with anosmia/hyposmia) has led to the identification of several genes that are critical to HPG axis function and olfactory development. Loss of function mutations in Kallmann syndrome-1 (*KAL-1*) (Franco et al., 1991; Legouis et al., 1991; Hardelin et al., 1993) and fibroblast growth factor receptor-1 (*FGFR1*) (Dode et al., 2003) are implicated in the X-linked and autosomal dominant forms of the disease, respectively. However, mutations in these two genes account for only 20% of patients with KS (Dode et al., 2006). Recently mutations in the prokineticin receptor-2 (*PROKR2*) and prokinetcin-2 (*PROK2*) genes were found in a cohort of KS patients (Dode et al., 2006). These genes encode a G-protein-coupled receptor (PROKR2) and its ligand (PROK2), and these findings add prokinetcin signaling to the list of genes known to be important for olfactory and HPG axis development. Of note, one of the patients in this series was heterozygous for both a *PROKR2* mutation and a *KAL1* mutation, suggesting a possible digenic mode of inheritance (Dode et al., 2006). In addition, mutations in nasal embryonic LHRH factor (*NELF*), which plays a role in migration of GnRH neurons and olfactory axon outgrowth (Kramer and Wray, 2000), have been implicated in the pathogenesis of KS (Miura et al., 2004). A heterozygous deletion in *NELF* has been reported as a synergistic component of digenic inheritance along with *FGFR1*, although it is less clear if mutations in *NELF* alone lead to KS (Pitteloud et al., 2007a).

It is important to recognize that the distinction among different abnormalities of pubertal development is not absolute. For example, mutations in *FGFR1* can cause both KS and IHH (without anosmia/hyposmia) (Pitteloud et al., 2006). A recent paper also identified a homozygous mutation in *PROK2* which caused both KS and normosmic IHH within the same kindred; it will be important to investigate this further and determine if this is a common finding among different IHH and KS families (Pitteloud et al., 2007b). It has also been reported that loss of function mutations in *FGFR1* can cause delayed puberty in members of IHH pedigrees (Pitteloud et al., 2005, 2006); however, it remains to be determined whether genetic variation in *FGFR1* is a common cause of delayed puberty in the general population. Causes of reversible IHH have also recently been reported (Raivio et al., 2007), further blurring the distinction between IHH and CDGP.

Finally, as mentioned above, further complexity derives from two groups' recent reports that the presentation of hypogonadotropic hypogonadism can result from the combination of mutations in different genes (*FGFR1* and *NELF*; *FGFR1* and *GNRHR*; *PROKR2* and *KAL1*) (Dode et al., 2006; Pitteloud et al., 2007a). These initial reports likely represent just the beginning of our understanding of the roles that multigenic inheritance and modifier genes may play in the phenotypic variability within IHH.

9.5.2.3 Leptin and other genes

The importance of nutrition in modulating the HPG axis is evidenced by the discovery that IHH can also result from defects in leptin (*LEP*) or the leptin receptor (*LEPR*) gene. These findings initially led to speculation that leptin was a trigger for pubertal onset (Mantzoros et al., 1997); however, the more widely held view is that leptin acts as a permissive factor in pubertal maturation (Plant and Barker-Gibb, 2004). Interestingly, there is recent evidence that leptin action on GnRH is modulated by kisspeptins, including (1) leptin receptors are present on KISS1 neurons, though not on GnRH neurons, and (2) leptin-deficient mice show decreased hypothalamic KISS1 mRNA levels (Smith et al., 2006). This important finding demonstrates that leptin may integrate energy balance and metabolic regulation into the control of pubertal timing by acting as a regulator of *KISS1* expression. The possibility that common variations in *LEP* and *LEPR* may contribute to variation of pubertal timing within the general population is interesting. However, common polymorphisms in *LEP* and *LEPR* were not associated with CDGP in a recent association study (Banerjee et al., 2006).

Other causes of IHH include mutations in genes that are critical to HPG development. This category includes the orphan nuclear receptors *DAX-1* (dosage-sensitive sex reversal adrenal congenital hypoplasia (DSS-AHC) critical region on the X chromosome) and steriodogenic factor-1 (*SF-1*). *DAX-1* is expressed in the hypothalamus, pituitary gonadotropes, adrenal cortex, and gonads. The role of *DAX-1* mutations in the pathogenesis of IHH is complex and appears to involve a combined hypothalamic and pituitary defect, as well as a defect in spermatogenesis (Habiby et al., 1996; Seminara et al., 1999). *SF-1* plays an important role in gonadotropes by regulating the transcription of several pituitary genes, including *GNRHR* and *LH-β* (Parker and Schimmer, 1997). Pituitary specific *SF-1* knockout mice showed no pituitary expression of LH or FSH and had no sexual maturation (Bakke et al., 2001); however, the role of *SF-1* in humans remains to be fully elucidated. Mutations in several pituitary transcription factors, including homeobox gene expressed in ES cells (HESX-1), LIM homeobox gene 3 (LHX3), and prophet of PIT1 (PROP1), can lead to combined pituitary hormone deficiencies that include IHH as a phenotype. Finally, prohormone convertase-1 (*PC-1*) has been associated with IHH and obesity, possibly as a result of defective processing of any of the neuropeptides or prohormones that are components of GnRH secretion (Jackson et al., 1997; Silveira et al., 2002a).

9.5.3 Genetic Variation in Normal Puberty

The information gained from the study of IHH and KS is critical to our understanding of the reproductive–endocrine axis, but the role that the genes (and pathways) identified in these disorders play in regulating the variation in pubertal in healthy adolescents remains unclear. CDGP, which probably represents the extreme end of normal pubertal timing, clusters in families, and has a strong genetic component (Reindollar and McDonough, 1981; Toublanc et al., 1991; Sperlich et al., 1995; Sedlmeyer et al., 2002; Sedlmeyer and Palmert, 2002). Less severe genetic variation (polymorphisms) in any of the above genes (or others that have yet to be identified) could help explain the variation in pubertal timing seen within the general population. This is illustrated in **Figure 1** which depicts a possible paradigm for understanding the genetics of delayed puberty. It is known that specific genes are involved in the pathogenesis of IHH and KS; however, the genes that regulate timing of puberty in the general population or in populations with CDGP have yet to be fully elucidated. It has already been noted that mutations in both *FGFR1* and *GNRHR* have been found in patients with delayed puberty who are members of families with IHH or KS but who themselves have no features of these disorders (Lin et al., 2006; Pitteloud et al., 2006, 2007a); however, whether genetic variation in these genes plays a role in modulating pubertal timing in the general population is not clear (Sedlmeyer et al., 2005).

We are actively investigating the role that variations in genes responsible for IHH and KS play in modulating pubertal timing. We have resequenced several candidate genes, including *GNRHR*, *GNRH*, *LEP*, and *LEPR*, in populations of individuals with late, but otherwise normal pubertal development (Sedlmeyer et al., 2005; Banerjee et al., 2006). No rare mutations have been identified in these genes thus far. We and others have also assessed the role that common, more subtle variations (single nucleotide polymorphisms or SNPs) in these genes may play in regulating pubertal timing in the normal population. Although analyses are ongoing, no evidence for association between SNPs in *GNRH*, *GNRHR* (Sedlmeyer et al., 2005; Gajdos et al., 2007); *LEP*, *LEPR* (Banerjee et al., 2006; Gajdos et al., 2007); or *KISS-1*, *GPR54*, *FGFR1* (Gajdos et al., 2007) and alterations in pubertal timing have been found to date. Thus, unless other genes, or combinations of genes, in these pathways modulate the timing of puberty in the general population, new regulators need to be identified and studied (Roth et al., 2007; Seminara, 2007). This need for new genetic investigation is perhaps not surprising since ~70% of

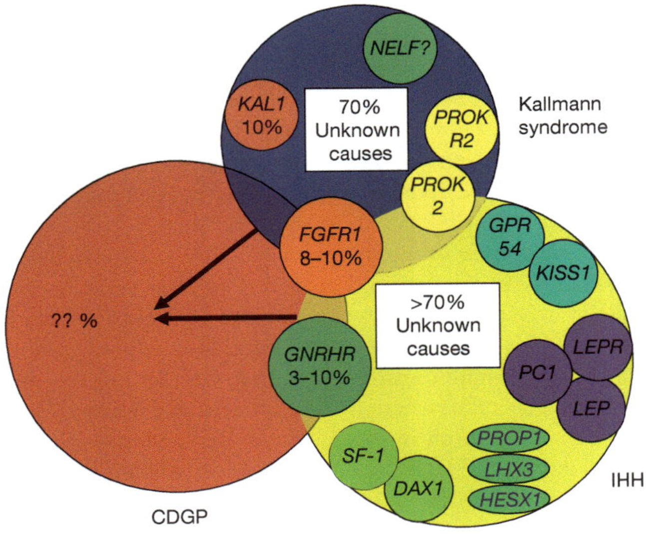

Figure 1 Genetic basis of delayed puberty. A possible paradigm for understanding the genetics of puberty is shown. Some genes underlying the pathogenesis of Kallmann syndrome (KS) and idiopathic hypogonadotropic hypogonadism (IHH) have been identified and are depicted with each diagnosis; less is known about the genetic basis of constitutional delay of growth and puberty (CDGP). There is overlap between the three clinical entities as illustrated by the overlapping circles. There is also likely overlap in their genetic bases, as has been reported for *FGFR1* (in KS, IHH, and CDGP), *GNRHR* (in IHH and possibly CDGP), and *PROK2* (in KS and IHH). Several groups of genes have been associated with particular phenotypes; for example, mutations in *LEP*, *LEPR*, and *PC1* have been found in cases of IHH and obesity, while mutations in *PROP1*, *LHX3*, and *HESX1* can lead to combined pituitary hormone deficiencies. Thus far, mutations in only *GNRHR* and *FGFR1* have been found in cases of delayed puberty without evidence of KS or permanent IHH. Thus, it is possible that variation in these genes may account for variation in pubertal timing in the general population, but this has yet to be proven experimentally. In both KS and IHH, >70% of the genetic cause is still unknown (see text for detailed discussion of each gene and its role in these clinical conditions).

the causes of IHH and KS remain unexplained (**Figure 1**). This is an important area of ongoing investigation because 50–80% of the variation in pubertal timing in the general population is believed to be genetically determined (Palmert and Hirschhorn, 2003; Parent et al., 2003).

9.5.4 Quantitative Trait Loci Associated with Timing of Puberty

Another approach to discovering genes that modulate pubertal timing is the identification of quantitative trait loci (QTLs) through genome-wide linkage analyses. As discussed previously, using these analyses to identify QTLs that modulate complex traits is difficult. Thus far, no genes that modulate the timing of puberty in the general population have been identified using linkage analysis. However, several recent studies have used linkage to investigate the genetic basis for variation in AAM in human populations (see **Table 2**). Guo et al. (2006) used a genome-wide linkage analysis of the AAM in 2461 Caucasian women and identified a statistically significant linkage signal at 22q13 (LOD 3.70), and two suggestive linkages at 22q11 and 11q23. Rothenbuhler et al. (2006) investigated 98 sister pairs for AAM and menarcheal weight-adjusted AAM. No significant linkage was found for AAM; however, the menarcheal weight-adjusted AAM genome scan revealed QTLs with strongly suggestive LOD scores at 16q21, 16q12, and 8p12. Long et al. (2005) found a suggestive peak associated with AAM at 6q25.3; however, this was not a genome-wide scan, but rather a linkage analysis of a genomic region containing a candidate gene on chromosome 6 (Long et al., 2005). It is important to note that none of these studies have been independently replicated and they have all described QTLs at different genomic locations (**Table 2**). It is not clear whether this lack of replication of significant linkage is due to genetic or environmental heterogeneity or perhaps statistical fluctuations. Thus, although progress has been made in investigating the genetic basis of pubertal timing using linkage analysis, replication of positive findings remains necessary.

Table 2 Quantatative trait loci (QTLs) associated with age at menarche (AAM) in linkage analyses

	Human chromosome																						
Trait (reference)	*1*	*2*	*3*	*4*	*5*	*6*	*7*	*8*	*9*	*10*	*11*	*12*	*13*	*14*	*15*	*16*	*17*	*18*	*19*	*20*	*21*	*22*	*X/Y*
AAM (Guo et al., 2006)																						22q11 (2.68) **22q13** (3.70)	
AAM (Long et al., 2005)						6q25.3 (2.01)																	
AAM (Rothenbuler et al., 2006)																							
AAM/weight SDS (Rothenbuler et al., 2006)								8p12 (2.18)								**16q12** (3.12) **16q21** (3.33) 16q21 (2.46)							
Age at VO (Nathan et al., 2006)												**12p11–12**											

Compiled data from several studies indicating significant QTLs (and LOD scores) associated with age at menarche. Those results with evidence of statistically significant linkage are shown in bold; areas with only suggestive linkage are in normal text. In Rothenbuler et al. the analyses for AAM found only nominally significant QTLs; however, when AAM was adjusted for weight standard deviation score (weight SDS) at menarche, several significant QTLs were identified. The bottom row represents data from mice showing a significant linkage with timing of vaginal opening (VO), a marker of pubertal onset in mice, on the region of mouse chromosome 6 that correlates with human chromosome 12p11–12.

These genome-wide linkage studies are important since they do not require *a priori* assumptions about causative genes or pathways and may lead to discovery of novel regulatory genes. However, it is sometimes difficult to assemble the large human populations needed for linkage studies. Thus, it is important to augment human studies with investigation of animal models. Genome-wide scans can be performed readily in mice, and by identifying genes, or even the chromosomal locations of QTLs, important information can be obtained and used to prioritize genes for candidate analysis in humans. For example, we have used chromosome substitution strains to identify a statistically significant QTL associated with timing of vaginal opening (a phenotypic marker of puberty in mice) on mouse chromosome 6, corresponding to human chromosome 12p11–12 (Nathan et al., 2006a). We have also identified chromosome 13 as harboring genes which regulate timing of vaginal opening and are working toward the identification of QTL(s) on this chromosome as well (Krewson et al., 2004). These and other studies (Ojeda et al., 2006; Roth et al., 2007) may eventually provide important clues as to candidate genes/pathways to investigate regarding the genetic basis of variation in pubertal timing within the normal human population.

9.6 Neuroendocrine Regulation of Pubertal Onset

We refer the reader to for a complete discussion of the neuroendocrine control of puberty. We emphasize that as the genes responsible for regulating pubertal timing are being elucidated, key components of the neuroendocrine control of pubertal onset will also be delineated. Investigation has thus far shown that the neuroendocrine control of puberty involves a complex network of genes and cell–cell interactions. The GnRH neurons are under both excitatory and inhibitory control, the balance of which controls the onset of puberty. In order for the pulsatile release of GnRH to increase, excitatory inputs to the neuronal network increase, while inhibitory tone decreases (Ojeda et al., 2006).

Gamma-aminobutyric acid (GABA) is likely the major neurotransmitter that inhibits GnRH secretion during prepubertal development (Mitsushima et al., 1994; Terasawa and Fernandez, 2001; Plant and Barker-Gibb, 2004). In addition, gene expression data indicate that neuropeptide Y (NPY) may play a role in the dampening of the HPG axis after infancy, as mRNA levels of NPY are higher in juvenile monkeys than neonatal animals (El Majdoubi et al., 2000). Other factors that likely inhibit GnRH secretion include endogenous opioids and melatonin, although these likely have limited influence on regulating the timing of puberty in the general population (Terasawa and Fernandez, 2001).

The initiation of puberty is prompted by an increase in glutamatergic neurotransmission (Ojeda et al., 2006) which increases GnRH secretion in animal models through its actions on *N*-methyl-D-aspartate and kainite receptors (Donoso et al., 1990; Claypool et al., 2000). Other stimulators of GnRH secretion include norepinephrine, dopamine, transforming growth factor-α (TGF-α) signaling via erbB1 receptors, neuregulin signaling via erbB4 receptors, galanin-like peptide, kisspeptins, and leptin (Terwilliger and Weiss, 1998; Prevot et al., 2003; Gottsch et al., 2004a,b; Plant and Barker-Gibb, 2004). It has been postulated that these signals act within a complex network facilitating communication between glial cells and the neurons of the hypothalamus. In addition, a hierarchical system of genes involved in the upstream regulation of these signals has been proposed (Ojeda et al., 2003, 2006; Roth et al., 2007), and a new transcriptional regulator of female puberty (enhanced at puberty (EAP1)) has recently been described (Heger et al., 2007; Roth et al., 2007).

9.7 Environmental Influences on Pubertal Timing

9.7.1 Obesity and the Relationship to Pubertal Timing

The relationship between nutritional status and pubertal onset has long been a focus of investigation (Parent et al., 2003; Plant and Barker-Gibb, 2004). With the recent increases in obesity among youth, one wonders if increased adiposity has contributed to the possible shifts toward earlier puberty. Several studies have reported associations between earlier breast development and menarche and indices of overweight, but it is unclear if this has led to a shift in pubertal timing within the general population (reviewed in Parent et al. (2003), Styne (2004), and Himes (2006)). Interestingly, overweight status in boys may, in some cases, be associated with later as opposed to earlier maturation (Wang, 2002; Nathan et al., 2006b).

9.7.2 Endocrine Disrupters and Environmental Influences

Synthetic and naturally occurring substances that alter normal reproductive development and function are known as endocrine-disrupting chemicals (EDCs). Several environmental agents, including polychlorinated biphenyls, organochlorine pesticides, and phthalates, as well as naturally occurring soy-based phytoestrogens, have been implicated as EDCs (Setchell et al., 1997; Gore, 2001; Rasier et al., 2006). Several individual examples of EDCs affecting sexual maturation have been reported (including an association between phthalates and premature thelarche in young girls (Colon et al., 2000), as well as an association between lavender and tea tree oil and prepubertal gynecomastia in boys (Henley et al., 2007)), and some authors have suggested that shifts toward earlier puberty could stem from EDC exposure (Herman-Giddens et al., 1997; Rasier et al., 2006). Murine studies have also indicated that exposure to EDCs can lead to advancement of puberty, reproductive endocrine abnormalities, and cancer (Rasier et al., 2006). This is an area of ongoing important research; however, no evidence for a widespread link between human EDC exposure and altered pubertal timing has yet been identified (Greim, 2004).

9.8 Behavior Related to Variations in Pubertal Timing

The behavioral and psychological changes that occur during puberty and adolescence have been extensively reviewed in the previous edition (Styne and Grumbach, 2002), as well as in several excellent review articles (Waylen and Wolke, 2004; Michaud et al., 2006; Patton and Viner 2007), and we direct the reader to these for a more detailed discussion. In this section, we briefly review several behavioral changes associated with pubertal development, particularly those related to variation in pubertal timing, and present a brief update of this field.

The onset of puberty brings not only the physical and biological changes of sexual maturity, but also the behavioral, psychological, cognitive, emotional, and social changes that comprise the adolescent stage of life. Adolescence can be divided into three stages: early adolescence, which encompasses the majority of the physical changes of puberty, maturing of abstract thinking, and increased peer influence on behavior; middle adolescence, during which increased independence is achieved, and executive function continues to mature; and late adolescence, which involves progression to adult roles in society. It should be noted that in our Westernized society, the length of time from achieving reproductive maturity in early adolescence to assuming a fully realized adult role has lengthened in many instances, such as an extended postsecondary education, thus lengthening the duration of the late adolescent stage (Waylen and Wolke, 2004; Gluckman and Hanson, 2006).

9.8.1 Psychosocial Changes of Puberty

There has been much interest regarding the interplay among the physical changes of puberty, hormonal changes, pubertal stage, and pubertal timing as well as the effect of these on behavior. As adolescents confront increasingly complex interactions with both their families and their peers, both positive and negative coping mechanisms can develop. Adolescence also brings with it the often turbulent emotions associated with emergence of sexuality, sexual attraction, and romantic relationships. Increases in self-harm, sexual activity, risk taking, substance use, mental health, and depression all occur during puberty and adolescence. Several aspects of pubertal timing can affect adolescents and are areas of increasing investigation (Michaud et al., 2006), including the effect of early versus late body transformations on self-image, parental and adult expectations, and peer interactions. In addition, hormonal changes during puberty may have a direct effect on brain development (Sisk and Zehr, 2005).

The psychosocial effects of the timing of puberty differ between boys and girls and have been studied by several authors among youth who experienced pubertal onset at the extremes of the normal population. In discussing these reports, it is important to emphasize that the vast majority of these studies pertain to the range of normal and do not directly pertain to clinically defined precocious or delayed puberty. Boys who experience puberty at a relatively early age may be more likely to engage in exploratory behaviors (sexual activity and substance use) and have been reported to have higher self-esteem (Michaud et al., 2006); conversely, boys who experience late or delayed puberty may have lower self-esteem and be less assertive, less popular, and later in engaging in sexual activity (Waylen and Wolke, 2004). Girls who enter puberty early may be more likely to have lower self-esteem, more emotional and behavioral problems, and earlier sexual activity (Michaud et al., 2006).

One recent cross-sectional survey of 12–15-year-olds in the US and Australia reported an average prevalence of deliberate self-harm of 3.7%, with more than twofold higher rates in females (5% in females vs. 2.2% in males). Despite older age being reported to have a protective effect against self-harm, four- to fivefold higher rates of self-harm were found among girls in later stages of puberty (4.6% during late puberty vs. 1% during pre/early puberty for definitive self-harm) (Patton et al., 2007). This apparent inconsistency raises the possibility that those who experience puberty at younger ages may be at increased risk of self-injurious behaviors (in this study, self-laceration and self-poisoning).

Earlier timing of puberty in both males and females has also been associated with earlier onset of sexual activity (Kaltiala-Heino et al., 2003; Johansson and Ritzen, 2005). A Swiss study reported increased sexual activity in early-maturing girls, as well as increased sexual activity and substance use in early-maturing boys (Michaud et al., 2006). The reverse is also true with late-maturing males and females engaging in sexual activity less often and at a later age. Interestingly, one study also reported increased rates of higher education for women who had undergone late pubertal development than those with early pubertal development, citing early engagement in sexual activity as a contributor to the decreased education achievement of the early maturers (Johansson and Ritzen, 2005).

Substance use is another exploratory behavior which has been studied and may be increased in association with earlier timing of puberty (Dick et al., 2000; Lanza and Collins, 2002; Bratberg et al., 2007). In the Swiss study, early-maturing boys had a greater risk of substance use than early-maturing girls (Michaud et al., 2006). As with sexual activity, later puberty may confer a lower risk of substance use.

The literature that addresses mental health in adolescence and pubertal timing has mainly focused on girls. Most studies agree that early-maturing females may suffer from mental health issues, specifically depression, in a higher proportion than normal or late-developing females (Graber et al., 1997; Kaltiala-Heino et al., 2003; Johansson and Ritzen, 2005). Some females who experience early menarche report higher rates of depression and depressive symptoms during adolescence, which may last until late adolescence (Ge et al., 2001; Stice et al., 2001). Limited data suggest that late-developing boys may be more likely to suffer from not only depression, but also more frequent delinquent behaviors, although, as previously noted, they also have been reported to be less assertive (Kaltiala-Heino et al., 2003; Graber et al., 2004; Waylen and Wolke, 2004).

For both genders, it is not clear that there are long-term effects of early puberty on mental health (Johansson and Ritzen, 2005). Pubertal development and adolescence are associated with increased rates of risky behaviors, and these may be increased in those who experience the changes of puberty at the extremes of normal pubertal timing. Whether the effects of early or late puberty persist into adulthood has been addressed in one study by Johansson and Ritzen (2005) who found that girls with early puberty engaged in norm-breaking behavior, including delinquency, as well as early sexual experiences at a higher rate. However, by adult age (27 and 43 years), there were "no differences in psychosocial adjustment between the early- and late-developed women" as well as no differences in quality of life (as assessed by self-image, satisfaction with family, work, leisure, and life). The authors concluded that the effects of early menarche on psychosocial problems seem to be limited to adolescence.

In reviewing the literature, it is important to note that the total number of studies is relatively small, that the study groups are often of limited size, and that appropriate control groups are not always included. It is also important to draw a distinction between the reported associations seen in a portion of early- or late-maturing adolescents and the population as a whole. Although adolescence is certainly a challenging time, we would not want to overemphasize self-image, mental health issues, or risk-taking behaviors since many (perhaps most) adolescents adapt reasonably well to the changes of puberty and mature with few adverse events.

9.8.2 Brain Development during Puberty

The development of the brain during adolescence has been an active area of research, specifically the study of the development of executive function and social cognition. It has been found that changes in brain structure occur during adolescence and into early adulthood, which may have implications for the changes in behavior that occur during puberty.

Multiple MRI studies have reported a consistent linear increase in white matter during adolescence, with several studies agreeing that this increase takes place in the frontal and parietal cortices (Pfefferbaum et al., 1994; Giedd et al., 1999, 2006; Sowell et al., 2001, 2003; Barnea-Goraly et al., 2005). Conversely, gray matter density reaches a peak at the onset of puberty as a result of increased synapses, before declining through early adulthood as a result of post-pubertal synaptic pruning (Giedd et al., 1999, 2006; Sowell et al., 2001). Given that the frontal lobes control executive function, it follows that the development of the frontal cortex during puberty would result in improvement in executive function abilities (e.g., decision making, multitasking, selective attention, and problem solving; Blakemore and Choudhury, 2006). Overall, the remodeling of the cortical and limbic structures that occurs during adolescence leads to the development of adult cognition, decision-making skills, and social behaviors. The effect of gonadal hormones on the organization of the brain during puberty is an area of current investigation, although the specific mechanism of action of these hormones in relation to development of behaviors or structural brain development is currently unknown (reviewed in Sisk and Zehr (2005)). It is also not known whether variations in timing of puberty affect brain development.

9.9 Conclusion

In conclusion, we emphasize that there is great variation in the timing of puberty in humans. The exact causes and mechanisms underlying this variation are still largely unknown, but new attempts to use genetics and genomics may inform our understanding of the spectrum of pubertal development. The effects of puberty and variation in pubertal timing on adolescent behavior have begun to be investigated; however, thus far studies are limited and not definitive. Further investigation into the mechanisms underlying variation in pubertal timing and the effects of this variation on behaviors are needed to provide a comprehensive understanding of this important life stage.

References

Altmuller J, Palmer LJ, Fischer G, Scherb H, and Wjst M (2001) Genomewide scans of complex human diseases: True linkage is hard to find. *American Journal of Human Genetics* 69(5): 936–950.

Ambrosone CB, Moysich KB, Furberg H, et al. (2003) CYP17 genetic polymorphism, breast cancer, and breast cancer risk factors. *Breast Cancer Research* 5(2): R45–R51.

Apter D, Butzow TL, Laughlin GA, and Yen SS (1993) Gonadotropin-releasing hormone pulse generator activity during pubertal transition in girls: Pulsatile and diurnal patterns of circulating gonadotropins. *Journal of Clinical Endocrinology and Metabolism* 76(4): 940–949.

Bakke M, Zhao L, and Parker KL (2001) Approaches to define the role of SF-1 at different levels of the hypothalamic–pituitary–steroidogenic organ axis. *Molecular and Cellular Endocrinology* 179(1–2): 33–37.

Banerjee I, Trueman JA, Hall CM, et al. (2006) Phenotypic variation in constitutional delay of growth and puberty: Relationship to specific leptin and leptin receptor gene polymorphisms. *European Journal of Endocrinology* 155(1): 121–126.

Barnea-Goraly N, Menon V, Eckert M, et al. (2005) White matter development during childhood and adolescence: A cross-sectional diffusion tensor imaging study. *Cerebral Cortex* 15(12): 1848–1854.

Bayley N and Pinneau SR (1952) Tables for predicting adult height from skeletal age: Revised for use with the Greulich–Pyle hand standards. *Journal of Pediatrics* 40(4): 423–441.

Bedecarrats GY, Linher KD, Janovick JA, et al. (2003a) Four naturally occurring mutations in the human GnRH receptor affect ligand binding and receptor function. *Molecular and Cellular Endocrinology* 205(1–2): 51–64.

Bedecarrats GY, Linher KD, and Kaiser UB (2003b) Two common naturally occurring mutations in the human gonadotropin-releasing hormone (GnRH) receptor have differential effects on gonadotropin gene expression and on GnRH-mediated signal transduction. *Journal of Clinical Endocrinology and Metabolism* 88(2): 834–843.

Beranova M, Oliveira LM, Bedecarrats GY, et al. (2001) Prevalence, phenotypic spectrum, and modes of inheritance of gonadotropin-releasing hormone receptor mutations in idiopathic hypogonadotropic hypogonadism. *Journal of Clinical Endocrinology and Metabolism* 86(4): 1580–1588.

Bhagavath B, Ozata M, Ozdemir IC, Bolu E, Bick DP, Sherins RJ, and Layman LC (2005) The prevalence of gonadotropin-releasing hormone receptor mutations in a large cohort of patients with hypogonadotropic hypogonadism. *Fertility and Sterility* 84(4): 951–957.

Blakemore SJ and Choudhury S (2006) Development of the adolescent brain: Implications for executive function and social cognition. *Journal of Child Psychology and Psychiatry* 47(3–4): 296–312.

Boot AM, van der Sluis IM, de Muinck Keizer-Schrama SM, van Meurs JB, Krenning EP, Pols HA, and Uitterlinden AG (2004) Estrogen receptor alpha gene polymorphisms and bone mineral density in healthy children and young adults. *Calcified Tissue International* 74(6): 495–500.

Boyar R, Finkelstein J, Roffwarg H, Kapen S, Weitzman E, and Hellman L (1972) Synchronization of augmented luteinizing hormone secretion with sleep during puberty. *New England Journal of Medicine* 287(12): 582–586.

Bratberg GH, Nilsen TI, Holmen TL, and Vatten LJ (2007) Perceived pubertal timing, pubertal status and the prevalence of alcohol drinking and cigarette smoking in early and late adolescence: A population based study of 8950 Norwegian boys and girls. *Acta Paediatrica* 96(2): 292–295.

Carlson CS, Eberle MA, Kruglyak L, and Nickerson DA (2004) Mapping complex disease loci in whole-genome association studies. *Nature* 429(6990): 446–452.

Caron P, Chauvin S, Christin-Maitre S, et al. (1999) Resistance of hypogonadic patients with mutated GnRH receptor genes to pulsatile GnRH administration. *Journal of Clinical Endocrinology and Metabolism* 84(3): 990–996.

Chalumeau M, Hadjiathanasiou CG, Ng SM, et al. (2003) Selecting girls with precocious puberty for brain imaging: Validation of European evidence-based diagnosis rule. *Journal of Pediatrics* 143(4): 445–450.

Claypool LE, Kasuya E, Saitoh Y, Marzban F, and Terasawa E (2000) *N*-methyl D,L-aspartate induces the release of luteinizing hormone-releasing hormone in the prepubertal and pubertal female rhesus monkey as measured by *in vivo* push-pull perfusion in the stalk-median eminence. *Endocrinology* 141(1): 219–228.

Clement K, Vaisse C, Lahlou N, et al. (1998) A mutation in the human leptin receptor gene causes obesity and pituitary dysfunction. (see comments). *Nature* 392(6674): 398–401.

Colon I, Caro D, Bourdony CJ, and Rosario O (2000) Identification of phthalate esters in the serum of young Puerto Rican girls with premature breast development. *Environmental Health Perspectives* 108(9): 895–900.

Comings DE, Muhleman D, Johnson JP, and MacMurray JP (2002) Parent–daughter transmission of the androgen receptor gene as an explanation of the effect of father absence on age of menarche. *Child Development* 73(4): 1046–1051.

Costa EM, Bedecarrats GY, Mendonca BB, Arnhold IJ, Kaiser UB, and Latronico AC (2001) Two novel mutations in the gonadotropin-releasing hormone receptor gene in Brazilian patients with hypogonadotropic hypogonadism and normal olfaction. *Journal of Clinical Endocrinology and Metabolism* 86(6): 2680–2686.

Darvasi A (1998) Experimental strategies for the genetic dissection of complex traits in animal models. *Nature Genetics* 18(1): 19–24.

de Roux N, Genin E, Carel JC, Matsuda F, Chaussain JL, and Milgrom E (2003) Hypogonadotropic hypogonadism due to loss of function of the KiSS1-derived peptide receptor GPR54. *Proceedings of the National Academy of Sciences of the United States of America* 100: 10972–10976.

de Roux N, Young J, Brailly-Tabard S, Misrahi M, Milgrom E, and Schaison G (1999) The same molecular defects of the gonadotropin-releasing hormone receptor determine a variable degree of hypogonadism in affected kindred. *Journal of Clinical Endocrinology and Metabolism* 84(2): 567–572.

de Roux N, Young J, Misrahi M, Genet R, Chanson P, Schaison G, and Milgrom E (1997) A family with hypogonadotropic hypogonadism and mutations in the gonadotropin-releasing hormone receptor. *New England Journal of Medicine* 337(22): 1597–1602.

Dick DM, Rose RJ, Viken RJ, and Kaprio J (2000) Pubertal timing and substance use: Associations between and within families across late adolescence. *Developmental Psychology* 36(2): 180–189.

Dode C, Levilliers J, Dupont JM, et al. (2003) Loss-of-function mutations in FGFR1 cause autosomal dominant Kallmann syndrome. *Nature Genetics* 33(4): 463–465.

Dode C, Teixeira L, Levilliers J, et al. (2006) Kallmann syndrome: Mutations in the genes encoding prokineticin-2 and prokineticin receptor-2. *PLoS Genetics* 2(10): e175.

Donoso AO, Lopez FJ, and Negro-Vilar A (1990) Glutamate receptors of the non-*N*-methyl-D-aspartic acid type mediate the increase in luteinizing hormone-releasing hormone release by excitatory amino acids *in vitro*. *Endocrinology* 126(1): 414–420.

Dunkel L, Alfthan H, Stenman UH, Tapanainen P, and Perheentupa J (1990) Pulsatile secretion of LH and FSH in prepubertal and early pubertal boys revealed by ultrasensitive time-resolved immunofluorometric assays. *Pediatric Research* 27(3): 215–219.

Dunning AM, Healey CS, Pharoah PD, et al. (1998) No association between a polymorphism in the steroid metabolism gene CYP17 and risk of breast cancer. *British Journal of Cancer* 77(11): 2045–2047.

El Majdoubi M, Sahu A, Ramaswamy S, and Plant TM (2000) Neuropeptide Y: A hypothalamic brake restraining the onset of puberty in primates. *Proceedings of the National Academy of Sciences of the United States of America* 97(11): 6179–6184.

Emans SJ and Biro F (1998) Secondary sexual characteristics and menses in young girls. *Pediatrics* 101(5): 949–950.

Eriksson AL, Suuriniemi M, Mahonen A, Cheng S, and Ohlsson C (2005) The COMT val158met polymorphism is associated with early pubertal development, height and cortical bone mass in girls. *Pediatric Research* 58(1): 71–77.

Farooqi IS, Jebb SA, Langmack G, et al. (1999) Effects of recombinant leptin therapy in a child with congenital leptin deficiency. (see comments). *New England Journal of Medicine* 341(12): 879–884.

Feigelson HS, Coetzee GA, Kolonel LN, Ross RK, and Henderson BE (1997) A polymorphism in the CYP17 gene increases the risk of breast cancer. *Cancer Research* 57(6): 1063–1065.

Finlay F and Jones R (2000) Precocious puberty. *Pediatrics* 106(1 pt 1): 162–163.

Franco B, Guioli S, Pragliola A, et al. (1991) A gene deleted in Kallmann's syndrome shares homology with neural cell adhesion and axonal path-finding molecules. *Nature* 353 (6344): 529–536.

Frayling TM (2007) Genome-wide association studies provide new insights into type 2 diabetes aetiology. *Nature Reviews Genetics* 8(9): 657–662.

Funes S, Hedrick JA, Vassileva G, et al. (2003) The KiSS-1 receptor GPR54 is essential for the development of the murine reproductive system. *Biochemical and Biophysical Research Communications* 312(4): 1357–1363.

Gajdos Z, DeLellis Henderson K, Butler J, et al. (2007) Is age at menarche in the Hawaii and Los Angeles Multiethnic Cohort influenced by common sequence variation in genes that cause hypogonadotropic hypogonadism? 89th Annual Meeting of the Endocrine Society, Toronto, Canada.

Ge X, Conger RD, and Elder GH, Jr. (2001) Pubertal transition, stressful life events, and the emergence of gender differences in adolescent depressive symptoms. *Developmental Psychology* 37(3): 404–417.

Giedd JN, Blumenthal J, Jeffries NO, et al. (1999) Brain development during childhood and adolescence: A longitudinal MRI study. *Nature Neuroscience* 2(10): 861–863.

Giedd JN, Clasen LS, Lenroot R, et al. (2006) Puberty-related influences on brain development. *Molecular and Cellular Endocrinology* 254–255: 154–162.

Glazier AM, Nadeau JH, and Aitman TJ (2002) Finding genes that underlie complex traits. *Science* 298(5602): 2345–2349.

Gluckman PD and Hanson MA (2006) Evolution, development and timing of puberty. *Trends in Endocrinology and Metabolism* 17(1): 7–12.

Gorai I, Tanaka K, Inada M, et al. (2003) Estrogen-metabolizing gene polymorphisms, but not estrogen receptor-alpha gene polymorphisms, are associated with the onset of menarche in healthy postmenopausal Japanese women. *Journal of Clinical Endocrinology and Metabolism* 88(2): 799–803.

Gore AC (2001) Environmental toxicant effects on neuroendocrine function. *Endocrine* 14(2): 235–246.

Gottsch ML, Clifton DK, and Steiner RA (2004) Galanin-like peptide as a link in the integration of metabolism and reproduction. *Trends in Endocrinology and Metabolism* 15(5): 215–221.

Gottsch ML, Clifton DK, and Steiner RA (2006) Kisspepeptin-GPR54 signaling in the neuroendocrine reproductive axis. *Molecular and Cellular Endocrinology* 254–255: 91–96.

Gottsch ML, Cunningham MJ, Smith JT, et al. (2004b) A role for kisspeptins in the regulation of gonadotropin secretion in the mouse. *Endocrinology* 145(9): 4073–4077.

Graber JA, Lewinsohn PM, Seeley JR, and Brooks-Gunn J (1997) Is psychopathology associated with the timing of pubertal development? *Journal of the American Academy of Child and Adolescent Psychiatry* 36(12): 1768–1776.

Graber JA, Seeley JR, Brooks-Gunn J, and Lewinsohn PM (2004) Is pubertal timing associated with psychopathology in young adulthood. *Journal of the American Academy of Child and Adolescent Psychiatry* 43(6): 718–726.

Greim HA (2004) The endocrine and reproductive system: Adverse effects of hormonally active substances? *Pediatrics* 113(supplement 4): 1070–1075.

Greulich WW and Pyle SI (1959) *Radiographic Atlas of Skeletal Development of the Hand and the Wrist.* Stanford, CA: Stanford University Press.

Grunt JA, Midyett LK, Simon SD, and Lowe L (2004) When should cranial magnetic resonance imaging be used in girls with early sexual development? *Journal of Pediatric Endocrinology and Metabolism* 17(5): 775–780.

Guo Y, Shen H, Xiao P, et al. (2006) Genomewide linkage scan for quantitative trait loci underlying variation in age at menarche. *Journal of Clinical Endocrinology and Metabolism* 91(3): 1009–1014.

Habiby RL, Boepple P, Nachtigall L, Sluss PM, Crowley WF, Jr., and Jameson JL (1996) Adrenal hypoplasia congenita with hypogonadotropic hypogonadism: Evidence that DAX-1 mutations lead to combined hypothalmic and pituitary defects in gonadotropin production. (see comments). *Journal of Clinical Investigation* 98(4): 1055–1062.

Haiman CA, Hankinson SE, Spiegelman D, et al. (1999) The relationship between a polymorphism in CYP17 with plasma hormone levels and breast cancer. *Cancer Research* 59(5): 1015–1020.

Han SK, Gottsch ML, Lee KJ, et al. (2005) Activation of gonadotropin-releasing hormone neurons by kisspeptin as a neuroendocrine switch for the onset of puberty. *Journal of Neuroscience* 25(49): 11349–11356.

Hardelin JP, Levilliers J, Blanchard S, et al. (1993) Heterogeneity in the mutations responsible for X chromosome-linked Kallmann syndrome. *Human Molecular Genetics* 2(4): 373–377.

Hardelin JP, Levilliers J, del Castillo I, et al. (1992) X chromosome-linked Kallmann syndrome: Stop mutations validate the candidate gene. *Proceedings of the National Academy of Sciences of the United States of America* 89(17): 8190–8194.

Heger S, Mastronardi C, Dissen GA, et al. (2007) Enhanced at puberty 1 (EAP1) is a new transcriptional regulator of the female neuroendocrine reproductive axis. *Journal of Clinical Investigation* 117(8): 2145–2154.

Henley DV, Lipson N, Korach KS, and Bloch CA (2007) Prepubertal gynecomastia linked to lavender and tea tree oils. *New England Journal of Medicine* 356(5): 479–485.

Herman-Giddens ME, Bourdony C, Slora E, and Wasserman R (2001a) Early puberty: A cautionary tale. *Pediatrics* 107(3): 609–610.

Herman-Giddens ME, Kaplowitz PB, and Wasserman R (2004) Navigating the recent articles on girls' puberty in Pediatrics: What do we know and where do we go from here? *Pediatrics* 113(4): 911–917.

Herman-Giddens ME, Slora EJ, Wasserman RC, Bourdony CJ, Bhapkar MV, Koch GG, and Hasemeier CM (1997) Secondary sexual characteristics and menses in young girls seen in office practice: A study from the Pediatric Research in Office Settings network. (see comments). *Pediatrics* 99(4): 505–512.

Herman-Giddens ME, Wang L, and Koch G (2001b) Secondary sexual characteristics in boys: Estimates from the national health and nutrition examination survey III, 1988–1994. *Archives of Pediatrics and Adolescent Medicine* 155(9): 1022–1028.

Himes JH (2006) Examining the evidence for recent secular changes in the timing of puberty in US children in light of increases in the prevalence of obesity. *Molecular and Cellular Endocrinology* 254–255: 13–21.

Hirschhorn JN and Daly MJ (2005) Genome-wide association studies for common diseases and complex traits. *Nature Reviews Genetics* 6(2): 95–108.

Hodges CA and Palmert MR (2007) Genetic regulation of the variation in pubertal timing. In: Pescovitz OH and Walvoord EC (eds.) *When Puberty is Precocious: Scientific and Clinical Aspects*, pp. 83–102. Totowa NJ: Humana Press.

Horikawa Y, Oda N, Cox NJ, et al. (2000) Genetic variation in the gene encoding calpain-10 is associated w th type 2 diabetes mellitus. *Nature Genetics* 26(2): 163–175.

Ilondo MM, Vanderschueren-Lodeweyckx M, Vlietinck R, Pizarro M, Malvaux P, Eggermont E, and Eeckels R (1982) Plasma androgens in children and adolescents. Part II. A longitudinal study in patients with hypopituitarism. *Hormone Research* 16(2): 78–95.

Irwig MS, Fraley GS, Smith JT, et al. (2004) Kisspeptin activation of gonadotropin releasing hormone neurons and regulation of KiSS-1 mRNA in the male rat. *Neuroendocrinology* 80(4): 264–272.

Jackson RS, Creemers JW, Ohagi S, et al. (1997) Obesity and impaired prohormone processing associated with mutations in the human prohormone convertase 1 gene. *Nature Genetics* 16(3): 303–306.

Jimenez-Sanchez G, Childs B, and Valle D (2001) Human disease genes. *Nature* 409(6822): 853–855.

Johansson T and Ritzen EM (2005) Very long-term follow-up of girls with early and late menarche. *Endocrine Development* 8: 126–136.

Jorm AF, Christensen H, Rodgers B, Jacomb PA, and Easteal S (2004) Association of adverse childhood experiences, age of menarche, and adult reproductive behavior: Does the androgen receptor gene play a role? *American Journal of Medical Genetics, B* 125(1): 105–111.

Kalantaridou SN and Chrousos GP (2002) Clinical review 148: Monogenic disorders of puberty. *Journal of Clinical Endocrinology and Metabolism* 87(6): 2481–2494.

Kaltiala-Heino R, Kosunen E, and Rimpela M (2003) Pubertal timing, sexual behaviour and self-reported depression in middle adolescence. *Journal of Adolescence* 26(5): 531–545.

Kaplan SL, Grumbach MM, and Aubert ML (1976) The ontogenesis of pituitary hormones and hypothalamic factors in the human fetus: Maturation of central nervous system regulation of anterior pituitary function. *Recent Progress in Hormone Research* 32: 161–243.

Kaplowitz PB and Oberfield SE (1999) Reexamination of the age limit for defining when puberty is precocious in girls in the United States: Implications for evaluation and treatment. Drug and Therapeutics and Executive Committees of the Lawson Wilkins Pediatric Endocrine Society. *Pediatrics* 104(4 pt 1): 936–941.

Kaplowitz PB, Slora EJ, Wasserman RC, Pedlow SE, and Herman-Giddens ME (2001) Earlier onset of puberty in girls: Relation to increased body mass index and race. *Pediatrics* 108(2): 347–353.

Karges B, Karges W, Mine M, Ludwig L, Kuhne R, Milgrom E, and de Roux N (2003) Mutation Ala(171)Thr stabilizes the gonadotropin-releasing hormone receptor in its inactive conformation, causing familial hypogonadotropic hypogonadism. *Journal of Clinical Endocrinology and Metabolism* 88(4): 1873–1879.

Kottler ML, Chauvin S, Lahlou N, et al. (2000) A new compound heterozygous mutation of the gonadotropin-releasing hormone receptor (L314X, Q106R) in a woman with complete hypogonadotropic hypogonadism: Chronic estrogen administration amplifies the gonadotropin defect. *Journal of Clinical Endocrinology and Metabolism* 85(9): 3002–3008.

Kramer PR and Wray S (2000) Novel gene expressed in nasal region influences outgrowth of olfactory axons and migration of luteinizing hormone-releasing hormone (LHRH) neurons. *Genes and Development* 14(14): 1824–1834.

Krewson TD, Supelak PJ, Hill AE, Singer JB, Lander ES, Nadeau JH, and Palmert MR (2004) Chromosomes 6 and 13 harbor genes that regulate pubertal timing in mouse chromosome substitution strains. *Endocrinology* 145: 4447–4451.

Kruglyak L and Nickerson DA (2001) Variation is the spice of life. *Nature Genetics* 27(3): 234–236.

Lai J, Vesprini D, Chu W, Jernstrom H, and Narod SA (2001) *CYP* gene polymorphisms and early menarche. *Molecular Genetics and Metabolism* 74(4): 449–457.

Lanza ST and Collins LM (2002) Pubertal timing and the onset of substance use in females during early adolescence. *Prevention Science* 3(1): 69–82.

Lapatto R, Pallais JC, Zhang D, et al. (2007) Kiss1-/-mice exhibit more variable hypogonadism than Gpr54-/-mice. *Endocrinology* 148(10): 4927–4936.

Layman LC, Cohen DP, Jin M, et al. (1998) Mutations in gonadotropin-releasing hormone receptor gene cause hypogonadotropic hypogonadism. *Nature Genetics* 18(1): 14–15.

Lee PA, Guo SS, and Kulin HE (2001) Age of puberty: Data from the United States of America. *Apmis* 109(2): 81–88.

Legouis R, Hardelin JP, Levilliers J, et al. (1991) The candidate gene for the X-linked Kallmann syndrome encodes a protein related to adhesion molecules. *Cell* 67(2): 423–435.

Lin L, Conway GS, Hill NR, Dattani MT, Hindmarsh PC, and Achermann JC (2006) A homozygous R262Q mutation in the gonadotropin-releasing hormone receptor presenting as constitutional delay of growth and puberty with subsequent borderline oligospermia. *Journal of Clinical Endocrinology and Metabolism* 91(12): 5117–5121.

Long JR, Xu H, Zhao LJ, et al. (2005) The oestrogen receptor alpha gene is linked and/or associated with age of menarche in different ethnic groups. *Journal of Medical Genetics* 42(10): 796–800.

Mantzoros CS, Flier JS, and Rogol AD (1997) A longitudinal assessment of hormonal and physical alterations during normal puberty in boys. V. Rising leptin levels may signal the onset of puberty. *Journal of Clinical Endocrinology and Metabolism* 82(4): 1066–1070.

Marshall WA (1974) Interrelationships of skeletal maturation, sexual development and somatic growth in man. *Annals of Human Biology* 1(1): 29–40.

Marshall WA and Tanner JM (1969) Variations in pattern of pubertal changes in girls. *Archives of Disease in Childhood* 44(235): 291–303.

Marshall WA and Tanner JM (1970) Variations in the pattern of pubertal changes in boys. *Archives of Disease in Childhood* 45(239): 13–23.

Marti-Henneberg C and Vizmanos B (1997) The duration of puberty in girls is related to the timing of its onset. *Journal of Pediatrics* 131(4): 618–621.

Messager S, Chatzidaki EE, Ma D, et al. (2005) Kisspeptin directly stimulates gonadotropin-releasing hormone release via G protein-coupled receptor 54. *Proceedings of the National Academy of Sciences of the United States of America* 102(5): 1761–1766.

Michaud PA, Suris JC, and Deppen A (2006) Gender-related psychological and behavioural correlates of pubertal timing in a national sample of Swiss adolescents. *Molecular and Cellular Endocrinology* 254–255: 172–178.

Midyett LK, Moore WV, and Jacobson JD (2003) Are pubertal changes in girls before age 8 benign? *Pediatrics* 111(1): 47–51.

Mitamura R, Yano K, Suzuki N, Ito Y, Makita Y, and Okuno A (1999) Diurnal rhythms of luteinizing hormone, follicle-stimulating hormone, and testosterone secretion before the onset of male puberty. *Journal of Clinical Endocrinology and Metabolism* 84(1): 29–37.

Mitsushima D, Hei DL, and Terasawa E (1994) Gamma-Aminobutyric acid is an inhibitory neurotransmitter restricting the release of luteinizing hormone-releasing hormone before the onset of puberty. *Proceedings of the National Academy of Sciences of the United States of America* 91(1): 395–399.

Miura K, Acierno JS, Jr., and Seminara SB (2004) Characterization of the human nasal embryonic LHRH factor gene, NELF, and a mutation screening among 65 patients with idiopathic hypogonadotropic hypogonadism (IHH). *Journal of Human Genetics* 49(5): 265–268.

Montague CT, Farooqi IS, Whitehead JP, et al. (1997) Congenital leptin deficiency is associated with severe early-onset obesity in humans. *Nature* 387(6636): 903–908.

Muscatelli F, Strom TM, Walker AP, et al. (1994) Mutations in the DAX-1 gene give rise to both X-linked adrenal hypoplasia congenita and hypogonadotropic hypogonadism. *Nature* 372(6507): 672–676.

Nathan BM, Hodges CA, Supelak PJ, Burrage LC, Nadeau JH, and Palmert MR (2006a) A quantitative trait locus on chromosome 6 regulates the onset of puberty in mice. *Endocrinology* 147: 5132–5138.

Nathan BM, Sedlmeyer IL, and Palmert MR (2006b) Impact of body mass index on growth in boys with delayed puberty. *Journal of Pediatric Endocrinology and Metabolism* 19(8): 971–977.

Navarro VM, Castellano JM, Fernandez-Fernandez R, et al. (2004a) Developmental and hormonally regulated messenger ribonucleic acid expression of KiSS-1 and its putative receptor, GPR54, in rat hypothalamus and potent luteinizing hormone-releasing activity of KiSS-1 peptide. *Endocrinology* 145(10): 4565–4674.

Navarro VM, Fernandez-Fernandez R, Castellano JM, et al. (2004b) Advanced vaginal opening and precocious activation of the reproductive axis by KiSS-1 peptide, the endogenous ligand of GPR54. *Journal of Physiology* 561 (pt 2): 379–386.

Neely EK, Hintz RL, Wilson DM, Lee PA, Gautier T, Argente J, and Stene M (1995) Normal ranges for immunochemiluminometric gonadotropin assays. *Journal of Pediatrics* 127(1): 40–46.

Ng SM, Kumar Y, Cody D, Smith CS, and Didi M (2003) Cranial MRI scans are indicated in all girls with central precocious puberty. *Archives of Disease in Childhood* 88(5): 414–418; discussion 414–418.

Ojeda SR, Lomniczi A, Mastronardi C, et al. (2006) Minireview: The neuroendocrine regulation of puberty: Is the time ripe for a systems biology approach? *Endocrinology* 147(3): 1166–1174.

Ojeda SR, Prevot V, Heger S, Lomniczi A, Dziedzic B, and Mungenast A (2003) Glia-to-neuron signaling and the neuroendocrine control of female puberty. *Annals of Medicine* 35(4): 244–255.

Ong KK, Potau N, Petry CJ, et al. (2004) Opposing influences of prenatal and postnatal weight gain on adrenarche in normal boys and girls. *Journal of Clinical Endocrinology and Metabolism* 89(6): 2647–2651.

Palmert MR and Boepple PA (2001) Variation in the timing of puberty: Clinical spectrum and genetic investigation. *Journal of Clinical Endocrinology and Metabolism B* 86(6): 2364–2368.

Palmert MR, Hayden DL, Mansfield MJ, et al. (2001) The longitudinal study of adrenal maturation during gonadal suppression: Evidence that adrenarche is a gradual process. *Journal of Clinical Endocrinology and Metabolism* 86(9): 4536–4542.

Palmert MR and Hirschhorn JN (2003) Genetic approaches to stature, pubertal timing, and other complex traits. *Molecular Genetics and Metabolism* 80(1–2): 1–10.

Parent AS, Teilmann G, Juul A, Skakkebaek NE, Toppari J, and Bourguignon JP (2003) The timing of normal puberty and the age limits of sexual precocity: Variations around the world, secular trends, and changes after migration. *Endocrine Reviews* 24(5): 668–693.

Parker KL and Schimmer BP (1997) Steroidogenic factor 1: A key determinant of endocrine development and function. *Endocrine Reviews* 18(3): 361–377.

Patton GC, Hemphill SA, Beyers JM, et al. (2007) Pubertal stage and deliberate self-harm in adolescents. *Journal of the American Academy of Child and Adolescent Psychiatry* 46(4): 508–514.

Patton GC and Viner R (2007) Pubertal transitions in health. *Lancet* 369(9567): 1130–1139.

Pfefferbaum A, Mathalon DH, Sullivan EV, Rawles JM, Zipursky RB, and Lim KO (1994) A quantitative magnetic resonance imaging study of changes in brain morphology from infancy to late adulthood. *Archives of Neurology* 51(9): 874–887.

Pitteloud N, Acierno JS, Jr., Meysing AU, Dwyer AA, Hayes FJ, and Crowley WF, Jr. (2005) Reversible Kallmann syndrome, delayed puberty, and isolated anosmia occurring in a single family with a mutation in the fibroblast growth factor receptor 1 gene. *Journal of Clinical Endocrinology and Metabolism* 90(3): 1317–1322.

Pitteloud N, Acierno JS, Jr., Meysing A, et al. (2006) Mutations in fibroblast growth factor receptor 1 cause both Kallmann syndrome and normosmic idiopathic hypogonadotropic hypogonadism. *Proceedings of the National Academy of Sciences of the United States of America* 103(16): 6281–6286.

Pitteloud N, Quinton R, Pearce S, et al. (2007a) Digenic mutations account for variable phenotypes in idiopathic hypogonadotropic hypogonadism. *Journal of Clinical Investigation* 117(2): 457–463.

Pitteloud N, Zhang C, Pignatelli D, et al. (2007b) Loss-of-function mutation in the prokineticin 2 gene causes Kallmann syndrome and normosmic idiopathic hypogonadotropic hypogonadism. *Proceedings of the National Academy of Sciences of the United States of America* 104(44): 17447–17452.

Plant TM (2006) The role of KiSS-1 in the regulation of puberty in higher primates. *European Journal of Endocrinology* 155(supplement 1): S11–S16.

Plant TM and Barker-Gibb ML (2004) Neurobiological mechanisms of puberty in higher primates. *Human Reproduction Update* 10(1): 67–77.

Pralong FP, Gomez F, Castillo E, et al. (1999) Complete hypogonadotropic hypogonadism associated with a novel inactivating mutation of the gonadotropin-releasing hormone receptor. *Journal of Clinical Endocrinology and Metabolism* 84(10): 3811–3816.

Prevot V, Rio C, Cho GJ, et al. (2003) Normal female sexual development requires neuregulin-erbB receptor signaling in hypothalamic astrocytes. *Journal of Neuroscience* 23(1): 230–239.

Quigley CA (2002) Editorial: The postnatal gonadotropin and sex steroid surge-insights from the androgen insensitivity syndrome. *Journal of Clinical Endocrinology and Metabolism* 87(1): 24–28.

Raivio T, Falardeau J, Dwyer A, et al. (2007) Reversal of idiopathic hypogonadotropic hypogonadism. *New England Journal of Medicine* 357(9): 863–873.

Rasier G, Toppari J, Parent AS, and Bourguignon JP (2006) Female sexual maturation and reproduction after prepubertal exposure to estrogens and endocrine disrupting chemicals: A review of rodent and human data. *Molecular and Cellular Endocrinology* 254–255: 187–201.

Reindollar RH and McDonough PG (1981) Delayed sexual development: Common causes and basic clinical approach. *Pediatric Annals* 10(5): 30–39.

Reiter EO and Lee PA (2001) Have the onset and tempo of puberty changed? *Archives of Pediatrics and Adolescent Medicine* 155(9): 988–989.

Remer T and Manz F (1999) Role of nutritional status in the regulation of adrenarche. *Journal of Clinical Endocrinology and Metabolism* 84(11): 3936–3944.

Risch N and Merikangas K (1996) The future of genetic studies of complex human diseases. *Science* 273(5281): 1516–1517.

Ritzen EM (2003) Early puberty: What is normal and when is treatment indicated? *Hormone Research* 60(supplement 3): 31–34.

Rosenfield RL, Bachrach LK, Chernausek SD, et al. (2000) Current age of onset of puberty. *Pediatrics* 106(3): 622–623.

Roth CL, Mastronardi C, Lomniczi A, et al. (2007) Expression of a tumor-related gene network increases in the mammalian hypothalamus at the time of female puberty. *Endocrinology* 148(11): 5147–5161.

Rothenbuhler A, Fradin D, Heath S, Lefevre H, Bouvattier C, Lathrop M, and Bougneres P (2006) Weight-adjusted genome scan analysis for mapping quantitative trait Loci for menarchal age. *Journal of Clinical Endocrinology and Metabolism* 91(9): 3534–3537.

Sachidanandam R, Weissman D, Schmidt SC, et al. (2001) A map of human genome sequence variation containing 1.42 million single nucleotide polymorphisms. *Nature* 409(6822): 928–933.

Sedlmeyer IL, Hirschhorn JN, and Palmert MR (2002) Pedigree analysis of constitutional delay of growth and maturation: Determination of familial aggregation and inheritance patterns. *Journal of Clinical Endocrinology and Metabolism* 87(12): 5581–5586.

Sedlmeyer IL and Palmert MR (2002) Delayed puberty: Analysis of a large case series from an academic center. *Journal of Clinical Endocrinology and Metabolism* 87(4): 1613–1620.

Sedlmeyer IL, Pearce CL, Trueman JA, et al. (2005) Determination of sequence variation and haplotype structure for the gonadotropin-releasing hormone (GnRH) and GnRH Receptor Genes: Investigation of role in pubertal timing. *Journal of Clinical Endocrinology and Metabolism* 90(2): 1091–1099.

Seminara SB (2007) Converging at puberty's hub. *Endocrinology* 148(11): 5145–5146.

Seminara SB, Achermann JC, Genel M, Jameson JL, and Crowley WF, Jr. (1999) X-linked adrenal hypoplasia congenita: A mutation in DAX1 expands the phenotypic spectrum in males and females. *Journal of Clinical Endocrinology and Metabolism* 84(12): 4501–4509.

Seminara SB, Messager S, Chatzidaki EE, et al. (2003) The GPR54 gene as a regulator of puberty. *New England Journal of Medicine* 349(17): 1614–1627.

Setchell KD, Zimmer-Nechemias L, Cai J, and Heubi JE (1997) Exposure of infants to phyto-oestrogens from soy-based infant formula. *Lancet* 350(9070): 23–27.

Shahab M, Mastronardi C, Seminara SB, Crowley WF, Ojeda SR, and Plant TM (2005) Increased hypothalamic GPR54 signaling: A potential mechanism for initiation of puberty in primates. *Proceedings of the National Academy of Sciences of the United States of America* 102(6): 2129–2134.

Silveira LF, MacColl GS, and Bouloux PM (2002a) Hypogonadotropic hypogonadism. *Seminars in Reproductive Medicine* 20(4): 327–338.

Silveira LF, Stewart PM, Thomas M, Clark DA, Bouloux PM, and MacColl GS (2002b) Novel homozygous splice acceptor site GnRH receptor (GnRHR) mutation: Human GnRHR knockout. *Journal of Clinical Endocrinology and Metabolism* 87(6): 2973–3977.

Sisk CL and Zehr JL (2005) Pubertal hormones organize the adolescent brain and behavior. *Frontiers in Neuroendocrinology* 26(3–4): 163–174.

Sklar CA, Kaplan SL, and Grumbach MM (1980) Evidence for dissociation between adrenarche and gonadarche: Studies in patients with idiopathic precocious puberty, gonadal dysgenesis, isolated gonadotropin deficiency, and constitutionally delayed growth and adolescence. *Journal of Clinical Endocrinology and Metabolism* 51(3): 548–556.

Smith JT, Acohido BV, Clifton DK, and Steiner RA (2006) KiSS-1 neurones are direct targets for leptin in the ob/ob mouse. *Journal of Neuroendocrinology* 18(4): 298–303.

Soderlund D, Canto P, de la Chesnaye E, Ulloa-Aguirre A, and Mendez JP (2001) A novel homozygous mutation in the second transmembrane domain of the gonadotrophin releasing hormone receptor gene. *Clinical Endocrinology (Oxf)* 54(4): 493–498.

Sowell ER, Peterson BS, Thompson PM, Welcome SE, Henkenius AL, and Toga AW (2003) Mapping cortical change across the human life span. *Nature Neuroscience* 6(3): 309–315.

Sowell ER, Thompson PM, Tessner KD, and Toga AW (2001) Mapping continued brain growth and gray matter density reduction in dorsal frontal cortex: Inverse relationships during postadolescent brain maturation. *Journal of Neuroscience* 21(22): 8819–8829.

Sperlich M, Butenandt O, and Schwarz HP (1995) Final height and predicted height in boys with untreated constitutional growth delay. *European Journal of Pediatrics* 154(8): 627–632.

Stanhope R (2003) Gonadotrophin-dependent [correction of dependant] precocious puberty and occult intracranial tumors: Which girls should have neuro-imaging? *Journal of Pediatrics* 143(4): 426–427.

Stavrou I, Zois C, Ioannidis JP, and Tsatsoulis A (2002) Association of polymorphisms of the oestrogen receptor alpha gene with the age of menarche. *Human Reproduction* 17(4): 1101–1105.

Stefansson H, Sarginson J, Kong A, et al. (2003) Association of neuregulin 1 with schizophrenia confirmed in a Scottish population. *American Journal of Human Genetics* 72(1): 83–87.

Stefansson H, Sigurdsson E, Steinthorsdottir V, et al. (2002) Neuregulin 1 and susceptibility to schizophrenia. *American Journal of Human Genetics* 71(4): 877–892.

Stice E, Presnell K, and Bearman SK (2001) Relation of early menarche to depression, eating disorders, substance abuse, and comorbid psychopathology among adolescent girls. *Developmental Psychology* 37(5): 608–619.

Strobel A, Issad T, Camoin L, Ozata M, and Strosberg AD (1998) A leptin missense mutation associated with hypogonadism and morbid obesity [news]. *Nature Genetics* 18(3): 213–215.

Styne DM (1994) Physiology of puberty. *Hormone Research* 41(supplement 2): 3–6.

Styne DM (2004) Puberty, obesity and ethnicity. *Trends in Endocrinology and Metabolism* 15(10): 472–478.

Styne DM and Grumbach MM (2002) Puberty in boys and girls. In: Pfaff DW, Arnold AP, Etgen AM, Fahrbach SE, and Rubin RT (eds.) *Hormones, Brain, and Behavior*, pp. 661–716. New York: Academic Press.

Sun SS, Schubert CM, Chumlea WC, et al. (2002) National estimates of the timing of sexual maturation and racial differences among US children. *Pediatrics* 110(5): 911–919.

Tanner JM, Whitehouse RH, Marshall WA, et al. (1975) *Assessment of Skeletal Maturity and Prediction of Adult Height: TW 2 Method.* New York: Academic Press.

Terasawa E and Fernandez DL (2001) Neurobiological mechanisms of the onset of puberty in primates. *Endocrine Reviews* 22(1): 111–151.

Terwilliger JD and Weiss KM (1998) Linkage disequilibrium mapping of complex disease: Fantasy or reality? *Current Opinion in Biotechnology* 9(6): 578–594.

Toublanc JE, Roger M, and Chaussain JL (1991) Etiologies of late puberty. *Hormone Research* 36(3–4): 136–140.

Towne B, Czerwinski SA, Demerath EW, Blangero J, Roche AF, and Siervogel RM (2005) Heritability of age at menarche in girls from the Fels Longitudinal Study. *American Journal of Physical Anthropology* 128(1): 210–219.

Waldhauser F, Weissenbacher G, Frisch H, and Pollak A (1981) Pulsatile secretion of gonadotropins in early infancy. *European Journal of Pediatrics* 137(1): 71–74.

Wang Y (2002) Is obesity associated with early sexual maturation? A comparison of the association in American boys versus girls. *Pediatrics* 110(5): 903–910.

Watanabe G and Terasawa E (1989) *In vivo* release of luteinizing hormone releasing hormone increases with puberty in the female rhesus monkey. *Endocrinology* 125(1): 92–99.

Waylen A and Wolke D (2004) Sex 'n' drugs 'n' rock 'n' roll: The meaning and social consequences of pubertal timing. *European Journal of Endocrinology* 151(supplement 3): U151–U159.

Weel AE, Uitterlinden AG, Westendorp IC, et al. (1999) Estrogen receptor polymorphism predicts the onset of natural and surgical menopause. *Journal of Clinical Endocrinology and Metabolism* 84(9): 3146–3150.

Wennink JM, Delemarre-van de Waal HA, Schoemaker R, Schoemaker H, and Schoemaker J (1990) Luteinizing hormone and follicle stimulating hormone secretion patterns in girls throughout puberty measured using highly sensitive immunoradiometric assays. *Clinical Endocrinology (Oxf)* 33(3): 333–344.

Weston A, Pan CF, Bleiweiss IJ, Ksieski HB, Roy N, Maloney N, and Wolff MS (1998) CYP17 genotype and breast cancer risk. *Cancer Epidemiology Biomarkers Prevention* 7(10): 941–944.

Wierman ME, Beardsworth DE, Crawford JD, et al. (1986) Adrenarche and skeletal maturation during luteinizing hormone releasing hormone analogue suppression of gonadarche. *Journal of Clinical Investigation* 77(1): 121–126.

Wolczynski S, Laudanski P, Jarzabek K, Mittre H, Lagarde JP, and Kottler ML (2003) A case of complete hypogonadotropic hypogonadism with a mutation in the gonadotropin-releasing hormone receptor gene. *Fertility and Sterility* 79(2)442–424.

Wu FC, Butler GE, Kelnar CJ, Huhtaniemi I, and Veldhuis JD (1996) Ontogeny of pulsatile gonadotropin releasing hormone secretion from midchildhood, through puberty, to adulthood in the human male: A study using deconvolution analysis and an ultrasensitive immunofluorometric assay. *Journal of Clinical Endocrinology and Metabolism* 81(5): 1798–1805.

Wu T, Mendola P, and Buck GM (2002) Ethnic differences in the presence of secondary sex characteristics and menarche among US girls: The Third National Health and Nutrition Examination Survey, 1988–1994. *Pediatrics* 110(4): 752–757.
Wyshak G and Frisch RE (1982) Evidence for a secular trend in age of menarche. *New England Journal of Medicine* 306(17): 1033–1035.
Xita N, Tsatsoulis A, Stavrou I, and Georgiou I (2005) Association of SHBG gene polymorphism with menarche. *Molecular Human Reproduction* 11(6): 459–462.
Zanaria E, Muscatelli F, Bardoni B, et al. (1994) An unusual member of the nuclear hormone receptor superfamily responsible for X-linked adrenal hypoplasia congenita. *Nature* 372(6507): 635–641.

10 The Biology of Sexual Orientation and Gender Identity

F J Sánchez, S Bocklandt, and E Vilain, UCLA School of Medicine, Los Angeles, CA, USA

Chapter Outline

Glossary

congenital adrenal hyperplasia An autosomal recessive condition caused by an abnormal adrenal steroidogenesis, often triggered by an enzymatic defect in the cortisol and aldosterone synthesis pathway. This condition results in an abnormal production of sex steroids and may alter sexual development.

disorders of sex development The updated name for the medical classification of conditions that used to be known as intersex (e.g., true hermaphroditism, pseudohermaphroditism, and XX sex reversal). These are congenital conditions in which there is an atypical development of chromosomal, gonadal, or anatomical sex.

epigenetics The study of regulatory mechanisms of gene expression that are stable over rounds of cell division, and sometimes between generations, but do not involve changes in the underlying DNA sequence of the organism. The molecular basis of epigenetics involves chemical modifications to DNA and the chromatin proteins that associate with it.

fraternal birth order effect An observed phenomenon in which older brothers increase the likelihood that later-born (maternally related) males will be homosexual. This is sometimes referred to as the older brother effect.

gender identity disorder A diagnosable mental disorder characterized by a strong cross-sex identification (in attitude, interests, and behaviors) and significant discomfort with one's birth sex. The disorder may impair psychological functioning and steps may be taken to alter one's primary and secondary sexual characteristics in order to be congruent with his or her sense of gender.

linkage analysis A gene-mapping technique in which genetic markers are typed in families to identify regions that are associated with a disease or trait of interest. Markers that are shared more often than expected by chance by family members sharing the trait of interest are assumed to be in close physical proximity to the gene(s) influencing this trait.

LOD score The LOD score (logarithm (base 10) of odds, also called logit by mathematicians) is a statistical metric often used for linkage analysis. The higher the LOD score, the more significant the finding. For instance, an LOD score greater than 3.0 means that the likelihood of observing the given data if the loci are not linked is less than 1 in 1000. MLOD is the maximum LOD score obtained for the various transmission models tested in the statistical analysis.

X-inactivation A process by which one of the two copies of the X chromosomes in female mammalian cells is inactivated. The inactive X chromosome is silenced by its packaging into transcriptionally inactive heterochromatin. X-inactivation occurs so that the female, with two X chromosomes, does not have twice as many X chromosome gene products as the male, which only possess a single copy of the X chromosome, a process called dosage compensation. The choice of which X chromosome will be inactivated is random in placental mammals such as mice and humans.

10.1 Introduction

Everyone's identity is multidimensional. Yet, two dimensions – sexual orientation and gender identity – play a critical and fundamental role in people's lives. Sexual orientation refers to one's sexual attraction, fantasies, and behaviors. Gender identity refers to an inner sense of maleness or femaleness, which may or may not be congruent with the sex assigned at birth.

While both of these discrete traits can significantly impact our lives, what contributes to their development is not fully understood. Complicating matters is that both of these are intertwined in popular culture. Within major social and political movements, individuals who are outside the mainstream for these two traits tend to organize together in order to secure basic human rights and privileges. Although this may be beneficial in terms of increasing the power of marginalized groups, it blurs the distinction between these two complex traits.

In order to avoid contributing to this confusion and to isolate our knowledge on both of these issues, we divide this chapter in two parts. First we focus on sexual orientation and then concentrate on gender identity in humans.

10.2 Sexual Orientation

Sexual attraction and sexual reproduction are essential for the survival of a species. The majority of species within the animal kingdom have two sexes, and organisms seek out opposite-sex mates for reproduction. While a substantial amount of knowledge exists on this behavior, less is known about organisms that deviate from this norm. That is, why do some male organisms seek to mate with only male organisms, and why do some female organisms seek to mate with only female organisms?

This question garners substantial media attention whenever biological studies are released addressing the issue. This section reviews most of this literature. While heterosexuality is the dominant sexual orientation in human cultures, this section focuses on homosexuality.

10.2.1 Defining and Describing Homosexuality

The majority of the population identifies as heterosexual – sexual attraction to someone of the opposite sex. However, some individuals do not. In popular culture, individuals with a nonheterosexual sexual orientation may adopt one of numerous identity labels (e.g., lesbian, gay, bisexual, and same-gender-loving), while some do not adopt such identities. Furthermore, because the term homosexuality was used as a diagnostic label when it was considered a mental disorder, many individuals reject the term. However, from a biological perspective, the term is descriptive of an organism that engages in same-sex behavior. Nevertheless, social trends and sociopolitical movements can complicate the study of sexual orientation because of varying definitions.

It is hard to estimate the prevalence of homosexuality. The popular perception is that 10% of the population is homosexual. Estimates in the general literature range between 1% and 14% (Sell et al., 1995; Voeller, 1990). Sex researchers estimate the rate to be 5–6% for men and 2–3% for women (Diamond, 1993). Part of the problem with these wide-ranging estimates is that measuring sexual orientation can be difficult.

Most research has relied on individuals self-identifying as nonheterosexual. However, this is problematic because it fails to capture people who may engage in nonheterosexual behavior but who do not identify as lesbian, gay, or bisexual. To try to account for this, some researchers have chosen to use the Kinsey scale (Kinsey et al., 1948/1975), – a scale that ranges from 0 (exclusively heterosexual) to 6 (exclusively homosexual) and that takes into account an individual's sexual attraction, sexual fantasies, sexual behavior, and self-identification. This measure is highly correlated in men where the distribution is largely bimodal (Diamond, 1993; Hamer et al., 1993). However, the distribution is more complex in women where the percentage of women that show exclusive same-sex attraction is lower than men, and many more women than men report erotic fantasies toward both sexes (Hu et al., 1995). Yet, use of the Kinsey scale has also been criticized because it relies on individuals responding accurately.

Illustrating the problem with self-reports are two objective studies of sexual arousal. It has been shown that men are category specific when it comes to their sexual arousal (Chivers et al., 2004, Safron et al., 2007). That is, men are either sexually aroused by a particular stimulus or they are not. Yet, one study demonstrated that a particular group of self-identified heterosexual men were sexually aroused by same-sex stimuli (Adams et al., 1996). Another study found that a group of self-identified bisexual men were only sexually aroused by same-sex stimuli and not by opposite-sex stimuli (Rieger et al., 2005). While studies on sexual arousal are criticized for narrowly defining sexual orientation (i.e., it ignores other aspects of sexual attraction such as emotionality), they highlight the limitations of self-report.

10.2.2 Theory

While measuring sexual orientation can be difficult, even more problematic is explaining what contributes to such a basic biological process – sexual attraction. Within the social sciences, a variety of theories exist as to why a homosexual identity may develop. They include early-childhood developmental disruptions (Freud, 1957), pleasurable same-sex experiences early in life (Churchill, 1967), and incest or unpleasant opposite-sex experiences (Cameron and Cameron, 1995). Some social theorists have argued that sexual orientation is meaningless and purely socially constructed (Foucault, 1978; Halperin, 1990). Yet, these theories are neither equally accepted nor empirically supported. Consequently, the social sciences have contributed little to our understanding of the development of homosexuality.

At the same time, the life sciences have struggled to make sense of this anomalous behavior – a behavior that seems to have little evolutionary purpose. Many different attempts have been made to determine if hormones, *in utero* anomalies, genetic variability, or other factors affect sexual orientation. Yet, few consistent results exist. Nevertheless, scientific research has shed light on possible biological underpinnings for this behavior, as described below.

10.2.3 The Biology of Sexual Orientation

Sexual orientation is a complex trait influenced by both environmental and biological factors. A variety of approaches have been utilized to isolate the influence of biological factors. We divide these approaches into three different sections: hormonal influences, correlational studies, and genetic studies.

10.2.3.1 Hormonal influences

The role that sex hormones play in homosexual behavior has been extensively researched in the biological study of sexual orientation. Initially, it was believed that homosexual men and women had abnormal levels of sex hormones (Meyer-Bahlburg, 1977). Based on this hypothesis, some attempts were made to induce heterosexual behavior in homosexual men by administering testosterone. Ironically, one study found that this treatment actually increased homosexual behavior (Glass and Johnson, 1944). While some studies did show that homosexual men had reduced plasma testosterone and impaired spermatogenesis (Kolodny et al., 1971, 1972), the majority found that gonadotropin and testosterone measurements in homosexual men did not differ from heterosexual men (e.g., Barlow et al., 1974; Pillard et al., 1974; Tourney and Hatfield, 1973). Furthermore, testosterone levels in homosexual women were found to be comparable to that in heterosexual women (Dancey, 1990; Downey et al., 1987; Gartrell et al., 1977).

A second line of hormonal research focused on the prenatal environment. It is well known that hormones secreted by the testes and ovaries during sexual differentiation of the fetus significantly influence the developing brain and body (see **Chapter 8, Gonadal Hormones and Sexual Differentiation of Human Brain and Behavior**). Given the role that organizational hormones play in subsequent behavior, could atypical variations of prenatal hormones lead to atypical sexual behavior?

Studies using animal models have attempted to demonstrate the behavioral effect of hormonal variations on a developing fetus (e.g., Dörner et al., 1975, Galdue, 1984). For instance, in guinea pigs, sex-atypical behavior was observed in offspring when hormonal levels were manipulated – either by gonadectomy or by testosterone injections – during the critical perinatal period of sensitivity (Phoenix et al., 1959). Furthermore, the social and sexual behavior of rats have been altered (e.g., inducing lordosis in male rats) by manipulating hormonal levels (e.g., Davidson, 1969; Larsson and Södersten, 1971; Paup et al., 1975; Quadagno et al., 1972).

While these studies seem to implicate the effect of atypical hormonal levels during sexual differentiation, these manipulations go beyond naturally occurring variations. In addition, the behavior analyzed in rats does not compare well to sexual behavior in humans. For instance, the increased frequency of lordosis in male rats does not provide a good model for human male homosexuality: a systematic choice of same-sex partners by the rodents would be comparable, yet this behavior has never been shown to be provoked by hormonal treatment.

One final approach at studying the influence of hormones on sexual orientation has been to study individuals who have been diagnosed with a disorder of sex development. Many of these disorders result from extreme variations of sex hormones during fetal development – an experimental manipulation that would be unethical to conduct on humans. Thus, individuals diagnosed with such a disorder may offer the closest estimation of how atypical hormone levels affect subsequent behavior.

Initially, studies looking at individuals with partial androgen insensitivity syndrome (Money and Ogunro, 1974), complete androgen insensitivity syndrome (Masica et al., 1971), and congenital adrenal hyperplasia (CAH; Lev-Ran, 1974; Money and Dalery, 1976) found that all subjects reported a heterosexual orientation congruent with their gender of rearing rather than their chromosomal sex. However, other reports on women with CAH showed an increased incidence of homosexual or bisexual behavior (e.g., Hines et al., 2004; Johannsen et al., 2006; Money et al., 1984; Zucker et al., 1996). Recently, Meyer-Bahlburg et al. (2008) found that sexual orientation varied depending on the severity of CAH: women who had been exposed to higher levels of androgens were more likely to be classified as homosexual or bisexual than women exposed to less androgen – which could account for some of the inconsistencies in earlier reports.

While these reports on women with CAH seem to implicate hormones during fetal development, it is misleading to generalize these findings to the general population. Women with CAH are exposed to androgen levels that sometimes exceed what men are exposed to *in utero*, and the elevated levels of androgen often lead to virilization or masculinization of the external female genitalia (Grumbach et al., 2003). Furthermore, the majority of women with CAH still report exclusively heterosexual fantasies and behaviors even if they exhibit more stereotypically masculine traits (Meyer-Bahlburg et al., 2008). Consequently, the relevance of these findings beyond women with CAH is unclear.

As a whole, studies looking at the role that hormones play in human sexual orientation have been inconsistent. The only evidence for a hormonal role comes from women with CAH. It seems that only gross hormonal variations seem to have an effect on sexual orientation for particular groups of people. While this does not completely negate the role that hormones play in sexual orientation, evidence has yet to show any clear causal connection in humans.

10.2.3.2 Correlational studies

Another approach to determine possible biological influences on sexual orientation is through correlational studies. This method attempts to assess the interdependence between biological variants and trait differences within a population. Often, the underlying assumption is that the measured traits are indicative of prenatal androgen exposure.

One area of correlational research has focused on the 2D:4D ratio – the ratio between the index finger and the ring finger – which has been shown to be sexually dimorphic. In men, 2D has been shown to be shorter than 4D, whereas for women the lengths are about equal (Manning et al., 1998; Phelps, 1952). Manning et al. (1998, p. 3003) suggested that this ratio reflects gonadal functioning during sexual differentiation where "high concentrations of fetal testosterone lead to a low 2D:4D ratio which therefore indicates high prenatal testicular activity."

Subsequently, several studies investigated the 2D:4D ratio in homosexual men and homosexual women. The first study found that this ratio was significantly altered toward a masculine ratio in the right hand of homosexual women when compared with heterosexual women (Williams et al., 2000). No difference was found between the homosexual and heterosexual men. Williams et al. (2000, p. 455) concluded that their results suggested that "events before birth...influence human sexual orientation" given that "all non-gonadal somatic sex differences in humans appear to be the result of fetal androgens that masculinize males" – a statement that now seems incongruent with evidence from nonhuman mammals (Dewing et al., 2003; Gatewood et al., 2006).

Since the Williams et al. study, the results have been contradictory and inconsistent (e.g., Robinson and Manning, 2000). For instance, a larger study failed to replicate the above observation when controlling for ethnicity (Lippa, 2003), while other small-scale studies confirmed the more male-typical ratio in homosexual women (e.g., Kraemer et al., 2006). McFadden et al. (2005) attempted to resolve the conflicting findings of five studies focused on sexual orientation; but, the reanalysis of the raw data failed to yield any significant findings beyond the established ratio between men and women.

While research continues on the 2D:4D ratio, one point of contention remains whether this finger length ratio is actually a measure of prenatal androgen exposure as suggested by Manning et al. (1998). For instance, it has been proposed that genetic differences may account for differences in 2D:4D between groups of men from different countries (Loehlin et al., 2006). In addition, studies of CAH women who have been exposed to high levels of prenatal androgen are contradictory depending on whether bone (Buck et al., 2003) or soft tissue of the fingers (Brown et al., 2002; Ökten et al., 2002) is measured. Consequently, the connection between prenatal hormones and finger length ratio has yet to be proven. Yet even if prenatal androgen levels influence the 2D:4D ratio, that does not rationalize the use of 2D:4D ratio as a surrogate measure of prenatal androgen levels because other factors likely influence the ratio, which could explain the differences found between specific groups of people.

In addition to the 2D:4D ratio, other correlational studies linking various sexually dimorphic anthropometric characteristics to sexual orientation have been published. For instance, sexual orientation has been correlated with fingerprint asymmetry (Hall and Kimura, 1994), age at which puberty is initiated (Bogaert et al., 2002), height and weight (Bogaert and Blanchard, 1996), and soft sounds emitted by the inner ear (known as otoacoustic emissions) as well as brain-wave patterns evoked by sounds (McFadden and Champlin, 2000; McFadden and Pasanen, 1999). However, results have been inconsistent (see Mustanski et al. (2002)). Ultimately, it has not been shown that variations in androgen levels within the physiological range influence most of these traits.

The fraternal birth order is one biological influence on male sexual orientation that has been replicated over a dozen times and is the best-established effect in sexual orientation research. Homosexual men have been found to have a greater number of older brothers when compared to heterosexual men (Blanchard and Bogaert, 1996, 1997). Each older brother increases the odds that a male will be homosexual by 33%. While this ratio may seem high, it is relative to the base rate of approximately 2.5% of the population being homosexual (Blanchard and Bogaert, 2004). However, based on a statistical model used in epidemiological studies, it has been estimated that 15% of homosexual men can attribute their sexual orientation to this effect (Cantor et al., 2002). Older sisters do not seem to affect the rate increase, and this trend has not been reported for homosexual women.

It was further shown that only biological older brothers and not older sisters influence sexual orientation (Bogaert, 2006). In addition, this trend is only true for right-handed men (Blanchard et al., 2006). It is uncertain why older biologically related brothers affect sexual orientation. The dominant theory – which lacks any molecular or biochemical evidence – is that the mother's immune system may be affecting her son's developing brain. That is, because male H–Y antigens are foreign to a woman's body, her immune system becomes more efficient at producing antibodies to attack a male fetus with each successive male conception (Blanchard and Bogaert, 1996; Blanchard and Klassen, 1997).

Another characteristic that has been found to correlate with sexual orientation is childhood gender nonconformity. By early childhood, most children achieve gender constancy. That is, they have adopted a gender identity – typically congruent with the sex assigned at birth – that will remain stable throughout life (Bem, 1989; Chauhan et al., 2005; Gouze and Nadelman, 1980). Moreover, their behaviors are typically congruent with their sex. However, some children exhibit interests and behaviors that are atypical for their sex. These children may cross-dress, associate mostly with children of the opposite sex (which is atypical for very young children), and will exhibit play behavior typical for the opposite sex (e.g., a boy playing house with girls vs. rough-and-tumble play with other boys).

While most individuals do not identify as nonheterosexual until early adulthood, there is some evidence that the gender-nonconforming behavior of these children may be a sign of their developing sexual identity (e.g., Drummond et al., 2008). For instance, Green (1976, 1985) followed a group of gender-conforming and -nonconforming boys over several years. When he compared the groups at late adolescence/early adulthood, he found that all of the gender-conforming boys had had exclusive or almost exclusively heterosexual fantasies, whereas 37% of the gender-nonconforming boys had had exclusive or almost exclusively homosexual fantasies, (Green et al., 1987). Furthermore, of the sexually active males, none of the gender-conforming boys had had sexual contact with other males, whereas 83% of the gender-nonconforming boys had had some sexual contact with other males (26.6% exclusively with males). While this neither means that all effeminate boys will grow up to be homosexual, nor that all homosexual men were effeminate as boys, early-childhood behavior may be predictive of how sex typical a male will be in adulthood.

Correlational studies on brain structure and function have also been conducted. Three particular brain regions have been correlated with homosexuality, the third interstitial nucleus of the anterior hypothalamus (INAH-3; LeVay, 1991), the anterior commissure (Allen and Gorski, 1992), and the argine vasopressin neuronal population of the suprachiasmatic nucleus (Swaab et al., 1997). In homosexual men, these regions were found to be significantly skewed in the female direction.

The most publicized of these brain findings was reported by LeVay (1991). He compared three groups of autopsied brains: (1) 18 homosexual men and one bisexual man; (2) 16 presumed heterosexual men; and (3) six presumed heterosexual women. He found that the first group of brains differed from their heterosexual counterparts: in particular, he found that the INAH-3 – a sexually dimorphic brain structure shown to be significantly smaller in women when compared to men – was smaller in the first group of brains when compared to the heterosexual men's brains.

As important as LeVay's finding was, it was harshly criticized because all the homosexual men and the bisexual man had died of AIDS complications (although seven of the controls had also died of AIDS), it was based on postmortem tissue, and because of the potential effect of unsystematic variability due to the subjects' different lifestyles. Furthermore, a subsequent postmortem study found that INAH-3 in homosexual men did not significantly differ from heterosexual men; however, it also did not significantly differ from heterosexual women (Byne et al., 2001).

Interestingly, a study on rams found a similar trend. Among the ram population, a significant number of rams (approximately 8–10%) seem to lack sexual interest in female ewes. These rams are known as shy breeders by some ranchers, but a study subsequently found that they were not shy at all. Instead, after an extensive series of sexual behavioral tests consisting of a variety of scenarios, these particular rams were found to be interested in only mounting other rams (Katz et al., 1988; Price et al., 1988). When the brains of these shy breeders were compared to controls, it was found that the sexually dimorphic nucleus in the preoptic area of the hypothalamus was comparable in size to that of the female (Roselli et al., 2004).

Some studies examining cognitive abilities in homosexuals have suggested that homosexual men may perform more similarly to heterosexual women. For instance, on tests measuring visuospatial abilities (e.g., mental rotation), homosexual men perform significantly worse than their heterosexual counterparts; and, on test measuring verbal ability (e.g., word production), homosexual men performed significantly better than heterosexual men (e.g., Neave et al., 1999; Wegesin, 1998). However, other researchers have failed to replicate such findings (e.g., Gladue and Bailey, 1995; Tuttle and Pillard, 1991). One hypothesis for this inconsistency is that the studies have sampled different types of homosexual men in relation to their gender role behavior. If studies are able to find difference between stereotypically masculine homosexual men and stereotypically feminine homosexual men, then this may support the idea that organizational hormones may have affected the developing fetuses differently.

Several other biological traits have been correlated with sexual orientation. These include scalp hair-whorl

(Klar, 2004) nonright-handedness (Lalumière et al., 2000; Lindesay, 1987; McCormick and Witelson, 1991), voice quality (Pierrehumbert et al., 2004), penis size (Bogaert and Hershberger, 1999), eye blinks (Rahman et al., 2003), body morphology (Johnson et al., 2007), facial features (Rule and Ambady, 2008), cerebral asymmetry (Rahman et al., 2008), gray-matter concentration (Ponseti et al., 2007), and brain responses to odors (Berglund et al., 2006; Savic et al., 2005; Savic and Lindström, 2008). While we do not know the physiological relevance of these differences, the biological evidence is mounting that sexual orientation may be in part influenced by one's biology.

10.2.3.3 Genetics studies

Scant evidence supports connecting differences in sexual orientation to variations in prenatal androgens. However, there is abundant evidence that sexual orientation has a strong genetic component.

Family studies have found a higher rate of homosexuality in both the siblings of homosexual men and homosexual women and the maternal uncles of homosexual men (Bailey and Bell, 1993; Bailey and Benishay, 1993; Bailey et al., 1995, 1999; Bailey and Pillard, 1991; Pattatucci and Hamer, 1995; Pillard, 1990; Pillard and Weinrich, 1986). For instance, brothers of homosexual men had a median rate for homosexuality of 9%, which is above the expected frequency (Bailey and Pillard, 1995). One limitation of family studies is that they are unable to separate genetic and environmental effects – something that is possible with twin studies.

Twin studies have yielded compelling evidence that support a large genetic component in the development of sexual orientation (Bailey et al., 2000; Bailey and Pillard, 1991; Kendler et al., 2000; Kirk et al., 2000). Although concordance rates of homosexuality in twins can vary between the studies as a result of different ascertainment methods, the remaining variance in sexual orientation is almost wholly a result of differences in the nonshared environment. This suggests that biological environments specific to each twin (e.g., intrauterine and intracellular processes including epigenetics) have a larger effect on sexual orientation than environmental variables, such as family dynamics, parental relationships, and school setting, which are assumed to affect both twins about equally.

Pedigree analysis and family linkage studies have also led to some profound results. In particular, Hamer et al. (1993) found a pattern of maternal loading for male homosexuality. In other words, there was an increased rate of homosexuality on the mother's side of the family compared to the father's side. In addition, an increase in maternal transmission of male homosexuality relative to paternal transmission has been reported (Rice et al., 1999b), though other studies failed to replicate this finding (Bailey et al., 1999; McKnight and Malcolm, 2000). Nevertheless, these studies suggested that homosexuality is an X-linked trait given the preferential maternal transmission.

Consequently, Hamer et al. (1993) carried out a linkage scan of the X chromosome and found evidence of significant linkage of male homosexuality to the Xq28 region. This result was replicated in a subsequent study from the same group based on a new sample set (Hu et al., 1995). However, a separate research team had inconclusive findings with regard to Xq28 linkage, though these results were not published in a peer-reviewed journal (see Sanders and Dawood (2003)). In addition, a fourth study from another independent group found no evidence of X-linkage (Rice et al., 1999a). A meta-analysis of the four studies was performed by Hamer (1999) that produced an estimated level of 64% Xq28 allele sharing between gay brothers versus the expected 50%. The meta-analysis also reported a statistically suggestive multiple scan probability (MSP) value of 0.00 003 (Sanders and Dawood, 2003). The exact gene involved has yet to be identified 15 years after the initial findings.

A recent study found a role for the X chromosome in male sexual orientation in a very different way (Bocklandt et al., 2006). Typically, male cells contain a single X chromosome while female cells contain two. In order to regulate the phenotypic expression of traits determined by genes on the X chromosome, the cells in a developing female embryo inactivate or silence one of its two X chromosomes. This process – known as dosage compensation – occurs at random and typically results with half the cells inactivating the paternal X chromosome and half the cells inactivating the maternal X chromosome. Once embryogenesis is complete, all the descending daughter cells will keep the same X chromosome inactive (Brown and Robinson, 2000).

By comparing mothers of homosexual sons with women without homosexual sons, it was shown that the number of women with extreme skewing of X-inactivation was significantly higher in mothers of homosexual men (13%) compared to controls (4%). For mothers with two or more homosexual sons, the incidence of extreme skewing was even higher (23%) (Bocklandt et al., 2006). Therefore, the cells in a mother with homosexual sons tended to inactivate either the maternal X chromosome or the paternal X chromosome at a statistically higher rate than would be expected by chance (i.e., 50/50).

It is still unclear whether this unusual X-inactivation is a direct influence on or a consequence of a mechanism influencing sexual orientation. The most probable explanation of the skewing is that sequence variations that influence sexual orientation in the child could affect the growth or survival of white blood cells or stem cells in the mother. As a result, cells that have inactivated one allele or the other are selected for. Arguably, the effect of the X chromosome gene(s) or mechanisms that influence sexual orientation in the sons is visible in the blood of their mothers. The unusual X chromosome methylation pattern in this sample of mothers of homosexual men strengthens the case for genes on the X chromosome playing a role in male sexual orientation.

Ultimately, the determination of sexual orientation is likely to involve many genes based on the complexity of the trait. A large number of these genes are expected to be autosomal rather than sex linked since the modest linkage levels reported for the X chromosome cannot explain more than a fraction of the heritability of male sexual orientation as determined by twin studies. Recently, the results of a genome-wide linkage scan to identify genes that influence sexual orientation were published (Mustanski et al., 2005). This study did not exclude families that displayed evidence of nonmaternal transmission and calculated separate logarithm of odds score (LOD score, which indicates the statistical significance of a linkage scan where a higher LOD means a more significant link between traits) for maternal, paternal, as well as combined transmission. Mustanski et al. found a maximum LOD (MLOD) score of 3.45 near D7S798 on 7q36. The second highest MLOD score was 1.96, near D8S505 on 8p12. Both had similar paternal and maternal contributions. Near marker D10S217 on 10q26, a maternal origin effect was observed. The MLOD score at this position was 1.81 for maternal meioses and no paternal contribution, which suggests the presence of a maternally expressed but paternally silenced imprinted gene for sexual orientation in 10q26. The relatively low MLOD scores from this linkage scan is an effect of the small sample size. The results of a much larger linkage scan on 1000 homosexual male sib pairs are expected shortly (A. Sanders, 2008, personal communication).

One of the most intriguing results from the Mustanski et al. (2005) study is the possibility that there is an imprinted gene on chromosome 10. Prior reports of maternal loading of male sexual orientation transmission were used to link the X chromosome to sexual orientation, but it could also indicate that epigenetic factors act on autosomal genes. Bocklandt and Hamer (2003) had earlier speculated that imprinting could affect human sexual orientation. If so, then some of the discordance in the sexual orientation of identical twins could be accounted for by epigenetic regulation. Even though identical twins have the same DNA sequence, DNA methylation could diverge leading to differences in gene expression during critical periods of development (Fraga et al., 2005). Consequently, the observed discordance for homosexuality in twins, which is currently attributed to each twin's unique environment, could very well be caused by the epigenetic environment.

10.3 Gender Identity

Gender plays a critical role in a person's development. Most people adopt a gender identity that is congruent with the sex assigned at birth; and this identity will remain constant throughout life (Chauhan et al., 2005; Gouze and Nadelman, 1980; Leonard and Archer, 1989). Certainly, the development of our gender identity and the specific attitudes and roles that are ascribed to such an identity are heavily influenced by social factors (Eagly and Wood, 1999; Wood and Eagly, 2002). Yet, what biological underpinnings affect the development of our gender identity? One approach to answer this question has been to study individuals who develop a cross-gender identity – in particular, transsexuals. This section focuses on this particular population.

10.3.1 Defining and Describing Transsexualism

Across time and across cultures there have been individuals who exhibit interests and behaviors that are markedly atypical for their gender. Many different terms have been used in regard to this group, including gender identity disorder (GID), transgender, and transsexual. Given that most people are not familiar with the transgender community and given the common misperceptions of this community, some time is spent describing issues pertinent to gender-variant individuals. Most of the following information is based on the United States (US) and other western societies. However, it should be noted that gender-related constructs vary cross-culturally (e.g., the *fa'afafine* of Samoa and *hijra* of India) and not all people fit or identify with the following descriptions (e.g., Tsoi, 1990).

10.3.1.1 Gender identity disorder

In the general population some people report discomfort with their gender identity. Many of these individuals – such as the above-described gender-nonconforming children – will eventually adopt a gender identity that is consistent with the sex assigned at birth. However, for others, this dysphoria over their gender may persist and cause clinically significant distress, which may lead to a diagnosis of GID (American Psychiatric Association, APA, 2000). The full criteria for GID can be reviewed in the DSM-IV-TR (APA, 2000). We briefly mention some of the key characteristics here.

Individuals who are diagnosed with GID report a strong and enduring identification with the gender opposite to the one they were assigned at birth. Children with GID may exhibit gender-nonconforming behavior, including cross-dressing and playing with children of the opposite sex. They may even believe that they are going to grow up to be the opposite sex only to be disappointed at puberty. Adults with GID also exhibit a similar pattern of behavior and may take extreme steps to alter their physical traits.

GID is a controversial diagnosis in part because some believe it pathologizes normal variation within sexual and gender expression (Lev, 2005; Zucker and Spitzer, 2005). It is important to remember that in order to have a diagnosable disorder, a person must experience gender dysphoria and not merely exhibit gender-nonconforming behaviors and attitudes. Consistent with other diagnosable mental disorders, the persons's gender dysphoria must cause "clinically significant distress or impairment in social, occupational, or other important areas of functioning" (Criterion D, APA, 2000, p. 579) and not simply be behaviors that conflict with societal norms "unless the deviance or conflict is a symptom of a dysfunction in the individual" (APA, 2000, pp. xxx–xxxi). Furthermore, individuals who have been medically diagnosed with a disorder of sex development do not fit within the GID criteria (Criterion C); though such persons can be diagnosed with GID–Not Otherwise Specified if they are experiencing gender dysphoria. Nevertheless, these diagnostic criteria are criticized for being too vague and subjective in practice (Wakenfield, 1997; Widiger and Clark, 2000) and that they are discriminatory against particular groups of people (Hill et al., 2005; Lev, 2005).

10.3.1.2 Transgender

Transgender is an umbrella term used to describe individuals whose gender identity and behavior significantly differs from the social norms for males and females (Davidson, 2007). This can include people who identify as a crossdresser, gender-queer (individuals who identify as both male and female or who identify as neither), and drag queens/kings (individuals who crossdress typically for theatrical purposes). Within popular culture, individuals who are transgender may be assumed to be nonheterosexual; however, this is not necessarily true. This confusion possibly stems from a lack of public awareness and because the transgender community is associated with the gay, lesbian, and bisexual community within sociopolitical movements.

Recently, some individuals within the transgender community have begun to adopt the label of intersex without any formal medical diagnosis. This may partly be a play on words and/or a personal belief that there is a biological basis for their own gender identity, which could fall under the group of intersex conditions (now formally known as disorders of sex development). While some individuals who are formally diagnosed with an intersex condition may associate with the transgender community, it is important to realize that the issues faced by transgender and intersex people are markedly different. Where transgender individuals are born with what would be seen as typical anatomy, intersex individuals are not. Furthermore, few individuals with intersex conditions experience the degree of gender identity issues reported by transgender individuals (Intersex Society of North America, nd).

10.3.1.3 Transsexualism

Within the transgender community, probably the most fascinating individuals are those who identify as transsexual – both male-to-female (MtF) transsexuals and female-to-male (FtM) transsexuals. These individuals report extreme distress over their sexual anatomy and may take drastic steps to alter their physical appearance so that it is congruent with their gender identity. While transsexualism is rare, there are conflicting statistics on just how prevalent transsexualism is. The estimates in peer-reviewed journals range from 1:100 000 for MtF and 1:400 000 for FtM to as high as 1:2900 for MtF and 1:8300 for

FtM (de Cuypere et al., 2007; Tsoi, 1988; van Kesteren et al., 1997; Wålinder, 1971).

In terms of treatment, the vast majority of transsexuals will take cross-sex hormones and will undergo facial procedures (e.g., electrolysis and rhinoplasty). Most will opt to have top surgery (i.e., masectomies in FtM and breast augmentation for MtF). Some will choose to have bottom surgery (i.e., phalloplasty for FtM and vaginoplasty for MtF). The degree to which one will alter his or her appearance can be affected by many factors, including economic limitations, medical conditions that prevent the performance of certain procedures, the fear of losing significant relationships, and a desire not to conduct an irreversible procedure.

Given the irreversibility of genital surgery, it is important that proper evaluation be conducted prior to surgery – especially given that there are reports of regret after this drastic procedure (e.g., Olsson and Möller, 2006). The biggest concern among patients is the effect that the surgery will have on their sexual functioning. In a survey of 55 transsexuals, de Cuypere et al. (2005) found that the majority were satisfied with their surgeries. However, 24.2% of the MtFs and 19.0% of the FtMs were sexually unsatisfied with 13.8% of the MtFs and 10.0% of the FtMs reporting a worsening in their sex life.

Far less drastic is the treatment of transsexuals with cross-sex hormones. Yet, the physical effect of hormone therapy is more dramatic for FtM transsexuals than for MtF transsexuals (Meyer et al., 1986). FtM transsexuals exhibit a marked increase in facial and body hair, the development of male-pattern baldness, a deepening of the voice, an increase in lean muscle mass, a cessation of menstrual periods, and an increase in sexual drive (Futterweit et al., 1986; Meyer et al., 1986; Oriel, 2000). Consequently, the physical appearance of FtM transsexuals is generally compatible with their new identity (Smith et al., 2005).

For MtF transsexuals, the effect of cross-sex hormones on their physical appearance is somewhat mixed. As a whole, the physiological changes are generally consistent across MtF transsexuals. These include a decrease in hair loss, slowdown in facial hair growth, reduction of facial acne, increase in nipple tenderness with some breast enlargement, and a decrease in sex drive (Meyer et al., 1986; Oriel, 2000). However, the subjective appraisal of the physical appearance of MtF transsexuals – when assessed by independent raters – is generally more believable for those who transition earlier in life than later (Smith et al., 2005).

10.3.1.4 Primary and secondary MtF transsexuals

The outward acknowledgment and adoption of a transsexual identity – sometimes termed coming out – is somewhat different between MtF and FtM transsexuals: in general, MtF transsexuals come out either in adolescence or at midlife. To differentiate these two groups, the literature may refer to MtF transsexuals as early-onset or primary transsexuals and late-onset or secondary transsexuals (Person and Ovesey, 1974a,b; Landén et al., 1998). This bimodal trend is not seen among FtM transsexuals (Smith et al., 2005).

If we examine these two separate groups, some characteristics seem to be pervasive within each group of MtF transsexuals. The primary transsexuals are typically from racial/ethnic minority communities, they exhibit stereotypically feminine interests and behaviors, and they are sexually attracted to men. The secondary transsexuals typically identify as Caucasian, exhibit stereotypically masculine interests and behaviors, are sexually attracted to women, report a history of being married to a woman, and tend to have children (Herman-Jeglinska et al., 2002). While these demographic characteristics are not exclusive of each group, this difference has been one of the most widely reported in peer-reviewed journals. Consequently, the route that leads to the development of a transsexual identity may be different between primary and secondary subtypes, which would reflect a different etiology for each group.

10.3.2 Theory

Many theories have been proposed about what contributes to the development of a transsexual identity. Within the social sciences, these theories range from poor parenting to the idea that transsexualism is an invented construct. For instance, theorists have suggested that transexualism is the result of an absent father (Stoller, 1979), a mother being enmeshed with her son (Loeb and Shane, 1982), or the parents wishing they had had a child of the opposite sex (Green, 1974). Meanwhile, others have argued that it is society's rigidly constructed rules on gender that create the problem versus transsexualism itself being a problem (Roy, 2001).

Likewise, some have tried to implicate biological factors for transsexualism. The most dominant theory in this realm is the brain sex theory. This theory suggests that MtF transsexuals have female brains trapped in a man's body and vice versa for FtM transsexuals. Support for this theory is limited and is discussed below.

Theories have also been proposed to explain the primary and secondary MtF transsexuals. One hypothesis is that because the secondary transsexuals are typically attracted to women, the gender dysphoria they experience may be more easily suppressed than for the primary transsexuals who are sexually attracted to men. In addition, because the secondary transsexuals are typically Caucasian men who are in stereotypically masculine occupations, there may be a fear of losing many of the privileges that come with such a status; whereas primary transsexuals – who are typically racial/ethnic minorities – are already marginalized within the mainstream US culture and thus have less social privilege to lose.

A controversial explanation for this difference has recently received substantial attention. Based on numerous studies conducted in the 1980s and 1990s and that were published in peer-reviewed journals, Blanchard (1989b, 1991, 1992, 2005) asserted that gender dysphoria may not be characteristic of both primary and secondary transsexuals. Blanchard (1985) believed that these two groups of MtF transsexuals could be classified based on whether they were homosexual or not prior to transitioning. The secondary or nonhomosexual (including asexual and bisexual) group was far more likely to report a history of transvestic-fetishism (sexual arousal when cross-dressing) and to be sexually aroused at the thought or image of themselves as a women (known as autogynephilia; Blanchard (1989a)) when compared to the primary or homosexual group. Consequently, the distress that nonhomosexual transitioners report may be precipitated by social mores and sexual prohibitions rather than GID. Such a trend in FtM transsexuals has not been reported, though recent data suggest that there may be a subtype of FtMs who also report a history of sexual arousal while crossdressing (Smith et al., 2005).

Regardless of the motive for transitioning, what does seem to differentiate transsexuals is their sexual orientation. Interestingly, many of the above noted biological trends among nontranssexual homosexual men and women have been shown among transsexuals. We now turn our attention to the limited peer-reviewed reports.

10.3.3 The Biology of Gender Identity

As with sexual orientation, several different approaches have been used to study transsexualism. These include studies on hormones, correlational studies, and genetic studies. Far less data exist on transsexualism as compared to homosexuality. What follows is based on this limited set of knowledge.

10.3.3.1 Hormonal influences

As with homosexuality, a predominant hypothesis was that hormonal levels in transsexuals were not in the typical range and/or they may have been exposed to atypical hormonal levels *in utero*. Early reports suggested that plasma testosterone was lower in MtF transsexuals when compared to heterosexual men (Starká et al., 1975) and higher in FtM transsexuals when compared to heterosexual women (Spiová and Starká, 1977). However, several subsequent studies failed to find significant differences in gonadotropin and sex steroid secretions between MtF transsexuals and control males (e.g., Spijkstra et al., 1988). In addition, clinical trials have reported that MtF transsexuals at baseline were in the normal range (e.g., Dittrich et al., 2005) even if the participants had had a prior history of cross-sex hormonal treatment (Meyer et al., 1986).

Aside from the aforementioned hormonal studies on animal models and women with disorders of sex development, a few studies attempted to test whether transsexuals responded differently to hormonal treatment when compared to control men. However, when MtF transsexuals who had no history of cross-sex hormonal treatment were administered gonadotropins and sex steroids, they responded no differently to luteinizing hormone or estrogen provocation when compared to control males (Goodman et al., 1985, Gooren, 1986). Thus, these studies did not find any organizational differences with these methods.

10.3.3.2 Correlational studies

As with sexual orientation, attempts have been made to find correlations between different traits among transsexuals. To date, two studies focusing on 2D:4D ratio in adult transsexuals have been published. When controlling for handedness, Schneider et al. (2006) found that heterosexual females, FtM transsexuals, and MtF transsexuals all had more feminized 2D:4D ratios than heterosexual men. However, FtM transsexuals did not differ from heterosexual females. Although Schneider et al. (2006, p. 268) concluded that their data "strongly support a role for biological factors in the etiology of [MtF] transsexualism," this claim seemed premature given the inconsistencies found in 2D:4D studies with homosexual samples and given the uncertain significance of finger length ratio. Furthermore, a subsequent study focused on both children and adults with a cross-sex identity found no difference (Wallien et al., 2008).

Given the relationship found between fraternal birth order and male homosexuality, some research has examined whether a birth-order effect is present among transsexuals. Interestingly, most of the reports show that the trend can be seen in transsexuals who would be characterized as primary transitioners and not the secondary transitioners (Blanchard et al., 1996; Blanchard and Sheridan, 1992; Poasa et al., 2004). However, some mixed results were found in a sample of Korean transsexuals (Zucker et al., 2006a).

As discussed earlier, some children will exhibit gender-atypical interests and behaviors. Is this an early sign of transsexualism? This topic has received recent media attention, which seems to be increasing public awareness of GID and transsexualism. One question that is addressed in public discussions related to childhood gender nonconformity is whether a parent should encourage or discourage the behavior. This is a difficult question to address because the relationship between these early behaviors and later development is unclear.

At this time, there is no empirically supported treatment for children with GID, and professionals must use their clinical judgment when devising a care plan. Where some suggest that children with GID must be actively treated to make their gender identity congruent with their birth sex (e.g., Zucker, 2001), others believe that children should be allowed to freely express themselves and begin the process of transitioning (Johnson, 2008; Reed, 2006). Consequently, conversations in both the public and professional forum over what interventions primary caregivers and mental health professions should provide to children have become extremely contentious.

What complicates addressing these treatment concerns is that GID in childhood often subsides by adulthood, and many homosexual men recall exhibiting gender-atypical behavior as boys (Bailey and Zucker, 1995; Zucker et al., 2006b). For instance, in the previously described longitudinal study by Green (1985), the majority of the gender-nonconforming boys – including many who repeatedly voiced wanting to be a girl – were classified as homosexual or bisexual at follow-up. Only one identified as transsexual. Two subsequent reports on males found a similar trend – the majority of gender-nonconforming boys later identified as nonheterosexual rather than transsexual (Cohen-Kettenis and Pfäfflin, 2003; Zucker and Bradley, 1995). However, these studies also found higher rates of transsexualism than did the Green study.

Altogether, published research examining the relationship between childhood gender nonconformity and later development shows that the majority of these children do not identify as transsexual in adulthood. Consequently, caution must be taken when considering how to respond to children with GID (see Meyer et al. (2001) and Zucker (2005)). This is especially true when it comes to considering treatment with cross-sex hormones and any surgical procedures.

Another area of correlational research that has received substantial attention has focused on the brain. Similar to the LeVay (1991) findings in homosexual men, Zhou et al. (1995) released an influential brain study on MtF transsexuals. When they compared the autopsied brains of six MtF transsexuals to the brains of heterosexual men, heterosexual women, and homosexual men, they found that the sexually dimorphic bed nucleus of the stria terminalis of the hypothalamus (BSTc) was of female size in the MtF transsexuals (Zhou et al., 1995; Kruijver et al., 2000) and male size in homosexual men. The research group followed up this report with a study showing that the MtF transsexuals had female-like neuronal density in the BSTc (Kruijver et al., 2000). From these two studies, the team concluded that "transsexualism may reflect a form of brain hermaphroditism such that this limbi nucleus itself is structurally sexually differentiated opposite to the transsexual's genetic and genital sex" (Kruijver et al., 2000, p. 2041).

These two findings seemed to support a hypothesis that is known as the brain sex theory. This theory suggests that transsexuals have brains consisting of the sexually dimorphic structures of the opposite sex. At face value, these two studies – which used the same autopsied MtF brains – seem to bolster this belief. However, the generalization of these findings is limited by the inherent pitfalls of postmortem studies, the relatively small sample size, and the fact that all the MtF transsexuals had had an extensive history of estrogen therapy, which has been shown to alter the brains of MF transsexuals toward female proportions (Hulshoff Pol et al., 2006).

Aside from brain autopsies, another approach has been to use neuropsychological tests that men and women tend to perform differently on. While the magnitude of the difference is small on most of these tests (e.g., vocabulary), some show marked differences (e.g., mental rotation; Hyde (2005)). Thus, some researchers have assessed if transsexuals show a cross-sex pattern on such tests. MtF transsexuals currently on cross-sex hormones were found to perform in a feminized direction on certain cognitive abilities (e.g., Cohen and Forget, 1995), but these findings are uncertain because of the effect of hormone treatment. Thus, several researchers tested MtF transsexuals prior to hormonal treatment and found that they had feminized performances on certain neurocogntive tasks, in particular, tests of visuospatial abilities (Cohen-Kettenis et al., 1998; van Gooren et al., 2002). As with other studies, reports have been conflicting and attempts to replicate previous findings are inconsistent (e.g., Miles et al., 1998, 2006; van Goozen et al., 1995).

Many other traits have been correlated with transsexualism. These include fingerprint asymmetry (Green and Young, 2000; Slabbekoorn et al., 2000), height and weight (Blanchard et al., 1995; Smith et al., 2005), nonright-handedness (Green and Young, 2001), and hyphothalamic activation to the odor of steroids secreted by men and women (Berglund et al., 2008). As with sexual orientation, we do not know the relevance of these differences. However, future research may help us better understand primary and secondary transsexuals.

10.3.3.3 Genetic studies

The body of literature on the genetic basis of transsexualism is extremely limited. Although there are reports of families where more than one member identify as transsexuals (e.g., Green, 2000), such reports are rare. **Table 1** presents a list of familial cases of transsexuals that have been reported in the peer-reviewed literature; however, additional cases have been reported elsewhere (e.g., Diamond and Hawk, 2004). Only a small number of twin case reports have been published – some discordant (Garden and Rothery, 1977; Green and Stoller, 1971; Segal, 2006), some concordant (Hyde and Kenna, 1977; Sadeghi and Fakhrai, 2000), and some unclear (Hepp et al., 2004). Given that no systematic twin study has been reported, it is impossible to separate genetic from environmental influences in the reported cases. Consequently, there is no support for a genetic basis of transsexualism.

A number of chromosomal abnormalities have been reported in transsexual individuals (Haberman et al., 1975; Buhrich et al., 1978; Snaith et al., 1991; Taneja et al., 1992; Turan et al., 2000; Mouaffak et al., 2008). In all cases, sex chromosomes were involved. The most common association was described with disomy Y (47, XYY). However, because of the relatively high frequency of sex chromosome aneuploidy (1 in 900 males for XYY; Nielsen and Wohlert, 1990) a statistically significant association with transsexualism has not been demonstrated.

Table 1 Family co-occurrence of transsexualism reported in peer-reviewed literature

Case type	*No.*	*Source*
MZ brothers concordant for MtF	1	Anchersen (1956)
	1	Hyde and Kenna (1977)
	1	Green (2000)
	1	Tsur et al. (1991)
Triplets: both males MtF	1	McKee et al. (1976)
Two brothers, nontwins, MtF	1	Ball (1981)
	1	Stoller and Baker (1972)
	1	Hore et al. (1973)
	3	Green (2000)
Three brothers, nontwins, MtF	1	Saba is et al. (1974)
MtF with GID dad	1	Green (2000)
MtF with GID/TV son	1	Green (2000)
MtF with GID sister	1	Green (2000)
FtM with TV dad	1	Green (2000)
Two sisters, nontwins, FtM	1	Green (2000)
	1	Joyce and Ding (1985)
MZ sisters concordant for FtM	1	Sadeghi and Fakhrai (2000)
	1	Knoblauch et al. (2007)

MZ, monozygotic (identical) twins; MtF, male-to-female transsexualism; FtM, female-to-male transsexualism; GID, gender identity disorder; TV, transvestite/crossdresser.

A small number of candidate genes have been studied for transsexualism. A recent study looked at a polymorphism in the gene coding for 5-α reductase and found no assocation in a sample of MtF and FtM transsexuals (Bentz et al., 2007). The same group did find a significant association between a single nucleotide polymorphism in the *CYP17* gene (which encodes the 17α-hydroxylase enzyme) in FtM transsexuals but not MtF transsexuals (Bentz et al., 2008). However, their sample size was small and they reported a significant difference in allele distribution between male and female controls as well, shedding doubts on these results.

A small study of MtF transsexuals in a Swedish population studied repeat sequences in or near the androgen receptor gene (*AR*), the estrogen receptor beta gene (ERβ), and the aromatase gene. They found an association between MtF transsexualism and a dinucleotide CA polymorphism in the ERβ gene ($p = 0.03$; Henningsson et al., 2005). However, a subsequent study consisting of 112 MtF transsexuals failed to replicate the Swedish findings, but found an association between transsexualism and a longer polymorphism for *AR* – a gene that plays a role in the masculinization of the brain (Hare et al., in press). Overall, our understanding of the genetic basis of transsexualism is less advanced when compared to the study of sexual orientation.

10.4 Conclusion

Sexual orientation and gender identity are two complex traits that play a critical role in human lives. This chapter has presented a representative sample of the biological research that has investigated how and why these traits develop. Results from hormonal, correlational, and genetic studies offer a glimpse at what might underlie these behaviors. While there are limitations to this research, the results of these important studies have helped determine where future research should go as methodology and technology continue to advance.

Indeed, social and environmental factors contribute to the development of sexual orientation and gender identity as well. In addition, sociopolitical forces influence the kind of research questions that are and are not explored by scientists. While some voice fear about the potential implications of isolating the biological factors contributing to these fundamental traits, others want to understand these complex traits, which are relevant not just to those who have been historically marginalized, but to all human beings.

References

Adams HE, Wright LW, Jr., and Lohr BA (1996) Is homophobia associated with homosexual arousal? *Journal of Abnormal Psychology* 105: 440–445.

Allen LS and Gorski RA (1992) Sexual orientation and the size of the anterior commissure in the human brain. *Proceedings of the National Academy of Sciences of the United States of America* 89: 7199–7202.

American Psychiatric Association (2000) *Diagnostic and Statistical Manual of Mental Disorders,* 4th edn. Washington, DC: Author.

Anchersen P (1956) Problems of transvestism. *Acta Psychiatrica Scandinavica* 33: 249–256.

Bailey JM and Bell AP (1993) Familiality of female and male homosexuality. *Behavior Genetics* 23: 313–322.

Bailey JM and Benishay DS (1993) Familial aggregation of female sexual orientation. *American Journal of Psychiatry* 150: 272–277.

Bailey JM, Bobrow D, Wolfe M, and Mikach S (1995) Sexual orientation of adult sons of gay fathers. *Developmental Psychology* 31: 124–129.

Bailey JM, Dunne MP, and Martin NG (2000) Genetic and environmental influences on sexual orientation and its correlates in an Australian twin sample. *Journal of Personality and Social Psychology* 78: 524–536.

Bailey JM and Pillard RC (1991) A genetic study of male sexual orientation. *Archives of General Psychiatry* 48: 1089–1096.

Bailey JM and Pillard RC (1995) Genetics of human sexual orientation. *Annual Review of Sex Research* 6: 126–150.

Bailey JM, Pillard RC, Dawood K, Miller MB, Farrer LA, Trivedi S, and Murphy RL (1999) A family history study of male sexual orientation using three independent samples. *Behavior Genetics* 29: 79–86.

Bailey JM and Zucker KJ (1995) Childhood sex-typed behavior and sexual orientation: A conceptual analysis and quantitative review. *Developmental Psychology* 31: 43–55.

Ball J (1981) Thirty years experience with transsexualism. *Australian and New Zealand Journal of Medicine* 15: 39–43.

Barlow LH, Abel GG, Blanchard EB, and Mevissakalian M (1974) Plasma testosterone levels and male homosexuality: A failure to replicate. *Archives of Sexual Behavior* 3: 571–575.

Bem SL (1989) Genital knowledge and gender constancy in preschool children. *Child Development* 60: 649–662.

Bentz E-K, Hefler LA, Kaufman U, Huber JC, Kolbus A, and Tempfer CB (2008) A polymorphism of the CYP17 gene related to sex steroid metabolism is associated with female-to-male but not male-to-female transsexualism. *Fertility and Sterility* 90: 56–59.

Bentz E-K, Schneeberger C, Hefler LA, van Trotsenburg M, Kaufmann U, Huber JC, and Tempfer CB (2007) A common polymorphism of the SRD5A2 gene and transsexualism. *Reproductive Science* 14: 705–709.

Berglund H, Lindström P, Dhejne-Helmy C, and Savic I (2008) Male-to-female transsexual show sex-atypical hypothalamus activation when smelling odours steroids. *Cerebral Cortex* 18: 1900–1908.

Berglund H, Lindström P, and Savic I (2006) Brain response to putative pheromones in lesbian women. *Proceedings of the National Academy of Sciences of the United States of America* 103: 8269–8274.

Blanchard R (1985) Typology of male-to-female transsexualism. *Archives of Sexual Behavior* 14: 247–261.

Blanchard R (1989a) The classification and labeling of nonhomosexual gender dysphoria. *Archives of Sexual Behavior* 18: 315–334.

Blanchard R (1989b) The concept of autogynephilia and the typology of male gender dysphoria. *Journal of Nervous and Mental Diseases* 177: 616–623.

Blanchard R (1991) Clinical observations and systematic studies of autogynephilia. *Journal of Sex and Marital Therapy* 17: 235–251.

Blanchard R (1992) Nonmonotonic relation of autogynephilia and heterosexual attraction. *Journal of Abnormal Psychology* 101: 271–276.

Blanchard R (2005) Early history of the concept of autogynephilia. *Archives of Sexual Behavior* 34: 439–446.

Blanchard R and Bogaert AF (1996) Homosexuality in men and number of older brothers. *American Journal of Psychiatry* 153: 27–31.

Blanchard R and Bogaert AF (1997) Additive effects of older brothers and homosexual brothers in the prediction of marriage and cohabitation. *Behavioral Genetics* 27: 45–54.

Blanchard R and Bogaert AF (2004) Proportion of homosexual men who owe their sexual orientation to fraternal birth order:

An estimate based on two national probability samples. *American Journal of Human Biology* 16: 151–157.
Blanchard R, Cantor JM, Bogaert AF, Breedlove SM, and Ellis L (2006) Interaction of fraternal birth order and handedness in the development of male homosexuality. *Hormones and Behavior* 49: 405–414.
Blanchard R, Dickey R, and Jones CL (1995) Comparison of height and weight in homosexual versus nonhomosexual male gender dysphorics. *Archives of Sexual Behavior* 24: 543–554.
Blanchard R and Klassen P (1997) H-Y antigen and homosexuality in men. *Journal of Theoretical Biology* 185: 373–378.
Blanchard R and Sheridan P (1992) Sibship size, sibling sex ratio, birth order, and parental age in homosexual and nonhomosexual gender dysphorics. *Journal of Nervous and Mental Disease* 180: 40–47.
Blanchard R, Zucker KJ, Cohen-Kettenis PT, Gooren LJG, and Bailey JM (1996) Birth order and sibling ratio in two samples of Dutch gender-dysphoric homosexual males. *Archives of Sexual Behavior* 25: 495–514.
Bocklandt S and Hamer DH (2003) Beyond hormones: A novel hypothesis for the biological basis of male sexual orientation. *Journal of Endocrinological Investigation* 26: 8–12.
Bocklandt S, Horvath S, Vilain E, and Hamer DH (2006) Extreme skewing of X chromosome inactivation in mothers of homosexual men. *Human Genetics* 118: 691–694.
Bogaert AF (2006) Biological versus nonbiological older brothers and men's sexual orientation. *Proceedings of the National Academy of Sciences of the United States of America* 103: 10771–10774.
Bogaert AF and Blanchard R (1996) Physical development and sexual orientation in men: Height, weight, and onset of puberty differences. *Personality and Individual Differences* 21: 77–84.
Bogaert AF, Friesen C, and Klentrou P (2002) Age of puberty and sexual orientation in a national probability sample. *Archives of Sexual Behavior* 31: 73–81.
Bogaert AF and Hershberger S (1999) The relation between sexual orientation and penile size. *Archives of Sexual Behavior* 28: 213–221.
Brown CJ and Robinson WP (2000) The causes and consequences of random and non-random X chromosome inactivation in human. *Clinical Genetics* 58: 353–363.
Brown WM, Finn CJ, Cooke BM, and Breedlove SM (2002) Differences in finger length between self-identified "butch" and "femme" lesbians. *Archives of Sexual Behavior* 31: 123–128.
Buck JJ, Williams RM, Hughes IA, and Acerini CL (2003) *In-utero* androgen exposure and 2nd to 4th digit length ratio – comparisons between healthy controls and females with classical congenital adrenal hyperplasia. *Human Reproduction* 18: 976–979.
Buhrich N, Barr R, and Lam-Po-Tang PR (1978) Two transsexuals with 47-XYY karyotype. *British Journal of Psychiatry* 133: 77–81.
Byne W, Tobet S, Mattiace L, et al. (2001) The interstitial nuclei of the human anterior hypothalamus: An investigation of variation within sex, sexual orientation and HIV status. *Hormones and Behavior* 40: 86–92.
Cameron P and Cameron K (1995) Does incest cause homosexuality? *Psychological Reports* 76: 611–621.
Cantor JM, Blanchard R, Paterson AD, and Bogaert AF (2002) How many gay men owe their sexual orientation to fraternal birth order? *Archives of Sexual Behavior* 31: 63–71.
Chauhan GS, Shastri J, and Mohite P (2005) Development of gender constancy in preschoolers. *Psychological Studies* 50: 62–71.
Chivers ML, Rieger G, Latty E, and Bailey JM (2004) A sex difference in the specificity of sexual arousal. *Psychological Science* 15: 736–744.
Churchill W (1967) *Homosexual Behavior among Males: A Cross-Cultural and Cross-Species Investigation.* New York: Hawthorn Books.
Cohen H and Forget H (1995) Auditory cerebral lateralization following cross-gender hormone therapy. *Cortex* 31: 565–573.
Cohen-Kettenis PT and Pfäfflin F (2003) *Transgenderism and Intersexuality in Childhood and Adolescence: Making Choices.* Thousand Oaks, CA: Sage.
Cohen-Kettenis PT, van Goozen SHM, Doorn CD, and Gooren LJG (1998) Cognitive ability and cerebral lateralisation in transsexuals. *Psychoneuroendocrinology* 23: 631–641.
Dancey CP (1990) Sexual orientation in women: An investigation of hormonal and personality variables. *Biological Psychiatry* 30: 251–264.
Davidson JM (1969) Effects of estrogen on the sexual behavior or male rats. *Endocrinology* 84: 1365–1372.
Davidson M (2007) Seeking refuge under the umbrella: Inclusion, exclusion, and organizing within the category transgender. *Sexuality Research and Social Policy* 4(4): 60–80.
de Cuypere G, T'Sjoen G, Beerten R, et al. (2005) Sexual and physical health after sex reassignment surgery. *Archives of Sexual Behavior* 34: 679–690.
de Cuypere G, van Hemelrijck M, Michel A, et al. (2007) Prevalence and demography of transsexualism in Belgium. *European Psychiatry* 22: 137–141.
Dewing P, Shi T, Horvath S, and Vilain E (2003) Sexually dimorphic gene expression in mouse brain precedes gonadal differentiation. *Molecular Brain Research* 118: 82–90.
Diamond M (1993) Homosexuality and bisexuality in different populations. *Archives of Sexual Behavior* 22: 291–310.
Diamond M and Hawk ST (2004) Concordance for gender identity among monozygotic and dizygotic twin pairs. Paper Presented at the 116th Annual Meeting of the American Psychological Association, Honolulu, HI, USA, 28 July–1 August.
Dittrich R, Binder H, Cupisti S, Hoffmann I, Beckmann MW, and Mueller A (2005) Endocrine treatment of male-to-female transsexuals using gonadotropin-releasing hormone agonist. *Experimental and Clinical Endocrinology and Diabetes* 113: 586–592.
Dörner G, Rhode W, Stahl F, Krell L, and Masius WG (1975) A neuroendocrine predisposition for homosexuality in men. *Archives of Sexual Behavior* 4: 1–8.
Downey J, Ehrhardt AA, Schiffman M, Dyrenfurth I, and Becker J (1987) Sex hormones in lesbian and heterosexual women. *Hormones and Behavior* 21: 347–357.
Drummond KD, Bradley SJ, Peterson-Badali M, and Zucker KJ (2008) A follow-up of girls with gender identity disorder. *Developmental Psychology* 44: 34–45.
Eagly AH and Wood W (1999) The origins of sex differences in human behavior: Evolved dispositions versus social roles. *American Psychologist* 54: 408–423.
Foucault M (1978) *The History of Sexuality,* vol.1. New York: Pantheon.
Fraga MF, Ballestar E, Paz MF, et al. (2005) Epigenetic differences arise during the lifetime of monozygotic twins. *Proceeding of the National Academy of Sciences of the United States of America* 102: 10604–10609.
Freud S (1957) Leonardo da Vince and a memory of his childhood. In: Strachey J (ed.) *Standard Edition of the Complete Psychological Works of Sigmund Freud.* London: Hogarth Press.

Futterweit W, Weiss RA, and Fagerstrom RM (1986) Endocrine evaluation of forty female-to-male transsexuals: Increased frequency of polycystic ovarian disease in female transsexualism. *Archives of Sexual Behavior* 15: 69–78.

Garden GM and Rothery DJ (1977) A female monozygotic twin pair discordant for transsexualism. Some theoretical implications. *British Journal of Psychiatry* 161: 852–854.

Gartrell NK, Loriaux DL, and Chase TN (1977) Plasma testosterone in homosexual and heterosexual women. *American Journal of Psychiatry* 134: 1117–1119.

Gatewood JD, Wills A, Shetty S, Xu J, Arnold AP, Burgoyne PS, and Rissman EF (2006) Sex chromosome complement and gonadal sex influence aggressive and parental behaviors in mice. *Journal of Neuroscience* 26: 2335–2342.

Gladue BA (1984) Neuroendocrine response to estrogen and sexual orientation. *Science* 22: 1496–1499.

Gladue BA and Bailey JM (1995) Spatial ability, handedness and human sexual orientation. *Psychoneuroendocrinology* 20: 487–497.

Glass SJ and Johnson RW (1944) Limitations and complications of organotherapy in male homosexuality. *Journal of Clinical Neuroendocrinology* 4: 550–554.

Goodman RE, Anderson DC, Bu'lock DE, Sheffield B, Lynch SS, and Butt WR (1985) Study of the effect of estradiol on gonadotrophin levels in untreated male-to-female transsexuals. *Archives of Sexual Behavior* 14: 141–146.

Gooren L (1986) The neuroendocrine response of LH to estrogen administration in heterosexual, homosexual and transsexual subjects. *Journal of Clinical Endocrinology and Metabolism* 63: 583–586.

Gouze KR and Nadelman L (1980) Constancy of gender identity for self and others in children between the ages of three and seven. *Child Development* 51: 275–278.

Green R (1974) *Sexual Identity Conflict in Children and Adults.* New York: Basic Books.

Green R (1976) One hundred and ten feminine and masculine boys: Behavioral contrasts and demographic similarities. *Archives of Sexual Behavior* 5: 425–446.

Green R (1985) Gender identity in childhood and later sexual orientation: Follow-up of 78 males. *American Journal of Psychiatry* 142: 339–341.

Green R (2000) Family coocurrence of 'gender dysphoria': Ten sibling or parent–child pairs. *Archives of Sexual Behavior* 29: 499–507.

Green R, Roberts CW, Williams K, Goodman M, and Mixon A (1987) Specific cross-gender behaviour in boyhood and later homosexual orientation. *British Journal of Psychiatry* 151: 84–88.

Green R and Stoller RJ (1971) Two monozygotic (identical) twin pairs discordant for gender identity. *Archives of Sexual Behavior* 1: 321–327.

Green R and Young R (2000) Fingerprint asymmetry in male and female transsexuals. *Personality and Individual Differences* 29: 993–942.

Green R and Young R (2001) Hand preference, sexual preference, and transsexualism. *Archives of Sexual Behavior* 30: 565–574.

Grumbach MM, Hugh IA, and Conte FA (2003) Disorders of sex differentiation. In: Larsen PR, Kronenberg HM, Melmed S, and Polonsky KS (eds.) *Williams Textbook of Endocrinology,* 10th edn., pp. 842–1002. Philadelphia, PA: W. B. Sanders.

Haberman M, Hollingsworth F, Falek A, and Michael RP (1975) Gender identity confusion, schizophrenia and a 47 XYY karyotype: A case report. *Psychoneuroendocrinology* 1: 207–209.

Hall J and Kimura D (1994) Dermatoglyphic asymmetry and sexual orientation in men. *Behavioral Neuroscience* 108: 1203–1206.

Halperin DM (1990) *One Hundred Years of Homosexuality and Other Essays on Greek Love.* New York: Routledge.

Hamer D (1999) Genetics and male sexual orientation. *Science* 285: 803a.

Hamer D, Hu S, Magnuson V, Hu N, and Pattatucci AM (1993) A linkage between DNA markers on the X chromosome and male sexual orientation. *Science* 261: 321–327.

Hare L, Bernard P, Sánchez FJ, Vilain E, Kennedy T, and Harley VR (in press) Androgen receptor (AR) repeat length polymorphism associated with male-to-female transsexualism. *Biological Psychiatry.*

Henningsson S, Westberg L, Nilsson S, et al. (2005) Sex steroid-related genes and male-to-female transsexualism. *Psychoneuroendocrinology* 30: 657–664.

Hepp U, Milos G, and Braun-Scharm H (2004) Gender identity disorder and anorexia nervosa in male monozygotic twins. *International Journal of Eating Disorders* 35: 239–243.

Herman-Jeglinska A, Grabowska A, and Dulko S (2002) Masculinity, femininity, and transsexualism. *Archives of Sexual Behavior* 31: 527–534.

Hill DB, Rozanski C, Carfagnini J, and Willoughby B (2005) Gender identity disorders in childhood and adolescence: A critical inquiry. *International Journal of Sexual Health* 19: 57–75.

Hines M, Brook C, and Conway GS (2004) Androgen and psychosexual development: Core gender identity, sexual orientation, and recalled childhood gender role behavior in women and men with congenital adrenal hyperplasia (CAH). *Journal of Sex Research* 41: 75–81.

Hore BD, Nicolle FV, and Calnan JS (1973) Male transsexualism: Two cases in a single family. *Archives of Sexual Behavior* 2: 317–321.

Hu S, Pattatucci A, Patterson C, et al. (1995) Linkage between sexual orientation and chromosome Xq28 in males but not females. *Nature Genetics* 11: 248–256.

Hulshoff Pol HE, Cohen-Kettenis PT, van Haren, et al. (2006) Changing your sex changes your brain: Influences of testosterone and estrogen on adult human brain structure. *European Journal of Endocrinology* 155: S107–S114.

Hyde C and Kenna JC (1977) A male monozygotic twin pair concordant for transsexualism, discordant for schizophrenia. *Acta Psychiatrica Scandinavica* 56: 265–275.

Hyde JS (2005) The gender similarities hypothesis. *American Psychologist* 60: 581–592.

Intersex Society of North America (nd) What's the difference between being transgender or transexual and having an intersex condition? http://www.isna.org/faq/transgender (accessed April 2008).

Johannsen TH, Ripa CPL, Mortensen EL, and Main KM (2006) Quality of life in 70 women with disorders of sex development. *European Journal of Endocrinology* 155: 877–885.

Johnson C (2008) Transgender teens: Doctors refine hormone, other therapies. *Foster Daily Democrat* http://www.fosters.com/apps/pbcs.dll/article?AID=/20080127/GJNEWS_01/205304745 (accessed June 2008).

Johnson KL, Gill S, Reichman V, and Tassinary LG (2007) Swagger, sway, and sexuality: Sexual orientation from body motion and morphology. *Journal of Personality and Social Psychology* 93: 321–334.

Joyce PR and Ding L (1985) Transsexual sisters. *Australian and New Zealand Journal of Psychiatry* 19: 188–189.

Katz LS, Price EO, Wallach SJR, and Zenchak JJ (1988) Sexual performance of rams reared with or without females after weaning. *Journal of Animal Science* 66: 1166–1173.

Kendler KS, Thornton LM, Gilman SE, and Kessler RC (2000) Sexual orientation in a U.S. national sample of twin and nontwin sibling pairs. *American Journal of Psychiatry* 157: 1843–1846.

Kinsey AC, Pomeroy WB, and Martin CE (1948/1975) *Sexual Behavior in the Human Male.* Bloomington, IN: Indiana University Press.

Kirk KM, Bailey JM, Dunne MP, and Martin NG (2000) Measurement models for sexual orientation in a community twin sample. *Behavior Genetics* 30: 345–356.

Klar AJS (2004) Excess of counterclockwise scalp hair-whorl rotation in homosexual men. *Journal of Genetics* 83: 251–255.

Knoblauch H, Busjahn A, and Wegener B (2007) Monozygotic twins concordant for female-to-male transsexualism: A case report. *Archives of Sexual Behavior* 36: 135–137.

Kolodny RC, Jacobs LS, Masters WH, Toro G, and Daughaday WH (1972) Plasma gonadotrophins and prolactins in male homosexuals. *Lancet* 2(776): 18–20.

Kolodny RC, Masters WH, Hendryx J, and Toro G (1971) Plasma testosterone and semen analysis in male homosexuals. *New England Journal of Medicine* 285: 1170–1174.

Kraemer B, Noll T, Delsignore A, Milos G, Schnyder U, and Hepp U (2006) Finger length ratio (2D:4D) and dimensions of sexual orientation. *Neuropsychobiology* 53: 210–214.

Kruijver FPM, Fernández-Guasti A, Fodor M, Kraan EM, and Swaab DF (2000) Male-to-female transsexuals have female neuron numbers in a limbic nucleus. *Journal of Clinical Endocrinology and Metabolism* 85: 2034–2041.

Landén M, Wålinder J, and Lundstöm B (1998) Clinical characteristics of a total cohort of female and male applicants for sex reassignment: A descriptive study. *Acta Psychiatrica Scandinavica* 97: 189–194.

Lalumière ML, Blanchard R, and Zucker KL (2000) Sexual orientation and handedness in men and women: A meta-analysis. *Psychological Bulletin* 126: 575–592.

Larsson K and Södersten P (1971) Lordosis behavior in male rats treated with estrogen in combination with tetrabenazine and nialamide. *Psychopharmacologia* 21: 13–16.

Leonard SP and Archer J (1989) A naturalistic investigation of gender constancy in three- to four-year-old children. *British Journal of Developmental Psychology* 7: 341–346.

Lev AI (2005) Disordering gender identity: Gender identity disorder in the DSM-IV-TR. *Journal of Psychology and Human Sexuality* 17: 35–69.

Lev-Ran A (1974) Sexuality and educational levels of women with the late treated adrenogenital syndrome. *Archives of Sexual Behavior* 3: 27–32.

LeVay S (1991) A difference in hypothalamic structure between heterosexual and homosexual men. *Science* 253: 1034–1037.

Lindesay J (1987) Laterality shift in homosexual men. *Neuropsychologia* 25: 965–969.

Lippa RA (2003) Are 2D:4D finger-length ratios related to sexual orientation? Yes for men, no for women. *Journal of Personality and Social Psychology* 85: 179–188.

Loeb L and Shane M (1982) The resolution of a transsexual wish in a five-year-old boy. *Journal of the American Psychoanalytic Association* 30: 419–434.

Loehlin JC, McFadden D, Medland SE, and Martin NG (2006) Population differences in finger-length ratio: Ethnicity or latitude? *Archives of Sexual Behavior* 35: 739–742.

McCormick CM and Witelson SF (1991) A cognitive profile of homosexual men compared to heterosexual men and women. *Psychoneuroendocrinology* 15: 459–473.

McFadden D and Champlin CA (2000) Comparison of auditory evoked potentials in heterosexual, homosexual, and bisexual males and females. *Journal of the Association for Research in Otolaryngology* 1: 89–99.

McFadden D, Loehlin JC, Breedlove SM, Lippa RA, Manning JT, and Rahman Q (2005) A reanalysis of five studies on sexual orientation and the relative length of the 2nd and 4th fingers (the 2D:4D ratio). *Archives of Sexual Behavior* 34: 341–356.

McFadden D and Pasanen EG (1999) Spontaneous otoacoustic emissions in heterosexuals, homosexuals, and bisexuals. *Journal of the Acoustical Society of America* 105: 2403–2413.

McKee EA, Roback HB, and Hallender MH (1976) Transsexualism in two male triplets. *American Journal of Psychiatry* 133: 334–337.

McKnight J and Malcolm J (2000) Is male homosexuality maternally linked? *Psychology Evolution Gender* 2: 229–239.

Manning JT, Scutt D, Wilson J, and Lewis-Jones DI (1998) The ratio of 2nd to 4th digit length: A predictor of sperm numbers and concentrations of testosterone, luteinizing hormone and oestrogen. *Human Reproduction* 13: 3000–3004.

Masica D, Money J, and Ehrhardt A (1971) Fetal feminization and female gender identity in the testicural feminization syndrome of androgen insensitivity. *Archives of Sexual Behavior* 1: 131–142.

Meyer W, III, Bockting WO, Cohen-Kettenis P, et al. (2001) The Harry Benjamin International Gender Dysphoria Association's Standards of Care for Gender Identity Disorders, sixth version. *Journal of Psychology and Human Sexuality* 13: 1–30.

Meyer WJ, III, Webb A, Stuart CA, Finkelstein JW, Lawrence B, and Walker PA (1986) Physical and hormonal evaluation of transsexual patients: A longitudinal study. *Archives of Sexual Behavior* 15: 121–138.

Meyer-Bahlburg HFL (1977) Sex hormones and male homosexuality: A comparative perspective. *Archives of Sexual Behavior* 6: 297–325.

Meyer-Bahlburg HFL, Dolezal C, Baker SW, and New MI (2008) Sexual orientation in women with classical or non-classical congenital adrenal hyperplasia as a function of degree of prenatal androgen excess. *Archives of Sexual Behavior* 37: 85–99.

Miles C, Green R, and Hines M (2006) Estrogen treatment on cognition, memory and mood in male-to-female transsexuals. *Hormones and Behavior* 50: 708–717.

Miles C, Green R, Sanders G, and Hines M (1998) Estrogen and memory in a transsexual population. *Hormones and Behavior* 34: 199–208.

Money J and Dalery J (1976) Latrogenic homosexuality: Gender identity in seven 46, XX chromosomal females with hyperadrenocortical hermaphroditism born with a penis. Three reared as boys, four reared as girls. *Journal of Homosexuality* 1: 357–371.

Money J and Ogunro C (1974) Behavioral sexology: Ten cases of genetic male intersexuality with impaired prenatal and pubertal androgenization. *Archives of Sexual Behavior* 3: 181–205.

Money J, Schwartz M, and Lewis V (1984) Adult erotosexual status and fetal hormonal masculinization and demasculinization 46, congenital virilizing adrenal hyperplasia and 46 XY androgen-insensitivity syndrome compared. *Psychoneuroendocrinology* 9: 405–414.

Mouaffak F, Gallarda T, Baup N, Olié JP, and Krebs MO (2008) Gender identity disorders and bipolar disorder associated with the ring Y chromosome. *American Journal of Psychiatry* 164: 1122–1123.

Mustanski BS, Chivers ML, and Bailey JM (2002) A critical review of recent biological research on human sexual orientation. *Annual Review of Sex Research* 13: 89–140.
Mustanski BS, Dupree MG, Nievergelt CM, Bocklandt S, Schork NJ, and Hamer DH (2005) A genomewide scan of male sexual orientation. *Human Genetics* 116: 272–278.
Neave N, Menaged M, and Weightman DR (1999) Sex difference in cognition: The role of testosterone and sexual orientation. *Brain and Cognition* 41: 245–262.
Nielsen J and Wohlert M (1990) Sex chromosome abnormalities found among 34,910 newborn children: Results from a 13-year incidence study in Arhus, Denmark. *Birth Defects Original Article Series* 26: 209–223.
Ökten A, Kalyoncu M, and Yaris N (2002) The ratio of second- and fourth-digit lengths and congenital adrenal hyperplasia due to 21-hydroxylase deficiency. *Early Human Development* 70: 47–54.
Olsson S and Möller A (2006) Regret after sex reassignment surgery in male-to-female transsexual: A long-term follow-up. *Archives of Sexual Behavior* 35: 501–506.
Oriel KA (2000) Medical care of transsexual patients. *Journal of the Gay and Lesbian Medical Association* 4: 185–194.
Pattatucci AML and Hamer DH (1995) Development and familiality of sexual orientation in females. *Behavior Genetics* 25: 407–420.
Paup DC, Mennin SP, and Gorski RA (1975) Androgen- and estrogen-induced copulatory behavior and inhibition of luteinizing hormone (LH) secretion in the male rat. *Hormones and Behavior* 6: 35–46.
Person E and Ovesey L (1974a) The transsexual syndrome in males: I. Primary transsexualism. *American Journal of Psychotherapy* 28: 4–20.
Person E and Ovesey L (1974b) The transsexual syndrome in males: II. Secondary transsexualism. *American Journal of Psychotherapy* 28: 174–193.
Phelps VR (1952) Relative index finger length as a sex-influenced trait in man. *American Journal of Human Genetics* 4: 72–89.
Phoenix CH, Goy RW, Gerall AA, and Young WC (1959) Organizing action of prenatally administered testosterone propionate on the tissue mediating mating behavior in the female guinea pig. *Endocrinology* 65: 369–382.
Pierrehumbert JB, Bent T, Munson B, Bradlow AR, and Bailey JM (2004) The influence of sexual orientation on vowel production (L). *Journal of the Acoustical Society of America* 116: 1905–1908.
Pillard RC (1990) The Kinsey scale: Is it familial? In: McWhirter DP, Sanders SA, and Reinisch JM (eds.) *Homosexuality/Heterosexuality: Concepts of Sexual Orientation. The Kinsey Institute Series*, pp. 88–100. New York: Oxford University Press.
Pillard RC, Rose RM, and Sherwood M (1974) Plasma testosterone levels in homosexual men. *Archives of Sexual Behavior* 3: 453–458.
Pillard RC and Weinrich JD (1986) Evidence of familial nature of male homosexuality. *Archives of General Psychiatry* 43: 808–812.
Poasa KH, Blanchard R, and Zucker KJ (2004) Birth order in transgendered males from Polynesia: A quantitative study of Samoan fa'afafine. *Journal of Sex and Marital Therapy* 30: 13–23.
Ponseti J, Siebner HR, Klöppel S, Wolff S, Granert O, Mehdorn HM, and Bosinski HA (2007) Homosexual women have less grey matter in perirhinal cortex than heterosexual women. *PLoS ONE* 2, doi:10.1371/journalpone.0000762.
Price EO, Katz LS, Wallach SJR, and Zenchak JJ (1988) The relationship of male–male mounting to the sexual preferences of young rams. *Applied Animal Behaviour Science* 21: 347–352.
Quadagno DM, Shryne J, Anderson C, and Gorski RA (1972) Influence of gonadal hormones on social, sexual, emergence, and open field behaviour in the rat (*Rattus norvegicus*). *Animal Behaviour* 20: 732–740.
Rahman Q, Cockburn A, and Govier E (2008) A comparative analysis of functional cerebral asymmetry in lesbian women, heterosexual women, and heterosexual men. *Archives of Sexual Behavior* 37: 566–571.
Rahman Q, Kumari V, and Wilson GD (2003) Sexual orientation-related differences in prepulse inhibition of the human startle response. *Behavioral Neuroscience* 117: 1096–1102.
Reed B (2006) Early medical treatment for transsexual people. *Gender Identity Research and Education Society.* http://www.gires.org.uk/earlymed.php (accessed June 2008).
Rice G, Anderson C, Risch N, and Ebers G (1999a) Male homosexuality: Absence of linkage to microsatellite markers at Xq28. *Science* 284: 665–667.
Rice G, Risch N, and Ebers G (1999b) Response: Genetics and male sexual orientation. *Science* 285: 803.
Rieger G, Chivers ML, and Bailey JM (2005) Sexual arousal patterns of bisexual men. *Psychological Science* 16: 579–584.
Robinson SJ and Manning JT (2000) The ratio of 2nd to 4th digit length and male homosexuality. *Evolution and Human Behavior* 21: 333–345.
Roselli CE, Larkin K, Schrunk JM, and Stormshak F (2004) Sexual partner preference, hypothalamic morphology and aromatase in rams. *Physiology and Behavior* 83: 233–245.
Roy WG (2001) *Making Societies: The Historical Construction of Our World.* Thousand Oaks, CA: Pine Forge Press.
Rule NO and Ambady N (2008). Brief exposures: Male sexual orientation is accurately perceived at 50 ms. *Journal of Experimental Social Psychology* 44: 1100–1105.
Sabalis RF, Frances A, Appenzeller SN, and Moseley WB (1974) The three sisters: Transsexual male siblings. *American Journal of Psychiatry* 131: 907–909.
Sadeghi M and Fakhrai A (2000) Transsexualism in female monozygotic twins: A case report. *Australian and New Zealand Journal of Psychiatry* 34: 862–864.
Safron A, Barch B, Bailey JM, Gitelman DR, Parrish TB, and Reber PJ (2007) Neural correlates of sexual arousal in homosexual and heterosexual men. *Behavioral Neuroscience* 121: 237–248.
Sanders AR and Dawood K (2003) Sexual orientation. In: *Nature Encyclopedia of Life Sciences.* London: Nature Publishing Group (doi: 10.1038/npg.els.0001480).
Savic I, Berglund H, and Lindström P (2005) Brain response to putative pheromones in homosexual men. *Proceedings of the National Academy of Sciences of the United States of America* 102: 7356–7361.
Savic I and Lindström P (2008) PET and MRI show differences in cerebral asymmetry and functional connectivity between homo- and heterosexual subjects. *Proceedings of the National Academy of Sciences of the United States of America* 105: 9403–9408.
Schneider HJ, Pickel J, and Stalla GK (2006) Typical female 2nd–4th finger length (2D:4D) ratios in male-to-female transsexuals – possible implications for prenatal androgen exposure. *Psychoneuroendocrinology* 31: 265–269.
Segal NL (2006) Two monozygotic twin pairs discordant for female-to-male transsexualism. *Archives of Sexual Behavior* 35: 347–358.

Sell RL, Wells JA, and Wypij D (1995) The prevalence of homosexual behavior and attraction in the United States, the United Kingdom and France: Results of national population-based samples. *Archives of Sexual Behavior* 24: 235–248.
Slabbekoorn D, van Goozen SHM, Sanders G, Gooren LJG, and Cohen-Kettenis PT (2000) The dermatoglyphic characteristics of transsexuals: Is there evidence for an organizing effect of sex hormones? *Psychoneuroendocrinology* 25: 365–375.
Smith YLS, van Goozen SHM, Kuiper AJ, and Cohen-Kettenis PT (2005) Transsexual subtypes: Clinical and theoretical significance. *Psychiatry Research* 137: 151–160.
Snaith RP, Penhale S, and Horsfield P (1991) Male-to-female transsexual with XYY karyotype. *Lancet* 337: 557–558.
Spijkstra JJ, Spinder T, and Gooren LJG (1988) Short-term patterns of pulsatile luteinizing hormone secretion do not differ between male-to-female transsexuals and heterosexual men. *Psychoneuroendocrinology* 13: 279–273.
Spiová L and Starká L (1977) Plasma testosterone values in transsexual women. *Archives of Sexual Behavior* 6: 477–481.
Starká L, Spiová I, and Hynie J (1975) Plasma testosterone in male transsexuals and homosexuals. *Journal of Sex Research* 11: 134–138.
Stoller RJ (1979) Fathers of transsexual children. *Journal of the American Psychoanalytic Association* 27: 837–866.
Stoller RJ and Baker HJ (1973) Two male transsexuals in one family. *Archives of Sexual Behavior* 2: 323–328.
Swaab DF, Zhou JN, Fodor M, and Hofman MA (1997) Sexual differentiation of the human hypothalamus: Differences according to sex, sexual orientation, and transsexuality. In: Ellis L and Ebertz L (eds.) *Sexual Orientation: Toward Biological Understanding*, pp. 129–150. Westport, CT: Praeger Publishers.
Taneja N, Ammini AC, Mohapatra I, Saxena S, and Kucheria K (1992) A transsexual male with 47, XYY karyotype. *British Journal of Psychiatry* 161: 698–699.
Tourney G and Hatfield LM (1973) Androgen metabolism in schizophrenics, homosexuals, and normal controls. *Biological Psychiatry* 6: 23–26.
Tsoi WF (1988) The prevalence of transsexualism in Singapore. *Acta Psychiatrica Scandinavica* 78: 501–504.
Tsoi WF (1990) Developmental profile of 200 male and 100 female transsexuals in Singapore. *Archives of Sexual Behavior* 19: 595–605.
Tsur H, Borenstein A, and Seidman DS (1991) Transsexualism. *Lancet* 333: 945–946.
Turan MT, Esel E, Dündar M, Candemir Z, Bastürk M, Sofuglu S, and Ozkul Y (2000) Female-to-male transsexual with 47, XXX karyotype. *Biological Psychiatry* 48: 1116–1117.
Tuttle GE and Pillard RC (1991) Sexual orientation and cognitive abilities. *Archives of Sexual Behavior* 20: 307–318.
van Kesteren PJM, Asscheman H, Megens JAJ, and Gooren JG (1997) Mortality and morbidity in transsexual subjects treated with cross-sex hormones. *Clinical Endocrinology* 47: 337–343.
van Goozen SHM, Cohen-Kettenis PT, Gooren LJG, Frijda NH, and van de Poll NE (1995) Gender differences in behaviour: Activating effects of cross-sex hormones. *Psychoneuroendocrinology* 20: 343–363.
van Goozen SHM, Slabbekoorn D, Gooren LJG, and Sanders G (2002) Organizing and activating effects of sex hormones in homosexual transsexuals. *Behavioral Neuroscience* 116: 982–988.
Voeller B (1990) Some uses and abuses of the Kinsey Scale. In: McWhirther DP, Sanders SA, and Reinisch JM (eds.) *Homosexuality/Heterosexuality: Concepts of Sexual Orientation*, pp. 32–38. New York: Oxford University Press.
Wakenfield JC (1997) Diagnosing *DSM-IV* – Part I: *DSM-IV* and the concept of disorder. *Behaviour Research and Therapy* 35: 633–649.
Wålinder J (1971) Incidence and sex ratio of transsexualism in Sweden. *British Journal of Psychiatry* 119: 195–196.
Wallien MSC, Zucker KJ, Steensma TD, and Cohen-Kettenis PT (2008) 2D:4D finger-length ratios in children and adults with gender identity disorder. *Hormones and Behavior*. 54: 450–454.
Wegesin DJ (1998) A neuropsychological profile of homosexual and heterosexual men and women. *Archives of Sexual Behavior* 27: 91–108.
Widiger TA and Clark LA (2000) Toward *DSM-V* and the classification of psychopathology. *Psychological Bulletin* 126: 946–963.
Williams TJ, Pepitone ME, Christensen SE, et al. (2000) Finger-length ratios and sexual orientation. *Nature* 404: 455–456.
Wood W and Eagly AH (2002) A cross-cultural analysis of the behavior of women and men: Implications for the origins of sex differences. *Psychological Bulletin* 128: 699–727.
Zhou JN, Hofman MA, Gooren LJ, and Swaab DF (1995) A sex difference in the human brain and its relation to transsexuality. *Nature* 378: 68–70.
Zucker KJ (2001) Gender identity disorder in children and adolescents. In: Gabbard GO (ed.) *Treatments of Psychiatric Disorders,* 3rd edn., pp. 2069–2094. Washington, DC: American Psychiatric Association.
Zucker KJ (2005) Gender identity disorder in children and adolescents. *Annual Review of Clinical Psychology* 1: 467–492.
Zucker KJ, Blanchard R, Kim T-K, Pae C-U, and Lee C (2006a) Birth order and sibling ratio in homosexual transsexual South Korean men: Effects of the male-preference stopping rule. *Psychiatric and Clinical Neurosciences* 61: 529–533.
Zucker KJ and Bradely SJ (1995) *Gender Identity Disorder and Psychosexual Problems in Children and Adolescence.* New York: Guilford Press.
Zucker KJ, Bradley SJ, Oliver G, Blake J, Fleming S, and Hood J (1996) Psychosexual development of women with congenital adrenal hyperplasia. *Hormones and Behavior* 30: 300–318.
Zucker KJ, Mitchell JN, Bradley SJ, Tkachuk J, Cantor JM, and Allin SM (2006b) The recalled childhood gender identity/gender role questionnaire: Psychometric properties. *Sex Roles* 54: 469–483.
Zucker KJ and Spitzer RL (2005) Was the gender identity disorder of childhood diagnosis introduced into the *DSM-III* as a backdoor maneuver to replace homosexuality? *Journal of Sex and Marital Therapy* 31: 31–41.

Further Reading

Manning JT, Barley L, Walton J, et al. (2000) The 2nd:4th digit ratio, sexual dimorphism, population differences, and reproductive success: Evidence for sexually antagonistic genes. *Evolution and Human Behavior* 21: 163–183.
Tsoi WF, Kok LP, and Long FY (1977) Male transsexualism in Singapore: A description of 56 cases. *British Journal of Psychiatry* 131: 405–409.

11 Sexual Orientation in Men and Women

L J Gooren, VU University Medical Center, Amsterdam, The Netherlands
W Byne, Mount Sinai School of Medicine, New York, NY, USA

Chapter Outline

Glossary

androphilia Erotosexual love directed to a man.
gender One's personal, social, and legal status as male or female.
gender identity One's sense of self as belonging to a particular gender category or one's sense of goodness of fit with one or more gender categories.
gynephilia Erotosexual love directed to a woman.
intersex A term which has replaced hermaphroditism and pseudohermaphroditism indicating that one or more biological variables of sex is discordant with the others or is incompletely differentiated.

11.1 History of the Concept of Homosexuality

Homosexual activity has been universal throughout history. In every historical period there is some record of homosexual activity although different societies, cultures, and belief systems have regarded these activities with a very wide range of attitudes (Bullough, 1976).

The phenomenon now called homosexuality did not have that name until it was coined by Karl-Maria Kertbeny in 1869:

> In addition to the normal sexual urge in men and women, Nature in her sovereign mood has endowed at birth certain male and female individuals with the homosexual urge, thus placing them in a sexual bondage which renders them physically and psychically incapable – even with the best intention – of normal erection. This urge creates in advance a direct horror of the opposite sex, and the victim of this passion finds it impossible to suppress the feeling which individuals of his own sex exercise upon him.
> (Bullough, 1976: 637)

Kertbeny believed that homosexuality was inborn and unchangeable, opposing the view that homosexual men committed sodomy out of mere wickedness. He argued that homosexual men were not by nature effeminate and also pointed out that many of the great heroes of history had engaged in homosexual behavior. Kertbeny was certainly a campaigner for homosexual rights. But unlike Ulrichs and Hirschfeld later, Kertbeny did not seek to use primarily biological arguments for the liberation of homosexuals. He made the point that the modern state should extend the principle of not intervening in the private lives of citizens to apply to homosexuals, too:

> To prove innateness ... is a dangerous double edged weapon. Let this riddle of nature be very interesting from the anthropological point of view. Legislation is not concerned whether this inclination is innate or not, legislation is only interested in the personal and social dangers associated with it ... Therefore we would not win anything by proving innateness beyond a shadow of doubt. Instead we should convince our opponents – with precisely the same legal notions used by them – that they do not have anything at all to do with this inclination, be it innate or intentional, since the state does not have the right to intervene in anything that occurs between two consenting persons older than fourteen, which does not affect the public sphere, nor the rights of a third party.

One could say, a remarkably modern position.

11.1.1 The Third Sex as Homosexuality

In contrast to Kertbeny, Ulrichs (1825–95) and later Hirschfeld (1868–1935) were not confident that they could successfully argue legal rights for homosexuals unless it became irrefutably clear that homosexuality was inborn and had a biological substrate. If proven, they believed, it would be gross injustice to prosecute homosexuals since their behaviors were not acts of free will but determined by their biological constitution. In a letter dated 22 September 1862 addressed to his sister, the German lawyer and scholar Karl Heinrich Ulrichs used the term third sex to refer to homosexuals (Kennedy, 1997). He explains that people like him are not men in the common sense of the word, since there is a decidedly female element in them. Although they have a male body, they are spiritually female in correspondence to the direction of their sexual love. Insisting on the inborn and natural character of such a sexual disposition, Ulrichs redefined sexuality within a triadic scope of sexual possibilities and concluded: we constitute a third sex.

11.1.2 Hirschfeld and the Concept of the Third Sex

Magnus Hirschfeld, who used the term third sex for the first time in 1899, made the concept popular in the twentieth century (Bullough, 2000). Since third sex – like the term Urning [Uranian], another of Ulrichs' terms – had no moral implications, it became widely used in self-descriptive contexts.

With third sex, Hirschfeld designated a whole range of intermediate expressions of sexuality that could not be readily classified using the male/female scheme. Although Hirschfeld did not employ the term very often, he conceded that its advantage over homosexuality consisted in the fact that it does not necessarily connote sexual acts.

Subscribing to the adage justice through science both Ulrichs and Hirschfeld believed that presenting homosexuality as a biologically determined entity, and therefore, a category beyond one's free will, would foster the social acceptance of homosexuals and confer protection from prosecution. Therefore, Hirschfeld developed a vivid interest in the biological experiments of Steinach.

11.1.3 The Hormonal Theories of Steinach

In considering how homosexuality might come about, Hirschfeld was influenced by the developments in endocrinology, especially by the research of the Viennese endocrinologist, Eugen Steinach (1861–1944) (Bullough, 1976). During the first decade of the twentieth century, Steinach performed transplantations of testes and ovaries in rats and guinea pigs. His research showed that these glands secrete hormones into the bloodstream that influence not only the animals' physical development but also their sexual behavior. These secretions, he argued, were responsible for the sexualization of the brain as male or female. He suggested that this sexualization occurs early in life, because he observed the most dramatic effects in guinea pigs when the transplantations were performed shortly after birth. Steinach developed the notion, partly under the influence of Hirschfeld's quest for biological theories of a third sex, that testicular secretions in homosexual men were abnormal and that they drove brain development in a female, rather than a male, direction. He even believed that he saw microscopic differences in the structure of the testis between homosexual and heterosexual men; these differences were not in the sperm-forming cells but in the interstitial cells, the cells that he had shown to be responsible for the secretion of testicular hormones.

True to his training as an experimentalist, Steinach tested his hypothesis by conducting transplants in humans. In 1917, he published a sensational report in the Jahrbuch that described the results of transplanting a testicle from a heterosexual man into an effeminate, passive homosexual man. According to the report, the man was totally cured; he was said to have lost all attraction to men and to have developed normal heterosexual feelings.

Steinach's experiment seemed to provide dramatic support for a biological explanation of homosexuality and for the concept that homosexual men have a female sexual brain. But Hirschfeld published the account of another man whose experience was quite different:

> After my wife gave her consent, I underwent a bilateral castration. The operation was performed by a well-known surgeon, with the understanding that it would be followed later by the implantation of a testicle from a heterosexual man. As I was already over forty, the initial operation didn't have any dramatic effects. My voice and facial hair weren't affected. My sex drive declined in strength but didn't change its direction. I did lose my body hair, though. A year later the testicle of a heterosexual man was implanted in my abdominal cavity. My body hair began to regrow, but six months later it disappeared again. My sex drive gradually declined until it finally disappeared, but it never changed its direction.

11.2 Paradigm of Biomedical Research into Homosexuality

When research tools in biomedicine improved, the principle guiding biomedical research of homosexuality remained faithful to the conceptualization of Ulrichs, as identifying female biological traits in male homosexuals and male biological traits in female homosexuals. Not stated anymore explicitly, the paradigm in this type of research has remained that expressions of sexuality, such as homosexuality, are variations on the fundamentals of male/female dichotomy and heterosexual interaction (Gorman, 1994; Gooren, 2006). Some researchers (Dorner, 1988) make the gross distinction of heterosexual versus nonheterosexual and do not differentiate between homosexuality and transgenderism, viewing them as variations only in the degree of expressions of femininity in males and of masculinity in females. This approach lacks thorough enquiry into the matter, since the two phenomena (sexual orientation and gender identity) are different.

Transsexualism involves identifying oneself as a member of the other gender or desiring to become the other gender. On the other hand, homosexuality usually entails no desire to become the other gender but merely signifies one's erotic interests are by and large confined to members of one's own sex and gender. The issue of femininity in male homosexuals has been reviewed recently establishing large degrees of fluidity in populations and eras (Sandfort, 2005). The exaggerated effeminacy encountered in some homosexuals is not rarely a caricature of women's manners, an attempt to amuse, and sometimes acted out only in the company of peers and not in other contexts. Even though effeminate homosexual men may see themselves as sharing some temperamental characteristics and interests with women and, therefore, identify with, not as, women, few express the desire to become women.

The terms homosexual and heterosexual are problematic in a variety of contexts, such as in transgendered and intersex subjects. One might ask, homosexual with respect to what – a particular variable of biological sex (e.g., gonads, external genitalia, and chromosomes) or perhaps homosexual with respect to gender role or identity? Androphilic, gynephilic, and bisexual are much more precise terms for sexual orientation and do not confound sexual orientation with sexual identity. Terms such as transgendered, gay, lesbian, bisexual, etc., describe identities, not orientations. Nevertheless, a reputable expert has labeled androphilic transsexualism, which usually has an early onset as homosexual transsexualism, and non-androphilic transsexualism, which usually has a later onset as nonhomosexual transsexualism in males (Blanchard, 2005). Transgender/transsexual phenomena are poorly understood and their typologies are a subject of current controversy among transgendered individuals as well as those who study them. These controversies likely reflect the inability of any single taxonomic schema to accommodate the multiplicity of presentations of transgenderism. Thus, many transsexuals do not recognize themselves in Blanchard's typology. First, the so-called homosexual male-to-female transsexuals may not view themselves as homosexual. If they interact with a man before sex-reassignment surgery, they may view themselves in their sexual interaction as women for the time being handicapped by a male anatomy, not as homosexual men interacting with another homosexual man. Their potential partners at that stage, though, are often homosexual men who are frustrated by the denial of manhood on the side of the transsexual. Post-sex-reassignment surgery these transsexuals have a neovagina and their partners are more likely to be heterosexual men and the sexual behavior of these transsexual subjects can no longer be labeled as homosexual. But the transsexuals do not think their sexual orientation has changed; they themselves never regarded themselves as homosexual. In fact, their orientation has not changed. It was androphilic prior to surgery and remains so afterward. Even before sex reassignment, transgendered males desiring to have sex with men do not regard themselves as homosexuals since they view themselves as female. Not rarely, they find labeling of their sexual behavior as homosexual insulting. This may be because it denies their being female. Alternatively, internalized antihomosexual attitudes may be one of the factors that motivates, at least some androphilic males, to seek sex reassignment to reconcile their sexual orientation to men with an adopted female gender identity (Hellman et al., 1981). That is, their defense mechanisms may not allow them to view themselves as homosexual because they view homosexuality as unacceptable. It is not possible to know the motivations of others with certainty, leaving much room for experts to disagree as to what factors motivate individuals to seek sex-reassignment surgery. A current matter of contentious debate concerns the motivations of Blanchard's nonhomoseuxal male-to-female transsexuals who are in the early phase usually, but not always, sexually oriented toward women. Even those who at that stage may see themselves as nonhomosexual may desire sex with men. Some have argued that such desires are not true androphilia, but instead, men are used as sexual props in order to heighten the nonhomosexual's fantasy of being a woman (analogous to the way donning feminine attire may intensify that fantasy). Postoperatively, one study found that about 50% experience a shift in sexual orientation to men (Lawrence, 2005). At this point, it remains controversial as to what has changed – their erotic desire or their narrative interpretation of that desire. As with the homosexual transsexuals of Blanchard's typology, this is another situation in which biologically male individuals may desire and engage in sexual intercourse with other biological males and in which, not only do the participants not see themselves as homosexuals, but also sexologists are divided as to whether or not they should be considered homosexual.

A similar reasoning applies to men who engage in sexual activities with men for financial reasons but do not derive any lust and fulfillment from these engagements. They do not regard themselves as homosexual. An important issue in defining homosexuality is whether the criterion should be sexual acts between two persons with the same body morphology or whether being attracted to/falling in love with persons with the same body morphology should also be a criterion. Currently, a distinction is generally made between homosexuality which is defined by erotic desire and homosexual behavior which does not necessarily require sexual desire for a same-sexed partner.

The conclusion must be that sexuality of the human species is characterized by the attribution of the meanings that human beings give to their sexuality which they signify with language. This adds a level of complexity to human sex research compared to animal sexology where only observable behaviors can be examined. It further underscores the need to use separate terminology for descriptions of sexual acts (homosexual or heterosexual), sexual orientation (androphilic, gynephilic, analophilic, and bisexual), and sexual identities. Moreover, identities are multilayered and reflect not only how a person regards the self in terms of biological status (e.g., male, female, or intersex), social category (e.g., man or woman), and sexual orientation (e.g., heterosexual, homosexual, bisexual, or analosexual), but with respect to the interactions across those individual layers (e.g., gay man, nongay homosexual, lesbian, male-to-female transsexual, or transsexual woman).

The assumption of female traits in male homosexuals and, vice versa – of male traits in female homosexuals – has a tradition since Ulrich and Hirschfeld, and its validity to guide biomedical research is up for review. Surprisingly, up to the present day this intersex hypothesis of homosexuality is used frequently without questioning – as if it were self-evident that homosexual behavior is an expression of cross-sex sexual differentiation of the substrate of sexual behavior, the brain. This is not to say that sexual behavior, including homosexual behavior, has no biological substrate. The following review, however, will show that the search for biological signs of femaleness in male homosexuals and for signs of maleness in female homosexuals has met with little success, especially with respect to homosexuality in men.

11.3 The Search for Cross-Sex Endocrine Findings in Homosexuals

Biomedical research following the intersex hypothesis of homosexuality got a major impetus when methods for accurate measurement of androgens and estrogens, often referred to as, respectively, male and female hormones (semantically significant!), became more widely available. At least 20 studies have measured peripheral levels of sex steroids in male homosexuals and at least five studies did so in female homosexuals. A few claimed to find differences between homosexuals and heterosexuals in the theorized direction: less male hormone and/or more female hormone in male homosexuals and vice versa in female homosexuals; however, some found differences in the opposite direction. These findings have been reviewed by several authors (Meyer-Bahlburg, 1977, 1979; Gooren, 1990) and must be dismissed as suffering from faulty design and interpretation. Moreover, with the present insights into hormonal action it is evident that the peripheral blood level of a hormone is merely one of the indices that interactively determine the strength of a hormonal signal. Its eventual biological effect will be co-determined by the properties of hormonal metabolic and converting enzymes, as well as receptors and postreceptor events, rendering the peripheral blood level not interpretable in the absence of knowledge of the other indices. In other words, in some men, hormone blood levels may be somewhat lower, but yet provide a stronger biological androgenic signal than in other men who have slightly higher hormone levels.

Capitalizing on these new insights, studies have been performed to analyze whether genetic properties of sex-steroid receptors in transsexuals and homosexuals differ, rendering them differently sensitive to the masculinizing action of sex steroids. In a recent study, sex-steroid-related genes were investigated in male-to-female transsexuals who appeared to have longer estrogen-receptor-beta (ERβ) repeat polymorphisms but no differences were found with regard to the androgen receptor and the aromatase gene (Henningsson et al., 2005). While one study found variations of the androgen-receptor gene associated with rating scale measures of masculinity and femininity (Macke et al., 1993) another study in monozygotic and dizygotic twins found no such relation (Loehlin et al., 2004). Further, gene properties coding for aromatase – the enzyme which converts androgens to estrogens – have not been observed to differ between homosexual and heterosexual men (DuPree et al., 2004). Finally, if sexual orientation were determined by developmental hormonalization, one might expect to find clinical manifestations of altered hormone profiles in homosexuals. These, however, usually turn out to be nonexistent or very subtle, and of dubious clinical relevance. Conversely, one might expect to see increased rates of homosexuality in individuals known to have experienced atypical hormonalization developmentally. In clinical practice, numerous patients are encountered with gross abnormalities of their hormonal profiles. As a rule this does not impact on either their gender identity or sexual orientation. The *a priori* likelihood that very subtle changes in hormonal profiles and action, not strong enough to be clinically manifest, would impact on sexual orientation is, therefore, not large. Somewhat more sophisticated, but – from historical and endocrine viewpoints – misguided, was the search for evidence of an estrogen positive-feedback signal on secretion of luteinizing hormone (LH) – the endocrine correlate of ovulation in women of fertile age. In lower mammals prenatal androgenization masculinizes behavior and simultaneously abolishes the capacity to display an estrogen positive-feedback signal. The estrogen positive-feedback signal in homosexual and transsexual males, if it were demonstrable, was thought to be proof of less-than-normal androgenization in the prenatal lives of men. Two studies claimed to demonstrate an estrogen positive feedback in homosexuals and transsexuals (Gladue et al., 1984; Dorner, 1988), but the designs of these studies were inadequate in the sense that they failed to demonstrate that all requirements of a true estrogen positive feedback were fulfilled. Replication studies by Gooren (1986a,b) demonstrated that in transsexual and homosexual males with their normal male hormonal milieus an estrogen positive feedback cannot be elicited. However, in transsexuals, having undergone sex reassignment and subsequently having a female hormonal milieu, the estrogen positive feedback became demonstrable. So, it appeared that the ambient endocrine milieu, rather than the prenatal endocrine history, determines whether an estrogen positive feedback can be elicited. Maybe an even more important case in point is that in primates, in contrast to lower mammals, prenatal androgen exposure of neuroendocrine structures does not abolish the capacity to respond with an estrogen positive feedback (Westfahl et al., 1984). Women with substantial prenatal androgen exposure, due to congentital virilizing adrenal hyperplasia, who are born with masculinized external genitalia are capable of having full-term pregnancies and offspring (Krone et al., 2001), and this was well known at least two decades prior to the first attempts to differentiate between homosexual and heterosexual men on the basis of positive-feedback responses. This fundamentally undermines the attempts to use the estrogen positive feedback as a telltale of prenatal brain androgenization in primates, including humans (Gooren, 1990). In conclusion, attempts to find signs of a female hormonal milieu in male-to-female transsexuals and male homosexuals and vice versa have been unsuccessful. It is disconcerting that so many faulty endocrine studies passed the review process in journals which reported these findings. A corollary of the expectation to find positive feedback in homosexual but not heterosexual men would be the absence of a positive-feedback response and, therefore, anovulatory infertility in lesbians. Regular menstruation and pregnancy in lesbians disproves this simple hypothesis. Sexology prides itself in being a multidisciplinary or an interdisciplinary science. The downside of this lofty ideal is that, compared to submissions to monodisciplinary journals, research papers in sexology spanning several disciplines are often insufficiently scrutinized for their scientific rigor.

11.4 The Prenatal Hormonal Hypothesis

According to this hypothesis, prenatal hormones act, primarily during embryonic and fetal development, to mediate the sexual differentiation not only of the internal and external genitalia but also of the brain. The sexually differentiated state of the brain then influences the subsequent expression of gender identity and sexual orientation (Gooren, 2006). Intersexuality results from variation in the normative course of somatic sexual differentiation, and homosexuality and bisexuality have been proposed to reflect variant sexual differentiation of hypothetical neural substrates that mediate sexual orientation. Similarly, transgenderism has been conjectured to reflect variant differentiation of hypothetical neural substrates that mediate gender identity. Some of the same hormones and hormonal receptors mediate the sexual differentiation of both the brain and the genitalia. Thus, the brains, as well as the genitalia, of intersexes may exhibit sexual differentiation that is intermediate between that of normatively developed males and females.

The prenatal hormonal hypothesis is widely used to account for the sexual orientation and gender identity that will manifest themselves postnatally. We wish to consider here whether, and when, in early development, hormones, particularly testosterone and its metabolites, might exert such organizing influences on sexual orientation.

It is therefore appropriate to review properties of prenatal testosterone physiology and some research findings with regard to sexual differentiation of the human brain.

11.4.1 Prenatal/Postnatal Testosterone Physiology

The literature on the biological aspects of homosexuality often refers to prenatal testosterone as the determinant of a male or female sexual differentiation of the brain (Gladue et al., 1984; Dorner, 1988). But, in the relevant literature, there is rarely a detailed description of the prenatal testosterone physiology. Sex differences in testosterone levels are not quantitatively significant throughout pregnancy but only between weeks 9 and 26 of pregnancy – the period of formation of the internal and external genitalia. Testosterone production by the human fetal testis is detectable at 9 weeks, and peaks between 14 and 17 weeks of prenatal development. The presence of testosterone is pivotal for prenatal male development of Wolffian ducts and the masculinization of external genitalia. Differentiation of fetal Leydig cells and early testosterone secretion depends on placental human chorionic gonadotropin (hCG) in the first and second trimesters of fetal life and on the LH of the fetus' own pituitary thereafter. Leydig cell number peaks at mid-gestation and then slightly decreases. Testosterone production by the human fetal testes, starting at approximately 8–9 weeks of prenatal development with differentiation of the Leydig cells, reaches a maximum between 14 and 19 weeks and subsequently declines sharply, so the serum concentrations of testosterone are not distinctly different in males and females in late pregnancy (Word et al., 1989). Testicular and serum levels of testosterone are closely correlated with hCG concentration such that the peak of fetal testicular-testosterone production coincides with the acme of concentrations of hCG in the circulation.

Current understanding of sexual differentiation of the internal/external genitalia is still in agreement with the classical experiments of Jost (1947). To a first approximation, an embryo, regardless of its genetic sex, develops along female lines provided it is not exposed to testicular secretions. The urogenital sinus and external genitalia are virilized by androgens. The internal reproductive tract is similar in male and female fetuses up to 56 days of prenatal development. The upper part of the Wolffian duct differentiates into the epididymis. The more caudal part of the Wolffian duct develops a layer of smooth muscle to become the vas deferens.

Up to approximately 72–74 days the urogenital sinus and external genitalia are not differentiated. Male development of the external genitalia starts by 72 days of prenatal development with a fusion of the labioscrotal folds to form the epithelial seam inducing the closure of the primary urethral groove. The next step is the extension of the urethral plate in the roof of the primary urethral groove to the tip of the penis. The development of the urethra is completed after approximately 98 days.

When chorionic gonadotropin declines in the third trimester, the hypothalamic–pituitary axis takes over (Rabinovici and Jaffe, 1990). Impaired pituitary secretion of LH in 46, XY fetuses (such as in idiopathic hypogonadotropic hypogonadism (IHH) or Kallmann syndrome, characterized by the agenesis of neurons producing LH-releasing hormone) does not result in major signs of impaired male sexual differentiation because the most important steps of somatic sexual differentiation would have already taken place during the fourth month, when Leydig cells are controlled by hCG. These newborns may present with cryptorchidism and micropenis, owing to reduced Leydig cell number and low testosterone production during the third trimester. Since 1975, there has been a postnatal test of sensitivity to testosterone so that in cases of an extreme degree of micropenis, boys with Kallmann syndrome have not been reassigned as girls (Grumbach, 2005).

Given that it is a rare disease, data on the incidence of IHH are limited. The estimated incidence varies from 1/10 000 to 1/30 000 (Fromantin et al., 1973; Filippi, 1986). Isolated GnRH deficiency occurs more commonly in men than in women. Based on the review of 250 consecutive cases seen at the Massachusetts General Hospital, the male:female ratio is 4:1.

Fetal testosterone levels have been lower than normal for male fetuses in the third trimester of pregnancy in subjects with IHH or Kallmann syndrome. In the male infant, serum FSH, LH, and testosterone levels increase during the second postnatal week, reach a maximum at 4–10 weeks, and decline to low levels by about 6 months of age (Grumbach, 2005). This postnatal testosterone surge occurring in boys depends on stimulation by their own gonadotropins, the production of which is lacking in boys with Kallmann syndrome (Grumbach, 2005). This surge often has been thought to be significant for male psychosexual differentiation; however, there is no evidence that men with Kallmann syndrome are overrepresented in samples of transsexuals. Transsexuals undergo a somatic screening before endocrine treatment and there is only one report of a male subject with Kallmann's syndrome who presented as a male-to-female transsexual (Meyenburg and Sigusch, 2001). There are no literature data on sexual orientation of men with Kallmann syndrome. Patients with Kallmann's syndrome are, in principle, fertile and those in a relationship with a female partner often desire to father children (Buchter et al., 1998). In the authors' opinion, Kallmann syndrome constitutes an example of a very plausible diminished degree of exposure of the brain to testosterone during the last trimester of pregnancy and during the first postnatal months. As indicated above, these men do no distinguish themselves significantly as homosexual and transsexual.

11.4.2 Impact of Prenatal Hormones on Sexual Orientation/Gender Identity: Lessons from Clinical Syndromes

The extent to which hormones contribute to sexual differentiation of the brain in humans remains to be determined (Collaer et al., 2004). In humans this information is to be obtained from experiments of nature: genetic and endocrine disorders that spontaneously occur in the fetus (Money, 1981; Gooren, 1990; Migeon and Wisniewski, 2000, 2003). The morphology of the brains of subjects with these conditions has not been studied. But overall, clinical observations support the hypothesis that in human prenatal development, effects of androgens on sexual-brain differentiation can be recognized, but these are not of the hormonal-robot type found in subprimate mammals, in which sex steroids, in the set of behaviors studied, typically exert on–off activational effects and straightforward organizational effects (some of which are simple on–off effects (e.g., defeminization of estrogen positive feedback) and some of which appear to be graded responses depending on the length and magnitude of androgen exposure (e.g., mounting) (Money, 1981; Gooren, 1990; Migeon and Wisniewski, 2000, 2003)). In addition, there are certainly other unidentified factors that modulate or override androgen effects on the central nervous system. For instance, male and female cells differ because of the differential effects of sex-chromosome genes expressed within the cells themselves (Arnold et al., 2003) and some genes on the X chromosome escape inactivation potentially resulting in sex differences in gene dosing (Xu et al., 2006).

11.4.3 Disorders of Sexual Differentiation

Human sexual differentiation is a multistep, sequentially interrelated process in which genetic information is translated into the phenotype of a person who subsequently establishes a gender identity and sexual orientation (Money, 1981; Gooren, 1990). The human embryo is initially bipotential with respect to genital development, and consequently, disorders in any of these steps can result in ambiguity of the genitalia (Hughes, 2001; Migeon and Wisniewski, 2003). The brain, as the substrate of sex-typical and sexual behaviors, might be initially bipotential in its development as well. Gender identity and sexual orientation are enduring and must therefore have some form of equally enduring neural substrate, though that substrate need not be the same in everyone with a particular identity and orientation. In theory, sexual orientation and gender identity could have rather idiosyncratic representations in the brains of different individuals.

As indicated earlier, brain research, mostly performed in lower mammals, demonstrates a significant role of prenatal and perinatal sex hormones in the sexual differentiation of brain and behavior (Baum, 2003; McEwen, 2001; Gorski, 2002; Swaab et al., 2002; Baum, 2006). Several studies suggest that such hormonal effects are present in humans as well, but the association is not sufficiently robust or absolute to draw firm conclusions (Money, 1981, 2002; Gooren, 1990; Migeon and Wisniewski, 2000, 2003; Meyer-Bahlburg, 2002).

The clinical syndromes that allow assessment of prenatal androgen effects on future gender identity/role, sexual orientation, and other behaviors are 46,XX subjects with congenital adrenal hyperplasia (CAH) (Money, 1991; Meyer-Bahlburg, 2001; Meyer-Bahlburg et al., 2003; Hrabovszky and Hutson, 2002; Warne, 2003), 46, XY subjects with hypoandrogenism (Meyer-Bahlburg, 1999; Ahmed et al., 2000; Migeon et al., 2002a; Melo et al., 2003), such as complete and partial androgen insensitivity or defective prenatal testosterone synthesis, and further, children with nonhormonally induced severe genital malformations such as cloacal exstrophy and penile agenesis/ablatio/micropenis who have been assigned to the female sex but whose prenatal androgen production/exposure has been similar to normal males (Wisniewski et al., 2001; Reiner, 2002; Reiner and Gearhart, 2004; Reiner and Kropp, 2004).

11.4.3.1 Complete androgen insensitivity

Children afflicted with the complete androgen insensitivity syndrome (CAIS) have a 46, XY karyotype and testes as gonads (Wisniewski et al., 2000; Wisniewski and Migeon, 2002; Hines et al., 2003). An abbreviated blind vaginal pouch is present, but no uterus or fallopian tubes. Because the external genitalia have a normal female appearance, the disorder of these patients is often unnoticed at birth. Surgical repair of an inguinal hernia containing a testis may reveal the condition. Hormonal puberty is, without intervention, feminizing due to the aromatization of endogenous androgens to estrogens. In cases of CAIS, sex assignment and rearing are almost invariably female. The differentiation of gender identity/role is feminine (Meyer-Bahlburg, 1999; Wisniewski et al., 2000; Boehmer et al., 2001; Migeon et al., 2002a,b; Wisniewski and Migeon, 2002; Ghali et al., 2003; Hines et al., 2003; Melo et al., 2003; Minto et al., 2003). This fact is theoretically important because it shows that the nature of the chromosomes and gonads *per se* does not dictate gender identity or social role; and further, that the absence of androgen sensitivity is associated with a female gender identity and a sexual orientation toward men. Another, theoretically important aspect of this condition is the high circulating levels of estradiol, derived from the elevated levels of testosterone production. In lower mammals prenatal estrogen exposure has behaviorally defeminizing and masculinizing effects, but in humans this apparently is not the case, as is also true for nonhuman primates (Rochira et al., 2005; Wallen, 2005). In adulthood, gender identity/role and sexuality conform to typical heterosexual feminine expectations.

11.4.3.2 Partial androgen resistance syndromes

The spectrum of phenotypes in 46, XY may include individuals with almost normal female external genitalia, children with ambiguous genitalia (perineoscrotal hypospadias, a microphallus, and cryptorchidism), and a normal male phenotype (Meyer-Bahlburg, 1999; Ahmed et al., 2000; Migeon et al., 2002a, b; Melo et al., 2003). There may be some (Ahmed et al., 2000; Ghali et al., 2003) relation between the nature of the androgen-receptor defect and the physical phenotype. At puberty, depending on the degree of androgen insensitivity, the development of male secondary sex characteristics may be compromised. Gynecomastia may develop as a result of estrogenic stimulation in the setting of compromised androgen action. Less severe cases may have either hypospadias or a normal male phenotype and normal male development at puberty with azoospermia.

There is considerable variability in expression of partial androgen insensitivity syndrome (AIS) (Meyer-Bahlburg, 1999; Ahmed et al., 2000; Boehmer et al., 2001; Migeon et al., 2002a, b; Melo et al., 2003). Minor deviations may go unnoticed or may be addressed surgically (e.g., hypospadias). In more severe cases, the child has ambiguous genitalia and decisions must be made about the most appropriate gender assignment. Since the mid-1950s surgery has often been carried out in infancy in order to minimize discrepancies between the assigned gender and external genital morphology. The wisdom of such surgeries in infancy has been called into question, most notably by adults who, in infancy, were recipients of such surgeries. They suggest that affected infants can be assigned to a gender and reared accordingly without surgery and its associated physical complications and ethical dilemmas (Consortium on the Management of Disorders of Sex Development, 2006).

In accordance with the optimal gender policy (Meyer-Bahlburg, 1999) with its emphasis on normal appearing external genitalia, over the past five decades, most affected individuals have been assigned female because the results of genitoplasty were deemed more satisfactory than phalloplasty.

While studies indicate relatively normal female development in most such individuals treated with castration and vaginoplasty, outcome studies have been limited and that conclusion must be interpreted with caution (Meyer-Bahlburg, 1999; Ahmed et al., 2000; Migeon et al., 2002a, b; Melo et al., 2003).

If a male sex of rearing has been chosen, genital surgeries have usually been initiated in childhood, with surgical correction of pubertal gynecomastia. The majority of 46, XY intersex patients with partial androgen insensitivity seem to develop an identity commensurate with the assigned gender and only rarely change their gender later (Meyer-Bahlburg, 2002). Childhood gender identity will, in most cases, continue into adolescence and adulthood, but patient-initiated gender change in intersex patients does seem to happen more often in adolescence and adulthood than in childhood (Gooren and Cohen-Kettenis, 1991). More female-assigned 46, XY patients initiate gender change to male than male-assigned 46, XY patients to female, possibly indicating the prenatal effects of androgens (Meyer-Bahlburg, 2001, 2002; Hrabovszky and Hutson, 2002). A recent review found that approximately 10% self-initiated sex reassignment (Mazur, 2005). Follow-up studies of adult sexual orientation are few. In a recent study of 15 affected individuals assigned as male, all identified as heterosexual men (Bouvattier et al., 2006).

11.4.3.3 5α-Reductase deficiency

5α-Dihydrotestosterone (DHT), the most potent natural androgen, is formed exclusively through 5α-reduction of testosterone by the enzyme 5α-reductase (Russell and Wilson, 1994; Mendonca, 2003). Affected individuals are born with labioscrotal folds and a clitoridean phallus. At puberty they become moderately virilized or remain eunuchoid with enlargement of the phallus. No breast development is seen.

The first reports claimed that these people were reared as girls during childhood, but after pubertal physical changes, took up life as men (Russell and Wilson, 1994). The interpretation offered was that against the backdrop of prenatal testosterone exposure, the pubertal surge of testosterone apparently induces a reversal of gender identity and role and generates a male sex drive (Russell and Wilson, 1994). Later studies showed that this interpretation probably needs some modification (Mendonca, 2003).

Local people are usually aware of the genital disorder of affected neonates and of their potential future male pubertal development. In a relatively recent study from Brazil (Mendonca, 2003), 25 of 26 affected with 5α-reductase type-2 deficiency were assigned at birth to the female sex and raised as girls. Thirteen changed to the male sex after puberty. This was associated with some virilization of the external genitalia. There was no straightforward relationship between the severity of the condition and change of gender (Mendonca, 2003). It is of note that both testosterone itself and DHT are capable of masculinizing the brain in nonhuman primates (Wallen, 2005), so prenatally there has been potentially a fair amount of brain masculinization in subjects with this condition.

11.4.3.4 17β-Hydroxysteroid dehydrogenase defiency

17β-Hydroxysteroid dehydrogenase 3 is involved in the terminal step in the synthesis of testosterone in the Leydig cell and of estradiol in the ovarian granulosa cell (Andersson et al., 1996). Subjects with an XY chromosomal pattern and testes affected with 17β-hydroxysteroid dehydrogenase 3 deficiency have more or less female external genitalia due to the lack of an effective androgenic stimulus at the time of the differentiation of the external genitalia (Andersson et al., 1996; Boehmer et al., 1999; Wilson, 1999). Such children are usually assigned to the female sex at birth and raised as girls (Wilson, 1999). A particular feature of this disorder is that testosterone production increases with time (due to a higher LH drive and alternative pathways of testosterone production), and subjects may have near-normal testosterone levels at the time of puberty, inducing substantial virilization. There are several reports of affected individuals raised as females who have changed their gender-role behavior from female to male at the time of expected puberty (Wilson, 1999). This is not universally the case but appears to happen in approximately 50% of reported cases in the literature (Andersson et al., 1996; Wilson, 1999). Children with 5α-reductase deficiency and 17β-hydroxysteroid dehydrogenase deficiency have genital ambiguity on the basis of deficient prenatal androgen exposure. It is the nature of these endocrine defects that biological activity of androgens at the time of puberty becomes stronger compared to prenatal life. Several studies document that approximately 50% of these children, originally assigned to the female sex, initiate a reassignment to the male sex, arguing in favor of the effects of prenatal androgen exposure on future gender identity. The latter two syndromes with a less-than-normal prenatal androgen exposure and a much stronger androgen exposure following puberty pose the theoretically interesting question whether androgen exposure in puberty is a contributor to sex reassignment. So, the least one can say is that prenatal androgen exposure is associated with an increased chance of later patient-initiated gender reassignment to male after initial female assignment in infancy or early childhood (Meyer-Bahlburg, 2005). Much less information is available on the sexual orientation of these subjects. Gender identity has its public expression but sexual orientation is not manifest unless disclosed by the individual and disclosure may be embarrassing for some individuals.

11.4.3.5 Congenital adrenal (virilizing) hyperplasia in women

CAH is a disorder occurring in both sexes involving undue/untimely exposure to androgens. Early reports indicated an overriding influence of the sex of assignment and rearing on the gender identity of CAH girls (Money, 1981, 1991). If CAH subjects were assigned as girls, they turned out to have a female gender identity, but with tomboyish behavior in play and activity and high energy expenditure – a marked masculine shift on the scale of sex dimorphic behavior likely due to prenatal and possibly postnatal androgen exposure (Money, 1991; Meyer-Bahlburg, 2001; Meyer-Bahlburg et al., 2003). Thus, there is robust evidence that prenatal androgenization of 46, XX fetuses leads to marked masculinization of later gender-related behavior, but gender-identity confusion/dysphoria is rare and does not support evidence for a direct determination of gender identity by prenatal androgens. This together with the potential for fertility as females, therefore, does not support a male-gender assignment at birth of even the most markedly masculinized females (Meyer-Bahlburg et al., 2004). This was also confirmed in a later study (Meyer-Bahlburg et al., 2006) of 63 women with classical CAH (42 with the salt wasting (SW) and 21 with the simple virilizing (SV) variant), 82 women with the nonclassical (NC) variant, and 24 related non-CAH sisters and female cousins as controls (COS). NC women showed a few signs of gender shifts in the masculine direction, SV women were intermediate, and SW women most severely affected. In terms of gender identity, two SW women were gender-dysphoric, and a third had changed to male in adulthood. All others identified as women.

We conclude that behavioral masculinization/defeminization is pronounced in SW-CAH women, slight but still clearly demonstrable in SV women, and probable, but still in need of replication in NC women (Meyer-Bahlburg et al., 2006).

Some are less content with life as women without having an explicit gender-identity disorder (Meyer-Bahlburg, 2001; Meyer-Bahlburg et al., 2003). In a report of older CAH subjects reared as girls, 37% rated themselves as homosexual or bisexual or they had fewer heterosexual experiences than the comparison group (Meyer-Bahlburg, 2001). It must be noted that this operational definition which lumps together different categories maximizes the number of nonheterosexuals.

This finding has been further corroborated (Meyer-Bahlburg et al., 2003; Gastaud et al., 2007). Another study was less affirmative in this regard (Money, 1991). Nevertheless, retrospective studies indicate that that there may be a decreased sexual interest and below-average engagement in heterosexual relationships among affected women. This may also be due to anatomical masculinization of the genitalia or inadequacies of vaginoplasty (Meyer-Bahlburg, 2001, 2003; Warne, 2003). Further, hirsutism may be a confounding factor. The likelihood of a gender change later in life in such females correlates with the presumed degree of prenatal androgen exposure, though the association is not very strong. Naturally, the degree of prenatal and postnatal androgen exposure also determines the extent of genital ambiguity, which, together with the postnatal biography and considerations of quality of life, may also be factors in a change of gender. Predictors of gender change are stigmatization, gonadectomy, and/or feminizing surgery after the age of 3. A relative absence of gender dysphoria in childhood does not preclude a gender change later in life. By contrast, those subjects assigned as boys due to the high degree of masculinization of their external genitalia successfully developed a male gender identity and role (Money, 1991), though patient-initiated reassignments to the female gender have been reported (Meyer-Bahlburg, 2001; Meyer-Bahlburg et al., 2003).

11.4.3.6 Cloacal exstrophy

Cloacal exstrophy is a disorder of embryogenesis, involving the genitourinary and intestinal tracts, affecting both genetic males and females. Its incidence is believed to be less than one in 400 000 births. In genetic males, the external genitalia are often grossly anomalous or absent. Boys born with cloacal exstrophy have normal testes and a presumed normal prenatal exposure to androgens. Genetic males with this disorder are often reassigned female, castrated to prevent emergence of masculine secondary sexual characteristics, and treated with estrogen at puberty to stimulate the development of female secondary sexual characteristics. In some studies, about 50% of those assigned to the female sex have been reported to evidence dissatisfaction with this and change to life as a male (Reiner, 2002; Reiner and Gearhart, 2004; Reiner and Kropp, 2004). In a series of ten affected 46, XY individuals between the ages of 4 and 14 who were castrated and unequivocally raised as girls from birth, all were close to the male range in attitudes, activities, and behaviors (Reiner, 2002). Three had declared themselves to be male (at ages 8, 9, and 12, respectively).

In a detailed clinical study of affected adolescents who were reared as males, all were described as exhibiting psychosexual dysfunction and anxiety leading to social and sexual developmental impairment (Reiner, 2004). Nevertheless, all retained a male gender identity. The interpretation is that prenatal androgen exposure underlies this change and is consistent with the notion that there is an increased risk of later patient-initiated gender reassignment to male after female assignment in infancy or early childhood in 46, XY subjects born with ambiguous genitalia with other etiologies (Meyer-Bahlburg, 2005). The latter probably also plays a significant role in the change of gender observed in boys with cloacal exstrophy. It is clear that androgen exposure in XY-subjects predisposes, but does not assure, a male gender identity. If the newborn has originally been assigned to the female sex, there is an approximately 40–50% chance of development of a female gender identity (Reiner, 2005). Schober et al. (2002) have commented that Reiner's results differ markedly from their own; however, no details were given.

After careful review, Meyer-Bahlberg concluded that the data indicate an increased risk of patient-initiated gender reassignment to male after female assignment in infancy or early childhood. Nevertheless, the findings were incompatible with the notion of a full determination of gender identity by prenatal androgens (Meyer-Bahlburg, 2005). The latter seemed to be confirmed in a series of 18 children born with a micropenis – 13 assigned as males and five as females – all subjects were satisfied with their sex of rearing in adulthood, though both men and women expressed dissatisfaction with their genital status (Wisniewski et al., 2001).

11.4.3.7 Summary of the findings in subjects with disorders of sexual differentiation

In summary, the evidence available to date permits the following conclusions: (1) the organizational effects of prenatal androgens are more noticeable in gender-role behavior than in gender identity; (2) gender identity can develop as female or male over wide variations of gender-role behavior; (3) there is suggestive, but not conclusive, evidence that a male gender identity/role occurs rather frequently in patients with a history of fully male-typical prenatal androgenization, such as in cloacal exstrophy; and (4) women with prenatal androgen exposure are often less interested in heterosexual intercourse and have inclinations to homosexual encounters. It is of note that in severe cases their conditions are hindrance to engage in heterosexual encounters (hirsutism, genital virilization, and sometimes inadequate vaginoplasties). The overall conclusion must be that prenatal exposure to testosterone in women predisposes to gynephilia (erotosexual attraction to women) but it is in no way deterministic. On the other hand, the fact gynephilia often emerges in individuals with known prenatal and/or perinatal and/or pubertal androgen exposure that is reduced relative to that of normal males argues that androgen deficiency is not a likely explanation for the lack of gynephilia in the majority of homosexual men who exhibit no somatic evidence of androgen deficiency.

11.5 Digit Ratios as Marker of Prenatal Testosterone

The ratio of the second-to-fourth finger length was first proposed as a marker for prenatal androgen action in 1998, and over 100 studies have been published testing the association between the digit ratio and prenatal androgens, or have employed digit ratios as a marker to investigate the association between prenatal androgens and a variety of outcomes, including behavior, fertility, and disease risks. The validity of digit ratios to serve as an adult marker of prenatal androgen action remains controversial (for review see McIntyre (2006)). In short: adult men tend to have longer ring fingers (fourth digits) than adult women relative to the lengths of other fingers (McIntyre, 2006). Most researchers have focused on the digit ratio of 2D:4D, the index and ring finger, respectively. In adults, the point-biserial correlation between sex and right hand, skin-surface 2D:4D is not high, even in racially homogenous samples ($r=0.22$, ~60% overlap between male and female distributions (Manning et al., 1998)). Furthermore, the large population and racial differences observed in both adults and children may introduce serious confounding (McIntyre, 2006).

CAH treated soon after birth has been the gold-standard method for studying effects of prenatal or perinatal androgens. One study found that 13 females and seven males with postnatally treated CAH had lower 2D:4D (but only on the right hands of females and left hands of males) than 44 female and 28 male relatives unaffected by CAH (Brown et al., 2002). Another study found that 27 females and nine males with postnatally treated CAH had lower 2D:4D (on both hands of females and only the right hands of males) than 52 female and 52 male age-matched controls unaffected by CAH (Okten et al., 2002). However, a study employing 2D:4D measured on radiographic films of the left hand failed to find a significant difference between 66 CAH females and 69 age-matched control females and 77 males, though 2D:4D was intermediate between the means for unaffected males and females (Buck et al., 2003). The 2D:4D ratio is not a robust marker of prenatal androgen exposure. It is also of note that there is a multitude of factors influencing bone growth prenatally. Recent endocrine insights are that that estrogens (derived from androgens) are probably more significant than androgens themselves in bone development of men (Vanderschueren et al., 2004). Several studies have attempted to link sexual orientation to 2D:4D ratio. As reviewed by McFadden et al. (2005), four studies found a correlation between 2D:4D and sexual orientation in men two of which were interpreted as suggestive of hypermasculinization (Robinson and Manning, 2000; Rahman and Wilson, 2003) and two of which were interpreted as evidence for hypomasculinization (McFadden and Shubel, 2002; Lippa, 2003). Of the two interpreted as hypermasclulinizing, one found a correlation only for the left hand (Robinson and Manning, 2000) and one only for the right (Rahman and Wilson, 2003). Of those suggestive of hypomasculinization, one found the effect bilaterally (Lippa, 2003) and one for the left hand only (McFadden and Shubel, 2002).

In women, at least three studies were interpreted as suggestive of hypermasculinizatin in lesbians. Of these, one found a correlation bilaterally (Rahman and Wilson, 2003) and two for the right hand (Williams et al., 2000; McFadden and Shubel, 2002). Several studies have found correlation with sexual orientation in only one sex but not in the other (Williams et al., 2000; Lippa, 2003) or in neither men nor women (McFadden et al., 2005; van Anders and Hampson, 2005).

Finally, it must be noted that the logic that led to the hypothesized link between 2D:4D and prenatal androgen exposure is flawed. According to Manning and Robinson (Manning and Robinson, 2003), the developmental links between fingers and 'gonads' led Manning et al. to suggest that patterns of digit differentiation, in particular the 2D:4D ratio, may reflect patterns of 'gonad' differentiation and therefore prenatal testosterone production. However, the work cited by Manning et al. linked genital, not gonadal, development with development of the digits. Specifically, some homeobox (*hox*) genes (within the hox a and hox d cluster) are essential for the development of the external genital tubercle and the development of the digits (Kondo et al., 1997). If these *hox* genes are absent, there will be no development of the genitals at all, regardless of testosterone. The *hox* genes are crucial for the genital bud to develop and then testosterone masculinizes it. Because gonads, not genitals, synthesize testosterone, the study cited by Manning et al. does not suggest any relationship between gonadal testosterone and digit formation. The fact that the premise of the hypothesis is flawed, however, would not undermine empirical observations derivative of the hypothesis. That is, empirical evidence could link prenatal testosterone and digit lengths, even if the mediating factor were not the *hox* genes as proposed by Manning and colleagues. That being said, however, as noted above substantiating empirical evidence supporting such a link is lacking.

The above research is largely based on statistical outcomes. While statistical analysis has been one of the major sources of progress in (biomedical) science, it is important to attribute proper weight to statistical significance. In statistical analysis a result is called significant if it is unlikely to have occurred by chance. A statistically significant difference simply means there is statistical evidence that there is a difference; it does not imply the difference is biologically relevant or is necessarily large, important, or is universally true. Both studies make inferences about prenatal androgen effects and their supposed androgenization of the brain (insufficient in homosexual men and excessive in homosexual women). The digit-marker theory uses a surrogate marker of prenatal androgenization. A recent study measured the 2D:4D index in human fetus with ages ranging between 9 and 40 weeks of gestation. The 2D:4D ratio was significantly higher in females compared to males ($p < 0.05$) and mean 2D:4D did not change with gestational age (Malas et al., 2006). Testosterone production in the male fetus starts at 8–9 weeks of gestation and peaks between 14 and 19 weeks. If, indeed, the 2D:4D index does not change with gestational age (Malas et al., 2006), it is difficult to regard 2D:4D index as a robust indicator of prenatal androgen exposure.

It is more appropriate to say that the 2D:4D index is sexually influenced rather than sexually dimorphic in recognition of the fact that the effect size is fairly small (2D:4D distributions of the two sexes overlap to a great degree), especially as compared to other sexually dimorphic traits such as height. Manning and colleagues have shown that 2D:4D ratios vary greatly between different ethnic groups (Manning et al., 2004). This variation is far larger than the differences between sexes, as Manning puts it, "There's more difference between a Pole and a Finn than a man and a woman." The variation has been reported to be related to number of sexual partners in heterosexual men (Honekopp et al., 2006), as well as to latitude, such that more northerly populations have higher digit ratios (Manning et al., 2004). Further, 2D:4D ratio declined between ages 6 and 8 years in a longitudinal sample (McIntyre et al., 2006), rendering data collection of 2D:4D ratios an even more frail base to make inferences on prenatal androgen exposure. In other words, every investigator seems to have found some association between 2D:4D and whatever other variables were available or selected to study but most findings remain unreplicated or have failed more than one test of independent replication. Moreover, even if one were to accept 2D:4D as a marker of prenatal androgen exposure, based on these data one might hypothesize that homosexuality results from excessive androgen exposure in both sexes.

11.6 The Fraternal Birth Order in Males

In diverse samples investigated by Blanchard and his collaborators, homosexual men have been found to have a greater number of older brothers than heterosexual men (Blanchard and Bogaert, 1997; Bogaert, 1998; Purcell et al., 2000; Blanchard and Ellis, 2001; Blanchard and Bogaert, 2004; Blanchard et al., 2005). It has been estimated that for men each older brother increases the relative risk of being homosexual by 33–48%, although these odds translate into population-probability estimates of only a few percent (Blanchard and Bogaert, 2004; Blanchard et al., 2005). Their study indicates this antecedent older-brother effect accounts for 15% (or 1:7) of homosexuality in men (Cantor et al., 2002; Blanchard and Bogaert, 2004) in the population and, therefore, certainly does not provide a universal hypothesis for the origins of homosexuality. The hypothesis advanced by these authors is that the late birth order, with more male siblings born earlier, could lead to a progressive immune response of the mother to products associated with a male which could impair brain masculinization in the fetus (Blanchard and Bogaert, 2004; Blanchard et al., 2005). The existence of such antimale antibodies is purely hypothetical as is the male-associated factor that would trigger their production. Androgens have been hypothesized in this regard (Whitehead, 2007); however, under usual circumstances they are not antigenic. The Y-linked minor histocompatibility locus (H-Y antigen) has also been hypothesized as the responsible factor. However, how such a mechanism could selectively impair only certain, less well-established, processes, such as the organization of the brain, and not others, like formation of the genitalia, is not explained by this hypothesis, and not even addressed by its proponents (Gooren and Kruijver, 2002). Nor does this theory explain why the majority of boys late in birth order do not become homosexual, even if their elder brother is homosexual. To the best of this author's knowledge, it has not been attempted to demonstrate antibodies against testosterone or proteins encoded on the Y-chromosome in women who have one or more homosexual sons with a late birth order; nor is this author aware of a clinical syndrome which is based on antibodies to testosterone or proteins encoded on the Y-chromosome (Whitehead, 2007). The formation of antibodies against a steroid hormone is an unlikely event. While the statistical association has been documented in numerous samples, the advanced hormonal explanation for the association lacks direct experimental support (Whitehead, 2007). Since its introduction as a theory, the fraternal birth-order theory has not progressed over time in substantiating the alleged deficient androgenization of boys late in the birth order.

To put the theoretical framework of prenatal brain androgenization in perspective, the known facts about prenatal testosterone levels and brain development and brain sex dimorphism will be discussed here to explore whether the broad and loose inferences about prenatal brain androgenization have any relationship with the known facts.

11.7 Hormonal Effects on the Developing Brain

Over the past three decades, sex differences have been confirmed in the size of several brain structures in a variety of laboratory animals. These findings have generated speculation concerning the existence of parallel differences in the human brain associated not only with sex, but also with sexual orientation.

Most of the structural sex differences identified involve specific cell groups within a broad region of the rodent hypothalamus that participates in regulating a variety of functions including sexually dimorphic copulatory behaviors. Like the sex differences in copulatory behaviors, several structural sex differences develop in response to sex differences in early androgen exposure: high androgen levels at the appropriate time lead to male-typical anatomy, whereas low levels lead to female-typical anatomy. Consequently, behavioral sex differences are thought to be mediated, at least in part, by structural differences.

The best-studied anatomical sex difference involves a cell group that straddles the medial preoptic and anterior regions of the hypothalamus – the sexually dimorphic nucleus of the preoptic area (SDN-POA). In the rat, this structure is 5–8 times larger in males than in females. Damage to the preoptic region decreases mounting behavior in laboratory animals (Arendash and Gorski, 1983), whereas electrical stimulation elicits mounting behavior (Perachio et al., 1979).

The belief that the SDN-POA participates in regulating male sex behavior in rats has led the search for a comparable nucleus in humans' sexual dimorphism of the brain.

11.7.1 Nucleus Intermedius

Sexual differentiation of the hypothalamus of the human brain is believed to take place around mid-pregnancy and thought to be related to the development of sexual orientation and gender identity. A life-span study has been conducted on nucleus intermedius of the human hypothalamus. This nucleus has been designated as the sexually dimorphic nucleus; however, it must be borne in mind that only one out of four laboratories that have measured this nucleus found it to be sexually dimorphic. A study of this nucleus in more than 100 subjects reported, however, that at the age of 2–4 years its cell number reaches a peak value, and that only after this age does sexual differentiation become manifest.

Furthermore, this nucleus did not differ in size or cell number between heterosexual and homosexual men, but was significantly larger in both groups of men than in age-matched women (Swaab and Hofman, 1988).

11.7.2 The Caudal Part of the Bed Nucleus of the Stria Terminalis

The human bed nucleus of the stria terminalis (BSTc) is sexually dimorphic in size and neuron number (Zhou et al., 1995; Kruijver et al., 2000; Chung et al., 2002). Both measures are larger in males. No relationship between these BSTc measures and sexual orientation was found, whereas a striking relationship with gender identity was observed (Zhou et al., 1995; Kruijver et al., 2000). Regardless of sexual orientation or adult endocrine status male-to-female transsexuals had a BSTc with a female size and neuron number. Conversely, the reversed pattern in the only available brain so far of a female-to-male transsexual was found (Kruijver et al., 2000). The functional implications of these sex findings which are yet to be independently replicated are still far from clear, but it is of note that in animals subdivisions of the BST have been implicated in the regulation of, for example, female reproductive (lordosis) and maternal behavior (Sheehan and Numan, 2002). The BSTc expresses both androgen receptors and estrogen receptors, with more nuclear ERβ expression of low intensity in young adult men (Kruijver et al., 2003) that might be related to sexually dimorphic functions. The same holds true for the observation that men have more BSTc neurons than women (Kruijver et al., 2000) (Chung et al., 2002).

Together with literature of differences in neuropeptide content (Zhou et al., 1995; Kruijver et al., 2000), these data (what data – the morphometric data?) clearly indicate that different subdivisions of the human BST are functionally different. Furthermore, there are androgen receptors, estrogen receptors, and progesterone receptors in the adult and developing BSTc, but on the basis of limited post-mortem studies in humans some investigators have concluded that in adulthood there is a lack of effect of changes in hormone levels on the size and cell number of the BSTc (Zhou et al., 1995; Kruijver et al., 2000). The neurodevelopmental study by Chung et al. (2002) indicates the BSTc area becomes only sexually dimorphic around young adulthood.

The group of Dr. Swaab of the Dutch Brain Research Institute in Amsterdam who performed the above studies concludes in a review paper that the period of manifest sexual differentiation of the human hypothalamus occurs between approximately 4 years of age and adulthood, thus much later than is generally presumed (Swaab et al., 2002). They theorize that the late sexual differentiation may be based upon processes that have already been programmed in mid-pregnancy or during the neonatal period but this remains conjectural. In any case there is no known biological action of testosterone which shows this temporal pattern of action.

11.7.3 Interstitial Nucleus of the Anterior Hypothalamus 3

Another structure, the third interstitial nucleus of the anterior hypothalamus (INAH3) is also considered a candidate for homology with the rat's SDN-POA. Three independent laboratories found this nucleus to be larger in presumed heterosexual men than in women (Allen et al., 1989; LeVay, 1991; Byne et al., 2001) and one laboratory found the nucleus to contain more neurons in men (Byne et al., 2001). By extrapolation from animal work, this sex difference is widely believed to reflect sex differences in early hormone exposure, although at present that hypothesis cannot be directly tested in humans. The AIDS epidemic has made it possible to study this nucleus in individuals whose medical records indicated homosexual behavior as the HIV risk factor. (Unless someone dies from complications of AIDS there is usually no documentation of sexual orientation in the medical records available for autopsy studies.) These studies suggest that the volume of INAH3 may be smaller in homosexual men than in heterosexual men, but the number of neurons within the nucleus does not vary with sexual orientation. The suggestion of volume reduction must be viewed skeptically for a variety of technical reasons. For example, tissue shrinks in the process of fixation for histological analysis. This shrinkage influences measures of volume but not measures of cell number. Thus, the finding of equal numbers of neurons in homosexual and heterosexual men may be a more reliable finding than the suggestion of a difference in the volume of the nucleus. Alternatively, the presence of an equal number of cells in a smaller volume could reflect decreased neuropil volume (e.g., synapses and dendrites) in homosexual men. Decreased neuropil could conceivably reflect low androgens either prenatally or at the time of death due to AIDS or the antiandrogenic effects of some medications used in the prophylaxis of particular opportunistic infections associated with

AIDS (Eil, 1992). The latter possibility should not be lightly dismissed since gay men with HIV infections are likely to be more compliant with their medical regimens than are heterosexual men whose HIV risk factor was intravenous drug abuse, and whose lives tend to be chaotic due to their drug addiction. Additional studies are needed to clarify these issues.

Speculation regarding the function of INAH3 has been based on the assumption that it is the homolog of the rat's SDN-POA. The finding that the size of the SDN-POA correlates positively with the frequency of mounting behavior displayed by male rats (Anderson et al., 1986) established the belief that the SDN-POA participates in regulating male sex behavior. That interpretation, however, is at odds with the fact, the rat's nucleus can be destroyed by electrolytic lesions without any discernible effect on mounting behavior (Arendash and Gorski, 1983). A better case can be made that the SDN-POA may inhibit the expression of a female mating posture by male rats. In rats in which intracerebral prostaglandin-E2 was manipulated, the volume of the SDN-POA was found to correlate with defeminization and not with masculinization of copulatory behaviors (Todd et al., 2005). Moreover, following lesions that include the SDN-POA, male rats can be induced to display the female copulatory response if they are also injected with estrogen and progesterone (Hennessey et al., 1986).

11.7.4 Other Neuroanatomical Studies

In addition to the hypothalamus, researchers have sought to identify variation with sex and sexual orientation in the brain commissure, the fiber bundles that connect the left and right hemispheres of the brain. The rationale underlying these studies was drawn from speculation regarding the neuroanatomical basis of statistical sex differences in the lateralization of particular cognitive functions within the brain. Several lines of investigation suggest that women compared to men are more likely to exhibit more bilaterally symmetric hemispheric activation while performing particular cognitive tasks. A popular hypothesis holds that this sex difference in brain activity reflects increased interhemispheric communication in women compared to men, and that this increased communication between the hemispheres is due to more interconnecting fibers between the hemispheres. Over 50 studies have examined the width or area of particular portions of the corpus callosum, the largest of the brains commissures, for sex differences (Bishop and Wahlsten, 1997). Popular accounts to the contrary, the majority of these studies has found no sexual dimorphism. Studies, controlling for variation in the callosum as a function of handedness and aging, however, have found differences in the isthmus of the corpus callossum which contains axons connecting right and left parietotemporal cortical regions. These regions mediate functions of language and spatial cognition which are asymmetrically represented and which according to some studies vary with both sex and sexual orientation. A recent study, yet to be subjected to replication attempts, found the isthmus to be larger in right-handed homosexual men compared to right-handed heterosexual men (Witelson et al., 2007).

In addition, several studies have examined the anterior commissure, a small bundle of fibers connecting portions of the left and right temporal lobes. These studies have produced conflicting results regarding variation with both sex and sexual orientation (Lasco et al., 2002).

The prenatal hormonal hypothesis figures prominently in a large number of research papers but upon critical review of research findings it is difficult, if not impossible, to relate sex steroids in a consistent way to the anatomical differences of the brain between the sexes that have been reported in the literature.

11.8 Conclusion

It is not an exaggeration to say that our understanding of the biological underpinnings of homosexuality has not progressed much.

It is remarkable that no study provides a clear operational definition of homosexuality. It may be obvious what homosexuality is, but scientific investigations need a clear working hypothesis to enable verification and replication which lies at the core of scientific enquiry.

The overwhelming majority of studies has searched for signs and symptoms of biological femaleness in homosexual men and biological maleness in homosexual women. Some studies have found statistically significant associations (prenatal androgen exposure, digit ratios) but the results of the studies were not very significant in providing a comprehensive theory why some people are attracted to the same sex, neither why the majority is not.

The search for femaleness in homosexual men and for maleness in homosexual women has a long tradition since Ulrichs formulated this concept in 1862.

It is bewildering that the concept has not been seriously questioned by those embarking on research on the biology of sexual orientation as if it were self-evident that homosexual men are failed men and homosexual women are malelike creatures. Time has come for a different paradigm. It is not improbable that broad research into the general principles of sexual orientation would enhance not only our understanding of homosexuality but also of heterosexuality. The latter, even though being the statistical norm, is as much in need of scientific enquiry as homosexuality is. Once we have an understanding of sexual orientation in general, it is probably a smaller step to comprehend why a majority is oriented to the opposite sex and a minority to the same sex. Sexual orientation not only implies attraction to one sex but also an incapacity to relate in a passionate way or an aversion to the other sex one is not oriented to as outlined by Kertbeny but ignored in further analysis of sexual orientation. To include the latter element might also help to develop a paradigm of sexual orientation.

As indicated above, almost all biomedical research into homosexuality has been put in the perspective of male/female brain differentiation and its disorders (Dorner, 1988). Hirschfeld and his contemporaries were fascinated by the research endeavors of Steinach showing that male/female copulatory patterns could be hormonally manipulated. This has inspired future generations of biomedical researchers into homosexuality. While their research data are solid and undoubtedly applicable to the species studied, facile extrapolations to the human species and even primates are not warranted (Wallen, 2005; Baum, 2006). Sex appropriate behavior, including copulation, in lower mammals is much more stereotyped than in primates, but even in lower mammals there is more variation that most researchers assume. A hallmark of gender identity/role and sexual orientation/interaction in the human species is the impressive inter-individual variation. This is probably best captured in John Money's concept of the lovemap (Money, 1986), defined as our own concept of an idealized lover and idealized erotic and sexual activity in our imagery or actually carried out. Later Money amplified the concept of lovemaps to gender maps and sexual orientation maps, and he defined homosexuality as a gender transposition of the idealized lover in the lovemap, with a large variety in idealized erotic and sexual activity in our imagery or actually carried out (Money, 1988). This concept accommodates the large inter-individual variation lived out by the human species.

Ulrichs and Hirschfeld assumed that understanding the biology of homosexuality would bring justice through science. History has proven them wrong. Research endeavors with a political mission are more likely to serve politics than science.

References

Ahmed SF, Cheng A, Dovey L, et al. (2000) Phenotypic features, androgen receptor binding, and mutational analysis in 278 clinical cases reported as androgen insensitivity syndrome. *Journal of Clinical Endocrinology and Metabolism* 85(2): 658–665.

Allen LS, Hines M, Shryne JE, and Gorski RA (1989) Two sexually dimorphic cell groups in the human brain. *Journal of Neuroscience* 9(2): 497–506.

Anderson RH, Fleming DE, Rhees RW, and Kinghorn E (1986) Relationships between sexual activity, plasma testosterone, and the volume of the sexually dimorphic nucleus of the preoptic area in prenatally stressed and non-stressed rats. *Brain Research* 370(1): 1–10.

Andersson S, Geissler WM, Wu L, et al. (1996) Molecular genetics and pathophysiology of 17 beta-hydroxysteroid dehydrogenase 3 deficiency. *Journal of Clinical Endocrinology and Metabolism* 81(1): 130–136.

Arendash GW and Gorski RA (1983) Effects of discrete lesions of the sexually dimorphic nucleus of the preoptic area or other medial preoptic regions on the sexual behavior of male rats. *Brain Research Bulletin* 10(1): 147–154.

Arnold AP, Rissman EF, and De Vries GJ (2003) Two perspectives on the origin of sex differences in the brain. *Annals of the New York Academy of Sciences* 1007: 176–188.

Baum MJ (2003) Activational and organizational effects of estradiol on male behavioral neuroendocrine function. *Scandinavian Journal of Psychology* 44(3): 213–220.

Baum MJ (2006) Mammalian animal models of psychosexual differentiation: When is translation to the human situation possible? *Hormones and Behavior* 50(4): 579–588.

Bishop KM and Wahlsten D (1997) Sex differences in the human corpus callosum: Myth or reality? *Neuroscience and Biobehavioral Reviews* 21(5): 581–601.

Blanchard R (2005) Early history of the concept of autogynephilia. *Archives of Sexual Behavior* 34(4): 439–446.

Blanchard R and Bogaert AF (1997) Additive effects of older brothers and homosexual brothers in the prediction of marriage and cohabitation. *Behavioral Genetics* 27(1): 45–54.

Blanchard R and Bogaert AF (2004) Proportion of homosexual men who owe their sexual orientation to fraternal birth order: An estimate based on two national probability samples. *American Journal of Human Biology* 16(2): 151–157.

Blanchard R, Cantor JM, Bogaert AF, Breedlove SM, and Ellis L (2005) Interaction of fraternal birth order and handedness in the development of male homosexuality. *Hormones and Behavior* 49: 405–414.

Blanchard R and Ellis L (2001) Birth weight, sexual orientation and the sex of preceding siblings. *Journal of Biosocial Science* 33(3): 451–467.

Boehmer AL, Brinkmann AO, Sandkuijl LA, et al. (1999) 17 Beta-hydroxysteroid dehydrogenase-3 deficiency: Diagnosis, phenotypic variability, population genetics, and worldwide distribution of ancient and de novo mutations. *Journal of Clinical Endocrinology and Metabolism* 84(12): 4713–4721.

Boehmer AL, Brüggenwirth H, van Assendelft C, et al. (2001) Genotype versus phenotype in families with androgen insensitivity syndrome. *Journal of Clinical Endocrinology and Metabolism* 86(9): 4151–4160.

Bogaert AF (1998) Birth order and sibling sex ratio in homosexual and heterosexual non-white men. *Archives of Sexual Behavior* 27(5): 467–473.

Bouvattier C, Mignot B, Lefèvre H, Morel Y, and Bougnères P (2006) Impaired sexual activity in male adults with partial androgen insensitivity. *Journal of Clinical Endocrinology and Metabolism* 91(9): 3310–3315.

Brown WM, Hines M, Fane BA, and Breedlove SM (2002) Masculinized finger length patterns in human males and females with congenital adrenal hyperplasia. *Hormones and Behavior* 42(4): 380–386.

Buchter D, Behre HM, Kliesch S, and Nieschlag E (1998) Pulsatile GnRH or human chorionic gonadotropin/human menopausal gonadotropin as effective treatment for men with hypogonadotropic hypogonadism: A review of 42 cases. *European Journal of Endocrinology* 139(3): 298–303.

Buck JJ, Williams RM, Hughes IA, and Acerini CL (2003) In-utero androgen exposure and 2nd to 4th digit length ratio-comparisons between healthy controls and females with classical congenital adrenal hyperplasia. *Human Reproduction* 18(5): 976–979.

Bullough V (1976) *Sexual Variance in Society and History*. Chicago, IL: University of Chicago Press.

Bullough V (ed.) (2000) *The Homosexuality of Men and Women*. (trans. Lombardi-Nash M) Buffalo, NY: Prometheus Books.

Bullough VL (1976) *Sexual Variance in Society and History*. New York: Wiley.

Byne W, Tobet S, Mattiace LA, et al. (2001) The interstitial nuclei of the human anterior hypothalamus: An investigation of variation with sex, sexual orientation, and HIV status. *Hormones and Behavior* 40(2): 86–92.

Cantor JM, Blanchard R, Paterson AD, and Bogaert AF (2002) How many gay men owe their sexual orientation to fraternal birth order? *Archives of Sexual Behavior* 31(1): 63–71.

Chung WC, De Vries GJ, and Swaab DF (2002) Sexual differentiation of the bed nucleus of the stria terminalis in humans may extend into adulthood. *Journal of Neuroscience* 22(3): 1027–1033.

Collaer ML, Tory HO, and Valkenburgh MD (2004) Do sex steroid hormones contribute to sexual differentiation of the human brain? In: Legato MJ (ed.) *Principles of Gender-Specific Medicine*, pp. 71–83. San Diego, CA: Elsevier Science.

Consortium on the Management of Disorders of Sex Development (2006) *Clinical Guidlines for the Management of Disorders of Sex Development in Childhood*. Rohnert Park, CA: Intersex Society of North America: http://www.dsdguidelines.org (accessed July 2008).

Dorner G (1988) Neuroendocrine response to estrogen and brain differentiation in heterosexuals, homosexuals, and transsexuals. *Archives of Sexual Behavior* 17(1): 57–75.

DuPree MG, Mustanski BS, Bocklandt S, Nievergelt C, and Hamer DH (2004) A candidate gene study of CYP19 (aromatase) and male sexual orientation. *Behavioral Genetics* 34(3): 243–250.

Eil C (1992) Ketoconazole binds to the human androgen receptor. *Hormone and Metabolic Research* 24(8): 367–370.

Filippi G (1986) Klinefelter's syndrome in Sardinia. Clinical report of 265 hypogonadic males detected at the time of military check-up. *Clinical Genetics* 30(4): 276–284.

Fromantin M, Gineste J, Didier A, and Rouvier J (1973) Impuberism and hypogonadism at induction into military service. Statistical study. *Problèmes Actuels d'Endocrinologie et de Nutrition* 16: 179–199.

Gastaud F, Bouvattier C, Duranteau L, et al. (2007) Impaired sexual and reproductive outcomes in women with classical forms of congenital adrenal hyperplasia. *Journal of Clinical Endocrinology and Metabolism* 92(4): 1391–1396.

Ghali SA, Gottlieb B, Lumbroso R, et al. (2003) The use of androgen receptor amino/carboxyl-terminal interaction assays to investigate androgen receptor gene mutations in subjects with varying degrees of androgen insensitivity. *Journal of Clinical Endocrinology and Metabolism* 88(5): 2185–2193.

Gladue BA, Green R, and Hellman RE (1984) Neuroendocrine response to estrogen and sexual orientation. *Science* 225 (4669): 1496–1499.

Gooren L (1986a) The neuroendocrine response of luteinizing hormone to estrogen administration in heterosexual, homosexual, and transsexual subjects. *Journal of Clinical Endocrinology and Metabolism* 63(3): 583–588.

Gooren L (1986b) The neuroendocrine response of luteinizing hormone to estrogen administration in the human is not sex specific but dependent on the hormonal environment. *Journal of Clinical Endocrinology and Metabolism* 63(3): 589–593.

Gooren L (1990) The endocrinology of transsexualism: A review and commentary. *Psychoneuroendocrinology* 15(1): 3–14.

Gooren L (2006) The biology of human psychosexual differentiation. *Hormones and Behavior* 50(4): 589–601.

Gooren L and Cohen-Kettenis PT (1991) Development of male gender identity/role and a sexual orientation towards women in a 46, XY subject with an incomplete form of the androgen insensitivity syndrome. *Archives of Sexual Behavior* 20(5): 459–470.

Gooren LJ and Kruijver FP (2002) Androgens and male behavior. *Molecular and Cellular Endocrinology* 198(1–2): 31–40.

Gorman MR (1994) Male homosexual desire: Neurological investigations and scientific bias. *Perspectives in Biology and Medicine* 38(1): 61–81.

Gorski RA (2002) Hypothalamic imprinting by gonadal steroid hormones. *Advances in Experimental Medicine and Biology* 511: 57–70; discussion 70–73.

Grumbach MM (2005) A window of opportunity: The diagnosis of gonadotropin deficiency in the male infant. *Journal of Clinical Endocrinology and Metabolism* 90(5): 3122–3127.

Hellman RE, Green R, Gray JL, and Williams K (1981) Childhood sexual identity, childhood religiosity, and homophobia as influences in the development of transsexualism, homosexuality, and heterosexuality. *Archives of General Psychiatry* 38(8): 910–915.

Hennessey AC, Wallen K, and Edwards DA (1986) Preoptic lesions increase the display of lordosis by male rats. *Brain Research* 370(1): 21–28.

Henningsson S, Westberg L, Nilsson S, et al. (2005) Sex steroid-related genes and male-to-female transsexualism. *Psychoneuroendocrinology* 30(7): 657–664.

Hines M, Ahmed SF, and Hughes IA (2003) Psychological outcomes and gender-related development in complete androgen insensitivity syndrome. *Archives of Sexual Behavior* 32(2): 93–101.

Honekopp J, Voracek M, and Manning JT (2006) 2nd to 4th Digit ratio (2D:4D) and number of sex partners: Evidence for effects of prenatal testosterone in men. *Psychoneuroendocrinology* 31(1): 30–37.

Hrabovszky Z and Hutson JM (2002) Androgen imprinting of the brain in animal models and humans with intersex disorders: Review and recommendations. *Journal of Urology* 168(5): 2142–2148.

Hughes IA (2001) Minireview: Sex differentiation. *Endocrinology* 142(8): 3281–3287.

Jost A (1947) Recherche sur la differenciacion sexuelle de l'embryon de lapin. *Archives d'Anatomie Microscopique et de Morphologie Experimentale* 36: 271–315.

Kennedy H (1997) Karl Heinrich Ulrichs: First theorist of homosexuality science and homosexualities. In: Rosario V (ed.) *Science and Homosexualities*, pp. 26–45. New York: Routledge.

Kondo T, Zakany J, Innis JW, and Duboule D (1997) Of fingers, toes and penises. *Nature* 390(6655): 29.

Krone N, Wachter I, Stefanidou M, Roscher AA, and Schwarz HP (2001) Mothers with congenital adrenal hyperplasia and their children: Outcome of pregnancy, birth and childhood. *Clinical Endocrinology (Oxf)* 55(4): 523–529.

Kruijver FP, Balesar R, Espila AM, Unmehopa UA, and Swaab DF (2003) Estrogen-receptor-beta distribution in the human hypothalamus: Similarities and differences with ER alpha distribution. *Journal of Comparative Neurology* 466(2): 251–277.

Kruijver FP, Zhou JN, Pool CW, Hofman MA, Gooren LJG, and Swaab DF (2000) Male-to-female transsexuals have female neuron numbers in a limbic nucleus. *Journal of Clinical Endocrinology and Metabolism* 85(5): 2034–2041.

Lasco MS, Jordan TJ, Edgar MA, Petito CK, and Byne W (2002) A lack of dimorphism of sex or sexual orientation in the human anterior commissure. *Brain Research* 936(1–2): 95–98.

Lawrence AA (2005) Sexuality before and after male-to-female sex reassignment surgery. *Archives of Sexual Behavior* 34(2): 147–166.

LeVay S (1991) A difference in hypothalamic structure between heterosexual and homosexual men. *Science* 253(5023): 1034–1037.

Lippa RA (2003) Are 2D:4D finger-length ratios related to sexual orientation? Yes for men, no for women. *Journal of Personality and Social Psychology* 85(1): 179–188.

Loehlin JC, Jonsson EG, Gustavsson JP, et al. (2004) Gender diagnosticity and androgen receptor gene CAG repeat sequence. *Twin Research* 7(5): 456–461.

Long DN, Wisniewski AB, and Migeon CJ (2004) Gender role across development in adult women with congenital adrenal hyperplasia due to 21-hydroxylase deficiency. *Journal of Pediatric Endocrinology and Metabolism* 17(10): 1367–1373.

Macke JP, Hu N, Hu S, et al. (1993) Sequence variation in the androgen receptor gene is not a common determinant of male sexual orientation. *American Journal of Human Genetics* 53(4): 844–852.

Malas MA, Dogan S, Evcil EH, and Desdicioglu K (2006) Fetal development of the hand, digits and digit ratio (2D:4D). *Early Human Development* 82(7): 469–475.

Manning JT and Robinson SJ (2003) 2nd to 4th Digit ratio and a universal mean for prenatal testosterone in homosexual men. *Medical Hypotheses* 61(2): 303–306.

Manning JT, Scutt D, Wilson J, and Lewis-Jones DI (1998) The ratio of 2nd to 4th digit length: A predictor of sperm numbers and concentrations of testosterone, luteinizing hormone and oestrogen. *Human Reproduction* 13(11): 3000–3004.

Manning JT, Stewart A, Bundred PE, and Trivers RL (2004) Sex and ethnic differences in 2nd to 4th digit ratio of children. *Early Human Development* 80(2): 161–168.

Mazur T (2005) Gender dysphoria and gender change in androgen insensitivity or micropenis. *Archives of Sexual Behavior* 34(4): 411–421.

McEwen BS (2001) Invited review: Estrogens effects on the brain: Multiple sites and molecular mechanisms. *Journal of Applied Physiology* 91(6): 2785–2801.

McFadden D, Loehlin JC, Breedlove S, Lippa R, Manning J, and Rahman Q (2005) A reanalysis of five studies on sexual orientation and the relative length of the 2nd and 4th fingers (the 2D:4D ratio). *Archives of Sexual Behavior* 34(3): 341–356.

McFadden D and Shubel E (2002) Relative lengths of fingers and toes in human males and females. *Hormones and Behavior* 42(4): 492–500.

McIntyre MH (2006) The use of digit ratios as markers for perinatal androgen action. *Reproductive Biology and Endocrinology* 4: 10.

McIntyre MH, Cohn BA, and Ellison PT (2006) Sex dimorphism in digital formulae of children. *American Journal of Physical Anthropology* 129(1): 143–150.

Melo KF, Mendonca BB, Billerbeck AEC, et al. (2003) Clinical, hormonal, behavioral, and genetic characteristics of androgen insensitivity syndrome in a Brazilian cohort: Five novel mutations in the androgen receptor gene. *Journal of Clinical Endocrinology and Metabolism* 88(7): 3241–3250.

Mendonca BB (2003) Male pseudohermaphroditism due to 5 alpha-reductase 2 deficiency; outcome of a brazilian cohort. *The Endocrinologist* 13: 201–204.

Meyenburg B and Sigusch V (2001) Kallmann's syndrome and transsexualism. *Archives of Sexual Behavior* 30(1): 75–81.

Meyer-Bahlburg HF (1977) Sex hormones and male homosexuality in comparative perspective. *Archives of Sexual Behavior* 6(4): 297–325.

Meyer-Bahlburg HF (1979) Sex hormones and female homosexuality: A critical examination. *Archives of Sexual Behavior* 8(2): 101–119.

Meyer-Bahlburg HF (1999) Gender assignment and reassignment in 46, XY pseudohermaphroditism and related conditions. *Journal of Clinical Endocrinology and Metabolism* 84(10): 3455–3458.

Meyer-Bahlburg HF (2001) Gender and sexuality in classic congenital adrenal hyperplasia. *Endocrinology Metabolism Clinics of North America* 30(1): 155–171viii.

Meyer-Bahlburg HF (2002) Gender assignment and reassignment in intersexuality: Controversies, data, and guidelines for research. *Advances in Experimental Medicine and Biology* 511: 199–223.

Meyer-Bahlburg HF (2005) Gender identity outcome in female-raised 46, XY persons with penile agenesis, cloacal exstrophy of the bladder, or penile ablation. *Archives of Sexual Behavior* 34(4): 423–438.

Meyer-Bahlburg HF, Baker SW, Dolezal C, Carlson AD, Obeid JS, and New MI (2003) Long-term outcome in congenital adrenal hyperplasia: Gender and sexuality. *The Endocrinologist* 13: 227–232.

Meyer-Bahlburg HF, Dolezal C, Baker SW, Carlson AD, Obeid JS, and New MI (2004) Prenatal androgenization affects gender-related behavior but not gender identity in 5–12-year-old girls with congenital adrenal hyperplasia. *Archives of Sexual Behavior* 33(2): 97–104.

Meyer-Bahlburg HF, Dolezal C, Baker SW, et al. (2006) Gender development in women with congenital adrenal hyperplasia as a function of disorder severity. *Archives of Sexual Behavior* 35(6): 667–684.

Migeon CJ and Wisniewski AB (2000) Human sex differentiation: From transcription factors to gender. *Hormone Research* 53(3): 111–119.

Migeon CJ and Wisniewski AB (2003) Human sex differentiation and its abnormalities. *Best Practice and Research Clinical Obstetrics and Gynaecology* 17(1): 1–18.

Migeon CJ, Wisniewski AB, Brown TR, et al. (2002a) 46, XY Intersex individuals: Phenotypic and etiologic classification, knowledge of condition, and satisfaction with knowledge in adulthood. *Pediatrics* 110(3): e32.

Migeon CJ, Wisniewski AB, Gearhart JP, et al. (2002b) Ambiguous genitalia with perineoscrotal hypospadias in 46, XY individuals: Long-term medical, surgical, and psychosexual outcome. *Pediatrics* 110(3): e31.

Minto CL, Liao KL, Conway GS, and Creighton SM (2003) Sexual function in women with complete androgen insensitivity syndrome. *Fertility and Sterility* 80(1): 157–164.
Money J (1981) The development of sexuality and eroticism in humankind. *Quarterly Review of Biology* 56(4): 379–404.
Money J (1986) *Lovemaps: Clinical Concepts of Sexual/Erotic Health and Pathology, Paraphilia, and Gender Transposition in Childhood, Adolescence, and Maturity*. New York: Irvington.
Money J (1988) *Gay, Straight, and In-between: The Sexology of Erotic Orientation*. New York: Oxford University Press.
Money J (1991) *Biographies of Gender and Hermaphroditism in Paired Comparisons*. Amsterdam: Elsevier Science.
Money J (2002) Amative orientation: The hormonal hypothesis examined. *Journal of Pediatric Endocrinology and Metabolism* 15(7): 951–957.
Okten A, Kalyoncu M, and Yaris N (2002) The ratio of second- and fourth-digit lengths and congenital adrenal hyperplasia due to 21-hydroxylase deficiency. *Early Human Development* 70(1–2): 47–54.
Perachio AA, Marr LD, and Alexander M (1979) Sexual behavior in male rhesus monkeys elicited by electrical stimulation of preoptic and hypothalamic areas. *Brain Research* 177(1): 127–144.
Purcell DW, Blanchard R, and Zucker KJ (2000) Birth order in a contemporary sample of gay men. *Archives of Sexual Behavior* 29(4): 349–356.
Rabinovici J and Jaffe RB (1990) Development and regulation of growth and differentiated function in human and subhuman primate fetal gonads. *Endocrine Reviews* 11(4): 532–557.
Rahman Q and Wilson GD (2003) Sexual orientation and the 2nd to 4th finger length ratio: Evidence for organising effects of sex hormones or developmental instability? *Psychoneuroendocrinology* 28(3): 288–303.
Reiner WG (2002) Gender identity and sex assignment: A reappraisal for the 21st century. *Advances in Experimental Medicine and Biology* 511: 175–189; discussion 189–197.
Reiner WG (2004) Psychosexual development in genetic males assigned female: The cloacal exstrophy experience. *Child and Adolescent Psychiatric Clinics of North America* 13(3): 657–674, ix.
Reiner WG (2005) Gender identity and sex-of-rearing in children with disorders of sexual differentiation. *Journal of Pediatric Endocrinology and Metabolism* 18(6): 549–553.
Reiner WG and Gearhart JP (2004) Discordant sexual identity in some genetic males with cloacal exstrophy assigned to female sex at birth. *New England Journal of Medicine* 350(4): 333–341.
Reiner WG and Kropp BP (2004) A 7-year experience of genetic males with severe phallic inadequacy assigned female. *Journal of Urology* 172(6 Pt. 1): 2395–2398.
Robinson SJ and Manning JT (2000) The ratio of 2nd to 4th digit length and male homosexuality. *Evolution and Human Behavior* 21(5): 333–345.
Rochira V, Granata AR, Madeo B, Zirilli L, Rossi G, and Carani C (2005) Estrogens in males: What have we learned in the last 10 years? *Asian Journal of Andrology* 7(1): 3–20.
Russell DW and Wilson JD (1994) Steroid 5 alpha-reductase: Two genes/two enzymes. *Annual Review of Biochemistry* 63: 25–61.
Sandfort TG (2005) Sexual orientation and gender: Stereotypes and beyond. *Archives of Sexual Behavior* 34(6): 595–611.
Schober JM, Carmichael PA, Hines M, Ransley PG, and Grady RW (2002) The ultimate challenge of cloacal exstrophy. *Journal of Urology* 167(1): 300–304.
Sheehan T and Numan M (2002) Estrogen, progesterone, and pregnancy termination alter neural activity in brain regions that control maternal behavior in rats. *Neuroendocrinology* 75(1): 12–23.
Swaab DF, Chung WC, Kruijver FPM, Hofman MA, and Ishunina TA (2002) Sexual differentiation of the human hypothalamus. *Advances in Experimental Medicine and Biology* 511: 75–100; discussion 100–105.
Swaab DF and Hofman MA (1988) Sexual differentiation of the human hypothalamus: Ontogeny of the sexually dimorphic nucleus of the preoptic area. *Brain Research – Developmental Brain Research* 44(2): 314–318.
Todd BJ, Schwarz JM, and McCarthy MM (2005) Prostaglandin-E2: A point of divergence in estradiol-mediated sexual differentiation. *Hormones and Behavior* 48(5): 512–521.
van Anders SM and Hampson E (2005) Testing the prenatal androgen hypothesis: Measuring digit ratios, sexual orientation, and spatial abilities in adults. *Hormones and Behavior* 47(1): 92–98.
Vanderschueren D, Vandenput L, Boonen S, Lindberg ML, Bouillon R, and Ohlsson C (2004) Androgens and bone. *Endocrine Reviews* 25(3): 389–425.
Wallen K (2005) Hormonal influences on sexually differentiated behavior in nonhuman primates. *Frontiers in Neuroendocrinology* 26(1): 7–26.
Warne G (2003) Congenital adrenal hyperplasia: Long-term outcome studies. *The Endcrinologist* 13: 179–181.
Westfahl PK, Stadelman HL, Horton LE, and Resko JA (1984) Experimental induction of estradiol positive feedback in intact male monkeys: Absence of inhibition by physiologic concentrations of testosterone. *Biology of Reproduction* 31(5): 856–862.
Whitehead NE (2007) An antiboy antibody? Re-examination of the maternal immune hypothesis. *Journal of Biosocial Science* 39(6): 905–921.
Williams TJ, Pepitone ME, Christensen SE, et al. (2000) Finger-length ratios and sexual orientation. *Nature* 404 (6777): 455–456.
Wilson JD (1999) The role of androgens in male gender role behavior. *Endocrine Reviews* 20(5): 726–737.
Wisniewski AB and Migeon CJ (2002) Long-term perspectives for 46, XY patients affected by complete androgen insensitivity syndrome or congenital micropenis. *Seminars in Reproductive Medicine* 20(3): 297–304.
Wisniewski AB, Migeon CJ, Gearhart JP, et al. (2001) Congenital micropenis: Long-term medical, surgical and psychosexual follow-up of individuals raised male or female. *Hormone Research* 56(1–2): 3–11.
Wisniewski AB, Migeon CJ, Meyer-Bahlburg HFL, et al. (2000) Complete androgen insensitivity syndrome: Long-term medical, surgical, and psychosexual outcome. *Journal of Clinical Endocrinology and Metabolism* 85(8): 2664–2669.
Witelson SF, Kigar DL, et al. (2007) Corpus callosum anatomy in right-handed homosexual and heterosexual men. *Archives of Sexual Behavior* 37: 857–863 (doi: 10.1007/s10508-007-9276-y).
Word RA, George FW, Wilson JD, and Carr BR (1989) Testosterone synthesis and adenylate cyclase activity in the early human fetal testis appear to be independent of human chorionic gonadotropin control. *Journal of Clinical Endocrinology and Metabolism* 69(1): 204–208.
Xu J, Watkins R, and Arnold AP (2006) Sexually dimorphic expression of the X-linked gene Eif2s3x mRNA but not protein in mouse brain. *Gene Expression Patterns* 6(2): 146–155.
Zhou JN, Hofman MA, Gooren LJ, and Swaab DF (1995) A sex difference in the human brain and its relation to transsexuality. *Nature* 378(6552): 68–70.

12 Sex Differences in Competitive Confrontation and Risk-taking

M Wilson and M Daly, McMaster University, Hamilton, ON, Canada
N Pound, Brunel University, Uxbridge, UK

Chapter Outline

Glossary

adaptation An anatomical, physiological, or behavioral attribute of an organism that has been designed by a history of natural selection to achieve some specific function that contributes to the organism's fitness.

fitness Fitness is the expected value, in the statistical sense, of a phenotypic design's success in promoting the relative replicative success of its bearer's genes, in competition with their alleles (alternative variants at the same genetic locus), in the environment(s) in which that phenotypic design evolved.

future discounting Future discounting is the rate at which one discounts the future; that is, the rate at which the subjective value of future consumption diminishes relative to the alternative of present consumption (or, if you like, the interest rate required to motivate foregoing consumption).

secondary sexual characteristic A secondary sexual characteristic is an attribute that differs between the sexes for reasons other than a direct sex-specific role in reproduction.

sexual selection Sexual selection is the component of natural selection (the nonrandom differential reproduction of types) that results from differential access to mates and their reproductive labors.

12.1 Introduction

We discuss variations in homicide as indicative of variations in competitive risk taking, interpreting prevalent conflict typologies and demographic patterns as reflections of evolved motivational and information-processing mechanisms that function to regulate competitive inclinations and actions. Connections are then drawn to research on future discounting and impulsivity, on the effects of inequity on violence, and on the bidirectional influences between circulating testosterone (T) levels and social experience. We argue that Darwinian theory, especially sexual selection theory, provides a framework that can both synthesize existing knowledge in these disparate domains and facilitate future discovery.

12.2 An Evolutionary Psychological Perspective

Evolutionary psychology is the pursuit of psychological science with explicit consideration of the fact that the psyche is, like the body, a product of evolutionary processes. More specifically, evolutionary psychologists assume that brains and minds are complexly structured as a result of a history of Darwinian selection, and that it is therefore enlightening to view the psyche and its functional components as adaptations.

The proposition that some attribute is an adaptation is a hypothesis about special-purpose design, and the value of such a hypothesis is that it suggests avenues of inquiry. As Mayr (1983) has noted, for example, generating hypotheses about what the heart or lungs or liver are for were essential first steps for investigating their physiology. Similarly, if psychological mechanisms and processes are characterized at an appropriate level of generality, then they too can be assumed to be for something (brightness compensation, memory consolidation, communication, courtship, etc.). In other words, the psychological constituents of human (and nonhuman) nature, like the anatomical and physiological elements thereof, exhibit adaptive design for the solution of particular recurrent problems faced by our ancestors.

Notions about adaptive design inevitably provide direction to the research enterprise. If we suppose, for example, that forgetting (the nonavailability of previously accessible information) constitutes repression of memories whose recall might interfere with effective action, our experimental hypotheses and methods will be very different than if we instead hypothesize that it is primarily an adaptive process of discarding obsolete information that has lost its predictive utility, or that it is merely a nonadaptive by-product of limited capacity. But if psychological research is almost invariably conducted in the shadow of the investigators' assumptions about what components of the brain/mind are for, then it is surely useful to scrutinize those assumptions, rather than letting them remain implicit, and to consider them in the light of what evolutionary biologists know about the process that creates functional organization in living creatures.

The Darwinian process of natural selection (including sexual selection) is of course the unwitting designer. Selection, systematic differentials in reproduction, and the proliferation of gene copies, is the force that created existing adaptations as solutions to the adaptive problems that were confronted by ancestral generations. Of course, not all aspects of contemporary organisms are adaptations; the colors of our internal organs, for example, have no functional significance in their own right, and are better understood as nonadaptive by-products of other adaptations. However, any complex, functional component of a living creature is almost certain to be an adaptation whose basic architecture and parameters have been shaped by a history of past selection, replicating better than alternatives because it dealt with some challenge better and thus promoted its bearers' fitness better.

Fitness is a technical term that has been used in several slightly different ways in evolutionary biology (Dawkins, 1982). Our meaning here is the same as its predominant contemporary meaning: the expected value, in the statistical sense, of a phenotypic design's success in promoting the relative replicative success of its bearer's genes, in competition with their alleles (alternative variants at the same genetic locus), in the environment(s) in which that phenotypic design evolved. Because psychological adaptations control behavioral outputs in systematic response to internal and extrinsic situational variables, we may expect that they have evolved to produce responses that are appropriately contingent, on both immediate and developmental timescales, on those aspects of the environment (including social variables) that were statistical predictors of the average fitness consequences of alternative courses of action in the past.

It should be clear from the previous paragraphs that evolutionary psychologists – like behavioral endocrinologists and neuroscientists, but unlike most social scientists – see no distinction in kind between psychological and biological influences and processes. We take it for granted that processes described at a psychological level (e.g., in terms of cognitions or traits or stimulus–response associations) are somehow instantiated in neural and hormonal mechanisms. We are not saying merely that psychological processes are engaged in continual reciprocal causal interactions with the sorts of physiological states and processes that are often referred to as biological influences on behavior: psychological processes are fundamentally biological processes. It is no surprise to an evolutionary psychologist, for example, that T should have a variety of subtle effects on information processing and behavior, both by virtue of action in the central nervous system itself and by virtue of other peripheral effects, nor that plasma T levels are themselves affected by social experience (see, e.g., McCaul et al. (1992) and Mazur et al. (1992)). Neither is it surprising to learn that an insult that evokes a surge of T in men raised in the honor culture of the US South, where retaliatory aggression is admired, has no such effect on men raised in the North, where retaliatory aggression is not admired (Cohen et al., 1996). It is often the case that psychophysiological adaptations are best characterized in terms of contingent responsiveness not only to immediate circumstances or stimuli but also to the cumulative consequences of experience, and there is no reason why the experiences in question should not include the assimilation and internalization of local cultural norms.

12.2.1 Decision-Making Adaptations

Evolutionists routinely model the costs and benefits of alternative decision rules concerning such matters as how many eggs a bird should lay before she starts incubation, or when a plant should stop putting all its energy into growth and start putting some into reproduction. We call these decisions because the organism has sufficient flexibility or plasticity to pursue alternative courses of action (e.g., either to continue growing taller, or else to terminate growth and flower instead), and because the particular course of action that is actually taken is complexly contingent on information available in the environment that imperfectly but usefully predicts relevant future conditions.

To a layperson, this use of the word decisions may sound metaphorical; a real decision is surely the product of deliberation by a conscious, intentional agent! But exactly how such real decisions differ from the thoughtless decision processes made by our peripheral organs, or by a plant, is less clear than it may first appear. Social and cognitive psychologists have shown that people do not necessarily enjoy privileged insight into the determinants of their own decisions, and that the phenomenology of deliberation and reasoned choice is often illusory and reconstructive. A genuine causal determinant of behavioral choice may be surreptitiously manipulated by an experimenter, while both observers and the subjects themselves confidently espouse coherent but manifestly false explanations of the subjects' choice of action (see, e.g., Nisbett and Wilson (1977), Nisbett and Ross (1980), and Kahneman et al. (1982)). For example, subjects in such experiments provide elaborate esthetic justifications for preferences that are actually governed by manipulated noise levels whose relevance goes unnoticed. Our reason for mentioning these results is not to show that human decision making and the mental procedures underlying it are illogical or dysfunctional; human inference and choice procedures apparently deal with naturalistic inputs very well (Cosmides and Tooby, 1996; Gigerenzer and Todd, 1999; Rode et al., 1999). Our point is simply that neither mental procedures themselves nor their logic and functionality are necessarily transparent to introspection. More generally, the phenomenology of reasoned consideration notwithstanding, we suggest that essentially all decision making necessarily relies on a great deal of inaccessible mental processing by complex devices designed by selection to generate inferences and choices on the basis of partial information and probabilistic cues.

Emotions are crucial, directive components of this complex evolved machinery, despite the popular misconception that emotion is the enemy of reason. This baseless prejudice has even invaded the social sciences (see Daly and Wilson (1997) for criminological examples and a critique of the presumption that emotional arousal during the commission of crimes is an impediment to effective action and that rational choice theories of crime must therefore be wrong). If it were truly the case that fear, anger, jealousy, and other emotional states interfered with our capacity to make decisions that further our interests, then we would have evolved to be affectless zombies. Clearly, we have not, and that very fact is testimony to the functionality of emotional states. More dramatic evidence to this effect is provided by people whose emotional mechanisms are operating abnormally, and who are therefore incapacitated (Nesse, 1990). Thus, rather than being impediments to dispassionate evaluation of alternatives, emotional states are best interpreted as functional operating modes whose specific attributes are design features facilitating effective response to the particular situations that arouse each of them (Nesse, 1990). In this light, an admonition such as "don't get mad, get even" is grounded in confusion; one of the things that "getting mad" exists for is "getting even."

Once the complexity of the psychophysiological machinery generating even our rational choices is recognized, it no longer seems metaphorical to speak of decisions about such things as when to ovulate, or when to undergo puberty, in the same language that we apply to processes that select among behavioral options. In all these cases, complex, imperfectly understood procedures, involving evolved information-processing machinery, the vestiges of individual experience, and contemporary extrinsic inputs, generate one choice rather than another. True, the deliberative homunculus of folk psychology seems not to be involved in deciding how much you will let your bone calcium be depleted during lactation, but that is not a problem, for he contributes nothing toward a genuine explanation of how you decide anything. Philosophically, evolutionary psychologists are committed materialists and mechanists, even if the mechanisms of which they speak are often characterized as cybernetic abstractions, such as decision rules or algorithms or contingent responsiveness, rather than in terms of identified neural circuitry or chemical titers.

12.2.2 Adaptation versus Pathology

Our own research has sought to illuminate the evolutionary psychology of risk acceptance and competition through the window afforded by the epidemiology of lethal violence. It may seem odd to suggest that such behavior can illuminate adaptation. Criminal violence is often ineffectual, and we will argue later that homicide is better seen as an assay of motives whose useful consequences typically fall short of lethality than as an adaptation in its own right. Moreover, violence is sometimes truly pathological: a product of alcohol-induced psychoses, delusions, organic defects, and so forth (see, e.g., Raine (1993), Aarsland et al. (1996), and Giancola and Zeichner (1995)). Nevertheless, the popular metaphor whereby violence itself is deemed a social pathology is weak and potentially misleading.

Perhaps because violence is abhorrent, and because it is a popular metaphor to call socially desirable outcomes healthy, violent behavior is often disparaged as sick. But the implication that violence *per se* is pathological is certainly false (Monahan and Splane, 1980; Cohen and Machalek, 1994). Pathologies are failures – due to mishap, senescent decline, or subversion by biotic agents with antagonistic interests – of anatomical, physiological, and psychological entities or processes, reducing their effectiveness in achieving the adaptive functions for which they evolved (Williams and Nesse, 1991; Nesse and Williams, 1994). To assert that violence is a pathology (and not simply that there are pathologies of violence) is to maintain that it is nothing more than a functionless manifestation of such failures, analogous to a fracture or a delusional psychosis. This cannot be correct, however, because people and other animals possess psychological and physiological machinery that is very clearly designed for the production and regulation of violence.

The evidence on this point is multifaceted. Most importantly, the typical elicitors of violent response are threats to survival and reproductive prospects, and its typical effect is to counter those threats. Animals (including people) react violently to usurpation of essential resources by rivals, and they direct their violence against those rivals (see, e.g., Archer (1988) and Huntingford and Turner (1987)). Moreover, such violent response is adaptively modulated. Behavioral ecologists have analyzed the cost–benefit structure of confrontational violence in terms of factors affecting the expected results of electing to fight and of escalating, and they have found that animals indeed exhibit contingent response to available cues of the probable costs and benefits of alternative actions (see, e.g., Pruett-Jones and Pruett-Jones (1994), Chase et al. (1994), Turner (1994), and Kvarnemo et al. (1995)). These analyses leave little doubt that violence is regulated with sensitivity to its probable consequences (Andersson, 1980; Clutton-Brock and Parker, 1995; Enquist and Leimar, 1990; Oliveira et al., 1998).

In addition to contextual appropriateness, the motivational states that produce violent action entail complex psychophysiological mobilization for effective agonistic action, and they induce postures appropriate for attack and defense. Some anatomical structures function solely or primarily as intraspecific weapons, and these often develop and regress seasonally or in delimited life stages, so that their availability tracks variations in the utility of violent action. Such weaponry is often sexually differentiated, in ways that make sense when one considers sex differences in the intensity of intrasexual competition, as discussed later. There are also neuroanatomical structures and chemical systems that are dedicated to the production of effective violence, and these too are likely to be sexually differentiated (Archer, 1988; Huntingford and Turner, 1987). Moreover, the magnitude of these sex differences in anatomical weaponry and intrasexual aggressive behavior is variable across species, and this variability is systematically related to aspects of the breeding system, especially its average level of polygamy (Daly and Wilson, 1983). All of these facts testify to the potency of Darwinian selection in shaping the anatomy and psychology of intrasexual aggression.

The misconception that human violence is mere pathology has been reinforced by its contemporary link to disadvantaged backgrounds that include real threats to healthy development, but this link is by no means universal. In face-to-face societies lacking strong central authority, like the foraging societies in which we evolved, violence was a recourse of highly successful men, too, and contributed to their success (see, e.g., Chagnon (1988, 1996) and Betzig (1986)). In modern nation-states, most people no longer rely on their own or their allies' violent capabilities, with the result that violence is indeed relatively often due to psychopathology, but such pathology is in the regulation of violence, not in its existence. Moreover, we must be alert to politically motivated attributions of pathology that serve to trivialize the grievances of the underprivileged. Violent offenders are disproportionately drawn from social strata with scant access to the opportunities and protective state services available to more fortunate citizens, leaving people in self-help circumstances much like those experienced by most of our ancestors. Violence in such circumstances may be deplorable, but it is not obviously a mistake in terms of the perpetrators' self-interest, and even in those people whose violence is counterproductive, the functional organization of its contingent controls remains evident.

12.2.3 Sexual Selection and Competition

In 1871, Darwin provided a significant addendum to his 1859 theory of evolution by natural selection: sexual selection. By natural selection, Darwin meant the nonrandom differential reproduction of types as a result of differential success in coping with challenges such as finding food, avoiding predators, and enduring climatic factors. Sexual selection covered the nonrandom differential reproduction of types as a result of differential access to mates and their reproductive labors. Darwin distinguished these two forces because they could sometimes act in opposition, and the distinction could help explain the evolution of attributes that have no apparent survival value. For example, sexual selection has produced conspicuous courtship structures, like the peacock's tail, that enhance mating success but apparently also increase mortality risk.

It is often useful to further partition sexual selection into effects of competition for mates among same-sex rivals versus effects exerted by the preferences of the opposite sex. Darwin noted that in most taxa, including the mammals, competition for access to mates occurs mainly (or, at least, most conspicuously) among males rather than females, but he could offer no explanation for this pattern. The issue was clarified by Bateman (1948), Williams (1966), and Trivers (1972), who pointed out that if one sex (e.g., males) provides only a minority of the resources necessary for producing viable young, then individuals of this less investing sex can increase their fitness by polygamous mating, and will have a higher ceiling on potential fitness than individuals of the more investing sex, whose fitness will be limited not by access to more mates but by the time and energy costs of raising offspring. Thus, the less investing sex is likely to have a higher fitness variance, as each successful polygamist consigns one or more rivals to unmated status; the less investing sex will therefore be sexually selected to compete for matings, which represent precious opportunities to parasitize the efforts of the sex that invests more. (There are cases, although not among the mammals, in which males are the more investing sex, and aggressive competition for access to mates occurs mainly among females (see, e.g., Gwynne and Bailey (1999)). More recent theorists have argued that sex differences in potential reproductive rate (Clutton-Brock and Vincent, 1991) or time allocation (Parker and Simmons, 1996) are more fundamental than the sex difference in parental investment emphasized by Bateman (1948) and Trivers (1972), but the essence of the theory is unchanged: male mammals have been subject to sexual selection favoring intrasexual competitive inclinations and abilities more than has been the case for females because (and to the degree that) a female must invest more time or energy than a male in order to reproduce, making access to females the crucial resource that limits a male's fitness.

The mate choice component of sexual selection sometimes acts in concert with same-sex competition, such as when females evaluate males on the basis of the size or symmetry of weapons, or when they actively select the winners. More often, however, female choice appears to favor somewhat different male design features than those that are useful for fighting (see, e.g., Moore and Moore (1999) and Norry et al. (1999)). Much research has been concerned with demonstrating that females respond to male attributes that are informative with respect to the male's genetic quality, especially his genotype's success in resisting locally prevalent strains of disease organisms. Some of these attributes, such as aerobic capacity, are selected for both by female choice and by their effects in male–male competition, but others, such as elaborate display organs or pheromones, probably function only in mate choice and not in intrasexual competition. In addition, females often select mates on the basis of the resources that they offer (e.g., territories, nest sites, gifts of food, and status) and although these various sorts of wealth are not necessarily traits that impart success in male–male competition, they are things that males compete to attain or monopolize. The effect is that mate choice by females selectively reinforces male abilities and inclinations to compete with one another, whether females attend directly to male performance in competitive confrontations or not.

What is responsible for variations in the intensity of intrasexual competition? A major factor is certainly the extent of male polygamy. Where male fitness variance is relatively large, the prize to the winners of aggressive competitions is also large, and there are large numbers of losers who are denied mating access; both these considerations favor the use of relatively extreme, reckless tactics. This principle is especially clear in comparisons across groups of related species such as seals or ungulates: the more polygamous the species (as indicated by the variance in male reproductive success or the numbers of females in the harems of the most successful males), the larger the sex differences in body size, weaponry, and violent combat, as well as in other forms of risk taking and in mortality schedules (see, e.g., Daly and Wilson (1983)). However, the same principle applies to variations within species, too. Unmated males have less to lose than those who already have a mate, for if the latter are contributing to the prospects of their progeny, or even just guarding the mates that they have acquired, then they are likely to evaluate the prospective costs and benefits of overt competition over access to a new female somewhat differently than unmated males, and are likely to be less reckless in the pursuit of new mating opportunities. In the human case, these considerations suggest that we should see not just sex differences in intrasexual competition, but differences in relation to the actors' marital and parental statuses as well, and we might also expect to see patterned variation as a function of life stage, cues of the intensity of local competition, and cues indicative of one's life prospects.

12.2.4 Homicide as an Assay of Competitive Confrontation and Risk Taking

How is the notion that intrasexual competition is a predictable consequence of sexual selection relevant to interpersonal violence? Homicide refers to lethal violence undertaken privately (i.e., excluding killing that is socially legitimized, such as in war). It can be looked on as a form of risk taking since many cases culminate confrontational conflicts in which both parties risked injury, death, or arrest, and even when killers do not put themselves in physical danger they risk dire legal consequences. Most homicides involve a victim and a killer who are unrelated men, more often acquaintances than strangers, and most are transparently competitive, although the contested resources are not necessarily material ones. The competitive element in robbery homicide is obvious: one party had something that the other wished to usurp. Also clearly competitive are jealous killings of sexual rivals, although the contested resource is now a person. These two conflict typologies are both extremely common, but there is a third variety that is even more prevalent, especially where homicide rates are relatively high, such as is the case in many US cities: disputes over intangible social resources like face and respect (Daly and Wilson, 1988). In the criminological literature, these escalated status disputes and insult-precipitated killings are called trivial altercations, but in their social context, these disputes are not trivial. The precipitating insult may seem petty, but it is usually a deliberate provocation (or is perceived to be), and hence a public challenge that cannot be shrugged off. It often takes the form of disparaging the challenged party's manhood: his nerve, strength, or savvy, or the virtue of his wife, girlfriend, or female relatives. If there is a disparity in social rank, the high-status man may be able to ignore a challenge without loss of face, but not when the two are approximate equals. The challenge itself is likely to have been issued in response to status-inappropriate behavior: offense was taken because one party was elevating himself by putting on superior airs or failing to show deference to those of slightly higher rank. In these regards, homicidal altercations among low-status men resemble killings in defense of personal honor that are celebrated as heroic in other social contexts, especially when the killer is of higher status.

One interesting question is why these contests must continue to a deadly end. Sometimes, there seems to be almost an agreement that the conflict will be resolved violently (see, e.g., Luckenbill (1977) and Toch (1969)), and it is often the eventual victim who forces the issue (Wolfgang, 1958). The eventual killer may announce and justify his deadly intentions both to his victims and to their audience. In these features, the homicidal altercation seems more like a formal duel (see, e.g., Baldick (1965) and Williams (1980)) than a senseless eruption of violence. The vigilant maintenance of a reputation for refusing to tolerate insult or disrespect is likely to have a deterrent effect on the probability of further disrespect, trespass or abuse, and such deterrence is particularly important when one cannot rely on dispassionate third parties such as the law to protect one's interests (Nisbett and Cohen, 1996). In a sociable species such as our own, where success and failure in competition have lasting social consequences mediated by rank and reputation, we expect an evolved inclination to display one's competitive risk-taking skills, and this should be especially characteristic of males (Wilson and Daly, 1985).

As would be expected from the combined considerations that most homicides entail competition between unrelated same-sex rivals and that intrasexual competition during human evolution was more intense in males than in females, the victims and killers in these competitive conflicts are overwhelmingly male. It is perhaps especially noteworthy that this is true even of the victims of robbery homicides, despite the fact that women are as likely as men to be victims of nonlethal robbery (Wilson and Daly, 1985). There may be several reasons for this, but at least part of the explanation seems to lie in the fact that robbery homicides often contain the same elements of competition and face that characterize altercations. This point is nicely made by Toch (1969) who has analyzed violent escalation in police–suspect interactions in terms of the stubborn aggressiveness of both parties when concerned to maintain face in front of witnesses.

Of course, most altercations do not end in death, and most robberies do not either. We use homicides as our assay of the intensity of intrasexual competition, despite their relative rarity and possible atypicality, because killing is recorded more reliably and with less bias than any lesser manifestation of interpersonal conflict. If readiness to participate in dangerous competition varies in relation to a demographic or circumstantial factor, then so should the rates of these homicides, regardless of whether killing is (or formerly was) adaptive in such contexts or is instead a by-product of aggressive motives that evolved to produce nonlethal effects. Buss (1999, 2000) has argued that the prevalence of homicidal fantasies, and the fact that many killings are intentional (rather than being accidents or slips), are reasons to doubt that homicide is a nonadaptive by-product of psychological adaptations and to conclude that what we are studying is the behavioral outputs of psychological adaptations for homicide. This argument is unpersuasive (at least to us) because neither fantasy nor intent is a criterion by which one can distinguish adaptations from by-products. People intend to watch their favorite TV shows or feed their cats more often than they intend to kill, for example, but that does not imply that these actions are adaptations in their own right rather than being by-products of adaptations. Similarly, whereas many male undergraduates indeed report that they have had homicidal fantasies, even higher proportions of the men in one study reported that they had fantasized that they were flying, that they were playing video games, and that they had won the lottery; obviously, human beings possess psychological adaptations for none of these things (Kai, unpublished BSc thesis). In any event, epidemiological analyses of homicide can be used to test evolutionary psychological hypotheses, as illustrated later, regardless of whether there are aspects of the human psyche that deserve to be considered adaptations for lethal action, and regardless of whether homicide typically promotes the killers' fitness or ever did so in human ancestry.

12.2.5 The Sex Difference in Human Intrasexual Competition and Violence

Both the intensity of social competition and the local homicide rate are hugely variable in time and space, but one difference is apparently universal: men kill unrelated men at vastly higher rates than women kill unrelated women, everywhere (**Table 1**).

Criminologists and other social scientists have offered a wide range of hypotheses to explain sex differences in the use of lethal violence. Many such theories attribute men's greater use of violence to local aspects of particular societies, and thus shed no light on the phenomenon's cross-cultural universality (see, e.g., Daly and Wilson (1988, pp. 149–157; 1990, pp. 86–87)). Others invoke the human male's greater size and strength, but although this might explain an asymmetry in the use of violence against the opposite sex, it can hardly be said to predict or account for the sex difference in **Table 1**; one might as readily have predicted that the group with the physical capacity to inflict the most damage (i.e., men) would be attacked least. The crucial omission from most such discussions of gendered behavior in the social sciences is of course some consideration of the effects of our sexual selective history. In any species in which the zero-sum game that partitions

Table 1 Numbers of same-sex nonrelative homicides in various studies for which information on sex and relationship of killer and victim were available[a]

	Homicides	
	Male	*Female*
Chicago 1965–89	9761	229
Detroit 1972	316	11
Miami 1980	358	0
Canada 1974–90	3881	94
England and Wales 1977–90	3087	108
Scotland 1953–74	143	5
Iceland 1946–70	10	0
Tzeltal (Mexico) 1938–65	15	0
Bison-Horn Maria (India) 1920–41	36	1[b]
Munda (India)	34	0
Oraon (India)	26	0
Bhil (India) 1971–75	50	1[b]
Tiv (Nigeria) 1931–49	74	1
BaSoga (Uganda) 1936–55	38	0
Gisu (Uganda) 1948–54	44	2
Banyoro (Uganda) 1936–55	9	1[b]
Alur (Uganda) 1945–54	33	1[b]
BaLuyia (Kenya) 1949–54	65	3[b]
JoLuo (Kenya) c. 79	22	2[b]
!Kung San (Botswana) 1920–55	12	0

[a]Data from Daly M and Wilson M (1988) *Homicide*. New York: de Gruyter, and unpublished data.
[b]Victim and killer were unrelated co-wives of a polygynous man in the lone female–female cases in the Maria, Bhil, Banyoro, and Alur samples, as well as in one of the three BaLuyia cases and one of two JoLuo cases. We include co-wife cases, despite otherwise excluding marital as well as genetic relatives, because unrelated co-wives represent a female analogue of male–male rivalries.

ancestry among males is played with different rules or parameters than the corresponding game among females, the selective process will favor different attributes, including psychological attributes, in the two sexes. Any attempt to explain the phenomena in **Table 1** that presupposes the psychological identity of the sexes is a nonstarter.

Sex-differential violence against same-sex antagonists appears to be one of many manifestations of the fact that the human male psyche has evolved to be more risk accepting in competitive situations than the female psyche (Wilson and Daly, 1985; Daly and Wilson, 1990, 1993). Our sex difference in competitive violence is not unique: other species in which the variance in fitness (and the risk of total reproductive failure) is greater for males than for females exhibit the same thing. The evidence that human beings evolved under chronic circumstances of a somewhat greater fitness variance among males than among females is abundant and consistent. Even in societies where people live by hunting and foraging, with much less disparity of wealth than is seen in agricultural societies or modern states, some men monopolize several women and thereby consign others to bachelorhood, with the result that variance in offspring production is greater among men than among women (Howell, 1979; Hewlett, 1988; Hill and Hurtado, 1995).

As noted earlier, the natural selective link between such a mating system and sex differences in competitive violence is well understood: basically, greater fitness variance selects for greater acceptance of risk in the pursuit of scarce means to the end of fitness (e.g., Daly and Wilson, 1988, pp. 163–168). Furthermore, reckless life-threatening risk proneness is especially likely to evolve where staying alive by opting out of competition promises to yield no fitness at all and is therefore the natural selective equivalent of death (Daly and Wilson, 1988; Rubin and Paul, 1979). It is not surprising to an evolutionist that men compete with one another more intensely and dangerously than do women, and that this sex difference apparently transcends cultural diversity.

12.2.6 Demography of Masculine Competitive and Risk-Taking Inclinations

For further insight about sources of variability in violence, beyond the ubiquitous sex difference, the best models are likely to be those that incorporate sources of variance in social competition and the likelihood of reproductive failure in ancestral environments, and propose cues by which actors detect and respond to these variations. We propose that dangerous competitive violence reflects the activation of a risk-prone mindset that is modulated by present and past cues of one's social and material success, and by some sort of mental model of the current local utility of competitive success both in general and in view of one's personal situation. Thus, sources of variability in addition to sex might include the potentially violent individual's age, material and social status, marital status, and parental status; local population parameters such as the sex ratio, prevalence of polygamy, and cohort sizes; and ecological factors that affect resource flow stability and expected life span. Only a few of these factors have been extensively studied with respect to their possible impacts on violence, and some have scarcely yet been considered at all.

Criminal violence exhibits an age pattern that is almost as dramatic and consistent as the sex difference: Offending rises rapidly after puberty, reaches a peak in young adulthood, and then slowly declines. **Figure 1** illustrates this pattern with homicide data.

This pattern is characteristic of other violent crimes such as robbery and sexual assault (see, e.g., Thornhill and Thornhill (1983) and Chilton (1987)), and of noncriminal confrontational risk taking as well (Hilton et al., 2000; Moffitt, 1993). Challenging criminologists, Hirschi and Gottfredson (1983, p. 55), to explain this striking pattern asserted that "the age distribution of crime is invariant across social and cultural conditions," and that it "cannot be accounted for by any variable or combination of variables currently available to criminology." Expanding on these assertions, Hirschi and Gottfredson (1986) argued that changes in employment status cannot explain why offending decreases with age,

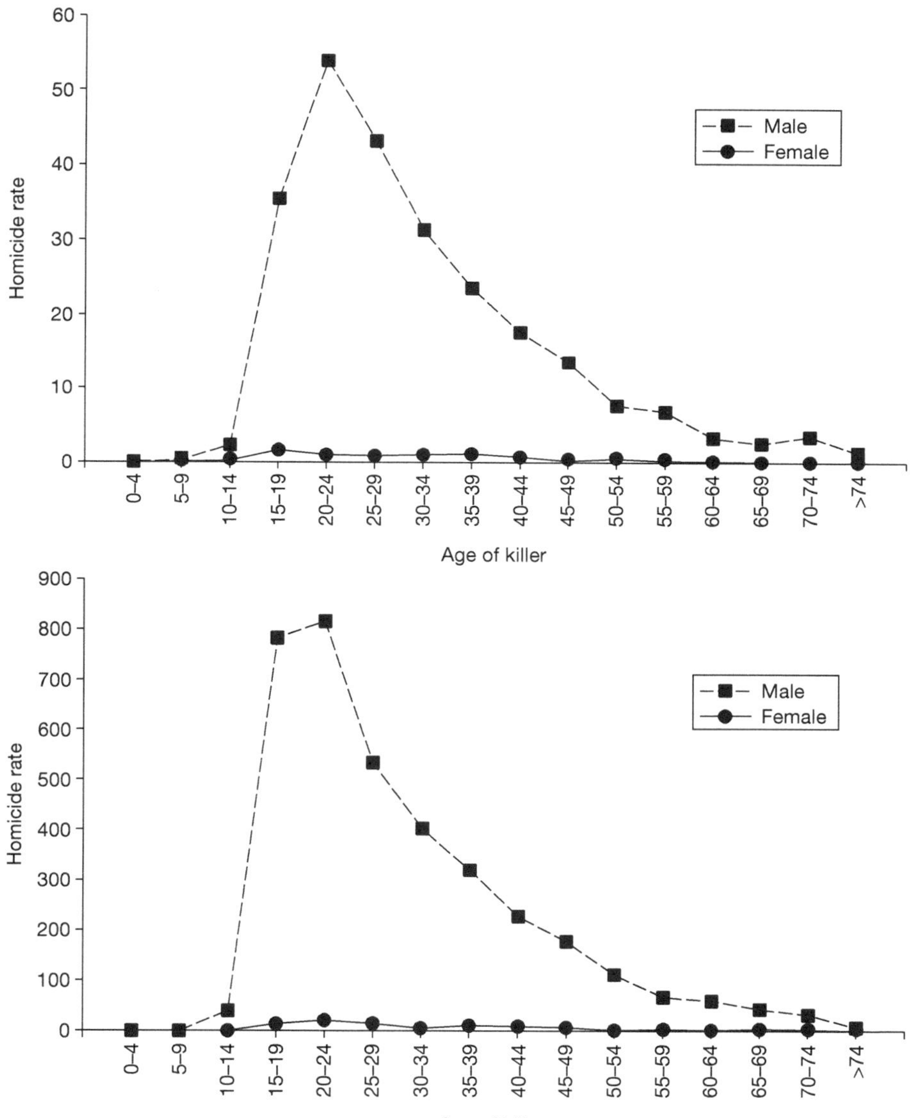

Figure 1 Age-specific homicide rates (homicides per million persons per annum) for men and women who killed an unrelated person of the same sex in Canada, 1974–92 (upper panel), and in Chicago, 1965–89 (lower panel). Data include all homicides known to police in which a killer was identified. Data from Wilson M and Daly M (1994) A lifespan perspective on homicidal violence: The young male syndrome. In Block CR and Block RL (eds.) *Proceedings of the 2nd Annual Workshop of the Homicide Research Working Group*, pp. 29–38. Washington, DC: National Institute of Justice.

citing evidence that working and nonworking teenagers incur similar arrest rates; that an age-related decline in the proportion of men who are unmated is likewise irrelevant, citing evidence that delinquent boys are more, not less, likely to have girlfriends than their nondelinquent age-mates; and that becoming a father plays no role either, although on this point they cited no evidence at all. Thus, they maintained, "Change in crime with age apparently cannot be explained... by change in the social situation of people over the course of life" (Hirschi and Gottfredson, 1986: 67), and social explanations having failed, these sociologists concluded that the age–crime curve must be a reflection of human biology.

The assumptions that a biological explanation is an alternative to a social one and that its domain of applicability is that which is invariant are of course quite at odds with the evolutionary perspective outlined earlier. If it were true that a man's likelihood of doing violence were unaffected by his material and social circumstances, this would be as surprising to an evolutionary biologist or psychologist as to a sociologist, since natural and sexual selection is expected to shape the psychology underlying risky competition in such a way as to modulate behavior in relation to one's circumstance. A married father has more to lose in an altercation than a childless bachelor of the same age, for example, and it is implausible that the human psyche should have evolved to ignore such factors. To say this is not to deny that there is an evolved human life course, such that young men are specialized, both physically and psychologically, for competitive risk acceptance; male muscle strength and aerobic capacity, for example, rise and fall in a pattern rather like that of the age–crime curve, even when effects of exercise are controlled, and various sorts of voluntary risk-taking rise and fall similarly (Daly and Wilson, 1990). What is scarcely plausible is that either competitive inclinations or the evaluation of risks should have evolved to follow a life span trajectory that is impervious to one's personal circumstances and associated cues of risk's costs and benefits.

So is Hirschi and Gottfredson's (1986) conclusion that social circumstance is irrelevant to the age–crime curve correct? Certainly not with respect to homicide: although employment status (**Figure 2**) and marital status (**Figure 3**) are poor predictors of offense rates by teenagers, they are very good predictors at subsequent ages, with the employed and the married much less likely than other men to become involved in the sorts of lethal disputes that we

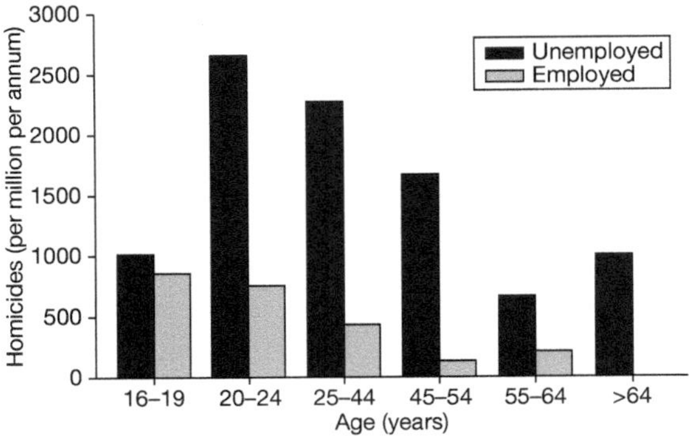

Figure 2 Age-specific homicide rates (homicides per million men per annum) for men who have killed an unrelated man, according to whether the killer was employed or unemployed for Detroit 1972. Data from Wilson M and Daly M (1985) Competitiveness, risk-taking and violence: The young male syndrome. *Ethology and Sociobiology* 6: 59–73.

interpret as competitive (Wilson and Daly, 1985; Daly and Wilson, 1990). A complication is that single men may differ from their married counterparts with respect to attributes that have causal impact on both their single status and their violence, but the notion that marriage is itself pacifying, and not simply a correlate of more relevant unmeasured variables, gains support from the observation that both divorced and widowed men apparently behave more like never-married men of the same age than like married ones (**Figure 3**). One could argue that the divorced are another selected subset who are more violent by disposition than those who stay married, but this sort of selection argument seems less plausible for the widowed. These data are also reminiscent of the longitudinal study by Mazur and Booth (1998), which indicated that T levels are reduced in married men, compared to same-age single men in the same profession (the US military), and that they rise again in those who divorce. We tentatively conclude that marriage's apparent pacifying effect on men really is an effect.

It should of course be noted that we have not explained away the age–crime curve by these analyses. Although employment status and marital status make a big difference, massive age effects persist in **Figures 2** and **3**. Whether fatherhood has pacifying effects over and above those of marriage we cannot say, as no substantial homicide data set contains the requisite information, and, of course, no one has yet assessed the simultaneous impacts of these several factors. Thus, exactly how much of the age-related variability in crime will eventually be attributed to

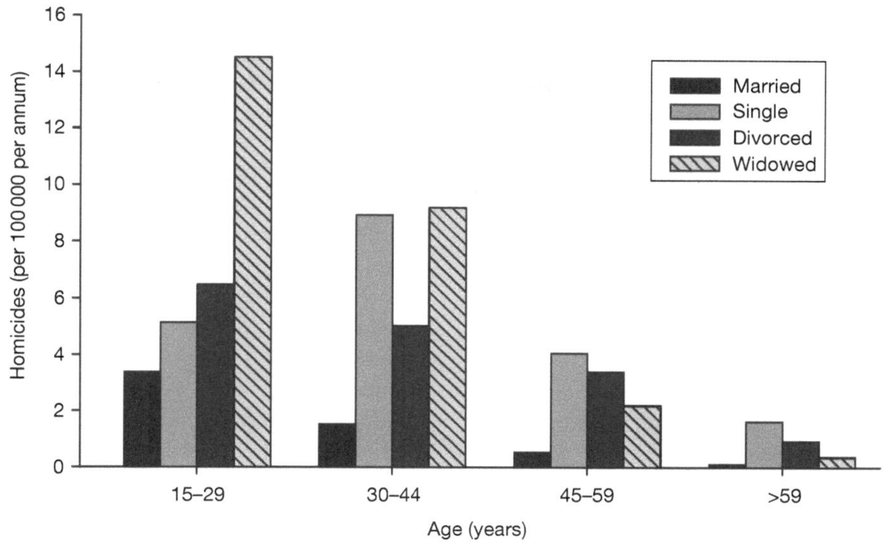

Figure 3 Age-specific homicide rates (homicides per million men per annum) for men killing unrelated men, according to the killer's marital status in Canada, 1974–90. Data from Daly M and Wilson M (2001) Risk-taking, intrasexual competition, and homicide. In: French JA, Kamil AC, and Leger DW (eds.) *Nebraska Symposium on Motivation, Vol. 47: Evolutionary Psychology and Motivation*, pp. 1–36. Lincoln, NE: University of Nebraska Press.

age-related changes in social and material circumstances, and whether an age–crime curve will persist when all correlated social factors are controlled is still unknown.

12.2.7 Discounting the Future

The rate at which one discounts the future is the rate at which the subjective value of future consumption diminishes relative to the alternative of present consumption (or, if you like, the interest rate required to motivate foregoing consumption). If A discounts more steeply than B, then A values present rewards more highly relative to future rewards than B, and is less tolerant of delay of gratification.

Psychologists, economists, and criminologists have reported that young adults, the poor, and criminal offenders all tend to discount the future relatively steeply. Wilson and Herrnstein (1985), for example, reviewed persuasive evidence that men who engage in predatory violence and other risky criminal activity have different time horizons than law-abiding men, weighing the near future relatively heavily against the long term. There are many terms for such a tendency, most of them pejorative: impulsivity, myopia, impatience, lack of self-control, and inability to delay gratification. Behind the use of such terms lies an implicit, false presumption that there is a single right answer to the adaptive problem of how one should value present rewards relative to more distal ones, independent of life stage and socioeconomic circumstance, and that steep discounting is therefore pathological. By contrast, an evolutionary perspective suggests that adjusting one's personal time horizons can be an adaptive response to predictive information about the stability of one's social order and ownership rights and one's expected longevity (Daly and Wilson, 1990, 2005; Rogers, 1994; Sozou, 1998; Daly and Wilson, 2005) as well as being a facultative response to one's current situation and perceived opportunities (Wilson and Daly, 2004; Van den Bergh et al., 2008).

Much of the literature on these matters equates impulsivity with low intelligence, a prejudicial view that is perhaps based on the premise that an ability to plan for a distant future is a hallmark of the cognitive skills that make humans unique. The trouble with this anthropocentric view is that the problem of how to discount the future confronts all creatures. Indeed, it is the same problem as that addressed by Fisher (1930) and by all subsequent life-history theorists: How should the future be weighted in deciding present allocations of effort (see, e.g., Clinton and LeBoeuf (1993), Grand (1999), and Roitberg et al. (1992))? The right answer depends on statistical expectations of the present and future reproductive payoffs of alternatives, which vary in relation to cues to which organisms – brainless creatures (and plants) as well as sophisticated cognizers – have evolved facultative responses. By this reasoning, what selects for delay of gratification is a high likelihood that present somatic effort can be converted to future reproduction, and rather than reflecting stupidity, short time horizons are predictable attributes of those with short life expectancies, those whose likely sources of mortality are independent of their actions, and those for whom the expected fitness returns of present striving are positively accelerated rather than exhibiting diminishing marginal returns.

How human beings and other animals discount the future has been described in some detail by experimental psychologists, but a fuller understanding of these processes requires an infusion of evolutionary insights. The most noteworthy conundrum concerns the shape of discount functions, which are often, perhaps typically, hyperbolic (Kirby and Herrnstein, 1995). Hyperbolic discounting is puzzling because it engenders predictable reversals of preference between alternative futures as time passes, and hence predictable regret of what will become bad decisions in retrospect, with the result that people and other animals will even work to erect impediments to their own anticipated future choices of action (Kirby and Herrnstein, 1995). Why are the psychological underpinnings of time preference such as to produce these seemingly maladaptive internal struggles? This question can only be addressed by interpreting the relevant decision processes as adaptations to the structure of problems in nature, and Kacelnik (1997) has provided a satisfying answer to this particular problem, by showing that hyperbolic discounting is an expected consequence of mental evaluations whose function is to maximize rate of return while foraging or otherwise investing time in a task with sporadic returns. The experiments that produce hyperbolic discounting require animals to choose between rewards, after alternative delays that are unaffected by what one does during the delay and are followed by obligate time-outs which make the longer delay the optimal choice. The natural world contains no such problems. Instead, animals must choose among alternative activities with different expected rates of return (e.g., prey encounter rates), and the opportunity to resume foraging after an interval is typically under one's own control. Thus, with a psychology adapted to the problem of allocating efforts among endeavors with sporadic payoffs, evolved decision makers treat delays as time invested in the task, and get it wrong only because of the artificiality of experimental situations.

Rogers (1994, 1997) has brought evolutionary reasoning to bear on the issue of optimal age-specific rates of future discounting, given the age-specific mortality and fertility schedules of human populations. His analysis suggests that people of both sexes should have evolved to have the shortest time horizons and to be maximally risk accepting in young adulthood. More specifically, his theoretical curve of age-specific optimal discount rates looks very much like the actual life span trajectory of reckless risk proneness that may be inferred from accidental death rates and homicide perpetration. This striking result seems paradoxical, given that we might expect indicators of a short or uncertain expected future life span to be responded to as cues favoring risk acceptance, but certain peculiarities of human life history and sociality, namely, gradually diminishing fertility long before death and a shifting allocation of familially controlled resources between personal reproductive efforts and descendants' reproductive efforts, dictate Rogers's counterintuitive result. One implication of this analysis is that a young adult peak in competitive risk taking may be a human peculiarity, while other species, including even our close relatives, the great apes, will exhibit monotonic increases in these tendencies into old age (Daly and Wilson, 2005). Comparative research on this question is needed.

As argued earlier, homicide and other criminal violence can be considered outcomes of steep future discounting and risky escalation of social competition. On the assumption that people are sensitive to social information predictive of their probable futures, Wilson and Daly (1997) hypothesized that homicide rates would vary as a function of local life expectancy, and tested this idea in Chicago, a city divided into 77 long-standing neighborhoods with relatively stable social and economic characteristics. In 1990, male life expectancy at birth in these neighborhoods ranged from 54.3 to 77.4 years, even with the effect of homicide as a cause of death removed, and this life expectancy proved to be the best available predictor of neighborhood-specific homicide rates (which ranged from 1.3 to 156 homicides per 100 000 persons per annum): the bivariate correlation between these variables was –0.88 (**Figure 4**).

Might people actually be tracking something akin to local life expectancy? At present, we cannot disentangle possible effects of low or uncertain life span from the closely linked problems of low and

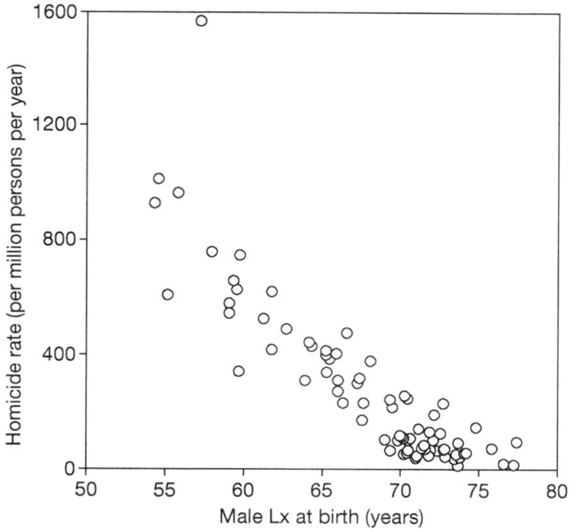

Figure 4 Neighborhood-specific homicide rates (homicides per million persons per annum) in relation to male life expectancy at birth (with effects of homicide mortality removed) for 77 community areas of Chicago, 1988–93. Data from Wilson M and Daly M (1997) Life expectancy, economic inequality, homicide, and reproductive timing in Chicago neighbourhoods. *British Medical Journal* 314: 1271–1274.

uncertain future resource flow, and other evolution-minded social scientists have emphasized the latter sort of uncertainty as the factor favoring short time horizons (Hill et al., 1997). However, it is conceivable that the human psyche computes something functionally equivalent to an estimate of the distribution of local life spans, based on the fates of salient others; if several family members and friends have died young, discounting the future would follow. Moreover, if this mortality appeared to be due to bad luck that was independent of the decedents' behavior, it would be all the more adaptive to respond by accepting risks in the pursuit of short-term advantage. If such inference processes exist, they are unlikely to be transparent to introspection, but they may be revealed in attitudes and expectations, and it is interesting in this regard that testimony of the US urban poor contains many articulate statements about the specter of early death, the unpredictability of future resources, and the futility of long-term planning (see, e.g., Hagedorn (1988) and Jankowski (1992)). An interesting question for psychological research is how mental models and subjective parameters in these domains develop over the life span; another is whether media representations (including fictitious ones) affect these inference processes and their development in the same way as real information about neighbors and relatives.

Local life expectancy appears to be predictive of future discounting in nonviolent domains, too, and we suggest that life expectancy cues may affect investments in the future through education, preventive health measures, and savings, as well as the timing of major transitions and life events. In Chicago, rates of truancy (school absenteeism for nonmedical reasons) are high where life expectancy is low, and somewhat surprisingly, this relationship is stronger for primary school truancy than for high school (Daly and Wilson, 2001); one interpretation is that parents' motivation to invest in education by enforcing school attendance varies in relation to cues of the chances that doing so will eventually pay off. Teenage pregnancy is another phenomenon in which perceptions of local life expectancy or life prospects may play a role: although early reproduction is widely deemed a social pathology, reflecting a failure to exercise choice, Burton (1990) and Geronimus (1992, 1996) report that poor teenage mothers are active decision makers who expressly wish to become mothers and grandmothers while still young and efficacious because they anticipate problems of early weathering, poor health, and a life course more compressed in time than that of more affluent people. Wilson and Daly (1997) interpret age-related fertility patterns in Chicago neighborhoods accordingly.

12.2.8 Inequity and Lethal Competitive Violence

Homicide rates are highly variable between times and places (see, e.g., Archer and Gartner (1984) and Eisner (2003)), and the discussion earlier suggests that much of this variance may be attributable to variance in the severity of male–male competition. When rewards are inequitably distributed and those at the bottom feel they have little to lose, escalated tactics of social competition, including violent tactics, become attractive. When the prize for competitive ascendancy is smaller, and even those at the bottom have something to lose, such tactics lose their appeal. One might therefore expect that income inequality will account for a large share of the variance in homicide rates.

This expectation is upheld. In cross-national analyses, the Gini index of income inequality (which ranges from 0 at perfect equity to 1 when all income accrues to the single wealthiest unit) consistently outperforms most other predictors of homicide rates, including various indices of average income or welfare, suggesting that it is relative rather than absolute deprivation that has the greater effect on levels of violent competition (Daly et al., 2001). Only the Gartner (1990) study disaggregated the overall homicide rate, and she found that income inequality predicts the rates at which adults, but not children, are killed, and is a stronger predictor of men's than of women's victimization. In general, the results of these cross-national studies are highly compatible with the proposition that homicide rates assay the local intensity of competitive conflict, especially among men.

Research on income inequality and homicide rates within, rather than between, nations is relatively scarce, but the results have been striking. Kennedy et al. (1996) found that the Gini index was significantly correlated with many components of mortality across the 50 US states in 1990, but with none more highly than homicide ($r = 0.73$). Blau and Blau (1982) found that income inequality accounted for more of the variance in homicide rates among 125 US cities than other measures including percent below the poverty line. At a still finer level, namely across the 77 Chicago neighborhoods, the correlation between the Gini index and the homicide rate is even higher ($r = 0.81$; Daly et al., 2001).

Despite this abundant evidence, the proposition that inequity *per se* is relevant has remained somewhat controversial because high income inequality is generally associated with a low level of economic development and average welfare, and even though the Gini index usually predicts the homicide rate better than other economic indicators, one might still argue that it is somehow just a better indicator of general prosperity. Analyses of Canadian homicide data (Daly et al., 2001) refute this hypothesis: the most inequitable Canadian provinces are the richest ones, in direct contrast to the correlations across US states and Chicago neighborhoods, but the positive association between the Gini index and homicide is undiminished ($r = 0.85$). Moreover, the data for the Canadian provinces and the United States lie almost exactly on the same regression line (**Figure 5**). These results suggest that the single variable of economic inequality is largely responsible for the notorious difference in these neighboring countries' homicide rates.

The dramatic association between inequitable access to resources and homicide in modern nation-states may not extend to traditional nonstate societies like those in which we evolved. Homicide rates in modern hunter–gatherer societies generally dwarf those of modern nation-states (Daly and Wilson, 1988),

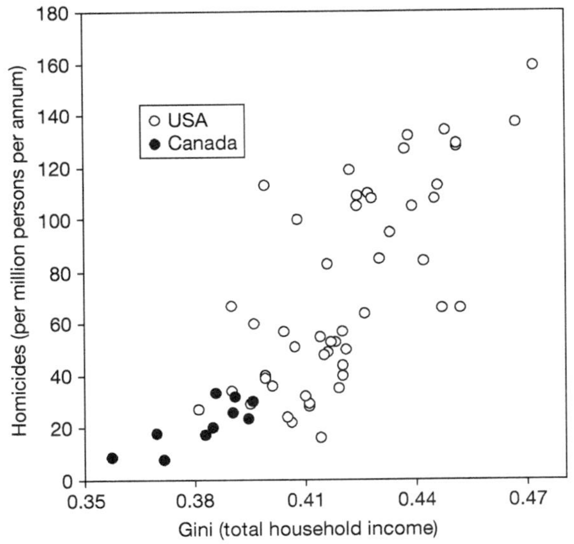

Figure 5 Homicide rates (homicides per million persons per annum) in relation to the Gini coefficient (based on total household income) for the 50 United States, 1990, and the ten provinces of Canada, 1988–92. Data from Daly M, Wilson M, and Vasdev S (2001) Income inequality and homicide rates in Canada and the United States. *Canadian Journal of Criminology* 43: 219–236.

even though economic (material) inequity is seldom extreme. One reason for high homicide rates in these relatively egalitarian societies is the absence of modern medicine, which makes a broader range of wounds life-threatening, but a possibly more important reason is the absence of police power and an effective system of disinterested third-party justice: without effective police and judiciary, a credible threat of personal or kin violence is a crucial social asset regardless of one's wealth or status, and as noted earlier, the familiar tendency for violence to be primarily a recourse of the disadvantaged disappears. Moreover, although an inequitable distribution of wealth is less conspicuous in face-to-face societies than in modern nation-states, inequitable access to marital and reproductive opportunities may actually be more extreme and highly salient. In any case, we would still expect that *ceteris paribus*, dangerous tactics of social competition, will be more attractive to those who have less to lose.

Our Chicago analyses, plus consideration of the social environments in which humans evolved, make us suspect that the social comparison processes mediating the effects of inequity probably operate at more local levels than considered in most criminological and economic research on income inequality's effects. The lives and deaths of personally known individuals are especially salient to one's mental model of one's life prospects. Nevertheless, it is also interesting to ask whether a more global perception that one lives in a winner-take-all society (Frank and Cook, 1995) inspires competitive escalation, and whether media portrayals (including even fictional ones) affect mental models of the rewards available to the winners.

Social scientists have long been interested in the socially undesirable effects of inequality. What an evolutionary psychological approach adds is the suggestion that inequality has its effects not simply by virtue of nonadaptive stress effects, but also by inspiring a rational escalation of costly tactics of social competition. This consideration complicates causal analysis, because it implies that the distribution of age-specific mortality is more than an outcome variable, having feedback effects on its own causal factors and hence on itself. Although we excluded homicide mortality from our analyses of the apparent effects of life expectancy, for example, in order to eliminate spurious autocorrelational effects, it is likely that local levels of homicidal violence affect expectations of future life, discount rates, and hence further violence. Inequality is also expected to affect expectations of future life and discount rates.

12.2.9 Making Sense of Individual Differences

We have stressed that theory and research in psychology, including evolutionary psychology, has been more concerned with characterizing panhuman nature than with human diversity. But diversity exists, and it too requires evolutionary explanation. Much behavioral heterogeneity depends, of course, on the contingent responsiveness of the adult psyche to differences in circumstance, but not all can be explained as resulting from a shared set of if–then rules in conjunction with individual differences in one's current situation. Some, instead, derives from enduring personality traits established early in development. In explaining and predicting others' behavior, people attribute traits such as a short temper or jealousy or fair-mindedness to particular acquaintances because such attributions appear to carry useful information about the reasons for past actions and the likelihoods of future actions. This appearance is no illusion, for although attributions can be excessive or misguided, they do improve prediction. That is, stable differences in personality do indeed account statistically for some substantial fraction of behavioral diversity (Kenrick and Funder, 1988).

Why this should be so is a conundrum from an evolutionary perspective. If selection favors an optimal species-typical design, it is challenging to address the question of individual differences in personality from an adaptationist perspective (Buss, 1991; Tooby and Cosmides, 1990). Why, for example, does violent aggressivity seem to be a personality variable? Why, in other words, should selection not have favored the optimal set of social psychological propensities (i.e., those that best promoted fitness in ancestral environments), with the result that everyone's behavioral repertoire would include more or less identical responses, including violent responses, to particular threats and challenges?

One plausible answer is that people are more alike in this regard than they appear, because human facultative responsiveness is largely developmental rather than just being a matter of contingent reaction to immediately present challenges. Consistent with this idea is the fact that experience with violence predicts its subsequent use (see, e.g., Coie et al. (1991) and Dodge et al. (1990a,b,c)); one reason why people are developmentally labile in this way could be that expertise in the use of violence raises its effectiveness and hence its appeal. Moreover, the information that is of relevance to adaptive decision making in dangerous confrontations has greater time depth than just the immediate situation: how short one's temper ought, ideally, to be, for example, depends on statistical attributes of one's social milieu that can only be induced (if at all) from cumulative experience over a long time.

There is substantial evidence that readiness to use violence is indeed developmentally labile, and some of this evidence suggests that this plasticity may be functional for the actors in the manner just suggested. In a cross-cultural analysis of child-rearing practices in nonstate societies, Low (1989) showed that in societies which have repeatedly engaged in war in recent history, parents and others strive to inculcate aggressivity, strength, skilled use of weapons, and tolerance of pain in boys more than is the case in nonwarring societies. Within Western industrialized nations, there is evidence that people with childhood experience of violence, whether as victims or as witnesses, are relatively likely to use violence (Dodge et al., 1990b; Widom, 1989). Longitudinal studies of juvenile delinquents and career criminals reveal a prior history of various social transgressions including violence (see, e.g., Wilson and Herrnstein (1985), Farrington (1991), Tonry et al. (1991), and Sampson and Laub (1993)).

Psychiatrists have identified a personality type that is particularly likely to engage in violent aggressivity: the antisocial personality (American Psychiatric Association, 1994), which is often diagnosed in both juvenile offenders and career criminals (Moffitt, 1993; Olweus et al., 1986). The diagnostic criteria for antisocial personality include a history of conduct disorder prior to age 15 and a continued pattern of disregard for and violation of the rights of others into adulthood (American Psychiatric Association, 1994, p. 649). There are apparently a number of reliable risk factors associated with the development and maintenance of antisocial personality, including poverty, maleness, early maturity, poor school performance, parental criminal history and psychopathology, and having a lone mother in *loco parentis* (see, e.g., Tonry et al. (1991) and Moffitt (1993)). These risk factors largely overlap the risk factors for juvenile deliquency and violent crime (see, e.g., Wilson and Herrnstein (1985), Farrington (1991), and Sampson and Laub (1993)); moreover, the same risk factors characterize many urban communities with high rates of violence (see, e.g., Krahn et al. (1986), Sampson (1991), and Coulton et al. (1995)). All these considerations suggest that this so-called disorder is largely a reflection of a facultative developmental response to indicators of the futility of developing a more prosocial personality.

Notwithstanding these reasons for invoking facultative development, however, there is also considerable evidence from twin and adoption studies that antisocial personality is substantially heritable (Carey, 1994; Carey and Gottesman, 1996; Lyons, 1996). Thus, the question remains: Why does genetic variability affecting traits like antisocial personality persist? Behavior geneticists seldom consider their findings in the context of Darwinian selection, so the question has scarcely been addressed (but see Rowe (1994)). The available answers appear to be few. One possibility is that selection was weak in ancestral generations, so the variability was effectively neutral with respect to fitness. This is especially plausible when the attributes in question develop in interaction between the effects of particular genotypes and particular novel aspects of modern environments; sensitivity to novel chemical pollutants, for example, can be highly heritable, but these differential effects of the relevant genotypes were of no selective relevance in ancestral environments in which those chemicals did not exist. Another possibility is that selection pressures have been heterogeneous in time and/or space, so that no single optimal phenotype could become universal across the population's whole range of environments (Williams, 1992). Finally, perhaps the most interesting possibility in the present context is that of frequency-dependent selection.

Alternative types are said to incur frequency-dependent selection when their respective fitnesses change as a function of their relative frequency (see, e.g., Heino et al. (1998)). In some animals, for example, there is a rare male mating advantage (Ehrman, 1972) such that whichever of two alternative types of male is rarer in a local population begins to be preferred by females, with the result that the rarer type outreproduces the commoner and neither is likely to go extinct. Mealey (1995) and accompanying commentaries discuss the possibility that sociopaths or psychopaths – an exploitative and often charming personality type apparently lacking empathy – might be maintained analogously in human populations, with their success as deceivers tending to be inversely correlated with their prevalence.

On this argument, psychopathology is not really a pathology at all, and we should not expect to see the signs of dysfunction that characterize unequivocally pathological states like autism, schizophrenia, or Down's syndrome. It is true that psychopaths are different from normal prosocial people in that they lack empathy, tolerate aversive stimuli, are more likely to use violence and coercion, and generally process emotional information differently (Harris et al., 2001; Quinsey et al., 1998; Williamson et al., 1991; Hare, 1996), but are they defective, or just different, albeit in an unpleasant way? Lalumière and Quinsey (1996); Quinsey et al. (1998), Lalumière et al. (1999, 2001), and Harris et al. (2001) argue that the specific attributes that characterize psychopaths are intelligible design features of a frequency-dependent alternative reproductive strategy that cheats by exploiting the prosocial inclinations of the majority. One indication that psychopaths may not be defective is that they do not show the other signs of developmental damage that often accompany other serious psychiatric disorders; they differ from groups with other clinical diagnoses, for example, in the fact that they are no more likely than normals to have been the victims of complications at their births (Harris et al., 2001), and in the fact that their bodies are just as bilaterally symmetrical as those of normals (Lalumière et al., 2001).

12.2.10 Testosterone and the Modulation of Confrontational Competitive Risk Taking

There has been considerable controversy about the links between circulating levels of plasma T and behavior. However, the largest published study of the issue indicates that high T levels are associated with a variety of antisocial behaviors: among 4462 US military veterans, criterion groups who reported antisocial acts such as childhood truancy, violent behavior, military AWOL, drug and alcohol abuse, on-the-job problems, and marital disruption each had on average higher T-levels than those who did not report the acts (Dabbs and Morris, 1990; Booth et al., 1985; Booth and Osgood, 1993; Booth and Dabbs, 1993; Mazur, 1998). Dabbs et al. (1987, 1995) have also reported that T levels in a sample of incarcerated men were related to the violence of the crime. An earlier study (Ehrenkranz et al., 1974) had shown that chronically aggressive men in prison for violent crimes had higher levels of plasma T than nonaggressive men in prison for nonviolent crimes. Other researchers, however, have found no significant differences in T between men charged with murder or assault and those charged with property crimes (Bain et al., 1987). Descriptions of the behavior and attitudes of the incarcerated men in Dabbs's studies suggest that it is not the type of crime *per se* that is associated with elevated T-levels, but whether it entailed aggressive confrontation; the men with higher T were also characterized as tough by their peers (Dabbs et al., 1987) and were known for violating prison rules, especially rules involving overt confrontation (Dabbs et al., 1995).

There are some inconsistencies in the literature on associations between circulating T levels and individual differences in confrontational and competitive behaviors. However, according to Archer (2006), these may be explicable in light of the challenge hypothesis – a conceptual framework proposed by Wingfield (1984) and Wingfield et al. (1990) for understanding how androgen levels in vertebrates are modulated by aspects of the social environment. Wingfield et al. (1990) noted that in various species, transitory elevations in male T levels can be induced by interactions with receptive females and male–male interactions that constitute challenges to a male's territorial or reproductive interests. They argued that the extent of such androgen responsiveness to social cues should vary systematically across species, and between individuals, according to factors such as mating system and degree of paternal investment. Moreover, they suggested that variation in the display of aggressive or confrontational behavior is likely to be more closely associated with T responsiveness to social modulation rather than absolute or baseline levels of T. Not all predictions made by Wingfield et al. (1990) have received empirical support, but evidence for social modulation of androgens, consistent with many aspects of the challenge hypothesis, has accumulated for nonhuman species (Oliveira, 2004; Hirschenhauser and Oliveira, 2006) and for men (Archer, 2006).

There is some evidence that men's T levels do increase after brief interactions with women (Roney et al., 2003, 2007) and after watching sexually arousing films (Pirke et al., 1974; Hellhammer et al., 1985; Stoléru et al.,1993, 1999). Moreover, there are numerous demonstrations that circulating T levels in men can be affected by social cues associated with involvement in competitive events. T-levels rise in anticipation of athletic contests such as tennis and wrestling matches, for example, and rise still higher in the aftermath of a victory, whereas there is no further rise or even a drop after a loss (Mazur and Lamb, 1980; Booth et al., 1989; Elias, 1981; Campbell et al., 1997). T responses to competition are not limited to physically demanding activities, but have also been seen in chess matches (Mazur et al., 1992), in a reaction time contest (Gladue et al., 1989), and even in games of chance with no element of skill (McCaul et al., 1992). The juice of victory may even be felt among spectators: Bernhardt et al. (1998) measured T-levels in fans watching the 1994 World Cup final on television, and found that supporters of the winning Brazilian side exhibited a postmatch increase while supporters of the losing Italian side exhibited a decrease. Even cichlid fish have been shown to experience vicarious endocrine responses, exhibiting increases in T levels after watching a dominance fight between conspecific males (Oliveira et al., 2001).

Not all studies find effects of competitive outcomes on T-levels, but even the exceptions may reflect interpretable variations in the meaningfulness of victory. There is some evidence that T increases in response to success in competitive encounters depend on a man experiencing mood elevation as a consequence of winning (McCaul et al., 1992) and the degree of mood elevation may depend on the salience of the competitive encounter for the individual. Gonzalez-Bono et al. (1999), for example, found no significant change in the average T-levels of winners or losers in a close professional basketball game, but they did find effects in the players who purportedly played the largest role. Similarly, Edwards et al. (2005) found that the percent increase in T seen after having won a soccer match was positively associated with self- and other-rated connectedness with teammates. Other competitions may be unarousing because the outcome is a foregone conclusion: Mazur et al. (1992) found postmatch T-changes in chess winners and losers only when opponents were of similar ability.

Of course, T-responses, like average T-levels, may also exhibit individual differences, and one relevant attribute (perhaps intermediate in stability between a personality trait and a situational response) is the individual's inclination to compete and/or dominate. Schultheiss et al. (1999), for example, found that scores on a personal power motivation scale were positively correlated with T-responses to victory in a race to connect a series of numbers. Similarly, Zumoff et al. (1984) reported that men who are intensely competitive and impatient had higher daytime T-levels than more placid men. Moreover, there is evidence that T levels are generally high in men in action-oriented occupations like firefighters and paramedics (Dabbs and Dabbs, 2000).

The challenge hypothesis suggests that variation in the display of aggressive or confrontational behavior is likely to be more closely associated with T responsiveness to social influences rather than with absolute or baseline levels of T. Few studies, however, have investigated individual differences in T responsiveness in response to social challenges. However, the previously mentioned study by Cohen et al. (1996) found that among University of Michigan students (presumably a relatively nonviolent group), T-levels rose within 20 min after a verbal insult in men hailing from southern states, known for espousing the honor culture, but not in those from the North (Cohen et al., 1996). More recently, it has been shown that circulating T-levels after an experience of competitive success (which was actually experimentally determined) is highest in men with relatively masculine facial structure (Pound et al., 2009).

One adaptationist interpretation of the increase in T after a win is that the winners in social competition will soon be challenged again, and that increased T-levels help prepare tissues for such challenges (Archer, 2006; Wingfield, 1984; Wingfield et al., 1990). A related, and perhaps more general, idea is that victory and defeat are informative as regards one's likelihood of further success or failure in the near future, so that winners should continue to be assertive both in male–male competition and in courtship, whereas losers are better advised to retrench and wait for better opportunities; the T-response may then be mediating the contingent modulation of sexual and aggressive motivation and initiative. Recent evidence suggests that, if experimentally allocated to a losing condition in a rigged competition, a man's T-response to the experience is predictive of the likelihood that he will choose to compete again (Mehta and Josephs, 2006).

It remains unclear through what proximate psychological mechanisms T might influence competitiveness and risk sensitivity. Recent evidence, however, suggests that acute administration of T (in women at least) can reduce attention to threatening stimuli (Van Honk et al., 2005), the magnitude of the fear-potentiated startle response (Hermans et al., 2006a), and empathetic behavior (Hermans et al., 2006b). Moreover, recent experimental evidence suggests that circulating T-levels are associated with cognitive processes that could have a proximate role in modulating confrontational and competitive behavior. Burnham (2007) measured T-levels in Harvard economics students playing the ultimatum game, a two-step procedure in which one party proposes a split of a windfall and the other then either accepts the offer, effectuating the proposed split, or rejects it, leaving both players with nothing. Men who rejected an anonymous peer's low and perhaps insulting offer of $5 out of $40 (the stingy condition) had higher T-levels, measured before playing the game, than those who accepted the offer. Unwillingness to accept outcomes that are perceived as unfair is likely to be related to a man's tendency to get involved in tit-for-tat disputes that could escalate into violent confrontations.

12.2.11 Testosterone as a Mediator of Mating Effort

Work in nonhuman species has indicated that testosterone may play a role in mediating life-history tradeoffs between reproductive and immune functions (Muehlenbein and Bribiescas, 2005) and that social modulation of T-levels may permit individual males to facultatively adjust their agonistic behavior according to their own competitive ability (Oliveira, 2004). Investigation of the links between T and behavior in humans has been complex and controversial not only because causal impacts are bidirectional, but also because there are simultaneous impacts on several analytically separable timescales.

At one extreme, stable personality differences must be invoked when interpreting findings such as the association between prison inmates' T-levels and the qualities of their past crimes (Dabbs et al., 1987, 1995). At the opposite extreme are the rapid T-responses to social experience reviewed earlier. Intermediate cases include perhaps the most important effects: changes in relatively chronic T-levels and behavioral practices over the life span and in relation to long-lasting social status attributes such as marriage and fatherhood.

As we might expect from the evidence that T is causally implicated in both sexual behavior and male–male competition, T-levels are low during most of prepubertal childhood (although there is a period of relatively high T-levels in infants, related to ongoing sexual differentiation), and they rise from puberty to a peak in young adulthood, then fall very slowly at about 1% per year (Liu et al., 2007), remaining above childhood levels into old age (Simon et al., 1992; Dai et al., 1981; Gray et al., 1991a,b; Field et al., 1994; Dabbs, 1990). This age pattern is reminiscent of the age–crime curve illustrated in **Figure 1** and other age-related patterns of risk taking, although the peak in youth is less dramatic, and one is tempted to infer that these changes in T-levels play a role in mediating mating effort. This term (Low, 1978) refers to the allocation of one's time, energy, and attention to pursuits, including both courtship and male–male competition, that might pay off in mating opportunities, and is contrasted with the alternative form of reproductive effort, namely parental effort. Of course, correspondence between T levels and the age–crime curve cannot be taken to imply causation. As noted previously, involvement in social situations entailing competition and challenges to status can cause transient elevations in T-levels, and therefore plausibly could also have chronic effects.

In a predominantly monogamous species with biparental care, such as *Homo sapiens*, we might expect individuals who have acquired mates to reallocate time and energy from mating effort into other pursuits, and it is interesting in this context that T-levels have been shown to vary systematically in relation to relationship status. A number of studies have shown that, in North America, partnered men (i.e. married or in a committed relationship) have lower T (see, e.g., Gray et al. (2002), Burnham et al. (2003), Gray et al. (2002, 2004), and van Anders and Watson (2006)) and similar findings have now been reported in other populations. For example, among the Ariaal of Northern Kenya men with wives have significantly lower T levels than unmarried men (Gray et al., 2007). A number of studies, however, have reported only marginal, nonsignificant, differences in T levels between partnered and unpartnered men (Gray et al., 2006; Sakaguchi et al., 2006, 2007).

In the large study of US military veterans mentioned earlier, not only was it the case that men with high T-levels were relatively likely to be single, or, if they did marry, to divorce (Booth and Dabbs, 1993), but there was also evidence that changing T-levels tracked marital status. The relevant data come from a 10-year longitudinal study of 2100 male US Air Force veterans (aged 32–68 years) whose T-levels were measured in 1982, 1985, 1987, and 1992: in a pattern reminiscent of the homicide data in **Figure 3**, the transition to marriage was associated with a fall in T-levels and divorce was associated with a resurgence (Mazur and Booth, 1998; Mazur and Michalek, 1998; but see Flinn et al. (1998)). A more recent (albeit much smaller) longitudinal study (van Anders and Watson, 2006), however, has suggested that differences in T levels between partnered and unpartnered men may arise as a consequence of stable individual differences in T levels that are systematically related to the likelihood of acquiring and maintaining partnered status (rather than being attributable to the effects of being partnered or not). If both elevated T-levels and homicide offending reflect the allocation of reproductive effort to mating effort, then it is not surprising that they exhibit parallel patterns associated with marital status, perhaps analogous to the seasonal changes in T-levels and mating competition that occur in some nonhuman species.

Recent reports indicate that married men's T-levels fall still further as parental effort is demanded: Gray et al. (2006) found that, in a sample of men from a Chinese university, fathers had lower T than both married men without children and unmarried men. Similarly, Storey et al. (2000) reported that T-levels in men were low around the time that their wives gave birth, especially in those fathers who scored high on indices of commitment to the pregnancy and the child. An obvious interpretation is that high T-levels interfere with paternal effort by inspiring greater allocation to mating effort instead, and are therefore adaptively reduced as part of getting into a paternally investing mode, an interpretation that is reinforced by parallel evidence in other biparental species, including some experimental evidence. In dark-eyed juncos, for example, Raouf et al. (1997) reported that males who were given T-implants reduced their levels of paternal care relative to males with control implants, and were more likely than controls to sire offspring in rivals' nests, presumably because they put more effort into the pursuit of extrapair matings. Remarkably, these T-implanted male juncos were also less likely to be cuckolded themselves, presumably because they also defended their exclusive access to their own mates more effectively; little wonder that they had to cut back on feeding the nestlings!

In further support of the testosterone–mating effort connection in men, there is some intriguing evidence that individual differences in men's T-levels are predictive of their reported numbers of sexual partners (Dabbs and Dabbs, 2000; Booth et al., 1999; Kemper, 1990). However, what these studies might really reflect is a tendency for men with high T-levels to brag or exaggerate, and even if the effect is genuine, the direction of causality is not clear. Certainly, copulation can sometimes boost T-levels in nonhuman species such as male rats, for example (Hart, 1983). In humans, however, although evidence suggests that brief interactions with women (Roney et al., 2003, 2007) and sexual arousal associated with viewing erotic films (Pirke et al., 1974; Hellhammer et al., 1985; Stoléru et al., 1993, 1999) can both cause acute increases in T-levels, experimental evidence indicates that they are unaffected by sexual activity and orgasm (Exton et al., 2001; Krüger et al., 2003). There is evidence, however, that T levels may rise in anticipation of sexual activity. For example, nearly 40 years ago in an anonymous report in *Nature* (Anonymous, 1970), an author described how his beard growth (an androgen-dependent process) slowed during periods of isolation on a remote island but increased again in the days immediately prior to returning to the mainland and resuming sexual activity with his partner. More recent experimental evidence suggests that, relative to men who have not abstained, T-levels are elevated immediately prior to sexual arousal in men who have undergone a period of several weeks sexual abstinence (Exton et al., 2001). Consequently, it is possible to speculate that differences in T-levels between partnered and unpartnered men could be caused by differential exposure to periods of anticipatory abstinence.

12.2.12 Testosterone's Costs and Honest Signaling

A male's level of circulating T affects not only his own behavioral decision making, but the social signals that he emits and the reactions that he elicits from others of both sexes. Female choice often favors males who maintain and somehow are able to display higher than average T-levels, although this may be less true of pair-forming biparental animal species than of species in which male fitness variance is high and paternal investment in young is nonexistent. Female meadow voles, for example, prefer intact high-T males over castrates on the basis of smell (Ferkin et al., 1997). In the black grouse, a bird in which females choose with whom to mate at a communal display ground and then nest alone without male help, T-levels (as well as seniority) predict which males will hold the advantageous central territories on the display ground and how many females will elect to copulate with them (Alatalo et al., 1996).

In red grouse, T-administration enhances the size (Mougeot et al., 2004) and redness (Mougeot et al., 2007) of the comb, but impairs T-cell-mediated immunity (Mougeot et al., 2004), and experimental reductions in nematode parasite load also increase comb redness (Martinez-Padilla et al., 2007). It has proven rather difficult to identify the phenotypic differences among displaying males by which female grouse choose males, but relevant cues have been discovered in many other bird species, with females showing sensitivity to detailed aspects of male secondary sexual traits or ornaments such as plumage brightness, long tails, and the size and coloration of featherless protruberances. (The evidence for fine female discriminations with respect to behavioral displays such as songs and aerobatic maneuvers is much scarcer, but this should not be considered evidence of their lesser importance. They are just harder to measure and to manipulate experimentally than morphological ornaments.) That the more elaborately ornamented males indeed get a disproportionate share of paternity has now been documented with DNA evidence in a number of avian species (Møller and Ninni, 1998).

The dependence of male secondary sexual traits on androgens, both for their initial development and for their maintenance, is widely documented in a range of vertebrates (see, e.g., Pinxten et al. (2000), Wingfield et al. (1997), and Owens and Short (1995)). But why should females prefer males whose sexual ornaments are both androgen-dependent and brighter or otherwise more elaborate than average? One hypothesis that has been popular with theoretical biologists is that these traits and associated female preferences are arbitrary products of a runaway process whereby selection for the trait and for the preference reinforce one another until costs of the trait stop the elaborative process. This model has never been as popular with field researchers, however, not just because it is hard to design a study that would demonstrate the process in nature, but because whenever preference criteria are carefully studied, they turn out not to be arbitrary, as the runaway model predicts, but to be carriers of useful information about the signaler. Accordingly, theories that posit that a female's mate choice criteria actually function to improve the quality of her offspring are in the ascendancy, and perhaps the most influential of these is that proposed by Hamilton and Zuk (1982).

Initially concerned only with birds, Hamilton and Zuk (1982) proposed that male display traits, especially plumage color and luster and the quality of songs, provide evidence of the male's current burden of disease microorganisms (pathogens), and of his past and present pathogen resistance (see also Møller et al. (1998)). One benefit of choosing a relatively pathogen-free male is reduced risk of disease transmission at the time of mating, but the function that Hamilton and Zuk stressed is more subtle: by mating with a male who is relatively resistant to whatever pathogen strains are currently, locally prevalent, a female recruits the genetic basis for that resistance for her offspring. This theory has now garnered extensive support in a wide range of vertebrate and invertebrate taxa, including comparative demonstrations that species with higher pathogen loads have more elaborate sexual ornaments than related species under lesser pathogen pressure (see, e.g., Read (1988)); demonstrations that the ornaments to which females attend indeed provide evidence of a male's disease history or current pathogen loads, and (in some cases) that females actually pay the most attention to the traits that are most informative in this regard (see, e.g., Zuk et al. (1990a,b)); and demonstrations that females who are permitted to choose their mates indeed produce offspring that are more disease resistant than those of females who are assigned a mate and cannot choose (Drickamer et al., 2000; Johnsen et al., 2000).

A long-standing puzzle about courtship displays is why, if a signal such as a bright ornament or a song of some particular quality is attractive to females, do all males not display it. The answer seems to be that selection on the recipients of signals favors attending only to honest signals that are genuinely informative and ignoring those that an impostor is capable of faking (Grafen, 1990; Zahavi and Zahavi, 1997; Johnstone, 1995). One consideration that apparently constrains the ability of low-quality males to cheat by displaying T-dependent ornaments is that T suppresses the immune system (Wichmann et al., 1997; Angele et al., 1998; Yao et al., 2003), making males with high T-levels more vulnerable to infections (see, e.g., Folstad and Karter (1992), Thompson et al. (1997), and Kacelnik and Norris (1998)). Males with high-energy reserves, no wounds, and genotypes that are relatively resistant to local pathogen strains can presumably afford the immunosuppression that developing and maintaining ornaments requires, when other males cannot. (Why ornaments should not then evolve to no longer be androgen dependent, so that males who can develop them without paying these costs can invade the population, is another puzzle. Perhaps this does happen to different ornaments in different evolving lineages, and females are then selected to ignore the signal, contributing to the taxonomic diversification of display features.)

The essence of the immunocompetence hypothesis is that males invest in the production of the secondary sexual traits that females prefer to the degree that they are of sufficient intrinsic quality and/or in sufficiently good condition to tolerate the associated costs, especially that of compromising their immune system (Folstad and Karter, 1992). The several elements of this argument have seldom been empirically demonstrated in a single system, but research on the red jungle fowl (*Gallus gallus*), the ancestor of the domestic hen, comes close. Females have been shown to prefer males with the largest and brightest ornaments, and to pay the greatest attention to the ornaments that respond most to the male's disease history (Ligon et al., 1990; Zuk et al., 1990a,b). One of the most important male attributes for female choice is the size of the comb, which has been shown to be positively correlated with T-levels (Zuk et al., 1995; Verhulst et al., 1999). Furthermore, immune response is negatively correlated with both T-levels and the size of the comb, a remarkable demonstration that the relatively vigorous and decorated cocks that hens prefer are indeed paying a price. However, relationships among T, immune response, and secondary sexual traits are not always this consistent with the immunocompetence hypothesis in other species (Hasselquist et al., 1999; Ros et al., 1997; Weatherhead et al., 1993) and evidence in support of the specific premise that developing T-dependent traits imposes an immunocompetence cost is mixed and needs to be interpreted with caution (Roberts et al., 2004). But the costs that keep T-dependent signals honest need not reflect direct immunosuppressive effects;

they may instead be mediated by other correlated factors, such as lipid stores and leptin levels (Alonso-Alvarez et al., 2007a), or by T's effects on such things as antioxidant activity (Alonso-Alvarez et al., 2007b), and carotenoid deposition (Blas et al., 2006). Ongoing studies should help identify important modulating variables and clarify whether the diverse findings to date can be encompassed under a single theory.

A difficulty in testing the immunocompetence hypothesis is that even where it is correct, different relationships between secondary sexual traits and immune response may be observed. High-quality males can sometimes afford T's negative effects on their immune systems precisely because they still perform better than those of low-quality males, in which case the correlation will still be positive. In such cases, it may be most informative to compare males after administering both exogenous T and an immune challenge, eliminating the confound between T-levels and male quality (see, e.g., Ros et al. (1997)). A correlation between ornamentation and immune response in males whose T-levels have not been experimentally manipulated is likely to reflect variation in allocation between current reproductive effort and its alternative, somatic effort (growth and maintenance), which can be considered an investment in future reproduction. Stickleback (*Gasterosteus aculeatus*) males invest in the pigmentation that makes them attractive to females (reproductive effort) when in good condition (indexed by lipid stores), as one might expect, but also do so when in especially poor condition or in the presence of a predator (Candolin, 1998, 1999); it appears that males in the best condition find the prospective fitness benefits of reproductive effort sufficient to cover the costs and those in somewhat poorer condition do not, but those in the worst condition give up on future reproduction altogether, in response to cues indicating that they are unlikely to get another chance (Candolin, 1998, 1999; Kokko, 1997). From a female perspective, such males who invest everything in a last-ditch mating effort are dishonest signalers, and selection may be expected to favor females who are able to detect quality cues that cannot be faked, so there is male–female conflict in these cases, and a potential for evolutionary arms races.

Males with high T-levels incur other costs in addition to immunosuppression. Their basal metabolic rate may be elevated (Buchanan et al., 2001), and the behavioral risk taking that creates mating opportunities can also increase the chance of injury in an aggressive contest and of falling victim to a predator (see, e.g., Daly et al. (1990)). The upshot is that a male whose blood T increases may gain in reproductive success, but risks shortening his life and there is evidence for this occurring in humans (Hamilton, 1948; Hamilton and Mestler, 1969; but see also Nieschlag et al. (1993)) and in nonhuman species (Wingfield et al., 1997; Dufty, 1989). Moreover, as we have already noted, T enhances mating effort at the expense of paternal investment (Raouf et al., 1997; Wingfield et al., 1997) so that even apart from the energetic and mortality costs, T may have simultaneous opposing effects on different components of fitness. The optimal level of circulating T for a male is clearly both species specific and a complex function of his individual situation and attributes.

12.3 Concluding Remarks

The psychophysiological controls and behavioral manifestations of intrasexual competition have undoubtedly been shaped by a history of sexual selection. This implies that perceptual and cognitive mechanisms subserving interpretation of social situations and decision making must be functionally integrated with endocrine, immune, and other physiological systems to produce coordinated and subtly modulated responses to cues of the utility of competitive and aggressive inclinations and actions, at least as they would have paid off in fitness in ancestral environments. We have proposed that dangerous competitive violence is modulated by cues of one's social and material circumstances, both absolute and relative, and by some sort of mental model of the current local utility of competitive success. An evolutionary psychological perspective can help direct research on psychophysiological mechanisms by identifying the sorts of contingent response that are likely to have been favored by selection, and is equally of relevance for the social sciences, by suggesting hypotheses about the probable impacts of social factors such as inequitable resource distributions.

Acknowledgments

MW and MD thank the Social Sciences and Humanities Research Council of Canada, the Natural Sciences and Engineering Research Council of Canada, the Harry Frank Guggenheim Foundation, and the Catherine T. and John D. MacArthur Foundation for financial support of their research. NP thanks the Ontario Graduate Scholarship fund, McMaster University, the Medical Research Council (UK), and the Brunel Research Initiative and Enterprise Fund.

References

Aarsland D, Cummings JL, Yenner G, and Miller B (1996) Relationship of aggressive behavior to other neuropsychiatric symptoms in patients with Alzheimer's disease. *American Journal of Psychiatry* 153: 243–247.

Alatalo RV, Höglund J, Lundberg A, Rintamäki PT, and Silverin B (1996) Testosterone and male mating success on the black grouse leks. *Proceedings of the Royal Society of London, B* 263: 1697–1702.

Alonso-Alvarez C, Bertrand S, and Sorci G (2007a) Energetic reserves, leptin and testosterone: A refinement of the immunocompetence handicap hypothesis. *Biology Letters* 3: 271–274.

Alonso-Alvarez C, Bertrand S, Faivre B, Chastel O, and Sorci G (2007b) Testosterone and oxidative stress: The oxidation handicap hypothesis. *Proceedings of the Royal Society of London, B* 274: 819–825.

American Psychiatric Association (1994) *Diagnostic and Statistical Manual of Mental Disorders (DSM-IV)*, 4th edn. Washington, DC: American Psychiatric Association.

Andersson M (1980) Why are there so many threat displays? *Journal of Theoretical Biology* 86: 773–781.

Angele MK, Ayala A, Cioffi WG, Bland KI, and Chaudry IH (1998) Testosterone: The culprit for producing splenocyte immune depression after trauma hemorrhage. *American Journal of Physiology Cell Physiology* 274: C1530–C1536.

Anonymous (1970) Effects of sexual activity on beard growth in man. *Nature* 226: 869–870.

Archer D and Gartner R (1984) *Violence and Crime in Cross-National Perspective*. New Haven, CT: Yale University Press.

Archer J (1988) *The Behavioural Biology of Aggression*. Cambridge: Cambridge University Press.

Archer J (2006) Testosterone and human aggression: An evaluation of the challenge hypothesis. *Neuroscience and Biobehavioral Reviews* 30: 319–345.

Bain J, Langevin R, Dickey R, and Ben-Aron M (1987) Sex hormones in murderers and assaulters. *Behavioral Sciences and the Law* 5: 95–101.

Baldick R (1965) *The Duel*. New York: Clarkson N. Potter.

Bateman AJ (1948) Intra-sexual selection in *Drosophila*. *Heredity* 2: 349–368.

Bernhardt PC, Dabbs JM, Jr., Fielden JA, and Lutter CD (1998) Testosterone changes during vicarious experiences of winning and losing among fans at sporting events. *Physiology and Behavior* 65: 59–62.

Betzig L (1986) *Despotism and Differential Reproduction: A Darwinian View of History*. New York: de Gruyter.

Blas J, Pérez-Rodríguez L, Bortolotti GR, Viñuela J, and Marchant TA (2006) Testosterone increases bioavailability of carotenoids: Insights into the honesty of sexual signaling. *Proceedings of the National Academy of Sciences of the United States of America* 103: 18633–18637.

Blau J and Blau P (1982) The cost of inequality. *American Sociological Review* 47: 114–129.

Booth A and Dabbs J, Jr. (1993) Testosterone and men's marriages. *Social Forces* 72: 463–477.

Booth A, Johnson DR, and Granger DA (1999) Testosterone and men's health. *Journal of Behavioral Medicine* 22: 1–19.

Booth A, Johnson D, White L, and Edwards J (1985) Predicting divorce and permanent separation. *Journal of Family Issues* 6: 331–346.

Booth A and Osgood D (1993) The influence of testosterone on deviance in adulthood. *Criminology* 31: 93–117.

Booth A, Shelley G, Mazur A, Tharp G, and Kittok R (1989) Testosterone and winning and losing in human competition. *Hormones and Behavior* 23: 556–571.

Buchanan KL, Evans MR, Goldsmith AR, Bryant DM, and Rowe LV (2001) Testosterone influences basal metabolic rate in male house sparrows: A new cost of dominance signalling? *Proceedings of the Royal Society of London, B* 268: 1337–1344.

Burnham T (2007) High-testosterone men reject low ultimatum game offers. *Proceedings of the Royal Society of London, B* 274: 2327–2330.

Burnham TC, Flynn Chapman JC, Gray PB, McIntyre M, Lipson SF, and Ellison PT (2003) Men in committed, romantic relationships have lower testosterone. *Hormones and Behavior* 44: 119–122.

Burton LM (1990) Teenage childbearing as an alternative life-course strategy in multigenerational black families. *Human Nature* 1: 123–143.

Buss DM (1991) Evolutionary personality psychology. *Annual Review of Psychology* 42: 459–491.

Buss DM (1999) *Evolutionary Psychology: The New Science of the Mind*. Needham Heights, MA: Allyn and Bacon.

Buss DM (2000) *The Dangerous Passion: Why Jealousy Is as Necessary as Love and Sex*. New York: The Free Press.

Campbell A, Muncer S, and Odber J (1997) Aggression and testosterone: Testing a bio-social model. *Aggressive Behavior* 23: 229–238.

Candolin U (1998) Reproduction under predation risk and the trade-off between current and future reproduction in the threespine stickleback. *Proceedings of the Royal Society of London, B* 265: 1171–1175.

Candolin U (1999) The relationship between signal quality and physical condition: Is sexual signalling honest in the three-spined stickleback? *Animal Behaviour* 58: 1261–1267.

Carey G (1994) Genetics and violence. In: Reiss AJ, Miczek KA, and Roth JA (eds.) *Understanding and Preventing Violence: Biobehavioral Influences*, vol. 2, pp. 21–58. Washington, DC: National Academy Press.

Carey G and Gottesman II (1996) Genetics and antisocial behavior: Substance versus sound bytes. *Politics and the Life Sciences* 15: 88–90.

Chagnon NA (1988) Life histories, blood revenge, and warfare in a tribal population. *Science* 239: 985–992.

Chagnon NA (1996) Chronic problems in understanding tribal violence and warfare. In: Bock GR and Goode JA (eds.) *CIBA Foundation Symposium 194: Genetics of Criminal and Antisocial Behaviour*, pp. 202–236. Chichester: Wiley.

Chase ID, Bartolomeo C, and Dugatkin LA (1994) Aggressive interactions and inter-contest interval: How long do winners keep winning? *Animal Behaviour* 48: 393–400.

Chilton R (1987) Twenty years of homicide and robbery in Chicago: The impact of the city's changing racial and age composition. *Journal of Quantitative Criminology* 3: 195–214.

Clinton WL and LeBoeuf BJ (1993) Sexual selection's effects on male life history and the pattern of male mortality. *Ecology* 74: 1884–1892.

Clutton-Brock TH and Parker GA (1995) Punishment in animal societies. *Nature* 373: 209–216.

Clutton-Brock TH and Vincent ACJ (1991) Sexual selection and the potential reproductive rates of males and females. *Nature* 351: 58–60.

Cohen D, Nisbett RE, Bowdle BF, and Schwarz N (1996) Insult, aggression, and the southern culture of honor: An experimental ethnography. *Journal of Personality and Social Psychology* 70: 945–959.

Cohen LE and Machalek R (1994) The normalcy of crime. From Durkheim to evolutionary ecology. *Rationality and Society* 6: 286–308.

Coie JD, Dodge KA, Terry R, and Wright V (1991) The role of aggression in peer relations: An analysis of aggression episodes in boys' play groups. *Child Development* 62: 812–826.

Cosmides L and Tooby J (1996) Are humans good intuitive statisticians after all? Rethinking some conclusions from the literature on judgment under uncertainty. *Cognition* 58: 1–73.

Coulton CJ, Korbin JE, Su M, and Chow J (1995) Community level factors and child maltreatment rates. *Child Development* 66: 1262–1276.

Dabbs JM, Jr. (1990) Age and seasonal variation in serum testosterone concentration among men. *Chronobiology International* 7: 245–249.

Dabbs J, Jr., Frady R, Carr T, and Besch N (1987) Saliva testosterone and criminal violence in young adult prison inmates. *Psychosomatic Medicine* 49: 174–182.

Dabbs J, Jr. and Morris R (1990) Testosterone, social class, and antisocial behavior in a sample of 4,462 men. *Psychological Science* 1: 209–211.

Dabbs JM and Dabbs MG (2000) *Heroes, Rogues, and Lovers. Testosterone and Behavior*. New York: McGraw-Hill.

Dai WD, Kuller LH, LaPorte RE, Gutal JP, Falvo-Gerard L, and Caggiula A (1981) The epidemiology of plasma testosterone levels in middle-aged men. *American Journal of Epidemiology* 114: 804–816.

Daly M and Wilson MI (1983) *Sex, Evolution and Behavior: Adaptations for Reproduction*, 2nd edn. Boston, MA: Willard Grant Press.

Daly M and Wilson M (1988) *Homicide*. New York: de Gruyter.

Daly M and Wilson M (1990) Killing the competition. *Human Nature* 1: 83–109.

Daly M and Wilson M (1993) Evolutionary psychology of male violence. In: Archer J (ed.) *Male Violence*, pp. 253–288. London: Routledge, Chapman and Hall.

Daly M and Wilson M (1997) Crime and conflict: Homicide in evolutionary perspective. *Crime and Justice* 22: 251–300.

Daly M and Wilson M (2001) Risk-taking, intrasexual competition, and homicide. In: French JA, Kamil AC, and Leger DW (eds.) *Nebraska Symposium on Motivation, Vol. 47: Evolutionary Psychology and Motivation,* pp. 1–36. Lincoln, NE: University of Nebraska Press.

Daly M and Wilson M (2005) Carpe diem: Adaptation and devaluing the future. *Quarterly Review of Biology* 80: 55–61.

Daly M, Wilson MI, Behrends PR, and Jacobs LF (1990) Characteristics of kangaroo rats (*Dipodomys merriami*) associated with differential predation risk. *Animal Behaviour* 40: 380–389.

Daly M, Wilson M, and Vasdev S (2001) Income inequality and homicide rates in Canada and the United States. *Canadian Journal of Criminology* 43: 219–236.

Dawkins R (1982) *The Extended Phenotype*. Oxford: Freeman.

Dodge KA, Bates JE, and Pettit GS (1990a) Mechanisms in the cycle of violence. *Science* 250: 1678–1683.

Dodge KA, Coie JD, Pettit GS, and Price JM (1990b) Peer status and aggression in boys' groups: Developmental and contextual analyses. *Child Development* 61: 1289–1309.

Dodge KA, Price JM, and Bachorowski J-A (1990c) Hostile attributional biases in severely aggressive adolescents. *Journal of Abnormal Psychology* 99: 385–392.

Drickamer LC, Gowaty PA, and Holmes CM (2000) Free female mate choice in house mice affects reproductive success and offspring viability and performance. *Animal Behaviour* 59: 371–378.

Dufty AM (1989) Testosterone and survival: A cost of aggressiveness. *Hormones and Behavior* 23: 185–193.

Edwards DA, Wetzel K, and Wyner DR (2005) Intercollegiate soccer: Saliva cortisol and testosterone are elevated during competition, and testosterone is related to status and social connectedness with teammates. *Physiology and Behaviour* 87: 135–143.

Ehrenkranz J, Bliss E, and Sheard MH (1974) Plasma testosterone: Correlation with aggressive behavior and social dominance in man. *Psychosomatic Medicine* 36: 469–475.

Ehrman L (1972) Genetics and sexual selection. In: Campbell B (ed.) *Sexual Selection and the Descent of Man 1871–1971*, pp. 105–135. Chicago, IL: Aldine.

Eisner M (2003) Long-term historical trends in violent crime. *Crime and Justice* 30: 83–142.

Dabbs J, Jr., Carr S, Frady R, and Riad J (1995) Testosterone, crime, and misbehavior among 692 male prison inmates. *Personality and Individual Differences* 18: 627–633.

Elias M (1981) Serum cortisol, testosterone, and testosterone-binding globulin responses to competitive fighting in human males. *Aggressive Behavior* 7: 215–224.

Enquist M and Leimar O (1990) The evolution of fatal fighting. *Animal Behaviour* 39: 1–9.

Exton MS, Kruger TH, Bursch N, Haake P, Knapp W, Schedlowski M, and Hartmann U (2001) Endocrine response to masturbation-induced orgasm in healthy men following a 3-week sexual abstinence. *World Journal of Urology* 19: 377–382.

Farrington DP (1991) Psychological contributions to the explanation of offending. *Issues in Criminological and Legal Psychology* 1: 7–19.

Ferkin MH, Sorokin ES, and Johnston RE (1997) Effect of prolactin on the attractiveness of male odors to females in meadow voles: Independent and additive effects with testosterone. *Hormones and Behavior* 31: 55–63.

Field AE, Colditz GA, Willett WC, Longcope C, and McKinlay JB (1994) The relation of smoking, age, relative weight, and dietary intake to serum adrenal steroids, sex hormones, and sex hormone-binding globulin in middle-aged men. *Journal of clinical Endocrinology and Metabolism* 79: 1310–1316.

Fisher RA (1930) *The Genetical Theory of Natural Selection*. Oxford: Oxford University Press (Clarendon).

Flinn M, Baerwald C, Decker S, and England B (1998) Evolutionary functions of neuroendocrine response to social environment. *Behavioral and Brain Sciences* 21: 372–374.

Folstad I and Karter AJ (1992) Parasites, bright males, and the immunocompetence handicap. *American Naturalist* 139: 603–622.

Frank RH and Cook PJ (1995) *The Winner-Take-All Society*. New York: Simon and Schuster.

Gartner R (1990) The victims of homicide: A temporal and cross-national review. *American Sociological Review* 55: 92–106.

Geronimus AT (1992) The weathering hypothesis and the health of African–American women and infants: Evidence and speculation. *Ethnicity and Disease* 2: 207–221.

Geronimus AT (1996) What teen mothers know. *Human Nature* 7: 323–352.

Giancola PR and Zeichner A (1995) An investigation of gender differences in alcohol-related aggression. *Journal of Studies on Alcohol* 56: 573–579.

Gigerenzer G and Todd PM (1999) *Simple Heuristics That Make Us Smart*. New York: Oxford University Press.
Gladue B, Boechler M, and McCaul K (1989) Hormonal response to competition in human males. *Aggressive Behavior* 15: 409–422.
Gonzalez-Bono E, Salvador A, Serrano MA, and Ricarte J (1999) Testosterone, cortisol, and mood in a sports team competition. *Hormones and Behavior* 35: 55–62.
Grafen A (1990) Biological signals as handicaps. *Journal of Theoretical Biology* 144: 517–546.
Grand TC (1999) Risk-taking behaviour and the time of life history events: Consequences of body size and season. *Oikos* 85: 467–480.
Gray A, Berlin JA, McKinlay JB, and Longcope C (1991a) An examination of research design effects on the association of testosterone and male aging: Results of a metaanalysis. *Journal of Clinical Epidemiology* 44: 671–684.
Gray A, Feldman HA, McKinlay JB, and Longcope C (1991b) Age, disease, and changing sex hormone levels in middle-aged men: Results of the Massachusetts male aging study. *Journal of clinical Endocrinology and Metabolism* 73: 1016–1025.
Gray PB, Campbell BC, Marlowe FW, Lipson SF, and Ellison PT (2004) Social variables predict between-subject but not day-to-day variation in the testosterone of US men. *Psychoneuroendocrinology* 29: 1153–1162.
Gray PB, Ellison PT, and Campbell BC (2007) Testosterone and marriage among Ariaal men of Northern Kenya. *Current Anthropology* 48: 750–755.
Gray PB, Kahlenberg SM, Barrett ES, Lipson SF, and Ellison PT (2002) Marriage and fatherhood are associated with lower testosterone in males. *Evolution and Human Behavior* 23: 193–201.
Gray PB, Yang CFJ, and Pope HG (2006) Fathers have lower salivary testosterone levels than unmarried men and married non-fathers in Beijing, China. *Proceedings of the Royal Society of London, B* 273: 333–339.
Gwynne DT and Bailey WJ (1999) Female–female competition in katydids: Sexual selection for increased sensitivity to a male signal? *Evolution* 53: 546–551.
Hagedorn JD (1988) *People and Folks*. Chicago, IL: Lake View Press.
Hamilton JB (1948) The role of testicular secretions as indicated by the effects of castration in man and by studies of pathological conditions and the short lifespan associated with maleness. *Recent Progress in Hormone Research* 3: 257–322.
Hamilton JB and Mestler GE (1969) Mortality and survival: Comparison of eunuchs with intact men and women in a mentally retarded population. *Journals of Gerontology* 24: 395–411.
Hamilton WD and Zuk M (1982) Heritable true fitness and bright birds: A role for parasites? *Science* 213: 384–387.
Hare RD (1996) Psychopathy: A clinical construct whose time has come. *Criminal Justice and Behavior* 23: 25–54.
Harris GT, Skilling TA, and Rice ME (2001) The construct of psychopathy. In: Tonry M (ed.) *Crime and Justice: An Annual Review of Research*, pp. 197–264. Chicago, IL: University of Chicago Press.
Hart BL (1983) Role of testosterone secretion and penile reflexes in sexual behavior and sperm competition in male rats: A theoretical contribution. *Physiology and Behavior* 31: 823–827.
Hasselquist D, March JA, Sherman PW, and Wingfield JC (1999) Is avian humoral immunocompetence suppressed by testosterone? *Behavioral Ecology and Sociobiology* 45: 167–175.
Heino M, Metz JAJ, and Kaitala V (1998) The enigma of frequency-dependent selection. *Trends in Ecology and Evolution* 13: 367–370.
Hellhammer DH, Hubert W, and Schürmeyer T (1985) Changes in saliva testosterone after psychological stimulation in men. *Psychoneuroendocrinology* 10: 77–81.
Hermans EJ, Putman P, and van Honk J (2006b) Testosterone administration reduces empathetic behavior: A facial mimicry study. *Psychoneuroendocrinology* 31: 859–866.
Hermans P, Putman J, Baas H, Koppeschaar J, and van Honk (2006a) A single administration of testosterone reduces fear-potentiated startle in humans. *Biological Psychiatry* 59: 872–874.
Hewlett BS (1988) Sexual selection and paternal investment among Aka pygmies. In: Betzig L, Borgerhoff Mulder M, and Turke P (eds.) *Human Reproductive Behaviour*, pp. 263–276. Cambridge: Cambridge University Press.
Hill EM, Ross LT, and Low BS (1997) The role of future unpredictability in human risk taking. *Human Nature* 8: 287–325.
Hill K and Hurtado AM (1995) *Ache Life History*. Hawthorne, NY: de Gruyter.
Hilton NZ, Harris GT, and Rice ME (2000) The functions of aggression by male teenagers. *Journal of Personality and Social Psychology* 79: 988–994.
Hirschenhauser K and Oliveira RF (2006) Social modulation of androgen levels in vertebrates: A meta-analysis of the challenge hypothesis. *Animal Behaviour* 71: 265–277.
Hirschi T and Gottfredson MR (1983) Age and the explanation of crime. *American Journal of Sociology* 89: 552–584.
Hirschi T and Gottfredson MR (1986) The distinction between crime and criminality. In: Hartnagel TF and Silverman (eds.) *Critique and Explanation*, pp. 55–70. New Brunswick, NJ: Transaction Books.
Howell N (1979) *The Demography of the Dobe !Kung*. New York: Academic Press.
Huntingford P and Turner A (1987) *Animal Conflict*. London: Chapman and Hall.
Jankowski MS (1992) *Islands in the Street*. Berkeley, CA: University of California Press.
Johnsen A, Anderson V, Sunding C, and Lifjeld JT (2000) Female bluethroats enhance offspring immunocompetence through extra-pair copulations. *Nature* 406: 296–299.
Johnstone RA (1995) Sexual selection, honest advertisement and the handicap principle, reviewing the evidence. *Biological Reviews* 70: 1–65.
Kacelnik A (1997) Normative and descriptive models of decision making, time discounting and risk sensitivity. In: Bock G and Cardew G (eds.) *Ciba Foundation Symposium 208: Characterizing Human Psychological Adaptations*, pp. 51–70. London: Wiley.
Kacelnik A and Norris S (1998) Signalling via testosterone: Communicating health and vigour. *Behavioral and Brain Sciences* 21: 378.
Kahneman D, Slovic P, and Tversky A (eds.) (1982) *Judgment under Uncertainty*. New York: Cambridge University Press.
Kemper TD (1990) *Social Structure and Testosterone*. New Brunswick, NJ: Rutgers University Press.
Kennedy BP, Kawachi I, and Prothrow-Stith D (1996) Income distribution and mortality: Cross sectional ecological study of the Robin Hood index in the United States. *British Medical Journal* 312: 1004–1007and 1194.
Kenrick DT and Funder DC (1988) Profiting from controversy: Lessons from the personality-situation debate. *American Psychologist* 43: 23–34.
Kirby KN and Herrnstein RJ (1995) Preference reversals due to myopic discounting of delayed reward. *Psychological Science* 6: 83–89.
Kokko H (1997) Evolutionary stable strategies of age-dependent sexual advertisement. *Behavioral Ecology and Sociobiology* 41: 99–107.

Krahn H, Hartnagel TF, and Gartrell JW (1986) Income inequality and homicide rates: Cross-national data and criminological theories. *Criminology* 24: 269–295.

Krüger TH, Haake P, Chereath D, et al. (2003) Specificity of the neuroendocrine response to orgasm during sexual arousal in men. *Journal of Endocrinology* 177: 57–64.

Kvarnemo C, Forsgren E, and Magnhagen C (1995) Effects of sex-ratio on intra-sexual and inter-sexual behavior in sand gobies. *Animal Behaviour* 50: 1455–1461.

Lalumière ML, Harris GT, and Rice ME (1999) Birth order and fluctuating asymmetry: A first look. *Proceedings of the Royal Society of London, B* 266: 2351–2354.

Lalumière ML, Harris GT, and Rice ME (2001) Psychopathy and developmental instability. *Evolution and Human Behavior* 22: 75–92.

Lalumière ML and Quinsey VL (1996) Sexual deviance, antisociality, mating effort, and the use of sexually coercive behaviors. *Personality and Individual Differences* 21: 33–48.

Ligon JD, Thornhill R, Zuk M, and Johnson K (1990) Male–male competition, ornamentation and the role of testosterone in sexual selection in red jungle fowl. *Animal Behaviour* 40: 367–373.

Liu PY, Beilin J, Meier C, et al. (2007) Age-related changes in serum testosterone and sex hormone binding globulin in Australian men: Longitudinal analyses of two geographically separate regional cohorts. *Journal of Clinical Endocrinology and Metabolism* 92: 3599–3603.

Low BS (1978) Environmental uncertainty and the parental strategies of marsupials and placentals. *American Naturalist* 112: 197–213.

Low BS (1989) Cross-cultural patterns in the training of children: An evolutionary perspective. *Journal of Comparative Psychology* 103: 311–319.

Luckenbill DF (1977) Criminal homicide as a situated transaction. *Social Problems* 25: 176–186.

Lyons MJ (1996) A twin study of self-reported criminal behaviour. In: Bock GR and Goode JA (eds.) *CIBA Foundation Symposium 194: Genetics of Criminal and Antisocial Behaviour*, pp. 61–69. Chichester: Wiley.

Martinez–Padilla J, Mougeot F, Perez Rodriguez L, and Bortolotti GR (2007) Nematode parasites reduce carotenoid–based signalling in male red grouse. *Biology Letters* 3: 161–164.

Mayr E (1983) How to carry out the adaptationist program? *American Naturalist* 121: 324–334.

Mazur A (1998) Marriage, divorce, and male testosterone. *Social Forces* 77: 315–330.

Mazur A and Booth A (1998) Testosterone and dominance in men. *Behavioral and Brain Sciences* 21: 353–363.

Mazur A, Booth A, and Dabbs J, Jr. (1992) Testosterone and chess competition. *Social Psychology Quartely* 55: 70–77.

Mazur A and Lamb T (1980) Testosterone, status, and mood in human males. *Hormones and Behavior* 14: 236–246.

Mazur A and Michalek J (1998) Marriage, divorce, and male testosterone. *Social Forces* 77: 315–330.

McCaul K, Gladue B, and Joppa M (1992) Winning, losing, mood, and testosterone. *Hormones and Behavior* 26: 486–504.

Mealey L (1995) The sociobiology of sociopathy: An integrated evolutionary model. *Behavioral and Brain Sciences* 18: 523–599.

Mehta PH and Josephs RA (2006) Testosterone change after losing predicts the decision to compete again. *Hormones and Behavior* 50: 684–692.

Moffitt T (1993) Adolescence-limited and life-course persistent antisocial behavior: A developmental taxonomy. *Psychological Review* 100: 674–701.

Møller AP, Dufva R, and Erritzøe J (1998) Host immune function and sexual selection in birds. *Journal of Evolutionary Biology* 11: 703–719.

Møller AP and Ninni P (1998) Sperm competition and sexual selection: A meta-analysis of paternity studies of birds. *Behavioral Ecology and Sociobiology* 43: 345–358.

Monahan J and Splane S (1980) Psychological approaches to criminal behavior. In: Bittner E and Messinger S (eds.) *Criminology Review Yearbook*, vol. 2, pp. 17–47. Beverly Hills, CA: Sage.

Moore AJ and Moore PJ (1999) Balancing sexual selection through opposing mate choice and male competition. *Proceedings of the Royal Society of London, B* 266: 711–716.

Mougeot F, Irvine JR, Seivwright L, Redpath SM, and Piertney S (2004) Testosterone, immunocompetence, and honest sexual signaling in male red grouse. *Behavioral Ecology* 15: 930–937.

Mougeot F, Perez-Rodriguez L, Martinez-Padilla J, Redpath S, and Leckie F (2007) Parasites, testosterone and honest carotenoid-based signaling of health. *Functional Ecology* 21: 886–898.

Muehlenbein MP and Bribiescas RG (2005) Testosterone-mediated immune functions and male life histories. *American Journal of Human Biology* 17: 527–558.

Nesse RM (1990) Evolutionary explanations of emotions. *Human Nature* 1: 261–289.

Nesse RM and Williams GC (1994) *Why We Get Sick*. New York: Random House.

Nieschlag E, Nieschlag S, and Behre HM (1993) Lifespan and testosterone. *Nature* 366: 215.

Nisbett RE and Cohen D (1996) *Culture of Honor: The Psychology of Violence in the South*. Boulder, CO: Westview Press.

Nisbett RE and Ross L (1980) *Human Inference: Strategies and Shortcomings of Social Judgment*. Englewood Cliffs, NJ: Prentice Hall.

Nisbett RE and Wilson TD (1977) Telling more than we can know: Verbal reports on mental processes. *Psychological Review* 84: 231–259.

Norry FM, Calcagno G, Vera MT, Manso F, and Vilardi JC (1999) Sexual selection on male morphology independent of male–male competition in the Mediterranean fruit fly (*Diptera: Tephritidae*). *Annals of the Entomological Society of America* 92: 571–577.

Oliveira RF (2004) Social modulation of androgens in vertebrates: Mechanisms and function. In: Slater PJB, Rosenblatt JS, Snowdon CT, and Roper TJ (eds.) *Advances in the Study of Behavior*, vol. 34, pp. 165–239. New York: Academic Press.

Oliveira RF, Lopes M, Carneiro LA, and Canario AVM (2001) Watching fights raises fish hormone levels: Cichlid fish wrestling for dominance induce an androgen surge in male spectators. *Nature* 409: 475.

Oliveira RF, Mcgregor PK, and Latruffe C (1998) Know thine enemy. Fighting fish gather information from observing conspecific interactions. *Proceedings of the Royal Society of London, B* 265: 1045–1049.

Olweus D, Block J, and Radke-Yarrows M (eds.) (1986) *Development of Antisocial and Prosocial Behavior.* New York: Academic Press.

Owens IPF and Short RV (1995) Hormonal basis of sexual dimorphism in birds: Implications for new theories of sexual selection. *Trends in Ecology and Evolution* 10: 44–47.

Parker GA and Simmons LW (1996) Parental investment and the control of sexual selection: Predicting the direction of sexual competition. *Proceedings of the Royal Society of London, B* 263: 315–321.

Pinxten R, de Ridder E, Balthazart J, Berghman L, and Eens M (2000) The effect of castration on aggression in the nonbreeding season is age-dependent in male European starlings. *Behaviour* 137: 647–661.

Pirke K, Kockott G, and Dittmar F (1974) Psychosexual stimulation and plasma testosterone in man. *Archives of Sexual Behavior* 3: 577–584.

Pound N, Penton-Voak IS, and Surridge (2009) Testosterone responses to competition in men are related to facial masculinity. *Proceedings of the Royal Society of London, B* 276: 153–159.

Pruett-Jones S and Pruett-Jones M (1994) Sexual competition and courtship disruptions: Why do male bowerbirds destroy each other's bowers? *Animal Behaviour* 47: 607–620.

Quinsey VL, Harris GT, Rice ME, and Cormier CA (1998) *Violent Offenders: Appraising and Managing Risk*. Washington, DC: American Psychological Association.

Raine A (1993) *The Psychopathology of Crime and Criminal Behavior as a Clinical Disorder*. San Diego, CA: Academic Press.

Raouf SA, Parker PG, Ketterson ED, Nolan V, Jr., and Ziegenfus C (1997) Testosterone affects reproductive success by influencing extra-pair fertilizations in male dark-eyed juncos (Aves: *Junco hyemalis*). *Proceedings of the Royal Society of London, B* 264: 1599–1603.

Read AF (1988) Sexual selection and the role of parasites. *Trends in Ecology and Evolution* 3: 97–105.

Roberts ML, Buchanan KL, and Evans MR (2004) Testing the immunocompetence handicap hypothesis: A review of the evidence. *Animal Behaviour* 68: 227–239.

Rode C, Cosmides L, Hell W, and Tooby J (1999) When and why do people avoid unknown probabilities in decisions under uncertainty? Testing some predictions from optimal foraging theory. *Cognition* 72: 269–304.

Rogers AR (1994) Evolution of time preference by natural selection. *American Economic Review* 84: 460–481.

Rogers AR (1997) The evolutionary theory of time preference. In: Bock G and Cardew G (eds.) *Ciba Foundation Symposium 208: Characterizing Human Psychological Adaptations*, pp. 231–252. London: Wiley.

Roitberg BD, Mangel M, Lalonde RG, Roitberg CA, van Alphen JJM, and Vet L (1992) Seasonal dynamic shifts in patch exploitation by parasitic wasps. *Behavioral Ecology* 3: 156–165.

Roney JR, Lukaszewski AW, and Simmons ZL (2007) Rapid endocrine responses of young men to social interactions with young women. *Hormones and Behavior* 52: 326–333.

Roney JR, Mahler SV, and Maestripieri D (2003) Behavioral and hormonal responses of men to brief interactions with women. *Evolution and Human Behavior* 24: 365–375.

Ros AFH, Groothuis TGG, and Apanius V (1997) The relation among gonadal steroids, immunocompetence, body mass, and behavior in young black-headed gulls (*Larus ridibundus*). *American Naturalist* 150: 201–219.

Rowe DC (1994) *The Limits of Family Influence: Genes, Experience, and Behavior*. New York: Guilford Press.

Rubin PH and Paul CW (1979) An evolutionary model of taste for risk. *Economic Inquiry* 17: 585–596.

Sakaguchi K, Oki M, Honma S, and Hasegawa T (2006) Influence of relationship status and personality traits on salivary testosterone among Japanese men. *Personality and Individual Differences* 41: 1077–1087.

Sakaguchi K, Oki M, Honma S, Uehara H, and Hasegawa T (2007) The lower salivary testosterone levels among unmarried and married sexually active men. *Journal of Ethology* 25: 223–229.

Sampson RJ (1991) Linking the micro- and macrolevel dimensions of community social organization. *Social Forces* 70: 43–64.

Sampson RJ and Laub JH (1993) *Crime in the Making: Pathways and Turning Points through Life*. Cambridge, MA: Harvard University Press.

Schultheiss OC, Campbell KL, and McClelland DC (1999) Implicit power motivation moderates men's testosterone responses to imagined and real dominance success. *Hormones and Behavior* 36: 234–241.

Simon D, Preziosi P, Barrett-Connor E, Roger M, Saint-Paul M, Nahoul K, and Papoz L (1992) The influence of aging on plasma sex hormones in men. *American Journal of Epidemiology* 135: 783–791.

Sozou PD (1998) On hyperbolic discounting and uncertain hazard rates. *Proceedings of the Royal Society of London, B* 265: 2015–2020.

Stoléru S, Ennaji A, Cournot A, and Spira A (1993) LH pulsatile secretion and testosterone blood levels are influenced by sexual arousal in human males. *Psychoneuroendocrinology* 18: 205–218.

Stoléru S, Grégoire MC, Gérard D, et al. (1999) Neuroanatomical correlates of visually evoked sexual arousal in human males. *Archives of Sexual Behavior* 28: 1–21.

Storey AE, Walsh CJ, Quinton RL, and Wynne-Edwards KE (2000) Hormonal correlates of paternal responsiveness in new and expectant fathers. *Evolution and Human Behavior* 21: 79–95.

Thompson CW, Hillgarth N, Leu M, and McClure HE (1997) High parasite load in house finches (*Carpodacus mexicanus*) is correlated with reduced expression of a sexually selected trait. *American Naturalist* 149: 270–294.

Thornhill R and Thornhill NW (1983) Human rape: An evolutionary analysis. *Ethology and Sociobiology* 4: 137–173.

Toch H (1969) *Violent Men: An Inquiry into the Psychology of Violence*. Chicago, IL: Aldine.

Tonry MH, Ohlin LE, and Farrington DP (1991) *Human Development and Criminal Behavior*. New York: Springer-Verlag.

Tooby J and Cosmides L (1990) On the universality of human nature and the uniqueness of the individual: The role of genetics and adaptation. *Journal of Personality* 58: 17–67.

Trivers RL (1972) Parental investment and sexual selection. In: Campbell B (ed.) *Sexual Selection and the Descent of Man. 1871–1971*, pp. 136–172. Chicago, IL: Aldine.

Turner GF (1994) The fighting tactics of male mouthbrooding cichlids: The effects of size and residency. *Animal Behaviour* 47: 655–662.

van Anders SM and Watson NV (2006) Relationship status and testosterone in North American heterosexual and non-heterosexual men and women: Cross-sectional and longitudinal data. *Psychoneuroendocrinology* 31: 715–723.

Van den Bergh B, Dewitte S, and Warlop L (2008) Bikinis instigate generalized impatience in intertemporal choice. *Journal of Consumer Research* 35: 85–97.

Van Honk J, Peper JS, and Schutter DJ (2005) Testosterone reduces unconscious fear but not consciously experienced anxiety Implications for the disorders of fear and anxiety. *Biological Psychiatry* 58: 218–225.

Verhulst S, Dieleman SJ, and Parmentier HK (1999) A tradeoff between immunocompetence and sexual ornamentation in domestic fowl. *Proceedings of the National Academy of Sciences of the United States of America* 96: 4478–4481.

Weatherhead PJ, Metz KJ, Bennett GF, and Irwin RE (1993) Parasite faunas, testosterone and secondary sexual traits in male red-winged blackbirds. *Behavioral Ecology and Sociobiology* 33: 13–23.

Wichmann MW, Ayala A, and Chaudry IH (1997) Male sex steroids are responsible for depressing macrophage immune function after trauma-hemorrhage. *American Journal of Physiology Cell Physiology* 273: C1335–C1340.

Widom CS (1989) The cycle of violence. *Science* 244: 160–166.
Williams GC (1966) *Adaptation and Natural Selection*. Princeton, NJ: Princeton University Press.
Williams GC (1992) *Natural Selection: Domains, Levels, and Challenges*. Oxford: Oxford University Press.
Williams GC and Nesse RM (1991) The dawn of Darwinian medicine. *Quarttely Review of Biology* 66: 1–22.
Williams JH (1980) *Dueling in the Old South. Vignettes of Social History*. College Station, TX: Texas A&M University Press.
Williamson S, Harpur TJ, and Hare RD (1991) Abnormal processing of affective words by psychopaths. *Psychophysiology* 28: 260–273.
Wilson JQ and Herrnstein RJ (1985) *Crime and Human Nature*. New York: Simon and Schuster.
Wilson M and Daly M (1985) Competitiveness, risk-taking and violence: The young male syndrome. *Ethology and Sociobiology* 6: 59–73.
Wilson M and Daly M (1994) A lifespan perspective on homicidal violence: The young male syndrome. In: Block CR and Block RL (eds.) *Proceedings of the 2nd Annual Workshop of the Homicide Research Working Group*, pp. 29–38. Washington, DC: National Institute of Justice.
Wilson M and Daly M (1997) Life expectancy, economic inequality, homicide, and reproductive timing in Chicago neighbourhoods. *British Medical Journal* 314: 1271–1274.
Wilson M and Daly M (2004) Do pretty women inspire men to discount the future? *Proceedings of the Royal Society of London, B* 271(supplement 4): S177–S179.
Wingfield JC (1984) Environmental and endocrine control of reproduction in the song sparrow, *Melospiza melodia*. *General and Comparative Endocrinology* 56: 417–424.
Wingfield JC, Hegner RE, Dufty AM, Jr., and Ball GF (1990) The challenge hypothesis: Theoretical implications for patterns of testosterone secretion, mating systems, and breeding strategies. *American Naturalist* 136: 829–846.
Wingfield JC, Jacobs J, and Hillgarth N (1997) Ecological constraints and the evolution of hormone–behavior interrelationships. *Annals of the New York Academy of Sciences* 807: 22–41.
Wolfgang ME (1958) *Patterns in Criminal Homicide*. Philadelphia, PA: University of Pennsylvania Press.
Yao G, Liang J, Han X, and Hou Y (2003) *In vivo* modulation of the circulating lymphocyte subsets and monocytes by androgen. *International Immunopharmacology* 3: 1853–1860.
Zahavi A and Zahavi A (1997) *The Handicap Principle*. New York: Oxford University Press.
Zuk M, Johnsen TS, and Maclarty T (1995) Endocrine–immune interactions, ornaments and mate choice in red jungle fowl. *Proceedings of the Royal Society of London, B* 260: 205–210.
Zuk M, Johnson K, Thornhill R, and Ligon JD (1990a) Parasites and male ornaments in free-ranging and captive red jungle fowl. *Behaviour* 114: 232–248.
Zuk M, Thornhill R, Ligon JD, and Johnson K (1990b) Parasites and mate choice in red jungle fowl. *American Zoologist* 30: 235–244.
Zumoff B, Rosenfeld RS, Friedman M, Byers SO, Rosen-man RH, and Hellman L (1984) Elevated daytime urinary excretion of testosterone glucuronide in men with the type A behavior pattern. *Psychosomatic Medicine* 46: 223–225.

Further Reading

Adkins-Regan E (2005) *Hormones and Animal Social Behaviour*. Princeton, NJ: University Press.
Campbell B (1972) *Sexual: Selection and the Descent of Man: 1871–1971*. Chicago, IL: Aldine.
Daly M and Wilson MI (1995) Discriminative parental solicitude and the relevance of evolutionary models to the analysis of motivational systems. In Gazzaniga M (eds.) *The Cognitive Neurosciences*, pp. 1269–1286. Cambridge, MA: MIT Press.
Darwin CR (1871) *The Descent of Man, and Selection in Relation to Sex*. London: John Murray.

13 Prolactin Actions in the Brain

D R Grattan, University of Otago, Dunedin, New Zealand
R S Bridges, Tufts University School of Veterinary Medicine, North Grafton, MA, USA

Chapter Outline

Glossary

JAK/STAT pathway Janus kinase/signal transducer and activator of transcription pathway. A signal transduction pathway characteristically activated by the cytokine family of receptors.

medial preoptic nucleus The region of the rostral hypothalamus adjacent to the third ventricle that is a focal area involved in the regulation of both maternal behavior and neuroendocrine feedback. In rodents, bilateral lesions of this area result in the elimination of ongoing maternal care.

neurogenesis The production or genesis of new neurons, which is not only common during development but can also occur throughout adulthood within specific regions of the adult brain.

suppressors of cytokine signaling (SOCS) A family of proteins that once activated act as inducable feedback regulators of the JAK/STAT pathway.

tuberoinfundibular dopamine neurons A specialized population of neuroendocrine dopamine neurons, in the arcuate nucleus in the mediobasal hypothalamus, that release dopamine into the pituitary portal blood to regulate PRL secretion from the anterior pituitary gland.

13.1 Introduction

Prolactin (PRL) was purified as an anterior pituitary hormone in 1931 (Riddle et al., 1931), and named based on several earlier studies identifying a lactotrophic factor in anterior pituitary extracts. While its name reflects its critical role to stimulate lactation, an incredibly wide range of functions of PRL have been identified since that time. Over 300 functions have been characterized, and these have been broadly grouped into six categories: water and electrolyte balance, growth and development, endocrinology and metabolism, immunoregulation and protection, reproduction, and brain and behavior (Bole-Feysot et al., 1998). This chapter focuses on the growing evidence that PRL plays an important role in numerous brain functions, ranging from the control of its own secretion through to regulation of complex behaviors. Central nervous system actions of PRL were first identified in the early 1960s, when PRL was implicated in brooding behavior in birds (Lehrman and Brody, 1961), nest-building behavior in rabbits (Zarrow et al., 1961, 1963), and fin-fanning behavior in some fish to induce fresh water flow over their eggs (Bluem and Fiedler, 1965). These observations provided early evidence that PRL was involved in reproductive behaviors, particularly those focused on nuturing eggs or offspring, in a manner clearly complementary to its established role in stimulating milk production in the mammary gland during lactation, in mammals. It was subsequently found that PRL exerted key actions in the brain to control its own secretion, by stimulating hypothalamic dopamine synthesis (Hokfelt and Fuxe, 1972) and turnover (Eikenburg et al., 1977; Annunziato and Moore, 1978) and promoting dopamine secretion into the pituitary portal blood (Gudelsky and Porter, 1979). In subsequent years, a large range of other functions of PRL in the brain have been identified, including regulation of oxytocin neurons, suppression of fertility, suppression of the stress response and reducing anxiety-related behaviors, stimulation of appetite and food intake, stimulation of myelination in the central nervous system, and generation of new neurons in the olfactory bulb. Thus, PRL is a pleiotropic hormone, exerting multiple different actions on brain and behavior (Grattan and Kokay, 2008).

Many of the actions of PRL in the brain are mediated by peripheral PRL, after entering the central nervous system through a carrier-mediated transport system (Walsh et al., 1987). Brain production of PRL has also been identified, however, and this may contribute to the brain actions of PRL under certain circumstances (Dutt et al., 1994). PRL receptors are widely expressed in the brain. Although they have a predominantly hypothalamic distribution, extrahypothalamic expression of PRL receptors has also been reported. Thus, PRL acts directly on neurons to influence their activity, thereby affecting brain function. There is also extensive evidence for PRL actions on glial cells, and this may also play an important role in brain function. In this chapter, we describe the neuroendocrine mechanisms controlling PRL secretion, and the factors controlling access of PRL to the brain. We also summarize the major changes in patterns of PRL secretion that occur under different physiological conditions, and describe how this impacts on brain function in each state. We will characterize the distribution of PRL receptors in the brain, and identify specific neuronal populations expressing PRL receptors. Finally, we systematically describe the neuroendocrine and behavioral consequences of PRL action in the brain.

13.2 Hypothalamic Control of PRL Secretion

13.2.1 PRL Secretion Is Inhibited by Dopamine from the Hypothalamus

When the pituitary gland is removed from the control of the hypothalamus, most hypophyseal hormone secretion ceases. Under these conditions, however, isolated pituitary cells are able to spontaneously release large amounts of PRL, suggesting that the hypothalamus tonically inhibits PRL secretion. This discovery originally came from the observation that the rat pituitary could maintain corpus luteum function even after the pituitary stalk had been severed, or after the transplantation of a pituitary gland under the kidney capsule (Everett, 1954). These observations suggested the spontaneous secretion of a luteotrophic substance from the pituitary gland. Furthermore, it was shown that ectopic pituitary grafts could maintain lactation, suggesting high levels of PRL secretion without the requirement for hypothalamic stimulation (Cowie et al., 1960). With the development of PRL radioimmunoassays in the early 1970s, it was confirmed that severing the pituitary stalk (Kanematsu and Sawyer, 1973) or lesions of the median eminence (Arimura et al., 1972) resulted in elevated PRL secretion. Transplanting a pituitary gland back under the median eminence of the hypothalamus, where the pituitary was re-vascularized by portal vessels, restored the normal, inhibitory control of PRL secretion (Nikitovitch-Winer and Everett, 1958). Thus, it was apparent that the hypothalamus secreted an inhibitory hormone into the portal blood to control PRL secretion. Research showing that dopamine inhibited PRL secretion both from isolated pituitary glands (Koch et al., 1970; MacLeod et al., 1970) and *in vivo* (MacLeod et al., 1970) suggested that dopamine might be involved in the regulation of PRL secretion. Dopamine was subsequently identified in the pituitary portal blood (Kamberi et al., 1970) and an elegant series of studies showed that changing levels of dopamine in the portal blood could account for much of the variation in PRL secretion from the anterior pituitary gland (Ben-Jonathan et al., 1977; Gibbs and Neill, 1978; De Greef and Neill, 1979; Ben-Jonathan et al., 1980).

Evidence from rodents suggests that PRL secretion is regulated by up to three populations of hypothalamic dopaminergic systems (DeMaria et al., 1999): the tuberoinfundibular (TIDA), tuberohypophyseal (THDA), and periventricular hypophyseal dopaminergic (PHDA) neurons. TIDA neurons arise from the dorsomedial arcuate nucleus and project to the external zone of the median eminence (Bjorklund et al., 1973). Dopamine diffuses into the capillaries of the pituitary portal blood vessels and is transported to the anterior pituitary where it acts on type 2 dopamine receptors on lactotrophs to tonically inhibit PRL secretion (Mansour et al., 1990). THDA neurons originate in the rostral arcuate nucleus and project in the hypothalamohypophyseal tract to the intermediate and neural lobes of the pituitary gland (Fuxe, 1964). PHDA neurons arise more rostrally in the periventricular nucleus, with axons terminating solely in the intermediate lobe of the pituitary gland (Goudreau et al., 1992). Dopamine from THDA and PHDA neurons contributes to basal regulation of PRL secretion, after transport to the anterior pituitary gland through short portal vessels from the neurohypophysis (Freeman et al., 2000).

13.2.2 Short-Loop Negative Feedback

Concurrent with studies establishing the role of dopamine as a physiological PRL-inhibiting factor, it was found that PRL selectively stimulated hypothalamic dopamine synthesis (Hokfelt and Fuxe, 1972) and turnover (Eikenburg et al., 1977; Annunziato and Moore, 1978) and promoted dopamine secretion into the pituitary portal blood (Gudelsky and Porter, 1979). These observations identified the afferent limb of a short-loop feedback mechanism regulating PRL secretion (see **Figure 1**). Multiple studies in the subsequent two decades have confirmed the reciprocal relationship between PRL and hypothalamic dopamine secretion. Dopamine

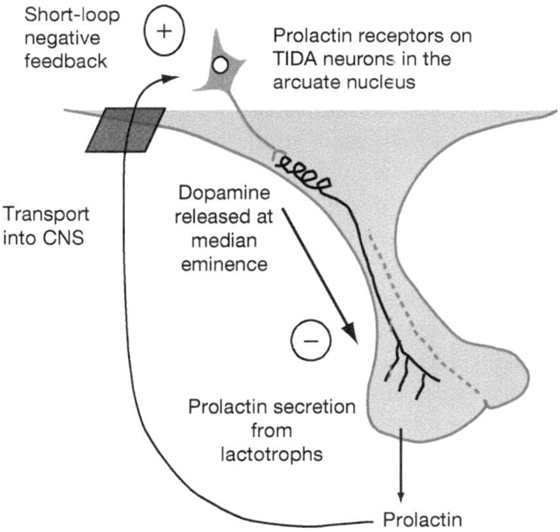

Figure 1 Diagrammatic representation of the neuroendocrine feedback pathway regulating prolactin secretion. Prolactin secretion is inhibited by dopamine released into the pituitary portal blood from hypothalamic dopamine neurons, including the tuberoinfundibular dopamine (TIDA) neurons. In turn, prolactin is transported into the brain and acts to stimulate activity of TIDA neurons and thereby inhibits its own secretion through short-loop negative feedback.

synthesis and release are increased in response to acute or chronic increases in PRL levels in blood (Demarest et al., 1984, 1986; Moore, 1987), while hypoprolactinemia results in the suppression of dopamine secretion (Arbogast and Voogt, 1991a), thus fulfilling criteria to be a homeostatic regulatory system. The action of PRL on hypothalamic dopamine neurons has become the best-characterized action of PRL in the brain.

PRL acts directly on hypothalamic dopamine neurons, which express PRL receptors (Lerant and Freeman, 1998; Kokay and Grattan, 2005). The PRL receptor is a member of the class 1 cytokine receptor superfamily (Bole-Feysot et al., 1998), with at least two isoforms of the PRL receptor molecule – a long form and one or more short forms – produced by alternative splicing of the PRL receptor gene. The different forms of the PRL receptor have identical extracellular portions, and hence are all able to bind PRL, but they differ in their ability to activate intracellular signaling pathways (see **Figure 2**). The long form appears to be the major isoform capable of activating most PRL-sensitive signal transduction pathways, including the JAK/STAT pathway, and is the most abundantly expressed in brain tissue. The short forms may also have a limited but important signaling role.

Most studies into PRL action on the TIDA neurons have used biochemical approaches, because it has been difficult to specifically isolate TIDA neurons for electrophysiological studies. While there have been a number of studies evaluating electrical activity of arcuate nucleus neurons, few have identified dopamine neurons specifically (Loose et al., 1990), and none of these has examined the effect of PRL on arcuate dopamine neurons. PRL has been shown to rapidly enhance the rate of spontaneous firing in arcuate neurons in a brain slice preparation (Haskins and Moss, 1983), but the specific neurons involved were unidentified. In putative tuberoinfundibular neurons, identified by antidromic stimulation from the median eminence, PRL was ineffective at altering neuronal firing rate (Nishihara and Kimura, 1989). The use of transgenic mice with fluorescently labeled dopamine neurons (e.g., tyrosine hydroxylase (TH)-green fluorescent protein (GFP); Sawamoto et al., 2001) might allow more specific electrophysiological studies to be undertaken in the near future.

Using biochemical indices of activity of TIDA neurons, the time course of PRL action is complex, with a rapid component of increased activity observed 2–4 h after PRL treatment (Selmanoff, 1985), and a delayed component seen approximately 12 h after PRL treatment (Demarest et al., 1984, 1986). The latter component requires protein synthesis (Johnston et al., 1980). These different time courses may reflect the multiple mechanism of PRL action in these neurons. PRL increases gene expression for TH, the rate-limiting enzyme in catecholamine synthesis (Arbogast and Voogt, 1991a), dependent on activation of the JAK/STAT pathway (see **Figure 2**). This effect is likely to account for the delayed component of PRL action. The specific requirement for STAT5b in mediating PRL action on TIDA neurons (Grattan et al., 2001; Ma et al., 2005a) strongly implicates the long form of the PRL receptor as the critical protein mediating PRL signal in this population of hypothalamic neurons, as this is the only form that can activate the JAK-STAT pathway. The long form of the PRL receptor mRNA has been specifically identified in TIDA neurons (Kokay and Grattan, 2005). PRL also causes rapid changes in serine phosphorylation of the TH protein (Ma et al., 2005b), resulting in increased enzyme activity. These effects of PRL are presumably mediated more acutely by a range of other signaling pathways, including protein kinase C and MAP kinase (Ma et al., 2005b). Finally, there is evidence that PRL stimulates the acute release of dopamine from isolated synaptosomes (Gregerson and Selmanoff, 1988). These latter mechanisms likely account for the rapid component of PRL action. Importantly, removal of endogenous PRL by bromocriptine treatment results in opposite changes on TH mRNA levels and TH phosphorylation (Arbogast and Voogt, 1995), demonstrating that basal activity of these neurons is dependent on circulating PRL.

The TIDA neurons represent a key point of hypothalamic regulation of PRL secretion. Many of the neuropeptides and hormones that have been reported to influence PRL secretion act through changing the activity of the TIDA neurons. For example, opioid peptides stimulate PRL secretion through inhibition of TIDA neurons (Enjalbert et al., 1979; Sagrillo and Voogt, 1991; Manzanares et al., 1993; Wagner et al., 1994; Callahan et al., 2000; Andrews and Grattan, 2003), while factors such as neurotensin inhibit PRL secretion by activating TIDA neurons (Tojo et al., 1986; Thomas et al., 1988a; Pan et al., 1992; Hentschel et al., 1998). Hormones, such as estrogen and progesterone, also influence PRL secretion, both by altering TH phosphorylation and influencing TH gene expression (Arbogast and Voogt, 1993, 1994; Liu and Arbogast, 2008).

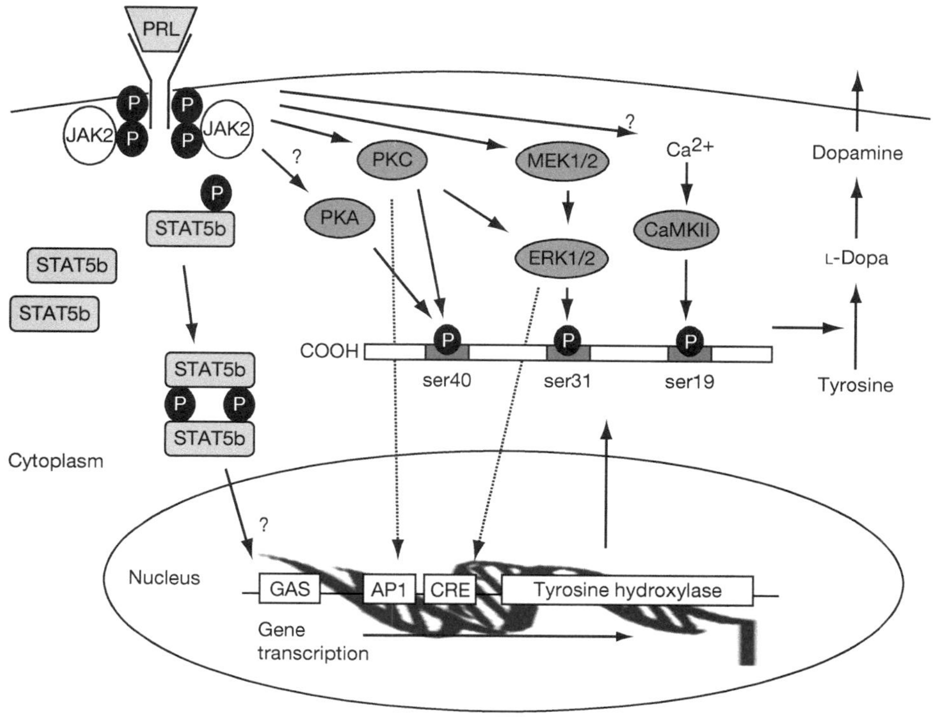

Figure 2 Prolactin signal transduction in TIDA neurons. The prolactin (PRL) receptor is a member of the class 1 cytokine receptor superfamily (Bole-Feysot et al., 1998). There are at least two isoforms of the prolactin receptor molecule, a long form and a short form, produced by alternative splicing of the prolactin receptor gene. The two forms of the prolactin receptor have identical extracellular portions, and hence both are able to bind prolactin, but they differ in their ability to activate intracellular signaling pathways. Binding of prolactin to its receptor induces receptor dimerization or a change in configuration of preformed dimers of receptor molecules (Sakal et al., 1997; Gadd and Clevenger, 2006), and then activation of multiple intracellular signaling proteins (Bole-Feysot et al., 1998; Freeman et al., 2000). Ligand interaction with the long form of the receptor leads to phosphorylation (P) of the receptor-associated Janus tyrosine kinase 2 (JAK2). JAK2 phosphorylates tyrosine residues on the prolactin receptor, resulting in recruitment of a family of latent cytoplasmic proteins known as signal transducer and activator of transcription (STAT) proteins. STAT molecules are also phosphorylated by JAK2, then forming homo- or heterodimers, translocating to the nucleus and binding to specific promoter sequences of target genes (Leonard and O'Shea, 1998). In the TIDA neurons, prolactin treatment induces nuclear translocation of STAT5 (Lerant et al., 2001; Cave et al., 2005; Ma et al., 2005a), and in particular, STAT5b is specifically required for the negative-feedback action of prolactin (Grattan et al., 2001). As only the long form of the receptor can activate the JAK-STAT pathway, involvement of STAT5b in the prolactin regulation of TIDA strongly implicates the long form of the receptor as the critical protein mediating prolactin signal, at least in this population of neuroendocrine neurons. Prolactin also regulates the phosphorylation of TH protein, and multiple signal transduction pathways have been implicated in this action, including protein kinases A (PKA) and C (PKC), extracellular regulated kinases 1 and 2 (ERK1/2), and calcium/calmodulin-dependent kinase (CamKII) (see Ma et al. (2005b)). While the short form of the receptor cannot activate the JAK-STAT pathway, it can mediate some actions of prolactin through these alternative pathways (Das and Vonderhaar, 1995). Hence, the short form of the receptor, which is expressed in the arcuate nucleus of the hypothalamus, may also play a role in regulating neuronal function.

The PRL receptor is a member of the class 1 cytokine receptor superfamily (Bole-Feysot et al., 1998). There are at least two isoforms of the PRL receptor molecule, a long form and a short form, produced by alternative splicing of the PRL receptor gene. The two forms of the PRL receptor have identical extracellular portions, and hence are both able to bind PRL, but they differ in their ability to activate intracellular signaling pathways. Binding of PRL to its receptor induces receptor dimerization or a change in configuration of preformed dimers of receptor molecules (Sakal et al., 1997; Gadd and Clevenger, 2006), and then activation of multiple intracellular signaling proteins (Bole-Feysot et al., 1998; Freeman et al., 2000). Ligand interaction with the long form of the receptor leads to phosphorylation (P) of the receptor-associated Janus tyrosine kinase 2 (JAK2). JAK2 phosphorylates tyrosine residues on the PRL receptor, resulting in recruitment of a family of latent cytoplasmic proteins known as signal transducer and activator of transcription (STAT) proteins. STAT molecules are also phosphorylated by JAK2, then forming homo- or heterodimers, translocating to the nucleus and binding to specific promoter sequences of target genes (Leonard and O'Shea, 1998). In the TIDA neurons, PRL treatment induces nuclear translocation of STAT5 (Lerant et al., 2001; Cave et al., 2005; Ma et al., 2005a), and in particular, STAT5b is specifically required for the negative-feedback action of PRL (Grattan et al., 2001). As only the long form of the receptor can activate the JAK-STAT pathway, involvement of STAT5b in the PRL regulation of TIDA neurons strongly implicates the long form of the receptor as the critical protein mediating PRL signal, at least in this population of neuroendocrine neurons. PRL also regulates the phosphorylation of TH protein, and multiple signal transduction pathways have been implicated in this action, including protein kinases A (PKA) and C (PKC), extracellular regulated kinases 1 and 2 (ERK1/2), and calcium/calmodulin dependent kinase (CamKII) (see Ma et al. (2005b)). While the short form of the receptor cannot activate the JAK-STAT pathway, it can mediate some actions of PRL through these alternative pathways (Das and Vonderhaar, 1995). Hence, the short form of the receptor, which is expressed in the arcuate nucleus of the hypothalamus, may also play a role in regulating neuronal function.

13.2.3 Role of a PRL-Releasing Factor

While the dopamine-mediated inhibition of PRL secretion is well established, there is much debate about the additional role of hypothalamic-releasing factors in stimulating secretion of PRL. Disinhibition induced by administration of a dopamine D2R antagonist can result in a release of PRL of similar magnitude to that seen during suckling (Andrews and Grattan, 2004). However, whether such a total withdrawal of dopamine ever occurs *in vivo* is unlikely (Martinez de la Escalera and Weiner, 1992). Where measured, the fall in portal blood dopamine may not account for the full release of PRL (de Greef et al., 1981). Hence, there have been many attempts to identify additional factors involved in stimulating PRL secretion. Lesions of the paraventricular nucleus (PVN) markedly impair the suckling-induced release of PRL (Kiss et al., 1986; Bodnar et al., 2002), suggesting that this nucleus plays a key role in stimulating PRL secretion. Similarly, removal of the posterior lobe of the pituitary blocks the suckling-induced release of PRL (Murai and Ben-Jonathan, 1987b). A range of putative candidates from the PVN and/or posterior pituitary have now been examined, but the evidence for a physiological PRL-releasing factor remains inconclusive. For example, passive immunization against thyrotropin-releasing hormone (TRH) blocked the suckling-induced PRL release (de Greef et al., 1987), but TRH release from the median eminence did not increase in response to suckling (Rondeel et al., 1988; Thomas et al., 1988b) and there is no concurrent suckling-induced release of thyroid-stimulating hormone (TSH) (Riskind et al., 1984; Vanhaasteren et al., 1996), suggesting that it is unlikely that TRH is a physiological regulator of PRL secretion during suckling. Similarly, both positive and negative evidence exist for vasopressin (Nagy et al., 1991) and oxytocin (Samson et al., 1986; Johnston and Negro-Vilar, 1988) as PRL-releasing factors during suckling. More recently, the dopamine derivative, salsolinol, has been proposed as a potential PRL-releasing factor (Toth et al., 2001) that is derived from the posterior lobe of the pituitary and involved in both the stress and suckling-induced PRL secretion (Bodnar et al., 2004). Clearly, the jury is still out as to whether one or more PRL-releasing factors are involved in the stimulation of PRL release, and if so, their identity and physiological role remains to be determined. There was much excitement in the field when a novel neuropeptide was identified and, based on *in vitro* actions, named PRL-releasing peptide (Hinuma et al., 1998). Unfortunately, based on a range of subsequent anatomical and physiological studies, this peptide is almost certainly not a physiological PRL-releasing factor (Watanobe et al., 2000; Samson et al., 2003).

13.3 Access of PRL to the Brain

13.3.1 Transport into the Central Nervous System

The clear role for PRL as a feedback regulator of hypothalamic dopamine neurons suggests that there must be a mechanism for PRL to enter the brain. Being a relatively large polypeptide hormone (197–199 amino acids), PRL would be expected to be excluded from nervous tissue by the specialized tight junctions between vascular endothelial cells in the brain which form the so-called blood–brain barrier

(BBB) (Nilsson et al., 1992). There is clear evidence, however, that systemic PRL gains access to the cerebrospinal fluid (CSF), from where it can diffuse to numerous brain regions (Nicholson et al., 1980). Systemic PRL appears to enter the CSF through a saturable, carrier-mediated process (Walsh et al., 1987), possibly involving PRL-binding sites in the choroid plexus (Walsh et al., 1978). The choroid plexus contains fenestrated capillaries, but maintains a blood–CSF barrier by means of continuous tight junctions between the choroid plexus epithelial cells (Nilsson et al., 1992). Hence, blood-borne PRL has ready access to the choroid plexus epithelial cells, which express high levels of both the short and long forms of the PRL receptor (Augustine et al., 2003). The precise mechanism that translocates PRL across this epithelial layer into the CSF, however, remains to be determined, and may be an important regulatory checkpoint. The arcuate nucleus/median eminence region potentially also has an incomplete BBB, so hormones may have more direct access to neurons in this region (Faouzi et al., 2007). Some early reports also suggest that hypothalamic neurons might be exposed to very high concentrations of PRL from the anterior pituitary by retrograde diffusion in the pituitary portal vessels (Oliver et al., 1977). PRL from this source could theoretically enter the brain through the fenestrated capillaries of the median eminence.

In addition to PRL's gaining access to the brain, there is evidence in both rats and humans that placental lactogens (PLs) enter the brain in significant amounts during pregnancy (Peake et al., 1983; Bridges et al., 1996). Specifically, using a push–pull perfusion technique to continuously collect CSF, it was shown that rat PL-I was the dominant lactogen detected in the CSF on day 12 of gestation in rats, whereas rat PL-II was most dominant on day 21 of pregnancy (Bridges et al., 1996). Thus, over the course of pregnancy the maternal brain is exposed to elevated concentrations of a range of lactogens, proteins that have the capacity to stimulate maternal behavior (Bridges et al., 1996).

13.3.2 The Brain Also Produces PRL

While there is extensive evidence of systemic PRL being transported into the brain, PRL of brain origin has also been reported. The role of brain PRL remains controversial, however, despite numerous reports of both the protein and its mRNA in brain tissue. The original observation of immunoreactive PRL in the hypothalamus (Fuxe et al., 1977) has been independently replicated by several groups (Emanuele et al., 1986; DeVito et al., 1987; DeVito, 1988; Shivers et al., 1989; Paut-Pagano et al., 1993; Suzuki and Handa, 2005). Importantly, all groups found that PRL-like immunoreactivity was not depleted by hypophysectomy, suggesting that it was unlikely to be of pituitary origin. Moreover, *de novo* PRL synthesis and PRL release could be detected from hypothalamic fragments, *in vitro* (DeVito et al., 1987, 1991, 1992a), strongly suggesting local production of PRL in neurons or glial cells. Subsequently, PRL mRNA was detected in brain tissue by reverse transcriptase polymerase chain reaction (RT-PCR) (DeVito et al., 1992a; Emanuele et al., 1992; Clapp et al., 1994), and very recently, by *in situ* hybridization in fetal sheep brains (Roselli et al., 2008). PRL expression in the brain has been co-localized with estrogen receptors (ERs) (Shivers et al., 1989; Suzuki and Handa, 2005), and appears to be stimulated by gonadal steroids (DeVito, 1989; Shivers et al., 1989; DeVito et al., 1992a; Emanuele et al., 1992; Torner et al., 1999). PRL mRNA in the brain is also increased by stress and during lactation (Torner et al., 2002; Torner et al., 2004). Although there is not universal agreement, PRL expression seems to be widespread in the hypothalamus and most studies report PRL or its mRNA in the medial preoptic, paraventricular, periventricular, and arcuate nuclei. The arcuate nucleus, indeed, appears to have the highest level of expression for the PRL gene within those putative central sites where PRL synthesis might be expected (as measured by real-time RT-PCR; Bridges et al., unpublished data). While it is not clear what levels of PRL might be released in the central nervous system, relative to levels entering the brain from the systemic circulation, it seems that some actions of PRL in the brain might be mediated by PRL acting as a neuropeptide, as well as a neuroendocrine hormone (Dutt et al., 1994).

13.4 PRL Receptor Expression in the Brain

13.4.1 High Levels of Expression of PRL Receptors in the Choroid Plexus

The distribution of PRL receptor expression in the brain has been described by studies using *in vitro* binding assays and autoradiography (Muccioli et al., 1988; Crumeyrolle Arias et al., 1993; Mustafa et al., 1994, 1995), immunohistochemistry (Roky et al., 1996;

Pi and Grattan, 1998b, 1999c), and *in situ* hybridization (Chiu and Wise, 1994, 1996; Bakowska and Morrell, 1997, 2003). In all studies, the highest level of PRL receptor mRNA detected was in the choroid plexus. Both the short and long forms are present in the choroid plexus (Pi and Grattan, 1998a, 1999a,b), and levels of both forms increase in response to estrogen (Pi and Voogt, 2002), during pregnancy and lactation (Sugiyama et al., 1996; Augustine et al., 2003), and in response to stress (Fujikawa et al., 1995). There is widespread speculation that the PRL receptor may be involved in the carrier-mediated transport of PRL into the CSF, and in particular, that this might be a function of the short form of the receptor. The specific mechanisms by which a cytokine receptor might function as a transepithelial transporter, however, are not clear. At least for the PRL receptor, the evidence for such a mechanism is only circumstantial. PRL and PRL receptor internalization is well documented (Lu et al., 2002), although this is often associated with receptor degradation (Swaminathan et al., 2008). However, evidence is accumulating that the carrier-mediated transport of the adipose-derived hormone, leptin, into the CSF could involve the short form of the leptin receptor, a related cytokine-family receptor. Transport is impaired in mice specifically lacking the short form of the leptin receptor (Kastin et al., 1999), and transfection of the short form of the receptor into an epithelial cell line confers enhanced ability to transport leptin across an epithelial layer (Hileman et al., 2000). The leptin receptor is regulated in the choroid plexus and brain microvessels in a manner consistent with changing levels of leptin transport into the brain (Hileman et al., 2002). Hence, at least conceptually, the short form of the PRL receptor may be acting in a similar fashion, and may provide a site of regulated access of PRL to the brain.

13.4.2 PRL Receptors Are Widespread in the Hypothalamus

Consistent with the discussion of PRL feedback, above, there is also clear evidence of PRL receptor expression in the arcuate nucleus. Indeed, double-label immunohistochemistry (Arbogast and Voogt, 1997; Lerant and Freeman, 1998; Grattan, 2001) and *in situ* hybridization (Kokay and Grattan, 2005) studies have established that most of the PRL receptor-containing neurons in the arcuate nucleus are TH positive, representing the TIDA neurons involved in negative-feedback regulation of PRL secretion. However, PRL receptor expression is much more widespread than simply the choroid plexus and arcuate nucleus of the hypothalamus. While there are some inconsistencies in the various studies, a relatively clear picture of PRL receptor distribution is beginning to emerge. The most thorough survey of the distribution of both short- and long-form PRL receptor mRNA has been provided by Bakowska and Morrell (1997, 2003), and these studies are summarized in **Table 1**. The relative distribution of the two isoforms is very similar, although the long form is present in most areas at higher levels than the short form. In the rostral hypothalamus, PRL receptor mRNA has been observed in the anteroventral periventricular nucleus, the median

Table 1 Distribution of prolactin receptor mRNA in the brain

Brain region	*Long form*	*Short form*
Choroid plexus	✓✓✓✓✓	✓✓✓
Telencephalon		
Piriform cortex	×	✓
Bed nucleus of the stria terminalis	✓✓	✓
Medial amygdala	✓✓	✓
Subfornical organ	✓✓	×
Preoptic area		
Anteroventral periventricular nucleus	✓✓✓	✓
Median preoptic nucleus	✓✓✓	✓
Medial preoptic area	✓✓	✓
Preoptic periventricular nucleus	✓✓	✓
Diencephalon		
Supraoptic nucleus	✓✓✓	✓
Paraventricular nucleus	✓✓✓	✓
Periventricular nucleus	✓✓	✓
Suprachiasmatic nucleus	×	✓
Dorsomedial hypothalamic nucleus	✓	✓
Ventromedial hypothalamic nucleus	✓✓✓	✓
Arcuate nucleus	✓✓✓✓	✓✓
Zona incerta	×	✓
Paraventricular nucleus of thalamus	×	✓
Mesencephalon		
Periaqueductal gray matter	✓	n.r.
Interpeduncular nucleus	✓	n.r.

n.r., not reported.
Data derived from Bakowska JC and Morrell JI (1997) Atlas of the neurons that express mRNA for the long form of the prolactin receptor in the forebrain of the female rat. *Journal of Comparative Neurology* 386: 161–177; Bakowska JC and Morrell JI (2003) The distribution of mRNA for the short form of the prolactin receptor in the forebrain of the female rat. *Brain Research – Molecular Brain Research* 116: 50–58; and Grattan and Kokay (unpublished).

preoptic nucleus, the medial preoptic nucleus, and the periventricular nucleus. Both forms are also detected in the magnocellular neurons of the supraoptic nucleus and PVN, and to a lesser extent in the parvocellular subdivisions of the PVN. In the mediobasal hypothalamus, both forms of the receptor are also expressed in the arcuate and ventromedial nuclei, particularly in the ventrolateral divisions of this latter nucleus. Outside the hypothalamus and preoptic areas, PRL receptor mRNA has also been observed in the bed nucleus of the stria terminalis and the medial amygdala, as well as in the subfornical organ, periaquaductal gray matter, and interpeduncular nucleus. The distribution of the long form of PRL receptor mRNA is illustrated in **Figure 3**.

There are only a few regions showing differential patterns of distribution of the short- and long-form mRNA. Only the short-form mRNA is found in the suprachiasmatic nucleus, an area in which PRL receptor protein has been reported in some (Crumeyrolle Arias et al., 1993; Roky et al., 1996), but not all (Pi and Grattan, 1998b) studies. Other areas in which the short form was found, without significant levels of the long form are the piriform cortex and the zona incerta (Bakowska and Morrell, 2003). The relative expression in the magnocellular hypothalamic nucleus is particularly interesting. While both the short and long form of the PRL receptor mRNA are found in these nuclei, the distribution is not identical. All cells appear to express the short-form mRNA, while only a subset contains the long form (Bakowska and Morrell, 1997, 2003). Consistent with these observations, in dual-label *in situ* hybridization studies, the long form of the PRL receptor has only been observed in oxytocin neurons in the supraoptic nucleus and PVN (Kokay

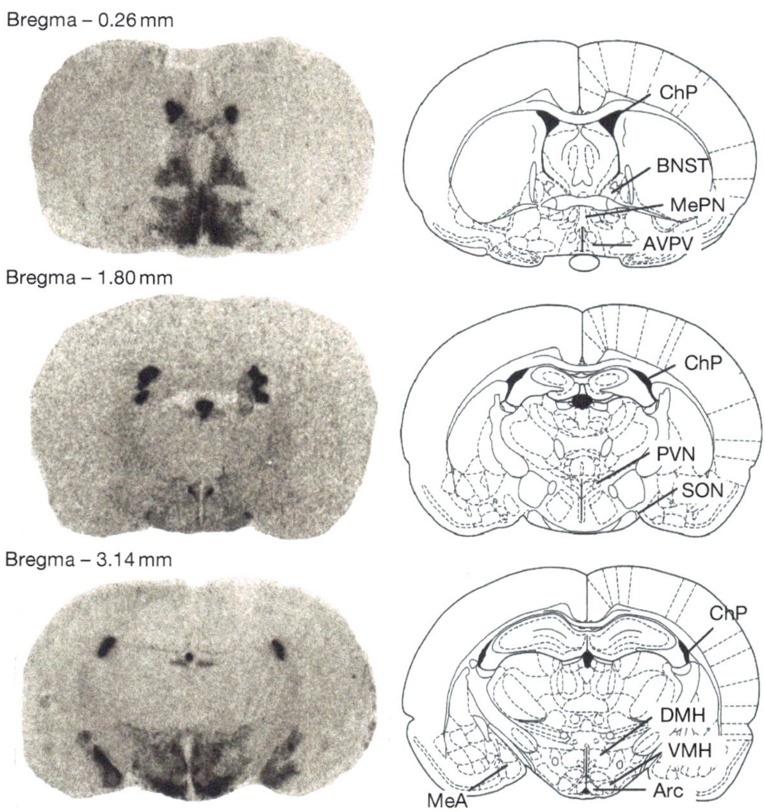

Figure 3 Distribution of prolactin receptor (long-form) mRNA in the brain. Left-hand images show autoradiograms of *in situ* hybridization for prolactin receptor mRNA in coronal sections of the rat brain. Black regions shown binding of a ^{32}P-labeled riboprobe identifying prolactin receptor mRNA. Three levels, with approximate coordinates, relative to bregma, are shown. Right-hand diagrams show images from a brain axis (Paxinos and Watson, 1997), with major nuclei that show prolactin receptor mRNA labeled. Arc, arcuate nucleus; AVPV, anteroventral periventricular nucleus; BNST, bed nucleus of the stria terminalis; ChP, choroid plexus; DMH, dorsomedial hypothalamic nucleus; MeA, medial amygdala; MePN, median preoptic nucleus; PVN, paraventricluar nucleus; and SON, supraoptic nucleus.

et al., 2006), and functional studies suggest a specific effect of PRL on oxytocin neurons (Ghosh and Sladek, 1995a,b; Kokay et al., 2006). In contrast, vasopressin neurons do not express the long form of the PRL receptor (Kokay et al., 2006), but may express the short form of the receptor as PRL receptor-like immunoreactivity has been identified on these cells (Mejia et al., 2003). The functional consequences of neurons expressing either or both of the short and long forms of the PRL receptor are uncertain. As described above, the two isoforms can activate distinct signal transduction pathways in neurons, and thus might exert different actions. One additional possibility is that the short form of the receptor may function as a negative regulator of long-form signaling. Because homodimers of the long form of the receptor are required to activate JAK/STAT signaling, the short form can function as a competitive inhibitor by forming inactive heterodimer receptor complexes with the long form (Perrot-Applanat et al., 1997). Such a mechanism has not yet been demonstrated to occur in neurons.

The distribution of the PRL receptor mRNA, described above, predominantly matches that which has been reported for the PRL receptor protein or PRL binding. Key areas of overlap include the hypothalamus, where PRL receptor protein has been observed in the arcuate, ventromedial, supraoptic, paraventricular, and peoptic areas (Crumeyrolle Arias et al., 1993; Roky et al., 1996; Pi and Grattan, 1998b). Some subtle differences between these studies may have occurred due to differences in experimental models. For example, Pi and Grattan (1999c) did not detect PRL receptor immunoreactivity in the PVN and ventromedial hypothalamic nucleus in diestrous animals, but clearly detected it during lactation. There are some notable exceptions, however. Some studies have reported dense immunostaining in cell bodies of the cerebral cortex, substantia nigra, and ventral tegmental area (VTA) (Roky et al., 1996), and this has not been supported by the mRNA studies. Furthermore, several studies reported immunostaining of neuronal processes in regions of the striatum and basal ganglia (Roky et al., 1996; Pi and Grattan, 1998b). While there is no PRL receptor mRNA in these regions, it remains possible that receptor proteins are transported into this region in neuronal processes.

The neurochemical identity of PRL receptor-expressing neurons has not been fully elucidated. Besides being expressed on TIDA neurons in the arcuate nucleus (Lerant and Freeman, 1998; Kokay and Grattan, 2005), PRL receptors are also expressed in enkephalin neurons (Kokay and Grattan, unpublished), as well as some as-yet-unidentified neurons in this nucleus. Despite evidence that PRL might regulate β-endorphin production (Panerai et al., 1980; Sarkar and Yen, 1985; Tong and Pelletier, 1992), only a few proopiomelanocortin (POMC) neurons express the long form of the PRL receptor (Kokay and Grattan, 2005). Neuropeptide Y (NPY) neurons in the arcuate nucleus do not appear to express PRL receptors (Chen and Smith, 2004), but there are a population of neurons in the dorsomedial hypothalamic nucleus that express NPY during lactation, and are directly regulated by PRL (Chen and Smith, 2004). In the ventromedial hypothalamic nucleus, PRL receptors are localized on enkephalin neurons (Kokay and Grattan, unpublished data). In addition, there are extensive PRL receptor-containing neurons in the medial preoptic nucleus and anteroventral periventricular nucleus. While a few of these neurons may be gonadotropin-releasing hormone (GnRH) neurons (Grattan et al., 2007a), most are of an as-yet-undetermined phenotype.

13.4.3 Regulation of PRL Receptor Expression in the Brain

Interestingly, the overall distribution of PRL receptor-containing neurons (**Figure 2**) strongly resembles the distribution of ERα in the mammalian brain (Krieger et al., 1976). There is significant evidence that PRL receptor expression is directly influenced by estradiol (Muccioli et al., 1991; Mustafa et al., 1995; Shamgochian et al., 1995; Lerant and Freeman, 1998; Pi et al., 2003) and progesterone (Bridges and Hays, 2005), and we have recently shown that some of the biological actions of PRL in the brain are dependent on estradiol (Anderson et al., 2008). PRL receptor expression in the brain is also regulated during pregnancy (Sugiyama et al., 1994; Bakowska and Morrell, 1997; Mann and Bridges, 2002; Augustine et al., 2003) and lactation (Pi and Grattan, 1999a,b,c; Pi and Voogt, 2000; Mann and Bridges, 2002). The mechanisms mediating increased expression of PRL receptors in the brain of pregnant and/or lactating rats are not known. PRL itself may mediate an upregulation of its receptors (Di Carlo and Muccioli, 1981; Muccioli and Di Carlo, 1994), as occurs in other tissues. It is probable that hormonal changes of pregnancy, such as the high progesterone levels, prolonged elevations of placental and then pituitary lactogens, and the increasing estrogen/progesterone ratio prior to parturition, all play a role in stimulating the expression of PRL receptors during pregnancy and lactation. The elevated PRL receptor expression during lactation appears to result from the combined stimulatory effects of suckling-induced hyperprolactinemia and the neural input from the suckling stimulus *per se* (Pi and Voogt, 2001).

In addition to these changes in the regulation of PRL receptor expression, the fact that a female experienced a prior pregnancy and lactation can also affect both basal PRL receptor expression as well as the response of the PRL receptor system to a PRL challenge. When comparing the level of expression of the long form of the PRL receptor in the brain of female rats that have previously raised a litter with that of females that have never given birth, it was shown that basal expression of the receptor was higher in the medial preoptic area (mPOA) and arcuate region of experienced females (Anderson et al., 2006c). Moreover, systemic PRL treatment further increased receptor expression in the (mPOA) of experienced females, while it failed to affect expression in the inexperienced controls (Anderson et al., 2006c). Thus, reproductive experience appears to be a potent regulator of neural PRL receptor expression. Whether this shift translates into either behavioral or neuroendocrine changes has not been established, but is a likely possibility.

13.5 Changes in Patterns of PRL Secretion

With transport of PRL into the CSF being well established, it is clear that changes in PRL in the peripheral circulation may influence brain function. Hence, in the following sections, we will summarize the key changes that occur in PRL secretion in different physiological conditions.

13.5.1 Estrous/Menstrual Cycle

PRL levels are low throughout most of the reproductive cycle in female mammals, except for a large preovulatory surge driven by chronic elevation in 17β-estradiol levels from the developing preovulatory follicle. Estradiol potently stimulates PRL secretion, both through actions on the pituitary gland as well as through hypothalamic mechanisms. Interestingly, in many species, the estradiol-induced PRL secretion is entrained to the light–dark cycle (Palm et al., 2001). In rats, estradiol induces a single afternoon PRL surge observed during the afternoon of proestrus (Smith et al., 1975), while in ovariectomized rats, chronic elevation in estradiol induces a daily afternoon PRL surge (Subramanian and Gala, 1976). The mechanism by which rising estradiol stimulates a PRL surge is not clear. Certainly, it is associated with a decrease in activity of TIDA neurons (Demarest et al., 1981; DeMaria et al., 1998) with a consequent decrease in dopamine in the portal blood (Ben-Jonathan et al., 1977). One or more PRL-releasing factors may also be involved (Freeman et al., 2000).

In humans, a similar midcycle surge of PRL, associated with the LH surge, has been observed in the follicular phase of the menstrual cycle in some (Vekemans et al., 1977; Khamad'ianov et al., 1980; Gonen and Casper, 1990; Subramanian et al., 1997), but not all (Franchimont et al., 1976; Werawatgoompa et al., 1981), studies. A PRL surge is also observed during an exogenous estrogen-induced LH surge (Messinis and Templeton, 1990). Thus, it would seem that the estrogen-induced PRL surge is common throughout a range of mammalian species.

13.5.2 Stress-Induced Changes in PRL Secretion

PRL is recognized as a stress hormone (Moore et al., 1987), and is acutely increased in response to a wide variety of stressors (Terkel et al., 1972; Gala, 1990). Stress-induced increase in PRL secretion is brief (5–10 min) and of low magnitude, and there is evidence that cortisol, which is concomitantly released in response to stress, may play a critical role in limiting PRL secretion (Gala, 1990). The stress-induced release of PRL appears to involve an acute suppression in activity of TIDA neurons (Demarest et al., 1985a,b), although several other studies have failed to observe this (Gala, 1990; Kehoe et al., 1992), perhaps due to the specific stressor used (Lookingland et al., 1990). Importantly, stress is able to induce a release of PRL even in the presence of exogenous dopamine, suggesting the involvement of a PRL-releasing factor (Shin, 1979). Lesions of the PVN of the hypothalamus (Minamitani et al., 1987) or removal of the neurointermediate lobe of the pituitary (Murai and Ben-Jonathan, 1987a) both abolish the stress-induced release of PRL, suggesting that a PRL-releasing factor might originate in one or both of these structures. A recent report suggested that salsolinol, a catecholamine derivative, may function as such a PRL-releasing factor during stress (Bodnar et al., 2004).

While the stress-induced increase in PRL is well established, it is less well known that during times of elevated PRL, such as during the proestrous PRL surge (Gala and Haisenleder, 1986; Poletini et al., 2006), the PRL surges of early pregnancy (Morehead and Gala, 1989; Morehead et al., 1990), or during lactation (Banky et al., 1994), stress in fact decreases PRL secretion. The stress-induced decrease in PRL during these surges is associated with increased activity of TIDA neurons (Gala, 1990). The differential response appears to be controlled by the different levels of ovarian steroid hormones present under different conditions (Poletini et al., 2006).

The physiological function of stress-induced PRL secretion is not known, although it has been suggested that elevated PRL may function to prevent overactivation of the stress response (Drago et al., 1989; Cook, 1997) or to contribute to stimulation of the immune system (Demoraes et al., 1995; Clevenger et al., 1998; Chikanza, 1999; Matera et al., 2000; Yu-Lee, 2002) as part of an integrated stress response.

13.5.3 Pregnancy

Pregnancy is unique with respect to lactogenic hormone action in the brain, in that as well as the maternal pituitary gland, the placenta provides an additional source of lactogenic hormones in the circulation of many species. In rodents, serum PRL levels vary throughout pregnancy and can be categorized into three distinct phases: early, mid-, and late pregnancy. The pattern of PRL secretion characteristic of early pregnancy is initiated by copulation (or copulomimetic stimuli). The mating stimulus induces a unique neuroendocrine reflex that generates twice-daily PRL surges that continue for 8–10 days (Butcher et al., 1972; Gunnet and Freeman, 1983). The PRL surges act to rescue the corpora lutea from the previous estrous cycle, establishing the ongoing progesterone secretion that is necessary to maintain early pregnancy (Gunnet and Freeman, 1983). In keeping with the feedback relationship discussed above, TIDA neurons also exhibit a semicircadian pattern of activity during early pregnancy, with reduced dopamine release observed during the PRL surges and elevated release between surges (McKay et al., 1982; Demarest et al., 1983d; Arbogast and Voogt, 1991b). The periodic reductions in dopamine secretion are required for the expression of PRL surges, while a feedback-induced activation of TIDA neurons terminates the surges and restores basal secretion of PRL. It should be noted, however, that a low-amplitude rhythm of TIDA activity persists even when PRL secretion is suppressed with bromocriptine (Demarest et al., 1983d). It is thought that the diurnal and nocturnal PRL surges are driven by a complex reciprocal interaction involving stimulation of PRL secretion by oxytocin in the portal blood followed (with an intrinsic delay) by feedback inhibition involving TIDA neurons. This rhythm is entrained to the light–dark cycles by a vasoactive intestinal peptide innervation of oxytocin neurons from the suprachiasmatic nucleus (Egli et al., 2004, 2006; Bertram et al., 2006; McKee et al., 2007).

The second phase of PRL secretion occurs around mid-pregnancy, when the PRL surges cease during days 8–10 of pregnancy (Smith and Neill, 1976) due to a sustained increase in dopamine release from TIDA neurons (McKay et al., 1982). These events are associated with the onset of PL secretion (Smith and Neill, 1976). PL is structurally homologous to PRL (Voogt et al., 1982; Voogt, 1984) and acts in a PRL-like manner on PRL receptors on TIDA neurons to stimulate dopamine synthesis and release, thus inhibiting PRL secretion from the maternal pituitary gland (Voogt et al., 1982; Demarest et al., 1983a,c; Tonkowicz and Voogt, 1983; Lee and Voogt, 1999). Two forms of PL have been identified in rats, PL-I (rPL-I) and -II (rPL-II). rPL-I is present from day 8 of pregnancy and peaks around day 13 (Robertson et al., 1982) while rPL-II is first detected around day 10 of pregnancy, peaking near parturition (Robertson and Friesen, 1981). Like PRL, both variants can enter the brain and bind to PRL receptors (Robertson et al., 1982), thus activating the short-loop feedback system inhibiting PRL secretion (Demarest et al., 1983a) and potentially regulate other PRL-sensitive neurons. As rPL-I is present earlier, it is most likely the signal that terminates early pregnancy PRL surges (Voogt et al., 1996), while rPL-II maintains this inhibition of PRL secretion during the remainder of pregnancy. Thus, from the time of cessation of the PRL surges until very near term, TIDA neuronal activity is high and maternal PRL secretion is suppressed (Andrews et al., 2001).

The third phase of PRL secretion during pregnancy occurs in a short period during very late pregnancy, in the 24–36 h prior to parturition, when there is a distinct change in the mechanisms controlling PRL secretion. Despite the continued presence of rPL-II, PRL levels begin to rise during the night preceding parturition, generating a large nocturnal surge of PRL (Grattan and Averill, 1990, 1991b). This antepartum PRL surge occurs at a time when PL levels remain high, suggesting a disruption in the normal negative feedback mechanisms controlling

PRL release. As treatment with bromocriptine (a dopamine agonist) completely abolishes the antepartum PRL surge (Grattan and Averill, 1995), it is clear that the disruption in PRL feedback does not occur at the level of the lactotrophs, but rather at the level of the TIDA neurons. This is confirmed by the fact that TIDA neuronal activity significantly decreases approximately 36 h prior to parturition even though PL levels are still elevated, and that no further change in TIDA neuronal activity is observed in response to the antepartum PRL surge (Andrews et al., 2001). Moreover, administration of ovine PRL (Fliestra and Voogt, 1997) or human PL (hPL) (Grattan and Averill, 1991b) fails to activate TIDA neurons or prevent the antepartum PRL surge. Similarly, hypothalamic implants of anterior pituitary tissue (that release PRL into the CSF, stimulating TIDA neurons (Gonzalez et al., 1989)) and, thus, reducing endogenous PRL secretion (Grattan and Averill, 1991a) or intraventricular transplants of rPL-secreting placental tumor cells are both unable to abolish the antepartum surge (Fliestra and Voogt, 1997). These data strongly suggest that TIDA neurons become unresponsive to both PL and PRL during late pregnancy, allowing hyperprolactinemia to develop (see below).

In other rodents, such as mice and hamsters, PL secretion is almost identical to the rat, with two major forms of PL secreted at specific times during gestation (Ogren and Telamantes, 1988). Many other species, including nonhuman primates, cows, sheep, and goats, show a gradual, but continuous, increase in placental lactogen in the blood starting relatively early in gestation and reaching a peak near parturition (Ogren and Telamantes, 1988). By contrast, there are some mammalian species, including pigs, rabbits, and dogs, that apparently do not express a PL (Talamantes et al., 1980). Interestingly, in these species, pituitary PRL secretion is elevated throughout the latter half of gestation and especially immediately prior to parturition (Cowie et al., 1980).

In humans, both PRL (Tyson et al., 1972; Schweizer et al., 1984) and hPL (Braunstein et al., 1980) are present in increasing amounts with advancing gestation. This is interesting, as one would expect that the placental hormone would feedback to induce an inhibition of PRL secretion, as in other species. It is possible that the TIDA neurons become insensitive to PRL relatively early during human gestation, although this has not been studied. Alternatively, it is possible that hPL does not function at the PRL receptors controlling dopamine activity in the human brain. Unlike PLs of other species, hPL is actually more closely related to human growth hormone than human PRL (Soares et al., 1998).

13.5.4 Suckling-Induced Release of PRL

Following parturition, PRL secretion becomes closely linked to the suckling stimulus, which can be considered the prototypical neuroendocrine signal to stimulate PRL release. Suckling of the nipple by the young produces a signal in afferent somatosensory neuronal pathways that is relayed to the hypothalamus and then pituitary, inducing PRL secretion. This is a classical neuroendocrine reflex, where the amount of PRL released is proportional to the intensity and duration of suckling (Neill and Nagy, 1994). This highly coordinated release of PRL is caused by a suckling-induced decrease of activity of TIDA neurons (Selmanoff and Wise, 1981; Demarest et al., 1983b; Selmanoff and Gregerson, 1985), resulting in profound suppression of TH mRNA levels (Wang et al., 1993) and reduced secretion of dopamine into the portal blood (Ben-Jonathan et al., 1980). One or more PRL-releasing factors may also be released (see Section 13.2.3). Following pup withdrawal, TH mRNA (Wang et al., 1993) and dopamine secretion (Ben-Jonathan et al., 1980) increase relatively rapidly to levels similar to that of nonpregnant animals. As with late pregnancy, TIDA neuronal activity remains low during lactation even during periods of elevated PRL secretion (Ben-Jonathan et al., 1980; Demarest et al., 1983b). Furthermore, even in the absence of suckling, treatment with PRL during lactation does not affect TIDA neuronal activity (Demarest et al., 1983b; Arbogast and Voogt, 1996), suggesting that as in late pregnancy, short-loop negative feedback regulation of PRL during lactation is impaired. The loss of activation of TIDA neuronal activity in response to PRL secretion during late pregnancy is a critical maternal adaptation, which allows the high levels of PRL that are required for milk production and maternal behavior to be maintained unencumbered by an autoregulatory feedback mechanism (Andrews, 2005).

13.5.5 Mechanisms Contributing to the Change in the Neuroendocrine Control of PRL Secretion during Late Pregnancy and Lactation

13.5.5.1 Change in PRL signal transduction in TIDA neurons

Mechanisms underlying the loss of responsiveness of TIDA neurons to PRL during late pregnancy and lactation have been reviewed recently (Grattan et al., 2008). This does not appear to be mediated by a suppression of PRL receptor expression on the TIDA neurons. In fact, there is a twofold increase in the number of neurons expressing PRL receptors in the arcuate nucleus during lactation (Pi and Grattan, 1999c). Similarly, levels of mRNA for both isoforms of PRL-R mRNA are increased in the arcuate nucleus throughout pregnancy, although this is not maintained into lactation (Augustine et al., 2003). Specifically within TIDA neurons, there is no evidence of a reduction in long-form PRL receptor mRNA in TIDA neurons during lactation, compared to nonpregnant females (Kokay and Grattan, 2005). Thus, it seems likely that the altered response to PRL is mediated by changes in the downstream signaling pathways activated by PRL in these neurons. As described above, in the TIDA neurons, PRL specifically induces phosphorylation and nuclear translocation of STAT5b (Lerant et al., 2001; Anderson et al., 2006a,b), and this action is required for the short-loop negative-feedback action of PRL (Grattan et al., 2001; Ma, 2005a). While PRL can readily activate STAT5 in TIDA neurons in nonpregnant animals, it is completely ineffective during lactation (Anderson et al., 2006a,b). These data suggest that the loss of response to PRL during late pregnancy and lactation is specifically mediated by a loss of PRL-induced phosphorylation and nuclear translocation of STAT5b.

Recent evidence suggests that suppressors of cytokine signaling (SOCS) proteins may be involved in the loss of PRL signaling within hypothalamic dopaminergic neurons during pregnancy (Anderson et al., 2006b) and lactation (Anderson et al., 2006a,b). SOCS proteins are induced in response to cytokine signaling and act as intracellular feedback regulators (Starr and Hilton, 1999). Eight members of the SOCS family (SOCS-1 to SOCS-7 and CIS) have been identified to date (Larsen and Ropke, 2002), and all have conserved SH2 domains in common with STAT proteins. Thus, SOCS proteins, like STATs, are able to bind the kinase domain of JAK, but they block access of STATs to receptor-binding sites to suppress PRL signaling (Yoshimura, 1998; Pezet et al., 1999; Kile et al., 2001). In ovariectomized rats, PRL induces SOCS-1, SOCS-3, and CIS but not SOCS-2 within the arcuate nucleus (Anderson et al., 2006b). This suggests that, as in response to other cytokines, these proteins are induced as a part of the PRL signal transduction response. During late pregnancy and lactation, levels of SOCS-1, SOCS-3, and CIS mRNA were all markedly elevated in the arcuate nucleus (Anderson et al., 2006b), consistent with the hypothesis that these proteins mediate the loss of PRL-induced activation of STAT5b in TIDA neurons (see **Figure 4**).

Interestingly, bromocriptine treatment to suppress PRL levels during late pregnancy or pup withdrawal to prevent suckling-induced increases in PRL during lactation both prevented the rise in SOCS mRNA levels in the arcuate nucleus (Anderson et al., 2006b), suggesting that PRL mediates the increase in SOCS mRNA at these times (Anderson et al., 2006a,b). The classical role of SOCS as a STAT-induced feedback regulator of JAK/STAT signaling, however, presents something of a paradox in interpreting the above data. If PRL-induced SOCS expression inhibits the activation of STAT5b, thereby rendering the neurons unresponsive to PRL, it would clearly be self-limiting. Thus, it seems that the increased SOCS expression cannot be solely dependent on PRL action through the JAK/STAT pathway, and alternative pathways must also be involved. For example, PRL is known to induce gene transcription by various MAPK signaling cascades in many other cellular systems (Piccoletti et al., 1994; Das and Vonderhaar, 1995; Rao et al., 1995), and there are some reports that MAPK can activate SOCS3 (Canfield et al., 2005). Alternatively, it is possible that other extracellular or hormonal signals present during late pregnancy and lactation might induce SOCS, either independently or synergistically with PRL. For example, estrogen, which is high during late pregnancy, has been reported to induce SOCS gene expression in other systems (Leong et al., 2004; Leung, 2004; Matthews et al., 2005).

13.5.5.2 Role of ovarian steroids in the regulation of PRL feedback during pregnancy and lactation

The adaptations to PRL action on TIDA neurons during late pregnancy are critically dependent on the changing levels of ovarian steroids present at this time. Estrogen levels increase throughout pregnancy, peaking near parturition (Bridges, 1984). In contrast, progesterone levels are high throughout pregnancy, before rapidly declining to near basal levels approximately 24–36 h prior to parturition (Bridges, 1984). As dopaminergic neurons within the arcuate nucleus express both the ER and progesterone receptor (PR) (Steyn et al., 2007), it is likely that ovarian steroids directly influence these neurons. In nonpregnant animals, estrogen decreases dopamine turnover within the arcuate nucleus and median eminence (Blum et al., 1987; Morrell et al., 1989;

Jones and Naftolin, 1990), suppresses the release of dopamine (Cramer et al., 1979), and inhibits the expression of TH in the arcuate nucleus (Jones and Naftolin, 1990; Arbogast and Voogt, 1993). In contrast, progesterone increases TIDA neuronal activity (Rance et al., 1981; DeMaria et al., 2000) and increases hypothalamic TH mRNA expression (Arbogast and Voogt, 1993, 1994). We have observed similar effects during late pregnancy, as premature withdrawal of progesterone in the presence of ongoing estrogen levels induced a decrease in TIDA neuronal activity, while progesterone treatments to prevent the antepartum decline in progesterone levels maintained the elevated TIDA neuronal activity, thus preventing the normal PRL surge at this time (Steyn et al., 2008).

To investigate the possibility that ovarian steroids might influence TIDA neurons by altering SOCS expression, we measured SOCS mRNA levels in the arcuate nucleus after manipulating ovarian steroid levels during late pregnancy. The decline of progesterone proved to be a key determining factor that allows both increased SOCS expression as well as the antepartum PRL surge (Steyn et al., 2008). Neither of these events occur when the fall in progesterone levels is prevented with exogenous progesterone (Steyn et al., 2008). These data suggest that the decline of progesterone during late pregnancy allows the induction of SOCS mRNA within the arcuate nucleus by PRL and/or estrogen, resulting in the loss of ability of PRL to activate STAT5b in these neurons, reducing dopamine synthesis and secretion and allowing expresion of the antepartum PRL surge (**Figure 4**).

13.5.5.3 A proposed model for the pregnancy-induced adaptation of the neuroendocrine control of PRL secretion

In examining the feedback regulation of PRL secretion during late pregnancy and lactation, several key

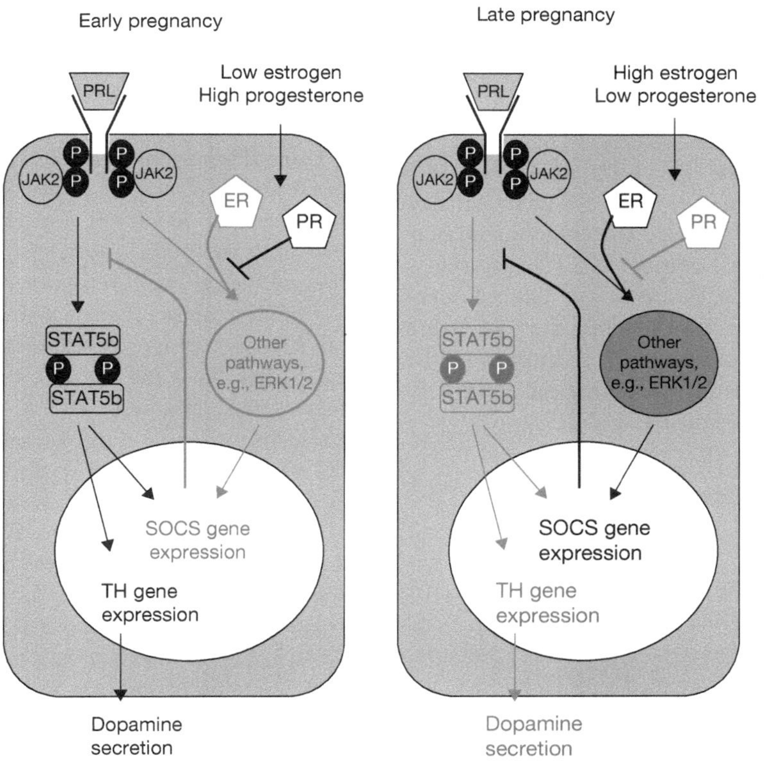

Figure 4 Model showing possible changes in prolactin signal transduction on TIDA neurons during late pregnancy (see text for details). Under normal conditions, such as early pregnancy, depicted on the left, prolactin stimulates dopamine secretion, acting via the JAK/STAT5b pathway. During late pregnancy, and persisting into lactation, the changing estrogen:progesterone ratio favors the prolactin-induced activation of SOCS expression, presumably mediated through alternative signal transduction pathways. Expression of SOCS inhibits signaling through the JAK/STAT pathway, resulting in reduced dopamine secretion. This allows hyperprolactinemia to occur, unimpeded by a feedback regulatory mechanism.

observations have been made: (1) that there is no reduction in the levels of long-form PRL receptor mRNA in TIDA neurons during late pregnancy and lactation, and, therefore, it is likely that the change in PRL feedback occurs downstream of the receptor; (2) that PRL signaling in TIDA neurons is mediated via STAT5b, and that PRL-induced activation of STAT5b in TIDA neurons is markedly reduced during late pregnancy (thereby contributing to the desensitization of TIDA neurons to PRL at this time); (3) that levels of specific SOCS mRNAs (known inhibitors of STAT signaling) increase in the arcuate nucleus during late pregnancy and lactation; (4) that both estrogen and PRL can stimulate SOCS mRNA expression in the arcuate nucleus, but that this is inhibited by the presence of progesterone; and (5) that the timing of ovarian steroid hormone changes, normally associated with late pregnancy, regulates the induction of the antepartum PRL surge and the increase of SOCS mRNA expression within the arcuate nucleus. These observations have led to the formulation of a hypothesis to explain the loss of sensitivity of TIDA neurons to PRL during late pregnancy. Withdrawal of progesterone, in the presence of elevated estrogen, promotes a PRL-induced increase of SOCS proteins within TIDA neurons, which then disrupt PRL signaling via the JAK/STAT5b pathway. The PRL effect on SOCS mRNA must be mediated by a non-STAT5b pathway, as activation of STAT5b is specifically inhibited during late pregnancy and lactation. As TH expression is dependent on STAT5b, this results in reduced dopamine synthesis and release, allowing the subsequent induction of the antepartum PRL surge and the continued hyperprolactinemia during lactation. This model is depicted in **Figure 4**.

13.6 Brain Actions of PRL in Mammals

PRL can be considered a pleiotropic hormone, exerting a wide range of actions in the body (Grattan and Kokay, 2008). The following section details the range of different functions influenced by PRL in the brain, and attempts to place these in the context of how these actions play a specific role in the physiology of the animals. Because many of these functions are influenced by PRL of systemic origin, the physiological functions are often most pronounced during pregnancy or lactation, when PRL levels in the blood are high.

13.6.1 Maternal Behavior

There is a strong body of findings that support a role for PRL in the central stimulation of the onset of maternal behavior. This work, primarily conducted in the rat, indicates that PRL and the related lactogenic hormones of pregnancy, rPLs, act via PRL receptors within the mPOA to promote a rapid expression of maternal care toward foster young.

The initial findings that demonstrated a role for PRL in the onset of maternal behavior utilized a rat preparation in which gonadectomized, virgin rats were treated sequentially with a systemically (Silastic capsules) administered steroid regimen that consisted of 10 days of progesterone exposure followed by exposure to estradiol. This steroid treatment stimulates a fast onset of maternal care toward foster young (Bridges et al., 1985) as well as elevates circulating PRL concentrations (Bridges and Ronsheim, 1990). Hormone-treated, behaviorally inexperienced virgin rats retrieve, group, and crouch over the test young with latencies of 1–2 days (Bridges, 1984). Moreover, when steroid-treated females are concurrently treated with bromocriptine to suppress endogenous PRL secretion, the onset of maternal behavior toward foster young is delayed from a latency of 1–2 days to 4–5 days (Bridges and Ronsheim, 1990).

Using this hormone paradigm, Bridges et al. (1990) established that central infusions of PRL (Bridges et al., 1990) as well as rPL-I or rPL-II (Bridges et al., 1996) into the mPOA stimulated maternal behavior. The mPOA is well recognized as a critical site controlling maternal behavior (Numan, 1994, 2006), and PRL receptors are expressed throughout this region (Bakowska and Morrell, 1997; Pi and Grattan, 1998b, 1999c; Bakowska and Morrell, 2003). When steroid-primed virgin rats that have their endogenous PRL secretion suppressed by SC injections of bromocriptine are given bilateral infusions of PRL into the mPOA beginning a day prior to the start of behavioral testing (at the time of progesterone removal and estradiol exposure) and continuing for an additional 2 days, a robust stimulation of maternal behavior occurs. As shown in **Figure 5**, PRL infusions (40 ng/side) stimulated a significant increase in responsiveness over a 6-day test period, reaching significance by the test session of day 2 of testing (Bridges et al., 1990). This effect appeared specific to the mPOA, since infusions of this amount of PRL into the ventricular system was without effect, whereas much higher doses infused into the lateral ventricle could stimulate a rapid onset of maternal behavior (Bridges et al., 1990). It is noted

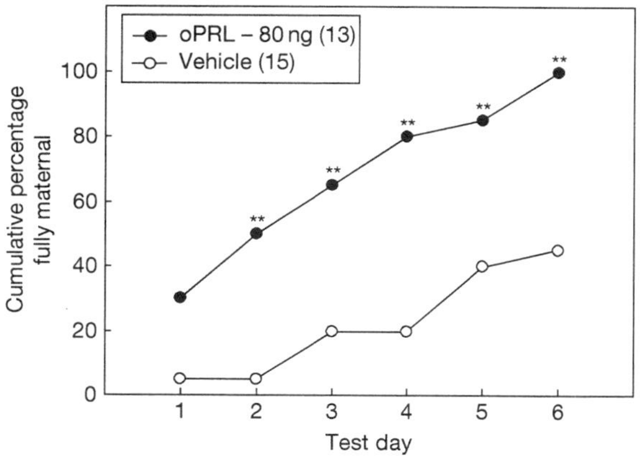

Figure 5 Stimulation of maternal behavior after bilateral infusions of oPRL (40 ng/side) into the mPOA region of steroid-primed virgin rats. Subjects were tested once daily for 1 h for responsiveness toward three foster young. A female was rated maternal if she retrieved, grouped, and crouched over all three test young during a test session. ** $P < 0.01$ for controls.

that the actions of PRL are dependent upon the presence of the combination of progesterone and estradiol, since infusions of PRL into the mPOA in the absence of steroids failed to affect maternal care.

The central site(s) of PRL stimulation of maternal behavior appear localized to the mPOA. The stimulatory steroid regimen of progesterone and estradiol alters expression of the long form of the PRL receptor in the mPOA (Bridges and Hays, 2005); progesterone treatment itself appears to reduce mRNA expression within the dorsal mPOA. Studies of PRL's actions in other brain regions where PRL receptors are abundant have not been systematically examined for an influence on maternal behavior. A study in the ventromedial nucleus of the hypothalamus (VMH) of the action of PRL on the onset of maternal behavior is suggestive, but not conclusive, of a stimulatory action of PRL (Bridges et al., 1999). Whereas latencies of PRL-infused rats were short, control infusion latencies were also unexpectedly short, indicative of a slight facilitation of behavior by the infusion and/or cannulation procedures (Bridges et al., 1999). Earlier work in mice has also implicated the hypothalamus as a possible site of PRL action in stimulating maternal behavior and nest building (Voci and Carlson, 1973). Likewise, in ring doves PRL stimulates parental-related behaviors when infused into the basal hypothalamus (Buntin et al., 1991; Buntin, 1996) further studies, however, are needed to parcel out the specific actions of PRL in these brain regions. Other neural regions which may regulate either the establishment or ongoing maternal care include the medial amygdala where PRL receptors are abundant (Bakowska and Morrell, 1997). Moreover, this region is known to mediate olfactory sensations associated with maternal behavior (Fleming et al., 1980, 1994).

Additional studies demonstrate that two other lactogenic hormones, recombinant rPL-I and purified rPL-II, are also able to stimulate a fast onset of maternal care in steroid-primed virgin rats (Bridges et al., 1996). Since CSF concentrations of these lactogens are elevated from day 11 of pregnancy until shortly prior to parturition (Bridges et al., 1996), these findings raise the intriguing possibility that the developing conceptus also provides an endocrine signal to the gestating maternal brain, which serves to prime the expectant mother to respond maternally at the time of birth. It is likely that some redundancy or safety system exists to insure that the new mother is maternal at birth. This system includes both sets of lactogenic hormones, PRL and the PLs.

The central PRL receptor system within the mPOA appears to mediate the actions of PRL in the onset of maternal behavior. Using the established steroid hormone regimen, it was found that bilateral mPOA infusions of the putative PRL receptor antagonist S179D-PRL delayed the onset of maternal behavior in virgin rats by about 2 days (Bridges et al., 2001). These findings provide support for an involvement of the PRL receptor in the regulation of the onset of maternal care in the rat. This role for the PRL receptor is further supported by studies in mice with null mutations for the PRL receptor (Lucas et al., 1998) in which deficits in maternal care in PRL receptor-knockout mice were found. Specifically, heterozygote mice displayed partial deficits in maternal care, whereas mice with complete deletions of the PRL receptor exhibited more severe deficits. Presumably, these effects are mediated through receptor deletions within the central nervous system.

Whereas the onset of maternal behavior in the rat is stimulated in part by PRL, the role for PRL in ongoing or established maternal care has not been as systematically explored. To date, evidence for a role for PRL in ongoing maternal care is limited. One possible developmental shift relative to ongoing maternal care that merits consideration is that a biochemical transition occurs in the regulation of maternal care from one being endocrine-based during an initial pregnancy to another being more actively mediated by a neural lactogenic system once maternal behavior is established. Recent studies of the effects of reproductive experience on the central PRL system indicate that a prior pregnancy and lactation result in a long-term upregulation of the PRL receptor system within the mPOA as well as other central sites of PRL action, that is, arcuate, PVN, etc. (Anderson et al., 2006c). Basal expression of the long form of the PRL receptor as well as the response of this system to a PRL challenge increases as a result of reproductive experience. Thus, it is an exciting, albeit unconfirmed, possibility that brain regions involved in maternal care, that is, mPOA, become more responsive to PRL or PRL-like signals once maternal care is initiated for the first time, thereby reducing the need for pituitary PRL or PLs in ongoing maternal care. Perhaps, central signals themselves are sufficient to maintain maternal responsiveness once maternal behavior is established in combination with shifts in sensory processing which may make stimuli received from young more salient. These possibilities are the topics of future studies.

13.6.2 Stress Response and Anxiety

PRL acts as an endogenous anxiolytic agent, able to induce dose-dependent suppression of anxiety behaviors (Donner et al., 2007) and inhibit the acute stress response (Carter and Lightman, 1987; Cook, 1997; Torner et al., 2001). Chronic PRL treatment blocks the stress-induced increase in corticotropin-releasing hormone (CRH) mRNA in the PVN, and reduces neuronal activation in response to stress, as measured by fos mRNA (Donner et al., 2007). The effects of PRL can be prevented by intracerebroventricular (ICV) administration of antisense oligonucleotides directed against the PRL receptor (Torner et al., 2001), demonstrating a direct action of PRL in the brain. This may be mediated by PRL receptors found in the PVN, described above, although specific expression of PRL receptors on CRH neurons has not been reported. Alternatively, because the amygdala is critically involved in anxiety and stress responses (Schulkin et al., 2005; Schulkin, 2006), PRL receptors in the medial amygdala could be involved in the anxiolytic actions of PRL.

PRL can also act in the brain (Fujikawa et al., 2004) to prevent the stress-induced formation of gastric ulcers (Drago et al., 1985, 1990). This effect is associated with increased PRL receptor expression in the choroid plexus (Fujikawa et al., 1995) and in the PVN of the hypothalamus (Fujikawa et al., 2004). Antisense-induced inhibition of PRL receptor expression in the PVN prevents the ability of PRL to block stress-induced changes in plasma calcium and gastric ulceration (Fujikawa et al., 2005). Thus, stress-induced changes in PRL secretion (described above), might have protective actions in the body as well as reducing anxiety associated with the stressor. In addition, stress has also been shown to induce brain PRL production in the PVN (Torner et al., 2004), suggesting that local release of PRL might compliment PRL in the systemic circulation.

Lactation is characterized by hyperprolactinemia, and may be the most physiologically relevant period when PRL might alter emotionality in the female. It is well documented that lactation is a stress hyporesponsive state (Stern et al., 1973; Lightman, 1992; Neumann et al., 1998; Lightman et al., 2001; Tilbrook and Clarke, 2006; Slattery and Neumann, 2008). These changes are thought to be protective for the fetus, preventing excessive exposure to high levels of maternal cortisol (Brunton and Russell, 2008b; Slattery and Neumann, 2008). PRL receptor expression is increased in the brain during lactation, and in particular, high levels are found in regions involved in the control of stress responses, including the PVN (Pi and Grattan, 1999c). Moreover, PRL synthesis in the brain is also increased during pregnancy and lactation (Torner et al., 2002, 2004). Antisense oligonucleotides directed against PRL receptors restore stress responsiveness during lactation (Torner and Neumann, 2002), suggesting that anxiolytic actions of PRL (Torner et al., 2001; Donner et al., 2007) may play a critical role in regulating stress responsiveness during lactation (Neumann, 2003).

13.6.3 Regulation of Oxytocin Neurons

A complex reciprocal relationship exists between PRL and oxytocin. Oxytocin is known to stimulate PRL secretion both *in vitro* (Lumpkin et al., 1983; Johnston and Negro-Vilar, 1988; Chadio and Antoni, 1993; Liu and Ben-Jonathan, 1994) and *in vivo* (Samson et al., 1986), and oxytocin antagonists can block PRL secretion (Johnston and Negro-Vilar, 1988; Samson et al., 1989; Mogg and Samson, 1990) suggesting an endogenous role for oxytocin in the control of PRL secretion. In particular, oxytocin appears to be important in stimulating PRL secretion at proestrus (Johnston and Negro-Vilar, 1988) and in early pregnancy (Arey and Freeman, 1990; Egli et al., 2004, 2006). PRL receptors are expressed in oxytocin neurons in both the supraoptic nucleus and PVN (Kokay et al., 2006), and PRL acutely inhibits the firing rate of these neurons in nonpregnant females (Townsend et al., 2005; Kokay et al., 2006). PRL also inhibits stress-induced oxytocin secretion (Carter and Lightman, 1987; Torner et al., 2002). It has been proposed that this inhibitory action of PRL on oxytocin neurons might reflect a feedback role, regulating the putative PRL-releasing role of oxytocin (White and Samson, 2006). In lactating animals, however, there is clear evidence that PRL stimulates expression of oxytocin mRNA (Ghosh and Sladek, 1995b; Popeski et al., 2003) and oxytocin secretion (Sarkar, 1989; Parker et al., 1991). Moreover, ICV administration of a PRL receptor antagonist impairs parturition (Nephew et al., 2007), consistent with a role for PRL in stimulating the normal firing pattern of oxytocin neurons at this time. Hence, PRL has a complex role in the regulation of oxytocin neurons, with specific effects dependent on the physiological state of these neurons. Indeed, it is possible that the changing actions of PRL on oxytocin neurons play an important role in mediating the adaptive changes in firing pattern of oxytocin neurons in different physiological states.

13.6.4 Regulation of Reproductive Behavior and Fertility

Hyperprolactinemia is the most common endocrine disorder of the hypothalamic–pituitary axis (Mah and Webster, 2002) and is well recognized as a major cause of reproductive dysfunction in males and females. Women with chronic hyperprolactinemia may present with infertility, galactorrhea, amenorrhoea, reduced libido, and even habitual abortion. Up to 40% of women presenting with secondary amenorrhoea have increased serum PRL levels (Evans et al., 1982; Meaney and O'Keane, 2002). Men may similarly present with infertility and galactorrhea, in addition to gynecomastia and complaints of decreased libido. In men, hyperprolactinemia is present in approximately 16% of patients with erectile dysfunction and 11% of men with oligospermia (De Rosa et al., 2003). Infertility and loss of libido are serious side effects of neuroleptic drugs (Knegtering et al., 2008), affecting up to 70% of patients, and representing one of the major causes of noncompliance with antipsychotic medication (Meaney and O'Keane, 2002).

PRL action on fertility is likely to be mediated in the hypothalamus. PRL reduces GnRH secretion into the portal blood (Weber et al., 1983; Koike et al., 1984, 1991; Sarkar et al., 1992) and suppresses both frequency and amplitude of luteinizing hormone (LH) pulses (Cohen-Becker et al., 1986; Fox et al., 1987; Park and Selmanoff, 1991; Grattan and Selmanoff, 1994). The action of PRL to suppress LH secretion is dependent on the presence of estradiol (Anderson et al., 2008), consistent with the idea that PRL and estradiol may act on the same populations of neurons. PRL can also act in the pituitary gland to suppress LH secretion (Smith, 1978, 1982; Cheung, 1983; Morel et al., 1994; Tortonese et al., 1998). In humans, pulsatile GnRH replacement can reverse the infertility induced by hyperprolactinaemia (Polson et al., 1986; Matsuzaki et al., 1994; Lecomte et al., 1997), suggesting a PRL-induced suppression of GnRH release is the primary cause of infertility. Such an effect might be mediated directly on GnRH neurons. The GnRH-secreting GT1 cell line expresses PRL receptors, and PRL suppresses release of GnRH from these cells (Milenkovic et al., 1994). However, we have recently reported that only a very small proportion of GnRH neurons express the PRL receptor *in vivo* (Grattan et al., 2007a). In contrast, large numbers of other neurons in the vicinity of GnRH neurons express PRL receptor mRNA, consistent with the hypothesis that PRL action might be mediated indirectly, through PRL-sensitive afferent neurons.

In addition to actions on GnRH neurons, PRL may impair reproduction through the regulation of reproductive behavior. Hyperprolactinemia causes erectile dysfunction (Doherty et al., 1990) and suppression of copulatory behavior in male rats (Bailey and Herbert, 1982; Doherty et al., 1985, 1989). While one study reported similar inhibitory effects on female reproductive behavior (Dudley et al., 1982), others reported that PRL may in fact enhance lordosis. Injection of PRL directly into the midbrain central gray matter stimulates lordosis (Harlan et al., 1983), while suppression of the proestrous PRL surge by bromocriptine reduced expression of this key reproductive behavior (Witcher and Freeman, 1985). The reasons for the differences in these studies are not clear.

In humans, there is a marked release of PRL following orgasm in both male and females (Brody and Kruger, 2006). Because of the suppressive effect of PRL on reproduction and libido, discussed above, it has been proposed that this prolonged postcoital elevation in PRL may provide a feedback system, reducing sexual drive and motivation during the refractory stage after intercourse (Kruger et al., 2002, 2005). This concept is supported by anecdotal studies, showing that in an individual male who exhibited spontaneous multiple orgasms (i.e., lacking a postcoital refractory stage), the postcoital rise in PRL was absent (Haake et al., 2002).

13.6.5 Neurotrophic Effects, Neurogenesis, and Glial Cell Function

PRL effects on neural development were first reported in studies investigating brain development in amphibians in the early 1970s (Hunt and Jacobson, 1970, 1971). In mammals, as well as regulating activity of TIDA neurons in the adult, studies in dwarf mice lacking PRL have shown that PRL is required for the normal development of the TIDA neurons (Romero and Phelps, 1993; Phelps et al., 1994; Phelps, 2004). PL can apparently substitute for PRL in this developmental function (Phelps and Horseman, 2000). Similarly, PRL replacement can rescue dopamine neurons in mice lacking endogenous PRL (Morgan and Besch, 1990), and although it was originally thought that exposure must occur through a specific developmental window (Romero and Phelps, 1995; Phelps et al., 2003), recent data suggest that long-term PRL treatment may be able to stimulate differentiation of dopamine neurons, even in the adult (Khodr et al., 2008).

Interest in the role of PRL in neural development has recently been revived, with the increased recognition that neurogenesis occurs in the mammalian brain throughout adulthood (Zhao et al., 2008), and the surprising observation that elevated PRL secretion seen during pregnancy in mice causes an increase in mitogenesis in the subventricular zone of the mother, resulting in increased incorporation of new neurons into the olfactory bulb (Shingo et al., 2003). Subsequently, similar observations have been made following enhanced PRL secretion induced by pheromone exposure (Mak et al., 2007; Larsen et al., 2008). It has been hypothesized that the PRL-induced changes in adult neurogenesis may play important roles in both mating (Mak et al., 2007) and maternal behaviors (Shingo et al., 2003; Larsen et al., 2008).

PRL has also been shown to stimulate mitogenesis in astrocyte (DeVito et al., 1992b, 1993, 1995; Mangoura et al., 2000) and oligodendrocyte (Gregg et al., 2007) populations. These glial cell actions might play a role in helping the brain recover following acute injury (DeVito et al., 1995). Acute brain injury has been shown to upregulate PRL expression in the brain (DeVito et al., 1995; Moderscheim et al., 2007), and PRL treatment enhances glial cell differentiation and function (DeVito et al., 1995; Moderscheim et al., 2007).

13.6.6 Appetite and Food Intake

A number of studies have documented the ability of systemic PRL administration to increase food intake in a variety of species (Buntin and Figge, 1988; Gerardo-Gettens et al., 1989; Noel and Woodside, 1993). Because elevated PRL can also affect the reproductive cycle, the increase in food intake in female rats following PRL treatment could be mediated through effects on gonadal steroids, which are known to affect food intake (Hervey and Hervey, 1967; Nance and Gorski, 1978; Sieck et al., 1978; Eckel, 2004; Santollo et al., 2007). However, ICV administration of PRL increases food intake without influencing estrous cyclicity (Noel and Woodside, 1993; Sauvé and Woodside, 2000; Naef and Woodside, 2007). Moreover, ICV administration of PRL also increases food intake in ovariectomized rats (Sauvé and Woodside, 1996, 2000). Together, these data argue for the ability of central PRL receptor activation to increase food intake independent of steroid hormone levels.

The phasic PRL secretion of early pregnancy is likely to contribute to the rapid increase in food intake seen in pregnant rats, both indirectly, by promoting progesterone secretion, and directly, through actions in the hypothalamus. As described above, PRL receptors are found in many of the nuclei involved in the homeostatic regulation of food intake, including the arcuate, ventromedial, and paraventricular nuclei, and these nuclei form potential targets for PRL action during pregnancy and lactation. In the arcuate nucleus, however, PRL receptors do not appear to be expressed in the NPY (Li et al., 1999; Chen and Smith, 2004) and POMC neurons (Kokay and Grattan, 2005) that regulate appetite. Hence, it seems likely that PRL acts downstream of the arcuate neurons. Consistent with this hypothesis, localized injections of PRL directly into the paraventricular nucleus stimulate food intake in a dose-dependent manner in female rats (Sauvé and Woodside, 2000).

In addition to stimulating appetite during pregnancy, PRL also facilitates food intake by inducing a state of leptin resistance. By day 14 of pregnancy in rats, leptin is unable to suppress food intake (Ladyman and Grattan, 2004, 2005). By contrast, pseudo-pregnant rats remain fully responsive to leptin, even when pseudo-pregnancy is maintained beyond 14 days by supplementing progesterone. In this model, however, chronic PRL infusion to mimic the chronic elevation in PL secretion characteristic of mid-pregnancy induces a leptin-resistant state similar to pregnant animals (Augustine and Grattan, 2008). In nonpregnant female rats, chronic ICV infusion of PRL prevents the reduction in body weight or food intake normally seen in response to a central injection of leptin (Naef and Woodside, 2007). The lack of a behavioral response to leptin is accompanied by a reduction in the ability of leptin to induce Fos expression and phosphorylation of STAT3 in both the PVN and ventromedial hypothalamic nucleus (Naef and Woodside, 2007). Taken together, these data suggest that the PRL surges during early pregnancy stimulate an initial orexigenic response through direct and indirect actions. The resultant hyperphagia is then maintained, in the face of rising leptin levels and a positive energy balance, by PL-induced leptin resistance.

PRL also contributes to the hyperphagia of lactation (Woodside, 2007). PRL receptors are expressed on a population of neurons in the dorsomedial hypothalamus during lactation, and PRL stimulates NPY expression in these neurons (Chen and Smith, 2004). Upregulation of NPY in this part of the hypothalamus, which is also seen in genetic models of obesity, augments the orexigenic effects of other NPY inputs to the PVN (Li et al., 1999). PRL is the major lactogenic hormone in the rat; thus, examining the effects of PRL on food intake by pharmacological manipulation in postpartum rats is confounded due to the energetic demands of increased milk production causing an increase in appetite, independent of any direct effects of PRL. To eliminate this confounding variable, Woodside used a model in which the galactophores connecting the mammary gland to the nipple were cut. Galactophore-cut rats continue to suckle their litters, but do not deliver milk, and have similar hormonal profiles to those of intact lactating rats. Thus, it is possible to study the specific contributions of suckling-induced PRL to the hyperphagia of lactation. When PRL levels are suppressed in galactophore-cut rats using the dopamine D_2 agonist, bromocriptine, food intake and body weight is decreased. Replacement of PRL in bromocriptine-treated galactophore-cut rats, by chronic intraventricular infusion, restores food intake without changing the length of lactational infertility suggesting that the orexigenic effects of PRL acting on PRL receptors within the brain make a major contribution to the hyperphagia of lactation (Woodside, 2007).

13.6.7 PRL and the Neurobiological Adaptation to Pregnancy and Lactation

Rather than multiple unrelated functions, the pleiotropic actions of PRL in the brain can be considered as a coordinated response to promote the neurobiological adaptation to pregnancy and lactation (**Figure 6**).

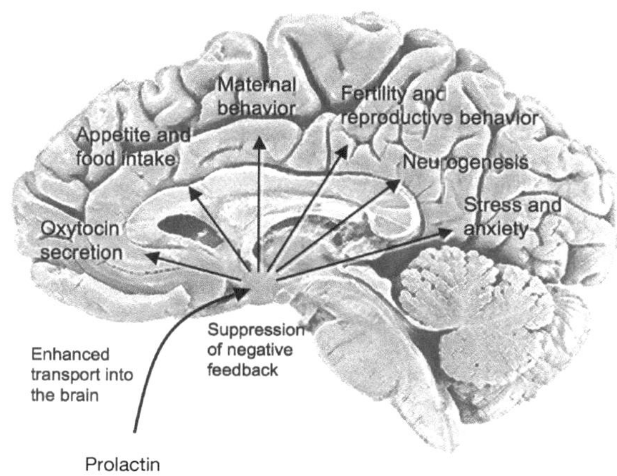

Figure 6 Diagram representing the range of functions influenced by prolactin in the maternal brain. The high levels of prolactin (and/or PL) present during pregnancy and lactation are well placed to coordinate a broad range of adaptive responses in the mother, facilitating the nuturing of the offspring.

The physiological changes of pregnancy put enormous new strains on the mother's body. The endocrine changes associated with pregnancy function to coordinate the multifactorial adaptive responses required for the mother to cope with these strains (Brunton and Russell, 2008b). As with many adaptive homeostatic responses, the brain plays a significant role in monitoring the changing physiological conditions and organizing appropriate responses. There are multiple adaptations to maintain high levels of lactogenic hormones throughout pregnancy and lactation. Initially, PL secretion overcomes the normal low levels of PRL maintained by negative feedback. Subsequently, short-loop negative feedback of PRL is suppressed, allowing hyperprolactinemia to develop (Grattan et al., 2008). Elevated PRL contributes to the establishment of maternal behavior (Bridges et al., 1985, 1990; Lucas et al., 1998), and the consequent suckling stimulus drives further PRL secretion. Moreover, brain PRL production is also increased during pregnancy and lactation (Torner et al., 2002, 2004) and may combine with PRL of systemic origin to regulate brain function. PRL receptor expression is also upregulated in the hypothalamus during pregnancy (Pi and Grattan, 1999c), suggesting that the hypothalamus is highly sensitive to PRL at this time. Thus, PRL (together with the related PLs) is well placed to provide an essential afferent signal to communicate the reproductive state to the maternal brain, driving the adaptive responses.

PRL stimulates food intake during pregnancy and lactation by increasing orexigenic drive and helping to establish leptin resistance in the mother (Grattan et al., 2007b; Naef and Woodside, 2007; Augustine and Grattan, 2008). These are key metabolic adaptations promoting energy storage in preparation for the demands of fetal growth and lactation (Ladyman and Grattan, 2005; Augustine et al., 2008). The antistress effects of PRL may contribute to the suppression of stress responses during late pregnancy and lactation (Torner and Neumann, 2002; Slattery and Neumann, 2008), an adaptive response to minimize the risk of adverse fetal programming from glucocorticoids (Brunton et al., 2006; Brunton and Russell, 2008a). PRL has also been shown to prevent stress-induced hyperthermia (Drago and Amir, 1984), and hence may be involved in the loss of the febrile response during late pregnancy (Spencer et al., 2008). Oxytocin neurons undergo marked changes in firing pattern during parturition and in lactation (Russell et al., 2001), and hence, the differential regulation by PRL may have important consequences for the functioning of these critical neurons during parturition and/ or milk ejection (Kokay et al., 2006). PRL-induced suppression of fertility (Grattan et al., 2007a), likewise, may have physiological importance in lactational infertility and birth spacing (McNeilly, 2001). Finally, PRL-induced neurogenesis during pregnancy (Shingo et al., 2003) may have important implications for maternal recognition of the offspring, contributing to the PRL-induced enhancement of maternal behavior.

13.7 Conclusion

In summary, PRL and the PLs, both of which bind to PRL receptors in the brain, alter a broad range of behavioral and homeostatic responses that are crucial for propagation of the species. The actions of PRL within the central nervous system are significantly altered by the reproductive state of the female, and perhaps the male, as reflected by hormone-induced shifts in neurochemical sensitivities. Moreover, developmental factors, such as prior experience, also appear capable of altering PRL receptor expression and regulating responsiveness to changing PRL levels. Dysfunction of this system can result in pathological conditions, including hyperphagia and emotional disturbances such as those reported during pseudopregnancy in humans (Sobrinho, 1993, 1998). Through the advance of knowledge generated by recent studies of the molecular actions of PRL and its transduction systems, future studies are now poised to elucidate the central mechanisms, that is, transduction mechanisms and genetic components, involved in controlling crucial behavioral and neuroendocrine responses. The central production of PRL and the reported stimulation of neurogenesis by this protein also open new avenues in understanding how the brain adapts over the course of development to shifts in endocrine and environmentally altered endocrine states. Once considered a hormone that primarily altered peripheral organ systems, that is, hepatic and mammary, prolactin's role in brain function has now become apparent, and its vital role in the adaptation of the female brain is now recognized.

References

Anderson GM, Beijer P, Bang AS, Fenwick MA, Bunn SJ, and Grattan DR (2006b) Suppression of prolactin-induced signal transducer and activator of transcription 5b signaling and induction of suppressors of cytokine signaling messenger ribonucleic acid in the hypothalamic arcuate nucleus of the rat during late pregnancy and lactation. *Endocrinology* 147: 4996–5005.

Anderson GM, Grattan DR, van den Ancker W, and Bridges RS (2006c) Reproductive experience increases prolactin responsiveness in the medial preoptic area and arcuate nucleus of female rats. *Endocrinology* 147: 4688–4694.

Anderson GM, Kieser DC, Steyn FJ, and Grattan DR (2008) Hypothalamic prolactin receptor messenger ribonucleic acid levels, prolactin signaling, and hyperprolactinemic inhibition of pulsatile luteinizing hormone secretion are dependent on estradiol. *Endocrinology* 149: 1562–1570.

Anderson ST, Barclay JL, Fanning KJ, Kusters DH, Waters MJ, and Curlewis JD (2006a) Mechanisms underlying the diminished sensitivity to prolactin negative feedback during lactation: Reduced STAT5 signalling and upregulation of cytokine-inducible SH2-domain-containing protein (CIS) expression in tuberoinfundibular dopaminergic neurons. *Endocrinology* 147: 1195–1202.

Andrews ZB (2005) Neuroendocrine regulation of prolactin secretion during late pregnancy: Easing the transition into lactation. *Journal of Neuroendocrinology* 17: 466–473.

Andrews ZB and Grattan DR (2003) Opioid receptor subtypes involved in the regulation of prolactin secretion during pregnancy and lactation. *Journal of Neuroendocrinology* 15: 227–236.

Andrews ZB and Grattan DR (2004) The roles of dopamine and the neurointermediate lobe of the pituitary in the regulation of prolactin secretion during late pregnancy in rats. *Journal of Neuroendocrinology* 16: 859–865.

Andrews ZB, Kokay IC, and Grattan DR (2001) Dissociation of prolactin secretion from tuberoinfundibular dopamine activity in late pregnant rats. *Endocrinology* 142: 2719–2724.

Annunziato L and Moore KE (1978) Prolactin in CSF selectively increases dopamine turnover in the median eminence. *Life Sciences* 22: 2037–2042.

Arbogast LA and Voogt JL (1991a) Hyperprolactinemia increases and hypoprolactinemia decreases tyrosine hydroxylase messenger ribonucleic acid levels in the arcuate nuclei, but not the substantia nigra or zona incerta. *Endocrinology* 128: 997–1005.

Arbogast LA and Voogt JL (1991b) Mechanisms of tyrosine hydroxylase regulation during pregnancy: Evidence for protein dephosphorylation during the prolactin surges. *Endocrinology* 129: 2575–2582.

Arbogast LA and Voogt JL (1993) Progesterone reverses the estradiol-induced decrease in tyrosine hydroxylase mRNA levels in the arcuate nucleus. *Neuroendocrinology* 58: 501–510.

Arbogast LA and Voogt JL (1994) Progesterone suppresses tyrosine hydroxylase messenger ribonucleic acid levels in the arcuate nucleus on proestrus. *Endocrinology* 135: 343–350.

Arbogast LA and Voogt JL (1995) Hypoprolactinemia decreases tyrosine hydroxylase activity in the tuberoinfundibular dopaminergic neurons acutely by protein dephosphorylation and chronically by changes in gene expression. *Endocrine* 3: 801–806.

Arbogast LA and Voogt JL (1996) The responsiveness of tuberoinfundibular dopaminergic neurons to prolactin feedback is diminished between early lactation and midlactation in the rat. *Endocrinology* 137: 47–54.

Arbogast LA and Voogt JL (1997) Prolactin (PRL) receptors are colocalized in dopaminergic neurons in fetal hypothalamic cell cultures: Effect of PRL on tyrosine hydroxylase activity. *Endocrinology* 138: 3016–3023.

Arey BJ and Freeman ME (1990) Oxytocin, vasoactive-intestinal peptide, and serotonin regulate the mating-induced surges of prolactin secretion in the rat. *Endocrinology* 126: 279–284.

Arimura A, Dunn JD, and Schally AV (1972) Effect of infusion of hypothalamic extracts on serum prolactin levels in rats treated with nembutal, CNS depressants or bearing hypothalamic lesions. *Endocrinology* 90: 378–383.

Augustine RA and Grattan DR (2008) Induction of central leptin resistance in hyperphagic pseudopregnant rats by chronic prolactin infusion. *Endocrinology* 149: 1049–1055.

Augustine RA, Kokay IC, Andrews ZB, Ladyman SR, and Grattan DR (2003) Quantitation of prolactin receptor mRNA in the maternal rat brain during pregnancy and lactation. *Journal of Molecular Endocrinology* 31: 221–232.

Augustine RA, Ladyman SR, and Grattan DR (2008) From feeding one to feeding many: Hormone-induced changes in bodyweight homeostasis during pregnancy. *Journal of Physiology* 586: 387–397.

Bailey DJ and Herbert J (1982) Impaired copulatory behaviour of male rats with hyperprolactinaemia induced by domperidone or pituitary grafts. *Neuroendocrinology* 35: 186–193.

Bakowska JC and Morrell JI (1997) Atlas of the neurons that express mRNA for the long form of the prolactin receptor in the forebrain of the female rat. *Journal of Comparative Neurology* 386: 161–177.

Bakowska JC and Morrell JI (2003) The distribution of mRNA for the short form of the prolactin receptor in the forebrain of the female rat. *Brain Research – Molecular Brain Research* 116: 50–58.

Banky Z, Nagy GM, and Halasz B (1994) Analysis of pituitary prolactin and adrenocortical response to ether, formalin or restraint in lactating rats: Rise in corticosterone, but no increase in plasma prolactin levels after exposure to stress. *Neuroendocrinology* 59: 63–71.

Ben-Jonathan N, Neill MA, Arbogast LA, Peters LL, and Hoefer MT (1980) Dopamine in hypophysial portal blood: Relationship to circulating prolactin in pregnant and lactating rats. *Endocrinology* 106: 690–696.

Ben-Jonathan N, Oliver C, Weiner HJ, Mical RS, and Porter JC (1977) Dopamine in hypophysial portal plasma of the rat during the estrous cycle and throughout pregnancy. *Endocrinology* 100: 452–458.

Bertram R, Egli M, Toporikova N, and Freeman ME (2006) A mathematical model for the mating-induced prolactin rhythm of female rats. *American Journal of Physiology – Endocrinology and Metabolism* 290: E573–E582.

Bjorklund A, Moore RY, Nobin A, and Stenevi U (1973) The organization of tubero-hypophyseal and reticulo-infundibular catecholamine neuron systems in the rat brain. *Brain Research* 51: 171–191.

Bluem V and Fiedler K (1965) Hormonal control reproductive behavior in some cichlid fish. *General and Comparative Endocrinology* 56: 186–196.

Blum M, McEwen BS, and Roberts JL (1987) Transcriptional analysis of tyrosine hydroxylase gene expression in the tuberoinfundibular dopaminergic neurons of the rat arcuate nucleus after estrogen treatment. *Journal of Biological Chemistry* 262: 817–821.

Bodnar I, Banky ZS, Toth BE, Nagy GM, and Halasz B (2002) Brain structures mediating the suckling stimulus-induced release of prolactin. *Journal of Neuroendocrinology* 14: 384–396.

Bodnar I, Mravec B, Kubovcakova L, et al. (2004) Stress-as well as suckling-induced prolactin release is blocked by a structural analogue of the putative hypophysiotrophic prolactin-releasing factor, salsolinol. *Journal of Neuroendocrinology* 16: 208–213.

Bole-Feysot C, Goffin V, Edery M, Binart N, and Kelly PA (1998) Prolactin (PRL) and its receptor: Actions, signal transduction pathways and phenotypes observed in PRL receptor knockout mice. *Endocrine Reviews* 19: 225–268.

Braunstein GD, Rasor JL, Engvall E, and Wade ME (1980) Interrelationships of human chorionic gonadotropin, human placental lactogen, and pregnancy-specific beta 1-glycoprotein throughout normal human gestation. *American Journal of Obstetrics and Gynecology* 138: 1205–1213.

Bridges RS (1984) A quantitative analysis of the roles of dosage, sequence, and duration of estradiol and progesterone exposure in the regulation of maternal behavior in the rat. *Endocrinology* 114: 930–940.

Bridges RS, DiBiase R, Loundes DD, and Doherty PC (1985) Prolactin stimulation of maternal behavior in female rats. *Science* 227: 782–784.

Bridges RS and Hays LE (2005) Steroid-induced alterations in mRNA expression of the long form of the prolactin receptor in the medial preoptic area of female rats: Effects of exposure to a pregnancy-like regimen of progesterone and estradiol. *Brain Research – Molecular Brain Research* 140: 10–16.

Bridges RS, Mann PE, and Coppeta JS (1999) Hypothalamic involvement in the regulation of maternal behaviour in the rat: Inhibitory roles for the ventromedial hypothalamus and the dorsal/anterior hypothalamic areas. *Journal of Neuroendocrinology* 11: 259–266.

Bridges RS, Numan M, Ronsheim PM, Mann PE, and Lupini CE (1990) Central prolactin infusions stimulate maternal behavior in steroid-treated, nulliparous female rats. *Proceedings of the National Academy of Sciences of the United States of America* 87: 8003–8007.

Bridges RS, Rigero BA, Byrnes EM, Yang LL, and Walker AM (2001) Central infusions of the recombinant human prolactin receptor antagonist, S179D-PRL, delay the onset of maternal behavior in steroid-primed, nulliparous female rats. *Endocrinology* 142: 730–739.

Bridges RS, Robertson MC, Shiu RPC, Friesen HG, Stuer AM, and Mann PE (1996) Endocrine communication between conceptus and mother: Placental lactogen stimulation of maternal behavior. *Neuroendocrinology* 64: 57–64.

Bridges RS and Ronsheim PM (1990) Prolactin (PRL) regulation of maternal behavior in rats: Bromocriptine treatment delays and PRL promotes the rapid onset of behavior. *Endocrinology* 126: 837–848.

Brody S and Kruger TH (2006) The post-orgasmic prolactin increase following intercourse is greater than following masturbation and suggests greater satiety. *Biological Psychology* 71: 312–315.

Brunton PJ, Bales J, and Russell JA (2006) Neuroendocrine stress but not feeding responses to centrally administered neuropeptide Y are suppressed in pregnant rats. *Endocrinology* 147: 3737–3745.

Brunton PJ and Russell JA (2008a) Attenuated hypothalamo-pituitary–adrenal axis responses to immune challenge during pregnancy: The neurosteroid opioid connection. *Journal of Physiology* 586: 369–375.

Brunton PJ and Russell JA (2008b) The expectant brain: Adapting for motherhood. *Nature Reviews Neuroscience* 9: 11–25.

Buntin JD (1996) Neural and hormonal control of parental behavior in birds. *Advances in Study of Behavior* 25: 161–213.

Buntin JD, Becker GM, and Ruzycki E (1991) Facilitation of parental behavior in ring doves by systemic or intracranial injections of prolactin. *Hormones and Behavior* 25: 424–444.

Buntin JD and Figge GR (1988) Prolactin and growth hormone stimulate food intake in ring doves. *Pharmacology, Biochemistry, and Behavior* 31: 533–540.

Butcher RL, Fugo NW, and Collins WE (1972) Semicircadian rhythm in plasma levels of prolactin during early gestation in the rat. *Endocrinology* 90: 1125–1127.

Callahan P, Klosterman S, Prunty D, Tompkins J, and Janik J (2000) Immunoneutralization of endogenous opioid peptides prevents the suckling-induced prolactin increase and the inhibition of tuberoinfundibular dopaminergic neurons. *Neuroendocrinology* 71: 268–276.

Canfield S, Lee Y, Schroder A, and Rothman P (2005) Cutting edge: IL-4 induces suppressor of cytokine signaling-3 expression in B cells by a mechanism dependent on activation of p38 MAPK. *Journal of Immunology* 174: 2494–2498.

Carter DA and Lightman SL (1987) Oxytocin responses to stress in lactating and hyperprolactinaemic rats. *Neuroendocrinology* 46: 532–537.

Cave BJ, Norman M, Flynn A, Townsend J, Wakerley JB, and Tortonese DJ (2005) Prolactin-induced activation of STAT5 within the hypothalamic arcuate nucleus. *NeuroReport* 16: 1423–1426.

Chadio SE and Antoni FA (1993) Specific oxytocin agonist stimulates prolactin release but has no effect on inositol phosphate accumulation in isolated rat anterior pituitary cells. *Journal of Molecular Endocrinology* 10: 107–114.

Chen P and Smith MS (2004) Regulation of hypothalamic neuropeptide Y messenger ribonucleic acid expression during lactation: Role of prolactin. *Endocrinology* 145: 823–829.

Cheung CY (1983) Prolactin suppresses luteinizing hormone secretion and pituitary responsiveness to luteinizing hormone-releasing hormone by a direct action at the anterior pituitary. *Endocrinology* 113: 632–638.

Chikanza IC (1999) Prolactin and neuroimmunomodulation: *In vitro* and *in vivo* observations. *Annals of the New York Academy of Sciences* 876: 119–130.

Chiu S and Wise PM (1994) Prolactin receptor mRNA localization in the hypothalamus by *in situ* hybridization. *Journal of Neuroendocrinology* 6: 191–199.

Chiu S and Wise PM (1996) Prolactin receptor gene expression in specific hypothalamic nuclei increases with age. *Journal of Gerontology Series A, Biological Sciences and Medical Sciences* 51: B220–B224.

Clapp C, Torner L, GutierrezOspina G, et al. (1994) The prolactin gene is expressed in the hypothalamic–neurohypophyseal system and the protein is processed into a 14-kDa fragment with activity like 16-kDa prolactin. *Proceedings of the National Academy of Sciences of the United States of America* 91: 10384–10388.

Clevenger CV, Freier DO, and Kline JB (1998) Prolactin receptor signal transduction in cells of the immune system. *Journal of Endocrinology* 157: 187–197.

Cohen-Becker IR, Selmanoff M, and Wise PM (1986) Hyperprolactinemia alters the frequency and amplitude of pulsatile luteinizing hormone secretion in the ovariectomized rat. *Neuroendocrinology* 42: 328–333.

Cook CJ (1997) Oxytocin and prolactin suppress cortisol responses to acute stress in both lactating and non-lactating sheep. *Journal of Dairy Research* 64: 327–339.

Cowie A, Forsyth I, and Hart I (1980) *Hormonal Control of Lactation.* New York: Springer.

Cowie AT, Tindal JS, and Benson GK (1960) Pituitary grafts and milk secretion in hypophysectomized rats. *Journal of Endocrinology* 21: 115–123.

Cramer OM, Parker CR, Jr., and Porter JC (1979) Estrogen inhibition of dopamine release into hypophysial portal blood. *Endocrinology* 104: 419–422.

CrumeyrolleArias M, Latouche J, Jammes H, Djiane J, Kelly PA, Reymond MJ, and Haour F (1993) Prolactin receptors in the rat hypothalamus: Autoradiographic localization and characterization. *Neuroendocrinology* 57: 457–466.

Das R and Vonderhaar BK (1995) Transduction of prolactin's (PRL) growth signal through both long and short forms of the PRL receptor. *Molecular Endocrinology* 9: 1750–1759.

De Greef WJ and Neill JD (1979) Dopamine levels in hypophysial stalk plasma of the rat during surges of prolactin secretion induced by cervical stimulation. *Endocrinology* 105: 1093–1099.

De Greef WJ, Plotsky PM, and Neill JD (1981) Dopamine levels in hypophysial stalk plasma and prolactin levels in peripheral plasma of the lactating rat: Effects of a simulated suckling stimulus. *Neuroendocrinology* 32: 229–233.

De Greef WJ, Voogt JL, Visser TJ, Lamberts SW, and van der Schoot P (1987) Control of prolactin release induced by suckling. *Endocrinology* 121: 316–322.

De Rosa M, Zarrilli S, DiSarno A, et al. (2003) Hyperprolactinemia in men: Clinical and biochemical features and response to treatment. *Endocrine* 20: 75–82.

Demarest KT, Duda NJ, Riegle GD, and Moore KE (1983a) Placental lactogen mimics prolactin in activating tuberoinfundibular dopaminergic neurons. *Brain Research* 272: 175–178.

Demarest KT, Johnston CA, and Moore KE (1981) Biochemical indices of catecholaminergic neuronal activity in the median eminence during the estrous cycle of the rat. *Neuroendocrinology* 32: 24–27.

Demarest KT, McKay DW, Riegle GD, and Moore KE (1983b) Biochemical indices of tuberoinfundibular dopaminergic neuronal activity during lactation: A lack of response to prolactin. *Neuroendocrinology* 36: 130–137.

Demarest KT, Moore KE, and Riegle GD (1983c) Role of a uterine-placental factor in the cessation of the semicircadian rhythm of tuberoinfundibular dopaminergic neuronal activity at midpregnancy in the rat. *Neuroendocrinology* 36: 409–414.

Demarest KT, Moore KE, and Riegle GD (1983d) Role of prolactin feedback in the semicircadian rhythm of

tuberoinfundibular dopaminergic neuronal activity during early pregnancy in the rat. *Neuroendocrinology* 36: 371–375.
Demarest KT, Moore KE, and Riegle GD (1985a) Acute restraint stress decreases dopamine synthesis and turnover in the median eminence: A model for the study of the inhibitory neuronal influences on tuberoinfundibular dopaminergic neurons. *Neuroendocrinology* 41: 437–444.
Demarest KT, Moore KE, and Riegle GD (1985b) Acute restraint stress decreases tuberoinfundibular dopaminergic neuronal activity: Evidence for a differential response in male versus female rats. *Neuroendocrinology* 41: 504–510.
Demarest KT, Riegle GD, and Moore KE (1984) Prolactin-induced activation of tuberoinfundibular dopaminergic neurons: Evidence for both a rapid 'tonic' and a delayed 'induction' component. *Neuroendocrinology* 38: 467–475.
Demarest KT, Riegle GD, and Moore KE (1986) The rapid 'tonic' and the delayed 'induction' components of the prolactin-induced activation of tuberoinfundibular dopaminergic neurons following the systemic administration of prolactin. *Neuroendocrinology* 43: 291–299.
DeMaria JE, Lerant AA, and Freeman ME (1999) Prolactin activates all three populations of hypothalamic neuroendocrine dopaminergic neurons in ovariectomized rats. *Brain Research* 837: 236–241.
DeMaria JE, Livingstone JD, and Freeman ME (1998) Characterization of the dopaminergic input to the pituitary gland throughout the estrous cycle of the rat. *Neuroendocrinology* 67: 377–383.
DeMaria JE, Livingstone JD, and Freeman ME (2000) Ovarian steroids influence the activity of neuroendocrine dopaminergic neurons. *Brain Research* 879: 139–147.
Demoraes MCL, Touraine P, Gagnerault MC, Savino W, Kelly PA, and Dardenne M (1995) Prolactin receptors and the immune system. *Annals of Endocrinology* 56: 567–570.
DeVito WJ (1988) Distribution of immunoreactive prolactin in the male and female rat brain: Effects of hypophysectomy and intraventricular administration of colchicine. *Neuroendocrinology* 47: 284–289.
DeVito WJ (1989) Immunoreactive prolactin in the hypothalamus and cerebrospinal fluid of male and female rats. *Neuroendocrinology* 50: 182–186.
DeVito WJ, Avakian C, Stone S, and Ace CI (1992a) Estradiol increases prolactin synthesis and prolactin messenger ribonucleic acid in selected brain regions in the hypophysectomized female rat. *Endocrinology* 131: 2154–2160.
DeVito WJ, Avakian C, Stone S, and Okulicz WC (1993) Prolactin-stimulated mitogenesis of cultured astrocytes is mediated by a protein kinase C-dependent mechanism. *Journal of Neurochemistry* 60: 832–842.
DeVito WJ, Connors JM, and Hedge GA (1987) Immunoreactive prolactin in the rat hypothalamus: *In vitro* release and subcellular localization. *Neuroendocrinology* 46: 155–161.
DeVito WJ, Okulicz WC, Stone S, and Avakian C (1992b) Prolactin-stimulated mitogenesis of cultured astrocytes. *Endocrinology* 130: 2549–2556.
DeVito WJ, Stone S, and Avakian C (1991) Stimulation of hypothalamic prolactin release by veratridine and angiotensin II in the female rat: Effect of ovariectomy and estradiol administration. *Neuroendocrinology* 54: 391–398.
DeVito WJ, Stone S, and Shamgochian M (1995) Prolactin induced expression of glial fibrillary acidic protein and tumor necrosis factor-alpha at a wound site in the rat brain. *Molecular and Cellular Endocrinology* 108: 125–130.
Di Carlo R and Muccioli G (1981) Changes in prolactin binding sites in the rabbit hypothalamus induced by physiological and pharmacological variations of prolactin serum levels. *Brain Research* 230: 445–450.
Doherty PC, Bartke A, Smith MS, and Davis SL (1985) Increased serum prolactin levels mediate the suppressive effects of ectopic pituitary grafts on copulatory behavior in male rats. *Hormone Behavior* 19: 111–121.
Doherty PC, Lane SJ, Pfeil KA, Morgan WW, Bartke A, and Smith MS (1989) Extra-hypothalamic dopamine is not involved in the effects of hyperprolactinemia on male copulatory behavior. *Physiology and Behavior* 45: 1101–1105.
Doherty PC, Wu DE, and Matt KS (1990) Hyperprolactinemia preferentially inhibits erectile function in adrenalectomized male rats. *Life Sciences* 47: 141–148.
Donner N, Bredewold R, Maloumby R, and Neumann ID (2007) Chronic intracerebral prolactin attenuates neuronal stress circuitries in virgin rats. *European Journal of Neuroscience* 25: 1804–1814.
Drago F and Amir S (1984) Effects of hyperprolactinaemia on core temperature of the rat. *Brain Research Bulletin* 12: 355–358.
Drago F, Continella G, Conforto G, and Scapagnini U (1985) Prolactin inhibits the development of stress-induced ulcers in the rat. *Life Sciences* 36: 191–197.
Drago F, D'Agata V, Iacona T, et al. (1989) Prolactin as a protective factor in stress-induced biological changes. *Journal of Clinical Laboratory Analysis* 3: 340–344.
Drago F, D'Agata V, Spadaro F, Grassi M, and Valerio C (1990) Prolactin as a protective factor in stress-induced gastric ulcers. *Annals of the New York Academy of Sciences* 597: 248–251.
Dudley CA, Jamison TS, and Moss RL (1982) Inhibition of lordosis behavior in the female rat by intraventricular infusion of prolactin and by chronic hyperprolactinemia. *Endocrinology* 110: 677–679.
Dutt A, Kaplitt MG, Kow LM, and Pfaff DW (1994) Prolactin, central nervous system and behavior: A critical review. *Neuroendocrinology* 59: 413–419.
Eckel LA (2004) Estradiol: A rhythmic, inhibitory, indirect control of meal size. *Physiology and Behavior* 82: 35–41.
Egli M, Bertram R, Sellix MT, and Freeman ME (2004) Rhythmic secretion of prolactin in rats: Action of oxytocin coordinated by vasoactive intestinal polypeptide of suprachiasmatic nucleus origin. *Endocrinology* 145: 3386–3394.
Egli M, Bertram R, Toporikova N, Sellix MT, Blanco W, and Freeman ME (2006) Prolactin secretory rhythm of mated rats induced by a single injection of oxytocin. *American Journal of Physiology – Endocrinology and Metabolism* 290: E566–E572.
Eikenburg DC, Ravitz AJ, Gudelsky GA, and Moore KE (1977) Effects of estrogen on prolactin and tuberoinfundibular dopaminergic neurons. *Journal of Neural Transmission* 40: 235–244.
Emanuele NV, Jurgens JK, Halloran MM, Tentler JJ, Lawrence AM, and Kelley MR (1992) The rat prolactin gene is expressed in brain tissue: Detection of normal and alternatively spliced prolactin messenger RNA. *Molecular Endocrinology* 6: 35–42.
Emanuele NV, Metcalfe L, Wallock L, et al. (1986) Hypothalamic prolactin: Characterization by radioimmunoassay and bioassay and response to hypophysectomy and restraint stress. *Neuroendocrinology* 44: 217–221.
Enjalbert A, Ruberg M, Arancibia S, Priam M, and Kordon C (1979) Endogenous opiates block dopamine inhibition of prolactin secretion *in vitro*. *Nature* 280: 595–597.
Evans WS, Cronin MJ, and Thorner MO (1982) Hypogonadism in hyperprolactinemia: Proposed mechanisms. In: Ganong WF and Martini L (eds.) *Frontiers in Neuroendocrinology*, vol. 7, pp. 77–122. New York: Raven Press.

Everett JW (1954) Luteotrophic function of autographs of the rat hypophysis. *Endocrinology* 54: 685–690.
Faouzi M, Leshan R, Bjornholm M, Hennessey T, Jones J, and Munzberg H (2007) Differential accessibility of circulating leptin to individual hypothalamic sites. *Endocrinology* 148: 5414–5423.
Fleming AS, Suh EJ, Korsmit M, and Rusak B (1994) Activation of Fos-like immunoreactivity in the medial preoptic area and limbic structures by maternal and social interactions in rats. *Behavioral Neuroscience* 108: 724–734.
Fleming AS, Vaccarino F, and Luebke C (1980) Amygdaloid inhibition of maternal behavior in the nulliparous female rat. *Physiology and Behavior* 25: 731–743.
Fliestra RJ and Voogt JL (1997) Lactogenic hormones of the placenta and pituitary inhibit suckling-induced prolactin (PRL) release but not the ante-partum PRL surge. *Proceedings of the Society for Experimental Biology and Medicine* 214: 258–264.
Fox SR, Hoefer MT, Bartke A, and Smith MS (1987) Suppression of pulsatile LH secretion, pituitary GnRH receptor content and pituitary responsiveness to GnRH by hyperprolactinemia in the male rat. *Neuroendocrinology* 46: 350–359.
Franchimont P, Dourcy C, Legros JJ, Reuter A, Vrindts-Gevaert Y, VanCauwenberge JR, and Gaspard U (1976) Prolactin levels during the menstrual cycle. *Clinical Endocrinology (Oxf)* 5: 643–650.
Freeman ME, Kanyicska B, Lerant A, and Nagy G (2000) Prolactin: Structure, function, and regulation of secretion. *Physiological Reviews* 80: 1523–1631.
Fujikawa T, Soya H, Tamashiro KL, et al. (2004) Prolactin prevents acute stress-induced hypocalcemia and ulcerogenesis by acting in the brain of rat. *Endocrinology* 145: 2006–2013.
Fujikawa T, Soya H, Yoshizato H, Sakaguchi K, Doh-Ura K, Tanaka M, and Nakashima K (1995) Restraint stress enhances the gene expression of prolactin receptor long form at the choroid plexus. *Endocrinology* 136: 5608–5613.
Fujikawa T, Tamura K, Kawase T, et al. (2005) Prolactin receptor knockdown in the rat paraventricular nucleus by a morpholino-antisense oligonucleotide causes hypocalcemia and stress gastric erosion. *Endocrinology* 146: 3471–3480.
Fuxe K (1964) Cellular localization of monoamines in the median eminence and the infundibular stem of some mammals. *Zeitschrift fuer Zellforschung und Mikroskopische Anatomie* 61: 710–724.
Fuxe K, Hokfelt T, Eneroth P, Gustafsson JA, and Skett P (1977) Prolactin-like immunoreactivity: Localization in nerve terminals of rat hypothalamus. *Science* 196: 899–900.
Gadd SL and Clevenger CV (2006) Ligand-independent dimerization of the human prolactin receptor isoforms: Functional implications. *Molecular Endocrinology* 20: 2734–2746.
Gala RR (1990) The physiology and mechanisms of the stress-induced changes in prolactin secretion in the rat. *Life Sciences* 46: 1407–1420.
Gala RR and Haisenleder DJ (1986) Restraint stress decreases afternoon plasma prolactin levels in female rats. Influence of neural antagonists and agonists on restraint-induced changes in plasma prolactin and corticosterone. *Neuroendocrinology* 43: 115–123.
Gerardo-Gettens T, Moore BJ, Stern JS, and Horwitz BA (1989) Prolactin stimulates food intake in a dose-dependent manner. *American Journal of Physiology* 256: R276–R280.
Ghosh R and Sladek CD (1995a) Prolactin modulates oxytocin mRNA during lactation by its action on the hypothalamo-neurohypophyseal axis. *Brain Research* 672: 24–28.
Ghosh R and Sladek CD (1995b) Role of prolactin and gonadal steroids in regulation of oxytocin mRNA during lactation. *American Journal of Physiology* 269: E76–E84.
Gibbs DM and Neill JD (1978) Dopamine levels in hypophysial stalk blood in the rat are sufficient to inhibit prolactin secretion *in vivo*. *Endocrinology* 102: 1895–1900.
Gonen Y and Casper RF (1990) Transient hyperprolactinemia is associated with a midcycle luteinizing hormone surge. *Fertility and Sterility* 54: 936–938.
Gonzalez HA, Kedzierski W, Aguila-Mansilla N, and Porter JC (1989) Hormonal control of tyrosine hydroxylase in the median eminence: Demonstration of a central role for the pituitary gland. *Endocrinology* 124: 2122–2127.
Goudreau JL, Lindley SE, Lookingland KJ, and Moore KE (1992) Evidence that hypothalamic periventricular dopamine neurons innervate the intermediate lobe of the rat pituitary. *Neuroendocrinology* 56: 100–105.
Grattan DR (2001) The actions of prolactin in the brain during pregnancy and lactation. *Progress in Brain Research* 133: 153–171.
Grattan DR and Averill RL (1990) Effect of ovarian steroids on a nocturnal surge of prolactin secretion that precedes parturition in the rat. *Endocrinology* 126: 1199–1205.
Grattan DR and Averill RL (1991a) Intrahypothalamic pituitary grafts elevate prolactin in the cerebrospinal fluid and attenuate prolactin release following ether stress. *Proceedings of the Society for Experimental Biology and Medicine* 196: 42–46 Issn: 0037-9727.
Grattan DR and Averill RL (1991b) Role of the placenta in the control of the ante-partum surge of prolactin in the rat. *Journal of Endocrinology* 130: 401–407.
Grattan DR and Averill RL (1995) Absence of short-loop autoregulation of prolactin during late pregnancy in the rat. *Brain Research Bulletin* 36: 413–416.
Grattan DR, Jasoni CL, Liu X, Anderson GM, and Herbison AE (2007a) Prolactin regulation of gonadotropin-releasing hormone neurons to suppress luteinizing hormone secretion in mice. *Endocrinology* 148: 4344–4351.
Grattan DR and Kokay IC (2008) Prolactin: A pleiotropic neuroendocrine hormone. *Journal of Neuroendocrinology* 20: 752–763.
Grattan DR, Ladyman SR, and Augustine RA (2007b) Hormonal induction of leptin resistance during pregnancy. *Physiology and Behavior* 91: 366–374.
Grattan DR and Selmanoff M (1994) Prolactin- and testosterone-induced inhibition of LH secretion after orchidectomy: Role of preoptic and tuberoinfundibular γ-aminobutyric acidergic neurones. *Journal of Endocrinology* 143: 165–174.
Grattan DR, Steyn FJ, Kokay IC, Anderson GM, and Bunn SJ (2008) Pregnancy-induced adaptation in the neuroendocrine control of prolactin secretion. *Journal of Neuroendocrinology* 20: 497–507.
Grattan DR, Xu JJ, McLachlan MJ, Kokay IC, Bunn SJ, Hovey RC, and Davey HW (2001) Feedback regulation of PRL secretion is mediated by the transcription factor, signal transducer, and activator of transcription 5b. *Endocrinology* 142: 3935–3940.
Gregerson KA and Selmanoff M (1988) Selective effects of hyperprolactinemia on *in vitro* dopamine release from median eminence synaptosomes. *Journal of Neuroscience* 8: 2477–2484.
Gregg C, Shikar V, Larsen P, Mak G, Chojnacki A, Yong VW, and Weiss S (2007) White matter plasticity and enhanced remyelination in the maternal CNS. *Journal of Neuroscience* 27: 1812–1823.

Gudelsky GA and Porter JC (1979) Release of newly synthesized dopamine into the hypophysial portal vasculature of the rat. *Endocrinology* 104: 583–587.
Gunnet JW and Freeman ME (1983) The mating-induced release of prolactin: A unique neuroendocrine response. *Endocrine Reviews* 4: 44–61.
Haake P, Exton MS, Haverkamp J, et al. (2002) Absence of orgasm-induced prolactin secretion in a healthy multi-orgasmic male subject. *International Journal of Impotence Research* 14: 133–135.
Harlan RE, Shivers BD, and Pfaff DW (1983) Midbrain microinfusions of prolactin increase the estrogen-dependent behavior, lordosis. *Science* 219: 1451–1453.
Haskins JT and Moss RL (1983) Differential effects of morphine, dopamine and prolactin administered ionophoretically on arcuate-ventromedial hypothalamic neurons. *Brain Research* 268: 185–188.
Hentschel K, Cheung S, Moore KE, and Lookingland KJ (1998) Pharmacological evidence that neurotensin mediates prolactin-induced activation of tuberoinfundibular dopamine neurons. *Neuroendocrinology* 68: 71–76.
Hervey E and Hervey GR (1967) The effects of progesterone on body weight and composition in the rat. *Journal of Endocrinology* 37: 361–381.
Hileman SM, Pierroz DD, Masuzaki H, Bjorbaek C, El-Haschimi K, Banks WA, and Flier JS (2002) Characterizaton of short isoforms of the leptin receptor in rat cerebral microvessels and of brain uptake of leptin in mouse models of obesity. *Endocrinology* 143: 775–783.
Hileman SM, Tornoe J, Flier JS, and Bjorbaek C (2000) Transcellular transport of leptin by the short leptin receptor isoform ObRa in Madin-Darby Canine kidney cells. *Endocrinology* 141: 1955–1961.
Hinuma S, Habata Y, Fujii R, et al. (1998) A prolactin-releasing peptide in the brain. *Nature* 393: 272–276.
Hokfelt T and Fuxe K (1972) Effects of prolactin and ergot alkaloids on the tubero-infundibular dopamine (DA) neurons. *Neuroendocrinology* 9: 100–122.
Hunt RK and Jacobson M (1970) Brain enhancement in tadpoles: Increased DNA concentration after somatotrophin or prolactin. *Science* 170: 342–344.
Hunt RK and Jacobson M (1971) Neurogenesis in frogs after early larval treatment with somatotropin or prolactin. *Developmental Biology* 26: 100–124.
Johnston CA, Demarest KT, and Moore KE (1980) Cycloheximide disrupts the prolactin-mediated stimulation of dopamine synthesis in tuberoinfundibular neurons. *Brain Research* 195: 236–240.
Johnston CA and Negro-Vilar A (1988) Role of oxytocin on prolactin secretion during proestrus and in different physiological or pharmacological paradigms. *Endocrinology* 122: 341–350.
Jones EE and Naftolin F (1990) Estrogen effects on the tuberoinfundibular dopaminergic system in the female rat brain. *Brain Research* 510: 84–91.
Kamberi IA, Mical RS, and Porter JC (1970) Prolactin-inhibiting activity in hypophysial stalk blood and elevation by dopamine. *Experientia* 26: 1150–1151.
Kanematsu S and Sawyer CH (1973) Elevation of plasma prolactin after hypophysial stalk section in the rat. *Endocrinology* 93: 238–241.
Kastin AJ, Pan WH, Maness LM, Koletsky RJ, and Ernsberger P (1999) Decreased transport of leptin across the blood–brain barrier in rats lacking the short form of the leptin receptor. *Peptides* 20: 1449–1453.
Kehoe L, Janik J, and Callahan P (1992) Effects of immobilization stress on tuberoinfundibular dopaminergic (TIDA) neuronal activity and prolactin levels in lactating and non-lactating female rats. *Life Sciences* 50: 55–63.
Khamad'ianov UR, Laricheva IP, and Smirnova LK (1980) Blood serum content of prolactin, follicle-stimulating hormone, luteinizing hormone, estradiol and progesterone in women with normal menstrual cycle. *Problemy Endokrinologii (Mosk)* 26: 32–37.
Khodr CE, Clark SM, Hurley DL, and Phelps CJ (2008) Long-term, homologous prolactin, administered through ectopic pituitary grafts, induces hypothalamic dopamine neuron differentiation in adult Snell dwarf mice. *Endocrinology* 149: 2010–2018.
Kile BT, Nicola NA, and Alexander WS (2001) Negative regulators of cytokine signaling. *International Journal of Hematology* 73: 292–298.
Kiss JZ, Kanyicska B, and Nagy GY (1986) The hypothalamic paraventricular nucleus has a pivotal role in regulation of prolactin release in lactating rats. *Endocrinology* 119: 870–873.
Knegtering H, van den Bosch R, Castelein S, Bruggeman R, Sytema S, and van Os J (2008) Are sexual side effects of prolactin-raising antipsychotics reducible to serum prolactin? *Psychoneuroendocrinology* 33: 711–717.
Koch Y, Lu KH, and Meites J (1970) Biphasic effects of catecholamines on pituitary prolactin release *in vitro*. *Endocrinology* 87: 673–675.
Koike K, Aono T, Miyake A, Tasaka K, Chatani F, and Kurachi K (1984) Effect of pituitary transplants on the LH-RH concentrations in the medial basal hypothalamus and hypophyseal portal blood. *Brain Research* 301: 253–258.
Koike K, Miyake A, Aono T, Sakumoto T, Ohmichi M, Yamaguchi M, and Tanizawa O (1991) Effect of prolactin on the secretion of hypothalamic GnRH and pituitary gonadotropins. *Hormone Research* 35: 5–12.
Kokay IC, Bull PM, Davis RL, Ludwig M, and Grattan DR (2006) Expression of the long form of the prolactin receptor in magnocellular oxytocin neurons is associated with specific prolactin regulation of oxytocin neurons. *American Journal of Physiology – Regulatory, Integrative and Comparative Physiology* 290: R1216–R1225.
Kokay IC and Grattan DR (2005) Expression of mRNA for prolactin receptor (long form) in dopamine and pro-opiomelanocortin neurones in the arcuate nucleus of non-pregnant and lactating rats. *Journal of Neuroendocrinology* 17: 827–835.
Krieger MS, Morrell JI, and Pfaff DW (1976) Autoradiographic localization of estradiol-concentrating cells in the female hamster brain. *Neuroendocrinology* 22: 193–205.
Kruger TH, Haake P, Hartmann U, Schedlowski M, and Exton MS (2002) Orgasm-induced prolactin secretion: Feedback control of sexual drive. *Neuroscience Biobehavioral Reviews* 26: 31–44.
Kruger TH, Hartmann U, and Schedlowski M (2005) Prolactinergic and dopaminergic mechanisms underlying sexual arousal and orgasm in humans. *World Journal of Urology* 23: 130–138.
Ladyman SR and Grattan DR (2004) Region-specific reduction in leptin-induced phosphorylation of signal transducer and activator of transcription-3 (STAT3) in the rat hypothalamus is associated with leptin resistance during pregnancy. *Endocrinology* 145: 3704–3711.
Ladyman SR and Grattan DR (2005) Suppression of leptin receptor messenger ribonucleic acid and leptin responsiveness in the ventromedial nucleus of the hypothalamus during pregnancy in the rat. *Endocrinology* 146: 3868–3874.
Larsen CM, Kokay IC, and Grattan DR (2008) Male pheromones initiate prolactin-induced neurogenesis and advance maternal behavior in female mice. *Hormones and Behavior* 53: 509–517.
Larsen L and Ropke C (2002) Suppressors of cytokine signalling: SOCS. *Acta Pathologica, Microbiologica et Immunologica* 110: 833–844.

Lecomte P, Lecomte C, Lansac J, Gallier J, Sonier CB, and Simonetta C (1997) Pregnancy after intravenous pulsatile gonadotropin-releasing hormone in a hyperprolactinaemic woman resistant to treatment with dopamine agonists. *European Journal of Obstetrics, Gynecology, and Reproductive Biology* 74: 219–221.

Lee Y and Voogt J (1999) Feedback effects of placental lactogens on prolactin levels and fos-related antigen immunoreactivity of tuberoinfundibular dopaminergic neurons in the arcuate nucleus during pregnancy in the rat. *Endocrinology* 140: 2159–2166.

Lehrman DS and Brody P (1961) Does prolactin induce incubation behavior in the ring dove? *Journal of Endocrinology* 22: 269–275.

Leonard WJ and O'Shea JJ (1998) Jaks and STATs: Biological implications. *Annual Review of Immunology* 16: 293–322.

Leong GM, Moverare S, Brce J, et al. (2004) Estrogen up-regulates hepatic expression of suppressors of cytokine signaling-2 and -3 *in vivo* and *in vitro*. *Endocrinology* 145: 5525–5531.

Lerant A and Freeman ME (1998) Ovarian steroids differentially regulate the expression of prolactin receptors in neuroendocrine dopaminergic neuron populations – a double-label confocal microscopic study. *Brain Research* 802: 141–154.

Lerant A, Kanyicska B, and Freeman ME (2001) Nuclear translocation of STAT5 and increased expression of Fos related antigens (FRAs) in hypothalamic dopaminergic neurons after prolactin administration. *Brain Research* 904: 259–269.

Leung KC (2004) Regulation of cytokine receptor signaling by nuclear hormone receptors: A new paradigm for receptor interaction. *DNA and Cell Biology* 23: 463–474.

Li C, Chen P, and Smith MS (1999) Neuropeptide Y and tuberoinfundibular dopamine activities are altered during lactation: Role of prolactin. *Endocrinology* 140: 118–123.

Lightman SL (1992) Alterations in hypothalamic–pituitary responsiveness during lactation. *Annals of the New York Academy of Sciences* 652: 340–346.

Lightman SL, Windle RJ, Wood SA, Kershaw YM, Shanks N, and Ingram CD (2001) Peripartum plasticity within the hypothalamo-pituitary–adrenal axis. *Progress in Brain Research* 133: 111–129.

Liu B and Arbogast LA (2008) Phosphorylation state of tyrosine hydroxylase in the stalk-median eminence is decreased by progesterone in cycling female rats. *Endocrinology* 149: 1462–1469.

Liu JW and Ben-Jonathan N (1994) Prolactin-releasing activity of neurohypophysial hormones: Structure-function relationship. *Endocrinology* 134: 114–118.

Lookingland KJ, Gunnet JW, Toney TW, and Moore KE (1990) Comparison of the effects of ether and restraint stress on the activity of tuberoinfundibular dopaminergic neurons in female and male rats. *Neuroendocrinology* 52: 99–105.

Loose MD, Ronnekleiv OK, and Kelly MJ (1990) Membrane properties and response to opioids of identified dopamine neurons in the guinea pig hypothalamus. *Journal of Neuroscience* 10: 3627–3634.

Lu JC, Scott P, Strous GJ, and Schuler LA (2002) Multiple internalization motifs differentially used by prolactin receptor isoforms mediate similar endocytic pathways. *Molecular Endocrinology* 16: 2515–2527.

Lucas BK, Ormandy CJ, Binart N, Bridges RS, and Kelly PA (1998) Null mutation of the prolactin receptor gene produces a defect in maternal behavior. *Endocrinology* 139: 4102–4107.

Lumpkin MD, Samson WK, and McCann SM (1983) Hypothalamic and pituitary sites of action of oxytocin to alter prolactin secretion in the rat. *Endocrinology* 112: 1711–1717.

Ma FY, Anderson GM, Gunn TD, Goffin V, Grattan DR, and Bunn SJ (2005a) Prolactin specifically activates STAT5b in neuroendocrine dopaminergic neurons. *Endocrinology* 146: 5112–5119.

Ma FY, Grattan DR, Goffin V, and Bunn SJ (2005b) Prolactin-regulated tyrosine hydroxylase activity and messenger ribonucleic acid expression in mediobasal hypothalamic cultures: The differential role of specific protein kinases. *Endocrinology* 146: 93–102.

MacLeod RM, Fontham EH, and Lehmeyer JE (1970) Prolactin and growth hormone production as influenced by catecholamines and agents that affect brain catecholamines. *Neuroendocrinology* 6: 283–294.

Mah PM and Webster J (2002) Hyperprolactinemia: Etiology, diagnosis, and management. *Seminars in Reproductive Medicine* 20: 365–374.

Mak GK, Enwere EK, Gregg C, Pakarainen T, Poutanen M, Huhtaniemi I, and Weiss S (2007) Male pheromone-stimulated neurogenesis in the adult female brain: Possible role in mating behavior. *Nature Neuroscience* 10: 1003–1011.

Mangoura D, Pelletiere C, Leung S, Sakellaridis N, and Wang DX (2000) Prolactin concurrently activates src-PLD and JAK/Stat signaling pathways to induce proliferation while promoting differentiation in embryonic astrocytes. *International Journal of Developmental Neuroscience* 18: 693–704.

Mann PE and Bridges RS (2002) Prolactin receptor gene expression in the forebrain of pregnant and lactating rats. *Brain Research – Molecular Brain Research* 105: 136–145.

Mansour A, Meador-Woodruff JH, Bunzow JR, Civelli O, Akil H, and Watson SJ (1990) Localization of dopamine D2 receptor mRNA and D1 and D2 receptor binding in the rat brain and pituitary: An *in situ* hybridization-receptor autoradiographic analysis. *Journal of Neuroscience* 10: 2587–2600.

Manzanares J, Wagner EJ, Moore KE, and Lookingland KJ (1993) Kappa opioid receptor-mediated regulation of prolactin and alpha-melanocyte-stimulating hormone secretion in male and female rats. *Life Sciences* 53: 795–801.

Martinez de la Escalera G and Weiner RI (1992) Dissociation of dopamine from its receptor as a signal in the pleiotropic hypothalamic regulation of prolactin secretion. *Endocrine Reviews* 13: 241–255 Issn: 0163-0769x.

Matera L, Mori M, Geuna M, Buttiglieri S, and Palestro G (2000) Prolactin in autoimmunity and antitumor defence. *Journal of Neuroimmunology* 109: 47–55.

Matsuzaki T, Azuma K, Irahara M, Yasui T, and Aono T (1994) Mechanism of anovulation in hyperprolactinemic amenorrhea determined by pulsatile gonadotropin-releasing hormone injection combined with human chorionic gonadotropin. *Fertility and Sterility* 62: 1143–1149.

Matthews J, Almlof T, Kietz S, Leers J, and Gustafsson JA (2005) Estrogen receptor-alpha regulates SOCS-3 expression in human breast cancer cells. *Biochemical and Biophysical Research Communications* 335: 168–174.

McKay DW, Pasieka CA, Moore KE, Riegle GD, and Demarest KT (1982) Semicircadian rhythm of tuberoinfundibular dopamine neuronal activity during early pregnancy and pseudopregnancy in the rat. *Neuroendocrinology* 34: 229–235.

McKee DT, Poletini MO, Bertram R, and Freeman ME (2007) Oxytocin action at the lactotroph is required for prolactin

surges in cervically stimulated ovariectomized rats. *Endocrinology* 148: 4649–4657.
McNeilly AS (2001) Lactational control of reproduction. *Reproduction, Fertility and Development* 13: 583–590.
Meaney AM and O'Keane V (2002) Prolactin and schizophrenia: Clinical consequences of hyperprolactinaemia. *Life Sciences* 71: 979–992.
Mejia S, Torner LM, Jeziorski MC, Gonzalez C, Morales MA, de la Escalera GM, and Clapp C (2003) Prolactin and 16K prolactin stimulate release of vasopressin by a direct effect on hypothalamo-neurohypophyseal system. *Endocrine* 20: 155–162.
Messinis IE and Templeton AA (1990) Prolactin secretion in women during the oestradiol-induced luteinizing hormone surge with or without progesterone. *Clinical Endocrinology (Oxf)* 32: 755–761.
Milenkovic L, D'Angelo G, Kelly PA, and Weiner RI (1994) Inhibition of gonadotropin hormone-releasing hormone release by prolactin from GT1 neuronal cell lines through prolactin receptors. *Proceedings of the National Academy of Sciences of the United States of America* 91: 1244–1247.
Minamitani N, Minamitani T, Lechan RM, Bollinger-Gruber J, and Reichlin S (1987) Paraventricular nucleus mediates prolactin secretory responses to restraint stress, ether stress, and 5-hydroxy-L-tryptophan injection in the rat. *Endocrinology* 120: 860–867.
Moderscheim TA, Gorba T, Pathipati P, Kokay IC, Grattan DR, Williams CE, and Scheepens A (2007) Prolactin is involved in glial responses following a focal injury to the juvenile rat brain. *Neuroscience* 145: 963–973.
Mogg RJ and Samson WK (1990) Interactions of dopaminergic and peptidergic factors in the control of prolactin release. *Endocrinology* 126: 728–735.
Moore KE (1987) Interactions between prolactin and dopaminergic neurons. *Biology of Reproduction* 36: 47–58.
Moore KE, Demarest KT, and Lookingland KJ (1987) Stress, prolactin and hypothalamic dopaminergic neurons. *Neuropharmacology* 26: 801–808.
Morehead MH and Gala RR (1989) The restraint stress-induced decrease of the nocturnal prolactin surge and the physiology of pseudopregnancy and pregnancy in the rat. *Life Sciences* 45: 207–215.
Morehead MH, Lookingland KJ, and Gala RR (1990) Stress-induced suppression of the prolactin afternoon surge in ovariectomized, estrogen-treated rats and the nocturnal surge in pseudopregnant rats are accompanied by an increase in median eminence dihydroxyphenylacetic acid concentrations. *Neuroendocrinology* 51: 208–212.
Morel G, Ouhtit A, and Kelly PA (1994) Prolactin receptor immunoreactivity in rat anterior pituitary. *Neuroendocrinology* 59: 78–84 Issn: 0028-3835.
Morgan WW and Besch KC (1990) Effect of prolactin replacement on the number of tyrosine hydroxylase expressing neurons in the arcuate nuclei of Ames dwarf and normal mice. *Neuroendocrinology* 52: 70–74.
Morrell JI, Rosenthal MF, McCabe JT, Harrington CA, Chikaraishi DM, and Pfaff DW (1989) Tyrosine hydroxylase mRNA in the neurons of the tuberoinfundibular region and zona incerta examined after gonadal steroid hormone treatment. *Molecular Endocrinology* 3: 1426–1433.
Muccioli G, Bellussi G, Ghe C, Pagnini G, and DiCarlo R (1988) Regional distribution and species variation of prolactin binding sites in the brain. *General and Comparative Endocrinology* 69: 399–405.
Muccioli G and DiCarlo R (1994) Modulation of prolactin receptors in the rat hypothalamus in response to changes in serum concentration of endogenous prolactin or to ovine prolactin administration. *Brain Research* 663: 244–250.
Muccioli G, Ghe C, and DiCarlo R (1991) Distribution and characterization of prolactin binding sites in the male and female rat brain: Effects of hypophysectomy and ovariectomy. *Neuroendocrinology* 53: 47–53.
Murai I and Ben-Jonathan N (1987a) Prolactin secretion in posterior pituitary lobectomized rats: Differential effects of 5-hydroxytryptophan and ether. *Brain Research* 420: 227–232.
Murai I and Ben-Jonathan N (1987b) Posterior pituitary lobectomy abolishes the suckling-induced rise in prolactin (PRL): Evidence for a PRL-releasing factor in the posterior pituitary. *Endocrinology* 121: 205–211.
Mustafa A, Nyberg F, Bogdanovic N, Islam A, Roos P, and Adem A (1994) Somatogenic and lactogenic binding sites in rat brain and liver: Quantitative autoradiographic localization. *Neuroscience Research* 20: 257–263.
Mustafa A, Nyberg F, Bogdanovic N, et al. (1995) Prolactin binding sites in rat brain and liver: Effects of long-term ovariectomy and ovarian steroids. *Neuroscience Letters* 200: 179–182.
Naef L and Woodside B (2007) Prolactin/leptin interactions in the control of food intake in rats. *Endocrinology* 148: 5977–5983.
Nagy GM, Gorcs TJ, and Halasz B (1991) Attenuation of the suckling-induced prolactin release and the high afternoon oscillations of plasma prolactin secretion of lactating rats by antiserum to vasopressin. *Neuroendocrinology* 54: 566–570.
Nance DM and Gorski RA (1978) Similar effects of estrogen and lateral hypothalamic lesions on feeding behavior of female rats. *Brain Research Bulletin* 3: 549–553.
Neill JD and Nagy GM (1994) Prolactin secretion and its control. In: Knobil E and Neill JD (eds.) *The Physiology of Reproduction,* 2nd edn., pp. 1833–1860. New York: Raven Press.
Nephew BC, Amico J, Cai HM, Walker AM, and Bridges RS (2007) Intracerebroventricular administration of the prolactin (PRL) receptor antagonist, S179D PRL, disrupts parturition in rats. *Reproduction* 134: 155–160.
Neumann ID (2003) Brain mechanisms underlying emotional alterations in the peripartum period in rats. *Depression and Anxiety* 17: 111–121.
Neumann ID, Johnstone HA, Hatzinger M, et al. (1998) Attenuated neuroendocrine responses to emotional and physical stressors in pregnant rats involve Adenohypophysial changes. *Journal of Physiology* 508: 289–300.
Nicholson G, Greeley GH, Jr., Humm J, Youngblood WW, and Kizer JS (1980) Prolactin in cerebrospinal fluid: A probable site of prolactin autoregulation. *Brain Research* 190: 447–457.
Nikitovitch-Winer M and Everett JW (1958) Functional restitution of pituitary grafts re-transplanted from kidney to median eminence. *Endocrinology* 63: 916–930.
Nilsson C, Lindvall-Axelsson M, and Owman C (1992) Neuroendocrine regulatory mechanisms in the choroid plexus-cerebrospinal fluid system. *Brain Research Reviews* 17: 109–138.
Nishihara M and Kimura F (1989) Postsynaptic effects of prolactin and estrogen on arcuate neurons in rat hypothalamic slices. *Neuroendocrinology* 49: 215–218.
Noel MB and Woodside B (1993) Effects of systemic and central prolactin injections on food intake, weight gain, and estrous cyclicity in female rats. *Physiology and Behavior* 54: 151–154.
Numan M (1994) A neural circuitry analysis of maternal behavior in the rat. *Acta Paediatrica* 397(supplement): 19–28.

Numan M (2006) Hypothalamic neural circuits regulating maternal responsiveness toward infants. *Behavioral and Cognitive Neuroscience Reviews* 5: 163–190.

Ogren L and Talamantes F (1988) Prolactins of pregnancy and their cellular source. *International Review of Cytology* 112: 1–65.

Oliver C, Mical RS, and Porter JC (1977) Hypothalamic–pituitary vasculature: Evidence for retrograde blood flow in the pituitary stalk. *Endocrinology* 101: 598–604.

Palm IF, van der Beek EM, Swarts HJ, et al. (2001) Control of the estradiol-induced prolactin surge by the suprachiasmatic nucleus. *Endocrinology* 142: 2296–2302.

Pan JT, Tian Y, Lookingland KJ, and Moore KE (1992) Neurotensin-induced activation of hypothalamic dopaminergic neurons is accompanied by a decrease in pituitary secretion of prolactin and alpha-melanocyte-stimulating hormone. *Life Sciences* 50: 2011–2017 Issn: 0024-3205.

Panerai AE, Sawynok J, LaBella FS, and Friesen HG (1980) Prolonged hyperprolactinemia influences beta-endorphin and Met-enkephalin in the brain. *Endocrinology* 106: 1804–1808.

Park SK and Selmanoff M (1991) Dose-dependent suppression of postcastration luteinizing hormone secretion exerted by exogenous prolactin administration in male rats: A model for studying hyperprolactinemic hypogonadism. *Neuroendocrinology* 53: 404–410.

Parker SL, Armstrong WE, Sladek CD, Grosvenor CE, and Crowley WR (1991) Prolactin stimulates the release of oxytocin in lactating rats: Evidence for a physiological role via an action at the neural lobe. *Neuroendocrinology* 53: 503–510.

Paut-Pagano L, Roky R, Valatx JL, Kitahama K, and Jouvet M (1993) Anatomical distribution of prolactin-like immunoreactivity in the rat brain. *Neuroendocrinology* 58: 682–695.

Paxinos G and Watson C (1997) *The Rat Brain in Stereotaxic Coordinates*. San Diego, CA: Academic Press.

Peake GT, Buckman MT, Davis LE, and Standefer J (1983) Pituitary and placentally derived hormones in cerebrospinal fluid during normal human pregnancy. *Journal of Clinical Endocrinology and Metabolism* 56: 46–52.

Perrot-Applanat M, Gualillo O, Pezet A, Vincent V, Edery M, and Kelly PA (1997) Dominant negative and cooperative effects of mutant forms of prolactin receptor (published erratum appears in *Molecular Endocrinology* 1997 Nov; 11(12): 1781). *Molecular Endocrinology* 11: 1020–1032.

Pezet A, Favre H, Kelly P, and Edery M (1999) Inhibition and restoration of prolactin signal transduction by suppressors of cytokine signalling. *Journal of Biological Chemistry* 274: 24497–24502.

Phelps CJ (2004) Postnatal regression of hypothalamic dopaminergic neurons in prolactin-deficient Snell dwarf mice. *Endocrinology* 145: 5656–5664.

Phelps CJ, Carlson SW, and Vaccarella MY (1994) Hypothalamic dopaminergic neurons in prolactin-deficient Ames dwarf mice: Localization and quantification of deficit by tyrosine hydroxylase immunocytochemistry. *Journal of Neuroendocrinology* 6: 145–152.

Phelps CJ and Horseman ND (2000) Prolactin gene disruption does not compromise differentiation of tuberoinfundibular dopaminergic neurons. *Neuroendocrinology* 72: 2–10.

Phelps CJ, Romero MI, and Hurley DL (2003) Prolactin replacement must be continuous and initiated prior to 21 d of age to maintain hypothalamic dopaminergic neurons in hypopituitary mice. *Endocrine* 20: 139–148.

Pi XJ and Grattan DR (1998a) Differential expression of the two forms of prolactin receptor mRNA within microdissected hypothalamic nuclei of the rat. *Brain Research – Molecular Brain Research* 59: 1–12.

Pi XJ and Grattan DR (1998b) Distribution of prolactin receptor immunoreactivity in the brain of estrogen-treated, ovariectomized rats. *Journal of Comparative Neurology* 394: 462–474.

Pi XJ and Grattan DR (1999a) Expression of prolactin receptor mRNA is increased in the preoptic area of lactating rats. *Endocrine* 11: 91–98.

Pi XJ and Grattan DR (1999b) Increased expression of both short and long forms of prolactin receptor mRNA in hypothalamic nuclei of lactating rats. *Journal of Molecular Endocrinology* 23: 13–22.

Pi XJ and Grattan DR (1999c) Increased prolactin receptor immunoreactivity in the hypothalamus of lactating rats. *Journal of Neuroendocrinology* 11: 693–705.

Pi XJ and Voogt JL (2000) Effect of suckling on prolactin receptor immunoreactivity in the hypothalamus of the rat. *Neuroendocrinology* 71: 308–317.

Pi XJ and Voogt JL (2001) Mechanisms for suckling-induced changes in expression of prolactin receptor in the hypothalamus of the lactating rat. *Brain Research* 891: 197–205.

Pi X and Voogt JL (2002) Sex difference and estrous cycle: Expression of prolactin receptor mRNA in rat brain. *Brain Research – Molecular Brain Research* 103: 130–139.

Pi X, Zhang B, Li J, and Voogt JL (2003) Promoter usage and estrogen regulation of prolactin receptor gene in the brain of the female rat. *Neuroendocrinology* 77: 187–197.

Piccoletti R, Maroni P, Bendinelli P, and Bernelli-Zazzera A (1994) Rapid stimulation of mitogen-activated protein kinase of rat liver by prolactin. *Biochemical Journal* 303: 429–433.

Poletini MO, Szawka RE, Franci CR, and Anselmo-Franci JA (2006) Ovarian steroids but not the locus coeruleus regulate stress-induced prolactin secretion in female rats. *Journal of Neuroendocrinology* 18: 938–948.

Polson DW, Sagle M, Mason HD, Adams J, Jacobs HS, and Franks S (1986) Ovulation and normal luteal function during LHRH treatment of women with hyperprolactinaemic amenorrhoea. *Clinical Endocrinology (Oxf)* 24: 531–537.

Popeski N, Amir S, Diorio J, and Woodside B (2003) Prolactin and oxytocin interaction in the paraventricular and supraoptic nuclei: Effects on oxytocin mRNA and nitric oxide synthase. *Journal of Neuroendocrinology* 15: 687–696.

Rance N, Wise PM, and Barraclough CA (1981) Negative feedback effects of progesterone correlated with changes in hypothalamic norepinephrine and dopamine turnover rates, median eminence luteinizing hormone-releasing hormone, and peripheral plasma gonadotropins. *Endocrinology* 108: 2194–2199.

Rao YP, Buckley DJ, and Buckley AR (1995) Rapid activation of mitogen-activated protein kinase and p21ras by prolactin and interleukin 2 in rat Nb2 node lymphoma cells. *Cell Growth Differentiation* 6: 1235–1244.

Riddle O, Bates RW, and Dykshorn SW (1931) The preparation, identification and assay of prolactin – a hormone of the anterior pituitary. *American Journal of Physiology* 105: 191–216.

Riskind PN, Millard WJ, and Martin JB (1984) Evidence that thyrotropin-releasing hormone is not a major prolactin-releasing factor during suckling in the rat. *Endocrinology* 115: 312–316.

Robertson MC and Friesen HG (1981) Two forms of rat placental lactogen revealed by radioimmunoassay. *Endocrinology* 108: 2388–2390.

Robertson MC, Gillespie B, and Friesen HG (1982) Characterization of the two forms of rat placental lactogen (rPL): rPL-I and rPL-II. *Endocrinology* 111: 1862–1866.

Roky R, Paut-Pagano L, Goffin V, Kitahama K, Valatx JL, Kelly PA, and Jouvet M (1996) Distribution of prolactin receptors in the rat forebrain. Immunohistochemical study. *Neuroendocrinology* 63: 422–429.
Romero MI and Phelps CJ (1993) Prolactin replacement during development prevents the dopaminergic deficit in hypothalamic arcuate nucleus in prolactin-deficient Ames dwarf mice. *Endocrinology* 133: 1860–1870.
Romero MI and Phelps CJ (1995) Prolactin replacement in adult dwarf mice does not reverse the deficit in tuberoinfundibular dopaminergic neuron number. *Endocrinology* 136: 3238–3246.
Rondeel JM, de Greef WJ, Visser TJ, and Voogt JL (1988) Effect of suckling on the *in vivo* release of thyrotropin-releasing hormone, dopamine and adrenaline in the lactating rat. *Neuroendocrinology* 48: 93–96.
Roselli CE, Bocklandt S, Stadelman HL, Wadsworth T, Vilain E, and Stormshak F (2008) Prolactin expression in the sheep brain. *Neuroendocrinology* 87: 206–215.
Russell JA, Douglas AJ, and Ingram CD (2001) Brain preparations for maternity – adaptive changes in behavioral and neuroendocrine systems during pregnancy and lactation. An overview. *Progress in Brain Research* 133: 1–38.
Sagrillo CA and Voogt JL (1991) Endogenous opioids mediate the nocturnal prolactin surge in the pregnant rat. *Endocrinology* 129: 925–930.
Sakal E, Elberg G, and Gertler A (1997) Direct evidence that lactogenic hormones induce homodimerization of membrane-anchored prolactin receptor in intact Nb2-11C rat lymphoma cells. *FEBS Letters* 410: 289–292.
Samson WK, Bianchi R, Mogg RJ, Rivier J, Vale W, and Melin P (1989) Oxytocin mediates the hypothalamic action of vasoactive intestinal peptide to stimulate prolactin secretion. *Endocrinology* 124: 812–819.
Samson WK, Lumpkin MD, and McCann SM (1986) Evidence for a physiological role for oxytocin in the control of prolactin secretion. *Endocrinology* 119: 554–560.
Samson WK, Taylor MM, and Baker JR (2003) Prolactin-releasing peptides. *Regulatory Peptides* 114: 1–5.
Santollo J, Wiley MD, and Eckel LA (2007) Acute activation of ER alpha decreases food intake, meal size, and body weight in ovariectomized rats. *American Journal of Physiology – Regulatory, Integrative and Comparative Physiology* 293: R2194–R2201.
Sarkar DK (1989) Evidence for prolactin feedback actions on hypothalamic oxytocin, vasoactive intestinal peptide and dopamine secretion. *Neuroendocrinology* 49: 520–524.
Sarkar DK, Frautschy SA, and Mitsugi N (1992) Pituitary portal plasma levels of oxytocin during the estrous cycle, lactation, and hyperprolactinemia. *Annals of the New York Academy of Sciences* 652: 397–410 Issn: 0077-8923.
Sarkar DK and Yen SS (1985) Hyperprolactinemia decreases the luteinizing hormone-releasing hormone concentration in pituitary portal plasma: A possible role for beta-endorphin as a mediator. *Endocrinology* 116: 2080–2084.
Sauvé D and Woodside B (1996) The effect of central administration of prolactin on food intake in virgin female rats is dose-dependent, occurs in the absence of ovarian hormones and the latency to onset varies with feeding regimen. *Brain Research* 729: 75–81.
Sauvé D and Woodside B (2000) Neuroanatomical specificity of prolactin-induced hyperphagia in virgin female rats. *Brain Research* 868: 306–314.
Sawamoto K, Nakao N, Kobayashi K, et al. (2001) Visualization, direct isolation, and transplantation of midbrain dopaminergic neurons. *Proceedings of the National Academy of Sciences of the United States of America* 98: 6423–6428.
Schulkin J (2006) Angst and the amygdala. *Dialogues in Clinical Neuroscience* 8: 407–416.
Schulkin J, Morgan MA, and Rosen JB (2005) A neuroendocrine mechanism for sustaining fear. *Trends in Neuroscience* 28: 629–635.
Schweizer FW, Kim MH, and Malarkey WB (1984) The diurnal variations of serum prolactin levels before and during pregnancy in normal and hyperprolactinemic patients. *American Journal of Obstetrics and Gynecology* 149: 367–371.
Selmanoff M (1985) Rapid effects of hyperprolactinemia on basal prolactin secretion and dopamine turnover in the medial and lateral median eminence. *Endocrinology* 116: 1943–1952.
Selmanoff M and Gregerson KA (1985) Suckling decreases dopamine turnover in both medial and lateral aspects of the median eminence in the rat. *Neuroscience Letters* 57: 25–30.
Selmanoff M and Wise PM (1981) Decreased dopamine turnover in the median eminence in response to suckling in the lactating rat. *Brain Research* 212: 101–115.
Shamgochian MD, Avakian C, Truong NH, Stone S, Tang KT, and DeVito WJ (1995) Regulation of prolactin receptor expression by estradiol in the female rat brain. *NeuroReport* 6: 2537–2541.
Shin SH (1979) Prolactin secretion in acute stress is controlled by prolactin releasing factor. *Life Sciences* 25: 1829–1835.
Shingo T, Gregg C, Enwere E, et al. (2003) Pregnancy-stimulated neurogenesis in the adult female forebrain mediated by prolactin. *Science* 299: 117–120.
Shivers BD, Harlan RE, and Pfaff DW (1989) A subset of neurons containing immunoreactive prolactin is a target for estrogen regulation of gene expression in rat hypothalamus. *Neuroendocrinology* 49: 23–27.
Sieck GC, Nance DM, and Gorski RA (1978) Estrogen modification of feeding behavior in the female rat: Influence of metabolic state. *Physiology and Behavior* 21: 893–897.
Slattery DA and Neumann ID (2008) No stress please! Mechanisms of stress hyporesponsiveness of the maternal brain. *Journal of Physiology* 586: 377–385.
Smith MS (1978) A comparison of pituitary responsiveness to luteinizing hormone-releasing hormone during lactation and the estrous cycle of the rat. *Endocrinology* 102: 114–120.
Smith MS (1982) Effect of pulsatile gonadotropin-releasing hormone on the release of luteinizing hormone and follicle-stimulating hormone *in vitro* by anterior pituitaries from lactating and cycling rats. *Endocrinology* 110: 882–891.
Smith MS, Freeman ME, and Neill JD (1975) The control of progesterone secretion during the estrous cycle and early pseudopregnancy in the rat: Prolactin, gonadotropin and steroid levels associated with rescue of the corpus luteum of pseudopregnancy. *Endocrinology* 96: 219–226.
Smith MS and Neill JD (1976) Termination at midpregnancy of the two daily surges of plasma prolactin initiated by mating in the rat. *Endocrinology* 98: 696–701.
Soares MJ, Muller H, Orwig KE, Peters TJ, and Dai G (1998) The uteroplacental prolactin family and pregnancy. *Biology of Reproduction* 58: 273–284.
Sobrinho LG (1993) The psychogenic effects of prolactin. *Acta Endocrinologica, Copenhagen* 129(supplement 1): 38–40.
Sobrinho LG (1998) Emotional aspects of hyperprolactinemia. *Psychotherapy and Psychosomatics* 67: 133–139.
Spencer SJ, Mouihate A, Galic MA, and Pittman QJ (2008) Central and peripheral neuroimmune responses: Hyporesponsiveness during pregnancy. *Journal of Physiology* 586: 399–406.

Starr R and Hilton DJ (1999) Negative regulation of the JAK/STAT pathway. *BioEssays* 21: 47–52.

Stern JM, Goldman L, and Levine S (1973) Pituitary–adrenal responsiveness during lactation in rats. *Neuroendocrinology* 12: 179–191.

Steyn FJ, Anderson GM, and Grattan DR (2007) Expression of ovarian steroid hormone receptors in tuberoinfundibular dopaminergic neurones during pregnancy and lactation. *Journal of Neuroendocrinology* 19: 788–793.

Steyn FJ, Anderson GM, and Grattan DR (2008) Hormonal regulation of suppressors of cytokine signaling (SOCS) messenger ribonucleic acid in the arcuate nucleus during late pregnancy. *Endocrinology* 149: 3206–3214.

Subramanian MG and Gala RR (1976) The influence of cholinergic, adrenergic, and serotonergic drugs on the afternoon surge of plasma prolactin in ovariectomized, estrogen-treated rats. *Endocrinology* 98: 842–848.

Subramanian MG, Kowalczyk CL, Leach RE, et al. (1997) Midcycle increase of prolactin seen in normal women is absent in subjects with unexplained infertility. *Fertility and Sterility* 67: 644–647.

Sugiyama T, Minoura H, Kawabe N, Tanaka M, and Nakashima K (1994) Preferential expression of long form prolactin receptor mRNA in the rat brain during the oestrous cycle, pregnancy and lactation: Hormones involved in its gene expression. *Journal of Endocrinology* 141: 325–333.

Sugiyama T, Minoura H, Toyoda N, Sakaguchi K, Tanaka M, Sudo S, and Nakashima K (1996) Pup contact induces the expression of long form prolactin receptor mRNA in the brain of female rats: Effects of ovariectomy and hypophysectomy on receptor gene expression. *Journal of Endocrinology* 149: 335–340.

Suzuki S and Handa RJ (2005) Estrogen receptor-beta, but not estrogen receptor-alpha, is expressed in prolactin neurons of the female rat paraventricular and supraoptic nuclei: Comparison with other neuropeptides. *Journal of Comparative Neurology* 484: 28–42.

Swaminathan G, Varghese B, Thangavel C, et al. (2008) Prolactin stimulates ubiquitination, initial internalization, and degradation of its receptor via catalytic activation of Janus kinase 2. *Journal of Endocrinology* 196: R1–R7.

Talamantes F, Ogren L, Markoff E, Woodard S, and Madrid J (1980) Phylogenetic distribution, regulation of secretion, and prolactin-like effects of placental lactogens. *Federation Proceedings* 39: 2582–2587.

Terkel J, Blake, and Sawyer CH (1972) Serum prolactin levels in lactating rats after suckling or exposure to ether. *Endocrinology* 91: 49–53.

Thomas GB, Cummins JT, Griffin N, and Clarke IJ (1988a) Effect and site of action of hypothalamic neuropeptides on prolactin release in sheep. *Neuroendocrinology* 48: 252–257.

Thomas GB, Cummins JT, Yao B, Gordon K, and Clarke IJ (1988b) Release of prolactin is independent of the secretion of thyrotrophin-releasing hormone into hypophysial portal blood of sheep. *Journal of Endocrinology* 117: 115–122.

Tilbrook AJ and Clarke IJ (2006) Neuroendocrine mechanisms of innate states of attenuated responsiveness of the hypothalamo-pituitary–adrenal axis to stress. *Frontier in Neuroendocrinology* 27: 285–307.

Tojo K, Kato Y, Kabayama Y, Ohta H, Inoue T, and Imura H (1986) Further evidence that central neurotensin inhibits pituitary prolactin secretion by stimulating dopamine release from the hypothalamus. *Proceedings of the Society for Experimental Biology and Medicine* 181: 517–522.

Tong Y and Pelletier G (1992) Prolactin regulation of pro-opiomelanocortin gene expression in the arcuate nucleus of the rat hypothalamus. *Neuroendocrinology* 56: 561–565.

Tonkowicz PA and Voogt JL (1983) Termination of prolactin surges with development of placental lactogen secretion in the pregnant rat. *Endocrinology* 113: 1314–1318.

Torner L, Maloumby R, Nava G, Aranda J, Clapp C, and Neumann ID (2004) *In vivo* release and gene upregulation of brain prolactin in response to physiological stimuli. *European Journal of Neuroscience* 19: 1601–1608.

Torner L, Nava G, Duenas Z, et al. (1999) Changes in the expression of neurohypophyseal prolactins during the estrous cycle and after estrogen treatment. *Journal of Endocrinology* 161: 423–432.

Torner L and Neumann ID (2002) The brain prolactin system: Involvement in stress response adaptations in lactation. *Stress* 5: 249–257.

Torner L, Toschi N, Nava G, Clapp C, and Neumann ID (2002) Increased hypothalamic expression of prolactin in lactation: Involvement in behavioural and neuroendocrine stress responses. *European Journal of Neuroscience* 15: 1381–1389.

Torner L, Toschi N, Pohlinger A, Landgraf R, and Neumann ID (2001) Anxiolytic and anti-stress effects of brain prolactin: Improved efficacy of antisense targeting of the prolactin receptor by molecular modeling. *Journal of Neuroscience* 21: 3207–3214.

Tortonese DJ, Brooks J, Ingleton PM, and McNeilly AS (1998) Detection of prolactin receptor gene expression in the sheep pituitary gland and visualization of the specific translation of the signal in gonadotrophs. *Endocrinology* 139: 5215–5223.

Toth BE, Homicsko K, Radnai B, et al. (2001) Salsolinol is a putative endogenous neuro-intermediate lobe prolactin-releasing factor. *Journal of Neuroendocrinology* 13: 1042–1050.

Townsend J, Cave BJ, Norman MR, Flynn A, Uney JB, Tortonese DJ, and Wakerley JB (2005) Effects of prolactin on hypothalamic supraoptic neurones: Evidence for modulation of STAT5 expression and electrical activity. *Neuro Endocrinology Letters* 26: 125–130.

Tyson JE, Hwang P, Guyda H, and Friesen HG (1972) Studies of prolactin secretion in human pregnancy. *American Journal of Obstetrics and Gynecology* 113: 14–20.

Vanhaasteren GAC, Vantoor H, Klootwijk W, et al. (1996) Studies on the role of TRH and corticosterone in the regulation of prolactin and thyrotrophin secretion during lactation. *Journal of Endocrinology* 148: 325–336.

Vekemans M, Delvoye P, L'Hermite M, and Robyn C (1977) Serum prolactin levels during the menstrual cycle. *Journal of Clinical Endocrinology and Metabolism* 44: 989–993.

Voci VE and Carlson NR (1973) Enhancement of maternal behavior and nest building following systemic and diencephalic administration of prolactin and progesterone in the mouse. *Journal of Comparative and Physiological Psychology* 83: 388–393.

Voogt JL (1984) Evidence for an inhibitory influence of rat placental lactogen on prolactin release *in vitro*. *Biology of Reproduction* 31: 141–147.

Voogt J, Robertson M, and Friesen H (1982) Inverse relationship of prolactin and rat placental lactogen during pregnancy. *Biology of Reproduction* 26: 800–805.

Voogt JL, Soares MJ, Robertson MC, and Arbogast LA (1996) Rat placental lactogen-I abolishes nocturnal prolactin surges in the pregnant rat. *Endocrine* 4: 233–238.

Wagner EJ, Manzanares J, Moore KE, and Lookingland KJ (1994) Neurochemical evidence that estrogen-induced suppression of kappa-opioid-receptor-mediated regulation of tuberoinfundibular dopaminergic neurons is prolactin-independent. *Neuroendocrinology* 59: 197–201.

Walsh RJ, Posner BI, Kopriwa BM, and Brawer JR (1978) Prolactin binding sites in the rat brain. *Science* 201: 1041–1043.

Walsh RJ, Slaby FJ, and Posner BI (1987) A receptor-mediated mechanism for the transport of prolactin from blood to cerebrospinal fluid. *Endocrinology* 120: 1846–1850.

Wang HJ, Hoffman GE, and Smith MS (1993) Suppressed tyrosine hydroxylase gene expression in the tuberoinfundibular dopaminergic system during lactation. *Endocrinology* 133: 1657–1663.

Watanobe H, Schioth HB, Wikberg JES, and Suda T (2000) Evaluation of the role for prolactin-releasing peptide in prolactin secretion induced by ether stress and suckling in the rat: Comparison with vasoactive intestinal peptide. *Brain Research* 865: 91–96.

Weber RF, de Greef WJ, de Koning J, and Vreeburg JT (1983) LH-RH and dopamine levels in hypophysial stalk plasma and their relationship to plasma gonadotrophins and prolactin levels in male rats bearing a prolactin- and adrenocorticotrophin-secreting pituitary tumor. *Neuroendocrinology* 36: 205–210.

Werawatgoompa S, Tankeyoon M, Kongseripong R, Chaniyom K, Virutamasen P, Dusitsin N, and Boonsiri B (1981) Hormonal changes in normal menstrual cycle of Thai women. *Contraception* 23: 301–313.

White MM and Samson WK (2006) Closing the link in short-loop prolactin feedback. *Trends in Endocrinology and Metabolism* 17: 217–218.

Witcher JA and Freeman ME (1985) The proestrous surge of prolactin enhances sexual receptivity in the rat. *Biology of Reproduction* 32: 834–839.

Woodside B (2007) Prolactin and the hyperphagia of lactation. *Physiology and Behavior* 91: 375–382.

Yoshimura A (1998) The CIS family: Negative regulators of JAK-STAT signaling. *Cytokine and Growth Factor Reviews* 9: 197–204.

Yu-Lee LY (2002) Prolactin modulation of immune and inflammatory responses. *Recent Progress in Hormone Research* 57: 435–455.

Zarrow MX, Farooq A, Denenberg VH, Sawin PB, and Ross S (1963) Maternal behaviour in the rabbit: Endocrine control of maternal nest building. *Journal of Reproduction and Fertility* 6: 375–383.

Zarrow MX, Sawin PB, Ross S, Denenberg VH, Crary D, Wilson ED, and Farooq A (1961) Maternal behaviour in the rabbit: Evidence for an endocrine basis of maternalnest building and additional data on maternal-nest building in the Dutch-belted race. *Journal of Reproduction and Fertility* 2: 152–162.

Zhao C, Deng W, and Gage FH (2008) Mechanisms and functional implications of adult neurogenesis. *Cell* 132: 645–660.

14 Growth Hormone and Insulin-Like Growth Factor-I: Effects on the Brain

Z Laron, Tel Aviv University, Tel Aviv, Israel

Chapter Outline

14.1 Introduction

Since the first edition of this chapter (Laron, 2002a), new experimental and clinical data on the influence of growth hormone (GH) and insulin-like growth factor I (IGF-I) have been reported. Following is an update, including for the year 2007.

The extraction and purification of pituitary GH (Li and Papkoff, 1956) followed by its recombinant biosynthesis in 1985 (Kaplan et al., 1986) as well as the biosynthesis of its effector hormone IGF-I (Niwa et al., 1986) enabled the initiation of studies of the role of GH and IGF-I on the development and functions of the nervous tissue.

The GH axis is an interactive mechanism with many players (**Figure 1**), including hormones, their receptors, and specific binding (carrier) proteins. Investigations in recent years have shown that pituitary GH secretion is regulated not only by GH-releasing hormone (GHRH) and somatostatin (SMS = GH-inhibiting hormone) but also by neurotransmitters and secretagogs (Merimee and Laron, 1996).

GH and its protein anabolic effector hormone IGF-I exert many metabolic actions affecting tissue, organ, and body growth, including the central and peripheral nervous tissue (Fernandez et al., 2007). Forthwith is presented a chapter on the presently known effects of these two hormones on the brain and its functions, both in man and in animal models.

14.2 The GHRH–GH–IGF-I Axis

The secretion of pituitary GH is regulated by GHRH, a GH-stimulating hormone, and SMS, a GH-inhibiting hormone. The role of Ghrelin, an additional GH secretagog, is as yet not completely elucidated (Kojima et al., 1999). All three are synthesized and secreted in the hypothalamus, being true neurohormones, and also by cells in the gastrointestinal tract where they act as paracrine hormones.

14.2.1 Growth Hormone-Releasing Hormone

The GHRH gene is a member of a large family of hormones and factors, which includes glucagon, secretin, vasoactive intestinal polypeptide (VIP), and others (Hofman and Pescovitz, 1999). Humans have a single copy of the GHRH gene (Mayo et al., 1985). It is localized on chromosome 20q12 and p11.23 (Perez Jurado et al., 1994) and has five exons.

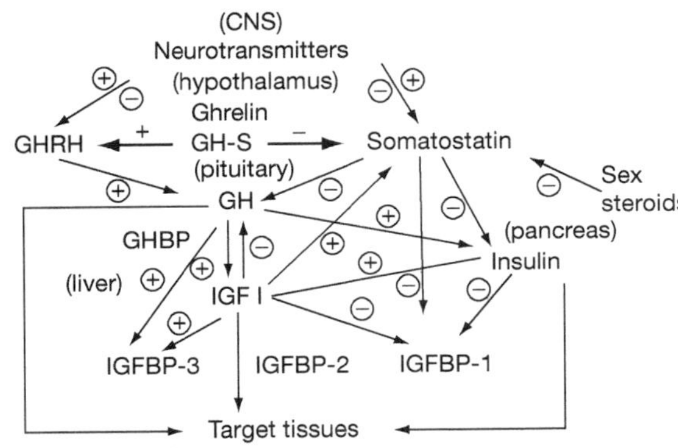

Figure 1 The GHRH–GH–IGF-I cascade. GH, growth hormone; GHRH, growth hormone-releasing hormone; GHBP, GH-binding protein; IGF-I, insulin-like growth factor I; IGFBP, IGF-binding protein.

GHRH is localized in the arcuate and ventromedial nuclei of the hypothalamus, and also in the gastrointestinal tract, including the pancreas (Bosman et al., 1984), testes, and placenta (Berry et al., 1992), and in other tissues as well as in tumors (Rivier et al., 1982; Asa et al., 1985). GHRH is secreted in three molecular forms GHRH 1–44NH2, GHRH 1–40-OH, and GHRH 1–37-OH (Guillemin et al., 1984). All three forms are biologically active, even as a 1–29 fragment (Laron, 1991).

14.2.2 Human GHRH Receptor

The GHRH-receptor (GHRH-R) gene is homologous to a subfamily of G-protein-coupled receptors, which include VIP, glucagon-like peptide-1 (GLP-1), secretin, glucagon, gastric inhibitory polypeptide (GIP), pituitary adenylate cyclase-activating polypeptide (PACAP), calcitonin, parathyroid hormone (PTH), and corticotropin-releasing hormone (CRH) (Mayo, 1992). The human GHRH-R gene is located on chromosome 7p13–p21 (Vamvakopoulos et al., 1994) and probably at p15 as well (Wajnrajch et al., 1994).

14.2.3 GH Secretagogs

During studies of the opioidal control of GH secretion several analogs of met-enkephalin were found to be potent GH secretagogs. Among them were GH-releasing peptide-6 (GHRP-6), and hexarelin (His-D2MeTRP-Ala-Trp-DPhe-Lys-NH2) (Laron, 1995). They act via a receptor unrelated to that of GHRH (Howard et al., 1996). The potent biologic action of the GHRPs and the identification of a specific receptor suggested the existence of a natural ligand.

14.2.4 Ghrelin

The natural ligand was identified as Ghrelin, an acylated peptide of 28 amino acids. Though isolated from the stomach, it is present in the hypothalamus (Kojima et al., 1999). It probably acts in synergism with GHRH to stimulate GH secretion (Kineman and Luque, 2007). In addition, Ghrelin releases appetite-stimulatory signals from the stomach (Asakawa et al., 2001). The orexigenic effect of Ghrelin does not seem to be GH dependent, but acts by increasing hypothalamic neuropeptide Y (NPY) (Shintani et al., 2001).

14.2.5 Somatostatin

SMS or SST is secreted in two forms, a peptide containing 14 amino acids, that is, the major form in the brain, and SMS with 28 amino acids, the major form in the gastrointestinal tract. In addition to inhibiting GH, SMS also inhibits thyroid-stimulating hormone (TSH), insulin, and glucagon secretion (Reichlin, 1983). Its neuroregulation is not clear; IGF-I was found to stimulate SMS (Gil-Ad et al., 1996) which seems part of the negative feedback mechanism of GH secretion (**Figure 1**).

14.2.6 Somatostatin Receptors

Five SMS receptors (SSTR1–5) are known (Shimon et al., 1997): SSTR1 and SSTR2 are the most abundant in the brain. SSTR4 is expressed in the hippocampus and, of all receptors expressed in the pituitary, SSTR2 and SSTR5 are the most abundant on somatotrophs.

14.2.7 Cortistatin

Cortistatin (CST-14) is a recently described neuropeptide which differs from SMS-14 by three-amino-acid residues (Rubinfeld and Shimon, 2006). Some of its actions not only overlap with those of SMS, such as GH suppression (Deghenghi et al., 2001), but it also has sleep- and locomotor-modulating activity.

14.2.8 Human GH

Human GH (hGH) consists of a cluster of five similar genes in the following order: 5′ hGH-1, (or –N), chorionic somatomammotropin pseudo gene (CSHP), chorionic somatomammotropin-1 (CSH-1); hGH-2 (or V) or placental GH which differs from the primary sequence of hGH-N by 13 amino acids (Frankenne et al., 1988) and replaces pituitary GH in the maternal circulation during the second-half of pregnancy. The hGH genes are located on the long arm of chromosome 17q22–24 (George et al., 1981). CSH-1 and CSH-2 encode chorionic somatomammotropin. Only hGH-1 has anabolic growth-promoting actions.

hGH is produced as a single-chain, 191-amino-acid 22-kDa protein (Lewis et al., 1980b). Alternate splicing of the second codon resulting in deletion of amino acids 31–46 yields a 20-kDa form (about 5–10% of pituitary hGH) (Lewis et al., 1980a). A small amount of 17-kDa hGH is formed as well (Cooke et al., 1988). The spectrum of biological activity of the 20-kDa hGH form is very similar to that of the 22-kDa hGH (Culler et al., 1988), although a less insulinotropic effect has occasionally been found.

14.2.9 GH Receptor

The GH receptor (GH-R) belongs to the family of cytokine receptors, which includes receptors for prolactin, erythropoietin, and interleukins (Kelly et al., 1994). The human GH-R gene cloned in 1987 by Leung et al. (1987) is a protein of 620 amino acids and consists of ten exons (Godowski et al., 1989). Exons 2–7 encode the extracellular domain, exon 8 the single membrane-spanning domain, and exons 9 and 10 the intracellular (cytoplasmatic) domain (Godowski et al., 1989). The human GH-R gene is located on the short arm of chromosome 5p–3.1 (Barton et al., 1989).

14.2.10 GH-Binding Protein

Thirty to fifty percent of the circulating GH is bound to GH-binding protein (GHBP) (Herington et al., 1986; Baumann et al., 1986). This GHBP was shown to be identical in structure with the extracellular hormone-binding domain of the GH-R (Leung et al., 1987). Quantitative measurements revealed that its serum concentrations change with age, being low in neonates and reaching maximal values in young adulthood (Silbergeld et al., 1989).

14.2.11 Insulin-Like Growth Factor I

The IGFs (IGF-I and IGF-II) are members of a family of insulin-related peptides, which include relaxin and several peptides isolated from lower invertebrates (Blundell and Humbel, 1980). IGF-I is a small peptide consisting of 70 amino acids with a molecular weight of 7649 Da (Rinderknecht and Humbel, 1978). The structural similarity to insulin explains the ability of IGF-I to bind (with low affinity) to the insulin receptor.

The IGF-I gene is encoded on the long arm of chromosome 12q22–23 (Mullis et al., 1991) and consists of six exons, including two leader exons (Rotwein, 1991).

14.2.12 IGF-Binding Proteins

The IGFs circulate in plasma, 99% complexed to a family of binding proteins (BPs) which modulate the availability of free IGF-I to various tissues. There are six BPs (Hwa et al., 1999). In humans almost 80% of circulating IGF-I is carried by insulin-like growth factor-binding protein-3 (IGFBP-3), a ternary complex consisting of one molecule of IGF-I, one molecule of IGFBP-3 and a molecule of an 88-kDa protein named acid-labile subunit (ALS) (Lewitt et al., 1994). IGFBP-1 is regulated by both insulin and IGF-I (Laron et al., 1992c), whereas IGFBP-3 is mainly regulated by GH and, to a lesser degree, by IGF-I (Kanety et al., 1993). In states of GH deficiency (GHD) and IGF-I deficiency, serum IGFBP-3 is low (Laron et al., 1992b) but IGFBP-1 is raised (Laron et al., 1992c) due to a negative insulin and IGF-I feedback mechanism.

14.2.13 IGF-I Receptor

The human IGF-I receptor (IGF-I-R; type 1 receptor) is the product of a single-copy gene spanning the end of the long arm of chromosome 15q25–26 (Abbott et al., 1992). The gene contains 21 exons and its organization resembles that of the structurally related insulin receptor (Seino et al., 1989). The type-1 IGF-R gene is expressed by virtually every tissue and cell type even during embryogenesis (Bondy et al., 1990) and on cancerous cells.

Like the insulin receptor, the IGF-I-R is a heterotetramer composed of two extracellular-spanning α-subunits and two transmembrane β-subunits. The α-subunits have the binding sites for IGF-I and are linked by disulfide bonds. The β-subunit has a short extracellular domain, a transmembrane, and an intracellular domain. The intracellular part contains a tyrosine kinase domain which constitutes the signal transduction mechanism. Similar to the insulin receptor, the IGF-I receptor undergoes ligand-induced autophoshorylation (Kato et al., 1994). The activated IGF-I receptor is capable of phosphorylating other tyrosine-containing substrates, such as insulin receptor substrate 1 (IRS-1), and continues a cascade of enzyme activations (LeRoith et al., 1995).

14.3 GH Crosses the Blood–Brain Barrier

In patients treated by hGH, it was found that following the hormone administration, the hGH levels in the cerebrospinal fluid (CSF) increase (Burman et al., 1996; Coculescu et al., 1998; Johansson et al., 1995; **Figure 2**). Elevated levels of hGH have been

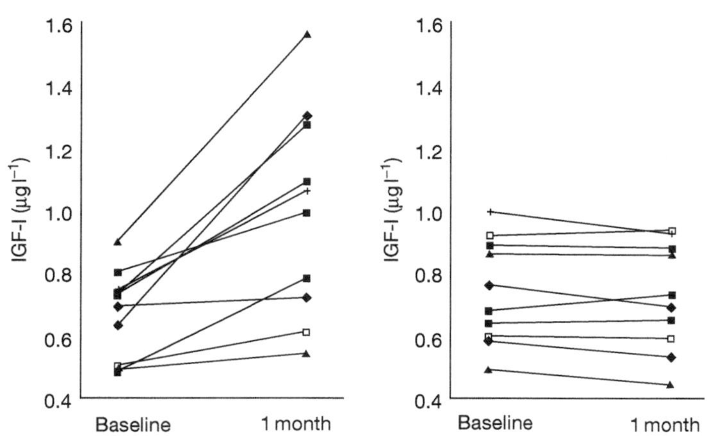

Figure 2 Effects of hGH (left) and placebo (right) on the hGH concentration in CSF of ten GH-deficient adult patients. Reproduced from Johansson JO, Larson N, Andersson M, et al. (1995) Treatment of growth hormone (GH)-deficient adults with recombinant human GH increases the concentration of GH in the cerebrospinal fluid and affects the neurotransmitters. *Neuroendocrinology* 61: 57–66.

reported in patients with acromegaly (Linfoot et al., 1970). The concentration of GH in the CSF is low (~5%) compared to that in the serum. The above observations prove that, in man, hGH crosses the blood–brain barrier (BBB). It is probable that the transport mechanism involves a gradient which is set by a receptor-mediated transport mechanism (transcytosis) (Partridge, 1986) involving the high density of GH receptors in the choroid plexus (Harvey et al., 1993; Lai et al., 1991). High doses of hGH may cause intracranial hypertension (pseudotumor cerebri) by overstimulation of the CSF-producing choroid plexus (Underwood, 1997; Coculescu, 1999). It remains to be established whether the effects evoked by exogenous administration of GH passing the BBB are direct effects of GH on the nervous tissue or effects resulting from the activation of paracrine IGF-I present in the central nervous system (CNS).

14.4 IGF-I Crosses the BBB

In vitro and *in vivo* experimental data demonstrated that IGF-I also crosses the BBB. The injection of ^{125}I-labeled IGF-I infused into the carotid artery of adult rats resulted in the detection of the ligand in the choroid plexus, median eminence, forebrain, and blood vessels (Reinhardt and Bondy, 1994; Pan and Kastin, 2000). IGF receptors are also present at the BBB (Rosenfeld et al., 1987). Also human ^{125}I-IGF-I injected subcutaneously (SC) in rats was found to pass into the CSF (Armstrong et al., 2000). As both insulin and IGF-I are present in the blood, they might cause reciprocal inhibition in the transport across the BBB (Yu et al., 2006) and in their actions in the brain.

Overexpression of IGF-I in transgenic mice results in enlarged brains (Ye et al., 1995) as well as in increased presence of IGF-I during postnatal brain development (Zhang and D'Ecole, 2004).

Recently, it was found that the concentrations of IGF-I, IGF-II, IGFBP-1, and IGFBP-3 in the CSF, from infants less than 6 months of age, were significantly higher than in older children and adults (Bunn et al., 2005). This may indicate a greater permeability of the BBB in the first months of life critical for neurodevelopment. Data from rats showed that uptake of circulating IGF-I into CSF is independent of receptor saturation or sequestration by its binding protein (Pulford and Ishii, 2001).

14.5 Expression of GH in the Central Nervous Tissue

The role of GH on nervous tissue, in addition to the known effects on the hypothalamus (Clark et al., 1988), has been proven by the demonstration of GH-R in various locations of the rat, rabbit, and human brain (Lai et al., 1991; Mustafa et al., 1994; Posner et al., 1994). Thus, Zhai et al. (1994) using ^{125}I-labeled GH in male Sprague-Dawley rats found high-density binding in the choroid plexus, hypothalamus, hippocampus, pituitary, and spinal cord, whereas a weaker binding was observed in the cortex (**Figure 3**). Identical findings were reported for the human brain (Nyberg, 2000). Binding was also found in the brainstem and in retinal ganglion cells (Lobie et al., 1993) as well as in glial cells and astrocytes, along with ependyma of the choroid plexus and pia mater. ^{125}I-labeled hGH showed very high binding on the choroid plexus of the rat (Lai et al., 1991; Zhai et al., 1994) and humans (Lai et al., 1993). Contrary to its ontogeny in the liver, GH-R mRNA in the CNS decreased with postnatal age in the rat and man (Lai et al., 1993), being high in the neonatal period similar to IGF-I-R expression (Pomerance et al., 1988). Maximal binding of GH in the rat brain was found between 12 and 18 weeks of age, followed by a

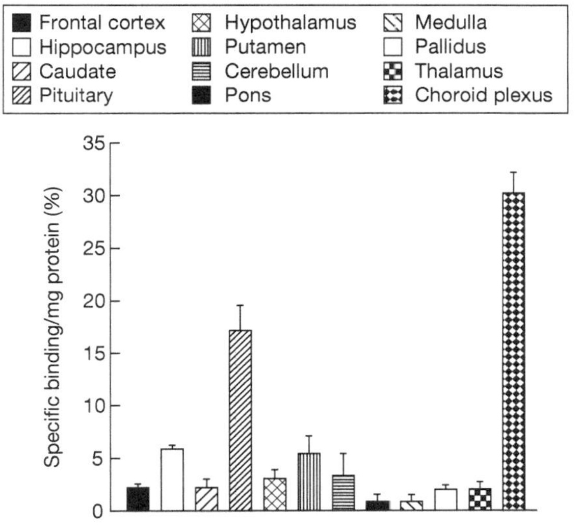

Figure 3 Specific binding of ^{125}I-hGH to membranes from different parts of the brain; values as m ± SD. Reproduced from Lai Z, Emtner M, Roos P, and Nyberg F (1991) Characterization of putative growth hormone receptors in the human choroid plexus. *Brain Research* 546: 222–226, with permission from Elsevier.

decline; at 53 weeks most of the GH binding had disappeared (Zhai et al., 1994). There is also transient GH immunoreactivity in the late fetal–early neonatal stage of the rat brain (Hojvat et al., 1982). The density of binding in human and rat brain is altered with aging (Nyberg, 1997).

The findings suggest a role for GH in neuronal maturation and glial cell formation (Ilkbahar et al., 1995). It may also be that loss of GH-R in the brain of aged people is connected with certain pathophysiological consequences. The apparent sex differences in GH binding, greater binding in females both in rats and humans (Mustafa et al., 1994), are probably due to the fact that hGH also binds to lactogenic receptors (Lai et al., 1991), as demonstrated by blocking the lactogenic sited by prolactin and reducing the binding (Nyberg and Burman, 1996).

The high concentration of GH-R in the choriod plexus has been associated with a role of this tissue in the transport of GH across the BBB (Coculescu, 1999; Nyberg and Burman, 1996). mRNA for the GH-R has been identified in the rat brain and pituitary (Lobie et al., 1993) as well as in glial, neuronal, and endothelial cells, as has mRNA for GHBP (Lobie et al., 1993; Burton et al., 1992).

Recently, GH-R expression has been described in a human glioblastoma cell line (Castro et al., 2000). The presence of the GH-R is probably involved in modulating paracrine IGF-I present in many nervous tissues as there is no evidence for a direct GH action on these tissues or cells. It has been suggested that GH may also be a substitute for metabolic enzymes generating neuroactive peptides, such as fragments of GH interacting with opiod receptors (Nyberg et al., 1989).

Lack of estrogens, as well as ovariectomy, decrease the somatogenic binding sites in the choroid plexus, but not in other brain areas (Nyberg, 2000). The observation that estradiol replacement restored the GH-R levels in the liver and not in the choroid plexus suggested that the GH-binding sites are differently regulated in various tissues (Mustafa et al., 1994).

14.6 Expression of IGF-I and Its Receptor Gene in the Nervous Tissue

IGF-I appears early in life and during the development of the rat CNS, it is maximally expressed at embryonic day 14 with a 3–4 factor decline until birth (Rotwein et al., 1988). In adult rats, IGF mRNA is 8–10 times more abundant in the cervicothoracic spinal cord and olfactory bulb than in the whole brain (Rotwein et al., 1988). Demonstration of local IGF-I synthesis using hybridization techniques in the neonatal and adult mouse and rat brain (hippocampus, olfactory bulb, and cerebellum) pointed to a paracrine system of IGF-I in the CNS (Bartlett et al., 1991; Werther et al., 1990; **Table 1**). IGF-I mRNA was also found in the choroid plexus (Marks et al., 1992). In spite of apparently constant levels of IGF-I mRNA between ages 2 and 30 months in the rat, there is a 36% decrease in the cortical IGF-I protein between 11 and 32 months of age, suggesting that IGF-I function is decreased with advancing age (Niblock et al., 1998). IGF-I and IGF-II gene expression was also described in the avian brain (Holzenberger and Lapointe, 2000). IGF-I was found in the cyst fluid of patients with craniopharyngioma but in lower concentrations than that of IGF-II (Zumkeller et al., 1991). The important effects of IGF-I in different periods of the CNS development have been reviewed by Anlar et al. (1999).

IGF-I distribution in the brain is reduced in Snell dwarf mice (Noguchi et al., 1987). (The Snell mouse is the Pit-1 transcription defect model lacking GH, prolactin, and TSH.) In contradistinction, IGF-I transgenic mice have an increased number of neurons and more abundant IGFBP-3 and -4 (Chrysis et al., 2001; Ye et al., 2003).

Of note is the fact that IGF-II is abundant in the CNS and pituitary, but its physiological function, in addition to an important role in the fetal brain, is yet to be elucidated (Bach and Bondy, 1992; Yokoyama et al., 1997).

14.7 IGFBPs in the Brain

Like the IGFs, IGFBPs are produced by many tissues. Their role is to transport (therefore the term carrier proteins) IGF to various tissues and to modulate the availability of free, biologically active IGF-I (Clemmons, 1998; Zapf et al., 1996). The mRNA expression profiles of BP-2, -4, -5, and -6 in the normal developing and adult CNS have been characterized and appear to have distinctive, nonoverlapping distributions (Ocrant, 1993). The CSF contains a mixture of BPs, probably transferred through the BBB. The choroid plexus was found to synthesize only IGFBP-2 (Ocrant et al., 1990). Astrocytes, C6 glial cells, and fetal neurons are also capable of synthesizing IGFBPs (Ocrant et al., 1989). The most frequently expressed BPs in the brain (cortex,

Table 1 IGF-I mRNA distribution in the mouse CNS

	Grading of distribution
Forebrain	
Cortical regions	
Cerebral cortex	+
Piriform cortex	+++
White matter	
Corpus callosum	−
External capsule	−
Lateral olfactory tract	−
Olfactory regions	
Interngranular cell layer	−
External plexiform layer	−
Mitral cell layer	+++
Glomerular layer	+++
Olfactory nerve layer	−
Hippocampal formation	
Stratum oriens	++
Stratum radiatum	++
Stratum lac.-mol.	++
Stratum pyramidial	−
Alveus and fimbra	−
Dentate gyrus	
Stratum polymorph	++
Basal ganglia	
Caudate putamen	+
Thalamic nuclei	
Ventral lateral	++
Ventral posterolateral	++
Lateral dorsal	−
Cerebellum	
Cortex	
Purkinje cell layer	+++
External granular cell layer	−
Internal granular cell layer	−
Molecular layer	−
White matter	++
Deep nuclei	++
Brainstem	
Inferior olive	+++
Vestibular nuclei	++
Cochlear nucleus	++
Spinal trigeminal nucleus	−

Note: Relative distribution of IGF-I mRNA in the neonatal mouse brain. The strength of hybridization signal is graded based on the number of grains per positive cell observed in dark- and bright-field micographs. +++, strongest signal; ++, moderate signal; +, weak signal; −, very weak hybridization signal.
Reproduced from Bartlett WP, Li X-S, Williams M, and Benkovic S (1991) Localization of insulin-like growth factor-1 mRNA in murine central nervous system during postnatal development. *Developmental Biology* 147: 239–250, with permission from Elsevier.

hippocampus, cerebellum, brainstem, leptomeninges, and myelin) are BP-2, and (Bondy and Lee, 1993; Cheng et al., 1996; Lee et al., 1993). IGFBP-2, -4, and -5 are found in the olfactory nerves (Bondy and Lee, 1993; Sullivan and Feldman, 1994) and retina (Lee et al., 1993). IGFBP-5 seems to play an important role in the brain (Ye and d'Ercole, 1998) and peripheral nerve cells (Cheng and Feldman, 1997). The mRNA expression of IGFBP-6 in rat brain, spinal cord, and sensory ganglia has been reported (Naeve et al., 2000). It is of note that little evidence of IGFBP-3, the major BP in peripheral circulation, has been forthcoming from the nervous tissue. From available studies it is also not clear whether the expression and localization of IGFBPs found in experimental animals can be related to the human nervous tissue. It has also been suggested that the IGFBPs may be involved in the so far unidentified action of IGF-II in the brain (Walter et al., 1999).

14.8 IGF as a Neurotropic and Antiapoptotic Factor

IGF-I has neurotropic activity, that is, it promotes the survival and differentiation of neuronal cells, including sensory ones (Carro et al., 2003; Feldman et al., 1997; Oorschot and McLennan, 1998), sympathetic neurons (Recio-Pinto et al., 1986; Sendtner, 1995), and motoneurons (Oorschot and McLennan, 1998; Rabinovsky et al., 2003). The ability to promote survival of neurons is partially due to the antiapoptotic effect of IGF-I (Tagami et al., 1997a; Mason et al., 2000; Zheng et al., 2002; Linseman et al., 2002). A strong neuronal effect of IGF-I has been found with sympathetic neurons exposed to high glucose (Russell and Feldman, 1999) or neuroblastoma cells exposed to hyperosmosis (Van Golen and Feldman, 2000).

Adolescent mice have high levels of neurogenesis compared to adults suggesting a dramatic loss of neurogenesis during the transition from adolescence to adult age (He and Crews, 2007). Transgenic mice overexpressing IGF-I with a cerebellar overgrowth have an increase in granule neuron number (Zhang and D'Ercole, 2004). On the other hand, IGF-I levels are decreased in neurodegeneratative diseases (Busiguina et al., 2000).

The above properties of IGF-I probably play an important role in injuries of the CNS (Russo et al., 2005), such as ischemic injuries (Johnston et al., 1996; Saatman et al., 1997). After hypoxia–ischemia, IGF-I mRNA and protein as well as IGFBP-2, BP-3, and BP-5 mRNA were upregulated (Gluckman et al., 1993). Recent studies in hypoxic rats showed that IGF-I administration reduced cortical damage only when given 2 h, but not 6 h, after the insult (Guan et al., 2000). It was also observed that rise of blood-borne IGF-I (by exercise or by exogenous IGF-I injection) leads to neuroprotective actions (Carro et al., 2000). IGF-I also improves the survival of neuroretinal cells *in vitro* under conditions of hypoxia (Seigel et al., 2000).

14.9 GH/IGF-I and Cerebral Myelinization

IGF-I is required for normal oligodendrocyte development and myelinization (Zegher et al., 2007) and has been shown to ameliorate demyelinization induced by tumor necrosis factor alpha (TNFα) in transgenic mice (Ye et al., 2007). GH-deficient mice present with cerebral hypomyelinization (Morisawa et al., 1989). Whether these properties may prove of therapeutic value in multiple sclerosis remains to be tried.

14.10 Effect of GH and IGF-I on Brain Development and Growth – Animal Studies

Studies of the Snell mouse and little mouse, models of dwarfed mice, have rendered interesting information on the role of GH on the developing brain. The Snell mouse (dw) (Snell, 1929) lacks GH, TSH, and prolactin caused by mutations in the transcription factor Pit-1 (Camper et al., 1990). The little mouse (lit) (Eicher and Breamer, 1976) is the counterpart of the human GHRH-R mutation (Godfrey et al., 1993; Lin et al., 1993) leading to isolated GHD (IGHD).

The weight of the brain of Snell mice was found to be up to two-thirds smaller than that of normal mice (Noguchi, 1996) and their corpus callosum contained a reduced number of myelinated fibers (Noguchi et al., 1983). Administration of bovine GH and thyroxine to the Snell mice restored the normal state but only if administered during the first 20 days of life. The Snell mice also have a below-normal concentration of gangliosides in the cerebrum (Sugisaki et al., 1987). From the above studies it was not clear whether the restricting effect on brain development and myelinization was due to GH or TSH deficiency or a combination of both hormones. The response was obtained by studies in the lit mouse (Godfrey et al., 1993; Lin et al., 1993) lacking only GH. In these mice the weight of the cerebellum was decreased and the pyramidal neurons were found to have short primary dendrites with sparse branching (Noguchi and Sugisaki, 1985). This state could be restored by GH administration during the first 20 postnatal days of life (Noguchi et al., 1988). The above experiments proved that IGHD in the prenatal period exerted specific defects on the murine brain and that these defects could be reversed by administration of GH during a critical time period in early life similar to the critical time known for the congenitally hypothyroid mouse (Noguchi and Sugisaki, 1984).

In transgenic mice which overexpressed IGF-I, the brains were 50–70% larger than those of controls and had an increased cell number and cell size (Carson et al., 1993). It remains to be established whether the effects of GHD or GH administration in the murine brain are due to a direct GH effect or are mediated by IGF-I, either by IGF-I transferred from the blood (Armstrong et al., 2000) or by the paracrine IGF-I widely distributed throughout the CNS (Werther et al., 1990; Bartlett et al., 1991). Carson et al. (1993) showed in transgenic mice that IGF-I increases brain growth and CNS myelinization, and Dentremont et al. (1999) reported that IGF-I differentially increases neuron number and growth in medullary nuclei during the early postnatal development in mice. Conversely, the number of oligodendrocytes and myelinization is reduced in IGF-I knockout (KO) mice as is the total brain size (Beck et al., 1995; Cheng et al., 1999).

Studies using IGF-I KO mice (Accilli et al., 1999), GH-R KO mice (Zhou et al., 1997), and GH or IGF-I transgenic mice (Ye et al., 1997) offer further information.

14.11 Additional Effects of IGF-I on the Central and Peripheral Nervous System

IGF-I acts most probably on neural stem cells which can multiply throughout life (Eriksson et al., 1998). Using SC IGF-I infusion in 50-day-old hypophysectomized rats, Aberg et al. (2000) demonstrated that IGF-I increases progenitor cell proliferation and selectively induces neurogenesis in the progeny of adult neural progenitor cells. Exogenously administered IGF-I also enhances regeneration of peripheral nerves, such as the sciatic nerve in mice after crush damage (Contreras et al., 1995). IGF-I also acts *in vitro* by increasing the growth rate and size of rat glial progenitor cells (McMorris and Duboic-Dalcq, 1988). Evidence that IGF-I regulates dendritic elaboration in cortical neurons was shown by its ability to increase the branching and extent of dendrites of pyramidal cells in slices of rat primary somatosensory cortex (Niblock et al., 2000). IGF-I was also found to stimulate chemotaxis of human neuroblasts (Puglianiello et al., 2000) and myelinization of both the CNS (Goddard et al., 1999) and the peripheral nervous system (Cheng et al., 1999). Carson et al. (1993) also found increased myelinization in the CNS of IGF-I transgenic mice. IGF-I also attenuates or prevents apoptosis in brain neurons after ischemia (Tagami et al., 1997a,b). It has been suggested that the potent antipoptotic activity of IGF-I is linked to an interaction with prostaglandins and proinflammatory cytokines (Lackey et al., 2000).

14.12 GH and IGF-I Effects on Brain Growth in Children

Head circumference is a measure of brain growth (Kaplan, 1987) increasing in size by inside pressure exercised by the developing brain. Head circumference measurements made in infants with either congenital IGHD (Laron et al., 1993; Zachmann et al., 1980) or in infants with hereditary isolated IGF-I deficiency, such as Laron syndrome (LS = primary GH insensitivity) (Laron, 2004) revealed subnormal head circumference in untreated children. A rapid catch-up growth was seen once hGH treatment was instituted in young children with IGHD (Laron et al., 1979, 1995) and during IGF-I replacement treatment in children with Laron syndrome (Laron et al., 1992a; Laron, 2008). Recent studies using magnetic resonance imaging (MRI) of the skull in a group of patients with Laron syndrome revealed various degrees of diffuse brain parenchymal damage (Kornreich et al., 2002) which differed with the type of the GH receptor mutations (Shevah et al., 2005).

Because of the marked retardation in skeletal age in untreated children with GH or IGF-I deficiency (Laron, 1999b), the catch-up in head circumference (i.e., brain growth) of untreated patients can last into pubertal age (Laron, 1999a). Nevertheless, as shown in the next section, there seems to be a critical period in early age in which brain development, growth, and intellectual advancement coincide. In children with either congenital IGHD or IGF-I receiving replacement treatment, the brain has a faster catch-up growth velocity and normalization of size than the body height or foot size (Silbergeld et al., 2006, 2007).

Recent studies by Attias and Laron (submitted) revealed that untreated or late-treated patients with Laron syndrome suffer from a sensorial acoustic hearing defect. Children treated with IGF-I before age 4 have a normal hearing.

In children who had received cranial irradiation in the dosage range 2700–4750 centi-Geigy, which reduces GH secretion, showed a decrease in the occipitofrontal head circumference (Clayton et al., 1987). The same authors noted that in the years following cranial irradiation, head growth is unaffected by GH therapy.

14.13 Effect of GH and/or IGF-I on Intellectual Performance

Studies on the mental and motor abilities in both patients with congenital and acquired isolated deficiencies of GH and/or IGF-I before and during respective replacement therapies are scant as the free availability of human GH since the 1960s leads to immediate initiation of therapy. Laron and Galatzer (1981) measured the intelligence quotient (IQ) using the Wechsler Intelligence Scale (WISC) (Wechsler, 1949/1974) and Stanford Binet Tests (Terman and Merill, 1960); in four children with IGHD due to hGH-N gene deletion (Laron et al., 1985) before and several years after hGH treatment. In the three children in whom treatment was initiated at a bone age of below 5 years, the IQ rose significantly from an IQ score of 80, 96, and 99 to 105, 125, and 112. In the boy in whom treatment was started during puberty the IQ score was 80 (low), and remained unchanged. In another study of 21 children with IGHD, performed during hGH treatment, for various lengths of time (Laron and Galatzer, 1980), it was found that the IQ deficits were moderate, but there were more patients in the very low score range (**Table 2**).

Abbott et al. (1982) reported psychological and emotional tests in five children (three males, two females, aged 4–16) with IGHD before and 1 year after intermittent GH treatment. The results, mixed with those in children with multiple pituitary hormone deficiencies (MPHD), showed a low average IQ of 88 ± 17 (using Wechsler, Beery DAP, and WRAP tests). The visuomotor integration skills were also low. GH treatment had no significant effect.

Meyer-Bahlburg et al. (1978) reported intelligence tests in 29 children with GHD; of these 14 presented with IGHD. The mean age was 11 years.

Children with multiple pituitary hormone deficiency (MPHD) had a lower IQ than those with IGHD. GH treatment did not affect IQ, but many patients were not tested before initiation of therapy. Of note is that in the 1960s and early 1970s GH was not injected daily and even then in lower doses than nowadays.

We studied a group of patients with Laron syndrome who have a congenital inability to generate IGF-I (Laron, 2004). The Wechsler intelligence tests for children or adults (Wechsler, 1949/1974, 1955), Stanford–Binet Tests of Intelligence (Terman and Merill, 1960), The Catell Infant Intelligence Scale (Catell, 1940) and the Bender Visual Motor Gestalt Test (Bender, 1948; Koppitz, 1964), as well as Human Figure Drawing Tests (Alexander et al., 1966) were performed on several occasions, at various ages in children, adolescents, and adults with this syndrome. The early study by Frankel and Laron (1968) of 17 children and adolescents with Laron syndrome revealed a lower IQ as a whole compared to normal controls and a distinct deficiency in visuomotor functioning. Galatzer et al. (1993) reexamined 12 of these patients in adult age (28.9 ± 8.6 years) and additional patients who had been tested at a mean age of 13.3 ± 5.2 years (range 7–19 years). The results revealed that 4 out of 18 patients had an IQ lower than 69, which means mental retardation; of these one female patient is institutionalized. The mean performance IQ on retesting was 86.6 ± 15.5 compared to 83.3 ± 17.52 before, that is, no change. However, the mean verbal IQ had risen from 81.8 ± 17.24 to 91.9 ± 16.91. Comparing the IQ distribution shows that in the original testing, none of the patients reached the upper quartile of the normal distribution (IQ > 110), while 3 out of 12 patients (one-fourths) patients, on retesting, scored in this range (Galatzer et al., 1993; **Table 3**).

Further investigations in patients with Laron syndrome showed that there was a great variability in IQ from normal to mental retardation and that there was a correlation with the degree of brain pathology (Kornreich et al., 2002) and the type of GH-R gene mutation (Shevah et al., 2005). These findings explain the normal performance on psychological tests in Ecuadorian patients with Laron syndrome reported by Kranzler et al. (1998) with an E180 GH receptor

Table 2 Distribution of IQ scores of 21 patients with isolated growth hormone deficiency (IGHD)

IQ	*Expected (%)*	*Verbal (%)*	*Performance (%)*	*Total (%)*
>130	2.2	0	5.3	4.8
120–129	6.7	10.5	10.5	4.8
110–119	16.1	10.5	10.5	9.5
90–109	50.0	42.1	42.1	38.1
80–89	16.1	31.6	5.3	23.8
70–79	6.7	0	15.8	4.8
<69	2.2	5.3	10.5	14.2

Reproduced from Laron Z and Galatzer A (1980) Aspects of brain development in children and adolescents with pituitary growth hormone deficiency. In: De Wied D and van Keep PA (eds.) *Hormones and the Brain*, pp. 293–302. Lancaster: MTP Press.

Table 3 Comparison of full-scale, verbal, and performance IQ distribution of patients with Laron syndrome, tested once during adolescence (I: $n = 18$) and repeated 10 years later as adults (II: $n = 12$)

Score	*Expected*	*Full-scale*		*Verbal*		*Performance*	
		I	*II*	*I*	*II*	*I*	*II*
>130	2.2						
120–129	6.7		8.3		8.3		
110–119	16.1		16.7		16.7		8.3
100–109	25.0	22.2		11.1		16.7	16.7
90–99	25.0	27.8	16.7	22.2	25.0	16.7	16.7
80–89	16.1	27.8	16.7	33.3	16.7	33.3	
70–79	6.7	5.5	41.6	16.7	33.3	11.1	58.3
≤69	2.2	27.8		16.7		22.2	

Reproduced from Laron Z (2002) Final height and psychosocial outcome of patients with Laron syndrome (primary GH resistance): A model of genetic IGF-I deficiency. In: Gilli G, Schell LM, and Benso L (eds.) *Human Growth from Conception to Maturity*, ch. 23, pp. 215–225. London: Smith-Gordon.

mutation in exon 6 (Berg et al., 1994). We found the same in one of our patients with the same mutation (Shevah and Laron, 2006) and a normal brain MRI.

Woods et al. (1997) reported an increased frequency of mental subnormalities in reviewing a group of 82 children with Laron syndrome belonging to many ethnic groups: European, Indian, Pakistan, Vietnam, Cambodia, etc. Woods et al. (1996) also reported a patient with IGF-I gene deletion who was severely mentally retarded.

It cannot be excluded that some of these differences are due to low socioeconomic class of the patients – a known phenomenon (Kennet and Cropley, 1970).

Stabler et al. (1998) reported that a 3-year treatment with GH of 72 children with GHD (peak <10 μg ml^{-1}) improved their IQ and few social scores. A positive link between IGF-I and intelligence was also demonstrated in a study of healthy 547 white boys ($n = 301$) and girl ($n = 246$) (the Avon study) who underwent the Wechsler Intelligence and Objective reading tests (Gunnell et al., 2005) and serum IGF-I determinations. The positive influence was seen in relation to the verbal rather than performance components. Whether psychosocial dwarfism is due to a hypothalamic-driven deficiency of GH or in food is uncertain (Stanhope et al., 1994).

14.14 Influence of Untreated and Treated GH and IGF-I Deficiency on Psychosocial Well-Being and Quality of Life

Untreated patients with GHD (rarely encountered in developed countries) and patients with Laron syndrome have many emotional problems, difficulties in vocational training and social life, resulting from their marked short stature, physical handicaps, and progressive obesity (Laron, 1999b) in addition to varying intellectual and school performance deficits (see Section 14.13; **Figure 4**). Only a few reach higher education; the majority, if employed, perform clinical or low-skilled manual jobs. Few are institutionalized. In our cohort of 66 dwarfed patients with Laron syndrome (58 adults), four males and six females are married and have children; one female is divorced, two females have a boyfriend, and some females and males have had no sexual relations. One untreated male congenital IGHD is married and has children, and so are two female patients. One can assume that many of the difficulties in adjustment and social integration are due to the direct and indirect damage of the hormone deficiency

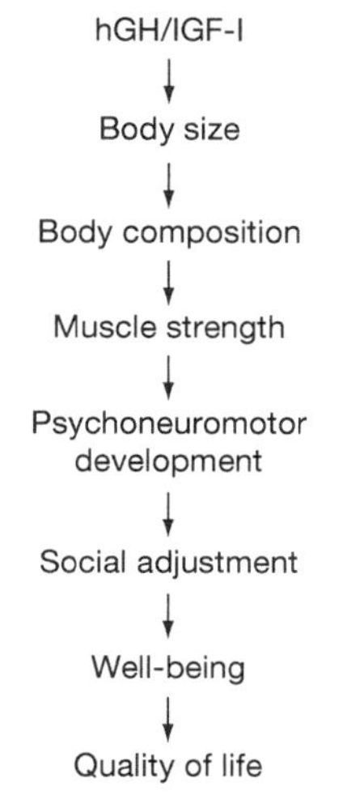

Figure 4 Effects of GH and IGF-I on social adjustment and quality of life. Reproduced from Laron Z (2002a) Growth hormone and insulin-like growth factor-I: Effects on the brain. In: Pfaff DW, Arnold AP, Etgen AM, Fahrbach SE, and Rubin RT (eds.) Hormones, Brain and Behavior, ch. 85, pp. 75–96. Amsterdam: Academic Press, with permission from Elsevier.

to the nervous system, largely encountered during the prenatal and early infantile critical periods. Little is known on the continuous effects of GH and IGF-I deficiency on the central and peripheral nervous tissue.

Several controversial studies on adult patients, with childhood-onset GHD who have been treated until final height and then stopped, have been published (Laron and Butenandt, 1993).

With the availability of unlimited amounts of biosynthetic hGH in the last decade, treatment of adults with GHD has been started in several countries (Laron and Butenandt, 1993; Carroll et al. 1998; Cummings and Merriam, 1999; Juul and Jorgensen, 2000). Due to uncertainty of the dose of hGH needed in adult age and the development of adverse effects (Abs et al., 1999; Holmes and Shalet, 1995), a very low replacement dose has been recommended (Growth Hormone Research Society, 1998). The majority of investigations performed in adults involved patients with MPHD, including GH, and middle-aged patients after operative pituitary ablation; thus, psychological deficits cannot be blamed with certainty on the GHD.

Certain studies suggested that GH replacement in adults with acquired GHD improved quality of life (QoL) (Sartorio et al., 1996). Studies by Baum et al. (1998) on 40 GHD men (age 24–64) and that by Florkowski et al. (1998) in 20 adult GHD patients (age 20–69), both studying QoL and psychological parameters, found no changes.

Almqvist et al. (1986) and Deijen et al. (1996, 1998) reported that GH replacement had beneficial effects on memory and cognitive functions, as did Van Dam (2006). In several studies QoL assessments were made by two self-rating prestructured questionnaires – The Nottingham Health Profile (NHP) and the Psychological General Well-Being Index (PGWB). In placebo-controlled investigations, McGauley (1989) and Bengtsson et al. (1993) described psychological improvement already after 6 months of hGH treatment. In other studies emotional reaction and social isolation was improved (Mardh et al., 1994). Koltowska-Haggstrom et al. (2006), having collected data from four countries, report that long-term GH replacement results in sustained improvement in QoL.

However, following a large group of patients, it became apparent that not all patients respond in the same way during long-term treatment (Wiren et al., 1998). It has been suggested that the beneficial effects on mood and cognitive function in GH treatment in adults are caused by affecting neurotransmitters of GH treatment in adults, such as elevating aspartate concentrations and lowering the dopamine metabolite homovanilic acid (HVA) in the CSF (Burman et al., 1993, 1996). On the other hand, considering the grading of passage of hGH and IGF-I through the BBB (Armstrong et al., 2000; Coculescu, 1999), it is difficult to conceive how the very low doses of hGH administered affect the CNS.

Arwert et al. (2005a) compared neuropsychological (mood states profile, short- and long-term memory, and associated learning task) and functional MRI (fMRI) during a memory task in 13 GH-treated (mean duration, 10.9 ± 6.74 years) childhood-onset GH-deficient adults (eight males and five females aged 27.3 ± 6.94 years) and matched controls. The authors concluded that there was no difference in the quality of performance in the working memory task between the two groups but untreated patients reported more fatigue. The fMRI showed that GH-deficient patients have a subnormal memory speed, but no impaired quality of memory performance, suggesting that GH/IGF-I contribute to prefrontal functioning.

It should be kept in mind that with advancing age the GH production and serum IGF-I levels progressively decrease and that adults are sensitive to even low doses of exogenous hGH or IGF-I – the former causing fluid retention, arthralgia, and carpal tunnel syndrome (Vance and Mauras, 1999) and the latter causing hypoglycemia, in addition (Laron, 2004).

Radcliffe et al. (2004) summarized the findings on QoL notes that open-label trials in adults are inconsistent and limited by potential placebo effects and patient self-selection. Similar conclusions were reached by Maruff and Falleti (2005) summarizing cross-sectional studies on cognitive function in GHD in adults, and those by Arwert et al. (2005a,b). The conflicting findings are mostly due to mixing patients with IGHD and MPHD, congenital versus acquired GHD as well as age and duration of therapy.

To ascertain the effects of hGH and IGF-I versus thyroid and other pituitary hormones, randomized studies with placebos in patients with childhood-onset IGHD, investigated before and during retreatment in adult age, are obviously needed. Also, cost–benefit analyses need to be done.

14.15 GH and IGF-I and the Aging Brain

There is much speculation as to whether aging individuals characterized by low circulating levels of GH and IGF-I could benefit from GH administration to improve their cognitive function (Aleman et al., 1999; Kalmijn et al., 2000). The results of existing studies are, so far, inconclusive as many factors change with aging, on the one hand, and influence GH/IGF-I secretion, on the other (Cherrier et al., 2004; Maruff and Falleti, 2005; Okereke et al., 2006).

Investigating the possible correlation between age-dependent decline in cognitive functions and the GH/IGF-I axis in 22 subjects aged 65–84. Rollero et al. (1998) found that serum IGF-I levels directly correlated with the Mini Mental State Examination (MMSE) scores. As there are many nutritional and hormonal changes during aging, which also directly or indirectly influence GH and IGF-I secretion (Fernandez et al., 2007; Garcia-Segura, 2007), it is difficult to determine to what degree GH/IGF-I influence the declining brain function in aging.

A recent metanalysis by Arwert et al. (2005b) concluded that it is not possible, at present, to definitely state that GH treatment is more efficacious than placebo to improve QoL in elderly people.

14.16 GH and IGF-I Effects on Memory in Mice

Of interest are some experiments in new animal models; however, it is not known to what degree they relate to man. Kinney et al. (2001) found that aging GH-R KO mice (Laron mice) with an absence of GH and IGF-I signaling have an improved long-term memory. Svensson et al. (2006) described that old mice with inactivated liver IGF-I (LI-IGF-I−/−), resulting in 80–85% reduction of circulating IGF-I, had a slower acquisition of spatial tasks than controls. Histochemical analysis revealed increased dynorphin and enkephalin immunoreactivity, astrocytosis, and metabotropic glutamate receptor, suggesting synapting dysfunction in these mice.

Administration of IGF-I to adult diabetic rats prevented deterioration of spatial memory (Lupien et al., 2003).

14.17 GH and IGF-I in Neurological Disorders

The relationship between CNS diseases and GH and/or IGF-I is not sufficiently clarified. Histological examination of brain tissue obtained at postmortem from 14 patients with multiple sclerosis (MS), age range 22–69, revealed the expression of IGF-I and IGF-BP-1 in active lesions in hypertrophic astrocytes, whereas IGF-IR was immunolocalized in macrophages (Gveric et al., 1999). In experimental studies, Cheng et al. (1999) showed that IGF-I is critical for long-term myelinization and Schwann cell attachment. In other experimental studies, Nieder et al. (2005) showed that IGF-I can play a protective role in a model of high-dose spinal cord irradiation.

Based on the property of neuroregeneration and neuroprotection of IGF-I (Aberg et al., 2006) IGF-I administration has been tried in head injuries (Hatton et al., 1997) or in laryngeal paralysis (Shiotani et al., 1998). The results are too few to judge.

Alterations of the IGF-I action have also been suggested in Huntington's disease (Humbert et al., 2002) and Parkinson's disease (Offen et al., 2001).

The lower than normal GH/IGF-I levels in blood or CSF in some diseases have initiated therapeutic clinical trials (Gasparini and Xu, 2003). So far, the use of GH or IGF-I in nonendocrine neurological and psychiatric disorders have not yielded encouraging results.

The use of GH or IGF-I in neuromuscular degenerative diseases deserves further trials, even if it is effective in only strengthening the muscular system and improving the mobility of the patients as shown in patients with HIV.

14.18 GH and IGF-I in Psychiatric Disorders

The reports on the relationship between GH secretion and depression are contradictory (Zalsman et al., 2006). Studies in children have found a blunted response to GH stimulation by clonidine (Jensen and Garfinkel, 1990) or insulin hypoglycemia (Ryan et al., 1994) and to GHRH (Dahl et al., 2000). These and similar reports have suggested that a blunted GH response may be a trait marker for depression and even a risk for suicide (Dahl et al., 1992). In adult patients with depression, both higher or lower GH secretion have been reported (Linkowski et al., 1987; Sakkas et al., 1998).

Panic patients, in contrast to depressed patients, have normal GH-axis activity (Abelson et al., 2005). Ten drug-free patients with schizophrenia (age range 23–37) submitted to 14 days of GHRH administration showed a normal response of GH (Mayerhoff et al., 1990). In brains of patients with Alzheimer's disease, Rivera et al. (2005) found a reduction in the IGF-I and insulin gene expression, resembling diabetes. Reduced GH, IGF-I, and insulin levels have also been registered in patients with amyotrophic lateral sclerosis (ALS) (Bilic et al., 2006). In contrast, children with autism were found to have increased levels of IGF-I, IGFBP-2, and -3, as well as a greater head circumference than controls (Mills et al., 2007).

It has been suggested that the changes of GH secretion are secondary to neurobiological changes in the CNS related to the disorder, such as dysregulation of serotonin (5-HT) neurotransmission (Bhagwagar et al., 2002).

14.19 Psychological Effects of GH Administration to Nongrowth Hormone-Deficient Short Children

Ross (2005) reported that absence of 1–7 years GH treatment affects cognitive functions and/or QoL outcomes in girls with Turner syndrome. Boulton et al. (1991) reported improved emotional attitude and perception during a 2-year GH treatment of short, normal children. Hokken-Koelega et al. (2005) studied IQ and psychosocial functioning in two trials with either 73 or 87 children born short for gestational age (SGA). Mean age at start of treatment was 7.4 years and the mean duration of hGH treatment was 8 years. Total IQ score and block design s-score increased significantly during the hGH treatment; in the normal verbal IQ score there was no change, but self-perception and behavior improved significantly.

Radcliffe et al. (2004) concluded that the evidence for improved QoL as a result of hGH treatment in children with idiopathic short stature is not sufficiently proven.

14.20 GH and IGF-I and Risk for Brain Malignancy

IGF-I has been described to stimulate mitogenic activity in the human neuroblastoma cell line SH-SY5Y (Sumantran and Feldman, 1993) but little is known relating GH/IGF-I and brain tumors. In a 5-year survey of 29 133 men (ages 50–69 years), in whom 22 patients with glioma were found (Lönn et al., 2007), an inverse association between glioma and serum IGF-I was registered. In another study, Glick and Unterman (1995) measured IGF-I and IGF-II in 26 tumors located adjacent to the ventricular system. In pituitary tumors, IGF-I levels in the CSF were elevated, whereas in other tumors (gliomas, meningiomas, medulloblastomas, and metastases) the IGF-I levels in the CSF were normal, but the IGF-II levels were elevated.

The evidence that IGF-I enhances cancer risk is not necessarily related to increased serum IGF-I levels, as in acromegaly and colon cancer (Loeper and Ezzat, 2008), but to increased sensitivity to IGF-I, as evidenced by an increased number of IGF-I receptors on cancer tissues (Werner et al., 1993). Such studies are missing for brain tumors.

14.21 Conclusions

Studies in animals and humans have unequivocally documented the important roles of GH and IGF-I on the CNS. **Figure 5** illustrates the various pathways by which GH and IGF-I exert their effects. Using animal or human models of genetically determined or acquired GH and/or IGF-I deficiencies have shown that both hormones seem to be needed for the normal early development of the brain and psychoneural functions. GH and IGF-I have direct effects and also seem to act in combination or by activation of a number of neurogenic molecules, such as neurogenin-1, suppressor of cytokine signaling-2 (SOCS-2; Turnley, 2005), and brain-derived neurotrophic factor (BDNF).

IGF-I has important roles in neuronal development, synaptogenesis, myelinization, and apoptosis, as well as on neuronal rescue and cell survival in acute trauma. The ability of IGF-I to regulate neurogenesis and myelinization may be of therapeutic value. Whether GH treatment can restore major intellectual functions in adults lacking hGH or aging remains controversial.

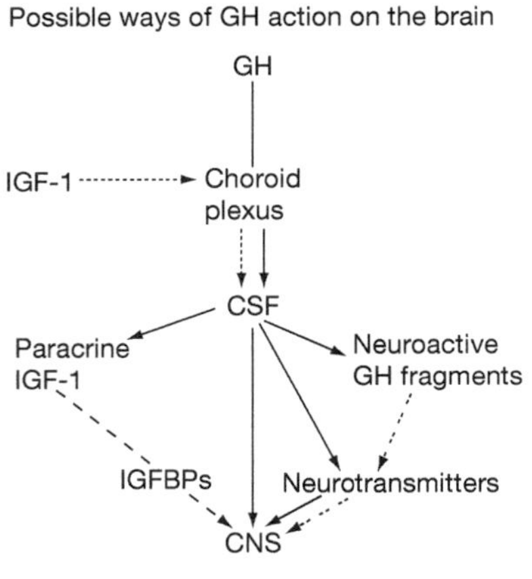

Figure 5 Possible ways of GH and IGF-I action on the brain. Reproduced from Larzon Z (2002a) Growth hormone and insulin-like growth factor-I: Effects on the brain. In: Pfaff DW, Arnold AP, Etgen AM, Fahrbach, SE, and Rubin RT (eds.) Hormones, Brain and Behavior, ch. 85, pp. 75–96. Amsterdam: Academic Press, with permission of Elsevier.

Acknowledgment

The author thanks Mrs. G. Waichman for technical assistance in the preparation of the manuscript.

References

Abbott AM, Bueno R, Pedrini MT, Murray JM, and Smith RJ (1992) Insulin-like growth factor I receptor gene structure. *Journal of Biological Chemistry* 267: 10759–10763.

Abbott D, Rotnem D, Genel M, and Cohen DJ (1982) Cognitive and emotional functioning in hypopituitary short-statured children. *Schizophrenia Bulletin* 8: 310–319.

Abelson JL, Curtis GC, and Uhde TW (2005) Twenty-four hour growth hormone secretion in patients with panic disorder. *Psychoneuroendocrinology* 30: 72–79.

Aberg MAI, Aberg D, Hedbacker H, Oscarsson J, and Eriksson PS (2000) Peripheral infusion of IGF-I selectively induces neurogenesis in the hippocampus. *Journal of Neuroscience* 20: 2896–2903.

Aberg ND, Brywe KG, and Isgaard J (2006) Aspects of growth hormone and insulin-like growth factor-I related to neuroprotection, regeneration, and functional plasticity in the adult brain. *Scientific World Journal* 6: 53–80.

Abs R, Bengtsson B-A, Hernberg-Stahl E, Monson JP, TauberJ-P, Wilton P, and Wuster C (1999) On behalf of the KIMS Study Group and the KIMS International Board, GH replacement in 1034 growth hormone deficient hypopituitary adults: Demographic and clinical characteristics, dosing and safety. *Clinical Endocrinology* 50: 703–713.

Accilli D, Nakae J, Kim JJ, Park B-C, and Rother KI (1999) Targeted gene mutations define the roles of insulin and IGF-I receptors in mouse embyronic development. *Journal of Pediatric Endocrinology and Metabolism* 12: 475–485.

Aleman A, Verhaar HJJ, de Haan EHF, et al. (1999) Insulin-like growth factor-I and cognitive function in healthy older men. *Journal of Clinical Endocrinology and Metabolism* 84: 471–475.

Alexander D, Erhardt AA, and Money J (1966) Defective figure drawing, geometric and human, in Turner's syndrome. *Journal of Nervous and Mental Disease* 142: 161–167.

Almqvist O, Thoren M, Saaf M, and Eriksson O (1986) Effects of growth hormone substitution on mental performance in adults with growth hormone deficiency: A pilot study. *Psychoneuroendocrinology* 11: 347–352.

Anlar B, Sullivan KA, and Feldman EL (1999) Insulin-like growth factor I and central nervous system development. *Hormone and Metabolic Research* 31: 120–125.

Armstrong CS, Wuarin L, and Ishii DN (2000) Uptake of circulating insulin-like growth factor-I into the cerebrospinal fluid of normal and diabetic rats and normalization of IGF-II mRNA content in diabetic rat brain. *Journal of Neuroscience Research* 59: 649–660.

Arwert LI, Deijen JB, and Drent ML (2005a) Effects of growth hormone deficiency and growth hormone treatment on quality of life in growth-deficient patients. In: Jorgensen JOL and Christiansen JS (eds.) *Growth Hormone Deficiency in Adults, Frontiers in Hormone Research,* vol. 33, pp. 196–208. Basel: Karger.

Arwert LI, Veltman DJ, Deijen JB, Sytze van Dam P, Delemarre-van de Waal HA, and Drent MJ (2005b) Growth hormone deficiency and memory functioning in adults visualized by functional magnetic resonance imaging. *Neuroendocrinology* 82: 32–40.

Asa SL, Kovacs J, Thorner MO, Leong DA, Rivier J, and Vale W (1985) Immunohistological localization of growth hormone-releasing hormone in human tumors. *Journal of Clinical Endocrinology and Metabolism* 60: 423–427.

Asakawa A, Inui A, Kaga T, et al. (2001) Ghrelin is an appetite-stimulatory signal from stomach with structural resemblance to motilin. *Gastroenterology* 120: 337–345.

Bach MA and Bondy CA (1992) Anatomy of the pituitary insulin-like growth fact or system. *Endocrinology* 131: 2588–2594.

Bartlett WP, Li X-S, Williams M, and Benkovic S (1991) Localization of insulin-like growth factor-1 mRNA in murine central nervous system during postnatal development. *Developmental Biology* 147: 239–250.

Barton DE, Foellmer BE, Wood WI, and Francke U (1989) Chromosome mapping of the growth hormone receptor gene in man and mouse. *Cytogenetic and Genome Research* 50: 137–141.

Baum HBA, Katznelson L, Sherman JC, et al. (1998) Effects of physiological growth hormone (GH) therapy on cognition and quality of life in patients with adult-onset GH deficiency. *Journal of Clinical Endocrinology and Metabolism* 83: 3184–3189.

Baumann G, Stolar MN, Amburn K, Barsano CP, and DeVries BC (1986) A specific GH-binding protein in human plasma: Initial characterization. *Journal of Clinical Endocrinology and Metabolism* 62: 134–141.

Beck KD, Powell-Braxton L, Widmer HR, Valverde J, and Hefti F (1995) IGF-I gene disruption results in reduced brain size, CNS hyopmyelination, and loss of hippocampal granule and striatal parvalbumin-containing neurons. *Neuron* 14: 717–730.

Bender L (1948) *Instruction for the Use of Visual Motor Gestalt Test.* New York: American Orthopsychiatric Association and Psychological Corp.

Bengtsson BA, Eden S, Lonn L, et al. (1993) Treatment of adults with growth hormone (GH) deficiency with recombinant human GH. *Journal of Clinical Endocrinology and Metabolism* 76: 309–317.

Berg MA, Peoples R, Perez-Jurado L, et al. (1994) Receptor mutations and haplotypes in growth hormone receptor deficiency: A global survey and identification of the Ecuadorean E180 splice mutation in an oriental Jewish patient. *Acta Paediatrica* 399(supplement): 112–114.

Berry SA, Srivastava CH, Rubin LR, Phipps WR, and Pescovitz OH (1992) Growth hormone-releasing hormone-like messenger ribonucleic acid and immunoreactive peptide are present in human testis and placenta. *Journal of Clinical Endocrinology and Metabolism* 75: 281–284.

Bhagwagar Z, Hafizi S, and Cowen PJ (2002) Cortisol modulation of 5-HT-mediated growth hormone release in recovered depressed patients. *Journal of Affective Disorders* 72: 249–255.

Bilic E, Bilic E, Rudan I, Kusec V, Zurak N, Delimar D, and Zagar M (2006) Comparison of the growth hormone, IGF-1 and insulin in cerebrospinal fluid and serum between patients with motor neuron disease and healthy controls. *European Journal of Neurology* 13: 1340–1345.

Blundell TL and Humbel RE (1980) Hormone families: Pancreatic hormones and homologous growth factors. *Nature* 287: 781–787.

Bondy C and Lee W-H (1993) Correlation between insulin-like growth factor (IGF)-binding protein 5 and IGF-I gene expression during brain development. *Journal of Neuroscience* 13: 5092–5104.

Bondy CA, Werner H, Roberts CT, Jr., and LeRoith D (1990) Cellular pattern of insulin-like growth factor I (IGF-I) and type I IGF receptor gene expression in early organogenesis: Comparison with IGF-II gene expression. *Molecular Endocrinology* 4: 1386–1398.

Bosman FT, Van Assche C, Nieuwenhuyzen Kruseman AC, Jackson S, and Lowry PJ (1984) Growth hormone releasing factor (GRF) immunoreactivity in human and rat gastrointestinal tract and pancreas. *Journal of Histochemistry and Cytochemistry* 32: 1139–1144.

Boulton TJ, Dunn SM, Quigley CA, Taylor JJ, and Thompson L (1991) Perceptions of self and short stature: Effects of two years of growth hormone treatment. *Acta Paediatrica Scandinavica* 377(supplement): 20–27; discussion 28.

Bunn RC, King WD, Winkler MK, and Fowlkes JL (2005) Early developmental changes in IGF-I, IGF-II, IGF binding protein-1, and IGF binding protein-3 concentration in the cerebrospinal fluid of children. *Pediatric Research* 58: 89–93.

Burman P, Hetta J, and Karlsson A (1993) Effect of growth hormone on brain neurotransmitters. *Lancet* 342: 1492–1493.

Burman P, Hetta J, Wide L, Mansson JE, Ekman R, and Karlsson FA (1996) Growth hormone treatment affects brain neurotransmitters and thyroxine. *Clinical Endocrinology (Oxf)* 44: 319–324.

Burton KA, Kabigting EB, Clifton DK, and Steiner RA (1992) Growth hormone receptor messenger ribonucleic acid distribution in the adult male rat brain and its colocalization in hypothalamic-somatostatin neurons. *Endocrinology* 44: 319–324.

Busiguina S, Fernandez AM, Barrios V, Clark R, Tolbert DL, Berciano J, and Torres-Aleman I (2000) Neurodegeneration is associated to changes in serum insulin-like growth factors. *Neurobiological Disorders* 7: 657–665.

Camper SA, Saunders TL, Katz RW, and Reeves RH (1990) The Pit-1 transcription factor gene is a candidate for the murine Snell dwarf mutation. *Genomics* 8: 586–590.

Carro E, Nunez A, Busiguina S, and Torres-Aleman I (2000) Circulating insulin-like growth factor I mediates effects of exercise on the brain. *Journal of Neuroscience* 20: 2926–2933.

Carro E, Trejo JL, Núñez A, and Torres-Aleman I (2003) Brain repair and neuroprotection by serum insulin-like growth factor I. *Molecular Neurobiology* 27: 153–162.

Carroll PV, Christ ER and the members of Growth Hormone, Research Society Scientific, Committee et al. (1998) Growth hormone deficiency in adulthood and the effects of growth hormone replacement. A review. *Journal of Clinical Endocrinology and Metabolism* 83: 382–395.

Carson MJ, Behringer RR, Brinster RL, and McMorris FA (1993) Insulin-like growth factor-1 increases brain growth and

central nervous system myelination in transgenic mice. *Neuron* 10: 729–740.
Castro JR, Costoya JA, Gallego R, Prieto A, Arce VM, and Senaris R (2000) Expression of growth hormone receptor in the human brain. *Journal of Neuroscience Letters* 281: 147–150.
Catell P (1940) *Infant Intelligence Scale*. New York: The Psychological Corporation.
Cheng HL and Feldman EL (1997) Insulin-like growth factor-I (IGF-I) and IGF binding protein protein-5 in Schwann cell differentiation. *Journal of Cell Physiology* 171: 161–167.
Cheng HL, Russell JW, and Feldman EL (1999) IGF-I promotes peripheral nervous system myelination. *Annals of the New York Academy of Sciences* 883: 124–130.
Cheng HL, Sullivan KA, and Feldman EL (1996) Immonunohistochemical localization of insulin-like growth factor binding protein-5 in the developing rat nervous system. *Developments in Brain Research* 92: 211–218.
Cherrier MM, Plymate S, Mohan S, et al. (2004) Relationship between testosterone supplementation and insulin-like growth factor-I levels and cognition in healthy older men. *Psychoneuroendocrinology* 29: 65–82.
Chrysis D, Calikoglu AS, Ye P, and d'Ercole AJ (2001) Insulin-like growth factor-I overexpression attenuates cerebellar apoptosis by altering the expression of Bcl family proteins in a developmentally specific manner. *Journal of Neuroscience* 21: 1481–1489.
Clark RG, Carlson MS, and Robinson ICAF (1988) Growth hormone secretion in the conscious rat: Negative feedback of GH on its own release. *Journal of Endocrinology* 119: 201–209.
Clayton PE, Shalet SM, Price DA, Surtees RAH, and Pearson D (1987) The role of growth hormone in stunted head growth after cranial irradiation. *Pediatric Research* 22: 402–404.
Clemmons DR (1998) Role of insulin-like growth factor binding proteins in controlling IGF actions. *Molecular and Cellular Endocrinology* 140: 19–24.
Coculescu M (1999) Blood–brain barrier for human growth hormone and insulin-like growth factor-I. *Journal of Pediatric Endocrinology and Metabolism* 12: 113–124.
Coculescu M, Cucu C, Catrina SB, et al. (1998) The blood–brain barrier permeability by human growth hormone (hGH) as reflected by hGH concentration in cerebrospinal fluid. *Journal of Endocrinology* 159(supplement): OC22 (abstract).
Contreras PC, Steffler C, Yu E, Callison K, Stong D, and Vaught JL (1995) Systemic administration of rhIGF-I enhanced regeneration after sciatic nerve crush in mice. *Journal of Pharmacology and Experimental Therapeutics* 274: 1443–1449.
Cooke NE, Ray J, Watson MA, Ester PA, Kuo BA, and Liebhaber SA (1988) Human growth hormone gene and the highly homologous growth hormone variant gene display different splicing patterns. *Journal of Clinical Investigation* 82: 270–275.
Culler FL, Kaufman S, Frigeri LG, and Jones KL (1988) Comparison of the acute metabolic effects of 22,000-dalton and 20,000-dalton growth hormone in human subjects. *Hormone and Metabolic Research* 20: 107–109.
Cummings DE and Merriam GR (1999) Growth hormone and growth hormone secretagogues in adults. In: Meikle AW (ed.) *Hormone Replacement Therapy,* pp. 61–90. Totowa, NJ: Humana Press.
Dahl RE, Birmaher B, Williamson DE, et al. (2000) Low growth hormone response to growth hormone-releasing hormone in child depression. *Biological Psychiatry* 48: 981–988.
Dahl RE, Ryan ND, Williamson DE, et al. (1992) Regulation of sleep and growth hormone in adolescent depression. *Journal of the American Academy of Child and Adolescent Psychiatry* 31: 615–621.
Deghenghi R, Avallone R, Torsello A, Mucciolo G, Ghigo E, and Locatelli V (2001) Growth hormone inhibiting activity of cortistatin in the rat. *Journal of Endocrinological Investigation* 24: RC1–RC3.
Deijen JB, de Boer H, Blok GJ, and van der Veen EA (1996) Cognitive impairments and mood disturbances in growth hormone deficient men. *Psychoneuroendocrinology* 21: 313–322.
Deijen JB, de Boer H, and van der Veen EA (1998) Cognitive changes during growth hormone replacement in adult men. *Psychoneuroendocrinology* 23: 45–55.
Dentremont KD, Ye P, D'Ercole AJ, and O'Kusky JR (1999) Increased insulin-like growth factor I (IGF-I) expression during early postnatal development differentially increases neuron number and growth in medullary nuclei of the mouse. *Brain Research – Developmental Brain Research* 114: 135–141.
Eicher EM and Breamer WG (1976) Inherited ateliotic dwarfism in mice; characteristics of mutation, Little, on chromosome 6. *Journal of Heredity* 67: 87–91.
Eriksson PS, Perfilieva E, Bjork-Eriksson T, Alborn AM, Nordborg C, Peterson DA, and Gage FH (1998) Neurogenesis in the adult human hippocampus. *Nature Medicine* 4: 1313–1317.
Feldman EL, Sullivan KA, Kim B, and Russell JW (1997) Insulin-like growth factors regulate neuronal differentiation and survival. *Neurobiological Disorders* 4: 201–214.
Fernandez S, Fernandez AM, Lopez-Lopez C, and Torres-Aleman I (2007) Emerging roles of insulin-like growth factor-I in the adult brain. *Growth Hormone and IGF Research* 17: 89–95.
Florkowski CM, Stevens I, Joyce P, Espiner EA, and Donald RA (1998) Growth hormone replacement does not improve psychological well-being in adult hypopituitarism: A randomized crossover trial. *Psychoneuroendocrinology* 23: 57–63.
Frankel JJ and Laron Z (1968) Psychological aspects of pituitary insufficiency in children and adolescents with special reference to growth hormone. *Israel Journal of Medical Sciences* 4: 953–961.
Frankenne F, Closset J, Gomez F, Scippo ML, Smal J, and Hennen G (1988) The physiology of growth hormones (GHs) in pregnant women and partial characterization of the placental GH variant. *Journal of Clinical Endocrinology and Metabolism* 66: 1171–1180.
Galatzer A, Aran O, Nagelberg N, Rubitzek J, and Laron Z (1993) Cognitive and psychosocial functioning of young adults with Laron syndrome. In: Laron Z and Parks JS (eds.) *Pediatric and Adolescent Endocrinology, Vol. 24: Lessons from Laron Syndrome (LS) 1966–1992*, pp. 53–60. Basel-New York: Karger.
Garcia-Segura LM, Diz-Chaves Y, Perez-Martin M, and Darnaudery M (2007) Estradiol, insulin-like growth factor-I and brain aging. *Psychoneuroendocrinology* 32: 557–561.
Gasparini L and Xu H (2003) Potential roles of insulin and IGF-1 in Alzheimer's disease. *Trends in Neuroscience* 26: 404–406.
George DL, Phillips JA, III, Francke U, and Seebur PH (1981) The genes for growth hormone and chorionic somatomammotropin are on the long arm of chromosome 17 in req21-qter. *Human Genetics* 57: 138–141.
Gil-Ad I, Koch Y, Silbergeld A, Dickerman Z, Kaplan B, Weizman A, and Laron Z (1996) Differential effect of insulin-like growth factor I (IGF-I) and growth hormone (GH) on hypothalamic regulation of GH secretion in the rat. *Journal of Endocrinological Investigation* 19: 542–547.

Glick RP and Unterman TG (1995) Radioimmunoassay of insulin-like growth factors I and II in the cerebrospinal fluid of patients with pituitary and other central nervous system tumors. *Neurosurgery* 36: 556–564.

Gluckman PD, Guan J, Beilharz EJ, et al. (1993) The role of the insulin-like growth factor system in neuronal rescue. *Annals of the New York Academy of Sciences* 692: 138–148.

Goddard DR, Berry M, and Butt AM (1999) *In vivo* actions of fibroblast growth factor-2 and insulin-like growth factor-I on oligodendrocyte development and myelination in the central nervous system. *Journal of Neuroscience Research* 57: 74–85.

Godfrey P, Rahal JO, Beamer WG, Copeland NG, Jenkins NA, and Mayo KE (1993) GHRH receptor of little mice contains a missense mutation in the extracellular domain that disrupts receptor function. *Nature Genetics* 4: 227–232.

Godowski PJ, Leung DW, Meacham LR, et al. (1989) Characterization of the human growth hormone receptor gene and demonstration of a partial gene deletion in two patients with Laron type dwarfism 8. *Proceedings of the National Academy of Sciences of the United States of America* 6: 8083–8088.

Growth Hormone Research Society (1998) Consensus guidelines for the diagnosis and treatment of adults with growth hormone deficiency: Summary statement of the Growth Hormone Research Society Workshop on adult growth hormone deficiency. *Journal of Clinical Endocrinology and Metabolism* 83: 379–381.

Guan J, Gunn AJ, Sirimanne ES, Tuffin J, Gunning MI, Clark R, and Gluckman PD (2000) The window of opportunity for neuronal rescue with insulin-like growth factor-1 after hypoxia-ischemia in rats is critically modulated by cerebral temperature during recovery. *Journal of Cerebral Blood Flow and Metabolism* 20: 513–519.

Guillemin R, Zeytin F, Ling N, et al. (1984) Growth hormone-releasing factor: Chemistry and physiology (41813). *Proceedings of the Society for Experimental Biology and Medicine* 175: 407–413.

Gunnell D, Miller LL, Rogers I, and Holly JMP the ALSPAC Study Team (2005) Association of insulin-like growth factor I and insulin-like growth factor-binding protein-3 with intelligence quotient among 8- to 9-year-old children in the Avon longitudinal study of parents and children. *Pediatrics* 116: e681–e686.

Gveric D, Cuzner ML, and Newcombe J (1999) Insulin-like growth factors and binding proteins in multiple sclerosis plaques. *Neuropathology and Applied Neurobiology* 25: 215–225.

Harvey S, Hull KL, and Fraser RA (1993) Growth hormone: Neurocrine and neuroendocrine perspectives. *Growth Regulation* 3: 161–171.

Hatton J, Rapp RP, Kudsk KA, et al. (1997) Intravenous insulin-like growth factor-I (IGF-I) in moderate-to-severe head injury: A phase II safety and efficacy trial. *Journal of Neurosurgery* 86: 779–786.

He J and Crews FT (2007) Neurogenesis decreases during brain maturation from adolescence to childhood. *Pharmacology, Biochemistry, and Behavior* 86: 327–333.

Herington AC, Ymer S, and Stevenson J (1986) Identification and characterization of specific binding proteins for growth hormone in normal human sera. *Journal of Clinical Investigation* 77: 1817–1823.

Hofman PL and Pescovitz OH (1999) Growth hormone releasing hormone – biological and molecular aspects. In: Handwerger S (ed.) *Molecular and Cellular Pediatric Endocrinology*, pp. 85–112. Totowa, NJ: Humana Press.

Hojvat S, Emanuele N, Baker G, Connick E, Kirsteins L, and Lawrence AM (1982) Growth hormone, thyroid stimulating hormone and luteinizing hormone-like peptides in the rodent brain: Non-parallel ontogenic development with pituitary counterparts. *Developmental Brain Research* 4: 427–434.

Hokken-Koelega A, van Pareren Y, and Arends N (2005) Effects of growth hormone treatment on cognitive function and head circumference in children born small for gestational age. *Hormone Research* 64(supplement 3): 95–99.

Holmes SJ and Shalet SM (1995) Which adults develop side-effects of growth hormone replacement? *Clinical Endocrinology* 43: 143–149.

Holzenberger M and Lapointe F (2000) Expression of insulin-like growth factor-I (IGF-I) and IGF-II in the avian brain: Relationship of *in situ* hybridization patterns with IGF type 1 receptor expression. *International Journal of Developmental Neuroscience* 18: 69–82.

Howard AD, Feighner SD, Cully DF, et al. (1996) A receptor in pituitary and hypothalamus that functions in growth hormone release. *Science* 273: 974–977.

Humbert S, Bryson E, Cordelières F, et al. (2002) The IGF-1/Akt pathway is neuroprotective in Huntington's disease and involves Huntingtin phosphorylation by Akt. *Developmental Cell* 2: 831–837.

Hwa V, Oh Y, and Rosenfeld RG (1999) The insulin-like growth factor binding protein (IGFBP) superfamily. *Endocrine Reviews* 20: 761–787.

Ilkbahar YN, Wu K, Thordarson G, and Talamantes F (1995) Expression and distribution of messenger ribonucleic acids for growth hormone (GH) receptor and GH-binding protein in mice during pregnancy. *Endocrinology* 136: 386–392.

Jensen JB and Garfinkel BD (1990) Growth hormone dysregulation in children with major depressive disorder. *Journal of the American Academy of Child and Adolescent Psychiatry* 29: 295–301.

Johansson JO, Larson N, Andersson M, et al. (1995) Treatment of growth hormone (GH)-deficient adults with recombinant human GH increases the concentration of GH in the cerebrospinal fluid and affects the neurotransmitters. *Neuroendocrinology* 61: 57–66.

Johnston BM, Mallard EC, Williams CE, and Gluckman PD (1996) Insulin-like growth factor-I is a potent neuronal rescue agent after hypoxic-ischemic injury in fetal lambs. *Journal of Clinical Investigation* 97: 300–308.

Juul A and Jorgensen JOL (2000) *Growth Hormone in Adults. Physiological and Clinical Aspects*. Cambridge: Cambridge University Press.

Kalmijn S, Janssen JAMJL, Pols HAP, Lamberts SWJ, and Breteler MMB (2000) A prospective study on circulating insulin-like growth factor I (IGF-I), IGF-binding proteins, and cognitive function in the elderly. *Journal of Clinical Endocrinology and Metabolism* 85: 4551–4555.

Kanety H, Karasik A, Klinger B, Silbergeld A, and Laron Z (1993) Long-term treatment of Laron type dwarfs with insulin-like growth factor I increases serum insulin-like growth factor-binding protein 3 in the absence of growth hormone activity. *Acta Endocrinologica* 128: 144–149.

Kaplan SL (1987) Growth: Normal and abnormal. In: Rudolph AM (ed.) *Pediatrics,* 18th edn., pp. 75–92. Norwalk, CT: Appleton & Lange.

Kaplan SL, August GP, Blethen SL, et al. (1986) Clinical studies with recombinant-DNA-derived methionyl human growth hormone in growth hormone deficient children. *Lancet* 327: 697–700.

Kato H, Faria TN, Stannard B, Roberts CT, Jr., and LeRoith D (1994) Essential role of tyrosine residues 1131, 1135, and 1136 of the insulin-like growth factor-I (IGF-I) receptor in IGF-I action. *Molecular Endocrinology* 8: 40–50.

Kelly PA, Goujon L, Sotiropoulos A, et al. (1994) The GH receptor and signal transduction. *Hormone Research* 42: 133–139.

Kennet KF and Cropley AJ (1970) Intelligence, family size and socio-economic status. *Journal of Biosocial Science* 2: 227–236.

Kineman RD and Luque RM (2007) Evidence that Ghrelin is as potent as growth hormone (GH)-releasing hormone (GHRH) in releasing GH from primary pituitary cell cultures of a nonhuman primate (*Papio anubis*), acting through intracellular signaling pathways distinct from GHRH. *Endocrinology* 148: 4440–4449.

Kinney BA, Coschigano KT, Kopchick JJ, Steger RW, and Bartke A (2001) Evidence that age-induced decline in memory retention is delayed in growth hormone resistant GH-R-KO (Laron) mice. *Physiology and Behavior* 72: 653–660.

Kojima M, Hosoda H, Date Y, Nakazato M, Matsuo H, and Kangawa K (1999) Ghrelin is a growth-hormone-releasing acylated peptide from stomach. *Nature* 402: 656–660.

Koltowska-Haggstrom M, Mattsson AF, Monson JP, et al. (2006) Does long-term GH replacement therapy in hypopituitary adults with GH deficiency normalise quality of life? *European Journal of Endocrinology* 155: 109–119.

Koppitz E (1964) *The Bender Gestalt Test for Young Children*. New York: Grune and Stratton.

Kornreich L, Horev G, Schwarz M, Karmazyn B, and Laron Z (2002) Craniofacial and brain abnormalities in Laron syndrome (primary growth hormone insensitivity). *European Journal of Endocrinology* 146: 499–503.

Kranzler JH, Rosenbloom AL, Martinez V, and Guevara-Aguirre J (1998) Normal intelligence with severe insulin-like growth factor I deficiency due to growth hormone receptor deficiency: A controlled study in a genetically homogeneous population. *Journal of Clinical Endocrinology and Metabolism* 83: 1953–1958.

Lackey BR, Gray SL, and Henricks DM (2000) Does the insulin-like growth factor system interact with prostaglandins and proinflammatory cytokines during neurodegeneration? *Proceedings of the Society for Experimental Biology and Medicine* 224: 20–27.

Lai Z, Emtner M, Roos P, and Nyberg F (1991) Characterization of putative growth hormone receptors in the human choroid plexus. *Brain Research* 546: 222–226.

Lai Z, Roos P, Zhai Q, Olsson Y, Fholenhag K, Larsson C, and Nyberg F (1993) Age-related reduction of human growth hormone-binding sites in the human brain. *Brain Research* 621: 260–266.

Laron Z (1991) Usefulness of the growth hormone-releasing hormone test regardless of which fragment is used (GHRH 1–44, 1–40 or 1–29). *Israel Journal of Medicine Science* 27: 343–345.

Laron Z (1995) Growth hormone secretagogues: Clinical experience and therapeutic potential. *Drugs* 50: 595–601.

Laron Z (1999a) The essential role of IGF-I. Lessons from the long-term study and treatment of children and adults with Laron syndrome. *Journal of Clinical Endocrinology and Metabolism* 84: 4397–4404.

Laron Z (1999b) Natural history of the classical form of primary growth hormone (GH) resistance (Laron syndrome). *Journal of Pediatric Endocrinology and Metabolism* 12: 231–249.

Laron Z (2002) Final height and psychosocial outcome of patients with Laron syndrome (primary GH resistance): A model of genetic IGF-I deficiency. In: Gill G, Scell LM, Benso L (eds.) *Human Growth from Conception to Maturity*, ch. 23. pp. 215–225. London: Smith-Gordon.

Laron Z (2002a) Growth hormone and insulin-like growth factor-I: Effects on the brain. In: Pfaff DW, Arnold AP, Etgen AM, Fahrbach SE, and Rubin RT (eds.) *Hormones, Brain and Behavior*, ch. 85, pp. 75–96. Amsterdam: Academic Press (Elsevier Science).

Laron Z (2004) Laron syndrome (primary growth hormone resistance or insensitivity). The personal experience 1958–2003. *Journal of Clinical Endocrinology and Metabolism* 89: 1031–1044.

Laron Z (2008) Insulin-like growth factor-I treatment of children with Laron syndrome (primary growth hormone insensitivity). *Pediatric Endocrinology Reviews* 5: 766–771.

Laron Z, Anin S, Klipper-Aurbach Y, and Klinger B (1992a) Effects of insulin-like growth factor on linear growth, head circumference, and body fat in patients with Laron-type dwarfism. *Lancet* 339: 1258–1261.

Laron Z and Butenandt O (1993) *Growth Hormone Replacement Therapy in Adults – Pros and Cons*. London-Tel Aviv: Freund Publishing House Ltd.

Laron Z and Galatzer A (1980) Aspects of brain development in children and adolescents with pituitary growth hormone deficiency. In: De Wied D and van Keep PA (eds.) *Hormones and the Brain*, pp. 293–302. Lancaster: MTP Press.

Laron Z and Galatzer A (1981) Effect of hGH on head circumference and IQ in isolated growth hormone deficiency. *Early Human Developmental* 5: 211–214.

Laron Z, Kelijman M, Pertzelan A, Keret R, Shoffner JM, and Parks JS (1985) Human growth hormone gene deletion without antibody formation or growth arrest during treatment – a new disease entity? *Israel Journal of Medicine Science* 21: 999–1006.

Laron Z, Klinger B, Blum WF, Silbergeld A, and Ranke MB (1992b) IGF binding protein-3 in patients with Laron-type dwarfism: Effect of exogenous rIGF-I. *Clinical Endocrinology* 36: 301–304.

Laron Z, Lazar L, Galatzer A, and Pertzelan A (1995) Growth hormone treatment accelerates head circumference growth in isolated GH deficiency. In: Hauspie R, Lindgren G, and Falkner F (eds.) *Essays on Auxology*, pp. 48–495. Hertfordshire, UK: Castlemead Publishers.

Laron Z, Lazar L, and Klinger B (1993) Growth hormone, insulin-like growth factor I and brain function. In: Castells S and Wisniewski KE (eds.) *Growth Hormone Treatment in Down's Syndrome*, pp. 151–161. Chichester, NY: Wiley.

Laron Z, Roitman A, and Kauli R (1979) Effect of human growth hormone therapy on head circumference in children with hypopituitarism. *Clinical Endocrinology* 10: 393–399.

Laron Z, Suikkairi AM, Klinger B, Silbergeld A, Pertzelan A, Seppala M, and Koivisto VA (1992c) Growth hormone and insulin-like growth factor regulate insulin-like growth factor-binding protein-1 in Laron type dwarfism, growth hormone deficiency and constitutional short stature. *Acta Endocrinologica* 127: 351–358.

Lee WH, Michels KM, and Bondy CA (1993) Localization of insulin-like growth factor binding protein-2 messenger RNA during postnatal brain development: Correlation with insulin-like growth factors I and II. *Neuroscience* 53: 251–265.

LeRoith D, Werner H, Beitner-Johnson D, and Roberts CT, Jr. (1995) Molecular and cellular aspects of the insulin-like growth factor I receptor. *Endocrine Reviews* 16: 143–163.

Leung DW, Spencer SA, Cachianes G, et al. (1987) Growth hormone receptor and serum binding protein: Purification, cloning and expression. *Nature* 330: 537–543.

Lewis UJ, Bonewald LF, and Lewis LJ (1980a) The 20,000 dalton variant of human growth hormone: Location of the amino acid deletions. *Biochemical and Biophysical Research Communications* 92: 511–516.

Lewis UJ, Singh RNP, Tutwiler GH, Siegel MB, van der Laan EF, and Van der Laan WP (1980b) Human growth hormone: A complex of proteins. *Recent Progress in Hormone Research* 36: 477–508.

Lewitt MS, Saunders H, Phuyal JL, and Baxter RC (1994) Complex formation by human insulin-like growth factor-binding protein-3 and human acid-labile subunit in growth hormone-deficient rats. *Endocrinology* 134: 2402–2409.

Li CH and Papkoff H (1956) Preparation and properties of growth hormone from human and monkey pituitary glands. *Science* 124: 1293–1294.

Lin S-C, Lin CR, Gukovsky I, Lusis AJ, Sawchenko PE, and Rosenfeld MG (1993) Molecular basis of the little mouse phenotype and implications for cell type-specific growth. *Nature* 364: 208–213.

Linfoot JA, Garcia JF, Wei W, Fink R, Sarin R, Born JL, and Lawrence JH (1970) Human growth hormone levels in cerebrospinal fluid. *Journal of Clinical Endocrinology and Metabolism* 31: 230–232.

Linkowski P, Mendelewicz J, Kerkhofs M, et al. (1987) 24-Hour profiles of adrenocorticotropin, cortisol, and growth hormone in major depressive illness: Effect of antidepressant treatment. *Journal of Clinical Endocrinology and Metabolism* 65: 141–152.

Linseman DA, Phelps RA, Bouchard RJ, Le SS, Laessig TA, McClure ML, and Heidenreich KA (2002) Insulin-like growth factor-i blocks Bcl-2 Interacting mediator of cell death (Bim) induction and intrinsic death signaling in cerebellar granule neurons. *Journal of Neuroscience* 22: 9287–9297.

Lobie PE, Garcia-Arragon J, Lincoln D, Barnard R, Wilcox JN, and Waters MJ (1993) Localization and ontogeny of growth hormone receptor gene expression in the central nervous system. *Developmental Brain Research* 74: 225–233.

Loeper S and Ezzat S (2008) Acromegaly: Re-thinking the cancer risk. *Reviews in Endocrine and Metabolic Disorders* 9: 41–58.

Lönn S, Inskip PD, Pollak MN, Weinstein SJ, Virtamo J, and Albanes D (2007) Glioma risk in relation to serum levels of insulin-like growth factors. *Cancer Epidemiology Biomarkers and Prevention* 16: 844–846.

Lupien SB, Bluhm EJ, and Ishii DN (2003) Systemic insulin-like growth factor-I administration prevents cognitive impairment in diabetic rats, and brain IGF regulates learning/memory in normal adult rats. *Journal of Neuroscience Research* 74: 512–523.

Mardh G, Lundin K, Borg G, Jonsson B, and Lindeberg A (1994) Growth hormone replacement therapy in adult hypopituitary patients with growth hormone deficiency: Combined data from 12 European placebo-controlled clinical trials. *Endocrinology and Metabolism* 1(supplement A): 43–49.

Marks JL, Porte D, Jr., and Baskin DG (1992) Localization of type I insulin-like growth factor receptor messenger RNA in the adult rat brain by *in situ* hybridization. *Molecular Endocrinology* 5: 1158–1168.

Maruff P and Falleti M (2005) Cognitive function in growth hormone deficiency and growth hormone replacement. *Hormone Research* 64(supplement 3): 100–108.

Mason JL, Ye P, Suzuki K, D'Ercole AJ, and Matsushima GK (2000) Insulin-like growth factor-1 inhibits mature oligodendrocyte apoptosis during primary demyelination. *Journal of Neuroscience* 20: 5703–5708.

Mayerhoff DI, Lieberman JA, Lemus CZ, Pollack S, and Schneider BS (1990) Growth hormone response to growth hormone-releasing hormone in schizophrenic patients. *American Journal of Psychiatry* 147: 1072–1074.

Mayo KE (1992) Molecular cloning and expression of pituitary-specific receptor for growth hormone-releasing hormone. *Molecular Endocrinology* 6: 1734–1744.

Mayo KE, Cerelli GM, Lebo RV, Bruce BD, Rosenfeld MG, and Evans RM (1985) Gene encoding human growth hormone-releasing factor precursor: Structure, sequence, and chromosomal assignment. *Proceedings of the National Academy of Sciences of the United States of America* 82: 63–67.

McGauley GA (1989) Quality of life assessment before and after growth hormone treatment in adults with growth hormone deficiency. *Acta Paediatrica Scandinavica* 356(supplement): 70–72.

McMorris FA and Duboic-Dalcq M (1988) Insulin-like growth factor I promotes cell proliferation and oligodendroglial commitment in rat glial progenitor cells developing *in vitro*. *Journal of Neuroscience Research* 21: 199–209.

Merimee T and Laron Z (1996) *Growth Hormone, IGF-I and Growth: New Views of Old Concepts. Modern Endocrinology and Diabetes,* vol. 4. London-Tel Aviv: Freund Publishing House Ltd.

Meyer-Bahlburg HFL, Feinmann JA, MacGillivray MH, and Aceto T, Jr. (1978) Growth hormone deficiency, brain development, and intelligence. *American Journal of Diseases of Childhood* 132: 565–572.

Mills JL, Hediger ML, Molloy CA, et al. (2007) Elevated levels of growth-related hormones in autism and autism spectrum disorder. *Clinical Endocrinology* 67: 230–237.

Morisawa K, Sugisaki T, Kanamatsu T, Aoki T, and Noguchi T (1989) Factors contributing to cerebral hypomyelination in the growth hormone deficient little mouse. *Neurochemical Research* 14: 173–177.

Mullis PE, Patel MS, Brickell PM, Hindmarsh PC, and Brook CD (1991) Growth characteristics and response to growth hormone therapy in patients with hypochondroplasia: Genetic linkage of the insulin-like growth factor I gene at chromosome 12q23 to the disease in a subgroup of these patients. *Clinical Endocrinology* 34: 265–274.

Mustafa A, Adem A, Roos P, and Nyberg F (1994) Sex differences in binding of human growth hormone to rat brain. *Neuroscience Research* 19: 93–99.

Naeve GA, Vana AM, Eggold JR, Verge G, Ling N, and Foster AC (2000) Expression of rat insulin-like growth factor binding protein-6 in the brain, spinal cord, and sensory ganglia. *Molecular Brain Research* 75: 185–197.

Niblock MM, Brunso-Bechtold JK, Lynch CD, Ingram RL, McShane T, and Sonntag WE (1998) Distribution and levels of insulin-like growth factor I mRNA across the life span in the Brown Norway X Fischer 344 rat brain. *Brain Research* 804: 79–86.

Niblock MM, Brunso-Bechtold JK, and Riddle DR (2000) Insulin-like growth factor I stimulated dendritic growth in primary somatosensory cortex. *Journal of Neuroscience* 201: 4165–4176.

Nieder C, Andratschke N, Price RE, Rivera B, and Kian Ang K (2005) Evaluation of insulin-like growth factor-1 for prevention of radiation-induced spinal cord damage. *Growth Factors* 23: 15–18.

Niwa M, Sato Y, Saito Y, et al. (1986) Chemical synthesis, cloning and expression of genes for human somatomedin C (insulin like growth factor I) and 59Val somatomedin C. *Annals of the New York Academy of Sciences* 469: 31–52.

Noguchi T (1996) Effects of growth hormone on cerebral development: Morphological studies. *Hormone Research* 45: 5–17.

Noguchi T, Kurata LM, and Sugisaki T (1987) Presence of a somatomedin C – Immunoreactive substance in the central nervous system: Immunohistochemical mapping studies. *Neuroendocrinology* 46: 277–282.

Noguchi T, Sekiguchi M, Sugisaki T, Tsukada Y, and Shimai K (1983) Faulty development of cortical neurons in the Snell dwarf cerebrum. *Developmental Brain Research* 10: 125–138.

Noguchi T and Sugisaki T (1984) Hypomyelination in the cerebrum of the congenitally hypothyroid mouse (*hyt*). *Journal of Neurochemistry* 42: 891–893.

Noguchi T and Sugisaki T (1985) Abnormal neuronal growth in the Little (lit) cerebrum. *Experimental Neurology* 89: 274–278.

Noguchi T, Sugisaki T, Nishikawa N, and Tsukada Y (1988) Restoration of microcephalic cerebrum with hypomyelination in the growth hormone-deficient mouse (*lit*): Stimulatory effects of GH restricted to the first 20 days of postnatal life. *Neurochemical Research* 13: 249–252.

Nyberg F (1997) Aging effects on growth hormone receptor binding in the brain. *Experimental Gerontology* 32: 521–528.

Nyberg F (2000) Growth hormone in the brain: Characteristics of specific brain targets for the hormone and their functional significance. *Frontiers in Neuroendocrinology* 21: 330–348.

Nyberg F and Burman P (1996) Growth hormone and its receptors in the central nervous system-location and functional significance. *Hormone Research* 45: 18–22.

Nyberg F, Nalen B, Fholenhag K, Fryklund L, and Albertsson-Wikland K (1989) Enzymatic release of peptide fragments from human growth hormone which displace 3H-dihydromorphine from rat brain opioid receptors. *Journal of Endocrinological Investigation* 12(supplement 2): 140.

Ocrant I (1993) Insulin-like growth factor binding proteins in nervous-tissue-derived cells. *Annals of the New York Academy of Sciences* 692: 44–50.

Ocrant I, Fay CT, and Parmelee JT (1990) Characterization of insulin-like growth factor binding proteins produced in the rat central nervous system. *Endocrinology* 127: 1260–1267.

Ocrant I, Pham H, Oh Y, and Rosenfeld RG (1989) Characterization of insulin-like growth factor binding proteins of cultured rat astroglial and neuronal cells. *Biochemical and Biophysical Research Communications* 159: 1316–1322.

Offen D, Shtaif B, Hadad D, Weizman A, Melamed E, and Gil-Ad I (2001) Protective effect of insulin-like-growth-factor-1 against dopamine-induced neurotoxicity in human and rodent neuronal cultures: Possible implications for Parkinson's disease. *Neuroscience Letters* 316: 129–132.

Okereke OI, Kang JH, Ma J, Gaziano JM, and Grodstein F (2006) Midlife plasma insulin-like growth factor I and cognitive function in older men. *Journal of Clinical Endocrinology and Metabolism* 91:4306–4312.

Oorschot DE and McLennan IS (1998) The trophic requirements of mature motoneurons. *Brain Research* 789: 315–321.

Pan W and Kastin AJ (2000) Interactions of IGF-I with the blood brain barrier *in vivo* and *in situ*. *Neuroendocrinology* 72: 171–178.

Partridge WM (1986) Receptor-mediated peptide transport through the blood–brain barrier. *Endocrine Reviews* 7: 314–330.

Perez Jurado LA, Phillips JA, III, Summar ML, et al. (1994) Genetic mapping of the human growth hormone-releasing factor gene (GHRF) using two intragenic polymorphism detected by PCR amplification. *Genomics* 20: 132–134.

Pomerance M, Gavaret JM, Jacquemin C, Matricon C, Toru-Delbauffe D, and Pierre M (1988) Insulin and insulin like growth factor I receptors during post-natal development of rat brain. *Developmental Brain Research* 42: 77–83.

Posner BI, Kelly PA, and Friesen HG (1994) Studies of insulin, growth hormone and prolactin binding: Tissue distribution, species variation and characterization. *Endocrinology* 95: 521–531.

Puglianiello A, Germani D, Rossi P, and Cianfarani S (2000) IGF-I stimulates chemotaxis of human neuroblasts. Involvement of type 1 IGF receptor, IGF binding proteins, phosphatidylinositol-3 kinase pathway and plasmin system. *Journal of Endocrinology* 165: 123–131.

Pulford BE and Ishii DN (2001) Uptake of circulating insulin-like growth factors (IGFs) into cerebrospinal fluid appears to be independent of the IGF receptors as well as IGF-binding proteins. *Endocrinology* 142: 213–220.

Rabinovsky ED, Gelir E, Gelir S, et al. (2003) Targeted expression of IGF-1 transgene to skeletal muscle accelerates muscle and motor neuron regeneration. *FASEB Journal* 17: 53–55.

Radcliffe DJ, Pliskin JS, Silvers JB, and Cuttler L (2004) Growth hormone therapy and quality of life in adults and children. *Pharmacoeconomics* 22: 499–524.

Recio-Pinto E, Rechler MM, and Ishii DN (1986) Effects of insulin, insulin-like growth factor-II, and nerve growth factor on neurite formation and survival in cultured sympathetic and sensory neurons. *Journal of Neuroscience* 6: 1211–1219.

Reichlin S (1983) Somatostatin. *New England Journal of Medicine* 309: 1495–1501.

Reinhardt RR and Bondy CA (1994) Insulin-like growth factors cross the blood brain–barrier. *Endocrinology* 135: 1753–1761.

Rinderknecht E and Humbel RE (1978) The amino acid sequence of human insulin like growth factor I and its structural homology, with proinsulin. *Journal of Biological Chemistry* 253: 2769–2776.

Rivera EJ, Goldin A, Fulmer N, Tavares R, Wands JR, and de la Monte SM (2005) Insulin and insulin-like growth factor expression and function deteriorate with progression of Alzheimer's disease: Link to brain reductions in acetylcholine. *Journal of Alzheimers Disease* 8: 247–268.

Rivier J, Spiess J, Thorner M, and Vale W (1982) Characterization of a growth hormone-releasing factor from a human pancreatic islet tumour. *Nature* 300: 276–278.

Rollero A, Murialdo G, Fonzi S, Garrone S, Gianelli MV, Gazzerro E, Barreca A, and Polleri A (1998) Relationship between cognitive function, growth hormone and insulin-like growth factor I plasma levels in aged subjects. *Neuropsychobiology* 38: 73–79.

Rosenfeld RG, Pham H, Keller BT, Borchard RT, and Pardridge WM (1987) Demonstration and structural comparison of receptors for insulin-like growth factor-I and-II (IGF-I and-II) in brain and blood–brain barrier. *Biochemical and Biophysical Research Communications* 149: 159–166.

Ross JL (2005) Effects of growth hormone on cognitive function. *Hormone Research* 64(supplement 3): 89–94.

Rotwein P (1991) Structure, evolution, expression and regulation of insulin-like growth factors I and II. *Growth Factors* 5: 3–18.

Rotwein P, Burgess SK, Milbrandt JD, and Krause JE (1988) Differential expression of insulin-like growth factor genes in rat central nervous system. *Neurobiology* 85: 265–269.

Rubinfeld H and Shimon I (2006) Cortistatin – a new neuroendocrine hormone? *Pediatic Endocrinology Review* 4: 106–110.

Russell JW and Feldman EL (1999) Insulin-like growth factor-I prevents apoptosis in sympathetic neurons exposed to high glucose. *Hormone and Metabolic Research* 31: 90–96.

Russo VC, Gluckman PD, Feldman EL, and Werther GA (2005) The insulin-like growth factor system and its pleiotropic functions in brain. *Endocrine Reviews* 26: 916–943.

Ryan ND, Dahl RE, Birmaher B, et al. (1994) Stimulatory tests of growth hormone secretion in prepubertal major depression: Depressed versus normal children. *Journal of the American Academy of Child and Adolescent Psychiatry* 33: 824–833.

Saatman KE, Contreras PC, Smith DH, et al. (1997) Insulin-like growth factor-1 (IGF-1) improves both neurological motor and cognitive outcome following experimental brain injury. *Experimental Neurology* 147: 418–427.

Sakkas PN, Soldatos CR, Bergiannaki JD, Paparrigopoulos TJ, and Stefanis CN (1998) Growth hormone secretion during sleep in male depressed patients. *Progress in Neuropsychopharmacology and Biological Psychiatry* 22: 467–483.

Sartorio A, Conn A, Molinari E, Riva G, Morabito F, and Faglia G (1996) Growth, growth hormone and cognitive functions. *Hormone Research* 45: 23–29.

Seigel GM, Chiu L, and Paxhia A (2000) Inhibition of neuroretinal cell death by insulin-like growth factor-1 and its analogs. *Molecular Vision* 6: 157–163.

Seino S, Seino M, Nishi S, and Bell GI (1989) Structure of the human insulin receptor gene and characterization of its promoter. *Proceedings of the National Academy of Sciences of the United States of America* 86: 114–118.

Sendtner M (1995) Molecular biology of neurotrophic factors. *Baillieres Clinical Neurology* 4: 575–591.

Shevah O, Kornreich L, Galatzer A, and Laron Z (2005) The intellectual capacity of patients with Laron syndrome (LS) differs with various molecular defects of the growth hormone receptor gene. Correlation with CNS abnormalities. *Hormone Metabolic Research* 37: 757–760.

Shevah O and Laron Z (2006) Genetic analysis of the pedigrees and molecular defects of the GH-receptor gene in the Israeli cohort of patients with Laron syndrome. *Pediatric Endocrinology Reviews* 3(supplement 3): 489–497.

Shimon I, Taylor JE, and Dong JZ (1997) Somatostatin receptor subtype specificity in human fetal pituitary cultures. Differential role of SSTR2 and SSTR5 for growth hormone, thyroid-stimulating hormone, and prolactin regulation. *Journal of Clinical Investigation* 99: 789–798.

Shintani M, Ogawa Y, Ebihara K, et al. (2001) Ghrelin, an endogenous growth hormone secretagogue, is a novel orexigenic peptide that antagonizes leptin action through the activation of hypothalamic neuropeptide Y/Y1 receptor pathway. *Diabetes* 50: 227–232.

Shiotani A, O'Malley BW, Coleman ME, Alila HW, and Flint PW (1998) Reinnervation of motor endplates and increased muscle fiber size after human insulin-like growth factor I gene transfer into the paralyzed larynx. *Human Gene Therapy* 9: 2039–2047.

Silbergeld A, Lazar L, Erster B, Keret R, Tepper R, and Laron Z (1989) Serum growth hormone binding protein activity in healthy neonates, children and young adults correlation with age, height and weight. *Clinical Endocrinology* 31: 295–303.

Silbergeld A, Lilos P, and Laron Z (2006) Comparative catch-up growth of height, foot and brain in children with Laron syndrome (LS) treated by IGF-I to that of children with IGHD treated by hGH. (Abstract #P03–610). 45th Annual Meeting of the European Society for Paediatric Endocrinology(ESPE), Rotterdam, The Netherlands, June 30–July 3, 2006. *Hormone Research* 65(supplement 4): 177.

Silbergeld A, Lilos P, and Laron Z (2007) Foot length, before and during insulin-like growth factor-I treatment of children with Laron syndrome compared to human growth hormone treatment of children with isolated growth hormone deficiency. *Journal of Pediatric Endocrinology and Metabolism* 20: 1325–1328.

Snell GD (1929) Dwarf, a new mendelian recessive character of the house mouse. *Proceedings of the National Academy of Sciences of the United States of America* 15: 733–734.

Stabler B, Siegel PT, Clopper RR, Stoppani CE, Compton PG, and Underwood LE (1998) Behavior change after growth hormone treatment of children with short stature. *Journal of Pediatrics* 133: 366–373.

Stanhope R, Wilks Z, and Hamil G (1994) Failure to grow: Lack of food or lack of love? *Professional Care of Mother and Child* 4: 234–237.

Sugisaki T, Noguchi T, Kanamatsu T, and Nishikawa N (1987) Alteration of cerebral ganglioside metabolism in developing Snell dwarf mice. *International Journal of Developmental Neuroscience* 5: 399–405.

Sullivan KA and Feldman EL (1994) Immunohistochemical localization of insulin-like growth factor II and insuln-like growth factor binding protein-2 during development in the rat. *Endocrinology* 135: 540–547.

Sumantran VN and Feldman EL (1993) Insulin-like growth factor I regulates c-myc and GAP-43 messenger ribonucleic acid expression in SH-SY5Y human neuroblastoma cells. *Endocrinology* 132: 2017–2023.

Svensson J, Diez M, Engel J, et al. (2006) Endocrine, liver-derived IGF-I is of importance for spatial learning and memory in old mice. *Journal of Endocrinology* 189: 617–627.

Tagami M, Ikeda K, Nara Y, Fujino H, Kubota A, Numano F, and Yamori Y (1997a) Insulin-like growth factor-1 attenuates apoptosis in hippocampal neurons caused by cerebral iscehmia and reperfusion in stroke-prone spontaneously hypertensive rats. *Laboratory Investigation* 76: 613–617.

Tagami M, Yamagata K, Nara Y, Fujino H, Kubota A, Numano F, and Yamori Y (1997b) Insulin-like growth factors prevent apoptosis in cortical neurons isolated from stroke-prone spontaneously hypertensive rats. *Laboratory Investigation* 76: 603–612.

Terman LM and Merill MA (1960) *Stanford Binet Intelligence Scale*. Boston, MA: Houghton Rifflin Company.

Turnley AM (2005) Growth hormone and SOCS2 regulation of neuronal differentiation: Possible role in mental function. *Pediatric Endocrinology Review* 2: 366–371.

Underwood LE (1997) Short stature. In: Bardin CW (ed.) *Current Therapy in Endocrinology and Metabolis*, pp. 66–67. Boston: Mosby.

Vamvakopoulos NC, Kunz J, Olberding U, et al. (1994) Mapping the human growth hormone-releasing hormone receptor (GHRHR) gene to the short arm of chromosome 7 (7p13-p21) near the epidermal growth factor receptor (EGFR) gene. *Genomics* 20: 338–340.

Vance ML and Mauras N (1999) Growth hormone therapy in adults and children. *New England Journal of Medicine* 341: 1206–1216.

Van Dam S (2006) Somatropin therapy and cognitive function in adults with growth hormone deficiency: A critical review. *Treaties in Endocrinology* 5: 159–170.

Van Golen CM and Feldman EL (2000) Insulin-like growth factor I is the key growth factor in serum that protects neuroblastoma cells from hyperosmotic-induced apoptosis. *Journal of Cell Physiology* 184: 24–32.

Wajnrajch MP, Chua SC, Green ED, and Leibel RL (1994) Human growth hormone-releasing hormone receptor (GHRHR) maps to a YAC at chromosome 7p15. *Mammalian Genetics* 5: 595.

Walter HJ, Berry M, Hill DJ, Cwyfan-Hughes S, Holly JMP, and Logan A (1999) Distinct sites of insulin-like growth factor (IGF)-II expression and localization in lesioned rat brain: Possible roles of IGF binding proteins (IGFBPs) in the mediation of IGF-II activity. *Endocrinology* 140: 520–522.

Wechsler D (1949/1974) *Wechsler Intelligence Scale for Children (WISC)*. New York: The Psychological Corporation.

Wechsler D (1955) *Wechsler Adult Intelligence Scale*. New York: The Psychological Corporation.

Werner H, Re GG, Drummond IA, et al. (1993) Increased expression of the insulin-like growth factor I receptor gene,

IGF1R, in Wilms tumor is correlated with modulation of IGF1R promoter activity by the WT1 Wilms tumor gene product. *Proceedings of the National Academy of Sciences of the United States of America* 90: 5828–5832.
Werther GA, Abate M, Hogg A, et al. (1990) Localization of insulin-like growth factor-I mRNA in rat brain by *in situ* hybridization – relationship to IGF-I receptors. *Molecular Endocrinology* 4: 773–778.
Wiren L, Bengtsson BA, and Johannsson G (1998) Beneficial effects of long-term GH replacement therapy on quality of life in adults with GH deficiency. *Clinical Endocrinology (Oxf)* 48: 613–620.
Woods KA, Camacho-Hubner C, Savage MO, and Clark AJL (1996) Intrauterine growth retardation and postnatal growth failure associated with deletion of the insulin-like growth factor I gene. *New England Journal of Medicine* 335: 1363–1367.
Woods KA, Dastot F, Preece MA, et al. (1997) Phenotype: genotype relationships in growth hormone insensitivity syndrome. *Journal of Clinical Endocrinology and Metabolism* 82: 3529–3535.
Ye P, Carson J, and D'Ercole AJ (1995) *In vivo* actions of insulin-like growth factor-I (IGF-I) on brain myelination: Studies of IGF-I and IGF binding protein-1 (IGFBP-1) transgenic mice. *Journal of Neuroscience* 15: 7344–7356.
Ye P and d'Ercole J (1998) Insulin-like growth factor I (IGF-I) regulates IGF binding protein-5 gene expression in the brain. *Endocrinology* 139: 65–71.
Ye P, Kollias G, and D'Ercole AJ (2007) Insulin like growth factor (IGF)-I ameliorates demyelination induced by tumor necrosis factor (TNF)-alpha in transgenic mice. *Journal of Neuroscience Research* 85: 712–772.
Ye P, Price W, Kassiotis G, Kolllias G, and D'Ercole AJ (2003) Tumor necrosis factor alpha regulation of insulin like growth factor I, type 1 IGF receptor, and IGF binding protein expression in cerebellum of transgenic mice. *Journal of Neuroscience Research* 71: 721–731.
Ye P, Umayahara Y, Ritter D, Bunting T, Auman H, Rotwein P, and D'Ercole AJ (1997) Regulation of insulin-like growth factor I (IGF-I) gene expression in brain of transgenic mice expressing an IGF-I-luciferase fusion gene. *Endocrinology* 138: 5466–5475.
Yokoyama S, Stefaneanu L, and Kovacs K (1997) Pituitary insulin like growth factors. *Endocrinology and Pathology* 3: 167–179.
Yu Y, Kastin AJ, and Pan W (2006) Reciprocal interactions of insulin and insulin like growth factor I in receptor mediated transport across the blood–brain barrier. *Endocrinology* 147: 2611–2615.
Zachmann M, Fernandez F, Tassinari D, Thakker R, and Prader A (1980) Anthropometric measuremin patients with growth hormone deficiency before treatment with human growth hormone. *European Journal of Pediatrics* 133: 277–282.
Zalsman G, Oquendo M, Greenhill L, Goldberg P, Kamali M, Martin A, and Mann JJ (2006) Neurobiology of depression in children and adolescents. *Child and Adolescent Psychiatric Clinics of North America* 15: 843–868.
Zapf J, Gosteli-Peter M, and Schmid C (1996) Insulin-like growth factor binding proteins (IGFBPs). In: Merimee T and Laron Z (eds.) *Growth Hormone, IGF-I and Growth: New Views of Old Concepts. Modern Endocrinology and Diabetes,* vol. 4, pp. 45–71. London-Tel Aviv: Freund Publishing House Ltd.
Zegher M, Popken G, Zhang J, et al. (2007) Insulin like growth factor type 1 receptor signaling in the cells of oligodentrocyte lineage is required for normal *in vivo* oligodendrocyte development and myelination. *Glia* 55: 400–411.
Zhai Q, Lai Z, Roos P, and Nyberg F (1994) Characterization of growth hormone binding sites in rat brain. *Acta Psychiatrica Scandinavica* 406(supplement): 92–95.
Zhang J and D'Ercole AJ (2004) Expression of Mcl-1 in cerebellar granule neurons is regulated by IGF-I in a developmentally specific fashion. *Developmental Brain Research* 152: 255–263.
Zheng WH, Kar S, and Quirion R (2002) Insulin-like growth factor-1-induced phosphorylation of transcription factor FKHRL1 is mediated by phosphatidylinositol 3-kinase/akt kinase and role of this pathway in insulin-like growth factor-1-induced survival of cultured hippocampal neurons. *Molecular Pharmacology* 62: 225–233.
Zhou Y, Xu BC, Maheshwari HG, et al. (1997) A mammalian model for Laron syndrome produced by targeted disruption of the mouse growth hormone receptor/binding protein gene (the Laron mouse). *Proceedings of the National Academy of Sciences of the United States of America* 94: 13215–13220.
Zumkeller W, Saaf M, Rahn T, and Hall K (1991) Demonstration of insulin-like growth factors I, II and heterogeneous insulin-like growth factor binding proteins in the cyst fluid of patients with craniopharyngioma. *Neuroendocrinology* 54: 196–201.

Further Reading

Laron Z (2002b) Molecular mutations in the human growth hormone axis. In: Eugster EA and Pescovitz OH (eds.) *Developmental Endocrinology From Research to Clinical Practice*, pp. 43–76. Totowa, NJ: The Humana Press Inc.
Mustafa A, Bogdanovic N, Nyberg F, et al. (2005) Effects of long-term ovariectomy and ovarian steroids on somatogenic binding sites in rat brain and liver. *Neuroscience Letters* 194: 193–196.

15 Neurosteroids: From Basic Research to Clinical Perspectives

C A Frye, University at Albany-State University of New York, Albany, NY, USA

Chapter Outline

Glossary

allopregnanolone Also known as 3α-hydroxy-5α-pregnan-20-one (3α,5α-THP), it is a neurosteroid and a metabolite of progesterone that has agonist-like actions at gamma-aminobutyric acid (GABA) receptors in the brain that modify response to stress and may influence expression of many behaviors.

autocrine signaling This is a form of signaling in which a cell secretes a hormone that binds to receptors on the same cell, thereby leading to changes in the cell.

chemoconvulsants These are drugs, which when administered, can induce seizures.

endocrine signaling This is a form of signaling in which a gland secretes a hormone that gets into circulation and thereby binds to receptors on distant cells, which leads to changes in the cell.

intracrine signaling This refers to hormones that act inside cells, such as steroid hormones that act through intracellular, nuclear cognate steroid receptors. Other actions of hormones occur via endocrine, autocrine, or paracrine effects by binding to receptors on cell surfaces.

neuroactive steroids These are formed in the central and peripheral nervous system from prohormones secreted by peripheral glands and typically have actions through nontraditional steroid receptor targets.

neurosteroids These are synthesized in the central and peripheral nervous system, in glial cells, *de novo* from cholesterol, independent of peripheral glands.

paracrine signaling This is a form of signaling in which a cell secretes a hormone that binds to receptors on an adjacent cell, thereby leading to changes in the cell.

progestins These are synthetic steroid hormones that can produce effects similar to progestogens.

progesterone It is the major naturally occurring human progestogen. It is a 21-carbon steroid hormone that when declining over the menstrual cycle leads to menstruation, and is increased during pregnancy to support gestation and embryogenesis.

progestogens These are other natural steroid hormones that produce effects similar to progesterone.

prohormones These are hormones that may have minimal effects by themselves but serve as a precursor to another hormone that may mediate more substantive effects.

15.1 Introduction

Steroid hormones play fundamental roles in the development and function of the central nervous system (CNS). Gonadal hormones, secreted by endocrine glands, exert effects during critical periods of development (typically pre- or perinatal) to organize or permanently change structure (such as sexually dimorphic areas of the hypothalamus) and/or function (hypothalamic–pituitary–gonadal (HPG) axis) of the brain. Also, steroid hormones have activational effects to mitigate temporary changes in brain function in the already-developed CNS. Gender differences and/or hormonal effects in the incidence and/or expression of neuropsychiatric and/or neurodegenerative disorders may reflect organizing and/or activating effects of steroids on CNS function, and imply that neuroendocrine factors may play an important role. In recent years, evidence has emerged that has broadened the scope of our understanding of some of the effects and mechanisms of steroids. In addition to classic endocrine effects, steroids have autocrine actions to influence cells that produce them, intracrine effects to mediate intracellular events, and paracrine (or neurostransmitter-like) effects to induce biological response in adjacent cells. The rapidity of some of these effects of steroids in the CNS is incongruous with classic genomic actions of steroids through binding to cognate, intracellular steroid receptors in hypothalamic or limbic regions to modulate transcription and translation (Pfaff et al., 1976; Osterlund et al., 2000; Shughrue et al., 1997). Steroids also have membrane and/or rapid signaling actions that cannot readily be explained by binding to cognate intracellular steroid receptors. This chapter discusses the role of steroids that are produced in the brain, and/or have rapid signaling actions, in neuropsychiatric, neurodevelopmental, neurodegenerative, and/or age-related disorders.

15.2 The Brain is an Endocrine Organ – Neurosteroidogenesis

The brain, similar to peripheral glands (gonads, adrenals, and placenta), is an endocrine organ. Coordinated actions of steroidogenic enzymes in neurons and glia in different parts of the CNS metabolize peripheral steroids to neuroactive products (neuroactive steroids) and/or produce steroids *de novo* in the brain, independent of peripheral gland secretion (neurosteroids). In general, neuro(active)steroids produce rapid effects on neuronal excitability and synaptic function that involve direct or indirect modulation of ion-gated or other neurotransmitter receptors and transporters rather than actions via classic, nuclear steroid hormone receptors.

15.2.1 The Discovery of Biosynthesis

Steroids that could be synthesized within the CNS and/or peripheral nervous system (PNS) were termed neurosteroids by Baulieu (1980, 1991) to differentiate them from steroids derived from peripheral organs (gonads, adrenals, and placenta). The initial discovery that steroids were synthesized in the brain came from observations that the levels of steroids, such as pregnenolone, dehydroepiandrosterone (DHEA), and their sulfate and lipoidal esters, were higher in the CNS and PNS, than in circulation. The same enzymes involved in steroidogenesis in peripheral glands were identified in the nervous system and found to be responsible for biosynthesis; albeit, they are typically expressed in brain at levels 2 to 5 orders of magnitude lower than in adrenal or gonadal tissues (Compagnone and Mellon, 2000; Furukawa et al., 1998). Further, central steroid concentrations can be much higher than in the circulation and these levels can persist after extirpation of peripheral glands (gonadectomy and/or adrenalectomy). Together, these findings demonstrated that steroids are synthesized in the nervous system and their localization in the CNS/PNS was neither due to peripheral synthesis, nor sequestration and/or accumulation in neural tissues (Baulieu, 1980, 1991; Majewska, 1992; Mellon, 1994; Paul and Purdy, 1992). Thus, early investigations of neurosteroids by Baulieu and colleagues, which they refer to so eloquently as being, "of the nervous system, by the nervous system, and for the nervous system," have been a profound discovery. Key information regarding the central production of neurosteroids follows.

15.2.2 Peripheral-Type Benzodiazepine Receptor Recognition Site

The peripheral-type benzodiazepine receptor (PBR) recognition site, which is essential for neurosteroidogenesis, binds cholesterol in nanomolar affinities. The PBR was first identified in 1977 as the binding site for diazepam in peripheral tissues; however, the most extensively investigated functions of PBRs are their roles in biosynthesis of steroids. The PBR is a high-affinity cholesterol-binding protein that helps to import cholesterol into the mitochrondria, whereupon it can be oxidized to pregnenolone by two proteins that initiate steroidogenesis – the steroidogenic acute regulatory (StAR) protein and cytochrome P450-dependent C27 side chain cleavage enzymes (P450scc), rate-limiting steps in steroid biosynthesis (King et al., 2004; Mellon and Deschepper, 1993; Papadopoulos et al., 2006).

15.2.3 Metabolic Pathways

The nervous system expresses all of the enzymes required for steroid biosynthesis that produce a variety of neurosteroids, such as progesterone (P), DHEA, 17β-estradiol (E_2), 3α,5β-34 tetrahydroprogesterone (pregnanolone), and 5α-reduced steroids, such as 3α,5α-tetrahydrodeoxycorticosterone (THDOC; Compagnone and Mellon, 2000). Steroid sulfates, such as DHEA-sulfate (DHEA-S) and pregnenolone sulfate, are also produced *de novo* in the primate and rodent brain (Ebner et al., 2006; Kríz et al., 2005; Liere et al., 2004; Weill-Engerer et al., 2002). It is beyond the scope of this chapter to review the effects and mechanisms of all neurosteroids. As such, this chapter focuses on the role of pregnane neurosteroids, which has been the most extensively investigated. We have recently reviewed the effects and mechanisms of and rostane neurosteroids (Frye, 2006, 2007).

Once pregnenolone is synthesized from cholesterol by StAR and P450scc, several 3-hydroxysteroid dehydrogenase (3-HSD) enzymes can convert the prohormones to neuroactive steroids which occur through the actions of 3α-reduced enzymes. Rodents have a single 3α-HSD pathway for steroidogenesis which mediates all reactions, whereas that of people is more complex. Key examples of neurosteroids and neuroactive steroids, which are derived from cholesterol and/or blood-borne precursors, respectively, are pregnenolone, which is converted to P, and its product, 5α-pregnane-3,20-dione-dihydroprogesterone (DHP), which can be converted to 3α-hydroxy-5α-pregnan-20-one (3α,5α-THP) and its 5β-stereoisomer (3α,5β-THP; see **Figure 1**). There is also an analogous pathway for biosynthesis of androgens. Testosterone (T), and its product, 5α-androstan-3,20-dione-dihydrotestosterone (DHT), can be converted to 5α-androstan-3α-17β-diol (3α-diol) and its 3β-stereoisomer (3β-diol).

15.2.4 Metabolic Enzymes

Whether estrogens, progestogens, and/or androgens are synthesized in the CNS/PNS depends upon many factors including timing during development, expression of metabolism enzymes, and substrates for them in particular tissues (brain regions), cell types (i.e., neurons vs. glia), and/or areas of cells (cell bodies, fibers, etc.). Neurosteroids are formed from biosynthesis of cholesterol by P450 and non-P450 enzymes (17β-HSD, aromatase, 3α-HSD, and 5α-reductase). Gene expression of P450c17 is

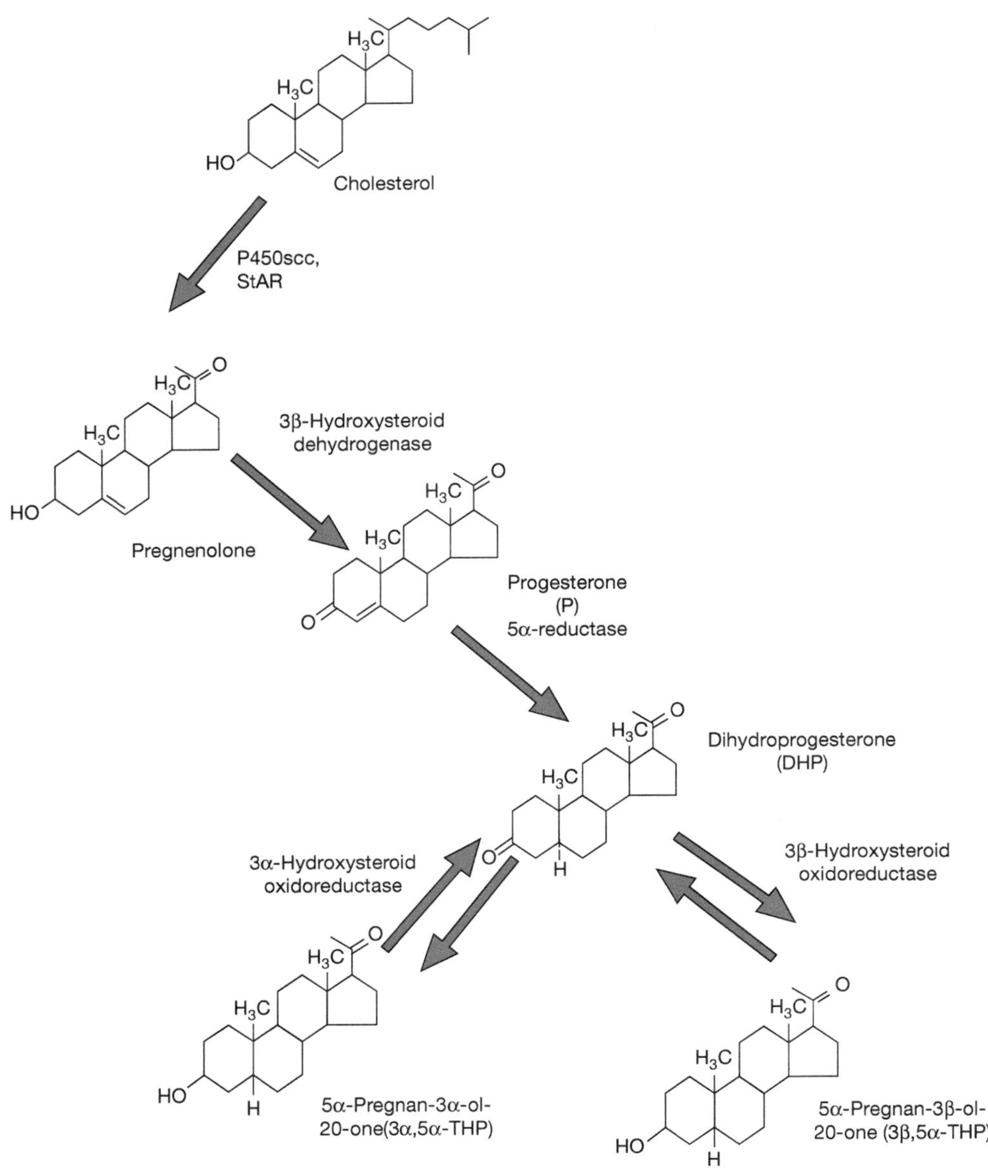

Figure 1 Pregnane neurosteroid metabolism/biosynthesis pathway.

developmentally regulated with early expression being different from that observed in the mature brain (Mellon, 2007). This implies that certain metabolites may play key roles that are relegated throughout development without continuous expression in adulthood. Further, there are developmental differences in the substrates that two 3α-HSDs in human fetal brain utilize, with one using DHP and another using DHT. Conservation in the localization of steroidogenic enzymes, which are involved in synthesizing neurosteroids, to specific brain regions within a variety of species, indicates that their function may be similar throughout evolution. Most steroidogenic enzymes have been found in limbic regions (cortex, hippocampus, basal ganglia, hypothalamus, and thalamus), cerebellum, tegmentum, tectum, pons, medulla, spinal cord, pituitary, and various PNS regions. Some investigations have attempted to co-localize steroidogenic enzymes in whole brain, neuronal, and/or glial cell cultures. Indeed, some steroidogenic enzymes are expressed only in neurons, or only in glia (type 1 astrocytes, Schwann cells), while other enzymes are expressed in both cell types. Investigations are still needed to elucidate how specific brain regions synthesize or metabolize steroids, which are essential to reveal the function of neurosteroids. Ascertaining the relative concentrations of peripheral versus central levels of 3α,5α-THP is of great interest to understanding the functional significance of neurosteroids (see Mellon (2007) for an expert review of this topic).

15.2.5 Patterns in Secretion

In people, variations in P and 3α,5α-THP levels have been examined. During the menstrual cycle, patterns in plasma levels of pregnenolone and 3α,5α-THP are similar to that of P. During the luteal phase, circulating concentrations of pregnenolone and 3α,5α-THP are two- to fourfold higher (2–4 $nmol\ L^{-1}$) than they are during the follicular phase (~1 $nmol\ L^{-1}$; Genazzani et al., 1998; Purdy et al., 1990; Sundström and Bäckström, 1998a,b; Wang et al., 1996). Throughout pregnancy, serum levels of pregnenolone, P, and 3α,5α-THP increase with gestation (Luisi et al., 2000) and peak late in the third trimester (50–100 $nmol\ L^{-1}$; Herbison, 2001; Luisi et al., 2000). The levels achieved during late pregnancy are within the range that can produce sedation (80–160 $nmol\ L^{-1}$; Sundström et al., 1999). Notably, within 1 h after delivery, maternal serum 3α,5α-THP levels are decreased significantly, albeit pregnenolone levels in serum do not decline as substantially until 1 day later.

There is evidence that P and 3α,5α-THP bioaccumulate in brain. Investigation of levels of P and 3α,5α-THP in the postmortem brain of pre- and postmenopausal women shows the expected patterns of menstrual variations and lower levels postclimacteric (Bixo et al., 1997; Purdy et al., 1991). 3α,5α-THP levels varied by region with the highest concentrations being seen in the midbrain and hypothalamus (14–21 $ng\ g^{-1}$; Bixo et al., 1997). These variations over reproductive events in circulating concentrations of prohormones for P, and its products, imply that there may be effects of peripheral steroid secretion on central biosynthesis. In addition, these variations indicate that neurosteroidogenesis may influence circulating steroid concentrations. Indeed, one of the rate-limiting factors in understanding more about the functional significance of neurosteroids lies within the challenge of being able to parse out the relative contributions of central versus peripheral endocrine glands. Irrespective of this, there is evidence that local biosynthesis and paracrine/autocrine effects of neurosteroids can precisely and rapidly alter neuronal function in a manner not achievable by neuroactive steroids derived from circulation. Some of these diverse effects are discussed below.

15.3 Actions of Neurosteroids

15.3.1 Nonclassical Actions of Neurosteroids

While steroids secreted by peripheral glands are typically considered to have their effects through classic nuclear steroid receptors, neurosteroids can have more immediate, rapid signaling effects. Neurosteroids can have actions through ion-channel-associated membrane receptors to elicit rapid changes in signaling that can occur within milliseconds to seconds. The most extensively investigated actions of neurosteroids are those at synaptic and extrasynaptic γ-aminobutyric acid type A ($GABA_A$) receptors, which can mediate phasic and tonic inhibition of neurons. The GABA transmitter system is the major inhibitory system in the mammalian CNS. In nanomolar concentrations, 5α-reduced neurosteroids (3α,5α-THP, 3α-diol, and THDOC) are positive modulators of $GABA_A$. These neurosteroids can allosterically modulate and directly activate $GABA_A$ to increase chloride channel currents and thereby lower neuronal excitability with 20- and 200-fold higher efficacy than benzodiazepines and barbiturates, respectively (Belelli and Lambert, 2005; Brot et al., 1997; Gee et al., 1995; Lambert et al., 2003; Morrow et al., 1987; Reddy, 2004; Weir et al., 2004). Pregnenolone sulfate and DHEA-S are negative modulators and antagonize $GABA_A$ channel activity (Majewska and Schwartz, 1987; Majewska et al., 1986, 1990). In addition to distinct neurosteroids exerting specific effects, the nature of these effects are also influenced by many other factors, including the concentration of the steroid and the subunit composition of the receptor.

15.3.2 Actions of Neurosteroids through $GABA_A$ Receptors

When GABA binds to its receptors, which consist of five co-assembled subunits that form ligand-gated chloride channels, the influx of chloride ions increases and thereby hyperpolarizes the postsynaptic membrane rendering the postsynaptic cell less prone to excitation (Mehta and Ticku, 1999). To date, 19 different $GABA_A$ receptor subunits, α1–6, β1–4, γ1–3, δ, ε, π, and ρ1 and 2, have been cloned (Whiting et al., 1999). $GABA_A$ receptors of different subunit composition vary in their anatomical distribution (Pirker et al., 2000) and physiological properties (Hevers and Lüddens, 1998). Although there is an enormous number of possible $GABA_A$ combinations, the most prevalent combination of $GABA_A$ receptors is the α1β2γ2 subunit combination (Whiting et al., 1999), which has α4/δ subunit co-localization. High densities of α4/δ-subunit-containing $GABA_A$s are responsible for mediating tonic inhibition by altering the efficacy of GABA as an agonist. Such receptor composition is found in the thalamus, striatum, hippocampal dentate gyrus, olfactory bulb, and layers 2–3 of the neocortex (Bianchi and Macdonald, 2003; Wohlfarth et al., 2002). Neurosteroids potentiate the action of GABA at the δ-subunit-containing $GABA_A$s by increasing the efficacy of the agonist, due to a change in the desensitization property of the receptors in the presence of the neurosteroid (Bianchi and Macdonald, 2003; Wohlfarth et al., 2002). Neurosteroids, in physiological concentrations, appear to solely target the tonic inhibition mediated by δ-subunit-containing $GABA_A$ to have a net effect to reduce network excitability (Stell et al., 2003).

15.3.3 Other Targets for Neurosteroids

Neurosteroids can also have actions through other nonsteroidal, ligand-gated, ion channel, and/or G-protein-coupled receptors. Neurosteroids can have actions through receptors specific for glycine, sigma type 1 (σ1), glutamate, serotonin, acetylcholine, oxytocin, and nicotinic/muscarinic receptors (Rupprecht and Holsboer, 1999). P, 3α,5α-THP, pregnanolone, pregnenolone sulfate, and DHEA-S are all negative modulators of glycine receptors. At sigma type 1 (σ1) receptors, DHEA, pregnenolone, and their sulfate ester are positive modulators, whereas P is a negative modulator. Actions of neurosteroids at glutamate receptors include negative modulatory effects of pregnenolone sulfate at AMPA and kainate receptors but positive modulatory effects at *N*-methyl-D-aspartic acid (NMDA) receptors. P and/or 3α,5α-THP have negative modulatory actions at receptors for norepinephrine, dopamine, serotonin, acetylcholine, oxytocin, as well as at nicotinic/muscarinic receptors. In addition to actions at these nonsteroidal receptors, neurosteroids also alter neuronal function membrane E_2 receptors (Chaban et al., 2004) and G-protein-coupled membrane progestin receptors (Zhu et al., 2003).

15.4 Neurosteroids Clinical Relevance

15.4.1 Neurosteroids and Neuronal Growth and Development

Neurosteroids may have an important role in maintaining normal growth and development and may help protect against fetal brain injury. Cell-culture experiments suggest that neurosteroids may increase the survival and differentiation of both neurons and glial cells (Marx et al., 2000). Also, neurosteroids can regulate synthesis of myelin proteins and thereby improve and/or facilitate myelinization (Baulieu and Schumacher, 2000; Chan et al., 1998; Koenig et al., 1995). Plasma 3α,5α-THP concentrations are increased during pregnancy and high levels have been reported in the circulation of the fetus (Bicikova et al., 2000, 2002). In sheep, central levels of 3α,5α-THP also increase throughout gestation and then decline at birth (Nguyen et al., 2004). Indeed, circulating levels of 3α,5α-THP in the fetal sheep may depend upon maternal levels (Nguyen et al., 2003a,b); however, central production may be somewhat independent as finasteride only reduced central, not plasma, 3α,5α-THP levels of fetal sheep (Yawno et al., 2007). 3α,5α-THP mediates CNS activity of fetal sheep. Inhibiting fetal production of 3α,5α-THP decreases sleep-like behavior and increases arousal (Crossley et al., 1997). Although levels of 3α,5α-THP decline in the fetus circulation and brain after delivery, stress can profoundly increase levels of 3α,5α-THP in the brain and may be protective. For example, hypoxia (Nguyen et al., 2004) or endotoxin (Billiards et al., 2006) markedly increases 3α,5α-THP production in the fetal sheep brain and attenuated the apoptosis produced by the damage (Yawno et al., 2007). Together, these findings suggest that the 3α,5α-THP in the fetal sheep brain may mediate the sleep-like state seen during pregnancy; the fetal brain has the capacity for 3α,5α-THP biosynthesis, which may mediate growth and/or protective effects of (dis)stress.

15.4.2 Neurosteroids and Gestation

In addition to direct effects on the fetus, progestogens may also have indirect effects on offspring due to their role in the successful maintenance of pregnancy. In rats, pregnancy is characterized by increased secretion of P by the corpora lutea and placenta, which maintain progestogen levels, including 3α,5α-THP. Both circulating and brain levels of P and 3α,5α-THP are high in pregnancy of rats; however, central P levels peak and begin to decline earlier than 3α,5α-THP (Concas et al., 1998). 3α,5α-THP peaks around gestation day (GD) 19, such that at term, around GD 21, 3α,5α-THP levels are at nadir prior to onset of parturition (Concas et al., 1998), such that the decline in 3α,5α-THP coincides with the onset of labor. This may reflect an inhibitory action of 3α,5α-THP on oxytocin neurons in late pregnancy (Brussaard and Herbison, 2000), which may be important in preventing preterm labor, since oxytocin is an important stimulator of uterine contractions at term (Antonijevic et al., 2000; Russell et al., 2003).

15.4.3 Neurosteroids and Preterm Birth

In humans, environmental stress caused by adverse life events in late gestation can have profound negative consequences on the developing brain that persist into adulthood. Stress manipulations in the current study occurred at a critical time during hippocampal formation in the rodent brain. Much neuronal specification and cell body migration occur in this late stage of gestation. In rodents, maternal stress during this period produces deficits in long-term potentiation and long-term depression in hippocampus, impairs spatial memory, increases anxiety and depressive behavior, and enhances susceptibility to addiction (Deminière et al., 1992; Walf and Frye, 2006; Weinstock, 2001, 2005). Among women, stress during pregnancy can result in preterm labor and a wide variety of physical and cognitive birth defects in offspring (Stewart et al., 1983, 1987; Lancefield et al., 2006; Nosarti et al., 2002, 2004). Thus, we have been investigating the role of inhibiting 3α,5α-THP production on birth outcomes and the potential negative impact on offspring of preterm birth.

15.4.4 Neurosteroids and Autism Spectrum Disorders

Autism spectrum disorders (ASDs) are neurodevelopmental disorders that are associated with deficits in social, cognitive, and/or affective behavior. There are gender differences in the incidence and/or expression of ASD, with boys being afflicted 4 times more often than girls. Differences in stress-related neurobiological factors, such as the adrenal hormone, cortisol, have also been reported in ASDs. Despite these findings which suggest that neuroendocrine factors may be involved in the etiology and/or expression of some ASDs, the possible involvement of neurosteroids in autism has been poorly studied. One study reported that nonmedicated subjects with autism have lower levels of plasma DHEA-S compared to normal controls (Strous et al., 2003b); however, another found no difference in DHEA-S levels between pre- and postpubertal males with autism and controls (Tordjman et al., 1995).

15.4.5 Neurosteroids and Drug Abuse

There are gender differences in some drug addictions. Although men may be more likely to become addicted, this may be accounted for by the greater opportunity or access they have to some drugs. Indeed, evidence is emerging that women may be more sensitive to effects of some drugs. Moreover, during adolescence, individuals may be much more sensitive to hormonal and/or stress effects, which may influence their vulnerability to drug effects. Given the role of 3α,5α-THP in stress and reward processes, we and others have begun to consider its role in some effects of drugs.

3α,5α-THP may have effects on reward processes (Frye, 2008b). In support, 3α,5α-THP mediates sex behavior of female rodents. Further, 3α,5α-THP can enhance conditioned place preference (CPP; Finn et al., 1997; Frye, 2008b). Among rats, 3α,5α-THP administration can dose-dependently increase the release of dopamine in the nucleus accumbens (Rougé-Pont et al., 2002). However, a conditioned place aversion has been demonstrated among rats administered 3α,5α-THP (Beauchamp et al., 2000). Moreover, mating increases 3α,5α-THP biosynthesis and CPP (Frye, 2008b; Frye and Rhodes, 2008a; Jenkins and Becker, 2003; Paredes and Vazquez, 1999). Thus, 3α,5α-THP has reinforcing properties that may underlie some effects of drugs of abuse. A discussion of how 3α,5α-THP may modify responses to two drugs of abuse, alcohol and cocaine, follows.

15.4.5.1 Neurosteroids and alcohol

Neurosteroids influence rewarding effects of alcohol in some models. As seen in both male and female adolescent humans (Torres and Ortega, 2003, 2004), in male and female rats (Morrow et al., 1999), ethanol increases 3α,5α-THP levels. P from the adrenals (Holzbauer et al., 1985) serves as a prohormone for neurosteroids in brain (see Barbaccia et al. (2001) and Mellon et al. (2001)). Acute administration of ethanol activates the HPA axis and increases plasma precursors for neurosteroids, resulting in elevated neurosteroids in the brain (Barbaccia et al., 1996; Morrow et al., 2001). Alcohol's effects on neurosteroidogenesis are blocked by adrenalectomy (Khisti et al., 2002, 2003) and/or gonadectomy (O'Dell et al., 2004).

Agonist-like actions of neurosteroids at $GABA_A$ receptors (Harrison and Simmonds, 1984; Majewska et al., 1986) may contribute to ethanol's GABAmimetic profile. In support, although 3α,5α-THP does not alter ethanol-induced CPP, other work emphasized a relationship of neurosteroid action to ethanol pharmacodynamics (Gabriel et al., 2004), discrimination for ethanol generalized to neurosteroids

(Ator et al., 1993; Bowen et al., 1999; Grant et al., 1996). In addition, neurosteroids increase ethanol consumption in a two-bottle preference test (Sinnott et al., 2002), as well as reinforcement and reinstatement of previously extinguished responding for ethanol (Janak et al., 1998; Nie and Janak, 2003). Thus, ethanol and neurosteroids may interact to influence $GABA_A$ receptors.

15.4.5.2 Neurosteroids and cocaine

Cocaine, another rewarding drug, can have more salient effects among females than males; however, progestogens can influence this (Chen and Kandel, 2002; Frye, 2008b; Koob and Kreek, 2007). When endogenous progestogen levels are low, women may experience enhanced effects of cocaine, whereas P administration reduces the effect of cocaine on rodents' CPP and psychomotor responses and peoples' desire to self-administer (Becker, 1999; Evans et al., 2002; Niyomchai et al., 2005; Quiñones-Jenab et al., 1999; Roberts et al., 1989; Russo et al., 2003; Sofuoglu et al., 2004). Cocaine administration to rats increases central levels of P and 3α,5α-THP (Frye, 2008b). Thus, cocaine and progestogens are related, as are progestogens and natural-rewarding sexual stimuli.

Progestogens and cocaine both influence HPA-stress-axis function. Biosynthesis of 3α,5α-THP occurs rapidly in response to stress and increases parasympathetic tone, in part, by decreasing HPA activation and B levels (Barbaccia et al., 2001). Females, compared to males, typically have greater HPA activation in response to stress; however, progestogens levels influence this. Decline in progestogen levels, or progestogen withdrawal, increases stress and anxiety, and high progestogen levels decrease stress and anxiety (Bäckström et al., 2003). Stress is a risk factor greatly associated with cocaine use and relapse (Amiel-Tison et al., 2004; Johnson et al., 2006; Latkin et al., 2007; McKay et al., 1995; O'Brien et al., 1998; Roxburgh et al., 2006; Sinha et al., 2005, 2003). Cocaine administration markedly increases plasma corticosterone levels, whereas administering P prior to cocaine suppresses cocaine-induced corticosterone increases (Frye, 2008b). Thus, differential HPA responses associated with variations in 3α,5α-THP may mediate some sex differences in cocaine's psychotropic effects.

15.4.6 Neurosteroids and Depression

Depression is a serious and widespread mental disorder that may be influenced by many factors including steroid hormones. Findings have suggested that progestogens may have effects upon depressive behaviors in both humans and animals. Throughout the life span, women experience varied and occasionally dramatic changes in their hormonal and reproductive cycles. Among some women, hormonal and/or reproductive events may influence the onset or expression of depression and/or anxiety disorders, such as premenstrual syndrome (PMS), premenstrual dysphoric disorder (PMDD), and postpartum depression, syndromes which occur when endogenous progestogen levels are low (Bäckström et al., 2003; Endicott et al., 1999; Glick and Bennett, 1981; Markou et al., 2005; Pearlstein et al., 2005; Rapkin et al., 2002). Compared to men, who do not experience profound changes in progestogens, women are more susceptible to depression and/or anxiety disorders. Among some women at menopause, reduced levels of 3α,5α-THP and other neurosteroids have been associated with depression and other mood disorders (Freeman et al., 2002; Girdler et al., 2001; Pearlstein, 1995). Thus, these clinical findings suggest that progestogens may be involved in the symptomology and/or etiology of depressive disorders.

15.4.6.1 Neurosteroids and depression – etiology

Clinical findings support a role of neurosteroids in the etiology of depression. Stressful life events can precipitate depression (Brown et al., 1994) and individuals with depression often have difficulties in coping with stress (Young et al., 2000). Some people with major depression have higher concentrations of corticotropin-releasing factor (CRF), cortisol, and/or impaired glucocortoid feedback to dexamethasone (Carroll et al., 1976, 1981; Halbreich et al., 1985; Nemeroff et al., 1988; Rubin et al., 1987b). Given that activation of the HPA can alter neurosteroid production, one must consider that clinical depression is associated with dysregulation of this axis. Some of the most compelling evidence that neurosteroids are involved in depression come from clinical use of finasteride, a 5α-reductase inhibitor, which decreases production of 3α,5α-THP (Townsend and Marlowe, 2004). Nineteen of 23 patients treated with finasteride for alopecia reported depressive symptoms (Altomare and Capella, 2002). Also, men with benign prostate hyperplasia are much more likely to develop depressive disorders when their treatment includes finasteride (Clifford and Farmer, 2002). Thus, decreasing 3α,5α-THP production may precipitate depressive symptoms in some individuals without depression.

15.4.6.2 Neurosteroids and depression – treatment

Moreover, the efficacy of some therapeutics to treat depressive and/or anxiety disorders is associated with the capacity of the therapeutics to increase levels of progestogens (Griffin and Mellon, 1999; Uzunov et al., 1996, 1998). Interestingly, recent clinical findings indicate that P's antidepressant effects may involve actions of 3α,5α-THP. Some patients who have depressive disorders have reduced plasma concentrations and/or cerebrospinal fluid levels of 3α,5α-THP (Romeo et al., 1998; Stahl, 1997; Uzunova et al., 1998). Administration of antidepressants, such as fluoxetine or fluvoxamine, normalizes decreased 3α,5α-THP concentrations concomitant with reducing depressive symptomology of depressed individuals (Uzunova et al., 2004, 2006). Thus, 3α,5α-THP may contribute to the antidepressant effects of some therapeutics.

Neurosteroids may also play a role in the therapeutic treatment of depression. Various treatments of depression alter neurosteroids. Antidepressants (Dubrovsky, 2006; Uzunova et al., 2004), sleep deprivation (Schüle et al., 2003), electroconvulsive therapy (Baghai et al., 2005), and transcraneal magnetic stimulation (Padberg et al., 2002) have been investigated for their capacity to alter neurosteroids and depression. Notably, measurable changes in plasma levels of neurosteroids are not as readily observable with nonpharmacological treatments of depression. However, common effects of therapeutic treatments may include direct central changes in neurosteroidogenesis or alterations in peripheral HPA function that may change steroid biosynthesis and thereby alter core symptoms of depression, such as anxiety, memory, sleep, and sexual function (Dubrovsky, 2006). Although the ability to assess central levels of neurosteroids in people limits the capacity to elucidate this, in animal models, 3α,5α-THP has effects in adult rodents to decrease depressive behavior (Khisti et al., 2000; discussed further below), aggression (Kavaliers, 1988), and CNS excitability (Landgren et al., 1987; Frye, 2008).

15.4.7 Neurosteroids and Anxiety

Given 3α,5α-THP's actions as a positive modulator at $GABA_A$ receptors, its role in various anxiety disorders has been investigated. Baseline plasma levels of 3α,5α-THP are normal in patients with generalized anxiety disorders and social phobia (Le Mellédo and Baker, 2002). Moreover, levels are within normal limits following administration of pentagastrin, a panic-inducing agent (Tait et al., 2002). In people with panic disorder, 3α,5α-THP levels are higher than normal (Brambilla et al., 2003; Ströhle et al., 2003), but were decreased by infusions of sodium lactate or cholecystokinintetrapeptide (CCK4) to induce panic attacks (Ströhle et al., 2003). Notably, in people without a history of panic attacks, levels of neurosteroids are not affected or are increased following CCK4 treatment (Eser et al., 2005; Zwanzger et al., 2004). In women with panic disorder and agoraphobia, perimenstrual, but not midluteal, 3α,5α-THP levels were significantly higher than among normal controls, and they correlated with their panic-phobic symptoms (Brambilla et al., 2003). In collaboration with Dr. John Casada, we have found that 3α,5α-THP levels among those with post-traumatic stress disorder (PTSD) are higher and correlated with anxiety state following exposure to PTSD-relevant stimuli (Frye and Rhodes, 2008b). However, there is emerging evidence that there are biphasic effects of 3α,5α-THP, such that high levels may in fact increase aggression, hostility, and/or negative mood (Andréen et al., 2005; Fish et al., 2001; Spalletta et al., 2005). A discussion of how 3α,5α-THP may produce mood dysregulation is as follows.

15.4.8 Neurosteroids and Mood Dysregulation

Most women experience some physical and/or psychological symptoms related to the cyclical variations in ovarian steroids; however, some women (~5%) experience such profound symptoms that they require medical treatment (Sveindottir and Bäckström 2000). PMDD is a clinical condition that is characterized by heightened sensitivity to fluctuations in steroids (Bäckström et al., 1983; Hammarbäck et al., 1989; Sanders et al., 1983). PMDD is defined by the occurrence of debilitating negative mood and physical symptoms that uniquely occur when hormone levels fluctuate during the luteal phase but are absent when there is luteal insufficiency (Bäckström et al., 1983; Endicott et al., 1999; Hammarbäck and Bäckström, 1988; Hammarbäck et al., 1991; Ramcharan et al., 1992; Sveindottir and Bäckström, 2000). For example, women with PMDD report improved mood with ovarian suppression, but negative mood is increased by administration of E_2 and/or P (Schmidt et al., 1998). Furthermore, mood is more negative when hormone levels are higher during the luteal phase in PMDD patients (Hammarbäck et al., 1989; Seippel and Bäckström, 1998). Indeed, we have also found that among some PMDD patients treated with selective serotonin reuptake inhibitors (SSRIs), who had clinical improvement induced by placebo or SSRI, had lower levels of 3α,5α-THP (Freeman et al., 2002). However, other studies have shown that women with PMDD report improved symptoms during luteal phases when 3α,5α-THP levels are higher (Wang et al., 1996) or higher anxiety and irritability when 3α,5α-THP levels are lower (Girdler et al., 2001). Indeed, we have seen this opposite pattern as well. Most studies show that absolute levels of E_2, P, and/or 3α,5α-THP do not differ among women with PMDD and controls (Epperson et al., 2002; Halbreich et al., 1993a; Rubinow et al., 1988; Schmidt et al., 1994; Sundström and Bäckström, 1998a; Wang et al., 1996, 2001), but some have reported significantly lower (Bicikova et al., 1998; Monteleone et al., 2000; Rapkin et al., 1997) or significantly higher concentrations of 3α,5α-THP in PMDD patients (Girdler et al., 2001). One explanation for some of the heterogeneity in the results is that some women with PMDD may have enhanced sensitivity of their $GABA_A$ receptor (Sundström et al., 1997a,b, 1998). Indeed, there is evidence for heterogeneity in responses to steroid hormones among women with PMDD and between women with PMDD and others.

15.4.9 Neurosteroids and Schizophrenia

3α,5α-THP may be important in the pathophysiology of schizophrenia, which is characterized by deficits in social, affective, and cognitive functioning (Shirayama et al., 2002). First, stress reactivity may underlie the etiology and/or manifestation of schizophrenia. Among people with schizophrenia, dysregulation of the HPA axis is common (Butzlaff and Hooley, 1998; Lukoff et al., 1984; Malla et al., 1990; Myin-Germeys et al., 2001; Norman and Malla, 1993; Read et al., 2001) and stress can precipitate psychiatric episodes related to schizophrenia (Read et al., 2001). Second, stress-induced 3α,5α-THP production can be disrupted in schizophrenia. A novel polymorphism and genetic mutation in the sequence encoding the gene for the mitochondrial benzodiazepine receptor, which is necessary for 3α,5α-THP biosynthesis in glial cells, has been demonstrated among some schizophrenics, and may create a predisposition to oversensitivity to stress (Kurumaji et al., 2000; Myin-Germeys et al., 2001; Read et al., 2001). In addition, social isolation (an animal model of schizophrenia) decreases 3α,5α-THP biosynthesis in the frontal cortex of male Swiss-Webster mice, compared to group-housed controls (Dong et al., 2001). Third, there is evidence that 3α,5α-THP metabolized in the brain from peripheral prohormones may reduce the incidence and/or expression of schizophrenia. Women, compared to men, typically have higher levels of 3α,5α-THP, are more likely to have late-onset schizophrenia, better prognosis, and therapeutic response to lower dosages of antipsychotics (Bloch et al., 2000). When 3α,5α-THP levels are low perimenstrually, first onset, or recurrence of psychotic episodes is more likely and more negative symptoms are reported (Hallonquist et al., 1993; Hendrick et al., 1996; Huber et al., 2001). After menopause, when 3α,5α-THP levels are lower, recurrence of psychiatric episodes increases. Fourth, effective pharmacotherapies for schizophrenia can alter 3α,5α-THP levels. The atypical antipsychotic drug, olanzapine, enhances social functioning and increases 3α,5α-THP levels (Frye and Seliga (2003) cited in Frye (2008b); Marx and Lieberman 1998; Marx et al., 2000). Together, these data suggest that schizophrenia may involve a reduced capacity to synthesize 3α,5α-THP in the brain, which may increase sensitivity to stress and thereby vulnerability to psychosis.

15.4.10 Neurosteroids, Aging, Menopause, and Hormone Therapy

Reproductive senescence is characterized by a decline in ovulatory cycles and consequent reduction in progestogen concentrations. For example, in the 6 years leading up to menopause, the frequency of ovulatory cycles decreases from about 60% to 10% (Rannevik et al., 1995). P is produced by the corpus luteum after ovulation. As such, reproductively aging women typically have serum P concentrations of less than 2 nmol L^{-1} postmenopausally. Indeed, 3α,5α-THP levels in plasma and brain of postmenopausal women are lower than that of premenopausal women during the luteal, but not follicular, phase (Bixo et al., 1997; Genazzani et al., 1998).

Decline in production of ovarian steroids with menopause results in physical and psychological symptoms among some women. As such, hormone replacement therapies, such as conjugated equine estrogen (CEE), are used by some women to manage minor climacteric symptoms, vasomotor symptoms, atrophic vaginitis, and/or mood (Campbell and Whitehead, 1977; Ditkoff et al., 1991; MacLennan et al., 2001; Wiklund et al., 1993). Regarding the latter, in healthy, surgically menopausal, nondepressed women, circulating E_2 levels and mood covary (Sherwin, 1988). Furthermore, among naturally menopausal women, those taking higher concentrations of E_2 reported having more energy and enhanced well-being (Sherwin and Gelfand, 1989).

The extent to which these effects are through direct actions of E_2 and/or its effects to enhance biosynthesis of progestogens is of interest. Administration of E_2-based hormone therapies increases 3α,5α-THP levels above non-HRT baseline, but levels are not increased as much as is seen with E_2 and progestin combination therapy (Andréen et al., 2005; Bernardi et al., 2003; Wihlbäck et al., 2005).

E_2-based hormone replacement therapies are typically augmented with progestins to minimize endometrial hyperplasia and/or increased risk of endometrial cancer (Voigt et al., 1991; Whitehead, 1978). However, this can cause cyclical mood changes among some women that can be analogous to those that they experienced during the perimenstruum (Andréen et al., 2003; Björn et al., 2000; Hammarbäck et al., 1985; Magos et al., 1986; Wihlbäck et al., 2001). Often, these negative mood symptoms gradually develop during progestogen phase and then peak during the beginning of the next E_2-only phase, concomitant with withdrawal from progestogens (Andréen et al., 2003; Björn et al., 2000). Many women report that they have discontinued hormone therapy report due to negative effects on mood (Björn and Bäckström, 1999). Interestingly, two factors seem to mitigate whether negative mood occurs. First, the most prominent increases in negative mood seem to be produced by natural progestogens that can more readily increase pregnenolone concentrations than do 19-nor derivatives (such as medoxyprogesterone acetate (MPA), nomegestrol, dihydrogesterone, or cyproteroneacetate; Wihlbäck et al., 2001). Second, prior adverse response to steroids is predictive of negative effects of hormone therapy with reproductive senescence. For example, among most women norethisteroneacetate (NETA) produces more negative mood than does MPA (Björn et al., 2000, 2002); however, both MPA and NETA provoke negative symptoms among women with a history of PMS (Björn et al., 2000; Odmark et al., 2004). Thus, the type of progestin may be less important than history of response to steroids.

15.4.11 Neurosteroids and Neurodegeneration

The role of neurosteroids for aging/changes in cognition across the life span is of interest. The effects of aging on cognitive performance have been well demonstrated. About 60% of aging individuals will experience some cognitive decline. Among these 60%, about one half of the them will experience benign senescent forgetfulness, a cognitive impairment that levels out and appears to be due to cell loss in a region of the hippocampus called the subiculum; the other half will likely experience Alzheimer's disease (AD). A discussion of the effects of neurosteroids on neurodegenerative processes follows.

15.4.11.1 Neurosteroids and seizure disorder

Epilepsy is a neurodegenerative disorder that afflicts 50 million people of all ages (Satzinger, 1994). Steroid hormones may have an important protective role in epilepsy. Infantile spasms, convulsive seizures that occur in the first year of life, may be due in part to withdrawal from steroids associated with pregnancy or delivery. Seizure disorders usually begin during childhood, when hormone levels are low. Many children with epilepsy experience changes in patterns of seizures with the onset of puberty or in early adolescent years when hormone levels are in flux. Some adolescent girls with epilepsy experience exacerbation of their seizures around menstruation, when their progestogen levels are low, and can benefit from P therapy to help control their seizure disorder. Given that drug therapies for management of seizure disorder have typically targeted the GABA neurotransmitter system and steroid hormones can also alter the GABA system, steroids represent an emerging therapeutic strategy for epilepsy.

In general, P has robust antiseizure effects (Selye, 1941). Among women, decreases in seizure number during the luteal phase correlate with the highest concentrations of P (Bäckström, 1976). P infusions that produce circulating concentrations akin to the luteal phase, suppress epileptiform spikes in some patients (Bäckström et al., 1984; Landgren et al., 1978). P therapy has been used effectively to treat some women with catamenial epilepsy (Herzog, 1995).

The role of neurosteroids in the pathophysiology of seizure disorder is relevant to other disorders for several reasons. First, steroids play a salient role in the etiology, expression, and treatment of seizure disorder, which may be relevant to other disorders. Second, the hippocampus is a primary target of actions of steroids in epilepsy and may also be relevant to neuropsychiatric disorders. Third, epilepsy is characterized by disruption of endocrine function and advanced aging, which may provide insight into the role of steroids in other age-related neurodegenerative diseases.

15.4.11.2 Neurosteroids and AD

AD is the most common cause of dementia in older persons. AD is characterized by deposition of senile plaques composed of aggregated Aβ-amyloid (Aβ)

40/42 peptides and neurofibrillary tangles (NFTs), due to aggregation of pathological tau proteins in vulnerable brain regions. Steroids may influence AD. Most research on steroids and AD has focused on E_2. E_2 can promote neuronal survival by increasing cellular defenses and antiapoptotic strategies (Bcl1-2, Bcl-x) in the face of diverse challenges. Neurons preexposed to E_2 have significantly less damage induced by excitotoxic glutamate, or free radical generators (hydrogen peroxide; Pike, 1999; Brinton, 2004; Brinton et al., 2000; Green et al., 2000). E_2 *in vitro* or *in vivo* reduces Aβ production (Pike, 1999), increases microglial uptake (Li et al., 2001), and influences AD neuropathologies (Brinton, 2004).

15.4.11.3 Neurosteroids and Niemann–Pick type C

Another neurodegenerative system in which neurosteroids may exert a profound role is in Niemann–Pick type C (NP-C), an autosomal recessive neurodegenerative disease caused by mutations in the lysosomal NPC1 protein, which results in accumulation of cholesterol esters in lysosomes. Research by Mellon and colleagues has demonstrated that brains of adult NP-$C^{2/2}$ mice contain much lower concentrations of neurosteroids compared with their normal littermates (reviewed in Mellon et al. (2008)). They also showed that the synthesis of 3α,5α-THP from P is much less in NP-$C^{2/2}$ mice than in wild-type mice and that expression of enzymes required for 3α,5α-THP synthesis is decreased in specific brain regions of NP-$C^{2/2}$ mice. Notably, these effects are observed before the onset of symptoms. Moreover, a single injection of 3α,5α-THP on postnatal day 7 can prevent neurodegeneration. Their findings suggest that decreased synthesis of neurosteroids may be a factor that ultimately leads to neurodegeneration in these animals.

15.4.12 Neurosteroids, Apoptosis, and Neurogenesis

The previous sections described the myriad effects of neurosteroids on functional processes across the life span. Of interest is how some of these effects may be related to steroids' trophic effects in the CNS. Indeed, gonadal hormones influence neural plasticity (Parducz et al., 2006). In general, E_2 can enhance cell proliferation and P can attenuate some of these effects (Galea, 2008; Tanapat et al., 2005). However, some neurosteroids can protect neuron cells from apoptosis (Charalampopoulos et al., 2004). For example, 3α,5α-THP protects sympathoadrenal medulla cells against apoptosis via antiapoptopic Bcl-2proteins, which can decrease the generation of reactive oxygen species (ROS) in the mitochondria (Jayanthi et al., 2004). Neurosteroids, such as 3α,5α-THP, can also enhance neurogenesis (Magnaghi et al., 2001; Schumacher et al., 2000, 2001; Young, 2002). 3α,5α-THP levels in the brain of Alzheimer's patients are decreased (Weill-Engerer et al., 2002). 3α,5α-THP dose-dependently increases cell proliferation in embryonic hippocampal neurons in culture (Wang et al., 2005).

One intriguing possibility that may underlie some of the common actions of 3α,5α-THP is its effects on neuroplasticity. One theory about some neuropsychiatric disorders is that they are associated with decreased neural plasticity in the hippocampus (depression and anxiety disorders) and/or the prefrontal cortex (schizophrenia), which may account for some of the differences in HPA feedback seen among individuals with these disorders. In support, some treatments may enhance neurogenesis which may be necessary for some of their therapeutic effects. For example, isoproterenol and serotonin increase glial fibrillary acidic protein (GFAP) mRNA levels, an index of glial cell differentiation, which may involve cyclic adenosine monophosphate (cAMP)-dependent increase in 5α-reductase expression. Evidence that progestogen activation of $GABA_A$ receptors is necessary to increase GFAP gene mRNA levels in type1 rat astrocytes includes that finasteride or $GABA_A$ receptor antagonists can abolish these effects (Koschel and Tas, 1993; Le Prince et al., 1991; Morita et al., 2006; Melcangi et al., 1992; Papodopoulos and Guarneri, 1994; Segovia et al., 1994; Shain et al., 1992). Moreover, adrenergic and serotonergic stimulation of glial cells enhances the expression of 5α-reductase genes (Morita et al., 2004), emphasizing the bidirectional and interactive nature of these effects.

A nonpharmacological paradigm reveals interactive therapeutic effects of neurogenesis and neurosteroidogenesis. Electroconvulsive seizures, which produce antidepressant effects, enhance hippocampal neurogenesis in the adult rat (Hellsten et al., 2002). We have previously found that seizures induced by chemoconvulsants enhanced activity of 5α-reductase enzymes and, thereby, neurosteroidogenesis (Frye and Rhodes, 2005). Indeed, it has been proposed that increases in neural proliferation and neuronal activity may lead to structural changes within the brain, and this neuroplasticity may be of some importance for some therapeutic effects (Hellsten et al., 2004). Further investigation of the role of neurosteroids for their functional effects and the extent to which these may be related to their effects on neuroplasticity is needed.

15.5 Conclusions

There is clear evidence that neurosteroids can play a role in hormonally and developmentally mediated behavioral processes, and that they may serve clinically relevant functions across the life span. However, understanding the role of neurosteroids can be a challenge because of their lability and the inherent challenges associated with defining neuroethologically relevant system variables. In order to further elucidate the role of neurosteroids, it may be necessary to think beyond heretofore-defined clinical syndromes and consider common underlying symptomology that may be explained, or influenced in part, by a role of neurosteroids. A truly productive translational approach will require careful elucidation and consideration of the common features of neurosteroids' role in basic animal research and in clinical syndromes. As such, continued evaluation of the relationship between neurosteroidogenic capacity, neuroprotection, neurogenesis, and therapeutic effects on basic biobehavioral processes across a number of paradigms may provide insight to some of the diverse underlying actions of 3α,5α-THP.

References

Altomare G and Capella GL (2002) Depression circumstantially related to the administration of finasteride for androgenetic alopecia. *Journal of Dermatology* 29: 665–669.

Amiel-Tison C, Cabrol D, Denver R, Jarreau PH, Papiernik E, and Piazza, PV (2004) Fetal adaptation to stress: Part II. Evolutionary aspects; stress-induced hippocampal damage; long-term effects on behavior; consequences on adult health. *Early Human Development* 78(2): 81–94.

Andréen L, Bixo M, Nyberg S, Sundström-Poromaa I, and Bäckström T (2003) Progesterone effects during sequential hormone replacement therapy. *European Journal of Endocrinology* 148: 571–577.

Andréen L, Sundström-Poromaa I, Bixo M, Andersson A, Nyberg S, and Bäckström T (2005) Relationship between allopregnanolone and negative mood in postmenopausal women taking sequential hormone replacement therapy with vaginal progesterone. *Psychoneuroendocrinology* 30: 212–224.

Antonijevic IA, Russell JA, Bicknell RJ, Leng G, and Douglas AJ (2000) Effect of progesterone on the activation of neurones of the supraoptic nucleus during parturition. *Journal of Reproduction and Fertility* 120: 367–376.

Ator NA, Grant KA, Purdy RH, Paul SM, and Griffiths RR (1993) Drug discrimination analysis of endogenous neuroactive steroids in rats. *European Journal of Pharmacology* 241: 237–243.

Bäckström T (1976) Epileptic seizures in women related to plasma estrogen and progesterone during the menstrual cycle. *Acta Neurologica Scandinavica* 54: 321–347.

Bäckström T, Andersson A, Andreé L, et al. (2003) Pathogenesis in menstrual cycle-linked CNS disorders. *Annals of the New York Academy of Sciences* 1007: 42–53.

Bäckström T, Sanders D, Leask R, Davidson D, Warner P, and Bancroft J (1983) Mood, sexuality, hormones, and the menstrual cycle. II. Hormone levels and their relationship to the premenstrual syndrome. *Psychosomatic Medicine* 45: 503–507.

Bäckström T, Zetterlund B, Blom S, and Romano M (1984) Effects of intravenous progesterone infusions on the epileptic discharge frequency in women with partial epilepsy. *Acta Neurologica Scandinavica* 69(4): 240–248.

Baghai TC, di Michele F, Schüle C, et al. (2005) Plasma concentrations of neuroactive steroids before and after electroconvulsive therapy in major depression. *Neuropsychopharmacology* 30: 1181–1186.

Barbaccia ML, Roscetti G, Trabucchi M, Mostallino MC, Concas A, Purdy RH, and Biggio G (1996) Time-dependent changes in rat brain neuroactive steroid concentrations and $GABA_A$ receptor function after acute stress. *Neuroendocrinology* 63: 166–172.

Barbaccia ML, Serra M, Purdy RH, and Biggio G (2001) Stress and neuroactive steroids. *International Review of Neurobiology* 46: 243–272.

Baulieu EE (1980) Steroid hormone receptors. *Exposés Annuels de Biochimie Médicale* 34: 1–25.

Baulieu EE (1991) Neurosteroids: A new function in the brain. *Biology of Cell* 71: 3–10.

Baulieu EE and Schumacher M (2000) Progesterone as a neuroactive neurosteroid, with special reference to the effect of progesterone on myelination. *Human Reproduction* 15: 1–13.

Beauchamp MH, Ormerod BK, Jhamandas K, Boegman RJ, and Beninger RJ (2000) Neurosteroids and reward: Allopregnanolone produces a conditioned place aversion in rats. *Pharmacology, Biochemistry, and Behavior* 67: 29–35.

Becker A, Schmitz M, and Grecksch G (2006) Kindling modifies morphine, cocaine and ethanol place preference. *Experimental Brain Research* 168(1–2): 33–40.

Belelli D and Lambert JJ (2005) Neurosteroids: Endogenous regulators of the GABA(A) receptor. *Nature Reviews. Neuroscience* 6: 565–575.

Bernardi F, Pieri M, Stomati M, et al. (2003) Effect of different hormonal replacement therapies on circulating allopregnanolone and dehydroepiandrosterone levels in postmenopausal women. *Gynecological Endocrinology* 17: 65–77.

Bianchi MT and Macdonald RL (2003) Neurosteroids shift partial agonist activation of GABA(A) receptor channels from low- to high-efficacy gating patterns. *Journal of Neuroscience* 23: 10934–10943.

Bicikova M, Dibbelt L, Hill M, Hampl R, and Starka L (1998) Allopregnanolone in women with premenstrual syndrome. *Hormone and Metabolic Research* 30: 227–230.

Bicíková M, Klak J, Hill M, Zizka Z, Hampl R, and Calda P (2002) Two neuroactive steroids in midpregnancy as measured in maternal and fetal sera and in amniotic fluid. *Steroids* 67(5): 399–402.

Bicíková M, Tallová J, Hill M, Krausová Z, and Hampl R (2000) Serum concentrations of some neuroactive steroids in women suffering from mixed anxiety-depressive disorder. *Neurochemical Research* 25(12):1623–1627.

Billiards SS, Nguyen PN, Scheerlinck JP, Phillips DJ, Canny BJ, Walker DW, and Hirst JJ (2006) Hypoxia potentiates endotoxin-induced allopregnanolone concentrations in the newborn brain. *Biology of the Neonate* 90: 258–267.
Bixo M, Andersson A, Winblad B, Purdy RH, and Bäckström T (1997) Progesterone, 5α-pregnane-3,20-dione and 3α-hydroxy-5α-pregnane-20-one in specific regions of the human female brain in different endocrine states. *Brain Research* 764: 173–178.
Björn I and Bäckström T (1999) Drug related negative side-effects is a common reason for poor compliance in hormone replacement therapy. *Maturitas* 32: 77–86.
Björn I, Bixo M, Nöjd KS, Collberg P, Nyberg S, Sundström-Poromaa I, and Bäckström T (2002) The impact of different doses of medroxyprogesterone acetate on mood symptoms in sequential hormonal therapy. *Gynecological Endocrinology* 16: 1–8.
Björn I, Bixo M, Strandberg-Nöjd K, Nyberg S, and Bäckström T (2000) Negative mood changes during hormone replacement therapy: A comparison between two progestogens. *American Journal of Obstetrics and Gynecology* 183: 1419–1426.
Bloch M, Schmidt PJ, Danaceau M, Murphy J, Nieman L, and Rubinow DR (2000) Effects of gonadal steroids in women with a history of postpartum depression. *American Journal of Psychiatry* 157: 924–930.
Bowen CA, Purdy RH, and Grant KA (1999) An investigation of endogenous neuroactive steroid-induced modulation of ethanol's discriminative stimulus effects. *Behavioural Pharmacology* 10: 297–311.
Brambilla F, Biggio G, Pisu MG, et al. (2003) Neurosteroid secretion in panic disorder. *Psychiatry Research* 118: 107–116.
Brinton RD (2004) Impact of estrogen therapy on Alzheimer's disease: A fork in the road? *CNS Drugs* 18: 405–422.
Brinton RD, Chen S, Hsieh D, Minaya J, and Montoya M (2000) The estrogen replacement therapy of the Women's Health Initiative promotes the cellular mechanisms of memory and neuronal survival in neurons vulnerable to Alzheimer's disease. *Maturitas* 1: 34–52.
Brot MD, Akwa Y, Purdy RH, Koob GF, and Britton KT (1997) The anxiolytic-like effects of the neurosteroid allopregnanolone: Interactions with GABA(A) receptors. *European Journal of Pharmacology* 325: 1–7.
Brown GW, Harris TO, Hepworth C, and Robinson R (1994) Clinical and psychosocial origins of chronic depressive episodes. II. A patient enquiry. *British Journal of Psychiatry* 165: 457–465.
Brussaard AB and Herbison AE (2000) Long-term plasticity of postsynaptic GABAA-receptor function in the adult brain: Insights from the oxytocin neurone. *Trends in Neuroscience* 23: 190–195.
Butzlaff RL and Hooley JM (1998) Expressed emotion and psychiatric relapse: A meta-analysis. *Archives of General Psychiatry* 55: 547–552.
Campbell S and Whitehead M (1977) Oestrogen therapy and the menopausal syndrome. *Clinical Obstetrics and Gynaecology* 4: 31–47.
Carroll BJ, Curtis GC, and Mendels J (1976) Neuroendocrine regulation in depression. II. Discrimination of depressed from nondepressed patients. *Archives of General Psychiatry* 33: 1051–1058.
Carroll BJ, Greden JF, Feinberg M, et al. (1981) Neuroendocrine evaluation of depression in borderline patients. *Psychiatric Clinics of North America* 4: 89–99.
Chaban VV, Lakhter AJ, and Micevych P (2004) A membrane estrogen receptor mediates intracellular calcium release in astrocytes. *Endocrinology* 145: 3788–3795.
Chan JR, Phillips LJ, II, and Glaser M (1998) Glucocorticoids and progestins signal the initiation and enhance the rate of myelin formation. *Proceedings of the National Academy of Sciences USA* 95(18): 10459–10464.
Charalampopoulos I, Tsatsanis C, Dermitzaki E, Alexaki VI, Castanas E, Margioris AN, and Gravanis A (2004) Dehydroepiandrosterone and allopregnanolone protect sympathoadrenal medulla cells against apoptosis via antiapoptotic Bcl-2 proteins. *Proceedings of the National Academy of Sciences of the United States of America* 101: 8209–8214.
Chen HH, Yang YK, Yeh TL, et al. (2004) Methamphetamine-induced conditioned place preference is facilitated by estradiol pretreatment in female mice. *The Chinese Journal of Physiology* 46(4): 169–174.
Clifford GM and Farmer RD (2002) Drug or symptom-induced depression in men treated with α 1-blockers for benign prostatic hyperplasia? A nested case-control study. *Pharmacoepidemiology and Drug Safety* 11: 55–61.
Compagnone NA and Mellon SH (2000) Neurosteroids: Biosynthesis and function of these novel neuromodulators. *Frontiers in Neuroendocrinology* 21: 1–56.
Concas A, Mostallino MC, Porcu P, et al. (1998) Role of brain allopregnanolone in the plasticity of gamma-aminobutyric acid type A receptor in rat brain during pregnancy and after delivery. *Proceedings of the National Academy of Sciences of the United States of America* 95: 13284–13289.
Crossley KJ, Nicol MB, Hirst JJ, Walker DW, and Thorburn GD (1997) Suppression of arousal by progesterone in fetal sheep. *Reproduction, Fertility and Development* 9(8): 767–773.
Deminière JM, Piazza PV, Guegan G, Abrous N, Maccari S, LeMoal M, and Simon H (1992) Increased locomotor response to novelty and propensity to intravenous amphetamine self-administration in adult offspring of stressed mothers. *Brain Research* 586: 135–139.
Ditkoff EC, Crary WG, Cristo M, and Lobo RA (1991) Estrogen improves psychological function in asymptomatic postmenopausal women. *Obstetrics and Gynecology* 18: 991–995.
Dong E, Matsumoto K, Uzunova V, et al. (2001) Brain 5α-dihydroprogesterone and allopregnanolone synthesis in a mouse model of protracted social isolation. *Proceedings of the National Academy of Sciences of the United States of America* 98: 2849–2854.
Dubrovsky B (2006) Neurosteroids, neuroactive steroids, and symptoms of affective disorders. *Pharmacology, Biochemistry and Behavior* 84: 644–655.
Ebner MJ, Corol DI, Havlíková H, Honour JW, and Fry JP (2006) Identification of neuroactive steroids and their precursors and metabolites in adult male rat brain. *Endocrinology* 147: 179–190.
Endicott J, Amsterdam J, Eriksson E, et al. (1999) Is premenstrual dysphoric disorder a distinct clinical entity? *Journal of Women's Health and Gender-Based Medicine* 8: 663–679.
Epperson CN, Haga K, Mason GF, et al. (2002) Cortical gamma-aminobutyric acid levels across the menstrual cycle in healthy women and those with premenstrual dysphoric disorder: A proton magnetic resonance spectroscopy study. *Archives of General Psychiatry* 59: 851–858.
Eser D, di Michele F, Zwanzger P, et al. (2005) Panic induction with cholecystokinin-tetrapeptide (CCK-4) increases plasma concentrations of the neuroactive steroid 3α, 5α tetrahydrodeoxycorticosterone (3α, 5α-THDOC) in healthy volunteers. *Neuropsychopharmacology* 30: 192–195.
Evans SM, Haney M, and Foltin RW (2002) The effects of smoked cocaine during the follicular and luteal phases of the menstrual cycle in women. *Psychopharmacology (Berl.)* 159(4): 397–406.

Finn DA, Phillips TJ, Okorn DM, Chester JA, and Cunningham CL (1997) Rewarding effect of the neuroactive steroid 3 alpha-hydroxy-5 alpha-pregnan-20-one in mice. *Pharmacology, Biochemistry, and Behavior* 56: 261–264.
Fish EW, Faccidomo S, DeBold JF, and Miczek KA (2001) Alcohol, allopregnanolone and aggression in mice. *Psychopharmacology (Berl)* 153: 473–483.
Freeman EW, Frye CA, Rickels K, Martin PA, and Smith SS (2002) Allopregnanolone levels and symptom improvement in severe premenstrual syndrome. *Journal of Clinical Psychopharmacology* 22: 516–520.
Frye CA (2006) Role of androgens in epilepsy. *Expert Review of Neurotherapeutics* 6: 1061–1075.
Frye CA (2007) Some rewarding effects of androgens may be mediated by actions of its 5α-reduced metabolite 3α-androstanediol. *Pharmacology, Biochemistry, and Behavior* 86: 354–367.
Frye CA (2008) Progestins influence motivation, reward, conditioning, stress, and/or response to drugs of abuse. *Pharmacology, Biochemistry, and Behavior* 86: 209–219.
Frye CA and Rhodes ME (2005) Estrogen-priming can enhance progesterone's anti-seizure effects in part by increasing hippocampal levels of allopregnanolone. *Pharmacology Biochemistry and Behavior* 81(4): 907–916.
Frye CA and Rhodes ME (2008a) The role of midbrain 3α,5α-THP in mediating exploration, anxiety, social and reproductive behavior. In: Ritsner MS and Weizman A (eds.) *Neuroactive Steroids in Brain: From Experiments to Psychopathology and Treatment*, pp 449–482. Berlin: Springer.
Frye CA and Rhodes ME (2008b) The role and mechanisms of steroid hormones to enhance approach behavior. In: Elliot A (ed.) *Handbook of Approach and Avoidance Motivation*, pp 109–126. Mahwah, NJ: LEA.
Frye CA and Seliga AM (2003) Olanzapine's effects to reduce fear and anxiety and enhance social interactions coincide with increased progestin concentrations of ovariectomized rats. *Psychoneuroendocrinology* 28(5): 657–673.
Furukawa A, Miyatake A, Ohnishi T, and Ichikawa Y (1998) Steroidogenic acute regulatory protein (StAR) transcripts constitutively expressed in the adult rat central nervous system: Colocalization of StAR, cytochrome P-450SCC (CYP XIA1), and 3β-hydroxysteroid dehydrogenase in the rat brain. *Journal of Neurochemistry* 71: 2231–2238.
Gabriel KI, Cunningham CL, and Finn DA (2004) Allopregnanolone does not influence ethanol-induced conditioned place preference in DBA/2J mice. *Psychopharmacology (Berl)* 176: 150–156.
Galea LA (2008) Gonadal hormone modulation of neurogenesis in the dentate gyrus of adult male and female rodents. *Brain Research Reviews* 57: 332–341.
Gee KW, McCauley LD, and Lan NC (1995) A putative receptor for neurosteroids on the $GABA_A$ receptor complex: The pharmacological properties and therapeutic potential of epalons. *Critical Reviews in Neurobiology* 9: 207–227.
Genazzani AR, Petraglia F, Bernardi F, et al. (1998) Circulating levels of allopregnanolone in humans: Gender, age, and endocrine influences. *Journal of Clinical Endocrinology and Metabolism* 83: 2099–2103.
Girdler SS, Straneva PA, Light KC, Pedersen CA, and Morrow AL (2001) Allopregnanolone levels and reactivity to mental stress in premenstrual dysphoric disorder. *Biological Psychiatry* 49: 788–797.
Glick ID and Bennett SE (1981) Psychiatric complications of progesterone and oral contraceptives. *Journal of Clinical Psychopharmacology* 1: 350–356.
Grant KA, Azarov A, Bowen CA, Mirkis S, and Purdy RH (1996) Ethanol-like discriminative stimulus effects of the neurosteroid 3 alpha-hydroxy-5 alpha-pregnan-20-one in female *Macaca fascicularis* monkeys. *Psychopharmacology* 124: 340–346.
Green PS, Simpkins JW, and Yang SH (2000) Neuroprotective effects of phenolic A ring oestrogens. *Novartis Foundation Symposium* 230: 202–213.
Griffin LD and Mellon SH (1999) Selective serotonin reuptake inhibitors directly alter activity of neurosteroidogenic enzymes. *Proceedings of the National Academy of Sciences of the United States of America* 96: 13512–13517.
Halbreich U, Bancroft J, Dennerstein L, et al. (1993) Menstrually related disorders: Points of consensus, debate, and disagreement. *Neuropsychopharmacology* 9: 13–15; discussion 17–29.
Halbreich U, Endicott J, and Lesser J (1985) The clinical diagnosis and classification of premenstrual changes. *Canadian Journal of Psychiatry* 30: 489–497.
Hallonquist JD, Seeman MV, Lang M, and Rector NA (1993) Variation in symptom severity over the menstrual cycle of schizophrenics. *Biological Psychiatry* 33: 207–209.
Hammarbäck S and Bäckström T (1988) Induced anovulation as treatment of premenstrual tension syndrome. A double-blind cross-over study with GnRH-agonist versus placebo. *Acta Obstetricia et Gynecologica Scandinavica* 67: 159–166.
Hammarbäck S, Bäckström T, Holst J, von Schoultz B, and Lyrenäs S (1985) Cyclical mood changes as in the premenstrual tension syndrome during sequential estrogen–progestagen postmenopausal replacement therapy. *Acta Obstetricia et Gynecologica Scandinavica* 64: 393–397.
Hammarbäck S, Bäckström T, and MacGibbon-Tyler B (1989) Diagnosis of pre-menstrual syndrome: Description and evaluation of a procedure for diagnosis and differential diagnosis. *Journal of Psychosomatic Obstetrics and Gynaecology* 10: 25–42.
Hammarbäck S, Ekholm UB, and Bäckström T (1991) Spontaneous anovulation causing disappearance of cyclical symptoms in women with the premenstrual syndrome. *Acta Endocrinologica (Copenhagen)* 125: 132–137.
Harrison NL and Simmonds MA (1984) Modulation of the GABA receptor complex by a steroid anaesthetic. *Brain Research* 323(2): 287–292.
Hellsten J, Wennström M, Bengzon J, Mohapel P, and Tingström A (2004) Electroconvulsive seizures induce endothelial cell proliferation in adult rat hippocampus. *Biological Psychiatry* 55: 420–427.
Hellsten J, Wennström B, Mohapel CT, Ekdahl J, Bengzon A, and Tingström A (2002) Electroconvulsive seizures increase hippocampal neurogenesis after chronic corticosterone treatment. *European Journal of Neuroscience* 16: 283–290.
Hendrick V, Altshuler LL, and Burt VK (1996) Course of psychiatric disorders across the menstrual cycle. *Harvard Review of Psychiatry* 4: 200–207.
Herbison AE (2001) Physiological roles for the neurosteroid allopregnanolone in the modulation of brain function during pregnancy and parturition. *Progress in Brain Research* 133: 39–47.
Herzog AG (1995) Progesterone therapy in women with complex partial and secondary generalized seizures. *Neurology* 45: 1660–1662.
Hevers W and Lüddens H (1998) The diversity of GABAA receptors. Pharmacological and electrophysiological properties of $GABA_A$ channel subtypes. *Molecular Neurobiology* 18: 35–86.
Holzbauer M, Birmingham MK, De Nicola AF, and Oliver JT (1985) *In vivo* secretion of 3 alpha-hydroxy-5 alpha-pregnan-20-one, a potent anaesthetic steroid, by the adrenal gland of the rat. *Journal of Steroid Biochemistry* 22: 97–102.
Huber TJ, Rollnik J, Wilhelms J, von zur Muhlen A, Emrich HM, and Schneider U (2001) Estradiol levels in psychotic disorders. *Psychoneuroendocrinology* 26: 27–35.

Janak PH, Redfern JE, and Samson HH (1998) The reinforcing effects of ethanol are altered by the endogenous neurosteroid, allopregnanolone. *Alcoholism, Clinical and Experimental Research* 5: 1106–1112.

Jayanthi S, Deng X, Noailles PA, Ladenheim B, and Cadet JL (2004) Methamphetamine induces neuronal apoptosis via cross-talks between endoplasmic reticulum and mitochondria-dependent death cascades. *FASEB Journal* 18: 238–251.

Jenkins WJ and Becker JB (2003) Female rats develop conditioned place preferences for sex at their preferred interval. *Hormones and Behavior* 43: 503–507.

Johnson SD, Striley C, and Cottler LB (2006) The association of substance use disorders with trauma exposure and PTSD among African American drug users. *Addictive Behaviors* 31: 2063–2073.

Kavaliers M (1988) Inhibitory influences of the adrenal steroid, 3α,5α-tetrahydrodeoxycorticosterone (correction of tetrahydroxycorticosterone) on aggression and defeat-induced analgesia in mice. *Psychopharmacology (Berlin)* 95: 488–492.

Khisti RT, Chopde CT, and Jain SP (2000) Antidepressant-like effect of the neurosteroid 3α-hydroxy-5α-pregnan-20-one in mice forced swim test. *Pharmacology, Biochemistry, and Behavior* 67: 137–143.

Khisti RT, Kumar S, and Morrow AL (2003) Ethanol rapidly induces steroidogenic acute regulatory protein expression and translocation in rat adrenal gland. *European Journal of Pharmacology* 473(2–3): 225–227.

Khisti RT, Penland SN, VanDoren MJ, Grobin AC, and Morrow AL (2002) GABAergic neurosteroid modulation of ethanol actions. *World Journal of Biological Psychiatry* 3(2): 87–95.

King SR, Matassa AA, White EK, et al. (2004) Oxysterols regulate expression of the steroidogenic acute regulatory protein. *Journal of Molecular Endocrinology* 32: 507–517.

Koenig HL, Schumacher M, Ferzaz B, et al. (1995) Progesterone synthesis and myelin formation by Schwann cells. *Science* 268: 1500–1503.

Koob G and Kreek MJ (2007) Stress, dysregulation of drug reward pathways, and the transition to drug dependence. *American Journal of Psychiatry* 164(8): 1149–1159.

Koschel K and Tas PWI (1993) Lysophosphatidic acid reverts the β-adrenergic agonist-induced morphological response in C6 rat glioma cells. *Experimental Cell Research* 206: 162–166.

Kríz L, Bicíková M, Hill M, and Hampl R (2005) Steroid sulfatase and sulfuryl transferase activity in monkey brain tissue. *Steroids* 70: 960–969.

Kurumaji A, Nomoto H, Yoshikawa T, Okubo Y, and Toru M (2000) An association study between two missense variations of the benzodiazepine receptor (peripheral) gene and schizophrenia in a Japanese sample. *Journal of Neural Transmission* 107: 491–500.

Lambert JJ, Belelli D, Peden DR, Vardy AW, and Peters JA (2003) Neurosteroid modulation of $GABA_A$ receptors. *Progress in Neurobiology* 71: 67–80.

Lancefield K, Nosarti C, Rifkin L, Allin M, Sham P, and Murray R (2006) Cerebral asymmetry in 14 year olds born very preterm. *Brain Research* 1093: 33–40.

Landgren S, Aasly J, Bäckström T, Dubrovsky B, and Danielsson E (1987) The effect of progesterone and its metabolites on the interictal epileptiform discharge in the cat's cerebral cortex. *Acta Physiologica Scandinavica* 131: 33–42.

Landgren S, Bäckström T, and Kalistratov G (1978) The effect of progesterone on the spontaneous interictal spike evoked by the application of penicillin to the cat's cerebral cortex. *Journal of the Neurological Sciences* 36: 119–133.

Latkin CA, Curry AD, Hua W, and Davey MA (2007) Direct and indirect associations of neighborhood disorder with drug use and high-risk sexual partners. *American Journal of Preventive Medicine* 32: S234–S2341.

LeMellédo JM and Baker GB (2002) Neuroactive steroids and anxiety disorders. *Journal of Psychiatry and Neuroscience* 27: 161–165.

LePrince G, Fages C, Rolland B, Nunez J, and Tardy M (1991) cAMP effect on the expression of GFAP and of its encoding mRNA in astroglial primary cultures. *Glia* 4: 322–326.

LiY, Liu L, Barger SW, Mrak RE, and Griffin WS (2001) Vitamin E suppression of microglial activation is neuroprotective. *Journal of Neuroscience Research* 66: 163–170.

Liere P, Pianos A, Eychenne B, et al. (2004) Novel lipoidal derivatives of pregnenolone and dehydroepiandrosterone and absence of their sulfated counterparts in rodent brain. *Journal of Lipid Research* 45: 2287–2302.

Luisi S, Petraglia F, Benedetto C, et al. (2000) Serum allopregnanolone levels in pregnant women: Changes during pregnancy, at delivery, and in hypertensive patients. *Journal of Clinical Endocrinology and Metabolism* 85: 2429–2433.

Lukoff D, Snyder K, Ventura J, and Nuechterlein KH (1984) Life events, familial stress, and coping in the developmental course of schizophrenia. *Schizophrenia Bulletin* 10: 258–292.

MacLennan A, Lester S, and Moore V (2001) Oral oestrogen replacement therapy versus placebo for hot flushes. *Cochrane Database Systemic Reviews* CD002978.

Magnaghi V, Cavarretta I, Galbiati M, Martini L, and Melcangi RC (2001) Neuroactive steroids and peripheral myelin proteins. *Brain Research. Brain Research Reviews* 37: 360–371.

Magos AL, Brewster E, Sing R, O'Dowd TM, and Studd JWW (1986) The effect of norethisterone in postmenopausal women on oestrogen therapy: A model for the premenstrual syndrome. *British Journal of Obstetrics and Gynaecology* 93: 1290–1296.

Majewska MD (1992) Neurosteroids: Endogenous bimodal modulators of the GABAA receptor. Mechanism of action and physiological significance. *Progress in Neurobiology* 38: 379–395.

Majewska MD, Demirgören S, and London ED (1990) Binding of pregnenolone sulfate to rat brain membranes suggests multiple sites of steroid action at the $GABA_A$ receptor. *European Journal of Pharmacology* 189: 307–315.

Majewska MD, Harrison NL, Schwartz RD, Barker JL, and Paul SM (1986) Steroid hormone metabolites are barbiturate-like modulators of the GABA receptor. *Science* 232: 1004–1007.

Majewska MD and Schwartz RD (1987) Pregnenolone-sulfate: An endogenous antagonist of the gamma-aminobutyric acid receptor complex in brain? *Brain Research* 404: 355–360.

Malla AK, Cortese L, Shaw TS, and Ginsberg B (1990) Life events and relapse in schizophrenia. A one year prospective study. *Social Psychiatry and Psychiatric Epidemiology* 25: 221–224.

Markou A, Duka T, and Prelevic GM (2005) Estrogens and brain function. *Hormones (Athens)* 4: 9–17.

Marx CE, Duncan GE, Gilmore JH, Lieberman JA, and Morrow AL (2000) Olanzapine increases allopregnanolone in the rat cerebral cortex. *Biological Psychiatry* 47: 1000–1004.

Marx CE and Lieberman JA (1998) Psychoneuroendocrinology of schizophrenia. *Psychiatric Clinics of North America* 21: 413–434.

McKay JR, Rutherford MJ, Alterman AI, Cacciola JS, and Kaplan MR (1995) An examination of the cocaine relapse process. *Drug and Alcohol Dependence* 38: 35–43.

Mehta AK and Ticku MK (1999) An update on $GABA_A$ receptors. *Brain Research. Brain Research Reviews* 29: 196–217.

Melcangi RC, Celotti F, Castano P, and Martini L (1992) Intracellular signaling systems controlling the 5α-reductase in glial cell cultures. *Brain Research* 585: 411–415.

Mellon SH (1994) Neurosteroids: Biochemistry, modes of action, and clinical relevance. *Journal of Clinical Endocrinology and Metabolism* 78: 1003–1008.

Mellon SH (2007) Neurosteroid regulation of central nervous system development. *Pharmacology and Therapeutics* 6: 107–124.

Mellon SH and Deschepper CF (1993) Neurosteroid biosynthesis: Genes for adrenal steroidogenic enzymes are expressed in the brain. *Brain Research* 3: 283–292.

Mellon SH, Gong W, and Schonemann MD (2008) Endogenous and synthetic neurosteroids in treatment of Niemann–Pick type C disease. *Brain Research Reviews* 57: 410–420.

Mellon SH and Vaudry H (2001) Biosynthesis of neurosteroids and regulation of their synthesis. *International Review of Neurobiology* 46: 33–78.

Monteleone P, Luisi S, Tonetti A, et al. (2000) Allopregnanolone concentrations and premenstrual syndrome. *European Journal of Endocrinology* 142: 269–273.

Morita K, Arimochi H, and Tsuruo Y (2004) Adrenergic activation of steroid 5α-reductase gene expression in rat C6 glioma cells: Involvement of cyclic AMP/protein kinase A-mediated signaling pathway. *Journal of Molecular Neuroscience* 22: 205–212.

Morita K, Hideki A, Hiroyuki I, and Song H (2006) Possible involvement of 5α-reduced neurosteroids in adrenergic and serotonergic stimulations of GFAP gene expression in rat C6 glioma cells. *Brain Research* 1085: 49–56.

Morrow AL, Janis GC, VanDoren MJ, Matthews DB, Samson HH, Janak PH, and Grant KA (1999) Neurosteroids mediate pharmacological effects of ethanol: A new mechanism of ethanol action? *Alcoholism, Clinical and Experimental Research* 23: 1933–1940.

Morrow AL, Suzdak PD, and Paul SM (1987) Steroid hormone metabolites potentiate GABA receptor-mediated chloride ion flux with nanomolar potency. *European Journal of Pharmacology* 142: 483–485.

Morrow AL, VanDoren MJ, Penland SN, and Matthews DB (2001) The role of GABAergic neuroactive steroids in ethanol action, tolerance and dependence. *Brain Research. Brain Research Reviews* 37: 98–109.

Myin-Germeys I, van Os J, Schwartz JE, Stone AA, and Delespaul PA (2001) Emotional reactivity to daily life stress in psychosis. *Archives of General Psychiatry* 58: 1137–1144.

Nemeroff CB (1988) The role of corticotropin-releasing factor in the pathogenesis of major depression. *Pharmacopsychiatry* 21: 76–82.

Nemeroff CB, Knight DL, Krishnan RR, et al. (1988) Marked reduction in the number of platelet-tritiated imipramine binding sites in geriatric depression. *Archives of General Psychiatry* 45(10): 919–923.

Nguyen PN, Billiards SS, Walker DW, and Hirst JJ (2003b) Changes in 5alpha-pregnane steroids and neurosteroidogenic enzyme expression in the perinatal sheep. *Pediatric Research* 53: 956–964.

Nguyen PN, Billiards SS, Walker DW, and Hirst JJ (2003a) Changes in 5alpha-pregnane steroids and neurosteroidogenic enzyme expression in fetal sheep with umbilicoplacental embolization. *Pediatric Research* 54: 840–847.

Nguyen PN, Ross Young I, Walker DW, and Hirst JJ (2004) Allopregnanolone in the brain and blood after disruption of the hypothalamic–pituitary–adrenal axis in fetal sheep. *Journal of Endocrinology* 182: 81–88.

Nie H and Janak PH (2003) Comparison of reinstatement of ethanol- and sucrose-seeking by conditioned stimuli and priming injections of allopregnanolone after extinction in rats. *Psychopharmacology (Berl)* 168: 222–228.

Niyomchai T, Akhavan A, Festa ED, Lin SN, Lamm L, Foltz R, and Quiñones-Jenab V (2006) Estrogen and progesterone affect cocaine pharmacokinetics in female rats. *Brain Research Bulletin* 68(5): 310–314.

Norman RM and Malla AK (1993) Stressful life events and schizophrenia. I: A review of the research. *British Journal of Psychiatry* 162: 161–166.

Nosarti C, Al-Asady MH, Frangou S, Stewart AL, Rifkin L, and Murray RM (2002) Adolescents who were born very preterm have decreased brain volumes. *Brain* 125: 1616–1623.

Nosarti C, Rushe TM, Woodruff PW, Stewart AL, Rifkin L, and Murray RM (2004) Corpus callosum size and very preterm birth: Relationship to neuropsychological outcome. *Brain* 127: 2080–2089.

Numan M and Insel TR (2003) Motivational models of the onset and maintenance of maternal behavior and maternal aggression. In: Bridges R (ed.) *The Neurobiology of Parental Behavior*, pp 69–106. Oxford: Springer.

O'Brien CP, Childress AR, Ehrman R, and Robbins SJ (1998) Conditioning factors in drug abuse: Can they explain compulsion? *Journal of Psychopharmacology* 12: 15–22.

O'Dell LE, Alomary AA, Vallée M, Koob GF, Fitzgerald RL, and Purdy RH (2004) Ethanol-induced increases in neuroactive steroids in the rat brain and plasma are absent in adrenalectomized and gonadectomized rats. *European Journal of Pharmacology* 484: 241–247.

Odmark IS, Bixo M, Jonsson B, and Bäckström T (2004) Long term effects of two different continuous combined regimens of hormone replacement therapy on well being. *Gynecological Endocrinology* 18: 305–317.

Osterlund MK, Gustafsson JA, Keller E, and Hurd YL (2000) Estrogen receptor β (ERβ) messenger ribonucleic acid (mRNA) expression within the human forebrain: Distinct distribution pattern to ERα mRNA. *Journal of Clinical Endocrinology and Metabolism* 85: 3840–3846.

Padberg F, di Michele F, Zwanzger P, et al. (2002) Plasma concentrations of neuroactive steroids before and after repetitive transcranial magnetic stimulation (rTMS) in major depression. *Neuropsychopharmacology* 27: 874–878 (erratum in: *Neuropsychopharmacology* 2003, 28: 610–611).

Papodopoulos V and Guarneri P (1994) Regulation of C6 glioma cell steroidogenesis by adenosine 3′,5′,cyclic monophosphate. *Glia* 75–78.

Papadopoulos V, Lecanu L, Brown RC, Han Z, and Yao ZX (2006) Peripheral-type benzodiazepine receptor in neurosteroid biosynthesis, neuropathology and neurological disorders. *Neuroscience* 138: 749–756.

Parducz A, Hajszan T, Maclusky NJ, et al. (2006) Synaptic remodeling induced by gonadal hormones: Neuronal plasticity as a mediator of neuroendocrine and behavioral responses to steroids. *Neuroscience* 138: 977–985.

Paredes RG and Vazquez B (1999) What do female rats like about sex? Paced mating. *Behavioural Brain Research* 105: 117–127.

Paul SM and Purdy RH (1992) Neuroactive steroids. *FASEB Journal* 6: 2311–2322.

Pearlstein TB (1995) Hormones and depression: What are the facts about premenstrual syndrome, menopause, and hormone replacement therapy? *American Journal of Obstetrics and Gynecology* 173: 646–653.

Pearlstein TB, Bachmann GA, Zacur HA, and Yonkers KA (2005) Treatment of premenstrual dysphoric disorder with a new drospirenone-containing oral contraceptive formulation. *Contraception* 72: 414–421.

Pfaff DW, Gerlach JL, McEwen BS, Ferin M, Carmel P, and Zimmerman EA (1976) Autoradiographic localization of hormone-concentrating cells in the brain of the female rhesus monkey. *Journal of Comparative Neurology* 170: 279–293.
Pike CJ (1999) Estrogen modulates neuronal Bcl-xl expression and β-amyloid-induced apoptosis: Relevance to Alzheimer's disease. *Journal of Neurochemistry* 72: 1552–1563.
Pirker S, Schwarzer C, Wieselthaler A, Sieghart W, and Sperk G (2000) GABA(A) receptors: Immunocytochemical distribution of 13 subunits in the adult rat brain. *Neuroscience* 101: 815–850.
Purdy RH, Morrow AL, Blinn JR, and Paul SM (1990) Synthesis, metabolism, and pharmacological activity of 3α-hydroxy steroids which potentiate GABA-receptor-mediated chloride ion uptake in rat cerebral cortical synaptoneurosomes. *Journal of Medicinal Chemistry* 33: 1572–1581.
Purdy RH, Morrow AL, Moore PH, Jr., and Paul SM (1991) Stress-induced elevations of gamma-aminobutyric acid type A receptor-active steroids in the rat brain. *Proceedings of the National Academy of Sciences of the United States of America* 88: 4553–4557.
Quiñones-Jenab V, Ho A, Schlussman SD, Franck J, and Kreek MJ (1999) Estrous cycle differences in cocaine-induced stereotypic and locomotor behaviors in Fischer rats. *Behavioural Brain Research* 101(10): 15–20.
Ramcharan S, Love EJ, Fick GH, and Goldfien A (1992) The epidemiology of premenstrual symptoms in a population-based sample of 2650 urban women: Attributable risk and risk factors. *Journal of Clinical Epidemiology* 45: 377–392.
Rannevik G, Jeppsson S, Johnell O, Bjerre B, Laurell-Borulf Y, and Svanberg L (1995) A longitudinal study of the perimenopausal transition: altered profiles of steroid and pituitary hormones, SHBG and bone mineral density. *Maturitas* 21: 103–113.
Rapkin AJ, Mikacich JA, Moatakef-Imani B, and Rasgon N (2002) The clinical nature and formal diagnosis of premenstrual, postpartum, and perimenopausal affective disorders. *Current Psychiatry Reports* 4: 419–428.
Rapkin AJ, Morgan M, Goldman L, Brann DW, Simone D, and Mahesh VB (1997) Progesterone metabolite allopregnanolone in women with premenstrual syndrome. *Obstetrics and Gynecology* 90: 709–714.
Read J, Perry BD, Moskowitz A, and Connolly J (2001) The contribution of early traumatic events to schizophrenia in some patients: A traumagenic neurodevelopmental model. *Psychiatry* 64: 319–345.
Reddy DS (2004) Pharmacology of catamenial epilepsy. *Methods and Findings in Experimental and Clinical Pharmacology* 26: 547–561.
Roberts DC, Bennett SA, and Vickers GJ (1989) The estrous cycle affects cocaine self-administration on a progressive ratio schedule in rats. *Psychopharmacology* 98: 408–411.
Romeo E, Ströhle A, Spalletta G, et al. (1998) Effects of antidepressant treatment on neuroactive steroids in major depression. *American Journal of Psychiatry* 155: 910–913.
Rougé-Pont F, Mayo W, Marinelli M, Gingras M, LeMoal M, and Piazza PV (2002) The neurosteroid allopregnanolone increases dopamine release and dopaminergic response to morphine in the rat nucleus accumbens. *European Journal of Neuroscience* 16: 169–173.
Roxburgh A, Degenhardt L, and Copeland J (2006) Posttraumatic stress disorder among female street-based sex workers in the greater Sydney area, Australia. *BMC Psychiatry* 6: 24.
Rubinow DR, Hoban MC, Grover GN, Galloway DS, Roy-Byrne P, Andersen R, and Merriam GR (1988) Changes in plasma hormones across the menstrual cycle in patients with menstrually related mood disorder and in control subjects. *American Journal of Obstetrics and Gynecology* 158: 5–11.
Rupprecht R and Holsboer F (1999) Neuroactive steroids: Mechanisms of action and neuropsychopharmacological perspectives. *Trends in Neuroscience* 22: 410–416.
Russell JA, Leng G, and Douglas AJ (2003) The magnocellular oxytocin system, the fount of maternity: Adaptations in pregnancy. *Frontiers in Neuroendocrinology* 24: 27–61.
Russo SJ, Festa ED, Fabian SJ, Gazi FM, Kraish M, Jenab S, and Quinones-Jenab V (2003) Gonadal hormones differentially modulate cocaine-induced conditioned place preference in male and female rats. *Neuroscience* 120: 523–533.
Sanders D, Warner P, Bäckström T, and Bancroft J (1983) Mood, sexuality, hormones and the menstrual cycle. I. Changes in mood and physical state: Description of subjects and method. *Psychosomatic Medicine* 45: 487–501.
Satzinger G (1994) Antiepileptics from gamma-aminobutyric acid. *Arzneimittelforschung* 44: 261–266.
Schmidt PJ, Nieman LK, Danaceau MA, Adams LF, and Rubinow DR (1998) Differential behavioral effects of gonadal steroids in women with and in those without premenstrual syndrome. *New England Journal of Medicine* 338: 209–216.
Schmidt PJ, Purdy RH, Moore PH, Jr., Paul SM, and Rubinow DR (1994) Circulating levels of anxiolytic steroids in the luteal phase in women with premenstrual syndrome and in control subjects. *Journal of Clinical Endocrinology and Metabolism* 79: 1256–1260.
Schüle C, di Michele F, Baghai T, et al. (2003) Influence of sleep deprivation on neuroactive steroids in major depression. *Neuropsychopharmacology* 28: 577–581.
Schumacher M, Akwa Y, Guennoun R, et al. (2000) Steroid synthesis and metabolism in the nervous system: Trophic and protective effects. *Journal of Neurocytology* 29: 307–326.
Schumacher M, Guennoun R, Mercier G, et al. (2001) Progesterone synthesis and myelin formation in peripheral nerves. *Brain Research. Brain Research Reviews* 37: 343–359.
Segovia J, Lawless GM, Tillakaratne NJ, Brenner M, and Tobin AJ (1994) Cyclic AMP decreases the expression of a neuronal marker (GAD67) and increases the expression of an astroglia marker (GFAP) in C6 cells. *Journal of Neurochemistry* 63: 1218–1225.
Seippel L and Backstrom T (1998) Luteal-phase estradiol relates to symptom severity in patients with premenstrual syndrome. *Journal of Clinical Endocrinology and Metabolism* 45: 1988–1992.
Selye H (1941) Anesthetic effects of steroid hormones. *Proceedings of the Society for Experimental and Biological Medicine* 46: 116–121.
Shain W, Bausback D, Fiero A, Madelian V, and Turner JN (1992) Regulation of receptor-mediated shape change in astroglial cells. *Glia* 5: 223–238.
Sherwin BB (1988) Affective changes with estrogen and androgen replacement therapy in surgically menopausal women. *Journal of Affective Disorders* 14: 177–187.
Sherwin BB and Gelfand MM (1989) A prospective one-year study of estrogen and progestin in postmenopausal women: Effects on clinical symptoms and lipoprotein lipids. *Obstetrics and Gynecology* 73: 759–766.
Shirayama Y, Hashimoto K, Suzuki Y, and Higuchi T (2002) Correlation of plasma neurosteroid levels to the severity of negative symptoms in male patients with schizophrenia. *Schizophrenia Research* 58: 69–74.
Shughrue PJ, Lane MV, and Merchenthaler I (1997) Regulation of progesterone receptor messenger ribonucleic acid in the rat medial preoptic nucleus by estrogenic and antiestrogenic

compounds: An *in situ* hybridization study. *Endocrinology* 138: 5476–5484.

Sinha R, Lacadie C, Skudlarski P, Fulbright RK, Rounsaville BJ, Kosten TR, and Wexler BE (2005) Ural activity associated with stress-induced cocaine craving: A functional magnetic resonance imaging study. *Psychopharmacology (Berl)* 183: 171–180.

Sinha R, Talih M, Malison R, Cooney N, Anderson GM, and Kreek MJ (2003) Hypothalamic–pituitary–adrenal axis and sympatho-adreno-medullary responses during stress-induced and drug cue-induced cocaine craving states. *Psychopharmacology (Berl)* 170: 62–72.

Sinnott RS, Phillips TJ, and Finn DA (2002) Alteration of voluntary ethanol and saccharin consumption by the neurosteroid allopregnanolone in mice. *Psychopharmacology* 162: 438–447.

Sofuoglu M, Mitchell E, and Kosten TR (2004) Effects of progesterone treatment on cocaine responses in male and female cocaine users. *Pharmacology, Biochemistry, and Behavior* 78: 699–705.

Spalletta G, Romeo E, Bonaviri G, Bernardi G, Caltagirone C, and di Michele F (2005) Preliminary evidence for an association between aggressive and hostile behaviour and $3\alpha,5\alpha$-tetrahydroprogesterone plasma levels in schizophrenia. *Journal of Psychiatry and Neuroscience* 30: 49–52.

Stahl SM (1997) Sex therapy in psychiatric treatment has a new partner: Reproductive hormones. *Journal of Clinical Psychiatry* 58: 468–469.

Stell BM, Brickley SG, Tang CY, Farrant M, and Mody I (2003) Neuroactive steroids reduce neuronal excitability by selectively enhancing tonic inhibition mediated by delta subunit-containing $GABA_A$ receptors. *Proceedings of the National Academy of Sciences of the United States of America* 100: 14439–14444.

Stewart AL, Reynolds EO, Hope PL, et al. (1987) Probability of neurodevelopmental disorders estimated from ultrasound appearance of brains of very preterm infants. *Developmental Medicine and Child Neurology* 29: 3–11.

Stewart AL, Thorburn RJ, Hope PL, Goldsmith M, Lipscomb AP, and Reynolds EO (1983) Ultrasound appearance of the brain in very preterm infants and neurodevelopmental outcome at 18 months of age. *Archives of Disease in Childhood* 58: 598–604.

Ströhle A, Romeo E, di Michele F, et al. (2003) Induced panic attacks shift gamma-aminobutyric acid type A receptor modulatory neuroactive steroid composition in patients with panic disorder: Preliminary results. *Archives of General Psychiatry* 60: 161–168.

Strous RD, Maayan R, Lapidus R, Stryjer R, Lustig M, Kotler M, and Weizman A (2003) Dehydroepiandrosterone augmentation in the management of negative, depressive, and anxiety symptoms in schizophrenia. *Archives of General Psychiatry* 60: 133–141.

Sundström I, Andersson A, Nyberg S, Ashbrook D, Purdy RH, and Bäckström T (1998) Patients with premenstrual syndrome have a different sensitivity to a neuroactive-steroid during the menstrual cycle compared to control subjects. *Neuroendocrinology* 67: 126–138.

Sundström I, Ashbrook D, and Bäckström T (1997) Reduced benzodiazepine sensitivity in patients with premenstrual syndrome: A pilot study. *Psychoneuroendocrinology* 22: 25–38.

Sundström I and Bäckström T (1998a) Patients with premenstrual syndrome have decreased saccadic eye velocity compared to control subjects. *Biological Psychiatry* 44: 755–764.

Sundström I and Bäckström T (1998b) Citalopram increases pregnanolone sensitivity in patients with premenstrual syndrome: An open trial. *Psychoneuroendocrinology* 23: 73–88.

Sundström I, Bäckström T, Wang M, Olsson T, Seippel L, and Bixo M (1999) Premenstrual syndrome, neuroactive steroids and the brain. *Gynecological Endocrinology* 13: 206–220.

Sundström I, Nyberg S, and Bäckström T (1997) Patients with premenstrual syndrome have reduced sensitivity to midazolam compared to control subjects. *Neuropsychopharmacology* 17: 370–381.

Sveindottir H and Bäckström T (2000) Prevalence of menstrual cycle symptom cyclicity and premenstrual dysphoric disorder in a random sample of women using and not using oral contraceptives. *Acta Obstetricia et Gynecologica Scandinavica* 79: 405–413.

Tait GR, McManus K, Bellavance F, Lara N, Chrapko W, and LeMellédo JM (2002) Neuroactive steroid changes in response to challenge with the panicogenic agent pentagastrin. *Psychoneuroendocrinology* 27: 417–429.

Tanapat P, Hastings NB, and Gould E (2005) Ovarian steroids influence cell proliferation in the dentate gyrus of the adult female rat in a dose- and time-dependent manner. *Journal of Comparative Neurology* 481: 252–265.

Tordjman S, Anderson GM, McBride PA, et al. (1995) Plasma androgens in autism. *Journal of Autism and Developmental Disorders* 25: 295–304.

Torres JM and Ortega E (2003) Alcohol intoxication increases allopregnanolone levels in female adolescent humans. *Neuropsychopharmacology* 28: 1207–1209.

Torres JM and Ortega E (2004) Alcohol intoxication increases allopregnanolone levels in male adolescent humans. *Psychopharmacology (Berl)* 172: 352–355.

Townsend KA and Marlowe KF (2004) Relative safety and efficacy of finasteride for treatment of hirsutism. *Annals of Pharmacotherapy* 38: 1070–1073.

Uzunov DP, Cooper TB, Costa E, and Guidotti A (1996) Fluoxetine-elicited changes in brain neurosteroid content measured by negative ion mass fragmentography. *Proceedings of the National Academy of Sciences of the United States of America* 93: 12599–12604.

Uzunova V, Sampson L, and Uzunov DP (2006) Relevance of endogenous 3α-reduced neurosteroids to depression and antidepressant action. *Psychopharmacology (Berl)* 186: 351361.

Uzunova V, Sheline Y, Davis JM, Rasmusson A, Uzunov DP, Costa E, and Guidotti A (1998) Increase in the cerebrospinal fluid content of neurosteroids in patients with unipolar major depression who are receiving fluoxetine or fluvoxamine. *Proceedings of the National Academy of Sciences of the United States of America* 95: 3239–3244.

Uzunova V, Wrynn AS, Kinnunen A, Ceci M, Kohler C, and Uzunov DP (2004) Chronic antidepressants reverse cerebrocortical allopregnanolone decline in the olfactory-bulbectomized rat. *European Journal of Pharmacology* 486: 31–34.

Voigt LF, Weiss NS, Chu J, Daling JR, McKnight B, and van Belle G (1991) Progestagen supplementation of exogenous oestrogens and risk of endometrial cancer. *Lancet* 338: 274–277.

Walf AA and Frye CA (2006) A review and update: Estrogen's effects, brain targets, and mechanisms for anxiety and depression behavior. *Neuropsychopharmacology* 31: 1097–1111.

Wang M, Bäckström T, Sundström I, et al. (2001) Neuroactive-steroids and central nervous system disorders. *International Review of Neurobiology* 46: 421–459.

Wang JM, Johnston PB, Ball BG, and Brinton RD (2005) The neurosteroid allopregnanolone promotes proliferation of rodent and human neural progenitor cells and regulates

cell-cycle gene and protein expression. *The Journal of Neuroscience* 25(19): 4706–4718.
Wang M, Seippel L, Purdy RH, and Bäckström T (1996) Relationship between symptom severity and steroid variation in women with premenstrual syndrome: Study on serum pregnenolone, pregnenolone sulfate, 5α-pregnane-3, 20-dione and 3α-hydroxy-5α-pregnan-20-one. *Journal of Clinical Endocrinology and Metabolism* 81: 1076–1082.
Weill-Engerer S, David JP, Sazdovitch V, et al. (2002) Neurosteroid quantification in human brain regions: Comparison between Alzheimer's and nondemented patients. *Journal of Clinical Endocrinology and Metabolism* 87: 5138–5143.
Weinstock M (2001) Alterations induced by gestational stress in brain morphology and behaviour of the offspring. *Progress in Neurobiology* 65: 427–451.
Weinstock M (2005) The potential influence of maternal stress hormones on development and mental health of the offspring. *Brain, Behavior, and Immunity* 19: 296–308.
Weir CJ, Ling AT, Belelli D, Wildsmith JA, Peters JA, and Lambert JJ (2004) The interaction of anaesthetic steroids with recombinant glycine and $GABA_A$ receptors. *British Journal of Anaesthesia* 92: 704–711.
Whitehead MI (1978) The effects of oestrogens and progestogens on the postmenopausal endometrium. *Maturitas* 1: 87–98.
Whiting PJ, Bonnert TP, McKernan RM, et al. (1999) Molecular and functional diversity of the expanding GABA-A receptor gene family. *Annals of New York Academy of Sciences* 868: 645–653.
Wihlbäck AC, Nyberg S, Bäckström T, Bixo M, and Sundström-Poromaa I (2005) Estradiol and the addition of progesterone increase the sensitivity to a neurosteroid in postmenopausal women. *Psychoneuroendocrinology* 30: 38–50.
Wihlbäck AC, Sundström-Poromaa I, Nyberg S, and Bäckström T (2001) Sensitivity to a neurosteroid is increased during addition of progestagen to postmenopausal hormone replacement therapy. *Neuroendocrinology* 73: 397–407.
Wiklund I, Karlberg J, and Mattsson LÅ (1993) Quality of life of postmenopausal women on a regimen of transdermal estradiol therapy: A double-blind placebo-controlled study. *American Journal of Obstetrics and Gynecology* 168: 824–830.
Wohlfarth KM, Bianchi MT, and Macdonald RL (2002) Enhanced neurosteroid potentiation of ternary GABA(A) receptors containing the delta subunit. *Journal of Neuroscience* 22: 1541–1549.
Yawno T, Yan EB, Walker DW, and Hirst JJ (2007) Inhibition of neurosteroid synthesis increases asphyxia-induced brain injury in the late gestation fetal sheep. *Neuroscience* 146: 1726–1733.
Young EA, Midgley AR, Carlson NE, and Brown MB (2000) Alteration in the hypothalamic–pituitary–ovarian axis in depressed women. *Archives of General Psychiatry* 57: 1157–1162.
Young LT (2002) Neuroprotective effects of antidepressant and mood stabilizing drugs. *Journal of Psychiatry and Neuroscience* 27: 8–9.
Zhu Y, Bond J, and Thomas P (2003) Identification, classification, and partial characterization of genes in humans and other vertebrates homologous to a fish membrane progestin receptor. *Proceedings of the National Academy of Sciences of the United States of America* 100: 2237–2242.
Zwanzger P, Eser D, Padberg F, et al. (2004) Neuroactive steroids are not affected by panic induction with 50 microg cholecystokinin-tetrapeptide (CCK-4) in healthy volunteers. *Journal of Psychiatric Research* 38: 215–217.

Further Reading

Baulieu EE, Bénavides J, Désarnaud F, et al. (2001) Progesterone synthesis and myelin formation in peripheral nerves. *Brain Research Reviews* 37: 343–359.
Carlsson ML (1998) Hypothesis: Is infantile autism a hypoglutamatergic disorder? Relevance of glutamate–serotonin interactions for pharmacotherapy. *Journal of Neural Transmission* 105: 525–535.
Engel SR and Grant KA (2001) Neurosteroids and behavior. *International Review of Neurobiology* 46: 321–348.
Erskine MS and Kornberg E (1992) Stress and ACTH increase circulating concentrations of 3 α-androstanediol in female rats. *Life Sciences* 51: 2065–2071.
Freeman EW, Sammel MD, Lin H, Gracia CR, and Kapoor S (2008) Symptoms in the menopausal transition: Hormone and behavioral correlates. *Obstetrics and Gynecology* 111: 127–136.
Frye CA (2008) Hormonal influences on seizures: Basic neurobiology. In: Cynthia Harding (ed.) *Epilepsy in Women – Scientific Management,* pp 28–79. Mahwah, NJ: LEA.
Frye CA and Rhodes ME (2008) Infusions of 3α,5α-THP to the VTA enhance exploratory, anti-anxiety, social, and sexual behavior and increase levels of 3α,5α-THP in midbrain, hippocampus, diencephalon, and cortex of female rats. *Behavioural Brain Research* 187: 88–99.
Genazzani AR, Stomati M, Bernardi F, et al. (2004) Conjugated equine estrogens reverse the effects of aging on central and peripheral allopregnanolone and β-endorphin levels in female rats. *Fertility and Sterility* 81: 757–766.
Kavaliers M and Kinsella DM (1994) Male preference for the odors of estrous female mice is reduced by the neurosteroid pregnenolone sulfate. *Brain Research* 682: 222–226.
Kehoe P, Mallinson K, McCormick CM, and Frye CA (2000) Central allopregnanolone is increased in rat pups in response to repeated, short episodes of neonatal isolation. *Developmental Brain Research* 124: 133–136.
Kellogg CK and Frye CA (1999) Endogenous levels of 5α-reduced progestins and androgens in fetal vs. adult rat brains. *Developmental Brain Research* 115: 17–24.
Kessler RC, McGonagle KA, Nelson CB, Hughes M, Swartz M, and Blazer DG (1994) Sex and depression in the National Comorbidity Survey. II: Cohort effects. *Journal of Affective Disorder* 30: 15–26.
McFarlane HG, Kusek GK, Yang M, Phoenix JL, Bolivar VJ, and Crawley JN (2008) Autism-like behavioral phenotypes in BTBR T+tf/J mice. *Genes, Brain, and Behavior* 7: 152–163.
Moy SS, Nadler JJ, Young NB, et al. (2007) Mouse behavioral tasks relevant to autism: Phenotypes of 10 inbred strains. *Behavioural Brain Research* 176: 4–20.
Numan M and Insel TR (2003) Motivational models of the onset and maintenance of maternal behavior and maternal aggression. In: Bridges R (ed.) *The Neurobiology of Parental Behavior*, pp. 69–106. Oxford: Springer.
Patchev VK and Almeida OF (1996) Gonadal steroids exert facilitating and "buffering" effects on glucocorticoid-mediated transcriptional regulation of corticotropin-releasing hormone and corticosteroid receptor genes in rat brain. *Journal of Neuroscience* 16: 7077–7084.
Patchev VK, Hassan AH, Holsboer DF, and Almeida OF (1996) The neurosteroid tetrahydroprogesterone attenuates the endocrine response to stress and exerts glucocorticoid-like effects on vasopressin gene transcription in the rat hypothalamus. *Neuropsychopharmacology* 15: 533–540.
Patchev VK, Shoaib M, Holsboer F, and Almeida OF (1994) The neurosteroid tetrahydroprogesterone counteracts

corticotropin-releasing hormone-induced anxiety and alters the release and gene expression of corticotropin-releasing hormone in the rat hypothalamus. *Neuroscience* 62: 265–271.

Rubin RT, Poland RE, Lesser IM, Martin DJ, Blodgett AL, and Winston RA (1987b) Neuroendocrine aspects of primary endogenous depression. III. Cortisol secretion in relation to diagnosis and symptom patterns. *Psychological Medicine* 17: 609–619.

Rubin RT, Poland RE, Lesser IM, Winston RA, and Blodgett AL (1987a) Neuroendocrine aspects of primary endogenous depression. I. Cortisol secretory dynamics in patients and matched controls. *Archives of General Psychiatry* 44: 328–336.

Schneider JS, Burgess C, Sleiter NC, Doncarlos LL, Lydon JP, O'Malley B, and Levine JE (2005) Enhanced sexual behaviors and androgen receptor immunoreactivity in the male progesterone receptor knockout mouse. *Endocrinology* 146: 4340–4348.

Schumacher M, Weill-Engerer S, Liere P, et al. (2003) Steroid hormones and neurosteroids in normal and pathological aging of the nervous system. *Progress in Neurobiology* 71: 3–29.

Strous RD, Maayan R, and Weizman A (2006) The relevance of neurosteroids to clinical psychiatry: From the laboratory to the bedside. *European Neuropsychopharmacology* 16: 155–169.

Young E, Carlson NE, and Brown MB (2001) Twenty-four-hour ACTH and cortisol pulsatility in depressed women. *Neuropsychopharmacology* 25: 267–276.

16 Brain Peptides: From Laboratory to Clinic

T D Geracioti, Jr., J R Strawn, N N Ekhator, and M Wortman, University of Cincinnati, Cincinnati, OH, USA
J Kaskow, University of Pittsburgh Medical Center, Pittsburgh, PA, USA

Chapter Outline

16.1 Introduction

This chapter focuses on clinically relevant neuropeptides of central nervous system (CNS) origin, including those hypothalamic peptides which subserve or regulate body temperature, osmolality, appetite, mood, libido, anger, circadian rhythms, and the response to stress. In addition, we review progress in the clinical development of therapeutics, with specific cortical and extrahypothalamic peptidergic targets, which is nevertheless emerging.

Neuropeptides – relatively short chains of amino acids – are potent and specific neuromodulators that play a role in both synaptic and nonsynaptic neuronal communication (diffusion or volume neurotransmission, see Agnati et al. (1995) and Zoli et al. (1999)). Peptide targets are tyically distant from the cell bodies from which they are released (Herkenham, 1987). The ability of peptides to diffuse or travel to distant sites, and to retain potency in small concentrations, permits them to entrain mass-sustained functions, such as mood, appetite, and libidinal tone that require the coordination of diverse neuronal pathways and multiple-organ systems. Significantly, numerous neuropeptides share functions, indicating the existence of signaling-system redundancies within the CNS. Studies with transgenic mice involving the deletion of neuropeptides have revealed this redundancy phenomenon; the importance of the brain necessitates that its critical mechanisms be subserved by multiple peptides capable of sharing the same function (Inui, 2000).

Despite the multiplicity of neuropeptides that have been discovered and characterized and despite the ever-quickening realization of the clinical promise of targeted modification of their signaling pathways, the routine clinical use of neuropeptidergic interventions to modify CNS function remains limited. In some cases, this is because clinical trials of available CNS-penetrating peptide analogs or antagonists have not been performed. In other cases, this is because proven therapies (such as chronic gonadotropin-releasing hormone (GnRH) agonist administration to induce a hypoandrogenic state and eliminate libido in pedophilia) are rarely utilized.

CNS neuropeptides have received abundant attention from clinical psychobiologists and psychiatrists for three decades (Prange et al., 1979). As progress is made in the identification of the neurobehavioral pathways and functions subserved by specific peptides, novel therapeutics based on the chemical structures of the peptides in question will be increasingly generated. CNS-specific peptide agonists and antagonists will likely have important clinical applications in the treatment of neurologic and psychiatric conditions. Furthermore, it is likely that modulation of gene expression with these therapeutics will provide the most power to manipulate the production and actions of proteins and neuropeptides.

The neuropeptides to be discussed in this chapter include the (1) hypothalamic-releasing hormones: growth-hormone-releasing hormone (GHRH), corticotropin-releasing hormone (CRH), somatostatin, thyrotropin-releasing hormone (TRH), and GnRH; (2) the proopiomelanocortins (POMCs): α-melanocyte-stimulating hormone (MSH), and adrenocorticotropin hormone (ACTH); (3) the gut-derived peptide cholecytokinin, which also plays an important role in the brain; (4) the opioids: β-endorphin, dynorphin, and leu- and met-enkephalin; (5) the neurohypophysial hormones vasopressin (VP) and oxytocin; and finally (6) several neuropeptides of emerging or expanding psychiatric interest, including substance P, neuropeptide Y (NPY), and orexin (hypocretin). We overview the basic biology of these peptides and consider their clinical relevance. As can be seen from the table of neuropeptides (**Table 1**), we have only scratched the surface of the field here.

16.2 Growth-Hormone-Releasing Hormone

GHRH is a growth-hormone-stimulating peptide which belongs to the secretin–glucagon family that includes vasoactive intestinal peptide (VIP). GHRH neurons are localized in the arcuate nucleus, the supraoptic nucleus (SON), and paraventricular nucleus (PVN) of the hypothalamus and the pituitary stalk. SON and PVN neurons secrete GHRH into the pituitary portal system from the median eminence. GHRH receptors are G-protein-coupled receptors and they signal via adenylate cyclase, phosphatidylinositol, the Ca^{2+} calmodulin system, and the arachadonic acid–eicosanoid system (Harvey, 1995). A number of GHRH receptor splice variants have been identified and several appear to be of clinical significance.

16.2.1 Regulation of GHRH

Regulation of GHRH is known to occur through several mechanisms (**Figure 1**). Hypothalamic GHRH stimulates growth hormone (GH) secretion from pituitary somatotropes. The balance between GHRH, a secretagog, and hypophyseal somatostatin, an inhibitor of GH release, appears to be the most important determinant of pituitary GH pulsatility. Hypothalamic somatostatin, in addition to its pituitary effects, may also dampen GHRH secretion directly (Fodor et al., 2006). Ghrelin (a GH-releasing peptide) stimulates GHRH (and increases pituitary sensitivity to its effects), while NPY, by stimulating somatostatin, inhibits GHRH (**Figure 1**) (see Smith (2005) and Fodor et al. (2006)).

16.2.2 Functions of GHRH

GHRH is the central effector of the GH/insulin-like growth factor-I axis. GHRH stimulates GH synthesis and release and the proliferation and differentiation of pituitary somatotropes, beginning in fetal life. In growing children, GHRH is secreted in a pulsatile manner about every 3 h and more frequently during sleep (Tapanainen et al., 1989). The pulsatile release of GHRH is age dependent and there is also a sexually dimorphic pattern, particularly at puberty when GHRH function becomes sexually dimorphic due to

Table 1 Table of neuropeptides

Activin
Adrenocorticotropin hormone (ACTH)
Agouti-related peptide
Amylin
Angiotensin
Apelin
Atrial naturetic peptide
Bombesin
Brain-derived neurotrophic factor (BDNF)
Brain natriuretic peptide
C-natriuretic peptide
Calcitonin
Calcitonin-gene-related peptide (CGRP)
Cholecystokinin (CCK)
Cocaine- and amphetamine-related transcript (CART)
Corticotropin-releasing hormone (CRH)
δ-Sleep-inducing peptide
Dynorphin
Endomorphin
β-Endorphin
Endothelins
Enkephalin
Follicle-stimulating hormone (FSH)
Galanin
Gastric inhibitory peptide
Gastrin
Gastrin-releasing peptide
Ghrelin
Glucagon
Glucagon-like peptide-1 (GLP-1)
Gonadotropin-releasing hormone (GnRH)
Growth hormone (GH)
Growth-hormone-releasing hormone (GHRH)
Insulin
Insulin growth factors
Kisspeptin
Leptin
β-Lipotropin
Lutenizing hormone (LH)
Met- and leu-enkephalin
Melanocyte-stimulating hormone (MSH)
Motilin
Nerve growth factor
Neurokinins
Neuromedins
Neuropeptide Y
Neuropeptide YY
Neurophysins
Neurotensin
Opioids
Orexins (hypocretins)
Oxytocin
Pituitary adenylate cyclase-activating polypeptide (PACAP)
Pancreatic polypeptide
Peptide histidine isoleucine
Peptide YY
Prolactin

Continued

Table 1 Continued

Secretin
SF-1
Somatostatin
Substance P
Thyrotropin-releasing hormone (TRH)
Thyroid-stimulating hormone (TSH)
Urocortin
Vasopressin
Vasoactive intestinal polypeptide (VIP)

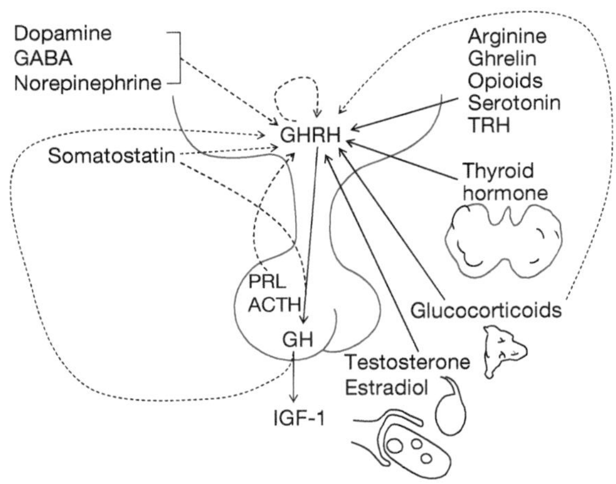

Figure 1 Clinically important physiologic regulators of growth-hormone-releasing hormone (GHRH) release. Positive effectors are indicated by solid arrows and negative effectors are denoted by dashed arrows.

testosterone. GHRH actions in adult or aging animals become less responsive to testosterone and the feedback of GH on GHRH neurons also declines considerably with aging (Veldhuis et al., 1997). In aging, GHRH receptors diminish.

16.2.3 Growth Hormone

GH, which is stimulated by the pulsatile secretion of GHRH, is an anabolic hormone which promotes growth, stimulates protein synthesis, glycogenolysis, and lipolysis. The GH receptor belongs to the hematopoietin superfamily which includes receptors for erythropoietin, the interleukins, the interferons, and several white-blood-cell stimulating factors. Insulin growth factor-I, which mediates some of theGH's effects, is among one of the factors that regulate GH gene expression at the transcriptional level (Bertherat et al., 1995).

16.2.4 Clinical Implications: Disease States with GHRH-Related Abnormalities

GHRH deficiency during development results in dwarfism or small stature. Mutations of both the GHRH gene and the GHRH receptor are characterized by GH deficiency and dwarfism (Maheshwari et al., 1998; Netchine et al., 1998; Salvatori et al., 1999; Hayashida et al., 2000; Hilal et al., 2008).

GHRH deficiency can also result from cranial irradiation and traumatic brain injury, or be part of a syndrome of hypothalamic–pituitary disease that involves deficiencies in other hypophysiotropic or hypohyseal hormones. Old age is a commonplace cause of GHRH deficiency.

While traumatic brain injury can, in 15% to >50% of cases, depending on the case series reported, cause insufficiencies in virtually all the hypothalalmic–pituitary hormone systems of known clinical relevance, it is the GHRH–GH system that is most likely to be affected (Kelly et al., 2000; Aimaretti et al., 2004; Klose et al., 2007; Tanriverdi et al., 2008). GHRH and GH deficiency have been seen in boxers, football players, martial artists, motorcycle riders, and so on. Preliminary data indicate that GH deficiency in patients with mild to moderate traumatic brain injuries improves over 1–3 years (about half the time), but may rarely worsen (Tanriverdi et al., 2008). Given the high prevalence of American soldiers in the Iraq theater returning home with head injuries, we predict (along with Stonesifer, 2008) that CNS-mediated hypoituitarism involving GHRH deficiency will be a source of significant psychiatric morbidity.

GHRH (often in combination with the amino acid arginine or with GH-releasing peptide) is used clinically as a provocative test of GH secretion in the diagnostic workup of suspected GH deficiency (see, e.g., Ho, 2007).

Extensive investigation of the GHRH–GH axis in patients with depression reveals consistent abnormalites in noradrenergic-mediated GH release, which probably occurs via GHRH-containing neurons (Dinan, 1998), although patients with both major depression and a full-blown anxiety disorder have more blunted GH responses to the centrally active adrenergic α-2 autoreceptor agonist clonidine (which inhibits noradrenergic transmission) than patients with depression alone (Cameron, 2006). Some patients with major depressive disorders also appear to have diminished GH secretion and/or diminished GH responses to GHRH and it is possible that normalization of the GHRH–GH hypothalamic–pituitary axis might have an antidepressant effect. Panic disorder – a syndrome involving the sudden onset of discrete episodes of crippling anxiety or anxiety equivalents, such as acute and apparently inexplicable somatic symptoms – may also occur in the context of a hyporesponsive hypothalamic-GH system (Uhde et al., 1992).

In contrast, acromegaly, a syndrome characterized by many dysmorphic changes, including bony and soft-tissue overgrowth, in most patients, is caused by excessive circulating GH, most often as a result of a GH-secreting pituitary tumor. A GHRH-producing tumor is nevertheless an important, if relatively infrequent, cause of acromegaly, and quantification of circulating GHRH is a necessary element in the clinical evaluation of the syndrome. Because of its insidious onset, the development of the dsymorphic signs of acromegaly is often not recognized by the afflicted individual or his/her friends and family. Consequently, acromegaly is frequently not diagnosed until after the development of one of its complications, such as diabetes mellitus, cardiovascular disease, and hyperthyroidism, or from symptoms related to tumor enlargement.

In anorexia nervosa, altered hypothalamic control of pituitary GH secretion exists, presumably involving GHRH, with patients showing basal GH hypersecretion (Garfinkel et al., 1975; Scacchi et al., 1997; Gianotti et al., 2000). It is not clear whether or not the hypersecretion of GH is a manifestation of the pathophysiology of the eating disorder or an attempt to compensate for food restriction by elaborating an anabolic, growth-promoting hormone. The GHRH–GH abnormalities in anorexia nervosa normalize with sustained weight restoration. In contrast to anorexia nervosa, obesity is related to low activity of the GHRH–GH system.

16.2.5 Clinical Implications: Therapeutics

Pharmacologic treatment of acromegaly – notwithstanding surgery and radiation – is used in the attempt to reduce GH hypersecretion. Administration of long-acting somatostatin analogs such as octreotide or a dopamine agonist, for example, bromocriptine, for this purpose shows varying degrees of success in reducing GH concentrations. Specific GH-receptor antagonists (Trainer et al., 2000) may eventually prove to be superior. The advent of potent GHRH antagonists also constitutes a new era of therapeutics (Schally and Varga, 1999). Evidence is accumulating that GHRH antagonists can be anticancer agents by blocking the growth of tumors for which GH, IGF-I, IGF-II, and/or GHRH are growth factors (Kovács et al., 2008). Dwarfism in children can be treated with GHRH administration and an acceleration of growth can be observed when it is given in a pulsatile manner, but growth-stimulation declines after a year of therapy (Vance and Thorner, 1988). Long-acting GHRH pharmaceutical preparations are in clinical development (Clemmons, 2007). GH itself has been used to treat not only dwarfism but also, increasingly, idiopathic short stature.

A potentially useful new class of synthetic and orally active peptide and nonpeptide GH secretagogs that act synergistically with and largely independently of GHRH – including GHRP-6, hexarelin, and others – may find clinical utility as an alternative to treatment with GHRH in states of GH deficiency (Ghigo et al., 1998). Other GH secretagogs, such as ghrelin, act via specific receptors in the pituitary and hypothalamus that are different from those upon which GHRH acts. The development of GH secretagogs by Smith (2005) led to the discovery of ghrelin.

Since GHRH (and GH) secretion declines with aging and senesence, it is possible that this phenomenon is intrinsically related to the aging process. Consequently, there has been speculation that administration of GH or stimulating its secretion may be beneficial in normal aging, and this has led to widespread off-label use of GH; however, there are still very few well-controlled studies of the effects and side effects of GH or GH secretagogs in aging (Cummings and Merriam, 1999). Some evidence does suggest that in individuals with age-related GH deficiency, GH replacement increases lean muscle mass, reduces abdominal adiposity, and improves cardiovascular health and feelings of general well-being. However, it is of interest that GH-receptor knockout mice live considerably longer than their normal littermates (Laron, 2002), similar to the life extension seen in animals subjected to caloric restriction. It is of additional interest that daily GHRH for 6 months improved cognition in a group ($n=89$) of healthy older adults with a mean age of 68 (Vitiello et al., 2006).

Doping with GH has become an increasing problem in sports during the last two decades. A few controlled studies have been performed with supraphysiological GH doses in athletes, which have shown conflicting results. Indeed, the chronic hypersecretion of GH in acromegaly shows that long-term elevations in circulating GH concentrations, although associated with increased muscle volume, do not beget increased strength or performance and actually worsen exercise tolerance and fatigue. Moreover, GH-related hyperglycemia, reduced insulin sensitivity, edema, and even bony overgrowth are among the potential serious side effects. However, although many believe that short-term administration of supraphysiological doses of GH may have a beneficial effect on athletic performance, scientific evidence for this claim – although clinical studies may not have accurately reflected real-world doping protocols – is lacking (Liu et al., 2008).

16.3 Gonadotropin-Releasing Hormone

GnRH was first isolated in 1971 (Schally et al., 1971; Amoss et al., 1971). GnRH neurons are present in loose networks rather than discrete nuclei and are found in the medial basal hypothalamus (MBH) and the arcuate nucleus of the hypothalamus. Most GnRH neurons project to the median eminence, the final common pathway from the brain to the anterior pituitary, but some GnRH fibers continue down the infundibular stalk to enter the posterior pituitary. Secretion of GnRH is pulsatile and the pattern of pulses is influenced by the pulse generator in the hypothalamus, specifically the MBH and arcuate nucleus (Knobil, 1989). The GnRH receptor is a surface receptor; the second-messenger system for GnRH is the phosphatidyl–inositol system and there is increasing evidence of functionally significant GnRH-receptor subtypes that could eventually lead to advances in GnRH analog design.

16.3.1 GnRH Regulation

A number of factors can modulate release of GnRH, such as ovarian steroids, neuropeptides, and neurotransmitters (**Figure 2**). In addition, GnRH also regulates its own release via either a short C- or ultrashort-loop-feedback mechanism. In the female, there is an abrupt and steep rise in GnRH secretion prior to ovulation.

16.3.2 Functions of GnRH

GnRH has important pituitary–gonadal effects, essentially involving the regulation of LH and FSH during reproductive life. GnRH also plays a role in the development of the pituitary–gonadal regulation during fetal life. The synthesis and secretion of gonadotropins is critically dependent on GnRH production by the fetal hypothalamus. Circulating GnRH acts on specific GnRH receptors in the gonads.

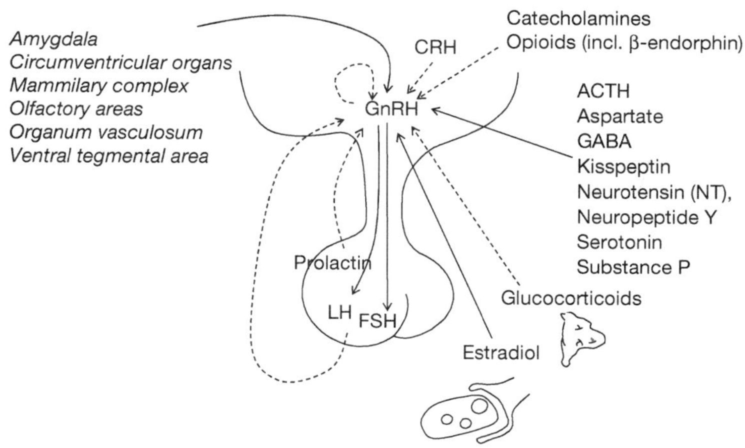

Figure 2 Clinically important physiologic regulators of gonadotropin-releasing hormone (GnRH) secretion. Positive effectors are indicated by solid arrows and negative effectors are denoted by dashed arrows. Structures and tracts are italicized.

16.3.3 Clinical Implications

Sexual development into a mature phenotype is dependent on adequate function of the hypothalamic–pituitary–gonadal axis; activation of the brain GnRH pulse generator is, therefore, critical for sexual maturation and reproduction, while developmental failure to activate GnRH results in sexual infantilism. This failure is a matter of degree (as only one example, Kallmann's syndrome – congenital gonadotropin deficiency with anosmia – has variable genetic underpinnings and variable phenotypic severity). In addition to puberty-related activation of the GnRH-controlled system, there is also a brief postnatal surge of GnRH (lasting only months) that provides a window for evaluation of suspected gonadotropin deficiency in infants (van Tijn et al., 2007). A number of GnRH-receptor mutations associated with loss of function have also been identified in recent years. If identified and corrected early, the negative sequelae of congenital gonadotropin insufficiency can be avoided and sexual, somatic, and psychological growth better promoted and timed. Adult-onset gonadotropin deficiency results in insufficiency of sex-hormone production with attendant deficits that may include fatigue, lack of vigor, impaired libido and sexuality, irritability, anxiety, and depressed mood. Among the causes of adult-onset GnRH deficiency are head trauma and aging.

In contrast to the hypogonadism and failure to undergo puberty that is associated with developmental lack of GnRH pulsing, early activation of GnRH is the most common cause of precocious puberty (see Carel and Leger (2008) for a review). Although precise norms for what constitutes precocious puberty can be debated, generally, this means puberty onset before the age of 8 years in girls and 9.5 years in boys; however, blacks enter puberty, on an average, sooner than whites.

Many and various synthetic GnRH agonists (and antagonists) are available for clinical application. The synthetic GnRH agonists or analogs are typically of greater potency and longer acting than veridical GnRH. Although short-term or pulsatile administration of a GnRH agonist stimulates the pituitary–gonadal axis, continuous or chronic use of GnRH agonists – or use of long-acting GnRH analogs – results in desensitization of pituitary GnRh receptors and supression of gonadal activity via inhibition of the synthesis and release of LH and FSH by the anterior pituitary. Developmental attempts are underway to generate a GnRH antagonist (see below) that produces immediate inhibition of LH and sex hormones (van Poppel and Nilsson, in press).

GnRH-agonist-induced suppression of pituitary–gondal activity is helpful in the treatment of some clinical syndromes whose pathophysiologies are dependent upon gonadal hormones. For example, GnRh agonists, such as leuprolide acetate, reduce symptoms of premenstrual syndrome (or premenstrual dysphoric disorder) – a syndrome characterized in part by emotional lability, depression, and irritability that occur during the late luteal phase (approximately 1 week before the onset of menses) and resolve upon or within 1 week of the commencement of menstruation. Leuprolide appears to be ineffective in those women who have premenstural excacerbations of ongoing or menstrual-cycle-independent symptoms of dysphoria (Freeman et al., 1997).

The side effects of continuous GnRH-agonist therapy include those directly related to gonadal suppression – including anovulation in women. In the female, chronic GnRH-induced hypoestrogenic side effects include vaginal dryness, bone demineralization, vasomotor changes, and headache. In the male, hypoandrogen side effects include erectile failure (Marumo et al., 1999). Psychiatric side effects include major anxiety and depressive syndromes (Warnock and Bundren, 1997). Typically, GnRH-agonist trials in the premenstrual syndrome are reseved for women who experience particularly severe or resistant symptoms, while specific serotonin reuptake inhibitors (SSRIs) are typically used as first-line agents because of their relatively benign side-effect profile and ease of use. One potential complication of suppression of sex hormones by long-acting or continuously administered GnRH-agonists is the initial testosterone surge that takes place after the initiation of treatment, but before inihibition of the sex hormone axis occurs (e.g., a concern in the treatment of sex-hormone-dependent cancers).

In men, the antiandrogen effects of GnRH agonists can be effectively used to treat paraphilias (historically called sexual perversions) by reducing or eliminating sexual fantasies, libido, and erectile function (Thibaut et al., 1996). Essentially, GnRH analogs are able to effect a reversible chemical castration. Monthly injections of the long-acting GnRH analog triptorelin, for example, reduce serum testosterone concentrations to 4% of baseline in men with sexual deviancies and abolish unwanted sexual behavior (Rosler and Witztum, 1998), and, in pedophiles, similar GnRH agonist injections reduced serum testosterone levels to castrate levels and nearly eliminated sexual urges toward children (Schober et al., 2005). We anticipate a much greater use of GnRH agonists in the treatment of sexual predators, and possibly – pending clinical research in the area – in violent predators, than what currently occurs.

Other applications of GnRH analogs include roles in various clinical conditions involving problems of reproduction. As noted, GnRH can be used to either stimulate or inhibit the pituitary–gonadal axis, such as in the treatment of either precocious or delayed puberty. Also, endometriosis, some uterine tumors, and prostate cancer can be treated with these analogs (Laufer and Rein, 1993; Ensom, 2000). GnRH is also helpful in increasing fertility in females (by inducing ovulation) if it is administered in a pulsatile manner. *In vitro* fertilization methods utilize GnRH analogs and ovarian steroids (Fahri et al., 1996) to induce ovulation.

GnRH antagonists block the ability of endogenous GnRH to bind to their pituitary receptors, thus immediately suppressing secretion of gonadotopins. The GnRH antagonists have a number of clinical uses, largely in the practice of obstetrics and gynecology at present – although, the ability of GnRH antagonists to rapidly suppress testosterone concentrations in men, with a single dose (Behre et al., 1992; Erb et al., 2000), suggests that their indications will increase. Indeed, GnRH antagonists may eventually replace the use of the GnRH agonists in many clinical situations (Felberbaum et al., 2000). Finally, it should be noted that the azoospermia engendered by GnRH antagonists may eventually make possible a hormonal contraceptive for males (Bremner et al., 1991; Bastias et al., 1993), although the hypoandrogenism associated with GnRH antagonism is problematic.

Finally, it should be noted that a number of single-gene mutations that directly or indirectly affect GnRH neuronal function are increasingly recognized as common causes of the disorders of puberty (see Herbison (2007) for a review).

16.4 Somatostatin

The initial evidence suggesting the existence of somatostatin (somatotropin-release-inhibiting factor) was from Kruhlich et al. (1968) who reported on a GH-release-inhibiting substance in hypothalamic extracts. Somatostatin was subsequently isolated, sequenced, and synthesized by Brazeau et al. (1973) and was found to not only inhibit GH, but also the release of many other hormones as well as the exocrine secretions of the GI tract (Zeng and Sachs, 1998).

16.4.1 Localization

In mammals, somatostatin is located in the periventricular preoptic area and the arcuate and ventromedial nuclei, and these neurons have long projections to the median eminence as well as limbic and brainstem structures. These projections overlap with those of GHRH-expressing neurons. Furthermore, these somatostatin-expressing neurons can be seen in the cerebral cortex and autonomic nervous system (Elde and Hokfelt, 1979), supporting a role of somatostatin in autonomic regulation. Somatostatin is also seen in the GI tract – in mucosal D cells, in myenteric neurons, and in D cells of the pancreas – and is involved in the secretion of insulin and glucagon.

16.4.2 Somatostatin Receptors

Somatostatin receptors have their highest concentrations in the cerebral cortex. Intermediate levels of expression can be observed in the hypothalamus, thalamus, amygdala, and hippocampus. Low amounts of expression can be observed in the mid- and hindbrain (Vecsei and Widerlov, 1990). Somatostatin receptors are also enriched in the anterior pituitary, *zona glomerulosa* of adrenal cortex, and exocrine pancreas (Patel, 1987). Five types of somatostatin receptors have been discovered (Epelbaum et al., 1994). The genes for all five of these have been cloned. All receptors bind somatostatin-14 and -28 with high affinity. In the pituitary, the somatotropes, thyrotropes, and lactotropes have a single class of binding site for somatostatin, whereas the brain has multiple subtypes. Receptor subtypes 1 and 2 are involved in the neuroendocrine regulation of GH (Beaudet et al., 1995). Subtype 3 is found in the motoneurons of the spinal cord and motor nuclei of the brainstem as well as in the sensory neurons of the spinal ganglia (Senaris et al., 1995).

16.4.3 Physiologic Effects

Somatostatin is secreted in a pulsatile manner, which is coordinated with GHRH (which has opposing or reciprocal effects), and inhibits the secretion of GH (and several other hormones).

Somatostatin influences numerous organismic functions, including regulation of body temperature, blood pressure, hunger/satiety, nociception, the acoustic startle resonse, learning, and memory. Analgesia can be induced, probably secondary to the interaction of somatostatin, with opioid receptors. Somatostatin increases rapid eye movement (REM) sleep (Toppila et al., 1996); it also has immunomodulatory and anti-inflammatory effects (Karalis et al., 1994; and see Guillemin (2004)).

16.4.4 Clinical Implications

The discovery that somatostatin inhibits pituitary secretion of GH and other hormones such as insulin and glucagon, forms the basis for its clinical use in acromegaly and to inhibit other hormone-secreting tumors. The emergence of various somatostatin analogs, with different receptor affinities and pharmacological properties, has led to the possibility of matching a specific analog to a given tumor.

Somatostatin generally inhibits a number of hormone-secreting, somatostatin-receptor-containing tumors, including, but not limited to, GH-secreting pituitary tumors, thyroid-stimulating hormone (TSH)-secreting tumors, and at least some prolactin (PRL)-secreting tumors (although the latter are preferentially treated with dopamine agonists). Long-acting forms of somatostatin – including octreotide (SMS 201–995), an octapeptide – are especially useful. Several other somatostatin analogs, including nonpeptide compounds, have been developed for the treatment of acromegaly. Induction of apoptosis has been described with high doses of another agent – lanreotide. These analogs are especially important since most patients with acromegaly escape from somatostatin-analog therapy with regard to both hormonal production and tumor growth; the mechanism behind the tachyphylaxis is not yet known.

In Alzheimer's disease (AD), somatostatin peptide levels are consistently decreased in the cerebrospinal fluid (CSF) and receptors are decreased in the brain; in this regard, it appears that certain genetic variants of the somatostatin gene may protect against the disorder (Vepsäläinen et al., 2007; and see Burgos-Ramos et al. (2008)). However, while pilot data are intriguing (Craft et al., 1999), clinical trials of somatostatin analogs in AD are lacking. There is also evidence that somatostatin is altered in patients with mood and cognitive disorders (Roca et al., 1999). Specifically, there are reports indicating that somatostatin-like immunoreactivity is decreased in the CSF of patients with major depression (Molchan et al., 1993; Kling et al., 1993).

16.5 Corticotropin-Releasing Hormone

CRH, or corticotropin-releasing factor (CRF), the principle CNS effector of the pituitary–adrenocortical response to stress, is a 41-amino-acid-containing neuropeptide which was isolated by Vale et al. (1981) and subsequently cloned (Shibahara et al., 1983). The structure of CRH varies from species to species but its biological activity does not. CRH from one species will stimulate ACTH secretion in another. CRH is found in highest concentration in the parvocellular neurons of the PVN of the hypothalamus. The axons of these neurons terminate in the external zone of the median eminence. CRH is released from these neurons into the portal capillary system. This CRH–tuberoinfundibular system stimulates ACTH secretion. Half of the parvocellular CRH neurons also express VP, which synergizes with CRH in its actions on ACTH (Ma and Aguilera, 1999). Outside of the hypothalamus, CRH is localized to the limbic system, including the amygdala, septum, bed nucleus of stria terminalis (BNST), the brainstem, and the spinal cord (Cummings et al., 1983).

16.5.1 CRH Regulation

CRH is regulated by a number of factors (**Figure 3**). To oversimplify, glucocorticoids negatively regulate CRH, while physical and emotional stressors positively regulate its expression (Ma and Aguilera, 1999). CRH and norepinephrine may normally participate in a positive, reverberating, feed-forward loop, wherein activation of one is associated with activation of the other (for human data see Wong et al. (2000) and Vythilingam et al. (2000)). Testosterone suppresses CRH promoter acitivy in the human hypothalamic PVN *in vitro* (Bao et al., 2006); clinical data suggest that testosterone may dampen the effects of CRH-induced cortisol release at the level of the adrenal cortex or through a mechanism that does not involve the suppression of ACTH secretion (Rubinow et al., 2005). CRH hypothalamic–pituitary–adrenal (HPA) responses vary with the type of stressor, the duration and pattern of the stressor, the age and sex of the individual, perinatal experience, and dominance position within the social group.

16.5.2 CNS CRH Circadian Rhythm

CRH is secreted on a circadian basis. In the human, CSF CRH concentrations show a diurnal rhythm; levels rise in the morning upon arousal and peak in the early morning with a decline over the remaining 24-h period (Kling et al., 1991; **Figure 4**). Thus, interestingly, the CRH circadian rhythm is roughly the opposite of the circadian rhythm of cortisol secretion from the adrenal cortex, even though CRH is the brain effector of adrenocortical cortisol release. In nocturnal animals, the situation is reversed. The hypothalamic suprachiasmatic nucleus (SCN) is thought to be one of the prime regulators of this clock.

16.5.3 The CRH Receptor

The CRH receptor is a seven-transmembrane protein coupled to G-proteins (Perrin and Vale, 1999). It is linked to cyclic adenosine monophosphate (cAMP) production. Brain and pituitary receptors are high-affinity receptors which share similar kinetic and pharmacologic characteristics. CRH receptors are located in the anterior and intermediate lobes of

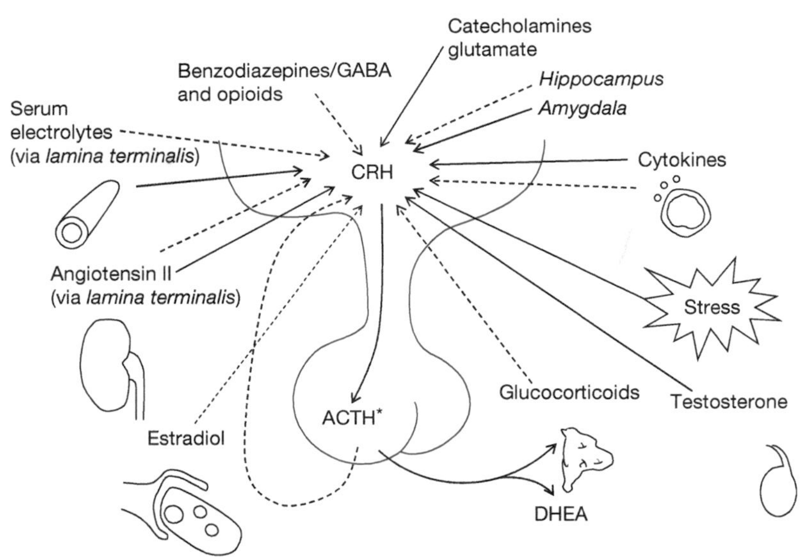

Figure 3 Clinically important physiologic regulators of corticotropin-releasing hormone (CRH) secretion. Structures and tracts are italicized. Asterisk denotes that in addition to adrenocorticotropin-releasing hormone (ACTH), other POMC products are released (see **Figure 5**).

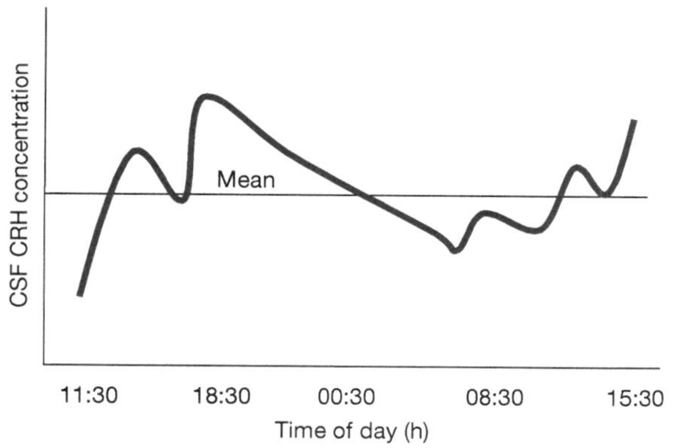

Figure 4 Diurnal variations in cerebrospinal fluid (CSF) corticotropin-releasing hormone (CRH) concentrations in six healthy volunteers over the course of 30h of continuous CSF sampling. The first sample was collected approximately 3 h after the insertion of a lumbar subarachnoid catheter. Data from Kling MA, DeBellis MD, O'Rourke DK, et al. (1994) Diurnal variation of cerebrospinal fluid immunoreactive corticotropin-releasing hormone levels in healthy volunteers. *Journal of Clinical Endocrinology and Metabolism* 79: 233–239.

the pituitary, olfactory bulb, cerebellum, cerebral cortex, and amygdala (Chen et al., 2000). In the hypothalamus, basal levels of expression of CRH receptors are low but levels of expression increase with stress.

CRH-receptor subtypes include the 1, 2-α, 2-β, and 2-γ. The CRH_1 receptor is localized in neocortical, cerebellar, and sensory structures, particularly in the lateral septum, ventromedial hypothalamus, and paraventricular hypothalamus; CRH_2 receptors are also present in human brain. Type-1 and-2 receptors differ in structure, pharmacology, and regulatory responses to stress (Perrin and Vale, 1999).

16.5.4 Physiologic Effects

CRH has numerous functions, but most importantly it coordinates the CNS response to stress. Its best-known role is that of stimulating pituitary-ACTH secretion – via stimulating cleavage of the POMC molecule, the precursor of ACTH – leading to adrenocortical release of glucocorticoids and dehydroepiandrosteroe (DHEA). Centrally, CRH initiates a stress response and inhibits appetitive and sexual behavior in favor of a fight-or-flight program. Central administration of CRH also causes behavioral activation; high doses lead to seizures. In addition, central administration of CRH can reinforce behavioral responses to stressors and lead to anxiogenic behaviors. CRH is also involved in the physiologic regulation of the autonomic nervous system. CRH can evoke epinephrine and norepinephrine secretion and can inhibit vagal-nerve activity following central administration.

16.5.5 Clinical Implications

Given that CRH is the principle CNS effector of the organismic response to stress, stress-related conditions, particularly depressive and anxiety syndromes, are the most clinically relevant to the this neuropeptide.

That many depressed patients are hypercortisolemic, perhaps 50%, is a well-replicated observation, leading to the hypothesis that the adrenocortical hyperactivity of depression is driven centrally by hypersecretion of CRH (Gold et al., 1984, 1986; Holsboer et al., 1984; Nemeroff et al., 1984; Wong et al., 2000). Among the other data supportive of a pathophysiologic role for CNS-derived CRH hypersecretion in major depression is the finding, largely from one group, that concentrations of CRH in CSF immediately after lumbar puncture are elevated in many depressed patients (Nemeroff et al., 1984; Banki et al., 1987; Arato et al., 1989), especially those with melancholic and/or psychotic features, hypercortisolemia, and nonsuppression of cortisol secretion after administration of the glucocorticoid dexamethasone. Additionally, successful treatment of depressed patients with the antidepressant fluoxetine (DeBellis et al., 1993) or amitriptyline (Heuser et al., 1998) is associated with a decrease in CSF CRH concentrations. Furthermore, CSF CRH levels decline in patients successfully treated with electroconvulsive therapy (Nemeroff et al., 1991; Kling et al., 1994). In this regard, CRH within the CNS may itself directly evoke a depressive-symptom complex consisting of loss of appetite, insomnia, and intense anxiety (see Gold et al. (1988), Chrousos and Gold (1992), and Steckler and Holsboer (1999) for reviews). Finally, we note that a pilot study recently revealed that genes that are involved in the activation of the CRH neuron were found to be overexpressed in the postmortem hypothalamus (PVN) of depressed patients (for the CRH_1 receptor, α-estogen receptor, arginine VP (AVP)-1 receptor, and for the GR type 1 – mineralocorticoid – receptor; Wang et al., 2008). Single-nucleotide polymorphisms have also been reported in the genes for the CRH_1 and AVP-1 receptors in patients with panic disorder (Keck et al., 2008).

Basal elevations in CNS CRH are also observed in post-traumatic stress disorder (PTSD) – a syndrome with a lifetime prevalence of around 7% in Americans and whose development is intrinsically related to the experience of overwhelming trauma – both by Bremner et al. (1997) in single-point post-lumbar puncture CSF samples and by our group in CSF samples collected serially over 6h via an indwelling catheter (Baker et al., 1999). Even emotional trauma in childhood correlates positively with CSF CRH levels in adulthood (Lee et al., 2005). However, acute presentation of traumatic imagery to patients with combat-related-PTSD results in lower mood ratings, increased anxiety, and, surprisingly, a decrease in CSF CRH concentrations, suggesting an acute stress-related inhibition or suppression of CRH secretion, if not increased brain uptake or metabolism of CRH (Geracioti et al., 2008). It will be of interest to pursue this finding, especially since hypoactive adrenal activity has been described in combat veterans with PTSD.

Major efforts are ongoing to develop orally available peptide and nonpeptide CRH-receptor antagonists that cross the blood–brain barrier for the treatment of depressive and anxiety disorders. Preliminary data in depressed patients at first suggested that the CRH-1-receptor antagonist R121919 does indeed reduce anxiety and depression and is well tolerated (Zobel et al., 2000), although a more recent trial of a CRH_1-receptor antagonist in major depression was negative (Binneman et al., 2008). Additional clinical trials of various CRH-receptor antagonists are still awaited by clinical psychiatrists and are ongoing for a number of conditions, including mood and anxiety disorders as well as general medical conditions (e.g., irritable-bowel syndrome).

Clinically, it should be kept in mind that about half of depressed patients are not hypercortisolemic. While the weight of available evidence suggests that a significant portion of depressed patients manifest CRH hypersecretion, particularly those with melancholia and/or psychotic symptoms, another substantial subgroup of depressed patients appear to have a hypoactive hypothalamic–pituitary–adrenocortical axis and pathologically diminished elaboration of CNS CRH (Geracioti et al., 1992, 1997; Kasckow et al., 2001). Often, these patients suffer from a depressive syndrome that is, despite its high prevalence, called atypical. The syndrome of the so-called atypical depression is characterized by hypersomnia, hyperphagia, and mood reactivity (Liebowitz et al., 1984), and these patients are prominently represented among the approximately 50% of depressed patients who are not hypercortisolemic.

In this regard, we used an indwelling subarachnoid catheter and continuously sampled CSF over 6h in a group of depressed patients, comprising mainly of atypical depressives (Geracioti et al., 1992, 1997). The patient group as a whole was eucortisolemic, with low-to-normal plasma ACTH concentrations. We found clearly diminished CSF CRH concentrations in the depressed patients relative to healthy volunteers, but, for the most part, retention of the normal diurnal concentration pattern. Furthermore, close inspection of the existing data on CSF CRH levels in depressed patients, obtained from single-point lumbar-puncture-derived samples (Nemeroff et al., 1984, 1991; Banki et al., 1987; Kling et al., 1991; DeBellis et al., 1993; Molchan et al., 1993; France et al., 1988; Pitts et al., 1995), shows that many patients had low CSF CRH concentrations, often more than one standard deviation below the mean level of the normal comparison subjects (Kling et al., 1991; Pitts et al., 1995). In addition, evidence of decreased CRH secretion has been reported in patients with chronic fatigue syndrome, a condition clinically reminiscent of atypical depression (Demitrack and Crofford, 1998). Furthermore, hyperphagic depression is often observable in Cushing's syndrome – a syndrome characterized by hypercortisolemia due to an ACTH-producing tumor, but, unlike the situation in melancholic depression, with normal negative-feedback suppression of CNS CRH by the high ambient glucocorticoid concentrations – when CSF CRH levels are decreased (Kling et al., 1991). Decreased CRH secretion has also been postulated to exist in patients with winter depression, a condition which includes anergia, hyperphagia, and hypersomnia – symptoms seen in atypical depression.

Abnormalities in CRH secretion or signal conduction – resulting in either the under- or over-amplification of the CRH signal – may also be a part of the pathophysiology of certain rheumatoid disorders (Baerwald et al., 2000; Eijsbouts and Murphy, 1999; Webster et al., 1998). Decrements of cortical CRH have been observed in the brains of patients with AD (Bissette, 1997) and diminished CSF CRH concentrations have also been inconsistently observed in demented patients (see Rehman (2002)). Indeed, there is accumulating evidence that CRH actions may be neuroprotective (see Bayatti and Behl (2005) for a review).

It is not surprising that CRH is like many other hormones relevant to clinical medicine in that both hyperactive and hypoactive physiologic derangements occur in nature. Thus, in addition to the promise of clinically available CRH antagonists, we predict that CRH analogs – or molecules whose mechanism of action results in the stimulation of CRH secretion – will eventually have a major place in the clinical armamentarium.

16.6 Thyrotropin-Releasing Hormone

TRH (**Figure 5**) was isolated by Guillemin and Schally and was shown by their group to consist of three amino acids – glutamine, histidine, and proline with molecular weight of 362 (Guillemin et al., 1982). It is an evolutionary conserved peptide; for instance, it can be found in insects (Dubois, 1980). TRH-containing neurons are also seen in arcuate and tuberomammillary nuclei. About 70% of TRH is found outside of the hypothalamus; in the spinal cord, TRH concentrations are comparable to those seen in the hypothalamus. TRH neurons are frequently encountered in the limbic system, including the amygdala, nucleus accumbens, and olfactory lobe (Hokfelt et al., 1989).

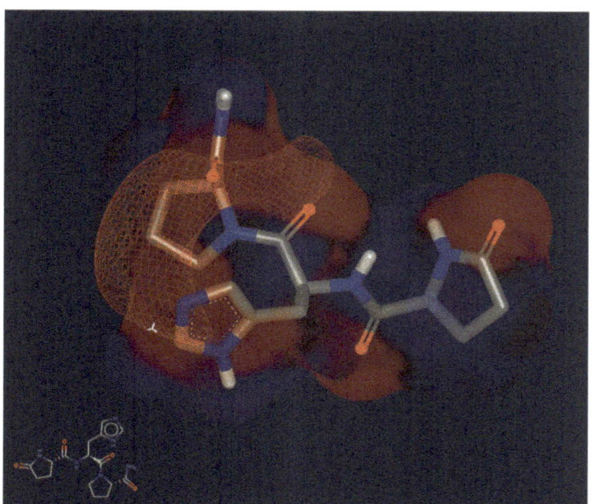

Figure 5 Human thyrotropin-releasing hormone (TRH) receptor. Three-dimensional (center) and two-dimensional (corner) representations of TRH. 3D coordinates were obtained by repeated conformational sampling to identify the global minimum followed by energy minimization in explicit solvent (water, 0.9% NaCl) at pH 7.4 and 300 K. TRH atoms (gray = carbon, red = oxygen, blue = nitrogen, and white = hydrogen) are visible through a solid, transparent molecular surface colored by electrostatic charge (red = negative, blue = positive, and white = neutral). Also displayed is a mesh electrostatic field that represents charge character (±) and magnitude surrounding the molecule. Figure prepared using VIDA (Openeye Scientific Software).

TRH is secreted in a diurnal rhythm wherein TRH mRNA in the hypothalamus peaks at 7 a.m. In addition to the brain, TRH is also expressed, among other places, in the GI tract and the placenta (Bajoria and Babawale, 1998; Walter and Kissel, 1994).

16.6.1 Regulation of TRH

TRH is synthesized from a larger preprohormone molecule that is post-translationally cleaved, processed, and packaged into secretory granules. TRH is inactivated by a highly specific TRH-degrading ectoenzyme.

There are numerous compounds and conditions which regulate TRH (**Figure 6**). Environmentally, acute cold exposure leads to an increase in hypothalamic TRH mRNA. A long negative-feedback loop exists between the hypothalamus and the pituitary–thyroid products, TSH, and thyroid hormones, to regulate TRH. Ultrashort loop self-regulation and short-loop feedback from agouti-related peptide (AgRP) and cocaine- and amphetamine-related transcript (CART; stimulatory effect) and from NPY and α-MSH (inhibitory effect) appear to be of importance (Fekete et al., 2002) (**Figure 6**). Available data indicate that cortisol suppresses TRH secretion in our species (Alkemade et al., 2005).

16.6.2 TRH Receptors

Several TRH receptors have been identified – α 1, β 1, and β 2. The first two are the most prevalent type (Bauer et al., 1999). The TRH receptor is expressed in thyrotropes and mammotropes (PRL-secreting cells) and also in a small fraction of somatotropes (GH-secreting cells). Activation of signaling pathways (which is beyond the scope of this chapter) results in increased TSH secretion and exocytosis.

16.6.3 TRH Function

TRH functions as a stimulator of TSH secretion. In humans, an intravenous (IV) dose of 75 ng min^{-1} leads to peak TSH levels in 20–30 min. TRH also stimulates PRL secretion and can also function as a neuromodulator or neurotransmitter. Administration of TRH to neurons can lead to early neuronal excitation or a late longer-lasting modulation. TRH also has numerous behavioral and vegetative effects. For instance, TRH antagonizes the effects of opioids, is involved in thermoregulation, and increases body temperature. Either central or peripheral injection

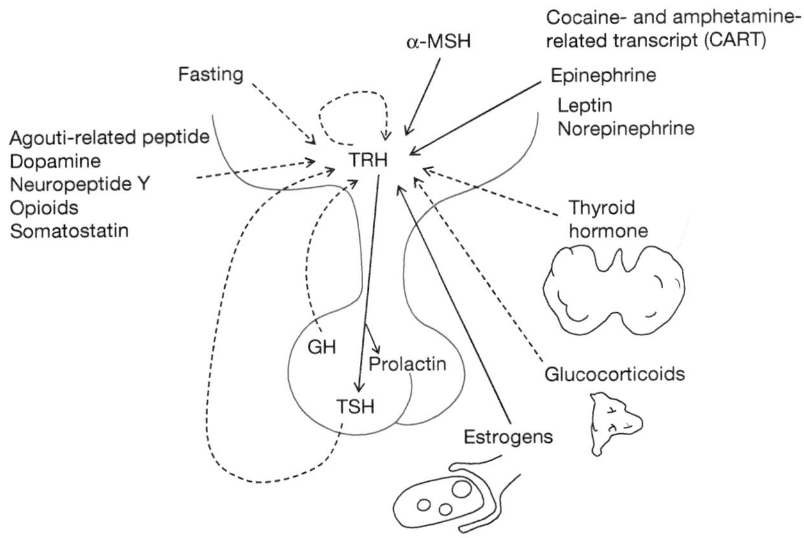

Figure 6 Clinically important regulators of thyrotropin-releasing hormone (TRH) secretion. Positive effectors are indicated by solid arrows and negative effectors are denoted by dashed arrows.

of TRH promotes thermogenesis; the site of action appears to be central although TRH only modestly crosses the blood–brain barrier in humans.

16.6.4 Clinical Implications

Abnormalities of thyroid function are of great clinical importance in psychiatry and medicine. Although TRH is the main CNS effector of the hypothalamic–pituitary–thyroid (HPT) axis, compared with evaluation of the pituitary–thyroid unit, TRH itself is much less often considered in the diagnostic work-up of a suspected HPT abnormality. Some exceptions exist. For example, patients with hypothyroidism may have TSH that possesses reduced biological activity. Oral TRH can restore TSH activity and thus, there may be a defect in TRH production in these patients (Beck-Peccoz et al., 1985). Because of its ability to cause the release of thyrotropin (TSH), PRL, and, under particular circumstances, other adenohypopyseal hormones, from the pituitary, TRH was used as a diagnostic tool for about 30 years. The recent introduction of an ultrasensitive TSH assay, able to clearly distinguish suppressed from unsuppressed TSH levels, has made the use of the TRH test obsolete in the diagnosis of classic hyperthyroidism. (Yet, it should be noted that because of the pulsatile nature of the release of TSH from the pituitary, along with the TSH diurnal rhythm, several single TSH concentrations – if not serial levels to capture the evening or sleep-related TSH surge – might be necessary to adequately characterize the TSH activity in the clinical situation.) However, the TRH test can still be useful in hyperthyroid patients who show inappropriate secretion of thyrotropin, allowing clinicians to make the distinction between TSH-secreting pituitary tumors (usually unresponsive) and the pituitary variant of resistance to thyroid hormone (PRTH) syndrome (always responsive; Faglia, 1998).

The direct therapeutic use of TRH has been attempted to treat depression with reported success. Intrathecal (spinal canal) administration of TRH has antidepressant-like effects. Intrathecal administration of a TRH analog (protirelin) can induce 2–3-day remissions of major depression more reliably than IV administration. Although clinically unwieldy, these remissions are rapid, occur within hours, and survive at least one night's sleep (Marangell et al., 1997; Sattin, 1998). A single 500-μg IV infusion at midnight to depressed patients induced a >50% reduction in depressive symptoms within 24h in bipolar-spectrum depressives (Szuba et al., 2005).

Administration of TRH or one of its analogs intrathecally might also slow the rate of deterioration of patients with amyotropic lateral sclerosis (Guiloff and Eckland, 1987; Munsat et al., 1989).

TRH analogs are also used to promote recovery from head-trauma-induced alterations in consciousness and from hemorrhagic shock.

Expanded uses of TRH in the neuropsychiatric clinical armamentarium are likely in the years ahead.

Thyroid hormone itself is required for normal brain development and function. Although a consideration of the major issue of clinical abnormalities in the secretion of thyroid hormone (and in its end-organ effects – or alterations thereof) is well beyond the scope of this chapter, it should be mentioned that major depression (or one of its variants) is often associated with one or another degree of hypothyroidism. Insufficient thyroid function is associated with mental slowing and other depression-like symptoms. However, paradoxically, severe hypothyroidism can manifest as an agitated, paranoid psychosis. Hyperthyroidism is frequently associated with anxiety symptoms. Therefore, HPT function should be carefully assessed in the work-up of patients presenting with these syndromes. Even minor thyroid abnormalities (such as compensated hypothyroidism or marginally normal thyroid function) may be problematic for the mood-disordered individual. Robust thyroid function appears to be necessary for antidepressant medication to be effective. Furthermore, triodothyronine (T3), usually at doses of 25–50 μg (sometimes higher), continues to be used successfully and routinely to rapidly augment antidepressants in the treatment of depressive disorders (Prange et al., 1984, Prange 1996). Supraphysiologic doses of thyroid hormone are occasionally successful in converting a treatment of a nonresponsive patient with manic depression into a responder. Finally, we note that chronic lithium treatment of bipolar affective disorder (manic depression and its variants) can lead to hypothyroidism. Lithium is concentrated by the thyroid and inhibits thyroidal iodine uptake. It also inhibits iodotyrosine coupling, alters thyroglobulin structure, and inhibits thyroid hormone secretion. The latter effect is critical to the development of hypothyroidism and goiter (Lazarus, 1998).

16.7 POMC-Derived Neuropeptides: Melanocortins

POMC is a 241-amino-acid prohormone that contains the entire sequences of several other hormones, called melanocortins (**Figure** 7). Cleavage of POMC yields ACTH 1–39, MSH 1–13, and shorter ACTH fragments, especially ACTH 4–10 (see Castro and Morrison (1997)). β-Lipotropin and ACTH are the predominant melanocortins in humans. α-MSH dominates in most other species (Tatro, 1996).

16.7.1 Tissue-Specific Processing of POMC

The POMC prohormone gives rise to a variety of neuropeptides, the type depending on the degree to which the prohormone is processed (**Figure 7**), which is in turn dependent upon the specific enzymes present in that tissue. For example, POMC is processed by corticotropes to ACTH 1–39, β-lipoprotein, and β-endorphin. These peptides are themselves processed into a multiplicity of smaller peptides of variable biological activity (see Solomon, 1999) for example).

16.7.2 Melanocyte-Stimulating Hormone

There are three forms of MSH – designated α, β, and γ. There are not large amounts of α-MSH in adult human pituitaries (and even less of β-MSH) although it is present in the brain. β-MSH, like α-MSH, is involved in melanogenesis and skin pigmentation (Spencer and Schallreuter, 2008). γ-MSH has little melanotropic activity; its physiologic function is not well known.

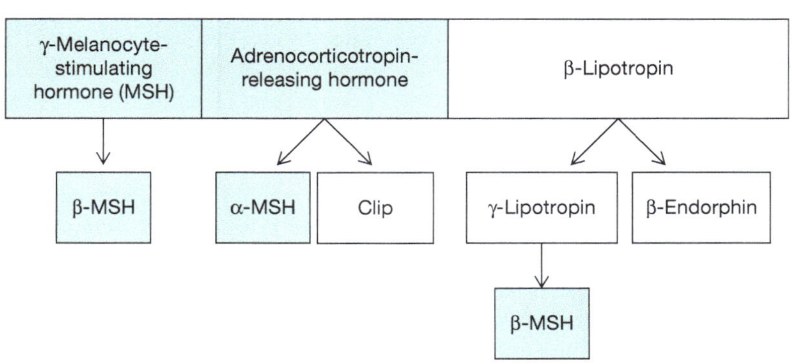

Figure 7 Schematic structure of proopiomelanocortin (POMC; simplified). Melanocortins are shown in light blue. For further details, please see Coll (2007).

16.7.3 Lipotropin

β-Lipotropin is a 91-amino-acid peptide and itself a precursor to several other peptides, including β-MSH and β-endorphin. In human CSF, we observed that β-lipotropin is more plentiful than β-endorphin (Baker et al., 1997) – although the physiologic role of β-lipotropin remains to be determined.

16.7.4 Distribution of POMC and Its Derived Peptides

POMC is expressed in hypothalamic arcuate nucleus and in the fibers that project from these cells to extrahypothalamic structures. These include the nucleus accumbens, reticular formation, periaqueductal gray, and the medial amygdala. We found that POMC is present in human CSF in quantities larger than its cleavage products, such as β-endorphin (Baker et al., 1997).

POMC-derived neuropeptides are also present in many hypothalamic nuclei, the pituitary stalk, and the median eminence (Krieger et al., 1980). ACTH 1–39, α-MSH, and β-lipotropins are expressed in the anterior pituitary corticotropes. They comprise 10–15% of adenohypophyseal cells. Lesser amounts are in the amygdala, mesencephalon, and septum.

Another group of POMC neurons includes cells within the nucleus of the tractus solitarius of the caudal medulla. These cells produce α-MSH, corticotropin-like intermediate lobe peptide (CLIP), β-endorphin, and very little ACTH and β-lipotropin. POMC also appears to be expressed in spinal motoneurons. Furthermore, POMC and its products are expressed in the gut, the pancreas, lymphocytes, macrophages, the spleen, the testis, ovary, adrenal medulla, placenta, lung, skin, and heart.

α-MSH has its highest level of expression in the hypothalamus and pineal gland; lower amounts are present in thalamus, brainstem, cerebrum, and cortex (Eberle, 1988). Melanotropes in the pituitary intermediate lobe also contain MSH and other derivatives of POMC.

16.7.5 Regulation of the POMC Gene and POMC-Derived Peptides

CRH is the primary positive regulator of anterior-pituitary ACTH secretion. Other peptides act synergistically with CRH to influence cleavage of POMC, including VP, angiotensin II, VIP, peptide histidine isoleucine (PHI), GHRH, and norepinephrine.

CRH also stimulates the release of MSH from the intermediate lobe of the pituitary. In addition, epinephrine released during stress releases both α-MSH and β-endorphin from pituitary melanotropes (Berkenbosch et al., 1983). Dopamine is perhaps the main MSH-inhibiting factor. Gamma-aminobutyric acid (GABA) and NPY also inhibit MSH release.

16.7.6 Melanocortin Receptors and Second Messengers

MSH-related receptors are localized to the septal area, septohypothalamic nucleus, the BNST, and the medial preoptic area based on MSH binding. Intermediate areas are seen in various hypothalamic regions. All these sites bind ACTH, MSH, and the MSH fragment – NDP–MSH (N-pLeu4, D-Phe7 MSH; Tatro, 1990).

Five melanocortin receptors (MC receptors or MCRs) have been localized and all of these signal through G-protein-coupled receptors which are, in turn, coupled to cAMP production. All five of these receptors have been cloned (Chhajlani et al., 1992; Gantz et al., 1993, 1994; Mountjoy et al., 1992, 1994; Roselli-Rehfuss et al., 1993; Adan and Gispen, 1997).

The type 1 and 2 MCRs are absent in the brain. The type 1 MCR is expressed mostly in melanocytes and melanoma cells; this is the receptor which binds MSH, leading to increased pigmentation (although agouti-signaling peptide also binds to the MC-1 receptor as an antagonist or inverse agonist, inhibiting pigmentation). The type 2 MCR binds to the ACTH receptor and this is localized in the adrenal cortex. ACTH activates all other MCRs (Hadley and Haskell-Luevano, 1999), while α-MSH activates not only MC1 receptors, but also the MC3 and MC4 receptors that are found in the CNS.

The type 3–5 MCRs are the three neuronal receptors. The type 3 MCR is expressed in the limbic system including the hypothalamus. The type 3 MCR is activated by ACTH 1–39 and also α, β, γ 1–3 MSH. The type 4 MCR, the MC4 (**Figure 8**), is found in every brain region, including the PVN. ACTH 1–39 and α-MSH equally activate the MC 4. The type 5 MCR is expressed in the brain as well as in many peripheral tissues such as skin, adrenal gland, adipose tissue, skeletal muscle, and lymphoid tissue (Chhajlani et al., 1993; Gantz et al., 1994; Griffon et al., 1994).

16.7.7 Functions of ACTH and MSH

Although MSH released from the pituitary is well known to stimulate melanin synthesis in skin

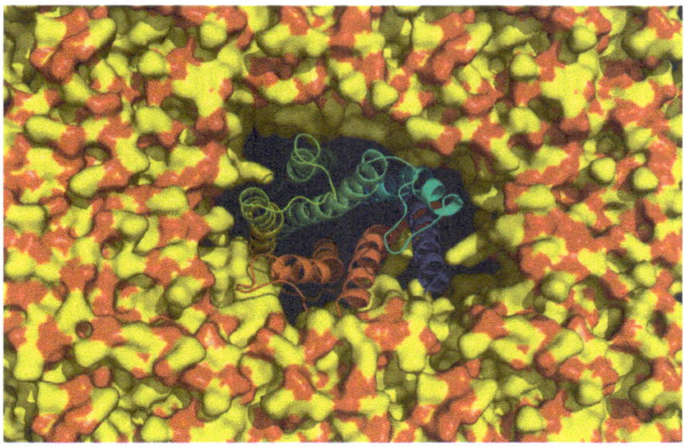

Figure 8 An extracellular view of an all-atom model of the human melanocortin 4 receptor embedded within a double-layer POPC membrane. The yellow and red surfaces represent the lipid and polar membrane constituents, respectively. The receptor is comprised of seven rainbow-colored transmembrane helices. This model was created based on homology to bovine rhodopsin (PDB ID 1f88) and subsequently used to identify biologically active small molecules with nanomolar AC50. Figure prepared using Pymol (Delano Scientific).

melanocytes (melanocytes are also present in hair and the eye), it also has important CNS effects, such as on energy balance and feeding (see below). In addition to the anorexigenic, appetite-impairing effects of MSH, its ability to stimulate penile erection in males and sexual desire and arousal in females (Hadley, 2005) has led to advances in neuroscience, pharmacology, and in the clinical treatment of human sexual disorders (see below).

Similarly, while ACTH is well known as a secretagog of cortisol from the adrenal cortex, it also is present in brain. α-MSH has long been regarded to function as a neurotransmitter in the spinal cord (Krivoy and Guillemin, 1961). In addition to its corticotropic role, ACTH 1–39 and 1–24 possess neurotropic activity. The critical sequences reside in residues 11–24. ACTH acts on the type 2 melanocortin receptor (MCR2) in the *zona fasciculata* to mediate its corticotropic activity. MSH binds to the type 1 MCR on melanocytes and melanoma cells to distribute melanin and darken color (Thody, 1999).

In higher vertebrates, data more than three decades old indicate that ACTH 1–10, 4–10, and 4–7 potently influence behavior, attention, and learning (DeWied and Jolles, 1982). ACTH peptide effects on learning, attention enhancement, or motivation have long been known to be independent of adrenal steroids (De Wied, 1969; Sandman and Kastin, 1977). ACTH and MSH are known to have facilitating influences on memory storage and there is evidence suggesting that this may involve neuromodulatory effects on norepinephrine (Gold and Delanoy, 1981). On the other hand, pathologically increased ACTH secretion – as is seen clinically, for example, in Cushing's syndrome – is associated with decreases in certain types of memory abilities (this may be largely a consequence of the associated chronic increases is glucocorticoid concentration). Intracerebroventricular injection of ACTH 1–24 leads to other behaviors not seen with peripheral administration of the peptide (Ferrari, 1958), including increases in grooming, yawning, and stretching. Interestingly, ACTH has been shown to promote the regeneration of crushed nerve (see Strand (2000)).

16.7.8 Other Effects of Melanocortins

The CNS–melanocortin system is a key regulator of food intake and body weight (see Seeley et al. (2004)). Increasing melanocortin signaling in the brain, via MC3 and MC4 receptors, by pharmacologic means or via genetic manipulation, causes reduction in food intake and leads to weight loss. α-MSH administered into the CNS of rodents robustly decreases food intake and leads to weight loss. Conversely, overexpression of AgRP, an antagonist of the CNS-enriched MC3 and MC4 receptors, causes hyperphagia and obesity. Knockout mice lacking the type 4 MCR mice are hyperphagic, obese, and develop hyperinsulinemia and hyperglycemia (Fan et al., 1997).

ACTH, ACTH 4–10, or α-MSH can cause pressor and cardioaccelerator effects, the tachycardia being due to increased sympathetic influences (Bohus et al., 1993). These cardiovascular effects are also observable during passive avoidance paradigms, suggesting that ACTH and MSH peptides act centrally to affect autonomic responses.

α-MSH and C-terminal peptide 11–13 (Lys-Pro-Val) inhibit inflammation when administered systemically; these are effects mediated by pathways involving both central and peripheral sites of action (Macaluso et al., 1994). ACTH also suppresses immune functions via glucocorticoids and also by acting directly on ACTH receptors on lymphocytes and monocytes.

ACTH and α-MSH can antagonize the analgesic effects of opioids. α-MSH plays a role in pigmenting skin via functioning as an agonist at the MC1 receptor, where AgRP is an antagonist.

16.7.9 Clinical Implications

The melanocortin system of is of wide general relevance to human brain function, including in its CNS regulation of the stress response, sexuality, appetitive behavior and body weight, mood and anxiety (see Chaki and Okuyama (2005)), pain tolerance, inflammatory responses, and cognition.

Mutations of the MC4 receptor have been detected in patients with obesity (Vaisse et al., 2000). POMC gene defects have also been described, leading to hyperphagia, light skin, bright-red hair, and adrenal insufficiency (Krude et al., 1998). Brain-penetrating MC3 and, especially, MC4-receptor agonists are major drug-developmental foci for the treatment of obesity (see, e.g., Adan and van Dijk (2006)). However, there are a number of enduring open questions about the suitability of MC angonists for use in human obesity; as the relationships between hypothalamic and extrahypothalamic brain melanocortin signaling are not well elucidated (neither is the relationship between the CNS and peripheral systems), the possibility of adverse effects remains to be fully explored (e.g., hyperpigmentation or untimely sexual arousal), and the potential compensatory response of AgRP remains to be determined (see Seeley et al. (2004)).

MC3/MC4 agonists are presently under development to promote sexual function in men and women. The MSH analog melanotan II – so named because the original development goal had been to synthesize a compound that would pharmaceutically generate skin tan (Hadley, 2005) – turned out to have unexpected, potent penile-erection-inducing effects (Hadley, 2005). Intranasal bremelanotide, an active metabolite of melanotan II, is presently in clinical trials in men and women, and showing promising results for inducing erection in men and sexual arousal in women (Safarinejad and Hosseini, 2008; Safarinejad, 2008). Various other melanocortinergic agents are under investigation for enhancement of sexual arousal (see Shadiack et al., 2007).

Addison's disease is a disorder in which there is diminished functioning of the adrenal glands (adrenal insufficiency) leading to decreased levels of circulating corticosteroids (mainly cortisol, and also DHEA). This may be due to adrenal failure, pituitary ACTH dysfunction, or deficiency in brain CRH (such as might rarely be acquired after severe brain injury). Psychiatric symptoms include apathy, fatigue, lack of initiative, and poverty of thought. Depression and psychomotor retardation may be evident. In acute crises, psychotic manifestations may emerge (Baker, 1997; Reiser and Reiser, 1985). These patients require glucocorticoid and mineralocorticoid replacement therapy. Adrenal insufficiency, whether primary (of the adrenal) or secondary (of the hypothalamic–pituitary unit), can range from mild – manifesting only during periods of stress wherein the lack of ability to mobilize an adrenal reserve results in psycho–physical fragmentation and inability to cope with physiologic and cognitive stresses – to severe and life threatening. It is the authors' experience that psychiatric manifestations, such as depression, fatigue, and stress intolerance, may be the sole presenting complaints in patients with low-grade adrenal insufficiency.

In opposition to adrenal insufficiency, Cushing's syndrome is a condition resulting from chronic excesses of circulating cortisol, usually from an ACTH-secreting tumor. Frequent clinical findings include weight gain, truncal obesity, striae, hypertension, glucose intolerance, and infections. Cushing's syndrome can be characterized by an excess of ACTH which can be due to a pituitary adenoma (Cushing's disease), an adrenal adenoma, an ACTH-secreting carcinoma of the lung or thyroid, or a hypothalamic CRH-secreting tumor. When the source of ACTH excess is derived from the pituitary, the various therapies used include transsphenoidal microsurgery to remove the pituitary, pituitary irradiation, or the administration of ACTH-suppressive drugs, such as cyproheptadine or bromocriptine. Adrenal enzyme inhibitors, such as ketoconazole, are often particularly effective. Antagonism of glucocorticoid receptors is presently limited by compensatory increases in ACTH (Gross et al., 2007). The best screening test for Cushing's syndrome is a 24-h urine collection with analysis for urinary-free cortisol examination of diurnal and circadian salivary cortisol concentrations. Various additional tests such as the dexamethasone suppression test, the CRH stimulation test, and corticotropin assays may be helpful diagnostically. Treatment should aim to cure the hypercortisolism and to eliminate any tumor that threatens the patient's health, while minimizing the chance of an endocrine deficiency or long-term dependence on medications. There are numerous psychiatric problems associated with Cushing's syndrome and these have been correlated with endocrine manifestations. These include depression, agitation, elated mood or signs of manic excitement, memory disturbances, or psychotic phenomena. (Kirk et al., 2000; Orth, 1995; Gold et al., 1995; Stratakis and Chrousos, 1995). Generalized and panic anxiety are also seen in the majority of patients (Loosen et al., 1992).

Other melanocortins which have been discussed above have been shown to have clinical relevance. For instance, immunoreactive β-lipotropin, one of the precursors of β-endorphin, is expressed in high amounts in human CSF and the levels detected are even higher than those of β-endorphin itself (Baker et al., 1997). The physiologic significance of this is unknown. Furthermore, in humans, various melanocortins have been shown to maintain alertness during long-term performance but have no direct effect on learning or memory. α-MSH has been demonstrated to have a beneficial effect on visual recall and there is also evidence that ACTH 4–10 may help retarded adults with their cognitive abilities. MSH has also been shown to help enhance attention (Kastin et al., 1971; Beckwith, 1988; Sandman et al., 1972).

16.8 Opioid Peptides

The three well-known families of opiate peptides include the endorphins, the enkephalins, and the dynorphins (the endomorphins remain less known). These are coded by three separate genes. The endorphins are derived from β-lipotropin (itself a POMC derivative). β-Lipotropin serves as a prohormone for α-, β-, and γ-endorphin. The two enkephalins, leu-enkephalin and met-enkephalin, are pentapeptides which are different at the C-terminal in having either leucine or methionine. Proenkephalin is the precursor molecule for both. The dynorphins are opiate peptides derived from a different precursor, prodynorphin, and this is sometimes called proenkephalin B. The endorphins are found in the cells of the intermediate lobe of the pituitary as well as in the anterior-pituitary cells that contain ACTH. The enkephalins are present in the intermediate and posterior pituitary but are absent from the anterior lobe. β-Endorphin and enkephalin are also expressed in different neuronal systems within the brain. Enkephalins are expressed not only in short interneurons, but can also be found in longer-projecting neuronal systems.

16.8.1 Prodynorphin (Proenkephalin B) and Dynorphin

Prodynorphin is a 30-kDa molecule that contains three sequences of leu-enkephalin and several other opiate peptides; these compounds are all C-terminal extensions of leu-enkephalin – α- and β-neoendorphin, dynorphin A (1–17), dynophin A (1–8), and dynorphin B (1–29). Tissue-specific processing of prodynorphin varies. Dynorphin is a heptadecapeptide with uncertain physiological significance, with both analgesic and cytotoxic effects possible (Nakazawa et al., 1985; Brugos and Hochhaus 2004; Hauser et al., 2005). Prodynorphin is distributed throughout the neuraxis. It is found in local and long-tract projections, often in parallel with proenkephalin systems.

16.8.2 Proenkephalin A

Proenkephalin was discovered in the adrenal gland (Lewis et al., 1980). Human brain proenkephalin has 267 amino acids. Proenkephalin neurons are more widespread in the brain than POMC-containing neurons. Proenkephalin is enriched in the cerebral cortex, basal ganglia, and the limbic telencephalic nuclei – including the nucleus accumbens, BNST, and amygdala. Enkephalin-containing neurons are also present in the hypothalamus, thalamus, spinal cord, and anterior and posterior pituitary. Peripheral localization includes the adrenal medulla and reproductive tissues.

16.8.3 Nociceptin

In addition to the three opiate systems, there also exists an opiate-like compound called nociceptin or ophanin FQ. This is a 17-amino-acid peptide similar to dynorphin which is the endogenous ligand for the opiate-receptor-like receptor (Mollerau et al., 1996). It has modest orexigenic properties (Olszewski and Levine 2004). This ligand differs from other opiate-receptor agonists in that it lacks the N-terminal tyrosine residue needed for agonist activity at the μ, δ, and κ opiate receptors (Reinscheid et al., 1996). Nociceptin causes analgesia when given intracerebroventricularly in mice and also inhibits electrically induced contractions of the mouse vas deferens; this is nonopiate mediated since opiate antagonists will not alter this effect (Berzetei-Gurske et al., 1996). Nociceptin is processed from a larger precursor molecule, prenociceptin. Nociceptin plays a role in the central control of water balance, regulation of arterial pressure, ataxia, loss of the righting reflex, and the spinal nococeptive-flexor reflex. Opiate antagonists do not affect any of these actions.

16.8.4 Endomorphin

There also exists another important, more recently discovered class of opioids, the endomorphins (Zadina et al., 1997, 1999). Endomorphin-1 (Tyr-Pro-Trp-Phe-NH2, EM-1) and endomorphin-2 (Tyr-Pro-Phe-Phe-NH2, EM-2) show the highest affinity and selectivity for the μ (morphine) opiate receptor of all the known endogenous opioids. The endomorphins have both analgesic and GI effects. At the cellular level, they activate G-proteins and likewise inhibit calcium currents. EM-1 is more widely and densely distributed throughout the brain than EM-2. Their distributions are consistent with a role in the modulation of diverse functions, including autonomic, neuroendocrine, and reward functions as well as in the modulation of responses to pain and stress.

16.8.5 Opiate-Receptor Distribution

Opiate binding was described as early as 1973 by Pert and Snyder (1973). The highest number of sites can be detected in the telencephalon, diencephalon, and the older limbic system (Atweh and Kuhar, 1983). In sensory systems, opiate receptors are expressed in structures relevant to pain and temperature transmission: the dorsal horn of the spinal cord and the substantia gelatinosa of the spinal cord and medulla. In addition, receptors are present in the neospinothalamic tract which has projections to the thalamus and to the somatosensory cortex for the perception of primary pain. Also, the pleospinothalamic tract projects to the periaqueductal gray in the reticular formation, then to the thalamus, and finally to limbic and subcortical regions involved in secondary or subjective pain. It is thought that these latter projections account for the subjective aspects of pain perception (see Besson and Chaouch (1987)). The μ-, δ-, and κ-receptors play an important role in the ascending pathways as well as in the periaqueductal gray, reticular nuclei, and raphe nuclei but in descending and thalamic systems, the μ- and κ-receptors play a predominant role (Mansour et al., 1995).

16.8.6 Role of Receptor Subtypes

The μ-, δ-, and κ-receptors have been cloned, and have the most important actions for endogenous opiates (Reisine and Bell, 1993). The μ- and δ-receptors bind enkephalins and endorphins and the κ-receptors bind dynorphin. Based on Pasternak (1988), there are two μ-subtypes. μ1 is supraspinal and has been implicated in PRL release, feeding, and acetylcholine release. μ2 is mostly spinal in localization and is implicated physiologically in respiratory depression, GI transit, brain dopamine turnover, feeding, and cardiovascular actions (Ding et al., 1996). High expression is present in the thalamus, striatum, locus ceruleus (LC), and nucleus of the solitary tract (Mansour et al., 1995). κ-Receptors bind dynorphin compounds with high affinity. They are present in the nucleus accumbens, substantia nigra, ventral tegmental area, and the nucleus of the solitary tract. The receptors comprise 10% of the brain opiate receptors and are the predominant hypothalamic opiate receptors (Mansour et al., 1995, 1996). Functions include antinociception, fluid homeostasis, hormonal regulation, nigrostriatal function, and visceral responses. δ-Receptors exhibit high affinity for enkephalin compounds. The gene for the receptor is located on chromosome 1p34 (Evans et al., 1992). Their distribution includes olfactory-related areas, the olfactory tubercle and the amygdala (Mansour et al., 1987). High concentrations of δ-receptors are expressed in the cortex, striatum, and the lateral reticular area. The type 1 δ-receptor is supraspinal in distribution and the type 2 δ-receptor has both a spinal and supraspinal distribution. σ-Receptors overlap with binding sites for nonopiate drugs of abuse, including phencyclidine, and antipsychotics such as haloperidol.

The σ-receptor was once considered to be an opioid-receptor subtype, but no longer. It is an interesting receptor that binds various psychotropic drugs, yet is beyond the scope of this chapter.

The gene structure of the three opiate receptors suggests that they belong to one gene family, although they are located on different chromosomes. Opiate receptors are G-protein-coupled receptors with seven transmembrane domains. μ-Receptors have high affinity for morphine and lower affinity for enkephalin. They are generally involved in sensory information processing. The orphan opiate-receptor-like or ORL1 receptor is a G-protein-coupled receptor which binds to nociceptin. This receptor shares 60% homology with opiate receptors.

16.8.7 Physiologic Roles of Opioids

Central and peripheral administration of enkephalin or β-endorphin in mammals results in antinociceptive effects ranging from an increased pain threshold to frank analgesia. Opiate receptor subtypes operate in a modality-specific manner; for instance, processing of thermonociceptive stimuli involves μ- and δ-receptors, but not κ-receptors (Schmauss and Yaksh, 1984). The enkephalins mediate analgesia at spinal and supraspinal levels, including the midbrain (Han and Terenius, 1982). In humans, β-endorphin causes analgesia and high CSF levels of endorphins correlate with the need for opiate analgesics and high pain thresholds. Physical exercise also leads to β-endorphin release and can affect blood pressure and altered pain thresholds after exercise (Hoffman et al., 1996). Analgesia mediated at the peripheral level is also engendered by opioids.

Morphine and endogenous opioids increase PRL levels, inhibit oxytocin, and thereby inhibit lactation. Morphine suppresses substance P in the CNS. Morphine and endogenous opioids increase GH secretion. Morphine and endogenous opiates inhibit gonadotropin release.

In addition, oxytocin release in animals can be inhibited by opioids: morphine and its derivatives inhibit suckling-induced and acetylcholine-induced oxytocin release.

Opioids, in general, stimulate feeding, while antagonists generally suppress it. Eating also increases plasma levels of β-endorphin. Opioids dampen cardiovascular and GI functions (e.g., causing chronic constipation in opioid addicts), decrease respiratory activity, and decrease locomotion. Chronic opiate administration also inhibits the immune system (Stefano et al., 1996).

Alcohol promotes release of opiod peptides, especially β-endorphin. High doses of morphine reduce ethanol consumption, while low doses increase ethanol consumption (Ulm et al., 1995). It is possible that an opiate imbalance may be involved in the reinforcing effects of ethanol (Swift, 1995) since opiate antagonists modestly reduce ethanol consumption in rats and humans (O'Malley, 1995; Guardia et al., 2002).

Opioids exhibit marked effects on mood and motivation. In many humans, opioids induce euphoria – although sometimes nausea and/or dysphoria can occur; conversely, opioids may be released during rewarding situations (Herz, 1997). The euphoric or mood-enhancing/anxiety-reducing effects of opioids are extremely reinforcing. Consequently, opioids – such as heroin or morphine among many others – are compounds that are frequently abused. Long-term administration causes tolerance and physical dependence. Repeated administration of opioids leads to decreases in opiate-receptor number (Cox, 1994). It has long been suggested that the opiate system in these effects is closely tied to dopamine reward pathways (Koob and Bloom, 1988; Stinus et al., 1986; Shippenberg and Elmer, 1998).

Opioid withdrawal is associated with diarrhea, weight loss, body-temperature dysregulation, pupil dilation, psychomotor agitation, wet-dog shakes, and teeth chattering. Place and taste aversion, withdrawal from others, and irritability are common during withdrawal (Olson et al., 1995).

16.8.8 Clinical Implications

As is well known, morphine and other opioids are used extensively to treat all types of pain syndromes; the μ-receptor is the principle target for these opioid analgesics. The potent, selective affinity of the endogenous endomorphins for the μ-receptor has suggested the possibility of analgesic effects without as many side effects as of the plant-derived opioids (e.g., morphine). Indeed, intense efforts have been made to develop endomorphine analogs for analgesia that have excellent bioavailability and tissue penetration (see Janecka et al. (2007) for a review). The use of these agents to treat pain does not necessarily lead to addiction and drug dependence. This is especially true with cancer pain in which it is recommended that oral around-the-clock dosages be utilized (Cherny, 2000). Correct diagnoses of common pain patterns, including breakthrough and incident pain, are essential components of pain-management strategies. The concept of rescue dosing in safe analgesic titration and management of breakthrough/incident pain is an additional key concept. Choice of a specific opioid is often less important than correct dosing, as side effects are similar among the commonly prescribed drugs and often include nausea.

Opioids are highly reinforcing agents which have a high abuse potential in vulnerable individuals (although there are chronically depressed individuals, some of them with PTSD, in whom the use of opioids apparently normalizes moods). The opioid agonist methadone is used to treat heroin addiction (Kreek, 2000). Methadone is longer-acting than heroin, is less associated with a subjective high, and is less potent. Nevertheless, in adequate doses it eliminates or greatly reduces the craving for heroin and prevents opioid withdrawal. It is also one of the few opioids that can be legally prescribed to treat opioid dependence. Naloxone is an opioid antagonist that has a variety of clinical uses, including the treatment of opioid overdose. Opioids produce inhibition at the chemoreceptors in the brainstem respiratory centers via μ-opioid receptors and in the medulla via μ- and δ-receptors (White and Irvine, 1999). Thus, respiratory depression is the mechanism of death in overdose. Two other effective opioid-agonist treatments have been developed: the even longer-acting opioid agonist l-α-acetylmethadol (LAAM), which has been approved for pharmacotherapy for opiate dependence (Kreek, 2000), and the more recent opioid partial agonist–antagonist buprenorphine–naloxone combination (suboxone®). A variety of studies, both laboratory based and clinical, have revealed the mechanisms of action of long-acting opioid agonists in treatment of opioid dependence, including prevention of disruption of molecular, cellular, and physiologic events and, in fact, allowing normalization of those functions disrupted by chronic heroin use (Shvartzman, 2000). Acquiring further understanding of heroin's effects on μ- and κ-opiate receptors has led to investigations of the partial μ-agonist buprenorphine in opiate maintenance. Evidence for hyperexcitability of LC noradrenergic neurons and excitatory amino acid inputs to the LC has guided the development of new trials of novel α2-adrenergic agonists or excitatory amino acid antagonists to alleviate opiate withdrawal. In addition, naltrexone is another opiate antagonist approved for the treatment of alcohol dependence (Volpicelli et al., 1995; Guardia et al., 2002); it is modestly effective and available to clinicians in an oral formulation and in long-acting intramuscular preparation which is released over 4 weeks (Garbutt et al., 2005).

A number of human μ-opioid receptor polymorphisms have been identified (Bond et al., 1998) and are associated with alterations in opioid-receptor binding and signal transduction. Opioid-receptor polymorphisms are associated with differential postoperative morphine need and pain control in orthopedic patients (Chou et al., 2006) and predict treatment response to naltrexone in alcohol dependence (Anton et al., 2008). It is anticipated that μ-opioid receptor genotyping will have additional clinical importance, such as in the assessment of vulnerability to opioid abuse and dependence, selection of analgesic regimens, and possibly in the formation of treatment plans for alcohol-dependent patients.

Finally, clinical experience reveals that opioids are sometimes markedly effective as antidepressant or anxiolytic agents. However, in the United States, use of potent opioids for these clinical purposes is illegal (with mandated restrictions to use for chronic pain or opioid dependence).

16.9 Oxytocin

Oxytocin was the first peptide hormone to have its structure elucidated and synthesized (du Vigneaud et al., 1953). Oxytocin is a nonpeptide with a ring structure formed by a sulfur bridge connecting two conserved cysteine residues and a C-terminal tripeptide. VP differs from oxytocin only by the substitution of amino acids in the third and eighth position. The third position in oxytocin is isoleucine and the eighth position is arginine. Neurophysins are small acidic proteins 93–95 residues long which are about 10 times the length of oxytocin and exist in 1:1 molar noncovalent complexes with oxytocin and VP. They exist as part of the precursor dimers or even as higher oligomers. Neurophysin 1 is associated with oxytocin while neurophysin 2 is associated with VP. The role of neurophysins is not clear; it has been suggested that they not only function as carriers, but also protect the active hormone from enzymatic destruction (Legros and Geenen, 1996).

The oxytocin gene is located on the same chromosome as VP. In the human, this is chromosome 20. Both molecules are oriented in opposite transcriptional directions. Gene expression of oxytocin is confined mostly to magnocellular neurons of the SON and PVN. The hypothalamus connects to the neurohypophysis via the supraopticohypophyseal tract from the SON and the paraventriculohypophyseal tract which arises from the PVN. The larger VP-containing cells are located more medially in the mid- and caudal regions of the PVN than the oxytocin-containing neurons (Ginsberg et al., 1994). The axons of both of these neurons terminate in the perivascular spaces of the neural lobe near blood capillaries and are surrounded by the pituicytes, the supporting cells. After crossing the basement membrane, the axons enter the systemic circulation. The medial preoptic area also contains oxytocin neurons; this area contains estrogens as well and is likely a site where these two hormones interact (Caldwell, 1992). The collateral axons of oxytocin and VP neurons from the SON and PVN magnocellular neurons also project to the spinal cord and interact with autonomic cardiovascular systems as well as other brain regions where they act as neuromodulators or neurotransmitters. Included in these pathways are projections of oxytocin fibers from the SON and PVN which innervate β-endorphin neurons in the arcuate nucleus of the hypothalamus (Csiffary et al., 1992). In addition, oxytocin-expressing neurons exhibit paracrine and autocrine actions with themselves or with other neurons in the hypothalamus. There are also numerous afferent inputs to the magnocellular system from the olfactory system, particularly the olfactory bulbs (Hatton et al., 1992). Oxytocin is also expressed in uterine epithelium and the ovary.

16.9.1 Processing and Metabolism of Oxytocin

Pro-oxytocin is the precursor molecule of oxytocin which not only includes oxytocin, but also includes neurophysin 1 (Mizuno and Matsuo, 1994). Oxytocin is stored in secretory granules and released by the classic regulated secretory pathway. During pregnancy, oxytocin mRNA becomes polyadenylated (Carter and Murphy, 1991). Oxytocin is inactivated by clearance through the kidney, liver, and intestine. There are brain degradative processes also. During pregnancy, a vasopressinase–oxytocinase enzyme destroys oxytocin.

16.9.2 Regulation of the Oxytocin Gene and Peptide

The 5′ flanking region of the oxytocin gene contains estrogen, thyroid hormone, and retinoid response elements. The estrogen-responsive element has been localized to nucleotides 168–156 (Burbach et al., 1992, 1995). Gene expression can be influenced by development, puberty, various reproductive states of the female, thyroid hormones, and plasma osmolarity (Burbach et al., 1992). The increase in oxytocin which occurs prior to puberty can be eliminated by gonadectomy and restored by treatment with estrogen or testosterone. Thyroid hormones are thought to be involved in developmental changes in oxytocin gene expression.

With regard to reproductive states in the female adult rat, oxytocin gene expression peaks at estrus and is lowest at metestrus; essentially, estrogen stimulates oxytocin synthesis and secretion also while progesterone inhibits these processes. This has been confirmed via microdialysis studies which measure oxytocin secretion in the ventromedial nucleus of sheep following treatment with estrogen and/or progesterone (Kendrick and Keverne, 1992). In the pregnant rat, oxytocin levels rise toward the end of gestation and peak just before and during parturition. During parturition, oxytocin release is increased following the initiation of uterine contractions and cervical dilation; those catecholaminergic neurons which interact with neurons in the nucleus tractus solitarius and ventromedial medulla and project to the preoptic area and hypothalamus likely account for this activation (Luckman, 1995). These changes are accompanied by increases in neurohypohyseal oxytocin levels and an increase in oxytocin receptors in the myometrium and mammary gland. Oxytocin gene expression in the uterus increases more than 150-fold during pregnancy; at term, oxytocin-gene expression in the uterus exceeds hypothalamic mRNA by a factor of 70 (Zingg et al., 1995). In addition, during parturition, oxytocin release from dendrites leads to plastic changes in the morphology of magnocellular neurons; synapses are remodeled to bring them closer together (Perlmutter et al., 1984).

With lactation, oxytocin is chronically secreted and pituitary stores become depleted. Nipple stimulation is associated with brief high-frequency electrical discharges of hypothalamic neurons which lead to oxytocin release from neurohypophyseal nerve endings (Bourgue and Renaud, 1990). Fos-immunostaining in the supraoptic rats is increased markedly during these events (Leng et al., 1993). Release of oxytocin leads to mammary-gland myoepithelial-cell contraction and consequent milk ejection.

Hyperosmolality associated with a 2% saline drink or dehydration leads to a two- to fourfold increase in oxytocin mRNA content of SON and PVN neurons. This can be demonstrated by quantitation of oxytocin mRNA as well as protein content in these areas (Sharp et al., 1991). In addition, oxytocin is known to produce natriuresis in mammals (Zingg et al., 1995).

It was briefly mentioned above that there are interactions between the oxytocin system and the opiate system. For instance, oxytocin nerve fibers project to β-endorphin cells in the arcuate nucleus to interact with pain systems. Progesterone is also involved in this mechanism; high progesterone concentrations induce the expression of β-endorphin secretion which, in turn, inhibits oxytocin release. In late gestation, progesterone levels drop, opiate secretion falls, and the inhibitory effect on oxytocin secretion is released.

16.9.3 Oxytocin Receptors

There is one type of oxytocin receptor. The region of the receptor which is crucial to receptor specificity (Chini et al., 1995) includes the interaction of position eight of oxytocin and a residue in the first extracellular loop. In the rat, high-affinity binding sites can be identified in the olfactory system, basal ganglia, limbic system, thalamus, hypothalamus, cortical regions, the brainstem, and spinal cord. A high density of oxytocin receptors are present in the ventromedial hypothalamus (Bale and Dorsa, 1995). Oxytocin receptors are also present in peripheral tissues such as the ovary, testis, and adrenal as well as in the uterus, mammary gland, liver, and fat cells. VP also binds with high affinity to oxytocin receptors (Tribollet et al., 1992).

Oxytocin receptors signal through G-protein-mediated mechanisms. Peripheral oxytocin receptors are linked to phosphatidyl inositol production, whereas the second-messenger systems in the CNS which couple to oxytocin receptors remain, to our knowledge, unknown. Like oxytocin, oxytocin-receptor numbers are regulated during development, suckling, mating, pregnancy, as well as by the steroids estrogen and testosterone (Bale and Dorsa, 1995; Zingg et al., 1998).

16.9.4 Behavioral Effects of Oxytocin

Oxytocin facilitates all aspects of reproductive behavior in mammals. All of these behaviors are dependent upon priming by gonadal steroids and, in turn, oxytocin plays a role in regulating the influences that gonadal steroids exert on such behaviors (Pedersen et al., 1992). Sexual and maternal behaviors in female rodents ultimately depend upon estrogen, but these behaviors can be facilitated by oxytocin. For instance, lordosis is enhanced by oxytocin and the brain area implicated in these effects is the ventromedial nucleus of the hypothalamus (McCarthy et al., 1994). Furthermore, maternal behavior involves anogenital licking of pups, retrieval of pups, response to pup ultrasonic vocalizations, and crouching in a suckling position. These behaviors require prior exposure to estrogen and can also be facilitated by oxytocin (Pedersen et al., 1985). Oxytocin has also been dubbed the satisfaction hormone. For instance, oxytocin can substitute for morphine in ameliorating morphine withdrawal and infusion of oxytocin into the brain produces feelings of satiety (Caldwell, 1992). Oxytocin is also one of the most potent stimuli which can induce penile erection in the rat. In addition, oxytocin shortens ejaculation latency and the postejaculation interval in rats.

Oxytocin has been implicated in monogymous behavior in prairie voles. Adult prairie voles form pair bonds following mating and oxytocin is released during this process. Essentially, oxytocin is required for partner-preference formation, which begins the process of pair bonding in males (Insel et al., 1995). Oxytocin infusions in female prairie voles can also hasten the formation of partner preference and increase maternal behavior.

Oxytocin has been implicated in other behaviors. Grooming is a response to stress, unfamiliar surroundings, and can be evoked by CRH, ACTH, AVP, and also oxytocin. Electrical stimulation of the PVN and infusion of oxytocin into this area will also increase self-grooming in the rat. This suggests that electrical stimulation may induce grooming by activation of the oxytocin system originating from the PVN (McCarthy and Altemus, 1997). Central oxytocin also plays a role in the mechanism of genital grooming which increases during parturition and copulation (Pedersen et al., 1988). Yawning can also be induced by oxytocin injections into the PVN (Melis and Argiolas, 1995). Furthermore, central administration of oxytocin enhances rat social behavior, increases grooming, and impairs memory consolidation and retrieval.

16.9.5 Clinical Implications

Of oxytocin's many actions, "its only irreplaceable role is to mediate milk let-down" (Leng et al., 2005) – at least from an evolutionary perspective, in women with dependent infants. Of interest is that the oxytocin-mediated milk let-down response can be initiated psychologically, through mental images or meanings. This observation supports a role for brain oxytocin in mediating human bonding, at least in women (CSF concentrations of oxytocin are higher in women than in men (Altemus et al., 1999), although the significance of this is unknown). Human experimental data on the affiliative and bonding actions of CNS oxytocin lag far behind those in lower animals, although the accumulating clinical data are consistent with affiliation-enhancing effects of the hormone. For example, clinical data suggest that intranasal oxytocin increases interpersonal trust (Kosfeld et al., 2005) and increases gazing at the eye region of human faces (Guastella et al., 2008). The clinical potential for oxytocin to promote pair bonding, maternal behavior, social attunement, and socialization in certain psychiatric syndromes, including autism, depression, PTSD, severe obsessive–compulsive disorder, and possibly even in some psychoses, is intriguing.

It is well established in humans that oxytocin is released into circulating blood during coitus and orgasm in both sexes, but secretion into CSF during an in-laboratory, masturbation-induced sexual-response cycle in males is not significant (Kruger et al., 2006). Although oxytocin signals may be more salient for female sexuality than for male sexuality, there are some data suggesting that oxytocin may increase sexual responsiveness in both sexes. In the way of anecdotal evidence, a 26-year-old woman used synthetic oxytocin spray to enhance lactation and several hours later noted intense sexual arousal, initiated sexual intercourse with her husband, and had enhanced uterine and vaginal contractions during orgasm (Anderson-Hunt and Dennerstein, 1995). Recently, intranasal oxytocin was used successfully to treat a previously treatment-refractory male patient with anorgasmia (Ishak et al., 2008), although the limited data to date on the experimental effects of intranasal oxytocin on the human sexual response in men are equivocal (Burri et al., 2008). Well-powered, well-designed, randomized studies of intranasal oxytocin in the treatment of anorgasmic and hypoarousal disorders are needed to elucidate the effects of the hormone on affiliative, bonding, and romantic behavior.

Although peripherally administered oxytocin has poor access to the CNS due to the blood–brain barrier, intranasal administration of the hormone permits better brain penetration. In the periphery, oxytocin is secreted into the circulation by healthy, cycling women in response to interpersonal, emotional stimuli (Turner et al., 1999).

Outside the CNS, intravenous oxytocin is commonly used to stimulate uterine contractions in pregnant women to induce labor. It is optimal that the peptide be administered in a pulsatile manner (Dawood, 1995). Scientists have also been working to design peptide analogs of oxytocin for the treatment of preterm labor (Goodwin and Zograbyan, 1998). The available tocolytics used currently include betamimetics, indomethacin, and atosiban. (ethanol also has this effect). These agents have been shown to decrease the risk of delivery in preterm labor within 7 days. However, tocolytics are not associated with improved perinatal outcomes (Gyetvai et al., 1999).

16.10 Vasopressin

VP – or AVP – is a nine-amino-acid-containing neuropeptide which has diverse physiologic actions (Buijs, 1987), including the regulation of fluid balance and in affiliation, social memory, aggression, and even depression. AVP is expressed in magnocellular neurons of the SON and PVN and these neurons project to the posterior pituitary to release AVP and VP-neurophysin into the peripheral circulation. Parvocellular neurons also secrete AVP into the portal system to exert central actions related to stress. Parvocellular neurons also project to the brainstem and spinal cord. AVP is also found in the SCN, BNST, and the medial amygdaloid nucleus; these VP neurons project to the lateral septum, ventral hippocampus, and the habenular area. The BNST AVP neurons also innervate the central amygdaloid nucleus. AVP expression in the BNST is greater in males and is also dependent on testosterone (De Vries et al., 1994). Peripherally, AVP can be found in the anterior pituitary, testis, pancreas, adrenal, ovary, and placenta (Strand, 1999). VP also exhibits a diurnal variation in the SCN, but this is not seen in the PVN or SON (Burbach et al., 1988).

16.10.1 AVP Precursor and Post-Translational Products

Propresorphysin is the precursor molecule for VP. AVP is part of the N-terminal and included in this molecule is a neurophysin and a C-terminal glycopeptide (Schmale et al., 1987).

16.10.2 AVP Receptors

There are at least three AVP receptors: V_{1a}, V_{1b}, and V_2, each of which is coupled to G-proteins (see Caldwell et al. (2008)). The V_{1a} is present in the brain, pituitary, liver, blood vessels, and kidney. These receptors in the brain are found in the SCN, the sigmoid, and arcuate nuclei as well in the lateral hypothalamic and dorsochiasmatic nuclei, the anterior central amygdala, the mesencephalic central gray, and the choroid plexus (de Kloet et al., 1990). The V_{1b} receptor is in the anterior pituitary and, here, it modulates secretion of ACTH and it can not only be seen in various peripheral tissues (Lolait et al., 1995), but is also found in the brain. Both V_{1a} and V_{1b} receptors utilize phosphatidyl inositol hydrolysis and mobilization of Ca^{2+}. The V_2 receptor is mainly expressed in kidney and mediates the antidiuretic function of AVP (Jard, 1998).

16.10.3 Physiologic Functions

There are numerous stimuli which can alter VP expression. Plasma hypertonicity and hypovolemia of the extracellular fluid are stimuli for VP release in the hypothalamus. There are also considerable neural inputs to the PVN neurons which can modulate this function; it has been estimated that there are about 2800 axon terminals which influence a single PVN neurosecretory neuron (Kiss et al., 1983). In the neurohypophysis, there is also evidence supporting therole of presynaptic dopamine in modulating VP release (Carter and Lightman, 1985). Loss of blood volume also stimulates VP release. The relevant neurocircuitry involved in processing this includes afferents from the mid- and hindbrain and the anterior wall of the third ventricle. Badoer and Merolli (1998) has demonstrated that magnocellular neurons as well as parvocellular neurons are activated following hemorrhage. Angiotensin and its receptor have also been implicated in this phenomenon. Interleukin-1, when administered, centrally increases AVP release as well (Landgraf et al., 1995).

VP acts on the kidney to conserve water. This action involves modulation of solute and water transport, vasoconstriction, stimulation of prostaglandin synthesis, inhibition of renin release, and mitogenic effects. V_1 and V_2 receptors have been implicated.

Importantly, VP is a direct ACTH secretagog (Rubin et al., 1999), second in importance only to CRH; VP also enhances CRH-mediated ACTH secretion in the pituitary and this is mediated through the parvocellular–median eminence route (Antoni, 1993). Other brain areas have implicated VP effects in stress as follows: (1) VP depresses locomotor activity and (2) VP stimulates the parasympathetic system to result in passive coping with stress (Bohus, 1993). AVP also appears to suppress leptin secretion (Rubin et al., 2003).

16.10.4 Behavioral Effects

Behavioral effects of VP have been studied over the past 30 years. VP improves the consolidation and retrieval of memory. It reverses memory loss, including amnesia resulting from high doses of epinephrine. VP actions in the amygdala also help with preserving memory; however, this antiamnestic effect is seen in males but not in females. Interestingly, VP 1–8 has the behavioral activities of VP, but lacks the pressor and antidiuretic actions (De Wied, 1983).

VP can facilitate the development of tolerance to morphine and ethanol. In addition, high doses of VP have an antinociceptive effect in rats and this can be blocked by a VP antagonist. VP also has an antipyretic effect. VP is implicated in hibernation and if injected into the lateral septum, hibernation can be interrupted (Hermes et al., 1993).

16.10.5 Clinical Implications of VP

The clinical availability of VP formulations makes it possible to explore the effects of this hormone on a spectrum of human behaviors, including mood, anxiety, social memory and recognition, male fidelity toward a mate, and aggression – all behaviors or states that have been implicated in VP's regulatory domain (see Caldwell et al. (2008)). VP may be given intranasally (by spray, dropper, or via cotton pledgets), or by injection. Complications include water intoxication and hyponatremia.

If AVP promotes mate pair bonding in men (Walum et al., 2008), might interventions that enhance AVP signaling prove useful in promoting romantic attachments in humans? Using a homeopathic approach, a salt load might have this effect (although not if ingested with high glucose concentrations). Intranasal VP might be of interest to study in this regard.

Deficiencies of VP (or end-organ resistance to VP) are a major cause of diabetes insipidus (see Bohus and deWied (1998)), a condition characterized by renal failure to reabsorb water, frequent voiding of dilute urine, great thirst, and often voracious appetite, but with emaciation and weakness.

Hyponatremia can occur in the syndrome of inappropriate antidiuretic hormone secretion, usually from neoplasms releasing VP. The conventional treatment of hyponatremia used to be fluid restriction and treatment of the underlying disorder. This kind of treatment has been unreliable, cumbersome, and difficult to comply with for the patient. VP V2 antagonists are currently under development for clinical use in the treatment of hyponatremia (Gross and Palm, 2000).

In the treatment of enuresis, desmopressin, a nasally administered form of VP, has been shown to rapidly reduce frequency of bed-wetting episodes, but this effect does not appear to be sustained after stopping treatment. Minor side effects of desmopressin in the trials performed include nasal irritation and nose bleeds. However, the risk of water intoxication associated with overdrinking before bedtime has also been reported (Glazener and Evans, 2000). Desmopressin stimulates endogenous release of factor VIII and von Willebrand factor, which have been implicated in bleeding disorders; it increases platelet adhesiveness and shortens bleeding time.

Although little recent clinical work has been reported in the area, VP is known to enhance memory. This effect is observable in patients with mild memory deficits, although not apparently in those with neurodegenerative disorders (Jolles, 1987). In healthy humans, VP also has memory-enhancing effects, as it also does in patients with diabetes insipidus (Beckwith et al., 1984; Van Ree et al., 1985) and interestingly, these effects are thought to be modulated through effects on attention and arousal (Beckwith et al., 1983). The size of VP neurons in the human supraoptic and PVNs increases in the age group 80–100 years, while the size of the SCN VP neurons decrease. This is more exaggerated in patients with Alzheimer's dementia (Swaab et al., 1987).

VP plays an important role in peripheral vasoconstriction, hypertension, and in several disease conditions with dilutional hyponatremia in edematous disorders, such as congestive heart failure, liver cirrhosis, syndrome of inappropriate antidiuretic hormone (SIADH), and nephrotic syndrome. A series of orally active nonpeptide antagonists against the VP-receptor subtypes has recently been synthesized and is now under intensive examination. Nonpeptide V1a-receptor-specific antagonists, OPC 21268 and SR 49059, nonpeptide V2-receptor-specific antagonists, SR 121463 A and VPA 985, and combined V1a-/V2-receptor antagonists, OPC 31260 and YM 087, have become available for clinical research. The term aquaretic drugs (aquaretics) has been coined for these drugs to highlight their different mechanisms compared with the saluretic diuretic furosemide. V1a-receptor antagonists might offer new therapeutic advantages in the treatment of vasoconstriction and hypertension. Combined V1a-/V2-receptor antagonists might be beneficial in the treatment of congestive heart failure (Mayinger and Hensen, 1999).

In a hospital- and population-based case-control, age-matched retrospective study, the incidence of human gastroduodenal ulceration is significantly higher in the normal population (in whom the release of VP is presumed to be intact) than in the VP-deficient one (central diabetes insipidus patients). It thus appears that endogenous VP plays an aggressive role in the development of gastroduodenal ulceration, especially that related to stress (Laszlo et al., 1998).

Finally, dysregulation of VP has been implicated in eating disorders (Gold et al., 1983, Demitrack et al., 1989, 1992) and obsessive–compulsive disorder (Altemus et al., 1992).

16.11 Cholecystokinin

In 1975, many years after its discovery in the GI tract, cholecystokinin (CCK) was noted to be enriched in the mammalian brain (Dockray, 1976), thus becoming one of the first in a long line of peptides known to have such distribution. In fact, CCK appears to be one of the most highly concentrated and widely dispersed neuropeptides in brain (Crawley, 1985).

The appetite-suppressing or satiety-inducing effects of CCK, first noted by Gibbs et al. (1973), physiologically involve communication of feeding-induced, CCK-mediated satiety signals from the periphery to the CNS via the vagal nerves and brainstem. However, centrally elaborated CCK is involved in the regulation of many other behaviors, including analgesia, memory and cognition, anxiety and panic, locomotion, and in the regulation of dopamine release and HPA-stress-system activity.

CCK is one of many neuropeptides that act on the intestine, pancreas, and gallbladder. Endocrine sources of CCK include endocrine cells of the intestinal mucosa, some pituitary cells, and adrenal medulla cells. In the CNS, it is present in the cerebral cortex, hypothalamus, and arcuate nucleus, among other loci. Low levels are also present in the spinal cord and cerebellum (Crawley, 1985). In the mesencephalon, CCK is co-localized in dopamine neurons which project to the limbic forebrain and the ventromedial hypothalamus.

CCK is also found in the peripheral nerves of the GI tract, including in the circular muscle layer of the colon and in the myenteric and submucosal plexuses, where it stimulates the release of acetylcholine and the contraction of smooth muscle. It is also found surrounding the islets of Langerhans in the pancreas.

16.11.1 Structure of CCK

CCK was isolated from porcine intestine by Mutt and Jorpes (1968). In common with gastrin is a C-terminal pentapeptide, Gly-Trp-Asp-Met-Phe-NH2. CCK binds to the CCK-gastrin receptor. There are a variety of sizes of CCK molecules (Rehfeld and Kruse-Larsen, 1978, 2000; Rehfeld et al., 1985; Liddle et al., 1985; Reeve et al., 1994). Variations include CCK-39, 33, 25, 18, 8, 7, and 5. In the nervous system, the predominant form is CCK-8. Heterogenous CCK molecular forms are also observed in human CSF. CCK-8 has been regarded to be the most common form of CCK in most assessments of human CSF (Rehfeld, 2000), although we found substantial quantities of an intermediate form of CCK, with a molecular weight of 1.66kDa, in this fluid – most likely CCK-12 (Geracioti et al., 1993). It is of interest that CCK-4, an anxiogenic peptide, when administered to humans, is relatively absent from human CSF. In any event, selectivity of CCK's action requires seven amino acids and sulfation of tyrosine at position seven from the C-terminal.

Precholecystokinin is a 115-amino-acid polypeptide which is cleaved to form pro-CCK and then CCK-58 (Eng et al., 1990). α-Amidation of the C-terminal follows and then O-sulfation of Tyr in the seventh position from the amidated C-terminus. Further cleavage at monobasic or dibasic residues leads to the smaller forms.

The CCK gene, found on chromosome 3, is regulated by feeding; decreased expression can be achieved by fasting and somatostatin. Somatostatin can lower CCK mRNA levels in the intestine (Liddle, 1994a,b).

16.11.2 Localization

CCK is one of the most prevalent peptides in the CNS, where it is commonly co-localized in neurons with dopamine, glutamate, or GABA (among other classical neurotransmitters and peptides). In the human, CCK mRNA is found in the neocortex, amygdala, claustrum, and hippocampus (Hokfelt et al., 1991). CCK levels in the cerebral cortex are high – comparable to those seen in the duodenum – with extensive distribution throughout cortical fields in the primate (Kritzer et al., 1987). Tissue-specific post-translational processing accounts for differences in the final form of CCK (Gubler et al., 1984). In some species, CCK mRNA positive neurons may be present in the dorsal root ganglia and in dorsal horn neurons, especially in the deeper layers of the substantia gelatinosa and lamina III. Other areas of expression include the periaqueductal gray, thalamus, and cortex, regions implicated in nociception (Schiffmann and Vanderhaeghen, 1991).

CCK is released from GI endocrine cells after a meal and circulating CCK concentrations remain elevated as long as there is food in the stomach. In the human, proteins, amino acids, and fats are the strongest stimulants of CCK release. In the rat and human, trypsin inhibits the release of CCK via a negative-feedback mechanism (Owyang et al., 1986). There also exists a pancreatic monitor peptide (Lu et al., 1989) and an intestinal CCK-releasing factor (Miyasaka et al., 1989) which can stimulate CCK release. Food must be present in the intestine for these mechanisms to be operative.

16.11.3 CCK Receptors

Two types of CCK receptors called CCK-A (or CCK-1) and CCK-B (or CCK-2) – both G-protein-linked receptors – have been identified (Wank, 1995) in addition to a gastrin receptor that is functionally similar to the CCK-B receptor. CCK-B receptors are predominantly brain receptors and appear to mediate CCK signals that are involved with memory, cognition, antianalgesia, and anxiety, while CCK-A receptors (which exist in a variety of forms) are predominantly in the peripheral alimentary system and more fundamentally convey digestive-tract-mediated CCK satiety signaling (Moran et al., 1993). However, CCK-B receptors have a role in the satiety response to feeding (e.g., by conveying the state of gastric distenion), while CCK-A receptors in the brain have a role, along with CCK-B receptors, in modulating the release or actions of dopamine (Bush et al., 1999; Wunderlich et al., 2000). CCK-A- and CCK-B-receptor subtypes are identical in the brain and GI system.

16.11.4 CCK Physiology

CCK is released from the intestine after eating and stimulates gallbladder contraction and pancreatic exocrine secretion. CCK can potentiate the action or release of several secretions from the pancreas. This includes the effects that secretin exerts on pancreatic bicarbonate release, the amino-acid-stimulated release of insulin β cells that have CCK receptors and glucagon, and pancreatic polypeptide release. CCK released after eating is a physiologic regulator of postprandial satiety in experimental animals (Gibbs et al., 1978) and in humans (Geracioti and Liddle, 1988). These peripheral actions of CCK are predominantly mediated via the CCK-A receptor and the vagus nerve. The nucleus tractus solitarius and area postrema are critical brain areas that must be intact in order for the peripherally generated satiety signals to be effective (Edwards and Ritter, 1981). Thus, the receptor antagonist devazepide acts on CCK-A receptors in the gallbladder to inhibit CCK actions (Liddle et al., 1989). CCK-A antagonists can lead to increases in food intake in laboratory animals (Brenner and Ritter, 1996).

CCK can also alter mobility of the GI tract – it reduces stomach motility and gastric emptying but stimulates intestinal motility. In the upper region of the gut, CCK regulates lower esophageal sphincter pressure and, via stimulation of CCK receptors that ring the gastric pylorus, slows the rate at which food is emptied from the stomach. In the human, CCK stimulates antropyloroduodenal motility through reactions with cholinergic neurons in this region. The net effect is increased intestinal mobility and transit of food through the intestine. CCK also induces relaxation of the sphincter of Oddi, thereby regulating the passage of pancreatic and biliary secretions into the duodenum. CCK-8 can cause gastric mucosal hyperemia. This is brought about by a vago-vagal reflex that involves acetylcholine, calcitonin-gene-related peptide, and nitric oxide (Heinemann et al., 1996). CCK has a trophic effect on the pancreas. It increases the secretory capacity of the gland by increasing cell number and size.

CCK also modulates nociception and can antagonize the effects of opioids (Han et al., 1986). In this regard, CCK exerts antagonistic effects on both the antinociceptive and GI-antimotility actions of morphine (Singh et al., 1996). CCK tetrapeptide (Trp-Met-ASP-Phe) induces panic attacks in humans upon IV injection (Rehfeld, 2000). It has been postulated that the anxiogenic effect of CCK may be through mesocorticolimbic dopamine neurons (Crawley, 1991). CCK in the brain is frequently intraneurally co-localized with dopamine. CCK modulates dopamine neurotransmission, either increasing or decreasing it depending on: the nature and location of the dopaminergic or CCK challenge, the specific area of brain dopaminergic activity assessed, and the species of experimental animal tested (Lanca et al., 1998; Reum et al., 1997; Kawai et al., 1997; You et al., 1998). Gene expression of CCK and CCK-B receptors is upregulated in dorsal root ganglia cells following axotomy (Hokfelt et al., 1994), implicating a role of this peptide in nerve injury.

CCK and CCK-B agonists stimulate the pituitary–adrenocortical axis (e.g., Abelson et al., 1997), probably via mediation by CNS CRH – at least in the rat (Kamilaris et al., 1992). In human beings, we found that a significant relationship exists between concentrations of CCK and CRH in CSF, especially in depressed patients (Geracioti et al., 1999) – a situation that is of potential relevance to the treatment of anxiety and depression (see below).

16.11.5 Clinical Implications

A variety of specific CCK-receptor agonists and antagonists exist or are under development (see, e.g., Martin-Martinez et al. (2000), Bellier et al. (2000), Ursini et al. (2000), Low et al. (2000), Gouldson et al. (2000), Gardner and Jensen (1984), and Liddle et al. (1989)). The clinical availability of these agents provides the tools to make imminently possible the understanding of the role of the various CNS CCK-receptor-subtype systems to (human) consciousness and behavior and its fragmentation. Some notable observations have already been made.

The panic-anxiety-inducing effects of the fragment CCK-4 were reportedly first observed as early as 1979 by Rehfeld and a colleague who injected the substance into each other via IV push (Rehfeld, 2000). Over a decade ago, the first of what have now become many clinical studies demonstrating this CCK-4 effect emerged (de Montigny, 1989; Bradwejn et al., 1990). Although CCK-4 may or may not have physiologic relevance in the human being, pharmacologically, CCK-4 appears to be a CCK-B-receptor agonist. Recent attempts to treat panic disorder with the CCK-B antagonists CCI-988 or L-365,260 for 6 weeks showed no effect (Kramer et al., 1995; Pande et al., 1999). Acute administration of CCI-998 also failed to attentuate meta-chlorophenylpiperazine (mCPP)-induced anxiety in patients with generalized anxiety disorder (Goddard et al., 1999). However, the panicogenic effects of the synthetic CCK-B agonist pentagastrin, which includes the entire amino-acid-sequence of CCK-4, were apparently blocked by pretreatment with 10–50 mg of L-365,260 (Lines et al., 1995) – doses less than half of those used without effect in the 6-week clinical trial in panic-disorder patients. Additionally, a preliminary study recently found that a CCK-B-receptor gene polymorphism was significantly associated with panic disorder (Kennedy et al., 1999). Interestingly, CCK-B agonists such as BC264 – which is supposedly devoid of anxiety-producing effects, at least in the rat – have been proposed as potential cognitive enhancers (Lena et al., 1999; Taghzouti et al., 1999).

Given the important role of dopamine in the pathophysiology and treatment of the psychoses, along with the well-documented ability of CCK to regulate dopamine release in the brain, a long-standing search for a role of CNS CCK in the pathophysiology and treatment of schizophrenia continues (see, e.g., Nair et al. (1982), Minato et al. (2007), Toirac et al. (2007)). CCK-A receptor polymorphisms may be more frequent in schizophrenic patients, especially those of the paranoid type, than in healthy controls (Tachikawa et al., 2000). In patients with schizophrenia, there is evidence that CCK levels are abnormally expressed. For instance, Bachus et al. (1997) reported that levels of CCK mRNA are decreased in entorhinal cortex and subiculum in schizophrenics relative to controls. Cellular analysis indicates that there is a decrease in density of CCK mRNA in labeled neurons. In so far as CCK is co-localized with GABA or glutamate neurons, these authors hypothesized that the interaction of these neurotransmitters may be important in the pathogenesis of schizophrenia. The search for an antipsychotic effect of CCK-based treatments has stalled after early attempts to treat psychosis with peripheral administration of CCK-8 or an analog thereof were unsuccessful (Hommer et al., 1985; Mattes et al., 1985; Peselow et al., 1987).

CCK has antiopioid effects; therefore, it is possible that blockade of CCK neurotransmisson could have pro-opioid effects. In this regard, under double-blind conditions, the nonspecific CCK antagonist proglumide increased the analgesic effect of morphine in some patients with chronic pain (McCleane, 1998).

Finally, patients with bulimia nervosa – characterized by binge eating and impaired postprandial satiety – have impaired postprandial secretion of

CCK (Geracioti and Liddle, 1988; Brambilla et al., 1995; Devlin et al., 1997) – a phenomenon which can be normalized in some bulimic patients treated with antidepressant drugs. Peripheral CCK influences brain function via vagal afferents that synapse in the brainstem. CCK administration itself increases feelings of satiety in humans (Wolkewitz et al., 1990). However, the decrease in feeding (mainly due to a decrease in meal size) engendered by pharmacologic administration of CCK cannot easily be sustained due to the emergence of compensatory consumption of more frequent, if smaller, meals. A more intermittent, rather than chronic, application would seem to be the best approach; this would seem to be well suited to a clinical trial in bulimia nervosa. However, the potential of CCK-A (CCK-1) agonists to have a clinical role in the treatment or control of obesity has thus far been unrewarding (Jordan et al., 2008).

16.12 Neuropeptides of Emerging or Expanding Psychiatric Interest

16.12.1 Substance P

Extracted over 75 years ago from equine brain and intestine (Von Euler and Gaddum, 1931), the 11-amino-acid compound identified as substance P is now known to subsume a number of major physiologic functions, including transmission of nociceptive and stressful signals within the CNS (Stout et al., 2001; Mantyh, 2002; Rupniak, 2002) and mediation of the mammalian nausea response. This peptide is one of the most abundant hormones in the mammalian brain and, in conjunction with neurokinin (NK) A, is encoded by the preprotachykinin A (PPTA or Tac1) gene (Krause et al., 1989).

Within the human CNS, the NK receptors to which substance P binds are distributed heterogeneously throughout the spinal cord, brainstem, limbic system prefrontal cortex, hippocampus, and thalamus (Cuello et al., 1976; Schoenen et al., 1985; Bouras et al., 1986; Bennett et al., 1986; Hargreaves, 2002). NK-1, NK-2, and NK-3 receptors have been identified in mammals and endogenous NKs interact with all three receptors though substance P binds with highest affinity to the NK-1 receptor in humans. Substance P exerts its cellular effects via G-protein-coupled NK receptors (Khawaja and Rogers, 1996) which can activate multiple second-messenger systems (see Ebner and Singewald (2006) for review).

16.12.2 Clinical Implications: Populations of Interest

Although substance P is now not routinely measured in clinical practice, CSF substance P concentrations have been found to be elevated in a number of psychiatric conditions, including PTSD and major depression (Geracioti et al., 2006). Analogously, postmortem studies of human brains by Stockmeier et al. (2002) have observed a decreased density of NK-1 receptors in the orbitofrontal cortex of individuals who died by suicide and individuals with major depression, compared with individuals who were reportedly psychiatrically normal which "raised the possibility of chronic hypersecretion of substance P and subsequent down-regulation of these NK-1 receptors" (Geracioti et al., 2006). In addition, CSF substance P has been found to be elevated in patients with fibromyalgia, chronic arthralgias, and chronic daily headaches, and concentrations of this compound are reportedly abnormal in asthma, nausea, inflammatory bowel syndrome, and urinary incontinence (see Alvaro and Di Fabio (2007)). Indeed, it will be of great interest to see the results of additional studies of substance P in patients with other conditions and, to this end, studies are currently underway to examine the dynamics of this system in patients with panic disorder and alcohol dependence as well as to further elucidate the role of substance P in postoperative and chemotherapy-induced nausea.

16.12.3 Clinical Implications: Diagnostic Testing

Despite a significant volume of literature on substance P in various disease states, the utility of these measures in routine clinical practice remains to be determined. In fact, we are unaware of any studies which have yet examined the relationship between central (e.g., CSF) and peripheral (e.g., plasma or saliva) measures of substance P.

16.12.4 Clinical Implications: Therapeutics

One substance P antagonist, aprepitant (MK-869), is currently Food and Drug Administration (FDA)-approved for use in the United States for acute and delayed chemotherapy-induced nausea and vomiting (Patel and Lindley, 2003). However, other clinically available drugs appear to reduce substance P release, including pregabalin (Fehrenbacher et al., 2003), which is used to treat fibromyalgia, diabetic peripheral neuropathic pain, postherpetic neuralgia, and adjunctive treatment of partial seizures – although use in psychiatric patients with mood disorder and borderline personality disorders is increasing.

Aprepitant acts as an antagonist at the NK-1 receptor and was reported to be an effective antidepressant in a double-blind, placebo-controlled trial (Kramer et al., 1998), although subsequent clinical trials failed to confirm this finding (Krishnan, 2002). However, another substance P antagonist, L-759274, did exhibit antidepressant effects in a controlled clinical trial (Kramer et al., 2004).

Interestingly, we recently noted pregabalin to be effective in treating chronic, treatment-refractory PTSD wherein substance P levels are similarly known to be elevated (Strawn et al., in press), though this finding is not entirely surprising. Finally, it is of interest that this compound has demonstrated efficacy in the treatment of generalized anxiety disorder in five double-blind placebo-controlled trials – an effect which may be related to substance P pathophysiology or to its effect on other panic-related neurotransmitters (Strawn and Geracioti, 2007).

Accumulating data indicate that substance P has anxiogenic effects; conversely, antagonism of the NK-1 receptor is anxiolytic (for a review, see Herpfer and Lieb (2005)). In this regard, the NK-1 antagonist GR205171 significantly reduced symptoms of social phobia during a 4-week trial (Furmark et al., 2005). In addition, patients treated with the NK-1 antagonist improved comparably to those who received the SSRI citalopram (Furmark et al., 2005). Currently, at least two additional trials of NK-1 antagonists are underway for the treatment of PTSD, including an NIMH-sponsored trial of GR205171 which is expected to be complete by 2009. It will be of great interest to determine the effectiveness of these compounds in the treatment of PTSD, especially given that CSF substance P levels in PTSD patients markedly increase during exposure to traumatic psychological stimuli, but not during exposure to neutral psychological stimuli (Geracioti et al., 2006).

16.12.5 Neuropeptide Y

In 1982, Tatemoto et al. (1982) discovered NPY, a 36-amino-acid peptide, from porcine brain, which is known to be widely distributed in a number of mammalian organ systems, including the central and peripheral nervous systems (Hendry, 1993; Sundler et al., 1993). In fact, in the human brain, this neuropeptide is thought to be among the most abundant of human neuropeptides, with concentrations that often exceed those of CCK and somatostatin (Adrian et al., 1983). In the human brain, NPY is found throughout the basal ganglia, amygdala, and nucleus accumbens, and NPY neuronal cell bodies are similarly located in the caudate, putamen, and the cortex (especially in layers V and VI) (Adrian et al., 1983). In addition, NPY is localized within the forebrain limbic structures, such as the hippocampus, amygdala, cortex, and the BDST, as well as within the hypothalamus (Hendry, 1993).

The gene encoding NPY is located on chromosome 7 in the human (Takeuchi et al., 1986) and its expression is influenced by a number of endocrine factors, through a variety of cell-specific mechanisms. For example, transcription of NPY is inhibited by leptin in hypothalamic neurons (Higuchi et al., 2005) and is upregulated by ovarian hormones (Peng et al., 1994) and ghrelin.

There are at least six NPY receptors (Y1–6) and all are coupled to G-proteins. However, the Y1, Y2, and Y5 receptors appear to have the greatest CNS significance. In the brain, concentrations of NPY receptors are found throughout the cortex, limbic system, hippocampus, nucleus accumbens, PVN, arcuate nucleus, and brainstem, though peripherally they are found in intestinal, cardiac, splenic, and kidney tissue (Strand, 1999; Silva et al., 2005).

16.12.6 Clinical Implications: Populations of Interest

Clinical and preclinical studies have implicated NPY in the pathophysiology of a multitude of diseases including epilepsy, eating disorders and obesity, anxiety disorders and affective disorders, and congestive heart failure.

With regard to the role of NPY in epilepsy, there are limited clinical studies of NPY concentrations in epileptic patients, though the extant data from lower animals strongly implicate this neuropeptide in the seizure pathophysiology. In particular, activation of Y2 and Y5 receptors and, interestingly, antagonism of central Y5 have been associated with anticonvulsant effects in a variety of animal seizure models (Meurs et al., 2007).

Given the recent epidemic of obesity and its general health consequences, significant efforts have been directed at understanding the relationship between NPY, eating dynamics, and modulators of metabolism, including leptin and α-MSH. In clinical studies of obese subjects, absolute CSF concentrations of NPY are not significantly different between nonobese and healthy women, although significant decreases in CSF NPY in the obese subjects were observed after weight loss (Nam et al., 2001). Interestingly, this study did not observe a correlation between CSF leptin and CSF NPY, a finding that has been observed in obesity and anorexia and had suggested that the feedback mechanism between leptin and NPY is lost in both obesity and anorexia nervosa (Baranowska et al., 2003).

In depressed patients, peripheral NPY concentrations have been found to be low following suicide attempts as compared to healthy control subjects (Westrin et al., 1999). Interestingly, CSF NPY concentrations increase following successful treatment with the selective serotonin receptor antagonist citalopram (Nikisch et al., 2005) and electroconvulsive therapy (Nikisch and Mathé, 2008).

With regard to anxiety disorders, CSF NPY levels are similar to healthy control subjects in patients with obsessive–compulsive disorder (Altemus et al., 1999). More recently, we observed strikingly low CSF NPY concentrations in unmedicated patients with chronic combat-related PTSD (Sah et al., 2009). Plasma NPY levels increase in response to major stress (Morgan et al., 2000), and may be reduced in patients with PTSD (Rasmusson et al., 2000). Yehuda and colleagues examined NPY concentrations in PTSD patients and found plasma NPY levels to be predicted by symptom improvement and lower combat exposure. In addition, the authors of this study noted a trend with regard to NPY and positive coping suggesting a resilience-modulating role associated with this peptide (Yehuda et al., 2006). Plasma NPY concentrations are similar in patients with social phobia and panic disorder (Stein et al., 1996). NPY concentrations in circulating blood do not reflect those in the CSF (Dötsch et al., 1997).

16.12.7 Clinical Implications: Therapeutics

Although currently, there are no clinically approved NPY analogs, antagonists, or agonists, CNS penetration of NPY can be achieved by intranasal application. Variety of intranasal NPY analogs potentially suitable for human are under development (e.g., see Hallschmid et al. (2004)). In addition, the Y5 antagonist MK-0557 has been studied in the treatment of obesity and results of a 1-year clinical trial suggest that this compound can produce modest weight loss (MacNeil, 2007), although formal reporting of the clinical trial results as well as additional trials of this compound in combination with sibutramine or orlistat have not been released. Additional trials of Y5 antagonists in the treatment of cognitive deficits associated with schizophrenia are currently ongoing. Thus, the possibility that the central NPY system could become a therapeutic target in any of the previously mentioned psychiatric conditions is of great interest.

16.12.8 Orexins (Hypocretins)

Since the discovery of the orexin (hypocretin) system, this neuropeptide system has become a focus of increasing clinical and basic science research owing to its importance in regulating sleep, wakefulness, and feeding. Over the last decade, before which orexins were completely unknown, impressive amounts of information have been quickly learned in studies spanning gene regulation to biochemistry to involvement in human disease. Nonetheless, the arousal-regulating properties of orexin and potential involvement in clinical neurologic syndromes are becoming apparent, while only a handful of studies have examined its role in psychiatric illness.

Owing to the simultaneous discovery by two different groups, one of which identified the peptides by searching for the endogenous ligands for orphan G-protein-coupled receptors (Sakurai et al., 1998), and the other of which discovered the peptides by subtractive PCR in hypothalamic extracts (De Lecea et al., 1998), this class has been called orexins (A and B) or hypocretins (1 and 2), respectively. Both orexin-A, a 33-amino-acid peptide (3562 Da), and orexin-B, a 28-amino-acid peptide (2937 Da), are produced from proteolytic processing of the prepro-orexin peptide (Nishino, 2003). The prepro-orexin gene, which includes two exons and one intron, is located on chromosome 17q21.

Two orexin receptors have been identified (orexin-1 and orexin-2 receptors) and both are G-protein-coupled receptors. The orexin-2 receptor is thought to be nonselective for orexin-A and orexin B, while the orexin-1 receptor is significantly more selective for orexin-A based on *in vitro* binding studies (Takeshi, 2006). The orexin receptors are expressed throughout the brain though they are extensively localized within the hypothalamus, hippocampus, raphe nuclei, basal ganglia, LC, and cortex. Also of note, orexin receptors are prominent in peripheral tissues such as intestine where they are affected by nutritional status and may, at least in part, regulate gut motility (Heinonen et al., 2008).

16.12.9 Clinical Implications: Special Populations

CSF levels of the orexin-A have been consistently noted to be low (or even undetectable) in patients with narcolepsy; this finding appears to be quite specific among sleep disorders and neurologic diseases (Mignot et al., 2002; Ripley et al., 2001).

CSF orexins have also been reported to be low in patients with intracerebral hemorrhage (Dohi et al., 2008).

There are few studies of orexins in psychiatric patients. However, CSF orexin-A levels are reportedly normal in patients with schizophrenia and major depression, although in the later group, there was a trend toward higher levels which decreased significantly following treatment with the selective serotonin-reuptake inhibitor (SSRI), sertraline, but not with the norepinephrine and dopamine reuptake-inhibiting bupropion (Salomon et al., 2003). This finding suggests, at least in depressed patients, that orexin-A may be influenced by serotonergic tone. Interestingly, the amplitudes of the normal diurnal CSF orexin-A concentration variation were less in patients with depression, suggesting that the pulsatility of release may be altered in this population (Salomon et al., 2003). Finally, our group recently observed low CSF levels of orexin-A, measured over hours, in patients with chronic, combat-related PTSD (Strawn et al., under review).

16.12.10 Clinical Implications: Diagnostics

There are no commercially and clinically available (FDA-approved) tests for orexin-A or-B in CSF (or any other biological matrix); a test might be particularly useful in the diagnosis of narcolepsy wherein low levels are highly sensitive and specific for the condition. However, recent data suggest that peripheral orexin-A highly correlates with CSF orexin-A, which raises the possibility that plasma concentrations could be used as surrogate measures for central concentrations of this compound (Strawn et al., under review). This might be particularly helpful as the narcolepsy diagnosis often is not made until years after onset.

16.12.11 Clinical Implications: Therapeutics

To date, there are no clinically available orexin-1 or 2 receptor antagonists or agonists, although there are reports of experimental orexin-1 antagonists which have been studied in lower animals (e.g., SB334867 in rats; Yamada et al., 2005). Should a centrally acting form of the drug be developed or another form of replacement therapy be made feasible, including gene therapy, there would be intense interest in its clinical effects in narcolepsy, as well as in depression and PTSD.

16.13 Concluding Remark

In summary, it should be evident from this chapter that neuropeptides play a multitude of roles in CNS function and behavior. The rapid recent development of molecular approaches to neuropeptide research coupled with advances in behavioral studies and clinical physiology and pharmacology will only further our understanding of the pathophysiology of numerous psychiatric and neurological disorders. More importantly, elucidating the role of these peptides will allow us to increasingly target interventions in the CNS. CNS-specific peptide-receptor agonists and antagonists and other selective modifications of peptide systems, such as through modification of genetic expression, will have increasing clinical importance in the years ahead.

References

Abelson JL, Khan S, Liberzon I, and Young EA (1997) HPA axis activity in patients with panic disorder: Review and synthesis of four studies. *Depress Anxiety* 24: 66–76.

Adan RA and Gispen WH (1997) Brain melanocortin receptors: From cloning to function. *Peptides* 18: 1279–1287.

Adan RAH and van Dijk G (2006) Melanocortin receptors as drug targets for disorders of energy balance. *CNS and Neurological Disorders Drug Targets* 5: 251–261.

Adrian TE, Allen JM, Bloom SR, et al. (1983) Neuropeptide Y distribution in human brain. *Nature* 306: 584–586.

Agnati LF, Zoli M, Stromberg I, and Fuxe K (1995) Intercellular communication in the brain: Writing versus volume transmission. *Neuroscience* 69: 711–726.

Aimaretti G, Ambrisio MR, Di Somma C, et al. (2004) Traumatic brain injury and subarchnoid haemorrhage are conditions at high risk for hypopituitarism: Screening study at 3months after brain injury. *Clinical Endocrinology* 61: 320–326.

Alkemade A, Unmehopa UA, Wiersinga WM, Swaab DF, and Fliers E (2005) Glucocorticoids decrease thyrotropin-releasing hormone messenger ribonucleic acid expression in the paraventricular nucleus of the human hypothalamus. *Journal of Clinical Endocrinology and Metabolism* 90: 323–327.

Altemus M, Jacobson KR, Debellis M, Kling M, Pigott T, Murphy DL, and Gold PW (1999) Normal CSF oxytocin and NPY levels in OCD. *Biological Psychiatry* 45: 931–933.

Altemus M, Pigott T, Kalogeras KT, Demitrack M, Dubbert B, Murphy DL, and Gold PW (1992) Abnormalities in the regulation of vasopressin and corticotropin releasing factor secretion in obsessive–compulsive disorder. *Archives of General Psychiatry* 49: 9–20.

Alvaro G and Di Fabio R (2007) Neurokinin 1 receptor antagonists – current prospects. *Current Opinion in Drug Discovery and Development* 10: 613–621.

Amoss M, Burgus R, Blackwell R, Vale W, Fellows R, and Guillemin R (1971) Purification, amino acid composition and N-terminus of the hypothalamic luteinizing hormone releasing factor (LRF) of ovine origin. *Biochemical and Biophysical Research Communications* 44: 205–210.

Anderson-Hunt M and Dennerstein L (1995) Oxytocin and female sexuality. *Gynecologic and Obstetric Investigation* 40: 217–221.

Antoni FA (1993) Vasopressinergic control of pituitary adrenocorticotropin secretion comes of age. *Frontiers of Neuroendocrinology* 14: 76–122.

Anton RF, Oroszi G, O'Malley S, Couper D, Swift R, Pettinati H, and Goldman D (2008) An evaluation of mu-opioid receptor (OPRM1) as a predictor of naltrexone response in the treatment of alcohol dependence: Results from the Combined Pharmacotherapies and Behavioral Interventions for Alcohol Dependence (COMBINE) study. *Archives of General Psychiatry* 65:135–144.

Arato M, Banki CM, Bissette G, and Nemeroff CB (1989) Elevated CSF CRH in suicide victims. *Biological Psychiatry* 25: 355–359.

Atweh SF and Kuhar MJ (1983) Distribution and physiological significance of opioid receptors in the brain. *British Medical Bulletin* 39: 47–52.

Bachus SE, Hyde TM, Herman MM, Egan MF, and Kleinman JE (1997) Abnormal cholecystokinin mRNA levels in entorhinal cortex of schizophrenics. *Journal of Psychiatric Research* 31: 233–256.

Badoer E and Merolli J (1998) Neurons in the hypothalamic paraventricular nucleus that project to the rostral ventrolateral medulla are activated by haemorrhage. *Brain Research* 791: 317–320.

Baerwald CG, Mok CC, Tickly M, et al. (2000) Corticotropin releasing hormone (CRH) promoter polymorphisms in various ethnic groups of patients with rheumatoid arthritis. *Zeitschrift fur Rheumatologie* 59: 29–34.

Bajoria R and Babawale M (1998) Ontogeny of endogenous secretion of immunoreactive-thyrotropin releasing hormone by the human placenta. *Journal of Clinical Endocrinology and Metabolism* 83: 4148–4155.

Baker JT (1997) Adrenal disorders. A primary care approach. *Lippincott's Primary Care Practice* 5: 527–536.

Baker DG, West SA, Nicholson WE, et al. (1999) Serial CSF corticotropin-releasing hormone levels and adrenocortical activity in combat veterans with posttraumatic stress disorder. *American Journal of Psychiatry* 156: 585–588.

Baker DG, West SA, Orth DN, et al. (1997) Cerebrospinal fluid and plasma beta-endorphin in combat veterans with post-traumatic stress disorder. *Psychoneuroendocrinology* 22: 517–529.

Bale TL and Dorsa DM (1995) Regulation of oxytocin receptor messenger ribonucleic acid in the ventromedial hypothalamus by testosterone and its metabolites. *Endocrinology* 136: 5135–5138.

Banki CM, Bissette G, Arato M, O'Connor L, and Nemeroff CB (1987) Cerebrospinal fluid corticotropin-releasing factor-like immunoreactivity in depression and schizophrenia. *American Journal of Psychiatry* 144: 873–877.

Bao AM, Fischer DF, Wu YH, et al. (2006) A direct androgenic involvement in the expression of human corticotropin-releasing hormone. *Molecular Psychiatry* 11: 567–576.

Baranowska B, Wolinska-Witort E, Wasilewska-Dziubinska E, Roguski K, Martynska L, and Chmielowska M (2003) The role of neuropeptides in the disturbed control of appetite and hormone secretion in eating disorders. *Neuroendocrinology Letters* 24: 431–434.

Bastias MC, Kamijo H, and Pavlou SN (1993) Sperm motion parameters after suppression of spermatogenesis with a gonadotropin-releasing hormone antagonist plus testosterone supplementation. *Fertility and Sterility* 59: 1261–1265.

Bauer K, Schomburg L, Heuer H, and Schafer MK (1999) Use instead thyrotopin releasing hormone (TRH), the TRH-receptor and the TRH-degrading ectoenzyme; three elements of a peptidergic signalling system. *Results and Problems in Cell Differentiation* 26: 13–42.

Bayatti N and Behl C (2005) The neuroprotective actions of corticotropin releasing hormone. *Ageing Research Reviews* 4: 258–270.

Beaudet A, Greenspun D, Raelson J, and Tannenbaum GS (1995) Patterns of expression of SSTR1 and SSTR2 somatostatin receptor subtypes in the hypothalamus of the adult rat: Relationship to neuroendocrine function. *Neuroscience* 65: 551–561.

Beck-Peccoz P, Amr S, Menezes-Ferreira MM, Faglia G, and Weintraub BD (1985) Decreased receptor binding of biologically inactive thyrotropin in central hypothyroidism. Effect of treatment with thyrotropin-releasing hormone. *New England Journal of Medicine* 312: 1085–1090.

Beckwith BE (1988) The melanotropins: Learning and memory. In: Hadley ME (ed.) *The Melanotropic Peptides*, vol. 2, pp. 43–85. Boca Raton, FL: CRC Press.

Beckwith BE, Couk DI, and Till TS (1983) Vasopressin analog influences the performance of males on a reaction time task. *Peptides* 4: 707–709.

Beckwith BE, Till RE, and Schneider V (1984) Vasopressin analog (DDAVP) improves memory in human males. *Peptides* 5: 819–822.

Behre HM, Klein B, Steinmeyer E, McGregor GP, Voigt K, and Nieschlag E (1992) Effective suppression of luteinizing hormone and testosterone by single doses of the new gonadotropin-releasing hormone antagonist cetrorelix (SB-75) in normal men. *Journal of Clinical Endocrinology and Metabolism* 75: 393–398.

Bellier B, Million ME, DaNascimento S, Meudal H, Kellou S, Maigret B, and Garbay C (2000) Replacement of glycine with dicarbonyl and related moieties in analogues of the C-terminal pentapeptide of cholecystokinin: CCK (2) agonists displaying a novel binding mode. *Journal of Medicinal Chemistry* 43: 3614–3623.

Bennett GW, Nathan PA, Wong KK, and Marsden CA (1986) Regional distribution of immunoreactive-thyrotrophin-releasing hormone and substance P, and indoleamines in human spinal cord. *Journal of Neurochemistry* 46: 1718–1724.

Berkenbosch F, Tilders FJ, and Vermes I (1983) Beta-adrenoceptor activation mediates stress-induced secretion of beta-endorphin-related peptides from intermediate but not anterior pituitary. *Nature* 305: 237–239.

Bertherat J, Bluet-Pajot MT, and Epelbaum J (1995) Neuroendocrine regulation of growth hormone. *European Journal of Endocrinology* 132: 12–24.

Berzetei-Gurske IP, Schwartz RW, and Toll L (1996) Determination of activity for nociceptin in the mouse vas deferens. *European Journal of Pharmacology* 302: R1–2.

Besson JM and Chaouch A (1987) Peripheral and spinal mechanisms of nociception. *Physiological Reviews* 67: 67–186.

Binneman B, Feltner D, Kolluri S, Shi Y, Qiu R, and Stiger T (2008) A 6-week randomized, placebo-controlled trial of CP-316,311 (a selective CRH1 antagonist) in the treatment of major depression. *American Journal of Psychiatry* 165: 617–620.

Bissette G (1997) Neuropeptides and Alzheimer's disease pathology. *Annals of the New York Academy of Sciences* 814: 17–29.

Bohus B (1993) Physiologic functions of vasopressin in behavioral and autonomic responses to stress. In: Burdoch JPH and De Wied D (eds.) *Brain Functions of Neuropeptides*, pp. 15–40. New York: Parthenon.

Bohus B and de Wied D (1998) The vasopressin deficient Brattleboro rats: A natural knockout model used in the search for CNS effects of vasopressin. *Progress in Brain Research* 119: 555–573.

Bond C, LaForge KS, Tian M, et al. (1998) Single-nucleotide polymorphism in the human mu opioid receptor gene alters β-endorphin binding and activity: Possible implications for opiate addiction. *Proceedings of the National Academy of Sciences of the United States of America* 95: 9608–9613.

Bouras C, Vallet PG, Dobrinov H, de St-Hilaire S, and Constantinidis J (1986) Substance P neuronal cell bodies in the human brain: Complete mapping by immunohistofluorescence. *Neuroscience Letters* 69: 31–36.

Bourgue CW and Renaud LP (1990) Electrophysiology of mammalian magnocellular vasopressin and oxytocin neurons. *Frontiers in Neuroendocrinology* 11: 183–212.

Bradwejn J, Koszycki D, and Meterissian G (1990) Cholecystokinin-tetrapeptide induces panic attacks in patients with panic disorder. *Canadian Journal of Psychiatry* 35: 83–85.

Brambilla F, Brunetta M, Draisci A, et al. (1995) T-lymphocyte concentrations of cholecystokinin-8 and beta-endorphin in eating disorders: II. Bulimia nervosa. *Psychiatry Research* 59: 51–56.

Brazeau P, Vale W, Burgus R, Ling N, Butcher M, Rivier J, and Guillemin R (1973) Hypothalamic polypeptide that inhibits the secretion of immunoreactive pituitary growth hormone. *Science* 179: 77–79.

Brenner LA and Ritter RC (1996) Type A CCK receptors mediate satiety effects of intestinal nutrients. *Pharmacology, Biochemistry, and Behavior* 54: 625–631.

Bremner WJ, Bagatell CJ, and Steiner RA (1991) Gonadotropin-releasing hormone antagonist plus testosterone: A potential male contraceptive. *Journal of Clinical Endocrinology and Metabolism* 73: 465–469.

Bremner JD, Licinio J, Darnell A, et al. (1997) Elevated CSF corticotropin-releasing factor concentrations in posttraumatic stress disorder. *American Journal of Psychiatry* 154: 624–629.

Buijs RM (1987) Vasopressin localization and putative functions in the brain. In: Gash DM and Boer GJ (eds.) *Vasopressin: Principles and Properties*, pp. 91–115. New York: Plenum Press.

Burbach JP, Adan RA, and de Bree FM (1992) Regulation of oxytocin gene expression and forms of oxytocin in the brain. *Annals of the New York Academy of Sciences* 652: 1–13.

Burbach JP, va Schaick HS, de Bree FM, Lopes da Silva S, and Adan RA (1995) Functional domains in the oxytocin gene for regulation of expression and biosynthesis of gene products. *Advances in Experimental Medicine and Biology* 395: 9–21.

Burbach JPH, Liu B, Seger MA, et al. (1988) Vasopressin gene regulation and gene products in the brain. In: Crowley AW, Jr., Liard JF, and Ausiello DA (eds.) *Vasopressive: Cellular and Integrative Functions*, pp. 295–300. New York: Raven Press.

Brugos B and Hochhaus G (2004) Metabolism of dynorphin A(1–13). *Pharmazie* 59: 339–343.

Burgos-Ramos E, Hervás-Aguilar A, Aguado-Llera D, et al. (2008) Somatostatin and Alzheimer's disease. *Molecular and Cellular Endocrinology* 286: 104–111.

Burri A, Heinrichs M, Schedlowski M, and Kruger TH (2008) The acute effects of intranasal oxytocin administration on endocrine and sexual function in males. *Psychoneuroendocrinology* 33: 591–600.

Bush DE, DeSousa NJ, and Vaccarino FJ (1999) Self-administration of intravenous amphetamine: Effect of nucleus accumbens CCKB receptor activation on fixed-ratio responding. *Psychopharmacology (Berl.)* 147: 331–334.

Caldwell JD (1992) Central oxytocin and female sexual behavior. *Annals of the New York Academy of Sciences* 652: 166–179.

Caldwell JD, Lee H-J, Macbeth AH, and Young WS (2008) Vasopressin: Behavioral roles of an original neuropeptide. *Progress in Neurobiology* 84:1–24.

Cameron OG (2006) Anxious-depressive comorbidity: Effects on HPA axis and CNS noradrenergic functions. *Essential Psychopharmacology* 7: 24–34.

Carel J-C and Leger J (2008) Precocious puberty. *New England Journal of Medicine* 358: 2366–2376.

Carter DA and Lightman SI (1985) Neuroendocrine control of vasopressin secretion. In: Baylis PH and Padfield PL (eds.) *The Posterior Pituitary: Hormone Secretion in Health and Disease*, pp. 53–118. New York: Marcel Dekker.

Carter DA and Murphy D (1991) Rapid changes in poly (A) tail length of vasopressin and oxytocin mRNAs form a common early component of neurohypophyseal peptide gene activation following physiological stimulation. *Neuroendocrinology* 53: 1–6.

Castro MG and Morrison E (1997) Post-translational processing of proopiomelanocortin in the pituitary and in the brain. *Critical Reviews in Neurobiology* 11: 35–57.

Chaki S and Okuyama S (2005) Involvement of melanocortin-4 receptor in anxiety and depression. *Peptides* 26:1952–1964.

Chen Y, Brunson KL, Muller MB, Cariaga W, and Baram TZ (2000) Immunocytochemical distribution of corticotropin-releasing hormone receptor type-1 (CRF(1))-like immunoreactivity in the mouse brain: Light microscopy analysis using an antibody directed against the C-terminus. *Journal of Comparative Neurology* 420: 305–323.

Cherny N (2000) New strategies in opioid therapy for cancer pain. *Journal of Oncology Management* 9: 8–15.

Chhajlani V, Muceniece R, and Wikberg JE (1993) Molecular cloning of a novel human melanocortin receptor. *Biochemical and Biophysical Research Communications* 195: 866–873.

Chini B, Mouillac B, Ala Y, et al. (1995) Molecular basis for agonist selectivity in the vasopressin/oxytocin receptor family. In: Ivell R and Russell JA (eds.) *Oxytocin*, pp. 321–328. New York: Plenum Press.

Chou WY, Yang LC, Lu HF, et al. (2006) Association of mu-opioid receptor gene polymorphism (A118G) with variations in morphine consumption for analgesia after total knee arthroplasty. *Acta Anaesthesiologica Scandinavica* 50: 787–792.

Chrousos GP and Gold PW (1992) The concepts of stress and stress system disorders: Overview of physical and behavioral homeostasis. *Journal of the American Medical Association* 267: 1244–1252.

Clemmons DR (2007) Long-acting forms of growth hormone-releasing hormone and growth hormone: Effects in normal volunteers and adults with growth hormone deficiency. *Hormone Research* 68: 178–181.

Coll AP (2007) Effects of pro-opiomelanocortin (POMC) on food intake and body weight: Mechanisms and therapeutic potential? *Clinical Science (London)* 113:171–182.

Cox BM (1994) Opiate drug tolerance: An introduction. *Regulatory Peptides* 54: 71–72.

Craft S, Asthana S, Newcomer JW, et al. (1999) Enhancement of memory in Alzheimer's disease with insulin and somatostatin, but not glucose. *Archives of General Psychiatry* 56: 1135–1140.

Crawley JN (1985) Comparative distribution of cholecystokinin and other neuropeptides. Why is this peptide different from all other peptides? *Annals of the New York Academy of Sciences* 448: 1–8.

Crawley JN (1991) Cholecystokinin–dopamine interactions. *Trends in Pharmacological Science* 12: 232–236.

Csiffary A, Ruttner Z, Toth Z, and Palkovits M (1992) Oxytocin nerve fibers innervate beta-endorphin neurons in the arcuate nucleus of the rat hypothalamus. *Neuroendocrinology* 56: 429–435.

Cuello AC, Polak JM, and Pearse AG (1976) Substance P: A naturally occurring transmitter in human spinal cord. *Lancet*. 2:1054–1056.

Cummings DE and Merriam GR (1999) Age-related changes in growth hormone secretion: Should the somatopause be treated? *Seminars in Reproductive Endocrinology* 17: 311–325.

Cummings S, Elde R, Ells J, and Lindall A (1983) Corticotropin-releasing factor immunoreactivity is widely distributed within the central nervous system of the rat: An immunohistochemical study. *Journal of Neuroscience* 3: 1355–1368.

Dawood MY (1995) Novel approach to oxytocin induction-augmentation of labor. Application of oxytocin physiology during pregnancy. *Advances in Experimental Medicine and Biology* 395: 585–594.

DeBellis MD, Geracioti TD, Jr., Gold PW, and Kling MA (1993) Fluoxetine reduces CSF SRH and AVP in major depression. *American Journal of Psychiatry* 150: 656–657.

de Kloet ER, Joels M, and Urban IJA (1990) Central neurohypophyseal hormone receptors and receptor mediated cellular responses. In: De Wied D (ed.) *Neuropeptides: Basics and Perspectives*, pp. 105–138. Amsterdam: Elsevier.

De Lecea L, Kilduff TS, Peyron C, et al. (1998) The hypocretins: Hypothalamus-specific peptides with neuroexcitatory activity. *Proceedings of the National Academy of Sciences of the United States of America* 95: 322–327.

de Montigny C (1989) Cholecystokinin tetrapeptide induces panic-like attacks in healthy volunteers. Preliminary findings. *Archives of General Psychiatry* 46: 511–517.

De Vries GJ, al-Shamma HA, and Zhou L (1994) The sexually dimorphic vasopressin innervation of the brain as a model for steroid modulation of neuropeptide transmission. *Annals of the New York Academy of Sciences* 743: 95–120.

De Wied D (1969) Effects of peptide hormones on behavior. In: Gangon WF and Martini L (eds.) *Frontiers in Endocrinology*, pp. 97–140. New York: Oxford University Press.

De Wied D (1983) Central actions of neurohypophysial hormones. *Progress in Brain Research* 60: 155–167.

De Wied D and Jolles J (1982) Neuropeptides derived from pro-opiocortin: Behavioral, physiological, and neurochemical effects. *Physiological Reviews* 62: 976–1059.

Demitrack MA, Lesem MD, Brandt HA, Pigott TA, Jimerson DC, Altemus M, and Gold PW (1989) Neurohypophyseal dysfunction: Implications for the pathophysiology of eating disorders. *Psychopharmacology Bulletin* 25: 439–443.

Demitrack MA, Kalogeras KT, Altemus M, Pigott TA, Listwak SJ, and Gold PW (1992) Plasma and cerebrospinal fluid measures of arginine vasopressin secretion in patients with bulimia nervosa and in healthy subjects. *Journal of Clinical Endocrinology and Metabolism* 74: 1277–1283.

Demitrack MA and Crofford LJ (1998) Evidence for and pathophysiologic implications of hypothalamic–pituitary–adrenal axis dysregulation in fibromyalgia and chronic fatigue syndrome. *Annals of the New York Academy of Sciences* 840: 684–697.

Devlin MJ, Walsh BT, Guss JL, Kissileff HR, Liddle RA, and Petkova E (1997) Postprandial cholecystokinin release and gastric emptying in patients with bulimia nervosa. *American Journal of Clinical Nutrition* 65: 114–120.

Dinan TG (1998) Psychoneuroendocrinology of depression. Growth hormone. *Psychiatric Clinics of North America* 21: 325–339.

Ding YQ, Kaneko T, Nomura S, and Mizuno N (1996) Immunohistochemical localization of mu-opioid receptors in the central nervous system of the rat. *Journal of Comparative Neurology* 367: 375–402.

Dockray GJ (1976) Immunochemical evidence of cholecystokinin-like peptides in brain. *Nature* 264: 568–570.

Dohi K, Ripley B, Fujiki N, et al. (2008) CSF orexin-A/hypocretin-1 concentrations in patients with intracerebral hemorrhage (ICH). *Regulatory Peptides* 145: 60–64.

Dötsch J, Adelmann M, Englaro P, et al. (1997) Relation of leptin and neuropeptide Y in human blood and cerebrospinal fluid. *Journal of Neurological Sciences* 151: 185–188.

du Vigneaud V, Ressler C, Swan JM, et al. (1953) The synthesis of an octapeptide amide with the hormonal activity of oxytocin. *Journal of the American Chemical Society* 75: 4879–4880.

Dubois MP (1980) Immunocytochemistry of polypeptide hormones: A review. *Acta Histochemica* 22: 141–177.

Eberle AN (1988) *The Melatropins: Chemistry, Physiology and Mechanisms of Action*. Basel, Switzerland: Basel and Karger.

Ebrier K and Singewald N (2006) The role of substance P in stress and anxiety responses. *Amino acids* 31: 251–272.

Edwards GL and Ritter RC (1981) Ablation of the area postrema causes exaggerated consumption of preferred foods in the rat. *Brain Research* 216: 265–276.

Eijsbouts AM and Murphy EP (1999) The role of the hypothalamic–pituitary–adrenal axis in rheumatoid arthritis. *Bailliere's Best Practice and Research. Clinical Rheumatology*. 13: 599–613.

Elde R and Hokfelt T (1979) Localization of hypophysiotropic peptides and other biologically active peptides within the brain. *Annual Review of Physiology* 41: 587–602.

Eng J, Li HR, and Yalow RS (1990) Purification of bovine cholecystokinin-58 and sequencing of its N-terminus. *Regulatory Peptides* 30: 15–19.

Ensom MH (2000) Gender-based differences and menstrual cycle-related changes in specific diseases: Implications for pharmacotherapy. *Pharmacotherapy* 20: 523–539.

Epelbaum J, Dournaud P, Fodor M, and Viollet C (1994) The neurobiology of somatostatin. *Critical Reviews in Neurobiology* 8: 25–44.

Erb K, Pechstein B, Schueler A, Engel J, and Hermann R (2000) Pituitary and gonadal endocrine effects and pharmacokinetics of the novel luteinizing hormone-releasing hormone antagonist teverelix in healthy men – a first-dose-in-humans study. *Clinical Pharmacology and Therapeutics* 67: 660–669.

Evans CJ, Keith DE, Jr., Morrison H, Magendzo K, and Edwards RH (1992) Cloning of a delta opioid receptor by functional expression. *Science* 258: 1952–1955.

Faglia G (1998) The clinical impact of the thyrotropin-releasing hormone test. *Thyroid* 8: 903–908.

Fan W, Boston BA, Kesterson RA, Hruby VJ, and Cone RD (1997) Role of melanocortinergic neurons in feeding and the agouti obesity syndrome. *Nature* 385: 165–168.

Farhi J, West C, Patel A, and Jacobs HS (1996) Treatment of anovulatory infertility: The problem of multiple pregnancy. *Human Reproduction* 11: 429–434.

Fehrenbacher JC, Taylor CP, and Vasko MR (2003) Pregabalin and gabapentin reduce release of substance P and CGRP from rat spinal tissues only after inflammation or activation of protein kinase C. *Pain* 105: 133–141.

Fekete C, Sarkar S, Rand WM, Harney JW, Emerson CH, Bianco AC, and Lechan RM (2002) Agouti-related protein (AGRP) has a central inhibitory action on the hypothalamic–pituitary–thyroid (HPT) axis: Comparisons between the effect of AGRP and neuropeptide Y on energy homeostatsis and the HPT axis. *Endocrinology* 143: 3846–3853.

Felberbaum RE, Ludwig M, and Diedrich K (2000) Clinical application of GnRH-antagonists. *Molecular and Cellular Endocrinology* 166: 9–14.

Ferrari W (1958) Behavioral changes in animals after intracisternal injection with adrenocorticotrophic hormone and melanocyte stimulating hormone. *Nature* 181: 925–926.

Fodor M, Kordon C, and Epelbaum J (2006) Anatomy of the hypophysiotropic somatostatinergic and growth hormone-releasing hormone system minireview. *Neurochemical Research* 31: 137–143.

France RD, Urban B, Krishnan RR, Bissett G, Banki CM, Nemeroff C, and Speilman FJ (1988) CSF corticotropin-releasing factor-like immunoactivity in chronic pain patients with and without major depression. *Biological Psychiatry* 23: 86–88.

Freeman EW, Sondheimer SJ, and Rickels K (1997) Gonadotropin-releasing hormone agonist in the treatment of premenstrual symptoms with and without ongoing dysphoria: A controlled study. *Psychopharmacology Bulletin* 33: 303–309.

Furmark T, Appel L, Michelgård A, et al. (2005) Cerebral blood flow changes after treatment of social phobia with the neurokinin-1 antagonist GR205171, citalopram, or placebo. *Biological Psychiatry* 58: 132–142.

Gantz I, Miwa H, Konda Y, et al. (1993) Molecular cloning, expression, and gene localization of a fourth melanocortin receptor. *Journal of Biological Chemistry* 268: 15174–15179.

Gantz I, Shimoto Y, Konda Y, Miwa H, Dickinson CJ, and Yamada T (1994) Molecular cloning, expression, and characterization of a fifth melanocortin receptor. *Biochemical and Biophysical Research Communications* 200: 1214–1220.

Garbutt JC, Kranzler HR, O'Malley SS, et al. (2005) Efficacy and tolerability of long-acting injectable naltrexone for alcohol dependence: A randomized controlled trial. *Journal of the American Medical Association* 293: 1617–1625.

Gardner JD and Jensen RT (1984) Cholecystokinin receptor antagonists. *American Journal of Physiology* 246: G471–G476.

Garfinkel PE, Brown GM, Stancer HC, and Moldofsky H (1975) Hypothalamic–pituitary function in anorexia nervosa. *Archives of General Psychiatry* 32: 739–744.

Geracioti TD, Jr., Baker DG, Kasckow JW, et al. (2008) Effects of trauma-related audiovisual stimulation on cerebrospinal fluid norepinephrine and corticotropin-releasing hormone concentrations in post-traumatic stress disorder. *Psychoneuroendocrinology* 33: 416–424.

Geracioti TD, Jr., Carpenter LL, Owens MJ, et al. (2006) Elevated cerebrospinal fluid substance P concentrations in post-traumatic stress disorder and depression. *American Journal of Psychiatry* 163: 637–643.

Geracioti TD, Jr., Ekhator NN, Nicholson WE, Arndt S, Loosen PT, and Orth DN (1999) Intra- and inter-individual correlations between cholecystokinin and corticotropin-releasing hormone concentrations in human cerebrospinal fluid. *Depression and Anxiety* 10: 77–80.

Geracioti TD, Jr. and Liddle RA (1988) Impaired cholecystokinin secretion in bulimia nervosa. *New England Journal of Medicine* 319: 683–688.

Geracioti TD, Jr., Loosen PT, and Orth DN (1997) Low cerebrospinal fluid corticotropin-releasing hormone concentrations in eucortisolemic depression. *Biological Psychiatry* 42: 166–174.

Geracioti TD, Jr., Nicholson WE, Orth DN, Ekhator NN, and Loosen PT (1993) Cholecystokinin in human cerebrospinal fluid: Concentrations, dynamics, molecular forms, and relationship to fasting and feeding in health, depression, and alcoholism. *Brain Research* 629: 260–268.

Geracioti TD, Jr., Orth DN, Ekhator NN, Blumenkopf B, and Loosen PT (1992) Serial cerebrospinal fluid corticotropin-releasing hormone concentrations in healthy and depressed humans. *Journal of Clinical Endocrinology and Metabolism* 74: 1325–1330.

Ghigo E, Arvat E, and Camanni F (1998) Orally active growth hormone secretagogues: State of the art and clinical perspective. *Annals of Medicine* 30: 159–168.

Gianotti L, Pincelli AL, Scacchi M, et al. (2000) Effects of recombinant human insulin-like growth factor I administration on spontaneous and growth hormone (GH)-releasing hormone-stimulated GH secretion in anorexia nervosa. *Journal of Clinical Endocrinology and Metabolism* 85: 2805–2809.

Gibbs J, Young RC, and Smith GP (1973) Cholecystokinin elicts satiety in rats with open gastric fistulas. *Nature* 245: 323–325.

Ginsberg SD, Hof PR, Young WG, and Morrison JH (1994) Noradrenergic innervation of vasopressin- and oxytocin-containing neurons in the hypothalamic paraventricular nucleus of the macaque monkey: Quantitative analysis using double-label immunohistochemistry and confocal laser microscopy. *Journal of Comparative Neurology* 341: 476–491.

Glazener CM and Evans JH (2000) Desmopressin for nocturnal enuresis in children. *Cochrane Database of Systematic Reviews* 2: CD002112.

Goddard AW, Woods SW, Money R, et al. (1999) Effects of the CCK(B) antagonist CI-988 on responses to mCPP in generalized anxiety disorder. *Psychiatry Research* 85: 225–240.

Gold PE and Delanoy RL (1981) ACTH modulation of memory storage processing. In: Martizner JL, Jr., Jensen RA, Messing RB, et al. (eds.) *Endogenous Peptides and Learning and Memory Processes*, pp. 79–98. New York: Academic Press.

Gold PW, Kaye W, Robertson GL, and Ebert M (1983) Abnormalities in plasma and cerebrospinal-fluid arginine vasopressin in patients with anorexia nervosa. *New England Journal of Medicine* 308: 1117–1123.

Gold PW, Chrousos G, Kellner C, et al. (1984) Psychiatric implications of basic and clinical studies with corticotropin-releasing factor. *American Journal of Psychiatry* 141: 619–627.

Gold PW, Goodwin FK, and Chrousos GP (1988) Clinical and biochemical manifestation of depression: Relation to the neurobiology of stress (part II of II parts). *New England Journal of Medicine* 319: 413–420.

Gold PW, Licinio J, Wong ML, and Chrousos GP (1995) Corticotropin releasing hormone in the pathophysiology of melancholic and atypical depression and in the mechanism of action of antidepressant drugs. *Annals of the New York Academy of Sciences* 771: 716–729.

Gold PW, Loriaux DL, Roy A, et al. (1986) Responses to corticotropin-releasing hormone in the hypercortisolism of depression and Cushing's disease. Pathophysiologic and diagnostic implications. *New England Journal of Medicine* 314: 1329–1335.

Goodwin TM and Zograbyan A (1998) Oxytocin receptor antagonists. Update. *Clinics in Perinatology* 25: 859–871.

Gouldson P, Legoux P, Carillon C, et al. (2000) The agonist SR 146131 and the antagonist SR 27897 occupy different sites on the human CCK(1) receptor. *European Journal of Pharmacology* 400: 185–194.

Griffon N, Mignon V, Facchinetti P, Diaz J, Schwartz JC, and Sokoloff P (1994) Molecular cloning and characterization of the rat fifth melanocortin receptor. *Biochemical and Biophysical Research Communications* 200: 1007–1014.

Gross BA, Mindea SA, Pick AJ, Chandler JP, and Batjer HH. (2007) Medical management of Cushing disease. *Neurosurgical Focus* 23: E10.

Gross P and Palm C (2000) The treatment of hyponatraemia using vasopressin antagonists. *Experimental Physiology* 85: S253–S257.

Guardia J, Caso C, Arias F, et al. (2002) A double-blind, placebo-controlled study of naltrexone in the treatment of alcohol-dependence disorder: Results from a multicenter clinical trial. *Alcoholism, Clinical and Experimental Research* 26: 1381–1387.

Guastella AJ, Mitchell PB, and Dadds MR (2008) Oxytocin increases gaze to the eye region of human faces. *Biological Psychiatry* 63: 3–5.

Gubler U, Chua AO, Hoffman BJ, Collier KJ, and Eng J (1984) Cloned cDNA to cholecystokinin mRNA predicts an identical preprocholecystokinin in pig brain and gut. *Proceedings of the National Academy of Sciences of the United States of America* 81: 4307–4310.

Guillemin R (2004) Neuroendocrine basis of human disease. *Annals of the New York Academy of Sciences* 1038: 131–137.

Guillemin R, Brazeau P, Bohlen P, Esch F, Ling N, and Wehrenberg WB (1982) Growth hormone-releasing factor from a human pancreatic tumor that caused acromegaly. *Science* 218: 585–587.

Guiloff RJ and Eckland DJ (1987) Observations on the clinical assessment of patients with motor neuron disease. Experience with a TRH analogue. *Neurologic Clinic* 5: 171–192.

Gyetvai K, Hannah ME, Hodnett ED, and Ohlsson A (1999). Tocolytics for preterm labor: A systematic review. *Obstetrics and Gynecology* 94: 869–877.

Hadley ME (2005) Discovery that a melanocortin regulates sexual functions in male and female humans. *Peptides* 26: 1687–1689.

Hadley ME and Haskell-Luevano C (1999) The proopiomelanocortin system. *Annals of the New York Academy of Sciences* 885: 1–21.

Hallschmid M, Benedict C, Born J, Fehm HL, and Kern W (2004) Manipulating central nervous mechanisms of food intake and body weight regulation by intranasal administration of neuropeptides in man. *Physiology and Behavior* 83: 55–64.

Han JS, Ding XZ, and Fan SG (1986) Cholecystokinin octapeptide (CCK-8): Antagonism to electroacupuncture analgesia and a possible role in electroacupuncture tolerance. *Pain* 27: 101–115.

Han JS and Terenius L (1982) Neurochemical basis of acupuncture analgesia. *Annual Review of Pharmacology and Toxicology* 22: 193–220.

Hargreaves R (2002) Imaging substance P receptors (NK1) in the living human brain using positron emission tomography. *Journal of Clinical Psychiatry* 63(supplement 11): 18–22.

Harvey S (1995) Growth hormone release: Mechanisms. In: Harvey S, Scanes CG, and Daughaday WH (eds.) *Growth Hormone*, pp. 87–95. Boca Raton, FL: CRC Press.

Hatton GI, Modney BK, and Salm AK (1992) Increases in dendritic bundling and dye coupling of supraoptic neurons after the induction of maternal behavior. *Annals of the New York Academy of Sciences* 652: 142–155.

Hauser KF, Aldrich JV, Anderson KJ, et al. (2005) Pathobiology of dynorphins in trauma and disease. *Frontiers in Bioscience* 10: 216–235.

Hayashida CY, Gonda RG, Ferrari C, et al. (2000) Familial growth hormone deficiency with mutated GHRH receptor gene: Clinical and hormonal findings in homozygous and heterozygous individuals from Itabaianinha. *European Journal of Endocrinology* 142: 557–563.

Heinemann A, Jocic M, Peskar BM, and Holzer P (1996) CCK-evoked hyperemia in rat gastric mucosa involves neural mechanisms and nitric oxide. *American Journal of Physiology* 270: G253–G258.

Heinonen MV, Purhonen AK, Mäkelä KA, and Herzig KH (2008) Functions of orexins in peripheral tissues. *Acta Physiologica (Oxf)* 192: 471–485.

Hendry SHC (1993) Organization of Neuropeptide Y neurons in the mammalian central nervous system. In: Colmers WF and Wahlestedt C (eds.) *The Biology of Neuropeptide Y and Related Peptides*, pp. 65–135. Totowa, NJ: Humana Press.

Herbison AE (2007) Genetics of puberty. *Hormone Research* 68 (supplement 5): 75–79.

Herkenham M (1987) Mismatches between neurotransmitter and receptor localizations in brain: Observations and implications. *Neuroscience* 23: 1–38.

Hermes ML, Kalsbeek A, Kirsch R, Buijs RM, and Pevet P (1993) Induction of arousal in hibernating European hamsters (*Cricetus cricetus L.*) by vasopressin infusion in the lateral septum. *Brain Research* 631: 313–316.

Herpfer I and Lieb K (2005) Substance P receptor antagonists in psychiatry: Rationale for development and therapeutic potential. *CNS Drugs* 19: 275–293.

Herz A (1997) Endogenous opioid systems and alcohol addiction. *Psychopharmacology (Berl.)* 129: 99–111.

Heuser I, Bissette G, Dettling, M., et al. (1998) Cerebrospinal fluid concentrations of corticotropin-releasing hormone, vasopressin, and somatostatin in depressed patients and healthy controls: Response to amitripyline treatment. *Depression and Anxiety* 8: 71–79.

Higuchi H, Hasegawa A, and Yamaguchi T (2005) Transcriptional regulation of neuronal genes and its effect on neural functions: Transcriptional regulation of neuropeptide Y gene by leptin and its effect on feeding. *Journal of Pharmacological Sciences* 98: 225–231.

Hilal L, Hejaji Y, Vie-Luton MP, et al. (2008) Usual phenotypic features in a patient with a novel splice mutation in the GHRHR gene. *Molecular Medicine* 14: 286–292.

Ho KKY (2007) Consensus guidelines for the diagnosis and treatment of adults with GH deficiency II: A statement of the GH Research Society in association with the European Society for Pediatric Endocrinology, Lawson Wilkins Society, European Society of Endocrinology, Japan Endocrine Society, and Endocrine Society of Australia. *European Journal of Endocrinology* 157: 695–700.

Hoffman P, Jonsdottir IH, and Thoren P (1996) Activation of different opioid systems by muscle activity and exercise. *News in Physiological Sciences* 11: 223–228.

Hokfelt T, Tsuruo Y, Ulfhake B, et al. (1989) Distribution of TRH-like immunoreactivity with special reference to coexistence with other neuroactive compounds. *Annals of the New York Academy of Sciences* 553: 76–105.

Hokfelt T, Cortes R, Schalling M, Ceccatelli S, Pelto-Huikko M, Persson H, and Villar MJ (1991) Distribution patterns of CCK and CCK mRNA in some neuronal and non-neuronal tissues. *Neuropeptides* 19: 31–43.

Hokfelt T, Zhang X, and Wiesenfeld-Hallin Z (1994) Messenger plasticity in primary sensory neurons following axotomy and its functional implications. *Trends in Neurosciences* 17: 22–30.

Holsboer F, Von Bardeleben U, Gerken A, Stalla GK, and Muller OA (1984) Blunted corticotropin and normal corisol response to human corticotropin-releasing factor in depression. *New England Journal of Medicine* 311: 1127.

Hommer DW, Pickar D, Crawley JN, Weingartner H, and Paul SM (1985) The effects of cholecystokinin-like peptides in schizophrenics and normal human subjects. *Annals of the New York Academy of Sciences* 448: 542–552.

Insel TR, Winslow JT, Wang ZX, Young L, and Hulihan TJ (1995) Oxytocin and the molecular basis of monogamy. *Advances in Experimental Medicine and Biology* 395: 227–234.

Inui A (2000) Transgenic approach to the study of body weight regulation. *Pharmacological Reviews* 52: 35–61.

Ishak WW, Berman DS, and Peters A (2008) Male anorgasmia treated with oxytocin. *Journal of Sexual Medicine* 5: 1022–1024.

Janecka A, Staniszewska R, and Fichna J (2007) Endomorphine analogs. *Current Medicinal Chemistry* 14: 3201–3208.

Jard S (1998) Vasopressin receptors: A historical survey. *Advances in Experimental Medicine and Biology* 449: 1–13.

Jolles J (1987) Vasopressin and human behavior. In: Gash DM and Boer GJ (eds.)*Vasopressin: Principles and Properties*, pp. 549–578. New York: Plenum Press.

Jordan J, Greenway FL, Leiter LA, et al. (2008) Stimulation of cholecystokinin-A receptors with GI181771X does not cause weight loss in overweight or obese patients. *Clinical Pharmacology and Therapeutics* 83: 281–287.

Kamilaris TC, Johnson EO, Calogero AE, Kalogeras KT, Bernardini R, Chrousos GP, and Gold PW (1992) Cholecystokinin-octapeptide stimulates hypothalamic–pituitary–adrenal function in rats: Role of corticotropin-releasing hormone. *Endocrinology* 130: 1764–1774.

Karalis K, Mastorakos G, Chrousos GP, and Tolis G (1994) Somatostatin analogues suppress the inflammatory reaction *in vivo*. *Journal of Clinical Investigation* 93: 2000–2006.

Kasckow JW, Baker D, and Geracioti TD, Jr. (2001) Corticotropin-releasing hormone in depression and post-traumatic stress disorder. *Peptides* 22: 845–851.

Kastin AJ, Miller LH, Gonzalez-Barcena D, Hawley WD, et al. (1971) Psycho-physiologic correlates of MSH activity in man. *Physiology and Behavior* 7: 893–896.

Kawai N, Takamatsu Y, Yamamoto H, et al. (1997) Effect of methamphetamine and dopamine receptor antagonists on cholecystokinin-like immunoreactivity in the rat medial prefrontal cortex. *Pharmacology, Biochemistry, and Behavior* 58: 517–523.

Keck ME, Kern N, Erhardt A, et al. (2008) Combined effects of exonic polymorphisms in CRHR1 and AVPR1B genes in a case/control study for panic disorder. *American Journal of Medical Genetics. Part B, Neuropsychiatric Genetics*. Apr 2 (doi: 10.1002/ajmg.b.30750).

Kelly DF, Gonzalo IT, Cohan P, Berman N, Swerdloff R, and Wang C (2000) Hypopituitarism following traumatic brain injury and subarachnoid hemorrhage: A preliminary report. *Journal of Neurosurgery* 93: 743–752.

Kendrick KM and Keverne EB (1992) Control of synthesis and release of oxytocin in the sheep brain. *Annals of the New York Academy of Sciences* 652: 102–121.

Kennedy JL, Bradwejn J, Koszycki D, King N, Crowe R, Vincent J, and Fourie O (1999) Investigation of cholecystokinin system genes in panic disorder. *Molecular Psychiatry* 4: 284–285.

Khawaja AM and Rogers DF (1996) Tachykinins: Receptor to effector. *International Journal of Biochemistry and Cellular Biology* 28: 721–738.

Kirk LF, Jr., Hash RB, Katner HP, and Jones T (2000) Cushing's disease: Clinical manifestations and diagnostic evaluation. *American Family Physician* 62: 1119–1127; 1133–1134.

Kiss JZ, Palkovits M, Zaborszky L, Tribollet E, Szabo D, and Makara GB (1983) Quantitative histological studies on the hypothalamic paraventricular nucleus in rats. II. Number of local and certain afferent nerve terminals. *Brain Research* 265: 11–20.

Kling MA, DeBellis MD, O'Rourke DK, et al. (1994) Diurnal variation of cerebrospinal fluid immunoreactive corticotropin-releasing hormone levels in healthy volunteers. *Journal of Clinical Endocrinology and Metabolism* 79: 233–239.

Kling MA, Geracioti TD, Jr., Licinio J, Michelson D, Oldfield EH, and Gold PW (1994) Effects of electroconvulsive therapy on the CRH–ACTH–cortisol system in melancholic depression: Preliminary findings. *Psychopharmacology Bulletin* 30: 489–494

Kling MA, Roy A, Doran AR, et al. (1991) Cerebrospinal fluid immunoreactive corticotropin-releasing hormone and adrenocorticotropin secretion in Cushing's disease and major depression: Potential clinical implications. *Journal of Clinical Endocrinology and Metabolism* 72: 260–271.

Kling MA, Rubinow PR, Doran AR,et al. (1993) Cerebrospinal fluid immunoreactive somatostatin concentrations in patients with Cushing's disease and major depression: Relationship to indices of corticotropin-releasing hormone and cortisol secretion. *Neuroendocrinology* 57: 79–88.

Klose M, Juul A, Poulsgaard L, Kosteljanetz M, Brennum J, and Feldt-Rasmussen U (2007) Prevalence and predictive factors of post-traumatic hypopituitarism. *Clinical Endocrinology* 67: 193–201.

Knobil E (1989) The electrophysiology of the GnRH pulse generator in the rhesus monkey. *Journal of Steroid Biochemisry* 33: 669–671.

Koob GF and Bloom FE (1988) Cellular and molecular mechanisms of drug dependence. *Science* 242: 715–723.

Kosfeld M, Heinrichs M, Zak PJ, Fischbacher U, and Fehr E (2005) Oxytocin increases trust in humans. *Nature* 435: 67306.

Kovács M, Schally AV, Varga JL, and Zarándi M (2008) Endocrine and antineoplastic actions of growth hormone-releasing hormone antagonists. *Current Medicinal Chemistry* 15: 314–321.

Kramer MS, Cutler NR, Ballenger JC, et al. (1995) A placebo-controlled trial of L-365,260, a CCKB anatagonist, in panic disorder. *Biological Psychiatry* 37: 462–466.

Kramer MS, Cutler N, Feighner J, et al. (1998) Distinct mechanism for antidepressant activity by blockade of central substance P receptors. *Science* 281:1640–1645.

Kramer MS, Winokur A, Kelsey J, et al. (2004) Demonstration of the efficacy and safety of a novel substance P (NK1) receptor antagonist in major depression. *Neuropsychopharmacology* 29: 385–392.

Krause JE, MacDonald MR, and Takeda Y (1989) The polyprotein nature of substance P precursors. *Bioessays* 10 (2–3): 62–69.

Kreek MJ (2000) Methadone-related opioid agonist pharmacotherapy for heroin addiction. History, recent molecular and neurochemical research and future in mainstream medicine. *Annals of the New York Academy of Sciences* 909: 186–216.

Krieger DT, Liotta AS, Brownstein MJ, and Zimmerman EA (1980) ACTH, beta-lipotropin, and related peptides in brain, pituitary, and blood. *Recent Progress in Hormone Research* 36: 277–344.

Krishnan KRR (2002) Clinical experience with substance P receptor (NK1) antagonists in depression. *Journal of Clinical Psychiatry* 63(supplement 11): 25–29.

Kritzer MF, Innis RB, and Goldman-Rakic PS (1987) Regional distribution of cholecystokinin receptors in primate cerebral

cortex determined by *in vitro* receptor autoradiography. *Journal of Comparative Neurology* 263: 418–435.
Krivoy W and Guillemin R (1961) On a possible role of b-melanocyte stimulating hormone (b-MSH) in the central nervous system of mammalia. *Endocrinology* 69: 170–175.
Krude H, Biebermann H, Luck W, Horn R, Brabant G, and Gruters A. (1998) Severe early-onset obesity, adrenal insufficiency, and red hair pigmentation caused by POMC mutatins in humans. *Nature Genetics* 19:155–157.
Krüger TH, Schiffer B, Eikermann M, Haake P, Gizewski E, and Schedlowski M (2006) Serial neurochemical measurement of cerebrospinal fluid during the human sexual response cycle. *European Journal of Neuroscience* 24: 3445–3452.
Kruhlich L, Dhariwal A, and McCann S (1969) Stimulatory and inhibitory effects of purified hypothalamic extracts on growth hormone release from rat pituitary *in vitro*. *Endocrinology* 83: 783–790.
Lança AJ, De Cabo C, Arifuzzaman AI, and Vaccarino FJ (1998) Cholecystokinergic innervation of nucleus accumbens subregions. *Peptides* 19: 859–868.
Landgraf R, Neumann I, Holsboer F, and Pittman QJ (1995) Interleukin-1 beta stimulates both central and peripheral release of vasopressin and oxytocin in the rat. *European Journal of Neuroscience* 7: 592–598.
Laron Z (2002) Effects of growth hormone and insulin-like growth factor 1 deficiency on ageing and longevity. *Novartis Foundation Symposium* 242: 125–137.
Laszlo F, Pavo I, Szepes Z, Varga CS, and Laszlo FA (1998) Deleterious action of vasopressin in gastroduodenal ulceration: Experimental and clinical observations. *Scandinavian Journal of Gastroenterology* 228: 62–67.
Laufer MR and Rein MS (1993) Treatment of abnormal uterine bleeding with gonadotropin-releasing hormone analogues. *Clinical Obstetrics and Gynecology* 36: 668–678.
Lazarus JH (1998) The effects of lithium therapy on thyroid and thyrotropin-releasing hormone. *Thyroid* 8: 909–913.
Lee R, Geracioti TD, Jr., Kasckow JW, and Coccaro EF (2005) Childhood trauma and personality disorder: Positive correlation with adult CSF corticotropin-releasing factor concentrations. *American Journal of Psychiatry* 162: 995–997.
Legros JJ and Geenen V (1996) Neurophysins in central diabetes insipidus. *Hormone Research* 45: 182–186.
Lena I, Simon H, Roques BP, and Dauge V (1999) Opposing effects of two CCK(B) agonists on the retrieval phase of a two-trial memory task after systemic injection in the rat. *Neuropharmacology* 38: 543–553.
Leng G, Caquineau C, and Sabatier N (2005) Regulation of oxytocin secretion. *Vitamins and Hormones* 71:27–58.
Leng G, Luckman SM, Dyball RE, Hamamura M, and Emson PC (1993) Induction of c-fos in magnocellular neurosecretory neurons. A link between electrical activity and peptide synthesis? *Annals of the New York Academy of Sciences* 689: 133–145.
Lewis RV, Stern AS, Kimura S, Rossier J, Stein S, and Udenfriend S (1980) An about 50,000-dalton protein in adrenal medulla: A common precursor of [Met]- and [Leu] enkephalin. *Science* 208: 1459–1461.
Liddle RA (1994a) Regulation of cholecystokinin gene expression in rat intestine. *Annals of the New York Academy of Sciences* 713: 22–31.
Liddle RA (1994b) Cholecystokinin. In: Walsh JH and Dockray GJ (eds.) *Gut Peptides: Biochemistry and Physiology*, pp. 175–216. New York: Raven Press.
Liddle RA, Gertz BJ, Kanayama S, Beccaria L, Coker LD, Turnbull TA, and Morita ET (1989) Effects of a novel cholecystokinin (CCK) receptor antagonist, MK-329, on gallbladder contraction and gastric emptying in humans. Implications for the physiology of CCK. *Journal of Clinical Investigation* 84: 1220–1225.
Liddle RA, Goldfine ID, Rosen MS, Taplitz RA, and Williams JA (1985) Cholecystokinin bioactivity in human plasma: Molecular forms, responses to feeding, and relationship to gallbladdar contraction. *Journal of Clinical Investigation* 69: 218–225.
Liebowitz MR, Quitkin FM, Stewart JW,et al. (1984) Phenelzine v imipramine in atypical depression. A preliminary report. *Archives of General Psychiatry* 41: 669–677.
Lines C, Challenor J, and Traub M (1995) Cholecystokinin and anxiety in normal volunteers: An investigation of the anxiogenic properities of pentagastrin and reversal by the cholecystokinin receptor subtype B antagonist L-365,260. *British Journal of Clinical Pharmacology* 39: 235–242.
Liu H, Bravata DM, Olkin I, et al. (2008) Systematic review: The effects of growth hormone on athletic performance. *Annals of Internal Medicine* 148: 747–758.
Lolait SJ, O'Carroll AM, Mahan LC, et al. (1995) Extrapituitary expression of the rat V1b vasopressin receptor gene. *Proceedings of the National Academy of Sciences of the United States of America* 92: 6783–6787.
Loosen PT, Chambliss B, Debold CR, Shelton R, and Orth DN (1992) Psychiatric phenomenology of Cushing's disease. *Pharmacopsychiatry* 25: 192–198.
Low CM, Black JW, Broughton HB, et al. (2000) Development of peptide 3D structure mimetics: Rational design of novel peptoid cholecystokinin receptor antagonists. *Journal of Medicinal Chemistry* 43: 3505–3517.
Lu L, Louie D, and Owyang C (1989) A cholecystokinin releasing peptide mediates feedback regulation of pancreatic secretion. *American Journal of Physiology* 256: G430–G435.
Luckman SM (1995) Fos expression within regions of the preoptic area, hypothalamus and brainstem during pregnancy and parturition. *Brain Research* 669: 115–124.
Ma XM and Aguilera G (1999) Differential regulation of corticotropin-releasing hormone and vasopressin transcription by glucocorticoids. *Endocrinology* 140: 5642–5650.
Macaluso A, McCoy D, Ceriani G, Watanabe T, Biltz J, Catania A, and Lipton JM (1994) Antiinflammatory influences of alpha-MSH molecules: Central neurogenic and peripheral actions. *Journal of Neuroscience* 14: 2377–2382.
MacNeil DJ (2007) NPY Y1 and Y5 receptor selective antagonists as anti-obesity drugs. *Current Topics in Medicinal Chemistry* 7: 1721–1733.
Maheshwari HG, Silverman BL, Dupuis J, and Baumann G (1998) Phenotype and genetic analysis of a syndrome caused by an inactivating mutation in the growth hormone-releasing hormone receptor: Dwarfism of Sindh. *Journal of Clinical Endocrinology and Metabolism* 83: 4065–4074.
Mansour A, Burke S, Pavlic RJ, Akil H, and Watson SJ (1996) Immunohistochemical localization of the cloned kappa 1 receptor in the rat CNS and pituitary. *Neuroscience* 71: 671–690.
Mansour A, Fox CA, Akil H, and Watson SJ (1995) Opioid-receptor mRNA expression in the rat CNS: Anatomical and functional implications. *Trends in Neurosciences* 18: 22–29.
Mansour A, Khachaturian H, Lewis ME, Akil H, and Watson SJ (1987) Autoradiographic differentiation of mu, delta, and kappa opioid receptors in the rat forebrain and midbrain. *Journal of Neuroscience* 7: 2445–2464.
Marangell LB, George MS, Callahan AM, et al. (1997) Effects of intrathecal thyrotropin-releasing hormone (protirelin) in refractory depressed patients. *Archives of General Psychiatry* 54: 214–222.

Martin-Martinez M, De La Figuera N, Latorre M, et al. (2000) Beta-turned dipeptoids as potent and selective CCK(1) receptor antagonists. *Journal of Medicinal Chemistry* 43: 3770–3777.

Marumo K, Baba S, and Murai M (1999) Erectile function and nocturnal penile tumescence in patients with prostate cancer undergoing luteinizing hormone-releasing hormone agonist therapy. *International Journal of Urology* 6: 19–23.

Mattes JA, Hom W, Rochford JM, and Orlosky M (1985) Ceruletide for schizophrenia: A double-blind study. *Biological Psychiatry* 20: 533–538.

McCarthy MM and Altemus M (1997) Central nervous system actions of oxytocin and modulation of behavior in humans. *Molecular Medicine Today* 3: 269–275.

McCarthy MM, Kleopoulos SP, Mobbs CV, and Pfaff DW (1994) Infusion of antisense oligodeoxynucleotides to the oxytocin receptor in the ventromedial hypothalamus reduces estrogen-induced sexual receptivity and oxytocin receptor binding in the female rat. *Neuroendocrinology* 59: 432–440.

Mantyh PW (2002) Neurobiology of substance P and the NK1 receptor. *Journal of Clinical Psychiatry* 63: 6–10.

Mayinger B and Hensen J (1999) Nonpeptide vasopressin antagonists: A new group of hormone blockers entering the scene. *Experimental and Clinical Endocrinology and Diabetes* 107: 157–165.

McCleane GJ (1998) The cholecystokinin antagonist proglumide enhances the analgesic efficacy of morphine in humans with chronic benign pain. *Anesthesia and Analgesia* 87: 1117–1120.

Melis MR and Argiolas A (1995) Nitric oxide donors induce penile erection and yawning when injected in the central nervous system of male rats. *European Journal of Pharmacology* 294: 1–9.

Meurs A, Clinckers R, Ebinger G, Michotte Y, and Smolders I (2007) Clinical potential of neuropeptide Y receptor ligands in the treatment of epilepsy. *Current Topics in Medicinal Chemistry* 7: 1660–1674.

Mignot E, Lammers GJ, Ripley B, et al. (2002) The role of cerebrospinal fluid hypocretin measurement in the diagnosis of narcolepsy and other hypersomnias. *Archives of Neurology* 59: 1553–1562.

Minato T, Tochigi M, Kato N, and Sasaki T (2007) Association study between the cholecystokinin A receptor gene and schizophrenia in the Japanese population. *Psychiatric Genetics* 17: 117–119.

Miyasaka K, Guan DF, Liddle RA, and Green GM (1989) Feedback regulation by trypsin: Evidence for intraluminal CCK-releasing peptide. *American Journal of Physiology* 257: G175–G181.

Mizuno K and Matsuo H (1994) Processing of peptide hormone precursors. In: Imura H (ed.) *The Pituitary Gland*, pp. 153–178. New York: Raven Press.

Molchan SE, Hill JL, Martinez RA, et al. (1993) CSF somatostatin in Alzheimer's disease and major depression: Relationship to hypothalamic–pituitary–adrenal axis and clinical measures. *Psychoneuroendocrinology* 18: 509–519.

Mollerau C, Simons MJ, Soularue P, Liners F, Vassart G, Meunier JC, and Parmentier M (1996) Structure, tissue distribution and chromosomal localization of the prepronociceptin gene. *Proceedings of the National Academy of Science of the United States of America* 93: 8666–8670.

Moran TH, Ameglio PJ, Peyton HJ, Schwartz GJ, and McHugh PR (1993) Blockade of type A, but not type B, CCK receptors postpones satiety in rhesus monkeys. *American of Journal of Physicology* 265: R620–R624.

Morgan CA, 3rd, Wang S, Southwick SM, Rasmusson A, Hazlett G, Hauger RL, and Charney DS (2000) Plasma neuropeptide-Y concentrations in humans exposed to military survival training. *Biological Psychiatry* 47: 902–909.

Mountjoy KG, Mortrud MT, Low MJ, Simerly RB, and Cone RD (1994) Localization of the melanocortin-4 receptor (MC4-R) in neuroendocrine and autonomic control circuits in the brain. *Molecular Endocrinology* 8: 1298–1308.

Mountjoy KG, Robbins LS, Mortrud MT, and Cone RD (1992) The cloning of a family of genes that encode the melanocortin receptors. *Science* 257: 1248–1251.

Munsat TL, Lechan R, Taft JM, Jackson IM, and Reichlin S (1989) TRH and diseases of the motor system. *Annals of the New York Academy of Sciences* 553: 388–398.

Mutt V and Jorpes JE (1968) Structure of porcine cholecystokinin-pancreozymin. 1. Cleavage with thrombin and with trypsin. *European Journal of Biochemistry* 6: 156–162.

Nair NP, Bloom DM, and Nestoros JN (1982) Cholecystokinin appears to have antipsychotic properties. *Progress in Neuro-Psychopharmacology and Biological Psychiatry* 6: 509–512.

Nakazawa T, Ikeda M, Kaneko T, and Yamatsu K (1985) Analgesic effects of dynorphin-A and morphine in mice. *Peptides* 6: 75–78.

Nam SY, Kratzsch J, Kim KW, Kim KR, Lim SK, and Marcus C (2001) Cerebrospinal fluid and plasma concentrations of leptin, NPY, and alpha-MSH in obese women and their relationship to negative energy balance. *Journal of Clinical Endocrinology and Metabolism* 86: 4849–4853.

Nemeroff CB, Bissette G, Akil H, and Fink M (1991) Neuropeptide concentrations in the cerebrospinal fluid of depressed patients treated with electroconvulsive therapy: Corticotropin-releasing factor, β-endorphin, and somatostatin. *British Journal of Psychiatry* 158: 59–63.

Nemeroff CB, Widerlov E, Bissette G, et al. (1984) Elevated concentrations of CSF corticotropin-releasing factor-like immunoreactivity in depressed patients. *Science* 226: 1342–1344.

Netchine I, Talon, P., Dastot, F., Vitaus, F., Goossens, M. and Amselem, S. (1998). Extensive phenotypic analysis of a family with growth hormone (GH) deficiency caused by a mutation in the GH-releasing hormone receptor gene. *Journal of Clinical Endocrinology and Metabolism* 83: 432–436.

Nikisch G, Agren H, Eap CB, Czernik A, Baumann P, and Mathé AA (2005) Neuropeptide Y and corticotropin-releasing hormone in CSF mark response to antidepressive treatment with citalopram. *International Journal of Neuropsychopharmacology* 8: 403–410.

Nikisch G and Mathé AA (2008) CSF monoamine metabolites and neuropeptides in depressed patients before and after electroconvulsive therapy. *European Psychiatry* (doi:10.1016/j.eurpsy.2008.03.003).

Nishino S (2003) The hypocretin/orexin system in health and disease. *Biological Psychiatry* 54: 87–95

Olson GA, Olson RD, and Kastin AJ (1995) Endogenous opiates: 1994. *Peptides* 16: 1517–1555.

Olszewski PK and Levine AS (2004) Characterization of influence of central nociceptin/orphanin FQ on consummatory behavior. *Endocrinology* 145: 2627–2632.

O'Malley SS (1995) Current strategies for the treatment of alcohol dependence in the United States. *Drug and Alcohol Dependence* 39: S3–S7.

Orth DN (1995) Cushing's syndrome. *New England Journal of Medicine* 332: 791–803.

Owyang C, Louie DS, and Tatum D (1986) Feedback regulation of pancreatic enzyme secretion. Suppression of cholecystokinin release by trypsin. *Journal of Clinical Investigation* 77: 2042–2047.

Pande AC, Greiner M, Adams JB, Lydiard RB, and Pierce MW (1999) Placebo-controlled trial of the CCK-B antagonist, CI-988, in panic disorder. *Biological Psychiatry* 46: 860–862.

Pasternak GW (1988) Multiple morphine and enkephalin receptors and the relief of pain. *Journal of the American Medical Association* 259: 1362–1367.

Patel YC (1987) Somatostatin. In: Ludecke DK and Tolis G (eds.) *Growth Hormone, Growth Factors and Acromegaly*, pp. 21–36. New York: Raven Press.

Patel L and Lindley C (2003) Aprepitant – a novel NK1-receptor antagonist. *Expert Opinion on Pharmacotherapy* 4: 2279–2296.

Pedersen CA, Caldwell JD, Drago F, Noonan LR, Peterson G, Hood LE, and Prange AJ, Jr. (1988) Grooming behavioral effects of oxytocin. Pharmacology, ontogeny, and comparisons with other nonapeptides. *Annals of the New York Academy of Sciences* 525: 245–256.

Pedersen CA, Caldwell JD, Johnson MF, Fort SA, and Prange AJ, Jr. (1985) Oxytocin antiserum delays onset of ovarian steroid-induced maternal behavior. *Neuropeptides* 6: 175–182.

Pedersen CA, Caldwell JD, Peterson G, Walker CH, and Mason GA (1992) Oxytocin activation of maternal behavior in the rat. *Annals of the New York Academy of Sciences* 652: 58–69.

Peng C, Gallin W, Peter RE, Blomqvist AG, and Larhammar D (1994) Neuropeptide-Y gene expression in the goldfish brain: Distribution and regulation by ovarian steroids. *Endocrinology* 134: 1095–1103.

Perlmutter LS, Tweedle CD, and Hatton GI (1984) Neuronal/glial plasticity in the supraoptic dendritic zone: Dendritic bundling and double synapse formation at parturition. *Neuroscience* 13: 769–779.

Perrin MH and Vale WW (1999) Corticotropin releasing factor receptors and their ligand family. *Annals of the New York Academy of Sciences* 885: 312–328.

Pert CB and Snyder SH (1973) Opiate receptor: Demonstration in nervous tissue. *Science* 179: 1011–1014.

Peselow E, Angrist B, Sudilovsky A, Corwin J, Siekierski J, Trent F, and Rotrosen J (1987) Double blind controlled trials of cholecystokinin octapeptide in neuroleptic-refractory schizophrenia. *Psychopharmacology* (*Berl.*) 91: 80–84.

Pitts AF, Samuelson SD, Meller WH., Bissette G, Nemeroff CB, and Kathol RG (1995) Cerebrospinal fluid corticotropin-releasing hormone, vasopressin, and oxytocin concentrations in treated patients with major depression and controls. *Biological Psychiatry* 38: 330–335.

Prange AJ, Jr. (1996) Novel uses of thyroid hormones in patients with affective disorders. *Thyroid* 6: 537–543.

Prange AJ, Jr., Loosen PT, Wilson IC, and Lipton MA (1984) The therapeutic use of hormones of the thyroid axis in depression. In: Post RM and Ballenger JC (eds.) *Neurobiology of Mood Disorders*, pp. 311–322. Baltimore, MD: Williams and Wilkins.

Rasmusson AM, Hauger RL, Morgan CA, Bremner JD, Charney D, and Southwick SM (2000) Low baseline and yohimbine-stimulated plasma neuropeptide Y (NPY) levels in combat-related PTSD. *Biological Psychiatry* 47: 526–539.

Reeve JR, Jr., Eysselein VE, Ho FJ, Chew P, Vigna SR, Liddle RA, and Evans C (1994) Natural and synthetic CCK-58. Novel reagents for studying cholecystokinin physiology. *Annals of the New York Academy of Sciences* 713: 11–21.

Rehfeld JF (2000) Cholecystokinin and panic disorder – three unsettled questions. *Regulatory Peptides* 93: 79–83.

Rehfeld JF, Hansen HF, and Marley PD (1985) Molecular forms of cholecystokinin in the brain and the relationship to neuronal gastrins. *Annals of the NewYork Academy of Sciences* 448: 11–23.

Rehfeld JF and Kruse-Larsen C (1978) Gastrin and cholecystokinin in human cerebrospinal fluid. Immunochemical determinations of concentrations and molecular heterogeneity. *Brain Research* 155: 19–26.

Rehman HU (2002) Role of CRH in the pathogenesis of dementia of Alzheimer's type and other dementias. *Current Opinion in Investigational Drugs* 3: 1637–1642.

Reinscheid RK, Ardati A, Monsma FJ, Jr., and Civelli O (1996) Structure-activity relationship studies on the novel neuropeptide orphanin FQ. *Journal of Biological Chemistry* 271: 14163–14168.

Reiser LW and Reiser MF (1985) Endocrine disorders. In: Kaplan HI and Sadock BJ (eds.) *Comprehensive Textbook of Psychiatry*, 4th edn., pp. 1171. Baltimore, MD: Williams and Wilkins.

Reisine T and Bell GI (1993) Molecular biology of opioid receptors. *Trends in Neurosciences* 16: 506–510.

Reum T, Schafer U, Marsden CA, Fink H, and Morgenstern R (1997) Cholecystokinergic increases extracellular dopamine overflow in the anterior nucleus accumbens via CCK (B) receptors in the VTA assessed by *in vivo* voltammetry. *Neuropeptides* 31: 82–88.

Ripley B, Overeem S, Fujiki N,et al. (2001) CSF hypocretin/orexin levels in narcolepsy and other neurological conditions. *Neurology* 57: 2253–2258.

Roca CA, Su TP, Elpern S, McFarland H, and Rubinow DR (1999) Cerebrospinal fluid somatostatin, mood, and cognition in multiple sclerosis. *Biological Psychiatry* 46: 551–556.

Roselli-Rehfuss L, Mountjoy KG, Robbins LS, et al. (1993) Identification of a receptor for gamma melanotropin and other proopiomelanocortin peptides in the hypothalamus and limbic system. *Proceedings of the National Academy of Sciences of the United States of America* 90: 8856–8860.

Rosler A and Witztum E (1998) Treatment of men with paraphilia with a long-acting analogue of gonadotropin-releasing hormone. *New England Journal of Medicine* 338: 416–422.

Rubin RT, Rhodes ME, and Czambel RK (2003) Plasma leptin suppression by arginine vasopressin in normal women and men. *Life Sciences* 72: 1209–1220.

Rubin RT, Sekula LK, O'Toole S, Rhodes ME, and Czambel RK (1999) Pituitary–adrenal cortical responses to low-dose physostigmine and arginine vasopressin administration in normal women and men. *Neuropsychopharmacology* 20: 434–446.

Rubinow DR, Roca CA, Schmidt PJ, et al. (2005) Testosterone suppression of CRH-stimulated cortisol in men. *Neuropsychopharmacology* 30: 1906–1912.

Rupniak NM (2002) New insights into the antidepressant actions of substance P (NK1 receptor) antagonists. *Canadian Journal of Physiology and Pharmacology* 80: 489–494.

Safarinejad MR (2008) Evaluation of the safety and efficacy of bremelanotide, a melanocortin receptor agonist, in female subjects with arousal disorder: A double-blind placebo-controlled, fixed dose, randomized study. *Journal of Sexual Medicine* 5: 887–897.

Safarinejad MR and Hosseini SY (2008) Salvage of sildenafil failures with bremelanotide: A randomized, double-blind, placebo controlled study. *Journal of Urology* 179: 1066–1071.

Sah R, Ekhator NN, Strawn JR, Sallee FR, Horn PS, Baker DG, and Geracioti TD, Jr. (2009) Low cerebrospinal fluid neuropeptide Y concentrations in posttraumatic stress disorder. *Biological Psychiatry (accepted)*.

Sakurai T, Amemiya A, Ishil M, et al. (1998) Orexins and orexin receptors: A family of hypothalamic neuropeptides and G protein-coupled receptors that regulate feeding behavior. *Cell* 92: 573–585.

Salomon RM, Ripley B, Kennedy JS, Johnson B, Schmidt D, and Zeitzer (2003) Diurnal variation of cerebrospinal fluid hypocretin-1 (orexin-A) levels in control and depressed subjects. *Biological Psychiatry* 54: 96–104.

Salvatori R, Hayashida CY, Aguiar-Oliveira MH, et al. (1999) Familial dwarfism due to a novel mutation of the growth hormone-releasing hormone receptor gene. *Journal of Clinical Endocrinology and Metabolism* 84: 917–923.

Sandman CA and Kastin AJ (1977) Pitutary peptide influences on attention and memory. In: Drucker-Colin and McGaugh JL (eds.) *Neurobiology of Sleep and Memory*, pp. 347–360. New York: Academic Press.

Sandman CA, Miller LH, Kastin AJ, and Schally AV (1972) Neuroendocrine influence on attention and memory. *Journal of Comparative and Physiological Psychology* 80: 54–58.

Sattin A (1998) A heuristic model of mental depression derived from basic and applied research on thyrotropin-releasing hormone. *Thyroid* 8: 957–962.

Scacchi M, Pincelli Al, Caumo A, Tomasi P, Delitala G, Baldi G, and Cavagnini F (1997) Spontaneous nocturnal growth hormone secretion in anorexia nervosa *Journal of Clinical Endocrinology and Metabolism* 82: 3225–3229.

Schally AV, Arimura A, Baba Y, Nair RM, Matsuo H, Redding TW, and Debeljuk L (1971) Isolation and properties of the FSH and LH-releasing hormone. *Biochemical and Biophysical Research Communications* 43: 393–399.

Schally AV and Varga JL (1999) Antagonist analogs of growth hormone-releasing hormone: New potential antitumor agents. *Trends in Endocrinology and Metabolism* 10: 383–391.

Schiffmann SN and Vanderhaeghen JJ (1991) Distribution of cells containing mRNA encoding cholecystokinin in the rat central nervous system. *Journal of Comparative Neurology* 304: 219–233.

Schmale H, Fehr S, and Richter D (1987) Vasopressin biosynthesis – from gene to peptide hormone. *Kidney International Supplement* 2: S8–S13.

Schmauss C and Yaksh TL (1984) *In vivo* studies on spinal opiate receptor systems mediating antinociception. II. Pharmacological profiles suggesting a differential association of mu, delta and kappa receptors with visceral chemical and cutaneous thermal stimuli in the rat. *Journal of Pharmacology and Experimental Therapeutics* 228: 1–12.

Schober JM, Kuhn PJ, Kovacs PG, Earle JH, Byrne PM, and Fries RA (2005) Leuprolide acetate suppresses pedophilic urges and arousability. *Archives of Sexual Behavior* 34: 691–705.

Schoenen J, Lotstra F, Vierendeels G, Reznik M, and Vanderhaeghen JJ (1985) Substance P, enkephalins, somatostatin, cholecystokinin, oxytocin, and vasopressin in human spinal cord. *Neurology* 35: 881–890.

Seeley RJ, Drazen DL, and Clegg DJ (2004) The critical role of the melanocortin system in the control of energy balance. *Annual Review of Nutrition* 24: 133–149.

Senaris RM, Schindler M, Humphrey PP, and Emson PC (1995) Expression of somatostatin receptor 3 mRNA in the motorneurones of the rat spinal cord, and the sensory neurones of the spinal ganglia. *Brain Research. Molecular Brain Research* 29: 185–190.

Shadiack AM, Sharma SD, Earle DC, Spana C, and Hallam TJ (2007) Melanocortins in the treatment of male and female sexual dysfunction. *Current Topics in Medicinal Chemistry* 7: 1137–1144.

Sharp FR, Sagar SM, Hicks K, Lowenstein D, and Hisanaga K (1991) c-fos mRNA, Fos, and Fos-related antigen induction by hypertonic saline and stress. *Journal of Neuroscience* 11: 2321–2331.

Shibahara S, Morimoto Y, Furutani Y, et al. (1983). Isolation and sequence analysis of the human corticotropin-releasing factor precursor gene. *EMBO Journal* 2: 775–779.

Shippenberg TS and Elmer GI (1998) The neurobiology of opiate reinforcement. *Critical Reviews in Neurobiology* 12: 267–303.

Shvartzman P (2000) Pharmacological treatment of cancer pain. *Israel Medical Association Journal* 2: 536–539.

Silva AP, Xapelli S, Grouzmann E, and Cavadas C (2005) The putative neuroprotective role of neuropeptide Y in the central nervous system. *Current Drug Targets. CNS and Neurological Disorders*. 4: 331–347.

Singh L, Field MJ, Hunter JC, Oles RJ, and Woodruff GN (1996) Modulation of the *in vivo* actions of morphine by the mixed CCKA/B receptor antagonist PD 142898. *European Journal of Pharmacology* 307: 283–289.

Smith RG (2005) Development of growth hormone secretagogues. *Endocrine Reviews* 26: 346–360.

Solomon S (1999) POMC-derived peptides and their biological action. *Annals of the New York Academy of Sciences* 885: 22–40.

Spencer JD and Schallreuter KU (2008) Regulation of pigmentation in human epidermal melanocytes by functional high affinity {beta}-MSH/MC4-R signalling. *Endocrinology* (doi:10.1210/en.2008-1212).

Steckler T and Holsboer F (1999) Corticotropin-releasing hormone receptor subtypes and emotion. *Biological Psychiatry* 46: 1480–1508.

Stefano GB, Scharrer B, Smith EM, et al. (1996) Opioid and opiate immunoregulatory processes. *Critical Reviews in Immunology* 16; 109–144.

Stein MB, Hauger RL, Dhalla KS, Chartier MJ, and Asmundson GJ (1996) Plasma neuropeptide Y in anxiety disorders: Findings in panic disorder and social phobia. *Psychiatry Research* 59: 183–188.

Stinus L, Nadaud D, Jauregui J, and Kelley AE (1986) Chronic treatment with five different neuroleptics elicits behavioral supersensitivity to opiate infusion into the nucleus accumbens. *Biological Psychiatry* 21: 34–48.

Stockmeier CA, Shi X, Konick L, et al. (2002) Neurokinin-1 receptors are decreased in major depressive disorder. *Neuroreport* 13: 1223–1227.

Stonesifer LD (2008) Mild traumatic brain injury in U.S. soldiers returning from Iraq. *New England Journal of Medicine* 358: 2178 (letter).

Stout SC, Owens MJ, and Nemeroff CB (2001) Neurokinin(1) receptor antagonists as potential antidepressants. *Annual Review of Pharmacology and Toxicology* 41: 877–906.

Strand FL (2000) David and Goliath – the slingshot that started the neuropeptide revolution. *European Journal of Pharmacology* 405: 3–12.

Strand FL (1999) Tachykinins: Substance P, neurokinin A and neurokinin B. In: Stevens CF (ed.) *Neuropeptides: Regulators of Physiological Processes*, pp. 383–430. Cambridge, MA: MIT Press.

Stratakis CA and Chrousos GP (1995) Neuroendocrinology and pathophysiology of the stress system. *Annals of the New York Academy of Sciences* 771: 1–18.

Strawn JR, Dowling BP, and Geracioti TD, Jr. (2008) Pregabalin treatment of posttraumatic stress disorder. *Journal of Clinical Psychopharmacology* 28: 596–597.

Strawn JR and Geracioti TD, Jr. (2007) The treatment of generalized anxiety disorder with pregabalin, an atypical anxiolytic. *Neuropsychiatric Disease and Treatment* 3: 237–243.

Strawn JR, Pyne-Geithman GJ, Ekhator NN, et al. (under review) Low cerebrospinal fluid and plasma orexin-A (hypocretin-1) concentrations in combat-related posttraumatic stress disorder.

Sundler F, Bottcher G, Ekblad E, and Hakanson R (1993) PP, PYY and NPY: Occurence and distribution in the periphery. In: Colmers WF and Wahlestedt C (eds.) *The Biology of Neuropeptide Y and Related Peptides*, pp. 170–183. Totowa, NJ: Humana Press.

Swaab DF, Roozendaal B, Ravid R, Velis DN, Gooren L, and Williams RS (1987) Suprachiasmatic nucleus in aging, Alzheimer's disease, transsexuality and Prader–Willi syndrome. *Progress in Brain Research* 72: 301–310.

Swift RM (1995) Effect of naltrexone on human alcohol consumption. *Journal of Clinical Psychiatry* 56: 24–29.

Szuba MP, Amsterdam JD, Fernaando AT, 3rd, Gary KA, Whybrow PC, and Winokur A (2005) Rapid antidepressant response after nocturnal TRH administration in patients with bipolar type I and bipolar type II major depression. *Journal of Clinical Psychopharmacology* 25: 325–330.

Tachikawa H, Harada S, Kawanishi Y, Okubo T, and Shiraishi H (2000) Novel polymorphisms of the human cholecystokinin A receptor gene: An association analysis with schizophrenia. *American Journal of Medical Genetics* 96: 141–145.

Taghzouti K, Lena I, Dellu F, Roques BP, Dauge V, and Simon H (1999) Cognitive enhancing effects in young and old rats of pBC264, a selective CCK(B)receptor agonist. *Psychopharmacology (Berl.)* 143: 141–149.

Takeshi S (2006) Orexin and orexin receptors. In: Nishino S and Sakurai T (eds.) *The Orexin/Hypocretin System: Physiology and Pathophysiology*, pp. 13–20. Totowa, NJ: Humana Press Inc.

Takeuchi T, Gumucio DL, Yamada T, et al. (1986) Genes encoding pancreatic polypeptide and neuropeptide Y are on human chromosomes 17 and 7. *Journal of Clinical Investigation* 77: 1038–1041.

Tanriverdi F, Ulutabanca H, Unluhizarci K, Selcuklu A, Casanueva FF, Kelestimur F (2008) Three years prospective investigation of anterior pituitary function after traumatic brain injury: A pilot study. *Clinical Endocrinology (Oxf)* 68(4): 573–579.

Tapanainen P, Rantala H, Leppäluoto J, Lautala P, Käär ML, and Knip M (1989) Nocturnal release of immunoreactive growth hormone-releasing hormone and growth hormone in normal children. *Pediatric Research* 26: 404–409.

Tatemoto K, Carlquist M, and Mutt V (1982) Neuropeptide Y – a novel brain peptide with structural similarities to peptide YY and pancreatic polypeptide. *Nature* 296: 659–660.

Tatro JB (1990) Melanotropin receptors in the brain are differentially distributed and recognize both corticotropin and alpha-melanocyte stimulating hormone. *Brain Research* 536: 124–132.

Tatro JB (1996) Receptor biology of the melanocortins, a family of neuroimmunomodulatory peptides. *Neuroimmunomodulation* 3: 259–284.

Thibaut F, Cordier B, and Kuhn JM (1996) Gonadotropin hormone releasing hormone agoist in cases of severe paraphilia: A lifetime treatment? *Psychoneuroendocrinology* 21: 411–419.

Thody AJ (1999) Alpha-MSH and the regulation of melanocyte function. *Annals of the New York Academy of Sciences* 885: 217–229.

Toirac I, Sanjuán J, Aguilar EJ, et al. (2007) Association between CCK-AR gene and schizophrenia with auditory hallucinations. *Psychiatric Genetics* 17: 47–53.

Toppila J, Asikainen M, Alanko L, Turek FW, Stenberg D, and Porkka-Heiskanen T (1996) The effect of REM sleep deprivation on somatostatin and growth hormone-releasing hormone gene expression in the rat hypothalamus. *Journal Sleep Research* 5: 115–122.

Trainer PJ, Drake WM, Katznelson L, et al. (2000) Treatment of acromegaly with the growth hormone-receptor antagonist pegvisomant. *New England Journal of Medicine* 342; 1171–1177.

Tribollet E, Dubois-Dauphin M, Dreifuss JJ, Barberis C, and Jard S (1992) Oxytocin receptors in the central nervous system. Distribution, development, and species differences. *Annals of the New York Academy of Sciences* 652: 29–38.

Turner RA, Altemus M, Enos T, Cooper B, and McGuinness T (1999) Preliminary research on plasma oxytocin in normal cycling women: Investigating emotion and interpersonal distress. *Psychiatry* 62: 97–113.

Uhde TW, Tancer ME, Rubinow DR, et al. (1992) Evidence for hypothalamo-growth hormone dysfunction in panic disorder: Profile of growth hormone (GH) responses to clonidine, yohimbine, caffeine, glucose, GRF and TRH in panic disorder patients versus healthy volunteers. *Neuropsychopharmacology* 6: 101–118.

Ulm RR, Volpicelli JR, and Volpicelli LA (1995) Opiates and alcohol self-administration in animals. *Journal of Clinical Psychiatry* 56: 5–14.

Ursini A, Capelli AM, Carr RA, et al. (2000) Synthesis and SAR of new 5-phenyl-3-ureido-1,5-benzodiazepines as cholecystokinin-B receptor antagonists. *Journal of Medicinal Chemistry* 43: 3596–3613.

Vaisse C, Clement K, Durand E, Hercberg S, Guy-Grand B, and Froguel P (2000) Melanocortin-4 receptor mutations are a frequent and heterogeneous cause of morbid obesity. *Journal of Clinical Investigation* 106: 253–262.

Vale W, Spiess J, Rivier C, and Rivier J (1981) Characterization of a 41-residue ovine hypothalamic peptide that stimulates secretion of corticotropin and beta-endorphin. *Science* 213: 1394–1397.

Van Poppel H and Nilsson S (2008) Tesetosterone surge: Rationale for gonadotropin-releasing hormone blockers? *Urology* 71: 1001–1006.

Van Ree JM, Hijman R, Jolles J, and De Wied D (1985) Vasopressin and related peptides: Animal and human studies. *Progress in Neuro-Psychopharmacology and Biological Psychiatry* 9: 551–559.

van Tijn DA, Schroor EJ, Delemarre-van de Waal HA, de Viflder JJM, and Vulsma T (2007) Early assessment of hypothalamic–pituitary–gonadal function in patiens with congenital hypothyroidism of central origin. *Journal of Clinical Endocrinology and Metabolism* 92: 104–109.

Vance ML and Thorner MO (1988) Some clinical considerations of growth hormone releasing hormone. *Frontiers in Neuroendocrinology* 10: 279–294.

Vecsei L and Widerlov E (1990) Preclinical and clinical studies with cysteamine and pantethine related to the central nervous system. *Progress in Neuro-Psychopharmacology and Biological Psychiatry* 14: 835–862.

Veldhuis JD, Iranmanesh A, and Weltman A (1997) Elements in the pathophysiology of diminished growth hormone (GH) secretion in aging humans. *Endocrine* 7: 41–48.

Vepsäläinen S, Helisalmi S, Koivisto AM, Tapaninen T, Hiltunen M, and Soininen H (2007) Somatostatin genetic variants modify the risk for Alzheimer's disease among Finnish patients. *Journal of Neurology* 254: 1504–1508.

Vitiello MV, Moe KE, Merriam GR, Mazzoni G, Buchner DH, and Schwartz RS (2006) Growth hormone releasing hormone improves the cognition of healthy older adults. *Neurobiology of Aging* 27: 318–323.

Volpicelli JR, Volpicelli LA, and O'Brien CP (1995) Medical management of alcohol dependence: Clinical use and limitations of naltrexone treatment. *Alcohol* 30: 789–798.

Von Euler US and Gaddum JH (1931) An unidentified depressor substance in certain tissue extracts. *Journal of Physiology* 72: 74–87.

Vythilingam M, Anderson GM, Owens MJ, et al. (2000) Cerebrospinal fluid corticotropin-releasing hormone in healthy humans: effects of yohimbine and naloxone. *Journal of Clinical Endocrinology and Metabolism* 85: 4138–4145.
Walter E and Kissel T (1994) Transepithelial transport and metabolism of thyrotropin-releasing hormone (TRH) in monolayers of a human intestinal cell line (Caco-2): Evidence for an active transport component? *Pharmaceutical Research* 11: 1575–1580.
Walum H, Westberg L, Henningsson S, et al. (2008) Genetic variation in the vasopressin receptor 1a gene (AVPR1A) associates with pair-bonding behavior in humans. *Proceedings of the National Academy of Sciences of the United States of America* 105: 14153–14156.
Wang SS, Kamphuis W, Juitinga I, Zhou JN, and Swaab DF (2008) Gene expression analysis in the human hypothalamus in depression by laser microdissection and real-time PCR: The presence of multiple receptor imbalances. *Molecular Psychiatry* 13: 786–799.
Wank SA (1995) Cholecystokinin receptors. *American Journal of Physiology* 269: G628–G646.
Warnock JK and Bundren JC (1997) Anxiety and mood disorders associated with gonadotropin-releasing hormone agonist therapy. *Psychopharmocology Bulletin* 33: 311–316.
Webster EL, Torpy DJ, Elenkov IJ, and Chrousos GP (1998) Corticotropin-releasing hormone and inflammation. *Annals of the New York Academy of Sciences* 840: 21–32.
Westrin A, Ekman R, and Träskman-Bendz L (1999) Alterations of corticotropin releasing hormone (CRH) and neuropeptide Y (NPY) plasma levels in mood disorder patients with a recent suicide attempt. *European Neuropsychopharmacology* 9: 205–211.
White JM and Irvine RJ (1999) Mechanisms of fatal opioid overdose. *Addiction* 94: 961–972.
Wolkewitz OM, Gertz B, Weingartner H, Beccaria L, Thompson K, and Liddle RA (1990) Hunger in humans induced by MK-329, a specific peripheral-type cholecystokinin receptor antagonist. *Biological Psychiatry* 28: 169–173.
Wong M-L, Kling MA, Munson PJ, et al. (2000) Pronounced and sustained central hypernoradrenergic function in major depression with melancholic features: Relation to hypercortisolism and corticotropin-releasing hormone. *Proceedings of the National Academy of Sciences of the United States of America* 97: 325–330.
Wunderlich GR, DeSousa NJ, and Vaccarino FJ (2000) Cholecystokinin modulates both the development and the expression of behavioral sensitization to amphetamine in the rat. *Psychopharmacology (Berl)* 151: 283–290.
Yehuda R, Brand S, and Yang RK (2006) Plasma neuropeptide Y concentrations in combat exposed veterans: Relationship to trauma exposure, recovery from PTSD, and coping. *Biological Psychiatry* 59: 660–663.
You ZB, Tzschentke TM, Brodin E, and Wise RA (1998) Electrical stimulation of the prefrontal cortex increases cholecystokinin, glutamate, and dopamine release in the nucleus accumbens: An *in vivo* microdialysis study in freely moving rats. *Journal of Neuroscience* 18: 6492–6500.
Zadina JE, Hackler L, Ge LJ, and Kastin AJ (1997) A potent and selective endogenous agonist for the mu-opiate receptor. *Nature* 386: 499–502.
Zadina JE, Martin-Schild S, Gerall AA, Kastin AJ, Hackler L, Ge LJ, and Zhang X (1999) Endomorphins: Novel endogenous mu-opiate receptor agonists in regions of high mu-opiate receptor density. *Annals of the New York Academy of Sciences* 897: 136–144.
Zeng N and Sachs G (1998) Properties of isolated gastric enterochromaffin-like cells. *Yale Journal of Biology and Medicine* 71: 233–246.
Zingg HH, Grazzini E, Breton C, et al. (1998) Genomic and non-genomic mechanisms of oxytocin receptor regulation. *Advances in Experimental Medicine and Biology* 449: 287–295.
Zingg HH, Rozen F, Breton C, et al. (1995) Gonadal steroid regulation of oxytocin and oxytocin receptor gene expression. *Advances in Experimental Medicine Biology* 395: 395–404.
Zobel AW, Nickel T, Kunzel HE, Ackl N, Sonntag A, Ising M, and Holsboer F (2000) Effects of the high-affinity corticotropin-releasing hormone receptor 1 antagonist R121919 in major depression: The first 20 patients treated. *Journal of Psychiatric Research* 34: 171–181.
Zoli M, Jansson A, Sykova E, Agnati LF, and Fuxe K (1999) Volume transmission in the CNS and its relevance for neuropsychopharmacology. *Trends in Pharmacological Sciences* 20: 142–150.

Further Reading

Alves SE, Akbari HM, Azmitia EC, and Strand FL (1993) Neonatal ACTH and corticosterone alter hypothalamic monoamine innervation and reproductive parameters in the female rat. *Peptides* 14: 379–384.
Argente J, Chowen JA, Zeitler P, Clifton DK and Steiner RA (1991) Sexual dimorphism of growth hormone-releasing hormone and somatostatin gene expression in the hypothalamus of the rat during development. *Endocrinology* 128: 2369–2375.
Banks WA and Kastin AJ (1988) Interactions between the blood–brain barrier and endogenous peptides: Emerging clinical implications. *American of Journal of the Medical Sciences* 295: 459–465.
Beckwith BE, Sandman CA, Hothersall D, and Kastin AJ (1977) Influence of neonatal injections of alpha-MSH on learning, memory and attention in rats. *Physiology and Behavior* 18: 63–71.
Beglinger C, Hildebrand P, Meier R, et al. (1992) A physiological role for cholecystokinin as a regulator of gastrin secretion. *Gastroenterology* 103: 490–495.
Bonavera JJ, Kalra PS, and Kalra SP (1996) L-arginine/nitric oxide amplifies the magnitude and duration of the luteinizing hormone surge induced by estrogen: Involvement of neuropeptide Y. *Endocrinology* 137: 1956–1962.
Burgus R, Dunn TF, Desiderio D, Ward DN, Vale W, and Guillemin R (1970) Characterization of ovine hypothalamic hypophysiotropic TSH-releasing factor. *Nature* 226: 321–325.
Clasadonte J, Poulain P, Beauvillain JC, and Prevot V (2008) Activation of neuronal nitric oxide release inhibits spontaneous firing in adult gonadotropin-releasing hormone neurons: A possible local synchronizing signal. *Endocrinology* 149: 587–596.
Clement-Jones V, Corder R, Smith R, Medbak S, Lowry PJ, Rees LH, and Besser GM (1982) Met-enkephalin and related peptides in man. *Advances in Biochemical Psychopharmacology* 33: 379–386.
Conn PM and Bowers CY (1996) A new receptor for growth hormone-release peptide. *Science* 273: 923.
Conn PM and Venter JC (1985) Radiation inactivation (target size analysis) of the gonadotropin-releasing hormone receptor: Evidence for a high molecular weight complex. *Endocrinology* 116: 1324–1326.
Dores RM, Steveson TC, and Price ML (1993) A view of the N-acetylation of alpha-melanocyte-stimulating hormone and beta-endorphin from a phylogenetic perspective. *Annals of the New York Academy of Sciences* 680: 161–174.

Drake CT, Terman GW, Simmons ML, Milner TA, Kunkel DD, Schwartzkroin PA, and Chavkin C (1994) Dynorphin opioids present in dentate granule cells may function as retrograde inhibitory neurotransmitters. *Journal of Neuroscience* 14: 3736–3750.

Ehrnborg C, Bengtsson BA, and Rosen T (2000) Growth hormone abuse. *Bailliere's Best Practice and Research. Clinical Endocrinology and Metabolism* 14: 71–77.

Hokfelt T, Pernow B, and Wahren J (2001) Substance P: A pioneer amongst neuropeptides. *Journal of Internal Medicine* 249: 27–40.

Fink G (1988) Gonadotropin secretion and its control. In: Knobil E and Neill JD (eds.) *The Physiology of Reproduction*, pp. 1349–1377. New York: Raven Press.

Frieboes RM, Murck H, Maier P, Schier T, Holsboer F, and Steiger A (1995) Growth hormone-releasing peptide-6 stimulates sleep, growth hormone, ACTH and cortisol release in normal man. *Neuroendocrinology* 61: 584–589.

Fritz MA and Speroff L (1982) The endocrinology of the menstrual cycle: The interaction of folliculogenesis and neuroendocrine mechanisms. *Fertility and Sterility* 38: 509–529.

Frohman LA, Downs TR, Heimer, EP, and Felix AM (1989) Dipeptidylpeptidase IV and trypsin-like enzymatic degradation of human growth hormone-releasing hormone in plasma. *Journal of Clinical Investigation* 83: 1533–1540.

Genazzani AR, Bernardi F, Monteleone P, Luisi S, and Luisi M (2000) Neuropeptides, neurotransmitters, neurosteroids, and the onset of puberty. *Annals of the New York Academy of Sciences* 900: 1–9.

Goffin V, Binart N, Clement-Lacroix P, et al. (1999) From the molecular biology of prolactin and its receptor to the lessons learned from knockout mice models. *Genetic Analysis* 15: 189–201.

Guldenaar SE and Swabb DF (1995) Estimation of oxytocin mRNA in the human paraventricular nucleus in AIDS by means of quantitative *in situ* hybridization. *Brain Research* 700: 107–114.

Healy ML, Gibney J, Russel-Jones DL, Pentecost C, Croos P, Sonksen PH, and Umpleby AM (2003) High dose growth hormone exerts an anabolic effect at rest and during exercise in endurance-trained athletes. *Journal of Clinical Endocrinology and Metabolism* 88: 5221–5226.

Hill K, Bucuvalas J, McClain C, Kryscio R, Martini RT, Alfaro MP, and Maloney M (2000) Pilot study of growth hormone administration during the refeeding of malnourished anorexia nervosa patients. *Journal of Child and Adolescent Psychopharmacology* 10: 3–8.

Hoffman AR and Crowley WF, Jr. (1982) Induction of puberty in men by long-term pulsatile administration of low-dose gonadotropin-releasing hormone. *New England Journal of Medicine* 307: 1237–1241.

Janson ET and Oberg K (1999) Somatostatin receptor ligands and their use in the treatment of endocrine disorders. *Current Pharmaceutical Design* 5: 693–705.

King JA, Hassan MF, Mehl AE, and Millar RP (1988) Gonadotropin-releasing hormone molecular forms in mammalian hypothalamus. *Endocrinology* 122: 2742–2752.

Kiss JZ, Mezey E, Cassell MD, Williams TH, Mueller GP, O'Donohue TL, and Palkovits M (1985) Topographical distribution of pro-opiomelanocortin-derived peptides (ACTH/beta-END/alpha-MSH) in the rat median eminence. *Brain Research* 329: 169–176.

Lam KS, Lee MF, Tam SP, and Srivastava G (1996) Gene expression of the receptor for growth-hormone-releasing hormone is physiologically regulated by glucocorticoids and estrogen. *Neuroendocrinology* 63: 475–480.

Lamberts SW, Hofland LJ, de Herder WW, Kwekkeboom DJ, Reubi JC, and Krenning EP (1993) Octreotide and related somatostatin analogs in the diagnosis and treatment of pituitary disease and somatostatin receptor scintigraphy. *Frontiers in Neuroendocrinology* 14: 27–55.

Lechan RM, Wu P, Jackson IM, Wolf H, Cooperman S, Mandel G, and Goodman RH (1986) Thyrotropin-releasing hormone precursor: Characterization in rat brain. *Science* 231: 159–161.

Lin XW, Otto CJ, and Peter RE (1998) Evolution of neuroendocrine peptide systems: Gonadotropin-releasing hormone and somatostatin. *Comparative Biochemistry and Physiology. Part C, Pharmacology, Toxicology and Endocrinology* 119: 375–388.

Lloyd KC (1994) Gut hormones in gastric function. *Bailliere's Clinical Endocrinology and Metabolism* 8: 111–136.

Lovejoy DA and Balment RJ (1999) Evolution and physiology of the corticotropin-releasing factor (CRF) family of neuropeptides in vertebrates. *General and Comparative Endocrinology* 115: 1–22.

Maeda K, Tsukamura H, Ohkura S, Kawakami S, Nagabukuro H, and Yokoyama A (1995) The LHRH pulse generator: A mediobasal hypothalamic location. *Neuroscience and Biobehavioral Reviews* 19: 427–437.

Mason AJ, Pitts SL, Nikolics K, Szonyi E, Wilcox JN, Seeburg PH, and Stewart TA (1986) The hypogonadal mouse: Reproductive functions restored by gene therapy. *Science* 234: 1372–1378.

Mathias S, Held K, Ising M, Weikel JC, Yassouridis A, Steiger A (2007) Systemic growth hormone-releasing hormone (GHRH) impairs sleep in healthy young women. *Psychoneuroendocrinology* 32: 1021–1027.

Muglia L, Jacobson L, Dikkes P, and Majzoub JA (1995) Corticotropin-releasing hormone deficiency reveals major fetal but not adult glucocorticoid need. *Nature* 373: 427–432.

Nair RM, Barrett JF, Bowers CY, and Schally AV (1970) Structure of porcine thyrotropin releasing hormone. *Current Therapeutic Research, Clinical and Experimental* 12: 394–401.

Newman CB (1999) Medical therapy for acromegaly. *Endocrinology and Metabolism Clinics of North America* 28: 171–190.

Nikolics K, Mason AJ, Szonyi E, Ramachandran J, and Seeburg PH (1985) A prolactin-inhibiting factor within the precursor for human gonadotropin-releasing hormone. *Nature* 316: 511–517.

Ortmann O and Diedrich K (1999) Pituitary and extrapituitary actions of gonadotrophin-releasing hormone and its analogues. *Human Reproduction* 14: 194–206.

Owens MJ and Nemeroff CB (1991) Physiology and pharmacology of corticotropin-releasing factor. *Pharmacological Reviews* 43: 425–473.

Owyang C (1996) Physiological mechanisms of cholecystokinin action on pancreatic secretion. *American Journal of Physiology* 271: G1–G7.

Patel YC and Galanopoulou A (1995) Processing and intracellular targeting of prosomatostatin-derived peptides: The role of mammalian endoproteases. *Ciba Foundation Symposium* 190: 26–40.

Perone MJ, Murray CA, Brown OA, et al. (1998) Procorticotrophin-releasing hormone: Endoproteolytic processing and differential release of its derived peptides within AtT20 cells. *Molecular and Cellular Endocrinology* 142: 191–202.

Plotnikoff NP and Kastin AJ (1976) Neuropharmacological tests with alpha-melanocyte stimulating hormone. *Life Sciences* 18: 1217–1222.

Qian A, Wang W, and Sanborn BM (1998) Evidence for the involvement of several intracellular domains in the coupling of oxytocin receptor to G alpha(q/11). *Cellular Signalling* 10: 101–105.

Ristic H and Isaac L (1994) Pharmacological characterization of dynorphin A (1–17)-induced effects on spinal cord-evoked potentials. *Journal of Pharmacology and Experimental Therapeutics* 270: 534–539.

Rivier J, Spiess J, Thorner M, and Vale W (1982) Characterization of a growth hormone-releasing factor from a human pancreatic islet tumour. *Nature* 300: 276–278.

Rodgers RJ and Johnson NJ (1995) Cholecystokinin and anxiety: Promises and pitfalls. *Critical Reviews in Neurobiology* 9: 345–369.

Salomon Y (1990) Melanocortin receptors: Targets for control by extracellular calcium. *Molecular and Cellular Endocrinology* 70: 139–145.

Sato M and Takahara J (1997) Gene regulation of growth hormone-releasing hormone and its receptor. *Endocrine Journal* 44: 765–774.

Sevarino KA, Goodman RH, Spiess J, Jackson IM, and Wu P (1989) Thyrotropin-releasing hormone (TRH) precursor processing. Characterization of mature TRH and non-TRH peptides synthesized by transfected mammalian cells. *Journal of Biological Chemistry* 264: 21529–21535.

Severini C, Improta G, Falconieri-Espamer G, Saalvadori S, and Erspamer V (2002) The tachykinin peptide family. *Pharmacological Reviews* 54: 285–322.

Sherwood NM (1976) Gonadotropin-releasing hormone surge in pro-oestrous rats. *Nature* 264: 461–463.

Sherwood NM, Lovejoy DA, and Coe IR (1993) Origin of mammalian gonadotropin-releasing hormones. *Endocrine Reviews* 14: 241–254.

Sherwood NM and Parker DB (1990) Neuropeptide families: An evolutionary perspective. *Journal of Experimental Zoology* 4: 63–71.

Shivers BD, Harlan RE, Morrell JI, and Pfaff DW (1983) Absence of oestradiol concentration in cell nuclei of LHRH-immunoreactive neurones. *Nature* 304: 345–347.

Silverman AJ (1988) The gonadotropin-releasing hormone (GnRH) neuronal systems: Immunocytochemistry. In: Knobil E and Neill JD (eds.) *The Physiology of Reproduction*, pp.1283–1304. New York: Raven Press.

Smith GP and Gibbs J (1985) The satiety effect of cholecystokinin. Recent progress and current problems. *Annals of the New York Academy of Sciences* 448: 417–423.

Tachikawa H, Harada S, Kawanishi Y, Okubo T, and Shiraishi H (1999) Novel polymorphism in the promoter and coding regions of the human cholecystokinin B receptor gene: An association analysis with schizophrenia. *American Journal of Medical Genetics* 88: 700–704.

Thorner MO and Vance ML (1988) Growth hormone. *Journal of Clinical Investigation* 82; 745–747.

Vaccarino FJ, Sovran P, Baird JP, and Ralph MR (1995) Growth hormone-releasing hormone mediates feeding-specific feedback to the suprachiasmatic circadian clock. *Peptides* 16: 595–598.

Vale W, Vaughan J, and Perrin MH (1997) Corticotropin-releasing factor (CRF) family of ligands and their receptors. *The Endocrinologist* 7: 3S–9S.

van der Helm-Hylkema H and de Wied D (1976) Effect of neonatally injected ACTH and ACTH analogues on eye-opening of the rat. *Life Sciences* 18: 1099–1104.

Van Megen HJ, den Boer JA, and Westenberg HG (1994) On the significance of cholecystokinin receptors in panic disorder. *Progress in Neuro-Psychopharmacology and Biological Psychiatry* 18: 1235–1246.

Vanecek J and Klein DC (1995) Melatonin inhibition of GnRH-induced LH release from neonatal rat gonadotroph: Involvement of $Ca2^{+}$ not cAMP. *American Journal of Physiology* 269: E85–E90.

Wang JB, Imai Y, Eppler CM, Gregor P, Spivak CE, and Uhl GR (1993) Mu opiate receptor: cDNA cloning and expression. *Proceedings of the National Academy of Sciences of the United States of America* 90: 10230–10234.

Wang PS, Huang SW, Tung YF, et al. (1994b) Interrelationship between thyroxine and estradiol on the secretion of thyrotropin-releasing hormone and dopamine into hypophysial portal blood in ovariectomized-thyroidectomized rats. *Neuroendocrinology* 59: 202–207.

17 Melatonin Actions in the Brain

A J Lewy, J Emens, J Songer, and J Rough, Oregon Health and Science University, Portland, OR, USA

Chapter Outline

Glossary

circadian phase position The clock time or circadian time of a specified point of a circadian rhythm.

circadian time (CT) The internal body clock time of a circadian, often chosen in relation to the 10 pg ml^{-1} plasma or 3 pg ml^{-1} saliva DLMO indicating circadian time (CT) 14.

DLMO Dim light melatonin onset, which is the time when melatonin levels rise in plasma or saliva and is the best marker for circadian phase position in humans.

PAD Phase angle difference, which is the time interval between circadian phase positions of two circadian rhythms, for example, between the DLMO and the midpoint of sleep.

zeitgeber Time giver, or circadian phase resetting stimulus.

zeitgeber time (ZT) The time interval between the time of a zeitgeber (light exposure, which is usually waketime, or time of melatonin administration) and the phase of a circadian rhythm (which is often the DLMO).

17.1 Hormones

Hormones are biological substances that function as messengers in the body. They are secreted by highly differentiated glandular cells and then travel via the bloodstream (or diffusion) to adjacent or distant sites where they have specific effects. Their basic chemical structures are either of amino acid derivatives or of steroids. Hormones are present in very small concentrations, and discrete mechanisms have developed to facilitate their actions. They act via specific receptors that can be intracellular, nuclear, or membrane bound. Hormone binding causes a conformational change in the receptor that then triggers a sequence of events unique to the hormone–receptor complex in a particular tissue. These may include stimulation of second messenger production (cAMP and phosphoinositol), activation of protein kinases, or stimulation of DNA transcription. The end result is a change in the functional activity of the target cell.

A key feature of all endocrine systems is that they are under some type of feedback control. In some systems, such as the thyroid axis, a hormone may participate in its own regulation: circulating thyroid hormone downregulates secretion of thyrotropin (TSH) which leads to decreased thyroxine secretion by the thyroid gland. Metabolites can also provide feedback, such as serum glucose which stimulates insulin secretion by the pancreas. Homeostatic conditions such as fluid volume and osmolality can also control hormonal secretion. For example, low circulating volume stimulates the renin–angiotensin–aldosterone system, and both high serum osmolality and low volume stimulate the secretion of vasopressin. In the case of parathyroid hormone, a cation, serum calcium, downregulates its secretion. Input from the nervous system is also an important regulatory mechanism of hormonal secretion.

17.2 Melatonin as a Neurohormone

Melatonin is perhaps more accurately classified as a neurohormone, at least in humans in whom it does not appear to have any reproductive effects. It is an indoleamine synthesized from serotonin in two steps (**Figure 1**). The conversion of serotonin to N-acetylserotonin is the rate-limiting step (Klein and Moore, 1970). Melatonin was first isolated from bovine pineal tissue by Lerner et al. (1959). It is lipophilic, a property that allows it to go just about anywhere in the body and to pass through cell and nuclear membranes.

The main site of melatonin secretion in humans is the pineal gland, although it may also be synthesized for local use in the retina, lacrimal gland, and gut (Pang et al., 1993). Melatonin circulates via the bloodstream (and the cerebrospinal fluid (CSF)) and then binds to specific seven-transmembrane, G-protein-coupled receptors at various sites, of

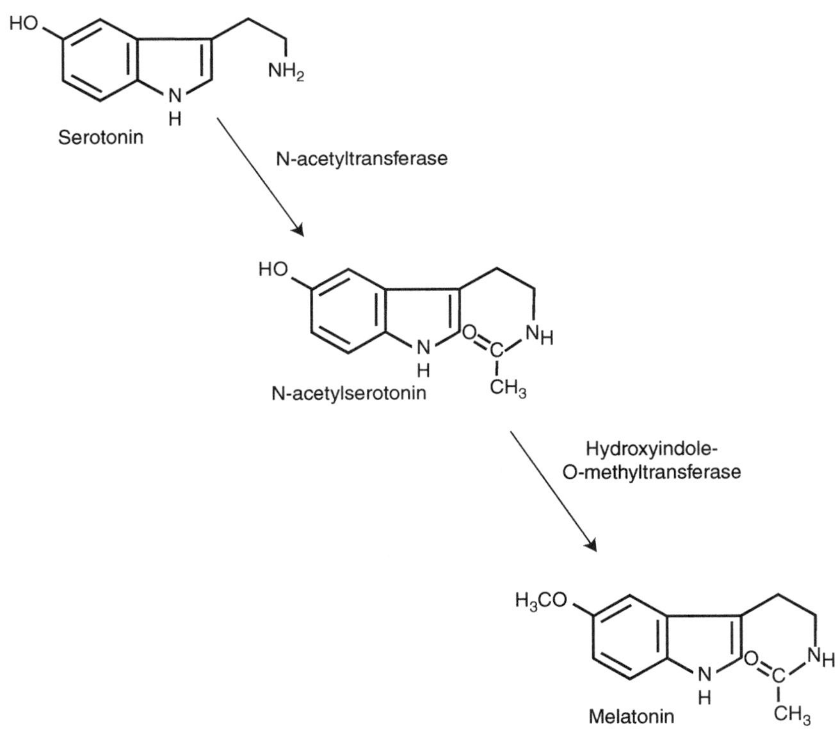

Figure 1 Synthesis of melatonin from serotonin. Adapted from Lewy AJ (1983b) Biochemistry and regulation of mammalian melatonin production. In: Relkin RM (ed.) *The Pineal Gland*, pp. 77–128. New York: Elsevier North-Holland, with permission from Elsevier.

which the suprachiasmatic nucleus (SCN) of the hypothalamus is the best characterized (Dubocovich et al., 1996). A nuclear receptor for melatonin has also been identified (Cardinali et al., 1997).

Melatonin is found diffusely in nature. While its function in vertebrates is the most clearly defined, it is also found in numerous plants and in lower phyla, including bacteria and fungi (Dubbels et al., 1995; Hardeland, 1999; Hattori et al., 1995; Murch et al., 1997). Melatonin production by the pineal is not influenced by circulating melatonin levels (Matsumoto et al., 1997), that is, there is no negative feedback inhibition, which is another reason why melatonin is better classified as a neurohormone in humans, rather than a hormone (which may be a more accurate classification in animals whose reproductive cycles are influenced by melatonin). This unique feature of melatonin is one reason why exogenous melatonin is considered safe.

As will be discussed in Section 17.3, melatonin's participation in the endogenous circadian system is its most clearly defined role. It has been proposed to have many other functions as well, but these are less well characterized. For example, it has been suggested that melatonin may be a gateway to puberty, because some studies have found a decrease in melatonin levels that coincides with early puberty; however, the studies in this area are contradictory, and the jury is still out on this question (Cavallo, 1991, 1992a,b; Cavallo and Dolan, 1996; Cavallo et al., 1992c; Molina-Carballo et al., 1996; Penny, 1982; Salti et al., 2000; Tetsuo et al., 1982; Waldhauser et al., 1988). A relationship has been proposed between melatonin and the reproductive axis in both men and women; however, here too, the jury is still out because of contradictory and inconsistent findings, despite the work of more than two decades. Multiple studies by several investigators have shown increased nocturnal melatonin levels in women with hypothalamic amenorrhea (Berga et al., 1988; Brzezinski et al., 1988; Kadva et al., 1998; Laughlin et al., 1991; Okatani and Sagara, 1994, 1995). Okatani and Sagara (1995) also found that exogenous estrogen significantly suppressed nocturnal melatonin levels in amenorrheic subjects. In men, however, the situation is less clear. Luboshitzky et al. (1995, 1996a,b, 1997, 2000) found that men with both idiopathic and congenital hypogonadotropic hypogonadism have high nocturnal melatonin levels which normalize with testosterone therapy. In contrast, they also found that men with primary hypogonadism (Klinefelter's syndrome) have low nocturnal melatonin levels which increase with exogenous testosterone (Luboshitzky et al., 1996c, 1997). Rajmil et al. (1997), on the other hand, found that men with primary hypogonadism have high nocturnal melatonin levels which decrease with testosterone therapy. These findings suggest that there is no simple classic feedback regulation of melatonin by testosterone. A recent review of this area reveals no publications that provide more specific information about a possible role of melatonin as a reproductive hormone. Therefore, humans are unique in this regard compared to other mammals.

17.3 Circadian Physiology

In virtually all vertebrates, circulating levels of melatonin are derived from the pineal gland with active production occurring only at night (Lewy et al., 1980b; Neuwelt and Lewy, 1983). The pineal gland measures 5–7 mm and lies in the center of the brain, (outside the blood–brain barrier), attached by a stalk to the roof of the third ventricle. Melatonin secretion is thought to represent the hormonal signal indicating nighttime darkness. In some nonmammalian vertebrates (including birds, reptiles, amphibians, and fish), the pineal gland itself contains photoreceptors involved in the regulation of melatonin secretion (Korf, 1994). This is in contrast to mammals where retinal, rather than pineal, photoreceptors are involved in entraining the endogenous circadian pacemaker (ECP) and in regulating melatonin synthesis and secretion (Korf, 1994).

In mammals, the site of the ECP or biological clock is the SCN of the hypothalamus. The SCN is a paired structure containing about 10 000 neurons situated in the anterobasal hypothalamus, just above the optic chiasm. The SCN receives external photic stimuli from the environment, maintaining circadian entrainment to the 24-h light/dark cycle. Photoreceptors in the retina send impulses to the SCN via a specific neural pathway, the retinohypothalamic tract, which is separate from the one that mediates vision (Moore and Lenn, 1972). Communication between the pineal gland and the SCN is bidirectional. The SCN sends signals to the pineal gland restricting the synthesis and secretion of melatonin to the appropriate 12-h interval of the 24-h day. The pineal gland communicates with the SCN by means of secretion of melatonin that is transmitted by diffusion or circulation to the SCN where it binds to specific receptors. Melatonin then acts on the SCN to help entrain circadian rhythms.

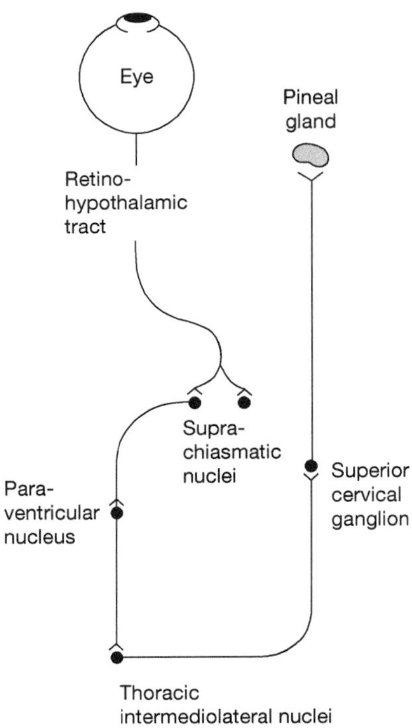

Figure 2 Schematic diagram depicting neuroanatomic regulation of mammalian melatonin production. Adapted from Lewy AJ (1983b) Biochemistry and regulation of mammalian melatonin production. In: Relkin RM (ed.) *The Pineal Gland*, pp. 77–128. New York: Elsevier North-Holland, with permission from Elsevier.

A neural pathway connects the SCN to the pineal gland (**Figure 2**). From the SCN, impulses are transmitted to the paraventricular nucleus (PVN) of the hypothalamus and then to the intermediolateral column of the upper thoracic cord. Preganglionic sympathetic fibers extend from the intermediolateral column to the superior cervical ganglion, adjacent to the internal carotid artery in the neck. Postganglionic sympathetic nerve fibers then innervate the pineal gland (Ariëns-Kappers, 1960). Norepinephrine is the neurotransmitter released from the postganglionic sympathetic fibers terminating on pinealocytes.

Neuroanatomical techniques such as *in vivo* neuronal tracings, immunohistochemistry, and *in situ* hybridization have demonstrated that the pineal gland has parasympathetic, in addition to sympathetic, innervation. Studies about the functional significance of the parasympathetic output, however, have shown conflicting results with regard to its effects on melatonin synthesis and release. Laitinen et al. (1995) found that carbachol, a parasympathomimetic, stimulates transpineal *in vivo* melatonin release from rat pineal gland in organ culture. In contrast, Drifhout et al. (1996) found, using transpineal *in vivo* microdialysis, that infusion of the muscarinic agonists carbechol and oxotremorine resulted in a decrease in melatonin release. Given these discrepancies, there is no clear evidence to date to support a role for parasympathetic neural input in the regulation of melatonin production or secretion.

Compared with all other endocrine glands, the pineal is unique, in that its sole regulator is neural input. Although other glands have sympathetic neural input or respond to circulating catecholamines, they differ from the pineal in that they are also controlled by other factors (such as hormones, cations, metabolites, or serum osmolality). In these glands, catecholamines affect the sensitivity of endocrine cells to feedback control by their usual regulators. An example of this dual regulation is the control of cortisol secretion in the adrenal gland. Cortisol secretion is primarily regulated by adrenocorticotropic hormone (ACTH) from the pituitary; however, its secretion is enhanced by electrical stimulation of adrenal nerves and is reduced by adrenal denervation (Edwards and Jones, 1993). In contrast, sympathetic neuronal stimulation is the only known regulator of melatonin secretion in the pineal gland.

Pinealocytes possess both beta-1 and alpha-1 adrenergic receptors. Norepinephrine stimulation (Axelrod and Zatz, 1977) leads to activation of N-acetyltransferase, the rate-limiting enzyme in melatonin synthesis (Klein and Moore, 1970). This action is potentiated by activation of alpha-1 adrenergic receptors (Klein et al., 1983). The nerve terminals of the sympathetic neurons terminating in the pineal gland have adrenergic receptors as well. These are alpha-2 adrenergic autoreceptors, and their activation leads to a decrease in melatonin production. Activation of the sympathetic fibers terminating in the pineal gland is tightly controlled by the SCN. Nonspecific sympathetic stimuli, collectively known as the fight or flight response, do not activate the pineal gland. This may be partly due to the active catecholamine-reuptake system present in the postganglionic sympathetic neurons innervating the pineal (Reiter, 1990). Indeed, Wetterberg (1979) studied two patients with high norepinephrine levels due to pheochromocytoma and found that they both had low (normal) daytime melatonin levels.

During the daily dark period, neural impulses from the SCN are transmitted via the above pathway to the pineal gland, where they stimulate melatonin synthesis and secretion. Melatonin is secreted both directly into the CSF and into the bloodstream; indeed, melatonin levels in the CSF may be much higher than those in blood. During the day, the SCN sends impulses that decrease otherwise on signal of the PVN. When not inhibited by the SCN, the PVN stimulates melatonin production and secretion (Pickard and Turek, 1983).

As expected, drugs that affect sympathetic activation can alter melatonin secretion. Beta-blocking agents, which interfere with beta-1 adrenergic stimulation in the pineal gland, cause a reduction of melatonin production (Wetterberg, 1978). Drugs that stimulate alpha-2 receptors, such as clonidine, will decrease melatonin output (Lewy et al., 1986), while drugs that block these receptors, such as yohimbine, increase melatonin production (Kennedy et al., 1995). Tricyclic antidepressants, which interfere with norepinephrine reuptake, increase melatonin production (Sack and Lewy, 1986). Only extreme physical exercise, such as high-altitude marathon races, will lead to increased melatonin production (Strassman et al., 1989). Drugs that affect melatonin production generally do so consistently throughout the night, so that they affect both the rise of melatonin production levels in the evening and the fall of levels in the morning, equally. This is important, because changes in the pattern of the melatonin secretory profile can lead to phase shifts of the ECP (see Section 17.7).

Even low levels of light exposure during the dark period (scotoperiod) can have an acute suppressive effect on melatonin production (Illnerová, 1978). This effect can only occur during the scotoperiod, because melatonin secretion is already stopped by the SCN during the daytime (photoperiod). In almost all species studied, there is no corresponding stimulatory effect of darkness during the day; that is, dark exposure during the daytime does not lead to an acute increase in melatonin production in humans. Thus, melatonin secretion is restricted to about 12 h during the night. The acute suppressant effect of light is critical for endogenous melatonin to augment entrainment of the ECP by the light/dark cycle (**Figure 3**). However, a more likely role for endogenous melatonin production will be discussed at the end of this chapter (however, see Section 17.12).

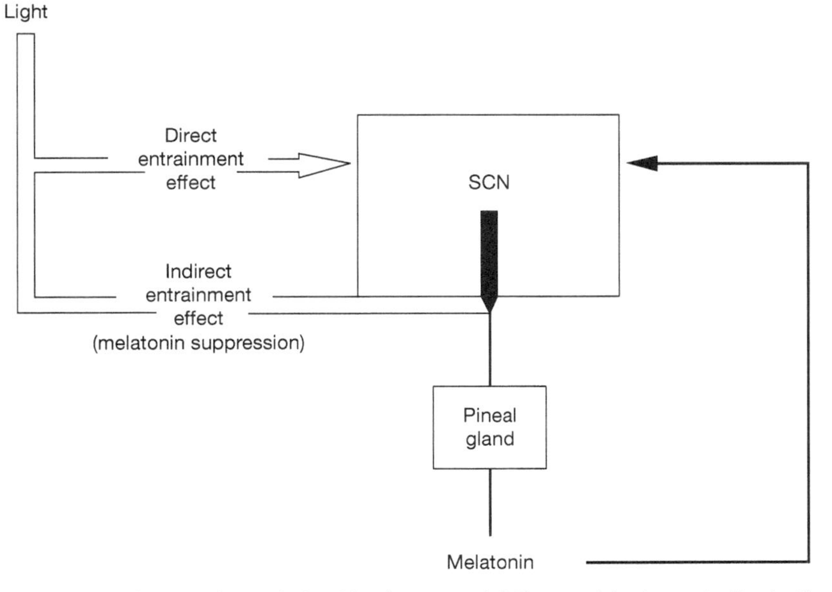

Figure 3 Schematic diagram of some of the relationships between nighttime melatonin production by the pineal gland, the light/dark cycle, and an endogenous circadian pacemaker located in the SCN. Acting on the SCN as described by the melatonin PRC (**Figure 4**) at any given time of the day or night, melatonin causes phase shifts opposite to those that light would cause (indicated by the opposing arrows). However, the suppressant effect of light pares the margins of the nighttime melatonin profile (vertical tapered arrow) and reduces endogenous melatonin's stimulation of the melatonin PRC at the day/night transitions. This second (indirect) pathway for entrainment by light is particularly significant during shifts of the light/dark cycle. Reprinted from Lewy AJ, Ahmed S, Jackson JML, and Sack RL (1992) Melatonin shifts circadian rhythms according to a phase-response curve. *Chronobiology International* 9: 380–392, with permission from Taylor & Francis Ltd.

17.4 Melatonin as a Phase Marker

Endogenous circadian rhythms are coordinated physiologic responses regulated by the SCN that occur cyclically during a 24-h period. Core body temperature, sleep propensity, and hormonal secretion are all linked to the SCN. The nadir of body temperature typically occurs around 3–5 a.m., and the peak occurs in the late afternoon. Cortisol levels generally reach a nadir around midnight and peak in the early morning. The secretion of TSH, prolactin, and growth hormone also has intrinsic circadian patterns, although sleep also influences their secretion (Czeisler and Klerman, 1999). Pathological processes that disrupt circadian rhythms can have deleterious effects.

The phase of the ECP is best assessed by monitoring melatonin levels. Other parameters such as body temperature and cortisol secretion can also be used, but both of these are more likely than melatonin to be affected by masking (interference by external stimuli such as activity, stress, and sleep deprivation). As discussed previously, melatonin levels are typically low during the day and generally begin to rise about 14 h after a person's habitual wake time. In normal individuals, the time of the melatonin onset (MO) is usually stable from day to day and can be used as a marker to evaluate a person's circadian phase. Studies often empirically use the time at which plasma melatonin levels reach $10\,pg\,ml^{-1}$ as the MO, although other operational definitions can be used as well. Because sampling for melatonin is usually done in sighted people under conditions of dim light (<10–30 lux) to avoid the suppressive effects of light on melatonin secretion, the MO under these conditions is called the dim light melatonin onset (DLMO; Lewy and Sack, 1989). A person whose plasma $DLMO_{10}$ occurs 14 h after wake time (2 h before bedtime) is said to be in phase. By frequently sampling overnight melatonin levels, it is also possible to map other points on the melatonin production profile. The time when melatonin levels start to fall (when melatonin synthesis stops) is called the synthesis offset (SynOff; Lewy et al., 1999), and the time when melatonin levels decrease below $10\,pg\,ml^{-1}$ is called the dim light melatonin offset (DLMOff). These phase markers could also be used to monitor a person's circadian rhythms, although the DLMOff may not be quite as reliable, as it is significantly influenced by the amplitude of melatonin secretion and melatonin half-life which can vary markedly between individuals. Also, measuring the DLMOff, and particularly the SynOff, requires waking subjects in the middle of the night, while the DLMO can usually be measured prior to sleep onset. For low secretors, the plasma $DLMO_2$ is used, which occurs on average about 1 h before the $DLMO_{10}$.

In addition to plasma, melatonin measurements can also be done in saliva (Carskadon et al., 1997; Sharkey and Eastman, 2000). Levels in saliva are generally about one-third of those in plasma when melatonin levels begin to rise. Saliva collection is less invasive than collection of blood and can easily be done at home. The corresponding salivary thresholds to the plasma $DLMO_{10}$ and $DLMO_2$ are $DLMO_3$ and $DLMO_{0.7}$, respectively. Because of the suppressant effect of ambient light, we and others have gradually reduced the recommended light intensity for assessing DLMOs over the years to 10–30 lux (too dim to read).

In the research lab, the DLMO has become the most frequently used marker for the phase of the ECP. The one remaining concern about the DLMO stems from studies in rodents suggesting that there are two weakly coupled oscillators responsible for the onset and offset of melatonin production. These studies have led to a model that attempts to explain how melatonin duration lengthens in the winter and shortens in the summer as it tracks the scotoperiod, or nighttime dark period. In this model, there is no need for an acute suppressant effect of light if there is an oscillator cued to dusk that controls the time of the melatonin onset and a separate oscillator cued to dawn that controls the time of the melatonin offset. Alternatively, we have proposed a clock-gate model in which the acute suppressant effect of light replaces the need for a second oscillator. In any event, humans (who do not have robust seasonal rhythms) do not usually have marked seasonal changes in the naturalistic endogenous melatonin profile. That is, the melatonin profile as it is measured under customary lighting is certainly not as robust as has been observed in many seasonally breeding mammals. In all likelihood, indoor lighting helps keep the melatonin duration from markedly changing across the year. Clearly, humans do not share the profound response to a half-hour change in the spring or fall photoperiod that regulates seasonal rhythms in some rodents.

In certain pathological conditions, for example, seasonal affective disorder, there may be a slight lengthening of melatonin duration in the winter, at least in some patients. Also, exotic experimental conditions can shift the onset and the offset (even during sampling under dim light) in opposite directions. However, it is generally the case that changes in the onset reflect changes in the offset. In some studies in which the offset has shifted in the same direction as the onset, but with a different magnitude, such evidence has been marshaled to support the two-oscillator model, when in fact a more parsimonious explanation is simply measurement variability. In any event, the finding that bright light suppresses melatonin production more than dim light originally led to the speculation that humans had circadian and seasonal rhythms cued to natural daylight relatively unperturbed by indoor light, when in fact this speculation may be more applicable to circadian than seasonal rhythms. Also, the precise relationship between intensities that suppress melatonin production and cause phase shifts is not known. In any event, circadian and not seasonal rhythms appear to be more relevant to human physiology.

17.5 Circadian Time

It is helpful to use the concept of circadian time (CT) when assessing phase. By convention, a person's wake time is called CT 0. On average, the $10\,pg\,ml^{-1}$ DLMO occurs at CT 14. Therefore, in studies the DLMO is often designated as CT 14 and then CTs can be assigned to other markers, accordingly. For example, if the core body temperature nadir is occurring 3 h after the DLMO, then it is occurring at CT 17. The most important use of CT is for scheduling bright-light exposure and melatonin administration, which will be explained in Sections 17.7 and 17.8.

17.6 Zeitgeber Time

Another important concept is zeitgeber time (ZT). The term zeitgeber (literally, time giver or time cue) refers to environmental variables that are capable of acting as circadian time cues. The light/dark cycle is the most important zeitgeber, but other stimuli such as melatonin can also function as zeitgebers. ZT is the temporal relation of the circadian rhythm, marked by the DLMO, to entraining signals such as dawn (or the first introduction of light, i.e. wake time). For example, a person whose DLMO occurs 13 h after wake time is said to have a DLMO ZT of 13. The ZT of the DLMO can provide information about the length of a person's circadian rhythm (tau). For example, a person whose DLMO occurs at ZT 13 most likely has a shortened tau, perhaps less than 24 h, while a person with a DLMO of ZT 15 most likely has an intrinsically long tau (Lewy et al., 2000).

17.7 Effects of Light on Circadian Rhythms

Lewy et al. (1980a) demonstrated that bright light could suppress melatonin secretion in humans. Prior to that time, studies done with ordinary-intensity room light had not been able to demonstrate a clear-cut suppressant effect of light on melatonin secretion (Akerstedt et al., 1979; Arendt, 1978; Wetterberg, 1978). The finding that light could suppress melatonin in humans implied that humans could have similar responses as other animals to sunlight or bright artificial light. This finding also suggested that bright light, but not ordinary-intensity room light, would be needed to markedly affect melatonin production. Brainard et al. (1985) further delineated the type of light that affects melatonin secretion. They originally found that blue-green light with a peak wavelength of 509 nm is most effective in suppressing secretion. This wavelength can be found in most white light sources and in regular fluorescent light. However, more recent studies suggest that the peak wavelength is actually closer to 480 nm (blue light; Brainard et al., 2001).

The discovery that bright light could suppress melatonin secretion (Lewy et al., 1987) allowed scientists to consider possible uses of light in the treatment of circadian rhythm disorders in humans. The first uses were in the treatment of depression. Kripke (1981) used bright light to treat nonseasonal depression. We identified winter depression and first treated it by extending the photoperiod with 3 h of 2000 lux light in the morning and in the late afternoon (Lewy et al., 1982). Shortly thereafter, Lewy et al. (1983) and Wever et al. (1983) showed that bright light could shift circadian rhythms in humans, although Wever did not postulate a phase response curve (PRC) to light for humans.

We postulated that humans have a PRC to light which is similar to that of animals, but that bright light, rather than ordinary room light, is necessary to optimally elicit the phase shifts. Lewy et al. (1984, 1985a) were able to demonstrate that light exposure could indeed be used in humans to shift the melatonin secretory profile. In these studies, the sleep/wake cycle was held constant while the light/dark cycle was manipulated. Alteration of the light/dark cycle led to a clear shift in the melatonin profile. These early studies were followed by numerous others that showed a shift in circadian rhythms in response to light exposure: light treatment at the end of the photoperiod delays the melatonin profile while light at the beginning of the photoperiod advances it (Czeisler et al., 1986, 1989;

Eastman and Miescke 1986; Honma and Honma, 1988; Lewy et al., 1987; Minors et al., 1991; Wever, 1989).

17.8 Effects of Melatonin on Circadian Rhythms

Attention then turned toward investigating whether exogenously administered melatonin had effects similar to those of light. Animal studies had already shown that melatonin was able to induce circadian rhythm phase shifts at both behavioral and cellular levels (Armstrong et al., 1986; Cassone et al., 1986a,b; Chesworth et al., 1987; Redman et al., 1983; Underwood, 1986). Arendt et al. (1985) and Claustrat and co-workers (Mallo et al., 1988) were the first to investigate the phase-shifting effects of melatonin in humans. Sack et al. (1987) undertook studies in which blind people were given melatonin to shift their circadian rhythms. This led to giving sighted human subjects melatonin on four consecutive days (Lewy et al., 1990a,b, 1991). These studies demonstrated that melatonin administered during the late afternoon or early evening would advance the DLMO, while melatonin given in the morning would delay it (**Figure 4**). These findings were subsequently replicated by other groups (Middleton et al., 1997; Zaidan et al., 1994). The results of Lewy et al. (1992) describe a melatonin PRC which makes it possible to accurately predict the phase-shifting effect of exogenously administered melatonin based on the time of administration.

The melatonin PRC reveals that there is an approximately 12-h interval during which exogenous melatonin will advance the endogenous pacemaker and a 12-h interval during which melatonin will delay it. Not surprisingly, the melatonin PRC is about 12 h out of phase with the light PRC (Lewy et al., 1998a). This would be expected because melatonin is thought to act as a chemical signal for darkness. The melatonin PRC is most helpful when it is expressed in CT rather than clock time. This allows the melatonin PRC to be used effectively in people whose circadian rhythms are out of phase with their environment, for example, immediately after transmeridian flight. The advance zone of the melatonin PRC is generally from CT 6 to CT 18 and the delay zone is from CT 18 to CT 6. This information can be used therapeutically in a number of clinical situations, as will be described in Section 17.11. **Figure 5** illustrates the fundamental ways in which bright light and melatonin should be scheduled according to clock times to correct circadian phase disorders.

17.9 Soporific Effects of Melatonin

In addition to phase-shifting effects, high doses of melatonin also have soporific effects. About 30%

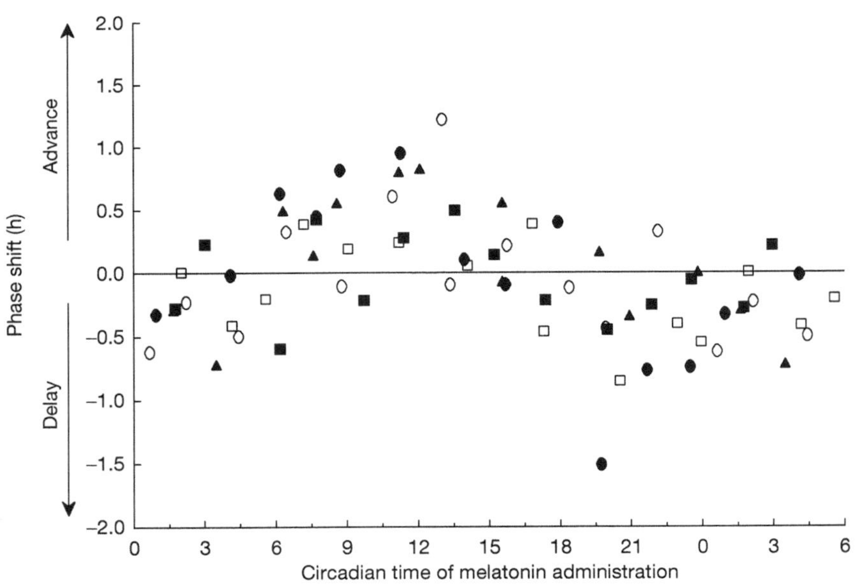

Figure 4 Individual PRCs from five subjects (four females; one male, λ) given melatonin (0.5 mg) for four consecutive days on 12 occasions. Phase advances (+) and delays (–) are plotted against circadian time (CT) of administration. Reproduced from Lewy AJ, Bauer VK, Ahmed S, et al. (1998a) The human phase response curve (PRC) to melatonin is about 12 hours out of phase with the PRC to light. *Chronobiology International* 15: 71–83, by permission of Taylor & Francis Inc.

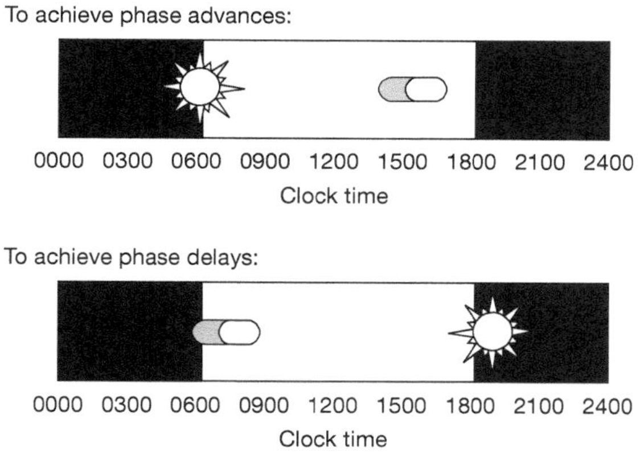

Figure 5 Phase-shifting effects of light and exogenous melatonin. In order to cause a phase advance, light should be scheduled in the morning and melatonin administered in the afternoon. In order to cause a phase delay, light should be scheduled in the evening and melatonin should be administered in the morning. Reproduced from Lewy AJ and Sack RL (1996) The role of melatonin and light in the human circadian system In: Buijs R, Kalsbeek A, Romijn A, Pennartz C, and Mirmiran M (eds.) *Progress in Brain Research: Hypothalamic Integration of Circadian Rhythms*, vol. 111, pp. 205–216. Amsterdam: Elsevier, with permission from Elsevier.

of people will develop sleepiness as a side effect of melatonin administration (>1 mg). Sack et al. (2000) performed a study in blind subjects whose circadian rhythms were close to 180° out of phase with their environment. The subjects were given either 10 mg of melatonin or placebo for four nights on an alternating basis. On the nights of melatonin administration, the subjects had significant improvement in total sleep time and sleep efficiency compared with placebo (Hughes et al., 1998b). In another study, Hughes and Badia (1997) administered either 1 , 10, or 40 mg of melatonin to a group of healthy young men after a 7-h nighttime sleep opportunity and then monitored sleep the next day during a 4-h sleep opportunity between noon and 4 p.m. All doses of melatonin significantly shortened sleep latency, increased total sleep time, and decreased amount of time awake after sleep onset compared with placebo. These studies demonstrate the direct sleep-promoting effects of melatonin (which are not as potent as those of sedatives/hypnotics).

17.10 Safety of Melatonin

In the United States, melatonin is easily available without a prescription, and millions of people have been taking it over the past few years. This is despite the fact that there are no published long-term safety data on the use of melatonin. If melatonin is taken incorrectly, it can potentially cause adverse consequences via its effects on sleep and circadian rhythms. For example, someone taking melatonin for jet lag could potentially retard the normalization of their circadian rhythms by taking melatonin at the wrong time of day. Another example would be the development of somnolence when melatonin is taken during the day. Besides predictable consequences such as these, there have been few, if any, reports of serious irreversible adverse effects from melatonin. It seems unlikely that melatonin, particularly at low doses, will have any harmful effects in humans.

In general, the collective published research indicates that melatonin is safe, and no significant irreversible toxicity has been demonstrated. People who choose to take melatonin should do so with the understanding that not much is known about the possible consequences, if any, of its long-term use. However, anecdotal reports thus far suggest that melatonin is safe to take long term.

17.11 Abnormalities in Circadian Rhythms

Numerous conditions exist in which circadian rhythm abnormalities lead to problems with either sleep or mood, or both.

17.11.1 Blindness

The circadian rhythms of blind people are often abnormal (Lewy, 1981; Lewy and Newsome, 1983). Many blind people have circadian rhythms that are not entrained to the 24-h day/night cycle with a period longer (or rarely, shorter) than 24 h; that is, they have free-running circadian rhythms. In these blind free-runners (BFRs), the circadian rhythms governed by the SCN, such as melatonin secretion, core body temperature, and sleep propensity, shift later and later (usually) each day. Because of their constantly drifting circadian rhythms, BFRs spend much of their lives out of phase. When they are out of phase, they often experience adverse effects, including nighttime insomnia and daytime somnolence. It is estimated that at least half of the approximately 200 000 totally blind people in the US have free-running circadian rhythms (Sack et al., 1998). Another subgroup of totally blind people appears to be entrained but at an abnormal phase: the MO may stably occur at any time of the day or night, and thus they are always out of phase with their environment. These people may have chronic problems with insomnia and daytime somnolence. Still other blind people may have normally and stably phased circadian rhythms.

Sack et al. (1987) published the first report of a BFR whose ECP was phase-shifted by melatonin. This was followed in 1991 by publication of the first clinical trial which demonstrated that exogenously administered melatonin can advance the circadian rhythms in free-running blind people (Sack et al., 1991). Five BFRs with an average tau of 24.71 h ± 0.19 h (range 24.55–25.10 h) were treated with 0.5 mg of melatonin for 3 weeks. Placebo was given for several weeks prior to active treatment. When the subjects' circadian rhythms were approaching normal phase, melatonin was started. Melatonin was continued for 3 weeks (the maximum treatment duration allowed by the Food and Drug Association (FDA) at that time), followed again by several weeks of placebo treatment. MOs were measured frequently throughout the study. Four of the five subjects had significant phase advances with melatonin treatment. Their average total phase advance over the 3 weeks of active treatment was 8.41 h.

Sack et al. (1992b) published a study that confirmed the high prevalence of circadian rhythm abnormalities in totally blind people. Out of 20 subjects tested with serial melatonin measurements, 11 subjects had free-running circadian rhythms and three had entrained, but out-of-phase, rhythms. Only three subjects appeared to have normally entrained circadian rhythms. However, it should be mentioned that it is difficult to rule out a free-running period close to 24 h in a blind person who may appear to be entrained; these people need to be studied over several months to verify entrainment or observe a very gradual drift.

More recently, we published work demonstrating that melatonin can be used successfully to treat circadian rhythm disorders in blind people (Sack et al., 2000). In this study, seven BFRs with endogenous circadian rhythms between 24.2 and 24.9 h were administered 10 mg of melatonin daily, 1 h before preferred bedtime. Treatment was again initiated when each subject's free-running rhythm was approaching a normal phase. Six of the seven subjects had their endogenous circadian rhythms entrained by the treatment. The seventh subject, who had an endogenous circadian period of 24.9 h, which was the longest in the group, had a shortening of his tau to at least 24.3 h but his melatonin rhythm was never entrained. Compared to values before treatment, the subjects overall were found to have greater sleep efficiency and less wake time after sleep onset. In this same study, three subjects had their melatonin doses gradually decreased to 0.5 mg per day after entrainment was achieved. These subjects were able to maintain their entrained circadian rhythms. Subsequently, Lewy et al. (2001) showed that a *de novo* dose of 0.5 mg can entrain BFRs; in fact, the one BFR who did not entrain to 10 mg (or to 20 mg) did entrain to 0.5 mg (Lewy et al., 2002). Our thinking is that high doses of melatonin spill over onto the wrong zone of the melatonin PRC and reduce the magnitude of its intended phase-shifting effect. Optimally, melatonin is taken at about 6 p.m., so that the endogenous MO occurs a few hours before bedtime. The few BFRs with intrinsic taus less than 24 h should take low-dose melatonin at awakening (Emens et al., 2006). In BFRs, Lewy et al. (2005) were able to demonstrate a dose–response curve for melatonin in the physiological range.

There is currently some debate as to how to best assess light perception in totally blind people – to rule out any possibility of subjective or objective perception. Light suppression by melatonin (Lewy et al., 1980a) has been used in this regard (Czeisler et al., 1995). We have raised several concerns about the way the melatonin suppression test (MST) is used for this purpose (Lewy, 2007). Blind people are obviously going to be subsensitive to light perception. Therefore, it is important to use the brightest, yet safest, light possible. Since blind people are normally exposed to outdoor light, collecting samples outdoors when their melatonin peak is occurring in the middle of the day can serve as the basis for such a test. Orange goggles, which block blue light transmission, should be able to reduce suppression by sunlight.

17.11.2 Advanced and Delayed Sleep Phase Syndromes

Sighted people can also have endogenous circadian rhythms that are out of phase with their environment and with their desired sleep time. People who suffer from advanced sleep phase syndrome (ASPS) have sleep times that are advanced in relation to normal clock time and in relation to their desired sleep time. These people feel excessively sleepy in the evening, fall asleep early and suffer from early morning awakening. This condition develops more commonly in the elderly (Lack, 1986; Pelayo et al., 1988; Thorpy et al., 1988; Weitzman et al., 1981). Delayed sleep phase syndrome (DSPS) is just the opposite. People with this syndrome have sleep delayed in relation to clock time and in relation to their desired sleep time. They are unable to fall asleep until late in the night, typically after 1 a.m., and they have difficulty awakening in the morning. This condition is more prevalent in the young (Thorpy et al., 1988).

Jones et al. (1999) reported on three families that each have multiple family members affected by familial advanced sleep phase syndrome (FASPS). In these families, FASPS segregates as a highly penetrant autosomal dominant trait. Compared to control subjects, affected family members showed 4-h advances in measures of sleep, including sleep onset, sleep offset, first slow-wave sleep, and first rapid eye movement sleep. The temperature rhythms and DLMOs of affected subjects were also advanced by 3–4 h. Data from five affected family members showed a mean wake time of 4:18 a.m. and mean DLMO of 5:31 p.m. With these data, we have calculated the ZT of the DLMO to be approximately 13.3. One affected subject was studied in temporal isolation for 3 weeks and was found to have a tau of 23.3 h, perhaps the shortest ever recorded in humans. As discussed in Section 17.6, the early ZT of the DLMO would be expected to correspond to a shortened tau, which is exactly what was found in this family.

To determine the genetic basis of this disorder, linkage analysis was performed in one large affected family (Toh et al., 2001). The FASPS gene was localized to an area near the telomere on chromosome 2q. A strong candidate gene (*hPER2*) mapped to the same locus. This gene is a human homolog of the period gene (*Per*) in *Drosophila*. Mutations in *Per* in the fly (Konopka and Benzer, 1971) produce a short tau phenotype. This would be consistent with the findings in these families since short tau mutants are usually phase-advanced (Pittendrigh and Daan, 1976). While more than one clock gene is probably involved in determining an individual's tau, perhaps the FASPS gene determines if tau is shorter than 24 h – a rare occurrence.

Some people may simply choose to have either advanced or delayed sleep patterns. In these people, there is no pathological sleep disorder. The diagnoses of ASPS or DSPS are only made when a person has attempted to normalize their sleep pattern to clock time (normal sleep time) and is unsuccessful in doing so.

Hughes and Lewy (1998a) and Lewy (1990) have proposed that there are three subtypes of circadian rhythm sleep disorders. In one subtype, both sleep and the other circadian rhythms are phase-shifted to the same extent. In the other two subtypes, sleep is not as phase-shifted as the other circadian rhythms or vice versa.

Both light and melatonin have been effectively used to treat circadian phase disorders in sighted people. Lewy et al. (1983, 1985b) successfully used morning light to treat a person with DSPS and subsequently treated a person with ASPS with evening light. Other studies then confirmed that light therapy could treat circadian phase sleep disorders (Joseph-Vanderpool et al., 1988, 1989; Singer and Lewy, 1989). The effective therapeutic light treatment duration has varied from 15 min to 4 h in these studies and the light intensity has varied from 2500 to 10 000 lux.

Melatonin therapy is another option for treating these disorders. Dahlitz et al. (1991) successfully treated patients with DSPS with melatonin. Five milligrams of melatonin was administered to subjects with DSPS each evening at 10:00 p.m. (5 h prior to usual sleep onset) for 4 weeks. The average sleep onset of the group advanced by 82 min. Studies by Tzischinsky et al. (1993) and Oldani et al. (1994) have shown similar results. Based on the melatonin PRC (Lewy et al., 1992, 1998a), ASPS should be treated with 0.5 mg of melatonin at each awakening during the night after 1 a.m. and at final awakening in the morning; DSPS should be treated with 0.5 mg of melatonin 8 h after usual weekday awakening, followed by 3 mg at bedtime.

17.11.3 Jet Lag

People who travel across multiple time zones are usually affected by jet lag, because their endogenous circadian rhythms are out of phase for the first few days in the new time zone. The first therapeutic use of melatonin for jet lag was by Arendt et al. (1986). Properly timed melatonin treatment and light exposure can be used to accelerate the synchronization of circadian rhythms to the new time zone after travel (Daan and Lewy, 1984; Lewy et al., 1995).

To facilitate phase advances, we recommend that people traveling between one and six time zones from west to east take melatonin at about 2:00 p.m. on the day prior to and day of departure. After arriving at their destination, they should then take melatonin at 2:00 p.m. corrected to the corresponding time in the new time zone. On days 2 and 3 following arrival, melatonin can be moved ideally 3 h earlier than the previous day. Because bright light has strong phase-shifting effects, these west to east travelers should seek out morning sunlight after arriving at their destination. With these measures, travelers can expect to normalize their ECP more quickly.

People traveling less than six time zones from east to west should take melatonin upon awakening the day before and the day of departure; then on the first day in the new time zone, they should take it at wake time in the pretravel time zone converted to the corresponding time in the destination time zone (or else at bedtime if this is not convenient). On days 2 and 3 after arrival, they should move the melatonin dose ideally 3 h later than the previous day. After arrival they should seek out sunlight late in the day.

For both eastward and westward travel crossing more than six time zones, the same guidelines as above can be used for melatonin administration; however, sunlight exposure must be avoided at certain times. Morning sunlight in this group of eastward travelers will actually cause phase delays and should therefore be avoided for the first few days after arrival. Similarly, westward travelers crossing over six time zones should avoid sunlight at the end of the day, which causes phase advances. Both groups will benefit from midday sunlight for the first few days after arriving at their destinations.

17.11.4 Shift Work

Shift workers can suffer physical and psychological consequences from constantly being out of phase. The effects of shift work may be especially hard to deal with as people age (Foret et al., 1981). People who permanently work day or evening shifts do not typically have chronobiological problems. The most significant problems are seen in those who do only part of their work at night. Almost all night-shift workers rotate shifts, if only between workdays and days off. In a typical week they may work five nights and then try to normalize their schedule to daytime hours on days off in order to spend time with family and friends. Other shift workers constantly rotate actual work schedules. They may, for example, do one or more day shifts, evening shifts, and night shifts within the same week. With these schedule disturbances, they may never fully synchronize with their environment, and consequently they often suffer from insomnia during their desired sleep hours and somnolence while at work. Shift workers can also have impaired productivity and work performance. For all of these reasons and because shift work is so prevalent in our society, shift work has been the subject of intense investigation.

Barnes et al. (1998) showed that socially isolated oil rig workers who were starting to work the night shift were able to phase-delay by approximately 90 min per day. At this rate, it would take 5–6 days before a person was fully adapted to the night-shift schedule. If, as is often the case, they then had an off-duty break of several days and reverted to a daytime-activity schedule, they would disrupt their rhythms just as they had achieved entrainment to the night-time schedule. Some people may take longer to shift. Sack et al. (1992a) found that seven of nine workers doing ten consecutive night shifts did not have an MO between 6 a.m. and noon as would be expected.

Both bright-light manipulation and melatonin can be used to help the circadian rhythm disturbances of shift workers. Czeisler et al. (1990) found that it is important for night workers (and thus daytime sleepers) to have complete darkness in their bedrooms in order to allow the appropriate phase delay. Eastman et al. (1994) showed that avoidance of morning bright light on the commute home from a night shift is also very important (her subjects wore dark welders' goggles). Sack et al. (2002) have found that 0.5 mg of melatonin taken before bedtime helps shiftworkers adapt. In addition to helping synchronize rhythms, nighttime bright light has been shown to increase alertness. Badia et al. (1991) provided pulsatile bright light during the night shift and found a corresponding increase in subjective alertness during the pulses. While high-intensity light leads to the most dramatic results, even light of moderate intensity can have rapid beneficial effects. Martin and Eastman (1998) compared the effects of exposure to 1230 lux versus 5700 lux for 3 h per night in nighttime workers. Although the effects of the higher-intensity light were more predictable, most workers responded to the lower-intensity light as well.

17.11.5 Seasonal Affective Disorder (Winter Depression)

Seasonal affective disorder (SAD) is another condition that appears to be related to endogenous circadian rhythms. SAD is a condition characterized by recurrent bouts of depression that occur predictably every fall or winter as the days become shorter and improve spontaneously every spring or summer as the days lengthen. All patients with SAD have depression, but they also commonly have increased sleep time, morning somnolence, hyperphagia (especially carbohydrate craving), weight gain, low energy, low motivation, and social withdrawal (Rosenthal et al., 1984).

Following the discovery that bright light could suppress melatonin secretion in humans, Lewy et al. (1980a) speculated that humans, like animals, might have biological rhythms that were cued to sunlight. Lewy et al. (1982) started experimenting with light treatment to determine if it could improve the symptoms of SAD. The first identified patient with winter depression was treated with 3 h of morning and late afternoon bright light with the goal of extending the photoperiod, based on the premise that the short winter days were responsible for precipitating his disorder. After 4 days, his symptoms began to resolve.

This case report paved the way for further research into the area of light treatment of SAD. One difficulty encountered in planning such studies was the question of what should be used as a placebo. It was necessary to choose a placebo that patients would expect to work as well as bright light. Some early studies used dim light; however, it is unlikely that this served as an adequate placebo, because of the obvious differences between the two treatments (James et al., 1985; Rosenthal et al., 1985a). Other studies used bright light at different times of day as a control (Lewy et al., 1987; Sack et al., 1990).

Because many SAD patients have morning hypersomnia, Lewy et al. (1983, 1984) predicted that these patients had delayed circadian rhythms. Based on this theory, Lewy et al. (1987) began treating SAD patients with morning bright light in order to cause a phase advance. One advantage of this type of treatment is that light at another time of day can serve as a control, since the effects of the two treatments can be directly compared. In these studies, evening bright light served as the control. In this early study, the subjects were not thought to have any greater expectation for the therapeutic effects of morning versus evening light, suggesting that evening light was in fact a plausible placebo. Indeed, at the time, even many researchers did not think that morning bright light would have superior antidepressant effects (James et al., 1985; Rosenthal et al., 1985a).

The study involved eight SAD patients and seven controls who were placed in dim light between 5 p.m. and 8 a.m. and slept between 10 p.m. and 6 a.m. After a baseline adaptation week, subjects were randomly assigned to 2 h of bright light either in the morning (6–8 a.m.) or in the evening (8–10 p.m.) for 1 week. They were then switched over to the other treatment for the second week. All subjects were treated with both morning and evening light during the third week. DLMO and depression assessments were done prior to treatment and during each week of the study.

The DLMOs of SAD patients were delayed compared to those of controls both at the beginning and at the end of the adaptation week, which confirmed our hypothesis that SAD is a phase-delay disorder. The fact that the DLMOs remained delayed even after the adaptation week demonstrates that it was not simply the pretreatment environment, such as later light exposure due to late awakening, which caused the DLMO to delay in affected patients. It should be noted, however, that in this and other studies, there has been a fair amount of overlap between the DLMOs of patients and controls. Therefore, a late DLMO is not sufficient criteria for a diagnosis of SAD. It may be, in fact, that the DLMOs of some people with SAD are not delayed compared to those of controls but rather that they are delayed when compared to their own DLMOs during the spring and summer when they are euthymic. This would be an ipsative, rather than a normative, pathological finding (Lewy et al., 1987).

In the above study, as expected, morning light treatment advanced the DLMOs and evening light treatment delayed them. Depression ratings significantly decreased after the week of morning light treatment compared to ratings at baseline and after evening light treatment (**Figure 6**). Depression ratings after the combination of morning and evening light were between those of the morning and evening groups. These results provide support for the hypothesis that treatment of SAD is dependent on phase-advancing circadian rhythms.

Sack et al. (1990) published a study that confirmed our earlier findings about the superiority of morning light. We studied eight patients with SAD and five controls and again found that the SAD patients had DLMOs that were significantly later than those of controls during the prebaseline comparison.

After a 1-week baseline period, subjects were randomized to 2 h of either morning (6–8 a.m.) or evening (7–9 p.m.) light for 1 week and then were crossed over to the alternate therapy for 1 week. All subjects were otherwise kept in dim light from 6 p.m. to 8 a.m. and were asked to sleep between 10 p.m. and 6 a.m. After exposure to morning light, SAD patients had DLMO phase advances of approximately 1 h and had significant decreases in all three measures of depression tested (Hamilton Depression Scale ratings (Ham-D ratings), Beck Depression Inventory self-ratings, and global depression ratings). Interestingly, in patients but not controls, there was no significant DLMO shift after evening light and only the Ham-D ratings were reduced (and to a significantly lesser

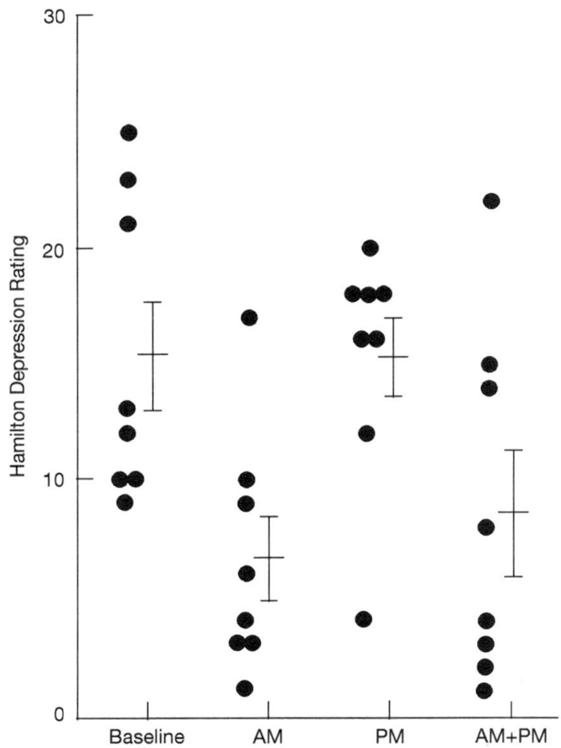

Figure 6 Individual and average 21-item Hamilton Depression Scale ratings (± SEM) for eight patients with winter depression for each of the 4 weeks of the study. An analysis of variance for repeated measures indicated a significant ($p = 0.026$) difference between treatments. Only the paired *t*-tests comparing the week of morning (a.m.) light and the baseline week ($p = 0.004$) and comparing the week of a.m. light and the week of evening (p.m.) light ($p = 0.045$) were significant. Average depression ratings (± SEM) for the seven normal control subjects were 3.0 ± 0.9 at baseline, 2.4 ± 0.3 (a.m. light), 6.1 ± 1.6 (p.m. light), and 4.3 ± 0.9 (a.m. + p.m. light). Reproduced from Lewy AJ, Sack RL, Miller S, and Hoban TM (1987) Antidepressant and circadian phase-shifting effects of light. *Science* 235: 352–354, with permission from AAAS.

extent than after morning treatment). It is possible that the antidepressant response to evening light may have been mainly due to the placebo component of light treatment. The lack of phase delay with evening light in this study compared to the previously discussed study by Lewy et al. (1987) may be due to the earlier time of evening melatonin administration in this study. SAD patients were allowed to keep their light boxes at the end of the study. Seven of the eight patients kept consistent logs of their light use after study termination, and all of them chose to use morning rather than evening light. Most patients stopped using their lights by the end of March.

Studies by other researchers have found either superiority of morning over evening light (Avery et al., 1991) or no difference between the two (Wirz-Justice et al., 1993). No studies have found evening light to be superior. A consensus seems to be building that SAD is best treated by morning light. Studies done by Eastman et al. (1998), Lewy et al. (1998c), and Terman et al. (1998) found morning light to be a more effective treatment of SAD. Bright-light exposure is the treatment of choice for SAD. A typical treatment will start with 1–2 h of light treatment up to 10 000 lux every morning. After they respond, patients can then decrease the duration of light exposure to less than 30 min per day (Lewy et al., 1988).

Although the studies mentioned above show a clear superiority of morning light for the treatment of SAD, it is noteworthy that light at any time of day still has a significant antidepressant effect. A possible energizing effect of light has been suggested (Lewy and Sack, 1986) although neither this nor any other specific mechanism has yet been proven. Therefore, this effect of light remains a placebo effect.

Studies on the effects of melatonin on the treatment of SAD have lagged behind studies of light therapy and, so far, very few have been published. Two studies in which melatonin was administered at times on the melatonin PRC that would not be expected to cause significant phase advances showed no benefit of melatonin in SAD (Rosenthal et al., 1985b; Wirz-Justice et al., 1990). A small pilot study showed melatonin to be effective in SAD (Lewy et al., 1998b).

In this study, five patients with SAD were treated with melatonin every afternoon (0.125 mg of melatonin administered at CT 8 and CT 12 based on patient wake time), and another group of five patients was treated with placebo. Patients were instructed to adhere to their usual sleep/wake cycles throughout the study. Depression was rated before and after treatment using the 29-item Structured Interview Guide for the Hamilton Depression Rating Scale-Seasonal Affective Disorder Version (SIGH-SAD). Using a 39% decrease in SIGH-SAD score as evidence of response (selected in order to categorize all five melatonin-treated patients as responders), only one of five placebo-treated patients responded after 2 weeks, a clear difference from the treatment group. These results support the phase shift hypothesis of SAD.

The optimal amount of phase advance seems to be about 1.5 h (Lewy et al., 2000a; Terman et al., 2001).

This shift can be achieved theoretically with either morning light, afternoon/evening melatonin, or a combination of the two. Combining the two may make the most sense for some people, because adding melatonin to light treatment will allow a shorter light-treatment duration which is often desirable. Using melatonin alone without light treatment is another option; however, this is not always feasible, because the doses of melatonin required to cause the necessary phase advance also cause somnolence in some patients with SAD who seem to be very sensitive to melatonin's soporific effect. This effect can often be eliminated by using very low doses of melatonin (e.g., 0.075–0.125 mg) given every few hours in the afternoon and evening; however, even these doses will be sedating in some people. Of note, it is possible to overly advance the ECP; a phase advance of greater than 1.5 h probably should be avoided.

It should be emphasized that to achieve the full therapeutic effect of the phase advance caused by morning light or evening melatonin, it is important that a person maintains the same sleep/wake cycle. That is, the ECP needs to be advanced in relation to the sleep/wake cycle (Lewy et al., 1990c, 1989).

Recently, Lewy et al. (2006) published a study that establishes the circadian basis of winter depression. The same study indicated that appropriately timed low-dose melatonin administration may be helpful. There appears to be a small subgroup of SAD patients who are phase-advanced and best benefit from melatonin administered in the morning so as to provide a corrective phase delay. However, most SAD patients should take low-dose melatonin in the afternoon/evening, at which time it will cause a therapeutic phase advance. The phase angle difference (PAD), or time interval, between the DLMO and the time of midsleep is on average 6 h in healthy controls (Lewy et al., 1998c; **Figure 7**).

PAD 6 appears to be the sweet spot for optimal mood. PAD 6 can also be used for phase typing. Individuals with PADs >6 can be considered to be phase-advanced and those with PADs ≤ 6 can be considered to be phase-delayed. Bright light and low-dose melatonin can then be scheduled at the appropriate times. In this study (Lewy et al., 2006), the prototypical phase-delayed patients who received low-dose melatonin in the afternoon/evening to provide a corrective phase advance showed a remarkably robust relationship between depression scores and circadian misalignment (**Figure 8**; replicated in Lewy et al. (2007a)). Indeed, this study may be the first to demonstrate a correlation between psychiatric

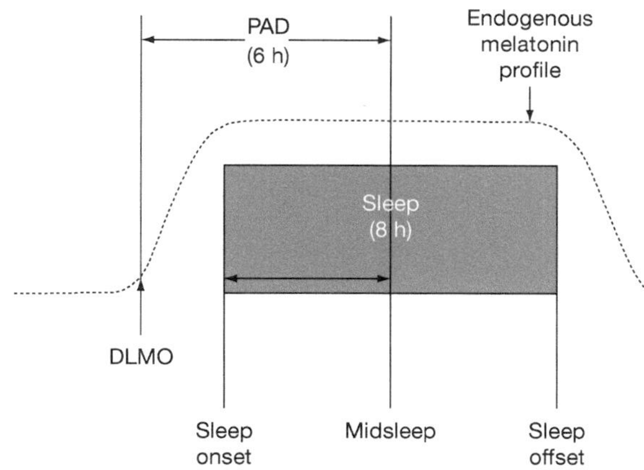

Figure 7 Schematic diagram of normal phase relationships (rounded to the nearest integer) between sleep phase markers, the 10 pg ml^{-1} plasma dim light melatonin onset (DLMO), (10) and the core body temperature minimum (Tmin) (27, 32) derived from historical controls. The present study used the DLMO/midsleep interval (phase angle difference, or PAD) of 6 h as the hypothesized therapeutic window for optimal circadian alignment. Sleep times were determined actigraphically. Plasma melatonin levels were obtained under dim light every 30 min in the evening. The operational definition of the DLMO is the interpolated time of continuous rise above the threshold of 10 pg m^{-1}; for example, if the melatonin level at 8 p.m. was 5 pg ml^{-1} and at 8:30 p.m. was 15 pg ml^{-1}, the DLMO would be 8:15 p.m. Reproduced from Lewy AJ, Lefler BJ, Emens JS, and Bauer VK (2006) The circadian basis of winter depression. *Proceedings of the National Academy of Sciences of the United States of America* 103: 7414–7419, copyright (2006) National Academy of Sciences, USA.

symptom severity and a physiological marker in the same subjects before and after treatment. Analysis of the pretreatment data in an extant data set (Lewy et al., 1998c) also revealed a significant parabolic correlation, with the minimum at PAD 6; once again, two-thirds of the patients were phase-delayed and the others were phase-advanced (Lewy et al., 2007a). This work is now being extended to other types of sleep and mood disorders, such as nonseasonal unipolar depression (Emens et al., 2008).

PAD 6 may also offer a way to create an animal model of circadian misalignment and perhaps depression. Restricted access to a running wheel and food availability can regulate a rodent's sleep/wake cycle. Since the melatonin rhythm is regulated by the light/dark cycle, the time of the melatonin onset can be precisely set in relation to midsleep. The behavior of the animal can then be studied, as well as myriad biological parameters that cannot be studied in humans.

The use of a dawn simulator is another possible way to provide phase advances. This method may be

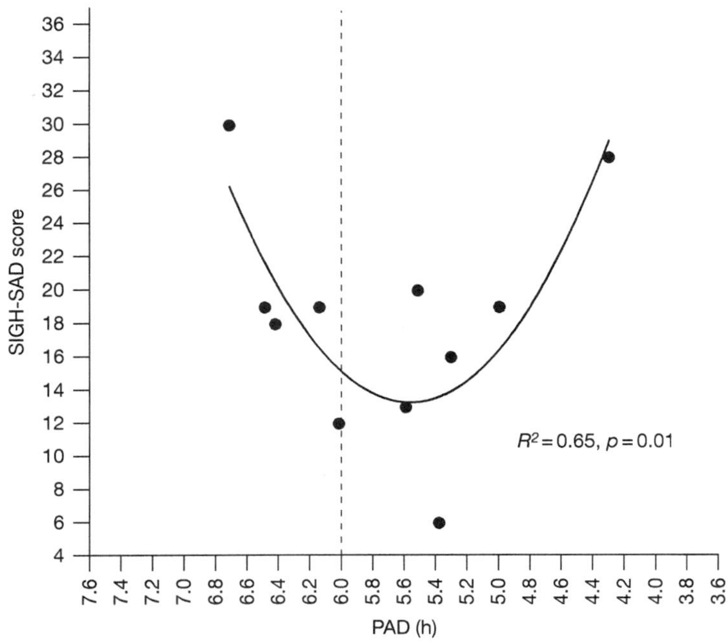

Figure 8 Posttreatment SIGH-SAD depression score as a function of PAD in seasonal affective disorder (SAD) patients who were phase-typed as delayed before treatment. They received PM melatonin to provide a phase advance of the DLMO with regard to mid sleep (i.e, to increase PAD). The parabolic curve indicates that PAD accounts for 65% of the variance in SIGH-SAD scores (F (2,8) = 7.57; minimum = 5.56). Four patients became overly phase-advanced (i.e, phase-shifted) through the therapeutic window, or sweet spot, of PAD 6, representing optimal circadian alignment and mood. Adapted from Lewy AJ, Lefler BJ, Emens JS, and Bauer VK (2006) The circadian basis of winter depression. *Proceedings of the National Academy of Sciences of the United States of America* 103: 7414–7419, copyright (2006) National Academy of Sciences, USA.

better tolerated in some people, because it does not require waking up early for treatment. The dawn simulator can be set to provide increasing light intensity starting very early in the morning, and it can cause a phase advance, even when the eyelids are closed during sleep (Avery et al., 1993; Terman et al., 1989). This is because, according to the light PRC, the greatest phase advances are achieved in the middle of the night. However, no test has yet been done without the presence of light exposure after awakening, that is, when the eyes have been closed throughout the light exposure.

17.12 Speculation on the Function of Endogenous Melatonin Production

As mentioned above, there is a dose–response curve for the phase-shifting effects of melatonin in the physiological range. In fact, BFRs have entrained to a daily dose of melatonin as low as 10 μg (**Figure 9**), which produces melatonin levels no greater than those produced endogenously. This further suggests that there is a circadian function for melatonin. In addition to the function discussed earlier, there may be a more important role for endogenous melatonin: endogenous melatonin from the mother may serve to entrain the third-trimester fetus and the suckling infant to her sleep/wake cycle (Lewy, 2007). After a few months of age, the light/dark cycle is able to entrain the infant.

17.13 A Possible Bioassay for Sensitivity to the Weak Zeitgebers Reveals a Gender Difference

Contrary to what was first thought, all BFRs studied to date under highly resolved circadian phase assessments (modally, about every 2 weeks for a complete circadian beat cycle) exhibit varying degrees of the same pattern of relative coordination to as-yet-unknown weak zeitgebers (Emens et al., 2005) which are probably related to social cues. These weak zeitgebers are insufficient for entraining most BFRs. However, there is a marked

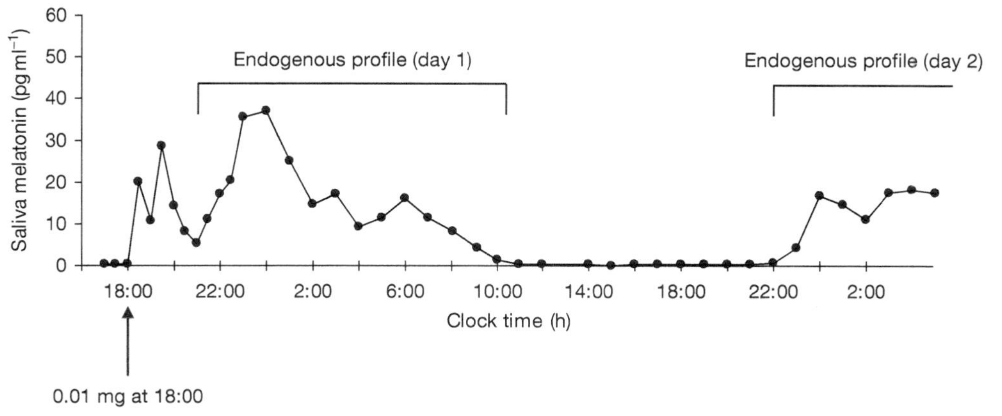

Figure 9 Salivary pharmacokinetic (PK) profile and endogenous melatonin levels in a BFR entrained to a daily dose of 0.01 mg of exogenous melatonin. The dose is given at 1800 (1 h earlier than usual in order to better discern its PK profile) and the endogenous melatonin offset (MO) occurs a few hours later. In the steady-state situation, there is no melatonin-free interval between the exogenous and endogenous profiles, maximizing the phase-shifting effects of this very low dose, which must produce a daily phase advance of least 0.1 h to entrain this subject. Although saliva levels are less reliable than plasma levels (at least for higher concentrations, when the three-fold difference between them becomes greater), there is no doubt that this dose is producing no more than a physiologic level for this subject. This finding suggests that very low doses may be therapeutic and that there may be a circadian function for endogenous melatonin production. Reproduced from Lewy AJ (2007) Melatonin and human chronobiology. Cold Spring Harbor Symposia on Quantitative Biology, vol. 72, pp. 626–636. Woodbury, NJ: Cold Spring Harbor Laboratory Press, with permission.

gender difference (Lewy, 2007; Lewy et al., 2007b). Using stricter operational definitions of entrainment versus free-running than previously used, about 25% of adult females appear to be naturally entrained to them, whereas none of 25 totally blind males appear to be entrained. This gender difference probably starts after puberty (in fact, there may be a higher proportion of prepubertal totally blind males who are naturally entrained than prepubertal females). Among adult BFRs, the amplitude of relative coordination is much greater in females than in males. Relative coordination amplitude and circadian status may thus be bioassays to assess physiologic sensitivity to the weak zeitgebers.

17.14 Summary

Progress has been made during the past 20 years elucidating the roles of the pineal gland and its primary secretory product, melatonin. Melatonin has been shown to be a hormone with a clearly defined physiologic role. The production and secretion of melatonin are regulated by the ECP in the SCN of the hypothalamus. Both light exposure and exogenous melatonin can influence the ECP and, consequently, melatonin secretion.

Normally phased endogenous rhythms are important in maintaining good physical and mental health. Disturbances in the ECP can lead to sleep and mood disturbances. Both melatonin and light can be used effectively to treat these rhythm disturbances. Ongoing work in this area will help refine the current therapeutic approaches and will hopefully find new ways to manipulate the body's rhythms to help maintain optimum health.

References

Akerstedt T, Fröberg JE, Friberg Y, and Wetterberg L (1979) Melatonin excretion, body temperature, and subjective arousal during 64 hours of sleep deprivation. *Psychoneuroendocrinology* 4: 219–225.

Arendt J (1978) Melatonin assays in body fluids. *Journal of Neural Transmission Supplement* 13: 265–278.

Arendt J, Aldhous M, and Marks V (1986) Aleviation of 'jet lag' by melatonin: Preliminary results of controlled double blind trial. *British Medical Journal* 292: 1170.

Arendt J, Bojkowski C, Folkard S, et al. (1985) Some effects of melatonin and the control of its secretion in humans. In: Evered D and Clark S (eds.) *Ciba Foundation Symposium 117: Photoperiodism, Melatonin and the Pineal*, pp. 266–283. London: Pitman.

Ariëns-Kappers J (1960) The development, topographical relations and innervation of the epiphysis cerebri in the albino rat. *Zeitschrift fuer Zellforschung und Mikroskopische Anatomie* 52: 163–215.

Armstrong SM, Cassone VM, Chesworth MJ, Redman JR, and Short RV (1986) Synchronization of mammalian circadian rhythms by melatonin. *Journal of Neural Transmission Supplement* 21: 375–394.

Avery DH, Bolte MA, Dager SR, Wilson LG, Weyer M, Cox GB, and Dunner DL (1993) Dawn simulation treatment of winter

depression: A controlled study. *American Journal of Psychiatry* 150: 113–117.
Avery D, Khan A, Dager S, Cohen S, Cox G, and Dunner D (1991) Morning or evening bright light treatment of winter depression? The significance of hypersomnia. *Biological Psychiatry* 29: 117–126.
Axelrod J and Zatz M (1977) The ß-adrenergic receptor and the regulation of circadian rhythms in the pineal gland. In: Litwack G (ed.) *Biochemical Actions of Hormones,* vol. 4, pp. 249–268. New York: Academic Press.
Badia P, Myers B, Boecker M, Culpepper J, and Harsh J (1991) Bright light effects on body temperature, alertness, EEG, and behavior. *Physiology and Behavior* 50: 583–588.
Barnes RG, Deacon SJ, Forbes MJ, and Arendt J (1998) Adaptation of the 6-sulphatoxymelatonin rhythm in shift-workers on offshore oil installations during a 2-week 12-h night shift. *Neuroscience Letters* 241: 9–12.
Berga SL, Mortola JF, and Yen SSC (1988) Amplification of nocturnal melatonin secretion in women with functional hypothalamic amenorrhea. *Journal of Clinical Endocrinology and Metabolism* 66: 242–244.
Brainard GC, Hanifin JP, Greeson JM, Byrne B, Glickman G, Gerner E, and Rollag MD (2001) Action spectrum for melatonin regulation in humans: Evidence for a novel circadian photoreceptor. *Journal of Neuroscience* 21(16): 6405–6412.
Brainard GC, Lewy AJ, Menaker M, et al. (1985) Effect of light wavelength on the suppression of nocturnal plasma melatonin in normal volunteers. *Annals of the New York Academy of Sciences* 453: 376–378.
Brzezinski A, Lynch HJ, Seibel MM, Deng MH, Nader TM, and Wurtman RJ (1988) The circadian rhythm of plasma melatonin during the normal menstrual cycle and in amenorrheic women. *Journal of Clinical Endocrinology and Metabolism* 66: 891–895.
Cardinali DP, Golombek DA, Rosenstein RE, Cutrera RA, and Esquifino AI (1997) Melatonin site and mechanism of action: Single or multiple? *Journal of Pineal Research* 23: 32–39.
Carskadon MA, Acebo C, Richardson GS, Tate BA, and Seifer R (1997) An approach to studying circadian rhythms of adolescent humans. *Journal of Biological Rhythms* 12: 278–289.
Cassone VM, Chesworth MJ, and Armstrong SM (1986a) Dose dependent entrainment of rat circadian rhythms by daily injection of melatonin. *Journal of Biological Rhythms* 1: 219–229.
Cassone VM, Chesworth MJ, and Armstrong SM (1986b) Entrainment of rat circadian rhythms by daily injection of melatonin depends upon the hypothalamic suprachiasmatic nuclei. *Physiology and Behavior* 36: 1111–1121.
Cavallo A (1991) Plasma melatonin rhythm in disorders of puberty: Interactions of age and pubertal stages. *Hormone Research* 37: 185–189.
Cavallo A (1992a) Melatonin secretion during adrenarche in normal human puberty and in pubertal disorders. *Journal of Pineal Research* 12: 71–78.
Cavallo A (1992b) Plasma melatonin rhythm in normal puberty: Interactions of age and pubertal stages. *Neuroendocrinology* 55: 372–379.
Cavallo A and Dolan LM (1996) 6-Hydroxymelatonin sulfate excretion in human puberty. *Journal of Pineal Research* 21: 225–230.
Cavallo A, Richards GE, and Smith ER (1992c) Relation between nocturnal melatonin profile and hormonal markers of puberty in humans. *Hormone Research* 37: 185–189.
Chesworth MJ, Cassone VM, and Armstrong SM (1987) Effects of daily melatonin injections on activity rhythms of rats in constant light. *American Journal of Physiology* 253: R101–R107.
Czeisler CA, Allan JS, Strogatz SH, et al. (1986) Bright light resets the human circadian pacemaker independent of the timing of the sleep–wake cycle. *Science* 233: 667–671.
Czeisler CA, Johnson MP, Duffy JF, Brown EN, Ronda JM, and Kronauer RE (1990) Exposure to bright light and darkness to treat physiologic maladaptation to night work. *New England Journal of Medicine* 322: 1253–1259.
Czeisler CA and Klerman EB (1999) Circadian and sleep-dependent regulation of hormone release in humans. *Recent Progress in Hormone Research* 54: 97–132.
Czeisler CA, Kronauer RE, Allan JS, Duffy JF, Jewett ME, Brown EN, and Ronda JM (1989) Bright light induction of strong (Type O) resetting of the human circadian pacemaker. *Science* 244: 1328–1333.
Czeisler CA, Shanahan TL, et al. (1995) Suppression of melatonin secretion in some blind patients by exposure to bright light. *New England Journal of Medicine* 332: 6–11.
Daan S and Lewy AJ (1984) Scheduled exposure to daylight: A potential strategy to reduce "jet lag" following transmeridian flight. *Psychopharmacology Bulletin* 20: 566–568.
Dahlitz M, Alvarez B, Vignau J, English J, Arendt J, and Parkes JD (1991) Delayed sleep phase syndrome response to melatonin. *Lancet* 337: 1121–1123.
Drifhout WJ, Grol CJ, and Westerink HC (1996) Parasympathetic inhibition of pineal indole metabolism by prejunctional modulation of noradrenaline release. *European Journal of Pharmacology* 308: 117–124.
Dubbels R, Reiter RJ, Klenke E, et al. (1995) Melatonin in edible plants identified by radioimmunoassay and by high performance liquid chromatography-mass spectrometry. *Journal of Pineal Research* 18: 28–31.
Dubocovich ML, Benloucif S, and Masana MI (1996) Melatonin receptors in the mammalian suprachiasmatic nucleus. *Behavioural Brain Research* 73: 141–147.
Eastman CI and Miescke KJ (1986) Bright light improves the entrainment of the circadian body temperature rhythm to a 26-hr sleep-wake schedule in humans. *Sleep Research* 15: 271.
Eastman CI, Stewart K, Mahoney MP, Liu L, and Fogg LF (1994) Dark goggles and bright light improve circadian rhythm adaptation to night-shift work. *Sleep* 17: 535–543.
Eastman CI, Young MA, Fogg LF, Liu L, and Meaden PM (1998) Bright light treatment of winter depression. *Archives of General Psychiatry* 55: 883–889.
Edwards AV and Jones CT (1993) Autonomic control of adrenal function. *Journal of Anatomy* 183: 291–307.
Emens JS, Lewy AJ, Lefler BJ, and Sack RL (2005) Relative coordination to unknown 'weak zeitgebers' in free-running blind individuals. *Journal of Biological Rhythms* 20(2): 159–167.
Emens J, Lewy AJ, Yuhas K, Jackman AR, and Johnson KP (2006) Melatonin entrains free-running blind individuals with circadian periods less than 24 hours. *Journal of Sleep and Sleep Disorders Research* 29: A62.
Emens J, Rough J, Arntz D, and Lewy A (2008) Circadian misalignment correlates with symptom severity in non-seasonal depression. *Sleep* (abstract supplement) 31: A314.
Foret J, Bensimon B, Benoit O, and Vieux N (1981) Quality of sleep as a function of age and shift work. In: Reinberg A, Vieux N, and Andlauer P (eds.) *Night and Shift Work: Biological and Social Aspects,* vol. 30, pp. 149–154. Oxford: Pergamon Press.
Hardeland R (1999) Melatonin and 5-methoxytryptamine in non-metazoans. *Reproduction, Nutrition, Development* 39: 399–408.

Hattori A, Migitaka H, Iigo M, et al. (1995) Identification of melatonin in plants and its effects on plasma melatonin levels and binding to melatonin receptors in vertebrates. *Biochemistry and Molecular Biology International* 35: 627–634.
Honma K and Honma S (1988) A human phase response curve for bright light pulses. *Japanese Journal of Psychiatry and Neurology* 42: 167–168.
Hughes RJ and Badia P (1997) Sleep-promoting and hypothermic effects of daytime melatonin administration in humans. *Sleep* 20: 124–131.
Hughes RJ and Lewy AJ (1998a) Light and melatonin treatment of circadian phase sleep disorders. In: Lam RW (ed.) *Seasonal Affective Disorder and Beyond: Light Treatment for SAD and Non-SAD Conditions*, pp. 221–252. Washington, DC: American Psychiatric Press.
Hughes RJ, Sack RL, and Lewy AJ (1998b) The role of melatonin and circadian phase in age-related sleep maintenance insomnia: Assessment in a clinical trial of melatonin replacement. *Sleep* 21: 52–68.
Illnerová H (1978) Melatonin in rat pineal gland and serum; rapid parallel decline after light exposure at night. *Neuroscience Letters* 9: 189–193.
James SP, Wehr TA, Sack DA, Parry BL, and Rosenthal NE (1985) Treatment of seasonal affective disorder with light in the evening. *British Journal of Psychiatry* 147: 424–428.
Jones CR, Campbell SS, Zone SE, et al. (1999) Familial advanced sleep-phase syndrome: A short-period circadian rhythm variant in humans. *Nature Medicine* 5: 1062–1065.
Joseph-Vanderpool JR, Kelly K, Schulz PM, Allen R, Souetre E, and Rosenthal NE (1988) Delayed sleep phase syndrome revisited: Preliminary effects of light and triazolam. *Sleep Research* 17: 381.
Joseph-Vanderpool JR, Rosenthal NE, Levendosky AA, et al. (1989) Phase-shifting effects of bright morning light as treatment for delayed sleep phase syndrome. *Sleep Research* 18: 422.
Kadva A, Djahanbakhch O, Monson J, Di WL, and Silman R (1998) Elevated nocturnal melatonin is a consequence of gonadotropin-releasing hormone deficiency in women with hypothalamic amenorrhea. *Journal of Clinical Endocrinology and Metabolism* 83: 3653–3662.
Kennedy SH, Gnam W, Ralevski E, and Brown GM (1995) Melatonin responses to clonidine and yohimbine challenges. *Journal of Psychiatry and Neuroscience* 20: 297–304.
Klein DC and Moore RY (1970) Indole metabolism in the pineal gland: A circadian rhythm in *N*-acetyltransferase. *Science* 169: 1093–1095.
Klein DC, Sugden D, and Weller JL (1983) Postsynaptic a-adrenergic receptors potentiate the ß-adrenergic stimulation of pineal serotonin *N*-acetyltransferase. *Proceedings of the National Academy of Sciences of the United States of America* 80: 599–603.
Konopka RJ and Benzer S (1971) Clock mutants of *Drosophila melanogaster*. *Proceedings of the National Academy of Sciences of the United States* 68: 2112–2116.
Korf HW (1994) The pineal organ as a component of the biological clock: Phyogenetic and ontogenetic considerations. *Annals of the New York Academy of Sciences* 31: 13–42.
Kripke DF (1981) Photoperiodic mechanisms for depression and its treatment. In: Perris C, Struwe G, and Jansson B (eds.) *Biological Psychiatry*, pp. 1249–1252. Amsterdam: Elsevier.
Lack LC (1986) Delayed sleep and sleep loss in university students. *Journal of American College Health* 35: 105–110.
Laitinen JT, Laitinen KS, and Kokkola T (1995) Cholinergic signaling in the rat pineal gland. *Cellular and Molecular Neurobiology* 15: 177–192.
Laughlin GA, Loucks AB, and Yen SSC (1991) Marked augmentation of nocturnal melatonin secretion in amenorrheic athletes, but not in cycling athletes: Unaltered by opioidergic or dopaminergic blockade. *Journal of Clinical Endocrinology and Metabolism* 73: 1321–1326.
Lerner AB, Case JD, and Heinzelman RV (1959) Structure of melatonin. *Journal of American Chemical Society* 81: 6084–6090.
Lewy AJ (1981) Human plasma melatonin studies: Effects of light and implications for biological rhythm research. In: Birau N and Schloot W (eds.) *Melatonin: Current Status and Perspectives, Vol. 29: Advances in the Biosciences*, pp. 397–400. Oxford: Pergamon Press.
Lewy AJ (1983) Biochemistry and regulation of mammalian melatonin production. In: Relkin RM (ed.) *The Pineal Gland*, pp. 77–128. New York: Elsevier North-Holland.
Lewy AJ (1990) Chronobiologic disorders, social cues and the light–dark cycle. *Chronobiology International* 7: 15–21.
Lewy AJ (2007) Melatonin and human chronobiology. *Cold Spring Harbor Symposia on Quantitative Biology*, vol. 72, pp. 626–636. Woodbury, NJ: Cold Spring Harbor Laboratory Press.
Lewy AJ, Ahmed S, Jackson JML, and Sack RL (1992) Melatonin shifts circadian rhythms according to a phase-response curve. *Chronobiology International* 9: 380–392.
Lewy AJ, Bauer VK, Ahmed S, et al. (1998a) The human phase response curve (PRC) to melatonin is about 12 hours out of phase with the PRC to light. *Chronobiology International* 15: 71–83.
Lewy AJ, Bauer VK, Cutler NL, and Sack RL (1998b) Melatonin treatment of winter depression: A pilot study. *Psychiatry Research* 77: 57–61.
Lewy AJ, Bauer VK, Cutler NL, et al. (1998c) Morning versus evening light treatment of patients with winter depression. *Archives of General Psychiatry* 55: 890–896.
Lewy AJ, Bauer VK, Hasler BP, Kendall AR, Pires LN, and Sack RL (2001) Capturing the circadian rhythms of free-running blind people with 0.5 mg melatonin. *Brain Research* 918: 96–100.
Lewy AJ, Bauer BK, Singer CW, Minkunas DV, and Sack RL (2000) Later circadian phase of plasma melatonin relative to usual waketime in older subjects. *Sleep* 23: A188.
Lewy AJ, Cutler NL, and Sack RL (1999) The endogenous melatonin profile as a marker for circadian phase position. *Journal of Biological Rhythms* 14: 227–236.
Lewy AJ, Emens JS, Lefler BJ, Yuhas K, and Jackman AR (2005) Melatonin entrains free-running blind people according to a physiological dose-response curve. *Chronobiology International* 22(6): 1093–1106.
Lewy AJ, Emens JS, Sack RL, Hasler BP, and Bernert RA (2002) Low, but not high, doses of melatonin entrained a free-running blind person with a long circadian period. *Chronobiology International* 19(3): 649–658.
Lewy AJ, Kern HA, Rosenthal NE, and Wehr TA (1982) Bright artificial light treatment of a manic-depressive patient with a seasonal mood cycle. *American Journal of Psychiatry* 139: 1496–1498.
Lewy AJ, Lefler BJ, Emens JS, and Bauer VK (2006) The circadian basis of winter depression. *Proceedings of the National Academy of Sciences of the United States of America* 103: 7414–7419.
Lewy AJ and Newsome DA (1983) Different types of melatonin circadian secretory rhythms in some blind subjects. *Journal of Clinical Endocrinology and Metabolism* 56: 1103–1107.
Lewy AJ, Rough J, Songer J, Mishra N, Yuhas K, and Emens J (2007a) The phase shift hypothesis for the circadian component of winter depression. *Dialogues in Clinical Neuroscience* 9: 291–300.

Lewy AJ and Sack RL (1986) Melatonin physiology and light therapy. *Clinical Neuropharmacology Supplement* 9: 196–198.
Lewy AJ and Sack RL (1989) The dim light melatonin onset (DLMO) as a marker for circadian phase position. *Chronobiology International* 6: 93–102.
Lewy AJ and Sack RL (1996) The role of melatonin and light in the human circadian system. In: Buijs R, Kalsbeek A, Romijn A, Pennartz C, and Mirmiran M (eds.) *Progress in Brain Research: Hypothalamic Integration of Circadian Rhythms,* vol. 111, pp. 205–216. Amsterdam: Elsevier.
Lewy AJ, Sack RL, Blood ML, Bauer VK, Cutler NL, and Thomas KH (1995) Melatonin marks circadian phase position and resets the endogenous circadian pacemaker in humans. In: Chadwick DJ and Ackrill K (eds.) *Ciba Foundation Symposium 183: Circadian Clocks and Their Adjustment*, pp. 303–321. New York: Wiley.
Lewy AJ, Sack RL, Fredrickson RH, Reaves M, Denney D, and Zielske DR (1983) The use of bright light in the treatment of chronobiologic sleep and mood disorders: The phase-response curve. *Psychopharmacology Bulletin* 19: 523–525.
Lewy AJ, Sack RL, and Latham JM (1990a) Circadian phase shifting of blind and sighted people with exogenous melatonin administration: Evidence for a phase response curve. *Society for Light Treatment and Biological Rhythms Abstracts* 2: 22.
Lewy AJ, Sack RL, and Latham JM (1990b) Exogenous melatonin administration shifts circadian rhythms according to a phase response curve. In: *The Vth Colloquium of the European Pineal Study Group,* Abstract 021. Guildford, UK.
Lewy AJ, Sack RL, and Latham JM (1991) Melatonin and the acute suppressant effect of light may help regulate circadian rhythms in humans. In: Arendt J and Pevét P (eds.) *Advances in Pineal Research,* vol. 5, pp. 285–293. London: John Libbey.
Lewy AJ, Sack RL, Miller S, and Hoban TM (1987) Antidepressant and circadian phase-shifting effects of light. *Science* 235: 352–354.
Lewy AJ, Sack RL, and Singer CM (1984) Assessment and treatment of chronobiologic disorders using plasma melatonin levels and bright light exposure: The clock-gate model and the phase response curve. *Psychopharmacology Bulletin* 20: 561–565.
Lewy AJ, Sack RL, and Singer CM (1985a) Immediate and delayed effects of bright light on human melatonin production: Shifting 'dawn' and 'dusk' shifts the dim light melatonin onset (DLMO). *Annals of the New York Academy of Sciences* 453: 253–259.
Lewy AJ, Sack RL, and Singer CM (1985b) Melatonin, light and chronobiological disorders. In: Evered D and Clark S (eds.) *Ciba Foundation Symposium: Photoperiodism, Melatonin and the Pineal,* vol. 117, pp. 231–252. London: Pittman.
Lewy AJ, Sack RL, and Singer CM (1990c) Bright light, melatonin, and winter depression: The phase-shift hypothesis. In: Shafii MA and Shafii SL (eds.) *Biological Rhythms, Mood Disorders, Light Therapy, and the Pineal Gland*, pp. 143–173. Washington, DC: American Psychiatric Press.
Lewy AJ, Sack RL, Singer CM, White DM, and Hoban TM (1988) Winter depression and the phase shift hypothesis for bright light's therapeutic effects: History, theory and experimental evidence. *Journal of Biological Rhythms* 3: 121–134.
Lewy AJ, Sack RL, Singer CM, White DA, and Hoban TM (1989) Winter depression: The phase angle between sleep and other circadian rhythms may be critical. In: Thompson C and Silversone T (eds.) *Seasonal Affective Disorder*, pp. 205–221. London: Clinical Neuroscience.
Lewy AJ, Siever LJ, Uhde TW, and Markey SP (1986) Clonidine reduces plasma melatonin levels. *Journal of Pharmacy and Pharmacology* 38: 555–556.
Lewy AJ, Tetsuo M, Markey SP, Goodwin FK, and Kopin IJ (1980a) Pinealectomy abolishes plasma melatonin in the rat. *Journal of Clinical Endocrinology and Metabolism* 50(1): 204–205.
Lewy AJ, Yuhas K, Emens J, et al. (2007b) Are the circadian rhythms of blind adult males less sensitive to social cues than females? *Sleep* 30(abstract supplement): A63.
Lewy AJ, Wehr TA, Goodwin FK, Newsome DA, and Markey SP (1980b) Light suppresses melatonin secretion in humans. *Science* 210: 1267–1269.
Luboshitzky R, Lavi S, Thuma I, Herer P, and Lavie P (1996a) Nocturnal secretory patterns of melatonin, luteinizing hormone, prolactin and cortisol in male patients with gonadotropin-releasing hormone deficiency. *Journal of Pineal Research* 21: 49–54.
Luboshitzky R, Lavi S, Thuma I, and Lavie P (1995) Increased nocturnal melatonin secretion in male patients with hypogonadotropic hypogonadism and delayed puberty. *Journal of Clinical Endocrinology and Metabolism* 80: 2144–2148.
Luboshitzky R, Lavi S, Thuma I, and Lavie P (1996b) Testosterone treatment alters melatonin concentrations in male patients with gonadotropin-releasing hormone deficiency. *Journal of Clinical Endocrinology and Metabolism* 81: 770–774.
Luboshitzky R, Shen-Orr Z, Ishai A, and Lavie P (2000) Melatonin hypersecretion in male patients with adult-onset idiopathic hypogonadotropic hypogonadism. *Experimental and Clinical Endocrinology and Diabetes* 108: 142–145.
Luboshitzky R, Wagner O, Lavi S, Herer P, and Lavie P (1996c) Decreased nocturnal melatonin secretion in patients with Klinefelter's syndrome. *Clinical Endocrinology* 45: 749–754.
Luboshitzky R, Wagner O, Lavi S, Herer P, and Lavie P (1997) Abnormal melatonin secretion in hypogonadal men: The effect of testosterone treatment. *Clinical Endocrinology* 47: 463–469.
Mallo C, Zaidan R, Faure A, Brun J, Chazot G, and Claustrat B (1988) Effects of a four-day nocturnal melatonin treatment on the 24 h plasma melatonin, cortisol and prolactin profiles in humans. *Acta Endocrinologica (Copen)* 119: 474–480.
Martin SK and Eastman CI (1998) Medium-intensity light produces circadian rhythm adaptation to simulated night-shift work. *Sleep* 21: 154–165.
Matsumoto M, Sack RL, Blood ML, and Lewy AJ (1997) The amplitude of endogenous melatonin production is not affected by melatonin treatment in humans. *Journal of Pineal Research* 22: 42–44.
Middleton B, Arendt J, and Stone BM (1997) Complex effects of melatonin on human circadian rhythms in constant dim light. *Journal of Biological Rhythms* 12: 467–477.
Minors DS, Waterhouse JM, and Wirz-Justice A (1991) A human phase-response curve to light. *Neuroscience Letters* 133: 36–40.
Molina-Carballo A, Muñoz-Hoyos A, Martin-García JA, Uberos-Fernández J, Rodriguez-Cabezas T, and Acuña-Castroviejo D (1996) 5-Methoxytryptophol and melatonin in children: Differences due to age and sex. *Journal of Pineal Research* 21: 73–79.
Moore RY and Lenn NJ (1972) A retinohypothalamic projectionin the rat. *Journal of Comprehensive Neurology* 146: 1–14.
Murch SJ, Simmons CB, and Saxena PK (1997) Melatonin in feverfew and other medicinal plants. *Lancet* 29: 1598–1599.
Neuwelt EA and Lewy AJ (1983) Disappearance of plasma melatonin after removal of a neoplastic pineal gland. *New England Journal of Medicine* 308: 1132–1135.

Okatani Y and Sagara Y (1994) Amplification of nocturnal melatonin secretion in women with functional secondary amenorrhoea: Relation to endogenous estrogen concentration. *Clinical Endocrinology* 41: 763–770.

Okatani Y and Sagara Y (1995) Enhanced nocturnal melatonin secretion in women with functional secondary amenorrhea: Relationship to opioid system and endogenous estrogen levels. *Hormone Research* 43: 194–199.

Oldani A, Ferini-Strambi L, Zucconi M, Stankov B, Fraschini F, and Smirne S (1994) Melatonin and delayed sleep phase syndrome: Ambulatory polygraphic evaluation. *NeuroReport* 6: 132–134.

Pang SF, Dubocovich ML, and Brown GM (1993) Melatonin receptors in peripheral tissues: A new area of melatonin research. *Biological Signals* 2: 177–180.

Pelayo RP, Thorpy MJ, and Glovinsky P (1988) Prevalence of delayed sleep phase syndrome in adolescents. *Sleep Research* 12: 239.

Penny R (1982) Melatonin excretion in normal males and females: Increase during puberty. *Metabolism* 31: 816–823.

Pickard GE and Turek FW (1983) The hypothalamic paraventricular nucleus mediates the photoperiodic control of reproduction but not the effects of light on the circadian rhythm of activity. *Neuroscience Letters* 43: 67–72.

Pittendrigh CS and Daan S (1976) A functional analysis of circadian pacemakers in nocturnal rodents: IV. Entrainment: Pacemaker as a clock. *Journal of Comparative Physiology* 106: 291–331.

Rajmil O, Puig-Domingo M, Tortosa F, Viader M, Patterson AG, Schwarzstein D, and de Leiva A (1997) Melatonin concentration before and during testosterone replacement in primary hypogonadic men. *European Journal of Endocrinology* 137: 48–52.

Redman J, Armstrong S, and Ng KT (1983) Free-running activity rhythms in the rat: Entrainment by melatonin. *Science* 219: 1089–1091.

Reiter RJ (1990) Pineal rhythmicity: Neural, behavioral, and endocrine consequences. In: Shafii M and Shafii SL (eds.) *Biological Rhythms, Mood Disorders, Light Therapy and the Pineal Gland*, pp. 39–66. Washington, DC: American Psychiatric Press.

Rosenthal NE, Sack DA, Carpenter CJ, Parry BL, Mendelson WB, and Wehr TA (1985) Antidepressant effects of light in seasonal affective disorder. *American Journal of Psychiatry* 142: 163–170.

Rosenthal NE, Sack DA, Gillin JC, et al. (1984) Seasonal affective disorder: A description of the syndrome and preliminary findings with light therapy. *Archives of General Psychiatry* 41: 72–80.

Rosenthal NE, Sack DA, James SP, Parry BL, Mendelson WB, Tamarkin L, and Wehr TA (1985b) Seasonal affective disorder and phototherapy. *Annals of the New York Academy of Sciences* 453: 260–269.

Sack RL, Blood ML, Hughes RJ, and Lewy AJ (1998) Circadian rhythm sleep disorders in the totally blind. *Journal of Visual Impairment and Blindness* 92: 145–161.

Sack RL, Blood ML, and Lewy AJ (1992a) Melatonin rhythms in night shift workers. *Sleep* 15: 434–441.

Sack R, Brandes R, Kendall A, and Lewy A (2000) Entrainment of free-running circadian rhythms by melatonin in blind people. *New England Journal of Medicine* 343: 1070–1077.

Sack RL, Blood ML, and Lewy AJ (2002) Melatonin rhythms in night shift workers. *Sleep* 15: 434–441.

Sack RL and Lewy AJ (1986) Desmethylimipramine treatment increases melatonin production in humans. *Biological Psychiatry* 21: 406–410.

Sack RL, Lewy AJ, Blood ML, Keith LD, and Nakagawa H (1992b) Circadian rhythm abnormalities in totally blind people: Incidence and clinical significance. *Journal of Clinical Endocrinology and Metabolism* 75: 127–134.

Sack RL, Lewy AJ, Blood ML, Stevenson J, and Keith LD (1991) Melatonin administration to blind people: Phase advances and entrainment. *Journal of Biological Rhythms* 6: 249–261.

Sack RL, Lewy AJ, and Hoban TM (1987) Free-running melatonin rhythms in blind people: Phase shifts with melatonin and triazolam administration. In: Rensing L, Ander Heiden U, and Mackey MC (eds.) *Temporal Disorder in Human Oscillatory Systems*, pp. 219–224. Heidelberg: Springer.

Sack RL, Lewy AJ, White DM, Singer CM, Fireman MJ, and Vandiver R (1990) Morning versus evening light treatment for winter depression; evidence that the therapeutic effects of light are mediated by circadian phase shifts. *Archives of General Psychiatry* 47: 343–351.

Salti R, Galluzzi F, Bindi G, Perfetto F, Tarquini R, Halberg F, and Cornelissen G (2000) Nocturnal melatonin patterns in children. *Journal of Clinical Endocrinology and Metabolism* 85: 2137–2144.

Sharkey KM and Eastman CI (2000) Phase-advancing human circadian rhythms with melatonin. *Society for Light Treatment and Biological Rhythms* 12.

Singer CM and Lewy AJ (1989) Case report: Use of the dim light melatonin onset in the treatment of ASPS with bright light. *Sleep Research* 18: 445.

Strassman RJ, Appenzeller O, Lewy AJ, Qualls CR, and Peake GT (1989) Increase in plasma melatonin, ß-endorphin, and cortisol after a 28.5-mile mountain race: Relationship to performance and lack of effect of Naltrexone. *Journal of Clinical Endocrinology and Metabolism* 69: 540–545.

Terman JS, Terman M, Lo E, and Cooper TB (2001) Circadian time of morning light administration and therapeutic response in winter depression. *Archives of General Psychiatry* 58: 69–75.

Terman M, Schlager D, Fairhurst S, and Perlman B (1989) Dawn and dusk simulation as a therapeutic intervention. *Biological Psychiatry* 25: 966–970.

Terman M, Terman JS, and Ross DC (1998) A controlled trial of timed bright light and negative air ionization for treatment of winter depression. *Archives of General Psychiatry* 55: 875–882.

Tetsuo M, Poth M, and Markey SP (1982) Melatonin metabolite excretion during childhood and puberty. *Journal of Clinical Endocrinology and Metabolism* 55: 311–313.

Thorpy MJ, Korman E, Spielman AJ, and Glovinsky PB (1988) Delayed sleep phase syndrome in adolescents. *Journal of Adolescent Health Care* 9: 22–27.

Toh KL, Jones CR, He Y, et al. (2001) An hPer2 phosphorylation site mutation in familial advanced sleep-phase syndrome. *Science* 291: 1040–1043.

Tzischinsky O, Dagan Y, and Lavie P (1993) The effects of melatonin on the timing of sleep in patients with delayed sleep phase syndrome. In: Touitou Y, Arendt J, and Pévet P (eds.) *Melatonin and the Pineal Gland-from Basic Science to Clinical Application*, pp. 251–254. New York: Elsevier.

Underwood H (1986) Circadian rhythms in lizards: Phase response curve for melatonin. *Journal of Pineal Research* 3: 187–196.

Vessely L and Lewy AJ (2002) Melatonin as a hormone and as a marker for circadian phase position in humans. In: Pfaff D, Arnold A, Etgen A, Fahrbach S, and Rubin R (eds.) *Hormones, Brain and Behavior,* vol. 5, pp. 121–141. San Diego, CA: Academic Press.

Waldhauser F, Weiszenbacher G, Tatzer E, Gisinger B, Waldhauser M, Schemper M, and Frisch H (1988) Alterations in nocturnal serum melatonin levels in humans with growth and aging. *Journal of Clinical Endocrinology and Metabolism* 66: 648–652.

Weitzman ED, Czeisler CA, Coleman RM, Spielman AJ, Zimmerman JC, and Dement W (1981) Delayed sleep phase syndrome: A chronobiological disorder with sleep-onset insomnia. *Archives of General Psychiatry* 38: 737–746.

Wetterberg L (1978) Melatonin in humans: Physiological and clinical studies (review). *Journal of Neural Transmission Supplement* 13: 289–294.

Wetterberg L (1979) Clinical importance of melatonin. *Progress in Brain Research* 52: 539–547.

Wever RA (1989) Light effects on human circadian rhythms. A review of recent Andechs experiments. *Journal of Biological Rhythms* 4: 161–186.

Wever R, Polasek J, and Wildgruber C (1983) Bright light affects human circadian rhythms. *European Journal of Physiology* 396: 85–87.

Wirz-Justice A, Graw P, Kräuchi K, Gisin B, Arendt J, Aldhous M, and Pöldinger W (1990) Morning or night-time melatonin is ineffective in seasonal affective disorder. *Journal of Psychiatric Research* 24: 129–137.

Wirz-Justice A, Graw P, Krauchi K, et al. (1993) Light therapy in seasonal affective disorder is independent of time of day or circadian phase. *Archives of General Psychiatry* 50: 929–937.

Zaidan R, Geoffriau M, Brun J, Taillard J, Bureau C, Chazot G, and Claustrat B (1994) Melatonin is able to influence its secretion in humans: Description of a phase-response curve. *Neuroendocrinology* 60: 105–112.

Further Reading

Arendt J, Skene DJ, Middleton B, Lockley SW, and Deacon S (1997) Efficacy of melatonin treatment in jet lag, shift work, and blindness. *Journal of Biological Rhythms* 12: 604–617.

Lewy A, Songer J, Yuhas K, and Emens J (2008) Melatonin regulation of sleep and circadian rhythms in humans. In: Squire LR (ed.) *Encyclopedia of Neuroscience*, pp. 893–900. Oxford: Academic Press.

Lewy AJ (1987) Treating chronobiologic sleep and mood disorders with bright light. *Psychiatric Annals* 17: 664–669.

Lewy AJ, Woods K, Kinzie J, Emens J, Songer J, and Yuhas K (2007c) DLMO/Mid-sleep interval of six hours phase types SAD patients and parabolically correlates with symptom severity. *Sleep* 30(abstract supplement): A63–A64.

18 Neuroendocrine–Immune Interactions: Implications for Health and Behavior

T W W Pace, C L Raison, and A H Miller, Emory University School of Medicine, Atlanta, GA, USA

Chapter Outline

Glossary

acquired immunity (aka adaptive or specific immunity) Elements of immune function that require specific recognition of foreign substances (pathogens) and allow for the establishment of memory cells that react more quickly upon future exposure to the same foreign substance, and rely on hematopoietically derived lymphocytic cells with the ability to specifically recognize a wide array of foreign substances.

clusters of differentiation Molecules expressed on the surface of immune cells, often ligands or receptors, that are commonly used as cell-type markers.

cytokines Proteins produced by immune cells that mediate inflammatory and immune responses and permit communication between immune cells (e.g., interleukin-1 and interleukin-6).

delayed-type hypersensitivity (DTH) reaction An immune reaction involving T-cell-dependent macrophages that results in tissue inflammation, and occurs in response to subcutaneous injection of antigen – an assay for cell-mediated immunity.

depressive-like behavior Behavior that occurs in the absence of cytokine-induced changes in motor activity, which might be more reflective of a sickness state.

enumerative tests of immune function Assessments of immune system function that utilize techniques such as flow cytometry to measure the number or percentages of various immune cell types based on the presence of cell-surface determinants or intracellular factors.

functional tests of immune function Immune system tests performed *in vitro* that evaluate the functional capacity of immune cells (e.g., proliferative assays, assays of natural killer cell activity, and mitogen-stimulated production of cytokines).

glucocorticoid resistance A state characterized by an overall decreased responsiveness of cells, organs, and physiological systems to the effects of glucocorticoids.

innate immunity A first line of defense involves attacking foreign substances rapidly in a nonspecific manner, without requirement for a specific antigen (or antibody-generating molecule) to be recognized, allows for a rapid response but occurs at the cost of not developing an immunological memory, uses relatively crude (nonspecific) pattern-recognition receptors referred to as Toll-like receptors to initiate and mobilize the response to infection and/or tissue damage and destruction.

Janus kinase/signal transducers and activators of transcription (Jak-STAT) A molecular signaling pathway activated when cytokines bind to type 1 or type 2 cytokine receptors, which in turn activate Janus kinase tyrosine kinases, with subsequent phosphorylation of signal transducers and activators of transcription (STAT) proteins that function as transcription factors in the cell nucleus.

mitogen-activated protein kinase (MAPK) cascade A molecular signaling cascade involving multiple individual signaling pathways that is activated by cytokines and/or antigens binding to receptors, all involving initial activation of the Ras protein leading to activation of various MAPKs (p38s JNKs, and ERKs), which in turn phosphorylate downstream enzyme and/or transcription factors.

natural killer cell activity (NKCA) An *in vitro* assay that tests the ability of endogenous natural killer cells (a subset of bone marrow-derived lymphocytes whose function is to kill microbe-infected cells) to destroy target cells, often immortalized cancer cells.

nuclear factor–kappa B (NFκB) A family of transcription factors activated when I kappa B kinase (IKK) phosphorylates IκB. IκB kinase is activated by cytokines and/or antigens binding to their receptors.

sickness behavior The behavioral response which accompanies activation of the innate immune response.

Th1 cytokine response A proinflammatory response involving the secretion of IFN-γ, IL-2, and TNF-α from lymphocytes that promotes cellular immunity and inhibits the Th2 response.

Th2 cytokine response A anti-inflammatory response involving secretion of IL-4, IL-10, and IL-13 from lymphocytes that promotes humoral immunity and inhibits the Th1 response.

18.1 Overview of the Immune System

Prior to consideration of the relationship among the nervous, endocrine, and immune systems, it is important to briefly review the general purpose and internal organization of the immune response. In vertebrates, the immune system has evolved a wide array of separate, but cooperative, mechanisms that serve to attack invading pathogens, destroy tumor cells, and remove and repair damaged tissue. The complexities of immune system functioning are beyond the scope of this chapter. Our attempt here is to provide a necessarily simplified overview of the major immune

Table 1 Divisions of the immune system: innate versus acquired

	Innate	*Acquired*
Physicochemical barriers	Skin, mucous membranes	Cutaneous and mucosal immune systems
Cells	Phagocytes (macrophages, neutrophils, and natural killer cells)	Lymphocytes (B- and T-cells)
Soluble mediators that affect other cells	Macrophage-derived cytokines, i.e., IL-1, IL-6, TNF-α, interferons	Lymphocyte-derived cytokines, i.e., IL-2, IL-4, IL-5, IL-6, IL-10, IFN-γ
Receptors	Toll-like receptors	T-cell receptors, B-cell receptors
Circulating molecules	Complement	Antibodies

IL, interleukin; IFN, interferon; TNF, tumor necrosis factor.

compartments and their integrated and carefully regulated interactions. More extensive reviews of immune system functioning are available elsewhere (Abbas and Lichtman, 2003; Beutler, 2004; Chaplin, 2006).

18.1.1 Innate versus Acquired Immunity

A central distinction within the immune system is between innate, also known as natural or nonspecific immunity, and acquired, or specific, immunity (**Table 1**). These two divisions subsume central tasks in protecting the self from encroachment. Innate immunity provides a first line of defense by attacking foreign substances rapidly in a nonspecific manner, without the requirement that a specific antigen (or antibody-generating molecule) be recognized. This nonselectivity allows for a rapid response, but occurs at the cost of not developing an immunological memory to speed response should a pathogen be encountered again. This immediate response is followed by activation of the acquired immune system, which responds more slowly but more selectively to the particular invading entity. An acquired immune response requires specific recognition of the foreign substance and allows for the establishment of memory cells that react far more quickly when the same foreign substance is again encountered. Finally, both innate and acquired immune systems include mechanisms to control and extinguish activation so that immune activity, with its attendant metabolic costs and danger of self-attack, does not become self-perpetuating.

Innate immunity begins with the skin and mucosal surfaces lining the gastrointestinal and respiratory tracts that provide a physical/chemical barrier to invasion by pathogens. When these surfaces are breached, phagocytic cells, such as macrophages and certain reticular cells, engulf and destroy foreign substances. Cells of the innate immune system use relatively crude (nonspecific) pattern-recognition receptors referred to as Toll-like receptors (TLRs) to initiate and mobilize the response to infectious agents and/or cellular components released as a function of tissue damage or destruction (Abbas and Lichtman, 2003). TLRs in turn are linked to fundamental inflammatory signaling pathways including nuclear factor kappa B (NF-κB) and mitogen-activated protein kinases (MAPKs), which when activated, stimulate the production of the innate immune cytokines interferon alpha (IFN-α), interleukin (IL)-1, IL-6, and tumor necrosis factor-alpha (TNF-α). Aside from the release of cytokines, activated cells of the innate immune system also release other inflammatory mediators including chemokines, which, in combination with the induction of adhesion molecules, attract and engage a multiplicity of immune cells at the site of tissue encroachment. IL-1, IL-6, and TNF-α are important in further orchestrating an inflammatory response at the site of pathogen invasion or tissue damage. In addition, IL-6 plays an essential role in stimulating the liver to produce a host of proteins known as acute-phase reactants that serve to both facilitate destruction of foreign substances and limit tissue damage from immune activation (**Figure 1**; Beutler, 2004). Examples of acute-phase reactants include C-reactive protein and haptoglobin. Inflammatory cytokines also have potent effects on the neuroendocrine system (especially the hypothalamic–pituitary–adrenal (HPA) axis) and the central nervous system (CNS) where they mediate many symptoms of illness, including fever, loss of appetite, social withdrawal, and sleep changes (Raison et al., 2006). Interestingly, as discussed below, evidence exists that innate immune cytokines can also be induced in response to stress in laboratory animals and humans and are increased in at least some patients with major depression. In addition, these cytokines can promote depressive-like changes in behavior and cognition (Raison et al., 2006). The behavioral response

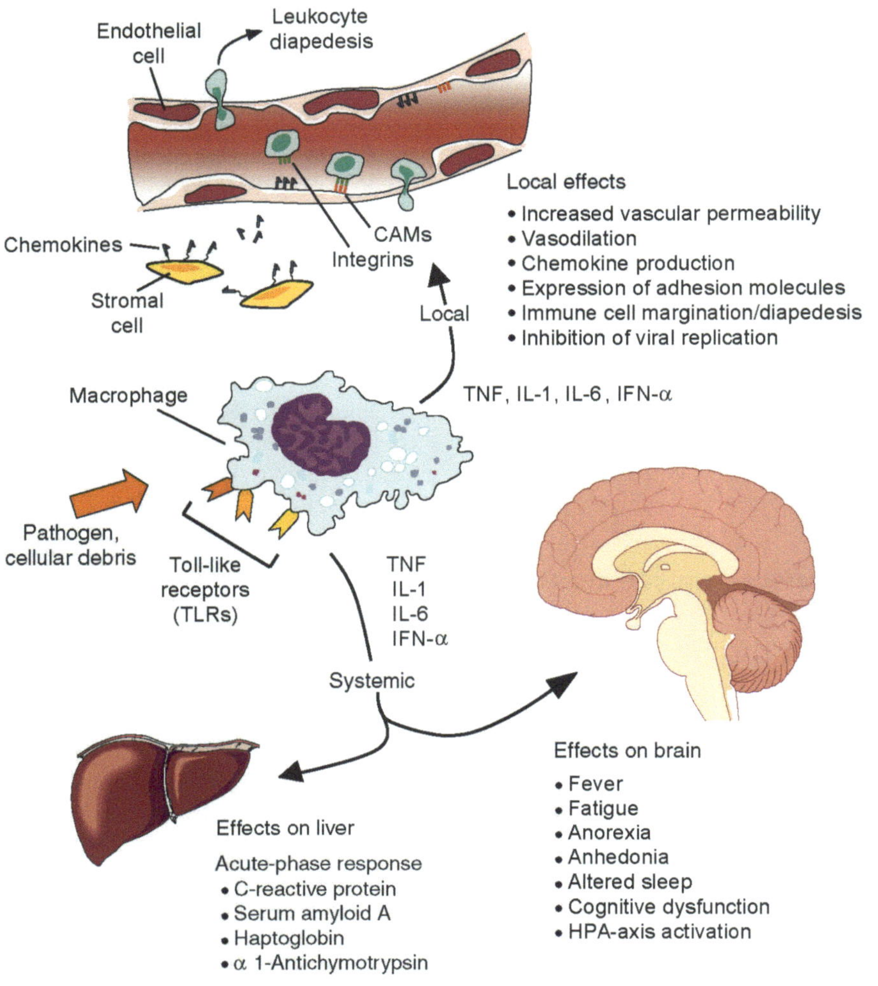

Figure 1 Innate immunity: Local and systemic responses to cytokine release secondary to tissue injury/infection. Macrophages release cytokines upon binding of pathogen or cellular debris to TLRs. Cytokines then act both locally and systemically to combat infection and/or tissue damage. Locally, cytokines stimulate the accumulation of inflammatory cells in several ways, including the activation of adjacent stroma cells to elicit other chemotactic cytokines, as well as the expression of cell-adhesion molecules (CAMs) by endothelial cells that bind inflammatory cells. Inflammatory cells then undergo diapedesis. Cytokines also act locally to inhibit viral replication. Systemically, cytokines induce the liver to produce acute phase proteins such as C-reactive protein. In the brain, proinflammatory cytokines – including interleukin (IL)-1, IL-6, TNF-α, and interferon (IFN)-α – activate the hypothalamic–pituitary–adrenal (HPA) axis and induce behavioral changes that subserve the metabolic demands of fever and inflammation. Reproduced from Cowles MK and Miller AH (2008) Stress, cytokines and depressive illness. In: Squire LR (ed.) *Encyclopedia of Neuroscience*, pp. 519–527. Oxford: Academic Press, with permission from Elsevier.

which accompanies activation of the innate immune response is referred to as sickness behavior and is believed to represent a reorganization of behavioral priorities that is in part geared to conserve energy resources for the attendant metabolic costs of fighting infection.

Acquired immunity relies on hematopoietically derived lymphocytic cells with the ability to specifically recognize an astoundingly wide array of foreign substances, while screening out lymphocytes that might react against self-molecules. These lymphocytes fall into two general categories. T-lymphocytes mature in the thymus and mediate cellular immunity, which is essential for protection against intracellular pathogens, such as viruses and mycobacteria. B-lymphocytes mature in the bone marrow and produce antibodies that are especially effective in neutralizing blood-borne and extracellular pathogens, such as parasites, viruses in replication phase, and many species of bacteria.

Unlike innate immunity, which responds in a general way to foreign substances, acquired immunity activates only cells that have some degree of specific affinity for features of the foreign substance. These features are known as antigens and typically include markers on tumor cells and molecules derived from pathogens, such as viral subunits, enzymes, and bacterial cell-wall glycoproteins. Whereas B-lymphocytes can bind directly to antigens, T-cells only recognize antigens when presented in conjunction with self-molecules known as major histocompatibility (MHC) molecules. Thus, whether and the degree to which an individual responds to a pathogen depends both on their arrays of randomly generated B- and T-cell receptors and their MHC makeup.

Lymphocytes involved in acquired immunity can be categorized by cell-surface markers known as clusters of differentiation (CD) into two primary groups: T-lymphocytes, marked by CD3, and B-lymphocytes marked by CD19. T-lymphocytes are further classified into helper/inducer (CD4), cytotoxic (CD8), and regulatory T-cells (CD25). T-helper cells play a central role in orchestrating acquired immunity via several mechanisms, including activating cytotoxic T-cells and B-cells and mediation of delayed-type hypersensitivity (DTH). Like phagocytic cells in the innate immune system, cytotoxic T-cells are able to attack and destroy foreign substances directly. However, unlike phagocytes, cytotoxic T-cells require activation by T-helper cells and must recognize a specific antigen presented in the context of an MHC molecule. Regulation of the acquired immune response is achieved in part by regulatory T-cells which are immunosuppressive in nature and serve to restrict immune responses to self-antigens. In addition, the inhibitory costimulatory molecule called programmed death-1 (PD-1) and its ligands have been shown to play an important role in regulating T-cell activation. Indeed, loss of PD-1 has been associated with an autoimmune diathesis in laboratory animals.

Like the innate immune system, acquired immunity utilizes soluble cytokine mediators. It is generally recognized that an acquired immune response develops along one of two lines, known as Th1 or Th2, based on the cytokine profiles produced by T-helper cells. A Th1 response is characterized by cytokines that promote cell-mediated inflammatory reactions, such as DTH. These cytokines include IL-2, IL-12, TNF-β, and IFN-γ. Cytokines generated by T-helper cells during a Th2 immune response include IL-4, IL-9, and IL-10. IL-6 is also produced by T-helper cells during a Th2-type acquired immune response (Beutler, 2004). The development of a Th2 response favors antibody production and can be associated with allergic responses.

An acquired immune system response consists of three phases: an induction phase in which the system detects the presence of antigen, an activation phase in which the presence of antigen triggers the expansion of antigen-specific T- and B-cells and an effector phase in which the foreign substance is cleared from the body. In this process, the acquired immune system utilizes and empowers many innate immune elements. As with other complicated physiological response systems, the immune response has built-in negative feedback elements that have evolved to limit immune reactivity once the pathogenic challenge has been met. Negative feedback elements that are intrinsic to the immune system include a number of cytokines, such as IL-10 and transforming growth factor-β (TGF-β). Other bodily systems, including the HPA axis, serve as important extrinsic immunomodulatory roles, and when functioning optimally, limit inflammation and immune system proliferation. A final feature of the acquired immune system that is central to its functioning is the formation of long-lived B- and T-lymphocytes that serve as memory cells for recognizing an antigen should it be encountered in the future.

18.1.2 Immune System Tests

In order to examine the effect of nervous and endocrine system factors on the immune system (especially in the context of clinical studies), various quantitative methods have been used, including both *in vitro* and *in vivo* assessments of immunity. *In vitro* tests of the immune system can be divided into enumerative and functional tests. Enumerative tests utilize techniques such as flow cytometry to measure the number and percentages of various immune system cell types based on the presence of cell-surface determinants, including CD markers, or the extent to which immune cells are positive for a particular intracellular marker of immune activation, including intracellular cytokine expression or evidence of activation of inflammatory signaling pathways (e.g., phosphorylated p38 as an indicator of p38 MAPK activation). Functional tests, as the name implies, evaluate the functional capacity of immune cells. The functional tests that have been most often employed include lymphocyte proliferative assays, assays of natural killer cell activity (NKCA), and mitogen-stimulated production of cytokines. These functional tests are performed *in vitro*, which means that the elements being studied are removed from their natural environment, including hormones, neurotransmitters, and cytokines. Nevertheless, *in vitro* studies using whole blood may circumvent some of the inherent limitations in such an approach.

In vivo tests seek to examine immune system status and immune responses *in situ*, which may better reflect functioning that is directly relevant to real world immune competence. These tests include antibody responses to antigenic challenge such as a vaccine, antibody titers to latent viruses, cutaneous DTH, and wound healing. In addition, plasma measures of circulating cytokines, and their soluble receptors as well as acute phase proteins, provide information on the status of the immune system *in situ*. Finally, activation of immunologic signaling pathways and a variety of immune mediators can be assessed by measuring relevant proteins and mRNAs in whole-cell, cytosolic, and nuclear extracts from immune cells using Western blot, electrophoretic mobility shift assay, DNA-binding enzyme-linked immunosorbent assay, reverse-transcriptase polymerase chain reaction, and RNA microarray analysis.

18.1.3 Regulation of the Immune Response

An effective immune response requires the cooperation of many components, often resulting in the augmentation of each component's contribution to the overall immune response. However, the simultaneous indiscriminate amplification of all aspects of the immune system would not be efficient and, in fact, could be disastrous. An overactive immune system may contribute to autoimmunity; furthermore, the inflammatory component of immune responses can be damaging if not controlled, as is seen in immune complex diseases and septic shock. In addition, prolonged periods of increased innate immune system activity (i.e., inflammation) has been shown to contribute to the development of a number of medical disorders, including heart disease, diabetes, and cancer (Aggarwal et al., 2006; Raison et al., 2006; Willerson and Ridker, 2004). Therefore, regulation of the immune response is necessary to insure that the response is energy efficient, focused on the infectious agent, counterbalanced in a fashion that does not cause short- or long-term self-damage, and reversible once the pathogen has been eliminated. The modes and the extent of immune system regulation are still poorly understood. Nevertheless, increasing evidence of neural–immune interactions indicates that extrinsic factors originating from the nervous and endocrine systems may play an important role in the regulation of the immune response.

18.2 Foundations of Neuroendocrine–Immune Interactions

Several seminal observations underlie the notion that the brain and immune system interact (**Table 2**).

Table 2 Foundations of nervous, endocrine, and immune system interactions

1. Expression of receptors for neurotransmitters, hormones, and neuropeptides on immune cells
2. Autonomic nervous system innervation of lymphoid tissues
3. Stress effects on immune function
4. Expression of cytokines and their receptors in the CNS
5. Influence of the immune system on neurotransmitter turnover, neuroendocrine function, and behavior

Table 3 Receptors for neurotransmitters, hormones, and peptides on immune cells

Neurotransmitters	*Hormones*	*Peptides*
Acetylcholine	Corticosteroids –	ACTH
Dopamine	glucocorticoids, mineralocorticoids	α-MSH
Histamine	Gonadal steroids –	AVP
Norepinephrine	estrogen, progesterone, testosterone	Calcitonin
Serotonin	Growth hormone	CGRP
	Prolactin	CRH
	Opioids	GHRH
	(endorphins,	GnRH
	enkephalins)	IGF-1
	Thyroid hormone	Melatonin
		NPY
		PTH
		Somatostatin
		Substance P
		TRH
		TSH
		VIP

ACTH, adrenocorticotropin c hormone; α-MSH, alpha-melanocyte stimulating hormone; AVP, arginine vasopressin; CGRP, calcitonin gene-related peptide; CRH, corticotropin-releasing hormone; GHRH, growth hormone-releasing hormone; GnRH, gonadotropin-releasing hormone; IGF-1, insulin-like growth factor-1; NPY, neuropeptide Y; PTH, parathyroid hormone; TRH, thyrotropin-releasing hormone; TSH, thyroid-stimulating hormone; VIP, vasoactive intestinal peptide.

First, immune cells have the capacity to receive signals from the nervous and endocrine systems by virtue of receptors for neurotransmitters, hormones, and neuropeptides. Receptors on immune cells that have been characterized include those for small-molecule neurotransmitters, gonadal and adrenal steroids, hypothalamic-releasing factors, and other neuropeptides (**Table 3**). In general, specific receptor densities vary among immune cell types and correlate with cell sensitivity to ligand.

In addition to harboring receptors for transmitters derived from or regulated by the nervous and endocrine systems, immune tissues are innervated by fibers derived from the autonomic nervous system (ANS) (**Table 2**; Nance and Sanders, 2007). Innervation by the sympathetic branch of the ANS has been best characterized. Typically, sympathetic nervous system (SNS) fibers enter immune tissues in association with the vascular supply. Deep in the tissue parenchyma of lymphoid tissues, these fibers are associated with vascular smooth muscle cells where they regulate vascular tone. However, electron microscopic images have revealed that sympathetic nerve endings as identified by tyrosine hydroxylase staining also occur in close approximation with lymphocytes and macrophages (**Figure 2**). Thus, the sympathetic branch of the ANS (e.g., via nervous innervation of lymphoid tissue) can influence the immune system either by changing the vascular tone and blood flow into immune organs, or by directly influencing immune cell function via locally released neurotransmitters, especially catecholamines (norepinephrine) and neuropeptides (e.g., neuropeptide Y, substance P, vasoactive intestinal peptide, calcitonin gene-related peptide, and corticotropin-releasing hormone (CRH)) that in turn interact with specific receptors on nearby immune cells (Nance and Sanders, 2007).

In addition to the SNS, there has been increasing interest in the role of the parasympathetic nervous system (PNS) in the regulation of immune responses. In particular, there have been a series of elegant studies indicating that stimulation of efferent vagus nerve fibers can inhibit cytokine and physiologic responses to endotoxin administration (Pavlov and Tracey, 2005; Tracey, 2002). These effects appear to be regulated in part by stimulation of the release of acetylcholine, which by binding to the $\alpha7$ subunit of the nicotinic acetylcholine receptor is able to inhibit activation of NFκB (the so-called cholinergic anti-inflammatory reflex). Given the stimulation of NFκB and proinflammatory cytokines by the SNS and catecholamines (discussed below), these data suggest that there may be a yin-yang influence of SNS and PNS on innate immune system activation that may be relevant to health. Indeed, alterations in heart-rate variability (which reflect the balance of PNS and SNS inputs to the heart) have been associated with cytokine (IL-6) concentrations in the peripheral blood (Sloan et al., 2007). It should be noted, however, that limited evidence exists that efferent vagal or parasympathetic fibers innervate immune organs, although the immunologic effects of activation of efferent parasympathetic pathways in other bodily tissues may be most relevant regarding the anti-inflammatory effects described above (Nance and Sanders, 2007; Pavlov and Tracey, 2004).

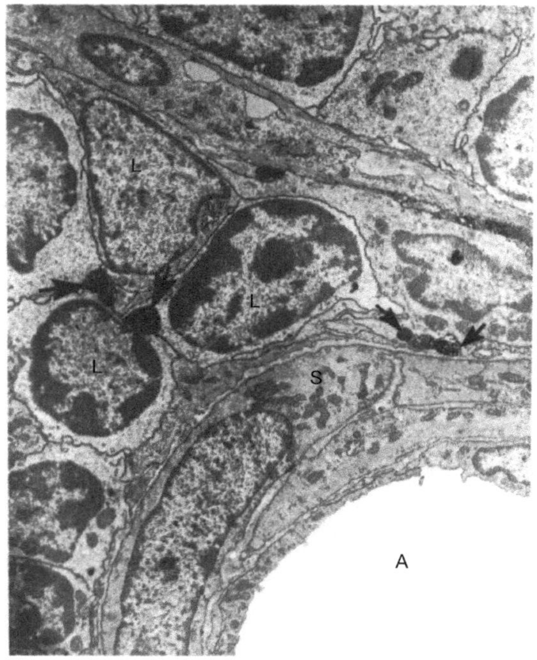

Figure 2 Sympathetic nervous system innervation of lymphoid tissue. Tyrosine hydroxylase-immunoreactive nerve processes (small arrowheads) in contact with the smooth muscle (S) of the central arteriole (A), and nerve processes (large arrowheads) in direct contact with lymphocytes (L) in the periarteriolar lymphatic sheath of the rat spleen. Transmission electron micrograph, X6732. Reproduced with permission from Denise L. Bellinger, Center for Neuroimmunology, Loma Linda University, Loma Linda, CA, and Suzanne Y. Stevens, Department of Neurobiology and Anatomy, University of Rochester School of Medicine and Dentistry, Rochester, NY

Finally, the potential role of cytokines in regulating the local release of transmitters from ANS fibers has become an active area of research.

Probably, the most extensive database documenting the existence of meaningful interactions among the nervous, endocrine and immune systems is that which examines the impact of stress on immune function (**Table 2**). A multitude of studies have demonstrated that stress is capable of altering virtually every aspect of the immune response. In general, the acute effects of stress enhance immune and inflammatory capacities, whereas chronic stress suppresses acquired immune responses, while potentially leading to chronic activation of innate immune responses. Stress effects are mediated in large part by the HPA axis and/or the SNS, which are regulated in part by CRH.

The demonstration of immune system effects on nervous and endocrine system function has documented that neuroendocrine–immune system interactions are bidirectional. Relevant pathways by which cytokines access the CNS have been described, and elucidation of the cytokine network within the CNS has been underway for the last several years. Finally, cytokines have been shown to influence multiple aspects of CNS function as well as behavior (**Table 2**).

18.3 Neuroendocrine Factors in Immune Regulation

While there are a multitude of factors that play relevant roles in immune regulation, much consideration has been given to the role of glucocorticoid hormones and the catecholamines.

18.3.1 Glucocorticoids

Of all the potential mediators of the nervous and endocrine systems on immune function, glucocorticoids have received the most attention. The effects of glucocorticoids, the final product of HPA-axis activation, on the immune system have been extensively characterized. In 1950, Philip Hench won the Nobel prize in medicine for his discovery of the role of glucocorticoids in modulating the expression of autoimmune disorders including rheumatoid arthritis and allergic responses (Hench, 1952). Since then, a massive literature has developed documenting the multiple immunologic effects of glucocorticoids and the biochemical and molecular mechanisms involved (for a summary, see McEwen et al. (1997)). In brief, glucocorticoids influence the immune response by (1) modulating cell-death pathways in immature and mature cell types (McEwen et al., 1997), (2) modulating the trafficking of immune cells throughout the body (Dhabhar, 2003; McEwen et al., 1997), (3) inhibiting the generation of products of the arachidonic acid pathway which mediate inflammation (Kozak et al., 2000), (4) inhibiting cytokine production and function through interaction of glucocorticoid receptors (GRs) with transcription factors (e.g., NFκB) which in turn regulate cytokine gene expression and/or the expression of cytokine-inducible genes (Rhen and Cidlowski, 2005), (5) modulating the Th1/Th2 phenotype of the immune response by inhibiting Th1 (cell-mediated) responses and enhancing Th2 (antibody) responses (McEwen et al., 1997), and (6) inhibiting T-cell-mediated and natural killer (NK)-cell-mediated cytotoxicity (McEwen et al., 1997).

Glucocorticoid action is a consequence of both the relative expression and activity of factors that modulate the access of glucocorticoids to their receptors, and the functioning of the receptors themselves. Although concentrations of circulating glucocorticoids are of immense importance, since the majority of circulating hormone is bound to corticosteroid-binding globulin (CBG), tissue concentrations of CBG are a critical determinant of the relative amount of free steroid that is available to passively diffuse into the cell. Once glucocorticoids enter the cell, the relative concentration of the enzymes, 11β-hydroxysteroid dehydrogenase (11β-HSD) type 1 and 11β-HSD type 2, will further determine access of hormone to receptor. 11β-HSD type 2 breaks down naturally occurring glucocorticoids (but not synthetic glucocorticoids) as they enter the cell, leaving the hormone in the form of inactive metabolites, whereas 11β-HSD type 1 does the reverse (Holmes et al., 2003). Interestingly, significant differences in 11β-HSD type 2 activity have been found among immune compartments, and there is a direct correlation between 11β-HSD type 2 activity and the preferential production of Th1-versus Th2-type cytokines by T-cells residing in a given tissue (Hennebold et al., 1996). Inhibition of 11β-HSD type 2 activity (which would enhance the available amount of hormone to the GR) leads to reduced Th1 responses and enhanced Th2 cytokine production (Hennebold et al., 1996), while mice deficient in 11β-HSD type 1 show delayed resolution of inflammation in experimentally induced arthritis (Chapman et al., 2006). Finally, the relative expression of the P-glycoprotein efflux pump (also referred to as the multiple drug-resistance pump-1 or

MDR-1) may modulate intracellular glucocorticoid hormone concentrations (Pariante, 2008). This pump actively extrudes synthetic glucocorticoids as well as cortisol and aldosterone. The contribution of MDR-1 to glucocorticoid action on immune cells has yet to be established, although patients with myasthenia gravis exhibit increased expression of MDR-1 (Richaud-Patin et al., 2004). Taken together, these data indicate that the microenvironment of the various immune compartments and the relative presence of factors that regulate glucocorticoid availability within the cell are critical in determining the impact of the neuroendocrine system on immune regulation.

One of the most important functions of glucocorticoid hormones is to protect the body against an overshoot of potentially damaging immune activation (Raison and Miller, 2003; Sapolsky et al., 2000). For example, neutralization of endogenous glucocorticoid function results in enhanced pathology and mortality in animals exposed to endotoxin (e.g., lipopolysaccharide (LPS)) and autoimmune-inducing stimuli (e.g., streptoccocal cell-wall antigen or myelin basic protein; Rivest, 2003; Sternberg et al., 1989). Similar enhanced pathology has been found to occur in the absence of glucocorticoids following viral infections. For example, mice infected with murine cytomegalovirus (MCMV) exhibit a robust plasma corticosterone response. The physiological relevance of this corticosterone surge has been established by rendering animals devoid of glucocorticoids (via adrenalectomy) which has been found to lead to death following MCMV infection (Ruzek et al., 1999; Silverman et al., 2005). Studies in cytokine-deficient and cytokine-neutralized mice have demonstrated that TNF-α is the primary mediator of this lethality (Ruzek et al., 1999). Of note, glucocorticoids have been found to restrain virally induced changes in anxiety behavior that are also driven by TNF-α (Silverman et al., 2007). Because glucocorticoids have been reported to inhibit IFN-γ and IL-2 expression (Th1) while enhancing IL-4 (Th2) (McEwen et al., 1997), these hormones may act to turn off immune responses that are no longer needed and/or selectively promote activation of other immune response components. Finally, it should be noted that some investigators have suggested that while glucocorticoids are anti-inflammatory in the periphery, they may support inflammatory responses in the brain. For example, animals exposed to chronic unpredictable stress and then administered LPS in the brain exhibited reduced evidence of CNS inflammation when administered the GR antagonist, RU486 (Sorrells and Sapolsky, 2007). Nevertheless, these data are in stark contrast to the devastating effects of RU486 on LPS-induced CNS inflammation in nonstressed animals (Nadeau and Rivest, 2003), and may reflect the contribution of glucocorticoids to immune-cell trafficking to the brain under conditions of chronic stress.

It should be noted that the ultimate impact of glucocorticoids is determined at the level of the GR, and therefore alterations involving GR expression and/or function related to environmental stressors and/or immune activation can lead to increased or decreased sensitivity of target immune tissues to glucocorticoid hormones irrespective of circulating or intracellular hormone concentrations (see Section 18.4.3.2) (Pace et al., 2007). Taken together, these data support the notion that glucocorticoids are immunomodulators, and the ultimate consequence of elevations in glucocorticoids is a function of the context within which they occur.

18.3.2 Catecholamines

Aside from glucocorticoids, probably the second most-studied pathway by which the CNS can influence the immune system is the SNS and the release of catecholamines.

Although early studies identified catecholamines as primarily inhibitory, more recent data have suggested that these transmitters have more complex effects on immune function (Nance and Sanders, 2007). As previously noted, SNS fibers are present in immune tissues, and immune cells have been shown to express receptors for catecholamines. The predominate receptor subtype expressed on T- and B-cells is the β2-adrenergic receptor, a G-protein-linked receptor that traditionally leads to increased cAMP and protein kinase A activation (Nance and Sanders, 2007). Nevertheless, other signal-transduction pathways shared by the immune system, such as MAPK pathways, are activated by β-adrenergic receptor binding , and it is the potential crosstalk between these pathways that is believed to explain the differential effects of norepinephrine on immune cell activity (Nance and Sanders, 2007). The β2 receptor is expressed on resting and activated B cells, naive CD4$^+$ T-cells, newly generated Th1 cells and Th1-cell clones. However, it is not expressed on newly generated Th2 cells or Th2 cell clones (Nance and Sanders, 2007). Consistent with these findings, norepinephrine has been found to enhance IL-12-induced differentiation of naive CD4$^+$ T-cells into Th1 cells and to promote production of IFN-γ by these cells (Kohm and Sanders, 2000). No effect was found on IL-4-induced Th2-cell differentiation. The effect of NE on Th1-type responses is also manifested by the ability of NE to help Th1 cells support B-cell antibody production. Interestingly, the above-noted NE effects are manifested only in Th1 cells derived from NE-exposed naive CD4$^+$ T-cells, underlining the importance of timing in the determination of the effects of catecholamines on immune function (Nance and Sanders, 2007).

Immune cells, including cells located within the spleen and thymus, as well as circulating peripheral blood mononuclear cells (PBMCs), also have been found to express α1-adrenergic receptors (Kavelaars, 2002). Interestingly, expression of α1-adrenergic receptors has been found to be higher in PBMCs from patients with juvenile idiopathic arthritis (JIA) compared to controls, suggesting that chronic inflammation promotes α1-adrenergic receptor expression. Moreover, stimulation of PBMCs collected from JIA patients with specific α1-adrenergic receptor agonists increased monocyte cytokine production, an effect not seen in cells from control subjects. Commensurate with these findings is the observation that treatment of cultured monocytes with LPS results in increased α1-adrenergic receptor expression (Kavelaars, 2002). Thus, α1-adrenergic receptors may play an important role in the relationship between stress-induced activation of SNS activity and the pathophysiology of disease states involving inflammatory excess.

In terms of *in vivo* studies, animal studies have demonstrated that surgical or chemical sympathectomy alters immune responses in rodents as well as attenuates stress-induced immune changes (Nance and Sanders, 2007). As might be anticipated by the pattern of nervous innervation of lymphoid tissues, abrogation of stress-induced immune changes by antagonizing the SNS is most apparent in solid immune tissues such as the spleen. For example, stress-induced suppression of splenic lymphocyte proliferation to polyclonal mitogens is not influenced by adrenalectomy but is markedly attenuated by β-adrenergic receptor antagonists (Cunnick et al., 1990). Based on this and other studies (Rabin, 1999), it is apparent that different neuroendocrine mechanisms can be operative in different immune compartments. In particular, immune responses in the peripheral blood seem to be more influenced by glucocorticoids, whereas immune responses in the spleen seem to be more sensitive to catecholamines. Further discussion of the role of catecholamines and the SNS in the effects of stress and psychiatric disorders on the immune system is provided below.

18.3.3 Corticotropin-Releasing Hormone

Given the importance of the HPA axis (glucocorticoids) and SNS (catecholamines) in modulating immune responses, a logical focus in investigating neuroendocrine–immune interactions has been CRH. CRH plays a pivotal role in regulating HPA and SNS activation. Interestingly, CRH is also capable of influencing immune cells directly.

The effects of CRH on a wide range of immune functions have been well characterized (for an overview, see Baigent (2001)). For example, in laboratory animals, intracerebroventricular (ICV) (Irwin et al., 1988) as well as intravenous (IV) (Baigent, 2001) administration of CRH has been shown to lead to suppression of NK activity . Furthermore, CRH inhibits *in vivo* and *in vitro* antibody formation, including the generation of an IgG response to immunization with keyhole limpet hemocyanin (Irwin, 1993). The influence of CRH on antibody responses is also apparent in CRH overproducing mice whose immune deficits are characterized by a profound decrease in the number of B-cells and severely diminished primary and memory antibody responses (Stenzel-Poore et al., 1996), likely secondary to alterations in HPA-axis function. In contrast, adrenalectomized CRH knockout mice display lower levels of inflammatory reaction to peripheral treatment with the immune stimulus carrageenin, suggesting that peripheral CRH also has important proinflammatory functions (Baigent, 2001).

Of note, in some cases, CRH has been found to stimulate lymphocyte proliferation (Baigent, 2001), as well as the release of proinflammatory cytokines in both laboratory animals and humans. Chronic ICV administration of CRH to rats led to induction of IL-1β mRNA in splenocytes, and acute IV infusion of CRH in humans led to an almost fourfold induction of IL-1α in the circulation (Labeur et al., 1995; Schulte et al., 1994). Both treatments also led to significant increases in the immunoregulatory cytokine, IL-2 (Labeur et al., 1995; Schulte et al., 1994). In addition, CRH has been found to induce the release of IL-1 and IL-6 from human mononuclear cells *in vitro* (Leu and Singh, 1992; Paez Pereda et al., 1995). CRH has also been shown to promote the release of adrenocorticotropic hormone (ACTH) from peripheral lymphocytes, suggesting that CRH may able to stimulate glucocorticoid synthesis and release via immune cells in the absence of the pituitary (Baigent, 2001; Blalock and Smith, 2007).

As for the neuroendocrine mechanisms by which CRH influences immune responses, activation of the SNS by ICV CRH has been found to be a major regulator of the effects of CRH on splenic NK activity. In particular, the sympathetic ganglionic blocker, chlorisandamine, was shown to reverse the inhibitory effects of ICV CRH on NK activity in the spleen (Irwin et al., 1988). The HPA axis is also involved in CRH immune effects as shown by Labeur et al. (1995) who demonstrated that the inhibitory effects of chronic ICV CRH on splenocyte-proliferative responses were eliminated by adrenalectomy.

CRH is additionally capable of directly modulating immune and inflammatory responses (Blalock and Smith, 2007). Local production of CRH has been demonstrated in inflammatory diseases such as arthritis, where it is proposed to act as a local proinflammatory agent (Nishioka et al., 1996). A recent study by Smith et al. (2006) demonstrated that CRH is able to enhance antigen-specific antibody responses by acting through the CRH receptor 1 subtype in cultured leukocytes, which in turn activates the NFκB pathway, one of the primary inflammatory signaling pathways. Finally, *in vivo* studies have shown that CRH can both inhibit (Irwin et al., 1988) and enhance (Jessop et al., 1997) NK cell activity and antibody production. Based on these capacities, as discussed below, CRH is well situated to play an important role in many of the effects of acute and chronic stress on immune function.

18.3.4 Other Factors

Given the explosion of data documenting the rich complexity of neuroimmune interactions, we should at least mention some of the factors other than glucocorticoids and catecholamines that have received attention, including the opioids, gonadal steroids, growth hormone, and prolactin.

Opioids have diverse effects on immune function *in vitro* and *in vivo* (Vallejo et al., 2004; Roy et al., 2006). Nevertheless, some of the most compelling findings are the profound suppression of a wide range of immune parameters after *in vivo* administration of morphine (Roy et al., 2006). Activation of central opioid receptors by morphine and μ-selective opioid receptor agonists has been consistently shown to inhibit peripheral immune functions, including NK activity, mitogen-induced lymphocyte proliferation, and phagocytic cell function (Roy et al., 2006). These effects appear to be mediated acutely by the SNS, whereas effects of more chronic administration appear to be mediated by the HPA axis (Vallejo et al., 2004). Direct effects of morphine on immune cell function are also involved. In addition, endogenous opioids appear to play a role in modulating immune responses following stress. For example, rats subjected to a footshock paradigm known to be associated with opioid analgesia exhibited decreased NK-cell activity that was prevented by administration of the opioid antagonist, naltrexone (Shavit et al., 1984). Interestingly, administration of the opioid, morphine, has also been found to activate glia and induce the production of proinflammatory cytokines, which in turn appear to contribute to opioid tolerance and hyperalgesia (Johnston et al., 2004).

Sexual dimorphism in immune function is an impressive example of the relevance of neuroendocrine–immune interactions in immune system regulation (Yu and Whitacre, 2004). Adult females have more exuberant antibody responses to immune challenge, reject transplanted tissues more vigorously, are more susceptible to allergies, and live longer than adult males (Bouman et al., 2005). In addition, of the approximately 40 autoimmune disorders, the majority is more common in women. In part, it is believed that these differences in immune responsiveness are secondary to a greater propensity in females to develop a Th1 response after infectious challenge or antigen exposure (except during pregnancy when a Th2 propensity prevails) (Bouman et al., 2005). The etiology of these sex differences in immune function is secondary to a combination of the influence of gonadal steroids on the development of nervous and immune system cells and tissues and direct effects of sexually dimorphic hormones, including estrogens, progesterone, androgens, prolactin, growth hormone, and insulin-like growth factor-1 on immune function (Martin, 2000). As noted previously, immune cells express receptors for these hormones, and transcripts for enzymes that synthesize gonadal steroids are expressed in immune tissues (Martin, 2000). Estrogens tend to promote Th1-type responses, while progesterone tends to promote Th2-type responses (Bouman et al., 2005). Testosterone exhibits anti-inflammatory and immunosuppressive properties as determined in animal models of autoimmunity (Bouman et al., 2005). During pregnancy, Th2-type immune responses prevail (possibly secondary to the increased influence of progesterone) and autoimmune disorders related to excessive Th1-like activity (multiple sclerosis and rheumatoid arthritis) improve (Bouman et al., 2005). Diseases related to Th2-like activity (systemic lupus erythematosis) are exacerbated during pregnancy (Bouman et al., 2005). Aside from direct effects on the immune system, gonadal steroids also have been shown to modulate the HPA-axis response to stress (Kudielka and Kirschbaum, 2005). Thus, gonadal hormones may also influence immune responses indirectly through effects on HPA-axis-immune pathways.

18.4 Role of Cytokines in the Regulation of the Neuroendocrine System and Behavior

Research lending support to the importance of interactions between the immune system and the CNS includes the discovery that cytokines are capable of exerting profound effects on the nervous and endocrine systems.

18.4.1 Pathways of Immune to Brain Signaling

Since cytokines do not freely cross the blood–brain barrier (BBB) under usual circumstances (i.e., in the absence of CNS infection or in cases of traumatic brain injury), considerable attention has been paid to how peripheral immune signals are transmitted to the brain. Several mechanisms have been proposed (**Table 4**). For example, it has been suggested that local (peripheral) production of proinflammatory cytokines can stimulate visceral afferent nerve fibers that communicate with the brain through the vagus nerve (Watkins et al., 1994). Vagus nerve stimulation by cytokines in turn modulates CNS function (e.g., CRH release from the hypothalamus) through interconnected neuronal circuits involving ascending catecholaminergic fibers (A2, C2 cell groups) of the nucleus of the solitary tract (NTS) that project to the parvocellular division of the paraventricular nucleus (PVN) of the hypothalamus (Dantzer et al., 2008).

Table 4 Pathways by which cytokines signal the brain

Stimulation of visceral afferent nerve fibers (e.g., vagus nerve)
Direct passage through leaky regions in the blood–brain barrier
Activation of intermediary cells in the vasculature and circumventricular organs
Active transport across the blood–brain barrier

Non-neural, or humoral, communication with the brain by cytokines also occurs. Circulating cytokines may enter the brain in regions where the BBB is leaky such as the circumventricular organs, allowing diffusion into the brain parenchyma and entry into cerebrospinal fluid (CSF)-flow pathways (Vitkovic et al., 2000). For example, a high dose of IL-1 administered peripherally causes the induction of early immediate gene, c-fos, in two such leaky regions, the area postrema and the vascular organ of the lamina terminalis (e.g., Brady et al., 1994; Ericsson et al., 1994). Cytokines can also communicate with the brain through intermediates without themselves entering the CNS parenchyma, for example, by acting on TLRs of cells located in the brain endothelium or choroid plexus and inducing the release of secondary messengers, including prostaglandins and nitric oxide, as well as proinflammatory cytokines (Quan et al., 1998). Of note, recent data indicate that NFκB may be critically involved in translating cytokine signals to the brain. Indeed, inhibition of NFκB signaling using a NFκB essential modulator (NEMO)-binding domain peptide, which blocks the association of NEMO with the IkB kinase (IKK) complex, reduced the induction of the early immediate gene, c-fos, in relevant brain regions and blocked the development of behavioral changes following peripheral administration of IL-1β (Nadjar et al., 2005). Active transport mechanisms for proinflammatory cytokines provide another way by which small quantities of proinflammatory cytokines may reach the brain and neuroendocrine regulatory circuits (Quan and Banks, 2007). Activation of cytokine signaling networks in the brain by peripheral stimuli are thought to require both vagal input and humoral input: early vagal input primes the brain to receive a humoral signal, whereas later humoral input produces increased expression of cytokines in the brain (Dantzer et al., 2008).

18.4.2 Cytokine Network in the Brain

Once cytokine signals access the brain, they can be transmitted and amplified in the context of the cytokine network within the brain. The principal cells in the brain to produce cytokines are microglial cells (Dheen et al., 2007). Receptors for proinflammatory cytokines have been located in the brain, including regions that play important roles in vegetative functions, emotion regulation, and memory. For example, IL-1β and its mRNA have been found in nerve cell bodies and nerve fibers within the hypothalamus, hippocampus, amygdala, and other regions in human and rodent brains (Dantzer et al., 2008). Furthermore, several cytokines derived from neural cells may play a role as intermediaries in communicating peripheral inflammatory signals to the brain. For example, peripheral injection of LPS and circulating cytokines such as IL-1 have been shown to induce neural cells within the hypothalamus and other brain regions to produce proinflammatory cytokines such as IL-1, IL-6, and TNF (Rivest, 2003). Among neural cells, activated glia (especially microglia) are rich sources of proinflammatory cytokines. Activated microglia as well as astrocytes can produce large quantities of IL-1, and proinflammatory cytokines are potent stimulants of glial activation (Dheen et al., 2007). Finally, nonimmunologic stressors (such as restraint stress) can induce cytokine expression in the brain (see below), suggesting that the behavioral and neuroendocrine response to stress may also involve cytokine-signaling pathways (Raison et al., 2006).

18.4.3 Impact of Cytokines on Nervous and Endocrine System Function

The impact of cytokines on CNS function has been examined in numerous *in vitro* and *in vivo* studies. Cytokines have been shown to influence virtually every pathophysiologic domain relevant to the regulation of behavior including neuroendocrine function as well as neurotransmitter metabolism, synaptic plasticity, and regional brain activity.

18.4.3.1 Cytokine effects on the HPA axis

Acting at the level of the hypothalamus, the pituitary, and the adrenal glands, immune system products, including IL-1, IL-6, TNF, leukemia inhibitory factor (LIF) and IL-2, appear to play a role in regulating the secretion of multiple hormones, most notably glucocorticoids (Silverman et al., 2005; Turnbull and Rivier, 1999). Based on a series of studies examining the mechanisms by which proinflammatory cytokines such as IL-1 and IL-6 lead to HPA-axis activation, a major final common pathway involves cytokine induction of CRH in the PVN of the hypothalamus (Silverman et al., 2005; Turnbull and Rivier, 1999). In addition to actions through CRH secretion, cytokines may regulate glucocorticoid release through multiple alternate pathways, including direct effects on the pituitary or adrenal glands (Silverman et al., 2005; Turnbull and Rivier, 1999). The simultaneous release of ACTH-like and β-endorphin-like products from immune cells in response to a variety of stimuli, including CRH, indicates that immunocytes (probably macrophages), like pituitary cells, are capable of transcribing the proopiomelanocortin gene, which is responsible for coding the precursor protein for ACTH and β-endorphin (Smith et al., 2006). Of note, other hormones found to be secreted by immunocytes include urocortin, somatostatin, vasoactive intestinal polypeptide, thyrotropin, and prolactin.

18.4.3.2 Cytokine effects on glucocorticoid receptors

As noted previously, the ultimate effect of glucocorticoid hormones on target immune tissues is determined at the level of the GR. There is a large body of data demonstrating that cytokines have significant effects on both GR number and function (Pace et al., 2007). In general, studies using whole-cell radioligand-binding techniques have found increased GR numbers following treatment with a host of cytokines, including IL-1, IL-2, and IL-4, whereas the majority of studies measuring the GR using a cytosolic radioligand-binding assay find GR to be decreased following treatment with the same group of cytokines (Miller et al., 1999). Studies on the effects of cytokines on the function of the GR have been somewhat more consistent with a number of studies demonstrating that a variety of cytokines can inhibit GR function (Pace et al., 2007).

A number of investigators have studied potential mechanisms of cytokine effects on the GR by examining the impact of cytokines on GR translocation from cytoplasm to nucleus and hormone-induced, GR-mediated gene transcription (**Figure 3**; e.g., Irusen et al., 2002; Pariante et al., 1999; Wang et al., 2004). For example, IL-1α was found to block GR translocation from the cytoplasm to the nucleus and significantly reduce dexamethasone (DEX)-induced GR-mediated gene transcription. These results suggest that proinflammatory cytokines like IL-1 may have direct effects on GR function that lead to resistance to glucocorticoids. Of note in this regard, studies in mice have demonstrated that activation of IL-1 in the context of social disruption stress is capable of leading to glucocorticoid resistance in splenic macrophages (Stark et al., 2001). This macrophage glucocorticoid resistance in turn was associated with a marked inflammatory response to viral infection that was associated with death in some animals (Quan et al., 2001). Interestingly, macrophage

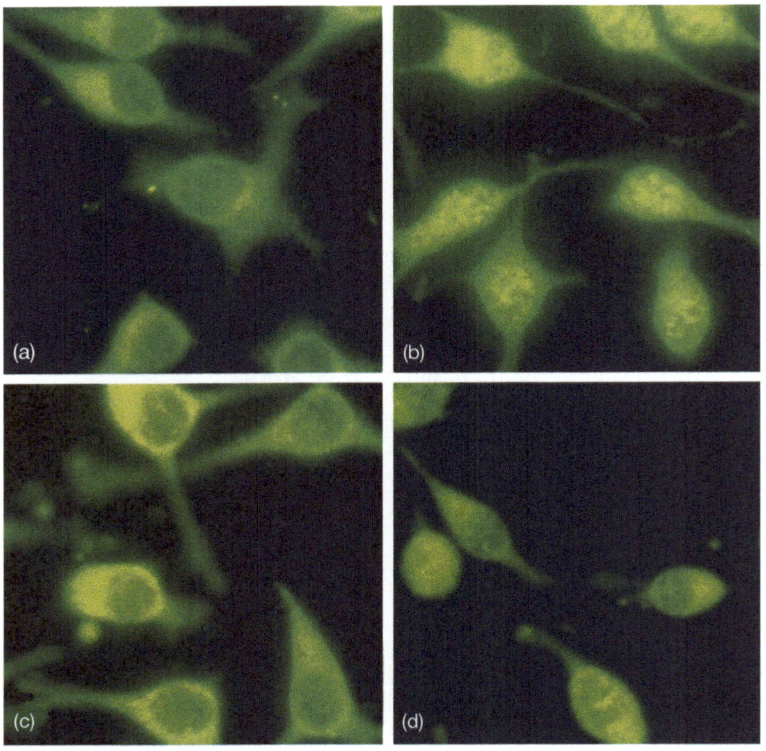

Figure 3 Interleukin (IL)-1α-reduced glucocorticoid receptor translocation. Inhibition of glucocorticoid receptor (GR) translocation after treatment with IL-1α and dexamethasone (DEX) was determined by immunostaining of GR. L929 cells were grown in steroid-free medium and treated with vehicle (a), DEX (10 nM) for 1.5 h (b), IL-1α (1000 U ml^{-1}) for 24h (c), or IL-1α (1000 U ml^{-1}) for 24h followed by IL-1α plus DEX (10 nM) for 1.5 h (d). GRs were immunostained using the anti-GR polyclonal antibody GR57. Note the primarily cytoplasmic staining in vehicle-treated cells and the increased cytoplasmic staining in cells treated with IL-1α alone. Note the increase in nuclear staining in DEX-treated cells, which is reduced in cells treated with IL-1 α plus DEX. Reprinted with permission from Pariante CM, Pearce BD, Pisell TL, Sanchez CI, Po C, Su C, and Miller AH (1999) The proinflammatory cytokine, interleukin-1α, reduces glucocorticoid receptor translocation and function. *Endocrinology* 140: 4359–4366. Copyright 1999, The Endocrine Society.

glucocorticoid resistance was eliminated in IL-1 knockout mice (Engler et al., 2008).

In terms of the mechanisms by which cytokines impair GR function, multiple inflammatory and immunoregulatory signaling pathways have been shown to play a role. Pathways that have received the most attention include those involving MAPK, NFκB, and STAT (**Figure 4**; Pace et al., 2007).

MAPKs such as the extracellular signal-related kinases (ERKs), jun amino-terminal kinases (JNK), and p38 activate nuclear transcription factors and promote proliferative and inflammatory activation in response to stress, infection, or other environmental stimuli. Multiple MAPK pathways have been shown to inhibit GR function, both by disrupting translocation of GR to the nucleus and by preventing normal GR function as a transcription factor (Pace et al., 2007). For example, constitutive JNK activity impairs DEX-induced expression of GR-mediated gene expression, while induction of p38 by IL-1 blocks translocation of GR to the nucleus as well as GR–DNA binding (Wang et al., 2005). In addition, both p38 and JNK pathways have been implicated in the inhibitory effects of TNF-α on GR function (Szatmary et al., 2004).

As noted previously, NFκB is a nuclear transcription factor that plays a pivotal role in mediating inflammatory and immune responses to proinflammatory cytokines. NFκB has long been recognized to interact with the GR at multiple levels, and NFκB has been shown to directly interact with GR in the nucleus through physical association, causing mutual repression of both GR and NFκB function (Rhen and Cidlowski, 2005).

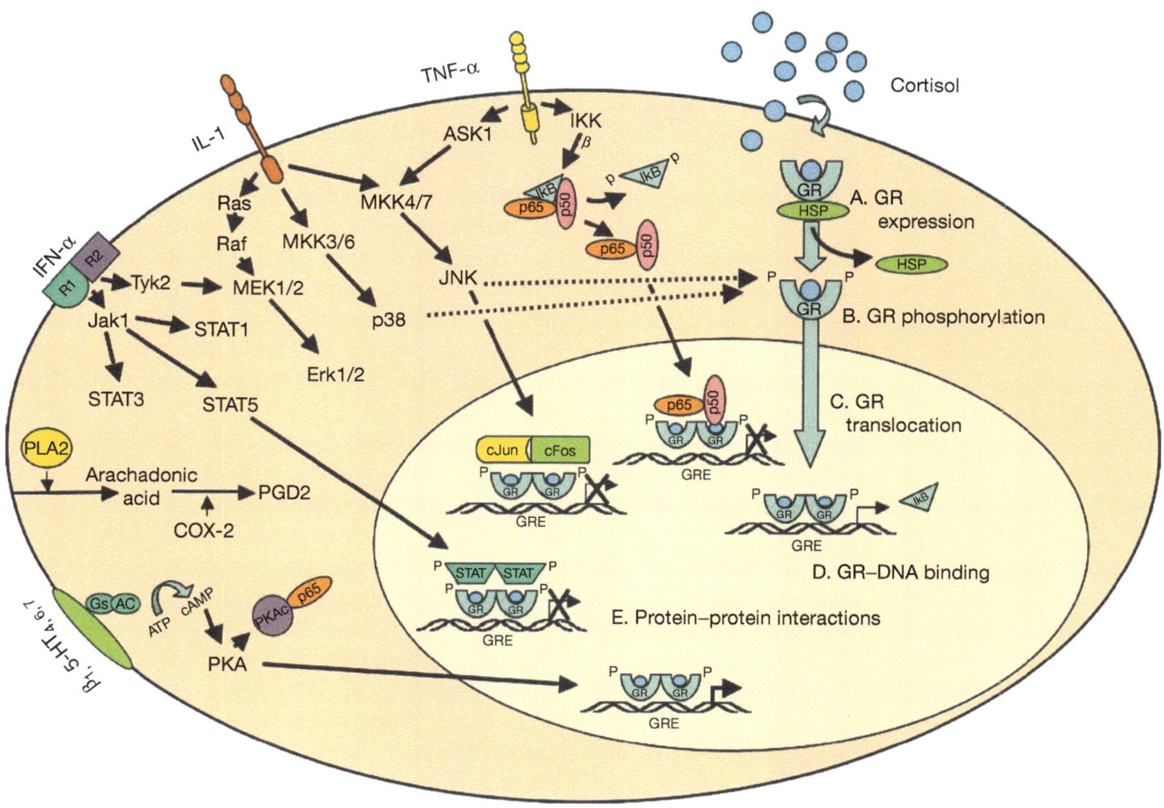

Figure 4 Interactions between cytokine and glucocorticoid receptor-signaling pathways. Selected cytokines and their signal-transduction pathways are depicted in simplified fashion to illustrate representative interactions between cytokine and glucocorticoid receptor (GR) signaling events. Cortisol binds to GR, resulting in dissociation of heat-shock protein (HSP) complexes and subsequent phosphorylation. GR then translocates to the nucleus where it dimerizes and either interacts with other transcription factors, or binds to glucocorticoid response elements (GREs) upstream of GR-regulated genes (e.g., inhibitor κB or IκB). TNF-α binds to its receptor and results in activation of IκB kinase β (IKKβ), which phosphorylates IκB, allowing NFκB (shown here as p65 and p50 Rel subunits) to translocate to the nucleus. Through protein–protein interactions, activated NFκB associates with GR, thus interfering with GR–DNA binding. IL-1 binds to its receptor initiating (1) mitogen-activated protein kinase (MAPK) kinase (MKK)4/7, which culminates in activation of Jun amino-terminal kinase (JNK); (2) MKK3/6, which culminates in activation of p38; and (3) Ras, which results in activation of the extracellular signal-related kinase (ERK)1/2. Of note, MKK4/7 activation of JNK can also occur through TNF-α receptor binding. As depicted by the dotted lines, both p38 and JNK can phosphorylate key GR residues, thereby disrupting nuclear translocation of GR. Interferon (IFN)-α binds to its receptor resulting in Janus kinase (Jak) phosphorylation, represented as Jak1 and tyrosine kinase (Tyk)2. Jak1 phosphorylates signal transducers and activators of transcription (STAT) proteins, including STAT1, STAT3, and STAT5. Tyk2 can also activate elements of the Ras signaling pathway, resulting in activation of Erk1/2. Activated STATs translocate to the nucleus, where they can interact with GR through protein–protein interactions, thereby interfering with GR–DNA binding. Phospholipids are hydrolyzed by phospholipase A2 (PLA2) to form arachidonic acid which is metabolized by cyclooxygenase (COX)-2 to produce prostaglandin D2 (PGD_2). Stimulation of serotonergic receptors 4, 6, or 7 (5-HT4, 6, 7) and β-adrenergic receptors (β1) induces a conformational change in G stimulatory (Gs) protein, which then activates adenylyl cyclase (AC). AC, in turn, converts adenosine triphosphate (ATP) to cyclic adenosine monophosphate (cAMP). cAMP then induces a conformational changes in protein kinase A (PKA), which translocates to the nucleus where it is able to enhance GR–DNA binding. In addition, the catalytic subunit of PKA (PKAc) interacts with p65, thereby inhibiting NFκB nuclear translocation. Reproduced from Pace TW, Hu F, and Miller AH (2007) Cytokine-effects on glucocorticoid receptor function: Relevance to glucocorticoid resistance and the pathophysiology and treatment of major depression. *Brain, Behavior, and Immunity* 21: 9–19, with permission from Elsevier.

Cytokine stimulation of Jak-STAT pathways also has been shown to influence GR function. Jak-STAT pathways are activated by stimulation of type I (e.g., IL-2, IL-4, IL-6, and IL-12) and type II (IFN-α/β, IFN-γ, and IL-10) cytokine receptors. The most-studied interactions between Jak-STAT and GR-signaling pathways have been protein–protein interactions between GR and STAT5 (for a recent review, see Rogatsky and

Ivashkiv (2006)). Immunoprecipitation studies have demonstrated that STAT5 and GR form complexes, and treatment of cultured cells with IL-2 and IFN-α, which increases STAT5 activity, has been shown to disrupt GR translocation and/or GR binding to DNA response elements in a number of studies (Pace et al., 2007). Less is known about GR interactions with other Jak or STAT proteins.

Another inflammatory signaling pathway that has been reported to interact with and modulate GR function is the phospholipase/cyclooxygenase (COX)/prostaglandin pathway. Cultured cells treated with compounds that block either COX-2 (nimesulide, celecoxib) or both COX-1 and COX-2 (ibuprofen) have been shown to enhance GR activity, including increased expression of GR-mediated genes and increased phosphorylation of GR (Pace et al., 2007). Interestingly, data suggest that inhibition of MAPK signaling pathways (p38) may be involved in the effects of COX-2 inhibition on GR function and may represent a final common target for several GR regulatory pathways as they relate to inflammation and immune regulation (Pace et al., 2007).

Finally, cytokines have been found to induce the expression of GRβ, the inert isoform of GR in humans, which has been associated with decreased sensitivity to glucocorticoids. For example, TNF-α and IL-1 have been shown to increase expression of hGRβ more than expression of GRα (Webster et al., 2001). In addition, hGRβ has been found to limit hGRα-dependent expression of glucocorticoid-sensitive genes (Lewis-Tuffin and Cidlowski, 2006).

18.4.3.3 Behavioral effects of cytokines

At high doses, human recombinant forms of cytokines such as IFN-α and IL-2 have been used for the treatment of viral and malignant illnesses, including chronic hepatitis B and C, malignant melanoma, and renal cell carcinoma (Capuron et al., 2002; Irwin and Miller, 2007; Raison et al., 2006). Although effective therapies for several of these illnesses, these cytokines are also known for a variety of constitutional and psychiatric complications that frequently limit the dose and duration of treatment. Similar behavioral changes are observed in laboratory animals administered these and other cytokines (Dantzer et al., 2008). The behavioral syndrome that develops as a function of cytokine administration is referred to as sickness behavior and can be divided into acute and chronic phases (Kent et al., 1992). The acute phase follows a similar pattern of signs and symptoms for different cytokines. Often described as a flu-like syndrome, the acute phase is characterized by fever, chills, myalgias, nausea, vomiting, and general malaise. Symptoms in the later phase resemble clinical depression in humans and include depressed mood, anxiety, psychomotor slowing, fatigue, cognitive dysfunction, and altered sleep (Raison et al., 2006). The resemblance of the latter syndrome to major depression has raised the specter that immunologic mechanisms may be involved in the pathophysiology of depression, especially in medically ill patients who have multiple sources of inflammation (and inflammatory cytokines) and high rates of depressive disorders (Irwin and Miller, 2007).

While it is widely known that treating humans or laboratory animals with cytokines leads to profound behavioral changes, it has been difficult to determine the extent to which depressive-like behaviors exist independently from sickness behaviors. Recent studies have begun to address this issue. Dantzer and colleagues have devised behavioral assessments and sampling strategies that provide a reasoned approach to differentiate sickness from depressive-like behavior in rodent models of cytokine-induced behavioral change (Frenois et al., 2007). For example, to be called depressive-like, altered performance in rodent tests of depressive-like behavior (e.g., the Porsolt forced swim test) must occur in the absence of cytokine-induced changes in motor activity, which might be more reflective of a sickness state (Dantzer et al., 2008).

18.4.3.3(i) Mechanisms of cytokine-induced depression/sickness behavior

There are a number of mechanisms by which cytokines may contribute to changes in behavior. These mechanisms may be related to either direct effects of the administered cytokine on the CNS, or the induction of the cytokine cascade that recruits a multiplicity of other cytokines and inflammatory mediators that in turn influence CNS function.

First, as noted above, cytokines have been shown to have potent stimulatory effects on the HPA axis, in large part through activation of CRH (Silverman et al., 2005; Turnbull and Rivier, 1999). CRH administration promotes the expression of behaviors in animals that are similar to those seen in patients suffering from depression/sickness behavior (including alterations in activity, appetite, and sleep) (Nemeroff, 1996). Moreover, patients with depression have been found to exhibit increased CRH activity as manifested by elevated concentrations of CRH in the CSF, increased mRNA in the PVN, a blunted ACTH response to CRH challenge (likely reflecting downregulation of pituitary CRH receptors), and downregulation of receptors for CRH in the frontal cortex of victims of suicide (presumably secondary to hypersecretion of CRH) (Nemeroff, 1996; Holsboer and Barden, 1996). In addition, as previously described, cytokines, including IL-1, may induce resistance of nervous, endocrine, and immune system tissues to circulating glucocorticoid hormones through direct inhibitory effects on the expression and/or function of GR (Pace et al., 2007). Glucocorticoid resistance has been repeatedly demonstrated in patients with depression (as reflected in nonsuppression on the DEX suppression test and the more sensitive DEX-CRH test (Holsboer, 2000), and may contribute to impaired feedback regulation of CRH and the release of proinflammatory cytokines (Raison and Miller, 2003).

Second, cytokines have been shown to alter the metabolism of monoamines, including norepinephrine, serotonin, and dopamine, all of which have been implicated in the pathophysiology of mood disorders (Dunn et al., 1999). For example, cytokines, including IFN-α, have been shown to reduce serum concentrations of L-tryptophan, likely secondary to induction of the enzyme, indoleamine 2,3 dioxygenase, which breaks down tryptophan into kynurenine and quinolinic acid (Dantzer, 2006). Tryptophan is the primary precursor of serotonin, and depletion of tryptophan has been associated with the precipitation of mood disturbances in vulnerable patients (Dantzer, 2006). Of note, quinolinic acid exerts agonistic effects on glutamate receptors, and thereby may contribute to neurotoxicity. Cytokine induction of MAPK pathways may also alter neurotransmitter availability by upregulating the expression and activity of monoamine reuptake pumps for serotonin, norepinephrine, and dopamine (Zhu et al., 2005). Indeed, reduced CSF concentrations of the serotonin metabolite, 5-hydroxyindoleacetic acid, was found to be associated with evidence of activation of p38 MAPK in PBMCs of rhesus monkeys exposed to early life abuse/neglect (Sanchez et al., 2007).

Third, cytokines appear to have significant effects on factors that influence synaptic plasticity including growth factors and neurogenesis. For example, intrahippocampal transplantation of neural precursor cells that overexpress IL-1 RA blocked chronic isolation-induced impairment in both memory and neurogenesis (Ben-Menachem-Zidon et al., 2007). Decreased hippocampal neurogenesis as a result of chronic unpredictable stress was also blocked in mice lacking the IL-1 receptor 1 (IL-1 R1) (Koo and Duman, 2008). Moreover, disruption of hippocampal-dependent memory as a result of exposure to social-isolation stress has been shown to critically involve inhibition of brain-derived neurotrophic factor (BDNF) by IL-1β (Barrientos et al., 2003).

Finally, regional brain activity in neural circuits relevant to behavior has been found to be altered by both cytokine (IFN-α) administration as well as immune activation following typhoid vaccination (Brydon et al., 2008; Capuron et al., 2005, 2007). Relevant brain regions that appear to be involved in cytokine-induced behavioral changes include the basal ganglia and the dorsal anterior cingulate cortex.

18.5 The Impact of Stress on the Immune System

The impact of stressors known to activate neuroendocrine outflow pathways on immune function is one of the most compelling demonstrations of meaningful interactions among the nervous, endocrine, and immune systems.

18.5.1 Acute Stress

In keeping with the concept that appropriate and time-limited activation of stress-response systems may be necessary for optimal functioning of the organism, data suggest that brief and/or mild stressors may enhance innate and acquired immune system functioning in ways that are likely to have beneficial effects on health, such as augmented DTH reactions, increased antigen-induced antibody production, increased circulating levels of proinflammatory cytokines, and enhanced macrophage function (Broug-Holub et al., 1998; Dhabhar, 2003; O'Connor et al., 2003; Silberman et al., 2003). For example, brief electric foot-and tailshock have been shown to increase alveolar macrophage production of inflammatory cytokines (Broug-Holub et al., 1998) and to augment nitric oxide release from macrophages stimulated *in vitro* with bacterial endotoxin. From an evolutionary perspective, enhanced macrophage function would be expected to serve an adaptive function, such as priming the immune system to cope with tissue trauma and/or infection. Evidence supporting this notion comes from a study showing that acute stress diminished the time required for resolution of inflammation induced by injection of bacteria into rodents (Deak et al., 1999).

Because innate immune mediators such as macrophages also play a central role in initiating antigen-mediated specific (or acquired) immune responses (i.e., cellular or humoral immunity), it might be predicted that evolutionary pressure would favor an enhancement of cellular or antibody-mediated immunity to complement augmented innate immunity following acutely stressful conditions. Significant evidence suggests that such an adaptive interaction exists in the context of acute and/or mild stressors. Dhabhar et al. (2000) have shown in a series of studies in rodents that a brief mild stressor, such as 2 h of physical restraint and/or shaking, administered prior to antigen challenge enhances cutaneous DTH, an antigen-specific reaction mediated by CD4 T-lymphocytes. The degree of enhancement correlates with increases in plasma corticosterone and appears to be mediated in part by corticosterone-induced alterations in leukocyte trafficking out of blood and spleen and into the skin and draining lymph nodes. Adrenalectomy abolishes this immune-enhancing effect of stress, suggesting that adrenal hormones play a central role in this process. Compatible with this notion is that administration of corticosterone at levels similar to those seen after acute stress mimics the immune enhancing effect of the stressor. Administration of epinephrine also enhances DTH, and the combined effect of corticosterone and epinephrine is additive. Stress-induced enhancement of DTH requires the local production of the T-cell-stimulating cytokine, IFN-γ (Dhabhar et al., 2000). More recent studies indicate that stress-induced increases in leukocyte trafficking may also enhance immunoprotective responses during wounding (e.g., trauma or surgery), infection and vaccination, although such stress-induced trafficking may contribute to pathology in the context of autoimmune or inflammatory disorders (Viswanathan and Dhabhar, 2005).

Acute stress in rodents has also been associated in several studies with enhanced humoral immunity as measured by antibody production in response to specific antigens (e.g., Silberman et al., 2003). This effect appears to be highly sensitive to the intensity of the stressor, given findings that mild stress (placement in a shock box) enhanced, but more severe stress (electric footshock) suppressed, antibody production when administered at the same time in relationship to immunization (Croiset et al., 1990; Zalcman et al., 1991). As with cellular immunity, significant evidence suggests that the ability of acute stress to enhance humoral immunity is mediated in part by neuroendocrine elements of the stress response. In rats, the mild stress of exposure to a novel environment enhances splenic antibody production to sheep red blood cells (RBCs) when the stress is delivered at the time of RBC immunization. This effect is blocked either by severing the splenic nerve, which is the chief source of SNS innervation of the spleen, or by pretreating the rats with a β-adrenergic antagonist (Croiset et al., 1990). A parallel involvement of the HPA axis is suggested by findings that pretreatment with an antibody to CRH blocks the ability of acute stress to enhance antibody production in the spleen (Berkenbosch et al., 1991).

In addition to effects on cell-mediated and humoral immune responses, challenge of rodents with acute stressors has been shown to alter circulating and tissue levels of innate immune cytokines. For example, exposure of rats to inescapable shock increases protein and gene expression of IL-1β in both the CNS and the periphery (e.g., O'Connor et al., 2003). These effects of stress appear to be related in part to activation of the SNS and binding of catecholamines to α- and β-adrenergic receptors. Indeed, administration of the α-adrenergic receptor antagonist, prazosin, was found to block stress-induced increases of proinflammatory cytokines in the blood of rats, while the β-adrenergic receptor blocker, propranolol, inhibited stress-induced cytokines in the brain (Johnson et al., 2005). Of note, antagonism of adrenergic α-receptors in mice blocked stress-induced increases in NFκB DNA binding, suggesting that catecholamines may activate cytokine expression in part through stimulation of relevant inflammatory-signaling pathways (Bierhaus et al., 2004).

As in studies of laboratory animals, many studies in humans document that a variety of acutely stressful situations produce short-term alterations in both enumerative and functional aspects of the immune system. Typical acute-stress paradigms include standardized laboratory tasks such as public speaking and/or performing mental arithmetic (Kirschbaum et al., 1993), as well as less controlled stressors such as first-time parachute jumping. Immune system responses are generally measured over minutes to several hours in order to investigate the extent to which a single laboratory stressor in humans is able to reliably affect long-term immune system functioning in ways similar to those seen in studies of laboratory animals (Rabin, 1999). Although there is limited literature prior to 2000 addressing the question as to whether or not acute laboratory-type stressors have relevant effects on more complex measures of immune function, in more recent years, studies have begun to investigate the topic, including ways in which acute stressor challenge influences production of antibodies in response to antigen. For example, acute stress has been found to enhance antibody production in response to vaccination, similar to effects seen in laboratory animals (Edwards et al., 2006; Viswanathan et al., 2005).

Table 5 lists immune changes most frequently associated with brief environmental stressors (such as parachute jumping) or laboratory stress (i.e., mental arithmetic and public speaking) in humans. Commonly reported enumerative changes include increased numbers of white blood cells, CD8 T-lymphocytes and NK cells, decreased numbers of T-cells and B-cells, a decrease in the ratio of CD4/CD8 cells, and changes in NK cells (Zorrilla et al., 2001). Of these changes, increases in the numbers of circulating NK cells and CD8 lymphocytes demonstrate the largest effect size and have been the most often reported. Interestingly, these increases are opposite to the decreases in NK and CD8 cell numbers observed in most studies of chronic stress in humans (Zorrilla et al., 2001).

In addition to effects on the trafficking of leukocytes between bodily compartments, acute stress appears to alter the *in vitro* functioning of lymphocytes as measured by proliferation in response to nonspecific mitogens and by the capacity of NK cells to lyse target tumor cells. Many studies and a

Table 5 Effects of acute stress, chronic stress, and major depression on immune parameters in humans

Immune variable	*Acute stress*	*Chronic stress*	*Major depression*
WBC	↑/↑	↑/↑	↑/↑
Lymphocytes	→	→	↓/↓
Monocytes	→	→	→
T-lymphocytes	→	↓/↓	→
B-lymphocytes	↓/↓	↓/↓	→
CD4+ cells	→	↓/↓	→
CD8+ cells	↑/↑	↓/↓	→
NK cells	↑/↑	↓/↓	↓/↓
CD4+/CD8+	↑	↓/↓	[illegible]
% B-cells	→	→	↓/↓
% T-cells	→	↓/↓	↓/↓
% CD4+	↓	→	↓/↓
% CD8+	↑	↓/↓	↓/↓
% CD4+/CD8+	→	→	→
% NK cells	?	→	→
NKCA (total)	↑/↑	↓/↓	↓/↓
NKCA (per cell)	→	↓	↓
Mitogen-induced proliferation	↓	↓/↓	↓/↓
Response to vaccine	↑	↓	↓
Antibody titers to EBV	?	↓/↓	?
Cytolytic T-cell response to antigen	?	↓	?
Wound healing	?	↓	?
Th1/Th2 balance	↓	↓	?
DTH	↑	→	↓
TNF-α	↑	?	↑
IL-1	↑	?	↑
IL-2	→	?	↑
IL-6	↑/↑	↑	↑
NFκB	↑	?	?
CRP	↑	↑	↑

↑/↑, positive effect confirmed by meta-analysis; ↑, majority of studies suggest positive effect; ↓/↓, negative effect confirmed by meta-analysis; ↓, majority of studies suggest negative effect; →, conflicting findings: ?, not enough data to suggest positive or negative relationship; WBC, white blood cells; NK cells, natural killer cells; CD4/CD8, ratio of CD4+ T-lymphocytes to CD8+ T-lymphocytes; NKCA, natural killer cell activity; EBV, Epstein–Barr Virus; Th1/Th2, the ratio of T-helper cell type 1 cytokines to T-helper cell type 2 cytokines; DTH, delayed type hypersensitivity; TNF, tumor necrosis factor; IL, interleukin; CRP, C-reactive protein; NFκB, nuclear factor kappa B.

large meta-analysis document that laboratory stressors suppress lymphocyte proliferation in response to a variety of mitogens, especially the nonspecific T-cell mitogens, concanavalin A (ConA) and phytohemagglutinin (PHA), and the nonspecific B-cell stimulus pokeweed mitogen (PWM) (Zorrilla et al., 2001). In this regard, laboratory stressors mirror many forms of chronic stress in humans, which have also been shown to reliably decrease mitogen-stimulated lymphocyte proliferation (Rabin, 1999). Interestingly, most, but not all, studies of short-term stress in laboratory animals also find a decrease in nonspecific lymphocyte proliferation. Why nonspecific lymphocyte proliferation in animals and humans does not demonstrate the type of enhancement seen in antigen-driven lymphocyte responses to acute stress is unknown. One possible explanation is that nonspecific lymphocyte responsiveness provides no survival advantage in the dangerous types of situations that activate stress response systems and hence has not been retained by natural selection. Nevertheless, the majority of the studies examining nonspecific lymphocyte responsiveness following stress in both laboratory animals and humans have sampled the peripheral blood from which relevant lymphocytes may have trafficked to other tissues where immunoprotective responses may be observed (Viswanathan and Dhabhar, 2005).

While studies in humans demonstrate consistent effects of acute stress on nonspecific lymphocyte proliferation, the effect of acute stress on the functional capacity of NK cells presents a more complicated scenario. Chronic stress tends to lower NKCA. However, several studies suggest that acute stressors potentiate NKCA (Zorrilla et al., 2001). It is attractive to speculate that the increased NKCA frequently reported in response to acute stress in humans reflects enhanced innate immune system functioning as observed in certain animal models of acute stress, but it should be remembered that most studies measure total NKCA per unit of blood, not activity per NK cell. Indeed, when NKCA is calculated on a per cell basis, it becomes far less clear that acute stress enhances NK cell functioning. For example, Naliboff et al. (1995) found that although a confrontational role-play stressor increased the number of circulating NK cells, it actually lowered NKCA when the increased numbers of NK cells were taken into account. Other studies suggest that acute stress enhances total NKCA, but has no effect on the cytolytic capacity of individual NK cells (e.g., Dopp et al., 2000). Even when an increase in per cell NKCA has been shown in response to acute stress, this effect is typically attenuated when compared to the concomitant increase in total NKCA. Thus, it seems that most reports of increased NKCA in response to acute stress reflect in large part stress-induced shifts in NK-cell trafficking.

Like in laboratory animals, acute stress in humans has also been associated with increases in circulating inflammatory markers, including IL-1α and -β, IL-6, TNF-α (Steptoe et al., 2007). For example, challenge with a psychosocial stressor consisting of a public-speaking task and a mental arithmetic task was found to increase plasma levels of IL-6 in men and women (e.g., Brydon et al., 2004; Pace et al., 2006). Circulating levels of anti-inflammatory factors such as IL-4 and IL-10 have been reported to decrease with acute stress challenge (Buske-Kirschbaum et al., 2007). Based on findings of increased stress-induced NFκB DNA binding in PBMCs, increases in proinflammatory cytokines during stress appear to be related in part to the activation NFκB signaling pathways (Bierhaus et al., 2004; Pace et al., 2006).

The effects of acute stress on immune measures in humans are mediated by neuroendocrine elements of the stress response. Most evidence points to a primary role for catecholamines, as opposed to glucocorticoids, in the rapid shifts in leukocyte trafficking that account for most immune alterations observed with laboratory stressors (Bierhaus et al., 2004; Johnson et al., 2005; Peters et al., 1999). Many immune parameters, including increased numbers of circulating NK and CD8 cells, decreased circulating CD4 cells, increased NKCA, and decreased mitogen-stimulated lymphocyte proliferation, correlate strongly with stress-induced cardiovascular changes that reflect heightened SNS activity (Benschop et al., 1998). Conversely, when subjects exposed to an acute stressor are divided between those who respond with high versus low SNS activity, only those with high SNS activity demonstrate immune system changes (Marsland et al., 2002). Blockade of adrenergic β- and/or α-receptors reliably attenuates or abolishes immune changes in response to acute stressors in humans (i.e., parachute jumping) (Klokker et al., 1997) as in laboratory rodents (Johnson et al., 2005). In addition, administration of catecholamines to humans transiently produces immune changes similar both qualitatively and quantitatively to those observed with acute stress (e.g., Benschop et al., 1996b). Catecholamines have been shown to decrease the number of adhesion molecules on lymphocytes (Rogers et al., 1999), suggesting that increased SNS activity may alter numbers of circulating lymphocytes by decreasing endothelial adhesion of these cells.

Because the HPA-axis response to stress occurs at a slower pace than the nearly instantaneous response of the SNS (Herbert and Cohen, 1993b), little is known about the contribution of cortisol to the transient immune changes observed in response to acute stress in humans. This is in contrast to animal models of acute stress, in which glucocorticoids have been shown to play a central role (e.g., Dhabhar and McEwen, 1999; McEwen et al., 1997). The little data available suggest that relatively late appearing effects may be less responsive to catecholaminergic manipulation than immune changes seen immediately after an acute stressor. It should be noted that because the SNS and HPA axis interact, manipulations that abolish the SNS reaction to acute stress can attenuate the stress-induced activation of the HPA axis (Benschop et al., 1996a). Thus, SNS manipulations such as β-adrenergic blockade may attenuate HPA-axis contributions to the effects of stress on immune parameters that would otherwise be recognized.

18.5.2 Chronic Stress

Like acute stress, chronic stress produces reliable alterations in enumerative and *in vitro* functional immune measures in humans (**Table 5**). Two influential meta-analyses found that chronic stress significantly increases the number of circulating white blood cells, while decreasing the number of circulating B-cells, T-cells, helper CD4 cells, cytotoxic/suppressor CD8 (Herbert and Cohen, 1993a; Zorrilla et al., 2001). The review by Herbert and Cohen (1993a) also reported a decreased number of NK cells and a decrease in the CD4/CD8 ratio as a result of chronic stress. When compared to nonsocial types of stress, interpersonal stressors more significantly increase the CD4/CD8 ratio and decrease the number of circulating B- and T-cells and the percentages of circulating CD4 and CD8 cells (Herbert and Cohen, 1993a). Compared with long-term stressors such as caring for a disabled spouse, naturalistic stress of short duration (i.e., exam taking) more sharply decreased the percentage of circulating CD4 cells. As is the case with CD8 cells, the number of NK cells increases in blood following acute stress, but decreases during conditions of chronic stress. In addition to these enumerative effects, the meta-analyses showed that chronic stress also reliably decreased several *in vitro* measures of lymphocyte functioning, including lymphocyte proliferation to the nonspecific mitogens, PHA and ConA, as well as NKCA (Herbert and Cohen, 1993a; Zorrilla et al., 2001).

In addition to enumerative and *in vitro* effects, chronic stress has been repeatedly shown to affect the functioning of both humoral and cell-mediated immunity (e.g., Burns et al., 2003; Kiecolt-Glaser et al., 1996; Li et al., 2007), although much more is known about stress effects on antibody production or cytotoxic T-cell functioning than about any direct health consequences that might stem from these stress-associated immune changes. The effects of naturalistic stressors such as exam taking or caregiving to a spouse with dementia on humoral immunity have been assessed through two principal methods: antibody titers to chronic latent viruses and antibody responses to vaccines. It is generally believed that increased antibodies to viruses such as Epstein–Barr virus (EBV) or herpes simplex-1 or-6 viruses (HSV-1, HSV-6) reflect a diminished capacity on the part of the immune system to maintain these chronic viruses in a latent state (Glaser and Kiecolt-Glaser, 1997). Therefore, findings that stress correlates with increased antibodies imply that the overall functioning of the immune system is compromised. In a meta-analysis, Herbert and Cohen (1993a) found a strong positive correlation (i.e., 0.528 for EBV and 0.848 for HSV-1) between all types of reported naturalistic stressors and antibody titers to EBV and HSV-1. However, a separate meta-analysis also published in 1993 only found a significant correlation between chronic stress and EBV titers, not HSV-1 titers (Rood et al., 1993). Similarly, Glaser et al. (1999) reported that exam stress in a group of West Point cadets increased EBV, but not HSV-1 antibody titers.

A number of studies have reported that chronic stress interferes with the body's ability to mount an antibody response to vaccines, with most of the work being done with the influenza and hepatitis B vaccines. For example, several studies report a decreased antibody response to influenza vaccine in geriatric subjects undergoing the stress of caring for a spouse or family member with dementia (Segerstrom et al., 2008). Because antibody response to vaccination is T-cell dependent, and because cellular immunity plays a central role in controlling latent viruses such as EBV, the findings of stress-associated decrements in these immune measures suggest that *in vivo* cellular immunity may also be impaired by chronic stress. Indeed, *in vitro* assessments of antigen-specific T-cell functioning seem to confirm such impairment. For example, both examination and caregiving stress have been reported to be associated with decrements in memory T-cell responses to latent virus antigens and to vaccines in terms of antigen-induced T-cell proliferation (Glaser and Kiecolt-Glaser, 1997, 1998). DTH is the standard test of *in vivo* cell-mediated immunity, and animal studies suggest that chronic stress impairs DTH (Dhabhar, 2003) as would be predicted by latent virus and vaccine-related findings in humans. However, little is known about the effect of chronic psychological stress on DTH in humans. Interestingly, two studies of long-term stress (bereavement and caregiving) found no effect on DTH, even while showing suppression of nonspecific T-cell proliferation (Bartrop et al., 1977) and other typical stress-related abnormalities in T-cell functional and enumerative measures (Pariante et al., 1997a). These discrepancies again point to the complexities involved in extrapolating from *in vitro* alterations in immune functioning to impairment of more complex *in vivo* immune responses. Nonetheless, chronic stress has been shown to affect wound healing, a complex process that involves aspects of both innate and acquired cellular immunity. For example, in dental students who volunteered to receive oral punch biopsies, healing took 40% longer during examination periods than during summer vacation, with no student showing faster healing during exams than vacation (Marucha et al., 1998). Chronic stress has also been shown to promote the development of ultraviolet light-induced squamous cell carcinoma in mice by suppressing type 1 cytokines (e.g., IFN-γ) and protective T-cells numbers, while increasing suppressor T-cells numbers (Saul et al., 2005).

As with acute stress exposure, challenge with chronic stress has also been shown to alter circulating and tissue levels of innate immune/inflammatory cytokines in both laboratory animals and humans. For example, rats exposed to chronic unpredictable stress exhibit increases in central as well as circulating concentrations of IL-1β and TNF-α (Grippo et al., 2005). Moreover, as noted above, the effects of chronic unpredictable stress in rats including depressive-like behavioral alterations (e.g., decreased sugar water consumption) and reduced hippocampal neurogenesis are abrogated in mice lacking IL-1 R1 (Koo and Duman, 2008), suggesting that IL-1 plays a significant role in chronic stress-induced changes in CNS function. A similar role of stress-induced IL-1 has been found in relation to the effects of social isolation stress on the expression of BDNF in the hippocampus (Barrientos et al., 2003). In humans, chronic caregivers have been shown to have elevated circulating concentrations of IL-1 relative to controls (von Kanel et al., 2006). Regarding potential neuroendocrine mechanisms of these effects, increased circulating concentrations of IL-6 in response to chronic exposure to high altitude were blocked by the α-adrenergic antagonist, prazosin, suggesting a role for catecholamines (Mazzeo et al., 2001).

18.5.3 Psychosocial Variables Mediating Neuroendocrine–Immune Interactions during Stress

We have highlighted the notion that stress (especially chronic stress) plays a central role in altering the immune system and thus represents a significant risk factor for the development of immunologic and possibly neuropsychiatric disorders (discussed below). However, even a moment's reflection upon human experience obviates so simple a notion.

Clearly, how a stimulus is perceived and the degree to which this perception activates stress-response elements has much to do with the psychological and physiological makeup of an organism, and will therefore vary tremendously between individuals. Thus, differences between individuals, or groups of individuals, will significantly affect relationships between a stressor and immune responses (Kemeny and Gruenewald, 2000). In fact, evidence suggests that both measures of immune suppression and activation correlate more strongly with an individual's perception and physiologic response to a stressor than to the experience of a stressful situation, *per se*. For example, although Vedhara and Nott (1996) observed no differences in *in vivo* cell-mediated immunity between a group of students in the stressful situation of taking final examinations and a carefully matched group of nonstudents, they noted that students with high subjective complaints of stress had significant reductions in DTH when compared to students with low subjective stress. Moreover, during examinations, students with low levels of anxiety showed increased NKCA, whereas students with high anxiety showed decreased NKCA (Borella et al., 1999). Yet another study found an association between perceived stress and low antibody titers to meningitis C conjugate vaccination, whereas no relationship was observed between number of stressful life events and the same antibody measure (Burns et al., 2002). Finally, women with regional breast cancer who displayed an early rapid decline in subjective distress after cancer diagnosis showed improved NK-cell cytotoxicity (Thornton et al., 2007).

An extensive literature exists concerning factors that moderate an individual's response to stress. Although numerous such factors have been putatively identified, this chapter focuses on the variables of social support and stressor controllability.

A number of studies have shown that social support influences both neuroendocrine and immune system functioning. This effect has been shown in laboratory animals and humans in the context of chronic and acute stress (DeVries et al., 2007; Uchino, 2006). For example, young monkeys who are removed from their natal social groups and who are housed alone show significantly increased levels of cortisol and attendant decreases in circulating T-cells (Gordon et al., 1992). Housing these animals with peer-age strangers does not modify the effect (Gust et al., 1992). However, allowing the separated animals to remain in the presence of a juvenile attachment figure (i.e., a friend) prevents these immune and neuroendocrine changes (Gust et al., 1996). Studies by DeVries et al. (2007) have shown that social contact also attenuates the negative effects of restraint stress on wound healing in Siberian hamsters. These positive effects of social contact appear to involve decreased restraint-induced glucocorticoid responses secondary to increased circulating concentrations of oxytocin, a neuropeptide that has been associated with pair bonding in voles.

In humans, several studies have reported an association between CD4 cells and social support in human immunodeficiency virus (HIV)-positive men (e.g., Uchino, 2006). In addition, the rate of progression from HIV infections to acquired immune deficiency syndrome virus (AIDS) has been associated with lower satisfaction with social support (Leserman et al., 2000). Associations have also been made on a number of separate studies between NKCA and social support (Uchino, 2006). For example, NKCA and social support were positively correlated in women with ovarian cancer (Lutgendorf et al., 2005). Social support has also been positively associated with antibody production following both hepatitis B and influenza vaccination (Uchino, 2006). In a classic study by Cohen et al. (1997), individuals with impoverished social support systems were more likely than people with more varied social ties to develop colds following exposure to a rhinovirus under laboratory conditions. Finally, higher levels of perceived social support were associated with Th1 dominance in male Japanese workers (Miyazaki et al., 2005).

Limited evidence suggests that the perception of whether a stressor is controllable or not and, not the actual stressor itself, may also modulate the effect of stress on immunity (Isowa et al., 2006; Peters et al., 1999). An influential early literature found that immune changes seen in response to electrical shock in rodents could be accounted for by whether an animal could escape the shock through its own efforts or not. For example, the ability to escape being shocked was protective against the decreased nonspecific mitogen-induced lymphocyte proliferation observed in animals without control of the stressor (Laudenslager et al., 1983; Millar et al., 1993). In humans, as in animals, controllability may play an important role in mediating the deleterious health effects of daily stress (Isowa et al., 2006; Nakata et al., 2000; Peters et al., 1999). For example, when the contributions of various aspects of the work place are assessed, job control is a central factor in accounting for immune alterations, especially the frequently reported finding of decreased numbers of circulating CD4 helper/inducer cells (Kawakami et al., 1997). It should be noted, however, that not all studies have found controllability effects on immune or neuroendocrine functioning. Tailshock in rats increases expression of CRH mRNA in the PVN of the hypothalamus, regardless of whether or not that shock is escapable (Helmreich et al., 1999). In addition, one study of acute stress in humans found that controllable stress decreased mitogen-induced lymphocyte proliferation and percentage of circulating monocytes, whereas uncontrollable stress had no effect on any immune measure (Weisse et al., 1990).

18.6 Neuroendocrine–Immune Interactions in Depression

18.6.1 Major Depression and Immune Parameters

An intimate connection between stress and mood disorders has been recognized for many years. Patients with major depression are characterized by persistent stress system hyperactivity (e.g., CRH hypersecretion, hypercortisolism, and increased catecholamine production; Raison and Miller, 2003) and, therefore, given the role of stress-related hormones in regulating the immune response, one would expect to see immune system changes in depression that mirror changes resulting from chronic stress. Much evidence suggests that this is the case, as can be seen from a comparison of immune system changes in chronic stress and depression, as listed in **Table 5**. Like chronic stress, major depression has been associated with alterations in enumerative immune measures, including increased numbers of circulating white blood cells and increased percentages of neutrophils and lymphocytes. Depression is also associated with an increased CD4/CD8 ratio and by altered *in vitro* measures of immune functioning, including decreased nonspecific mitogen-stimulated lymphocyte proliferation and NKCA (Irwin, 1999; Irwin and Miller, 2007; Zorrilla et al., 2001). Also reported to be associated with major depression are decreased percentages of lymphocytes including T-cells, B-cells, T-helper cells, and T-cytotoxic cells as well as a decrease in circulating numbers of NK cells (Zorrilla et al., 2001). However, not all studies have obtained results consistent with these findings, suggesting that clinical and biological factors may be important for interpreting immune changes in depressed patients (Irwin and Miller, 2007).

Since depressed patients represent a heterogeneous group, some researchers have proposed that the changes in immune functioning in depression are more driven by the characteristics of the patients and their illness than any process central to the pathophysiology of depression itself (Irwin and Miller, 2007). Investigators have considered factors such as age, hospitalization status, psychiatric comorbidity, alcohol and smoking history, and profiles of symptoms, including sleep disturbance, appetite change, and presence of melancholia or atypical depression (Irwin and Miller, 2007). Support for this idea comes from a study of depressed subjects showing that of six symptom clusters, only motor retardation and sleep disturbance correlated with decreased NKCA (Cover and Irwin, 1994). No correlation was seen for anxiety/somatization, weight loss, cognitive disturbance, or diurnal mood variation. Evidence also suggests that depressed patients with melancholia (nonreactive mood, diurnal mood variation, insomnia, and anorexia) are more likely than nonmelancholic patients to exhibit immune alterations, including suppression of *in vivo* cell-mediated immunity (as assessed by DTH reactions; Hickie and Lloyd, 1995; Lloyd et al., 1992; Maes et al., 1992), as well as increases in circulating inflammatory markers (Kaestner et al., 2005). Earlier age of onset of major depression has been linked to decreased NKCA, suggesting that alterations in NK-cell function may be a phenotypic feature of depression that begins earlier in life (Frank et al., 2002). Interestingly, Zaharia et al. (2000) found that patients with dysthymia, a condition less severe than major depression by definition, showed greater suppression of nonspecific mitogen-induced lymphocyte proliferation than patients with major depression. Dysthymic patients often have a longer illness course, raising the possibility that chronicity may be a more potent mediator of functional immune changes than symptom severity.

Sleep disturbance is a central feature of depression, and several converging lines of evidence suggest that sleep disturbance may significantly contribute to depression-related immune alterations, especially changes in NK-cell number and function (Irwin and Miller, 2007). For example, in depressed subjects, NKCA correlates with both subjective insomnia and polysomnographic findings of decreased total sleep time and sleep efficiency (defined as time asleep/time in bed; Cover and Irwin, 1994; Irwin et al., 1992). Moreover, these sleep disturbances equally affect NKCA in subjects without depression. For example, nondepressed insomniacs and healthy subjects who are sleep deprived exhibit decreased NKCA (Irwin and Miller, 2007). These findings suggest that sleep status affects immune function independent of the presence of a mood disorder.

18.6.2 Depression and Immune Activation

The past 15 years have seen a wealth of studies investigating the possibility that in addition to exhibiting evidence of suppression of certain aspects of the immune response, some patients with major depression demonstrate immune system activation

Table 6 Immune activation in major depression

Increased baseline plasma concentrations of interleukin (IL)-1β receptor antagonist, IL-6, soluble IL-6 receptor, IL-8, soluble IL-2 receptor
Increased psychosocial stress-induced plasma concentrations of IL-6 and peripheral blood mononuclear levels of nuclear factor kappa B
Increased cerebrospinal fluid concentration of IL-1β
Increased mitogen-induced production of IL-1β, IL-6, and interferon-γ
Increased plasma concentrations of neopterin
Activation of circulating T-cells as assessed by cell-surface markers
Increased plasma concentration of soluble CD8
Increased plasma concentrations of acute-phase reactants serum amyloid A, C-reactive protein, and haptoglobin
Increased plasma concentrations of prostaglandin E2 and thromboxane B2
Decreased plasma concentrations of zinc
Decreased omega-3 fatty acid/omega-6 fatty acid ratio

(Maes, 1999). Findings consistent with immune system activation are presented in **Table 6** and include elevated blood concentrations of various cytokines (especially those elaborated from an activated innate immune response), increased *in vitro* stimulated cytokine production (for a review, see Irwin and Miller (2007), Pace et al. (2007), and Raison et al. (2006)), increased markers of T-cell activation (Maes et al., 1993), and increased plasma concentrations of acute phase proteins, chemokines, and adhesion molecules (Irwin and Miller, 2007; Kling et al., 2007; Raison et al., 2006; Zorrilla et al., 2001). In addition, studies have reported that patients with major depression have elevated concentrations of cytokines (including IL-1β) in the cerebrospinal fluid (Levine et al., 1999), consistent with studies from the animal literature demonstrating that stress increases IL-1β in the brain (Nguyen et al., 1998). Also consistent with immune system activation in depression are findings of decreased serum zinc and a decrease in the ratio of omega-3 to omega-6 polyunsaturated fatty acids (Kiecolt-Glaser et al., 2007; Levenson, 2006). Inflammation reliably reduces zinc and causes a shift in the balance of fatty acids, favoring omega-3 at the expense of omega-6. Of note, increased inflammatory markers (e.g., plasma IL-6) in depressed patients are apparent across the circadian cycle (Alesci et al., 2005), suggesting that patients with major depression exhibit persistent innate immune system activation. Positive correlations between severity of depressive symptoms and plasma cytokine levels have also been described (Alesci et al., 2005; Miller et al., 2002a; Thomas et al., 2005). Of note, as with data on immunosuppression, not all studies have confirmed immune system activation in depressed patients. For example, no alterations in blood cytokine concentrations were found in depressed or schizophrenic subjects when age, body mass index, gender, smoking habits, ongoing or recent infections, or prior medication were taken into account (Haack et al., 1999).

In addition to increased basal innate immune system activation, patients with major depression may also show enhanced stress-induced inflammatory responses. For example, relative to healthy controls, male patients with major depression and increased early life stress were found to exhibit enhanced innate immune responses to psychosocial stress, as reflected by increased plasma IL-6 concentrations and increased NFκB DNA binding in PBMCs (Pace et al., 2006) (**Figure 5**). Of note, other studies have found a

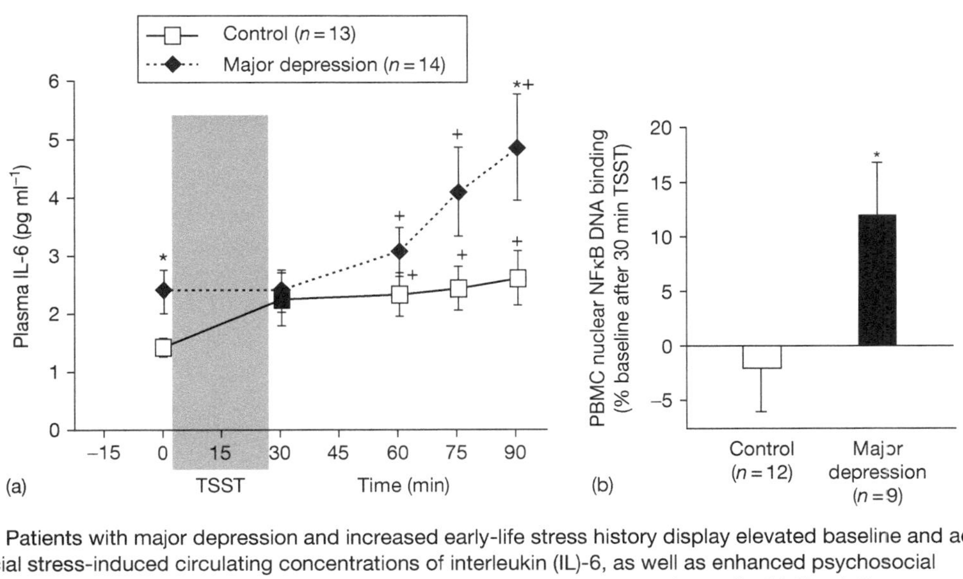

Figure 5 Patients with major depression and increased early-life stress history display elevated baseline and acute psychosocial stress-induced circulating concentrations of interleukin (IL)-6, as well as enhanced psychosocial stress-induced nuclear factor κB (NFκB) DNA binding in peripheral blood mononuclear cells. (a) Circulating concentrations of IL-6 before and 30, 60, 75, and 90min after the start of the trier social stressor test (TSST) in healthy controls vs. patients with current major depression and increased history of early-life stress. * vs. control group at the given timepoint, $p \leq 0.05$; + vs. 0min within the same group $p < 0.025$. (b) Percent change in nuclear NFκB DNA binding from before to 30 min after the start of the TSST (ΔNFκB) in a subset of the same participants. * vs. control group, $p \leq 0.05$. Reprinted with permission from Pace TW, Mletzko TC, Alagbe O, Musselman DL, Nemeroff CB, and Miller AH (2006) Increased stress-induced inflammatory responses in male patients with major depression and increased early life stress. *American Journal of Psychiatry* 163: 1630–1633. Copyright 2006, American Psychiatric Association.

relationship between early life stress and immune activation (Danese et al., 2007), especially as it relates to depressed patients (Danese et al., 2008).

In addition to direct evidence of immune system activation in patients with major depression, other lines of evidence strongly suggest that immune system activation may play an etiologic role in certain depressive syndromes. For example, it is increasingly recognized that immune system activation in humans and animals induces a physiological and behavioral syndrome that resembles depression (Dantzer et al., 2008; Raison et al., 2006; Smith, 1991). As noted above, this constellation of immune-mediated signs and symptoms has been referred to as sickness behavior (Dantzer, 2006). Much evidence has accrued that sickness behavior is mediated primarily by cytokines. Indeed, administration of cytokines to humans and animals produces a similar state, and blocking cytokine action using cytokine antagonists including IL-1ra, alpha melanocyte stimulating hormone, insulin-like growth factor and IL-10 prevents the development of sickness behavior in laboratory animals (Bluthe et al., 1992; Dantzer et al., 2008). Taken together, these findings demonstrate that cytokines are able to induce a syndrome virtually indistinguishable from idiopathic major depression. Moreover, these data provide powerful circumstantial evidence that the elevated cytokine levels reported in major depression may represent, not just an epiphenomenon, but a contributor to the pathogenesis of the disorder.

The similarity of depression and sickness behavior suggests that people with medical illnesses may have higher rates of major depression than people who are medically healthy in part related to contributions from immune activation (Irwin and Miller, 2007; Raison et al., 2006; Yirmiya et al., 2000). Activation of the innate immune response is common in the context of many medical illnesses, and contributes to much medical morbidity, from septic shock in bacteremia to cachexia in patients with cancer (Anker et al., 1999; Inui, 1999). Relevant to behavioral consequences of illness, work by our group found increased plasma concentrations of IL-6 in cancer patients with a diagnosis of depression (Musselman et al., 2001b). That immune activation may mediate the high rate of depression in medically ill patients is further suggested by the especially high rate of major depression seen in autoimmune disorders and other diseases in which immune activation is a cardinal feature (Raison et al., 2006).

18.7 Model for Neuroendocrine–Immune Interactions in Clinical Disease

18.7.1 A Neuroendocrine Diathesis Model of Inflammation

Immune and stress response systems interact via a complicated circuitry composed of both feed-forward excitatory and feedback inhibitory loops. Under normal conditions, inhibitory elements such as glucocorticoids and anti-inflammatory cytokines limit immune system activation to levels that are appropriate for clearing the initial antigenic stimulus (Beutler, 2004; Chaplin, 2006). However, it is becoming increasingly recognized that in certain conditions such as chronic stress these inhibitory feedback loops may become less efficient, allowing for chronic immune activation and possibly suppression of certain aspects of the immune response (especially acquired immune responses) (**Figure 6**; Raison et al., 2006). Glucocorticoid inhibition of inflammatory signaling involves GR antagonism of the NFκB pathway, and NFκB has been regarded as a lynchpin in the inflammatory signaling cascade leading to the production of proinflammatory cytokines (McKay and Cidlowski, 1998). In this regard, it is interesting that major depression, some autoimmune disorders, chronic fatigue syndrome, and some chronic pain states have been associated with deficits in glucocorticoid activity. Many patients with major depression demonstrate glucocorticoid resistance, suggesting that the high levels of cortisol observed in the disorder may be a compensatory mechanism for decreased glucocorticoid signaling (Pariante, 2004; Pariante and Miller, 2001; Raison and Miller, 2003). Interestingly, as noted above, innate immune cytokines have been shown to impair GR function in cultured cells, and antidepressant medications are known to reverse this condition (Pace et al., 2007; Pariante et al., 1999). These findings have led to the suggestion that increased cytokine activity may contribute to the pathogenesis of at least some types of major depression, especially severe or treatment-resistant depression, or in depressions where resistance to glucocorticoids is prominent (Dantzer et al., 2008; Irwin and Miller, 2007; Pace et al., 2007; Raison et al., 2006; Raison and Miller, 2003). As discussed previously, elevated levels of innate immune cytokines in the periphery can access the CNS through multiple routes, and once cytokine signals access the cytokine network within the brain, they

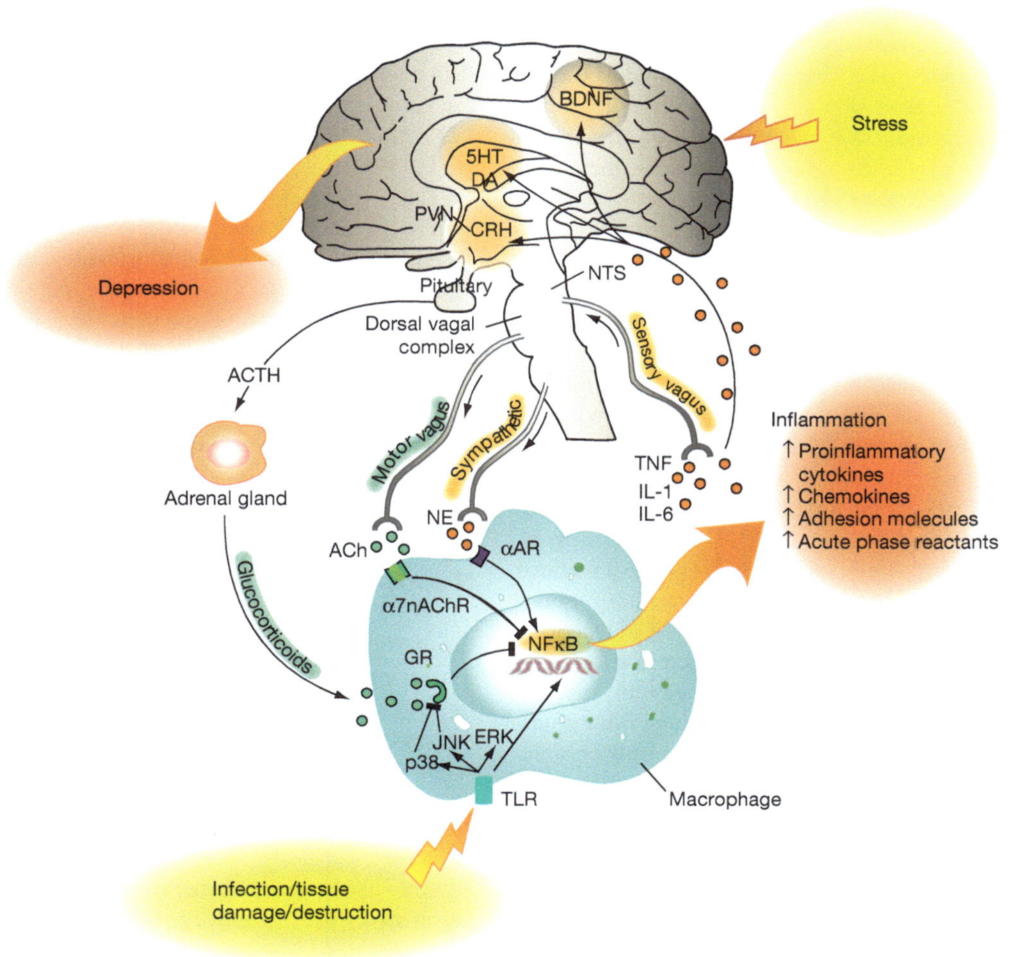

Figure 6 Stress and bidirectional neural–immune interactions. (1) Activation of NFκB through Toll-like receptors (TLRs) during immune challenge leads to an inflammatory response including (2) the release of proinflammatory cytokines TNF-α, IL-1, and IL-6. (3) These cytokines, in turn, access the brain via leaky regions in the blood–brain barrier, active transport molecules, and afferent nerve fibers (e.g., sensory vagus), which relay information through the nucleus tractus solitarius (NTS). (4) Once in the brain, cytokine signals participate in pathways known to be involved in the development of major depression , including: (i) altered metabolism of relevant neurotransmitters such as serotonin (5HT) and dopamine (DA); (ii) activation of CRH in the paraventricular nucleus (PVN) and the subsequent production and/or release of ACTH and glucocorticoids (cortisol); and (iii) disruption of synaptic plasticity through alterations in relevant growth factors (e.g., brain-derived neurotrophic factor (BDNF)). (5) Exposure to environmental stressors promotes activation of inflammatory signaling (NFκB) thoroughly increased outflow of proinflammatory sympathetic nervous system responses (release of norepinephrine (NE), which binds to the α (αAR) and β (βAR) adrenoceptors) (6) Stressors also induce withdrawal of inhibitory motor vagal input (release of acetylcholine (Ach), which binds to the α7 subunit of the nicotinic acetylcholine receptor (α7nAChR)). (7) Activation of the mitogen-activated protein kinase pathways, including p38 and Jun amino-terminal kinase (JNK), inhibit the function of glucocorticoid receptors (GRs), thereby releasing NKκB from negative regulation by glucocorticoids released as a result of HPA axis in response to stress. Reproduced from Raison CL, Capuron L, and Miller AH (2006) Cytokines sing the blues: Inflammation and the pathogenesis of major depression. *Trends in Immunology* 27: 24–31, with permission from Elsevier.

can influence numerous pathways relevant to depression, including neurotransmitter metabolism, neuroendocrine function, synaptic plasticity, and regional brain activity.

Although major depression may be characterized by reduced glucocorticoid signaling at the receptor level accompanied by increased circulating levels of glucocorticoids, somatic pain syndromes (such as fibromyalgia and chronic fatigue syndrome (CFS)) as well as autoimmune disorders (such as rheumatoid arthritis) appear to be characterized by reduced glucocorticoid transmission at the level of the hormone (i.e., adrenal production of cortisol), given that many patients with these disorders demonstrate decreased basal and/or stimulated cortisol production (Heim et al., 2000). In addition, exposure to early life stress (such as neglect or sexual abuse) has been associated with hyposecretion of cortisol in women during non-stressful situations (Heim et al., 2004; Penza et al., 2003). These findings may be related to hyporesponsiveness of the adrenal gland, which has been identified during the ACTH stimulation test in individuals with a history of early life stress (Heim et al., 2001). Of note, early life stress has been found to be a significant risk factor for the development of major depression (Kendler et al., 2002). Hypocortisolism in any of the above would be expected to have the effect of diminishing negative feedback upon the production of innate immune cytokines, even if glucocorticoid sensitivity is normal (or even increased).

Although no single explanatory system can yet fully account for the divergent immune alterations that have been reported in depression and chronic stress, evidence exists that activation of a positive feed-forward loop composed of innate immune cytokines, CRH and the SNS may suppress certain aspects of immune system functioning, especially acquired immune responses such as mitogen-induced lymphocyte proliferation (while activating the innate immune response) (Maes et al., 1994). For example, studies have demonstrated that both CRH and IL-1 administered ICV can inhibit lymphocyte proliferation while elevating proinflammatory cytokine production in the brain and periphery (see above). Activation of the SNS reliably suppresses NKCA in splenocytes, an effect reversed by blockade of adrenergic β-receptors (Hellstrand et al., 1985). Kubera et al. (1998) reported that mice exposed to chronic mild stress, which is a well-accepted animal model for depression, develop increased production of IL-1 and decreased mitogen-induced lymphocyte proliferation and NKCA. It has been suggested, in fact, that immune suppression especially involving acquired immune responses seen in depression may develop in response to elevated cytokine activity. Evidence for this comes from research demonstrating that immune activation, as assessed by elevated serum levels of PGE2 and other markers of an inflammatory response, correlates with a decrease in mitogen-induced lymphocyte proliferation in patients with major depression (Maes et al., 1995). Moreover, in research on laboratory animals, it has been shown that administration of the IL-1 receptor antagonist can reverse the effects of stress on antibody formation to Keyhole Limpet Hemocyanin antibody (Fleshner et al., 1998).

Taken together, these data suggest that increased activity in a circuit composed of inflammatory cytokines, CRH, and the SNS may provide a mechanism to account for at least a portion of the immune activation and suppression reported in patients with depression and chronic stress. Such a pattern would be consistent with inadequate HPA-axis inhibitory regulation of this circuitry, as would occur with hypocortisolemia or glucocorticoid resistance (**Figure 6**).

18.8 Therapeutic Implications of Neuroendocrine–Immune Interactions

Given the model presented above, there are several important sites where therapeutic interventions could be targeted to disrupt the potential consequences of a degenerative inflammatory cascade. These include targeting behavior, the neuroendocrine system, and the immune system.

18.8.1 Behavioral Interventions in Immunologic Disorders

One obvious approach to ameliorating the effects of chronic stress and depression on the immune system is to reduce stress levels and treat psychological distress (e.g., depression) via psychosocial interventions. Psychosocial interventions include psychoeducational approaches (including stress management), cognitive/behavioral interventions, psychotherapy (including group and individual therapy), and meditation. Spiegel et al. (1989) and Fawzy et al. (1993) treated patients with cancer using supportive group psychotherapy under controlled conditions and found improved outcomes in the intervention group compared to those receiving standard treatment. These outcomes included lengthened survival in patients with metastatic breast cancer and decreased recurrence in patients with malignant melanoma. Nevertheless, not all studies have reproduced these findings. For example, two studies, using cognitive behavior therapy (CBT) in patients with metastatic breast cancer, found no improved survival in the intervention group (Cunningham et al., 1998;

Edelman et al., 1999). Moreover, in a recent replication of his original study, Spiegel was only able to reproduce his findings in a small subset of patients whose breast cancers were estrogen receptor negative (Spiegel et al., 2007). However, it remains to be determined if more powerful treatment effects would be observed if studies targeted patients with psychosocial risk factors, including depressive symptoms, reduced social support, or impaired coping strategies.

The efficacy of meditation practices to optimize immune and endocrine parameters has also been investigated. For example, training in mindfulness meditation was associated with enhanced antibody responses to influenza vaccination (Davidson et al., 2003). In addition, preliminary work from our group suggests that practicing compassion mediation may reduce resting and stress-induced circulating concentrations of IL-6, as well as stress-induced feelings of distress.

18.8.2 Neuroendocrine Interventions in Immunologic Disorders

Some of the most exciting possibilities in addressing the pathophysiology of disorders of immune dysregulation involve targeting neuroendocrine system participants in this process. As noted above, proinflammatory cytokines have the capacity to disrupt glucocorticoid feedback inhibition through direct effects on the GR. A number of studies have established that a variety of antidepressants have the capacity to enhance GR translocation and function both *in vivo* and *in vitro*. Indeed, early studies using the antidepressant desipramine (DMI) demonstrated that DMI was capable of both increasing DEX-induced GR-mediated gene transcription and increasing GR translocation (even in the absence of DEX) (Pariante et al., 1997b). Other antidepressants, including clomipramine, fluoxetine, paroxetine, and citalopram, have also exhibited these effects on the GR (Pariante et al., 2001). It has been demonstrated that the mechanism of antidepressant effects on GR function is related in part to their inhibitory effects on the multidrug resistance pump (making more hormone available for GR activation); however, other antidepressant-induced signal transduction pathways including cAMP appear to be involved. Indeed, the phosphodiesterase inhibitor, rolipram, which increases intracellular concentrations of cAMP has been shown to enhance GR function *in vitro*, while exhibiting both anti-inflammatory and antidepressant effects *in vivo* (Miller et al., 2002b). Of relevance to interactions between cytokines and GR, data also indicate that antidepressants exhibit the capacity to inhibit cytokine production both *in vitro* and *in vivo* (Kenis and Maes, 2002). For example, exposure of mixed glial culture to amitriptyline led to a decrease in LPS-stimulated release of both IL-1β and TNF-α (Obuchowicz et al., 2006). Nevertheless, whether these effects of antidepressants on cytokines are related to their effects on the GR have yet to be established.

In terms of studies with laboratory animals, a number of studies have shown that *in vivo* treatment with a range of antidepressant agents, including both tricyclic antidepressants and selective serotonin reuptake inhibitors, is capable of enhancing glucocorticoid-mediated feedback inhibition and increasing GR protein and/or mRNA in key brain regions including the hippocampus (which has been shown to mediate an inhibitory influence on CRH in the PVN) (Holsboer and Barden, 1996; Pariante and Miller, 2001). Moreover, antidepressant medications have been associated with the resolution of HPA-axis alterations in patients with depression (Holsboer and Barden, 1996). Thus, antidepressants may facilitate feedback inhibition pathways and thereby enhance negative feedback of endogenous glucocorticoids on cytokine responses. Consistent with this notion is data demonstrating that antidepressant pretreatment for 2 weeks substantially reduces the development of depression and neurotoxicity in humans during administration of the innate immune cytokine, IFN-α (Musselman et al., 2001a).

18.8.3 Immune Interventions in Behavioral Disorders

Given the integral role of proinflammatory cytokines in the inflammatory glucocorticoid cascade, there is no question that therapeutic agents that directly target inflammatory mediators are a logical choice for approaching pathologic interactions among the nervous, endocrine, and immune systems. Antagonists for IL-1 and TNF-α are already available for clinical usage, including IL-1 receptor antagonist (anakinra) for IL-1, and etanercept (blocks TNF receptors) and infliximab (TNF-α antibodies) for TNF-α (Irwin and Miller, 2007). Recent data from a randomized, double-blind, placebo-controlled trial demonstrated that etanercept, a drug which antagonizes the activity of TNF-α, significantly improved depressive symptoms in patients with psoriasis (Tyring et al., 2006). Of note, the improvement in depressive symptoms was independent of reductions in disease activity (skin lesions or joint pain) in these patients.

Anti-inflammatory agents (e.g., nonsteroidal anti-inflammatory drugs such as celecoxib) have also been examined, and have shown promise in a recent study as an adjunct to conventional antidepressant medication (Muller et al., 2006). Agents that target NFκB signaling, including natural compounds such as curcumin (the active element in the Indian spice turmeric), have also shown antidepressant effects in studies with laboratory animals (Xu et al., 2005a,b). Omega-3 fatty acids have demonstrated anti-inflammatory and immunosuppressive effects via the inhibition of prostaglandin and cytokine production (Harbige, 1998). Finally, since immune activation can lead to reduced availability of tryptophan through the induction of IDO (Dantzer et al., 2008), dietary tryptophan supplementation or inhibition of IDO could be considered as potential treatments to disrupt the consequences of inflammation on behavior. Indeed, blockade of IDO has been shown to resolve behavioral changes in mice administered LPS (Dantzer et al., 2008).

References

Abbas AK and Lichtman AH (2003) *Cellular and Molecular Immunology,* 5th edn. Philadelphia, PA: Saunders.

Aggarwal BB, Shishodia S, Sandur SK, Pandey MK, and Sethi G (2006) Inflammation and cancer: How hot is the link? *Biochemical Pharmacology* 72: 1605–1621.

Alesci S, Martinez PE, Kelkar S, et al. (2005) Major depression is associated with significant diurnal elevations in plasma interleukin-6 levels, a shift of its circadian rhythm, and loss of physiological complexity in its secretion: Clinical implications. *Journal of Clinical Endocrinology and Metabolism* 90: 2522–2530.

Anker SD, Ponikowski PP, Clark AL, et al. (1999) Cytokines and neurohormones relating to body composition alterations in the wasting syndrome of chronic heart failure. *European Heart Journal* 20: 683–693.

Baigent SM (2001) Peripheral corticotropin-releasing hormone and urocortin in the control of the immune response. *Peptides* 22: 809–820.

Barrientos RM, Sprunger DB, Campeau S, Higgins EA, Watkins LR, Rudy JW, and Maier SF (2003) Brain-derived neurotrophic factor mRNA downregulation produced by social isolation is blocked by intrahippocampal interleukin-1 receptor antagonist. *Neuroscience* 121: 847–853.

Bartrop RW, Luckhurst E, Lazarus L, Kiloh LG, and Penny R (1977) Depressed lymphocyte function after bereavement. *Lancet* 1: 834–836.

Ben-Menachem-Zidon O, Goshen I, Kreiesl T, Ben-Menahem Y, Reinhartz E, Ben-Hur T, and Yirmiya R (2007) Intrahippocampal transplantation of transgenic neural precursor cells overexpressing interleukin-1 receptor antagonist blocks chronic isolation-induced impairment in memory and neurogenesis. *Neuropsychopharmacology* 33: 2251–2262.

Benschop RJ, Geenen R, Mills PJ, et al. (1998) Cardiovascular and immune responses to acute psychological stress in young and old women: A meta-analysis. *Psychosomatic Medicine* 60: 290–296.

Benschop RJ, Jacobs R, Sommer B, Schurmeyer TH, Raab JR, Schmidt RE, and Schedlowski M (1996a) Modulation of the immunologic response to acute stress in humans by beta-blockade or benzodiazepines. *FASEB Journal* 10: 517–524.

Benschop RJ, Rodriguez-Feuerhahn M, and Schedlowski M (1996b) Catecholamine-induced leukocytosis: Early observations, current research, and future directions. *Brain, Behavior, and Immunity* 10: 77–91.

Berkenbosch F, Wolvers DA, and Derijk R (1991) Neuroendocrine and immunological mechanisms in stress-induced immunomodulation. *Journal of Steroid Biochemistry and Molecular Biology* 40: 639–647.

Beutler B (2004) Innate immunity: An overview. *Molecular Immunology* 40: 845–859.

Bierhaus A, Humpert PM, and Nawroth PP (2004) NF-kappaB as a molecular link between psychosocial stress and organ dysfunction. *Pediatric Nephrology* 19: 1189–1191.

Blalock JE and Smith EM (2007) Conceptual development of the immune system as a sixth sense. *Brain, Behavior, and Immunity* 21: 23–33.

Bluthe RM, Dantzer R, and Kelley KW (1992) Effects of interleukin-1 receptor antagonist on the behavioral effects of lipopolysaccharide in rat. *Brain Research* 573: 318–320.

Borella P, Bargellini A, Rovesti S, Pinelli M, Vivoli R, Solfrini V, and Vivoli G (1999) Emotional stability, anxiety, and natural killer activity under examination stress. *Psychoneuroendocrinology* 24: 613–627.

Bouman A, Heineman MJ, and Faas MM (2005) Sex hormones and the immune response in humans. *Human Reproduction Update* 11: 411–423.

Brady LS, Lynn AB, Herkenham M, and Gottesfeld Z (1994) Systemic interleukin-1 induces early and late patterns of c-fos mRNA expression in brain. *Journal of Neuroscience* 14: 4951–4964.

Broug-Holub E, Persoons JH, Schornagel K, Mastbergen SC, and Kraal G (1998) Effects of stress on alveolar macrophages: A role for the sympathetic nervous system. *American Journal of Respiratory Cell and Molecular Biology* 19: 842–848.

Brydon L, Edwards S, Mohamed-Ali V, and Steptoe A (2004) Socioeconomic status and stress-induced increases in interleukin-6. *Brain, Behavior, and Immunity* 18: 281–290.

Brydon L, Harrison NA, Walker C, Steptoe A, and Critchley HD (2008) Peripheral inflammation is associated with altered substancia nigra activity and psychomotor slowing in humans. *Biological Psychiatry* 63: 1022–1029.

Burns VE, Carroll D, Drayson M, Whitham M, and Ring C (2003) Life events, perceived stress and antibody response to influenza vaccination in young, healthy adults. *Journal of Psychosomatic Research* 55: 569–572.

Burns VE, Drayson M, Ring C, and Carroll D (2002) Perceived stress and psychological well-being are associated with antibody status after meningitis C conjugate vaccination. *Psychosomatic Medicine* 64: 963–970.

Buske-Kirschbaum A, Kern S, Ebrecht M, and Hellhammer DH (2007) Altered distribution of leukocyte subsets and cytokine production in response to acute psychosocial stress in patients with psoriasis vulgaris. *Brain, Behavior, and Immunity* 21: 92–99.

Capuron L, Gumnick JF, Musselman DL, Lawson DH, Reemsnyder A, Nemeroff CB, and Miller AH (2002) Neurobehavioral effects of interferon-alpha in cancer patients: Phenomenology and paroxetine responsiveness of symptom dimensions. *Neuropsychopharmacology* 26: 643–652.

Capuron L, Pagnoni G, Demetrashvili M, Woolwine BJ, Nemeroff CB, Berns GS, and Miller AH (2005) Anterior

cingulate activation and error processing during interferon-alpha treatment. *Biological Psychiatry* 58: 190–196.
Capuron L, Pagnoni G, Demetrashvili MF, et al. (2007) Basal ganglia hypermetabolism and symptoms of fatigue during interferon-alpha therapy. *Neuropsychopharmacology* 32: 2384–2392.
Chaplin DD (2006) Overview of the human immune response. *Journal of Allergy and Clinical Immunology* 117: S430–S435.
Chapman KE, Coutinho A, Gray M, Gilmour JS, Savill JS, and Seckl JR (2006) Local amplification of glucocorticoids by 11beta-hydroxysteroid dehydrogenase type 1 and its role in the inflammatory response. *Annals of the New York Academy of Sciences* 1088: 265–273.
Cohen S, Doyle WJ, Skoner DP, Rabin BS, and Gwaltney JM, Jr. (1997) Social ties and susceptibility to the common cold. (see comments). *Journal of the American Medical Association* 277: 1940–1944.
Cover H and Irwin M (1994) Immunity and depression: Insomnia, retardation, and reduction of natural killer cell activity. *Journal of Behavioral Medicine* 17: 217–223.
Cowles MK and Miller AH (2008) Stress, cytokines and depressive illness. In: Squire LR (ed.) *Encyclopedia of Neuroscience*, pp. 519–527. Oxford: Academic Press.
Croiset G, Heijnen CJ, and de Wied D (1990) Passive avoidance behavior, vasopressin and the immune system. A link between avoidance latency and immune response. *Neuroendocrinology* 51: 156–161.
Cunnick JE, Lysle DT, Kucinski BJ, and Rabin BS (1990) Evidence that shock-induced immune suppression is mediated by adrenal hormones and peripheral beta-adrenergic receptors. *Pharmacology, Biochemistry and Behavior* 36: 645–651.
Cunningham AJ, Edmonds CV, Jenkins GP, Pollack H, Lockwood GA, and Warr D (1998) A randomized controlled trial of the effects of group psychological therapy on survival in women with metastatic breast cancer. *Psycho-Oncology* 7: 508–517.
Danese A, Moffitt TE, Pariante CM, Ambler A, Puoulton R, and Caspi A (2008) Elevated inflammation levels in depressed adults with a history of childhood maltreatment. *Archives of General Psychiatry* 65: 409–415.
Danese A, Pariante CM, Caspi A, Taylor A, and Poulten R (2007) Childhood maltreatment predicts adult inflammation in a life-course study. *Proceedings of the National Academy of Sciences of the United States of America* 104: 1319–1324.
Dantzer R (2006) Cytokine, sickness behavior, and depression. *Neurologic Clinics* 24: 441–460.
Dantzer R, O'Connor JC, Freund GG, Johnson RW, and Kelley KW (2008) From inflammation to sickness and depression: When the immune system subjugates the brain. *Nature Reviews Neuroscience* 9: 46–56.
Davidson RJ, Kabat-Zinn J, Schumacher J, et al. (2003) Alterations in brain and immune function produced by mindfulness meditation. *Psychosomatic Medicine* 65: 564–570.
Deak T, Nguyen KT, Fleshner M, Watkins LR, and Maier SF (1999) Acute stress may facilitate recovery from a subcutaneous bacterial challenge. *Neuroimmunomodulation* 6: 344–354.
DeVries AC, Craft TK, Glasper ER, Neigh GN, and Alexander JK (2007) 2006 Curt P. Richter award winner: Social influences on stress responses and health. *Psychoneuroendocrinology* 32: 587–603.
Dhabhar FS (2003) Stress, leukocyte trafficking, and the augmentation of skin immune function. *Annals of the New York Academy of Sciences* 992: 205–217.
Dhabhar FS and McEwen BS (1999) Enhancing versus suppressive effects of stress hormones on skin immune function. *Proceedings of the National Academy of Sciences of the United States of America* 96: 1059–1064.
Dhabhar FS, Satoskar AR, Bluethmann H, David JR, and McEwen BS (2000) Stress-induced enhancement of skin immune function: A role for gamma interferon. *Proceedings of the National Academy of Sciences of the United States of America* 97: 2846–2851.
Dheen ST, Kaur C, and Ling EA (2007) Microglial activation and its implications in the brain diseases. *Current Medicinal Chemistry* 14: 1189–1197.
Dopp JM, Miller GE, Myers HF, and Fahey JL (2000) Increased natural killer-cell mobilization and cytotoxicity during marital conflict. *Brain, Behavior, and Immunity* 14: 10–26.
Dunn AJ, Wang J, and Ando T (1999) Effects of cytokines on cerebral neurotransmission. Comparison with the effects of stress. *Advances in Experimental Medicine and Biology* 461: 117–127.
Edelman S, Lemon J, Bell DR, and Kidman AD (1999) Effects of group CBT on the survival time of patients with metastatic breast cancer. *Psycho-Oncology* 8: 474–481.
Edwards KM, Burns VE, Reynolds T, Carroll D, Drayson M, and Ring C (2006) Acute stress exposure prior to influenza vaccination enhances antibody response in women. *Brain, Behavior, and Immunity* 20: 159–168.
Engler H, Bailey MT, Engler A, Stiner-Jones LM, Quan N, and Sheridan JF (2008) Interleukin-1 receptor type 1-deficient mice fail to develop social stress-associated glucocorticoid resistance in the spleen. *Psychoneuroendocrinology* 33: 108–117.
Ericsson A, Kovacs KJ, and Sawchenko PE (1994) A functional anatomical analysis of central pathways subserving the effects of interleukin-1 on stress-related neuroendocrine neurons. *Journal of Neuroscience* 14: 897–913.
Fawzy FI, Fawzy NW, Hyun CS, Elashoff R, Guthrie D, Fahey JL, and Morton DL (1993) Malignant melanoma. Effects of an early structured psychiatric intervention, coping, and affective state on recurrence and survival 6 years later. *Archives of General Psychiatry* 50: 681–689.
Fleshner M, Nguyen KT, Cotter CS, Watkins LR, and Maier SF (1998) Acute stressor exposure both suppresses acquired immunity and potentiates innate immunity. *American Journal of Physiology* 275: R870–R878.
Frank MG, WieselerFrank JL, Hendricks SE, Burke WJ, and Johnson DR (2002) Age at onset of major depressive disorder predicts reductions in NK cell number and activity. *Journal of Affective Disorders* 71: 159–167.
Frenois F, Moreau M, O'Connor J, et al. (2007) Lipopolysaccharide induces delayed FosB/DeltaFosB immunostaining within the mouse extended amygdala, hippocampus and hypothalamus, that parallel the expression of depressive-like behavior. *Psychoneuroendocrinology* 32: 516–531
Glaser R and Kiecolt-Glaser JK (1997) Chronic stress modulates the virus-specific immune response to latent herpes simplex virus type 1. *Annals of Behavioral Medicine* 19: 78–82.
Glaser R and Kiecolt-Glaser JK (1998) Stress-associated immune modulation: Relevance to viral infections and chronic fatigue syndrome. *American Journal of Medicine* 105: 35S–42S.
Glaser R, Kiecolt-Glaser JK, Marucha PT, MacCallum RC, Laskowski BF, and Malarkey WB (1999) Stress-related changes in proinflammatory cytokine production in wounds. *Archives of General Psychiatry* 56: 450–456.
Gordon TP, Gust DA, Wilson ME, Ahmed-Ansari A, Brodie AR, and McClure HM (1992) Social separation and reunion affects immune system in juvenile rhesus monkeys. *Physiology and Behavior* 51: 467–472.
Grippo AJ, Francis J, Beltz TG, Felder RB, and Johnson AK (2005) Neuroendocrine and cytokine profile of chronic mild

stress-induced anhedonia. *Physiology and Behavior* 84: 697–706.
Gust DA, Gordon TP, Brodie AR, and McClure HM (1996) Effect of companions in modulating stress associated with new group formation in juvenile rhesus macaques. *Physiology and Behavior* 59: 941–945.
Gust DA, Gordon TP, Wilson ME, Brodie AR, Ahmed-Ansari A, and McClure HM (1992) Removal from natal social group to peer housing affects cortisol levels and absolute numbers of T cell subsets in juvenile rhesus monkeys. *Brain, Behavior, and Immunity* 6: 189–199.
Haack M, Hinze-Selch D, Fenzel T, Kraus T, Kuhn M, Schuld A, and Pollmacher T (1999) Plasma levels of cytokines and soluble cytokine receptors in psychiatric patients upon hospital admission: Effects of confounding factors and diagnosis. *Journal of Psychiatric Research* 33: 407–418.
Harbige LS (1998) Dietary n-6 and n-3 fatty acids in immunity and autoimmune disease. *Proceedings of the Nutrition Society* 57: 555–562.
Heim C, Ehlert U, and Hellhammer DH (2000) The potential role of hypocortisolism in the pathophysiology of stress-related bodily disorders. *Psychoneuroendocrinology* 25: 1–35.
Heim C, Newport DJ, Bonsall R, Miller AH, and Nemeroff CB (2001) Altered pituitary–adrenal axis responses to provocative challenge tests in adult survivors of childhood abuse. (see comment). *American Journal of Psychiatry* 158: 575–581.
Heim C, Plotsky PM, and Nemeroff CB (2004) Importance of studying the contributions of early adverse experience to neurobiological findings in depression. *Neuropsychopharmacology* 29: 641–648.
Hellstrand K, Hermodsson S, and Strannegard O (1985) Evidence for a beta-adrenoceptor-mediated regulation of human natural killer cells. *Journal of Immunology* 134: 4095–4099.
Helmreich DL, Watkins LR, Deak T, Maier SF, Akil H, and Watson SJ (1999) The effect of stressor controllability on stress-induced neuropeptide mRNA expression within the paraventricular nucleus of the hypothalamus. *Journal of Neuroendocrinology* 11: 121–128.
Hench PS (1952) The reversibility of certain rheumatic and non-rheumatic conditions by the use of cortisone or of the pituitary adrenocorticotropic hormone. *Annals of Internal Medicine* 36: 1–25.
Hennebold JD, Ryu SY, Mu HH, Galbraith A, and Daynes RA (1996) 11 Beta-hydroxysteroid dehydrogenase modulation of glucocorticoid activities in lymphoid organs. *American Journal of Physiology* 270: R1296–R1306.
Herbert TB and Cohen S (1993a) Depression and immunity: A meta-analytic review. *Psychological Bulletin* 113: 472–486.
Herbert TB and Cohen S (1993b) Stress and immunity in humans: A meta-analytic review. *Psychosomatic Medicine* 55: 364–379.
Hickie I and Lloyd A (1995) Are cytokines associated with neuropsychiatric syndromes in humans? *International Journal of Immunopharmacology* 17: 677–683.
Holmes MC, Yau JL, Kotelevtsev Y, Mullins JJ, and Seckl JR (2003) 11 Beta-hydroxysteroid dehydrogenases in the brain: Two enzymes two roles. *Annals of the New York Academy of Sciences* 1007: 357–366.
Holsboer F (2000) The corticosteroid receptor hypothesis of depression. *Neuropsychopharmacology* 23: 477–501.
Holsboer F and Barden N (1996) Antidepressants and hypothalamic–pituitary–adrenocortical regulation. *Endocrine Reviews* 17: 187–205.
Inui A (1999) Cancer anorexia-cachexia syndrome: Are neuropeptides the key? *Cancer Research* 59: 4493–4501.
Irusen E, Matthews JG, Takahashi A, Barnes PJ, Chung KF, and Adcock IM (2002) p38 Mitogen-activated protein kinase-induced glucocorticoid receptor phosphorylation reduces its activity: Role in steroid-insensitive asthma. *Journal of Allergy and Clinical Immunology* 109: 649–657.
Irwin M (1993) Brain corticotropin-releasing hormone- and interleukin-1 beta-induced suppression of specific antibody production. *Endocrinology* 133: 1352–1360.
Irwin M (1999) Immune correlates of depression. *Advances in Experimental Medicine and Biology* 461: 1–24.
Irwin M, Hauger RL, Brown M, and Britton KT (1988) CRF activates autonomic nervous system and reduces natural killer cytotoxicity. *American Journal of Physiology* 255: R744–R747.
Irwin M, Smith TL, and Gillin JC (1992) Electroencephalographic sleep and natural killer activity in depressed patients and control subjects. *Psychosomatic Medicine* 54: 10–21.
Irwin MR and Miller AH (2007) Depressive disorders and immunity: 20 Years of progress and discovery. *Brain, Behavior, and Immunity* 21: 374–383.
Isowa T, Ohira H, and Murashima S (2006) Immune, endocrine and cardiovascular responses to controllable and uncontrollable acute stress. *Biological Psychology* 71: 202–213.
Jessop DS, Harbuz MS, Snelson CL, Dayan CM, and Lightman SL (1997) An antisense oligodeoxynucleotide complementary to corticotropin-releasing hormone mRNA inhibits rat splenocyte proliferation *in vitro*. *Journal of Neuroimmunology* 75: 135–140.
Johnson JD, Campisi J, Sharkey CM, Kennedy SL, Nickerson M, Greenwood BN, and Fleshner M (2005) Catecholamines mediate stress-induced increases in peripheral and central inflammatory cytokines. *Neuroscience* 135: 1295–1307.
Johnston IN, Milligan ED, Wieseler-Frank J, et al. (2004) A role for proinflammatory cytokines and fractalkine in analgesia, tolerance, and subsequent pain facilitation induced by chronic intrathecal morphine. *Journal of Neuroscience* 24: 7353–7365.
Kaestner F, Hettich M, Peters M, et al. (2005) Different activation patterns of proinflammatory cytokines in melancholic and non-melancholic major depression are associated with HPA axis activity. *Journal of Affective Disorders* 87: 305–311.
Kavelaars A (2002) Regulated expression of alpha-1 adrenergic receptors in the immune system. *Brain, Behavior, and Immunity* 16: 799–807.
Kawakami N, Tanigawa T, Araki S, Nakata A, Sakurai S, Yokoyama K, and Morita Y (1997) Effects of job strain on helper-inducer (CD4+CD29+) and suppressor-inducer (CD4+ CD45RA+) T cells in Japanese blue-collar workers. *Psychotherapy and Psychosomatics* 66: 192–198.
Kemeny ME and Gruenewald TL (2000) Affect, cognition, the immune system and health. *Progress in Brain Research* 122: 291–308.
Kendler KS, Gardner CO, and Prescott CA (2002) Toward a comprehensive developmental model for major depression in women. *American Journal of Psychiatry* 159: 1133–1145.
Kenis G and Maes M (2002) Effects of antidepressants on the production of cytokines. *International Journal of Neuropsychopharmacology* 5: 401–412.
Kent S, Bluthe RM, Kelley KW, and Dantzer R (1992) Sickness behavior as a new target for drug development. *Trends in Pharmacological Sciences* 13: 24–28.
Kiecolt-Glaser JK, Belury MA, Porter K, Beversdorf DQ, Lemeshow S, and Glaser R (2007) Depressive symptoms, omega-6: Omega-3 fatty acids, and inflammation in older adults. *Psychosomatic Medicine* 69: 217–224.
Kiecolt-Glaser JK, Glaser R, Gravenstein S, Malarkey WB, and Sheridan J (1996) Chronic stress alters the immune response to influenza virus vaccine in older adults. *Proceedings of the*

National Academy of Sciences of the United States of America 93: 3043–3047.
Kirschbaum C, Pirke KM, and Hellhammer DH (1993) The 'trier social stress test' – a tool for investigating psychobiological stress responses in a laboratory setting. *Neuropsychobiology* 28: 76–81.
Kling MA, Alesci S, Csako G, et al. (2007) Sustained low-grade pro-inflammatory state in unmedicated, remitted women with major depressive disorder as evidenced by elevated serum levels of the acute phase proteins C-reactive protein and serum amyloid A. *Biological Psychiatry* 62: 309–313.
Klokker M, Secher NH, Madsen P, Pedersen M, and Pedersen BK (1997) Adrenergic beta 1- and beta 1 + 2-receptor blockade suppress the natural killer cell response to head-up tilt in humans. *Journal of Applied Physiology* 83: 1492–1498.
Kohm AP and Sanders VM (2000) Norepinephrine: A messenger from the brain to the immune system. *Immunology Today* 21: 539–542.
Koo JW and Duman RS (2008) IL-1beta is an essential mediator of the antineurogenic and anhedonic effects of stress. *Proceedings of the National Academy of Sciences of the United States of America* 105: 751–756.
Kozak W, Kluger MJ, Tesfaigzi J, Kozak A, Mayfield KP, Wachulec M, and Dokladny K (2000) Molecular mechanisms of fever and endogenous antipyresis. *Annals of the New York Academy of Sciences* 917: 121–134.
Kubera M, Basta-Kaim A, Holan V, et al. (1998) Effect of mild chronic stress, as a model of depression, on the immunoreactivity of C57BL/6 mice. *International Journal of Immunopharmacology* 20: 781–789.
Kudielka BM and Kirschbaum C (2005) Sex differences in HPA axis responses to stress: A review. *Biological Psychology* 69: 113–132.
Labeur MS, Arzt E, Wiegers GJ, Holsboer F, and Reul JM (1995) Long-term intracerebroventricular corticotropin-releasing hormone administration induces distinct changes in rat splenocyte activation and cytokine expression. *Endocrinology* 136: 2678–2688.
Laudenslager ML, Ryan SM, Drugan RC, Hyson RL, and Maier SF (1983) Coping and immunosuppression: Inescapable but not escapable shock suppresses lymphocyte proliferation. *Science* 221: 568–570.
Leserman J, Petitto JM, Golden RN, et al. (2000) Impact of stressful life events, depression, social support, coping, and cortisol on progression to AIDS. *American Journal of Psychiatry* 157: 1221–1228.
Leu SJ and Singh VK (1992) Stimulation of interleukin-6 production by corticotropin-releasing factor. *Cellular Immunology* 143: 220–227.
Levenson CW (2006) Zinc: The new antidepressant? *Nutrition Reviews* 64: 39–42.
Levine J, Barak Y, Chengappa KN, Rapoport A, Rebey M, and Barak V (1999) Cerebrospinal cytokine levels in patients with acute depression. *Neuropsychobiology* 40: 171–176.
Lewis-Tuffin LJ and Cidlowski JA (2006) The physiology of human glucocorticoid receptor beta (hGRbeta) and glucocorticoid resistance. *Annals of the New York Academy of Sciences* 1069: 1–9.
Li J, Cowden LG, King JD, et al. (2007) Effects of chronic stress and interleukin-10 gene polymorphisms on antibody response to tetanus vaccine in family caregivers of patients with Alzheimer's disease. *Psychosomatic Medicine* 69: 551–559.
Lloyd A, Hickie I, Hickie C, Dwyer J, and Wakefield D (1992) Cell-mediated immunity in patients with chronic fatigue syndrome, healthy control subjects and patients with major depression. *Clinical and Experimental Immunology* 87: 76–79.
Lutgendorf SK, Sood AK, Anderson B, et al. (2005) Social support, psychological distress, and natural killer cell activity in ovarian cancer. *Journal of Clinical Oncology* 23: 7105–7113.
Maes M (1999) Major depression and activation of the inflammatory response system. *Advances in Experimental Medicine and Biology* 461: 25–46.
Maes M, Meltzer HY, Stevens W, Calabrese J, and Cosyns P (1994) Natural killer cell activity in major depression: Relation to circulating natural killer cells, cellular indices of the immune response, and depressive phenomenology. *Progress in Neuropsychopharmacology and Biological Psychiatry* 18: 717–730.
Maes M, Scharpe S, Meltzer HY, Bosmans E, Suy E, Calabrese J, and Cosyns P (1993) Relationships between interleukin-6 activity, acute phase proteins, and function of the hypothalamic–pituitary–adrenal axis in severe depression. *Psychiatry Research* 49: 11–27.
Maes M, Scharpe S, Verkerk R, et al. (1995) Seasonal variation in plasma L-tryptophan availability in healthy volunteers. Relationships to violent suicide occurrence. *Archives of General Psychiatry* 52: 937–946.
Maes M, Stevens W, Peeters D, et al. (1992) A study on the blunted natural killer cell activity in severely depressed patients. *Life Sciences* 50: 505–513.
Marsland AL, Bachen EA, Cohen S, Rabin B, and Manuck SB (2002) Stress, immune reactivity and susceptibility to infectious disease. *Physiology and Behavior* 77: 711–716.
Martin JT (2000) Sexual dimorphism in immune function: The role of prenatal exposure to androgens and estrogens. *European Journal of Pharmacology* 405: 251–261.
Marucha PT, Kiecolt-Glaser JK, and Favagehi M (1998) Mucosal wound healing is impaired by examination stress. *Psychosomatic Medicine* 60: 362–365.
Mazzeo RS, Donovan D, Fleshner M, Butterfield GE, Zamudio S, Wolfel EE, and Moore LG (2001) Interleukin-6 response to exercise and high-altitude exposure: Influence of alpha-adrenergic blockade. *Journal of Applied Physiology* 91: 2143–2149.
McEwen BS, Biron CA, Brunson KW, et al. (1997) The role of adrenocorticoids as modulators of immune function in health and disease: Neural, endocrine and immune interactions. *Brain Research – Brain Research Reviews* 23: 79–133.
McKay LI and Cidlowski JA (1998) Cross-talk between nuclear factor-kappa B and the steroid hormone receptors: Mechanisms of mutual antagonism. *Molecular Endocrinology* 12: 45–56.
Millar DB, Thomas JR, Pacheco ND, and Rollwagen FM (1993) Natural killer cell cytotoxicity and T-cell proliferation is enhanced by avoidance behavior. *Brain, Behavior, and Immunity* 7: 144–153.
Miller AH, Pariante CM, and Pearce BD (1999) Effects of cytokines on glucocorticoid receptor expression and function. Glucocorticoid resistance and relevance to depression. *Advances in Experimental Medicine and Biology* 461: 107–116.
Miller GE, Stetler CA, Carney RM, Freedland KE, and Banks WA (2002) Clinical depression and inflammatory risk markers for coronary heart disease. *American Journal of Cardiology* 90: 1279–1283.
Miller AH, Vogt G, and Pearce BD (2002b) The phosphodiesterase type 4 inhibitor, rolipram, enhances glucocorticoid receptor function. *Neuropsychopharmacology* 27: 939–948.
Miyazaki T, Ishikawa T, Nakata A, et al. (2005) Association between perceived social support and Th1 dominance. *Biological Psychology* 70: 30–37.
Muller N, Schwarz MJ, Dehning S, et al. (2006) The cyclooxygenase-2 inhibitor celecoxib has therapeutic

effects in major depression: Results of a double-blind, randomized, placebo controlled, add-on pilot study to reboxetine. *Molecular Psychiatry* 11: 680–684.
Musselman DL, Lawson DH, Gumnick JF, et al. (2001) Paroxetine for the prevention of depression induced by high-dose interferon alfa. (see comments). *New England Journal of Medicine* 344: 961–966.
Musselman DL, Miller AH, Porter MR, et al. (2001b) Higher than normal plasma interleukin-6 concentrations in cancer patients with depression: Preliminary findings. *American Journal of Psychiatry* 158: 1252–1257.
Nadeau S and Rivest S (2003) Glucocorticoids play a fundamental role in protecting the brain during innate immune response. *Journal of Neuroscience* 23: 5536–5544.
Nadjar A, Bluthe RM, May MJ, Dantzer R, and Parnet P (2005) Inactivation of the cerebral NFkappaB pathway inhibits interleukin-1beta-induced sickness behavior and c-Fos expression in various brain nuclei. *Neuropsychopharmacology* 30: 1492–1499.
Nakata A, Araki S, Tanigawa T, et al. (2000) Decrease of suppressor-inducer (CD4+ CD45RA) T lymphocytes and increase of serum immunoglobulin G due to perceived job stress in Japanese nuclear electric power plant workers. *Journal of Occupational and Environmental Medicine* 42: 143–150.
Naliboff BD, Solomon GF, Gilmore SL, Fahey JL, Benton D, and Pine J (1995) Rapid changes in cellular immunity following a confrontational role-play stressor. *Brain, Behavior, and Immunity* 9: 207–219.
Nance DM and Sanders VM (2007) Autonomic innervation and regulation of the immune system (1987–2007). *Brain, Behavior, and Immunity* 21: 736–745.
Nemeroff CB (1996) The corticotropin-releasing factor (CRF) hypothesis of depression: New findings and new directions. (see comment). *Molecular Psychiatry* 1: 336–342.
Nguyen KT, Deak T, Owens SM, Kohno T, Fleshner M, Watkins LR, and Maier SF (1998) Exposure to acute stress induces brain interleukin-1beta protein in the rat. *Journal of Neuroscience* 18: 2239–2246.
Nishioka T, Kurokawa H, Takao T, Kumon Y, Nishiya K, and Hashimoto K (1996) Differential changes of corticotropin releasing hormone (CRH) concentrations in plasma and synovial fluids of patients with rheumatoid arthritis (RA). *Endocrine Journal* 43: 241–247.
O'Connor KA, Johnson JD, Hansen MK, WieselerFrank JL, Maksimova E, Watkins LR, and Maier SF (2003) Peripheral and central proinflammatory cytokine response to a severe acute stressor. *Brain Research* 991: 123–132.
Obuchowicz E, Kowalski J, Labuzek K, Krysiak R, Pendzich J, and Herman ZS (2006) Amitriptyline and nortriptyline inhibit interleukin-1 release by rat mixed glial and microglial cell cultures. *International Journal of Neuropsychopharmacology* 9: 27–35.
Pace TW, Hu F, and Miller AH (2007) Cytokine-effects on glucocorticoid receptor function: Relevance to glucocorticoid resistance and the pathophysiology and treatment of major depression. *Brain, Behavior, and Immunity* 21: 9–19.
Pace TW, Mletzko TC, Alagbe O, Musselman DL, Nemeroff CB, and Miller AH (2006) Increased stress-induced inflammatory responses in male patients with major depression and increased early life stress. *American Journal of Psychiatry* 163: 1630–1633.
Paez Pereda M, Sauer J, Perez Castro C, Finkielman S, Stalla GK, Holsboer F, and Arzt E (1995) Corticotropin-releasing hormone differentially modulates the interleukin-1 system according to the level of monocyte activation by endotoxin. *Endocrinology* 136: 5504–5510.
Pariante CM (2004) Glucocorticoid receptor function *in vitro* in patients with major depression. *Stress* 7: 209–219.
Pariante CM (2008) The role of multi-drug resistance P-glycoprotein in glucocorticoid function: Studies in animals and relevance in humans. *European Journal of Pharmacology* 583: 263–271.
Pariante CM, Carpiniello B, Orru MG, et al. (1997a) Chronic caregiving stress alters peripheral blood immune parameters: The role of age and severity of stress. *Psychotherapy and Psychosomatics* 66: 199–207.
Pariante CM, Makoff A, Lovestone S, Feroli S, Heyden A, Miller AH, and Kerwin RW (2001) Antidepressants enhance glucocorticoid receptor function *in vitro* by modulating the membrane steroid transporters. *British Journal of Pharmacology* 134: 1335–1343.
Pariante CM and Miller AH (2001) Glucocorticoid receptors in major depression: Relevance to pathophysiology and treatment. *Biological Psychiatry* 49: 391–404.
Pariante CM, Pearce BD, Pisell TL, Owens MJ, and Miller AH (1997b) Steroid-independent translocation of the glucocorticoid receptor by the antidepressant desipramine. *Molecular Pharmacology* 52: 571–581.
Pariante CM, Pearce BD, Pisell TL, Sanchez CI, Po C, Su C, and Miller AH (1999) The proinflammatory cytokine, interleukin-1alpha, reduces glucocorticoid receptor translocation and function. *Endocrinology* 140: 4359–4366.
Pavlov VA and Tracey KJ (2004) Neural regulators of innate immune responses and inflammation. *Cellular and Molecular Life Sciences* 61: 2322–2331.
Pavlov VA and Tracey KJ (2005) The cholinergic anti-inflammatory pathway. *Brain, Behavior, and Immunity* 19: 493–499.
Penza KM, Heim C, and Nemeroff CB (2003) Neurobiological effects of childhood abuse: Implications for the pathophysiology of depression and anxiety. *Archives of Women's Mental Health* 6: 15–22.
Peters ML, Godaert GL, Ballieux RE, et al. (1999) Immune responses to experimental stress: Effects of mental effort and uncontrollability. *Psychosomatic Medicine* 61: 513–524.
Quan N, Avitsur R, Stark JL, et al. (2001) Social stress increases the susceptibility to endotoxic shock. *Journal of Neuroimmunology* 115: 36–45.
Quan N and Banks WA (2007) Brain-immune communication pathways. *Brain, Behavior, and Immunity* 21: 727–735.
Quan N, Whiteside M, and Herkenham M (1998) Time course and localization patterns of interleukin-1beta messenger RNA expression in brain and pituitary after peripheral administration of lipopolysaccharide. *Neuroscience* 83: 281–293.
Rabin B (1999) *Stress, Immune Function, and Health: The Connection*. New York: Wiley.
Raison CL, Capuron L, and Miller AH (2006) Cytokines sing the blues: Inflammation and the pathogenesis of major depression. *Trends in Immunology* 27: 24–31.
Raison CL and Miller AH (2003) When not enough is too much: The role of insufficient glucocorticoid signaling in the pathophysiology of stress-related disorders. *American Journal of Psychiatry* 160: 1554–1565.
Rhen T and Cidlowski JA (2005) Antiinflammatory action of glucocorticoids – new mechanisms for old drugs. *New England Journal of Medicine* 353: 1711–1723.
Richaud-Patin Y, Vega-Boada F, Vidaller A, and Llorente L (2004) Multidrug resistance-1 (MDR-1) in autoimmune disorders IV. P-glycoprotein overfunction in lymphocytes from myasthenia gravis patients. *Biomedicine and Pharmacotherapy* 58: 320–324.

Rivest S (2003) Molecular insights on the cerebral innate immune system. *Brain, Behavior, and Immunity* 17: 13–19.
Rogatsky I and Ivashkiv LB (2006) Glucocorticoid modulation of cytokine signaling. *Tissue Antigens* 68: 1–12.
Rogers CJ, Brissette-Storkus CS, Chambers WH, and Cameron JL (1999) Acute stress impairs NK cell adhesion and cytotoxicity through CD2, but not LFA-1. *Journal of Neuroimmunology* 99: 230–241.
Rood YR, Bogaards M, Goulmy E, and Houwelingen HC (1993) The effects of stress and relaxation on the *in vitro* immune response in man: A meta-analytic study. *Journal of Behavioral Medicine* 16: 163–181.
Roy S, Wang J, Kelschenbach J, Koodie L, and Martin J (2006) Modulation of immune function by morphine: Implications for susceptibility to infection. *Journal of Neuroimmune Pharmacology* 1: 77–89.
Ruzek MC, Pearce BD, Miller AH, and Biron CA (1999) Endogenous glucocorticoids protect against cytokine-mediated lethality during viral infection. *Journal of Immunology* 162: 3527–3533.
Sanchez MM, Alagbe O, Felger JC, et al. (2007) Activated p38 MAPK is associated with decreased CSF 5-HIAA and increased maternal rejection during infancy in rhesus monkeys. *Molecular Psychiatry* 12: 895–897.
Sapolsky RM, Romero LM, and Munck AU (2000) How do glucocorticoids influence stress responses? Integrating permissive, suppressive, stimulatory, and preparative actions. *Endocrine Reviews* 21: 55–89.
Saul AN, Oberyszyn TM, Daugherty C, et al. (2005) Chronic stress and susceptibility to skin cancer. *Journal of the National Cancer Institute* 97: 1760–1767.
Schulte HM, Bamberger CM, Elsen H, Herrmann G, Bamberger AM, and Barth J (1994) Systemic interleukin-1 alpha and interleukin-2 secretion in response to acute stress and to corticotropin-releasing hormone in humans. *European Journal of Clinical Investigation* 24: 773–777.
Segerstrom SC, Schipper LJ, and Greenberg RN (2008) Caregiving, repetitive thought, and immune response to vaccination in older adults. *Brain, Behavior, and Immunity.* 22: 744–752.
Shavit Y, Lewis JW, Terman GW, Gale RP, and Liebeskind JC (1984) Opioid peptides mediate the suppressive effect of stress on natural killer cell cytotoxicity. *Science* 223: 188–190.
Silberman DM, Wald MR, and Genaro AM (2003) Acute and chronic stress exert opposing effects on antibody responses associated with changes in stress hormone regulation of T-lymphocyte reactivity. *Journal of Neuroimmunology* 144: 53–60.
Silverman MN, Macdougall MG, Hu F, Pace TW, Raison CL, and Miller AH (2007) Endogenous glucocorticoids protect against TNF-alpha-induced increases in anxiety-like behavior in virally infected mice. *Molecular Psychiatry* 12: 408–417.
Silverman MN, Pearce BD, Biron CA, and Miller AH (2005) Immune modulation of the hypothalamic–pituitary–adrenal (HPA) axis during viral infection. *Viral Immunology* 18: 41–78.
Sloan RP, McCreath H, Tracey KJ, Sidney S, Liu K, and Seeman T (2007) RR interval variability is inversely related to inflammatory markers: The CARDIA study. *Molecular Medicine* 13: 178–184.
Smith RS (1991) The macrophage theory of depression. *Medical Hypotheses* 35: 298–306; erratum appears in *Medical Hypotheses* 1991 Oct; 36(2): 178.
Smith EM, Gregg M, Hashemi F, Schott L, and Hughes TK (2006) Corticotropin releasing factor (CRF) activation of NF-kappaB-directed transcription in leukocytes. *Cellular and Molecular Neurobiology* 26: 1021–1036.
Sorrells SF and Sapolsky RM (2007) An inflammatory review of glucocorticoid actions in the CNS. *Brain, Behavior, and Immunity* 21: 259–272.
Spiegel D, Bloom JR, Kraemer HC, and Gottheil E (1989) Effect of psychosocial treatment on survival of patients with metastatic breast cancer. (see comments). *Lancet* 2: 888–891.
Spiegel D, Butler LD, Giese-Davis J, et al. (2007) Effects of supportive-expressive group therapy on survival of patients with metastatic breast cancer: A randomized prospective trial. *Cancer* 110: 1130–1138.
Stark JL, Avitsur R, Padgett DA, Campbell KA, Beck FM, and Sheridan JF (2001) Social stress induces glucocorticoid resistance in macrophages. *AJP – Regulatory, Integrative and Comparative Physiology* 280: R1799–R1805.
Stenzel-Poore MP, Duncan JE, Rittenberg MB, Bakke AC, and Heinrichs SC (1996) CRH overproduction in transgenic mice: Behavioral and immune system modulation. *Annals of the New York Academy of Sciences* 780: 36–48.
Steptoe A, Hamer M, and Chida Y (2007) The effect of acute psychological stress on circulating inflammatory factors in humans: A review and meta-analysis. *Brain, Behavior, and Immunity* 21: 901–912.
Sternberg EM, Hill JM, Chrousos GP, Kamilaris T, Listwak SJ, Gold PW, and Wilder RL (1989) Inflammatory mediator-induced hypothalamic–pituitary–adrenal axis activation is defective in streptococcal cell wall arthritis-susceptible Lewis rats. *Proceedings of the National Academy of Sciences of the United States of America* 86: 2374–2378.
Szatmary Z, Garabedian MJ, and Vilcek J (2004) Inhibition of glucocorticoid receptor-mediated transcriptional activation by p38 mitogen-activated protein (MAP) kinase. *Journal of Biological Chemistry* 279: 43708–43715.
Thomas AJ, Davis S, Morris C, Jackson E, Harrison R, and O'Brien JT (2005) Increase in interleukin-1beta in late-life depression. *American Journal of Psychiatry* 162: 175–177.
Thornton LM, Andersen BL, Crespin TR, and Carson WE (2007) Individual trajectories in stress covary with immunity during recovery from cancer diagnosis and treatments. *Brain, Behavior, and Immunity* 21: 185–194.
Tracey KJ (2002) The inflammatory reflex. *Nature* 420: 853–859.
Turnbull AV and Rivier CL (1999) Regulation of the hypothalamic–pituitary–adrenal axis by cytokines: Actions and mechanisms of action. *Physiological Reviews* 79: 1–71.
Tyring S, Gottlieb A, Papp K, et al. (2006) Etanercept and clinical outcomes, fatigue, and depression in psoriasis: Double-blind placebo-controlled randomised phase III trial. (see comment). *Lancet* 367: 29–35.
Uchino BN (2006) Social support and health: A review of physiological processes potentially underlying links to disease outcomes. *Journal of Behavioral Medicine* 29: 377–387.
Vallejo R, deLeon-Casasola O, and Benyamin R (2004) Opioid therapy and immunosuppression: A review. *American Journal of Therapeutics* 11: 354–365.
Vedhara K and Nott K (1996) The assessment of the emotional and immunological consequences of examination stress. *Journal of Behavioral Medicine* 19: 467–478.
Viswanathan K, Daugherty C, and Dhabhar FS (2005) Stress as an endogenous adjuvant: Augmentation of the immunization phase of cell-mediated immunity. *International Immunology* 17: 1059–1069.
Viswanathan K and Dhabhar FS (2005) Stress-induced enhancement of leukocyte trafficking into sites of surgery or immune activation. *Proceedings of the National Academy of Sciences of the United States of America* 102: 5808–5813.
Vitkovic L, Konsman JP, Bockaert J, Dantzer R, Homburger V, and Jacque C (2000) Cytokine signals propagate through the

brain. *Molecular Psychiatry* 5: 604–615; erratum appears in *Molecular Psychiatry* 2001 Mar; 6(2): 249.
von Kanel R, Dimsdale JE, Mills PJ, Ancoli-Israel S, Patterson TL, Mausbach BT, and Grant I (2006) Effect of Alzheimer caregiving stress and age on frailty markers interleukin-6, C-reactive protein, and D-dimer. *Journals of Gerontology, Series A* 61: 963–969.
Wang X, Wu H, Lakdawala VS, Hu F, Hanson ND, and Miller AH (2005) Inhibition of Jun N-terminal kinase (JNK) enhances glucocorticoid receptor-mediated function in mouse hippocampal HT22 cells. *Neuropsychopharmacology* 30: 242–249.
Wang X, Wu H, and Miller AH (2004) Interleukin-1 alpha-induced activation of p38 mitogen-activated kinase inhibits glucocorticoid receptor function. *Molecular Psychiatry* 9: 65–75.
Watkins LR, Wiertelak EP, Goehler LE, et al. (1994) Neurocircuitry of illness-induced hyperalgesia. *Brain Research* 639: 283–299.
Webster JC, Oakley RH, Jewell CM, and Cidlowski JA (2001) Proinflammatory cytokines regulate human glucocorticoid receptor gene expression and lead to the accumulation of the dominant negative beta isoform: A mechanism for the generation of glucocorticoid resistance. *Proceedings of the National Academy of Sciences United States of America* 98: 6865–6870.
Weisse CS, Pato CN, McAllister CG, Littman R, Breier A, Paul SM, and Baum A (1990) Differential effects of controllable and uncontrollable acute stress on lymphocyte proliferation and leukocyte percentages in humans. *Brain, Behavior, and Immunity* 4: 339–351.
Willerson JT and Ridker PM (2004) Inflammation as a cardiovascular risk factor. *Circulation* 109: II2–II10.
Xu Y, Ku BS, Yao HY, Lin YH, Ma X, Zhang YH, and Li XJ (2005a) Antidepressant effects of curcumin in the forced swim test and olfactory bulbectomy models of depression in rats. *Pharmacology, Biochemistry and Behavior* 82: 200–206.
Xu Y, Ku BS, Yao HY, Lin YH, Ma X, Zhang YH, and Li XJ (2005b) The effects of curcumin on depressive-like behaviors in mice. *European Journal of Pharmacology* 518: 40–46.
Yirmiya R, Pollak Y, Morag M, et al. (2000) Illness, cytokines, and depression. *Annals of the New York Academy of Sciences* 917: 478–487.
Yu CY and Whitacre CC (2004) Sex, MHC and complement C4 in autoimmune diseases. *Trends in Immunology* 25: 694–699.
Zaharia MD, Ravindran AV, Griffiths J, Merali Z, and Anisman H (2000) Lymphocyte proliferation among major depressive and dysthymic patients with typical or atypical features. *Journal of Affective Disorders* 58: 1–10.
Zalcman S, Henderson N, Richter M, and Anisman H (1991) Age-related enhancement and suppression of a T-cell-dependent antibody response following stressor exposure. *Behavioral Neuroscience* 105: 669–676.
Zhu CB, Carneiro AM, Dostmann WR, Hewlett WA, and Blakely RD (2005) p38 MAPK activation elevates serotonin transport activity via a trafficking-independent, protein phosphatase 2A-dependent process. *Journal of Biological Chemistry* 280: 15649–15658.
Zorrilla EP, Luborsky L, McKay JR, et al. (2001) The relationship of depression and stressors to immunological assays: A meta-analytic review. *Brain, Behavior, and Immunity* 15: 199–226.

PART II

ENDOCRINOLOGICALLY IMPORTANT BEHAVIORAL SYNDROMES

19 Diseases of Hypothalamic Origin

J D Carmichael and G D Braunstein, Cedars-Sinai Medical Center, Los Angeles, CA, USA

Chapter Outline

19.1 Anatomy

The hypothalamus is located at the base of the brain below the thalamus, above the pituitary, and is one of the major structures of the diencephalon. The borders of the hypothalamus are somewhat arbitrary, but are defined anteriorly by the anterior margin of the optic chiasm and lamina terminalis and posteriorly by the posterior margins of the mamillary bodies and posterior commisure. The lateral borders vary and are less well defined. They include the optic tracts, internal capsule, pes pedunculi, globus pallidus, and ansa lenticularis (Daniel and Prichard, 1975). On the ventral surface, between the optic chiasm and mamillary bodies lay the tuber cinereum and infundibulum, from which the pituitary stalk originates. The third ventricle lies in the center of the hypothalamus, and connects to the fourth ventricle through the aqueduct of Sylvius and to the lateral ventricles through the foramen of Monro. The weight is approximately 2.5 g, and overall size approximately $1.5 \times 1.5 \times 1.3$ cm (Daniel and Treip, 1977).

The hypothalamus can be subdivided anatomically into three zones in the medial to lateral direction (periventricular, medial, and lateral) and four regions in the anterior to posterior direction (preoptic, supraoptic, tuberal, and mamillary) (Boshes, 1969; Kirgis and Locke, 1972; Bruesch, 1984). Within these areas, the hypothalamus is composed of groups of nerve cell bodies that form distinct nuclei (Martin and Reichlin, 1987; Braunstein, 2002) (**Figures 1** and **2**). Afferent and efferent nerve fibers provide complex interconnections among hypothalamic nuclei as well as neural structures beyond the hypothalamus, including the cerebral cortex, thalamus, limbic system, midbrain, and spinal regions (Martin and Reichlin, 1987).

19.2 Hypothalamic Functions

The hypothalamus integrates stimuli from a wide array of sources to control multiple behavioral, metabolic, and autonomic functions. It is responsible for homeostatic mechanisms such as water metabolism, temperature regulation, appetite control, sleep–wake cycles, circadian cycles, and autonomic functions. It also has a role in integrating memory, behavior, and emotion. Finally, the hypothalamus is essential in the regulation of anterior pituitary hormone production and release and, with this function, plays a critical role in the integration between neural and hormonal activity.

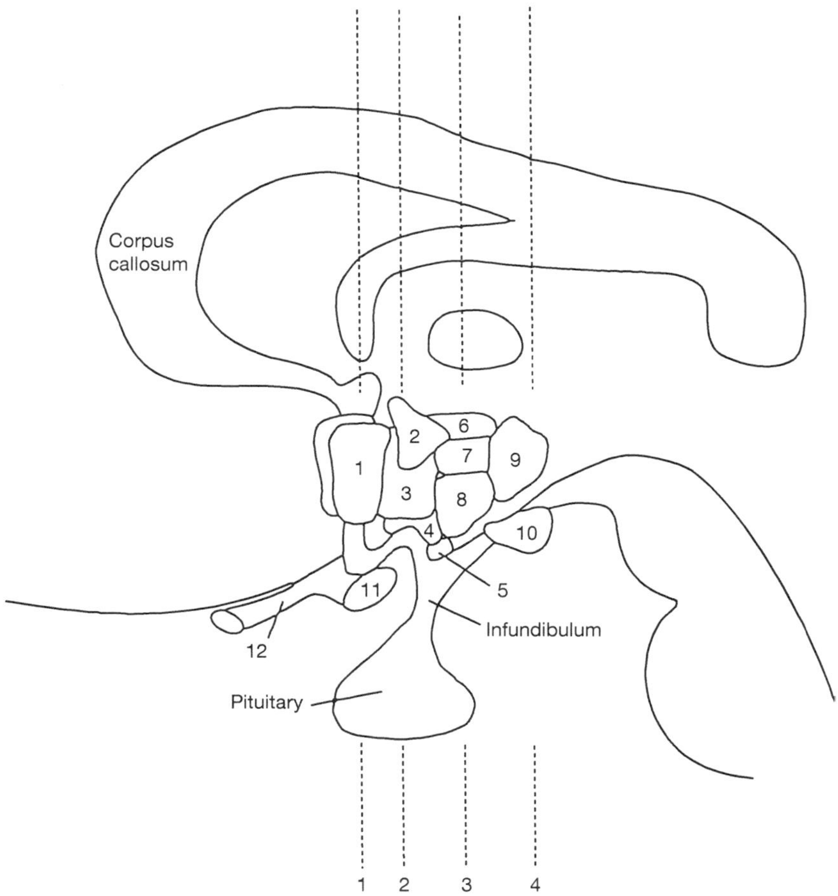

Figure 1 Schematic representation of lateral brain section demonstrating hypothalamic nuclei. Dashed lines represent the frontal (coronal) section planes illustrated in **Figures 2** and **3**. 1, preoptic nucleus; 2, paraventricular nucleus; 3, anterior hypothalamic areas; 4, supraoptic nucleus; 5, arcuate nucleus; 6, dorsal hypothalamic area; 7, dorsomedial nucleus; 8, ventromedial nucleus; 9, posterior hypothalamic area; 10, mamillary body; 11, optic chiasm; 12, optic nerve. Reproduced from Braunstein GD (2002) The hypothalamus. In: Melmed, S (ed.) *The Pituitary*, pp. 317–348. Cambridge, MA: Blackwell Scientific, with permission from Wiley-Blackwell Publishing Ltd.

19.2.1 Water Metabolism

The regulation of fluid balance and serum osmolality is crucial to survival. The integration of thirst and free-water excretion is maintained by nuclei in the medial part of the anterior hypothalamus. Arginine vasopressin (AVP) is a nine-amino-acid peptide that is synthesized in the magnocellular nerve cell bodies of the supraoptic and paraventricular nuclei (Verbalis, 2003). The peptide is enzymatically cleaved from its carrier protein, neurophysin II, as it is axonally transported to the synaptic terminal in the posterior pituitary. It is stored there with neurosecretory granules until it is secreted into the bloodstream. Increases in serum osmolarity and decreases in blood volume stimulate release of the AVP. Osmoreceptors located in the circumventricular organs of the lamina terminalis sense small changes in the plasma osmolality and rapidly stimulate AVP release and thirst (Carmel, 1980). In the euvolemic state, AVP release is under tonic inhibition by baroreceptors in the left atrium, carotid sinus, and aortic arch. Extracellular volume changes of approximately 5–10% reduce this tonic inhibition, resulting in AVP secretion (Antunes-Rodrigues et al., 2004). AVP release is much more sensitive to changes in osmotic stimuli, though the magnitude of AVP response to changes in blood volume is much greater (Verbalis, 2003). AVP acts on the V_2

receptor of the distal tubule of the collecting ducts of the nephron to increase water permeability through insertion of aquaporin-II water channels, allowing for water reabsorption. Combined reabsorption of free water and ingestion of fluids in response to thirst stimuli restore volume and serum osmolality.

19.2.2 Temperature Regulation

The preoptic anterior hypothalamic area (POA), dorsomedial nucleus (DMN) of the hypothalamus, periaqueductal gray matter of the midbrain, and the nucleus raphe pallidus in the medulla have integral roles in the central regulation of body temperature

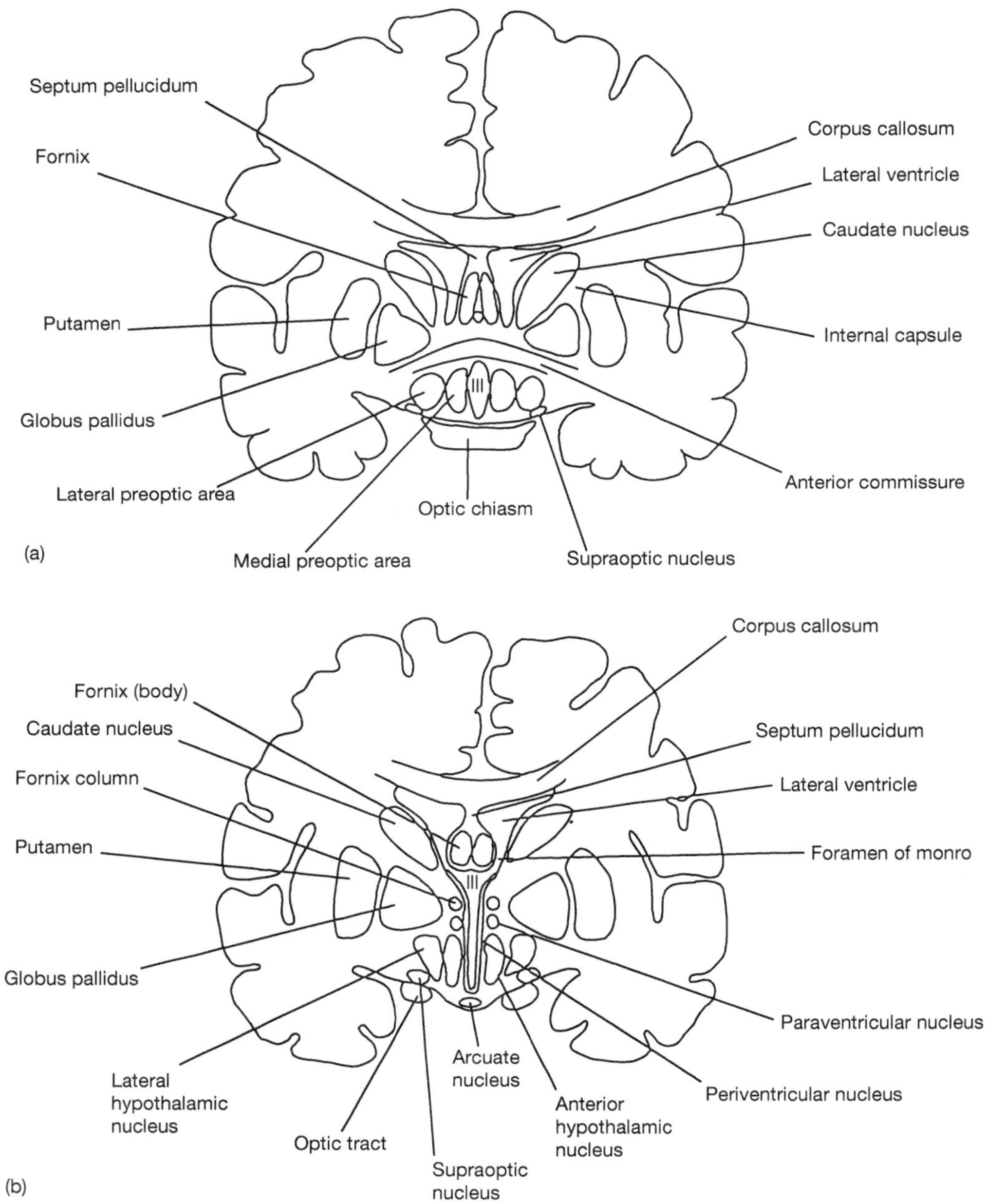

Figure 2 Continued on next page

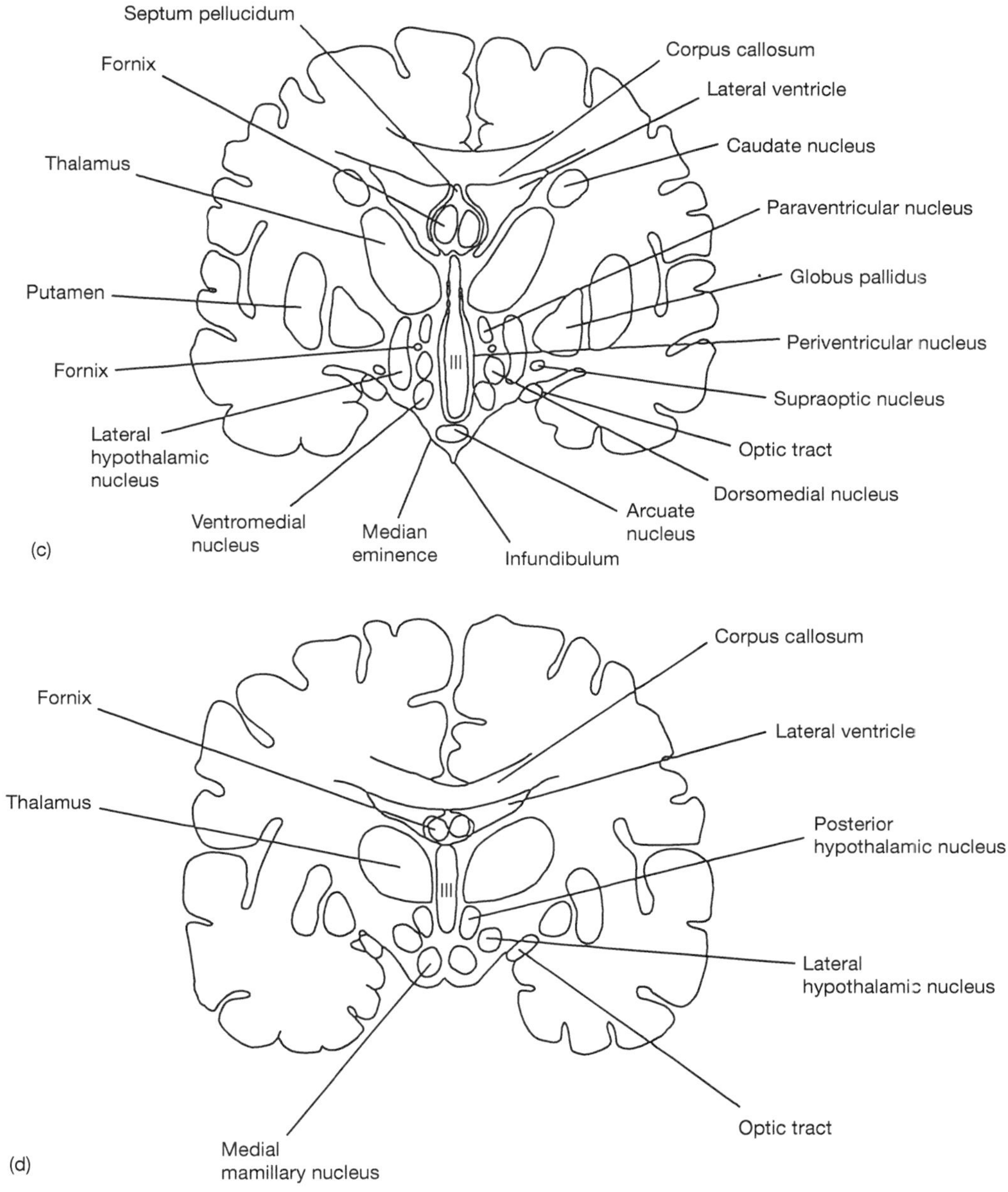

Figure 2 (a–d) Frontal (coronal) sections of the hypothalamic regions. Part (a) represents the preoptic region (frontal section plane 1 in **Figure 1**); part (b) represents the supraoptic region (frontal section plane 2 in **Figure 1**); part (c) represents the tuberal region (frontal section plane 3 in **Figure 1**); part (d) represents the mamillary region (frontal section plane 4 in **Figure 1**). Reproduced from Braunstein GD (2002) The hypothalamus. In: Melmed S (ed.) *The Pituitary*, pp. 317–348. Cambridge, MA: Blackwell Scientific, with permission from Wiley-Blackwell Publishing Ltd.

(Benarroch, 2007). The POA contains warm-sensitive, cold-sensitive, and temperature-insensitive neurons, with the warm-sensitive neurons holding the critical role in temperature homeostasis. Warm-sensitive neurons are spontaneously active, and a rise in temperature of the hypothalamus causes inactivation of inhibitory currents, resulting in increased activity of these neurons. Depending on the level of activity, these warm-sensitive neurons trigger either heat-loss or heat-gain mechanisms. Core temperature is maintained within a narrow zone, fluctuating between 0.2 and 0.5 °C. Superficial peripheral thermosensory neurons, the majority of which are cold sensitive, react to changing temperature with increased activity, allowing for rapid reactions to environmental changes (Romanovsky, 2007). Warm-sensitive peripheral sensors are deeper in the epidermis and their signals travel by unmyelinated c-fibers. Core-body temperature is sensed by deep body neurons in the esophagus, stomach, and large intra-abdominal veins. Afferent thermoeffective signals are sent to the brain through a spino-reticulo-hypothalamic pathway, in which lamina-I neurons project to the reticular formation, incorporating sensory input from multiple sources (Romanovsky, 2007). The neurons of the spino-reticulo-hypothalamic pathway project to hypothalamic structures via the periventricular stratum or the median forebrain bundle. Efferent thermoeffector pathways are activated causing temperature-adaptive mechanisms of vasomotor tone, shivering, and nonshivering thermogenesis to appropriately respond to temperature change (Plum and Van Uitert, 1978; Carmel, 1980; Romanovsky, 2007). Activation of warm-sensitive neurons in the hypothalamus results in heat dissipation mechanisms including skin vasodilatation and sweating, whereas heat production through shivering, nonshivering thermogenesis, and alterations in vasomotor tone are predominantly mediated by inhibitory signals from warm-sensitive neurons (Nagashima et al., 2000).

19.2.3 Appetite Control

Despite wide fluctuations in caloric intake, most healthy adults maintain a steady weight over many years due to a balance of caloric intake and energy expenditure (Edholm, 1977). Early theories regarding peripheral signaling to a central regulator of energy intake and body weight were proposed, with experimental models demonstrating changes in eating behavior with ablation of precise areas of the rat hypothalamus (Coll et al., 2007). Lesions in the ventromedial, paraventricular, and DMN induced hyperphagia, whereas lesions in the lateral hypothalamus caused anorexia and weight loss (King, 2006). A great deal of progress has been made in our understanding of the peripheral signaling involved in hunger, satiety, and energy expenditure. Circulating peptides produced by adipose tissue, the gastrointestinal tract, and the endocrine pancreas have a profound effect on appetite behavior through actions on the hypothalamus, brainstem, and afferent autonomic nerves. These peripheral signals and the central regulation of appetite and energy balance may also be intricately linked to circadian rhythms (Kalra et al., 1999).

Studies of mouse models of obesity (the *ob/ob* and *db/db* mice) led to the discovery of leptin, a peptide produced by adipose tissue (Zhang et al., 1994). Further work elaborated the role of leptin in satiety signaling and demonstrated receptors for leptin on neurons in the arcuate nucleus (Schwartz et al., 2000; Cone, 2005). Leptin acts on two sets of neurons in the arcuate nucleus: one set expressing orexigenic peptides, agouti-related peptide (AgRP), and neuropeptide Y (NPY); and the second set expressing cocaine and amphetamine-related transcript (CART) and proopiomelanocortin (POMC). Both sets of neurons project to melanocortin 4 receptor (MCR4)-expressing neurons in the hypothalamus and elsewhere in the brain (Schwartz et al., 2000; Coll et al., 2007). Other hypothalamic factors affecting appetite regulation include melanin-concentrating hormone, orexins, endocannabinoids, brain-derived neurotrophic factor, and nesfatin-1 (Schwartz et al., 2000; Dhillo, 2007).

Signals from the gastrointestinal tract include cholecystokinin, a gut peptide that signals satiety through vagal stimulation of appetite centers in the brainstem. Peptide YY is a gut peptide that has local effects on motility and suppresses appetite. Ghrelin is the endogenous ligand for the growth hormone secretagog (GHS) receptor. It is produced by the stomach, and serum levels have been found to rise prior to meal initiation, supporting a hypothesis that ghrelin is involved in preparing the body for incoming nutrients (Higgins et al., 2007). Obestatin, glucagon-like peptide-1, and oxyntomodulin are recently found proteins whose roles in appetite regulation have yet to be fully determined (Coll et al., 2007). Pancreatic hormones, well known for their roles in glucose homeostasis, may also play a role in appetite regulation. Insulin and pancreatic polypeptide are clearly related to caloric intake, but their precise role in appetite and satiety signaling has yet to be determined.

19.2.4 Sleep–Wake Cycle and Circadian Rhythm Control

The regulation of sleep and wakefulness involves an ascending arousal system, sleep-promoting neurons in the anterior hypothalamus, and a hypothalamic switch to regulate the activation of one center while inhibiting the other. The ascending arousal system originates in the rostral pons and traverses through the midbrain reticular formation, and is known as the ascending reticular-activating system (Saper et al., 2005). This system has two major branches, the first of which ascends to the thalamus and activates the thalamic relay neurons, which in turn transmit information to the cerebral cortex. This branch is active during wakefulness and REM sleep. The second branch bypasses the thalamus and activates neurons in the hypothalamus, basal forebrain, and cerebral cortex. Sleep-promoting neurons are located in the ventrolateral preoptic nucleus (VLPO). The VLPO has inhibitory connections with major centers in the hypothalamus and brainstem involved in wakefulness and is most active during sleep (Szymusiak et al., 2007). Orexin-producing neurons exert a balancing influence on sleep and wakeful states such that these states are stabilized and maintained without intermediate transition states (Saper et al., 2001). The suprachiasmatic nucleus, using inputs from the retina during the day and melatonin secretion from the pineal gland at night, regulates the circadian rhythm of sleep as well as other diurnal functions (Aschoff, 1979; Swaab et al., 1993; Saper et al., 2005).

19.2.5 Regulation of Visceral (Autonomic) Function

The hypothalamus has an important role in integrating sympathetic and parasympathetic autonomic nervous system activity. Stimulation of the posteromedial hypothalamus results in activation of the thoracolumbar autonomic response and the fight-or-flight response. Heart rate and cardiac output increases, blood pressure rises, and there is vasoconstriction of the α-adrenergic receptors on visceral vascular beds and vasodilatation of the β-adrenergic responsive blood vessels in skeletal muscle (Sano et al., 1970). Stimulation of the POA leads to an increased vagal and autonomic response (bradycardia, hypotension, increased blood flow to the visceral vascular beds, and decreased blood flow to skeletal muscle) (Sano et al., 1970; Plum and VanUitert, 1978; Carpenter and Sutin, 1983). The hypothalamus has also been ascribed a role in autonomic functions such as micturition, defecation, and penile erection (Boshes, 1969; Dong and Swanson, 2006a,b).

19.2.6 Emotional Expression and Behavior

The ventromedial nucleus (VMN) has been found to play a role in integrating cortical input with regard to behavior. Both humans and animals with lesions of the VMN exhibit rage and aggressive, often violent behavior (Plum and VanUitert, 1978). This behavior is referred to as sham rage and distinguishes it from voluntary or cortical rage. Electrical stimulation of the medial or posterior hypothalamus results in the sensation of fear or horror, while reduced activity is found with destructive lesions of these areas (Spiegel and Wycis, 1968; Sano et al., 1970; Schvarcz et al., 1972). Lesions of the limbic system in the caudal hypothalamus have been associated with aggressive, hypersexual behavior (Poeck and Pilleri, 1965; Frohman et al., 2002).

19.2.7 Memory

Memory is a complex process requiring an intact group of neurological structures termed the Papez circuit: the cerebral cortex, the hippocampal cortex, the fornix, mamillary bodies of the hypothalamus, the thalamus, and cingulate cortex (Brewer et al., 2007). Research has demonstrated a relationship between sleep and memory, with slow wave sleep contributing to declarative memory formation (Gais and Born, 2004). The hypothalamic–pituitary–adrenal (HPA) axis is also linked to memory. In addition, increased acute and chronic stress and glucocorticoid administration have been found to have adverse consequences on certain types of learning and memory (Wolf, 2003).

19.2.8 Control of Anterior Pituitary Function

Several hypophysiotropic hormones are synthesized and secreted by the hypothalamus, some of which stimulate while others inhibit anterior pituitary function. These neurons converge at the median eminence where nerve terminals secrete the hypophysiotropic hormones into the hypothalamic-hypophyseal portal circulation system. The hormones bind to receptors on the pituitary and regulate pituitary hormone production and secretion.

The hypophysiotropic hormones are produced in neurons that are located in a variety of areas in the hypothalamus. Gonadotropin-releasing-hormone (GnRH)-producing neurons are concentrated in the medial basal hypothalamus and preoptic areas (Pelletier, 1980). Thyrotropin-releasing hormone (TRH) neurons are found in the suprachiasmatic, preoptic medial, and paraventricular nuclei (Hokfelt et al., 1975). Corticotropin-releasing-hormone (CRH) neurons have been localized to the paraventricular nucleus (Pelletier et al., 1983). Growth hormone-releasing hormone-producing neuron cell bodies are located in the arcuate nucleus, as are the neurons synthesizing somatostatin (Desy and Pelletier, 1977; Pelletier et al., 1986). Dopaminergic neurons, which exert tonic inhibition of prolactin-secreting lactotrophs, are found predominantly in the arcuate nucleus, with smaller concentrations in the dorsomedial, ventromedial, periventricular, paraventricular nuclei, and median forebrain bundle (Palkovits et al., 1974). Integral to the control of anterior pituitary function is the feedback to the hypothalamus (and pituitary) by peripheral hormones which have a regulatory role on the production and secretion of the hypophysiotropic hormones.

19.3 Pathophysiological Principles

With the large number of important physiological functions that depend on the integrity of the hypothalamus, the high concentration of discreet nuclei and tracts, and the small size of the hypothalamus, one would anticipate that lesions of the hypothalamus would present with a great variety of syndromes. Despite the diversity of findings from patient to patient, many general principles regarding the manifestations of hypothalamic disease can be illustrated by careful clinico-pathological correlation.

1. Diseases of differing etiology can give rise to identical clinical signs and symptoms and can involve the hypothalamus. Bauer correlated the clinical signs and symptoms of 60 patients with autopsy results of hypothalamic lesions of varying locations and etiologies (Bauer, 1954, 1959). Despite the diversity of these cases, a number of similarities were found: 78% had neuro-ophthalmologic abnormalities (13% as the primary manifestation), 75% had pyramidal tract or sensory nerve involvement, 65% had headaches, 62% showed extrapyramidal cerebellar signs, and 40% exhibited recurrent vomiting. In his series, 40% manifested precocious puberty, 35% diabetes insipidus, 32% hypogonadism, 30% somnolence, 28% dysthermia, 25% and obesity or emaciation. This general uniformity of presentation is related to both the discreet location of the nuclei devoted to hypothalamic function and the often large lesions interrupting multiple tracts and nuclei. However, there are some pathological processes that have consistent and discreet manifestations of disease such as the gliosis of the supraoptic and paraventricular nuclei that occurs in familial or idiopathic diabetes insipidus and the consistent finding of precocious puberty and gelastic seizures with hypothalamic hamartomas.

2. Lesions may cause specific manifestations of hypothalamic disease without direct involvement of the site responsible for that specific function. The hypothalamic nuclei are densely located among the various tracts that integrate various afferent and efferent signals. Lesions of the hypothalamus are rarely discreet, and more often affect multiple areas of the hypothalamus.

3. Patients with hypothalamic involvement of systemic diseases usually have evidence of disease in nonhypothalamic areas. Many systemic illnesses can involve the hypothalamus, and, in fact, may cause the presenting symptom. Systemic diseases such as sarcoidosis, histiocytosis, and acute leukemia may have isolated lesions in the hypothalamus, but these are rare exceptions. Focal granulomatous disease has been described in the hypothalamus, but these lesions are usually accompanied by histiocytic infiltration of bone. Tuberculous meningitis, neurosyphilis, and viral infections may present with signs and symptoms of hypothalamic disease, but these illnesses are not usually limited to the hypothalamus.

4. Clinical manifestations can be related to the rate of progression of the disease. Rapidly progressing lesions of the hypothalamus often cause symptoms early on in the course of the disease, while more indolent lesions may be asymptomatic for long periods of time. Presumably, the slow growth allows for adaptive mechanisms to compensate for the deficits caused by the lesion. Acute insults such as trauma or hemorrhage tend to cause decreased consciousness, hyperthermia, diabetes insipidus, and acute endocrine alterations, which may be transient if the patient survives the injury. Chronic lesions affecting neuroendocrine function are usually not reversible.

5. Most lesions causing symptoms have bilateral involvement of the hypothalamus. Most hypothalamic functions are controlled by one or more paired nuclei. Unilateral lesions may cause symptoms, but in most cases bilateral disease is required to sufficiently compromise the hypothalamic function. Involvement of multiple areas (metastatic tumors and granulomatous disease), diseases in the third ventricle (colloid cysts), enlargement of the third ventricle (pinealomas, germ cell tumors, midbrain gliomas, and aqueductal stenosis), or midline lesions invading the floor of the hypothalamus (craniopharyngiomas, optic gliomas, and pituitary adenomas) are more likely to result in clinical signs and symptoms of hypothalamic disease than lesions involving more lateral portions of the hypothalamus.

6. Lesions of hypothalamic nuclei may result in different symptoms depending on whether the lesion has a stimulatory or destructive effect on the nuclei. Precocious puberty may result from stimulatory

lesions in the tuberal area, while destructive lesions may lead to hypogonadism. Similarly, stimulation of the preoptic nucleus results in hypothermia, whereas destruction causes hyperthermia.

7. Clinical manifestations of hypothalamic disease depend on the age of onset. The hypothalamus of neonates is immature and changes morphologically with age. Accordingly, the functionality of the hypothalamus differs from that of older individuals. Diseases that affect the neonatal hypothalamus present with different symptoms than the same disease in the same location later in life. Gliomas causing the diencephalic syndrome of infancy also cause loss of weight despite adequate intake of food coupled with euphoric behavior. Surviving individuals dramatically undergo a change to weight gain, obesity, and irritable behavior (Plum and VanUitert, 1978). Manifestations of gonadotropin deficiency are clearly different depending on the age of onset, with prepubertal patients failing to develop secondary sexual characteristics, while adults retain these characteristics. Growth hormone (GH) deficiency in childhood results in short stature, whereas in adults the manifestations are related to metabolic and body compositional changes. GH excess dramatically differs in its presentation of gigantism in childhood, and acromegaly in adults.

19.4 Manifestations of Hypothalamic Disease

Using the general principles outlined above, and relying on careful pathologic studies of patients with hypothalamic diseases, a topographic map of the hypothalamus correlating clinical findings to the anatomic sites of lesions can be constructed (McLean, 1934; Rothballer and Dugger, 1955; White and Hain, 1959; Reeves and Plum, 1969; Fox et al., 1970; Lewin et al., 1972; Kamalian et al., 1975; Celesia et al., 1981; Haugh and Markesbery, 1983; Schwartz et al., 1986; Pinkney et al., 2002) (**Figure 3** and **Table 1**).

19.4.1 Disorders of Water Metabolism

19.4.1.1 Central diabetes insipidus

Central diabetes insipidus (DI) results from the complete or partial lack of secretion of AVP. The absence of AVP action in the distal tubules and collecting ducts of the nephron prevents the reabsorption of free water, resulting in the inability to adequately concentrate urine. Polyuria ensues, ranging from 2 l a day in partial DI to 10–15 l a day in complete DI, triggering the thirst mechanism prompting water-seeking behavior and polydipsia (Rose and Post, 2001). When the thirst mechanism is intact, and the patient has access to water and can drink, the plasma osmolality can be maintained within the normal range. When the osmoreceptors are damaged, altering the patients' ability to perceive thirst, or if the patient is unable to take a sufficient quantity of water, hypernatremia can develop with rapid mental status changes ranging from lethargy to stupor to coma. Patients with partial deficiency of AVP may release enough of the hormone to maintain normal water balance, and concentrate their urine to an osmolarity above the serum osmolarity. Both partial and complete DI are characterized by an increase in urine osmolarity after the administration of exogenous vasopressin, in contrast to normal individuals who maximally concentrate their urine with endogenous vasopressin secretion in response to dehydration.

Lesions involving the magnocellular neurons of the supraoptic and paraventricular nuclei or the supraoptico-hypophyseal tracts, which terminate in the pituitary stalk and posterior pituitary, can cause DI. A variety of hypothalamic lesions commonly cause DI, including primary or metastatic tumors, infiltrative diseases, vascular causes, infections, and trauma. Transient DI can occur following resection of a pituitary adenoma resulting in manipulation or resection of the pituitary stalk, or by acute reversible hypothalamic lesions (Bauer, 1954, 1959).

In normal individuals, the posterior pituitary can be identified by its hyperintense signal on sagittal T1-weighted imaging. The absence of this signal is a nonspecific finding consistent with central DI. A thickened infundibulum or pituitary stalk suggests an infiltrative cause for DI. The thickening of the pituitary stalk in association with autoimmune or inflammatory disease correlates with histological features of lymphocyte and plasma cell infiltration, fibrosis, and necrosis (Maghnie, 2003).

A spectrum of pathologic lesions results in DI (Blotner, 1958; Randall et al., 1961; Scherbaum et al., 1986; Maghnie et al., 2000). Idiopathic DI comprises the largest proportion of patients and is commonly associated with vasopressin-cell antibodies and markers of autoimmunity (Scherbaum et al., 1986; Maghnie et al., 2000, 2006). Loss of magnocellular nerve bodies and gliosis is also seen in the supraoptic and paraventricular nuclei (Bergeron et al., 1991). These findings are also present in patients with

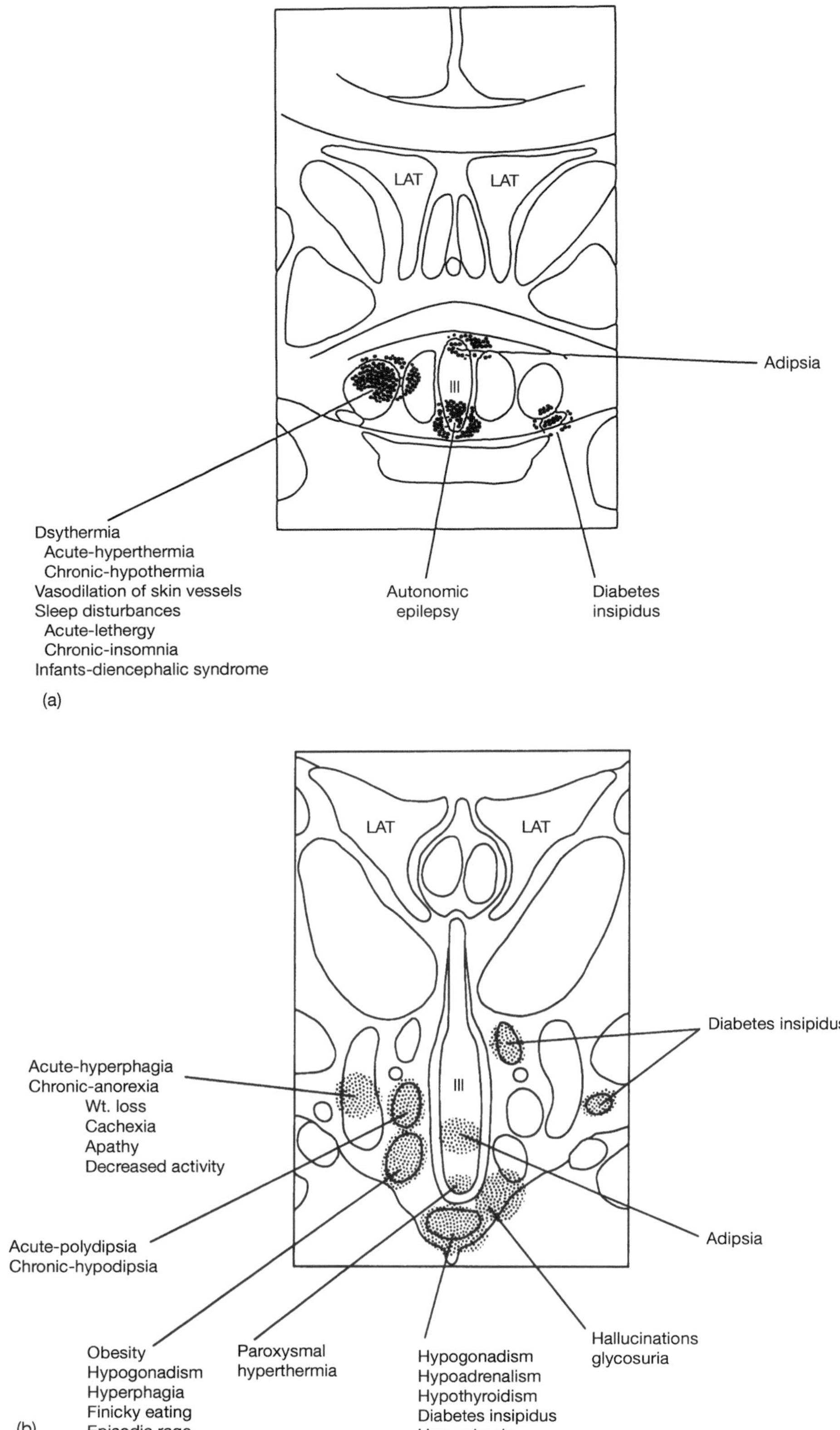

Figure 3 Continued on next page

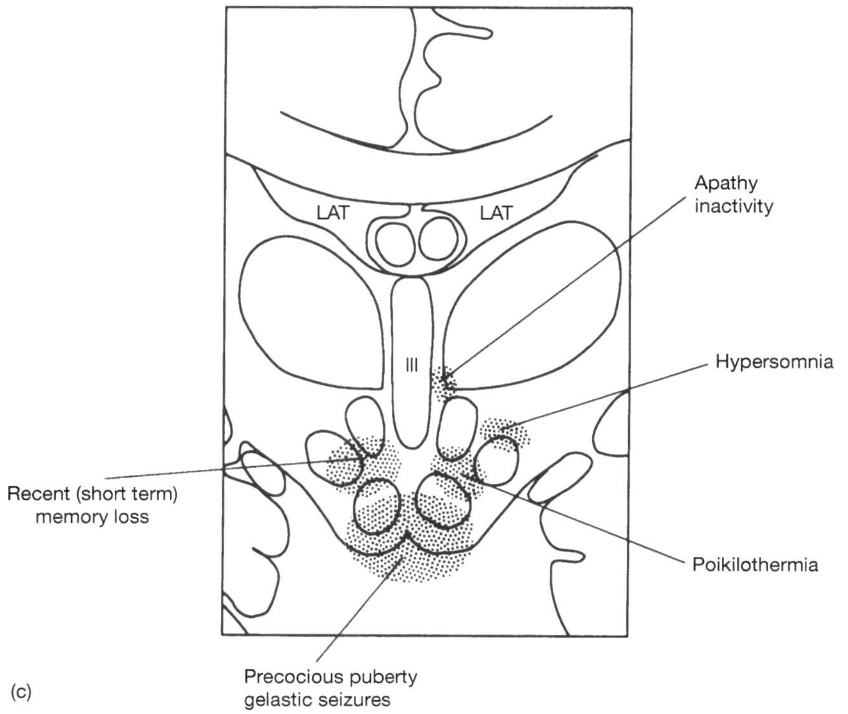

Figure 3 (a–c) Clinical findings associated with hypothalamic lesions located at various anatomic sites. Clinicopathological correlation based on multiple studies (McLean, 1934; Rothballer and Dugger, 1955; White and Hain, 1959; Reeves and Plum, 1969; Fox et al., 1970; Lewin et al., 1972; Kamalian et al., 1975; Celesia et al., 1981; Haugh and Markesbery, 1983; Schwartz et al., 1986; Pinkney et al., 2002): (a) corresponds to region depicted in **Figure 2(a)**; (b) corresponds to region depicted in **Figure 2(c)**; (c) corresponds to section depicted in **Figure 2(d)**. Adapted from Braunstein GD (2002) The hypothalamus. In: Melmed S (ed.) *The Pituitary*, pp. 317–348. Cambridge, MA: Blackwell Scientific, with permission from Wiley-Blackwell Publishing Ltd.

familial central DI, but lack the antibodies to the vasopressin cells. In these cases, the disease is caused by a mutation in the vasopressin pro-hormone gene resulting in a deficiency of vasopressin (Macleod et al., 1997). Multiple different mutations have been identified with most mutations occurring within the neurophysin domain of the precursor. Almost all are inherited in an autosomal dominant form. The DIDMOAD syndrome (Wolfram's syndrome) refers to central DI, diabetes mellitus, optic atrophy, and deafness (Rimon and Phillips, 1996; Maghnie, 2003; Minton et al., 2003). It is caused by a loss-of-function mutation in the gene for wolframin, a protein with unknown function found in the endoplasmic reticulum.

DI is frequently seen in specific hypothalamic lesions such as suprasellar germinomas (85%) (Imura et al., 1987; Buchfelder et al., 1989), pineal germinomas (40%) (Jennings et al., 1985; Imura et al., 1987), chronic disseminated histiocytosis (50%) (Stosel and Braunstein, 1991), and sarcoidosis (58%) (Delaney, 1977; Vesely et al., 1977; Stuart et al., 1978; Jawadi et al., 1980). DI may also be found in patients with septo-optic dysplasia (23%) (Arslanian et al., 1984; Izenberg et al., 1984; Margalith et al., 1984; 1985), pinealomas (18%) (Imura et al., 1987), hypothalamic gliomas (17%) (Roberson and Till, 1974; Borit and Richardson, 1982), and craniopharyngiomas (14%) (Jennings et al., 1985).

19.4.1.2 Adipsic or essential hypernatremia

Damage to the osmoreceptors in the anterior medial and anterior preoptic regions causes impaired or absent thirst. Chronic, inadequate fluid intake can lead to hypernatremia in this condition. The hypernatremia seen in this syndrome often reaches dangerously high levels. Likewise, ingestion of large amounts of fluid can result in hyponatremia. These patients have normal extracellular fluid volume and maintain normal blood pressure, pulse, and glomerular filtration rate. Due to the proximity of the osmoreceptors responsible for thirst regulation and AVP

Table 1 Hypothalamic functions, the nuclei or regions involved with the specific functions and the disorders resulting from stimulatory or destructive lesions in the regions

Function	*Nuclei or region involved*	*Disorders*
Water metabolism	Supraoptic (n), paraventricular (n) Circumventricular organs (r)	Diabetes mellitus Essential hypernatremia SIADH
Temperature regulation	Preoptic anterior hypothalamic (r) Posterior hypothalamus (r)	Hyperthermia Hypothermia Poikilothermia
Appetite control	Ventromedial (n) (satiety center) Lateral hypothalamic (r) (feeding center)	Hypothalamic obesity Cachexia Anorexia nervosa Diencephalic syndrome Diencephalic glycosuria
Sleep–wake cycle and circadian rhythm	Ventrolateral preoptic anterior hypothalamic (r) (sleep center) Posterior hypothalamic (r) including tuberomamillary (n) (arousal center) Suprachiasmatic (n)	Somnolence Reversal of sleep–wake cycle Akinetic mutism Coma
Visceral (autonomic) function	Posterior medial (r) (sympathetic region) Preoptic anterior hypothalamus (r) (parasympathetic region)	Sympathetic activation Parasympathetic activation
Emotional expression and behavior	Ventromedial (n) Medial and posterior hypothalamus (r) Caudal hypothalamic (r)	Sham rage Fear or horror Apathy Hypersexual behavior
Memory	Ventromedial (n), mamillary bodies	Short-term memory loss
Control of anterior pituitary function	Arcuate (n) Preoptic (n) Suprachiasmatic (n) Paraventricular (n) Neovascular zone (median eminence)	Hyperfunction syndromes Hypofunction syndromes

McLean (1934), Dott (1938), Riddoch (1938), Bauer (1954, 1959), Rothballer and Dugger (1955), White and Hain (1959), Boshes (1969), Reeves and Plum (1969), Fox et al. (1970), Sano et al. (1970), Lewin et al. (1972), Kamalian et al. (1975), Plum and Van Uitert (1978), Carmel (1980), Frohman (1980), Garnica et al. (1980), Celesia et al. (1981), Haugh and Markesbery (1983), Schwartz et al. (1986), Braunstein (2002, 2006), Thompson et al. (2003).
Reproduced from Braunstein GD (2006) Hypothalamic syndromes. In: DeGroot LJ and Jameson JL (eds.) *Endocrinology*, 5th edn., p. 377. Philadelphia, PA: Elsevier, with permission from Elsevier.

release, many patients have altered thirst and partial or complete DI (Ball et al., 1997). However, since baroreceptor control of AVP release remains intact, fluctuations in volume can suppress or stimulate vasopressin release. Generally, patients are able to tolerate hypernatremia and stay relatively asymptomatic with chronic elevations. When levels are above 160 mEq l^{-1}, patients may develop fatigue, lethargy, and weakness. Hypothalamic endocrinopathies are present with 71% exhibiting anterior pituitary hormone deficiencies and 43% having obesity (DeRubertis et al., 1974). Essential hypernatremia has been reported with suprasellar germinomas, histiocytosis, sarcoidosis, craniopharyngiomas, ruptured aneurysms, optic nerve gliomas, pineal tumors, trauma, hydrocephalus, cysts, inflammatory conditions, and toluene exposure (DeRubertis et al., 1974; Teelucksingh et al., 1991; Ball et al., 1997). Some children have been found to have essential hypernatremia without a structural hypothalamic defect (Hayek–Peake syndrome). This syndrome is characterized by hypernatremia, hypodipsia, obesity, hyperprolactinemia, hypothyroidism, hyperlipidemia, and growth hormone deficiency (GHD) (Hayek and Peake, 1982). Treatment is challenging and relies on careful monitoring and regulation of weight, urine output, and water intake (Ball et al., 1997).

19.4.1.3 Syndrome of inappropriate secretion of antidiuretic hormone

Syndrome of inappropriate secretion of antidiuretic hormone (SIADH) is characterized by inappropriate urinary concentration ($U_{osm} > 100$ mosm kg^{-1} H_2O)

and decreased effective osmolality of the extracellular compartment ($P_{osm} < 275$ mosm kg^{-1} H_2O) in a clinically euvolemic patient (Ellison and Berl, 2007). This results in hyponatremia in light of continued urinary sodium excretion of varying severity and hypouricemia. Other potential causes of euvolemic hypoosmolality such as hypothyroidism, adrenal insufficiency, or diuretic use are absent.

The syndrome may be the result of nervous system disorders such as mass lesions, inflammatory diseases, degenerative demyelinating diseases, or head trauma or surgery as well as tumors producing ectopic AVP (Imura et al., 1987). Multiple drugs of different classes have also been found to cause SIADH. Vasopressin production can also be initiated by pain from major surgery in addition to infections of the respiratory system.

Clinical signs and symptoms are related to the severity and/or rapidity of changes in the serum sodium. Common symptoms for mild hyponatremia include nausea, vomiting, headache, dizziness, and lethargy. Mental confusion, seizures and coma may develop with sodium concentrations less than 120 mEq l^{-1} especially when changes are rapid.

Treatment of hyponatremia in SIADH depends upon the duration of hyponatremia and the presence of neurological symptoms. The mainstay of treatment of asymptomatic hyponatremia due to SIADH is fluid restriction. Additional treatments such as salt (hypertonic saline) administration, loop diuretics, demeclocycline, or lithium are reserved for severe, symptomatic, or resistant forms of SIADH. Of note, correction of serum sodium at too rapid of a rate may result in central pontine myelinolysis. New medications with vasopressin receptor antagonist activity have been shown to be effective in raising serum sodium and may have a potential role in treatment of severe or chronic SIADH patients (Saito et al., 1997; Decaux, 2001).

19.4.1.4 Cerebral salt wasting

Cerebral salt wasting is another potential cause of hyponatremia. It is found in patients with central nervous system (CNS) disease or in the postoperative setting. The laboratory findings are identical to SIADH, and the distinction between these two entities is often difficult to make. With cerebral salt wasting, the major distinguishing factor is the volume status that results from hypovolemic activation of ADH release. The putative mechanisms for the salt wasting and hypovolemia are disruption of sympathetic nervous system input to the kidney and increased levels of a central natriuretic factor (Palmer, 2003). Distinction between SIADH and cerebral salt wasting is important because their respective management differs. Fluid restriction in a patient with cerebral salt wasting can result in disastrous hypovolemia and hypotension.

19.4.1.5 Reset osmostat

Resetting of the central osmostat may occur in euvolemic and hypovolemic hyponatremia. A reset osmostat is also seen in quadriplegia, psychosis, tuberculosis, and chronic malnutrition (Robertson et al., 1982). Patients usually have a serum sodium between 125 and 135 mEq l^{-1} that remains stable despite variations in water and sodium intake. Water excretion is normal in a reset osmostat as opposed to impaired water excretion seen in SIADH. Attempts to correct hyponatremia in these cases are unnecessary and are unlikely to be effective.

19.4.2 Dysthermia

19.4.2.1 Hyperthermia

Acute injury to the preoptic area and anterior hypothalamus from intracranial hemorrhage, neurosurgery, or trauma may produce a profound hyperthermia with temperature elevations up to 41°C. Tachycardia and loss of consciousness may accompany the hyperthermia and the fever rarely persists longer than a few days. With acute injury to the hypothalamus, heat production may continue while heat-dissipating mechanisms are lost. Of note, the tachycardia associated with the hyperthermia in these patients is not increased to the same extent as patients with febrile reactions to infection or inflammation (Bauer, 1959; Boshes, 1969; Plum and VanUitert, 1978; Carmel, 1980).

Neuroleptic malignant syndrome (NMS) is an idiosyncratic complication of treatment with high potency neuroleptics such as haloperidol, atypical neuroleptics, for example, risperidone and olanzapine, non-neuroleptic drugs including metoclopramide, prochlorperazine, promethazine, or after sudden withdrawal of dopamine agonists (Rusyniak and Sprague, 2005, 2006). Medications with more potent antidopaminergic activity cause a higher incidence of this syndrome to occur (Horn et al., 1988). The pathophysiology is largely unknown; however, blockade of basal ganglia D2 receptors may have a role. Antidopaminergic activity activates heat generation through muscle contraction, impairs heat dissipation through hypothalamic injury, and inhibits diaphoresis through peripheral anticholinergic effects. Hypothalamic damage has been seen at autopsy in the preoptic medial and tuberal nuclei (Horn et al., 1988). In addition to the hyperthermia, symptoms include lead pipe rigidity, altered mental status ranging from agitation to stupor and coma, and autonomic dysfunction (Rusyniak and Sprague, 2005, 2006). The clinical presentation may vary, but symptoms can progress from mental status changes to muscle rigidity to autonomic instability and then hyperthermia (Velamoor et al., 1994). Treatment of NMS consists of removal of the offending drug, supportive care, and the use of dantrolene and bromocriptine.

Hyperthermia is also evident in the serotonin syndrome and sympathomimetic syndrome. Acute intoxication with cocaine, methamphetamine, and 3,4-methylenedioxymethamphetamine (MDMA) can cause the serotonin syndrome. Therapeutic use of stimulants and antidepressants has also been associated with the serotonin syndrome. Most cases present with altered mental status, autonomic instability, and abnormal neuromuscular activity (Rusyniak and Sprague, 2006).

Malignant hyperthermia is characterized by hyperthermia, hypotension, and muscle rigidity during and after anesthesia in genetically susceptible individuals. Symptoms are a consequence of increased calcium release in skeletal muscle, subsequent uncoupling of oxidative phosphorylation and excess cellular metabolism (Rusyniak and Sprague, 2006). Treatment with dantrolene causes complete and sustained relaxation of skeletal muscle and inhibits intracellular calcium release.

Chronic hyperthermia is due to lesions in the tuberoinfundibular region. In Bauer's series, 10% of patients exhibited chronic hyperthermia (Bauer, 1954, 1959). This sustained hyperthermia is thought to be secondary to several mechanisms such as loss of heat dissipation mechanisms, stimulation of heat production, and elevation of the set point (Plum and VanUitert, 1978). Paradoxically, vasoconstriction of the peripheral vasculature occurs which presents as cold, clammy extremities.

Paroxysmal hyperthermia is thought to be of hypothalamic origin due to accompanying signs and symptoms. This condition is characterized by brief, sporadic episodes of hyperthermia, shaking chills, hypertension, vomiting, and peripheral vasoconstriction. Episodes are brief in duration and resolution is accompanied by vasodilatation and diaphoresis.

19.4.2.2 Hypothermia

Chronic central hypothermia is usually associated with large lesions of the posterior hypothalamus, or the entire hypothalamus. Lesions causing defects in thermoregulation usually cause other hypothalamic defects as well. Destruction of the thermoregulatory mechanisms results in an inability to generate heat through shivering and vasoconstriction. Multiple lesions have been noted to cause such thermoregulatory defects: neoplasms (craniopharyngiomas and neuroblastoma); infections (poliomyelitis and neurosyphilis); sarcoidosis; multiple sclerosis; Wernicke's encephalopathy glial scarring; Parkinson's disease; and traumatic or vascular injury (Wolff et al., 1964; Delaney, 1977; Martin and Reichlin, 1987; Sandyk et al., 1987; Haak et al., 1990; Edwards et al., 1996; White et al., 1996). Drugs including barbiturates and alcohol may cause a defect in heat-maintenance mechanisms resulting in chronic hypothermia (Martin and Reichlin, 1987).

Spontaneous periodic hypothermia is a rare syndrome that has been attributed to autonomic seizure activity (also referred to as diencephalic autonomic epilepsy) and is characterized by periodic hypothermia with normal body temperature regulation mechanisms (Martin and Reichlin, 1987). Cases have been associated with agenesis of the corpus callosum (Shapiro's syndrome) and with the absence of systemic disease or structural hypothalamic lesions (Shapiro et al., 1969; Fox et al., 1973; Kloos, 1995). Accompanying symptoms of autonomic nervous system activity such as sweating, vasodilatation, nausea, vomiting, lacrimation, salivation, and bradycardia may occur (Martin and Reichlin, 1987). Mentation may be altered during or after the episode and shivering may be seen with the return of normal body temperature. The frequency of episodes varies from hours to years, and episodes may be prolonged lasting hours to weeks. Inability to warm these patients has demonstrated the maintenance of an altered temperature set point in those affected with periodic hypothermia (Kloos, 1995). While it has been proposed that these episodic alterations in autonomic function are related to seizure activity, definitive evidence of seizure activity has not been documented and the cause still remains unknown. Other possible mechanisms for spontaneous periodic hypothermia include degenerative processes, neurochemical dysfunction, and irritative mechanisms (Kloos, 1995).

19.4.2.3 Poikilothermia

This condition is characterized by core temperatures that follow the ambient temperature due to loss of function of heat conservation as well as heat-dissipation mechanisms. Compared to the tight regulation of core body temperature in normal individuals, these patients' core temperatures vary more than 2 °C (MacKenzie, 1997). Patients become hyperthermic in hot conditions and hypothermic in cold conditions without experiencing any discomfort. They make no effort to change their environment to alter their core temperature. This condition not only has been found in patients with large anterior and posterior hypothalamic destruction, but has also been reported in subjects with multiple sclerosis and Wernicke's encephalopathy (Bauer, 1954, 1959; Plum and VanUitert, 1978).

19.4.3 Disorders of Caloric Balance

19.4.3.1 Hypothalamic obesity

A great deal has been learned regarding the neuroendocrine mechanisms of appetite regulation. In many cases, this knowledge has come from studies on the effects of structural lesions and genetic defects in hypothalamic obesity. Structural damage to the hypothalamus commonly results in obesity, either as a consequence of the tumor or its treatment. Space-occupying lesions such as craniopharyngiomas, other hypothalamic tumors, aneurysms, inflammatory and infiltrative diseases all have been known to be associated with obesity (Pinkney et al., 2002). In a series of 212 patients with craniopharyngiomas, 125 of whom had hypothalamic involvement, body mass index (BMI) was higher in those with hypothalamic lesions at diagnosis, and was more progressive during follow-up (Muller et al., 2003). Lesions causing obesity tend to be large, but careful analysis of patients with discreet lesions reveals that bilateral destruction of the ventromedial nucleus results in obesity, as is the case with experimental animals (Bauer, 1954, 1959; White and Hain, 1959; Boshes, 1969; Reeves and Plum, 1969; Bray and Gallagher, 1975; Celesia et al., 1981). Hypothalamic obesity is also associated with gene defects. Prader–Willi syndrome (Mann and Bartolomei, 1999), mutations affecting the leptin gene (Montague et al., 1997), the leptin receptor (Clement et al., 1998), the melanocortin 4 receptor (Yeo et al., 1998), and proopiomelanocortin (POMC) (Krude et al., 1998) are associated with obesity.

Hyperphagia and food-seeking behavior are characteristic of hypothalamic obesity, and have been described in patients with destructive lesions, children with Prader–Willi syndrome, and defects in leptin and its receptor. Patients with extreme cases exhibit food stealing and constant foraging. Their obesity is not due to hyperphagia alone, however, as hyperinsulinemia (greater than age- and weight-matched controls), a lower resting metabolic rate, reduced activity, and deficiencies of GH, thyroid stimulating hormone (TSH), and GnRH may also contribute to excessive weight gain in patients with hypothalamic disease (Pinkney et al., 2002).

19.4.3.2 Hypothalamic cachexia in adults

Destructive lesions of the lateral hypothalamus, with or without involvement of the ventromedial nucleus lead to anorexia and emaciation (White and Hain, 1959; Reeves and Plum, 1969; Kamalian et al., 1975). Lateral hypothalamic lesions result in a rapidly progressive weight loss, decreased appetite and food intake, muscle wasting, and lethargy, leading to cachexia and death. Neoplasms are the most common etiology, but cysts and multiple sclerosis have also been described (White and Hain, 1959; Kamalian et al., 1975). In Bauer's series of patients with hypothalamic tumors, 18% had substantial weight loss, 7% had anorexia, and 8% were bulimic (Bauer, 1954, 1959).

Cachexia is a common finding in patients with cancer and chronic disease. Recent research regarding the neuropeptides involved in the hypothalamic control of satiety and hunger suggests that cytokines produced by tumors and inflammation interact with the hypothalamic signaling of hunger, producing cachexia. This appears to be a multifactorial process involving leptin, melanocortin and orexin, and multiple cytokines (Ramos et al., 2004; Mitch, 2005).

19.4.3.3 Diencephalic syndrome of infancy

Diencephalic syndrome (DS) is a rare cause of failure to thrive in infancy. It is characterized by profound emaciation, an absence of subcutaneous adipose tissue despite normal caloric intake, normal linear growth, euphoria, hyperkinesis, and hyperalertness (Russell, 1951). The syndrome is associated with tumors of the anterior hypothalamus (80%), usually low-grade hypothalamic or optic gliomas affecting the ventral medial nuclei (Burr et al., 1976; Plum and VanUitert, 1978; Carmel, 1980). Other tumors such as ependymomas, gangliogliomas, and dysgerminomas have also been described in these patients (Burr et al., 1976). Infants appear normal at birth and develop normally during the first 3–12 months. They then begin to lose weight and subcutaneous fat, but continue normal linear growth. Signs of hyperactivity, cheerful affect, and an alert appearance, secondary to eyelid retraction (Collier sign) emerge (Burr et al., 1976). In addition, nystagmus, vomiting, pallor, optic atrophy, and tremors may also be present (Burr et al., 1976; Poussaint et al., 1997). Children are often brought to clinical attention for failure to thrive and diagnosis can be delayed due to the rarity of this syndrome (Huber et al., 2007). Endocrine abnormalities such as elevated GH levels that paradoxically rise in response to glucose, low insulin-like growth factor (IGF)-I levels, and absent diurnal variation of serum cortisol concentrations occur (Frohman et al., 1980). While elevated GH levels are not specific to this disease, GH dysregulation may explain the loss of subcutaneous fat seen in these patients. The low IGF-I levels are consistent with peripheral GH resistance; however, maintenance of linear growth differentiates this diagnosis from other GH resistance states (anorexia nervosa, chronic illness, and oncologic processes) (Fleischman et al., 2005). Most children succumb to the tumor and emaciation by 2 years of age; however, those that survive beyond this age often maintain their appetite, gain weight, and become obese. Mood changes, somnolence, and precocious puberty may also develop (Burr et al., 1976; Carmel, 1980, 1985). The protean signs of this syndrome clearly demonstrate the relationship between hypothalamic manifestations and the age of patient and development of the hypothalamus.

19.4.3.4 Anorexia nervosa

The etiology for anorexia nervosa is unknown. The diagnosis is made based on criteria of loss of weight to less than 85% of expected weight, fear of becoming overweight, a distorted body image, and amenorrhea in postmenarcheal women for greater than 3 months. Multiple endocrine abnormalities are present in anorexia nervosa, perhaps due to adaptive mechanisms designed to facilitate survival during prolonged starvation. The overwhelming majority of the patients are young, Caucasian women.

Amenorrhea is one of the diagnostic criteria, and is central in its etiology. The circadian pattern of gonadotropin secretion is identical to that of the prepubertal stage, with low levels of luteinizing hormone (LH), follicle-stimulating hormone (FSH), and circulating estradiol (Thomas and Rebar, 1990). There is a diminished response to GnRH stimulation; however, when low-dose pulses are administered, gonadotropin release normalizes (Stoving et al., 1999b). With weight gain, nocturnal secretory pulses resume, eventually progressing to an adult circadian pattern of pulsatile LH secretion. Menstrual periods resume in most patients with weight gain.

The HPA axis is also disturbed in patients with anorexia nervosa. Though plasma cortisol levels are elevated, the circadian rhythm is preserved and the response to dexamethasone suppression is partial, mimicking a pseudo-Cushing's state (Munoz and Argente, 2002). Adrenocorticotrophic hormone (ACTH) levels are normal, but the ACTH response to CRH administration is blunted, suggesting an elevated level of CRH secretion. Abnormalities in the HPA axis return to normal with as little as 10% weight gain (Munoz and Argente, 2002).

In patients with anorexia nervosa, spontaneous GH levels are elevated and IGF-I levels are low as compared to healthy subjects (Stoving et al., 1999a). Responses to various stimuli have been heterogeneous. Most responses to growth-hormone-releasing hormone (GHRH) administration have been elevated, while responses to insulin, clonidine, and hexarelin are low (Munoz and Argente, 2002). Detailed analysis of GH secretory dynamics demonstrates an increase in pulse amplitude, frequency of peaks, and duration of peaks (Misra et al., 2003). Possible etiologies of the altered GH–IGF-I axis include decreased IGF-I production secondary to malnutrition, or GH resistance (Stoving et al., 1999a). Both GH-binding proteins and IGF-I-binding protein-3 are low in anorexia nervosa, most likely reflecting the poor nutritional state (Munoz and Argente, 2002). Increases in weight normalize IGF-I levels, GH secretory dynamics, and binding protein abnormalities (Munoz and Argente, 2002).

In addition, thyroid secretion is abnormal in anorexia nervosa with most patients presenting with low tri-iodothyronine (T3) levels, normal or low levels of thyroxine (T4), and normal TSH levels, similar to findings in nonthyroidal illness or euthyroid sick syndrome. As with nonthyroidal illness there is a preferential deiodination of T4 to reverse-T3 (rT3), causing depletion of circulating T3. Thyroid hormonal abnormalities also return to normal with weight gain.

Eating behavior is regulated by the hypothalamus through the integration of peripheral signals of hunger and satiety that modulate energy expenditure and autonomic function. Not surprisingly, neuropeptide levels involved in peripheral signaling of appetite are altered in anorexia nervosa. Peptide YY (an intestinally derived anorexic peptide) and ghrelin (a gut peptide with GH secretory properties) have been found to be elevated (Misra et al., 2005, 2006). In contrast, leptin, an adipocytokine expressed by adipose tissue, has been found to be low in patients with anorexia nervosa (Misra et al., 2005). These patients may have additional derangement of hypothalamic function, including hyperprolactinemia with galactorrhea, poikilothermia, and partial DI (Mecklenberg et al., 1974). Although no etiology has been found for anorexia nervosa, the alterations in hypothalamic regulation of appetite and neuroendocrine disturbances appear to be related primarily to altered body composition, body weight, and malnutrition.

19.4.3.5 Diencephalic glycosuria

Injury to the hypothalamus may cause elevations in cortisol, growth hormone, and catecholamines, all of which have insulin counterregulatory effects. Lesions specific to the tuberoinfundibular region can have transient hyperglycemia and glycosuria (Clark, 1938; Boshes, 1969).

19.4.4 Sleep–Wake Cycle Circadian Abnormalities

In 1918, Baron Constatin von Economo first described a viral encephalitis characterized by excessive sleep that specifically affected areas of the brain that would eventually be confirmed to regulate sleep and wakefulness (Saper et al., 2005). Patients slept excessively for many weeks, only arising to eat and drink. These patients were found to have lesions at the junction of the midbrain and diencephalon, and it was proposed that an arousal mechanism in this area of the brain was responsible for maintaining wakefulness. Structural lesions of the hypothalamus commonly cause a disturbance in the sleep–wake cycle, usually resulting in somnolence. In Bauer's series of patients with proven hypothalamic disease 30% suffered from somnolence, while in 10% it was the presenting feature (Bauer, 1954, 1959). Lesions affecting the posterior hypothalamus can result in symptoms ranging from drowsiness to coma, depending on the size and acuity of the lesion (Plum and VanUitert, 1978). Lesions causing insomnia are more rare, but are generally located more anteriorly in the preoptic nuclei (Clark, 1938). Lesions involving the suprachiasmatic nuclei cause more alterations in circadian rhythm, with daytime somnolence and nighttime wakefulness (Cohen and Albers, 1991).

Disruptions of the circadian system have been described without structural lesions of the hypothalamus. These sleep disorders occur when the internal timing mechanism is altered or when there is misalignment between sleep and the usual 24-h environmental and social cues (Lu and Zee, 2006). Delayed initiation of sleep, involuntary patterns of sleep initiation and early waking, alternating insomnia and hypersomnolence, irregular sleep–wake patterns, shift-work disorders, and jet lag have all been associated with temporary or chronic abnormalities in the circadian sleep–wake rhythm (Lu and Zee, 2006).

Narcolepsy is defined as excessive daytime sleepiness with cataplexy with or without abnormal REM sleep phenomena such as sleep paralysis and hypnagogic hallucinations (Nishino, 2007). While a hypothalamic cause has been theorized, the pathophysiology has only recently been described. Using animal models, genes encoding for the hypocretin/orexin ligand and its receptor were identified simultaneously by two independent research groups (de Lecea et al., 1998; Sakurai et al., 1998). This neuropeptide is also involved in other neuroendocrine functions, but has been identified as a key modulator of sleep–wake-state stability (Saper et al., 2005). The underlying etiology behind the loss of function of hypocretin/orexin has yet to be fully elucidated. Most patients with narcolepsy lack genetic mutations in pertinent genes, as would be suspected since most cases are sporadic and not familial (Nishino, 2007). Low levels of hypocretin-1 have been found in most patients with narcolepsy however, and consideration has been made for an autoimmune process, with strong associations with HLA-DR2 in these patients with narcolepsy (Nishino, 2007).

19.4.5 Behavioral Abnormalities

Spontaneous rage reactions characterized by emotional lability, agitation, and aggressive and destructive behavior have been reported with lesions involving the ventromedial nuclei (Plum and Van Uitert, 1978; Carmel, 1980; Haugh and Markesbery, 1983). The episodes are usually accompanied by activation of the autonomic nervous system with symptoms of tachycardia, blood pressure elevation, diaphoresis, and pupillary dilatation. Similar sham rage reactions are also seen in patients with lesions of the medial temporal lobes or orbitofrontal cortex (Plum and VanUitert, 1978).

Apathy, somnolence, hypoactivity, and general indifference have been described in patients with medial posterior hypothalamic lesions or destruction of the mamillary bodies. Akinetic mutism and unresponsiveness to vocal and auditory stimuli have also been noted in patients with these types of lesions (Boshes, 1969; Plum and VanUitert, 1978).

Patients with Wernicke's encephalopathy and its related syndrome, Korsakoff's psychosis, may have characteristic findings of confabulation and short-term memory deficits, both of which are associated with lesions of the mamillary bodies, periaqueductal gray matter, and thalamus.

Sexual dysfunction is commonly seen in hypothalamic disorders. In Bauer's series, hypogonadism was present in 32% of patients, with the majority of lesions located near the floor of the third ventricle and involving the anterior hypothalamus, ventromedial nuclei, and tuberoinfundibular regions (Bauer, 1954, 1959). Symptoms include amenorrhea in women and decreased libido and impotence in men. Lesions of the hypothalamus may cause hypogonadism directly through interruption of GnRH secretion or indirectly by other means (e.g., hyperprolactinemia).

Hypersexual paraphilias may accompany lesions of the caudal hypothalamus, as well as the limbic system and the medial temporal lobe (Fenzi et al., 1993; Frohman et al., 2002). The Kleine–Levin syndrome is characterized by recurrent episodes of somnolence, cognitive disturbances, a sensation of detachment from reality, eating disorders (megaphagia), depression, irritability, and hypersexuality (Arnulf et al., 2005). Adolescent boys are most often affected. The spectrum of symptoms points toward a hypothalamic etiology, but aside from rare reports of hypothalamic abnormalities on pathological examination, no unifying cause has been found (Arnulf et al., 2005).

19.4.6 Diencephalic Epilepsy

Any seizure activity arising from the hypothalamus is broadly defined as diencephalic epilepsy and can include periodic hypothermia, periodic hyperthermia, and other autonomic activity. Gelastic seizures are epileptic events characterized by bouts of inappropriate laughter. They are seen frequently in children with hypothalamic hamartomas of the tuber cinereum and other lesions near the floor of the third ventricle and extending into the mamillary region (Breningstall, 1985). Usually laughter-like vocalization is combined with facial contraction in the form of a smile. A subset of these seizures, called dacrystic seizures, has a crying quality and accompanying facial features similar to a grimace. Both types of seizures may occur in the same patient and even in the same seizure event. Generally, children do not lose consciousness although the seizures may progress to partial complex or generalized seizures (Takeuchi and Handa, 1985; Harvey and Freeman, 2007). Autonomic features such as flushing, tachycardia, and altered respiration may be present (Cerullo et al., 1998). The diagnosis is made by establishing recurrent laughter in the absence of an appropriate context, other associated signs of seizure activity, and ictal or inter-ictal electroencephalogram (EEG) abnormalities (Gascon and Lombroso, 1971; Breningstall, 1985; Sharma, 1987).

19.5 Disordered Control of Anterior Pituitary Function

19.5.1 Hyperfunction Syndromes

19.5.1.1 Precocious puberty

Pubertal development that occurs in children below the age of 8 years in girls and 9 years in boys has been considered early or precocious pubertal development. There is a significant degree of racial variability noted in the age of onset for normal pubertal development, as well as a general trend noted over the past few decades toward children going through puberty at a younger age (Sun et al., 2002). Classification of precocious puberty is generally divided into central and peripheral forms. Central forms are characterized by both breast and pubic hair maturation in females and both testicular enlargement and pubic hair development in males. These children have early activation of the normal pubertal process that is mediated by the hypothalamic–pituitary–gonadal axis. Peripheral forms of precocity are due to excess gonadal hormones produced independent of gonadotropins or GnRH. When sexual characteristics are consistent with the child's gender, they are termed isosexual as opposed to virilization in girls or feminization in boys, which is known as contrasexual. Causes of central precocity are divided into two general categories, idiopathic and organic. Organic causes are related to CNS lesions, either congenital or acquired. There is a female predominance in central precocious puberty and it is much more common for girls to have the idiopathic form (Chemaitilly et al., 2001). While the idiopathic form is far more common, several different CNS lesions have been reported to cause precocious puberty. In Bauer's series of known hypothalamic lesions, precocious puberty was noted in 40% of cases, most often due to neoplasms (60%). Tumors were generally located in the posterior hypothalamus, at or near the mamillary bodies, or hamartomas in the tuber cinereum (20%) (Bauer, 1954, 1959). CNS irradiation has also been reported to cause early puberty especially at lower doses (Ogilvy-Stuart et al., 1994). **Table 2** lists the causes of precocious puberty.

The normal onset of puberty is the result of a change in stimulatory actions of glutamatergic neurotransmitters and suppressive effects of gamma-aminobutyric acid neurotransmission (Muir, 2006). This change results in increased pulsatile secretion of GnRH from the arcuate nucleus and the subsequent secretion of LH and FSH from the pituitary. LH and FSH then cause production of gonadal steroids and the development of secondary sexual characteristics. Patients with idiopathic precocious puberty go through the same sequence but have an earlier activation of the hormonal cascade. Patients with inflammatory conditions or increased intracranial pressure may also have early activation due to mass effect or pressure on the hypothalamus. Tumors such as germinomas may secrete human chorionic gonadotropin (hCG), which directly stimulates the gonads to secrete sex hormones. Early production of sex steroids due to congenital adrenal hyperplasia or McCune–Albright polyostotic fibrous dysplasia syndrome may prime the hypothalamus for subsequent early pubertal activation. Hypothalamic hamartomas and gliomas have the potential to secrete GnRH and may directly cause pubertal changes, or may indirectly cause early activation of the hypothalamus through mass effect (Judge et al., 1977; Hochman et al., 1981). Hypothyroidism can cause galactorrhea due to TRH-induced hyperprolactinemia as well as precocious puberty (Van Wyk–Grumbach syndrome) which is amenable to treatment with thyroid hormone replacement (Chattopadhyay et al., 2003).

Table 2 Causes of central precocious puberty

Idiopathic	*Neoplasms*
Congenital abnormalities	Optic nerve glioma
Hypothalamic hamartoma	Hypothalamic glioma
Arachnoid cyst	Neurofibroma
Myelomeningocele	Astrocytoma
Aqueductal stenosis with hydrocephalus	Ependymoma
Tuberous sclerosis	Infundibuloma
Congenital optic nerve hypoplasia	Pinealoma
Congenital adrenal hyperplasia	Neuroblastoma
McCune–Albright syndrome	Craniopharyngioma
Angioma cavernosum	Germinoma
	Medulloblastoma
Inflammatory conditions	Hemispheric tumor
Tuberculosis	Cerebral stem tumor
Sarcoidosis	Meningioma
Meningoencephalitis	
Subdural hematoma	
Primary hypothyroidism	

Gross (1940), Weinberger and Grant (1941), Banna (1976), Balagura et al. (1979), Margalith et al. (1984, 1985), Laue et al. (1985), Gillett and Symon (1987), Shankar and Pescovitz (1995), Chemaitilly et al. (2001).
Adapted from Braunstein GD (2002) The hypothalamus. In: Melmed, S (ed.) *The Pituitary*, pp. 317–348. Cambridge, MA: Blackwell Scientific, with permission from Wiley-Blackwell Publishing Ltd.

19.5.1.2 Acromegaly

Acromegaly is most often due to excessive GH production from a pituitary adenoma. Rarely, acromegaly is caused by excessive production of GHRH, or even more rarely by ectopic secretion of GH (Melmed et al., 1985; Losa and von Werder, 1997; Beuschlein et al., 2000). Eutopic hypersecretion of GHRH is found in tumors of the hypothalamus arising from cells having the physiologic capability to secrete GHRH. Hypothalamic hamartomas, gangliocytomas, and ganglioneuromas have been associated with pituitary adenomas or hyperplasia in patients with acromegaly (Asa et al., 1980, 1984; Saeger et al., 1994; Losa and von Werder, 1997). Ectopic production of GHRH originates from tumors outside the hypothalamus. Carcinoid tumors, pancreatic islet cell tumors, adrenal adenoma, pheochromocytoma, and lung carcinoma have been found to secrete GHRH resulting in excessive GH secretion from the pituitary, manifesting as acromegaly (Losa and von Werder, 1997). In contrast to normal subjects, when GHRH is administered as a continuous infusion, GH reserves are not depleted in subjects manifesting acromegalic symptoms (Losa et al., 1984). Furthermore, the sustained release of GH and GHRH is resistant to the negative feedback of elevated IGF-I levels seen in patients with acromegaly (Berelowitz et al., 1981). Most commonly, the elevated levels of GHRH cause pituitary hyperplasia with retention of normal sinusoidal architecture (Losa and von Werder, 1997). Prolonged exposure to elevated levels of GHRH has given rise to GH-secreting adenomas, supporting the hypothesis that somatotroph hyperplasia may be an intermediate stage in pituitary tumorigenesis (Frohman et al., 1980). However, the monoclonal nature of pituitary adenomas argues against this hypothesis (Alexander et al., 1990; Herman et al., 1990; Jacoby et al., 1990).

19.5.1.3 Cushing's disease

Pituitary ACTH-dependent hypersecretion of cortisol is referred to as Cushing's disease. It is commonly associated with adrenal hyperplasia and a pituitary adenoma, although as many as 40% of patients with central ACTH secretion will not have evidence of a pituitary adenoma on magnetic resonance imaging (MRI) (Newell-Price et al., 2006). Stepwise diagnostic testing is required to establish central ACTH-dependent disease from other forms of Cushing's syndrome. In addition, further testing is required to differentiate this entity from excessive ectopic CRH production. Reported sources of ectopic CRH production include pheochromocytoma, malignant gastrinoma, ganglioneuroblastoma, medullary thyroid carcinoma, bronchial and thymic carcinoid, small cell lung carcinoma, and prostate carcinoma (Morris et al., 2006). A hypothalamic etiology for Cushing's disease had been considered with the hypersecretion of ACTH. An altered set point for feedback inhibition at the hypothalamus causing hypersecretion of ACTH is thought to be one hypothetical etiology for Cushing's disease (Biller, 1994). Patients with Cushing's disease have a diminished responsiveness to glucocorticoid suppression, although they are more responsive than patients with ectopic production of ACTH (Newell-Price et al., 1998). They also retain responsiveness to exogenous CRH administration (Nieman et al., 1993). In opposition to this hypothesis is the finding of monoclonal cells in ACTH-secreting pituitary adenomas, and the scarcity of corticotroph hyperplasia found in patients with Cushing's disease (Biller et al., 1992; Biller, 1994; Faglia and Spada, 1995).

19.5.1.4 Hyperprolactinemia

Lactotrophs are the sole source for prolactin secretion and are under tonic inhibition from dopamine that is synthesized and secreted by the hypothalamus. Physiologic causes of hyperprolactinemia include pregnancy, nipple stimulation, and stress. Pathologic causes include pituitary lactotroph adenomas, various drugs, and any hypothalamic or pituitary diseases causing interruption of normal dopaminergic suppression. Amenorrhea and galactorrhea in women, and impotence and diminished libido in men are variably present. Hypogonadotropic hypogonadism is found in these patients most often due to interruption of the pulsatile secretion of GnRH and gonadotropins, and also due to direct effects on the ovary which inhibit folliculogenesis and aromatase activity (Horseman and Greerson, 2006). Inhibition of dopaminergic suppression generally causes mild elevation of prolactin (less than 70 ng ml^{-1}), although occasionally levels rise to up to 150 ng ml^{-1} (Kapcala et al., 1980). Levels higher than 200 ng ml^{-1} are almost always due to lactotroph adenomas. A large series of craniopharyngiomas reported hyperprolactinemia in 55% of adults (Karavitaki et al., 2005). In a series of patients with various hypothalamic tumors, hyperprolactinemia was found in 36% of craniopharyngiomas, 79% of suprasellar germinomas, and 14% of patients harboring pineal germinoma (Jennings et al. 1985; Imura et al., 1987). A recent review of histologically proven nonfunctioning adenomas revealed hyperprolactinemia in 39% of patients with prolactin levels <100 ng ml^{-1} in 99% of patients (Karavitaki et al., 2006b). As with other pituitary adenomas, a hypothalamic etiology has been considered as the underlying causative factor for hyperplasia and neoplasia when dopaminergic deficiency is present. The monoclonal origin of pituitary adenomas seems to indicate that pituitary tumors arise from a single mutated cell (Alexander et al., 1990; Herman et al., 1990; Jacoby et al., 1990). Data regarding the monoclonal origin of prolactinomas are more scarce than those of other tumors due to the effectiveness of medical therapy in treating these tumors (Spada et al., 2005).

19.5.2 Hypofunction Syndromes

19.5.2.1 Acquired hypogonadotropic hypogonadism

Several etiologies can cause hypogonadotropic hypogonadism, including hypothalamic tumors, infiltrative diseases, infections, hemorrhage, trauma, surgery or radiation to the sellar region, hyperprolactinemia, critical illness, chronic systemic illness, diabetes mellitus, obesity, and drugs including glucocorticoids, opiates, and GnRH analogs. Hypogonadotropic hypogonadism is commonly seen in patients with hypothalamic tumors. In Bauer's series, almost one-third of patients had hypogonadism, with 84% of the lesions located in the floor of the third ventricle involving the tuberoinfundibular and more anterior regions of the hypothalamus (Bauer, 1954, 1959).

Depending on the age of onset, the manifestations of hypogonadotropic hypogonadism differ. Prepubertal onset results in delayed puberty or a lack of pubertal onset entirely. Both boys and girls lack secondary sexual characteristics, and have diminished pubic and axillary hair development. A concomitant deficiency in androgens causes an absence of axillary hair growth. Males do not experience voice deepening, muscle development, and facial or chest hair growth. The testicles and penis remain small. Females will not experience menstruation, breast development, uterine enlargement, or vaginal cornification and mucus development. If GH is unaffected, individuals may continue to have linear growth due to the lack of closure of the epiphyseal plates. These children develop a eunuchoidal habitus in which the upper (crown to pubis) to lower segment (pubis to floor) ratio is less than 1, and the arm span exceeds total height by 5 cm or more. In adult males, hypogonadism may be detected by a lack of libido or erectile dysfunction. They also develop testicular atrophy, low semen volume and decreased sperm count, loss of muscle mass, and gradual loss of pubic and body hair. Adult premenopausal females experience amenorrhea and may have symptoms of menopause. Postmenopausal women do not experience symptoms directly related to the loss of gonadotropins. The characteristic laboratory findings are reduced levels of gonadal steroids with low or inappropriately normal LH and FSH. The response to a GnRH stimulation test may be low or normal but usually increases if the patient receives pulses of priming doses of GnRH every 90 min.

19.5.2.2 Congenital GnRH deficiency (idiopathic hypogonadotropic hypogonadism)

This form of hypogonadism is characterized by a functional absence of GnRH secretion from the hypothalamus or defect in the GnRH receptor, leading to deficiencies in LH and FSH. Congenital deficiency of GnRH is predominantly found in males with a 5:1 male to female ratio. When low LH, FSH, and sex steroids levels are accompanied by negative radiological studies of the hypothalamus and pituitary area, the diagnosis of idiopathic hypogonadotropic hypogonadism is made. In neonates, clinical presentation is marked by micropenis with or without cryptorchidism in males, due to a lack of normal neonatal testosterone production. Female neonates have no obvious presenting signs. During childhood, anosmia or midline cranial defects may reveal a diagnosis; however, low or undetectable gonadal hormones normally characterize this phase of life. At puberty, both males and females fail to undergo sexual maturation and do not manifest secondary sexual characteristics. However, linear growth and adrenal androgen activity is normal in these patients.

Multiple genetic defects are associated with congenital hypogonadotropic hypogonadism. The majority of cases of idiopathic hypogonadotropic hypogonadism are due to mutations in three genes: anosmin-1 (KAL1), fibroblast growth factor-1 receptor (FGFR1), and gonadotropin-releasing-hormone receptor (GnRHR) (Layman, 2007). The X-linked recessive form of Kallman's syndrome is due to a mutation in the KAL1 gene on the Xp22.3 region of the X-chromosome. The defect has been identified as altered expression of anosmin, a neural cell adhesion molecule, which results in impaired migration of the GnRH neurons to the ventral hypothalamus. Ten to fifteen percent of sporadic cases of anosmic hypogonadal male patients have mutations in the KAL1 gene, whereas 30–60% of familial cases are due to defects in KAL1 (Layman, 2007). Kallman's syndrome is characterized by hypogonadism accompanied by one or more nongonadal congenital anomalies, including anosmia, color blindness, midline facial abnormalities, urogenital tract defects, and hearing loss. Mutations of the FGFR1, can cause both anosmic and normosmic forms of hypogonadism and occur in approximately 7–10% of patients with autosomal dominant congenital hypogonadotropic hypogonadism (Layman, 2007). Interactions have been proposed between the products of the KAL1 and FGFR1 genes. Increased levels of anosmin in females may explain the male predominance of hypogonadism in the autosomal FGFR1 mutation (Gonzalez-Martinez et al., 2004). Loss of function mutations in the GnRH receptor comprise approximately 5% of patients with idiopathic hypogonadotropic hypogonadism (Layman, 2007). Both autosomal recessive forms and sporadic cases have been described in these normosmic patients. Mutations in genes for leptin, the leptin receptor, G-protein-coupled receptor-54 (GPR54), and dosage-sensitive sex reversal-adrenal hypoplasia congenita critical region on the X-chromosome, gene 1 (DAX1) have all been described in association with idiopathic hypogonadotropic hypogonadism as well (Layman, 2007).

19.5.2.3 Growth hormone deficiency

GHD can result from both acquired and congenital causes. Acquired etiologies include perinatal and postnatal traumatic brain injury, infection, CNS tumors of the hypothalamic–pituitary axis, cranial irradiation, chemotherapy, pituitary infarction, and psychosocial deprivation. Congenital causes include many genetic defects involved in pituicyte differentiation such as homeobox, embryonic stem cell expressed 1 (HESX1), LIM homeobox protein 3 (LHX3), LIM homeobox protein 4 (LHX4), prophet of PIT-1 (PROP-1), and POU domain, class 1, transcription factor 1 (PIT-1) (POU1F1) (Dattani and Hindmarsh, 2006). Structural defects of the brain such as agenesis of the corpus collosum, septo-optic dysplasia, holoprosencephaly, encephalocele, and hydrocephalus can cause congenital GHD (Dattani and Preece, 2004). The most common etiology for congenital GHD is idiopathic. Patients with congenital GHD have low birth weight, a higher rate of breech presentation, and perinatal asphyxia. Perinatal hypoglycemia may be severe, especially when associated with ACTH and TSH deficiency. Postnatal growth is abnormal and becomes apparent within the first few months of life (Pena-Almazan et al., 2001). Early phenotypic features include immature facies with a prominent forehead and a characteristic depression of the bridge of the nose. Cleft lip, cleft palate, and a single central incisor may also accompany GHD. Both acquired and congenital forms demonstrate short stature and low growth velocity for age. Untreated, these children achieve only 70% of their predicted height, with deficits averaging 38 cm in males and 33 cm in females (Dattani and Preece, 2004).

The diagnosis of GHD is reliant on testing, but the assessment also includes growth patterns, medical history, and characteristic physical findings to both make the diagnosis and determine if biochemical testing is warranted. Single serum GH measurement is not useful, but in children, measurement of IGF-I and insulin-like growth factor binding protein-3 (IGFBP-3) serves as useful screening tests (Rosenfeld et al., 1995). In children, there are many options for stimulation testing available, each of them having their own limitations. General consensus is that two stimulation tests are required for diagnosis (Raiti et al., 1967). The diagnosis of GHD in adults is less reliant on auxological findings, as linear growth is no longer an issue. GH secretion is usually the first pituitary hormone affected by hypothalamic–pituitary insults. Testing is completely reliant on stimulation testing, of which the insulin tolerance test (ITT) and GHRH-arginine stimulation test appear to be the most sensitive and specific (Biller et al., 2002). IGF-I measurement is generally not helpful as a screening tool, as up to 50% of patients with severe GHD have normal IGF-I levels (Biller et al., 2002). However, a low IGF-I in the setting of multiple pituitary deficiencies is a strong indicator of GHD (Hartman et al., 2002). Treatment of children prior to closure of the epiphyses results in improvement in linear growth, while treatment of both children and adults improves metabolic and body composition parameters (Molitch et al., 2006).

19.5.2.4 Hypothalamic hypoadrenalism

The most common cause of tertiary hypoadrenalism is the withdrawal of glucocorticoids after a prolonged course or high-dose treatment. Large cumulative doses of glucocorticoids cause suppression of CRH production and secretion, leading to decreased ACTH secretion, decreased cortisol secretion, and eventual atrophy of the zona reticularis and zona fasciculata of the adrenal gland as well as the corticotroph cells in the pituitary. A similar pathophysiology results from cure of Cushing's syndrome by removal of the source of autonomous production of ACTH or cortisol. Radiation, trauma, and lesions of the hypothalamus, including tumors and infiltrative diseases, can interfere with the normal secretion of CRH and result in adrenal insufficiency (Schmiegelow et al., 2003). Several rare causes for isolated ACTH deficiency have been demonstrated, including autoimmune disorders, trauma, and genetic defects (Andrioli et al., 2006).

Tertiary adrenal insufficiency can be fatal if not recognized. Hypoglycemia may be severe especially when accompanied by GH deficiency. Children may also manifest hypotension, although electrolyte abnormalities are usually absent due to intact aldosterone production. Patients generally do well in unstressed situations, but may become symptomatic during times of physiologic stress. Symptoms of nausea and vomiting, extreme fatigue, abdominal pain, hypotension, hypoglycemia, and fever may be seen in acute adrenal crisis. Hyperpigmentation, encountered in primary adrenal insufficiency, is not seen in central causes of adrenal insufficiency.

19.5.2.5 Hypothalamic hypothyroidism

Tertiary hypothyroidism is a common consequence of hypothalamic tumors, including craniopharyngiomas, suprasellar germinomas, and septo-optic dysplasia. Infiltrative lesions, trauma, infection, and radiation may also cause central hypothyroidism (Fahlbusch et al., 1979; Izenberg et al., 1984; Margalith et al., 1984, 1985; Imura et al., 1987; Buchfelder et al., 1989; Seminara et al., 1998). By definition, these patients have low circulating thyroxine. Serum TSH concentrations may be low, normal, or high. Normal or high levels seen in hypothalamic hypothyroidism reflect decreased biologic activity due to diminished glycosylation, and retained immunoactivity (Beck-Peccoz et al., 1985; Persani et al., 2000). TRH administration does not help distinguish hypothalamic hypothyroidism from pituitary hypothyroidism, as both may have delayed responses or blunted responses (Mehta et al., 2003). A TSH peak 60 min or more after TRH administration, compared to 20–30 min in normal subjects, is characteristic of hypothalamic hypothyroidism (Snyder et al., 1974). Clinical symptoms are similar to those of primary hypothyroidism, but may be less severe and include thyroid gland atrophy, lethargy, cold intolerance, dry skin, constipation, hypothermia, bradycardia, and weight gain.

19.6 Specific Hypothalamic Disorders

19.6.1 Prader–Willi Syndrome

Prader–Willi syndrome is a genetic disease first described in 1956 by Prader, Labhart, and Willi. Major features include neonatal hypotonia, feeding problems with poor suckling, poor weight gain in infancy which rapidly changes to rapid weight gain leading to central obesity, hyperphagia, developmental delay, short stature, hypogonadism, and characteristic facial features (Prader et al., 1956; Holm et al., 1993; Goldstone, 2004). The frequency of published diagnostic criteria in patients with Prader–Willi is found in **Table 3**. This complex syndrome arises from the lack of expression of paternally inherited genes on chromosome 15q11-q13 (Rimoin and Schimke, 1971; Sapienza and Hall, 1995; Cassidy, 1997; Cassidy and Schwartz, 1998). Lack of expression is due to paternal deletion of 15q11-q13 in 75% of cases, maternal unipaternal disomy in 22%, imprinting errors in 3%, and paternal chromosomal translocation in less than 1% of cases (Goldstone, 2004). The prevalence of the syndrome is estimated to be 1 in 10 000–16 000 live-born infants (Bray et al., 1983).

Clinical manifestations begin before birth and include decreased fetal movement, increased incidence of breach presentation, prematurity, and low birth weight. Fetal hypotonia and poor feeding may be present, often necessitating enteral feeding. Affected children demonstrate multiple abnormalities, including narrowing of the bitemporal diameter of the cranium, strabismus, almond-shaped palpebral fissures, micrognathism, and small hands and feet (Rimoin and Schimke, 1971; Bray et al., 1983; Sapienza and Hall,

Table 3 Prevalence of diagnostic features of Prader–Willi syndrome in 163 patients

Feature	*Percentage*
Major criteria[a]	
Neonatal hypotonia	93
Feeding problems in infancy	86
Excessive weight gain	82
Facial features	88[b]
Hypogonadism	71
Developmental delay	98
Hyperphagia	84[b]
Minor criteria[a]	
Decreased fetal activity	70
Behavioral problems	78
Sleep disturbance	76[b]
Short stature	76
Hypopigmentation	73[b]
Small hands and/or feet	89
Narrow hands/straight ulnar borders	82[b]
Eye abnormalities	66
Thick viscous saliva	89[b]
Articulation defects	80[b]
Skin picking	83[b]

[a]Diagnosis established with: 5 points with ≥4 points from major criteria strongly suggest PWS in children ≤3 years old; 8 points with ≥5 points from major criteria are indicative in older individuals; and major criteria: 1 point each; minor criteria: 0.5 points each.
[b]Data from 90 patients. Bray et al. (1983), Ho m et al. (1993), and Gunay-Aygun et al., (2001).

1995; Cassidy, 1997; Burman et al., 2001). Children usually become overweight by the age of 4 due to compulsive eating habits and insatiable appetite. Obesity is progressive with age, often reaching levels of morbid obesity (Hall and Smith, 1972).

Short stature appears in about 90% of affected individuals and can be attributed to GH deficiency. Though decreased GH secretion is usually seen as a consequence of simple obesity, the obesity associated with Prader–Willi syndrome is due to dysfunction of either the hypothalamus or pituitary. In contrast to children with simple obesity, serum IGF-I levels are reduced and GH responses to GH secretagogs remain blunted despite administration of a cholinesterase inhibitor (Burman et al., 2001).

Hypogonadism is present in a majority of individuals as early as the prenatal period as evidenced by micropenis and cryptorchidism in male infants and labial hypoplasia in female infants (Bray et al., 1983; Sapienza and Hall, 1995). Puberty is delayed or nonexistent, although some reports have been made of precocious puberty (Kauli et al., 1978; Bray et al., 1983; Vanelli et al., 1984). Many patients experience premature adrenarche with development of axillary and pubic hair, but no progression beyond this stage (Garty et al., 1982). When present, menses are usually irregular (39%) and probably related to aromatization of androgens to estrogen by fat tissue in obese patients (Bray et al., 1983; Greenswag, 1987).

The hypogonadism associated with Prader–Willi syndrome appears to hypothalamic in origin. Most individuals demonstrate a poor response to GnRH that eventually improves with prolonged clomiphene or GnRH administration (Bray et al., 1983). Though most hypogonadism appears to be central, rare cases of pregnancy and primary testicular failure have been reported (Burman et al., 2001).

Despite multiple hypothalamic abnormalities, the thyroid and adrenal axes appear intact. Responses to TRH and provocative stimuli to ACTH and cortisol are normal as are prolactin concentrations (Bray et al., 1983).

Treatment is directed at behavior modification as well as correction of endocrine deficiencies (Greenswag and Alexander, 1995). GH replacement has resulted in improvements in linear growth as well as in body composition with reduction in fat mass and an increase in muscle mass. Sex hormone replacement therapy is a controversial topic and prospective data are lacking. Theoretical benefits include improvement of bone mineral density, prevention of osteoporosis, and development of secondary sexual characteristics (Burman et al., 2001). Repair of cryptorchidism improves detection of testicular malignancies, but is inconsistently performed (Greenswag, 1987). Treatment of comorbidities associated with obesity such as diabetes, hypertension, and hyperlipidemia should be addressed using standard methods.

19.6.2 Septo-Optic Dysplasia

Septo-optic dysplasia (De Morsier syndrome) is a heterogeneous condition defined by a combination of optic nerve hypoplasia, midline neurological malformations (such as agenesis of the corpus collosum and absence of the septum pellucidum), and pituitary–hypothalamic dysfunction (Polizzi et al., 2006). Several other dysmorphic features may be found in patients with the disease, such as midline facial abnormalities (cleft palate, hypertelorism, depressed nasal bridge, and agenesis of the olfactory nerves), musculoskeletal system (syndactyly and joint contractures), and genitalia (micropenis). This rare disorder has an estimated incidence of 1 in 10 000 live births (Patel et al., 2006). Most cases are sporadic, but occasional familial cases have been reported. Mutations in the HESX1 homeobox gene have been implicated as a possible genetic cause (Kelberman and Dattani, 2007). The etiology is most likely multifactorial, with several etiologies including various genetic abnormalities and *in utero* injuries (Polizzi et al., 2006). Associations have also been made with fetal alcohol syndrome, preeclampsia, gestational diabetes, and maternal drug use (Margalith et al., 1984; Garcia et al., 2006). Affected children may have developmental delay, perinatal hypoglycemia, seizures, cerebral palsy, and variable visual acuity (Siatkowski et al., 1997; Garcia et al., 2006).

Endocrinopathies associated with septo-optic dysplasia range from isolated GH deficiency to panhypopituitarism with DI. The most common endocrine abnormality is GH deficiency causing short stature followed by ACTH deficiency, DI, and hypothyroidism (Arslanian et al., 1984; Izenberg et al., 1984; Margalith et al., 1985). Abnormal TRH testing is more commonly seen than clinical hypothyroidism. Hypogonadism is infrequently reported most likely due to the fact that the disease is diagnosed in childhood, when most patients are too young to detect hypogonadotropic hypogonadism. However, there have been reports of sparing of the gonadotropic axis despite multiple pituitary hormone deficiencies, as well as cases of precocious puberty (Huseman et al., 1978; Nanduri and Stanhope, 1999). This may be a reflection of the hypothalamic locus of the endocrinopathy, as well as the relatively late migration of GnRH neurons to the hypothalamus *in utero* after the development of a midline defect (Nanduri and Stanhope, 1999). The hypothalamic locus has been confirmed by autopsy results demonstrating the absence of the supraoptic and paraventricular nuclei, hypoplasia of the posterior pituitary, ependymal scars around the third ventricle, and a normal anterior pituitary (Roessmann et al., 1987). Features of septo-optic dysplasia are listed in **Table 4**.

19.6.3 Psychosocial Short Stature

Children with this rare form of short stature exhibit growth failure and appetite disturbance in association with emotional deprivation (Powell et al., 1967a,b). Onset of the growth failure is usually before the age of 2, with dietary changes manifesting often after 3 years of age (Skuse et al., 1996, Gilmour et al., 2001). Most are exposed to a stressful home life with a social history of divorced or separated parents resulting in a disturbed parent–child relationship

Table 4 Clinical, anatomic, and biochemical features of septo-optic-pituitary dysplasia

Feature	*Number abnormal/number studied*	*Percentage abnormal*
Anatomic		
Hypoplastic optic nerves	122/122	100
Absent septum pellucidum	91/154	59
Agenesis of corpus callosum	7/109	6
Clinical: nonendocrine		
Visual problems	63/87	72
Mental retardation	60/106	57
Cerebral palsy	29/106	27
Nystagmus	75/122	61
Seizures	19/106	18
Neonatal jaundice	10/109	9
Neonatal hypoglycemia	12/115	10
Clinical: endocrine		
Short stature	41/116	35
Decrease growth rate	41/122	34
Diabetes insipidus	34/185	18
Precocious puberty	7/106	7
Endocrine testing		
Human growth hormone deficiency	135/181	75
Adrenocorticotropic hormone deficiency	68/175	39
Thyroid-stimulating hormone deficiency	42/179	23
Abnormal thyrotropin-releasing hormone test	11/73	15
Gonadotropin deficiency	5/66	8
Hyperprolactinemia	5/79	6
Multiple hormone deficiencies	70/166	42

Arslanian et al. (1984), Izenberg et al. (1984), Margalith et al. (1984, 1985), Yukizane et al. (1990), and Birkebaek et al. (2003).
Adapted from Braunstein GD (2002) The hypothalamus. In: Melmed, S (ed.) *The Pituitary*, pp. 317–348. Cambridge, MA: Blackwell/Scientific, with permission from Wiley-Blackwell Publishing Ltd.

(Powell et al., 1967a,b). The syndrome is characterized by multiple features implicating hypothalamic dysfunction, including abnormal responses to GH stimulation, polydipsia, and polyphagia (Gohlke et al., 2002). Affected children have been noted to eat 2–3 times the amount of food compared to unaffected siblings. They often exhibit bizarre eating behavior such as food stealing, hoarding, pica, eating from garbage bins, drinking from toilet bowls, and wandering at night in search of food or water. Despite hyperphagia, they are not obese and may even have low weight, leading to a workup for malnutrition, which is not present. The lack of weight gain may be related to increased activity and higher caloric expenditure (Skuse et al., 1996). Unusual behavior and emotional retardation are frequent characteristics of this syndrome (Powell et al., 1967a,b).

Since the original description, subtypes have been delineated based on the type of appetite disturbance and GH status (Gohlke et al., 2002). The first subtype describes hyperphagic patients with reversible GH insufficiency. These patients exhibit a rapid rebound growth spurt when removed from the stressful environment and show minimal response to GH therapy. The second subtype consists of a heterogeneous group of nonhyperphagic patients who do not exhibit growth when removed from the stressful environment and has a variable response to GH treatment. With this subtype, other causes for short stature besides the adverse environment have been excluded in these children. An anorexic form is the last subtype with characteristics of failure to thrive, depression, and normal GH status but a response to GH treatment. These children seem to have a form of idiopathic hypopituitarism.

Removal of the child from the adverse environment improves the symptoms of polyuria, polyphagia, and bizarre food-related behavior. Growth resumes, and in many patients, an accelerated, rebound growth phase follows (Gohlke et al., 2002). Normalization of GH responses to stimuli is seen. However, symptoms return when the child is again placed in the toxic environment. The etiology of this syndrome is unclear but appears to be primarily related to an adverse physiologic reaction to extreme psychological stressors.

19.6.4 Pseudocyesis

Pseudocyesis, or false pregnancy, represents a woman's delusional belief that she is pregnant coupled with objective signs and symptoms of pregnancy. It is usually seen in women with an intense desire to conceive, or a fear of pregnancy, but may be related to neuroendocrine changes associated with major depressive disorder. A spectrum of signs and symptoms has been described, including amenorrhea, progressive abdominal distension, breast engorgement with or without galactorrhea, morning sickness, sensation of fetal movements, uterine enlargement, and labor pains (Yen et al., 1976; Zuber and Kelly, 1984; Small, 1986; Paulman and Sadat, 1990; Bray et al., 1991). Hyperprolactinemia is present in the majority of women and may be a contributing factor to the physical signs and symptoms, by causing a persistent corpus luteum in affected individuals. Elevated levels of LH may also be present as well as a contributing factor. A variety of therapies have been used, but most women respond to revealing that they are not pregnant with abdominal imaging, coupled with counseling and education, and treatment of any underlying depression. In some cases, signs and symptoms may persist for months to years (Paulman and Sadat, 1990).

19.7 Neoplasms Involving the Hypothalamus

19.7.1 Hypothalamic Hamartoma

Hypothalamic hamartomas are rare non-neoplastic heterotopic nodules of the tuber cinereum, commonly presenting with gelastic seizures, central precocious puberty, and developmental delay (Diebler and Ponsot, 1983; Comite et al., 1984; Breningstall, 1985; Takeuchi and Handa, 1985; Valdueza et al., 1994). Size ranges from 5 to 50 mm in diameter, but is usually between 10 and 30 mm in diameter. They primarily consist of neurons and glial cells. Myelinated and unmyelinated fibers have also been identified and may connect the hypothalamic nuclei. Hypothalamic hamartomas may be due to tissue displacement as the ventral aspect of the neuraxis approaches the anterior tip of the notochord in the fifth or sixth week of gestation (Beningfield et al., 1988).

Most patients present with symptoms before the age of 2. Precocious puberty is found in 63% and seizures in 61% of patients. The majority of seizures are gelastic in nature (90%) (Nguyen et al., 2003). These seizures consist of unprovoked laughter, do not impair consciousness, and may progress to more generalized forms of epilepsy. The gelastic seizures seen commonly in this disease have been shown to arise from intrinsic epileptogenic activity in the hamartoma itself. Origin of the seizure activity has been confirmed by EEG recordings made directly from the hamartomas (Munari et al., 1995). Developmental delay is common in patients with early onset seizures, and deterioration of cognitive abilities coincides with late onset of seizure activity (Nguyen et al., 2003). Varying degrees of cognitive deficiencies, behavioral abnormalities, and psychiatric disturbances have been observed in affected children. Other manifestations include a high rate of obesity and, more rarely, acromegaly and pituitary deficiencies have been reported (Arita et al., 2005).

Clinical signs of precocious puberty include deepening of the voice and enlarged testes and penis in males, breast development and menses in females, and muscular development, pubic hair, and adolescent personality in both sexes (Starceski et al., 1990; Arita et al., 2005). In contrast to idiopathic precocious puberty, pubertal development due to hypothalamic hamartomas tends to occur much earlier with initial manifestations occurring before age 2 in 82% of cases (Arita et al., 2005). The mechanism of precocious puberty in these patients is controversial. Some hamartomas may stimulate gonadotroph secretion through production of GnRH or transforming growth factor-α. Alternatively, abnormal neuronal networks originating from the hamartomas may stimulate release of GnRH from normal cells (Sharma, 1987). A third hypothesis suggests that mechanical compression of the hypothalamus interrupts the normal inhibition of gonadotropin-secreting cells (Arita et al., 2005).

Seizures are generally not controlled by medication and while surgical resection targeted at the hamartoma may be successful, the perioperative and postoperative mortality and morbidity have been high in the past (Regis et al., 2006). Gamma-knife surgery has shown improvement in over 60% of patients and carries far less morbidity than conventional surgical techniques.

Treatment of precocious puberty with surgery has yielded mixed results with recent attempts showing marked improvement. Medical treatment with continuous GnRH stimulation with long-acting GnRH agonists is effective, but is expensive and may be required chronically. In general, the mode of treatment depends on the major manifestations of the disease. Medical therapy is the treatment of choice for precocious puberty, while early surgical intervention with gamma knife is the preferred treatment for seizure activity (Hochman et al., 1981; Sato et al., 1985; Arita et al., 2005).

The Pallister–Hall syndrome consists of hypothalamic hamartoma, polydactyly, panhypopituitarism, imperforate anus, and other visceral anomalies (Clarren et al., 1980; Hall et al., 1980; Biesecker et al., 1996; Biesecker and Graham, 1996). The hypopituitarism is manifested by cryptorchidism and micropenis in affected males and adrenal insufficiency, which can be lethal in the perinatal period. Craniofacial abnormalities are felt to be secondary to disruption of midline development. Features include short nose, flat nasal bridge, low set and rotated ears, cleft palate and uvula, buccal frenula, bifid epiglottis, and cleft larynx. Limb abnormalities including polydactyly, syndactyly, shortened limbs, and nail dysplasia are also present. The inheritance is autosomal dominant and is associated with a frameshift mutation of the GLI-Kruppel family member 3 (GLI3) gene, located on chromosome 7p13 (Biesecker et al., 1996; Kang et al., 1997; Boudreau et al., 2005). While initial reports described six infants who died in the neonatal period, other cases have been described since with much less severe disease (Arita et al., 2005).

19.7.2 Germ Cell Tumor

Germ cell tumors are a heterogeneous group of neoplasms that occur in both children and adults. The tumors are broadly classified into germinomatous (seminomas; 65% of intracranial germ cell tumors) and nongerminomatous forms, which include embryonal carcinoma, yolk sac tumor, choriocarcinoma, teratoma, and mixed germ cell tumors (Jennings et al., 1985). The incidence of these tumors varies worldwide constituting 0.4–3.4% of primary CNS tumors in Western countries with a much higher incidence in Asian countries. Germ cell tumors frequently arise in the pineal and suprasellar region, with a twofold greater predilection for the pineal region; however, between 5% and 10% of patients present with tumors in both locations (Jennings et al., 1985). The tumors also occur in other areas of the brain and tend to arise in the midline, and may include the fourth ventricle, basal ganglia, and thalamus. Germ cell tumors arising from the basal ganglia or thalamus are more commonly the germinomatous form (Packer et al., 2000). Germinomatous and nongerminomatous forms also differ in their gender ratio, age of onset, and prognosis. Males are affected twice as frequently as females for all germ cell tumors, with even higher male-to-female ratio for the nongerminomatous forms (Russell and Rubensten, 1984). Suprasellar germinomas are more frequently seen in females, whereas males more frequently harbor tumors in the pineal region (Packer et al., 2000). Age of diagnosis for germ cell tumors is 10–12 years with nongerminomatous forms frequently diagnosed earlier in life (Packer et al., 2000).

Clinical features of suprasellar germ cell tumors vary depending on the location of the lesion. The clinical features are listed in **Table 5**. DI, visual field abnormalities, and anterior pituitary deficiencies are present in patients who have anatomic involvement of the suprasellar space including the median eminence, optic chiasm, and third ventricle. However, patients with suprasellar disease can be asymptomatic, or have isolated DI or hypopituitarism, which may delay the diagnosis. Tumors arising from the pineal region tend to present with more neurological signs due to obstruction of the third ventricle and aqueduct of Sylvius, including hydrocephalus, paralysis of upward gaze (Parinaud's sign) obtundation, pyramidal tract signs, ataxia, anterior pituitary deficiencies, and DI (Jennings et al., 1985). Precocious puberty is a rare manifestation of germ cell tumors, occurring in approximately 5%, and occurring more frequently in choriocarcinomas (Simson et al., 1968; Sklar et al., 1981; Jennings et al., 1985; Navarro et al., 1985; Imura et al., 1987; Buchfelder et al., 1989). Symptoms may develop due to pressure on the median eminence. In some children, hCG and LH levels may be elevated which may explain the development of precocious puberty. In males, Leydig cell stimulation in the testes produces androgens, resulting in the pubertal symptoms; however, sperm production is absent because FSH production is suppressed. Females rarely present with precocious puberty, most likely due to the requirement of FSH for ovarian estradiol production (Kubo et al., 1977).

Diagnosis often requires biopsy of the intracranial tumor, though MRI and CT are highly sensitive. The presence of calcification of the pineal region may aid in the diagnosis, since 70% of pineal germ cell tumors have calcifications (Smirniotopoulos et al., 1992). Tumor markers may be present in the serum and cerebrospinal fluid (CSF) though they are more sensitive and reliable in the latter. Alpha-fetoprotein has been found when endodermal sinus tumors are present. Human chorionic gonadotropin levels are elevated in choriocarcinoma and mixed germ cell tumors (Packer et al., 2000).

Table 5 Common clinical and biochemical features of patients with various neoplasms of the hypothalamus

Features	*Percentage of suprasellar germ cell tumors*	*Percentage of optic chiasm/ hypothalamic gliomas*	*Percentage of craniopharyngiomas*	*Percentage of suprasellar arachnoid cysts*
Hydrocephalus		32	19	86
Headache	47	20	69	24
Papilledema		76	19	20
Nausea/vomiting	16	15	30	8
Brainstem/thalamic compression				70[a]
Ataxia		16	5	28
Seizure		8		
Spasticity				55
Tremor				20
Head-bobbing				13[a]
Hypothalamic–pituitary dysfunction				21
Optic nerve involvement		54		36
Visual acuity/field defect	87	66	77	28
Growth failure	27	22	28	
Diabetes insipidus	85	15	23	
Associated neurofibromatosis		25[a]		
Diencephalic syndrome		21		
Precocious puberty	6	34	6	17
Multiple hormone deficiency	64	6	83	
Gonadotropin deficiency	88	14	82	
Adrenocorticotropic hormone deficiency	67		68	22
Abnormal growth hormone dynamics	98			
Growth hormone deficiency			88	7
Anterior pituitary dysfunction		35	91	
Hypothyroidism	81	6	39	
Hyperprolactinemia	76		40	

[a]Unique characteristic of the disease.
Simson et al. (1968), Roberson and Till (1974), Banna (1976), Korsgaard et al. (1976), Takeuchi et al. (1978), Fahlbusch et al. (1979), Kjellberg (1979), Sklar et al. (1981), Borit and Richardson (1982), Hoffman et al. (1982), Rush et al. (1982), Randall et al. (1984), Helcl and Petraskova (1985), Imura et al. (1987), Packer et al. (1988), Buchfelder et al. (1989), Wen et al. (1989), Pierre-Kahn et al. (1990), Rodriguez et al. (1990), Sanford and Muhlbauer (1991), Aida et al. (1993), Nishio et al. (1993), Rappaport (1993), Seminara et al. (1998), Singhal et al. (2002), Shinoda et al. (2004), and Karavitaki et al. (2005).

The prognosis for germ cell tumors is highly dependent on the histological subtype of the tumor. In general, germinomas carry an excellent prognosis, with 5-year survival rates and cure above 90% (Packer et al., 2000). Nongerminomatous forms including mixed germ cell tumors and embryonal cell carcinomas have a poorer prognosis with survival rates ranging between 40% and 70% (Packer et al., 2000).

Treatment recommendations differ between germinomatous and nongerminomatous forms of the disease. For germinomas, local radiation therapy with doses of 4000 cGy results in an excellent cure rate (Jennings et al., 1985). Chemotherapy causes tumor shrinkage and may have a role as adjunctive therapy to radiation, allowing a lower dose of radiation to be delivered. The respective roles of chemotherapy and debulking surgery in overall survival in these tumors have yet to be established. Treatment of nongerminomatous tumors is less clear and treatment may vary among the different subtypes. Craniospinal radiation is essential in the nongerminomatous forms of the disease; however, it is rarely curative as a single modality. An evolving role for chemotherapy as an adjunctive therapy is developing. Due to the rarity of this disease, an effective, uniform treatment regimen has yet to be developed (Packer et al., 2000).

19.7.3 Optic Chiasm and Hypothalamic Glioma

Optic pathway gliomas are a subset of low-grade astrocytomas, accounting for 4–6% of all CNS tumors in the pediatric age group (Keles et al., 2004). These tumors grow as infiltrative lesions and often extend into adjacent structures, most commonly the hypothalamus. A majority of cases involve the chiasm or hypothalamus. Tumors of the optic chiasm and hypothalamus are often indistinguishable due to infiltration of both structures and thus grouped together. Fifteen to twenty percent of patients with neurofibromatosis (NF) type 1 have optic gliomas on MRI scans, but only 1–5% become symptomatic (Tseng and Haas-Kogan, 2004). Patients with optic pathway gliomas and NF type 1 seem to have a better overall prognosis than those without NF. Chiasmatic gliomas tend to have a more aggressive course with invasion of the hypothalamus and by causing hydrocephalus. In children less than 5 years of age, hypothalamic and optic gliomas behave more aggressively. Optic pathway gliomas usually present with some form of visual loss. Pathologically, optic tract gliomas are low-grade lesions, mostly pilocytic astrocytomas. In non-NF patients, tumors are confined to the optic nerve without involvement of the meninges. Optic nerve gliomas present with unilateral visual loss, proptosis, papilledema, strabismus, and nystagmus (MacCarty et al., 1970; Roberson and Till, 1974; Rogol, 1981; Borit and Richardson, 1982; Rush et al., 1982; Helcl and Petraskova, 1985; Laue et al., 1985; Packer et al., 1988; Rodriguez et al., 1990; Alshail et al., 1997). Hypothalamic gliomas are more associated with endocrine disturbances, including precocious puberty and DI as well as visual field abnormalities, hydrocephalus, and the diencephalic syndrome of infancy. The clinical features are summarized in **Table 5**. Diagnosis is usually made by MRI and surgical biopsy is not needed for confirmation. Treatment of these tumors has been controversial. Symptomatic patients may be treated with surgery as the initial therapy. Many indolent tumors may be observed without intervention until symptoms manifest. Radiation therapy may be used to inhibit tumor growth and delay time to recurrence. Chemotherapy has been advocated as a means of delaying the use of radiation in children less than 5 years of age. Patients with intraorbital optic gliomas have a good prognosis, whereas hypothalamic gliomas carry a worse prognosis regardless of histology. There is significant morbidity associated with surgical resection of hypothalamic gliomas. Sequelae include immediate endocrinologic or neurologic deficits. Treatment with radiation also carries with its significant side effects of hypothalamic dysfunction and cognitive deficits. The presence of NF may improve the overall prognosis, though studies are conflicting.

19.7.4 Craniopharyngioma

Craniopharyngiomas are rare tumors that arise from epithelial cell rests along the path of the craniopharyngeal duct. They have a histologically benign appearance, but often behave in a malignant fashion with an unpredictable growth pattern and a tendency to infiltrate neural structures in the parasellar region.

These tumors account for 5.6–15% of intracranial tumors in children and 2–5% of all primary intracranial neoplasms (Karavitaki et al., 2006a). They have a bimodal age distribution, with incidence rates in children peaking between 5 and 14 years and in adults between 50 and 74 years of age with no differences between genders (Randall et al., 1984; Wen et al., 1989; Bunin et al., 1998). Most tumors are cystic or mixed lesions (84–99%) (Banna, 1976; Karavitaki et al., 2006a). Two main histological subtypes have been identified – the adamantinomatous and papillary type, but mixed or transitional forms have also been described (Karavitaki et al., 2006a).

Presenting features and prognosis may vary depending on the age of the patient, the location of the tumor, and the size. The clinical features are described in **Table 5**. Children present more often with headache (78%) nausea and vomiting (54%), and papilledema (29%), due to hydrocephalus and increased intracranial pressure (Banna, 1976; Kjellberg, 1979; Rogol, 1981; Sanford and Muhlbauer, 1991; Seminara et al., 1998; Karavitaki et al., 2005). Visual field defects and decreased visual acuity are also commonly observed. Multiple hormone deficiencies are frequently found with growth failure due to GH deficiency being most common. DI, ACTH, and TSH deficiency are also commonly reported. However, endocrine abnormalities are rarely the presenting complaint (Sklar, 1994). Excessive somnolence and sleep–wake-cycle disturbances are more frequently seen in children than in adults (Banna, 1976; Randall et al., 1984; Wen et al., 1989; Yasargil et al., 1990; Seminara et al., 1998). Common presenting features in adulthood include progressive visual loss and bitemporal hemianopsia (Banna, 1976;

Wen et al., 1989; Yasargil et al., 1990). Other common manifestations include headache, personality changes, cognitive decline, nausea and vomiting, and weight gain (Banna, 1976; Wen et al., 1989; Yasargil et al., 1990). Anterior pituitary dysfunction such as deficiencies in GH, gonadotropins, ACTH, and TSH is frequently seen in the majority of patients. Hyperprolactinemia and DI are also common. Differences in the presenting manifestations between children and adults may be primarily related to the increased incidence of hydrocephalus in children with headache, nausea, vomiting, cranial nerve palsies, and papilledema more commonly seen in children than adults. A large retrospective study found no significant difference in the other presenting manifestations (Karavitaki et al., 2005).

Therapy of craniopharyngiomas is a controversial topic with multiple treatments available, each tailored to the individual presentation. Goals of treatment should include reduction of symptoms, prevention of recurrence, minimalization of treatment-induced morbidity, including neuroendocrine, visual, behavioral, and cognitive sequelae, and improvements in survival and quality of life. Surgical resection combined with aspiration of cystic contents is the first-line therapy. Residual tumor can be treated with external beam radiation. Alternative therapies have been proposed including intracystic irradiation or chemotherapy, stereotactic radiotherapy or radiosurgery, and systemic chemotherapy. Predictors for recurrence have not been established and are highly variable among the different treatment modalities. Long-term morbidity is substantial and is related to both tumor invasiveness and effects of treatment. The severity of radiation therapy sequelae is dose dependent. Partial or complete hypopituitarism is seen in the overwhelming majority of patients with multiple pituitary deficiencies seen in 54–100% (Karavitaki et al., 2006a). Restoration of preexisting hormone deficiencies after treatment is not common. Residual visual compromise is common and is directly related to presence of visual symptoms at diagnosis and radiation dose. Hypothalamic damage may be extensive and endocrine sequelae include obesity (26–61%), DI with or without thirst disturbance, and a variety of neuroendocrine dysfunction (Karavitaki et al., 2006a). Hypothalamic morbidity is related to age of presentation, symptoms at diagnosis, aggressiveness of surgical intervention, and radiation dose (Karavitaki et al., 2006a). Patients are significantly affected by neuropsychological and cognitive decline, which adversely affect their quality of life.

19.7.5 Suprasellar Meningioma

Meningiomas that arise from the diaphragma sellae, tuberculum sellae, and planum sphenoidale often encroach upon, but do not invade the hypothalamus (Kadis et al., 1979). Diaphragma sellae meningiomas are subclassified into three types, based on the site of origination (Kinjo et al., 1995). Most patients present between the age of 40 and 60 years of age with a female predominance of 3 to 1 (Jen and Lee, 1997). Due to the proximity of the optic tracts, several neuro-ophthalmic defects have been observed. Eighty percent of patients present with progressive loss of vision in one eye and over 90% have objective evidence of diminished visual acuity. In addition to color perception and visual fields possibly being affected, pallor of the optic disks and an afferent papillary defect may be present. Other presenting signs and symptoms include presence of the Foster–Kennedy syndrome and abnormal extraocular movements. Headache is the second most common symptom presenting in over 50% of patients (Finn and Mount, 1974). It is usually nonlocalizing, but some patients may complain of retro-orbital pain on the side of the lesion. Confusion, memory loss, and decline in cognitive function may also be observed. Endocrine abnormalities have been found in 22% of patients, most commonly hypogonadism, hypothyroidism, hyperprolactinemia, and DI (Finn and Mount, 1974). Such patients with diaphragma sellae meningiomas have been noted to have higher rates of endocrine defects (Kinjo et al., 1995). Presence of estrogen receptors in these tumors may cause an increase in size during pregnancy or during the menstrual cycle and may explain the fluctuation of signs and symptoms. Surgical resection is the treatment of choice for symptomatic lesions with varying approaches depending on the location and structures involved. The overall morbidity and mortality is low and improvement in vision is seen in 40–80% (Chi and McDermott, 2003). Postoperatively, vision stabilizes in 18–40% and worsens in 17–20% (Chi and McDermott, 2003). The degree of visual defect, age of the patient, and duration of visual symptoms prior to diagnosis are predictive of surgical outcome.

19.7.6 Suprasellar Arachnoid Cyst

Arachnoid cysts are congenital collections of CSF contained within the arachnoid membrane and subarachnoid space. Some cysts remain stable in size, while others enlarge, causing symptoms of mass effect. Sixty to ninety percent of all patients with arachnoid cysts are children with a male-to-female predominance of approximately 2:1 (Pradilla and Jallo, 2007). Symptoms are related to hydrocephalus due to obstruction of the foramen of Monro. The clinical features are described in **Table 5**. Endocrine dysfunction includes growth hormone and ACTH deficiency as well as precocious puberty in these children. Expansion of suprasellar arachnoid cysts can cause visual field defects in approximately 30% of patients (Hoffman et al., 1982). Treatment of symptomatic cysts is surgical decompression. A variety of approaches have been developed over the years with varying morbidity. Cystoperitoneal or ventriculoperitoneal shunting may be required for recurrence. Percutaneous ventriculocystostomy and endoscopic fenestration have both been shown to be effective techniques with low morbidity (Pierre-Kahn et al., 1990; Kirollos et al., 2001).

19.7.7 Colloid Cyst of the Third Ventricle

Colloid cysts are benign cysts that are located at the anterior roof of the third ventricle or, rarely, in the area of the septum pellucidum. Most case series report a preponderance of males affected, while some reports have an equal distribution between the sexes (Hellwig et al., 2003). The age of presentation for patients usually ranges from 30 to 60 (Hellwig et al., 2003). There are three major clinical presentations that these patients exhibit. Most patients will present with signs and symptoms of increased intracranial pressure, with headache (72%), nausea and vomiting (32%), and papilledema (21%). Others will present with a fluctuating or progressive dementia (22%), gait disturbance (12%), and incontinence (2%), reminiscent of normal pressure hydrocephalus (Hellwig et al., 2003). Another subset of patients will present with intermittent headache, vomiting, and visual disturbances followed by a loss of consciousness for a variable period (21%) (Hernesniemi and Leivo, 1996). The headache is often associated with a sudden onset with frontal localization and usually precipitated by positional changes of the head. The pain intensity rises rapidly and the patient develops nausea and vomiting and then loses consciousness. This classical presentation is due to the cyst acting as a ball valve, obstructing the foramen of Monro or aqueduct of Sylvius. Drop attacks (0.5%) and sudden death have both been reported and presumed to be due to the acute obstruction of the flow of CSF (Kelly, 1951; Little and MacCarty, 1974). Microsurgical excision of these benign cysts has been the standard therapy. The morbidity and mortality rates correlate with the surgical approach used. The use of rigid endoscopic techniques appears to minimize the complications of surgery (Hellwig et al., 2003).

19.8 Infiltrative Disorders

19.8.1 Neurosarcoidosis

Sarcoidosis affects the CNS in approximately 5–15% of individuals and has a poor prognosis (Bihan et al., 2007). Males and females are equally affected and over 80% of neurosarcoidosis have evidence of systemic disease, especially pulmonary involvement. Lesions of hypothalamus tend to occur in the area of the subependymal region around the third ventricle as well as the posterior pituitary. Neuroendocrine abnormalities include DI in 17–90% and hyperprolactinemia in 3–32% (Bihan et al., 2007). Patients may present with other hypothalamic dysfunction, including SIADH, thirst abnormalities, sleep disturbance, personality changes, morbid obesity, and temperature dysregulation (Porter et al., 2003). In addition to hypothalamic infiltration, cranial nerves may be affected as well as the leptomeninges (Porter et al., 2003; Bihan et al., 2007). Hypothalamic hypopituitarism is a very common manifestation of hypothalamic granulomas. GHD (92%), hypogonadotropic hypogonadism (79%), ACTH deficiency (58%), and hypothyroidism (6%) have been seen in patients with neurosarcoidosis (Vesely et al., 1977; Nakao et al., 1978; Stuart et al., 1978; Jawadi et al., 1980). TRH testing reveals a rise in TSH and gonadotropin secretion with GnRH infusion, indicating an intact anterior pituitary.

Optimal treatment for neurosarcoidosis has yet to be established; however, most patients are treated with corticosteroids as first-line therapy. Cytotoxic drugs and immunomodulators are used as adjuvant therapy to continued steroid treatment (Hoitsma et al., 2004). Recovery of hypothalamic function is usually poor even with treatment.

19.8.2 Histiocytosis

Langerhans cell histiocytosis (LCH) is a rare proliferative disease characterized by granulomatous infiltration of dendritic cells. It is predominantly a disease of childhood, but can present at any age.

Hypothalamic involvement is common, and usually accompanies chronic, disseminated, multifocal disease. It rarely presents as a unifocal mass lesion involving the hypothalamic–pituitary axis. Hand–Schüller–Christian disease describes the triad of calvarial bone disease, exophthalmos, and DI. DI is the most common endocrine manifestation and affects 15–50% of all patients with LCH (Stosel and Braunstein, 1991; Kaltsas et al., 2000; Nanduri et al., 2000). DI usually presents within 5 years of diagnosis and can also be the presenting feature of the disease. When accompanied by structural abnormalities, DI is usually accompanied by anterior pituitary hormone deficits (Makras et al., 2007). Anterior pituitary deficiencies occur in up to 20% of individuals, with GHD being most common and resulting in growth retardation (Braunstein and Kohler, 1972). Gonadotropin, adrenocorticotropin, and thyrotropin deficiency have also been described (Makras et al., 2007). Hypothalamic infiltration can also result in neuropsychiatric and behavioral disorders as well as autonomic and metabolic disturbances. Depending on the region of the hypothalamus affected, granulomatous infiltration can result in disturbances in sleep, appetite control, temperature regulation, and behavior. Obesity secondary to increased appetite and hyperphagia is the most common hypothalamic abnormality (Amato et al., 2006). Cognitive decline, memory impairment, and adipsia may also result from hypothalamic disease (Kaltsas et al., 2000). Low-dose radiation and chemotherapy have been used to treat the disease, but have not been successful in restoring hypothalamic–pituitary function (Stosel and Braunstein, 1991; Makras et al., 2007).

19.8.3 Leukemia

DI is a rare clinical feature of acute leukemia. The pathogenesis may be related to leukemic infiltration or small vessel thrombosis of the posterior pituitary gland, pituitary stalk, or hypothalamus. DI may be a presenting sign, and association with monosomy 7 and abnormalities of chromosome 3 in acute myelogenous leukemia has been observed (Nieboer et al., 2000; Lavabre-Bertrand et al., 2001; Breccia et al., 2002; Keung et al., 2002; Otrock et al., 2006). Of those leukemic patients with DI, 72% present with the acute myelogenous form, 14% with the acute lymphocytic form, 10% with the chronic myelogenous form, and 3% with the chronic lymphocytic form (Ra'anani et al., 1994). Despite chemotherapy, DI is usually persistent.

19.8.4 Paraneoplastic Syndrome

Idiopathic hypothalamic dysfunction has been described in tumors of neural crest origin without metastasis. Symptoms include alteration in sleep–wake cycle, hyperphagia, obesity, polyuria, central respiratory depression, and aggressive behavioral changes. Although few cases have been described, a paraneoplastic syndrome characterized by lymphocytic and histiocytic infiltration of the hypothalamus has been reported (Ouvrier et al., 1995; Nunn et al., 1997; Sirvent et al., 2003).

19.9 Cranial Irradiation

With advances in therapy for childhood cancer, more children are surviving into adulthood after treatment. The endocrine effects of radiation to the hypothalamic–pituitary region have been well known for decades. Cranial irradiation increases the incidence of growth failure, gonadotropin deficiency, adrenal insufficiency, and central hypothyroidism in a dose-dependent fashion. Multiple hormonal deficiencies are seen at doses >30 Gy (Rutter and Rose, 2007).

Recent data emphasize the magnitude of health-related sequelae and the severity of the health problems facing these patients. A large retrospective cohort study comparing the health status of 10 397 survivors of childhood cancer to siblings demonstrated the increased relative risk for any chronic (RR: 3.1; 95% CI: 3.1–3.6), severe or life-threatening health conditions (RR: 7.0; 95% CI: 5.8–8.5) with cranial irradiation (Oeffinger et al., 2006). Analysis of a subset of these children demonstrated one or more endocrinopathies in 43% of survivors (Gurney et al., 2003). Hypothyroidism, GHD, and hypogonadism were reported with increasing frequency depending on the treatment modality, with significantly more cases in children treated with radiation with and without chemotherapy compared to surgery alone (Gurney et al., 2003). Children with low-grade gliomas in various parts of the brain have a much higher rate of endocrinopathy when treated with radiation compared to surgery alone (88% vs. 0%) (Benesch et al., 2006). GHD seems to be more prevalent in younger patients as compared to adults (Mechanick et al., 1986; Samaan et al., 1987). In a large study examining endocrine sequelae of 166 patients (65 of whom were studied prospectively), irradiation of the hypothalamus (50 Gy) and pituitary (57 Gy) for nasopharyngeal carcinoma and paranasal sinus tumors resulted in hormonal deficits indicative of hypothalamic dysfunction in 69% compared to 40% with primary pituitary dysfunction (Samaan et al., 1987). Incidence of endocrine dysfunction was highest in the first 5 years following therapy, but late effects were seen during more prolonged follow-up periods. Other consequences of radiation therapy related to hypothalamic insult include an increased rate of obesity, alteration in the sleep/wake cycle, behavioral abnormalities, and cognitive decline (Gurney et al., 2003; Sarkissian, 2005; Benesch et al., 2006; Kelsey and Marks, 2006; Oeffinger et al., 2006; Rutter and Rose, 2007). Risk factors for the development of sequelae of radiation therapy include dose of radiation, the interval over which the radiation is delivered, and the age of the patient. Development of these sequelae may be delayed after treatment and long-term surveillance is required (Samaan et al., 1987).

Hypothalamic dysfunction is not limited to patients treated in childhood. Adults treated with cranial irradiation for nonpituitary tumors also demonstrate a higher incidence of hypopituitarism (41%) compared to radiation naive matched controls (Agha et al., 2005). Hyperprolactinemia (32%) and deficiencies of GH (32%), ACTH (21%), gonadotropins (27%), and TSH (9%) were noted in radiation-exposed patients. ACTH and LH deficiencies are seen in greater frequency in adults (Mechanick et al., 1986; Samaan et al., 1987). The degree of hypopituitarism and GH deficiency was directly associated with the length of time from radiation treatment and the effective dose (Agha et al., 2005).

19.10 Traumatic Brain Injury

Traumatic brain injury (TBI) is the leading cause of death and disability in young adults. In industrialized nations, 180–250 persons per 100 000 per year die or are hospitalized as a result of TBI (Bondanelli et al., 2005). Post-traumatic hypopituitarism has long been known to be a sequela of TBI, but it was thought to be very rare. Recently, several studies have drawn attention to the high prevalence of pituitary defects following TBI. In 2000, a review of the literature dating back to 1942 summarized findings in 367 patients with post-traumatic hypopituitarism. These patients had deficiencies of gonadotropins (100%), ACTH (53%), TSH (44%), GH (24%), and hyperprolactinemia (48%) (Benvenga et al., 2000). Since then, a number of studies have sought to systematically assess the frequency of hypopituitarism following TBI. While the definition of abnormality, the degree of traumatic injury, and diagnostic methodology have varied from study to study, it has become clear that hypopituitarism after brain injury is a common consequence with a prevalence of at least 25% (Agha and Thompson, 2006).

The mechanism of TBI to hypothalamic–pituitary function stems from the anatomy and fragile vasculature supplying the anterior pituitary. The major blood supply is from the hypophyseal portal circulation. The inferior hypophyseal artery, which arises from the internal carotid artery, supplies a small portion of the adenohypophysis and the posterior lobe. Damage may be caused by compression of the hypothalamus and pituitary due to edema, increased intracranial pressure, skull fracture, and/or hemorrhage. Hypoxia and direct mechanical injury to the pituitary stalk or hypothalamus are also causes of pituitary deficiencies. In cases of TBI, autopsy results have revealed injury to the hypothalamus, pituitary gland, or stalk.

Acute alterations in neuroendocrine function following TBI have been demonstrated in various trials, but the findings have been inconsistent. One prospective trial demonstrated low GH (18%), low cortisol (16%), and responses to glucagon stimulation (80%) (Agha et al., 2004a,b). Secondary hypogonadism (80%), TSH deficiency (2%), and hyperprolactinemia (50%) were also found in the acute phase of TBI. After the acute phase, recovery of anterior pituitary function is seen in most patients by 6 months. More severe deficiencies in the acute phase are associated with lower rates of recovery. However, some patients with normal responses in the acute phase develop late onset deficiencies, diagnosed at 6–12 months (Agha and Thompson, 2006).

In a recent systematic review, anterior pituitary hormone deficiencies were found in 15–50% of TBI patients and 38–55% in cases of subarachnoid hemorrhage. For TBI, 809 patients were studied showing GH deficiency ranging from 6% to 33%, gonadotropin deficiency from 2% to 20%, ACTH deficiency from 0% to 19%, TSH deficiency from 1% to 10%, and multiple deficiencies from 4% to 12% (Schneider et al., 2007). For subarachnoid hemorrhage, 102 patients were studied showing GH deficiency ranging from 12% to 36%, gonadotropin deficiency from 0% to 13%, ACTH deficiency from 6% to 40%, TSH deficiency from 3% to 9%, and multiple deficiencies from 6% to 13% (Schneider et al., 2007).

An optimal strategy for screening patients for hypopituitarism for TBI has recently been suggested

(Ghigo et al., 2005). Screening of patients with moderate or severe brain injury should be done with hormonal testing if clinical indicated with follow-up testing at 3 months and 1 year. When patients are seen more than 12 months after the injury, they should undergo baseline hormonal testing. Provocative testing may be required to further assess hormonal deficiencies.

Ongoing studies are required to adequately assess the benefits of individual hormone replacement in patients in the recovery phase of TBI. Replacement of sex steroids and GH should be done on a case-by-case basis. Persistent adrenal insufficiency and central hypothyroidism should be replaced when encountered more than 3–6 months after the initial injury.

19.11 Critical Illness

Advances in the overall management of critically ill patients have led to increased survival and present physicians with the complex task of managing acute and chronic phases of critical illness as well as recovery. Critical illness is accompanied by alterations in the hypothalamic–pituitary axes, and acute illness differs in its endocrine derangements from prolonged critical illness. GH levels quickly rise immediately during acute critical illness, with elevated peaks and high interpulse secretion (Vanhorebeek et al., 2006). GH resistance develops with levels of IGF-I, IGFBP-3, and GH-binding protein decreasing. During prolonged illness, GH levels are lower, but the pulsatile secretion is altered such that pulses are absent and basal secretion is elevated compared to normal. As with noncritical illness, the thyroid axis is altered in critical illness. Acutely, T3 levels decline and rT3 levels increase due to de-iodination of circulating T4 to rT3. TSH levels and T4 levels briefly rise and return to normal, although T4 levels may become low (Vanhorebeek and Van den Berghe, 2006). When critical illness is prolonged, TSH pulsatility is diminished and levels of T4 and T3 are low. Lower levels of TSH, T4, and T3 and higher levels of rT3 have been associated with mortality in critical illness (Vanhorebeek et al., 2006). In the acute phase, testosterone levels drop dramatically and LH levels rise (Van den Berghe, 2003). When illness is prolonged, testosterone levels remain low, and gonadotropins diminish resulting in hypogonadotropic hypogonadism. Prolactin levels are elevated during acute illness, but suppressed during the prolonged phase. Cortisol levels are usually elevated during times of physiological stress, and an inability to respond to stress with appropriately elevated levels of cortisol has been described as relative adrenal insufficiency (Vanhorebeek et al., 2006). However, low total cortisol levels may result from hypoproteinemia, with low cortisol-binding globulin, while serum levels of free cortisol remain normal (Hamrahian et al., 2004). Cortisol levels slowly decline during the prolonged phase of critical illness, reaching normal levels in the recovery phase. Of these abnormalities, replacement of cortisol has been associated with a reduction in mortality (Marik, 2003).

References

Agha A, Rogers B, Sherlock M, O'Kelly P, Tormey W, Phillips J, and Thompson CJ (2004a) Anterior pituitary dysfunction in survivors of traumatic brain injury. *Journal of Clinical Endocrinology and Metabolism* 89: 4929–4936.

Agha A, Sherlock M, Brennan S, et al. (2005) Hypothalamic–pituitary dysfunction after irradiation of nonpituitary brain tumors in adults. *Journal of Clinical Endocrinology and Metabolism* 90: 6355–6360.

Agha A and Thompson CJ (2006) Anterior pituitary dysfunction following traumatic brain injury (TBI). *Clinical Endocrinology (Oxf)* 64: 481–488.

Agha A, Thornton E, O'Kelly P, Tormey W, Phillips J, and Thompson CJ (2004b) Posterior pituitary dysfunction after traumatic brain injury. *Journal of Clinical Endocrinology and Metabolism* 89: 5987–5992.

Aida T, Abe H, Fujieda K, and Matsuura N (1993) Endocrine functions in children with suprasellar germinoma. *Neurologia Medico Chirurgica (Tokyo)* 33: 152–157.

Alexander JM, Biller BM, Bikkal H, Zervas NT, Arnold A, and Klibanski A (1990) Clinically nonfunctioning pituitary tumors are monoclonal in origin. *Journal of Clinical Investigation* 86: 336–340.

Alshail E, Rutka JT, Becker LE, and Hoffman HJ (1997) Optic chiasmatic–hypothalamic glioma. *Brain Pathology* 7: 799–806.

Amato MC, Elias LL, Elias J, Santos AC, Bellucci AD, Moreira AC, and DeCastro M (2006) Endocrine disorders in pediatric-onset Langerhans cell histiocytosis. *Hormone and Metabolic Research* 38: 746–751.

Andrioli M, Giraldi FP, and Cavagnini F (2006) Isolated corticotrophin deficiency. *Pituitary* 9: 289–295.

Antunes-Rodrigues J, de Castro M, Elias LL, Valenca MM, and McCann SM (2004) Neuroendocrine control of body fluid metabolism. *Physiological Reviews* 84: 169–208.

Arita K, Kurisu K, Kiura Y, Iida K, and Otsubo H (2005) Hypothalamic hamartoma. *Neurologia Medico Chirurgica (Tokyo)* 45: 221–231.

Arnulf I, Seitzer JM, File J, Farber N, and Mignot E (2005) Kleine–Levin syndrome: A systematic review of 186 cases in the literature. *Brain* 128: 2763–2776.

Arslanian SA, Rothfus WE, Foley TP, Jr., and Becker DJ (1984) Hormonal, metabolic, and neuroradiologic abnormalities associated with septo-optic dysplasia. *Acta Endocrinologica (Copenh)* 107: 282–288.

Asa SL, Bilbao JM, Kovacs K, and Linfoot JA (1980) Hypothalamic neuronal hamartoma associated with pituitary growth hormone cell adenoma and acromegaly. *Acta Neuropathologica (Berl)* 52: 231–234.

Asa SL, Scheithauer BW, Bilbao JM, et al. (1984) A case for hypothalamic acromegaly: A clinicopathological study of six patients with hypothalamic gangliocytomas producing growth hormone-releasing factor. *Journal of Clinical Endocrinology and Metabolism* 58: 796–803.

Aschoff J (1979) Circadian rhythms: General features and endocrinological aspects. In: Kreiger DT (ed.) *Endocrine Rhythms*, pp. 1–61. New York: Raven Press.

Balagura S, Shulman K, and Sobel EH (1979) Precocious puberty of cerebral origin. *Surgical Neurology* 11: 315–326.

Ball SG, Vaidja B, and Baylis PH (1997) Hypothalamic adipsic syndrome: Diagnosis and management. *Clinical Endocrinology (Oxf)* 47: 405–409.

Banna M (1976) Pathology and clinical manifestations. In: Hankinson J and Banna M (eds.) *Pituitary and Parapituitary Tumours*, pp. 13–58. London: W. B. Saunders.

Bauer HG (1954) Endocrine and other clinical manifestations of hypothalamic disease: A survey of 60 cases, with autopsies. *Journal of Clinical Endocrinology and Metabolism* 14: 13–31.

Bauer HG (1959) Endocrine and metabolic conditions related to pathology in the hypothalamus: A review. *Journal of Nervous and Mental Disease* 128: 323–338.

Beck-Peccoz P, Amr S, Menezes-Ferreira MM, Faglia G, and Weintraub BD (1985) Decreased receptor binding of biologically inactive thyrotropin in central hypothyroidism. Effect of treatment with thyrotropin-releasing hormone. *New England Journal of Medicine* 312: 1085–1090.

Benarroch EE (2007) Thermoregulation: Recent concepts and remaining questions. *Neurology* 69: 1293–1297.

Benesch M, Lackner H, Sovinz P, et al. (2006) Late sequela after treatment of childhood low-grade gliomas: A retrospective analysis of 69 long-term survivors treated between, 1983 and 2003. *Journal of Neurooncology* 78: 199–205.

Beningfield SJ, Bonnici F, and Cremin BJ (1988) Magnetic resonance imaging of hypothalamic hamartomas. *British Journal of Radiology* 61: 1177–1180.

Benvenga S, Campenni A, Ruggeri RM, and Trimarchi F (2000) Clinical review 113: Hypopituitarism secondary to head trauma. *Journal of Clinical Endocrinology and Metabolism* 85: 1353–1361.

Berelowitz M, Szabo M, Frohman LA, Firestone S, Chu L, and Hintz RL (1981) Somatomedin-C mediates growth hormone negative feedback by effects on both the hypothalamus and the pituitary. *Science* 212: 1279–1281.

Bergeron C, Kovacs K, Ezrin C, and Mizzen C (1991) Hereditary diabetes insipidus: An immunohistochemical study of the hypothalamus and pituitary gland. *Acta Neuropathologica (Berl)* 81: 345–348.

Beuschlein F, Strasburger CJ, Siegerstetter V, et al. (2000) Acromegaly caused by secretion of growth hormone by a non-Hodgkin's lymphoma. *New England Journal of Medicine* 342: 1871–1876.

Biesecker LG, Abbott M, Allen J, et al. (1996) Report from the workshop on Pallister–Hall syndrome and related phenotypes. *American Journal of Medical Genetics* 65: 76–81.

Biesecker LG and Graham JM, Jr. (1996) Pallister–Hall syndrome. *Journal of Medical Genetics* 33: 585–589.

Bihan H, Christozova V, Dumas JL, et al. (2007) Sarcoidosis: Clinical, hormonal, and magnetic resonance imaging (MRI) manifestations of hypothalamic–pituitary disease in 9 patients and review of the literature. *Medicine (Baltimore)* 86: 259–268.

Biller BM (1994) Pathogenesis of pituitary Cushing's syndrome. Pituitary versus hypothalamic. *Endocrinology and Metabolism Clinics of North America* 23: 547–554.

Biller BM, Alexander JM, Zervas NT, Hedley-Whyte ET, Arnold A, and Klibanski A (1992) Clonal origins of adrenocorticotropin-secreting pituitary tissue in Cushing's disease. *Journal of Clinical Endocrinology and Metabolism* 75: 1303–1309.

Biller BM, Samuels MH, Zagar A, et al. (2002) Sensitivity and specificity of six tests for the diagnosis of adult GH deficiency. *Journal of Clinical Endocrinology and Metabolism* 87: 2067–2079.

Birkebaek NH, Patel L, Wright NB, et al. (2003) Endocrine status in patients with optic nerve hypoplasia: Relationship to midline central nervous system abnormalities and appearance of the hypothalamic–pituitary axis on magnetic resonance imaging. *Journal of Clinical Endocrinology and Metabolism* 88: 5281–5286.

Blotner H (1958) Primary or idiopathic diabetes insipidus: A system disease. *Metabolism* 7: 191–200.

Bondanelli M, Ambrosio MR, Zatelli MC, DeMarinis L, and degli Uberti EC (2005) Hypopituitarism after traumatic brain injury. *European Journal of Endocrinology* 152: 679–691.

Borit A and Richardson EP, Jr. (1982) The biological and clinical behaviour of pilocytic astrocytomas of the optic pathways. *Brain* 105: 161–187.

Boshes B (1969) Syndromes of the diencephalon. In: Vinken PJ and Bruyn GW (eds.) *Handbook of Clinical Neurology, Vol. 2: Localization in Clinical Neurology*, pp. 432–468. Amsterdam: North-Holland.

Boudreau EA, Liow K, Frattali CM, et al. (2005) Hypothalamic hamartomas and seizures: Distinct natural history of isolated and Pallister–Hall syndrome cases. *Epilepsia* 46: 42–47.

Braunstein GD (2002) The hypothalamus. In: Melmed S (ed.) *The Pituitary*, pp. 317–348. Cambridge, MA: Blackwell Scientific.

Braunstein GD (2006) Hypothalamic syndromes. In: DeGroot LJ and Jameson JL (eds.) *Endocrinology*, pp. 373–386. Philadelphia, PA: Elsevier Saunders.

Braunstein GD and Kohler PO (1972) Pituitary function in Hand–Schüller–Christian disease. Evidence for deficient growth-hormone release in patients with short stature. *New England Journal of Medicine* 286: 1225–1229.

Bray GA, Dahms WT, Swerdloff RS, Fiser RH, Atkinson RL, and Carrel RE (1983) The Prader–Willi syndrome: A study of 40 patients and a review of the literature. *Medicine (Baltimore)* 62: 59–80.

Bray GA and Gallagher TF, Jr. (1975) Manifestations of hypothalamic obesity in man: A comprehensive investigation of eight patients and a reveiw of the literature. *Medicine (Baltimore)* 54: 301–330.

Bray MA, Muneyyirci-Delale O, Kofinas GD, and Reyes FI (1991) Circadian, ultradian, and episodic gonadotropin and prolactin secretion in human pseudocyesis. *Acta Endocrinologica (Copenh)* 124: 501–509.

Breccia M, Petti MC, Ottaviani E, Mancini M, D'Elia GM, Mecarocci S, and Alimena G (2002) Diabetes insipidus as first manifestation of acute myeloid leukaemia with EVI-1-positive, 3q21q26 syndrome and T cell-line antigen expression: What is the EVI-1 gene role? *British Journal of Haematology* 118: 438–441.

Breningstall GN (1985) Gelastic seizures, precocious puberty, and hypothalamic hamartoma. *Neurology* 35: 1180–1183.

Brewer JB, Gabrieli JD, Preston AR, Vaidya CJ, and Rosen AC (2007) Memory. In: Goetz CG (ed.) *Textbook of Clinical Neurology*, pp. 63–76. Philadelphia, PA: Saunders Elsevier.

Bruesch SR (1984) Anatomy of the human hypothalamus. In: Givens JR, Kitabchi AE, and Robertson JT (eds.) *The Hypothalamus*, pp. 1–16. Chicago, IL: Year Book Medical Publishers.

Buchfelder M, Fahlbusch R, Walther M, and Mann K (1989) Endocrine disturbances in suprasellar germinomas. *Acta Endocrinologica (Copenh)* 120: 337–342.

Bunin GR, Surawicz TS, Witman PA, Preston-Martin S, Davis F, and Bruner JM (1998) The descriptive epidemiology of craniopharyngioma. *Journal of Neurosurgery* 89: 547–551.

Burman P, Ritzen EM, and Lindgren AC (2001) Endocrine dysfunction in Prader–Willi syndrome: A review with special reference to GH. *Endocrine Reviews* 22: 787–799.

Burr IM, Slonim AE, Danish RK, Gadoth N, and Butler IJ (1976) Diencephalic syndrome revisited. *Journal of Pediatrics* 88: 439–444.

Carmel PW (1980) Surgical syndromes of the hypothalamus. *Clinical Neurosurgery* 27: 133–159.

Carmel PW (1985) Tumours of the third ventricle. *Acta Neurochirurgica (Wien)* 75: 136–146.

Carpenter MB and Sutin J (1983) *Human Neuroanatomy.* Baltimore, MD: Williams and Wilkins.

Cassidy SB (1997) Prader–Willi syndrome. *Journal of Medical Genetics* 34: 917–923.

Cassidy SB and Schwartz S (1998) Prader–Willi and Angelman syndromes. Disorders of genomic imprinting. *Medicine (Baltimore)* 77: 140–151.

Celesia GG, Archer CR, and Chung HD (1981) Hyperphagia and obesity. Relationship to medial hypothalamic lesions. *Journal of the American Medical Association* 246: 151–153.

Cerullo A, Tinuper P, Provini F, Contin M, Rosati A, Marini C, and Cortelli P (1998) Autonomic and hormonal ictal changes in gelastic seizures from hypothalamic hamartomas. *Electroencephalography and Clinical Neurophysiology* 107: 317–322.

Chattopadhyay A, Kumar V, and Marulaiah M (2003) Polycystic ovaries, precocious puberty and acquired hypothyroidism: The Van Wyk and Grumbach syndrome. *Journal of Pediatric Surgery* 38: 1390–1392.

Chemaitilly W, Trivin C, Adan L, Gall V, Sainte-Rose C, and Brauner R (2001) Central precocious puberty: Clinical and laboratory features. *Clinical Endocrinology (Oxf)* 54: 289–294.

Chi JH and McDermott MW (2003) Tuberculum sellae meningiomas. *Neurosurgical Focus* 14(6): e6.

Clark LG (1938) The hypothalamus in man. In: Le Gros Clark WE, Beattie J, Riddoch G, and Dott NM (eds.) *The Hypothalamus. Morphological, Functional, Clinical and Surgical Aspects*, pp. 59–68. London: Oliver and Boyd.

Clarren SK, Alvord EC, Jr., and Hall JG (1980) Congenital hypothalamic hamartoblastoma, hypopituitarism, imperforate anus, and postaxial polydactyly – a new syndrome? Part II: Neuropathological considerations. *American Journal of Medical Genetics* 7: 75–83.

Clement K, Vaisse C, Lahlou N, et al. (1998) A mutation in the human leptin receptor gene causes obesity and pituitary dysfunction. *Nature* 392: 398–401.

Cohen RA and Albers HE (1991) Disruption of human circadian and cognitive regulation following a discrete hypothalamic lesion: A case study. *Neurology* 41: 726–729.

Coll AP, Farooqi IS, and O'Rahilly S (2007) The hormonal control of food intake. *Cell* 129: 251–262.

Comite F, Pescovitz OH, Rieth KG, et al. (1984) Luteinizing hormone-releasing hormone analog treatment of boys with hypothalamic hamartoma and true precocious puberty. *Journal of Clinical Endocrinology and Metabolism* 59: 888–892.

Cone RD (2005) Anatomy and regulation of the central melanocortin system. *Nature Neuroscience* 8: 571–578.

Daniel PM and Prichard MML (1975) Studies of the hypothalamus and the pituitary gland. *Acta Endocrinologica* 80: 1–216.

Daniel PM and Treip CS (1977) The pathology of the hypothalamus. *Journal of Clinical Endocrinology and Metabolism* 6: 3–19.

Dattani MT and Hindmarsh PC (2006) Growth hormone deficiency in children. In: DeGroot LJ and Jameson JL (eds.) *Endocrinology,* vol. 1, pp. 733–754. Philadelphia, PA: Elsevier Saunders.

Dattani M and Preece M (2004) Growth hormone deficiency and related disorders: Insights into causation, diagnosis, and treatment. *Lancet* 363: 1977–1987.

Decaux G (2001) Long-term treatment of patients with inappropriate secretion of antidiuretic hormone by the vasopressin receptor antagonist conivaptan, urea, or furosemide. *American Journal of Medicine* 110: 582–584.

Delaney P (1977) Neurologic manifestations in sarcoidosis: Review of the literature, with a report of 23 cases. *Annals of Internal Medicine* 87: 336–345.

de Lecea L, Kilduff TS, Peyron C, et al. (1998) The hypocretins: Hypothalamus-specific peptides with neuroexcitatory activity. *Proceedings of the National Academy of Sciences of the United States of America* 95: 322–327.

DeRubertis FR, Michelis MF, and Davis BB (1974) Essential hypernatremia. Report of three cases and review of the literature. *Archives of Internal Medicine* 134: 889–895.

Desy L and Pelletier G (1977) Immunohistochemical localization of somatostatin in the human hypothalamus. *Cell Tissue Research* 184: 491–497.

Dhillo WS (2007) Appetite regulation: An overview. *Thyroid* 17: 433–445.

Diebler C and Ponsot G (1983) Hamartomas of the tuber cinereum. *Neuroradiology* 25: 93–101.

Dong HW and Swanson LW (2006a) Projections from bed nuclei of the stria terminalis, dorsomedial nucleus: Implications for cerebral hemisphere integration of neuroendocrine, autonomic, and drinking responses. *Journal of Comparative Neurology* 494: 75–107.

Dong HW and Swanson LW (2006b) Projections from bed nuclei of the stria terminalis, magnocellular nucleus: Implications for cerebral hemisphere regulation of micturition, defecation, and penile erection. *Journal of Comparative Neurology* 494: 108–141.

Dott NM (1938) Surgical aspects of the hypothalamus. In: Le Gros Clark,WE, Beattie J, Riddoch G, and Dott NM (eds.) *The Hypothalamus. Morphological, Functional, Clinical, and Surgical Aspects*, pp. 1131–1185. London: Oliver and Boyd.

Edholm OG (1977) Energy balance in man studies carried out by the Division of Human Physiology, National Institute for Medical Research. *Journal of Human Nutrition* 31: 413–431.

Edwards S, Lennox G, Robson K, and Whiteley A (1996) Hypothermia due to hypothalamic involvement in multiple sclerosis. *Journal of Neurology, Neurosurgery, and Psychiatry* 61: 419–420.

Ellison DH and Berl T (2007) Clinical practice. The syndrome of inappropriate antidiuresis. *New England Journal of Medicine* 356: 2064–2072.

Faglia G and Spada A (1995) The role of the hypothalamus in pituitary neoplasia. *Bailliere's Clinical Endocrinology and Metabolism* 9: 225–242.

Fahlbusch R, Muller OA, and von Werder K (1979) Functional endocrinological disturbances in parasellar processes. *Acta Neurochirurgica. Supplementum (Wien)* 28: 456–460.

Fenzi F, Simonati A, Crosato F, Ghersini L, and Rizzuto N (1993) Clinical features of Kleine–Levin syndrome with localized encephalitis. *Neuropediatrics* 24: 292–295.

Finn JE and Mount LA (1974) Meningiomas of the tuberculum sellae and planum sphenoidale. A review of 83 cases. *Archives of Ophthalmology* 92: 23–27.

Fleischman A, Brue C, Poussaint TY, et al. (2005) Diencephalic syndrome: A cause of failure to thrive and a model of partial growth hormone resistance. *Pediatrics* 115: e742–e748.

Fox RH, Davies TW, Marsh FP, and Urich H (1970) Hypothermia in a young man with an anterior hypothalamic lesion. *Lancet* 2: 185–188.

Fox RH, Wilkins DC, Bell JA, et al. (1973) Spontaneous periodic hypothermia: Diencephalic epilepsy. *British Medical Journal* 2: 693–695.

Frohman LA (1980) Clinical aspects of hypothalamic disease. In: Motta M (ed.) *The Endocrine Function of the Brain*, pp. 419–446. New York: Raven Press.

Frohman EM, Frohman TC, and Moreault AM (2002) Acquired sexual paraphilia in patients with multiple sclerosis. *Archives of Neurology* 59: 1006–1010.

Frohman LA, Szabo M, Berelowitz M, and Stachura ME (1980) Partial purification and characterization of a peptide with growth hormone-releasing activity from extrapituitary tumors in patients with acromegaly. *Journal of Clinical Investigation* 65: 43–54.

Gais S and Born J (2004) Declarative memory consolidation: Mechanisms acting during human sleep. *Learning and Memory* 11: 679–685.

Garcia ML, Ty EB, Taban M, David Rothner A, Rogers D, and Traboulsi EI (2006) Systemic and ocular findings in 100 patients with optic nerve hypoplasia. *Journal of Child Neurology* 21: 949–956.

Garnica AD, Netzloff ML, and Rosenbloom AL (1980) Clinical manifestations of hypothalamic tumors. *Annals of Clinical and Laboratory Science* 10: 474–485.

Garty B, Shuper A, Mimouni M, Varsano I, and Kauli R (1982) Primary gonadal failure and precocious adrenarche in a boy with Prader–Labhart–Willi syndrome. *European Journal of pediatrics* 139: 201–203.

Gascon GG and Lombroso CT (1971) Epileptic (gelastic) laughter. *Epilepsia* 12: 63–76.

Ghigo E, Masel B, Aimaretti G, et al. (2005) Consensus guidelines on screening for hypopituitarism following traumatic brain injury. *Brain Injury* 19: 711–724.

Gillett GR and Symon L (1987) Hypothalamic glioma. *Surgical Neurology* 28: 291–300.

Gilmour J, Skuse D, and Pembrey M (2001) Hyperphagic short stature and Prader–Willi syndrome: A comparison of behavioural phenotypes, genotypes and indices of stress. *British Journal of Psychiatry* 179: 129–137.

Gohlke BC, Frazer FL, and Stanhope R (2002) Body mass index and segmental proportion in children with different subtypes of psychosocial short stature. *European Journal of Pediatrics* 161: 250–254.

Goldstone AP (2004) Prader–Willi syndrome: Advances in genetics, pathophysiology and treatment. *Trends Endocrinology and Metabolism* 15: 12–20.

Gonzalez-Martinez D, Kim SH, Hu Y, et al. (2004) Anosmin-1 modulates fibroblast growth factor receptor 1 signaling in human gonadotropin-releasing hormone olfactory neuroblasts through a heparan sulfate-dependent mechanism. *Journal of Neuroscience* 24: 10384–10392.

Greenswag LR (1987) Adults with Prader–Willi syndrome: A survey of 232 cases. *Developmental Medicine and Child Neurology* 29: 145–152.

Greenswag LR and Alexander RC (1995) *Management of Prader–Willi Syndrome.* New York: Springer.

Gross RE (1940) Neoplasms producing endocrine disturbances in childhood. *American Journal of Diseases of Children* 59: 579–628.

Gunay-Aygun M, Schwartz S, Heeger S, O'Riordan MA, and Cassidy SB (2001) The changing purpose of Prader–Willi syndrome clinical diagnostic criteria and proposed revised criteria. *Pediatrics* 108: E92.

Gurney JG, Kadan-Lottick NS, Packer RJ, et al. (2003) Endocrine and cardiovascular late effects among adult survivors of childhood brain tumors: Childhood Cancer Survivor Study. *Cancer* 97: 663–673.

Haak HR, van Hilten JJ, Roos RA, and Meinders AE (1990) Functional hypothalamic derangement in a case of Wernicke's encephalopathy. *Netherlands Journal of Medicine* 36: 291–296.

Hall JG, Pallister PD, Clarren SK, et al. (1980) Congenital hypothalamic hamartoblastoma, hypopituitarism, imperforate anus and postaxial polydactyly – a new syndrome? Part I: Clinical, causal, and pathogenetic considerations. *American Journal of Medical Genetics* 7: 47–74.

Hall DM and Smith DW (1972) Prader–Willi syndrome. *Journal of Pediatrics* 81: 286–293.

Hamrahian AH, Oseni TS, and Arafah BM (2004) Measurements of serum free cortisol in critically ill patients. *New England Journal of Medicine* 350: 1629–1638.

Hartman ML, Crowe BJ, Biller BM, Ho KK, Clemmons DR, and Chipman JJ (2002) Which patients do not require a GH stimulation test for the diagnosis of adult GH deficiency? *Journal of Clinical Endocrinology and Metabolism* 87: 477–485.

Harvey AS and Freeman JL (2007) Epilepsy in hypothalamic hamartoma: Clinical and EEG features. *Seminars in Pediatric Neurology* 14: 60–64.

Haugh RM and Markesbery WR (1983) Hypothalamic astrocytoma. Syndrome of hyperphagia, obesity, and disturbances of behavior and endocrine and autonomic function. *Archives of Neurology* 40: 560–563.

Hayek A and Peake GT (1982) Hypothalamic adipsia without demonstrable structural lesion. *Pediatrics* 70: 275–278.

Helcl F and Petraskova H (1985) Gliomas of visual pathways and hypothalamus in children – a preliminary report. *Acta Neurochirurgica. Supplementum (Wien)* 35: 106–110.

Hellwig D, Bauer BL, Schulte M, Gatscher S, Riegel T, and Bertalanffy H (2003) Neuroendoscopic treatment for colloid cysts of the third ventricle: The experience of a decade. *Neurosurgery* 52: 525–533;discussion 532–533.

Herman V, Fagin J, Gonsky R, Kovacs K, and Melmed S (1990) Clonal origin of pituitary adenomas. *Journal of Clinical Endocrinology and Metabolism* 71: 1427–1433.

Hernesniemi J and Leivo S (1996) Management outcome in third ventricular colloid cysts in a defined population: A series of 40 patients treated mainly by transcallosal microsurgery. *Surgical Neurology* 45: 2–14.

Higgins SC, Gueorguiev M, and Korbonits M (2007) Ghrelin, the peripheral hunger hormone. *Annals of Medicine* 39: 116–136.

Hochman HI, Judge DM, and Reichlin S (1981) Precocious puberty and hypothalamic hamartoma. *Pediatrics* 67: 236–244.

Hoffman HJ, Hendrick EB, Humphreys RP, and Armstrong EA (1982) Investigation and management of suprasellar arachnoid cysts. *Journal of Neurosurgery* 57: 597–602.

Hoitsma E, Faber CG, Drent M, and Sharma M (2004) Neurosarcoidosis: A clinical dilemma. *Lancet Neurology* 3: 397–407.

Hokfelt T, Fuxe K, Johansson O, Jeffcoate S, and White N (1975) Distribution of thyrotropin-releasing hormone (TRH) in the central nervous system as revealed with immunohistochemistry. *European Journal of Pharmacology* 34: 389–392.

Holm VA, Cassidy SB, Butler MG, Hanchett JM, Greenswag LR, Whitman BY, and Greenberg F (1993) Prader–Willi syndrome: Consensus diagnostic criteria. *Pediatrics* 91: 398–402.

Horn E, Lach B, Lapierre Y, and Hrdina P (1988) Hypothalamic pathology in the neuroleptic malignant syndrome. *American Journal of Psychiatry* 145: 617–620.

Horseman ND and Greerson KA (2006) Prolactin. In: DeGroot LJ and Jameson JL (eds.) *Endocrinology,* vol. 1, pp. 309–322. Philadelphia, PA: Elsevier Saunders.

Huber J, Sovinz P, Lackner H, Mokry M, Eder H, and Urban C (2007) Diencephalic syndrome: A frequently delayed diagnosis in failure to thrive. *Klinische Padiatrie* 219: 91–94.

Huseman CA, Kelch RP, Hopwood NJ, and Zipf WB (1978) Sexual precocity in association with septo-optic dysplasia and hypothalamic hypopituitarism. *Journal of Pediatrics* 92: 748–753.

Imura H, Kato Y, and Nakai Y (1987) Endocrine aspects of tumors arising from suprasellar, third ventricular regions. *Progress in Experimental Tumor Research* 30: 313–324.

Izenberg N, Rosenblum M, and Parks JS (1984) The endocrine spectrum of septo-optic dysplasia. *Clinical Pediatrics (Phila)* 23: 632–636.

Jacoby LB, Hedley-Whyte ET, Pulaski K, Seizinger BR, and Martuza RL (1990) Clonal origin of pituitary adenomas. *Journal of Neurosurgery* 73: 731–735.

Jawadi MH, Hanson TJ, Schemmel JE, Beck P, and Katz FH (1980) Hypothalamic sarcoidosis and hypopituitarism. *Hormone Research* 12: 1–9.

Jen SL and Lee LS (1997) Suprasellar meningiomas: Analysis of 32 cases. *Zhonghua Yi Xue Za Zhi (Taipei)* 59: 7–14.

Jennings MT, Gelman R, and Hochberg F (1985) Intracranial germ-cell tumors: Natural history and pathogenesis. *Journal of Neurosurgery* 63: 155–167.

Judge DM, Kulin HE, Page R, Santen R, and Trapukdi S (1977) Hypothalamic hamartoma: A source of luteinizing hormone-releasing factor in precocious puberty. *New England Journal of Medicine* 296: 7–10.

Kadis GN, Mount LA, and Ganti SR (1979) The importance of early diagnosis and treatment of the meningiomas of the planum sphenoidale and tuberculum sellae: A retrospective study of 105 cases. *Surgical Neurology* 12: 367–371.

Kalra SP, Dube MG, Pu S, Xu B, Horvath TL, and Kalra PS (1999) Interacting appetite-regulating pathways in the hypothalamic regulation of body weight. *Endocrine Reviews* 20: 68–100.

Kaltsas GA, Powles TB, Evanson J, et al. (2000) Hypothalamo-pituitary abnormalities in adult patients with langerhans cell histiocytosis: Clinical, endocrinological, and radiological features and response to treatment. *Journal of Clinical Endocrinology and Metabolism* 85: 1370–1376.

Kamalian N, Keesey RE, and ZuRhein GM (1975) Lateral hypothalamic demyelination and cachexia in a case of malignant multiple sclerosis. *Neurology* 25: 25–30.

Kang S, Allen J, Graham JM, Jr., et al. (1997) Linkage mapping and phenotypic analysis of autosomal dominant Pallister–Hall syndrome. *Journal of Medical Genetics* 34: 441–446.

Kapcala LP, Molitch ME, Post KD, Biller BJ, Prager RJ, Jackson IM, and Reichlin S (1980) Galactorrhea, oligo/amenorrhea, and hyperprolactinemia in patients with craniopharyngiomas. *Journal of Clinical Endocrinology and Metabolism* 51: 798–800.

Karavitaki N, Brufani C, Warner JT, et al. (2005) Craniopharyngiomas in children and adults: Systematic analysis of 121 cases with long-term follow-up. *Clinical Endocrinology (Oxf)* 62: 397–409.

Karavitaki N, Cudlip S, Adams CB, and Wass JA (2006a) Craniopharyngiomas. *Endocrine Reviews* 27: 371–397.

Karavitaki N, Thanabalasingham G, Shore HC, et al. (2006b) Do the limits of serum prolactin in disconnection hyperprolactinaemia need re-definition? A study of 226 patients with histologically verified non-functioning pituitary macroadenoma. *Clinical Endocrinology (Oxf)* 65: 524–529.

Kauli R, Prager-Lewin R, and Laron Z (1978) Pubertal development in the Prader–Labhart–Willi syndrome. *Acta Paediatrica Scandinavica* 67: 763–767.

Kelberman D and Dattani MT (2007) Genetics of septo-optic dysplasia. *Pituitary* 10: 393–407.

Keles GE, Banerjee A, Puri D, and Bergen MS (2004) Supratentorial gliomas. In: Gupta N, Banerjee A, and Haas-Kogan D (eds.) *Pediatric CNS Tumors*, pp. 1–26. New York: Springer.

Kelly R (1951) Colloid cysts of the third ventricle; analysis of twenty-nine cases. *Brain* 74: 23–65.

Kelsey CR and Marks LB (2006) Somnolence syndrome after focal radiation therapy to the pineal region: Case report and review of the literature. *Journal of Neurooncology* 78: 153–156.

Keung YK, Buss D, Powell BL, and Pettenati M (2002) Central diabetes insipidus and inv(3)(q21q26) and monosomy 7 in acute myeloid leukemia. *Cancer Genetics and Cytogenetics* 136: 78–81.

King BM (2006) The rise, fall, and resurrection of the ventromedial hypothalamus in the regulation of feeding behavior and body weight. *Physiology and Behavior* 87: 221–244.

Kinjo T, al-Mefty O, and Ciric I (1995) Diaphragma sellae meningiomas. *Neurosurgery* 36: 1082–1092.

Kirgis HD and Locke W (1972) Anatomy and embryology. In: Locke W and Schally AV (eds.) *The Hypothalamus and Pituitary in Health and Disease*, pp. 3–21. Springfield, MA: Charles C. Thomas.

Kirollos RW, Javadpour M, May P, and Mallucci C (2001) Endoscopic treatment of suprasellar and third ventricle-related arachnoid cysts. *Child's Nervous System* 17: 713–718.

Kjellberg RN (1979) Craniopharyngiomas. In: Tindall GT and Collins WF (eds.) *Clinical Management of Pituitary Disorders*, pp. 373–388. New York: Raven Press.

Kloos RT (1995) Spontaneous periodic hypothermia. *Medicine (Baltimore)* 74: 268–280.

Korsgaard O, Lindholm J, and Rasmussen P (1976) Endocrine function in patients with suprasellar and hypothalamic tumours. *Acta Endocrinologica (Copenh)* 83: 1–8.

Krude H, Biebermann H, Luck W, Horn R, Brabant G, and Gruters A (1998) Severe early-onset obesity, adrenal insufficiency and red hair pigmentation caused by POMC mutations in humans. *Nature Genetics* 19: 155–157.

Kubo O, Yamasaki N, Kamijo Y, Amano K, Kitamura K, and Demura R (1977) Human chorionic gonadotropin produced by ectopic pinealoma in a girl with precocious puberty. Case report. *Journal of Neurosurgery* 47: 101–105.

Laue L, Comite F, Hench K, Loriaux DL, Cutler GB, Jr., and Pescovitz OH (1985) Precocious puberty associated with neurofibromatosis and optic gliomas. Treatment with luteinizing hormone releasing hormone analogue. *American Journal of Diseases of Children* 139: 1097–1100.

Lavabre-Bertrand T, Bourquard P, Chiesa J, et al. (2001) Diabetes insipidus revealing acute myelogenous leukaemia with a high platelet count, monosomy 7 and abnormalities of chromosome 3: A new entity? *European Journal of Haematology* 66: 66–69.

Layman LC (2007) Hypogonadotropic hypogonadism. *Endocrinology and Metabolism Clinics of North America* 36: 283–296.

Lewin K, Mattingly D, and Millis RR (1972) Anorexia nervosa associated with hypothalamic tumour. *British Medical Journal* 2: 629–630.

Little JR and MacCarty CS (1974) Colloid cysts of the third ventricle. *Journal of Neurosurgery* 40: 230–235.

Losa M, Bock L, Schopohl J, Stalla GK, Muller OA, and von Werder K (1984) Growth hormone releasing factor infusion does not sustain elevated GH-levels in normal subjects. *Acta Endocrinologica (Copenh)* 107: 462–470.

Losa M and von Werder K (1997) Pathophysiology and clinical aspects of the ectopic GH-releasing hormone syndrome. *Clinical Endocrinology (Oxf)* 47: 123–135.
Lu BS and Zee PC (2006) Circadian rhythm sleep disorders. *Chest* 130: 1915–1923.
MacCarty CS, Boyd AS, Jr., and Childs DS, Jr. (1970) Tumors of the optic nerve and optic chiasm. *Journal of Neurosurgery* 33: 439–444.
MacKenzie MA (1997) Pathophysiology and clinical implications of human poikilothermia. *Annals of the New York Academy of Sciences* 813: 738–740.
Macleod JF, Kouvacs L, and Gaskill MB (1997) Familial neurohypopyseal diabetes insipidus associated with a signal peptide mutation. *Journal of Clinical Endocrinology and Metabolism* 77: 599A–599G.
Maghnie M (2003) Diabetes insipidus. *Hormone Research* 59(supplement 1): 42–54.
Maghnie M, Cosi G, Genovese E, et al. (2000) Central diabetes insipidus in children and young adults. *New England Journal of Medicine* 343: 998–1007.
Maghnie M, Ghirardello S, DeBellis A, et al. (2006) Idiopathic central diabetes insipidus in children and young adults is commonly associated with vasopressin-cell antibodies and markers of autoimmunity. *Clinical Endocrinology (Oxf)* 65: 470–478.
Makras P, Alexandraki KI, Chrousos GP, Grossman AB, and Kaltsas GA (2007) Endocrine manifestations in Langerhans cell histiocytosis. *Trends in Endocrinology and Metabolism* 18: 252–257.
Mann MR and Bartolomei MS (1999) Towards a molecular understanding of Prader–Willi and Angelman syndromes. *Human Molecular Genetics* 8: 1867–1873.
Margalith D, Jan JE, McCormick AQ, Tze WJ, and Lapointe J (1984) Clinical spectrum of congenital optic nerve hypoplasia: Review of 51 patients. *Developmental Medicine and Child Neurology* 26: 311–322.
Margalith D, Tze WJ, and Jan JE (1985) Congenital optic nerve hypoplasia with hypothalamic–pituitary dysplasia. A review of 16 cases. *American Journal of Diseases of Children* 139: 361–366.
Marik PE (2003) Steroids and drotrecogin alfa (activated) for severe sepsis. *Chest* 124: 2033–2034.
Martin JB and Reichlin S (1987) *Clinical Neuroendocrinology.* Philadelphia, PA: F. A. Davis.
McLean AJ (1934) Autonomic epilepsy. *Archives of Neurology* 32: 189–197.
Mechanick JI, Hochberg FH, and LaRocque A (1986) Hypothalamic dysfunction following whole-brain irradiation. *Journal of Neurosurgery* 65: 490–494.
Mecklenberg RS, Loriaux DL, Thompson RH, Anderson AE, and Lipsett MB (1974) Hypothalamic dysfunction in patients with anorexia. *Medicine* 53: 147–159.
Mehta A, Hindmarsh PC, Stanhope RG, Brain CE, Preece MA, and Dattani MT (2003) Is the thyrotropin-releasing hormone test necessary in the diagnosis of central hypothyroidism in children. *Journal of Clinical Endocrinology and Metabolism* 88: 5696–5703.
Melmed S, Ezrin C, Kovacs K, Goodman RS, and Frohman LA (1985) Acromegaly due to secretion of growth hormone by an ectopic pancreatic islet-cell tumor. *New England Journal of Medicine* 312: 9–17.
Minton JA, Rainbow LA, Ricketts C, and Barrett TG (2003) Wolfram syndrome. *Reviews in Endocrine and Metabolic Disorders* 4: 53–59.
Misra M, Miller KK, Almazan C, Worley M, Herzog DB, and Klibanski A (2005) Hormonal determinants of regional body composition in adolescent girls with anorexia nervosa and controls. *Journal of Clinical Endocrinology and Metabolism* 90: 2580–2587.
Misra M, Miller KK, Bjornson J, et al. (2003) Alterations in growth hormone secretory dynamics in adolescent girls with anorexia nervosa and effects on bone metabolism. *Journal of Clinical Endocrinology and Metabolism* 88: 5615–5623.
Misra M, Miller KK, Tsai P, et al. (2006) Elevated peptide YY levels in adolescent girls with anorexia nervosa. *Journal of Clinical Endocrinology and Metabolism* 91: 1027–1033.
Mitch WE (2005) Cachexia in chronic kidney disease: A link to defective central nervous system control of appetite. *Journal of Clinical Investigation* 115: 1476–1478.
Molitch ME, Clemmons DR, Malozowski S, Merriam GR, Shalet SM, Vance ML, and Stephens PA (2006) Evaluation and treatment of adult growth hormone deficiency: An endocrine society clinical practice guideline. *Journal of Clinical Endocrinology and Metabolism* 91: 1621–1634.
Montague CT, Farooqi IS, Whitehead JP, et al. (1997) Congenital leptin deficiency is associated with severe early-onset obesity in humans. *Nature* 387: 903–908.
Morris DG, Grossman AB, and Nieman LK (2006) Cushing's syndrome. In: DeGroot LJ and Jameson JL (eds.) *Endocrinology,* vol. 1, pp. 429–464. Philadelphia, PA: Elsevier Saunders.
Muir A (2006) Precocious puberty. *Pediatrics in Review* 27: 373–381.
Muller HL, Faldum A, Etavard-Gorris N, Gebhardt U, Oeverink R, Kolb R, and Sorensen N (2003) Functional capacity, obesity and hypothalamic involvement: Cross-sectional study on 212 patients with childhood craniopharyngioma. *Klinische Pädiatrie* 215: 310–314.
Munari C, Kahane P, Francione S, et al. (1995) Role of the hypothalamic hamartoma in the genesis of gelastic fits (a video-stereo-EEG study). *Electroencephalography and Clinical Neurophysiology* 95: 154–160.
Munoz MT and Argente J (2002) Anorexia nervosa in female adolescents: Endocrine and bone mineral density disturbances. *European Journal of Endocrinology* 147: 275–286.
Nagashima K, Nakai S, Tanaka M, and Kanosue K (2000) Neuronal circuitries involved in thermoregulation. *Autonomic Neuroscience* 85: 18–25.
Nakao K, Noma K, Sato B, Yano S, Yamamura Y, and Tachibana T (1978) Serum prolactin levels in eighty patients with sarcoidosis. *European Journal of Clinical Investigation* 8: 37–40.
Nanduri VR, Bareille P, Pritchard J, and Stanhope R (2000) Growth and endocrine disorders in multisystem Langerhans' cell histiocytosis. *Clinical Endocrinology (Oxf)* 53: 509–515.
Nanduri VR and Stanhope R (1999) Why is the retention of gonadotrophin secretion common in children with panhypopituitarism due to septo-optic dysplasia? *European Journal of Endocrinology* 140: 48–50.
Navarro C, Corretger JM, Sancho A, Rovira J, and Morales L (1985) Paraneoplasic precocious puberty. Report of a new case with hepatoblastoma and review of the literature. *Cancer* 56: 1725–1729.
Newell-Price J, Bertagna X, Grossman AB, and Nieman LK (2006) Cushing's syndrome. *Lancet* 367: 1605–1617.
Newell-Price J, Trainer P, Besser M, and Grossman A (1998) The diagnosis and differential diagnosis of Cushing's syndrome and pseudo-Cushing's states. *Endocrine Reviews* 19: 647–672.
Nguyen D, Singh S, Zaatreh M, Novotny E, Levy S, Testa F, and Spencer SS (2003) Hypothalamic hamartomas: Seven cases and review of the literature. *Epilepsy and Behavior* 4: 246–258.
Nieboer P, Vellenga E, Adriaanse R, and van de Loosdrecht AA (2000) Central diabetes insipidus preceding acute myeloid

leukemia with t(3;12)(q26;p12). *Netherlands Journal of Medicine* 56: 45–47.
Nieman LK, Oldfield EH, Wesley R, Chrousos GP, Loriaux DL, and Cutler GB, Jr. (1993) A simplified morning ovine corticotropin-releasing hormone stimulation test for the differential diagnosis of adrenocorticotropin-dependent Cushing's syndrome. *Journal of Clinical Endocrinology and Metabolism* 77: 1308–1312.
Nishino S (2007) Clinical and neurobiological aspects of narcolepsy. *Sleep Medicine* 8: 373–399.
Nishio S, Inamura T, Takeshita I, Fukui M, and Kamikaseda K (1993) Germ cell tumor in the hypothalamo-neurohypophysial region: Clinical features and treatment. *Neurosurgical Review* 16: 221–227.
Nunn K, Ouvrier R, Sprague T, Arbuckle S, and Docker M (1997) Idiopathic hypothalamic dysfunction: A paraneoplastic syndrome? *Journal of Child Neurology* 12: 276–281.
Oeffinger KC, Mertens AC, Sklar CA, et al. (2006) Chronic health conditions in adult survivors of childhood cancer. *New England Journal of Medicine* 355: 1572–1582.
Ogilvy-Stuart AL, Clayton PE, and Shalet SM (1994) Cranial irradiation and early puberty. *Journal of Clinical Endocrinology and Metabolism* 78: 1282–1286.
Otrock ZK, Salti I, Merheb M, and Taher AT (2006) Diabetes insipidus and thrombocytosis as the presenting symptoms of acute myeloblastic leukemia with monosomy 7. *American Journal of Hematology* 81: 152–153.
Ouvrier R, Nunn K, Sprague T, McLean C, Arbuckle S, Hopkins I, and North K (1995) Idiopathic hypothalamic dysfunction: A paraneoplastic syndrome? *Lancet* 346: 1298.
Packer RJ, Cohen BH, and Cooney K (2000) Intracranial germ cell tumors. *Oncologist* 5: 312–320.
Packer RJ, Sutton LN, Bilaniuk LT, et al. (1988) Treatment of chiasmatic/hypothalamic gliomas of childhood with chemotherapy: An update. *Annals of Neurology* 23: 79–85.
Palkovits M, Brownstein M, Saavedra JM, and Axelrod J (1974) Norepinephrine and dopamine content of hypothalamic nuclei of the rat. *Brain Research* 77: 137–149.
Palmer BF (2003) Hyponatremia in patients with central nervous system disease: SIADH versus CSW. *Trends in Endocrinology and Metabolism* 14: 182–187.
Patel L, McNally RJ, Harrison E, Lloyd IC, and Clayton PE (2006) Geographical distribution of optic nerve hypoplasia and septo-optic dysplasia in Northwest England. *Journal of Pediatrics* 148: 85–88.
Paulman PM and Sadat A (1990) Pseudocyesis. *Journal of Family Practice* 30: 575–576.
Pelletier G (1980) Localization of active peptices in brain. In: Motta M (ed.) *Endocrine Functions of the Brain*, pp. 155–169. New York: Raven Press.
Pelletier G, Desy L, Cote J, Lefevre G, and Vaudry H (1986) Light-microscopic immunocytochemical localization of growth hormone-releasing factor in the human hypothalamus. *Cell Tissue Research* 245: 461–463.
Pelletier G, Desy L, Cote J, and Vaudry H (1983) Immunocytochemical localization of corticotropin-releasing factor-like immunoreactivity in the human hypothalamus. *Neuroscience Letters* 41: 259–263.
Pena-Almazan S, Buchlis J, Miller S, Shine B, and MacGillivray M (2001) Linear growth characteristics of congenitally GH-deficient infants from birth to one year of age. *Journal of Clinical Endocrinology and Metabolism* 86: 5691–5694.
Persani L, Ferretti E, Borgato S, Faglia G, and Beck-Peccoz P (2000) Circulating thyrotropin bioactivity in sporadic central hypothyroidism. *Journal of Clinical Endocrinology and Metabolism* 85: 3631–3635.
Pierre-Kahn A, Capelle L, Brauner R, Sainte-Rose C, Renier D, Rappaport R, and Hirsch JF (1990) Presentation and management of suprasellar arachnoid cysts. Review of 20 cases. *Journal of Neurosurgery* 73: 355–359.
Pinkney J, Wilding J, Williams G, and MacFarlane I (2002) Hypothalamic obesity in humans: What do we know and what can be done? *Obesity Reviews* 3: 27–34.
Plum F and VanUitert R (1978) Nonendocrine disease and disorders of the hypothalamus. In: Reichlin S, Baldessarini RJ, and Martin JB (eds.) *The Hypothalamus*, pp. 415–473. New York: Raven Press.
Poeck K and Pilleri G (1965) Release of hypersexual behavior due to a lesion in the limbic system. *Acta Neurologica Scandinavica* 41: 233–244.
Polizzi A, Pavone P, Iannetti P, Manfre L, and Ruggieri M (2006) Septo-optic dysplasia complex: A heterogeneous malformation syndrome. *Pediatric Neurology* 34: 66–71.
Porter N, Beynon HL, and Randeva HS (2003) Endocrine and reproductive manifestations of sarcoidosis. *Quarterly Journal of Medicine* 96: 553–561.
Poussaint TY, Barnes PD, Nichols K, Anthony DC, Cohen L, Tarbell NJ, and Goumnerova L (1997) Diencephalic syndrome: Clinical features and imaging findings. *American Journal of Neuroradiology* 18: 1499–1505.
Powell GF, Brasel JA, and Blizzard RM (1967) Emotional deprivation and growth retardation simulating idiopathic hypopituitarism. Part I: Clinical evaluation of the syndrome. *New England Journal of Medicine* 276: 1271–1278.
Powell GF, Brasel JA, Raiti S, and Blizzard RM (1967) Emotional deprivation and growth retardation simulating idiopathic hypopituitarism. Part II: Endocrinologic evaluation of the syndrome. *New England Journal of Medicine* 276: 1279–1283.
Prader A, Labhart A, and Willi H (1956) Ein syndrome von adipositas, kleinwuchs, kryptorchismus, und oligophrenie nach myotonieartigen zustand im neugeborenenalter. *Schweizerische Medizinische Wochenschrift* 86: 1260–1261.
Pradilla G and Jallo G (2007) Arachnoid cysts: Case series and review of the literature. *Neurosurgical Focus* 22: E7.
Ra'anani P, Shpilberg O, Berezin M, and Ben-Bassat I (1994) Acute leukemia relapse presenting as central diabetes insipidus. *Cancer* 73: 2312–2316.
Raiti S, Davis WT, and Blizzard RM (1967) A comparison of the effects of insulin hypoglycaemia and arginine infusion on release of human growth hormone. *Lancet* 2: 1182–1183.
Ramos EJ, Suzuki S, Marks D, Inui A, Asakawa A, and Meguid MM (2004) Cancer anorexia–cachexia syndrome: Cytokines and neuropeptides. *Current Opinion in Clinical Nutrition and Metabolic Care* 7: 427–434.
Randall RV, Clark EC, and Dodge HWJ (1961) Scientific exhibit: Diabetes insipidus. Current concepts in the production of antidiuretic hormone. Clinical and experimental observations. *Postgraduate Medicine* 29: 97–107.
Randall RV, Laws ERJ, and Abboud CF (1984) Clinical presentations of craniopharyngiomas. A brief review of 300 cases. In: Givens JR, Kitabchi AE, and Robertson JT (eds.) *The Hypothalamus*, pp. 321–333. Chicago, IL: Year Book Medical Publishers.
Rappaport ZH (1993) Suprasellar arachnoid cysts: Options in operative management. *Acta Neurochirurgica (Wien)* 122: 71–75.
Reeves AG and Plum F (1969) Hyperphagia, rage, and dementia accompanying a ventromedial hypothalamic neoplasm. *Archives of Neurology* 20: 616–624.
Regis J, Scavarda D, Tamura M, et al. (2006) Epilepsy related to hypothalamic hamartomas: Surgical management with

special reference to gamma knife surgery. *Child's Nervous System* 22: 881–895.

Riddoch G (1938) Clinical aspects of hypothalamic derangement. In: Le GrosClark,WE, Beattie J, Riddoch G, and Dott NM (eds.) *The Hypothalamus. Morphological, Functional, Clinical and Surgical Aspects*, pp. 101–130. London: Oliver and Boyd.

Rimoin DL and Schimke RN (1971) *Genetic Disorders of the Endocrine Glands.* St. Louis, MO: C. V. Mosby.

Rimon DL and Phillips JA, III (1996) Genetic disorders of the pituitary gland. In: Rimoin DL, Conner JM, and Pyeritz RI (eds.) *Emery and Rimoin's Principles and Practice of Medical Genetics,* vol. 1, pp. 1331–1364. New York: Churchill Livingstone.

Roberson C and Till K (1974) Hypothalamic gliomas in children. *Journal of Neurology, Neurosurgery and Psychiatry* 37: 1047–1052.

Robertson GL, Aycinena P, and Zerbe RL (1982) Neurogenic disorders of osmoregulation. *American Journal of Medicine* 72: 339–353.

Rodriguez LA, Edwards MS, and Levin VA (1990) Management of hypothalamic gliomas in children: An analysis of 33 cases. *Neurosurgery* 26: 242–246; discussion 246–247.

Roessmann U, Velasco ME, Small EJ, and Hori A (1987) Neuropathology of septo-optic dysplasia (de Morsier syndrome) with immunohistochemical studies of the hypothalamus and pituitary gland. *Journal of Neuropathology and Experimental Neurology* 46: 597–608.

Rogol AD (1981) Pituitary and parapituitary tumors of childhood and adolescence. In: Givens JR, Kitabchi AE, and Robertson JT (eds.) *Hormone-Secreting Pituitary Tumors*, pp. 349–375. Chicago, IL: Year Book Medical Publishers.

Romanovsky AA (2007) Thermoregulation: Some concepts have changed. Functional architecture of the thermoregulatory system. *American Journal of Physiology: Regulatory, Integrative and Comparative Physiology* 292: R37–R46.

Rose BD and Post TW (eds.) (2001) *Clinical Physiology of Acid–Base and Electrolyte Disorders,* pp. 751–754. New York: McGraw-Hill.

Rosenfeld RG, Albertsson-Wikland K, Cassorla F, et al. (1995) Diagnostic controversy: The diagnosis of childhood growth hormone deficiency revisited. *Journal of Clinical Endocrinology and Metabolism* 80: 1532–1540.

Rothballer AB and Dugger GS (1955) Hypothalamic tumor. Correlation between symptomatology, regional anatomy, and neurosecretion. *Neurology* 5: 160–177.

Rush JA, Younge BR, Campbell RJ, and MacCarty CS (1982) Optic glioma. Long-term follow-up of 85 histopathologically verified cases. *Ophthalmology* 89: 1213–1219.

Russell A (1951) A diencephalic syndrome of emaciation in infancy and childhood. *Archives of Disease in Childhood* 26: 274.

Russell DS and Rubensten LJ (1984) *Pathology of Tumors of the Nervous System.* Baltimore, MD: Williams and Wilkins.

Rusyniak DE and Sprague JE (2005) Toxin-induced hyperthermic syndromes. *Medicine Clinics of North America* 89: 1277–1296.

Rusyniak DE and Sprague JE (2006) Hyperthermic syndromes induced by toxins. *Clinical Laboratory Medicine* 26: 165–184.

Rutter MM and Rose SR (2007) Long-term endocrine sequelae of childhood cancer. *Current Opinion in Pediatrics* 19: 480–487.

Saeger W, Puchner MJ, and Ludecke DK (1994) Combined sellar gangliocytoma and pituitary adenoma in acromegaly or Cushing's disease. A report of 3 cases. *Virchows Archiv* 425: 93–99.

Saito T, Ishikawa S, Abe K, et al. (1997) Acute aquaresis by the nonpeptide arginine vasopressin (AVP) antagonist OPC-31260 improves hyponatremia in patients with syndrome of inappropriate secretion of antidiuretic hormone (SIADH). *Journal of Clinical Endocrinology and Metabolism* 82: 1054–1057.

Sakurai T, Amemiya A, Ishii M, et al. (1998) Orexins and orexin receptors: A family of hypothalamic neuropeptides and G protein-coupled receptors that regulate feeding behavior. *Cell* 92: 573–585.

Samaan NA, Schultz PN, Yang KP, Vassilopoulou-Sellin R, Maor MH, Cangir A, and Goepfert H (1987) Endocrine complications after radiotherapy for tumors of the head and neck. *Journal of Laboratory and Clinical Medicine* 109: 364–372.

Sandyk R, Iacono RP, and Bamford CR (1987) The hypothalamus in Parkinson disease. *Italian Journal of Neurological Sciences* 8: 227–234.

Sanford RA and Muhlbauer MS (1991) Craniopharyngioma in children. *Neurologic Clinics* 9: 453–465.

Sano K, Mayanagi Y, Sekino H, Ogashiwa M, and Ishijima B (1970) Results of stimulation and destruction of the posterior hypothalamus in man. *Journal of Neurosurgery* 33: 689–707.

Saper CB, Chou TC, and Scammell TE (2001) The sleep switch: Hypothalamic control of sleep and wakefulness. *Trends in Neuroscience* 24: 726–731.

Saper CB, Scammell TE, and Lu J (2005) Hypothalamic regulation of sleep and circadian rhythms. *Nature* 437: 1257–1263.

Sapienza C and Hall JG (1995) Genetic imprinting in human disease. In: Scriver CR, Beaudet AR, and Sly WS (eds.) *The Metabolic and Molecular Basis of Injerited Disease,* vol. 1, pp. 437–458. New York: McGraw-Hill.

Sarkissian V (2005) The sequelae of cranial irradiation on human cognition. *Neuroscience Letters* 382: 118–123.

Sato M, Ushio Y, Arita N, and Mogami H (1985) Hypothalamic hamartoma: Report of two cases. *Neurosurgery* 16: 198–206.

Scherbaum WA, Wass JA, Besser GM, Bottazzo GF, and Doniach D (1986) Autoimmune cranial diabetes insipidus: Its association with other endocrine diseases and with histiocytosis X. *Clinical Endocrinology (Oxf)* 25: 411–420.

Schmiegelow M, Feldt-Rasmussen U, Rasmussen AK, Lange M, Poulsen HS, and Muller J (2003) Assessment of the hypothalamo-pituitary–adrenal axis in patients treated with radiotherapy and chemotherapy for childhood brain tumor. *Journal of Clinical Endocrinology and Metabolism* 88: 3149–3154.

Schneider HJ, Kreitschmann-Andermahr I, Ghigo E, Stalla GK, and Agha A (2007) Hypothalamo-pituitary dysfunction following traumatic brain injury and aneurysmal subarachnoid hemorrhage: A systematic review. *Journal of the American Medical Association* 298: 1429–1438.

Schvarcz JR, Driollet R, Rios E, and Betti O (1972) Stereotactic hypothalamotomy for behavior disorders. *Journal of Neurology, Neurosurgery and Psychiatry* 35: 356–359.

Schwartz WJ, Busis NA, and Hedley-Whyte ET (1986) A discrete lesion of ventral hypothalamus and optic chiasm that disturbed the daily temperature rhythm. *Journal of Neurology* 233: 1–4.

Schwartz MW, Woods SC, Porte D, Jr., Seeley RJ, and Baskin DG (2000) Central nervous system control of food intake. *Nature* 404: 661–671.

Seminara SB, Hayes FJ, and Crowley WF, Jr. (1998) Gonadotropin-releasing hormone deficiency in the human (idiopathic hypogonadotropic hypogonadism and Kallmann's syndrome): Pathophysiological and genetic considerations. *Endocrine Reviews* 19: 521–539.

Shankar RR and Pescovitz OH (1995) Precocious puberty. *Advances in Endocrinology and Metabolism* 6: 55–89.
Shapiro WR, Williams GH, and Plum F (1969) Spontaneous recurrent hypothermia accompanying agenesis of the corpus callosum. *Brain* 92: 423–436.
Sharma RR (1987) Hamartoma of the hypothalamus and tuber cinereum: A brief review of the literature. *Journal of Postgraduate Medicine* 33: 1–13.
Shinoda J, Sakai N, Yano H, Hattori T, Ohkuma A, and Sakaguchi H (2004) Prognostic factors and therapeutic problems of primary intracranial choriocarcinoma/germ-cell tumors with high levels of HCG. *Journal of Neurooncology* 66: 225–240.
Siatkowski RM, Sanchez JC, Andrade R, and Alvarez A (1997) The clinical, neuroradiographic, and endocrinologic profile of patients with bilateral optic nerve hypoplasia. *Ophthalmology* 104: 493–496.
Simson LR, Lampe I, and Abell MR (1968) Suprasellar germinomas. *Cancer* 22: 533–544.
Singhal S, Birch JM, Kerr B, Lashford L, and Evans DG (2002) Neurofibromatosis type 1 and sporadic optic gliomas. *Archives of Disease in Childhood* 87: 65–70.
Sirvent N, Berard E, Chastagner P, Feillet F, Wagner K, and Sommelet D (2003) Hypothalamic dysfunction associated with neuroblastoma: Evidence for a new paraneoplastic syndrome? *Medical and Pediatric Oncology* 40: 326–328.
Sklar CA (1994) Craniopharyngioma: Endocrine abnormalities at presentation. *Pediatric Neurosurgery* 21(supplement 1): 18–20.
Sklar CA, Grumbach MM, Kaplan SL, and Conte FA (1981) Hormonal and metabolic abnormalities associated with central nervous system germinoma in children and adolescents and the effect of therapy: Report of 10 patients. *Journal of Clinical Endocrinology and Metabolism* 52: 9–16.
Skuse D, Albanese A, Stanhope R, Gilmour J, and Voss L (1996) A new stress-related syndrome of growth failure and hyperphagia in children, associated with reversibility of growth-hormone insufficiency. *Lancet* 348: 353–358.
Small GW (1986) Pseudocyesis: An overview. *Canadian Journal of Psychiatry* 31: 452–457.
Smirniotopoulos JG, Rushing EJ, and Mena H (1992) Pineal region masses: Differential diagnosis. *Radiographics* 12: 577–596.
Snyder PJ, Jacobs LS, Rabello MM, Sterling FH, Shore RN, Utiger RD, and Daughaday WH (1974) Diagnostic value of thyrotrophin-releasing hormone in pituitary and hypothalamic diseases. Assessment of thyrotrophin and prolactin secretion in 100 patients. *Annals of Internal Medicine* 81: 751–757.
Spada A, Mantovani G, and Lania A (2005) Pathogenesis of prolactinomas. *Pituitary* 8: 7–15.
Spiegel EA and Wycis HT (1968) Multiplicity of subcortical localization of various functions. *Journal of Nervous and Mental Disease* 147: 45–48.
Starceski PJ, Lee PA, Albright AL, and Migeon CJ (1990) Hypothalamic hamartomas and sexual precocity. Evaluation of treatment options. *American Journal of Diseases of Children* 144: 225–228.
Stosel H and Braunstein GD (1991) Endocrine abnormalities associated with Langerhans' cell histiocytosis. *Endocrinologist* 1: 393–397.
Stoving RK, Hangaard J, Hansen-Nord M, and Hagen C (1999a) A review of endocrine changes in anorexia nervosa. *Journal of Psychiatric Research* 33: 139–152.
Stoving RK, Veldhuis JD, Flyvbjerg A, et al. (1999b) Jointly amplified basal and pulsatile growth hormone (GH) secretion and increased process irregularity in women with anorexia nervosa: Indirect evidence for disruption of feedback regulation within the GH-insulin-like growth factor I axis. *Journal of Clinical Endocrinology and Metabolism* 84: 2056–2063.
Stuart CA, Neelon FA, and Lebovitz HE (1978) Hypothalamic insufficiency: The cause of hypopituitarism in sarcoidosis. *Annals of Internal Medicine* 88: 589–594.
Sun SS, Schubert CM, Chumlea WC, et al. (2002) National estimates of the timing of sexual maturation and racial differences among US children. *Pediatrics* 110: 911–919.
Swaab DF, Hofman MA, Lucassen PJ, Purba JS, Raadsheer FC, and Van de Nes JA (1993) Functional neuroanatomy and neuropathology of the human hypothalamus. *Anatomy and Embryology (Berl)* 187: 317–330.
Szymusiak R, Gvilia I, and McGinty D (2007) Hypothalamic control of sleep. *Sleep Medicine* 8: 291–301.
Takeuchi J and Handa H (1985) Pubertas praecox and hypothalamic hamartoma. *Neurosurgical Review* 8: 225–231.
Takeuchi J, Handa H, and Nagata I (1978) Suprasellar germinoma. *Journal of Neurosurgery* 49: 41–48.
Teelucksingh S, Steer CR, Thompson CJ, Seckl JR, Douglas NJ, and Edwards CR (1991) Hypothalamic syndrome and central sleep apnoea associated with toluene exposure. *QJM* 78: 185–190.
Thomas MA and Rebar RW (1990) The endocrinology of anorexia nervosa and bulimia nervosa. *Current Opinion in Obstetrics and Gynecology* 2: 831–836.
Thompson HJ, Tkacs NC, Saatman KE, Raghupathi R, and McIntosh TK (2003) Hyperthermia following traumatic brain injury: A critical evaluation. *Neurobiology of Disease* 12: 163–173.
Tseng BS and Haas-Kogan D (2004) Neurocutaneous syndromes and associated CNS tumors. In: Gupta N, Banerjee A, and Haas-Kogan D (eds.) *Pediatric CNS Tumors*, pp. 183–203. New York: Springer.
Valdueza JM, Cristante L, Dammann O, et al. (1994) Hypothalamic hamartomas: With special reference to gelastic epilepsy and surgery. *Neurosurgery* 34: 949–958; discussion 958.
Van den Berghe G (2003) Endocrine evaluation of patients with critical illness. *Endocrinology and Metabolism Clinics of North America* 32: 385–410.
Vanelli M, Bernasconi S, Caronna N, Virdis R, Terzi C, and Giovannelli G (1984) Precocious puberty in a male with Prader–Labhart–Willi syndrome. *Helvetica Paediatrica Acta* 39: 373–377.
Vanhorebeek I, Langouche L, and Van den Berghe G (2006) Endocrine aspects of acute and prolonged critical illness. *Nature Clinical Practice Endocrinology and Metabolism* 2: 20–31.
Vanhorebeek I and Van den Berghe G (2006) The neuroendocrine response to critical illness is a dynamic process. *Critical Care Clinics* 22: 1–15.
Velamoor VR, Norman RM, Caroff SN, Mann SC, Sullivan KA, and Antelo RE (1994) Progression of symptoms in neuroleptic malignant syndrome. *Journal of Nervous and Mental Disease* 182: 168–173.
Verbalis JG (2003) Disorders of body water homeostasis. *Best Practice and Research Clinical Endocrinology and Metabolism* 17: 471–503.
Vesely DL, Maldonodo A, and Levey GS (1977) Partial hypopituitarism and possible hypothalamic involvement in sarcoidosis: Report of a case and review of the literature. *American Journal of Medicine* 62: 425–431.
Weinberger LM and Grant FC (1941) Precocious puberty and tumors of the hypothalamus. *Archives of Internal Medicine* 67: 762–792.
Wen BC, Hussey DH, Staples J, Hitchon PW, Jani SK, Vigliotti AP, and Doornbos JF (1989) A comparison of the roles of surgery and radiation therapy in the management of

craniopharyngiomas. *International Journal of Radiation Oncology, Biology, Physics* 16: 17–24.
White LE and Hain RF (1959) Anorexia in association with a destruction lesion of the hypothalamus. *American Medical Association: Archives of Pathology* 68: 275–281.
White KD, Scoones DJ, and Newman PK (1996) Hypothermia in multiple sclerosis. *Journal of Neurology, Neurosurgery and Psychiatry* 61: 369–375.
Wolf OT (2003) HPA axis and memory. *Best Practice and Research Clinical Endocrinology and Metabolism* 17: 287–299.
Wolff SM, Adler RC, Buskirk ER, and Thompson RH (1964) A syndrome of periodic hypothalamic discharge. *American Journal of Medicine* 36: 956–967.
Yasargil MG, Curcic M, Kis M, Siegenthaler G, Teddy PJ, and Roth P (1990) Total removal of craniopharyngiomas. Approaches and long-term results in 144 patients. *Journal of Neurosurgery* 73: 3–11.
Yen SS, Rebar RW, and Quesenberry W (1976) Pituitary function in pseudocyesis. *Journal of Clinical Endocrinology and Metabolism* 43: 132–136.
Yeo GS, Farooqi IS, Aminian S, Halsall DJ, Stanhope RG, and O'Rahilly S (1998) A frameshift mutation in MC4R associated with dominantly inherited human obesity. *Nature Genetics* 20: 111–112.
Yukizane S, Kimura Y, Yamashita Y, Matsuishi T, Horikawa H, Ando H, and Yamashita F (1990) Growth hormone deficiency of hypothalamic origin in septo-optic dysplasia. *European Journal of Pediatrics* 150: 30–33.
Zhang Y, Proenca R, Maffei M, Barone M, Leopold L, and Friedman JM (1994) Positional cloning of the mouse obese gene and its human homologue. *Nature* 372: 425–432.
Zuber T and Kelly J (1984) Pseudocyesis. *American Family Physician* 30: 131–134.

20 Stress and Anxiety Disorders

E A Young, S N Garfinkel, and I Liberzon, University of Michigan School of Medicine, Ann Arbor, MI, USA

Chapter Outline

20.1 Introduction

20.1.1 Stress, Fear, and Anxiety

This chapter reviews the laboratory and clinical findings of interaction between stress exposure, fear, and anxiety, with particular emphasis on the possible role of stress in the generation of anxiety symptoms and anxiety disorders. Careful examination of the interrelationships between stress, anxiety, and fear reveals an often-confusing picture due to both the degree of conceptual overlap and the liberal use of these definitions in the literature. To minimize potential confusion, we will be using these concepts in the following manner: stress represents an interaction between a particular type of environmental stimuli (stressors) and a number of specific stress response systems (namely hypothalamic–pituitary–adrenal (HPA) axis and/or catecholamines). Anxiety and fear, on the other hand, constitute a set of behavioral, cognitive, and physiologic responses to threatening situations or uncertainty. While fear often constitutes a normal response to a well-defined threat, anxiety is often dissociated from the external stimulus, and is not necessarily associated with a particular physiological response. From these definitions one can appreciate the potential for conceptual overlap and confusion, stemming from two sources: (1) anxiety and fear can be a part of the stress response, and

(2) anxiety and fear, in turn, can constitute a component of a potential stressor. However, in this chapter we focus on the long-lasting effects of stress on anxiety symptoms and behaviors, usually examining the relationships between stress exposure and symptoms of anxiety that are dissociated in time.

An additional important distinction is between normal or adaptive fear and anxiety and the pathological conditions. While the character of behavioral cognitive and autonomic responses might not differ between the normal and the pathological conditions, the context in which they occur, their intensity and the degree of their effects on overall behavior, defines the extent of the pathology. This phenomenological overlap, at times, leads to erroneous assumptions of identical neurophysiology underlying both normal and pathological anxiety. For example, animal models of normal fear demonstrated a central role of the amygdaloid complex in the expression of fear and anxiety. Interestingly, when researchers modeled abnormal fear, or pathological anxiety, additional or extra-amygdaloid neuroanatomical regions have often been implicated as well as the central gray or lateral hypothalamus. These regions exchange projections with the amygdaloid complex and it is possible that abnormal input or abnormal modulation of amygdaloid activity, originating from these regions, is involved in abnormal or pathological fear. However, it is also possible that abnormal function of these regions independent of the amygdaloid complex activity is involved in the generation of pathological anxiety. A better understanding of pathological anxiety and valid animal models is needed in order to empirically test these competing hypotheses.

The existing overlap between depression and pathological anxiety further complicates the picture and contributes to overlapping definitions. The role of stress in the generation of depression has been described extensively (Brown and Harris, 1978; Brown et al., 1994; Frank et al., 1994), and often, prominent anxiety symptoms are found in depressed patients. However, depression without anxiety symptoms can also be associated with stress, while abnormal anxiety can occur without obvious link to stress exposure – as exemplified by simple phobias. These observations suggest that more than a single mechanism might be involved both in the generation of pathological anxiety and in the effects of stress on fear and mood regulation. A particularly interesting example of stress/anxiety interaction is the field of post-traumatic stress. Post-traumatic stress disorder (PTSD) is, per definition, a stress disorder (induced by trauma) and the clinical picture includes multiple manifestations of pathological anxiety (among other symptoms). Furthermore, laboratory findings in PTSD also suggest changes in hormonal systems involved in the stress response. Traditionally, stress studies have primarily been focused on the investigation of particular neuroendocrine axes, while studies of fear and anxiety focused on cognitive and psychophysiologic responses, in humans, and on behavioral responses, in animals. A combination of these diverse modalities and both clinical and basic science approaches have contributed to substantial growth of knowledge in these fields lately, and even more integrative research will be needed in the future to further elucidate complex interactions between these systems.

20.1.2 Anxiety Disorders and Stressful Events – Is There a Connection? The Role of Life Events

Since the early idea of stress and the description of the general adaptational syndrome by Selye, the association of psychiatric disorders with stress has persisted. The first effort to measure life events, and to relate them to onset, severity, and/or course of illness, was by Holmes and Rahe (1967). There has been a substantial amount of research on the effects of social factors, stress, and, specifically, life events on the occurrence of depression (Brown and Harris, 1978, 1989; Finlay-Jones, 1981; O'Connell and Mayo, 1988; Paykel, 1994). The preponderance of the data supports a role for life events in the occurrence of depression. There has been less research on the role of life events in people with anxiety disorders, although some research supports a role in anxiety disorders other than PTSD. Finlay-Jones and Brown (1981) reported that life events associated with danger were associated with anxiety symptomatology, while those associated with loss were associated with depression; subsequent research has provided some support for this finding (Miller and Ingham, 1983; Torgesen, 1985; Deadman et al., 1989), although not all studies agree (Eaton and Ritter, 1988). This group (Brown et al., 1993; Brown and Harris, 1993) has proposed a model, based on their data, in which childhood abuse and neglect lead to increased risk for both depression and anxiety, while recent stressful life events lead to depression.

Other investigators have also addressed this question in anxiety. Raskin et al. (1982) found that, in comparison to patients with generalized anxiety disorder, panic disorder patients reported more grossly disturbed childhood environment. Faravelli (1985) reported that panic disorder patients showed a large increase in significant life events in the month before panic onset. Roy-Byrne and Uhde (1988) found that the occurrence of either loss or separation in panic disorder patients raised the likelihood of the subsequent occurrence of a depressive episode in these panic disorder patients, but did not influence the severity of the preexisting panic disorder.

20.2 Description of Basic Stress and Anxiety Systems

20.2.1 Stress-Response Systems: Stress and HPA-Axis Regulation

Stress activates secretion of a number of hormones, but the main stress hormone system is the HPA axis. Stress-sensitive systems in multiple areas of the brain are activated by stress and integrated at the hypothalamus resulting in a hormonal cascade leading to cortisol secretion by the adrenals. Neurons in the paraventricular nucleus (PVN) of the hypothalamus synthesize corticotropin-releasing factor (CRF), the lead hormone in this cascade, which is secreted into the hypophyseal portal system via the median eminence (Swanson et al., 1983). In man, CRF is believed to be the primary secretagog driving pituitary corticotropes to release adrenocorticotropic hormone (ACTH). The majority of stressors that activate CRF secretion in humans are physiological/hormonal, such as exercise, insulin-induced hypoglycemia, and infection, while evidence that psychological stressors activate CRF secretion in humans is inconsistent between individuals (Hellhammer and Wade, 1993). One exception is novelty, which many studies suggest activates the HPA axis in humans (Mason, 1968). CRF release stimulates the secretion of ACTH from pituitary corticotropes, which, in turn, stimulates the secretion of cortisol from the adrenal cortex in a feedforward cascade. Glucocorticoid secretion is tightly controlled and limited by the negative feedback effects of glucocorticoids at both pituitary and brain sites. The ability of glucocorticoids to inhibit their own release has formed the basis for challenge studies such as the dexamethasone suppression test. Negative feedback of glucocorticoids on CRF and ACTH secretion can occur very rapidly, within 5–10 min, and provides real-time inhibition to limit the stress response and prevent oversecretion of glucocorticoids (Keller-Wood and Dallman, 1985). In addition to stress as an activator of CRF/ACTH/cortisol secretion, intrinsic rhythmic elements in the suprachiasmatic nucleus (SCN) drive secretion from the HPA axis in a circadian pattern. In man, the circadian rhythm in ACTH and cortisol secretion is entrained to the wake/sleep cycle (Krieger, 1979). The ACTH secretion is pulsatile in nature with the trough of secretion occurring in the evening and early night and the peak of secretion occurring just before awakening. Active secretion continues throughout the morning and early afternoon.

20.2.1.1 Links between HPA axis and noradrenergic function in animal studies

Emerging studies on stress may provide a neurobiological mechanism to explain the coactivation of both the HPA axis and sympathetic nervous system (SNS). Basic science studies on the biology of stress have suggested a central role for CRF in the coordination and integration of the stress response throughout the brain (Dunn and Berridge, 1990; Butler and Nemeroff, 1990; Koob et al., 1993). While the role of CRF from the PVN of the hypothalamus as the releasing factor for ACTH is well established (Plotsky et al., 1989), a number of studies in rodents suggest that CRF outside the PVN nucleus appears to mediate the general stress response, including the behaviors of decreased sleep, anorexia, inhibition of sexual receptivity, altered gastrointestinal (GI) motility, decreased locomotion increased startle reflex, and decreased exploratory behavior in novel environments (Dunn and Berridge, 1990; Butler and Nemeroff, 1990; Koob et al., 1993). Additionally, a number of behavioral effects of stress have been demonstrated to be reversed by central administration of alpha-helical CRF (9–41), a CRF antagonist (Koob et al., 1993). Following the initial isolation and sequencing of CRF by Vale et al. (1981), Brown et al. (1982) demonstrated that injection of CRF activated the SNS. While it was long known that stress activated the locus ceruleus (LC), the studies by Valentino (1989) demonstrating direct effects of CRF on LC neurons were particularly critical for understanding the role of CRF in mediating arousal. Subsequent studies by Aston-Jones et al. (1991) have demonstrated that the main afferent fibers to the LC arise from the nucleus paragigantocellularis (PGi) and that these neurons contain CRF. Thus, these anatomical data provide the mechanism for the LC production of arousal/anxiety behavior following CRF administration. Furthermore, studies by Plotsky (1987) found noradrenergic stimulation resulted in secretion of CRF into the hypophyseal portal blood. Consequently, it is possible that stimulation of LC noradrenergic outflow can result in the activation of the HPA axis. Finally, studies examining the effects of the HPA axis on the LC have demonstrated that cortisol may inhibit LC activity; an increase in tyrosine hydroxylase mRNA levels in LC following adrenalectomy and decreased SNS activation following increases in circulating plasma glucocorticoid levels have been reported (Mc Ewen, 1995). These studies on stress suggest an underlying mechanism by which activation of these two stress systems is linked, which is dependent upon the actions of CRF. The model of the interactions between these systems is shown in **Figure 1**.

Consequently, one can conceptualize two different but related CRF systems, the PVN/HPA-axis system and the PGi/LC system. In depression, there is clear evidence of HPA-axis activation indicating CRF hypersecretion from the PVN and suggesting extra-PVN CRF hypersecretion as a model of depression (Butler and Nemeroff, 1990). However, animal studies suggest that central CRF administration is also an excellent model of anxiety (Butler and Nemeroff, 1990). Central CRF administration can intensify anxiety symptoms in anxiogenic situations and these behavioral effects are blocked by benzodiazepines and alcohol (Britton et al., 1985; Thatcher-Britton and Koob1986). Thus, this CRF model of depression and anxiety would predict that HPA-axis activation and central noradrenergic activation would be linked. This is certainly the case in studies of mammals other than humans, but the situation is less clear with man. The majority of stressors that activate cortisol secretion in humans are physiological stressors, such as exercise, insulin-induced hypoglycemia, and infection, while evidence that psychological stressors activate cortisol secretion in humans is inconsistent between individuals (Hellhammer and Wade, 1993). One exception is novelty, which many studies suggest activates the HPA axis in humans (Mason, 1968). Certainly, evidence from examining anxiety states including reactions to exposure to phobic objects and to precipitated panic attacks suggest that these clearly psychologically stressful events are not necessarily accompanied by activation of the HPA axis, despite profound changes in heart rate and blood pressure, physiological measures dependent upon activation of the SNS including catecholamine secretion. Consequently, the circumstances under which these two stress systems are activated in a coordinated fashion in humans are not entirely clear. Exposure to extreme stressors like captivity or natural disasters (Rahe et al., 1990; Davidson and Baum, 1986) does appear to activate both systems, leading to increases in both urinary free cortisol (UFC) secretion and urinary norepinepherine (NE) excretion. However, in a study of a general epidemiological-based population sample, Young and Breslau (2004a) found no correlation between basal 24-h UFC secretion and basal 24-h urinary catecholamine secretion. But in a study examining both systems in depressed and anxious patients, we observed a significant correlation between ACTH reactivity to a stressor and growth hormone (GH) response to clonidine (an index of central noradrenergic receptor sensitivity) in normal subjects and those with anxiety but not in depressed subjects (Young et al., 2004). Because the majority of both clinical and nonclinical human studies have focused on one or the other of these systems, we know little about how they interact in humans. This may be salient to psychiatric disorders, where existing linkages, or defective linkages, between these two stress-response systems could have pathophysiological significance.

20.2.2 Anxiety and Fear – Neural Pathways

Elucidation of the neuroanatomic pathways and neurochemical functioning is crucial to the understanding of fear, abnormal anxiety, and stress/anxiety interaction. Not surprisingly, some of the same areas that have been implicated in the regulation of the stress response have also been found to play a central role in anxiety

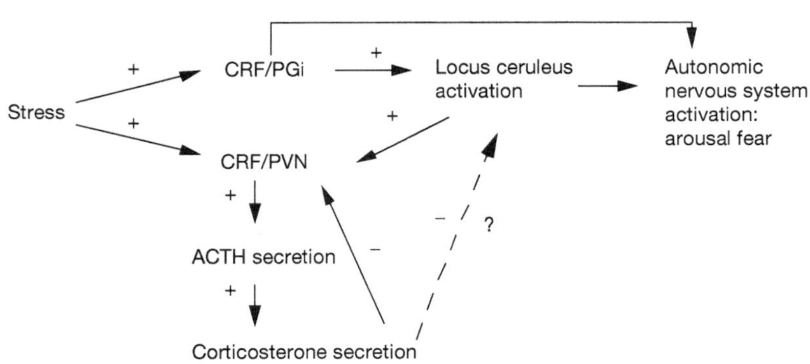

Figure 1 Rat HPA/automatic activation model.

and fear-related behaviors. Furthermore, neurotransmitter systems, like CRF and norepinephrine that are present in these regions, are involved in both the regulation of the stress response and in expression of anxiety and fear. Animal experiments and human studies have implicated a number of limbic/paralimbic regions, such as the anterior cingulate cortex, amygdaloid complex, bed nucleus of stria terminalis (BNST), temporal poles, inferior orbital cortex, and ventromedial cortex, as having a role in the regulation of fear and anxiety. In addition, structures like the hypothalamus, periaqueductal gray (PAG), and monoaminergic brainstem nuclei like LC, that have a role in arousal regulation, have been also implicated in the generation of anxiety. While detailed discussion of these regions and their involvement in stress and anxiety is beyond the scope of a single chapter and the comprehensive picture is very complex, the converging evidence suggests that the neurocircuitry that involves three groups of neuroanatomical structures are central to anxiety/stress interaction. These can be grouped into (1) limbic regions of ventral forebrain (e.g., amygdaloid complex, BNST, and hypothalamus) (2) limbic and paralimbic cortical areas, and (3) midbrain structures and monoaminergic nuclei of the brainstem.

The central role of the amygdala in fear-related behaviors has been firmly established and appears to be preserved across different species. Amygdaloid lesions in monkeys result in a loss of appropriate fearful responses (Weiskrantz, 1956). Direct stimulation of the amygdala in animals elicits dramatic behavioral responses, including fear, rage, and aggression (Kling and Brothers, 1992), and electrophysiological experiments show that it performs a mediating role in the fear-conditioning response (Davis, 1992). Davis and colleagues have demonstrated the central role of the amygdaloid structures like the central nucleus in fear-potentiated startle (Davis, 1986, 1998; Walker and Davis, 1997) which has been suggested as a model of anticipatory anxiety in humans (Davis, 1992). Finally, functional neuroimaging data in humans demonstrate amygdaloid activation in tasks associated with experience of fear, disgust, anger, and anxiety. More recent work also implicated additional ventral forebrain regions like BNST in the modulation of nonspecific anxiety (Davis, 1998). Since the BNST has cytoarchitectural similarities with the amygdala and is even considered an extended amygdala region, this further supports the role of the amygdaloid complex in fear/anxiety modulation. Interestingly, in addition to extensive projections that the BNST receives from the amygdala, it also serves as a relay for hippocampal projections to the hypothalamus involved in the regulation of stress response via fast negative feedback (Cullinan et al., 1993). This provides a potential neuroanatomical site for the interaction between stress and anxiety systems.

Limbic/paralimbic cortical regions like the anterior cingulate, temporal poles, orbitofrontal cortex, and medial prefrontal cortex (mPFC) send projections to the amygdaloid complex, and they have been implicated in mood and anxiety regulation in multiple imaging and EEG studies (Davidson et al., 1999; Kimbrell et al., 1999; Price, 1999). Some of the evidence suggests that cortical involvement in mood and anxiety regulation is lateralized, with right hemispheric activation playing a more prominent role in the generation of negative emotions and anxiety (Davidson et al., 1999). Since fear and anxiety in humans have obvious cognitive components, and can be further modulated using cognitive input, the role of inhibiting the amygdaloid response in providing cognitive context for fear/anxiety has been hypothesized for these cortical regions.

Finally, the midbrain structures and monoaminergic brainstem nuclei have been implicated in fear-related behaviors and abnormal anxiety. The PAG exchanges projections with forebrain limbic structures and both animal and human studies suggest that stimulation of this region produces extreme fear and escape-related behaviors that might represent panic-like state (Behbehani, 1995; Graeff et al., 1993). The LC, that contains up to 50% of all CNS noradrenergic cell bodies, sends direct projections to hypothalamus, amygdala, and cortex, and virtually every area in the amygdala receives at least modest noradrenergic input (Waterhouse et al., 1983; Jones et al., 1977; Swanson and Hartman, 1975; Freedman et al., 1975). The amygdala, in turn, sends direct projections to other areas like the hypothalamus and brainstem nuclei that are involved in the expression of fear or anxiety-related symptoms and behaviors. LeDoux et al. (1988) suggested that two pathways from the amygdala orchestrate various aspects of the response to fear: the amygdala–central gray pathway, that orchestrates behavioral responses, and the amygdala–lateral hypothalamus pathway, that orchestrates autonomic components of the fear response.

Interestingly, this more comprehensive description of the neuroanatomical and neurophysiological pathways involved in the fear response, also suggested the possible involvement of a number of additional neurotransmitter systems like cholecystokinin (CCK) and Substance P that were not traditionally associated with anxiety or fear responses.

20.2.3 The HPA Axis in Panic Disorder and Other Anxiety Disorders

The HPA axis has been examined in a number of anxiety disorders. Panic disorder and PTSD have been the most extensively studied of the anxiety disorders. Despite the profound psychological changes induced by either spontaneous or lactate-induced panic attack, there is no evidence that panic attacks *per se* are accompanied by cortisol secretion (Cameron et al., 1987; Abelson and Cameron, 1994). Furthermore, across a number of studies, the overall incidence of cortisol nonsupression with dexamethasone challenge is 17% in panic disorder (13 studies), while the incidence for major depression is 50% (Heninger, 1990). Studies of basal secretion of anxiety disorders are mixed (see Abelson et al. (2007) for review) but overall data do not support basal abnormalities in cortisol secretion in panic disorder. However, reactivity may be exaggerated leading to some increased cortisol in some basal studies (Abelson et al., 2007).

Studies with CRF challenge in panic disorder patients have demonstrated a decreased integrated ACTH response in comparison to controls in some studies (Roy-Byrne et al., 1986; Holsboer et al., 1987) but a normal response in others (Brambilla et al., 1992; Abelson et al., 2007). Similar to CRF challenge studies in depressed patients, baseline plasma cortisol was increased in patients with panic disorder who demonstrated blunted CRF responses. Cortisol secretion has not been demonstrated to accompany either spontaneously occurring panic attacks (Cameron et al., 1987) or lactate-induced panic attacks. Finally, studies by Nesse et al. (1985) which examined the hormonal response to exposure to a phobic object in patients with small animal phobias demonstrated only an extremely small cortisol response to this exposure, despite significant anxiety and subjective distress. Overall, data from anxiety disorders other than PTSD suggest normal HPA-axis function in these disorders.

20.2.4 The HPA Axis in PTSD

In general, the effects of repeated stress are of sensitization of the HPA axis to stressors, leading to a greater hormonal stress response over time, and increase in baseline cortisol (Dallman, 1993). Thus, it was expected that PTSD patients would show HPA-axis abnormalities similar to that seen in depressed patients or chronically stressed animals. However, that has clearly not been the case. An initial report by Mason et al. (1986) found that UFC excretion was lower in the PTSD than major depression patients, but that UFC excretion was similar between PTSD and paranoid schizophrenic patients. All patients were on psychotropic medications. After this initial report, Halbreich et al. (1988) noted that in patients with major depression, endogenous subtype, those who also met criteria for PTSD demonstrated significantly lower baseline and post-dexamethasone plasma cortisol than depressed patients without PTSD. Furthermore, none of the PTSD patients were dexamethasone nonsuppressors. Both alcohol abuse and chronic pain were present in this PTSD sample. Pitman and Orr (1990) found increased UFC excretion in outpatient PTSD veterans compared to combat controls without PTSD. In contrast, Yehuda (reviewed in Yehuda (2002)) reported decreased UFC excretion in PTSD veterans compared to normal controls. A clear difference between these two studies is the use of normal controls versus combat controls. No study of veterans has compared UFC excretion in PTSD patients, combat controls, and normal controls not exposed to combat. The one study on response to CRF challenge, in veterans with PTSD, showed normal to increased plasma cortisol at the time of the CRH challenge (Smith et al., 1989). More recent studies have examined response to low-dose dexamethasone in PTSD veterans, veterans exposed to combat without PTSD, and normal controls and found enhanced feedback to dexamethasone in veterans who met criteria for PTSD. The presence of co-morbid major depression did not alter the picture. Combat-exposed control veterans demonstrated normal supression compared to noncombat normal subjects (Yehuda, 2002). The above study suggest that the abnormalities seen in PTSD are not a consequence of exposure to trauma *per se*, but either a reflection of the underlying disorder of PTSD or a preexisting condition that may predispose to PTSD. Although Yehuda has linked the enhanced suppression of cortisol to low-dose dexamethasone in PTSD veterans to increased numbers of glucocorticoid receptors in lymphocytes, increased numbers of glucocorticoid receptors were also seen in combat-exposed veterans without PTSD. Furthermore, lower plasma cortisol would produce decreased occupancy of glucocorticoid receptors and thus, increased numbers of glucocorticoid receptors would be seen in receptor-binding assays. The most significant issue of the above studies remains the nature of the sample – veterans who are all males who also demonstrate substantial co-morbid Axis I and Axis II disorders, particularly substance abuse, and may be very different endocrinologically and psychiatrically than women exposed to trauma.

In order to address some of the concerns about the ability to extrapolate from male veterans with significant past substance abuse to civilian populations including women, Yehuda (2002) has examined UFC excretion in holocaust survivors. In this case, three groups of subjects have been studied: Holocaust subjects with PTSD, Holocaust subjects without PTSD, and age-matched normal subjects without exposure to the Holocaust. Again, these studies have shown reduced UFC excretion in subjects with PTSD, compared to normal subjects and subjects exposed to trauma without PTSD. While these studies are promising with regard to replicating the work with veterans and extending it to individuals of both genders and reducing the problems with co-morbid substance abuse, there are still problems with this population. The elderly nature of the population, the extremely long time since exposure to the trauma, the young age of the subjects at the time of trauma (often children and adolescents), which may result in different adaptations than would be observed in an adult, and the problems in classifying individuals who met criteria for PTSD in the past but who are now well, complicate interpretation of these data.

A number of studies have sought to address this problem by using nonveteran subjects recruited from clinics and the community. The majority of these studies has examined women with childhood sexual abuse. While some studies have demonstrated increased UFC (Lemieux and Coe, 1995), others have demonstrated similar plasma cortisol (Rasmusson et al., 2001) and still others have found lower cortisol and enhanced suppression to dexamethasone (Stein et al., 1997b). A recent Dutch study of chronic PTSD in civilian trauma found lower 9 a.m. plasma cortisol in PTSD subjects (Olff et al., 2006). A comprehensive study examining cortisol production rate over 24 h in ten subjects with PTSD and ten age-, sex-matched controls found normal production rate, normal 24-h plasma cortisol, normal saliva cortisol, and normal lymphocyte GR receptor number and affinity (Wheler et al., 2006). However, UFC was lower in PTSD subjects, but this was not confirmed by gas–liquid chromatography (GLC)/mass spectrometry analysis of the urine. The issue of co-morbid depression in the PTSD population is not clear, with most studies including co-morbid individuals and few analyzing the data by the absence or presence of co-morbid depression. Exceptions are the studies of Heim et al. (2000, 2001) focusing upon childhood abuse and major depressive disorder (MDD), which examined multiple HPA-axis challenges in the same subjects. These studies found an effect of early abuse (with co-morbid PTSD in 11/13 subjects) and MDD on stress reactivity, with both an increased ACTH and cortisol response to the stressor, compared to either controls or depressed patients without childhood abuse. In the same subjects, they found a blunted response to CRH challenge in MDD patients, with or without childhood abuse, but an increased response to CRH in abused patients without MDD. The abused subjects also showed a blunted cortisol response to ACTH 1–24. Thus, childhood abuse produced enhanced pituitary response with counter-regulatory adrenal adaptations, a change compatible with low or normal basal cortisol. Furthermore, they found lower cortisol and enhanced feedback to low-dose dexamethasone in the same subjects (Newport et al., 2004). These data were analyzed by the presence or absence of PTSD as the primary diagnosis and again found enhanced feedback in PTSD patients.

Epidemiological-based samples in adults have focused upon natural disasters and have generally examined exposure with high and low-PTSD symptoms (Davidson and Baum, 1986; Anisman et al., 2001), but without diagnostic information. One exception was the study of Maes et al. (1998) which looked at PTSD subjects recruited from community disasters and demonstrated increased UFC in PTSD. In general, community-based studies suggest that exposure to disaster increases plasma (Fukuda et al., 2000) and saliva cortisol (Anisman et al., 2001) and UFC (Davidson and Baum, 1986). Studies examining motor-vehicle-accident survivors (Hawk et al., 2000) found no difference in cortisol between those with and without PTSD 6 months later. Studies of male and female adults with exposure to mixed traumas have found either no effect of PTSD on basal cortisol (Young and Breslau, 2004a,b; Young et al., 2004; Kellner et al., 2002) or elevated basal cortisol (Atmaca et al., 2002; Lindley et al., 2004). Of relevance to the findings of increased basal cortisol in PTSD is our finding of increased saliva cortisol with exposure to trauma in the past year (Young et al., 2004) so that cases of PTSD with recent trauma exposure may show increased saliva cortisol. Furthermore, co-morbid depression, along with PTSD, resulted in increased saliva cortisol (Young and Breslau, 2004b; Young et al., 2004) and the majority of studies of trauma and PTSD included subjects with co-morbid depression, and commonly, subjects have both disorders.

In addition to the issue of exposure to trauma, the persistence of the neuroendocrine changes following recovery from PTSD is unclear. In an early study, Yehuda reported that Holocaust survivors with past, but not current, PTSD demonstrated normal UFC, while later studies of offspring of the Holocaust survivors (Yehuda, 2002) suggested that changes in cortisol may persist beyond the duration of the symptoms, and thus may be a marker of underlying vulnerability to PTSD. The large analysis by Boscarino (1996), of cortisol data from several thousand combat veterans, showed a very small effect of current PTSD on basal cortisol, but a very clear effect of combat exposure, with increasing levels of severity of combat exposure associated with increasingly lower cortisol. In this study, lifetime PTSD was not associated with lower cortisol. But another recent study found elevated cortisol in women with partner-violence-related lifetime PTSD (Inslicht et al., 2006).

Studies examining the response to low-dose dexamethasone in PTSD veterans, combat-exposed veterans without PTSD, and normal controls found enhanced feedback to dexamethasone in veterans with PTSD, whether or not co-morbid MDD was present; combat-exposed controls demonstrated normal suppression compared to noncombat normal subjects (Yehuda, 2002). Similar enhanced suppression to dexamethasone has been found in the Holocaust survivors with PTSD and their offspring (Yehuda, 2002). This enhanced suppression to dexamethasone has been found in studies looking at either plasma or saliva cortisol (Yehuda, 2002). In Yehuda's (2002) studies, as well as the report by Stein et al. (1997), the enhanced suppression is also paired with low baseline cortisol. Lindley et al. (2004) examined a treatment-seeking nonveteran PTSD population and found elevated basal cortisol and normal suppression to dexamethasone in subjects with PTSD. Kellner examined response to low-dose dexamethasone in anxiety disorders and found a normal response in both PTSD patients and patients with panic disorders (Kellner et al., 2002). Few other groups have utilized the low-dose dexamethasone suppression test to determine whether this is a replicable finding in PTSD and whether it is present in other anxiety disorders. More recent studies continue to show variable results (deKloet et al., 2007; Olff et al., 2006; Inslicht et al., 2006).

Activational challenges have generally used CRF challenge. An initial CRF challenge study in combat-related PTSD showed normal to increased plasma cortisol at time of challenge (Smith et al., 1989) and a decreased ACTH response in subjects with high baseline cortisol. A study by Rasmusson et al. (2001) examined women with history of childhood abuse who met criteria for PTSD. Women with PTSD showed enhanced cortisol response to CRF and to exogenous ACTH infusion, as well as a trend toward higher 24-h UFC. Interestingly, all the women with PTSD had either past or current major depression, so co-morbidity was the rule. In the study by Heim et al. (2001) examining response to CRF in women with major depression, with and without childhood abuse, 14 of 15 childhood abuse MDD patients also met criteria for PTSD. This group, with co-morbid MDD and PTSD, demonstrated a blunted ACTH response to CRF challenge, similar to that observed in MDD alone without PTSD. The abused groups also demonstrated lower baseline and stimulated cortisol in response to CRF challenge as well as following ACTH infusion. These same groups of women showed a significantly greater HPA-axis response to the trier social stress test (TSST), despite smaller responses to CRF (Heim et al., 2000).

Several additional studies have evaluated response to stressors. An early study by us using combat noise versus white noise in male veterans with PTSD showed elevated basal and postprovocation cortisol compared to combat controls but no evidence of a difference between the combat and white noise days. A study by Bremner et al. (2003) of PTSD subjects of both sexes used a stressful cognitive challenge and found elevated basal saliva cortisol and continuing higher cortisol for 60 min postchallenge. Eventually the saliva cortisol of the PTSD group returned to the same level as controls, raising the issue of whether the basal samples were truly basal or influenced by the anticipation of the challenge. Similar data were found in the study of Elzinga et al. (2003) using trauma scripts in women with childhood abuse and PTSD versus abused-no PTSD. In that study, saliva cortisol was again significantly elevated at baseline, increased in response to the challenge (while controls showed no response), and then greatly decreased following the stressor, compatible with the idea that basal levels already reflected exaggerated stress sensitivity in this group. Using a 1-min cold pressor test a very recent report (Santa Ana et al., 2006) compared the plasma ACTH and cortisol response in PTSD subjects with either childhood trauma or adult trauma to controls and saw lower basal cortisol in the childhood abuse group. However, their data do not support an actual change in ACTH or cortisol in response to the stressor in any group; so it is difficult to take their findings as reflecting differences in the stress response in subjects with PTSD. Together the existing stress data suggest an exaggerated stress response in PTSD.

Furthermore, the challenge studies suggest that the picture is complicated in PTSD with co-morbid depression; the findings of some studies look like depression while others look quite different, for example, showing a smaller response to ACTH infusion when MDD patients show an augmented response. Finally, one study by Yehuda et al. (2002), of combat veterans with PTSD, demonstrated greater rebound ACTH secretion, compared to controls, following administration of metyrapone in the morning, indicating that increased CRF drive is present in the morning and is normally restrained by cortisol feedback (Yehuda, 2002). The other two studies examining metyrapone challenge in PTSD found a normal ACTH response to afternoon or overnight metyrapone as well as a normal response to cortisol infusion in PTSD subjects and panic disorder subjects (Kellner et al., 2004; Kanter et al., 2001). In summary, these data suggest that there may be no simple relationship between PTSD and specific HPA-axis abnormalities. The lack of consistent direction of findings suggests either no changes or that differences in populations, rather than disorder, account for the variability in PTSD.

20.4 The Sympthetic Nervous System in Anxiety Disorders

20.4.1 Central Noradrenergic Regulation in Anxiety Disorders

GH response to clonidine has been widely used as a marker of central noradrenergic activity in psychiatric disorders. Clonidine is a selective α_2-adrenergic receptor partial agonist which reduces central noradrenergic outflow by activation of presynaptic receptors at noradrenergic reuptake sites. It releases GH through direct agonistic activity at postsynaptic sites. Blunted GH responses to clonidine are thought to reflect subsensitivity (downregulation) of these postsynaptic α_2-adrenergic receptors (Siever et al., 1982). Downregulation presumably occurs in response to chronic, excessive noradrenergic outflow from the LC, which is thought to play a role in anxiety states (Uhde et al., 1992). The finding of blunted GH responses to GH-releasing hormone (GHRH), and possibly to other challenges, in panic patients has raised the possibility of a more generalized dysregulation of the hypothalamic–pituitary–somatrophic (HPS) axis, in panic disorder patients, that may not reflect a specific noradrenergic defect (Uhde et al., 1992). However, the blunted response to clonidine is the only abnormality that has been replicated in panic disorder; panic disorder patients have normal GH responses to apomorphine (Pichot et al., 1995), exercise, and pyridostigmine and panic disorder patients have normal 24-h GH secretion (Abelson et al., 2005). These findings, and basic science evidence (Devesa et al., 1991), continue to support the assumption that the GH response to clonidine provides a probe of the central noradrenergic system and that the blunted response seen in panic disorder patients reflects a specific noradrenergic abnormality. Our recent study of social phobia, major depression, and their co-morbid condition found (1) a blunted GH response in the pure anxiety group; (2) a normal GH response in pure depression; and (3) a mixed response in the co-morbid group, such that those persons with predominant anxiety symptom demonstrated the same blunting as the pure anxiety group, while those individuals with predominant depression appeared like the pure depression group. (Cameron et al., 2004). We found no relationship of the GH response to dimensional measures of anxiety. These studies led us to conclude that excessive noradrenergic activation is a marker of an anxiety disorder and not of depression. Previous studies demonstrating decreased GH response to noradrenergic challenge in depression were likely affected by prior tricyclic exposure.

20.4.2 Other Noradrenergic Markers in Panic Disorders

In addition to the blunted GH response to clonidine, there is other evidence of both central and peripheral noradrenergic abnormalities in these disorders. Catecholamine and 3-methoxy-4-hydroxyphenylglycol (MHPG) levels appear to be either normal or mildly elevated in panic patients. However, there is no evidence that panic attacks are accompanied by peripheral secretion of catecholamines (Abelson and Cameron, 1994). Challenges with other pharmacological agents in addition to clonidine, primarily yohimbine, have been used, but nonspecific anxiogenic effects have made results of pharmacological challenges difficult to interpret. This problem is also true for the β-adrenergic agonist isoproterenol (as well as for a number of nonadrenergic challenge agents, such as caffeine and carbon dioxide). Results of adrenergic receptor binding on blood cells have been inconsistent, although in some groups, decreased platelet α_2-adrenergic receptor binding has been consistently observed (Cameron et al., 1984, 1990). In addition, most data suggest decreases in β-adrenergic receptor function of lymphocytes. Studies with depression and co-morbid panic disorder demonstrated that unlike panic disorder patients, patients with major depression without panic disorder demonstrated an increase in platelet α_2-adrenergic binding, while patients with co-morbid depression plus panic disorder demonstrated a decrease in platelet α_2 adrenergic binding that was even greater than the pure panic disorder patients, suggesting that the effects of noradrenergic hyperactivity predominate in the co-morbid state (Grunhaus et al., 1990). Generalized anxiety disorder and PTSD appear similar to panic disorder, although less well studied, while another obsessive–compulsive disorder (OCD) does not show evidence of noradrenergic hyperactivity. In general, measures under truly basal conditions are typically normal, but panic disorder patients may be more reactive to provocative stimuli than are normal subjects. This is true not only for anxiogenic stimuli, but also for physiologic stimuli, such as change in posture. There is evidence, from some studies, of differences between panic disorder patients and control subjects in the hemodynamic and catecholamine responses to standing, suggesting an abnormality in systemic autonomic reactivity as reflected in the control of vascular tone (Abelson and Cameron, 1994).

20.4.3 Peripheral Sympathetic Nervous System Function in PTSD

While increased adrenergic activation would appear to underlie the pathophysiology of several anxiety disorders, alterations in peripheral catecholamine systems has been difficult to demonstrate for any anxiety disorder, (Abelson and Cameron, 1994). The situation is similar in PTSD. Electrophysiological studies examining baseline heart rate, blood pressure, and galvanic skin response have demonstrated no consistent alterations in veterans with PTSD. In contrast, challenge studies have demonstrated exaggerated autonomic reactivity in response to various combat-related stimuli but not in response to non-trauma-related stimuli. Pitman's studies (Pitman et al., 1987) have used scripts of the trauma situation, lending support for the idea that memories of the trauma can activate these physiological parameters. These studies provide indirect evidence of a hyperactive SNS system in PTSD patients, but suggest the abnormality may be present episodically, dependent upon environmental cues, specifically cues associated with the trauma (Pitman et al., 1987; Murburg et al., 1994; Hamner et al., 1994). Given that these autonomic measures are regulated in opposing directions by SNS and parasympathetic nervous system input, and that increase in a parameter can result from an increase in adrenegic tone or a decrease in vagal tone, more direct measures of SNS activity are necessary. SNS activity can be evaluated by measurement of plasma epinepherine (Epi) and NE or quantitation of urinary excretion. Three published studies have examined 24-h urinary catecholamines in PTSD veterans, with conflicting results. The study by Kosten et al. (1987) and Yehuda et al. (1992) demonstrated an increase in urinary Epi and NE, in comparison to other psychiatric disorder controls (Kosten) or normal controls (Yehuda), while Pitman and Orr (1990) found no difference in PTSD veterans compared to combat controls.

Again the issue of the nature of the control group is critical, and it may be that exposure to trauma itself alters urinary Epi and NE. This possibility is supported by the studies of Davidson and Baum (1986), demonstrating increases in urinary NE in a civilian population exposed to the Three Mile Island explosion compared to individuals 80 miles away, as well as the studies of Rahe et al. (1990) demonstrating increased urinary catecholamine excretion in the American hostages shortly after they were freed from Iran. Our own studies in a population-based sample (Young and Breslau, 2004a) found increased 24-h urinary Epi, NE, and dopamine in subjects with lifetime PTSD, whether or not the PTSD was accompanied by depression, but normal urinary catecholamine secretion in subjects with pure depression. Persons exposed to trauma without PTSD demonstrated significantly lower Epi and dopamine than either unexposed or PTSD persons. Urinary NE was the same as in unexposed but lower than in PTSD persons. Recency of trauma exposure had no effect on catecholamine secretion. These data lead us to conclude that increased urinary catecholamines found in general-population studies of those exposed to trauma were likely confounded by the failure to separate those with PTSD from those exposed without PTSD.

In contrast to the data from urinary studies, plasma studies have found no elevation in baseline plasma catecholamines in PTSD patients compared to normal controls (McFall et al., 1990; Blanchard et al., 1991; Hamner et al., 1994; Southwick et al., 1993; Murburg et al., 1994) but increased catecholamine response to trauma-related stimuli were reported by Murburg et al. (1994). Interestingly, the PTSD patients in the Murburg study did not have an exaggerated plasma catecholamine response to a stressful, but trauma-unrelated, stimulus. These findings again suggest that the increases in catecholamine secretion may be occurring sporadically throughout the 24 h, perhaps in response to specific trauma-related cues. In summary, there are discrepancies between conclusions from urinary measures and plasma measures that could result from episodic activation of catecholamine secretion which are captured by urinary measures but missed by plasma measures, or it could be that the urinary measures represent a type I error because of the small sample sizes in these studies. Furthermore, the role of exposure to trauma versus PTSD symptoms has not been examined in the urinary measures.

20.5 Modeling Stress/Anxiety Interaction in Animals

Establishment of a valid animal model of a disorder or disease is one of the central steps in defining pathophysiological process, examining etiology, and developing effective treatment. The interspecies differences make this task difficult, and in particular, in conditions that are: (1) heterogeneous, (2) most likely multifactorial, and (3) characterized by humans-specific symptomatology, for example, cognitive/emotional symptoms. It is not surprising, therefore, that while extensive work has been done on the neuroanatomy and neurophysiology of fear, and a number of well-validated animal fear paradigms have been established, the establishment of animal models of anxiety disorders and stress/anxiety interaction is in relatively early stages of development. A number of animal models have attempted to emulate both the symptoms associated with anxiety disorders and the interaction between stress and anxiety; however, no single model adequately mirrors the full range of affective and cognitive symptoms associated with anxiety disorders or comprehensively addresses the issue of stress/anxiety interaction. Historically, the validity of these models has been based on presumed specificity of behavioral profiles and/or on the pharmacological actions – namely, the sensitivity of the behavioral measure to benzodiazepine compounds. This validation is weakened, however, by: (1) the recognized heterogeneity of anxiety symptoms across various anxiety disorders, (2) high degree of co-morbidity between anxiety and depression and the overlap in symptom profiles, and finally (3) the discovery of pharmacologic agents (selective serotonin reuptake inhibitors) that do not have characteristic anxiolytic properties in acute administration but are effective in treatment of anxiety disorders. In spite of the progress made in modeling specific processes, since disease-specific laboratory findings are yet to be identified in anxiety disorders, the majority of proposed models still rely on face validity of environmental manipulation, and on an exaggerated expression of normal anxiety in animals.

20.5.1 Modeling Fear versus Modeling Abnormal Anxiety

Common outcome measures used in these studies were behaviors that reflect the animal's fearfulness in threatening situations that supposedly reflect anxiety symptoms. This assumption ignores, however, the possibility that pathological anxiety and normal fear might not share the same mechanisms. For example, severe and uncontrollable stress in animals is considered similar in some aspects to psychological trauma in humans, and if an animal exposed to this type of stressor exhibits exaggerated startle response or exaggerated avoidance of open spaces, this can be proposed as a model of abnormal anxiety. This interpretation has both the obvious limitation of relying on face validity of environmental manipulation, and the assumption that exaggerated normal response is similar to abnormal or pathological anxiety that we see in anxiety disorders. It is possible, however, that pathological anxiety involves inherently different mechanisms than exaggerated normal fear. Some of the more recent animal models, however, attempted to combine both behavioral and neuroendocrine characteristics reported in anxiety disorders. This approach might provide better validated models in the future, especially when abnormal or altered neuroendocrine characteristics, that have been associated with anxiety, can be demonstrated in animals that exhibit an excessive amount of anxiety or fear.

20.5.2 Behavioral Test versus Models of Anxiety Disorders

It is important to keep in mind the distinction between models of abnormal anxiety and the behavioral paradigms that elicit specific aspects of fear-related behaviors. While the former attempts to model a pathological condition (like panic disorder, generalized anxiety, or PTSD) or particular pathophysiologic process (effect of early stress on anxiety in later life), the latter attempts to identify and isolate different types of basic processes underlying fear-related behaviors (inhibitory avoidance, condition fear conditioning, one-way escape, freezing, etc.).

Initial attempts have been made to tie these basic fear processes with a particular anxiety disorder, for example, connecting inhibitory avoidance to generalized anxiety disorder, one-way escape to panic attacks (Graeff et al., 1998), and anxiety-potentiated startle to PTSD (Davis and Shi, 1999). Some of these procedures like elevated plus maze (or its derivatives – elevated T or elevated O) or acoustic startle are more widely adopted as standard testing paradigms since they allow both assessment of the relevant underlying processes (conditioned fear, unconditioned fear, anxious avoidance, and escape) and comparison of the behavioral results across different models. These procedures, in combination with sensitivity to anxiolytic agents and specific biochemical assays, are used in more recent studies to validate animal models of anxiety.

Multiple factors play an important role in the development of anxiety disorders, including genetic predisposition, developmental vulnerabilities, and environmental exposure. Therefore, modeling anxiety disorders can involve manipulation of any one, or the combination, of these factors. An exhaustive, or even a comprehensive, review of animal models of anxiety is beyond the scope of a single chapter; therefore, we focus selectively on some of the more recent and interesting models that combine stress exposure and anxiety. A number of promising genetic (knockout models or selection of anxious breeds), developmental (intervening during vulnerable period of development), neuroanatomical/neurophysiological (lesion or stimulation studies), or stressor-specific (exposure to a particular type of stressor) models that produce pathological anxiety or specific neuroendocrine or neurophysiological changes are selected.

20.5.3 Effects of Stressful Exposure on Endocrine and Behavioral Variables

20.5.3.1 Stressor characteristics

Since fear-related behaviors, both in animals and humans, are subject to reinforcement, conditioning, and extinction, a number of animal models attempted to elicit anxiety symptoms in animals by exposing them to various types of environmental manipulations or stressors. For example, fear conditioning in animals can induce fearful response to an otherwise nonaversive stimulus, processes that might be similar in some aspects to the one that is found in simple phobia. However, simple fear conditioning is not effective in animal models of anxiety disorders, like panic disorder or PTSD. These observations contributed to the development of experimental paradigms where different types of stress exposure were used to elicit abnormal anxiety and fear in animals, focusing on particular characteristics of stress exposure (type, duration, repetitiveness, controllability, etc.). Exposure to a natural predator, social defeat, conflict, and the single prolonged stress (SPS) paradigm are some of the stressor-specific models that were proposed for anxiety disorders. Other stressor-specific stress models like uncontrollable stress or learned helplessness have also been proposed as potential models for anxiety disorders; however, the behavioral profile of the animals and the HPA-axis changes suggested that these conditions mimic depression-like states better (Yehuda and Antelman, 1993).

Adamec and Shallow (1993) reported that 5-min exposure to a natural predator (cat) produces long-lasting effects on anxiety-like behaviors in elevated plus maze, in Lewis rat (high-emotionality strain). Blockade of cholecystokin B (CCKB) receptors, that are implicated in anxiety, prevented these effects. Berton et al. (1999) suggested that social defeat in the resident–intruder paradigm, followed by continued exposure to dominant animal (separated by a transparent divider that allows visual and olfactory contact) produces long-lasting behavioral metabolic and endocrine changes. These animals were more anxious in elevated plus maze, and demonstrated both altered resting corticosterone levels and corticosterone response to forced swim test. Interestingly, fluoxetine pretreatment abolished these behavioral and HPA-axis changes. The SPS model developed in our laboratory (Liberzon et al., 1997) emphasized continued prolonged stress exposure to various stressors and prevention of habituation by eliminating any contact following stress exposure. Animals exposed to the SPS paradigm developed enhanced, fast glucocorticoid feedback of the HPA axis, which is characteristic of HPA abnormality reported in PTSD, as well as changes in glucocorticoid receptors in the hippocampus (Liberzon et al., 1999a). Earlier works by Van Dijken et al. (1993) and Antelman et al. (1988) have also suggested that a single exposure to the stressor can lead to persistent behavioral and endocrine changes. Thus, the existing evidence suggests that while most chronic stressors produce a depression-like picture in animals, in order to elicit a long-lasting anxiety-like picture (animal model of anxiety), a single session of severe stress that does not support habituation might be more effective. Furthermore, it is possible that species-specific psychological stressors might be as, or even more effective, than traditional physiologic stressors.

20.5.4 Summary of Animal Models

The initial works on animal models of stress/anxiety interaction have already produced a number of tangible results that have contributed to our current understanding of the etiology, pathophysiology, and treatment of anxiety. Recognition of the roles of serotonin receptors and CRF-binding protein in anxiety regulation, identification of the particularly vulnerable developmental stage that is highly sensitive to maternal stress and maternal behavior (and anxiety), identification of relevant characteristics of stress exposure (e.g., duration, repetition, and type), and identification of neuroanatomical regions (DMH) that are central to panic symptomatology have all been elucidated by these animal models. Even more importantly, these studies have highlighted the urgent need to better define and operationalize concepts like stress and anxiety, and to develop more sophisticated methodologies combining psychological, behavioral, and neurophysiological aspects. For example, while a few models have been established using anxious strains of animals, we are yet to find a model that used controlled manipulations on two different levels – for example, knockout animals (genetic model) that were exposed to maternal separation during development (developmental model) or social defeat (environmental model) in the adulthood. As mentioned before, the current view of the anxiety disorders often involves both developmental or genetic predisposition and environmental exposure, and if this is true, models that combine manipulations of two or more different levels might be the next logical step.

20.6 Imaging the Fear and Anxiety Pathways

In recent years, a major transformation in the concepts of mental function occurred, substituting older dualistic models, of mental versus physical or psychological versus biological, with a more integrative view of function and structure interaction. These developments also offered a new conceptualization of psychiatric disorders, which were traditionally seen as psychological or functional in nature. PTSD, for example, has been considered as a preeminently functional disorder; however, multiple lines of evidence have demonstrated stress-related neurobiological changes and neuroanatomical sequelae of traumatic exposure that might underlie symptoms of PTSD, as well as lead to increased vulnerability of PTSD following trauma (e.g., Gilbertson et al., 2002). Accordingly, an increasing number of investigators are searching for structural or functional neuroanatomical abnormalities in other anxiety disorders. One of the disorders that received a lot of attention with respect to structural and functional neuroanatomy, and that is also classified in DSM IV as an anxiety disorder, is OCD. The centrality and the prominence of abnormal cognitive processes in OCD, however, initiated many scientists to question whether OCD should be considered more appropriately as a disorder of thought and cognition, or seen as separate from the entity of other anxiety disorders. Since the consideration of the neuroanatomical circuitry of cognitive processing is clearly beyond the scope of this chapter, it will focus primarily on the findings in other anxiety disorders.

20.6.1 Structural Neuroimaging in PTSD and Anxiety Disorders – Is Cortisol Bad for Your Hippocampus?

Magnetic resonance imaging (MRI) enabled examination of the neuroanatomy of small central nervous system (CNS) structures that were previously inaccessible with computerized tomography (CT). In particular, MRI reveals structural details of medial temporal lobe (TL) areas relevant to anxiety, PTSD, and other psychiatric disorders, including the hippocampal formation, parahippocampal gyrus, and amygdaloid nuclei. Surprisingly, little structural neuroanatomical work has been done in anxiety disorders. Some reports suggested increased number of nonspecific MRI abnormalities in panic disorder (Dantendorfer et al., 1996), and to date only limited structural changes have been reported in panic patients. Gray matter volume differences have been found using voxel-based morphometry, with bilateral putamen decreases in panic subjects that were negatively correlated with duration of illness (Yoo et al., 2005). In addition, smaller TL volumes were reported in panic patients, both bilaterally and in the left TL, with no differences in the hippocampus (Vythilingam et al., 2000). Other studies using volumetric MRI reported no difference in the size of TLs in panic patients and controls, but did demonstrate smaller left-sided and right-sided amygdala volumes (Massana, 2003). Significant findings that have been reported so far focus largely on PTSD.

Bremner et al. (1995) reported that the right hippocampus of PTSD patients was approximately 8% smaller than that of a matched control group unilaterally. In a follow-up study of 17 PTSD patients with histories of childhood sexual abuse, reduced hippocampal size was found, except that the significant finding occurred in the left hippocampus (Bremner et al., 1997). Gurvetz et al. (1996) found a reduction in both right and left hippocampal size, with the PTSD groups as much as 30% smaller than the controls. Stein et al. (1997) found 5% reduction in left hippocampal volume in women who reported sexual victimization in childhood. Negative findings concerning hippocampal abnormalities associated with PTSD have also been reported (De Bellis et al., 2001). Schuff et al. (2001) found no differences in hippocampal volumes in veterans with PTSD and Bonne et al. (2001) found no hippocampal volume differences between subjects with and without PTSD at either 1 week or 6 months following traumatic events. It has been speculated that this short duration might not be long enough to produce detectable structural changes (e.g., Sala et al., 2004). Whereas, studies that had positive findings tended to include participants who had chronic PTSD with unremitting symptoms present over a period of years or even decades (Bremner et al., 1995, 1997a; Gurvits et al., 1996; Stein et al., 1997a).

Based on the important work of Sapolsky (1996), who has demonstrated in animal studies that high cortisol levels can be neurotoxic to hippocampal neurons, it has been argued that reduced volume of the medial TL in PTSD patients reflects response to environmental stress and subsequent cortisol secretion. This interpretation was further supported by smaller hippocampal volume findings in patients with Cushing's disorder (Starkman et al., 1992). Arguing from this data, some investigators suggested that the smaller hippocampal volume in PTSD could result from excessive cortisol secretion and subsequent neurotoxicity. This argument, however, failed to account for the asymmetrical volume loss reported, or for the low 24-h cortisol secretion reported in studies of PTSD (Yehuda et al., 1995). Finally, work in twins has helped to differentiate acquired signs from predisposing factors in PTSD. Gilbertson et al. (2002) investigated monozygotic twins discordant for combat exposure and PTSD status. It was found that individuals with severe PTSD and their trauma-unexposed co-twin had significantly smaller hippocampi than twin pairs without PTSD. The severity of the disorder of the PTSD patient was found to correlate with the hippocampal volume of both the PTSD patient and their trauma-unexposed, identical co-twin, serving as compelling evidence that smaller hippocampi constitute a risk factor for the development of PTSD.

20.6.2 Functional Imaging of Stress/Anxiety States

20.6.2.1 Imaging of fear in normal controls

In contrast to structural neuroimaging, functional neuroimaging depicts the brain as it carries out a particular activity, such as neuronal metabolism, blood flow, or neurotransmission, and provides considerably greater information about the function of the specific region. Current technologies include positron emission tomography (PET), single photon emission computed tomography (SPECT), both of which use radionuclide-labeled molecules, and functional magnetic resonance imaging (fMRI), which typically uses the paramagnetic properties of deoxyhemoglobin to mark blood flow changes. Recent neuroimaging work has begun to identify brain regions involved in the regulation of negative emotions like fear and anxiety. In humans viewing pictures of aversive visual stimuli, such as facial mutilation, cerebral perfusion increases in the amygdala, particularly on the left side (Irwin et al., 1996; Breiter et al., 1996; Taylor et al., 1998). Neuroimaging studies focused on the neuroanatomy of fear in control subjects strongly support the central role of the amygdala in this process. Amygdaloid activation has been implicated in the production of conditioned fear responses (Knight et al., 2005), and observed in response to unseen fear-conditioned stimuli (Morris et al., 1999), and to linguistic threat (Isenberg et al., 1999). As the putative site for attaching emotional valence, the amygdala appears to be central to the ability of sensory input to elicit emotional memories, and these data support the idea that dysfunction in the amygdala might be involved in the symptomatology of anxiety and stress disorders like phobia and PTSD.

The role of cortical regions in fear/anxiety responses is less well understood than that of the amygdala, which may reflect the fact they appear to mediate more cognitive components, for which animal models are not readily available. However, some initial cortico-amygdaloid pathway had been implicated in fear regulation. Work done by our group demonstrated that appraising emotional stimuli leads to decreased activation of the bilateral insula and right amygdala (Taylor et al., 2003). Research by Ochsner et al. (2002)

investigated how explicit reappraisal of negative emotional scenes leads to decreases in negative affect accompanied by increases in activation observed in the lateral and medial prefrontal regions, and decreased activation in the amygdala and orbitofrontal regions.

20.6.2.2 Functional neuroimaging in anxiety disorders

Both symptom provocation studies during blood-flow activation and PET receptor imaging have been employed to examine the possible role of limbic brain regions and their neurochemistry in the pathophysiology of anxiety disorders. For example, in OCD, a comparison of the cerebral blood flow (CBF) patterns while the subject views the contaminated, compared to the uncontaminated, stimulus shows activation in the orbitofronal cortex, anterior cingulate, basal ganglia, thalamus, and lateral frontal cortex (Rauch et al., 1994; Chen et al., 2004). Patients with simple phobias to small animals, such as spiders, show blood-flow increases in similar cortical regions, including the anterior cingulate, left orbitofrontal cortex, and right temporal pole, as well as the left thalamus (Rauch et al., 1995). Wik et al. (1997) reported activation of subcortical regions in animal-phobic patients involving the amygdala, thalamus, and striatum. In symptom provocation studies of social phobics, the findings more consistently point toward the involvement of the amygdaloid region, upon exposure to human faces (Birbaumer et al., 1998). In a study that investigated the anticipatory anxiety of a public-speaking task, social phobics showed greater fMRI activity in subcortical (ventral striatum and pons), limbic (amygdala region), and lateral anterior paralimbic belt (insula and temporal pole) regions and reduced activity in cortical regions (cingulate/PFC; Lorberbaum et al., 2004). In a different study, increased subjective anxiety in the social phobics during public speaking was accompanied by enhanced regional CBF (rCBF) in the amygdaloid complex (Tillfors et al., 2001). Bishop et al. (2004) have investigated the degree to which amydgala responsivity is affected by attentional focus, and reported that anxiety may interact with attentional focus to determine the magnitude of the amygdala response (Bishop et al., 2004). To date, only limited fMRI studies have been performed in panic disorder patients. In one small study using directed imagery, activations in the orbitofrontal cortex and cingulate cortex in panic disorder patients, relative to normal controls, was found for high-anxiety situations (Bystritsky et al., 2001). In another small study of fear processing, patients with panic disorder ($N = 6$), demonstrated significantly greater activation in the posterior cingulate cortex than normal subjects in response to threat words. In addition, panic disorder patients also had enhanced activation for threat-related words in the left dorsolateral PFC (Maddock et al., 2003).

In a more recent study, 22 subjects with panic disorder, on medication, had decreased rCBF in right superior temporal regions relative to healthy comparison subjects. In addition, this decrease was found to be correlated with the symptom severity of individuals' panic disorder (Lee et al., 2006). Another study investigating differences in rCBF and rCBF asymmetry index values between 22 panic disorder patients and 19 normal comparison subjects using SPECT, found panic to be associated with decreases in perfusion in the bilateral frontal regions and a relative increase in perfusion in the right medial and superior frontal regions. In addition, significant positive correlations between scores on the panic and agoraphobia scale and rCBF asymmetry index values of the parietal, superior temporal, and lateral temporal regions in the panic disorder patients were observed (Eren et al., 2003).

A few studies, to date, have examined rCBF or metabolism in panic disorder patients. In studies examining metabolic rate changes, Nordahl et al. (1990, 1998) reported right/left asymmetries (decreased L/R ratio) in hippocampal and posterior–inferior prefrontal glucose metabolism in both unmedicated and medicated panic patients as compared to controls. Bisaga et al. (1998) reported an increase in left hippocampal and parahippocampal glucose metabolism in six women with panic disorder and a decrease in the right inferior parietal and right superior temporal brain regions. Based on the anxiolytic properties of benzodiazepine compounds and their interaction with GABAergic transmission, a number of investigators examined the distribution of benzodiazepine receptors *in vivo* in panic disorder patients using specific PET and SPECT ligands. Malizia et al. (1998) reported an overall general decrease in flumazenil binding by PET imaging in panic disorder patients across the whole brain, as compared to the control group, with a greater decrease in the orbitofrontal cortex and right insula. Others reported decrease in left hippocampal and parahippocampal regions using SPECT imaging and iomazenil, while Abadie et al. (1999) found no relationships between anxiety and flumazenil binding in the brain, and no difference between flumazenil binding in the brain of control subjects and anxiety patients. While these are preliminary studies and the findings emerging are not always consistent, they do provide support for an emerging functional neuroanatomy of anxiety disorders.

20.6.2.3 Functional neuroimaging in PTSD

20.6.2.3(i) Symptom provocation

Both generic reminders of traumatic experience, such as battle footage from movies, and imagery induced by personalized scripts of the traumatic event reliably elicit exaggerated skin conductance and heart-rate responses or enhanced plasma catecholamine secretion in PTSD patients. With recent progress in the functional neuroanatomy of emotions, the exploration of rCBF changes associated with the specific psychophysiologic responses characteristic of PTSD, likely to represent some of the PTSD-related pathophysiology, became possible. Rauch et al. (1996), were the first to examine eight PTSD subjects using [^{15}O] water PET. A number of limbic and paralimbic regions were activated by traumatic imagery in this study: medial (posterior) orbitofrontal cortex, insular cortex, anterior temporal pole, and medial temporal cortex – all on the right.

Shin et al. (1997) used PET activation to study groups of combat veterans and combat controls (seven subjects each) with combat-related, neutral, and emotionally negative (but combat-unrelated) pictures, with verbal descriptions. Activation in the anterior cingulate was present in PTSD during combat imagery compared with neutral pictures, and activation in the right amygdaloid regions in PTSD was found in the comparison of combat imagery to the neutral condition. Although the amygdala was among the predicted results, interpretation was not straightforward, since a relative deactivation also occurred for normal controls in some comparisons. In a study using survivors of childhood sexual abuse, the authors demonstrated greater activation in the orbitofrontal cortex and anterior temporal poles following exposure to personal account of trauma, in survivors of abuse with PTSD. Furthermore, the PTSD group also had smaller activation of the anterior cingulate region as compared to trauma controls (Shin et al., 1999). Bremner et al. (1999) also examined response to trauma-related stimuli in Vietnam veterans with PTSD and reported differential responses in the mPFC (area 25) and anterior cingulate (area 24). We have studied PTSD patients, normal and combat controls, using SPECT imaging and found activation within the left amygdaloid region of the PTSD patients and no activation in this region in control subjects (Liberzon et al., 1999b). All three groups did show activation in the rostral anterior cingulate and mPFC. This site was very near a focus of activity reported in association with anxiety symptoms after yohimbine infusion, both in healthy controls and in panic disorder patients (Woods et al., 1988). If activation of the mPFC is associated with PTSD/anxiety symptom generation, these findings might also implicate noradrenergic mechanisms in PTSD symptomatology. This area of the brain is also associated with emotional regulation, including the recall of emotional experiences (Reiman et al., 1997). It is possible, therefore, that this medial frontal cortex activation may be associated with the processing of arousal component of meaningful stimuli, in general. A number of researchers have reported exaggerated response of the amygdala to masked fearful faces, in PTSD (e.g., Rauch et al., 2000; Armony et al., 2005). This activation in the amygdaloid region occurred in the absence of the activation in the anterior cingulate/medial prefrontal region, suggesting that the exaggerated activation in the amygdala might be independent of the diminished activation in the anterior cingulate. Others, however, found an association between increased amygdale responses and decreased mPFC responses to overt fearful faces relative to happy faces, in PTSD (Shin et al., 2005). A growing number of researchers have employed functional connectivity analyses to assess interregional covariations in brain activity in PTSD. This method reflects the growing awareness that complex cognitive and emotional processes depend on interactions of distributed brain networks. Using traumatic script-driven imagery, functional connectivity maps were found to differ in subjects with PTSD exhibiting dissociative responses relative to trauma-exposed control subjects without PTSD. Comparison of functional connectivity maps revealed that dissociated PTSD subjects showed greater covariation than the control subjects in the right insula, left parietal lobe, right middle frontal gyrus, superior temporal gyrus, and right cuneus (Lanius et al., 2005).

Investigators have also used correlational analyses to examine relationships between the activation patterns and the symptom severity. These have tended to implicate the same subset of limbic and cortical regions that were identified in earlier studies. In personalized script-driven imagery, rCBF in the brainstem, insula, and hippocampus correlated with flashback intensity (Osuch et al., 2001). Another script-driven and imagery study in Vietnam veterans (Shin et al., 2004) found that symptom severity, as determined by CAPS, was positively related to rCBF in the right amygdala and negatively related to rCBF in the medial frontal gyrus. Another approach is to investigate the time course of neural responses in PTSD, as compared to controls, particularly in regard to traumatic stimuli. Hendler and colleagues investigated whether neural responses to repeated versus novel presentations of stimuli differed in PTSD. Repeated presentations induced a greater decline in blood-oxygen-level-dependent (BOLD) signal in the lateral occipital cortex in the non-PTSD control group relative to the PTSD group, indicating a propensity for sustained neuronal responding in PTSD individuals in response to traumatic stimuli (Hendler et al., 2001).

Recent work by our group investigated changes in corticolimbic blood flow during script-driven imagery. During [0–15] H20 PET scanning, 16 combat veterans with PTSD, 15 combat veterans without PTSD, and 14 healthy aged-matched noncombat control subjects recalled emotional and neutral autobiographical events. This design allowed the differentiation of changes resulting from trauma from changes specifically associated with PTSD. Interestingly, when comparing traumatic/stressful to neutral scripts, significant deactivation patterns were found in the medial frontal cortex and cingulate cortex, with PTSD subjects deactivating the rACC to a greater extent than the control groups, while both control groups deactivated the ventromedial prefrontal cortex (vmPFC). In addition, the traumatic/stressful versus neutral comparison revealed that normal control subjects activated the amygdala to a greater extent than the two trauma groups, both with and without PTSD (Britton et al., 2005).

20.6.2.3(ii) Pharmacological challenge

In addition to psychological challenge paradigms, pharmacological challenge procedures that activate stress-response systems have also been used in PET imaging of anxiety. Yohimbine, an α-adrenergic agonist, elicits anxiety and trauma-related symptoms in PTSD patients (Southwick et al., 1997). Bremner et al. (1997) used PET [^{18}F]-2-fluoro-deoxy-D-glucose (FDG) and yohimbine infusion in a group of PTSD patients and normal controls. Yohimbine administration produced more anxiety, panic, and flashbacks in PTSD subjects, and the authors also noted a differential effect of yohimbine on brain metabolism in three out of seven hypothesized regions (orbitofrontal cortex, temporal cortex, and postcentral gyrus). These findings supported the *a priori* hypothesis of a more sensitive response to catecholamine secretion in some brain regions in subjects with PTSD. However, there was no association between anxiety symptoms and changes in brain metabolism. One of the major challenges in these types of pharmacological studies, however, is the interpretation of results. The pharmacological agent might exert a direct effect on rCBF or brain metabolism, unrelated to symptoms of interest. In addition, the pharmacological probes are seldom symptom-specific, and they elicit a set of different symptoms, making a connection between a specific symptom and rCBF change problematic. On the other hand, the potential pharmacological specificity of these probes provides additional information regarding the possible neurochemical processes underlying the observed blood-flow changes. A recent PET study performed by our group used the μ-opioid receptor radiotracer [^{11}C] carfentanil to examine the μ-opioid neurotransmitter system, implicated in responses to stress and the suppression of pain, in PTSD patients and two non-PTSD control groups, with and without combat exposure. Trauma-exposed groups, relative to noncombat controls, demonstrated decreased μ-opioid receptor-binding potential BP2 in the rostral component of extended amygdala system (SLEA and NAc), nucleus accumbens, and dorsal frontal and insular cortex, while in the orbitofrontal cortex, higher BP2 was observed. In addition, μ-opioid receptor binding in the combat-exposed subjects without PTSD, relative to PTSD patients, was lower in the amygdala but higher in the orbitofrontal cortex (Liberzon et al., 2007).

20.6.2.3(iii) Future directions in imaging of anxiety

The structural and functional neuroanatomy of human emotional responses are clearly in their early stages of development. A network of regions, also including medial TL structures, mPFC and orbitofrontal cortex, thalamus, and anterior cingulate, have been implicated in human response to stress and anxiety. Alterations of function in the amygdaloid, anterior cingulate, and other regions might have relevance to pathological processes in PTSD and other anxiety disorders; however, specificity of these findings remains to be elucidated. Future studies examining the neurochemical substrates underlying these abnormalities – PET receptor imaging and magnetic resonance spectroscopy (MRS) – should prove to be very helpful in clarifying the pathophysiology of abnormal responses to stress and trauma. *In vivo* receptor-imaging studies that identify regionally specific neurochemistry, in combination with activation paradigms, can provide the crucial link between abnormal functioning of a particular region and underlying neurophysiologic changes. The absence of receptor ligands relevant to anxiety symptomatology has prevented this type of investigation until now; however, the recent synthesis of specific anxiety-relevant PET ligands makes this strategy possible. The use of combined receptor ligands and rCBF in the same individual will contribute greatly to defining the neural pathways and neurotransmitters involved in normal anxiety as well as anxiety disorders.

References

Abadie P, Boulenger JP, Benali K, Barre L, Zarifian E, and Baron JC (1999) Relationships between trait and state anxiety and the central benzodiazepine receptor: A PET study. *European Journal of Neuroscience* 11: 1470–1478.

Abelson JL and Cameron OG (1994) Adrenergic dysfunction in anxiety disorders. In: Cameron OG (ed.) *Adrenergic Dysfunction and Psychobiology*, pp. 365–401. Washington, DC: American Psychiatric Press.

Abelson JL, Curtis GC, and Uhde TW (2005) Twenty-four hour growth hormone secretion in patients with panic disorder. *Psychoneuroendocrinology* 30: 72–79.

Abelson JL, Khan S, Liberzon I, and Young EA (2007) HPA axis activity in patients with panic disorder: Review and synthesis of four studies. *Depression and Anxiety* 24: 66–76.

Adamec RE and Shallow T (1993) Lasting effects on rodent anxiety of a single exposure to a cat. *Physiology and Behavior* 54: 101–109.

Anisman H, Griffiths J, Matheson K, Ravindran AV, and Merali Z (2001) Posttraumatic stress symptoms and salivary cortisol levels. *American Journal of Psychiatry* 158: 1509–1511.

Antelman SM, Knopf S, Kocan D, Edwards DJ, Ritchie JC, and Nemeroff CB (1988) One stressful event blocks multiple actions of diazepam for up to at least a month. *Brain Research* 445: 380–385.

Armony JL, Corbo V, Clement MH, and Brunet A (2005) Amygdala response in patients with acute PTSD to masked and unmasked emotional facial expressions. *American Journal of Psychiatry* 162(10): 1961–1963.

Aston-Jones G, Shipley MT, Chouvet G, et al. (1991) Afferent regulation of locus coeruleus neurons: Anatomy, physiology and pharmacology. *Progress in Brain Research* 88: 47–75.

Atmaca M, Kuloglu M, Tezcan E, Onal S, and Ustundag B (2002) Neopterin levels and dexamethasone suppression test in posttraumatic stress disorder. *European Archives of Psychiatry and Clinical Neuroscience* 252: 161–165.

Behbehani MM (1995) Functional characteristics of the midbrain periaqueductal gray. *Progress in Neurobiology* 46: 575–605.

Berton O, Durand M, Aguerre S, Mormede P, and Chaouloff F (1999) Behavioral, neuroendocrine and serotonergic consequences of single social defeat and repeated fluoxetine pretreatment in the Lewis rat strain. *Neuroscience* 92: 327–341.

Birbaumer N, Grodd W, Diedrich O, et al. (1998) fMRI reveals amygdala activation to human faces in social phobics. *NeuroReport* 9: 1223–1226.

Bisaga A, Katz JL, Antonini A, Wright CE, Margouleff C, Gorman JM, and Eidelberg D (1998) Cerebral glucose metabolism in women with panic disorder. *American Journal of Psychiatry* 155: 1178–1183.

Bishop SJ, Duncan J, and Lawrence AD (2004) State anxiety modulation of the amygdala response to unattended threat-related stimuli. *Journal of Neuroscience* 24(46): 10364–10368.

Blanchard EB, Kolb LC, Prins A, Gates S, and McCoy GC (1991) Changes in plasma norepinepherine in combat related stimuli among Vietnam veterans with PTSD. *Journal of Nervous and Mental Disease* 179: 371–373.

Bonne O, Brandes D, Gilboa A, et al. (2001) Longitudinal MRI study of hippocampal volume in trauma survivors with PTSD. *American Journal of Psychiatry* 158: 1248–1251.

Boscarino JA (1996) Posttraumatic stress disorder, exposure to combat, and lower plasma cortisol among Vietnam veterans: Findings and clinical implications. *Journal of Consulting and Clinical Psychology* 64: 191–201.

Brambilla F, Bellodi L, Perna G, et al. (1992) Psychoimmunoendocrine aspects of panic disorder. *Neuropsychobiology* 26: 12–22.

Breiter HC, Etcoff NL, Whalen PJ, et al. (1996) Response and habituation of the human amygdala during visual processing of facial expression. *Neuron* 17: 875–887.

Bremner JD, Randall P, Scott TM, et al. (1995) MRI-based measurement of hippocampal volume in patients with combat-related posttraumatic stress disorder. *American Journal of Psychiatry* 152: 973–981.

Bremner JD, Randall P, Vermetten E, et al. (1997) Magnetic resonance imaging-based measurement of hippocampal volume in posttraumatic stress disorder related to childhood physical and sexual abuse – a preliminary report. *Biological Psychiatry* 41: 23–32.

Bremner JD, Staib LH, Kaloupek D, Southwick SM, Soufer R, and Charney DS (1999) Neural correlates of exposure to traumatic pictures and sound in Vietnam combat veterans with and without posttraumatic stress disorder: A positron emission tomography study. *Biological Psychiatry* 45: 806–816.

Bremner JD, Vythilingam M, Vermetten E, et al. (2003) Cortisol response to a cognitive stress challenge in posttraumatic stress disorder (PTSD) related to childhood abuse. *Psychoneuroendocrinology* 28: 733–735.

Britton JC, Phan KL, Taylor SF, Fig LM, and Liberzon I (2005) Corticolimbic blood flow in posttraumatic stress disorder during script-driven imagery. *Biological Psychiatry* 57(8): 832–840.

Britton KT, Morgan J, Rivier J, Vale W, and Koob GF (1985) Chlorodiazepoxide attenuates response suppression induced by corticotropin-releasing factor in the conflict test. *Psychopharmacology* 86: 170–174.

Brown GW (1993) Life events and affective disorder: Replications and limitations. *Psychosomatic Medicine* 55: 248–259.

Brown GW and Harris T (1978) *Social Origins of Depression: A Study of Psychiatric Disorder in Women.* New York: Free Press.

Brown GW and Harris TO (1989) Depression. In: Brown GW and Harris TO (eds.) *Life Events and Illness*, pp. 49–93. New York: Guilford Press.

Brown GW and Harris TO (1993) Aetiology of anxiety and depressive disorders in an inner-city population. 1. Early adversity. *Psychological Medicine* 23: 143–154.

Brown GW, Harris TO, and Eales MJ (1993) Aetiology of anxiety and depressive disorders in an inner-city population. 2. Comorbidity and adversity. *Psychological Medicine* 23: 155–165.

Brown GW, Harris TO, and Hepworth C (1994) Life events and endogenous depression. A puzzle reexamined. *Archives of General Psychiatry* 51: 525–534.

Brown MR, Fisher LA, Spiess J, Rivier C, Rivier J, and Vale W (1982) Corticotropin releasing factor: Actions on sympathetic

nervous system and metabolism. *Endocrinology* 111: 928–931.
Butler PD and Nemeroff CB (1990) Corticotropin releasing factor as a possible cause of comorbidity in anxiety and depressive disorders. In: Maser JD and Cloninger CR (eds.) *Comorbidity of Mood and Anxiety Disorders*, pp. 413–435. Washington, DC: American Psychiatric Press.
Bystritsky A, Pontillo D, Powers M, Sabb FW, Craske MG, and Bookheimer SY (2001) Functional MRI changes during panic anticipation and imagery exposure. *NeuroReport* 12(18): 3953–3957.
Cameron OG, Abelson J, and Young EA (2004) Anxious and depressive disorders and their comorbidity: Effect on central nervous system noradrenergic function. *Biological Psychiatry* 56: 875–883.
Cameron OG, Lee MA, Curtis GC, and McCann DS (1987) Endocrine and physiological changes during 'spontaneous' panic attacks. *Psychoneuroendocrinology* 12: 321–331.
Cameron OG, Smith CB, Hollingsworth PJ, Neese RM, and Curtis GC (1984) Platelet alpha 2-adrenergic receptor binding and plasma catecholamines. Before and during imipramine treatment in patients with panic anxiety. *Archives of General Psychiatry* 41: 1144–1148.
Cameron OG, Smith CB, Lee MA, Holligsworth PJ, Hill EM, and Curtis GC (1990) Adrenergic status in anxiety disorders: Platelets alpha 2-adrenergic receptor binding, blood pressure, pulse, and plasma catecholamines in panic and generalized anxiety disorder patients and in normal subjects. *Biological Psychiatry* 28: 3–20.
Chen XL, Xie JX, Han HB, Cui YH, and Zhang BQ (2004) MR perfusion-weighted imaging and quantitative analysis of cerebral hemodynamics with symptom provocation in unmedicated patients with obsessive–compulsive disorder. *Neuroscience Letters* 370(2–3): 206–211.
Cullinan WE, Herman JP, and Watson SJ (1993) Ventral subicular interaction with the hypothalamic paraventricular nucleus: Evidence for a relay in the bed nucleus of the stria terminalis. *Journal of Comparative Neurology* 332: 1–20.
Dallman MF (1993) Stress update: Adaptation of the hypothalamaic–pituitary–adrenal axis to chronic stress. *Trends in Endocrinology and Metabolism* 4: 62–69.
Dantendorfer K, Prayer D, Kramer J, et al. (1996) High frequency of EEG and MRI brain abnormalities in panic disorder. *Psychiatry Research* 68: 41–53.
Davidson LM and Baum A (1986) Chronic stress and posttraumatic disorders. *Journal of Clinical and Consulting Psychology* 54: 303–308.
Davidson RJ, Abercrombie H, Nitschke JB, and Putnam K (1999) Regional brain function, emotion and disorders of emotion. *Current Opinion in Neurobiology* 9: 228–234.
Davis M (1986) Pharmacological and anatomical analysis of fear conditioning using the fear-potentiated startle paradigm. *Behavioral Neuroscience* 100: 814–824.
Davis M (1992) The role of the amygdala in conditioned fear. In: Aggleton JP (ed.) *The Amygdala: Neurobiological Aspects of Emotion, Memory and Mental Dysfunction*, pp. 255–306. New York: Wiley-Liss.
Davis M (1998) Are different parts of the extended amygdala involved in fear versus anxiety? *Biological Psychiatry* 44: 1239–1247.
Davis M and Shi C (1999) The extended amygdala: Are the central nucleus of the amygdala and the bed nucleus of the stria terminalis differentially involved in fear versus anxiety? *Annals of the New York Academy of Sciences* 877: 281–291.
Deadman JM, Dewey MJ, Owens RG, Leinster SJ, and Slade PD (1989) Danger and loss in breast cancer. *Psychological Medicine* 19: 677–681.
De Bellis MD, Hall J, Boring AM, Frustaci K, and Mortiz G (2001) A pilot longitudnal study of hippocampal volumes in pediatric maltreatment-related posttraumatic stress disorder. *Biological Psychiatry* 50: 305–309.
de Kloet CS, Vermetten E, Heijnen CJ, Geuze E, Lentjes EG, and Westenberg HG (2007) Enhanced cortisol suppression in response to dexamethasone administration in traumatized veterans with and without posttraumatic stress disorder. *Psychoneuroendocrinology* 32(3): 215–226.
Devesa J, Diaz MJ, Tresguerres JAF, Arce V, and Lima L (1991) Evidence that alpha2-adrenergic pathways play a major role in growth hormone (GH) neuroregulation: Alpha2-adrenergic agonism counteracts the inhibitory effect of muscarinic cholinergic receptor blockade on the GH response to GH-releasing hormone, while alpha2-adrenergic blockade diminishes the potentiating effect of increased cholinergic tone on such stimulation in normal men. *Journal of Clinical Endocrinology and Metabolism* 73: 252–256.
Dunn AJ and Berridge CW (1990) Physiological and behavioral responses to corticotropin-releasing factor administration: Is CRF a mediator of anxiety or stress response. *Brain Research Reviews* 15: 71–100.
Eaton WW and Ritter C (1988) Distinguishing anxiety and depression with field survey data. *Psychological Medicine* 18: 155–166.
Elzinga BM, Schmahl CG, Vermetten E, van Dyck R, and Bremner JD (2003) Higher cortisol levels following exposure to traumatic reminders in abuse-related PTSD. *Neuropsychopharmacology* 28(9): 1656–1665.
Eren I, Tukel R, Polat A, Karaman R, and Unal S (2003) Evaluation of regional cerebral blood flow changes in panic disorder with Tc99m-HMPAO SPECT. *Psychiatry Research: Neuroimaging* 123(2): 135–143.
Faravelli C (1985) Life events preceding the onset of anxiety disorder. *Journal of Affective Disorders* 9: 103–105.
Finlay-Jones R (1981) Showing that life events are a cause of depression – a review. *Australian and New Zealand Journal of Psychiatry* 15: 229–238.
Finlay-Jones R and Brown GW (1981) Types of stressful life event and the onset of anxiety and depressive disorders. *Psychological Medicine* 11: 803–815.
Frank E, Anderson B, Reynolds CF, Ritenour A, and Kupfer DJ (1994) Life events and the research diagnostic criteria endogenous subtype. A confirmation of the distinction using the Bedford College methods. *Archives of General Psychiatry* 51: 519–524.
Freedman R, Foote SL, and Bloom FE (1975) Histochemical characterization of a neocortical projection of the nucleus locus coeruleus in the squirrel monkey. *Journal of Comparative Neurology* 164: 209–231.
Fukuda S, Morimota K, Kanae M, and Maruyama S (2000) Effect of the Hanshin–Awaji earthquake on posttraumatic stress, lifestyle changes, and cortisol levels of victims. *Archives of Environmental Health* 55: 121–125.
Gilbertson MW, Shenton ME, Ciszewski A, et al. (2002) Smaller hippocampal volume predicts pathologic vulnerability to psychological trauma. *Nature Neuroscience* 5(11): 1242–1247.
Graeff FG, Netto CF, and Zangrossi H, Jr. (1998) The elevated T-maze as an experimental model of anxiety. *Neuroscience and Biobehavioral Reviews* 23: 237–246.
Graeff FG, Silveira MC, Nogueira RL, Audi EA, and Oliveira RM (1993) Role of the amygdala and periaqueductal gray in anxiety and panic. *Behavioural Brain Research* 58: 123–131.
Grunhaus LJ, Cameron OG, Pande AC, Haskett RF, Hollingsworth PJ, and Smith CB (1990) Comorbidity of panic

disorder and major depressive disorder: Effects on platelet alpha2-adrenergic receptors. *Acta Psychiatrica Scandinavica* 81: 216–219.

Gurvits TV, Shenton ME, Hokama H, et al. (1996) Magnetic resonance imaging study of hippocampal volume in chronic, combat-related posttraumatic stress disorder. *Biological Psychiatry* 40: 1091–1099.

Halbreich U, Olympia J, Glogowski J, Carson S, Axelrod S, and Yeh CM (1988) The importance of past psychological trauma and pathophysiological process as determinants of current biologic abnormalities. *Archives of General Psychiatry* 45: 293–294.

Hamner MB, Diamond BI, and Hitri A (1994) Plasma norepinephrine and MHPG responses to exercise stress in PTSD. In: Murburg MM (ed.) *Cathecolamine Function in Posttraumic Stress Disorder: Emerging Concepts*, pp. 221–232. Washington, DC: American Psychiatric Press.

Hawk LW, Dougall AL, Ursano RJ, and Baum A (2000) Urinary catecholamines and cortisol in recent-onset posttraumatic stress disorder after motor vehicle accidents. *Psychosomatic Medicine* 62(3): 423–434.

Heim C, Newport DJ, Bonsall R, Miller AH, and Nemeroff CB (2001) Altered pituitary–adrenal axis responses to provocative challenge tests in adult survivors of childhood abuse. *American Journal of Psychiatry* 158: 575–581.

Heim C, Newport DJ, Heit S, et al. (2000) Pituitary–adrenal and autonomic responses to stress in women after sexual and physical abuse in childhood. *Journal of the American Medical Association* 284: 592–597.

Hellhammer DH and Wade S (1993) Endocrine correlates of stress vunerability. *Psychotherapy and Psychosomatics* 60: 8–17.

Hendler T, Rotshtein P, and Hadar U (2001) Emotion-perception interplay in the visual cortex: The eyes follow the heart. *Cellular and Molecular Neurobiology* 21(6): 733–752.

Heninger GR (1990) A biological perspective on comorbidity of major depressive disorder and panic disorder. In: Maser JD and Cloninger CR (eds.) *Comorbidity of Mood and Anxiety Disorders.* Washington, DC: American Psychiatric Press.

Holmes TH and Rahe RH (1967) The social readjustment rating scale. *Journal of Psychosomatic Research* 11: 213–218.

Holsboer F, von Bardeleben U, Buller R, Heuser I, and Steiger A (1987) Stimulation response to corticotropin-releasing hormone (CRH) in patients with depression, alcoholism and panic disorder. *Hormone and Metabolic Research* 16(supplement): 80–88.

Inslicht SS, Marmar CR, Neylan TC, et al. (2006) Increased cortisol in women with intimate partner violence-related posttraumatic stress disorder. *Annals of the New York Academy of Sciences* 1071: 428–429.

Irwin W, Davidson RJ, Lowe MJ, Mock BJ, Sorenson JA, and Turski PA (1996) Human amygdala activation detected with echo-planar functional magnetic resonance imaging. *NeuroReport* 7: 1765–1769.

Isenberg N, Silbersweig D, Engelien A, et al. (1999) Linguistic threat activates the human amygdala. *Proceedings of the National Academy of Sciences of the United States of America* 96: 10456–10459.

Jones BE, Halaris AE, McIlhany M, and Moore RY (1977) Ascending projections of the locus coeruleus in the rat. I. Axonal transport in central noradrenaline neurons. *Brain Research* 127: 1–21.

Kanter ED, Wilkinson CW, Radant AD, et al. (2001) Glucocorticoid feedback sensitivity and adrenocortical responsiveness in posttraumatic stress disorder. *Biological Psychiatry* 50: 238–245.

Keller-Wood ME and Dallman MF (1985) Corticosteroid inhibition of ACTH secretion. *Endocrine Reviews* 5: 1–24.

Kellner M, Baker DG, Yassouridis A, Bettinger S, Otte C, Naber D, and Wiedemann K (2002) Mineralocorticoid receptor function in patients with posttraumatic stress disorder. *American Journal of Psychiatry* 159(11): 1938–1940.

Kellner M, Otte C, Yassouridis A, Schick M, Jahn H, and Wiedemann K (2004) Overnight metyrapone and combined dexamethasone/metyrapone tests in post-traumatic stress disorder: Preliminary findings. *European Neuropsychopharmacology* 14: 337–339.

Kimbrell TA, George MS, Parekh PI, et al. (1999) Regional brain activity during transient self-induced anxiety and anger in healthy adults. *Biological Psychiatry* 46: 454–465.

Kling AS and Brothers L (1992) The amygdala and social behavior. In: Aggleton JP (ed.) *The Amygdala: Neurolbiological Aspects of Emotion, Memory and Mental Dysfunction*, pp. 353–377. New York: Wiley-Liss.

Knight DC, Nguyen HT, and Bandettini PA (2005) The role of the human amygdala in the production of conditioned fear responses. *NeuroImage* 26(4): 1193–1200.

Koob GF, Heinrichs SC, Pich EM, et al. (1993) The role of corticotropin-releasing factor in behavioral responses to stress. *Ciba Foundation Symposium* 172: 277–289.

Kosten TR, Mason JW, Giller EL, Ostroff RB, and Harkness L (1987) Sustained urinary norepinephine and epinephrine elevation in post-traumatic stress disorder. *Psychoneuroendocrinology* 12: 13–20.

Krieger D (1979) Rhythms in CRH, ACTH and corticosteroids. *Endocrine Reviews* 1: 1–23.

Lanius RA, Williamson PC, Bluhm RL, et al. (2005) Functional connectivity of dissociative responses in posttraumatic stress disorder: A functional magnetic resonance imaging investigation. *Biological Psychiatry* 57(8): 873–884.

LeDoux JE, Iwata J, Cicchetti P, and Reis DJ (1988) Different projections of the central amygdaloid nucleus mediate autonomic and behavioral correlates of conditioned fear. *Journal of Neuroscience* 8: 2517–2529.

Lee YS, Hwang J, Kim SJ, et al. (2006) Decreased blood flow of temporal regions of the brain in subjects with panic disorder. *Journal of Psychiatric Research* (6): 528–534.

Lemieux AM and Coe CL (1995) Abuse-related posttraumatic stress disorder: Evidence for chronic neuroendocrine activation in women. *Psychosomatic Medicine* 57: 105–115.

Liberzon I, Krstov M, and Young EA (1997) Stress–restress: Effects on ACTH and fast feedback. *Psychoneuroendocrinology* 22: 443–453.

Liberzon I, Lopez JF, Flagel SB, Vazquez DM, and Young EA (1999a) Differential regulation of hippocampal glucocorticoid receptors mRNA and fast feedback: Relevance to post-traumatic stress disorder (in process citation). *Journal of Neuroendocrinology* 11: 11–17.

Liberzon I, Taylor SF, Amdur R, et al. (1999b) Brain activation in PTSD in response to trauma-related stimuli. *Biological Psychiatry* 45: 817–826.

Liberzon I, Taylor SF, Phan KL, et al. (2007) Altered central mu-opioid receptor binding after psychological trauma. *Biological Psychiatry* 61(9): 1030–1038.

Lindley SE, Carlson EB, and Benoit M (2004) Basal and dexamethasone suppressed salivary cortisol concentrations in a community sample of patients with posttraumatic stress disorder. *Biological Psychiatry* 55: 940–945.

Lorberbaum JP, Kose S, Johnson MR, et al. (2004) Neural correlates of speech anticipatory anxiety in generalized social phobia. *NeuroReport* 15(18): 2701–2705.

Maddock RJ, Buonocore MH, Kile SJ, and Garrett AS (2003) Brain regions showing increased activation by

threat-related words in panic disorder. *NeuroReport* 14(3): 325–328.
Maes M, Lin A, Bonaccorso S, et al. (1998) Increased 24-hour urinary cortisol excretion in patients with post-traumatic stress disorder and patients with major depression, but not in patients with fibromyalgia. *Acta Psychiatrica Scandinavica* 98: 328–335.
Malizia AL, Cunningham VJ, Bell CJ, Liddle PF, Jones T, and Nutt DJ (1998) Decreased brain GABA(A)-benzodiazepine receptor binding in panic disorder: Preliminary results from a quantitative PET study. *Archives of General Psychiatry* 55: 715–720.
Mason JW (1968) A review of psychoendocrine research on the pituitary adrenal cortical system. *Psychosomatic Medicine* 30: 576–607.
Mason JW, Giller EL, Kosten TR, Ostroff RB, and Podd L (1986) Urinary free-cortisol levels in posttraumatic stress disorder patients. *Journal of Nervous and Mental Disease* 174: 145–149.
Massana G (2003) Amygdalar atrophy in panic disorder patients detected by volumetric magnetic resonance imaging. *NeuroImage* 19(1): 80–90.
McEwen BS (1995) Adrenal steroid action on brain: Dissecting the fine line between protection and damage. In: Friedman MJ, Charney DS, and Deutch AY (eds.) *Neurobiological and Clinical Consequences of Stress: From Normal Adaptation to PTSD*, pp. 135–147. Philadelphia, PA: Lippincott-Raven.
McFall M, Murburg M, Ko G, and Veith RC (1990) Autonomic response to stress in Vietnam combat veterans with post-traumatic stress disorder. *Biological Psychiatry* 27: 1165–1175.
Miller PM and Ingham JG (1983) Dimensions of experience. *Psychological Medicine* 13: 417–429.
Morris JS, Ohman A, and Dolan RJ (1999) A subcortical pathway to the right amygdala mediating unseen fear. *Proceedings of the National Academy of Sciences of the United States of America* 96: 1680–1685.
Murburg MM, McFall ME, and Veith RC (1994) Basal sympathoadrenal function in patients with PTSD and depression. In: Murburg MM (ed.) *Cathecolamine Function in Posttraumatic Stress Disorder: Emerging Concepts*, pp. 175–188. Washington, DC: American Psychiatric Press.
Nesse RM, Curtis GC, Thyer BA, McCann DS, Huber-Smith MJ, and Knopf RF (1985) Endocrine and cardiovasular responses during phobic anxiety. *Psychosomatic Medicine* 47: 320–332.
Newport DJ, Heim C, Bonsall R, Miller AH, and Nemeroff CB (2004) Pituitary–adrenal responses to standard and low-dose dexamethasone suppression tests in adult survivors of child abuse. *Biological Psychiatry* 55(1): 10–20.
Nordahl TE, Semple WE, Gross M, et al. (1990) Cerebral glucose metabolic differences in patients with panic disorder. *Neuropsychopharmacology* 3: 261–272.
Nordahl TE, Stein MB, Benkelfat C, et al. (1998) Regional cerebral metabolic asymmetries replicated in an independent group of patients with panic disorders. *Biological Psychiatry* 44: 998–1006.
Ochsner KN, Bunge SA, Gross JJ, and Gabrieli JDE (2002) Rethinking feelings: An fMRI study of the cognitive regulation of emotion. *Journal of Cognitive Neuroscience* 14(8): 1215–1229.
O'Connell RA and Mayo JA (1988) The role of social factors in affective disorders: A review. *Hospital and Community Psychiatry* 39: 842.
Olff M, Guzelcan Y, de Vries GJ, Assies J, and Gersons BP (2006) HPA- and HPT-axis alterations in chronic posttraumatic stress disorder. *Psychoneuroendocrinology* 31: 1220–1230.
Osuch EA, Benson B, Geraci M, Podell D, Herscovitch P, McCann UD, and Post RM (2001) Regional cerebral blood flow correlated with flashback intensity in patients with posttraumatic stress disorder. *Biological Psychiatry* 50(4): 246–253.
Paykel ES (1994) Life events, social support and depression. *Acta Psychiatrica Scandinavica* 377: 50–53.
Pichot W, Hansenne M, Gonzalez-Mareno A, and Ansseau M (1995) Growth hormone response to aopmorphine in panic disorder: Comparison with major depression and normal controls. *European Archives of Psychiatry and Clinical Neuroscience* 245: 306–308.
Pitman RG and Orr SP (1990) Twenty-four hour urinary cortisol and catecholamine excretion in combat-related posttraumatic stress disorder. *Biological Psychiatry* 27: 245–247.
Pitman RK, Orr SP, Forgue DF, de Jong JB, and Claiborn JM (1987) Psychophysiologic assessment of posttraumatic stress disorder imagery in Vietnam combat veterans. *Archives of General Psychiatry* 44: 970–975.
Plotsky PM (1987) Facilitation of immunoreactive corticotropin-releasing factor secretion into the hypophyseal-portal circulation after activation of catechoaminergic pathways or central norepinepherine injection. *Endocrinology* 121: 924–930.
Plotsky PM, Cunningham ET, and Widmaier EP (1989) Catecholaminergic modulation of corticotropin-releasing factor and adrenocorticotropin secretion. *Endocrine Reviews* 10: 437–458.
Price JL (1999) Prefrontal cortical networks related to visceral function and mood. *Annals of the New York Academy of Sciences* 877: 383–396.
Rahe RH, Karson S, Howard NS, Rubin RT, and Poland RE (1990) Psychological and physiological assessment on American hostages freed from captivity in Iran. *Psychosomatic Medicine* 52: 1–16.
Raskin M, Peeke HV, Dickman W, and Pinsker H (1982) Anxiety and generalized anxiety disorders. Developmental antecedents and precipitants. *Archives of General Psychiatry* 39: 687–689.
Rasmusson AM, Lipschitz DS, Wang S, et al. (2001) Increased pituitary and adrenal reactivity in premenopausal women with posttraumatic stress disorder. *Biological Psychiatry* 50: 965–977.
Rauch SL, Jenike MA, Alpert NM, Baer L, Breiter HC, Savage CR, and Fischman AJ (1994) Regional cerebral blood flow measured during symptom provocation in obsessive–compulsive disorder using oxygen 15-labeled carbon dioxide and positron emission tomography. *Archives of General Psychiatry* 51: 62–70.
Rauch SL, Savage CR, Alpert NM, et al. (1995) A positron emission tomographic study of simple phobic symptom provocation. *Archives of General Psychiatry* 52: 20–28.
Rauch SL, van der Kolk BA, Fisler RE, et al. (1996) A symptom provocation study of posttraumatic stress disorder using positron emission tomography and script-driven imagery. *Archives of General Psychiatry* 53: 380–387.
Rauch SL, Whalen PJ, Shin LM, et al. (2000) Exaggerated amygdala response to masked facial stimuli in posttraumatic stress disorder: A functional MRI study. *Biology of Psychiatry* 47(9): 769–776.
Reiman EM, Lane RD, Ahern GL, et al. (1997) Neuroanatomical correlates of externally and internally generated human emotion. *American Journal of Psychiatry* 154: 918–925.
Roy-Byrne PP and Uhde TW (1988) Exogenous factors in panic disorder: Clinical and research implications. *Journal of Clinical Psychiatry* 49: 56–61.

Roy-Byrne PP, Uhde TW, Post RM, Galucci W, Chrosous GP, and Gold PW (1986) The corticotropin-releasing-hormone stimulation test in patients with panic disorder. *American Journal of Psychiatry* 148: 896–899.

Sala M, Perez J, Soloff P, di Nemi SU, Caverzasi E, Soares JC, and Brambilla P (2004) Stress and hippocampal abnormalities in psychiatric disorders. *European Neuropsychopharmacology* 14(5): 393–405.

Santa Ana EJ, Saladin ME, Back SE, et al. (2006) PTSD and the HPA axis: Differences in response to the cold pressor task among individuals with child vs. adult trauma. *Psychoneuroendocrinology* 31: 501–509.

Sapolsky RM (1996) Why stress is bad for your brain. *Science* 273: 749–750.

Schuff N, Neylan TC, Lenoci MA, Du A, Weiss DS, Marmar CR, and Weine MW (2001) Decreased hippocampal *N*-acetylaspartate in the absence of atrophy in posttraumatic stress disorder. *Biological Psychiatry* 50(12): 952–959.

Shin LM, Kosslyn SM, McNally RJ, et al. (1997) Visual imagery and perception in posttraumatic stress disorder. A positron emission tomographic investigation. *Archives of General Psychiatry* 54: 233–241.

Shin LM, McNally RJ, Kosslyn SM, et al. (1999) Regional cerebral blood flow during script-driven imagery in childhood sexual abuse-related PTSD: A PET investigation. *American Journal of Psychiatry* 156: 575–584.

Shin LM, Orr SP, Carson MA, et al. (2004) Regional cerebral blood flow in the amygdala and medial prefrontal cortex during traumatic imagery in male and female Vietnam veterans with PTSD. *Archives of General Psychiatry* 61(2): 168–176.

Shin LM, Wright CI, Cannistraro PA, et al. (2005) A functional magnetic resonance imaging study of amygdala and medial prefrontal cortex responses to overtly presented fearful faces in posttraumatic stress disorder. *Archives of General Psychiatry* 62(3): 273–281.

Siever LJ, Uhde TW, Silberman EK, et al. (1982) Growth hormone response to clonidine as a probe of noradrenergic receptor responsiveness in affective disorder patients and controls. *Psychiatry Research* 6: 171–183.

Smith MA, Davidson J, Ritchie JC, Kudler H, Lipper S, Chappell P, and Nemeroff CB (1989) The corticotropin-releasing hormone test in patients with posttraumatic stress disorder. *Biological Psychiatry* 26: 349–355.

Southwick SM, Krystal JH, Morgan CA, et al. (1993) Abnormal noradrenergic function in posttraumatic stress disorder. *Archives of General Psychiatry* 50: 266–274.

Southwick SM, Morgan CA, Bremner AD, Grillon CG, Krystal JH, Nagy LM, and Charney DS (1997) Noradrenergic alterations in posttraumatic stress disorder. *Annals of the New York Academy of Sciences* 821: 125–141.

Starkman MN, Gebarski SS, Berent S, and Schteingart DE (1992) Hippocampal formation volume, memory dysfunction, and cortisol levels in patients with Cushing's syndrome. *Biological Psychiatry* 32: 756–765.

Stein MB, Koverola C, Hanna C, Torchia MG, and McClarty B (1997a) Hippocampal volume in women victimized by childhood sexual abuse. *Psychological Medicine* 27: 951–959.

Stein MB, Yehuda R, Koverola C, and Hanna C (1997b) Enhanced dexamethasone suppression of plasma cortisol in adult women traumatized by childhood sexual abuse. *Biological Psychiatry* 42: 680–686.

Swanson LW and Hartman BK (1975) The central adrenergic system. An immunofluorescence study of the location of cell bodies and their efferent connections in the rat utilizing dopamine-beta-hydroxylase as a marker. *Journal of Comparative Neurology* 163: 467–505.

Swanson LW, Sawchenko PE, Rivier J, and Vale WW (1983) Organization of ovine corticotropin-releasing factor immunoreactive cells and fibers in the rat brain. An immunohistochemical study. *Neuroendocrinology* 36: 165–186.

Taylor SF, Liberzon I, Fig LM, Decker LR, Minoshima S, and Koeppe RA (1998) The effect of emotional content on visual recognition memory: A PET activation study. *NeuroImage* 8: 188–197.

Taylor SF, Phan KL, Decker LR, and Liberzon I (2003) Subjective rating of emotionally salient stimuli modulates neural activity. *NeuroImage* 18(3): 650–659.

Thatcher-Britton K and Koob G (1986) Alcohol reverses the proconflict effect of corticotropin-releasing factor. *Regulatory Peptides* 16: 315–320.

Tillfors M, Furmark T, Marteinsdottir I, et al. (2001) Cerebral blood flow in subjects with social phobia during stressful speaking tasks: A PET study. *American Journal of Psychiatry* 158(8): 1220–1226.

Torgersen S (1985) Developmental differentiation of anxiety and affective neuroses. *Acta Psychiatrica Scandinavica* 71: 304–310.

Uhde TW, Tancer ME, Rubinow DR, et al. (1992) Evidence for hypothalamic-growth hormone dysfunction in panic disorder: Profile of growth hormone (GH) responses to clonidine, yohimbine, caffeine, glucose, GRF and TRH in panic disorder patients versus healthy volunteers. *Neuropsychopharmacology* 6: 101–118.

Vale W, Spiess J, Rivier J, and Rivier C (1981) Characterization of a 41-residue ovine hypothalamic peptide that stimulates secretions of corticotropin and beta-endorphin. *Science* 213: 1394–1397.

Valentino RJ (1989) Corticotropin-releasing factor: Putative neurotransmitter in the noradrenergic nucleus locus coeruleus. *Psychoparmacological Bulletin* 25: 306–311.

van Dijken HH, de Goeij DC, Sutanto W, Mos J, de Kloet ER, and Tilders FJ (1993) Short inescapable stress produces long-lasting changes in the brain–pituitary–adrenal axis of adult male rats. *Neuroendocrinology* 58: 57–64.

Vythilingam M, Anderson ER, Goddard A, Woods SW, Staib LH, Charney DS, and Bremner JD (2000) Temporal lobe volume in panic disorder – a quantitative magnetic resonance imaging study. *Psychiatry Research: Neuroimaging* 99(2): 75–82.

Walker DL and Davis M (1997) Double dissociation between the involvement of the bed nucleus of the stria terminalis and the central nucleus of the amygdala in startle increases produced by conditioned versus unconditioned fear. *Journal of Neuroscience* 17: 9375–9383.

Waterhouse BD, Lin CS, Burne RA, and Woodward DJ (1983) The distribution of neocortical projection neurons in the locus coeruleus. *Journal of Comparative Neurology* 217: 418–431.

Weiskrantz L (1956) Behavioral changes associated with ablation of the amygdaloid complex in monkeys. *Journal of Child Psychology and Psychiatry* 49: 381–391.

Wheler GH, Brandon D, Clemons A, Riley C, Kendall J, Loriaux DL, and Kinzie JD (2006) Cortisol production rate in posttraumatic stress disorder. *Journal of Clinical Endocrinology and Metabolism* 91(9): 3486–3489.

Wik G, Fredrikson M, and Fischer H (1997) Evidence of altered cerebral blood-flow relationships in acute phobia. *International Journal of Neuroscience* 91: 253–263.

Woods SW, Koster K, Krystal JK, Smith EO, Zubal IG, Hoffer PB, and Charney DS (1988) Yohimbine alters regional cerebral blood flow in panic disorder. *Lancet* 2: 678.

Yehuda R (2002) Current status of cortisol findings in post-traumatic stress disorder. *Psychiatric Clinics of North America* 25: 341–368.

Yehuda R and Antelman SM (1993) Criteria for rationally evaluating animal models of posttraumatic stress disorder. *Biological Psychiatry* 33: 479–486.
Yehuda R, Kahana B, Binder-Brynes K, Southwick SM, Mason JW, and Giller EL (1995) Low urinary cortisol excretion in Holocaust survivors with posttraumatic stress disorder. *American Journal of Psychiatry* 152: 982–986.
Yoo HK, Kim MJ, Kim SJ, et al. (2005) Putaminal gray matter volume decrease in panic disorder: An optimized voxel-based morphometry study. *European Journal of Neuroscience* 22(8): 2089–2094.
Young EA and Breslau N (2004a) Cortisol and catecholamines in posttraumatic stress disorder: A community study. *Archives of General Psychiatry* 61: 394–401.
Young EA and Breslau N (2004b) Saliva cortisol in a community sample with posttraumatic stress disorder. *Biological Psychiatry* 56: 205–9.
Young EA, Tolman R, Witkowski K, and Kaplan G (2004) Salivary cortisol and PTSD in a low income community sample of women. *Biological Psychiatry* 55: 621–626.

Further Reading

Baxter L, Jr., Phelps ME, Mazziotta JC, Guze BH, Schwartz JM, and Selin CE (1987) Local cerebral glucose metabolic rates in obsessive–compulsive disorder. A comparison with rates in unipolar depression and in normal controls. *Archives of General Psychiatry* 44: 211–218.
Jones BE and Moore RY (1977) Ascending projections of the locus coeruleus in the rat. II. Autoradiographic study. *Brain Research* 127: 25–53.
LaBar KS, Gatenby JC, Gore JC, LeDoux JE, and Phelps EA (1998) Human amygdala activation during conditioned fear acquisition and extinction: A mixed-trial fMRI study. *Neuron* 20: 937–945.
Schneider F, Weissa U, Kesslera, et al. (1999) Subcortical correlates of differential classical conditioning of aversive emotional reactions in social phobia. *Biological Psychiatry* 45: 863–871.

21 Mood Disorders

R T Rubin, VA Greater Los Angeles Healthcare System, Los Angeles, CA, USA
B J Carroll, Pacific Behavioral Research Foundation, Carmel, CA, USA

Published by Elsevier Inc.

Chapter Outline

Glossary

ACTH – adrenocorticotropic hormone; corticotropin A hormone produced in the pituitary gland that stimulates the adrenal gland to secrete steroid hormones.

AVP – arginine vasopressin A hormone produced in the hypothalamic area of the brain that stimulates ACTH secretion and increases blood pressure and water retention by the kidney.

cortisol A hormone released from the adrenal cortex (the classical stress hormone) that promotes glucose mobilization, and catabolic and anti-inflammatory tissue responses. Cortisol is the primary adrenal glucocorticoid hormone in man.

CRH – corticotropin-releasing hormone A hormone produced in the hypothalamic area of the brain that stimulates the pituitary gland to secrete ACTH.

Cushing syndrome Hyperactivity of the pituitary–adrenocortical system mainly caused by pituitary or adrenal hormone-secreting tumors or by treatment with steroid medications.

glucocorticoid A steroid hormone with metabolic effects, primarily on energy balance.

leptin A hormone produced in fat cells that inhibits food intake.

melatonin A hormone produced in the pineal gland that is regulated by the light–dark cycle and influences sleep and other endocrine processes such as reproduction in seasonally breeding animals.

neuroactive steroid A steroid hormone produced in or outside the nervous system that has activity within the nervous system.

neuromodulator A compound that alters the sensitivity of neurons to neurotransmitters.

neurosteroid A steroid hormone produced in the nervous system and is active there.

neurotransmitter Compound that is released by a neuron and conveys its signal across a synapse to another neuron, thereby exciting or inhibiting the activity of the second neuron.

prolactin A hormone produced in the pituitary gland that primarily stimulates the secretion of milk in nursing mothers.

somatostatin A hormone produced in the hypothalamic area of the brain that inhibits the secretion of growth hormone and TSH by the pituitary gland.

steroid hormones Class of hormones, of a particular molecular structure, that can have diverse effects on energy metabolism, salt and water balance, male and female sex characteristics and behaviors, and central nervous system activity.

synapse; synaptic cleft Space between the processes of two neurons, across which neurotransmitters convey signals from one neuron to the other.

transporter A domain of proteins within a cell membrane that mediates drug uptake or efflux.

TRH – thyrotropin-releasing hormone A hormone produced in the hypothalamic area of the brain that stimulates the pituitary gland to secrete TSH and prolactin.

TSH – thyroid-stimulating hormone; thyrotropin A hormone produced in the pituitary gland that stimulates the thyroid gland to secrete thyroid hormones.

21.1 Introduction

Mood disorders are abnormal states of feeling, primarily excessive sadness or elation, that (1) are sustained over a period of weeks or longer, (2) represent a clear difference from a person's usual feeling state, and (3) in many instances, are recurrent throughout the individual's lifetime. During a mood episode, a variety of affective displays such as sadness, anger, and panic may be seen. The neuroendocrine research strategy in mood disorders arose because of the existence of important functional relationships among the limbic system sites that control mood, affect, and neuroendocrine function (Rubin and Mandell, 1966). The two essential themes have been that patterns of dysregulated neuroendocrine activity may serve as proxy evidence of limbic system dysfunction in disordered mood states, and that study of neurotransmitter influences on neuroendocrine functions in patients compared with control subjects may be informative vis-à-vis hypothesized neurotransmitter bases of the mood disorders. Abnormal secretion patterns of several pituitary hormones and their target endocrine gland hormones have been noted in mood disorders, and, conversely, clinical mood episodes commonly occur in some primary endocrine conditions, for example, secondary depression in Cushing's disease and hypothyroidism. Moreover, both hormonal augmentation and hormonal blockade or suppression have been studied as treatments of primary and secondary mood disorders. This chapter reviews these neuroendocrine aspects of mood disorders.

21.1.1 Classification

Patients with unipolar depression display only depressive phases. In patients with bipolar disorder, both manic and depressive phases occur. Clinical depression may be mild (dysthymia) or severe (major depression), episodic or chronic, and with melancholic or psychotic or atypical features. Likewise, elevated mood states may be mild (hypomania) or severe (mania) and may display psychotic features. The boundaries between mild depression or mild hypomania and depressive or hypomanic or cyclothymic temperament are uncertain. On follow-up, many subjects with initially subclinical symptoms will eventually display fully symptomatic clinical episodes.

21.1.2 Diagnostic Criteria and Depressive Subtypes

These are subjects of ongoing debate, with multiple and competing diagnostic systems based on *ad hoc* consensus criteria. For this chapter, we will adopt the criteria and subtypes of the *Diagnostic and Statistical Manual of Mental Disorders*, Fourth Edition (DSM-IV) (American Psychiatric Association, 1994), with the caveat that these current criteria do not necessarily map onto earlier constructs, such as endogenous depression (Taylor and Fink, 2006). Even the term melancholia has undergone significant changes in definition between 1980 (DSM-III) and 1994 (DSM-IV) (Rubin et al., 2002; Taylor and Fink, 2006). As a rule, neuroendocrine changes are associated with more severe and classically melancholic depressive episodes. In principle, the debates about depressive subtypes can be informed by examination of biological markers, including neuroendocrine measures. Unfortunately, few studies have adopted the pluridiagnostic approach needed to help the field choose one diagnostic system over another (Carroll, 1989). There is no certainty that patients diagnosed by DSM-III or DSM-IV criteria will be clinically uniform from one research center to another. This problem is a serious confound in studies of neuroendocrine function and other biomarkers, and is responsible for frequent nonconfirmations of findings among centers.

21.1.3 Genetics

The inheritance patterns of both major depression and bipolar illness suggest contributions from many genes, and an influence of environmental factors on the genetic predisposition (diathesis) to develop the clinical syndromes. Genetic profiling of prefrontal cortex in patients with major depression suggests "a rich profile of dysregulated genes" (Kang et al., 2007), but which of these are etiologic and which are epiphenomenal remains to be determined.

The gene for the serotonin (5-HT) transporter contains an *s/l* polymorphism (variable number of tandem nucleotide repeats) in the promoter region (5HTTLPR). The long form in Caucasians is more active than the short form, and the opposite is the case in some Asian populations (Japan and Korea). Caucasians with the short form of 5HTTLPR appear to be at greater risk for depression following stressful experiences (Wilhelm et al., 2006) and are less responsive to 5-HT uptake-inhibiting antidepressants (Laje and McMahon, 2007). There is only a weak overall relationship between the 5HTTLPR polymorphism and the occurrence of major depression, however; the polymorphism appears to confer vulnerability to developing depression following less severe stresses. Similarly, a single nucleotide polymorphism in the norepinephrine (NE) transporter (NET G1287A) is associated with preferential response to the NE uptake-inhibiting antidepressant, nortriptyline (Kim et al., 2006). At present, there is no consensus on which other genes may be consistently dysregulated across cohorts of unipolar depressed or bipolar patients. Whole-genome studies offer some promise for clarification of this area of study.

21.1.4 Epidemiology

The lifetime prevalence of major depression is 10–25% for women and 5–12% for men, and the lifetime prevalence of bipolar disorder is approximately 1–2% in both sexes. Prior to puberty, the incidence of depression is similar in boys and girls, but the incidence increases in girls during puberty, likely related to their changing estrogen and testosterone concentrations (Angold et al., 1999), such that a 2:1 female:male ratio prevails throughout adulthood. Premenstrual, postpartum, and perimenopausal depressive symptoms of varying severity occur in many women, usually related to hypoestrogenism or estrogen withdrawal (Sichel et al., 1995). Severe postpartum psychosis is commonly a precursor of later bipolar disorder (Sit et al., 2006). In contrast to puberty, there does not appear to be a major change in the incidence of depression in women at the time of menopause (Pearlstein et al., 1997).

21.1.5 Neurocircuitry of Depression

Functional imaging studies confirm that frontal-striatal and frontal-limbic circuits are activated in depressed patients. Each of the three major cortico–striato–thalamo–cortical circuits is implicated in mood disorders. These circuits originate in the anterior cingulate cortex (motivation, drive, and incentive functions), the orbitofrontal cortex (executive control of behavioral inhibitory functions), and the dorsolateral prefrontal cortex (cognitive flexibility and set-shifting). The major striatal component of these circuits is the ventral striatum (nucleus accumbens), rather than the caudate-putamen; their thalamic component is the mediodorsal thalamic nucleus (Mink, 2003). Neurotransmitter inputs from the brainstem that regulate these circuits include NE, 5-HT, dopamine (DA), and acetylcholine (ACh). Limbic system sites that link to these circuits include the amygdala, hippocampus, and hypothalamus (Robbins and Everitt, 2003). These links provide the anatomical basis for the neuroendocrine research strategy in mood disorders. For instance, the amygdala stimulates HPA-axis activity through the hypothalamus, while the hippocampus has an inhibitory influence on HPA-axis activity.

Clinical neuroimaging studies confirm hyperactivity in these emotional circuits. Activated regions include the amygdala, subgenual anterior cingulate cortex (SACC; after correction for volume reduction), medial orbital cortex, left ventrolateral prefrontal cortex (PFC), lateral orbital cortex, anterior insular cortex, and medial thalamus (Drevets, 2000). Consistent findings were reported by Mayberg et al. (2005), who found that clinical improvement after 3 months' continuous deep-brain stimulation of the SACC was accompanied by reduced activity in the SACC, medial frontal cortex, orbital frontal cortex, anterior insula, and hypothalamus. In a study of patients with Bipolar II depression, activated areas included bilateral ventral striatum, left orbitofrontal cortex, left parahippocampal gyrus, and left posterior cingulate cortex (Mah et al., 2007). Neuroendocrine correlates of these brain-imaging findings give proof of principle of the neuroendocrine research strategy in mood disorders. For example, activity in the left amygdala correlated significantly with plasma cortisol elevations in depressed patients (Drevets et al., 2002). Moreover, Aihara et al. (2007) found that resolution of depression was accompanied not only by resolution of HPA-axis dysregulation, but also by normalization of regional glucose metabolism in prefrontal cortical, limbic, and paralimbic regions.

21.1.6 Neurotransmitter and Neuromodulator Function

Neurotransmitters and neuromodulators act on both neurons and microglia (Pocock and Kettenmann, 2007) and influence many central nervous system (CNS) functions (Ordway et al., 2002), as shown in **Figure 1**. NE, 5-HT, DA, and ACh have received the most attention as possible neurochemical substrates of the affective disorders. In the simplest formulation, deficiencies of NE and 5-HT neurotransmission and excessive cholinergic neurotransmission in the limbic system and hypothalamus have been postulated to underlie depression. The hypotheses usually involve a balance between neurotransmitters; for example, an excess of cholinergic transmission relative to noradrenergic transmission may underlie depression, and the converse may underlie mania. The defect may be anywhere in the chain of events that includes transmitter synthesis, release, metabolism, receptor activation, and the postreceptor metabolic cascade. In addition, some neurotransmitter receptors exist as homodimers, higher-order homomultimers, and heteromers (e.g., the dopamine D1–D2 receptor heteromer), which can modulate both pre- and postsynaptic neurotransmission (Ferre et al., 2007), adding another

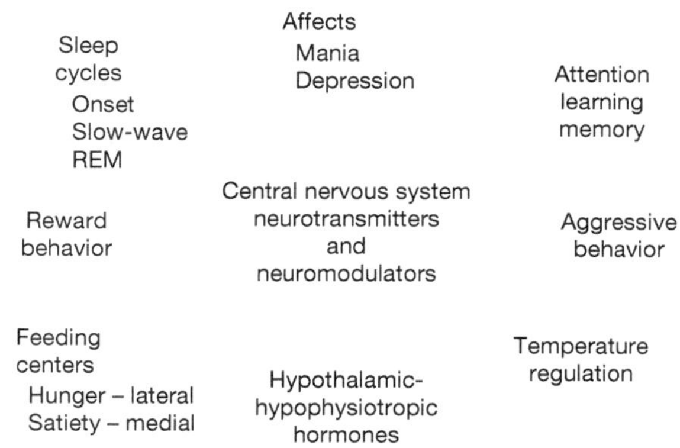

Figure 1 Neurotransmitters and neuromodulators influence many CNS functions, including (clockwise from top left) cognition (attention, learning, memory), affective tone, aggressive and impulsive behavior, temperature regulation, anterior pituitary hormone releasing and inhibiting factors (hypothalamic–hypophysiotropic hormones), hunger and satiety, and sleep–wake cycles. They can act on both neurons and microglia.

level of complexity to the concept of neurotransmitter dysregulation in affective disorders. In addition, many neuromodulators (e.g., neurotrophins, steroid hormones, and thyroid hormone) affect synaptic transmission and neuronal activity. Glucocorticoid hormones, in excess, accelerate nerve conduction velocity and slow synaptic transmission time. These effects may contribute to the disordered information processing occurring in Cushingoid states and psychotic depression.

Much of the empirical evidence supporting one or another neurotransmitter hypothesis derives from pharmacological treatment studies; for example, most antidepressants block the presynaptic transporter of NE and/or 5-HT, thereby increasing their concentrations in the synaptic cleft and, ostensibly, their activation of postsynaptic receptors. The enduring neurotransmitter hypotheses of affective disorders additionally have a body of experimental evidence in both animals and humans, including neuroendocrine data, to support them.

21.1.6.1 Acetylcholine and norepinephrine

Supporting data for the cholinergic/adrenergic balance hypothesis come from several physiological and pharmacological domains (Janowsky and Overstreet, 1995). For example, physostigmine, a cholinesterase inhibitor that increases CNS ACh levels, will quickly interrupt a manic episode and produce an anergic, depression-like state. It has a similar depressogenic effect in normal individuals. Organophosphorous insecticides, which are irreversible cholinesterase inhibitors, can produce depressive symptoms (Gershon and Shaw, 1961). Drugs that activate cholinergic mechanisms, such as arecoline, produce neuroendocrine and polysomnographic changes indistinguishable from those occurring in major depression (Dilsaver, 1986). Conversely, sympathomimetic compounds are behaviorally activating and can exacerbate a manic state (Borgerding et al., 2007). Compared to normal subjects, patients with major depression appear to have a heightened sensitivity to cholinomimetic drugs administered in experimental challenge paradigms, in that they have greater pupillary constriction (miosis), they go into rapid eye movement (REM) sleep more quickly, and they have a greater secretion of pituitary–adrenocortical-axis hormones and growth hormone (Dilsaver, 1986). The cholinergic hypersensitivity of depressed patients and the depressogenic response of manic patients to physostigmine are mediated by muscarinic cholinergic receptors, in that they are all blocked by atropine.

A deficiency of CNS NE neurotransmission in depression and an excess in mania represent the other pole of the cholinergic/adrenergic balance hypothesis (van Moffaert and Dierick, 1999). NE and its metabolites in cerebrospinal fluid (CSF) and urine have been reported as variably decreased in depressed patients and elevated in manic patients. The most persuasive evidence of impaired NE turnover in depression comes from a study of brain arteriovenous concentration differences for NE and its metabolites (Lambert et al., 2000). In patients with refractory depression, this measure of NE turnover was reduced by over 75%. In comparison, more modest reductions of 5HT and DA turnover occurred. A pharmacological argument for the involvement of NE neurotransmission in depression is that treatment with almost all antidepressants leads to downregulation of β-adrenergic receptors in the CNS, and the time course of this downregulation, over several weeks, parallels the time course of clinical improvement in depressive symptoms. However, some antidepressants do not have this receptor effect. In addition, the downregulation of β-adrenergic receptors by antidepressants that block the NE transporter also occurs in normal animals, and thus may be nothing more than a local adaptation to increased synaptic NE concentrations.

21.1.6.2 Serotonin

A deficiency of 5-HT neurotransmission has been implicated in some patients with major depression, one reflection of which is reduced CSF concentrations of the 5-HT metabolite, 5-hydroxyindoleacetic acid (5-HIAA) (Stockmeier, 1997). Depletion of tryptophan, the dietary amino acid precursor of 5-HT, can precipitate depressive symptoms in some at-risk individuals and some depressives who have responded to a 5-HT transporter inhibitor (Heninger, 1995). However, tryptophan depletion does not cause relapse in depressed patients who have responded to electroconvulsive therapy (ECT), or in bipolar patients who have responded to lithium (Cassidy et al., 1997, 1998). Positron emission tomography (PET) studies have shown elevated 5-HT transporter binding in untreated unipolar and bipolar depressed patients and normal binding in recovered unipolar patients, compared to control subjects (Cannon et al., 2007; Bhagwagar et al., 2007). Altered 5-HT neurotransmission has also been proposed as a neurochemical substrate in a variety of other psychiatric disorders, such as obsessive–compulsive disorder (OCD), premenstrual dysphoria, social phobia, post-traumatic stress disorder (PTSD), and pathological aggression, as mentioned earlier. 5-HT uptake-inhibiting drugs, which are most often used as antidepressants (Nelson, 1999), are therapeutically efficacious in these other disorders as well (see Rubin et al. (2002)).

21.1.6.3 Dopamine

CNS dopaminergic systems have been implicated in the pathophysiology of depression, based on the hypothesis that the mesocorticolimbic dopaminergic system functions as a reward pathway and modulates goal-directed behavior (Willner, 1995). Hyperfunction of mesocorticolimbic dopaminergic neurotransmission might result in manic behavior, and, conversely, hypofunction of this system might result in the loss of pleasure (anhedonia), loss of interest, and lack of motivation that occur in major depression. Reduced CSF concentrations of the DA metabolite, homovanillic acid (HVA), in depressed patients with psychomotor retardation and suicide attempters support this hypothesis (Brown and Gershon, 1993).

21.1.6.4 Other neuroendocrine peptides

Hypothalamic peptides such as corticotropin-releasing hormone (CRH), thyrotropin-releasing hormone (TRH), and growth hormone-inhibiting hormone (somatostatin) have extrahypothalamic distributions in the brain and behavioral effects. Altered concentrations of these peptides have been found in the CSF of depressed patients, suggesting pervasive alterations in neurotransmitter and neuropeptide modulator function in the CNS (Plotsky et al., 1995). The endocrine functions of these peptides will be discussed in later sections.

21.1.6.5 Brain-derived neurotrophic factor

A deficiency of brain-derived neurotrophic factor (BDNF) has been implicated in the etiology of depression. BDNF mRNA and protein concentrations, and trkB, the BDNF receptor, are reduced by about 50% in the brains of suicide victims (Dwivedi et al., 2003). Plasma BDNF concentrations in depressed patients are lower than in controls and increase with antidepressant treatment (Gönül et al., 2005; Aydemir et al., 2006; Piccinni et al., 2008). In animals, acute and chronic stress can decrease BDNF levels in the hippocampus, and chronic antidepressant treatment increases hippocampal BDNF in a time course consistent with clinical improvement in patients (Duman and Monteggia, 2006; Monteggia et al., 2007). There is also synergism between CNS BDNF and 5-HT systems: 5-HT uptake-inhibiting antidepressants enhance BDNF gene expression, and BDNF promotes the differentiation and survival of 5-HT neurons (Martinowich and Lu, 2008). Hypothetically, lack of hippocampal BDNF leads to neuronal loss and some clinical symptoms of depression (e.g., memory deficits), and increased BDNF following antidepressant treatment permits hippocampal neuronal regeneration and clinical improvement. In at least one mouse strain, however, the behavioral effects of the 5-HT uptake-inhibiting antidepressant, fluoxetine, are not dependent on hippocampal neurogenesis (Holick et al., 2008).

21.1.6.6 Neurosteroids and neuroactive steroids

Neuroactive steroids are those steroid hormones that have CNS activity and include most metabolically active steroids. Neurosteroids are considered to be those that are also synthesized in the glia and neurons of the CNS (Bennaroch, 2007). Prominent among the latter are progesterone (P), tetrahydro-P (THP; allopregnanolone), dehydroepiandrosterone (DHEA) and DHEA-sulfate (DHEA-S), estradiol, and testosterone. Neuroactive steroids have multiple effects, including on neuronal proliferation, differentiation, migration, survival, and synaptogenesis. Effects can be opposing; for example, estradiol promotes dentate gyrus neurogenesis, whereas P inhibits it (Bennaroch, 2007).

Several of these steroids have been implicated in affective disorders (for review, see Dubrovsky (2006)). Memory deficits can occur in severe depression, and THP, an allosteric modulator of GABA-A receptors, can disrupt memory in animal models. In contrast, DHEA-S is memory-enhancing, and some effectiveness of DHEA-S in the treatment of depression has been proposed. Several neuroactive steroids affect long-term potentiation, a proposed model for associative learning, and sleep processes, which are often interrupted in affective disorders (Dubrovsky, 2006).

21.2 Hypothalamic–Pituitary–Adrenocortical Axis

The most prominent and well-documented neuroendocrine change in major depression is overactivity of the HPA axis, as reflected by increased CRH and arginine vasopressin (AVP) expression in the paraventricular nucleus (PVN) of the hypothalamus, increased circulating adrenocorticotropic hormone (ACTH) and cortisol concentrations, increased CSF cortisol concentrations, increased urine free cortisol (UFC) excretion, and cortisol resistance to dexamethasone (DEX) suppression (see Rubin et al. (2002) for references). The HPA-axis overactivity is mild to moderate, occurs in 30–50% of major depressives, and occurs throughout the 24 h, even when patients are asleep. By measures such as cortisol production rate and UFC excretion, the highest values occurring in severe major depression overlap with the lowest values occurring in Cushing's disease.

21.2.1 Secretion of Adrenocorticotropic Hormone and Cortisol in Depression

In major depression, both the peak and the nadir in circulating cortisol concentrations are elevated (Rubin et al., 1987a), but overall there appears to be little reduction in the amplitude of the circadian rhythm, nor is its timing significantly shifted (**Figure 2**). A feature of the HPA-axis hyperactivity in major depression that distinguishes it from Cushing's disease and Cushing's syndrome is that the circadian rhythms of ACTH and cortisol are preserved. In contrast, in most Cushing's cases the autonomously functioning basophilic pituitary adenoma or adrenal tumor generates relatively constant production of ACTH and/or cortisol throughout the 24 h. The constant tissue exposure to elevated glucocorticoid concentrations likely produces the overt clinical stigmata in Cushing's patients (Newell-Price et al., 1998). There may be some chemical changes in major

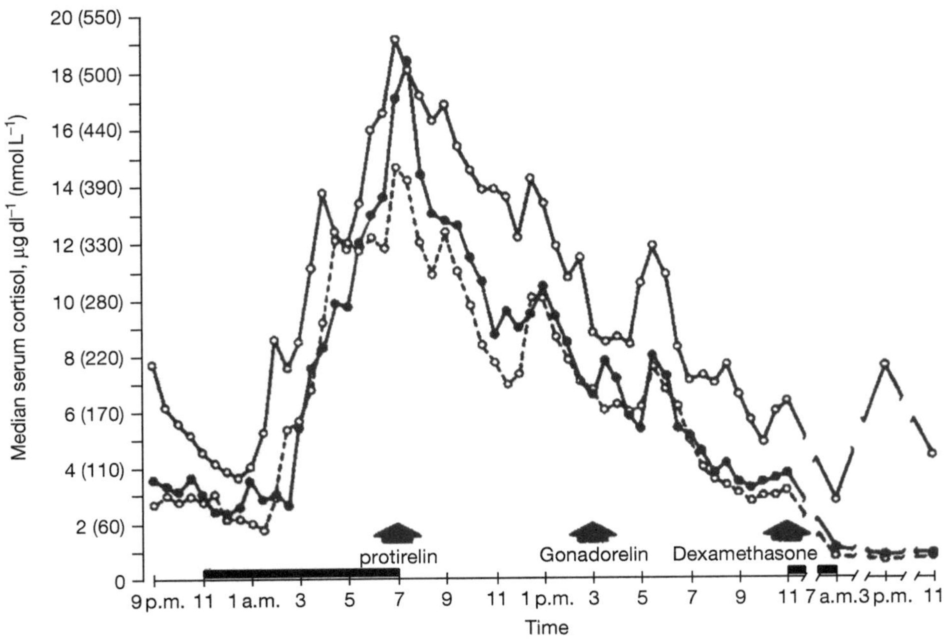

Figure 2 Median serum cortisol concentrations at time of each blood sampling for 15 endogenous depressed patients who were post-DEX cortisol escapers (open circles with solid line), 25 patients who were suppressors (closed circles), and 40 matched normal control subjects (broken line). Times of sleep (lights out) are marked by horizontal bars. Times of administration of protirelin (TRH; 100 μg IV), gonadorelin (LHRH; 100 μg IV), and DEX elixir (1 mg orally) are marked by arrows. Protirelin and gonadorelin had no effect on serum cortisol concentrations. Increased pre-DEX serum cortisol concentrations are evident in post-DEX cortisol escapers, compared with suppressors and controls, at all times of night and day. Reprinted with permission from *Archives of General Psychiatry* 44: 332, 1987. Copyright 1987, American Medical Association.

depressives secondary to their increased cortisol production, such as increased serum sodium and decreased serum potassium (Reus, 1984), but the physical changes characteristic of Cushing's patients are absent in major depression. Another important reason for the absence of clinical stigmata of Cushing's disease in severe depression is that depressed patients maintain normal plasma concentrations of corticosteroid-binding globulin (CBG), whereas plasma CBG is severely reduced in Cushing's disease, with a resultant disproportionate increase in circulating free cortisol (Schlechte et al., 1986). Thus, plasma-free cortisol levels in depression are higher than normal, but well below those in Cushing's disease (Carroll et al., 1976). Nevertheless, even nonobese patients with major depression and bipolar depression, but not minor depression, do manifest an important metabolic sign of Cushing's disease – insulin resistance (Hung et al., 2007), and bipolar patients have a high incidence of type II diabetes mellitus (Cassidy et al., 1999). Atypical antipsychotic drugs should be used cautiously in these patients, because these drugs can lead to the same metabolic disturbances.

Carroll et al. (2007) studied HPA-axis regulation in hypercortisolemic, severely depressed patients by measuring plasma ACTH and cortisol every 10 min for 24 h. The activity of the central HPA-axis pulse generator was normal, as judged by circadian acrophase and nadir of ACTH and cortisol, the circadian amplitude of each hormone, the pulse frequency of each hormone, and the entropic orderliness of ACTH secretion. Basal and pulsatile ACTH secretions were increased, whereas the plasma half-life of ACTH was shortened, so that mean 24-h plasma ACTH concentrations were not significantly elevated. Thus, the central HPA-axis overdrive consisted primarily of an elevated burst mass per pulse of ACTH, with an additional contribution from increased basal ACTH secretion, the latter possibly related to increased anterior pituitary volume in depression (see below). There was a normal number of cortisol pulses but diminished regularity (high approximate entropy), impaired linkage of cortisol to ACTH, normal circadian rhythmicity, increased burst mass of cortisol per pulse, and a 60% increase of total cortisol secretion, with a normal plasma cortisol half-life. These results indicate increased central drive of ACTH secretion, but no abnormality of limbic system programming of the HPA axis. They also are consistent with increased, ACTH-independent secretion of cortisol, possibly related to adrenal hypertrophy (see below).

21.2.2 Secretion of Corticotropin-Releasing Hormone in Depression

Corticotropin-releasing hormone (CRH) is produced by the parvocellular neurons of the hypothalamic PVN and is released from their axon terminals in the external layer of the median eminence, in proximity to the capillaries of the pituitary portal circulation, which carry CRH and other hypothalamic hormones to the anterior pituitary gland. CRH receptors are present on pituitary corticotrophs. CRH is the primary stimulus to ACTH secretion, with variable contributions from AVP, catecholamines, and angiotensin-II (Orth, 1992; De Souza and Grigoriadis, 2002). CRH is the lead hormone in the HPA axis, and its secretion is influenced by a number of factors, including age, food intake, nicotine intake and withdrawal, and the stress of lumbar puncture (Geracioti et al., 1997).

Studies of CRH concentrations in the CSF in major depression have yielded varying results. CSF CRH concentrations do not follow plasma ACTH or cortisol circadian rhythms, and the CSF concentration of CRH appears to be independent of the HPA axis (Wong et al., 2000). Rather, CSF CRH may reflect activation of non-HPA CRH pathways in the brain (Vythilingam et al., 2000). Some studies of CSF CRH in depression report elevated concentrations, and others do not (Geracioti et al., 1997; Mitchell, 1998). The best evidence comes from Wong et al. (2000), who sampled plasma and CSF over 24 h and found no elevation of CSF CRH concentrations in hypercortisolemic, melancholic depressed patients. In aggregate, the studies point to mild elevations of CSF CRH in a small percentage of patients with major depression. This finding has not been linked to any particular depressive subtype, and it is not specific to depression. For example, elevated CSF CRH concentrations are reported in alcohol withdrawal, PTSD, and schizophrenia. Several small studies suggest that CSF CRH concentrations are somewhat reduced after successful treatment of depression with antidepressant drugs or ECT, but not with vagus nerve stimulation. In these studies, however, the effect was small and simple regression to the mean cannot be excluded. In addition, an initial association between CRH-binding protein gene polymorphisms and major depression was not replicated in a subsequent study (Van Den Eede et al., 2007).

In addition to the CRH receptors present on pituitary corticotrophs, CRH-containing cell bodies and saturable, high-affinity CRH receptors are distributed throughout the cerebral cortex and limbic system (Orth, 1992; De Souza and Grigoriadis, 2002), which suggests that CRH acts as a CNS neuromodulator separate from its hormonal action at the anterior pituitary. Increased activity of CRHergic pathways to the locus ceruleus (LC) (van Bockstaele et al., 1998) stimulates NE release from the LC and may underlie some symptoms of major depression, such as anxiety (Gold et al., 1988). Elevated tissue CRH concentrations have been reported in the LC and prefrontal cortex of depressed suicides (Bissette et al., 2003; Merali et al., 2006). Chronic stress leads to downregulation of CRH-1 receptors in some brain regions (Fuchs and Flugge, 1995), and there is evidence of this downregulation in suicide brains (Merali et al., 2004). The specificity of this finding is unknown.

21.2.3 Secretion of Arginine Vasopressin in Depression

Arginine vasopressin (AVP) plays a prominent role in stress-mediated HPA-axis responses. Circulating AVP is elevated in depression, but the plasma concentration of vasopressin neurophysin 1 is markedly reduced (Laruelle et al., 1990). Desmopressin (ddAVP) reverses the blunted ACTH response to CRH in this illness (Dinan et al., 1999; Dinan and Scott, 2005; Landgraf, 2006), and, by itself, produces higher ACTH and cortisol responses than in normal control subjects, suggesting enhanced sensitivity of V3 receptors in depression (Dinan et al., 2004). Premenopausal female depressives were found to have greater ACTH and AVP responses to cholinergic challenge (low-dose physostigmine) compared to matched female controls, whereas the reverse was true for male depressives and matched male controls (Rubin et al., 1999). The relative contributions of CRH and AVP in driving the HPA-axis hyperactivity that is characteristic of 30–50% of major depressives remain to be determined. Pitts et al. (1995) found that CSF AVP was not elevated in depressed patients and that it did not differ between DEX suppression test (DST) suppressors and nonsuppressors. DST status is also not related to the ACTH and cortisol responses to AVP (Carroll et al., 1993) or to the plasma concentrations of vasopressin neurophysin 1 (Laruelle et al., 1990). These results are consistent with the fact that the majority of circulating AVP and vasopressin neurophysin 1 derives from the posterior pituitary.

21.2.4 Perturbation Tests of HPA-Axis Function in Depression

21.2.4.1 Dexamethasone suppression test

DEX, a potent synthetic glucocorticoid, primarily binds to glucocorticoid receptors (GRs) on anterior pituitary corticotrophs and, by feedback inhibition, suppresses ACTH and cortisol secretion. The degree and duration of suppression depends on a balance among the amount of DEX administered, its pharmacokinetics in a given subject, and the degree of that individual's suprapituitary drive of the HPA axis (primarily CRH and AVP secretion into the pituitary portal circulation, as discussed above). Low-dose and high-dose DEX suppression tests (DSTs) have been used for the differential diagnosis of Cushing's disease (Newell-Price et al., 1998), and a low-dose DST has been used as a marker of HPA-axis hyperactivity in affective disorders in an attempt to aid differential diagnosis and to follow the course of treatment (see Rubin et al. (2002) for references).

The most widely used low-dose protocol in affective disorders has been the administration of DEX, 1 mg orally, at 11 p.m. or midnight, followed by serum cortisol determinations at intervals over the following 24 h. Early studies used 2 and 1.5 mg doses, but the compromise between sensitivity versus specificity of the test is best with the 1.0-mg dose. In normal individuals, cortisol remains suppressed to very low levels for the full 24 h. In contrast, 40–70% of patients with major depression show cortisol nonsuppression or early escape from suppression (a positive DST) during the 24 h following DEX administration. Positive DST results are found most often in patients with severe depression, especially those with psychotic features or melancholic features, and in cases of mixed bipolar disorder. Many studies suggest that only about 10% of patients with milder depressions and most other psychiatric illnesses show cortisol nonsuppression. Longitudinal studies in rapidly cycling bipolar disorder show abnormal DST results during or slightly preceding the depressive phases and normal results in the manic phases (Greden et al., 1982). Abnormal DST results also occur often in anorexia nervosa, dementia, and acutely psychotic patients.

The DST has an interesting history, beginning with careful delineation of its sensitivity, specificity, and positive predictive value as an ancillary diagnostic test for severe or melancholic major depression. Unfortunately, psychiatrists began using the DST as a screening test, rather than as an ancillary diagnostic test after clinical examination suggests a high likelihood of major depression. The use of the DST as a screening test created much confusion when nondepressed patients had positive DSTs, based on the less-than-perfect specificity of the test (Rubin and Poland, 1984; Carroll, 1985) and led to its demise a few years after its adoption by psychiatrists in clinical practice, especially after the importance of controlling for plasma DEX concentrations was appreciated.

The DST has also been used to follow the course of treatment in patients who had a positive DST while depressed. In successfully treated patients, the DST gradually becomes normal (**Figure 3**), and such patients tend to remain in remission longer than patients who may show clinical improvement but still have an abnormal DST. The use of the DST to follow the adequacy of treatment of depression still appears to hold promise (Kin et al., 1997). An abnormal DST in patients with unipolar depression is associated with a large drug versus placebo response difference approximating 40%, whereas this difference is only about 10% in depressed patients with a normal DST (Brown, 2007). These findings result

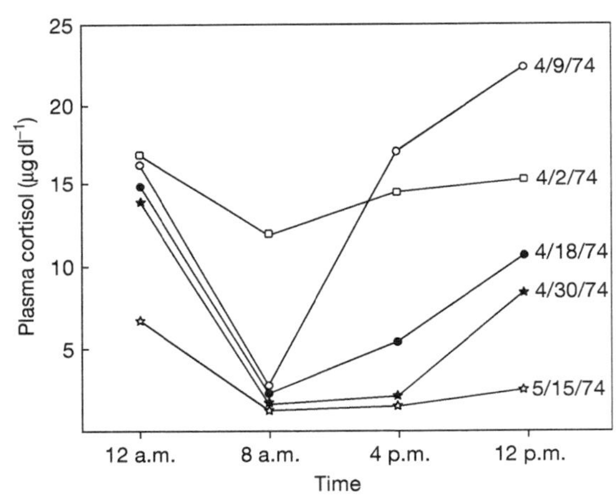

Figure 3 Plasma cortisol responses to DEX administration (2 mg orally) given just after 12 a.m. blood draw in a 49-year-old woman hospitalized with agitated, unipolar depression. During the first week, placebo medication was given, followed by antidepressants after the 4/9/74 DST. Gradual return of a very abnormal test to full, 24-h cortisol suppression is evident across the 6 weeks of serial testing. Reprinted with permission from *Archives of General Psychiatry* 33: 1041, 1976. Copyright 1976, American Medical Association.

from a low rate of response to placebo (about 20%) in patients with an abnormal DST compared with a placebo response rate over 50% in depressed patients with a normal DST. The DST provides better discrimination of placebo responders and nonresponders than either depression subtype or depression severity. Similar differences appear to hold for response rates to psychotherapy alone in depressed patients (Brown, 2007). Likewise, an abnormal DST during an index episode constitutes an approximately four- to eightfold risk factor for eventual suicide, for switch to bipolar disorder, and for re-hospitalization after treatment of the index episode (Coryell, 1990; Unden and Aperia, 1994; Coryell and Schlesser, 2001). The prediction of completed suicide by DST status appears strongest for inpatients with manifest suicidality. Death from all causes is also associated with DST status in the index depressive episode. On the other hand, the DST does not predict nonlethal suicidality and it is not informative in nondepressed psychiatric populations at risk of suicide (Coryell et al., 2006; Jokinen et al., 2007). The DST may be a more powerful predictor of completed suicide in depression than the customary clinical measures such as age, male sex, or past suicide attempts (Coryell and Schlesser, 2001). Notwithstanding the present demise of the DST as a diagnostic test, it remains a very informative biological marker in depression (Fink, 2005).

21.2.4.2 CRH stimulation test

ACTH responses to exogenous CRH administration are often blunted in depressed patients compared to controls (Gold et al., 1986; Holsboer et al., 1986; Rubin et al., 1996). Cortisol responses may not be similarly reduced, suggesting increased sensitivity or capacity of the adrenal cortex to endogenous ACTH. ACTH and other pro-opiomelanocortin (POMC)-derived peptide responses to CRH in depressives can be enhanced by pretreatment with metyrapone, which, by inhibiting cortisol synthesis, interrupts cortisol negative feedback to the pituitary and hippocampus, thus implicating increased circulating cortisol as the main factor influencing the reduced corticotroph response in depression (von Bardeleben et al., 1988; Ur et al., 1992; Young et al., 1994). The CRH test has also been administered following pretreatment with DEX, which results in larger, rather than smaller, ACTH and cortisol response differences in depressed patients compared to normal controls (Heuser et al., 1994). The combined DEX-CRH test has also been used to predict the possibility of relapse in remitted depressed patients (Zobel et al., 1999), similar to the use of the DST itself. The combined DEX-CRH test is subject to the same caveats concerning DEX plasma concentrations and drug interferences as the original DST.

21.2.4.3 ACTH stimulation test

Exogenous $ACTH_{1-24}$ administration has been used as a direct test of adrenocortical responsiveness in depression. Increased cortisol responses to direct ACTH stimulation, as well as normal cortisol responses in the face of reduced ACTH responses following CRH, as noted above (e.g., Amsterdam et al., 1987; Jaeckle et al., 1987) suggest increased adrenocortical responsiveness to ACTH. These studies were done with the standard clinical protocol that uses a supramaximal stimulation dose of 250 μg $ACTH_{1-24}$, thus testing maximal adrenal secretory capacity. Several studies using much lower, more physiologic doses of $ACTH_{1-24}$ have also been conducted, in an attempt to measure adrenal sensitivity to stimulation, rather than maximal secretory capacity (reviewed in Rubin et al. (2006a)). Although the methodologies varied considerably, the majority of studies found no significant difference in cortisol responses between depressed patients and control subjects, suggesting there is no adrenocortical hypersensitivity to ACTH stimulation in major depression. Likewise, Carroll et al. (2007) demonstrated that adrenocortical sensitivity to endogenous ACTH is normal in hypercortisolemic depressives.

21.2.4.4 Serotonergic stimulation

5-HT input to the hypothalamus is an important stimulus to CRH release. Of the many 5HT receptors, the 5-HT1A receptor appears dominant in this regard (Dinan, 1997). Stimulation of these receptors in humans by azaspirones such as ipsapirone activates the HPA axis and induces hypothermia. These responses appear to be attenuated in major depression (Lesch et al., 1990a). High basal cortisol levels were present in these patients, so that their impaired HPA response may have been due to glucocorticoid-induced subsensitivity of postsynaptic 5-HT1A receptors or defective postreceptor signaling pathways.

Chronic treatment with the tricyclic antidepressant, amitriptyline, caused further impairment in 5-HT1A-mediated hypothermia in these patients (Lesch et al., 1990b), supporting the concept that effective antidepressant treatment downregulates 5-HT1A receptors. Treatment of patients with OCD with the 5-HT uptake-inhibiting antidepressant, fluoxetine, yielded similar findings, in that the ability of ipsapirone to induce hypothermia and ACTH/cortisol release was attenuated (Lesch et al., 1991).

21.2.5 Pituitary and Adrenal Volumetric Studies in Depression

Stimulation of the anterior pituitary by CRH and AVP, and, in turn, stimulation of the adrenal cortex by ACTH, can be expected to result in some hypertrophy of the target endocrine glands. Studies of this possibility have been conducted in depressed patients, by both direct determination of gland weight in postmortem samples and by CT and MRI imaging.

21.2.5.1 Pituitary gland

Using MRI, Krishnan et al. (1991) reported significantly greater pituitary volume in patients with major depression compared to age- and sex-matched controls, and Axelson et al. (1992) further suggested a significant relationship between pituitary volume and 10 p.m. post-DEX cortisol. In addition, anterior pituitary volume is significantly increased, by 25%, in adolescent patients with major depression (MacMaster and Kusumakar, 2004). These findings suggest a dynamic hypertrophy of the pituitary gland in conditions of increased HPA-axis activity, as suggested also by Pariante et al. (2005), who observed increased anterior pituitary volume in nondepressed patients with acute psychosis. In treatment-naive pediatric patients with major depressive disorder, anterior pituitary volume is increased and the sex difference in this measure (boys smaller than girls) is lost (MacMaster et al., 2006). In contrast, Sassi et al. (2001) found no increase of anterior pituitary volume in unipolar depression. On balance, the evidence for increased anterior pituitary volume in depression is credible and is consistent with the concept of central HPA overdrive, but it is not a specific finding.

21.2.5.2 Adrenal gland

Because adrenal hypertrophy follows chronic stimulation by ACTH (Orth et al., 1992), adrenal gland volume and adrenal weight have been determined in major depressives or victims of suicide in whom depression prior to death was inferred (reviewed in Rubin et al. (1996)). Larger adrenals in patients before treatment, compared to controls, were found in most of these studies, but the results were not entirely consistent. Rubin et al. (1995) did find a return of adrenal volume to control values in successfully treated major depressives, indicating that the enlargement was reversible. Of note, in every study in which HPA function was assessed, there was no significant relationship between adrenal gland size and hormone measures. These were all cross-sectional studies, however; longitudinal studies in the same patients in and out of episode might reveal a better correlation between adrenal size and HPA-axis activity. The adrenocortical hypertrophy in depression likely is another indication of increased activity in the entire HPA axis. It is also noteworthy that reversible hypertrophy of the adrenal cortex is a classical neuroendocrine response to chronic stress.

21.2.6 Glucocorticoid Receptor Function in Depression

The hippocampus contains high concentrations of mineralocorticoid receptors (MRs) and GRs. MRs have high affinity for glucocorticoids and are almost saturated at low, basal circulating glucocorticoid levels. On the other hand, GRs are low-affinity receptors and become occupied, along with MRs, at high-stress levels of glucocorticoids (de Kloet et al., 1998). The actions of glucocorticoids on higher CNS functions are complex and depend on the concentration of steroid and the duration of any perturbation of its circadian rhythm. Generally, it is considered that MRs operate in a proactive mode, determining the sensitivity of HPA-axis responses to stress, while GRs operate in a reactive mode, terminating the stress response and facilitating recovery from stress. At lower steroid levels, activation of MRs maintains neuronal excitability, such that excitatory input to the hippocampus results in steady excitatory output of the CA1 region, whereas activation of GRs at higher steroid levels depresses CA1 output. Behaviorally, activation of MRs affects integration of sensory information, interpretation of environmental experiences, and execution of appropriate behavioral reactions, whereas activation of GRs facilitates information storage (memory) and elimination of inadequate behavioral responses (de Kloet et al., 1998). When stimulation of the two receptor types become chronically imbalanced, for example, in cases of prolonged HPA-axis hyperactivity or exogenous glucocorticoid administration, effects on cognition can turn from adaptive to maladaptive, with resultant decrements in cognitive function (for review see Herbert et al. (2006)). There is some evidence that polymorphisms in the *GR* gene can influence the HPA-axis response to stress differentially in normal men and women (Kumsta et al., 2007).

Studies in depressed patients suggest that MR function is normal (Young et al., 2003) or even upregulated (Mason and Pariante, 2006). Impaired GR signaling has been implicated in the HPA-axis hyperactivity of major depression, as well as in some psychopathological aspects of the illness. Antidepressant drugs may also relieve depression by increasing the sensitivity of GRs (Pariante, 2006).

21.2.7 Effects of Antidepressants on the HPA Axis

Antidepressants that block the transporters for 5-HT and NE increase HPA-axis activity following their acute administration (Holsboer and Barden, 1996). Chronic administration of antidepressants to depressed patients, however, results in reversion to normal of increased HPA-axis activity when there is complete or nearly complete remission of the depressive episode. Increased concentrations of hippocampal MRs and GRs occur transiently between 2 and 5 weeks following the start of antidepressant treatment (Holsboer and Barden, 1996; Okugawa et al., 1999), which is roughly the time course of clinical improvement of depressive symptoms and reduction of HPA-axis hyperactivity, suggesting that increased corticosteroids might be contributory to some dimensions of depressive symptomatology, such as memory and concentration difficulties (Holsboer and Barden, 1996). Support for this hypothesis comes from transgenic mice with impaired GR function, which show behavioral changes suggestive of cognitive impairment and which improve with chronic antidepressant treatment (Montkowski et al., 1995).

21.2.8 CRH-Receptor Antagonists in the Treatment of Depression

As discussed above, CRH, in addition to being a major stimulus to ACTH secretion, is distributed widely throughout the CNS. There are two CRH receptors, CRH1 and CRH2, encoded by two distinct genes. CRH-binding protein is co-localized with CRH in many brain areas and has high affinity for the peptide (McCarthy et al., 1999). In animals, increased CRH-binding sites in the amygdala and hindbrain occur in response to maternal deprivation, along with increased anxiety-like behavior; increased CRH stimulation of NE secretion from the LC has been proposed as one explanation for this behavior.

Increased CRH neuronal activity has been hypothesized as underlying the anxiety frequently occurring in major depression. CRH-receptor antagonists reduce experimental anxiety in primates (Habib et al., 2000) and are being studied as putative treatments for depression. In one early phase II study, the CRH-receptor antagonist R121919 was tolerated, but, unexpectedly it did not alter basal or CRH-stimulated HPA-axis activity. Clinical improvement was dose-dependent, but, because the study had no placebo or active control, no conclusion as to efficacy is possible (Zobel et al., 2000). Moreover, increased hepatic enzyme activity in some patients prompted discontinuation of further development of this compound.

21.2.9 Cortisol Synthesis Inhibitors and Glucocorticoid Receptor Antagonists in the Treatment of Depression

Suppression of glucocorticoid synthesis in major depression has been proposed as a useful adjunct in patients resistant to treatment with antidepressant drugs, and steroid synthesis inhibitors (ketoconazole, metyrapone, and aminoglutethimide) have been used both alone and as adjunctive treatments to standard antidepressants (Wolkowitz and Reus, 1999). Glucocorticoid synthesis inhibitors alone have a moderate antidepressant effect, but not sufficient to produce acceptable clinical remission. In addition, hypercortisolemic patients, rather than normocortisolemic patients, are the responders (Wolkowitz and Reus, 1999). On the other hand, a glucocorticoid synthesis inhibitor may be a useful adjunct in depressed patients with HPA-axis hyperactivity who do not fully respond to standard antidepressant treatment. Therapeutic concerns are the side effects of the specific inhibitors, as well as the physiological consequences of chronic steroid suppression. Moreover, repeated administration of a cortisol synthesis inhibitor like metyrapone soon leads to massive secretion of ACTH and a breakthrough of enzyme inhibition, so that cortisol levels return to normal or even exceed basal levels (Veldhuis et al., 2001; Holger et al., 2004). This effect is counterproductive to the therapeutic goal of reducing tissue exposure to cortisol in depression. Thus, there is little prospect of successfully treating depression with cortisol synthesis inhibitors. These agents are useful in clinical endocrinology primarily in cases of Cushing's syndrome due to non-ACTH-dependent adrenal tumors.

An alternative approach is to block the GR. Mifepristone (RU486), a progesterone receptor antagonist at low doses and GR antagonist at higher doses, has been proposed in the treatment of major depression (Murphy, 1997). Its use for the treatment of the psychotic subtype of major depression, however, has shown no convincing effect, for either the depressive or the psychotic symptoms (Rubin and Carroll, 2004; Carroll and Rubin, 2007). Three recent phase III trials of the drug also failed to demonstrate efficacy of RU486 in psychotic depression.

21.3 Hypothalamic–Pituitary–Thyroid Axis

Thyroid hormones are fundamental to normal brain development and regulate neuronal growth and synaptogenesis (Oppenheimer and Schwartz, 1997). TRH, the hypothalamic tripeptide that stimulates the anterior pituitary to secrete TSH, and prolactin as well (Leong et al., 1983), is produced primarily from the parvocellular region of the PVN of the hypothalamus (Guldenaar et al., 1996), but is also expressed in other areas of the brain. There is extensive intraneuronal co-localization of TRH with other neuroactive substances, including 5-HT, substance P, neuropeptide Y (NPY), and DA (Hökfelt et al., 1989). TRH synthesis and release are regulated primarily by the thyroid hormone, triiodothyronine (T_3), through negative feedback. 5-HT inhibits TRH release (Morley et al., 1981), and DA, via the D2 receptor, has a stimulatory effect on TRH (Lewis et al., 1987).

Pituitary TSH synthesis and release is controlled by TRH in concert with negative feedback by thyroid hormones and the inhibitory influences of DA (Pourmand et al., 1980) and somatostatin (Weeke et al., 1975). TSH stimulates glandular growth and the synthesis and release of the thyroid hormones, thyroxine (T_4) and T_3. In normal humans, both T_4 and T_3 are released from the thyroid gland, but most plasma T_3 is derived from peripheral tissues by deiodination of T_4. Both T_4 and T_3 are bound firmly but reversibly to several plasma proteins, the principal one being thyroxine-binding protein (TBG).

21.3.1 Basal Thyroid Function in Depression

About 20–30% of patients with mood disorders exhibit some form of thyroid dysfunction. Increased circulating concentrations of total and/or free T_4, although still within the normal range, and which regress after successful treatment, are the most frequently reported abnormalities (Chopra et al., 1990). There is also a loss of the normal nocturnal TSH rise between midnight and 3 a.m. (reviewed in Rubin et al. (1987b)).

Nearly 5–10% of the normal population are thyroid-antibody positive, and 15% of depressed patients have evidence of autoimmune thyroiditis, with the highest incidence occurring in women over 60 years of age (Rosenthal et al., 1987). Autoimmune thyroiditis, with marginally elevated basal TSH levels and exaggerated TSH responses to TRH indicative of subclinical hypothyroidism, is overrepresented in major depression (Kraus et al., 1997). Subtle neuropsychological abnormalities have been shown in subjects with borderline TSH elevation (Manzoni et al., 1993), and subclinical hypothyroidism contributes to treatment resistance in depression (Hickie et al., 1996). Patients with frank hypothyroidism and euthyroid or subclinically hypothyroid depressives share many clinical features, including fatigue, dysphoric mood, poor memory, apathy, and cognitive dysfunction.

21.3.2 TRH Stimulation of TSH in Depression

Administration of TRH initially was undertaken in depression as a possible therapeutic measure (reviewed in Rubin et al. (2002)). Later studies suggested that the TSH response to TRH, in the context of normal basal T_4 and T_3 levels, was blunted in about 25% of depressives compared to normal subjects. In thyrotoxicosis, the TSH response to TRH is flat and circulating T_3 is elevated, but depressed patients do not manifest other metabolic features of hyperthyroidism.

21.3.3 Relationship to the HPA Axis

There is a considerable evidence to suggest that HPT dysregulation may arise as a consequence of altered HPA-axis activity in depression (reviewed in Rubin et al. (2002)). Glucocorticoids inhibit TSH secretion, and an inverse relationship between nocturnal cortisol and TSH levels has been reported in depressed patients. A strong positive correlation has also been noted between the TSH response to TRH and the ACTH response to CRH in depressives, suggesting a common regulator of both systems, but there is no significant association of abnormal DST results with blunting of the TSH response to TRH in depression.

As indicated earlier, successful antidepressant treatment normalizes HPA-axis hyperactivity in major depression. Antidepressant treatment also reduces plasma T_4 levels, with responders having a more pronounced reduction in total and free T_4 than nonresponders, and normalizes the TSH responses to TRH, indicating that a blunted TSH response to TRH is state- rather than trait-dependent. It has been suggested that the normalization of subtle dysregulation of the HPT axis may be part of the mechanism of action of antidepressant therapy, but it is more likely that the reversal of HPT irregularities occurs secondary to normalization of HPA-axis alterations (reviewed in Rubin et al. (2002)).

21.3.4 Diagnostic and Prognostic Utility of the TRH Stimulation Test

The demonstration of blunting of the TSH response to TRH in depression led to exploration of its potential utility in the diagnosis and subtyping of depressive disorders, but results have been disappointing (reviewed in Rubin et al. (2002)). As a diagnostic tool, the TRH stimulation test has a low specificity for depression, as indicated by blunted TSH responses in an array of other psychiatric disorders, including alcohol dependence, PTSD, premenstrual syndrome, OCD, panic disorder, and Alzheimer's disease, as well as nonspecific suicidality. A change in the TSH response to TRH from blunted to normal also does not predict a positive response to antidepressant therapy (reviewed in Rubin et al. (2002)).

21.3.5 Adjuvant Therapy with Thyroid Hormones

Because, as noted above, there is considerable overlap between the symptoms of affective disorders and those of hypothyroidism as well as hyperthyroidism, it is important to determine the thyroid status of both depressed and manic patients and to ensure they are euthyroid during treatment of the affective disturbance. In particular, patients treated long term with lithium need to be checked regularly for possible drug-related hypothyroidism. In euthyroid patients, as well, thyroid hormones are used as adjuvant therapy to antidepressants (reviewed in Joffe (2002)).

21.3.5.1 Acceleration of antidepressant effect

The full therapeutic response to antidepressants usually requires 3–4 weeks of treatment, which prompted the initial use of thyroid hormones to accelerate the clinical response. An acceleration of antidepressant effect by T_3 was initially seen in female, but not male, depressed patients, but others could not replicate this finding. These studies were limited by small sample sizes and inadequate doses of antidepressants.

21.3.5.2 Augmentation of antidepressant effect

Up to 30% of patients fail to respond to antidepressant treatment, and thyroid hormone augmentation (low-dose T_3) has been studied as a means of converting antidepressant nonresponders into responders. In early studies a response rate of 25–90% has been reported; however, very few of the early studies were double-blind, and an assessment of basal thyroid state of the patients was often not included (reviewed in Joffe (2002) and Rubin et al. (2002)). As indicated earlier, lithium is also used as an adjunct to antidepressant treatment in partial responders and nonresponders; there appears to be no difference in the clinical response to tricyclic antidepressants between augmentation with lithium or augmentation with T_3. There are also reports of the beneficial response of adjuvant therapy with T_3 after failure to respond to fluoxetine alone. T_3 as an adjuvant to ECT has also been shown to increase clinical responsiveness. In regard to the relationship between basal thyroid hormone status and response to T_3 augmentation, eventual responders to augmentation were found to have, prior to any antidepressant treatment, lower circulating TSH, higher circulating T_4, and a greater free thyroxine index (FTI) compared to augmentation nonresponders. Another strategy for augmentation with thyroid hormones has been the use of mega-doses of T_4 in refractory depression (Bauer et al., 1998) and for preventing relapses (Bauer et al., 2002), but this strategy remains controversial. Variability in response to T_4 may be related to a reduction of CSF transthyretin, the T_4-transporter, in depressed patients (Sullivan et al., 2006).

21.3.5.3 Mode of action of thyroid hormone augmentation

Most studies of thyroid hormone augmentation have used T_3 rather than T_4. T_3 appears to be significantly more effective than T_4 in depressed patients unresponsive to tricyclic antidepressants (reviewed in Joffe (2002)). T_3, the active thyroid hormone in the brain, is derived in the CNS from T_4 by brain type II 5′ deiodinase, an enzyme that is inhibited by cortisol (Hindall and Kaplan, 1988). Such inhibition would be accompanied by an elevation in 3,3′5′-triiodo-L-thyronine (rT_3) levels, which has been reported in the CSF of unipolar depressed subjects. It remains possible that elevated cortisol levels, in severely depressed patients, impair the intracerebral conversion of T_4 to T_3, so that T_3 augmentation serves to restore thyroid hormone homeostasis in the CNS.

21.4 Growth Hormone (Somatotropin)

Growth hormone (GH) containing cells (somatotrophs) account for about 40% of anterior pituitary cells. The circadian secretion of GH has been well characterized. For most of the day, except following meals, circulating GH levels are less than 5 mU l^{-1}. In pubertal children and young adults, prominent GH secretion occurs early in the sleep period, in association with stage 3–4 or deep sleep. In contrast, REM sleep inhibits GH release. A variety of stressors, both physical (e.g., exercise) and psychological, can elevate GH. Two hypothalamic peptides regulate GH: GH-releasing hormone (GHRH) and GH-inhibiting hormone (somatostatin). These, in turn, are regulated by several neurotransmitter systems. In addition, ghrelin, a peptide hormone produced by the stomach, has strong GH-releasing activity, and lesser activity as an ACTH and prolactin secretagog (van der Lely et al., 2004).

GH promotes the production of insulin-like growth factors (IGFs) by the liver. Through negative feedback, GH secretion is inhibited by the IGFs and by GH itself. Estrogen enhances the secretion of GH, but oral estrogen can impair the metabolic action of GH in the liver, leading to a fall in IGF-1 production and fat oxidation (Leung et al., 2004).

21.4.1 Regulation of GH Secretion

GHRH-containing neurons are found in the arcuate nucleus of the hypothalamus, and nerve fibers from this nucleus project almost entirely to the median eminence, suggesting that its principal role is the regulation of GH secretion. A smaller group of fibers projects to the periventricular region of the hypothalamus, close to where somatostatin neurons are located, raising the possibility that GHRH neurons also modulate somatostatinergic activity. Axons of somatostatin-containing neurons in the periventricular nucleus of the anterior hypothalamus also terminate in the median eminence. Fibers from these neurons also project to the arcuate and ventromedial nuclei, where they have synaptic connections with GHRH neurons, allowing for direct interaction between these opposing hormones on GH secretion (Meister et al., 1990).

21.4.2 Basal GH Secretion in Depression

Circadian studies of GH secretion in depression have reported no difference in mean 24-h or nocturnal or awake serum GH concentrations, diurnal GH hypersecretion but normal nocturnal secretion, and reduced sleep-related GH release (reviewed in Rubin et al. (2002)). In aggregate, the studies suggest that basal nocturnal GH secretion may be reduced in major depression, diurnal GH secretion is not abnormal, and the nocturnal decrease may be a trait phenomenon, that is, persisting into clinical remission. The nocturnal decrease may be related to the relative lack of stage IV sleep in depression during the first half of the sleep period.

21.4.3 Monoamines and GH Secretion in Depression

21.4.3.1 Norepinephrine

In man, at least half of the NE-containing neurons in the CNS are localized in the LC. Their axons innervate the entire cortex and spinal cord. Considerable evidence indicates that α_2-adrenergic receptors in the hypothalamus control the release of GHRH. In rat, dog, and rhesus monkey, the stimulation of GH secretion by the α_2-agonist, clonidine, is abolished by the α_2-antagonist, yohimbine. GHRH-induced GH release was unaffected by this antagonist, whereas concomitant blockade of α_1- and α_2-adrenergic receptors by phentolamine abolished the GHRH-induced GH rise. Blunted GH release in response to clonidine challenge occurs in depressed patients, suggesting a subsensitivity or downregulation of α_2-adrenergic receptors (reviewed in Rubin et al. (2002)).

Desipramine, a tricyclic antidepressant, is a potent inhibitor of presynaptic NE uptake and elevates GH levels when given acutely to healthy volunteers. Laakmann et al. (1986) examined desipramine-induced GH release following pretreatment with methysergide (5-HT receptor blocker), propranolol (β-adrenergic receptor blocker), phentolamine, yohimbine, and prazosin (α_2-adrenergic receptor blocker). Methysergide and prazosin had no effect on the GH response, propranolol enhanced the response, and phentolamine and yohimbine attenuated the response. These results support the concept that α_2-adrenergic receptors enhance, and β-adrenergic receptors inhibit, GH release.

In regard to GH responses to adrenergic challenge during treatment of depression, Checkley et al. (1986) examined α_2-adrenergically mediated GH responses to clonidine throughout the course of desipramine treatment. After 1 week of treatment, the GH response to clonidine was enhanced, but it was reduced after the second and third weeks of treatment. This accords with the concept of slow downregulation of α_2-adrenergic receptors during antidepressant treatment. No alteration in α_1-adrenergic tone was noted, as measured by the responsiveness of pupil diameter to phenylephrine. Depressed patients treated with the 5-HT uptake inhibitor, fluoxetine, also showed a significant decrease in clonidine-stimulated GH release, suggesting that fluoxetine can alter α_2-adrenergic responses indirectly (O'Flynn et al., 1991).

The attenuation of α_2-adrenergically mediated GH release appears to be too nonspecific to be useful in the differential diagnosis of major depression. Similar abnormalities have been reported in mania, panic disorder, and irritable bowel syndrome, in the absence of depressive symptoms (Rubin et al., 2002).

21.4.3.2 Dopamine

Apomorphine, a DA agonist, has been used to stimulate GH release. Both a blunted GH release in response to apomorphine and no difference in response between depressed patients and normal controls have been reported, with the weight of the evidence suggesting no abnormality of DA regulation of GH in major depression (Rubin et al., 2002).

21.4.3.3 Serotonin

Tryptophan and 5-hydroxytryptophan (5-HTP), the precursors of 5-HT, have been used as stimuli for GH release, although their specificity as 5-HT challenge agents has been questioned (Van Praag et al., 1986). There is little evidence to support altered 5-HT regulation of GH release in depression.

21.4.3.4 Acetylcholine

The cholinergic control of GH release is probably largely under muscarinic influence. Oxotremorine, a muscarinic cholinergic agonist, induces GH release, and atropine, a cholinergic antagonist, blocks this response (Casanueva, 1993). In contrast, nicotine did not alter resting GH concentrations, and pretreatment with the nicotinic receptor blocker, mecamylamine, did not counteract the GH-releasing effect of the cholinesterase inhibitor, eserine (physostigmine). ACh inhibits the release of somatostatin from hypothalamic slices (Ghigo et al., 1997), which appears to be the mechanism of action of muscarinic agonists.

Pyridostigmine, an acetylcholinesterase inhibitor, increases GH levels, in man, in a dose-dependent manner. Enhanced GH responses to pyridostigmine have been found in depressed patients, supporting the hypothesis of cholinergic supersensitivity in major depression (O'Keane et al., 1992). Similarly, enhanced GH responses to low-dose physostigmine occurred in young adult depressed women compared to matched controls, but not in young adult depressed men, suggesting that cholinergic supersensitivity in major depression is a sexually diergic phenomenon (Rubin et al., 2003).

21.4.3.5 Gamma-aminobutyric acid

Hypothalamic GABA receptors are involved in the modulation of GH secretion. Baclofen, a GABA-B agonist, readily crosses the blood–brain barrier and stimulates the secretion of GH. The GH response of depressed patients to baclofen has been reported as significantly blunted, especially in patients who were DST nonsuppressors (reviewed in Rubin et al. (2002)). This may indicate a relationship between a GABA-B receptor abnormality and HPA-axis dysfunction in major depression.

21.4.4 Glucocorticoids and GH Secretion in Depression

Glucocorticoids have a biphasic effect on GH release, being initially stimulating and then inhibiting. The mechanism is likely through acute inhibition of somatostatin, and then increased somatostatinergic tone (Thakore and Dinan, 1994). The disturbances in the GH axis in depression thus can be viewed as secondary to hyperactivity of the HPA axis. In other clinical states characterized by hypercortisolemia, such as Cushing's disease, there is a blunting of GH responses to all commonly used stimuli including GHRH, a situation similar to that occurring in depression. There are also data suggesting a state-dependent normalization of the GH response with recovery from depression (Thakore and Dinan, 1995).

21.4.5 Peptide-Stimulated GH Secretion in Depression

21.4.5.1 Growth hormone-releasing hormone

Studies of GHRH-stimulated GH secretion in depression have produced conflicting findings – some having reported a blunted GH response and others normal or enhanced responses (reviewed in Skare et al. (1994)). Most studies have not reported baseline cortisol or IGF-1 levels, making data interpretation difficult.

21.4.5.2 *Corticotropin-releasing hormone*

CRH inhibits spontaneous pulsatile GH secretion and attenuates stimulation by GHRH (Rosen et al., 1994), most likely on the basis of increased somatostatin tone. However, one study suggested a significant GH response to CRH in depressed patients (Lesch et al., 1988). The GH increase was not correlated with baseline ACTH or cortisol or with the CRH-induced ACTH or cortisol responses. It may be that increased production of CRH in depression results in a paradoxical GH response to CRH, but the mechanism is unclear.

21.4.5.3 *Thyrotropin-releasing hormone*

As with CRH, GH release in response to TRH has been reported in depression, with no responses occurring in healthy subjects. After clinical recovery, depressed patients have been reported to show no response to TRH, suggesting a state-dependent phenomenon. Other studies, however, have found no difference in GH responses to TRH between major depressives and normal controls (reviewed in Rubin et al. (1990)).

21.5 Hypothalamic–Pituitary–Gonadal Axis

Whereas menopause in women is a time of rapid decline of gonadal steroid hormone production and arrest of reproductive capacity, men do not experience such an abrupt hormonal shift (Kaufman and Vermeulen, 2005). A discussion of alterations in gonadotropin and sex steroid hormone secretion in affective disorders therefore needs to consider premenopausal female, postmenopausal female, and male patients as separate groups, because of their clearly different hormonal physiologies. While some early studies suggested there may be reduced serum luteinizing hormone (LH), estradiol, and testosterone concentrations in depressed patients, compared to normal controls, the findings were not consistent in either female or male patients. Methodological concerns in some of these early studies included imprecise diagnostic terminology, lack of control groups, and small sample sizes. More recent studies generally have attended to these methodological issues, and their findings are correspondingly more robust.

21.5.1 Depressed Men

Findings in this group of patients have been both negative and positive, the latter including lower serum basal LH and testosterone, with no difference in FSH or estradiol; reduced circulating testosterone (and increased cortisol), compared to testosterone and cortisol levels following successful treatment; and decreased diurnal and nocturnal plasma testosterone, with no difference in plasma gonadotropins (reviewed in Rubin et al. (2002)). In aggregate, these studies indicate that there may be somewhat reduced circulating testosterone in depressed men, the reduction being correlated with severity of depression and perhaps with increased circulating glucocorticoids as well, but the rest of the HPG axis is most often normal.

21.5.2 Premenopausal Depressed Women

Studies of basal serum LH, FSH, and estradiol, and of LH and FSH responses to low doses of gonadotropin-releasing hormone (GnRH) in this group of patients generally have been negative, suggesting that premenopausal depressed women do not have a consistent dysregulation of their HPG axis (reviewed in Rubin et al. (2002)). Variation in the nucleotide composition of the estrogen-receptor alpha gene, however, may predispose to premenstrual dysphoric disorder (Huo et al., 2007). Postpartum depression may be different in that, following childbirth, there are marked changes in the maternal hormonal milieu. Nevertheless, evidence for an etiological role for these hormone changes in postpartum depression is scanty.

21.5.3 Peri/Postmenopausal Depressed Women

In contrast to these negative findings in premenopausal patients, significantly lower LH concentrations have been reported in non-estrogen-replaced postmenopausal depressives compared to postmenopausal controls, but other studies have found either no difference in circulating gonadotropins or gonadotropin responses to GnRH stimulation or, in perimenopausal patients, increased responses to GnRH (reviewed in Rubin et al. (2002)). There were also consistent positive correlations between basal and stimulated LH and ratings of depression severity in the patients. Taken together, these data, while seemingly discrepant, may be consistent in the context of the patients' ages and number of years following menopause. Depressed women early in their transition to menopause may have higher HPG-axis activity than do comparable normal controls, which becomes subnormal after several years of menopausal status. This suggests that the depressive state might confer less ability to normally regulate the HPG axis, no matter where in the menopausal process the patient is when her depressive episode occurs.

21.5.4 Gonadal Steroid Pharmacotherapy

Estrogen replacement in postmenopausal women has a number of salutary effects, both physical and mental. Cognitive performance and memory are enhanced when estrogen supplementation is given to estrogen-deficient women (Sherwin, 2003). Estrogen enhances serotonergic neurotransmission by increasing the density of 5-HT2A-binding sites (Fink et al., 1998; Joffe and Cohen, 1998) and inhibits monoamine oxidase activity (Chakravorty and Halbreich, 1997). A number of studies support the use of estrogen as part of the overall treatment of depression in estrogen-deficient women secondary to childbirth, the peri/postmenopause, and antiestrogen treatment (reviewed in Rubin et al. (2002)).

Androgens improve libido in postmenopausal women as well as in hypogonadal men (Epperson et al., 1999) and can be a useful adjunct in the treatment of depression in hypogonadal men (Seidman, 2006). In a few cases of severe hypogonadism, testosterone treatment alone has alleviated depressive symptoms. To avoid untoward side effects, such as increased risk of estrogen-sensitive malignancies in women and testosterone-induced erythrocytosis, prostate hypertrophy, irritability, and occasional psychotic episodes in men, gonadal steroid treatment should be reserved for those patients with demonstrated hormone deficiencies. In contrast to the qualified utility of estrogen and testosterone as adjuncts in the treatment of affective disorders, progesterone appears to be of little or no benefit in the treatment of premenstrual or postpartum depressive symptomatology.

21.6 Prolactin

Lactotrophs, the prolactin secreting cells, account for approximately 20% of the anterior pituitary cell mass. The breast is the primary target site for prolactin, but a role in immunoregulation has also been proposed (Horseman and Gregerson, 2006). Prolactin secretion is pulsatile, and there is a circadian rhythm of serum concentrations, with peak levels occurring during the mid-sleep cycle and the nadir occurring during the day. Hypothalamic DA is the dominant regulator of prolactin release, having a major inhibitory role (Ben-Jonathan and Hnasko, 2001). DA is released from the tuberoinfundibular portion of the hypothalamus into the pituitary portal circulation and acts on D2 receptors on the lactotrophs. Other putative inhibitors include GABA and somatostatin.

5-HT is an important stimulus for prolactin release, probably acting via an intermediate peptide such as TRH, vasoactive intestinal peptide, or vasopressin (reviewed in Rubin et al. (2002)). Both 5-HT1A and 5-HT2A/2C receptors have been implicated. In humans, intravenous (IV) l-tryptophan, a 5-HT precursor, elevates prolactin concentrations in a dose-dependent manner and is thought to be mediated by 5-HT1 receptors. Fenfluramine, which brings about the release and inhibits the uptake of 5-HT, also produces a dose-dependent increase in prolactin in humans, probably mediated by 5-HT, because pretreatment with cyproheptadine, a 5-HT and histamine antagonist, blocks the response.

21.6.1 Basal Prolactin Secretion in Depression

Several studies have examined spontaneous prolactin release in depressed patients, with varying results, including diurnal elevation, nocturnal decrease, and no change (reviewed in Rubin et al. (1989b) and Nicholas et al. (1998)). In one study (Mai et al., 1985), no differences were found between the male patients and controls, but major differences occurred among the premenopausal versus the postmenopausal female patients and controls, emphasizing the need to control for menstrual status in studies of prolactin release.

21.6.2 Prolactin Responses to Serotonergic Challenges in Depression

While, as indicated, no consistent differences in basal prolactin levels have been revealed in depression, studies based on the pharmacological stimulation of prolactin release have been more fruitful (reviewed in Rubin et al. (2002)). Prolactin secretion in response to IV tryptophan challenge has been consistently blunted in depression, particularly in those who had not lost weight. The 5-HT precursor, 5-HTP, also has been used as a serotonergic probe; depressed women with melancholia had enhanced prolactin responses to oral 5-HTP in comparison to healthy controls. Such a difference was not observed in male patients.

Many studies have reported blunted prolactin secretion in response to racemic d,l-fenfluramine challenge. A blunted prolactin response has also been related to a history of suicide. The specificity of d,l-fenfluramine for the serotonergic system, however, has been questioned, because the l-isomer possesses catecholaminergic activity. Some studies, therefore, have made use of the d-isomer, d-fenfluramine, which is more selective for the serotonergic system. Again, a blunted prolactin response has been found, but there have also been negative reports. Challenge studies using the 5-HT agonists mCPP and MK-212 have failed to detect any blunting of prolactin release in depression. Some, but not all, challenge studies using the 5HT1A agonist, buspirone, however, did find differences in prolactin response. A tentative conclusion is that, overall, 5-HT regulation of prolactin in depression is normal, but that certain receptor subtypes, for example, 5HT1A, might be subsensitive, perhaps owing to increased glucocorticoids, as suggested for 5HT1A regulation of the HPA axis. With reference to prolactin, however, such putative 5HT1A subsensitivity appears to be compensated for by other 5-HT pathways.

21.6.3 Prolactin Secretion Following Treatment of Depression

Prolactin responses to d,l-fenfluramine are significantly increased following treatment with imipramine, amitriptyline, clomipramine, fluoxetine, and ECT, irrespective of therapeutic outcome (reviewed in Rubin et al. (2002)). The blunting of serotonergically mediated prolactin release may provide a state marker of depression: When patients clinically improve, the blunted prolactin response becomes normal. However, unlike HPA-axis disturbances, a trend toward normalization of the prolactin response is not always associated with clinical recovery.

21.7 Melatonin

Melatonin, formed by O-methylation of 5-HT, is secreted by the pineal gland. Melatonin has received the most attention as a regulator of certain biological rhythms. Melatonin as a natural remedy for sleep disturbance, particularly jet lag, has captured public interest, although its effect is questionable. Nocturnal melatonin secretion provides a measure of the phase of the circadian pacemaker and of noradrenergic receptor sensitivity at the pinealocyte. Although some studies of depressed subjects have found a reduction in nocturnal melatonin-secretion amplitude (reviewed in Rubin et al. (1992)), the large variance in melatonin rhythms in depressed subjects undermines its usefulness as a marker for circadian disturbances.

21.7.1 Melatonin and Seasonal Affective Disorder

Melatonin is suppressed by light treatment given at night, which forms the rationale for the treatment of seasonal affective disorder (SAD) by light therapy. SAD is a subtype of depression in which depressed mood most often develops during the autumn and winter – when the length of darkness increases – and remits during the spring and summer. Photoperiodic time measurement is mediated by melatonin, which led to the theory that SAD was related to abnormal melatonin function. Bright light given in the morning and evening can suppress melatonin, with resulting improvement in depressed mood, and there are reports that depressed patients with winter SAD have abnormal sensitivity to light, as assessed by light-induced suppression of melatonin secretion at night (Nathan et al., 1999). However, not all studies have demonstrated a link between melatonin suppression and clinical improvement (Partonen et al., 1997). Studies of light therapy in SAD suggest that bright light exposure should be given immediately on awakening (Lewy et al., 2006).

21.7.2 Relationship to the HPA Axis

An interaction between the frequently observed hyperactivity of the HPA axis in depression and low circulating melatonin has been hypothesized, in support of which is the demonstration that CRH inhibits melatonin secretion in healthy volunteers, but other studies have not confirmed this possibility and have even suggested increased melatonin secretion (reviewed in Rubin et al. (1992)).

21.8 Other Neuroendocrine Peptides

21.8.1 Opioid Peptides

The endogenous opioid peptides – endorphins, metenkephalins, and dynorphins – are located in the brain, spinal cord, autonomic ganglia, and enteric nervous system. They act on μ-, κ-, and δ-opioid receptors, respectively. The opioidergic system exerts a predominantly inhibitory influence on the HPA axis (Zis et al., 1989). Administration of morphine reduces plasma cortisol concentrations in normal subjects and in depressed patients. However, DST-positive depressed patients also display abnormal early escape from morphine suppression of cortisol (Zis et al., 1985). Like the DST, this abnormality was reversed by successful treatment. Naloxone, an opiate receptor antagonist, blocks the inhibitory influence of endogenous opioids and produces an increase in circulating ACTH and cortisol in normal subjects (Zis et al., 1989). The ACTH and cortisol responses to naloxone challenge are considered a marker of central opioid tone. The naloxone test has been used in a variety of clinical settings, in the diagnosis of central adrenal insufficiency and Cushing's syndrome. Burnett et al. (1999) examined ACTH and cortisol release in response to an infusion of naloxone, 0.125 $\mu g\ kg^{-1}$, in depressed outpatients and healthy controls; the patients released significantly less cortisol than did the controls, as well as less ACTH. Depressed patients thus may have decreased central opioid tone, possibly mediated by subsensitive μ-opioid receptors. There is also evidence that the sexually diergic prolactin response to morphine is impaired in depressed patients (Zis et al., 1989). Together, these studies suggest impaired central opioid tone in depression as a state phenomenon.

21.8.2 Substance P

Substance P is a tachykinin, which acts on NK1 receptors. It has a role in pain modulation, control of vomiting, and cardiovascular responses to stress. Initial clinical trials suggesting that NK1 antagonists are effective in the treatment of depression and emesis, but not pain, have not been supported (McLean, 2005; Chahl, 2006). The role of substance P in the endocrine disturbances occurring in depression has not been adequately explored.

21.8.3 Arginine Vasopressin

Chronic stress in animals is associated with a shift from predominantly CRH to predominantly AVP regulation of the HPA axis. AVP acts on V1A, V1B, and V2 receptors, with V1B receptors being predominant on anterior pituitary corticotrophs. ddAVP has greater affinity for V2 than V1 receptors. Dinan et al. (1999) found that ACTH responses to CRH were blunted in depressed patients compared to healthy controls, but when CRH was combined with ddAVP, ACTH release was similar in the two groups of subjects. These findings suggest enhanced V1B receptor responsiveness in depression. Other studies indicate that basal plasma AVP may be elevated in depression, the extent of increase being positively correlated with psychomotor retardation (Hebb et al., 2005). A difference in the AVP responses to cholinergic challenge between male and premenopausal female patients with major depression has also been reported, the women having greater responses than the men (Rubin et al., 1999). In normal subjects, the opposite was found, normal men having greater responses than normal women (Rubin et al., 2006b).

Regarding treatment, Devanand et al. (1998) studied 55 medication-free depressed patients receiving unilateral or bilateral ECT. Post-ECT increases in AVP and oxytocin were not predictive of the patients' clinical response to treatment.

21.8.4 Neurotensin and NPY

Neurotensin stimulates ACTH secretion in the rat (Malendowicz et al., 1991), but, to date, no reports of neurotensin profiles in depression appear to have been published. NPY has been examined in patients with major depression (reviewed in Karl and Herzog (2007)); patients with a history of suicide attempts had the lowest plasma concentrations. DEX decreased plasma NPY concentrations in healthy subjects but not in depressed patients, and suicidal patients showed a significant positive correlation between plasma NPY and plasma cortisol. Bipolar patients had reduced levels of NPY mRNA expression in the prefrontal cortex. NPY has shown antidepressant properties in animal studies, and a deficiency thus might play a role in depressive illness, although this remains speculative.

21.8.5 Cholecystokinin and Endogenous Opioids

There appears to be a reciprocal neurochemical relationship between cholecystokinin (CCK), which is anxiogenic and memory-enhancing, and endogenous opioids, which are anxiolytic and amnestic, as indicated above (Hebb et al., 2005). High CSF CCK concentrations have been reported in major depressives who had made one or more suicide attempts (Lofberg et al., 1998); there were no significant correlations between CSF CCK and measures of HPA-axis activity or between CSF CCK and CSF 5-HIAA or HVA. Decreased serum β-endorphin concentrations have also been reported in patients with severe depression and anxiety. In animal studies, antidepressants have been shown to promote CNS opioidergic activity and to block the anxiogenic effect of administered CCK, further implicating these peptides in the neurochemistry of depressive disorders (Hebb et al., 2005).

21.8.6 Leptin

Leptin, the protein product of the *ob* gene, plays a role in inhibiting food intake. Plasma leptin concentrations have been reported as higher in female depressed patients compared to normal female controls, but not higher in male patients compared to male controls (Esel et al., 2005). In contrast, brain leptin, determined by arteriovenous difference, was low in depressed patients, and leptin mRNA was not detectable in the brains of suicide victims, whereas it was present in the brains of nonsuicide donors (Eikelis et al., 2006). The apparently reciprocal relationship between circulating leptin and brain leptin remains to be confirmed and its mechanism and importance delineated.

21.9 Summary

Neuroendocrine studies are one of several ways to examine CNS function in psychiatric disorders – through the neuroendocrine window to the brain. The influence of a host of neurotransmitters and neuromodulators on the neurosecretory cells of the hypothalamus, as well as autonomic innervation of the target endocrine glands, provides the neurophysiological rationale for these studies. Although the hormonal outcome measures are indirect, the information gained from neuroendocrine studies can advance our understanding of two major aspects of affective disorders: their underlying neurochemical mechanisms, and their diagnosis and treatment through the use of endocrine tests to provide ancillary clinical information.

With reference to underlying neurochemical mechanisms, as indicated in Section 21.1, dysregulation of several neurotransmitters has been implicated in the etiology of major depression, and continued investigations are revealing interesting new information. For example, as indicated in Section 21.2, opposite male and female differences in responses of this axis to cholinergic stimulation have been reported in normal subjects versus patients with major depression (Rubin et al., 1999), suggesting that premenopausal women normally have less responsive CNS cholinergic systems than men, but that CNS neurochemical changes underlying the onset of a major depressive episode (or at least concomitant with the episode) may render their cholinergic systems more sensitive compared to depressed men. The roles of estrogen and other gonadal steroids in this switch in cholinergic sensitivity in premenopausal women, and the contribution of this switch in sensitivity to the 2:1 greater incidence of depression in women during their childbearing years, are important questions yet to be answered.

Forty years ago, Mason (1968), through careful experimentation, established the concept of coordinated endocrine responses to stress – immediate catabolic hormone secretion, to provide energy for rapidly coping with a threat, and delayed anabolic hormone secretion, to rebuild energy stores. Our modern techniques for measurement of the secretion of specific hormones, combined with molecular biologic studies of mechanisms of their secretion, for example, measurement of hormone ribonucleic acids (RNAs) and proteins in specific brain areas, afford remarkably refined methods of study compared to Mason's collecting plasma and urine from monkeys placed in restraining chairs and subjected to various environmental manipulations. With these more sophisticated methods, it is highly likely that Mason's concept of organization of psychoendocrine mechanisms into coordinated, physiologically meaningful responses will be validated and considerably refined.

With reference to psychiatric diagnostics, the establishment of a clinically useful biological test involves several difficult steps. As mentioned in Section 21.1, anchor points of psychiatric diagnosis, against which biological correlates are established, can be moving targets, melancholic depression being a good example. Does a test relate better to a syndrome in its totality, for example, major depression; to a diagnostic subtype, for example, melancholic or psychotic major depression; or to a specific dimension of the syndrome, for example, the agitation/anxiety component of major depression? And, what is the ultimate utility of this relationship for clinical decisions about patient management – does a test have the requisite sensitivity, specificity, and positive and/or negative predictive value to be a screening test for a given psychiatric diagnosis (highly unlikely and not shown for any biological test to date)? Or can its utility be defined more narrowly, that is, a clinically useful sensitivity, specificity, and positive and/or negative predictive value in specified Bayesian contexts, possibly in combination with other tests, or conversion of an initially abnormal test to normal as a hallmark of successful treatment? This is a long developmental road for a clinically useful biological marker of psychiatric disorders, no different from the clinical electroencephalogram or the glucose tolerance test in general medicine (Rubin and Poland, 1984; Carroll, 1989).

Finally, it should be noted that the focus of this chapter on affective disorders has been primarily on major depression – bipolar (manic-depressive) illness and other affective disorders have been given somewhat short shrift. Part of the problem is that bipolar patients have been studied less frequently, because of their scarcity compared to major depressives and their more unpredictable cooperation with study requirements when in a manic phase. In addition, a number of their neuroendocrine responses are similar to those of patients with major depression, such as having a similar, or in some studies an even higher, incidence of HPA-axis hyperactivity, including an abnormal DST. Mainly, though, the issue has been one of relative numbers of published studies on the different diagnostic categories.

Acknowledgment

This work is supported by National Institute of Mental Health research grant MH28380 (to R.T.R.).

References

Aihara M, Ida I, Yuuki N, et al. (2007) HPA axis dysfunction in unmedicated major depressive disorder and its normalization by pharmacotherapy correlates with alteration of neural activity in prefrontal cortex and limbic/paralimbic regions. *Psychiatry Research: Neuroimaging* 155: 245–256.

American Psychiatric Association (1994) *Diagnostic and Statistical Manual of Mental Disorders,* 4th edn., pp. 317–391. Washington, DC: American Psychiatric Association.

Amsterdam JD, Maislin G, Winokur A, Kling M, and Gold P (1987) Pituitary and adrenocortical responses to the ovine corticotropin releasing hormone in depressed patients and healthy volunteers. *Archives of General Psychiatry* 44: 775–781.

Angold A, Costello EJ, Erkanli A, and Worthman CM (1999) Pubertal changes in hormone levels and depression in girls. *Psychological Medicine* 29: 1043–1053.

Axelson DA, Doraiswamy PM, Boyko OB, et al. (1992) *In vivo* assessment of pituitary volume with magnetic resonance imaging and systematic stereology: Relationship to dexamethasone suppression test results in patients. *Psychiatry Research* 44: 63–70.

Aydemir C, Yalcin ES, Aksaray S, Kisa C, Yildirim SG, Uzbay T, and Goka E (2006) Brain-derived neurotrophic factor (BDNF) changes in the serum of depressed patients. *Progress in Neuropsychopharmacology and Biological Psychiatry* 30: 1256–1260.

Bauer A, Berghofer A, Bschor T, et al. (2002) Supraphysiological doses of L-thyroxine in the maintenance treatment of prophylaxis resistant affective disorders. *Neuropsychopharmacology* 27: 620–628.

Bauer M, Hellweg R, Graf KJ, and Baumgartner A (1998) Treatment of refractory depression with high-dose thyroxine. *Neuropsychopharmacology* 18: 444–455.

Ben-Jonathan N and Hnasko R (2001) Dopamine as a prolactin inhibitor. *Endocrine Reviews* 22: 724–763.

Bennaroch EE (2007) Neurosteroids: Endogenous modulators of neuronal excitability and plasticity. *Neurology* 68: 945–947.

Bhagwagar Z, Murthy N, Selvaraj S, et al. (2007) 5-HTT binding in recovered depressed patients and healthy volunteers: A positron emission tomography study with [^{11}C]DASB. *American Journal of Psychiatry* 164: 1858–1865.

Bissette G, Klimek V, Pan J, Stockmeier C, and Ordway G (2003) Elevated concentrations of CRF in the locus coeruleus of depressed subjects. *Neuropsychopharmacology* 28: 1328–1335.

Borgerding LA, Lineberry TW, and Bostwick JM (2007) Creatine kinase elevation associated with bipolar disorder and ephedrine abuse. *Psychosomatics* 48: 348–349.

Brown AS and Gershon S (1993) Dopamine and depression. *Journal of Neural Transmission* 91: 75–109.

Brown WA (2007) Treatment response in melancholia. *Acta Psychiatrica Scandinavica* 115(supplement 433): 125–129.

Burnett FE, Scott LV, Weaver MG, Medbak SH, and Dinan TG (1999) The effect of naloxone on adrenocorticotropin and cortisol release: Evidence for a reduced response in depression. *Journal of Affective Disorders* 53: 263–268.

Cannon DM, Ichise M, Rollis D, et al. (2007) Elevated serotonin transporter binding in major depressive disorder assessed using positron emission tomography and [^{11}C]DASB: Comparison with bipolar disorder. *Archives of General Psychiatry* 64(2): 201–208.

Carroll BJ (1985) Dexamethasone suppression test: A review of contemporary confusion. *Journal of Clinical Psychiatry* 46: 13–24.

Carroll BJ (1989) Diagnostic validity and laboratory studies: Rules of the game. In: Robbins LN and Barrett JE (eds.) *The Validity of Psychiatric Diagnosis*, pp. 229–245. New York: Raven Press.

Carroll BJ, Cassidy F, Naftolowitz D, et al. (2007) Pathophysiology of hypercortisolism in depression. *Acta Psychiatrica Scandinavica* 115(supplment 433): 90–103.

Carroll BJ, Curtis GC, and Mendels J (1976) Cerebrospinal fluid and plasma free cortisol concentrations in depression. *Psychological Medicine* 6: 235–244.

Carroll BJ and Rubin RT (2007) Mifepristone in psychotic depression? *Biological Psychiatry* 63, doi:10.1016/j.biopsych.2007.03.032.

Carroll BT, Meller WH, Kathol RG, Gehris TL, Carter JL, Samuelson SD, and Pitts AF (1993) Pituitary–adrenal axis response to arginine vasopressin in patients with major depression. *Psychiatry Research* 46: 119–126.

Casanueva FF (1993) Physiology of growth hormone secretion in man. *Journal of Clinical Endocrinology and Metabolism* 76: 168–171.

Cassidy F, Ahearn E, and Carroll BJ (1999) Elevated frequency of diabetes mellitus in hospitalized manic-depressive patients. *American Journal of Psychiatry* 156: 1417–1420.

Cassidy F, Murry E, and Carroll BJ (1998) Tryptophan depletion in recently manic patients treated with lithium. *Biological Psychiatry* 43: 230–232.

Cassidy F, Murry E, Weiner RD, and Carroll BJ (1997) Lack of relapse with tryptophan depletion following successful treatment with ECT. *American Journal of Psychiatry* 154: 1151–1152.

Chahl LA (2006) Tachykinins and neuropsychiatric disorders. *Current Drug Targets* 7: 993–1003.

Chakravorty SG and Halbreich U (1997) The influence of estrogen on monoamine oxidase activity. *Psychopharmacology Bulletin* 33: 229–233.

Checkley SA, Corn TH, Glass IB, Thompson C, Franey C, and Arendt J (1986) Neuroendocrine and other studies of the mechanism of antidepressant action of desipramine. *Ciba Foundation Symposium* 123: 126–147.

Chopra IJ, Solomon DH, and Huang T-S (1990) Serum thyrotropin in hospitalized psychiatric patients: Evidence for hyperthyrotropinaemia as measured by an ultrasensitive thyrotropin assay. *Metabolism* 39: 538–543.

Coryell W (1990) DST abnormality as a predictor of course in major depression. *Journal of Affective Disorders* 19: 163–169.

Coryell W and Schlesser M (2001) The dexamethasone suppression test and suicide prediction. *American Journal of Psychiatry* 158: 748–753.

Coryell W, Young E, and Carroll BJ (2006) Hyperactivity of the hypothalamic–pituitary–adrenal axis and mortality in major depressive disorder. *Psychiatry Research* 142: 99–104.

de Kloet ER, Vreugdenhil E, Oitzl MS, and Joels M (1998) Brain corticosteroid receptor balance in health and disease. *Endocrine Reviews* 19: 269–301.

De Souza EB and Grigoriadis DE (2002) Corticotropin-releasing factor: Physiology, pharmacology, and role in central nervous system disorders. In: Davis KL, Charney D, Coyle JT, and Nemeroff C (eds.) *Neuropsychopharmacology: The Fifth Generation of Progress*, ch. 7, pp. 91–107. Philadelphia, PA: Lippincott Williams and Wilkins.

Devanand DP, Lisanby S, Lo ES, Fitzsimons L, Cooper TB, Halbreich U, and Sackeim HA (1998) Effects of electroconvulsive therapy on plasma vasopressin and oxytocin. *Biological Psychiatry* 44: 610–616.

Dilsaver SC (1986) Pathophysiology of cholinoceptor supersensitivity in affective disorders. *Biological Psychiatry* 21: 813–829.

Dinan TG (1997) Serotonin and the regulation of hypothalamic–pituitary–adrenal axis function. *Life Sciences* 58: 1683–1693.

Dinan TG, Lavelle E, Scott LV, Newell-Price J, Medbak S, and Grossman AB (1999) Desmopressin normalizes the blunted adrenocorticotropin response to corticotropin-releasing hormone in melancholic depression: Evidence of enhanced vasopressinergic responsivity. *Journal of Clinical Endocrinology and Metabolism* 84: 2238–2240.

Dinan TG, O'Brien S, Lavelle E, and Scott LV (2004) Further neuroendocrine evidence of enhanced vasopressin V3 receptor responses in melancholic depression. *Psychological Medicine* 34: 169–172.

Dinan TG and Scott LV (2005) Anatomy of melancholia: Focus on hypothalamic–pituitary–adrenal axis overactivity and the role of vasopressin. *Journal of Anatomy* 207: 259–264.

Drevets WC (2000) Neuroimaging studies of mood disorders. *Biological Psychiatry* 48: 813–829.

Drevets WC, Price JL, Bardgett ME, Reich T, Todd RD, and Raichle ME (2002) Glucose metabolism in the amygdala in depression: Relationship to diagnostic subtype and plasma cortisol levels. *Pharmacology, Biochemistry, and Behavior* 71: 431–447.

Dubrovsky B (2006) Neurosteroids, neuroactive steroids, and symptoms of affective disorders. *Pharmacology, Biochemistry, and Behavior* 84: 644–655.

Duman RS and Monteggia LM (2006) A neurotrophic model for stress-related mood disorders. *Biological Psychiatry* 59: 1116–1127.

Dwivedi Y, Rizavi HS, Conley RR, Roberts RC, Tamminga CA, and Pandey GN (2003) Altered gene expression of brain-derived neurotrophic factor and receptor tyrosine kinase B in postmortem brain of suicide subjects. *Archives of General Psychiatry* 60: 804–815.

Eikelis N, Esler M, Barton D, Dawood T, Wiesner G, and Lambert G (2006) Reduced brain leptin in patients with major depressive disorder and in suicide victims. *Molecular Psychiatry* 11: 800–801.

Epperson CN, Wisner KL, and Yamamoto B (1999) Gonadal steroids in the treatment of mood disorders. *Psychosomatic Medicine* 61: 676–697.

Esel E, Özsoy S, Tutus A, et al. (2005) Effects of antidepressant treatment and of gender on serum leptin levels in patients with major depression. *Progress in Neuropsychopharmacology and Biological Psychiatry* 29: 565–570.

Ferre S, Ciruela F, Woods AS, Lluis C, and Franco R (2007) Functional relevance of neurotransmitter receptor heteromers in the central nervous system. *Trends in Neurosciences* 30: 440–446.

Fink G, Sumner BE, Rosie R, Grace O, and Quinn JP (1998) Sex steroid control of mood, mental state and memory. *Clinical and Experimental Pharmacology and Physiology* 25: 764–775.

Fink M (2005) Should the dexamethasone suppression test be resurrected? *Acta Psychiatrica Scandinavica* 112: 245–249.

Fuchs E and Flugge G (1995) Modulation of binding sites for corticotropin-releasing hormone by chronic psychosocial stress. *Psychoneuroendocrinology* 20: 33–51.

Geracioti TD, Jr., Loosen PT, and Orth DN (1997) Low cerebrospinal fluid corticotropin-releasing hormone concentrations in eucortisolemic depression. *Biological Psychiatry* 42: 166–174.

Gershon S and Shaw FH (1961) Psychiatric sequelae of chronic exposure to organophosphorous insecticides. *Lancet* 1: 1371–1374.

Ghigo E, Arvat E, Muccioli G, and Camanni F (1997) Growth hormone releasing peptides. *European Journal of Endocrinology* 136: 445–460.

Gold PW, Goodwin FK, and Chrousos GP (1988) Clinical and biochemical manifestations of depression. Relation to the neurobiology of stress. *New England Journal of Medicine* 319: 413–420.

Gold PW, Loriaux DL, Roy A, et al. (1986) Responses to corticotropin-releasing hormone in the hypercortisolism of depression and Cushing's disease: Pathophysiologic and diagnostic implications. *New England Journal of Medicine* 314: 1329–1335.

Gönül AS, Akdeniz F, Taneli F, Donat O, Eker Ç, and Vahip S (2005) Effect of treatment on serum brain-derived neurotrophic factor levels in depressed patients. *European Archives of Psychiatry and Clinical Neuroscience* 255: 381–386.

Greden JF, deVigne JP, Albala AA, Tarika J, Buttenheim M, Eiser A, and Carroll BJ (1982) Serial dexamethasone suppression tests among rapid cycling bipolar patients. *Biological Psychiatry* 17: 455–462.

Guldenaar SE, Veldkamp B, Bakker O, Wiersinga WM, Swaab DF, and Fliers E (1996) Thyrotropin-releasing hormone gene expression in the human hypothalamus. *Brain Research* 743: 93–101.

Habib KE, Weld KP, Rice KC, et al. (2000) Oral administration of a corticotropin-releasing hormone receptor antagonist significantly attenuates behavioral, neuroendocrine, and autonomic responses to stress in primates. *Proceedings of*

the National Academy of Sciences of the United States of America 97: 6079–6084.
Hebb ALO, Poulin J-F, Roach SP, Zacharko RM, and Drolet G (2005) Cholecystokinin and endogenous opioid peptides: Interactive influence on pain, cognition, and emotion. *Progress in Neuropsychopharmacology and Biological Psychiatry* 29: 1225–1238.
Heninger GR (1995) The role of serotonin in clinical disorders. In: Bloom FE and Kupfer DJ (eds.) *Psychopharmacology: The Fourth Generation of Progress*, pp. 471–482. New York: Raven Press.
Herbert J, Goodyer IM, Grossman AB, et al. (2006) Do corticosteroids damage the brain? *Journal of Neuroendocrinology* 18: 393–411.
Heuser IJ, Yassourdis A, and Holsboer F (1994) The combined dexamethasone/CRH test: A refined laboratory test for psychiatric disorders. *Journal of Psychiatric Research* 28: 341–356.
Hickie I, Bennett B, Mitchell P, Wilhelm K, and Orlay W (1996) Clinical and subclinical hypothyroidism in patients with chronic and treatment-resistant depression. *Australian and New Zealand Journal of Psychiatry* 30: 246–252.
Hindall JT and Kaplan MM (1988) Inhibition of thyroxine 5′-deiodination type II in cultured human placental cells by cortisol, insulin, 3,5′-cyclic adenosine monophosphate and butyrate. *Metabolism – Clinical and Experimental* 37: 664–668.
Hökfelt T, Tsuruo Y, Ulfhake B, et al. (1989) Distribution of TRH-like immunoreactivity with special reference to co-existence with other neuroactive compounds. *Annals of the New York Academy of Sciences* 553: 76–105.
Holger J, Schick M, Kiefer F, Kellner M, Yassouridis A, and Wiedemann K (2004) Metyrapone as additive treatment in major depression: A double-blind and placebo-controlled trial. *Archives of General Psychiatry* 61: 1235–1244.
Holick KA, Lee DC, Hen R, and Dulawa SC (2008) Behavioral effects of chronic fluoxetine in BALB/cJ mice do not require adult hippocampal neurogenesis or the serotonin 1A receptor. *Neuropsychopharmacology* 33: 406–417.
Holsboer F and Barden N (1996) Antidepressants and hypothalamic–pituitary–adrenocortical regulation. *Endocrine Reviews* 17: 187–205.
Holsboer F, Gerken A, von Bardeleben U, Grimm W, Beyer H, Müller OA, and Stalla GK (1986) Human corticotropin-releasing hormone in depression. *Biological Psychiatry* 21: 601–611.
Horseman ND and Gregerson KA (2006) Prolactin. In: DeGroot LJ and Jameson JL (eds.) *Endocrinology,* 5th edn., pp. 309–322. Philadelphia, PA: Elsevier Saunders.
Hung Y-J, Hsieh C-H, Chen Y-J, et al. (2007) Insulin sensitivity, proinflammatory markers and adiponectin in young males with different subtypes of depressive disorder. *Clinical Endocrinology* 67: 784–789.
Huo L, Straub RE, Roca C, et al. (2007) Risk for premenstrual dysphoric disorder is associated with genetic variation in ESR1, the estrogen receptor alpha gene. *Biological Psychiatry* 62: 925–933.
Jaeckle RS, Kathol RG, Lopez JF, Meller WH, and Krummel SJ (1987) Enhanced adrenal sensitivity to exogenous cosyntropin (ACTH alpha 1–24) stimulus in major depression. Relationship to dexamethasone suppression test results. *Archives of General Psychiatry* 44: 233–240.
Janowsky DS and Overstreet DH (1995) The role of acetylcholine mechanisms in mood disorders. In: Bloom FE and Kupfer DJ (eds.) *Psychopharmacology: The Fourth Generation of Progress*, pp. 945–969. New York: Raven Press.
Joffe H and Cohen LS (1998) Estrogen, serotonin, and mood disturbance: Where is the therapeutic bridge? *Biological Psychiatry* 44: 798–811.
Joffe RT (2002) Hypothalamic–pituitary–thyroid axis. In: Pfaff D, Arnold AP, Etgen AM, Fahrbach SE, Moss RL, and Rubin RT (eds.) *Hormones, Brain and Behavior,* vol. 4, ch. 82, pp. 867–881. New York: Academic Press.
Jokinen J, Carlborg A, Martensson B, Forslund K, Nordstrom AL, and Nordstrom P (2007) DST non-suppression predicts suicide after attempted suicide. *Psychiatry Research* 150: 297–303.
Kang HJ, Adams DH, Simen A, et al. (2007) Gene expression profiling in postmortem prefrontal cortex of major depressive disorder. *Journal of Neuroscience* 27: 13329–13340.
Karl T and Herzog H (2007) Behavioral profiling of NPY in aggression and neuropsychiatric disorders. *Peptides* 28: 326–333.
Kaufman JM and Vermeulen A (2005) The decline of androgen levels in elderly men and its clinical and therapeutic implications. *Endocrine Reviews* 26: 833–876.
Kim H, Lim S-W, Kim S, Kim J-W, Chang YH, Carroll BJ, and Kim DK (2006) Monoamine transporter gene polymorphisms and antidepressant response in Koreans with late-life depression. *Journal of the American Medical Association* 296: 1609–1618.
Kin NMKNY, Nair NPV, Amin M, et al. (1997) The dexamethasone suppression test and treatment outcome in elderly depressed patients participating in a placebo-controlled multicenter trial involving moclobemide and nortriptyline. *Biological Psychiatry* 42: 925–931.
Kraus RP, Phoenix E, Edmonds MW, Nicholson IR, Chandarana PC, and Tokmakejian S (1997) Exaggerated TSH responses to TRH in depressed patients with normal baseline TSH. *Journal of Clinical Psychiatry* 58: 266–270.
Krishnan KR, Doraiswamy PM, Lurie SN, et al. (1991) Pituitary size in depression. *Journal of Clinical Endocrinology and Metabolism* 72: 256–259.
Kumsta R, Entringer S, Koper JW, van Rossum EFC, Hellhammer DH, and Wüst S (2007) Sex-specific associations between common glucocorticoid receptor gene variants and hypothalamus–pituitary–adrenal axis responses to psychosocial stress. *Biological Psychiatry* 62: 863–869.
Laakmann G, Zygan K, Schoen H-W, Weiss A, Wittmann M, Meissner R, and Blaschke D (1986) Effect of receptor blockers (methysergide, propranolol, phentolamine, yohimbine and prazosin) on desimipramine-induced pituitary hormone stimulation in humans. I: Growth hormone. *Psychoneuroendocrinology* 11: 447–461.
Laje G and McMahon FJ (2007) The pharmacogenetics of major depression: Past, present, and future. *Biological Psychiatry* 62: 1205–1207.
Lambert G, Johansson M, Agren H, and Friberg P (2000) Reduced brain norepinephrine and dopamine release in treatment-refractory depressive illness. *Archives of General Psychiatry* 57: 787–793.
Landgraf R (2006) The involvement of the vasopressin system in stress-related disorders. *CNS and Neurological Disorders – Drug Targets* 5: 167–179.
Laruelle M, Seghers A, Goffinet S, Bouchez S, and Legros JJ (1990) Plasmatic vasopressin neurophysin in depression: Basic levels and relations with HPA axis. *Biological Psychiatry* 27: 1249–1263.
Leong DA, Frawley LS, and Neill JD (1983) Neuroendocrine control of prolactin secretion. *Annual Review of Physiology* 45: 109–127.
Lesch KP, Disselkamp-Tietze J, and Schmidtke A (1990a) 5-HT1A receptor function in depression: Effect of chronic amitriptyline treatment. *Journal of Neural Transmission* 80: 157–161.
Lesch KP, Hoh A, Schulte HM, Osterheider M, and Müller T (1991) Long-term fluoxetine treatment decreases 5-HT1A receptor responsivity in obsessive-compulsive disorder. *Psychopharmacology* 105: 415–420.

Lesch KP, Laux G, Schulte HM, Pfuller H, and Beckmann H (1988) Abnormal responsiveness of growth hormone to human corticotropin-releasing hormone in major depressive disorder. *Journal of Affective Disorders* 14: 245–250.
Lesch KP, Mayer S, Disselkamp-Tietze D, Hoh A, Wiesmann M, Osterheider M, and Schulte HM (1990b) 5-HT1A receptor responsivity in unipolar depression: Evaluation of ipsapirone-induced ACTH and cortisol secretion in patients and controls. *Biological Psychiatry* 28: 620–628.
Leung K-C, Johannson G, Leong GM, and Ho KKY (2004) Estrogen regulation of growth hormone action. *Endocrine Reviews* 25: 693–721.
Lewis BM, Dieguez C, Lewis MD, and Scanlon MF (1987) Dopamine stimulates release of thyrotrophin-releasing hormone from perfused intact hypothalamus via hypothalamic D2 receptors. *Journal of Endocrinology* 115: 419–424.
Lewy AJ, Emens J, Jackman A, and Yuhas K (2006) Circadian uses of melatonin in humans. *Chronobiology International* 23: 403–412.
Lofberg C, Agren H, Harro J, and Oreland L (1998) Cholecystokinin in CSF from depressed patients: Possible relations to severity of depression and suicidal behaviour. *European Neuropsychopharmacology* 8: 153–157.
MacMaster FP and Kusumakar V (2004) MRI study of the pituitary gland in adolescent depression. *Journal of Psychiatric Research* 38: 231–236.
MacMaster FP, Russell A, Mirza Y, et al. (2006) Pituitary volume in treatment-naive pediatric major depressive disorder. *Biological Psychiatry* 60: 862–866.
Mah L, Zarate CA, Jr., Singh J, Duan YF, Luckenbaugh DA, Manji HK, and Drevets WC (2007) Regional cerebral glucose metabolic abnormalities in bipolar II depression. *Biological Psychiatry* 61: 765–775.
Mai FM, Shaw BF, Jenner MR, Wielgosz G, and Giles D (1985) Nocturnal prolactin secretion in depression. *British Journal of Psychiatry* 147: 314–317.
Malendowicz LK, Lesniewska B, Miskowiak B, Nussdorfer GG, and Nowak M (1991) Effects of neurotensin on the pituitary–adrenocortical axis of intact and dexamethasone-suppressed rats. *Experimental Pathology* 43: 205–211.
Manzoni F, Del Gerra P, Caraccion N, Prunetti CA, Pucci E, Luisi M, and Baschieri L (1993) Subclinical hypothyroidism: Neurobehavioral features and beneficial effect of l-thyroxine treatment. *Clinical Investigation* 71: 367–371.
Martinowich K and Lu B (2008) Interaction between BDNF and serotonin: Role in mood disorders. *Neuropsychopharmacology Reviews* 33: 73–83.
Mason BL and Pariante CM (2006) The effects of antidepressants on the hypothalamic–pituitary–adrenal axis. *Drug News and Perspective* 19: 603–608.
Mason JW (1968) Organization of psychoendocrine mechanisms. *Psychosomatic Medicine* 30: 565–575.
Mayberg HS, Lozano AM, Voon V, et al. (2005) Deep brain stimulation for treatment-resistant depression. *Neuron* 45: 651–660.
McCarthy JR, Heinrichs SC, and Grigoriadis DE (1999) Recent advances with the CRF1 receptor: Design of small molecule inhibitors, receptor subtypes and clinical indications. *Current Pharmaceutical Design* 5: 289–315.
McLean S (2005) Do substance P and the NK1 receptor have a role in depression and anxiety? *Current Pharmaceutical Design* 11: 1529–1547.
Meister B, Scanlon MS, and Hökfelt T (1990) Occurrence of galanin-like immunoreactivity in growth hormone-releasing factor (GRF)-containing neurons of the monkey (*Macaca fascicularis*) infundibular nucleus and median eminence. *Neuroscience Letters* 119: 136–139.
Merali Z, Du L, Hrdina P, Palkovits M, Faludi G, Poulter MO, and Anisman H (2004) Dysregulation in the suicide brain: mRNA expression of corticotropin-releasing hormone receptors and GABA(A) receptor subunits in frontal cortical brain region. *Journal of Neuroscience* 24: 1478–1485.
Merali Z, Kent P, Du L, et al. (2006) Corticotropin-releasing hormone, arginine vasopressin, gastrin-releasing peptide, and neuromedin B alterations in stress-relevant brain regions of suicides and control subjects. *Biological Psychiatry* 59: 594–602.
Mink JW (2003) The basal ganglia. In: Squire LR, Bloom FE, McConnell SK, Roberts JL, Spitzer NC, and Zigmond MJ (eds.) *Fundamental Neuroscience,* 2nd edn., pp. 815–839. Amsterdam: Academic Press.
Mitchell AJ (1998) The role of corticotropin releasing factor in depressive illness: A critical review. *Neuroscience and Biobehavioral Reviews* 22: 635–651.
Monteggia LM, Luikart B, Barrot M, et al. (2007) Brain-derived neurotrophic factor conditional knockouts show gender differences in depression-related behaviors. *Biological Psychiatry* 61: 187–197.
Montkowski A, Barden N, Wotjak C, et al. (1995) Long-term antidepressant treatment reduces behavioural deficits in transgenic mice with impaired glucocorticoid receptor function. *Journal of Neuroendocrinology* 7: 841–845.
Morley JE, Brammer GL, Sharp B, Yamada T, Yuwiler A, and Hershman JM (1981) Neurotransmitter control of hypothalamic–pituitary–thyroid function in rats. *European Journal of Pharmacology* 70: 263–271.
Murphy BE (1997) Antiglucocorticoid therapies in major depression: A review. *Psychoneuroendocrinology* 22(supplement 1): S125–S132.
Nathan PJ, Burrows GD, and Norman TR (1999) Melatonin sensitivity to dim white light in affective disorders. *Neuropsychopharmacology* 21: 408–413.
Nelson JC (1999) A review of the efficacy of serotonergic and noradrenergic reuptake inhibitors for treatment of major depression. *Biological Psychiatry* 46: 1301–1308.
Newell-Price J, Trainer P, Besser M, and Grossman A (1998) The diagnosis and differential diagnosis of Cushing's syndrome and pseudo-Cushing's states. *Endocrine Reviews* 19: 647–672.
Nicholas L, Dawkins K, and Golden RN (1998) Psychoneuroendocrinology of depression: Prolactin. *Psychiatric Clinics of North America* 21: 341–358.
O'Flynn K, O'Keane V, Lucey JV, and Dinan TG (1991) Effect of fluoxetine on noradrenergic mediated growth hormone release: A double-blind, placebo controlled study. *Biological Psychiatry* 30: 377–382.
O'Keane V, O'Flynn K, Lucey IV, and Dinan TG (1992) Pyridostigmine-induced growth hormone responses in healthy and depressed subjects: Evidence for cholinergic supersensitivity in depression. *Psychological Medicine* 22: 55–60.
Okugawa G, Omori K, Suzukawa J, Fujiseki Y, Kinoshita T, and Inagaki C (1999) Long-term treatment with antidepressants increases glucocorticoid receptor binding and gene expression in cultured rat hippocampal neurons. *Journal of Neuroendocrinology* 11: 887–895.
Oppenheimer JH and Schwartz HL (1997) Molecular basis of thyroid hormone – brain development. *Endocrine Reviews* 18: 462–475.
Ordway GA, Klimek V, and Mann JJ (2002) Neurocircuitry of mood disorders. In: Davis KL, Charney D, Coyle JT, and Nemeroff C (eds.) *Neuropsychopharmacology: The Fifth Generation of Progress*, ch. 73, pp. 1051–1074. Philadelphia, PA: Lippincott Williams and Wilkins.
Orth DN (1992) Corticotropin-releasing hormone in humans. *Endocrine Reviews* 13: 164–191.

Orth DN, Kovacs WJ, and DeBold CR (1992) The adrenal cortex. In: Wilson JD and Foster DW (eds.) *Williams Textbook of Endocrinology,* 8th edn., pp. 489–619. Philadelphia, PA: W.B. Saunders.

Pariante CM (2006) The glucocorticoid receptor: Part of the solution or part of the problem? *Journal of Psychopharmacology* 20: 79–84.

Pariante CM, Dazzan P, Danese A, et al. (2005) Increased pituitary volume in antipsychotic-free and antipsychotic-treated patients of the AEsop first-onset psychosis study. *Neuropsychopharmacology* 30: 1923–1931.

Partonen T, Vakkuri O, and Lonnqvist J (1997) Suppression of melatonin secretion by bright light in seasonal affective disorder. *Biological Psychiatry* 42: 509–513.

Pearlstein T, Rosen K, and Stone AB (1997) Mood disorders and menopause. *Endocrinology and Metabolism Clinics of North America* 26: 279–294.

Piccinni A, Marazziti D, Catena M, et al. (2008) Plasma and serum brain-derived neurotrophic factor (BDNF) in depressed patients during 1 year of antidepressant treatments. *Journal of Affective Disorders* 105: 279–283.

Pitts AF, Samuelson SD, Meller WH, Bissette G, Nemeroff CB, and Kathol RG (1995) Cerebrospinal fluid corticotrophin-releasing hormone, vasopressin, and oxytocin concentrations in treated patients with major depression and controls. *Biological Psychiatry* 38: 330–335.

Plotsky PM, Owens MJ, and Nemeroff CB (1995) Neuropeptide alterations in mood disorders. In: Bloom FE and Kupfer DJ (eds.) *Psychopharmacology: The Fourth Generation of Progress*, pp. 971–981. New York: Raven Press.

Pocock JM and Kettenmann H (2007) Neurotransmitter receptors on microglia. *Trends in Neurosciences* 30: 527–531.

Pourmand M, Rodriguez-Arnao MD, Weightman DR, Hall R, Cook DB, Lewis M, and Scanlon MF (1980) Domperidone: A novel agent for the investigation of anterior pituitary function in man. *Clinical Endocrinology* 12: 211–215.

Reus VI (1984) Diagnosis and treatment in endocrinology and psychiatry: From Cushing's syndrome to disorder to mood. In: van Dyke C, Temoshok L, and Zegans LS (eds.) *Emotions in Health and Illness: Applications to Clinical Practice*, pp. 23–34. New York: Grune & Stratton.

Robbins TW and Everitt BJ (2003) Motivation and reward. In: Squire LR, Bloom FE, McConnell SK, Roberts JL, Spitzer NC, and Zigmond MJ (eds.) *Fundamental Neuroscience,* 2nd edn., pp. 1109–1126. Amsterdam: Academic Press.

Rosen T, Johannsson G, Hallgren P, Bengtsson BA, and Rafaelson E (1994) Beneficial effects of 12 months replacement therapy with recombinant human growth hormone to growth hormone deficient adults. *Endocrinology and Metabolism* 1: 55–66.

Rosenthal MJ, Hunt WC, Garry PJ, and Goodwin JS (1987) Thyroid failure in the elderly: Microsomal antibodies as discriminant for therapy. *Journal of the American Medical Association* 258: 209–213.

Rubin RT, Abbasi SA, Rhodes ME, and Czambel RK (2003) Growth hormone responses to low-dose physostigmine administration: Sexual diergism between major depressives and matched controls. *Psychological Medicine* 33: 655–665.

Rubin RT and Carroll BJ (2004) Mifepristone (RU486) in the treatment of psychotic depression: Re-evaluation of published data. *Neuropsychopharmacology* 29(supplement 1): S203–S204.

Rubin RT, Dinan TG, and Scott LV (2002) The neuroendocrinology of affective disorders. In: Pfaff D, Arnold AP, Etgen AM, Fahrbach SE, Moss RL, and Rubin RT (eds.) *Hormones, Brain and Behavior,* vol. 5, ch. 97, pp. 467–514. New York: Academic Press.

Rubin RT, Heist K, McGeoy SS, Hanada K, and Lesser IM (1992) Neuroendocrine aspects of primary endogenous depression XI. Serum melatonin measures in patients and matched control subjects. *Archives of General Psychiatry* 49: 558–567.

Rubin RT and Mandell AJ (1966) Adrenal cortical activity in pathological emotional states: A review. *American Journal of Psychiatry* 123: 387–400.

Rubin RT, Miller TH, Rhodes ME, and Czambel RK (2006a) Adrenal cortical responses to low- and high-dose ACTH1–24 administration in major depressives vs. matched controls. *Psychiatry Research* 143: 43–50.

Rubin RT, O'Toole SM, Rhodes ME, Sekula LK, and Czambel RK (1999) Hypothalamo-pituitary–adrenal cortical responses to low–dose physostigmine and arginine vasopressin administration: Sex differences between major depressives and matched controls. *Psychiatry Research* 89: 1–20.

Rubin RT, Phillips JJ, McCracken JT, and Sadow TF (1996) Adrenal gland volume in major depression: Relationship to basal and stimulated pituitary–adrenal cortical axis function. *Biological Psychiatry* 40: 89–97.

Rubin RT, Phillips JJ, Sadow TF, and McCracken JT (1995) Adrenal gland volume in major depression: Increase during the depressive episode and decrease with successful treatment. *Archives of General Psychiatry* 52: 213–218.

Rubin RT and Poland RE (1984) The dexamethasone suppression test in depression: Advantages and limitations. In: Burrows GD, Norman TR, and Maguire KP (eds.) *Biological Psychiatry: Recent Studies*, pp. 76–83. London: John Libbey.

Rubin RT, Poland RE, and Lesser IM (1990) Neuroendocrine aspects of primary endogenous depression X: Serum growth hormone measures in patients and matched control subjects. *Biological Psychiatry* 27: 1065–1082.

Rubin RT, Poland RE, Lesser IM, Winston RA, and Blodgett AL (1987a) Neuroendocrine aspects of primary endogenous depression. I. Cortisol secretory dynamics in patients and matched controls. *Archives of General Psychiatry* 44: 328–336.

Rubin RT, Poland RE, Lesser IM, and Martin DJ (1987b) Neuroendocrine aspects of primary endogenous depression. IV. Pituitary–thyroid axis activity in patients and matched control subjects. *Psychoneuroendocrinology* 12: 333–347.

Rubin RT, Poland RE, Lesser IM, and Martin DJ (1989b) Neuroendocrine aspects of primary endogenous depression V. Serum prolactin measures in patients and matched control subjects. *Biological Psychiatry* 25: 4–21.

Rubin RT, Rhodes ME, Miller TH, Jakab RL, and Czambel RK (2006b) Sequence of pituitary–adrenal cortical hormone responses to low-dose physostigmine administration in young adult women and men. *Life Sciences* 79: 2260–2268.

Sassi RB, Nicoletti M, Brambilla P, et al. (2001) Decreased pituitary volume in patients with bipolar disorder. *Biological Psychiatry* 50: 271–280.

Schlechte J, Sherman B, and Pfohl B (1986) A comparison of adrenal cortical function in patients with depressive illness and Cushing's disease. *Hormone Research* 23: 1–8.

Seidman SN (2006) Normative hypogonadism and depression: Does 'andropause' exist? *International Journal of Impotence Research* 18: 415–422.

Sherwin BB (2003) Estrogen and cognitive functioning in women. *Endocrine Reviews* 24: 133–151.

Sichel DA, Cohen LS, Robertson LM, Ruttenberg A, and Rosenbaum JF (1995) Prophylactic estrogen in recurrent postpartum affective disorder. *Biological Psychiatry* 38: 814–818.

Sit D, Rothschild AJ, and Wisner KL (2006) A review of postpartum psychosis. *Journal of Women's Health* 15: 352–368.

Skare SS, Dysken MW, and Billington CJ (1994) A review of GHRH stimulation test in psychiatry. *Biological Psychiatry* 36: 249–265.
Stockmeier CA (1997) Neurobiology of serotonin in depression and suicide. *Annals of the New York Academy of Sciences* 836: 220–232.
Sullivan GM, Mann JJ, Oquendo MA, Lo ES, Cooper TB, and Gorman JM (2006) Low cerebrospinal fluid transthyretin levels in depression: Correlations with suicidal ideation and low serotonin function. *Biological Psychiatry* 60: 500–506.
Taylor MA and Fink M (2006) *Melancholia.* New York: Cambridge University Press.
Thakore JH and Dinan TG (1994) Growth hormone secretion: The role of glucocorticoids. *Life Sciences* 55: 1083–1099.
Thakore JH and Dinan TG (1995) Effect of fluoxetine on dexamethasone-induced growth hormone release in depression: A double-blind placebo controlled study. *American Journal of Psychiatry* 152: 616–618.
Unden F and Aperia B (1994) Major affective disorders: Prospective clinical course during a 5-year to 7-year follow-up in relation to neuroendocrine function tests. *Nordic Journal of Psychiatry* 48: 131–137.
Ur E, Dinan TG, O'Keane VO, et al. (1992) Effect of metyrapone on the pituitary–adrenal axis in depression: Relation to dexamethasone suppressor status. *Neuroendocrinology* 56: 533–538.
van Bockstaele EJ, Colago EEO, and Valentino RJ (1998) Amygdaloid corticotropin-releasing factor targets locus coeruleus dendrites: Substrate for the co-ordination of emotional and cognitive limbs of the stress response. *Journal of Neuroendocrinology* 10: 743–757.
Van Den Eede F, Venken T, Del-Favero J, et al. (2007) Single nucleotide polymorphism analysis of corticotropin-releasing factor-binding protein gene in recurrent major depressive disorder. *Psychiatry Research* 153: 17–25.
van der Lely AJ, Tschöp M, Heiman ML, and Ghigo E (2004) Biological, physiological, pathophysiological, and pharmacological aspects of ghrelin. *Endocrine Reviews* 25: 426–457.
van Moffaert M and Dierick M (1999) Noradrenaline (norepinephrine) and depression. *CNS Drugs* 12: 293–305.
Van Praag HM, Lemus C, and Kahn R (1986) The pitfalls of serotonin precursors as challengers in hormonal probes of central serotonin activity. *Psychopharmacology Bulletin* 22: 565–570.
Veldhuis JD, Iranmanesh A, Naftolowitz D, Tatham N, Cassidy F, and Carroll BJ (2001) Corticotropin secretory dynamics in humans under low glucocorticoid feedback. *Journal of Clinical Endocrinology and Metabolism* 86: 5554–5563.
Von Bardeleben U, Stalla GK, Muller OA, and Holsboer F (1988) Blunting of ACTH response to human CRH in depressed patients is avoided by metyrapone pretreatment. *Biological Psychiatry* 24: 782–786.
Vythilingam M, Anderson GM, Owens MJ, et al. (2000) Cerebrospinal fluid corticotropin-releasing hormone in healthy humans: Effects of yohimbine and naloxone. *Journal of Clinical Endocrinology and Metabolism* 85: 4138–4145.
Weeke J, Hansen AP, and Lundaek K (1975) Inhibition by somatostatin of basal levels of serum thyrotropin (TSH) in normal men. *Journal of Clinical Endocrinology and Metabolism* 41: 168–171.
Wilhelm K, Mitchell PB, Niven H, et al. (2006) Life events, first depression onset and the serotonin transporter gene. *British Journal of Psychiatry* 188: 210–215.
Willner P (1995) Dopaminergic mechanisms in depression and mania. In: Bloom FE and Kupfer DJ (eds.) *Psychopharmacology: The Fourth Generation of Progress*, pp. 921–931. New York: Raven Press.
Wolkowitz OM and Reus VI (1999) Treatment of depression with antiglucocorticoid drugs. *Psychosomatic Medicine* 61: 698–711.
Wong M-L, Kling MA, Munson PJ, et al. (2000) Pronounced and sustained central noradrenergic function in major depression with melancholic features: Relation to hypercortisolism and corticotropin-releasing hormone. *Proceedings of the National Academy of Sciences of the United States of America* 97: 325–330.
Young EA, Haskett RF, Weinberg VM, Watson SJ, and Akil H (1994) Increased evening activation of the hypothalamic–pituitary–adrenal axis in depressed patients. *Archives of General Psychiatry* 51: 701–707.
Young EA, Lopez JF, Murphy-Weinberg V, Watson SJ, and Akil H (2003) Mineralocorticoid receptor function in major depression. *Archives of General Psychiatry* 60: 24–28.
Zis AP, Haskett RF, Albala AA, Carroll BJ, and Lohr NE (1985) Opioid regulation of hypothalamic–pituitary–adrenal function in depression. *Archives of General Psychiatry* 42: 383–386.
Zis AP, Remick RA, Clark CM, Goldner E, Grant BEK, and Brown GM (1989) Effect of morphine on cortisol and prolactin secretion in anorexia nervosa and depression. *Clinical Endocrinology* 30: 421–427.
Zobel AW, Nickel T, Kunzel HE, Ackl N, Sonntag A, Ising M, and Holsboer F (2000) Effects of the high-affinity corticotropin-releasing hormone receptor 1 antagonist R121919 in major depression: The first 20 patients treated. *Journal of Psychiatric Research* 34: 171–181.
Zobel AW, Yassourdis A, Frieboes RM, and Holsboer F (1999) Prediction of medium-term outcome by cortisol response to the combined dexamethasone-CRH test in patients with remitted depression. *American Journal of Psychiatry* 156: 949–951.

Further Reading

Carroll BJ and Rubin RT (2006) Is mifepristone useful in psychotic depression? *Neuropsychopharmacology* 31: 2793–2794.
Holsboer F (2000) The corticosteroid receptor hypothesis of depression. *Neuropsychopharmacology* 23: 477–501.
Holsboer F, Spengler D, and Heuser I (1992) The role of corticotropin-releasing hormone in the pathogenesis of Cushing's disease, anorexia nervosa, alcoholism, affective disorders and dementia. *Progress in Brain Research* 93: 385–417.
Raadsheer FC, Hoogendijk WJG, Stam FC, Tilders FJH, and Swaab DF (1994) Increased numbers of corticotropin-releasing hormone expressing neurons in the hypothalamic paraventricular nucleus of depressed patients. *Neuroendocrinology* 60: 436–444.
Singh V, Muzina DJ, and Calabrese JR (2005) Anticonvulsants in bipolar disorder. *Psychiatric Clinics of North America* 28: 301–323.
Young EA, Carlson NE, and Brown MB (2001) Twenty-four-hour ACTH and cortisol pulsatility in depressed women. *Neuropsychopharmacology* 25: 267–276.

22 Premenstrual Dysphoric Disorder

B L Parry, S Nowakowski, and L F Martinez, University of California, San Diego, La Jolla, CA, USA
S L Berga, Emory University, Atlanta, GA, USA

Chapter Outline

Glossary

estradiol (17β-estradiol) Mislabeled the female hormone, it is also present in males; it represents the major estrogen in humans. Estradiol has not only a critical impact on reproductive and sexual functioning, but also affects other organs including bone structure.

follicle-stimulating hormone (FSH) Synthesized and secreted by gonadotropes in the anterior pituitary gland. FSH regulates development, growth, pubertal maturation, and reproductive processes in humans. FSH and luteinizing hormone (LH) act synergistically in reproduction.

luteinizing hormone (LH) Produced by the anterior pituitary gland; in females, an acute rise of LH – the LH surge – triggers ovulation, while in males it stimulates Leydig cell production of testosterone.

premenstrual dysphoric disorder (PMDD) A severe form of premenstrual syndrome (a collection of physical, psychological, and emotional symptoms related to a woman's menstrual cycle of sufficient severity to interfere with some aspects of life), afflicting 3–8% of women. It is a mood disorder associated with the luteal phase of the menstrual cycle.

selective serotonin reuptake inhibitors (SSRIs) A class of antidepressants used to treat depression, anxiety disorders, and some personality disorders. As the first class of psychotropic drugs to be rationally designed, SSRIs are the most widely prescribed antidepressants in many countries.

thyroid-stimulating hormone (TSH) A peptide hormone synthesized and secreted by thyrotrope cells in the anterior pituitary gland which regulates the endocrine function of the thyroid gland.

22.1 Introduction

Premenstrual mood changes have been described since the time of Hippocrates. Frank (1931) more specifically identified physical, psychological, and behavioral changes corresponding to monthly changes in reproductive hormones. Public awareness of the concurrent change in mood with phases of the menstrual cycle has increased markedly over the past two decades. The media has disseminated a wide body of information shaping a societal notion of what is popularly termed premenstrual syndrome (PMS). Definitions for PMS have varied widely. Typically, PMS has been defined by a large number of symptoms, including irritability, tension, fatigue, dysphoria, distractibility, impaired motor coordination, changes in eating and sleeping, and libido changes, which occur in the late luteal phase and remit after the beginning of menstruation. Although a majority of women may experience some premenstrual symptoms, and up to 50% of women may experience many symptoms together, comprising a syndrome, more rigorous criteria are defined in the *Diagnostic and Statistical Manual of Mental Disorders*, previously termed late luteal phase dysphoric disorder (LLPDD) in DSM-III-R and currently as premenstrual dysphoric disorder (PMDD) in DSM-IV (American Psychiatric Association, 1994).

22.2 Diagnostic Issues

Many women are thought to experience disruptions in mood or physical complaints during the menstrual cycle. Thus, the diagnosis of PMS has been one of controversy among psychiatrists, psychologists, gynecologists, and sociologists. One reason for the debate among researchers may be due to ambiguity resulting from the failure to distinguish normal from pathological premenstrual mood disturbance. Unlike other psychiatric syndromes at present, the remission and relapse of premenstrual symptoms is connected to a physiological process – the menstrual cycle. Every woman does not experience a pathological, premenstrual mood disorder, however. Thus, the differentiation between the normal and pathological is important in terms of investigative research, interdisciplinary discourse, as well as treatment options.

Premenstrual symptoms are currently included in the appendix of the DSM-IV under the heading, premenstrual dysphoric disorder (PMDD), and classified as depressive disorder, not otherwise specified in the main body of the text. The criteria for this disorder are more stringent and delineate increased specificity and severity of premenstrual symptoms American Psychiatric Association, 1994. In this definition of the disorder, the (1) nature, (2) severity, and (3) timing of the symptoms are critical in establishing the diagnosis. First, symptoms must be primarily psychological, cognitive, or affective in nature (i.e., the predominant symptoms are depression, anxiety, or irritability as opposed to breast swelling or tenderness, abnormal bloating, weight gain, or other somatic symptoms). Second, the symptoms must be of sufficient severity to disrupt social or occupational functioning. Third, the timing of the symptoms in relation to the phases of the menstrual cycle needs to be established such that the symptoms reach their peak intensity during the premenstrual or late luteal phase of the menstrual cycle and remit shortly after the onset of menses in the follicular phase. With this latter criteria, symptoms cannot be merely a premenstrual exacerbation of another underlying disorder that would not be expected to remit after menses.

22.2.1 Clinical Phenomenology

Symptoms of PMDD include the common to major depressive disorders, such as depressed mood, eating and sleeping disturbances, decreased interest in usual activities, lethargy, difficulty concentrating, feelings of hopelessness, or anxiety. Women with PMDD may also experience increased sensitivity to rejection, irritability, a sense of being out of control, and concomitant physical symptoms, such as headaches, breast tenderness, and bloating. In aggregate, the symptoms suggest increased central and peripheral reactivity or sensitivity to the customary sex steroid excursions that accompany an ovulatory menstrual cycle.

Besides emotional distress, the symptoms must cause impairment in occupational or social functioning. Also, the mood disturbances occur during the luteal phase and remit during the follicular phase; this pattern of mood fluctuation must be documented over at least two consecutive cycles to confirm the diagnosis. Retrospective ratings are notoriously unreliable, likely due to an expectation bias. Thus, an accurate diagnosis rests on prospectively charting important symptoms on a daily basis across two menstrual intervals. In the United States, approximately 3–5% of women are thought to experience symptoms that meet criteria for PMDD. Over time, untreated PMDD may become progressively more severe, and the episodes of dysphoria may extend in duration. Consistent with a model of kindling and behavioral sensitization (Post et al., 1986), women who suffer recurrent episodes of PMDD that are not treated are vulnerable to developing a major depressive disorder over time.

PMDD causes significant morbidity and may be linked to affective illness. The incidence of PMDD has been estimated at 5%, although 20–80% of women report some mood, cognitive, and behavioral disturbances associated with their menstrual cycle (Hamilton et al., 1984). In the premenstrual phase, women report symptoms such as depression, anxiety, irritability, and difficulty concentrating, as well as sleep, appetite, and energy disturbances (Dalton, 1964). These symptoms become severe enough to disrupt normal functioning in work and interpersonal relationships in some, and have resulted in psychosis and suicidal depressions in others (Dalton, 1964; Endo et al., 1978). An increasing number of studies indicate that PMDD is related to the mood disorders, which forms the rationale for its categorization under mood disorders, depression not otherwise specified in DSM-IV (De Ronchi et al., 2005; Kornstein et al., 2005; Hallman, 1986; Mackenzie et al., 1986; Halbreich and Endicott, 1985; Wetzel et al., 1975; Schuckit et al., 1975), although some authors argue that it is a distinct diagnostic entity (Landen and Eriksson, 2003). Although neurotransmitter, neuroendocrine, and chronobiological disturbances have been implicated in mood disorders and in PMDD, no single theory has been well substantiated, nor no single treatment shown to be consistently effective with rigorous testing.

In addition to disturbances in mood, women with PMDD may experience physical and cognitive deficits. Cognitive deficits in major depression have been well documented and often include disturbances in attention, forgetfulness, psychomotor retardation, and proneness to confusion. In some cases, these cognitive disturbances may be secondary to mood disturbances. In PMDD, however, the role reproductive hormone fluctuation may have on memory and cognitive functioning has recently been the focus of several studies. Estrogen (E_2) receptors in the brain could mediate estrogen-induced vasodilatation that could affect the blood supply of the brain. E_2 receptors are located in areas of the brain involved with cognition, for example, cerebral cortex, basal forebrain, and hippocampus. Endogenous fluctuation in E_2 associated with the menstrual cycle has been associated with cognitive changes such that specific cognitive skills are elevated at times during the cycle when estrogen levels are high. In healthy women without psychiatric illness, performance on neuropsychological tests has been found to vary with the monthly fluctuation in reproductive hormones. High levels of estrogen, as experienced in the late follicular phase, may elevate performance on automatization abilities, perceptual-motor speed, mental arithmetic, and verbal memory (Hampson and Kimura, 1988; Hampson, 1990; Klaiber et al., 1982). Lower levels of estrogen, as experienced during the menstrual phase, may be associated with higher performance on visuospatial tasks. The majority of these effects has been subtle and unable to be replicated; this may be due to difficulties in documentation and measurement of ovulation, plasma hormone levels, mood symptoms, and individual variation in cycle length. Furthermore, cyclic performance changes are minor and are unlikely to interfere with normal functioning. These data suggested the hypothesis that cycle-related changes in performance may be exaggerated in women with PMDD. Comparative studies in animals have provided physiological evidence for the distinctiveness of cognitive change associated with reproductive hormone cyclicity. In rats, dendritic spine density within the hippocampus, an area of the brain associated with memory and learning, changes rapidly and systematically with estrus (Woolley and McEwen, 1992). These effects may represent an interaction of hormones upon brain substrate and/or cholinergic, serotoninergic, or gabamin GABAergic neuromodulatory systems implicated in cognition.

Relatively few studies have focused on cognitive functioning in PMDD. Women who have PMDD, like individuals with depression, often report a subjective sense of altered cognitive functioning during symptomatic phases: decreased concentration, motor inefficiency, forgetfulness, and indecisiveness are among the most frequent complaints (Diener et al., 1992). Keenan et al. (1992) found that women meeting DSM-IV criteria for PMDD exhibited non-phase-dependent deficits in encoding and retrieval of verbal material. In a follow-up study, Keenan et al. (1995) replicated the general findings that women with PMDD exhibit subtle yet persistent verbal memory deficits. Similarly, several authors have reported that premenstrual women shown compromised inhibitory capacity on executive function tasks primarily medicated by the frontal lobe (Brugger et al., 1993; Keenan et al., 1992). Resnick et al. (1998) found subtle psychomotor slowing in the late follicular phase when compared with late luteal phase performance in women with and without PMDD. However, the authors did not find discernable group differences in cognitive functioning between women with and without PMDD. This lack of group differences has been found in several other studies (Man et al., 1999; Morgan et al., 1996; Morgan and Rapkin, 2002; Ussher and Wilding, 1991). These equivocal findings may be due to methodological weaknesses, failure to objectively assess type and intensity of PMDD symptoms, to categorize women properly, limitations in testing instruments, and/or inclusion of potential moderating variables, such as sleep. They may also reflect the insensitivity of the instruments used or be a function of ceiling effects.

22.2.2 Relationship to Depression

Depression, one manifestation of which may be PMDD, is a major mental health problem in women. Women, as compared to men, have a greater lifetime risk for depression, and the risk for depression appears to be increasing each generation (Gershon et al., 1987; Weissman et al., 1984). Women predominate with respect to unipolar depression (Weissman and Klerman, 1977), the depressive subtype of bipolar illness (Angst, 1978), and cyclical forms of affective illness such as rapid cycling manic-depressive illness (Dunner et al., 1977) and seasonal affective disorder (Rosenthal et al., 1984). In addition, events associated with the reproductive cycle are capable of provoking affective changes in predisposed individuals. Examples include depression associated with oral contraceptives (Parry and Rush, 1979), the luteal phase of the menstrual cycle (Dalton, 1964), the postpartum period (Brockington and Kumar, 1982), and menopause (Angst, 1978; Winokur, 1973; Weissman, 1979; Schmidt et al., 2004; Freeman et al., 2006; Cohen et al., 2006). Sex differences in the rates of depression begin to appear in adolescence (Weissman et al., 1987), a time of major change in the neuroendocrine reproductive axis. Thus, the fluctuation of ovarian steroids during specific phases of the reproductive cycle may bear some relationship to the particular vulnerability of women for mood disorders. The reproductive hormones could exert their effects on mood directly or indirectly by their effect on neurotransmitter (McEwen and Parsons, 1982), neuroendocrine (Meites et al., 1979), brain-derived neurotrophic factor, protein kinase C (Payne, 2003), or circadian systems (Albers et al., 1981b; Wehr, 1984), all of which have been implicated in the pathogenesis of affective illness. One clinical model for studying the relationship of ovarian hormones and mood disorders is the affective changes associated with the menstrual cycle (PMDD). One scientific advantage of studying PMDD is that the mood and behavioral changes are recurrent and predictable and thus can be studied prospectively and longitudinally. Since PMDD criteria were not available when many of the studies described below were done, unless otherwise specified, the older term, premenstrual syndrome or PMS, will be used.

22.2.3 Risk Factors, Inheritance and Relationship to Other Mood Disorders

22.2.3.1 Mood disorders

Increasing evidence suggests that women with a lifetime risk for major depressive disorders are more likely to have premenstrual depression, and alternatively, women with PMS may later develop major mood disorders (Mackenzie et al., 1986; Halbreich and Endicott, 1985; Wetzel et al., 1975; Schuckit et al., 1975).

22.2.3.2 Familial factors

Certain studies (Widholm, 1979; van den Akker et al., 1987) suggest that women whose mothers reported premenstrual tension were more likely (70% of daughters of affected mothers vs. 37% of daughters of unaffected mothers) to develop premenstrual mood symptoms. Also, concordance rates for symptoms of PMS are significantly higher in monozygotic twins (93%) compared to dizygotic twins (44%) or sibling controls (31%) (Dalton et al., 1987). Glick et al. (1993), however, did not find an increased incidence of the disorder in a population sampled in more detail without preselected probands.

22.2.3.3 Other reproductive-related mood disorders

Women with PMS are at greater risk for menopausal symptoms (Freeman et al., 2004b; Becker et al., 2007) and postpartum depression (Chuong and Burgos, 1995). Brockington et al. (1988) have suggested that the postpartum period is a particularly vulnerable time period for the development or exacerbation of premenstrual depression. Some anecdotal evidence suggests an increased incidence of PMS in women who have used oral contraceptives (Parry et al., 1991; Altshuler et al., 1995).

22.2.3.4 Age

Consistent with the model of kindling and behavioral sensitization (see below), premenstrual mood disturbances tend to become more severe with increasing age (Post et al., 1986). The probable trigger is rapid changes in ovarian sex steroids. It is interesting to note that as the ovary ages, its secretion of estradiol (E_2) and progesterone (P) alters. During the late perimenopausal years, ovarian secretion becomes erratic with precipitous changes in E_2 (Santoro et al., 1996). These changes may also reflect aging of the brain's neural machinery.

22.2.4 Cultural Aspects

Estimates of the worldwide prevalence of what historically has been referred to as PMS have ranged from 3% to 95% due to a lack of diagnostic precision among researchers (American Psychiatric Association, 1994; Banerjee et al., 2000; Di Giulio and Reissing, 2006; Halbreich et al., 2003; Richardson, 1995; Takeda et al., 2006). A limited number of studies have examined symptoms of PMDD in non-Western societies. This research has been plagued by a number of important methodological concerns and questionable validity of instruments that have been translated for use in populations other than those in which they were developed. In addition, the socio-political position of women in other cultures and their societal beliefs and expectations about menstruation are difficult to quantify and thus have not been taken into account. Despite these difficulties, cross-cultural studies are of interest in considering the relative importance of biological versus sociological factors in the etiology of PMDD. To date, every culture assessed for PMS has reported negative mood changes and physical discomfort associated with the period prior to menstruation, although the frequency, number, type, and nature of symptoms reported may vary across cultures (Chandra and Chaturvedi, 1989; Chang et al., 1995; Monagle et al., 1993; Janiger et al., 1972; Meaden et al., 2005; Most et al., 1981; Rasheed and Al-Sowielem, 2003; Shye and Jaffe, 1991; Lu, 2001; Snowden and Christian, 1983).

The largest cross-cultural study on perceptions of menstrual bleeding was performed by the World Health Organization (WHO) in 1981 on 5300 parous women from 14 cultural groups (Snowden and Christian, 1983). Although no clear pattern of perimenstrual symptom type or frequency emerged, the most frequently reported physical and affective symptoms were back pain, abdominal pain, irritability, lethargy, and depression (Chang et al., 1995; Merikangas et al., 1993). In the premenstrual phase, symptom reporting of physical discomfort ranged from 43% to 70% of women, and perceived mood change ranged from 13% to 71% of women. Eriksson et al. (2002) found that women in developing countries tended to report shifts in mood associated with the onset of menstruation or throughout the menstrual cycle. Gravidity or parity may affect the expression of these symptoms.

22.3 Etiology

Various disciplines have conceptualized different etiological precipitants for premenstrual mood disturbances. Biological theorists maintain that premenstrual symptoms originate from the biological reactivity or sensitivity to reproductive hormones. Social theorists assert that the social, patriarchal environment elicits symptoms. Psychological theorists suggest that premenstrual symptoms arise from the perception and maladaptive responses of the individual to the environment. Regardless of the etiological model, hormonal changes associated with the menstrual cycle are likely to serve as contributory factors to mood disruption; physiological changes are intrinsically involved in the temporal sequence of the menstrual cycle.

22.3.1 Biomedical Model

Investigators and clinicians have recognized the spectrum of premenstrual symptoms that range from mild to severe; those with severe symptoms meet criteria for PMDD. Rather than psychosomatic origins, researchers have investigated physiological underpinnings to the disorder that are expressed in mood and behavior, that is, it is conceptualized as a somatopsychic illness.

Research in this area began with the hypothesis that the absolute amount of estrogen and progesterone (e.g., P deficiency, estrogen excess) were responsible for mood disturbances. Inconsistencies in research findings have called this hypothesis into question. However, the rate of hormonal change may play an important role in the pathogenesis of the disorder (Halbreich et al., 1986). Sex differences in the rates of major depressive disorder begin at puberty, a time of major change in the neuroendocrine reproductive axis. Women also predominate with respect to cyclic mood disorders, such as in rapid cycling bipolar disorder and seasonal affective disorder. In addition, women are at increased risk for experiencing depression during the postpartum period and the peri-menopause, which are both times of rapid changes in reproductive hormones.

Estrogen and P variations are likely to exert their influence on mood indirectly rather than directly. Research has suggested that possible abnormal fluctuations in neurotransmitter systems, prolactin, mineralocorticoids, prostaglandins (PGs), and endogenous opiates correspond to changes in the menstrual cycle. The central nervous system (CNS) is thus a nontraditional target tissue for sex steroids. Hormones may prime the brain to gate attention to certain stimuli or alter responses to those stimuli. They also may heighten awareness to nonverbal cues such as olfactory cues. In PMDD, these heightened responses may be exaggerated.

Many neuroendocrine theories have been proposed for PMDD. However, no single theory has proven consistently true, and until the recent trails of selective serotonin reuptake inhibitors, no single treatment was consistently effective (Parry and Rausch, 1988). Although a pattern of cyclic mood changes occurring in association with the menstrual cycle has been noted since the time of Hippocrates, it was not until 1931 that Frank first proposed a hormonal etiology (Frank, 1931). Frank attributed menstrually related mood symptoms to an excess production and decreased excretion of female hormones. With the burgeoning of radioimmunoassay techniques, a multitude of hormonal systems have been implicated in the pathogenesis of PMS. These hormonal systems include the gonadal steroids (in particular, a disturbed estrogen–progesterone ratio), prolactin, cortisol, thyroid hormones, PGs, endorphins, and the biogenic amines.

In this chapter, the focus is on more recent studies that document prospective ratings of symptoms and that include control groups. Also, particular controversial studies which may not have included all the aforementioned criteria are also discussed. The studies of biological correlates of PMDD have been divided into six sections: (1) gonadal steroids and gonadotropins; (2) neurovegetative signs; (3) neuroendocrine measures; (4) serotonin and other neurotransmitters; (5) β-endorphin; and (6) other biologic factors, including PGs, vitamins, electrolytes, and carbon dioxide. Note: earlier studies often used the terminology PMS before PMDD criteria were available. In the interim, some studies used the DSM-III-R criteria of late luteal phase dysphoric disorder (LLPDD).

Before reviewing gonadal steroids, a brief summary of endogenous cyclical changes in these hormones associated with the menstrual cycle is delineated below.

22.3.2 Neuroendocrine Control of the Menstrual Cycle

Menstrual cyclicity is the direct result of ovarian cyclicity. Ovarian cyclicity starts with the development of a cohort of follicles, one of which will become dominant. The follicles are composed of an oocyte surrounded by granulosa cells which, in turn, are surrounded by theca cells. Follicular development is initiated by the hypothalamic release of gonadotropin-releasing hormone (GnRH) at a pulse frequency of about once every 90 min. GnRH stimulates the release of the pituitary gonadotropins, luteinizing harmone (LH) and follicle stimulating harmone (FSH). In turn, LH stimulates ovarian theca cells to synthesize and secrete androgens, while FSH induces granulosa cell development, including the enzyme aromatase, which converts the thecally produced androgens to estrogens. In the presence of a constant GnRH pulse interval of 90 min, the secretion of LH and FSH will be regulated primarily by E_2 feedback at the level of the pituitary. Rising levels of E_2 suppress FSH, thereby limiting the number of follicles that will develop mature oocytes. When E_2 concentrations exceed a critical threshold and remain elevated for at least 36 h, which approximates the pattern one fully mature follicle produces, an LH surge is triggered and ovulation ensues approximately 36 h later. Thereafter, granulosa cells transform into progesterone-secreting luteal cells and the ovulated follicle is then referred to as the corpus luteum.

22.3.2.1 Gonadal steroids/gonadotropins

Backstrom et al. (2003) and Rubinow and Schmidt (2006) reviewed the role of gonadal steroid regulation of mood as exemplified in PMS. Recently,

Huo et al. (2007) reported that the risk for PMDD was associated with genetic variation in ESR1, the estrogen receptor-α gene. Hsiao et al. (2004), however, found no correlation of depression and anxiety to plasma estrogen and progesterone levels in PMDD patients. Rubinow et al. (1988) examined E_2, P, FSH, LH, testosterone-E_2-binding globulin, dihydroepiandrosterone sulfate, dihydrotestosterone, prolactin, and cortisol in 17 women with prospectively confirmed PMS and nine control subjects. The diagnosis was made by obtaining daily ratings for two cycles. If there was a 30% increase in symptoms in the premenstrual compared to the follicular phase, then a diagnosis of PMS was made providing the schedule for affective disorders and schizophrenia (SADS) interview did not reveal concomitant psychiatric illness. Blood samples were drawn at 8 a.m. during the early, mid-, and late follicular phases, and the early, mid-, and late luteal phases. There were no diagnosis-related changes in any of the hormones. The authors suggest the need for dynamic rather than baseline measures to examine biological differences in PMS patients.

Hammarback et al. (1989) examined 18 PMS patients (no controls) diagnosed using daily ratings for two cycles, but control women were not included. Blood samples for E_2, P, FSH, and LH were taken daily in the luteal phase. Increased E_2 and P levels were associated with increased symptomatology. Increased FSH levels were inversely related to symptoms of breast swelling and tenderness. The authors suggest that the relationship between E_2, P, and FSH may be important in the production of PMS symptoms.

Watts et al. (1985) measured E_2, P, FSH, LH, cortisol, prolactin, thyroid-stimulating hormone (TSH), and testosterone in 35 PMS patients and controls diagnosed by daily prospective ratings for 2 months. The time of blood samples were taken between 08:30 and 17:00 h (the study did not control for possible circadian variation), weeks 1–4 of the menstrual cycle. Ovulation was determined by ultrasound. PMS patients were found to have earlier ovulation, possibly a longer luteal phase, and increased cortisol levels. There was some suggestion of a phase advance of the E_2 peak in PMS patients versus controls; however, the levels of these and other measured hormones were not different between the two groups.

Halbreich et al. (1986) examined the rate of change in gonadal hormones in relationship to PMS symptomatology. Seventeen patients with prospectively confirmed PMS from daily ratings using the premenstrual assessment form (PAF) and the SADS, had blood samples drawn between 8:00 and 10:00 a.m. every other day for E_2 and P levels. Clinical assessment was used in determining the most symptomatic versus the least symptomatic individuals. A faster rate of decline of P was associated with increased symptomatology with a time large of 4–7 days between hormonal decline and presentation of symptoms in PMS patients. Only three control subjects were studied.

Backstrom et al. (1985) studied seven PMS patients diagnosed using 1 month of prospective ratings and seven controls who had undergone hysterectomy. Their methods included surgical removal of the corpus luteum. The luteal phase of PMS patients was associated with decreased P and FSH levels and increased E_2 levels. The authors suggest that these findings may implicate increased inhibin levels in PMS. Many of the women had fibroids and other medical reasons for hysterectomy. Because of the difficulty of performing the techniques involved, the study has not been replicated.

Smith et al. (2003), examining the response to transcranial magnetic stimulation of the motor cortex in nine women with PMS and 14 control subjects, found that in the luteal, but not the follicular phase, PMS women showed relative facilitation, indicating an abnormal brain response to P, perhaps mediated by gamma aminobutyric acid (GABA).

Dysregulation of the neuroactive P metabolite, allogregnanolone, also has been reported in PMDD (Girdler et al., 2001). Allopregnanolone decreased in association with symptom improvement in women with severe PMS treated with a gonadotropin-releasing hormone agonist or placebo (Oberlander et al., 2006). Other neuroactive steroids also may be involved in menstrual cycle-related CNS disorders (N-Wihlback et al., 2006). These hormones are higher in the luteal phase in women with bipolar illness or major depressive disorder (Hardoy et al., 2006). The depressogenic effects of P administration may be mediated by the effect on allopregnanolone and GABA concentrations, which is modified by a history of depression (Klatzkin et al., 2006).

Summary. Although the studies are variable, the majority of well-controlled studies do not support the hypothesis that PMDD is associated with aberrations in E_2, P, FSH, or LH in PMS patients compared with asymptomatic control subjects. Studies that investigate GnRH agonists generally show an improvement in mood but variable results when E_2 and P necessarily are added back to prevent the effects of a chemical oophorectomy (menopause) on cardiovascular, reproductive, and skeletal systems

(Mortola et al., 1991; Schmidt et al., 1991; Brown et al., 1994; Mezrow et al., 1994). A more parsimonious interpretation suggested by these data, when taken in aggregate, is that women with PMDDs are more reactive or sensitive to customary ovarian sex steroid excursions. Although not studied in PMDD patients, waking and sleep episodes can affect LH secretion, an area which to date has not been systematically investigated (Hall et al., 2005).

22.3.2.2 Neurovegetative signs and psychophysiological responses

Both-Orthman et al. (1988) examined appetite changes in 21 PMS patients diagnosed by 3 months of self-ratings and 13 control subjects diagnosed using 2 months of ratings. Based on the premenstrual assessment form (PAF), in PMS patients, increased appetite correlated with depressed mood. These findings led the authors to suggest links between PMS and atypical depression and implicate the serotonergic system.

Parry et al. (2006) reviewed studies of sleep, biological rhythms and women's mood during the menstrual cycle. Although some studies suggested lighter, more disturbed sleep premenstrually and increased rapid eye movement (REM) and decreased REM latency in depressed women, the marked inconsistency in the results of these studies is likely the result of the variability of methods employed (e.g., lack of definitive diagnostic criteria, lack of hormonal measures to define menstrual cycle phase, small sample sizes, concomitant use of medications).

Mauri et al. (1988) examined self-reports of sleep in the premenstrual phase in 14 PMS patients diagnosed using prospective assessments and 26 control subjects. Their methods included using two retrospective questionnaires, the postsleep inventory (PSI) and the premenstrual tension syndrome (PMTS) form, by Steiner (a yes/no questionnaire). PMS patients reported increased sleep disturbances in the luteal phase. Sleep disturbances discriminated between PMS patients and control subjects with 82% accuracy. However, the questionnaires were based on subjective, retrospective reports.

Parry et al. (1989b) measured sleep electroencephalogram (EEG), temperature, and activity during the menstrual cycle in eight PMS patients and eight control subjects screened using 2 months of daily ratings and weekly Hamilton and Beck depression ratings. Sleep EEG recordings were made 2 times weekly for the duration of one menstrual cycle; activity was measured daily using a wrist actogram, and temperature was measured by means of a nocturnal indwelling rectal probe. PMS patients had more stage 2 and less REM sleep compared to control subjects. There were no significant differences between the groups with regard to daily activity measurements. PMS patients, however, had earlier nocturnal temperature minima compared to control subjects at all menstrual cycle phases. Both groups had increased awakenings in the late luteal phase. The sleep changes, though different in patients versus controls, did not parallel sleep changes characteristic of patients with major depressive disorders such as shortened REM latency or decreased delta sleep.

Baker et al. (2007) found that women with severe PMS reported a significantly poorer subjective sleep quality during the late luteal phase, but there was no evidence for polysomnographic sleep disturbances. Both women with PMS and normal controls had increased wakefulness after sleep onset and increased sigma power (14–15 Hz) during nonrapid eye movement sleep in the luteal compared with the follicular phase. Trait markers of decreased δ and increased θ incidence occurred in women with PMS irrespective of menstrual cycle phase.

In examining the role of naps in women with PMS versus controls, LaMarche et al. (2007) found that both groups of women had less slow-wave sleep and more lighter stages of sleep (stage 2) at night as well as higher daytime and nocturnal temperature during the late luteal phase. Women with symptoms were sleepier and less alert during the late luteal phase, but no significant group differences were found in nap or nocturnal sleep characteristics.

Van den Akker and Steptoe (1989) examined psychophysiological responses (heart rate, skin conductance, and electromyogram) in 16 women reporting severe premenstrual symptoms and in eight control women, but found no marked differences in resting autonomic activity.

Landen et al. (2004) examined heart-rate variability in PMDD and found that PMDD patients had reduced vagal tone compared to normal controls in the nonsymptomatic follicular phase, perhaps mediated by dopamine or other monoamines.

Summary. Although the studies need to be replicated in a larger number of patients, the initial findings support differences in neurovegetative signs and symptoms (sleep and appetite) in PMS patients versus controls during the menstrual cycle.

22.3.2.3 Neuroendocrine

Thyroid. In a very controversial study using retrospective questionnaires, Brayshaw and Brayshaw

(1987) identified 20 patients with PMS and 12 without PMS. The authors performed thyrotropin-releasing hormone (TRH) infusions and then treated symptomatic patients with thyroxine (Synthroid). They claimed that PMS patients showed increased TSH responses to TRH and that 100% of their PMS patients responded to thyroxine. The main criticism of the study is that PMS patients were not diagnosed using prospective ratings and that the subject population included patients with thyroid disorder, affective disorder, and anorexia, not PMS. Also, specific outcome measures were not delineated.

In contrast, Roy-Bryne et al. (1987) did not find group or follicular–luteal differences in TSH and prolactin levels after TRH infusion in 14 women with prospectively confirmed PMS and controls (documented by daily ratings and 30% change criteria).

Casper et al. (1989) also found no differences in TSH or prolactin response to TRH during either follicular or luteal phases in 15 PMS patients and 19 controls subjects that were selected using every third-day ratings for one (controls) to two (patients) menstrual cycles.

Parry et al. (1991) examined eight PMS patients who completed daily ratings for several cycles and found that there were normal TSH but enhanced prolactin responses to TRH administered in the follicular and luteal menstrual cycle phases when compared to published normal control values. In addition, in this study, cerebrospinal fluid (CSF) samples for 3-methoxy-4 hydroxyphenylglycol (MHPG; a metabolite of norepinephrine), homovanillic acid (HVA; a metabolite of dopamine), 5-hydroxyindoleacetic acid (5-HIAA; a metabolite of serotonin), GABA, β-endorphin, and PGs were obtained from PMS patients in an asymptomatic follicular and a symptomatic luteal menstrual cycle phase. There were significant increases in CSF MHPG in the premenstrual compared to the follicular phase. Follicular and luteal phase dexamethasone suppression tests (DST) were performed in subsequent months after initial circadian hormone profiles of cortisol were obtained. Baseline cortisol levels showed significant increases in the late follicular phase, probably an estrogen effect. Sixty-two percent of the patients showed nonsuppression to dexamethasone. This abnormality, however, occurred in both follicular and luteal menstrual cycle phases. In a study examining the effects of sleep deprivation, Parry et al. (1996) found that TSH circadian rhythms occurred earlier in PMDD compared to normal control subjects. Girdler et al. (2004) reported increased conversion of T_4 to T_3 and increased binding of thyroid hormones in women with PMDD with previous sexual abuse.

Stress hormones: cortisol, prolactin, and glucocorticoids. Other groups have examined cortisol differences in PMS patients and normal controls. Haskett et al. (1984) examined urinary free cortisol (UFC) and a dexamethasone suppression test (DST) in PMS patients. Forty-two PMS women were selected on the basis of self-report scales and clinical interviews obtained during follicular and luteal phases (no daily ratings; no controls). One milligram of dexamethasone was administered, and UFC (24 h) obtained on cycle day 9 and 26. There was no cortisol hypersecretion and there was normal 4 p.m. suppression of cortisol after DST. No changes in UFC between follicular and luteal phases were found. The authors suggest that PMS is not a model for endogenous depression.

Roy-Byrne et al. (1986) also examined the DST in 11 women with prospectively confirmed PMS (daily ratings for 2 months). No follicular–luteal differences in DST results were found in either PMS patients or controls.

Steiner et al. (1984) examined the circadian profile of prolactin, growth hormone, and cortisol in two women with PMS and in two controls assessed by A Moos Menstrual Distress Questionnaire – Today Form (MDQ-T). Blood samples were obtained at 30-min intervals for 24 h in follicular and luteal phases. There were increased prolactin levels in the luteal phase in both PMS patients and controls and normal growth hormone and cortisol concentrations. The small sample size limits the interpretation of the findings.

Parry et al. (1994) found that the peak of the cortisol circadian rhythm was delayed significantly in the late luteal compared to the mid-follicular phase in normal controls, but not in women with PMDD. Altered timing, but not amplitude disturbances also were observed in PMDD versus control women in a study of sleep deprivation effects (Parry et al., 2000). In these circadian rhythm studies, prolactin peak and amplitude were higher and acrophase earlier in PMDD compared with normal controls (Parry et al., 1994, 2000).

Bloch et al. (1998), however, found no differences between PMS and control groups in β-endorphin, adrenocorticotrophic hormone (ACTH) or cortisol. Rosenstein et al. (1996) also found no differences in arginine vasopressin (AVP) or ACTH in PMS patients versus controls, although AVP concentrations were lower throughout the menstrual cycle in symptomatic PMS patients compared to normal controls or PMS patients during asymptomatic cycles. Other investigators observed lower evening cortisol levels in the premenstrual phase in women who were significantly more depressed (Odber et al., 1998). In response to exercise during the luteal phase, women with PMS fail to show the normal increased hypothalamic–pituitary–adrenal (HPA)-axis response mediated by P, but do not display HPA-axis abnormalities characteristic of major depression (Roca et al., 2003). Stoddard et al. (2007) found that the beneficial effects of exercise on premenstrual distress were mediated by the effects on estrone glucuronide and pregnanediol glucuronide.

Reid et al. (1986) performed a 5-h oral glucose tolerance test (GTT) in six PMS women (assessed by Steiner questionnaire but not daily ratings) and five normal control subjects. Glucose tolerance did not differ between follicular and luteal phases nor between normal and PMS patients. Also, there were no differences in glucose, insulin, or glucagon responses to naloxone. Denicoff et al. (1990) also examined GTTs in 11 women with prospectively confirmed PMS during follicular and luteal phases. Although patients experienced hypoglycemic symptoms, they were not specific to the luteal phase and did not resemble their PMS symptoms.

Altemus et al. (1997) found that healthy women may have reduced glucocorticoid feedback regulation in the mid-luteal phase of the menstrual cycle resulting in higher levels of cortisol or enhanced activation of central stress-response systems. No significant differences were found between women with premenstrual mood disorders and normal women in the HPA-axis function at difference phases of the cycle. Women with PMDD had brief, transient, heightened cortisol responses to ovine (o) corticotropin-releasing hormone (CRH) stimulation, blunted cortisol responses to serotonergic agonists in the late luteal phase, and lower evening plasma cortisol levels. This finding suggests that physiological dysfunction between the HPA and the hypothalamic–pituitary–ovarian (HPO) axis for women with PMDD may be complex and include a variety of other systems in its interaction.

Melatonin. Parry et al. (1990) examined the melatonin circadian profile in eight PMS patients (documented by 2 months of daily ratings) and eight age-matched normal control subjects during the early follicular, late follicular, mid-luteal, and late luteal menstrual cycle phases. When compared to normal control subjects, PMS patients showed significantly lower levels of melatonin and a significant phase advance of the melatonin offset at all menstrual cycle phases. These findings, which suggest chronobiologic disturbances in PMDD, now have been replicated in a larger sample size (Parry et al., 1997b). In contrast, investigators examining 12–24 h samples of the urinary metabolite, 6-sulfatoxy melatonin, found no menstrual cycle phase or group differences in patients with PMS versus controls (McIntyre and Morse, 1990; Hamilton et al., 1988). It may be an altered relationship of melatonin to sleep and temperature rhythms (internal desynchronization) during the menstrual cycle that precipitates premenstrual symptoms (Shinohara et al., 2000).

Summary. There are no consistent findings with respect to thyroid, cortisol, prolactin, or glucose abnormalities in PMS patients versus controls. The differences in melatonin secretion between PMDD patients and controls are suggestive of differences in the circadian clock and parallel findings in serotonin metabolism (see below). As an aberrant brain response to P may be mediated by GABA, and serotonin and GABA are linked, the effects of serotonin on the clock may be mediated in part by P altering serotonin input to the clock via its effect on GABA.

22.3.2.4 Neurotransmitters: Serotonin, norepinephrine, and GABA

An increasing database links abnormalities of the serotonin neurotransmitter system to both PMDD and major depressive disorder, differentiating patients from healthy control subjects with regard to deficiencies in this system. Investigators have found that women with PMS have reduced platelet uptake of serotonin 1 week before menstruation, low whole-blood serotonin during the last 10 days of the menstrual cycle, and an abnormal response to tryptophan loading in the late luteal phase. Furthermore, selective serotonin reuptake inhibitors (e.g., fluoxetine, sertraline, or paroxetine) have been found to be efficacious in the treatment of this disorder. Recent work by Steiner et al. (1997) has shown that fluoxetine can be effective if administered only during the luteal phase; further work is necessary to corroborate their findings. This medication regimen may be preferable to some women who do not wish to take chronic medication for a periodic condition.

22.3.2.4(i) Baseline studies

Ashby et al. (1988) examined serotonergic mechanisms and monoamine oxidase (MAO) in PMS patients who were diagnosed with daily visual analog scales

(VASs) for anxiety for two cycles, the requirement that there be a 30% increase in symptoms premenstrually compared to postmenstrually (30% criteria) (Hamilton et al., 1984) and a Minnesota multiple personality inventory (MMPI). Blood samples were obtained premenstrually (days 1–9 before menses) and postmenstrually (days 5–9) for platelet uptake and content of 5-hydroxytryptamine (5-HT), MAO, and tryptophan. The uptake (V_{max}) and content of serotonin decreased in PMS patients premenstrually compared to controls. MAO decreased postmenstrually compared to premenstrually. There were no significant changes in tryptophan. This study examined PMS anxiety, and sample size was not described. As the authors discuss, their findings implicate changes in serotonergic function in PMS. In a follow-up report, Ashby et al. (1990) found that plasma obtained from PMS patients caused less stimulation of 5-HT uptake compared to plasma from the control group.

Taylor et al. (1984) also examined serotonin levels and platelet uptake in 16 PMS patients assessed by the Moos Menstrual Distress Questionnaire. Blood levels were drawn in pre- and postmenstrual phases. The uptake of serotonin was significantly lower during the premenstrual phase. There were no differences in affinity (K_m) values. The study used no controls, the screening of subjects was not described, and the cycle phase was not documented. The findings, however, are consistent with those of Ashby et al. (1988).

Rapkin et al. (1987) examined whole-blood serotonin in 14 PMS subjects and 13 age-matched controls selected by symptom diaries for 1 month, the profile of mood states (POMS), and the 30% criteria (Hamilton et al., 1984). Blood samples were obtained during the late luteal and premenstrual cycle phases. Serotonin levels in PMS patients were lower during the last 10 days of the cycle. The time of day when samples were collected was not specified, however, and diaries were obtained for only 1 month.

Malmgren et al. (1987) examined platelet serotonin uptake and pyridoxine (B_6) in 19 women with PMS and 19 age-matched controls who completed the Moos Menstrual Distress (MDQ) and Spielberger Anxiety questionnaires on cycle days 5–7 and 25–27. Blood sampling was done during pre- and postmenstrual phases. There were stable number (V_{max}) and affinity (K_m) K_m values at both menstrual cycle phases. While there were no group differences, lower V_{max} values occurred in spring. There was no effect of B_6. The limit of this study was that daily ratings were not obtained.

Rojansky et al. (1991) examined imipramine receptor binding and serotonin uptake in platelets of women with premenstrual changes (PMCs). Although the 18 subjects with PMCs showed large interindividual variability with no consistent pattern during the late symptomatic versus the early asymptomatic luteal phase, their imipramine receptor binding was lower compared to the nine control subjects prior to the development of symptoms in the luteal phase and was similar to the control subjects during the symptomatic phase. The findings implicate a preexistent vulnerability of impaired gonadal hormone modulation of the serotonergic system in this disorder.

Ashby et al. (1992) examined inhibition of serotonin uptake in rat brain synaptosomes by plasma from six women with and six women without PMS. Plasma obtained premenstrually from subjects with PMS inhibited synaptosomal uptake of 5-HT to a greater degree than plasma from the control group. In contrast to the control subjects in whom inhibition of uptake was significantly greater during the postmenstrual versus the premenstrual interval, there was no difference in the magnitude of the inhibition produced by plasma in the pre- and postmenstrual intervals in the PMS patients. The findings suggest that the plasma from PMS patients and controls contains endogenous factors that differentially inhibit synaptosomal 5-HT uptake.

In one of the few negative studies, Veeninga and Westenberg (1992) investigated in 38 women who met DSM-III-R criteria for LLPDD and 18 control subjects, the 5-HT uptake kinetics of platelets in the premenstrual (day 26) and postmenstrual (day 4) phase of the cycle. In addition, plasma samples for cortisol and β-endorphin were obtained before and after oral administration of 200 mg of 5-hydoxytryptophan (5-HTP) to LLPDD and control subjects in both menstrual cycle phases. Subjects with LLPDD did not exhibit menstrual cycle phase differences in 5-HT uptake and content. During the premenstrual phase, their results did not differ from those of normal control subjects. Group differences were not observed in neuroendocrine responses to 5-HTP stimulation in either the premenstrual or postmenstrual phase. The findings from this study do not support a specific role for 5-HT in the pathophysiology of LLPDD.

More recent studies continue to support the role of serotoninergic dysfunction in PMDD whether assessed by plasma 5-HIAA (Clayton et al., 2006), binding of [H]paroxetine to serotonin uptake sites (Bixo et al., 2001), changes in brain serotonin precursor trapping (Eriksson et al., 2006), or by the effects of a serotonin receptor antagonist, metergoline (Roca et al., 2002), although many of the studies continue to find the dysfunction a trait, rather than a state, marker. A study of the serotonin transporter, tryptophan hydoxylase, and MAO A gene polymorphisms found no association, however, between any genotype and PMDD versus control groups and no significant allelic distribution profiles in either clinical category (Magnay et al., 2006). Nonetheless, an association between the presence of PMDD, family history, and 5-HTTLPR (serotonin transporter promoter gene) long/short allele heteroozygozity was found in women with seasonal affective disorder (Praschak-Rieder et al., 2002).

22.3.2.4(ii) Challenge studies

Challenge studies have the potential to be more revealing. Bancroft et al. (1991) studied the prolactin and growth hormone responses to L-tryptophan infusions in 13 women with, and 13 women without, premenstrual depression. In depressed women, both responses were blunted pre- and postmenstrually. The prolactin response was blunted premenstrually in both groups. As other studies have found, the findings suggest that women who experience premenstrual depression may have neuroendocrine abnormalities throughout the menstrual cycle. The premenstrual phase, with its attendant neuroendocrine changes, represents a vulnerable time period for the expression of symptoms in predisposed women.

Rasgon et al. (2000) also examined neuroendocrine responses to an intravenous L-tryptophan challenge administered 2 times a week during 1 month to five subjects with prospectively documented PMS and five age- and body mass-matched subjects. Whole-blood serotonin response to the L-tryptophan challenge was blunted in the luteal phase of the menstrual cycle in subjects with PMS compared to controls. Cortisol, but not prolactin, levels were higher at baseline in the luteal phase in women with PMS, whereas neither postchallenge cortisol nor prolactin levels differed between groups. The results support previously reported findings in tryptophan handling in women with PMS.

In a subsequent study, Bancroft and Cook (1995) examined the prolactin and cortisol response to d-fenfluramine in 17 women with premenstrual depression and 14 controls. In contrast to their earlier findings with intravenous L-tryptophan challenge, the d-fenfluramine challenge failed to show any difference in neuroendocrine response between women with premenstrual depression and controls, suggesting that 5-HTP_2 receptor function is unaltered in this disorder.

In contrast, FitzGerald et al. (1997) administered the serotonin-releasing drug dl-fenfluramine versus placebo to nine women with PMDD and 11 healthy female volunteers in the luteal phase of the menstrual cycle. Compared with the normal subjects, the women with PMDD had a significantly blunted prolactin response to fenfluramine, suggesting that the disorder is associated with serotonergic deficiency.

Yatham et al. (1993) studied prolactin responses to buspirone challenges in seven women with LLPDD and in seven healthy controls. Women with LLPDD had blunted responses during the follicular phase, suggesting that 5-HT_{1A} receptor subsensitivity is a trait rather than a state marker for LLPDD.

FitzGerald et al. (1997) found blunted serotonin response to fenfluramine challenge in the luteal phase in nine women with PMDD versus 11 control women. Su et al. (1997) investigated the behavioral and endocrine responses to the serotonergic agonist m-chlorophenylpiperazine (m-CPP) in ten patients with prospectively documented PMS and in ten healthy controls. In PMS patients, m-CPP administration during the luteal phase resulted in improvement in PMS symptoms. Plasma cortisol and ACTH responses to m-CPP were blunted in both menstrual cycle phases in PMS patients compared to controls. The findings provide support for the acute efficacy of m-CPP in the treatment of PMS and additional evidence for dysregulation of serotonin control of the HPA axis in PMS. They do not provide support, however, for luteal-specific serotonergic dysfunction in this disorder, suggesting that the serotonin system is a modulating, rather than a causal factor in PMS.

22.3.2.4(iii) CSF studies

Eriksson et al. (1994) examined CSF levels of monoamine metabolites in 13 women with LLPDD and in 13 control subjects. Neither in the follicular phase nor in the luteal phase did the mean concentration of CSF monoamine metabolites (including levels of the serotonin metabolite 5-HIAA) differ from corresponding values in the control group. Neither in the LLPDD group nor in the control group did the values differ in the follicular versus the luteal phase. The intraindividual variations of dopamine and serotonin metabolites, however, were smaller in the LLPDD versus the control group and the ratio of dopamine to serotonin metabolites was lower in the

LLPDD group. These findings are in part similar to some of the findings by Parry et al. (1991) who also found no menstrual cycle-phase differences in CSF metabolites of serotonin (5-HIAA) in women with PMS, but altered ratios of serotonin and dopamine (HVA) metabolites.

22.3.2.4(iv) Summary

With the exception of Malmgren's study (see Malmgren et al. (1987)) in which daily ratings were not obtained, and the Veeninga and Westenberg 1992 study, most other baseline studies of serotonin in PMS patients versuscontrol subjects show a consistent decrease in the V_{max} or levels of serotonin premenstrually. Studies with larger sample sizes in well-diagnosed PMDD patients and controls are needed to replicate these findings, but the results to date show a consistent trend. The studies of CSF metabolites of serotonin (Eriksson et al., 1994; Parry et al., 1991), however, do not indicate altered serotonin function in PMDD subjects during the symptomatic luteal versus the asymptomatic follicular, menstrual cycle phase, nor significant differences between patient and control groups. Challenge studies tend to indicate that serotonergic dysfunction is more of a trait, rather than a state marker (Kouri and Halbreich, 1997) but may reflect diversified serotonergic systems that are selectively affected by fluctuations in gonadal hormones. Therapeutic ranges of gonadal hormones, serotonin and melatonin, in ratios with appropriate relative rates of change may be required to elicit functional behavioral responses. In further support of the role of serotonergic dysfunction in PMDD is that patients respond better to treatment with serotonergic (sertraline), rather than noradrenergic (desipramine) antidepressants (Freeman et al., 1999), although one study did report that higher α_2-adrenergic receptor density in the follicular phase predicted more severe luteal symptoms and correlated positively with symptom severity in PMDD patients (Gurguis et al., 1998).

Increasing evidence points to the role of the GABA system in PMDD. Epperson et al. (2002) examined nine women with PMDD and 14 healthy controls with serial proton magnetic resonance spectroscopic measures of occipital cortex GABA levels across the menstrual cycle. In contrast to the healthy controls in whom there was a reduction in GABA levels across the menstrual cycle, in women with PMDD, there was an increase in GABA levels from the follicular phase to the mid and late luteal phases and a reduction in follicular cortical GABA levels. Further support for the role of GABA in PMDD derives from studies of the benzodiazepine site antagonist flumazenil, which induced much greater panic response in women with PMDD than in comparison to subjects (Le Melledo et al., 2000).

22.3.2.5 β-Endorphin

Chuong et al. (1985) examined neuropeptide levels in 20 PMS patients and 20 controls. Patients completed the MDQ and daily diaries for 3 months while controls did so for 1 month. Blood samples were collected every 2–3 days for 1 month for β-endorphin. β-Endorphin levels were lower in PMS patients than in controls. In PMS patients, luteal levels were lower than follicular levels. There were no changes in neurotensin, human pancreatic peptide (HPP), vasointestinal peptide (VIP), gastrin, or bombesin. Although peripheral measures were taken and circadian effects were not assessed, the authors suggested that β-endorphin may be a state marker for PMS.

Facchinetti et al. (1987) also examined plasma β-endorphin in 11 PMS patients and 8 controls who completed the MDQ every 2 days. Blood samples were collected every 2–3 days for 1 month for β-endorphin and β-lipotropin hormone (β-LPH). PMS patients showed a decrease in β-endorphin premenstrually and during menses. There were normal follicular values. There were no changes in controls and there were no changes in β-LPH during the menstrual cycle. Although there were no daily ratings, the investigators did do prospective assessments using the MDQ every other day. The authors implicated the failure of central opioid tone premenstrually, although no assessment of central tone was made.

Tulenheimo et al. (1987) examined plasma β-endorphin immunoreactivity in 12 PMS patients and 14 controls based on daily records (0–3 severity). Morning blood samples were collected at mid- and late follicular, early and late luteal, and premenstrual cycle phases. No differences in E_2, P, LH, or cortisol were found between groups. β-Endorphin levels were lower in PMS patients versus controls in the luteal phase. There were no menstrual cycle phase differences. In this study, the investigators found similar trends to other studies of lowered β-endorphin of PMS patients versus controls in the luteal phase.

Giannini and Martin (1989) reported that of 53 women with LLPDD (DSM-III-R criteria), 21 had significant declines in serum β-endorphin on the 20th day of the cycle that was associated with increased anxiety, physical discomfort, decreased concentration, and increased caloric consumption. The limitations of this study, however, were that there were no control subjects, peripheral serum levels of β-endorphin do not reflect central opioid tone, the sampling interval was infrequent, and 32 women had no β-endorphin decline.

In contrast, Bloch et al. (1998) found no differences between β-endorphin, ACTH, or cortisol in ten PMS versus ten controls.

Summary. Although differences in assay sensitivity and circadian variability need to be assessed, the studies show a consistent trend of decreased β-endorphin in PMS patients versus controls in the luteal phase. The major limitation of these studies is that plasma β-endorphin is a peripheral measure. Measurement of β-endorphin in CSF in PMS patients did not decline premenstrually (Parry et al., 1991).

22.3.2.6 Other (PGs, CCK, alpha asymmetry, brain metabolic changes, acupuncture, vitamins, electrolytes, and CO_2 inhalation)

Jakubowicz et al. (1984) examined the use of mefenamic acid, a PG synthetase inhibitor, in PMS patients. Eighty patients were treated with mefenamic acid, 500mg 3times daily. In 19 patients, PGs were measured for three cycles. Subjects were selected using a daily symptoms checklist for one cycle. Blood samples were obtained every 3 days. Although 86% of patients improved with mefenamic acid versus placebo, there were no changes in PGs during the menstrual cycle and PGs were lower in patients versus controls.

The role of cholecystokinin (CCK) was reported by Le Melledo et al. (1999) who found that 18 women with PMDD compared to 21 normal controls showed a greater anxiety and panic response to CCK-4. In studies of frontal α-symmetry, Baehr et al. (2004) found that in five women with PMDD, but not in five normal controls, asymmetry scores fell into the negative range during the luteal phase. Menstrual cycle-related brain metabolic changes were found using H magnetic resonance spectroscopy in a pilot study of five women with PMDD versus six control subjects (Rasgon et al., 2001). The beneficial effects of acupuncture in treating 77.8% of women with PMS, as compared to 5.9% treated with placebo, may be attributed to its effects on serotonergic and opioidergic neurotransmission (Habek et al., 2002).

Mira et al. (1988) examined vitamins and trace elements in PMS. Thirty-eight patients with PMS and 23 controls completed prospective symptom reports for three cycles. Samples were collected during midfollicular and premenstrual cycle phases for magnesium, zinc, and vitamins A, E, and B_6. No differences between groups were found during the cycle for any of the nutritional parameters.

A more recent and larger study reported the efficacy of calcium supplements (TUMS) in prospectively diagnosed patients with PMDD (Thys-Jacobs et al., 1998). These findings may reflect that calcium is an important cofactor for neural transmission.

Varma (1984) examined hormones and electrolytes in 25 PMS patients and ten controls selected by daily visual analog scales (VASs). Blood samples were obtained on days 3, 7, 11, 15, 19, 24, and 27 of each cycle. No differences in sodium or potassium were found between PMS and control subjects and there were no menstrual cycle-phase differences. Although there was a slight increase in cortisol in the luteal phase in the most severely affected women with PMS, levels were still in the normal range. No group differences were found for prolactin, FSH, LH, E_2, or P, although a slight increase in the E_2:P ratio was noted in the PMS patients in the luteal phase.

Women with LLPDD have been found to be more sensitive to the anxiolytic properties of carbon dioxide (CO_2) inhalation (doublebreath or rebreathing) as well as lactate infusion than have symptomatic controls (Harrison et al., 1989b). None of the control women developed intense anxiety or panic attacks, while over half of the LLPDD women did so. These findings suggest that patients with LLPDD and anxiety disorders may have a shared vulnerability.

Summary and interpretation. The studies do not support PG, nutritional (vitamin), or electrolyte disturbances in PMS patients. The work of CO_2 inhalation suggests biological differences between patients with LLPDD and normal controls and perhaps a shared vulnerability of patients with LLPDD and those with anxiety or panic disorders.

22.3.3 Chronobiological Hypotheses

Besides antidepressant medication, sleep deprivation and exposure to bright light at critical times of the day (light therapy) have been found to be beneficial in reducing symptoms. The efficacy of these types of treatment is hypothesized to involve chronobiological mechanisms. Patients with major depressive disorders may have disruptions in the internal regulation of biological rhythms, a phenomenon termed internal desynchronization. In this condition, circadian output rhythms (such as REM sleep propensity, temperature, cortisol, and melatonin) are not in phase with the external light–dark cycle and not in phase with each other. The oscillators regulating these circadian rhythms may be internally desynchronized with respect to the sleep–wake cycle and externally desynchronized with respect to the light/dark cycle. External cues, or zeitgebers, such as bright light, play an important role in the modulation and synchronization of these circadian rhythms. Light (200–500 lux) has the capacity to acutely suppress melatonin secretion, and to shift (>200 lux) circadian rhythms. One of the best markers for the central clock in humans (when light conditions are controlled) is circulating melatonin concentrations. Reports by Lewy et al. (1981) and Nurnberger et al. (1988) have shown that affectively ill patients are supersensitive to the suppressive effects of light on melatonin secretion. Other, but not all, investigators have observed a decreased amplitude of melatonin secretion and higher nocturnal core body temperatures (Beck-Friis and Wetterberg, 1984; Claustrat et al., 1984; Brown et al., 1987; Mendlewicz et al., 1980; Thompson et al., 1988; Rubin et al., 1992; Avery, 1987). One hypothesis suggests that depressed patients may receive inadequate exposure to daytime light.

There may be some differences between men and women in the regulation or expression of circadian rhythms. As compared to men, menstruating women have a shorter free-running period (i.e., length of time of a rhythm in an environment free of external cues, such as light), longer sleep duration, and lower amplitude in body temperature that varies with phases of the menstrual cycle (Wever, 1988). These findings suggest that reproductive hormones (including testosterone) have an effect on circadian rhythm amplitude and synchronization. Indeed, comparative studies in rodents have shown that estrogen advances and P delays circadian rhythms (Albers et al., 1981a,b; Morin et al., 1977). Estrogen also appears to enhance synchronization between circadian oscillators (Thomas and Armstrong, 1989) and affect entrainment pathways to light (Davis et al., 1983).

Thus, the fluctuation of reproductive hormones during the menstrual cycle may alter circadian rhythmicity. An instability of circadian rhythms that may result from changing reproductive hormonal levels may then put some women at increased risk for depression. Chronobiologic hypotheses contend that disrupted circadian rhythmicity, mimicking the malaise of jet lag, may be one factor contributing to depression. Women with PMDD have been reported to have decreased melatonin amplitudes (Parry et al., 1990, 1997a), high nocturnal body temperatures (Parry et al., 1997b; Severino et al., 1991), and disturbances in prolactin, TSH, and cortisol circadian rhythms (Parry et al., 1994, 1996). Also, the onset and duration of the melatonin rhythm may be disturbed, possibly due to an abnormal response to light (Parry et al., 1997c). Light therapy and sleep deprivation (Parry and Wehr, 1987; Parry et al., 1989a, 1993, 1995, 1997a) may be beneficial for women with PMDD because these treatments may help to realign the underlying circadian clocks.

Since the menstrual cycle is an inherent biological rhythm, our laboratory has been testing the hypothesis that disturbances in the timing or regulation of this infradian rhythm may predispose women to mood disturbances.

The phase-advance hypothesis of affective disorder postulates that the central neural oscillator regulating the circadian rhythms of REM sleep, temperature, cortisol, and melatonin is shifted earlier (phase-advanced) with respect to the sleep–wake cycle (Kripke et al., 1978; Wehr and Goodwin, 1980). Our preliminary (Parry et al., 1990) and now replicated work (Parry et al., 1997a) suggests that when compared to normal control subjects, women with PMDD have significantly lower integrated melatonin concentrations and an earlier offset time of melatonin secretion. If melatonin is a marker for the phase and amplitude of central circadian oscillators, then our data might suggest that these circadian oscillators are not only advanced, but also dampened, at all menstrual cycle phases in women with this disorder.

The amplitude of a circadian rhythm may reflect the relative strength of the underlying pacemaker. Aschoff stated over a decade ago that the stability of a circadian system is positively correlated with its amplitude; circadian systems with labile phases can be expected to have smaller amplitudes. If phase lability is symptomatic of a certain illness, then measures taken to improve amplitude are likely to be beneficial (Aschoff, 1983). The corollary is also likely to be true; the circadian system of individuals with low amplitude rhythms may be expected to be easier to manipulate. This phenomenon may be reflected in the adaptability to shift work; individuals with a low-amplitude core temperature rhythm adapt best

(Reinberg et al., 1984). Also, in PMDD patients, lower-amplitude rhythms may be manipulated or buttressed with bright light with therapeutic benefits (Parry et al. 1989a, 1993). The amplitude of a circadian rhythm may be its most relevant parameter because it determines internal stability (Wever, 1988). The larger the amplitude of the rhythm, the smaller the cycle-to-cycle variability, and the more resistant is the rhythm to hormonal or environmental perturbations, including those of the menstrual or light–dark cycle.

If phase lability is symptomatic for certain psychiatric illness (Aschoff, 1983), then it is likely that measures that improve the amplitude would be beneficial. In PMDD patients, the low amplitude of melatonin secretion, which suggests dampened central circadian oscillators, may predispose to circadian desynchrony and thus possibly to mood shifts. Rapid shifts in mood are considered one of the characteristic features of PMDD (1994). Similar predispositions to circadian phase lability and mood disturbances may occur in elderly depressed patients who also exhibit low amplitudes of circadian oscillators as measured by melatonin (Sack et al., 1986). Indeed, PMDD symptoms tend to increase with increasing age (Golub, 1988). A reduction in the amplitude of melatonin and other circadian rhythms was reported in one study to be the most relevant circadian abnormality observed in depressed patients (Souetre et al., 1989).

As Souetre et al. (1989) suggested, a reduction in the nocturnal rise of melatonin may alter the coupling processes between the pineal gland and other endocrine functions. As a result, a clear neuroendocrine signal may not be provided to peripheral clocks. Thus, melatonin may serve to entrain internal circadian and neuroendocrine systems and their relationship to the temporal structure of the external environment. If depression is conceptualized as a form of weakened or altered entrainment, then melatonin may serve to restore these normal coupling processes. Manipulating other major synchronizers or *zeitgebers*, such as sleep or light, may be beneficial for depression by helping to reinforce endogenous coupling processes.

In summary, preliminary data demonstrate that women with PMDD manifest chronobiological abnormalities of melatonin secretion. The fact that the symptoms of such patients respond to specific treatments that affect circadian physiology, such as sleep deprivation (Parry and Wehr, 1987; Parry et al., 1995) and bright light (Parry et al., 1989a, 1993), suggests that circadian system abnormalities may contribute to the pathogenesis of PMDD, and that correcting such disturbances may result in clinical remission. Further studies examining the interrelationship of circadian and neuroendocrine systems in menstrually related mood disorders would help elucidate the psychobiological relationships comprising this multifaceted disorder.

22.3.4 Summary

Research efforts aimed at elucidating the pathophysiology of PMDD have attempted to identify disturbances of ovarian and other peripheral glandular secretion. To date, there are few data to support the notion that PMDD is due to abnormalities of ovarian hormone secretion. The alternative hypothesis, that PMDD results from the impact of ovarian cyclicity upon vulnerable central processes underlying mood and behavior, is supported by the finding that (1) there are disturbances in central neuroregulation as evidenced by phase advances in temperature and melatonin secretion; (2) oophorectomy (both surgical and medical) interrupts the cyclic symptomatology; (3) gonadal steroidal replacement doses that approximate ovarian secretion provokes affective symptomatology in predisposed individuals; and (4) women with PMDD appear to be at high risk for the development of major mood disorders. If the primary disturbance underlying PMDD is central, then interventions directed at central processes, such as the interaction of sex steroids with GABA, may be more therapeutic.

22.3.5 Emergence of a Biopsychosocial Model

Since the relatively recent invention of radioimmunoassay and other more sensitive techniques for assessing ligands of interest, biomedical research has clearly dominated the area of PMS. Because the symptoms of PMS are associated with hormonal fluctuations during the menstrual cycle, cultural and psychological phenomena associated with the disorder have often been ignored in scientific research, which has resulted in a poor integration of the findings. In the field of depression and other affective disorders, multidimensional theoretical models have emerged and contributed greatly to understanding the experience of depression for an individual because experience alters the underlying neural substrates. In addition, exploring the interaction of biological, psychological, and social processes has expanded the range of treatment options and provided efficacious alternatives to medication to people who suffer from depression. Consistently, psychotherapeutic and pharmacological combined treatment strategies offer more benefit than either treatment intervention alone, especially in the psychosomatic and somato-psychic disorders. For similar reasons, an integrative approach could greatly benefit research and clinical options in the area of premenstrual mood disturbances.

In the area of affective illness, a social *zeitgeber* (challenge or stressor) theory of mood disturbance has been one unifying hypothesis linking biological and psychosocial models (Ehlers et al., 1988). Social *zeitgebers* are personal relationships, social demands, or tasks that entrain biological rhythms. Similar to the effect of light, social events are external regulators that synchronize circadian rhythms. For instance, marriage, birth of a child, divorce, loss of a job, etc., are social events that disrupt natural mealtimes, sleeping times, and times of activity, which affect the stability of biological rhythms. In individuals predisposed to affective disruption, disturbances in the biological clock may develop into a state of ongoing desynchronization as observed in major depression. Social rhythm disruptions also have been implicated in the onset of manic episodes. Ehlers and her colleagues suggested that both interpersonal psychotherapy and cognitive behavior therapy, two empirically supported treatments for depression, address the regularity of social routine. In addition, they state that the model is likely to be influenced by personality factors, gender, social support, coping, genetic/familial loading, and past treatment experience. A tenet of this conceptualization is that social behaviors alter biological outputs.

Adoption of the social *zeitgeber* model for premenstrual mood disturbances may potentially elucidate the interaction of important psychosocial and biological etiological variables. Socially, women may be particularly prone to respond more dramatically to interpersonal conflicts and stresses because they may rely on social *zeitgebers*, or social relationships, more than men. Carol Gilligan (1982) proposed that women may view relationships in terms of interconnections and caring, and men in terms of hierarchy and power. Developmental research has documented that girls may be more likely to engage in prosocial behavior than boys. On self-report measures, women are often more empathetic and nurturing than men. The emphasis that women may place on connection with others possibly serves as a vulnerability to a mood disturbance; interpersonal disruptions may have a greater impact on women as compared to men. At the same time, this increased need for social support may be more adaptive in the long term; women may cope better than men to stress, such as to a loss or a death. Therefore, the increased variation in mood may be more adaptive by enhancing the mechanism of homeostasis, thus protecting women against other types of long-standing illness and increasing longevity.

Entrainment of social *zeitgebers* may potentially serve as a mechanism for the treatment of PMDD or other mood disorders. Preliminary findings from Frank and her colleagues have indicated that interpersonal psychotherapy may be more efficacious for women than cognitive behavior therapy. Animal and human studies have also shown that responses to social interactions can lead to fluctuations in whole-blood serotonin levels. Therefore, women may be able to regulate serotonin function by seeking appropriately rewarding social interactions; the inability to seek or receive the social interactions could lead to further physiologic dysregulation via circadian rhythm disruption and/or reduction of serotonin, manifesting as symptoms of PMDD. These correlations between behavior and neuroendocrinology, in which behavioral phenotypes are a clue to underlying neurobiology, have been explored within a framework of functional hypothalamic amenorrhea. For some time, researchers have noted that psychosocial variables, such as exercise, personality traits, and environmental stress, have the potential to induce ovarian acyclicity by inhibiting the release of GnRH through mediating variables, such as CRH, endogenous opioids, dopamine, and TSH. Further work in this area could potentially elucidate important somato-psychic interactions, which are key to understanding the etiology of PMDD.

22.4 Treatment

PMDD has a range of psychiatric and physical symptoms. Most women who seek psychiatric help for this disorder present with symptoms of premenstrual depression, anxiety, and/or irritability. Although until recently no consistently safe and effective treatments were available, a number of treatment strategies currently exist that target these symptoms and appear beneficial in treating them (Altshuler et al., 1995). The SSRIs, fluoxetine and sertraline, have been approved by the Food and Drug Administration (FDA) for the treatment of PMDD.

A careful medical history will reveal the presence of any physical conditions. If a physical disorder (e.g., hypothyroidism) exists, it should be treated first to determine whether alleviation of the medical problem also relieves the premenstrual symptoms. The clinician should obtain a careful psychiatric history, as well as prospective ratings of a patient's symptoms for 2–3 months. Such assessments can help rule out premenstrual exacerbation of another psychiatric condition such as a major depressive disorder that may be confused with PMDD but requires a different treatment approach. Prospective ratings allow easy comparison of symptom severity between the week prior to menses and the week after menses. They can be made with a simple 10-mm VAS that the patient completes at approximately the same time each day for each symptom present (Aitken, 1969). To meet criteria, symptoms should increase (at least 30%) in the luteal phase and remit in the follicular phase.

If the patient meets PMDD criteria after two consecutive cycles, she should receive supportive counseling in conjunction with other treatment strategies. Helping a woman to see that the timing of her symptoms is predictable can help her plan ways to reduce stress in her work or personal life at the appropriate times. This approach leads to a decreased sense of vulnerability and an increased sense of control over her life. Bringing in spouses or partners, explaining the condition, validating the symptoms, and helping the couple to adjust their lifestyle to reduce stress during premenstrual times can be very helpful.

If depressive symptoms are mild, vitamins (B_6, A, and E), minerals (calcium, magnesium, and zinc), evening primrose oil, or a diuretic may be tried (Altshuler et al., 1995). The mechanisms of actions of these treatments are thought to exert their effects by acting as important coenzymes in neurotransmitter pathways such as norepinephrine or serotonin, by affecting PG or fatty acid metabolism, or by altering the mineralcorticoid system, respectively. Support for their efficacy is more than anecdotal, but studies are often inconsistent (see, for more extensive review, Altshuler et al. (1995)). Patients often prefer these agents over psychotropic strategies, since they need be taken only for part of the cycle and do not carry the stigma of psychotropic medication. Vitamin B_6 should be given in doses of 25 mg to 100 mg day^{-1} (starting with 25–50 mg day^{-1}) from midcycle to the onset of menses over two or three cycles. The daily dose should not exceed 100 mg day^{-1} to avoid potential development of peripheral neuropathy. Optivite, calcium, or magnesium have had positive effects on mood, anxiety, and irritability in at least half of the very few double-blind studies that have been reported (Altshuler et al., 1995), although a recent study found no evidence of magnesium deficiency, and magnesium treatment was not superior to placebo in the mitigation of mood symptoms in women with PMDD (Khine et al., 2006). Diuretics should be tried if a woman has significant weight gain premenstrually along with psychiatric symptoms. A starting dose of one tablet of the potassium-sparing combination diuretic (hydrochlorothiazide 25 mg/triamterene 50 mg) or its equivalent from the onset of symptoms to the onset of menses is usually effective. Even the potassium-sparing diuretics may cause potassium loss, so electrolytes must be monitored. Bromocriptine, a dopamine agonist that lowers prolactin, a hormone that is associated with mastodynia, in dosages of 1.25–7.5 mg day^{-1} in divided doses, may be an alternative if breast pain is a significant concomitant symptom. In a patient who could not tolerate the side effects of SSRIs, St. John's Wort (Hypericum Perforatum, 900 mg day^{-1}) was beneficial (Huang and Tsai, 2003).

If symptoms of depression are moderate to severe, the physician should consider instituting treatment with a psychotropic drug or a hormone. Most trials have assessed patients only after two or three cycles of treatment and do not comment on response during the first cycle. On average, studies have noted improvement after 2–4 months of treatment. For a patient a trial should occur over two or three menstrual cycles.

For antidepressants, the SSRIs such as fluoxetine (Menkes et al., 1992; Steiner et al., 1995; Stone et al., 1991; Wood et al., 1992), sertraline (Yonkers et al., 1997), or clomipramine (Sundblad et al., 1992, 1993) appear to be highly effective. Nortriptyline in dosages of 50–125 mg to achieve therapeutic levels may also be beneficial (Harrison et al., 1989a). Fluoxetine in doses of 20–40 mg day^{-1} (started at 20 mg day^{-1} for two or three cycles and then increased) may be given continuously throughout the cycle. Doses of sertraline of 50–100 mg are most effective (Yonkers et al., 1997). Some studies, particularly with the SSRIs such as fluoxetine which has a long half-life, have shown that luteal phase administration only is effective in relieving symptoms and minimizing side effects (Steiner et al., 1997; Freeman, 2004). Intermittent luteal phase dosing of sertraline also has been found to be effective and well tolerated (Halbreich et al., 2002), but postmenstrual symptoms limit response (Freeman et al., 2004a). Low doses (25–50 mg day^{-1}) can produce significant improvement in mood over two menstrual cycles (Kornstein et al., 2006). When the medication is discontinued, however, there is rapid recurrence in moderate to severe symptoms in up to 77% of women (Freeman et al., 2004c). Continuous and intermittent doses of paroxetine controlled release (12.5 and 25 mg daily) also have been shown to be effective and well tolerated in PMDD (Cohen et al., 2004b; Steiner et al., 2005; Landen et al., 2007). Irritability, affect lability, and mood swings respond more rapidly to the SSRIs, allowing for intermittent treatment, than do somatic symptoms (Landen et al., 2007). The differential time course efficacy on dysphoric and physical symptoms of the intermittent dosing of fluoxetine in PMDD may be attributed in part to synaptic effects, effects on neuroactive steroids or mediation through spinal–thalamic pathways (Tamayo et al., 2004). Treatment can improve work capacity within a month (Steiner et al., 2003). Weekly dosing of enteric-coated 90 mg fluoxetine given once or twice a cycle also may provide safe and effective treatment of PMDD (Miner et al., 2002). Women with severe PMDD may respond better to luteal phase dosing than symptom-onset dosing of escitalopram (Freeman et al., 2005). Continuous and intermittent dosing of venlafaxine has been reported to be efficacious and well tolerated, although withdrawal reactions have been associated with low-dose treatment (Freeman et al., 2001b; Cohen et al., 2004a; Hsiao and Liu, 2004). Clomipramine may be effective in doses of 25–75 mg day^{-1} from midcycle to the onset of menses or throughout the cycle.

Hormonal options include GnRH agonists, danazol and E_2, but regular use of these medications cannot be recommended until more is known about their safety in long-term use. GnRH agonists are still experimental and are not yet approved for clinical use in this context. They can induce a menopausal state and the associated hypoestrogenism predisposes to osteoporosis and heart disease. They potentially could be utilized for short-term stabilization or to help make diagnoses, although better methods are available. Depressive mood symptoms increase in women treated with GnRH agonist therapy for endometriosis, thought to be mediated by the decline in estrogen levels. Symptoms respond to sertraline treatment (Warnock et al., 1998). In studies of using GnRH agonists to treat PMS, the addition of hormonal add-back therapy does not necessarily reduce the efficacy of the GnRH agonist alone (Wyatt et al., 2004) and may prevent bone-loss (Mitwally et al., 2002). In women who have not had a hysterectomy, estrogen should be administered with progestogens to reduce the risk of endometrial hyperplasia. Premenstrual symptoms may recur, however, in association with progestogen use (Watson et al., 1989). Additional studies are necessary to assess the extent to which progestogens counteract estrogen's beneficial effects. Fortunately, another alternative exists to oral or topical progestagens, a progestin-containing IUD, which can be employed to protect against endometrial hyperplasia. The serum levels of progestin are negligible in women using a progestin IUD. The adverse effects of estrogen include nausea, weight gain, breast tenderness, and headache. High estrogen levels may provoke symptoms in women with PMDD (Schmidt et al., 1991). Those of danazol include hirsutism, acne, weight gain, and nausea. These strategies often are poorly tolerated. P does not appear to have a beneficial effect over placebo in most studies, and it may exacerbate depressive symptoms in women with a history of depression. Oral contraceptives have not been adequately studied for PMDD, but they, too, may exacerbate depressive symptoms in women with a history of depression and should be used cautiously, if at all, for such patients (Parry and Rush, 1979). Newer agents, such as a combination of drosperinone and ethinyl estradiol, have been reported to be efficacious (Freeman et al., 2001a; Pearlstein et al., 2005; Yonkers et al., 2005; Sangthawan and Taneepanichskul, 2005). The beneficial effects of oral contraceptives may be mediated by effects on neurosteroids (Kurshan and Neill Epperson, 2006). Clinicians should discuss both psychotropic and hormonal options with patients. If a patient prefers to try a hormonal strategy prior to an antidepressant trial, a specialist should be consulted.

If anxiety is a prominent symptom, with dysphoria occurring secondarily, using the anxiolytics alprazolam or buspirone may be worthwhile. Alprazolam, unlike nortryptyline or fluoxetine, which for the most part have been prescribed throughout the cycle, can be given during days 12–28 of the cycle. Dosage should start at 0.25 mg day^{-1} and be increased as necessary (in divided daily doses) to relieve symptoms. The total dose should not exceed 4 mg day^{-1}. With the onset of menses, the dosage should be tapered by 25% per day (Harrison et al., 1990). Although measures of abuse liability were not increased following the acute administration of alprazolam for PMDD, Evans et al. (1998) found it did not improve negative premenstrual mood. For patients who cannot tolerate the abrupt tapering, a longer-acting benzodiazepine such as clonazepam may be helpful, although no studies have ascertained its efficacy in treating PMDD. Benzodiazepines used during symptomatic days may also reduce symptoms of anxiety, however, they may exacerbate depression in some patients. Because patients may have difficulties with withdrawal symptoms or tolerance, benzodiazepines should be considered with caution. In women with a history of drug abuse or dependence, these agents should be prescribed only after carefully weighing other options. Buspirone can be used throughout the cycle, or from midcycle to the onset of menses, with a starting dose of 5 mg orally 3 times per day (Rickels et al., 1989).

For patients with recurrent suicidal depression, lithium should be considered. Lithium, taken from midcycle to the onset of menses, has been studied as a treatment option for PMDD. Although several case reports suggest the efficacy of lithium for relieving tension, irritability, insomnia, restlessness, depression, and edema, three controlled studies (Mattsson and von Schoultz, 1974; Singer et al., 1974; Steiner et al., 1980) which included two double-blind studies, one prospectively, one retrospectively diagnosed, have produced negative results. Lithium may target suicidal symptom through its effect on the protein kinase C, phosphotidyl inositol pathway, just as tamoxifen, an antagonist to this pathway, may induce suicidal symptoms in predisposed women.

The use of alternative therapies remains to be systematically investigated (Domoney et al., 2003).

22.4.1 The Future

More recently, there appears to be a paradigm shift occurring in the way psychiatric syndromes are approached. Several recent trials have suggested that a combination of therapies may yield better acute and chronic results that either a single agent or behavioral intervention alone. In some cases, there has been synergism between behavioral strategies and psychotropic agents. A combination of therapies may also prove best for women with PMDD, although investigators have yet to determine which combinations offer the most promise. With cessation of treatment, most patients relapse within a few months. Further work is needed.

References

Aitken RC (1969) Measurement of feelings using visual analogue scales. *Proceedings of the Royal Society of Medicine* 62: 989–993.

Albers HE, Gerall AA, and Axelson JF (1981a) Effect of reproductive state on circadian periodicity in the rat. *Physiology and Behavior* 26: 21–25.

Albers HE, Gerall AA, and Axelson JF (1981b) Effect of reproductive state on circadian system of the rat. *Annual Review of Pharmacology and Toxicology* 22: 555–598.

Altemus M, Redwine L, Leong YM, Yoshikawa T, Yehuda R, Detera-Wadleigh S, and Murphy DL (1997) Reduced sensitivity to glucocorticoid feedback and reduced glucocorticoid receptor mRNA expression in the luteal phase of the menstrual cycle. *Neuropsychopharmacology* 17: 100–109.

Altshuler LL, Hendrick V, and Parry B (1995) Pharmacological management of premenstrual disorder. *Harvard Review of Psychiatry* 2: 233–245.

American Psychiatric Association (1994) *Diagnostic and Statistical Manual of Mental Disorders,* 4th edn. Washington, DC: American Psychiatric Association.

Angst J (1978) The course of affective disorders. II. Typology of bipolar manic-depressive illness. *Archiv für Psychiatrie und Nervenkrankheiten* 226: 65–73.

Aschoff J (1983) Disorders of the circadian system as discussed in psychiatric research. In: Wehr TA and Goodwin FK (eds.) *Circadian Rhythms in Psychiatry*, pp. 33–39. Pacific Grove, CA: Boxwood Press.

Ashby CR, Jr., Carr LA, Cook CL, Steptoe MM, and Franks DD (1988) Alteration of platelet serotonergic mechanisms and monoamine oxidase activity in premenstrual syndrome. *Biological Psychiatry* 24: 225–233.

Ashby CR, Jr., Carr LA, Cook CL, Steptoe MM, and Franks DD (1990) Alteration of 5-HT uptake by plasma fractions in the premenstrual syndrome. *Journal of Neural Transmission. General Section* 79: 41–50.

Ashby CR, Jr., Carr LA, Cook CL, Steptoe MM, and Franks DD (1992) Inhibition of serotonin uptake in rat brain synaptosomes by plasma from patients with premenstrual syndrome. *Biological Psychiatry* 31: 1169–1171.

Avery DH (1987) REM sleep and temperature regulation in affective disorder. In: Halaris A (ed.) *Chronobiology and Psychiatric Disorders*, pp. 75–101. Elsevier: New York.

Backstrom T, Andreen L, Birzniece V, et al. (2003) The role of hormones and hormonal treatments in premenstrual syndrome. *CNS Drugs* 17: 325–342.

Backstrom T, Smith S, Lothian H, and Baird DT (1985) Prolonged follicular phase and depressed gonadotrophins following hysterectomy and corpus lute-ectomy in women with premenstrual tension syndrome. *Clinical Endocrinology (Oxf)* 22: 723–732.

Baehr E, Rosenfeld P, Miller L, and Baehr R (2004) Premenstrual dysphoric disorder and changes in frontal alpha asymmetry. *International Journal of Psychophysiology* 52: 159–167.

Baker FC, Kahan TL, Trinder J, and Colrain IM (2007) Sleep quality and the seep electroencephalogram in women with severe premenstrual syndrome. *Sleep* 30: 1283–1291.

Bancroft J and Cook A (1995) The neuroendocrine response to d-fenfluramine in women with premenstrual depression. *Journal of Affective Disorders* 36: 57–64.

Bancroft J, Cook A, Davidson D, Bennie J, and Goodwin G (1991) Blunting of neuroendocrine responses to infusion of L-tryptophan in women with perimenstrual mood change. *Psychological Medicine* 21: 305–312.

Banerjee N, Roy KK, and Takkar D (2000) Premenstrual dysphoric disorder – a study from India. *International Journal of Fertility and Women's Medicine* 45: 342–344.

Becker D, Orr A, Weizman A, Kotler M, and Pines A (2007) Depressed mood through women's reproductive cycle: Correlation to mood at menopause. *Climacteric* 10: 46–50.

Beck-Friis J and Wetterberg L (1984) A possible low melatonin syndrome in depressed patients. *Annals of Clinical Research* 11: 163–170.

Bixo M, Allard P, Backstrom T, Mjorndal T, Nyberg S, Spigset O, and Sundstrom-Poromaa I (2001) Binding of [3H]paroxetine to serotonin uptake sites and of [3H]lysergic acid diethylamide to 5-HT2A receptors in platelets from women with premenstrual dysphoric disorder during gonadotropin releasing hormone treatment. *Psychoneuroendocrinology* 26: 551–564.

Bloch M, Schmidt PJ, Su TP, Tobin MB, and Rubinow DR (1998) Pituitary–adrenal hormones and testosterone across the menstrual cycle in women with premenstrual syndrome and controls. *Biological Psychiatry* 43: 897–903.

Both-Orthman B, Rubinow DR, Hoban MC, Malley J, and Grover GN (1988) Menstrual cycle phase-related changes in appetite in patients with premenstrual syndrome and in control subjects. *American Journal of Psychiatry* 145: 628–631.

Brayshaw ND and Brayshaw DD (1987) Premenstrual syndrome and thyroid dysfunction. *Integrated Psychiatry* 5: 179–193.

Brockington IF, Kelly A, Hall P, and Deakin W (1988) Premenstrual relapse of puerperal psychosis. *Journal of Affective Disorders* 14: 287–292.

Brockington IF and Kumar R (1982) *Motherhood and Mental Illness.* London: Academic Press.

Brown CS, Ling FW, Andersen RN, Farmer RG, and Arheart KL (1994) Efficacy of depot leuprolide in premenstrual syndrome: Effect of symptom severity and type in a controlled trial. *Obstetrics and Gynecology* 84: 779–786.

Brown RP, Kocsis JH, Caroff S, Amsterdam J, Winokur A, Stokes P, and Frazer A (1987) Depressed mood and reality disturbance correlate with decreased nocturnal melatonin in depressed patients. *Acta Psychiatrica Scandinavica* 76: 272–275.

Brugger P, Milicevic A, Regard M, and Cook ND (1993) Random-number generation and the menstrual cycle: Preliminary evidence for a premenstrual alteration of frontal lobe functioning. *Perceptual and Motor Skills* 77: 915–921.

Casper RF, Patel-Christopher A, and Powell AM (1989) Thyrotropin and prolactin responses to thyrotropin-releasing hormone in premenstrual syndrome. *Journal of Clinical Endocrinology and Metabolism* 68: 608–612.

Chandra PS and Chaturvedi SK (1989) Cultural variations of premenstrual experience. *International Journal of Social Psychiatry* 35: 343–349.

Chang AM, Holroyd E, and Chau JP (1995) Premenstrual syndrome in employed Chinese women in Hong Kong. *Health Care Women International* 16: 551–561.

Chuong CJ and Burgos DM (1995) Medical history in women with premenstrual syndrome. *Journal of Psychosomatic Obstetrics and Gynecology* 16: 21–27.

Chuong CJ, Coulam CB, Kao PC, Bergstralh EJ, and Go VL (1985) Neuropeptide levels in premenstrual syndrome. *Fertility and Sterility* 44: 760–765.

Claustrat B, Chazot G, Brun J, Jordan D, and Sassolas G (1984) A chronobiological study of melatonin and cortisol secretion in depressed subjects: Plasma melatonin, a biochemical marker in major depression. *Biological Psychiatry* 19: 1215–1228.

Clayton AH, Keller AE, Leslie C, and Evans W (2006) Exploratory study of premenstrual symptoms and serotonin variability. *Archives of Women's Mental Health* 9: 51–57.

Cohen LS, Soares CN, Lyster A, Cassano P, Brandes M, and Leblanc GA (2004a) Efficacy and tolerability of premenstrual use of venlafaxine (flexible dose) in the treatment of premenstrual dysphoric disorder. *Journal of Clinical Psychopharmacology* 24: 540–543.

Cohen LS, Soares CN, Vitonis AF, Otto MW, and Harlow BL (2006) Risk for new onset of depression during the menopausal transition: The Harvard study of moods and cycles. *Archives of General Psychiatry* 63: 385–390.

Cohen LS, Soares CN, Yonkers KA, Bellew KM, Bridges IM, and Steiner M (2004b) Paroxetine controlled release for premenstrual dysphoric disorder: A double-blind, placebo-controlled trial. *Psychosomatic Medicine* 66: 707–713.

Dalton K (1964) *The Premenstrual Syndrome.* Springfield, IL: Charles C. Thomas.

Dalton K, Dalton ME, and Guthrie K (1987) Incidence of the premenstrual syndrome in twins. *British Medical Journal (Clinical Research Edition)* 295: 1027–1028.

Davis FC, Darrow JM, and Menaker M (1983) Sex differences in the circadian control of hamster wheel-running activity. *American Journal of Physiology* 244: R93–R105.

De Ronchi D, Ujkaj M, Boaron F, Muro A, Piselli M, and Quartesan R (2005) Symptoms of depression in late luteal phase dysphoric disorder: A variant of mood disorder? *Journal of Affective Disorders* 86: 169–174.

Denicoff KD, Hoban C, Grover GN, and Rubinow DR (1990) Glucose tolerance testing in women with premenstrual syndrome. *American Journal of Physiology* 147: 477–480.

Di Giulio G and Reissing ED (2006) Premenstrual dysphoric disorder: Prevalence, diagnostic considerations, and controversies. *Journal of Psychosomatic Obstetrics and Gynecology* 27: 201–210.

Diener D, Greenstein FL, and Turnbough PD (1992) Cyclical variation in digit-span and visual-search performance in women differing in the severity of their premenstrual symptoms. *Perceptual and Motor Skills* 74: 67–76.

Domoney CL, Vashisht A, and Studd JW (2003) Premenstrual syndrome and the use of alternative therapies. *Annals of the New York Academy of Sciences* 997: 330–340.

Dunner DL, Patrick V, and Fieve RR (1977) Rapid cycling manic depressive patients. *Comparative Psychiatry* 18: 561–566.

Ehlers CL, Frank E, and Kupfer DJ (1988) Social zeitgebers and biological rhythms. A unified approach to understanding the etiology of depression. *Archives of General Psychiatry* 45: 948–952.

Endo M, Daiguji M, Asano Y, Yamashita I, and Takahashi S (1978) Periodic psychosis recurring in association with menstrual cycle. *Journal of Clinical Psychiatry* 39: 456–466.

Epperson CN, Haga K, Mason GF, et al. (2002) Cortical gamma-aminobutyric acid levels across the menstrual cycle in healthy women and those with premenstrual dysphoric disorder: A proton magnetic resonance spectroscopy study. *Archives of General Psychiatry* 59: 851–858.

Eriksson E, Alling C, Andersch B, Andersson K, and Berggren U (1994) Cerebrospinal fluid levels of monoamine metabolites. A preliminary study of their relation to menstrual cycle phase, sex steroids, and pituitary hormones in healthy women and in women with premenstrual syndrome. *Neuropsychopharmacology* 11: 201–213.

Eriksson E, Andersch B, Ho HP, Landen M, and Sundblad C (2002) Diagnosis and treatment of premenstrual dysphoria. *Journal of Clinical Psychiatry* 63(supplement 7): 16–23.

Eriksson O, Wall A, Marteinsdottir I, et al. (2006) Mood changes correlate to changes in brain serotonin precursor trapping in women with premenstrual dysphoria. *Psychiatry Research* 146: 107–116.

Evans SM, Haney M, Levin FR, Foltin RW, and Fischman MW (1998) Mood and performance changes in women with premenstrual dysphoric disorder: Acute effects of alprazolam. *Neuropsychopharmacology* 19: 499–516.

Facchinetti F, Martignoni E, Petraglia F, Sances MG, Nappi G, and Genazzani AR (1987) Premenstrual fall of plasma beta-endorphin in patients with premenstrual syndrome. *Fertility and Sterility* 47: 570–573.

FitzGerald M, Malone KM, Li S, et al. (1997) Blunted serotonin response to fenfluramine challenge in premenstrual

dysphoric disorder. *American Journal of Psychiatry* 154: 556–568.
Frank RC (1931) The hormonal causes of premenstrual tension. *Archives of Neurology and Psychiatry* 26: 1053–1057.
Freeman EW (2004) Luteal phase administration of agents for the treatment of premenstrual dysphoric disorder. *CNS Drugs* 18: 453–468.
Freeman EW, Kroll R, Rapkin A, et al. (2001) Evaluation of a unique oral contraceptive in the treatment of premenstrual dysphoric disorder. *Journal of Women's Health and Gender-Based Medicine* 10: 561–569.
Freeman EW, Rickels K, Sondheimer SJ, and Polansky M (1999) Differential response to antidepressants in women with premenstrual syndrome/premenstrual dysphoric disorder: A randomized controlled trial. *Archives of General Psychiatry* 56: 932–939.
Freeman EW, Rickels K, Sondheimer SJ, Polansky M, and Xiao S (2004a) Continuous or intermittent dosing with sertraline for patients with severe premenstrual syndrome or premenstrual dysphoric disorder. *American Journal of Psychiatry* 161: 343–351.
Freeman EW, Rickels K, Yonkers KA, Kunz NR, McPherson M, and Upton GV (2001b) Venlafaxine in the treatment of premenstrual dysphoric disorder. *Obstetrics and Gynecology* 98: 737–744.
Freeman EW, Sammel MD, Lin H, and Nelson DB (2006) Associations of hormones and menopausal status with depressed mood in women with no history of depression. *Archives of General Psychiatry* 63: 375–382.
Freeman EW, Sammel MD, Rinaudo PJ, and Sheng L (2004b) Premenstrual syndrome as a predictor of menopausal symptoms. *Obstetrics and Gynecology* 103: 960–966.
Freeman EW, Sondheimer SJ, Rickels K, and Martin PG (2004c) A pilot naturalistic follow-up of extended sertraline treatment for severe premenstrual syndrome. *Journal of Clinical Psychopharmacology* 24: 351–353.
Freeman EW, Sondheimer SJ, Sammel MD, Ferdousi T, and Lin H (2005) A preliminary study of luteal phase versus symptom-onset dosing with escitalopram for premenstrual dysphoric disorder. *Journal of Clinical Psychiatry* 66: 769–773.
Gershon ES, Hamovit JH, Guroff JJ, and Nurnberger JI (1987) Birth-cohort changes in manic and depressive disorders in relatives of bipolar and schizoaffective patients. *Archives of General Psychiatry* 44: 314–319.
Giannini JA and Martin DM (1989) B-endorphin related symptoms in late luteal disorder. New Research Abstract NR227. In: American Psychiatric Association Meeting, San Francisco, CA.
Gilligan C (1982) *In a Different Voice.* Cambridge, MA: Harvard University Press.
Girdler SS, Straneva PA, Light KC, Pedersen CA, and Morrow AL (2001) Allopregnanolone levels and reactivity to mental stress in premenstrual dysphoric disorder. *Biological Psychiatry* 49: 788–797.
Girdler SS, Thompson KS, Light KC, Leserman J, Pedersen CA, and Prange AJ, Jr. (2004) Historical sexual abuse and current thyroid axis profiles in women with premenstrual dysphoric disorder. *Psychosomatic Medicine* 66: 403–410.
Glick H, Endicott J, and Nee J (1993) Premenstrual changes: Are they familial? *Acta Psychiatrica Scandinavica* 88: 149–155.
Golub S (1988) A developmental perspective. In: Gise LH (ed.) *The Premenstrual Syndromes,* vol. 2, pp. 7-20. New York: Churchill Livingstone.
Gurguis GN, Yonkers KA, Phan SP, Blakeley JE, Williams A, and Rush AJ (1998) Adrenergic receptors in premenstrual dysphoric disorder. I: Platelet alpha 2 receptors: Gi protein coupling, phase of menstrual cycle, and prediction of luteal phase symptom severity. *Biological Psychiatry* 44: 600–609.
Habek D, Habek JC, and Barbir A (2002) Using acupuncture to treat premenstrual syndrome. *Archives of Gynecology and Obstetrics* 267: 23–26.
Halbreich U, Bergeron R, Yonkers KA, Freeman E, Stout AL, and Cohen L (2002) Efficacy of intermittent, luteal phase sertraline treatment of premenstrual dysphoric disorder. *Obstetrics and Gynecology* 100: 1219–1229.
Halbreich U, Borenstein J, Pearlstein T, and Kahn LS (2003) The prevalence, impairment, impact, and burden of premenstrual dysphoric disorder (PMS/PMDD). *Psychoneuroendocrinology* 28(supplement 3): 1–23.
Halbreich U and Endicott J (1985) Relationship of dysphoric premenstrual changes to depressive disorders. *Acta Psychiatrica Scandinavica* 71: 331–338.
Halbreich U, Endicott J, Goldstein S, and Nee J (1986) Premenstrual changes and changes in gonadal hormones. *Acta Psychiatrica Scandinavica* 74: 576–586.
Hall JE, Sullivan JP, and Richardson GS (2005) Brief wake episodes modulate sleep-inhibited luteinizing hormone secretion in the early follicular phase. *Journal of Clinical Endocrinology and Metabolism* 90: 2050–2055.
Hallman J (1986) The premenstrual syndrome – an equivalent of depression? *Acta Psychiatrica Scandinavica* 73: 403–411.
Hamilton JA, Gallant SA, and Pinkel S (1988) Urinary 6-hydroxymelatonin in menstruating women. *Biological Psychiatry* 24: 845–852.
Hamilton JA, Parry BL, Alagna SW, Blumenthal SJ, and Herz E (1984) Premenstrual mood changes: A guide to evaluation and treatment. *Psychiatric Annals* 14: 426–435.
Hammarback S, Damber JE, and Backstrom T (1989) Relationship between symptom severity and hormone changes in women with premenstrual syndrome. *Journal of Clinical Endocrinology and Metabolism* 68: 125–130.
Hampson E (1990) Estrogen-related variations in human spatial and articulatory-motor skills. *Psychoneuroendocrinology* 15: 97–111.
Hampson E and Kimura D (1988) Reciprocal effects of hormonal fluctuations on human motor and perceptual-spatial skills. *Behavioral Neuroscience* 102: 456–459.
Hardoy MC, Serra M, Carta MG, Contu P, Pisu MG, and Biggio G (2006) Increased neuroactive steroid concentrations in women with bipolar disorder or major depressive disorder. *Journal of Clinical Psychopharmacology* 26: 379–384.
Harrison WM, Endicott J, and Nee J (1989a) Treatment of premenstrual depression with nortriptyline: A pilot study. *Journal of Clinical Psychiatry* 50: 136–139.
Harrison WM, Endicott J, and Nee J (1990) Treatment of premenstrual dysphoria with alprazolam. A controlled study. *Archives of General Psychiatry* 47: 270–275.
Harrison WM, Sandberg D, Gorman JM, Fyer M, Nee J, Uy J, and Endicott J (1989b) Provocation of panic with carbon dioxide inhalation in patients with premenstrual dysphoria. *Psychiatry Research* 27: 183–192.
Haskett RF, Steiner M, and Carroll BJ (1984) A psychoendocrine study of premenstrual tension syndrome. A model for endogenous depression? *Journal of Affective Disorders* 6: 191–199.
Hsiao CC, Liu CY, and Hsiao MC (2004) No correlation of depression and anxiety to plasma estrogen and progesterone levels in patients with premenstrual dysphoric disorder. *Psychiatry and Clinical Neurosciences* 58: 593–599.
Hsiao MC and Liu CY (2004) Withdrawal reactions associated with low-dose venlafaxine treatment in a patient with premenstrual dysphoric disorder. *Journal of Clinical Psychiatry* 65: 1147–1148.

Huang KL and Tsai SJ (2003) St. John's wort (Hypericum perforatum) as a treatment for premenstrual dysphoric disorder: Case report. *International Journal of Psychiatry Medicine* 33: 295–297.

Huo L, Straub RE, Roca C, et al. (2007) Risk for premenstrual dysphoric disorder is associated with genetic variation in ESR1, the estrogen receptor alpha gene. *Biological Psychiatry* 62: 925–933.

Jakubowicz DL, Godard E, and Dewhurst J (1984) The treatment of premenstrual tension with mefenamic acid: Analysis of prostaglandin concentrations. *British Journal of Obstetrics and Gynaecology* 91: 78–84.

Janiger O, Riffenburgh R, and Kersh R (1972) Cross cultural study of premenstrual symptoms. *Psychosomatics* 13: 226–235.

Keenan PA, Lindamer LA, and Jong SK (1995) Menstrual phase-independent retrieval deficit in women with PMS. *Biological Psychiatry* 38: 369–377.

Keenan PA, Stern RA, Janowsky DS, and Pedersen CA (1992) Psychological aspects of premenstrual syndrome. I: Cognition and memory. *Psychoneuroendocrinology* 17: 179–187.

Khine K, Rosenstein DL, Elin RJ, Niemela JE, Schmidt PJ, and Rubinow DR (2006) Magnesium (Mg) retention and mood effects after intravenous Mg infusion in premenstrual dysphoric disorder. *Biological Psychiatry* 59: 327–333.

Klaiber EL, Broverman DM, Vogel W, Kennedy JA, and Nadeau CJ (1982) Estrogens and central nervous system function: Electroencephalography, cognition, and depression. In: Friedman RC (ed.) *Behavior and the Menstrual Cycle*, pp. 267–290. New York: Marcel Dekker.

Klatzkin RR, Morrow AL, Light KC, Pedersen CA, and Girdler SS (2006) Associations of histories of depression and PMDD diagnosis with allopregnanolone concentrations following the oral administration of micronized progesterone. *Psychoneuroendocrinology* 31: 1208–1219.

Kornstein SG, Harvey AT, Rush AJ, et al. (2005) Self-reported premenstrual exacerbation of depressive symptoms in patients seeking treatment for major depression. *Psychological Medicine* 35: 683–692.

Kornstein SG, Pearlstein TB, Fayyad R, Farfel GM, and Gillespie JA (2006) Low-dose sertraline in the treatment of moderate-to-severe premenstrual syndrome: Efficacy of 3 dosing strategies. *Journal of Clinical Psychiatry* 67: 1624–1632.

Kouri EM and Halbreich U (1997) State and trait serotonergic abnormalities in women with dysphoric premenstrual syndromes. *Psychopharmacology Bulletin* 33: 767–770.

Kripke DF, Mullaney DJ, Atkinson M, and Wolf S (1978) Circadian rhythm disorders in manic-depressives. *Biological Psychiatry* 13: 335–351.

Kurshan N and NeillEpperson C (2006) Oral contraceptives and mood in women with and without premenstrual dysphoria: A theoretical model. *Archives of Women's Mental Health* 9: 1–14.

Lamarche LJ, Driver HS, Wiebe S, Crawford L, and De Koninck JM (2007) Nocturnal sleep, daytime sleepiness, and napping among women with significant emotional/behavioral premenstrual symptoms. *Journal of Sleep Research* 16: 262–268.

Landen M and Eriksson E (2003) How does premenstrual dysphoric disorder relate to depression and anxiety disorders? *Depression and Anxiety* 17: 122–129.

Landen M, Nissbrandt H, Allgulander C, Sorvik K, Ysander C, and Eriksson E (2007) Placebo-controlled trial comparing intermittent and continuous paroxetine in premenstrual dysphoric disorder. *Neuropsychopharmacology* 32: 153–161.

Landen M, Wennerblom B, Tygesen H, et al. (2004) Heart rate variability in premenstrual dysphoric disorder. *Psychoneuroendocrinology* 29: 733–740.

LeMelledo JM, Merani S, Koszycki D, et al. (1999) Sensitivity to CCK-4 in women with and without premenstrual dysphoric disorder (PMDD) during their follicular and luteal phases. *Neuropsychopharmacology* 20: 81–91.

LeMelledo JM, VanDriel M, Coupland NJ, Lott P, and Jhangri GS (2000) Response to flumazenil in women with premenstrual dysphoric disorder. *American Journal of Psychiatry* 157: 821–823.

Lewy AJ, Wehr TA, Goodwin FK, Newsome DA, and Rosenthal NE (1981) Manic-depressive patients may be supersensitive to light. *Lancet* 1: 383–384.

Lu ZJ (2001) The relationship between menstrual attitudes and menstrual symptoms among Taiwanese women. *Journal of Advanced Nursing* 33: 621–628.

Mackenzie TB, Wilcox K, and Baron H (1986) Lifetime prevalence of psychiatric disorders in women with perimenstrual difficulties. *Journal of Affective Disorders* 10: 15–19.

Magnay JL, Ismail KM, Chapman G, Cioni L, Jones PW, and O'Brien S (2006) Serotonin transporter, tryptophan hydroxylase, and monoamine oxidase A gene polymorphisms in premenstrual dysphoric disorder. *American Journal of Obstetrics and Gynecology* 195: 1254–1259.

Malmgren R, Collins A, and Nilsson CG (1987) Platelet serotonin uptake and effects of vitamin B6-treatment in premenstrual tension. *Neuropsychobiology* 18: 83–86.

Man MS, MacMillan I, Scott J, and Young AH (1999) Mood, neuropsychological function and cognitions in premenstrual dysphoric disorder. *Psychological Medicine* 29: 727–733.

Mattsson B and von Schoultz B (1974) A comparison between lithium, placebo and a diuretic in premenstrual tension. *Acta Psychiatrica Scandinavica* 255(supplement): 75–84.

Mauri M, Reid RL, and MacLean AW (1988) Sleep in the premenstrual phase: A self-report study of PMS patients and normal controls. *Acta Psychiatrica Scandinavica* 78: 82–86.

McEwen BS and Parsons B (1982) Gonadal steroid action on the brain: Neurochemistry and neuropharmacology. *Annual Review of Pharmacology and Toxicology* 22: 555–598.

McIntyre IM and Morse C (1990) Urinary 6-sulphatoxy melatonin levels within the menstrual cycle and in patients with premenstrual syndrome. *Psychoneuroendocrinology* 15: 233–236.

Meaden PM, Hartlage SA, and Cook-Karr J (2005) Timing and severity of symptoms associated with the menstrual cycle in a community-based sample in the Midwestern United States. *Psychiatry Research* 134: 27–36.

Meites J, Bruni JF, VanVugt DA, and Smith AF (1979) Relation of endogenous opioid peptides and morphine to neuroendocrine functions. *Life Sciences* 24: 1325–1336.

Mendlewicz J, Branchey L, Weinberg U, Branchey M, Linkowski P, and Weitzman ED (1980) The 24 hour pattern of plasma melatonin in depressed patients before and after treatment. *Communications in Psychopharmacology* 4: 49–55.

Menkes DB, Taghavi E, Mason PA, Spears GF, and Howard RC (1992) Fluoxetine treatment of severe premenstrual syndrome. *British Medical Journal* 305: 346–347.

Merikangas KR, Foeldenyi M, and Angst J (1993) The Zurich Study. XIX. Patterns of menstrual disturbances in the community: Results of the Zurich Cohort Study. *European Archives of Psychiatry and Clinical Neuroscience* 243: 23–32.

Mezrow G, Shoupe D, Spicer D, Lobo R, Leung B, and Pike M (1994) Depot leuprolide acetate with estrogen and progestin add-back for long-term treatment of premenstrual syndrome. *Fertility and Sterility* 62: 932–937.

Miner C, Brown E, McCray S, Gonzales J, and Wohlreich M (2002) Weekly luteal-phase dosing with enteric-coated fluoxetine 90 mg in premenstrual dysphoric disorder: A randomized, double-blind, placebo-controlled clinical trial. *Clinical Therapeutics* 24: 417–433.

Mira M, Stewart PM, and Abraham SF (1988) Vitamin and trace element status in premenstrual syndrome. *American Journal of Clinical Nutrition* 47: 636–641.

Mitwally MF, Gotlieb L, and Casper RF (2002) Prevention of bone loss and hypoestrogenic symptoms by estrogen and interrupted progestogen add-back in long-term GnRH-agonist down-regulated patients with endometriosis and premenstrual syndrome. *Menopause* 9: 236–241.

Monagle L, Dan A, Krogh V, Jossa F, Farinaro E, and Trevisan M (1993) Perimenstrual symptom prevalence rates: An Italian–American comparison. *American Journal of Epidemiology* 138: 1070–1081.

Morgan M and Rapkin A (2002) Cognitive flexibility, reaction time, and attention in women with premenstrual dysphoric disorder. *Journal of Gender-Specific Medicine* 5: 28–36.

Morgan M, Rapkin AJ, D'Elia L, Reading A, and Goldman L (1996) Cognitive functioning in premenstrual syndrome. *Obstetrics and Gynecology* 88: 961–966.

Morin LP, Fitzgerald KM, and Zucker I (1977) Estradiol shortens the period of hamster circadian rhythms. *Science* 196: 305–307.

Mortola JF, Girton L, and Fischer U (1991) Successful treatment of severe premenstrual syndrome by combined use of gonadotropin-releasing hormone agonist and estrogen/progestin. *Journal of Clinical Endocrinology and Metabolism* 72: 252A–252F.

Most AF, Woods NF, Dery GK, and Most BM (1981) Distress associated with menstruation among Israeli women. *International Journal of Nursing Studies* 18: 61–71.

Nurnberger JI, Jr., Berrettini W, Tamarkin L, Hamovit J, Norton J, and Gershon E (1988) Supersensitivity to melatonin suppression by light in young people at high risk for affective disorder. A preliminary report. *Neuropsychopharmacology* 1: 217–223.

N-Wihlback AC, Sundstrom-Poromaa I, and Backstrom T (2006) Action by and sensitivity to neuroactive steroids in menstrual cycle related CNS disorders. *Psychopharmacology (Berl)* 186: 388–401.

Oberlander TF, Warburton W, Misri S, Aghajanian J, and Hertzman C (2006) Neonatal outcomes after prenatal exposure to selective serotonin reuptake inhibitor antidepressants and maternal depression using population-based linked health data. *Archives of General Psychiatry* 63: 898–906.

Odber J, Cawood EH, and Bancroft J (1998) Salivary cortisol in women with and without perimenstrual mood changes. *Journal of Psychosomatic Research* 45: 557–568.

Parry BL, Berga SL, Kripke DF, Klauber MR, Laughlin GA, Yen SS, and Gillin JC (1990) Altered waveform of plasma nocturnal melatonin secretion in premenstrual depression. *Archives of General Psychiatry* 47: 1139–1146.

Parry BL, Berga SL, Mostofi N, Klauber MR, and Resnick A (1997a) Plasma melatonin circadian rhythms during the menstrual cycle and after light therapy in premenstrual dysphoric disorder and normal control subjects. *Journal of Biological Rhythms* 12: 47–64.

Parry BL, Berga SL, Mostofi N, Sependa PA, Kripke DF, and Gillin JC (1989a) Morning versus evening bright light treatment of late luteal phase dysphoric disorder. *American Journal of Psychiatry* 146: 1215–1217.

Parry BL, Cover H, Mostofi N, LeVeau B, Sependa PA, Resnick A, and Gillin JC (1995) Early versus late partial sleep deprivation in patients with premenstrual dysphoric disorder and normal comparison subjects. *American Journal of Psychiatry* 152: 404–412.

Parry BL, Gerner RH, Wilkins JN, et al. (1991) CSF and endocrine studies of premenstrual syndrome. *Neuropsychopharmacology* 5: 127–137.

Parry BL, Hauger R, LeVeau B, Mostofi N, Cover H, Clopton P, and Gillin JC (1996) Circadian rhythms of prolactin and thyroid-stimulating hormone during the menstrual cycle and early versus late sleep deprivation in premenstrual dysphoric disorder. *Psychiatry Research* 62: 147–160.

Parry BL, Hauger R, Lin E, LeVeau B, Mostofi N, Clopton PL, and Gillin JC (1994) Neuroendocrine effects of light therapy in late luteal phase dysphoric disorder. *Biological Psychiatry* 36: 356–364.

Parry BL, Javeed S, Laughlin GA, Hauger R, and Clopton P (2000) Cortisol circadian rhythms during the menstrual cycle and with sleep deprivation in premenstrual dysphoric disorder and normal control subjects. *Biological Psychiatry* 48: 920–931.

Parry BL, LeVeau B, Mostofi N, Naham HC, Loving R, Clopton P, and Gillin JC (1997b) Temperature circadian rhythms during the menstrual cycle and sleep deprivation in premenstrual dysphoric disorder and normal comparison subjects. *Journal of Biological Rhythms* 12: 34–46.

Parry BL, Mahan AM, Mostofi N, Klauber MR, Lew GS, and Gillin JC (1993) Light therapy of late luteal phase dysphoric disorder: An extended study. *American Journal of Psychiatry* 150: 1417–1419.

Parry BL, Martinez LF, Maurer EL, Lopez AM, Sorenson DL, and Meliska CJ (2006) Sleep rhythms and women's mood. Part I: Menstrual cycle, pregnancy and postpartum. *Sleep Medicine Reviews* 10: 129–144.

Parry BL, Mendelson WB, Duncan WC, Sack DA, and Wehr TA (1989) Longitudinal sleep EEG, temperature, and activity measurements across the menstrual cycle in patients with premenstrual depression and in age-matched controls. *Psychiatry Research* 30: 285–303.

Parry BL and Rausch JL (1988) Evaluation of biologic research. In: Gise LH (ed.) *The Premenstrual Syndromes,* vol. 2, pp. 46–58. New York: Churchill Livingstone.

Parry BL and Rush AJ (1979) Oral contraceptives and depressive symptomatology: Biologic mechanisms. *Comprehensive Psychiatry* 20: 347–358.

Parry BL, Udell C, Elliott JA, et al. (1997c) Blunted phase-shift responses to morning bright light in premenstrual dysphoric disorder. *Journal of Biological Rhythms* 12: 443–456.

Parry BL and Wehr TA (1987) Therapeutic effect of sleep deprivation in patients with premenstrual syndrome. *American Journal of Psychiatry* 144: 808–810.

Payne JL (2003) The role of estrogen in mood disorders in women. *International Review of Psychiatry* 15: 280–290.

Pearlstein TB, Bachmann GA, Zacur HA, and Yonkers KA (2005) Treatment of premenstrual dysphoric disorder with a new drospirenone-containing oral contraceptive formulation. *Contraception* 72: 414–421.

Post RM, Rubinow DR, and Ballenger JC (1986) Conditioning and sensitisation in the longitudinal course of affective illness. *British Journal of Psychiatry* 149: 191–201.

Praschak-Rieder N, Willeit M, Winkler D, et al. (2002) Role of family history and 5-HTTLPR polymorphism in female seasonal affective disorder patients with and without premenstrual dysphoric disorder. *European Neuropsychopharmacology* 12: 129–134.

Rapkin AJ, Edelmuth E, Chang LC, Reading AE, McGuire MT, and Su TP (1987) Whole-blood serotonin in premenstrual syndrome. *Obstetrics and Gynecology* 70: 533–537.

Rasgon N, McGuire M, Tanavoli S, Fairbanks L, and Rapkin A (2000) Neuroendocrine response to an intravenous L-tryptophan challenge in women with premenstrual syndrome. *Fertility and Sterility* 73: 144–149.

Rasgon NL, Thomas MA, Guze BH, Fairbanks LA, Yue K, Curran JG, and Rapkin AJ (2001) Menstrual cycle-related brain metabolite changes using 1H magnetic resonance spectroscopy in premenopausal women: A pilot study. *Psychiatry Research* 106: 47–57.

Rasheed P and Al-Sowielem LS (2003) Prevalence and predictors of premenstrual syndrome among college-aged women in Saudi Arabia. *Annals of Saudi Medicine* 23: 381–387.

Reid RL, Greenaway-Coates A, and Hahn PM (1986) Oral glucose tolerance during the menstrual cycle in normal women and women with alleged premenstrual hypoglycemic attacks: Effects of naloxone. *Journal of Clinical Endocrinology and Metabolism* 62: 1167–1172.

Reinberg A, Andlauer P, DePrins J, Malbecq W, Vieux N, and Bourdeleau P (1984) Desynchronization of the oral temperature circadian rhythm and intolerance to shift work. *Nature* 308: 272–274.

Resnick A, Perry W, Parry B, Mostofi N, and Udell C (1998) Neuropsychological performance across the menstrual cycle in women with and without premenstrual dysphoric disorder. *Psychiatry Research* 77: 147–158.

Richardson JT (1995) The premenstrual syndrome: A brief history. *Social Science and Medicine* 41: 761–767.

Rickels K, Freeman E, and Sondheimer S (1989) Buspirone in treatment of premenstrual syndrome. *Lancet* 1: 777.

Roca CA, Schmidt PJ, Altemus M, Deuster P, Danaceau MA, Putnam K, and Rubinow DR (2003) Differential menstrual cycle regulation of hypothalamic–pituitary–adrenal axis in women with premenstrual syndrome and controls. *Journal of Clinical Endocrinology and Metabolism* 88: 3057–3063.

Roca CA, Schmidt PJ, Smith MJ, Danaceau MA, Murphy DL, and Rubinow DR (2002) Effects of metergoline on symptoms in women with premenstrual dysphoric disorder. *American Journal of Psychiatry* 159: 1876–1881.

Rojansky N, Halbreich U, Zander K, Barkai A, and Goldstein S (1991) Imipramine receptor binding and serotonin uptake in platelets of women with premenstrual changes. *Gynecologic and Obstetric Investigation* 31: 146–152.

Rosenstein DL, Kalogeras KT, Kalafut M, Malley J, and Rubinow DR (1996) Peripheral measures of arginine vasopressin, atrial natriuretic peptide and adrenocorticotropic hormone in premenstrual syndrome. *Psychoneuroendocrinology* 21: 347–359.

Rosenthal NE, Sack DA, Gillin JC, et al. (1984) Seasonal affective disorder. A description of the syndrome and preliminary findings with light therapy. *Archives of General Psychiatry* 41: 72–80.

Roy-Byrne PP, Rubinow DR, Gwirtsman H, Hoban MC, and Grover GN (1986) Cortisol response to dexamethasone in women with premenstrual syndrome. *Neuropsychobiology* 16: 61–63.

Roy-Byrne PP, Rubinow DR, Hoban MC, Grover GN, and Blank D (1987) TSH and prolactin responses to TRH in patients with premenstrual syndrome. *American Journal of Psychiatry* 144: 480–484.

Rubin RT, Heist EK, McGeoy SS, Hanada K, and Lesser IM (1992) Neuroendocrine aspects of primary endogenous depression. XI. Serum melatonin measures in patients and matched control subjects. *Archives of General Psychiatry* 49: 558–567.

Rubinow DR, Hoban MC, Grover GN, Galloway DS, Roy-Byrne P, Andersen R, and Merriam GR (1988) Changes in plasma hormones across the menstrual cycle in patients with menstrually related mood disorder and in control subjects. *American Journal of Obstetrics and Gynecology* 158: 5–11.

Rubinow DR and Schmidt PJ (2006) Gonadal steroid regulation of mood: The lessons of premenstrual syndrome. *Frontiers in Neuroendocrinology* 27: 210–216.

Sack RL, Lewy AJ, Erb DL, Vollmer WM, and Singer CM (1986) Human melatonin production decreases with age. *Journal of Pineal Research* 3: 379–388.

Sangthawan M and Taneepanichskul S (2005) A comparative study of monophasic oral contraceptives containing either drospirenone 3 mg or levonorgestrel 150 microg on premenstrual symptoms. *Contraception* 71: 1–7.

Santoro N, Brown JR, Adel T, and Skurnick JH (1996) Characterization of reproductive hormonal dynamics in the perimenopause. *Journal of Clinical Endocrinology and Metabolism* 81: 1495–1501.

Schmidt PJ, Haq N, and Rubinow DR (2004) A longitudinal evaluation of the relationship between reproductive status and mood in perimenopausal women. *American Journal of Psychiatry* 161: 2238–2244.

Schmidt PJ, Nieman LK, Grover GN, Muller KL, Merriam GR, and Rubinow DR (1991) Lack of effect of induced menses on symptoms in women with premenstrual syndrome. *New England Journal of Medicine* 324: 1174–1179.

Schuckit MA, Daly V, Herrman G, and Hineman S (1975) Premenstrual symptoms and depression in a university population. *Diseases of Nervous System* 36: 516–517.

Severino SK, Wagner DR, Moline ML, Hurt SW, Pollak CP, and Zendell S (1991) High nocturnal body temperature in premenstrual syndrome and late luteal phase dysphoric disorder. *American Journal of Psychiatry* 148: 1329–1335.

Shinohara K, Uchiyama M, Okawa M, Saito K, Kawaguchi M, Funabashi T, and Kimura F (2000) Menstrual changes in sleep, rectal temperature and melatonin rhythms in a subject with premenstrual syndrome. *Neuroscience Letters* 281: 159–162.

Shye D and Jaffe B (1991) Prevalence and correlates of perimenstrual symptoms: A study of Israeli teenage girls. *Journal of Adolescent Health* 12: 217–224.

Singer K, Cheng R, and Schou M (1974) A controlled evaluation of lithium in the premenstrual tension syndrome. *British Journal of Psychiatry* 124: 50–51.

Smith MJ, Adams LF, Schmidt PJ, Rubinow DR, and Wassermann EM (2003) Abnormal luteal phase excitability of the motor cortex in women with premenstrual syndrome. *Biological Psychiatry* 54: 757–762.

Snowden R and Christian B (1983) *Patterns and Perceptions of Menstruation, a World Health Organization International Collaborative Study in Egypt, India, Indonesia, Jamaica, Mexico, Pakistan, Philippines, Republic of Korea, United Kingdom and Yugoslavia.* New York: Croom Helm, Long and Canberra and St. Martin's Press.

Souetre E, Salvati E, Belugou JL, et al. (1989) Circadian rhythms in depression and recovery: Evidence for blunted amplitude as the main chronobiological abnormality. *Psychiatry Research* 28: 263–278.

Steiner M, Brown E, Trzepacz P, et al. (2003) Fluoxetine improves functional work capacity in women with premenstrual dysphoric disorder. *Archives of Women's Mental Health* 6: 71–77.

Steiner M, Haskett RF, Carroll BJ, Hays SE, and Rubin RT (1984) Circadian hormone secretory profiles in women with severe premenstrual tension syndrome. *British Journal of Obstetrics and Gynaecology* 91: 466–471.

Steiner M, Haskett RF, Osmun JN, and Carroll BJ (1980) Treatment of premenstrual tension with lithium carbonate. A pilot study. *Acta Psychiatrica Scandinavica* 61: 96–102.

Steiner M, Hirschberg AL, Bergeron R, Holland F, Gee MD, and Van Erp E (2005) Luteal phase dosing with paroxetine controlled release (CR) in the treatment of premenstrual dysphoric disorder. *American Journal of Obstetrics and Gynecology* 193: 352–360.

Steiner M, Korzekwa M, Lamont J, and Wilkins A (1997) Intermittent fluoxetine dosing in the treatment of women with premenstrual dysphoria. *Psychopharmacology Bulletin* 33: 771–774.

Steiner M, Steinberg S, Stewart D, et al. (1995) Fluoxetine in the treatment of premenstrual dysphoria. Canadian Fluoxetine/ Premenstrual Dysphoria Collaborative Study Group. *New England Journal of Medicine* 332: 1529–1534.

Stoddard JL, Dent CW, Shames L, and Bernstein L (2007) Exercise training effects on premenstrual distress and ovarian steroid hormones. *European Journal of Applied Physiology* 99: 27–37.

Stone AB, Pearlstein TB, and Brown WA (1991) Fluoxetine in the treatment of late luteal phase dysphoric disorder. *Journal of Clinical Psychiatry* 52: 290–293.

Su TP, Schmidt PJ, Danaceau M, Murphy DL, and Rubinow DR (1997) Effect of menstrual cycle phase on neuroendocrine and behavioral responses to the serotonin agonist m-chlorophenylpiperazine in women with premenstrual syndrome and controls. *Journal of Clinical Endocrinology and Metabolism* 82: 1220–1228.

Sundblad C, Hedberg MA, and Eriksson E (1993) Clomipramine administered during the luteal phase reduces the symptoms of premenstrual syndrome: A placebo-controlled trial. *Neuropsychopharmacology* 9: 133–145.

Sundblad C, Modigh K, Andersch B, and Eriksson E (1992) Clomipramine effectively reduces premenstrual irritability and dysphoria: A placebo-controlled trial. *Acta Psychiatrica Scandinavica* 85: 39–47.

Takeda T, Tasaka K, Sakata M, and Murata Y (2006) Prevalence of premenstrual syndrome and premenstrual dysphoric disorder in Japanese women. *Archives of Women's Mental Health* 9: 209–212.

Tamayo JM, Gomez G, Barrios R, Holguin J, and Adrianzen C (2004) Differential time course efficacy on dysphoric and physical symptoms of the intermittent dosing of fluoxetine in the premenstrual dysphoric disorder. *Journal of Clinical Psychopharmacology* 24: 469–471.

Taylor DL, Mathew RJ, Ho BT, and Weinman ML (1984) Serotonin levels and platelet uptake during premenstrual tension. *Neuropsychobiology* 12: 16–18.

Thomas EM and Armstrong SM (1989) Effect of ovariectomy and estradiol on unity of female rat circadian rhythms. *American Journal of Physiology* 257: R1241–R1250.

Thompson C, Franey C, Arendt J, and Checkley SA (1988) A comparison of melatonin secretion in depressed patients and normal subjects. *British Journal of Psychiatry* 152: 260–265.

Thys-Jacobs S, Starkey P, Bernstein D, and Tian J (1998) Calcium carbonate and the premenstrual syndrome: Effects on premenstrual and menstrual symptoms. Premenstrual Syndrome Study Group. *American Journal of Obstetrics and Gynecology* 179: 444–452.

Tulenheimo A, Laatikainen T, and Salminen K (1987) Plasma beta-endorphin immunoreactivity in premenstrual tension. *British Journal of Obstetrics and Gynecology* 94: 26–29.

Ussher JM and Wilding JM (1991) Performance and state changes during the menstrual cycle, conceptualised within a broad band testing framework. *Social Science and Medicine* 32: 525–534.

van den Akker O and Steptoe A (1989) Psychophysiological responses in women reporting severe premenstrual symptoms. *Psychosomatic Medicine* 51: 319–328.

van den Akker OB, Stein GS, Neale MC, and Murray RM (1987) Genetic and environmental variation in menstrual cycle: Histories of two British twin samples. *ACTA Geneticae Medicae et Gemellologiae (Roma)* 36: 541–548.

Varma TR (1984) Hormones and electrolytes in premenstrual syndrome. *International Journal of Gynecology and Obstetrics* 22: 51–58.

Veeninga AT and Westenberg HG (1992) Serotonergic function and late luteal phase dysphoric disorder. *Psychopharmacology (Berl)* 108: 153–158.

Warnock JK, Bundren JC, and Morris DW (1998) Depressive symptoms associated with gonadotropin-releasing hormone agonists. *Depression and Anxiety* 7: 171–177.

Watson NR, Studd JW, Savvas M, Garnett T, and Baber RJ (1989) Treatment of severe premenstrual syndrome with oestradiol patches and cyclical oral norethisterone. *Lancet* 2: 730–732.

Watts JF, Butt WR, LoganEdwards R, and Holder G (1985) Hormonal studies in women with premenstrual tension. *British Journal of Obstetrics and Gynaecology* 92: 247–255.

Wehr TA (1984) Biological rhythms and manic-depressive illness. In: Post RM and Ballenger JC (eds.) *The Neurobiology of Mood Disorders*, pp. 190–206. Baltimore, MD: Williams and Wilkins.

Wehr TA and Goodwin FK (1980) Desynchronization of circadian rhythms as a possible source of manic-depressive cycles. *Psychopharmacology Bulletin* 16: 1920.

Weissman MM (1979) The myth of involutional melancholia. *Journal of the American Medical Association* 242: 742–744.

Weissman MM, Gammon GD, John K, Merikangas KR, Warner V, Prusoff BA, and Sholomskas D (1987) Children of depressed parents. Increased psychopathology and early onset of major depression. *Archives of General Psychiatry* 44: 847–853.

Weissman MM and Klerman GL (1977) Sex differences and the epidemiology of depression. *Archives of General Psychiatry* 34: 98–111.

Weissman MM, Leaf PJ, Holzer CE, III, Myers JK, and Tischler GL (1984) The epidemiology of depression. An update on sex differences in rates. *Journal of Affective Disorders* 7: 179–188.

Wetzel RD, Reich T, McClure JN, Jr., and Wald JA (1975) Premenstrual affective syndrome and affective disorder. *British Journal of Psychiatry* 127: 219–221.

Wever RA (1988) Order and disorder in human rhythmicity possible relations to mental disorders. In: Kupfer DJ, Monk TH, and Barchas JD (eds.) *Biological Rhythms and Mental Disorders*, pp. 253–346. New York: Guildford Press.

Widholm O (1979) Dysmenorrhea during adolescence. *Acta Obstetricia et Gynecologica Scandinavica* 87(supplement): 61–66.

Winokur G (1973) Depression in the menopause. *American Journal of Psychiatry* 130: 92–93.

Wood SH, Mortola JF, Chan YF, Moossazadeh F, and Yen SS (1992) Treatment of premenstrual syndrome with fluoxetine: A double-blind, placebo-controlled, crossover study. *Obstetrics and Gynecology* 80: 339–344.

Woolley CS and McEwen BS (1992) Estradiol mediates fluctuation in hippocampal synapse density during the estrous cycle in the adult rat. *Journal of Neuroscience* 12: 2549–2554.

Wyatt KM, Dimmock PW, Ismail KM, Jones PW, and O'Brien PM (2004) The effectiveness of GnRHa with and without 'add-back' therapy in treating premenstrual syndrome: A meta analysis. *British Journal of Obstetrics and Gynaecology* 111: 585–593.

Yatham LN (1993) Is 5HT1A receptor subsensitivity a trait marker for late luteal phase dysphoric disorder? A pilot study. *Canadian Journal of Psychiatry* 38: 662–664.

Yonkers KA, Brown C, Pearlstein TB, Foegh M, Sampson-Landers C, and Rapkin A (2005) Efficacy of a new low-dose oral contraceptive with drospirenone in premenstrual dysphoric disorder. *Obstetrics and Gynecology* 106: 492–501.

Yonkers KA, Halbreich U, Freeman E, et al. (1997) Symptomatic improvement of premenstrual dysphoric disorder with sertraline treatment. A randomized controlled trial. Sertraline Premenstrual Dysphoric Collaborative Study Group. *Journal of the American Medical Association* 278: 983–988.

23 Post-Traumatic Stress Disorder

R Yehuda and C Sarapas, James J. Peters VA Medical Center, Bronx, NY, USA

Published by Elsevier Inc.

Chapter Outline

Glossary

adrenal insufficiency A failure by the adrenal gland to produce adequate amounts of cortisol in response to stress.

amplitude-to-mesor ratio In analysis of an oscillation (for instance, diurnal variation in levels of a hormone), the ratio of the total range of variation (amplitude) to the mean of the oscillation (mesor – midline estimating statistic of rhythm). In endocrinology, the amplitude-to-mesor ratio is an index of a hormone's dynamic range.

endocrine challenge test Any of a number of tests used to assess the responsiveness of an endocrine axis, wherein a hormone, synthetic hormone agonist, or other compound is administered and changes from baseline in the concentrations of hormones of interest are observed.

glucocorticoid receptor (GR) A steroid hormone receptor with high affinity to cortisol. Once bound to cortisol, the GR both promotes transcription of certain genes and represses the transcription of others.

negative feedback inhibition A homeostatic control mechanism wherein a biologic response to a change in conditions (e.g., pH or circulating levels of a hormone) is initiated; those conditions are monitored by receptors; and the response is terminated before conditions fall outside of a normal range. Enhanced negative feedback inhibition occurs when the response is terminated before homeostasis is fully restored (for instance, due to hypersensitive receptors).

23.1 Introduction

Post-traumatic stress disorder (PTSD) was originally described in 1980 as an anxiety disorder that can occur in persons who experience fear, helplessness, or horror following a traumatic event (American Psychiatric Association, 1980). This condition is characterized by the presence of three distinct but co-occurring symptom clusters: reexperiencing, avoidance, and hyperarousal symptoms. Reexperiencing symptoms describe spontaneous, often insuppressible intrusions of the traumatic memory in the form of images, nightmares, or flashbacks that are accompanied by an intense physiological distress similar to that experienced when the event actually occurred. Avoidance symptoms involve restricting thoughts and keeping away from other reminders of the event, including, in the extreme, developing amnesia for distressing aspects of the event. More generalized symptoms signifying emotional withdrawal are also included in this category. Hyperarousal symptoms such as insomnia, irritability, impaired concentration, hypervigilance, and increased startle responses concern more physiological manifestations of trauma exposure. Criteria for PTSD are met when these symptoms co-occur for at least 1 month and lead to impairments in social, occupational, or interpersonal functioning. **Table 1** lists the diagnostic criteria for PTSD (American Psychiatric Association, 2000). An important observation, from epidemiologic studies, has been that only a minority of persons exposed to extreme trauma develop this disorder (Kessler et al., 1995). This fact established that while trauma exposure is a necessary immediate precipitant of PTSD, exposure alone may not explain either the development of PTSD or its chronicity.

Table 1 DSM-IV-TR diagnostic criteria for PTSD

A. The person has been exposed to a traumatic event in which both of the following were present:
 (1) the person experienced, witnessed, or was confronted with an event or events that involved actual or threatened death or serious injury, or a threat to the physical integrity of self or others
 (2) the person's response involved intense fear, helplessness, or horror

B. The traumatic event is persistently reexperienced in one (or more) of the following ways:
 (1) recurrent and intrusive distressing recollections of the event, including images, thoughts, or perceptions
 (2) recurrent distressing dreams of the event
 (3) acting or feeling as if the traumatic event were recurring (includes a sense of reliving the experience, illusions, hallucinations, and dissociative flashback episodes, including those that occur on awakening or when intoxicated)
 (4) intense psychological distress at exposure to internal or external cues that symbolize or resemble an aspect of the traumatic event
 (5) physiological reactivity on exposure to internal or external cues that symbolize or resemble an aspect of the traumatic event

C. Persistent avoidance of stimuli associated with the trauma and numbing of general responsiveness (not present before the trauma), as indicated by three (or more) of the following:
 (1) efforts to avoid thoughts, feelings, or conversations associated with the trauma
 (2) efforts to avoid activities, places, or people that arouse recollections of the trauma
 (3) inability to recall an important aspect of the trauma
 (4) markedly diminished interest or participation in significant activities
 (5) feeling of detachment or estrangement from others
 (6) restricted range of affect (e.g., unable to have loving feelings)
 (7) sense of a foreshortened future (e.g., does not expect to have a career, marriage, children, or a normal life span)

D. Persistent symptoms of increased arousal (not present before the trauma), as indicated by two (or more) of the following:
 (1) difficulty falling or staying asleep
 (2) irritability or outbursts of anger
 (3) difficulty concentrating
 (4) hypervigilance
 (5) exaggerated startle response

E. Duration of the disturbance (symptoms in criteria B, C, and D) is more than 1 month.

F. The disturbance causes clinically significant distress or impairment in social, occupational, or other important areas of functioning.

Specify if: Acute: if duration of symptoms is less than 3 months
 Chronic: if duration of symptoms is 3 months or more

Specify if: With delayed onset: if onset of symptoms is at least 6 months after the stressor

Reprinted with permission from the Diagnostic and Statistical Manual of Mental Disorders, 4th edn., Text Revision, (copyright 2000). American Psychiatric Association.

The impairment resulting from PTSD is frequently augmented by poor coping strategies, substance abuse, co-occurring mood and anxiety disorders, and lack of social support (Keane et al., 1988; Kessler et al., 1995; Johnsen et al., 2002; Schumm et al., 2006). However, the disability associated with PTSD is not confined to the psychological domain, nor are co-morbidities exclusively related to other mental illnesses. PTSD has been demonstrated to increase risk for numerous medical conditions, such as hypertension, diabetes, metabolic syndrome, and immune and pain disorders (Beckham et al., 1998; Boscarino, 2004; Vieweg et al., 2006; Trief et al., 2006), that are classically underpinned by abnormalities of the hypothalamic–pituitary–adrenal (HPA) axis and other endocrine systems.

Interestingly, however, the study of the neuroendocrinology of PTSD has highlighted endocrine alterations that have not classically been associated with either stress or stress-related disease. The most infamous of these findings – low cortisol levels – has been subjected to much discussion and scrutiny, likely because it has been a counterintuitive result given modern interpretations of the damaging effects of stress hormones. The initial observation of low cortisol in a disorder precipitated by extreme stress directly contradicted the emerging and popular formulation of hormonal responses to stress, the glucocorticoid cascade hypothesis (Sapolsky et al., 1986), which was emerging as a cogent rationale for anti-glucocorticoid treatments in depression and other psychiatric disorders thought to be driven by hypercortisolism. However, the finding of low cortisol in PTSD is part of a growing body of neuroendocrine data providing evidence of insufficient glucocorticoid signaling in stress-related neuropsychiatric disorders (Raison and Miller, 2003).

That cortisol levels are low in PTSD is particularly noteworthy when considered in the context of findings of corticotropin-releasing factor (CRF) levels, which appear to be elevated. Further, as reviewed in detail below, PTSD is also associated with increased cortisol suppression in response to dexamethasone (DEX) administration possibly resulting from increased responsiveness of glucocorticoid receptor (GR). In contrast, studies of acute and chronic stress and depression have demonstrated increased CRF and cortisol levels and reduced cortisol suppression to DEX, and GR responsiveness.

This chapter summarizes findings of the HPA axis in PTSD. Majority of neuroendocrine studies of PTSD demonstrate alterations consistent with an enhanced negative feedback inhibition of cortisol on the pituitary, and/or an overall hyperresponsiveness of other target tissues (adrenal gland, hypothalamus, etc.) to hormonal and other regulatory signals. Recent data, however, suggest that some of the cortisol alterations associated with PTSD may reflect pre-existing risk factors. The significance of these observations will also be discussed.

23.2 Cortisol Levels in PTSD

23.2.1 Twenty-Four-Hour Urinary Excretion of Cortisol

Mason et al. (1986) first reported that the mean 24-h urinary excretion of cortisol was significantly lower (but well within the normal endocrinological range) in Vietnam combat veterans with PTSD compared to psychiatric patients in four other diagnostic groups. The authors expressed surprise at the lower cortisol levels since "certain clinical features such as depression and anxiety [in PTSD] might have been expected to be associated with increased activity of the pituitary–adrenal cortical system." In this initial study, lower cortisol levels were described as sustained since they had remained low across four separate longitudinal observations. However, in a follow-up study examining cortisol levels during a hospitalization designed to discuss highly traumatic material in the context of specialized PTSD psychotherapy, individual fluctuations in cortisol levels were noted relating to different stages of illness and psychological coping strategies (Mason et al., 2001, 2002).

The stability of cortisol levels in PTSD remains an unresolved issue. However, the first longitudinal study of urinary cortisol levels recently completed in Holocaust survivors has shed light on this issue. The first wave of data collection demonstrated lower cortisol levels in Holocaust survivors with, as compared to those without, PTSD (Yehuda et al., 1995b). When the sample was studied 10 years later, there was a high correlation between cortisol levels at time 1 and time 2 ($r(28) = 0.53, p < 0.01$) in the entire sample. However, higher cortisol levels at time 2 than time 1 were only found in those whose PTSD had remitted at a 10-year follow-up, while those who had developed or failed to recover from PTSD had decreased cortisol at follow-up (Yehuda et al., 2007b). These findings are the first suggestion that cortisol levels continue to decline in PTSD in the absence of a remission in clinical state.

Table 2 Summary of data from studies of 24-h urinary cortisol excretion in adults with PTSD

Author(s) (year)	*Trauma survivors with PTSD*		*Trauma w/out PTSD*		*Normal comparison*		*Psychiatric comparison*	
	Cortisol μg day^{-1} (*n*)		Cortisol μg day^{-1} (*n*)		Cortisol μg day^{-1} (*n*)		Cortisol μg day^{-1} (*n*)	
Mason et al. (1986)[a]	33.3	(*n* = 9)					48.5	(*n* = 35)
Kosten et al. (1990)[a]	50.0	(*n* = 11)			55.0	(*n* = 28)	70.0	(*n* = 18)
Pitman and Orr (1990)[b]	107.3	(*n* = 20)	80.5	(*n* = 15)				
Yehuda et al. (1990)[a]	40.9	(*n* = 16)			62.8	(*n* = 16)		
Yehuda et al. (1993a)[a]	38.6	(*n* = 8)					69.4	(*n* = 32)
Yehuda et al. (1995a)[a]	32.6	(*n* = 22)	62.7	(*n* = 25)	51.9	(*n* = 15)		
Lemieux and Coe (1995)[b]	111.8	(*n* = 11)	83.1	(*n* = 8)	87.8	(*n* = 9)		
Maes et al. (1998)[b]	840.0	(*n* = 10)			118	(*n* = 17)	591.0	(*n* = 10)
Thaller et al. (1999)[a]	130.9	(*n* = 34)			213.9	(*n* = 17)		
Baker et al. (1999)	84.4	(*n* = 11)			76.2	(*n* = 12)		
De Bellis et al. (1999)[b]	57.3	(*n* = 18)			43.6	(*n* = 24)	56.0	(*n* = 10)
Yehuda et al. (2000)[a]	48.3	(*n* = 22)			65.1	(*n* = 15)		
Rasmusson et al. (2001)	42.8	(*n* = 12)			34.6	(*n* = 8)		
Glover and Poland (2002)[a,c]	9.8	(*n* = 14)	16.5	(*n* = 7)	12.8	(*n* = 8)		
Otte et al. (2005)	52	(*n* = 20)	43	(*n* = 16)				
Bierer et al. (2006)[a]	46.3	(*n* = 32)	72.2	(*n* = 10)				
Wheler et al. (2006)[a, d]	50.6	(*n* = 10)	73.9	(*n* = 10)				
Yehuda et al. (2007a)	30.8	(*n* = 17)	39.1	(*n* = 11)				
Simeon et al. (2007)[e]	62	(*n* = 32)			62	(*n* = 56)	83	(*n* = 44)
Lemieux et al. (2008)	22.2	(*n* = 11)	48.8	(*n* = 13)	82.2	(*n* = 12)		

[a]Denotes findings in which cortisol levels were significantly lower than comparison subjects, or, in the case of Kosten et al. from depression only.
[b]Denotes findings in which cortisol levels were significantly higher than comparison subjects.
[c]Results are from a 12-h rather than 24-h urine collection and are expressed as $\mu g^{-1} 2\,h^{-1}$.
[d]Cortisol levels were significantly lower in PTSD as measured by radioimmunoassay but not by gas chromatography-mass spectroscopy.
[e]No means reported in the text; data estimated from the figures provided.

Table 2 presents published studies on 24-h urinary cortisol levels in PTSD. The majority of these have found evidence of low cortisol in PTSD, but group differences are not always present between subjects with and without PTSD. This is particularly true when samples are from a general civilian population of persons exposed to a wide range of traumatic events, such as in the subjects obtained by Young and Breslau (2004). Since there are numerous sources of potential variability in such studies related to individual differences, samples heterogeneous with respect to age, gender, trauma type, PTSD chronicity, and developmental stage at which trauma occurred may obscure overall group differences. It is more likely that low cortisol levels are observed in studies examining subjects who are similar in the above characteristics. It is also clear from **Table 2** that enthusiasm for urinary cortisol studies may have been diminished in the last decade. In particular, studies using radioimmunoassay (RIA) to measure cortisol levels from 24-h urinary samples often provide an estimate of total glucocorticoids (free cortisol + inactive metabolites), which may result in different conclusions about cortisol activity. Indeed, Wheler et al. (2006) demonstrated low urinary cortisol in PTSD when using RIA, but found no differences in free cortisol when measured using gas chromatography-mass spectroscopy (GCMS), a finding recently replicated in our laboratory (unpublished observations). It is, of course, more likely the case that urinary cortisol studies have appeared less frequently in the literature because of the logistic difficulties in ensuring completeness of at-home collections. Indeed, studies of cortisol in PTSD in the last decade have tended to use salivary and plasma samples. For a review of cortisol findings in PTSD, see Yehuda (2002).

23.2.2 Single-Time-Point Estimates of Basal Cortisol

Although assessments of single plasma and salivary cortisol levels have become increasingly popular given the relative ease in acquiring samples, these studies have also been hampered by methodological problems and mixed findings. Because of moment-to-moment fluctuations in cortisol levels related to transient stressors (including routine venipuncture) the use of a single sampling of cortisol, particularly at a set time of the day, may not represent an appropriate method for estimating cortisol levels. Variability in single sampling estimates of cortisol may also reflect individual variation in sleep cycles. Given that alterations in basal levels are in the normal endocrinologic range and may be subtle, a very large sample size may be required to discern group differences using single, or even very few, time points during the day. This point is particularly well illustrated in a report by Boscarino of low cortisol in a large epidemiologic sample of over 2000 Vietnam veterans with PTSD compared to those without PTSD, which implies that to consistently observe low morning cortisol would require an extremely large sample size (Boscarino, 1996). The magnitude of difference between PTSD and non-PTSD subjects at 8.00 a.m. was very modest, at just 4%. Cortisol levels were significantly lower in combat veterans with very high exposure (17.9 $\mu g\, day^{-1}$) compared to those with no or low exposure (19.1 $\mu g\, day^{-1}$).

23.2.3 Circadian Rhythm of Cortisol

The best methodology for precise estimates of cortisol levels over the diurnal cycle is one in which subjects are examined under controlled conditions to ensure that complete and representative data have been obtained. In an initial study of circadian parameters in PTSD, blood samples were collected from Vietnam combat veterans with PTSD, subjects (largely veterans) with major depression, and non-psychiatric comparison subjects – every 30 min over a 24-h period under carefully controlled laboratory conditions (Yehuda et al., 1990). Mean basal cortisol release was significantly lower in the PTSD group, with reductions particularly prominent in the late evening and early morning hours. By the time of awakening, however, cortisol release was comparable in PTSD subjects and age-matched controls. PTSD subjects displayed a greater dynamic range of cortisol, as reflected in an increased amplitude-to-mesor ratio. That is, although the cortisol peak among individuals without PTSD was not statistically different from the peak among individuals with PTSD, the lower trough and longer period spent at the nadir in the latter group resulted in a decreased mesor. The amplitude-to-mesor ratio provides an estimate of the signal-to-noise ratio of the system. In contrast, depressed patients showed a less dynamic circadian release of cortisol, reflected in an increased mesor of cortisol release over the 24-h cycle, a decreased amplitude-to-mesor ratio, and an elevated trough. The main implication of these findings is the potential for a greater reactivity of the HPA axis in PTSD.

In a second study, cortisol levels were obtained every 15 min over a 24-h period from a sample of 52 women with and without a history of early childhood sexual abuse and PTSD (Bremner et al., 2003a). Those with PTSD again had significantly lower cortisol levels, this time in the afternoon and evening hours. A more recent study by Bremner et al. (2007) also found lower cortisol, in women with PTSD and early childhood sexual abuse, between noon and 8:00 p.m, compared both to women with abuse but no PTSD and healthy women without history of abuse. Another recent study specifically examined the cortisol-awakening response (the increase in cortisol generally seen in the first hour after waking up) in battered women with PTSD (Johnson et al., 2007). Those with more severe PTSD symptoms showed a greater increase of cortisol after wake up, but cortisol levels were negatively correlated with the duration of abuse. The finding that symptomatology and trauma chronicity may have different effects on cortisol may help explain why findings on cortisol rhythmicity have sometimes been discrepant.

23.2.4 Cortisol Levels in Response to Stress

The circadian rhythm findings demonstrating a dynamic range of cortisol release in PTSD presented a biological scenario to support the idea that the HPA axis may be maximally responsive to stress-related cues in persons who develop PTSD. This would be in contrast to major depressive disorder, which might reflect a condition of minimal responsiveness to the environment. In support of this, Liberzon et al. (1999) observed increased cortisol (but not increased corticotropin (ACTH)) in response to combat sounds versus white noise in combat veterans with PTSD compared to controls. Elizinga et al. (2003) also observed that women with PTSD related to childhood abuse had substantially higher salivary cortisol levels in response to hearing scripts related to their childhood experiences compared to controls who heard scripts of other people's traumatic stories. Similarly, Bremner et al. (2003b) also observed an increased salivary cortisol response in anticipation of a cognitive challenge test in women with PTSD related to childhood abuse, relative to controls. The authors suggest that although cortisol levels were found low at baseline, there did not appear to be an impairment in the cortisol response to stressors in PTSD. These studies demonstrate transient increases in cortisol levels, which are consistent with the notion of a more generalized HPA-axis reactivity in PTSD. An important implication of heightened cortisol reactivity under stressful conditions is that it challenges the idea that cortisol levels are low as a result of a reduced capacity of the adrenal to synthesize cortisol. Rather, lower ambient cortisol levels appear to reflect other regulatory considerations (as described below).

23.2.5 Cortisol as a Pretraumatic Risk Factor

Although cortisol findings in PTSD were initially interpreted as reflecting pathophysiology of this disorder resulting from trauma exposure and/or chronic symptoms, data from prospective longitudinal studies raised the possibility that low cortisol levels reflect pretraumatic predictors of PTSD rather than a consequence of trauma exposure. Low cortisol levels in the immediate aftermath of a motor vehicle accident were found to predict the development of PTSD in a group of 35 accident victims consecutively presenting to an emergency room (Yehuda et al., 1998). Delahanty et al. (2000) also reported that low cortisol levels in the immediate aftermath of a trauma contributed to the prediction of PTSD symptoms at 1 month. Similarly, in a sample of 115 persons who survived a natural disaster, cortisol levels were similarly found to be lowest in those with highest PTSD scores at 1 month post-trauma. However, cortisol levels were not predictive of symptoms at 1 year (Anisman et al., 2001). Lower morning, but higher evening cortisol levels were observed in 15 subjects with more severe PTSD symptoms 5 days following a mine accident in Lebanon, compared to 16 subjects with lower levels of PTSD symptoms (Aardal-Eriksson et al., 2001).

In a study examining the cortisol response in the acute aftermath of rape, low cortisol levels were associated with prior rape or assault, themselves risk factors for PTSD (Resnick et al., 1995), but not with the development of PTSD *per se*. A *post hoc* analysis of the data reported in a study by Yehuda et al. (1998) confirmed that low cortisol levels were associated with prior trauma exposure in this group as well (McFarlane et al. personal communication).

These findings suggest that cortisol levels might have been abnormally low even before trauma exposure in survivors who develop PTSD, therefore representing a preexisting risk factor, at least in some persons (e.g., see Pervanidou et al. (2007), who found that elevated evening cortisol within 24 h of a motor vehicle accident predicted PTSD at 6 months in children and adolescents). Consistent with this, low cortisol levels in adult children of Holocaust survivors were specifically associated with the risk factor of parental PTSD, as shown by both 24-h urine collection (Yehuda et al., 2001) and by plasma samples taken every 30 min over 24 h (Yehuda et al., 2007c). The findings suggest that low cortisol may contribute to secondary biological alterations which ultimately lead to the development of PTSD. Interestingly, the risk factor of parental PTSD in offspring of Holocaust survivors was also associated with an increased incidence of traumatic childhood antecedents in the above study. Thus, low cortisol levels may be present in those who have experienced an adverse event early in life, and then remain different from those not exposed to early adversity. Although there might reasonably be HPA-axis fluctuations in the aftermath of stress, and even differences in the magnitude of such responses compared to those not exposed to trauma early in life, HPA-axis parameters would subsequently recover to their (abnormal) pre-stress baseline. In line with these cortisol findings, maternal (but not paternal) PTSD was significantly associated with offspring having PTSD themselves, suggesting that epigenetic mechanisms such as prenatal glucocorticoid programming of HPA axis responsiveness may be involved in transmission of PTSD risk (Yehuda et al., 2008).

Low cortisol levels may impede the process of biological recovery from stress, resulting in a cascade of alterations that lead to intrusive recollections of the event, avoidance of reminders of the event, and symptoms of hyperarousal. This failure may represent an alternative trajectory to the normal process of adaptation and recovery after a traumatic event. Additionally, it is possible that there is an active process of adaptation and attempt at achieving homeostasis in the period following a trauma before the development of PTSD and that PTSD symptoms are themselves determined by biological responses, rather than the opposite.

PTSD may arise from any number of circumstances, one of which may be the hormonal milieu at the time of trauma, which may itself reflect an interaction of pre- and peritraumatic influences. These responses may be further modified in the days and weeks preceding it by a variety of other influences. For example, under normal circumstances, CRF and ACTH are activated in response to stress, and ultimately culminate in cortisol release, which negatively feeds back to keep the stress response in check. An attenuated cortisol response to an acute trauma might initially lead to a stronger activation of the pituitary due to increased CRF stimulation in synergy with other neuropeptides, such as arginine vasopressin, resulting in a high-magnitude ACTH response. This, in turn, could lead to a greater necessity by the pituitary for negative feedback inhibition in order to achieve regulation, for instance, via a progressive decline in the ACTH/cortisol ratio, facilitated by accommodations in the sensitivity of GRs and other central neuromodulators, ultimately leading to an exaggerated negative feedback inhibition. Affecting these hormonal responses might also be the demands made by post-traumatic factors. Although such a model is hypothetical, it is consistent with the adaptational process of allostatic load described by McEwan and Seeman (1999): physiologic systems accommodate to achieve homeostasis based on already existing predispositions to stress responses. Thus, the neuroendocrine response to the trauma of a person with lower pretrauma cortisol may be fundamentally different from that of someone with a greater adrenal capacity and higher ambient cortisol levels.

One of the most compelling lines of evidence supporting the hypothesis that lower cortisol levels may be an important pathway to the development of PTSD lies in a study by Schelling et al. (2001), who administered stress doses of hydrocortisone during septic shock and evaluated the effects of this treatment on the development of PTSD and traumatic memories in a randomized, double-blind study. High, but physiologic, stress doses of hydrocortisone were associated with reduced PTSD symptoms compared to the group that received saline. Recent research has further demonstrated the therapeutic implications of the low cortisol–PTSD link – patients with PTSD reported lower intrusive symptoms after treatment with low-dose cortisol for 1 month (de Quervain and Margraf, 2008).

23.3 CRF and ACTH Release in PTSD: Baseline Studies

23.3.1 Corticotropin-Releasing Factor

PTSD is associated with a unique profile in that despite low ambient cortisol levels, CRF levels appear to be increased. Several published reports have examined the concentration of CRF in cerebrospinal fluid (CSF) in PTSD (Bremner et al., 1997, Baker et al., 1999, Sautter et al., 2003). Elevated CRF levels were not found to be correlated with 24-h urinary cortisol release (Baker et al., 1999). A recent report found that basal plasma CRF levels were elevated in veterans with PTSD, compared not only to healthy controls but also to veterans without PTSD matched for time and place of deployment (de Kloet et al., 2008a). This indicates that elevated CRF is likely a marker of PTSD *per se*, rather than of simple trauma exposure.

23.3.2 Adrenocorticotropin Hormone

If CRF is hypersecreted in PTSD but cortisol is low, it becomes of great interest to examine the pituitary gland in PTSD, both with respect to baseline ACTH release, and response of ACTH to CRF and related challenge. Among the difficulties in assessing pituitary activity under basal conditions is the fact that the pituitary is subject to multiple positive and negative feedback influences, making its activity difficult to interpret. The pituitary's ACTH release receives both CRF stimulation from the hypothalamus and inhibition from negative feedback of adrenal corticosteroids, thus baseline ACTH levels may appear to be normal even though the pituitary gland is receiving excessive stimulation from CRF. In some studies, ACTH levels in PTSD patients were reported to be comparable to nonexposed subjects. Majority of studies have not reported detectable differences in ACTH levels between PTSD and comparison subjects, even when cortisol levels obtained from the same sample were found to be significantly lower (Kellner et al., 2000; Hocking et al., 1993, Yehuda et al., 1996a, Kanter et al., 2001 Neylan et al., 2003).

Lower cortisol levels in the face of normal ACTH levels can reflect decreased adrenal output. Yet in classic adrenal insufficiency, ACTH release is usually elevated over normal levels. Thus, in PTSD, there may be an additional component of feedback on the pituitary acting to depress ACTH, with the net effect that levels appear normal rather than elevated. Indeed, elevations in ACTH would be expected not only from a reduced adrenal output but also from increased CRF stimulation (Baker et al., 1999, Bremner et al., 1997). Alternatively, the adrenal output in PTSD may be relatively decreased, but not substantially enough to affect ACTH levels. In any event, normal ACTH levels in PTSD, in the context of the other findings, suggests a more complex model of the regulatory influences on the pituitary than adrenal insufficiency.

23.4 Endocrine Challenge Findings Implicating CRF Hypersecretion in PTSD

23.4.1 The Metyrapone Stimulation Test

Metyrapone prevents adrenal steroidogenesis by blocking the conversion of 11-deoxycortisol to cortisol, thereby releasing the pituitary gland from the negative feedback inhibition. A sufficiently high dose of metyrapone (such that an almost complete suppression of cortisol is achieved) allows a direct examination of pituitary release of ACTH without the potentially confounding effects of differing ambient cortisol levels. If metyrapone is administered in the morning, when HPA axis activity is relatively high, maximal pituitary activity can be achieved, facilitating an evaluation of group differences in pituitary capability.

The administration of 2.5 g metyrapone in the morning resulted in a similar and almost-complete reduction in cortisol levels (and consequent removal of negative feedback inhibition) in both PTSD and normal subjects, but in a higher increase in ACTH and 11-deoxycortisol in Vietnam combat veterans with PTSD compared to nonexposed subjects (Yehuda et al., 1996a). Neither pituitary nor adrenal insufficiency would likely result in an increased ACTH response to removal of negative feedback inhibition, since the former would be associated with an attenuated ACTH response, and reduced adrenal output would not necessarily affect the ACTH response. The increased ACTH response is most easily explained by increased suprapituitary activation. However, a sufficiently strong negative feedback inhibition would account for the augmented ACTH response, even in the absence of hypothalamic CRF hypersecretion.

Kanter et al. (2001) failed to find evidence for an exaggerated negative feedback inhibition using a different type of metyrapone stimulation paradigm. In this study, a lower dose of metyrapone was administered over a 3-h period (750 mg at 7.00 a.m. and 10.00 a.m.), and rather than simply examining the ACTH response to this manipulation, the cortisol was infused intravenously, allowing the effects of negative feedback inhibition to be evaluated more systematically. Under conditions of enhanced negative feedback inhibition, the introduction of cortisol following metyrapone administration should result in a greater suppression of ACTH in PTSD. However, no significant differences were observed in the ACTH response to cortisol infusion between PTSD and comparison subjects, although there was a nonsignificant trend, $p = 0.10$, for such a reduction. The authors concluded that their findings provided evidence of subclinical adrenocortical insufficiency. However, the dose of metyrapone used did not accomplish a complete suppression of cortisol (but produced a more robust suppression of cortisol in comparison subjects). Thus, the lack of ACTH reduction may have been caused by a floor effect, rather than by a lack of reactivity of the system. The endogenous cortisol present may already have been high enough to suppress ACTH secretion in the PTSD group. Interestingly, although metyrapone did not result in as great a decline in cortisol in PTSD, it did result in the same level of cortisol inhibition, implying differences in the activity of the enzyme 11-β-hydroxylase, which merits further investigation.

A third study used metyrapone to evaluate CRF effects on sleep: 750 mg of metyrapone was administered at 8.00 a.m. every 4 h for 16 h, and cortisol, 11-deoxycortisol, and ACTH levels were measured at 8.00 a.m. the following morning. All three of these were increased in the PTSD group relative to the controls, suggesting that the same dose of metyrapone did not produce the same degree of adrenal suppression of cortisol synthesis. Under these conditions, it is difficult to evaluate the true effect on ACTH and 11-deoxycortisol, which depends on achieving complete cortisol suppression, or at least the same degree of cortisol suppression in both groups. The endocrine response to metyrapone in this study does not support the model of reduced adrenal capacity; this model predicts a large ratio of ACTH to cortisol release, but the mean ACTH/cortisol ratio prior to metyrapone was no different in PTSD than in controls. On the other hand, the mean ACTH/cortisol ratio postmetyrapone was (nonsignificantly) lower, suggesting, if anything, an exaggerated negative feedback rather than reduced adrenal capacity (Neylan et al., 2003).

23.4.2 CRF Challenge Findings

Infusion of exogenous CRF increases ACTH levels, and provides a test of pituitary sensitivity. In several studies in major depression, the ACTH response to CRF was shown to be blunted, reflecting a reduced sensitivity of the pituitary to CRF (e.g., Nerozzi et al., 1988). This finding has been widely interpreted as reflecting a downregulation of pituitary CRF receptors secondary to CRF hypersecretion, but may also reflect increased cortisol inhibition of ACTH secondary to hypercortisolism (Yehuda and Nemeroff, 1994).

An initial study demonstrated a similarly blunted ACTH response to CRF in PTSD (Smith et al., 1989). However, although the authors noted a uniform blunting of the ACTH response, this did not always occur in the context of hypercortisolism. Furthermore, although the ACTH response was significantly blunted, the cortisol response was not (the area under the curve for cortisol was 38% less than controls, but this was not statistically significant). Bremner et al. (2003a) also observed a blunted ACTH response to CRF in women with PTSD as a result of early childhood sexual abuse.

In contrast, Rasmusson et al. (2001) recently reported an augmented ACTH response to CRF in 12 women with PTSD compared to 11 healthy controls. In the same subjects, the authors also performed a neuroendocrine challenge with 250 μg of Cosyntropin (ACTH) to determine the response of the pituitary gland to this maximally stimulating dose. Women with PTSD demonstrated an exaggerated cortisol response to ACTH compared to healthy subjects. Basal assessments did not reveal group differences in 24-h urinary cortisol levels, basal plasma cortisol, or ACTH levels. The authors concluded that their findings suggested an increased reactivity of both the pituitary and adrenal glands in PTSD.

However, this explanation seems unlikely since the ACTH response to CRF was 87% greater in the subjects with PTSD, but the cortisol response was only 35% higher. Thus, the more marked increase in ACTH in PTSD subjects was not accompanied by a comparable stimulation of cortisol, suggesting a reduced adrenal capacity or an enhanced inhibition of cortisol. But this would contradict the finding of an increased cortisol response to Cosyntropin in the same patients. The observation of an increased ACTH response to CRF would be compatible with a study by Heim et al. (2001), who examined this response in abused women with and without major depressive disorder compared with nonabused depressed women and comparison subjects. Abused women without depression showed an augmented ACTH response to CRF, but a reduced cortisol response to ACTH compared to other groups. It is possible that low cortisol levels resulting from this early trauma may also be influenced by PTSD-related alterations (i.e., increased GR responsiveness and increased responsiveness of negative feedback inhibition).

23.4.3 Cholecystokinin Tetrapeptide Challenge Findings

Cholecystokinin tetrapeptide (CCK-4) is a potent stimulator of ACTH. Kellner et al. (2000) administered a 50-μg bolus of CCK-4 to subjects with PTSD and found substantially attenuated elevations of ACTH, which occurred despite comparable ACTH levels at baseline. Cortisol levels were lower in PTSD at baseline, but rose to a comparable level in PTSD and control subjects after CCK-4. However, the rate of decline from the peak was faster, leading to an overall lower total cortisol surge. The attenuated ACTH response to CCK-4 is compatible with the idea of CRF overdrive in PTSD and is a test similar to the CRF stimulation test described below. That less ACTH can produce a similar activation of the adrenal gland but a more rapid decline of cortisol is also consistent with a more sensitive negative feedback inhibition secondary to increased GR activity at the pituitary. Although the comparatively greater effect on cortisol relative to ACTH is also compatible with an increased sensitivity of the adrenals to ACTH (rather than of the pituitary to negative feedback), this explanation only accounts for the greater rise of cortisol following CCK-4, and not for its more rapid decline after peaking.

23.5 The Dexamethasone Suppression Test and Glucocorticoid Receptors in PTSD

23.5.1 The Dexamethasone Suppression Test

Results using the DEX suppression test (DST) have presented a more consistent view of reduced cortisol suppression in response to DEX administration. The DST provides a direct test of the effects of GR activation in the pituitary on ACTH secretion, and cortisol levels following DEX administration are thus interpreted as an estimate of the strength of negative feedback inhibition, provided that the adrenal response to ACTH is not altered. There are several hundred published studies reporting on the use of the DST in depression, generally reporting that approximately 40–60% of patients with major depression fail to suppress cortisol levels below $5.0\,\mu g\,100\,dl^{-1}$ in response to 1.0 mg of DEX (Ribeiro et al., 1993). Nonsuppression of cortisol results from a reduced ability of DEX to exert negative feedback inhibition on the release of CRF and ACTH (Holsboer, 2000).

The initial DST studies in PTSD using the 1-mg dose of DEX did not consider the possibility of a hypersuppression to DEX, and tested the hypothesis that patients with PTSD might show a nonsuppression of cortisol similar to patients with major depressive disorder. A large proportion of the PTSD subjects studied also met criteria for major depression. Four (Dinan et al., 1990; Halbreich et al., 1989; Kosten et al., 1990; Reist et al., 1995) out of five (Kudler et al., 1987) of the earlier studies noted that PTSD did not appear to be associated with cortisol nonsuppression, using the established criterion of 5 μg 100 ml^{-1} at 4.00 p.m.

Although the 1-mg DST studies primarily focused on evaluating failure of normal negative feedback inhibition, Halbreich et al. (1989) noted that post-DEX cortisol levels in the PTSD group were lower than in subjects with depression and even those in comparison subjects. The mean post-DEX cortisol levels were $0.96 \pm 0.63\ \mu g\,dl^{-1}$ in PTSD compared to $3.72 \pm 3.97\ \mu g\,dl^{-1}$ in depression and $1.37 \pm 0.95\ \mu g\,dl^{-1}$ in comparison subjects, raising the possibility that the 1-mg dose produced a floor effect in the PTSD group.

When lower doses of DEX are administered, 0.50-mg and 0.25-mg doses, a cortisol hypersuppression can clearly be observed, as indicated in **Table 3**.

Results from these studies, in **Table 3**, are expressed as the extent of cortisol suppression, evaluated by the quotient of 8.00 a.m. post-DEX cortisol to 8.00 a.m. baseline cortisol. Expressing the data in this manner accounts for individual differences in baseline cortisol levels and allows for a more precise characterization of the strength of negative feedback inhibition as a continuous rather than as a dichotomous variable. Whereas studies of major depression emphasize the 4.00 p.m. post-DEX value as relevant to the question of nonsuppression (Stokes et al., 1984), studies of PTSD have been concerned with the degree to which DEX suppresses negative feedback at the level of the pituitary, rather than the question of early escape from the effects of DEX.

Following the initial studies of cortisol suppression in response to DEX, there has been some debate about whether DST hypersuppression reflects trauma exposure in psychiatric patients, or PTSD *per se*. Díaz-Marsá et al. (2007) found a similarly enhanced cortisol suppression to DEX in eating disorder patients with childhood trauma compared to patients without history of trauma. The authors attribute the finding to trauma exposure. On the other hand, Grossman et al. (2003) examined the cortisol response to 0.50-mg DEX in a sample of personality-disordered subjects and found that cortisol hypersuppression was related to the comorbid presence of PTSD, but not to trauma exposure.

Yehuda et al. (2004b) observed cortisol hypersuppression following 0.50 mg DST in subjects with PTSD, both with and without comorbid depression, but noted that hypersuppression was particularly prominent in subjects with depression if there had been a prior traumatic experience. Thus, cortisol hypersuppression in response to DEX appears to be associated with PTSD, but in subjects with depression, it may be present as a result of early trauma and, possibly, past PTSD (Yehuda et al., 2004b).

Table 3 Cortisol suppression to DEX in PTSD and comparison subjects

Author (year)	*DEX dose day^{-1}*	*PTSD (%) supp (n)*		*Comparison (%) supp (n)*	
Yehuda et al. (1993b)[a]	0.5	87.5	($n = 21$)	68.3	($n = 12$)
Stein et al. (1997)[a]	0.5	89.1	($n = 13$)	80.0	($n = 21$)
Yehuda et al. (1995b)[a]	0.5	90.0	($n = 14$)	73.4	($n = 14$)
Yehuda et al. (1995b)[a]	0.25	54.4	($n = 14$)	36.7	($n = 14$)
Kellner et al. (1997)[b]	0.50	90.1	($n = 7$)		
Yehuda et al. (2002)[a, c]	0.50	89.9	($n = 17$)	77.9	($n = 23$)
Grossman et al. (2003)[a, d]	0.50	83.6	($n = 16$)	63.0	($n = 36$)
Newport et al. (2004)[a, e]	0.50	92.3	($n = 16$a)	77.78	($n = 19$)
Yehuda et al. (2004a)[a]	0.50	82.5	($n = 19$)	68.9	($n = 10$)
Lange et al. (2005)[a, f]	0.50	85.7	($n = 12$)	65.5	($n = 9$)
Griffin et al. (2005)	0.50	85.0	($n = 42$)	75.0	($n = 14$)

[a]Significantly more suppressed than controls.
[b]No control group was studied.
[c]PTSD group only includes subjects without depression; subjects with both PTSD and MDD ($n = 17$) showed a percent suppression of 78.8, which differs from our previous report (Yehuda et al. 1993b) in younger combat veterans.
[d]Comparison subjects were those with personality disorders but without PTSD.
[e]It is impossible from this chapter to get the correct mean for the actual 15 subjects with PTSD. These 16 subjects had MDD, but 15/16 also had PTSD, so this group also contains one subject who had been exposed to early abuse with past, but not current PTSD.
[f]This study compared borderline personality disorder patients with and without comorbid PTSD.

23.5.2 The Combined DEX/CRF Test

The response to CRF challenge after pretreatment with DEX has been investigated in PTSD – an approach used frequently in affective disorders but only recently in PTSD. In the first such study, no differences in cortisol or ACTH were found the day after 11.00 p.m. self-administration of DEX, either before or after CRF administration, although patients with history of childhood abuse actually had a significantly higher ACTH response to DEX (Muhtz et al., 2007). Heim et al. (2008) found a similarly enhanced ACTH response to DEX/CRF in depressed men abused in childhood over both healthy controls and nonabused depressed men, suggesting that this particular finding may be a marker for childhood trauma rather than an associate of specific psychopathology. Using the combined DEX/CRF challenge in women with borderline personality disorder with and without PTSD related to sustained childhood abuse, Rinne et al. (2002) demonstrated that chronically abused patients had a significantly enhanced ACTH and cortisol response to the DEX/CRF challenge compared with nonabused subjects, suggesting a hyperresponsiveness of the HPA axis.

Ströhle et al. (2008) examined effects of the DEX/CRF test while controlling for potential confounds to a greater degree than previous studies – patients in the PTSD group were not medicated and had no history of childhood trauma. More consistent with previous challenge studies, PTSD patients showed a blunted ACTH response to CRF and a hypersuppression of cortisol to DEX compared to healthy controls. In de Kloet et al. (2008b), there were no significant differences between PTSD and controls in response to DEX/CRF. A subgroup of PTSD patients with comorbid depression did exhibit a blunted ACTH response, one possible explanation being that there are biological subtypes within people diagnosed with PTSD, associated with different patterns of HPA-axis responsiveness.

23.5.3 Glucocorticoid Receptors

Type II GRs are expressed in ACTH- and CRF-producing cells of the pituitary, hypothalamus, and hippocampus, and mediate most systemic glucocorticoid effects, particularly those related to stress responsiveness (deKloet et al., 1991). Low circulating levels of a hormone or neurotransmitter can result in increased numbers of available receptors (Sapolsky et al., 1984), improving response capacity and facilitating homeostasis. However, alterations in the number and sensitivity of both type I (mineralocorticoid) and type II GRs can also significantly influence HPA-axis activity, particularly by modulating the strength of negative feedback, thereby regulating hormone levels (Svec, 1985; Holsboer et al., 1995).

Lymphocyte and brain GRs have been found to share similar regulatory and binding characteristics (Lowy, 1989). A greater number of 8.00 a.m., but not 4.00 p.m., mononuclear leukocyte (presumably lymphocyte) type II GRs was reported in Vietnam veterans with PTSD compared to a normal comparison group (Yehuda et al., 1991). Subsequently, Yehuda et al. (1993a) reported an inverse relationship between 24-h urinary cortisol excretion and lymphocyte GR number in PTSD and depression (with low cortisol and increased receptor levels in PTSD vs. elevated cortisol and reduced receptor levels in depression). Although it is not clear whether alterations in GR number reflect an adaptation to low cortisol levels or some other alteration, the observation of an increased number of lymphocyte GRs provided the basis for the hypothesis of an increased negative feedback inhibition of cortisol secondary to increased receptor sensitivity.

Following the administration of 0.25-mg dose of DEX, it was possible to observe that the cortisol response was accompanied by a concurrent decline in the number of cytosolic lymphocyte receptors (Yehuda et al., 1996b). This finding contrasts with that of a reduced decline in the number of cytosolic lymphoycte receptors in major depression, implying that the reduced cortisol levels following DEX administration may reflect an enhanced negative feedback inhibition in PTSD (Gormley et al., 1985).

Observations regarding the cellular immune response in PTSD are also consistent with enhanced GR responsiveness in the periphery. In one study, beclomethasone-induced vasoconstriction was increased in female PTSD subjects compared to healthy, non-trauma-exposed comparison subjects (Coupland et al., 2003). Similarly, an enhanced delayed-type hypersensitivity of skin-test responses was observed in women who survived childhood sexual abuse versus those who did not (Altemus et al., 2003). Because immune responses, like endocrine ones, can be multiply regulated, these studies provide only indirect evidence of GR responsiveness. However, the observation that PTSD patients showed increased expression of the receptors in all lymphocyte subpopulations, despite both a relatively lower quantity of intracellular GR (as determined by flow cytometry), and lower ambient cortisol levels

(Gotovac et al., 2003), suggests more convincingly that these findings are due to an enhanced sensitivity of the GR to glucocorticoids. Furthermore, Kellner et al. (2002) reported an absence of alterations of the mineralocorticoid receptor in PTSD as investigated by examining the cortisol and ACTH response to spironolactone following CRF stimulation.

Finally, one study provided a demonstration of an alteration in target tissue sensitivity in glucocorticoids using an *in vitro* paradigm. Mononuclear leukocytes isolated from the blood of 26 men with PTSD and 18 men without PTSD were incubated with a series of concentrations of DEX to determine the rate of inhibition of lysozyme activity. A portion of cells was frozen for the determination of GR. Subjects with PTSD showed evidence of a greater sensitivity to glucocorticoids as reflected by a significantly lower mean lysozyme $IC_{50\text{-}DEX}$. The lysozyme $IC_{50\text{-}DEX}$ was significantly correlated with age at exposure to the first traumatic event in subjects with PTSD. The number of cytosolic GR was correlated with age at exposure to the focal traumatic event (Yehuda et al., 2004a).

23.5.4 Effects of Exogenous Cortisol Administration

Although the above studies provide information about cortisol negative feedback inhibition and GR responsiveness in the periphery, the extent to which they reflect changes in central glucocorticoid responsiveness is not clear. To address this question, 17.5 mg of hydrocortisone succinate was administered in a single intravenous (IV) dose and the ACTH response to this dose was quantitated. The ACTH response to hydrocortisone is conceptually similar to the cortisol response to DEX, but DEX does not cross the blood–brain barrier, and acts mostly at the level of the pituitary. A greater ACTH decline was observed in combat veterans with, compared to those without, PTSD, implying that both peripheral and central GR are more responsive (Yehuda et al., 2006). Importantly, this single dose was found to affect memory performance (Yehuda et al., 2007a) and alter glucose metabolism (Yehuda et al., in press), thus suggesting that enhanced glucocorticoid responsiveness may contribute to PTSD pathophysiology.

23.6 Putative Models of HPA-Axis Alterations in PTSD

Cortisol levels are most often found to be abnormally low in PTSD, but can also be similar to or greater than those in comparison subjects. Findings of changes in circadian rhythm suggest that there may be regulatory influences that result in a greater dynamic range of cortisol release over the diurnal cycle in PTSD. Together, they suggest that although cortisol levels may be generally lower, the adrenal gland is certainly capable of producing adequate amounts of cortisol in response to challenge.

The model of enhanced negative feedback inhibition is compatible with the idea that there may be transient elevations in cortisol, but suggests that when present, these increases will be shorter-lived due to a more efficient containment of ACTH release as a result of enhanced GR activation. Thus, there may be chronic or transient elevations in CRF in PTSD which stimulate pituitary release of ACTH, which, in turn, stimulates the adrenal release of cortisol. However, increased GR number and sensitivity can result in reduced cortisol levels under ambient conditions (Yehuda et al., 1996b). In contrast to explanatory models of endocrinopathy, which identify specific and usually singular primary alterations in endocrine organs and/or regulation, the model of enhanced negative feedback inhibition in PTSD is, in large part, descriptive. It currently offers little explanation for why some individuals show such alterations of the HPA axis following exposure to traumatic experiences while others do not. It nonetheless represents an important development in the field of neuroendocrinology of PTSD by accounting for a substantial proportion of the findings observed, and can be put to further hypothesis testing.

23.7 Conclusions

HPA-axis alterations in PTSD seem to be complex, and may be associated with different aspects of the disorder, including preexisting risk factors. For the findings to coalesce into an integrative neuroendocrine hypothesis of PTSD, the following would have to be demonstrated:

1. Some features of the HPA axis may be altered prior to the exposure to a focal trauma;
2. Components of the HPA axis are not uniformly regulated (e.g., circadian rhythm patterns, tonic cortisol secretion, negative feedback inhibition, and the cortisol response to stress are differentially mediated);
3. The system is dynamic, and may therefore show transient increases or hyperresponsivity under certain environmental conditions;

4. Other regulatory influences may affect HPA-axis regulation in PTSD; and probably (though not necessarily),
5. There may be different biologic variants of PTSD with relatively similar phenotypic expressions, as is the case with major depressive disorder.

The wide range of findings on the neuroendocrinology of PTSD underscores the important observation made by Mason et al. (1986): HPA-axis response patterns in PTSD are fundamentally in the normal range and do not reflect endocrinopathy. In endocrinologic disorders, where there is typically a lesion in one or more target tissues or biosynthetic pathways, endocrine methods can usually isolate the problem with appropriate tests and subsequently obtain rather consistent results. In psychiatric disorders, neuroendocrine alterations may be subtle; consequently, it is not always likely that standard tools for assessing endocrine alterations will reveal all the alterations consistent with a neuroendocrine explanation of the pathology in tandem. Additionally, disparate results may be observed within the same patient group owing to a stronger compensation or re-regulation of the HPA axis following challenge.

The next generation of studies should aim to apply more rigorous neuroendocrine tests to the investigation of PTSD, in consideration of appropriate developmental issues, the longitudinal course of the disorder, and individual differences that affect relevant processes. No doubt such studies will require a closer examination of a wide range of biologic responses, including the cellular and molecular mechanisms involved in adaptation to stress, and an understanding of the relationship between the endocrine findings and other identified biologic alterations in PTSD.

Acknowledgments

This work was supported by MH 49555, MH 55-7531, and MERIT review funding.

References

Aardal-Eriksson E, Eriksson TE, and Thorell L (2001) Salivary cortisol, posttraumatic stress symptoms and general health in the acute phase and during 9-month follow-up. *Biological Psychiatry* 50: 986–993.

Altemus M, Cloitre M, and Dhabhar FR (2003) Enhanced cellular immune responses in women with PTSD related to childhood abuse. *American Journal of Psychiatry* 160: 1705–1707.

American Psychiatric Association (1980) *Diagnostic and Statistical Manual of Mental Disorders*, 3rd edn. Washington, DC: American Psychiatric Association.

American Psychiatric Association (2000) *Diagnostic and Statistical Manual of Mental Disorders*, 4th edn. Washington, DC: American Psychiatric Association.

Anisman H, Griffiths J, Matheson K, Ravindran AV, and Merali Z (2001) Posttraumatic stress symptoms and salivary cortisol levels. *American Journal of Psychiatry* 158: 1509–1511.

Baker DG, West SA, Nicholson WE, et al. (1999) Serial CSF corticotropin-releasing hormone levels and adrenocortical activity in combat veterans with posttraumatic stress disorder (published erratum appears in *American Journal of Psychiatry* (1999) 156: 986). *American Journal of Psychiatry* 156: 585–588.

Beckham JC, Moore SD, Feldman ME, Hertzberg MA, Kirby AC, and Fairbank JA (1998) Health status, somatization, and severity of posttraumatic stress disorder in Vietnam combat veterans with posttraumatic stress disorder. *American Journal of Psychiatry* 155: 1565–1569.

Bierer LM, Tischler L, Labinsky E, Cahill S, Foa E, and Yehuda R (2006) Clinical correlates of 24-h cortisol and norepinephrine excretion among subjects seeking treatment following the World Trade Center attacks on 9/11. *Annals of the New York Academy of Sciences* 1071: 514–520.

Boscarino JA (1996) Posttraumatic stress disorder, exposure to combat, and lower plasma cortisol among Vietnam veterans: Findings and clinical implications. *Journal of Consulting and Clinical Psychology* 64: 191–201.

Boscarino JA (2004) Posttraumatic stress disorder and physical illness: Results from clinical and epidemiologic studies. *Annals of the New York Academy of Sciences* 1032: 141–153.

Bremner D, Vermetten E, and Kelley ME (2007) Cortisol, dehydroepiandrosterone, and estradiol measured over 24 hours in women with childhood sexual abuse-related posttraumatic stress disorder. *Journal of Nervous and Mental Disease* 195: 919–927.

Bremner JD, Licinio J, Darnell A, et al. (1997) Elevated CSF corticotropin-releasing factor concentrations in posttraumatic stress disorder. *American Journal of Psychiatry* 154: 624–629.

Bremner JD, Vythilingam M, Anderson G, et al. (2003a) Assessment of the hypothalamic–pituitary–adrenal axis over a 24-hour diurnal period and in response to neuroendocrine challenges in women with and without childhood sexual abuse and posttraumatic stress disorder. *Biological Psychiatry* 54: 710–718.

Bremner JD, Vythilingma M, Vermetten E, et al. (2003b) Cortisol response to a cognitive stress challenge in posttraumatic stress disorder related to childhood abuse. *Psychoneuroendocrinology* 28: 733–750.

Coupland NJ, Hegadoren KM, and Myrholm J (2003) Increaed beclomethasone-induced vasoconstriction in women with posttraumatic stress disorder. *Journal of Psychiatric Research* 37: 221–228.

De Bellis MD, Baum AS, Birmaher B, et al. (1999) A.E. Bennett Research Award. Developmental traumatology. Part I: Biological stress systems. *Biological Psychiatry* 45: 1259–1270.

de Kloet ER, Joels M, Oitzl M, and Sutanto W (1991) Implication of brain corticosteroid receptor diversity for the adaptation syndrome concept. *Methods and Achievements in Experimental Pathology* 14: 104–132.

de Kloet CS, Vermetten E, Geuze E, Lentjes EG, Heijnen CJ, Stalla GK, and Westenberg HG (2008a) Elevated plasma corticotrophin-releasing hormone levels in veterans with

posttraumatic stress disorder. *Progress in Brain Research* 167: 287–291.
de Kloet CS, Vermetten E, Lentjes E, et al. (2008b) Differences in the response to the combined DEX-CRH test between PTSD patients with and without co-morbid depressive disorder. *Psychoneuroendocrinology* 33: 313–320.
Delahanty DL, Raimonde AJ, and Spoonster E (2000) Initial posttraumatic urinary cortisol levels predict subsequent PTSD symptoms in motor vehicle accident victims. *Biological Psychiatry* 48: 940–947.
de Quervain DJ and Margraf J (2008) Glucocorticoids for the treatment of post-traumatic stress disorder and phobias: A novel therapeutic approach. *European Journal of Pharmacology* 583: 365–371.
Díaz-Marsá M, Carrasco JL, Basurte E, Pastrana JI, Sáiz-Ruiz J, and López-Ibor JJ (2007) Findings with 0.25 mg dexamethasone suppression test in eating disorders: Association with childhood trauma. *CNS Spectrum* 12: 675–680.
Dinan TG, Barry S, Yatham LN, Mobayed M, and Brown I (1990) A pilot study of a neuroendocrine test battery in posttraumatic stress disorder. *Biological Psychiatry* 28: 665–672.
Glover D and Poland R (2002) Urinary cortisol and catecholamines in mothers of child cancer survivors with and without PTSD. *Psychoneuroendocrinology* 27: 805–819.
Gormley GJ, Lowy MT, Reder AT, Hospelhorn VD, Antel JP, and Meltzer HY (1985) Glucocorticoid receptors in depression: Relationship to the dexamethasone suppression test. *American Journal of Psychiatry* 142: 1278–1284.
Griffin MG, Resick PA, and Yehuda R (2005) Enhanced cortisol suppression following dexamethasone administration in domestic violence survivors. *American Journal of Psychiatry* 162: 1192–1199.
Grossman R, Yehuda R, New A, et al. (2003) Dexamethasone suppression test findings in subjects with personality disorders: Associations with posttraumatic stress disorder and major depression. *American Journal of Psychiatry* 160: 1291–1298.
Halbreich U, Olympia J, Carson S, Glogowski J, Yeh CM, Axelrod S, and Desu MM (1989) Hypothalamo–pituitary–adrenal activity in endogenously depressed posttraumatic stress disorder patients. *Psychoneuroendocrinology* 14: 365–370.
Heim C, Mletzko T, Purselle D, Musselman DL, and Nemeroff CB (2008) The dexamethasone/ corticotropin-releasing factor test in men with major depression: Role of childhood trauma. *Biological Psychiatry* 63: 398–405.
Heim C, Newport DJ, Bonsall R, Miller AH, and Nemeroff CB (2001) Altered pituitary–adrenal axis responses to provocative challenge tests in adult survivors of childhood abuse. *American Journal of Psychiatry* 158: 575–581.
Holsboer F (2000) The corticosteroid receptor hypothesis of depression. *Neuropsychopharmacology* 23: 477–501.
Holsboer F, Lauer CJ, Schreiber W, and Krieg JC (1995) Altered hypothalamic–pituitary–adrenocortical regulation in healthy subjects at high familial risk for affective disorders. *Neuroendocrinology* 62: 340–347.
Johnsen BH, Eid J, Laberg JC, and Thayer JF (2002) The effect of sensitization and coping style on post-traumatic stress symptoms and quality of life: Two longitudinal studies. *Scandinavian Journal of Psychology* 43: 181–188.
Johnson DM, Delahanty DL, and Pinna K (2007) The cortisol awakening response as a function of PTSD severity and abuse chronicity in sheltered battered women. *Journal of Anxiety Disorders* 22: 793–800.
Kanter ED, Wilkinson CW, Radant AD, et al. (2001) Glucocorticoid feedback sensitivity and adrenocortical responsiveness in posttraumatic stress disorder. *Biological Psychiatry* 50: 238–245.
Keane TM, Gerardi RJ, Lyons JA, and Wolfe J (1988) The interrelationship of substance abuse and posttraumatic stress disorder: Epidemiological and clinical considerations. *Recent Developments in Alcoholism* 6: 27–48.
Kellner M, Baker DG, and Yehuda R (1997) Salivary cortisol in Operation Desert Storm returnees. *Biological Psychiatry* 42: 849–850.
Kellner M, Baker DG, Yassouridis A, Bettinger S, Otte C, Naber D, and Wiedemann K (2002) Mineralocorticoid receptor function in patients with posttraumatic stress disorder. *American Journal of Psychiatry* 159: 1938–1940.
Kellner M, Wiedemann K, Yassouridis A, Levengood R, Guo LS, Holsboer F, and Yehuda R (2000) Behavioral and endocrine response to cholecystokinin tetrapepetide in patients with posttraumatic stress disorder. *Biological Psychiatry* 47: 107–111.
Kessler RC, Sonnega A, Bromet E, Hughes M, and Nelson CB (1995) Posttraumatic stress disorder in the National Comorbidity Survey. *Archives of General Psychiatry* 52: 1048–1060.
Kosten TR, Wahby V, Giller E, and Mason J (1990) The dexamethasone suppression test and thyrotropin-releasing hormone stimulation test in posttraumatic stress disorder. *Biological Psychiatry* 28: 657–664.
Kudler H, Davidson J, Meador K, Lipper S, and Ely T (1987) The DST and posttraumatic stress disorder. *American Journal of Psychiatry* 14: 1058–1071.
Lange W, Wulff H, Berea C, Beblo T, Saavedra AS, Mensebach C, Wingenfeld K, and Driessen M (2005) Dexamethasone suppression test in borderline personality disorder – effects of posttraumatic stress disorder. *Psychoneuroendocrinology* 30: 919–923.
Lemieux AM and Coe CL (1995) Abuse-related posttraumatic stress disorder: Evidence for chronic neuroendocrine activation in women. *Psychosomatic Medicine* 57: 105–115.
Lemieux A, Coe CL, and Carnes M (2008) Symptom severity predicts degree of T cell activation in adult women following childhood maltreatment. *Brain, Behavior and Immunity* 22: 994–1003.
Liberzon I, Abelson JL, Flagel SB, Raz J, and Young EA (1999) Neuroendocrine and psychophysiologic responses in PTSD: A symptom provocation study. *Neuropsychopharmacology* 21: 40–50.
Lowy MT (1989) Quantification of type I and II adrenal steroid receptors in neuronal, lymphoid and pituitary tissues. *Brain Research* 503: 191–197.
Maes M, Lin A, Bonaccorso S, et al. (1998) Increased 24-hour urinary cortisol excretion in patients with post-traumatic stress disorder and patients with major depression, but not in patients with fibromyalgia. *Acta Psychiatrica Scandinavica* 98: 328–335.
Mason JW, Giller EL, Kosten TR, Ostroff RB, and Podd L (1986) Urinary free-cortisol levels in posttraumatic stress disorder patients. *Journal of Nervous and Mental Disease* 174: 145–159.
Mason JW, Wang S, Yehuda R, Lubin H, Johnson D, Bremner JD, Charney D, and Southwick S (2002) Marked lability in urinary cortisol levels in subgroups of combat veterans with posttraumatic stress disorder during an intensive exposure treatment program. *Psychosomatic Medicine* 64: 238–246.
Mason JW, Wang S, Yehuda R, Riney S, Charney DS, and Southwick SM (2001) Psychogenic lowering of urinary

cortisol levels linked to increased emotional numbing and a shame-depressive syndrome in combat-related posttraumatic stress disorder. *Psychosomatic Medicine* 63: 387–401.
McEwan BS and Seeman T (1999) Protective and damaging effects of mediators of stress: Elaborating and testing the concepts of allostasis and allostatic load. *Annals of the New York Academy of Sciences* 896: 30–47.
Muhtz C, Wester M, Yassouridis A, Wiedemann K, and Kellner M (2007) A combined dexamethasone/corticotropin-releasing hormone test in patients with chronic PTSD – first preliminary results. *Journal of Psychiatric Research* 42: 689–693.
Nerozzi D, Bersani G, Melia E, Magnani A, Antonozzi I, and Frajese G (1988) Corticotropin-releasing factor and adrenal function in major depression. *Journal of Endocrinological Investigation* 11: 697–701.
Newport DJ, Heim C, Bonsall R, Miller AH, and Nemeroff CB (2004) Pituitary–adrenal responses to standard and low-dose dexamethasone suppression tests in adult survivors of child abuse. *Biological Psychiatry* 55: 10–20.
Neylan TC, Lenoci M, Maglione ML, et al. (2003) Delta sleep response to metyrapone in posttraumatic stress disorder. *Neuropsyceopharmacology* 28: 1666–1676.
Otte C, Lenoci M, Metzler T, Yehuda R, Marmar CR, and Neylan TC (2005) Hypothalamic–pituitary–adrenal axis activity and sleep in posttraumatic stress disorder. *Neuropsychopharmacology* 30: 1173–1180.
Pervanidou P, Kolaitis G, Charitaki S, et al. (2007) Elevated morning serum interleukin (IL)-6 or evening salivary cortisol concentrations predict posttraumatic stress disorder in children and adolescents six months after a motor vehicle accident. *Psychoneuroendocrinology* 32: 991–999.
Pitman RK and Orr SP (1990) Twenty-four hour urinary cortisol and catecholamine excretion in combat-related posttraumatic stress disorder. *Biological Psychiatry* 27: 245–247.
Raison CL and Miller AH (2003) When not enough is too much: The role of insufficient glucocorticoid signaling in the pathophysiology of stress related disorders. *American Journal of Psychiatry* 160: 1554–1565.
Rasmusson AM, Lipschitz DS, Wang S, Hu S, Vojvoda D, Bremner JD, Southwick SM, and Charney D (2001) Increased pituitary and adrenal reactivity in premenopausal women with posttraumatic stress disorder. *Biological Psychiatry* 12: 965–977.
Reist C, Kauffmann ED, Chicz-Demet A, Chen CC, and Demet EM (1995) REM latency, dexamethasone suppression test, and thyroid releasing hormone stimulation test in posttraumatic stress disorder. *Progress in Neuropsychopharmacology and Biological Psychiatry* 19: 433–443.
Resnick HS, Yehuda R, Pitman RK, and Foy DW (1995) Effect of previous trauma on acute plasma cortisol level following rape. *American Journal of Psychiatry* 152: 1675–1677.
Ribeiro SC, Tandon R, Grunhaus L, and Greden JF (1993) The DST as a predictor of outcome in depression: A meta-analysis. *American Journal of Psychiatry* 150: 1618–1629.
Rinne T, deKloet ER, Wouters L, Goekoop JG, DeRijk RH, and van de Brink W (2002) Hyperresponsiveness of hypothalamic–pituitary–adrenal axis to combined dexamethasone/corticotropin-releasing hormone challenge in female borderline personality disorder subjects with a history of sustained childhood abuse. *Biological Psychiatry* 52: 1102–1112.
Sapolsky RM, Krey LC, and McEwen BS (1984) Stress down-regulates corticosterone receptors in a site-specific manner in the brain. *Endocrinology* 114: 287–292.
Sapolsky RM, Krey LC, and McEwen BS (1986) The neuroendocrinology of stress and aging: The glucocorticoid cascade hypothesis. *Endocrine Reviews* 7: 284–301.
Sautter FJ, Bisette G, Wiley J, et al. (2003) Corticotropin-releasing factor in posttraumatic stress disorder with secondary psychotic symptoms, nonpsychotic PTSD, and health control subjects. *Biological Psychiatry* 54: 1382–1388.
Schelling G, Briegel J, Roozendaal B, Stoll C, Rothenhausler HB, and Kapfhammer HP (2001) The effect of stress doses of hydrocortisone during septic shock on posttraumatic stress disorder in survivors. *Biological Psychiatry* 50: 978–985.
Schumm JA, Briggs-Phillips M, and Hobfoll SE (2006) Cumulative interpersonal traumas and social support as risk and resiliency factors in predicting PTSD and depression among inner-city women. *Journal of Traumatic Stress* 19: 825–836.
Simeon D, Knutelska M, Yehuda R, Putnam F, Schmeidler J, and Smith LM (2007) Hypothalamic–pituitary–adrenal axis function in dissociative disorders, post-traumatic stress disorder, and healthy volunteers. *Biological Psychiatry* 61: 966–967.
Smith MA, Davidson J, Ritchie JC, Kudler H, Lipper S, Chappell P, and Nemeroff CB (1989) The corticotropin-releasing hormone test in patients with posttraumatic stress disorder. *Biological Psychiatry* 26: 349–355.
Stein MB, Yehuda R, Koverola C, and Hanna C (1997) Enhanced dexamethasone suppression of plasma cortisol in adult women traumatized by childhood sexual abuse. *Biological Psychiatry* 42: 680–686.
Stokes PE, Stoll PM, Koslow SH, Maas JW, Davis JM, Swann AC, and Robins E (1984) Pretreatment DST and hypothalamic–pituitary–adrenocortical function in depressed patients and comparison groups. A multicenter study. *Archives of General Psychiatry* 41: 257–267.
Ströhle A, Scheel M, Modell S, and Holsboer F (2008) Blunted ACTH response to dexamethasone suppression-CRH stimulation in posttraumatic stress disorder. *Journal of Psychiatric Research* 42: 1185–1188.
Svec F (1985) Glucocorticoid receptor regulation. *Life Sciences* 36: 2359–2366.
Thaller V, Vrkljan M, Hotujac L, and Thakore J (1999) The potential role of hypocortisolism in the pathophysiology of PTSD and psoriasis. *Collegium Antropologicum* 23: 611–619.
Trief PM, Ouimette P, Wade M, Shanahan P, and Weinstock RS (2006) Post-traumatic stress disorder and diabetes: Co-morbidity and outcomes in a male veterans sample. *Journal of Behavioral Medicine* 29: 411–418.
Vieweg WV, Julius DA, Fernandez A, Tassone DM, Narla SN, and Pandurangi AK (2006) Posttraumatic stress disorder in male military veterans with comorbid overweight and obesity: Psychotropic, antihypertensive, and metabolic medications. *Primary Care Companion to the Journal of Clinical Psychiatry* 8: 25–31.
Wheler GH, Brandon D, Clemons A, Riley C, Kendall J, Loriaux DL, and Kinzie JD (2006) Cortisol production rate in posttraumatic stress disorder. *Journal of Clinical Endocrinology and Metabolism* 91: 3486–3489.
Yehuda R (2002) Current status of cortisol findings in post-traumatic stress disorder. *Psychiatric Clinics of North America* 25: 341–368, vii.
Yehuda R, Bell A, Bierer LM, and Schmeidler J (2008) Maternal, not paternal, PTSD is related to increased risk for PTSD in offspring of Holocaust survivors. *Journal of Psychiatric Research* 42: 1104–1111.
Yehuda R, Bierer LM, Schmeidler J, Aferiat DH, Breslau I, and Dolan S (2000) Low cortisol and risk for PTSD in adult

offspring of holocaust survivors. *American Journal of Psychiatry* 157: 1252–1259.
Yehuda R, Boisoneau D, Lowy MT, and Giller EL (1995a) Dose-response changes in plasma cortisol and lymphocyte glucocorticoid receptors following dexamethasone administration in combat veterans with and without posttraumatic stress disorder. *Archives of General of Psychiatry* 52: 583–593.
Yehuda R, Boisoneau D, Mason JW, and Giller EL (1993a) Glucocorticoid receptor number and cortisol excretion in mood, anxiety, and psychotic disorders. *Biological Psychiatry* 34: 18–25.
Yehuda R, Golier J, Yang RK, and Tischler L (2004a) Enhanced sensitivity to glucocorticoids in peripheral mononuclear leukocytes in posttraumatic stress disorder. *Biological Psychiatry* 55: 1110–1116.
Yehuda R, Halligan SL, Golier J, Grossman R, and Bierer LM (2004b) Effects of trauma exposure on the cortisol response to dexamethasone administration in PTSD and major depressive disorder. *Psychoneuroendocrinology* 29: 389–404.
Yehuda R, Halligan SL, and Grossman R (2001) Childhood trauma and risk for PTSD: Relationship to intergenerational effects of trauma, parental PTSD and cortisol excretion. *Development and Psychopathology* 13: 733–753.
Yehuda R, Halligan SL, Grossman R, Golier JA, and Wong C (2002) The cortisol and glucocorticoid receptor response to low dose dexamethasone administration in aging combat veterans and holocaust survivors with and without posttraumatic stress disorder. *Biological Psychiatry* 52: 393–403.
Yehuda R, Harvey PD, Buchsbaum M, Tischler L, and Schmeidler J (2007a) Enhanced effects of cortisol administration on working memory in PTSD. *Neuropsychopharmacology* 32: 2581–2591.
Yehuda R, Harvey PD, Golier J, et al. (in press) ^{18}FDG-PET response to cortisol in aging veterans with post-traumatic stress disorder. *Journal of Neuropsychiatry and Clinical Neurosciences*.
Yehuda R, Kahana B, Binder-Brynes K, Southwick S, Mason JW, and Giller EL (1995b) Low urinary cortisol excretion in Holocaust survivors with posttraumatic stress disorder. *American Journal of Psychiatry* 152: 982–986.
Yehuda R, Levengood RA, Schmeidler J, Wilson S, Guo LS, and Gerber D (1996a) Increased pituitary activation following metyrapone administration in post-traumatic stress disorder. *Psychoneuroendocrinology* 21: 1–16.
Yehuda R, Lowy MT, Southwick S, Shaffer D, and Giller EL (1991) Lymphocyte glucocorticoid receptor number in posttraumatic stress disorder. *American Journal of Psychiatry* 148: 499–504.
Yehuda R, Morris A, Labinsky E, Zemelman S, and Schmeidler J (2007b) Ten-year follow-up study of cortisol levels in aging holocaust survivors with and without PTSD. *Journal of Traumatic Stress* 20: 757–761.
Yehuda R and Nemeroff CB (1994) Neuropeptide alterations in affective and anxiety disorders. In: DenBoer JA and Sisten A (eds.) *Handbook on Depression and Anxiety: A Biological Approach*, pp. 543–571. New York: Marcel Dekker.
Yehuda R, Shalev AY, and McFarlane AC (1998) Predicting the development of posttraumatic stress disorder from the acute response to a traumatic event. *Biological Psychiatry* 44: 1305–1313.
Yehuda R, Southwick SM, Krystal JH, Bremner D, Charney DS, and Mason JW (1993b) Enhanced suppression of cortisol following dexamethasone administration in posttraumatic stress disorder. *American Journal of Psychiatry* 150: 83–86.
Yehuda R, Southwick SM, Nussbaum G, Wahby V, Giller EL, and Mason JW (1990) Low urinary cortisol excretion in patients with posttraumatic stress disorder. *Journal of Nervous and Mental Disease* 178: 366–369.
Yehuda R, Teicher MH, Seckl JR, Grossman RA, Morris A, and Bierer LM (2007) Parental posttraumatic stress disorder as a vulnerability factor for low cortisol trait in offspring of holocaust survivors. *Archives of General Psychiatry* 64: 1040–1048.
Yehuda R, Teicher MH, Trestman RL, Levengood RA, and Siever LJ (1996b) Cortisol regulation in posttraumatic stress disorder and major depression: A chronobiological analysis. *Biological Psychiatry* 40: 79–88.
Yehuda R, Yang RK, Buchsbaum M, and Golier JA (2006) Alterations in cortisol negative feedback inhibition as examined using the ACTH response to cortisol administration in PTSD. *Psychoneuroendocrinology* 31: 447–451.
Young EA and Breslau N (2004) Cortisol and catecholamines in posttraumatic stress disorder: an epidemiologic community study. *Archives of General Psychiatry* 61: 394–401.

24 Anorexia Nervosa and Bulimia Nervosa

G J Paz-Filho and J Licinio, University of Miami Miller School of Medicine, Miami, FL, USA

This chapter is a revision of the previous edition chapter by Andre B. Negrão and Julio Licinio, Volume 5, pp. 515–530, © 2002 Elsevier Inc.

Chapter Outline

Glossary

actigraphy Method of monitoring human rest and activity cycles through a portable actigraph unit.

adipocytokine Adipocyte-derived cytokine.

standardized mortality ratios (SMRs) Ratio of observed deaths to expected deaths in a population.

ultradian Biologic variations or rhythms occurring in cycles more frequent than every 24 h.

24.1 Overview

Anorexia nervosa (AN) and bulimia nervosa (BN) belong to a group of eating disorders characterized by pathological alterations in food intake, attributed to overevaluation of shape and weight. Patients with eating disorders judge their self-worth in terms of their shape and weight, and their ability to control them. Rigid restriction of food intake, which can alternate with loss of control in the drive to eat, is the most striking behavioral change seen in patients with AN and BN, respectively. Excessive fear of becoming overweight and body image disturbances are the psychological features most clearly present in both disorders. Both are complex psychiatric disorders with social, psychological, and biological processes playing a role in the etiopathogenesis. Those disorders are an important cause of physical and psychosocial morbidity in adolescent girls and young adult women, occurring less frequently in men and older adults.

Patients with AN pursue a disproportionately low body weight, which is achieved through severe restrictions in quantity and quality of food intake. Primary or secondary amenorrhea is one of the diagnostic criteria, although many other endocrine disturbances may be seen during the course of the disorder. In contrast, purging behavior is more prominent in patients with BN, after uncontrollably eating large quantities of food in a defined period (binge eating). Because of binge eating, bulimic patients often refer themselves as failed anorexics. Eating disorders appear to have increased in incidence in the general population at a rate of eight new cases per 100 000 per year for AN, and 12 new cases per 100 000 per year for BN in young females, between the last century and the 1970s (Hoek and Van Hoeken, 2003). More recent studies are discordant, and show either a decrease (Currin et al., 2005; Van Son et al., 2006; Keel et al., 2006) or an increase (Hudson et al., 2007; Hay et al., 2008) in prevalence of eating disorders over the last 20 years.

It is particularly worrisome that eating disorders (particularly AN) may result in death due to suicide, starvation, electrolyte imbalances, and ipecac abuse. Mortality outcome data from case series show that the standardized mortality ratios (SMR) for AN ranges from 1.36 among females 20 years following treatment, to 30.5 among females at <1 year following treatment (Birmingham et al., 2005; Eckert et al., 1995; Herzog et al., 1996; Lee et al., 2003; Moller-Madsen et al., 1996; Patton, 1988). In a 22-year follow-up conducted in Great Britain among 168 cases of AN, SMR was 1.36 in England and 4.71 in Scotland (Crisp et al., 1992). In a recent review of the literature, the mortality rate for AN was estimated to be about 4% per decade of follow-up (Signorini et al., 2007). Across BN studies, either no participants were deceased or the SMR was not significantly different from the rate expected in the population matched by age and sex (Fichter and Quadflieg, 2004; Herzog et al., 2000; Keel et al., 2003; Patton, 1988). After 10-year follow-up prospective cohort studies conducted among AN patients in Göteborg, Sweden (Ivarsson et al., 2000; Nilsson et al., 1999; Rastam et al., 2003; Wentz et al., 2001, 2000), 6% of the patients still presented AN and 4% still fulfilled diagnostic criteria for BN. According to an operationalized global outcome score, 49% had a good outcome, 41% an intermediate outcome, and 10% a poor outcome. The AN group was significantly more likely than the comparison group to have a personality disorder, obsessive–compulsive disorder, and Asperger's syndrome or autism-spectrum disorder, but not other anxiety disorders. The AN group also had a higher lifetime prevalence of depression. Regular menses were present in 65% after the 10-year follow-up (Berkman et al., 2007). Among patients with BN, one case series study with healthy comparisons showed that at 12 years of follow-up, 41% had a psychiatric disorder in the month before assessment, half of the patients had suffered from a lifetime mood disorder or major depression, and 36% had suffered from an anxiety or substance-use disorder (Fichter and Quadflieg, 2004). Besides these psychological and psychiatric outcomes, AN and BN may lead to dermatologic, gastrointestinal, cardiovascular, pulmonary, nutritional, osteo-metabolic, and endocrine medical complications (Mitchell and Crow, 2006). Morbidity in eating disorders is often long lasting and causes a lifelong impairment in patients' personal and family lives, as well as in professional development.

Historically, the clinical presentation of AN was attributed to pituitary insufficiency. Although there are several endocrine disturbances associated with AN or BN, it is now clear that there is no evidence for primary pituitary dysfunction. In fact, most of the physiological abnormalities found in eating disorders are secondary to a cluster of dieting, weight loss, and behavior directed to lose weight. This cluster leads to a dysfunction in hypothalamic centers, which can be reversed once there is recovery of healthy eating habits and normal weight. The still undefined pathophysiology of eating disorders is being unveiled by the recent discovery of signaling molecules originating within the brain, adipose tissue, and gastrointestinal tract, which are involved in controlling food intake and energy expenditure. Although it is well accepted that eating disorders result from the combination of biological, cultural, and social factors, it is unlikely that the initial trigger will be identified in the near future. Moreover, it is still unclear whether the endocrine alterations seen in eating disorders are a component of disease causation or whether they are an epiphenomenon – a consequence of starvation.

For decades, such complex interactions of hormones, brain, and behavior have stimulated clinicians and researchers to further investigate the underlying biological mechanisms in AN and BN. In this chapter, we describe the clinical aspects of anorexia and bulimia and review the current status of knowledge on the endocrine abnormalities found in these disorders.

24.2 Clinical Presentation

24.2.1 Anorexia Nervosa

AN has the highest mortality of any psychiatric disorder (Hoek, 2006). In AN, the intense fear of being or becoming fat drives patients to a continuous pursuit of weight loss. This pursuit is successful as the result of a severe and selective restriction of food intake, with the exclusion of foods viewed as fattening. Patients continue to view themselves as overweight and have a heightened desire to lose more weight, although they may be hungry and thin. Besides this distortion of self-image, other psychological processes – such as asceticism, competitiveness, and a wish to punish themselves – may be present (Beumont, 2002). AN is often accompanied by compulsive, strenuous exercise, which can contribute to the patients' low weight. Symptoms of depression and anxiety disorders, irritability, lability of mood, impaired concentration, loss of sexual appetite, social withdrawal, and obsessive features are frequent and also appear in other semistarvation states. These symptoms tend to worsen as patients lose weight and improve upon regaining weight, but the underlying mechanisms are still unknown. In addition, a subgroup of patients presents purging behavior, with or without binge eating.

According to the *Diagnostic and Statistical Manual of Mental Disorders*, Fourth Edition (DSM-IV) of the American Psychiatric Association, the diagnosis of AN requires fulfillment of all four diagnosis criteria: (1) refusal to maintain weight within a normal range for height and age (more than 15% below ideal body weight); (2) fear of weight gain; (3) severe body image disturbance; and (4) presence of primary or secondary amenorrhea (American Psychiatry Association, 1994). In patients with AN, self-esteem is regulated to an excessive degree by their views on body shape and weight, associated with distorted body image perception and denial of the medically adverse outcomes of extreme weight loss. Although amenorrhea is a diagnostic criterion, it is unreliable given the prevalent use of birth-control pills to regulate menses in individuals with AN. Moreover, there do not appear to be meaningful differences between individuals with AN who do and do not menstruate (Watson and Andersen, 2003).

The DSM-IV specifies two subtypes of AN: binge eating/purging type and restrictive type, according to the presence or absence of those behaviors. Prevalence rates for AN are generally described as ranging from 0.3% to 1.0% among females (Hoek and Van Hoeken, 2003; Hudson et al., 2007), with males being affected about one-tenth as frequently (Hudson et al., 2007; Lucas et al., 1991). Epidemiological studies reveal that the peak age of onset is between 15 and 19 years old (Lucas et al., 1991). However, there are reports of increasing presentations in prepubertal children (Gowers et al., 1991) and in mid- to late life (Beck et al., 1996; Boast et al., 1992; Bowler, 1992; Inagaki et al., 2002). In the last decade, the notion of AN as a Western, culture-bound syndrome, afflicting predominantly upper-middle-class white females (Dolan et al., 1990) is gradually being dispelled with increasing literature from East Asian countries and India (Huon et al., 2002; Khandelwal et al., 1995; Lee et al., 2005; Pike and Mizushima, 2005; Tareen et al., 2005).

Out of all psychiatric disorders, AN has the highest mortality rate (Hoek, 2006), mostly due to suicide. The course and outcome of AN is extremely variable. Early age of onset is an important positive prognostic factor for AN. Significant predictors of chronic AN (intermediate or poor outcome) include an extreme compulsive drive to exercise (Strober et al., 1997), a history of poor social relationships preceding onset of illness (Strober et al., 1997), and unsatisfactory evaluation scores concerning hypochondriasis, paranoia, and psychopathic deviance (Dancyger et al., 1997). In the UK, predictors of death included a weight less than 35 kg at presentation and more than one inpatient admission (Patton, 1988). Other predictors of mortality include greater severity of alcohol use disorders, greater severity of substance use disorders, poor social adjustment, and low Global Assessment of Functioning scores (Patton, 1988).

Multiple organ systems are affected by AN. A variety of dermatologic abnormalities can be seen in patients with AN as well as BN. Those abnormalities include xerosis (dry skin), lanugo-like body hair on the back, abdomen, and forearms, telogen effluvium (hair loss and a positive hair-pull test), acne, carotenoderma (yellowing of the skin because of excessive ingestion of carotenoid-rich vegetables), acrocyanosis, pruritis, purpura (caused by thrombocytopenia), stomatitis, nail dystrophy, and presence of Russell's sign (formation of scar/callus over the dorsal surface of the hand, as the hand is used to stimulate the gag reflex to induce vomiting). Massive gastric dilatation and necrosis can occur in patients with AN who have episodes of binge eating. Other gastrointestinal disorders include delayed gastric emptying and constipation (Chial et al., 2002). Patients with AN have increased risk for heart abnormalities like arrhythmias, bradycardia, shift to the right of the QRS axis (i.e., the average direction of electrical activity during ventricular depolarization), enhancement of the QT/RR slope, and diminished amplitudes of the QRS complex and T wave. Cases of ampulla cardiomyopathy (characterized by extensive akinesis of the apical region with hypercontraction of the basal segment of the ventricle; Ohwada et al. (2005)) and spontaneous pneumomediastinum (Sundararaghavan et al., 2005) have also been reported in patients with AN. In addition, osteopenia is a frequent, early, and serious complication in patients with AN. There is a threefold increase in long-term risk of fractures later in life for people with a history of AN (Lucas et al., 1999). Other medical complications include amenorrhea, failure to develop sexual characters, decreased fertility, poorer birth outcomes, and dehydration.

24.2.2 Bulimia Nervosa

BN shares many clinical and biological features with AN. The main difference is that attempts to restrict food intake are punctuated by repeated episodes of binges. In most cases, binge eating is followed by compensatory self-induced vomiting or laxative misuse, but there is a subgroup who do not purge. Body weight is generally normal, as opposed to the weight of patients with AN. According to the DSM-IV criteria, a patient with BN must present the following: (1) episodes of binge eating with a sense of loss of control; (2) binge eating is followed by compensatory behavior of the purging type (self-induced vomiting, laxative abuse, and diuretic abuse) or nonpurging type (excessive exercise, fasting, or strict diets); (3) binges and the resulting compensatory behavior must occur a minimum of 2 times per week for 3-months; and (4) dissatisfaction with body shape and weight (American Psychiatry Association, 1994). Stressful situations seem to trigger binge-eating episodes in these patients, which generate avoidance of these situations. Consequently, patients have increased feelings of frustration and dissatisfaction toward themselves, leading to further binge eating. Self-esteem is closely and momentarily related to their ability to maintain a diet, which is completely compromised by a new episode of binge eating.

Prevalence of BN is greater than that of AN, between 1% and 3% (Kreipe and Birndorf, 2000; Ben-Tovim et al., 1989). The ratio of female patients to male patients ranges from 10:1 to 20:1 (Woodside et al., 2001). Previous history of AN, obesity (self and familial), early menarche, and substance abuse are risk factors for BN (Fairburn and Harrison, 2003). Across several studies, depression was the only consistent factor relating to worse BN outcomes. A substantial proportion of individuals continue to suffer from eating disorders over time but BN was not associated with increased mortality risk. In a 12-year follow-up case series study with a comparison group, 67% of the group patients who had received inpatient treatment for BN were recovered and had no eating disorder diagnosis (Fichter and Quadflieg, 2004). In another study, 73% had achieved a full recovery after 7 years. However, 35% of women who had previously met criteria for being fully recovered had relapsed, and no significant predictor factors of recovery could be identified (Herzog et al., 1999). Approximately half of the patients suffer from a lifetime mood disorder or major depression and 36% suffer from an anxiety or substance use disorder (Fichter and Quadflieg, 2004). Mortality rates are much lower in BN than in AN, which are not significantly different from the rate expected in the population matched by age and sex (Fichter and Quadflieg, 2004; Herzog et al., 2000; Keel et al., 2003; Patton, 1988). Most of the medical outcomes are caused by the purgative behavior, such as enlargement of salivary glands, abrasion of the dorsum of the hands, dental erosion, esophagitis, gastric dilation, severe constipation, and electrolyte abnormalities. Generally, bone density is normal unless there is a history of AN. Menses are generally irregular, but fertility is restored upon remission. Repeated abuse of ipecac can cause cardiomyopathy and muscle weakness (Ho et al., 1998).

24.3 Hormonal Findings

Physiologically, anorexia and bulimia are states of nutritional deprivation that have many endocrine outcomes, as an effort to preserve essential body functions. Eating disorders, voluntary starvation, and constitutional thinness have many endocrine changes in common that have been described in numerous studies. More recently, better understanding of the physiology of leptin, the antistarvation hormone, as well as other substances involved in the regulation of food intake have brought to light new knowledge regarding the etiopathogenesis and the treatment of these disorders. Herein we describe the endocrine changes associated with AN and BN. **Table 1** summarizes the common findings in AN and BN.

24.3.1 Reproductive System

The presence of a critical fat mass is crucial for the initiation of menstruation (Frisch and Revelle, 1970). Upon the discovery of leptin, now this link between adipose tissue and reproduction is clear. Leptin regulates the ultradian oscillations in the levels of LH and estradiol, and the nocturnal rise in leptin determines the change in nocturnal LH. A leptin level below 2 mg l^{-1} has been invoked to represent the critical threshold value for amenorrhea (Audi et al., 1998; Kopp et al., 1997). Hypoleptinemia is responsible for inhibiting the hypothalamic–pituitary–gonadal (HPG) axis, leading to amenorrhea. It is suggested that this mechanism is mediated through an

Table 1 Common endocrine alterations in anorexia and bulimia

	Anorexia	*Bulimia*
Reproductive system	Hypogonadotropic hypogonadism	Hypogonadotropic hypogonadism or eugonadism
Thyroid gland	Low T3 syndrome Delayed TSH response to TRH	Low T3 syndrome (purge phase) Normal or high T3 (binge phase) Normal or delayed TSH response to TRH
Adrenal gland	Hypercortisolism	Normo- or hypercortisolism
Somatotropic axis	Increased GH Decreased IGF-I	Increased GH Decreased IGF-I
Bone metabolism	Decreased bone mineral density	Normal bone mineral density
Leptin	Decreased	Varies with patterns of food intake
Glucose homeostasis	Increased insulin sensitivity Decreased insulin secretion	Normal insulin sensitivity Normal insulin secretion
Resistin	Normal	Normal
Adiponectin	Increased	Normal
Ghrelin	Increased	Increased or normal
PYY	Increased	Increased or normal
NPY	Increased	Increased

Adapted from Negrao AB and Licinio J (2002) Anorexia nervosa and bulimia nervosa. In: Pfaff DW, Arnold AP, Etgen AM, Fahrbach SE, and Rubin RT (eds) *Hormones, Brain and Behavior*, 1st edn., vol. 5, pp. 515–530. San Diego, CA: Academic Press, with permission from Elsevier.

upregulation of cocaine and amphetamine-regulated transcript (CART), which reduces the secretion of the hypothalamic gonadotropin-releasing hormone (Audi et al., 1998; Chan and Mantzoros, 2005; Di Carlo et al., 2002; Kopp et al., 1997). Hypoleptinemia may also affect the reproductive axis directly at the level of the pituitary and the ovary (Cioffi et al., 1997; Chan and Mantzoros, 2001; Munoz and Argente, 2002).

The most remarkable outcome of AN and BN is hypogonadism due to alterations in the HPG axis. These alterations manifest as amenorrhea, oligomenorrhea, delay in the onset of puberty, arrested growth, infertility, insufficient weight gain during pregnancy, and low-birth-weight infants (Becker et al., 1999; Fairburn and Harrison, 2003). In AN, gonadotropin-releasing hormone pulsatility is altered due to reduced leptin, excessive exercise, alterations in neurotransmitters (such as melanocortin and 5-hydroxytryptamine), and, possibly, psychological factors (Golden, 2003; Chan and Mantzoros, 2005; Munoz and Argente, 2002). Luteinizing hormone (LH) and follicle stimulating hormone (FSH) are reduced, resembling secretory patterns of prepubertal children (Di Carlo et al., 2002; Tomova et al., 2007). Nevertheless, leptin plays an important role in reproduction, as its low levels observed in patients with eating disorders may contribute to explaining their hypogonadism. In women with hypothalamic hypogonadism due to strenuous exercise or low weight, recombinant leptin treatment increased mean LH levels and LH pulse frequency after 2 weeks, and increased maximal follicular diameter, the number of dominant follicles, ovarian volume, and estradiol levels over a period of 3 months (Welt et al., 2004). Pituitary response to luteinizing hormone-releasing hormone (LHRH) is impaired but can be normalized with low-dose or pulsatile treatment (Beumont et al., 1978). Interestingly, in approximately 20% of cases, amenorrhea may even precede significant weight loss (Golden et al., 1997). In BN, changes in the reproductive system are less evident. Approximately 50% present amenorrhea and/or anovulatory cycles, which are also due to the reduction in LH pulse frequency and amplitude (Resch et al., 2004a,b).

24.3.2 Thyroid Gland

Hormonal changes seen in patients with AN are the same as those observed in other starvation disorders: low T3, T4 in low-normal range, and normal concentrations of thyroid-stimulating hormone (TSH; low-T3 or euthyroid sick syndrome). Most patients with AN show a delayed or a hyporesponsive TSH response to thyrotropin-releasing hormone (TRH). Moreover, TRH cerebrospinal fluid (CSF) is decreased. After weight gain, T4, T3, free T3, and TSH increase significantly, whereas rT3 decreases (Tamai et al., 1986). In patients with BN, thyroid function varies according to the binge eating–fasting cycle of the disorder. During the binging phase of the illness, patients have low total T3 levels. After 7 weeks without binge eating or purging, patients have low total T3, total T4, free T3, free FT4, rT3, and thyroid-binding globulin. Bulimics in the binging phase of the illness showed a positive correlation between caloric intake and TSH values, suggesting that food binging may stimulate thyroid activity (Altemus et al., 1996). TSH responses to TRH are normal or delayed in most of the patients (Kiyohara et al., 1988).

24.3.3 Adrenal Gland

Patients with eating disorders have a mild increase in cortisol and urinary free cortisol due to increased frequency of secretory bursts at the level of the hypothalamus, and to increased cortisol half-life (Misra et al., 2004b). Corticotropin-releasing hormone (CRH) is increased, adrenocorticotropic hormone (ACTH) is inappropriately normal (Licinio et al., 1996), and dehydroepiandrosterone sulfate (DHEAS) is low, as a reflection of reduced ovarian function and low androstenedione production (Gordon et al., 2002). Administration of recombinant leptin did not change serum cortisol in women with hypothalamic hypogonadism (Welt et al., 2004). Once body weight normalizes, hypercortisolism resolves, although the blunted ACTH response can persist for 6 months.

24.3.4 Growth Hormone

Patients with AN display extensive changes of the hypothalamic–pituitary function with activation of the growth hormone-releasing hormone–growth hormone (GHRH–GH) axis. Low circulating insulin-like growth factor I (IGF-I) levels are associated with an enhanced GH production rate and a highly disordered mode of somatotropin release. GH secretion appears to be refractory to the hypoglycemia, while hyperglycemia is unable to blunt the response of GH to GHRH. Conversely, the influence of amino acids on GH release appears to be conserved in these patients. Patients with AN have an increase in GH peak frequency and valley hormonal concentrations, a decrease in IGF-I production, a reduction in central somatostatin tone, and hypersecretion of ghrelin (Scacchi et al., 1997). While plasma IGF-I concentrations are low, IGF-II circulating levels are normal, insulin-like growth factor-binding protein-3 (IGFBP-3) is low, and IGFBP-1 and IGFBP-2 are significantly elevated (Scacchi et al., 2003). Not only are these abnormalities likely due to the lack of negative IGF-I feedback, but also to a primary hypothalamic alteration with increased frequency of GHRH discharges and decreased somatostatinergic tone. It is questionable whether the IGF-I levels are falsely low due to a reduction of the bound fraction, with an increase in the free fraction, or due to peripheral GH resistance. Given the reversal of the above alterations following weight recovery, these abnormalities can be seen as secondary, and possibly adaptive, to nutritional deprivation.

24.3.5 Bone Metabolism

Low bone mineral density (BMD) is a consequence of many of the clinical features of AN: (1) estrogen deficiency, which increases bone resorption and decreases bone formation; (2) IGF-I deficiency and GH resistance, which decrease bone metabolic effects of GH; (3) low DHEAS, which leads to possible loss of IGF-I-mediated anabolic and anti-osteolytic features; (4) hypercortisolemia, which decreases bone formation; (5) low free testosterone, which decreases bone formation and increases bone resorption; (6) increased osteoprotegerin, which is indicative of a compensatory suboptimal bone remodeling process; (7) reduced leptin, which may centrally regulate bone formation and resorption; (8) undernutrition, which independently contributes to bone formation; (9) increased peptide YY (PYY), which may reduce osteoblastic activity; and (10) hyperadrenalism, which is associated with increased bone resorption (Jayasinghe et al., 2008). Women with AN have increased risk for osteoporosis and bone fractures, and weight gain is not sufficient to increase and normalize BMD. In patients with BN, BMD is frequently normal.

24.3.6 Leptin

Leptin, the product of the *ob* gene, is an adipocyte-derived hormone with key effects on reproduction (Bluher and Mantzoros, 2007; Chan and Mantzoros, 2005), glucose homeostasis (Brennan and Mantzoros, 2007; Ceddia, 2005), bone formation (Cock and Auwerx, 2003; Karsenty, 2006), tissue remodeling (Cock and Auwerx, 2003; Karsenty, 2006), inflammation (Otero et al., 2005), as well as on other elements of the endocrine (Ahima, 2006; Chan and Mantzoros, 2005; Chan et al., 2003) and immune systems (Lago et al., 2007). The most important and acknowledged function of leptin resides in the regulation of energy expenditure and food intake, through its actions on the arcuate nucleus of the hypothalamus. In humans, functional mutations of the leptin gene result in morbid obesity, hypogonadism, and other metabolic dysfunctions. In our studies, replacement with recombinant methionyl human leptin-induced weight loss (as fat body mass), as a result of a substantial increase in physical activity energy expenditure (measured by actigraphy) and a decrease in energy intake, with dietary changes regarding macro- and micronutrient composition (Licinio et al., 2004, 2007a,b; Williamson et al., 2005). Hypogonadism was reversed in all adult patients. One patient had a diagnosis of type 2 diabetes before treatment, and normalization of glucose levels was achieved as the patient lost weight. We collected blood continuously for 24 h at 7-min intervals, and we showed that leptin increased the 24-h average concentrations of LH, testosterone, and cortisol. This suggests that leptin may influence the GnRH–gonadotropin (Licinio et al., 2007a,b; Williamson et al., 2005) and the hypothalamic–pituitary–adrenal (HPA) axis (Licinio et al., 1997, 1996). Before treatment, we also showed that TSH pulsatility and circadian rhythmicity were disrupted (Mantzoros et al., 2001).

Leptin concentrations correlate with the amount of fat mass, with lower levels in lean individuals (Considine et al., 1996). Females have higher leptin levels than males as the result of the higher percentage of body fat, but also due to a direct influence of sex hormones on leptin secretion (Wabitsch et al., 1997). After a 2.5-day fast, leptin concentrations in healthy females drop by 75% (Bergendahl et al., 1999, 2000).

In starvation, leptin levels drop as patients lose weight. Replacement with exogenous leptin resulted in recovery of levels of energy expenditure, muscle work efficiency, sympathetic nervous system (SNS) tone, T3 and T4 (Rosenbaum et al., 2005). AN patients share physiological, endocrinological, and psychological features with healthy subjects who have lost a substantial amount of body weight. For example, AN patients, too, have reduced basal metabolic rate, T3 and T4. Amenorrhea, bradycardia, and hypothermia are somatic symptoms of AN, which also occur in healthy subjects upon substantial weight loss. Patients with AN have hypoleptinemia, as well as subnormal leptin concentrations in the CSF. However, the CSF to plasma leptin ratio is higher in patients than in controls, suggesting increased efficiency of leptin transport to the brain (Mantzoros et al., 1997). Upon weight regain, some patients may develop relative hyperleptinemia (Hebebrand et al., 1997; Wabitsch et al., 2001). Moreover, normalization of plasma and CSF leptin can occur even before normal weight is reached (Mantzoros et al., 1997). Therefore, AN leads to a long-term destabilization of leptin secretion, during which both relative hypoleptinemia and hyperleptinemia can occur (Hebebrand et al., 2007). However, these observations must be carefully analyzed, since these patients with AN also have high levels of the leptin-soluble receptor (Misra et al., 2004a; Monteleone et al., 2002), which may interfere with leptin assays. The degree of hypoleptinemia in acute AN is an indicator of the severity of the disorder, reflecting the low fat mass and revealing the neuroendocrine adaptations to semistarvation.

Leptin is the signal of adiposity that regulates the HPG axis, and its low levels observed in patients with AN lead to disruptions in the HPG axis, as previously discussed. Achievement of a weight approximately 90% of standard body weight has been shown to lead to resumption of menses in 85% of adolescent AN patients within 6 months (Golden et al., 1997).

Hypoleptinemia also plays an important role in determining low BMD in patients with AN. Leptin-deficient *ob/ob* mice demonstrate an increase in bone resorption, possibly due to upregulation of the neuropeptide CART (Elefteriou, 2005; Elmquist and Strewler, 2005). Moreover, since leptin seems to play a role in cognition (Harvey et al., 2005), hypoleptinemia could also have implications for AN patients, particularly among severely emaciated patients whose cognitive functions are reduced. Hypoleptinemia could also have an effect on sleep, as both *ob/ob* mice and patients with AN show an increase of wakefulness after sleep onset, a higher number of arousals, and a reduction of slow-wave sleep and slow-wave activity (Laposky et al., 2006; Nobili et al., 2004). In animal models, hypoleptinemia can induce hyperactivity, and it has been proposed that hyperactivity seen in AN patients is attributed to low leptin levels as well (Exner et al., 2000; Hebebrand et al., 2003; Holtkamp et al., 2003).

In patients with BN, as well as in normal individuals, leptin levels are normal since most of these patients have a normal weight (Ferron et al., 1997). However, binges and disruptions of mealtimes alters the diurnal pattern of leptinemia (**Figure 1**; Taylor et al., 1999).

24.3.7 Glucose Homeostasis

Studies reporting changes of insulin sensitivity in AN provide rather contradictory results. Although the majority of studies has found increased insulin sensitivity (Delporte et al., 2003; Misra et al., 2004a; Fukushima et al., 1993; Dostalova et al., 2007), either decreased (Pannacciulli et al., 2003) or unchanged (Castillo et al., 1985) insulin sensitivity has been reported as well. This increase in insulin sensitivity can be explained by the hyperadiponectinemia found in AN

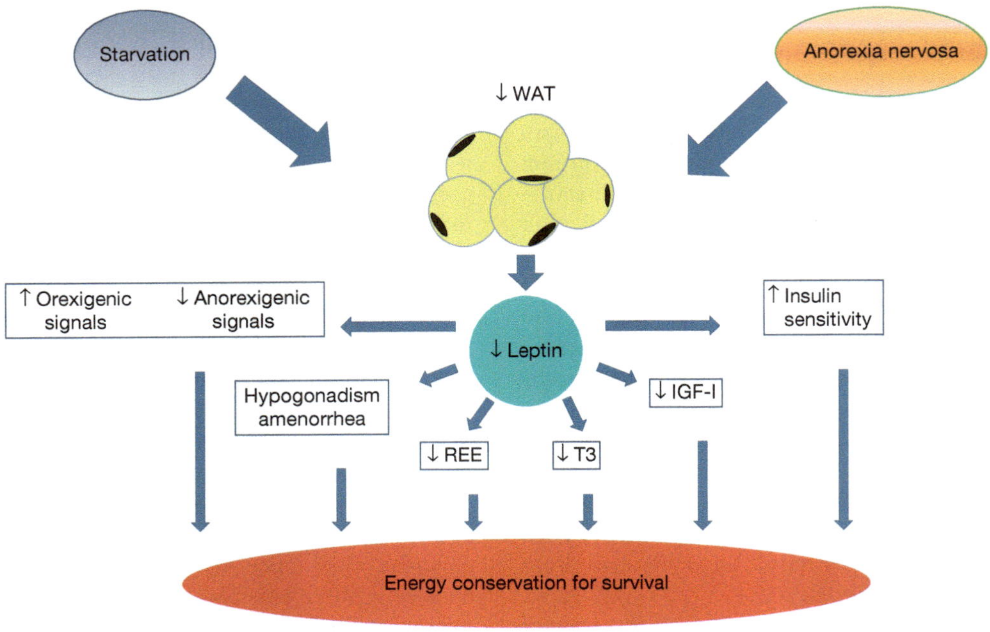

Figure 1 Hypoleptinemia is a common finding in starvation and in anorexia nervosa, leading to endocrine alterations. These alterations take place as an evolutionary effort to conserve energy for survival. In spite of the activation of signals to increase food intake mediated by low leptin levels, anoretic patients have decreased food intake. Other unknown signals interacting with psychological inputs override leptin-mediated signal transduction in anorexia nervosa. This is a key difference between anorexia and starvation. REE, resting energy expenditure; WAT, white adipose tissue.

patients since this adipocytokine has insulin-sensitizing effects (Yamauchi et al., 2001). However, it has also been hypothesized that hyperadiponectinemia is just a compensatory mechanism for the reduced insulin-stimulated glucose metabolism in AN (Pannacciulli et al., 2003). An excess of GH, cortisol, and plasma free fatty acid levels, together with beta-cell dysfunction associated with AN, may also account for the altered glucose tolerance (Dostalova et al., 2007). Hypoleptinemia may also contribute to the increase in insulin sensitivity. This hypothesis was confirmed when our leptin-deficient patients developed higher insulin resistance, assessed by euglycemic hyperinsulinemic clamps, when leptin replacement was initiated (unpublished data). In addition, patients with AN present decreased insulin secretion (Fukushima et al., 1993; Nozaki et al., 1994). Restoration of normal body weight is associated with normalization of virtually all measures, and the existing literature offers no conclusive evidence for a syndrome-specific impairment of carbohydrate metabolism in AN (Casper, 1996). In BN patients who have normal glucose tolerance and no family history of diabetes, normal insulin secretion, normal insulin sensitivity, and reduced glucose effectiveness are observed (Taniguchi et al., 1997).

24.3.8 Other Endocrine Systems

Neuropeptide Y (NPY) is the most potent endogenous stimulant of feeding behavior within the central nervous system (CNS; Kaye, 1996). Kalra et al. (1991) showed that NPY is dynamically secreted in the paraventricular nucleus (PVN) and is associated with increased appetite in rats submitted to food restriction. After feeding, NPY release rapidly decreases. Anorectic patients also have higher levels of NPY in the CSF, secondary to malnutrition and as a result of an ineffective homeostatic mechanism to stimulate feeding. After long-term weight restoration, levels of NPY normalize in those patients (Kaye et al., 1989). These findings were replicated in a study by Gendall et al. (1999) showing that NPY, PYY, and leptin concentrations in the CSF are normal in women who recover from AN and BN (Gendall et al., 1999). Oswiecimska et al. (2005) demonstrated that treatment of anoretic patients leads to the normalization of NPY levels, in a manner that is not commensurate with increasing body weight and body mass index (BMI). In patients with BN, plasma NPY levels are significantly higher as compared with AN and with a control group (Baranowska et al., 2001).

PYY is an intestinal hormone which acts as a satiety factor (Batterham and Bloom, 2003; Batterham et al., 2002, 2003). PYY belongs to the pancreatic polypeptide (PP) family, which includes PP and NPY. In a study including young anoretic women, circulating PYY levels were significantly elevated in AN and, given the controversially discussed anorectic effect of PYY, could theoretically contribute to that syndrome (Pfluger et al., 2007). Other studies did not confirm that finding, either at baseline or after a meal (Stock et al., 2005; Otto et al., 2007).

While adiponectin levels are elevated in AN, resistin levels appear to be unchanged. Resistin is a cytokine produced by the adipose tissue linked to inflammation, but circulating resistin does not appear to be related to the nutritional status (Housova et al., 2005). In a study that correlated gene expression in subcutaneous tissue with blood levels of adipocytokines in AN patients, hypoleptinemia was correlated with decreased expression of the leptin gene. However, resistin mRNA expression was increased, whereas resistin levels were normal (Dolezalova et al., 2007). Possibly, resistin levels are in fact increased in patients with AN, when interstitially measured by microdialysis (Dostalova et al., 2006).

Ghrelin is an upstream regulator of the orexigenic peptides NPY and agouti-related peptide (AgRP) and acts as a natural antagonist to leptin's effects on NPY/AgRP expressing neurons, resulting in an increase in feeding and body weight. Plasma ghrelin levels are elevated in lean patients with AN (Ariyasu et al., 2001; Hotta et al., 2004; Nakai et al., 2003). Infusion of ghrelin to patients with AN and patients with constitutional thinness led to the partial recovery of original body weight in some of them (Miljic et al., 2006). Anoretic patients were significantly less hungry, suggesting that patients with AN are resistant to the orexigenic effects of ghrelin in comparison with healthy control individuals. In another study, however, patients with restrictive AN exhibited increased hunger after the administration of ghrelin, which was in a proportion similar to that seen in normal individuals (Broglio et al., 2004).

Obestatin, which counteracts ghrelin action on feeding, is derived from the same propeptide as ghrelin. Its role on feeding and drinking behavior is yet unclear, but it reportedly reduces appetite (Zhang et al., 2005; Zizzari et al., 2007) and slows gastrointestinal motility (Gourcerol et al., 2006). Both obestatin and ghrelin are increased in AN and decreased in obesity, suggesting that obestatin is a nutritional marker reflecting body adiposity and insulin resistance (Nakahara et al., 2008; Harada et al., 2008).

24.4 Multifactorial Etiology

24.4.1 Functional Studies

Patients with eating disorders frequently show excessive cognitive preoccupation with food, often in an obsessive manner. Compared to normal controls, patients with AN respond differently to food stimuli. Such responses include increased interference to food words, high recognition for food words (irrespective of being hungry or satiated), and negative valence of food stimuli. Functional magnetic imaging resonance (fMRI) is a new technology based on the increase in blood flow to the local vasculature that accompanies neural activity in the brain. This increase results in a corresponding local reduction in deoxyhemoglobin because the increase in blood flow occurs without an increase of similar magnitude in oxygen extraction. Thus, deoxyhemoglobin serves as the natural contrast, source of the signal for fMRI. Despite the fact that fMRI still needs further development and standardization, it is being used to evaluate many psychiatric disorders, such as AN and BN. Studies found that patients with AN and BN have decreased activation in the inferior parietal lobe (IPL) and the occipital cortex (Uher et al., 2003, 2004; Santel et al., 2006) and increased activation in the medial prefrontal cortex (MPC) when viewing food pictures (Uher et al., 2003, 2004). It has been found that the IPL is associated with satiation, so it is expected that, in a satiated state, AN patients show weaker IPL activation to food stimuli than controls. The occipital cortex is a visual area, and it is also expected that AN patients show less activation to the visual stimuli. Finally, MPC is commonly associated with cognitive control mechanisms. Thus, one expects that AN patients have more activation in these areas when exposed to food, as part of a higher degree of self-control against food intake.

Positron emission tomography (PET) is based on the variations in cerebral glucose metabolism, which correlates with neuronal activity. It is another modality of neuroimaging that has shown hypometabolism in the frontal region of AN patients (Delvenne et al., 1996), and a common parietal cortex dysfunction in patients with AN and BN (Delvenne et al., 1999). Another study showed hypometabolism in the MPC (Takano et al., 2001). PET and single photon emission computed tomography (SPECT) have also been shown to be valuable instruments to evaluate the role of the serotoninergic (5-HT) system on the modulation of appetite. There is strong evidence that the serotoninergic pathways also contribute to the pathophysiology of eating disorders (Kaye et al., 2005). Studies suggest that disturbances of 5-HT activity occur during the ill state and persist after recovery. AN patients have higher than normal concentrations of 5-HT metabolites, at concentrations 50% greater than those found in the ill AN state. Acutely ill AN patients have an abnormal hormonal response to 5-HT-specific challenges in most studies. Neuroimaging studies reinforce the importance of the 5-HT system, showing that recovered pure-restrictive-type AN participants have reduced 5-HT2A receptor activity in the subgenual and pregenual cingulate cortex, and the mesial temporal and parietal cortical areas (Frank et al., 2002). Recovered bulimic-type AN women have reduced 5-HT2A receptor activity relative to controls in the subgenual cingulate, mesial temporal, lateral temporal, parietal, and occipital cortical regions (Bailer et al., 2004). Another study showed that ill AN patients have reduced 5-HT2A activity in the left frontal, bilateral parietal, and occipital cortex (Audenaert et al., 2003).

Another modality of neuroimaging is magnetic resonance spectroscopy, which evaluates the regional metabolic changes. A recent study found that AN patients had significantly lower levels of *N*-acetyl-aspartate (NAA, a marker of neuronal functionality), glutamine/glutamate (Glx, an index of energy metabolism), and myo-inositol (mI, a membrane constituent and a cell messenger with an osmoregulatory function) than the control group in the frontal gray matter, and that these metabolites tend to increase at follow-up after weight recovery (Castro-Fornieles et al., 2007). It is questioned whether these changes are attributed to the pathophysiology of the disease or merely to the nutritional deficits.

Although data from neuroimaging studies are striking, nonconcordant results have been found due to the existence of many confounding factors in the studies. Because of those factors, larger and well-designed studies are required.

24.4.2 Genetics

AN is a highly familial disorder, whereas heritability in BN is not so clear (Fairburn and Harrison, 2003). The relative risk for AN in family members of probands with AN is 11.3 (Strober et al., 2000). In twin studies, heritability was between 48% and 76% (Bulik et al., 2006; Klump et al., 2001; Kortegaard et al., 2001; Wade et al., 2000). Recently, the Genetics of Anorexia Nervosa (GAN) Collaborative Study provided detailed context for the recently completed genotyping of the first 200 families by the Center for Inherited Disease Research (CIDR; Kaye et al., 2008). This group plans to identify the genomic regions that might harbor predisposing genes for AN. This aim will be reached through genome-wide linkage studies in roughly 400 families. Previous reports from the Price Foundation Genetic Study of Anorexia Nervosa (which is part of the GAN Collaborative Study) revealed several regions of suggestive and significant linkage in AN (Bergen et al., 2003; Devlin et al., 2002; Grice et al., 2002). Findings from association studies indicate some role for the serotonin system in the development of AN (Gorwood et al., 2003; Klump and Gobrogge, 2005). A recent meta-analysis proposed that the 1438A allele of the 5-HT2A gene is significantly associated but not linked (according to a transmission disequilibrium test on a large sample of trios) with AN (Gorwood et al., 2003). Other candidates are the norepinephrine transporter gene (Urwin et al., 2002), the catechol-*O*-methyl transferase gene (Frisch et al., 2001), the brain-derived neurotrophic gene (BDNF; Hashimoto et al., 2005), the estrogen receptor beta gene (Nilsson et al., 2004), and the melanocortin-4 receptor gene (Branson et al., 2003). Additional research is still needed for other candidate genes involved in the neurotransmitter system (e.g., dopamine, glutamate, and opioids) and also in the regulation of feeding and energy expenditure (e.g., agouti-related protein, NPY, leptin, and reproductive hormones). Genomic regions on chromosomes 1 and 10 are also likely to harbor susceptibility genes for AN as well as a range of eating pathologies. Of particular interest, chromosome 10 has also been implicated in obesity (Hager et al., 1998). Moreover, the linkage peak in this chromosome is glutamicacid decarboxylase 2 (GAD2) which has been linked to impulsive eating in obese individuals (Boutin et al., 2003). Many candidate genes have been studied, but few have been replicated in other studies.

24.5 Endocrine Treatment

The treatment of AN and BN is complex and consists of various types of pharmacological approaches, often combined with psychotherapeutic interventions, such as behavioral, cognitive-behavioral, interpersonal and family therapies. Even new technologies, such as the internet and remote text messaging, play a role in the treatment (Ljotsson et al., 2007; Robinson et al., 2006; Myers et al., 2004). The treatment of these disorders has been reviewed elsewhere (American Psychiatric Association, 2006; Abbate Daga et al., 2004; Attia and Schroeder, 2005; Bulik et al., 2007; Fairburn, 2005; Fairburn and Harrison, 2003; Fisher, 2006; Lilly, 2003; Shapiro et al., 2007; Taylor et al., 2007; Wilson and Shafran, 2005; Yager and Andersen, 2005). In this section, we briefly focus on the endocrine treatments.

Menses are usually restored when patients achieve a weight approximately 90% of the median weight for age and height. Once the treatment goal weight is achieved, restoration of the hypothalamic–pituitary–ovarian function can be monitored by measuring serum estradiol levels. An estradiol level (110 pmol l^{-1} (30 pg ml^{-1}) is predictive of resumption of menses within 3–6 months (relative risk 4.6, 95% confidence interval 1.9–11.2; Golden et al., 1997).

As previously discussed, low BMD is caused by diverse factors, and the degree of BMD reduction is directly related to the duration of amenorrhea and the degree of malnutrition (Golden, 2007). Weight gain and resumption of menses is associated with an increase in BMD, which does not return to normal. Hormone replacement therapy (HRT) increases BMD in perimenopausal women, but the same effect is not obtained in AN (Liu and Lebrun, 2006). Thus, treatment of low BMD in AN consists of weight restoration, resumption of spontaneous menstruation, and calcium and vitamin D supplementation. Moreover, patients should be advised to engage carefully monitored weight-bearing exercise, such as walking, jogging, dancing, and jumping activities, which increase bone mass. Treatments with IGF-I (Grinspoon et al., 2002), dehydroepiandrosterone (Gordon et al., 2002), r-metHuLeptin (Lamarca and Volpe, 2004), and bisphosphonates (Miller et al., 2004; Golden et al., 2005) may all increase BMD, although these are limited to clinical trials. Finally, adolescents should be discouraged from smoking, drinking carbonated beverages, or consuming alcohol, because these activities reduce BMD (Golden, 2007).

Although patients with AN have increased GH levels, their levels of IGF-I are decreased, perhaps due to a decreased ability of GH to stimulate secretion of IGF–I. Based on this possibility, a trial evaluated the effects of recombinant human GH (rhGH) in patients with AN. In that trial, rhGH restored the GH–IGF-I axis, increased BMI, food intake, T3, and in addition normalized other parameters of malnutrition, indicating that rhGH administration may be a feasible modality of treatment (Hashizume et al., 2007).

24.6 Conclusion

AN and BN are psychiatric and endocrine disorders that determine several endocrine changes due to starvation. Most of these changes are a protective mechanism against the absence of food, which leads to severe and irreversible consequences if not treated adequately. Patients with eating disorders, particularly anoretic patients, are human models of starvation; a better understanding of the pathophysiology of the processes involved in malnutrition may facilitate the development of novel approaches to conditions that involve similar mechanisms, such as obesity and insulin resistance. In both starvation and AN, appetitive signals favor eating and weight gain. A key difference between AN and starvation however is that in starvation once food is available, eating is increased, leading to weight gain, but in contrast in AN, particularly the restrictive subtype, patients will not eat in the context of food availability and of biological signals that should lead them to eat. The elucidation of the complex psychological, neuronal, and endocrine signal transduction pathways that override such signals is a fascinating area for future research in psychiatry and neuroendocrinology.

References

Abbate Daga G, Gianotti L, Mondelli V, Quartesan R, and Fassino S (2004) The psychopharmacotherapy of anorexia nervosa: Clinical, neuroendocrine and metabolic aspects. *Journal of Endocrinological Investigation* 27: 975–981.

Ahima RS (2006) Adipose tissue as an endocrine organ. *Obesity (Silver Spring)* 14(supplement 5): 242S–249S.

Altemus M, Hetherington M, Kennedy B, Licinio J, and Gold PW (1996) Thyroid function in bulimia nervosa. *Psychoneuroendocrinology* 21: 249–261.

Ariyasu H, Takaya K, Tagami T, et al. (2001) Stomach is a major source of circulating ghrelin, and feeding state determines plasma ghrelin-like immunoreactivity levels in humans. *Journal of Clinical Endocrinology and Metabolism* 86: 4753–4758.

American Psychiatry Association (1994) *Diagnostic and Statistical Manual of Mental Disorders*, 4th edn. Washington, DC: American Psychiatry Association.

American Psychiatric Association (2006) Practice guidelines for the treatment of patients with eating disorders, third edition. *American Journal of Psychiatry* 163: 4–54.

Attia E and Schroeder L (2005) Pharmacologic treatment of anorexia nervosa: Where do we go from here? *International Journal of Eating Disorders* 37(supplement): S60–S63; discussion S87–S89.

Audenaert K, Van Laere K, Dumont F, et al. (2003) Decreased 5-HT2a receptor binding in patients with anorexia nervosa. *Journal of Nuclear Medicine* 44: 163–169.

Audi L, Mantzoros CS, Vidal-Puig A, Vargas D, Gussinye M, and Carrascosa A (1998) Leptin in relation to resumption of

menses in women with anorexia nervosa. *Molecular Psychiatry* 3: 544–547.
Bailer UF, Price JC, Meltzer CC, et al. (2004) Altered 5-HT(2A) receptor binding after recovery from bulimia-type anorexia nervosa: Relationships to harm avoidance and drive for thinness. *Neuropsychopharmacology* 29: 1143–1155.
Baranowska B, Wolinska-Witort E, Wasilewska-Dziubinska E, Roguski K, and Chmielowska M (2001) Plasma leptin, neuropeptide Y (NPY) and galanin concentrations in bulimia nervosa and in anorexia nervosa. *Neuroendocrinology Letters* 22: 356–358.
Batterham RL and Bloom SR (2003) The gut hormone peptide YY regulates appetite. *Annals of the New York Academy of Sciences* 994: 162–168.
Batterham RL, Cohen MA, Ellis SM, et al. (2003) Inhibition of food intake in obese subjects by peptide YY3-36. *New England Journal of Medicine* 349: 941–948.
Batterham RL, Cowley MA, Small CJ, et al. (2002) Gut hormone PYY(3–36) physiologically inhibits food intake. *Nature* 418: 650–654.
Beck D, Casper R, and Andersen A (1996) Truly late onset of eating disorders: A study of 11 cases averaging 60 years of age at presentation. *International Journal of Eating Disorders* 20: 389–395.
Becker AE, Grinspoon SK, Klibanski A, and Herzog DB (1999) Eating disorders. *New England Journal of Medicine* 340: 1092–1098.
Ben-Tovim DI, Subbiah N, Scheutz B, and Morton J (1989) Bulimia: Symptoms and syndromes in an urban population. *Australian and New Zealand Journal of Psychiatry* 23: 73–80.
Bergen AW, Van Den Bree MB, Yeager M, et al. (2003) Candidate genes for anorexia nervosa in the 1p33–36 linkage region: Serotonin 1D and delta opioid receptor loci exhibit significant association to anorexia nervosa. *Molecular Psychiatry* 8: 397–406.
Bergendahl M, Evans WS, Pastor C, Patel A, Iranmanesh A, and Veldhuis JD (1999) Short-term fasting suppresses leptin and (conversely) activates disorderly growth hormone secretion in midluteal phase women – a clinical research center study. *Journal of Clinical Endocrinology and Metabolism* 84: 883–894.
Bergendahl M, Iranmanesh A, Evans WS, and Veldhuis JD (2000) Short-term fasting selectively suppresses leptin pulse mass and 24-hour rhythmic leptin release in healthy midluteal phase women without disturbing leptin pulse frequency or its entropy control (pattern orderliness). *Journal of Clinical Endocrinology and Metabolism* 85: 207–213.
Berkman ND, Lohr KN, and Bulik CM (2007) Outcomes of eating disorders: A systematic review of the literature. *International Journal of Eating Disorders* 40: 293–309.
Beumont PJV (2002) Clinical presentation of anorexia nervosa and bulimia nervosa. In: Brownell BK and Fairburn CG (eds.) *Eating Disorders and Obesity: A Comprehensive Handbook*, vol. 1, ch. 29, pp. 162–170. New York: Guildford Press.
Beumont PJ, Abraham SF, Argall WJ, and Turtle JR (1978) Plasma gonadotrophins and LHRH infusions in anorexia nervosa. *Australian and New Zealand Journal of Psychiatry* 8: 509–514.
Birmingham CL, Su J, Hlynsky JA, Goldner EM, and Gao M (2005) The mortality rate from anorexia nervosa. *International Journal of Eating Disorders* 38: 143–146.
Bluher S and Mantzoros CS (2007) Leptin in reproduction. *Current Opinion in Endocrinology, Diabetes and Obesity* 14: 458–464.
Boast N, Coker E, and Wakeling A (1992) Anorexia nervosa of late onset. *British Journal of Psychiatry* 160: 257–260.
Boutin P, Dina C, Vasseur F, et al. (2003) GAD2 on chromosome 10p12 is a candidate gene for human obesity. *PLoS Biology* 1: E68.
Bowler C (1992) Late-onset anorexia nervosa. *British Journal of Psychiatry* 160: 717.
Branson R, Potoczna N, Kral JG, Lentes KU, Hoehe MR, and Horber FF (2003) Binge eating as a major phenotype of melanocortin 4 receptor gene mutations. *New England Journal of Medicine* 348: 1096–1103.
Brennan AM and Mantzoros CS (2007) Leptin and adiponectin: Their role in diabetes. *Current Diabetes Reports* 7: 1–2.
Broglio F, Gianotti L, Destefanis S, et al. (2004) The endocrine response to acute ghrelin administration is blunted in patients with anorexia nervosa, a ghrelin hypersecretory state. *Clinical Endocrinology (Oxf)* 60: 592–599.
Bulik CM, Berkman ND, Brownley KA, Sedway JA, and Lohr KN (2007) Anorexia nervosa treatment: A systematic review of randomized controlled trials. *International Journal of Eating Disorders* 40: 310–320.
Bulik CM, Sullivan PF, Tozzi F, Furberg H, Lichtenstein P, and Pedersen NL (2006) Prevalence, heritability, and prospective risk factors for anorexia nervosa. *Archives of General Psychiatry* 63: 305–312.
Casper RC (1996) Carbohydrate metabolism and its regulatory hormones in anorexia nervosa. *Psychiatry Research* 62: 85–96.
Castillo M, Scheen A, Lefebvre PJ, and Luyckx AS (1985) Insulin-stimulated glucose disposal is not increased in anorexia nervosa. *Journal of Clinical Endocrinology and Metabolism* 60: 311–314.
Castro-Fornieles J, Bargallo N, Lazaro L, Andres S, Falcon C, Plana MT, and Junque C (2007) Adolescent anorexia nervosa: Cross-sectional and follow-up frontal gray matter disturbances detected with proton magnetic resonance spectroscopy. *Journal of Psychiatric Research* 41: 952–958.
Ceddia RB (2005) Direct metabolic regulation in skeletal muscle and fat tissue by leptin: Implications for glucose and fatty acids homeostasis. *International Journal of Obesity (Lond)* 29: 1175–1183.
Chan JL, Heist K, Depaoli AM, Veldhuis JD, and Mantzoros CS (2003) The role of falling leptin levels in the neuroendocrine and metabolic adaptation to short-term starvation in healthy men. *Journal of Clinical Investigation* 111: 1409–1421.
Chan JL and Mantzoros CS (2001) Leptin and the hypothalamic–pituitary regulation of the gonadotropin–gonadal axis. *Pituitary* 4: 87–92.
Chan JL and Mantzoros CS (2005) Role of leptin in energy-deprivation states: Normal human physiology and clinical implications for hypothalamic amenorrhoea and anorexia nervosa. *Lancet* 366: 74–85.
Chial HJ, Mcalpine DE, and Camilleri M (2002) Anorexia nervosa: Manifestations and management for the gastroenterologist. *American Journal of Gastroenterology* 97: 255–269.
Cioffi JA, Van Blerkom J, Antczak M, Shafer A, Wittmer S, and Snodgrass HR (1997) The expression of leptin and its receptors in pre-ovulatory human follicles. *Molecular Human Reproduction* 3: 467–472.
Cock TA and Auwerx J (2003) Leptin: Cutting the fat off the bone. *Lancet* 362: 1572–1574.
Considine RV, Sinha MK, Heiman ML, et al. (1996) Serum immunoreactive-leptin concentrations in normal-weight and obese humans. *New England Journal of Medicine* 334: 292–295.
Crisp AH, Callender JS, Halek C, and Hsu LK (1992) Long-term mortality in anorexia nervosa. A 20-year follow-up of the St George's and Aberdeen cohorts. *British Journal of Psychiatry* 161: 104–107.
Currin L, Schmidt U, Treasure J, and Jick H (2005) Time trends in eating disorder incidence. *British Journal of Psychiatry* 186: 132–135.

Dancyger IF, Sunday SR, Eckert ED, and Halmi KA (1997) A comparative analysis of minnesota multiphasic personality inventory profiles of anorexia nervosa at hospital admission, discharge, and 10-year follow-up. *Comparative Psychiatry* 38: 185–191.

Delporte ML, Brichard SM, Hermans MP, Beguin C, and Lambert M (2003) Hyperadiponectinaemia in anorexia nervosa. *Clinical Endocrinology (Oxf)* 58: 22–29.

Delvenne V, Goldman S, De Maertelaer V, and Lotstra F (1999) Brain glucose metabolism in eating disorders assessed by positron emission tomography. *International Journal of Eating Disorders* 25: 29–37.

Delvenne V, Goldman S, De Maertelaer V, Simon Y, Luxen A, and Lotstra F (1996) Brain hypometabolism of glucose in anorexia nervosa: Normalization after weight gain. *Biological Psychiatry* 40: 761–768.

Devlin B, Bacanu SA, Klump KL, et al. (2002) Linkage analysis of anorexia nervosa incorporating behavioral covariates. *Human Molecular Genetics* 11: 689–696.

Di Carlo C, Tommaselli GA, De Filippo E, et al. (2002) Menstrual status and serum leptin levels in anorectic and in menstruating women with low body mass indexes. *Fertility and Sterility* 78: 376–382.

Dolan B, Lacey JH, and Evans C (1990) Eating behaviour and attitudes to weight and shape in British women from three ethnic groups. *British Journal of Psychiatry* 157: 523–528.

Dolezalova R, Lacinova Z, Dolinkova M, et al. (2007) Changes of endocrine function of adipose tissue in anorexia nervosa: Comparison of circulating levels versus subcutaneous mRNA expression. *Clinical Endocrinology (Oxf)* 67: 674–678.

Dostalova I, Kunesova M, Duskova J, Papezova H, and Nedvidkova J (2006) Adipose tissue resistin levels in patients with anorexia nervosa. *Nutrition* 22: 977–983.

Dostalova I, Smitka K, Papezova H, Kvasnickova H, and Nedvidkova J (2007) Increased insulin sensitivity in patients with anorexia nervosa: The role of adipocytokines. *Physiological Research* 56: 587–594.

Eckert ED, Halmi KA, Marchi P, Grove W, and Crosby R (1995) Ten-year follow-up of anorexia nervosa: Clinical course and outcome. *Psychological Medicine* 25: 143–156.

Elefteriou F (2005) Neuronal signaling and the regulation of bone remodeling. *Cellular and Molecular Life Sciences* 62: 2339–2349.

Elmquist JK and Strewler GJ (2005) Physiology: Do neural signals remodel bone? *Nature* 434: 447–448.

Exner C, Hebebrand J, Remschmidt H, et al. (2000) Leptin suppresses semi-starvation induced hyperactivity in rats: Implications for anorexia nervosa. *Molecular Psychiatry* 5: 476–481.

Fairburn CG (2005) Evidence-based treatment of anorexia nervosa. *International Journal of Eating Disorders* 37(supplement): S26–S30; discussion S41–S42.

Fairburn CG and Harrison PJ (2003) Eating disorders. *Lancet* 361: 407–416.

Ferron F, Considine RV, Peino R, Lado IG, Dieguez C, and Casanueva FF (1997) Serum leptin concentrations in patients with anorexia nervosa, bulimia nervosa and non-specific eating disorders correlate with the body mass index but are independent of the respective disease. *Clinical Endocrinology (Oxf)* 46: 289–293.

Fichter MM and Quadflieg N (2004) Twelve-year course and outcome of bulimia nervosa. *Psychological Medicine* 34: 1395–1406.

Fisher M (2006) Treatment of eating disorders in children, adolescents, and young adults. *Pediatrics in Review* 27: 5–16.

Frank GK, Kaye WH, Meltzer CC, Price JC, Greer P, Mcconaha C, and Skovira K (2002) Reduced 5-HT2A receptor binding after recovery from anorexia nervosa. *Biological Psychiatry* 52: 896–906.

Frisch A, Laufer N, Danziger Y, et al. (2001) Association of anorexia nervosa with the high activity allele of the COMT gene: A family-based study in Israeli patients. *Molecular Psychiatry* 6: 243–245.

Frisch RE and Revelle R (1970) Height and weight at menarche and a hypothesis of critical body weights and adolescent events. *Science* 169: 397–399.

Fukushima M, Nakai Y, Taniguchi A, Imura H, Nagata I, and Tokuyama K (1993) Insulin sensitivity, insulin secretion, and glucose effectiveness in anorexia nervosa: A minimal model analysis. *Metabolism* 42: 1164–1168.

Gendall KA, Kaye WH, Altemus M, Mcconaha CW, and La Via MC (1999) Leptin, neuropeptide Y, and peptide YY in long-term recovered eating disorder patients. *Biological Psychiatry* 46: 292–299.

Golden NH (2003) Eating disorders in adolescence and their sequelae. *Best Practice and Research Clinical Obstetrics and Gynaecology* 17: 57–73.

Golden NH (2007) Eating disorders in adolescence: What is the role of hormone replacement therapy? *Current Opinion in Obstetrics and Gynecology* 19: 434–439.

Golden NH, Iglesias EA, Jacobson MS, et al. (2005) Alendronate for the treatment of osteopenia in anorexia nervosa: A randomized, double-blind, placebo-controlled trial. *Journal of Clinical Endocrinology and Metabolism* 90: 3179–3185.

Golden NH, Jacobson MS, Schebendach J, Solanto MV, Hertz SM, and Shenker IR (1997) Resumption of menses in anorexia nervosa. *Archives of Pediatrics and Adolescent Medicine* 151: 16–21.

Gordon CM, Grace E, Emans SJ, et al. (2002) Effects of oral dehydroepiandrosterone on bone density in young women with anorexia nervosa: A randomized trial. *Journal of Clinical Endocrinology and Metabolism* 87: 4935–4941.

Gorwood P, Kipman A, and Foulon C (2003) The human genetics of anorexia nervosa. *European Journal of Pharmacology* 480: 163–170.

Gourcerol G, Million M, Adelson DW, et al. (2006) Lack of interaction between peripheral injection of CCK and obestatin in the regulation of gastric satiety signaling in rodents. *Peptides* 27: 2811–2819.

Gowers SG, Crisp AH, Joughin N, and Bhat A (1991) Premenarcheal anorexia nervosa. *Journal of Child Psychology and Psychiatry* 32: 515–524.

Grice DE, Halmi KA, Fichter MM, et al. (2002) Evidence for a susceptibility gene for anorexia nervosa on chromosome 1. *American Journal of Human Genetics* 70: 787–792.

Grinspoon S, Thomas L, Miller K, Herzog D, and Klibanski A (2002) Effects of recombinant human IGF-I and oral contraceptive administration on bone density in anorexia nervosa. *Journal of Clinical Endocrinology and Metabolism* 87: 2883–2891.

Hager J, Dina C, Francke S, et al. (1998) A genome-wide scan for human obesity genes reveals a major susceptibility locus on chromosome 10. *Nature Genetics* 20: 304–308.

Harada T, Nakahara T, Yasuhara D, et al. (2008) Obestatin, acyl ghrelin, and des-acyl ghrelin responses to an oral glucose tolerance test in the restricting type of anorexia nervosa. *Biological Psychiatry* 63: 245–247.

Harvey J, Shanley LJ, O'Malley D, and Irving AJ (2005) Leptin: A potential cognitive enhancer? *Biochemical Society Transactions* 33: 1029–1032.

Hashimoto K, Koizumi H, Nakazato M, Shimizu E, and Iyo M (2005) Role of brain-derived neurotrophic factor in eating disorders: Recent findings and its pathophysiological implications. *Progress in Neuropsychopharmacology and Biological Psychiatry* 29: 499–504.

Hashizume K, Suzuki S, Komatsu A, et al. (2007) Administration of recombinant human growth hormone normalizes GH–IGF1 axis and improves malnutrition-related disorders in patients with anorexia nervosa. *Endocrine Journal* 54: 319–327.

Hay PJ, Mond J, Buttner P, and Darby A (2008) Eating disorder behaviors are increasing: Findings from two sequential community surveys in south Australia. *PLoS ONE* 3: e1541.

Hebebrand J, Blum WF, Barth N, et al. (1997) Leptin levels in patients with anorexia nervosa are reduced in the acute stage and elevated upon short-term weight restoration. *Molecular Psychiatry* 2: 330–334.

Hebebrand J, Exner C, Hebebrand K, et al. (2003) Hyperactivity in patients with anorexia nervosa and in semistarved rats: Evidence for a pivotal role of hypoleptinemia. *Physiology and Behavior* 79: 25–37.

Hebebrand J, Muller TD, Holtkamp K, and Herpertz-Dahlmann B (2007) The role of leptin in anorexia nervosa: Clinical implications. *Molecular Psychiatry* 12: 23–35.

Herzog DB, Dorer DJ, Keel PK, et al. (1999) Recovery and relapse in anorexia and bulimia nervosa: A 7.5-year follow-up study. *Journal of the American Academy of Child and Adolescent Psychiatry* 38: 829–337.

Herzog DB, Field AE, Keller MB, West JC, Robbins WM, Staley J, and Colditz GA (1996) Subtyping eating disorders: Is it justified? *Journal of the American Academy of Child and Adolescent Psychiatry* 35: 928–936.

Herzog DB, Greenwood DN, Dorer DJ, et al. (2000) Mortality in eating disorders: A descriptive study. *International Journal of Eating Disorders* 28: 20–26.

Ho PC, Dweik R, and Cohen MC (1998) Rapidly reversible cardiomyopathy associated with chronic ipecac ingestion. *Clinical Cardiology* 21: 780–783.

Hoek HW (2006) Incidence, prevalence and mortality of anorexia nervosa and other eating disorders. *Current Opinion in Psychiatry* 19: 389–394.

Hoek HW and Van Hoeken D (2003) Review of the prevalence and incidence of eating disorders. *International Journal of Eating Disorders* 34: 383–396.

Holtkamp K, Herpertz-Dahlmann B, Mika C, et al. (2003) Elevated physical activity and low leptin levels co-occur in patients with anorexia nervosa. *Journal of Clinical Endocrinology and Metabolism* 88: 5169–5174.

Hotta M, Ohwada R, Katakami H, Shibasaki T, Hizuka N, and Takano K (2004) Plasma levels of intact and degraded ghrelin and their responses to glucose infusion in anorexia nervosa. *Journal of Clinical Endocrinology and Metabolism* 89: 5707–5712.

Housova J, Anderlova K, Krizova J, et al. (2005) Serum adiponectin and resistin concentrations in patients with restrictive and binge/purge form of anorexia nervosa and bulimia nervosa. *Journal of Clinical Endocrinology and Metabolism* 90: 1366–1370.

Hudson JI, Hiripi E, Pope HG, Jr., and Kessler RC (2007) The prevalence and correlates of eating disorders in the National Comorbidity Survey Replication. *Biological Psychiatry* 61: 348–358.

Huon GF, Mingyi Q, Oliver K, and Xiao G (2002) A large-scale survey of eating disorder symptomatology among female adolescents in the People's Republic of China. *International Journal of Eating Disorders* 32: 192–205.

Inagaki T, Horiguchi J, Tsubouchi K, Miyaoka T, Uegaki J, and Seno H (2002) Late onset anorexia nervosa: Two case reports. *International Journal of Psychiatry in Medicine* 32: 91–95.

Ivarsson T, Rastam M, Wentz E, Gillberg IC, and Gillberg C (2000) Depressive disorders in teenage-onset anorexia nervosa: A controlled longitudinal, partly community-based study. *Comprehensive Psychiatry* 41: 398–403.

Jayasinghe Y, Grover SR, and Zacharin M (2008) Current concepts in bone and reproductive health in adolescents with anorexia nervosa. *BJOG* 115: 304–315.

Kalra SP, Dube MG, Sahu A, Phelps CP, and Kalra PS (1991) Neuropeptide Y secretion increases in the paraventricular nucleus in association with increased appetite for food. *Proceedings of the National Academy of Sciences of the United States of America* 88: 10931–10935.

Karsenty G (2006) Convergence between bone and energy homeostases: Leptin regulation of bone mass. *Cell Metabolism* 4: 341–348.

Kaye WH (1996) Neuropeptide abnormalities in anorexia nervosa. *Psychiatry Research* 62: 65–74.

Kaye WH, Berrettini WH, Gwirtsman HE, Gold PW, George DT, Jimerson DC, and Ebert MH (1989) Contribution of CNS neuropeptide (NPY, CRH, and beta-endorphin) alterations to psychophysiological abnormalities in anorexia nervosa. *Psychopharmacology Bulletin* 25: 433–438.

Kaye WH, Bulik CM, Plotnicov K, et al. (2008) The genetics of anorexia nervosa collaborative study: Methods and sample description. *International Journal of Eating Disorders* 41: 289–300.

Kaye WH, Frank GK, Bailer UF, and Henry SE (2005) Neurobiology of anorexia nervosa: Clinical implications of alterations of the function of serotonin and other neuronal systems. *International Journal of Eating Disorders* 37(supplement): S15–S19; discussion S20–S21.

Keel PK, Dorer DJ, Eddy KT, Franko D, Charatan DL, and Herzog DB (2003) Predictors of mortality in eating disorders. *Archives of General Psychiatry* 60: 179–183.

Keel PK, Heatherton TF, Dorer DJ, Joiner TE, and Zalta a K (2006) Point prevalence of bulimia nervosa in 1982, 1992, and 2002. *Psychological Medicine* 36: 119–127.

Khandelwal SK, Sharan P, and Saxena S (1995) Eating disorders: An Indian perspective. *International Journal of Social Psychiatry* 41: 132–146.

Kiyohara K, Tamai H, Kobayashi N, and Nakagawa T (1988) Hypothalamic–pituitary–thyroidal axis alterations in bulimic patients. *American Journal of Clinical Nutrition* 47: 805–809.

Klump KL and Gobrogge KL (2005) A review and primer of molecular genetic studies of anorexia nervosa. *International Journal of Eating Disorders* 37(supplement): S43–S48; discussion S87–S89.

Klump KL, Miller KB, Keel PK, Mcgue M, and Iacono WG (2001) Genetic and environmental influences on anorexia nervosa syndromes in a population-based twin sample. *Psychological Medicine* 31: 737–740.

Kopp W, Blum WF, Von Prittwitz S, et al. (1997) Low leptin levels predict amenorrhea in underweight and eating disordered females. *Molecular Psychiatry* 2: 335–340.

Kortegaard LS, Hoerder K, Joergensen J, Gillberg C, and Kyvik KO (2001) A preliminary population-based twin study of self-reported eating disorder. *Psychological Medicine* 31: 361–365.

Kreipe RE and Birndorf SA (2000) Eating disorders in adolescents and young adults. *Emergency Medicine Clinics of North America* 84: 1027–1049, viii–ix.

Lago F, Dieguez C, Gomez-Reino J, and Gualillo O (2007) Adipokines as emerging mediators of immune response and inflammation. *Nature Clinical Practice Rheumatology* 3: 716–724.

Lamarca A and Volpe A (2004) Recombinant human leptin in women with hypothalamic amenorrhea. *New England Journal of Medicine* 351: 2343; author reply 2343.

Laposky AD, Shelton J, Bass J, Dugovic C, Perrino N, and Turek FW (2006) Altered sleep regulation in leptin-deficient mice. *AJP – Regulatory, Integrative and Comparative Physiology* 290: R894–R903.

Lee S, Chan YY, and Hsu LK (2003) The intermediate-term outcome of Chinese patients with anorexia nervosa in Hong Kong. *American Journal of Psychiatry* 160: 967–972.
Lee HY, Lee EL, Pathy P, and Chan YH (2005) Anorexia nervosa in Singapore: An eight-year retrospective study. *Singapore Medical Association* 46: 275–281.
Licinio J, Caglayan S, Ozata M, et al. (2004) Phenotypic effects of leptin replacement on morbid obesity, diabetes mellitus, hypogonadism, and behavior in leptin-deficient adults. *Proceedings of the National Academy of Sciences of the United States of America* 101: 4531–4536.
Licinio J, Mantzoros C, Negrao AB, et al. (1997) Human leptin levels are pulsatile and inversely related to pituitary–adrenal function. *Nature Medicine* 3: 575–579.
Licinio J, Milane M, Thakur S, et al. (2007a) Effects of leptin on intake of specific micro- and macronutrients in a woman with leptin gene deficiency studied off and on leptin at stable body weight. *Appetite* 49: 594–599.
Licinio J, Ribeiro L, Busnello JV, et al. (2007b) Effects of leptin replacement on macro- and micronutrient preferences. *International Journal of Obesity (Lond)* 31: 1859–1863.
Licinio J, Wong ML, and Gold PW (1996) The hypothalamic–pituitary–adrenal axis in anorexia nervosa. *Psychiatry Research* 62: 75–83.
Lilly RZ (2003) Bulimia nervosa. *British Medical Journal* 327: 380–381.
Liu SL and Lebrun CM (2006) Effect of oral contraceptives and hormone replacement therapy on bone mineral density in premenopausal and perimenopausal women: A systematic review. *British Journal of Sports Medicine* 40: 11–24.
Ljotsson B, Lundin C, Mitsell K, Carlbring P, Ramklint M, and Ghaderi A (2007) Remote treatment of bulimia nervosa and binge eating disorder: A randomized trial of Internet-assisted cognitive behavioural therapy. *Behaviour Research and Therapy* 45: 649–661.
Lucas AR, Beard CM, O'fallon WM, and Kurland LT (1991) 50-Year trends in the incidence of anorexia nervosa in Rochester, Minn.: A population-based study. *American Journal of Psychiatry* 148: 917–922.
Lucas AR, Melton LJ, III, Crowson CS, and O'Fallon WM (1999) Long-term fracture risk among women with anorexia nervosa: A population-based cohort study. *Mayo Clinic Proceedings* 74: 972–977.
Mantzoros C, Flier JS, Lesem MD, Brewerton TD, and Jimerson DC (1997) Cerebrospinal fluid leptin in anorexia nervosa: Correlation with nutritional status and potential role in resistance to weight gain. *Journal of Clinical Endocrinology and Metabolism* 82: 1845–1851.
Mantzoros CS, Ozata M, Negrao AB, et al. (2001) Synchronicity of frequently sampled thyrotropin (TSH) and leptin concentrations in healthy adults and leptin-deficient subjects: Evidence for possible partial TSH regulation by leptin in humans. *Journal of Clinical Endocrinology and Metabolism* 86: 3284–3291.
Miljic D, Pekic S, Djurovic M, et al. (2006) Ghrelin has partial or no effect on appetite, growth hormone, prolactin, and cortisol release in patients with anorexia nervosa. *Journal of Clinical Endocrinology and Metabolism* 91: 1491–1495.
Miller KK, Grieco KA, Mulder J, et al. (2004) Effects of risedronate on bone density in anorexia nervosa. *Journal of Clinical Endocrinology and Metabolism* 89: 3903–3906.
Misra M, Miller KK, Almazan C, et al. (2004a) Hormonal and body composition predictors of soluble leptin receptor, leptin, and free leptin index in adolescent girls with anorexia nervosa and controls and relation to insulin sensitivity. *Journal of Clinical Endocrinology and Metabolism* 89: 3486–3495.
Misra M, Miller KK, Almazan C, et al. (2004b) Alterations in cortisol secretory dynamics in adolescent girls with anorexia nervosa and effects on bone metabolism. *Journal of Clinical Endocrinology and Metabolism* 89: 4972–4980.
Mitchell JE and Crow S (2006) Medical complications of anorexia nervosa and bulimia nervosa. *Current Opinion in Psychiatry* 19: 438–443.
Moller-Madsen S, Nystrup J, and Nielsen S (1996) Mortality in anorexia nervosa in Denmark during the period 1970–1987. *Acta Psychiatrica Scandinavica* 94: 454–459.
Monteleone P, Fabrazzo M, Tortorella A, Fuschino A, and Maj M (2002) Opposite modifications in circulating leptin and soluble leptin receptor across the eating disorder spectrum. *Molecular Psychiatry* 7: 641–646.
Munoz MT and Argente J (2002) Anorexia nervosa in female adolescents: Endocrine and bone mineral density disturbances. *European Journal of Endocrinology* 147: 275–286.
Myers TC, Swan-Kremeier L, Wonderlich S, Lancaster K, and Mitchell JE (2004) The use of alternative delivery systems and new technologies in the treatment of patients with eating disorders. *International Journal of Eating Disorders* 36: 123–143.
Nakahara T, Harada T, Yasuhara D, et al. (2008) Plasma obestatin concentrations are negatively correlated with body mass index, insulin resistance index, and plasma leptin concentrations in obesity and anorexia nervosa. *Biological Psychiatry* 64: 252–255.
Nakai Y, Hosoda H, Nin K, Ooya C, Hayashi H, Akamizu T, and Kangawa K (2003) Plasma levels of active form of ghrelin during oral glucose tolerance test in patients with anorexia nervosa. *European Journal of Endocrinology* 149: R1–R3.
Negrao AB and Licinio J (2002) Anorexia nervosa and bulimia nervosa. In: Pfaff DW, Arnold AP, Etgen AM, Fahrbach SE, and Rubin RT (eds.) *Hormones, Brain and Behavior*, 1st edn., vol. 5, pp. 515–530. San Diego, CA: Academic Press.
Nilsson EW, Gillberg C, Gillberg IC, and Rastam M (1999) Ten-year follow-up of adolescent-onset anorexia nervosa: Personality disorders. *Journal of the American Academy of Child and Adolescent Psychiatry* 38: 1389–1395.
Nilsson M, Naessen S, Dahlman I, Linden Hirschberg A, Gustafsson JA, and Dahlman-Wright K (2004) Association of estrogen receptor beta gene polymorphisms with bulimic disease in women. *Molecular Psychiatry* 9: 28–34.
Nobili L, Baglietto MG, Beelke M, et al. (2004) Impairment of the production of delta sleep in anorectic adolescents. *Sleep* 27: 1553–1559.
Nozaki T, Tamai H, Matsubayashi S, Komaki G, Kobayashi N, and Nakagawa T (1994) Insulin response to intravenous glucose in patients with anorexia nervosa showing low insulin response to oral glucose. *Journal of Clinical Endocrinology and Metabolism* 79: 217–222.
Ohwada R, Hotta M, Kimura H, Takagi S, Matsuda N, Nomura K, and Takano K (2005) Ampulla cardiomyopathy after hypoglycemia in three young female patients with anorexia nervosa. *Internal Medicine* 44: 228–233.
Oswiecimska J, Ziora K, Geisler G, and Broll-Waska K (2005) Prospective evaluation of leptin and neuropeptide Y (NPY) serum levels in girls with anorexia nervosa. *Neuroendocrinology Letters* 26: 301–304.
Otero M, Lago R, Lago F, Casanueva FF, Dieguez C, Gomez-Reino JJ, and Gualillo O (2005) Leptin, from fat to inflammation: Old questions and new insights. *FEBS Letters* 579: 295–301.
Otto B, Cuntz U, Otto C, Heldwein W, Riepl RL, and Tschop MH (2007) Peptide YY release in anorectic patients after liquid meal. *Appetite* 48: 301–304.
Pannacciulli N, Vettor R, Milan G, et al. (2003) Anorexia nervosa is characterized by increased adiponectin plasma levels and reduced nonoxidative glucose metabolism.

Journal of Clinical Endocrinology and Metabolism 88: 1748–1752.
Patton GC (1988) Mortality in eating disorders. *Psychological Medicine* 18: 947–951.
Pfluger PT, Kampe J, Castaneda TR, et al. (2007) Effect of human body weight changes on circulating levels of peptide YY and peptide YY3-36. *Journal of Clinical Endocrinology and Metabolism* 92: 583–588.
Pike KM and Mizushima H (2005) The clinical presentation of Japanese women with anorexia nervosa and bulimia nervosa: A study of the eating disorders inventory-2. *International Journal of Eating Disorders* 37: 26–31.
Rastam M, Gillberg C, and Wentz E (2003) Outcome of teenage-onset anorexia nervosa in a Swedish community-based sample. *European Child and Adolescent Psychiatry* 12(supplement 1): 178–190.
Resch M, Szendei G, and Haasz P (2004a) Bulimia from a gynecological view: Hormonal changes. *Journal of Obstetrics and Gynaecology* 24: 907–910.
Resch M, Szendei G, and Haasz P (2004b) Eating disorders from a gynecologic and endocrinologic view: Hormonal changes. *Fertility and Sterility* 81: 1151–1153.
Robinson S, Perkins S, Bauer S, Hammond N, Treasure J, and Schmidt U (2006) Aftercare intervention through text messaging in the treatment of bulimia nervosa – feasibility pilot. *International Journal of Eating Disorders* 39: 633–638.
Rosenbaum M, Goldsmith R, Bloomfield D, et al. (2005) Low-dose leptin reverses skeletal muscle, autonomic, and neuroendocrine adaptations to maintenance of reduced weight. *Journal of Clinical Investigation* 115: 3579–3586.
Santel S, Baving L, Krauel K, Munte TF, and Rotte M (2006) Hunger and satiety in anorexia nervosa: fMRI during cognitive processing of food pictures. *Brain Research* 1114: 138–148.
Scacchi M, Ida Pincelli A, and Cavagnini F (2003) Nutritional status in the neuroendocrine control of growth hormone secretion: The model of anorexia nervosa. *Frontiers in Neuroendocrinology* 24: 200–224.
Scacchi M, Pincelli AI, Caumo A, Tomasi P, Delitala G, Baldi G, and Cavagnini F (1997) Spontaneous nocturnal growth hormone secretion in anorexia nervosa. *Journal of Clinical Endocrinology and Metabolism* 82: 3225–3229.
Shapiro JR, Berkman ND, Brownley KA, Sedway JA, Lohr KN, and Bulik CM (2007) Bulimia nervosa treatment: A systematic review of randomized controlled trials. *International Journal of Eating Disorders* 40: 321–336.
Signorini A, De Filippo E, Panico S, De Caprio C, Pasanisi F, and Contaldo F (2007) Long-term mortality in anorexia nervosa: A report after an 8-year follow-up and a review of the most recent literature. *European Journal of Clinical Nutrition* 61: 119–122.
Stock S, Leichner P, Wong AC, Ghatei MA, Kieffer TJ, Bloom SR, and Chanoine JP (2005) Ghrelin, peptide YY, glucose-dependent insulinotropic polypeptide, and hunger responses to a mixed meal in anorexic, obese, and control female adolescents. *Journal of Clinical Endocrinology and Metabolism* 90: 2161–2168.
Strober M, Freeman R, Lampert C, Diamond J, and Kaye W (2000) Controlled family study of anorexia nervosa and bulimia nervosa: Evidence of shared liability and transmission of partial syndromes. *American Journal of Psychiatry* 157: 393–401.
Strober M, Freeman R, and Morrell W (1997) The long-term course of severe anorexia nervosa in adolescents: Survival analysis of recovery, relapse, and outcome predictors over 10–15 years in a prospective study. *International Journal of Eating Disorders* 22: 339–360.
Sundararaghavan S, Pitts TY, Suarez WA, and Johnstone C (2005) Chest pain among adolescents with anorexia nervosa. *Pediatric Emergency Care* 21: 603–605.
Takano A, Shiga T, Kitagawa N, Koyama T, Katoh C, Tsukamoto E, and Tamaki N (2001) Abnormal neuronal network in anorexia nervosa studied with I-123-IMP SPECT. *Psychiatry Research* 107: 45–50.
Tamai H, Mori K, Matsubayashi S, et al. (1986) Hypothalamic–pituitary–thyroidal dysfunctions in anorexia nervosa. *Psychotherapy and Psychosomatics* 46: 127–131.
Taniguchi A, Nakai Y, Fukushima M, et al. (1997) Intravenous glucose tolerance test-derived glucose effectiveness in bulimia nervosa. *Metabolism* 46: 484–486.
Tareen A, Hodes M, and Rangel L (2005) Non-fat-phobic anorexia nervosa in British South Asian adolescents. *International Journal of Eating Disorders* 37: 161–165.
Taylor CB, Franko DL, Neumark-Sztainer D, Paxton SJ, and Shapiro JR (2007) Public-health approach to eating disorders. *Lancet* 369: 1928.
Taylor AE, Hubbard J, and Anderson EJ (1999) Impact of binge eating on metabolic and leptin dynamics in normal young women. *Journal of Clinical Endocrinology and Metabolism* 84: 428–434.
Tomova A, Makker K, Kirilov G, Agarwal A, and Kumanov P (2007) Disturbances in gonadal axis in women with anorexia nervosa. *Eating and Weight Disorders* 12: e92–e97.
Uher R, Brammer MJ, Murphy T, Campbell IC, Ng VW, Williams SC, and Treasure J (2003) Recovery and chronicity in anorexia nervosa: Brain activity associated with differential outcomes. *Biological Psychiatry* 54: 934–942.
Uher R, Murphy T, Brammer MJ, et al. (2004) Medial prefrontal cortex activity associated with symptom provocation in eating disorders. *American Journal of Psychiatry* 161: 1238–1246.
Urwin RE, Bennetts B, Wilcken B, et al. (2002) Anorexia nervosa (restrictive subtype) is associated with a polymorphism in the novel norepinephrine transporter gene promoter polymorphic region. *Molecular Psychiatry* 7: 652–657.
Van Son GE, Van Hoeken D, Bartelds AI, Van Furth EF, and Hoek HW (2006) Time trends in the incidence of eating disorders: A primary care study in the Netherlands. *International Journal of Eating Disorders* 39: 565–569.
Wabitsch M, Ballauff A, Holl R, Blum WF, Heinze E, Remschmidt H, and Hebebrand J (2001) Serum leptin, gonadotropin, and testosterone concentrations in male patients with anorexia nervosa during weight gain. *Journal of Clinical Endocrinology and Metabolism* 86: 2982–2988.
Wabitsch M, Blum WF, Muche R, et al. (1997) Contribution of androgens to the gender difference in leptin production in obese children and adolescents. *Journal of Clinical Investigation* 100: 808–813.
Wade TD, Bulik CM, Neale M, and Kendler KS (2000) Anorexia nervosa and major depression: Shared genetic and environmental risk factors. *American Journal of Psychiatry* 157: 469–471.
Watson TL and Andersen AE (2003) A critical examination of the amenorrhea and weight criteria for diagnosing anorexia nervosa. *Acta Psychiatrica Scandinavica* 108: 175–182.
Welt CK, Chan JL, Bullen J, et al. (2004) Recombinant human leptin in women with hypothalamic amenorrhea. *New England Journal of Medicine* 351: 987–997.
Wentz E, Gillberg IC, Gillberg C, and Rastam M (2000) Ten-year follow-up of adolescent-onset anorexia nervosa: Physical health and neurodevelopment. *Developmental Medicine and Child Neurology* 42: 328–333.
Wentz E, Gillberg C, Gillberg IC, and Rastam M (2001) Ten-year follow-up of adolescent-onset anorexia nervosa: Psychiatric disorders and overall functioning scales. *Journal of Child Psychology and Psychiatry* 42: 613–622.

Williamson DA, Ravussin E, Wong ML, et al. (2005) Microanalysis of eating behavior of three leptin deficient adults treated with leptin therapy. *Appetite* 45: 75–80.
Wilson GT and Shafran R (2005) Eating disorders guidelines from NICE. *Lancet* 365: 79–81.
Woodside DB, Garfinkel PE, Lin E, Goering P, Kaplan AS, Goldbloom DS, and Kennedy SH (2001) Comparisons of men with full or partial eating disorders, men without eating disorders, and women with eating disorders in the community. *American Journal of Psychiatry* 158: 570–574.
Yager J and Andersen AE (2005) Clinical practice. Anorexia nervosa. *New England Journal of Medicine* 353: 1481–1488.
Yamauchi T, Kamon J, Waki H, et al. (2001) The fat-derived hormone adiponectin reverses insulin resistance associated with both lipoatrophy and obesity. *Nature Medicine* 7: 941–946.
Zhang JV, Ren PG, Avsian-Kretchmer O, Luo CW, Rauch R, Klein C, and Hsueh AJ (2005) Obestatin, a peptide encoded by the ghrelin gene, opposes ghrelin's effects on food intake. *Science* 310: 996–999.
Zizzari P, Longchamps R, Epelbaum J, and Bluet-Pajot MT (2007) Obestatin partially affects ghrelin stimulation of food intake and growth hormone secretion in rodents. *Endocrinology* 148: 1648–1653.

25 Aging and Alzheimer's Disease

S J Lupien, Université de Montréal, Montreal, QC, Canada
C Lord, McMaster University Women's Health Concerns Clinic, Hamilton, ON, Canada
S Sindi, McGill University, Montreal, QC, Canada
C W Wilkinson, Geriatric Research Education and Clinical Center, VA Puget Sound Health Care System, Seattle, WA, USA and University of Washington, Seattle, WA, USA
A J Fiocco, University of California, San Francisco, CA, USA

Chapter Outline

Glossary

alzheimer's disease A progressive and irreversible neurodegenerative disorder that is characterized by the proliferation of senile plaques and neurofibrillary tangles, resulting in extensive memory and functional impairment.

apolipoprotein E gene The APOE gene is mapped to chromosome 19 and is polymorphic with three different alleles, $\varepsilon 2$, $\varepsilon 3$, and $\varepsilon 4$, with the $\varepsilon 4$ allele increasing risk for the development of Alzheimer's disease.

choline acetyltransferase Neuronal enzyme that joins acetyl-CoA and choline to form acetylcholine, a neurotransmitter implicated in learning and memory.

gonadal hormones Sex steroids secreted by the gonads that not only influence the development of reproductive organs but also cross the blood–brain barrier to influence brain function through various mechanisms. The hormone produced by the testis is testosterone and hormones produced by the ovaries include estrogen and progesterone.

hippocampus A sensitive and malleable brain structure that is important for declarative learning and memory.

hypothalamic–pituitary–adrenal axis A closed-loop stress-sensitive system that is necessary for survival of the organism, involving the production of hormones from the hypothalamus, the pituitary, and the adrenal glands.

25.1 Introduction

Alzheimer's disease (AD) is a progressive neurodegenerative and irreversible disorder that was first characterized by Dr. Alois Alzheimer in 1906 (Hodges, 2006). Among all of the known types of dementia, AD is the most common. According to estimates offered by the Alzheimer's Disease Facts and Figures 2007 report (published by the Alzheimer's Association), 5.1 million Americans currently have AD. Furthermore, given the population size of the baby boomer generation, the United States Census Bureau (July 2005) has estimated that the prevalence of AD will increase up to 16 million by 2050, unless more efficient prevention strategies are implemented. Currently, over 24 million people worldwide live with dementia, and by 2040, this number may rise up to 81 million cases (Ferri et al., 2005). The average number of survival years following the onset of AD is reportedly between 3 and 10 years, with standard deviations of up to 6 years (Heyman et al., 1996; Wolfson et al., 2001). The number of survival years may depend on the patient's age and the severity of the disorder (Bonsignore and Heun, 2003).

25.1.1 Diagnosis of AD

Although varying forms of dementia are currently diagnosed according to criteria stipulated in the fourth edition of the *Diagnostic and Statistical Manual of Mental Disorders* (DSM-IV), AD is most often diagnosed based on the criteria developed by the National Institute of Neurologic and Communicative Disorders and Stroke – Alzheimer's Disease and Related Disorders Association (NINCD-ADRDA) (McKhann et al., 1984).

A diagnosis of AD is classified as definite, probable, or possible. A definite diagnosis may only be given when clinical symptoms are accompanied by histologic evidence, once a postmortem analysis has been conducted (Small et al., 1997). A probable diagnosis consists of clinical symptoms without histologic confirmation, and a possible diagnosis is given when clinical symptoms without histologic evidence are atypical, yet no alternative diagnosis can be given (Cummings, 2004). The NINCD-ADRDA diagnostic criteria have an accuracy rate of 79% and a sensitivity value of 86% (Hogervorst et al., 2000).

In addition to tests of mental and cognitive status, modern brain imaging techniques such as magnetic resonance imaging (MRI) and positron emission tomography (PET) allow for the identification of atrophy and functional abnormalities of various brain regions that are involved in AD pathology. Such techniques are proving to be valuable tools that allow for the detection of individuals who are at risk for AD development (Jagust et al., 2006; Cummings, 2004). Moreover, once patients have already been diagnosed with AD, MRI scans allow for tracking the progression of the disease in a longitudinal manner (Thompson et al., 2007). However, despite the value of these tools, brain imaging techniques have been reported to predict cognitive decline and brain atrophy that are not specific to AD but are more general in detecting other forms of dementia (Grossman et al., 2006; Jagust et al., 2006).

Blood assays can be useful in examining endocrine profiles and detecting dysregulation associated with AD (Kawas, 2003). Examination of cerebrospinal fluid (CSF) can detect classical markers of AD, including increased levels of CSF β-amyloid (Aβ) peptides, hyperphosphorylated tau, and total tau levels. Although such measures can be useful to identify increased risk, they are neither specific to serve as diagnostic markers nor are they adequately sensitive to capture the longitudinal progression of the disease (Thompson et al., 2007; Grossman et al., 2006).

25.1.2 Pathophysiology of AD

In AD, neuronal loss or alterations in neuronal function are observed in several brain regions, including limbic structures such as the hippocampus, parahippocampal gyrus, entorhinal cortex, and amygdala. Also, neuronal alterations have been observed in the frontal, temporal, parietal, and occipital association cortices (Selkoe, 2001). The neuropathological hallmarks of AD include the presence of senile plaques (SPs) and neurofibrillary tangles (NFTs). SPs contain extracellular deposits of Aβ, which is an essential pathologic marker of the disease and is found in the form of amyloid fibrils. Aβ-filled SPs are located in areas of the brain that are important for learning and memory, including neocortical areas and limbic structures such as the hippocampus (Lee et al., 2005; Master and Beyreuther, 2005; Selkoe, 2001). NFTs are filamentous bundles of hyperphosphorylated tau proteins and are mostly found in the hippocampus (Selkoe, 2001; Masters, 2005). The hippocampus and the entorhinal cortex are medial temporal lobe structures that not only play an important role in learning and memory (Braak and Braak, 1991) but also are found to be among the first structures of the brain to be affected by AD (Jack et al., 1997; Xu et al., 2000; Dickerson et al., 2001). Although atrophy is commonly noted in these structures, whole brain volume is relatively preserved (Wolf et al., 2004; Pantel, 2002). The volumes of the hippocampus and entorhinal cortex have also been shown to predict decline in memory performance (Grundman et al., 2003; Jagust et al., 2006).

25.1.3 Clinical Features of AD

Among the typical neuropsychological features observed during the early stages of AD is an impairment of episodic memory (Greene et al., 1996). Patients with AD also suffer from deficits in various attentional processes, including selective attention (Pignatti et al., 2005), divided attention (Baddeley et al., 2001) and sustained attention, or vigilance (Perry and Hodges, 1999). Furthermore, language capacities gradually deteriorate (Price et al., 1993), and visuospatial deficits emerge (Kirk and Kertesz, 1991; Perry et al., 2000). As the disease continues to progress into its later stages, motor and sensory abnormalities develop (McKhann et al., 1984). Clinical manifestations may also include aphasia, apraxia, and disorientation, as well as a decline in judgment and executive function capacities (Small et al., 1997).

In addition to cognitive deficits, patients with AD may be identified as having psychiatric and behavioral symptoms, including depressive symptoms, sleep disruption, apathy, agitation, and aggression (Hodges, 2006; Jost and Grossberg, 1996; Mega et al., 1996; Neumann et al., 2001; Starkstein et al., 2006).

25.1.4 Stages of AD

The course of AD is progressive in nature with notable changes in clinical features and functional performance from early to late stages of the disease. While many studies broadly refer to these stages as mild, moderate, and severe, the Functional Assessment Staging procedure provides a more detailed description of functional level across seven stages of the disease (Reisberg, 1988; Sclan and Reisberg, 1992): Stage 1: No cognitive impairment, where there is no evidence of objective or subjective cognitive decline. Stage 2: Very mild decline, where subjective deficits may not be apparent to family members or during a medical exam, including deficits in recalling names or object locations. Stage 3: Mild cognitive decline, where memory deficits start becoming apparent and there is a possible diagnosis of early-stage AD. During this stage, work performance may decline, problems are encountered when traveling or driving in unfamiliar locations, and skills related to planning and organizing may also decrease. Stage 4: Moderate cognitive decline, also referred to as mild or early-stage AD. During this stage, there is an observable decrease in the capacity to perform complex and challenging daily tasks, including paying bills and purchasing grocery items. Such difficulties are often accompanied by reduced knowledge of recent events and a decrease in memory for personal history. Stage 5: Moderately severe cognitive decline, also referred to as moderate or mid-stage AD. During this stage, more memory deficits appear with regard to details, such as one's phone number or address, and assistance is required with daily tasks, such as choosing appropriate clothing, and reminders for routine activities, such as bathing. Stage 6: Severe cognitive decline, also known as moderately severe or mid-stage AD. During this stage, frequent assistance is needed for all daily tasks and sleep–wake cycles may become dysregulated. Awareness of one's surrounding declines, along with a substantial decrease in memory of recent events and activities. Stage 7: Very severe cognitive decline, also known as severe or late-stage AD. During this stage, overall capacities are significantly decreased with regard to responding to the surrounding environment. Speech may be restricted to short sentences and then to single words. Reflexes no longer function normally, and the abilities to sit up, smile, and swallow are lost.

25.1.5 Mild Cognitive Impairment: Between Norm and Pathology

On a continuum of cognitive functioning in old age, mild cognitive impairment (MCI) is a construct that overlaps with normal age-related cognitive decline on one end of the continuum and the earliest stages of AD on the other end (Petersen, 2004). Although MCI patients show impairment beyond that observed in normal cognitive aging, they do not meet the criteria for dementia (Gauthier et al., 2006). Furthermore, MCI-associated functional impairments are quite subtle and difficult to differentiate from problems experienced by healthy older adults (Petersen, 2004).

The criteria for MCI involve three main domains: (1) reporting of subjective memory complaints by the individual, (2) the presence of objective memory impairment, and (3) general decline in intellectual functioning, including nonmemory cognitive domains (Petersen, 2004). Importantly, MCI proves to be a valuable clinical entity that allows for the identification of individuals who are at increased risk of developing dementia (Ames, 2006). As shown by longitudinal data, participants diagnosed with MCI develop AD at a rate of about 12% per year, while age-matched controls develop AD at a rate of 1–2% per year (Petersen et al., 2001). Furthermore, MRI studies show that participants with MCI present atrophy of the hippocampus and the entorhinal cortex when compared to control participants (Jack et al., 1999; Pennanen et al., 2004). However, the atrophy detected in MCI patients is not as severe as that observed among AD patients. Interestingly, smaller volumes of the hippocampus and entorhinal cortex are associated with a more rapid progression rate from MCI to AD (Korf et al., 2004; Jack et al., 1999).

In addition to the presence of decreased levels of hippocampal metabolism that is observed in AD patients (Mosconi et al., 2005), MCI patients display higher levels of the pathological hallmarks of AD, especially high density of NFT and increased levels of amyloid deposits in the brain (Mufson et al., 1999; Price and Morris, 1999; Kordower et al., 2001; Mitchell et al., 2002).

Despite the fact that the pathological hallmarks of AD have all been characterized, including SP, NFT, and neuronal loss, little is known about the etiology and pathogenesis of the disease. Over the past few decades, a number of hormones have been investigated in relation to the pathophysiology of AD.

25.2 Hormones, Aging, and AD

In addition to neurotransmitters and neuropeptides, the brain produces hormones (or neurohormones) as well as receptors that are able to recognize peripherally secreted hormones. Hormones are very sensitive to environmental influences and, once produced in the periphery, many of these hormones can easily cross the blood–brain barrier (BBB) and affect and/or modify behavior or cognition by binding to various receptor types in the brain. Over the past few decades, research has shown that a number of hormones and their respective systems are dysregulated in the course of cognitive aging and AD. In the next section, we summarize what is known about each of these hormones.

25.2.1 A Brief History on Hormones and AD

The observation that there is an age-associated reduction in concentrations of several hormones, such as estrogen, testosterone, dehydroepiandrosterone (DHEA), and growth hormone, led to the inference that these alterations of neuroendocrine systems might contribute to the pathophysiology of AD. The fact that AD occurs more frequently in women than in men has also contributed to the neuroendocrine involvement hypothesis. However, this sex difference is still a matter of debate.

A causal relationship between hormones and AD remains to be established since attempts to treat AD with hormone supplementation have been disappointing. Also, the assumption that aging is inevitably associated with a reduction in hormone levels has been proven to be inconsistent for some hormones and incorrect for other hormones. However, despite these drawbacks, the hormonal involvement in AD is still under investigation and evidence is accumulating supporting the complex role of hormones in the pathophysiology of AD.

25.3 Gonadal Hormones

The biosynthesis of steroids in the gonads (ovaries and testes) is stimulated by a cascade of hormonal events that begins at puberty and terminates at menopause in women, but continues until the end of the life cycle in men. Briefly, the cascade of events begins in the hypothalamus, which releases gonadotropin-releasing hormones (GnRH) in a pulsatile fashion and stimulates the anterior pituitary to secrete two gonadotropins, luteinizing hormone (LH) and follicle-stimulating hormone (FSH), which, in turn, stimulate the secretion of gonadal hormones (estrogen and progesterone in women and testosterone in men). In women, sex steroids have feedback effects on the secretion of LH, FSH, and GnRH, thus forming a closed-loop system of the hypothalamic–pituitary–gonadal (HPG) axis (Ferrari et al., 2000).

Men and women differ greatly in their reproductive senescence. While both men and women encounter a decrease in sex steroid levels with age, the rate at which this change occurs is drastically different between the sexes. In men, hormonal changes are subtle, with testosterone and estrogen levels declining gradually and linearly with age (Vermeulen et al., 2002; Lamberts et al., 1997). In contrast, women encounter hormonal milestones that can occur abruptly and have a profound influence on endogenous gonadal hormone levels, such as menarche, pregnancy, breast-feeding, and menopause. The postmenopausal years (Vermeulen et al., 2002), following the cessation of ovarian production of sex steroids, are characterized by very low levels of progesterone and estrogen (Lamberts et al., 1997; Speroff and Fritz, 2005). These low levels of gonadal steroids, in turn, give rise to a shift in the balance of the HPG-axis feedback loop, resulting in an increase in the production of gonadotropins. In serum, a three- to fourfold increase in LH and a four- to 18-fold increase in FSH are seen in women (Chakravarti et al., 1976) whereas a two- to threefold increase in both LH and FSH characterize senescence in men (Neaves et al., 1984; McEwen, 2002; McEwen and Alves, 1999). These sex differences in rate and intensity of hormonal change are central to the hypothesis regarding the role of sex steroids in the neuroendocrine etiology of AD.

25.3.1 Gonadal Hormones and Neuroprotection

There are many ways by which sex steroids and gonadotropins may alter AD risk. Steroids are small lipophilic molecules that readily cross the BBB in their unbound state and act on the brain through several mechanisms and interactions throughout life.

25.3.1.1 Estrogen neuroprotection

The classical genomic pathway of estrogen involves intracellular receptors, namely estrogen receptors (ERs) alpha and beta. These ERs are expressed in large quantity with great overlap between the two subtypes, in both males and females of different species, in structures involved in hormonal regulation and sexual behavior, such as the hypothalamus, the preoptic area, and the anterior pituitary. Interestingly, they are also highly concentrated in areas involved in learning and memory, such as the amygdala, the hippocampus, the basal forebrain, and the frontal cortex (Osterlund et al., 2000a,b; McEwen, 2002).

It has been observed that estrogen may impact brain metabolic state by enhancing glucose transport and cerebral blood flow, which are altered in AD (Maki and Resnick, 2000; Greene, 2000; Eberling et al., 2000). *In vivo* evidence suggests that estrogen can protect neuronal loss that is characteristic of AD by promoting neurogenesis, synaptic plasticity, dentritic spine density, and connectivity (McEwen, 2002; Behl, 2002). Recently, it was shown that acute treatment with the right moderate dose of estrogen did not stimulate neurogenesis in animals that were in a prolonged state of estrogen deficiency, supporting a time-dependent effect of estrogen on neurogenesis (Tanapat et al., 2005). This has important clinical implications if we believe that there is a critical window of opportunity for protective estrogen action on the brain to take place.

Estrogens have neurotrophic properties and also have a profound influence on the activity of other growth factors, such as brain-derived neurotrophic factors (BDNF) and nerve growth factor (NGF) (Toran-Allerand, 2004; Scharfman and Maclusky, 2006). Recent results demonstrate that plasma BDNF levels are influenced by both endogenous and exogenous estrogen levels, with higher BDNF levels observed during high-estrogen periods across the menstrual cycle, lower BDNF levels following low-estrogen periods during menopause, and restored BDNF levels in postmenopausal women using hormone therapy (HT).

In vitro evidence suggests that estrogen might act as a chemical shield by protecting neurons against a wide range of insults such as Aβ toxicity and glutamate neurototoxicity (Petanceska et al., 2000). Bhavnani (2003) demonstrated that different compounds found in the conjugated equine estrogen (CEE) preparation, often used as HT, were all neuroprotective against glutamate-induced neurotoxicity in hippocampal cultured cells. However, this effect was found to be dose-dependent and differed with different compound potencies (Bhavnani, 2003).

Studies on estrogen-induced enhancement of cholinergic enzymes have highlighted the nonreproductive actions of estrogen on the brain that are essential for cognitive function (Luine, 1985). Different levels of estrogen seen during the estrous cycle and during ovariectomy/estrogen add-back studies have revealed an estrogen induction of choline acetyltransferase (ChAT), an enzyme involved in the synthesis of acetylcholine, in the basal forebrain, cortex, and hippocampus of female rats (Luine, 1985). In humans, it was reported that long-term HT can enhance cholinergic function in postmenopausal women and that it might be related to duration of estrogen use (Van Amelsvoort et al., 2003).

25.3.1.2 Testosterone neuroprotection

Interest in the influence of testosterone as a neuroprotective agent in AD is fairly new. Androgen receptors (ARs) that are found in several brain regions, such as the thalamus, hippocampus, and cerebral cortex, are often co-localized with ERs (Simerly et al., 1990). Testosterone is also converted into estrogen by aromatase. Thus, for men and women, testosterone can influence cerebral function directly or indirectly through conversion from testosterone to estrogen.

Similar to estrogen findings, studies have shown that testosterone can directly regulate Aβ levels and attenuate Aβ toxicity both *in vitro* and in humans (Xu et al., 2000; Gandy et al., 2001; Gillett et al., 2003). Testosterone is also able to prevent heat shock-induced hyperphosphorylation of tau protein in the brain (Papasozomenos, 1997), which, in turn, may prevent the formation of NFT.

In vitro and animal studies suggest that activation of AR results in several genomic and nongenomic changes involved in neuron viability and neuronal resistance to insult. For example, testosterone has the ability to block neuronal apoptosis (Lue et al., 1999) and increase the production of NGF (Pike et al., 2006; Bates et al., 2005).

25.3.2 Gonadal Hormones and Risk of AD

25.3.2.1 Estrogen and risk

25.3.2.1(i) Endogenous estrogen-related risk factors

It is hypothesized that the rapid estrogen-deprived state brought on by menopause renders women at a greater risk than men in AD development. However, to date, results have been mixed. Some authors speculate that the sex difference in AD development can be attributed to longevity, with women living longer than men and therefore increasing their chance of exhibiting pathological cognitive decline. However, epidemiological studies have reported that women suffer from AD about twice as much as men, regardless of the longevity factor (Launer et al., 1999; Group, 1994; Andersen et al., 1999). Other factors such as incidence of (other) dementia, duration of illness, and education level might contribute to blur the results and explain the mixed findings in the field of estrogen and AD (Baum, 2005).

Endogenous estrogen exposure factors, such as age at menarche, age at menopause, length of reproductive period, and number of children, are variables recently added to the field. An early study by Paganini-Hill and Henderson (1994) paved the way to endogenous estrogen exposure markers by reporting that the risk of AD decreases with an increase in endogenous estrogen exposure. The authors reported an increased risk with increasing age at menarche, but failed to find a relationship with age at menopause and number of children (Paganini-Hill and Henderson, 1994).

Studies that have assessed menopause and number of pregnancies have reported that having children (multiparous) is positively correlated with AD diagnosis (Ptok et al., 2002), and number of pregnancies is negatively correlated with age of AD onset (Sobow and Koszewska, 2003). In terms of menopause, Sobow and Koszewska (2003) reported a positive association between age at menopause and age of AD onset. Furthermore, a recent study demonstrated that women who undergo an oophorectomy (surgically induced menopause) are at increased risk of cognitive impairment or dementia and that this risk increases with younger age of surgery (Rocca et al., 2007).

Geerlings et al. (2001) were the first to investigate lifelong endogenous exposure to estrogen in relation to incidence of dementia. Endogenous exposure was measured by reproductive period, which was defined as age at natural menopause minus age at menarche. The authors reported that women with a longer reproductive period displayed greater risk of dementia. However, this association was only found in women who were carriers of a genetic risk marker of AD, namely carriers of the apolipoprotein E ε4 allele, and was not observed in non-ε4 carriers (Geerlings et al., 2001).

Overall, there is a need to consider endogenous estrogen exposure in the etiology of AD in women. Further studies are needed to better understand the role of punctual dramatic increases of sex steroids (e.g., sharp rise during pregnancy) on the brain in contrast to lifelong constant fluctuating levels (e.g., fluctuations in nulliparous women during menstrual cycle).

25.3.2.1(ii) Exogenous estrogen-related risk factors

For over a century, ovarian steroids have been used as HT to alleviate symptoms associated with menopause (Speroff and Fritz, 2005). HT can be comprised of either a combination of both estrogens and progestogens (EPT) or estrogens only (ET). With continual conversations regarding the health risk–benefit of HT, the medical community received a shock in 2002, following the publication and media coverage of the Women's Health Initiative (WHI) Study results that stated HT causes more harm than benefit to women's health (Rossouw et al., 2002). Soon after release of the WHI report, HT began to be considered a dangerous drug, giving rise to widespread controversies that brought the future of HT to a crossroad. Consequently, an important critical reevaluation of past scientific results took place and new variables were introduced to better understand the effect of HT on health and the brain. Variables such as HT duration, time of HT initiation, type of HT used, and HT regiment (cyclic vs. constant) are now taken into account and will be discussed further in Section 25.3.4.

Before publication of the Women's Health Initiative Memory Study (WHIMS) dementia results (Shumaker et al., 2004; Shumaker et al., 2003), the majority of studies that investigated the influence of HT on cognitive function suggested that HT lowers the risk for AD in postmenopausal women (Henderson et al., 2005; Zandi et al., 2002; Fillit, 2002; Yaffe et al., 1998; Kawas et al., 1997; Paganini-Hill and Henderson, 1994). The WHIMS was a randomized, double-blind, placebo-controlled clinical trial with the primary endpoint being incidence of dementia. Participants were postmenopausal women aged 65–79, who were randomly assigned to receive CEE (ET trial), a combination of CEE and medroxyprogesterone acetate (MPA; women with an intact uterus; EPT trial), or placebo. In comparing the EPT group to the placebo group, results showed that 66% of EPT versus 34% of placebo users were diagnosed with probable dementia, thus doubling the hazard ratio in the EPT group (HR 2.05). In comparing ET to placebo, it was reported that 60% of ET and 40% of placebo users were diagnosed with probable dementia, thus increasing the HR by half (HR 1.49). Incidence rates for probable dementia were similar for both the EPT trial and the ET trial. Contrary to what was expected, HT therapy (ET and EPT) did not reduce incidence of dementia but rather increased the risk. The authors concluded that the use of HT in the prevention of dementia in women 65 years of age or older is not recommended. However, because the WHIMS did not include women under 65, and previous studies reported positive effects of HT in the prevention of AD in younger women (Yaffe et al., 1998; Fillit, 2002; Henderson et al. 2005), not only could the WHIMS results not be generalized to younger postmenopausal groups, but it also contributed to the critical window of opportunity theory. This critical period is around early menopausal years, during which HT could reduce risk of AD. However, once older women are deprived of sex steroids for several years, and have thus passed the window of opportunity, the use of HT may have detrimental effects and increase the risk of AD

(Sherwin, 2007; Resnick and Henderson, 2002). Indeed, this assumption falls in line with the aforementioned animal study that reported a time-dependent effect of estrogen on neurogenesis (Tanapat et al., 2005). Recently, in an attempt to assess the importance of timing and age at HT initiation, the WHIMS data were revised, taking into account prior use of HT. The results demonstrated that prior HT users had significantly lower risk of dementia and AD (Henderson et al., 2007). Thus, women who began treatment at an earlier age experienced greater benefit compared to women who started therapy in later years. Altogether, these findings point to a time-dependent protective effect of estrogen on risk for AD.

25.3.2.2 *Testosterone and risk*

Testosterone circulates in the blood, bounded to the sex hormone-binding globulin (SHBG), which is an enzyme that is known to increase with age in men. Common testosterone measurements are total testosterone, free (unbound) testosterone, or free androgen index (FAI; based on the ratio of total testosterone to SHBG). Aging in men is accompanied by a linear, but individually variable, decline of circulating free testosterone levels and FAI. However, testosterone levels below normal values occur only in a minority of elderly men from 7% of men between the ages of 40 and 60, to 20% in men between the ages of 60 and 80, and 35% in men over 80 years old (Lamberts et al., 1997; Kaufman and Vermeulen, 2005; Vermeulen and Kaufman, 1995). The role of declining testosterone in the etiology of AD has recently gained considerable attention. A handful of cross-sectional studies consistently report lower testosterone concentrations or FAI in men diagnosed with AD (Hogervorst et al. 2004, 2003, 2001; Paoletti et al. 2004). Although limited in number, the recent results tend to show that low testosterone is a predecessor of AD. Results from the Baltimore Longitudinal Study of Aging (BLSA) support this view by demonstrating that men who were diagnosed with AD had lower free testosterone levels, compared to controls, years before a diagnosis of AD was made (Moffat et al., 2004). Another study assessing levels of testosterone in brain tissue demonstrated that brain levels of testosterone were significantly lower in a severe AD group and in an early-stage AD group compared to healthy controls (Rosario et al., 2004). These two studies suggest that testosterone decline precedes AD onset and support the view of low testosterone as being a risk factor and not just a consequence of the disease pathology in men (Hogervorst et al., 2001; Paoletti et al., 2004). Although these studies have been limited to men, one study demonstrated that women with normal global cognitive function, as determined by the Mini Mental State Exam (MMSE, score above 23), had significantly higher total and free testosterone levels compared to women who scored below normal on the MMSE (score below 23; Barrett-Connor and Goodman-Gruen, 1999).

25.3.3 Gonadotropins

A potential consequence of age-related decline in gonadal hormones is increased levels of gonadotropins (LH and FSH). Recently, it was hypothesized that low levels of sex steroids could be attributed to elevated LH and/or FSH levels. This hypothesis emerged from the observation that patients with AD have a twofold increase in gonadotropin levels compared to healthy controls (Short et al., 2001; Bowen et al., 2000). However, the effects of increased gonadotropins due to the loss of the negative feedback function of sex steroids are somewhat unexplored. In accordance with the speculated implication of gonadotropins in AD development, elevated LH in vulnerable neurons of AD patients was observed when compared to nondemented controls (Bowen et al., 2002). Furthermore, it has been demonstrated that increased gonadotropin levels in aged women are only found in women with AD (Short et al., 2001) and significant elevations in gonadotropins in men only occur in the later stages of life (Morley et al., 1997). This coincides with new results showing a significant relationship between gonadotropins and AD in older old men (i.e., mean age of 80) and not in younger old men (i.e., mean age of 65; Hogervorst et al., 2003, 2004).

Overall, the sex difference found in age-related elevations in gonadotropin levels, along with the decrease in sex steroids observed after menopause in women and gradual decline in men, support the sex-difference hypothesis in the prevalence of AD and highlight the importance of the critical window of opportunity for HT in women.

25.3.4 Gonadal Hormones: Prevention and Treatment

25.3.4.1 *Estrogen*

For over 20 years now, researchers have investigated the potential use of HT as a therapeutic agent to prevent, delay, or slow the progression of AD. Unfortunately, no consensus has been met. However, based on the results, HT seems to be associated with a reduced risk in AD development. An important meta-analysis assessing the effect of HT in the risk of developing dementia reported a 29% reduced risk in HT users (Yaffe et al., 1998). These findings have since been replicated (for review, Fillit (2002) and Pinkerton and Henderson (2005)). Furthermore, the importance of prior HT use was highlighted in a study that reported older women with a history of HT use of more than 10 years had a fivefold lower risk of developing AD (Zandi et al., 2002). Along the same line, taking age into account, Henderson et al. (2005) stratified their analyses across three age groups (50–63, 64–71, and 72–99) and reported that HT was significantly associated with reduced risk in the youngest group only. Together, these findings suggest that the protective effect of HT might be confined to a younger subgroup of women or women who begin HT at an earlier age, thus confirming the window of opportunity hypothesis.

In terms of HT as a treatment for AD symptomatology, most clinical trials suggest that it does not affect AD symptoms or disease progression. However, most of these trials were limited by small sample size, very short treatment duration, and variance in stage of the disease (Yaffe et al., 1998; Pinkerton and Henderson, 2005; Fillit, 2002). The results from the aforementioned prevention studies suggest that age, initial timing, and HT duration are crucial factors in determining treatment efficacy and thus additional studies should investigate the effect of HT on disease progression while taking these impact factors into account. However, one cannot rule out the fact that HT may only be effective as a preventative agent and may not be effective in slowing disease progression once the mechanisms involved in AD have been put into motion and neuronal injury has been incurred (Fillit, 2002).

A recent randomized placebo-controlled study investigated the selective estrogen receptor modulator (SERM) raloxifene in risk of AD development. The study concluded that compared to placebo, a raloxifene dose of 120 mg day^{-1} for 3 years, but not 60 mg day^{-1}, is associated with a 27% reduced risk of MCI or dementia (Yaffe et al., 2005).

Based on the above findings, more investigations are necessary to examine different forms, dosages, and route of administration, along with the relevance of endogenous estrogen exposure and timing, as well as duration of exogenous estrogen exposure in the treatment and prevention of AD.

25.3.4.2 Testosterone

Recent investigations, although limited in number, point toward a positive effect of testosterone supplementation on cognitive performance and brain function in elderly men (Moffat, 2005; Driscoll and Resnick, 2007). In a pilot study, Tan and Pu (2003) demonstrated that in a small group of men newly diagnosed with AD, a 12-month testosterone supplementation regime significantly improved scores on tests of cognitive function compared to the placebo-treated group who showed impairment. Similarly, in a randomized study assessing a 6-week testosterone supplementation regime in 32 men diagnosed with AD or MCI, it was found that men in the treatment group exhibited improvements on a spatial memory and constructional abilities test compared to men in the placebo-control group. Furthermore, the treatment group displayed increased testosterone and estradiol levels compared to the control group (Cherrier et al., 2005). In a study by Lu et al. (2006), while no improvement in cognitive function was observed following a 6-month placebo-control treatment regime, it was reported that AD patients receiving testosterone exhibited greater improvement in quality of life, compared to healthy controls and placebo-control AD patients (Lu et al., 2006). These preliminary results are encouraging, but studies with larger samples sizes are required before clinical decisions related to AD prevention and/or treatment can be made.

25.4 Adrenal Hormones

The hypothalamic–pituitary–adrenal (HPA) axis is a stress-sensitive system that is central to stress resistance. A functional HPA system, beginning with the synthesis and release of corticotropin-releasing hormones (CRH) from the medial paraventricular nucleus (PVN) of the hypothalamus, followed by the synthesis and release of adrenocorticotropic hormones (ACTH) from the pituitary, culminating in the production of glucocorticoids (GCs: namely cortisol in humans) and catecholamines from the adrenals, is necessary for survival of the organism. Following activation of the system, and once the perceived stressor has subsided, feedback loops are put into motion at various levels of the system (i.e., from the adrenal gland to the hypothalamus and other brain regions) in order to shut the HPA axis down and return to homeostasis.

As part of the feedback loop, GCs cross the BBB, where the steroid not only acts on the hypothalamus, but further acts on suprahypothalamic regions that contain GC receptors, including the amygdala, frontal cortex, and hippocampus (Diorio et al., 1993;

Mizoguchi et al., 2003; McEwen et al., 1968; Herman et al., 2005). The hippocampus is a very sensitive and malleable brain structure that is important for declarative learning and memory (Manns and Eichenbaum, 2006). The hippocampus is also involved in inhibiting most aspects of HPA activity, including the nadir and peak phase of the circadian rhythm of cortisol, as well as the onset and termination of the stress response (Feldman and Weidenfeld, 1999; Jacobson and Sapolsky, 1991; Born and Fehm, 1998). Indeed, damage or atrophy of the hippocampus has been found to impair the organism's ability to shut off the HPA axis once the stressor has subsided, leading to prolonged HPA activation (Herman and Cullinan, 1997; Jacobson and Sapolsky, 1991). Given the brain regions involved in the HPA axis, notably the hippocampus, dysregulation of the HPA system has been investigated in relation to risk and pathology of AD.

25.4.1 Glucocorticoids

Under basal conditions, cortisol secretion exhibits a 24-h circadian profile in which cortisol concentrations rise in the early morning to reach a circadian peak, followed by a decline in cortisol to reach a circadian trough in the evening and nocturnal period, which is then followed by an abrupt elevation after the first few hours of sleep. Cortisol secretion also exhibits seasonal variations, with higher cortisol levels secreted during long photoperiod months (e.g., May) compared to shorter photoperiod months (e.g., November; Arsenault-Lapierre et al., 2008).

Circulating GCs bind with high affinity to two GC receptor subtypes: the mineralocorticoid (or type I) and the GC (or type II) receptors. Although both receptor types have been implicated in mediating GC-feedback effects (Reul and De Kloet, 1985), there are two major differences between type I and type II GC receptors. First, type I receptors bind GC with an affinity that is about 6–10 times higher than that of type II receptors. This differential affinity results in a striking difference in occupation of the receptor types under different conditions and time of day. The second major difference between these two receptor types is related to their distribution throughout the brain. Type I receptors are present exclusively in the limbic system, with a preferential distribution in the hippocampus, parahippocampal gyrus, and entorhinal and insular cortices. In contrast, type II receptors are present in both subcortical (PVN and other hypothalamic nuclei, the hippocampus, and parahippocampal gyrus) and cortical structures, with a preferential distribution in the prefrontal cortex (Diorio et al., 1993; Meaney and Aitken, 1985; McEwen et al., 1968, 1986).

It is suggested that exposure to high levels of GCs over the life span plays a significant role in cognitive aging (Sapolsky et al., 1984). However, whether cortisol secretion increases with age is still somewhat controversial. Numerous studies have reported that overall basal cortisol levels generally do not change across the life span in healthy older populations (Baranowska et al., 2007; Boscaro et al., 1998; West et al., 1961; Jensen and Blichert-Toft, 1971; Sherman et al., 1985; Waltman et al., 1991). Still, several other researchers have reported higher evening (Touitou et al., 1982; Van Cauter et al., 1996; Raff et al., 1999; Van Cauter et al., 2000) or morning levels of basal cortisol (Bélanger et al., 1994), as well as a phase advance in their diurnal rhythm (Sherman et al., 1985; Van Cauter et al., 1996). Discrepancies in findings may be attributed to cortisol sampling method (multiple samples vs. one time sample), time of cortisol measurement (diurnal and seasonal difference), as well as differences in gender and age ranges of the population sample under investigation. It is also possible that even within clinically healthy populations, subpopulations of the elderly may have subsyndromal abnormalities that are related, either as cause or effect, to increases in basal cortisol concentration, for example, sleep fragmentation (Van Cauter et al., 2000) or decreased bone mineral density (Raff et al., 1999).

It has been proposed that increased HPA-axis activity is not a necessary consequence of aging *per se*, but is significantly more prevalent in aged rats selected for spatial memory deficits than in cognitively unimpaired aged rats (Landfield et al., 1978; Issa et al., 1990). These results are further confirmed in human cross-sectional and longitudinal research that show age-related increases in cortisol levels are specific to aged populations exhibiting signs of cognitive impairment (Dori et al., 1994; Giubilei et al., 2001; Lee et al., 2007; Rasmuson et al., 2001; Spada et al., 2001). Lupien et al. (1994, 1998) performed a longitudinal study in which they assessed 24-h serum cortisol levels, in a population of 92 older adults for a period of 4 years. Results revealed that about 30% of the older adult population exhibited increased secretion of cortisol over years. Furthermore, studies by this group revealed that aged adults displaying increased secretion of total cortisol over years presented both memory impairment (Lupien et al., 1994) and smaller volume of the hippocampus (Lupien et al., 1998), when compared to aged adults with normal secretion of cortisol over years. Also, controlling for seasonal effects, a recent study by Arsenault-Lapierre et al. (2008) showed a stepwise increase in basal salivary cortisol secretion, with normal elderly controls secreting less cortisol compared to MCI patients who in turn secreted lower cortisol levels than AD patients. Given these findings in normal healthy older adults, it is conceivable that dysregulation of the HPA-axis system leads to decreased cognitive function in old age, which may ultimately lead to the development of AD.

25.4.1.1 GCs and risk of AD

Numerous studies have reported altered HPA-axis activity, noted by increased GC secretion and decreased HPA-axis responsivity in AD patients (Gurevich et al., 1990; Rasmuson et al., 2001; Swanwick et al., 1998; Umegaki et al., 2000a). As previously mentioned, one of the first neuronal regions to undergo degeneration is the hippocampus (Dickerson et al., 2001; Jack et al., 1997; Xu et al., 2000), which is noted to have a high volume of GC receptors (McEwen et al., 1968; Seckl et al., 1991). Based on these observations, Sapolsky et al. (1986) put forth the GC cascade hypothesis. As the hippocampus and HPA axis are in a closed feedback loop, increased exposure to stress hormones may result in hippocampal cell loss, which may induce hypercortisolemia, which, in turn, may lead to further degeneration of the hippocampus (Sapolsky et al., 1986).

Many studies reporting hypercortisolism in AD have attributed the elevated GC levels to hippocampal atrophy associated with the disease. However, studies have shown that GC levels may predict AD progress (Csernansky et al., 2006; Umegaki et al., 2000a). In a 4-year longitudinal study by Csernansky et al. (2006), GC levels and cognitive scores were assessed in mild AD patients and healthy controls. Results showed that patients with moderate dementia secreted higher GC levels than healthy control older adults. More importantly, the authors reported that elevated cortisol levels in AD patients, but not healthy controls, predicted greater increase in clinical symptomatology and quicker decline in cognitive function (Csernansky et al., 2006). This finding showed that elevated GC secretion in AD is not simply a result of the disease process, but serves as a risk factor for decline in cognitive capacity.

Indeed, it has been suggested that elevated GC levels may impose negative effects on the brain by reducing hippocampal neuronal ability to survive a variety of insults, including AD-related neurotoxicity. In line with this, animal models of AD have shown that GCs increase Aβ and tau pathology, thus accelerating the formation of SPs and NFTs in the AD brain (Green et al., 2006).

However, very recently a new hypothesis was put forward by Landfield et al. (2007) in order to explain the association between GCs and AD. As stated above, the original GC hypothesis of brain aging and AD proposed that chronic exposure to GCs promotes hippocampal aging and AD. This hypothesis involves a very important notion, specifically that the expression of most target biomarkers of brain aging should be regulated in the same direction (increased or decreased) by both GCs and aging. In order to test this hypothesis, Landfield et al. (2007) used microarray analyses to identify a panel of hippocampal gene-expression changes that were age dependent and corticosterone dependent. The results obtained were inconsistent with the GC hypothesis of brain aging since the authors found that a majority of biomarker genes were regulated in opposite directions by aging and GCs, particularly inflammatory and astrocyte-specific genes. Consequently, the results obtained by Landfield et al. (2007) suggest that, in the brain, GCs and aging interact in more complex ways that depend on the cell type. Therefore, the authors now propose a new version of the GC-brain aging hypothesis. It is now suggested that aging not only selectively increases GC efficacy in some cell types (e.g., neurons where it enhances catabolic processes), but also selectively decreases GC efficacy in other cell types (e.g., astrocytes where there is a weakening GC anti-inflammatory activity). Finally, they proposed that the efficacy of GCs in terms of brain aging may be mediated in part by cell type, with specific shifts in the antagonistic action of GC and insulin. The latter antagonistic action may be of great relevance for AD pathogenesis.

25.4.1.2 GCs: Prevention and treatment

Despite strong evidence showing damaging effects of GCs on the aging brain, clinical trials were designed in the late 1990s in order to test the potentially positive effects of exogenous GCs on the course of AD. The rationale for these clinical trials was based on evidence showing that there are a number of inflammatory mechanisms (such as activation of the complement cascade; upregulation of a number of acute-phase proteins, cytokines, and chemokines) that are intimately involved in the development of AD (Akiyama et al., 2000; McGeer and McGeer, 2001). Based on these findings, it was conceivable that a neuroinflammatory process exacerbates AD pathology. Consequently, it was suggested that anti-inflammatory drugs (such as exogenous GC) may be useful in either delaying onset or slowing the progression of AD (Pratico and Trojanowski, 2000).

Thus far, only a few epidemiological studies exist on the association between GC use and AD. Breitner et al. (1994) performed a twin study, revealing that the onset of AD was inversely associated with prior concomitant use of GCs or ACTH (Breitner et al., 1994). In a second study by this group, no significant difference was reported in AD risk after exposure to GC in high-risk siblings (Breitner et al., 1995). This lack of association between GC use and AD was also found in The Canadian Study of Health and Aging (1994), which was a large population-based study.

Yet, despite the negative effects of GCs reported on the human brain, and despite the lack of association reported between GC use and AD, in the early 1990s the Multicenter Trial of Prednisone in Alzheimer's Disease was established (Aisen and Davis, 1997). A total of 138 participants were randomly assigned to receive either placebo or an initial dose of 20 mg of prednisone (a synthetic GC), which was then tapered, after 4 weeks, to 10mg and continued for a year. Results from this study became available Aisen and Pasinetti (1998) and in Aisen et al. (2000). The results showed that there were no differences in performance on the cognitive subscale of the Alzheimer's Disease Assessment Scale. More importantly, the results showed that the prednisone-treated participants showed a behavioral decline compared with the placebo group, a result that goes along with previous studies that have shown that GCs can have damaging effects on the human brain.

In 2002, based on the hypothesis that high levels of GCs may be detrimental for the human brain, Belanoff et al. (2002) started a clinical trial in AD, testing the efficacy of mifepristone, a GC antagonist, in decelerating the rate of cortisol-related cognitive decline in participants with mild-to-moderate AD. In this study, controls and AD subjects were randomly assigned to receive either placebo or 200 mg of mifepristone on a daily basis for 6 months. The results of this study have not yet been published.

25.4.2 Dihydroepiandrosterone

Dihydroepiandrosterone (free form=DHEA) and dihydroepiandrosterone sulfate (DHEA-S) are steroids synthesized mostly in the adrenal glands, but may also be synthesized in the brain (neurosteroid). DHEA is the precursor to testosterone, estrone, and estradiol. Despite the fact of being a neurosteroid, it has been observed that DHEA does not have high affinity for the AR.

DHEA interacts with other hormones and neurotransmitters. It is known for its powerful anti-GC action and has been suggested to protect the brain from the potential damaging effects of GCs on the aging brain (Blauer et al., 1991; Kimonides et al., 1998). It is reported that DHEA can also increase the effects of glutamate (Debonnel et al., 1996) and decrease the effects of gamma-aminobutyric acid (GABA; Majewska et al., 1990). Moreover, the neurosteroid has neuroprotective properties against oxidative stress and damage induced by Aβ synthesis (Bastianetto et al., 1999; Kimonides et al., 1998).

25.4.2.1 DHEA and risk of AD

A vast majority of studies found that DHEA and DHEA-S decline with age, illness, and stress. By the age of 80, DHEA and DHEA-S levels are found to be 20% of the levels found in 20-year-olds. Some qualify DHEA supplementation as an antiaging agent, improving well-being and counteracting the suggested neurotoxic exacerbation of declining DHEA levels in old age. In some countries, supplements are sold in large quantities over-the-counter despite the scarce and conflicting scientific evidence on the beneficial effects of DHEA.

It was observed that levels of DHEA and DHEA-S are lower in AD patients compared to healthy controls (Genedani et al., 2004; Ferrari et al., 2000; Armanini et al., 2003; Bernardi et al., 2000). Hillen et al. (2000) measured DHEA-S levels prior to AD diagnosis (incident cases) and found that individuals who developed AD displayed the lowest DHEA levels. This population-based prospective study supports the role of DHEA-S as a risk factor for AD (Hillen et al., 2000). However, the specificity of this result to AD has been questioned since reduced concentrations of serum DHEA-S have also been observed in cerebrovascular dementia. Furthermore, a number of studies found no association between DHEA-S levels and AD (Bo et al., 2006; Carlson et al., 1999; De Bruin et al., 2002; Nasman et al., 1991; Schneider et al., 1992). It has been hypothesized that these results might reflect a common neurodegenerative phenomenon (Yanase et al., 1996). However, studies have failed to find a link between Parkinson disease, a well-known neurodegenerative disorder, and DHEA-S levels (Genedani et al., 2004).

25.4.2.2 Dihydroepiandrosterone: Prevention and treatment

Few studies exist on DHEA supplementation in aging and dementia. In healthy aged individuals no significant effect of DHEA, compared with placebo, has been observed on cognitive functions (Grimley Evans et al., 2006). To date, only one study has assessed the effects of DHEA administration in AD patients, reporting no improvement in cognitive performance or overall change in severity of cognitive impairment following treatment (Wolkowitz et al., 2003).

25.4.3 Catecholamines

Catecholamines act both as hormones and as neurotransmitters. The focus here will be narrowed to the consideration of epinephrine (EPI) and norepinephrine (NE), at the expense of dopamine (DA), which does not function as a classical hormone. Although catecholamine biosynthesis is discussed in greater detail in other chapters, it is necessary to briefly identify the enzymes involved. The catecholamines are derived from the amino acid tyrosine, which is oxidized to dihydroxyphenylalanine (L-DOPA) by the rate-limiting enzyme tyrosine hydroxylase (TH). Decarboxylation to DA is followed by oxidation to NE by DA β-hydroxylase. EPI is synthesized by the methylation of NE by phenylethanolamine *N*-methyltransferase (PNMT), primarily in the chromaffin cells of the adrenal medulla and in adrenergic neurons of the rostral ventrolateral medulla oblongata in the caudal brainstem.

25.4.3.1 Epinephrine

EPI is the primary catecholamine secreted by the human adrenal medulla, at a ratio of approximately 4:1 relative to NE (Wang et al., 1999; Saar et al., 1980), a ratio that varies considerably across species. EPI is rapidly released in response to stress as part of the fight-or-flight response.

25.4.3.1(i) EPI and aging

In humans, there is no change, or a slight decline, in basal plasma EPI concentrations with increasing age (Franco-Morselli et al., 1977; Esler et al., 1995; Mazzeo et al., 1997; Penev et al., 2005; Seals and Esler, 2000). This relative constancy of plasma EPI is the result of a significant age-related reduction in secretion concurrent with marked decreases in clearance of EPI (Seals and Esler, 2000; Esler et al., 1995; Mazzeo et al., 1997). Increases in the secretion of plasma EPI in response to mental stress, isometric stress, and dynamic exercise have been found to be markedly attenuated (33–44%) in older men compared to young men (Esler et al., 1995). However, submaximal exercise has also been reported to elicit similar increases in EPI secretion in both old and young participants (Mazzeo et al., 1997). The mechanisms responsible for age-related decreases in resting and stimulated EPI secretions have not been identified, but may include decreased sympathetic nervous system (SNS) stimulation of the adrenal medulla, reduced responsiveness to stimulation, or deficits in synthesis and/or storage of EPI (Seals and Esler, 2000). One study has provided evidence from human chromaffin cells *in vitro* showing that uncoupling of EPI secretion from depolarization-mediated activation of calcium channels may occur in cells from older participants (Elhamdani et al., 2002).

25.4.3.1(ii) EPI and cognition

An interesting and potentially important aspect of age-related alterations of EPI function is the changing relationship between EPI and glucose that may be related to cognitive impairment in aging. EPI raises plasma glucose concentrations both by stimulating glucose production and limiting glucose utilization. A marked reduction in glycemic sensitivity to EPI has been observed in older individuals by Marker et al. (1998). Infusion of EPI resulted in significantly higher plasma levels in old, compared to young, individuals, confirming slower EPI clearance in aging. Despite elevated plasma EPI concentrations, increases in plasma glucose were markedly attenuated in old compared to young groups. This effect was attributed to desensitization of cellular responses to catecholamines, resulting from increased SNS activity in older participants.

Findings of attenuated EPI responses to stress, together with the reduced effectiveness of EPI in raising plasma glucose levels in older individuals have implications for learning and memory because of demonstrated effects of both EPI and glucose on cognitive function. It was first demonstrated more than 20 years ago that peripheral EPI administration, following appetitive or avoidance training, enhances later memory retention in both young and old rats (Sternberg et al., 1985) despite the very poor penetration of EPI through the BBB (Oldendorf, 1971). This finding has been replicated many times in rats, and the phenomenon has been reproduced in human studies (Cahill and Alkire, 2003). Glucose administered peripherally at training has the same effect on memory retention, suggesting that the effects of EPI may be mediated, at least in part, via alterations of blood glucose levels (Hall and Gold, 1986).

EPI secretion is suggested to be responsible for improved retention of memories associated with stress or emotional arousal. However, the influence of increases in plasma EPI on retention appears to depend on the specific nature and intensity of the emotional arousal or stress and the type of memory that is measured. In human studies, administration of propranolol, which blocks β-adrenergic receptors both peripherally and centrally, is found to impair recognition (Cahill et al., 1994) and short- and long-term declarative memory (Maheu et al., 2004) for emotionally arousing, but not neutral, stimuli. However, when an unrelated cognitive stressor is presented prior to a neutral narrative, neither short-nor long-term declarative memory for narrative material is affected by propranolol administration (Maheu et al., 2005). There have been suggestions that the effects of EPI on retention can be described by an inverted U-shaped dose–response curve.

A combination of anatomical, pharmacological, and electrophysiological approaches has led to evidence for the mediation of EPI effects on learning and memory by afferent nerve fibers in the vagus nerve that project to the nucleus of the solitary tract, where they synapse on neurons innervating the locus ceruleus (LC) to activate noradrenergic pathways to the hippocampus and amygdala. Peripheral endings of the vagus nerve are densely embedded with β-adrenergic receptors that bind EPI, and injections of EPI elicit a significant increase in vagal afferent nerve firing in rats that can be blocked completely by sotalol, a β-adrenergic receptor antagonist (Miyashita and Williams, 2006). Indeed, vagus nerve stimulation has been shown to enhance recognition memory in human participants (Clark et al., 1999).

25.4.3.1(iii) EPI and risk of AD

Attempts have been made to link age-related decreases in EPI's efficacy to stimulate glucose mobilization to altered cholinergic function in AD. In response to stress, acetylcholine is released in the hippocampus, and this cholinergic activation has been associated with increases in plasma glucose and catecholamine concentrations. Degeneration of the hippocampal cholinergic system is one of the most robust pathological features of AD, and there is one report of decreased plasma EPI levels in AD patients relative to vascular dementia patients and cognitively normal control participants (Umegaki et al., 2000b). However, other groups have not found evidence of decreased plasma EPI concentrations in AD (Pascualy et al., 2000; Peskind et al., 1998; Petrie et al., 2001). Although peripheral EPI concentrations do not appear to change in AD, responsiveness to EPI may be reduced, possibly at peripheral β-adrenergic receptors and at other points along the multi-synaptic pathway ultimately leading to decreased NE release in the forebrain.

25.4.3.2 Norepinephrine

The other major source of circulating catecholamines, in addition to the adrenal medulla, is the postganglionic noradrenergic neurons of the SNS that release NE that diffuses into the plasma compartment and is referred to as spillover. NE from the SNS plays a critical role in homeostasis, particularly with regard to blood pressure and cardiovascular responses, and the regulation of body temperature (Seals and Esler, 2000). It also ensures rapid and appropriate responses of skeletal muscle, the heart, and other organs in fight-or-flight situations. Because plasma NE is prevented from crossing the BBB and because its affinity for β_2-adrenergic receptors involved in glucose regulation and afferent signaling via the vagus nerve is much lower than that of EPI, its direct influence on the CNS, and thus on cognition and behavior, is modest. However, it serves, when interpreted carefully, as a valuable index of SNS activation.

25.4.3.2(i) NE and aging

In humans, basal plasma NE concentrations have been reported to increase 10–15% per decade (Ziegler et al., 1976; Goldstein et al., 1983). Although clearance rates are reportedly reduced with age, the major factor responsible for higher NE levels appears to be greater basal SNS activity and thus greater spillover (Marker et al., 1994). Early studies reported age-related increases in venous plasma NE in response to acute stress (Young et al., 1980; Barnes et al., 1982), but in studies in which arterial NE concentrations and spillover are determined in response to a variety of stressors, including isometric and dynamic exercise, orthostatis, cognitive challenge, local cold stimulation, and hypoxia, SNS responsiveness in older adults is not found to differ from young (Esler et al., 1995; Mazzeo et al., 1997).

In the CNS, the major source of NE is the LC in the rostral pons, supplying approximately 70% of NE in the brain (Moore and Bloom, 1979) and providing the exclusive source of noradrenergic input to the hippocampus and prefrontal cortex (Leranth and Hajszan, 2007; Morilak et al., 2005). In terms of neuronal loss, the LC is one of the cell groups most intensively affected by both normal aging and AD (Lyness et al., 2003; Palmer and DeKosky, 1993). Indeed, human LC neuronal loss with normal aging is well established (Chan-Palay and Asan, 1989; Vijayashankar and Brody, 1979; Lohr and Jeste, 1988; Manaye et al., 1995), but the findings of age-related losses are not unanimous (Ohm et al., 1997; Kubis et al., 2000; Mouton et al., 1994). Nonetheless, the occurrence of losses of noradrenergic LC neurons in AD is unequivocal (Matthews et al., 2002; German et al., 1992; Leverenz et al., 2001; Zarow et al., 2003).

In addition to LC neuronal loss, when inputs to terminal fields involved in learning and memory are considered, particularly those in the hippocampus and cortex, the picture of age- and AD-related changes in noradrenergic function becomes much more complex. Indeed, findings from animal literature suggest that LC projections to the dentate gyrus and frontal cortex of the rat are lost with increasing age between 7 and 15 months, at which time the LC neurons begin to give rise to axonal branchings, reinnervating the target areas (Ishida et al., 2001, 2000; Shirokawa et al., 2000). It is suggested that the plastic changes in the axon terminals that occur with age may compensate for the loss of LC innervations in the aged brain.

25.4.3.2(ii) NE and AD

The evidence for compensatory axonal sprouting and additional compensatory responses to losses of innervation from the LC is even more compelling in studies of human noradrenergic function in AD. Several changes in the LC noradrenergic system consistent with compensation for neuronal losses in AD have been identified: (1) increased TH expression in the remaining LC neurons: (2) sprouting of dendrites into the peri-LC dendritic zone, as determined by binding to α_2-adrenergic receptors and NE-binding sites; (3) sprouting of axonal projections to the hippocampus as determined by normal or elevated (depending on cell layer) α_1- and α_2-adrenergic receptor binding; and (4) sprouting of axonal projections to the prefrontal cortex as shown by maintenance of α_1- and α_2-adrenergic-binding sites despite significant neuronal loss (Szot et al., 2006, 2007).

Associations between AD, and in particular, aggressive behavior in AD, have frequently been reported in conjunction with altered α-adrenergic binding, but the methodologies and the nature of the findings are highly variable. Frontal cortex α_2 binding using tissue homogenates has been found to be unaltered (Matthews et al., 2002) or decreased (Meana et al., 1992; Kalaria and Andorn, 1991) in AD. Using a slice-binding approach, α_2 binding in - cerebellar cortex has been reported to be greater in aggressive AD patients (Russo-Neustadt and Cotman, 1997), and greater in LC projection areas in patients with dementia with Lewy bodies, usually associated with frequent agitation, than in AD or control subjects (Leverenz et al., 2001). In other studies using membrane-binding homogenate preparations, increased α_1 binding (as opposed to α_2 binding) was found to be associated with AD, or more specifically, with aggressive behavior in AD (Sharp et al., 2007).

Loss of LC neurons has also been linked to the pathogenesis of AD by evidence that LC degeneration potentiates inflammatory responses and amyloid plaque formation (Heneka et al., 2006). In addition to the loss of noradrenergic LC neurons, a connection has been postulated between adrenergic neurons of the C1 region of the pons/medulla and AD. PNMT activity and PNMT protein are reportedly decreased in areas of the AD brain that are affected by the disease, but not in the cerebellum. Also, the magnitude of the decrease in PNMT in the brain is correlated with the severity of the dementia (Burke et al., 1987). Further studies have shown that the decrease in PNMT activity in adrenergic projections to the LC is due to retrograde degeneration of these neurons and is significantly correlated with the extent of LC neuronal loss (Burke et al., 1988). In advanced AD, PNMT activity and protein in the adrenergic axon terminals diminish further, but PNMT in the cell bodies in the adrenergic neurons of the C1 region of the medulla are reported to be increased above control levels, suggesting failure of appropriate axonal transport (Burke et al., 1990). A relationship has been postulated between decreased brain PNMT and homocysteinemia, common in AD. Elevated levels of homocysteine have been reported in plasma, CSF, and brain of AD patients, although definitive links to AD have not been established (Boldyrev and Johnson, 2007).

Hyperhomocysteinemia can increase S-adenosylhomocysteine (SAH), a potent methyltransferase inhibitor that acts on both PNMT and catechol-*O*-methyltransferase (COMT), which is involved in the metabolism of catecholamines. SAH is found to be significantly (26%) higher in the prefrontal cortex of AD patients compared to normal controls (Kennedy et al., 2004). Furthermore, brain homogenates from AD patients significantly inhibit an exogenous methyltransferase *in vitro* that has shown to be inversely correlated with MMSE scores. PNMT and COMT activities are reportedly decreased by 30% in AD brains, and PNMT activity correlates significantly with cognitive function, age of AD onset, and choline acetyltransferase activity, but is negatively correlated with NFTs. These links, although only correlative, suggest that increased SAH and decreased PNMT activity may be related to disease progression and cognitive impairment in AD. In addition, genetic analysis has shown a significant association between two polymorphisms in the 5′ flanking (promoter) region of the human PNMT gene and early-onset, but not late-onset AD (Mann et al., 2001). In a recent study, the entire human PNMT gene was resequenced using DNA from 120 subjects, and in a luciferase reporter assay, the haplotype containing the two polymorphisms above showed a decrease of approximately 30% in reporter gene activity when compared with the wild type. This finding gives functional significance to the relationship of the PNMT polymorphisms to AD (Ji et al., 2005).

A better understanding of the several apparent links between catecholaminergic function, both in the periphery and the CNS, may provide insight into new approaches for the treatment of AD. The ability of surviving LC neurons to sprout and to reinnervate terminal regions after neuronal loss suggests that discovering a mechanism to aid the LC in compensating for neuronal loss might at least slow the progression of AD. Aging-related changes in the capacity of peripheral EPI to influence memory retention and links between PNMT expression in the CNS and AD suggest that increasing the synthesis of and/or the responsiveness of neural systems to EPI may have benefits, at least in delaying the cognitive losses inherent in AD.

25.5 Insulin

Insulin is an anabolic polypeptide hormone that is mainly secreted by β cells in the pancreas. Insulin plays a key role in regulating energy metabolism and in maintaining stable blood glucose levels (Davis and Grannar, 1996). Glucose is a hydrophyillic molecule that is able to cross the BBB (Gould and Bell, 1990).

Insulin is capable of crossing the BBB, reaching its receptors in various regions of the brain (Banks, 2004; Schulingkamp et al., 2000), namely the cortex and the hippocampus (Bondy and Cheng, 2004; Jacobson and Sapolsky, 1991). Importantly, insulin plays a critical role in the function of the hippocampus, and is involved in learning and memory (Park, 2001; Zhao and Alkon, 2001). Studies show that intravenous insulin administration is associated with an enhancement in memory performance (Craft et al., 2003). Furthermore, it has been reported that endogenous insulin, but not glucose predicts memory facilitation (Craft et al., 1999a).

25.5.1 Insulin and Cognition

In the field of memory and aging, it is well established that chronic increases in cortisol are associated with decreases in hippocampal volume, and consequent decline in hippocampal-dependent memory performance (Lupien et al., 1994, 1998). Recent evidence suggests that diabetes and insulin resistance are associated with loss of hippocampal volume (Convit et al., 2003). Furthermore, participants with type II diabetes show deficits in declarative memory performance in addition to reductions in hippocampal volume (Convit, 2005; Gold et al., 2007; Hendrickx et al., 2005).

It is suggested that cortisol levels may have an impact on the relationship between glucose, insulin, and memory performance (Convit, 2005). Indeed, elevated cortisol has been shown to be associated with insulin resistance and a decrease in insulin secretion (Rosmond, 2003). Further, type II diabetic patients tend to show increased basal cortisol levels and overactive HPA-axis activity (Bruehl et al., 2007; Lee et al., 1999).

Recently, Convit (2005) proposed a theoretical model to explain the mechanisms by which insulin resistance has a negative impact on the hippocampus when it is combined with a dysregulated HPA axis resulting in subsequent cognitive impairment. Indeed, it has been shown that glycemic control plays an important role in mediating the relationship between high cortisol levels and deficits in memory performance in type II diabetics (Bruehl et al., 2007).

25.5.2 Insulin and Diabetes: Risk for AD

Type II diabetes is the most prevalent form of hyperglycemia and the most common form of diabetes among older adults (Adult Treatment Panel III, 2001). The progress from norm to diabetic state begins with insulin resistance (noninsulin-dependent diabetes mellitus), where cells' sensitivity to insulin is decreased and consequently does not sufficiently respond to circulating insulin levels. The pancreas attempts to compensate by increasing levels of insulin production, which leads to hyperinsulinemia. However, sustained over long periods of time, the pancreas loses its capacity to produce insulin which results in the development of type II diabetes (Hendrickx et al., 2005; Kahn, 1994; Sacks and Mcdonald, 1996).

Interstingly, type II diabetes is associated with decreased verbal and visual memory performance (Messier, 2005; Schnaider Beeri et al., 2004; Strachan et al., 1997). Prospective studies have shown that participants with diabetes or prediabetes (i.e., elevated fasting glucose levels) exhibit cognitive decline and poorer performance on cognitive tasks compared to participants with normal glucose metabolism (Gregg et al., 2000; Kanaya et al., 2004; Yaffe et al., 2004). Compared to younger adults, older adults may be more vulnerable to the deleterious effects of type II diabetes on specific congnitive domains including reasoning, psychomotor speed, and attention (Awad et al., 2004; Ryan and Geckle, 2000; Yaffe et al., 2004).

A recent review of population-based studies reported that patients with diabetes mellitus are at an enhanced risk of developing dementia (Biessels et al., 2006) and those with type II diabetes have a higher risk of developing AD (Arvanitakis et al., 2004; Schnaider Beeri et al., 2004). In a population-based sample of diabetic older adults, MRI scans revealed an atrophy of the hippocampus and amygdala, when compared to a healthy control group of older adults (Den Heijer et al., 2003). Based on existing evidence from epidemiological studies, it has been suggested that type II diabetes on its own may not be sufficient to increase the risk for AD, but instead, hyperinsulinemia (occuring in the early stages of the development of diabetes) may hold a more important role in predicting the relationship (Qiu and Folstein, 2006). Indeed, longitudinal studies have shown that hyperinsulinemia increases the risk for the development of dementia and AD (Kuusisto et al., 1997; Luchsinger et al., 2004).

Insulin dysregultion has also been examined in AD populations. A recent community-based study showed that prevalence rates of type II diabetes were higher among patients with AD, when compared to healthy older adults (Janson et al., 2004). Also, AD patients are more likely to be hyperinsulinemic and hyperglycaemic compared to healthy nondemented controls (Razay and Wilcock, 1994). Overall, AD patients show a similar insulin profile as nondemented patients with insulin resistance, with elevated levels of plasma insulin and lower CSF insulin levels (Craft et al., 1999b).

In measuring glucose, it is reported that AD patients are presented with decreased levels of glucose uptake when compared to healthy controls (Herholz et al., 2002; Silverman et al., 2001). Patients with early onset AD are reported to have more pronounced hypometabolism of glucose when compared to late-onset patients (Kim et al., 2005). Furthermore, among patients with age-associated cognitive decline, those who exhibited a reduction in glucose metabolism and decreased levels of glucose uptake were at a higher risk for developing AD, evidenced by a significant decrease in performance on tests of cognitive function (Hunt et al., 2007).

Insulin-degrading enzyme (IDE) is a metalloendopeptidase that is involved in breaking down amyloid-forming proteins and in cleaving several small peptides, including insulin, insulin-like growth factors I and II, glucagon, calcitonin, and amylin, which are linked to type II diabetes (Farris et al., 2003; Qiu and Folstein, 2006). IDE is expressed in several brain regions in varying degrees of concentration; in the cerebellum, IDE levels are higher than those found in the hippocampus and cortex (Caccamo et al., 2005). Evidence for the interaction of insulin and Aβs has emerged from findings that show when insulin levels are too high, they are capable of substantially inhibiting the degradation of certain Aβ derivatives, while the reverse is similar to a lesser extent (Hoyer, 2002; Qiu and Folstein, 2006; Perez et al., 2000). Although IDE has been shown to decrease with increasing age, especially in the hippocampus, its activity has been demonstrated to be further reduced in the AD brain, which may predict an augmentation in Aβ levels due to a decreased capacity to degrade Aβ (Perez et al., 2000).

25.5.3 Insulin: Prevention and Treatment

25.5.3.1 Nonpharmacological interventions

Recent evidence demonstrates that it is never too late to make lifestyle modifications that would improve both cognitive and physical health (Kramer et al., 2005). Programs that target lifestyle changes in order to incorporate healthier diets and increase physical activity have proven to be effective in regulating blood glucose levels (Tuomilehto et al., 2001). In a randomized control trial, 522 overweight middle-aged individuals who presented with impaired glucose intolerance were randomly assigned to the intervention or control group. The intervention group involved weight reductions through reduced consumption of fat, and an augmentation of physical activity and fiber intake. The risk for diabetes among the intervention group was reduced by 58%.

Longitudinal studies (ranging 5–8 years) show that physical activity in aged individuals reduces the risk for cognitive impairment, dementia, and AD onset (Abbott et al., 2004; Barnes et al., 2003; Larson et al., 2006b; Yaffe et al., 2001). Further, brain imaging studies show that fitness training predicts higher levels of anterior white matter and gray matter in various regions of the brain, including the anterior cingulate, superior temporal lobe, and medial frontal gyrus (Colcombe et al., 2004; Kramer and Erickson, 2007). fMRI data also show that exercise activity in older adults heightens activation of the left dorsolateral cortex which is involved in working memory (Kramer, 2007). Interestingly, the mechanism by which exercise is associated with better cognitive performance is through its modulation of IGF-1, which has neuroprotective properties (Bondy and Cheng, 2004).

In terms of nutrition, consuming a diet that is high in fat may contribute toward metabolic dysregulation and has been shown to be associated with insulin resistance and to have an impact on cognitive performance (Greenwood and Winocur, 2005). Preliminary evidence suggests that diets with a lower glycaemic index may be associated with enhanced insulin sensitivity and lower insulin concentration levels among other positive outcomes (Rizkalla et al., 2004).

25.5.3.2 Pharmacological interventions

Many individuals with type II diabetes require medications that ameliorate dysregulated glycemic control associated with insulin resistance. While some pharmacologic interventions increase insulin levels by supplementing with exogenous insulin, others reduce insulin resistance as they improve insulin sensitivity (Malinowski and Bolesta, 2000). For patients with diabetes who also have insulin resistance, it is suggested that reducing insulin resistance as opposed to increasing insulin may reduce the risk for AD (Qiu and Folstein, 2006). Indeed, medications that stimulate the pancreas to produce more insulin in diabetic patients have been shown to increase the risk for AD (Qiu and Folstein, 2006).

Treatments that target glycemic regulation have proven to be successful in reducing cognitive decline, especially when utilized during the earlier stages of AD and MCI. In a sample of participants with mild AD or MCI, Watson et al. (2005) assessed the effectiveness of a 6-month treatment regime with rosiglitazone (PPAR-γ agonist) which increases insulin sensitivity. Results showed that participants who were treated with rosiglitazone performed better on memory and attention tasks when compared to participants who were assigned to the placebo condition (Watson et al., 2005).

25.6 Melatonin

Melatonin (5-methoxy-*N*-acetyltryptamine) is an endocrine hormone that is produced in the pineal gland and is regulated by the suprachiasmatic nucleus (SCN) of the hypothalamus. The biosynthetic pathway of melatonin secretion begins with L-tryptophan, which is converted into serotonin (5-HT), which is then metabolized to *N*-acetyl-5-hydroxytryptamine, and ends in the synthesis of melatonin.

The SCN is synchronized by the light–dark cycle and follows a circadian rhythm, thus acting as the body's natural internal clock. Light impulses are sent from the external environment through the retinohypothalamic pathway to the SCN, and the SCN then sends signals through the peripheral SNS to the pineal gland. As a result, the secretion of melatonin follows a 24-h circadian rhythm, with low levels of melatonin produced during the day and a rise in melatonin levels during the night. Although light inhibits the secretion of melatonin, it is important to note that the circadian rhythm of melatonin has an endogenous origin, evidenced by the presence of circadian rhythm of melatonin in blind individuals (Klerman et al., 2001). Also, melatonin can further manipulate the SCN through feed-forward and feed-backward mechanisms.

Centrally involved in sleep regulation, melatonin also serves as the body's chronological pacemaker by signaling the time of day as well as time of year. Consequently, as melatonin secretion is associated with length of nocturnal hours, seasonal changes are commonly observed with greater melatonin levels produced during winter months and lower levels produced during summer months.

25.6.1 Melatonin and Aging

As humans age, there is a change in the sleep–wake pattern, noted by increased nocturnal awakening, decreased sleep efficiency, and increased napping (Bliwise et al., 2005; Huang et al., 2002). Mirroring these changes are age-related changes in melatonin production and rhythm. While some studies suggest no change in melatonin secretion with age (Zeitzer et al., 1999), many studies have shown that increased age is associated with a flattening of the circadian profile as well as a shift in peak timing, thus producing lower levels of melatonin during nocturnal hours (Duffy et al., 2002; Graham and Mclachlan, 2004; Magri et al., 2004). In AD patients, a change in diurnal rhythm is also noted during daytime hours, with an increase in melatonin levels during the day (Ohashi et al., 1999). While the exact reason for this change is still under investigation, it has been suggested that change in circadian melatonin is a result of the age-related change in SCN regulation (Hofman, 2000; Wu et al., 2006). Also, it should be noted that interindividual differences do exist within the population and therefore, not all older adults experience a drastic change in melatonin production (Mahlberg et al., 2006). However, it is suggested that this change in timing and intensity of the melatonin peak may lead to adverse effects on physiological and cognitive function (Magri et al., 2004).

25.6.2 Melatonin Deficiency and Risk of AD

Apart from its role in biological circadian rhythm, melatonin has also been noted for its neuroprotective properties. Melatonin acts as a free radical scavenger and further acts as an indirect antioxidant by augmenting the activity of important antioxidant enzymes.

Melatonin has been found to protect the human brain against age-induced insult through various avenues. Indeed, low levels of melatonin, in both serum and CSF, have been reported in aged individuals and even more so in patients suffering from dementia (Zhou et al., 2003; Ferrari et al., 2000). Also, studies have reported a decline in melatonin receptors in the pineal gland, SCN, and cortex of AD patients (Wu et al., 2007; Brunner et al., 2006). It has been suggested that deficiency in melatonin production serves as a risk for the development of pathological cognitive decline by compromising the hormones' neuroprotective properties (Zhu et al., 2004, 2005).

As mentioned previously, two characteristic hallmarks of the AD brain include NFTs and SPs. It has been shown that melatonin regulates protein kinase and protein phosphatase activity which, in turn, regulate the production of tau, the protein responsible for producing NFTs in the human brain. Drug studies have shown that agents that induce tau phosphorylation, such as wortmannin and isoproterenol, are inhibited by melatonin (Liu and Wang, 2002; Wang et al., 2004, 2005b). Furthermore, it has been reported that melatonin not only inhibits these agents, but further suppresses the oxidative stress produced by these agents (Deng et al., 2005).

In addition to preventing the formation of NFT by attenuating tau phosphorylation, melatonin is reported to attenuate the formation of SPs in the brain. Indeed, melatonin is found to regulate the metabolism of amyloid precursor protein (APP), which is a precursor to Aβ peptide. Aggregation of Aβ in the brain is responsible for neuronal dysfunction and degeneration commonly associated with SP (Selkoe, 2002, 2004; Walsh et al., 2002). Thus, by regulating APP metabolism, melatonin diminishes the formation of Aβ peptide in the brain, and thus decreases the formation of SPs.

In addition to its direct effect on Aβ formation, melatonin has been reported to protect cells from Aβ-induced oxidative stress, inflammation, and degradation of the cholinergic system (Rosales-Corral et al., 2003; Feng et al., 2004; Shen et al., 2007, 2002a,b). Thus, with a decrease in melatonin in the aging brain comes a decrease in the hormone's ability to protect the human brain from the various aforementioned insults that lead to deterioration of cognitive function and pathological cognitive decline. For this reason, studies have assessed the efficacy of melatonin as a treatment option to slow the progression of cognitive deterioration.

25.6.3 Melatonin: Prevention and Treatment

Clinical trial studies suggest that administration of melatonin may serve as a good treatment for dementia symptomatology (Frank and Gupta, 2005). Indeed, animal studies show that melatonin is necessary for hippocampal synaptic plasticity and memory processes (Larson et al., 2006a; El-Sherif et al., 2003). Also, experimental models of AD pathology show that melatonin may decrease oxidative injury and apoptosis, inhibit Aβ deposition and fiber formation, prevent loss of cholinergic signaling, and improve cognitive function (Lahiri et al., 2005; Cheng et al., 2006).

In humans, it has been shown that melatonin administration improves sleep disturbance in old age (Pandi-Perumal et al., 2005; Haimov et al., 1995; Brzezinski, 1997). Also, a study by Peck et al. (2004) showed that exogenous melatonin administration not only improved sleep, but also improved cognitive function in nondemented older adults. The authors of this study further reported that improvement in cognitive function was independent from improvement in sleep quality. Thus, melatonin treatment may be considered a useful tool in treating age-related cognitive decline.

In patient populations, melatonin administration has been found to improve scores on tests of cognitive function as well as depressive symptomatology in older adults with MCI (Furio et al., 2007). Hormone treatment has also been found to diminish afternoon agitation, or sundowning, in AD patients (Volicer et al., 2001; Cardinali et al., 2002) and improve cognitive function (Brusco et al., 2000; Cardinali et al., 2002). In a study by Brusco et al. (2000), AD patients treated with melatonin over 22–35 months failed to show progression of cognitive and behavioral deterioration associated with the disease during the time of treatment. Overall, these findings suggest that melatonin administration may be valuable in the treatment and prevention of neurodegenerative disease. However, caution must be taken when considering melatonin treatment as additional case-control and longitudinal studies are required to evaluate safety as well as proper dosing and administration over the long term (Avery et al., 1998).

25.7 Genes, Hormones, and AD

No chapter on cognitive aging and AD is complete without a discussion on genetics. Indeed, the impact of genetics on the aging process has been widely assessed for decades. In this section, we shed light on some of the genetic variants that have been found to either directly or indirectly influence the aforementioned hormones with respect to cognitive aging.

25.7.1 Glucocorticoid Receptor Polymorphism

To date, very few studies have assessed the impact of genetic polymorphism of the glucocorticoid receptor (GR) and mineralcorticoid receptor (MR) gene on cognitive function in old age. In a longitudinal prospective study by Kuningas et al. (2007), the impact of MR and GR genetic variants on cognitive performance was assessed in a group of nondemented older adults above the age of 85. Overall, no genetic contribution was detected for either MR or GR polymorphisms. It was suggested that genetic polymorphism of MR and GR are specific to stress reactivity and do not influence receptor activity in basal nonstressful conditions (Kuningas et al., 2007).

In contrast, van Rossum et al. (2006) reported that the ER22/23EK allele was associated with a decreased risk for pathological cognitive decline, with lower presence and progression of white-matter lesions observed in ER22/23EK carriers compared to noncarriers. Although no group differences were found on tests of memory, the ER22/23EK carriers outperformed the noncarriers on a psychomotor speed test (Van Rossum and Lamberts, 2004).

25.7.2 Apolipoprotein E Gene and Hormone Modulation

The apolipoprotein E gene (APOE) is currently the most widely assessed polymorphism with regard to AD development as it is found to play a significant role in the pathophysiology of AD (Saunders, 2000; Strittmatter et al., 1993; Kamboh, 1995). In humans, APOE is coded by three different alleles: ε2, ε3, and ε4. Notably, the ε4 allele is found to have a gene-dose effect in the risk for AD (Kamboh, 1995; Poirier et al., 1993; Saunders, 2000; Strittmatter et al., 1993). As well, it has been reported that the presence of one ε4 allele significantly affects the trajectory from intact cognition, to mild cognitive impairment, to dementia status (Tyas et al., 2007; Payami et al., 1997).

Interestingly, recent studies have shown APOE to modulate the impact of certain hormones on cognitive function in old age. As mentioned above, Geerlings et al. (2001) reported a modulating effect of APOE in the relationship between lifelong estrogen exposure and risk of dementia.

APOE is also reported to moderate the relationship between insulin resistance and AD (Craft et al., 1998). Studies have shown that possession of the ε4 allele may protect individuals with hyperinsulinemia from AD risk (Kuusisto et al., 1997). Compared to AD ε4 carriers, AD non-ε4 carriers have been reported to display higher plasma insulin levels and lower ratios of CSF to plasma insulin (Craft et al., 1998). However, more recent studies report contrary findings, with hyperinsulinemic ε4-carriers displaying higher risk for AD (Luchsinger et al., 2004).

25.7.3 COMT Gene

The COMT gene has recently been studied in cognitive aging. COMT is an enzyme, located in the kidneys, liver, and CNS, that catalyzes the degradation of catecholamines and the methylation of catechol estrogens (Yavich et al., 2007).

The most widely assessed polymorphism of the COMT gene is the Val[158]Met polymorphism. Val and Met enzyme variants differ in their activity level, with the Val variant exhibiting a three- to fourfold increase in enzyme activity compared to the Met variant (Lachman et al., 1996; Chen et al., 2004). To this end, the Val/Val genotype has been associated with less efficient cognitive processing and poorer cognitive performance compared to the Met/Met genotype (Bruder et al., 2005; Caldu et al., 2007; Rosa et al., 2004; Malhotra et al., 2002). In terms of cognitive aging, a longitudinal study by de Frias et al. (2005) revealed that over a 5-year period, Val/Val carriers declined in executive functioning whereas Met/Met carriers remained stable over years (de Frias et al., 2005). However, in a more recent study, although baseline differences on executive function between Val/Val and Met/Met carriers were reported, no significant group differences on cognitive change over a 4-year period were found (Starr et al., 2007).

Interestingly, it is postulated that the COMT gene may play a role in the estrogen–AD link through its estrogen-metabolizing role. Indeed, it has been shown that the Val/Val genotype is related to lower estrogen levels than the Met/Met genotype (Worda et al., 2003). Although there are no studies reporting a direct link between AD (without psychosis) and COMT polymorphism, a study by Wang et al. (2005a) found a synergistic relationship between the COMT and APOE gene. Specifically, it was reported that carriers of the Val/Val genotype in conjunction with APOE ε4 allele exhibited increased risk in the development of AD. It was suggested that the potential protective effect of estrogen in ε4 carriers is reduced in Val/Val carriers compared to Met carriers due to increased metabolism of catechol estrogens (Wang et al., 2005a).

25.7.4 Estrogen Receptor Genes

As mentioned above, estrogen exerts its effects on the aging brain, notably through nuclear receptors including the estrogen receptor 1 alpha (ER-α) and the estrogen receptor 2 beta (ER-β). Of the two receptors, the ER-α gene (ESR1) has received the most attention regarding cognitive function in old age. Specifically, the *Pvu*II and the *Xba*I polymorphisms are reportedly implicated in cognitive impairment in older nondemented postmenopausal women and increase the risk for AD (Yaffe et al., 2002).

Unfortunately, studies reporting a role of the ESR1 gene in risk for AD development have been inconsistent. For one, studies are mixed regarding which variant of the polymorphism serves as the risk allele (Corbo et al., 2006; Olsen et al., 2006; Mattila et al., 2000; Yaffe et al., 2002; Porrello et al., 2006). The second inconsistency relates to whether the ESR1 gene poses a direct or indirect effect on AD onset. While some have suggested a direct role of ESR1 on AD risk (Brandi et al., 1999), others have reported an indirect link (Porrello et al., 2006). However, one common thread among the majority of these studies is a synergistic effect of ESR1 in conjunction with the APOE gene (Brandi et al.,1999; Corbo et al., 2006; Mattila et al., 2000; Porrello et al., 2006). Indeed, it has been shown that activation of specific ERs may mediate APOE levels in the hippocampus and thus reduce the risk of pathological cognitive decline (Csernansky et al., 2006).

Overall, there is no single gene that can explain cognitive decline and AD development in relation to hormonal dysregulations. Instead, there are many genetic variants, and interactions among genetic variants that may increase the risk for pathological cognitive decline.

25.8 Conclusion

Altogether, this chapter shows that there are various hormones involved in AD, although no hormone taken in isolation can explain the origin of the disorder, nor its course. From this report, it becomes clear that various interactions between hormones and their respective systems are at play in AD.

Another factor that is important to mention is that almost all hormones in the body respond to environmental demands (e.g., stress). Further studies trying to delineate the modulating impact of hormones as a potential bridge between the environment and the brain may thus prove important in elucidating part of the pathophysiology of AD.

Acknowledgments

Dr. Lupien is supported by a Senior Investigator Research Chair Award by the Canadian Institute of Health Research (CIHR) Institute of Gender and Health, Dr. Lord is supported by the Fonds de la recherche en santé du Québec (FRSQ) Fellowship Award, Dr. Wilkinson is supported by the US Department of Veterans Affairs, and Dr. Fiocco is supported by the Canadian Institute of Health Research Fellowship Award from the Institute of Aging.

References

Abbott RD, White LR, Ross GW, Masaki KH, Curb JD, and Petrovitch H (2004) Walking and dementia in physically capable elderly men. *Journal of the American Medical Association* 292: 1447–1453.

Aisen PS and Davis KL (1997) The search for disease-modifying treatment for Alzheimer's disease. *Neurology* 48: S35–S41.

Aisen PS, Davis KL, Berg JD, et al. (2000) A randomized controlled trial of prednisone in Alzheimer's disease. Alzheimer's disease cooperative study. *Neurology* 54: 588–593.

Aisen PS and Pasinetti GM (1998) Glucocorticoids in Alzheimer's disease. The story so far. *Drugs Aging* 12: 1–6.

Akiyama H, Barger S, Barnum S, et al. (2000) Inflammation and Alzheimer's disease. *Neurobiology of Aging* 21: 383–421.

Ames D (2006) For debate: Is mild cognitive impairment a clinically useful concept? *International Psychogeriatrics* 18: 393–414.

Andersen K, Launer LJ, Dewey ME, et al. (1999) Gender differences in the incidence of AD and vascular dementia: The EURODEM Studies. EURODEM Incidence Research Group. *Neurology* 53: 1992–1997.

Armanini D, Vecchio F, Basso A, et al. (2003) Alzheimer's disease: Pathophysiological implications of measurement of plasma cortisol, plasma dehydroepiandrosterone sulfate, and lymphocytic corticosteroid receptors. *Endocrine* 22: 113–118.

Arsenault-Lapierre G, Chertkow H, and Lupien S (2008) Seasonal effects on cortisol secretion in normal aging, mild cognitive impairment and Alzheimer's disease. *Neurobiology of Aging* [Epub ahead of print].

Arvanitakis Z, Wilson RS, Bienias JL, Evans DA, and Bennett DA (2004) Diabetes mellitus and risk of Alzheimer disease and decline in cognitive function. *Archives in Neurology* 61: 661–666.

Avery D, Lenz M, and Landis C (1998) Guidelines for prescribing melatonin. *Annals of Medicine* 30: 122–130.

Awad N, Gagnon M, and Messier C (2004) The relationship between impaired glucose tolerance, type 2 diabetes, and cognitive function. *Journal of Clinical and Experimental Neuropsychology* 26: 1044–1080.

Baddeley AD, Baddeley HA, Bucks RS, and Wilcock GK (2001) Attentional control in Alzheimer's disease. *Brain* 124: 1492–508.

Banks WA (2004) The source of cerebral insulin. *European Journal of Pharmacology* 490: 5–12.

Baranowska B, Wolinska-Witot E, Bik W, Baranowska-Bik A, Martynska L, and Chmielowska M (2007) Evaluation of neuroendocrine status in longevity. *Neurobiology of Aging* 28: 774–783.

Barnes DE, Yaffe K, Satariano WA, and Tager IB (2003) A longitudinal study of cardiorespiratory fitness and cognitive function in healthy older adults. *Journal of the American Geriatrics Society* 51: 459–465.

Barnes RF, Raskind M, Gumbrecht G, and Halter JB (1982) The effects of age on the plasma catecholamine response to mental stress in man. *Journal of Clinical Endocrinology and Metabolism* 54: 64–69.

Barrett-Connor E and Goodman-Gruen D (1999) Cognitive function and endogenous sex hormones in older women. *Journal of the American Geriatrics Society* 47: 1289–1293.

Bastianetto S, Ramassamy C, Poirier J, and Quirion R (1999) Dehydroepiandrosterone (DHEA) protects hippocampal cells from oxidative stress-induced damage. *Brain Research – Molecular Brain Research* 66: 35–41.

Bates KA, Harvey AR, Carruthers M, and Martins RN (2005) Androgens, andropause and neurodegeneration: Exploring the link between steroidogenesis, androgens and Alzheimer's disease. *Cellular and Molecular Life Sciences* 62: 281–292.

Baum LW (2005) Sex, hormones, and Alzheimer's disease. *Journals of Gerontology Series A: Biological Sciences and Medical Sciences* 60: 736–743.

Behl C (2002) Oestrogen as a neuroprotective hormone. *Nature Reviews Neuroscience* 3: 433–442.

Bélanger A, Candas B, Dupont A, Cusan L, Diamond P, Gomez JL, and Labrie F (1994) Changes in serum concentrations of conjugated and unconjugated steroids in 40- to 80-year-old men. *Journal of Clinical Endocrinology and Metabolism* 79: 1086–1090.

Belanoff JK, Rothschild AJ, Cassidy F, DeBattista C, Baulieu EE, Schold C, and Schatzberg AF (2002) An open label trial of C-1073 (mifepristone) for psychotic major depression. *Biological Psychiatry* 52(5): 386–392.

Bernardi F, Lanzone A, Cento RM, et al. (2000) Allopregnanolone and dehydroepiandrosterone response to corticotropin-releasing factor in patients suffering from Alzheimer's disease and vascular dementia. *European Journal of Endocrinology* 142: 466–471.

Bhavnani BR (2003) Estrogens and menopause: Pharmacology of conjugated equine estrogens and their potential role in the prevention of neurodegenerative diseases such as Alzheimer's. *Journal of Steroid Biochemistry and Molecular Biology* 85: 473–482.

Biessels GJ, Staekenborg S, Brunner E, Brayne C, and Scheltens P (2006) Risk of dementia in diabetes mellitus: A systematic review. *Lancet Neurology* 5: 64–74.

Blauer KL, Poth M, Rogers WM, and Bernton EW (1991) Dehydroepiandrosterone antagonizes the suppressive effects of dexamethasone on lymphocyte proliferation. *Endocrinology* 129: 3174–3179.

Bliwise DL, Ansari FP, Straight LB, and Parker KP (2005) Age changes in timing and 24-hour distribution of self-reported sleep. *American Journal of Geriatric Psychiatry* 13: 1077–1082.

Bo M, Massaia M, Zannella P, et al. (2006) Dehydroepiandrosterone sulfate (DHEA-S) and Alzheimer's dementia in older subjects. *International Journal of Geriatric Psychiatry* 21: 1065–1070.

Boldyrev AA and Johnson P (2007) Homocysteine and its derivatives as possible modulators of neuronal and non-neuronal cell glutamate receptors in Alzheimer's disease. *Journal of Alzheimer's Disease* 11: 219–228.

Bondy CA and Cheng CM (2004) Signaling by insulin-like growth factor 1 in brain. *European Journal of Pharmacology* 490: 25–31.

Bonsignore M and Heun R (2003) Mortality in Alzheimer's disease. *Dementia and Geriatric Cognitive Disorders* 15: 231–236.

Born J and Fehm HL (1998) Hypothalamus–pituitary–adrenal activity during human sleep: A coordinating role for the limbic hippocampal system. *Experimental and Clinical Endocrinology and Diabetes* 106: 153–163.

Boscaro M, Poaletta A, Scarpa E, Barzon L, Fusaro P, Fallo F, and Sonino N (1998) Age-related changes in flucocorticoid fast feedback inhibition of adrenocorticotropin in man. *Journal of Clinical Endocrinology and Metabolism* 83: 1380–1383.

Bowen RL, Isley JP, and Atkinson RL (2000) An association of elevated serum gonadotropin concentrations and Alzheimer disease? *Journal of Neuroendocrinology* 12: 351–354.

Bowen RL, Smith MA, Harris PL, et al. (2002) Elevated luteinizing hormone expression colocalizes with neurons

vulnerable to Alzheimer's disease pathology. *Journal of Neuroscience Research* 70(3): 514–518.
Braak H and Braak E (1991) Neuropathological stageing of Alzheimer-related changes. *Acta Neuropathology* 82: 239–259.
Brandi ML, Becherini L, Gennari L, et al. (1999) Association of the estrogen receptor alpha gene polymorphisms with sporadic Alzheimer's disease. *Biochemical and Biophysical Research Communications* 265: 335–358.
Breitner JC, Gau BA, Welsh KA, Plassman BL, Mcdonald WM, Helms MJ, and Anthony JC (1994) Inverse association of anti-inflammatory treatments and Alzheimer's disease: Initial results of a co-twin control study. *Neurology* 44: 227–232.
Breitner JC, Welsh KA, Helms MJ, et al. (1995) Delayed onset of Alzheimer's disease with nonsteroidal anti-inflammatory and histamine H2 blocking drugs. *Neurobiology of Aging* 16: 523–530.
Bruder GE, Keilp JG, Xu H, Shikhman M, Schori E, Gorman JM, and Gilliam TC (2005) Catechol-*O*-methyltransferase (COMT) genotypes and working memory: Associations with differing cognitive operations. *Biological Psychiatry* 58: 901–907.
Bruehl H, Rueger M, Dziobek I, et al. (2007) Hypothalamic–pituitary–adrenal axis dysregulation and memory impairments in type 2 diabetes. *Journal of Clinical Endocrinology and Metabolism* 92: 2439–2445.
Brunner P, Sozer-Topcular N, Jockers R, et al. (2006) Pineal and cortical melatonin receptors MT1 and MT2 are decreased in Alzheimer's disease. *European Journal of Histochemistry* 50: 311–316.
Brusco LI, Marquez M, and Cardinali DP (2000) Melatonin treatment stabilizes chronobiologic and cognitive symptoms in Alzheimer's disease. *Neuroendocrinology Letters* 21: 39–42.
Brzezinski A (1997) Melatonin in humans. *New England Journal of Medicine* 336: 186–195.
Burke WJ, Chung HD, Huang JS, et al. (1988) Evidence for retrograde degeneration of epinephrine neurons in Alzheimer's disease. *Annals of Neurology* 24: 532–536.
Burke WJ, Chung HD, Marshall GL, Gillespie KN, and Joh TH (1990) Evidence for decreased transport of PNMT protein in advanced Alzheimer's disease. *Journal of the American Geriatrics Society* 38: 1275–1282.
Burke WJ, Chung HD, Nakra BR, Grossberg GT, and Joh TH (1987) Phenylethanolamine *N*-methyltransferase activity is decreased in Alzheimer's disease brains. *Annals of Neurology* 22: 278–280.
Caccamo A, Oddo S, Sugarman MC, Akbari Y, and Laferla FM (2005) Age- and region-dependent alterations in Abeta-degrading enzymes: Implications for Abeta-induced disorders. *Neurobiology of Aging* 26: 645–654.
Cahill L and Alkire MT (2003) Epinephrine enhancement of human memory consolidation: Interaction with arousal at encoding. *Neurobiology of Learning and Memory* 79: 194–198.
Cahill L, Prins B, Weber M, and Mcgaugh JL (1994) Beta-adrenergic activation and memory for emotional events. *Nature* 371: 702–704.
Caldu X, Vendrell P, Bartres-Faz D, et al. (2007) Impact of the COMT Val(108/158) Met and DAT genotypes on prefrontal function in healthy subjects. *NeuroImage* 37: 1437–1444.
Canadian Study of Health Aging (1994) Risk factors for Alzheimer's disease in Canada. *Neurology* 44: 2073–2080.
Cardinali DP, Brusco LI, Liberczuk C, and Furio AM (2002) The use of melatonin in Alzheimer's disease. *Neuroendocrinology Letters* 23(supplement 1): 20–23.
Carlson LE, Sherwin BB, and Chertkow HM (1999) Relationships between dehydroepiandrosterone sulfate (DHEAS) and cortisol (CRT) plasma levels and everyday memory in Alzheimer's disease patients compared to healthy controls. *Hormones and Behavior* 35: 254–263.
Chakravarti S, Collins WP, Forecast JD, Newton JR, Oram DH, and Studd JW (1976) Hormonal profiles after the menopause. *British Medical Journal* 2: 784–787.
Chan-Palay V and Asan E (1989) Quantitation of catecholamine neurons in the locus coeruleus in human brains of normal young and older adults and in depression. *Journal of Comparative Neurology* 287: 357–372.
Chen J, Lipska BK, Halim N, et al. (2004) Functional analysis of genetic variation in catechol-O-methyltransferase (COMT): Effects on mRNA, protein, and enzyme activity in postmortem human brain. *American Journal of Human Genetics* 75: 807–821.
Cheng Y, Feng Z, Zhang QZ, and Zhang JT (2006) Beneficial effects of melatonin in experimental models of Alzheimer disease. *Acta Pharmacologica Sinica* 27: 129–139.
Cherrier MM, Matsumoto AM, Amory JK, et al. (2005) Testosterone improves spatial memory in men with Alzheimer disease and mild cognitive impairment. *Neurology* 64: 2063–2068.
Clark KB, Naritoku DK, Smith DC, Browning RA, and Jensen RA (1999) Enhanced recognition memory following vagus nerve stimulation in human subjects. *Nature Neuroscience* 2: 94–98.
Colcombe SJ, Kramer AF, Erickson KI, et al. (2004) Cardiovascular fitness, cortical plasticity, and aging. *Proceedings of the National Academy of Sciences of the United States of America* 101: 3316–3321.
Convit A (2005) Links between cognitive impairment in insulin resistance: An explanatory model. *Neurobiology of Aging* 26 (supplement 1): 31–35.
Convit A, Wolf OT, Tarshish C, and de Leon MJ (2003) Reduced glucose tolerance is associated with poor memory performance and hippocampal atrophy among normal elderly. *Proceedings of the National Academy of Sciences of the United States of America* 100(4): 2019–2022.
Corbo RM, Gambina G, Ruggeri M, and Scacchi R (2006) Association of estrogen receptor alpha (ESR1) PvuII and XbaI polymorphisms with sporadic Alzheimer's disease and their effect on apolipoprotein E concentrations. *Dementia and Geriatric Cognitive Disorders* 22: 67–72.
Craft S, Asthana S, Cook DG, et al. (2003) Insulin dose-response effects on memory and plasma amyloid precursor protein in Alzheimer's disease: Interactions with apolipoprotein E genotype. *Psychoneuroendocrinology* 28: 809–822.
Craft S, Asthana S, Newcomer JW, et al. (1999a) Enhancement of memory in Alzheimer disease with insulin and somatostatin, but not glucose. *Archives of General Psychiatry* 56: 1135–1140.
Craft S, Asthana S, Schellenberg G, et al. (1999b) Insulin metabolism in Alzheimer's disease differs according to apolipoprotein E genotype and gender. *Neuroendocrinology* 70: 146–152.
Craft S, Peskind E, Schwartz MW, Schellenberg GD, Raskind M, and Porte D, Jr. (1998) Cerebrospinal fluid and plasma insulin levels in Alzheimer's disease: Relationship to severity of dementia and apolipoprotein E genotype. *Neurology* 50: 164–168.
Csernansky JG, Dong H, Fagan AM, Wang L, Xiong C, Holtzman DM, and Morris JC (2006) Plasma cortisol and progression of dementia in subjects with Alzheimer-type dementia. *American Journal of Psychiatry* 163: 2164–2169.
Cummings JL (2004) Alzheimer's disease. *New England Journal of Medicine* 351: 56–67.

Davis SN and Grannar DK (1996) Insulin, oral hypoglycemic agents, and the pharmacology of the endocrine pancreas. In: Hardman JG, Gilman AG, and Limbird LE (eds.) *Goodman and Gilman's the Pharmacological Basis of Therapeutics*, 9th edn., pp. 1487–1518. New York: McGraw-Hill.

Debonnel G, Bergeron R, and de Montigny C (1996) Potentiation by dehydroepiandrosterone of the neuronal response to *N*-methyl-D-aspartate in the CA3 region of the rat dorsal hippocampus: An effect mediated via sigma receptors. *Journal of Endocrinology* 150: S33–S42.

De Bruin VM, Vieira MC, Rocha MN, and Viana GS (2002) Cortisol and dehydroepiandosterone sulfate plasma levels and their relationship to aging, cognitive function, and dementia. *Brain and Cognition* 50: 316–323.

de Frias CM, Annerbrink K, Westberg L, Eriksson E, Adolfsson R, and Nilsson LG (2005) Catechol O-methyltransferase Val158Met polymorphism is associated with cognitive performance in nondemented adults. *Journal of Cognitive Neuroscience* 17: 1018–1025.

Den Heijer T, Vermeer SE, Van Dijk EJ, Prins ND, Koudstaal PJ, Hofman A, and Breteler MM (2003) Type 2 diabetes and atrophy of medial temporal lobe structures on brain MRI. *Diabetologia* 46: 1604–1610.

Deng YQ, Xu GG, Duan P, Zhang Q, and Wang JZ (2005) Effects of melatonin on wortmannin-induced tau hyperphosphorylation. *Acta Pharmacologica Sinica* 26: 519–526.

Dickerson BC, Goncharova I, Sullivan MP, et al. (2001) MRI-derived entorhinal and hippocampal atrophy in incipient and very mild Alzheimer's disease. *Neurobiology of Aging* 22: 747–754.

Diorio D, Viau V, and Meaney MJ (1993) The role of the medial prefrontal cortex (cingulate gyrus) in the regulation of hypothalamic–pituitary–adrenal responses to stress. *Journal of Neuroscience* 13: 3839–3847.

Dori D, Casale G, Solerte SB, Fioravanti M, Migliorati G, Cuzzoni G, and Ferrari E (1994) Chrono-neuroendocrinological aspects of physiological aging and senile dementia. *Chronobiologia* 21: 121–126.

Drafta D, Schindler AE, Stroe E, and Neacsu E (1982) Age-related changes of plasma steroids in normal adult males. *Journal of Steroid Biochemistry* 17: 683–687.

Driscoll I and Resnick SM (2007) Testosterone and cognition in normal aging and Alzheimer's disease: An update. *Current Alzheimer Research* 4: 33–45.

Duffy JF, Zeitzer JM, Rimmer DW, Klerman EB, Dijk DJ, and Czeisler CA (2002) Peak of circadian melatonin rhythm occurs later within the sleep of older subjects. *American Journal of Physiology–Endocrinology and Metabolism* 282: E297–E303.

Eberling JL, Reed BR, Coleman JE, and Jagust WJ (2000) Effect of estrogen on cerebral glucose metabolism in postmenopausal women. *Neurology* 55: 875–877.

El-Sherif Y, Tesoriero J, Hogan MV, and Wieraszko A (2003) Melatonin regulates neuronal plasticity in the hippocampus. *Journal of Neuroscience Research* 72: 454–460.

Elhamdani A, Palfrey CH, and Artalejo CR (2002) Ageing changes the cellular basis of the fight-or-flight response in human adrenal chromaffin cells. *Neurobiology of Aging* 23: 287–293.

Esler M, Kaye D, Thompson J, et al. (1995) Effects of aging on epinephrine secretion and regional release of epinephrine from the human heart. *Journal of Clinical Endocrinology and Metabolism* 80: 435–442.

Farris W, Mansourian S, Chang Y, et al. (2003) Insulin-degrading enzyme regulates the levels of insulin, amyloid beta-protein, and the beta-amyloid precursor protein intracellular domain *in vivo*. *Proceedings of the National Academy of Sciences of the United States of America* 100: 4162–4167.

Feldman S and Weidenfeld J (1999) Glucocorticoid receptor antagonists in the hippocampus modify the negative feedback following neural stimuli. *Brain Research* 821: 33–37.

Feng Z, Chang Y, Cheng Y, Zhang BL, Qu ZW, Qin C, and Zhang JT (2004) Melatonin alleviates behavioral deficits associated with apoptosis and cholinergic system dysfunction in the APP 695 transgenic mouse model of Alzheimer's disease. *Journal of Pineal Research* 37: 129–136.

Ferrari E, Arcaini A, Gornati R, et al. (2000) Pineal and pituitary–adrenocortical function in physiological aging and in senile dementia. *Experimental Gerontology* 35: 1239–1250.

Ferri CP, Prince M, Brayne C, et al. (2005) Global prevalence of dementia: A Delphi consensus study. *Lancet* 366: 2112–2117.

Fillit HM (2002) The role of hormone replacement therapy in the prevention of Alzheimer disease. *Archives of Internal Medicine* 162: 1934–1942.

Franco-Morselli R, Elghozi JL, Joly E, Di Giuilio S, and Meyer P (1977) Increased plasma adrenaline concentrations in benign essential hypertension. *British Medical Journal* 2: 1251–1254.

Frank B and Gupta S (2005) A review of antioxidants and Alzheimer's disease. *Annals of Clinical Psychiatry* 17: 269–286.

Furio AM, Brusco LI, and Cardinali DP (2007) Possible therapeutic value of melatonin in mild cognitive impairment: A retrospective study. *Journal of Pineal Research* 43: 404–409.

Gandy S, Almeida OP, Fonte J, et al. (2001) Chemical andropause and amyloid-beta peptide. *Journal of the American Medical Association* 285: 2195–2196.

Gauthier S, Reisberg B, Zaudig M, et al. (2006) Mild cognitive impairment. *Lancet* 367: 1262–1270.

Geerlings MI, Ruitenberg A, Witteman JC, et al. (2001) Reproductive period and risk of dementia in postmenopausal women. *Journal of the American Medical Association* 285: 1475–1481.

Genedani S, Rasio G, Cortelli P, et al. (2004) Studies on homocysteine and dehydroepiandrosterone sulphate plasma levels in Alzheimer's disease patients and in Parkinson's disease patients. *Neurotoxicity Research* 6: 327–332.

German DC, Manaye KF, White CL, III, et al. (1992) Disease-specific patterns of locus coeruleus cell loss. *Annals of Neurology* 32: 667–676.

Gillett MJ, Martins RN, Clarnette RM, Chubb SA, Bruce DG, and Yeap BB (2003) Relationship between testosterone, sex hormone binding globulin and plasma amyloid beta peptide 40 in older men with subjective memory loss or dementia. *Journal of Alzheimer's Disease* 5: 267–269.

Giubilei F, Patacchioli FR, Antonini G, et al. (2001) Altered circadian cortisol secretion in Alzheimer's disease: Clinical and neuroradiological aspects. *Journal of Neuroscience Research* 66: 262–265.

Gold SM, Dziobek I, Sweat V, et al. (2007) Hippocampal damage and memory impairments as possible early brain complications of type 2 diabetes. *Diabetologia* 50: 711–719.

Goldstein DS, Lake CR, Chernow B, et al. (1983) Age-dependence of hypertensive–normotensive differences in plasma norepinephrine. *Hypertension* 5: 100–104.

Gould GW and Bell GI (1990) Facilitative glucose transporters: An expanding family. *Trends in Biochemical Sciences* 15: 18–23.

Graham D and Mclachlan A (2004) Declining melatonin levels and older people. How old is old? *Neuroendocrinology Letters* 25: 415–418.

Green KN, Billings LM, Roozendaal B, Mcgaugh JL, and Laferla FM (2006) Glucocorticoids increase amyloid-beta and tau pathology in a mouse model of Alzheimer's disease. *Journal of Neuroscience* 26: 9047–9056.
Greene JD, Baddeley AD, and Hodges JR (1996) Analysis of the episodic memory deficit in early Alzheimer's disease: Evidence from the doors and people test. *Neuropsychologia* 34: 537–551.
Greene RA (2000) Estrogen and cerebral blood flow: A mechanism to explain the impact of estrogen on the incidence and treatment of Alzheimer's disease. *International Journal of Fertility and Women's Medicine* 45: 253–257.
Greenwood CE and Winocur G (2005) High-fat diets, insulin resistance and declining cognitive function. *Neurobiology of Aging* 26(supplement 1): 42–45.
Gregg EW, Yaffe K, Cauley JA, Rolka DB, Blackwell TL, Narayan KM, and Cummings SR (2000) Is diabetes associated with cognitive impairment and cognitive decline among older women? Study of Osteoporotic Fractures Research Group. *Archives of Internal Medicine* 160: 174–180.
Grimley Evans J, Malouf R, Huppert F, and VanNiekerk JK (2006) Dehydroepiandrosterone (DHEA) supplementation for cognitive function in healthy elderly people. *Cochrane Database of Systematic Reviews,* doi: 10.1002/14651858. CD006221.
Grossman H, Bergmann C, and Parker S (2006) Dementia: A brief review. *Mount Sinai Journal of Medicine* 73: 985–992.
Group CSOHAAW (1994) Canadian study of health and aging: Study methods and prevalence of dementia. *Canadian Medical Association Journal* 150: 899–913.
Grundman M, Jack CR, Jr., Petersen RC, et al. (2003) Hippocampal volume is associated with memory but not monmemory cognitive performance in patients with mild cognitive impairment. *Journal of Molecular Neuroscience* 20: 241–248.
Gurevich D, Siegel B, Dumlao M, Perl E, Chaitin P, Bagne C, and Oxenkrug G (1990) HPA axis responsivity to dexamethasone and cognitive impairment in dementia. *Progress in Neuropsychopharmacology and Biological Psychiatry* 14: 297–308.
Haimov I, Lavie P, Laudon M, Herer P, Vigder C, and Zisapel N (1995) Melatonin replacement therapy of elderly insomniacs. *Sleep* 18: 598–603.
Hall JL and Gold PE (1986) The effects of training, epinephrine, and glucose injections on plasma glucose levels in rats. *Behavioral and Neural Biology* 46: 156–167.
Henderson VW, Benke KS, Green RC, Cupples LA, and Farrer LA (2005) Postmenopausal hormone therapy and Alzheimer's disease risk: Interaction with age. *Journal of Neurology, Neurosurgery and Psychiatry* 76: 103–105.
Henderson VW, Espeland MA, Hogan PE, et al. (2007) Prior use of hormone therapy and incident Alzheimers disease in the womens health initiative memory study. *59th American Academy of Neurology Annual Meeting.* Boston, MA.
Hendrickx H, McEwen BS, and Ouderaa F (2005) Metabolism, mood and cognition in aging: The importance of lifestyle and dietary intervention. *Neurobiology of Aging* 26(supplement 1): 1–5.
Heneka MT, Ramanathan M, Jacobs AH, et al. (2006) Locus ceruleus degeneration promotes Alzheimer pathogenesis in amyloid precursor protein 23 transgenic mice. *Journal of Neuroscience* 26: 1343–1354.
Herholz K, Salmon E, Perani D, et al. (2002) Discrimination between Alzheimer dementia and controls by automated analysis of multicenter FDG PET. *Neuro Image* 17: 302–316.
Herman JP and Cullinan WE (1997) Neurocircuitry of stress: Central control of the hypothalamo-pituitary–adrenocortical axis. *Trends in Neuroscience* 20: 78–84.
Herman JP, Ostrander MM, Mueller NK, and Figueiredo H (2005) Limbic system mechanisms of stress regulation: Hypothalamo-pituitary–adrenocortical axis. *Progress in Neuropsychopharmacology and Biological Psychiatry* 29: 1201–1213.
Heyman A, Peterson B, Fillenbaum G, and Pieper C (1996) The consortium to establish a registry for Alzheimer's disease (CERAD). Part XIV: Demographic and clinical predictors of survival in patients with Alzheimer's disease. *Neurology* 46: 656–660.
Hillen T, Lun A, Reischies FM, Borchelt M, Steinhagen-Thiessen E, and Schaub RT (2000) DHEA-S plasma levels and incidence of Alzheimer's disease. *Biology of Psychiatry* 47: 161–163.
Hodges JR (2006) Alzheimer's centennial legacy: Origins, landmarks and the current status of knowledge concerning cognitive aspects. *Brain* 129: 2811–2822.
Hofman MA (2000) The human circadian clock and aging. *Chronobiology International* 17: 245–259.
Hogervorst E, Bandelow S, Combrinck M, and Smith AD (2004) Low free testosterone is an independent risk factor for Alzheimer's disease. *Experimental Gerontology* 39: 1633–1639.
Hogervorst E, Barnetson L, Jobst KA, Nagy Z, Combrinck M, and Smith AD (2000) Diagnosing dementia: Interrater reliability assessment and accuracy of the NINCDS/ADRDA criteria versus CERAD histopathological criteria for Alzheimer's disease. *Dementia and Geriatric Cognitive Disorders* 11: 107–113.
Hogervorst E, Combrinck M, and Smith AD (2003) Testosterone and gonadotropin levels in men with dementia. *Neuroendocrinology Letters* 24: 203–208.
Hogervorst E, Williams J, Budge M, Barnetson L, Combrinck M, and Smith AD (2001) Serum total testosterone is lower in men with Alzheimer's disease. *Neuroendocrinology Letters* 22: 163–168.
Hoyer S (2002) The brain insulin signal transduction system and sporadic (type II) Alzheimer disease: An update. *Journal of Neural Transmission* 109: 341–360.
Huang YL, Liu RY, Wang QS, Van Someren EJ, Xu H, and Zhou JN (2002) Age-associated difference in circadian sleep–wake and rest-activity rhythms. *Physiology and Behavior* 76: 597–603.
Hunt A, Schonknecht P, Henze M, Seidl U, Haberkorn U, and Schroder J (2007) Reduced cerebral glucose metabolism in patients at risk for Alzheimer's disease. *Psychiatry Research* 155: 147–154.
Ishida Y, Shirokawa T, Miyaishi O, Komatsu Y, and Isobe K (2000) Age-dependent changes in projections from locus coeruleus to hippocampus dentate gyrus and frontal cortex. *European Journal of Neuroscience* 12: 1263–1270.
Ishida Y, Shirokawa T, Miyaishi O, Komatsu Y, and Isobe K (2001) Age-dependent changes in noradrenergic innervations of the frontal cortex in F344 rats. *Neurobiology of Aging* 22: 283–286.
Issa AM, Rowe W, Gauthier S, and Meaney MJ (1990) Hypothalamic–pituitary–adrenal activity in aged, cognitively impaired and cognitively unimpaired rats. *Journal of Neuroscience* 10: 3247–3254.
Jack CR, Jr., Petersen RC, Xu YC, et al. (1997) Medial temporal atrophy on MRI in normal aging and very mild Alzheimer's disease. *Neurology* 49: 786–794.
Jack CR, Jr., Petersen RC, Xu YC, et al. (1999) Prediction of AD with MRI-based hippocampal volume in mild cognitive impairment. *Neurology* 52: 1397–1403.

Jacobson L and Sapolsky R (1991) The role of the hippocampus in feedback regulation of the hypothalamic–pituitary–adrenocortical axis. *Endocrine Reviews* 12: 118–134.

Jagust W, Gitcho A, Sun F, Kuczynski B, Mungas D, and Haan M (2006) Brain imaging evidence of preclinical Alzheimer's disease in normal aging. *Annals of Neurology* 59: 673–681.

Janson J, Laedtke T, Parisi JE, O'brien P, Petersen RC, and Butler PC (2004) Increased risk of type 2 diabetes in Alzheimer disease. *Diabetes* 53: 474–481.

Jensen HK and Blichert-Toft M (1971) Investigation of pituitary–adrenocortical function in the elderly during standardized operations and postoperative intravenous metyrapone test assessed by plasma cortisol, plasma compound S and eosinophil cell determinations. *Acta Endocrinologica (Copenh)* 67: 495–507.

Ji Y, Salavaggione OE, Wang L, Adjei AA, Eckloff B, Wieben ED, and Weinshilboum RM (2005) Human phenylethanolamine *N*-methyltransferase pharmacogenomics: Gene re-sequencing and functional genomics. *Journal of Neurochemistry* 95: 1766–1776.

Jost BC and Grossberg GT (1996) The evolution of psychiatric symptoms in Alzheimer's disease: A natural history study. *Journal of the American Geriatrics Society* 44: 1078–1081.

Kahn CR (1994) Banting lecture. Insulin action, diabetogenes, and the cause of type II diabetes. *Diabetes* 43: 1066–1084.

Kalaria RN and Andorn AC (1991) Adrenergic receptors in aging and Alzheimer's disease: Decreased alpha 2-receptors demonstrated by [^{3}H]p-aminoclonidine binding in prefrontal cortex. *Neurobiology of Aging* 12: 131–136.

Kamboh MI (1995) Apolipoprotein E polymorphism and susceptibility to Alzheimer's disease. *Human Biology* 67: 195–215.

Kanaya AM, Barrett-Connor E, Gildengorin G, and Yaffe K (2004) Change in cognitive function by glucose tolerance status in older adults: A 4-year prospective study of the Rancho Bernardo study cohort. *Archives of Internal Medicine* 164: 1327–1333.

Kaufman JM and Vermeulen A (2005) The decline of androgen levels in elderly men and its clinical and therapeutic implications. *Endocrine Reviews* 26: 833–876.

Kawas CH (2003) Clinical practice. Early Alzheimer's disease. *New England Journal of Medicine* 349: 1056–1063.

Kawas C, Resnick S, Morrison A, et al. (1997) A prospective study of estrogen replacement therapy and the risk of developing Alzheimer's disease: The Baltimore Longitudinal Study of Aging. *Neurology* 48: 1517–1521.

Kennedy BP, Bottiglieri T, Arning E, Ziegler MG, Hansen LA, and Masliah E (2004) Elevated S-adenosylhomocysteine in Alzheimer brain: Influence on methyltransferases and cognitive function. *Journal of Neural Transmission* 111: 547–567.

Kim EJ, Cho SS, Jeong Y, et al. (2005) Glucose metabolism in early onset versus late onset Alzheimer's disease: An SPM analysis of 120 patients. *Brain* 128: 1790–1801.

Kimonides VG, Khatibi NH, Svendsen CN, Sofroniew MV, and Herbert J (1998) Dehydroepiandrosterone (DHEA) and DHEA-sulfate (DHEAS) protect hippocampal neurons against excitatory amino acid-induced neurotoxicity. *Proceedings of the National Academy of Sciences of the United States of America* 95: 1852–1857.

Kirk A and Kertesz A (1991) On drawing impairment in Alzheimer's disease. *Archives of Neurology* 48: 73–77.

Klerman EB, Zeitzer JM, Duffy JF, Khalsa SB, and Czeisler CA (2001) Absence of an increase in the duration of the circadian melatonin secretory episode in totally blind human subjects. *Journal of Clinical Endocrinology and Metabolism* 86: 3166–3170.

Kordower JH, Chu Y, Stebbins GT, DeKosky ST, Cochran EJ, Bennett D, and Mufson EJ (2001) Loss and atrophy of layer II entorhinal cortex neurons in elderly people with mild cognitive impairment. *Annals of Neurology* 49: 202–213.

Korf ES, Wahlund LO, Visser PJ, and Scheltens P (2004) Medial temporal lobe atrophy on MRI predicts dementia in patients with mild cognitive impairment. *Neurology* 63: 94–100.

Kramer AF, Colcombe SJ, Mcauley E, Scalf PE, and Erickson KI (2005) Fitness, aging and neurocognitive function. *Neurobiology of Aging* 26(supplement 1): 124–127.

Kramer AF and Erickson KI (2007) Effects of physical activity on cognition, well-being, and brain: Human interventions. *Alzheimer's and Dementia* 3: S45–S51.

Kubis N, Faucheux BA, Ransmayr G, et al. (2000) Preservation of midbrain catecholaminergic neurons in very old human subjects. *Brain* 123(pt 2): 366–373.

Kuningas M, DeRijk RH, Westendorp RG, Jolles J, Slagboom PE, and VanHeemst D (2007) Mental performance in old age dependent on cortisol and genetic variance in the mineralocorticoid and glucocorticoid receptors. *Neuropsychopharmacology* 32: 1295–1301.

Kuusisto J, Koivisto K, Mykkanen L, et al. (1997) Association between features of the insulin resistance syndrome and Alzheimer's disease independently of apolipoprotein E4 phenotype: Cross sectional population based study. *British Medical Journal* 315: 1045–1049.

Lachman HM, Morrow B, Shprintzen R, et al. (1996) Association of codon 108/158 catechol-*O*-methyltransferase gene polymorphism with the psychiatric manifestations of velo-cardio-facial syndrome. *American Journal of Medical Genetics* 67: 468–472.

Lahiri DK, Chen DM, Lahiri P, Bondy S, and Greig NH (2005) Amyloid, cholinesterase, melatonin, and metals and their roles in aging and neurodegenerative diseases. *Annals of the New York Academy of Sciences* 1056: 430–449.

Lamberts SW, VanDen Beld AW, and VanDer Lely AJ (1997) The endocrinology of aging. *Science* 278: 419–424.

Landfield PW, Blalock EM, Chen KC, and Porter NM (2007) A new glucocorticoid hypothesis of brain aging: Implications for Alzheimer's disease. *Current Alzheimer Research* 4: 205–212.

Landfield PW, Waymire JC, and Lynch G (1978) Hippocampal aging and adrenocorticoids: Quantitative correlations. *Science* 202: 1098–1102.

Larson J, Jessen RE, Uz T, Arslan AD, Kurtuncu M, Imbesi M, and Manev H (2006a) Impaired hippocampal long-term potentiation in melatonin MT2 receptor-deficient mice. *Neuroscience Letters* 393: 23–26.

Larson EB, Wang L, Bowen JD, Mccormick WC, Teri L, Crane P, and Kukull W (2006b) Exercise is associated with reduced risk for incident dementia among persons 65years of age and older. *Annals of Internal Medicine* 144: 73–81.

Launer LJ, Andersen K, Dewey ME, et al. (1999) Rates and risk factors for dementia and Alzheimer's disease: Results from EURODEM pooled analyses. EURODEM Incidence Research Group and Work Groups. European Studies of Dementia. *Neurology* 52: 78–84.

Lee BK, Glass TA, McAtee MJ, Wand GS, Bandeen-Roche K, Bolla KI, and Schwartz BS (2007) Association of salivary cortisol with cognitive function in the Baltimore memory study. *Archives of General Psychiatry* 64: 810–818.

Lee HG, Castellani RJ, Zhu X, Perry G, and Smith MA (2005) Amyloid-beta in Alzheimer's disease: The horse or the cart? Pathogenic or protective? *International Journal of Experimental Pathology* 86: 133–138.

Lee ZS, Chan JC, Yeung VT, et al. (1999) Plasma insulin, growth hormone, cortisol, and central obesity among young Chinese type 2 diabetic patients. *Diabetes Care* 22: 1450–1457.

Leranth C and Hajszan T (2007) Extrinsic afferent systems to the dentate gyrus. *Progress in Brain Research* 163: 63–799.
Leverenz JB, Miller MA, Dobie DJ, Peskind ER, and Raskind MA (2001) Increased alpha 2-adrenergic receptor binding in locus coeruleus projection areas in dementia with Lewy bodies. *Neurobiology and Aging* 22: 555–561.
Liu SJ and Wang JZ (2002) Alzheimer-like tau phosphorylation induced by wortmannin *in vivo* and its attenuation by melatonin. *Acta Pharmacologica Sinica* 23: 183–187.
Lohr JB and Jeste DV (1988) Locus ceruleus morphometry in aging and schizophrenia. *Acta Psychiatrica Scandinavica* 77: 689–697.
Lu PH, Masterman DA, Mulnard R, et al. (2006) Effects of testosterone on cognition and mood in male patients with mild Alzheimer disease and healthy elderly men. *Archives of Neurology* 63: 177–185.
Luchsinger JA, Tang MX, Shea S, and Mayeux R (2004) Hyperinsulinemia and risk of Alzheimer disease. *Neurology* 63: 1187–1192.
Lue YH, Hikim AP, Swerdloff RS, et al. (1999) Single exposure to heat induces stage-specific germ cell apoptosis in rats: Role of intratesticular testosterone on stage specificity. *Endocrinology* 140: 1709–1717.
Luine VN (1985) Estradiol increases choline acetyltransferase activity in specific basal forebrain nuclei and projection areas of female rats. *Experimental Neurology* 89: 484–490.
Lupien S, Lecours AR, Lussier I, Schwartz G, Nair NP, and Meaney MJ (1994) Basal cortisol levels and cognitive deficits in human aging. *Journal of Neuroscience* 14: 2893–2903.
Lupien SJ, De Leon M, De Santi S, et al. (1998) Cortisol levels during human aging predict hippocampal atrophy and memory deficits. *Nature Neuroscience* 1: 69–73.
Lyness SA, Zarow C, and Chui HC (2003) Neuron loss in key cholinergic and aminergic nuclei in Alzheimer disease: A meta-analysis. *Neurobiology of Aging* 24: 1–23.
Maes M, Calabresse J, Lee M, and Meltzer HY (1994) Effects of age on spontaneous cortisolaemia of normal volunteers and depressed patients. *Psychoneuroendocrinology* 19: 79–84.
Magri F, Sarra S, Cinchetti W, Guazzoni V, Fioravanti M, Cravello L, and Ferrari E (2004) Qualitative and quantitative changes of melatonin levels in physiological and pathological aging and in centenarians. *Journal of Pineal Research* 36: 256–261.
Maheu FS, Joober R, Beaulieu S, and Lupien SJ (2004) Differential effects of adrenergic and corticosteroid hormonal systems on human short – and long-term declarative memory for emotionally arousing material. *Behavioral Neuroscience* 118: 420–428.
Maheu FS, Joober R, and Lupien SJ (2005) Declarative memory after stress in humans: Differential involvement of the beta-adrenergic and corticosteroid systems. *Journal of Clinical Endocrinology and Metabolism* 90: 1697–1704.
Mahlberg R, Walther S, Kalus P, et al. (2006) Pineal calcification in Alzheimer's disease: An *in vivo* study using computed tomography. *Neurobiology of Aging* 29(2): 203–209.
Majewska MD, Demirgoren S, Spivak CE, and London ED (1990) The neurosteroid dehydroepiandrosterone sulfate is an allosteric antagonist of the GABAA receptor. *Brain Research* 526: 143–146.
Maki PM and Resnick SM (2000) Longitudinal effects of estrogen replacement therapy on PET cerebral blood flow and cognition. *Neurobiology of Aging* 21: 373–383.
Malhotra AK, Kestler LJ, Mazzanti C, Bates JA, Goldberg T, and Goldman D (2002) A functional polymorphism in the COMT gene and performance on a test of prefrontal cognition. *American Journal of Psychiatry* 159: 652–654.
Malinowski JM and Bolesta S (2000) Rosiglitazone in the treatment of type 2 diabetes mellitus: A critical review. *Clinical Therapeutics* 22: 1151–1168; discussion 1149–1150.
Manaye KF, Mcintire DD, Mann DM, and German DC (1995) Locus coeruleus cell loss in the aging human brain: A non-random process. *Journal of Comparative Neurology* 358: 79–87.
Mann MB, Wu S, Rostamkhani M, Tourtellotte W, MacMurray J, and Comings DE (2001) Phenylethanolamine *N*-methyltransferase (PNMT) gene and early-onset Alzheimer disease. *American Journal of Medical Genetics* 105: 312–316.
Manns JR and Eichenbaum H (2006) Evolution of declarative memory. *Hippocampus* 16: 795–808.
Marker JC, Clutter WE, and Cryer PE (1998) Reduced epinephrine clearance and glycemic sensitivity to epinephrine in older individuals. *American Journal of Physiology* 275: E770–E776.
Marker JC, Cryer PE, and Clutter WE (1994) Simplified measurement of norepinephrine kinetics: Application to studies of aging and exercise training. *American Journal of Physiology* 267: E380–E387.
Masters CL and Beyreuther K (2005) The neuropathology of Alzheimer's disease in the year 2005. In: Beal F, Lang AE, and Ludolph A (eds.) *Neurodegenerative Diseases: Neurobiology Pathogenesis and Therapeutics*, pp. 433–440. Cambridge: Cambridge University Press.
Matthews KL, Chen CP, Esiri MM, Keene J, Minger SL, and Francis PT (2002) Noradrenergic changes, aggressive behavior, and cognition in patients with dementia. *Biological Psychiatry* 51: 407–416.
Mattila KM, Axelman K, Rinne JO, et al. (2000) Interaction between estrogen receptor 1 and the epsilon4 allele of apolipoprotein E increases the risk of familial Alzheimer's disease in women. *Neuroscience Letters* 282: 45–48.
Mazzeo RS, Rajkumar C, Jennings G, and Esler M (1997) Norepinephrine spillover at rest and during submaximal exercise in young and old subjects. *Journal of Applied Physiology* 82: 1869–1874.
McEwen B (2002) Estrogen actions throughout the brain. *Recent Progress in Hormone Research* 57: 357–384.
McEwen BS and Alves SE (1999) Estrogen actions in the central nervous system. *Endocrine Reviews* 20: 279–307.
McEwen BS, DeKloet ER, and Rostene W (1986) Adrenal steroid receptors and actions in the nervous system. *Physiological Reviews* 66: 1121–1188.
McEwen BS, Weiss JM, and Schwartz LS (1968) Selective retention of corticosterone by limbic structures in rat brain. *Nature* 220: 911–912.
McGeer EG and McGeer PL (2001) Innate immunity in Alzheimer's disease: A model for local inflammatory reactions. *Molecular Interventions* 1: 22–29.
McKhann G, Drachman D, Folstein M, Katzman R, Price D, and Stadlan EM (1984) Clinical diagnosis of Alzheimer's disease: Report of the NINCDS-ADRDA Work Group under the auspices of Department of Health and Human Services Task Force on Alzheimer's Disease. *Neurology* 34: 939–944.
Meana JJ, Barturen F, Garro MA, Garcia-Sevilla JA, Fontan A, and Zarranz JJ (1992) Decreased density of presynaptic alpha 2-adrenoceptors in postmortem brains of patients with Alzheimer's disease. *Journal of Neurochemistry* 58: 1896–1904.
Meaney MJ and Aitken DH (1985) [^{3}H]Dexamethasone binding in rat frontal cortex. *Brain Research* 328: 176–180.
Mega MS, Cummings JL, Fiorello T, and Gornbein J (1996) The spectrum of behavioral changes in Alzheimer's disease. *Neurology* 46: 130–135.
Messier C (2005) Impact of impaired glucose tolerance and type 2 diabetes on cognitive aging. *Neurobiology of Aging* 26 (supplement 1): 26–30.
Mitchell TW, Mufson EJ, Schneider JA, et al. (2002) Parahippocampal tau pathology in healthy aging, mild

cognitive impairment, and early Alzheimer's disease. *Annals of Neurology* 51: 182–189.
Miyashita T and Williams CL (2006) Epinephrine administration increases neural impulses propagated along the vagus nerve: Role of peripheral beta-adrenergic receptors. *Neurobiology of Learning and Memory* 85: 116–124.
Mizoguchi K, Ishige A, Aburada M, and Tabira T (2003) Chronic stress attenuates glucocorticoid negative feedback: Involvement of the prefrontal cortex and hippocampus. *Neuroscience* 119: 887–897.
Moffat SD (2005) Effects of testosterone on cognitive and brain aging in elderly men. *Annals of the New York Academy of Sciences* 1055: 80–92.
Moffat SD, Zonderman AB, Metter EJ, Kawas C, Blackman MR, Harman SM, and Resnick SM (2004) Free testosterone and risk for Alzheimer disease in older men. *Neurology* 62: 188–193.
Moore RY and Bloom FE (1979) Central catecholamine neuron systems: Anatomy and physiology of the norepinephrine and epinephrine systems. *Annual Review of Neuroscience* 2: 113–168.
Morilak DA, Barrera G, Echevarria DJ, Garcia AS, Hernandez A, Ma S, and Petre CO (2005) Role of brain norepinephrine in the behavioral response to stress. *Progress in Neuropsychopharmacology and Biological Psychiatry* 29: 1214–1224.
Morley JE, Kaiser F, Raum WJ, et al. (1997) Potentially predictive and manipulable blood serum correlates of aging in the healthy human male: Progressive decreases in bioavailable testosterone, dehydroepiandrosterone sulfate, and the ratio of insulin-like growth factor 1 to growth hormone. *Proceedings of the National Academy of Sciences of the United States of America* 94: 7537–7542.
Mosconi L, Tsui WH, DeSanti S, et al. (2005) Reduced hippocampal metabolism in MCI and AD: Automated FDG-PET image analysis. *Neurology* 64: 1860–1867.
Mouton PR, Pakkenberg B, Gundersen HJ, and Price DL (1994) Absolute number and size of pigmented locus coeruleus neurons in young and aged individuals. *Journal of Chemical Neuroanatomy* 7: 185–190.
Mufson EJ, Chen EY, Cochran EJ, Beckett LA, Bennett DA, and Kordower JH (1999) Entorhinal cortex beta-amyloid load in individuals with mild cognitive impairment. *Experimental Neurology* 158: 469–490.
Nasman B, Olsson T, Backstrom T, Eriksson S, Grankvist K, Viitanen M, and Bucht G (1991) Serum dehydroepiandrosterone sulfate in Alzheimer's disease and in multi-infarct dementia. *Biological Psychiatry* 30: 684–690.
Neaves WB, Johnson L, Porter JC, Parker CR, Jr., and Petty CS (1984) Leydig cell numbers, daily sperm production, and serum gonadotropin levels in aging men. *Journal of Clinical Endocrinology and Metabolism* 59: 756–763.
Neumann PJ, Araki SS, Arcelus A, et al. (2001) Measuring Alzheimer's disease progression with transition probabilities: Estimates from CERAD. *Neurology* 57: 957–964.
Ohashi Y, Okamoto N, Uchida K, Iyo M, Mori N, and Morita Y (1999) Daily rhythm of serum melatonin levels and effect of light exposure in patients with dementia of the Alzheimer's type. *Biological Psychiatry* 45: 1646–1652.
Ohm TG, Busch C, and Bohl J (1997) Unbiased estimation of neuronal numbers in the human nucleus coeruleus during aging. *Neurobiology of Aging* 18: 393–399.
Oldendorf WH (1971) Brain uptake of radiolabeled amino acids, amines, and hexoses after arterial injection. *American Journal of Physiology* 221: 1629–1639.
Olsen L, Rasmussen HB, Hansen T, et al. (2006) Estrogen receptor alpha and risk for cognitive impairment in postmenopausal women. *Psychiatric Genetics* 16: 85–88.
Osterlund MK, Gustafsson JA, Keller E, and Hurd YL (2000a) Estrogen receptor beta (ERbeta) messenger ribonucleic acid (mRNA) expression within the human forebrain: Distinct distribution pattern to ERalpha mRNA. *Journal of Clinical Endocrinology and Metabolism* 85: 3840–3846.
Osterlund MK, Keller E, and Hurd YL (2000b) The human forebrain has discrete estrogen receptor alpha messenger RNA expression: High levels in the amygdaloid complex. *Neuroscience* 95: 333–342.
Paganini-Hill A and Henderson VW (1994) Estrogen deficiency and risk of Alzheimer's disease in women. *American Journal of Epidemiology* 140: 256–261.
Palmer AM and DeKosky ST (1993) Monoamine neurons in aging and Alzheimer's disease. *Journal of Neural Transmission, General Section* 91: 135–159.
Pandi-Perumal SR, Zisapel N, Srinivasan V, and Cardinali DP (2005) Melatonin and sleep in aging population. *Experimental Gerontology* 40: 911–925.
Pantel J, Schönknecht P, Essig M, Amann M, Eysenbach K, Schad LR, and Schröder J (2002) Progressive medial temporal lobe changes in Alzheimer's disease revealed by quantitative MRI: Potential use for monitoring of drug-related changes. *Drug Developmental Research* 56: 51–56.
Paoletti AM, Congia S, Lello S, et al. (2004) Low androgenization index in elderly women and elderly men with Alzheimer's disease. *Neurology* 62: 301–303.
Papasozomenos SC (1997) The heat shock-induced hyperphosphorylation of tau is estrogen-independent and prevented by androgens: Implications for Alzheimer disease. *Proceedings of the National Academy of Sciences of the United States of America* 94: 6612–6617.
Park CR (2001) Cognitive effects of insulin in the central nervous system. *Neuroscience and Biobehavioral Reviews* 25: 311–323.
Pascualy M, Petrie EC, Brodkin K, Peskind ER, Wilkinson CW, and Raskind MA (2000) Hypothalamic pituitary adrenocortical and sympathetic nervous system responses to the cold pressor test in Alzheimer's disease. *Biological Psychiatry* 48: 247–254.
Payami H, Grimslid H, Oken B, et al. (1997) A prospective study of cognitive health in the elderly (Oregon Brain Aging Study): Effects of family history and apolipoprotein E genotype. *American Journal of Human Genetics* 60: 948–956.
Peck JS, Legoff DB, Ahmed I, and Goebert D (2004) Cognitive effects of exogenous melatonin administration in elderly persons: A pilot study. *American Journal of Geriatric Psychiatry* 12: 432–436.
Penev P, Spiegel K, Marcinkowski T, and VanCauter E (2005) Impact of carbohydrate-rich meals on plasma epinephrine levels: Dysregulation with aging. *Journal of Clinical Endocrinology and Metabolism* 90: 6198–6206.
Pennanen C, Kivipelto M, Tuomainen S, et al. (2004) Hippocampus and entorhinal cortex in mild cognitive impairment and early AD. *Neurobiology of Aging* 25: 303–310.
Perez A, Morelli L, Cresto JC, and Castano EM (2000) Degradation of soluble amyloid beta-peptides 1–40, 1–42, and the Dutch variant 1–40Q by insulin degrading enzyme from Alzheimer disease and control brains. *Neurochemical Research* 25: 247–255.
Perry RJ and Hodges JR (1999) Attention and executive deficits in Alzheimer's disease. A critical review. *Brain* 122(pt 3): 383–404.
Perry RJ, Watson P, and Hodges JR (2000) The nature and staging of attention dysfunction in early (minimal and mild) Alzheimer's disease: Relationship to episodic and semantic memory impairment. *Neuropsychologia* 38: 252–271.

Peskind ER, Elrod R, Dobie DJ, et al. (1998) Cerebrospinal fluid epinephrine in Alzheimer's disease and normal aging. *Neuropsychopharmacology* 19: 465–471.
Petanceska SS, Nagy V, Frail D, and Gandy S (2000) Ovariectomy and 17beta-estradiol modulate the levels of Alzheimer's amyloid beta peptides in brain. *Neurology* 54: 2212–2217.
Petersen RC (2004) Mild cognitive impairment as a diagnostic entity. *Journal of Internal Medicine* 256: 183–194.
Petersen RC, Stevens JC, Ganguli M, Tangalos EG, Cummings JL, and DeKosky ST (2001) Practice parameter: Early detection of dementia: Mild cognitive impairment (an evidence-based review). Report of the Quality Standards Subcommittee of the American Academy of Neurology. *Neurology* 56: 1133–1142.
Petrie EC, Peskind ER, Dobie DJ, Veith RC, and Raskind MA (2001) Plasma catecholamine and cardiovascular responses to physostigmine in Alzheimer's disease and aging. *Psychoneuroendocrinology* 26: 147–164.
Pignatti R, Rabuffetti M, Imbornone E, Mantovani F, Alberoni M, Farina E, and Canal N (2005) Specific impairments of selective attention in mild Alzheimer's disease. *Journal of Clinical and Experimental Neuropsychology* 27: 436–448.
Pike CJ, Rosario ER, and Nguyen TV (2006) Androgens, aging, and Alzheimer's disease. *Endocrine* 29: 233–241.
Pinkerton JV and Henderson VW (2005) Estrogen and cognition, with a focus on Alzheimer's disease. *Seminars in Reproductive Medicine* 23: 172–179.
Poirier J, Davignon J, Bouthillier D, Kogan S, Bertrand P, and Gauthier S (1993) Apolipoprotein E polymorphism and Alzheimer's disease. *Lancet* 342: 697–699.
Porrello E, Monti MC, Sinforiani E, et al. (2006) Estrogen receptor alpha and APOEepsilon4 polymorphisms interact to increase risk for sporadic AD in Italian females. *European Journal of Neurology* 13: 639–644.
Pratico D and Trojanowski JQ (2000) Inflammatory hypotheses: Novel mechanisms of Alzheimer's neurodegeneration and new therapeutic targets? *Neurobiology of Aging* 21: 441–445; discussion 451–453.
Price BH, Gurvit H, Weintraub S, Geula C, Leimkuhler E, and Mesulam M (1993) Neuropsychological patterns and language deficits in 20 consecutive cases of autopsy-confirmed Alzheimer's disease. *Archives in Neurology* 50: 931–937.
Price JL and Morris JC (1999) Tangles and plaques in nondemented aging and preclinical Alzheimer's disease. *Annals of Neurology* 45: 358–368.
Ptok U, Barkow K, and Heun R (2002) Fertility and number of children in patients with Alzheimer's disease. *Archives of Women's Mental Health* 5: 83–86.
Qiu WQ and Folstein MF (2006) Insulin, insulin-degrading enzyme and amyloid-beta peptide in Alzheimer's disease: Review and hypothesis. *Neurobiology of Aging* 27: 190–198.
Raff H, Raff JL, Duthie EH, Wilson CR, Sasse EA, Rudman I, and Mattson D (1999) Elevated salivary cortisol in the evening in healthy elderly men and women: Correlation with bone mineral density. *Journals of Gerontology: Series A: Biological Sciences and Medical Sciences* 54: M479–M483.
Rasmuson S, Andrew R, Nasman B, Seckl JR, Walker BR, and Olsson T (2001) Increased glucocorticoid production and altered cortisol metabolism in women with mild to moderate Alzheimer's disease. *Biological Psychiatry* 49: 547–552.
Razay G and Wilcock GK (1994) Hyperinsulinaemia and Alzheimer's disease. *Age and Ageing* 23: 396–399.
Reisberg B (1988) Functional assessment staging (FAST). *Psychopharmacology Bulletin* 24: 653–659.
Resnick SM and Henderson VW (2002) Hormone therapy and risk of Alzheimer disease: A critical time. *Journal of the American Medical Association* 288: 2170–2172.
Reul JM and DeKloet ER (1985) Two receptor systems for corticosterone in rat brain: Microdistribution and differential occupation. *Endocrinology* 117: 2505–2511.
Rizkalla SW, Taghrid L, Laromiguiere M, et a . (2004) Improved plasma glucose control, whole-body glucose utilization, and lipid profile on a low-glycemic index diet in type 2 diabetic men: A randomized controlled trial. *Diabetes Care* 27: 1866–1872.
Rocca WA, Bower JH, Maraganore DM, et al. (2007) Increased risk of cognitive impairment or dementia in women who underwent oophorectomy before menopause. *Neurology* 69: 1074–1083.
Rosa A, Peralta V, Cuesta MJ, Zarzuela A, Serrano F, Martinez-Larrea A, and Fananas L (2004) New evidence of association between COMT gene and prefrontal neurocognitive function in healthy individuals from sibling pairs discordant for psychosis. *American Journal of Psychiatry* 161: 1110–1112.
Rosales-Corral S, Tan DX, Reiter RJ, Valdivia-Velazquez M, Martinez-Barboza G, Acosta-Martinez JP, and Ortiz GG (2003) Orally administered melatonin reduces oxidative stress and proinflammatory cytokines induced by amyloid-beta peptide in rat brain: A comparative, *in vivo* study versus vitamin C and E. *Journal of Pineal Reserach* 35: 80–84.
Rosario ER, Chang L, Stanczyk FZ, and Pike CJ (2004) Age-related testosterone depletion and the development of Alzheimer disease. *Journal of the American Medical Association* 292: 1431–1432.
Rosmond R (2003) Stress induced disturbances of the HPA axis: A pathway to type 2 diabetes? *Medical Science Monitor* 9: RA35–RA39.
Rossouw JE, Anderson GL, Prentice RL, et al. (2002) Risks and benefits of estrogen plus progestin in healthy postmenopausal women: Principal results From the Women's Health Initiative randomized controlled trial. *Journal of the American Medical Association* 288: 321–333.
Russo-Neustadt A and Cotman CW (1997) Adrenergic receptors in Alzheimer's disease brain: Selective increases in the cerebella of aggressive patients. *Journal of Neuroscience* 17: 5573–5580.
Ryan CM and Geckle M (2000) Why is learning and memory dysfunction in type 2 diabetes limited to older adults? *Diabetes/Metabolism Research and Reviews* 16: 308–315.
Saar N, Jackson R, Bachmann A, and Gordon R (1980) Effect of sampling site and conditions on plasma levels of noradrenaline, adrenaline and dopamine *Progress in Biochemistry and Pharmacology* 17: 90–97.
Sacks DB and Mcdonald JM (1996) The pathogenesis of type II diabetes mellitus. A polygenic disease. *American Journal of Clinical Pathology* 105: 149–156.
Sapolsky RM, Krey LC, and McEwen BS (1984) Glucocorticoid-sensitive hippocampal neurons are involved in terminating the adrenocortical stress response. *Proceedings of the National Academy of Sciences of the United States of America* 81: 6174–6177.
Sapolsky RM, Krey LC, and McEwen BS (1986) The neuroendocrinology of stress and aging: The glucocorticoid cascade hypothesis. *Endocrine Reviews* 7: 284–301.
Saunders A (2000) Apolipoprotein E and Alzheimer disease: An update on genetic and functional analyses. *Journal of Neuropathology and Experimental Neurology* 59: 751–758.
Scharfman HE and Maclusky NJ (2006) Estrogen and brain-derived neurotrophic factor (BDNF) in hippocampus: Complexity of steroid hormone-growth factor interactions in the adult CNS. *Frontiers in Neuroendocrinology* 27: 415–435.
SchnaiderBeeri M, Goldbourt U, Silverman JM, et al. (2004) Diabetes mellitus in midlife and the risk of dementia three decades later. *Neurology* 63: 1902–1927.

Schneider LS, Hinsey M, and Lyness S (1992) Plasma dehydroepiandrosterone sulfate in Alzheimer's disease. *Biological Psychiatry* 31: 205–208.
Schulingkamp RJ, Pagano TC, Hung D, and Raffa RB (2000) Insulin receptors and insulin action in the brain: Review and clinical implications. *Neuroscience and Biobehavioral Reviews* 24: 855–872.
Sclan SG and Reisberg B (1992) Functional assessment staging (FAST) in Alzheimer's disease: Reliability, validity, and ordinality. *International Psychogeriatrics* 4(supplement 1): 55–69.
Seals DR and Esler MD (2000) Human ageing and the sympathoadrenal system. *Journal of Physiology* 528: 407–417.
Seckl JR, Dickson KL, Yates C, and Fink G (1991) Distribution of glucocorticoid and mineralocorticoid receptor messenger RNA expression in human postmortem hippocampus. *Brain Research* 561: 332–337.
Selkoe DJ (2001) Alzheimer's disease: Genes, proteins, and therapy. *Physiological Reviews* 81: 741–766.
Selkoe DJ (2002) Alzheimer's disease is a synaptic failure. *Science* 298: 789–791.
Selkoe DJ (2004) Cell biology of protein misfolding: The examples of Alzheimer's and Parkinson's diseases. *Nature Cell Biology* 6: 1054–1061.
Sharp SI, Ballard CG, Chen CP, and Francis PT (2007) Aggressive behavior and neuroleptic medication are associated with increased number of alpha1-adrenoceptors in patients with Alzheimer disease. *American Journal of Geriatric Psychiatry* 15: 435–437.
Shen YX, Wei W, and Xu SY (2002a) Protective effects of melatonin on cortico-hippocampal neurotoxicity induced by amyloid beta-peptide 25–35. *Acta Pharmacologica Sinica* 23: 71–76.
Shen YX, Xu SY, Wei W, Sun XX, Liu LH, Yang J, and Dong C (2002b) The protective effects of melatonin from oxidative damage induced by amyloid beta-peptide 25–35 in middle-aged rats. *Journal of Pineal Research* 32: 85–89.
Shen Y, Zhang G, Liu L, and Xu S (2007) Suppressive effects of melatonin on amyloid-beta-induced glial activation in rat hippocampus. *Archives of Medical Research* 38: 284–290.
Sherman B, Wysham C, and Pfohl B (1985) Age-related changes in the circadian rhythm of plasma cortisol in man. *Journal of Clinical Endocrinology and Metabolism* 61: 439–443.
Sherwin BB (2007) The critical period hypothesis: Can it explain discrepancies in the oestrogen-cognition literature? *Journal of Neuroendocrinology* 19: 77–81.
Shirokawa T, Ishida Y, and Isobe KI (2000) Age-dependent changes in axonal branching of single locus coeruleus neurons projecting to two different terminal fields. *Journal of Neurophysiology* 84: 1120–1122.
Short RA, Bowen RL, O'brien PC, and Graff-Radford NR (2001) Elevated gonadotropin levels in patients with Alzheimer disease. *Mayo Clinic Proceedings* 76: 906–909.
Shumaker SA, Legault C, Kuller L, et al. (2004) Conjugated equine estrogens and incidence of probable dementia and mild cognitive impairment in postmenopausal women: Women's Health Initiative Memory Study. *Journal of the American Medical Association* 291: 2947–2958.
Shumaker SA, Legault C, Rapp SR, et al. (2003) Estrogen plus progestin and the incidence of dementia and mild cognitive impairment in postmenopausal women: The Women's Health Initiative Memory Study: A randomized controlled trial. *Journal of the American Medical Association* 289: 2651–2662.
Silverman DH, Small GW, Chang CY, et al. (2001) Positron emission tomography in evaluation of dementia: Regional brain metabolism and long-term outcome. *Journal of the American Medical Association* 286: 2120–2127.
Simerly RB, Chang C, Muramatsu M, and Swanson LW (1990) Distribution of androgen and estrogen receptor mRNA-containing cells in the rat brain: An *in situ* hybridization study. *Journal of Comparative Neurology* 294: 76–95.
Small GW, Rabins PV, Barry PP, et al. (1997) Diagnosis and treatment of Alzheimer disease and related disorders. Consensus statement of the American Association for Geriatric Psychiatry, the Alzheimer's Association, and the American Geriatrics Society. *Journal of the American Medical Association* 278: 1363–1371.
Sobow TM and Koszewska I (2003) Modulation of age at onset in late-onset sporadic Alzheimer's disease by estrogen-related factors:the age at menopause and number of pregnancies. *German Journal of Psychiatry* 6: 49–55.
Spada RS, Cento RM, Proto C, et al. (2001) Twenty-four-hour urinary free cortisol levels in vascular dementia and in Alzheimer's disease. *Archives of Gerontology and Geriatrics* 7(supplement): 363–367.
Speroff L and Fritz AF (2005) *Clinical Gynecologic Endocrinology and Infertility.* Philadelphia, PA: Lippincott Williams & Wilkins.
Starkstein SE, Jorge R, Mizrahi R, and Robinson RG (2006) A prospective longitudinal study of apathy in Alzheimer's disease. *Journal of Neurology, Neurosurgery and Psychiatry* 77: 8–11.
Starr JM, Fox H, Harris SE, Deary IJ, and Whalley L (2007) COMT genotype and cognitive ability: A longitudinal aging study. *Neurosciences Letters* 421: 57–61.
Sternberg DB, Martinez JL, Jr., Gold PE, and Mcgaugh JL (1985) Age-related memory deficits in rats and mice: Enhancement with peripheral injections of epinephrine. *Behavioral and Neural Biology* 44: 213–220.
Strachan MW, Deary IJ, Ewing FM, and Frier BM (1997) Is type II diabetes associated with an increased risk of cognitive dysfunction? A critical review of published studies. *Diabetes Care* 20: 438–445.
Strittmatter WJ, Weisgraber KH, Huang DY, et al. (1993) Binding of human apolipoprotein E to synthetic amyloid beta peptide: Isoform-specific effects and implications for late-onset Alzheimer disease. *Proceedings of the National Academy of Sciences of the United States of America* 90: 8098–8102.
Swanwick GR, Kirby M, Bruce I, Buggy F, Coen RF, Coakley D, and Lawlor BA (1998) Hypothalamic–pituitary–adrenal axis dysfunction in Alzheimer's disease: Lack of association between longitudinal and cross-sectional findings. *American Journal of Psychiatry* 155: 286–289.
Szot P, White SS, Greenup JL, Leverenz JB, Peskind ER, and Raskind MA (2006) Compensatory changes in the noradrenergic nervous system in the locus ceruleus and hippocampus of postmortem subjects with Alzheimer's disease and dementia with Lewy bodies. *Journal of Neuroscience* 26: 467–478.
Szot P, White SS, Greenup JL, Leverenz JB, Peskind ER, and Raskind MA (2007) Changes in adrenoreceptors in the prefrontal cortex of subjects with dementia: Evidence of compensatory changes. *Neuroscience* 146: 471–480.
Tan RS and Pu SJ (2003) A pilot study on the effects of testosterone in hypogonadal aging male patients with Alzheimer's disease. *Aging Male* 6(1): 13–17.
Tanapat P, Hastings NB, and Gould E (2005) Ovarian steroids influence cell proliferation in the dentate gyrus of the adult female rat in a dose- and time-dependent manner. *Journal of Comparative Neurology* 481: 252–265.
Thompson PM, Hayashi KM, Dutton RA, et al. (2007) Tracking Alzheimer's disease. *Journal of Comparative Neurology* 1097: 183–214.

Toran-Allerand CD (2004) Estrogen and the brain: Beyond ER-alpha and ER-beta. *Experimental Gerontology* 39: 1579–1586.

Touitou Y, Sulon J, Bogdan A, et al. (1982) Adrenal circadian system in young and elderly human subjects: A comparative study. *Journal of Endocrinology* 93: 201–210.

Tuomilehto J, Lindstrom J, Eriksson JG, et al. (2001) Prevention of type 2 diabetes mellitus by changes in lifestyle among subjects with impaired glucose tolerance. *New England Journal of Medicine* 344: 1343–1350.

Tyas SL, Salazar JC, Snowdon DA, et al. (2007) Transitions to mild cognitive impairments, dementia, and death: Findings from the Nun Study. *American Journal of Epidemiology* 165: 1231–1238.

Umegaki H, Ikari H, Nakahata H, et al. (2000a) Plasma cortisol levels in elderly female subjects with Alzheimer's disease: A cross-sectional and longitudinal study. *Brain Research* 881: 241–243.

Umegaki H, Tamaya N, Shinkai T, and Iguchi A (2000b) The metabolism of plasma glucose and catecholamines in Alzheimer's disease. *Experimental Gerontology* 35: 1373–1382.

Van Amelsvoort T, Murphy DG, Robertson D, Daly E, Whitehead M, and Abel K (2003) Effects of long-term estrogen replacement therapy on growth hormone response to pyridostigmine in healthy postmenopausal women. *Psychoneuroendocrinology* 28: 101–112.

Van Cauter E, Leproult R, and Kupfer DJ (1996) Effects of gender and age on the levels and circadian rhythmicity of plasma cortisol. *Journal of Clinical Endocrinology and Metabolism* 81: 2468–2473.

van Cauter E, Leproult R, and Plat L (2000) Age-related changes in slow wave sleep and REM sleep and relationship with growth hormone and cortisol levels in healthy men. *Journal of the American Medical Association* 284: 861–868.

van Rossum EF, de Jong FJ, Koper JW, et al. (2006) Glucocorticoid receptor variant and risk of dementia and white matter lesions. *Neurobiology of Aging* 29(5): 716–723.

van Rossum EF and Lamberts SW (2004) Polymorphisms in the glucocorticoid receptor gene and their associations with metabolic parameters and body composition. *Recent Progress in Hormone Research* 59: 333–357.

Vermeulen A and Kaufman JM (1995) Ageing of the hypothalamo-pituitary–testicular axis in men. *Hormone Research* 43: 25–28.

Vermeulen A, Kaufman JM, Goemaere S, and VanPottelberg I (2002) Estradiol in elderly men. *Aging Male* 5: 98–102.

Vijayashankar N and Brody H (1979) A quantitative study of the pigmented neurons in the nuclei locus coeruleus and subcoeruleus in man as related to aging. *Journal of Neuropathology and Experimental Neurology* 38: 490–497.

Volicer L, Harper DG, Manning BC, Goldstein R, and Satlin A (2001) Sundowning and circadian rhythms in Alzheimer's disease. *American Journal of Psychiatry* 158: 704–711.

Walsh DM, Klyubin I, Fadeeva JV, et al. (2002) Naturally secreted oligomers of amyloid beta protein potently inhibit hippocampal long-term potentiation *in vivo*. *Nature* 416: 535–539.

Waltman C, Blackman MR, Chrousos GP, Riemann C, and Harman SM (1991) Spontaneous and glucocorticoid-inhibited adrenocorticotropic hormone and cortisol secretion are similar in healthy young and old men. *Journal of Clinical Endocrinology and Metabolism* 73: 495–502.

Wang DL, Ling ZQ, Cao FY, Zhu LQ, and Wang JZ (2004) Melatonin attenuates isoproterenol-induced protein kinase A overactivation and tau hyperphosphorylation in rat brain. *Journal of Pineal Research* 37: 11–16.

Wang PN, Liu HC, Liu TY, Chu A, Hong CJ, Lin KN, and Chi CW (2005a) Estrogen-metabolizing gene COMT polymorphism synergistic APOE epsilon4 allele increases the risk of Alzheimer disease. *Dementia and Geriatric Cognitive Disorders* 19: 120–125.

Wang Y, Fice DS, and Yeung PK (1999) A simple high-performance liquid chromatography assay for simultaneous determination of plasma norepinephrine, epinephrine, dopamine and 3,4-dihydroxyphenyl acetic acid. *Journal of Pharmaceutical and Biomedical Analysis* 21: 519–525.

Wang XC, Zhang J, Yu X, Han L, Zhou ZT, Zhang Y, and Wang JZ (2005b) Prevention of isoproterenol-induced tau hyperphosphorylation by melatonin in the rat. *Sheng Li Xue Bao* 57: 7–12.

Watson GS, Cholerton BA, Reger MA, et al. (2005) Preserved cognition in patients with early Alzheimer disease and amnestic mild cognitive impairment during treatment with rosiglitazone: A preliminary study. *American Journal of Geriatric Psychiatry* 13: 950–958.

West CD, Brown H, Simons EL, Carter DB, Kumagai LF, and Englert E Jr. (1961) Adrenocortical function and cortisol metabolism in old age. *Journal of Clinical Endocrinology and Metabolism* 21: 1197–1207.

Wolf H, Julin P, Gertz HJ, Winblad B, and Wahlund LO (2004) Intracranial volume in mild cognitive impairment, Alzheimer's disease and vascular dementia: Evidence for brain reserve? *International Journal of Geriatric Psychiatry* 19: 995–1007.

Wolfson C, Wolfson DB, Asgharian M, M'lan CE, Ostbye T, Rockwood K, and Hogan DB (2001) A reevaluation of the duration of survival after the onset of dementia. *New England Journal of Medicine* 344: 1111–1116.

Wolkowitz OM, Kramer JH, Reus VI, et al. (2003) DHEA treatment of Alzheimer's disease: A randomized, double-blind, placebo-controlled study. *Neurology* 60: 1071–1076.

Worda C, Sator MO, Schneeberger C, Jantschev T, Ferlitsch K, and Huber JC (2003) Influence of the catechol-*O*-methyltransferase (COMT) codon 158 polymorphism on estrogen levels in women. *Human Reproduction* 18: 262–266.

Wu YH, Fischer DF, Kalsbeek A, et al. (2006) Pineal clock gene oscillation is disturbed in Alzheimer's disease, due to functional disconnection from the master clock. *FASEB Journal* 20: 1874–1876.

Wu YH, Zhou JN, VanHeerikhuize J, Jockers R, and Swaab DF (2007) Decreased MT1 melatonin receptor expression in the suprachiasmatic nucleus in aging and Alzheimer's disease. *Neurobiology of Aging* 28: 1239–1247.

Xu Y, Jack CR, Jr., O'brien PC, et al. (2000) Usefulness of MRI measures of entorhinal cortex versus hippocampus in AD. *Neurology* 54: 1760–1767.

Yaffe K, Barnes D, Nevitt M, Lui LY, and Covinsky K (2001) A prospective study of physical activity and cognitive decline in elderly women: Women who walk. *Archives of Internal Medicine* 161: 1703–1708.

Yaffe K, Blackwell T, Kanaya AM, Davidowitz N, Barrett-Connor E, and Krueger K (2004) Diabetes, impaired fasting glucose, and development of cognitive impairment in older women. *Neurology* 63: 658–663.

Yaffe K, Krueger K, Cummings SR, et al. (2005) Effect of raloxifene on prevention of dementia and cognitive impairment in older women: The Multiple Outcomes of Raloxifene Evaluation (MORE) randomized trial. *American Journal of Psychiatry* 162: 683–690.

Yaffe K, Lui LY, Grady D, Stone K, and Morin P (2002) Estrogen receptor 1 polymorphisms and risk of cognitive impairment in older women. *Biological Psychiatry* 51: 677–682.

Yaffe K, Sawaya G, Lieberburg I, and Grady D (1998) Estrogen therapy in postmenopausal women: Effects on cognitive function and dementia. *Journal of the American Medical Association* 279: 688–695.

Yanase T, Fukahori M, Taniguchi S, et al. (1996) Serum dehydroepiandrosterone (DHEA) and DHEA-sulfate (DHEA-S) in Alzheimer's disease and in cerebrovascular dementia. *Endocrine Journal* 43: 119–123.

Yavich L, Forsberg MM, Karayiorgou M, Gogos JA, and Mannisto PT (2007) Site-specific role of catechol-*O*-methyltransferase in dopamine overflow within prefrontal cortex and dorsal striatum. *Journal of Neuroscience* 27: 10196–10209.

Young JB, Rowe JW, Pallotta JA, Sparrow D, and Landsberg L (1980) Enhanced plasma norepinephrine response to upright posture and oral glucose administration in elderly human subjects. *Metabolism* 29: 532–539.

Zandi PP, Carlson MC, Plassman BL, Welsh-Bohmer KA, Mayer LS, Steffens DC, and Breitner JC (2002) Hormone replacement therapy and incidence of Alzheimer disease in older women: The Cache County Study. *Journal of the American Medical Association* 288: 2123–2129.

Zarow C, Lyness SA, Mortimer JA, and Chui HC (2003) Neuronal loss is greater in the locus coeruleus than nucleus basalis and substantia nigra in Alzheimer and Parkinson diseases. *Archives of Neurology* 60: 337–341.

Zeitzer JM, Daniels JE, Duffy JF, Klerman EB, Shanahan TL, Dijk DJ, and Czeisler CA (1999) Do plasma melatonin concentrations decline with age? *American Journal of Medicine* 107: 432–436.

Zhao WQ and Alkon DL (2001) Role of insulin and insulin receptor in learning and memory. *Molecular and Cellular Endocrinology* 177: 125–134.

Zhou JN, Liu RY, Van Heerikhuize J, Hofman MA, and Swaab DF (2003) Alterations in the circadian rhythm of salivary melatonin begin during middle-age. *Journal of Pineal Research* 34: 11–16.

Zhu LQ, Wang SH, Ling ZQ, Wang Q, Hu MQ, and Wang JZ (2005) Inhibition of melatonin biosynthesis activates protein kinase A and induces Alzheimer-like tau hyperphosphorylation in rats. *Chinese Academy of Medical Sciences* 20: 83–87.

Zhu LQ, Wang SH, Ling ZQ, Wang DL, and Wang JZ (2004) Effect of inhibiting melatonin biosynthesis on spatial memory retention and tau phosphorylation in rat. *Journal of Pineal Research* 37: 71–77.

Ziegler MG, Lake CR, and Kopin IJ (1976) Plasma noradrenaline increases with age. *Nature* 261: 333–335.

26 Genetic Defects of Female Sexual Differentiation

A B Dessens, Sophia Children's Hospital/Erasmus MC, Rotterdam, The Netherlands
M B C M Cools, Ghent University, Ghent, Belgium
A Richter-Unruh, Endokrinologikum MC, Bochum, Germany
L H J Looijenga and J A Grootegoed, Erasmus MC, Rotterdam, The Netherlands
S L S Drop, Sophia Childen's Hospital/Erasmus MC, Rotterdam, The Netherlands

Chapter Outline

Glossary

chimerism The existence of two or more cell lines of different genetic origin in one individual, often intermingled in a given tissue.

disorders of sex development (DSD) Congenital conditions in which development of chromosomal, gonadal, or anatomical sex is atypical.

gender identity The sense of self as being male or female.

gender role behavior Behavior that is culturally associated with gender or shows sex difference.

gonadal sex reversal The chromosomal sex is in disagreement with the gonadal sex.

gonadoblastoma This is an *in situ* neoplastic lesion, consisting of large germ cells, similar to those of seminoma, and small immature Sertoli/granulosa cells; Leydig cells surrounding this lesion may or may not be present.

mosaicism The presence of two or more subtypes of cell lines in one individual, having different karyotypes (such as 46,XY and 45,X) or showing some other genetic or epigenetic difference, such as inactivation of the X chromosome of either paternal or maternal origin in 46,XX cell lines.

sexual orientation This is defined in terms of the sex of preferred erotic partners, both in fantasy and in behavior.

streak gonads Functionless gonadal tissue in which all germ cells are lost.

undifferentiated gonadal tissue Tissue that is not differentiated either into testicular or ovarian direction, but containing germ cells.

26.1 Introduction

In October 2005, the Lawson Wilkins Pediatric Endocrine Society (LWPES) and the European Society for Paediatric Endocrinology (ESPE) organized a consensus meeting on intersex. As a result, a consensus statement on the management of individuals with various disorders of sex development (DSD) was published (Hughes et al., 2006). Advances in identification of molecular genetic causes of abnormal sex development with increased awareness of ethical issues and patient advocacy concerns necessitated the reexamination of nomenclature. First of all, the umbrella term DSD was proposed and defined as a congenital condition in which development of chromosomal, gonadal, or anatomical sex is atypical.

In the consensus statement, the terms female and male pseudohermaphroditism have been replaced by 46,XX DSD and 46,XY DSD, respectively. As this chapter deals with genetic defects in female differentiation, the focus will be on 46,XX DSD. The second and third root classifications of 46,XX DSD disorders are given in **Table 1** (Hughes et al., 2007). In addition, the term sex chromosome DSD refers to Klinefelter and Turner syndromes and variants, as well as various forms of mosaicism or chimerism of the sex chromosomes. The latter may lead to abnormal gonadal development, such as ovotesticular DSD, and are incorporated into this chapter.

We provide a brief outline of what is known and unknown about ovarian and female development. In addition, defects in female differentiation as a result of sex chromosomal mosaicism are discussed. Subsequently, various causes of maternal and fetal androgen excess syndromes are reviewed. As indicated, the agenesis syndromes of uterus and vagina form a separate entity within the 46,XX DSD category. Finally, we summarize relevant points regarding the independent effects of genetic factors and androgens on brain and behavior.

Table 1 Causes of 46,XX DSD

A: Disorders of gonadal (ovarian) development
1. Gonadal dysgenesis
2. Ovotesticular DSD
3. Testicular DSD
(e.g., *SRY+*, *dup SOX9*, *RSP01*
B: Androgen excess
1. *Fetal*
3β-hydroxysteroid dehydrogenase 2 *HSD3B2*, 21-hydroxylase (*CYP21A2*), P450 oxidoreductase (*POR*), 11β-hydroxylase (*CYP11B1*)
Glucocorticoid receptor mutations
2. *Fetoplacental*
Aromatase (*CYP 19*) deficiency
Oxidoreductase (*POR*) deficiency
3. *Maternal*
Maternal virilizing tumors (e.g., luteomas)
Androgenic drugs
C: Other
1. Syndromic associations (e.g., cloacal anomalies)
2. Müllerian agenesis/hypoplasia (e.g., MRKH)
3. Labial adhesions

Reproduced from Hughes IA, Nihoul-Fekete C, Thomos B, and Cohen-Kettenis PT (2007) Consequences of the ESPE/LWPES guidelines for diagnosis and treatment of disorders of sex development. *Best Practice and Research Clinical Endocrinology and Metabolism* 21: 351–365, with permission from Elsevier.

26.2 Ovarian and Female Development

26.2.1 Primary Sex Determination: Sex Chromosomes Dictate Gonadal Sex

Virtually all vertebrates – from fish, amphibians, and reptiles to birds and mammals – live their individual lives either as female or male. Only some fish species hang on to a hermaphroditic reproduction strategy (Mackiewicz et al., 2006). To become either female or male implies that at some point in embryonic development sex determination pathways need to be activated, starting within the fetal gonads. The indifferent fetal gonadal structures are considered bipotential, because they can develop into either ovaries or testes. These indifferent gonads originate during vertebrate embryogenesis under the influence of specific genes, and dysregulation of genes involved in their formation, such as WT1 (Wilms tumor 1) and SF1 (steroidogenic factor 1), will impact on both female and male development (Wilhelm and Englert, 2002). Once the indifferent gonads are established and have become populated by germline cells with an extragonadal, even an extraembryonic, origin (McLaren and Lawson, 2005), gonadal sex determination can occur.

In many nonmammalian vertebrate species, gonadal sex determination is controlled by environmental temperature, in the absence of any genetic difference between females and males. However, next to temperature-dependent sex determination, genetic mechanisms have also evolved (for a review, see Ezaz et al. (2006)). In its simplest form, this is found among vertebrates in the medaka fish. Male medaka fish carry the gene DMY which activates male gonadal sex determination, whereas females lack this gene. This is the only genetic difference between medaka females and males (Matsuda et al., 2007). The DMY gene is residing on what is called the Y chromosome, and the homologous chromosome of this pair which lacks *DMY* is named the X chromosome.

Placental mammals would not be able to cope with temperature-dependent sex determination, as they develop within a uterine environment of 37 °C. Hence, placental mammals also rely on a genetic mechanism for primary sex determination. Until the late 1950s, it was thought that mammals become female if they carry two X chromosomes, as opposed to becoming a male when there is only one X chromosome. From analysis of Turner syndrome females (45,XO) and Klinefelter syndrome males (47,XXY) it has become clear that it is not the number of X chromosomes which determines sex, but rather the presence of the Y chromosome (Ford et al., 1959; Jacobs and Strong, 1959). The mammalian zygote, containing a paternal pronucleus with either an X or Y of paternal origin, next to the maternal pronucleus always containing a maternal X, has already undergone the first step of primary sex determination. The chromosomal (genetic) sex is fixed at fertilization. However, in an early human embryo, before week 7 after conception, the gonads and all the structures which will develop into future female or male internal and external genitalia are present in an undifferentiated form, indistinguishable between the two sexes.

An exciting series of hypotheses and discoveries finally led to the identification of the *SRY* gene (sex-determining region on the Y chromosome gene), encoding the SRY protein which acts as the key factor determining development as a male (Berta et al., 1990; Sinclair et al., 1990). Translocation of the *SRY* gene to an X chromosome (or to one of the autosomes) or a mutational defect in the encoded SRY protein, will result, respectively, in 46,XX female-to-male or 46,XY male-to-female gonadal sex reversal, meaning that the chromosomal sex is in disagreement with the gonadal sex (Hawkins et al., 1991). Skewing of X inactivation toward the paternal X carrying the translocated *SRY* gene can lead to an ambiguous phenotype (Kusz et al., 1999). The above fact was discovered based on findings for human sex-reversed individuals. Then, shortly after the discovery of the SRY gene, it was also shown that loss of *SRY* gene function in the mouse results in XY male-to-female sex reversal (Lovell-Badge and Robertson, 1990), and that transgenic expression of SRY from an autosomal *SRY* transgene can lead to female-to-male sex reversal of an XX mouse embryo (Koopman et al., 1991). This latter sex reversal was quite impressive, generating mice, which behaved like real males, although they had small testes, lacking complete spermatogenesis which requires transcription of some genes on the Y. Still, there must be other differences between XX and XY gonadal males (and XX and XY gonadal females), independent from all sex differences evoked by gonadal hormones. As compared to medaka fish, where females and males differ only for one *DMY* gene, human X and Y chromosomes have evolved in a quite bizarre fashion, becoming widely different, with many biological implications, which are still not fully understood (Graves, 2006). There is an increasing amount of evidence pointing to a direct role of sex chromosomal genes with regard to sex differences (Arnold, 2004; Glickman et al., 2005; Migeon, 2007).

The mammalian SRY protein is not similar to the medaka DMY but serves the same purpose: to activate a cascade of events leading to male gonadal sex determination. Gonadal sex differentiation starts with expression of the *SRY* gene, in human embryos around week 7 after conception. The SRY protein binds to specific DNA binding sites, causing bending of the DNA and resulting in changes of gene expression. The SOX9 gene is immediately downstream of SRY action, and induced transcription of SOX9/Sox9 can result in testis formation even in the absence of SRY, both in the mouse and humans (Kojima et al., 2007; Qin and Bishop, 2005). However, *SRY* is a male-specific gene, whereas the SOX9 protein is involved in several developmental processes, both in females and males (Barrionuevo et al., 2006).

When SRY is expressed, in pre-Sertoli cells (Albrecht and Eicher, 2001), the indifferent fetal gonads embark on a tissue differentiation pathway leading to the formation of testis tubules. SRY expression triggers cellular signaling events, which include migration of mesonephric cells into the fetal gonad (Capel et al., 1999). In the absence of SRY, this migration of mesonephric cells does not occur, and the female gonadal sex-determination pathway takes place (see **Figure 1**).

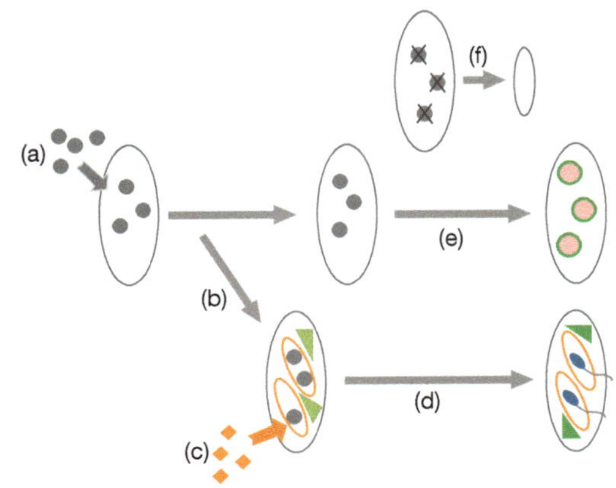

Figure 1 Gonadal sex determination and development. (a) The indifferent bipotential gonads become populated by primordial germ cells. (b) Shortly thereafter, SRY triggers a cascade of events (c) which includes migration of somatic cells from the adjacent mesonephros into the gonads. These somatic cells contribute to formation of the wall of the testicular tubules (orange), in which the germline cells become enclosed, together with the pre-Sertoli cells. (d) In these tubules, the germline cells do not enter meiotic prophase, until spermatogenesis together with steroidogenesis by Leydig cells (green) are activated by gonadotropic stimulation at puberty. The first generation of steroidogenic Leydig cells (light green) develop already in the fetal testis, and this is independent of the presence of germline cells. Also during adult reproductive life, steroidogenesis by Leydig cells is not primarily dependent on spermatogenesis. (e) In the absence of SRY, the gonadal environment is conducive to entry of the germline cells into meiotic prophase, with retinoic acid acting as an inducing compound. When meiotic prophase has generated diplotene stage oocytes enclosed by pre-granulodsa cells in primordial follicles, these oocytes start to function as signaling centers for complete ovarian development, including development of steroidogenic theca cell layers (green) surrounding growing follicles, throughout adult reproductive life. (f) If the gonads do not become populated by primordial germ cells, or when no primordial follicles can be formed or remain present, the ovary does not develop or maintain steroidogenic activity, and the tissue collapses.

26.2.2 Ovarian Development: Orchestrated by Ovary-Determining Genes?

The mammalian testis-determining gene *SRY* is highly sex-specific. A long-standing question is whether there are any known genes which are specifically required for differentiation of the bipotential gonads to become ovaries? An interesting variety of candidate ovary-determining genes have been extensively studied, as described in several comprehensive recent reviews (Kim and Capel, 2006; Ottolenghi et al., 2007a,b; Wilhelm et al., 2007a,b; Yao, 2005). If there are any new candidate ovary-determining genes to be discovered, these might be identified by analysis of transcriptional programs, studying differential gene expression patterns of female and male gonads in the mouse, at E10.5–E11.5 of embryonic development, around the period that the *Sry* gene and then the *Sox9* gene are first expressed (Beverdam and Koopman, 2006; Nef et al., 2005).

DAX1 (dosage-sensitive sex reversal–adrenal hypoplasia congenital–critical region of the X chromosome gene 1) was long considered to encode a promising candidate. The gene is located on the X chromosome and has no homolog on Y. Moreover, duplication of *DAX1* can result in male-to-female sex reversal of XY individuals. Its possible role as an antitestis factor, antagonizing SRY action (Swain et al., 1998), was challenged by the observation that DAX1 is required for testis determination in a mouse strain that is susceptible to sex reversal because of an altered *Sry* gene (Meeks et al., 2003). Furthermore, other studies in mice provided evidence that the *Dax1* gene is not strictly required for ovarian development (Yu et al., 1998).

The encoded protein plays some quite important and dosage-dependent role(s) in early gonadal development, both ovary and testis, but it is not the elusive ovary-determining factor.

Whereas DAX1 is a nuclear transcription factor, the *WNT4* gene encodes an intercellular signaling molecule that is involved in several developmental processes, including roles in sexual development. Even before gonadal sex differentiation occurs, WNT4 acts in the development of the Müllerian ducts, which later develop into the fallopian tubes, uterus, and upper vagina, in female embryos. In addition, *WNT4* knockout (KO) mice show partial female-to-male sex reversal. It appears that *WNT4* is involved in balancing the bipotential gonad between either ovary or testis formation (Kim and Capel, 2006). *WNT4* inhibits the formation of testis-specific vascularization in the ovary, but it is not considered an ovary-determining factor.

At this point in time, *FOXL2* is perhaps the most promising candidate gene to play a leading role in ovary development. The encoded protein contains a so-called forkhead box, a protein domain binding to DNA. This domain is also known as the winged helix, based on the butterfly-like appearance of the loops in its protein structure. FOXL2 is implicated in XX gonadal female-to-male sex reversal, and genetic inactivation of the *FOXL2* gene in mice demonstrates a role in granulosa cell differentiation and maintenance of the ovary. This inactivation does not cause a defect in early ovary formation, so that FOXL2 cannot be named an ovary-determining factor. However, FOXL2 appears to have an ongoing role to maintain ovary function by acting on follicle formation and development, thereby counteracting sex reversal activities (Ottolenghi et al., 2005, 2007a,b; Uhlenhaut and Treier, 2006; Wilhelm et al., 2007a,b).

Ovary development from the bipotential gonads can be viewed as a pathway which unfolds in the absence of SRY action. Some genes implicated in female-to-male gonadal sex reversal normally seem to suppress testis development, rather than to exert a primary ovary-determining effect. Perhaps the situation can be best described as follows.

In XX embryos, the developing fetal gonads become populated with mitotic germline cells, forming germ cell cysts which, after some time, undergo a transition toward meiotic prophase (Pepling and Spradling, 2001). The transition to meiotic prophase is activated by retinoic acid signaling (Bowles et al., 2006; Koubova et al., 2006). When these future oocytes approach toward the end of meiotic prophase, an interaction with somatic gonadal cells becomes evident, leading to the formation of primordial follicles in which diplotene-stage prophase oocytes become arrested, enclosed by a layer of future granulosa cells. These primordial follicles form the reservoir of oocytes from which follicles are recruited for growth throughout adult reproductive life. Primordial and growing follicles can be considered as organizers of ovarian tissue structure, also recruiting somatic cells to become steroidogenic theca cells. Without germline cells, chromosomal female gonads will never develop as ovaries, but rather turn into functionless streak tissue. If the germline cells are unable to organize the surrounding cells into follicular structures, a dysgenetic gonad, in which undifferentiated gonadal tissue persists, will develop. In this process, the sex chromosome content of the germline cells is not the key factor. In mouse, XO and even XY germline cells can also form follicles, if these cells are located in a fetal gonad where they are stimulated or admitted to enter meiotic prophase (McLaren, 1995). However, there is increased early loss of such germline cells, as found for oocytes in XO mice (Burgoyne and Baker, 1985). Similarly, in human Turner syndrome, it is most likely an early loss of XO germline cells which results in the formation of a streak gonad. Since ovary development fails when no germline cells are present in the gonads, or when follicle development or maintenance is dysregulated in any other manner (**Figure 1**), it can be suggested that perhaps we should classify several genes that are critically involved in follicle formation and maintenance as ovary-determining genes.

Ovary development fails when no germline cells are present in the gonads, or when follicle development or maintenance is dysregulated in any other manner (**Figure 1**). Oocytes arrested in the diplotene stage of meiotic prophase, in primordial follicles, act as signaling centers. As early as day E13.5, these oocytes express the transcription factor folliculogenesis-specific basic helix–loop–helix factor in the germline alpha (FIGalpha, encoded by the gene Figla). Known target genes for this transcription factor in the oocyte are the genes encoding the zona pellucida proteins ZP1, 2, and 3 (Liang et al., 1997). Moreover, FIGalpha was shown to be essential for recruitment of pre-granulosa cells to form the primordial follicle (Soyal et al., 2000). The transcription factor FOXL2 is found in the pre-granulosa cells, where it plays a role in their differentiation associated with primordial follicle formation (Schmidt et al., 2004; Uda et al., 2004). Together, FIGalpha in the oocyte and FOXL2 in the pre-granulosa cells are critical factors in ovary development. For certain, there are many other critical factors, further downstream in the follicular and ovarian development pathway, such as growth differentiation factor 9 (GDF9) which is also originating from the oocyte acting as signaling centers (Dong et al., 1996). If any ovary-determining gene remains to be identified, it will be acting upstream of FIGalpha and FOXL2.

Nature has found a very effective way to curtail ovary formation in male mammals by expressing the *SRY* gene. Through its actions, the SRY protein forces the indifferent fetal gonads to embark on a tissue differentiation pathway leading to the formation of testis tubules. It turns out that retinoic acid, which signals entry into meiotic prophase in the early ovary, is metabolically inactivated in the fetal testis (Bowles et al., 2006; Koubova et al., 2006). The germline cells present in the primitive gonads will remain in a mitotic stage, not entering into the prophase of meiosis, while they are enclosed in developing testis tubules in close association with the pre-Sertoli cells.

In XX–XY chimeric mouse testis, Sertoli cells are predominantly XY (Palmer and Burgoyne, 1991). Whether human gonads in cases of 45,X/46,XY mosaicism and variants will develop as testes, ovaries, ovotestis, or streak gonads is most likely determined by the percentage of cells carrying a Y chromosome and by the germline karyotypes, although additional mechanisms may add to the complexity of this situation (see Section 26.3.2). Sertoli cells expressing SRY may exert a dominant effect, driving gonadal determination pathways toward testis due to the activation of a cascade of testis-determining genes, such as *SOX9* (see above). Below a threshold of Y-positive cells, 45,X/46,XY gonads will most often develop as a streak, because continuation of ovarian development requires the continuous presence of oocytes, which in humans is incompatible with 45,X or 46,XY germline karyotypes.

26.2.3 Secondary Sex Determination: Gonadal Hormones and the Sexual Phenotype

Gonadal sex determination is followed by sex-specific gonadal hormone production, which then evokes the most evident differences between females and males. Like in all other vertebrates, differences between females and males are effectuated by gonadal hormone production, first during fetal development and then around puberty – when steroid hormone production by the gonads is activated through awakening of the brain–gonadal endocrine axis, leading to the development of dimorphic secondary sexual characteristics.

The first hormone to be produced by the fetal testis is the anti-Müllerian hormone (AMH). Being secreted by the immature Sertoli cells in the testis tubules, AMH reaches the Müllerian ducts, leading to their regression through apoptosis. Hence, males will not develop fallopian tubes, uterus, and upper vagina (Josso et al., 2006). Next, the fetal Leydig cells in the testis interstitial tissue will engage in steroidogenesis, producing testosterone. This androgenic steroid hormone acts on the Wolffian ducts, which are thereby stimulated to develop into epididymides and vasa deferentes (Hannema and Hughes, 2007). Lack of steroidogenesis in the fetal ovary – where the primordial follicles have not yet reached the adult reproductive stage of recruiting cells to become steroidogenic theca cells – will result in regression of the Wolffian ducts in gonadal females, due to lack of stimulation by testosterone. Extragonadal conversion of testosterone to dihydrotestosterone by the enzyme 5α-reductase generates an androgen with higher potency, capable of stimulating prostate development and virilizing the external genitalia. Normal XX gonadal females are not exposed to AMH and androgenic steroid hormones at the fetal stage, so that female development of the internal and external genitalia, with the upper vagina being connected to the vaginal orifice, can take place. The hormones from the fetal testes overrule this female pathway, by counteracting the development of the Müllerian ducts and by effectuating growth of the penis and fusion of folds.

Dysregulation of secondary sex determination of females will occur if the fetal ovaries, or an extragonadal fetal source, produce any of the hormones promoting male sex differentiation. Most importantly, this is found in mutational dysregulation of steroidogenesis by the adrenal glands, leading to improper production of androgens in XX gonadal female fetuses (as discussed below).

Whereas SRY is a male-specific factor, and with the exception of Y-chromosomal genes implicated in spermatogenesis and perhaps some other male-specific functions, all genes involved in the subsequent primary and secondary male sex determination pathway are not male-specific. This holds true for the first downstream gene for SRY action, the autosomal gene *SOX9*. In the mouse, transgenic expression of *Sox9* evokes XX female-to-male sex reversal (Vidal et al., 2001), but *Sox9* is also expressed in normal females at a low level and even in undifferentiated gonads, and is involved in chondrogenesis in both sexes (Barrionuevo et al., 2006). Similarly, all hormones produced by the testes and promoting male sex differentiation are not male-specific and can exert various actions. This is certainly true for AMH and testosterone. The difference between females and males concerns developmental timing and dosage of hormone production, as well as conversion rates to other hormones. During adult reproductive life, ovarian follicles produce AMH, which may play a role in fine-tuning the recruitment of follicles for growth (Durlinger et al., 2002). It is quite remarkable that this hormone, which is highly destructive for Müllerian ducts at the fetal stage and which, therefore, should not at all be produced by fetal XX gonads, becomes a paracrine regulator in the adult ovary. Similarly, testosterone is not produced by the fetal ovary, but growing follicles in the adult reproductive ovary recruit steroidogenic theca cells to produce testosterone as precursor for conversion to estradiol (E_2) by the enzyme CYP19A1.

26.2.4 Sex Differentiation of the Brain: Genes versus Hormones

Experimental research in nonhuman species has shown that manipulating androgens during comparable developmental periods influences fundamental processes of sex differentiation in the body and the brain. Consequently, many behavioral sex differences are also established by these hormones (De Vries and Simerley, 2002; Phoenix et al., 1959). During the last two decades, advances in genetics have broadened the focus of research on sexual differentiation to include the many genes involved in the sexual differentiation of the gonads and the brain (Arnold, 2002; Arnold et al., 2004).

It had already been demonstrated that some regions in the central nervous system (CNS) and some behaviors can be fully sex-reversed by treating females pre- or perinatally with testosterone, or by preventing the action of pre- or perinatal testicular hormones in males. The development and actions of these nervous system regions and these behaviors, therefore, seem fully dependent on the organizational actions of gonadal hormones (Swaab, 2007).

However, not all sex differences could be explained by actions of gonadal hormones and, in some cases, sex differences precede the onset of gonadal secretions, suggesting that these sex differences may be caused by other sexually dimorphic signals, such as sex chromosomes and genes (Arnold et al., 2004).

By interfering with the *Sry* gene responsible for testis differentiation, it became possible to produce gonadally sex-reversed mice; XY mice with ovaries and XX mice with testes. Mice with different complements of sex chromosomes (XX vs. XY) but with the same gonadal sex can be compared to determine whether the sex chromosomes influence specific traits directly, independent from gonadal hormone production (De Vries et al., 2002; Gatewood et al., 2006). XY^- mice (with the *Sry* gene deleted from the Y chromosome) develop ovaries, like normal XX females. Both XY^- and XX mice carrying a *Sry* transgene on an autosome, XY^-*Sry* and XX *Sry* mice, respectively, develop testes like normal XY males. In these genetically manipulated mice, numerous sexual phenotypes have been measured in the brain. Different behaviors including male reproductive behavior and social investigation behavior were recorded. Moreover, the morphology of various sexually dimorphic CNS regions, including the cerebral cortex, hypothalamus, septum, and spinal cord, was studied. All measurements were made in gonadectomized adult mice treated with equal levels of testosterone, or in gonadally intact newborn mice. For most of these variables, XX *Sry* and XY^-*Sry* mice with testes are more masculine than XX and XY^- mice with ovaries. The sex chromosomes induce no group differences in mice with the same gonadal sex, indicating that gonadal hormones are responsible for the induction of these sex differences. However, for some sex differences a sex chromosomal effect has been demonstrated. For instance, the lateral septum of XY normal males and XY^- gonadal females contained more dopamine neurons, as compared to XX normal females and XX *Sry* gonadal males, indicating an effect of XX versus XY on brain, regardless of the gonadal sex (Arnold and Burgoyne, 2004; De Vries et al., 2002; Gatewood et al., 2006; Xu et al., 2002).

26.3 Sex Chromosomal Disorders of Sex Development and Female Development

26.3.1 Incidence and Origin of 45,X/46,XY Mosaicism

Sex chromosome mosaicism (45,X/46,XY and variants) occurs with an estimated incidence of around 1.5/10 000 (Chang et al., 1990) and may be due to loss of the Y chromosome through anaphase lag or to interchromosomal rearrangements with final loss of a structurally abnormal Y chromosome. Mosaicism always arises from errors in mitosis after fertilization has occurred, in contrast to chimerism (the existence of two or more cell lines of different genetic origin in one individual), which arises at or immediately after fertilization (Grumbach et al., 2003). Spontaneous abortion occurs in over 99% of 45,X conceptuses, as opposed to much higher survival rates in embryos mosaic for a second cell line (45,X/46,XX or 45,X/46,XY), which led to the hypothesis that all liveborn 45,X individuals are, in fact, hidden mosaics (Hook and Warburton, 1983). This hypothesis has not been confirmed by molecular genetic studies with X or Y specific probes (Chu and Connor, 1995; Fernandez-Garcia et al., 2000; Gicquel et al., 1992; Grumbach et al., 2003; Kocova et al., 1993; Larsen et al., 1995), although the possibility of prenatal mosaicism with postnatal loss of a structurally abnormal sex chromosome cannot be ruled out.

26.3.2 Phenotypic Spectrum of 45,X/46,XY Mosaicism

The distribution of Y-bearing cell lines in mosaic patients may be markedly different in various tissues of the body (Cools et al., 2007; Petrusevska et al., 1996). As a result, the patient's individual phenotype may range from a normal fertile male to a normal fertile female, but is frequently characterized by ambiguity of internal and/or external genitals, with no correlations found between the phenotypic appearance and the peripheral blood karyotype (Chang et al., 1990; Grumbach et al., 2003; Telvi et al., 1999).

A study on the phenotypic appearance of 76 prenatally diagnosed 45,X/46,XY fetuses revealed a normal male appearance in 72 of them (95%), three males (4%) were diagnosed with hypospadias, and only one patient (1%) with clitoromegaly was assigned a female gender (Chang et al., 1990). Other anomalies encountered in this patient series included: atrial septal defect, ventricular septal defect, Meckel's diverticulum, inguinal hernia, and scrotal hydrocoele. The authors concluded that in prenatally diagnosed 45,X/46,XY mosaicism, most individuals are phenotypically normal males and that the degree of mosaicism in amniotic fluid does not correlate with the degree of genital or gonadal abnormality and thus should not be considered a predictor of phenotype in genetic counseling.

However, if diagnosed postnatally, 45,X/46,XY mosaicism is associated with marked phenotypic variability, including females with Turner syndrome features, children with ambiguous genitals, and normal males (Grumbach et al., 2003; Telvi et al., 1999). Again, no correlation is found between the proportion of 45,X/46,XY cell lines in peripheral blood or fibroblasts and the observed phenotype.

Short stature is a common finding in these patients and appears to be responsive to growth hormone treatment, in males as well as in females, similar to 45,X Turner patients treated with growth hormone (Richter-Unruh et al., 2004). Other Turner stigmata encountered in 45,X/46,XY mosaicism include renal abnormalities, aortic coarctation, cubitus valgus, and low posterior hairline. Mild mental retardation, autism, and facial dysmorphism may also be part of the phenotypic spectrum (Richter-Unruh et al., 2004; Telvi et al., 1999).

Genital ambiguity is typically associated with the presence of testicular tissue, which will be associated with fetal production of AMH and testosterone, causing partial activation of male development pathways, whereas in the case of a total lack of gonadal development and fetal gonadal hormone production (streak gonads), an unambiguously female phenotype will develop. However, genital ambiguity may be discrete, isolated proximal hypospadias in otherwise normal males being the most frequent finding (Chang et al. (1990), Telvi et al. (1999), and personal observations). Mild clitoromegaly in phenotypic females also warrants for the presence of testicular tissue. Therefore, a low threshold to perform karyotyping on peripheral blood, and if possible additionally on skin fibroblasts, in isolated hypospadias is appropriate; a combination of hypospadias and uni- or bilateral chryptorchidism always requires further investigation. We observed a case of penoscrotal hypospadias associated with unilateral cryptorchidsm. Blood karyotype was 46,XY. Following the finding of a utriculus cyst, which is considered to be a Müllerian duct remnant, karyotyping was additionally performed on skin fibroblasts and gonadal tissue, and revealed 45,X/46,XY mosaicism in both tissues (unpublished observation).

26.3.3 Gonadal Histology, Tumor Risk, and Fertility

45,X/46,XY patients diagnosed with Turner syndrome tend to have streak gonads (Grumbach et al. (2003), Telvi et al. (1999), and personal observations), defined as functionless gonadal tissue in which all germ cells are lost. However, spontaneous puberty, fertility, and pregnancy have been reported in rare cases (Dumic et al., 2008; Landin-Wilhelmsen et al., 2004). Uni- or bilateral dysgenetic testes are a consistent finding in patients with genital ambiguity in this group (Chang et al. (1990), Grumbach et al. (2003), Telvi et al. (1999), and personal observations), while presence of ovotestes has also been observed (Chang et al., 1990).

In mice, it was found that whether the gonads will develop as testes, ovaries, or ovotestes is determined by the percentage of cells carrying a Y chromosome. Sertoli cells expressing SRY exert a dominant effect, driving gonadal determination pathways toward testis development, due to the activation of a cascade of testis-determining genes. Lack of testis development in mice gonads can occur when the percentage of Y-positive cells falls below a specific threshold, estimated at around 30%. In this case, 45,X/46,XY gonads most often will develop as an ovary (Burgoyne et al., 1988; Koopman et al., 1991), with possible loss of germ cells due to increased apoptosis afterward (see Section 26.2.2).

Proper testis and ovarian development in men depends on even more complex, probably locally active processes.

The authors of this chapter and others performed detailed studies on the gonadal differentiation patterns in 23 gonads from 16 different patients with 45,X/46,XY karyotype as opposed to six gonads from three 46,XX/46,XY patients and one 46,XX/47,XXY patient (Cools et al. (2007, 2006a,b,c), Pena-Alonso et al. (2005), Petrusevska et al. (1996), and unpublished data). Within most gonads, a combination of various differentiation patterns was found, ranging from normal testes, testis tubules in a background of stromal tissue, undifferentiated gonadal tissue (i.e., tissue that is not differentiated either into testicular nor into ovarian direction, containing germ cells) (Cools et al., 2006a,b,c), normal ovaries (defined by the presence of germ cells that are all enclosed in primordial or growing follicles), and streak gonadal tissue (which may be the end stage of an ovary or of an undifferentiated region in which all germ cells are lost; **Figure 1**). Sometimes, scarce primitive testicular differentiation can be found within overall streak tissue. A normal ovary was only encountered unilaterally in two patients from the second group (one 46,XX/46,XY patient and one 46,XX/47,XXY patient); the other gonads in this group consisted of one gonad with testis and testis in a background of stromal tissue, two ovotestes, and one streak. No (pre) malignant lesion was found in this group. In the 23 gonads from the 45,X/46,XY group, no ovarian tissue (ovary or ovotestis) was encountered, underscoring the importance of an XX cell line for proper ovarian development. 8/23 gonads (35%) were testes, of which two gonads displayed the abnormal pattern of testis tubules in a background of stromal tissue. 3/23 (13%) gonads contained only undifferentiated gonadal tissue, 4/23 (17%) gonads consisted of testicular and undifferentiated gonadal tissue. Furthermore, a streak was found on five occasions (22%) and three gonadoblastomas (see below) (13%) were encountered, two in combination with streak gonadal tissue, one in combination with undifferentiated and testicular tissue. All patients revealing some degree of testicular differentiation were born with ambiguous genitalia, as well as some patients with undifferentiated gonadal tissue and two of three patients in whom a gonadoblastoma was found. *In situ* hybridization studies with X and Y specific probes visualized in individual gonadal cells in this patient series clearly demonstrated that factors other than the absolute number of Y-positive cells must be involved in gonadal differentiation into the male or female direction (Cools et al., 2007). As described in Section 26.2.2, the *SOX9* gene is activated immediately downstream of SRY, and *FOXL2* plays a leading role in ovary development. We demonstrated that local relative expression levels of *SOX9* and *FOXL2* determine the fate of the X/XY gonad (Hersmus et al., 2008).

Dysgenetic gonads are characterized by an increased risk for the development of the so-called type II germ cell tumors (carcinoma *in situ*, gonadoblastoma, and their invasive counterparts (non) seminoma and dysgerminoma) in the presence of (part of) the Y chromosome (Cools et al., 2006a,b,c; Oosterhuis and Looijenga, 2005). The *TSPY* gene, encoding the so-called testis-specific protein on the Y chromosome, is located close to the centromeric region and has been identified as the candidate gene for the gonadoblastoma-susceptible region on the Y chromosome (GBY), required for malignant transformation, which is likely associated with induction of proliferation (Cools et al., 2006a,b,c; Kersemaekers et al., 2005; Lau et al., 2000; Lau, 1999; Page, 1987). Consecutively, it was shown that immature germ cells, blocked in their maturation due to an inappropriate or unsupportive environment, are at highest risk. This is most clearly demonstrated by the presence of the OCT3/4 (POU5F1) protein, which under physiological conditions is only present in primordial germ cells and gonocytes, but is lost upon further maturation (Cools et al., 2006a,b,c, 2005; de Jong et al., 2005; Honecker et al., 2004; Looijenga et al., 2003; Rajpert-De Meyts et al., 2004; Stoop et al., 2005). Thus, gonads with the lowest degree of differentiation (undifferentiated gonadal tissue) are prone to malignant transformation, whereas in highly differentiated testicular tissue, the risk is considerably lower, even in the presence of TSPY. No TSPY expression is noticed in the ovaries or ovarian parts of ovotestes from 46,XX/46,XY patients. Hence, it is anticipated that no germ cell tumors will arise here (Cools et al., 2006a,b,c). Gonadoblastoma is believed to be the *in situ* malignant lesion arising from undifferentiated gonadal tissue, whereas carcinoma *in situ* originates from (dysgenetic) testes (Cools et al., 2006a,b,c, 2005). Thus, in 45,X/46,XY mosaic patients, gonadoblastoma and carcinoma *in situ* are both encountered (Grumbach et al., 2003; Pena-Alonso et al., 2005; Slowikowska-Hilczer et al., 2003). However, due to the higher differentiation grade of testicular tissue, as compared to undifferentiated gonadal tissue, gonadoblastoma is the predominant lesion in these patients (Cools et al., 2006a,b,c). For the same reasons, one could speculate that tumor development is probably more likely in 45,X/46,XY mosaicism as compared to 46,XX/46,XY chimerism or mosaicism (personal observations). Overall, the risk for germ cell tumor development in Y+ mosaic patients is estimated at 12% (Cools et al., 2006a,b,c) although it is to be expected that specific subgroups can be identified with a higher and lower risk.

26.3.4 Diagnosis and Treatment

The diagnosis is established by demonstrating a 45,X/46,XY karyotype in peripheral blood and/or skin fibroblast culture and/or gonadal tissue (Grumbach et al., 2003). As stated above, a low threshold to perform blood karyotyping or eventually fibroblast or gonadal karyotyping is appropriate, even in isolated hypospadias or clitoral hypertrophy. Hypospadias associated with cryptorchidism must always be further examined for the presence of hidden mosaicism. The presence of functional testicular tissue can be demonstrated by an hCG test for the evaluation of Leydig cell function, whereas plasma AMH and inhibin B levels are indicative of Sertoli cell function. Obtained serum levels of AMH, inhibin B, and testosterone must be interpreted in view of the patient's age (Crofton et al., 2002; Grumbach et al., 2003; Rey et al., 1999). Ultrasonography is indicative for the presence or absence of Müllerian structures but the sensitivity of this technique is highly dependent on the examinator's experience. Laparoscopic exploration of internal gonadal structures is usually mandatory. The decision on sex of rearing must be made after a thorough evaluation of prenatal androgen exposure, the possibility to make genitalia consistent with the chosen sex, gonadal function, and fertility chances (Hughes et al., 2006).

In case of male gender assignment, testicles should be brought into a palpable, if possible low scrotal position, to allow follow-up for the development of malignancy by physical examination and ultrasound. In this context, the presence of bilateral microlithiasis might be a specific point of interest. If it is impossible to bring the testes into a scrotal position, they should be removed and testicular prostheses should be placed. Hypospadias should be corrected. The possibility of phalloplastic surgery, analogous to phalloplasty in female-to-male transsexuals, offers new perspectives for patients with micropenis (Hoebeke et al., 2003; Selvaggi et al., 2007).

In case of female gender assignment, removal of a testis, ovotestis, or streak is mandatory to avoid malignancy and eventually the production of cross-sex hormones. Clitoris reduction should be performed by an experienced surgeon, only in cases of severe clitoromegaly. The ideal timing for vaginoplasty (early in life vs. during puberty) is controversial. Vaginal dilatation should not be undertaken until puberty (Bondy and Turner Syndrome Study, 2007; Grumbach et al., 2003; Hughes et al., 2006). Detailed analysis of gonadal biopsy material for the presence of premalignant changes in DSD gonads (Cools et al., 2006a,b,c, 2005; Honecker et al., 2004; Kersemaekers et al., 2005; Stoop et al., 2005) allowed avoidance of gonadectomy in some cases. However, this technique is limited due to the fact that a biopsy is only partially representative for the gonad as a whole. We observed a case of 46,XX/47,XXY mosaicism, born with ambiguous genitalia, who was assigned a female gender. Laparoscopy and bilateral gonadal biopsies at 2 years of age revealed the likely presence of a normal ovary and hemiuterus on one side and a testicle on the other. After removal of the testicle, which turned out to be an ovotestis after complete evaluation, we performed hCG testing in combination with AMH and inhibin B screening. The results confirmed the absence of remaining functional testicular tissue; hence, it was decided to leave the ovary in place. However, based on this information, we are unable to exclude the presence of undifferentiated parts in the remaining gonad, containing a high risk for the development of gonadoblastoma. The need for hormonal replacement therapy (androgens or estrogens) depends on the capacity of the remaining gonad(s) to produce sex hormones.

26.4 Disorders of Androgen Excess

The placenta acts in a protective way by converting androgens to estrogens (Wang et al., 2005) but androgenization of the female fetus will occur if the capacity of the placental CYP19A1 is exceeded.

In female fetuses, exposure to androgens during the critical 9–14 gestational weeks period may lead to variable degrees of masculinization. Expression of the androgen receptor in the female genitalia diminishes after the first trimester, and becomes limited to the clitoris, which thus maintains the potential for androgen-dependent growth (Hanley and Arlt, 2006; Shapiro et al., 2000). Maternal hyperandrogenemia beyond the critical period of the 9th–14th week will lead to clitoral hypertrophy without affecting the labioscrotal folds.

Androgen excess can be of maternal origin (see below) but results most often from adrenal androgen production of the fetus.

26.4.1 Fetal Origin

Goto et al. (2006) have studied early adrenal function and its significance for androgen-mediated sex development. As early as 50–52 days post-conception (dpc), there is already expression of all key elements of steroidogenesis in the fetal adrenal gland, including STAT, CYPIIAI, CYP17AI, CYP21A2, CYPIIBI, and CYPIIB2. It was demonstrated that, at the same time, early cortisol biosynthesis is facilitated by transient adrenal expression of nerve growth factor IB (NGFIB) involved in HSD3B2 regulation (Goto et al., 2006). Moreover, the important elements of the hypothalamus–pituitary–adrenal feedback system, such as pituitary expression of adrenocorticotropic hormone (ACTH) and the glucocorticoid receptor (GR), are also present at that stage. There is compelling evidence for the human adrenal as a source of active androgens during fetal development. Thus, in the physiological situation, cortisol is capable of attenuating ACTH stimulation of the adrenal and subsequent upregulation of androgen production, thereby acting as a safeguard mechanism for normal female sexual development (**Figure 2**). Support for the loss-of-feedback model is provided by clinical evidence, from the use of the potent synthetic glucocorticoid (GC) dexamethasone, which crosses the placenta, in women carrying a fetus with CYP21 deficiency. If administered by 8 weeks of gestation, dexamethasone can restore normal female differentiation of the external genitalia through resumption of adequate negative feedback at the anterior pituitary (Forest, 2004).

26.4.1.1 21-Hydroxylase deficiency

The clinical manifestations of congenital adrenal hyperplasia (CAH) indicate the significance of the distinct steroidogenic enzymes in the process of normal sex differentiation. As extensively reviewed elsewhere, the adrenal gland plays a key role in human development (Krone et al., 2007b; Merke and Bornstein, 2005; Speiser and White, 2003). Steroid 21-hydroxylase (CYP21A2) deficiency is most common and accounts for about 90–95% of all cases. The classical form of CYP21A2 deficiency occurs with an incidence of 1:7000 to 1:15 000 (Speiser and White, 2003). CYP21A2 is almost exclusively expressed in the adrenal cortex and facilitates two key steps in human GC and mineralocorticoid biosynthesis: the conversion of progesterone to 11-desoxycorticosterone and 17-hydroxyprogesterone to 11-desoxycortisol (**Figures 3** and **4**).

26.4.1.2 11-Beta hydroxylase deficiency

In most populations, 11-beta hydroxylase (CYP11B1) deficiency comprises approximately 5–8% of cases of CAH and occurs in 1:100 000 births. The clinical manifestations are similar to CYP21A2 deficiency, that is, masculinization of the external genitalia in the 46,XX female individual. In addition, hypertension is seen in about two-thirds of the patients, probably resulting from increased levels of corticosterone or its metabolites having mineralocorticoid activity (White et al., 1994). 11-Desoxycortisol and desoxycorticosterone are not effectively converted to cortisol and corticosterone, respectively.

The *CYP11B1* gene is located on chromosome 8q21–22, and gene mutations in patients with the classic form fully abolish enzymatic activity (Curnow et al., 1993).

26.4.1.3 Steroidogenic acute regulatory protein mutations

Mutations in the gene encoding for the steroidogenic acute regulatory protein (StAR) are the cause of a very rare form of CAH termed lipoid CAH (Bose et al., 1996). About 30 46,XX patients have been described (Bhangoo et al., 2007). Steroidogenesis is deficient in both adrenals and gonads and, therefore, the disorder is characterized by adrenal insufficiency and female external genitalia in both sexes. In 46,XX individuals the ovary is initially spared from damage as steroidogenesis is delayed until the time of puberty. In the course of time, the patients develop irregular menses, anovulatory cycles and/or ovarian

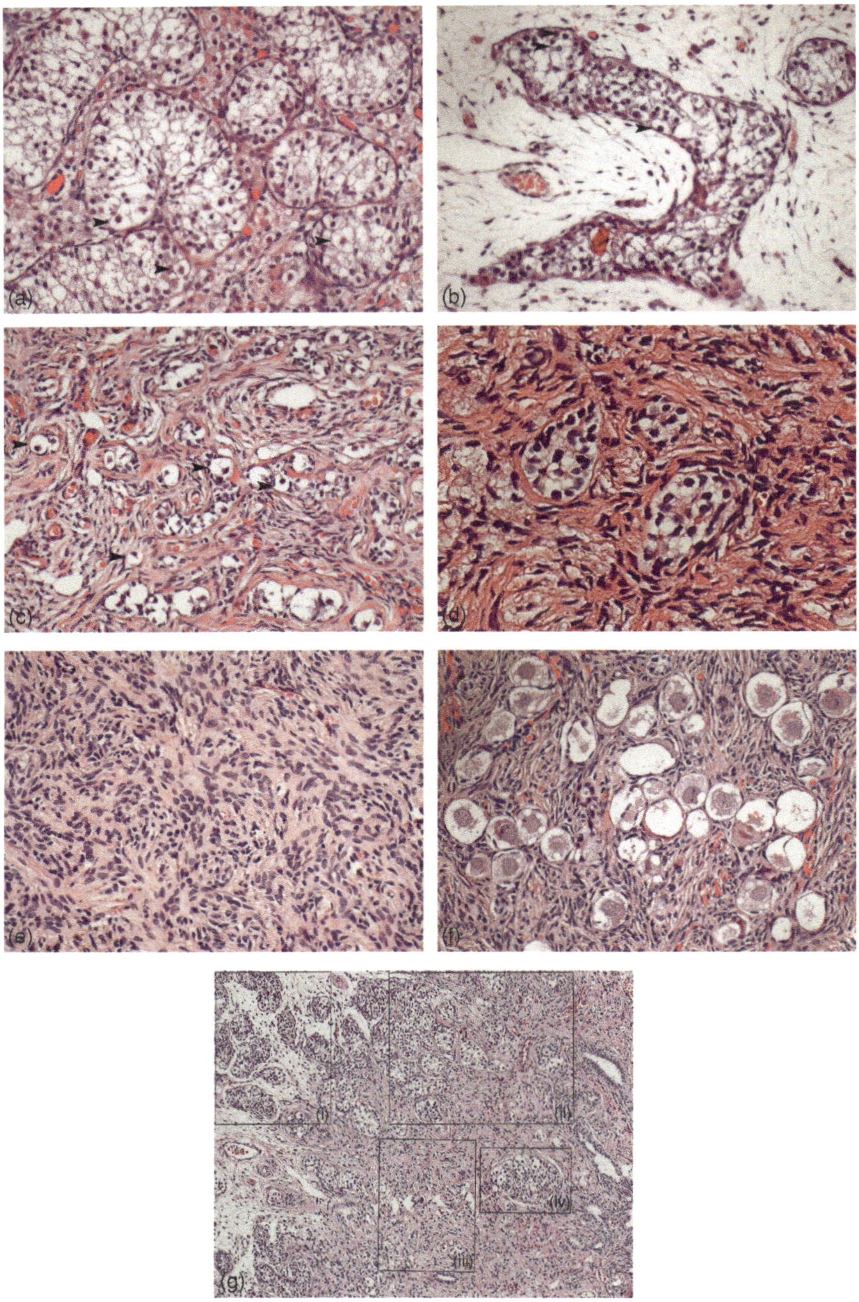

Figure 2 (a) Area with normal testis differentiation. The testis tubules contain germ cells (arrows). Patient karyotype: 45,X, inv(5) (q22q33.2); 46,X i(Y9), inv(5) (q22q33.2). HE staining, 200X. (b) Area with markedly dysgenetic testis tubules. The testis tubules contain germ cells (arrows). Same patient as IA. HE staining, 200X. (c) Area with undifferentiated gonadal tissue. Germ cells (arrows) line up with Sertoli/granulosa cells to form cord-like structures that are not differentiated into testis tubules, nor do they form ovarian follicular structures. Patient karyotype: 45,X/46,XY. HE staining, 200×. (d) Testis tubules devoid of germ cells in a background of ovarian stroma. Patient karyotype 45,X/46,XY. HE staining, 400×. (e) Area with streak gonadal tissue. Same patient as ID. HE staining, 200×. (f) Ovarian differentiation pattern, showing germ cells enclosed in primordial follicles. Patient karyotype: 46,XX/47,XXY. HE staining, 200×. (g) Combination of various gonadal differentiation patterns within a limited area: (i) Testicular tissue. (ii) Testis differentiation in a background of ovarian stroma. (iii) Undifferentiated gonadal tissue. (iv) Gonadoblastoma. Patient karyotype: 45,X/46,XY. HE staining, 100×.

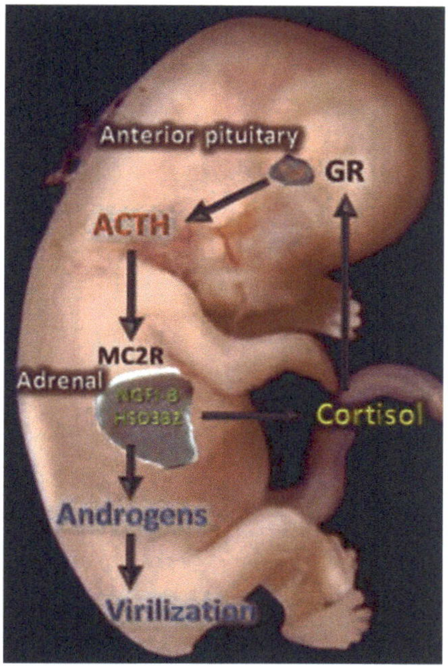

Figure 3 Human early adrenal function and its implications for androgen-mediated development according to Goto et al. (2006). Early cortisol biosynthesis is facilitated by transient adrenal expression of NGFI-B and HSD3B2. The subsequent increase in cortisol production increases the feedback to the anterior pituitary corticotroph, diminishes adrenocorticotrophic hormone (ACTH) production, and thereby reduces the ACTH-driven androgen secretion. GR, glucocorticoid receptor; MC2R, Melanocortin 2 (ACTH) receptor. This mechanism acts a safeguard during the crucial period of female sexual development. Reprinted from Krone N, Hanley NA, and Arlt W (2007b) Age-specific changes in sex steroid biosynthesis and sex development. *Best Practice and Research Clinical Endocrinology and Metabolism* 21: 393–401, with permission from Elsevier.

cysts, progressive ovarian failure, and hypergonadotropic hypogonadism (Bhangoo et al., 2005; Bose et al., 1997; Chen et al., 2005).

26.4.1.4 17-Alpha-hydroxylase and 21-hydroxylase deficiency

In 1985, a new clinical entity was described with apparent combined 17-alpha-hydroxylase (CYP17A1) and CYP21A2 deficiency (Peterson et al., 1985). Initially it was suggested that this disorder would be caused by concurrent mutations in the *CYP17A1* and *CYP21A2* genes, but subsequent studies failed to reveal any mutations in these genes.

More recently, two study groups reported independently that this disease was caused by inactivating mutations in P450 oxidoreductase (POR), a key electron donor enzyme providing electrons to all P450 (CYP) enzymes (Arlt et al., 2004; Dhir et al., 2007; Fluck et al., 2004; Krone et al., 2007a).

It is of specific interest that 46,XX individuals with POR deficiency present with ambiguous genitalia. In this case of 46,XX DSD there is increased exposure to androgens during fetal development, but postnatal androgen levels are normal or even low (Arlt, 2007; Arlt et al., 2004). This apparent contradiction has been proposed to be explained by the existence of an alternative pathway in human androgen synthesis, present only in the human fetal and early neonatal life and circumventing the classical sex steroid biosynthetic pathway (see **Figure 1**). After birth, the alternative pathway is no longer active, possibly due to a change in expression pattern of 5 alpha-reductase type 1 versus type 2.

The *POR* gene is located on the long arm of chromosome 7 (7q11.2). It consists of 15 exons spanning a region of 32.9 kb. The cDNA consists of 2043 bp encoding for a protein of 681 amino acids. A variety of inactivating mutations have been reported (Arlt, 2007).

26.4.1.5 CYP17A1/17,20-lyase deficiency

Cytochrome p450 CYP17A1 encoded by the *CYP17A1* gene, catalyzes two separate enzymatic reactions during steroid hormone biosynthesis. It hydroxylates progesterone or pregnenolone at the 17 carbon, and it converts 21-carbon steroids to 19-carbon steroids such as androstenedione or DHEA by its 17,20-lyase activity. CYP17A1 deficiencies are caused by mutations of the *CYP17A1* gene and account for approximately 1% of all cases of CAH. Its incidence is estimated at about 1:50 000 newborns. Since cloning of the *CYP17* gene, over 40 different mutations have been reported (Costa-Santos et al., 2004a,b; Hahm et al., 2004). Lack of gene transcription or mutational loss of enzymatic activity may cause steroid hormone-based disorders such as hypertension and hypokalemia. In a 46,XX individual, the occurrence of pubertal arrest or primary amenorrhoea depends on enzymatic rest function, as 5–8% of 17,20-lyase activity may be sufficient for pubertal development (Costa-Santos et al., 2004b) .

26.4.1.6 Glucocorticoid resistance

GC resistance (GRe) is an autosomal recessive or dominant condition characterized by generalized, partial, target-tissue insensitivity to GCs. Compensatory elevations in circulating ACTH concentrations

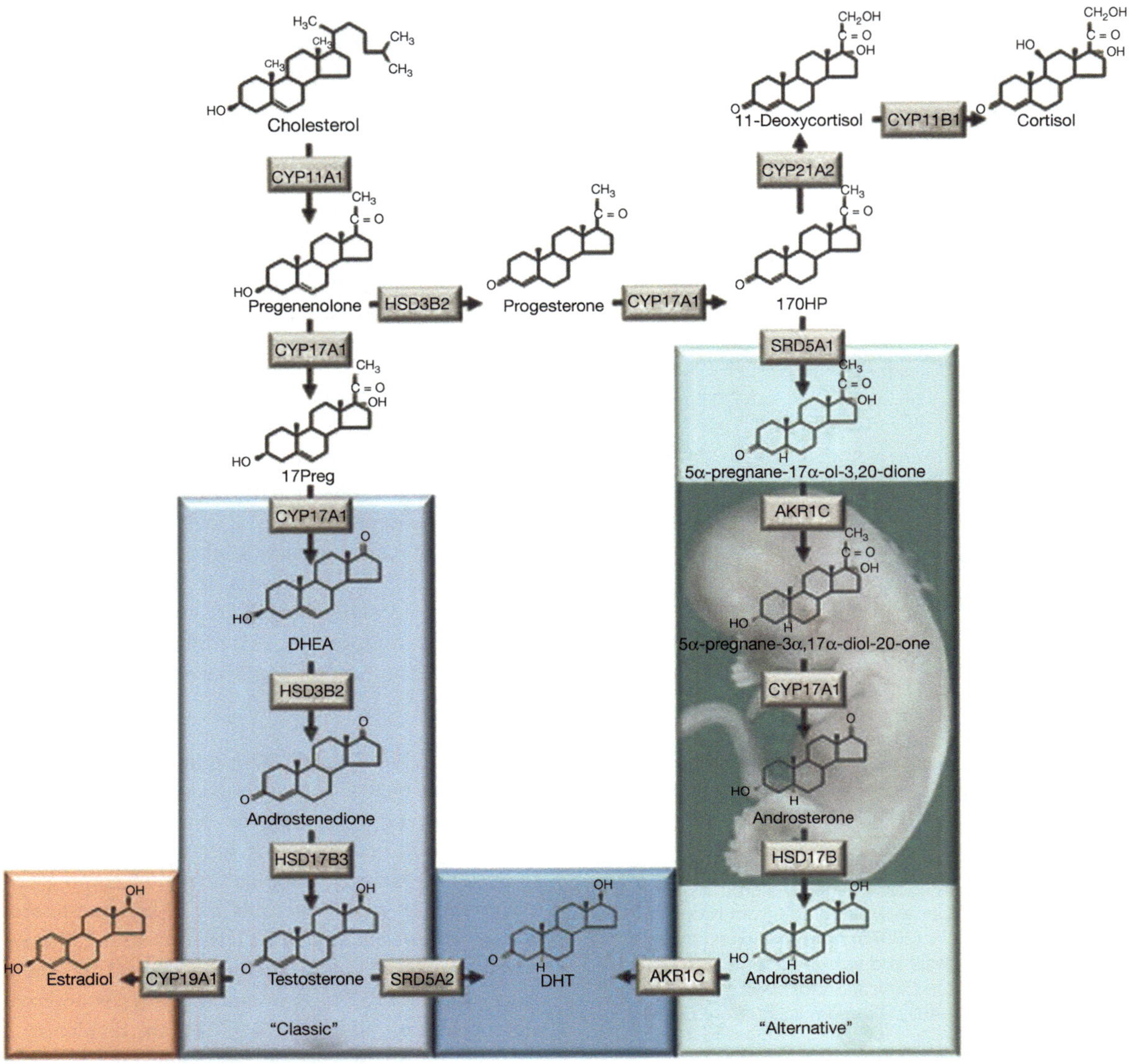

Figure 4 Classic and alternative pathways of steroidogenesis as proposed by Krone et al. (2007b). The biosynthetic pathways toward androgens are thought to be relevant to the testis and the adrenal cortex, whereas cortisol biosynthesis is confined to the adrenal. The classical androgen pathway is boxed in blue. The alternative pathway is given in light blue and is thought to be present in human fetal life only. The functionally active end product of both pathways is dihydrotestosterone (DHT). The pink box illustrates possible inactivation of testosterone (and androstenedione) toward estrogens by P450 aromatase (CYP19A1). Most importantly, 17-hydroxyprogesterone (17OHP) can act as the first metabolite for the alternative pathway. The pathway toward human mineralocorticoids is not shown. CYP11A1, P450 side-chain cleavage enzyme; HSD3B2, 3β-hydroxysteroid dehydrogenase type 2; CYP17A1, 17α-hydroxylase; CYP21A2, 21-hydroxylase; CYP11B1, 11β-hydroxylase; HSD17B3, 17β-hydroxysteroid dehydrogenase type 3; CYP19A1: P450 aromatase; SRD5A2, 5α-reductase type 2; SRD5A1, 5α-reductase type 1; AKR1C, aldo-keto reductase with 3α-hydroxysteroid dehydrogenase activity. Reproduced from Krone N, Hanley NA, and Arlt W (2007b) Age-specific changes in sex steroid biosynthesis and sex development. *Best Practice and Research Clinical Endocrinology and Metabolism* 21: 393–401, with permission from Elsevier.

lead to increased secretion of cortisol and adrenal steroids with mineralocorticoid and/or androgenic activity, but no clinical evidence of hypercortisolism but with resistance of the hypothalamic–pituitary–adrenal axis to dexamethasone suppression. As a consequence, patients present with signs of adrenal overproduction of mineralocorticoids (hypertension and hypokalemic alkalosis) and, in females, of androgens (ambiguous genitalia in newborn, precocious puberty, acne, hirsutism, infertility, male-pattern hair loss, and menstrual irregularities) (van Rossum and Lamberts, 2006). The molecular basis of GRe in several kindreds has been ascribed to mutations in the GR which belongs to the nuclear receptor superfamily (Hollenberg et al., 1985). Two GR isoforms were identified from alternative splicing, a ligand-binding isoform, GRα and a dominant negative non-ligand-binding isoform, GRβ. Inactivating mutations have been described within the ligand-binding and DNA-binding domain of the receptor as well a 4-bp deletion in exon 6 (Charmandari et al., 2007, 2006, 2004, 2005; Hurley et al., 1991; Karl et al., 1993, 1996; Kino et al., 2002; Mendonca et al., 2002) but in a number of patients the cause of GC resistance has not yet been elucidated (Huizenga et al., 2000). Interindividual variation in GC-sensitivity can be partly explained by polymorphisms in the *GR* gene (Russcher et al., 2005). The ER22/23EK and N363S polymorphisms have been described to be associated with lower and higher GC sensitivity, respectively. There are no recommendations for gender assignment for markedly virilized 46,XX females due to *GR* polymorphisms. Evidence supports a current recommendation to raise 46,XX newborn due to CAH as females; therefore, these girls may also be assigned a female gender (Clayton et al., 2002; Dessens et al., 2005). GRe is very rare, and especially therapeutic experience in childhood in this disorder is limited and experimental (Malchoff et al., 1994).

26.4.2 Fetoplacental Origin

26.4.2.1 *Aromatase deficiency*

CYP19A1 (15q21.2) is a cytochrome P450 enzyme ($P450_{arom}$) that plays a crucial role in the biosynthesis of estrogens (C18 steroids) from androgens (C19 steroids) in all vertebrate species. In humans, CYP19 expression is regulated by different tissue-specific promoters in the placenta, ovary, breast, bone, adipose tissue, vascular endothelium, and brain, resulting in systemic (gonadal/ovarian) and local (extragonadal) estrogen production in these tissues.

During pregnancy, androgens produced by the fetal adrenal gland, then sulfated and aromatized by the placenta, are the major sources of circulating estrogens. In the absence of $P450_{arom}$, the androgenic steroids (androstenedione, testosterone, and 16α-hydroxydehydroepiandrosterone) cannot be converted to estrone (E_1), estradiol (E_2), and estriol (E_3), and, therefore, large quantities of androstenedione and testosterone are transferred to the maternal and fetal circulation which results in an apparent rare cause of 46,XX disorder of sexual development (46, XX DSD) with masculinization of the urogenital sinus and external genitalia and in a progressive virilization of the mother during pregnancy which resolves after delivery. Importantly, about 1% of the normal placental activity of the wild-type $P450_{arom}$ seems to be enough to prevent virilization of the mother. Therefore, in the affected females so far, the exposure to a variable excess of androgens resulted in a distinct virilization of the mother and in masculinization of the fetal external genitalia (see **Table 2**). Furthermore, the lower the CYP19A1 activity, the greater was the degree of masculinization of the external genitalia at birth (Mullis et al., 1997). Later on, no estrogens can be produced resulting in delayed or absence of puberty, tall stature, and reduced bone mineral density.

The human *CYP19* gene encoding for the P450 enzyme consists of nine exons (exons 2 to 10). Upstream of exon 2 a number of alternative first exons is located that contain different tissue-specific promoters, although the translated protein product remains the same in all tissues. Since 1992, 15 well-documented cases harboring a mutation of the *CYP19* gene have been reported, including eight females (Belgorosky et al., 2003; Carani et al., 1997; Conte et al., 1994; Deladoey et al., 1999; Harada et al., 1992; Herrmann et al., 2002; Lin et al., 2007; Ludwig et al., 1998; Morishima et al., 1995; Mullis et al., 1997; Portrat-Doyen et al., 1996) (Richter-Unruh unpublished observations). These individuals were all born with ambiguous genitalia and assigned female in infancy. Plastic surgery and hormone replacement in puberty and adulthood seem to be absolutely vital. Therapeutic experience in childhood is limited because of the rareness of disease.

26.4.3 Maternal Origin

26.4.3.1 *Luteoma of pregnancy*

Pregnancy luteomas are distinctive hyperplastic lesions arising in pregnancy, mimicking ovarian tumors (Feldmann et al., 1991; Jenkins et al., 1968; Schmitt et al., 1990). Most often, they are asymptomatic and are only discovered incidentally, for instance, during a cesarean section or postpartum tubal sterilization; but maternal pregnancy luteomas may be responsible for virilization of both, a newborn and the mother (Spitzer et al., 2007; Wang et al., 2005). It is of interest to note that pregnancy luteomas constitute a benign condition that tends to regress spontaneously after delivery. As a rare occurrence,

Table 2 Genotype–phenotype correlation in 46,XX DSD due to aromatase deficiency

*Genotype**	*Phenotype*	*Virilization of the mother*	*$P450_{arom}$ activity (% of normal activity)*	*Publication*
Splice site exon/intron 6: insert of 29 amino acids at the donor (homoz.)	46,XX-DSD, clitoris hypertrophy, complete fusion of the labios crotal fold	Virilization during the third trimester	0.3%	Harada et al. (1992)
Exon 10: R435C (heteroz.) and C437Y (heteroz.)	46,XX-DSD, Prader stage III–IV	No virilization	1.1	Conte et al. (1994)
Exon 9, V370M (homoz.)	46,XX-DSD, Prader V	At 12th week	Not determined	Ludwig et al. (1998)
Exon 9 R375C (homoz.)	46,XX-DSD, Prader IV–V	At the 5th month	0.2	Morishima et al. (1995)
Exon 10, R457X (homoz.)	46,XX-DSD, Prader IV	Not reported	Not determined	Portrat-Doyen et al. (1996)
Splice site exon/intron 3 (heteroz.) and exon 9: P408X (heteroz.)	46,XX-DSD, Prader V	At 12th week	0	Mullis et al. (1997)
Exon 5: donor splice site (heteroz.) and exon 9: G412X (heteroz.)	46,XX-DSD, Prader IV	At second trimester	Not determined	Belgorosky et al. (2003)
Exon 3: M85R (heteroz.) and donor splice site exon 6 (heteroz.)	46,XX-DSD, Prader IV	At 12th week	0	Richter-Unruh, unpublished observations

a 46,XX girl has been described with marked virilization not only due to a maternal luteoma but also due to fetal CYP21A2 deficiency (Warmann et al., 2000). Similarly, virilization may very rarely occur as a result of a late onset non-salt-losing CYP21A2 deficiency of the mother apparently sufficiently mild to allow a normal menstrual cycle and pregnancy (Kai et al., 1979). Few reports relate recurrent maternal virilization to hyperthecosis or polycystic ovarian syndrome (PCOS) (Ben-Chetrit and Greenblatt, 1995; Holt et al., 2005; Sarlis et al., 1999).

26.5 Müllerian Agenesis/Hypoplasia Syndromes

The Mayer–Rokitansky–Küster–Hauser syndrome (MRKH) is a congenital malformation of the female internal genitalia resulting from an interrupted embryonic development of the Müllerian ducts. It occurs in 1:4000 female live births (Oppelt et al., 2006). The molecular genetic basis has not been elucidated. No mutations of the genes encoding AMH or the AMH type 2 receptor (AMHR2) have been found in MRKH patients (Zenteno et al., 2004).

However, in one patient with features of MRKH and additional androgen excess a *WNT4* mutation was established (Biason-Lauber et al., 2004). In a subsequent study of the same authors, out of six patients with different degrees of Müllerian abnormalities one additional patient was found with a novel loss-of-function dominant negative *WNT4* gene mutation also exhibiting androgen excess (Biason-Lauber et al., 2007). Thus, it was suggested that these rare patients may represent a clinical entity distinct from the classical MRKH syndrome. In fact, the typical MRKH syndrome is limited to vaginal aplasia and uterine hypoplasia in 64% of the cases. In the atypical form there may be malformations in the ovary or renal system. When additional associated malformations of the renal system and skeleton are present, the term Müllerian duct aplasia, renal aplasia, and cervicothoracic somite dysplasia (MURCS) has been proposed (Oppelt et al., 2006). McKusick–Kaufman syndrome (MKKS) is another rare syndrome comprised of multiple congenital anomalies, including postaxial polydactyly, congenital heart malformation, and hydrometrocolpos (Behera et al., 2005). It is more frequent in females and is inherited in an autosomal-recessive pattern. The most commonly reported genital malformations include distal vaginal agenesis and transverse vaginal membranes, giving rise to hydrometrocolpos. Uterine anomalies are more characteristic of a similar syndrome, Bardet–Biedl syndrome (BBS), having significant overlap with MKKS, because they are allelic to each other. BBS is characterized by retinitis pigmentosa, postaxial polydactyly, learning disability, childhood central obesity, and hypogenitalism. A wide variety of genital anomalies have been reported in women with BBS, including hypoplastic uterus, vaginal atresia, and hypoplastic fallopian tubes and ovaries. Absence of the vaginal and urethral orifice has also been reported (Behera et al., 2005).

26.6 Effects of Gonadal Steroids on Brain and Behavior

26.6.1 Role of Pre- and Postnatal Androgen Exposure

Experimental research in nonhuman species has shown that manipulating androgens during comparable developmental periods influences fundamental processes of neural development and differentiation, including cell survival neural connectivity and neurochemical characterization (De Vries and Simerley, 2002). During a critical period, the developing neuroendocrine system and the brain are sensitive to influences of gonadal hormones. In this period, the amount of circulating androgens will sculpt the developing fetal brain in male or female direction. The amounts of circulating perinatal androgens will also determine future sex dimorphic behaviors; low levels of androgens lead to demasculinization or feminization of behaviors in both males and females, whereas high levels of androgens lead to masculinization or defeminization of behaviors (Wallen, 2005). Studies in various species revealed that gonadal hormones have some behavioral influence in all species studied to date, although specific aspects of the effects, such as the range of behaviors influenced, the time when hormones are influential, or the specific hormones involved, differ from species to species (Arnold, 2002; De Vries and Simerley, 2002; Wallen, 2005). Most studies on the organizational effects of steroids on sexual differentiation of behavior have been carried out in altricial species that require a long period of nursing and who are born prior to complete neural differentiation; guinea pigs and nonhuman primates are the only precocial mammalian species whose sexual differentiation has been extensively studied. From these studies we know that precocial and altricial species differ in the timing of sexual differentiation and in the role of estrogenic metabolites of androgens in sexual differentiation (Wallen, 2005).

In human beings, naturally occurring abnormalities in prenatal gonadal hormone secretion are seen in patients with DSD. Most studies have been carried out in 46,XX females with CAH as CAH is the most common disorder of 46,XX DSD. Studies in 46,XX females with CAH on gender role behavior, aggression, and activity level, revealed that it is likely that several of the hormonal mechanisms that establish different types of sex dimorphic behaviors in animals are also applicable to human beings. However, the neural systems that regulate these altered behaviors are largely unknown.

26.6.2 Effects of Androgens on Sexuality

Human sexuality is considered multidimensional and includes at least three components: (1) gender role behavior, that is, behavior that is culturally associated with gender or that shows sex difference, (2) sexual orientation, defined in terms of the sex of preferred erotic partners, both in fantasy and in behavior, and (3) gender identity, defined as the sense of self as being male or female (Hines, 2004).

26.6.2.1 Gender role behavior

Studies on gender role behavior revealed that CAH females are more masculine in their gender role behavior than nonaffected females (Berenbaum, 1999; Collaer and Hines, 1995; Hines, 2004). For example, young CAH girls more often like boys as playmates, like to play with boy's toys, and like games that are typical for boys, such as games that involve a lot of rough-and-tumble play and require a high activity level. Adolescent girls and adult women show elevated interest in male-typical activities and careers and reduced interest in female-typical activities and careers (Berenbaum, 1999; Collaer and Hines, 1995; Meyer-Bahlburg et al., 2006). Dittmann et al. (1990) observed that adult females with the salt-wasting type of CAH were less interested in having their own children compared to simple virilizing CAH females and their nonaffected sisters. The degree of masculinity in behavior is related to the type of CAH. Females with the salt-wasting type of CAH produce larger amounts of androgens compared to girls with the simple virilizing type. In general, salt-wasting CAH females are born with more severely masculinized external genitalia. These salt-wasting CAH females are more masculine in their behavior compared to the simple-virilizing CAH females (Berenbaum et al., 2000; Collaer and

Hines, 1995). Adult females with the late onset or nonclassical type of CAH do not differ from unaffected females in their gender role behavior (Meyer-Bahlburg et al., 2006). The masculine preferences of CAH females are not influenced by parental permissiveness or encouragement of masculine behavior (Berenbaum and Hines, 1993; Pasterski et al., 2005; Slijper, 1984) as studies that compared the gender role behavior of CAH girls with the gender role behavior of nonaffected sisters and nonaffected control girls demonstrated that nonaffected sisters, and nonaffected controls girls had equal feminine scores on the rating scales, whereas the scores of the CAH girls were significantly more masculine (Pasterski et al., 2005; Slijper, 1984). Most parents in the study of Pasterski et al. (2005) encouraged feminine behavior but encouragement could not override their daughter's interest in boy's toys.

These findings in the gender role behavior in CAH females correspond well with findings in non-primate and primate research. It, therefore, can be concluded that also gender role behavior in humans and the brain structures that are involved are probably programmed by prenatal androgens. In non-CAH boys, testosterone peaks are between week 12 and 18 and from week 34 of gestational age into the first 3 months after birth. In males and females with CAH exposure to excessive amounts of androgens may start from week 6 when the adrenal fetal cortex starts to secrete steroids (Nimkarn and New, 2006). As CAH males do not differ from non-CAH boys with respect to masculinity of their gender role behavior, these findings suggest that programming of the human brain for human gender role behavior does not start before week 12 but probably takes place in these two peak periods. The fair correlation between the severity of masculinization of the external genitalia and masculinization of gender role behavior in CAH females suggests that the establishment of gender role behavior will start during the first testosterone peak. During this peak, testes of males and adrenals of CAH females probably produce sufficient androgens to establish male gender role behavior. Studies in children who had been prenatally exposed to mild increases of adrenal androgens and in females with nonclassical CAH, in whom the excessive production of androgens started postnatally, indicate that the hormonal masculinization of these kind of gender role behaviors takes place prenatally (Berenbaum et al., 2000; Meyer-Bahlburg et al., 2006).

26.6.2.2 Sexual orientation and sexual functioning

At present, 19 studies on sexual orientation in adult CAH women have been carried out. (Dessens et al., 2005; Meyer-Bahlburg et al., 2008). The total numbers of women who participated in these studies has been 604; of these females, 193 had the salt-wasting type of CAH, 128 had the simple-virilizing type, and 95 had the nonclassical type. Of the remaining 188 patients, no specification of the CAH diagnosis had been reported. Two studies (Ehrhardt et al., 1968; Lev-Ran, 1974) included females who had been diagnosed in adolescence or adulthood; treatment had started upon diagnosis. The women in these studies all probably had the simple-virilizing type as women with the salt-wasting type probably would not have survived. The ages of the women who participated in these studies ranged from 16 to 71. In eight studies, a control group had been included (Meyer-Bahlburg et al., 2008). Six studies reported decreased general sexual activity and decreased heterosexual activity and interest (Dittmann et al., 1990; Hines, 2004; Kuhnle and Bullinger, 1997; Slijper et al., 1992; Wisniewski et al., 2004; Zucker et al., 1996). In other studies, which included variables on fantasy and experiences, the number of women who reported bisexual or homosexual fantasy and experience was increased compared to findings in control groups (Dittmann et al., 1990; Ehrhardt, 1979; Ehrhardt et al., 1968; Gastaud et al., 2007; Guth et al., 2006; Hines et al., 2004; May et al., 1996; Meyer-Bahlburg et al., 2008; Money et al., 1984; Stikkelbroeck et al., 2003; Zucker et al., 1996). However, compared to the reported bisexual and homosexual romantic and erotic imagery, the percentages of women who were homosexually active or considered themselves lesbians was much smaller (Meyer-Bahlburg et al., 2008). The findings in these studies indicate that the majority of CAH women are heterosexual, but bisexual and homosexual orientation is increased (Meyer-Bahlburg et al., 2008). The study of Meyer-Bahlburg et al. (2008) is the first study that included a large number of females with nonclassical CAH. This study revealed that women with all types of CAH showed increased percentages of homosexual crushes, love, and genital sex and homosexual orientation. With respect to homosexual orientation, females with salt-wasting CAH scored highest in homosexual direction, followed by the group of females with simple-virilizing CAH. Females with the nonclassical type had lowest scores in the homosexual direction.

The CAH females in this sample did not differ from controls with respect to reaching the psychosexual milestones.

At this point it can be concluded that females with CAH display a decreased heterosexual activity and interest and increased rates on homosexual fantasy and experience. The number of females that consider themselves lesbian is small. Homosexual orientation is most frequently seen among females with the salt-wasting type, whereas homosexual orientation seems less frequent among females with the nonclassical form of CAH, suggesting a dose–response relationship of androgen excess on sexual orientation. As described in Section 26.6.2, behavioral-endocrinology research in nonprimate mammals demonstrated a profound organizational influence of sex hormones during early developmental periods on later mating behavior and sexual orientation. Perinatal androgens and estrogens (derived by aromatization of androgens within brain cells) support the development of masculine behavior whereas estrogens support the defeminization of behavior, followed by activating effects of sex-specific hormones from puberty onward (Arnold, 2002; Wallen and Baum, 2002). The findings on sexual activity and homosexual orientation in CAH females have been explained by prenatal androgen exposure (Zucker et al., 1996). In females with the nonclassical type of CAH, a mild androgen excess may already be present during the fetal stage. Meyer-Bahlburg et al. (2008) give two alternative explanations for these findings: (1) in females, homosexual orientation can be prenatally established in the developing brain by mild androgen excess that will be insufficient to affect the genitalia and (2) in females, homosexual orientation can be established postnatally by mild but persistent androgens excess. Bearman and Brückner (2002) proposed an alternative explanation. These authors suggest an interaction of social and biological factors: the associated delay of romantic practice in adolescence, perhaps also supported by less stereotypic feminine leisure-time activities and, for some women, relative isolation in the peer group, might increase the change into a bisexual/homosexual development rather than, exclusively, the direct effect of androgens on brain circuits that regulate sexual behavior.

These explanations do not explain the observed decreased sexual activity. Recently, Hines (2004) proposed that the pattern of hormonal exposure caused by CAH decreases interest in men (defeminization) as sexual partners without increasing interest in women (masculinization). With respect to the observed decreased heterosexual activity, alternative explanations can be offered. Decreased heterosexual activity can be a consequence of insufficient GC replacement or treatment interruptions, both leading to virilization. Overtreatment will lead to obesity. Both virilization and obesity may reduce attractivity to men and romantic approaches by men. Decreased heterosexual activity can also be explained as a consequence of postponement of romantic engagement in adolescence. Postponing romantic engagement may lead to a small developmental delay that may be difficult to catch up later in life. Young adults who lack these experiences may feel embarrassed by their lack of experiences and hesitate to contact a potential partner.

Despite early genital feminization, surgery does not always produce good results. In two studies (Minto et al., 2003; Mulaikal et al., 1987) females reported nonsensitivity and inability to reach orgasm. Moreover, the introitus was not adequate for sexual intercourse, and there was discomfort or pain during intercourse. Crouch et al. (2008) observed that penetration difficulties and low intercourse frequency were related to feminizing genital surgery. This study revealed a linear relationship between impairment to sensitivity and severity of sexual difficulties. So lack of sexual experience or heterosexual experience could result from vaginal inadequacy or pain at intercourse. In addition, body image concerns and doubts whether her vagina is adequate for intercourse may inhibit the adolescent girl to date and experiment with romantic and sexual relations.

26.6.2.3 Gender identity

Studies on gender identity in CAH females revealed that the vast majority (about 93%) of females have a female gender identity (Dessens et al., 2005; Gupta et al., 2006; Julka et al., 2006; Reiner, 2005; Richter-Appelt et al., 2005). A small group of patients report an ambiguous gender identity (0.5%), a male gender identity (1%), or gender dysphoria (4%). Only 1.8% had chosen to change their social gender role and to live as a man. However, from these figures we can also conclude that compared to nonaffected women, CAH women have an increased risk to develop a severe gender dysphoria: 7 out of 388 CAH women compared to 1 out of 30 400 diagnosed male-to-female transsexuals (Bakker et al., 1993). Contrary to the findings in gender role behavior, no relationship has been found between genital masculinization at birth (which is presumed to reflect the amounts of prenatally produced androgens) and the change in social gender role (Meyer-Bahlburg et al.,

1996; Zucker et al., 1996). No factors leading to social gender role change could be specified (Meyer-Bahlburg et al., 1996).

At present, the relationship between the degree of prenatal androgen exposure in 46,XX females and gender identity remains less clear than the relationship between prenatal exposure of androgens and gender role behavior. The brain structures involved in gender identity seem less sensitive to the influence of prenatal androgens compared to gender role behavior. From this, one might assume that gender identity is established early in development, before week 6–7 and that only genetic factors are involved. However, the finding that CAH females have an increased risk to develop an ambiguous gender identity or gender dysphoria suggests some involvement of gonadal hormones in the establishment of gender identity. The 46,XX CAH females who changed their social gender role into male had different degrees of masculinization of their external genitalia at birth (Prader stages 2–4) (Meyer-Bahlburg et al., 1996; Zucker et al., 1996). The external genitalia develop in weeks 9–14 of gestational age. From this, one might assume that involvement of androgens in the establishment of gender identity might be during the second peak of testosterone production in non-CAH males (gestational week 34–3 months of age), when the external genitalia have already developed.

26.6.3 Roles of Androgens on Activity

Normative studies of children show higher activity levels in boys than in girls. A study in a large group of 46,XX CAH females and their unaffected sisters aged 6–11 revealed that CAH girls are more active than their sisters, in particular, the younger CAH girls were most active (Pasterski et al., 2007). These results suggest that prenatal exposure to elevated levels of androgen also masculinizes activity level.

26.6.4 Roles of Androgens on Aggression

Rodent males, as well as men, are generally more aggressive than females, and this sex difference has been linked to prenatal (organizational) and pubertal and postpubertal (activational) actions of androgens (Simon, 2002). Studies on aggressive behavior in children, adolescents, and adults produced mixed results, which might be due to methodological limitations, such as small sample sizes, retrospective reporting, reliance on parental reports, and not including control groups (Berenbaum and Resnick, 1997; Ehrhardt et al., 1968; Ehrhardt and Baker, 1974; Money and Schwartz, 1976; Pasterski et al., 2007), and studied 113 American and British CAH girls and boys between the age of 3 and 11 as well as their unaffected sisters and brothers and found that the mothers of these children reported more aggressive behaviors of their CAH daughters compared to their nonaffected daughters. There were no differences regarding aggression between CAH boys and their brothers. As in studies on toy and playmate preference, these studies showed that early influence of androgens on aggression is manifest regardless of postpubertal hormone activation. The locus for hormonal influences could be in the medial amygdaloid nucleus (MA). High density of the androgen receptor has been found in the MA, and the MA exhibits structural and sexual dimorphism both in rodents and in humans (Pasterski et al., 2007).

26.6.5 Role of Androgens on Cognitive Capacities

Overall intelligence does not differ for males and females, but males are typically somewhat better in visuospatial skills, in particular tasks requiring mental rotation through 3-D space, spatial perception tasks, and quantitative problem solving. Females are better in verbal or associational fluency, perceptual speed, and arithmethic computation. It has been assumed that sex differences in visuospatial skills could be induced by the higher levels of prenatally produced androgens in males. This hypothesis has been tested in CAH females. These studies revealed that CAH females got equivalent intellectual abilities compared to well-matched control females (Collaer and Hines, 1995). With respect to verbal skills and perceptual speed accuracy CAH females received similar scores as controls. CAH females perform better on spatial tasks than do controls, suggesting that early androgens affect the development of visuospatial cognition (Puts et al., 2008). Johannsen et al. (2006) observed impaired intellectual abilities in adult CAH women. Decreased scores were particularly evident in salt-wasting CAH. Low IQ scores were related to hyponatremic crises. Several explanations have been given for these findings such as prenatal exposure to low levels of GCs, high levels of androgens, and suboptimal hormone replacement resulting in hyponatremia. Episodes of hypotension and hyponatremia may cause injury to the brain. To mothers who have an increased risk for giving birth to a CAH child, dexamethasone is prescribed during pregnancy in order to prevent prenatal genital masculinization. Follow-up studies up to 12 years of age on cognitive and motor development did not document any adverse effects of early dexamethasone treatment (Meyer-Bahlburg et al., 2004). However, Hirvikoski et al. (2007) suggest that prenatal dexamethasone treatment might be associated with previously not described long-term effects on verbal working memory and on certain aspects of self-perception that could be related to poorer verbal working memory.

26.6.6 Role of Prenatally Elevated Amounts of Estrogens on Behavior

Estrogens, perinatally administered to genetic female rodents, promote male-typical neural and behavioral differentiation. Female rodents perinatally treated with estrogens show enhanced male typical behavior and decreased female typical behaviors (Collaer and Hines, 1995). In the past, synthetic estrogens have been prescribed to pregnant women. Studies on sexual orientation in females prenatally exposed to the diethylstilbestrol (DES) suggested an increased bisexuality or homosexuality (Ehrhardt et al., 1985; Meyer-Bahlburg et al., 1995). However, the effects of these hormones are not universal; despite the increase in bisexuality and homosexuality, the majority of DES-exposed women are heterosexual. These findings in DES-exposed females show similarities with the findings in CAH females and indicate a defeminizing or masculinizing effect on sexual orientation. No effects of prenatal exposure of DES have been found on the development of gender identity.

A shift in male direction have been found in girls who had been prenatally exposed to androgenic progestins; on an aggression inventory, exposed girls more often choose physical aggressive responses than control girls did (Reinisch, 1981).

Prenatal masculinization is also seen in 46,XX females born with CYP19A1 deficiency. (see Section 26.4.2.1.). There are no reports of studies of the condition on the behavior and cognitive capacities of these females (Jones et al., 2007).

26.6.7 Concluding Remarks

Men and women consistently show differences in certain sexual behaviors, activity level, and aggression. Animal research has demonstrated that the action of prenatal gonadal hormones plays an important role in the establishment of these so-called sex dimorphic behaviors. Investigations in girls and women who have been prenatally exposed to excessive amounts of androgens or estrogens revealed that these hormonal changes cause masculinization of gender role behavior, activity level, and aggression. These findings correspond well with results in animal research. The findings on gender role behavior, activity level, and aggression in CAH females indicate that the excess of adrenal androgens are potent enough to change these behaviors into a male direction. This finding suggests that, during fetal programming, these three types of behavior are equally sensitive to steroid action. The observed dose–response relationship indicates that androgens have a strong influence on the establishment of these kind of behaviors.

CAH women show decreased heterosexual activity and increased rates on homosexual fantasy and experience. The majority of CAH women consider themselves heterosexual, but the proportion of women who consider themselves lesbian is raised compared to the population mean. These findings suggest that in many CAH women the prenatal exposure to adrenal androgens lead to defeminization of sexual activity in adulthood. As homosexual orientation is most frequently seen among women who had been prenatally exposed to the highest amounts of androgens, it can be assumed that for masculinization of sexual orientation a more intensive exposure to prenatal androgens is needed: large amounts of androgens over prolonged period of time, which is probably the case in those females with the salt-wasting type and with the most severely masculinized external genitalia. However, alternative explanations for the observed decreased sexual activity in these women have been formulated and these hypotheses need to be tested before it can be concluded that masculinized or defeminized sexual activity is caused by prenatal androgen action or, alternatively, by vaginal inadequacy.

The majority of CAH females identify themselves as females, but raised proportions of women experience gender dysphoria or identify themselves as males. Interpretation of the role of prenatal androgens on gender identity in these females remains most difficult. From these findings in CAH females, it seems likely that prenatal androgens contribute to the establishment of a defeminized or masculinized gender identity, but the present data do not give us a clue how the defeminization of masculinization of gender identity will be established. We do not know why some women develop gender identity problems whereas others do not; variables in which gender dysphoric women differ from nongender dysphoric women have not been detected yet.

The findings on gender role behavior, sexual orientation, and gender identity suggest that these three aspects of sexuality are different types of sex-specific behaviors and that the establishment of these aspects of sexuality are separate processes.

References

Albrecht KH and Eicher EM (2001) Evidence that sry is expressed in pre-sertoli cells and sertoli and granulosa cells have a common precursor. *Developmental Biology* 240: 92–107.

Arlt W (2007) P450 oxidoreductase deficiency and Antley–Bixler syndrome. *Reviews in Endocrine and Metabolic Disorders* 8: 301–307.

Arlt W, Walker EA, Draper N, et al. (2004) Congenital adrenal hyperplasia caused by mutant P450 oxidoreductase and human androgen synthesis: Analytical study. *Lancet* 363: 2128–2135.

Arnold AP (2002) Concepts of genetic and hormonal induction of vertebrate sexual differentiation in the twentieth century, with special reference to the brain. In: Pfaff DW, Arnold AP, Etgen AM, Fahrback SE, and Rubin RT (eds.) *Hormones, Brain and Behaviour,* vol. IV, pp. 105–135. Amsterdam: Academic Press (Elsevier Science).

Arnold AP (2004) Sex chromosomes and brain gender. *Nature Reviews Neuroscience* 5: 701–708.

Arnold AP and Burgoyne PS (2004) Are XX and XY brain cells intrinsically different? *Trends in Endocrinology and Metabolism* 15: 6–11.

Arnold AP, Xu J, Grisham W, Chen X, Kim YH, and Itoh Y (2004) Minireview: Sex chromosomes and brain sexual differentiation. *Endocrinology* 145: 1057–1062.

Bakker A, van Kesteren PJ, Gooren LJ, and Bezemer PD (1993) The prevalence of transsexualism in The Netherlands. *Acta Psychiatrica Scandinavica* 87: 237–238.

Barrionuevo F, Taketo MM, Scherer G, and Kispert A (2006) Sox9 is required for notochord maintenance in mice. *Developmental Biology* 295: 128–140.

Bearman PS and Brückner H (2002) Opposite-sex twins and adolescents same-sex attraction. *American Journal of Sociology* 107: 1179–1205.

Behera M, Couchman G, Walmer D, and Price TM (2005) Mullerian agenesis and thrombocytopenia absent radius syndrome: A case report and review of syndromes associated with Mullerian agenesis. *Obstetrical and Gynecological Survey* 60: 453–461.

Belgorosky A, Pepe C, Marino R, Guercio G, Saraco N, Vaiani E, and Rivarola MA (2003) Hypothalamic–pituitary–ovarian axis during infancy, early and late prepuberty in an aromatase-deficient girl who is a compound heterocygote for two new point mutations of the CYP19 gene. *Journal of Clinical Endocrinology and Metabolism* 88: 5127–5131.

Ben-Chetrit A and Greenblatt EM (1995) Recurrent maternal virilization during pregnancy associated with polycystic ovarian syndrome: A case report and review of the literature. *Human Reproduction* 10: 3057–3060.

Berenbaum SA (1999) Effects of early androgens on sex-typed activities and interests in adolescents with congenital adrenal hyperplasia. *Hormones and Behavior* 35: 102–110.

Berenbaum SA, Duck SC, and Bryk K (2000) Behavioral effects of prenatal versus postnatal androgen excess in children with 21-hydroxylase-deficient congenital adrenal hyperplasia. *Journal of Clinical Endocrinology and Metabolism* 85: 727–733.

Berenbaum SA and Hines M (1993) Early androgens are related to childhood sex-typed toy preferences. *Psychological Science* 3: 203–206.

Berenbaum SA and Resnick SM (1997) Early androgen effects on aggression in children and adults with congenital adrenal hyperplasia. *Psychoneuroendocrinology* 22: 505–515.

Berta P, Hawkins JR, Sinclair AH, Taylor A, Griffiths BL, Goodfellow PN, and Fellous M (1990) Genetic evidence equating SRY and the testis-determining factor. *Nature* 348: 448–450.

Beverdam A and Koopman P (2006) Expression profiling of purified mouse gonadal somatic cells during the critical time window of sex determination reveals novel candidate genes for human sexual dysgenesis syndromes. *Human Molecular Genetics* 15: 417–431.

Bhangoo A, Buyuk E, Oktay K, and Ten S (2007) Phenotypic features of 46,XX Females with StAR protein mutations. *Pediatric Endocrinology Reviews* 5: 633–641.

Bhangoo A, Gu WX, Pavlakis S, Anhalt H, Heier L, Ten S, and Jameson JL (2005) Phenotypic features associated with mutations in steroidogenic acute regulatory protein. *Journal of Clinical Endocrinology and Metabolism* 90: 6303–6309.

Biason-Lauber A, DeFilippo G, Konrad D, Scarano G, Nazzaro A, and Schoenle EJ (2007) WNT4 deficiency – a clinical phenotype distinct from the classic Mayer-Rokitansky-Kuster-Hauser syndrome: A case report. *Human Reproduction* 22: 224–229.

Biason-Lauber A, Konrad D, Navratil F, and Schoenle EJ (2004) A WNT4 mutation associated with Mullerian-duct regression and virilization in a 46,XX woman. *New England Journal of Medicine* 351: 792–798.

Bondy CA and Turner Syndrome Study Group (2007) Care of girls and women with Turner syndrome: A guideline of the Turner Syndrome Study Group. *Journal of Clinical Endocrinology and Metabolism* 92: 10–25.

Bose HS, Pescovitz OH, and Miller WL (1997) Spontaneous feminization in a 46,XX female patient with congenital lipoid adrenal hyperplasia due to a homozygous frameshift mutation in the steroidogenic acute regulatory protein. *Journal of Clinical Endocrinology and Metabolism* 82: 1511–1515.

Bose HS, Sugawara T, Strauss JF, III, and Miller WL (1996) The pathophysiology and genetics of congenital lipoid adrenal hyperplasia. International Congenital Lipoid Adrenal Hyperplasia Consortium. *New England Journal of Medicine* 335: 1870–1878.

Bowles J, Knight D, Smith C, et al. (2006) Retinoid signaling determines germ cell fate in mice. *Science* 312: 596–600.

Burgoyne PS and Baker TG (1985) Perinatal oocyte loss in XO mice and its implications for the aetiology of gonadal dysgenesis in XO women. *Journal of Reproduction and Fertility* 75: 633–645.

Burgoyne PS, Buehr M, Koopman P, Rossant J, and McLaren A (1988) Cell-autonomous action of the testis-determining gene: Sertoli cells are exclusively XY in XX–XY chimeric mouse testes. *Development* 102: 443–450.

Capel B, Albrecht KH, Washburn LL, and Eicher EM (1999) Migration of mesonephric cells into the mammalian gonad depends on Sry. *Mechanisms of Development* 84: 127–131.

Carani C, Qin K, Simoni M, et al. (1997) Effect of testosterone and estradiol in a man with aromatase deficiency. *New England Journal of Medicine* 337: 91–95.

Chang HJ, Clark RD, and Bachman H (1990) The phenotype of 45,X/46,XY mosaicism: An analysis of 92 prenatally

diagnosed cases. *American Journal of Human Genetics* 46: 156–167.
Charmandari E, Kino T, Ichijo T, Jubiz W, Mejia L, Zachman K, and Chrousos GP (2007) A novel point mutation in helix 11 of the ligand-binding domain of the human glucocorticoid receptor gene causing generalized glucocorticoid resistance. *Journal of Clinical Endocrinology and Metabolism* 92: 3986–3990.
Charmandari E, Kino T, Ichijo T, Zachman K, Alatsatianos A, and Chrousos GP (2006) Functional characterization of the natural human glucocorticoid receptor (hGR) mutants hGRalphaR477H and hGRalphaG679S associated with generalized glucocorticoid resistance. *Journal of Clinical Endocrinology and Metabolism* 91: 1535–1543.
Charmandari E, Kino T, Souvatzoglou E, Vottero A, Bhattacharyya N, and Chrousos GP (2004) Natural glucocorticoid receptor mutants causing generalized glucocorticoid resistance: Molecular genotype, genetic transmission, and clinical phenotype. *Journal of Clinical Endocrinology and Metabolism* 89: 1939–1949.
Charmandari E, Raji A, Kino T, Ichijo T, Tiulpakov A, Zachman K, and Chrousos GP (2005) A novel point mutation in the ligand-binding domain (LBD) of the human glucocorticoid receptor (hGR) causing generalized glucocorticoid resistance: The importance of the C terminus of hGR LBD in conferring transactivational activity. *Journal of Clinical Endocrinology and Metabolism* 90: 3696–3705.
Chen X, Baker BY, Abduljabbar MA, and Miller WL (2005) A genetic isolate of congenital lipoid adrenal hyperplasia with atypical clinical findings. *Journal of Clinical Endocrinology and Metabolism* 90: 835–840.
Chu CE and Connor JM (1995) Molecular biology of Turner's syndrome. *Archives of Disease in Childhood* 72: 285–286.
Clayton PE, Miller WL, Oberfield SE, Ritzen EM, Sippell WG, Speiser PW, and Group ELCW (2002) Consensus statement on 21-hydroxylase deficiency from the European Society for Paediatric Endocrinology and the Lawson Wilkins Pediatric Endocrine Society. *Hormone Research* 58: 188–195.
Collaer ML and Hines M (1995) Human behavioral sex differences: A role for gonadal hormones during early development? *Psychological Bulletin* 118: 55–107.
Conte FA, Grumbach MM, Ito Y, Fisher CR, and Simpson ER (1994) A syndrome of female pseudohermaphrodism, hypergonadotropic hypogonadism, and multicystic ovaries associated with missense mutations in the gene encoding aromatase (P450arom). *Journal of Clinical Endocrinology and Metabolism* 78: 1287–1292.
Cools M, Boter M, van Gurp R, Stoop H, et al. (2007) Impact of the Y-containing cell line on histological differentiation patterns in dysgenetic gonads. *Clinical Endocrinology (Oxf)* 67: 184–192.
Cools M, Drop SL, Wolffenbuttel KP, Oosterhuis JW, and Looijenga LH (2006a) Germ cell tumors in the intersex gonad: Old paths, new directions, moving frontiers. *Endocrine Reviews* 27: 468–484.
Cools M, Honecker F, Stoop H, et al. (2006b) Maturation delay of germ cells in fetuses with trisomy 21 results in increased risk for the development of testicular germ cell tumors. *Human Pathology* 37: 101–111.
Cools M, Stoop H, Kersemaekers AM, et al. (2006c) Gonadoblastoma arising in undifferentiated gonadal tissue within dysgenetic gonads. *Journal of Clinical Endocrinology and Metabolism* 91: 2404–2413.
Cools M, van Aerde K, Kersemaekers AM, et al. (2005) Morphological and immunohistochemical differences between gonadal maturation delay and early germ cell neoplasia in patients with undervirilization syndromes. *Journal of Clinical Endocrinology and Metabolism* 90: 5295–5303.
Costa-Santos M, Kater CE, Auchus RJ, and Brazilian Congenital Adrenal Hyperplasia Multicenter Study Group (2004a) Two prevalent CYP17 mutations and genotype-phenotype correlations in 24 Brazilian patients with 17-hydroxylase deficiency. *Journal of Clinical Endocrinology and Metabolism* 89: 49–60.
Costa-Santos M, Kater CE, Dias EP, and Auchus RJ (2004b) Two intronic mutations cause 17-hydroxylase deficiency by disrupting splice acceptor sites: Direct demonstration of aberrant splicing and absent enzyme activity by expression of the entire CYP17 gene in HEK-293 cells. *Journal of Clinical Endocrinology and Metabolism* 89: 43–48.
Crofton PM, Evans AE, Groome NP, Taylor MR, Holland CV, and Kelnar CJ (2002) Inhibin B in boys from birth to adulthood: Relationship with age, pubertal stage, FSH and testosterone. *Clinical Endocrinology (Oxf)* 56: 215–221.
Crouch NS, Liao LM, Woodhouse CR, Conway GS, and Creighton SM (2008) Sexual function and genital sensitivity following feminizing genitoplasty for congenital adrenal hyperplasia. *Journal of Urology* 179: 634–638.
Curnow KM, Slutsker L, Vitek J, et al. (1993) Mutations in the CYP11B1 gene causing congenital adrenal hyperplasia and hypertension cluster in exons 6, 7, and 8. *Proceedings of the National Academy of Sciences of the United States of America* 90: 4552–4556.
de Jong J, Stoop H, Dohle GR, et al. (2005) Diagnostic value of OCT3/4 for pre-invasive and invasive testicular germ cell tumours. *Journal of Pathology* 206: 242–249.
De Vries GJ, Rissman EF, Simerly RB, et al. (2002) A model system for study of sex chromosome effects on sexually dimorphic neural and behavioral traits. *Journal of Neuroscience* 22: 9005–9014.
De Vries GJ and Simerley RB (2002) Anatomy, development and function of sexually dimorphic neural circuits in the mammalian brain. In: Pfaff DW, Arnold AP, Etgen A, Fahrback SE, and Rubin RT (eds.) *Hormones, Brain and Behaviour,* vol. IV, pp. 137–191. Amsterdam: Academic Press.
Deladoey J, Fluck C, Bex M, Yoshimura N, Harada N, and Mullis PE (1999) Aromatase deficiency caused by a novel P450arom gene mutation: Impact of absent estrogen production on serum gonadotropin concentration in a boy. *Journal of Clinical Endocrinology and Metabolism* 84: 4050–4054.
Dessens AB, Slijper FM, and Drop SL (2005) Gender dysphoria and gender change in chromosomal females with congenital adrenal hyperplasia. *Archives of Sexual Behavior* 34: 389–397.
Dhir V, Ivison HE, Krone N, Shackleton CH, Doherty AJ, Stewart PM, and Arlt W (2007) Differential inhibition of CYP17A1 and CYP21A2 activities by the P450 oxidoreductase mutant A287P. *Molecular Endocrinology* 21: 1958–1968.
Dittmann RW, Kappes MH, Kappes ME, Borger D, Stegner H, Willig RH, and Wallis H (1990) Congenital adrenal hyperplasia. I: Gender-related behavior and attitudes in female patients and sisters. *Psychoneuroendocrinology* 15: 401–420.
Dong J, Albertini DF, Nishimori K, Kumar TR, Lu N, and Matzuk MM (1996) Growth differentiation factor-9 is required during early ovarian folliculogenesis. *Nature* 383: 531–535.
Dumic M, Lin-Su K, Leibel NI, et al. (2008) Report of fertility in a woman with a predominantly 46,XY karyotype in a family with multiple disorders of sexual development. *Journal of Clinical Endocrinology and Metabolism* 93: 182–189.
Durlinger AL, Gruijters MJ, Kramer P, et al. (2002) Anti-Mullerian hormone inhibits initiation of primordial follicle growth in the mouse ovary. *Endocrinology* 143: 1076–1084.

Ehrhardt AA (1979) Psychosocial adjustment in adolescence in patients with congenital abnormalities of their sex organs. In: Vallet HL and Porter IH (eds.) *Genetic Mechanisms of Sexual Development*, pp. 473–483. New York: Academic Press.

Ehrhardt AA and Baker SW (1974) Fetal androgens, human central nervous system differentiation, and behavior sex differences. In: Friedman RC, Richart RM, and van der Wiele RL (eds.) *Sex Differences and Behavior*, pp. 33–52. New York: Wiley.

Ehrhardt AA, Epstein R, and Money J (1968) Fetal androgens and female gender identity in the early-treated adrenogenital syndrome. *Johns Hopkins Medical Journal* 122: 160–167.

Ehrhardt AA, Meyer-Bahlburg HF, Rosen LR, Feldman JF, Veridiano NP, Zimmerman I, and McEwen BS (1985) Sexual orientation after prenatal exposure to exogenous estrogen. *Archives of Sexual Behavior* 14: 57–77.

Ezaz T, Stiglec R, Veyrunes F, and Marshall Graves JA (2006) Relationships between vertebrate ZW and XY sex chromosome systems. *Current Biology* 16: R736–R743.

Feldmann A, Placke J, Brunnert K, and Schulz H (1991) (Pregnancy luteoma, a differential diagnostic problem) Schwangerschaftsluteome, ein differentialdiagnostisches Problem. *Geburtshilfe und Frauenheilkunde* 51: 146–147.

Fernandez-Garcia R, Garcia-Doval S, Costoya S, and Pasaro E (2000) Analysis of sex chromosome aneuploidy in 41 patients with Turner syndrome: A study of 'hidden' mosaicism. *Clinical Genetics* 58: 201–208.

Fluck CE, Tajima T, Pandey AV, et al. (2004) Mutant P450 oxidoreductase causes disordered steroidogenesis with and without Antley-Bixler syndrome. *Nature Genetics* 36: 228–230.

Ford CE, Jones KW, Polani PE, De Almeida JC, and Briggs JH (1959) A sex-chromosome anomaly in a case of gonadal dysgenesis (Turner's syndrome). *Lancet* 1: 711–713.

Forest MG (2004) Recent advances in the diagnosis and management of congenital adrenal hyperplasia due to 21-hydroxylase deficiency. *Human Reproduction Update* 10: 469–485.

Gastaud F, Bouvattier C, Duranteau L, Brauner R, Thibaud E, Kutten F, and Bougneres P (2007) Impaired sexual and reproductive outcomes in women with classical forms of congenital adrenal hyperplasia. *Journal of Clinical Endocrinology and Metabolism* 92: 1391–1396.

Gatewood JD, Wills A, Shetty S, Xu J, Arnold AP, Burgoyne PS, and Rissman EF (2006) Sex chromosome complement and gonadal sex influence aggressive and parental behaviors in mice. *Journal of Neuroscience* 26: 2335–2342.

Gicquel C, Cabrol S, Schneid H, Girard F, and LeBouc Y (1992) Molecular diagnosis of Turner's syndrome. *Journal of Medical Genetics* 29: 547–551.

Glickman SE, Short RV, and Renfree MB (2005) Sexual differentiation in three unconventional mammals: Spotted hyenas, elephants and tammar wallabies. *Hormones and Behavior* 48: 403–417.

Goto M, Piper Hanley K, Marcos J, et al. (2006) In humans, early cortisol biosynthesis provides a mechanism to safeguard female sexual development. *Journal of Clinical Investigation* 116: 953–960.

Graves JA (2006) Sex chromosome specialization and degeneration in mammals. *Cell* 124: 901–914.

Grumbach MM, Hughes IA, and Conte FA (2003) Disorders of sex differentiation. In: Larsen PR, Kronenberg HM, Melmed S, and Polonsky KM (eds.) *Williams Textbook of Endocrinology*, pp. 842–1002. Philadelphia, PA: Saunders (Elsevier).

Gupta DK, Shilpa S, Amini AC, Gupta M, Aggarwal G, Deepika G, and Kamlesh K (2006) Congenital adrenal hyperplasia: Long-term evaluation of feminizing genitoplasty and psychosocial aspects. *Pediatric Surgery International* 22: 905–909.

Guth LJ, Witchel RI, Witchel SF, and Lee PA (2006) Relationships, sexuality, gender identity, gender roles and self concept of individuals who have congenital and adrenal hyperplasia: A quantitative investigation. *Journal of Gay and Lesbian Psychotherapy* 10: 57–75.

Hahm JR, Jung TS, Byun SY, Lee YN, Lee KH, and Kim DR (2004) Functional characterization of mutant CYP17 genes isolated from a 17 alpha-hydroxylase/17,20-lyase-deficient patient. *Metabolism* 53: 1527–1531.

Hanley NA and Arlt W (2006) The human fetal adrenal cortex and the window of sexual differentiation. *Trends in Endocrinology and Metabolism* 17: 391–397.

Hannema SE and Hughes IA (2007) Regulation of Wolffian duct development. *Hormone Research* 67: 142–151.

Harada N, Ogawa H, Shozu M, Yamada K, Suhara K, Nishida E, and Takagi Y (1992) Biochemical and molecular genetic analyses on placental aromatase (P-450AROM) deficiency. *Journal of Biological Chemistry* 267: 4781–4785.

Hawkins JR, Koopman P, and Berta P (1991) Testis-determining factor and Y-linked sex reversal. *Current Opinion in Genetics and Development* 1: 30–33.

Herrmann BL, Saller B, Janssen OE, et al. (2002) Impact of estrogen replacement therapy in a male with congenital aromatase deficiency caused by a novel mutation in the CYP19 gene. *Journal of Clinical Endocrinology and Metabolism* 87: 5476–5484.

Hersmus R, Kalfa N, deLeeuw B, et al. (2008) FOXL2 and SOX9 as parameters of female and male gonadal differentiation in patients with various forms of Disorders of Sex Development (DSD). *Journal of Pathology* 215: 31–38.

Hines M (2004) *Brain Gender.* New York: Oxford University Press.

Hines M, Brook C, and Conway GS (2004) Androgen and psychosexual development: Core gender identity, sexual orientation and recalled childhood gender role behavior in women and men with congenital adrenal hyperplasia (CAH). *Journal of Sex Research* 41: 75–81.

Hirvikoski T, Nordenstrom A, Lindholm T, Lindblad F, Ritzen EM, Wedell A, and Lajic S (2007) Cognitive functions in children at risk for congenital adrenal hyperplasia treated prenatally with dexamethasone. *Journal of Clinical Endocrinology and Metabolism* 92: 542–548.

Hoebeke P, deCuypere G, Ceulemans P, and Monstrey S (2003) Obtaining rigidity in total phalloplasty: Experience with 35 patients. *Journal of Urology* 169: 221–223.

Hollenberg SM, Weinberger C, Ong ES, et al. (1985) Primary structure and expression of a functional human glucocorticoid receptor cDNA. *Nature* 318: 635–641.

Holt HB, Medbak S, Kirk D, Guirgis R, Hughes I, Cummings MH, and Meeking DR (2005) Recurrent severe hyperandrogenism during pregnancy: A case report. *Journal of Clinical Patholology* 58: 439–442.

Honecker F, Stoop H, de Krijger RR, ChrisLau YF, Bokemeyer C, and Looijenga LH (2004) Pathobiological implications of the expression of markers of testicular carcinoma *in situ* by fetal germ cells. *Journal of Pathology* 203: 849–857.

Hook EB and Warburton D (1983) The distribution of chromosomal genotypes associated with Turner's syndrome: Livebirth prevalence rates and evidence for diminished fetal mortality and severity in genotypes associated with structural X abnormalities or mosaicism. *Human Genetics* 64: 24–27.

Hughes IA, Houk C, Ahmed SF, Lee PA, Group LC, and Group EC (2006) Consensus statement on management of intersex disorders. *Archives of Disease in Childhood* 91: 554–563.

Hughes IA, Nihoul-Fekete C, Thomas B, and Cohen-Kettenis PT (2007) Consequences of the ESPE/LWPES guidelines for diagnosis and treatment of disorders of sex development. *Best Practice and Research Clinical Endocrinology and Metabolism* 21: 351–365.

Huizenga NA, de Lange P, Koper JW, et al. (2000) Five patients with biochemical and/or clinical generalized glucocorticoid resistance without alterations in the glucocorticoid receptor gene. *Journal of Clinical Endocrinology and Metabolism* 85: 2076–2081.

Hurley DM, Accili D, Stratakis CA, et al. (1991) Point mutation causing a single amino acid substitution in the hormone binding domain of the glucocorticoid receptor in familial glucocorticoid resistance. *Journal of Clinical Investigation* 87: 680–686.

Jacobs PA and Strong JA (1959) A case of human intersexuality having a possible XXY sex-determining mechanism. *Nature* 183: 302–303.

Jenkins ME, Surana RB, and Russell-Cutts CM (1968) Ambiguous genitals in a female infant associated with luteoma of pregnancy. Report of a case. *American Journal of Obstetrics and Gynecology* 101: 923–928.

Johannsen TH, Ripa CP, Reinisch JM, Schwartz M, Mortensen EL, and Main KM (2006) Impaired cognitive function in women with congenital adrenal hyperplasia. *Journal of Clinical Endocrinology and Metabolism* 91: 1376–1381.

Jones ME, Boon WC, McInnes K, Maffei L, Carani C, and Simpson ER (2007) Recognizing rare disorders: Aromatase deficiency. *Nature Clinical Practice Endocrinology and Metabolism* 3: 414–421.

Josso N, Picard JY, Rey R, and diClemente N (2006) Testicular anti-Mullerian hormone: History, genetics, regulation and clinical applications. *Pediatric Endocrinology Reviews* 3: 347–358.

Julka S, Bhatia V, Singh U, et al. (2006) Quality of life and gender role behavior in disorders of sexual differentiation in India. *Journal of Pediatric Endocrinology and Metabolism* 19: 879–888.

Kai H, Nose O, Iida Y, Ono J, Harada T, and Yabuuchi H (1979) Female pseudohermaphroditism caused by maternal congenital adrenal hyperplasia. *Journal of Pediatrics* 95: 418–420.

Karl M, Lamberts SW, Detera-Wadleigh SD, et al. (1993) Familial glucocorticoid resistance caused by a splice site deletion in the human glucocorticoid receptor gene. *Journal of Clinical Endocrinology and Metabolism* 76: 683–689.

Karl M, Lamberts SW, Koper JW, et al. (1996) Cushing's disease preceded by generalized glucocorticoid resistance: Clinical consequences of a novel, dominant-negative glucocorticoid receptor mutation. *Proceedings of the Association of American Physicians* 108: 296–307.

Kersemaekers AM, Honecker F, Stoop H, et al. (2005) Identification of germ cells at risk for neoplastic transformation in gonadoblastoma: An immunohistochemical study for OCT3/4 and TSPY. *Human Pathology* 36: 512–521.

Kim Y and Capel B (2006) Balancing the bipotential gonad between alternative organ fates: A new perspective on an old problem. *Developmental Dynamics* 235: 229–300.

Kino T, Vottero A, Charmandari E, and Chrousos GP (2002) Familial/sporadic glucocorticoid resistance syndrome and hypertension. *Annals of the New York Academy of Sciences* 970: 101–111.

Kocova M, Siegel SF, Wenger SL, Lee PA, and Trucco M (1993) Detection of Y chromosome sequences in Turner's syndrome by Southern blot analysis of amplified DNA. *Lancet* 342: 140–143.

Kojima Y, Hayashi Y, Mizuno K, et al. (2007) Up-regulation of SOX9 in human sex-determining region on the Y chromosome (SRY)-negative XX males. *Clinical Endocrinology (Oxf)* 68: 791–799.

Koopman P, Gubbay J, Vivian N, Goodfellow P, and Lovell-Badge R (1991) Male development of chromosomally female mice transgenic for Sry. *Nature* 351: 117–121.

Koubova J, Menke DB, Zhou Q, Capel B, Griswold MD, and Page DC (2006) Retinoic acid regulates sex-specific timing of meiotic initiation in mice. *Proceedings of the National Academy of Sciences of the United States of America* 103: 2474–2479.

Krone N, Dhir V, Ivison HE, and Arlt W (2007a) Congenital adrenal hyperplasia and P450 oxidoreductase deficiency. *Clinical Endocrinology (Oxf)* 66: 162–172.

Krone N, Hanley NA, and Arlt W (2007b) Age-specific changes in sex steroid biosynthesis and sex development. *Best Practice and Research Clinical Endocrinology and Metabolism* 21: 393–401.

Kuhnle U and Bullinger M (1997) Outcome of congenital adrenal hyperplasia. *Pediatric Surgery International* 12: 511–515.

Kusz K, Kotecki M, Wojda A, et al. (1999) Incomplete masculinisation of XX subjects carrying the SRY gene on an inactive X chromosome. *Journal of Medical Genetics* 36: 452–456.

Landin-Wilhelmsen K, Bryman I, Hanson C, and Hanson L (2004) Spontaneous pregnancies in a Turner syndrome woman with Y-chromosome mosaicism. *Journal of Assisted Reproduction and Genetics* 21: 229–230.

Larsen T, Gravholt CH, Tillebeck A, Larsen H, Jensen MB, Nielsen J, and Friedrich U (1995) Parental origin of the X chromosome, X chromosome mosaicism and screening for hidden Y chromosome in 45,X Turner syndrome ascertained cytogenetically. *Clinical Genetics* 48: 6–11.

Lau Y, Chou P, Iezzoni J, Alonzo J, and Komuves L (2000) Expression of a candidate gene for the gonadoblastoma locus in gonadoblastoma and testicular seminoma. *Cytogenetic and Genome Research* 91: 160–164.

Lau YF (1999) Gonadoblastoma, testicular and prostate cancers, and the TSPY gene. *American Journal of Human Genetics* 64: 921–927.

Lev-Ran A (1974) Sexuality and educational levels of women with the late-treated adrenogenital syndrome. *Archives of Sexual Behavior* 3: 27–32.

Liang L, Soyal SM, and Dean J (1997) FIGalpha, a germ cell specific transcription factor involved in the coordinate expression of the zona pellucida genes. *Development* 124: 4939–4947.

Lin L, Ercan O, Raza J, et al. (2007) Variable phenotypes associated with aromatase (CYP19) insufficiency in humans. *Journal of Clinical Endocrinology and Metabolism* 92: 982–990.

Looijenga LH, Stoop H, de Leeuw HP, et al. (2003) POU5F1 (OCT3/4) identifies cells with pluripotent potential in human germ cell tumors. *Cancer Research* 63: 2244–2250.

Lovell-Badge R and Robertson E (1990) XY female mice resulting from a heritable mutation in the primary testis-determining gene, Tdy. *Development* 109: 635–646.

Ludwig M, Beck A, Wickert L, Bolkenius U, Tittel B, Hinkel K, and Bidlingmaier F (1998) Female pseudohermaphroditism associated with a novel homozygous G-to-A (V370-to-M) substitution in the P-450 aromatase gene. *Journal of Pediatric Endocrinology and Metabolism* 11: 657–664.

Mackiewicz M, Tatarenkov A, Turner BJ, and Avise JC (2006) A mixed-mating strategy in a hermaphroditic vertebrate. *Proceedings of the Royal Society B: Biological Sciences* 273: 2449–2452.

Malchoff CD, Reardon G, Javier EC, Rogol AD, McDermott P, Loriaux DL, and Malchoff DM (1994) Dexamethasone therapy for isosexual precocious pseudopuberty caused by generalized glucocorticoid resistance. *Journal of Clinical Endocrinology and Metabolism* 79: 1632–1636.

Matsuda M, Shinomiya A, Kinoshita M, et al. (2007) DMY gene induces male development in genetically female (XX) medaka fish. *Proceedings of the National Academy of Sciences of the United States of America* 104: 3865–3870.

May B, Boyle M, and Grant D (1996) A comparative study of sexual experiences: Women with diabetes and women with congenital adrenal hyperplasia. *Journal of Health Psychology* 1: 479–492.

McLaren A (1995) Germ cells and germ cell sex. *Philosophical Transactions of the Royal Society B: Biological Sciences* 35: 229–233.

McLaren A and Lawson KA (2005) How is the mouse germ-cell lineage established? *Differentiation* 73: 435–437.

Meeks JJ, Weiss J, and Jameson JL (2003) Dax1 is required for testis determination. *Nature Genetics* 34: 32–33.

Mendonca BB, Leite MV, de Castro M, et al. (2002) Female pseudohermaphroditism caused by a novel homozygous missense mutation of the GR gene. *Journal of Clinical Endocrinology and Metabolism* 87: 1805–1809.

Merke DP and Bornstein SR (2005) Congenital adrenal hyperplasia. *Lancet* 365: 2125–2136.

Meyer-Bahlburg HF, Dolezal C, Baker SW, Carlson AD, Obeid JS, and New MI (2004) Cognitive and motor development of children with and without congenital adrenal hyperplasia after early-prenatal dexamethasone. *Journal of Clinical Endocrinology and Metabolism* 89: 610–614.

Meyer-Bahlburg HF, Dolezal C, Baker SW, Ehrhardt AA, and New MI (2006) Gender development in women with congenital adrenal hyperplasia as a function of disorder severity. *Archives of Sexual Behavior* 35: 667–684.

Meyer-Bahlburg HF, Dolezal C, Baker SW, and New MI (2008) sexual orientation in women with classical or non-classical congenital adrenal hyperplasia as a function of degree of prenatal androgen excess. *Archives of Sexual Behavior* 37: 85–99.

Meyer-Bahlburg HFL, Ehrhardt AA, Rosen LR, Gruen RS, Veridiano NP, and Vann FH (1995) Prenatal estrogens and the development of homosexual orientation. *Developmental Psychology* 31: 12–21.

Meyer-Bahlburg HF, Gruen RS, New MI, et al. (1996) Gender change from female to male in classical congenital adrenal hyperplasia. *Hormones and Behavior* 30: 319–332.

Migeon BR (2007) Why females are mosaics, X-chromosome inactivation, and sex differences in disease. *Gender Medicine* 4: 97–105.

Minto CL, Liao LM, Woodhouse CR, Ransley PG, and Creighton SM (2003) The effect of clitoral surgery on sexual outcome in individuals who have intersex conditions with ambiguous genitalia: A cross-sectional study. *Lancet* 361: 1252–1257.

Money J and Schwartz M (1976) Fetal androgens in the early treated adrenogenital syndrome of 46,XX hermaphroditism: Influences on assertive ad aggressive types of behaviour. *Aggressive Behavior* 2: 19–30.

Money J, Schwartz M, and Lewis VG (1984) Adult erotosexual status and fetal hormonal masculinization and demasculinization: 46,XX congenital virilizing adrenal hyperplasia and 46,XY androgen-insensitivity syndrome compared. *Psychoneuroendocrinology* 9: 405–414.

Morishima A, Grumbach MM, Simpson ER, Fisher C, and Qin K (1995) Aromatase deficiency in male and female siblings caused by a novel mutation and the physiological role of estrogens. *Journal of Clinical Endocrinology and Metabolism* 80: 3689–3698.

Mulaikal RM, Migeon CJ, and Rock JA (1987) Fertility rates in female patients with congenital adrenal hyperplasia due to 21-hydroxylase deficiency. *New England Journal of Medicine* 316: 178–182.

Mullis PE, Yoshimura N, Kuhlmann B, Lippuner K, Jaeger P, and Harada H (1997) Aromatase deficiency in a female who is compound heterozygote for two new point mutations in the P450arom gene: Impact of estrogens on hypergonadotropic hypogonadism, multicystic ovaries, and bone densitometry in childhood. *Journal of Clinical Endocrinology and Metabolism* 82: 1739–1745.

Nef S, Schaad O, Stallings NR, et al. (2005) Gene expression during sex determination reveals a robust female genetic program at the onset of ovarian development. *Developmental Biology* 287: 361–377.

Nimkarn S and New MI (2006) Prenatal diagnosis and treatment of congenital adrenal hyperplasia owing to 21-hydroxylase deficiency. *Nature Clinical Practice Endocrinology and Metabolism* 3: 405–413.

Oosterhuis JW and Looijenga LH (2005) Testicular germ-cell tumours in a broader perspective. *Nature Reviews Cancer* 5: 210–222.

Oppelt P, Renner SP, Kellermann A, et al. (2006) Clinical aspects of Mayer–Rokitansky–Kuester–Hauser syndrome: Recommendations for clinical diagnosis and staging. *Human Reproduction* 21: 792–797.

Ottolenghi C, Omari S, Garcia-Ortiz JE, et al. (2005) Foxl2 is required for commitment to ovary differentiation. *Human Molecular Genetics* 14: 2053–2062.

Ottolenghi C, Pelosi E, Tran J, et al. (2007a) Loss of Wnt4 and Foxl2 leads to female-to-male sex reversal extending to germ cells. *Human Molecular Genetics* 16: 2795–2804.

Ottolenghi C, Uda M, Crisponi L, Omari S, Cao A, Forabosco A, and Schlessinger D (2007b) Determination and stability of sex. *Bioessays* 29: 15–25.

Page DC (1987) Hypothesis: A Y-chromosomal gene causes gonadoblastoma in dysgenetic gonads. *Development* 101 (supplement): 151–155.

Palmer SJ and Burgoyne PS (1991) *In situ* analysis of fetal, prepuberal and adult XX–XY chimaeric mouse testes: Sertoli cells are predominantly, but not exclusively, XY. *Development* 112: 265–268.

Pasterski V, Hindmarsh P, Geffner M, Brook C, Brain C, and Hines M (2007) Increased aggression and activity level in 3- to 11-year-old girls with congenital adrenal hyperplasia (CAH). *Hormones and Behavior* 52: 368–374.

Pasterski VL, Geffner ME, Brain C, Hindmarsh P, Brook C, and Hines M (2005) Prenatal hormones and postnatal socialization by parents as determinants of male-typical toy play in girls with congenital adrenal hyperplasia. *Child Development* 76: 264–278.

Pena-Alonso R, Nieto K, Alvarez R, et al. (2005) Distribution of Y-chromosome-bearing cells in gonadoblastoma and dysgenetic testis in 45,X/46,XY infants. *Modern Pathology* 18: 439–445.

Pepling ME and Spradling AC (2001) Mouse ovarian germ cell cysts undergo programmed breakdown to form primordial follicles. *Developmental Biology* 234: 339–351.

Peterson RE, Imperato-McGinley J, Gautier T, and Shackleton C (1985) Male pseudohermaphroditism due to multiple defects in steroid-biosynthetic microsomal mixed-function oxidases. A new variant of congenital adrenal hyperplasia. *New England Journal of Medicine* 313: 1182–1191.

Petrusevska R, Beudt U, Schafer D, et al. (1996) Distribution of marker-Y chromosome containing cells in different tissues of a Turner mosaic patient with mixed gonadal dysgenesis. *Clinical Genetics* 49: 261–266.

Phoenix CH, Goy RW, Gerall AA, and Young WC (1959) Organizing action of prenatally administered testosterone propionate on the tissues mediating mating behavior in the female guinea pig. *Endocrinology* 65: 369–382.

Portrat-Doyen S, Forest MG, Nicolino M, Morel Y, and Chatelain PC (1996) Female pseudohermaphroditism (FPH) resulting from aromatase (P450arom) deficiency associated with a novel mutation (R457) in the CYP19 gene. *Hormone Research* 46(supplement): 14.

Puts DA, McDaniel MA, Jordan CL, and Breedlove SM (2008) Spatial ability and prenatal androgens: Meta-analyses of congenital adrenal hyperplasia and digit ratio (2D:4D) studies. *Archives of Sexual Behavior* 37: 100–111.

Qin Y and Bishop CE (2005) Sox9 is sufficient for functional testis development producing fertile male mice in the absence of Sry. *Human Molecular Genetics* 14: 1221–1229.

Rajpert-DeMeyts E, Hanstein R, Jorgensen N, Graem N, Vogt PH, and Skakkebaek NE (2004) Developmental expression of POU5F1 (OCT-3/4) in normal and dysgenetic human gonads. *Human Reproduction* 19: 1338–1344.

Reiner WG (2005) Gender identity and sex-of-rearing in children with disorders of sexual differentiation. *Journal of Pediatric Endocrinology and Metabolism* 18: 549–553.

Reinisch JM (1981) Prenatal exposure to synthetic progestins increases potential for aggression in humans. *Science* 211: 1171–1173.

Rey RA, Belville C, Nihoul-Fekete C, et al. (1999) Evaluation of gonadal function in 107 intersex patients by means of serum antimullerian hormone measurement. *Journal of Clinical Endocrinology and Metabolism* 84: 627–631.

Richter-Appelt H, Discher C, and Gedrose B (2005) Gender identity and recalled gender related childhood play-behaviour in adult individuals with different forms of intersexuality. *Anthropologischer Anzeiger* 63: 241–256.

Richter-Unruh A, Knauer-Fischer S, Kaspers S, Albrecht B, Gillessen-Kaesbach G, and Hauffa BP (2004) Short stature in children with an apparently normal male phenotype can be caused by 45,X/46,XY mosaicism and is susceptible to growth hormone treatment. *European Journal of Pediatritics* 163: 251–256.

Russcher H, Smit P, van den Akker EL, et al. (2005) Two polymorphisms in the glucocorticoid receptor gene directly affect glucocorticoid-regulated gene expression. *Journal of Clinical Endocrinology and Metabolism* 90: 5804–5810.

Sarlis NJ, Weil SJ, and Nelson LM (1999) Administration of metformin to a diabetic woman with extreme hyperandrogenemia of nontumoral origin: Management of infertility and prevention of inadvertent masculinization of a female fetus. *Journal of Clinical Endocrinology and Metabolism* 84: 1510–1512.

Schmidt D, Ovitt CE, Anlag K, Fehsenfeld S, Gredsted L, Treier AC, and Treier M (2004) The murine winged-helix transcription factor Foxl2 is required for granulosa cell differentiation and ovary maintenance. *Development* 131: 933–942.

Schmitt K, Tulzer W, and Hohenauer L (1990) (Luteoma of the mother as a possible cause of external female pseudohermaphroditism) Luteom der mutter als mogliche ursache eines pseudohermaphroditismus femininus externus beim neugeborenen. *Klinische Pädiatrie* 202: 117–119.

Selvaggi G, Monstrey S, Ceulemans P, T'Sjoen G, De Cuypere G, and Hoebeke P (2007) Genital sensitivity after sex reassignment surgery in transsexual patients. *Annals of Plastic Surgery* 58: 427–433.

Shapiro E, Huang HY, and Wu XR (2000) Uroplakin and androgen receptor expression in the human fetal genital tract: Insights into the development of the vagina. *Journal of Urology* 164: 1048–1051.

Simon NG (2002) Hormonal processes in the development and expression of aggressive behaviour. In: Pfaff DW, Arnold AP, Etgen AM, Fahrbach SE, and Rubin RT (eds.) *Hormones, Brain and Behaviour,* vol. IV, pp. 339–392. Amsterdam: Academic Press (Elsevier Science).

Sinclair AH, Berta P, Palmer MS, et al. (1990) A gene from the human sex-determining region encodes a protein with homology to a conserved DNA-binding motif. *Nature* 346: 240–244.

Slijper FM (1984) Androgens and gender role behaviour in girls with congenital adrenal hyperplasia (CAH). *Progress in Brain Research* 61: 417–422.

Slijper FME, Van derKamp HJ, Brandenburg H, and de Muinck Keizer-Schrama SMBF (1992) Evaluation of psychosexual development in young women with congenital adrenal hyperplasia. A pilot study. *Journal of Sex Education and Therapy* 18: 200–207.

Slowikowska-Hilczer J, Romer TE, and Kula K (2003) Neoplastic potential of germ cells in relation to disturbances of gonadal organogenesis and changes in karyotype. *Journal of Andrology* 24: 270–278.

Soyal SM, Amleh A, and Dean J (2000) FIGalpha, a germ cell-specific transcription factor required for ovarian follicle formation. *Development* 127: 4645–4654.

Speiser PW and White PC (2003) Congenital adrenal hyperplasia. *New England Journal of Medicine* 349: 776–788.

Spitzer RF, Wherrett D, Chitayat D, Colgan T, Dodge JE, Salle JL, and Allen L (2007) Maternal luteoma of pregnancy presenting with virilization of the female infant. *Journal of Obstetrics and Gynaecology Canada* 29: 835–840.

Stikkelbroeck NM, Beerendonk CC, Willemsen WN, et al. (2003) The long term outcome of feminizing genital surgery for congenital adrenal hyperplasia: Anatomical, functional and cosmetic outcomes, psychosexual development, and satisfaction in adult female patients. *Journal of Pediatric and Adolescent Gynecology* 16: 289–296.

Stoop H, Honecker F, Cools M, de Krijger R, Bokemeyer C, and Looijenga LH (2005) Differentiation and development of human female germ cells during prenatal gonadogenesis: An immunohistochemical study. *Human Reproduction* 20: 1466–1476.

Swaab DF (2007) Sexual differentiation of the brain and behavior. *Best Practice and Research Clinical Endocrinology and Metabolism* 21: 431–444.

Swain A, Narvaez V, Burgoyne P, Camerino G, and Lovell-Badge R (1998) Dax1 antagonizes Sry action in mammalian sex determination. *Nature* 391: 761–767.

Telvi L, Lebbar A, Del Pino O, Barbet JP, and Chaussain JL (1999) 45,X/46,XY mosaicism: Report of 27 cases. *Pediatrics* 104: 304–308.

Uda M, Ottolenghi C, Crisponi L, et al. (2004) Foxl2 disruption causes mouse ovarian failure by pervasive blockage of follicle development. *Human Molecular Genetics* 13: 1171–1181.

Uhlenhaut NH and Treier M (2006) Foxl2 function in ovarian development. *Molecular Genetics and Metabolism* 88: 225–234.

van Rossum EF and Lamberts SW (2006) Glucocorticoid resistance syndrome: A diagnostic and therapeutic approach. *Best Practice and Research Clinical Endocrinology and Metabolism* 20: 611–626.

Vidal VP, Chaboissier MC, de Rooij DG, and Schedl A (2001) Sox9 induces testis development in XX transgenic mice. *Nature Genetics* 28: 216–217.

Wallen K (2005) Hormonal influences on sexually differentiated behavior in nonhuman primates. *Frontiers of Neuroendocrinology* 26: 7–26.

Wallen K and Baum MJ (2002) Masculinization and defeminization in altricial and precocial mammals: Comparative aspects of steroid hormone action. In: Pfaff DW, Arnold AP, Etgen AM, Fahrback SE, and Rubin RT (eds.) *Hormones, Brain and Behaviour,* vol. IV, pp. 105–135. Amsterdam: Academic Press (Elsevier Science).

Wang YC, Su HY, Liu JY, Chang FW, and Chen CH (2005) Maternal and female fetal virilization caused by pregnancy luteomas. *Fertility and Sterility* 84: 509.

Warmann S, Roth C, Gluer S, and Fuchs J (2000) Congenital adrenal hyperplasia associated with maternal pregnancy luteoma and the Antley-Bixler syndrome. *Journal of Pediatric Surgery* 35: 528–530.

White PC, Curnow KM, and Pascoe L (1994) Disorders of steroid 11 beta-hydroxylase isozymes. *Endocrine Reviews* 15: 421–438.

Wilhelm D and Englert C (2002) The Wilms tumor suppressor WT1 regulates early gonad development by activation of Sf1. *Genes and Development* 16: 1839–1851.

Wilhelm D, Hiramatsu R, Mizusaki H, Widjaja L, Combes AN, Kanai Y, and Koopman P (2007a) SOX9 regulates prostaglandin D synthase gene transcription *in vivo* to ensure testis development. *Journal of Biological Chemistry* 282: 10553–10560.

Wilhelm D, Palmer S, and Koopman P (2007b) Sex determination and gonadal development in mammals. *Physiological Reviews* 87: 1–28.

Wisniewski AB, Migeon CJ, Malouf MA, and Gearhart JP (2004) Psychosexual outcome in women affected by congenital adrenal hyperplasia due to 21-hydroxylase deficiency. *Jouranl of Urology* 171: 2497–2501.

Xu J, Burgoyne PS, and Arnold AP (2002) Sex differences in sex chromosome gene expression in mouse brain. *Human Molecular Genetics* 11: 1409–1419.

Yao HH (2005) The pathway to femaleness: Current knowledge on embryonic development of the ovary. *Molecular and Cellular Endocrinology* 230: 87–93.

Yu RN, Ito M, Saunders TL, Camper SA, and Jameson JL (1998) Role of Ahch in gonadal development and gametogenesis. *Nature Genetics* 20: 353–357.

Zenteno JC, Carranza-Lira S, and Kofman-Alfaro S (2004) Molecular analysis of the anti-Mullerian hormone, the anti-Mullerian hormone receptor, and galactose-1-phosphate uridyl transferase genes in patients with the Mayer–Rokitansky–Kuster–Hauser syndrome. *Archives of Gynecology and Obstetrics* 269: 270–273.

Zucker KJ, Bradley SJ, Oliver G, Blake J, Fleming S, and Hood J (1996) Psychosexual development of women with congenital adrenal hyperplasia. *Hormones and Behavior* 30: 300–318.

27 Genetic Defects of Male Sexual Differentiation

Y-S Zhu and J Imperato-McGinley, Weill Medical College of Cornell University, New York, NY, USA

Chapter Outline

Glossary

androgen receptor It is an intracellular protein that specifically binds to androgens and mediates androgen action.

coregulator A nuclear receptor coregulator is defined as a protein that interacts with nuclear receptors to enhance transactivation (coactivators) or reduce transactivation (corepressors) of target genes, but does not significantly affect the basal transcription rate.

dihydrotestosterone (DHT) A derivative of testosterone formed by the conversion of testosterone in the action of 5α-reductase isozymes.

disorders in male sexual differentiation A syndrome in which 46,XY individuals have testicular tissue, but the external genitalia and other sexual characteristics are so ambiguously developed that the sex of the individual is uncertain.

gender identity The sense of being male or female; it is the self-awareness of knowing one's sex.

gender role The expression of one's gender identity to the public; it is manifested by one's behavior or actions.

5α-reductase isozymes Microsomal NADPH-dependent proteins that reduce the double bond at the 4–5 position of a variety of C_{19} and C_{21} steroids including testosterone. Two 5α-reductase isozymes, type 1 and 2, are identified in humans.

testosterone (T) A male hormone that is produced by the testes and induces and maintains male secondary sexual characteristics.

27.1 Introduction

Any defect in the biosynthesis, metabolism, and action of androgens can result in disorders in male sexual differentiation. In many animal species, androgens have also been shown to be essential for sexual differentiation of the brain and for maintaining sexually dimorphic behavior throughout life.

The principles of sex determination, genetic and hormonal, in humans have proven to be similar to other mammals. However, the study of hormonal influence on sexual dimorphic differences in the nervous system in humans and sex differences in behavior is still an emerging field.

This chapter reviews subjects with enzyme defects in testosterone (T) biosynthesis, metabolism, and action, providing evidence for the role of androgens in male sexual differentiation and in gender and cognitive function. In particular, studies of subjects with inherited defects in male sexual differentiation, where numbers of patients with the same genetic defect have been studied, address the complex interaction of nature versus nurture.

27.2 Embryology of Male Sexual Differentiation and Development

27.2.1 Formation of the Bipotential Gonad

An undifferentiated gonad is present in both male and female fetuses. At the fifth week of fetal life, the undifferentiated gonad begins to develop when a thickened area of coelomic epithelium (germinal epithelium) appears on the medical aspect of the mesonephros. This is followed by proliferation of germinal epithelium cells and underlying mesenchyme to form a gonadal ridge and by development of primary sex cords migrating from the epithelium into the mesenchyme. The gonad, at this stage, consists mainly of mesodermal cells of coelomic epithelial origin (Arey, 1965).

The primordial germ cells become visible among the endodermal cells of the wall of the yolk sac near the origin of the allantois, during the third week of fetal life (Everett, 1943; Mintz and Russell, 1957). As the yolk sac folds, a portion of it is incorporated into the embryo, enabling the germ cells to migrate along the dorsal mesentery to the gonadal ridges.

The germ cells are spherical, with large vesicular nuclei and abundant cytoplasm. During the fifth week of fetal life, they multiply by mitosis during migration into the underlying mesenchyme, and at the end of the sixth week, the bipotential gonad is formed.

Thus, coelomic epithelium, underlying mesenchyme, and primordial germ cells form the undifferentiated gonad. Primordial germ cells become spermatogonia in the male and ova in the female; sex cords become either seminiferous tubules or primary ovarian follicles; and mesenchymal cells form either Leydig cells, or theca and stromal cells (Zachmann et al., 1972; Imperato-McGinley, 1983).

27.2.2 Testicular Differentiation

Testicular cords evolve from the primary sex cords of the indifferent gonad at approximately 7weeks of gestation. The Sertoli cells within the cords enlarge, become contiguous, and engulf the germ cells. The seminiferous cords interconnect and form a network of solid cords – the rete testes – which connect with the mesonephric tubules at the third month of gestation. By the sixth month, a lumen develops which is continuous with the mesonephric tubules and which ultimately develops into the ductuli efferentes.

Fetal Leydig cells are apparent by 8weeks of fetal life and completely fill the interstitial spaces of the developing testes at 3months of gestation (Jirasek, 1971; Imperato-McGinley, 1983).

27.2.3 Ovarian Differentiation

The ovary differentiates from the bipotential gonad at approximately 50days of gestation, when ovarian epithelial cells cluster into groups forming medullary cords which contain primitive granulosa cells and primordial oogonia. The prophase of meiosis begins at day 50–55, the leptotene stage, at day 60, the pachytene stage at approximately day 80; at day 90 the oogonia enter the diplotene stage (Bercu and Schulman, 1980).

As the oogonia differentiate, the primitive granulosa cells organize around the oogonia in a single layer constituting the primordial follicle, which becomes separated by connective tissue forming the primary follicles. At approximately 5months of gestation, the ovary has the maximum number of oogonia and oocytes of approximately 7million which diminish at birth to approximately 2million (Block, 1953; Baker, 1963; Imperato-McGinley, 1983).

27.2.4 Ductal Differentiation

The Wolffian ducts (mesonephric ducts) and Müllerian ducts (paramesonephric ducts) are present early in gestation, and appear at 25–30 days and 40–48 days of gestation, respectively. In the male fetus, the Müllerian ducts regress at about 8.5weeks of gestation; differentiation of the Wolffian ducts follow to form the epididymides, vas deferens, seminal vesicles, and ejaculatory ducts, a process which is completed by 12 weeks of gestation.

In the female, the Wolffian ducts regress at approximately 75days (50-mm stage), while the Müllerian ducts form the fallopian tubes, uterus, and upper portion of the vagina. The cervix becomes recognizable at this stage, but the uterus is not completely formed until the 150-mm stage (Imperato-McGinley, 1983).

27.2.5 Differentiation of the External Genitalia

The external genitalia in fetuses of both sexes also develop from common primordia: consisting of the urogenital tubercle, urogenital swellings, and urogenital folds. In the male fetus, external genital virilization begins shortly after Wolffian ductal differentiation; the urogenital tubercle becomes the glans penis, the urogenital folds become the shaft of the penis, and the urogenital swellings become the scrotum. The urogenital sinus forms the prostate, bulbourethral glands, and the prostatic and membranous portion of the urethra. The prostate develops from endodermal buds in the urethral lining; at 10 weeks, the endodermal buds grow into the mesenchyme to form the muscular and connective tissue components of the prostate. The entire process of male external sexual differentiation is completed by 14 weeks of gestation. Descent of the testes and growth of genitalia occur during the last two trimesters of pregnancy (Wilson, 1978).

In the female fetus at approximately 10½ weeks of gestation, the urogenital tubercle becomes the clitoris, the urogenital swellings become the labia majora, and the urogenital folds become the labia minora (Imperato-McGinley, 1983).

27.3 The Genetic and Hormonal Control of Male Sexual Differentiation

Multiple genetic and hormonal factors are involved in male sexual differentiation. For the purpose of this chapter, the hormonology of male sexual differentiation is described in detail, while the genetics of male sex determination is summarized.

27.3.1 The Genetic Control of Testicular Differentiation

The major step in male sex determination is the establishment of sex-specific gonads – the testes. The Y chromosome is critical for male sex determination. It is known that the *SRY* (sex-determining region-Y chromosome) gene located on the Y chromosome is the master switch for testis determination (Goodfellow and Lovell, 1993; Wilhelm et al., 2007). Based on genetic analyses in humans and animals, multiple genes, in addition to the *SRY* gene, are involved in the complex process of testicular differentiation and development (see **Figure 1**). These genes interact directly or indirectly, work in concert, and synchronize testicular differentiation and development.

SRY is a transcriptional factor that contains a HMG-box DNA-binding domain (DBD). The expression of *SRY* gene in the bipotential gonad is regulated by Wilms' tumor suppressor gene (WT1) and steroidogenic factor-1 (SF-1). WT1, SF-1, as well as LIM1 and Lhx9 are important in the development of the bipotential gonad, and a mutation in any of these genes will result in gonadal dysgenesis. They are expressed earlier than SRY in embryogenesis and may control *SRY* gene expression. WT1 is a transcription factor with tumor suppressor activity and is predominately expressed in the embryonic kidneys and gonads. It mediates the mesenchymal–epithelial transition and differentiation during morphogenesis of the kidneys and gonads. SF-1, an orphan nuclear receptor, is expressed in the bipotential gonad and the developing testis. The expression of SF-1 may be regulated by Lhx9, a member of the Limb Homeobox domain gene family, while the protein activity can be inhibited by DAX-1, an orphan nuclear receptor that influences testicular differentiation probably by interfering with SF-1 function. Overexpression due to gene duplication of either DAX1 or WNT4 antagonizes testis formation, giving rise to XY females.

Once the *SRY* gene is expressed, the expression of SOX-9, a SRY-like HMG Box gene, is upregulated in the developing testis. Overexpression of SOX-9 in humans and mice results in XX sex reversal, suggesting that SOX-9 is at the starting point of a cascade of genes responsible for testis determination. At the time *SRY* is turned on, other genes such as *DHH* and *FGF9* are switched on, while *DAX1* and *WNT4* are downregulated. Other factors such as the transcriptional factors DMRT1 and ARX, ATRX (a helicase), and DHH (Desert Hedge Hog), WNT4, and fibroblast growth factor 9 (FGF9, cell signal molecule) also work coordinately with SRY, and together with SOX-9, to direct testicular differentiation and

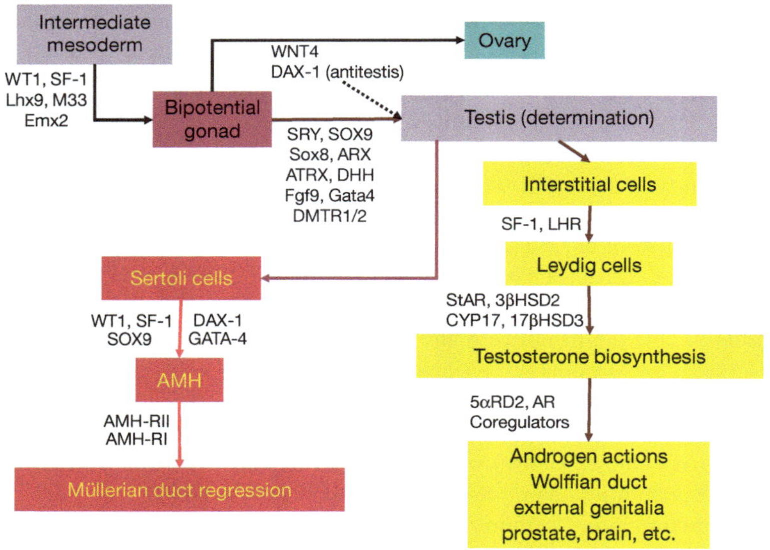

Figure 1 Schematic illustration of genetic and hormonal control of male sex determination and differentiation.

development (see **Figure 1**). For a full discussion of genetic control of testicular determination, the reader can refer to several recent review articles (Wilhelm et al., 2007; Maclaughlin and Donahoe, 2004; Kucinskas and Just, 2005).

27.3.2 Testicular Function

27.3.2.1 Testosterone production

Testicular differentiation occurs between weeks 7 and 8 of intrauterine life. At approximately 8 weeks of gestation following seminiferous tubule formation, the Leydig cells differentiate, increase in number, and secrete T. High serum levels of maternal chorionic gonadotropin coincide with the histological appearance of Leydig cells (Pelliniemi and Niemi, 1969) and T secretion (Siiteri and Wilson, 1974). In both animals and humans, the initiation of testicular T secretion coincides with male urogenital tract differentiation (Siiteri and Wilson, 1974).

Placental secretion of human chorionic gonadotropin (hCG) initiates fetal Leydig cell function. hCG appears in the maternal circulation 10 days to 2 weeks after ovulation, peaking at 10–12 weeks of gestation. Thereafter, serum hCG levels fall rapidly, with a more gradual decrease after 16 weeks (Clements et al., 1976; Takagi et al., 1977). Fetal pituitary gonadotropins, luteinizing hormone (LH) and follicular stimulating hormone (FSH), appear in fetal serum at 11–12 weeks of gestation and reach peak levels at 16–20 weeks of gestation. FSH levels are strikingly higher in female fetuses at mid-gestation, possibly reflecting decreased negative feedback from ovarian secretions.

hCG/LH acts on hCG/LH receptors that have been demonstrated in the fetal testis of humans and animals (Frowein and Engel, 1974; Huhtaniemi et al., 1977; Catt et al., 1975) to stimulate generation of cyclic 3′,5′-adenosine monophosphate (cAMP) and T (Huhtaniemi et al., 1977). Mutations in the hCG/LH receptor in 46,XY subjects results in defects in Leydig cell development, T biosynthesis, and consequently, male sexual differentiation (Stavrou et al., 1998). Serum T levels peak at 14–16 weeks (Siiteri and Wilson, 1974) in the male fetus, coinciding with the time when Leydig cells constitute half the testis volume, and with peak activity of steroidogenic enzymes (Pelliniemi and Niemi, 1969). Levels of maternal and fetal serum hCG decline by week 18 of gestation, and the Leydig cells begin to involute despite higher fetal pituitary gonadotropin levels (Imperato-McGinley, 1983).

27.3.2.2 Anti-Müllerian hormone

In classic experiments, when T was implanted into the peritoneal cavity of castrated fetal rabbits (male and female), Jost (1953) demonstrated that Wolffian differentiation and masculinization of the external genitalia occurred, while inhibition of Müllerian ductal differentiation did not occur. Thus, Jost postulated that two secretions from the fetal testes were essential for male development: (1) T and (2) Müllerian inhibiting substance (MIS). Subsequent studies with the antiandrogen cyproterone acetate also demonstrated that blocking androgen action in the male rabbit caused lack of differentiation of the Wolffian system and external genitalia, while inhibition of the Müllerian anlage proceeded normally (Neumann et al., 1970; Imperato-McGinley, 1983).

Anti-Müllerian hormone (AMH), or MIS, was ultimately isolated and found to be a high-molecular-weight glycoprotein (Josso and Picard, 1976). It is a member of the transforming growth factor-β (TGF-β) family and is secreted by the Sertoli cells shortly after testicular differentiation. Although it continues to be secreted through the perinatal period (2 years) (Donahoe et al., 1977), responsiveness to the hormone, however, is present only during the critical period of male sexual differentiation. The gene for AMH has five exons and is located on chromosome 19p13.2–p13.3 (Cohen-Haguenauer et al., 1987). AMH binds to the type II AMH receptor. However, the presence of type I receptor is required for AMH signaling (Massague, 1996). The expression of AMH is upregulated by WT1, GATA-4, and SF-1, while it is inhibited by DAX-1. Studies to date indicate that the normal development of Sertoli cells is dependent on FGF9, as demonstrated in animal studies (Colvin et al., 2001). In summary, both T and AMH must be secreted by the testes at a critical period *in utero* for normal male phenotypic development.

27.3.3 Enzymes and Genes Involved in Testosterone Biosynthesis

There are multiple enzymatic reactions involved in the biosynthesis of T from cholesterol (see **Figure 2**). Gene defects affecting any of these enzymatic reactions in the T biosynthetic pathway impair T production and can result in incomplete masculinization of the male genitalia. Moreover, since all steroid hormones need cholesterol as a precursor, defects in the gene, DHCR7 that encodes 7-dehydrosterol reductase and catalyzes the last step of endogenous cholesterol biosynthesis (Nowaczyk et al., 2001), and the StAR

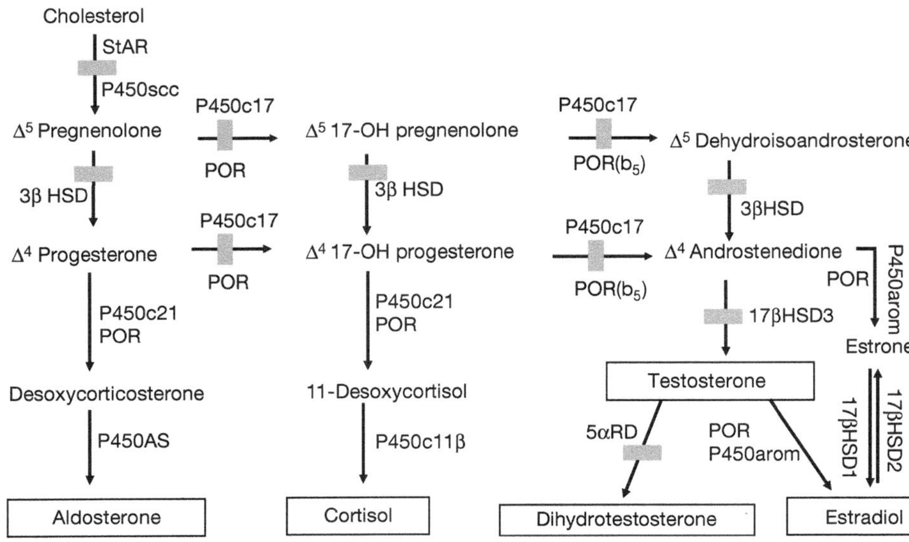

Figure 2 Pathway of steroid biosynthesis and enzymatic defects (shaded boxes) related to disorders of male sexual differentiation. SCC, side-chain cleavage; OH, hydroxylase; HSD, hydroxysteroid dehydrogenase; RD, reductase; Arom, aromatase; POR, P450 oxidoreductase.

protein, which is involved in cholesterol delivery to the mitochondria, have been described as causes of failure of masculinization (see **Figure 2**).

27.3.3.1 StAR protein

All steroid hormone biosynthesis utilizes cholesterol as a common precursor. Cholesterol, located in the outer mitochondrial membrane, must be delivered to the inner mitochondrial membrane, the site of the enzyme P450scc that converts cholesterol to pregnenolone – the first enzymatic reaction in steroidogenesis. This delivery is an assisted process and requires *de novo* protein synthesis (Garren et al., 1965). StAR, a steroidogenic acute regulatory protein, appears to be the regulator involved in cholesterol delivery to the mitochondrial enzyme P450scc (Stocco, 2000). How the StAR protein acts to transfer cholesterol across the membrane is currently obscure.

StAR has been purified and cloned (Clark et al., 1994). The gene is located on chromosome 8p11.2, spans 8 kb and consists of seven exons interrupted by six introns. Mutations in the StAR gene cause congenital lipoid adrenal hyperplasia, a rare, autosomal recessive, and potentially lethal condition resulting from a severe disruption of all adrenal and gonadal steroid biosynsthesis (Stocco, 2000; Imperato-McGinley, 1983; Lin et al., 1995). Affected individuals have low plasma levels of all steroid hormones, present with large adrenals, are yellowish in color and have high levels of cholesterol and cholesterol esters. In general, if hormonal replacement therapy does not occur, they will die from the severe glucocorticoid and mineralocorticoid deficiency. Affected 46,XY genetic males have inguinal testes, Wölffian differentiation, and no Müllerian structures with either female external genitalia or severe ambiguity of the external genitalia due to lack of T biosynthesis. As expected, affected 46,XX females have normal female internal and external accessory sex organs. An identical phenotype has been observed in StAR knockout (KO) mice (Caron et al., 1997), further demonstrating the critical role of StAR in steroidogenesis.

27.3.3.2 Cholesterol 20,22-desmolase

The P450scc gene is located on chromosome 15q23–q24 and is at least 20 kb in length (Sparkes et al., 1991). It is responsible for the conversion of cholesterol to pregnenolone, the rate-limiting step in all steroid biosynthesis and involves three biochemical reactions: 20α-hydroxylation, 22-hydroxylation, and side-chain cleavage of cholesterol. Cytochrome P450scc is a mitochondrial mixed-function oxidase mediating the three reactions. Mutations in this gene have not been reported to date.

27.3.3.3 3β-Hydroxysteroid dehydrogenases

Two genes, encoding isoenzymes 3β-hydroxysteroid dehydrogenases (3β-HSD) I and II, have been identified. These genes are located on chromosome 1 at locus p11–p13 (Berube et al., 1989; Lorence et al., 1990). They contain four exons and encode proteins comprising 371 and 372 amino acids, respectively, and have a 93.5% identity. Type I is expressed in peripheral tissues, particularly the liver and placenta, and type II is expressed in the adrenals and gonads.

Mutations in the type II 3β-HSD gene are linked to impaired steroid formation with decreased enzyme activity affecting both adrenals and gonads, resulting in congenital adrenal hyperplasia (CAH) and males being born with ambiguous genitalia (Bongiovanni, 1961; Simard et al., 2005). Male patients with classical 3β-HSD deficiency usually present with either perineal hypospadias or perineoscrotal hypospadias; female patients may have mild virilization of the external genitalia, thought to be due to increased peripheral conversion of dehydroepiandrosterone (DHEA) to T by type I 3β-HSD. In addition to decreased production of T, this enzyme deficiency results in decreased cortisol production resulting in elevated corticotropin secretion and increased production of Δ^5-steroids, particularly DHEA and DHEA-sulfate (DHEAS). Aldosterone production is also decreased. In severe cases, death occurs in early infancy due to marked salt-wasting. In milder cases, there is greater ability to synthesize cortisol, aldosterone, and T. Plasma levels of pregnenolone and 17α-hydroxypregnenolone are elevated and the urinary steroid excretion pattern shows a predominance of steroids with a Δ^5-3β-hydroxy configuration, which increases after corticotropin stimulation. Paradoxically, the Δ^4 steroid, plasma 17α-hydroxyprogesterone (17OHP), and its urinary metabolite pregnanetriol may be moderately elevated. This is attributed to the hepatic conversion of plasma Δ^5-17α-hydroxypregnenolone (Δ^5-17OHP) to Δ^4-17OHP (27) by type I 3β-HSD. Also, DHEA can be converted to androstenedione (Δ4) by the same isozyme. The classical form of 3β-HSD deficiency can be divided into the salt-wasting or non-salt-wasting forms. The salt-wasting form is usually diagnosed during the first few months of life due to insufficient biosynthesis of aldosterone and consequent salt loss. Whereas, the non-salt-wasting form may be diagnosed later on in childhood in the presence of family history of death during early infancy, hypospadias in males, or failure to gain weight (Simard et al., 2005).

27.3.3.4 17α-Hydroxylase/17,20-desmolase

The human 17α-hydroxylase/17,20-desmolase (P450c17) gene is 6.6-kb long and located on chromosome 10 q24–q25 (Sparkes et al., 1991). It is a gene with two sites of activity that converts pregnenolone to 17α-hydroxypregnenolone and progesterone to 17OHP in the adrenals and gonads, respectively. It also catalyzes the 17,20-desmolase or lyase reaction converting the 17α-hydroxy compounds to DHEA and Δ4, respectively. Thus, the enzyme has two different and distinct catalytic sites. Unlike P450scc that is located in mitochondria, P450c17 is located in the endoplasmic reticulum. The differential expression of P450c17 enzymatic activity in various tissues directs pregnenolone to its final steroid biosynthesis pathway. When P450c17 is absent or deficient, such as in the zona glomerolosa, the C-21 17-deoxysteroids, such as aldosterone, are produced. When 17α-hydroxylase activity of P450c17 is present, C-21 17-hydroxysteroids such as cortisol are produced. When both 17α-hydroxylase and 17,20-lyase activities of P450c17 are present, C-19 precursors of sex steroids are produced. Gene defects in this enzyme may result in males born with ambiguous genitalia (Miller, 2002). A mutation in P450c17 may abolish either one enzymatic activity, or both 17α-hydroxylase and 17,20-lysase activities (Miller, 2002; Brooke et al., 2006).

With few exceptions (New, 1970; Jones et al., 1982) genetic males with 17α-hydroxylase deficiency are phenotypically female and will seek medical attention due to lack of pubertal development, inguinal hernias (testis), or presence of hypertension (Imperato-McGinley, 1996). It is interesting that genetic females with this condition do not develop breasts (Kershnar et al., 1976; Peterson et al., 1985), whereas genetic males can present with gynecomastia (New, 1970; Jones et al., 1982). The etiology of gynecomastia in the male is unclear, since plasma and urinary estrogens are low. Adrenals have been reported to be increased in size (Imperato-McGinley, 1996; Honour et al., 1981).

The 17,20-lyase (desmolase) deficiency was initially diagnosed in two XY cousins (1.8 and 2.2 years) from consanguineous parents and in their gonadectomized aunt (18 years), a male pseudohermaphrodite (Zachmann et al., 1972). The children had ambiguous genitalia with inguinal gonads and no evidence of Müllerian structures. The aunt also had ambiguous genitalia with inguinal and abdominal gonads. A laparatomy showed rudimentary Müllerian structures.

The typical features of the combined 17-hydroxylase/17,20-lyase deficiency include hypertension, hypokalemia, lack of pubertal development, sexual differentiation disorders in 46,XY subjects, suppressed plasma rennin activity, low plasma aldosterone concentration, and increased plasma 11-deoxycorticosterone and corticosterone levels. However, genotype–phenotype correlation studies indicated a remarkable variation in the clinical and biochemical presentation of this disorder (Rosa et al., 2007).

27.3.3.5 17β-Hydroxysteroid dehydrogenase

The 17β-hydroxysteroid dehydrogenase (17βHSD) isozymes catalyze oxidoreduction of hydroxyl/keto groups of androgens and estrogens and plays an important role in sex steroid biosynthesis and metabolism. To date, 14 mammalian 17βHSD isozymes have been identified (**Table 1**) and are designated as types 1 through 14 according to the chronologic order of cloning (Lukacik et al., 2006). Twelve of them, except 17βHSD6 and 17βHSD9, have been identified in humans (**Table 1**). All of these isozymes, except 17βHSD5, which is an aldo–keto reductase, belong to the short-chain dehydrogenases/reductases (SDR) family. These isozymes have different chromosomal location, tissue expression, subcellular location, substrate and catalytic preference, and low homology in amino acid sequences as summarized in **Table 1**.

17βHSD3, cloned by Andersson and his colleagues using expression cloning (Geissler et al., 1994), is the primary isozyme catalyzing T biosynthesis from Δ4, the last step in T biosynthesis in the testes (Andersson and Moghrabi, 1997). The 17βHSD3 gene is composed of 11 exons, encodes a 310-amino-acid protein and is localized to chromosome 9q22.

The type 3 isozyme, importantly, appears to be mainly expressed in the testes and is responsible for T biosynthesis during the critical period of sexual differentiation. The 46,XY individuals homozygous for 17βHSD3 gene defects are born with undermasculinization of the genitalia. The clinical description is discussed later in this chapter (see Section 27.4).

The type 1 isozyme preferentially catalyzes the reduction of estrone to estradiol, and to a minor extent Δ4-androstenedione to T (Labrie et al., 1997). This isozyme is predominately expressed in the ovaries, breast tissue, placenta, and prostate, and it is normal in subjects with 17βHSD deficiency (Geissler et al., 1994). Type 2 and 4 isozymes are mainly involved in the inactivation of sex steroids. Mutations in 17βHSD4 causes a severe form of Zellweger-like syndrome (Moller et al., 2001). Type 5 isozyme is the only human 17βHSD that belongs to the aldo–keto reductase family. Similar to 17βHSD3, it may be involved in T biosynthesis in extratesticular tissues (Zang et al., 1995). Type 6 and 9 isozymes have been cloned in the rodent (Biswas and Russell, 1997), but not yet in humans. Type 7 isozyme is mainly expressed in the ovary during pregnancy and may play an important role in the maintenance of pregnancy (Krazeisen et al., 1999). Recently, a closely related gene with 96% sequence identity and similar enzymatic properties as the type 7 isozyme has been cloned (Liu et al., 2005). Type 8 17βHSD isozyme is widely expressed, mainly in the liver and gonadal tissues, and catalyzes the conversion of estradiol to estrone and, to a lesser extent, T to Δ4. 17βHSD10 isozyme is a homotetrameric protein located in mitochondria (He and Yang, 2006). It is an oxidative enzyme with broad substrate specificities, involving steroid hormone inactivation and catabolic pathways of straight- and branched-chain short hydroxyacyl coenzyme A. This isozyme also exhibits 3αHSD activity converting 5α-androstanediol and allopregnanolone into 5α-dihydrotestosterone (DHT) and 5α-dihydroprogesterone, respectively, enabling cells to produce DHT in the absence of T. It is expressed in a variety of tissues, for example, prostate, brain, liver, and heart. Mutations in this gene are responsible for 2-methyl-3-hydroxybutyryl-CoA dehydrogenase deficiency, resulting in neurologic abnormalities, including psychomotor retardation and loss of mental and motor skills (Ofman et al., 2003). Type 11 17βHSD catalyzes the conversion of 5α-androstanediol to androsterone, and its function in human physiology remains to be established. Similar to types 1 and 7, human type 12 17βHSD is an estrogen-specific enzyme, catalyzing the transformation of estrone to estradiol (Luu-The et al., 2006), although it may also involve fatty acid metabolism. It is widely expressed in various tissues, including ovary, breast, liver, testis, and placenta. The co-expression of three 17βHSDs in ovarian, mammary, and placental tissues may reflect an optimal control of estradiol biosynthesis. Type 13 isozyme is originally cloned from human liver cDNA, and it is expressed in the liver, kidney, ovary, testis, and brain (Liu et al., 2007). Its substrate specificity and functional activity remain to be investigated. Type 14 isozyme is a recently cloned nicotinamide adenine dinucleotide (NAD^+)-dependent estradiol dehydrogenase that catalyzes the oxidation of estradiol to estrone, which regulates the intracellular levels of estrogens (Jansson et al., 2006).

27.3.3.6 P450 oxidoreductase

Steroid biosynthesis and metabolism involve multiple cytochrome P450 proteins that require P450

Table 1 Comparison of human 17βHSD isozymes

	Size (AA)	*Gene location*	*Cellular location*	*Substrate preference*	*Preferred cofactor*	*Catalytic preference*	*Tissue distribution*
Type 1	327	17q21	Cytosol	C_{18}	NADH NADPH	Reduction	Ovary, placenta, prostate, breast, brain
Type 2	387	16q23.3	ER	C_{19}, C_{18}, C_{21}	NAD^+	Oxidation	Ubiquitous, liver, prostate, placenta endometrium
Type 3	310	9q22	ER	C_{19}, C_{18}	NADPH	Reduction	Testis
Type 4	736	5q23.1	PXS	C_{18}	NAD^+	Oxidation	Ubiquitous
Type 5	323	10p15.1	Cytosol	C_{19}	NADPH	Reduction	Liver, prostate, brain
Type 7	341	10p11.2	Cytosol	C_{18}	NADPH	Reduction	Ovary, testis, breast, brain, prostate, liver
Type 8	270	6p21.32	?	C_{18}, C_{19}	NAD^+	Oxidation	Ubiquitous
Type 10	261/252	Xp11.2	Mitochondria	C_{18}, C_{19}, C_{21}, fatty acid	NAD^+	Oxidation	Liver, brain, prostate, testis
Type 11	300	4q22.1	ER	C_{19}	NAD^+	Oxidation	Ovary, testis, adrenal
Type 12	312	11p11.2	ER	C_{18}, fatty acid?	NADPH	Reduction	Ovary, breast, liver, kidney, adrenal, testis, placenta
Type 13	300	4q22.1	Cytosol	?	NAD^+?	Oxidation	Liver, ovary, testis, brain
Type 14	270	19q13.33	Cytosol	C_{18}	NAD^+	Oxidation	Brain, kidney

Human 17βHSD6 and 17βHSD9 have not been cloned. AA, amino acid numbers; ER, endoplasmic reticulum; PXS, peroxisomes.

oxidoreductase (POR) in the reaction as shown in **Figure 2** (Lu et al., 1969). In the endoplasmic reticulum, a single nicotinamide adenine dinucleotide phosphate (NADPH) POR supplies electrons to all microsomal P450s for catalytic activity. In addition to P450s, POR also supplies electrons to heme oxygenase, fatty acid desaturase and elongase, squalene monoxygenase, cytochrome b5, and sterol reductase (Fluck et al., 2007). POR is a flavoprotein that contains both flavin mononucleotide and flavin adenine dinucleotide as cofactors, which allow it to donate electrons directly from NADPH to all microsomal P450 enzymes (Fluck et al., 2007).

A single copy gene with a size of approximately 50 kb for human POR is located in chromosome 7 (7q11.2). It has 16 exons and encodes a 680-amino-acid protein (Shephard et al., 1989). Mutations in this gene have been associated with various diseases, including apparent combined P450C17 and P450C21 deficiency, amenorrhea and disordered steroidogenesis, CAH (Peterson et al., 1985), and Antley–Bixler syndrome (Fluck et al., 2004, 2007). So far, 32 mutations have been identified, and most of them are missense mutations, and related to disordered steroidogenesis (Human Gene Mutation Database, accessed in December 2007).

POR deficiency is a new form of CAH, first described in 2004 (Fluck et al., 2004) in a newborn 46,XX Japanese girl with craniosynothosis, hypertelorism, mid-face hypoplasia, radio-humeral synostosis, and arachnodactyly. The clinical and biochemical characteristics were previously described in the literature by Peterson et al. (1985) as the so-called mixed function oxidase disease, as the steroid profile typically suggests combined deficiencies of steroid 21-hydroxylase and 17α-hydroxylase/17,20-lyase activities. The clinical spectrum is very variable and ranges from affected males with ambiguous genitalia to adrenal insufficiency, the Antley–Bixler skeletal malformation syndrome to individuals with polycystic ovary syndrome features (Scott and Miller, 2008). Unlike the common single enzyme defects in adrenal steroidogenesis, which cause either virilization in girls or undervirilization in boys, POR deficiency can cause abnormal genital development in both sexes. Boys with POR deficiency are often undervirilized, while girls are frequently virilized, but with no postnatal progression. The origin of this virilization may be due to an impairment of P450aro (aromatase) activity, a POR-dependent microsomal P450, resulting in a decrease in estrogen production. Another hypothesis is that fetal and maternal virilization in POR deficiency is due to increased gonadal DHEA synthesis through a backdoor pathway, in which 21-carbon steroid precursors are 5α reduced and converted to, bypassing the conventional precursors, Δ4 and T (Auchus, 2004).

27.3.4 Androgens and Target-Organ Responsiveness

27.3.4.1 The enzyme 5α-reductase-2

Steroid 5α-reductase (5αRD) isozymes are microsomal NADPH-dependent proteins that reduce the double bond at the 4–5 position of a variety of C_{19} and C_{21} steroids including T (**Figure 3**). Two 5αRD isozymes, type 1 and type 2, encoded by two different genes, SRD5A1 and SRD5A2, are identified in the mammal (Andersson and Russell, 1990; Andersson et al., 1991; Labrie et al., 1992; Russell and Wilson, 1994; Zhu et al., 1998). The characteristics of two human 5αRDs are listed in **Table 2**. Both human 5αRD2 and 5αRD1 genes have five exons and four introns and encode highly hydrophobic 254- and 259-amino-acid proteins with a molecular weight of approximately 28.4 and 29.5 kDa, respectively. SRD5A2 is mapped to the short arm of chromosome 2p23, and SRD5A1 to chromosome 5p15. There is approximately a 50% homology in amino acid compositions between human type 1 and type 2 isozymes. The type 2 isozyme has a much higher affinity than type 1 isozyme for substrates such as T, and finasteride, a 5αRD2 inhibitor, while the type 1 isozyme has low affinity and high capacity for T and is less sensitive to finasteride. The apparent K_m (3–10 μM) for the NADPH cofactor is similar for both isozymes.

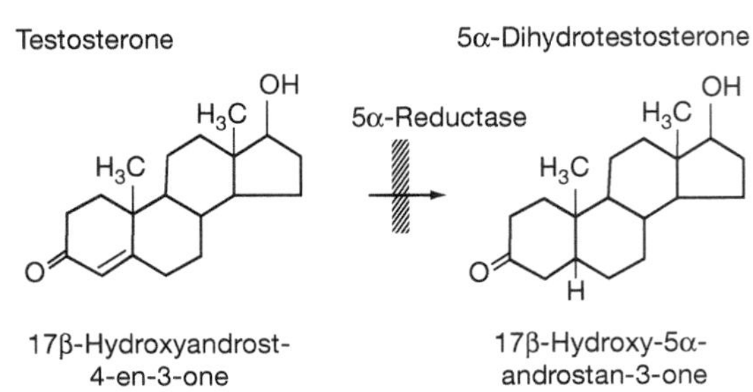

Figure 3 The 5α-reductases convert testosterone to DHT. Both type 1 and 2 isozymes can perform this conversion. The defect (indicated as shaded box) in 5α-reductase type 2 isozyme causes ambiguous genitalia in affected 46,XY subjects. Adapted from Imperato-McGinley J and Peterson RE (1976) Male pseudohermaphroditism: Complexities of male sexual development. *American Journal of Medicine* 61(2): 251–272.

Table 2 Comparison of human 5α-reductase type 1 and type 2

	Type 1	*Type 2*
Gene symbol	*SRD5A1*	*SRD5A2*
Chromosome location	5p15	2p23
Gene structure	5 exons, 4 introns	5 exons, 4 introns
Size	259 amino acids, $M_r = 29\,462$	254 amino acids, $M_r = 28\,398$
pH optima	Neutral to basic	Acidic or neutral
Tissue distribution	Liver, nongenital skin, prostate, brain, breast	Prostate, epididymis, seminal vesicle, genital skin, uterus, liver, breast, hair follicle, placenta
Prostate level	Low	High
Prostate cell distribution	Epithelial cells	Stromal and epithelial cells
Brain expression	High	Absent/low
Substrate (T) affinity	$K_m = 1–5\ \mu M$	$K_m = 0.004–1\ \mu M^a$
Mutation in 5α-reductase deficiency	Normal	Mutated
Finasteride inhibition	$K_i \geq 300$ nM	$K_i = 3–5$ nM

[a]The K_m of 5α-reductase-2 for testosterone (T) is dependent on the assay condition (Thigpen et al., 1993a).
Adapted from Wilson JD, Griffin JE, and Russell DW (1993) Steroid 5α-reductase 2 deficiency. *Endocrine Reviews* 14: 577–593; Russell DW and Wilson JD (1994) Steroid 5 alpha-reductase: Two genes/two enzymes. *Annual Review of Biochemistry* 63: 25–61; Zhu YS, Katz MD, and Imperato-McGinley J (1998) Natural potent androgens: Lessons from human genetic models. *Bailliere's Clinical Endocrinology and Metabolism* 12: 83–113; and Wiebe JP (2006) Progesterone metabolites in breast cancer. *Endocrine-Related Cancer* 13: 717–738.

The functional domains of 5αRD2 have been deduced from *in vitro* mutagenesis–transfection analysis in cultured mammalian cells (Russell and Wilson, 1994; Wigley et al., 1994; Can et al., 1998) of natural mutations of 5αRD2 gene and mutagenesis analysis of the 5αRD1 isozyme (Thigpen and Russell, 1992). Mutations affecting NADPH binding map to the carboxyl-terminal of the isozyme which appears to be the cofactor-binding domain even though consensus adenine dinucleotide-binding sequences have not been identified. In contrast, 5αRD2 mutations that affect substrate (T) binding appear to be located at both ends of the protein. However, the amino acid determinants of the substrate-binding domain appear to be located mainly at the amino terminal of the protein (Russell and Wilson, 1994).

5αRD2 is expressed in the external genital tissues early in gestation (Thigpen et al., 1993b). In adulthood, its expression in the prostate, genital skin, epididymis, seminal vesicle, and liver is relatively high, while it is quite low in other tissues. This isozyme also appears to be expressed in the ovary and hair follicles (Eicheler et al., 1994, 1995).

Mutations in the 5αRD2 gene are responsible for defects in male sexual differentiation due to 5αRD deficiency (Andersson et al., 1991; Can et al., 1998; Zhu and Imperato, 2004c). 5αRD activity is present in the external genital anlage prior to prostate and external genital differentiation, but not present in the Wolffian duct, at the time of epididymal, vas deferens, and seminal vesicle differentiation in human fetuses (Siiteri and Wilson, 1974). Studies of the affected subjects with 5αRD2 deficiency and decreased DHT production (see Section 27.4) have delineated specific actions for the two androgens *in utero*: T mediates Wolffian ductal differentiation forming the vas deferens, epididymides, and seminal vesicles, while DHT mediates male external genital and prostate differentiation (Imperato-McGinley et al., 1974; Imperato-McGinley and Zhu, 2002a). It has been shown that 5αRD activity in the external genitalia tissues and prostate of 5αRD2-deficient 46,XY individuals is decreased (Wilson et al, 1993).

5αRD1 is detected at birth in the liver and nongenital skin and is present throughout life. Its expression in embryonic tissues, however, is quite low. In adulthood, it is expressed in nongenital skin, liver, and certain brain regions; whereas its presence in the prostate, genital skin, epididymis, seminal vesicle, testis, adrenals, and kidney is low. The function of 5αRD1 in human physiology remains to be defined, although it appears to be involved in parturition and fetal survival in the rodent (Mahendroo et al., 1996, 1997).

27.3.4.2 The androgen receptor

To initiate androgen action at target areas, T enters the cell and either binds to an intracellular receptor or is metabolized to DHT depending upon the target tissue. Both T and DHT bind to the same high-affinity intracellular androgen receptor (AR) protein. The binding of androgen on the AR results in an AR conformational change that promotes the dissociation of chaperone proteins and facilitates receptor dimerization, nuclear transportation, phosphorylation, and DNA binding (Aranda and Pascual, 2001). Upon the recruitment of coregulators and general transcription factors, the transcription of a target gene is either induced or inhibited, and ultimately, leads to a change in androgen-target gene expression, and cellular or biological structures and functions (see **Figure 4**; Zhu, 2005). Recently, a number of coregulators, which can interact with AR at various domains to either enhance or reduce androgen–AR action on target gene transcription, have been identified (Heinlein and Chang, 2002). Furthermore, AR can be activated without ligands by other hormones, such as IGF-1 (Culig et al., 1994). Although significant progress has been made in the last decade in understanding androgen–AR action, the process of androgen binding to AR to the alteration of target gene transcription is not fully understood.

The AR is a member of the nuclear steroid receptor superfamily and a ligand-dependent nuclear transcription factor which was cloned in 1988 (Chang et al., 1988; Lubahn et al., 1988). It has approximately 910–919 amino acids encoded by the AR gene located in Xq11–12 (see **Figure 5**). The AR gene is a single-copy X-chromosomal gene that spans approximately 90 kb of genomic DNA. The encoding region of AR gene comprises 8 exons separated by 7 introns. Like other steroid receptors, the AR is a single polypeptide comprised of relatively distinct domains (**Figure 5**; Quigley et al., 1995): an amino-terminal A/B domain, a DBD (domain C), a hinge region (domain D), and a ligand-binding domain (LBD, domain E/F).

The large amino-terminal domain encoded by exon 1 contains a transactivation domain, activation function 1 (AF-1), and is involved in transcriptional activation of target genes. This domain plays a role in AR functions by intramolecular and/or intermolecular interaction with other coregulators (Heinlein and Chang, 2002). The N-terminal transactivation domain also contains three highly polymorphic direct repeats of amino acid residues: each containing glutamine,

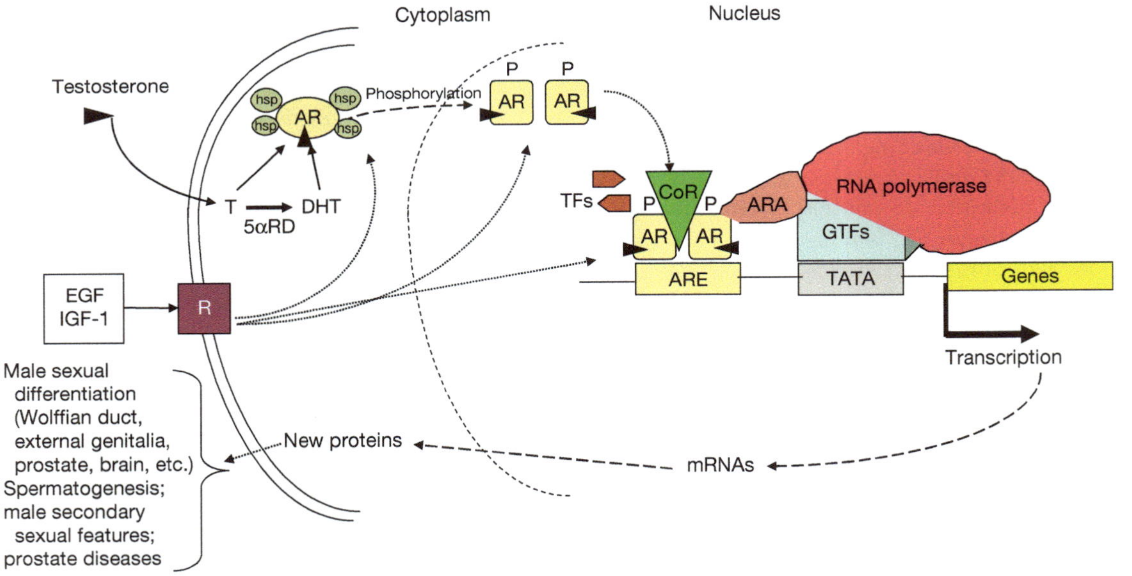

Figure 4 Illustrations of the molecular events of androgen–AR action in a target cell. When testosterone (T) enters the cell, it can be converted to dihydrotestosterone (DHT) in the cell by 5α-reductases (5αRD). Both T and DHT bind to androgen receptor (AR), resulting in a conformational change in AR and translocation of the receptor complex to nucleus. This complex interacts with androgen response element (ARE) on the target gene and regulates gene expression in concert with coregulators (CoR), transcription factors (TF), and the general transcription complex. The changes in androgen-target proteins in the cell eventually affect cellular structure and function related to male sexual differentiation, physiology, and pathophysiology. The function of AR can be activated or modified by nonligand factors such as growth factors, EGF and IGF-1. GTFs, general transcription factors; ARA, androgen-receptor-associated proteins; P, phosphorylation; TFs, transcription factors; hsp, heat shock protein. Adapted from Zhu YS (2005) Molecular basis of steroid action in the prostate. *Cell Science Reviews* 1: 27–55.

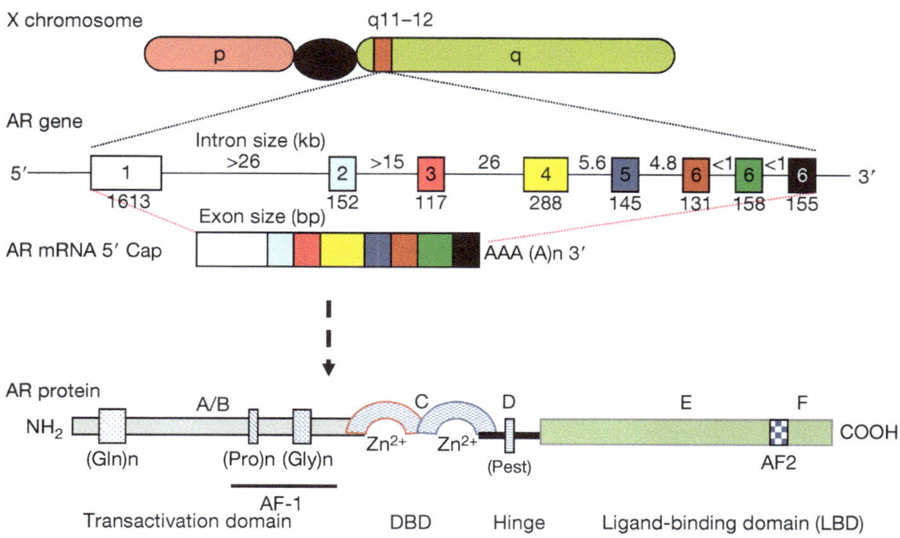

Figure 5 An illustration of the AR gene location, gene structure, and protein domains. (Top) The location of AR gene at the q11–12 of X chromosome. (Middle) The AR gene and its mRNA. The AR gene consists of 8 exons (box) and 7 introns (line), and the size of each exon and introns is indicated. (Bottom) The AR protein. The domains of AR are indicated. Relative positions of glutamine (Gln), proline (Pro), and glycine (Gly) repeats within the N-terminal domain are shown by the indicated boxes. The transactivation function domains, AF-1 and AF-2, are located within the N-terminal domain and ligand-binding domain, respectively. Two zinc fingers in the DNA-binding domain and a PEST sequence in the hinge region are indicated. Adapted from Zhu YS (2005) Molecular basis of steroid action in the prostate. *Cell Science Reviews* 1: 27–55.

proline, and glycine residues, respectively. The expansion of the size of the glutamine homopolymeric segment is related to spinal and bulbar muscular atrophy (Kennedy's disease) (La et al., 1991); conversely, the shortening of the length of glutamine and/or glycine repeats may be related to prostate cancer incidence (Hsing et al., 2000; Edwards et al., 1999; Lee and Chang, 2003), although the data are not conclusive. The mechanisms of how changes in the homopolymeric segments result in the pathological outcome are unclear, although it has been shown that expression of long-tract polyglutamine AR in neuronal cells results in aggregate formation (Piccioni et al., 2002) and progressive toxicity (Avila et al., 2003) within the cells.

The DBD encoded by exon 2 and 3 contains two zinc finger motifs that are hallmarks of all nuclear steroid receptors and is the most highly conserved region among steroid receptors (Freedman, 1992). The domain is responsible for the specific interaction between AR and its cognate DNA of target genes by interacting with the major groove of the DNA duplex. It is worthy to note that despite their exquisite functional specificity in the physiological context, receptors for androgens, glucocorticoids, progesterone, and mineralocorticoids can recognize the same DNA-response element in both *in vitro* binding assays and in functional analyses using transient cell transfection (Freedman, 1992; Evans, 1988). The answer to this nonspecific paradox may be related to the differential recruitment of various coregulators upon the ligand–receptor interaction, although further studies are required.

The carboxyl-terminus of the AR is the LBD, encoded by the 3′-portion of exon 4 and exons 5–8, and is responsible for the specific high-affinity ligand binding. The carboxyl-terminus also contains subdomains involved in dimerization and transcriptional activation (Jenster et al., 1991). The second transactivation function (AF-2) domain of AR resides within the LBD (Gronemeyer, 1991). Upon ligand binding, this AF-2 domain can interact with coregulators, such as coactivators, to affect AR function (Heinlein and Chang, 2002; Glass et al., 1997).

Between the DBD and the steroid-binding domain is the hinge region that contains the nuclear translocation signal (Jenster et al., 1991). This hinge region of all known mammalian ARs also contains a PEST (praline, glutamate, serine, and threonine-rich) sequence, which may function in proteasome-mediated AR turnover (Sheflin et al., 2000).

Despite the fact that T and DHT are active via the same AR, they produce distinct biological responses although the reasons for this are unclear. DHT has been reported to bind to the AR more avidly than T (Wilbert et al., 1983), and the DHT–receptor complex is more efficiently transformed to the DNA-binding state than is the T–receptor complex, although the differences are not enough to account for the distinct biologic responses (Kovacs et al., 1984). Whether the interaction of AR with T and DHT produces a differential recruitment of coregulators, and consequently a differential biological action, is a plausible mechanism to explain this phenomenon remains to be elucidated. Mutations in the AR result in the syndrome of androgen insensitivity (Quigley et al., 1995), which is discussed later in this article.

27.3.5 Summary

In summary, for development of the male phenotype, the testes differentiate and function at a critical period *in utero* (8–14 weeks). Acting locally, AMH secreted by the Sertoli cells suppresses the Müllerian anlage, while T produced by the Leydig cells induces Wolffian ductal differentiation. T circulates and enters the cells of the external genital anlage, where it is converted by the enzyme 5αRD2 to DHT resulting in prostate and male external genital differentiation (**Figure 1**). T and/or DHT are also responsible, at least in part, for the dimorphic differentiation and maintenance of neuronal structure and functions in the central nervous system (Swaab and Fliers, 1985; Beauchet, 2006; Zitzmann et al., 2001; Gooren and Kruijver, 2002).

27.4 Disorders of Male Sexual Differentiation Due to Defects in Androgen Production or Action

Three basic categories: disorders of testicular differentiation, disorders of testicular function, and disorders of function at androgen-dependent target areas can result in either incomplete masculinization of the genitalia or a female phenotype in 46,XY individuals (see **Table 3**).

The syndromes of 17βHSD3 deficiency, 5αRD2 deficiency, and androgen insensitivity are chosen for detailed discussion in this chapter. Studies of these syndromes lend support to the role of hormones in gender identity formation and cognitive function.

27.4.1 17βHSD3 Deficiency

27.4.1.1 The clinical syndrome of 17βHSD3 deficiency

The first case of a 46,XY male with ambiguous genitalia due to 17βHSD3 deficiency, also known as 17-ketosteroid reductase deficiency, was described in 1965 (Neher and Kahnt, 1965; Saez et al., 1971). Since then many other cases have been described. The largest pedigree of 17βHSD3 deficiency is found in an inbred Arab kindred from the Gaza Strip (Eckstein et al., 1989; Rosler et al., 1992). Affected 46,XY subjects have testes and male Wolffian duct-derived structures (epididymides, vasa deferentia, seminal vesicles, and ejaculatory ducts). Their external genitalia are frequently severely ambiguous at birth with most having pseudovaginal perineoscrotal hypospadias. Consequently, many are raised as girls (Saez et al., 1971; Rosler et al., 1992; Knorr et al., 1973; Imperato-McGinley et al., 1979c; Lee et al., 2007). Less severe defects in external genital masculinization, including micropenis, have been reported (Ulloa-Aguirre et al., 1985; Castro-Magana et al., 1993).

At puberty, substantial facial and body hair develops with genital growth (Rosler et al., 1992; Peterson and Imperato, 1984). If an infant is born with severe ambiguity of the genitalia at birth, the defect is frequently not noted until virilization occurs at puberty (Knorr et al., 1973; Virdis et al., 1978). The phenotype in adulthood is variable in that some patients can develop significant gynecomastia, while most do not.

In 46,XY subjects with 17βHSD3 deficiency, the severely ambiguous external genitalia at birth are enigmatic when compared with the significant virilization that occurs with puberty. For the fetus to have significant ambiguity of the external genitalia, theoretically, there would have to be little to no peripheral conversion of Δ4 by 17βHSD isozymes to T and DHT during early gestation (Imperato-McGinley et al., 1979c). In adulthood, however, peripheral 17βHSD activity, such as type 5 and 7, is known to be intact. An alternate possibility for lack of intrauterine virilization of the external genitalia could be due to the fact that fetal exposure to Δ4 is removed by the placenta by aromatization to estrogen, leaving little substrate available for extratesticular conversion to T and DHT (Goebelsmann et al., 1975).

Table 3 Classification of genetic defects associated with sexual determination and differentiation in 46,XY subjects

Gene	*Locus*	*Protein and function*	*Inheritance*	*Associated mutant phenotype*
Disorders of testicular differentiation and development				
WT1	11p13	TF	AD	Frasier syndrome, Denys–Drash syndrome with Wilms' tumor
SRY	Yp11	HMG-TF	Y	Gonadal dysgenesis
SOX9	17q24	HMG-TF	AD	Campomelic dysplasia, gonadal dysgenesis, or XY sex reversal
SF-1	9q33	TF, nuclear receptor	AD/AR	Gonadal and adrenal dysgenesis
DHH	12q13.1	Signal molecule	AR	Gonadal dysgenesis
ATRX	Xq13.3	Helicase	X	Gonadal dysgenesis, α-thalassaemia, mental retardation
DAX1	Xp21.3	TF, nuclear receptor	Duplication	Gonadal dysgenesis, CAH
ARX	Xp22.13	TF	X	Gonadal dysgenesis, X-linked lissencephaly, epilepsy
DMRT1	9p24.3	TF	Deletion	Gonadal dysgenesis, mental retardation
Disorders of testicular functions				
hCG/LH receptor	2p21	G-protein-coupled receptor	AR	Leydig cell dysgenesis
MIS or AMH type II receptor	12q12–13	Serine threonine kinase receptor	AR	Persistent Müllerian duct syndrome
MIS or AMH	19p13	Secreted protein, signaling molecule	AR	Persistent Müllerian duct syndrome
StAR	8p11.2	Steroidogenic acute regulatory protein	AR	Congenital lipoid adrenal hyperplasia
POR	7q11.2	P450 oxidoreductase	AR	Various phenotypes with mixed features of 21-hydroxylase, 17α-hydroxylase/17,20-lysase deficiency and DMSD
CYP17	10q24.3	17α-Hydroxylase/17,20-desmolase	AR	CAH, DMSD
HSD3B2	1p13.1	3β-Hydroxysteroid dehydrogenase type II	AR	CAH, primary adrenal failure, DMSD
HSD17B3	9q22	17β-Hydroxysteroid dehydrogenase-3	AR	DMSD
DHCR7	11q12–13	7-Dehydroxysterol reductase	AR	Smith–Lemli–Opitz syndrome
Disorders of function at the androgen-dependent target areas				
SRD5A2	2p23	5α-Reductase-2	AR	DMSD (5α-reductase-2 deficiency)
Androgen receptor	Xq11–12	Nuclear receptor, TF	X	DMSD (androgen insensitivity)
Co-activator(s)	?	Co-activator(s)	?	DMSD (androgen insensitivity)

AD, autosomal dominant; AR, autosomal recessive; CAH, congenital adrenal hyperplasia; DMSD, disorders in male sexual differentiation; HMG, high mobility group; TF, transcription factor; X, X-linked; Y, Y-linked.

27.4.1.2 Biochemical characterization of 17βHSD3 deficiency

In affected males with 17βHSD3 deficiency circulating plasma Δ4 is elevated and plasma T is low to low-normal with an elevated Δ4/T ratio. Plasma LH is elevated, while plasma FSH is normal (Goebelsmann et al., 1973) or elevated (Imperato-McGinley et al., 1979c; Givens et al., 1974). An alteration of the T/E2 ratio may be related to the development of gynecomastia in some patients with this condition (Imperato-McGinley et al., 1979c).

Plasma Δ4 and T levels are not significantly changed with suppression or stimulation of the adrenals, indicating the testes as the major source of the circulating Δ4. The testicular origin is confirmed by the findings that spermatic vein Δ4 levels are abnormally higher than T levels (Rosler and Kohn, 1983), and *in vitro* studies of testicular tissue demonstrate impaired conversion of Δ4 to T (Goebelsmann et al., 1975). hCG stimulation to enhance the enzyme defect is effectively used in the clinical diagnosis of this syndrome in the newborn and in prepubertal children (Ulloa-Aguirre et al., 1985; Levine et al., 1980; Berthezene et al., 1979).

The conversion of Δ4 to T in peripheral tissues is normal (cultured skin fibroblasts and erythrocytes) (Goebelsmann et al., 1975), and metabolic clearance rate studies reveal that approximately 90% of plasma T emanates from the extragonadal conversion of Δ4 to T in affected subjects; whereas the normal conversion is less than 1% (Horton and Tait, 1966). Plasma DHT can be normal in affected subjects despite low T, suggesting that a significant amount of DHT is produced via peripheral conversion from Δ4 (Can et al., 1998; Imperato-McGinley et al., 1979c; Goebelsmann et al., 1975). These findings correlate with the newly identified tissue distribution, catalytic activity, and substrate preference of the type 3 17βHSD isozyme and the other isozymes (see **Table 1**).

27.4.1.3 The molecular genetics of 17βHSD3 deficiency

Genetic screening of 17βHSD-deficient subjects reveal that the defect is due to mutations in the type 3 isozyme gene (Geissler et al., 1994; Can et al., 1998; Lee et al., 2007; Andersson et al., 1996b). Thus far, more than 25 different mutations of the 17βHSD3 gene have been detected (Lee et al., 2007; Andersson et al., 1996a; Zhu et al., 1998; Bertelloni et al., 2006, Human Gene Mutation Database, accessed in December 2007; **Figure 6**). The majority are missense mutations, inherited as homozygous mutations or compound heterozygous mutations. Missense mutations have been detected in seven of the 11 exons of 17βHSD3 gene, clustering in exon VIII, IX, and X. Other mutations include three splice junction alterations (325+4, 326−1, and 655−1), a single nucleotide polymorphism (G289S), a single amino acid deletion (ΔT187), and a small deletion (Δ777–783). Recently, large rearrangements of 17βHSD3 gene with a duplication of exon 2 or exons 3–10 have been reported in patients with 17βHSD3 deficiency (Michel-Calemard et al., 2005). Mutations in 17βHSD3 gene that inactivate enzymatic activity (Geissler et al., 1994; Andersson et al., 1996) cause impairment of T biosynthesis during the critical period of male sexual differentiation and result in incomplete masculinization.

Several mutations have been found in more than one family from different areas of the world. The 655–1 guanine to adenine (G→A) mutation has been identified in affected subjects from Syria, New York, and Turkey (Can et al., 1998; Andersson et al., 1996). The 325+4 adenine to thymidine (A→T) mutation is present in five families from the United States and in one family from Germany. The R80Q mutation is present in a large kindred from the Gaza Strip and in two affected subjects from Brazil. It should be noted that the 17βHSD3 deficiency is genetically heterogeneous and the phenotypic expression in the patients does not correlate with a particular 17βHSD3 gene mutation (Lee et al., 2007; Zhu et al., 1998).

27.4.2 5α-Reductase-2 Deficiency

27.4.2.1 The clinical syndrome of 5αRD2 deficiency

The syndrome of steroid 5αRD2 deficiency was first described, clinically and biochemically, in 1974 in a large Dominican kindred (**Figure 7**; Imperato-McGinley et al., 1974), and in two siblings from Dallas (Walsh et al., 1974). Two other large cohorts located in New Guinea (Imperato-McGinley et al., 1991) and Turkey (Akgun et al., 1986; Imperato-McGinley et al., 1987) have been described and

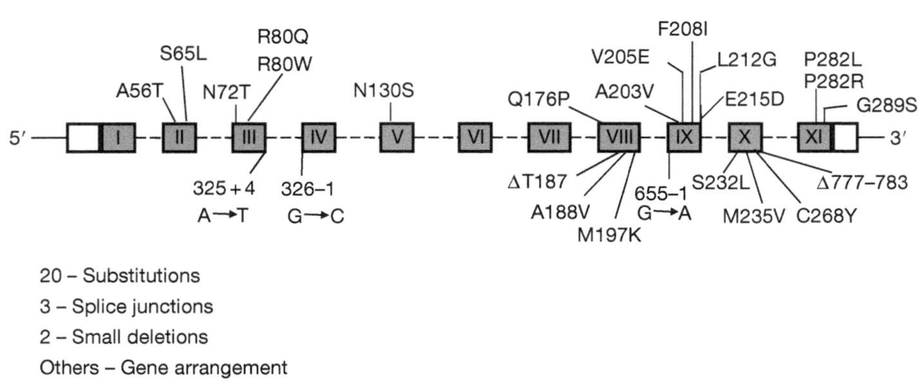

Figure 6 An illustration of gene mutations in human 17βHSD3 gene.

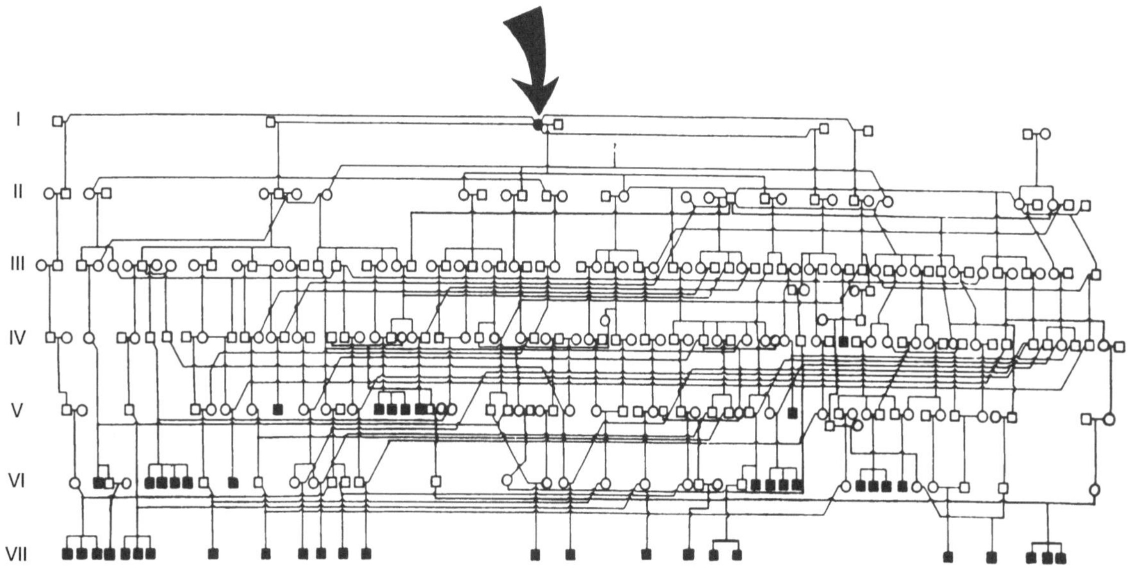

Figure 7 Pedigree of a kindred with 5α-reductase-2 deficiency. Blank circle denotes normal female; blank square denotes normal male; solid square denotes subject with 5α-reductase-2 deficiency. Adapted from Imperato-McGinley J (1996) Male pseudohermaphroditism. In: Adashi EY, Rock JA, and Rosenwaks Z (eds.) *Reproductive Endocrinology, Surgery, and Technology*, pp. 936–955. Philadelphia, PA: Lippincott-Raven Publishers.

many other cases identified (Wilson et al., 1993; Thigpen et al., 1992b; Zhu and Imperato, 2004b).

These authors had the unique opportunity of following affected members of the Dominican kindred for over 30 years, evaluating patients from childhood into adulthood. This has enabled these authors to make clinical observations providing information that is relevant to the biology of T and DHT in humans. The critical action of DHT in the development of the external genitalia and prostate is observed in patients with 5αRD2 deficiency (**Figure 8**). Most affected 46,XY subjects, homozygous for 5αRD2 deficiency, have a clitorus-like phallus, a severely bifid scrotum, pseudovaginal perineoscrotal hypospadias, and a rudimentary prostate (Imperato-McGinley and Zhu, 2002a; Zhu et al., 1998; Imperato-McGinley et al., 1974). Due to severe undermasculinization, affected males are often assigned a female gender at birth and are reared as girls (Imperato-McGinley et al., 1974, 1979a; Saenger et al., 1978; Cohen-Kettenis, 2005). On occasion, more masculinized subjects have been described who lack a separate vaginal opening (Imperato-McGinley et al., 1980), or who have penile hypospadias (Carpenter et al., 1990), or even a penile urethra (Ng et al., 1990). Wolffian ductal differentiation to form the seminal vesicles, vasa deferentia, epididymides, and ejaculatory ducts is normal. No Müllerian structures are present. Cryptorchidism is frequently described and the testes can be in the abdomen, but are more commonly found in the inguinal canal (Imperato-McGinley et al., 1974).

Pubertal changes include increased muscle mass and deepening of the voice (Imperato-McGinley et al., 1974; Imperato-McGinley and Zhu, 2002a). Prominent muscle mass is found in Dominican, New Guinean, and Turkish subjects (Imperato-McGinley et al., 1974, 1991; Akgun et al., 1986). Affected subjects are as tall as their unaffected male siblings, and there is no gynecomastia in adulthood (Imperato-McGinley and Zhu, 2002a; Peterson et al., 1977). With puberty, the phallus grows and the scrotum becomes rugated and hyperpigmented. Descent of the inguinal testes into the scrotum at the time of puberty has been observed in some patients (Cai et al., 1994). Libido is intact, and affected subjects have erections (Imperato-McGinley and Zhu, 2002a). From the clinical findings it appears that libido, increased muscle mass, deepening of the voice, male sexual function, and spermatogenesis are primarily T mediated (Imperato-McGinley and Zhu, 2002a).

Although patients are generally oligo- or azoospermic, normal sperm concentrations have been reported in patients with descended testes (Cai et al., 1994; Cantu et al., 1976; Katz et al., 1997). One

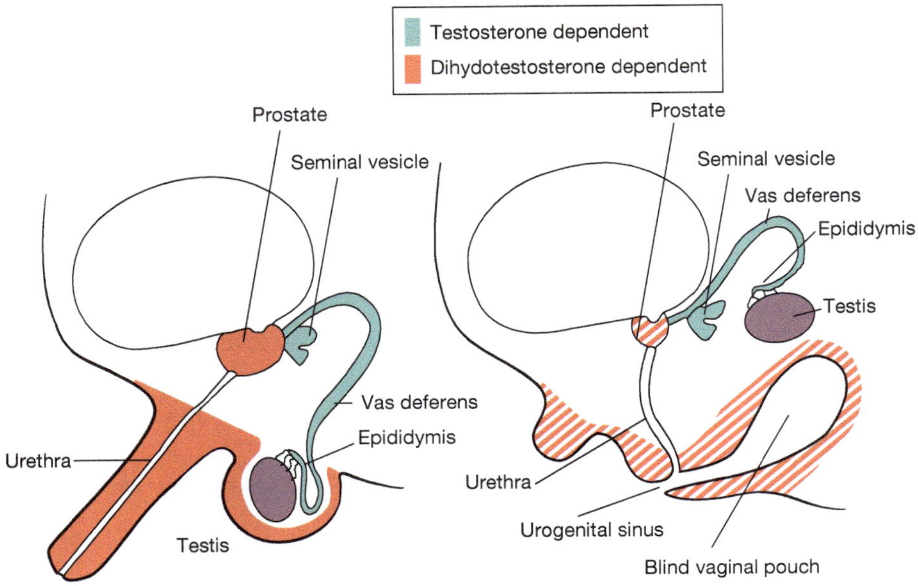

Figure 8 Illustration of the specific role of testosterone and DHT in male sexual differentiation *in utero*. Adapted from Imperato-McGinley, Guerrero L, Gautier T, and Peterson RE (1974) Steroid 5α-reductase deficiency in man: An inherited form of male pseudohermaphroditism. *Science* 186: 1213–1216.

patient from the Dominican kindred successfully fathered three children (two pregnancies) via intrauterine insemination of his wife with his sperm (Katz et al., 1997), and two brothers from Sweden have also been reported to have fathered children (Nordenskjold and Ivarsson, 1998).

46,XY individuals with 5αRD2 deficiency have less facial and body hair than their normal male relatives in adulthood, and male-pattern baldness has never been observed (Imperato-McGinley and Zhu, 2002a; Imperato-McGinley et al., 1974). Although the type 1, rather than the type 2 isozyme appears to predominate in the scalp (Thigpen et al., 1993b; Schweikert et al., 1974; Harris et al., 1992), baldness may be related to 5αRD2 expression in hair follicles (Eicheler et al., 1994, 1995) and/or to circulating DHT, which is mainly dependent on 5αRD2 activity.

Although males with inherited 5αRD deficiency rarely have significant acne, nonetheless, they produce normal amounts of sebum, suggesting that sebum production is under the control of the 5αRD1 isozyme (Imperato-McGinley et al., 1993) or, alternatively, is a T- and not a DHT-dependent process.

Affected males have nonpalpable prostates on rectal examination (Imperato-McGinley et al., 1974). On transrectal ultrasound and MRI visualization, the prostate was found to be rudimentary with prostatic volumes one-tenth the size of age-matched, normal controls (Imperato-McGinley and Zhu, 2002a; Zhu, 2005; Imperato-McGinley et al., 1992b). These findings have been confirmed by others (Mendonca et al., 1996), providing clinical evidence that prostate differentiation and growth is mediated largely by DHT. Consequently, 5αRD inhibition as treatment for benign prostatic hypertrophy evolved largely from the clinical observation that adult male subjects with 5αRD2 deficiency have rudimentary prostates due to lifelong DHT deficiency.

27.4.2.2 Biochemical characterization of 5αRD2 deficiency

The biochemical characteristics of 5αRD2 deficiency are characterized by (Zhu and Imperato-McGinley, 2004c; Imperato-McGinley and Zhu, 2002a; Imperato-McGinley et al., 1974): (1) normal-to-elevated levels of plasma T; (2) decreased levels of plasma DHT; (3) an increased T- to -DHT ratio at baseline in adults or following hCG stimulation in children; (4) decreased plasma and urinary 3α-androstanediol glucuronide, a major metabolite of DHT; (5) decreased conversion of T to DHT *in vivo*, with conversion ratios of T to DHT of $<1\%$ (normal $>7\%$); (6) reduced 5αRD activity in genital tissue slices and cultured fibroblasts; (7) normal metabolic clearance rates of T and DHT; and (8) a global defect in steroid 5α-reduction with decreased urinary 5α-reduced metabolites of C19 steroids in addition to T (androgens) and C21 steroids, that is, cortisol, corticosterone, 11β-hydroxy-androstenedione, and Δ4. However, despite the global steroid defect in 5α-reduction, only the defective reduction of T appears to have clinical significance (Imperato-McGinley and Zhu, 2002a).

Mean plasma levels of LH can be twice the normal despite elevated plasma T (Canovatchel et al., 1994). Pulsatility studies demonstrate an increase in LH pulse amplitude, with normal LH frequency, defining a role for DHT in the negative feedback control of LH (Canovatchel et al., 1994).

27.4.2.3 Molecular genetics of 5αRD2 deficiency

Approximately 61 mutations in the 5αRD2 gene have been identified to date (**Figure 9**), including the mutations affecting three large kindreds (Dominican, New Guinean, and Turkish kindreds) ((Zhu and Imperato, 2004c; Zhu et al., 1998; Baldinotti et al., 2008), and Human Gene Mutation Database, accessed in December 2007). A New Guinean kindred of affected males who were biochemically characterized as having 5αRD deficiency participated in the first genetic study (Andersson et al., 1991; Imperato-McGinley et al., 1991). A large deletion of more than 20 kb in the *SRD5A2* gene was demonstrated (Andersson et al., 1991). The Dominican kindred mutation is a missense mutation in exon 5, substituting a thymidine for a cytosine resulting in a substitution of the nonpolar amino acid tryptophan for the basic, polar amino acid arginine at position 246 of the isozyme (Thigpen et al., 1992a; Cai et al., 1996). In the Turkish kindred, a single base deletion in exon 5 of the 5αRD2 gene has been detected (Can et al., 1998), resulting in a frameshift at amino-acid-position 251 with an addition of 23 amino acids at the carboxyl-terminal of this 254-amino-acid isozyme. The mutation results in a complete loss of enzymatic activity.

The 61 mutations identified in the 5αRD2 gene, so far, are found throughout all five exons of the gene, and range from a single-point defect to a gene deletion. These mutations include 50 missense mutations, six small deletions, three splicing junction alterations, one single nucleotide insertion, and a large deletion involving the entire gene. As a consequence of the mutations, several types of enzymatic dysfunction occur, including impaired binding of substrate and cofactor to the isozyme; blocked formation of a functional isozyme (deletion, nonsense mutation, or splice-junction alterations); and an unstable isozyme (Thigpen et al., 1992b). Certain mutations can affect either substrate or cofactor binding, suggesting that these areas may be distinct (Russell and Wilson, 1994; Wigley et al., 1994). Approximately 35% of patients with 5αRD2 deficiency from different families worldwide have been found to be either

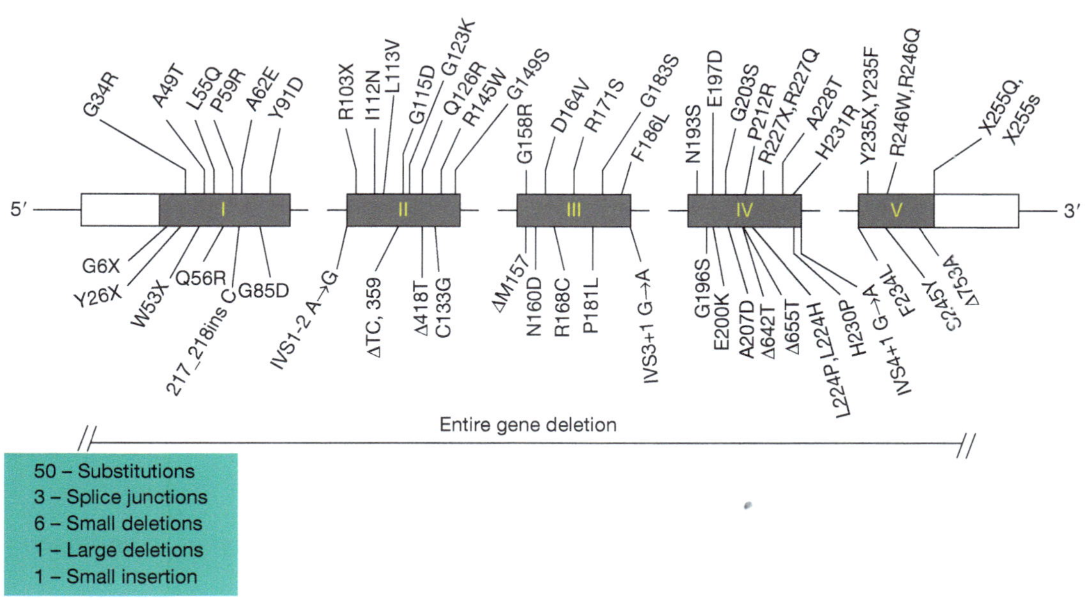

Figure 9 An illustration of gene mutations in human 5αRD2 gene.

compound, or inferred compound, heterozygotes with different allelic mutations, suggesting that the carrier frequency is higher than previously suggested (Thigpen et al., 1992b).

In the Dominican (Imperato-McGinley et al., 1974; Cai et al., 1996) and Turkish kindreds (Can et al., 1998), pedigree studies combined with geographic isolation demonstrate significant inbreeding indicating a founder effect. Mutations have also been found to be shared among different ethnic groups and may be due to hot spots or areas of vulnerability within the gene (Zhu et al., 1998; Thigpen et al., 1992b). In the Turkish kindred with the existence of a 5αRD2 and 17βHSD3 gene defect, some affected subjects described are homozygous for either 5αRD2 or 17βHSD3 and heterozygous for the other mutation (Can et al., 1998).

27.4.3 Androgen Insensitivity Syndrome

Androgen insensitivity syndrome can be classified as complete androgen insensitivity syndrome (CAIS) or partial androgen insensitivity syndrome (PAIS) (Quigley et al., 1995; Zhu and Imperato-McGinley, 2004a). CAIS was initially described in 1817 when an autopsy of a 21-year-old amenorrheic female revealed testes, excellent breast development, and lack of axillary, pubic, and body hair (Steglehner, 1817). In 1953, the condition was described in detail and named testicular feminization (Morris, 1953). Wilkins (1957) postulated that the syndrome was secondary to androgen insensitivity due to the fact that when he administered T intramuscularly to a patient with complete androgen insensitivity, failure of nitrogen retention or virilization was noted.

27.4.3.1 The androgen insensitivity syndrome

46,XY subjects with complete androgen insensitivity are phenotypic females. Consequently, the condition is rarely suspected before puberty, unless inguinal or labial masses are noted and found to be testes. During puberty, there is breast development, but absent or scant development of pubic and axillary hair.

A short vagina with an absent cervix and uterus in a hairless female with breast development strongly suggests the diagnosis of complete androgen insensitivity. Although subjects with complete androgen insensitivity have testes and normal androgen secretion, the external genitalia are female with an absent or hypoplastic Wolffian ductal system. Normal testicular secretion and responsiveness to AMH occurs *in utero*, resulting in Müllerian-derived structures that are absent or rudimentary.

Rarely, 46,XY individuals with 17α-hydroxylase deficiency can also have absent or decreased secondary sexual hair and develop gynecomastia at puberty (Imperato-McGinley and Canovatchel, 1992a).

Subjects with partial or incomplete forms of androgen insensitivity have a spectrum of clinical phenotypes including: gynecomastia with severely ambiguous genitalia, mild hypospadias, gynecomastia alone, normal genitalia with decreased body hair in adulthood, and infertility alone (Quigley et al., 1995; Zhu and Imperato, 2004a; Wilson, 1985).

27.4.3.2 The biochemical characterization of androgen insensitivity syndrome

In subjects with CAIS, plasma T levels are usually normal to high-normal. Plasma LH is increased despite the normal to high-normal levels of T, correlating well with the histological findings of Leydig cell hyperplasia (Judd et al., 1974; Imperato-McGinley et al., 1982; Imperato-McGinley and Canovatchel, 1992a). These findings are consistent with androgen unresponsiveness at the level of the hypothalamus and/or pituitary. Since LH levels are not in the castrate range, a supporting role for the negative feedback of estrogen on the hypothalamus and/or pituitary is suggested. FSH is normal to elevated (Judd et al., 1974; Imperato-McGinley et al., 1982).

Urinary estrogens range from high male to low female levels, with the testes producing substantial amounts of estradiol (Kelch et al., 1972; MacDonald et al., 1979). An unopposed estrogen effect, resulting from increased estrogen together with androgen unresponsiveness, is the theoretic explanation for breast development occurring during puberty (Imperato-McGinley et al., 1982).

Similar to individuals with CAIS, affected subjects with PAIS may have elevated LH and T levels (Quigley et al., 1995; Zhu and Imperato, 2004a). Total 17β-estradiol produced and secreted by the testes can be greater than in patients with CAIS (MacDonald et al., 1979). Despite increased estrogen production, the degree of feminization during puberty is not as marked as in testicular feminization. This may indicate a less severe androgen and estrogen imbalance at the cellular level.

27.4.3.3 Molecular genetics of androgen insensitivity syndrome

It has long been known from human pedigree studies that androgen insensitivity syndrome is maternally

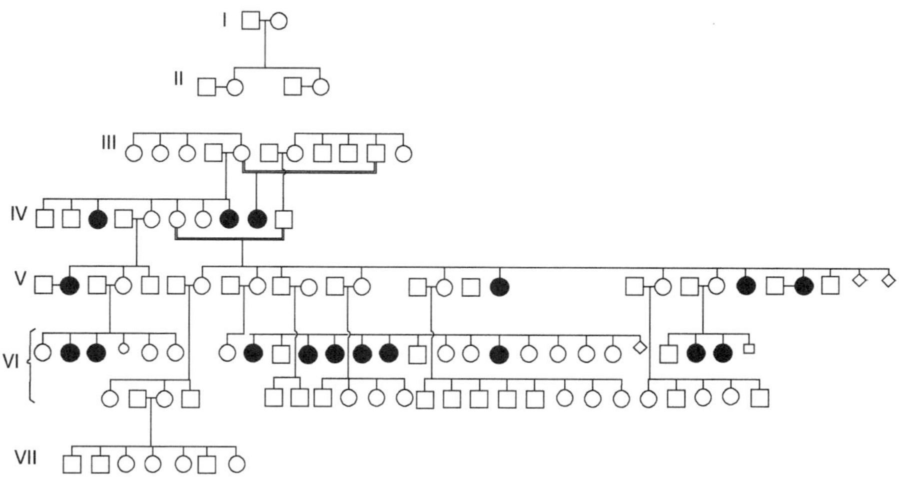

Figure 10 Pedigree of a kindred with complete androgen insensitivity. Blank circle denotes normal female; blank square denotes normal male; and solid circle denotes affected subject with complete androgen insensitivity. Adapted from Imperato-McGinley J, Peterson RE, Gauiter T, et al. (1982) Hormonal evaluation of a large kindred with complete androgen insensitivity: Evidence for secondary 5α-reductase deficiency. *Journal of Clinical Endocrinology and Metabolism* 54: 931–941.

transmitted to males only, suggesting X-linked recessive inheritance (**Figure 10**). With the cloning of AR gene (Chang et al., 1988; Lubahn et al., 1988), it was confirmed that mutations in AR gene located between Xq11 and Xq12 (Trapman et al., 1988; Brown et al., 1989; Imperato-McGinley et al., 1990) are responsible for the syndrome (Quigley et al., 1995; Zhu and Imperato, 2004a; Zhu et al., 1999).

To date, more than 295 different mutations throughout all eight exons in the AR gene have been reported, including the mutation for the largest pedigree of CAIS ((Zhu et al., 1999; Quigley et al., 1995; Zhu and Imperato, 2004a; Hiort et al., 1996), Human Gene Mutation Database and AR gene mutation database) are available online. These mutations range from a single point mutation to an entire gene deletion and can result in various AR dysfunction, including impaired androgen binding; impaired DNA binding; impaired cofactor interaction; blocked formation of a functional receptor (deletions, nonsense mutations, splice-junction alterations); decreased AR expression, and an unstable androgen–receptor complex. Depending upon the type of dysfunction, various degrees of functional impairment of the AR occur (Quigley et al., 1995; Brinkmann, 2001). Generally, mutations that delete the entire AR gene, or interrupt the AR open-reading frame, blocking the formation of a functional receptor resulted from premature termination, aberrant splicing, or deletion of partial or complete exon segments are associated with CAIS. This is because AR DBD and hormone-binding domain, two critical domains for AR function, are located at the carboxyl terminus of the AR protein. Therefore, defects truncating the receptor protein at any point during its synthesis result in the removal of a portion of one or both of these important functional domains, leading to a complete loss of AR function. A recent study using a transgenic mouse model has further demonstrated that the genomic actions of the AR, which regulates target gene expression, are required for normal male sexual differentiation (Notini et al., 2005).

The majority of AR mutations, approximately 67%, related to androgen insensitivity are located in the AR LBD. Most of these mutations are single nucleotide substitutions and cause defects in androgen binding. Approximately 15% of AR mutations are located in the DBD, which affect the interaction of the androgen–AR complex with the androgen-response DNA element of the androgen-target genes. Some AR mutations affect AR expression, resulting in decreased AR protein levels and decreased androgen-binding capacity in the genital skin fibroblasts of affected individuals. These mutant ARs may have subtle differences in function on selected target genes when analyzed by *in vitro* transfection. However, the levels of AR in these individuals are significantly decreased due to mutations that alter AR gene transcription, translation, post-translational processes, or stability (Avila et al., 2002).

In most patients with CAIS, a mutation in AR gene can be identified. However, a CAIS patient with a normal AR gene has been reported, which may be evidence of a possible coactivator defect based on the analysis of AR signal transduction (Adachi et al., 2000). This factor appears to be a specific AR coactivator, but it remains to be characterized. Furthermore, a KO of the SRC-1 gene (a general coactivator of steroid receptors) in the mouse has been shown to cause a partial resistance to sex hormones including androgens (Xu et al., 1998). However, to date, the SRC-1 gene mutation has not been identified in patients with androgen insensitivity syndrome.

In general, PAIS is usually related to AR gene mutations resulting in more subtle changes in receptor function. A variety of AR gene mutations have been identified with a wide spectrum of defects in receptor function (Quigley et al., 1995; Zhu and Imperato-McGinley, 2004a). However, there is no specific correlation between a specific AR gene mutation and a distinct clinical phenotype, especially in subjects with PAIS. Actually, a range of phenotypes have been described in 11 members of one family (Wilson et al., 1974), indicating that a single mutant gene, variably expressed, can be a factor in the variant phenotypic forms of partial androgen insensitivity.

27.5 Gender Identity Development

Gender identity is the sense of being male or female; it is the self-awareness of knowing one's sex. Gender role is the public expression of one's gender identity. It is manifested by one's behavior or actions as male, female, or ambivalent. The basic question of the relative influence of genetic factors, hormonal factors, and environmental factors in the determination of gender identity is the subject of continuous debate.

27.5.1 Social Theory in Gender Development

The theory of sexual neutrality at birth was proposed by Money et al. (1955b), which stated that sexuality in man was undifferentiated at birth. It became differentiated as either masculine or feminine with various social experiences, particularly, the sex of rearing. This concept was later expanded to recognize that human male and female infants expressed sexually dimorphic behavior from birth. However, it was postulated that such behavior could be incorporated into either a male or female gender identity pattern and consequently was not the exclusive property of either sex (Money and Ehrhardt, 1972; Money and Ogunro, 1974). The hypothesis was tested by matching subjects with ambiguous genitalia that were chromosomally, gonadally, and otherwise diagnostically the same. The matched pairs were said to differ only in their sex assignment and sex of rearing. The studies showed that the gender identity of the individuals was in accordance with the sex of rearing and not with the chromosomal or gonadal sex. This led to the conclusion that the sex of rearing, that is, social experiences, predominated in establishing gender identity in man (Money and Ogunro, 1974).

These studies were flawed since the methodology for adequate hormone evaluation was not available at the time. Consequently, subjects were not matched for their similar hormonal milieu – a critical factor. Exposure to androgens *in utero* and postnatally, therefore, was assumed but not documented to be similar for the matched pairs of subjects. In addition, castration and sex hormone therapy were usually initiated to coincide with the sex of rearing interrupting the natural hormonal sequence of events (Money and Ogunro, 1974; Money et al., 1955a). Thus, the issue of nature (i.e., androgen) versus nurture (i.e., sex of rearing) in the determination of a male gender identity was not resolved by these data mainly due to incomplete hormonal characterization (Imperato-McGinley et al., 1981).

27.5.2 Hormone-Influence Theory in Gender Development

Our initial studies of subjects with 5αRD2 and DHT deficiency (see Section 27.4) brought into question the theory of socially acquired gender (Imperato-McGinley et al., 1974, 1979a, 1981; Imperato-McGinley and Zhu, 2002b). These studies revealed that despite an unambiguous female sex of rearing, a male gender identity evolved followed by gender role change from female to male with or following puberty in many subjects. This unique condition, with normal T production and response, demonstrated for the first time the importance of androgen (T) exposure prenatally and/or postnatally in male gender identity formation in man. Based on these data, we proposed what is now known as the hormone-influence theory in gender identity development (Imperato-McGinley and Zhu, 2002a; Imperato-McGinley et al., 1979a).

This concept is supported by the famous John/Joan/John story (Diamond and Sigmundson, 1997) – a boy whose penis was accidently ablated at the age of 8 months during circumcision by cautery. This boy was subsequently raised as a girl, and orchidectomy and constructive surgery to fashion a full vagina were performed approximately within 1 year to facilitate feminization. Initially this individual was described as developing into a normally functioning female. However, the individual was later found to reject this sex of rearing. When the psychiatrist interviewed him, he chose to live as a male, was given male sex hormone injections, a mastectomy, and a phalloplasty. He married a woman at the age of 25 and adopted her children. He ultimately committed suicide which has been attributed partially if not completely to his anguished history (Byne, 2006).

Recent observations in 46,XY subjects with cloacal extrophy, a disorder of embryogenesis involving the genitourinary and intestinal tracts, provide further evidence to support the hormone-influence theory (Reiner and Gearhart, 2004; Reiner, 2004, 2005). The external genitalia in 46,XY individuals with cloacal extrophy are often grossly anomalous or absent; however, testicular function is normal. These individuals are often assigned as female at birth, castrated to prevent a male puberty, and treated with estrogen at puberty to stimulate the development of female secondary sexual characteristics. In a study that assessed 16 genetic males with cloacal extrophy, 14 underwent neonatal assignment to the female sex socially, legally, and surgically; two were assigned as male (Reiner and Gearhart, 2004). Eight of the 14 subjects assigned to the female sex declared themselves male during the course of 34–98 months of follow-up and ranged in age from 5 to 16, whereas the two who were raised as males remained male. All 16 subjects had moderate-to-marked interests and attitudes that were considered typical of males. These data collectively suggest, as we have postulated, that: (1) androgen exposure prenatally plays a significant role in male gender identity determination; (2) both environmental and hormonal factors act to determine male gender identity; and (3) gender identity is not fixed at 18 months to 3–4 years, but is continually expanding through childhood and adolescence, becoming firmly fixed thereafter.

27.5.3 Genetic Factors on Gender Development

To date, there is very little information concerning the influence of genetic factors in gender identity. There are a few reports of chromosomal abnormalities in transsexual individuals (Swaab, 2004). However, no evidence can be found that the chromosomal change has a direct effect on gender identity or sexual orientation. Rather, the influence is indirect through determination of the embryonic gonad and its hormonal secretion and action (Gooren, 2006). It is well documented that 46,XY individuals with CAIS due to AR mutations are phenotypically female, and the published literature does not contain any reports of affected individuals changing to a male gender identity (Zhu and Imperato, 2004a; Byne, 2006; Gooren, 2006; Meyer-Bahlburg, 1965). This suggests that androgen action, but not the Y chromosome itself, is the primary factor in controlling male gender development.

27.6 Gender Identity in Specific Inherited Disorders Affecting Androgen Biosynthesis and Androgen Actions

27.6.1 Gender Identity in Subjects with 5αRD2 Deficiency

Subjects with steroid 5αRD deficiency have a unique biochemical defect as discussed in detail earlier in this chapter. The biosynthesis of T and the peripheral action of T is normal and thus prenatal and postnatal T exposure of the brain proceeds as in the normal male. However, because of deficient 5αRD2 enzymatic activity which impairs conversion of T to DHT, decreased DHT production occurs *in utero*, resulting in genital ambiguity of affected male fetuses. Many affected subjects are believed to be female at birth and consequently are raised as girls. At puberty, however, significant virilization occurs under the influence of normal plasma T levels (Zhu and Imperato, 2004c; Imperato-McGinley and Zhu, 2002a; Imperato-McGinley et al., 1974; Peterson et al., 1977). Thus, 5αRD2-deficient subjects provide a unique opportunity to evaluate the relative influences of T and the sex of rearing in the determination of gender identity in man.

Studies of a large Dominican kindred revealed that there were 38 known subjects (adults and children) with a 5αRD deficiency from 23 interrelated families spanning four generations (Zhu and Imperato-McGinley, 2004c; Imperato-McGinley et al., 1974; Peterson et al., 1977). Interviews concerning the psychosexual development of affected subjects were carried out in two villages (Imperato-McGinley et al., 1979a, 1981). Historical data were obtained by interviewing both adult affected male subjects who were raised as females and those who were raised as males. Social practices within villages were also investigated in order to discern the influence of cultural factors on the gender of the subjects (see **Table 4**). Psychosexual data were obtained from 18 of 19 adult subjects (Imperato-McGinley et al., 1979a, 1981). Sixteen of the 18 affected subjects successfully changed to a male gender identity and a male gender role. The change occurred either during puberty or in the postpubertal period with an average age of 16. In some, a gender role change was delayed until they felt confident of their ability to defend themselves if it were deemed necessary (Imperato-McGinley et al., 1974, 1979a, 1981).

In affected individuals, libido is intact with normal erections (Imperato-McGinley et al., 1974; Imperato-McGinley and Zhu, 2002a). The ages of initiation of morning erections, nocturnal emissions, masturbation, and first sexual intercourse were compared to those raised as girls and those raised as boys and were not strikingly different. The time of first sexual intercourse was 15–18 for affected subjects raised as girls, 15–17 for those raised as boys, and 14–16 for male controls in the village (Imperato-McGinley et al., 1979a, 1981).

Table 4 Subjects with 5α-reductase deficiency who were unambiguously raised as females

Subject	*Age – 1978 (years)*	*Age of gender role change*
Village A		
1	59 (59)[a]	16
2	58	17
3	58	24
4	55	No change, male gender identity, and female gender role
5	53	Puberty
6	50	15
7	49	16
8	45 (74)[a]	21
9	44	16
10	40	15
11	36	15
15	24	16
Village B		
24	40	20
25	37	No change, female gender identity, and female gender role
26	35	Left town at 20 (no follow-up)
27	28	14
28	26	15
29	26	15
30	23	17

[a]Deceased.

Adapted from Imperato-McGinley J, Peterson RE, Gautier T, and Sturla E (1981) The impact of androgens on the evolution of male gender identity. In: Kogan SJ and Hafez ESE (eds.) *Pediatric Andrology*, pp. 99–108. The Hague: Martinus Nijhoff.

Four patterns of male sexual behavior differentiation, as described by Diamond (1965), were evaluated: sexual patterns (sex-related behavior) – which in the male includes direct aggressiveness, assertiveness, large motor activity, occupation; sexual gender identity – the sex to which an individual ascribes; sexual object of choice – the sex of the individual chosen as an erotically interesting partner; and sexual mechanisms – the features of sexual expression over which an individual has little control, which for males includes the ability to obtain and maintain an erection and to achieve orgasm. The affected subjects (with one exception) perform male work such as farmers, miners, or woodsmen in a society where there is a definite division of labor according to sex. They chose females as their sexual object of choice (**Table 4**). Fifteen of the 16 subjects who changed to a male gender role either live with women in common-law marriages or lived in common-law relationships in the past. They enjoy their role as head of the household with some choosing women who have children from previous unions. The adequacy of sexual intercourse depends upon the severity of the chordee and the size of the phallus.

Although these subjects behave as males, they experienced certain insecurities due to their abnormal appearing genitalia, and they view themselves as incomplete males. They also fear ridicule by members of the opposite sex and feel quite anxious about forming new sexual relationships.

Presently the villagers either raise their children as males from birth or change the sex of rearing to male when the problem is recognized. Some families will opt to raise affected subjects as females despite complete knowledge of what will transpire with puberty (Imperato-McGinley et al., 1979a, 1981).

The observation of the social and psychosexual development of New Guinean subjects with 5αRD2 deficiency provides another example of this phenomenon (Imperato-McGinley et al., 1991a; Gajdusek, 1977, 1989; Farquhar and Gajdusek, 1981). In the past, New Guinean 5αRD subjects were raised as girls until puberty, whereupon they made the transition to male (Gajdusek, 1977). Today, however, as in the Dominican kindred the condition is usually recognized in childhood, if not at birth, and the affected subjects are raised as male.

It is believed by some that New Guinean subjects with 5αRD deficiency are regarded as a third sex and change to a male gender role to adapt to their male-admiring society (Herdt and Davidson, 1988). It has been the experience of Carlton Gadjusek as well as our own experience that these subjects clearly regard themselves as male. To support this, three affected New Guinean 5αRD2 patients specifically requested and obtained genital correction in adulthood so that they could be, as they stated, complete men (Imperato-McGinley et al., 1991a; Fratinanni and Imperato-McGinley, 1994).

Gender change in subjects with 5αRD2 deficiency has been noted in affected subjects from many other countries including Turkey, Mexico, Cyprus, Algeria, Italy, Lebanon, Brazil, Pakistan, Saudi Arabia, UAE, Sweden, and other Dominican subjects not members of the large kindred (Zhu and Imperato, 2004c; Wilson et al., 1993; Zhu et al., 1998; Imperato-McGinley et al., 1979a). In a recent analysis, it has reported that roughly 63% of affected subjects who were raised as female at birth have changed gender role from female to male at age 12 or older (Cohen-Kettenis, 2005). However, this report may underestimate the numbers of gender role change as: (1) affected subjects under 12 years old were excluded from the analysis. Some of these subjects had already changed gender role to male, and some may later want to live as men and (2) some young affected subjects who live as female at the time reported (12–18 years old) may have a male gender identity role, but for societal reasons may be unable to live as a male. Our study in the Dominican kindred indicates that over 80% of affected subjects who were raised as a female at birth successfully changed to a male gender identity and a male gender role (Imperato-McGinley et al., 1979a, 1981).

Subjects with 5αRD deficiency demonstrate that despite a female sex of rearing, the majority of subjects change to a male gender identity and a male gender role during or following puberty. In the communities described, these events generally occurred without physician intervention and/or other social factors which might act to interrupt the sequence of events. There was no surgical correction of the external genitalia to coincide with the female sex assignment or the administration of female hormone therapy. The parental attitude of the Dominican subjects when the change was occurring was one of amazement, confusion, and finally acceptance. The embarrassment and possible harassment afterward by other villagers was the cause of major anxiety leading some to hesitate changing gender role for a time. However, the pressures were generally not strong enough to prevent the change to a male gender role from ultimately occurring.

We have continued to observe the behavioral characteristics of subjects for over 20 years and have no doubts as to their male behavioral pattern in adulthood. We also evaluated a 65-year-old male subject with 5αRD2 deficiency born in southern Italy and raised as a girl. The subject emigrated to the United States, and at age 16, without medical or psychologic intervention, realized a male gender identity with puberty. He has a male gender identity and has lived publically for many years with a woman but because of family pressure continues to retain a female gender role (Imperato-McGinley et al., 1980).

From these studies it can be postulated that in normal males, the sex of rearing and androgen imprinting of the brain act in concert to determine the expression of the male gender (**Figure 11**). These subjects demonstrate that in a *laissez-faire* environment, when the sex of rearing (female) is discordant with the T-mediated biologic sex, the biologic sex will prevail if puberty occurs. It appears that the extent of T exposure of the brain *in utero*, possibly in the early postnatal period and at puberty has greater impact in determining male gender identity than a female sex of rearing. It also appears that gender identity is not fixed at 18 months to 3–4 years, but is continually

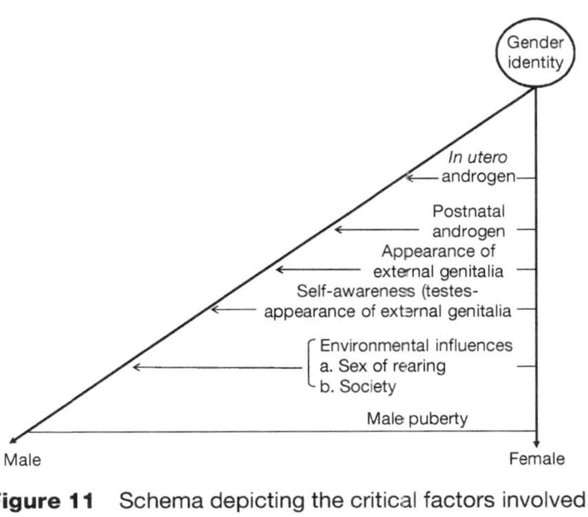

Figure 11 Schema depicting the critical factors involved in the evolution of a male gender identity in man. Adapted from Imperato-McGinley J, Peterson RE, Gautier T, and Sturla E (1981) The impact of androgens on the evolution of male gender identity. In: Kogan SJ and Hafez ESE (eds.) *Pediatric Andrology*, pp. 99–108. The Hague: Martinus Nijhoff.

expanding through childhood and adolescence becoming firmly fixed thereafter.

Theoretically, masculinization of the brain occurs under the influence of T-mediated puberty, and a male gender identity develops, overriding the female sex of rearing. Androgens therefore appear to act as inducers and activators in evolution of male gender identity in man (Imperato-McGinley et al., 1979a,b, 1981). In theory, however, it is unclear which androgen, T or DHT, mediates these processes since both type 1 and 2 5αRD isozymes are present in the brain, and since type 1 isozyme is normal in subjects with 5αRD2 deficiency it is possible that T is acting by conversion to DHT in the brain by the isozyme of 5αRD1 (see Section 27.3).

In 46,XY subjects who have inadequate T production or action, it is understandable that gender identity might coincide with the sex of rearing. Adequate androgen imprinting may not have occurred *in utero* and/or neonatally, and therefore the sex of rearing is predominant. Such individuals are a testimonial to the malleability of human beings in gender identity acquisition, but they do not approximate the normal events.

There are other reported cases of male sexual differentiation disorders, who have successfully changed gender from female to male occurring after the proposed critical period (Imperato-McGinley et al., 1979b; Reiner and Gearhart, 2004; Chapman et al., 1951; Burns et al., 1960; Berg and Leeds, 1963; Stoller, 1964). In many cases, the change in gender role occurred during adolescence. Although adequate hormonal evaluation was not available in these cases, they also challenge both the theory of the immutability of gender identity after the age of 3 or 4 and the sex of rearing as the major factor in determining male gender identity (Imperato-McGinley et al., 1979a).

In summary, environmental or sociocultural factors are not solely responsible for the formation of a male gender identity. Androgens make a strong and definite contribution. Analogous to the induction of a male phenotype from an inherent female phenotype, the formation of a male gender identity in man also appears to be at least partially induced by androgens from an undifferentiated and/or inherently female nervous system (**Figure 11**; Imperato-McGinley et al., 1979a, 1981).

In children with 5αRD2 deficiency, the most desirable situation is to raise the child as a male, since it is well documented that affected subjects can identify and behave as males in adulthood, marry, and father children despite a female sex of rearing (Zhu and Imperato-McGinley, 2004b; Katz et al., 1997; Imperato-McGinley, 1997). This necessitates early diagnosis of the condition followed by surgical correction of the external genitalia with correction of cryptorchidism if necessary. As most patients have perineoscrotal hypospadias and a small phallus, surgical correction is difficult. However, surgical correction of the genitalia is facilitated following enlargement of the phallus with topical administration of DHT cream (Carpenter et al., 1990; Imperato-McGinley, 1997). With successful genital repair, the parents can be assured that (1) the child will have a male puberty with normal male psychosexual development; (2) he will be equal in height to the normal males in his family; (3) there will be growth of the genitalia; (4) an increase in muscle mass and deepening of the voice; (5) gynecomastia will not be a concern; and (6) fertility is possible if cryptorchidism is corrected early, or through intrauterine insemination or *in vitro* fertilization (Katz et al., 1997; Nordenskjold and Ivarsson, 1998, and our unpublished data).

A serious debate involves management of subjects who are raised as females and diagnosed as having 5αRD2 deficiency in the peripubertal and postpubertal period. After careful psychiatric evaluation, subjects will often be found to identify as males and should be encouraged to take their place as males in society. It should be noted that pseudovaginal perineoscrotal hypospadias have been successfully repaired in adulthood in subjects with this condition. There are other subjects who identify as male but who may change gender role with time only if they can successfully deal with the social pressures of family, friends, etc. (unpublished). Thus, whether or not a gender role change from female to male will occur in an individual with 5αRD2 deficiency at this time is obviously dependent upon specific social and cultural factors, which might either consciously or subconsciously suppress or foster the gender change. All these factors must be considered by the patient's physician, as well as by the psychiatrist, working together with the patient and the family (Zhu and Imperato-McGinley, 2004b; Imperato-McGinley, 1997).

27.6.2 Gender Identity in Subjects with 17βHSD3 Deficiency

The first case of a documented change in gender from female to male in a subject born and raised in the United States with 17βHSD3 deficiency was in 1979 (Imperato-McGinley et al., 1979c). Since then, gender change from female to male has been reported in other subjects with 17βHSD3 deficiency

(Can et al., 1998; Rosler et al., 1992; Rosler and Kohn, 1983; Imperato-McGinley et al., 1987; Cohen-Kettenis, 2005; Mendonca et al., 2000), including subjects from a large Arab kindred in the Gaza Strip (Rosler and Kohn, 1983). Twenty-three affected subjects from 14 sibships, with ten sibships related to this large inbred cohort extending over eight generations, were identified. Affected male subjects with 17βHSD3 deficiency were described who were born with severe ambiguity of the genitalia; they were raised as girls and attended girls' schools. With puberty, they developed male secondary sexual characteristics with phallic enlargement and abundant body and facial hair, consequent to the peripheral conversion of Δ4 to T by other 17βHSD isozymes (Andersson and Moghrabi, 1997; Labrie et al., 1997; Can et al., 1998; Zhu and Imperato-McGinley, 2004b). Some exhibited aggressive behavior that led to their dismissal from school. Eventually, six spontaneously changed gender role and some began working in physically demanding jobs. This was done on their own initiative, some without parental consent or supportive psychiatric help. The affected subjects are capable of having erections with ejaculations. Other individuals have continued in their female gender role. Three subjects were castrated after diagnosis at the decision of the physicians. The adult subjects living as females stated that they would have preferred to have been raised as males. None of those living as females married. An 82-year-old subject from the kindred has a female gender role, but apparently has a male gender identity. This individual was working as a farm laborer and was proud that she was stronger and more productive than her male colleagues (Rosler and Kohn, 1983). A recent analysis of the published cases with 17βHSD3 deficiency shows that roughly 64% of subjects raised as female change their gender from female to male spontaneously or after consulting with a physician and psychiatrist peripubertally (Cohen-Kettenis, 2005). These data suggest that a biologic force can override the female sex of rearing.

It is known that conversion of Δ4 to T is possible in the human brain (Steckelbroeck et al., 1999), and except for the 17βHSD3 isozyme which is deficient in these subjects, other 17βHSD isozymes are normally expressed in the brains of both humans and animals (Martel et al., 1992; Stoffel-Wagner et al., 1999; Takeyama et al., 2000; and see **Table 1**). Thus, alternate pathways for T formation via other 17βHSD isozymes are present in the brain and extragonadal tissues in patients with 17βHSD3 deficiency and, theoretically, can cause androgen imprinting of the brain *in utero*.

27.7 Sex Differences in Cognitive Function and Laterality

Studies over the years, including two major meta-analyses, of sex differences in spatial abilities have indicated that spatial abilities of males on certain tasks are consistently superior than those of females (Maccoby and Jacklin, 1974; Linn and Petersen, 1986; Voyer et al., 1995), while females outperform males on verbal fluency and perceptual speed tasks (Gooren and Kruijver, 2002). A recent study by Voyer et al. (1995) has analyzed a variety of spatial abilities, and a number of tests showed significant sex differences with male advantage that were stable across age. These analyses included Mental Rotation tasks, the PMA Spatial Relations subtest, and the Rod-and-Frame test. Gender difference with male superiority in mathematics performance on SAT has remained constant (Feingold, 1988). Astur et al. (1998) have shown, by using a computerized version of the Morris Water Task, that males navigated to the hidden platform better than females across a variety of measures. This male superiority in visual–spatial tasks may be due to the fact that these tasks are mainly right hemispheric functions (Bogen, 1969), and males appear to have more consistent right hemispheric lateralization than females (Ray et al., 1976).

There is a trend for greater right ear superiority in men than women (Bryden, 1988). The auditory testing of laterality produces the most robust effects with the dichotic consonant–vowels (CV) syllable pairs tasks the most reliable approach to testing of this type (Voyer and Flight, 2000). It seems reasonable to suggest that the sex differences described above are linked to differences in functional organization of the brain.

Using echo-planar functional magnetic resonance imaging (fMRI), Shaywitz et al. (1995) and Pugh et al. (1996) for the first time objectively demonstrated sex differences in the functional organization of the brain. Male and female subjects who were right handed were imaged during letter recognition, rhyme, and semantic tasks. During phonological (rhyme) tasks, brain activation was found to be lateralized to the left inferior frontal gyrus regions in males, while in females it was more diffuse, involving both the left and right inferior frontal gyrus. Using a complex, three-dimensional, virtual reality maze analysis with fMRI, Gron et al. (2000) revealed a sex difference in the brain with a distinct activation of the left hippocampus in males, while a consistently activation of the right parietal and right prefrontal cortex occurred in females. Recently, using fMRI, we found a sex difference in brain activation with a mental rotation task (**Figure 12**; Butler et al., 2006). During mental rotation test, men activated primary sensory cortices as well as regions involved in implicit learning (basal ganglia) and mental imagery (precuneus), while women have greater activity in the dorsomedial prefrontal and other high-order heteromodal association cortices (Butler et al., 2006). Moreover, only women demonstrated ventral anterior cingulated cortex suppression and inverse functional connectivity with dorsal anterior cingulated cortex (Butler et al., 2007). Taken together, these studies have begun to provide objective evidence to support a sex difference in the functional organization of the brain.

The reasons for the sex difference in cognitive abilities and brain functions are not fully understood. One of the hypotheses is that prenatal/postnatal androgen exposure is responsible for the male-patterned cognitive abilities and functional organization of the brain, although there is no doubt that many additional factors may affect these cognitive abilities such as education, experience, environment, and cultural background (Gooren and Kruijver, 2002) (see Section 27.7.2 for further discussion).

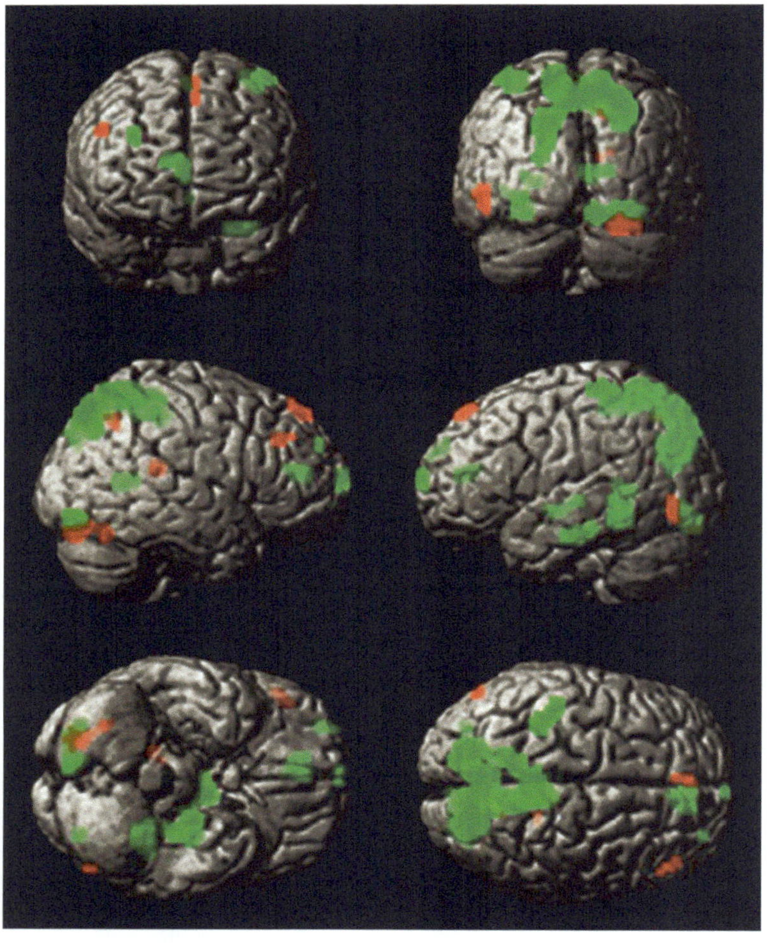

Figure 12 Network of brain regions active in inverse correlation with activity in left parieto-insular vestibular cortex (PIVC; x, −33, y, −3, z, 9) during mental rotation in men (green) and women (red). Results are displayed at a threshold of $P < 0.05$, corrected for multiple comparisons over the whole brain. Note extensive activation of parietal–occipital cortices in men, but not women. Adapted from Butler T, Imperato-McGinley J, Pan H, et al. (2006) Sex differences in mental rotation: Top–down versus bottom–up processing. *Neuro Image* 32: 445–456.

27.7.1 Cognitive Abilities in Androgen-Insensitive Subjects

Subjects with complete androgen insensitivity are phenotypic and psychosexual females who undergo breast development at puberty without secondary sexual hair.

Using the Spanish Version of the Wechsler Intelligence Scale (EIWA), we evaluated the relationship between androgen unresponsiveness and cognitive abilities with particular attention to subtests of visual–spatial ability (Imperato-McGinley et al., 1991). Subjects were matched for genetic and sociocultural factors. This was purposefully done to negate the possible influence of these factors on test performance. The study is unique in that the affected subjects, as well as control males and females as mentioned above, are all members of a large kindred with this condition (**Figure 10**). The control males were brothers and nephews of androgen-insensitive subjects. Control females included female siblings of androgen-insensitive subjects; they all had children, none of whom is known to be affected. Other female controls were daughters of female siblings of androgen-insensitive subjects, daughters of a normal brother, and nieces of androgen-insensitive subjects.

General intellectual scores for assessment of cognitive abilities, reported as Full Scale Intelligence Quotient (FIQ), Verbal Intelligence Quotient (VIQ), and Performance Intelligence Quotient (PIQ), were analyzed and compared, as well as the 11 subtests of the EIWA. Comparison was also made utilizing Cohen's cognitive factorial clusters in assessing the Wechsler Adult Intelligence Scale (WAIS) subtests (Cohen, 1957), which divided subtest scores into three factors: verbal comprehension (VC), perceptual organization (PO), and freedom from distractibility (FD). VC includes the verbal abilities subtests of information, comprehension, similarities, and vocabulary. PO includes the visuospatial subtests of digit symbol, picture completion, picture arrangement, block design, and object assembly, and FD includes the subtests of arithmetic and digit span.

Compared to the matched normal male and female controls, subjects with CAIS have no significant difference in FIQ, VIQ, and VIQ-performance IQ. However, significant differences are observed among these three groups in PIQ (see **Table 5**).

Examination of the univariable tests of significance indicated that there were significant differences on seven subtests of the WAIS using Cohen's factorial clusters. Multiple comparisons using the Newman–Keuls' method at the 0.05 level indicated that males were superior to females on the FD subtest of arithmetic and the PO subtests of block design, picture completion, and object assembly. Males had significantly higher scores than androgen-insensitive subjects on all seven of the subtests. Androgen-insensitive subjects scored lower than normal females on the subtests of digit symbol, block design, picture arrangement, and picture completion.

Since eight CAIS subjects were from two large families within the kindred, their scores were also compared to those of their female siblings. In one family, three CAIS subjects and three female siblings were tested, and in the second family, five CAIS and five female siblings were tested. There were no significant differences between CAIS subjects and their female siblings in this subsample on either age or education. The two groups were similar in their overall VIQ, although they differed in PIQ. With a three-way analysis of variance with androgen sensitivity and family as between-subjects factors, and the four subsets (digit symbol, picture completion, block design, and picture arrangement) as a within-subjects factor, an overall significant difference between CAIS subjects and their female siblings was found with no test by androgen insensitivity interaction, indicating that the group difference was consistent across all four subtests. Thus, CAIS subjects also show specific deficits on PIQ and four perceptual organization subtests when compared to the small group of their own female siblings. Interestingly, a left-handed CAIS subject and two elderly CAIS subjects demonstrated the same general basic test pattern as the right-handed subjects.

Since androgen-insensitive subjects are raised as females and have a totally female psychosexual orientation, their cognitive performance, therefore, may reflect their sex roles, as a reflection of social experience and values placed upon them. However, this consideration does not explain their significantly lower overall performance when compared to control females from the same kindred on the perceptual organization factor and subtests of spatial ability. This exaggerated female pattern of performance suggests an effect of androgen unresponsiveness.

Similar observations have been reported by other investigators (Masica et al., 1969; Collaer and Hines, 1995; Perlman, 2000). Subjects with CAIS have a modest but consistent and significant tendency toward superiority of verbal over space-form abilities using the WAIS. Performance-perceptual scores are poorer than both male and female controls.

Table 5 Comparison of cognitive functions among CAIS and age-matched males and females – Mean age-scaled scores of the Spanish version of the WAIS

Measure	*Group*		
	Males (n = 9)	*Females (n = 25)*	*CAIS (n = 10)*
Full IQ	115.2	108.7	104.1
Verbal IQ	116.8	111.1	109.4
Performance IQ	112.1[a]	105.2	98.2
Verbal IQ-performance IQ	4.7	5.8	11.2
Cohen's factorial clusters			
Verbal comprehension	12.7	12.4	12.2
Information	12.4	11.6	11.1
Similarities	12.6	12.8	11.7
Comprehension	12.7	12.2	12.6
Vocabulary	13.2	13.0	13.2
Perceptual organization	12.2[a,b]	10.7[c]	9.0
Digit symbol	11.8[a]	11.3[c]	8.6
Block design	12.7[a,b]	11.0[c]	9.3
Picture arrangement	11.6[a]	10.3[c]	8.2
Picture completion	11.9[a,b]	10.1[c]	8.5
Object assembly	13.0[a,b]	10.9	10.2
Freedom from destructibility	13.1[a,b]	10.8	10.3
Digit span	12.3[a]	10.8	9.9
Arithmetic	13.9[a,b]	10.9	10.6
Verbal comprehension Perceptual organization	0.5[a]	1.7	3.2

[a]$p < 0.05$ compared to CAIS.
[b]$p < 0.05$ compared to females.
[c]$p < 0.05$ compared to CAIS.
CAIS, complete androgen insensitive subjects.
Adapted from Imperato-McGinley J, Pichardo M, Gautier T, Voyer D, and Bryden MP (1991b) Cognitive abilities in androgen-insensitive subjects: Comparison with control males and females from the same kindred. *Clinical Endocrinology (Oxf)* 34: 341–347.

27.7.2 Other Studies of Cognitive Function in Hypogonadal Males

Men with idiopathic hypogonadotrophic hypogonadism and low T have also been found to have higher VIQ than Performance IQ on the EIWA (Bobrow et al., 1971). However, due to the small sample size of the VIQ investigation by Bobrow et al. (1971), statistical comparisons were not performed. Idiopathic hypogonadotrophic hypogonadal men show a visuospatial deficit in impaired performance on the Wechsler Block Design subtest as well as on the Space Relations subtest of the Differential Aptitude Test, the Embedded Figures Test, and the Rod-and-Frame Test (Buchsbaum and Henkin, 1980). Hier et al. (1982) reported that the severity of the hormonal defect in adult men with idiopathic hypogonadotrophic hypogonadism correlated with the degree of the visuospatial deficit; and treatment with T in adulthood did not improve performance on these tests. In contrast, they found that men who became hypogonadal following normal puberty had no visuospatial deficits, suggesting the importance of early androgen exposure before or at puberty in spatial proficiency. In other studies, however, these differences in cognitive abilities were not observed between men with hypogonadotrophic hypogonadism and control subjects (Buchsbaum and Henkin, 1980; Cappa et al., 1988). In another study, a subset of this category of patients showed abnormal spatial attention on a reaction time task, although the overall reaction time was faster than controls (Kertzman et al., 1990). A recent analysis suggests that low endogenous levels of T may be related to reduced cognitive ability; while T substitution in hypogonadal and elderly men may improve some aspects of cognitive ability such as visuospatial function (Beauchet, 2006). Further well-designed clinical studies are necessary to obtain conclusive information.

With the advance in imaging techniques, the effects of T on neurological activity in different regions of the brain have been analyzed. Using ^{18}F-deoxyglucose positron emission tomography, Zitzmann et al. (2001) studied cerebral glucose metabolism during a standardized mental rotation task in six hypogonadal men.

Each patient performed the test before and during treatment with T substitution. During T substitution, four patients exhibited improved visuospatial performance, which corresponded with enhanced cerebral glucose metabolism in corresponding brain regions such as right inferior occipital gyrus, right inferior frontal gyrus, right middle temporal gyrus, and left primary visual cortex during the test. Azad et al. (2003) showed that cerebral perfusion determined by single photon emission computed tomography was increased in the midbrain and the superior frontal gyrus after 3–5 weeks of T substitution in seven men with hypogonadism; and increased perfusion was still observed in the midbrain and midcingulate gyrus at 12–14weeks of treatment. Finally, Park et al. (2001) demonstrated differences in regional brain activation among 12 eugonadal sexually potent men and two hypogonadal impotent men in response to visual erotic stimulation using fMRI. Activation of cerebral cortices in the two men with hypogonadism was low but was partially restored following T substitution (Park et al., 2001). Taken together, these data indicate that exposure to androgen displays specific effects on brain functions related to cognitive ability in addition to its actions on sexual morphological dimorphism in the brain, although the biological effects of androgen in the brain are far from fully understood.

27.8 Conclusion

Natural human genetic models with a deficiency in androgen production or action provide an invaluable tool to evaluate the importance of androgen in the male sexual differentiation and in the development of male gender identity/role, and male-patterned behavior and cognitive function. Studies over the last three decades demonstrate that an androgen, most probably T, plays a critical role in the imprinting of male-patterned cognitive function and behavior, as well as in the development of male gender identity/role during a critical period. Although genetic factors may play a role in sexual morphological and functional dimorphism in the brain in animal studies, there is no such evidence in humans. 46,XY subjects with complete androgen insensitivity due to AR mutations have a female gender identity, female-patterned behavior, and cognitive function. These androgen actions appear to be mediated via androgens interacting with the AR and not via conversion of androgen to estrogen interacting with estrogen receptors as demonstrated in animal studies (Goy and McEwen, 1980; Ogawa et al., 2000). This concept is further supported by the fact that male subjects with an estrogen receptor mutation (Smith et al., 1994), or aromatase mutation (Carani et al., 1997; Bilezikian et al., 1998), resulting in deficiency in estrogen production and action, have male gender identity and role. The fact that the majority of subjects with 5αRD2 deficiency or 17βHSD3 deficiency, who were raised as female, changed their gender identity and role to male during or after puberty suggests that the development of gender identity is evolving throughout childhood and is flexible until puberty.

It should be realized that androgen is not the sole factor in controlling male cognitive ability and male gender identity/role development. Social or environmental factors work with the endocrine milieu together to determine the ultimate outcome. However, the critical period of androgen exposure of the brain for the development of male cognitive function and male gender identity and role is uncertain, and the biological effects of androgen in the brain are far from fully understood. Recent developments using advanced imaging technology to analyze neurological activity in humans have given objective information concerning androgen action and brain function, and it will continue to provide further insight on androgen effects on the sexual dimorphism of brain functions.

References

Adachi M, Takayanagi R, Tomura A, et al. (2000) Androgen-insensitivity syndrome as a possible coactivator disease. *New England Journal of Medicine* 343: 856–862.

Akgun S, Ertel N, Imperato-McGinley J, Sayli B, and Shackleton CHL (1986) Familial male pseudohermaphroditism in a Turkish village due to 5α-reductase deficiency. *American Journal of Medicine* 81: 267–274.

Andersson S, Berman DM, Jenkins EP, and Russell DW (1991) Deletion of steroid 5α-reductase-2 gene in male pseudohermaphroditism. *Nature* 354: 159–161.

Andersson S, Geissler WM, Wu L, et al. (1996a) Molecular genetics and pathophysiology of 17 beta-hydroxysteroid dehydrogenase 3 deficiency. *Journal of Clinical Endocrinology and Metabolism* 81: 130–136.

Andersson S and Moghrabi N (1997) Physiology and molecular genetics of 17β-hydroxysteroid dehydrogenases. *Steroids* 62: 143–147.

Andersson S and Russell DW (1990) Structural and biochemical properties of cloned and expressed human and rat steroid 5α-reductases. *Proceedings of the National Academy of Sciences of the United States of America* 87: 3640–3644.

Andersson S, Russell DW, and Wilson JD (1996b) 17β-Hydroxysteroid dehydrogenase 3 deficiency. *Trends in Endocrinology and Metabolism* 7: 121–126.

Aranda A and Pascual A (2001) Nuclear hormone receptors and gene expression. *Physiological Reviews* 81: 1269–1304.

Arey L (1965) *Developmental Anatomy: A Textbook and Laboratory Manual of Embryology*. Philadelphia, PA: Saunders.

Astur RS, Ortiz ML, and Sutherland RLA (1998) Characterization of performance by men and women in a virtual Morris water task: A large and reliable sex difference. *Behavioural Brain Research* 93: 185–190.

Auchus RJ (2004) The backdoor pathway to dihydrotestosterone. *Trends in Endocrinology and Metabolism* 15: 432–438.

Avila DM, Allman DR, Gallo JM, and McPhaul MJ (2003) Androgen receptors containing expanded polyglutamine tracts exhibit progressive toxicity when stably expressed in the neuroblastoma cell line, SH-SY 5Y. *Experimental Biology and Medicine (Maywood)* 228: 982–990.

Avila DM, Wilson CM, Nandi N, Griffin JE, and McPhaul MJ (2002) Immunoreactive AR and genetic alterations in subjects with androgen resistance and undetectable AR levels in genital skin fibroblast ligand-binding assays. *Journal of Clinical Endocrinology and Metabolism* 87: 182–188.

Azad N, Pitale S, Barnes WE, and Friedman N (2003) Testosterone treatment enhances regional brain perfusion in hypogonadal men. *Journal of Clinical Endocrinology and Metabolism* 88: 3064–3068.

Baker TG (1963) A quantitative and cytological study of germ cells in human ovaries. *Proceedings of the Royal Society of London, Series B* 158: 417–433.

Baldinotti F, Majore S, Fogli A, et al. (2008) Molecular characterization of 6 unrelated Italian patients with 5 {alpha}-reductase type 2 deficiency. *Journal of Andrology* 29: 20–28.

Beauchet O (2006) Testosterone and cognitive function: Current clinical evidence of a relationship. *European Journal of Endocrinology* 155: 773–781.

Bercu BB and Schulman JD (1980) Genetics of abnormalities of sexual differentiation and of female reproductive failure. *Obstetrical and Gynecological Survey* 35: 1–11.

Berg I and Leeds MB (1963) Change of assigned sex at puberty. *Lancet* 2: 1216.

Bertelloni S, Maggio MC, Federico G, Baroncelli G, and Hiort O (2006) 17Beta-hydroxysteroid dehydrogenase-3 deficiency: A rare endocrine cause of male-to-female sex reversal. *Gynecological Endocrinology* 22: 488–494.

Berthezene F, Forest MG, de-Peretti E, Chevallier M, Claustrat B, and Dubernard JM (1979) Study of a new case of male pseudohermaphroditism due to 17-ketosteroid reductase deficiency. *Annals of Endocrinology* 40: 549–550.

Berube D, Luu-The V, Lachance Y, et al. (1989) Assignment of the human 3β-hydroxysteroid dehydrogenase gene (HSDB3) to the p13 band of chromosome 1. *Cytogenetics and Cell Genetics* 52: 199–200.

Bilezikian JP, Morishima A, Bell J, and Grumbach MM (1998) Increased bone mass as a result of estrogen therapy in a man with aromatase deficiency. *New England Journal of Medicine* 339: 599–603.

Biswas MG and Russell DW (1997) Expression cloning and characterization of oxidative 17β- and 3α-hydroxysteroid dehydrogenases from rat and human prostate. *Journal of Biological Chemistry* 272: 15959–15966.

Block E (1953) Quantitative morphological investigation of follicular system in newborn female infants. *Acta Anatomica* 17: 201–206.

Bobrow NA, Money J, and Lewis VG (1971) Delayed puberty, eroticism, and sense of smell: A psychological study of hypogonadotropinism, osmatic and anosmatic (Kallmann's syndrome). *Archives of Sexual Behavior* 1: 329–344.

Bogen JE (1969) The other side of the brain. II. An appositional mind. *Bulletin of the Los Angeles Neurological Society* 34: 135–162.

Bongiovanni AM (1961) Unusual steroid pattern in congenital adrenal hyperplasia: Deficiency of 3β-hydroxy dehydrogenase. *Journal of Clinical Endocrinology and Metabolism* 21: 860–862.

Brinkmann AO (2001) Molecular basis of androgen insensitivity. *Molecular and Cellular Endocrinology* 179: 105–109.

Brooke AM, Taylor NF, Shepherd JH, et al. (2006) A novel point mutation in P450c17 (CYP17) causing combined 17alpha-hydroxylase/17,20-lyase deficiency. *Journal of Clinical Endocrinology and Metabolism* 91: 2428–2431.

Brown CJ, Goss SJ, Lubahn DB, et al. (1989) Androgen receptor locus on the human X chromosome: Regional localization to Xq11–12 and description of a DNA polymorphism. *American Journal of Human Genetics* 44: 264–269.

Bryden MP (1988) An overview of the dichotic listening procedure and its relation to cerebral organization. In: Hughdal K (ed.) *Handbook of Dichotic Listening*, pp. 1–45. Toronto, ON: Wiley.

Buchsbaum MS and Henkin RI (1980) Perceptual abnormalities in patients with chromatin negative gonadal dysgenesis and hypogonadotropic hypogonadism. *International Journal of Neuroscience* 11: 201–209.

Burns E, Segaloff A, and Carrera GM (1960) Reassignment of sex: Report of 3 cases. *Journal of Urology* 84: 126.

Butler T, Imperato-McGinley J, Pan H, et al. (2006) Sex differences in mental rotation: Top–down versus bottom–up processing. *Neuro Image* 32: 445–456.

Butler T, Imperato-McGinley J, Pan H, et al. (2007) Sex specificity of ventral anterior cingulate cortex suppression during a cognitive task. *Human Brain Mapping* 28: 1206–1212.

Byne W (2006) Developmental endocrine influences on gender identity: Implications for management of disorders of sex development. *Mount Sinai Journal of Medicine* 73: 950–959.

Cai L-Q, Fratianni CM, Gautier T, and Imperato-McGinley J (1994) Dihydrotestosterone regulation of semen in male pseudohermaphrodites with 5 alpha-reductase-2 deficiency. *Journal of Clinical Endocrinology and Metabolism* 79(2): 409–414.

Cai L-Q, Zhu YS, Katz MD, et al. (1996) 5α-Reductase-2 gene mutation in the Dominican Republic. *Journal of Clinical Endocrinology and Metabolism* 81: 1730–1735.

Can S, Zhu YS, Cai LQ, et al. (1998) The identification of 5α-reductase-2 and 17β-hydroxysteroid dehydrogenase-3 gene defects in male pseudohermaphrodites from a Turkish kindred. *Journal of Clinical Endocrinology and Metabolism* 83: 560–569.

Canovatchel WJ, Volquez D, Huang S, Wood E, Lesser ML, Gautier T, and Imperato-McGinley J (1994) Luteinizing hormone pulsatility in subjects with 5-alpha-reductase deficiency and decreased dihydrotestosterone production. *Journal of Clinical Endocrinology and Metabolism* 78: 916–921.

Cantu JM, Hernandez-Montes H, del Castillo V, Cortés-Gallegos V, Sandoval R, Armendares S, and Parra A (1976) Potential fertility in incomplete male pseudohermaphroditism type 2. *Revista de Investigacion Clinica* 28: 177–182.

Cappa SF, Guariglia C, Papagno C, et al. (1988) Patterns of lateralization and performance levels for verbal and spatial tasks in congenital androgen deficiency. *Behavioral Brain Research* 31: 177–183.

Carani C, Qin K, Simoni M, et al. (1997) Effect of testosterone and estradiol in a man with aromatase deficiency. *New England Journal of Medicine* 337: 91–95.

Caron KM, Soo SC, Wetsel WC, Stocco DM, Clark BJ, and Parker KL (1997) Targeted disruption of the mouse gene encoding steroidogenic acute regulatory protein provides insights into congenital lipoid adrenal hyperplasia. *Proceedings of the National Academy of Sciences of the United States of America* 94: 11540–11545.

Carpenter TO, Imperato-McGinley J, Boulware SD, Weiss RM, Shackleton CHL, and Wilson JD (1990) Variable expression of 5 alpha-reductase deficiency: Presentation with male phenotype in a child of Greek origin. *Journal of Clinical Endocrinology and Metabolism* 71(2): 318–322.

Castro-Magana M, Angulo M, and Uy J (1993) Male hypogonadism with gynecomastia caused by late-onset deficiency of testicular 17-ketosteroid reductase. *New England Journal of Medicine* 328: 1297–1301.

Catt KJ, Dufau ML, Neaves WB, Walsh PC, and Wilson JD (1975) LH–hCG receptors and testosterone content during differentiation of the testis in the rabbit embryo. *Endocrinology* 97: 1157–1165.

Chang C, Kokontis J, and Liao S (1988) Molecular cloning of human and rat complementary DNA encoding androgen receptors. *Science* 240: 324–326.

Chapman AH, Saslow G, and Watson F (1951) Pseudohermaphroditism, a medical, social, and psychiatric case study. *Psychosomatic Medicine* 13: 212.

Clark BJ, Wells J, King SR, and Stocco DM (1994) The purification, cloning, and expression of a novel luteinizing hormone-induced mitochondrial protein in MA-10 mouse Leydig tumor cells. Characterization of the steroidogenic acute regulatory protein (StAR). *Journal of Biological Chemistry* 269: 28314–28322.

Clements J, Reyes E, Winter JSD, and Faiman C (1976) Studies on human sexual development. III. Fetal pituitary and serum, and amniotic fluid concentrations of LH, CG, and FSH. *Journal of Clinical Endocrinology and Metabolism* 42: 9–19.

Cohen J (1957) The factorial structure of the WAIS between early adulthood and old age. *Journal of Consulting Psychology* 21: 451–457.

Cohen-Haguenauer O, Picard JY, Mattei MG, et al. (1987) Mapping of the gene for anti-Mullerian hormone to the short arm of human chromosome 19. *Cytogenetic and Genome Research* 44: 2–6.

Cohen-Kettenis PT (2005) Gender change in 46,XY persons with 5alpha-reductase-2 deficiency and 17beta-hydroxysteroid dehydrogenase-3 deficiency. *Archives of Sexual Behavior* 34: 399–410.

Collaer ML and Hines M (1995) Human behavioral sex differences: A role for gonadal hormones during early development? *Psychological Bulletin* 118: 55–107.

Colvin JS, Green RP, Schmahl J, Capel B, and Ornitz DM (2001) Male-to-female sex reversal in mice lacking fibroblast growth factor 9. *Cell* 104: 875–889.

Culig Z, Hobisch A, Cronauer MV, et al. (1994) Androgen receptor activation in prostatic tumor cell lines by insulin-like growth factor-I, keratinocyte growth factor, and epidermal growth factor. *Cancer Research* 54: 5474–5478.

Diamond MA (1965) Critical evaluation of the ontogeny of human sexual behavior. *Quarterly Review of Biology* 40: 147–175.

Diamond M and Sigmundson HK (1997) Sex reassignment at birth. Long-term review and clinical implications. *Archives of Pediatrics and Adolescent Medicine* 151: 298–304.

Donahoe PK, Ito Y, Morikawa Y, and Henderson WH (1977) Mullerian inhibiting substance in human testes after birth. *Journal of Pediatric Surgery* 12: 323–330.

Eckstein B, Cohen S, Farkas A, and Rosler A (1989) The nature of the defect in familial male pseudohermaphroditism in Arabs of Gaza. *Journal of Clinical Endocrinology and Metabolism* 68: 477–485.

Edwards SM, Badzioch MD, Minter R, et al. (1999) Androgen receptor polymorphisms: Association with prostate cancer risk, relapse and overall survival. *International Journal of Cancer* 84: 458–465.

Eicheler W, Dreher M, Hoffmann R, Happle R, and Aumuller G (1995) Immunohistochemical evidence for differential distribution of 5 alpha-reductase isoenzymes in human skin. *British Journal of Dermatology* 133: 371–376.

Eicheler W, Tuohimaa P, Vilja P, Adermann K, Forssmann WG, and Aumuller G (1994) Immunocytochemical localization of human 5 alpha-reductase 2 with polyclonal antibodies in androgen target and non-target human tissues. *Journal of Histochemistry and Cytochemistry* 42: 667–675.

Evans RM (1988) The steroid and thyroid hormone receptor superfamily. *Science* 240: 889–895.

Everett NB (1943) Observational and experimental evidences relating to origin and differentiation of definitive germ cells in mice. *Journal of Experimental Zoology* 92: 49–91.

Farquhar J and Gajdusek DC (1981) *Kuru: Early Letters and Field Notes from the Collection of D. Carleton Gajdusek*. New York: Raven Press.

Feingold A (1988) Cognitive gender differences are disappearing. *American Psychologist* 43: 95–103.

Fluck CE, Nicolo C, and Pandey AV (2007) Clinical, structural and functional implications of mutations and polymorphisms in human NADPH P450 oxidoreductase. *Fundamental and Clinical Pharmacology* 21: 399–410.

Fluck CE, Tajima T, Pandey AV, et al. (2004) Mutant P450 oxidoreductase causes disordered steroidogenesis with and without Antley–Bixler syndrome. *Nature Genetics* 36: 228–230.

Fratinanni CM and Imperato-McGinley J (1994) The syndrome of 5α-reductase deficiency. *Endocrinologist* 4: 302–314.

Freedman LP (1992) Anatomy of the steroid receptor zinc finger region. *Endocrine Reviews* 13: 129–145.

Frowein J and Engel W (1974) Constitutivity of the HCG-receptor protein in the testis of rat and man. *Nature (London)* 249: 377–379.

Gajdusek DC (1977) Urgent opportuniistic observations: The study of changing, transient and disappearing phenomena of medical interest in disrupted primitive human communities. In: *Health and Disease in Tribal Societies, CIBA Symposium*, vol. 49, pp. 69–102. Amsterdam: Elsevier Excerpta Medica.

Gajdusek DC (1989) *Journals 1957–76,* 18 vols. (with specific field data on Simbari male pseudohermaphrodites), vol. 18. Bethesda, MD: National Institutes of Health.

Garren LD, Ney RL, and Davis WW (1965) Studies on the role of protein synthesis in the regulation of corticosterone production by ACTH *in vivo*. *Proceedings of the National Academy of Sciences of the United States of America* 53: 1450.

Geissler WM, Davis DL, Wu L, et al. (1994) Male pseudohermaphroditism caused by mutations of testicular 17 beta-hydroxysteroid dehydrogenase 3 (see comments). *Nature Genetics* 7: 34–39.

Givens JR, Wiser WL, Summitt RL, Kerber IJ, Andersen RN, Pittaway DE, and Fish SA (1974) Familial male pseudohermaphroditism without gynecomastia due to deficient testicular 17-ketosteroid reductase activity. *New England Journal of Medicine* 291: 938–944.

Glass CK, Rose DW, and Rosenfeld MG (1997) Nuclear receptor coactivators. *Current Opinion in Cell Biology* 9: 222–232.

Goebelsmann U, Hall TD, Paul WL, and Stanczyk FZ (1975) *In vitro* steroid metabolic studies in testicular 17β-reduction deficiency. *Journal of Clinical Endocrinology and Metabolism* 41: 1136–1143.

Goebelsmann U, Horton R, Mestman JH, et al. (1973) Male pseudohermaphroditism due to testicular 17β-hydroxysteroid dehydrogenase deficiency. *Journal of Clinical Endocrinology and Metabolism* 36: 867–879.

Goodfellow PN and Lovell-Badge R (1993) SRY and sex determination in mammals. *Annual Review of Genetics* 27: 71–92.

Gooren L (2006) The biology of human psychosexual differentiation. *Hormones and Behavior* 50: 589–601.

Gooren LJ and Kruijver FP (2002) Androgens and male behavior. *Molecular and Cellular Endocrinology* 198: 31–40.

Goy RW and McEwen BS (1980) *Sexual Differentiation of the Brain*. Cambridge, MA: MIT Press.

Gron G, Wunderlich AP, Spitzer M, Tomczak R, and Riepe MW (2000) Brain activation during human navigation: Gender-different neural networks as substrate of performance. *Nature Neuroscience* 3: 404–408.

Gronemeyer H (1991) Transcription activation by estrogen and progesterone receptors. (review). *Annual Review of Genetics* 25: 89–123.

Harris G, Azzolina B, Baginsky W, et al. (1992) Identification and selective inhibition of an isozyme of steroid 5 alpha-reductase in human scalp. *Proceedings of the National Academy of Sciences of the United States of America* 89: 10787–10791.

He XY and Yang SY (2006) Roles of type 10 17beta-hydroxysteroid dehydrogenase in intracrinology and metabolism of isoleucine and fatty acids. *Endocrine, Metabolic and Immune Disorders – Drug Targets* 6: 95–102.

Heinlein CA and Chang C (2002) Androgen receptor (AR) coregulators: An overview. *Endocrine Reviews* 23: 175–200.

Herdt GH and Davidson J (1988) The Sambia 'turnim-man': Sociocultural and clinical aspects of gender formation in male pseudohermaphrodites with 5-alpha-reductase deficiency in Papua New Guinea. *Archives of Sexual Behavior* 17: 33–56.

Hier DB and Crowley WF, Jr. (1982) Spatial ability in androgen-deficient men. *New England Journal of Medicine* 306: 1202–1205.

Hiort O, Sinnecker GH, Holterhus PM, Nitsche EM, and Kruse K (1996) The clinical and molecular spectrum of androgen insensitivity syndromes. *American Journal of Medical Genetics* 63: 218–222.

Honour J, Millar G, Roitman E, and Shackleton CHL (1981) Steroid excretion in urine during suppression and stimulation of adrenals in the 17α-hydroxylase deficiency syndrome. *Journal of Clinical Endocrinology and Metabolism* 52: 1039–1042.

Horton R and Tait JF (1966) Androstenedione production and interconversion rates measured in peripheral blood and studies on the possible site of its conversion to testosterone. *Journal of Clinical Investigation* 45: 301–313.

Hsing AW, Gao YT, Wu G, et al. (2000) Polymorphic CAG and GGN repeat lengths in the androgen receptor gene and prostate cancer risk: A population-based case-control study in China. *Cancer Research* 60: 5111–5116.

Huhtaniemi IT, Korenbrot CC, and Jaffe RB (1977) HCG binding and stimulation of testosterone biosynthesis in the human fetal testis. *Journal of Clinical Endocrinology and Metabolism* 44: 963–967.

Imperato-McGinley J (1983) Sexual differentiation: Normal and abnormal. In: Martini L and James VHT (eds.) *Current Topics in Experimental Endocrinology, Vol. 5: Fetal Endocrinology and Me*, pp. 231–307. New York: Academic Press.

Imperato-McGinley J (1996) Male pseudohermaphroditism. In: Adashi EY, Rock JA, and Rosenwaks Z (eds.) *Reproductive Endocrinology, Surgery, and Technology*, pp. 936–955. Philadephia, PA: Lippincott-Raven Publishers.

Imperato-McGinley J (1997) 5 Alpha-reductase-2 deficiency. *Current Therapy in Endocrinology and Metabolism* 6: 384–387.

Imperato-McGinley J, Akgun S, Ertel NH, Sayli B, and Shackleton CHL (1987) The coexistence of male pseudohermaphrodites with 17-ketosteroid reductase deficiency and 5α-reductase deficiency within a Turkish kindred. *Clinical Endocrinology (Oxf)* 27: 135–143.

Imperato-McGinley J and Canovatchel WJ (1992a) Complete androgen insensitivity: Pathophysiology, diagnosis and management. *Trends in Endocrinology and Metabolism* 3: 75–81.

Imperato-McGinley J, Gautier T, Yee B, Cai L-Q, Epstein J, and Pochi P (1993) The androgen control of sebum production: Studies of subjects with dihydrotestosterone deficiency and complete androgen insensitivity. *Journal of Clinical Endocrinology and Metabolism* 76: 524–533.

Imperato-McGinley J, Gautier T, Zirinsky K, et al. (1992b) Prostate visualization studies in males homozygous and heterozygous for 5α-reductase deficiency. *Journal of Clinical Endocrinology and Metabolism* 75: 1022–1026.

Imperato-McGinley J, Guerrero L, Gautier T, and Peterson RE (1974) Steroid 5α-reductase deficiency in man: An inherited form of male pseudohermaphroditism. *Science* 186: 1213–1216.

Imperato-McGinley J, Ip NY, Gautier T, et al. (1990) DNA linkage analysis and studies of the androgen receptor gene in a large kindred with complete androgen insensitivity. *American Journal of Medical Genetics* 36(1): 104–108.

Imperato-McGinley J, Miller M, Wilson JD, Peterson RE, Shackleton CHL, and Gajdusek DC (1991a) A cluster of male pseudohermaphrodites with 5 alpha-reductase deficiency in Papua New Guinea. *Clinical Endocrinology (Oxf)* 34: 293–298.

Imperato-McGinley J and Peterson RE (1976) Male pseudohermaphroditism: Complexities of male sexual development. *American Journal of Medicine* 61(2): 251–272.

Imperato-McGinley J, Peterson RE, Gautier T, and Sturla E (1979a) Androgens and the evolution of male-gender identity among male pseudohermaphrodites with 5α-reductase deficiency. *New England Journal of Medicine* 300: 1233–1237.

Imperato-McGinley J, Peterson RE, Gautier T, and Sturla E (1979b) Male pseudohermaphroditism secondary to 5α-reductase deficiency: A model for the role of androgens in both the development of the male phenotype and the evolution of a male gender identity. *Journal of Steroid Biochemistry and Molecular Biology* 11: 637–645.

Imperato-McGinley J, Peterson RE, Gautier T, and Sturla E (1981) The impact of androgens on the evolution of male gender identity. In: Kogan SJ and Hafez ESE (eds.) *Pediatric Andrology*, pp. 99–108. The Hague: Martinus Nijhoff.

Imperato-McGinley J, Peterson RE, Gautier T, et al. (1982) Hormonal evaluation of a large kindred with complete androgen insensitivity: Evidence for secondary 5α-reductase deficiency. *Journal of Clinical Endocrinology and Metabolism* 54: 931–941.

Imperato-McGinley J, Peterson RE, Leshin M, Cooper G, Draghi S, Berenyi M, and Wilson JD (1980) Steroid 5α-reductase deficiency in a 65-year old male pseudohermaphrodite: The natural history, ultrastructure of the testes and evidence for inherited enzyme heterogeneity. *Journal of Clinical Endocrinology and Metabolism* 50: 15–22.

Imperato-McGinley J, Peterson RE, Stoller R, and Goodwin WE (1979c) Male pseudohermaphroditism secondary to 17β-hydroxysteroid dehydrogenase deficiency: Gender role change with puberty. *Journal of Clinical Endocrinology and Metabolism* 49: 391–395.

Imperato-McGinley J, Pichardo M, Gautier T, Voyer D, and Bryden MP (1991b) Cognitive abilities in androgen-insensitive subjects: Comparison with control males and females from the same kindred. *Clinical Endocrinology (Oxf)* 34: 341–347.

Imperato-McGinley J and Zhu YS (2002a) Androgens and male physiology – the syndrome of 5α-reductase-2 deficiency. *Molecular Cellular Endocrinology* 198: 51–59.

Imperato-McGinley J and Zhu YS (2002b) Gender and behavior in subjects with genetic defects in male sexual differentiation. In: Pfaff DW, Arnold AP, Etgen AM, Fahrbach SE, and Rubin RT (eds.) *Hormones, Brain, Behavior,* 1st edn., pp. 303–345. Orlando, FL: Academic Press.

Jansson AK, Gunnarsson C, Cohen M, Sivik T, and Stal O (2006) 17Beta-hydroxysteroid dehydrogenase 14 affects estradiol levels in breast cancer cells and is a prognostic marker in estrogen receptor-positive breast cancer. *Cancer Research* 66: 11471–11477.

Jenster G, van der Korput HA, van Vroonhoven C, Van der Kwast TH, Trapman J, and Brinkmann AO (1991) Domains of the human androgen receptor involved in steroid binding, transcriptional activation, and subcellular localization. *Molecular Endocrinology* 5: 1396–1404.

Jirasek JE (1971) *Development of the Genital System and Male Pseudohermaphroditism*. Baltimore, MD: John Hopkins Press.

Jones HW, Jr., Lee PA, Rock JA, Archer DF, and Migeon CJ (1982) A genetic male patient with 17α-hydroxylase deficiency. *Obstetrics and Gynecology* 59: 254–259.

Josso N and Picard JY (1976) The antimullerian hormone. *Recent Progress in Hormone Research* 33: 117–167.

Jost A (1953) Problems of fetal endocrinology: The gonadal and hypophyseal hormones. *Recent Progress in Hormone Research* 8: 379–418.

Judd HL, Hamilton CR, Barlow JJ, Yen SSC, and Kliman B (1974) Androgen and gonadotropin dynamics in testicular feminization syndrome. *Journal of Clinical Endocrinology and Metabolism* 38: 1143–1146.

Katz MD, Kligman I, Cai LQ, et al. (1997) Paternity by intrauterine insemination with sperm from a man with 5alpha-reductase-2 deficiency. *New England Journal of Medicine* 336: 994–997.

Kelch RP, Jenner MR, Weinstein R, Kaplan SL, and Grumbach MM (1972) Estradiol and testosterone secretion by human, simian and canine testes in males with hypogonadism and in male pseudohermaphrodites with the feminizing testes syndrome. *Journal of Clinical Investigation* 51: 824–829.

Kershnar AK, Borut D, Kogut MD, Biglieri EG, and Schambelan M (1976) Studies in a phenotypic female with 17α-hydroxylase deficiency. *Journal of Pediatrics* 89: 395–400.

Kertzman C, Robinson DL, Sherins RJ, Schwankhaus JD, and McClurkin JW (1990) Abnormalities in visual spatial attention in men with mirror movements associated with isolated hypogonadotropic hypogonadism. *Neurology* 40: 1057–1063.

Knorr D, Bidlingmaier F, and Engelhardt D (1973) Reifenstein's syndrome, a 17beta-hydroxysteroid-oxydoreductase deficiency? *Acta Endocrinologica* 173(supplement): 37.

Kovacs WJ, Griffin JE, Weaver DD, Carlson BR, and Wilson JDA (1984) Mutation that causes lability of the androgen receptor under conditions that normally promote transformation to the DNA-binding state. *Journal of Clinical Investigation* 73: 1095–1104.

Krazeisen A, Breitling R, Imai K, Fritz S, Moller G, and Adamski J (1999) Determination of cDNA, gene structure and chromosomal localization of the novel human 17beta-hydroxysteroid dehydrogenase type 7(1). *FEBS Letters* 460: 373–379.

Kucinskas L and Just W (2005) Human male sex determination and sexual differentiation: Pathways, molecular interactions and genetic disorders. *Medicina (Kaunas.)* 41: 633–640.

La Spada AR, Wilson EM, Lubahn DB, Harding AE, and Fischbeck KH (1991) Androgen receptor gene mutations in X-linked spinal and bulbar muscular atrophy. *Nature* 352: 77–79.

Labrie F, Luu-The V, Lin SX, Labrie C, Simard J, Breton R, and Belanger A (1997) The key role of 17β-hydroxysteroid dehydrogenases in sex steroid biology. *Steroids* 62: 148–158.

Labrie F, Sugimoto Y, Luu-The V, et al. (1992) Structure of human type II 5 alpha-reductase gene. *Endocrinology* 131: 1571–1573.

Lee DK and Chang C (2003) Endocrine mechanisms of disease: Expression and degradation of androgen receptor: Mechanism and clinical implication. *Journal of Clinical Endocrinology and Metabolism* 88: 4043–4054.

Lee YS, Kirk JM, Stanhope RG, et al. (2007) Phenotypic variability in 17beta-hydroxysteroid dehydrogenase-3 deficiency and diagnostic pitfalls. *Clinical Endocrinology (Oxf)* 67: 20–28.

Levine LS, Lieber E, Pang S, and New MI (1980) Male pseudohermaphroditism due to 17-ketosteroid reductase deficiency diagnosed in the newborn period. *Pediatric Research* 14: 480.

Lin D, Sugawara T, Strauss JF, III, et al. (1995) Role of steroidogenic acute regulatory protein in adrenal and gonadal steroidogenesis. *Science* 267: 1828–1831.

Linn MC and Petersen AC (1986) A meta-analysis of gender differences in spatial ability: Implications for mathematics and science achievement. In: Hyde JS and Linn MC (eds.) *The Psychology of Gender: Advances Through Meta-Analysis*, pp. 67–101. Baltimore, MD: Johns Hopkins University Press.

Liu H, Robert A, and Luu-The V (2005) Cloning and characterization of human form 2 type 7 17beta-hydroxysteroid dehydrogenase, a primarily 3beta-keto reductase and estrogen activating and androgen inactivating enzyme. *Journal of Steroid Biochemistry and Molecular Biology* 94: 173–179.

Liu S, Huang C, Li D, et al. (2007) Molecular cloning and expression analysis of a new gene for short-chain dehydrogenase/reductase 9. *Acta Biochimica Polonica* 54: 213–218.

Lorence MC, Corbin CJ, Kamimura N, et al. (1990) Structural analysis of the gene encoding human 3β-hydroxysteroid dehydrogenase/delta5–4-isomerase. *Molecular Endocrinology* 4: 1850–1855.

Lu AY, Junk KW, and Coon MJ (1969) Resolution of the cytochrome P-450-containing omega-hydroxylation system of liver microsomes into three components. *Journal of Biological Chemistry* 244: 3714–3721.

Lubahn DB, Joseph DR, Sullivan PM, Willard HF, French FS, and Wilson EM (1988) Cloning of human androgen receptor complementary DNA and localization to the X chromosome. *Science* 240: 327–330.

Lukacik P, Kavanagh KL, and Oppermann U (2006) Structure and function of human 17beta-hydroxysteroid dehydrogenases. *Molecular and Cellular Endocrinology* 248: 61–71.

Luu-The V, Tremblay P, and Labrie F (2006) Characterization of type 12 17beta-hydroxysteroid dehydrogenase, an isoform of type 3 17beta-hydroxysteroid dehydrogenase responsible for estradiol formation in women. *Molecular Endocrinology* 20: 437–443.

Maccoby EE and Jacklin CN (1974) *The Psychology of Sex Differences*. Stanford, CA: Stanford University Press.

MacDonald PC, Madden JD, Brenner PF, Wilson JD, and Siiteri PK (1979) Origin of estrogen in normal men and women with testicular feminization. *Journal of Clinical Endocrinology and Metabolism* 49: 905–916.

Maclaughlin DT and Donahoe PK (2004) Sex determination and differentiation. *New England Journal of Medicine* 350: 367–378.

Mahendroo MS, Cala KM, Landrum CP, and Russell DW (1997) Fetal death in mice lacking 5α-reductase type 1 caused by estrogen excess. *Molecular Endocrinology* 11: 917–927.

Mahendroo MS, Cala KM, and Russell DW (1996) 5α-Reduced androgens play a key role in murine parturition. *Molecular Endocrinology* 10: 380–392.

Martel C, Rheaume E, Takahashi M, et al. (1992) Distribution of 17 beta-hydroxysteroid dehydrogenase gene expression and activity in rat and human tissues. *Journal of Steroid Biochemistry and Molecular Biology* 41: 597–603.

Masica DN, Money J, Ehrhardt AA, and Lewis VG (1969) IQ, fetal sex hormones and cognitive patterns: Studies in the testicular feminizing syndrome of androgen insensitivity. *Johns Hopkins Medical Journal* 124: 34–43.

Massague J (1996) TGFbeta signaling: Receptors, transducers, and mad proteins. *Cell* 85: 947–950.

Mendonca BB, Inacio M, Arnhold IJ, et al. (2000) Male pseudohermaphroditism due to 17 beta-hydroxysteroid dehydrogenase 3 deficiency. Diagnosis, psychological evaluation, and management. *Medicine (Baltimore)* 79: 299–309.

Mendonca BB, Inacio M, Costa EM, et al. (1996) Male pseudohermaphroditism due to steroid 5alpha-reductase 2 deficiency. Diagnosis, psychological evaluation, and management. *Medicine (Baltimore)* 75: 64–76.

Meyer-Bahlburg HF (1965) Gender assignment and reassignment in 46,XY pseudohermaphroditism and related conditions. *Journal of Clinical Endocrinology and Metabolism* 84: 3455–3458.

Michel-Calemard L, Charmandari E, Acherman J, Skordis N, and Morel Y (2005) Large rearrangements of the HSD17B3 gene: A new type of mutation responsible for 17β-HSD deficiency. *Hormone Research* 64: 123–124.

Miller WL (2002) Androgen biosynthesis from cholesterol to DHEA. *Molecular and Cellular Endocrinology* 198: 7–14.

Mintz B and Russell ES (1957) Gene-induced embryological modifications of primordial germ cells in the mouse. *Journal of Experimental Zoology* 134: 207–237.

Moller G, van Grunsven EG, Wanders RJ, and Adamski J (2001) Molecular basis of D-bifunctional protein deficiency. *Molecular and Cellular Endocrinology* 171: 61–70.

Money J and Ehrhardt AA (1972) *Man and Woman, Boy and Girl: The Differentiation and Dimorphism of Gender Identity from Conception to Maturity*. Baltimore, MD: Johns Hopkins University Press.

Money J, Hampson JG, and Hampson JL (1955a) An examination of some basic sexual concepts: The evidence of human hermaphroditism. *Bulletin of the Johns Hopkins Hospital* 97: 301–319.

Money J, Hampson JG, and Hampson JL (1955b) Hermaphroditism: Recommendations concerning assignment of sex, change of sex and psychological management. *Bulletin of the Johns Hopkins Hospital* 97: 284–300.

Money J and Ogunro C (1974) Behavioral sexology: Ten cases of genetic male intersexuality with impaired prenatal and pubertal androgenization. *Archives of Sexual Behavior* 3: 181–205.

Morris JM (1953) The syndrome of testicular feminization in male pseudohermaphrodites. *American Journal of Obstetrics and Gynecology* 65: 1192–1211.

Neher R and Kahnt FW (1965) Gonadal steroid biosynthesis *in vitro* in four cases of testicular ferminization, androgens in normal and pathological conditions. 101: 130. (Excerpta Medica. International Congress Series Amsterdam.)

Neumann J, von Berswordt-Wallrabe R, Elger W, Steinbeck H, Hahn JD, and Kramer M (1970) Aspects of androgen-dependent events as studied by antiandrogens. *Recent Progress in Hormone Research* 26: 337–410.

New MI (1970) Male pseudohermaphroditism due to 17α-hydroxylase deficiency. *Journal of Clinical Investigation* 49: 1930–1941.

Ng WK, Taylor NF, Hughes IA, et al. (1990) 5α-Reductase deficiency without hypospadias. *Archives of Disease in Childhood* 65: 1166–1167.

Nordenskjold A and Ivarsson SA (1998) Molecular characterization of 5 alpha-reductase type 2 deficiency and fertility in a Swedish family. *Journal of Clinical Endocrinology and Metabolism* 83: 3236–3238.

Notini AJ, Davey RA, McManus JF, Bate KL, and Zajac JD (2005) Genomic actions of the androgen receptor are required for normal male sexual differentiation in a mouse model. *Journal of Molecular Endocrinology* 35: 547–555.

Nowaczyk MJ, Nakamura LM, and Waye JS (2001) DHCR7 and Smith–Lemli–Opitz syndrome. *Clinical and Investigative Medicine* 24: 311–317.

Ofman R, Ruiter JP, Feenstra M, et al. (2003) 2-Methyl-3-hydroxybutyryl-CoA dehydrogenase deficiency is caused by mutations in the HADH2 gene. *American Journal of Human Genetics* 72: 1300–1307.

Ogawa S, Krebs CJ, Zhu YS, and Pfaff DW (2000) Genes participating in reproductive behaviors. In: Pfaff DW, Berrettini WH, Joh TH, and Maxson SC (eds.) *Genetic Influences on Neural and Behavioral Functions*, pp. 417–430. Boca Raton, FL: CRC Press LLC.

Park K, Seo JJ, Kang HK, Ryu SB, Kim HJ, and Jeong GW (2001) A new potential of blood oxygenation level dependent (BOLD) functional MRI for evaluating cerebral centers of penile erection. *International Journal of Impotence Research* 13: 73–81.

Pelliniemi LJ and Niemi M (1969) Fine structure of the human foetal testis. I. The interstitial tissue. *Zeitschrift fuer Zellforschung und Mikroskopische Anatomie* 99: 507–522.

Perlman SM (2000) Cognitive abilities of children with hormone abnormalities: Screening by psychoeducational tests. *Journal of Learning Disabilities* 6: 21–29.

Peterson RE and Imperato-McGinley J (1984) Male pseudohermaphroditism due to inherited deficiencies of testosterone biosynthesis. In: Serio M, Motta M, Zanisi M, and Martini L (eds.) *Sexual Differentiation: Basic and Clinical Aspects*, pp. 301–319. New York: Raven Press.

Peterson RE, Imperato-McGinley J, Gautier T, and Shackleton CHL (1985) Male pseudohermaphroditism due to multiple defects in steroid-biosynthetic microsomal mixed-function oxidases. *New England Journal of Medicine* 313: 1182–1191.

Peterson RE, Imperato-McGinley J, Gautier T, and Sturla E (1977) Male pseudohermaphroditism due to steroid 5α-reductase deficiency. *American Journal of Medicine* 62: 170–191.

Piccioni F, Pinton P, Simeoni S, et al. (2002) Androgen receptor with elongated polyglutamine tract forms aggregates that alter axonal trafficking and mitochondrial distribution in motor neuronal processes. *FASEB Journal* 16: 1418–1420.

Pugh KR, Offywitz BA, Shaywitz SE, et al. (1996) Auditory selective attention: An fMRI investigation. *Neuro Image* 4: 159–173.

Quigley CA, De Bellis A, Marschke KB, el-Awady MK, Wilson EM, and French FS (1995) Androgen receptor defects: Historical, clinical, and molecular perspectives. *Endocrine Reviews* 16: 271–321.

Ray WJ, Morell M, and Frediani AW (1976) Sex differences and lateral specialization of hemispheric functioning. *Neuropsychologia* 14: 391–394.

Reiner WG (2004) Psychosexual development in genetic males assigned female: The cloacal exstrophy experience. *Child and Adolescent Psychiatric Clinics of North America* 13: 657–674, ix.

Reiner WG (2005) Gender identity and sex-of-rearing in children with disorders of sexual differentiation. *Journal of Pediatric Endocrinology and Metabolism* 18: 549–553.

Reiner WG and Gearhart JP (2004) Discordant sexual identity in some genetic males with cloacal exstrophy assigned to female sex at birth. *New England Journal of Medicine* 350: 333–341.

Rosa S, Duff C, Meyer M, et al. (2007) P450c17 deficiency: Clinical and molecular characterization of six patients. *Journal of Clinical Endocrinology and Metabolism* 92: 1000–1007.

Rosler A, Belanger A, and Labrie F (1992) Mechanisms of androgen production in male pseudohermaphroditism due to 17 beta-hydroxysteroid dehydrogenase deficiency. *Journal of Clinical Endocrinology and Metabolism* 75: 773–778.

Rosler A and Kohn G (1983) Male pseudohermaphroditism due to 17β-hydroxysteroid dehydrogenase deficiency: Studies on the natural history of the defect and effect of androgens on gender role. *Journal of Steroid Biochemistry and Molecular Biology* 19: 663–674.

Russell DW and Wilson JD (1994) Steroid 5 alpha-reductase: Two genes/two enzymes. *Annual Review of Biochemistry* 63: 25–61.

Saenger P, Goldman AS, Levine LS, et al. (1978) Prepubertal diagnosis of steroid 5 alpha-reductase deficiency. *Journal of Clinical Endocrinology and Metabolism* 46: 627–634.

Saez JM, De Peretti E, Morera AM, David M, and Bertrand J (1971) Familial male pseudohermaphroditism with gynecomastia due to a testicular 17-ketosteroid reductase defect. I. Studies *in vivo*. *Journal of Clinical Endocrinology and Metabolism* 32: 604–610.

Schweikert HU and Wilson JD, III (1974) Regulation of human hair growth by steroid hormones. I. Testosterone metabolism in isolated hairs. *Journal of Clinical Endocrinology and Metabolism* 38: 811–819.

Scott RR and Miller WL (2008) Genetic and clinical features of P450 oxidoreductase deficiency. *Hormone Research* 69: 266–275.

Shaywitz BA, Shaywitz SE, Pugh KR, et al. (1995) Sex differences in the functional organization of the brain for language. *Nature* 373: 607–609.

Sheflin L, Keegan B, Zhang W, and Spaulding SW (2000) Inhibiting proteasomes in human HepG2 and LNCaP cells increases endogenous androgen receptor levels. *Biochemical and Biophysical Research Communications* 276: 144–150.

Shephard EA, Phillips IR, Santisteban I, West LF, Palmer CN, Ashworth A, and Povey S (1989) Isolation of a human cytochrome P-450 reductase cDNA clone and localization of the corresponding gene to chromosome 7q11.2. *Annals of Human Genetics* 53: 291–301.

Siiteri P and Wilson JD (1974) Testosterone formation and metabolism during male sexual differentiation in the human embryo. *Journal of Clinical Endocrinology and Metabolism* 38: 113–125.

Simard J, Ricketts ML, Gingras S, Soucy P, Feltus FA, and Melner MH (2005) Molecular biology of the 3beta-hydroxysteroid dehydrogenase/delta5-delta4 isomerase gene family. *Endocrine Reviews* 26: 525–582.

Smith EP, Boyd J, Frank GR, et al. (1994) Estrogen resistance caused by a mutation in the estrogen-receptor gene in a man. *New England Journal of Medicine* 331: 1056–1061.

Sparkes RS, Klisak I, and Miller WL (1991) Regional mapping of genes encoding human steroidogenic enzymes: P450ssc to 15q23–q24, andrenodoxin to 11q22; andrenodoxin reductase to 17q24–q25; and P450c17 to 10q24–q25. *DNA and Cell Biology* 10: 359–365.

Stavrou SS, Zhu YS, Cai L-Q, Katz MD, Herrera C, DeFillo-Ricart M, and Imperato-McGinley J (1998) A novel mutation of the human luteinizing hormone receptor in 46XY and 46XX sisters. *Journal of Clinical Endocrinology and Metabolism* 83: 2091–2098.

Steckelbroeck S, Stoffel-Wagner B, Reichelt R, Schramm J, Bidlingmaier F, Siekmann L, and Klingmuller D (1999) Characterization of 17beta-hydroxysteroid dehydrogenase activity in brain tissue: Testosterone formation in the human temporal lobe. *Journal of Neuroendocrinology* 11: 457–464.

Steglehner (1817) *De Hermaphroditorum Natura* (Nature of Hermaphroditism). Leipzig.

Stocco DM (2000) The role of the StAR protein in steroidogenesis: Challenges for the future. *Journal of Endocrinology* 164: 247–253.

Stoffel-Wagner B, Watzka M, Steckelbroeck S, Schramm J, Bidlingmaier JF, and Klingmuller D (1999) Expression of 17beta-hydroxysteroid dehydrogenase types 1, 2, 3 and 4 in the human temporal lobe. *Journal of Endocrinology* 160: 119–126.

Stoller RJ (1964) A contribution to the study of gender identity. *International Journal of Psychoanalysis* 45: 220–226.

Swaab DF (2004) Sexual differentiation of the human brain: Relevance for gender identity, transsexualism and sexual orientation. *Gynecological Endocrinology* 19: 301–312.

Swaab DF and Fliers EA (1985) Sexually dimorphic nucleus in the human brain. *Science* 228: 1112–1115.

Takagi S, Yoshida T, Tsubata K, Ozaki H, Fuju TK, Nomura Y, and Sawada M (1977) Sex differences in fetal gonadotropins and androgens. *Journal of Steroid Biochemistry* 8: 609–620.

Takeyama J, Suzuki T, Hirasawa G, et al. (2000) 17Beta-hydroxysteroid dehydrogenase type 1 and 2 expression in the human fetus. *Journal of Clinical Endocrinology and Metabolism* 85: 410–416.

Thigpen AE, Cala KM, and Russell DW (1993a) Characterization of Chinese hamster ovary cell lines expressing human steroid 5 alpha-reductase isozymes. *Journal of Biological Chemistry* 268: 17404–17412.

Thigpen AE, Davis DL, Gautier T, Imperato-McGinley J, and Russell DW (1992a) The molecular basis of steroid 5 alpha-reductase deficiency in a large Dominican kindred. *New England Journal of Medicine* 327: 1216–1219.

Thigpen AE, Davis DL, Milatovich A, et al. (1992b) The molecular genetics of steroid 5α-reductase 2 deficiency. *Journal of Clinical Investigation* 90: 799–809.

Thigpen AE and Russell DW (1992) Four amino acid segments in steroid 5α-reductase-1 confers insensitivity to finasteride, a competitive inhibitor. *Journal of Biological Chemistry* 267: 8577.

Thigpen AE, Silver RI, Guileyardo JM, Casey ML, McConnell JD, and Russell DW (1993b) Tissue distribution and ontogeny of

steroid 5α-reductase isozyme expression. *Journal of Clinical Investigation* 92: 903–910.
Trapman J, Klaassen P, Kuiper G, et al. (1988) Cloning, structure, and expression of a cDNA encoding the human androgen receptor. *Biochemical and Biophysical Research Communications* 153: 241–248.
Ulloa-Aguirre A, Bassol S, Poo J, Mendez JP, Mutchinick O, Robles C, and Perez-Palacios G (1985) Endocrine and biochemical studies in a 46,XY phenotypically male infant with 17-ketosteroid reductase deficiency. *Journal of Clinical Endocrinology and Metabolism* 60: 639–643.
Virdis R, Saenger P, Senior B, and New MI (1978) Endocrine studies in a pubertal male pseudohermaphrodite with 17-ketosteroid reductase deficiency. *Acta Endocrinologica* 87: 212–224.
Voyer D and Flight J (2000) Reliability of a dichotic consonant-vowel pairs task using an ABX procedure. *Brain and Cognition* 43: 417–421.
Voyer D, Voyer S, and Bryden MP (1995) Magnitude of sex differences in spatial abilities: A meta-analysis and consideration of critical variables. *Psychological Bulletin* 117: 250–270.
Walsh PC, Madden JD, Harrod MJ, Goldstein JL, MacDonald PC, and Wilson JD (1974) Familial incomplete male pseudohermaphroditism, type 2. Decreased dihydrotestosterone formation in pseudovaginal perineoscrotal hypospadias. *New England Journal of Medicine* 291: 944–949.
Wiebe JP (2006) Progesterone metabolites in breast cancer. *Endocrine-Related Cancer* 13: 717–738.
Wigley WC, Prihoda JS, Mowszowicz I, Mendonca BB, New MI, Wilson JD, and Russell DW (1994) Natural mutagenesis study of the human steroid 5 alpha-reductase 2 isozyme. *Biochemistry* 33: 1265–1270.
Wilbert DM, Griffen JE, and Wilson JD (1983) Characterization of the cytosol androgen receptor of the human prostate. *Journal of Clinical Endocrinology and Metabolism* 56: 113–120.
Wilhelm D, Palmer S, and Koopman P (2007) Sex determination and gonadal development in mammals. *Physiological Reviews* 87: 1–28.
Wilkins L (1957) *The Diagnosis and Treatment of Endocrine Disorders in Childhood and Adolescence*, 2nd edn. Springfield, IL: Charles C. Thomas.
Wilson JD (1978) Sexual differentiation. *Annual Review of Physiology* 40: 270–306.
Wilson JP (1985) Mutations that impair androgen action. *Trends in Genetics* 1: 335–339.
Wilson JD, Griffin JE, and Russell DW (1993) Steroid 5α-reductase 2 deficiency. *Endocrine Reviews* 14: 577–593.
Wilson JD, Harrod MJ, Goldstein JL, Hemsell DL, and MacDonald PC (1974) Familial incomplete male pseudohermaphroditism type 1: Evidence for androgen resistance and variable clinical manifestations in a family with the Reifenstein syndrome. *New England Journal of Medicine* 290: 1097–1103.
Xu J, Qiu Y, DeMayo FJ, Tsai SY, Tsai MJ, and O'Malley BW (1998) Partial hormone resistance in mice with disruption of the steroid receptor coactivator-1 (SRC-1) gene. *Science* 279: 1922–1925.
Zachmann M, Vollmin JA, Hamilton W, and Prader A (1972) Steroid 17,20-desmolase deficiency: A new cause of male pseudohermaphroditism. *Clinical Endocrinology (Oxf.)* 1: 369–385.
Zang Y, Dufort I, Soucy P, Labrie F, and Luu-The V (1995) Cloning and expression of human type V 17β-hydroxysteroid dehydrogenase. Program and Abstracts, The Endocrine Society 77th Annual Meeting 622 (pp. 3–614).
Zhu YS (2005) Molecular basis of steroid action in the prostate. *Cell Science Reviews* 1: 27–55.
Zhu YS, Cai LQ, Cordero JJ, Canovatchel WJ, Katz MD, and Imperato-McGinley JA (1999) Novel mutation in the CAG triplet region of exon 1 of androgen receptor gene causes complete androgen insensitivity syndrome in a large kindred. *Journal of Clinical Endocrinology and Metabolism* 84: 1590–1594.
Zhu YS and Imperato-McGinley J (2004a) Androgen insensitivity syndrome. In: Martini L (ed.) *Encyclopedia of Endocrine Diseases*, pp. 214–220. San Diego, CA: Elsevier.
Zhu YS and Imperato-McGinley J (2004b) Male pseudohermaphroditism due to 5α-reductase-2 deficiency. In: Sciarra JJ (ed.) *Gynecology and Obstetrics,* vol. 5. CD version. Chicago, IL: Lippincott Williams & Wilkins.
Zhu YS and Imperato-McGinley J (2004c) Pseudohermaphroditism, male, due to 5α-reductase-2 deficiency. In: Martini L (ed.) *Encyclopedia of Endocrine Diseases*, pp. 131–135. San Diego, CA: Elsevier.
Zhu YS, Katz MD, and Imperato-McGinley J (1998) Natural potent androgens: Lessons from human genetic models. *Bailliere's Clinical Endocrinology and Metabolism* 12: 83–113.
Zitzmann M, Weckesser M, Schober O, and Nieschlag E (2001) Changes in cerebral glucose metabolism and visuospatial capability in hypogonadal males under testosterone substitution therapy. *Experimental and Clinical Endocrinology and Diabetes* 109: 302–304.

Relevant Website

http://www.mcgill.ca – McGill University, androgen database.

28 Assisted Reproduction in Infertile Women

L Baor, Tel-Aviv, Israel

Chapter Outline

28.1 Socio-Cultural Norms Regarding Parenthood and Infertility

In contemporary society parenting is perceived as the central developmental task of midlife. Societal and cultural norms constantly emphasize the importance of motherhood to the female role (Miall, 1986). Bearing and raising children constitute such behavior (Tulandi et al., 1981). Society values genetic motherhood. A leading infertility specialist wrote that the desire for a biological child is "an intrinsic part of the female nature and function" (Kentenich, 1989). For women who internalize this culturally formed notion, the inability to achieve this goal can be devastating. Because femininity is often perceived as synonymous with motherhood, childless women are excluded of the most central element of their gender identity and personal integrity (Miall, 1986).

Israel is an example for a paternalist society where procreation norms are imperative and childbirth is the highest personal priority. In such circumstances, being childless constitutes an unbearable stigma, and bears a wide range of deleterious psychological and social consequences for the couple, but mainly for women. Therefore, there is little desire to be single, rich, and childless as it is often the case in the USA (Remennick, 2000). The high social value of fertility in Israel is reflected in public and healthcare policies, whereby the Health Insurance Law provides for an unlimited number of *in vitro* fertilization (IVF) or other relevant infertility treatments until two children are born.

In modern society, young women who invest their younger years in education and career, erroneously take fertility for granted. Therefore, the use of contraception is prevailed, postponing pregnancy until certain life conditions are met. When the appropriate time for pregnancy comes, some couples may find that the goal of parenthood is beyond their control and out of their reach. This trend is reflected in an increasing incidence of age-related infertile couples.

Infertility is commonly defined as the failure to conceive after 12 months of unprotected regular sexual intercourse. Estimates of its prevalence in industrialized countries are 16–26% (Malin et al., 2001). The use of assisted reproductive technologies (ARTs) is frequently the end-stage treatment after many childless years.

The combination of presumably granted fertility coupled with cultural definitions linking fertility adulthood and sexuality suggests that infertility can deeply affect an individual's self-esteem, competence, adequacy, and efficacy (Mazor, 1984; Cook, 1987). Where childbearing is a societal and cultural imperative, childless couples feel that the main prism through which they perceive and define themselves, or judged by others, is the prism of infertility. Being different and deprived, they are determined to seek a solution. In this respect, ARTs are considered to be the ultimate solution, salivating the infertile couple from the social stigma, technically bypassing the infertility problem and helping the couple to fulfill their goal in life.

28.2 Assisted Reproductive Technologies

According to the Society for Assisted Reproductive Technology (SART), ART includes *in vitro* fertilization-embryo transfer (IVF-ET), gamete intrafallopian transfer (GIFT), zygote intrafallopian transfer (ZIFT), tubal embryo transfer (TET), and frozen embryo transfer (FET). These techniques also apply to oocyte donation and gestational carriers. These are relatively new procedures for couples who are unable to conceive by other methods.

IVF, GIFT, ZIFT, and TET are very similar procedures although there are a few significant differences. During IVF-ET, ZIFT, and TET, the oocytes and sperm are combined in a culture dish in the laboratory. Fertilization and very early embryo development occur outside the body, rather than in the fallopian tube. Once early embryo development is recognized, the embryos are transferred either into the uterus (IVF-ET) or in the fallopian tube (ZIFT, TET). Since most programs have seen no significant difference in success rates, they usually perform IVF-ET because it is less expensive and does not require laparoscopy and general anesthesia. In addition, IVF-ET is the only procedure available for women with damaged fallopian tubes.

28.2.1 ART Medications

28.2.1.1 GnRH agonists

Gonadotropin-releasing hormone (GnRH) is a hormone produced in the brain that indirectly stimulates ovarian function. Agonists of GnRH are synthetic forms of this hormone which do not directly induce follicle development or ovulation but which have become very important in ART therapy.

The major disadvantage of GnRH agonists is that most patients require more medication for ovarian stimulation. Occasionally, patients require adjustments in dosage of GnRH agonists or may respond better to treatment with GnRH antagonists.

28.2.1.2 Mechanism of action

Agonists of GnRH initially stimulate the pituitary gland to release all the stored gonadotropins (luteinizing hormone (LH) and follicle-stimulating hormone (FSH) – the hormones that normally stimulate ovarian function). Over the course of a week to 10 days, GnRH analogs suppress the production of any new LH and FSH. This effect appears to prevent the ovaries from receiving mixed signals from the patient's own LH and FSH and from the medications that we administer to stimulate follicle development. The result for many patients is a more synchronized development of mature oocytes.

28.2.2 Gonadotropins

In order to increase likelihood of pregnancy through ART, multiple oocytes must be produced. This is accomplished through the administration of gonadotropins-hormonal medications that directly stimulate the ovaries. Stimulation can be achieved with a variety of drug regimens. They replace a woman's own LH and FSH which are normally produced by the pituitary gland. This process ensures uniform purity and potency. Because the dose of hormones used in ART is greater than what the body normally produces, the ovaries typically develop more than one oocyte as occurs in a natural cycle.

Gonadotropins act directly on the ovary to stimulate the growth of follicles (the structures in ovaries which contain eggs). Granulosa cells within the follicles grow and develop which cause the follicles to enlarge and fill with follicular fluid. These developing follicles can be counted and measured using transvaginal ultrasound. As the follicles grow, they produce increasing amounts of estrogen, which can be measured with a laboratory blood test.

28.2.3 ART Procedure

Every cycle of ART involves multiple steps, and each occurs at a specific time during a 4- to 6-week period. The procedure begins in the month preceding the actual ART cycle.

28.2. 3.1 Cycle preceding ART cycle

The cycle preceding ART cycle is as follows:

1. start of oral contraceptives or documentation of ovulation (midluteal); and
2. start of GnRH agonist.

28.2.3.2 ART cycle

ART cycle is as follows:

1. baseline pelvic ultrasound on cycle day 2;
2. ovarian stimulation with gonadotropins;
3. monitoring follicle development with ultrasound and serum hormone levels;
4. hCG administration;
5. transvaginal oocyte retrieval;
6. embryo transfer;
7. progesterone supplementation;
8. hormonal studies and pregnancy test; and
9. follow-up consultation.

28.3 Psychological Reaction to Infertility

Psychosocial studies consistently show that infertility and ART are highly stressful and potentially lead to significant negative consequences, especially when treatment fails. More so than most medical problems, infertility is recently regarded as a couples' issue. Although among diagnosed infertile couples, 35% of the infertile diagnoses are with the woman's reproductive system, 35% with the man's, and 20% with both (Leiblum et al., 1987), in most cultures the inability to produce is regarded as a female problem. Women also serve the object of fertility treatments, even if the problem actually lies on the male side. A large body of research has proved that women show significantly higher levels of psychosocial distress than men (Abbey et al., 1994; Demyttenaere et al., 1998; Glover et al., 1999). They often tend to seek infertility treatment more than men and at a greater frequency than their partners, and are more committed to treatment. The majority of the medical tests and treatments involve the female's reproductive system; thus, women's life may be more disrupted than their partners (Abby et al., 1994).

The prime focus on conception affects every facet of the couple's psychological functioning: self-esteem, confidence, health, close relationship, security, and hope. Prevalent emotional responses include anxiety, diminished self-worth and quality of life, grief, frustration, depression, anger, guilt, shock or denial, and anxiety (Menning, 1980; Mahlstedt, 1985).

28.3.1 Loss of Relationship with Spouse

Many studies of coping with life stressors, including health problems, found that women and men cope differently. Women use more emotion-focused coping strategies, while men tend to use problem-focused coping. Wives' greater use of these emotion-focused coping strategies may be the result of their greater psychological distress, that lead to avoidant coping, seeking social support, escape, or avoid distraction and tension destruction (Jordan and Revenson, 1999). These differences in coping styles may result in conflict, communication problems, disagreements over medical treatment, lack of empathy, and differential investment in the infertility process (Andrews et al., 1991). Some individuals report increased anger, hostility, or resentment toward their spouse as a result of feeling lack of understanding and emotional support, or feeling that the other partner is not equally committed to having children (Mahlstedt, 1985). Because the couple may be unable to fulfill each other's needs and thereby experience the loss of closeness and increased isolation, for many infertile couples, end of a relationship may be either an actual or an unspoken threat. However, some individuals experience improved couple communication, increased intimacy, love, and support from their partners (Connolly et al., 1993; Menning, 1980).

28.3.2 Loss of Sexual Satisfaction

Since the course of treatment often requires scheduled sexual intercourse, the need to have sex on demand often makes it a chore rather than a pleasure. The evaluation of fertility is perceived as a constant intrusion into the most intimate aspects of life. For example, the initial interview focuses on the couple's sexual performance with questions regarding frequency sexual intercourses, coital techniques, premarital sexual behavior, discomfort with intercourse, etc. Such investigation can cause sexual dysfunction, purposeful avoidance of coitus or impotence. In addition, fertility tests, such as recorded body temperatures, serve as a constant reminder of the need to have sex on schedule, and also become a sexual report card to be scrutinized and critiqued by the medical staff (Keye, 1984). In turn, these experiences may reduce the feeling of being desired and lead to avoiding spontaneous sexual activity, failure to function on demand, and ultimately may cause sexual dissatisfaction (Mahlstedt, 1985; Abby, 2000).

28.3.3 Loss of Relationship within the Social Network

Couples report feeling socially unworthy or isolated, and jealousy, resentful, and envious of people who have achieved a successful pregnancy or are celebrating the birth of their biological child. Such feelings are frequently directed toward people who have children, mainly biologically similar family members (Connolly et al., 1993; Dunkle-Schetter and Lobel, 1991). Moreover, many couples are unwilling to confide in anyone their infertility problems because of embarrassment or the discomfort associated with discussing intimate issues. While understandable, this secrecy increases the likelihood that infertile couples may become the target of painful comments, leading the couple to choose to withdraw from their social contacts, including their family (Miall, 1986; Abbey et al., 1991).

28.3.4 Loss of Health

Those who pride themselves on taking care of their bodies may find the diagnosis of infertility incompatible with their previous concept of well-being. Uncomfortable, and sometimes unsuccessful treatments, as well as side effects of therapy, can make the couple feel very vulnerable and damage their body image as well. The diagnostic and medical procedures involved with infertility may lead couples to feel somehow damaged or defective. This feeling of being impaired often spreads to their overall sense of self-worth and body image (Keye, 1984).

28.3.5 Loss of Status and/or Prestige

Childlessness may be considered as a form of deviance in a society that eminently values parenthood. Some couples feel that their value to society is reduced by their inability to procreate. They may also feel that infertility compromises their individual sexual identities in the sense that the masculine and feminine roles might be questioned because infertility and virility are closely associated (Mahlstedt, 1985).

28.3.6 Loss of Self-Esteem

Self-esteem has defined as "the extent to which one prizes, values, approves, or likes one-self" (Blascovich and Tomaka, 1991). In adults, self-esteem is enhanced by the accomplishment of basic personal responsibilities as well as by the sense of how such activities are perceived by others. Because motherhood is traditionally perceived as the central role of women, procreation is more significant to a woman's identity than to man's, consequently leading to higher rate of evidence that women's self-esteem is more affected by having fertility problems (Danulik, 1997). Most people view the ability to reproduce as a central aspect of their personal identity. Thus, infertility is perceived as a personal failure that diminishes self-esteem (Matthews and Matthews, 1986). For some couples, the failure to conceive is likely to diminish their own pride, as they face daily external and internal reminders of the inability to accomplish their own expectations. In this case, however, feelings of failure and inadequacy are not limited to reproductive dysfunction; rather, they extend to sexual desirability, and to physical attractiveness and productivity in other spheres (Mahlstedt, 1985).

28.3.7 Loss of Confidence and/or Control

Self-directed individuals who are in control of their lives make goals for education, marriage, and careers are convinced that through hard work they are almost always able to obtain these goals. Facing the diagnosis of infertility, they realize that despite the costs in terms of money, time, persistence, commitment to a schedule, and sacrifice to self and marriage, the diagnosis cannot be changed and their goal of being parents is beyond their reach. Those unworthy efforts may result in distress, frustration, and hopelessness, followed by a feeling, for the first time, that they loose their sense of control over significant aspects of their lives (Keye, 1984). Because ART procedures demand high commitment and adherence, control of one's own life, bodily functions, and other daily activities are transferred to professionals, thus accentuating the sense of incompetence and failure. Due to this inevitable side effect, some couples feel loss of control over their ability to predict and plan for the future (Mahlstedt, 1985).

28.3.8 Loss of Security

Couples may become extremely stressful and insecure about the financial burden of repeated appointments, operations, and medications. Moreover, it is impossible to predict how many cycles are required to conceive in a given case; and, in fact, no upper limit really exists to the amount an infertile couple may invest in their pursuit for a baby. Job security and promotion are often affected by the investment in infertility treatments. Some individuals turn down promotions, relocation, or career changes, in order to keep in touch with their physicians and to continue the liberty to miss work for medical appointments.

On a deeper level, loss of security may extend to feelings regarding fairness and predictability of life. Couples often struggle to determine why they deserve being infertile and how they could have prevented it, fearing that if they could be afflicted by infertility, they could be afflicted with everything else (Menning, 1980).

28.3.9 Loss of Hope

Because ART treatments are tied to the menstrual cycle, couples begin each month feeling hopeful that the current week will bring its wishful pregnancy. Unfortunately, very often this hopefulness turns into hopelessness (Olshansky, 1988). The couple faces again the cruel fact that we are infertile. When infertility treatment is prolonged and unproductive, despair, exhaustion, and frustration are not uncommon, and couples become emotionally and financially depleted. The anticipatory anxiety from the next failure and the uncertainty whether they should start a new treatment cycle may lead to a radical change in the perspective of their future and in the perception of controlling life goals (Dunkle-Schetter and Lobel, 1991).

28.4 Multiple Pregnancy as a Side Effect of ART

ART is inevitable in modern infertility therapy from two main reasons. First, infertile patients who are under pressure to conceive cannot ignore the mere availability and effectiveness of a technology that solves most infertility problems. Second, the goal of parenthood usually justifies just every available means. Indeed, women who embark on the journey of infertility treatment manifest high commitment to achieve pregnancy at almost any personal cost, irrespective of the losses described above (Remennick, 2000). However, ART is associated with consistently increased rate of twin pregnancies and births (Blickstein and Keith, 2001; Martin and Park, 1999). Between 1980 and 1997, the number of twin births in the United States increased 52%. In a similar manner, between 1985 and 2001 in Israel, each year a consistent increase of 6.5% in the rate of multiple births was detected (Blickstein and Baor, 2004).

As early as 1995, infertile patients expressed a desire for multiples (Gleicher et al., 1995). Several articles note that the duration of infertility and the wish to have a multiple pregnancy are correlated, suggesting that ART couples, in fact, express their desperate wish for an instant family. For example, Murdoch (1997) evaluated 150 replies to a questionnaire regarding the ideal outcome of IVF treatment. Of the infertile respondents, 69% considered a multiple pregnancy as an ideal outcome. Ryan et al. (2004) studied the desired outcome of IVF treatment, with a possible reply of having one to four plus babies. In this sample, 20.3% of the patients listed a multiple birth as the most desired outcome. The majority of respondents (94%) ranked twins as their most desired outcome.

Researchers have assumed that young women may be more likely to underestimate the physical and economic resources needed to raise multiples, and are therefore less concerned about a multiple birth.

Whereas couples may consider themselves fortunate to have two or even three babies, the dream of a family does not include handicapped children, nor does it include an unhealthy mother – both a potential outcome of a multiple pregnancy. Indeed, the FIGO Committee for the Ethical aspects of Human Reproduction and Women's Health (FIGO, 2001) claimed "multiple pregnancy has very serious implications for the mother and her offspring, for the family and the community, and for the health service resources." Multiple gestation increases the risk of pregnancy-induced hypertension, preeclampsia, anemia, antepartum and postpartum hemorrhage, and maternal death. ART also contributes to the population of preterm, low-birth-weight, and very-low-birth-weight infants (Kinzler et al., 2000).

28.5 Psychological Reaction to Multiple Parenthood

Once pregnant, irrespective of the mode of conception or number of expected newborn babies, couples face a stage of transition to parenthood. This transition involves conditions that characterized by stress, negatively affecting (1) their feelings, as well as their mental and psychological well-being (Cowan and Cowan, 1992); (2) their relationship and marital satisfaction (Lewis and Spanier, 1979l; Glenn and McLanahan, 1982); and (3) parenting behaviors, ranging from physical and emotional neglect to aggressive behavior, the most extreme being child abuse (Holms and Rahe, 1967; Russell, 1974).

When emotional and psychological resources are diminished after years of infertility, couples usually consider their pregnancy as premium, and not to be lost. These fears and anxiety, much accentuated among parents of twins (Holditch-Davis et al., 1999; Groothuis et al., 1982; Nelson and Martine, 1985; Hay et al., 1990), contribute to a high level of stress, and add to the stress already present from the infertility treatments (Mushin et al., 1985; Raoul-Duval et al., 1993).

According to the cognitively oriented stress-and-coping model suggested by Lazarus and Folkman (1984), stress and coping are viewed as related constructs bearing on the interchange between a person's resources and the demands of the environment in which these resources may be taxed or exceeded. This view identifies primary and secondary cognitive appraisals, coping strategies, and the availability of support resources as central mediators in potentially stress-related responses, affecting immediate and long-term adjustment. They are also integrally related to influence one another in the process of adjustment.

The central role of the coping resources in understanding the mother's experience of stress, thus requiring an extended reference. Coping resources are both personally and socially originated. Sense of coherence (SOC) is one of the substantial personal resources. As the major variable in Antonovsky's salutogenic approach (Antonovsky, 1987), SOC means "a global orientation that expresses the extent to which one has a pervasive, enduring through dynamic feeling of confidence that (1) the stimuli, deriving from ones internal and external environments in the course of living are structured, predictable and explicable; (2) the resources are available to one to meet the demands posed by these stimuli; and (3) these demands are challenges, worthy of investment and engagement" (Antonovsky, 1987: 19).

Maternal self-efficacy, another fundamental personal resource, is defined as beliefs or judgments about the mother's competency or ability to be successful in the parenting role. Bandura (1989) postulated that an individual's sense of self-efficacy operates to reduce perceptions of and reactions to stress. In studies of the transition to parenthood, this concept has been measured by maternal self-efficacy.

Of the social resources, perceived social support encompasses the supportive provisions potentially available within the individual's social network (Pierce et al., 1996). Research on mothers and infants confirm the positive correlation between social support and adaptive maternal behavior. Maternal social support was related to greater parenting satisfaction, more positive behavioral interactions with babies (Crockenberg, 1988). The more intimate aspect of social support relates to the perceived marital quality. Spouse support is defined as the extent to which a person receives high levels of warmth, encouragement, and assistance in the interaction with the partner (Simons and Johnson, 1996).The effect of children on their parents' psychological well-being and marital relationship has been well studied in the past 30 years and have shown time and again, that the presence of children in the family and the experience of parenting stress lowers the level of marital happiness (Belsky and Pensky, 1988; Lavee et al., 1996).

Such coping resources are especially needed for the task of parenting multiples, because once pregnancy is over, families of twins often cannot imagine the realities of parenting twins and may have difficult adjustment (Golombok et al., 1995). During early infancy, difficulties in coping with the often unsynchronized sleeping, feeding, and crying patterns of two (or more) infants lead to fatigue and exhaustion, and feeling guilt associated with inadequate and unequal attention to both children (Sandbank, 1988). Isolation is common because mothers prefer to stay at home due to the complicated logistics involved in taking two (or more) infants out. The financial burden of raising multiples is higher than the cost of raising the same number of singletons and is commonly identified as a source of stress, with an attendant increase of child abuse (Groothuis et al., 1982). It is thus not surprising that mothers of multiples are particularly vulnerable to depression in comparison to singleton mothers (Thorpe et al., 1991). Overall, higher rates of parenting stress are found in ART parents of multiples compared to non-ART parents (Baor et al., 2004). Moreover, in a study that investigated the experience of stress among IVF mothers of twins, in comparison to spontaneously conceived mothers of twins, it was found that IVF mothers' coping resources were depleted in comparison to those of the comparison group of spontaneously conceived twins, and were related to the experience of stress. The higher the coping resources, the lower the parenting stress experience. Maternal efficacy had the highest contribution to the variance, followed by sense of coherence and perceived social support. In addition, maternal employment and infant health status contributed to parental stress (Baor and Soskolni, unpublished).

28.6 Parenting Preterm Multiples

The potential risks of complications in a pregnancy after infertility treatments (Andrews et al., 1993; Rufat et al., 1994; Van Balen et al., 1996) are best illustrated in the findings of Bergh et al. (1999) that the rate of multiples born before 37 weeks is 30.3% (compared to 6.3% in general population); the rate of low birth weight (<2500 g) among multiples is 27.4% (compared to 4.6% among controls); and perinatal mortality is 1.9% (1.1% in controls). Obviously, these issues add to the sense of stress of these couples (McMahon et al., 1997; Van Balen et al., 1996).

Parents often perceive the birth of a preterm infant as a major crisis. Undoubtedly, IVF parents, who timidly expected their perfect child, would consider the preterm birth of their multiples a great crisis, thus experiencing guilt and stress as if they were responsible for this outcome of pregnancy (Easterbrook, 1988). Mothers of preterm infants experience more severe levels of psychological distress during the neonatal period compared to mothers of full-term infants, and depression and anxiety noted at the time of discharge from hospital (Pederson et al., 1987). Moreover, the care of two or more preterm infants, or the care of a child (or children) with special needs, always challenges parents. The challenges increase by the difficulty of caring for other children of the same age, but with very different needs (Bryan, 1992).

Unrealistic expectations of IVF parents from their children can add stress that endangers the adjustment to parenthood of multiples (Belsky et al., 1986). When the children do not meet the expectations, mainly in cases of preterm birth and its sequelae, the adjustment to parenthood and parent–child interactions is delayed (Easterbrook, 1988; Belsky, 1986). These parental stresses reflect in increased incidence of child abuse and neglect, as has been described among infants who were born preterm or suffered significant neonatal illness (Harrison and Magill-Evance, 1996).

28.7 Perinatal Death

The risk of stillbirth and neonatal deaths in twins is more than twice and fivefold higher than as in singletons, respectively. Moreover, the bereavement that follows the death of a fetus or an infant is unique in the case of multiples: the parents may be confused by the apparent contradiction in their feelings of rejoicing in the new life and the simultaneous grief for the dead co-twin. Complex bereavement behavior of the parents results in idealization of the dead twin and possibly alienation from the survivor (Bryan, 1999).

28.8 Epilog

This chapter discusses the psychosocial experience of women undergoing ARTs trying to fulfill their natural and biological wish to become a mother. Going from the beginning of trying to conceive, through the exhausting and resource depleting fertility treatments and up to the experience of motherhood, evidences of the deleterious effects of ART can be found. Therefore, it can be concluded that first-time IVF mothers, especially mothers of premature babies, should be considered as a high-risk population. Ongoing consultation is needed to assist these mothers in decreasing the stress they experience, which might adversely affect their children's development and well-being.

At the organizational sphere, professional workers, trained in twin parenthood issues, are needed in order to meet the above recommendations. Furthermore, special services for these mothers should become an integral part of the public health and social services.

References

Abbey A, Andrews FM, and Halman LJ (1991) The importance of social relationships for infertile couple's well-being. In: Dunkle-Schetter AC (ed.) *Infertility: Perspectives from Stress and Coping Research Stanton*, ch. 4, pp. 61–86. New York: Plenum Press.

Abbey A, Andrews FM, and Halman LJ (1994) Infertility and parenthood: Does becoming a parent increase well being? *Journal of Consulting and Clinical Psychology* 62(2): 398–403.

Abby A (2000) Adjusting to infertility. In: Harvey JH and Miller ED (eds.) *Loss and Trauma: General and Close Relationship Perspective*, pp. 331–344. New York: Burner-Routledge.

Abby A, Andrews FM, and Halman LJ (1994) Psychosocial predictors of life quality: How are they affected by infertility, gender, and parenthood? *Journal of Family Issues* 15: 253–271.

Andrews FM, Abbey A, and Halman J (1991) Stress from infertility, marriage factors, and subjective well-being of wives and husbands. *Journal of Health and Social Behavior* 32: 238–253.

Andrews M, Mausher S, Levy D, et al. (1993) An analysis of the obstetric outcome of 125 consecutive pregnancies conceived *in vitro* and resulting in 100 deliveries. *American Journal of Obstetrics and Gynecology* 154: 848–854.

Antonovsky A (1987) *Unraveling the Mystery of Health: How People Manage Stress and Stay Well*, ch. 2, pp. 15–35. San Francisco, CA: Jossey-Bass.

Bandura A (1989) Regulation of cognitive processes through perceived self-efficacy. *Developmental Psychology* 25: 729–735.

Baor L, Bar-David J, and Blickstein I (2004) Psychological resource depletion of parents of twins after assisted versus spontaneous reproduction. *International Journal of Fertility and Women's Medicine* 49: 13–24.

Belsky J (1986) Transition to parenthood. *Medical Aspects of Human Sexuality* 20: 56–59.

Belsky J and Pensky E (1988) Marital change across the transition to parenthood. In: Palkovitz R and Sussman MB (eds.) *Transitions to Parenthood*, pp. 133–156. New York: Haworth Press.

Belsky J, Ward MJ, and Rovine M (1986) Parental expectations, postnatal experience and the transition to parenthood. In: Ashmore RD and Brodzinsky DM (eds.) *Thinking About the Family: Views of Parents and Children*, pp. 119–145. Hillsdale, NJ: Lawrence Erlbaum Associates.

Bergh T, Ericson A, Hillensjo T, Nygren KG, and Wennerholm UB (1999) Deliveries and children born after *in-vitro* fertilization in Sweden 1982–95: A retrospective cohort study. *Lancet* 354: 1579–1585.

Blascovich J and Tomaka J (1991) Measures of self esteem. In: Robinson JP, Shaver PR, and Wrightsman LS (eds.) *Measures of Personality and Psychological Attitudes*, pp. 115–160. New York: Academic Press.

Blickstein I and Baor L (2004) Multiple birth in Israel. In: Blickstein I and Keith LG (eds.) *Multiple Pregnancy: Epidemiology, Gestation and Perinatal Outcome,* 2nd edn., pp. 1–3. London: Taylor & Francis.

Blickstein I and Keith L (2001) The spectrum of iatrogenic multiple pregnancy. In: Blickstein I and Keith L (eds.) *Iatrogenic Multiple Pregnancy: Clinical Implications*, pp. 1–7. New York: The Partenon Publishing Group.

Bryan EM (1992) The disabled twin. In: Bryan E (ed.) *Twins and Higher Multiple Births. A Guide to Their Nature and Nurture*, pp. 165–170. London: Edward Arnold.

Bryan EM (1999) The death of a twin. In: Sandbank AC (ed.) *Twin and Triplet Psychology*, pp. 186–200. New York: Routledge.

Connolly KJ, Edelmann RJ, Bartlett H, Cook ID, Lenton E, and Pike S (1993) An evaluation of counseling for couples undergoing treatment for IVF. *Human Reproduction* 8(8): 1332–1338.

Cook E (1987) Characteristics of the biopsychosocial crisis of infertility. *Counseling and Development* 65: 465–470.

Cowan CP and Cowan PA (1992) *When Partners Become Parents: The Big Life Change for Couples*, pp. 24–45. New York: Basic Books.

Crockenberg SB (1988) Social support and parenting. In: Fitzgerald H, Lester B, and Yogman M (eds.) *Theory and Research in Behavioral Pediatrics*, pp. 141–174. New York: Plenum.

Danulik JC (1997) Gender and infertility. In: Leiblum SR (ed.) *Infertility: Psychological Issues and Counseling Strategies*, pp. 103–125. New York: Wiley.

Demyttenaere K, Bonte L, Gheldof M, Vervaeke M, Meuleman C, Vanderschuerem D, and D'Hooghe T (1998) Coping style and depression level influence outcome in *in vitro* fertilization. *Fertility and Sterility* 69: 1026–1033.

Dunkle-Schetter C and Lobel M (1991) Psychological reaction to infertility. In: Stanton A and Dunkle-Schetter C (eds.) *Infertility: Perspectives from Stress and Coping Research*, ch. 3, pp. 29–57. New York: Plenum Press.

Easterbrook MA (1988) Effects of infant risk status on the transition to parenthood. In: Michaels GY and Goldberg WA (eds.) *The Transition to Parenthood: Current Theory and Practice*, pp. 176–208. New York: Cambridge University Press.

FIGO Committee for the Ethical Aspects of Human Reproduction and Women's Health (2001) Ethical guidelines in the prevention of iatrogenic multiple pregnancy. *European Journal of Obstetrics and Gynecology and Reproductive Biology* 96: 209–210.

Gleicher N, Campbell DP, Chan CL, Karande V, Rao R, Balin M, and Pratt D (1995) The desire for multiple births in couples with infertility problems contradicts present practice patterns. *Human Reproduction* 10: 1079–1084.

Glenn ND and McLanahan S (1982) Children and marital happiness: A further specification of a relationship. *Journal of Marriage and Family* 44: 63–72.

Glover L, Hunter M, Richerds JM, Katz M, and Abel PD (1999) Development of the infertility adjustment scale. *Fertility and Sterility* 72(4): 623–628.

Golombok S, Cook R, Bish A, and Murray C (1995) Families created by the new reproductive technologies: Quality of parenting and social and emotional development of the children. *Child Development* 66: 285–298.

Groothuis JR, Altemeier WA, Robarge JP, O'Connor S, Sandler H, Vietze P, and Lustig JV (1982) Increased child abuse in families with twins. *Pediatrics* 70: 769–773.

Harrison MJ and Magill-Evance J (1996) Mother and father interactions over the first year with term and preterm infants. *Research in Nursing and Health* 19: 451–459.

Hay DA, Gleeson C, Lorden B, Mitchell D, and Paton L (1990) What information should the multiple birth family receive before, during and after the birth? *Acta Genetica Medica et Gemellol* 39: 213–222.

Holditch-Davis D, Roberts D, and Sandelovski M (1999) Early parental interaction with and perceptions of multiple birth infants. *Journal of Advanced Nursing* 30: 200–210.

Holms T and Rahe R (1967) The social readjustment rating scale. *Journal of Psychosomatic Research* 11: 213–218.

Jordan C and Revenson TA (1999) Gender differences in coping with infertility: A meta-analysis. *Journal of Behavioral Medicine* 22(4): 341–358.

Kentenich H (1989) Psychological guidance of IVF patients. *Human Reproduction* 4(supplement): 17–22.

Keye WR, Jr. (1984) Psychosexual responses to infertility. *Clinical Obstetrics and Gynecology* 27(3): 760–766.

Kinzler WL, Anath CV, and Vintzileos AM (2000) Medical and economic effects of twin gestations. *Journal of the Society for Gynecologic Investigation* 7: 321–327.

Lavee Y, Sharlin S, and Katz R (1996) The effect of parenting stress on marital quality: An integrated mother–father model. *Journal of Family Issues* 17: 114–135.

Lazarus RS and Folkman S (1984) *Stress, Appraisal and Coping*. New York: Springer.

Leiblum SR, Kemmann E, and Lane MK (1987) The psychological concomitants of *in vitro* fertilization. *Journal of Psychosomatic Obstetrics and Gynecology* 6: 165–178.
Lewis RA and Spanier GB (1979) Theorizing about the quality and stability of marriage. In: Burr WR, Hill RF, Nye I, and Reiss IL (eds.) *Contemporary Theories About the Family,* vol. 1, pp. 269–294. New York: Free Press.
Mahlstedt PP (1985) The psychological component of infertility. *Fertility and Sterility* 3: 335–346.
Malin M, Hemmink E, Räikkönen O, Sihvo S, and Perälä ML (2001) What do women want? Women's experiences of infertility treatment. *Social Science and Medicine* 53(1): 123–133.
Martin JA and Park MM (1999) Trends in twin and triplet births: 1980–1997. *National Vital Statistics Reports* 47: 1–17.
Matthews R and Matthews AM (1986) Infertility and involuntary childlessness: The transition to non-parenthood. *Journal of Marriage and Family* 8: 641–649.
Mazor MD (1984) Emotional reactions to infertility. In: Mazor MD and Simons H (eds.) *Infertility: Medical, Emotional and Social Considerations*, pp. 23–35. New York: Human Science Press.
McMahon CA, Ungerer JA, Tennant C, and Saunders D (1997) Psychosocial adjustment and the quality of the mother–child relationship at four months postpartum after conception by *in vitro* fertilization. *Fertility and Sterility* 68: 492–500.
Menning BE (1980) The emotional needs of infertile couples. *Fertility and Sterility* 34: 313–319.
Miall CE (1986) The stigma of involuntary childlessness. *Social Problems* 33: 268–282.
Murdoch A (1997) Triplets and embryo transfer policy. *Human Reproduction* 12: 88–92.
Mushin D, Spensley J, and Barreda-Hanson M (1985) Children of I.V.F. *Clinical Gynecology* 12: 865–876.
Nelson MHB and Martine CA (1985) Increased child abuse in twins. *Child Abuse and Neglect* 9: 501–505.
Olshansky EF (1988) Responses to high technology infertility treatment. *IMAGE* 20: 128–131.
Pederson DR, Bento S, Chance GW, Evans B, and Fox AM (1987) Maternal emotional responses to preterm birth. *American Journal of Orthopsychiatry* 57: 15–21.
Pierce GR, Sarason IG, and Sarason BR (1996) Coping and social support. In: Zeidner M and Endler N (eds.) *Handbook of Coping: Theory, Research, Applications*, pp. 434–451. New York: Wiley.
Raoul-Duval A, Bertrand-Servais M, and Frydman R (1993) Comparative prospective study of the psychological development of children born by *in vitro* fertilization and their mothers. *Journal of Psychosomatic Obstetrics and Gynecology* 14: 117–126.
Remennick L (2000) Childless in the land of imperative motherhood: Stigma and coping among infertile Israeli women. *Sex Roles* 43: 821–841.
Rufat P, Oliviennes F, DeMouzon J, Dehan M, and Frydman R (1994) Task force on the outcome of pregnancies and children conceived by *in vitro* fertilization (France: 1987 to 1989). *Fertility and Sterility* 61: 324–330.
Russell C (1974) Transition to parenthood. *Journal of Marriage and Family* 36: 294–302.
Ryan GL, Zhang SH, Dokras A, Syrop CH, and Van Voorhis BJ (2004) The desire of infertile patients for multiple births. *Fertility and Sterility* 81: 500–504.
Sandbank AC (1988) The effect of twins on family relationships. *Acta Genetica Medica et Gemellol* (*Roma*) 37: 161–171.
Simons RL and Johnson CA (1996) Social network and marital support as mediators and moderators of the impact of economic pressure on parental behavior. In: Pierce GR, Sarason BR, and Sarason IG (eds.) *Handbook of Social Support and the Family*, pp. 269–287. New York: Plenum.
Thorpe K, Golding J, MacGillivray I, and Greenwood R (1991) Comparison of prevalence of depression in mothers of twins and mothers of singletons. *British Medical Journal* 302: 875–878.
Tulandi T, Bull R, Cook R, and McInnes R (1981) Regrettable pregnancy after infertility. *Infertility* 4: 321–326.
Van Balen F, Naaktgeboren N, and Trimbos-Kemper TC (1996) *In vitro* fertilization: The experience of treatment, pregnancy and delivery. *Human Reproduction* 11: 95–98.

Further Reading

Berg BJ and Wilson JF (1991) Psychological functioning across stages of treatment in infertility. *Journal of Behavioral Medicine* 14: 11–26.

29 Transsexualism

R A Allison, CIGNA Medical Group of Arizona, Phoenix, AZ, USA

Chapter Outline

Glossary

gender One's personal, social, and legal status as a male or female.

gender binary The concept that there are only two genders, male and female.

gender dysphoria Persistent feeling of discomfort with one's biological sex and/or gender role.

gender expression The visual cues which identify a person, in the eyes of others, as belonging to a particular gender.

gender identity One's sense of self as belonging to a particular gender.

gender identity disorder Synonymous with transsexualism – a term considered pejorative by some transsexual persons.

genderqueer A new term, favored by young persons, describing those who do not fit a gender binary concept.

gender role The attribution by society of a person as male or female.

sex Physical identification of a person as man or woman according to external and internal sexual structures.

transsexual A person who has a persistent desire to live and be accepted as a member of the opposite sex, including hormonal and surgical treatment.

transvestite A person who wears clothing of the opposite sex but does not desire permanent changes including surgery.

29.1 Historical Perspective

Persons who live as members of the sex opposite their birth have been recognized and accepted in many societies throughout history. In the twentieth century, medical and surgical treatments were developed which allowed such persons to achieve a more satisfactory physical congruity with their chosen gender.

The first recorded example of surgery to reassign a person from male to female appears to be the case of Einar Wegener, who had surgery in Berlin in the 1920s to remove penis and testicles (Hoyer, 1933). Wegener took the name of Lili Elbe and was issued a Danish passport in that name.

Hirschfeld (1910) published *The Transvestite: An Investigation into the Erotic Impulse of Disguise.* He specifically distinguished the condition he called transvestism from homosexuality. For years, however, the psychiatric community did not recognize a difference between transvestism and transsexualism. Indeed, it was not a psychiatrist, but an endocrinologist, Harry Benjamin, who first used the term transsexualism in a medical presentation at the New York Academy of Medicine in 1953. This presentation followed the widely publicized report of the sex-change operation in Denmark which converted George Jorgensen into Christine. Benjamin (1966) deserves credit for his staunch support of transsexual patients despite strong opposition from psychoanalytic practicioners in the 1950s and 1960s. In 1966, his book, *The Transsexual Phenomenon*, became the standard text for doctors as well as for transsexual persons anxious to learn more about themselves.

The following decade saw the establishment of gender clinics in many university medical centers in the United States, led by the Johns Hopkins University. Such clinics included psychologists and psychiatrists; physicians who prescribed hormone treatment and provided medical care; and often surgeons who performed a range of genital and nongenital operations for transgender persons. Unfortunately, opposition to medical and surgical treatment of transsexual persons continued, primarily from psychoanalysts. Socarides (1978) referred to transsexualism as a psychosis. Meyer-Bahlburg (1982) claimed "Transsexualism is closely linked to perversions."

In the *Archives of General Psychiatry*, Meyer and Reter (1979) published 'Sex reassignment: Follow-up', which evaluated a small number of persons who had undergone surgery at Johns Hopkins. Out of 100 male-to-female patients, 34 had surgery but only 11 were available for follow-up, compared to 28 of 66 who did not have surgery. The study showed no differences in the two groups in terms of psychological, professional, residence stability, or global evaluation. Although such numbers are too small to have statistical significance, the authors concluded there was no benefit from surgery with regard to life outcomes. At least partially in response to this publication, many of the gender clinics eventually closed.

The management of transsexual patients continued, however, and the Harry Benjamin International Gender Dysphoria Association was established in that same year, 1979, to provide professional standards and support for those involved in the treatment of transsexual persons. The organization (now called the World Professional Association for Transgender Health (WPATH)) has established the Standards of Care for Gender Identity Disorders. These standards contain a general discussion of hormone therapy, including the types of hormone treatment used and their effects. Many physicians who provide medical and surgical care to transsexual persons are members of WPATH.

Estimates of the prevalence of transsexualism have changed over time. Pauly's initial estimate of 1 in 100 000 for male-to-female transsexualism was contrasted with Walinder's estimation from Sweden of 1 in 37 000 (Pauly, 1968; Walinder, 1968). Walinder acknowledged that these numbers were low, since they reflected only persons who had entered the medical system for treatment at that time. Eklund et al. (1988) reported a series of three surveys in the Netherlands, with increasing prevalence each time: in 1980, 1 in 45 000 (male to female); in 1983, 1 in 26 000; and in 1986, 1 in 18 000. The changes again are due to increasing numbers of transsexual persons coming out and seeking treatment. Bakker, also in the Netherlands in 1993, found a prevalence of 1 in 11 900 for male-to-female persons and 1 in 30 400 for female-to-male persons.

Conway, in 2007, used the estimated number of sex-reassignment surgeries performed annually on United States residents to calculate an overall prevalence of at least 1 in 2 500 and probably larger.

29.2 Terminology

29.2.1 Transsexual versus Gender Identity Disorder

The diagnosis of transsexualism was introduced in the third edition of the *Diagnostic and Statistical Manual of Mental Disorders, Third Edition (DSM-III)* in 1980 for individuals who demonstrated at least 2 years of continuous interest in transforming the sex of their bodies and their social gender status. In 1994, the DSM-IV committee replaced the diagnosis of transsexualism with gender identity disorder. Depending on their age, those with a strong and persistent cross-gender identification and a persistent discomfort with their sex or a sense of inappropriateness in the gender role of that sex were to be diagnosed as Gender Identity Disorder of Childhood (302.6), Adolescence, or Adulthood (302.85). For persons who did not meet these criteria, Gender Identity Disorder Not Otherwise Specified (GIDNOS) (302.6) was to be used.

There will likely never be a consensus among transsexual persons and healthcare providers regarding a single nomenclature. For purposes of clarity and simplicity, the term transsexual will be used in this review.

29.2.2 Transsexualism versus Crossdressing

The term crossdressing has replaced transvestism in common use and refers to persons who dress in clothing of the opposite gender but do not live full time in that gender. Crossdressing may or may not have an erotic component. Transsexual persons may or may not have a history of crossdressing behavior prior to beginning gender transition.

The term drag queen (or its counterpart, drag king) usually refers to a person who dresses as a member of the opposite sex for entertainment purposes, but does not seek sex reassignment, and may feel comfortable in the birth gender.

29.2.3 Transsexual versus Transgender

The term transgender at one time was used by persons who considered themselves to have a gender identity which was not congruent with their birth sex, but who did not desire full surgical change of sex. These persons often live full time in their desired gender role and may or may not take crossgender hormones.

In recent years the term transgender has taken on a broader interpretation and is used to encompass the spectrum of gender diversity, including crossdressers and transsexual persons. The transgender community is then a very diverse group with areas of similarity and areas of difference which may result in conflict.

29.2.4 Primary versus Secondary

The stage of life at which gender transition occurs is sometimes used to divide transsexual persons into two categories. Persons who never establish an adult gender role in their birth gender may refer to themselves as primary transsexuals or primaries, and to those who transition later in life as secondaries. Sometimes the distinction is used by primary transsexuals to imply their greater authenticity, compared to persons who establish careers in the birth gender role (and perhaps marry and have children). Of course, this terminology is not favored by persons who did transition later in life, many of whom state they were aware of their gender issues from an early age but did not live in an environment where transition was feasible.

Some sexologists have adopted this binary division to describe transsexual persons, particularly male to female. Seeing the difference in a sexual aspect, they call early transitioners homosexual transsexuals, implying that these young people are not women, but men who transition to have sex with men. They designate later transitioners by the term autogynephilia, by which they mean that these persons have autoerotic fantasies of themselves as women and ultimately fulfill these fantasies through medical and surgical transition.

This idea was advanced by Bailey (2003) in *The Man Who Would Be Queen: The Science of Gender-Bending and Transsexualism.* The book was published in 2003 and immediately became a source of great controversy. It was denounced as unscientific by transgender persons and by other sex researchers, including leaders of the Kinsey Institute and the World Professional Organization for Transgender Health.

While individual transsexual persons do report diverse reasons for beginning transition, there is general agreement that a division into two separate clinical syndromes based on age at beginning does not accurately or completely categorize all such persons.

29.2.5 Sexual Orientation versus Gender Identity

A person may be attracted to persons of the opposite sex (heterosexual); persons of the same sex (homosexual); persons of either sex (bisexual); or may have no sexual desire at all (asexual). Transsexual persons are no different. A male-to-female transsexual may be attracted to men prior to surgery and remain attracted to men after surgery. Did this person change her orientation from gay male to straight female? Or is she, then and now, attracted to the same people? Talk of sexual orientation may become confusing and perhaps irrelevant where transsexual people are concerned. This easily observable fact continues to confuse sexologists, who still use terms such as homosexual transsexual and nonhomosexual transsexual as if sexual activity were the motivation for transition.

Sexual orientation and gender identity are quite independent of one another.

29.3 Hormone Treatment of Transsexual Persons

The care of the transsexual person is usually a multidisciplinary effort, involving a behavioral health professional as well as medical and surgical providers. The role of the behavioral health professional is first to confirm that the diagnosis of transsexualism is correct and that transition is appropriate for the individual. The WPATH Standards are very helpful in this regard. Once the diagnosis is established and the patient has been followed for an appropriate length of time, it is appropriate to begin medical treatment with crossgender hormones.

At this time, hormone treatment of transsexual persons in the United States is off-label, meaning hormones are not approved by the Food and Drug Administration for such purposes. It is legal however, and physicians who prescribe hormones usually obtain detailed informed consent from the patient concerning the effects of treatment. Since the effects of hormone treatment will be irreversible in many instances, it is critical for the patient to be certain of his or her intentions.

The beginning of hormone therapy usually coincides somewhat closely with the beginning of living full time in the gender of choice. This is often called the real-life experience. According to the WPATH Standards of Care, it is recommended to begin hormone therapy after (1) a period of at least 3 months living in the new gender role or (2) an ongoing relationship with the behavioral health professional of at least 3 months. For many persons, the real-life experience is a time of increased psychological, social, and financial stress. Loss of family support, social relationships, employment, and savings produce anxiety. The physical and emotional changes which occur with hormone administration may relieve some of the anxiety.

29.4 Male-to-Female Hormone Treatment

To achieve feminization of the biological male, the use of estrogen is supplemented with agents which counteract the effects of testosterone. Estrogen itself has an antiandrogenic effect through suppression of pituitary gonadotropin, but significantly greater androgen suppression is achieved by adding an agent that suppresses the production or action of testosterone.

Estrogen may be administered in several ways. It can be taken orally, where it is absorbed through the intestine and metabolized by the liver. A side effect of this hepatic stimulation produces increased amounts of substances which promote blood clotting, so that increased risk of thrombotic disease (especially venous thrombosis or pulmonary embolism) is a concern. Other routes of administration which do not involve the liver include transdermal patches and intramuscular injection.

Some patients, and some physicians, have recommended cycling estrogen on a monthly basis to simulate the female menstrual cycle. There is no evidence that such cycling conveys any physiologic benefits. On the contrary, omission of estrogen for several days each month may allow breakthrough testosterone production in persons who have not yet had orchiectomy.

Progesterone is sometimes prescribed as an addition to estrogen treatment, and some transsexual persons feel it augments breast growth and body fat redistribution. There are no studies which confirm these impressions. Progesterone tends to negate the beneficial effect of estrogen on the production of high-density lipoprotein (HDL) cholesterol. It may also be given as an oral preparation or an intramuscular injection.

Several types of androgen antagonists are used in the treatment of transsexual persons who have not had orchiectomy. The most common, and least expensive, is spironolactone. This drug is felt to reduce testosterone production and block receptor sites for dihydrotestosterone (DHT), the active metabolite. Spironolactone is a potassium-sparing diuretic and may produce hypotension, dehydration, and hyponatremia. Monitoring of serum electrolytes and renal function must be performed in persons taking this drug.

Finasteride and dutasteride act by inhibiting conversion of testosterone to DHT. The resulting inhibition of testosterone action can be very beneficial with respect to body hair (reduced) and scalp hair (stimulated). These drugs are safe, with no major side effects.

Cyproterone acetate also blocks the DHT receptor site. This drug must be given orally twice a day to be most effective. Weakness and fatigue are common side effects. Cyproterone is not currently approved for prescription in the United States.

Leuprolide is a gonadotropin-releasing-hormone analog which produces testosterone suppression through gonadotropin inhibition. Unlike most of the other drugs, the antiandrogenic effects of leuprolide are reversible, at least up to a year of administration. Some doctors have used leuprolide as single therapy, without estrogen, in young transsexual persons to delay puberty and allow time to be certain of the diagnosis.

29.4.1 Effects of Hormone Treatment in Male-to-Female Transsexual Persons

Estrogen will begin to produce visible effects after several weeks to months after beginning administration. The most obvious early effect is breast enlargement. The degree of enlargement is usually modest, but may be as great as a biologic female, especially in persons whose female relatives tend to have large breasts. Sensitivity and tenderness also are noticed within a few weeks of starting treatment. Enlargement of areolae and nipples takes much longer, if it ever occurs. After 2 years of estrogen therapy, it is unlikely that significant further breast growth will be seen.

Estrogen causes redistribution of body fat, with an increase in subcutaneous fat generally. The veins in the arms and hands may become less visible. Fat redistribution in the buttocks may or may not be noticeable, and when it occurs it may take years.

Profound effects on male sexual organs are seen with continuing estrogen administration. Testicular atrophy is progressive and is often irreversible within 6–12 months of continuous treatment. Male-to-female persons should be advised that they will become sterile, and prior to beginning hormone treatment they should be offered the option of sperm cryopreservation for possible future fertility. The penis usually shows progressive diminution in size, and most (but not all) persons lose their ability to have erections. The prostate almost always shrinks with prolonged estrogen treatment, but this effect is reversible until orchiectomy is performed.

The mental and emotional effects of estrogen administration to male-to-female persons are not easily measurable, but almost every such person will report marked improvement in feeling of well-being and relief of anxiety. These emotional benefits are the usual reason for persons to continue estrogen treatment for many years after sex reassignment.

29.4.2 Limitations of Estrogen Therapy

There are a number of masculine characteristics which will not be changed by estrogen administration. Obviously, once adult bone growth has occurred, it will not be reversed. Significant changes in height do not occur, nor will the bones of the hands or feet become smaller. It is possible that some shrinkage of the muscles of the hands or feet may cause the person to perceive a slight size difference.

Estrogen, accompanied by antiandrogen treatment, will stop scalp hair loss and, in some cases, will promote regrowth if hair loss has not been too extensive. Its effect on body hair is variable. Many people have significant reduction in hair on the arms, legs, and even chest; while others do not achieve such results and must undergo other means of hair removal. The one area where estrogen does not change the male hair pattern is the beard. Nearly all male-to-female persons must rely on electrolysis for removal of hair from face and neck.

Estrogen will not produce changes in the voice. Once the vocal chords have reached male size, they will not shrink. Likewise, the thyroid cartilage (Adam's apple) remains prominent and must be reduced through surgery. Most surgical procedures on the vocal chords themselves have not yielded consistently good results, although some individual cases have been impressive. Speech therapy and vocal training are safer and more reliable ways to achieve a female voice.

29.4.3 Side Effects of Estrogen Therapy

Venous thromboembolism has already been mentioned as a potential side effect of oral estrogen administration. This complication is most likely with ethinyl estradiol or conjugated estrogens, and less so with oral estradiol. It does not seem to occur with transdermal or injectable estrogen.

Immobility, particularly bed rest after surgery, increases the risk of venous thromboembolism. For this reason, surgeons who perform sex reassignment surgery instruct their patients to withhold estrogen for a period of time (usually 2 weeks) prior to surgery. Most surgeons now have their patients ambulating earlier than in past years, which also reduces the risk.

The risk of arterial disease (heart attack or stroke) is difficult to determine in relation to estrogen. These diseases are so common in the population that it would be difficult to attribute their occurrence to any single factor. Risk factors such as hypertension, high cholesterol, diabetes, and cigarette smoking are seen in many male-to-female persons and should be managed aggressively. Persons who smoke and take estrogen are at especially high risk. The use of aspirin for its antithrombotic effect is encouraged in persons with all these risk factors.

Breast cancer appears extremely rare in male-to-female persons; but as more persons continue estrogen for many years, it is wise to continue surveillance, as any other woman, with breast self-examination and mammography.

One complication which has been reported is prolactinoma (adenoma of the pituitary gland). This was first thought to be due to excessively high doses of estrogen, but has been reported in male-to-female persons after years of normal-dose estrogen. It seems prudent to occasionally monitor serum prolactin levels as part of routine laboratory examination.

Carcinoma of the prostate has never been reported when orchiectomy is performed prior to age 40. Many male-to-female persons begin hormone therapy and have surgery after age 40, however, and several cases of prostate cancer have been reported, but the numbers are quite low relative to the number of persons undergoing hormone treatment.

29.5 Female-to-Male Hormone Treatment

The administration of testosterone to female-to-male transsexual persons produces masculinizing effects which are consistent and profound. Testosterone is almost always given as either an intramuscular injection or a transdermal patch or gel. The injections can be long lasting, but serum levels of testosterone diminish with long intervals between injections, so many persons choose smaller quantities, injected more frequently.

29.5.1 Effects of Testosterone Therapy

Testosterone will produce irreversible deepening of the voice after as little as 6 weeks of therapy. Lean body mass increases as body fat decreases (except abdominal fat). Beard growth occurs within months, and many persons will begin to exhibit male-pattern scalp hair loss. The skin becomes more oily and acne is common.

All female-to-male persons will experience clitoral enlargement with testosterone, but only a minority will have growth sufficient for intercourse. Libido is usually increased soon after starting testosterone treatment.

29.5.2 Limitations of Testosterone Therapy

Testosterone will not reduce the size of the breasts, and mastectomy is often the first operation the female-to-male person undergoes. For some, it will be their only surgery. Many female-to-male persons, aware of the less-than-ideal results of phalloplasty, choose not to have genital surgery, and often will not have removal of uterus or ovaries either. In most, but not all, persons, testosterone will cause cessation of menses. It does not reduce the risk of ovarian, uterine, or cervical cancer, nor is there evidence that it increases such risk. Several cases of ovarian cancer have been reported in female-to-male persons. Transition from female to male does not eliminate the need for annual pelvic examination, unless a person has had total hysterectomy. Neither does it eliminate the need for breast self-examination. Even when mastectomy has been performed, there have been cases of carcinoma in residual breast tissue.

29.6 The Social and Emotional Challenges of Gender Transition

In a society which expected its citizens to fill one of two specific and exclusive gender roles, the coordination of nature and nurture created a learning experience which is designed to teach children the skills needed to become a successful adult man or woman. Stereotypically, boys were encouraged in sports and in outdoor activities such as hunting and fishing. Girls were taught cooking, sewing, and homemaking skills, all designed to prepare them for their role as wife and mother. Times have changed somewhat: more women work outside the home, more fathers provide child care, and gender roles overlap in many families. Yet the binary culture persists, and in many places there is little tolerance for the assertive, athletic girl or the boy teased as a sissy. Children who feel themselves outside the gender binary may hide these feelings in an attempt to be accepted by peers and adults. The result is that they are socialized in their birth gender, and miss the experiences of girlhood or boyhood which would have made the eventual transition easy and natural.

A transsexual person who recognizes the need to transition at an early age – perhaps even as a teenager – may recover some socialization in the desired gender. Such fortunate young people may even be spared the unwanted physical changes of puberty in the birth gender. To succeed in transition as a teenager, of course, the support of parents is essential, and many persons still lack this support. As a result, they may leave home without good education or career skills and find themselves without the financial resources to achieve a successful transition. It is a sad paradox: they are young enough to obtain excellent results from hormone therapy, but lack the funds to pay for the hormones, much less sex reassignment surgery.

Contrast the persons who transition later in life. Here, adult gender roles assume greater importance. Someone transitioning from female to male may have had career success prior to transition, but often at the lower salaries paid to women compared to men. The relative lack of financial resources is intensified by the higher costs of female-to-male sex reassignment. Many are limited to mastectomy as their only affordable procedure, while others can finance hysterectomy but not the very expensive phalloplasty.

An adult transitioning from male to female may have had the opportunity to accumulate savings which will finance transition. Such years of saving come at a price: the hardening of male physical characteristics which will not be fully reversed by any hormone therapy, but only by cosmetic surgery. The deep voice is of great concern and impairs acceptance as a woman for many. Existing voice surgery procedures have not produced consistently impressive results, and voice therapy may be the best option.

Psychological adaptation to transition can be a challenge. An adult who has enjoyed professional success as a male may be accustomed to so-called male privilege and to having requests promptly answered. This privilege changes with assumption of the female role and the change is traumatic. While many report positive psychological effects of hormone treatment, the psychological adjustment to the absence of male privilege may impair the future career possibilities for a male-to-female transsexual person.

29.7 Conclusion

The dysphoria or discomfort which transsexual persons report as a constant part of their lives may be resolved through the transition process. A very important early step in this resolution is the beginning of crossgender hormone therapy. Hormone-induced changes occur rapidly after the start of therapy and are more profound with longer duration of treatment. The types of hormone therapy, and the expected results of each type, have been discussed. Further information is available from the WPATH Standards of Care for Gender Identity Disorders.

References

Bailey JM (2003) *The Man Who Would Be Queen: The Science of Gender-Bending and Transsexualism.* Washington, DC: Joseph Henry Press.

Benjamin H (1966) *The Transsexual Phenomenon.* New York: Julian Press.

Eklund PL, Gooren LJ, and Bezemer PD (1988) Prevalence of transsexualism in The Netherlands. *The British Journal of Psychiatry* 152: 638–640.

Hoyer N (1933) *Man into Woman: An Authentic Record of a Change of Sex.* New York: Dutton.

Meyer JK and Reter DJ (1979) Sex reassignment: Follow-up. *Archives of General Psychiatry* 36: 1010–1015.

Meyer-Bahlburg HFL (1982) Hormones and psychosexual differentiation: Implications for the management of intersexuality, homosexuality, and transsexuality. In: Bancroft J. (ed.) *Diseases of Sex and Sexuality. Clinics in Endocrinology and Metabolism* 11(3): 681–701.

Pauly I (1968) Current status of the change of sex operation. *Journal of Nervous and Mental Disease* 147: 460–471.

Socarides CW (1978) Transsexualism and psychosis. *International Journal of Psychoanalytic Psychotherapy* 7: 373–384.

Walinder J (1968) Transsexualism: Definition, prevalence, and sex distribution. *Acta Psychiatrica Scandinavica* 44: 255.

Further Reading

Ettner R, Monstrey S, and Eyler AE (2007) *Principles of Transgender Medicine and Surgery.* New York: The Haworth Press.

Feldman J (2007) Preventive care of the transgendered patient: An evidence-based approach. In: Ettner R, Monstrey S, and Eyler AE (eds.) *Principles of Transgender Medicine and Surgery.* New York: The Haworth Press.

Olyslager F and Conway L (2007) *On the Calculation of the Prevalence of Transsexualism.* Presented at the 2007 WPATH Symposium, Chicago.

van Kesteren PJ, Asscheman H, Megens JA, and Gooren LJ (1997) Mortality and morbidity in transsexual subjects treated with cross-sex hormones. *Clinical Endocrinology* 47: 337–342.

Relevant Websites

http://www.nctequality.org – National Center for Transgender Equality.

http://www.glma.org – The Gay and Lesbian Medical Association: Transgender Health Resources.

http://www.hrc.org – The Human Rights Campaign: Transgender Issues.

www.wpath.org – The World Professional Organization for Transgender Health.

30 Disorders of Salt and Fluid Balance

T Lenhard, University Clinic of Heidelberg, Heidelberg, Germany
M Bettendorf, University Clinic of Heidelberg, Heidelberg, Germany
S Schwab, University Clinic Erlangen-Nürnberg, Erlangen, Germany

Chapter Outline

Glossary

aldosterone The major mineralocorticoid that is synthesized in the *zona glomerulosa* of the adrenal gland. Biosynthesis uses progesterone, a cholesterolderivate, as central precursor also for cortisol, androgens, and estrogens. The lipophilic aldosterone molecule signals through cytoplamatic mineralocorticoid receptors and induces sodium reabsorption in the kidney.

aquaporin A family of transmembrane proteins forming a water-specific channel that facilitates water transport across membranes independently of solute transport. Aquaporins are highly conserved along evolution and can be divided into two functional groups: one solely permeable for water (aquaporins) and a second for additional small organic molecules, for example, glycerol (aquaglyceroporins). The transport rate for nonprotonated water molecules is very fast: one water molecule per nanosecond. Aquaporins play a crucial role in maintaining organ-specific fluids, for example, in the kidney (urine), in the brain (cerebrospinal fluid), or in the eye chamber.

beer potomania syndrome It is a rare renal excretion disorder seen in alcoholic patients who consume high amounts of beer (which has low salt content) and who eat little food. A diet poor in salt and protein (urea source) results in reduced excretion of urinary solutes that limits the ability to excrete free water.

chaperones Specialized proteins that facilitate the correct folding of immature proteins. The best-investigated protein chaperone is the heat shock protein-60 (hsp60). In addition, some organic molecules can also act as chaperones.

G-proteins Guanosine triphosphate (GTP)-binding proteins (G-proteins) are a family of relay proteins that functionally couple the largest family of transmembrane cell-surface receptors (G-protein-linked receptors, e.g., beta-adrenergic receptors, natriuretic peptide receptors, vasopressin receptors, and many more) to enzymes or ion channels and thus take part in downstream signaling and cellular responses to external stimuli.

neuroleptic malignant syndrome (NMS) It is a rare and potentially fatal complication of neuroleptic drugs and was described for the first time as akinetic hypertonic syndrome by Delay and colleagues after the introduction of the neuroleptics in 1960. Clinical criteria (*Diagnostic and Statistical Manual of Mental Disorders*, Fourth Edition) are muscle rigidity and elevated body temperature associated with an antipsychotic medication. These criteria must be accompanied by two or more additional symptoms, such as consciousness disturbances, autonomic dysregulation (tachycardia, hypertension, diaphoresis, and incontinence), evidence of muscle injury, or elevated leukocyte level.

proteasome A large adenosine triphosphate (ATP)-dependent cytoplamatic protease complex degrades damaged or unneeded proteins after attachment of those proteins with the small marker protein ubiquitin. In some inherited diseases (e.g., nephrogenic diabetes insipidus) the proteasome degrades mutated and falsely folded proteins. The proteasome also regulates physiological processes by degradation as, for example, antigen processing, gene transcription, cell-cycle progression, inflammatory responses, or circadian rhythm.

30.1 Physiology of Salt and Fluid Balance

The kidneys maintain water and salt homeostasis and extracellular fluid equilibrium. They regulate the volume and osmolality of the extracellular and vascular fluids via resorption and excretion of ions and water. The hormones involved in regulating water and salt equilibrium include argenine vasopressin (AVP), aldosterone, and natriuretic peptides. A disturbance in water and salt equilibrium at any level can cause severe changes in an individual's state of health. The pathological process for water and/or salt imbalance can be located within the kidney or in the brain (of the regulatory centers) but also in between as caused, for example, by paraneoplastic syndromes in some cancers. Disturbances in water and/or salt balance affect the functional metabolism of almost all organs but especially heart, muscle, and the brain. Among these, the brain is particularly sensitive to changes in water and sodium homeostasis. Symptoms extend from mild behavioral and cognitive deficits to life-threatening epileptic seizure with lethal brain edema. This section gives an introduction to the physiology of normal renal function and the hormonal and neural regulation of water and salt balance. Then a brief overview of the general symptoms of disturbed water volume and salt balance is presented. In subsequent sections, the different disorders that cause salt and water imbalance are described.

30.1.1 Salt and Fluid Balance in the Kidney: Normal Conditions

30.1.1.1 Structure of the nephron

The smallest functional and histological unit of the kidney is the nephron. The nephron consists of the Bowman capsule, containing the glomerular apparatus as a functional unit located within the renal cortex. The Bowman capsule joins into the proximal tubule, which is followed by the thin descending loop of Henle, the thin ascending loop of Henle, the thick ascending Henle tubule, the *macula densa*, and the distal tubule and connecting tubule to the collecting duct (**Figure 1**; Alpert and Hebert, 2007, pp. 479–627; Constanzo, 2006, pp. 235–299). The glomerular apparatus consists of a glomerular afferent and efferent arteriole. A specialized capillary system lies in between, forming the connecting unit between the blood stream and the urinary tubule system. The glomerular apparatus produces the primary urine. The glomerular filtration rate (GFR) of approximately 90 $ml\,min^{-1}$ is kept stable at a blood pressure ranging from 70 to 180 mmHg by an autoregulative capacity of the kidney. In addition, renal perfusion is regulated by nerve terminals of the sympathetic nervous system (SNS; vasoconstrictive: norepinephrine; vasodilatative: dopamine) and paracrine signals from hormones (vasoconstrictive: angiotensin II, 5-HT, endothelin, thromboxane; vasodilative: natriuretic pertides, NO, $PGE_2/-I_2$). The filtrated molecular mass is determined by the specific ultrastructure of the glomeruli. The capillary endothelium is fenestrated (50–100 nm). Podocytes and endothelial cells form a specialized basal membrane (300 nm in diameter; contains laminin, fibronectin, different isoforms of collagen type IV) on the basolateral endothelial site. The small foot-like processes of the podocytes form small slits of 5 nm and are surrounded by a negatively charged glycocalix (main protein: podocalixin, 144 kDa). This construction facilitates the filtration of small molecules but retains molecules with mass >65 kDa (filterability: water, urea, glucose, 100%; myoglobin with a mass of 16 kDa, 75%; albumin with a mass of 66 kDa, <1%). In the case of plasma proteins, the effective retention rate is greater due to their negative charges. The mesangium cell (MC) represents a third cell type of the glomerular apparatus. MCs lie between the capillary clew to the side of the glomerular pole and mediate mechanical stability. In addition, MCs synthesize renin in specilized cells, so-called epitheloid cells that are in close contact with the *macula densa* cells of a feed-loop-back structure of the distal tubules (for a more detailed description of the *macula densa*, see Section 30.1.2).

Two types of nephrons exist within the renal cortex, the superficial and the deep nephrons. The latter ones also form the juxtamedullar glomeruli close to the border of renal cortex and medulla. Only their efferent arterioles (only 10% in total!) contribute to renal medulla perfusion, whereas 90% of efferent glomerular arterioles perfuse only the renal cortex. The mechanisms by which kidney concentrates the primary filtrated urine is not solely a result of active transport mechanisms from tubule fluid back to the blood, but rather a result of a complex system using different ion gradients and changes in osmolality to induce secondary transport mechanisms. Different parts of the tubular system participate in reabsorbing different molecules from the primary urine. Quantitatively, water and Na^+ ion are reabsorbed to the greatest extent among K^+, Mg^{2+}, Ca^{2+}, Cl^-, HCO_3^-, urea, creatinine, glucose, amino acids, small peptides, and others. It should be mentioned that the kidney is also involved in active excretion of various molecules (e.g., urate, ammonium ions, and oxalate) or drugs (e.g., furosemide), in regulating acid–base equilibrium via buffered proton excretion, and has endocrinal capacities (secretion of rennin and erythropoietin; converting 25-hydroxycholecalciferol to 1,25-hydroxycholecalciferol). In this section we only address in detail the issue of how the kidney manages water and sodium balance. For more elaborate descriptions of the other processes we refer the reader to textbooks of renal physiology (Alpert and Hebert, 2007).

30.1.1.2 Mechanisms of urine concentration

The kidney reabsorbs 99% of sodium and water from the primary filtrate. For active transport mechanisms to cope with that effort, the metabolic energy expenditure would be extremely high because at the beginning the osmolality of the primary filtrate is much the same as that of plasma. The active extraction of sodium from the primary filtrate and maintaining passive water equilibrium would cost more than 300 times the energy than the kidney in reality needs. Instead, the urine is concentrated

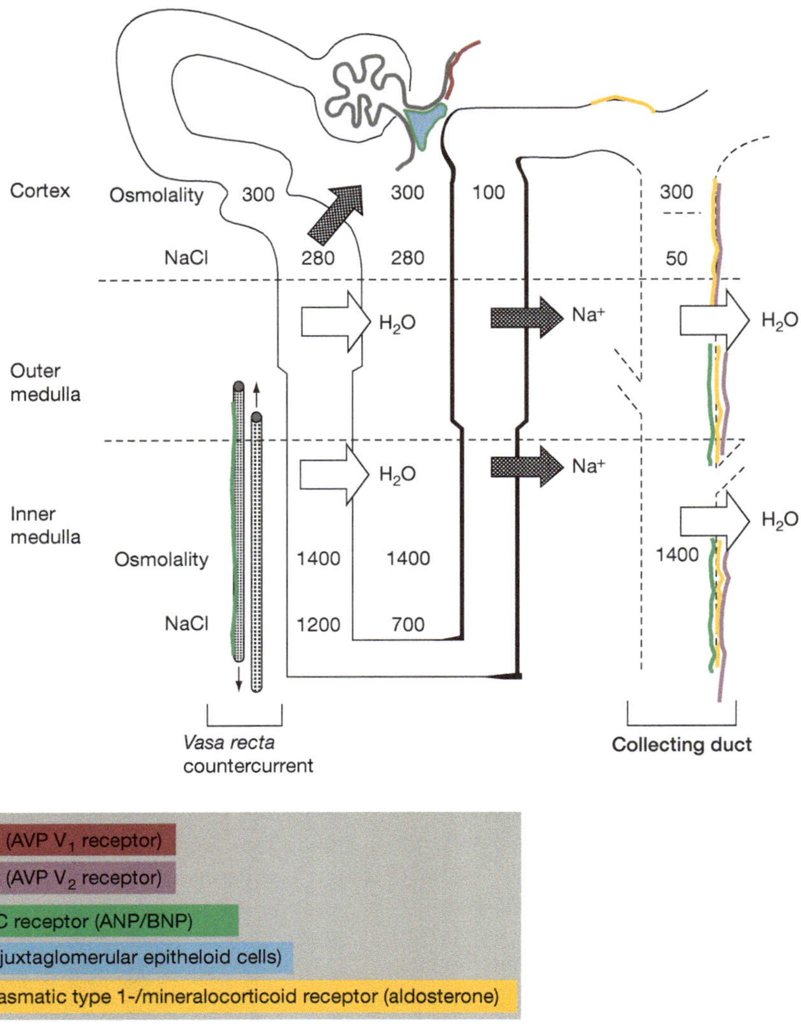

Figure 1 Schematic illustration of the model for the renal concentration mechanism and its hormonal regulation. Shown is a schematic drawing of the nephron structure and its relationship to the different renal zones. Thick lines indicate barriers for water permeability along the thin ascending and thick ascending loop of Henle. Arrows indicate relative magnitude of water (white arrows) and Na^+ fluxes (gray arrows) in the different parts of the nephron. The countercurrent principle is schematically depicted with the afferent *vasa recta* and the efferent *vasa recta* (dotted in gray). Note, that the contribution of urea to the osmolality in the different segments is neglected. In color are depicted the working points of regulative hormones and the regional expression pattern of its receptors along the tubular system, respectively, with permission. Modified with permission from Morello JP and Bichet DG (2001) Nephrogenic diabetes insipidus. *Annual Review of Physiology* 63: 607–630.

by a complex and subsequent interaction of primary filtrate with the different tubular parts and by a specifically formed medullary vessel system, the *vasa rectae*. The latter is involved in concentrating the urine via the so-called countercurrent mechanism. The countercurrent mechanism operates as the result of an exceptional histological co-location of the *vasa rectae* that are formed by a parallelizing course of the afferent and efferent vessels; this establishes a countercurrent blood flow in the vicinity of the tubular system (see also **Figure 1**). Via this mechanism in the *vasa rectae*, an increasing osmotic gradient is built up from the cortex to the inner zone of the medulla. Thus, the total urine concentration process is a result of a combined effect of alternating active and secondary active transport mechanisms with subsequent passive water and ion diffusion.

30.1.1.2(i) *Sodium reabsorption*

Sixty-five percent of Na^+ collected by glomerular filtration is reabsorbed in the proximal tubule, 25% within the ascending loop of Henle, 7% in the distal tubule, and 3% in the connecting tubule and collecting duct, respectively. In general, the driving force for Na^+ reabsorption is the concentration gradient between the tubular filtrate, the cytoplasm of the tubular cells, and the interstitium. This gradient is generated by the Na^+/K^+-ATPase that is located in the basolateral membranes of all tubular cells. The underlying mechanisms for Na^+ reabsorption differ among the specific cell types along the tubular system.

1. *Proximal tubule.* Na^+ reabsorption is mediated through a co-transport (also symport) with glucose, amino acids, and acid anions and an antiport with protons. The proton recycling system is provided by the enzyme carboanhydrase (catalyzes: $H_2O+CO_2 \rightarrow H^+ + HCO_3^-$). Passive paracellular H_2O flow stimulates the transport of further Na^+ to the interstitium, which is referred to as solvent drag.
2. The thin descending loop of Henle is not involved in Na^+ reabsorption. Its contribution to water diffusion is described in the next section.
3. The thin ascending loop of Henle has a passive paracellular diffusion selective for cations (Na^+, others) driven by a specific chloride permeability over the cell.
4. The thick ascending loop of Henle transports Na^+ ions via a $Na^+/K^+/2Cl^-$co-transporter (NKCC2). K^+ and Cl^- ions leave the cell at the basolateral membrane via specific ion channels. A portion of K^+ ions diffuses back to the tubular lumen via an apical K^+ channel. This mechanism establishes an electrochemical gradient with an excess of negative charges in the interstitium that provides the driving force for a paracellular transport of cations (Na^+, others).
5. The distal tubules reabsorb Na^+ via a Na^+/Cl^- symport where, as for all secondary transport processes that have already been described, the driving force is provided again by the basolateral Na^+/K^+-ATPase.
6. The collecting duct is permeable for Na^+ via specific Na^+ channels (epithelial-specific Na^+ channel, ENaC) in the apical membrane of the epithelium. In return, K^+ ions leave the cytoplasm via the apical membrane in the direction of the tubular lumen. This mechanism is regulated by aldosterone that induces channel expression and sorting to the apical plasma membrane. Aldosterone-mediated regulative effects are described in detail in Section 30.1.2.

30.1.1.2(ii) *Water diffusion*

Similar to sodium, 99% of the primary filtrated water (*c.* $180 l d^{-1}$) is reabsorbed along the tubular system: 65% within the proximal tubules, 18% within the descending loop of Henle/the distal tubules, and 10% along the collecting ducts. In the state of antidiuresis only 1% of water is excreted with the urine.

Due to the transcellular resorption of Na^+ and other osmotic, active molecules, the osmolality decreases and water flows both transcellularly and paracellularly in the proximal tubule. The paracellular transport causes the solvent drag, as already mentioned. The transcellular water flow is facilitated by members of a channel protein family, the aquaporins (AQP) (King et al., 2004), which span both the apical and the basolateral membrane (for more detailed description of AQP family proteins, see **Figure 2**). In the descending part of the loop of Henle, AQP1 is expressed both in the basolateral and apical membranes and thus facilitates transcellular water diffusion. The degree of urine concentration depends on the length of the loop of Henle, and this is associated with the increasing osmotic gradient to the deep part of medulla as the result of the countercurrent of the *vasa rectae*. The thin and thick ascending loops of Henle are highly impermeable to water because they lack water channel expression. The urine osmolality is fine-tuned at the connecting tubules and the collecting duct, where the architecture of the epithelium is different. In contrast to the tubular epithelia, the collecting duct cells do not allow a paracellular water transport due to tight junctions. In the collecting duct epithelium, water transport is a rather mediated controlled process. Again members of the AQP family, AQP3/AQP4, mediate transcellular water transport. Both are constitutively expressed within the basolateral membrane (see also **Figure 2(c)**). A third aquaporin, AQP2, is exceptionally expressed on the apical (luminal) membrane under the control of vasopressin (AVP). AVP regulates AQP2 expression via V_2 receptor (AVPR2) binding and a G-protein-mediated cAMP/PKA cascade by externalizing AQP2 from the endosome to the membrane. As a consequence, the transcellular water transport increases, again driven by the osmotic gradient from the countercurrent mechanism (Morello and Bichet, 2001).

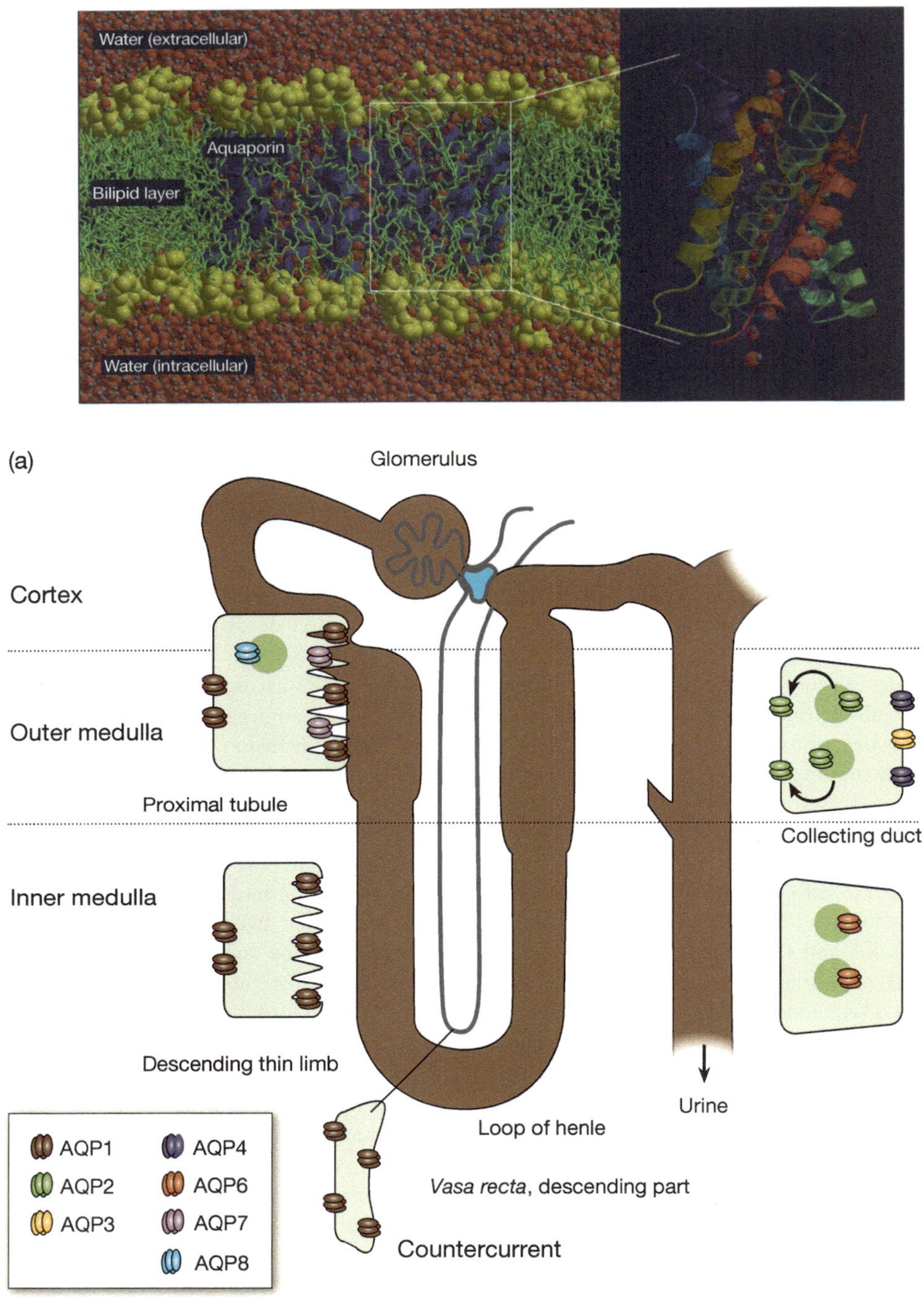

Figure 2 Aquaporin protein family. Aquaporins belong to a family of channel proteins and can be divided into two groups on the basis of their permeability characteristics. The members of the first group are only permeable for water (AQP0-2, -4, -5, -6, -8) whereby members of the second group (aquaglyceroporins; AQP3, -7, -9, -10) are permeable for water and for small organic molecules, particularly glycerol, as well. (a) All aquaporins share a unique protein structure as tetramers whereas each monomer consists of six membrane spanning helices forming its own water channel (right); (a, left) demonstrates an AQP tetramer complex integrated within a phospholipid bilayer. Water molecules are depicted as red balls flow pathing the water channel in a single trail. Aquaporins play a crucial role wherever specialized fluids of a compartment (of an organism and an organ/tissue, respectively)

30.1.2 Regulation of Fluid and Salt Balance

Extracellular volume (ECV) regulation is directly linked to Na^+ and water homeostasis since the Na^+ ions account for approximately 95% of the extracellular osmolality under physiological conditions. Glucose and blood urea nitrogen (BUN) participate minor, other molecules are negligible. Osmolality can be approximately calculated as follows: osmolality (mOsm/kg H_2O) $= 2 \times Na^+$ $(\text{mmol}l^{-1}) + \text{glucose}/18(\text{mgdl}^{-1}) + \text{BUN}/2.8(\text{mgdl}^{-1})$. It has to be mentioned that glucose and BUN may contribute significantly to the extracellular osmolality in diabetes mellitus and renal insufficiency, respectively. In general, many pathological states or therapeutic interventions can interfere with the ECV as mentioned, for example, treatment of brain edema with osmotic solutions or, mechanical ventilation with enhanced positive end-expiratory pressure. Three hormonal systems contribute mainly to the regulation of the salt and water homeostasis: AVP is the principal hormone for regulation of water balance, whereas salt balance is regulated by the renin–angiotensin–aldosteron system (RAAS) and counter-regulated by natriuretic peptides (see also **Figure 1**).

AVP is released from the posterior pituitary gland where the AVP-secreting axons terminate (Jennes et al., 2009). AVP acts, as already described in Section 30.1.1.2, on the collecting duct via externalization of apical AQP2. This translocation between the endosome and the apical membrane AQP2 is regulated by short-term AVP pulses since AVP has a plasma half-time clearance of 5 min. AVP binding and activation of AVPR2 activates an intracellular signaling cascade involving an increase in cyclic adenosine monophosphate (cAMP) and subsequent protein kinase A (PKA) activation. PKA promotes phosphorylation of AQP2 as an essential step for regulating the movement of the AQP2-containing vesicles to the apical plasma membrane. This signaling cascade is described in more detail in **Figure 3**. Continuously increased AVP, however, induces transcriptional activity of the AQP2 gene and thus enhances translocatable AQP2 units in collecting duct epithelial cells. The AVP effects can be antagonized at least in part by prostaglandin E_2 and by luminal calcium ion concentrations. An apical calcium/polycation receptor protein expressed in the terminal portion of the inner medullary collecting duct has been shown (in rats) to reduce AVP-mediated water permeability as a function of luminal calcium ion concentrations (Morello and Bichet, 2001).

The RAAS is the major regulatory system for controlling Na^+ homeostasis of the extracellular fluids and, as a result, of ECV and blood pressure. The RAAS consists of a cascade of mediators and is regulated by a feedback-loop mechanism. Renin is the initial activator. Once released, renin cleaves angiotensinogen to angiotensin I. This is then converted into angiotensin II by its converting enzyme (ACE) via passage through the pulmonary capillary system. Renin is synthesized in specialized MCs, the so-called epitheloid cells, in the vicinity of the *macula densa* of the distal tubule and the afferent arteriole next to the glomerular pole. The MCs, together with the epitheloid cells and the specialized *macula densa* epithelial cells of the distal tubule, form the juxtaglomerular apparatus as the anatomical substrate for the autoregulatory mechanism in the kidney.

have to be built and a steady state has to be maintained, such as in the eye (eye chambers, lens), the brain (cerebrospinal fluid), or the kidney. In addition, aquaporins play a role in water homeostasis in the lung, secretory glands, and more. In the brain, a rapid regulation of water and ion changes upon neuronal activation is fundamental since the brain is subject to a rigid physical constraint that is imposed by the neurocranium as well as by the tight brain capillar endothelium forming the blood–brain barrier. AQP4 is expressed to a high degree on astrocytes that play a crucial role in water homeostasis of the brain. AQP4 expression is clustered on specialized astroctic processes (endfeet) in connection to the pial surface and the basolateral side of brain capillary endothelium as well as being close to synaptic clefts. Studies from AQP4 nullmutant mice revealed a crucial role for the pathophysiology of brain edema (water intoxication, ischemic stroke). The ability of the kidney to control whole-body water balance strongly correlates with aquaporin function. Aquaporins are highly expressed on the epithelium of the nephron and on the kidney capillary system. Each nephron segment has a specific water permeability that corresponds with the presence or absence of a given aquaporin. (b) Shows a schematic drawing of the segmental AQP expression of the nephron. The water of the primary glomerular filtrate is reabsorbed besides the parecellular and passive solvent drag mechanism (see also Section 30.1.1.2.1) transcellularly through AQP1 in the proximal tubule and descending thin loop of Henle. AQP7 and AQP8 are also present in the proximal tubule. The endothelial cells of the *Vasa recta* also express AQP1, which facilitates the removal of water to maintain hypertonicity in the interstitium. The regulatory function of AVP on water reabsorbtion and the involvement of AQP2, AQP3, and AQP4 at the epithelium of the collecting duct are described in more detail in **Figure 3. (a) Courtesy: De Groot B, Max Planck Institute for Biophysical Chemistry, Göttingen, Germany.**

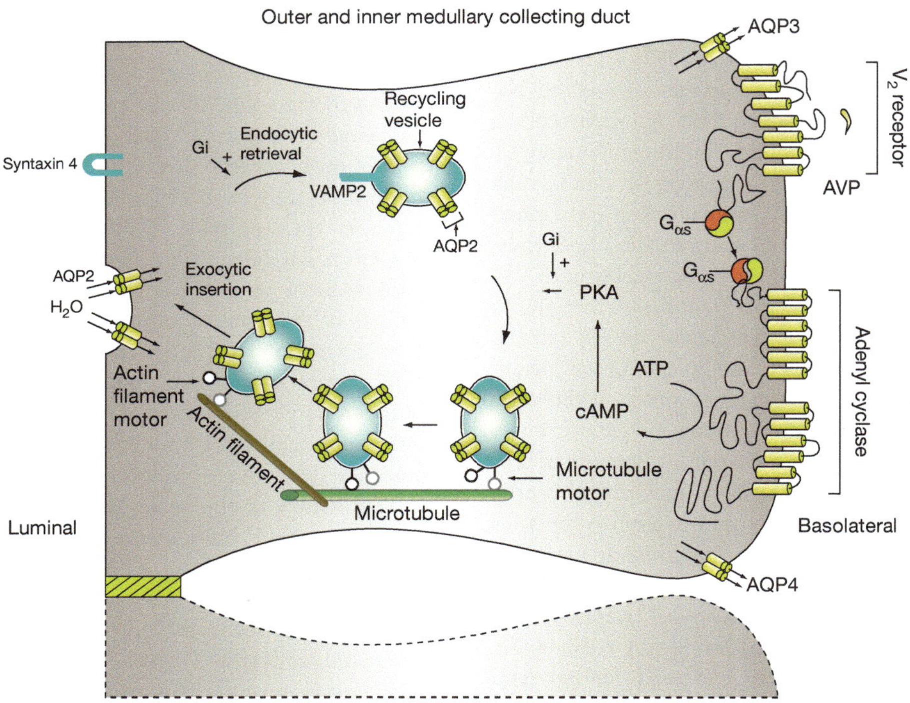

Figure 3 Molecular regulation of diuresis by AVP at the collecting duct. Depicted is an epithelial cell of the collecting duct. The yellow-shaded box indicates the tight junction at the apical site. AVP binds to the AVPR2 at the basolateral membrane. This activates a G-protein subunit and the adenyl cyclase, produces cAMP and as a consequence activates PKA. Upon activation of PKA, AQP2 reserves stored on cytoplasmatic vesicles become phosphorylated and are recruited to the apical membrane via tubulin- and actin-associated transport. Once externalized, AQP2 facilitates water flux into the cell. AQP3 and AQP4 channels that are constitutively expressed at the basolateral membrane allow flow in the direction of the interstitial space. The driving force of the transcellular water diffusion is the cortico-medullar osmotic gradient. Modified from Morello JP and Bichet DG (2001) Nephrogenic diabetes insipidus. *Annual Review of Physiology* 63: 607–630, with permission from the *Annual Review of Physiology*, Volume 61 © by Annual Review (www.annualreviews.org).

The autoregulatory mechanism consists of a myogenic reaction and a tubulo-glomerular feedback. A decrease in afferent arteriole blood pressure is sensed by the epitheloid cells, and prorenin stored in vesicles is cleaved to renin and secreted. Renin secretion is also stimulated by norepinephrine (β_1-receptors) and dopamine (D1-receptors) via sympathetic nerve terminals and by circulating epinephrine. The tubulo-glomerular feedback involves the epithelial cells of the *macula densa.* When renal arterial blood pressure increases, renal blood flow and GFR increase, too. *Macula densa* epithelial cells respond to the increased delivered load of solutes and water by secreting vasoactive molecules that constrict afferent arterioles. This in turn inhibits renin secretion. Which components of the tubular fluid are sensed (Na^+, others, or total osmolality) and what molecules mediate vasoconstriction (adenosine, protsaglandins, or kinins) are still unknown, however. Another negative regulator of renin is, via negative feedback, angiotensin II binding to its angiotensin II AT1-receptor.

Angiotensin II is the main regulatory mediator of the RAAS. Thus, it triggers the sensation of thirst, increases AVP secretion of the pituitary gland, enhances Na^+ reabsorption in the proximal tubule, and stimulates secretion of aldosterone in the *zona glomerulosa* of the adrenal gland.

Aldosterone acts on the connecting tubule and the collecting duct. It binds to the mineralocorticoid receptor (MR), also referred to as cytoplasmatic type-1 receptor. MR belongs to a superfamily of steroid/thyroid/retinoid/orphan (STRO) receptors (Hu and Funder, 2006). STRO/MRs are intracellular receptors and operate as ligand-activated transcription factors to regulate gene expression. MRs are expressed as epithelial receptors in the kidney and also in nonepithelial tissues such as the brain.

MR binds not only aldosterone but also glucocorticoids (cortisol, corticosterone). Steroid metabolites play a role in regulating aldosterone activity at the receptor level (Funder, 2007). Aldosterone binding mediates two temporally different responses at the tight epithelium of the connecting tubule and collecting duct: a fast (<10 min) and a slow response (>30 min up to hours). The slow response is a classical genomic response and involves gene transcription and protein synthesis. The main gene targets are the Na^+/K^+-ATPase, the ENaC, and some mitochondrial enzymes of the citrate cycle. The latter serves as a source for the required ATP turnover of the Na^+/K^+-ATPase. Thus, aldosterone binding first increases recruitment both Na^+/K^+-ATPase and ENaC from cytoplasmatic storage (within 30–60 min) and, second, transcription of the genes and externalization to the basolateral membranes (Na^+/K^+-ATPase) and apical membranes (ENaC), respectively. The fast response is also mediated by aldosterone but independently of MRs. A second, currently unknown receptor is postulated that acts through a nongenomic pathway. Interestingly, the fast response also increases K^+ in the collecting tubule and consecutively in urine before Na^+ reabsorption starts (**Figure 4**; Wehling, 2005).

Aldosterone effects can be antagonized by natriuretic peptides. These peptides and their sources are described in more detail in Section 30.3.2. Brain natriuretic peptide (BNP) and atrial natriuretic peptide (ANP) represent relevant candidates for counter-regulatory effects. ANP is mainly synthesized and constitutively stored in secretory vesicles in specialized myoendocrine cells in the right atrium of the heart. An increase in atrial pressure due to an increase in circulating vascular volume distends these cells and induces ANP secretion. Natriuretic peptides bind to specific transmembrane receptors (A-type for ANP and BNP) on their effector cell. These receptors are membrane-bound guanylate cyclases and mediate their intracellular response via cGMP. The receptors are expressed on afferent arterioles of the glomerular apparatus, on the *vasa rectae*, on the epithelium of the collecting duct, on the aldosterone-synthesizing cells, and on the renin-synthesizing epitheloid cells (see also **Figure 1**). Thus, ANP relaxes the afferent arterioles and subsequently decreases blood pressure. ANP directly antagonizes the aldosterone effects at the collecting duct and inhibits synthesis of the key mediators of RAAS. As a consequence of all these actions, ANP increases the GFR and excretion of Na^+ and secondarily also water. Whether physiological levels of BNP also play a role in aldosteron antagonism is unclear. Its role in the pathophysiology of cerebral salt-wasting syndrome (CSWS) is described in Section 30.3. For the sake of completeness, it has to be mentioned that besides A-type receptor, there exist two additional receptors. The C-type receptor seems to act as a clearing receptor for all natriuretic peptides and the B-type receptor binds C-type natriuretic peptide (CNP). CNP acts as a parakrine endothelium-derived hyperpolarizing factor and mediates vascular relaxation.

30.1.3 Symptoms of Disturbed Salt and Water Balance

In this section we briefly describe the general symptoms of hyponatremia and excessive water loss, irrespective of the underlying etiology. Symptoms or clinical findings that are more specific for underlying pathomechanisms of a given salt or water disturbance are described in the disease-specific sections for diabetes insipidus, CSWS, and syndrome of inappropriate antidiuresis (SIAD).

30.1.3.1 Hyponatremia

Central neuronal function is very sensitive to changes in extracellular Na^+ concentration. Neuronal disturbances occur not only as a function of the absolute extracellular Na^+ concentration but also as a function of the kinetics of these changes. Severe hyponatremia (below 120 mmol l^{-1}), especially if the condition develops rapidly within a period of 48 h, causes serious symptoms such as confusion, hallucinations, epileptic seizure, symptoms of elevated intracranial pressure and concomitant coma, decerebrate posture, and respiratory arrest leading to death. The severity of symptoms is not only related to acuity and severity of hyponatremia, but also to age. Futhermore, the threshold of plasma sodium levels at which symptoms occur seems to differ between genders and tends to be higher in women than in men. Although women and men are equally likely to develop hyponatremia, women have a 25-fold higher risk to die or receive permanent brain injury (Lien and Shapiro, 2007).

Mild symptoms of hyponatremia include headache, cognitive decline (concentrating and memory deficits), muscle cramps, and generalized weakness. Patients with chronic hyponatremia may be asymptomatic initially, but mild cognitive deficits and a significantly greater risk for falls, especially in elderly

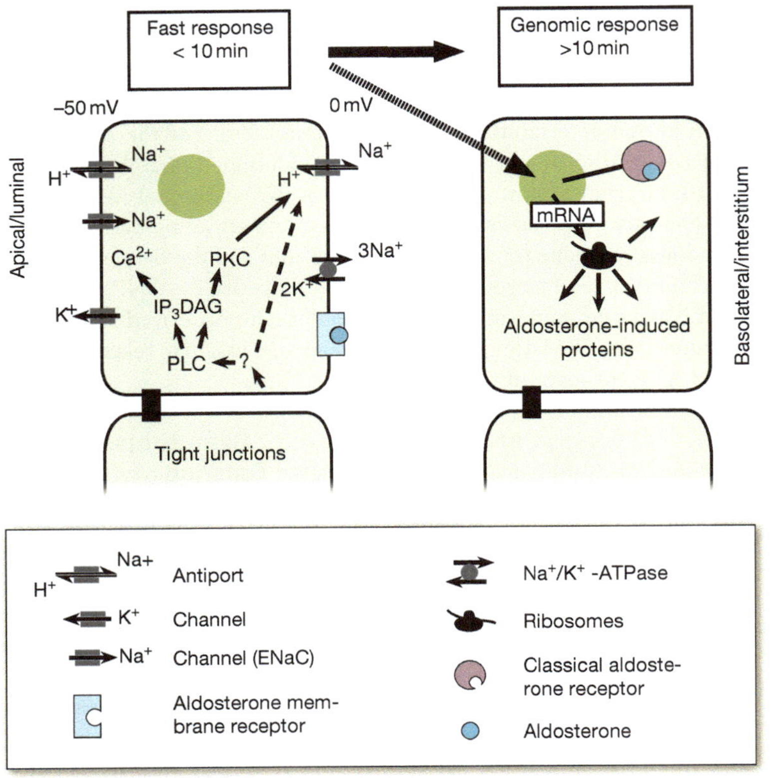

Figure 4 Nongenomic and genomic responses to aldosterone at the connecting tubule and collecting duct. Two-step model of nongenomic aldosterone action. Initial steps of mineralocorticoid action (activation of the Na^+/H^+ exchanger) and subsequent effects on Na^+/K^+ fluxes and Na^+/K^+-ATPase activation/production are shown schematically, including involvement of a still orphan membrane receptor for mineralocorticoids. This receptor could represent either the classic type-I-receptor or an unrelated novel structure. The genomic response involves the classical cytoplasmic type-1-receptor (MR). Upon its activation a *de novo* synthesis of mineralocorticoid response element regulated gene products is initiated (eg., Na^+/k^+-ATPase). IP_3, inositol-1-,-4,-5-trisphosphate; DAG, diacyl glycerol; PLC, phospholipase C; PKC, protein kinase C; IPYK, intermediate tyrosine kinase. Modified from Wehling M (2005) Effects of aldosterone and mineralocorticoid receptor blockade on intracellular electrolytes. *Heart Failure Reviews* 10: 39–46.

patients, are also evident. These apparently milder symptoms may occur more frequently in the elderly and seem to be underdiagnosed. However, their significance for quality of life and relevant side effects are underestimated.

30.1.3.2 Excessive renal loss of water

Polyuria is the central symptom of uncontrolled loss of water if the kidneys have lost their ability to concentrate the urine. The water content in humans with respect to lean body mass amounts to 72–74% independent of gender. Including the adipose tissue, water content varies with gender and age (infants, 75%; men, 65%; women, 50%). Within the body, water accumulates in two main spaces: the intracellular space as intracellular volume (ICV, 60–65% of body water) and the extracellular space as ECV (35–40% of body water). The latter can be further divided into the intercellular space (interstitium, 75% of ECV), the blood plasma volume of the vascular system (25% of ECV), and the transcellular fluids (e.g., cerebrospinal fluid or peritoneal fluid). A clinical significant variable is the effective arterial blood volume (EABV) since EABV defines organ perfusion. A healthy human can only resist thirst for a few days; otherwise, he will die of dehydration. The 24-h water balance of a 70-kg adult consists of 1.2l water intake from fluids, 0.9l from food, and 0.3l from oxidation water, and a loss of 1.4 l via urine, 0.9l via perspiration (mainly lung), and 0.1l via feces. The water lost via glomerular filtration amounts to $180 \mathrm{l\,day}^{-1}$; 90% is constitutively reabsorbed but 10% is under control of AVP in the collecting duct. In a state of antidiuresis <1% of the primarily filtrated water is excreted. Considering these concentrating mechanisms to keep water homeostasis, it is evident that even a minor decrease in osmoregulatory system functioning as disturbed AVP function, or loss of function in critical water transport mechanisms mediated by aquaporins cause severe water loss. Affected patients suffer from polyuria (urine volume >3l per 24h or >40ml kg^{-1} body weight), polydipsia (chronic excessive thirst), and hyposthenuria. Hyposthenuria is defined as secretion of urine of low specific gravity, due to inability of the tubules of the kidneys to produce concentrated urine. The triad of polyuria in conjunction with hyposthenuria and polydipsia are the cardinal symptoms of diabetes insipidus; however, it also occurs following excessive water ingestion in diabetes insipidus. DI patients can lose – depending on the nature of the underlying defect and the degree of severity – up to 40l urine per day. In infants, polyuria and concomitant exsiccation represent an urgent, life-threatening situation since infants do not tolerate fluid loss nearly as well as adults. In infants/children polyuria is defined as more than 3–5ml kg^{-1} body weight and per-hour urine production.

30.2 Diabetes Insipidus

Diabetes insipidus can be divided into two main groups – a central decrease/loss in AVP secretion, termed central diabetes insipidus (CDI), and a decrease/loss of function within the kidney itself to concentrate urine, termed nephrogenic diabetes insipidus (NDI) – with respect to the anatomical level of the underlying pathological process. Causes of secondary NDI must be distinguished from those of primary NDI since damage to the medullary system secondary to an underlying disease can also cause a loss of AVP function. Irrespective of the underlying pathomechanism and the anatomical level, the clinical manifestations are like those for polyuria, hyposthenuria, and polydipsia. In the following section, we describe the different pathomechanisms of nephrogenic and CDI. At the end of this section we address the issue of how to confirm diagnosis and how to treat these two totally different entities.

30.2.1 Nephrogenic Diabetes Insipidus

As already described in Section 30.1, concentration of the urine in terms of water reabsorption depends on an unregulated part (proximal tubule/descending loop of Henle) and the regulated part of the collecting duct. All of the known cases of primary NDI and hereditary NDI, respectively, are caused by defects of either loss of function of AVP signaling or loss of aquaportin function. Hereditary NDI can be devided into X-linked, autosomal recessive, and autosomal dominant NDI. Bartter's syndrome as a third entity with polyuria and further symptoms is closely related and also described in this section. Secondary NDI is caused by an aquired condition and can be divided into three different groups: electrolyte disturbances, lithium-induced, and pathology of renal parenchyma.

30.2.1.1 Aquaporin-associated nephrogenic diabetes insipidus

30.2.1.1(i) AQP1 pathology

The unregulated part of water reabsorption, accounting for 90% of GFR, is accomplished by passive paracellular and transcellular diffusion, the latter being mediated by AQP1 (see also **Figure 1(c)**). Both types of water flows are induced by the osmolality gradient between the tubular lumen and the interstitium. Clinically relevant defects of this aspect of urine concentration are not known although mutations in the AQP1 gene with a subclinical phenotype has been described in humans. An extracellular epitope of AQP1 has been encoded as the minor blood-group antigen Colton. Seven Colton-null families have been described worldwide. Three of them have homozygous mutations in the AQP1 gene (two homozygous exon deletions with frame shift, one homozygous missense mutation) and show low *in vitro* osmotic water permeability of red blood cells. To examine the hypothesis that AQP1 mutations would induce defects in water homeostasis under stress conditions, renal function was examined in two of these AQP1 null-mutant humans. Under normal conditions, these individuals have normal renal concentration capacities and normal renal baseline values. They also demonstrated normal increases in AVP levels and serum osmolality when water intake was restricted. Under stress conditions, however, both had a limited capacity to concentrate urine. Their maximal urine osmolality was less than half as concentrated as in normal individuals. Studies from Aqp1 knockout mice show a luminal hypotonic urine in the proximal tubules, increased amiotic fluid volume, and reduced osmotic water permeability of the peritoneal barrier but an overall normal behavioral and reproductive phenotype. It seems that mice would be more affected by a loss of AQP1 function than humans. This might be due to the fact that mice can concentrate urine to greater than 3000 $mOsm\,kg^{-1}$ as compared to humans, whose capacity is restricted to maximal $1200\,mOsm\,kg^{-1}$. Possibly, mice urine concentration depends more on AQP1 than in humans or humans have an additional, not yet known compensatory mechanism. Site-directed mutagenesis experiments of AQP1 revealed an add-on permeability for small molecules (urea, glycerol, and ammonia) and protons. Whether those mutations also play a role in human pathology is not known (Beitz et al., 2006).

30.2.1(ii) AQP2 pathology

The regulated part of urine concentration and water reabsorption is mediated via AQP2, AQP3, and AQP4 under regulation of AVP. The molecular mechanisms of AVP-regulated water reabsorption at the collecting duct have already been described in detail in Section 30.1 (see also **Figure 3**). AQP3 and AQP4 are expressed constitutively at the basolateral side of the collecting duct epithelium, whereas AQP2 is expressed on the apical side. Only in the deep part of renal medulla, the most hypertonic part of the kidney, is AQP2 also expressed on the basolateral side. The importance of this pattern is not known. *In vitro* experiments show that both chronic AVP exposure and hypertonic solutes – as an AVP-independent mechanism – can induce basolateral AQP2 gene transcription. An AQP2 defect reported by Deen and colleagues was the first example of a clinically important water channel defect in 1994. Autosomal-dominant and autosomal-recessive mutations of AQP2 have been described to cause hereditary NDI and account for 10% of all hereditary NDI. The pathophysiological consequences of dominant versus recessive mutations are considerably different. Dominant mutations always lead to chimeric tetramers consisting of normal and mutated monomers that cause a trafficking defect and AQP2 can no longer be externalized to the apical membrane. Recessive mutations cause a misfolding of all monomers and, as a consequence, degradation of inoperable proteins inside of the proteasome. Therefore, normal function is maintained in heterozygous individuals of recessive mutations, given that the mutated allele is silenced or gene products are degraded and monomers from the healthy allele form normal tetramers.

AQP2 dysfunction is also thought to cause acquired NDI in some cases. Chronic lithium exposure, chronic hypokalemia, or transient urinary obstruction can decrease AQP2 levels. Lithium, which is the drug of choice for bipolar disorders, causes NDI in 10–20% of individuals treated with this drug. Chronic lithium therapy seems to downregulate total AQP2 expression and the amount of externalized protein in the apical membrane. Furthermore, lithium seems to alter the ultrastructure of the collecting duct epithelium. In contrast, excessively high levels of AQP2 have been described under conditions of chronic fluid retention such as in congestive heart failure, liver cirrhosis, and pregnancy (Schrier and Martin, 1998; Robben et al., 2006).

30.2.1(iii) AQP3 and AQP4 pathology

AQP3 and AQP4 are both expressed along the collecting duct epithelium on the basolateral membrane. AQP3 is expressed along the whole collecting duct, whereas AQP4 is restricted to the inner medullary collecting duct. Aqp3 knockout mice suffer from severe polyuria, but surprisingly, an unexplained decrease in AQP2 expression could be observed in these animals too. AQP3 mutations have also been described in humans. Since AQP3 belongs to the aquaglyceroporins, it transports water, urea, and glycerol (see also **Figure 1**). AQP3 mutations cause a selective defect in glycerol transport but water-transport capacities are normal in these individuals (Roudier et al., 2002). Aqp4 knockout mice have a mild urine concentration defect but an overall normal phenotype and reproductive behavior.

Humans with AQP4 mutations have not been identified so far. For the sake of completeness the phenotype of AQP4 knockout mice is described shortly. These mice have a mild urine concentration defect but an overall normal phenotype and reproductive behavior. Closer analysis suggests that AQP4 is involved in brain edema pathology and cell migration of activated astrocytes, for example, after brain damage (Nicchia et al., 2003).

30.2.1.2 AVP V_2 receptor defects: X-linked nephrogenic diabetes insipidus

X-linked NDI is a rare disorder with approximately 0.9 cases per 100 000 male live births and is caused by various mutations in the AVP V_2 receptor (AVPR2) gene. In some populations, for example, in Canada, the incidence is much higher, which is attributed to the founder effect (Morello and Bichet, 2001). The AVPR2 gene consists of three exons and two small introns. The gene encodes for a protein belonging to the seven-transmembrane, G-protein-coupled receptor family (**Figure 3**). More than 150 mutations within the AVPR2 gene have been identified to cause X-linked NDI. Different types of mutations such as frame shift deletions or insertions, in-frame insertions or deletions, nonsense mutations, and splice-variant mutations have been reported. Twice as many mutations are found in the transmembrane helices as in the extra- or intracellular domains. At the cellular level, these various mutations cause three different types of mutations, which were originally classified from lipoprotein receptor mutations, and grouped according to their function and subcellular location. According to this classification, AVPR2 mutations cause either type 1 mutant receptors that reach the cell surface but display impaired ligand binding or impaired intracellular signal transduction (despite normal AVP binding; e.g., G-protein binding) or type 2 mutant receptors that display defective intracellular transport and sorting to the cell membrane. Thus, these receptors do not reach the cell surface. Type 2 mutations account for the vast majority of naturally occurring AVPR2 mutants that cause NDI. Type 3 mutations display ineffectively transcribed receptor genes (Morello and Bichet, 2001; Nguyen et al., 2003; Robben et al., 2006).

Affected males have a renal concentration defect whose central feature is a nonresponsiveness to AVP or AVP analog administration. The clinical presentation of X-linked NDI individuals exceeds simple polyuria. Individuals present with severe polyuria, hypernatremia, and hyperthermia, with renal insufficiency and mental retardation developing in the further course. The latter is presumably a consequence of repeated episodes of severe dehydration.

30.2.1.3 Other forms of hereditary nephrogenic diabetes insipidus

In the thick ascending loop of Henle, the tubular fluid is diluted via co-transport of $Na^+/K^+/2Cl^-$ (NKCC2 co-transporter) induced by the basolateral Na^+/K^+-ATPase. K^+ in part flows back to the luminal side via the potassium channels and Cl^- leaves the cell via specific chloride channels (chloride channel-kidney b (CLC-Kb)) at the basolateral side. This opposing flow of negative Cl^- and positive K^+ ions establishes a negative charge in the interstitium that triggers paracellular diffusion of cations (see also Section 30.1). Different mutations in these three channel proteins cause a defect in tubular urine concentration and are responsible for Bartter's syndrome. Patients with Bartter's syndrome present with salt wasting, hypokalemic alkalosis, and deficits in the concentrating and diluting capacity of the kidney (Morello and Bichet, 2001).

30.2.1.4 Nongenetic causes of nephrogenic diabetes insipidus

A more general definition of NDI is that it comprises all renal pathologies causing polyuria that does not respond to AVP. In terms of this general definition, several other underlying diseases/abnormalities can cause NDI, such as hypercalcemia, hypokalemia, hyperglycemia/glucosuria, drugs (as already mentioned lithium and others), and in a broader definition also pathologic processes in the kidney that produce defective countercurrent functions, such as renal failure due to intestinal nephritis, medullary damage due to sickle-cell anemia, amyloidosis, sarcoidosis, and more (Magner and Halperin, 1987; Garofeanu et al., 2005).

30.2.2 Central Diabetes Insipidus

AVP-synthesizing neurons are arranged in pairs in the paraventricular nuclei in the wall of the third ventricle and in the paired supraoptic nuclei above the lateral extremes of the optic chiasm. The axons project as a fiber track via the pituitary stalk to the posterior pituitary gland and store AVP in their terminals. AVP storage provides maximum antidiuretic capacity for nearly 10days and can maintain fluid balance for approximately 1month. The thirst center consists of a small group of neurons lying rostral to the supraoptic nuclei and is responsible for our awareness of the need for fluids. These neurons express angiotensin-II receptors. Angiotensin II is the most important mediator of thirst and is converted from angiotensin I upon renin secretion in the kidney (see also Section 30.1). The cell biology of AVP cells and central regulation of AVP are described elsewhere (Jennes et al., 2009). CDI is a heterogenous condition. Pathophysiology is mainly caused by local destruction or functional impairment of AVP-producing neurons. This involves tumors, trauma from surgery or accidents, local autoimmune, inflammatory, or vascular diseases. Rarely, genetic defects of AVP synthesis also contribute. Thirty to fifty percent of all CD1 cases are considered idiopathic (Maghnie, 2003; Ghirardello et al., 2005).

30.2.2.1 Destruction of AVP-producing neurons

CDI, especially in children and young adults, is mainly caused by destruction or neurodegeneration of the supraoptic and paraventricular neurons of the hypothalamus. The different causes include destruction by infiltration or compressive growth of local tumors or metastases such as germinoma, craniopharyngeoma, or Langerhans cell histiocytosis. Likewise, local inflammatory, autoimmune, or vascular disease can also cause CDI, as can trauma or pituitary surgery. Different cerebral and cranial malformations are associated with the risk of developing CDI. The highest risk of approximately 35–50% is with optic nerve hypoplasia, absence of corpus callosum, and holoprosencephalus. At minor risk are individuals with unilateral optic nerve hypoplasia (15–20%) and other malformations. These cranial malformations often present not only with DI but also with additional endocrinopathies of the different hypothalamic–pituitary axes such as growth hormone deficiency, neonatal hypoglycemia, or hypogonadotropic hypogonadism. Frequently, multiple pituitary hormone deficiency can be seen in combined malformations of the optic nerve and the absence of septum pelucidum (Traggiai and Stanhope, 2002).

30.2.2.2 Autoimmune pathology

In the group of idiopathic CDI, an autoimmune etiology seems to play a major role since circulating autoantibodies against AVP-producing cells (AVPc-Abs) can be demonstrated in 37% of these patients. There is an overlap in the autoimmune pathology to Langerhans cell histiocytosis since circulating AVPc-Abs can also be detected in more than 50% of patients with this disease. In rare cases, AVPc-Abs can also be detected in germinoma. The fact that only one-third of patients with idiopathic CDI have circulating AVPc-Abs suggests an additional pathomechanism such as a T-cell-mediated immune response. In accordance with such a postulated mechanism, a lymphocytic infundibulo-neurohypophysitis has recently been demonstrated. The Clinical Case 1 shown in **Figure 5** shows typical MRI abnormalities in the vicinity of the pituitary in a case of bioptic proven lymphocytic infundibulo-neurohypophysitis (Maghnie et al., 2006).

30.2.2.3 Familial neurohypophyseal diabetes insipidus

Another cause for what was formerly considered idiopathic CDI is a familial neurohypophyseal diabetes insipidus. Members of these families share mutations in the vasopressin pro-hormone gene, leading to a defective pre-hormone and a deficiency in AVP. More than 35 mutations have been described, most of them within the neurophysin domain of the precursor. Neurophysin is the binding protein for AVP in the producing neurons. All except one are inherited in an autosomal-dominant mode. Typically, symptoms of diabetes insipidus do not develop in these patients until late infancy but then it does progressively. In the solely autosomal-recessive form of familial CDI, symptoms are secondary due to a loss of function of the AVP. In the majority of the autosomal-dominant forms, however, the mechanisms leading to CDI are more complex and involve dominant-negative effects of the defective AVP or defective neurophysin over the physiological protein (derived from the residual healthy allele). Non-functioning monomers can form higher-order complexes of AVP and neurophysin. These protein complexes can accumulate and cause a secondary degeneration in the magnocellular neurons (Maghnie, 2003; Christensen and Rittig, 2006).

> ***Clinical Case 1***: Improvement of MRI abnormalities in a case of lymphocytic infundibulo-neurohypophysitis treated with dexamethasone. A female patient presents with CDI and amenorrhea at admission. **Figures 5(a)** and **5(c)** show saggital and coronal T1-weighted MRI scans at admission demonstrating absence of the physiological evident posterior pituitary hyperintensity (arrowhead) and thick pituitary stalk (arrow). A lymphocytic infundibulo-neurohypophysitis could be proven by pituitary stalk biopsy. **Figures 5(b)** and **5(d)** show MRI scans from the same patient after a short treatment with deaxamethasone. Normalization of pituitary stalk (arrow) is evident. Courtesy: Prof. Mohammad Maghnie, Department of Pediatrics IRCCS, Giannina Gaslini-University of Genova, Italy.

30.2.2.4 Primary polydipsia

Excessive fluid intake suppresses secretion of AVP and induces polyuria in otherwise healthy individuals in terms of water and fluid regulation systems. This phenomenon is called primary polydipsia. Patients with primary polydipsia in general show normal sodium concentrations despite high fluid intake. Low-normal sodium or rarely, severe hyponatremia and low osmolality may be observed. It should be mentioned that plasmatic $Na^+(PNa^+)$, serum osmolality, and urine osmolality can differ depending on the osmolality of the particular ingested fluid. The causes of primary polydipsia include habitually increased fluid intake and also psychogenic and psychiatric disorders such as obsessive–compulsive disorders or psychosis. Some drugs have been reported to induce primary polydipsia, for example,

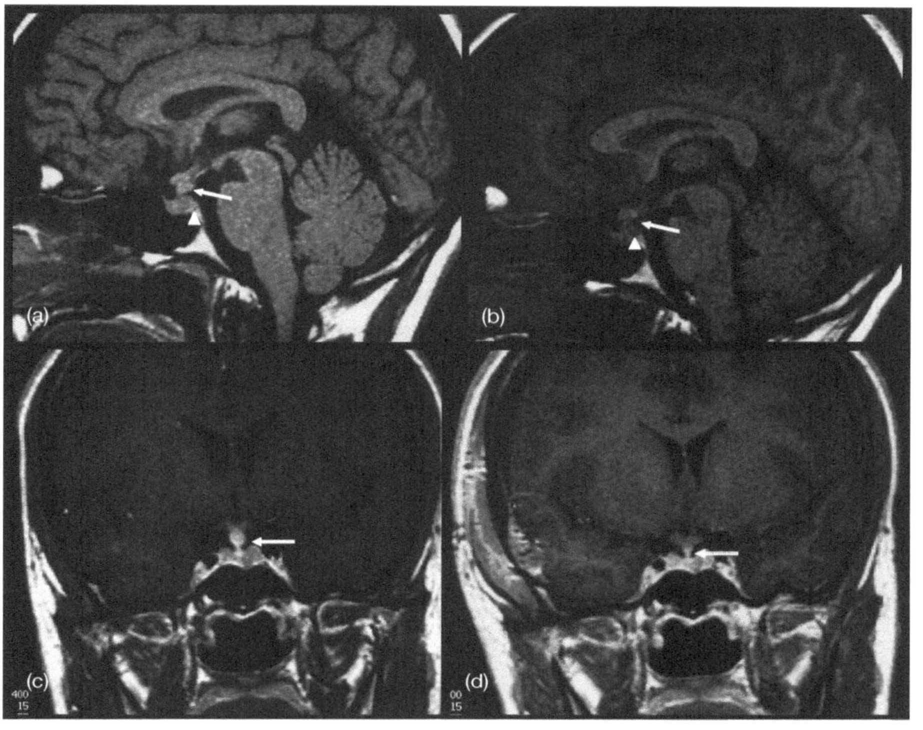

Figure 5 Improvement of MRI abnormalities in a case of lymphocytic infundibulo-neurohypophysitis treated with dexamethasone. Reproduced by permission of Maghnie M, Genova, Italy.

verapamil. In very rare cases, a lesion of neurons generating thirst within the hypothalamus can cause primary polydipsia. A patient with a hypothalamic defect suffering from celiac disease has been described, for example. The polyuria and polydipsia in primary polydipsia is reversible and strictly associated with improvement in the underlying disease (Maghnie, 2003).

30.2.3 Diagnostic Management of Polydipsia and Polyuria

The age at onset, the pattern of fluid intake, the occurrence of mental retardation, and the familial history represent important initial information for developing an individual diagnostic strategy. A familial history suggests the diagnosis of X-linked or autosomal-dominant NDI. Identifying the molecular defect in NDI is of immediate clinical significance because early diagnosis and adequate treatment can prevent physical and mental retardation as a result of frequent episodes of severe dehydration. Therefore, in known families with hereditary NDI, prenatal diagnostic studies (chorionic villus sample, amnion cell culture, and cord blood) are recommended (Morello and Bichet, 2001).

In more ambiguous cases of polyuria, one must first determine whether diabetes insipidus indeed represents the underlying pathology of the presenting symptoms and if so, at which level the pathology is present. Therefore, the existence of primary polyuria has to be excluded first (as a rule by normal or slightly decreased sodium and osmolality but also by behavioral observation). Further and general diagnostic procedures should comprise documentation of polyuria/polydipsia with a 24-h urine collection, balancing of fluid intake and urine output, analysis of serum and urine electrolytes, and osmolality. A general examination of kidney function is recommended. In cases of milder symptoms, fluid intake and output should be analyzed in more detail and should be combined with weight analysis (Maghnie, 2003; Ghirardello et al., 2007).

The next step to verify suspicion of diabetes insipidus is a water-deprivation test. A 6- to 8-h water deprivation test is sufficient to confirm the diagnosis, except in cases of primary polydipsia where longer dehydration is sometimes required. The test should be stopped if weight loss exceeds more than 5% of initial weight or if thirst becomes intolerable. To distinguish between central and nephrogenic diabetes insipidus, an AVP test should be performed. In cases of CDI, a marked increase in antidiuresis can be measured. In cases of NDI, AVP is ineffective.

AQP2 has been used to distinguish between NDI and CDI since AQP2 is excreted with the urine in the kidney. AQP2 is constitutively expressed and also excreted in diabetes insipidus to a basal degree. AQP2 levels do not change in diabetes insipidus under dehydration. In CDI AVP treatment increases excreted AQP2, whereas in NDI, AVP stimuli failed to increase AQP2. Thus, determining AQP2 in the urine can help in the differential diagnosis of CDI versus NDI.

Once the pathogenic mechanism of diabetes insipidus has been clarified, further investigations are mandatory to determine the nature of CDI and NDI, respectively. In cases of NDI, besides genetic/hereditary causes (which in general are suggested by the early onset and familial history) the nongenetic causes must also be taken into account, especially in adults. In CDI cases, advanced diagnostics on the cause of brain pathology should be performed, including brain MRI and tumor markers. MRI in particular represents the diagnostic method of choice for evaluating hypothalamic–pituitary-related endocrine alterations due to its ability to provide strongly contrasted, high-resolution images. With this technique the anatomy of the hypothalamus and the pituitary gland can be studied precisely, for example, to distinguish between the anterior and posterior pituitary lobes. Pituitary hyperintensity in the posterior part of the sella on MRI is considered to be a clear marker of neurohypophyseal functional integrity (see also **Figure 5**). Patients with suspected idiopathic CDI must be subjected to sequential imaging (e.g., 6–12 monthly) since a specific pituitary pathology may develop within the course. **Figure 6** summarizes an algorithm for diagnosis and management of CDI (Maghnie, 2003; Ghirardello et al., 2007).

30.2.4 Treatment Options for Diabetes Insipidus

In CDI the decrease or loss of AVP, irrespective of the underlying etiology, can be substituted by AVP analogs. The drug of choice is desmopressin (ddAVP®) which is a synthetic analog of AVP and is selective for AVPR2. Desmopressin can be optionally administered orally, intranasally, or parenterally. The required doses can vary (e.g., 2–40 μg intranasally) and should be divided into three doses per day. Always a low dose should be used initially and increased stepwise. A side effect that has to be considered is an overdosing, which causes a dilutional hyponatremia as in SIAD (see Sections 30.3.2). Caution should be exercised in patients under multidrug therapy because of the potential to develop complex water and salt disturbances and the risk of developing pontine myelinolysis (see also Section 30.3.3). Despite the availability of AVP analogs, in most cases of CDI, sole substitution of water is sufficient to improve symptoms and should always be attempted first.

Treatment of NDI is disproportionately more complicated since substitution of AVP is obviously not

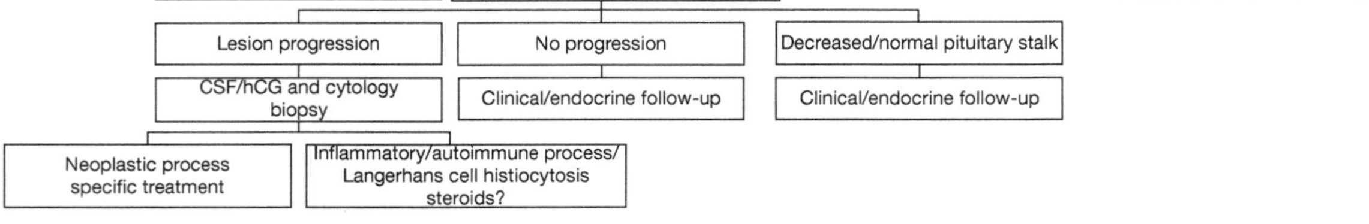

Figure 6 Algorithm for the diagnosis and management of central diabetes insipidus. Gd-DTPA, gadolinium diethylene triamine pentaacetic acid; CSF, cerebrospinal fluid; hCG, human chorionic gonadotropin. Reproduced with permission from Maghnie M (2003) Diabetes insipidus. *Hormone Research* 59 (supplement): 42–54.

effective. Only secondary NDI due to drug side effects (e.g., lithium), related metabolic disorders, or other underlying diseases as congestive heart failure may improve by drug withdrawal or treating the underlying disorder to reverse renal resistance to AVP.

The symptoms of polyuria in primary NDI can be decreased by 30–40% of urine excretion by administering thiazide diuretics in combination with salt restriction. Treatment can be attempted with high doses of a combination of hydrochlorothiazide and amiloride. Nevertheless, a lifelong, abundant water intake cannot be circumvented. It is difficult to achieve a normal growth curve despite the aforementioned treatment. Therefore, water/fluids should be offered every 2 h day and night, and weight growth curve and temperature should be monitored very closely. The voluminous amounts of water which are necessary to equilibrate fluid balance in NDI causes gastrointestinal reflux. Many affected boys suffer from vomiting due to these large amounts of water and from esophagitis. Treatment with H2-receptor antagonists and with metoclopramide (or domperidone) improves these symptoms. If an early diagnosis and adequate treatment can be ensured, even patients with NDI can reach a normal life span and mental development (Cheetham and Baylis, 2002; Rivkees et al., 2007).

Future treatment options for X-linked NDI may rise from a class of compounds called chemical chaperones. These molecules are capable of reversing the number of intracellular misfolded proteins. Among those molecules, glycerol and other polyols have been reported to stabilize protein conformations, to increase *in vitro* protein refolding, and to increase the kinetics of oligomeric receptor protein assembly. Such chemical chaperones may also facilitate *in vivo* mutant proteins into conformations that resemble the wild-type protein and allow them to escape from proteasome degradation. The first step in this direction has been made. Morello et al. could show in elegant *in vitro* experiments that the nonpeptide AVPR2 antagonist SR121463 can function as a chemical chaperone on AVPR2 mutations. SR121463 was able to restore functional AVPR2 in cells transfected with different human AVRP2 mutations, as demonstrated by an increase in intracellular cAMP levels upon AVP stimulation. This therapeutic strategy can only work in type 2 mutations with misfolded and retained receptors; however, type 2 mutations account for the majority of X-linked NDI. Chaperones do not work in type 1 and 3 mutations since type 1 mutations are properly directed to the surface and type 3 lead to early mRNA degradation (Morello et al., 2000; Robben et al., 2006; Bernier et al., 2006; Robben and Deen, 2007).

30.3 Dysregulation of Salt and Fluid Balance in Brain Disease

Hyponatremia frequently develops in patients on neurologic intensive care units; this important electrolyte imbalance can complicate the course of acute neurologic disorders. Two pathophysiological mechanisms have been suggested to cause noniatrogenic, central hyponatremia: CSWS and SIAD originally termed as inappropriate secretion of antidiuretic hormone (for details see Section 30.3.2). Other causes of hyponatremia such as adrenal insufficiency (aldosterone deficiency), pseudohyponatremia, iatrogenic hyponatremia caused, for example, by thiazide diuretics or rare disorders such as beer potomanial syndrome have to be taken into account and excluded before the treatment of central hyponatremia.

Both entities, CSWS and SIAD, are associated with hyponatremia and reduced serum osmolality but the underlying conditions with regard to the volume state are opposite. CSWS is a state of volume depletion, whereas SIAD is a eu- to mild hypervolemic state. This section describes the main clinical presentation, the different etiologies, and the pathophysiology of CSWS and SIAD. Our goal is to elaborate on the differences between the two entities even though overlapping clinical presentations may occur. We have attempted to discuss those intersecting points but also address open-ended questions. At the end of this section we have introduced a step that should help to move toward successive analysis of the cause of hyponatremia. Finally, we go into detail with regard to deducing the management of CSWS and SIAD, respectively.

30.3.1 Cerebral Salt-Wasting Syndrome

CSWS was first described by Peters et al. (1950) in an article reporting on three patients with neurological disorders who presented with hyponatremia. Detailed analysis of its origin at that time showed clinical signs of volume depletion and renal sodium wasting without disturbance of the pituitary–adrenal axis. However, since Schwartz et al. (1957) introduced SIAD as the cause of hyponatremia in two patients with bronchogenic carcinoma, the concept of an inappropriate secretion of ADH as the central pathomechanism has been adopted to nearly all cases of hyponatremia associated with central nervous system (CNS) disease. From then on, CSWS was viewed as an extremely rare disorder or a misnomer of SIAD. In recent years, however, CSWS has again received increasing attention as a distinct entity of hyponatremia, especially in the field of neurosurgery, probably because it is seen more often in these patients. Thus, for example, hyponatremia occurs in more than 30% of patients with subarachnoid hemorrhage (SAH) (Harrigan, 1996).

30.3.1.1 Clinical presentation of CSWS

The main clinical feature of hyponatremia in CSWS is a volume-depleted state. Patients show clinical signs of reduced ECV as orthostatic dysregulation with reflex tachycardia, low blood pressure, reduced skin turgor, and increased urine volume. Estimation of ECV using clinical signs may help distinguish CSWS from other causes (SIAD, iatrogenic hyponatremia), but often fails especially in patients on neurological intensive care units. The reasons for this are the delayed onset of CSWS in the course (mean is approximately tenth day) and the fact that these patients require intensive care with increased crystalloid infusion, co-administration of diuretics, and the need for mechanical ventilation. The latter implies increased positive end-expiratory pressure, which leads to decreased right atrial reverse blood flow and subsequent reallocation of body fluid between the extra- and intravascular compartments. Invasive measurement to evaluate central venous pressure (CVP) may establish the diagnosis earlier. CVP is reduced in CSWS ($\leq$2 mmHg) but normal or slightly enhanced in SIAD.

Plasma osmolality in general is decreased (normal values: 281–297 mOsm kg^{-1} H_2O). Further laboratory examination of the conditions in CSWS shows a renal sodium loss in combination with water loss in CSWS. Thus, 24-h urine collection shows enhanced sodium excretion with a urine sodium concentration above 20 mmol l^{-1}. Consistent with that is a high urine volume and high urine osmolality (>200 mOsm kg^{-1}). Furthermore, urine osmolality is always greater than blood osmolality. However, many of these urinary changes can also be seen in SIAD and cannot be applied to distinguish without any doubt between the two entities (Harrigan, 1996; Maesaka et al., 1999; Rabinstein and Wijdicks, 2003).

To distinguish CSWS from SIAD, Measaka et al. and others have suggested analyzing changes in uric acid metabolism since a renal urate transport abnormality in CSWS patients is different from that in SIAD. Both groups show a decrease in serum uric acid (SUr) and an increase in fractional excretion of uric acid (FEUr). In SIAD, SUr and FEUr return to normal after PNa^+ is corrected by restricting water intake. The mechanism of increased FEUr in SIAD is directly mediated by AVP since AVP stimulates uric acid secretion in the tubular system via AVPR1. In hyponatremia caused by CSWS, however, water restriction leads to pseudonormalization of sodium but SUr and FEUr remain unchanged. SUr values below 4 mg dl^{-1} and FEua fractions greater than 10% provide evidence of underlying CSWS and are not consistent with SIAD. It is worthy of mention that these urate changes can precede onset of hyponatremia in CSWS.

Some authors recommend analyzing changes in the level of BUN to distinguish CSWS from SIAD since a high BUN-to-creatinine ratio has been described as a clinical paradigm for a prerenal state mainly caused by volume depletion. Tendentially, the BUN-to-creatinine ratio is increased in CSWS and decreased in SIAD. However, analysis of the BUN-to-creatinine ratio is of limited value since changes can be observed in a consensual way in both CSWS and SIAD (Maesaka et al., 1999).

Since the central feature of CSWS is reflected in a reduced ECV that in turn represents the major clinical difference between the two entities, an inappropriate secretion of AVP could therefore not be observed in CSWS. Instead, two factors, BNP and ANP, have been found to be associated and pathophysiologically involved in volume depletion and sodium loss. Berendes et al. (1997) could demonstrate that only SAH patients develop a CSWS accompanied with elevated levels of BNP in SAH patients as compared to a control group of tumor patients after surgery (**Figure 7**). Several other reports have demonstrated increased circulating natriuretic peptides not only in CSWS in SAH patients but also in other pathological states. It seems that natriuretic peptides may play a major role in the pathophysiology of CSWS, which is described below. Therefore, measuring of blood BNP and/or ANP may help establish the diagnosis of CSWS. In addition, aldosterone, renin, and AVP levels may be additional helpful laboratory values to complete the diagnostic examinations in otherwise unclear cases. It must be mentioned that the significance of BNP and ANP is poor in the presence of coexisting congestive heart failure or sepsis (Christ-Crain and Müller, 2007). Then, it is impossible to

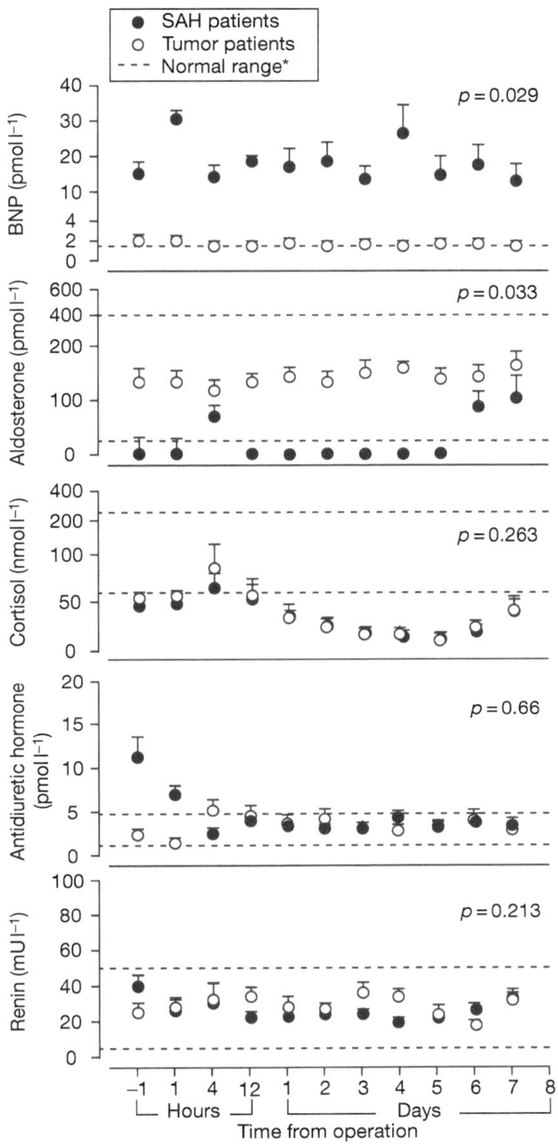

Figure 7 BNP levels in patients with SAH. Mean plasma concentration of BNP, aldosterone, cortisol, AVP, and renin in patients with SAH compared to those with tumors. Reproduced from Berendes E, Walter M, Cullen P, et al. (1997) Secretion of brain natriuretic peptide in patients with aneurysmal subarachnoid haemorrhage. *Lancet* 349: 245–249, with permission from Elsevier.

determine whether ANP/BNP originate in the heart or brain respectively.

In summary, the diagnosis of CSWS has to be confirmed carefully by integrating clinical signs and laboratory examinations. None of the clinical signs of volume depletion alone can be considered as proof of CSWS. For example, orthostatic hypertension is often seen in critically ill patients since long periods of bedrest impair the baroreceptor reflex. Furthermore, measuring CVP is not necessarily sufficient since cardiac stimulation with positive inotropic substances lowers CVP but enhances ECV. Physical examinations in general also fail in judging the impact of existing volume depletion in terms of CSWS since, for example, elderly patients frequently present with dehydration or patients on chronic diuretic therapy show also symptoms of mild volume depletion. Hence, CSWS is a complex diagnosis and can only arise from a differentiated investigation. An algorithm for the differential diagnosis of hyponatremia is described in Section 30.3.3.

30.3.1.2 Etiology of CSWS

Analysis of the cause of CSWS has not been investigated systematically in the past. Data are derived mainly from single case reports or case studies that have included only few patients. Only for SAH have larger groups of patients been investigated in more detail in the past. Prospective studies are rare to date. Summarizing all published data of suspected CSWS, the condition seems to be associated with different pathological states of the brain. These conditions can be divided into four groups:

1. CSWS as a complication of brain injury;
2. CSWS in the course of brain surgery, especially surgery of the skull base;
3. CSWS in association with brain infections; and
4. CSWS due to other rare conditions.

30.3.1(i) Brain injury

SAH is a life-threatening variant of intracranial bleeding and leads to death in approximately 23% of the cases. About 80% of nontraumatic SAH are caused by the rupture of a preexisting arterial a-neurysm. In the course of SAH, fluid and electrolyte imbalances are common. Among them, hyponatremia is the most common abnormality and occurs in 35% of these patients, most frequently seen between day 2 and 10 after onset of the bleeding. This onset closely parallels the period of cerebral vasospasm – one of the most dreaded complications of SAH that leads to ischemic stroke, consecutive acceleration of brain edema, and, last but not least, to death. The risk of developing hyponatremia is significantly increased with large amounts of blood deposits in the subarachnoid space, enlargement of the third ventricle (regardless of the size of the lateral ventricles), or presence of suprasellar or intraventricular blood on the initial computed tomography scan. Hyponatremia also seems to be significantly more frequently associated with aneurysms of the anterior communicating artery of *Circulus velisii* and the ophthalmic artery. Additional risk factors are emerging epileptic seizures and exacerbation of brain edema. The exact pathophysiology of these interrelations is unknown and some hypotheses are discussed in Section 30.3.1.3, Growing evidence suggests that particularly in SAH, hyponatremia is more frequently, if not exclusively, due to CSWS (Wijdicks et al., 1985; Harrigan, 1996; Shaffrey et al., 1996, pp. 1264–1271; Harrigan, 2001; Rabinstein and Wijdicks, 2003).

Hyponatremia can also be observed, but less frequently than in SAH, in other kinds of brain injury. These include traumatic head injury, intracerebral hemorrhage (ICH), and ischemic stroke. In contrast to SAH, the incidence and the exact underlying pathomechanism of hyponatremia has not been systematically studied in these conditions. Most knowledge from those patients is derived from single case reports. In traumatic head injury, hyponatremia develops several days after the injury, often in the course of disease after an initial period of hypernatremia due to secondary dehydration. The underlying mechanism for hyponatremia here is more likely a combination of hormonal water retention and sodium wasting. Elevated ADH levels can occur and reflect a more appropriate or regulatory response to decreased vascular volume. In some individual cases of pediatric head injury and pediatric stroke, CSWS with high levels of natriuretic peptides has been documented (Berger et al., 2002; Palmer, 2003; von Bismarck et al., 2006).

30.3.1.2(ii) Brain surgery

It should not come as a surprise that neurosurgical interventions can also produce hyponatremia in postoperative patients. Like in head injury or stroke, data from those patients have not been investigated systematically and are based mainly on single case reports. Most of these cases were associated with surgery of the skull base such as pituitary surgery, acoustic neurinoma surgery, craniopharyngioma surgery, or remodeling in craniosynostosis. The majority of the cases were also associated with CSWS.

Hyponatremia occurs particularly often after transsphenoidal resection of pituitary tumors. Frequency has been reported to range between 8% and 35 % and up to 20% with associated symptoms of hyponatremia. However, the dynamics of hyponatremia established under these conditions are different than in SAH. Immediately after surgery, a reversible CDI with polyuria can be observed in one-third of the cases due to manipulation of the pituitary stalk. This is followed by an oligouric interphase after which hyponatremia can develop with a latency of 1–2 weeks after surgery. This hyponatremia is independent of high levels of AVP and is characterized by elevated levels of natriuretic peptides (Shaffrey et al., 1996, pp. 1264–1271; Levine et al., 2001; Palmer, 2003; von Bismarck et al., 2006).

30.3.1.2(iii) Brain infections

As for neurosurgery and head injury, data from brain infections are also derived from single case reports and involve the following pathogens: *Coccidioides immitis* (Coccidiomycosis), *Cryptococcus neoformans*, *Herpes simplex virus*, *Mycobacterium tubulerculosis*, *Listeria monocytogenes* (see also the ***Clinical Case 3*** and **Figure 10**), and unspecified bacterial and viral meningitis. Among these different pathogens, tubulerculous meningitis accounts for more than half of the reported cases of hyponatremia. Interestingly, tubulerculous meningitis typically becomes manifest as brainstem meningoencephalits. The impact of this observation is discussed in the next section. The contribution of the two entities, CSWS and SIAD, respectively, to hyponatremia in brain infection is inconsistent but tends to be over-represented by CSWS. However, the significance of available data is limited since most reports are derived from children and only few from adults (von Bismarck et al., 2006).

30.3.1.2(iv) Others

Hyponatremia as a complication of brain pathology other than those described above is rare. In Guillain–Barré syndrome (GBS), mild hyponatremia can develop in the course of disease. Hyponatremia in these patients was more common if mechanical ventilation was required and appeared approximately at day 10 after intubation. Fluid restriction normalized sodium levels, suggesting SIAD as the underlying cause. Only in a subgroup of GBS patients with severe autonomic dysregulation and extremely high blood pressures were enhanced ANP levels present, which could be a hint for underlying CSWS (Ropper et al., 1991).

Pharmacogenic hyponatremia has been classically ascribed to SIAD and will be described in Section 30.3.2. However, a case of CSWS associated with neuroleptic medication and with coincident neuroleptic malignant syndrome has been reported (see also

***Clinical Case 2*:** CSWS in the course of a psychotic disorder treated with neuroleptics. A 57-year-old woman with the first psychotic episode was treated with the atypical neuroleptic olanzapine for 2 weeks. At this time she developed a malignant neuroleptic syndrome and severe hyponatremia. Clinically, she presented with consciousness disturbances with sopor and intermittent agitation. Physical examination showed tachycardia, reduced turgor of the skin, and pyrexia (40°C). Neurological examination revealed severe rigor of the neck and the limbs. CVP was low (2 mmHg) at admission. Laboratory findings showed severe hyponatremia (109 mM) with low plasma tonicity (235 mOsm kg^{-1} H_2O). The nature of hyponatriemia showed enhanced levels of brain-natriuretic peptide (BNP) but normal levels of vasopressin (AVP). The clinical signs and laboratory findings are summarized in (**Table 1**) and show the typical pattern of CSWS.

Treatment of the patient involved discontinuing of antipsychotic medication, sodium and fluid substitution, and datrolene intravenosly.

Table 1 Clinical and laboratory findings of CSWS in a case of NMS

Parameter	*Patient*	*Normal values*
Body temperature	40.0	36.5–37.5°C
Blood pressure (MAP)[a, b]	89	<115 mmHg
Heart rate[a, b, c]	100	50–80 min^{-1}
Water excretion	6,6	Approx. <3,0 L/24 h
Central venous pressure	2	2–10 mmHg
Plasma sodium	110	135–145 mM
Serum osmolality	239	281–297 mOsm kg^{-1} H_2O
Serum uric acid		<6 mg dl^{-1}
At admission	7	
In the course (lowest)	1,3	
Brain natriuretic factor		<334 ng l^{-1}
At admission	982	
In the course	457	
Vasopressin	2,7	2–8 ngl^{-1}
Leucocyte count		4–10 nl^{-1}
At admission	14,56	
In the course (highest)	15,02	
C-reactive protein		<5 mgl^{-1}
At admission	6,7	
In the course (highest)	64,7	

[a]MAP, mean arterial pressure.
[b]Values as mean of first 24 h.
[c]Resting pulse.
Lenhard T, Külkens S, and Schwab S (2007) Cerebral saltwasting syndrome in a patient with neuroleptic malignant syndrome. *Archives of Neurology* 64:122–125.

the *Clinical Case 2* and **Table 1**). For the sake of completeness it should be mentioned that a renal salt-wasting syndrome also exists in addition to CSWS. Causes are, among others, bone fractures in elderly patients, toxic nephropathies after chemotherapy (e.g., cisplatin), and some inherited renal tubular disorders such as sodium/glucose co-transporter defects (Cao et al., 2002; Calado et al., 2006; Maesaka et al., 2007).

In summary, CSWS can occur as a complication of cerebral diseases, mainly in SAH, where it has been studied most intensively, but also in neurosurgery patients, brain infections, and a small number of other conditions. Interestingly, if we take a closer look at the anatomical level that causes CSWS in brain diseases, there is striking evidence for involvement of the brainstem and diencephalic structures. This is supported by the observation that the risk for developing CSWS is high if a lesion or injury is close to basal brain structures, such as deep basal aneurysms, tubulerculous meningitis, or pituitary tumor surgery. The implication of these brain structures in the pathophysiology of CSWS is described in the next section.

30.3.1.3 Pathophysiological concepts of CSWS

The mechanism by which brain disease leads to salt wasting in the kidney is still only partially understood. The most likely process involves dysregulation of neural input and/or dysregulation of natriuretic peptide expression. Both lead to an increased urinary Na^+ excretion that decreases effective aterial blood volume (EABV) and, as a result, reduces the ECV. An astonishing finding is that despite reduced ECV and hyponatremia with subsequent blood hypoosmolality, the renin–angiotensin–aldosterone system fails to respond. Renin and aldosterone levels remain normal during developing CSWS. This is because NP secretion is, in a sense, inappropriate. NPs have inhibitory effects on the RAAS that are described in more detail below. On the other hand, reduced ECV induces a baroreceptor stimulus to release AVP, which in turn is often slightly elevated at the beginning of CSWS. This might be why CSWS was misdiagnosed as SIAD in the past but, in this context, AVP secretion must be judged as appropriate. It is not known why AVP levels return to normal in the course of the disease even though CSWS persists.

Six natriuretic active peptides have been described: ANP, BNP, CNP, the very recently discovered dendroaspis natriuretic peptide (DNP), and urodilatin (cleaved variant derived from pro-ANP). It is not clear whether all of them contribute to the pathophysiology of CSWS but a good amount of the data speak in favor of BNP and ANP. Their receptors are membrane-bound guanylate cyclases and are expressed on smooth muscle cells of the renal vessels (afferent glomerular arterioles, afferent *vasa recta*) and on epithelial cells of the collecting duct (see also Section 30.1 and **Figure 2**; Calado et al., 2006; Alpert and Hebert, 2007, pp. 947–979).

The pathomechanism by which NPs mediate sodium loss in the kidney is not fully understood. As NP secretion in CSWS is inappropriate, it antagonizes the RAAS. Visualizing the working points of ANP/BNP along the nephron and the renal vascular system, these peptides enhance first of all the GFR and perfusion of the renal medulla. Second, NPs directly inhibit water and sodium reabsorption at the collecting duct and thus antagonize aldosterone effects. Third, NPs suppress the RAAS by inhibiting renin excretion at the juxtaglomerular apparatus and aldosterone excretion in the adrenal medulla. The lack of aldosterone rise in response to reduced ECV also explains the absence of simultaneous renal K^+ wasting in CSWS. In addition, an increased sympathetic input additionally inhibits renin excretion in CSWS leading to an inhibited RAAS. The possible impact of the SNS on CSWS in general is described below.

As already mentioned, among the NPs, BNP/ANP seems to be one of the most prominent candidates to mediate renal salt wasting. Berendes et al. presented SAH patients after aneurysma clipping and compared them with controls suffering primarily from brain tumors after craniotomy. They could show a clear rise only in BNP in the SAH group that was accompanied by a slight aldosterone suppression but unchanged renin and cortisol levels (see also **Figure 7**). A slight AVP rise at the beginning of CSWS returns to normal levels in the course of disease. All SAH patients showed an increase in urine volume and increased Na^+ excretion but this was not found in any of the tumor patients (Wijdicks et al., 1991; Kurokawa et al., 1996; von Bismarck, 2006).

What is the source of NP in CSWS? Both, BNP and ANP, for example, are found in the cardiac ventricles (BNP) and artrium (ANP), respectively, and in the brain. In the heart a general stress response may trigger BNP/ANP excretion, whereas a disease-specific brain-associated BNP/ANP stimulation could have greater impact. As described above, the effect of circulating NPs at the nephron is well documented, whereas their intrinsic function within the CNS is less well understood.

Neurons in the vicinity of the third ventricle and the ventral part of the median preoptic nuclei express Na^+-sensing receptors. Hypertonic solutions increase local and plasma levels of ANP. The brain-derived expression of BNP is mainly localized in the hypothalamus. The brain region-specific expression pattern of BNP and a postulated stress-induced cardiac BNP excretion indicate a close functional connection with the SNS. Interestingly, a third source of BNP is even the adrenal medulla, a structure also abundantly innervated by sympathetic nerve terminals. The close connection of NP to the sympathoadrenal system (SAS) is further supported by the observation that deafferentation of the carotid baroreceptorreflex and of renal innervation decreases plasma ANP, irrespective of a hypertonic volume expansion and by the observation that CSWS can occur in the course of a neuroleptic malignant syndrome (NMS) (see also the ***Clinical Case** 2*).

NMS is a condition that is strongly associated with SNS dysregulation. Central regulators of the SNS are the lateral and posterior hypothalamus. The hypothalamus projects dopaminergic neurons to the spinal intermediolateral column, where the preganglionic sympathetic neurons arise. These neurons have been identified as SNS central command neurons and express high levels of BNP. Stimulation of these nuclei increases adrenal nerve activity and promotes organ-specific regulation of sympathetic activity. NMS is also a state of higher-order dopaminergic dysregulation and is caused by anti-dopaminergic drugs. Accordingly, it can cause symptoms of dysregulation in the autonomous nervous system, including diaphoresis, hyperthermia, incontinence, forced diuresis, and dehydration. As already described above, centrally acting ANP can mediate natriuresis and therefore require intact central aminergic system function. Haloperidol as a strong D2-receptor antagonist, for example, can increase the level of ANP and that in itself can induce natriuresis. Like in the case of NMS, an SNS dysregulation, possibly also via dopaminergic dysregulation, might also represent the central pathomechnism in other causes of CSWS, as for example is evident for SAH. Thus, especially in SAH patients, many features argue for a severely dysregulated stress response since those patients often require disproportionately high doses of analgosedatives to reach a sufficient sedation depth and those patients are at risk of developing cerebral vasospasms. Furthermore, many of the underlying pathological processes (aneurysm rupture, tubulerculous meningitis, and pituitary surgery) develop close to the hypothalamic region as already mentioned in Section 30.3.1.2. Enlargement of the third ventricle as predictor for CSWS in SAH as well as observed in other pathological stages (see as an example the ***Clinical Case 3*** and **Figure 10**) also points to the hypothamalimc surroundings as the critical anatomical brain structure for CSWS pathology (Gurrera, 1999; Antunes-Rodrigues et al., 2004; Lenhard et al., 2007).

In summary, CSWS is marked by increased renal Na^+ and water excretion. The basic pathophysiological process includes an enhanced glomerular filtration rate and attenuated reabsorption in the distal tubulus and collecting duct. Circulating natriuretic peptides and disturbed sympathetic renal innervation seem to be responsible for these renal effects. A higher-order brain sympathadrenal dysregulation may play a critical role for the underlying pathology.

30.3.2 Syndrome of Inappropriate Antidiuresis

SIAD is an imbalance of sodium and body fluid characterized by hpotonic hyponatremia with impaired water excretion. In contrast to CSWS, the volume state is euvolemic to mild hypervolemic. It is not caused by renal insufficiency, adrenal insufficiency, or other subsequent stimuli that trigger vasopressin (AVP) secretion. SIAD was first described by Schwartz, Bartter, and colleagues in 1957 and originally termed as syndrome of inappropriate antidiuretic hormone secretion in two patients with bronchogenic carcinoma. As already described in Section 30.3, the concept of SIAD was then proposed as a general pathomechanism of CNS-disease-associated hyponatremia. Both initially described protagonists suffered from severe hyponatremia at hospital admission. Both were treated with high saline intake and with deoxycorticosterone acetate (DOCA), a mineralocorticoid agonist. Nervertheless, PNa^+ decreased on balance in the course of disease and both patients died despite high saline infusion. The sodium balance remained negative with continuous renal sodium loss despite progressive drop of PNa^+ and osmolality. At autopsy, bronchogenic carcinoma with extensive metastatic status was found in both patients. Taking these findings together, a cancer-associated syndrome of inappropriate antidiuretic hormone secretion should be the likely pathomechanism of hyponatremia in both patients.

It is certainly difficult to form a critical opinion regarding the exact underlying condition of these two patients after more than 50 years. Taking a closer look, in fact, in one patient the condition suggests cancer-derived paraneoplastic excretion of AVP that led to the severe hyponatremia. Consistent with an SIAD diagnosis was a tendency for slight weight gain and the absence of clinical signs of volume depletion. Furthermore, a progressive hyponatremia and a drop of osmolality could be observed despite mineralocorticoid treatment with DOCA. SIAD as a consequence of paraneoplastic AVP/AVP-like peptides excretion from small cell lung cancer is a well-known etiology nowadays as will be elaborated in the next paragraph. It should be pointed out that Schwartz and Bartter did not measure AVP levels in the patients and natriuretic peptides were discovered years after the original publication.

In the second patient, the condition cannot be traced as clearly. This patient probably suffered more from CSWS than from SIAD. In support of this hypothesis is the fact that the second patient seemed to suffer from volume depletion with weight loss, high urine Na^+, and high urine volume. Furthermore, he sowed a positive if only brief response to DOCA with a transient PNa+ rise and a subsequent weight gain. Furthermore, he sowed a positive if only brief response to DOCA with a transient PNa^+ rise and a subsequent weight gain. Furthermore, the analysis of FEUr under conditions of water restriction could have advanced a more differentiated diagnosis. The critical review of both originally published cases with today's knowledge is a good example of the difficulty in diagnosing hyponatremia in terms of clearly differentiating the underlying cause – CSWS versus SIAD.

30.3.2.1 Pathophysiology of SIAD

In Section 30.3 we described causes of hyponatremia in association with brain lesions, which have been attributed to CSWS. Nevertheless, hyponatremia is frequently diagnosed and the criteria for CSWS do not fit. SIAD is diagnosed by exclusion and must be distinguished from other symptomatic causes of hyponatremia such as those associated with a hypervolemic state in congestive heart failure, liver cirrhosis, or nephrosis. For the differential diagnosis of hyponatremia, see Section 30.3.3. Robertson (2006) provided a new definition for SIAD that is independent of AVP levels. Upon closer analysis of patients with a euvolemic hypotonic hyponatremia, different types were established, one of which is not associated with

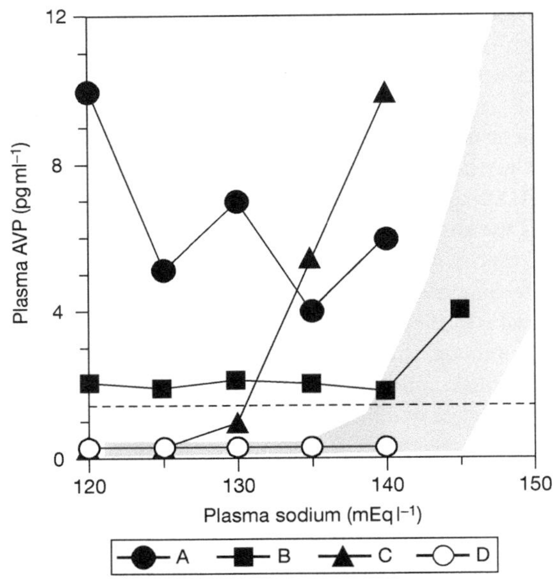

Figure 8 Osmoregulation in patients with SIAD. Different types of inappropriate secretion of AVP (depicted in capital letters) can be distinguished in patients. The shaded area represents physiological (appropriate) AVP response to increasing plasma sodium concentrations. Modified from Robertson GL (2006) Regulation of arginine vasopression in the syndrome of inappropriate antidiuresis. *American Journal of Medicine* 119(supplement 1): S36–S42, with permission from Elsevier.

an inappropriate AVP secretion. Therefore, the term SIAD is more precise. **Figure 8** illustrates the different types of inappropriate antidiuresis (Robertson, 2006; Ellison and Berl, 2007). Accordingly, four types of states of inappropriate antidiuresis could be distinguished with regard to AVP response. Types A, B, and C show an inappropriate secretion of AVP, but in type D, AVP cannot be detected. The physiological response of AVP secretion as a function of blood sodium concentration is composed of a very low basal secretion rate below 135 mmol l^{-1} PNa^{+} and an exponential secretion with further sodium rise. In the following, the four types of SIAD are described in detail.

Type A shows high, fluctuating AVP levels with no relationship to osmolality or sodium intake/infusion. Type A accounts for approximately 30% of SIAD cases and is not exclusively, but mostly, seen in patients with malignant tumors. The source in those patients is mostly ectopic AVP and not eutopic. Interestingly, the large AVP plasma fluctuations are not accompanied by changes in urine osmolality because the kidney has reached maximum concentration capacity under these conditions.

In type B, AVP levels are slightly and inappropriately elevated even though the sodium concentration and plasma osmolality would not necessarily demand an increase in AVP. However, if the sodium concentration increases to high normal levels, the AVP response is then appropriate and rises in conjunction with the osmotic stimulus. Type B also accounts for approximately 30% of SIAD cases. The source of AVP is also more ectopic rather than eutopic; it is also associated with malignancy and with local processes at the neurohypophysis as well.

Type C shows a pattern similar to the physiological response with low to nondetectable AVP and a robust response to rising osmolality, but with one critical difference: it seems that the central setpoint is shifted to lower osmolality and the response is then inappropriate. Type C again accounts for 30% of SIAD. The underlying pathomechanism is unclear. However, type C is associated with malignant tumors, too, whereby ectopic AVP secretion seems to be irrelevant. Probably, factors other than those produced by AVP influence eutopic AVP secretion via osmocenter afferent regulation.

The fourth type, type D, differs from the other types in terms of being inappropriate because AVP cannot be detected in these patients. Nevertheless, urine is not diluted in these individuals even if they receive a high water load. The reason for this defect is not known. A hint for a possible underlying pathomechnism comes from a kind of inherited SIAD for which a mutation in the AVPR2 could be demonstrated. This mutation causes a constantly active intracellular signal, also independently of AVP binding. However, this mutation cannot account for the sporadic cases. Possible mechanisms that may cause a type D constellation could be an acquired postreceptor defect in AQP2 water channel trafficking, affected AQP2 function (e.g., missorting of AQP2 to the membrane and resorting to endosomes, respectively or autoimmune processes), and other unknown AVP-like factors. A possible autoimmune mechanism is imaginable since, for example, an AQP4 autoimmunogenic mechanism has been described quite recently as cause of a brain demyelinating disease (neuromyelitis optica) (Halperin et al., 2001; Ellison and Berl, 2007; Wingerchuk et al., 2007).

30.3.2.2 Conditions favoring SIAD

The causes of SIAD are manifold. They can be divided into malignant tumors, pulmonary disease, and infection, CNS disorders, and drug-associated conditions. The main tumor entity that has been described to be associated with SIAD is SCLC. SCLC mostly contributes to ectopic AVP secretion as a paraneoplastic syndrome. Sometimes those tumors produce more than one hormonally active substance (e.g., ACTH). Other tumors include mesothelioma, gastric cancer, cancers of the genitourinary tract, lymphoma, and sarcomas.

Pulmonary diseases associated with SIAD are infections of different etiologies, asthma, cystic fibrosis, and respiratory failure associated with positive pressure breathing. Cerebral disorders can also cause SIAD. Among them several can also produce CSWS, such as GBS, brain tumors, mass bleedings, and brain infections.

Drugs play a major role as the cause of SIAD, especially psychoactive substances, such as tricyclic antidepressives, serotonin reuptake inhibitors, neuroleptic drugs, and antiepileptics (e.g., carbamazepine), but also narcotics, nonsteroidal antalgetics, cytostatics (vinorelbine, cyclophosphamide, ifosfamide), and others. Of particular interest is SIAD caused by antidepressants and neuroleptics because both substance classes interfere with aminergic neurotransmitter levels. The hypothalamus has both intrinsic dopaminergic neurons and dopaminergic projections from distant regions. Dopamine receptors are also expressed on the neurons of the nucleus paraventricularis that synthesize AVP. The hypothalamus is likewise intensely innervated with serotonin via projections from the raphe nuclei. It is still not known why antidopaminergic drugs can cause SIAD – an euvolemic to mild hypervolemic state – in some patients and CSWS – a hypovolemic state – in others (see also the ***Clinical Case*** *2*) (Robertson, 2006; Ellison and Berl, 2007).

30.3.3 Clinical Differentiation and Treatment of Hyponatremia

30.3.3.1 Diagnosis of CSWS and SIAD

Since CSWS reflects volume depletion and SIAD fluid overload as opposite key features and the deduced are also the opposite, an exact differential diagnosis is indispensable. For SAH with cerebral vasospasm, a so-called Triple-H-Therapy has been established. That means hemodilution and forced hypervolemia are required in addition to elevating the blood pressure. In those patients a misdiagnosis of CSWS and a treatment with water restriction, falsely assuming that SIAD is the underlying pathology of hyponatremia, would have far-ranging consequences culminating in excess vasospasms and, as the ultimate consequence, death.

The following section guides the reader through a diagnostic algorithm with several decision trees to distinguish CSWS, SIAD, and other possible causes of hyponatremia from each other. A respective algorithm for the differential diagnosis of hyponatremia is depicted in **Figure 9** (Albanese et al., 2001; Rabinstein and Wijdicks, 2003; Palmer, 2003).

If a patient with a brain lesion presents with hyponatremia, the first recommended diagnostic test is evaluation of serum osmolality. If serum osmolality is in the normal range, pseudohyponatremia (e.g., hyperlipidemia and hyperproteinemia) or hyperglycemia must be taken into account. If serum osmolality is below normal, the next step should be to evaluate urine osmolality and urine sodium from a single urine specimen. If the urine osmolality is appropriately low (below $100\,\mathrm{mOsm\,kg^{-1}}$ H_2O) as well as urine sodium ($<20\,\mathrm{mmol\,l^{-1}}$), the underlying pathophysiology is hypotonic hyperhydration as it can occur in primary polydipsia or in beer potomania syndrome. The latter is a rare renal excretion disorder seen in alcoholic patients who consume high amounts of beer (which has low salt content) and eat little food. A diet poor in salt and protein (urea source) results in reduced excretion of urinary solutes that limits the ability to excrete free water.

If the urine osmolality is inappropriately high (greater than $200\,\mathrm{mOsm\,kg^{-1}}$ H_2O) as well as urine sodium ($>20\,\mathrm{mmol\,l^{-1}}$), the status of extracellular fluid volume should be estimated. Clinical signs have been already described in detail in Section 30.3. Shortly summarized, these are orthostatic dysregulation with low blood pressure and reflex tachycardia, low CVP, and low blood volume. Simultaneously, thyroid and adrenal dysfunctions should be excluded (hypothyroidism and hypoaldosteronism). If the clinical signs indicate a volume-depleted state, a CSWS is most probable and fluid therapy can be started. If the estimation of the volume state is doubtful, urine should be collected for a 24-h period. Then, the sodium excretion and fraction of excreted urate should be determined (calculation of FEUr, see **Figure 9**). In both SIAD and CSWS, FEUr is increased above 10% and serum urate (SUr) is decreased. In order to clearly distinguish CSWS from SIAD, water intake should be restricted over 1 (maximum 2) days. In the case of SIAD, sodium, SUr, and FEUr return to normal levels, whereas in CSWS, FEUr remains elevated and SUr decreased. In SAH patients, this diagnostic procedure should be performed with caution since restricting water intake can aggravate cerebral vasospasms. In SAH, the

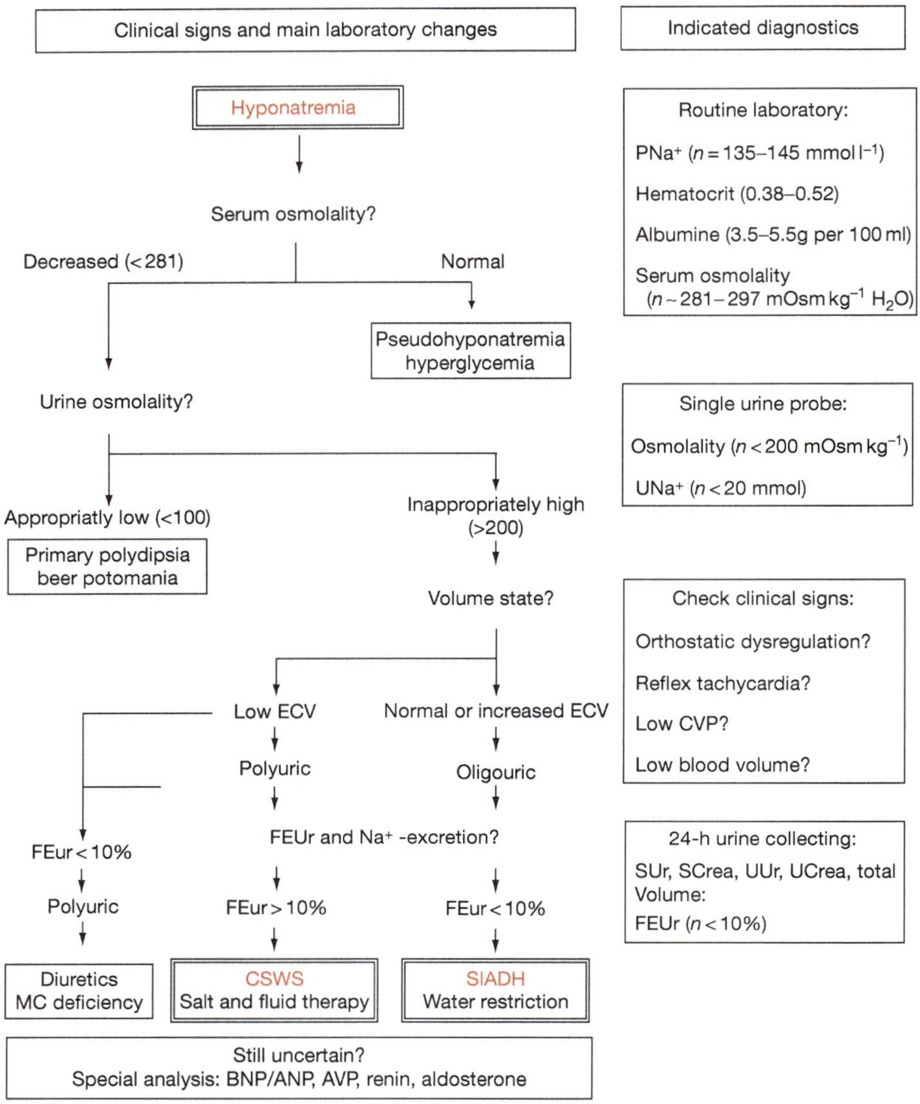

Figure 9 Approach to the management of patients with hyponatremia. Formula for the calculation of fractional excretion of urate: FEUr (%) = [UUr x SCrea]/[SUr x UCrea] x 100, with UUr, urine urate concentration; SCrea, serum creatinine concentration; SUr, serum urate concentration; UCrea, urine creatinine concentration (concentrations all as mg dl^{-1} or as $mmol l^{-1}$). BNP, brain natriuretic peptide; ANP, atrial natriuretic peptide; AVP, arginine vasopressin; MC, mineralocorticoid. Note that diuretics does not increase or decrease (furosemide) FEUr. MC deficiency is accompanied with increased potassium.

estimation of CSWS should only be based on laboratory values without restricting water intake and on clinical signs. Especially in the case of SAH, analysis of blood circulating BNP, AVP, renin, and aldosterone may provide additional help to reach a decision. In CSWS, BNP and ANP should be elevated, whereas levels of renin and aldosterone stay normal. The pathomechanisms for the inappropriate nonresponse of the RAAS have been described in Section 30.3. AVP levels can be elevated to a mild extent at the beginning of CSWS but return to normal in the course of the disease.

30.3.3.2 Therapy of hyponatremia in CSWS and SIAD

The therapy of hyponatremia should always be carried out with care and PNa^+ should be compensated slowly since pontine myelinolysis and also

Marchiafava-Bignami syndrome, which, however, occurs considerably less frequently, are dreaded complications. On the other hand, the degree of urgency for compensation treatment for hyponatremia depends on the speed at which hyponatremia developed and the severity of clinical presentation. The need for consequent and efficient but controlled compensation is much higher in patients in whom hyponatremia has developed in a short time and is accompanied by severe symptoms such as consciousness disturbances, seizures, and global brain edema. In patients with moderate hyponatremia and who present with only mild symptoms or with chronically developed hyponatremia, more elaborate diagnostic examinations should precede.

Before starting compensation of hyponatremia with sodium, one should consider total Na^+ distribution within the body. The initially measured PNa^+ may not be a steady-state value. Thus, after seizure, PNa^+ can transiently rise by about 10 mmol l^{-1}. Venous Na^+ concentration differs from arterial concentration; to the latter the brain is exposed. Furthermore, the risk of osmotic demyelination is strongly associated with water balance, and a diuretic state must be avoided. This implies that in CSWS, Na^+ compensation should always be given in conjunction with fluid. Many experts recommend concomitant use of furosemide (Adrogue and Madias, 2000; Halperin and Kamel, 2007). In the following, the principal management of hyponatremia irrespective of the underlying pathophysiology is described. The specific aspects of the treatment of CSWS and SIAD, respectively, are described thereafter.

In practice, management of compensation of hyponatremia starts by estimating the steady state of PNa^+ (=expected normal value=target value). The next step is to calculate the total amount of Na^+ to be substituted. A practical approach to calculate the total Na^+ requirement and accordingly to calculate the infusion rate of infusates is depicted in the following formulas:

$$\Delta[Na^+](\text{mmol}) = \text{TBW} \times [PNa^+_{\text{desired}}(\text{mmol}\,l^{-1}) - PNa^+_{\text{actual}}(\text{mmol}\,l^{-1})] \quad [1]$$

and

$$\text{Infusion rate}\ (\text{ml}\,h^{-1}) = Na^+_{\text{reqired}}(\text{mmol}\,l^{-1}) \times 1000/[Na^+]_{\text{infusate}}(\text{mmol}\,l^{-1}) \times \text{time(h)} \quad [2]$$

where $\Delta[Na^+]$ represents the required sodium and TBW represents the total body water, expressed in liters, and is approximately estimated with body weight × 0.6 (men)/× 0.5 (women) (children × 0.6; elderly women × 0.45/-men × 0.5).

Note that the maximum correction should not exceed 8–10 mmol l^{-1} in 24 h.

Adrogue and Madias recently proposed a formula to prove effectiveness and safety in managing hyponatremia. According to this formula, the anticipated change in the patient's plasma sodium concentration as a result of administration of 1 l of any infusate can be calculated by

$$\Delta[Na^+] = ([Na^+]_{inf} - [Na^+]_{plasma})/(\text{TBW} + 1) \quad [3]$$

where $\Delta[Na+]$ represents the expected change in sodium, $[Na^+]_{inf}$ the sodium content of infusion solution, and $[Na^+]_{plasma}$ the actual sodium concentration in the patient's plasma, all expressed in mmol l^{-1}. If the infused solution contains potassium in addition, a K^+-corrected formula is given as

$$\Delta[Na^+] = \{([Na^+]_{inf} + [K^+]_{inf}) - [Na^+]_{plasma}\}/(\text{TBW} + 1) \quad [4]$$

where $[K^+]_{inf}$ represents the potassium concentration of the infusate. The Na^+ content of the most commonly used infusates is depicted in **Table 2**.

Even if the required Na^+ to correct hyponatremia is calculated, PNa^+ must be monitored every 2–3 h and the infusion must be adjusted as needed. If PNa^+ exceeds above upper limits, one should consider administering the AVP analog desmopressin and 5% glucose solution to lower PNa^+. Case reports suggest that it may be possible to reverse symptoms of osmotic demyelination in this way.

If a patient presents with severe symptoms of hyponatremia (seizure, coma, and impending or manifest brain edema), this is a dilemma and the situation raises a conflict for therapy: first, immediate danger from high intracranial pressure, and second, a risk of provoking an osmotic demyelination. Nevertheless, the situation demands urgent action.

Table 2 Sodium content of established infusates

Infusate	*Infusate Na^+(mmol l^{-1})*
5% NaCl	855
3% NaCl	513
0.9% NaCl	154
Lactate Ringer's	130
0.45% NaCl	77
0.2% NaCl	34
5% Dextrose in water	0

***Clinical Case 3*:** CSWS in a patient with meningoencephalitis from *Listeria monocytogenes* improved under treatment with fludrocortisone. A 38-year-old woman had suffered from severe headache and arthralgia for several days. At the day of admission, she presented with a progressive disturbance of consciousness and a right-sided hemiparesis. Mechanical ventilation was mandatory. Lumbar puncture showed a xanthochromic CSF with 347 cells/μl (mixed pleocytosis), elevated lactate (10.1 mmol l^{-1}), decreased glucose (9 mg dl^{-1}), and enhanced protein (2.6 g l^{-1}). *Listeria monocytogenes* were raised from cultured CSF. Despite specific antibiotic treatment she developed an internal hydrocephalus and multiple ischemic infarctions, most probably due to a *Listeria*-associated vasculitis. The pattern of intracerebral lesions is depicted in **Figures 10(a)–10(c)**. In addition, consistent with vasculitis, transcranial doppler ultrasound revealed abnormally elevated bloodflow velocities.

During the course she developed a severe hyponatremia (117 mmol l^{-1} at lowest). Laboratory evaluation revealed a CSWS with increased renal sodium excretion, decreased CVP, decreased serum osmolality, decreased blood urate levels, enhanced FEUr (29%), and elevated BNP levels but normal renin and aldosterone levels. The consciousness disturbance was associated with blood sodium concentrations. The patient was first treated with hydrocortisone because of the coexisting vasculitis and with both sodium and fluid replacement. Due to its partially mineralocorticoid effect, hydrocortisone led to normalization of sodium levels. With reduction of hydrocortisone, hyponatremia reappeared. Additional treatment with fludrocortisone resolved the sodium imbalance, plasma tonicity, and consciousness disturbance completely. **Figures 10(d)–10(f)** summarize the laboratory alterations in the course of the CSWS.

Table 3 Course of CSWS under treatment with mineralocorticoids

Parameter	*Patient*	*Normal values*
Renal sodium excretion	512	120–260 mmol per 24 h
FEUrate1	28	<10%
Aldosterone	6.1	2–14 ng l^{-1}
Renin	11	2.4–29 mU l^{-1}
BNP		<153 ng l^{-1}
Baseline	625	
In the course of hydrocortisone	155	
Before fludrocortisone	1126	

Accordingly, one should infuse 3 mmol of hypertonic NaCl (3%) per liter of total body water in the first hour. The rapid infusion causes a faster Na^+ increase in the arteries than in the veins. This means that in brain, as compared to muscle, for example, cells will shrink faster as a consequence of higher blood flow per unit organ weight. Afterward, an infusion rate of 1–2 mmol per liter and hour is recommended; even if the hyponatremia has been present for more than 24 h, the maximal correction should also be kept at 8–10 mmol l^{-1} and 24 h. The risk for osmotic demyelination rises with a negative water balance. Thus, a water diuresis should be avoided by administering furosemide as co-treatment. If a seizure could not be interrupted, factors others than hyponatremia maintaining seizure should also be taken into account.

The approaches described above should be reserved for severe hyponatremia. In asymptomatic or mildly symptomatic cases, correction of salt and water imbalance is less urgent and a more differentiated diagnostic approach can be taken. In SIAD patients, water restriction alone may be sufficient. In CSWS patients, fluid infusion and augmented oral salt intake may also show sufficient benefit. A differentiated diagnostic approach regarding the origin of hyponatremia and adequate treatment is always at least as important as correcting sodium levels.

Although not well studied, CSWS tends to be transient in most cases, with evidence of resolving after 3–4 weeks. Sometimes, however, the condition persists, obviously because brain damage is irreversible or because the underlying disease cannot be adequately treated. For CSWS in those cases, treatment with aldosterone agonists may be justified. Two substances are suitable: hydrocortisone or fludrocortisone. Fludrocortsione is a pure MR agonist and has no glucocorticoid effects. Hydrocortisone has a 2:1 ratio of glucocorticoid to mineralocorticoid effects and can be used if an additional glucocorticoid effect is desired (e.g., coexisting autoimmune disease, vasculitis). The ***Clinical Case 3*** described in ***Clinical Case 3*** and **Figure 10** illustrates the complex course of disease in a female patient with CSWS due to *Listeria meningoecephalitis* and ultimate improvement of CSWS under treatment with mineralocorticoid stimulation.

Similar to CSWS, the only definitive therapy of SIAD is eliminating the underlying causes. The central treatment approach for non-life-threatening SIAD is water restriction. However, therapy of a chronic SIAD, if the condition causing it cannot be eliminated, requires a specific management since water restriction alone may not be sufficient. Often hyponatremia is not only of long but rather of unclear duration if the underlying cause could not be isolated. This is the case if, for example, cancer of

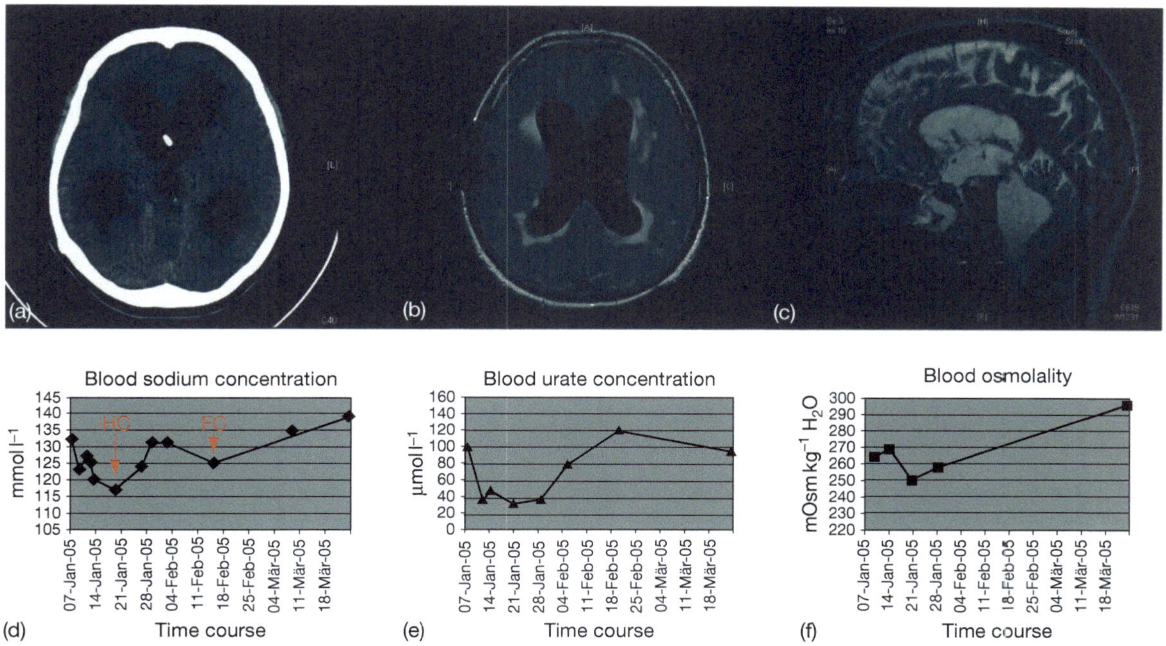

Figure 10 Brain imaging and sodium imbalance in the course of *Listeria meningoencephalitis*. Axial CT-scan (a), axial fluid-attenuated inversion-recovery (FLAIR) (b), and sagittal T2-weighted (c) MRI scans show internal hydrocephalus with ballooned lateral, third and fourth ventricles and consecutive subependymal CSF edema. Note that especially the third and fourth ventricle is enlarged excessively. Enlargement of the third ventricle is significantly associated with CSWS (see also Section 30.3). (d-f) Show plasma urate, plasma sodium, and serum osmolality before and in the course of treatment with hydrocortisone (HC) and fludrocortisone (FC), respectively. Unpublished data from Th. Lenhard and St. Schwab, Department of Neurology, University of Heidelberg, Germany.

unknown origin is assumed that has reached clinical significance only with SIAD. One option is treatment with AVP antagonists such as demeclocycline (Decclomycin®, Wyeth) or conivaptan (Vaprisol®, Astellas Pharma). Demeclocycline (300–600 mg twice a day) reduces urinary sololality and increases plasma sodium levels but has a risk of nephrotoxicity. Conivaptan can raise plasma sodium levels to a sufficient degree. As a nonselective AVPR1/2-antagonist, it can cause hypotension due to its vasodilatative effect. Side effects can be observed in 50% of the treated patients, so it should only be used in hospitalized and moderately affected patients. Drugs with a high potential for the future as oral AVP antagonists have been developed in recent years. Candidates, such as tolvaptan (Otsuka Pharma), are not yet clinically available but have been successively tested in clinical trials. Thus, tolvaptan could elevate sodium levels significantly in long-term therapy and improve mental outcome too (Oh, 2007).

In summary, severe hyponatremia presenting with seizure, coma, and brain edema should be treated immediately with sodium replacement using highly concentrated Na^+ solutions (3%) irrespective of the underlying pathogenesis. Water diuresis should be avoided by administering furosemide as co-treatment to prevent osmotic demyelination. In mild or asymptomatic hyponatremia, correction of salt and water imbalance should be slower and one should pay more attention to diagnosis of the underlying pathology and determine specific treatment.

References

Adrogue HJ and Madias N (2000) Hypernatremia. *New England Journal of Medicine* 342: 1493–1499.

Albanese A, Hindmarsh P, and Stanhope R (2001) Management of hyponatraemia in patients with acute cerebral insults. *Archives of Disease in Childhood* 85: 246–251.

Alpert RJ and Hebert SC (2007) *Seldin and Giebisch's The Kidney: Physiology and Pathophysiology*, 4th edn. St. Louis, MO: Elsevier.

Antunes-Rodrigues J, de Castro M, Elias LL, Valença MM, and McCann SM (2004) Neuroendocrine control of body fluid metabolism. *Physiological Reviews* 84: 169–208.

Beitz E, Wu B, Holm LM, Schultz JE, and Zeuthen T (2006) Point mutations in the aromatic/arginine region in aquaporin 1 allow passage of urea, glycerol, ammonia, and protons. *Proceedings of the National Academy of Sciences of the United States of America* 103: 269–274.

Berendes E, Walter M, Cullen P, et al. (1997) Secretion of brain natriuretic peptide in patients with aneurysmal subarachnoid haemorrhage. *Lancet* 349: 245–249.

Berger TM, Kistler W, Berendes E, Raufhake C, and Walter M (2002) Hyponatremia in a pediatric stroke patient: Syndrome of inappropriate antidiuretic hormone secretion or cerebral salt wasting? *Critical Care Medicine* 30: 792–795.

Bernier V, Morello JP, Zarruk A, et al. (2006) Pharmacologic chaperones as a potential treatment for X-linked nephrogenic diabetes insipidus. *Journal of the American Society of Nephrology* 17: 232–243.

Calado J, Loeffler J, Sakallioglu O, Gok F, Lhotta K, Barata J, and Rueff J (2006) Familial renal glucosuria: SLC5A2 mutation analysis and evidence of salt-wasting. *Kidney International* 69: 852–855.

Cao L, Joshi P, and Sumoza D (2002) Renal salt-wasting syndrome in a patient with cisplatin-induced hyponatremia: Case report. *American Journal of Clinical Oncology* 25: 344–346.

Cheetham T and Baylis PH (2002) Diabetes insipidus in children: Pathophysiology, diagnosis and management. *Paediatric Drugs* 4: 785–796.

Christ-Crain M and Müller B (2007) Biomarkers in respiratory tract infections: Diagnostic guides to antibiotic prescription, prognastic markers and mediators. *European Respiratory Journal* 30: 556–573.

Christensen JH and Rittig S (2006) Familial neurohypophyseal diabetes insipidus: An update. *Seminars in Nephrology* 26: 209–223.

Constanzo LS (2006) *Physiology,* 3rd edn., ch. 5, pp. 235–299. Philadelphia, PA: Saunders Elsevier.

Ellison DH and Berl T (2007) The syndrome of inappropriate antidiuresis. *New England Journal of Medicine* 356: 2064–2072.

Funder JW (2007) The role of aldosterone and mineralocorticoid receptors in cardiovascular disease. *American Journal of Cardiovascular Drugs* 7: 151–157.

Garofeanu CG, Weir M, Rosas-Arellano MP, Henson G, Garg AX, and Clark WF (2005) Causes of reversible nephrogenic diabetes insipidus: A systematic review. *American Journal of Kidney Diseases* 45: 626–637.

Ghirardello S, Garrè ML, Rossi A, and Maghnie M (2007) The diagnosis of children with central diabetes insipidus. *Journal of Pediatric Endocrinology and Metabolism* 20: 359–375.

Ghirardello S, Malattia C, Scagnelli P, and Maghnie M (2005) Current perspective on the pathogenesis of central diabetes insipidus. *Journal of Pediatric Endocrinology and Metabolism* 18: 631–645.

Gurrera RJ (1999) Sympathoadrenal hyperactivity and the etiology of neuroleptic malignant syndrome. *American Journal of Psychiatry* 156: 169–180.

Halperin ML, Bichet DG, and Oh MS (2001) Integrative physiology of basal water permeability in the distal nephron: Implications for the syndrome of inappropriate secretion of antidiuretic hormone. *Clinical Nephrology* 56: 339–345.

Halperin ML and Kamel KS (2007) A new look at an old problem: Therapy of chronic hyponatremia. *Nature Clinical Practice Nephrology* 3: 2–3.

Harrigan MR (1996) Cerebral salt wasting syndrome: A review. *Neurosurgery* 38: 152–160.

Harrigan MR (2001) Cerebral salt wasting syndrome. *Critical Care Clinics* 17: 125–138.

Hu X and Funder JW (2005) The evolution of mineralocorticoid receptors. *Molecular Endrocrinology* 20: 1471–1478.

Jennes L, Ulloa-Aguirre A, Janovick JA, Adjan VV, and Conn PM (2009) The gonadotropin-releasing hormone and its receptor. In Pfaff DW, Arnold AP, Etgen AE, Fahrbach SE, and Rubin RT (eds), *Hormones, Brain and Behavior,* 2nd edn, vol. 3, pp. 1645–1668. San Diego: Academic Press.

King LS, Kozono D, and Agre P (2004) From structure to disease: The evolving tale of aquaporin biology. *Nature Reviews Molecular Cell Biology* 5: 687–698.

Kurokawa Y, Uede T, Ishiguro M, Honda O, Honmou O, Kato T, and Wanibuchi M (1996) Pathogenesis of hyponatremia following subarachnoid hemorrhage due to ruptured cerebral aneurysm. *Surgical Neurology* 46: 500–507.

Lenhard T, Külkens S, and Schwab S (2007) Cerebral salt-wasting syndrome in a patient with neuroleptic malignant syndrome. *Archives of Neurology* 64: 122–125.

Levine JP, Stelnicki E, Weiner HL, Bradley JP, and McCarthy JG (2001) Hyponatremia in the postoperative craniofacial pediatric patient population: A connection to cerebral salt wasting syndrome and management of the disorder. *Plastic and Reconstructive Surgery* 108: 1501–1508.

Lien Y-HH and Shapiro JI (2007) Hyponatremia: Clinical diagnosis and management. *American Journal of Medicine* 120: 653–658.

Maesaka JK, Gupta S, and Fishbane S (1999) Cerebral salt-wasting syndrome: Does it exist? *Nephron* 82: 100–109.

Maesaka JK, Miyawaki N, Palaia T, Fishbane S, and Durham JH (2007) Renal salt wasting without cerebral disease: Diagnostic value of urate determinations in hyponatremia. *Kidney International* 71: 822–826.

Maghnie M (2003) Diabetes insipidus. *Hormone Research* 59 (supplement): 42–54.

Maghnie M, Ghirardello S, DeBellis A, et al. (2006) Idiopathic central diabetes insipidus in children and young adults is commonly associated with vasopressin-cell antibodies and markers of autoimmunity. *Clinical Endocrinology (Oxf)* 65: 470–478.

Magner PO and Halperin ML (1987) Polyuria – a pathophysiological approach. *North American Medical* 15: 2987–2997.

Morello JP and Bichet DG (2001) Nephrogenic diabetes insipidus. *Annual Review of Physiology* 63: 607–630.

Morello JP, Salahpour A, Laperrière A, et al. (2000) Pharmacological chaperones rescue cell-surface expression and function of misfolded V2 vasopressin receptor mutants. *Journal of Clinical Investigation* 105: 887–895.

Nguyen MK, Nielsen S, and Kurtz I (2003) Molecular pathogenesis of nephrogenic diabetes insipidus. *Clinical and Experimental Nephrology* 7: 9–17.

Nicchia GP, Frigeri A, Liuzzi GM, and Svelto M (2003) Inhibition of aquaporin-4 expression in astrocytes by RNAi determines alteration in cell morphology, growth, and water transport and induces changes in ischemia-related genes. *FASEB Journal* 17: 1508–1510.

Oh MS (2007) Management of hyponatremia and clinical use of vasopressin antagonists. *American Journal of the Medical Sciences* 333: 101–105.

Palmer BF (2003) Hyponatremia in patients with central nervous system disease: SIADH versus CSW. *Trends in Endocrinology and Metabolism* 14: 182–187.

Peters JP, Welt LG, Sims EA, Orloff J, and Needham J (1950) A salt-wasting syndrome associated with cerebral disease. *Transactions of the Association of American Physicians* 63: 57–64.

Rabinstein AA and Wijdicks EF (2003) Hyponatremia in critically ill neurological patients. *Neurologist* 9: 290–300.

Rivkees SA, Dunbar N, and Wilson TA (2007) The management of central diabetes insipidus in infancy:

Desmopressin, low renal solute load formula, thiazide diuretics. *Journal of Pediatric Endocrinology and Metabolism* 20: 459–469.
Robben JH and Deen PM (2007) Pharmacological chaperones in nephrogenic diabetes insipidus: Possibilities for clinical application. *BioDrugs* 21: 157–166.
Robben JH, Knoers NV, and Deen PM (2006) Cell biological aspects of the vasopressin type-2 receptor and aquaporin 2 water channel in nephrogenic diabetes insipidus. *American Journal of Physiology – Renal Physiology* 291: F257–F270.
Robertson GL (2006) Regulation of arginine vasopressin in the syndrome of inappropriate antidiuresis. *American Journal of Medicine* 119(supplement 1): S36–S42.
Ropper AH, Wijdicks EF, and Truax BT (1991) *Guillain-Barre' Syndrome*. Philadelphia, PA: FA Davis Company.
Roudier N, Ripoche P, Gane P, Le Pennec PY, Daniels G, Cartron JP, and Bailly P (2002) AQP3 deficiency in humans and the molecular basis of a novel blood group system, GIL. *Journal of Biological Chemistry* 277: 45854–45859.
Schrier RW and Martin PY (1998) Recent advances in the understanding of water metabolism in heart failure. *Advances in Experimental Medicine and Biology* 449: 415–426.
Schwartz WB, Bennett W, Curelop S, and Bartter FC (1957) A syndrome of renal sodium loss and hyponatremia probably resulting from inappropriate secretion of antidiuretic hormone. *American Journal of Medicine* 23: 529–542.
Shaffrey ME, Shaffrey C, and Lanzino G (1996) Nonoperative treatment of aneurysmal subarachnoid hemorrhage. In: Youmans JR (ed.) *Neurological Surgery,* 4th edn., pp. 1264–1271. Philadelphia, PA: W. B. Saunders.
Traggiai C and Stanhope R (2002) Endocrinopathies associated with midline cerebral and cranial malformations. *Journal of Pediatrics* 140: 252–255.
von Bismarck P, Ankermann T, Eggert P, Claviez A, Fritsch MJ, and Krause MF (2006) Diagnosis and management of cerebral salt wasting (CSW) in children: The role of atrial natriuretic peptide (ANP) and brain natriuretic peptide (BNP). *Child's Nervous System* 22: 1275–1281.
Wehling M (2005) Effects of aldosterone and mineralocorticoid receptor blockade on intracellular electrolytes. *Heart Failure Reviews* 10: 39–46.
Wijdicks EF, Ropper AH, Hunnicutt EJ, Richardson GS, and Nathanson JA (1991) Atrial natriuretic factor and salt wasting after aneurysmal subarachnoid hemorrhage. *Stroke* 22: 1519–1524.
Wijdicks EF, Vermeulen M, ten Haaf JA, Hijdra A, Bakker WH, and van Gijn J (1985) Volume depletion and natriuresis in patients with a ruptured intracranial aneurysm. *Annals of Neurology* 18: 211–216.
Wingerchuk DM, Lennon VA, Lucchinetti CF, Pittock SJ, and Weinshenker BG (2007) The spectrum of neuromyelitis optica. *Lancet Neurology* 6: 805–815.

Further Reading

Deen PM, Verdijk MA, Knoers NV, Wieringa B, Monnens LA, van Os CH, and van Oost BA (1994) Requirement of human renal water channel aquaporin-2 for vasopressin-dependent concentration of urine. *Science* 264: 92–95.
Funder JW (2000) Aldosterone and mineralocorticoid receptors: Orphan questions. *Kidney International* 57: 358–363.
Hobbs HH, Russell DW, Brown MS, and Goldstein JL (1990) The LDL receptor locus in familial hypercholesterolemia: Mutational analysis of a membrane protein. *Annual Review of Genetics* 24: 133–170.
King LS, Choi M, Fernandez PC, Cartron JP, and Agre P (2001) Defective urinary-concentrating ability due to a complete deficiency of aquaporin-1. *New England Journal of Medicine* 345: 175–179.
Lang F (2006) Mechanisms and significance of cell volume regulations. In: Ronco C (ed.) *Contributions to Nephrology,* vol. 152, pp. 1–8. Basel: Karger.
Mizobuchi M, Kunishige M, Kubo K, Komatsu M, Bando H, and Saito S (1994) Syndrome of inappropriate secretion of ADH (SIADH) due to small cell lung cancer with extremely high plasma vasopressin level. *Internal Medicine* 33: 501–504.
Moller K, Larsen FS, Bie P, and Skinhoj P (2001) The syndrome of inappropriate secretion of antidiuretic hormone and fluid restriction in meningitis – how strong is the evidence? *Scandinavian Journal of Infectious Diseases* 33: 13–26.
Papadopoulos MC and Verkman AS (2007) Aquaporin-4 and brain edema. *Pediatric Nephrology* 22: 778–784.
Potter LR, Abbey-Hosch S, and Dickey DM (2006) Natriuretic peptides, their receptors, and cyclic guanosine monophosphate-dependent signaling functions. *Endocrine Reviews* 27: 47–72.
Preston GM, Smith BL, Zeidel ML, Moulds JJ, and Agre P (1994) Mutations in aquaporin-1 in phenotypically normal humans without functional CHIP water channels. *Science* 265: 1585–1587.
Salvatoni A, Maghnie M, Lorini R, and Marni E (1990) Hyponatremia and seizures during desmopressin acetate treatment in hypothyroidism. *Journal of Pediatrics* 116: 835–836.
Sanghvi SR, Kellerman PS, and Nanovic L (2007) Beer potomania: An unusual cause of hyponatremia at high risk of complications from rapid correction. *American Journal of Kidney Diseases* 50: 673–680.
Verkman AS, Binder DK, Bloch O, Auguste K, and Papadopoulos MC (2006) Three distinct roles of aquaporin-4 in brain function revealed by knockout mice. *Biochimica et Biophysica Acta* 1758: 1085–1093.

31 Diabetes Mellitus and Neurocognitive Dysfunction

C M Ryan, University of Pittsburgh School of Medicine, Pittsburgh, PA, USA

Chapter Outline

Glossary

effect size It estimates the strength of the relationship between two variables (e.g., the correlation coefficient, *r*; or, when comparing two groups, the standardized magnitude of the between-group difference, e.g., Cohen's *d*). This provides a standard metric for determining whether observed relationships or differences are meaningful in a practical sense: $d = (\text{mean}_1 - \text{mean}_2)/[(SD_1 + SD_2)/2]$. According to Cohen, $d = 0.2$ is indicative of a small effect (corresponding to the 58th percentile); $d = 0.5$ is a medium effect (~69th percentile); and $d \geq 0.8$ is a large effect (~79th percentile).

executive cognitive processes A complex set of cognitive skills or strategies used by people to respond adaptively and efficiently to novel situations. Such abilities include planning or looking ahead, monitoring behavior, and using feedback to modify behavior in a purposive and efficient manner. These skills are largely considered to be mediated by frontal brain systems.

isoelectric EEG A reduction in the amplitude of the EEG recording until it is flat and activity can no longer be detected, that is, there is electrocerebral silence.

microangiopathy A disease process affecting the small blood vessels. In diabetes, this often occurs relatively early in the course of the disease and appears to be a consequence of glucose-induced changes in the endothelial cells lining the blood vessels. The subsequent formation of glycoproteins on their surface, as well as changes in the basement membrane, lead to a thickening, and weakening of the walls of the blood vessels, ultimately slowing blood flow. This is especially common in the retina and kidney and is the primary cause of diabetic retinopathy and nephropathy.

31.1 Introduction

Even before the discovery of insulin, clinicians had been intrigued by the possibility that the metabolic disorder, diabetes mellitus, could disrupt normal cognitive processes (Miles and Root, 1922). With the introduction of insulin treatment for diabetes, an additional risk factor for cognitive morbidity was identified: iatrogenic hypoglycemia. Insulin brought the elevated blood glucose levels characteristic of diabetes closer to the normal range and thereby led to dramatic improvements in health and longevity. Nevertheless, if glycemic levels fell too low, the resulting neuroglycopenia led to a decline in mental efficiency and slowed neurophysiological activity, sometimes eventuating in seizures or coma, and ultimately, in widespread neuronal damage.

A large body of research now demonstrates that diabetes, its complications, and its management can induce transient or permanent neurocognitive abnormalities that are largely a consequence of structural and functional changes occurring within the central nervous system (CNS). Although earlier writers attributed virtually all diabetes-related cognitive dysfunction to the adverse effects of severe and/or recurrent hypoglycemia on the CNS, beliefs about the underlying pathophysiology have shifted dramatically within the past 10 years. New evidence suggests that chronic hyperglycemia and its associated metabolic and vacular complications underlie most of the functional and structural CNS changes seen in many children and adults with diabetes mellitus. Delineating these neurocognitive phenotypes and identifying the metabolic and disease-related events that likely initiate brain dysfunction are the primary aims of this chapter.

31.2 Clinical Syndromes of Diabetes Mellitus

Understanding the differences between the two major forms of diabetes is critical for distinguishing diabetes-related neurocognitive phenotypes and for identifying both the putative risk factors and the biological mechanisms associated with their development.

31.2.1 Type 1 Diabetes

Type 1, or insulin-dependent, diabetes is usually diagnosed during childhood or adolescence, is characterized by autoimmune destruction of pancreatic beta cells, and is often signaled by a metabolic crisis (ketoacidosis) triggered by excessively high blood glucose levels (Daneman, 2006). Because of their complete inability to synthesize and release insulin endogenously, patients with type 1 diabetes must inject themselves several times a day with exogenous insulin and regulate both diet and exercise patterns to normalize carbohydrate metabolism. Too much insulin or a failure to balance self-administered insulin with food intake and activity level will induce acute hypoglycemia ($<3.5\ \text{mmol l}^{-1}$; $<65\ \text{mg dl}^{-1}$). In contrast, excessive carbohydrate consumption and/or inadequate insulin can lead to abnormally high ($>14\ \text{mmol l}^{-1}$; $>250\ \text{mg dl}^{-1}$) blood glucose levels.

Chronic hyperglycemia is associated with damage to both small and large blood vessels and greatly increases the risk of developing clinically significant diabetes-related biomedical complications (Donaghue et al., 2007; Pambianco et al., 2006). Microvascular damage affects the retina relatively early in the course of the disease and can lead to impaired vision or blindness (Antonetti et al., 2006). When it occurs within the glomerular loops of the kidney, diabetes-associated microangiopathy increases the risk of developing end-stage renal disease (DCCT/EDIC Research Group, 2003). Neuropathy is also a common early complication of type 1 diabetes. Patients may experience painful and/or reduced sensations in their extremities secondary to peripheral neuropathy (Said, 2007; Sugimoto et al., 2000), or may develop impotence, cardiac arrhythmias, or loss of urinary bladder sensation as a consequence of autonomic neuropathy (Nofzinger, 1997; Vinik et al., 2003; Vinik and Ziegler, 2007). Macrovascular disease tends to develop only after many years of diabetes and poor metabolic control and results in a greatly increased risk of heart attacks, stroke, and peripheral vascular disease – particularly in the individual with a history of hypertension, hypercholesterolemia, and/or smoking (Orchard et al., 2006).

Treatment is directed at maintaining good metabolic control by avoiding excessively high or low blood glucose levels. Diabetic patients' degree of metabolic control over the past 2–3 months can be estimated by measuring the glycosylated fractions of hemoglobin (hemoglobin A_{1c}; HbA1c) (Goldstein et al., 1995). The greater the degree and duration of chronic hyperglycemia, the higher this percentage, and the more out of control the patient. To avoid the vascular complications of diabetes, patients are now advised to maintain tight control and keep blood glucose levels as close as possible to the normal range (HbA1c $< 7\%$) by taking multiple daily injections of insulin, frequently monitoring blood glucose levels, and adjusting insulin doses accordingly (DCCT Research Group, 1993; DCCT/EDIC Research Group, 2002).

31.2.2 Type 2 Diabetes

Also known as non-insulin-dependent diabetes, type 2 diabetes was, until recently, predominantly a disease of middle-aged and elderly adults, although it is becoming increasingly common among obese adolescents (Shaw, 2007). In the US, more than 90% of all diabetes cases are of this form, which differs from type 1 diabetes in pathophysiology, etiology, and prevalence (Stumvoll et al., 2005). Onset is insidious over a number of years, with formal diagnosis often occurring only after the appearance of biomedical complications. Two different metabolic events underlie its development: the occurrence of insulin resistance followed by progressive impairment of beta cell function. Both reduce the bioavailability of insulin and ultimately lead to chronic hyperglycemia and its associated micro- and macrovascular complications. Insulin resistance is a loss of sensitivity to the glucose-lowering effects of insulin at peripheral cell receptor sites in muscle and liver. The pancreas initially compensates for the resulting increase in blood glucose levels by releasing more insulin, but over time, progressive deterioration in beta cell function and a parallel decline in insulin secretion ultimately results in chronically higher circulating blood glucose levels and a diagnosis of type 2 diabetes. Obesity, inactivity, and smoking are known to trigger insulin resistance in genetically susceptible individuals, although the exact physiological mechanisms underlying this process remain incompletely understood (Raz et al., 2005). Poor metabolic control has the same effect as previously described for those with type 1 diabetes: a greatly increased risk of stroke, heart attacks, kidney disease, blindness, neuropathy, and foot problems as a consequence of micro- and macroangiopathy secondary to chronic hyperglycemia.

31.3 Neurocognitive Phenotypes

Early neuropathological studies suggested that diabetes could eventuate in what has been characterized as a diabetic encephalopathy (Reske-Nielsen et al., 1965). In their seminal report of 16 poorly controlled, middle-aged adults who subsequently came to autopsy, Reske-Nielsen and associates noted profound neurological abnormalities and widespread neuropathology. After 17–36 years of diabetes, these patients showed evidence of cerebral angiopathy with enormously thickened walls in the arterioles and in the basement membrane of capillaries, accompanied by extensive gliosis in gray matter, basal ganglia, brainstem and cerebellum, as well as diffusely distributed degeneration of myelin sheaths and axon cylinders in the cranial nerves, optic chiasm, and white matter. All were blind, or nearly blind, and although all had neurological signs that ranged from reflex abnormalities to profound dementia, few details are available about the exact nature of their cognitive impairment because no formal neuropsychological assessment was conducted. The presence of both micro- and macrovascular disease made it impossible to determine whether the brain tissue abnormalities were due solely to ischemia secondary to vascular disease, or whether they were a direct effect of diabetes-associated metabolic abnormalities, including hypoglycemia.

Other researchers have also used the term diabetic encephalopathy occasionally (Dejgaard et al., 1991), but few have subsequently reported such profound brain damage associated with diabetes in young and middle-aged adults (Cotton et al., 2005). To some extent that may be a consequence of the dramatic improvements made in treating diabetes and reducing its vascular complications, although it may also reflect a failure, on the part of modern researchers, to enroll their most poorly controlled patients into research studies. Nevertheless, it is now apparent from more than 100 clinical and epidemiological studies that a subset of patients with diabetes may develop subclinical cognitive dysfunction characterized by mild to moderately severe neuropsychological deficits as well as abnormalities on structural and neurophysiological measures of CNS integrity. This cognitive dysfunction is rarely totally disabling from an academic or vocational perspective, but it appears to take the edge off many diabetic patients' level of functioning, rendering them somewhat less efficient mentally than would have been expected had they not developed diabetes. For those reasons, it has been suggested that the term diabetic encephalopathy be replaced by diabetes-associated cognitive decline (Mijnhout et al., 2006) – a term which is more akin to the concept of mild cognitive impairment as it has been used to characterize the functional changes occurring during the transition from normal aging to dementia (Gauthier et al., 2006; Van den Berg et al., 2005).

Diabetes is an unusual chronic disease insofar as it affects individuals across the life span, and its medical characteristics and neurocognitive manifestations vary according to the age of the patient, as do the biomedical risk factors associated with these cognitive complications (Biessels et al., 2008; Ryan, 1997). Studies of young and middle-aged adults with a childhood onset of type 1 diabetes illuminate the possible role played by micro- and macrovascular disease, as these individuals begin to manifest multiple complications. On the other hand, studies of children with type 1 diabetes provide insights into the effects of diabetes and its treatment on the developing nervous system, in the absence of clinically significant biomedical complications. Because type 2 diabetes is most common in individuals over the age of 60 who are beginning to show normal age-related neurocognitive changes and who are also developing disorders, like hypertension, that may independently disrupt cognition (Reitz et al., 2007; Waldstein et al., 2005), studies of this population yield information on possible interrelationships among vascular, glycemic, and neurodegenerative variables (Drachman, 2006). For each of these three patient groups one can identify neurocognitive phenotypes based on neuropsychological, electrophysiological, cerebrovascular, and neuroimaging evaluations.

31.3.1 Adults with Type 1 Diabetes

31.3.1.1 Cognitive manifestations

A distinctive, circumscribed pattern of mild cognitive dysfunction characterizes adults with type 1 diabetes. A systematic meta-analysis of data from 31 studies published in English between 1980 and 2004 compared the neuropsychological test scores of diabetic and nondiabetic adults on seven broad cognitive domains (general intelligence, language, attention, learning and memory, psychomotor speed, cognitive flexibility, and visual perception) (Brands et al., 2005). Effect sizes (Cohen's d) were also computed by dividing the between-group difference by the pooled standard (Cohen, 1988). Diabetic subjects, who were 18–50 years of age and in relatively good health, performed significantly more poorly on measures of intelligence, attention, psychomotor speed, cognitive flexibility, and visual perception (see **Table 1**). In contrast, no differences were found on measures of language or learning and memory. This work makes three notable points. First, differences in cognition between diabetic and nondiabetic adults are modest, at best, with effect sizes ranging from 0.3 to 0.8 standard deviation units. Second, not all cognitive domains are affected. Learning and memory skills, which are generally considered to be a reliable marker of early brain dysfunction (Winblad et al., 2004), are preserved in this patient population, despite 20 or more years of diabetes. Third, with only one exception (crystalized intelligence), most of the tasks on which diabetic patients perform worse than their nondiabetic counterparts – including tests requiring visual perception, attention, mental flexibility, or fluid intelligence – require rapid responding. That is, mental slowing appears to be the fundamental deficit associated with type 1 diabetes in adulthood (Ryan, 2005).

Table 1 Cognitive characteristics of adults with type 1 diabetes, based on a meta-analysis of published papers

Domain	*Effect size*	*Significance*	*Total N*	*Studies*
Overall cognition	0.40	$p < 0.001$	660	16
Intelligence				
Crystalized	0.80	$p < 0.01$	276	5
Fluid	0.50	$p < 0.01$	168	4
Language	0.05	NS	144	4
Attention				
Visual	0.40	$p < 0.001$	195	5
Sustained	0.30	$p < 0.01$	217	3
Learning and memory				
Working memory	0.10	NS	244	8
Verbal learning	0.20	NS	204	5
Verbal delayed memory	0.30	NS	157	3
Visual learning	0.10	NS	187	5
Visual delayed memory	0.10	NS	157	4
Psychomotor speed	0.60	$p < 0.05$	368	8
Cognitive flexibility	0.50	$p < 0.001$	364	9
Visual perception	0.40	$p < 0.001$	202	5

Standardized effect sizes (Cohen's *d*) for each cognitive domain reflect differences between diabetic and nondiabetic patients.
Adapted from Brands AMA, Biessels G-J, De Haan EHF, Kappelle LJ, and Kessels RPC (2005) The effects of type 1 diabetes on cognitive performance: A meta-analysis. *Diabetes Care* 28: 726–735.

The single published study of older adults with type 1 diabetes found similarly circumscribed results (Brands et al., 2006). Despite being more than 60 years old on average, and having diabetes for more than 34 years, diabetic patients performed as well as demographically similar nondiabetic subjects on measures of abstract reasoning, attention and executive function, and learning and memory. Impairments were restricted to tasks requiring rapid information processing and visuoconstructional skills, and the magnitude of those effects was, again, modest ($d = 0.3$–0.5). Although more studies of older adults with type 1 diabetes are needed, extant research finds no support for the view that global cognitive deterioration is common in the older adult who has experienced many years of chronic hyperglycemia and has developed multiple biomedical complications (Brands et al., 2006; Ryan, 2005).

31.3.1.2 Electrophysiological changes

Slowed neural processing, as demonstrated by measures of brainstem auditory evoked and visual-evoked potentials (BAEPs; VEPs), event-related potentials (ERPs), and electroencephalographic (EEG) recordings, is frequently seen in adults with type 1 diabetes. Those findings are consistent with the psychomotor slowing noted on neuropsychological assessment. Moreover, because they are seen most frequently in diabetic adults who are in poor metabolic control and/or manifest peripheral neuropathy, the presence of neural slowing has been interpreted as evidence that diabetes induces a central neuropathy that parallels the peripheral neuropathy that occurs relatively early in the course of diabetes (Dejgaard et al., 1991; Ryan et al., 1992).

Both animal (Biessels et al., 1999; Manschot et al., 2003) and human studies (Durmus et al., 2004; Lingenfelser et al., 1993; Virtaniemi et al., 1993; Ziegler et al., 1991) using BAEP measures have found differences between diabetic and nondiabetic subjects to be greatest in the latencies of wave III (indicative of activity in the superior olivary complex and the lateral lemniscus) or wave V (generated in axons and nuclei of the lateral lemniscus (Di Leo et al., 1997)). The most reliable measure of central transmission (wave interpeak I–V) is often (Durmus et al., 2004; Virtaniemi et al., 1993) but not invariably (Di Leo et al., 1997), prolonged in diabetic adults, as compared to healthy controls. Although effect sizes tend to be medium ($d = 0.6$–0.7), the actual degree of slowing is quite modest, ranging between 2% and 10% (Lingenfelser et al., 1993; Virtaniemi et al., 1993), unless subjects have clinically significant complications. Under those circumstances, effect sizes can be quite large ($d > 1.6$); diabetic subjects with complications have latencies that are between 13% (wave V) and 19% (wave III) longer than those without complications (Bayazit et al., 2000).

VEP latencies are also significantly longer in adults with type 1 diabetes (Parisi and Uccioli, 2001), and the presence of microvascular complications like retinopathy (Parisi et al., 1997) or neuropathy (Gregori et al., 2006; Várkonyi et al., 2002) is associated with additional slowing. To some extent, this slowing may be modifiable: strict metabolic control for even a very brief period of time (3 days) was found to reduce, but not completely normalize, VEP latencies in one small study (Ziegler et al., 1994a). On the other hand, a 10-day period of intensified insulin therapy with corresponding improvements in metabolic control did not lead to any meaningful reductions in BAEP latencies in a small group of diabetic young adults, suggesting that factors other than metabolic control (e.g., development of microvascular complications) may contribute to the slowed neural transmission in auditory pathways (Virtaniemi et al., 1995).

Electroencephalograms have been recorded only infrequently in adults with diabetes, but one recent study demonstrated widely distributed neural slowing when the EEG is measured at rest during euglycemia (Brismar et al., 2002). Despite being in good metabolic control and having no clinically significant complications, young adults with diabetes had abnormal EEG power spectra indicative of a loss of fast oscillations (alpha, beta, and gamma activity) in the temporal regions, with decreases in beta power also evident in the anterior temporal and occipital regions. Increases in slow wave delta–theta activity were also present and were especially prominent in frontal regions. Although no relationship was found between a past history of hypoglycemia and any EEG parameter, correlations between decreases in peripheral nerve conduction velocities and increases in slow activity (particularly in the parieto-temporal region) were found, consistent with the central neuropathy hypothesis. Unfortunately, because neuropsychological assessments were not conducted, it is not possible to determine whether those individuals with the greatest degree of EEG slowing were also those who were slowest on cognitive tasks requiring rapid information processing.

Event-related potentials have also been used to assess CNS integrity in adults with type 1 diabetes (Pozzessere et al., 1991). This task measures cortical changes that occur when the brain is activated during the performance of a complex cognitive task (e.g., paying attention to an infrequently occurring stimulus) and hence is quite different from either EEG or evoked potential paradigms where neural measures reflect either a resting state or a response to a simple sensory stimulus. In the largest study to date, P300 waveforms were recorded from 108 diabetic adults – approximately half of whom had previously experienced one or more hypoglycemic comas, while the remainder had no serious hypoglycemic events (Kramer et al., 1998). While there was no difference in P300 latency (or amplitude) between the two patient subgroups, the diabetic subjects as a whole tended to have P300 latencies that were significantly longer (344 ± 22 ms vs. 332 ± 21 ms) than the healthy reference group. According to those authors, a delay of this magnitude is comparable to approximately 10 years of normal aging and is indicative of subclinical CNS dysfunction – somewhat surprising, they argue, given the otherwise good health and excellent metabolic control of their diabetic subjects. Duration of diabetes was the strongest predictor of longer latencies; on the other hand, there was no effect of severe hypoglycemia on brain function.

31.3.1.3 Cerebrovascular outcomes

Cerebral blood flow (CBF), especially in the frontal and frontotemporal brain regions, is altered significantly by type 1 diabetes, with both decreases and increases in degree of perfusion (Jiménez-Bonilla et al., 2001; Keymeulen et al., 1995; MacLeod et al., 1994; Vázquez et al., 1999). In one of the largest studies to use single photon emission computerized tomography (SPECT), 82% of the middle-aged sample of diabetic subjects manifested hypoperfusion in one or more regions of interest, compared to only 10% of controls; similarly, 58% of the diabetic subjects manifested hyperperfusion, compared to 20% of controls (Quirce et al., 1997). The SPECT results indicated that essentially all brain regions showed changes in perfusion, with the greatest effects occurring in the cerebellum, frontal, and frontotemporal regions. None of those subjects had significant cardiovascular complications of diabetes but many had evidence of microvascular disease (most notably retinopathy), leading the authors to conclude that this cerebrovascular damage occurs in association with the microvascular complications of diabetes.

Additional support for that view comes from a small study that used positron emission tomography (PET) to measure cerebral glucose metabolism in young adults newly diagnosed with type 1 diabetes, and in diabetic patients with and without peripheral neuropathy (Ziegler et al., 1994b). Only those patients with complications showed significant reductions in regional cerebral metabolism rate (rCMR), although there was a significant relationship between rCMR, diabetes duration and age, but not HbA1c. Although BAEPs and ERPs were also assessed, no relationship between rCMR and any electrophysiological measure was noted, but subgroup sample sizes were extremely small.

Cerebrovascular reactivity, in response to an acetazolamide challenge, is also reduced in adults with type 1 diabetes. This effect appears to be limited to those with a longer duration of diabetes (>10 years) and higher rates of microvascular complications. In contrast, the middle cerebral artery blood flow velocities of diabetic patients with a shorter disease duration were comparable to those of healthy control subjects (Fülesdi et al., 1997). Measures of cerebrovascular reserve capacity were also reduced by nearly half in the long-duration diabetic subgroup as compared to the other two groups. These findings provide additional support for the view that diabetes is associated with pathological changes in the cerebral microvasculature – particularly in the brain resistance arterioles – and the responsible pathophysiological mechanisms are likely similar or identical to those processes underlying the development of diabetic microvascular complications.

31.3.1.4 Brain structure anomalies

Adults with a childhood onset of diabetes have marked reductions in brain matter density, as demonstrated by voxel-based morphometry (VBM) – a highly sensitive, semi-automated quantitative MRI-based methodology to evaluate brain structure. Compared to demographically similar healthy control subjects, diabetic young adults showed an asymmetrical reduction in gray matter density, with significantly less gray matter in the left temporal gyrus, left angular gyrus, left medial frontal gyrus, left inferior parietal lobule, and left thalamus (Musen et al., 2006). Only one region (superior temporal gyrus) was affected in the right hemisphere. The magnitude of these effects was modest, with an approximately 4–5% reduction in gray matter density. These regions are especially important for memory, language processing, and attention. Results from neuropsychological testing revealed that although the diabetic sample earned scores within the normal range, they performed significantly more poorly than controls on tests of vocabulary knowledge, verbal learning, and attention and executive functioning, as one might expect, given the pattern of structural neuroimaging results. The strongest predictor of gray matter density was degree of chronic hyperglycemia: higher lifetime HbA1c values were consistently correlated with lower gray matter density values in multiple brain regions, as was retinopathy severity. Higher frequency of severe recurrent hypoglycemic events was associated with reduced gray matter density in a single brain region, the cerebellum, which has been linked to higher-order executive cognitive processes (Schmahmann and Sherman, 1998).

Using a similar methodology, a small case/control study has provided additional support for the view that clinically significant microangiopathy predicts reductions in gray matter density. Patients with diabetic proliferative retinopathy had modest, but statistically reliable reductions in gray matter density in the left middle frontal gyrus, the right inferior frontal gyrus, the right occipital lobe, and the left cerebellum; those without retinopathy had gray matter values that were comparable to a nondiabetic comparison group (Wessels et al., 2006b). These authors speculate that the reduction in cortical density may be a consequence of microinfarctions because several of the affected regions (inferior frontal gyrus, occipital lobe) are watershed areas of the medial and posterior cerebral arteries and are particularly vulnerable to ischemia in individuals with vascular systems that are compromised secondary to hyperglycemia-induced microangiopathy.

When VBM techniques were used to estimate white matter fraction in a small sample of diabetic patients, not only did those with clinically significant retinopathy have less white matter volume than those without retinopathy, but also there was a strong relationship between degree of white matter loss and performance on measures of attention, executive function, and speed of information processing (Wessels et al., 2007). On the other hand, clinically meaningful white matter disease was absent in a very large cohort of young adults with type 1 diabetes (Weinger et al., 2008) – many of whom had previously been found to have a reduction in gray matter density (Musen et al., 2006), but those null results may have been a consequence of a relatively insensitive qualitative rating system to measure white matter lesions and/or the presence of minimal microangiopathy (van Harten et al., 2006).

Other neuroimaging studies have found either no MRI abnormalities or very small effects, but because they did not use VBM, these differences may reflect variations in methodology and/or limited statistical power because of much smaller study samples. One study of middle-aged patients with microvascular complications showed evidence of mild cerebral atrophy (~3.1% reduction in volume after adjusting for intracranial vault size), but there was no evidence of hippocampal atrophy nor any clinically significant white matter disease (Lobnig et al., 2005). In contrast, a study of older adults with type 1 diabetes found rates of cortical and subcortical atrophy and white matter lesions that were comparable to healthy comparison subjects (Brands et al., 2006). Given the data reported by Musen and her colleagues in their study of young adults, these latter findings are quite surprising, since Brands and her colleagues evaluated individuals who were significantly older (M = 61 vs. 32 years), and had a longer duration of diabetes (M = 34 vs. 20 years). It may be that in older adults, the presence of CNS changes associated with the process of normal aging impedes researchers' ability to identify brain anomalies that can be attributed specifically to diabetes (Drachman, 2006).

31.3.1.5 Alterations in brain metabolites

Proton magnetic resonance spectroscopy (^{1}H-MRS) has provided another means of evaluating the integrity of the CNS and has successfully demonstrated the extent to which cerebral metabolism is adversely affected by diabetes of long duration. Compared to healthy control subjects, middle-aged diabetic adults with retinopathy manifested significant increases in choline-containing compounds (Cho – a marker of myelin metabolism and other phospholipid cell membranes) that were particularly evident in white matter and thalamus. Myo-inositol (mI – a marker of glial proliferation or activation) was also increased in white matter, whereas levels of glucose were increased in all measured brain locations (Mäkimattila et al., 2004). Correlational analyses demonstrated that duration of poorer metabolic control (HbA1c months) was inversely correlated with levels of *N*-acetylaspartate and -glutamate (NAA; NAG – markers of neuronal integrity) and Cho in white matter, and Cho in deep gray matter. These findings have been interpreted as evidence of axonal injury, demyelination, and other neuronal pathology, as well as increased membrane proliferation secondary to gliosis that may be mediated by cerebrovascular changes secondary to chronic hyperglycemia. On the other hand, an earlier study of diabetic adults failed to find evidence of cerebral metabolic abnormalities despite high rates of cortical atrophy, but because that study focused on the effects of recurrent hypoglycemia, patients with significant microvascular complications had been excluded (Perros et al., 1997).

31.3.2 Children and Adolescents with Type 1 Diabetes

31.3.2.1 Cognitive manifestations

Neuropsychological profiles of children with diabetes differ, depending on the age when diabetes was diagnosed. Those diagnosed early in life, within the first 7 years, have a greatly elevated risk of manifesting severe cognitive deficits that affect a broad array of cognitive domains (Biessels et al., 2008; Northam et al., 2001, 2006; Rovet et al., 1987; Ryan et al., 1985b; Ryan, 2006b). In contrast, those diagnosed after that early critical period show very mild and highly circumscribed dysfunction that is virtually identical to what has been described for adults with type 1 diabetes. That is, children and adolescents with a later onset of diabetes earn somewhat lower scores than their nondiabetic counterparts on measures of intelligence, sustained attention, visuoperceptual skills, and psychomotor speed and cognitive flexibility (Desrocher and Rovet, 2004; Kaufman et al., 1999; Perantie et al., 2008). Academic achievement tends to be significantly poorer in children with diabetes, as measured by formal tests (Kaufman et al., 1999; Kovacs et al., 1992; McCarthy et al., 2003; Rovet et al., 1993; Ryan et al., 1985a), as well as by school grades (Dahlquist et al., 2007). As yet, there have been no published meta-analyses of these pediatric findings, but effect sizes are modest at best (e.g., $d = {\sim}4$ for verbal intelligence (Perantie et al., 2008); $d = {\sim}5$ for psychomotor speed (Ryan et al., 1984)) and are comparable to what has been reported for adults with type 1 diabetes (Brands et al., 2005).

Unlike adults with type 1 diabetes, children and adolescents with a later onset of disease may manifest mild memory impairments, but this finding remains controversial (Ryan, 1999). One group has sometimes (Hershey et al., 1999, 2003), but not invariably (Perantie et al., 2008), found later onset diabetic subjects performing more poorly on certain types of memory tasks, particularly those that have significant spatial component. In general, when memory dysfunction is found, it tends to be quite limited in scope (Kaufman et al., 1999).

On the other hand, diagnosis of diabetes early in life is associated with more severe cognitive dysfunction affecting virtually all cognitive domains, although the specific pattern of impairment varies somewhat, depending on the child's age at assessment. Younger children show performance decrements primarily on visuospatial tasks (e.g., copying complex designs, solving jigsaw puzzles, or reproducing patterns with blocks), and this is more common in girls than in boys (Rovet et al., 1987). By adolescence, impairments are evident in all cognitive domains, with boys and girls equally affected on tests of attention, mental efficiency, learning, memory, problem solving, eye–hand coordination, and general intelligence (Northam et al., 2001; Ryan et al., 1985b; Ryan, 2006b). Note that deficits on measures of learning and memory, and problem solving are far more prominent in children with an earlier, as compared to a later, onset of disease. Children and adolescents with an earlier onset of diabetes also have difficulty focusing attention (Hagen et al., 1990) and in selecting the correct target from among several similar choices on a complex visual attention test (Rovet and Alverez, 1997). In contrast, their use of other information-processing strategies (e.g., effective short-term memory rehearsal strategies) appears to be normal (Wolters et al., 1996).

Not all adolescents diagnosed early in life show cognitive deficits, but according to one estimate, nearly 25% of the adolescents diagnosed before 5 years of age meet criteria for clinically significant impairment, as compared to only 6% of adolescents diagnosed after age 5, or to 6% of a healthy community sample (Ryan et al., 1985b). Moreover, when followed over time, they are more likely to show elevated rates of delayed cognitive development – particularly on measures of intelligence (Northam et al., 2001; Schoenle et al., 2002). In a number of studies, investigators have stratified their samples in terms of severe hypoglycemic events (Bjørgaas et al., 1997; Hannonen et al., 2003) rather than age at onset, and have attributed cognitive dysfunction in these youngsters to severe hypoglycemia. However, severe hypoglycemia and onset age are almost invariably confounded. Both epidemiologic (Barkai et al., 1998) and cross-sectional (Lteif and Schwenk, 1999) studies demonstrate that children or adolescents who developed diabetes in the first 5–6 years of life also have a greatly increased rate of severe hypoglycemia as compared to those with a later onset of diabetes (e.g., 45% vs. 13%). Risk of severe hypoglycemia is increased further when more intensive therapeutic regimens are used to lower HbA1c values (Jones and Davis, 2003), and this effect is magnified in those with an earlier onset of diabetes (Wagner et al., 2005).

This age at onset phenomenon is not limited to children or adolescents but can be seen in adults diagnosed with diabetes early in life. Young adults who developed diabetes before 7 years of age earned lower performance IQ scores, and performed more poorly on cognitive tests assessing information-processing speed when compared to diabetic patients diagnosed after that age (Ferguson et al., 2005). Of even greater significance is the observation that an early onset of diabetes was associated with abnormalities in brain structure. MRI scans showed higher rates of mild-to-moderate ventricular atrophy (61% vs. 20%), as well as somewhat higher rates of small punctate white matter lesions within the hippocampus (14% vs. 2%). Brain volume was correlated with cognitive test performance, suggesting that an early onset of diabetes disrupts cognition by interfering with normal brain development.

Neurocognitive abnormalities emerge relatively early in the course of diabetes and may be detectable within 2–3 years of diagnosis. In the largest, longest prospective pediatric study to date, a representative sample of 90 newly diagnosed diabetic youngsters and 84 healthy children drawn from the community were followed over a 6-year period. No between-group differences were evident at study entry (Northam et al., 1995), but 2 years later, those children diagnosed before age 4 manifested a developmental delay insofar as their scores on both the Wechsler Vocabulary and Block Design subtests improved less over time, relative to either children with a later diabetes onset or to community control subjects (Northam et al., 1998). After 6 years of follow-up, diabetic children – regardless of age at diagnosis – performed worse than their nondiabetic peers on measures of intelligence, attention, processing speed, long-term memory, and executive skills. Children with an early age at onset were particularly affected and performed significantly worse on measures of attention and executive function as compared to those with a somewhat later onset of diabetes (Northam et al., 2001). Whether these children with an early onset of diabetes are merely experiencing developmental delays and will subsequently catch up during adolescence, or whether their performance will continue to deteriorate throughout adolescence and early adulthood, remains to be determined.

Most other studies assessing neurocognitive outcomes in diabetic children have not enrolled subjects at, or shortly after, diagnosis and so it has not been possible to determine unequivocally when diabetes-related CNS changes begin to occur. Nevertheless, those few studies that followed newly diagnosed children over time have found early changes on measures of psychomotor speed (Hershey et al., 1999) and intelligence (Kovacs et al., 1992; Schoenle et al., 2002). There is no comparable information currently available on the emergence of cognitive dysfunction in adults with either type 1 or type 2 diabetes.

31.3.2.2 Electrophysiological changes

Slowed neural activity – at rest and in response to sensory stimuli – is frequently seen in diabetic children. Relative to healthy comparison subjects, children and adolescents with at least a 2-year history of diabetes had greatly increased BAEP and VEP latencies (Seidl et al., 1996). Not only were effect sizes large (Cohen's $d = 1–1.2$), but for 37% of the sample, results were considered to be clinically pathological as well. In contrast, children with a briefer duration of diabetes had latencies comparable to nondiabetic controls. Subsequent research with somewhat older diabetic adolescents has demonstrated that abnormalities may be present shortly after diagnosis, but can be reversed with reductions in extent of chronic hyperglycemia (Verrotti et al., 2000). At diagnosis, HbA1c levels were 9.4% and P100 VEP latencies were significantly delayed. After 6 months of intensive treatment, HbA1c fell to 7.2% and VEP parameters were normalized. These very rapid changes in central neural transmission efficiency are most likely a consequence of acute metabolic changes within brainstem pathways that are secondary to a reduction in systemic glucose levels. Because the subjects were not followed over an extended period of time, however, it is impossible to determine whether after 2 or 5 years of diabetes, their VEP latencies would increase, relatively to healthy controls, as reported by Seidl et al. (1996).

Electroencephalography has also demonstrated evidence of altered neural activity during rest. Compared to age-matched healthy control subjects, diabetic adolescents in very good metabolic control (HbA1c = 6.7%) showed a marked increase in slow activity (delta and theta) and a reduction in alpha peak frequency, with these changes greatest in frontal regions (Hyllienmark et al., 2005). Both of those central changes were associated with peripheral nerve conduction slowing. Fast activity (alpha, beta, gamma) was also decreased, with the greatest declines seen bilaterally in posterior temporal regions. This pattern of results is remarkably similar – both in distribution and in magnitude – to what was previously described in well-controlled diabetic adults (mean HbA1c = 6.9%) who were studied using an identical methodology, but were significantly older (range: 21–41 years of age), with a longer duration of diabetes (Brismar et al., 2002). In the pediatric study, a history of severe hypoglycemia was correlated with poor metabolic control, and both were associated with significant increases in slow activity and decreases in alpha peak frequency. These results are consistent with data from other studies of children (Bjørgaas et al., 1996) and adults (Howorka et al., 2000). In contrast, no association was found between any biomedical parameter and the reduction in power of the faster frequencies. Because there was also no relationship between these posterior temporal fast activity changes and the more anterior increases in delta/theta power, the authors suggest that there may be two, qualitatively different, types of EEG abnormalities in diabetic children, with only the more anterior system clearly linked to diabetes-related biomedical variables.

Clinically significant EEG abnormalities are also common in children and adolescents with diabetes. In one study, 26% of the diabetic children showed pathological EEG patterns, as compared to 7% of the healthy control subjects (Soltész and Acsádi, 1989). Abnormalities were most likely in whose children with an earlier age at onset and multiple episodes of severe hypoglycemia. Similar results have been reported in a number of earlier studies (Eeg-Olofsson, 1977; Gilhaus et al., 1973; Haumont et al., 1979).

Although most writers have considered the occurrence of severe hypoglycemia to trigger EEG abnormalities, the obverse may also be true: children who have EEG abnormalities at the time of diabetes diagnosis may be more likely to subsequently experience one or more episodes of severe hypoglycemia. In a study where all pediatric patients received a routine EEG recording shortly after diagnosis (median age = 7.5 years), those who later experienced a hypoglycemic seizure or coma and had a follow-up EEG (median age = 13.3 years) were more likely to manifest EEG abnormalities at diagnosis, as compared to those who did not experience hypoglycemia but were assessed over the same time period (22% vs. 3%; Tupola et al., 1998). The etiology of these abnormal baseline EEGs remains poorly understood, but they could have a genetic or perinatal origin, rather than being due to diabetes-related metabolic disturbances. These very intriguing data suggest that having a preexisting brain injury or anomaly, as indexed by an abnormal EEG recording, may increase an individual's vulnerability to subsequently manifesting a severe neurological response (i.e., seizure or coma) during a hypoglycemic event.

31.3.2.3 Cerebrovascular outcomes

SPECT has been used in a single study to measure cerebral blood flow in children with diabetes. Compared to a small group of healthy controls, children with diabetes manifested lower levels of cerebral perfusion bilaterally, particularly in basal ganglia and frontal regions. Parietal and temporal regions were also affected, but to a lesser extent (Salem et al., 2002). This regional hypoperfusion pattern is similar to that previously reported in adults with type 1 diabetes (Quirce et al., 1997). Although several adult studies also noted relationships between degree of hypoperfusion and duration of diabetes (Rodriguez et al., 1993) or HbA1c (Keymeulen et al., 1995), no such association was found in this pediatric study, but disease duration was relatively brief (mean = 7 years). A neuropsychological assessment was also undertaken but there were no obvious relationships between cerebral hypoperfusion and cognitive performance. These data suggest that diabetes leads to reductions in regional cerebral blood flow in children and adolescents, but those cerebrovascular changes appear to be subclinical, and may have little impact on cognition.

31.3.2.4 Brain structure anomalies

Despite the very extensive neuroimaging literature on adults with diabetes (van Harten et al., 2006), there has been relatively little research on the structural integrity of the CNS in diabetic children. In the largest study to date, MRI scans were acquired from 108 diabetic children, 7–17 years of age and from 51 age-matched healthy nondiabetic sibling control subjects, and VBM was used to examine differences in gray and white matter volumes (Perantie et al., 2007). Unlike diabetic adults who showed small reductions in gray matter density (Musen et al., 2006), children with diabetes had gray and white matter volumes that were comparable to healthy controls. Nevertheless, subgroup analyses of the diabetic subjects showed that both severe hypoglycemia and chronic hyperglycemia were associated with regional differences in gray and white matter density. Children who experienced one or more episodes of severe hypoglycemia had smaller gray matter volumes in the left temporal–occipital region, as compared to diabetic subjects without severe hypoglycemia. Whether the left hemisphere is particularly vulnerable to hypoglycemia remains to be determined, but marked lateralization has been noted in several reports (Auer et al., 1989; Chalmers et al., 1991; Foster and Hart, 1987; Jarjour et al., 1995; Wayne et al., 1990).

Less cortical volume was also associated with chronic hyperglycemia in children (Perantie et al., 2007). This was particularly common in right posterior regions, including the right cuneus and precuneus – two areas also found to be sensitive to higher lifetime HbA1c values in adults (Musen et al., 2006). Higher HbA1c values were also correlated with less white matter volume in the right superior parietal region; a similar, but weaker relationship, was noted in the left superior parietal regions (Perantie et al., 2007). Somewhat paradoxically, increases in gray matter volumes were evident in the right prefrontal region. Since gray matter normally decreases in this area during later childhood and early adolescence (Shaw et al., 2006), this observation may reflect a delay in normal neurodevelopmental processes. Despite indications that both hyperglycemia and severe hypoglycemia may be associated with regional changes in gray and white matter in children with diabetes, these effects appear to be quite subtle: as a group, the diabetic subjects did not differ from their nondiabetic counterparts. The functional implications of these findings also remain unknown insofar as the authors have not reported relationships between these neuroimaging data and neuropsychological test performance.

A study limited to children who developed diabetes early in life (before 6 years age) has also noted structural changes, including a greatly elevated prevalence of clinically significant brain abnormalities (Ho et al., 2008). Within that entire diabetic sample (mean age = 10.0 years; mean age at onset = 3.0 years), 29% showed structural abnormalities, the most common being mesial temporal sclerosis (MTS), which was found in 16% of the patients. Although a healthy comparison group was not included, the rate of MTS in the nondiabetic pediatric population is extremely low – 0.77% (Ng et al., 2006). A past history of severe hypoglycemia was unrelated to the occurrence of MTS, but did predict cerebral gray and white matter volume. Subjects who had at least one severe hypoglycemic seizure or coma had gray matter volumes that were markedly smaller than those with no severe hypoglycemia (724 cm^3 vs. 764 cm^3; $d = 0.5$), regardless of when the severe hypoglycemic event occurred – that is, before or after the age of 6. Reduced white matter volumes were also associated with severe hypoglycemia, but the timing of the hypoglycemic event was critical. Those who experienced their first seizure before 6 years of age had smaller white matter volumes than those whose seizures occurred after that age (490 cm^3 vs. 531 cm^3; $d = 0.6$), whereas that latter group did not differ from those without a history of severe hypoglycemia (white matter volume: 518 cm^3). Whether structural changes of this magnitude would also be evident in children with a later onset of diabetes cannot be determined, because by design, this study was restricted to children with an early disease onset. It is also noteworthy that these structural changes, like those reported by Perantie et al. (2007), appeared relatively soon after diagnosis. The mean disease duration in this sample was approximately 7 years; in the Perantie study, it was 6 years.

31.3.2.5 Alterations in brain metabolites

Consistent with evidence of early structural CNS damage are data demonstrating that brain metabolites are altered to some extent in children with diabetes. Compared to healthy control subjects, diabetic children in very poor metabolic control (mean HbA1c = 11.9%) showed reductions in NAA in two brain regions – pons and posterior parietal white matter, with reduced choline values also evident in the pons. Brain structure, as assessed by MRI, was within normal limits, although more sensitive VBM techniques were not used (Sarac et al., 2005). Similar reductions in NAA have been reported previously in parietal white matter of diabetic adults in poor metabolic control (Kreis and Ross, 1992) and in hyperglycemic rats (Biessels et al., 2001). The implications of these findings are, unfortunately, complicated by the growing body of literature indicating that neurometabolic profiles may be exquisitely sensitive to transient metabolic derangements. For example, following an episode of severe hypoglycemia, three children with diabetes had reductions in NAA in frontal and temporal regions, and in the basal ganglia. Six months later, there was a marked increase in these values, although they were still somewhat lower than values from healthy comparison subjects (Rankins et al., 2005). Whether similar transient changes might occur following prolonged hyperglycemia, with or without ketoacidosis, remains to be determined.

31.3.3 Adults with Type 2 Diabetes

31.3.3.1 Cognitive manifestations

Neurocognitive research on type 2 diabetes was motivated initially by the belief that diabetes is a form of accelerated aging (Kent, 1976). Because many of the degenerative disorders associated with the normal aging process appear at a somewhat younger age in adults diagnosed with type 2 diabetes, it seemed reasonable to evaluate cognitive function in the diabetic adult, given the sensitivity of processes like attention, learning, memory, and psychomotor efficiency to normal age-related decline (Verhaeghen and Salthouse, 1997). Early studies found unequivocal evidence that older adults (e.g., 65+ years of age) with diabetes manifested marked difficulty learning new information and retaining it over a period of time, as compared to their nondiabetic peers (Mooradian et al., 1988; Perlmuter et al., 1984; Reaven et al., 1990). These observations have been replicated subsequently by numerous investigators (for reviews, see Allen et al. (2004), Awad et al. (2004), Stewart and Liolitsa (1999), and Strachan et al. (1997)), and it is now clear that in addition to learning and memory deficits, older adults with type 2 diabetes perform more poorly than nondiabetic peers on measures of attention, abstract reasoning and problem solving, and psychomotor efficiency. It is the presence of the mnestic deficits, however, that best differentiates the type 2 cognitive phenotype from that characteristic of type 1 diabetes.

Cognitive effect sizes are modest, at best, and similar to what has been reported in studies of adults with type 1 diabetes. A recent study comprised of 119 patients with type 2 diabetes (mean age = 66 years; mean diabetes duration = 8.7 years; mean HbA1c = 6.9%) and 55 age-matched healthy comparison subjects (Brands et al., 2007b) found that the diabetic subjects performed worst on measures of information processing speed ($d = 0.50$) and memory ($d = 0.43$), followed by tests of attention and executive function ($d = 0.37$). No statistically significant effects were evident on visuoconstructional ($d = 0.22$) or abstract reasoning ($d = 0.20$) tasks. Poorer cognitive function was associated with higher HbA1c values, but neither duration of diabetes nor severity of microvascular complications (peripheral neuropathy) influenced performance (Manschot et al., 2008).

Both the pattern and the magnitude of these cognitive test results are analogous to that characteristically seen in normal aging (Tisserand and Jolles, 2003; Verhaeghen and Salthouse, 1997) and are consistent with the hypothesis that type 2 diabetes, at least in older adults, may be associated with a premature aging of the brain. Additional evidence of premature aging that may ultimately eventuate into dementia comes from a series of prospective studies in which older diabetic subjects were followed for 2–8 years. Relative to nondiabetic adults, diabetic patients showed a greater rate of decline in cognitive function, as measured by tests of psychomotor efficiency (Fontbonne et al., 2001). Although there is no complete agreement, many studies have also noted an increased risk of dementia that typically range from 1.2 to 2.3 for Alzheimer's disease and 2.2 to 3.4 for vascular dementia (for reviews, see Biessels et al.

(2006), Cole et al., (2007), Cukierman et al., (2005), and Haan (2006)). In addition to affecting cognitive test performance, diabetes is associated with poorer quality of life. Rates of functional disability (e.g., ability to do housework efficiently; walk two to three blocks) are doubled, with a yearly incidence of 9.8% in older women with diabetes, as compared to 4.8% in those without diabetes (Gregg et al., 2002), and poorer performance of activities of daily living in homebound adults was correlated with scores on cognitive measures of visuospatial and executive function (Qiu et al., 2006).

Elderly adults with type 2 diabetes (mean age = 73 years; mean duration = 13.8 years; mean HbA1c = 7.7) show a similar pattern of results, with similar effect sizes (all approximately 0.4) on measures of executive functioning, memory, and mental and motor speed (van Harten et al., 2007). However, after statistically adjusting for blood pressure values, the memory effects now failed to reach the conventionally accepted level of statistical significance. This finding is not unexpected, given the fact that in older adults, hypertension influences memory functions independently (Korf et al., 2004; Raz et al., 2003; Reitz et al., 2007; Waldstein et al., 2005) and may also act synergistically with elevated blood glucose levels to disrupt cognition in adults with type 2 diabetes (Elias et al., 1997; Hassing et al., 2004).

Learning and memory deficits also appear to be absent in very old adults with diabetes. Results from a community-based cohort first studied at the age of 85 years and then followed for 5 years demonstrated that cognitive impairments were limited to tasks requiring mental flexibility and psychomotor speed (van den Berg et al., 2006). Cognitive function was not related to either HbA1c or blood pressure, but baseline cognition scores were associated with a past history of stroke. Both diabetic and nondiabetic subjects showed marked age-related declines over time, but there was no evidence of an accelerated rate of cognitive decline in the diabetic subjects. Their failure to find a robust relationship between diabetes and cognition in the very old may reflect measurement issues (e.g., floor effects), selective survival, high rates of comorbid conditions in both diabetic and nondiabetic subjects (e.g., marked vascular disease), or possible protective effects of elevated blood glucose values that could be analogous to the protective effects of elevated blood pressure values noted in very old adults (Qiu et al., 2005).

In reviewing the very large number of recent studies evaluating cognition in patients with type 2 diabetes, it becomes apparent that there is much variability in the magnitude of results across different studies (Arvanitakis et al., 2006; Asimakopoulou et al., 2002; Crooks et al., 2003; Grodstein et al., 2001; Korf et al., 2004; Kumari and Marmot, 2005; Lindeman et al., 2001; Luchsinger et al., 2007; Munshi et al., 2006; Verdelho et al., 2007). Although duration of diabetes and/or metabolic control usually correlate with cognitive test scores, occasionally one finds either no differences between diabetic and nondiabetic groups (Cosway et al., 2001) or an atypical pattern of results (Ryan and Geckle, 2000). The basis for this variability remains poorly understood but it is likely to reflect, at least in part, the large number of potentially confounding comorbid conditions that are associated with type 2 diabetes, including blood pressure, age, mood state (e.g., depression), diet, triglyceride levels, cerebrovascular disease, degree of insulin resistance or hyperinsulinemia, body mass index, and genetic factors (e.g., APOE 4; Awad et al., 2004; Brands et al., 2004; Bruce et al., 2008, 2005; Greenwood et al., 2003; Kalmijn et al., 1995; Watari et al., 2006).

31.3.3.2 Electrophysiological changes

Like younger adults with type 1 diabetes, older adults with type 2 diabetes show marked neural slowing on recordings of VEPs, BAEPs, ERPs, and resting EEGs, despite the fact that the type 2 patients have typically been diagnosed for a considerably shorter time than their type 1 counterparts. Prolonged VEP latencies are common in adults with type 2 diabetes and emerge relatively early in the course of the disease (Pozzessere et al., 1988). Not only are these effects particularly prominent in individuals with peripheral neuropathy, consistent with the central neuropathy hypothesis (Várkonyi et al., 2002), but they are also greater in patients with type 2, as compared to type 1 diabetes (Gregori et al., 2006). This phenomenon is also quite stable over time. Patients in good metabolic control who were followed over period of more than 4 years showed evidence of significant VEP abnormalities at both baseline and follow-up, but remarkably, did not appear to worsen over time, despite the fact that mean HbA1c values increased (from 6.7% to 7.5%).

Auditory-evoked potentials are also affected in older type 2 patients, with longer latencies seen in waves I, III, and V, as well as in wave I–V inter-peak latencies, and increased rates of clinically significant pathology (Durmus et al., 2004; Kurita et al., 1995; Nakamura et al., 1991). One of the few efforts to compare type 1 and type 2 diabetic subjects demonstrated a relatively greater prolongation of wave III and wave V latencies in those subjects with type 2 diabetes, although the magnitude of these effects is quite small – approximately 3–4% longer. Latencies were also more likely to be prolonged in older, as compared to younger subjects, and in those with a longer duration of diabetes (>10 years). Neither the presence of neuropathy, diabetes duration, nor level of metabolic control was correlated with prolonged latencies for either the type 1 or type 2 subjects, but this may reflect the relatively small number of subjects studied (Durmus et al., 2004).

Event-related potential latencies are consistently delayed in both older (Hissa et al., 2002) and younger (Dey et al., 1995; Kurita et al., 1996) adults with type 2 diabetes; latency increases ranged from approximately 4% to 11%. Having diabetes appears to be sufficient to induce these changes since neither degree of metabolic control nor duration of diabetes were associated with P300 latencies. Although the presence of microvascular complications in very poorly controlled patients has sometimes been linked to P300 latencies (Kurita et al., 1996), the magnitude of those effects appears to be miniscule ($d = 0.2$).

EEG recordings, only rarely obtained from patients with type 2 diabetes, show evidence of slowing that is similar to what has been reported previously for children and adults with type 1 diabetes. In what may be the only paper on this topic, older adults with type 2 diabetes showed reductions in alpha activity, particularly over the parietal area, and greater slow wave activity as indicated by increases in theta power recorded at central and parietal electrodes (Mooradian et al., 1988). Delta power was also increased somewhat but the very small sample size (13 diabetic and 8 control subjects) limits the sensitivity and generalizability of these findings.

31.3.3.3 Cerebrovascular outcomes

Altered cerebral blood flow is a well-established finding in older adults with type 2 diabetes. One early study, using SPECT techniques, demonstrated large reductions (15%, on average) in CBF in both hemispheres in all brain regions studied (Nagamachi et al., 1994). Of particular interest is their observation of a linear relationship between diabetes severity (defined here in terms of treatment, ranging from insulin – indicative of greatest severity – to oral medication, to diet) and CBF. For example, those treated with only dietary changes had CBF values in the frontal lobe that were significantly higher than those who had more serious diabetes and were treated with insulin.

More recent research, using continuous arterial spin labeling MRI techniques, has also reported overall reductions in CBF and CO_2 reactivity in older diabetic patients (mean age = 60 years), as compared to healthy control subjects, and this was correlated with structural changes within the CNS as well as with diabetes-related indicators of chronic hyperglycemia (Last et al., 2007). For all subjects, CBF in the parietal–occipital region was higher than in either the frontal or temporal regions, but diabetic subjects had significantly lower CBF values than controls when studied at rest and during CO_2 rebreathing, but not during hyperventilation. Diabetic subjects also had marked cortical and subcortical atrophy, as indexed by smaller gray and white matter volumes, and larger cerebrospinal fluid (CSF) volumes, all of which were associated with markers of chronic hyperglycemia (retinopathy, hypertension, and elevated HbA1c values). Especially noteworthy is the very strong relationship between brain structure parameters and cerebral blood flow. Although CBF measured at rest was highly correlated ($r = 0.77$) with regional gray matter volume for both diabetic and nondiabetic subjects, the diabetic subjects manifested a steeper CBF decline in frontal regions. Within the diabetic group, retinopathy, hypertension, and elevated HbA1c values were associated with lower CBF during hyper- and hypocapnia, particularly within the temporal region. These results, which are similar to that previously reported on patients with type 1 diabetes (Jiménez-Bonilla et al., 2001; Sabri et al., 2000; Vázquez et al., 1999), provide the strongest evidence to date that there are robust, and complex interrelationships among cortical and subcortical atrophy, cerebral perfusion, and biological markers of chronic hyperglycemia.

Additional evidence of diabetes-associated abnormalities in the cerebral microvasculature comes from measures of cerebral blood flow velocity and cerebrovascular resistance. Transcranial Doppler ultrasonography has demonstrated that adults with type 2 diabetes often, but not inevitably, show reductions in blood flow velocities in the middle cerebral arteries and increased cerebrovascular reactivity. Slower blood flow velocities were associated with poorer metabolic control, higher C-reactive protein, elevated systolic blood pressure, and volume of periventricular white matter hyperintensities (Novak et al., 2006). Whether these CBF changes are a consequence or a cause of the increased white matter (Novak et al., 2006) and decreased gray matter volumes (Last et al., 2007) seen on neuroimaging remain to be determined. The underlying pathophysiological processes appear to be strongly linked to the metabolic and vascular changes associated with chronic hyperglycemia, particularly in individuals with a long duration of diabetes (Fülesdi et al., 1999), multiple diabetes-related complications, and markers of inflammation like elevated levels of C-reactive protein and fibrinogen (Petrica et al., 2007). The failure to find a relationship between impairments in cerebrovascular reserve capacity and severity of peripheral neuropathy suggests that very different mechanisms may be responsible for these two phenomena, both of which are correlated with a longer duration of diabetes and poorer metabolic control (Hidasi et al., 2002).

31.3.3.4 Brain structure anomalies

Cerebral atrophy and white matter lesions have been reported frequently in studies of older adults with type 2 diabetes, but the details vary considerably, depending on the nature of the patient population and the neuroimaging techniques employed (van Harten et al., 2006).

In what may be the most sophisticated approach to date, automated evaluation of brain MR images revealed type 2 diabetes to be associated with a smaller volume of gray matter (~22-ml reduction), with greater subcortical atrophy (~7-ml increase in lateral ventricle volume), and larger white matter lesion volume (~57% increase) in a large sample of diabetic and nondiabetic adults recruited from the community (Jongen et al., 2007). Structural changes were significantly (and unexpectedly) more prominent in women than in men, and were associated with higher HbA1c values and older age, but were unrelated to hypertension, diabetes duration, or hypercholesterolemia. An earlier qualitative analysis of MRI data from essentially this same subject sample also showed that poorer cognitive function, particularly on measures of information processing speed and abstract reasoning, was correlated with degree of cortical atrophy and white matter lesions (Manschot et al., 2006). In turn, severity of cortical atrophy was positively associated with the presence of micro- and macrovascular complications, and negatively associated with the use of lipid-lowering medications (Manschot et al., 2007).

When identical neuroimaging and neurocognitive assessment parameters (Brands et al., 2006; Manschot et al., 2006) were used to compare 40 type 2 diabetic subjects carefully matched to 40 type 1 subjects on the basis of age (M = 61 years), gender, and estimated IQ, significantly greater cortical atrophy and more deep white matter lesions were found in those with type 2 diabetes, with effect sizes ranging from 0.5 to 0.66 (Brands et al., 2007a). What makes this especially noteworthy is that as a group, the type 2 subjects were in better metabolic control, had diabetes for a significantly shorter period of time (7 vs. 34 years), and had lower rates of clinically significant microvascular disease (laser-treated retinopathy: 8% vs. 38%). Because they had higher rates of macrovascular disease and more atherosclerotic risk factors (e.g., hypercholesterolemia, triglycerides, hypertension, and higher body mass index), the authors suggest that the pathophysiological processes underlying the development of brain anomalies in patients with type 2 diabetes may be qualitatively different from those associated with type 1 diabetes. Although their work strongly implicates atherosclerotic risk factors, studies of somewhat older subjects with pre-diabetes have suggested that impaired glucose and/or insulin regulation may also play a contributory role (Convit et al., 2003; den Heijer et al., 2003).

Other research has also reported relatively high rates of cerebral atrophy associated with diabetes; while these rates increased with increasing age, the diagnosis of diabetes appeared to have an additive rather than a synergistic effect on those rates. One early MRI study compared 159 diabetic and 2566 age-matched nondiabetic adults and found consistently higher rates of atrophy within the diabetic sample, but the magnitude of the between-group differences was quite similar regardless of age band. Of those 50–59 years of age, 41% of diabetic subjects, but only 18% of nondiabetic controls, showed cerebral atrophy; in the seventh decade rates were 60% versus 37%, and in the eighth decade rates were 92% versus 66% (Araki et al., 1994). The probability of manifesting atrophy was also associated with a diagnosis of hypertension and at least in relatively healthy diabetic patients – that is, those treated with diet or oral medications rather than insulin, atrophy is apparently limited to those who are also hypertensive. Normotensive diabetic subjects and hypertensive nondiabetic subjects did not manifest an elevated risk of severe cortical atrophy (Schmidt et al., 2004). Other studies have also found a relationship between hypertension and brain volume in patients with type 2 diabetes (Last et al., 2007), but this is not invariably so (Jongen et al., 2007).

The hippocampus is also affected in patients with type 2 diabetes, and may, in fact, show some of the earliest diabetes-associated CNS structural changes.

Diabetic adults 45–70 years of age (M = 59 years) without clinical evidence of microvascular complications had significantly smaller hippocampal volumes than age-matched healthy controls (5.4 vs. 6.2 cm^3; $d = 1.4$) but did not otherwise differ in number of white matter hyperintensities, or in brain volumes within frontal or temporal regions (Gold et al., 2007). Those diabetic subjects, all of whom had a milder form of the disease (i.e., were not being treated with insulin) and a relatively brief duration of diabetes (M = 6 years), also performed more poorly on measures of immediate memory, which were, in turn, weakly correlated with hippocampal volume (r=0.25). The best biomedical predictor of hippocampal atrophy was HbA1c, which explained 33% of the variance in multivariate modeling; neither hypertension nor dyslipidemia had an impact on outcomes.

Because only the hippocampus showed any volumetric reductions in this study, the authors suggest that the structure – which has not been evaluated specifically in most other recent neuroimaging studies (Akisaki et al., 2006; Jongen et al., 2007; Last et al., 2007; Manschot et al., 2006; Musen et al., 2006) – may be unusually vulnerable to the metabolic and vascular changes associated with diabetes. This possibility is consistent with other data demonstrating the sensitivity of the hippocampus to metabolic events like hypoglycemia (Auer, 2004) and is reinforced by the very strong relationship between hippocampal volume and metabolic control that was noted in relatively young older adults with no evidence of white matter hyperintensities or other brain anomalies (Gold et al., 2007). Since an earlier study of elderly nondiabetic adults with impaired glucose tolerance noted a similar pattern of results (Convit et al., 2003), it is plausible that the CNS changes found in adults with type 2 diabetes may be largely a consequence of the metabolic and microvascular changes associated with insulin resistance and glucose dysregulation, and to alterations in glucose transport into brain structures (Convit, 2005; Kumari et al., 2000).

Additional support for that possibility comes from several large population-based studies demonstrating the presence of hippocampal atrophy and neuropathological changes in elderly adults (mean age >70 years) with diabetes or with insulin resistance but no diabetes diagnosis. Subjects with type 2 diabetes who were followed in the Rotterdam Study had smaller hippocampal and amygdalar volumes than age-matched nondiabetic participants (den Heijer et al., 2003). Diabetes was associated with more vascular disease, but degree of atrophy was unrelated to severity of micro- and macrovascular disease or its risk factors. Because the nondiabetic adults in that study who had evidence of insulin resistance also showed smaller amygdalar volumes, the authors speculated that these structural changes may, at least in part, be a consequence of a dysfunction in insulin-signaling pathways. According to their model, the amygdala is particularly vulnerable because it has a lower density of insulin receptors than structures like the hippocampus (Schulinghamp et al., 2000).

Significant hippocampal atrophy has also been found in older adults with diabetes who were enrolled in the Honolulu–Asia Aging Study. Not only did the diabetic subjects have a twofold increased risk of hippocampal atrophy compared to those without diabetes (Korf et al., 2006), but also an ancillary autopsy study of a subset of participants demonstrated a synergism between diabetes and the presence of the APOE ε 4 allele (Peila et al., 2002). Participants with both conditions had higher numbers of neuritic plaques in the hippocampus, more neurofibrillary tangles in the hippocampus and cortex, and a greatly elevated risk of cerebral amyloid angiopathy as compared to either condition alone. Cognitive evaluations also indicated that the diagnosis of type 2 diabetes greatly increased the risk of manifesting Alzheimer's disease or vascular dementia, particularly in those carrying the APOE ε 4 allele.

31.3.3.5 Alterations in brain metabolites

Like adults with type 1 diabetes (Kreis and Ross, 1992; Mäkimattila et al., 2004), individuals with type 2 diabetes show marked elevations in certain brain metabolites, as demonstrated with ^{1}H-MRS. Changes in myo-inositol (mI) concentrations are most prominent, particularly in frontal white matter, with increases ranging from 16% (left hemisphere) to 26% (right hemisphere) in relatively healthy patients in good metabolic control (HbA1c = 7.1; Ajilore et al., 2007). Although not associated with glycosylated hemoglobin values, the mI values were correlated with scores on the Cerebrovascular Risk Factor Scale, suggesting that this frontal gliosis is secondary to cerebrovascular changes.

Other investigators, studying insulin-treated type 2 patients who were in poorer metabolic control (HbA1c = 8.0), have also found mI concentrations elevated in both white and gray matter, as well as increases in gray matter choline (Cho) values (Geissler et al., 2003). Again, there was no correlation between brain metabolites and HbA1c values, but there was a strong association with complications. Those subjects with peripheral neuropathy had higher Cho as well as higher white matter mI concentrations, as compared to those without neuropathy. In addition, there were strong correlations between neuropathy, white matter abnormalities, and duration of diabetes, consistent with the view that changes in brain metabolites reflect gliosis that may, in turn, be secondary to osmolarity changes and/or cerebral deposition of amylin.

31.3.4 Diabetes-Associated Neurocognitive Phenotypes: One or Many?

Reviewing the literature on the functional and structural CNS characteristics of people with diabetes suggests that there are actually few differences across discrete diabetic patient populations. Common to virtually all patients is slowed processing. Regardless of age or disease type, individuals with diabetes perform more poorly than their healthy peers on cognitive tasks requiring rapid responding, show longer central nerve conduction latencies in response to sensory or cognitive stimulation, and manifest slower brain wave activity when resting quietly. Also evident are reductions in cerebral blood flow and shifts in regional CBF patterns, abnormal levels of brain metabolites, higher rates of anomalies in brain structure, and somewhat less cerebral volume when evaluated with neuroimaging measures.

CNS sequelae of diabetes appear to be modest, at best, and tend not to become progressively larger in magnitude despite increasing duration of disease. Note that the effect size – the standardized difference between groups of age-matched diabetic and nondiabetic research subjects – is similar, regardless of how long the individual has had diabetes. Diabetic children, young adults with a childhood onset of diabetes, or older adults with diabetes, all typically manifest effect sizes ranging from 0.4 to 0.6, despite the fact that adults with onset in childhood or adolescence have had the disease for 20 or 30 years longer than the children. Moreover, the appearance of these neurocognitive sequelae seems to occur relatively early in the course of the disease. Although there has been no systematic effort to prospectively evaluate changes in brain function annually or biannually following diagnosis, a number of studies have reported differences between healthy comparison subjects and patients with a diabetes duration ranging from 2 to 6 years. Whether changes emerge even earlier than that remains unstudied.

Mental slowing may be the most ubiquitous manifestation of diabetes, but is not the only one. Both children and adults also perform more poorly on measures of crystalized intelligence, manifesting less knowledge about the world than their healthy peers. On the other hand, learning and memory impairments do not appear to be universal. While they are common in older adults with type 2 diabetes (who may have had diabetes for only a few years), and in children with an early onset of type 1 diabetes, they are only infrequently seen in children and young and middle-aged adults with a later onset of type 1 diabetes. Any etiologic model must not only explain why the pattern of cognitive dysfunction is so circumscribed, but must account for its relatively early appearance following diagnosis, its apparent minimally progressive nature over time, and its association with the metabolic abnormalities associated with diabetes.

31.4 Biomedical Risk Factors

Two types of disease-related variables (iatrogenic hypoglycemia and chronic hyperglycemia) have often been linked to the appearance of neurocognitive dysfunction in patients with diabetes. Establishing strong brain–behavior relationships with either of these putative risk factors has, however, been stymied by multiple measurement problems. Not only is there disagreement on the definition of hypoglycemia – in terms of relying on biochemical or clinical criteria (Ryan et al., 2005), but also with very few exceptions, determining the number, timing, and duration of such events across one's lifetime is almost always an unreliable activity, dependent as it is on the patient's (or parent's) ability to recall events that may have happened years earlier. Operationalizing chronic hyperglycemia also presents a significant challenge. The glycosylated hemoglobin assay provides an estimate of metabolic control over a period of 2–3 months, but because metabolic control may fluctuate extensively within that time period in response to changes in treatment regimens, overall health, and mood state, a single HbA1c value cannot accurately capture glycemic variability. Currently, there is no standard procedure for aggregating HbA1c data over multiple time points. This issue becomes even problematic in studies of type 2 patients because diabetes onset is not always readily recognized by the patient or physician in a timely fashion, leading to an underestimation of disease duration. For that reason, researchers have begun to use as surrogate markers of chronic hyperglycemia the appearance of clinical endpoints, like retinopathy, which have well-accepted diagnostic criteria (Boulton et al., 2004; Fong et al., 2004). Unfortunately, there is not a perfectly isomorphic relationship between chronic hyperglycemia and complications. Often they appear only after a number of years of diabetes and their occurrence may be influenced by temporal variations in metabolic control (DCCT/EDIC Research Group, 2000) and multiple, as yet unknown, genetic and other constitutional factors (Costacou et al., 2006; Gallego et al., 2007; Nathan, 1996).

31.4.1 Hypoglycemia

Hypoglycemia is a consequence of treatment of type 1 diabetes and, increasingly, is found to occur in patients with type 2 diabetes who are managed with insulin or oral medication regimens (Zammitt and Frier, 2005). From a pathophysiological perspective, it is quite plausible to attribute brain dysfunction to one or more episodes of hypoglycemia. Glucose normally serves as the primary, if not the sole, oxidative substrate for brain energy metabolism and because the brain neither stores nor produces sizable quantities of glucose, it must obtain it continuously from the systemic circulation via the process of facilitative diffusion across the blood–brain barrier (Nehlig, 1997). Falling blood glucose levels initiate a cascade of neurochemical events that lead to brain energy failure secondary to neuroglycopenia. Early signs include a slowing of the EEG which gradually becomes isoelectric. At the same time, changes in ion pump activity disrupt ion homeostasis, which in turn leads to the cellular influx of calcium and the occurrence of intracellular alkalosis. As the excitatory amino acids aspartate and, to a lesser extent, glutamate are released into the interstitial space of the brain, they bind to dendrites and ultimately induce nerve cell death (Auer, 2004; Auer and Siesjö, 1993).

31.4.1.1 CNS effects of extended episodes of profound hypoglycemia

It is important to keep in mind that these complex neurochemical changes induce significant structural CNS damage only when the level of hypoglycemia is profound – that is, when blood glucose falls to approximately 1–1.5 mmol l^{-1} (18–27 mg dl^{-1}), persists for an extensive period of time, and is accompanied by coma and/or seizure. Multiple case reports have noted that even a single episode of profound hypoglycemia can eventuate in a significant degree of cortical damage in frontal and/or temporal regions, as well as frequent involvement of basal ganglia, hippocampus, and brainstem (Aoki et al., 2004; Auer et al., 1989; Boeve et al., 1995; Chalmers et al., 1991; Fujioka et al., 1997; Jung et al., 2005; Mori et al., 2006; Perros et al., 1994). The resulting tissue degeneration is characterized by selective neuronal loss and a proliferation of astrocytic glial cells (Fujioka et al., 1997). Profound hypoglycemia may also be accompanied by elevated serum levels of neuron-specific enolase and protein S-100, two markers of neuronal injury (Strachan et al., 1999). In most instances, however, this CNS damage has been found in older adults who experienced profound hypoglycemia for several hours, had a form of brittle diabetes, with wildly fluctuating blood glucose levels, and/or had some other condition that could affect the CNS (e.g., long history of chronic alcohol abuse). Profound hypoglycemia of this sort is, however, remarkably infrequent and is quite atypical of most adults with diabetes.

31.4.1.2 Do single or recurrent episodes of less severe hypoglycemia have neurocognitive sequelae?

Far more prevalent are episodes of severe hypoglycemia. Multiple definitions of severe hypoglycemia have been offered, but the two most common are characterized clinically in terms of either the presence of transient neurological sequelae (seizure or coma), or in the absence of such signs, by an inability to treat the hypoglycemia without external assistance. Ordinarily, this corresponds to blood glucose values that are less than 3 mmol l^{-1} (55 mg dl^{-1}). In adults, the annual prevalence of severe hypoglycemia, defined as requiring external assistance, ranges from 30% to 40%, although the distribution is highly skewed, with 5% of patients accounting for more than 50% of the episodes (Strachan, 2007).

Animal research provides no evidence that recurrent bouts of severe hypoglycemia, without coma or an isoelectric EEG, necessarily produce structural brain changes in adult (Auer and Siesjö, 1993) or immature rats (Malone et al., 2006). Several smaller cross-sectional human studies suggested that repeated episodes of moderately severe hypoglycemia were associated with mild, relatively circumscribed neurocognitive dysfunction (Deary et al., 1993; Gold et al., 1993; Hershey et al., 2004). Most larger studies have either failed to find strong support for that view (Brands et al., 2005; Brismar et al., 2007; DCCT/EDIC Research Group, 2007; Ferguson et al., 2003;

Kramer et al., 1998; Strudwick et al., 2005; Wysocki et al., 2003) or have found weak relationships but have been unable to disentangle the possible contribution of recurrent hypoglycemia in patients who may also have highly variable levels of metabolic control over an extended period of time (Hyllienmark et al., 2005; Northam et al., 2001).

The most compelling evidence that recurrent hypoglycemia is not associated with an increased risk of enduring cognitive dysfunction comes from the Diabetes Control and Complications Trial (DCCT), and its follow-up natural history study, the Epidemiology of Diabetes Interventions and Complications (EDIC). As part of that clinical trial and its long-term follow-up, cognitive functioning was assessed over a period of 18 years, on average, in 1144 type 1 diabetic patients who at the time of their final cognitive evaluation ranged in age from 29 to 62 years ($M = 46$ years). Rates of severe hypoglycemia, defined here as including a seizure or coma, were high, with a total of 1355 episodes reported in 453 subjects over the duration of this study. Nevertheless, no relationship was found between the cumulative number of severe hypoglycemic episodes and changes in performance on a comprehensive neuropsychological test battery (DCCT/EDIC Research Group, 2007).

It would be inaccurate to conclude from the DCCT/EDIC follow-up results that severe hypoglycemia is entirely harmless. It is not. Transient reductions in mental efficiency have been well established in both children and adults (Ryan et al., 2005; Warren and Frier, 2005), are known to be sufficient to impair complex cognitive skills, including those associated with driving ability (Cox et al., 2000, 2003) and academic achievement (Gschwend et al., 1995), and may contribute to the dead in bed phenomenon (Sovik and Thordarson, 1999; Weston and Gill, 1999). As noted above, a single episode of profound hypoglycemia that persists for an extended period of time can eventuate in permanent CNS sequelae. Nevertheless, most recent studies on the effects of moderately severe single or recurrent episodes of hypoglycemia have not found evidence of long-term CNS sequelae. According to the selfish brain model (Peters et al., 2004), the CNS has a remarkable series of energy monitoring systems, with multiple short- and long-term feedback control processes to ensure that changes in glucose availability can be quickly identified by the organism and compensated for by a self-allocation process that leads to increases in food intake (or other behaviors) and/or by utilization of substrates like lactate or ketones. Such a system may thereby protect the brain from permanent structural damage when blood glucose values fall within very broad parameters. However, when blood glucose drops below some critical threshold value, compensatory mechanisms may fail and neuronal necrosis (and other CNS changes) may occur. This intriguing model of neuroprotection is consistent with the extant neurocognitive data, but much additional research is clearly needed.

31.4.2 Chronic Hyperglycemia

High blood glucose levels are the defining metabolic characteristic of both type 1 and type 2 diabetes. Patients in poor metabolic control not only have abnormally high blood glucose values over an extended period of time, but they also have a greatly increased risk of developing biomedical micro- and macrovascular complications like retinopathy, neuropathy, nephropathy, and cardiovascular disease (DCCT Research Group, 1993; UK Prospective Diabetes Study Group, 1998). A growing body of literature now supports the view that poor metabolic control is also associated with the development of cognitive complications (Cukierman et al., 2005), the severity of which may be ameliorated, at least to some limited extent, by improving metabolic control (DCCT/EDIC Research Group, 2007; Ryan et al., 2006), although this possibility remains controversial (Areosa Sastre and Grimley Evans, 2004).

31.4.2.1 Clinically significant microvascular complications predict cognitive impairment

The most direct support for a relationship between chronic hyperglycemia and cognition comes from longitudinal studies in which the onset and severity of microvascular complications as well as changes in cognitive performance are assessed over time. In one of the largest studies to date, young and middle-aged adults with a childhood onset of type 1 diabetes were followed over a 7-year period and compared to a group of demographically similar nondiabetic adults. Like healthy control subjects, those diabetic patients without proliferative retinopathy at either time point showed no change in mental efficiency. On the other hand, those who had proliferative retinopathy at the beginning of the study or who developed it during the 7-year follow-up period showed a significant decline in mental efficiency ($d = 0.56$ and 0.50, respectively), as assessed by tests of psychomotor efficiency (Ryan et al., 2003). Statistical modeling demonstrated that the risk of cognitive slowing was predicted by four variables: the presence or development of proliferative retinopathy, the presence of autonomic neuropathy, elevated systolic blood pressure, and longer duration of diabetes. This model identified, with 83% accuracy, subjects who showed significant cognitive declines over time and it explained 53% of the variance.

Cross-sectional research on young adults with type 1 diabetes has also demonstrated that the presence of background retinopathy is associated with white matter abnormalities on MRI and poorer performance on measures of attention, fluid intelligence, and information-processing efficiency (Ferguson et al., 2003). Additional support for an association between retinopathy and both functional and structural brain anomalies comes from a series of smaller case-control studies comparing groups of type 1 diabetic adults with, and without, retinopathy. Not only did those subjects with retinopathy have greater focal cortical atrophy on structural MRI (Wessels et al., 2006b), but they also manifested abnormal brain activation patterns on fMRI while performing a working memory task during an episode of experimentally induced hypoglycemia (Wessels et al., 2006a). Other microvascular complications, particularly peripheral neuropathy, have also been found to be associated with functional and structural changes in the CNS (Brismar et al., 2007; Dejgaard et al., 1991; Geissler et al., 2003; Gregori et al., 2006; Hyllienmark et al., 2005; Ryan et al., 1992), and as detailed above in Section 31.2, a wide range of studies have found relationships between measures of brain integrity, diabetes duration, and estimates of metabolic control in patients with either type 1 or type 2 diabetes.

31.4.2.2 Retinopathy as a surrogate marker of cerebral microangiopathy

Associations between retinopathy and neurocognitive dysfunction are not limited to people with diabetes. Results from the Atherosclerosis Risk in Communities Study (Wong et al., 2002) show that middle-aged nondiabetic adults with retinal microaneurysms also manifest a pattern of cognitive decline that is characterized by psychomotor slowing and is analogous to what has been reported in diabetic adults with clinically significant retinopathy (Ryan et al., 2003). Presence of retinal microaneurysms also predicts an increased risk of cerebral atrophy (Wong et al., 2003) and subclinical cerebral infarction (Cooper et al., 2006), with these effects being especially prominent when blood pressures are elevated. Cerebral small vessel disease is also associated with a decline in information-processing speed (Prins et al., 2005) and may precede development of white matter lesions (Kwa et al., 2002) in older adults without diabetes or dementia (for review, see Jellinger (2007)).

Because of the well-established homology between the retinal and cerebral microvasculatures (Patton et al., 2005), examination of retinal arterioles with techniques like digitized fundus photography can provide a noninvasive assessment of the cerebral microcirculation (Longstreth et al., 2006; Patton et al., 2006; Wong et al., 2001). It is likely, then, that in diabetic patients in poor metabolic control, the presence of retinopathy serves as a surrogate marker of structural changes within the brain microvasculature. The resulting cerebral hypoperfusion may contribute to the development of brain abnormalities by disrupting the efficient delivery of glucose and other nutrients to neural tissue (De La Torre, 2004; Ryan, 2006a). Associations reported between brain function and diabetic complications other than retinopathy (e.g., peripheral neuropathy) may reflect the fact that multiple microvascular complications tend to appear contemporaneously (Orchard et al., 1990) and have a common origin – that is, a consequence of diabetic microangiopathy occurring in multiple organ systems (Barr et al., 2006; Orchard et al., 1997). The extensive use of peripheral neuropathy as a marker of microangiopathy in early studies may be a consequence of its ease in measurement, although it now appears to be a less direct and reliable measure of microvascular disease than retinopathy (Cheung and Wong, 2008; Maser et al., 1992).

31.4.2.3 Chronic hyperglycemia may interfere with normal brain development

Similar vascular mechanisms may be responsible for the development of neurocognitive abnormalities in diabetic children, insofar as there is now evidence of cerebral hypoperfusion (Salem et al., 2002), retinopathy, and preclinical kidney dysfunction (Alibrahim et al., 2006; Donaghue et al., 2005), peripheral neuropathy (Karabouta et al., 2008; Karsidag et al., 2005), and endothelial abnormalities (Clarkson et al., 1996; Järvisalo et al., 2004) appearing within several years of diagnosis. In the special case of individuals with an early onset of type 1 diabetes, the possibility exists that the appearance of very early vascular changes, or more likely, the direct neurotoxic effects of elevated brain glucose levels, may adversely affect normal brain maturational process which are particularly labile during the first 5–6 years of life (Chugani, 1998; Huttenlocher and Dabholkar, 1997). According to this so-called diathesis or vulnerability hypothesis, the resulting brain abnormalities would disrupt cognition to a greater extent than would be seen in individuals who developed diabetes after that early critical period (Ryan, 2006b). This is consistent with a large body of neurobehavioral data, as well as with animal research demonstrating that continuous exposure to elevated glucose levels during early development induces structural changes within the CNS (most notably, reductions in cell size in the cortex and hippocampus and a reduction in myelinated processes; Malone et al., 2006).

31.5 Pathophysiological Mechanisms

Two broad classes of mechanistic models have been offered to explain the development of both biomedical and neurocognitive complications. The first of these emphasizes the toxicity of glucose to various tissue types and organ systems, including the vasculature and the CNS, whereas the second focuses on insulin dysregulation and its effect on multiple signaling pathways involved in learning and memory (and many other processes), particularly in individuals with type 2 diabetes and a past or current history of hyperinsulinemia.

31.5.1 Glucose Toxicity

A number of interrelated biochemical mechanisms of glucotoxicity have been identified as the putative basis for the development of both biomedical and neurocognitive diabetes-related complications, and recent review papers describe these mechanisms in more detail than is possible in this chapter (Brownlee, 2005; Sheetz and King, 2002; Tomlinson and Gardiner, 2008). The aldose reductase (polyol or sorbitol pathway) theory is most closely linked to the neurocognitive data and emerges largely from early work on the relationship between chronic hyperglycemia and changes in peripheral nerve conduction velocity. When intracellular blood glucose levels are chronically elevated, aldose reductase converts the glucose to sorbitol which is subsequently oxidized to fructose. The intracellular accumulation of sorbitol is associated with the depletion of myo-inositol, which in turn leads to reductions in Na^+/K^+ ATPase activity at the plasma membrane level, thereby disrupting intracellular metabolism and ultimately reducing nerve conduction velocity. Whether changes in the polyol pathway also underlie the development of the mental and motor slowing or central neuropathy that is commonly found in people with type 1 or type 2 diabetes is certainly plausible, but not yet proven.

Intracellular accumulation of advanced glycation end-products (AGEs) has also been offered as an explanation for what many consider to be premature aging – that is, the early emergence of cardio- and cerebrovascular changes in both children (Jakus and Rietbrock, 2004) and adults with diabetes (Singh et al., 2001). During normal aging, AGE develops as the product of the nonenzymatic glycation of proteins and not only affects the structural components of various tissues, including endothelial cells and myelin, but also modifies intracellular proteins, particularly those regulating gene transcription (Brownlee, 2005; Ramasamy et al., 2005). These oxidation products tend to accumulate in the hippocampus during the course of normal aging (Li et al., 1995) and may play a role in the pathogenesis of neuronal damage and the deposition of amyloid that is characteristic of Alzheimer's disease and normal aging (Münch et al., 1998). Cellular dysfunction associated with chronic hyperglycemia has also found to be mediated by changes within the protein kinase C pathway (Kuo et al., 2005) and the reactive oxygen intermediate pathway (Sheetz and King, 2002) but clear linkages between these molecular changes and changes within the CNS remain poorly understood (Klein and Waxman, 2003).

31.5.2 Hyperglycemia, Insulin Dysregulation, and Brain Dysfunction

By definition, diabetes is a disorder of insulin dysregulation. Patients with type 1 diabetes must obtain all insulin via injections, most typically as boluses several times a day, although increasingly, the use of subcutaneous continuous infusion pumps more closely mimics the natural pattern of insulin release. Although most patients with type 2 diabetes continue to secrete some insulin endogenously, the development of insulin resistance along with beta cell dysfunction leads to abnormally elevated insulin levels, and in fact this hyperinsulinemia may occur for several years prior to formal diagnosis of diabetes. Increasing evidence now suggests that excessive insulin levels may independently be associated with CNS dysfunction (for excellent reviews, see Cole et al. (2007), Craft and Watson (2004), Li and Hölscher (2007), Park (2001), and Woods et al. (2006)).

Rising with increasing age, systemic insulin levels are known to be strong predictors of cognitive impairment in adults without diabetes (Stolk et al., 1997). Elderly men with the highest insulin levels made 25% more errors on the mini-mental state examination than men with the lowest levels, and there was an interaction between age and insulin levels such that hyperinsulinemic men who were more than 75 years of age tended to make more errors than younger men with similar levels (Kalmijn et al., 1995). Elevated insulin levels also interact with blood pressure elevations to interfere with cognitive function, and small, but statistically significant correlations have been reported between fasting insulin and cognitive test performance (Kuusisto et al., 1993). When those subjects were subsequently reevaluated approximately 3.5 years later, elevated insulin levels were associated with a greatly increased risk of dementia (Kuusisto et al., 1997), a finding that has subsequently been replicated in other cohorts (Luchsinger et al., 2004). Hyperinsulinemia in the absence of diagnosed diabetes has been found in middle-aged adults to predict baseline performance on measures of memory, psychomotor efficiency, and verbal fluency, as well as to predict declines in cognitive performance over a 6-year period (Young et al., 2006). At least one study has suggested that the pattern of cognitive impairment associated with insulin resistance includes subcortical features, consistent with cerebrovascular pathology (Geroldi et al., 2005).

Exactly how high levels of circulating insulin disrupt brain function and adversely affect cognitive functioning remains incompletely understood (Gerozissis, 2003). Insulin receptors are widely distributed throughout the brain (Hopkins and Williams, 1997; Unger and Betz, 1998), and both brain insulin levels, as well as the density of these receptors, normally decrease with advancing age (Frölich et al., 1998). Insulin crosses the blood–brain barrier and enters the brain via a receptor-mediated active transport system (Baskin et al., 1987). Hyperinsulinemia is associated with a downregulation of this active transport process, leading to less availability of insulin to the CNS (Schwartz et al., 1992, 1994). Because brain insulin levels are particularly low in individuals with Alzheimer's disease, it has been hypothesized that the cognitive deterioration characteristic of that disorder, particularly the changes in learning and memory function, may be a consequence of alterations in insulin receptor activity (Biessels and Kappelle, 2005; Craft et al., 1999, 1998; Watson and Craft, 2004; Zhao et al., 2004). Although several intriguing explanatory theories of potential linkages between neurocognitive abnormalities in diabetic patients and insulin dysregulation exist (e.g., Convit, 2005; Sima et al., 2004), compelling empirical support is lacking at this time.

31.6 Diabetes and Brain Dysfunction: Some Final Thoughts

Diabetes mellitus is not a disorder of the CNS, but both the disease and its medical management can have a significant impact on brain structure and function. An enormous volume of research on this topic has emerged, especially during the past 10 years, to demonstrate that neurocognitive complications are a common occurrence in individuals with diabetes and these are more likely to be associated with a history of chronically elevated blood glucose levels than with recurrent bouts of moderately severe hypoglycemia. The focus of this chapter has been on delineating neurobehavioral phenotypes in an effort to clarify, for both the basic scientist and the clinical researcher, exactly what phenomena need to be explained by pathophysiological models.

From this reviewer's perspective, it appears as if there are three overlapping neurocognitive phenotypes that are characteristic of diabetes. Common to all are reductions in speed of information processing. Not only do diabetic patients of virtually any age manifest reductions in mental efficiency on cognitive tests that require rapid responding, but they also show slowed brain activity as measured at rest or in response to sensory or cognitive stimuli. Brain structure is also affected, with limited reductions in cortical gray matter density, increases in white matter anomalies, abnormal levels of brain metabolites, and changes in the rate and distribution of cerebral blood flow. These effects occur early in the course of diabetes and do not appear to worsen measurably over time, unless the individual develops significant micro- and macrovascular disease or other comorbid conditions, like severe hypertension, that can independently affect brain function. Even then, however, the severity of the cognitive dysfunction is modest; clinically significant impairments are seen only infrequently. Although many with this phenotype also perform more poorly on measures of academic achievement and crystallized intelligence, other cognitive domains – including language skills, problem solving, and memory – are typically unaffected.

A second phenotype, most commonly seen in individuals with type 2 diabetes, includes all of those features along with mild to moderately severe deficits on measures of learning and memory that are correlated with degree of hippocampal atrophy. Often, but not invariably, test performances requiring attention and abstract reasoning are also affected. Because the mnestic and hippocampal changes are not commonly found in most patients with type 1 diabetes, there is a growing belief that these may be related to hyperinsulinemia or to other forms of insulin dysregulation which are characteristic of type 2 diabetes. Both clinical and epidemiological studies are now beginning to explicitly examine structural and functional brain changes in individuals who have hyperinsulinemia. The challenge in studying brain function in patients with type 2 diabetes is that many also have multiple other comorbid conditions that can independently affect cognition, making it difficult to appropriately attribute neurocognitive phenomena to diabetes *per se*.

Children and young adults diagnosed with type 1 diabetes within the first several years of life manifest all of these features as well. However, not only are more cognitive domains affected, but also the magnitude of these effects tends to be somewhat larger and, in many instances, may frequently meet criteria for clinically significant impairment. It is likely, but as yet unproven, that this third neurocognitive phenotype is a direct consequence of diabetes-associated metabolic derangements disrupting normal brain development. Although earlier work attributed such changes to the effects of recurrent hypoglycemia – which occurs frequently in very young children – there is growing evidence from both human and animal studies that these changes may be triggered by the glucose toxicity associated with increased levels of brain glucose that occur during periods of chronic hyperglycemia.

There are many reasons to expect diabetes-associated neurocognitive dysfunction to progress in magnitude and in severity over time, particularly if glucotoxicity and cerebrovascular abnormalities underlie its development. Yet for the vast majority of diabetic patients, there is little convincing evidence at the present time to suggest that cognition deteriorates slowly with increasing duration of diabetes. Moreover, it does not appear that most people who have lived with diabetes for many years ultimately develop dementia. What is it that protects most people with diabetes from developing a classic encephalopathy? This issue has been almost entirely ignored. Not only is research needed to address that question, but as the worldwide epidemic of diabetes continues, also more research is required to identify effective biomedical interventions that can reverse, and prevent, neurocognitive complications in children and adults with diabetes mellitus.

References

Ajilore O, Haroon E, Kumaran S, et al. (2007) Measurement of brain metabolites in patients with type 2 diabetes and major depression using proton magnetic resonance spectroscopy. *Neuropsychopharmacology* 32: 1224–1231.

Akisaki T, Sakurai T, Takata T, et al. (2006) Cognitive dysfunction associates with white matter hyperintensities and subcortical atrophy on magnetic resonance imaging of the elderly diabetes mellitus Japanese elderly diabetes intervention trial (J-EDIT). *Diabetes/Metabolism Research and Reviews* 22: 376–384.

Alibrahim E, Donaghue KC, Rogers S, Hing S, Jenkins AJ, Chan A, and Wong TY (2006) Retinal vascular caliber and risk of retinopathy in young patients with type 1 diabetes. *Ophthalmology* 113: 1499–1503.

Allen KV, Frier BM, and Strachan MWJ (2004) The relationship between type 2 diabetes and cognitive dysfunction: Longitudinal studies and their methodological limitations. *European Journal of Pharmacology* 490: 169–175.

Antonetti DA, Barber AJ, Bronson SK, et al. (2006) Diabetic retinopathy: Seeing beyond glucose-induced microvascular disease. *Diabetes* 55: 2401–2411.

Aoki T, Sato T, Hasegawa K, Ishizaki R, and Saiki M (2004) Reversible hyperintensity lesion on diffusion-weighted MRI in hypoglycemic coma. *Neurology* 63: 392–393.

Araki Y, Nomura M, Tanaka H, Yamamoto H, Yamamoto T, Tsukaguchi I, and Nakamura H (1994) MRI of the brain in diabetes mellitus. *Neuroradiology* 36: 101–103.

Areosa Sastre A and Grimley Evans J (2004) Effect of the treatment of type II diabetes mellitus on the development of cognitive impairment and dementia. *Cochrane Database of Systematic Reviews* 4.

Arvanitakis Z, Wilson RS, Li Y, Aggarwal NT, and Bennett DA (2006) Diabetes and function in different cognitive systems in older individuals without dementia. *Diabetes Care* 29: 560–565.

Asimakopoulou KG, Hampson SE, and Morrish NJ (2002) Neuropsychological functioning in older people with type 2 diabetes: The effect of controlling for confounding factors. *Diabetic Medicine* 19: 311–316.

Auer RN (2004) Hypoglycemic brain damage. *Metabolic Brain Disease* 19: 169–175.

Auer RN, Hugh J, Cosgrove E, and Curry B (1989) Neuropathologic findings in three cases of profound hypoglycemia. *Clinical Neuropathology* 8: 63–68.

Auer RN and Siesjö BK (1993) Hypoglycaemia: Brain neurochemistry and neuropathology. *Baillière's Clinical Endocrinology and Metabolism* 7(3): 611–625.

Awad N, Gagnon M, and Messier C (2004) The relationship between impaired glucose tolerance, type 2 diabetes, and cognitive function. *Journal of Clinical and Experimental Neuropsychology* 26: 1044–1080.

Barkai L, Vámosi I, and Lukács K (1998) Prospective assessment of severe hypoglycaemia in diabetic children and adolescents with impaired and normal awareness of hypoglycaemia. *Diabetologia* 41: 898–903.

Barr ELM, Wong TY, Tapp RJ, Harper CA, et al. (2006) Is peripheral neuropathy associated with retinopathy and albuminuria in individuals with impaired glucose metabolism? The 1999–2000 AusDiab. *Diabetes Care* 29: 1114–1116.

Baskin DG, Figlewicz DP, Woods SC, Porte D, and Dorsa DM (1987) Insulin in the brain. *Annual Review of Physiology* 49: 335–347.

Bayazit Y, Yilmaz M, Kepekçi Y, Mumbuç S, and Kanlikama M (2000) Use of the auditory brainstem response testing in the clinical evaluation of the patients with diabetes mellitus. *Journal of Neurological Sciences* 181: 29–32.

Biessels G-J, Braun KPJ, de Graaf RA, van Eijsden P, Gispen WH, and Nicolay K (2001) Cerebral metabolism in streptozotocin-diabetic rats: An *in vivo* magnetic resonance spectroscopy study. *Diabetologia* 44: 346–353.

Biessels G-J, Cristino NA, Rutten G-J, Hamers FPT, Erkenlens DW, and Gispen WH (1999) Neurophysiological changes in the central and peripheral nervous system of streptozotocin-diabetic rats: Course of development and effects of insulin treatment. *Brain* 122: 757–768.

Biessels GJ, Deary IJ, and Ryan CM (2008) Cognition and diabetes: A lifespan perspective. *Lancet Neurology* 7: 184–190.

Biessels GJ and Kappelle LJ (2005) Increased risk of Alzheimer's disease in type II diabetes: Insulin resistance of the brain or insulin-induced amyloid pathology? *Biochemical Society Transactions* 33: 1041–1044.

Biessels G-J, Staekenborg S, Brunner E, and Scheltens P (2006) Risk of dementia in diabetes mellitus: A systematic review. *Lancet Neurology* 5: 64–74.

Bjørgaas M, Gimse R, Vik T, and Sand T (1997) Cognitive function in type 1 diabetic children with and without episodes of hypoglycaemia. *Acta Paediatrica* 86: 148–153.

Bjørgaas M, Sand T, and Gimse R (1996) Quantitative EEG in type 1 diabetic children with and without episodes of severe hypoglycemia: A controlled, blind study. *Acta Neurologica Scandinavica* 93: 398–402.

Boeve BF, Bell DG, and Noseworthy JH (1995) Bilateral temporal lobe MRI changes in uncomplicated hypoglycemic coma. *Canadian Journal of Neurological Sciences* 22: 56–58.

Boulton AJ, Malik RA, Arezzo JC, and Sosenko JM (2004) Diabetic somatic neuropathies. *Diabetes Care* 27: 1458–1486.

Brands AMA, Biessels G-J, De Haan EHF, Kappelle LJ, and Kessels RPC (2005) The effects of type 1 diabetes on cognitive performance: A meta-analysis. *Diabetes Care* 28: 726–735.

Brands AMA, Biessels GJ, Kappelle LJ, et al. (2007a) Cognitive functioning and brain MRI in patients with type 1 and type 2 diabetes mellitus: A comparative study. *Dementia and Geriatric Cognitive Disorders* 23: 343–350.

Brands AMA, Kessels RPC, Biessels GJ, et al. (2006) Cognitive performance, psychological well-being, and brain magnetic resonance imaging in older patients with type 1 diabetes. *Diabetes* 55: 1800–1806.

Brands AMA, Kessels RPC, De Haan EHF, Kappelle AC, and Biessels G-J (2004) Cerebral dysfunction in type 1 diabetes: Effects of insulin, vascular risk factors, and blood-glucose levels. *European Journal of Pharmacology* 490: 159–168.

Brands AMA, Van den Berg E, Manschot SM, Biessels GJ, Kappelle LJ, De Haan EHF, and Kessels RPC (2007b) A detailed profile of cognitive dysfunction and its relation to psychological distress in patients with type 2 diabetes mellitus. *Journal of the International Neuropsychological Society* 13: 288–297.

Brismar T, Hyllienmark L, Ekberg K, and Johansson B-L (2002) Loss of temporal lobe beta power in young adults with type 1 diabetes mellitus. *Neuroreport* 13: 2469–2473.

Brismar T, Maurex L, Cooray G, et al. (2007) Predictors of cognitive impairment in type 1 diabetes. *Psychoneuroendocrinology* 32: 1041–1051.

Brownlee M (2005) The pathobiology of diabetic complications: A unifying mechanism. *Diabetes* 54: 1615–1625.

Bruce DG, Davis WA, Casey GP, et al. (2008) Predictors of cognitive impairment and dementia in older people with diabetes. *Diabetologia* 51: 241–248.

Bruce DG, Davis WA, Starkstein SE, and Davis TME (2005) A prospective study of depression and mortality in patients with type 2 diabetes: The Fremantle Diabetes study. *Diabetologia* 48: 2532–2539.

Chalmers J, Risk MTA, Kean DM, Grant R, Ashworth B, and Campbell IW (1991) Severe amnesia after hypoglycemia. *Diabetes Care* 14: 922–925.

Cheung N and Wong TY (2008) Diabetic retinopathy and systemic vascular complications. *Progress in Retinal and Eye Research* 27: 161–176.

Chugani HT (1998) A critical period of brain development: Studies of cerebral glucose utilization with PET. *Preventive Medicine* 27: 184–188.

Clarkson P, Celermajer DS, Donald AE, et al. (1996) Impaired vascular reactivity in insulin-dependent diabetes mellitus is related to disease duration and low density lipoprotein cholesterol levels. *Journal of the American College of Cardiology* 28: 573–579.

Cohen J (1988) *Statistical Power Analysis for the Behavioral Sciences*, 2nd edn. Hillsdale, NJ: Lawrence Erlbaum.

Cole AR, Astell A, Green C, and Sutherland C (2007) Molecular connexions between dementia and diabetes. *Neuroscience and Biobehavioral Reviews* 31: 1046–1063.

Convit A (2005) Links between cognitive impairment in insulin resistance: An explanatory model. *Neurobiology of Aging* 26S: S31–S35.

Convit A, Wolf OT, Tarshish C, and DeLeon MJ (2003) Reduced glucose tolerance is associated with poor memory performance and hippocampal atrophy among normal elderly. *Proceedings of the National Academy of Sciences of the United States of America* 100: 2019–2022.

Cooper LS, Wong TY, Klein R, et al. (2006) Retinal microvascular abnormalities and MRI-defined subclinical cerebral infarction: The Atherosclerosis Risk in Communities study. *Stroke* 37: 82–86.

Costacou T, Chang Y-F, Ferrell RE, and Orchard TJ (2006) Identifying genetic susceptibilities to diabetes-related complications among individuals at low risk of complications: An application of tree-structured survival analysis. *American Journal of Epidemiology* 164: 862–872.

Cosway R, Strachan MWJ, Dougall A, Frier BM, and Deary IJ (2001) Cognitive function and information processing in type 2 diabetes. *Diabetic Medicine* 18: 803–810.

Cotton F, Kamoun S, Rety-Jacob F, Tran-Minh VA, Nighoghossian N, and Hermier M (2005) Acute hypertensive encephalopathy with widespread small-vessel disease at MRI in a diabetic patient: Pathogenetic hypotheses. *Neuroradiology* 47: 599–603.

Cox DJ, Gonder-Frederick LA, Kovatchev BP, Julian DM, and Clarke WL (2000) Progressive hypoglycemia's impact on driving simulation performance. *Diabetes Care* 23: 163–170.

Cox DJ, Penberthy JK, Zrebiec JF, et al. (2003) Diabetes and driving mishaps: Frequency and correlations from a multinational survey. *Diabetes Care* 26: 2329–2334.

Craft S, Asthana S, Newcomer JW, et al. (1999) Enhancement of memory in Alzheimer disease with insulin and somatostatin, but not glucose. *Archives of General Psychiatry* 56: 1135–1140.

Craft S, Peskind E, Schwartz MW, Schellenberg GD, Raskind M, and Porte D (1998) Cerebrospinal fluid and plasma insulin levels in Alzheimer's disease: Relationship to severity of dementia and apolipoprotein E genotype. *Neurology* 50: 164–168.

Craft S and Watson GS (2004) Insulin and neurodegenerative disease: Shared and specific mechanisms. *Lancet Neurology* 3: 169–178.

Crooks VC, Buckwalter JG, and Petitti DB (2003) Diabetes mellitus and cognitive performance in older women. *Annals of Epidemiology* 13: 613–619.

Cukierman T, Gerstein HC, and Williamson JD (2005) Cognitive decline and dementia in diabetes – systematic overview of prospective observational studies. *Diabetologia* 48: 2460–2469.

Dahlquist G, Källén B, and Swedish Childhood Diabetes Study Group (2007) School performance in children with type 1 diabetes: A population-based register study. *Diabetologia* 50: 957–964.

Daneman D (2006) Type 1 diabetes. *Lancet* 367: 847–858.

DCCT Research Group (1993) The effect of intensive treatment of diabetes on the development and progression of long-term complications in insulin-dependent diabetes mellitus. *New England Journal of Medicine* 329: 977–986.

DCCT/EDIC Research Group (2000) Retinopathy and nephropathy in patients with type 1 diabetes four years after a trial of intensive therapy. *New England Journal of Medicine* 342: 381–389.

DCCT/EDIC Research Group (2002) Effect of intensive therapy on the microvascular complications of type 1 diabetes. *Journal of the American Medical Association* 287: 2563–2569.

DCCT/EDIC Research Group (2003) Sustained effect of intensive treatment of type 1 diabetes mellitus on development and progression of diabetic nephropathy: The Epidemiology of Diabetes Interventions and Complications (EDIC) study. *Journal of the American Medical Association* 290: 2159–2167.

DCCT/EDIC Research Group (2007) Long term effects of diabetes and its treatment on cognitive function. *New England Journal of Medicine* 356: 1842–1852.

De La Torre JC (2004) Alzheimer's disease is a vasocognopathy: A new term to describe its nature. *Neurological Research* 26: 517–524.

Deary I, Crawford J, Hepburn DA, Langan SJ, Blackmore LM, and Frier BM (1993) Severe hypoglycemia and intelligence in adult patients with insulin-treated diabetes. *Diabetes* 42: 341–344.

Dejgaard A, Gade A, Larsson H, Balle V, Parving A, and Parving H (1991) Evidence for diabetic encephalopathy. *Diabetic Medicine* 8: 162–167.

den Heijer T, Vermeer SE, van Dijk EJ, Prins ND, Koudstaal PJ, Hofman A, and Breteler MMB (2003) Type 2 diabetes and atrophy of medial temporal lobe structures on brain MRI. *Diabetologia* 46: 1604–1610.

Desrocher M and Rovet J (2004) Neurocognitive correlates of type 1 diabetes in childhood. *Child Neuropsychology* 10: 36–52.

Dey J, Misra A, Desai NG, Mahapatra AK, and Padma MV (1995) Cerebral function in a relatively young subset of NIDDM patients. *Diabetologia* 38: 251.

Di Leo MAS, DiNardo W, Cercone S, et al. (1997) Cochlear dysfunction in IDDM patients with subclinical peripheral neuropathy. *Diabetes Care* 20: 824–828.

Donaghue KC, Chiarelli F, Trotta D, Allgrove J, and Dahl-Jorgensen K (2007) Microvascular and macrovascular complications. *Pediatric Diabetes* 8: 163–170.

Donaghue KC, Craig ME, Chan AKF, et al. (2005) Prevalence of diabetes complications 6 years after diagnosis in an incident cohort of childhood diabetes. *Diabetic Medicine* 22: 711–718.

Drachman DA (2006) Aging of the brain, entropy, and Alzheimer disease. *Neurology* 67: 1340–1352.

Durmus C, Yetiser S, and Durmus O (2004) Auditory brainstem evoked responses in insulin-dependent (ID) and non-insulin-dependent (NID) diabetic subjects with normal hearing. *International Journal of Audiology* 43: 29–33.

Eeg-Olofsson O (1977) Hypoglycemia and neurological disturbances in children with diabetes mellitus. *Acta Psychiatrica Scandinavica* (supplement 270): 91–95.

Elias PK, Elias MF, D'Agostino RB, Cupples LA, Wilson PW, Silbershatz H, and Wolf PA (1997) NIDDM and blood pressure as risk factors for poor cognitive performance. *Diabetes Care* 20: 1388–1395.

Ferguson SC, Blane A, Perros P, et al. (2003) Cognitive ability and brain structure in type 1 diabetes: Relation to microangiopathy and preceding severe hypoglycemia. *Diabetes* 52: 149–156.

Ferguson SC, Blane A, Wardlaw JM, Frier BM, Perros P, McCrimmon RJ, and Deary IJ (2005) Influence of an early-onset age of type 1 diabetes on cerebral structure and cognitive function. *Diabetes Care* 28: 1431–1437.

Fong DS, Aiello LP, Ferris FL, and Klein R (2004) Diabetic retinopathy. *Diabetes Care* 27: 2540–2553.

Fontbonne A, Ducimetiere P, Berr C, and Alperovitch A (2001) Changes in cognitive abilities over a 4-year period are unfavorably affected in elderly diabetic subjects: Results of the epidemiology of vascular aging study. *Diabetes Care* 24: 366–370.

Foster JW and Hart RG (1987) Hypoglycemic hemiplegia: Two cases and a clinical review. *Stroke* 18: 944–946.

Frölich L, Blum-Degen D, Bernstein H-G, et al. (1998) Brain insulin and insulin receptors in aging and sporadic Alzheimer's disease. *Journal of Neural Transmission* 105: 423–438.

Fujioka M, Okuchi K, Hiramatsu K, Sakaki T, Sakaguchi S, and Ishii Y (1997) Specific changes in human brain after hypoglycemic injury. *Stroke* 28: 584–587.

Fülesdi B, Limburg M, Bereczki D, et al. (1997) Impairment of cerebrovascular reactivity in long-term type 1 diabetes. *Diabetes* 46: 1840–1845.

Fülesdi B, Limburg M, Bereczki D, et al. (1999) Cerebrovascular reactivity and reserve capacity in type II diabetes mellitus. *Journal of Diabetes Complications* 13: 191–199.

Gallego PH, Wiltshire E, and Donaghue KC (2007) Identifying children at particular risk of long-term diabetes complications. *Pediatric Diabetes* 8(supplement 6): 40–48.

Gauthier S, Reisberg B, Zaudig M, et al. (2006) Mild cognitive impairment. *Lancet* 367: 1262–1270.

Geissler A, Fründ R, Schölmerich J, Feuerbach S, and Zietz B (2003) Alterations of cerebral metabolism in patients with diabetes mellitus studied by proton magnetic resonance spectroscopy. *Experimental and Clinical Endocrinology and Diabetes* 111: 421–427.

Geroldi C, Frisoni GB, Paolisso G, et al. (2005) Insulin resistance in cognitive impairment: The InCHIANTI study. *Archives of Neurology* 62: 1067–1072.

Gerozissis K (2003) Brain insulin: Regulation, mechanisms of actions and functions. *Cell and Molecular Biology* 23: 1–25.

Gilhaus KH, Daweke H, Lülsdorf HG, Sachsse R, and Sachsse B (1973) EEG-Veränderungen bei diabetischen Kindern. *Deutsche Medizinische Wochenschrift* 98: 1449–1454.

Gold AE, Deary IJ, and Frier BM (1993) Recurrent severe hypoglycaemia and cognitive function in type 1 diabetes. *Diabetic Medicine* 10: 503–508.

Gold SM, Dziobek I, Sweat V, et al. (2007) Hippocampal damage and memory impairments as possible early brain complications of type 2 diabetes. *Diabetologia* 50: 711–719.

Goldstein DE, Little RR, Lorenz RA, Malone JI, Nathan D, and Peterson CM (1995) Tests of glycemia in diabetes. *Diabetes Care* 18: 896–909.
Greenwood CE, Hebblethwaite S, Kaplan RJ, and Jenkins DJA (2003) Carbohydrate-induced memory impairment in adults with type 2 diabetes. *Diabetes Care* 26: 1961–1966.
Gregg EW, Mangione CM, Cauley JA, et al. (2002) Diabetes and incidence of functional disability in older women. *Diabetes Care* 25: 61–67.
Gregori B, Galié E, Pro S, Clementi A, and Accornero N (2006) Luminance and chromatic visual evoked potentials in type 1 and type 2 diabetes: Relationships with peripheral neuropathy. *Neurological Sciences* 27: 323–327.
Grodstein F, Chen J, Wilson RS, and Manson JE (2001) Type 2 diabetes and cognitive function in community-dwelling elderly women. *Diabetes Care* 24: 1060–1065.
Gschwend S, Ryan C, Atchinson J, Arslanian S, and Becker D (1995) Effects of acute hyperglycemia on mental efficiency and counterregulatory hormones in adolescents with insulin-dependent diabetes mellitus. *Journal of Pediatrics* 126: 178–184.
Haan MN (2006) Therapy insight: Type 2 diabetes mellitus and the risk of late-onset Alzheimer's disease. *Nature Clinical Practice Neurology* 2: 159–166.
Hagen JW, Barclay CR, Anderson BJ, Feeman DJ, Segal SS, Bacon G, and Goldstein GW (1990) Intellective functioning and strategy use in children with insulin-dependent diabetes mellitus. *Child Development* 61: 1714–1727.
Hannonen R, Tupola S, Ahonen T, and Riikonen R (2003) Neurocognitive functioning in children with type-1 diabetes with and without episodes of severe hypoglycemia. *Developmental Medicine and Child Neurology* 45: 262–268.
Hassing LB, Hofer SM, Nilsson SE, Berg S, Pedersen NL, McClearn G, and Johansson B (2004) Comorbid type 2 diabetes mellitus and hypertension exacerbates cognitive decline: Evidence from a longitudinal study. *Age and Ageing* 33: 355–361.
Haumont D, Dorchy H, and Pelc S (1979) EEG abnormalities in diabetic children: Influence of hypoglycemia and vascular complications. *Clinical Pediatrics* 18: 750–753.
Hershey T, Bhargava N, Sadler M, White NH, and Craft S (1999) Conventional vs. intensive diabetes therapy in children with type 1 diabetes: Effects on memory and motor speed. *Diabetes Care* 22: 1318–1324.
Hershey T, Lillie R, Sadler M, and White NH (2003) Severe hypoglycemia and long-term spatial memory in children with type 1 diabetes mellitus: A retrospective study. *Journal of the International Neuropsychology Society* 9: 740–750.
Hershey T, Lillie R, Sadler M, and White NH (2004) A prospective study of severe hypoglycemia and long-term spatial memory in children with type 1 diabetes. *Pediatric Diabetes* 5: 63–71.
Hidasi E, Káplár M, Diószeghy P, Bereczki D, Csiba L, Limburg M, and Fülesdi B (2002) No correlation between impairment of cerebrovascular reserve capacity and electrophysiologically assessed severity of neuropathy in noninsulin-dependent diabetes mellitus. *Journal of Diabetes Complications* 16: 228–234.
Hissa MN, D'Almeida JA, Cremasco F, and de Bruin VM (2002) Event related P300 potentials in NIDDM patients without cognitive impairment and its relationship with previous hypoglycemic episodes. *Neuroendocrinology Letters* 23: 226–230.
Ho MS, Weller NJ, Ives FJ et al. (2008) High prevalence of structural CNS abnormalities in early onset type 1 diabetes. *Journal of Pediatrics* 153(3): 385–390.
Hopkins DFC and Williams G (1997) Insulin receptors are widely distributed in human brain and bind human and porcine insulin with equal affinity. *Diabetic Medicine* 14: 1044–1050.
Howorka K, Pumprla J, Saletu B, Anderer P, Krieger M, and Schabmann A (2000) Decrease of vigilance assessed by EEG-mapping in type 1 diabetic patients with history of recurrent severe hypoglycaemia. *Psychoneuroendocrinology* 25: 85–105.
Huttenlocher PR and Dabholkar AS (1997) Regional differences in synaptogenesis in human cerebral cortex. *Journal of Comparative Neurology* 387: 167–178.
Hyllienmark L, Maltez J, Dandenell A, Ludviggson J, and Brismar T (2005) EEG abnormalities with and without relation to severe hypoglycaemia in adolescents with type 1 diabetes. *Diabetologia* 48: 412–419.
Jakus V and Rietbrock N (2004) Advanced glycation end-products and the progress of diabetic vascular complications. *Physiological Research* 53: 131–142.
Jarjour IT, Ryan CM, and Becker DJ (1995) Regional cerebral blood flow during hypoglycaemia in children with IDDM. *Diabetologia* 38: 1090–1095.
Järvisalo MJ, Raitakari M, Toikka JO, et al. (2004) Endothelial dysfunction and increased arterial intima-media thickness in children with type 1 diabetes. *Circulation* 109: 1750–1755.
Jellinger K (2007) The enigma of vascular cognitive disorder and vascular dementia. *Acta Neuropathologica (Berl)* 113: 349–388.
Jiménez-Bonilla JF, Quirce R, Hernández A, et al. (2001) Assessment of cerebral perfusion and cerebrovascular reserve in insulin-dependent diabetic patients without central neurological symptoms by means of ^{99m}Tc-HMPAO SPET with acetazolamide. *European Journal of Nuclear Medicine* 28: 1647–1655.
Jones TW and Davis EA (2003) Hypoglycemia in children with type 1 diabetes: Current issues and controversies. *Pediatric Diabetes* 4: 143–150.
Jongen C, Van der Grond J, Kappelle LJ, Biessels GJ, Viergever MA, Pluim JPW, and Utrecht Diabetic Encephalopathy Study Group (2007) Automated measurement of brain and white matter lesion volume in type 2 diabetes mellitus. *Diabetologia* 50: 1509–1516.
Jung SL, Kim BS, Lee KS, Yoon KH, and Byun JY (2005) Magnetic resonance imaging and diffusion-weighted imaging changes after hypoglycemic coma. *Journal of Neuroimaging* 15: 193–196.
Kalmijn S, Feskens EJM, Launer LJ, Stijnen T, and Kromhout D (1995) Glucose intolerance, hyperinsulinaemia, and cognitive function in a general population of elderly men. *Diabetologia* 38: 1096–1102.
Karabouta Z, Barnett S, Shield JPH, Ryan FL, and Crowne EC (2008) Peripheral neuropathy is an early complication of type 2 diabetes in adolescence. *Pediatric Diabetes* 9: 110–114.
Karsidag S, Morah S, Sargin M, Salman S, Karsidag K, and Us O (2005) The electrophysiological findings of subclinical neuropathy in patients with recently diagnosed type 1 diabetes. *Diabetes Research and Clinical Practice* 67: 211–219.
Kaufman FR, Epport K, Engilman R, and Halvorson M (1999) Neurocognitive functioning in children diagnosed with diabetes before age 10 years. *Journal of Diabetes Complications* 13: 31–38.
Kent S (1976) Is diabetes a form of accelerated aging? *Geriatrics* 31: 140–154.
Keymeulen B, Jacobs A, de Metz K, de Sadeleer C, Bossuyt A, and Somers G (1995) Regional cerebral hypoperfusion in long-term type 1 (insulin-dependent) diabetic patients: Relation to hypoglycaemic events. *Nuclear Medicine Communications* 16: 10–16.
Klein JP and Waxman SG (2003) The brain in diabetes: Molecular changes in neurons and their implications for end-organ damage. *Lancet Neurology* 2: 548–554.

Korf ESC, White LR, Scheltens P, and Launer LJ (2004) Midlife blood pressure and the risk of hippocampal atrophy: The Honolulu Asia Aging study. *Hypertension* 44: 29–34.
Korf ESC, White LR, Scheltens P, and Launer LJ (2006) Brain aging in very old men with type 2 diabetes. *Diabetes Care* 29: 2268–2274.
Kovacs M, Goldston D, and Iyengar S (1992) Intellectual development and academic performance of children with insulin-dependent diabetes mellitus: A longitudinal study. *Developmental Psychology* 28: 676–684.
Kramer L, Fasching P, Madl C, et al. (1998) Previous episodes of hypoglycemic coma are not associated with permanent cognitive brain dysfunction in IDDM patients on intensive insulin treatment. *Diabetes* 47: 1909–1914.
Kreis R and Ross BD (1992) Cerebral metabolic disturbances in patients with subacute and chronic diabetes mellitus: Detection with proton MR spectroscopy. *Radiology* 184: 123–130.
Kumari M, Brunner E, and Fuhrer R (2000) Minireview: Mechanisms by which the metabolic syndrome and diabetes impair memory. *Journal of Gerontology: Biological Sciences* 55A: B228–B232.
Kumari M and Marmot M (2005) Diabetes and cognitive function in a middle-aged cohort: Findings from the Whitehall II study. *Neurology* 65: 1597–1603.
Kuo H-K, Yen C-J, Chang C-H, Kuo C-K, Chen J-H, and Sorond F (2005) Relation of C-reactive protein to stroke, cognitive disorders, and depression in the general population: Systematic review and meta-analysis. *Lancet Neurology* 4: 371–380.
Kurita A, Katayama K, and Mochio S (1996) Neurophysiological evidence for altered higher brain functions in NIDDM. *Diabetes Care* 19: 361–364.
Kurita A, Mochio S, and Isogai Y (1995) Changes in auditory P300 event-related potentials and brainstem evoked potentials in diabetes mellitus. *Acta Neurologica Scandinavica* 92: 319–323.
Kuusisto J, Koivisto K, Mykkänen L, et al. (1993) Essential hypertension and cognitive function: The role of hyperinsulinemia. *Hypertension* 22: 771–779.
Kuusisto J, Koivisto K, Mykkänen L, et al. (1997) Association between features of the insulin resistance syndrome and Alzheimer's disease independently of apolipoprotein E4 phenotype: Cross sectional population based study. *British Medical Journal* 315: 1045–1049.
Kwa VIH, van der Stam J, Tijimes N, and Vrooland JL (2002) Retinal arterial changes correlate with cerebral small-vessel disease. *Neurology* 2002: 1536–1540.
Last D, Alsop DC, Abduljalil AM, et al. (2007) Global and regional effects of type 2 diabetes on brain tissue volumes and cerebral vasoreactivity. *Diabetes Care* 30: 1193–1199.
Li L and Hölscher C (2007) Common pathological processes in Alzheimer disease and type 2 diabetes: A review. *Brain Research Review* 56: 384–402.
Li JL, Surini M, Catsicas S, Kawashima E, and Bouras C (1995) Age-dependent accumulation of advanced glycosylation end products in human neurons. *Neurobiology of Aging* 16: 69–76.
Lindeman RD, Romero LJ, LaRue A, et al. (2001) A biethnic community survey of cognition in participants with type 2 diabetes, impaired glucose tolerance, and normal glucose tolerance. *Diabetes Care* 24: 1567–1572.
Lingenfelser T, Buettner UW, Sommerwerck U, Renn W, Eggstein M, and Jakober B (1993) Neurophysiological impairments in IDDM patients during euglycemia and hypoglycemia. *Diabetes Care* 16: 1438–1444.
Lobnig BM, Krömeke O, Optenhostert-Porst C, and Wolf OT (2005) Hippocampal volume and cognitive performance in long-standing type-1 diabetic patients without macrovascular complications. *Diabetic Medicine* 23: 32–39.
Longstreth WT, Marino-Larson EK, Klein R, Wong TY, Sharrett AR, Lefkowitz D, and Manolio T (2006) Associations between findings on cranial magnetic resonance imaging and retinal photography in the elderly: The Cardiovascular Health study. *American Journal of Epidemiology* 165: 78–84.
Lteif AN and Schwenk WF (1999) Type 1 diabetes mellitus in early childhood: Glycemic control and associated risk of hypoglycemic reactions. *Mayo Clinic Proceedings* 74: 211–216.
Luchsinger JA, Reitz C, Patel B, Tang M-X, Manly J, and Mayeux R (2007) Relation of diabetes to mild cognitive impairment. *Archives of Neurology* 64: 570–575.
Luchsinger JA, Tang M-X, Shea S, and Mayeux R (2004) Hyperinsulinemia and risk of Alzheimer disease. *Neurology* 63: 1187–1192.
MacLeod KM, Hepburn DA, Deary IJ, Goodwin GM, Dougall N, Ebmeier KP, and Frier BM (1994) Regional cerebral blood flow in IDDM patients: Effects of diabetes and recurrent severe hypoglycaemia. *Diabetologia* 37: 257–263.
Mäkimattila S, Malmberg-Cèder K, Häkkinen A-M, et al. (2004) Brain metabolic alterations in patients with type 1 diabetes-hyperglycemia-induced injury. *Journal of Cerebral Blood Flow and Metabolism* 24: 1393–1399.
Malone JI, Hanna SK, and Saporta S (2006) Hyperglycemic brain injury in the rat. *Brain Research* 1076: 9–15.
Manschot SM, Biessels GJ, DeValk HW, et al. (2007) Metabolic and vascular determinants of impaired cognitive performance and abnormalities on brain magnetic resonance imaging in patients with type 2 diabetes. *Diabetologia* 50: 2388–2397.
Manschot SM, Biessels GJ, Rutten GEHM, Kessels RPC, Gispen WH, and Kappelle LJ, and Utrecht Diabetic Encephalopathy Study Group (2008) Peripheral and central neurologic complications in type 2 diabetes mellitus: No association in individual patients. *Journal of Neurological Sciences* 264: 157–162.
Manschot SM, Brands AMA, van der Grond J, et al. (2006) Brain magnetic resonance imaging correlates of impaired cognition in patients with type 2 diabetes. *Diabetes* 55: 1106–1113.
Manschot SM, Gispen WH, Kappelle LJ, and Biessels GJ (2003) Nerve conduction velocity and evoked potential latencies in steptozotocin-diabetic rats: Effects of treatment with an angiotensin converting enzyme inhibitor. *Diabetes/Metabolism Research and Reviews* 19: 469–477.
Maser RE, Becker DJ, Drash AL, Ellis D, Kuller LH, Greene DA, and Orchard TJ (1992) Pittsburgh Epidemiology of Diabetes Complications study: Measuring diabetic neuropathy follow-up study results. *Diabetes Care* 15: 525–527.
McCarthy AM, Lindgren S, Mengeling MA, Tsalikian E, and Engvall JC (2003) Factors associated with academic achievement in children with type 1 diabetes. *Diabetes Care* 26: 112–117.
Mijnhout GS, Scheltens P, Diamant M, et al. (2006) Diabetic encephalopathy: A concept in need of a definition. *Diabetologia* 49: 1447–1448.
Miles WR and Root HF (1922) Psychologic tests applied to diabetic patients. *Archives of Internal Medicine* 30: 767–777.
Mooradian AD, Perryman K, Fitten J, Kavonian GD, and Morley JE (1988) Cortical function in elderly non-insulin dependent diabetic patients. Behavioral and electrophysiologic studies. *Archives of Internal Medicine* 148: 2369–2372.
Mori F, Nishie M, Houszen H, Yamaguchi J, and Wakabayashi K (2006) Hypoglycemic encephalopathy with extensive lesions in the cerebral white matter. *Neuropathology* 26: 147–152.
Münch G, Schinzel R, Loske C, et al. (1998) Alzheimer's disease – synergistic effects of glucose deficit, oxidative stress and advanced glycation endproducts. *Journal of Neural Transmission* 105: 439–461.

Munshi M, Grande L, Hayes M, et al. (2006) Cognitive dysfunction is associated with poor diabetes control in older adults. *Diabetes Care* 29: 1794–1799.
Musen G, Lyoo IK, Sparks CR, et al. (2006) Effects of type 1 diabetes on gray matter density as measured by voxel-based morphometry. *Diabetes* 55: 326–333.
Nagamachi S, Nishikawa T, Ono S, et al. (1994) Regional cerebral blood flow in diabetic patients: Evaluation by N-isopropyl-123I-IMP with SPECT. *Nuclear Medicine Communications* 15: 455–460.
Nakamura Y, Takahashi M, Kitaguti M, Imaoka H, Kono N, and Tarui S (1991) Abnormal brainstem evoked potentials in diabetes mellitus: Evoked potential testings and magnetic resonance imaging. *Electromyography and Clinical Neurophysiology* 31: 243–249.
Nathan DM (1996) The pathophysiology of diabetic complications: How much does the glucose hypothesis explain? *Annals of Internal Medicine* 124(supplement 1S-II): 86–89.
Nehlig A (1997) Cerebral energy metabolism, glucose transport and blood flow: Changes with maturation and adaptation to hypoglycaemia. *Diabetes and Metabolism* 23: 18–29.
Ng YT, McGregor AL, Duane DC, Jahnke HK, Bird CR, and Wheless JW (2006) Childhood mesial temporal sclerosis. *Journal of Child Neurology* 21: 512–517.
Nofzinger EA (1997) Sexual dysfunction in patients with diabetes mellitus: The role of a central neuropathy. *Seminars in Clinical Neuropsychiatry* 2: 31–39.
Northam EA, Anderson PJ, Jacobs R, Hughes M, Warne GL, and Werther GA (2001) Neuropsychological profiles of children with type 1 diabetes 6 years after disease onset. *Diabetes Care* 24: 1541–1546.
Northam E, Anderson P, Wether G, Adler R, and Andrewes D (1995) Neuropsychological complications of insulin dependent diabetes in children. *Child Neuropsychology* 1: 74–87.
Northam EA, Anderson PJ, Werther GA, Warne GL, Adler RG, and Andrewes D (1998) Neuropsychological complications of IDDM in children 2 years after disease onset. *Diabetes Care* 21: 379–384.
Northam EA, Rankins D, and Cameron FJ (2006) Therapy Insight: The impact of type 1 diabetes on brain development and function. *Nature Clinical Practice Neurology* 2: 78–86.
Novak V, Last D, Alsop DC, et al. (2006) Cerebral blood flow velocity and periventricular white matter hyperintensities in type 2 diabetes. *Diabetes Care* 29: 1529–1534.
Orchard TJ, Costacou T, Kretowski A, and Nesto RW (2006) Type 1 diabetes and coronary artery disease. *Diabetes Care* 29: 2528–2538.
Orchard TJ, Dorman JS, Maser RE, et al. (1990) Prevalence of complications in IDDM by sex and duration: Pittsburgh Epidemiology of Complications study II. *Diabetes* 39: 1116–1124.
Orchard TJ, Forrest KY-Z, Ellis D, and Becher DJ (1997) Cumulative glycemic exposure and microvascular complications in insulin-dependent diabetes mellitus: The glycemic threshold revisited. *Archives of Internal Medicine* 157: 1851–1856.
Pambianco G, Costacou T, Ellis D, Becker DJ, Klein R, and Orchard TJ (2006) The 30-year natural history of type 1 diabetes complications: The Pittsburgh Epidemiology of Diabetes Complications study experience. *Diabetes* 55: 1463–1469.
Parisi V and Uccioli L (2001) Visual electrophsiological responses in persons with type 1 diabetes. *Diabetes/Metabolism Research and Reviews* 17: 12–18.
Parisi V, Uccioli L, Monticone G, et al. (1997) Electrophysiological assessment of visual function in IDDM patients. *Electroencephalography and Clinical Neurophysiology* 104: 171–179.
Park CR (2001) Cognitive effects of insulin in the central nervous system. *Neuroscience and Biobehavioral Reviews* 25: 311–323.
Patton N, Aslam T, MacGillivray T, Pattie A, Deary IJ, and Dhillon B (2005) Retinal vascular image analysis as a potential screening tool for cerebrovascular disease: A rationale based on homology between cerebral and retinal microvasculatures. *Journal of Anatomy* 206: 319–348.
Patton N, Aslam TM, MacGillivray T, et al. (2006) Retinal image analysis: Concepts, applications and potential. *Progress in Retinal and Eye Research* 25: 99–127.
Peila R, Rodriguez BL, and Launer LJ (2002) Type 2 diabetes, *APOE* gene, and the risk for dementia and related pathologies: The Honolulu-Asia study. *Diabetes* 51: 1256–1262.
Perantie DC, Lim A, Wu J, et al. (2008) Effects of prior hypoglycemia and hyperglycemia on cognition in children with type 1 diabetes mellitus. *Pediatric Diabetes* 9: 95–97.
Perantie DC, Wu J, Koller JM, et al. (2007) Regional brain volume differences associated with hyperglycemia and severe hypoglycemia in youth with type 1 diabetes. *Diabetes Care* 30: 2331–2337.
Perlmuter LC, Hakami MK, Hodgson-Harrington C, Gingsberg J, Katz J, Singer DE, and Nathan DM (1984) Decreased cognitive function in aging non-insulin-dependent diabetic patients. *American Journal of Medicine* 77: 1043–1048.
Perros P, Deary IJ, Sellar RJ, Best JJK, and Frier BM (1997) Brain abnormalities demonstrated by magnetic resonance imaging in adult IDDM patients with and without a history of recurrent severe hypoglycemia. *Diabetes Care* 20: 1013–1018.
Perros P, Sellar RJ, and Frier BM (1994) Chronic pontine dysfunction following insulin-induced hypoglycemia in an IDDM patient. *Diabetes Care* 17: 725–727.
Peters A, Schweiger U, Pellerin L, et al. (2004) The selfish brain: Competition for energy resources. *Neuroscience and Biobehavioral Reviews* 28: 143–180.
Petrica L, Petrica M, Vlad A, et al. (2007) Cerebrovascular reactivity is impaired in patients with non-insulin dependent diabetes mellitus and microangiopathy. *Wiener Klinische Wochenschrift* 119: 365–371.
Pozzessere G, Rizzo PA, Valle E, et al. (1988) Early detection of neurological involvement in IDDM and NIDDM: Multimodal evoked potentials versus metabolic control. *Diabetes Care* 11: 473–480.
Pozzessere G, Valle E, DeCrignis S, et al. (1991) Abnormalities of cognitive functions in IDDM revealed by P300 event-related potential analysis. *Diabetes* 40: 952–958.
Prins ND, van Dijk EJ, den Heijer T, et al. (2005) Cerebral small-vessel disease and decline in information processing speed, executive function and memory. *Brain* 128: 2034–2041.
Qiu C, Winblad B, and Fratiglioni L (2005) The age-dependent relation of blood pressure to cognitive function and dementia. *Lancet Neurology* 4: 487–499.
Qiu WQ, Price LL, Hibberd P, et al. (2006) Executive dysfunction in homebound older people with diabetes mellitus. *Journal of American Geriatric Society* 54: 496–501.
Quirce R, Carril JM, Jiménez-Bonilla JF, et al. (1997) Semi-quantitative assessment of cerebral blood flow with ^{99m}Tc-HMPAO SPET in type 1 diabetic patients with no clinical history of cerebrovascular disease. *European Journal of Nuclear Medicine* 24: 1507–1513.
Ramasamy R, Vannucci SJ, Yan SSD, Herold K, and Schmidt AM (2005) Advanced glycation end products and RAGE: A common thread in aging, diabetes,

neurodegeneration, and inflammation. *Glycobiology* 15: 16R–28R.
Rankins D, Wellard RM, Cameron F, McDonnell C, and Northam E (2005) The impact of acute hypoglycemia on neuropsychological and neurometabolite profiles in children with type 1 diabetes. *Diabetes Care* 28: 2771–2773.
Raz I, Eldor R, Cernea S, and Shafrir E (2005) Diabetes: Insulin resistance and derangements in lipid metabolism. Cure through intervention in fat transport and storage. *Diabetes/Metabolism Research and Reviews* 21: 3–14.
Raz N, Rodrigue KM, and Acker JD (2003) Hypertension and the brain: Vulnerability of the prefrontal regions and executive functions. *Behavioral Neuroscience* 117: 1169–1180.
Reaven GM, Thompson LW, Nahum D, and Haskins E (1990) Relationship between hyperglycemia and cognitive function in older NIDDM patients. *Diabetes Care* 13: 16–21.
Reitz C, Tang M-X, Manly J, Mayeux R, and Luchsinger JA (2007) Hypertension and the risk of mild cognitive impairment. *Archives of Neurology* 64: 1734–1740.
Reske-Nielsen E, Lundbaek K, and Rafaelsen OJ (1965) Pathological changes in the central and peripheral nervous system of young long-term diabetics. *Diabetologia* 1: 232–241.
Rodriguez G, Nobili F, Celestino MA, et al. (1993) Regional cerebral blood flow and cerebrovascular reactivity in IDDM. *Diabetes Care* 16: 462–468.
Rovet J and Alverez M (1997) Attentional functioning in children and adolescents with IDDM. *Diabetes Care* 20: 803–810.
Rovet JF, Ehrlich RM, Czuchta D, and Akler M (1993) Psychoeducational characteristics of children and adolescents with insulin-dependent diabetes mellitus. *Journal of Learning Disabilities* 26: 7–22.
Rovet JF, Ehrlich RM, and Hoppe MG (1987) Intellectual deficits associated with the early onset of insulin-dependent diabetes mellitus in children. *Diabetes Care* 10: 510–515.
Ryan CM (1997) Effects of diabetes mellitus on neuropsychological functioning: A lifespan perspective. *Seminars in Clinical Neuropsychiatry* 2: 4–14.
Ryan CM (1999) Memory and metabolic control in children. *Diabetes Care* 22: 1242–1244.
Ryan CM (2005) Diabetes, aging, and cognitive decline. *Neurobiology of Aging* 26S: S21–S25.
Ryan CM (2006a) Diabetes and brain damage: More (or less) than meets the eye? *Diabetologia* 49: 2229–2233.
Ryan CM (2006b) Why is cognitive dysfunction associated with the development of diabetes early in life? The diathesis hypothesis. *Pediatric Diabetes* 7: 289–297.
Ryan CM, Freed MI, Rood JA, Cobitz AR, Waterhouse BR, and Strachan MWJ (2006) Improving metabolic control leads to better working memory in adults with type 2 diabetes. *Diabetes Care* 29: 345–351.
Ryan CM and Geckle MO (2000) Circumscribed cognitive dysfunction in middle-aged adults with type 2 diabetes. *Diabetes Care* 23: 1486–1493.
Ryan CM, Geckle MO, and Orchard TJ (2003) Cognitive efficiency declines over time in adults with type 1 diabetes: Effects of micro- and macrovascular complications. *Diabetologia* 46: 940–948.
Ryan C, Gurtunca N, and Becker D (2005) Hypoglycemia: A complication of diabetes therapy in children. *Pediatric Clinics of North America* 52: 1705–1733.
Ryan C, Longstreet C, and Morrow LA (1985a) The effects of diabetes mellitus on the school attendance and school achievement of adolescents. *Child: Care, Health, and Development* 11: 229–240.
Ryan C, Vega A, and Drash A (1985b) Cognitive deficits in adolescents who developed diabetes early in life. *Pediatrics* 75: 921–927.
Ryan C, Vega A, Longstreet C, and Drash L (1984) Neuropsychological changes in adolescents with insulin-dependent diabetes mellitus. *Journal of Consulting and Clinical Psychology* 52: 335–342.
Ryan CM, Williams TM, Orchard TJ, and Finegold DN (1992) Psychomotor slowing is associated with distal symmetrical polyneuropathy in adults with diabetes mellitus. *Diabetes* 41: 107–113.
Sabri O, Hellwig D, Schreckenberger M, et al. (2000) Influence of diabetes mellitus on regional cerebral glucose metabolism and regional cerebral blood flow. *Nuclear Medicine Communications* 21: 19–29.
Said G (2007) Diabetic neuropathy – a review. *Nature Clinical Practice Neurology* 3: 331–340.
Salem MAK, Matta LF, Tantawy AAG, Hussein M, and Gad GI (2002) Single photon emission tomography (SPECT) study of regional cerebral blood flow in normoalbuminuric children and adolescents with type 1 diabetes. *Pediatric Diabetes* 3: 155–162.
Sarac K, Akinci A, Alkan A, Baysal T, and Özcan C (2005) Brain metabolite changes on proton magnetic resonance spectroscopy in children with poorly controlled type 1 diabetes. *Neuroradiology* 47: 562–565.
Schmahmann JD and Sherman JC (1998) The cerebellear cognitive affective syndrome. *Brain* 121: 561–579.
Schmidt R, Launer LJ, Nilsson L-G, et al. (2004) Magnetic resonance imaging of the brain in diabetes: The Cardiovascular Determinants of Dementia (CASCADE) study. *Diabetes* 53: 687–692.
Schoenle EJ, Schoenle D, Molinari L, and Largo RH (2002) Impaired intellectual development in children with type 1 diabetes: Association with HbA_{1c}, age at diagnosis, and sex. *Diabetologia* 45: 108–114.
Schulinghamp RJ, Pagano TC, Hung D, and Raffa RB (2000) Insulin receptors and insulin action in the brain: Review and clinical implications. *Neuroscience and Biobehavioral Reviews* 24: 855–872.
Schwartz MW, Figlewicz DP, Baskin DG, Woods SC, and Porte D (1992) Insulin in the brain: A hormonal regulator of energy balance. *Endocrine Review* 13: 387–414.
Schwartz MW, Figlewicz DP, Baskin DG, Woods SC, and Porte D (1994) Insulin and the central regulation of energy balance: Update 1994. *Endocrine Reviews* 2: 109–113.
Seidl R, Birnbacher R, Hauser E, Gernert G, Freilinger M, and Schober E (1996) Brainstem auditory evoked potentials and visually evoked potentials in young patients with IDDM. *Diabetes Care* 19: 1220–1224.
Shaw J (2007) Epidemiology of childhood type 2 diabetes and obesity. *Pediatric Diabetes* 8(supplement 9): 7–15.
Shaw P, Greenstein D, Lerch J, et al. (2006) Intellectual ability and cortical development in children and adolescents. *Nature* 440: 676–679.
Sheetz MJ and King GL (2002) Molecular understanding of hyperglycemia's adverse effects for diabetic complications. *Journal of the American Medical Association* 288: 2579–2588.
Sima AF, Kamiya H, and GuoLi Z (2004) Insulin-C-peptide, hyperglycemia, and central nervous system complications in diabetes. *European Journal of Pharmacology* 490: 187–197.
Singh R, Barden A, Mori T, and Beilin L (2001) Advanced glycation end-products: A review. *Diabetologia* 44: 129–146.
Soltész G and Acsádi G (1989) Association between diabetes, severe hypoglycemia, and electroencephalographic abnormalities. *Archives of Disease in Childhood* 64: 992–996.
Sovik O and Thordarson H (1999) Dead-in-bed syndrome in young diabetic patients. *Diabetes Care* 22(supplement 2): B40–B42.

Stewart RJ and Liolitsa D (1999) Type 2 diabetes mellitus, cognitive impairment and dementia. *Diabetic Medicine* 16: 93–112.
Stolk RP, Breteler MMB, Ott A, Pols HAP, Lamberts SWJ, Grobbee DE, and Hofman A (1997) Insulin and cognitive function in an elderly population: The Rotterdam Study. *Diabetes Care* 20: 792–795.
Strachan MWJ (2007) Frequency and morbidity of severe hypoglycaemia in insulin-treated diabetic patients. In: Frier BM and Fisher M (eds.) *Hypoglycaemia in Clinical Diabetes,* 2nd edn., pp. 49–81. Chichester: Wiley.
Strachan MWJ, Abraha HD, Sherwood RA, et al. (1999) Evaluation of serum markers of neuronal damage following severe hypoglycaemia in adults with insulin-treated diabetes mellitus. *Diabetes/Metabolism Research and Reviews* 15: 5–12.
Strachan MWJ, Deary IJ, Ewing FME, and Frier BM (1997) Is type 2 (non-insulin dependent) diabetes mellitus associated with an increased risk of cognitive dysfunction? *Diabetes Care* 20: 438–445.
Strudwick SK, Carne C, Gardiner J, Foster JK, Davis EA, and Jones TW (2005) Cognitive functioning in children with early onset type 1 diabetes and severe hypoglycemia. *Journal of Pediatrics* 147: 680–685.
Stumvoll M, Goldstein BJ, and van Haeften TW (2005) Type 2 diabetes: Principles of pathogenesis and therapy. *Lancet* 365: 1333–1346.
Sugimoto K, Murakawa Y, and Sima AAF (2000) Diabetic neuropathy – a continuing enigma. *Diabetes/Metabolism Research and Reviews* 16: 408–433.
Tisserand DJ and Jolles J (2003) On the involvement of prefrontal networks in cognitive ageing. *Cortex* 39: 1107–1128.
Tomlinson DR and Gardiner NJ (2008) Glucose neurotoxicity. *Nature Reviews: Neuroscience* 9: 36–45.
Tupola S, Rajantie J, and Mäenpää J (1998) Severe hypoglycaemia in children and adolescents during multiple-dose insulin therapy. *Diabetic Medicine* 15: 695–699.
UK Prospective Diabetes Study Group (1998) Intensive blood-glucose control with sulphonylureas or insulin compared with conventional treatment and risk of complications in patients with type 2 diabetes (UKPDS 33). *Lancet* 352: 837–853.
Unger JW and Betz M (1998) Insulin receptors and signal transduction proteins in the hypothalamo-hypophyseal system: A review on morphological findings and functional implications. *Histology and Histpathology* 13: 1215–1224.
van den Berg E, DeCraen AJM, Biessels GJ, Gusselkloo J, and Westendorp RGJ (2006) The impact of diabetes mellitus on cognitive decline in the oldest old: A prospective population-based study. *Diabetologia* 49: 2015–2023.
Van den Berg E, Kessels RPC, De Haan EHF, Kappelle LJ, and Biessels GJ (2005) Mild impairments in cognition in patients with type 2 diabetes mellitus: The use of the concepts MCI and CIND. *Journal of Neurology, Neurosurgery and Psychiatry* 76: 1466–1467.
van Harten B, De Leeuw F-E, Weinstein HC, Scheltens P, and Biessels GJ (2006) Brain imaging in patients with diabetes: A systematic review. *Diabetes Care* 29: 2539–2548.
van Harten B, Oosterman J, Muslimovic D, Potter van Loon B-J, Scheltens P, and Weinstein HC (2007) Cognitive impairments and MRI correlates in the elderly patients with type 2 diabetes mellitus. *Age and Ageing* 36: 164–170.
Várkonyi TT, Petõ T, Dégi R, et al. (2002) Impairment of visual evoked potentials: An early manifestation of diabetic neuropathy? *Diabetes Care* 25: 1161–1162.
Vázquez LA, Amado JA, Carcía-Unzueta MT, et al. (1999) Decreased plasma endothelin-1 levels in asymptomatic type 1 diabetic patients with regional cerebral hypoperfusion assessed by Spect. *Journal of Diabetes Complications* 13: 325–331.
Verdelho A, Madureira S, Ferro JM, et al. (2007) Differential impact of cerbral white matter changes, diabetes, hypertension and stroke on cognitive performance among non-disabled elderly. The LADIS study. *Journal of Neurology, Neurosurgery and Psychiatry* 78: 1325–1330.
Verhaeghen P and Salthouse TA (1997) Meta-analysis of age-cognition relations in adulthood: Estimates of linear and nonlinear age effects and structural models. *Psychological Bulletin* 122: 231–249.
Verrotti A, Lobefalo L, Trotta D, et al. (2000) Visual evoked potentials in young persons with newly diagnosed diabetes: A long-term follow-up. *Developmental Medicine and Child Neurology* 42: 240–244.
Vinik AI, Maser RE, Mitchell BD, and Freeman R (2003) Diabetic autonomic neuropathy. *Diabetes Care* 26: 1553–1579.
Vinik AI and Ziegler D (2007) Diabetic cardiovascular autonomic neuropathy. *Circulation* 115: 387–397.
Virtaniemi J, Kuusisto J, Karjalainen L, Karjalainen S, and Laakso M (1995) Improvement of metabolic control does not normalize auditory brainstem latencies in subjects with insulin-dependent diabetes mellitus. *American Journal of Otolaryngology* 14: 413–418.
Virtaniemi J, Laakso M, Kärjä J, Nuutinen J, and Karjalainen S (1993) Auditory brainstem latencies in type 1 (insulin-dependent) diabetic patients. *American Journal of Otolaryngology* 14: 413–418.
Wagner VM, Grabert M, and Holl RW (2005) Severe hypoglycaemia, metabolic control and diabetes management in children with type 1 diabetes in the decade after the Diabetes Control and Complications Trial – a large scale multicentre study. *European Journal of Pediatrics* 164: 73–79.
Waldstein SR, Brown JRP, Maier KJ, and Katzel LI (2005) Diagnosis of hypertension and high blood pressure levels negatively affect cognitive function in older adults. *Annals of Behavioral Medicine* 29: 174–180.
Warren RE and Frier BM (2005) Hypoglycaemia and cognitive function. *Diabetes, Obesity and Metabolism* 7: 493–503.
Watari K, Letamendi A, Elderkin-Thompson V, Haroon E, Miller J, Darwin C, and Kumar A (2006) Cognitive function in adults with type 2 diabetes and major depression. *Archives of Clinical Neuropsychology* 21: 787–796.
Watson GS and Craft S (2004) Modulation of memory by insulin and glucose: Neuropsychological observations in Alzheimer's disease. *European Journal of Pharmacology* 490: 97–113.
Wayne EA, Dean HJ, Booth F, and Tenebein M (1990) Focal neurologic deficits associated with hypoglycemia in children with diabetes. *Journal of Pediatrics* 117: 575–577.
Weinger K, Jacobson AM, Musen G, Lyoo IK, Ryan CM, Jimerson DC, and Renshaw PF (2008) The effects of type 1 diabetes on cerebral white matter. *Diabetologia* 51: 417–425.
Wessels AM, Rombouts SARB, Remijnse PL, et al. (2007) Cognitive performance in type 1 diabetes patients is associated with cerebral white matter volume. *Diabetologia* 50: 1763–1769.
Wessels AM, Rombouts SARB, Simsek S, et al. (2006a) Microvascular disease in type 1 diabetes alters brain activation: A functional magnetic resonance imaging study. *Diabetes* 55: 334–340.
Wessels AM, Simsek S, Remijnse PL, et al. (2006b) Voxel-based morphometry demonstrates reduced gray matter density on brain MRI in patients with diabetic retinopathy. *Diabetologia* 49: 2474–2480.
Weston PJ and Gill GV (1999) Is undetected autonomic dysfunction responsible for sudden death in type 1 diabetes

mellitus? The 'dead in bed' syndrome revisted. *Diabetic Medicine* 16: 626–631.
Winblad B, Palmer K, Kivipelto M, et al. (2004) Mild cognitive impairment - beyond controversies, towards a consensus: Report of the International Working Group on mild cognitive impairment. *Journal of Internal Medicine* 256: 240–246.
Wolters CA, Yu SL, Hagen JW, and Kail R (1996) Short-term memory and strategy use in children with insulin-dependent diabetes mellitus. *Journal of Consulting and Clinical Psychology* 64: 1397–1405.
Wong TY, Klein R, Klein BEK, Tielsch JM, Hubbard L, and Nieto FJ (2001) Retinal microvascular abnormalities and their relationship with hypertension, cardiovascular disease, and mortality. *Survey of Ophthalmology* 46: 59–80.
Wong TY, Klein R, Sharrett AR, et al. (2002) Retinal microvascular abnormalities and cognitive impairment in middle-aged persons: The Atherosclerosis Risk in Communities study. *Stroke* 33: 1487–1492.
Wong TY, Mosley TH, Klein R, et al. (2003) Retinal microvascular changes and MRI signs of cerebral atrophy in healthy, middle-aged people. *Neurology* 61: 806–811.
Woods SC, Benoit SC, and Clegg DJ (2006) The brain–gut–islet connection. *Diabetes* 55(supplement 2): S114–S121.
Wysocki T, Harris MA, Mauras N, Fox L, Taylor A, Jackso SC, and White NH (2003) Absence of adverse effects of severe hypoglycemia on cognitive function in school-aged children with diabetes over 18 months. *Diabetes Care* 26: 1100–1105.
Young SE, Mainous AG, and Carnemolla M (2006) Hyperinsulinemia and cognitive decline in a middle-aged cohort. *Diabetes Care* 29: 2688–2693.
Zammitt NN and Frier BM (2005) Hypoglycemia in type 2 diabetes: Pathophysiology, frequency, and effects of different treatment modalities. *Diabetes Care* 28: 2948–2961.
Zhao W-Q, Chen H, Quon MJ, and Alkon DL (2004) Insulin and the insulin receptor in experimental mode s of learning and memory. *European Journal of Pharmacology* 490: 71–81.
Ziegler D, Hübinger A, and Gries FA (1991) Changes in brainstem auditory evoked potentials during insulin-induced hypoglycaemia in type 1 diabetic patients. *Diabetic Medicine* 8: 805–811.
Ziegler O, Guerci B, Algan M, Lonchamp P, Weber W, and Drouin P (1994a) Improved visual evoked potential latencies in poorly controlled diabetic patients after short-term strict metabolic control. *Diabetes Care* 17: 1141–1147.
Ziegler D, Langen K-J, Herzog H, Kuwert T, Mühlen H, Feinendegen LE, and Gries AF (1994b) Cerebral glucose metabolism in type 1 diabetic patients. *Diabetic Medicine* 11: 205–209.

Relevant Website

http://www.selfish-brain.org – The Selfish Brain Theory.

32 Alcohol Abuse: Endocrine Concomitants

E S Ginsburg, Brigham and Women's Hospital, Boston, MA, USA
N K Mello and J H Mendelson[†], McLean Hospital and Harvard Medical School, Boston, MA, USA

Chapter Outline

[†]Deceased.

Glossary

adrenocorticotropic hormone Secreted from anterior pituitary and stimulates adrenal cortisol secretion.

anovulation Absence of ovulation, with or without menstrual periods.

aromaotase Enzyme which converts androgens to estrogens; present in many tissues including fat, liver, and muscle

dehydrogenase Enzyme that removes a hydroxyl group (OH).

estradiol The primary ovarian steroid.

gonadotropins Follicle-stimulating hormone and luteinizing hormone; responsible for follicular development and ovulation.

hypogonadal Having low ovarian or testicular hormone secretion.

hyperprolactinemia Elevated serum prolactin levels.

inhibin It is produced by granulose cells and responsible in part for negative feedback between ovaries and pituitary FSH secretion.

isocaloric It is made equal in caloric content to the test drink/substance.

luteal The period of time (usually 2 weeks) between ovulation and the subsequent menstrual period.

mitogen It stimulates cell division.

opioid Substances which bind to opioid receptors.

progesterone The hormone produced by the ovaries after ovulation, during the luteal phase of the menstrual cycle. Also produced by placenta.

prolactin It is produced by the anterior pituitary, under negative feedback control by dopamine.

teratogen It induces birth defects.

32.1 Introduction

Alcohol abuse and alcohol dependence are associated with many disorders of reproductive function in women. Amenorrhea, anovulation, luteal-phase dysfunction, and hyperprolactinemia may occur in alcohol-dependent women and alcohol abusers. However, luteal-phase dysfunction, anovulation, and persistent hyperprolactinemia have also been observed in social drinkers who were studied under clinical research ward conditions. Tolerance may occur to alcohol's adverse effects on reproductive function since alcohol-dependent women do become pregnant. The reproductive consequences of alcohol abuse and alcohol dependence by women who conceive may culminate in the teratogenicity, evidenced in the fetal alcohol syndrome (FAS).

There has been surprisingly little research on alcohol's effects on reproductive function in women (Cicero, 1980; Mello, 1980, 1988; van Thiel and Gavaler, 1982) since it has long been known that alcohol abuse may induce a gonadal dysfunction in men.

This chapter focuses on alcohol's effects on reproductive hormones in both women and men. Acute effects of alcohol on the hypothalamic–pituitary–gonadaladrenal axis in normal women have been studied following stimulation with opioid antagonists, synthetic luteinizing hormone-releasing hormone (LHRH), and human chorionic gonadotropin (hCG). The impact of alcohol ingestion on hormone levels in postmenopausal women and its implications in the pathogenesis of breast cancer are reviewed. Alcohol's effects in animal studies are also described. The possible implications of alcohol's disruptive effects on pituitary, gonadal, and adrenal hormones in the pathogenesis of fetal dysmorphologies and developmental impairments are considered.

32.2 Alcohol and Reproductive System Dysfunction in Women

32.2.1 Overview of Effects of Alcohol on Reproductive Function

32.2.1.1 Anovulation and luteal-phase dysfunction in alcoholic women

Women who abuse alcohol may have anovulatory cycles or luteal-phase dysfunction (Hugues et al., 1980; Mello, 1988; Mello et al., 1989; Moskovic, 1975; Välimäki et al., 1990b) and report more menstrual cycle abnormalities than age-matched controls (Becker et al., 1989; Jones-Saumty et al., 1981). These disorders impair fertility by preventing pregnancy or by increasing the risk for spontaneous abortion.

Anovulation (failure to ovulate) is inferred from the absence of a mid-cycle gonadotropin surge and subsequent progesterone increase during the luteal phase of the menstrual cycle. Luteal-phase dysfunction is defined either as a short luteal-phase defect (8 days or less from ovulation to menses) or an inadequate luteal phase (when progesterone levels are abnormally low or endometrial biopsy results show a delay in endometrial development of >2 days on two separate cycles). (diZerega and Hodgen, 1981; Goodman and Hodgen, 1983; Sherman, 1984; Stouffer, 1990). In the normal cycle, the luteal phase lasts for approximately 14 days after ovulation. During this time the corpus luteum develops from the postovulatory follicle, producing peak progesterone levels about 7 days after ovulation, then regresses if pregnancy does not occur. The corpus luteum is essential for maintaining the progesterone secretion needed for endometrial development and support of early pregnancy.

Spontaneous abortion may be associated with luteal-phase defects. The relevance of luteal-phase defects in the general population is difficult to determine, in part, because of problems of accurate diagnosis (Balasch and Vanrell, 1987; McNeely and Soules, 1988). Progesterone is secreted in a pulsatile pattern and, therefore, conclusions drawn from single samples of progesterone may be misleading (Filicori et al., 1984; McNeely and Soules, 1988; Soules et al., 1988). Moreover, it has been difficult to define the specific parameters of luteal function necessary for initiation and maintenance of pregnancy (Stouffer, 1990).

There is only one report of the endocrine evaluation of alcohol-dependent women with anovulatory cycles or luteal-phase defects (Hugues et al., 1980). In this study four anovulatory women had scanty menses and intermittent amenorrhea. Clomiphene administration induced a significant increase in luteinizing hormone (LH) and estradiol (E_2). Three of the women had pancreatitis and one had cirrhosis (Hugues et al., 1980). Six women with luteal-phase inadequacy had mild oligomenorrhea and low luteal progesterone levels. Luteal-phase hCG administration increased progesterone levels to above 10 ng ml^{-1} in three of six women. Another small study of 20 women undergoing detoxification found lower levels of E_2, progesterone, and progesterone metabolites than matched controls. After detoxification the differences resolved (Hill et al., 2005).

32.2.1.2 Anovulation and luteal-phase defects in social drinkers

Luteal-phase dysfunction and anovulation have also been induced in healthy, well-nourished, socially drinking women who consumed alcohol during residence on a clinical research ward for 35 days (Mendelson and Mello, 1988). After a 7-day alcohol-free baseline period they could self-administer alcohol for 21 consecutive days. They could earn alcohol (beer, wine, or distilled spirits) or money (50 cents) for simple operant tasks. Points earned for alcohol and for money were not interchangeable (Mello et al., 1990; Mendelson and Mello, 1988). During the study, women were classified as heavy, social, or occasional alcohol users on the basis of the number of drinks they consumed during the 3 consecutive weeks of alcohol availability.

Five women who consumed an average of 7.8 (±0.69) drinks per day were classified as heavy drinkers while 12 women who consumed an average of 3.84 (±0.19) drinks per day were classified as moderate social drinkers. Nine women who consumed 1.22 (±0.21) drinks per day were designated occasional drinkers. These drinking patterns were consistent with each subject's self-report of alcohol use before admission to the clinical research ward. The heavy, social, and occasional alcohol users reported an average drinking history of 7.5, 6.6, and 6.9 years, respectively.

During the 21 days of alcohol availability, average peak blood alcohol levels measured in the moderate and heavy drinkers ranged from 109 (±16) to 199 (±13) mg dl^{-1}. Peak blood alcohol levels measured in the occasional drinkers averaged 48 (±10) to 87 (±22) mg dl^{-1} (Mendelson and Mello, 1988).

In one study, 60% of heavy drinkers and 50% of social drinkers who consumed more than three drinks per day had menstrual cycle disruption (Mendelson and Mello, 1988). Three heavy drinkers and one moderate social drinker, who consumed between 4.24 and 8.24 drinks per day, had persistent hyperprolactinemia (above 20 ng ml^{-1}) during at least 7 of the 21 days of alcohol consumption. Prolactin levels were significantly elevated in the moderate drinker within 5 days after initiation of drinking and peaked during a normal LH surge on study day 16. This woman's prolactin levels remained elevated throughout the luteal phase and after cessation of drinking.

One woman who drank an average of 4.10 ($\pm$0.77) drinks per day did not ovulate until the 28th day of her menstrual cycle. Three moderate social drinkers who consumed between 3.48 and 4.05 drinks per day had anovulatory cycles, as documented by absent LH surges on daily blood monitoring.

Alcohol-related menstrual cycle disorders appear related to the amount of alcohol consumed. There was no menstrual-cycle dysfunction or hormonal abnormalities in occasional drinkers or in two of the moderate social drinkers who consumed less than an average of three drinks per day. However, five of ten social drinkers (who drank more than three drinks per day) and three of the five heavy drinkers had significant derangements of the menstrual and hormonal function. The contrast between occasional and heavy social drinkers suggests that menstrual dysfunction in these otherwise healthy women was due to alcohol and not due to living conditions on the research ward *per se* (Mendelson and Mello, 1988). However, alcohol did not always cause menstrual-cycle abnormalities. The woman who consumed the most alcohol (10 $\pm$ 0.69 drinks per day) had normal cycles and prolactin levels, perhaps due to alcohol tolerance (Mendelson and Mello, 1988). Reproductive tolerance for alcohol can be inferred from the fact that many alcohol abusers and alcohol-dependent women become pregnant (Halmesmäki et al., 1987; Hollstedt et al., 1983; Teoh et al., 1992).

However, a stratified household sample of 917 women in the general population found a strong association between alcohol consumption and several menstrual disorders, including dysmenorrhea, heavy menstrual flow, and premenstrual discomfort. The incidence of these disorders increased as reported drinking levels rose; women who consumed six or more drinks each day at least 5 times a week had elevated rates of gynecological surgery (Wilsnack et al., 1984).

32.2.1.3 Amenorrhea

A number of conditions other than alcohol abuse are associated with amenorrhea. Severe weight loss, such as anorexia nervosa (Sherman, 1984), and obesity may also lead to amenorrhea (Frisch, 1982). Liver, renal, or thyroid disease, polycystic ovaries, and pituitary adenomas are also associated with amenorrhea (Shearman, 1985). In otherwise healthy women, amenorrhea may be associated with weight reduction, or athletic pursuits (Bullen et al., 1985; Frisch, 1982; Frisch and McArthur, 1974; McArthur et al., 1980; Sherman, 1984). It is unlikely that a single course is responsible for all amenorrhea.

A higher frequency of irregular menstrual cycles (erratic or abnormal time span between cycles and/or menstrual flow) has been reported in women who abuse alcohol in comparison with age-matched control women (Becker et al., 1988; Jones-Saumty et al., 1981). Chronic alcohol intake in rats leads to disruption of estrous cyclicity, similar to the menstrual irregularities seen in humans. Hormonal changes evidenced with 2 months of treatment included decreases in serum insulin-like growth factor (serum IGF), and with 2 weeks of use, temporary increases in circulating E_2 (Emanuele et al., 2001).

Women who abuse alcohol may also have persistent hyperprolactinemia even during alcohol abstinence (Teoh et al., 1992; Välimäki et al., 1990b). Early menopause has also been reported in association with alcohol abuse (Gavaler, 1985, 1988). These data have been derived primarily from endocrine evaluations and medical histories taken at the time of admission for treatment of alcohol-related problems. However, alcohol-dependent women often have medical disorders such as liver disease or pancreatitis that may be complicated by malnutrition (Mendelson et al., 1986a). It is therefore difficult to attribute the menstrual-cycle disorders solely to alcohol abuse *per se*. Animal models of alcohol abuse and dependence (Eskay et al., 1981; Gavaler et al., 1980; Krueger et al., 1983; Mello, 1988; Mello et al., 1983a, 1988; Sanchis et al., 1985; van Thiel et al., 1978) and data obtained from healthy social drinkers (Mello et al., 1988) indicate that similar reproductive disturbances occur in women with other medical conditions (Mello, 1988; Mello et al., 1989).

Amenorrhea has often been reported to occur in women who abuse alcohol (Hugues et al., 1980; Moskovic, 1975; Ryback, 1977; Seki, 1988; Seki et al., 1991a,b; van Thiel, 1984) and may persist for months or years (Hugues et al., 1980; Moskovic, 1975; Ryback, 1977; Seki, 1988; Seki et al., 1991a,b; Välimäki et al., 1984). To the best of our knowledge, there have been no longitudinal studies of amenorrheic women who abuse alcohol to determine if tolerance to its effects develops over time, and menstrual cycles return. Case reports suggest that amenorrhea may remit during alcohol abstinence, but menstrual cycles were not evaluated for normalcy, and endocrine testing was not reported (Ryback, 1977; Seki, 1988; Seki et al., 1991a).

There is minimal literature on endocrine parameters in women with alcohol-abuse-related amenorrhea. Twenty-two women admitted for liver disease or pancreatitis were studied in Finland ($n = 9$), and Paris ($n = 13$) (Hugues et al., 1980; Välimäki et al., 1984), and 23 women admitted for treatment of alcohol abuse were studied in Japan (Seki, 1988; Seki et al., 1991a). The Japanese women were not cirrhotic, but had hepatitis or fatty liver. Eight of the 22 European alcoholic amenorrheic women had normal circulating E_2 levels, and evidenced a positive E_2 response to stimulation with clomiphene or hCG (Hugues et al., 1980). However, the other amenorrheic alcoholic women had endocrine profiles similar to menopausal women as appropriate, since they were >45 years old. Their hormonal parameters were compared to postmenopausal controls. Fourteen of the European women had lower levels of estrogen and higher LH and follicle-stimulating hormone (FSH) levels than normal controls. The amenorrheic Japanese women also had low estrogen levels and high FSH levels, consistent with menopause or premenopause (Seki, 1988; Seki et al., 1991a). Not surprisingly, estrogen levels were higher in women who had withdrawal bleeding after provera than those who remained amenorrheic after the treatment ($29.7 \pm 12\,\mathrm{pg\,ml^{-1}}$ vs. $22.3 \pm 9.4\,\mathrm{pg\,ml^{-1}}$; Seki, 1988; Seki et al., 1991a). Such low E_2 levels could suggest that an element of hypothalamic amenorrhea not only may have occurred, but may also have only reflected impending menopause. Of note, hyperprolactinemia was prevalent in these women due to liver dysfunction. However, ovarian pathology has been reported in postmortem studies of alcohol-dependent women and alcohol-dependent rhesus monkeys and rats (Jung and Russfield, 1972; Mello et al., 1983a; van Thiel et al., 1978). Abnormally low estrogen levels could therefore reflect either impairment of ovarian function or disruption of gonadotropin secretory activity due to hyperprolactinemia, or both. The relative contribution of ovarian and hypothalamic factors to alcohol-related amenorrhea remains undetermined.

Animal models have been helpful in studying alcohol's effects under controlled conditions because neuroendocrine control of menstrual cycles in female rhesus monkeys is similar to human females (Knobil, 1974, 1980). Amenorrhea has been reported in alcohol-dependent women who also have liver disease (cirrhosis, pancreatitis, hepatitis, or fatty liver) (Hugues et al., 1980; Moskovic, 1975; Seki, 1988; Välimäki et al., 1984). Likewise, chronic alcohol self-administration resulted in amenorrhea, atrophy of the uterus, and decreased ovarian mass in otherwise healthy female macaque monkeys (Mello et al., 1983a). Daily self-administration of high doses of alcohol (2.9–$4.4\,\mathrm{g\,kg^{-1}\,day^{-1}}$) caused amenorrhea that persisted for 84 to more than 200 days (Mello, 1988; Mello et al., 1983a). Amenorrheic monkeys developed blood alcohol levels ranging from 266 to $438\,\mathrm{mg\,dl^{-1}}$ immediately following alcohol self-administration (Mello et al., 1983a).

These blood alcohol levels are similar to those observed in alcohol-dependent men during periods of intoxication (Mello and Mendelson, 1972). Monkeys that self-administered relatively lower doses of alcohol (1.3 and $1.6\,\mathrm{g\,kg^{-1}\,day^{-1}}$) for 119 and 173 days, respectively, continued to have ovulatory menstrual cycles (Mello et al., 1983a).

32.2.2 Effects of Alcohol on Hypothalamic, Pituitary, Gonadal, and Adrenal Hormones

32.2.2.1 Provocative tests of hormonal function

Provocative testing has been utilized for studying alcohol's effects on specific components of the hypothalamic–pituitary–ovarian axis, as well as for assessing pituitary–adrenal function (Rebar, 1986; Yen et al., 1985; Yen, 1983). For example, LHRH can be used to directly stimulate pituitary release of LH and FSH (Filicori, 1986; Yen, 1999) so that the effects of alcohol on LHRH stimulation of gonadotropin secretion may be examined (Mello, 1988; Mello et al., 1986a,b, 1989; Mendelson et al., 1989; Phipps et al., 1987).

Opioid antagonists also stimulate release of pituitary gonadotropins. The mechanism is thought to be antagonism of endogenous opioid peptides which mediate the inhibitory regulation of endogenous LHRH in the hypothalamus (Mendelson et al., 1979, 1986b; Mirin et al., 1976; Morley et al., 1980; Yen et al., 1985; Boyadjieva and Sarkar, 1997a). Two opioid antagonists, naloxone and naltrexone, stimulate hypothalamic release of endogenous LHRH which is followed by pituitary release of LH, FSH, and prolactin (Yen et al., 1985). However, one disadvantage of the short-acting narcotic antagonist, naloxone, is that it appears only to be effective during the late follicular and luteal phase of the menstrual cycle (Yen et al., 1985). In contrast, naltrexone, a long-acting opioid antagonist, acts to stimulate FSH and LH secretion during the early follicular phase in women (Mendelson et al., 1986b). Naltrexone administration in women also stimulates release of ACTH and cortisol (Mendelson et al., 1986b).

hCG stimulates release of ovarian steroids and can be used to evaluate the effects of alcohol on ovarian function (Teoh et al., 1990). Due to its relatively long half-life, hCG also can be used in primate models to simulate the endocrine milieu of early pregnancy (Ottobre and Stouffer, 1984; Wilks and Noble, 1983). In early pregnancy, endogenous chorionic gonadotropin, secreted by the placenta, prolongs the life span of the corpus luteum until placental progesterone and estrogen secretion become adequate for pregnancy maintenance (Klopper, 1985; Murad and Haynes, 1985a). Data obtained in studies of alcohol's effects on synthetic LHRH-, hCG-, or opioid antagonist-stimulated hormones are described as follows.

32.2.2.2 Follicular phase

32.2.2.2(i) Luteinizing hormone-releasing hormone, luteinizing hormone, and follicle-stimulating hormone

In studies carried out in Europe, amenorrheic alcoholic women had no significant difference in the LH and FSH response to LHRH stimulation (100 μg) as compared to normal controls (Hugues et al., 1980; Välimäki et al., 1984). LHRH (100 μg) also stimulated a rapid increase in LH and FSH in amenorrheic Japanese women aged 20–40 (Seki, 1988). Interestingly, the magnitude of the LHRH-stimulated increase in gonadotropins was significantly higher in women with less severe amenorrhea (Seki, 1988; Seki et al., 1991a). The normal gonadotropin response to exogenous LHRH stimulation suggests that the anterior pituitary may not be the primary site of alcohol's action causing amenorrhea in alcoholic women.

Hypothalamic amenorrhea, as well as several other disorders of reproductive function in nonalcoholic women, are associated with abnormal LH secretion (Crowley et al., 1985; Santoro et al., 1986b). Hypothalamic amenorrhea is often associated with low-frequency LH pulses or an apulsatile LH secretion (Crowley et al., 1985; Santoro et al., 1986b). Administration of pulsatile LHRH restored ovulatory menstrual cycles in women with hypothalamic amenorrhea, primary amenorrhea, and other endocrine disorders involving abnormal LHRH secretion (Conn and Crowley, 1991; Crowley et al., 1985; Hammond et al., 1979; Hurley et al., 1984; Leyendecker and Wildt, 1984; Santoro et al., 1986a). It is possible that alcohol-induced amenorrhea also reflects aberrant pituitary secretion of gonadotropins, but this has not yet been studied.

32.2.2.2(ii) Mechanisms of follicular-phase dysruption

Alcohol appears to prevent synthetic LHRH stimulation of FSH during the follicular phase of the menstrual cycle in normal female rhesus monkeys (Mello et al., 1986a). However, alcohol does not attenuate LHRH-stimulated increases in FSH in human males and ovariectomized rhesus females (Mello et al., 1986b; Phipps et al., 1987). In contrast, LH increased significantly within 15 min after LHRH stimulation when blood alcohol levels averaged 184–276 mg dl^{-1}. In addition, both FSH and LH increased significantly after LHRH stimulation when an isocaloric sucrose drink was substituted for alcohol (Mello et al., 1986a). It may be appropriate to hypothesize that if alcohol inhibits FSH responsiveness to endogenous LHRH during the follicular phase, this could result in menstrual cycle irregularities commonly seen in alcohol-dependent females.

In ovariectomized rhesus females, alcohol did not suppress LHRH-stimulated FSH, indicating that ovarian steroids have a modulatory role in FSH suppression (Mello et al., 1986b). After administration of LHRH, LH and FSH increased significantly in ovariectomized females when blood alcohol levels averaged 242 and 296 mg dl^{-1}(Mello et al., 1986b). The ovarian peptide, inhibin, suppresses FSH but has no effect on LH (Channing et al., 1985). In the normal human menstrual cycle, inhibin levels are inversely related to FSH during the mid- to late-follicular phase and function as a major regulator of FSH secretion from the pituitary (McLachlan et al., 1987). It is possible that alcohol suppresses LHRH-stimulated FSH by stimulating inhibin secretion; however, there are as yet no data on alcohol's effects on inhibin.

It is important to emphasize that clinical studies have not consistently implicated disruption of FSH secretion in the occurrence of luteal-phase deficiencies. Experimental data show that administration of rapid LHRH pulses during the follicular phase induced a luteal-phase deficiency in normal women (Soules et al., 1987). Endogenous high LH pulse frequency has been documented during the early follicular phase. In one study a significantly higher

LH pulse frequency during the early follicular phase distinguished luteal-phase defect patients from controls, whose FSH levels did not differ. This LH pulse frequency (12.8 ± 1.4 pulses per 12 h) persisted throughout the follicular phase in the patients, whereas LH pulse frequency in controls increased from 8.2 ± 0.7 pulses per 12 h to 15 pulses per 12 h during the late follicular phase. LH pulse amplitude was also lower in some patients than in controls (Soules, 1989a). These findings suggest that abnormally high-frequency LH pulses during the early follicular phase may induce impaired follicular development and thereby contribute to luteal-phase defects.

Alcohol did not alter stimulation of LH by synthetic LHRH or naloxone in follicular-phase rhesus females or in rhesus males even though peak blood alcohol levels ranged from 200 to above 300 ng dl^{-1} (Mello et al., 1985, 1986a). These findings were subsequently replicated in mid-luteal-phase women (Mendelson et al., 1987). Alcohol failed to attenuate LH after exogenous stimulation of the hypothalamus or pituitary by LHRH and naloxone. Under some conditions, alcohol actually augmented the LH response as compared to placebo. In the follicular phase, both women given the opioid antagonist naltrexone and ovariectomized rhesus females given synthetic LHRH showed a significant enhancement of LH after alcohol in comparison to placebo control conditions (Mello et al., 1986b; Teoh et al., 1988).

32.2.2.3 Amenorrhea and gonadotropin secretory activity

It is possible that alcohol suppresses hypothalamic release of endogenous LHRH with concomitant suppression of gonadotropin secretory activity. When hypothalamic release of endogenous LHRH was disrupted in ovariectomized rhesus monkeys by lesions of the hypothalamic arcuate nucleus and the median eminence, LH and FSH secretion were abolished. Pulsatile administration of synthetic LHRH restored LH and FSH secretory patterns, whereas continuous administration of LHRH did not, showing that the pulsatility of LHRH is vital for normal LH and FSH secretion to occur (Knobil, 1974, 1980; Knobil and Hotchkiss, 1988).

Clinical human data suggest that primary amenorrhea and secondary hypothalamic amenorrhea are also associated with the suppression of gonadotropin secretory activity (Berga et al., 1989; Conn and Crowley, 1991; Crowley et al., 1985; Reame et al., 1985; Santoro et al., 1986b). Low-frequency LH pulses were most commonly associated with secondary hypothalamic amenorrhea, but low-amplitude LH pulses were also sometimes observed (Berga et al., 1989; Conn and Crowley, 1991; Crowley et al., 1985; Reame et al., 1985; Santoro et al., 1986b). The most profound hypoestrogenism was associated with a complete absence of LH pulses and low LH levels (Crowley et al., 1985; Santoro et al., 1986b). Normal ovulatory function can be restored in amenorrheic women by pulsatile intravenous (IV) infusion of synthetic LHRH (Conn and Crowley, 1991; Crowley et al., 1985; Hurley et al., 1984; Leyendecker and Wildt, 1983; Santoro et al., 1986a,b).

In female macaque monkeys, trained to self-administer alcohol, average LH levels were significantly lower during amenorrheic cycles (16.9 ± 1.2 to 24 ± 1.4 ng ml^{-1}) than during nonalcoholic control cycles (28 ± 1.2 to 30 ± 2.2 ng ml^{-1}) (Mello, 1988). These data are consistent with human studies and support the hypothesis that amenorrhea may be related to suppression of gonadotropin levels. However, there have been no systematic studies to confirm or refute the hypothesis that alcohol-induced amenorrhea reflects abnormal gonadotropin secretion. It is still unknown if alcohol suppresses normal pulsatile gonadotropin secretion through a direct effect on hypothalamic LHRH release or by other mechanisms, such as stimulation of prolactin or corticotropin-releasing factor (CRF).

32.2.2.3(i) Possible mechanisms underlying anovulation and luteal-phase dysfunction

Factors leading to alcohol-related anovulation and luteal-phase dysfunction are poorly understood, but include systemic diseases and exercise disorders (Bullen et al., 1985; McNeely and Soules, 1988; Stouffer, 1990).

Although FSH is not the sole determinant of folliculogenesis (follicular growth), adequate FSH levels are necessary for a normal follicular phase (Goodman and Hodgen, 1983; Ross, 1985). FSH suppression during the follicular phase has been shown to delay follicular maturation and ovulation, or result in luteal-phase dysfunction (diZerega and Hodgen, 1981; diZerega and Wilks, 1984; Goodman and Hodgen, 1983; Wilks et al., 1977). Studies of folliculogenesis in the primate suggest that dominant follicle recruitment occurs during menstrual cycle days 1–4; a single follicle is selected during days 5–7; and the follicle achieves dominance during cycle days 8–12 (diZerega and Hodgen, 1981; Goodman and Hodgen, 1983; Hodgen, 1982). Inhibins, ovarian peptides, titrate FSH levels during folliculogenesis (Goodman and Hodgen, 1983; Soules et al., 1989b). If alcohol intoxication suppresses FSH directly or modulates inhibin, thereby downregulating FSH secretion, this could lead to anovulation or luteal-phase dysfunction (Mello, 1988; Mello et al., 1989).

32.2.2.3(ii) Mechanisms of alcohol-related pituitary and gonadal hormone stimulation

The physiological basis for the alcohol-induced augmentation of LH, prolactin, and E_2 is unclear. Alcohol administration may have a direct effect on endogenous LHRH, thereby increasing pituitary LH release. Increased E_2 levels after alcohol administration could enhance the LH response to LHRH stimulation, as seen in mid-cycle; the LH surge is dependent on the periovulatory increase in E_2 (Karsch et al., 1973). LHRH-stimulation leads to higher LH levels after alcohol (165 $mIU\,ml^{-1}$) than after placebo drink (105 $mIU\,ml^{-1}$) (Mello et al., 1986b), but, unfortunately, E_2 levels were not measured. E_2 pretreatment in ovariectomized monkeys also leads to increases in LHRH-stimulated LH secretion (Krey et al., 1973). The pretreatment also appears to increase pituitary sensitivity to LHRH stimulation in normal and hypogonadal women (Jaffe and Keye, 1974; Lasley et al., 1975) and in the intact diestrous rat (Arimura and Schally, 1971). Consequently, if alcohol administration increases E_2 levels in ovariectomized monkeys (Mello et al., 1986b), the E_2 may in turn sensitize the pituitary to produce an augmented LH response to LHRH stimulation. Ovariectomy reduces circulating E_2 by approximately 60%; however, estrogens are also produced through peripheral conversion of androgens to estrogens (Gavaler, 1985, 1988; Ross, 1985).

32.2.2.4 Effects of alcohol on ovarian hormones during the follicular phase

There is an evidence that increased E_2 levels during the early follicular phase suppresses FSH, inhibits preovulatory follicular growth, and prolongs the follicular phase (Dierschke et al., 1985, 1987; Zeleznik, 1981). After 6, 12, and 24 h of exposure to E_2 during the mid-follicular phase (day 6 or 7 of the menstrual cycle), luteal-phase defects were consistently observed (Dierschke et al., 1985). An increase of about 30 $pg\,ml^{-1}$ in E_2 significantly reduced FSH concentrations and prolonged the follicular phase (Zeleznik, 1981).

Acute alcohol intoxication appears to stimulate, not suppress, gonadotropins and ovarian steroids under a variety of experimental conditions (Mello, 1988; Mello et al., 1989). Clinical studies in normal follicular-phase female research volunteers have shown significant increases in E_2 levels during alcohol intoxication as compared to after placebo ingestion. Acute alcohol administration (0.695 $g\,kg^{-1}$) induced a significant increase of 19.5 ($\pm$4.1) $pg\,ml^{-1}$ in E_2 levels (Mendelson et al., 1988). Plasma E_2 reached peak levels within 25 min after initiation of drinking when blood alcohol levels were relatively low (mean 34 $mg\,dl^{-1}$). These data are shown in **Figure 1**. Plasma samples were taken every 5 min to detect an alcohol-related increase in plasma E_2 during the ascending phase of the blood alcohol curve. Prior studies used 20-min integrated sample collection procedures (Mendelson et al., 1981) and no significant changes in plasma E_2 levels were detected during the ascending, peak, or descending phase of the blood alcohol curve.

Figure 2 illustrates alcohol enhancement of naltrexone-stimulated E_2 levels in early follicular-phase women (Teoh et al., 1988). E_2 increased during the rising phase of the blood alcohol curve. Blood alcohol levels (123 $\pm$ 4.3 $mg\,dl^{-1}$) peaked 90 min after alcohol consumption began. In addition, following opioid antagonist stimulation with naloxone, E_2 levels increased significantly after alcohol as compared to after placebo ingestion (Mendelson et al., 1987). After LHRH stimulation, E_2 increased significantly more after alcohol than after placebo in both early follicular ($p < 0.0001$) and mid-luteal-phase women ($p < 0.01$) (Mendelson et al., 1987; Teoh et al., 1988). Alcohol-induced increases in E_2 shown in **Figures 1** and **2** are one illustration of the fact that alcohol intoxication may increase hormone levels.

Cohort data also support alcohol effects on circulating E_2 levels in women in the follicular phase of the menstrual cycle, and these elevations are also associated with higher breast-cancer risk. Blood samples were taken from 18 521 women, and 197 cases of breast cancer were diagnosed within the next 5–8 years. Women in the highest versus the lowest quartile of follicular-phase total and free E_2 had statistically significant higher risks of breast cancer: relative risk (RR) 2.1, confidence interval (CI) 1.1–4.1 ($p = 0.08$ for trend) and 2.4, CI 1.3–4.5 ($p = 0.01$), respectively. Trends were most notable for women with invasive breast cancer, and with estrogen- and progesterone-positive tumors (Eliassen et al., 2007). These data support the hypothesis that the E_2 elevations seen after alcohol ingestion may lead to a subsequent increase in breast cancer risk.

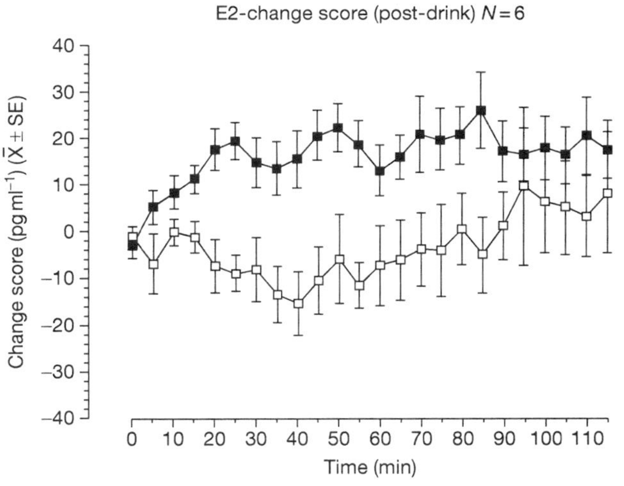

Figure 1 Change in estradiol (E_2) levels for 3 h after the onset of alcohol (closed square) or placebo (open square) ingestion in premenopausal women in the follicular phase of the menstrual cycle. Each data point is the mean ± SEM of six women. Reproduced from Ginsburg ES, Mello NK, and Mendelson JH (2002) Alcohol abuse: Endocrine concomitants. In: Pfaff DW, Arnold AP, Etgen AM, Fahrbach SE, and Rubin RT (eds.) *Hormones, Brain, and Behavior,* vol. 5, pp. 747–780. San Diego, CA: Academic Press, with permission from Elsevier.

32.2.2.5 Luteal phase

32.2.2.5 (i) Effects of alcohol on ovarian hormones during the luteal phase

Animal data support an impact of alcohol on luteal-phase function. Alcohol-related increases in E_2 during the luteal phase may lead to the functional equivalent of a luteal-phase defect. Administration of estrogen and progesterone capsules to rhesus females on luteal days 2–6 resulted in low progesterone levels, with menses occurring 5–6 days earlier than in control cycles (Hutchison et al., 1987). In monkeys with hypothalamic lesions or hypothalamic pituitary stalk transection where ovulatory menstrual cycles were restored by pulsatile administration of synthetic LHRH, E_2 administration did not result in premature luteal regression (Hutchison et al., 1987). These data suggest that estrogen's effects on the corpus luteum are mediated by the hypothalamic–pituitary axis in intact monkeys. In addition, in rhesus monkeys, alcohol (2.5 g kg^{-1}) significantly increased E_2 within 150–210 min under basal (nonstimulated) conditions in mid-luteal phase (Mello et al., 1983b).

A study of 60 premenopausal women sampled serum E_2 on the same day of the mid-luteal phase at the same time of day, a year apart, and samples were stored and analyzed simultaneously. E_2 levels were significantly higher in alcohol drinkers when the values of the two samples were averaged, with a dose–response effect seen (Muti et al., 1998). Mean (±SD) E_2 levels were 334.5 ± 104.0 in abstainers and 393.9 ± 124 in drinkers. The mean alcohol intake in drinkers was 91.4 ± 68.8 g per week.

In humans, increases in E_2 levels during acute alcohol intoxication were heightened following gonadotropin stimulation by naloxone and naltrexone (Mendelson et al., 1987; Teoh et al., 1988). The alcohol-related augmentation of opioid antagonist-stimulated E_2 levels was 45–50 pg ml^{-1}. Such E_2 levels are equivalent to those shown to selectively suppress FSH secretion in clinical studies (40–50 pg ml^{-1}) in the mid-follicular phase of the menstrual cycle (Marshall et al., 1983). When a fairly low dose of ethanol was administered to healthy female volunteers in the luteal phase of the menstrual cycle, resulting in mean maximal ethanol levels of only 33.5 mg dl^{-1}, and E_2 and androgen levels were measured 5 times over the following 24 h, a gradual decrease in steroid levels, prolactin, and FSH was seen, but was no different in controls (Becker et al., 1988).

In the European Prospective Investigation into Cancer and Nutrition cohort study, the largest cross-sectional study evaluating alcohol intake and circulating sex steroid concentrations (Rinaldi et al., 2006), women who consumed more than 25 g alcohol per day in the luteal phase of the menstrual cycle had dehydroepiandrosterone sulfate (DHEAS), androstenedione, and testosterone and estrone concentrations positively associated with alcohol consumption. Interestingly, no association with E_2 or sex-hormone-binding globulin (SHBG) was seen, unlike in other studies.

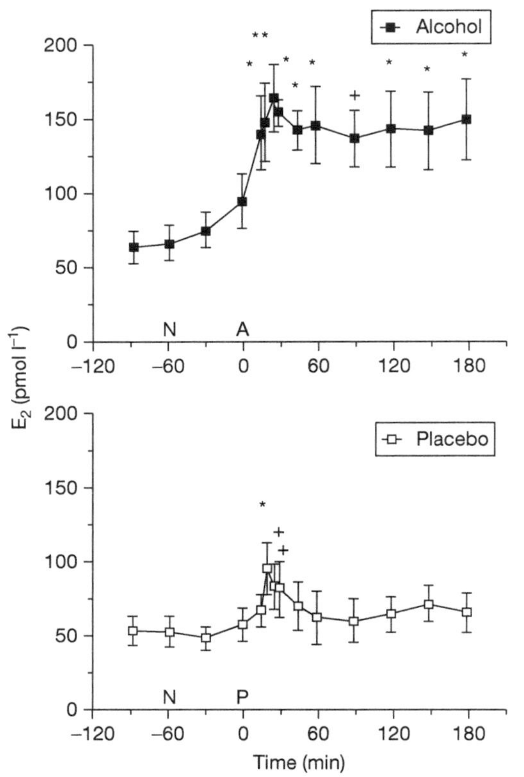

Figure 2 Mean (±SE) plasma and E_2 concentration before and after the administration of nahrexone (N) and alcohol (A; top panel) and after administration of nahrexone and placebo (P; bottom panel) in seven normal women. Plus, $p < 0.05$; asterisk, $p < 0.01$ (increase above baseline). Reproduced from Ginsburg ES, Mello NK, and Mendelson JH (2002) Alcohol abuse: Endocrine concomitants. In: Pfaff DW, Arnold AP, Etgen AM, Fahrbach SE, and Rubin RT (eds.) *Hormones, Brain, and Behavior,* vol. 5, pp. 747–780. San Diego, CA: Academic Press, with permission from Elsevier.

The authors hypothesized that the lack of an E_2 effect may have been due to the measurements being performed only during one menstrual cycle. Although menstrual-cycle-phase data were collected, variations in circulating ovarian steroid levels during various phases of the menstrual cycle were not reported separately. Because E_2 levels vary dramatically throughout the menstrual cycle, with a high mid-cycle peak, this may well have been part of reason that no effect of alcohol was shown in this study.

32.2.2.5(ii) Alcohol effects on circulating estrogens in oral contraceptive users

The results of acute alcohol ingestion (0.5 g kg^{-1}), ingested over 30 min on circulating estrogen levels, was also studied in 30 premenopausal women using oral contraceptives (OCs) and 40 mid-cycle controls (Sarkola et al., 1999). In contraceptive pill users, by 45 min after drinking began, E_2 levels increased, and estrone levels declined, thereby causing an increase in the E_2 to estrone ratio. Progesterone levels decreased in OC users from 45 min through 120 min after drinking began. No changes in E_2, estrone, their ratio, or progesterone was seen in the mid-cycle group. However, when data from another group of non-OC users were combined with this group, a decrease in progesterone levels over time was documented. Prolactin levels increased in both the groups. The authors concluded that alcohol decreased E_2 catabolism through an increase in the hepatic NADH:NAD ratio (reduced form of nicotinamide adenine dinucleotide:nicotinamide adenine dinucleotide). Prolactin elevations were felt to be due to acute hypothalamic changes in opioids and dopamine levels, but supportive data for this conclusion were not given. In premenopausal women using OCs, 0.5 g kg^{-1} of ethanol increases circulating E_2 levels and lowers progesterone levels. There was a significant increase in the estradiol:estrone ratio, indicating that the effect most likely included an increase in the hepatic NADH:NAD ratio leading to decreased E_2 catabolism. In ten women, 0.34–1.02 g kg^{-1} ethanol had no effect on circulating estrogen or progestin levels. Both OC and non-OC users had transient prolactin elevations felt to be due to hypothalamic opioid or dopamine levels (Sarkola et al., 1999).

32.2.2.5(iii) Mechanisms concerning alcohol-induced increases in circulating estrogens

There are several possibilities for the mechanism of the alcohol-stimulated increase in plasma E_2 after naloxone and nahrexone stimulation (Mendelson et al., 1987, 1988; Teoh et al., 1988). Alcohol may increase E_2 production or decrease E_2 metabolism. Because intrahepatic ethanol metabolism decreases NAD availability for other coupled oxidative reactions (Cronholm and Sjövall, 1968; Cronholm et al., 1969; Murono and Fisher-Simpson, 1984, 1985), this might reduce the rate of oxidation of E_2 to estrone and result in elevated E_2 levels (Mendelson et al., 1987, 1988). If this is the case, hepatic and gonadal oxidative metabolism of steroids may become rate limiting during alcohol metabolism when relatively low blood alcohol concentrations (45 mg dl^{-1} or 10 mmol l^{-1}) may saturate human alcohol dehydrogenase isoenzymes and decrease the NAD:NADH ratio. This in turn could decrease the rate of oxidation of E_2 to estrone and result in increased E_2 levels (Teoh et al., 1988).

Alcohol is thought to create tissue damage more rapidly in women than in men. Acetaldehyde, the primary metabolite of alcohol, is higher in cycling women and in those taking OCs than in men, and levels of acetaldehyde are associated with estrogen levels (Eriksson et al., 1996), indicating a role for estrogen in the tissue damage caused by alcohol.

32.2.2.5(iv) Significance of E_2 elevations in premenopausal women

Studies indicate that postmenopausal (Schairer et al., 2000) and premenopausal (Grabrick et al., 2000) estrogen use may increase breast cancer risk in subgroups of women. There is also evidence that breast cancer risk increases when hormone and alcohol use occurs simultaneously (Colditz et al., 1990; Gapstur et al., 1995).

Genetic influences are also important with regard to alcohol effects on E_2 elevations and breast cancer induction. A case control study of 117 moderate alcohol consumers with breast cancer, 107 healthy controls, and 111 age-matched women with alcohol-related diseases such as pancreatitis were compared with respect to acute elevations in E_2 after moderate alcohol consumption at a dose of $0.225\,g\,kg^{-1}$. E_2 levels rose 27–38%, particularly during mid-cycle in premenopausal women. Interestingly, women with breast cancer were more likely to have the alcohol dehydrogenase 1C allele (coding for rapid ethanol metabolism and higher acetaldehyde levels, an established carcinogen) at 62% versus 41.9% in age-matched alcoholic controls ($p = 0.0035$) (Coutelle et al., 2004). The alcohol dehydrogenase 3 (ADH3) genotype, coding for rapid metabolism of alcohol to acetaldehyde, is increased among women who drink alcohol and develop breast cancer. The Long Island Breast Cancer Study Project compared 1047 women with breast cancer with 1101 controls, and questionnaires and ADH status were compared. Fast metabolizers who consumed 15–30 g per day (1–2 drinks) had an odd ratio (OR) of 2.0, CI 1.1–3.5, compared to OR of 1.5, CI 0.9–2.4, and OR of 1.3, CI 0.5–3.5, in intermediate and slow metabolizers. Fast metabolizers had 2.3 times the risk of breast cancer than slow or intermediate metabolizers who did not drink (Terry et al., 2006).

A Kaiser Permanente cross-sectional study of 218 premenopausal women interested in evaluating genetic polymorphisms found that women with higher levels of alcohol consumption had higher circulating mean E_2 levels, during the menstrual cycle ($p = 0.09$); however, accurate documentation of ovulation, which may raise mean levels dramatically due to the mid-cycle peak, was not done. Androstenedione (secreted primarily by the ovaries) levels were also significantly elevated in drinkers. Two polymorphisms (CYP 1B1 L432V and S453N) were associated with increased luteal E_2 levels ($p = 0.04$). Even with the weakness of some aspects of the study design, these data may strengthen the acute ingestion data previously discussed (Garcia-Closas et al., 2002).

Therefore, acute and chronic elevations in E_2 levels may be a part of the mechanism through which alcohol could potentially increase breast cancer risk, and in turn these elevations appear to be in part-related genetic differences in alcohol metabolism.

32.2.3 Corticotropin-Releasing Factor

Acute alcohol drinking, as well as stress, can stimulate release of CRF and adrenocorticotropic hormone (ACTH), leading to increased levels of cortisol (Redei et al., 1986; Rivier et al., 1986; Rivier and Vale, 1984). In rats, this response to alcohol is more pronounced, with females secreting more ACTH and cortisol to the same alcohol dose than do males. Castrated males supplemented with E_2 have responses comparable to females (Rivier, 1993a; Ogilvie and Rivier, 1997). Elevations in ACTH occur with gene activation in the paraventricular nucleus of the hypothalamus (Ogilvie et al., 1997). In contrast, chronic alcohol feeding of rats with a diet containing 6.4% alcohol does not increase ACTH or corticosterone levels (Ogilvie et al., 1997). There are complex differences between responses to placebo or $0.5\,g\,kg^{-1}$ ethanol and stress on cortisol and ACTH, with differences between subjects with family history of alcohol dependence and those without. Low-risk subjects had less anxiety in response to alcohol and stress than high-risk background subjects, indicating possible genetic differences in responses to drinking and stress (Dai et al., 2007).

This may contribute to the increased incidence of amenorrhea seen in alcoholic women because alcohol stimulates CRF, ACTH, and adrenal hormones which, in turn, may suppress gonadotropin secretory activity and lead to amenorrhea. Administration of synthetic CRF inhibits pulsatile release of LH and FSH in ovariectomized rhesus females (Olster and Ferin, 1987). Interestingly, administration of ACTH and cortisol does not have the same effect (Xiao and Ferin, 1988).

In a study of eight male and female normal controls, acute administration of 95 ml kg^{-1} ethanol did not increase ACTH or cortisol levels unless gastrointestinal side effect such as nausea or vomiting occurred, and this effect (i.e., ACTH stimulation) appeared to be mediated by vasopressin (Inder et al., 1995a). There appeared to be a different pattern of response in chronic human alcoholics; as compared to sex-matched controls, their incremental ACTH response to naloxone and corticotrophin-releasing hormone (CRH) was blunted, though the cortisol response was not different (Inder et al., 1995b). In a rat model, synthetic CRF administration suppresses endogenous LHRH measured in portal blood (Petraglia et al., 1987). These data suggest that CRF-induced suppression of LH and FSH occurs, but is a central effect mediated through the hypothalamic–pituitary axis and not through adrenal stimulation (Xiao and Ferin, 1988). The relationship between alcohol-related increases in CRF and amenorrhea in alcohol-dependent women remains to be determined.

32.2.3.1 *Mechanisms of alcohol effects on the pituitary–adrenal axis*

In rat pituitary cultures, low doses of alcohol (up to 20 mM) tend to inhibit CRF-stimulated ACTH secretion, whereas higher doses are stimulatory. The effects can be modulated by addition of adenosine and urate, indicating that purine metabolism may be involved in the mechanism of ethanol stimulation of cortisol release at the level of the pituitary (Szabo et al., 1999). There is also evidence that in the rat, the dose-dependent increases in CRF, ACTH, and corticosterone levels, are modulated by vasopressin secretion which interacts with CRF to stimulate pituitary ACTH release (Rivier and Lee, 1996). In rats exposed to an alcohol diet over 7–10 days, the hypothalamic–pituitary–adrenal-axis response to alcohol is blunted by increased nitric oxide levels, which inhibit vasopressin stimulation of ACTH release (Ogilvie et al., 1998). In rats, alcohol feeding blunted the ACTH response to interleukin-1 (IL-1) by day 35 postnatally, but the effect is more pronounced at 45–55 days old (Lee and Rivier, 1994). Blunting of ACTH response to IL-1, but not cortisol response, is also seen in oophorectomized rats fed a 10-day alcohol diet (Lee and Rivier, 1993b). A similar blunting effect was seen with acute administration of 1.5 g kg^{-1} ethanol 30 min to 4 h before IL-1 administration, but not with administration of 0.65 g kg^{-1} (Rivier, 1993b).

32.2.4 Prolactin

Postpartum lactation and pituitary adenomas may both cause hyperprolactinemic amenorrhea and other disruptions of menstrual-cycle function such as luteal-phase defects (Buchanan and Tredway, 1979; Martin and Reichlin, 1987; Sauder et al., 1984; Tolis, 1980). However, hyperprolactinemia does not invariably cause amenorrhea (Buchanan and Tredway, 1979; Martin and Reichlin, 1987; Sauder et al., 1984; Tolis, 1980). Alcoholic women with liver disease can have amenorrhea with normal prolactin levels (Välimäki et al., 1984). In one study, alcoholic women (aged 23–40) reported amenorrhea of 3–12 months duration and their basal prolactin levels averaged only 10.6 ($\pm$1.1) ng ml^{-1} (Välimäki et al., 1984). Hyperprolactinemia unaccompanied by amenorrhea has also been reported in alcoholic women (aged 18–46) during abstinence (Välimäki et al., 1990b), and in healthy social drinkers during daily consumption of between 4.24 and 8.24 drinks per day (Mendelson and Mello, 1988).

Hyperprolactinemia is commonly seen in alcoholic women (Seki, 1988; Seki et al., 1991a,b; Teoh et al., 1992). In a Japanese study, 22 of 23 women admitted for alcoholism treatment had prolactin levels over 25 ng ml^{-1} on admission (normal up to 20 ng ml^{-1}) (Seki, 1988; Seki et al., 1991a). The women, who were 20–40 years old, met *Diagnostic and Statistical Manual of Mental Disorders*, Third Edition, Revised (DSM-III-R) criteria for alcoholism, and reported drinking an average of 84.1 g of alcohol daily for at least 7 years. Ten had hepatitis and the rest had fatty liver, but none had cirrhosis. All had oligomenorrhea ($n = 2$) or amenorrhea ($n = 21$) of 7–38 months duration, but their reproductive history was not described (Seki, 1988; Seki et al., 1991a,b). Six women had prolactin levels above 100 ng ml^{-1} (115–184 ng ml^{-1}). These are extremely high levels that may be seen in conjuction with pituitary macroadenomas, and ten had prolactin levels above 50 ng ml^{-1} (59–97 ng ml^{-1}). Of the remaining seven women, six had prolactin levels ranging between 27 and 38 ng ml^{-1} (Seki, 1988; Seki et al., 1991b). Prolactin levels returned to normal after up to 3 months of treatment (Seki, 1988). However, since other factors may cause amenorrhea, it is unclear whether this contributes to or causes their amenorrhea; 19 of these Japanese alcoholic women who were amenorrheic were found to have hyperprolactinemia on admission to the hospital (Seki, 1988; Seki et al., 1991b) with prolactin levels ranging from 37 to 184 ng ml^{-1} (mean 92.9 $\pm$ 9.9 ng ml^{-1}). In contrast, a European study reported that nine amenorrheic alcoholic women had a normal prolactin response to thyrotropin-releasing hormone (TRH) stimulation (Välimäki et al., 1984).

Sixteen alcoholic women had persistent hyperprolactinemia during 6 weeks of treatment in a clinical research ward (Välimäki et al., 1990b). They reported an average daily alcohol intake of 170 g for the past 2–16 years, but none had cirrhosis. Fourteen of the 16 women (aged 18–46) evidenced regular menstrual cycles associated with normal patterns of gonadotropin and ovarian steroid secretion; only two were anovulatory. Thirteen women had been pregnant.

Of 30 reported pregnancies, 16 were completed successfully, two were terminated by spontaneous abortions, and 12 by therapeutic abortions. Their alcohol abuse did not appear to have caused significant reproductive dysregulation or infertility.

Six of 12 alcohol-dependent women admitted to a Massachusetts hospital for the treatment had hyperprolactinemia, with prolactin levels ranging from 22.3 to 87.5 ng ml^{-1}. They reported a 7- to 33-year history of regular drinking of 75.7–247.2 g of alcohol per day (Teoh et al., 1992). Even in polysubstance abusers, regular menstrual cycles may occur; in a subgroup of six, five reported regular menses and four had live births despite a history of abuse of alcohol, cocaine, opiates, marijuana, and amphetamines (Teoh et al., 1992). This small sample of socially and economically disadvantaged women (Lex et al., 1990) illustrates the relative resilience of the reproductive system.

32.2.4.1 Hyperprolactinemia and alcohol-related amenorrhea

Animal models are helpful in teasing out the relationship between amenorrhea and prolactin in alcoholism. In one amenorrheic alcohol-dependent macaque monkey, during chronic high-dose alcohol self-administration (3.4 g kg^{-1} day^{-1}) prolactin levels increased from 16.5 to 63 ng ml^{-1}. Immunocytochemical examination of the anterior pituitary showed hyperplasia of the lactotrophs (Mello et al., 1983a). These data suggested that hyperprolactinemia might contribute to alcohol-induced amenorrhea. However, this hypothesis was not confirmed in subsequent studies (Mello, 1988). Four other amenorrheic cycles (85–194 days) were examined. Although prolactin levels were intermittently elevated above 20 ng ml^{-1}, average prolactin levels during the amenorrheic cycles (14.7 ± 1.8 to 19.6 ± 1.5 ng ml^{-1}) did not differ significantly from prolactin levels during normal ovulatory menstrual cycles when no alcohol was available (19.7 ± 0.63 ng ml^{-1}) (Mello, 1988).

Galactorrhea developed in one monkey during a 97-day amenorrheic cycle when alcohol self-administration averaged 3.35 g kg^{-1} day^{-1}. However, prolactin values averaged 19.6 ($\pm$1.5) ng ml^{-1} and ranged between 5.7 and 29.5 ng ml^{-1} – lower levels than one would expect to be associated with amenorrhea in humans. Galactorrhea and breast enlargement were first observed on cycle day 25 and persisted through cycle day 74 despite the fact that levels were not markedly elevated. Days 27–45 were associated with unusually high levels of alcohol self-administration ranging between 4.68 and 9.24 g kg^{-1} day^{-1} but prolactin values ranged from only 5.7 to a high of 24 ng ml^{-1} during this period. Clinical data indicate that galactorrhea is not always associated with hyperprolactinemia (Buchanan and Tredway, 1979; Edwards and Feek, 1981; Tolis et al., 1974). Galactorrhea with normal prolactin levels may reflect induction of prolactin receptors with an enhanced end-organ response. Overall, these data suggest that hyperprolactinemia probably is not the primary mechanism underlying alcohol-induced amenorrhea in the female macaque monkey model (Mello, 1988), leading to question its role as a human model, or alternatively, to question the role of hyperprolactinemia in human alcohol-associated amenorrhea.

32.2.4.2 Acute effects of alcohol on prolactin

The effects of acute alcohol intoxication on circulating prolactin levels have been studied in normal human subjects, but resulted in conflicting data. In a Finnish study of normal female volunteers, acute alcohol administration during the mid-luteal phase of the menstrual cycle significantly decreased basal prolactin levels over the first 4 h of observation (Välimäki et al., 1983). In Japan, however, the administration of 1.2 g kg^{-1} of alcohol during the luteal phase of the menstrual cycle was followed by a 150% increase in prolactin above baseline, and prolactin levels remained elevated throughout the 180-min sampling period (Seki et al., 1991b). Blood alcohol levels remained at intoxicated levels, above 100 mg dl^{-1}, for 40–180 min after alcohol ingestion (Seki et al., 1991b). An American study showed that a comparable alcohol dose given during mid-follicular phase had no effect on prolactin when blood alcohol levels averaged 88 mg dl^{-1}, except in women who complained of nausea and vomiting (Mendelson et al., 1981). Another study in mid-cycle controls and women using OCs found that 0.5 g kg^{-1} ethanol led to elevations of prolactin in both groups lasting up to 60 min after drinking began (Sarkola et al., 1999).

When prolactin secretion is stimulated by provocative testing, alcohol tends to augment increases in prolactin levels. In the mid-luteal phase, acute alcohol administration increased prolactin stimulation by naloxone (Mendelson et al., 1987). However, in the early follicular phase, higher mean peak blood alcohol levels ($123 \pm 4.3\,\text{mg}\,\text{dl}^{-1}$) did not enhance naltrexone-stimulated prolactin levels which were not different than after placebo drink was consumed (Teoh et al., 1988). Prolactin levels increased within 30 min in the mid-luteal phase when alcohol and hCG were administered simultaneously, whereas hCG and placebo administration did not lead to prolactin elevations (Teoh et al., 1990). The mechanism behind the stimulatory effects of combined alcohol, hCG, and naloxone on prolactin (but not after naltrexone stimulation) is unknown. It is possible, however, that an antecedent alcohol-related increase in E_2 may have affected the prolactin response. Following hCG and alcohol administration, E_2 increased first, followed 20 min later by prolactin (Teoh et al., 1990). Because E_2 is known to decrease the sensitivity of the pituitary lactotrophs to dopamine suppression (Prior et al., 1987), an alcohol-stimulated E_2 increase may modulate lactotroph sensitivity to the inhibitory effects of dopamine, resulting in prolactin elevations.

32.2.4.3 Luteal-phase dysfunction and prolactin abnormalities: Possible mechanisms

Luteal-phase dysfunction may also be associated with either pathologic increases or decreases in prolactin (McNeely and Soules, 1988). Ten to twenty percent of patients diagnosed with luteal-phase defects have been found to be hyperprolactinemia ($20–40\,\text{ng}\,\text{ml}^{-1}$) (McNeely and Soules, 1988). Low prolactin levels secondary to administration of a dopamine agonist were also associated with low levels of progesterone during the luteal phase in one study (Schulz et al., 1978). As discussed earlier, both hyperprolactinemia and decreased prolactin levels may occur during alcohol intoxication, depending on the dose and duration of alcohol intake and the conditions of gonadotropin stimulation. Alcohol-related changes in prolactin may contribute to luteal-phase defects observed in social drinkers and alcohol-dependent women (Hugues et al., 1980; Mendelson et al., 1988).

32.3 Alcohol Effects in Postmenopausal Women

It is possible that alcohol affects the age of menopause onset, although numbers are small. In a study of 983 women, 150 were postmenopausal and received no hormone replacement therapy (HRT), 277 were perimenopausal, with erratic menstrual cycles, and 378 were premenopausal. There was a significant association between alcohol consumption and E_2 levels. Moreover, moderate alcohol consumption was associated with delayed menopause.

Unfortunately, this sample may well have been biased, since current HRT users ($n = 178$) were excluded from analysis and they may have represented a group with earlier onset of menopause (Torgerson et al., 1997).

32.3.1 Alcohol Effects in Postmenopausal Women Not on HRT

32.3.1.1 Acute alcohol effects on the hypothalamic–pituitary–gonadal or adrenal axis

There is only one study we are aware of evaluating the effect of acute alcohol ingestion on circulating hormone levels in postmenopausal women. This blinded, crossover study evaluated the effect of $0.7\,\text{g}\,\text{kg}^{-1}$ ethanol given as 40% alcohol (Absolut vodka) or an isocaloric placebo drink to fasting postmenopausal women over a 15-min period (**Figure 3**). Blood alcohol levels peaked 40–60 min after drinking began. LH pulse frequency was no different after alcohol as compared with placebo ingestion, nor were pooled mean levels different (unpublished data). Circulating E_2 and estrone levels were not significantly different after alcohol as compared to after placebo ingestion (Ginsburg et al., 1996). ACTH and prolactin levels were not assessed in this study. A very small study of 30 g ETOH per day to healthy postmenopausal women over a period of 3 weeks found no changes in E_2 or testosterone levels; however, it did show increases in DHEAS of 16.5%, CI 8.0–24.9% (Sierksma et al., 2004).

In vitro evidence indicates that acute alcohol ingestion increases basal as well as prostaglandin E1 (PGE1)-stimulated hypothalamic β-endorphin release, apparently via regulation of cyclic adenosine monophosphate (cAMP) with the opposite effect occurring with chronic exposure (Boyadjieva et al., 1997; Boyadjieva and Sarkar, 1997b). The stimulatory

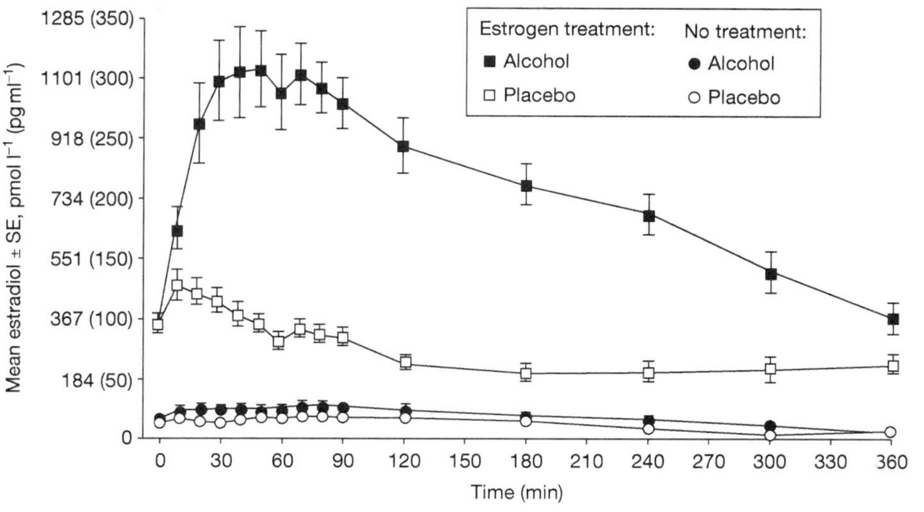

Figure 3 Circulating estradiol levels after alcohol and placebo ingestion in postmenopausal women. Each data point is the mean (±SE) of 12 women. Reproduced from Ginsburg ES, Mello NK, and Mendelson JH (2002) Alcohol abuse: Endocrine concomitants. In: Pfaff DW, Arnold AP, Etgen AM, Fahrbach SE, and Rubin RT (eds.) *Hormones, Brain, and Behavior,* vol. 5, pp. 747–780. San Diego, CA: Academic Press, with permission from Elsevier.

effect of alcohol on hypothalamic cultures may be mediated by acetaldehyde, its primary metabolite (Reddy et al., 1995; Reddy and Sarkar, 1993).

32.3.1.2 Chronic alcohol effects on the hypothalamic–pituitary–gonadal or adrenal axis

There is evidence to suggest that chronic alcohol consumption has a cumulative effect on circulating E_2 levels in postmenopausal women. Mean E_2 levels were higher in a cohort of 164 postmenopausal alcohol drinkers consuming from 0.1 to 28 drinks weekly, than in abstainers (Gavaler et al., 1993; Gavaler and van Theil, 1992). E_2 levels of drinkers averaged 162.6 pmol l^{-1} as compared to 100.8 pmol l^{-1} in abstainers. In addition, a study of 61 postmenopausal Japanese women found that after adjustment for age, height, and body mass index, alcohol consumption was positively associated with circulating E_2 levels, as well as DHEAS levels (Nagata et al., 1997).

Additional recent evidence in larger cross-sectional studies also supports the hypothesis that elevations in circulating estrogen levels do occur in women who consume alcohol chronically. A cross-sectional study of 17 357 Dutch women evaluated the relationship between alcohol consumption and hormone levels using a food frequency questionnaire. Women consuming more than 25 g of alcohol daily had significantly higher levels of estrone and E_2 than those who did not, and DHEAS levels also were non-significantly higher (Onland-Moret et al., 2005). In a follow-up study of 1291 postmenopausal women, those who consumed more than 25 g alcohol per day, had serum estrone, DHEAS, testosterone, and free testosterone levels that were 10–20% higher (Rinaldi et al., 2006). These data are supported by the finding that oophorectomized rats fed 5.5% ethanol in their drinking water over 10 weeks had significantly greater uterine weights than controls, an effect that is estrogen dependent (Gavaler et al., 1987). Not all animal models demonstrate increases in estrogen exposure in association with long-term alcohol use. A study of 46 ovariectomized monkeys who were trained to drink the equivalent of two drinks per day (0.5 g kg^{-1}), 5 days per week for 6 months, did not have more cell proliferation in the breast or endometrium in reponse to alcohol or placebo ingestion (Shively et al., 2004).

Interestingly, a study evaluating the effects of physiologic doses of E_2 on ovariectomized and sham-operated rats found that physiologic replacement doses of E_2 increased self-administration of ethanol, administered as a 10% ethanol solution, with no decrease in water consumption. Supraphysiologic doses of E_2 had no further effect on alcohol consumption. These data indicate that E_2 modulates alcohol consumption patterns (Ford et al., 2004).

Although postmenopausal women who consume alcohol do not consistently evidence estrogen elevations, pronounced enough to lead to uterine bleeding, they have higher breast cancer risk if alcohol is consumed regularly (Longnecker, 1993; Longnecker et al., 1988, 1995; Royo-Bordonada et al., 1997; Willett et al., 1987). Recent cohort studies also show that women who consume alcohol are more likely to develop breast cancer, and these cancers are more likely to be estrogen and progesterone receptor positive (Zhang et al., 2007; McDonald et al., 2004).

Whether the mechanism for this documented alcohol-associated increased breast cancer risk is due to elevations in E_2, or due to other mechanisms such as genetic differences in alcohol metabolism, is unknown.

32.3.2 Alcohol Effects in Postmenopausal Women on Estrogen Replacement Therapy

32.3.2.1 Acute alcohol effects: Gonadotropin and ovarian steroid hormones

As noted previously, postmenopausal women given $0.7\,g\,kg^{-1}$ ethanol or isocaloric placebo evidence no significant change in circulating E_2 levels (**Figure 3**; Ginsburg et al., 1996). However, when women using oral E_2 replacement therapy (ERT) were given the same alcohol dose, there was a threefold increase in circulating E_2 levels that peaked at $265 \pm 27\,pg\,ml^{-1}$ at 50 min after drinking began and persisted for 6 h. As seen in women not using ERT, blood alcohol levels peaked 40–60 min after drinking began. No difference in mean or pulsatile LH was seen over the course of the study (ES Ginsburg, NK Mello, JH Mendelson, et al., unpublished data). Estrone levels were lower after alcohol than after placebo ingestion, indicating that a change in redox potential with an increase in NADH may have interfered in part with E_2 metabolism.

However, the decline in estrone was not enough to account for the 300% increase in circulating E_2 levels. The mechanism for this phenomenon is not fully understood. Alcohol ingestion appears to have a less-pronounced impact on circulating E_2 levels in women using transdermal ERT. When $0.95\,g\,kg^{-1}$ ethanol was given to postmenopausal women using transdermal E_2, levels increased to 40% over baseline, in the same time frame as occurred in oral ERT users (**Figure 4**; Ginsburg et al., 1995a). Another study using the transdermal E_2 patch model showed that the half-life of E_2 is prolonged by 54% (Ginsburg et al., 1998).

32.3.2.2 Chronic alcohol effects: Estrogen and breast cancer

Although it is fairly clear that alcohol increases breast cancer risk (Smith-Warner et al., 1998), there is also evidence that women who use both alcohol and postmenopausal estrogen replacement have higher

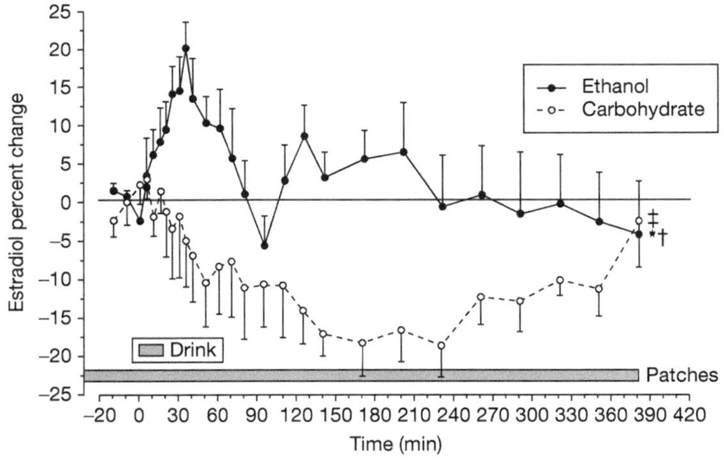

Figure 4 Percent changes in serum estradiol concentrations (mean ± SEM) before and after alcohol (closed circle) and placebo (open circle) ingestion. Asterisk indicates estradiol significantly higher after alcohol vs. placebo ingestion ($p = 0.038$); plus indicates estradiol significantly higher after alcohol vs. baseline ($p < 0.01$); and dagger indicates change in estradiol after placebo ingestion not significant. Reproduced from Ginsburg ES, Mello NK, and Mendelson JH (2002) Alcohol abuse: Endocrine concomitants. In: Pfaff DW, Arnold AP, Etgen AM, Fahrbach SE, and Rubin RT (eds.) *Hormones, Brain, and Behavior,* vol. 5, pp. 747–780. San Diego, CA: Academic Press, with permission from Elsevier.

breast cancer risks than would be expected by the use of either one alone (Colditz et al., 1990; Gapstur et al., 1995; Nielsen and Gronbaek., 2008). There is evidence that the increased risk of breast cancer associated with drinking is more prevalent in postmenopausal women, and is associated with long-term consumption of more than 30 g (two drinks) of alcohol daily. It has been hypothesized that alcohol-induced hyperinsulinemia leads to activation of insulin-like growth factor-1 (IGF 1) receptor in breast tissue, which in turn acts as a mitogen and transforming agent, and leads to an estrogen-independent, and more aggressive cancer cell type (Enger et al., 1999; Gapstur et al., 1995; Stoll, 1999).

A prospective cohort study involving 51 847 women in the Swedish Mammography Cohort evaluated the relationship between alcohol and postmenopausal breast cancer. Women self-reported alcohol consumption as part of the cohort study from 1987–97, and through 2004, 1188 cases of invasive breast cancer with known estrogen and progesterone receptor (ER and PR) status were identified. Alcohol intake was positively associated with an increased risk of ER-positive breast cancers, with the highest level of intake ($>10\,g\,day^{-1}$) compared to nondrinkers having an RR of 1.35 (CI 1.02 = 1.80). The risk of ER- and PR-positive breast cancer was increased even more in postmenopausal women who used hormones. There was no association between alcohol intake and the risk of ER-negative tumors (Suzuki et al., 2005). These findings are supported by additional studies as well (Klug et al., 2006).

Together, these studies confirm that there is an interaction between alcohol consumption and postmenopausal hormone use. Women at genetically increased risk of breast cancer due to the mutations in the alcohol dehydrogenase gene have marked elevations of E_2 after alcohol consumption and an increased risk of breast cancer, supporting that the mechanism of alcohol's increase in E_2 levels may induce breast cancer development. This hypothesis is further confirmed by studies showing that breast cancer risk is highest in postmenopausal women who both drink and take postmenopausal estrogen.

32.4 Implications of Stimulatory Effects of Alcohol on Pituitary and Gonadal Hormones

Alcohol may have either stimulatory or suppressive effects on pituitary and gonadal hormones, depending on the dose and duration of alcohol administration and conditions of gonadotropin stimulation. This seeming contradiction greatly complicates analysis of the mechanisms by which alcohol intoxication induces derangements of the menstrual cycle (Mello, 1988; Mello et al., 1989). The complex interrelationships between the hypothalamic–pituitary–ovarian axis and the hypothalamic–pituitary–adrenal axis and the incomplete evaluation of alcohol effects on them preclude any simplistic conclusion that alcohol acts primarily at one specific target site. It remains to be determined how alcohol affects the integration and regulation of these systems and causes disorders of reproductive function.

32.5 Implications of Alcohol-Induced Changes in Maternal Reproductive Hormones for Pregnancy and Fetal Growth and Development

Although alcohol may disrupt menstrual cycle dynamics, chronic alcoholic women can conceive and their children may be afflicted with the congenital anomalies, abnormal brain function, and retardation of growth and development of the FAS. Described behavioral disorders range from hyperactivity to mental retardation (Hannigan et al., 1992; Streissguth, 1986; Streissguth et al., 1991a; Federal Drug Administration, 1981; West et al., 1990).

Alcohol diffuses freely across the placental barrier; therefore, the fetus is exposed to the same dose of alcohol as the mother. It is not completely understood what range of alcohol intake is likely to produce adverse consequences for the fetus. This is important because in past reports most women (60%) of child-bearing age use some alcohol (Alcohol and Health, 1983), and confirmation of pregnancy often does not occur until midway into the first trimester, which is the period of organogenesis. Another unresolved question is the impact of duration of alcohol use on risk for fetal abnormalities. Although tolerance for alcohol develops in many systems, there are data suggesting that use during later pregnancy is more likely to be associated with a severe FAS than early pregnancy (Abel, 1984; Harrigan et al., 1992).

It has been estimated that one to three of every 1000 live births in the general population is afflicted with some variant of alcohol-related impairment (Abel and Sokol, 1987; Alcohol and Health, 1990). Among alcoholic women, the prevalence of the FAS has been estimated at 21–29 per 1000 births (Abel and Sokol, 1987).

The pathogenesis of fetal dysmorphologies and behavioral impairments associated with the FAS is unknown (Abel, 1984; Anderson, 1981; Smith and Asch, 1987). Fetal hypoxia, secondary to collapse of umbilical circulation, may be induced within 10–15 min after IV alcohol administration ($3\,g\,kg^{-1}$) to pregnant monkeys during the third trimester (120–147 days of gestation) (Mukherjee and Hodge, 1982). It is theorized that alcohol intoxication in pregnant women could produce recurrent episodes of fetal hypoxia with resulting damage to the developing nervous system (Mukherjee and Hodge, 1982). It has not been possible to isolate the effects of any toxins on a single component of the maternal–placental–fetal unit.

It is not definitively known whether alcohol's effects on maternal reproductive hormones influence pregnancy outcome and affect fetal growth and development (Alcohol and Health, 1983; Anderson, 1981; Beeley, 1986; Bowden et al., 1983; Clarren et al., 1988; Clarren and Bowden, 1982; Hutchings, 1989; Randall and Noble, 1980; Streissguth et al., 1980). Unfortunately, little information exists on alcohol's effects on the endocrine system during pregnancy (Anderson, 1981). A study suggests that E_2 levels are higher in pregnant women who drink than in those who do not when measured at 26 weeks gestation (Petridou et al., 1992). A study of 339 pregnant women from the Child Health Development Study cohort was done to determine the effects of alcohol on serum hormone levels. This cohort was monitored from 1959 to 1966, before risks of drinking during pregnancy were widely known. The amounts of alcohol consumed in the cohort were 0.2–0.5, 0.6–2.0, and 2.1–12.5 oz ethanol per week, or two drinks per week in the middle group and one drink per day in the highest intake group. There was no change in serum E_2 levels in association with the amount of alcohol consumed, but women in the middle and highest intake group had lower circulating testosterone levels than those in the lowers group; the significance of this is unknown (Stevens et al., 2005).

We know of only one assessment of pituitary and gonadal hormone levels during pregnancy in alcoholic mothers who gave birth to FAS infants (Halmesmäki et al., 1987). Increased levels of prolactin and low levels of E_2 and progesterone were reported during pregnancy weeks 16–24 when levels were measured (Halmesmäki et al., 1987). Prolactin levels were signicantly higher (mean $70\,ng\,ml^{-1}$, $p < 0.025$) in alcoholic women than in abstinent control women at the same gestational age (24 weeks).

However, alcoholic women who gave birth to normal infants had similar prolactin levels to those who had FAS infants (Halmesmäki et al., 1987). The importance of elevated prolactin levels in pregnant women who consume alcohol in terms of fetal growth and development are unclear. These findings are, however, consistent with the alcohol-induced enhancement of prolactin secretion after naloxone and LHRH stimulation discussed earlier (Mendelson et al., 1987; Phipps et al., 1987).

Endocrinologic alteration is probably only one of many factors leading to the occurrence of the FAS. In Finland, 5 of 20 alcoholic women studied from week 16 of gestation until term had children with FAS (Välimäki et al., 1990a). Alcoholic women had abnormal changes in serum lipids and lipoproteins during pregnancy. The usual increases in low-density lipoproteins were attenuated in alcoholic women, especially in mothers of FAS infants. Alcoholic women had accentuated increases in very low-density lipoprotein (VLDL) cholesterol as well, as compared to controls. FAS mothers continued to drink throughout pregnancy (range 21–105 drinks per week, or 294–1470 g per week). Nine alcoholic women with healthy babies reduced alcohol consumption to 3–10 drinks per week after counseling (Välimäki et al., 1990a). Whether these alcohol-related changes in serum lipids and lipoproteins contributed to the development of FAS, or are associated with etiologic factors is unknown.

32.5.1 Ovarian Steroid Hormones and Teratogenesis

There have been no studies of alcohol's effects on maternal hormones during early pregnancy. In the first trimester organogenesis occurs, so the developing fetus is especially vulnerable to drug-induced malformations (Beeley, 1986). Maternal drug use also may impair fetal growth and development during the second and third trimesters as well. The central nervous system (CNS) continues to develop throughout pregnancy and is therefore vulnerable to toxicities throughout (Beeley, 1986). A prospective study of 650 nonalcoholic women indicates that alcohol exposure (0.84–1.28 drinks per day) during the first and second month of the first trimester was associated with morphological abnormalities in the newborn (Day et al., 1989). An increased rate of decreased head circumference, length, and low birth weight was observed in children whose mothers drank one or more drinks per day (Day et al., 1989). Similar results were reported in a prospective study of 202 women conducted in France, where three drinks per day or more during the first trimester was associated with craniofacial anomalies and two cases of FAS, with a direct relationship seen between the incidence of abormalities and the amount of alcohol consumed (Rostand et al., 1990).

Alcohol-induced increases in plasma E_2 occur rapidly when blood alcohol levels exceed 20–30 mg dl^{-1}, when substrate saturation of hepatic alcohol dehydrogenase isoenzymes occurs (Mendelson et al., 1988). The stability of such E_2 changes as a function of frequency of drinking is unknown, however. If so, an alcohol-induced stimulation of E_2 might be maximal during the first trimester when hCG levels are high. Although diethylstillbestrol (DES), an extremely potent estrogen, has been accepted as a teratogen (Murad and Hayens, 1985b; Nora et al., 1976; Schardein, 1980) when administered in the first trimester of pregnancy, the lack of an increased rate of fetal malformations in women undergoing assisted reproductive technologies, with multiple follicular development, might argue against the role of elevated E_2 levels in the induction of FAS or other alcohol-induced abnormalities.

hCG is secreted by the syncytiotrophoblast during early pregnancy, reaching maximal levels at about 10 weeks gestation (Casey et al., 1985). In normal pregnancy, hCG maintains the corpus luteum during the luteal phase by stimulating ovarian-progesterone secretion until the placenta produces adequate progesterone. The luteoplacental shift in progesterone production occurs between 6 and 8 weeks after conception (Brodie and Wentz, 1989; Solomon, 1988; Talamantes and Ogren, 1988). In an attempt to simulate alcohol's effects on ovarian steroid secretion under conditions similar to that of early pregnancy, we administered hCG to normal healthy women (Teoh et al., 1990). Five thousand units of hCG (Profasi) intramuscularly and alcohol (0.7 g kg^{-1}) or placebo were administered simultaneously under double-blind conditions. Peak blood alcohol levels averaged 124 ($\pm$11) mg dl^{-1}, 45 min after alcohol and hCG administration (Teoh et al., 1990). Within 10 min after alcohol and hCG administration, E_2 levels rose significantly by 55–91 pg ml^{-1} and remained elevated for 240 min.

In contrast, after hCG and placebo–alcohol administration, E_2 gradually decreased and was significantly below baseline 240 min after hCG (Teoh et al., 1990). These findings supported earlier observations that E_2 increased to about 60% above baseline after alcohol plus opiate antagonist stimulation (Mendelson et al., 1987; Teoh et al., 1988) and LHRH stimulation (Mendelson et al., 1989). As discussed earlier, alcohol-related increases in E_2 may be due to decreased E_2 metabolism or increased E_2 secretion, or both.

In contrast to its effects on E_2, alcohol appeared to blunt hCG stimulation of progesterone (Teoh et al., 1990). After placebo–alcohol and hCG administration, progesterone gradually increased to levels significantly above baseline within 210 min. If alcohol suppresses progesterone levels during the first trimester of pregnancy when chorionic gonadotropin levels are high, this could potentially increase the risk for spontaneous abortion, since progesterone maintains the endometrium of early pregnancy.

If a competitive progesterone antagonist is given within 24 days after conception, abortion usually occurs (Itskovitz and Hodgen, 1988; Peyron et al., 1993). After naltrexone stimulation alcohol also suppressed progesterone levels as compared to placebo (Teoh et al., 1988). Alcoholic mothers of FAS infants had lower progesterone levels during gestational weeks 16–24 than control women (Halmesmäki et al., 1987); whether this contributes to the FAS is unknown. The mechanism of alcohol's suppression of progesterone is unclear. However, intrahepatic alcohol metabolism could reduce NAD availability for oxidation of pregnenolone to progesterone and thereby reduce progesterone levels (Teoh et al., 1988).

32.5.2 Hypothalamic–Pituitary–Adrenal Factors in Teratogenesis

Alcohol-induced changes in hypothalamic–pituitary–adrenal as well as hypothalamic–pituitary–gonadal function may be significant factors in the pathogenesis of the FAS (Anderson, 1981; Redei et al., 1986). Alcohol-related stimulation of adrenocortical activity leading to increases in corticosteroid secretion could produce adverse effects on the fetus.

As early as the 11th day of gestation alcohol stimulates an increase in basal corticosterone levels, stress-induced corticosterone, and adrenal weight in pregnant rats (Weinberg and Bezio, 1987). This may help explain the persistent increases in responsiveness of the hypothalamic–pituitary–adrenal axis following an ethanol challenge in rats exposed to ethanol *in utero* (Taylor et al., 1981, 1982, 1984; Lee and Rivier, 1993a). In addition, the patterns of response depend on the time course of alcohol exposure (Weinberg et al., 1996).

Alcohol and stress both stimulate CRF release which in turn may inhibit gonadotropin secretion from the pituitary (Olster and Ferin, 1987; Redei et al., 1986; Rivier et al., 1986; Rivier and Vale, 1984; Xiao and Ferin, 1988). CRF antagonist administration reversed the inhibitory action of stress on pulsatile LH release in rats (Rivier et al., 1986). CRF administration inhibited LH release in ovariectomized monkeys, whereas ACTH and cortisol did not affect LH secretory activity (Xiao and Ferin, 1988). Therefore, a direct central effect of alcohol on CRF secretion may compromise release of gonadotropins which are essential for maintenance of pregnancy due to their effects in stimulating ovarian progesterone secretion.

Alcohol stimulation of the hypothalamic–pituitary–adrenal axis may directly affect gonadal function. Cortisol directly suppresses plasma testosterone levels without altering LH or prolactin levels in normal men (Cumming et al., 1983). Alcohol-induced increases in cortisol may be one reason for the suppression of testosterone synthesis in alcoholic men. There are as yet no controlled clinical studies evaluating alcohol effects on the interactions of the hypothalamic–pituitary–adrenal and the hypothalamic–pituitary–gonadal axis in women.

32.5.3 Alcohol Use and Spontaneous Abortion

The adverse effects of moderate alcohol use on pregnancy are also suggested by data on abortion (Harlap and Shiono, 1980; Kline et al., 1980), but there is no consensus about the dose and frequency of drinking that increases miscarriage risk (Halmesmäki et al., 1987, 1989). In primates, alcohol exposure once a week before week 5 of gestation consistently induces abortion (Clarren et al., 1988).

A prospective survey of 32 019 women at their first antenatal visit to a Kaiser Hospital Clinic between 1974 and 1977 found a total of 1503 spontaneous abortions during this period. During the first trimester (5–14 weeks) there were 714 abortions, and during the second trimester (15–27 weeks) 789 abortions occurred. The overall rate of abortion was 14.4%, and 2.6% were during the second trimester. Self-reports of drinking indicated that women who had one or more drinks daily had higher spontaneous abortion rates, primarily during the second trimester, than women who were abstinent or occasional drinkers. The RR for second-trimester spontaneous abortion was 1.03 for occasional drinkers, 1.98 for those who consumed one to two drinks per day, and 3.53 for more than three drinks per day (Harlap and Shiono, 1980).

Alcohol did not significantly increase first-trimester miscarriage risk. The RR of first-trimester miscarriage was 1.12 in occasional drinkers, 1.15 for one to two drinks per day, and 1.16 for three drinks or more daily. These alcohol effects appeared to be independent of cigarette-smoking effects (Harlap and Shiono, 1980).

Other investigators have confirmed associations between spontaneous abortion and moderate drinking (Kline et al., 1980). One study surveyed 616 women who had spontaneous abortions and 632 women who delivered later than 28 weeks gestation for drinking frequency. Through a logistic regression model, it was estimated that 25% of women who drink twice a week will miscarry as compared to 14% of women who drink less frequently. This should be taken in the context of the fact that the overall clinical miscarriage risk in the general population has been estimated to be approximately 20%. One ounce of absolute alcohol was estimated to be the minimum harmful dose, and the type of alcohol was not significant (Kline et al., 1980).

However, studies conducted in Finland suggest that one or two drinks (13 g of alcohol) per week do not increase spontaneous abortion risk (Halmesmäki et al., 1987, 1989). Eighty women who were hospitalized just prior to spontaneous abortions (6–23 weeks) were compared with 81 age-matched pregnant controls at the time of ultrasound and delivery who ultimately had healthy babies (Halmesmaki et al., 1989), and controls were studied after ultrasound examination and again after delivery. The husbands in each group drank four to five drinks per week (Halmesmaki et al., 1989). Fifty-eight percent of women in both groups continued to drink during pregnancy. Fifteen (19%) and sixteen (20%) women in the abortion and control groups reported alcohol abstinence before and after pregnancy, respectively.

During early pregnancy (mean 10.7 ± 4.5 weeks), the spontaneous abortion group drank about one drink per week (16 g of alcohol, range 3–120 g) and the control group about one-half a drink per week (8 g of alcohol, range 3–23 g) (Halmesmaki et al., 1989). Although the average alcohol consumption reported by control women did not differ statistically from that of aborters, the range (3–23 g vs. 3–120 g) appears different. The impact of sporadic episodes of nine to ten drinks per week on the developing fetus is unknown. The investigators concluded that two or fewer drinks per week in the first trimester of pregnancy is not a significant risk factor for miscarriage (Halmesmaki et al., 1989). In a retrospective Swedish study, chronic alcoholism was also not associated with higher rates of spontaneous abortion (Hollstedt et al., 1983).

32.5.4 Alcohol and Reproductive System Development

Alcohol exposure of the fetus and the prepubertal female rodent may alter secretory patterns of pituitary hormones needed for normal reproductive function. During the week before birth, alcohol ($11.59 \pm 0.37\,g\,kg^{-1}\,day^{-1}$) caused lower postovariectomy levels of LH than pair-fed controls (Handa et al., 1985). In addition, there was diminished LH response to E_2 stimulation (Handa et al., 1985).

Alcohol exposure can delay the onset of puberty in female rodents (Dees and Skelley, 1990) which evidenced delayed puberty, which in turn was associated with lower levels of LH and growth hormone (GH) than seen in controls. However, FSH levels did not differ (Dees and Skelley, 1990; Dees et al., 1990). Elevations in hypothalamic growth-hormone-releasing hormone (GHRH) were interpreted to suggest that alcohol treatment decreased secretion of releasing hormone (Dees et al., 1990). These data suggest that alcohol may disrupt reproductive development in the prenatal period, puberty, and adulthood. The extent to which similar mechanisms occur in humans remains to be determined.

32.5.5 Alcohol Abuse and Teratogenesis: The FAS

It may seem surprising now that the 1973 report (Jones et al., 1973) defining FAS as a specific pattern of dysmorphologies and growth retardation in the children of alcoholic women was met with skepticism. Controversy centered around the fact that FAS mothers often used other drugs (Sokol et al., 1980), were malnourished, and had poor medical care during pregnancy. It was argued that any of these factors alone or in combination with alcohol abuse could lead to birth defects, and similar patterns of malformation can occur independently of alcohol abuse (Beeley, 1986). Attributing birth defects specifically to alcohol required controlling for malnutrition and polydrug abuse, a condition impossible to achieve in pregnant women. Animal models therefore were needed.

32.5.5.1 Animal models of FAS

Current evidence suggests that alcohol is a teratogen in several species studied under controlled laboratory conditions. Moreover, prenatal alcohol exposure appears to have similar effects in humans and animals (Driscoll et al., 1990; Randall et al., 1990). Because the reproductive biology of rhesus monkeys is so similar to that of human females (Goodman and Hodgen, 1983; Knobil, 1980), primate models have been extensively used.

Alcohol administration to pregnant macaques only once a week led to fetal malformations and behavioral retardation in the infants (Bowden et al., 1983; Clarren et al., 1988, 1990; Clarren and Bowden, 1982). In earlier studies, a high dose of alcohol (2.5–$4.1\,g\,kg^{-1}$) was administered to four pregnant females by gavage once a week from 40 days postconception to delivery. Within 2 h after alcohol administration, average blood alcohol levels ranged from 240 to $256\,mg\,dl^{-1}$ and 338 to $415\,mg\,dl^{-1}$, respectively (Bowden et al., 1983). There was one spontaneous abortion and three live births. The infants were followed for 6 months and compared with age- and sex-matched controls. One infant prenatally exposed to a high alcohol dose ($4.1\,g\,kg^{-1}$) had neurologic, developmental, and facial anomalies similar to those seen in human FAS infants. This infant showed profound retardation and cerebral asymmetry, minimal cortical organization, and hydrocephalus. The other two infants exposed to $2.5\,g\,kg^{-1}$ of alcohol prenatally, showed no abnormalities in the female but hyperkinesis, developmental retardation, and abnormal brain morphology in the male (Bowden et al., 1983).

These macaque infants differed from human FAS infants in two respects: (1) all were abnormally large at birth, whereas humans tend to have low birth weights, and (2) none showed malformations of heart, kidney, or other organs often seen in human FAS. However, the absence of organ malformation is probably because maternal alcohol administration did not begin until the end of the organogenic period. The size of the macaque infants may have reflected adequate nutritional status of the mothers and only once weekly alcohol intake, whereas alcoholic women usually drink daily (Bowden et al., 1983).

In a second series of studies, 54 pregnant macaque monkeys were given weekly doses of alcohol (or isocaloric and isovolemic sucrose control solution) from the 1st or the 5th week of gestation until term (Clarren et al., 1988). Abortion was induced by alcohol doses of 2.5 g kg^{-1} or more given from the 1st week of gestation. Therefore, the final dose regimens were 1.8 g kg^{-1} from week 1 of gestation (group 1) or 2.5 g kg^{-1} from days 33 to 46 days post-conception (group 2). Sixteen of twenty-eight infants had facial dysmorphia, growth deficiency, and CNS dysfunction. Group 1 infants had more consistent and severe cognitive abnormalities than group 2 infants even though the latter were exposed to higher alcohol doses. At 6 months of age, group 2 infants were more cognitively intact and showed less evidence of delayed development than group 1 infants. Facial dysmorphia occurred in only one group 1 infant, suggesting that behavioral teratogenic effects may occur independently of facial dysmorphia (Clarren et al., 1988). Mean maternal blood alcohol levels of 140 mg dl^{-1} and above were associated with developmental retardation in 10 of 12 animals. Microcephaly occurred in the single surviving infant in the 4.1 g kg^{-1} group (Clarren et al., 1988). These data indicate that alcohol-induced fetal impairments are related to alcohol levels, and how early chronic exposure begins during pregnancy (Clarren et al., 1988).

This study found that weekly alcohol exposure leading to average blood alcohol levels of 115, 140, and 249 mg dl^{-1} resulted in behavioral abnormalities in infants (Clarren et al., 1990). Exposure to these alcohol levels from week 1 of gestation to the end of pregnancy had a more severe impact on the fetus than higher average blood alcohol levels (260–540 mg dl^{-1}) obtained only after the 5th week of gestation (Clarren et al., 1990). However, infants sacrificed after 6 months had minimal dose-dependent structural brain abnormalities in the caudate nucleus by electron-micrographic analyses (Clarren et al., 1990). Seven infants had unilateral ocular anomalies similar to those in human FAS. The investigators concluded that prenatal exposure to alcohol may result in abnormal behavioral development in the presence of normal physical features and only subtle neuroanatomic and neurochemical alterations (Clarren et al., 1990).

32.5.5.2 Possible mechanisms of FAS

Prenatal alcohol exposure in a pregnant rat model indicates that an alcohol diet blunted pituitary ACTH and β-endorphin but not corticosterone release in response to peritoneal IL-1 injection, but there was no effect of injection of CRF (Lee and Rivier, 1993b). Pair-fed pups also demonstrated a decrease in pituitary responsiveness. These data indicate that alcohol interferes with IL-1β stimulation of pro-opiomelanacorticotrophic (POMC)-related peptides. Tumor necrosis factor α (TNFα) is neuroprotective, preventing neuronal excitement, but this effect is antagonized by low levels of alcohol in cultured murine cortical neurons (Gahring et al., 1999).

Interestingly, human first-trimester trophoblast cell line expresses high levels of granulocyte colony stimulating factor and IL-6. This may indicate that alcohol may modulate the cytokines the developing fetus is exposed to which could have adverse effects (Svinarich et al., 1998).

As previously discussed, alcohol and its metabolite acetaldehyde stimulate β-endorphin release from hypothalamic neurons in pituitary cultures. A study of cultured fetal hypothalamic cells found that chronic alcohol treatment desensitizes β-endorphin secreting neurons due to decreased cellular function, but not cell death. However, chronic acetaldehyde exposure reduces β-endorphin neurotransmission due to cell death. Pretreatment with cAMP appears to partly protect against these effects (Boyadjieva et al., 1997).

Glutamate, an excitatory neurotransmitter, is increased by ethanol. In an elegant study, 2 or 4 g kg^{-1} maternal body weight of ethanol was administered into the parasaggital cortex of near-term fetal lambs. Ethanol or saline was administered in divided doses over 5 h. A dose-dependent increase in glutamate resulted. As glutamate has been implicated in neuronal development, such increases may play a role in the development of the FAS (Reynolds et al., 1995). Ethanol has been implicated in alterations in glial cell development. *In vitro* studies of cortical astrocytes show multiple effects on DNA, RNA, and protein synthesis, including glial fibrillary acidic protein which is involved with glial growth (Guerri and Renau-Piqueras, 1997). It was shown that in 7-day-old Sprague-Dawley rats, alcohol administration caused neurodegeneration consistent with apoptosis, the same effect that could be found with other glutamate receptor blockers during the period of synaptogenesis (Ikonomidou et al., 1999). The mechanism of this was further elucidated in a study of pregnant rats administered various doses of alcohol. It was found that the apoptotic response required a minimum of 4 h of sustained blood alcohol levels of 200 mg dl^{-1} (Ikonomidou et al., 2000). In addition, neuronal populations had transient periods of vulnerability during which synaptogenesis occurred. Therefore, the weights of ethanol-treated fetal rat forebrains were significantly less than those of saline-treated rats, indicating that sigificant loss of brain tissue is associated with maternal alcohol abuse. The apoptotic effects are thought to be due to the blocking action of alcohol at the *N*-methyl D-aspartate (NMDA) glutamate receptors and its positive modulatory action at gamma-aminobutyric acid-A (GABA-A) receptors. In fact, NMDA antagonists (Ikonomidou et al., 2000), GABAergic agents, and ethanol all affect the immature brain during the period of synaptogenesis, which occurs during the last 3 months of gestation in humans (Ikonomidou et al., 2000).

Another effect of intrauterine alcohol exposure of the fetus is a decrease in fetal breathing movements. In fact, fetal blood alcohol levels are directly associated with fetal plasma PGE2 and cerebrospinal fluid concentrations, and inversely related to the incidence of fetal breathing movements. The alcohol effect on breathing suppression is blocked by the prostaglandin synthatase inhibitor indomethacin, confirming the causative PGE2 link (Brien and Smith, 1991).

There is also evidence that alcohol has a direct effect on fetal testes. Male fetuses of alcohol users compared to nonusers have lower amniotic fluid androstenedione, testosterone, and E_2 as well as lower ultimate birth weights (Westney et al., 1991).

In vitro evidence supports direct effects of alcohol on glial development. Alcohol treatment of cultures of glial cells from the cerebral cortex, a primary area affected in the FAS, shows that alcohol affects DNA, RNA, protein synthesis, decreases mitoses, alters the content and distribution of cytoskeletal proteins, causes oxidative stress, and decreases growth factor secretion by the glial cells. *In vivo* studies indicate that alcohol interferes with myelinogenesis and causes abnormal glial development (Guerri and Renau-Piqueras, 1997). In addition, cultures of fetal hypothalmic neurons indicate that alcohol causes neurotoxicity in these β-endorphin neurons during early neuronal differentiation through effects on cAMP control of apoptosis in these neurons (De et al., 1994). Children exposed to alcohol *in utero* are more likely to have behavioral and learning problems when followed longer term. In one study of 70 children, there was a relationship between the need for special education and behavioral problems, and the duration of alcohol exposure (Autti-Ramo, 2000). Even early drinking of alcohol prior to the 5th month of pregnancy was associated with learning disabilities, antisocial behavior, and school difficulty in a study of 464 children followed to age 14 (Olson et al., 1997; Streissguth et al., 1997; Streissguth, 1991b).

Although it has been accepted that alcohol, in sufficient doses, is teratogenic, the extent to which the FAS is specific to alcohol remains unclear. A number of other drugs are also fetotoxic in humans. Drug abuse during pregnancy can result in a combination of low birth weight, delayed development, and brain malformations similar to those reported after alcohol abuse (Hutchings, 1989). The minimum alcohol intake associated with physical or behavioral abnormalities of the FAS has not yet been established. Therefore, it is recommended that women contemplating pregnancy abstain from alcohol use (Henderson et al., 2007).

32.5.6 Polydrug Abuse

Abuse of cocaine and alcohol during the first trimester is associated with an increased risk for abortion (Smith and Smith, 1990). It appears that the combined effects of several drugs may have more significant consequences than use of a single drug, especially when both drugs have similar effects (Kreek, 1987, 1991).

32.6 Effects of Alcohol on Hormone Function in Men

32.6.1 Testosterone

In alcoholic men, impotence, testicular atrophy, gynecomastia, and decreased libido are associated with low testosterone levels, reflecting suppression of testicular testosterone synthesis (Boyden and Pamenter, 1983; Chiao and van Thiel, 1983; Cicero, 1980, 1982; Ellingboe and Varanelli, 1979; Mendelson et al., 1979; Noth and Walter, 1984; van Thiel, 1983, 1984; van Thiel and Gavaler, 1982; Wright et al., 1992; Sierskma et al., 2004). Alcohol appears to inhibit testosterone biosynthesis by direct toxic effects on the testicular Leydig cells (Chiao and van Thiel, 1983; Cicero, 1982; Ellingboe and Varanelli, 1979; van Thiel et al., 1983), perhaps also through its primary metabolite acetaldehyde, but the extent to which hypothalamic and pituitary factors contribute remains controversial (Clarren et al., 1990; Ellingboe, 1987; Välimäki et al., 1990c; van Thiel and Gavaler, 1982; Wright et al., 1992). However, when alcohol was adminstered to normal male volunteers over a

4-week period, LH levels and serum testosterone levels decreased, indicating that a central mechanism was also involved. Testosterone metabolic clearance rate was increased by alcohol, most likely due to a combination of decreased plasma SHBG binding and increased hepatic reductase activity (Gordon et al., 1976). The fact that the men maintained adequate nutrition and had stable weights indicated that nutritional deficiency did not play a role. A complicating factor associated with low testosterone levels is the fairly high prevalence of impotence in these men (Farnsworth et al., 1978). However, a study of 531 healthy Singaporean Chinese men aged 29–72 evaluated lifestyle parameters and measured sex hormones. A multivariate analysis adjusting for age and other confounders found that alcohol, along with exercise and smoking were related to higher androgen levels and lower body fat. The mechanism for these findings was unclear, but may be due to interactions between several factors (Goh et al., 2007).

Despite the hormonal derangements associated with alcohol use, it is not associated with the occurrence of prostate cancer (Tavani et al., 1994).

Exacerbating the direct suppression of alcohol intake on testosterone secretion is the increase of SHBG seen in noncirrhotic alcoholics following acute alcohol ingestion, but during the first days of abstinence. A study of 21 male alcoholics and 21 controls demonstrated that an increase in SHBG occurred, but gradually declined over the first 10 days, and free testosterone levels were normalized by day 10 (Iturriaga et al., 1999). Another study of 42 men with alcoholic cirrhosis and 21 healthy controls found that elevated SHBG was present in 71%, and serum estrone, E_2, FSH, and prolactin were also elevated in cases as compared to controls (Bahnsen et al., 1981). The increase in estrogen levels in alcoholics during withdrawal and 3 weeks of abstinence may be due in part to increased aromatase conversion of androstenedione to E_2 (Heinz et al., 1995).

Despite the significant hormonal derangements described in chronically alcoholic men, male alcoholics without hepatic or gonadal failure do not appear to have significant differences in circulating androgens or estrogens or obvious problems with sexual function. In a study of 45 alcoholic men and 30 healthy volunteers, the sexual function and erection scores were similar. Four of 45 alcoholic men, however, reported losing erections during sexual activity (Gumus et al., 1998).

However, stimulation testing has unmasked a disparity between the acute and chronic effects of alcohol on the hypothalamic–pituitary–gonadal axis and the hypothalamic–pituitary–adrenal axis.

32.6.2 Gonadal Steroids and Provocative Testing

In human males, alcohol significantly increased testosterone levels in comparison to placebo control conditions after LHRH stimulation (500 μm) (Phipps et al., 1987). In rhesus males, alcohol failed to block a naloxone-stimulated increase in testosterone even though blood alcohol levels ranged between 283 and 373 mg dl^{-1} (Mello et al., 1985).

Following hCG injection, ingestion of 200 g day^{-1} ethanol led to lower testosterone levels in 10 healthy men as compared to alcoholic men, though it was lower in both prior to drink ingestion (Bertello et al., 1986). There was no difference in E_2 levels after alcohol in alcoholic or normal men. Abstinence from drinking was followed by increased testosterone responses to hCG administration (Gatti et al., 1985).

32.6.2.1 Luteinizing hormone-releasing hormone/follicle-stimulating hormone/luteinizing hormone

When 1.5 g kg^{-1} was administered to eight healthy male volunteers, with blood samples taken every 20 min over 3 h, no change in LH or FSH secretion or pulsatility was detected (Välimäki et al., 1990c). In 14 alcoholic men with cirrhosis, mean testosterone levels were lower than in age-matched controls (Distiller et al., 1976). In addition, in five men with testicular atrophy, there was also an exaggerated FSH response to LHRH, which paralleled an exaggerated testosterone response to hCG administration. One study of noncirrhotic chronic alcoholic men found that mean 24-h mean levels of serum LH and FSH were elevated, with higher pulse amplitudes but longer interpulse intervals than in controls (Iranmanesh et al., 1988). Serum testosterone, E_2, and estrone levels were normal; however, free E_2 and testosterone levels were high. Interestingly, there was diminshed pituitaly LH and FSH secretion in response to IV LHRH. Together, these data suggest that alcohol causes pituitary suppression, compounding direct testicular effects. The combination of pituitary suppression, decrease in testicular testosterone synthesis and secretion, and an increase in hepatic SHBG combined led to hypogonadism in these patients (Farnsworth et al., 1978; Lester and van Theil, 1977).

A similar hypothesis can be advanced to account for the alcohol-induced enhancement of LHRH-stimulated testosterone levels in males (Phipps et al., 1987). Acute alcohol administration may increase hepatic blood flow (Castenfors et al., 1960; Mendeloff, 1954; Stein et al., 1963). Ethanol catabolism causes a prompt and dramatic increase in the hepatic NADH:NAD ratio (Forsander et al., 1958; Slater et al., 1964).

Increased testosterone levels after alcohol and concomitant gonadotropin stimulation may therefore be due in part to increased hepatic and gonadal conversion of precursor steroids such as androstenedione to testosterone as a consequence of increased NADH:NAD ratios during intrahepatic ethanol catabolism. Following acute alcohol ingestion ($1.5\ g\ kg^{-1}$), testosterone levels decrease by an average of 23% in healthy male volunteers (Välimäki et al., 1990c). However, studies have also found that acute alcohol ingestion ($0.8–1.5\ g\ kg^{-1}$ body weight) suppresses testosterone secretion more so after strenuous exercise (Heikkonnen et al., 1996). In healthy social drinkers, alcohol intake was unrelated to random levels of testosterone (Sparrow et al., 1980). When a fairly low dose of alcohol ($0.3\ g\ kg^{-1}$) was administered to normal men, there was no change in plasma testosterone or unconjugated E_2 or estrone; however, conjugated estrogens increased in parallel to blood alcohol levels, possibly also implicating a change in hepatic redox potential (Andersson et al., 1986, 1987).

Clomiphene citrate, an estrogen agonist–antagonist, has also been used to evaluate pituitary–gonadal function in male alcoholics (Martinez-Riera et al., 1995). When 63 male alcoholics and 15 controls were treated with 200 mg clomiphene citrate for 8 days, LH and FSH increased, and androstenedione and E_2 increased and testosterone levels decreased in male alcoholics with and without cirrhosis more so than in controls. Since clomiphene has no adrenal effect, this indicates that the elevated androstenedione levels were of testicular origin, and that increased aromatization was implicated.

In summary, chronic alcoholism leads to hypogonadism due to suppressive effects on testicular testosterone production, suppression of pituitary LH release, and elevation of SHBG. In men with testicular atrophy, hypogonadism is not reversible even with prolonged abstinence (van Thiel, 1983).

32.6.2.2 CRH/adrenocorticotropic hormone/cortisol

In a study of 297 Japanese men, alcohol ingestion and smoking history was noted in a study of cortisol levels, which were found to be lower in smokers but unrelated to alcohol ingestion (Handa et al., 1994). When $0.75\ g\ kg^{-1}$ ethanol was given to 14 healthy men, there was no difference in mean plasma ACTH or cortisol levels over the 180-min study period (Waltman et al., 1993). When the same alcohol dose or placebo was administered before IV CRH, the plasma response of ACTH and cortisol were blunted after alcohol compared to placebo. When ACTH was administered, however, there was no difference in cortisol levels after alcohol as compared to after placebo ingestion, indicating that alcohol may cause blunting of the hypothalamic–pituitary–adrenal axis at the pituitary. Another similar study evaluated the effect of alcohol ($0.75\ g\ kg^{-1}$) or placebo over 2 h. Interestingly, the response of ACTH to CRH evidenced a higher peak in seven men with alcoholic fathers than in 16 controls who did not have alcoholic fathers (Waltman et al., 1994).

Family history negative men had blunted ACTH responses after alcohol as compared to placebo, though family history positive men had similar responses during alcohol and placebo. There were no differences in cortisol responses to ACTH after alcohol and placebo between groups.

32.6.2.3 Adrenocorticotropic hormone

In male rats fed an alcohol diet for 10 days, the response of ACTH to immune signals such as IL-1β and endotoxin may be related to nitric oxide formation (Rivier, 1995). In 11 male abstaining alcoholics as compared to 10 healthy controls, there was a blunted response of ACTH and also cortisol and norepinephrine in response to IV CRF. This effect normalized after 12 weeks of abstinence (Ehrenreich et al., 1997). In a study of 18 sons of alcoholics and 18 sons of nonalcoholics, given $0.75\ ml\ kg^{-1}$ and $1.1\ ml\ kg^{-1}$ of ethanol, ACTH levels were lower in the sons of alcoholics than in controls after the higher alcohol dose. This suggests that cortisol changes following alcohol ingestion in this population is due in part to a difference in pituitary responsiveness (Schuckit et al., 1988). Acute alcohol ingestion ($1.5\ g\ kg^{-1}$) in eight human male volunteers increased cortisol levels by an average of 36% during the 3 h of sampling (Välimäki et al., 1990c).

32.6.2.4 Prolactin

Alcohol stimulates prolactin secretion in human males (Mendelson et al., 1987; Phipps et al., 1987; Teoh et al., 1990). In a study of seven normal male research subjects given 43% ethanol as whiskey and water over 30 min, and monitored over 3 h (Ida et al., 1992), both alcohol and prolactin levels peaked 60 min after the start of drinking, with testosterone levels falling at 30 min. When the same dose of alcohol was given to nine subjects over 7 days the same findings occurred; however, testosterone levels were suppressed at both 30 and 60 min. ACTH and cortisol levels were unaffected in both the acture and chronic alcohol administration arms. A study of eight healthy male subjects found that prolactin levels rose after 1.0 $g\,kg^{-1}$ body weight but not after 0.5 kNg was administered (Soyka et al., 1991). Concurrently measured epinephrine and norepinephrine were unchanged, indicating that the ethanol effects on prolactin were not stress mediated.

Interestingly, the serum prolactin response may vary depending on the family history of the person tested. A study of 44 nonalcoholic young men with an alcoholic first-degree relative, and four controls with no family history of alcoholism, indicates that prolactin levels increased in all subjects but returned to baseline later in those with a family history (150 vs. 90 min). These data suggest a possible genetic component linking prolactin response and alcoholism (Schuckit et al., 1983).

32.6.3 Thyroid Hormones

Chronic alcoholic men have been found to have profoundly disturbed thyroid regulation, and when TRH stimulation testing is done, a sick euthyroid picture arises, with low triiodothyronine (T3), high reverse triiodothyronine (rT3), and normal thyroxine (T4), as well as increased thyroid-hormone-binding capacity with decreased T3 uptake, and increased T4 binding globulin. In addition, thyroid-stimulating hormone (TSH) response to TRH was blunted in 31% of subjects (Loosen et al., 1983).

32.6.4 Mechanisms of Alcohol-Related Hormonal Changes in Men

Acute alcohol administration to normal men has been shown to induce small, but statistically significant, increases in prolactin levels but the biological significance of these prolactin elevations is unclear (Ellingboe et al., 1980; Välimäki et al., 1984).

Alternatively, since the LHRH-stimulated increase in LH preceded the increase in testosterone both in human and macaque males (Mello et al., 1985; Phipps et al., 1987), it is possible that this elevation in LH levels was sufficient to stimulate testosterone during alcohol intoxication. These data suggest that acute alcohol intoxication has minimal effects on hypothalamic–pituitary function (Mello et al., 1985). The mechanism of LHRH changes may be related to changes in hypothalamic neurons. In rat hypothalamic neurons in culture, 50 mM-alcohol treatment inhibits adenosine uptake, increasing extracellular adenosine activating membrane adenosine receptors, cAMP production, and β-endorphin secretion (Boyadjieva and Sarkar, 1999). Chronic alcohol treatment desensitizes the adenylate cyclase system in this cell population (De et al., 1999). The mechanism of these changes appears to involve calcium channels (Simasko et al., 1999).

32.7 Conclusions

Acute and chronic alcohol use and abuse causes multiple derangements of hypothalamic, pituitary, ovarian, and testicular function in women and men. The perturbations of ovarian hormones may lead to luteal-phase dysfunction or amenorrhea, resulting in impaired fertility. In men, hypogonadism may result in gynecomastia and impotence. In alcohol abusers, hepatic dysfunction may occur, leading to hyperestrogenic manifestations in men which include gyneomastia, and spider angiomata, and impaired coagulation and ability to metabolize medication in both men and women. Tragically in pregnancy the FAS has lifelong ramifications on the children of alcohol using women. Alcohol clearly interacts with estrogen in postmenopausaul estrogen users. Alcohol use in both premenopausal women and postmenopausal women appears to raise breast cancer risk. More research is needed to evaluate potential interactions of alcohol with medications and commonly used dietary supplements.

References

Abel EL (1984) Prenatal effects of alcohol. *Drug and Alcohol Dependence* 14: 1–10.
Abel EL and Sokol RJ (1987) Incidence of fetal alcohol syndrome and economic impact of FAS-related anomalies. *Drug and Alcohol Dependence* 19: 51–70.
Alcohol and Health (1983) *Fifth Special Report to Congress*. Washington, DC: US Government Printing Office.

Alcohol and Health (1990) *Seventh Special Report to Congress*. Washington, DC: US Government Printing Office.

Anderson RA (1981) Endocrine balance as a factor in the etiology of the fetal alcohol syndrome. *Neurobehavioral Toxicology and Teratology* 3: 89–104.

Andersson SH, Cronholm T, and Sjövall J (1986) Effects of ethanol on the levels of unconjugated and conjugated androgens and estrogens in plasma of men. *Journal of Steroid Biochemistry* 24: 1193–1198.

Andersson SH, Cronholm T, and Sjövall J (1987) Effects of ethanol on conjugated gonadal hormones in plasma. *Alcohol and Alcoholism, Supplement* 1: 529–531.

Arimura A and Schally V (1971) Augmentation of pituitary responsiveness to LH-releasing hormone (LHRH) by estrogen. *Proceedings of the Society for Experimental Biology and Medicine* 136: 290–298.

Autti-Ramo I (2000) Follow-up of children exposed to alcohol in utero. *Developmental Medicine and Child Neurology* 42: 406–411.

Bahnsen M, Gluud C, Johnsen SG, et al. (1981) Pituitary–testicular function in patients with alcoholic cirrhosis of the liver. *European Journal of Clinical Investigation* 11: 473–479.

Balasch J and Vanrell J (1987) Corpus luteum insufficiency and fertility: A matter of controversy. *Human Reproduction* 2: 557–567.

Becker U, Gluud C, Bennett P, et al. (1988) Effect of alcohol and glucose infusion on pituitary–gonadal hormones in normal females. *Drug and Alcohol Dependence* 22: 141–149.

Becker U, Tonnesen H, Kaas-Claesson N, et al. (1989) Menstrual disturbances and fertility in chronic alcoholic women. *Drug and Alcohol Dependence* 24: 75–82.

Beeley L (1986) Adverse effects of drugs in the first trimester of pregnancy. *Clinical Obstetrics and Gynecology* 13: 177–195.

Berga SL, Mortola JF, Girton L, et al. (1989) Neuroendocrine aberrations in women with functional hypothalamic amenorrhea. *Journal of Clinical Endocrinology and Metabolism* 68: 301.

Bertello P, Guirioli L, Gatti G, et al. (1986) Short-term ethanol ingestion can affect the testicular response to single-dose human chorionic gonadotropin in normal subjects. *Journal of Endocrinological Investigation* 9: 249–252.

Bowden DM, Weatherbee PS, Clarren SK, et al. (1983) A periodic dosing model of fetal alcohol syndrome in the pig-tailed macaque (*Macaca nemestrina*). *American Journal of Primatolology* 4: 143–157.

Boyadjieva NI, Reddy BV, and Sarker DK (1997) Forskolin delays the ethanol-induced desensitization of hypothalamic beta-endorphin neurons in primary cultures. *Alcoholism: Clinical and Experimental Research* 21: 477–482.

Boyadjieva NI and Sarkar DK (1997a) The role of cAMP in ethanol-regulated beta-endorphin relase by hypothalamic neurons. *Alcoholism: Clinical and Experimental Research* 21: 728–731.

Boyadjieva NI and Sarkar DK (1997b) Effects of ethanol on basal and prostaglandin EI-induced increases in beta-endorphin release and intracellular cAMP levels in hypothalamic cells. *Alcoholism: Clinical and Experimental Research* 21: 1005–1009.

Boyadjieva NI and Sarkar DK (1999) Effects of ethanol on basal and adenosine-induced increases in endorphin release and intracellular cAMP levels in hypothalamic cells. *Brain Research* 2: 112–118.

Boyden TW and Pamenter RW (1983) Effects of ethanol on the male hypothalamic–pituitary–gonadal axis. *Endocrine Reviews* 4: 389–395.

Brien JF and Smith GN (1991) Effects of alcohol (ethanol) on the fetus. *Journal of Developmental Physiology* 15: 21–32.

Brodie B and Wentz AC (1989) An update on the clinical relevance of luteal phase inadequacy. *Seminars in Reproductive Medicine* 7: 138–154.

Buchanan GC and Tredway DR (1979) Hyperpro-lactinemia and ovulatory dysfunction. In: Hafez ESE (ed.) *Human Ovulation*, pp. 255–277. Amsterdam: Elsevier Biomedical Press.

Bullen BA, Skinnar GS, Beitins IZ, et al. (1985) Induction of menstrual disorders by strenuous exercise in untrained women. *New England Journal of Medicine* 2: 1349–1353.

Casey ML, MacDonald PC, and Simpson ER (1985) Endocrinological changes of pregnancy. In: Wilson JD and Foster DW (eds.) *Williams Textbook of Endocrinology*, pp. 422–437. Philadelphia, PA: Saunders.

Castenfors H, Hultman E, and Josephson B (1960) Effect of intravenous infusions of ethyl alcohol on estimated hepatic blood flow in man. *Journal of Clinical Investigation* 39: 776–781.

Channing CP, Gordon WL, Liu WK, et al. (1985) Physiology and biochemistry of ovarian inhibin. *Proceedings of the Society for Experimental Biology and Medicine* 178: 339–361.

Chiao YB and van Thiel DH (1983) Biochemical mechanisms that contribute to alcohol-induced hypogonadism in the male. *Alcoholism: Clinical and Experimental Research* 6: 207–215.

Cicero TJ (1980) Common mechanisms underlying the effects of ethanol and the narcotics on neuroendocrine function. In: Mello NK (ed.) *Advances in Substance Abuse: Behavioral and Biological Research*, pp. 201–254. Greenwich, CT: JAI Press.

Cicero TJ (1982) Alcohol-indeced deficits in the hypothalamic–pituitary–luteinizing hormone axis in the male. *Alcoholism: Clinical and Experimental Research* 6: 207–215.

Clarren SK, Astley SJ, and Bowden DM (1988) Physical anomalies and developmental delays in nonhuman primate infants exposed to weekly doses of ethanol during gestation. *Tetratology* 37: 561–569.

Clarren SK, Astley SJ, Bowden DM, et al. (1990) Neuroanatomic and neurochemical abnormalities in non-human primate infants exposed to weekly doses of ethanol during gestation. *Alcoholism: Clinical and Experimental Research* 14: 674–683.

Clarren SK and Bowden DM (1982) Fetal alcohol syndrome: A new primate model for binge drinking and its relevance to human ethanol teratogenesis. *Journal of Paediatrics* 101: 819–824.

Colditz GA, Stampfer MJ, Willett WC, et al. (1990) Prospective study of estrogen replacement therapy and risk of breast cancer in postmenopausal women. *Journal of American Medical Association* 264: 2648–2653.

Conn PM and Crowley WEJ (1991) Gonadotropin-releasing hormone and its analogues. *New England Journal of Medicine* 324: 93–103.

Coutelle C, Hohn B, Benesova M, et al. (2004) Risk factors in alcohol associated breast cancer: Alcohol dehydrogenase polymorphism and estrogens. *International Journal of Oncology* 25: 1127–1132.

Cronholm T and Sjövall J (1968) Effect of ethanol on the concentrations of solvolyzable plasma steroids. *Steroids* 152: 233–236.

Cronholm T, Sjövall J, and Sjövall K (1969) Ethanol induced increase of the ratio between hydroxy- and ketosteroids in human pregnancy plasma. *Steroids* 13: 671–678.

Crowley WF, Jr., Filicori M, Spratt DI, et al. (1985) The physiology of gonadotropin-releasing hormone (GnRH) secretion in men and women. *Recent Progress in Hormone Research* 41: 473–526.

Cumming DC, Quigley ME, and Yen SSC (1983) Acute suppression of circulating testosterone levels by cortisol

in men. *Journal of Clinical Endocrinology and Metabolism* 57: 671–673.
Dai X, Thavundayil J, Santella S, and Gianoulakis C (2007) Response of the HPA axis to alcohol and stress as a function of alcohol dependence and family history of alcoholism. *Psychoneuroendocrinology* 32: 293–305.
Day N, Jasperse D, Richardson G, et al. (1989) Prenatal exposure to alcohol: Effect on infant growth and morphologic characteristics. *Pediatrics* 84: 536–541.
De A, Boyadjieva NI, Pastorcic M, et al. (1994) Cyclic AMP and ethanol interact to control apoptosis and differentiation in hypothalamic beta-endorphin neurons. *Journal of Biological Chemistry* 269: 697–705.
De A, Boyadjieva NI, and Sarkar DK (1999) Effects of ethanol on alpha-adrenergic and beta-adrenergic agonist stimulated beta-endorphin release and cAMP prodution in hypothalamic cells in primary cultures. *Alcoholism: Clinical and Experimental Research* 23: 46–51.
Dees WL and Skelley CW (1990) Effects of ethanol during the onset of puberty. *Neuroendocrinology* 51: 64–69.
Dees WL, Skelley CW, Hiney JK, et al. (1990) Actions of ethanol on hypothalamic and pituitary hormones in prepubertal female rats. *Alcohol* 7: 21–25.
Dierschke DJ, Hutz RJ, and Wolf RC (1985) Induced follicular atresia in rhesus monkeys: Strength-duration relationships of the estrogen stimulus. *Endocrinology (Baltimore)* 117: 1397–1403.
Dierschke DJ, Hutz RJ, and Wolf RC (1987) Atretogenic action of estrogen in rhesus monkeys: Effects of repeated treatment. *American Journal of Primatolology* 12: 251–261.
Distiller LA, Sagel J, Dubowitz B, et al. (1976) Pituitary–gonadal function in men with alcoholic cirrhosis of the liver. *Hormone and Metabolic Research* 8: 461–465.
diZerega GS and Hodgen GD (1981) Luteal phase dysfunction infertility: A sequel to aberrant folliculogenesis. *Fertility and Sterility* 35: 489–499.
diZerega GS and Wilks JW (1984) Inhibition of the primate ovarian cycle by a porcine follicular fluid protein(s). *Fertility and Sterility* 41: 1094–1100.
Driscoll CD, Streissguth AP, and Riley EP (1990) Prenatal alcohol exposure: Comparability of effects in humans and animal models. *Neurotoxicology and Teratology* 12: 231–237.
Edwards CRW and Feek CM (1981) Prolactinoma: A question of rational treatment. *British Medical Journal* 283: 1561–1562.
Ehrenreich H, Schuck J, Stender N, et al. (1997) Endocrine and hemodynamic effects of stress versus systemic CRF in alcoholics during early and medium term abstinence. *Alcoholism: Clinical and Experimental Research* 21: 1285–1293.
Eliassen AH, Missmer SA, Tworoger SS, et al. (2007) Endogenous steroid hormone concentrations and risk of breast cancer among premenopausal women. *Journal of the National Cancer Institute* 99: 409–410.
Ellingboe J (1987) Acute effects of ethanol on sex hormones in non-alcoholic men and women. *Alcohol, Supplement* 1: 109–116.
Ellingboe J, Mendelson JH, Kuehnle JC, et al. (1980) Effect of acute ethanol ingestion on integrated plasma prolactin levels in normal men. *Pharmacology, Biochemistry, and Behavior* 12: 297–301.
Ellingboe J and Varanelli CC (1979) Ethanol inhibits testosterone biosynthesis by direct action on Leydig cells. *Research Communications in Chemical Pathology and Pharmacology* 24: 87–102.
Emanuele NV, LaPaglia N, Steiner J, Kirsteins L, and Emanuele MA (2001) Effect of chronic ethanol exposure on female rat reproductive cyclicity and hormone secretion. *Alcoholism: Clinical and Experimental Research* 25: 1025–1029.
Enger SM, Ross RK, Paganini-Hill A, et al. (1999) Alcohol consumption and breast cancer oestrogen and progesterone receptor status. *British Journal of Cancer* 79: 1308–1314.
Eriksson CJ, Fukunaga T, Sarkola T, et al. (1996) Estrogen-related acetaldehyde elevation in women during alcohol in-toxication. *Alcoholism: Clinical and Experimental Research* 20: 1192–1195.
Eskay RL, Ryback RS, Goldman M, et al. (1981) Effect of chronic ethanol administration on plasma levels of LH and the estrous cycle in the female rat. *Alcoholism: Clinical and Experimental Research* 5: 204–206.
Farnsworth WE, Cavanaugh AH, Brown JR, et al. (1978) Factors underlying infertility in the alcoholic. *Archives in Andrology* 1: 193–195.
Federal Drug Administration (1981) Surgeon general's advisory on alcohol and pregnancy. *FDA Drug Bulletin* 11: 9–10.
Filicori M, Butler JP, and Crowley WM, Jr. (1984) Neuroendocrine regulation of the corpus luteum in the human. *Journal of Clinical Investigation* 73: 1638–1647.
Filicori M, Santoro N, Merriam GR, et al. (1986) Characterization of the physiological pattern of episodic gonadotropin secretion throughout the human menstrual cycle. *Journal of Clinical Endocrinology and Metabolism* 62: 1136–1144.
Ford MM, Eldridge JC, and Sampson HH (2004) Determination of an estradiol dose-response relationship in the modulation of ethanol intake. *Alcoholism: Clinical and Experimental Research* 28: 20–28.
Forsander O, Raiha N, and Sumalainen H (1958) Alko-holoxydation und bildung von acetoacetat in normaler und glykogenarmer intaker rattenleber. *Hoppe-Seyler's Zeitschrift fur Physiologische Chemie* 312: 243–248.
Frisch RE (1982) Fatness, puberty, menstrual periodicity and fertility. In: Vaitukaitis JL (ed.) *Clinical Reproductive Neuroendocrinology*, pp. 105–135. New York: Elsevier Biomedical.
Frisch RE and McArthur JW (1974) Menstrual cycles: Fatness as a determinant of minimum weight for height necessary for their maintenance or onset. *Proceedings of the Society for Experimental Biology and Medicine* 175: 487–490.
Gahring LC, Carlson NG, Wieggel WA, et al. (1999) Alcohol blocks TNF alpha but not other cytokine-mediated neuroprotection to NMDA. *Alcoholism: Clinical and Experimental Research* 23: 1571–1579.
Gapstur SM, Potter JD, Drinkard C, et al. (1995) Synergistic effect between alcohol and estrogen replacement therapy on risk of breast cancer differs by estrogen/ progesterone receptor status in the Iowa Women's Health Study. *Cancer Epidemiology Biomarkers and Prevention* 4: 313–318.
Garcia-Closas M, Herbstman J, Schiffman M, Glass A, and Dorgan JF (2002) Relationship between serum hormone concentrations, reproductive history, alcohol consumption and genetic polymorphisms in premenopausal women. *International Journal of Cancer* 102: 172–178.
Gatti G, Bertello P, Gurioli L, et al. (1985) Plasma testosterone, estradiol, progesterone and 17-hydroxy-progesterone responses to single-dose stimulation with human chorionic gonadotropin in chronic male alcoholics. *Journal of Endocrinological Investigation* 8: 313–319.
Gavaler JS (1985) Effects of alcohol on endocrine function in post-menopausal women: A review. *Journal of Studies on Alcohol* 46: 495–516.
Gavaler JS (1988) Effects of moderate consumption of alcoholic beverages on endocrine function in postmenopausal women: Bases for hypotheses. In: Galanter M, Begleiter H,

Deitrich DW, et al. (eds.) *Recent Developments in Alcoholism*, pp. 229–251. New York: Plenum Press.

Gavaler JS, Deal SR, van Thiel DH, et al. (1993) Alcohol and estrogen levels in postmenopausal women: The spectrum of effect. *Alcoholism: Clinical and Experimental Research* 17: 786–790.

Gavaler JS, Rosenblum E, and van Thiel DH (1987) Exposure dependent effects of ethanol on serum estradiol and uterus mass in sexually mature oophorectomized rats: A model for bilaterally ovariectomized-postmenopausal women. *Journal of Studies on Alcohol* 48: 295–303.

Gavaler JS and van Theil DH (1992) The association between moderate alcohol beverage consumption and serum estradiol and testosterone levels in normal postmenopausal women; relationship to the literature. *Alcoholism: Clinical and Experimental Research* 16: 87–92.

Gavaler JS, van Thiel DH, and Lester R (1980) Ethanol: A gonadal toxin in the mature rat of both sexes. *Alcoholism: Clinical and Experimental Research* 4: 271–276.

Ginsburg ES, Gao X, Shea BF, and Barbieri RL (1998) Half-life of estradiol in postmenopausal women using transdermal estradiol. *Gynecologic and Obstetric Investigation* 45: 45–48.

Ginsburg ES, Mello NK, and Mendelson JH (2002) Alcohol abuse: Endocrine concomitants. In: Pfaff DW, Arnold AP, Etgen AM, Fahrbach SE, and Rubin RT (eds.) *Hormones, Brain and Behavior.* Vol. 5, pp. 747–780. San Diego, CA: Academic Press.

Ginsburg ES, Mello NK, Mendelson JH, et al. (1996) Effects of alcohol ingestion on estrogens in postmenopausal women. *Journal of American Medical Association* 276: 1747–1751.

Ginsburg ES, Walsh BW, and Gao X (1995) The effects of ethanol on the clearance of estradiol in postmenopausal women. *Fertility and Sterility* 63: 1227–1230.

Goh VH, Tong TY, Mok HP, and Said B (2007) Interactions among age, adiposity, bodyweight, lifestyle factors and sex steroid hormones in healthy Singaporean Chinese men. *Asian Journal of Andrology* 9: 611–621.

Goodman AL and Hodgen GD (1983) The ovarian triad of the primate menstrual cycle. *Recent Progress in Hormone Research* 39: 1–67.

Gordon GG, Altman K, Southren AL, et al. (1976) Effect of alcohol (ethanol) administration on sex-hormone metabolism in normal men. *New England Journal of Medicine* 295: 793–797.

Grabrick DM, Hartmann LC, Cerhan JR, et al. (2000) Risk of breast cancer with oral contraceptive use in women with family history of breast cancer. *Journal of American Medical Association* 14: 1791–1798.

Guerri C and Renau-Piqueras J (1997) Alcohol, astroglia and brain development. *Molecular Neurobiology* 16: 65–81.

Gumus B, Yigitoglu MR, Lekili M, et al. (1998) Effect of long-term alcohol abuse on male sexual function and serum gonadal hormone levels. *International Urology and Nephrology* 30: 755–759.

Halmesmäki E, Autti I, Granström M-L, et al. (1987) Estradiol, progesterone, prolactin, and human chorionic gonadotropin in pregnant women with alcohol abuse. *Journal of Clinical Endocrinology and Metabolism* 64: 153–156.

Halmesmäki E, Välimäki M, Roine R, et al. (1989) Maternal and paternal alcohol consumption and miscarriage. *British Journal of Obstetrics and Gynecology* 96: 188–191.

Hammond C, Wiebe R, Haney A, et al. (1979) Ovulation induction with luteinizing hormone-releasing hormone in amenorrheic infertile women. *American Journal of Obstetrics and Gynecology* 135: 924–939.

Handa K, Kono S, Ishii H, et al. (1994) Relationship of alcohol consumption and smoking to plasma cortisol and blood pressure. *Journal of Human Hypertension* 8: 891–894.

Handa RJ, McGivern RF, Noble ESP, et al. (1985) Exposure to alcohol *in utero* alters the adult patterns of luteinizing hormone secretion in male and female rats. *Life Sciences* 37: 1683–1690.

Hannigan JH, Welch RA, and Sokol RJ (1992) Recognition of fetal alcohol syndrome and alcohol-related birth defects. In: Mendelson JH and Mello NK (eds.) *Third Edition of the Medical Diagnosis and Treatment of Alcoholism*, pp. 639–667. New York: McGraw-Hill.

Harlap S and Shiono PH (1980) Alcohol, smoking, and incidence of spontaneous abortions in the first and second trimester. *Lancet* 2: 173–176.

Heikkonnen E, Ylikahri R, Roine R, et al. (1996) The combined effect of alcohol and physical exercise on serum testosterone, luteinizing hormone and cortisol in males. *Alcoholism: Clinical and Experimental Research* 20: 711–716.

Heinz A, Rommelspacher H, Graf KJ, et al. (1995) Hypothalamic–pituitary–gonadal axis, prolactin, and cortisol in alcoholics during withdrawal and after three weeks of abstinence in comparison with healthy control subjects. *Psychiatry Research* 56: 81–95.

Henderson J, Gray R, and Brocklehurst P (2007) Systematic review of effects of low-moderate prenatal alcohol exposure on pregnancy outcome. *British Journal of Obstetrics and Gynaecology* 114: 243–253.

Hill M, Popov P, Havlikova H, et al. (2005) Altered profiles of serum neuroactive steroids in premenopausal women treated for alcohol addiction. *Steroids* 70: 515–524.

Hodgen GD (1982) The dominant ovarian follicle. *Fertility and Sterility* 38: 281–300.

Hollstedt C, Dahlgren L, and Rydberg U (1983) Outcome of pregnancy in women treated at an alcohol clinic. *Acta Psychiatrica Scandinavica* 667: 236–248.

Hugues JN, Coste T, Perret G, et al. (1980) Hypothalamo-pituitary ovarian function in 31 women with chronic alcoholism. *Clinical Endocrinology (Oxford)* 12: 543–551.

Hurley DM, Brian R, Outch J, et al. (1984) Induction of ovulation and fertility in amenorrheic women by pulsatile low-dose gonadotropin-releasing hormone. *New England Journal of Medicine* 310: 1069.

Hutchings DE (1989) *Prenatal Abuse of Licit and Illicit Drugs.* New York: New York Academic Science.

Hutchison JS, Kubik CJ, Nelson PB, et al. (1987) Estrogen induces premature luteal regression in rhesus monkeys during spontaneous menstrual cycles, but not in cycles driven by exogenous gonadotropin-releasing hormone. *Endocrinology (Baltimore)* 121: 466–474.

Ida Y, Tsujimara S, Nakamaura K, et al. (1992) Efffects of acute and repeated alcohol ingestion on hypothalamic–pituitary–gonadal and hypothalamic–pituitary–adrenal function in normal males. *Drug and Alcohol Dependence* 31: 57–64.

Ikonomidou C, Bittigau P, Ishimaru MJ, et al. (2000) Ethanol-induced apoptotic neurodegeneration and fetal alcohol syndrome. *Science* 287: 1056–1060.

Ikonomidou C, Bosch F, Miksa M, et al. (1999) Blockade of NMDA receptors and apoptotic neurodegeneration in the developing brain. *Science* 283: 70–74.

Inder WJ, Joyce PR, Ellis MJ, et al. (1995a) The effects of alcoholism on the hypothalamic–pituitary–adrenal axis: Interaction with endogenous opioid peptides. *Clinical Endocrinology (Oxford)* 43: 283–290.

Inder WJ, Joyce PR, Wells JE, et al. (1995b) The acute effects of oral ethanol on the hypothalamic–pituitary–adrenal axis in normal human subjects. *Clinical Endocrinology (Oxford)* 42: 65–71.

Iranmanesh A, Veldhuis JD, Samojlik E, et al. (1988) Alterations in the pulsatile properties of gonadotroin

secretion in alcoholic men. *Journal of Andrology* 9: 207–214.
Itskovitz J and Hodgen GD (1988) Endocrine basis for the initiation, maintenance and termination of pregnancy in humans. *Psychoneuroendocrinology* 13: 155–170.
Iturriaga H, Lioi X, and Valladares L (1999) Sex hormone binding globulin in non-cirrhotic alcoholic patients during early withdrawal and after long abstinence. *Alcohol and Alcoholism* 34: 903–909.
Jaffe RB and Keye WR (1974) Estradiol augmentation of pituitary responsiveness to gonadotropin-releasing hormone in women. *Journal of Clinical Endocrinology and Metabolism* 39: 850–855.
Jones KL, Smith DW, Ulleland CN, et al. (1973) Pattern of malformation in offspring of chronic alcoholic mothers. *Lancet* 1: 1267–1271.
Jones-Saumty DJ, Fabian MS, and Parsons OA (1981) Medical status and cognitive functioning in alcoholic women. *Alcoholism: Clinical and Experimental Research* 5: 372–377.
Jung Y and Russfield AB (1972) Prolactin cells in the hypophysis of cirrhotic patients. *Archives of Pathology* 94: 265–269.
Karsch EJ, Weick RF, Butler WR, et al. (1973) Induced LH surges in the rheusus monkey: Strength-duration characteristics of the estrogen stimulus. *Endocrinology (Baltimore)* 92: 1740–1747.
Kline J, Stein Z, Shrout P, et al. (1980) Drinking during pregnancy and spontaneous abortion. *Lancet* 2: 176–180.
Klopper A (1985) Steroids in pregnancy. In: Shearman RP (ed.) *Clinical Reproductive Endocrinology*, pp. 209–223. Edinburgh: Churchill-Livingstone.
Klug TL, Bageman E, Ingvar C, Rose C, and Jernstrom H (2006) Moderate coffee and alcohol consumption improves the estrogen metabolite profile in adjuvant treated breast cancer patients: A pilot study comparing pre and postoperative levels. *Molecular Genetics and Metabolism* 89: 381–389.
Knobil E (1974) On the control of gonadotropin secretion in the rhesus monkey. *Recent Progress in Hormone Research* 30: 1–46.
Knobil E (1980) The neuroendocrine control of the menstrual cycle. *Recent Progress in Hormone Research* 36: 53–88.
Knobil E and Hotchkiss J (1988) The menstrual cycle and its neuroendocrine control. In: Knobil E, Neill J, Ewing LL, Greenwald GS, Markert CL, and Pfaff DW (eds.) *The Physiology of Reproduction*, pp. 1971–1994. New York: Raven Press.
Kreek MJ (1987) Multiple drug abuse patterns and medical consequences. In: Meltzer H (ed.) *Psychopharmacology: The Third Generation of Progress*, pp. 1597–1604. New York: Raven Press.
Kreek MJ (1991) Multiple drug abuse patterns: Recent trends and associated medical consequences. In: Melloa NK (ed.) *Advances in Substance Abuse: Behavioral and Biological Research*, pp. 91–112. London: Jessica Kingsley Publishers.
Kreuger WA, Walter JB, and Rudeen PK (1983) Estrous cyclicity in rat-fed ethanol diet for 4 months. *Pharmacology, Biochemistry, and Behavior* 19: 583–585.
Krey LC, Butler WR, Weiss G, et al. (1973) Influence of endogenous and exogenous gonadal steroids on the action of synthetic RF in the rhesus monkey. *Excerpta Medica International Congress Series* 263: 39–47.
Lasley BL, Wang CF, and Yen SSC (1975) The effects of estrogen and progesterone on the functional capacity of the gonadotrophs. *Journal of Clinical Endocrinology and Metabolism* 1: 820–831.
Lee S and Rivier C (1993a) Prenatal alcohol exposure blunts interleukin-1-induced ACTH beta-endorphin secretion by immature rats. *Alcoholism: Clinical and Experimental Research* 17: 940–945.
Lee S and Rivier C (1993b) Effect of exposure to an alcohol diet for 10 days on the ability of interleukin-1 beta to release ACTH and corticosterone in the adult ovariectomized female rat. *Alcoholism: Clinical and Experimental Research* 17: 1009–1013.
Lee S and Rivier C (1994) Effect of postnatal exposure of female rats to an alcohol diet: Influence of age and circulating sex steroids. *Alcoholism: Clinical and Experimental Research* 18: 998–1003.
Lester R and van Theil DH (1977) Gonadal function in chronic alcoholic men. *Advances in Experimental Medicine and Biology* 85A: 399–413.
Lex BW, Teoh SK, Lagomasino I, et al. (1990) Characteristics of women receiving mandated treatment for alcohol or polysubstance dependence in Massachusetts. *Drug and Alcohol Dependence* 25: 13–20.
Leyendecker G and Wildt L (1983) Control of gonadotropin secretion in women. In: Norman RL (ed.) *Neuroendocrine Aspects of Reproduction*, pp. 295–323. New York: Academic Press.
Leyendecker G and Wildt L (1984) Pulsatile administration of GnRH in hypothalamic amenorrhea. *Upsala Journal of Medical Sciences* 89: 19.
Longnecker MP (1993) Do hormones link alcohol with breast cancer? *Journal of the National Cancer Institute* 85: 692–693.
Longnecker MP, Berlin JA, Orza MJ, et al. (1988) A metaanalysis of alcohol consumption in relation to breast cancer. *Journal of American Medical Association* 260: 652–656.
Longnecker MP, Newcomb PA, Mittendorf R, et al. (1995) Risk of breast cancer in relation to lifetime alcohol consumption. *Journal of the National Cancer Institute* 87: 923–929.
Loosen PT, Wilson IC, Dew BW, et al. (1983) Thyrotropin-releasing hormone (TRH) in abstinent alcoholic men. *American Journal of Psychiatry* 140: 1145–1149.
Marshall JC, Case GD, Valk TW, et al. (1983) Selective inhibition of follicle-stimulating hormone secretion by estradiol. *Journal of Clinical Investigation* 71: 248–257.
Martin JB and Reichlin S (1987) Regulation of prolactin secretion and its disorders. In: Martin JB and Reichlin S (eds.) *Clinical Neuroendocrinology*, pp. 201–231. Philadelphia, PA: Davis.
Martinez-Riera A, Santolaria-Fernandez F, Gonzalez Reimers E, et al. (1995) Alcoholic hypogonadism: Hormonal response to clomiphene. *Alcohol* 12: 581–587.
McArthur JW, Bullne BA, Beitins IZ, et al. (1980) Hypothalamic amenorrhea in runners of normal body composition. *Endocrine Research Communications* 7: 13–25.
McDonald JA, Mandel MG, Marchbanks PA, et al. (2004) Alcohol exposure and breast cancer: Results of the women's contraceptive and reproductive experiences study. *Cancer Epidemiology Biomarkers and Prevention* 13: 2106–2116.
McLachlan RI, Robertson DM, Healy DL, et al. (1987) Circulating immunoreactive inhibin levels during the normal human menstrual cycle. *Journal of Clinical Endocrinology and Metabolism* 65: 954.
McNeely MJ and Soules MR (1988) Diagnosis of luteal phase deficiency: A critical review. *Fertility and Sterility* 50: 1–15.
Mello NK (1980) Some behavioral and biological aspects of alcohol problems in women. In: Kalant OJ (ed.) *Alcohol and Drug Problems in Women: Research Advances in Alcohol and Drug Problems*, pp. 263–312. New York: Plenum Press.
Mello NK (1988) Effects of alcohol abuse on reproductive function in women. In: Galanter M, Beglieter H Dietrich RA, et al. (eds.) *Recent Developments in Alcoholism*, pp. 223–227. New York: Plenum Press.
Mello NK, Bree MR, Mendelson JH, et al. (1983a) Alcohol self-administration disrupts reproductive function in female macaque monkeys. *Science* 221: 677–679.

Mello NK, Ellingboe J, Bree MP, et al. (1983b) Alcohol effects on estradiol in female macaque monkey. *NIDA Research Monographs* 43: 210–216.

Mello NK and Mendelson JH (1972) Drinking patterns during work-contingent and noncontingent alcohol acquisition. *Psychosomatics* 34: 139–164.

Mello NK, Mendelson JH, Bree MP, et al. (1985) AST: Alcohol effects on LH and testosterone in male Macaque monkeys. *Journal of Pharmacology and Experimental Therapeutics* 223: 588–596.

Mello NK, Mendelson JH, Bree MP, et al. (1986a) Alcohol effects on LHRH-stimulated LH and FSH in female rhesus monkeys. *Journal of Pharmacology and Experimental Therapeutics* 236: 590–595.

Mello NK, Mendelson JH, Bree MP, et al. (1986b) Alcohol effects on LHRH stimulated LH and FSH in ovariectomized female rhesus monkeys. *Journal of Pharmacology and Experimental Therapeutics* 239: 693–700.

Mello NK, Mendelson JH, King NW, et al. (1988) Alcohol self-administration by female Macaque monkey: A model for study of alcohol dependence, hyperprolactinemia and amenorrhea. *Journal of Studies on Alcohol* 49: 551–560.

Mello NK, Mendelson JH, and Teoh SK (1989) Neuroendocrine consequences of alcohol abuse in women. In: Hutchings DE (ed.) *Prenatal Abuse of Licit and Illicit Drugs*, pp. 211–240. New York: New York Academy of Sciences.

Mello NK, Mendelson JH, Palmieri S, et al. (1990) Operant acquisition of alcohol by women. *Journal of Pharmacology and Experimental Therapeutics* 253: 237–245.

Mendeloff AI (1954) Effects of intravenous infusions of ethanol upon estimated hepatic blood flow in man. *Journal of Clinical Investigation* 33: 1298–1302.

Mendelson JH, Babor TF, Mello NK, et al. (1986a) Alcoholism and prevalence of medical and psychiatric disorders. *Journal of Studies on Alcohol* 47: 361–366.

Mendelson JH, Ellingboe J, Kuehnle JC, et al. (1979) Effects of naltrexone on mood and neuroendocrine function in normal adult males. *Psychoneuroendocrinology* 3: 231–236.

Mendelson JH, Lukas SE, Mello NK, et al. (1988) Acute alcohol effects on plasma estradiol levels in women. *Psychopharmacology (Berlin)* 94: 464–467.

Mendelson JH and Mello NK (1988) Chronic alcohol effects on anterior pituitary and ovarian hormones in healthy women. *Journal of Pharmacology and Experimental Therapeutics* 245: 407–412.

Mendelson JH, Mello NK, Cristofaro P, et al. (1986b) Use of naltrexone as a provocative test for hypothalamic–pituitary-hormone function. *Pharmacology, Biochemistry, and Behavior* 24: 309–313.

Mendelson JH, Mello NK, Cristofaro P, et al. (1987) Alcohol effects on naloxone-stimulated luteinizing hormone, prolactin, and estradiol in women. *Journal of Studies on Alcohol* 48: 287–294.

Mendelson JH, Mello NK, and Ellingboe J (1981) Acute alcohol intake and pituitary gonadal hormones in normal human females. *Journal of Pharmacology and Experimental Therapeutics* 218: 23–26.

Mendelson JH, Mello NK, Teoh SK, et al. (1989) Alcohol effects on luteinizing hormone-releasing hormone-stimulated anterior pituitary and gonadal hormones in women. *Journal of Pharmacology and Experimental Therapeutics* 250: 902–909.

Mirin SM, Mendelson JH, Ellingboe J, et al. (1976) Acute effects of heroin and naltrexone on testosterone and gonadotropin secretion: A pilot study. *Psychoneuroendocrinology* 1: 359–369.

Morley JE, Baranetsky NG, Wingert TD, et al. (1980) Endocrine effects of naloxone-induced opiate receptor blockade. *Journal of Clinical Endocrinology and Metabolism* 50: 252–357.

Moskovic S (1975) Effect of chronic alcohol intoxication on ovarian dysfunction. *Srpski Arhiv za Celokupno Lekarstvo* 103: 751–758.

Mukherjee AB and Hodge GD (1982) Maternal ethanol exposure induces transient impairment of umbilical circulation and fetal hypoxia in monkeys. *Science* 218: 700–702.

Murad F and Haynes RC (1985a) Adenohypophyseal hormones and related substances. In: Gilman AG, Goodman LS, Rail TW, and Murad F (eds.) *The Pharmacological Basis of Therapeutics,* 7th edn., pp. 1362–1368. New York: Macmillian.

Murad F and Haynes RC (1985b) Estrogens and progestins. In: Gilman AG, Goodman LS, Rail TW, and Murad F (eds.) *The Pharmacological Basis of Therapeutics,* 7th edn., pp. 1412–1439. New York: Macmillan.

Murono EP and Fisher-Simpson V (1984) Ethanol directly increases dihydrotestosterone conversion to 5α-androstan-3α, 17β-diol and 5α-androstan-3β, 17β-diol in rat Leydig cells. *Biochemical and Biophysical Research Communications* 121: 558–565.

Murono EP and Fisher-Simpson V (1985) Ethanol directly stimulates dihydro-testosterone conversion to 5α-androstan-3α, 17β-diol and 5α-androstan-3β, 17β-diol in rat liver. *Life Sciences* 36: 1117–1124.

Muti F, Trevisan M, Micheli A, et al. (1998) Alcohol consumption and total estradiol in premenopausal women. *Cancer Epidemiology* 7: 189–193.

Nagata C, Kabuto M, Takatsuka N, et al. (1997) Associations of alcohol, height, and reproductive factors with serum hormone concentrations in postmenopausal Japanese women. Steroid hormones in Japanese postmenopausal women. *Breast Cancer Research Treatment* 44: 235–241.

Nielsen NR and Gronbaek M (2008) Interactions between intakes of alcohol and postmenopausal hormones on risk of breast cancer. *International Journal of Cancer* 122: 1109–1113.

Nora JJ, Nora AH, Perinchiev AG, et al. (1976) Congenital abnormalities and first-trimester exposure to progestogen/oestrogen. *Lancet* 1: 313–314.

Noth RH and Walter RMJ (1984) The effects of alcohol on the endocrine system. *Medical Clinics of North America* 68: 133–146.

Ogilvie KM, Lee S, and Rivier C (1997) Effect of three different modes of alcohol administration on the activity of the rat hypothalamic–pituitary–adrenal axis. *Alcoholism: Clinical and Experimental Research* 21: 467–476.

Ogilvie KM, Lee S, Weiss B, et al. (1998) Mechanisms mediating the influence of alcohol on the hypothalamic–pituitary–adrenal axis responses to immune and nonimmune signals. *Alcoholism: Clinical and Experimental Research* 22: 243S–247S.

Ogilvie KM and Rivier C (1997) Gender difference in hypothalamic–pituitary–adrenal axis response to alcohol in the rat: Activational role of gonadal steroids. *Brain Research* 766: 19–28.

Olson HC, Streissguth AP, Sampson PD, et al. (1997) Association of prenatal alcohol exposure with behavioral and learning problems in early adolescence. *Journal of the American Academy of Child and Adolescent Psychiatry* 36: 1187–1194.

Olster DH and Ferin N (1987) Corticotropin-releasing hormone inhibits gonadotropin secretion in the ovariectomized rhesus monkey. *Journal of Clinical Endocrinology and Metabolism* 65: 262–267.

Onland-Moret NC, Peeters PH, van der Schouw YT, Grobbee DE, and van Gils CH (2005) Alcohol and endogenous sex steroid levels in postmenopausal women: A cross sectional study. *Journal of Clinical Endocrinology and Metabolism* 90: 1414–1419.

Ottobre JS and Stouffer RL (1984) Persistent versus transient stimulation of the macaque corpus luteum during prolonged exposure to human chorionic gonadotropin: A function of age of the corpus luteum. *Endocrinology (Baltimore)* 114: 2175–2182.

Petraglia F, Sutton S, Vale W, et al. (1987) Corticotropin-releasing factor decreases plasma LH levels in female rats by inhibiting gonadotropin-releasing hormone release into hypophysial–portal circulation. *Endocrinology (Baltimore)* 120: 1083–1088.

Petridou E, Katsouyanni K, Spanos E, Skalkidis Y, Pangiotopoulou K, and Trichopoulos D (1992) Pregnancy estrogens in relation to coffee and alcohol intake. *Annals of Epidemiology* 2(3): 241–247.

Peyron R, Aubeny E, Targosz V, et al. (1993) Early termination of pregnancy with mifepristone (R4 486) and the orally active prostaglanolin misoprostol. *New England Journal of Medicine* 328: 1509–1513.

Phipps WR, kukas SE, Mendelson JH, et al. (1987) Acute ethanol administration enhances plasma testosterone levels following gonadotropin stimulation in men. *Journal of Endocrinology* 101: 33–39.

Prior JC, Cox TA, Fairholm D, et al. (1987) Testosterone-related exacerbation of a prolactin-producing macroadenoma: Possible role for estrogen. *Journal of Clinical Endocrinology and Metabolism* 64: 391–394.

Randall CL, Ekblad U, and Anton RE (1990) Perspectives on the pathophysiology of fetal alcohol syndrome. *Alcoholism: Clinical and Experimental Research* 14: 807–812.

Randall CL and Noble E (1980) Alcohol abuse and fetal growth and development. In: Mello NK (ed.) *Advances in Substance Abuse: Behavioral and Biological Research,* vol. 1, pp. 327–367. Greenwich, CT: JAI Press.

Reame NE, Sauder ED, Case GD, et al. (1985) Pulsatile gonadotropin secretion in women with hypothalamic amenorrhea: Evidence that reduced frequency of gonadotropin-releasing hormone secretion is the mechanism of persistent anovulation. *Journal of Clinical Endocrinology and Metabolism* 61: 851–858.

Rebar RW (1986) Practical evaluation of hormonal status. In: Yen SSC and Jaffe RB (eds.) *Reproductive Endocrinology*, pp. 683–733. Philadelphia, PA: Saunders.

Reddy BV, Boyadjieva N, and Sarkar DK (1995) Effect of ethanol, propanol, butanol and cathalase enzyme blockers on beta-endorphin secretion from primary cultures of hypothalamic neurons: Evidence for a mediatory role of acetaldehyde in ethanol stimulation of beta-endorphin release. *Alcoholism: Clinical and Experimental Research* 19: 339–344.

Reddy BV and Sarkar DK (1993) Effect of alcohol, acetaldehyde, and salsolinol on beta-endorphin secretion from the hypothalamic neurons in primary cultures. *Alcoholism: Clinical and Experimental Research* 17: 1261–1267.

Redei E, Branch BJ, and Taylor AN (1986) Direct effect of ethanol on adrenocorticotropin (ACTH) release *in vitro*. *Journal of Pharmacology and Experimental Therapeutics* 237: 59–64.

Reynolds JD, Penning DH, Dexter F, et al. (1995) Dose-dependent effects of acute *in vivo* ethanol exposure on extracellular glutamate concentration in the cerbral cortex of near-term sheep. *Alcoholism: Clinical and Experimental Research* 19: 1447–1453.

Rinaldi S, Peeters PH, Bezemer ID, et al. (2006) Relationship of alcohol intake and sex steroid concentrations in blood in pre- and post-menopausal women: The European prospective investigation into cancer and nutrition. *Cancer Causes and Control* 17: 1033–1043.

Rivier C (1993a) Female rats release more corticosterone than males in response to alcohol: Influence of circulating sex steroids and possible consequences for blood alcohol levels. *Alcoholism: Clinical and Experimental Research* 17: 854–859.

Rivier C (1993b) Acute interactions between cytokines and alcohol on ACTH and corticosterone secretion in the rat. *Alcoholism: Clinical and Experimental Research* 17: 946–950.

Rivier C (1995) Adult male rats exposed to an alcohol diet exhibit a blunted adrenocorticotropic hormone response to immune or physical stress: Possible role of nitric oxide. *Alcoholism: Clinical and Experimental Research* 19: 1474–1479.

Rivier C and Lee S (1996) Acute alcohol administration stimulates the activity of hypothalamic neurons that express corticotropin-releasing factor and vasopressin. *Brain Research* 726: 1–10.

Rivier C, Rivier J, and Vale W (1986) Stress-induced inhibition of reproductive functions: Role of endogenous corticotropin-releasing factor. *Science* 31: 914–919.

Rivier C and Vale W (1984) Influence of corticotropin-releasing factor (CRF) on reproductive functions in the rat. *Endocrinology (Baltimore)* 114: 914–919.

Ross GT (1985) Disorders of the ovary and female reproductive tract. In: Wilson JD and Foster DW (eds.) *Williams Textbook of Endocrinology,* 7th edn., pp. 206–258. Philadelphia, PA: Saunders.

Rostand A, Kaminski M, Lelong N, et al. (1990) Alcohol use in pregnancy, craniofacial features, and fetal growth. *Journal of Epidemiology and Community Health* 44: 302–306.

Royo-Bordonada MA, Martin-Moreno JM, Guallar E, et al. (1997) Alcohol intake and risk of breast cancer: The euramic study. *Neoplasma* 44: 150–156.

Ryback RS (1977) Chronic alcohol consumption and menstruation. *Journal of American Medical Association* 238: 2143.

Sanchis R, Esquifino A, and Guerri C (1985) Chronic ethanol intake modifies estrous cyclicity and alters prolactin and LH levels. *Pharmacology, Biochemistry, and Behavior* 23: 221–224.

Santoro N, Filicori M, and Crowley J (1986a) Hypogonadotropic disorders in men and women: Diagnosis and therapy with pulsatile gonadotropin-releasing hormone. *Endocrine Reviews* 7: 11–23.

Santoro N, Wierman ME, and Filicori M (1986b) In-travenous administration of pulsatile gonadotropin-releasing hormone in hypothalamic amenorrhea: Effects of dosage. *Journal of Clinical Endocrinology and Metabolism* 62: 109–116.

Sarkola T, Makisalo H, Fukunaga T, et al. (1999) Acute effect of alcohol on estradiol, estrone, progesterone, prolactin, cortisol and luteinizing hormone in premenopausal women. *Alcoholism: Clinical and Experimental Research* 23: 976–982.

Sauder S, Frager M, Case G, et al. (1984) Abnormal patterns of pulsatile luteinizing hormone secretion in women with hyperprolactinemia and amenorrhea: Responses to bromocriptine. *Journal of Clinical Endocrinology and Metabolism* 59: 941–948.

Schairer C, Lubin J, Troisi R, et al. (2000) Menopausal estrogen and estrogen-progestin replacement therapy and breast cancer risk. *Journal of American Medical Association* 4: 485–491.

Schardein JL (1980) Cognenital abnormalities and hormones during pregnancy: A clinical review. *Teratology* 22: 251–270.

Schuckit MA, Parker DC, and Rossman LR (1983) Ethanol-related prolactin responses and risk for alcoholism. *Biological Psychiatry* 18: 1153–1159.
Schuckit MA, Risch SC, and Gold EO (1988) Alcohol, ACTH, and the family history of alcoholism. *American Journal of Psychiatry* 145: 1391–1395.
Schulz KD, Geiger W, del Pozo E, et al. (1978) Pattern of sexual steroids, prolactin and gonadotropic hormones during prolactin inhibition in normally cycling women. *American Journal of Obstetrics and Gynecology* 132: 561.
Seki M (1988) A physiopathological study on ovarian dysfunction in female patients with alcoholism. *Fukuoka Acta Medica* 79: 738–748.
Seki M, Yoshida K, and Okamura Y (1991a) Hormones in amenorrheic women with alcoholics. *Japanese Journal of Fertility and Sterility* 36: 630–635.
Seki M, Yoshida K, and Okamura Y (1991b) A study on hyperprolactinemia in female patients with alcoholics. *Japanese Journal of Alcohol Studies and Drug Dependence* 26: 49–59.
Sherman B (1984) Hypothalamic control of the menstrual cycle: Implications for the study of anorexia nervosa. In: Brown G, Koslow S, and Reichlin S (eds.) *Neuroendocrinology and Psychiatric Disorders*, pp. 315–323. New York: Raven Press.
Shearman RP (1985) Secondary amenorrhea. In: Shearman RP (ed.) *Clinical Reproductive Endocrinology*, pp. 493–507. New York: Churchill-Livingstone.
Shively CA, Register TC, Grant KA, Johnson JL, and Cline JM (2004) Effects of social status and moderate alcohol consumption on mammary gland and endometrium of surgically menopausal monkeys. *Menopause* 11: 389–399.
Sierksma A, Sarkola T, Eriksson CJ, van der Gaag MS, Grobbee DE, and Hendriks HF (2004) Effect of moderate alcohol consumption on plasma dehydroepiandrosterone sulfate, testosterone, and estradiol levels in middle-aged men and postmenopausal women: A diet-controlled intervention study. *Alcoholism: Clinical and Experimental Research* 28: 780–785.
Simasko SM, Boyadjieva N, De A, et al. (1999) Effect on ethanol on calcium regulation in rat fetal hypothalamic cells in culture. *Brain Research* 3: 89–96.
Slater TF, Sawyer BC, and Strauli UD (1964) Changes in liver nucleotide concentrations in experimental liver injury. *Biochemical Journal* 93: 267–270.
Smith C and Asch R (1987) Drug abuse and reproduction. *Fertility and Sterility* 48: 355–373.
Smith CG and Smith MT (1990) Substance abuse and reproduction. *Seminars in Reproduction and Endocrinology* 8: 55–64.
Smith-Warner SA, Spiegelman D, Yaun SS, et al. (1998) Alcohol and breast cancer in women: A pooled analysis of cohort studies. *Journal of American Medical Association* 279: 535–540.
Sokol R, Miller S, and Reed G (1980) Alcohol abuse during pregnancy. *Clinical Experimental Research* 4: 135–145.
Solomon S (1988) The placenta as an endocrine organ: Steroids. In: Knobil E and Neill J (eds.) *The Physiology of Reproduction*, pp. 2085–2091. New York: Raven Press.
Soules MR, Clifton DK, Bremner W, and Steiner R (1987) Corpus luteum insufficiency induced by a rapid gonadotropin-releasing hormone-induced gonadotropin secretion pattern in the follicular phase. *Journal of Clinical Endocrinology and Metabolism* 65: 457–464.
Soules MR, Clifton DK, Cohen DL, Bremner WJ, and Steiner RA (1989a) Luteal phase deficiency: Abnormal gonadotropin and progesterone secretion patterns. *Journal of Clinical Endocrinology and Metabolism* 69: 813–820.
Soules MR, Clifton DK, Steiner RA, Cohen NL, and Bremner WJ (1988) The corpus luteum: Determinants of progesterone secretion in the normal menstrual cycle. *Obstetrics and Gynecology* 71: 659.
Soules MR, McLachlan RI, Marit EK, et al. (1989b) Luteal phase deficiency: Characterization of reproductive hormones over the menstrual cycle. *Journal of Clinical Endocrinology and Metabolism* 69: 804–812.
Soyka M, Gorig E, and Naber D (1991) Serum prolactin increase induced by ethanol-a dose-dependent effect not related to stress. *Psychoneuroendocrinology* 16: 441–446.
Sparrow D, Bosse R, and Rowe JW (1980) The influence of age, alcohol consumption, and body build on gonadal function in men. *Journal of Clinical Endocrinology and Metabolism* 51: 508–512.
Stein SW, Lieber CS, Leery CM, et al. (1963) The effect of ethanol upon systemic and hepatic blood flow in man. *American Journal of Clinical Nutrition* 13: 68–74.
Stoll BA (1999) Alcohol intake and late-stage promotion of breast cancer. *European Journal of Cancer* 35: 1653–1658.
Stouffer RL (1990) Corpus luteum function and dysfunction. *Clinical Obstetrics and Gynecology* 33: 668–689.
Streissguth AP (1986) The behavioral teratology of alcohol: Performance, behavior, and intellectual deficits in prenatally exposed children. In: West J (ed.) *Alcohol and Brain Development*, pp. 3–44. New York: Oxford University Press.
Streissguth AP, Aase JM, Clarren SK, et al. (1991a) Fetal alcohol syndorme in adolescents and adults. *Journal of American Medical Association* 265: 1961–1967.
Streissguth AP, Grant TM, Barr HM, et al. (1991b) Cocaine and the use of alcohol and other drugs during pregnancy. *American Journal of Obstetrics and Gynecology* 164: 1239–1243.
Streissguth AP, Landesman-Dwyer S, Martin JC, et al. (1980) Teratogenic effects of alcohol in humans and laboratory animals. *Science* 209: 353–361.
Streissguth AP, Sampson PD, Barr HM, Bookstein EL, and Theide K (1997) Association of prenatal alcohol exposure with behavioral and learning problems in early adolescence. *Journal of the American Academy of Child and Adolescent Psychiatry* 36: 1187–1194.
Suzuki R, Ye W, Rylander-Rudqvist T, Saji S, Colditz GA, and Wolk A (2005) Alcohol and postmenopausal breast cancer risk defined by estrogen and progesterone receptor status: A prospective cohort study. *Journal of the National Cancer Institute* 97: 1563–1564.
Svinarich DM, DiCerbo JA, Zaher EM, et al. (1998) Ethanol-induced expression of cytokines in a first-trimester trophoblast cell line. *American Journal of Obstetrics and Gynecology* 179: 470–475.
Szabo J, Bruckner G, Medveczky I, et al. (1999) Ethanol's effect on rat pituitary adrenal axis is prevented by purine metabolic pathway inhibitors. *Proceedings of the Society for Experimental Biology and Medicine* 220: 112–118.
Talamantes F and Ogren L (1988) The placenta as an endocrine organ: Polypeptides. In: Knobil E and Neill J (eds.) *The Physiology of Reproduction*, pp. 2093–2144. New York: Raven Press.
Tavani A, Negri E, Franceschi S, et al. (1994) Alcohol consumption and risk of prostate cancer. *Nutrition and Cancer* 21: 24–31.
Taylor AN, Branch BJ, Liu S, et al. (1981) Fetal exposure to ethanol enhances pituitary–adrenal and temperature responses to ethanol in adult rats. *Alcoholism: Clinical and Experimental Research* 5: 237–246.
Taylor AN, Branch BJ, Liu S, et al. (1982) Long-term effects of fetal ethanol exposure on pituitarty–adrenal responses to stress. *Pharmacology, Biochemistry, and Behavior* 16: 585–590.
Taylor AN, Nelson LR, Branch BJ, et al. (1984) Altered stress responsiveness in adult rats exposed to ethanol *in*

utero: Neuroendocrine mechanisms. *Ciba Foundation Symposium* 1095: 47–65.

Teoh SK, Lex BW, Mendelson JH, et al. (1992) Hyperprolactinemia and macrocytosis in women with alcohol and polysubstance dependence. *Journal of Studies on Alcohol* 53: 176–182.

Teoh SK, Mendelson JH, Mello NK, et al. (1988) Alcohol effects on naltrexone-induced stimulation of pituitary, adrenal and gonadal hormones during the early follicular phase of the menstrual cycle. *Journal of Clinical Endocrinology and Metabolism* 66: 1181–1186.

Teoh SK, Mendelson JH, Mello NK, et al. (1990) Alcohol effects on hCG stimulated gonadal hormones in women. *Journal of Pharmacology and Experimental Therapeutics* 254: 407–411.

Terry MB, Gammon MD, Zhang FF, et al. (2006) ADH3 genotype, alcohol intake and breast cancer risk. *Carcinogenesis* 27: 840–847.

Tolis G, Somma M, van Campenhout J, et al. (1974) Prolactin secretion in sixty-five patients with galactorrhea. *American Journal of Obstetrics and Gynecology* 118: 91–101.

Torgerson DJ, Thomas E, Campbell MK, et al. (1997) Alcohol consumption and age of maternal menopause are associated with menopause onset. *Maturitas* 26: 21–25.

Välimäki M, Halmesmaki E, Keso L, et al. (1990a) Serum lipids and lipoproteins in alcoholic women during pregnancy. *Metabolism – Clinical and Experimental* 39: 486–493.

Välimäki M, Harkonen M, and Ylikahri R (1983) Acute effects of alcohol on female sex hormones. *Alcoholism: Clinical and Experimental Research* 7: 289–293.

Välimäki M, Pelkonen R, Harkonen M, et al. (1990b) Pituitary–gonadal hormones and adrenal androgens in noncirrhotic female alcoholics after cessation of alcohol intake. *European Journal of Clinical Investigating* 20: 177–181.

Välimäki M, Pelkonen R, Salaspuro M, et al. (1984) Sex hormones in amenorrheic women with alcoholic liver disease. *Journal of Clinical Endocrinology and Metabolism* 59: 133–138.

Välimäki M, Tuominen VA, Huhtaniemi I, et al. (1990c) The pulsatile secretion of gonadotropins and growth hormone, and the biological activity of luteinizing hormone in men acutely intoxicated with ethanol. *Alcoholism: Clinical and Experimental Research* 14: 928–931.

van Thiel DH (1983) Ethanol: Its adverse effects upon the hypothalamic–pituitary–gonadal axis. *Journal of Laboratory and Clinical Medicine* 101: 21–23.

van Thiel DH (1984) Ethyl alcohol and gonadal function. Physiology in medicine. *Hospital Practice* 19: 152–158.

van Thiel DH and Gavaler JS (1982) The adverse–effects of ethanol upon hypothalamic–pituitary–gonadal function in males and females compared and contrasted. *Alcoholism: Clinical and Experimental Research* 6: 179–185.

van Thiel DH, Gavaler JS, Cobb CF, et al. (1983) Ethanol, a Leydig cell toxin: Evidence obtained *in vivo* and *in vitro*. *Pharmacology, Biochemistry, and Behavior, Supplement* 18: 317–323.

van Thiel DH, Gavaler JS, and Lester R (1978) Alcohol-induced ovarian failure in the rat. *Journal of Clinical Investigation* 61: 624–632.

Waltman C, Blevins LS, Jr., Boyd G, et al. (1993) The effects of mild ethanol intoxication on the hypothalamic–pituitary–adrenal axis in nonalcoholic men. *Journal of Clinical Endocrinology and Metabolism* 77: 518–522.

Waltman C, McCaul ME, Wand GS, et al. (1994) Adreno-corticotropin responses following administration of ethanol and ovine corticotropin-releasing hormone in the sons of alcoholics and control subjects. *Alcoholism: Clinical and Experimental Research* 18: 826–830.

Weinberg J and Bezio S (1987) Alcohol-induced changes in pituitary–adrenal activity during pregnancy. *Alcoholism: Clinical and Experimental Research* 11: 274–280.

Weinberg J, Taylor AN, and Gianoulakis C (1996) Fetal ethanol exposure: Hypothalamic–pituitary–adrenal and beta-endorphin responses to repeated stress. *Alcoholism: Clinical and Experimental Research* 20: 122–131.

West JR, Goodlett CR, and Brandt JP (1990) New approaches to research on the long term consequences of prenatal exposure to alcohol. *Alcoholism: Clinical and Experimental Research* 14: 684–689.

Westney L, Bruney R, Ross B, et al. (1991) Evidence that gonadal hormone levels in amniotic fluid are decreased in males born to alcohol users in humans. *Alcohol and Alcoholism* 26: 403–407.

Wilks JW, Hodgen GD, and Ross GT (1977) Anovulatory menstrual cycles in rhesus monkeys: The significance of serum, follicle stimulating hormone/luteinizing hormone ratios. *Fertility and Sterility* 28: 1094–1101.

Wilks JW and Noble AS (1983) Steroidogenic responsiveness of the monkey corpus luteum to exogenous chorionic gonadotropin. *Endocrinology (Baltimore)* 112: 1256–1266.

Willett WC, Stampfer MJ, Colditz GA, et al. (1987) Moderate alcohol consumption and the risk of breast cancer. *New England Journal of Medicine* 316: 1174–1180.

Wilsnack SC, Klassen AD, and Wilsnack RW (1984) Drinking and reproductive dysfunction among women in a 1981 national survey. *Alcoholism: Clinical and Experimental Research* 89: 451–458.

Wright HI, Gavaler JS, Tabasco-Minguillan J, et al. (1992) Endocrine effects of alcohol abuse in males. In: Mendelson JH and Mello NY (eds.) *Medical Diagnosis and Treatment of Alcoholism*, pp. 341–362. New York: McGraw-Hill.

Xiao E and Ferin M (1988) The inhibitory action of corticotropin-releasing hormone on gonadotropin secretion in the ovariectomized rhesus monkey is not mediated by adrenocorticotropic hormone. *Biology of Reproduction* 38: 763–767.

Yen SSC (1983) Clinical applications of gonadotropin-releasing hormone and gonadotropin-releasing hormone analogs. *Fertility and Sterility* 39: 257–266.

Yen SSC (1999) The human menstrual cycle: Neuroen-docrine regulation. In: Yen SSC, Jaffe RB, and Barbieri RL (eds.) *Reproductive Endocrinology,* 4th edn., pp. 191–217. Philadelphia, PA: Saunders.

Yen SSC, Quigley ME, Reid RL, et al. (1985) Neuroen-docrinology of opioid peptides and their role in the control of gonadotropin and prolactin secretion. *American Journal of Obstetrics and Gynecology* 152: 485–493.

Zeleznik AJ (1981) Premature elevation of systemic estradid reduces serum levels of follicle-stimulating hormone and lengthens the follicular phase of the menstrual cycle in rhesus monkeys. *Endocrinology (Baltimore)* 109: 352–355.

Zhang SM, Lee IM, Manson JE, Cook NR, Willett WC, and Buring JE (2007) Alcohol consumption and breast cancer risk in the Women's Health Study. *American Journal of Epidemiology* 65: 667–676.

Further Reading

Gavaler JS (1992) Alcohol effects in postmenopausal women. In: Mendelson JH and Mello NK (eds.) *Third Edition of the Medical Diagnosis and Treatment of Alcoholism.* New York: McGraw-Hill.

Smith CG (1991) Marijuana and other drug effects on reproductive hormones in the primate. In: Mello NK (ed.) *Advances in Substance Abuse: Behavioral and Biological Research,* vol. 4, pp. 113–137. London: Jessica Kingsley Publishers.

Tolis G (1980) Prolactin: Physiology and pathology. In: Krieger DT and Hughes JC (eds.) *Neuroendocrinology: Interrelationships of the Body's Two Major Integrative Systems in Normal Physiology and in Clinical Disease*, pp. 321–328. Sunderland, MA: Sinauer Associates.

33 Effects of Smoking on Hormones, Brain, and Behavior

T Sidhartha, UT Southwestern Medical Center, Dallas, TX, USA
R E Poland, The Research and Education Institute for Texas Health Resources, Arlington, TX, USA
U Rao, UT Southwestern Medical Center, Dallas, TX, USA

Chapter Outline

33.1 Introduction

Nicotine is the main biologically active component of tobacco smoke, and one of the most addictive substances known to man. Individuals who initiate smoking end up smoking far more cigarettes and for a longer duration than originally intended. Many physical and psychological complications result due to the constant presence of nicotine and other components of cigarette smoke in the human body. Smoking harms nearly every organ of the body, causing many diseases and reducing the health of smokers in general. The adverse health effects from cigarette smoking account for an estimated 438000 deaths, or nearly one of every five deaths, each year in the United States (CDC, 2005). More deaths are caused each year by tobacco use than by all deaths from human immunodeficiency virus (HIV), illegal drug use, alcohol use, motor vehicle injuries, suicides, and murders, combined (McGinnis and Foege, 1993). Cigarette smoking is an addiction with major health consequences and although significant progress has been made in recent times in understanding the pathophysiology and treatment of smoking, effective interventions remain limited.

Numerous toxic compounds have been identified in cigarette smoke, including polycyclic aromatic hydrocarbons, nitroso compounds, and aromatic amines (Lofroth, 1989). The most well-studied component of tobacco smoke though is nicotine, and its effects on many biological systems are well studied and understood. In this chapter, the effects of smoking, and in particular of nicotine on the endocrine system, will be described. There is now ample evidence that smoking influences the functioning of the human endocrine system although the exact mechanisms and the extent of these influences are not well understood. Because of the intricate and interdependent relationships that the endocrine system has with other systems, smoking influences many of these systems both directly and indirectly. In this chapter, the interaction of smoking with various endocrine axes is discussed, with a special focus on the effects of this interaction on brain and behavior.

33.2 Hypothalamic–Pituitary–Adrenal Axis

The hypothalamic–pituitary–adrenal (HPA) axis is comprised of the hypothalamus, the pituitary, and the adrenal cortices. Neurosecretory cells within the paraventricular nucleus (PVN) of the hypothalamus secrete corticotropin-releasing hormone (CRH) and arginine vasopressin (AVP) into the primary capillaries of the microportal circulatory system of the pituitary stalk. CRH and AVP stimulate the release of adrenocorticotropic hormone (ACTH) from the anterior lobe of the pituitary. Cortisol, a major end product of the HPA axis, is released from the adrenal cortex in response to ACTH. Cortisol secretion reflects the peripheral activity of the HPA axis. This activity is driven by diurnal and metabolic inputs, as well as by stress responses. In humans, cortisol secretion peaks in the morning, about the time of awakening, and declines gradually through the waking hours to achieve a daily minimum during the first half of the sleep cycle (Czeisler et al., 1976). The morning burst of cortisol secretion is driven by the action of clock genes in the suprachiasmatic nucleus of the hypothalamus, initiating neuronal signals to the PVN (Linkowski et al., 1993). Specialized PVN neurons respond to these signals. This diurnal pattern is modulated throughout the day by metabolic inputs arising in relation to blood glucose levels, among others (Van Cauter et al., 1992). Cortisol also helps to regulate its own secretion by exerting a negative feedback to the pituitary and hypothalamus. In addition to the pituitary and hypothalamus, glucocorticoid receptors are also found in higher brain regions, including the hippocampus, other areas of the limbic system, and the prefrontal cortex (Sanchez et al., 2000). Worth noting is that a number of these brain regions have been implicated in mood and other psychiatric disorders (Harrison, 2002).

In several ways, smoking interferes with the regulatory aspects of the HPA axis, and some of the underlying mechanisms for this is discussed in this section. In addition, the HPA response to acute and chronic smoking differs in some ways, and this area is also reviewed. In addition, since HPA abnormalities are frequently reported in individuals with mood disorders, and a significant proportion of individuals suffering from mood disorders smoke, the possible ways in which smoking might influence the HPA axis in mood disorders is reviewed. It is widely appreciated that the HPA axis is of central importance to the individual in dealing with the stresses of life, be they physical, psychological, or social. Smoking and stress are intricately related and therefore the HPA response to stress in smokers is described and its implications for the development and maintenance of nicotine addiction is explored.

33.2.1 Acute Response of the HPA Axis to Smoking

Smoking causes an acute rise in plasma ACTH levels. Mendelson et al. (2005) studied the HPA response to smoking in individuals with nicotine dependence (ND) by rapid sampling of venous blood every 2 min after cigarette smoking. Plasma ACTH levels increased significantly above baseline within 12 min and reached peak levels within 20 min. ACTH increases were significantly correlated with increases in plasma nicotine. Cortisol and dehydroepiandrosterone (DHEA) levels increased significantly within 20 min and reached peak levels within 60 and 30 min, respectively. These changes were observed with high, but not with low, nicotine content cigarettes. Other investigators also found changes in HPA hormones only after smoking high nicotine content cigarettes, thereby suggesting that it is nicotine in cigarette smoke that is responsible for these changes (Kirschbaum et al., 1992; Seyler et al., 1986).

The acute HPA response to smoking in chronic smokers has been compared to the response observed in individuals who are nonsmokers. Gossain et al. (1986) reported that cortisol response to smoking was significantly higher among chronic smokers compared to nonsmokers. Kirschbaum et al. (1994) found a similar trend. The dose–response relationship between nicotine and HPA hormones has also been explored in nonsmokers. Newhouse et al. (1990) employed a continuous injection of different nicotine concentrations (0.125, 0.25, and 0.5 $\mu g\,kg^{-1}\,min^{-1}$) in 11 healthy nonsmokers for 1 h. ACTH, cortisol, and prolactin concentrations in plasma increased with increasing doses of nicotine. In addition, self-reported mood and anxiety symptoms also showed dose-dependent responses. Anxiety increased, and mood decreased, with increasing nicotine doses. To the best of our knowledge, this was the first time that a clear temporal association between mood and smoking-induced HPA changes were shown. Stimulation of the HPA axis by the administration of nicotine through the intranasal route has also been reported. Pomerleau (1992) demonstrated that intranasal application of 0.05-, 1.00-, and 2.00-mg nicotine resulted in dose-dependent activation of physiological and endocrine responses. The route of nicotine administration seems to influence the extent of the endocrine responses. Benowitz et al. (2002) compared smoking, nasal spray, and transdermal application and found that, with comparable nicotine doses, endocrine responses were highest when subjects smoked.

Mendelson et al. (2008) reported that HPA hormone and subjective mood elevations occurred after smoking one high-nicotine-content cigarette. However, although smoking more cigarettes at 1-h intervals produced elevations in HPA hormones, the peak levels did not increase. Nicotine levels increased in a cumulative fashion, but subjective positive mood and cortisol responses decreased after the third cigarette. The high subjective mood and hormonal response to the first cigarette are consistent with the usual observation that after a period of abstinence, the first cigarette is salient and most reinforcing. However, it is not clear if the decrease in cortisol and subjective effects after the third cigarette reflected the development of acute tolerance to nicotine or relief of nicotine withdrawal effects after overnight abstinence.

Influence of age or gender on HPA responses to smoking has not been explored in great detail. Studies that recruited both male and female smokers have not reported differences between them. In animal studies, though, significant gender differences have been found with respect to HPA responses to nicotine (Rhodes et al., 2001). Rats given intraperitoneal injections of 0, 0.03, 0.1, 0.3, or 0.5 $mg\,kg^{-1}$ of nicotine showed a sex-specific activation pattern. In male rats, the AVP response was higher, while female rats secreted more ACTH and corticosterone. Furthermore, ACTH and corticosterone responses to nicotine vary with the estrous cycle of female rats, showing highest increases in the proestrous and estrous phases (Rhodes et al., 2004).

33.2.2 HPA Axis in Chronic Smokers

Based upon the findings of elevated HPA response after acute smoking, it is reasonable to assume that chronic smokers would have elevated basal levels of HPA hormones. However, findings, to date, have been inconclusive in this regard. del Arbol et al. (2000) did not find elevated ACTH levels in smokers compared to nonsmokers but they did find elevated cortisol levels in subjects who smoked more than 20 cigarettes per day. A few studies (with small sample sizes) have failed to find increased levels of cortisol in chronic smokers (Benowitz et al., 1984; Gossain et al., 1986; Kirschbaum et al., 1994). Yeh and Barbieri (1989) did not find any differences in 24-h urinary free cortisol concentrations. One study with a larger community sample of middle-aged men found that the basal cortisol levels in smokers were increased as compared to nonsmokers, but only by 5% (Field et al., 1994), and another study with a smaller sample size found similar results (al'Absi et al., 2003). In contrast, in a Japanese study of 297 middle-aged men, morning cortisol levels were found to be lower in smokers compared to nonsmokers (Handa et al., 1994).

Steptoe and Ussher (2006) reported on comparison of salivary cortisol profiles, of smokers and nonsmokers, over the course of the day. In a sample that had 15 smokers and 152 nonsmokers, cortisol levels were significantly higher among smokers after adjustment for effects of age, sex, and body mass index. This study had a better design as cortisol levels were measured multiple times during the day and common confounding factors were controlled. Taken together, the variable findings in these studies might be due to differences in study design, such as different times of data collection, and therefore, it is difficult to make reasonable conclusions. Future studies with a larger sample of smokers will be helpful in gaining a better understanding of the effects of chronic smoking on HPA response.

33.2.3 Mechanism of HPA Activation by Nicotine

The exact mechanism of HPA activation following nicotine administration is debatable. Cam et al. (1979) demonstrated that hypophysectomy can lead to a reduction in ACTH release from the pituitary. Weidenfeld et al. (1989) showed that HPA activation after intravenous nicotine administration can be blocked by hypothalamic lesions in the PVN.

Other researchers demonstrated that nicotine perfusion led to ACTH release in isolated mouse brains (Marty et al., 1985), but not in isolated pituitary preparations (Matta et al., 1987). These results indicate that the HPA axis is stimulated by nicotine at the hypothalamus or at a higher level in the nervous system.

Matta et al. (1998) demonstrated that ACTH release by nicotine was blocked by α_1- and α_2-adrenergic receptor blockade in the hypothalamus. Using brainstem lesions, they found that activation of the nucleus tractus solitarius and the ventromedullary regions correlated with activation in the PVN as measured by cFos mRNA activation. They also demonstrated that dose-dependent release of epinephrine (which correlated with ACTH release) occurred with nicotine stimulation. These results strongly suggest that brainstem noradrenergic regions, which project to the PVN, play an important role in mediating nicotine stimulation of the HPA axis. Other investigators favor direct actions of nicotine at the hypothalamic level. Fuxe et al. described three types of nicotinic-binding sites in the hypothalamus (Fuxe et al., 1989). Nicotine mimics the effects of acetylcholine at selected central nicotinic cholinergic receptors, and thus, nicotine could activate the HPA axis by binding to the nicotinic receptors at various sites in the brain, including the PVN (Rosecrans and Karin, 1998).

With chronic nicotine use, there is ample evidence that desensitization of the nicotinic receptors takes place. In rat studies, a single injection of nicotine (0.5 $mg\,kg^{-1}$) completely abolished the response to a subsequent nicotine challenge for 1 h (Sharp and Beyer, 1986). Hypothalamic cFos expression and norepinephrine (NE) release in the PVN were significantly reduced in response to repeated nicotine injections (Matta et al., 1995; Sharp and Matta, 1993). As mentioned above, ACTH has not been found to be elevated in chronic smokers, and this might be due to the desensitization of central nicotinic cholinergic receptors (Fuxe et al., 1989).

In human studies, stimulation of the HPA axis by acute smoking in chronic smokers has been demonstrated consistently, whereas elevated basal levels of cortisol in smokers has not been a consistent finding. Studies examining the effects of acute smoking in chronic smokers include an overnight abstinence phase before the morning challenge, and this could account for the observed activation. One human study did demonstrate that by the third cigarette, HPA activation had reached its peak and cortisol levels had already started to decline (Mendelson et al., 2008). Taken together, these findings support the hypothesis that nicotine, following either acute or chronic administration or exposure, induces nicotinic receptor desensitization for varying periods of time.

33.2.4 Smoking, Mental Illness, and the HPA Axis

The association of elevated cortisol levels in depressed individuals was the first indication that HPA hormones might play a role in regulating emotions and behavior (Board et al., 1956). Since then, elevation of cortisol has been found to be one of the most consistent biological findings associated with depression (Young, 2004; Young et al., 2002). The HPA has also been implicated in the pathophysiology of mood, anxiety, and the psychotic disorders, although the association with depression is the most studied and robust.

33.2.4.1 Smoking, depression, and the HPA axis

Several studies have shown that HPA hyperactivity, as manifested by hypersecretion of CRH, increased cortisol levels in plasma, urine, and saliva exaggerated cortisol responses to ACTH, and enlarged pituitary and adrenal glands, occur in individuals suffering from severe mood disorders (Young, 2004). Approximately, 50% of the individuals suffering from depression also smoke (Hughes et al., 1986). Compared to nonsmokers, regular smokers report more depressive symptoms (Anda et al., 1990), more frequent and severe episodes of depression (Glassman et al., 1993), and higher rates of suicidal ideation and attempts (Breslau et al., 2005; Malone et al., 2003). Smokers with a history of depression who abstain from smoking are also significantly more likely to develop a new episode of major depression (Glassman et al., 2001). Along with unemployment and chronic medical illness as potential risk factors, smoking is one of the strongest correlates of current major depressive disorder, with smokers over 2.5 times more likely to be depressed than those who have never smoked (Wilhelm et al., 2003).

As discussed previously, there is sufficient evidence to suggest strong associations between depression and cigarette smoking, depression and the HPA axis, and between HPA axis and cigarette smoking. Despite strong evidence to support a link among these three factors, there are very few studies that have examined them simultaneously. In a longitudinal study of depressed and nondepressed adolescents with no prior evidence of smoking, elevated HPA activity at baseline increased the risk of smoking during prospective follow-up. Reciprocally, onset of regular smoking increased the risk for first episode of depression in controls and recurrent depressive episodes in the depressed cohort (Rao et al., 2007).

It is possible that, in depressed individuals, nicotine plays the role of stabilizing the HPA axis. Pomerleau et al. (2004) compared the severity of withdrawal symptoms and HPA sensitivity of women smokers who had a history of depression with those who had no such history. Smokers with a history of depression were found to have more severe dysphoric symptoms during withdrawal compared to smokers who had no history of depression. All participants evinced cortisol suppression in response to dexamethasone during both conditions, but the degree of suppression did not differ as a function of depression history. In history-positive smokers, however, ACTH levels trended toward overall elevation and showed almost no suppression during abstinence; thus, exacerbation of HPA dysregulation in history-positive smokers during smoking abstinence could not be ruled out. Thus, when smokers with a history of depression stop smoking, there may be a rebound effect on the HPA axis resulting in a loss of feedback inhibition of cortisol. Resumption of smoking has been found to alleviate depressive/dysphoric symptoms (Fagerstrom et al., 1990; Glassman, 1993). It is also speculated that chronic exposure to nicotine elicits changes in the brain that are depressogenic and that smokers are protected from the consequences of these changes, while they continue to smoke, by the antidepressant (AD) properties of nicotine (Balfour and Ridley, 2000; Djuric et al., 1999). Individuals in whom psychological and/or physiological adaptability is compromised, nicotine may serve to maintain homeostasis in critical systems such as the HPA axis; for such people, nicotine use may constitute a coping strategy for meeting the challenges of daily living (Pomerleau et al., 2004).

33.2.4.2 Schizophrenia, smoking, and the HPA axis

Previous epidemiological studies have shown that individuals suffering from schizophrenia smoke more frequently (50%) compared with the general population (23%) (Lasser et al., 2000). They also have a lower quit rate compared to the general population (27.5% vs. 42%) (Lasser et al., 2000). In addition they extract more nicotine from each cigarette smoked, presumably by deeper inhalation (Olincy et al., 1997). The high level of smoking has been proposed as a form of self-medication to alleviate symptoms of their illness including depression, anxiety, anhedonia, or amotivation (Glassman, 1993; Olincy et al., 1997; Tung et al., 1990). It is postulated that the high rates of smoking found in individuals suffering from schizophrenia can be explained by the ameliorating effects of nicotine on attentional abnormalities and improvement in cognitive deficits. Others have proposed that smoking alleviates symptoms of nicotine withdrawal or neuroleptic-induced side effects (Dalack et al., 1999; Goff et al., 1992; Nisell et al., 1995).

Both clinical and biological data indicate that schizophrenia patients are impaired in their biological response to stress (Jansen et al., 2000; Nuechterlein et al., 1994). This is associated with dysregulated HPA axis (Jakovljevic et al., 1998) and a blunted cortisol response to the stress of speaking in public (Jansen et al., 1998, 2000). DHEA is a major circulating steroid and serves as a precursor for both androgenic and estrogenic steroids. Its sulfated form (DHEA-S) is the most abundant steroid found in the body (Kroboth et al., 1999). It is considered both a neurosteroid, being produced in the brain, as well as a neuroactive steroid, produced in the adrenals and having its effect on the brain (Baulieu and Robel, 1998). Brain DHEA(S) (Corpechot et al., 1981) levels exceed their respective concentrations in plasma. Neurosteroidogenesis in the brain is independent of the peripheral production (Corpechot et al., 1981).

DHEA has been demonstrated to have memory-enhancing effects in rodents (Sujkovic et al., 2007) and in some human studies (Hirshman et al., 2003). It has been hypothesized that DHEA has neuroprotective effects on cognition (Wolf and Kirschbaum, 1999). DHEA and DHEA-S may have antistress properties by acting as an endogenous restraint against corticosterone. DHEA-S blocks the neurotoxic effects of corticosterone on hippocampal cells (Kimonides et al., 1999). DHEA also protects neurons against glutamate and β-amyloid-protein toxicity (Cardounel et al., 1999), excitatory amino acid-induced neurotoxicity (Kimonides et al., 1998), and numerous other insults resulting in oxidative stress.

These findings suggest the possibility that cortisol/DHEA and/or cortisol/DHEA-S ratios would be found elevated in schizophrenia patients as a result of stress associated with the illness. In a study by Ritsner et al. (2004), schizophrenia in patients demonstrated significantly higher levels of state and trait anxiety, anger expression index, and emotional and somatic self-reported distress scores. Cortisol/DHEA and cortisol/DHEA-S ratios were significantly higher in schizophrenia patients than in healthy comparison subjects. In another study by the same researchers the association of cortisol/DHEA-S ratios with antipsychotic response was studied and it was found that responders had significantly higher serum cortisol levels and cortisol/DHEA(S) ratios compared with nonresponders (Ritsner et al., 2005). Thus, these data provide evidence that elevated serum cortisol and cortisol/DHEA(S) ratios may serve as markers of biological mechanisms that are involved in responsivity of schizophrenia patients to antipsychotic treatment. Finally, in a recent study, cortisol/DHEA(S) ratios were measured sequentially during treatment with antipsychotics (Ritsner et al., 2007). Despite clinical improvement during the study period, cortisol/DHEA(S) molar ratios were found persistently elevated as compared to healthy controls. Elevated serum cortisol/DHEA(S) molar ratios were attributed to trait-anxiety and age rather than to clinical symptoms. The findings may indicate persistent dysfunction of the HPA axis in schizophrenia that is independent of the patients' clinical state (Ritsner et al., 2007).

33.2.4.3 Anxiety disorders, smoking and the HPA axis

In contrast to the fairly robust association seen with depression, the association of the HPA axis with anxiety disorders that are not caused by traumatic stressors is not certain. For instance, some findings in adults suggested increased HPA-axis activation in panic disorder (Schreiber et al., 1996) while other findings did not support this hypothesis (Curtis et al., 1997). Evidence has been found for lower cortisol levels in both adult and adolescent patients with posttraumatic stress disorder (Goenjian et al., 2003; Yehuda et al., 2006).

In the few studies that investigated the association between anxiety problems and cortisol levels in children and adolescents, findings were as inconclusive as in adults. Kagan and colleagues found that basal cortisol levels were higher in inhibited, than in uninhibited, young children (Kagan et al., 1987). In another study, Martel et al. (1999) did not find differences in basal cortisol levels of social phobic adolescent girls versus matched controls. In a recent study with 1700 participants, associations between cortisol levels and current anxiety problems were not found (Greaves-Lord et al., 2007). However, individuals with persistent anxiety problems had higher morning cortisol levels and a higher cortisol awakening response. Thus, it is possible that only persistent, and not current, anxiety problems are associated with higher HPA-axis activity. Alterations in HPA-axis activity might underlie persistent anxiety problems, or result from the stress accompanied by persistent anxiety problems (Greaves-Lord et al., 2007).

Smokers often report improvement in anxiety by smoking during stressful periods (Gilbert et al., 1989). They also report increased anxiety in the immediate period after smoking cessation (Giannakoulas et al., 2003; Jorenby et al., 1996), and studies suggest that smoking may be an effort to avoid this withdrawal symptom (Brown et al., 2001; Pomerleau et al., 2000). Individuals with a past history of smoking often relapse during periods of increased stress (reviewed in next section) and anxiety. Smoking is more common in those who suffer from anxiety disorders (Amering et al., 1999), and this is one of the reasons for arguing that chronic smoking probably plays a role in the development of anxiety (McCabe et al., 2004). It has also been reported that anxiety decreases approximately 1 week after smoking cessation (West and Hajek, 1997). The above findings, although in some ways contradictory, suggest that smoking definitely plays a role in modulating anxiety in smokers.

33.2.5 HPA Response to Stress in Smokers

Numerous studies have found that increased stress is a risk factor for smoking initiation (Byrne and Mazanov, 2003; Koval et al., 2000), as well as the transition to regular smoking (Orlando et al., 2001; Siqueira et al., 2000). Adolescent smokers commonly report stress reduction and calming influence of smoking as motives for smoking (Dozois et al., 1995; Nichter et al., 1997). Finally, individuals who fail to quit smoking, or relapse after a short period of smoking cessation, report higher levels of stress during abstinence compared to those who maintain abstinence (Cohen and Lichtenstein, 1990). Thus, stress is intricately linked to all aspects of smoking, from initiation to abstinence. It is well known that the HPA axis is involved in the physiological response to stress. In view of these associations, it is likely that the associations between smoking and stress are mediated by the HPA axis.

The close association of stress and smoking was clearly demonstrated by Rose et al. (1983). In this study, cigarette smokers were exposed to three conditions within a single session: stage-fright anxiety, monotonous concentration, and a relaxation control.

One cigarette was lit during the second 10-min half of each condition, and smoking topography (number of puffs and cumulative volume smoked) was continuously recorded. Subjects smoked significantly more in the two task conditions than during relaxation, supporting the hypothesis that anxiety-provoking and attention-demanding situations elicit smoking. Younger subjects increased their smoking more than older subjects during stage-fright, and females responded more than males to the concentration task. One interpretation of this finding is that stress may attenuate the effects of nicotine intake (including those related to reinforcement), and smokers therefore increase their smoking intensity in an effort to overcome this attenuation in nicotine's effects. This could, thus, contribute to the development of tolerance to nicotine's actions. A second possibility is that nicotine-induced elevations of corticosterone act to prevent a prolonged response to nicotine.

The HPA-axis response to stress itself appears to be altered in smokers. Kirschbaum et al. (1993) compared the physiological response of smokers and nonsmokers to a saline injection, hCRH injection, an exhaustive ergometry session (bicycle ergometry until exhaustion was reached), and a psychosocial stress test (Trier Social Stress Test; Kirschbaum et al., 1993). The stress test involved speaking and performing mental arithmetic in front of an audience. The HPA response to psychosocial stress was found to be blunted in smokers, with mean cortisol levels reaching only to a third of what was observed in nonsmokers. Responses to the CRH test were slightly, but nonsignificantly, lower in smokers, and ergometry failed to elicit a cortisol response in all subjects. The same group of investigators compared the hormone responses to injection of CRH following bicycle ergometry and psychological stress in ten habitual smokers and ten nonsmokers (Kirschbaum et al., 1994). Nonsmokers had stronger response to the different stimuli compared to smokers, but due to the small sample size the differences reached statistical significance only for growth hormone (GH) responses following ergometry and salivary cortisol responses after psychological stress. Rohleder et al. also demonstrated blunted cortisol responses in smokers to the Trier Social Stress Test compared to same sex nonsmokers (Rohleder and Kirschbaum, 2006).

The blunted HPA-axis response to stress in smokers might be related to the elevated levels of corticosterone associated with chronic smoking. This role would be consistent with a large body of evidence that corticosterone serves a general protective function of attenuating exaggerated or prolonged responses to exogenous agents that disturb homeostasis (Munck et al., 1984). In this case, elevated cortisol in response to one stressor (smoking) does not necessarily protect against the source of the stress but protects the body from mounting an exaggerated response to other stressors (psychological stress). It is also relevant that from the limited number of studies reviewed above, it appears that the HPA axis in smokers has a decreased sensitivity to psychological stress, but not physiological stressors. The differential sensitivity of the HPA axis to these two kinds of stressors could be due to the different brain pathways recruited by them to activate the HPA. We have already seen that HPA is activated by nicotine via direct actions of nicotine at the hypothalamic level and via the locus ceruleus (LC), which projects to the PVN. Brain regions higher than the hypothalamus (reviewed below) are also involved in the stress response and these regions are likely involved in triggering the HPA-axis response to psychological stress. Nicotine might be causing its effects on the HPA-axis response by preferentially acting on these regions. Or it could be that at a certain concentration in the brain, nicotine is more effective in the higher brain regions compared to the hypothalamus and brainstem.

It is clear, at least in animal studies, that the relationship between nicotine and at least one HPA-axis hormone, corticosterone, is bidirectional. Not only does nicotine affect the HPA axis and circulating corticosterone, but corticosterone can either directly or indirectly modulate nicotine's behavioral and physiological actions (Caggiula et al., 1998). It is now known that adrenalectomy increases nicotine responsiveness in female mice without altering blood or brain levels of nicotine, and there is a strong genetic influence on the extent of the change in responsiveness (Pauly et al., 1988, 1990b). In these studies, responsiveness was measured relative to nicotine-induced decreases in Y-maze crossings, heart rate, and body temperature, and increases in startle response. It has also been shown that chronic administration of corticosterone in mice produced subsensitivity to injections of nicotine (Pauly et al., 1990a) and that corticosterone withdrawal resulted in rapid reversal of the reduced sensitivity within 1–3 days (Grun et al., 1995).

The above discussion highlights the close association among HPA axis, stress, and smoking. These interactions are of particular importance in the effort to understand the biological mechanisms of nicotine addiction.

33.2.6 HPA Changes Associated with Nicotine Addiction

Although the interactions between the HPA axis and the addiction-related effects of drugs are poorly understood, in the recent past, a fair amount of attention has been paid to the possible mechanisms and significance of these interactions. The role of HPA-axis activation in cocaine addiction has been explored in various ways and may form the framework for similar studies with nicotine. HPA-axis activation is thought to contribute to drug abuse at several phases of the addictive process (Goeders, 2002). Pervasive alterations of HPA-axis stress responsivity in relation to drug exposure and addiction have been demonstrated. Acute administration of cocaine causes an HPA-axis response, leading to increased cortisol secretion (Broadbear et al., 2004). In animals, corticosterone seems to be crucial for the acquisition of cocaine use since self-administration does not occur unless this stress hormone is increased above a critical reward threshold (Goeders and Clampitt, 2002; Goeders and Guerin, 1996b). Increasing circulating levels of corticosterone also augments sensitivity to low doses of cocaine, suggesting that exposure to stress can increase individual vulnerability to cocaine (Goeders, 2002). Drugs affecting the synthesis and/or secretion of corticosterone decrease ongoing, low-dose cocaine self-administration (Goeders and Guerin, 1996a; Goeders et al., 1998). Hormonal studies using an experimenter-administered cocaine binge model and an escalation self-administration model have revealed large increases in ACTH and corticosterone in rats during an acute binge, with attenuation during the chronic binge stage and a reactivation of the HPA axis during acute withdrawal (Koob and Kreek, 2007).

Animal studies have also suggested a prominent role for CRH in addiction. Rapid cannabinoid withdrawal causes release of CRH in widespread brain regions, precipitating a systemic stress reaction (Rodriguez de Fonseca et al., 1997). Increases in ethanol self-administration during abstinence have been observed in animals that have been exposed to ethanol vapors sufficient to induce dependence (Rimondini et al., 2002). Intracerebroventricular administration of a competitive CRF antagonist can reduce ethanol self-administration during acute withdrawal and protracted abstinence in a dose-dependent manner (Valdez et al., 2002). These results suggest that, during the development of ethanol dependence, there is a recruitment of CRF activity, in the rat, of motivational significance that can persist into protracted abstinence. During short-term and protracted abstinence, human alcoholics showed a blunted cortisol response to CRF (Bailly et al., 1989). An elevation of cerebro spinal fluid (CSF) CRF from lumbar samples in human alcoholics during acute withdrawal has been observed (Adinoff et al., 1996).

Several human studies have also indicated that the HPA axis is intricately involved in the process of nicotine addiction. As reviewed in the previous section HPA-axis response is altered in smokers, and the inhibitory effect of cortisol on nicotine effects might be contributing to nicotine tolerance. It is quite likely that the HPA-axis response during smoking cessation is involved in the mechanisms of relapse. Comparison of cortisol levels during *ad libitum* smoking and abstinence in chronic smokers has led to mixed results. Some studies have shown greater levels during abstinence (Hughes et al., 1988), and others showed no differences (Pickworth et al., 1996). Frederick et al. (1998) found a positive association between the magnitude of cortisol drop from pre- to 2 weeks post-treatment and emotional distress after initial abstinence (Frederick et al., 1998). al'Absi et al. (2005) compared the HPA-axis response to a socially salient stressor (public speaking) among smokers who continued to smoke *ad libitum*, smokers who abstained from smoking, and nonsmokers. They found that only nonsmokers showed significant cortisol responses to the stressor. Smokers, regardless of whether they were smoking *ad libitum* or abstaining, showed attenuated systolic blood pressure responses. It is possible that chronic HPA-axis activation caused by smoking results in a dysregulated central modulation of adrenocortical and sympathetic responses to acute stress.

The same group of researchers conducted a study where salivary cortisol and mood reports were collected during 24-h *ad libitum* smoking and during the first 24 h of abstinence (al'Absi et al., 2004). Participants who relapsed in the first week of abstinence showed a greater decline in cortisol levels on the abstinence day compared to the *ad libitum* day. Participants who relapsed reported greater craving for cigarettes and overall distress during the first 24-h period of abstinence than those who maintained abstinence. These findings support the hypothesis that early relapse is associated with exaggerated adrenocortical and mood perturbations during the first 24 h of abstinence. In a subsequent study, the same group demonstrated that male smokers who relapsed within 4 weeks of smoking cessation, in comparison with those who maintained abstinence, showed attenuated ACTH and cortisol responses to a stressor within 24h of the cessation attempt (al'Absi et al., 2005). Relapsers showed reduced blood pressure responses to stress, exaggerated withdrawal symptoms, and mood deterioration after smoking cessation. Rao et al. (2007) evaluated the contribution of HPA-axis function and environmental stress to the development of smoking in adolescents. Both elevated HPA-axis activity and recent stressful life events predicted the onset of smoking during prospective follow-up in adolescents, and youngsters who had combination of both were at highest risk.

Smokers have substantially higher serum levels of DHEA, DHEA-S, and androstenedione (al'Absi et al., 2003; Baron et al., 1995). ACTH-stimulated androstenedione and DHEA levels appear to be higher in smokers (Hautanen et al., 1993). These steroids can be synthesized in the adrenals, and therefore, data demonstrating higher DHEA, DHEA-S, and androstenedione levels in smokers (and increased DHEA and androstenedione responses to ACTH) suggest a potential upregulation of the HPA axis in subjects who smoke. Consistent with this possibility, DHEA levels appear to decrease after smoking cessation (Oncken et al., 2002). Rasmusson et al. (2006) demonstrated that a decrease in plasma DHEA/cortisol ratio during 8days of abstinence from smoking was associated with relapse over the following week. DHEA-S levels during the abstinence period have been inversely correlated with negative affect and craving measures, and may predict severity of ND (Marx et al., 2006). Based on the above findings, DHEA has been suggested as a potential therapeutic agent to facilitate smoking cessation (Marx et al., 2006).

There has been a lot of progress in our understanding but it remains to be seen if the enhanced neurobiological understanding of the association of the stress response system with nicotine addiction will lead to the development of potent therapeutic agents. Pharmacological agents that can influence the HPA axis can be potentially useful for management of addiction. At the same time, optimal use of behavioral and psychological interventions (e.g., stress management techniques) which can modify the individual stress response continue to be an integral part of the management of addiction.

33.2.6.1 Brain regions involved in nicotine addiction and regulation of HPA axis

The emerging view of the commonalities among addictions is promoted by research showing that addictions involve common alterations in motivational systems within the brain. One critically important region of the brain mediating addiction is the mesolimbic dopamine (DA) system consisting of midbrain DA cells, primarily in the ventral tegmental area (VTA) and the nucleus accumbens (NAcc), which receives a dense projection from the VTA. All drugs of abuse appear to act directly at the VTA and/or NAcc to cause increases in DA levels (Nestler, 2001). Saal et al. (2003) demonstrated that *in vivo* administration of five different drugs of abuse with very different molecular mechanisms of action all elicit an enhancement of strength at excitatory synapses on midbrain DA neurons. Two psychoactive but therapeutic and nonaddictive medications (fluoxetine and carbamazepine) did not cause such a change. This degree of specificity suggests that this *in vivo*, drug-induced synaptic plasticity in DA neurons is an important component of the neural circuit adaptations that contribute to core features of addiction (Saal et al., 2003).

One of the major biological responses to acute stress is increased secretion of glucocorticoids and expression of glucocorticoid receptors (GRs). Saal et al. (2003) demonstrated that acute stress led to enhanced synaptic strength in mice midbrain dopaminergic neurons. They also demonstrated that *in vivo* administration of five different drugs of abuse, with very different molecular mechanisms of action, all elicit an enhancement of strength at excitatory synapses on midbrain dopaminergic neurons. Administering a GR antagonist (RU486), prior to stress exposure, blocked the synaptic change. On the other hand GR blockade did not influence cocaine-induced synaptic changes in DA neurons. Thus, GR blockade prevents the stress-enhancement of DA neuron excitability, although it does not prevent the drug-induced effect on this excitability. This suggests that stress and drugs of abuse may initiate their effects in different ways but that they both act on brain DA systems as a common pathway to self-administration (Saal et al., 2003).

Psychological stressors gain their influence because of how we interpret them in relation to our long-term plans and expectations about the world (Lazarus and Folkman, 1984). It is noteworthy that cortisol is quite responsive to acute psychological distress, suggesting that the source of HPA-axis activation, in such cases, must involve connections from the limbic system and prefrontal cortex to the hypothalamus. In addition, the presence of GRs above the hypothalamus suggests that higher brain centers are involved in the modulation of the stress response. It has been observed that during periods of psychological stress, the diurnal pattern of cortisol secretion is overridden by signals to the hypothalamus that originate in the limbic system. The signals arise in the amygdala and the bed nuclei of the stria terminalis (BNST). These brain structures are known to be activated by stimuli that convey information having survival value (LeDoux, 1993). The amygdala, therefore, stands at the center of a neural network that generates approach and avoidance reactions to innate and learned stimuli (Rolls and Stringer, 2001). Outputs from the amygdala and BNST interact with nearby structures, such as the NAcc, that, in turn, communicate extensively with the prefrontal cortex (Herman et al., 2003). The BNST also provide the primary inputs to the PVN that, in turn, generate an HPA-axis response to psychological stress. These frontal–limbic processes, therefore, form the neurophysiological mechanism through which psychological events generate cortisol responses (Lovallo, 2006; Lovallo et al., 2000). These central influences are augmented during periods of psychological stress through noradrenergic inputs that ascend from the LC in the brainstem to activate the cerebral cortex and limbic system (Pacak et al., 1995). The stress response is further integrated across the central nervous system by an extensive system of CRH-secreting neurons found in the cerebral cortex and limbic system (Lovallo, 2006; Petrusz et al., 1985).

The above findings indicate that the limbic system response to emotional stimuli and HPA-axis responses to stress are both of interest in relation to drug intake, addiction vulnerability, and potential for relapse in humans. Finally, indirect evidence from studies in humans supports the notion that some neuroendocrine responses to smoking are mediated by forebrain β-endorphin opioid mechanisms (Gilbert et al., 1992; Gorelick et al., 1988). Experimental data from rats suggest that, while acute nicotine administration stimulates release of β-endorphin from forebrain neurons (Boyadjieva and Sarkar, 1997; Davenport et al., 1990), chronic nicotine administration inhibits proopiomelanocortin gene expression and thereby, probably, biosynthesis of β-endorphin and other opiomelanocortins (Rasmussen, 1998). It can, therefore, be reasonably hypothesized that diminished forebrain β-endorphin biosynthesis, in response to long-term nicotine exposure by chronic smoking, could potentiate the self-administration of nicotine, in order to induce acute release of the available β-endorphin, minimizing the opioid withdrawal that would otherwise occur due to tonically decreased β-endorphin synthesis (Rasmussen, 1998).

33.2.7 Nicotinic Acetylcholinergic Receptors

Nicotine acts on various brain regions including the regions implicated in the regulation of the HPA axis by binding with nicotinic acetylcholinergic receptors (nAChRs). In previous sections, the associations between smoking HPA axis and some mental disorders were reviewed. In this section, the possible role of the nAChR in mediating these associations will be discussed.

The nAChR is a ligand-gated ion channel with a pentameric structure. It belongs to the superfamily of ligand-gated ion channels that includes gamma-aminobutyric acid (GABA), glycine, and 5-hydroxytryptamine (5-HT3) serotonin receptors. It is widely distributed in the brain and is controlled by acetylcholine and nicotine agonists. Several types of nAChRs have been identified. They vary based on the different subunit combinations. nAChR subunits can be separated into five major categories (α, β, δ, ϵ, and γ), and 11 different types of subunits have been described in mammals (α2–α9, β2–β4). The α4β2 and the α7 types are the most prevalent in the brain. Nicotine has a strong affinity for the α4β2 type and low affinity for the α7 type. The nAChRs not only exist on neuronal cell bodies and dendrites, but are also located on axon terminals and are involved in multiple neurotransmitter release, including ACh, DA, GABA, glutamate, NE, and serotonin (Dani and Bertrand, 2007). Postsynaptic nAChRs contribute a small minority of fast excitatory transmission, and nonsynaptic nAChRs modulate many neurotransmitter systems by influencing neuronal excitability.

33.2.7.1 Smoking, anxiety, and nicotinic acetylcholinergic receptors

Animal studies have suggested that, in a limited dose range, nicotine may have anxiolytic properties (Brioni et al., 1993; Cao et al., 1993), but higher doses are clearly anxiogenic (Cheeta et al., 2001; Irvine et al., 2001). Altered anxiety has been observed in several lines of mice with mutations in nicotinic receptor subunits. Nicotinic receptors containing α4 subunits likely modulate anxiety, as increased anxiety was observed in mice lacking the α4 subunit (Ross et al., 2000), as well as in mice expressing the Leucine9'Serine mutation in the same subunit

(Labarca et al., 2001). A decrease in anxiety was observed in mice lacking the β4 subunit (Salas et al., 2003). The opposing effects of nicotinic subunit deletion on anxiety support the possibility that the divergent effects of nicotine on anxiety may be mediated by different populations of nicotinic receptors.

Another type of nAChR, the beta-3 (β3) containing receptors, has the highest affinity for nAChRs studied to date. β3-Null mutant mice demonstrated decreased anxiety compared to the wild type and, interestingly, levels of the stress hormone, corticosterone, were significantly higher in the β3-null mutant mice at baseline and following exposure to stress (Booker et al., 2007). As discussed above, one of the ways that nicotine acts on the HPA axis is via the action of brainstem adrenergic neurons, such as those in the nucleus tractus solitatrius (NTS) which project to the PVN. High levels of β3 mRNA have not been detected in the NTS, but they are highly expressed in the LC (Cui et al., 2003). This area provides another source of noradrenergic input to the PVN (Swanson and Sawchenko, 1983) and, accordingly, β3 receptors may alter NE release via this pathway. Noradrenergic neurons from the locus ceruleus also project to the hippocampus as part of the 'extrahypothalamic' pathway through which corticosterone is increased in response to stress. Deletion of the β3 subunit significantly reduces nicotinic modulation of hippocampal NE release (Azam and McIntosh, 2006). Therefore, β3-containing nicotinic receptors in the LC–hippocampal pathway may mediate the enhanced HPA-axis response to stress noted above (Booker et al., 2007).

Alternatively, dopaminergic neurons project from the VTA/substantia nigra to the central nucleus of the amygdala (Fallon and Moore, 1978) and specifically target CRF neurons (Eliava et al., 2003), suggesting that dopaminergic transmission may modulate the HPA axis via CRF release from the amygdala. β3 is highly expressed in the VTA/substantia nigra and β3-containing nicotinic receptors modulate the release of DA in the nigrostriatal pathway (Cui et al., 2003). It is possible that β3-containing receptors modulate DA release in the amygdala in a similar manner and subsequently regulate corticosterone levels via CRF release.

33.2.7.2 Nicotinic acetylcholinergic receptors and schizophrenia

Schizophrenia is a syndrome and endophenotypes associated with this syndrome have been particularly useful for understanding specific biological mechanisms including genetics. A few measurable sensory deficits have been proposed as possible endophenotypes of schizophrenia. Individuals with schizophrenia have diminished suppression of auditory-evoked response (P50) to repeated stimuli and abnormal smooth pursuit eye movements (SPEM). Nicotine normalizes both these deficits (Adler et al., 1993). Relatives of individuals with schizophrenia also have poor P50 suppression (Clementz et al., 1998). It is important to note that clozapine normalizes the P50 ratio coincident with improvement in the clinical symptoms of schizophrenia (Nagamoto et al., 1999). Clozapine, which releases ACh in the hippocampus (Shirazi-Southall et al., 2002), may thereby indirectly act on the nicotinic cholinergic receptors to normalize the P50 ratio, as people with schizophrenia also decrease the amount of cigarettes they smoke while taking this medication (McEvoy et al., 1999). Administration of high-dose nicotine with mecamylamine (10mg), a high-affinity α4β2-nicotinic cholinergic receptor antagonist, still produces improvement in P50 suppression in schizophrenia (Freedman et al., 1994). Thus, since nicotine is a nonselective agonist and mecamylamine is blocking the high-affinity receptors, the improvement in suppression appears mediated through the low-affinity α7-nicotinic cholinergic receptors.

Genetic studies have provided another line of evidence for involvement of the α7-cholinergic receptor in the P50 auditory-evoked potential deficit. Linkage to the P50 deficit in families suffering from schizophrenia has been found at chromosome 15q14 (Freedman et al., 1997; Leonard et al., 1998). Polymorphisms in the core promoter region of the α7-nAChR gene occur more frequently in schizophrenic patients than in controls, and the presence of one polymorphism (2-bp deletion in exon 6) in controls was associated with failure to inhibit P50 (Leonard et al., 2002; Raux et al., 2002). Although the role of α7-nAChRs in auditory sensory processing has been documented extensively, there is evidence that the α4β2 subunit is also involved, particularly in relation to the acoustic startle response (Owens et al., 2003). A naturally occurring single nucleotide polymorphism in the α4 subunit was associated with nicotine-induced alteration in acoustic startle response (Tritto et al., 2002). In addition to the P50 deficits, there is a wealth of evidence for SPEM abnormalities in adult patients with schizophrenia and in their unaffected relatives (Kathmann et al., 2003; Ross et al., 2002). The SPEM deficits remain stable with treatment and across clinical states (Thaker et al., 1999).

People with schizophrenia also have abnormalities in expression and regulation of central nAChRs. Postmortem studies have shown reduced [125I] α-bungarotoxin binding in the hippocampus in schizophrenic patients, a ligand selective for the α7-nicotinic receptor subtype (Freedman et al., 1995). Decreased α7-receptor binding has also been demonstrated in the thalamus, cingulated cortex, and the frontal lobe regions (Court et al., 1999; Marutle et al., 2001).

Thus, there is a fair amount of data that suggests neurochemical and genetic factors contribute significantly to the observed association between smoking and schizophrenia. Based on the findings described above, drugs that interact with nAChRs are potential agents for pharmacological management of schizophrenia. Agonists at the α7-nicotinic cholinergic receptor are being investigated for this purpose and may become useful clinical tools in the not too distant future (Olincy and Stevens, 2007).

33.2.7.3 Nicotinic acetylcholinergic receptors and depression

In earlier sections, the links between smoking, depression, and the HPA axis have been described. It is speculated that chronic exposure to nicotine elicits changes in the brain that are depressogenic and that smokers are protected from the consequences of these changes, while they continue to smoke, by the AD properties of nicotine (Balfour and Ridley, 2000; Salin-Pascual and Drucker-Colin, 1998). These findings suggest that the association between tobacco smoking and depression is complex and pernicious. While much uncertainty remains about bio-behavioral mechanisms that might link smoking and depression, nicotine is known to have important effects on central acetylcholine receptors and catecholamines (Pomerleau and Pomerleau, 1984), both of which have been shown to play a role in the etiology of depression. Nicotine is known to affect brain regions that influence mood and well-being (Pomerleau and Rosecrans, 1989). Data from a twin study suggest that a genetically influenced common physiologic substrate may be associated with predisposition for depressive illness and smoking (Kendler et al., 1993). It is plausible, therefore, that the same genetic variations in brain neurotransmitter systems that influence the vulnerability to depression also increase the probability of smoking by enhancing the degree to which nicotine provides reinforcement via normalization of depressed affect (Kendler et al., 1993).

The theory that cholinergic systems are involved in depression has existed for decades (Janowsky et al., 1972). Strong evidence supports the presence of exaggerated responses (including behavioral, sleep, and neuroendocrine) to cholinergic agents in patients with affective disorders, relative to controls (Janowsky et al., 1994). For example, physostigmine, an indirectly acting cholinergic agonist, increases heart rate and blood pressure and produces symptoms of dysphoria, irritability, anxiety, and depression when administered to normal volunteers. When physostigmine was administered to patients with depression, symptoms of negative affect were more pronounced (Janowsky and Risch, 1984). Reduced latency to rapid eye movement (REM) (Anda et al., 1990) sleep and cortisol hypersecretion are consistent findings associated with depression, and depressed patients show shortening of REM latency and cortisol secretion to a greater extent in response to the administration of cholinergic agonists (Janowsky and Overstreet, 1995). Cholinergic hypersensitivity may also be a marker of genetic predisposition to mood disorders, with unaffected relatives of depressed patients also demonstrating exaggerated behavioral, sleep, and neuroendocrine responses to cholinergic agonists (for a review, see Janowsky et al. (1994)).

In experimental studies, cessation of continuous nicotine infusion produces withdrawal signs, and these withdrawal signs can also be precipitated by mecamylamine, a noncompetitive nAChR antagonist, dihydro-β-erythroidine, the α4β2 nAChR antagonist, and methyllycaconitine, an α7 antagonist, in animals showing evidence of ND (Damaj et al., 2003; Malin et al., 1998). The Flinders Sensitive Line (FSL) rat has been used as an animal model of depression. These animals, selectively bred for their hyper-responsiveness to cholinergic stimulation, demonstrate depression-related behaviors and physiology. FSL rats showed evidence of increased neuronal nAChR expression, and the most prominent change measured was in α4β2 density (Tizabi et al., 2000). Interestingly, the increased nAChR binding observed following chronic nicotine administration was less marked in FSL animals than in control animals, suggesting a potential mechanism of ND in depression. A study in humans demonstrated higher frequency of the nonfunctional variant (characterized by 2-bp deletion in exon 6) of the partially duplicated α7-nAChR gene in patients with depression than in controls (Lai et al., 2001).

Putative associations among depression, tobacco smoking, and nAChR function are best understood through the mechanisms of AD drug actions (Shytle et al., 2002). Bupropion, an atypical AD agent, has been found to be effective for reducing tobacco smoking in broad groups of patients (Covey et al., 2000; Tonstad, 2002). Bupropion is a relatively weak reuptake inhibitor of DA and NE, with no direct action on serotonergic neurotransmission (Cooper et al., 1994). Experimental studies have shown that bupropion blocks the activation of α3β2-, α4β2-, and α7-nAChRs by nicotine with some degree of selectivity, with the highest potency at α3β2 sites (Fryer and Lukas, 1999b; Slemmer et al., 2000). The functional blockade of the nAChRs was noncompetitive. Given the relative selectivity of bupropion at inhibiting α3β2 receptors, which are implicated in nicotine-induced DA release from midbrain DA neurons (and thereby enhance its rewarding properties), the therapeutic efficacy of bupropion is partly explained by these findings. In addition to bupropion, nortryptyline has shown efficacy in achieving smoking cessation (Hughes et al., 2007). Nortryptyline also inhibits nAChRs at therapeutic concentrations (Fryer and Lukas, 1999a; Shytle et al., 2002). In contrast, clinical trials with selective serotonin-receptor reuptake inhibitors (SSRI) have not been very beneficial (Hughes et al., 2007). In summary, these data suggest that altered nAChR function may be involved in the pathophysiology of both depression and ND. Some AD drugs are helpful in reducing tobacco smoking in addition to alleviating depressive symptoms. One potential mechanism of action of these drugs is antagonism at nAChRs.

Sensory processing measures might serve as potential tools for assessing nAChR function in relation to depression and ND. As reviewed above, research in schizophrenia (another psychiatric disorder with high incidence of tobacco smoking and which also shows some links with mood disorders) has benefited significantly from these measures. Although the data are limited, an altered P50-evoked potential was demonstrated in adult patients with unipolar and bipolar depression, and linkage to the 15q13–14 locus was detected in bipolar disorder (Baker et al., 1987, 1990; Edenberg et al., 1997). Impaired performance in SPEM was also observed in adult patients with depression and in their unaffected relatives (Abel et al., 1991; Kathmann et al., 2003; Mahlberg et al., 2001), and susceptibility loci for the illness were mapped to 6p21 in addition to 15q13–14 (Ginns et al., 1996; Zubenko et al., 2002). Although the findings are limited, these data suggest that the high incidence of tobacco use in patients with mood disorders may have an underlying genetic diathesis. Neurophysiological studies (specifically P50 and SPEM measures) might provide a means for identifying the links between the different facets of clinical phenotypes and the genes involved in their expression. Such knowledge potentially will be helpful in developing more specific preventive and treatment interventions for the different subgroups of patients.

In summary, there appears to be an intricate and complex relationship among brain nicotinic systems, smoking, and neuropsychiatric disorders. However, the exact relationship remains obscure. The additional involvement of the HPA axis on these systems and behaviors further complicates the picture. Nonetheless, based on extant data, it would appear that subtype-selective nicotine agonists or antagonists, or even gene therapy, might someday be useful for the treatment or prevention of various neuropsychiatric disorders.

33.2.8 Smoking and Other Pituitary Hormones

Acute smoking also leads to increases in the plasma levels of GH and AVP (Seyler et al., 1986; Wilkins et al., 1982). Levels of thyroid-stimulating hormone (TSH), luteinizing hormone (LH), and follicle stimulating hormone (FSH) do not increase significantly (Winternitz and Quillen, 1977). Chronic smoking leads to inhibition of prolactin secretion. Increased DA release secondary to activation of nicotinic receptors on the tubero-infundibular DA neurons by nicotine has been proposed as a mechanism for this action (Fuxe et al., 1989).

33.3 Thyroid Hormone

Cigarette smoking has multiple effects on the thyroid gland, and its association with certain thyroid diseases is quite well established. For example, there is evidence to suggest that smoking is a risk factor for Graves' hyperthyroidism, and especially Graves' ophthalmopathy (Krassas and Wiersinga, 2006; Shine et al., 1990; Thornton et al., 2007). The more severe the eye disease, the stronger is the association. In sharp contrast, smoking has been found to be negatively associated with thyroid cancer (Bufalo et al., 2006; Guignard et al., 2007; Mack et al., 2003).

The effects of smoking on thyroid function are less clear. In normal adults, small increases in serum triiodothyronine (T3) and thyroglobulin concentrations may occur with smoking (Utiger, 1998). This effect could be due to sympathetic activation by nicotine. In a few studies, TSH levels were reported to be slightly lower in smokers (Asvold et al., 2007; Christensen et al., 1984; Ericsson and Lindgarde, 1991; Fisher et al., 1997), whereas other studies refute this finding (Gu et al., 2007; Karakaya et al., 1987). With respect to overt hypothyroidism, the literature is again inconclusive. Some studies have shown an increased incidence (Nystrom et al., 1993; Vestergaard et al., 2002), some found no association (Prummel and Wiersinga, 1993; Vestergaard, 2002), and one study found a decreased incidence (Asvold et al., 2007) of hypothyroidism in smokers. In addition, it has been proposed that smoking probably reduces thyroid secretion in patients with subclinical hypothyroidism and exacerbates the peripheral effects of thyroid deficiency in overt hypothyroidism (Utiger, 1998). A study in women with hypothyroidism showed that subjects diagnosed with subclinical hypothyroidism who were also smokers had higher serum TSH concentrations and a higher ratio of T3 to free thyroxine than nonsmokers. However, in the same study, in patients with overt hypothyroidism, smokers and nonsmokers had similar thyroid hormone concentrations, but smokers had more severe symptoms and signs of hypothyroidism (Muller et al., 1995). There is some evidence to suggest that smoking is associated with Hashimoto's thyroiditis, but its association with the resulting hypothyroidism is unclear (Vestergaard, 2002).

A common presentation of thyroid disorder is goiter. The prevalence of nontoxic goiter is higher in smokers than nonsmokers, and this has a greater bias toward women than men (Christensen et al., 1984; Ericsson and Lindgarde, 1991; Hegedus et al., 1985; Vestergaard, 2002). Tobacco smoke contains several toxins such as thiocyanate, and this compound has been shown to be a potential goitrogen (Fukayama et al., 1992). Thiocyanate, which has a half-life of more than 6 days, inhibits iodide transport and organification, as well as increases the efflux of iodide from the gland. In the presence of iodine deficiency, thiocyanate can cause goiter. After a meta-analysis showed that smoking was not associated with toxic nodular goiter in women (Vestergaard, 2002), a recent large cohort study with data on 800000 smokers supported the presence of this association (Galanti et al., 2005). There was generally no clear trend for risk of goiter with regard to amount of cigarette consumption. As diffuse goiter is often seen in patients with Graves' disease, an increase in sympathetic activity in smokers may promote the development of thyrotoxicosis in these predisposed individuals.

The weak stimulatory effects of smoking observed in normal adults have also been seen in infants of smoking parents (Meberg and Marstein, 1986). 2,3-hydroxypyridine, which is present in cigarette smoke, inhibits thyroxine deiodination by limiting iodothyronine deiodinase activity. This effect may temporarily and mildly elevate serum thyroxine levels as a result of its deiodinase-altering activity prior to decreasing the levels. Other investigators have found that infants of parents who smoke have higher cord concentrations of serum thyroglobulin and thiocyanate at birth and at 1 year of age than infants of nonsmoking parents. However, no differences were observed in thyroid hormone levels (Gasparoni et al., 1998). Smoking during pregnancy was also reported to cause neonatal thyroid enlargement (Chanoine et al., 1991).

Mental health has been found to be intricately linked to functioning of the thyroid gland. Thyroid hormone deficiency as well as excess have been known to precipitate and exacerbate mental illness, especially mood and anxiety disorders. In addition, the prevalence of smoking in the mentally ill population is exceptionally high (Dixon et al., 2007; Leonard et al., 2001). These findings suggest the possibility of a link among smoking, thyroid status, and mental health, but to the best of our knowledge, this has not been studied yet. Examining the thyroid status as a mediating variable between smoking and psychiatric disorders could result in a better understanding of this association. It has been reported that smoking during adolescence is associated with a higher incidence of mood and anxiety disorders later in life (Brook et al., 1998; Clark et al., 2007). It will be interesting to explore whether this is mediated by subtle alterations in the thyroid status.

33.4 Sex Hormones

Estrogens are mostly protein bound in the circulation. They are strongly bound to sex hormone-binding globulin (SHBG), loosely to albumin, and about 1–3% is the free, unbound fraction. Concentrations of SHBG are higher in smokers and this lowers the biologically active fraction of estrogens (Cassidenti et al., 1990; Daniel et al., 1992). High serum SHBG would also be expected to raise the total levels of estrogen, but several earlier studies indicated that estrogen levels were lower in women who smoked (MacMahon et al., 1982; Sterzik et al., 1996; Westhoff et al., 1996). Recent studies have not found this association (Gallicchio et al., 2006; Law et al., 1997). Smoking also influences the hepatic metabolism of estrogens. It stimulates the 2-hydroxylation pathway of estradiol metabolism which leads to increased production of 2-hydroxyestradiol (Michnovicz et al., 1986). These compounds have minimal potency compared to estrogens, and are rapidly cleared from the circulation. It seems likely that smoking has an overall antiestrogenic effect even though basal serum levels of total estrogens in smokers have been found to be normal in several studies (Spangler, 1999; Tansavatdi et al., 2004).

The antiestrogenic effects of smoking are supported by several other observations. Windham et al. (1999) found that women who smoke more than 20 cigarettes per day had shorter menstrual cycle length than nonsmokers. Women smokers, in general, have a higher chance of having irregular menstrual cycles and length of menstruation (Hornsby et al., 1998; Windham et al., 1999). A negative association between smoking and time to conception has been demonstrated (Hughes and Brennan, 1996). The age of menopause is reduced and menopausal symptoms, such as hot flashes, are more common among smokers (Celentano et al., 2003; Harlow and Signorello, 2000; Schindler, 2006). Smoking has been shown to have effects on ovarian function. Ingredients of cigarette smoke and nicotine produce a direct inhibition of granulosa cell aromatase activity (Byrne et al., 1991; Shiverick and Salafia, 1999). Alkaloid contents of cigarettes also inhibit progesterone synthesis, both by inhibiting progesterone synthesis and by causing less specific cytotoxic effects (Gocze and Freeman, 2000; Gocze et al., 1999). Young women smokers taking gonadotropins for infertility have higher mean basal serum FSH levels and require a higher mean dose of gonadotropins for ovarian stimulation compared to nonsmokers (Van Voorhis et al., 1996).

Certain diseases that are estrogen dependent are found to be less common among smokers. Young women who smoke have a lower chance of suffering from hyperemesis gravidarum, uterine fibroids, and endometriosis (Spangler, 1999). Endometrial cancer is estrogen dependent and a lower prevalence is seen among women who smoke (Baron, 1996; Reeves et al., 2007). On the other hand, even though breast tissue is estrogen responsive, breast cancer incidence is not lower in smokers. These data suggest that the carcinogens in cigarette smoke along with genetic susceptibility had a stronger effect than the antiestrogenic effects of smoking (Breast Cancer Family Registry, 2008; Tansavatdi et al., 2004).

The antiestrogenic property of smoking is an important consideration in women who are on oral contraceptives or hormone replacement therapy (HRT). Smokers are more likely to have spotting or bleeding compared to nonsmokers, and this might lead to discontinuation of contraceptive use and increased risk of unwanted pregnancy (Rosenberg et al., 1996). The antiestrogenic effect of smoking may also impair the efficacy of the oral contraceptive pill (Rosenberg et al., 1996). The efficacy of oral HRT is decreased in smokers (Mueck and Seeger, 2003), and this interferes with the beneficial effects of HRT. Higher doses might be needed when treating postmenopausal women with osteoporosis or other postmenopausal symptoms. Increasing the dose of oral estrogen is not recommended as it results in the production of toxic estrogen conjugates, such as catechol estrogens and 16α-hydroxyoestrone, which have been implicated in breast cancer (Tanko and Christiansen, 2004). The transdermal route bypasses the liver allowing a lower dosage of estrogen to be used, and this route should be considered in women who are unable to stop smoking (Mueck and Seeger, 2003).

Maternal smoking during pregnancy is known to be associated with adverse pregnancy and fetal outcomes, including low birth weight, intrauterine growth retardation, premature delivery, spontaneous abortion, placental abruption, placenta praevia, perinatal mortality, and ectopic pregnancy, especially in older mothers (Shiverick and Salafia, 1999). During early pregnancy, smoking is associated with significantly lower levels of estriol, estradiol, human chorionic gonadotrophin, and human placental lactogen in the mother, and there appears to be a decline in these values with increasing cigarette consumption (Bernstein et al., 1989; Shiverick and Salafia, 1999). In addition, placental microsomes of smokers have increased 2- and 4-hydroxylation of estradiol (Juchau et al., 1982). These two effects may explain certain adverse effects of smoking. It has also been proposed that the smoking-induced corpus luteal deficiency could underlie the increase in early pregnancy termination observed in smokers (Shiverick and Salafia, 1999).

In males, most of the early studies examining the association between smoking and serum testosterone levels were inconclusive. Most of the circulating total testosterone is inactive as it is tightly bound to SHBG

(65–80%), whereas the biologically active fraction circulates either free (1–3%) or loosely bound to albumin (20–40%). Similar to women smokers, high SHBG is also found in men who smoke (Field et al., 1994), and it was postulated that the effects of smoking on the sex hormones, particularly in males, are secondary to increase in the SHBG levels (English et al., 2001). However, Svartberg et al. (2003) found a positive association between testosterone and smoking even after adjusting for SHBG though other plasma proteins were not taken into account. A recent large study with more than 3000 participants found that total and free serum testosterone levels were increased in male smokers (Svartberg and Jorde, 2007). Estrogen has been traditionally thought to be cardioprotective while androgens have been considered to be atherogenic, and the finding of increased levels of androgens in smokers suggests that smoking increases cardiovascular risk in multiple ways.

Vine et al. (1994), in their meta-analysis, found that sperm count is reduced by 13–17% in male smokers, but even then fertility appears to be quite resistant to deleterious effects of tobacco smoke (Bonde and Storgaard, 2002). *In utero* exposure to constituents of tobacco smoke could damage the fetal gonads, but experimental evidence is lacking (Jensen et al., 1998).

33.5 Smoking and Insulin Resistance

Several prospective studies have shown that smoking increases the risk of diabetes in both men and women. The risk of developing diabetes among smokers has ranged from 1.4 in women (Rimm et al., 1993) to 1.7 in men (Manson et al., 2000), after controlling for other covariates. Smoking mostly causes an increased risk for type 2 diabetes because type 1 diabetes is relatively rare in the age groups studied (Eliasson, 2003). This is consistent with several lines of evidence which suggest that smoking may contribute to the development of insulin resistance rather than the destruction of pancreatic beta cells.

Attvall et al. (1993) demonstrated that, among healthy smokers, acute smoking impairs insulin action due to lower peripheral glucose uptake. In cross-sectional studies, researchers have demonstrated that the measures of insulin sensitivity were significantly lower (10–40%) in smokers compared to nonsmokers (Eliasson et al., 1997; Facchini et al., 1992). Tobacco users are hyperinsulinemic and relatively glucose intolerant when compared with nonsmokers (Frati et al., 1996). Even though smoking is associated with insulin resistance, a consistent effect on HbA1c in individuals with type 2 diabetes mellitus has not been reported (McCulloch et al., 2002). In contrast, in a recent, large cross-sectional analysis, smoking habits were also correlated with HbA1c levels in nondiabetic individuals after adjustments for confounding factors (Sargeant et al., 2001). An improvement in insulin sensitivity and increase in high-density lipoprotein cholesterol occurs after cessation of smoking (Eliasson et al., 1997). In type 1 diabetic subjects, insulin requirements have also been found to be increased in smokers (Madsbad et al., 1980).

The reduced insulin sensitivity seen in smokers could be due to the increase in counter-regulatory hormones such as GH, cortisol, and catecholamines, all of which raise blood glucose levels. Increased glucagon levels have also been shown after acute smoking in men with type 1 diabetes mellitus although substantial changes in insulin sensitivity were not observed in these patients despite the rise in counter-regulatory hormones (Helve et al., 1986). Other investigators have shown that smoking in patients with insulin-dependent diabetes not only elicits higher GH, AVP, and cortisol responses than in normal subjects but also enhances the counter-regulatory responses to insulin-induced hypoglycaemia (Chiodera et al., 1997). These effects probably play a role in the pathogenesis of diabetic complications because increased cortisol and AVP cause an increase in blood pressure, and thus their enhanced secretion in smokers might contribute to cardiovascular, cerebrovascular, and renal diseases. Sonksen et al. (1993) have also suggested that hypersecretion of GH could be linked to the development of diabetic microangiopathy.

From the above discussion, it can be reasonably concluded that cigarette smoking can complicate the course and management of diabetes and possibly even play a causative role. In diabetes care, smoking cessation is of utmost importance to improve glycemic control and prevent complications (Eliasson, 2003).

33.6 Smoking and Osteoporosis

A community-based longitudinal study of men and women, aged more than 60 at study enrollment, found that smoking was associated, independently of calcium intake or body mass, with 5–8% lower bone mass density (BMD) in both men and women. A Norwegian cohort study followed 34856 adults aged 50 or older for 3 years, and found an increased risk of hip fracture in both female (risk ratio (RR) 1.5, 95% confidence interval (CI) 1.0–2.4) and male smokers (RR 1.8, 95% CI 1.2–2.9) compared with nonsmokers (Forsen et al., 1994). In this study, the effect of smoking on fracture risk appeared largely independent of BMD. Other researches have also found bone mass to be lower among postmenopausal women (Baheiraei et al., 2005; MacInnis et al., 2003), but in premenopausal women the results have been inconsistent. Some studies have reported decreased BMD in the premenopausal women who smoke (Jones and Scott, 1999; Mazess and Barden, 1991), and others have not found this association (Bainbridge et al., 2004). It should be noted that no bone mass differences have been found between former and never-smokers (Gerdhem and Obrant, 2002).

Although there are fewer data regarding the effects of smoking on bone health in men, smoking appears to be a significant risk factor for bone loss (Slemenda et al., 1992). After adjusting for potentially confounding variables, current male smokers had a 7.3% reduction in lumbar spine BMD compared with nonsmokers (Egger et al., 1996). There is some evidence to suggest that the skeletal effects of smoking may even be more pronounced in men (Bakhireva et al., 2004; Izumotani et al., 2003). There is also an increased risk of bone fracture in smokers; this effect is strongest in women, with a twofold increased risk of fracture for current smokers compared with current nonsmokers (Burns et al., 2003).

Part of this detrimental effect of smoking on bone metabolism is mediated by an adverse influence on sex-steroid metabolism, and in particular by an estrogen-lowering effect (Tanko and Christiansen, 2004). Menopause is associated with low estrogen levels and smoking, with its antiestrogenic effect, can contribute to an increased risk of osteoporosis and fractures in this age group. In addition, greater incidence in the postmenopausal periods may be partly explained by a greater cumulative tobacco exposure in older smokers or a greater sensitivity to smoking-induced bone loss (MacInnis et al., 2003). As mentioned above, some studies have found an increased risk of fractures in smokers independent of its association with lower bone mass. The mechanisms of this association are not clear but it could be due to the effect of smoking on lowering body mass (Jensen, 1986). Smokers are generally leaner than the rest of the population (Jones and Scott, 1999), and this could be related to the appetite-suppressing effect of smoking (Rasmussen, 1998). Serum 25-hydroxyvitamin D levels are lower in smokers than in nonsmokers (Valimaki et al., 2003). The reasons for this association are unclear but it might be related to increased hepatic metabolism induced by smoking. Smoking also seems to dampen the bone-protective effects of nutritional calcium in postmenopausal women (Sirola et al., 2003). Finally, at least part of the negative influence of smoking on bone mass is explained by weight and physical activity differences between smokers and nonsmokers (Nguyen et al., 1994).

Thus, smoking appears to exert a negative effect on bone mass and this influence appears independent of other risk factors for fracture, such as age, sex, weight, and menopausal status. Mechanisms underlying smoking-induced bone loss are not well understood. Nonetheless, along with pharmacological therapy, encouragement of lifestyle alterations, including smoking cessation, should be a major component of any bone therapeutic program (Wong et al., 2007).

33.7 Summary

Smoking leads to various physiological and psychological consequences in humans. The endocrine changes caused by smoking play an important role in mediating these consequences. The interaction of smoking with some endocrine hormone systems such as the HPA axis is well researched and thus we know more about it than most of the other endocrine systems. The close association that changes in the HPA axis has with the various stages of nicotine addiction opens up the possibility of finding pharmacological agents which can help with management of this problem. Smoking-induced changes in the functioning of other hormone systems are an area of active research. At the minimum, it appears that all hormonal systems in humans have a basal rhythm which is interrupted to some degree by the chronic presence of nicotine in the body. Therefore, smoking cessation should be a component in the management plan for most of the diseases affecting the endocrine system.

Acknowledgments

This work was supported in part by NIH grants R01 DA14037; R01 DA15131; R01 DA17804; R01 DA17805, R01 MH62462, R01 MH68391 and by Sarah M. and Charles E. Seay Endowment to UT Southwestern Medical Center.

References

Abel LA, Friedman L, Jesberger J, Malki A, and Meltzer HY (1991) Quantitative assessment of smooth pursuit gain and catch-up saccades in schizophrenia and affective disorders. *Biological Psychiatry* 29: 1063–1072.

Adinoff B, Kiser JM, Martin PR, and Linnoila M (1996) Response of dehydroepiandrosterone to corticotropin-releasing hormone stimulation in alcohol-dependent subjects. *Biological Psychiatry* 40: 1305–1307.

Adler LE, Hoffer LD, Wiser A, and Freedman R (1993) Normalization of auditory physiology by cigarette smoking in schizophrenic patients. *American Journal of Psychiatry* 150: 1856–1861.

al'Absi M, Hatsukami D, and Davis GL (2005) Attenuated adrenocorticotropic responses to psychological stress are associated with early smoking relapse. *Psychopharmacology (Berl)* 181(1): 107–117.

al'Absi M, Hatsukami D, Davis GL, and Wittmers LE (2004) Prospective examination of effects of smoking abstinence on cortisol and withdrawal symptoms as predictors of early smoking relapse. *Drug and Alcohol Dependence* 73: 267–278.

al'Absi M, Wittmers LE, Erickson J, Hatsukami D, and Crouse B (2003) Attenuated adrenocortical and blood pressure responses to psychological stress in *ad libitum* and abstinent smokers. *Pharmacology, Biochemistry, and Behavior* 74: 401–410.

Amering M, Bankier B, Berger P, Griengl H, Windhaber J, and Katschnig H (1999) Panic disorder and cigarette smoking behavior. *Comprehensive Psychiatry* 40: 35–38.

Anda RF, Williamson DF, Escobedo LG, Mast EE, Giovino GA, and Remington PL (1990) Depression and the dynamics of smoking. A national perspective. *Journal of the American Medical Association* 264: 1541–1545.

Asvold BO, Bjoro T, Nilsen TI, and Vatten LJ (2007) Tobacco smoking and thyroid function: A population-based study. *Archives of Internal Medicine* 167: 1428–1432.

Attvall S, Fowelin J, Lager I, Von Schenck H, and Smith U (1993) Smoking induces insulin resistance – a potential link with the insulin resistance syndrome. *Journal of Internal Medicine* 233: 327–332.

Azam L and McIntosh JM (2006) Characterization of nicotinic acetylcholine receptors that modulate nicotine-evoked [^{3}H] norepinephrine release from mouse hippocampal synaptosomes. *Molecular Pharmacology* 70: 967–976.

Baheiraei A, Pocock NA, Eisman JA, Nguyen ND, and Nguyen TV (2005) Bone mineral density, body mass index and cigarette smoking among Iranian women: Implications for prevention. *BMC Musculoskeletal Disorders* 6: 34.

Bailly D, Dewailly D, Beuscart R, et al. (1989) Adrenocorticotropin and cortisol responses to ovine corticotropin-releasing factor in alcohol dependence disorder. Preliminary report. *Hormone Research* 31: 72–75.

Bainbridge KE, Sowers M, Lin X, and Harlow SD (2004) Risk factors for low bone mineral density and the 6-year rate of bone loss among premenopausal and perimenopausal women. *Osteoporosis International* 15: 439–446.

Baker N, Adler LE, Franks RD, et al. (1987) Neurophysiological assessment of sensory gating in psychiatric inpatients: Comparison between schizophrenia and other diagnoses. *Biological Psychiatry* 22: 603–617.

Baker NJ, Staunton M, Adler LE, et al. (1990) Sensory gating deficits in psychiatric inpatients: Relation to catecholamine metabolites in different diagnostic groups. *Biological Psychiatry* 27: 519–528.

Bakhireva LN, Barrett-Connor E, Kritz-Silverstein D, and Morton DJ (2004) Modifiable predictors of bone loss in older men: A prospective study. *American Journal of Preventive Medicine* 26: 436–442.

Balfour DJ and Ridley DL (2000) The effects of nicotine on neural pathways implicated in depression: A factor in nicotine addiction? *Pharmacology, Biochemistry, and Behavior* 66: 79–85.

Baron JA (1996) Beneficial effects of nicotine and cigarette smoking: The real, the possible and the spurious. *British Medical Bulletin* 52: 58–73.

Baron JA, Comi RJ, Cryns V, Brinck-Johnsen T, and Mercer NG (1995) The effect of cigarette smoking on adrenal cortical hormones. *Journal of Pharmacology and Experimental Therapeutics* 272: 151–155.

Baulieu EE and Robel P (1998) Dehydroepiandrosterone (DHEA) and dehydroepiandrosterone sulfate (DHEAS) as neuroactive neurosteroids. *Proceedings of the National Academy of Sciences of the United States of America* 95: 4089–4091.

Benowitz NL, Hansson A, and Jacob P, 3rd (2002) Cardiovascular effects of nasal and transdermal nicotine and cigarette smoking. *Hypertension* 39: 1107–1112.

Benowitz NL, Kuyt F, and Jacob P, 3rd (1984) Influence of nicotine on cardiovascular and hormonal effects of cigarette smoking. *Clinical Pharmacology and Therapeutics* 36: 74–81.

Bernstein L, Pike MC, Lobo RA, Depue RH, Ross RK, and Henderson BE (1989) Cigarette smoking in pregnancy results in marked decrease in maternal hCG and oestradiol levels. *British Journal of Obstetrics and Gynaecology* 96: 92–96.

Bjornerem A, Straume B, Midtby M, et al. (2004) Endogenous sex hormones in relation to age, sex, lifestyle factors, and chronic diseases in a general population: The Tromso Study. *Journal of Clinical Endocrinology and Metabolism* 89: 6039–6047.

Board F, Persky H, and Hamburg DA (1956) Psychological stress and endocrine functions; Blood levels of adrenocortical and thyroid hormones in acutely disturbed patients. *Psychosomatic Medicine* 18: 324–333.

Bonde JP and Storgaard L (2002) How work-place conditions, environmental toxicants and lifestyle affect male reproductive function. *International Journal of Andrology* 25: 262–268.

Booker TK, Butt CM, Wehner JM, Heinemann SF, and Collins AC (2007) Decreased anxiety-like behavior in beta3 nicotinic receptor subunit knockout mice. *Pharmacology, Biochemistry, and Behavior* 87: 146–157.

Boudarene M, Legros JJ, and Timsit-Berthier M (2002) Study of the stress response: Role of anxiety, cortisol and DHEAs. *L'Encephale* 28: 139–146.

Boyadjieva NI and Sarkar DK (1997) The secretory response of hypothalamic beta-endorphin neurons to acute and chronic nicotine treatments and following nicotine withdrawal. *Life Sciences* 61: PL59–PL66.

Breast Cancer Family Registry (2008) Smoking and risk of breast cancer in carriers of mutations in BRCA1 or BRCA2 aged less than 50 years. *Breast Cancer Research and Treatment* 109: 67–75.

Breslau N, Schultz LR, Johnson EO, Peterson EL, and Davis GC (2005) Smoking and the risk of suicidal behavior: A prospective study of a community sample. *Archives of General Psychiatry* 62: 328–334.

Brioni JD, O'Neill AB, Kim DJ, and Decker MW (1993) Nicotinic receptor agonists exhibit anxiolytic-like effects on the elevated plus-maze test. *European Journal of Pharmacology* 238: 1–8.

Broadbear JH, Winger G, and Woods JH (2004) Self-administration of fentanyl, cocaine and ketamine: Effects on the pituitary–adrenal axis in rhesus monkeys. *Psychopharmacology* 176: 398–406.

Brook JS, Cohen P, and Brook DW (1998) Longitudinal study of co-occurring psychiatric disorders and substance use. *Journal of the American Academy of Child and Adolescent Psychiatry* 37: 322–330.

Brown RA, Kahler CW, Zvolensky MJ, Lejuez CW, and Ramsey SE (2001) Anxiety sensitivity: Relationship to negative affects smoking and smoking cessation in smokers with past major depressive disorder. *Addictive Behaviors* 26: 887–899.

Bufalo NE, Leite JL, Guilhen AC, Morari EC, Granja F, Assumpcao LV, and Ward LS (2006) Smoking and susceptibility to thyroid cancer: An inverse association with CYP1A1 allelic variants. *Endocrine-Related Cancer* 13: 1185–1193.

Buisson B and Bertrand D (2001) Chronic exposure to nicotine upregulates the human (alpha)4(beta)2 nicotinic acetylcholine receptor function. *Journal of Neuroscience* 21: 1819–1829.

Burns L, Ashwell M, Berry J, et al. (2003) UK Food Standards Agency Optimal Nutrition Status Workshop: Environmental factors that affect bone health throughout life. *British Journal of Nutrition* 89: 835–840.

Byrne B, Cunningham S, Igoe D, Conroy R, and McKenna TJ (1991) Sex steroids, adiposity and smoking in the pathogenesis of idiopathic hirsutism and polycystic ovary syndrome. *Acta Endocrinologica* 124: 370–374.

Byrne DG and Mazanov J (2003) Adolescent stress and future smoking behaviour: A prospective investigation. *Journal of Psychosomatic Research* 54: 313–321.

Caggiula AR, Donny EC, Epstein LH, et al. (1998) The role of corticosteroids in nicotine's physiological and behavioral effects. *Psychoneuroendocrinology* 23: 143–159.

Cam GR, Bassett JR, and Cairncross KD (1979) The action of nicotine on the pituitary–adrenal cortical axis. *Archives Internationales de Pharmacodynamie et de Therapie* 237: 49–66.

Cao W, Burkholder T, Wilkins L, and Collins AC (1993) A genetic comparison of behavioral actions of ethanol and nicotine in the mirrored chamber. *Pharmacology, Biochemistry, and Behavior* 45: 803–809.

Cardounel A, Regelson W, and Kalimi M (1999) Dehydroepiandrosterone protects hippocampal neurons against neurotoxin-induced cell death: Mechanism of action. *Proceedings of the Society for Experimental Biology and Medicine Society for Experimental Biology and Medicine (New York)* 222: 145–149.

Cassidenti DL, Vijod AG, Vijod MA, Stanczyk FZ, and Lobo RA (1990) Short-term effects of smoking on the pharmacokinetic profiles of micronized estradiol in postmenopausal women. *American Journal of Obstetrics and Gynecology* 163: 1953–1960.

CDC (2005) Annual smoking-attributable mortality, years of potential life lost, and productivity losses – United States, 1997–2001. *MMWR* 54: 625–628.

Celentano E, Galasso R, Berrino F, et al. (2003) Correlates of age at natural menopause in the cohorts of EPIC-Italy. *Tumori* 89: 608–614.

Chanoine JP, Toppet V, Bourdoux P, Spehl M, and Delange F (1991) Smoking during pregnancy: A significant cause of neonatal thyroid enlargement. *British Journal of Obstetrics and Gynaecology* 98: 65–68.

Cheeta S, Irvine EE, Kenny PJ, and File SE (2001) The dorsal raphe nucleus is a crucial structure mediating nicotine's anxiolytic effects and the development of tolerance and withdrawal responses. *Psychopharmacology* 155: 78–85.

Chiodera P, Volpi R, Capretti L, et al. (1997) Abnormal effect of cigarette smoking on pituitary hormone secretions in insulin-dependent diabetes mellitus. *Clinical Endocrinology* 46: 351–357.

Christensen SB, Ericsson UB, Janzon L, Tibblin S, and Melander A (1984) Influence of cigarette smoking on goiter formation, thyroglobulin, and thyroid hormone levels in women. *Journal of Clinical Endocrinology and Metabolism* 58: 615–618.

Clark C, Haines MM, Head J, et al. (2007) Psychological symptoms and physical health and health behaviours in adolescents: A prospective 2-year study in East London. *Addiction (Abingdon)* 102: 126–135.

Clementz BA, Geyer MA, and Braff DL (1998) Poor P50 suppression among schizophrenia patients and their first-degree biological relatives. *American Journal of Psychiatry* 155: 1691–1694.

Cohen S and Lichtenstein E (1990) Perceived stress, quitting smoking, and smoking relapse. *Health Psychology* 9: 466–478.

Cooper BR, Wang CM, Cox RF, Norton R, Shea V, and Ferris RM (1994) Evidence that the acute behavioral and electrophysiological effects of bupropion (Wellbutrin) are mediated by a noradrenergic mechanism. *Neuropsychopharmacology* 11: 133–141.

Corpechot C, Robel P, Axelson M, Sjovall J, and Baulieu EE (1981) Characterization and measurement of dehydroepiandrosterone sulfate in rat brain. *Proceedings of the National Academy of Sciences of the United States of America* 78: 4704–4707.

Court J, Spurden D, Lloyd S, et al. (1999) Neuronal nicotinic receptors in dementia with Lewy bodies and schizophrenia: Alpha-bungarotoxin and nicotine binding in the thalamus. *Journal of Neurochemistry* 73: 1590–1597.

Covey LS, Sullivan MA, Johnston JA, Glassman AH, Robinson MD, and Adams DP (2000) Advances in non-nicotine pharmacotherapy for smoking cessation. *Drugs* 59: 17–31.

Cui C, Booker TK, Allen RS, et al. (2003) The beta3 nicotinic receptor subunit: A component of alpha-conotoxin MII-binding nicotinic acetylcholine receptors that modulate dopamine release and related behaviors. *Journal of Neuroscience* 23: 11045–11053.

Curtis GC, Abelson JL, and Gold PW (1997) Adrenocorticotropic hormone and cortisol responses to corticotropin-releasing hormone: Changes in panic disorder and effects of alprazolam treatment. *Biological Psychiatry* 41: 76–85.

Curtis L, Buisson B, Bertrand S, and Bertrand D (2002) Potentiation of human alpha4beta2 neuronal nicotinic acetylcholine receptor by estradiol. *Molecular Pharmacology* 61: 127–135.

Czeisler CA, Ede MC, Regestein QR, Kisch ES, Fang VS, and Ehrlich EN (1976) Episodic 24-hour cortisol secretory patterns in patients awaiting elective cardiac surgery. *Journal of Clinical Endocrinology and Metabolism* 42: 273–283.

Dalack GW, Becks L, Hill E, Pomerleau OF, and Meador-Woodruff JH (1999) Nicotine withdrawal and psychiatric symptoms in cigarette smokers with schizophrenia. *Neuropsychopharmacology* 21: 195–202.

Damaj MI, Kao W, and Martin BR (2003) Characterization of spontaneous and precipitated nicotine withdrawal in the mouse. *Journal of Pharmacology and Experimental Therapeutics* 307: 526–534.

Dani JA and Bertrand D (2007) Nicotinic acetylcholine receptors and nicotinic cholinergic mechanisms of the central nervous system. *Annual Review of Pharmacology and Toxicology* 47: 699–729.

Daniel M, Martin AD, and Drinkwater DT (1992) Cigarette smoking, steroid hormones, and bone mineral density in young women. *Calcified Tissue International* 50: 300–305.

Davenport KE, Houdi AA, and Van Loon GR (1990) Nicotine protects against mu-opioid receptor antagonism by beta-funaltrexamine: Evidence for nicotine-induced release of endogenous opioids in brain. *Neuroscience Letters* 113: 40–46.

del Arbol JL, Munoz JR, Ojeda L, et al. (2000) Plasma concentrations of beta-endorphin in smokers who consume different numbers of cigarettes per day. *Pharmacology, Biochemistry, and Behavior* 67: 25–28.

Dixon L, Medoff DR, Wohlheiter K, et al. (2007) Correlates of severity of smoking among persons with severe mental illness. *American Journal on Addictions* 16: 101–110.

Djuric VJ, Dunn E, Overstreet DH, Dragomir A, and Steiner M (1999) Antidepressant effect of ingested nicotine in female rats of Flinders resistant and sensitive lines. *Physiology and Behavior* 67: 533–537.

Dozois DN, Farrow JA, and Miser A (1995) Smoking patterns and cessation motivations during adolescence. *International Journal of the Addictions* 30: 1485–1498.

Edenberg HJ, Foroud T, Conneally PM, et al. (1997) Initial genomic scan of the NIMH genetics initiative bipolar pedigrees: Chromosomes 3, 5, 15, 16, 17, and 22. *American Journal of Medical Genetics* 74: 238–246.

Egger P, Duggleby S, Hobbs R, Fall C, and Cooper C (1996) Cigarette smoking and bone mineral density in the elderly. *Journal of Epidemiology and Community Health* 50: 47–50.

Eliasson B (2003) Cigarette smoking and diabetes. *Progress in Cardiovascular Diseases* 45: 405–413.

Eliasson B, Attvall S, Taskinen MR, and Smith U (1997) Smoking cessation improves insulin sensitivity in healthy middle-aged men. *European Journal of Clinical Investigation* 27: 450–456.

Eliava M, Yilmazer-Hanke D, and Asan E (2003) Interrelations between monoaminergic afferents and corticotropin-releasing factor-immunoreactive neurons in the rat central amygdaloid nucleus: Ultrastructural evidence for dopaminergic control of amygdaloid stress systems. *Histochemistry and Cell Biology* 120: 183–197.

English KM, Pugh PJ, Parry H, Scutt NE, Channer KS, and Jones TH (2001) Effect of cigarette smoking on levels of bioavailable testosterone in healthy men. *Clinical Science (Lond)* 100: 661–665.

Ericsson UB and Lindgarde F (1991) Effects of cigarette smoking on thyroid function and the prevalence of goitre, thyrotoxicosis and autoimmune thyroiditis. *Journal of Internal Medicine* 229: 67–71.

Facchini FS, Hollenbeck CB, Jeppesen J, Chen YD, and Reaven GM (1992) Insulin resistance and cigarette smoking. *Lancet* 339: 1128–1130.

Fagerstrom KO, Heatherton TF, and Kozlowski LT (1990) Nicotine addiction and its assessment. *Ear, Nose, and Throat Journal* 69: 763–765.

Fallon JH, Koziell DA, and Moore RY (1978) Catecholamine innervation of the basal forebrain II: Amygdala, suprarhinal cortex, and entorhinal cortex. *Journal of Comparative Neurology* 180(3): 509–532.

Feldman HA, Johannes CB, McKinlay JB, and Longcope C (1998) Low dehydroepiandrosterone sulfate and heart disease in middle-aged men: Cross-sectional results from the Massachusetts male aging study. *Annals of Epidemiology* 8: 217–228.

Field AE, Colditz GA, Willett WC, Longcope C, and McKinlay JB (1994) The relation of smoking, age, relative weight, and dietary intake to serum adrenal steroids, sex hormones, and sex hormone-binding globulin in middle-aged men. *Journal of Clinical Endocrinology and Metabolism* 79: 1310–1316.

Fisher CL, Mannino DM, Herman WH, and Frumkin H (1997) Cigarette smoking and thyroid hormone levels in males. *International Journal of Epidemiology* 26: 972–977.

Forsen L, Bjorndal A, Bjartveit K, Edna TH, Holmen J, Jessen V, and Westburg G (1994) Interaction between current smoking, leanness, and physical inactivity in the prediction of hip fracture. *Journal of Bone and Mineral Research* 9: 1671–1678.

Frati AC, Iniestra F, and Ariza CR (1996) Acute effect of cigarette smoking on glucose tolerance and other cardiovascular risk factors. *Diabetes Care* 19: 112–118.

Frederick SL, Reus VI, Ginsberg D, Hall SM, Munoz RF, and Ellman G (1998) Cortisol and response to dexamethasone as predictors of withdrawal distress and abstinence success in smokers. *Biological Psychiatry* 43: 525–530.

Freedman R, Adler LE, Bickford P, et al. (1994) Schizophrenia and nicotinic receptors. *Harvard Review of Psychiatry* 2: 179–192.

Freedman R, Coon H, Myles-Worsley M, et al. (1997) Linkage of a neurophysiological deficit in schizophrenia to a chromosome 15 locus. *Proceedings of the National Academy of Sciences of the United States of America* 94: 587–592.

Freedman R, Hall M, Adler LE, and Leonard S (1995) Evidence in postmortem brain tissue for decreased numbers of hippocampal nicotinic receptors in schizophrenia. *Biological Psychiatry* 38: 22–33.

Fryer JD and Lukas RJ (1999a) Antidepressants noncompetitively inhibit nicotinic acetylcholine receptor function. *Journal of Neurochemistry* 72: 1117–1124.

Fryer JD and Lukas RJ (1999b) Noncompetitive functional inhibition at diverse, human nicotinic acetylcholine receptor subtypes by bupropion, phencyclidine, and ibogaine. *Journal of Pharmacology and Experimental Therapeutics* 288: 88–92.

Fukayama H, Nasu M, Murakami S, and Sugawara M (1992) Examination of antithyroid effects of smoking products in cultured thyroid follicles: Only thiocyanate is a potent antithyroid agent. *Acta Endocrinologica* 127: 520–525.

Fuxe K, Andersson K, Eneroth P, Harfstrand A, and Agnati LF (1989) Neuroendocrine actions of nicotine and of exposure to cigarette smoke: Medical implications. *Psychoneuroendocrinology* 14: 19–41.

Galanti MR, Granath F, Cnattingius S, Ekbom-Schnell A, and Ekbom A (2005) Cigarette smoking and the risk of goitre and thyroid nodules amongst parous women. *Journal of Internal Medicine* 258: 257–264.

Gallicchio L, Miller SR, Visvanathan K, Lewis LM, Babus J, Zacur H, and Flaws J (2006) Cigarette smoking, estrogen levels, and hot flashes in midlife women. *Maturitas* 53: 133–143.

Gasparoni A, Autelli M, Ravagni-Probizer MF, Bartoli A, Regazzi-Bonora M, Chirico G, and Rondini G (1998) Effect of passive smoking on thyroid function in infants. *European Journal of Endocrinology* 138: 379–382.

Gerdhem P and Obrant KJ (2002) Effects of cigarette-smoking on bone mass as assessed by dual-energy X-ray absorptiometry and ultrasound. *Osteoporosis International* 13: 932–936.

Giannakoulas G, Katramados A, Melas N, Diamantopoulos I, and Chimonas E (2003) Acute effects of nicotine withdrawal syndrome in pilots during flight. *Aviation, Space, and Environmental Medicine* 74: 247–251.

Gilbert DG, Meliska CJ, Williams CL, and Jensen RA (1992) Subjective correlates of cigarette-smoking-induced elevations of peripheral beta-endorphin and cortisol. *Psychopharmacology* 106: 275–281.

Gilbert DG, Robinson JH, Chamberlin CL, and Spielberger CD (1989) Effects of smoking/nicotine on anxiety, heart rate, and lateralization of EEG during a stressful movie. *Psychophysiology* 26: 311–320.

Ginns EI, Ott J, Egeland JA, et al. (1996) A genome-wide search for chromosomal loci linked to bipolar affective disorder in the Old Order Amish. *Nature Genetics* 12: 431–435.
Glassman AH (1993) Cigarette smoking: Implications for psychiatric illness. *American Journal of Psychiatry* 150: 546–553.
Glassman AH, Covey LS, Dalack GW, Stetner F, Rivelli SK, Fleiss J, and Cooper TB (1993) Smoking cessation, clonidine, and vulnerability to nicotine among dependent smokers. *Clinical Pharmacology and Therapeutics* 54: 670–679.
Glassman AH, Covey LS, Stetner F, and Rivelli S (2001) Smoking cessation and the course of major depression: A follow-up study. *Lancet* 357: 1929–1932.
Gocze PM and Freeman DA (2000) Cytotoxic effects of cigarette smoke alkaloids inhibit the progesterone production and cell growth of cultured MA-10 Leydig tumor cells. *European Journal of Obstetrics, Gynecology, and Reproductive Biology* 93: 77–83.
Gocze PM, Szabo I, and Freeman DA (1999) Influence of nicotine, cotinine, anabasine and cigarette smoke extract on human granulosa cell progesterone and estradiol synthesis. *Gynecological Endocrinology* 13: 266–272.
Goeders NE (2002) Stress and cocaine addiction. *Journal of Pharmacology and Experimental Therapeutics* 301: 785–789.
Goeders NE and Clampitt DM (2002) Potential role for the hypothalamo-pituitary–adrenal axis in the conditioned reinforcer-induced reinstatement of extinguished cocaine seeking in rats. *Psychopharmacology* 161: 222–232.
Goeders NE and Guerin GF (1996a) Effects of surgical and pharmacological adrenalectomy on the initiation and maintenance of intravenous cocaine self-administration in rats. *Brain Research* 722: 145–152.
Goeders NE and Guerin GF (1996b) Role of corticosterone in intravenous cocaine self-administration in rats. *Neuroendocrinology* 64: 337–348.
Goeders NE, Peltier RL, and Guerin GF (1998) Ketoconazole reduces low dose cocaine self-administration in rats. *Drug and Alcohol Dependence* 53: 67–77.
Goenjian AK, Pynoos RS, Steinberg AM, Endres D, Abraham K, Geffner ME, and Fairbanks (2003) Hypothalamic–pituitary–adrenal activity among Armenian adolescents with PTSD symptoms. *Journal of Traumatic Stress* 16: 319–323.
Goff DC, Henderson DC, and Amico E (1992) Cigarette smoking in schizophrenia: Relationship to psychopathology and medication side effects. *American Journal of Psychiatry* 149: 1189–1194.
Gorelick DA, Rose J, and Jarvik ME (1988) Effect of naloxone on cigarette smoking. *Journal of Substance Abuse* 1: 153–159.
Gossain VV, Sherma NK, Srivastava L, Michelakis AM, and Rovner DR (1986) Hormonal effects of smoking. II: Effects on plasma cortisol, growth hormone, and prolactin. *American Journal of the Medical Sciences* 291: 325–327.
Greaves-Lord K, Ferdinand RF, Oldehinkel AJ, Sondeijker FE, Ormel J, and Verhulst FC (2007) Higher cortisol awakening response in young adolescents with persistent anxiety problems. *Acta Psychiatrica Scandinavica* 116: 137–144.
Grun EU, Pauly JR, Bullock AE, and Collins AC (1995) Corticosterone reversibly alters brain alpha-bungarotoxin binding and nicotine sensitivity. *Pharmacology, Biochemistry, and Behavior* 52: 629–635.
Gu XL, Mao JY, Shan ZY, et al. (2007) Epidemiological study of the effects of smoking cigarette on thyroid gland. *Zhonghua Liu Xing Bing Xue Za Zhi = Zhonghua Liuxingbingxue Zazhi* 28: 53–56.
Guignard R, Truong T, Rougier Y, Baron-Dubourdieu D, and Guenel P (2007) Alcohol drinking, tobacco smoking, and anthropometric characteristics as risk factors for thyroid cancer: A countrywide case-control study in New Caledonia. *American Journal of Epidemiology* 166: 1140–1149.
Handa K, Kono S, Ishii H, Shinchi K, Imanishi K, and Arakawa K (1994) Relationship of alcohol consumption and smoking to plasma cortisol and blood pressure. *Journal of Human Hypertension* 8: 891–894.
Harlow BL and Signorello LB (2000) Factors associated with early menopause. *Maturitas* 35: 3–9.
Harrison PJ (2002) The neuropathology of primary mood disorder. *Brain* 125: 1428–1449.
Hautanen A, Manttari M, Kupari M, Sarna S, Manninen V, Frick MH, and Adlercreutz H (1993) Cigarette smoking is associated with elevated adrenal androgen response to adrenocorticotropin. *Journal of Steroid Biochemistry and Molecular Biology* 46: 245–251.
Hegedus L, Karstrup S, Veiergang D, Jacobsen B, Skovsted L, and Feldt-Rasmussen U (1985) High frequency of goitre in cigarette smokers. *Clinical Endocrinology* 22: 287–292.
Helve E, Yki-Jarvinen H, and Koivisto VA (1986) Smoking and insulin sensitivity in type I diabetic patients. *Metabolism: Clinical and Experimental* 35: 874–877.
Herman JP, Figueiredo H, Mueller NK, Ulrich-Lai Y, Ostrander MM, Choi DC, and Cullinan WE (2003) Central mechanisms of stress integration: Hierarchical circuitry controlling hypothalamo-pituitary–adrenocortical responsiveness. *Frontiers in Neuroendocrinology* 24: 151–180.
Hirshman E, Wells E, Wierman ME, Anderson B, Butler A, Senholzi M, and Fisher J (2003) The effect of dehydroepiandrosterone (DHEA) on recognition memory decision processes and discrimination in postmenopausal women. *Psychonomic Bulletin and Review* 10: 125–134.
Hornsby PP, Wilcox AJ, and Weinberg CR (1998) Cigarette smoking and disturbance of menstrual function. *Epidemiology (Cambridge, MA)* 9: 193–198.
Hughes EG and Brennan BG (1996) Does cigarette smoking impair natural or assisted fecundity? *Fertility and Sterility* 66: 679–689.
Hughes JR, Arana G, Amori G, Stewart F, and Workman R (1988) Effect of tobacco withdrawal on the dexamethasone suppression test. *Biological Psychiatry* 23: 96–98.
Hughes JR, Hatsukami DK, Mitchell JE, and Dahlgren LA (1986) Prevalence of smoking among psychiatric outpatients. *American Journal of Psychiatry* 143: 993–997.
Hughes JR, Stead LF, and Lancaster T (2007) Antidepressants for smoking cessation. *Cochrane Database of Systematic Reviews* 3: CD000031 (doi: 10.1002/14651858).
Iancu I, Tchernihovsky E, Maayan R, et al. (2007) Circulatory neurosteroid levels in smoking and non-smoking chronic schizophrenia patients. *European Neuropsychopharmacology* 17: 541–545.
Irvine EE, Cheeta S, Marshall M, and File SE (2001) Different treatment regimens and the development of tolerance to nicotine's anxiogenic effects. *Pharmacology, Biochemistry, and Behavior* 68: 769–776.
Izumotani K, Hagiwara S, Izumotani T, Miki T, Morii H, and Nishizawa Y (2003) Risk factors for osteoporosis in men. *Journal of Bone and Mineral Metabolism* 21: 86–90.
Jakovljevic M, Muck-Seler D, Pivac N, and Crncevic Z (1998) Platelet 5-HT and plasma cortisol concentrations after dexamethasone suppression test in patients with different time course of schizophrenia. *Neuropsychobiology* 37: 142–145.
Janowsky D and Overstreet D (1995) The role of acetylcholine mechanisms in mood disorders. In: Bloom F and Kupfer D (eds.) *Psychopharmacology: The Fourth Generation of Progress*, pp. 945–956. New York: Raven Press.

Janowsky DS, el-Yousef MK, Davis JM, and Sekerke HJ (1972) A cholinergic–adrenergic hypothesis of mania and depression. *Lancet* 2: 632–635.
Janowsky DS, Overstreet DH, and Nurnberger JI, Jr. (1994) Is cholinergic sensitivity a genetic marker for the affective disorders? *American Journal of Medical Genetics* 54: 335–344.
Janowsky DS and Risch SC (1984) Cholinomimetic and anticholinergic drugs used to investigate an acetylcholine hypothesis of affective disorders and stress. *Drug Development Research* 4: 125–142.
Jansen LM, Gispen-de Wied CC, Gademan PJ, De Jonge RC, van der Linden JA, and Kahn RS (1998) Blunted cortisol response to a psychosocial stressor in schizophrenia. *Schizophrenia Research* 33: 87–94.
Jansen LM, Gispen-de Wied CC, and Kahn RS (2000) Selective impairments in the stress response in schizophrenic patients. *Psychopharmacology* 149: 319–325.
Jensen GF (1986) Osteoporosis of the slender smoker revisited by epidemiologic approach. *European Journal of Clinical Investigation* 16: 239–242.
Jensen TK, Henriksen TB, Hjollund NH, et al. (1998) Adult and prenatal exposures to tobacco smoke as risk indicators of fertility among 430 Danish couples. *American Journal of Epidemiology* 148: 992–997.
Jones G and Scott FS (1999) A cross-sectional study of smoking and bone mineral density in premenopausal parous women: Effect of body mass index, breastfeeding, and sports participation. *Journal of Bone and Mineral Research* 14: 1628–1633.
Jorenby DE, Hatsukami DK, Smith SS, Fiore MC, Allen S, Jensen J, and Timorthy BB (1996) Characterization of tobacco withdrawal symptoms: Transdermal nicotine reduces hunger and weight gain. *Psychopharmacology* 128: 130–138.
Juchau MR, Namkung J, and Chao ST (1982) Mono-oxygenase induction in the human placenta. Interrelationships among position-specific hydroxylations of 17 beta-estradiol and benzo[a]pyrene. *Drug Metabolism and Disposition: The Biological Fate of Chemicals* 10: 220–224.
Kagan J, Reznick JS, and Snidman N (1987) The physiology and psychology of behavioral inhibition in children. *Child Development* 58: 1459–1473.
Karakaya A, Tuncel N, Alptuna G, Kocer Z, and Erbay G (1987) Influence of cigarette smoking on thyroid hormone levels. *Human Toxicology* 6: 507–509.
Kathmann N, Hochrein A, Uwer R, and Bondy B (2003) Deficits in gain of smooth pursuit eye movements in schizophrenia and affective disorder patients and their unaffected relatives. *American Journal of Psychiatry* 160: 696–702.
Kendler KS, Neale MC, MacLean CJ, Heath AC, Eaves LJ, and Kessler RC (1993) Smoking and major depression. A causal analysis. *Archives of General Psychiatry* 50: 36–43.
Kimonides VG, Khatibi NH, Svendsen CN, Sofroniew MV, and Herbert J (1998) Dehydroepiandrosterone (DHEA) and DHEA-sulfate (DHEAS) protect hippocampal neurons against excitatory amino acid-induced neurotoxicity. *Proceedings of the National Academy of Sciences of the United States of America* 95: 1852–1857.
Kimonides VG, Spillantini MG, Sofroniew MV, Fawcett JW, and Herbert J (1999) Dehydroepiandrosterone antagonizes the neurotoxic effects of corticosterone and translocation of stress-activated protein kinase 3 in hippocampal primary cultures. *Neuroscience* 89: 429–436.
Kirschbaum C, Scherer G, and Strasburger CJ (1994) Pituitary and adrenal hormone responses to pharmacological, physical, and psychological stimulation in habitual smokers and nonsmokers. *Clinical Investigator* 72: 804–810.
Kirschbaum C, Strasburger CJ, and Langkrar J (1993) Attenuated cortisol response to psychological stress but not to CRH or ergometry in young habitual smokers. *Pharmacology, Biochemistry, and Behavior* 44: 527–531.
Kirschbaum C, Wust S, and Strasburger CJ (1992) 'Normal' cigarette smoking increases free cortisol in habitual smokers. *Life Sciences* 50: 435–442.
Koob G and Kreek MJ (2007) Stress, dysregulation of drug reward pathways, and the transition to drug dependence. *American Journal of Psychiatry* 164: 1149–1159.
Koval JJ, Pederson LL, Mills CA, McGrady GA, and Carvajal SC (2000) Models of the relationship of stress, depression, and other psychosocial factors to smoking behavior: A comparison of a cohort of students in grades 6 and 8. *Preventive Medicine* 30: 463–477.
Krassas GE and Wiersinga W (2006) Smoking and autoimmune thyroid disease: The plot thickens. *European Journal of Endocrinology* 154: 777–780.
Kroboth PD, Salek FS, Pittenger AL, Fabian TJ, and Frye RF (1999) DHEA and DHEA-S: A review. *Journal of Clinical Pharmacology* 39: 327–348.
Labarca C, Schwarz J, Deshpande P, et al. (2001) Point mutant mice with hypersensitive alpha 4 nicotinic receptors show dopaminergic deficits and increased anxiety. *Proceedings of the National Academy of Sciences of the United States of America* 98: 2786–2791.
Lai IC, Hong CJ, and Tsai SJ (2001) Association study of nicotinic-receptor variants and major depressive disorder. *Journal of Affective Disorders* 66: 79–82.
Lapchak PA and Araujo DM (2001) Preclinical development of neurosteroids as neuroprotective agents for the treatment of neurodegenerative diseases. *International Review of Neurobiology* 46: 379–397.
Lasser K, Boyd JW, Woolhandler S, Himmelstein DU, McCormick D, and Bor DH (2000) Smoking and mental illness: A population-based prevalence study. *Journal of the American Medical Association* 284: 2606–2610.
Law MR, Cheng R, Hackshaw AK, Allaway S, and Hale AK (1997) Cigarette smoking, sex hormones and bone density in women. *European Journal of Epidemiology* 13: 553–558.
Lazarus RS and Folkman S (1984) Stress, Appraisal and Coping. New York: Springer.
LeDoux JE (1993) Emotional memory systems in the brain. *Behavioural Brain Research* 58: 69–79.
Leonard S, Adler LE, Benhammou K, et al. (2001) Smoking and mental illness. *Pharmacology, Biochemistry, and Behavior* 70: 561–570.
Leonard S, Gault J, Hopkins J, et al. (2002) Association of promoter variants in the alpha 7 nicotinic acetylcholine receptor subunit gene with an inhibitory deficit found in schizophrenia. *Archives of General Psychiatry* 59(12): 1085–1096.
Leonard S, Gault J, Moore T, et al. (1998) Further investigation of a chromosome 15 locus in schizophrenia: Analysis of affected sibpairs from the NIMH Genetics Initiative. *American Journal of Medical Genetics* 81: 308–312.
Linkowski P, Van Onderbergen A, Kerkhofs M, Bosson D, Mendlewicz J, and Van Cauter E (1993) Twin study of the 24-h cortisol profile: Evidence for genetic control of the human circadian clock. *American Journal of Physiology* 264: E173–E181.
Lofroth G (1989) Environmental tobacco smoke: Overview of chemical composition and genotoxic components. *Mutation Research* 222: 73–80.
Lovallo WR (2006) Cortisol secretion patterns in addiction and addiction risk. *International Journal of Psychophysiology* 59: 195–202.
Lovallo WR, Dickensheets SL, Myers DA, Thomas TL, and Nixon SJ (2000) Blunted stress cortisol response in abstinent

alcoholic and polysubstance-abusing men. *Alcoholism, Clinical and Experimental Research* 24: 651–658.
MacInnis RJ, Cassar C, Nowson CA, et al. (2003) Determinants of bone density in 30- to 65-year-old women: A co-twin study. *Journal of Bone and Mineral Research* 18: 1650–1656.
Mack WJ, Preston-Martin S, Dal Maso L, et al. (2003) A pooled analysis of case-control studies of thyroid cancer: Cigarette smoking and consumption of alcohol, coffee, and tea. *Cancer Causes Control* 14: 773–785.
MacMahon B, Trichopoulos D, Cole P, and Brown J (1982) Cigarette smoking and urinary estrogens. *New England Journal of Medicine* 307: 1062–1065.
Madsbad S, McNair P, Christensen MS, Christiansen C, Faber OK, Binder C, and Transbol I (1980) Influence of smoking on insulin requirement and metbolic status in diabetes mellitus. *Diabetes Care* 3: 41–43.
Mahlberg R, Steinacher B, Mackert A, and Flechtner KM (2001) Basic parameters of saccadic eye movements: Differences between unmedicated schizophrenia and affective disorder patients. *European Archives of Psychiatry and Clinical Neuroscience* 251: 205–210.
Malin DH, Lake JR, Upchurch TP, Shenoi M, Rajan N, and Schweinle WE (1998) Nicotine abstinence syndrome precipitated by the competitive nicotinic antagonist dihydro-beta-erythroidine. *Pharmacology, Biochemistry, and Behavior* 60: 609–613.
Malone KM, Waternaux C, Haas GL, Cooper TB, Li S, and Mann JJ (2003) Cigarette smoking, suicidal behavior, and serotonin function in major psychiatric disorders. *American Journal of Psychiatry* 160: 773–779.
Manson JE, Ajani UA, Liu S, Nathan DM, and Hennekens CH (2000) A prospective study of cigarette smoking and the incidence of diabetes mellitus among US male physicians. *American Journal of Medicine* 109: 538–542.
Martel FL, Hayward C, Lyons DM, Sanborn K, Varady S, and Schatzberg AF (1999) Salivary cortisol levels in socially phobic adolescent girls. *Depression and Anxiety* 10: 25–27.
Marty MA, Erwin VG, Cornell K, and Zgombick JM (1985) Effects of nicotine on beta-endorphin, alpha MSH, and ACTH secretion by isolated perfused mouse brains and pituitary glands, *in vitro*. *Pharmacology, Biochemistry, and Behavior* 22: 317–325.
Marutle A, Zhang X, Court J, Piggott M, Johnson M, Perry R, and Nordberg A (2001) Laminar distribution of nicotinic receptor subtypes in cortical regions in schizophrenia. *Journal of Chemical Neuroanatomy* 22: 115–126.
Marx CE, Trost WT, Shampine L, Behm FM, Giordano LA, Massing MW, and Rose JE (2006) Neuroactive steroids, negative affect, and nicotine dependence severity in male smokers. *Psychopharmacology* 186: 462–472.
Matta SG, Beyer HS, McAllen KM, and Sharp BM (1987) Nicotine elevates rat plasma ACTH by a central mechanism. *Journal of Pharmacology and Experimental Therapeutics* 243: 217–226.
Matta SG, Fu Y, Valentine JD, and Sharp BM (1998) Response of the hypothalamo-pituitary–adrenal axis to nicotine. *Psychoneuroendocrinology* 23: 103–113.
Matta SG, McCoy JG, Foster CA, and Sharp BM (1995) Nicotinic agonists administered into the fourth ventricle stimulate norepinephrine secretion in the hypothalamic paraventricular nucleus: An *in vivo* microdialysis study. *Neuroendocrinology* 61: 383–392.
Mazess RB and Barden HS (1991) Bone density in premenopausal women: Effects of age, dietary intake, physical activity, smoking, and birth–control pills. *American Journal of Clinical Nutrition* 53: 132–142.
McCabe RE, Chudzik SM, Antony MM, Young L, Swinson RP, and Zolvensky MJ (2004) Smoking behaviors across anxiety disorders. *Journal of Anxiety Disorders* 18: 7–18.
McCulloch P, Lee S, Higgins R, McCall K, and Schade DS (2002) Effect of smoking on hemoglobin A1c and body mass index in patients with type 2 diabetes mellitus. *Journal of Investigative Medicine* 50: 284–287.
McEvoy JP, Freudenreich O, and Wilson WH (1999) Smoking and therapeutic response to clozapine in patients with schizophrenia. *Biological Psychiatry* 46: 125–129.
McGinnis JM and Foege WH (1993) Actual causes of death in the United States. *Journal of the American Medical Association* 270: 2207–2212.
Meberg A and Marstein S (1986) Smoking during pregnancy: Effects on the fetal thyroid function. *Acta Paediatrica Scandinavica* 75: 762–766.
Mendelson JH, Goletiani N, Sholar MB, Siegel AJ, and Mello NK (2005) Effects of low- and high-nicotine cigarette smoking on mood states and HPA axis in men. *Neuropsychopharmacology* 30(9): 1751–1763.
Mendelson JH, Goletiani N, Sholar MB, Siegel AJ, and Mello NK (2008) Effects of smoking successive low- and high-nicotine cigarettes on hypothalamic–pituitary–adrenal axis hormones and mood in men. *Neuropsychopharmacology* 33: 749–760.
Michnovicz JJ, Hershcopf RJ, Naganuma H, Bradlow HL, and Fishman J (1986) Increased 2-hydroxylation of estradiol as a possible mechanism for the anti-estrogenic effect of cigarette smoking. *New England Journal of Medicine* 315: 1305–1309.
Mueck AO and Seeger H (2003) Smoking, estradiol metabolism and hormone replacement therapy. *Arzneimittel-Forschung* 53: 1–11.
Muller B, Zulewski H, Huber P, Ratcliffe JG, and Staub JJ (1995) Impaired action of thyroid hormone associated with smoking in women with hypothyroidism. *New England Journal of Medicine* 333: 964–969.
Munck A, Guyre PM, and Holbrook NJ (1984) Physiological functions of glucocorticoids in stress and their relation to pharmacological actions. *Endocrine Reviews* 5: 25–44.
Nagamoto HT, Adler LE, McRae KA, et al. (1999) Auditory P50 in schizophrenics on clozapine: Improved gating parallels clinical improvement and changes in plasma 3-methoxy-4-hydroxyphenylglycol. *Neuropsychobiology* 39: 10–17.
Nestler EJ (2001) Molecular basis of long-term plasticity underlying addiction. *Nature Reviews* 2: 119–128.
Newhouse PA, Sunderland T, Narang PK, Mellow AM, Fertig JB, Lawlor BA, and Murphy DL (1990) Neuroendocrine, physiologic, and behavioral responses following intravenous nicotine in nonsmoking healthy volunteers and in patients with Alzheimer's disease. *Psychoneuroendocrinology* 15: 471–484.
Nguyen TV, Kelly PJ, Sambrook PN, Gilbert C, Pocock NA, and Eisman JA (1994) Lifestyle factors and bone density in the elderly: Implications for osteoporosis prevention. *Journal of Bone and Mineral Research* 9: 1339–1346.
Nichter M, Nichter M, Vuckovic N, Quintero G, and Ritenbaugh C (1997) Smoking experimentation and initiation among adolescent girls: Qualitative and quantitative findings. *Tobacco Control* 6: 285–295.
Nisell M, Nomikos GG, and Svensson TH (1995) Nicotine dependence, midbrain dopamine systems and psychiatric disorders. *Pharmacology and Toxicology* 76: 157–162.
Nuechterlein KH, Dawson ME, Ventura J, et al. (1994) The vulnerability/stress model of schizophrenic relapse: A longitudinal study. *Acta Psychiatrica Scandinavica* 89 (supplement 382): 58–64.
Nystrom E, Bengtsson C, Lapidus L, Petersen K, and Lindstedt G (1993) Smoking – a risk factor for hypothyroidism. *Journal of Endocrinological Investigation* 16: 129–131.
Olincy A and Stevens KE (2007) Treating schizophrenia symptoms with an alpha7 nicotinic agonist, from mice to men. *Biochemical Pharmacology* 74: 1192–1201.

Olincy A, Young DA, and Freedman R (1997) Increased levels of the nicotine metabolite cotinine in schizophrenic smokers compared to other smokers. *Biological Psychiatry* 42: 1–5.

Oncken C, Prestwood K, Cooney JL, Unson C, Fall P, Kulldorff M, and Raisz LG (2002) Effects of smoking cessation or reduction on hormone profiles and bone turnover in postmenopausal women. *Nicotine and Tobacco Research* 4: 451–458.

Orlando M, Ellickson PL, and Jinnett K (2001) The temporal relationship between emotional distress and cigarette smoking during adolescence and young adulthood. *Journal of Consulting and Clinical Psychology* 69: 959–970.

Owens JC, Balogh SA, McClure-Begley TD, et al. (2003) Alpha 4 beta 2* nicotinic acetylcholine receptors modulate the effects of ethanol and nicotine on the acoustic startle response. *Alcoholism: Clinical and Experimental Research* 27(12): 1867–1875.

Pacak K, Palkovits M, Kvetnansky R, Yadid G, Kopin IJ, and Goldstein DS (1995) Effects of various stressors on *in vivo* norepinephrine release in the hypothalamic paraventricular nucleus and on the pituitary–adrenocortical axis. *Annals of the New York Academy of Sciences* 771: 115–130.

Pauly JR, Grun EU, and Collins AC (1990a) Chronic corticosterone administration modulates nicotine sensitivity and brain nicotinic receptor binding in C3H mice. *Psychopharmacology* 101: 310–316.

Pauly JR, Ullman EA, and Collins AC (1988) Adrenocortical hormone regulation of nicotine sensitivity in mice. *Physiology and Behavior* 44: 109–116.

Pauly JR, Ullman EA, and Collins AC (1990b) Strain differences in adrenalectomy-induced alterations in nicotine sensitivity in the mouse. *Pharmacology, Biochemistry, and Behavior* 35: 171–179.

Petrusz P, Merchenthaler I, Maderdrut JL, and Heitz PU (1985) Central and peripheral distribution of corticotropin-releasing factor. *Federation Proceedings* 44: 229–235.

Pickworth WB, Baumann MH, Fant RV, Rothman RB, and Henningfield JE (1996) Endocrine responses during acute nicotine withdrawal. *Pharmacology, Biochemistry, and Behavior* 55: 433–437.

Pomerleau CS, Marks JL, and Pomerleau OF (2000) Who gets what symptom? Effects of psychiatric cofactors and nicotine dependence on patterns of smoking withdrawal symptomatology. *Nicotine and Tobacco Research* 2: 275–280.

Pomerleau OF (1992) Nicotine and the central nervous system: Biobehavioral effects of cigarette smoking. *American Journal of Medicine* 93: 2S–7S.

Pomerleau OF and Pomerleau CS (1984) Neuroregulators and the reinforcement of smoking: Towards a biobehavioral explanation. *Neuroscience and Biobehavioral Reviews* 8: 503–513.

Pomerleau OF, Pomerleau CS, Snedecor SM, Gaulrapp S, Brouwer RN, and Cameron OG (2004) Depression, smoking abstinence and HPA function in women smokers. *Human Psychopharmacology* 19: 467–476.

Pomerleau OF and Rosecrans J (1989) Neuroregulatory effects of nicotine. *Psychoneuroendocrinology* 14: 407–423.

Prummel MF and Wiersinga WM (1993) Smoking and risk of Graves' disease. *Journal of the American Medical Association* 269: 479–482.

Rasmussen DD (1998) Effects of chronic nicotine treatment and withdrawal on hypothalamic proopiomelanocortin gene expression and neuroendocrine regulation. *Psychoneuroendocrinology* 23: 245–259.

Rasmusson AM, Wu R, Paliwal P, Anderson GM, and Krishnan-Sarin S (2006) A decrease in the plasma DHEA to cortisol ratio during smoking abstinence may predict relapse: A preliminary study. *Psychopharmacology* 186: 473–480.

Rao U, Hammen C, London ED, and Poland RE (2007) Vulnerability for cigarette smoking in adolescents: A diathesis-stress model. Presented at the 13th Annual Meeting of the Society for Research on Nicotine and Tobacco. Austin, TX. February 21–24.

Raux G, Bonnet-Brilhault F, Louchart S, et al. (2002) The 2-bp deletion in exon 6 of the alpha 7-like nicotinic receptor subunit gene is a risk factor for the P50 sensory gating deficit. *Molecular Psychiatry* 7(9): 1006–1011.

Reeves GK, Pirie K, Beral V, Green J, Spencer E, and Bull D (2007) Cancer incidence and mortality in relation to body mass index in the Million Women Study: Cohort study. *British Medical Journal* 333, doi:10.1136/bmj.39367.495995.AE.

Rhodes ME, Kennell JS, Belz EE, Czambel RK, and Rubin RT (2004) Rat estrous cycle influences the sexual diergism of HPA axis stimulation by nicotine. *Brain Research Bulletin* 64: 205–213.

Rhodes ME, O'Toole SM, Czambel RK, and Rubin RT (2001) Male–female differences in rat hypothalamic–pituitary–adrenal axis responses to nicotine stimulation. *Brain Research Bulletin* 54: 681–688.

Rimm EB, Manson JE, Stampfer MJ, et al. (1993) Cigarette smoking and the risk of diabetes in women. *American Journal of Public Health* 83: 211–214.

Rimondini R, Arlinde C, Sommer W, and Heilig M (2002) Long-lasting increase in voluntary ethanol consumption and transcriptional regulation in the rat brain after intermittent exposure to alcohol. *FASEB Journal* 16: 27–35.

Ritsner M, Gibel A, Maayan R, et al. (2005) Cortisol/dehydroepiandrosterone ratio and responses to antipsychotic treatment in schizophrenia. *Neuropsychopharmacology* 30: 1913–1922.

Ritsner M, Gibel A, Maayan R, Ratner Y, Ram E, Modai I, and Weizman A (2007) State and trait related predictors of serum cortisol to DHEA(S) molar ratios and hormone concentrations in schizophrenia patients. *European Neuropsychopharmacology* 17: 257–264.

Ritsner M, Maayan R, Gibel A, Strous RD, Modai I, and Weizman A (2004) Elevation of the cortisol/dehydroepiandrosterone ratio in schizophrenia patients. *European Neuropsychopharmacology* 14: 267–273.

Rodriguez de Fonseca F, Carrera MR, Navarro M, Koob GF, and Weiss F (1997) Activation of corticotropin-releasing factor in the limbic system during cannabinoid withdrawal. *Science (New York)* 276: 2050–2054.

Rohleder N and Kirschbaum C (2006) The hypothalamic–pituitary–adrenal (HPA) axis in habitual smokers. *International Journal of Psychophysiology* 59: 236–243.

Rolls ET and Stringer SM (2001) A model of the interaction between mood and memory. *Network (Bristol)* 12: 89–109.

Rose JE, Ananda S, and Jarvik ME (1983) Cigarette smoking during anxiety-provoking and monotonous tasks. *Addictive Behaviors* 8: 353–359.

Rosecrans JA and Karin LD (1998) Effects of nicotine on the hypothalamic–pituitary-axis (HPA) and immune function: Introduction to the sixth nicotine round table satellite, American society of addiction medicine nicotine dependence meeting, November 15, 1997. *Psychoneuroendocrinology* 23: 95–102.

Rosenberg MJ, Waugh MS, and Stevens CM (1996) Smoking and cycle control among oral contraceptive users. *American Journal of Obstetrics and Gynecology* 174: 628–632.

Ross RG, Olincy A, Mikulich SK, et al. (2002) Admixture analysis of smooth pursuit eye movements in probands with schizophrenia and their relatives suggests gain and leading saccades are potential endophenotypes. *Psychophysiology* 39(6): 809–819.

Ross SA, Wong JY, Clifford JJ, et al. (2000) Phenotypic characterization of an alpha 4 neuronal nicotinic acetylcholine receptor subunit knock-out mouse. *Journal of Neuroscience* 20: 6431–6441.

Saal D, Dong Y, Bonci A, and Malenka RC (2003) Drugs of abuse and stress trigger a common synaptic adaptation in dopamine neurons. *Neuron* 37: 577–582.

Salas R, Pieri F, Fung B, Dani JA, and De Biasi M (2003) Altered anxiety-related responses in mutant mice lacking the beta4 subunit of the nicotinic receptor. *Journal of Neuroscience* 23: 6255–6263.

Salin-Pascual RJ and Drucker-Colin R (1998) A novel effect of nicotine on mood and sleep in major depression. *Neuro Report* 9: 57–60.

Sanchez MM, Young LJ, Plotsky PM, and Insel TR (2000) Distribution of corticosteroid receptors in the rhesus brain: Relative absence of glucocorticoid receptors in the hippocampal formation. *Journal of Neuroscience* 20: 4657–4668.

Sargeant LA, Khaw KT, Bingham S, et al. (2001) Cigarette smoking and glycaemia: The EPIC-Norfolk Study. European prospective investigation into cancer. *International Journal of Epidemiology* 30: 547–554.

Schindler AE (2006) Climacteric symptoms and hormones. *Gynecological Endocrinology* 22: 151–154.

Schreiber W, Lauer CJ, Krumrey K, Holsboer F, and Krieg JC (1996) Dysregulation of the hypothalamic–pituitary–adrenocortical system in panic disorder. *Neuropsychopharmacology* 15: 7–15.

Seyler LE, Jr., Pomerleau OF, Fertig JB, Hunt D, and Parker K (1986) Pituitary hormone response to cigarette smoking. *Pharmacology, Biochemistry, and Behavior* 24: 159–162.

Sharp BM and Beyer HS (1986) Rapid desensitization of the acute stimulatory effects of nicotine on rat plasma adrenocorticotropin and prolactin. *Journal of Pharmacology and Experimental Therapeutics* 238: 486–491.

Sharp BM and Matta SG (1993) Detection by *in vivo* microdialysis of nicotine-induced norepinephrine secretion from the hypothalamic paraventricular nucleus of freely moving rats: Dose-dependency and desensitization. *Endocrinology* 133: 11–19.

Shine B, Fells P, Edwards OM, and Weetman AP (1990) Association between Graves' ophthalmopathy and smoking. *Lancet* 335: 1261–1263.

Shirazi-Southall S, Rodriguez DE, and Nomikos GG (2002) Effects of typical and atypical antipsychotics and receptor selective compounds on acetylcholine efflux in the hippocampus of the rat. *Neuropsychopharmacology* 26: 583–594.

Shiverick KT and Salafia C (1999) Cigarette smoking and pregnancy I: Ovarian, uterine and placental effects. *Placenta* 20: 265–272.

Shytle RD, Silver AA, Lukas RJ, Newman MB, Sheehan DV, and Sanberg PR (2002) Nicotinic acetylcholine receptors as targets for antidepressants. *Molecular Psychiatry* 7: 525–535.

Siqueira L, Diab M, Bodian C, and Rolnitzky L (2000) Adolescents becoming smokers: The roles of stress and coping methods. *Journal of Adolescent Health* 27: 399–408.

Sirola J, Kroger H, Honkanen R, Sandini L, Tuppurainen M, Jurvelin JS, and Saarikoski S (2003) Smoking may impair the bone protective effects of nutritional calcium: A population-based approach. *Journal of Bone and Mineral Research* 18: 1036–1042.

Slemenda CW, Christian JC, Reed T, Reister TK, Williams CJ, and Johnston CC, Jr. (1992) Long-term bone loss in men: Effects of genetic and environmental factors. *Annals of Internal Medicine* 117: 286–291.

Slemmer JE, Martin BR, and Damaj MI (2000) Bupropion is a nicotinic antagonist. *Journal of Pharmacology and Experimental Therapeutics* 295: 321–327.

Sonksen PH, Russell-Jones D, and Jones RH (1993) Growth hormone and diabetes mellitus. A review of sixty-three years of medical research and a glimpse into the future? *Hormone Research* 40: 68–79.

Spangler JG (1999) Smoking and hormone-related disorders. *Primary Care* 26: 499–511.

Steptoe A and Ussher M (2006) Smoking, cortisol and nicotine. *International Journal of Psychophysiology* 59: 228–235.

Sterzik K, Strehler E, De Santo M, Trumpp N, Abt M, Rosenbusch B, and Schneider A (1996) Influence of smoking on fertility in women attending an *in vitro* fertilization program. *Fertility and Sterility* 65: 810–814.

Sujkovic E, Mileusnic R, Fry JP, and Rose SP (2007) Temporal effects of dehydroepiandrosterone sulfate on memory formation in day-old chicks. *Neuroscience* 148: 375–384.

Svartberg J and Jorde R (2007) Endogenous testosterone levels and smoking in men. The fifth Tromso study. *International Journal of Andrology* 30: 137–143.

Svartberg J, Midtby M, Bonaa KH, Sundsfjord J, Joakimsen RM, and Jorde R (2003) The associations of age, lifestyle factors and chronic disease with testosterone in men: The Tromso study. *European Journal of Endocrinology* 149: 145–152.

Swanson LW and Sawchenko PE (1983) Hypothalamic integration: Organization of the paraventricular and supraoptic nuclei. *Annual Review of Neuroscience* 6: 269–324.

Tanko LB and Christiansen C (2004) An update on the antiestrogenic effect of smoking: A literature review with implications for researchers and practitioners. *Menopause (New York)* 11: 104–109.

Tansavatdi K, McClain B, and Herrington DM (2004) The effects of smoking on estradiol metabolism. *Minerva Ginecologica* 56: 105–114.

Thaker GK, Ross DE, Buchanan RW, Adami HM, and Medoff DR (1999) Smooth pursuit eye movements to extraretinal motion signals: Deficits in patients with schizophrenia. *Psychiatry Research* 88(3): 209–219.

Thornton J, Kelly SP, Harrison RA, and Edwards R (2007) Cigarette smoking and thyroid eye disease: A systematic review. *Eye (London)* 21: 1135–1145.

Tizabi Y, Rezvani AH, Russell LT, Tyler KY, and Overstreet DH (2000) Depressive characteristics of FSL rats: Involvement of central nicotinic receptors. *Pharmacology, Biochemistry, and Behavior* 66: 73–77.

Tonstad S (2002) Use of sustained-release bupropion in specific patient populations for smoking cessation. *Drugs* 62 (supplement 2): 37–43.

Tritto T, Stitzel JA, Marks MJ, Romm E, and Collins AC (2002) Variability in response to nicotine in the LSxSS RI strains: Potential role of polymorphisms in alpha 4 and alpha 6 nicotinic receptor genes. *Pharmacogenetics* 12(3): 197–208.

Tung CS, Grenhoff J, and Svensson TH (1990) Nicotine counteracts midbrain dopamine cell dysfunction induced by prefrontal cortex inactivation. *Acta Physiologica Scandinavica* 138: 427–428.

Utiger RD (1998) Effects of smoking on thyroid function. *European Journal of Endocrinology* 138: 368–369.

Valdez GR, Roberts AJ, Chan K, Davis H, Brennan M, Zorrilla EP, and Koob GF (2002) Increased ethanol self-administration and anxiety-like behavior during acute ethanol withdrawal and protracted abstinence: Regulation by corticotropin-releasing factor. *Alcoholism, Clinical and Experimental Research* 26: 1494–1501.

Valera S, Ballivet M, and Bertrand D (1992) Progesterone modulates a neuronal nicotinic acetylcholine receptor. *Proceedings of the National Academy of Sciences of the United States of America* 89: 9949–9953.
Valimaki MJ, Laitinen KA, Tahtela RK, Hirvonen EJ, and Risteli JP (2003) The effects of transdermal estrogen therapy on bone mass and turnover in early postmenopausal smokers: A prospective, controlled study. *American Journal of Obstetrics and Gynecology* 189: 1213–1220.
Van Cauter E, Shapiro ET, Tillil H, and Polonsky KS (1992) Circadian modulation of glucose and insulin responses to meals: Relationship to cortisol rhythm. *American Journal of Physiology* 262: E467–E475.
Van Voorhis BJ, Dawson JD, Stovall DW, Sparks AE, and Syrop CH (1996) The effects of smoking on ovarian function and fertility during assisted reproduction cycles. *Obstetrics and Gynecology* 88: 785–791.
Vestergaard P (2002) Smoking and thyroid disorders – a meta-analysis. *European Journal of endocrinology* 146: 153–161.
Vestergaard P, Rejnmark L, Weeke J, et al. (2002) Smoking as a risk factor for Graves' disease, toxic nodular goiter, and autoimmune hypothyroidism. *Thyroid* 12: 69–75.
Vine MF, Margolin BH, Morrison HI, and Hulka BS (1994) Cigarette smoking and sperm density: A meta-analysis. *Fertility and Sterility* 61: 35–43.
Weidenfeld J, Bodoff M, Saphier D, and Brenner T (1989) Further studies on the stimulatory action of nicotine on adrenocortical function in the rat. *Neuroendocrinology* 50: 132–138.
West R and Hajek P (1997) What happens to anxiety levels on giving up smoking? *American Journal of Psychiatry* 154: 1589–1592.
Westhoff C, Gentile G, Lee J, Zacur H, and Helbig D (1996) Predictors of ovarian steroid secretion in reproductive-age women. *American Journal of Epidemiology* 144: 381–388.
Wilhelm K, Mitchell P, Slade T, Brownhill S, and Andrews G (2003) Prevalence and correlates of DSM-IV major depression in an Australian national survey. *Journal of Affective Disorders* 75: 155–162.
Wilkins JN, Carlson HE, Van Vunakis H, Hill MA, Gritz E, and Jarvik ME (1982) Nicotine from cigarette smoking increases circulating levels of cortisol, growth hormone, and prolactin in male chronic smokers. *Psychopharmacology* 78: 305–308.
Windham GC, Elkin EP, Swan SH, Waller KO, and Fenster L (1999) Cigarette smoking and effects on menstrual function. *Obstetrics and Gynecology* 93: 59–65.
Winternitz WW and Quillen D (1977) Acute hormonal response to cigarette smoking. *Journal of Clinical Pharmacology* 17: 389–397.
Wolf OT and Kirschbaum C (1999) Actions of dehydroepiandrosterone and its sulfate in the central nervous system: Effects on cognition and emotion in animals and humans. *Brain Research* 30: 264–288.
Wong PK, Christie JJ, and Wark JD (2007) The effects of smoking on bone health. *Clinical Science (Lond)* 113: 233–241.
Woolf NJ (1991) Cholinergic systems in mammalian brain and spinal cord. *Progress in Neurobiology* 37: 475–524.
Yeh J and Barbieri RL (1989) Twenty-four-hour urinary-free cortisol in premenopausal cigarette smokers and nonsmokers. *Fertility and Sterility* 52: 1067–1069.
Yehuda R, Brand SR, Golier JA, and Yang RK (2006) Clinical correlates of DHEA associated with post-traumatic stress disorder. *Acta Psychiatrica Scandinavica* 114: 187–193.
Young AH (2004) Cortisol in mood disorders. *Stress (Amsterdam)* 7: 205–208.
Young AH, Gallagher P, and Porter RJ (2002) Elevation of the cortisol–dehydroepiandrosterone ratio in drug-free depressed patients. *American Journal of Psychiatry* 159: 1237–1239.
Zhou FM, Wilson CJ, and Dani JA (2002) Cholinergic interneuron characteristics and nicotinic properties in the striatum. *Journal of Neurobiology* 53: 590–605.
Zubenko GS, Hughes HB, Stiffler JS, Zubenko WN, and Kaplan BB (2002) Genome survey for susceptibility loci for recurrent, early-onset major depression: Results at 10cM resolution. *American Journal of Medical Genetics* 114: 413–422.

34 Cocaine, Hormones and Behavior

N K Mello and J H Mendelson, Harvard Medical School, Belmont, MA, USA

Chapter Outline

34.1 Introduction

Cocaine abuse and dependence continues to be one of the nation's most serious drug-abuse problems (SAMHSA, 2006; Mendelson and Mello, 2008) and the associated social and economic costs include a number of adverse effects on health (Mendelson and Mello, 2008). The latest data from the Drug Abuse Warning Network (DAWN, 2007) indicate that cocaine was the most frequently cited illicit drug in emergency-room visits. Moreover, cocaine abuse is not only restricted to a small group of polydrug abusers, but also affects the general population. In 2006, the National Survey on Drug Abuse and Health estimated that over 2.4 million people aged 12 and older used cocaine during the past year and 1.5 million used cocaine during the past month (SAMHSA, 2006). A major federal program to develop medications for cocaine-abuse treatment is underway (Vocci et al., 2005; Vocci and Ling, 2005), but thus far the available pharmacotherapies have not been as effective as methadone, levo-alpha acetyl methadol (LAAM), and buprenorphine for the treatment of opioid dependence (Rawson et al., 1991; Tutton and Crayton, 1993; Mendelson and Mello, 1996; Montoya et al., 2004; Mendelson and Mello, 2008).

This chapter focuses on clinical studies of the interactions between cocaine and the hypothalamic–pituitary–gonadal (HPG) axis and the hypothalamic–pituitary–adrenal (HPA) axis. Cocaine-related changes in the HPG- and HPA-axis hormones have broad implications for normal reproductive and immune functions as well as behavior. There is emerging evidence that cocaine's perturbation of anterior pituitary, gonadal, and adrenal hormones may influence its reinforcing properties (Mendelson et al., 1989a, 1992a, 2002; Goeders, 1997, 2002a,b; Evans, 2007). Moreover, the importance of studying possible sex differences in response to drugs is increasingly recognized (IOM, 2001; Cahill, 2006; Wetherington, 2007; Becker et al., 2008). Accordingly, some illustrative studies of the role of sex in modulating cocaine's neuroendocrine and behavioral effects are described.

This chapter describes the effects of cocaine on the HPA-axis hormones, adrenocorticotropic hormone (ACTH), and cortisol or corticosterone, as well as on the gonadotropins and gonadal steroid hormones; the interactions between cocaine, sex, and gonadal steroid hormones; and disruptive effects of chronic cocaine exposure on reproductive function, and some possible mechanisms by which cocaine and other abused drugs disrupt the menstrual cycle in women and compromise reproductive function in men. Cocaine's effects on prolactin were summarized in an earlier version of this chapter (Mello and Mendelson, 2002) and have not been updated here due to space limitations. We conclude that an improved understanding of the interactions between cocaine, the neuroendocrine system, and behavior may clarify some aspects of the neurobiology of psychostimulant abuse, and suggest new approaches to medication-based treatment.

34.2 Cocaine's Effects on ACTH and Cortisol/Corticosterone

34.2.1 Background

The HPA axis is the major hormonal system that integrates physiological responses to stress. Hypothalamic corticotropin-releasing hormone (CRH) regulates the pulsatile release of ACTH from the anterior pituitary. It has been long known that CRH activation of ACTH release and the subsequent increase in cortisol secretion from the adrenal is essential for a prompt cardiovascular-, respiratory-, gastrointestinal-, and immune-system response to stress. CRH is secreted by neurons in the basal hypothalamus and stimulates CRH receptors on anterior pituitary corticotropes to secrete ACTH, which in turn, stimulates cortisol or corticosterone release from the adrenal cortex. Both CRH and ACTH secretion are under negative-feedback control by cortisol in humans and in nonhuman primates, and by corticosterone in rodents. CRH cannot be measured in peripheral circulation, but increases in CRH can be inferred from increases in plasma ACTH. A number of neuronal systems are involved in the regulation of CRH secretion, and noradrenergic and adrenergic activity may increase the pulsatile release of CRH. Serotonergic and dopaminergic systems may be involved in both stimulation and inhibition of CRH secretion, and endogenous opioid agonists also have been shown to inhibit CRH secretion (Yen, 1999a). In addition to CRH secretion from the basal hypothalamus, CRH neurons are widely distributed throughout the central nervous system. These multiple CRH systems in the brain appear to regulate processes associated with the perception of pain, affective states, learning, arousal, and motivation (Chrousos and Gold, 1992, 1998; Yen, 1999a).

Cocaine, similar to stress, modulates HPA-axis activity. Cocaine may also disrupt normal immune–neuroendocrine interactions and immune function (see Reichlin (1993) and Besedovsky and Del Rey (1996) for review). There is increasing evidence that dysregulation of the HPA axis (i.e., sustained hyper- or hypoactivity) may increase vulnerability to depression, and a number of other psychiatric disorders, as well as cardiovascular disease (Chrousos and Gold, 1992, 1998; Heinrichs et al., 1995; Nemeroff, 1998; Holsboer, 2003). Cocaine's suppressive effects on immune function may amplify the risk for HIV (human immunodeficiency virus) infection, in part, through stimulation of the HPA axis (Schoenbaum et al., 1989; NIDA, 1991; Steel and Haverkos, 1992). For example, ACTH may inhibit macrophage activation, synthesis of IgG, and interferon-γ, and CRH may have a stimulatory effect on lymphocyte and monocyte proliferation and activation (Reichlin, 1993). Cocaine stimulation of cortisol may contribute to the suppression of a proinflammatory cytokine, interleukin-6 (Halpern et al., 2003). Corticotropin, acting through the adrenal cortex and the secretion of glucocorticoids, may suppress proliferation of lymphocytes and the secretion of inflammatory mediators (Jain et al., 1991; Pequegnat et al., 1992; Reichlin, 1993). Moreover, glucocorticoids directly stimulate the transcription of the HIV virus, *in vitro* (Markham et al., 1986; Soudeyns et al., 1993), and, therefore, may increase susceptibility to acquired immune deficiency syndrome (AIDS; Pequegnat et al., 1992). Thus, cocaine abusers may be at enhanced risk for HIV infection (Chaisson et al., 1989) due, in part, to cocaine-related activation of the HPA axis and subsequent immunosuppression.

Cocaine stimulates ACTH and cortisol secretion in humans and in rhesus monkeys, and ACTH and corticosterone release in rats (see Mello and Mendelson (1997, 2002, in press) for review). Although the exact mechanisms underlying cocaine's effects on the HPA axis remain to be clarified, it appears that cocaine-related stimulation of ACTH (and by inference CRH) is modulated by several interacting neurotransmitter systems. CRH release is regulated, in part, by dopamine and serotonin, and antagonists that are selective for dopamine receptors or 5-HT receptors attenuate cocaine-induced stimulation of ACTH (Levy et al., 1991; Borowsky and Kuhn, 1991a). Moreover, both dopamine and 5-HT receptor agonists stimulate ACTH release in rats (Borowsky and Kuhn, 1991a; Van de Kar et al., 1992; Levy et al., 1994; Baumann et al., 1995a). The complex relationships between cocaine, dopamine, serotonin, and the HPA axis have been reviewed elsewhere (Levy et al., 1994; Koob and LeMoal, 2006). Some illustrative studies of the acute and chronic effects of cocaine on ACTH and corticosterone in rats, and ACTH and cortisol in rhesus monkeys and humans, are described below.

34.2.2 Clinical Studies of the Acute Effects of Cocaine on ACTH and Cortisol

Cocaine administration is usually followed by an increase in ACTH and a subsequent increase in cortisol in human males. However, the route of cocaine administration and the rate of increase in plasma cocaine levels determine the time course and magnitude of the ACTH response. Intravenous (IV) cocaine administration consistently results in rapid increases in ACTH levels (Teoh et al., 1994a; Mendelson and Mello, 1998; Sholar et al., 1998; Elman et al., 1999). When cocaine was administered intranasally, a change in ACTH was not detected, but there was a significant increase in cortisol levels, which were maximal 60 min after cocaine administration (Heesch et al., 1995). Clinical studies of the effects of cocaine on basal levels of ACTH and cortisol, and on pulsatile release patterns of ACTH are described below.

34.2.2.1 Acute effects of cocaine on basal levels of ACTH and cortisol

The first study of cocaine's acute effects on ACTH was reported in 1992 (Mendelson et al., 1992b). Eighteen men who met *Diagnostic and Statistical Manual of Mental Disorders, Third Edition, Revised* (DSM-III-R) criteria for cocaine and opioid dependence were studied in a clinical research ward and were drug free for 6 days before cocaine exposure. Each subject served as his own control during placebo- and cocaine-administration conditions. ACTH levels were measured before and after IV administration of cocaine (30 mg over 1 min) or placebo. Baseline levels of ACTH were equivalent under both conditions and there were no significant changes in plasma ACTH levels after placebo administration. However, within 5 min after IV cocaine administration, ACTH levels increased significantly and remained significantly above baseline levels for 45 min. **Figure 1** shows plasma cocaine levels and plasma ACTH levels in six men with a history of cocaine abuse before and after IV cocaine administration (0.2 mg kg^{-1} over 1 min; Sholar et al., 1998). Peak plasma cocaine levels were measured at 6 ($\pm$ 1.4) min after IV injection and were coincident with the peak increase in plasma ACTH levels, which occurred 7.3 ($\pm$ 1.2) min after cocaine administration. ACTH increases were significantly correlated with increases in plasma cocaine levels (Sholar et al., 1998). Cardiovascular and subjective-effect measures paralleled increases in plasma cocaine and ACTH. Significant increases in reports of high, euphoria, and good were reported at 5 min after cocaine injection. Taken together, these data are consistent with the hypothesis that the reinforcing properties of cocaine may be related to cocaine-induced stimulation of endogenous CRH in the brain, as discussed later in this section (Mendelson et al., 1989a, 1992a).

Figure 1 also shows that cortisol increased significantly within 16 min after cocaine administration and reached peak levels within 30 min, that is, about 22 min after peak levels of ACTH were measured (Sholar et al., 1998). A similar time course of increases in ACTH and cortisol was measured in cocaine-dependent subjects after administration of 0.6 mg kg^{-1} IV cocaine (Elman et al., 1999). ACTH increased significantly within 3 min and reached peak levels within 10 min after IV cocaine

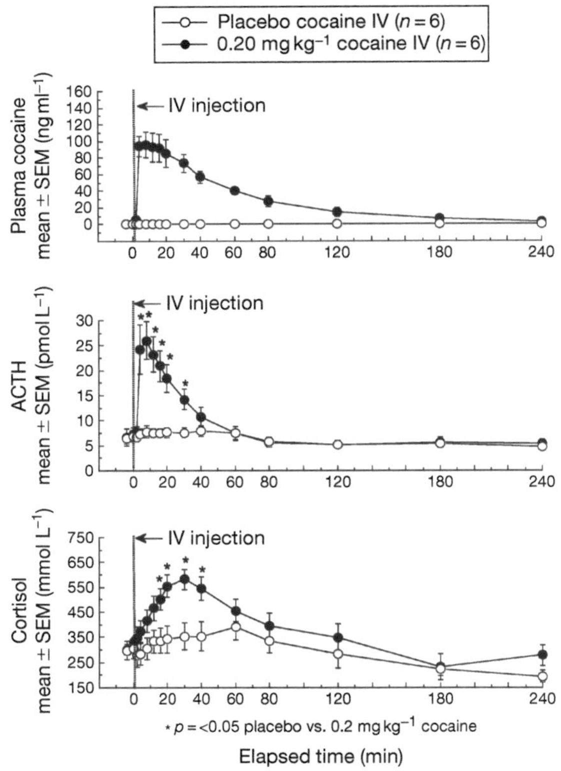

Figure 1 Covariance between ACTH and plasma cocaine levels in men. Plasma cocaine levels (ng ml^{-1}) are shown in row 1 and ACTH levels (pmol L^{-1}) and cortisol (mmol L^{-1}) are shown in rows 2 and 3. The time of sample collection is shown on the abscissa. The time of cocaine injection (0.4 mg kg^{-1}, IV) is indicated by a vertical line in each panel. Asterisks: significantly different from placebo ($P < 0.05$). Reproduced with permission from Sholar MB, Mendelson JH, Mello NK, et al. (1998) Concurrent pharmacokinetic analysis of plasma cocaine and adrenocorticotropic hormone in men. *Journal of Clinical Endocrinology and Metabolism* 83: 966–968. Copyright 1998, The Endocrine Society.

administration. Cortisol increased significantly within 10 min and reached peak levels within 30 min, and persistent elevations in both ACTH (30 min) and cortisol (120 min) were measured (Elman et al., 1999). These data are consistent with previous reports that cocaine stimulates cortisol release in men (Wilkins et al., 1992; Baumann et al., 1995; Heesch et al., 1995). A significant increase in plasma cortisol levels was found in cocaine-dependent men following IV infusion of 40 mg of cocaine (Baumann et al., 1995). However, the magnitude and duration of cortisol increases were not cocaine-dose-dependent when the effects of 10 or 90 mg of IV cocaine were examined in male cocaine abusers (Wilkins et al., 1992). Cocaine also increased plasma cortisol levels in cocaine-naive men after intranasal cocaine administration (2 mg kg^{-1}) in comparison to intranasal saline (Heesch et al., 1995).

In preclinical studies, chronic exposure to cocaine does not affect the ACTH response to an acute cocaine challenge (Borowsky and Kuhn, 1991b; Levy et al., 1992; Laviola et al., 1995) or a CRH challenge (Torres and Rivier, 1992c). Thus, it was surprising to find evidence of tolerance to cocaine's effects on ACTH in humans (Mendelson et al., 1998). Men who were dependent on both cocaine and opioids were compared with occasional cocaine abusers and matched for age and body mass index (BMI). The cocaine- and opioid-dependent men reported using cocaine for 9 years and opioids for 13 years, whereas the occasional cocaine abusers reported using cocaine on five or ten occasions during the previous year. All men were drug free at the time of the study, and the cocaine- and opioid-dependent subjects had lived in a clinical research ward under drug-free conditions for at least 9 days before the study. **Figure 2** shows the effects of 0.4 mg kg^{-1} IV cocaine on plasma cocaine levels and ACTH levels in these two groups. The time course and peak levels of plasma cocaine were almost identical, but peak levels of ACTH were significantly higher in the occasional cocaine users than in the cocaine- and opioid-dependent men. Reports of subjective euphoria and high as well as increases in heart rate were also significantly higher in the occasional cocaine users than in the cocaine- and opioid-dependent men. Although the contribution of concurrent opioid dependence to the apparent tolerance to cocaine's physiologic, neuroendocrine, and subjective effects cannot be determined with certainty, these findings suggest that tolerance to cocaine's effects may occur as a function of chronic cocaine dependence (Mendelson et al., 1998).

34.2.2.2 Acute effects of cocaine on pulsatile release of ACTH

The effects of cocaine on pulsatile secretion of ACTH in men examined under controlled clinical-research-ward conditions were very similar to those in male rhesus monkeys (Teoh et al., 1994c; Sarnyai et al., 1996). Eight men with concurrent cocaine and opioid dependence were drug free for at least 6 days before a cocaine or placebo challenge. Following an overnight fast, a challenge dose of cocaine (30 mg, IV) or placebo was administered under single-blind conditions in a randomized order on two study days. Blood samples were collected at 2-min intervals for 76 min during baseline and for an additional 76 min

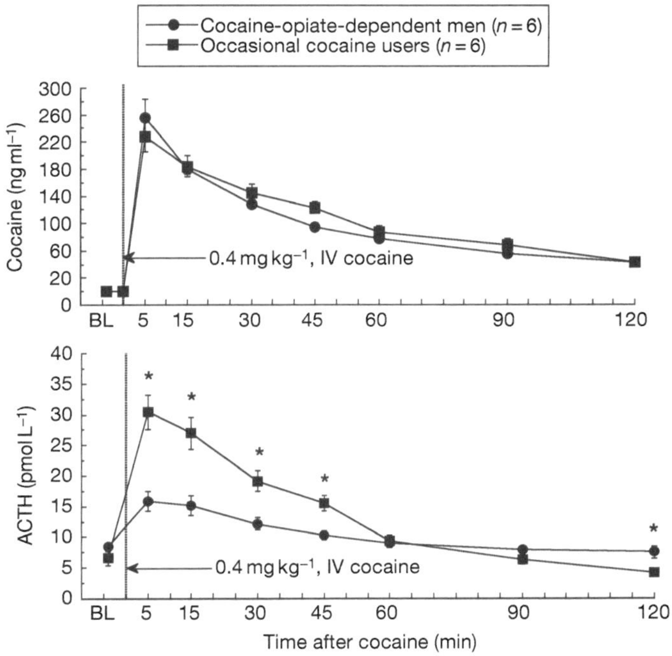

Figure 2 A comparison of cocaine's effects in occasional cocaine users and in cocaine- and opioid-dependent users. Plasma cocaine levels (ng ml^{-1}) after IV injection of 0.4 mg kg^{-1} cocaine ($\bar{x} \pm$ S.E.) are shown in row 1. ACTH levels (pmol L^{-1}) are shown in row 2. The time of cocaine injection is indicated by the vertical dotted lines. Precocaine baseline (BL) values and the times of postcocaine sample collection are shown on the abscissa. Each data point is based on six cocaine- and opiate-dependent men shown as black circles ($\bar{x} \pm$ S.E.) and six occasional cocaine users shown as black squares ($\bar{x} \pm$ S.E.). Asterisks indicate significant changes from baseline levels ($P < 0.05$). These data were adapted from Mendelson JH, Sholar M, Mello NK, Teoh SK, and Sholar J W (1998) Cocaine tolerance: behavioral, cardiovascular, and neuroendocrine function in men. *Neuropsychopharmacology* 18: 263–271.

following IV cocaine or placebo administration. Peak plasma cocaine levels of 313.8 ± 46.5 ng ml^{-1} were detected within 2 min after cocaine administration, and ACTH increased rapidly to peak levels within 8 min after cocaine administration. Cocaine-induced increases in ACTH peak amplitude persisted for the duration of the study, but ACTH-pulse frequency was not altered by cocaine.

Although the precise mechanism(s) of the action of cocaine on ACTH is unknown, clinical and preclinical studies suggest that CRH is critical for this effect. Hypothalamic CRH regulates the amplitude modulation of micropulsatile ACTH secretion, but the frequency of ACTH micropulses may reflect an intrinsic secretory rhythm of corticotrope cells in the anterior pituitary (Carnes et al., 1990). Intrinsic pulsatility of ACTH release from isolated human pituitary has also been demonstrated *in vitro* (Gambacciani et al., 1987). Immunoneutralization and receptor blockade of CRH completely abolished the effects of cocaine on ACTH and corticosterone in rats (Rivier and Vale, 1987; Sarnyai et al., 1992a). In men, cocaine increased the amplitude (a CRH-dependent component) of ACTH pulses, but not the frequency (a CRH-independent component) of pulsatile ACTH release (Teoh et al., 1994c). Taken together, these findings suggest that cocaine increases ACTH-pulse amplitude by stimulation of CRH release from the hypothalamus in men with a history of cocaine and opioid dependence.

34.2.3 Clinical Studies of Chronic Cocaine Effects on ACTH and Cortisol

The extent to which chronic cocaine abuse may cause persistent alterations of HPA-axis function is unclear. The pulsatile release of ACTH and cortisol in male cocaine abusers (Mendelson et al., 1989b; Teoh et al., 1994c) did not differ from data obtained in normal controls (Iranmanesh et al., 1990). Yet, as shown earlier in **Figure 2**, cocaine- and opioid-dependent men appeared to be tolerant to the neuroendocrine, cardiovascular, and subjective effects of cocaine in comparison to occasional cocaine users (Mendelson et al., 1998).

Opioid agonists alter basal and CRH-induced ACTH secretion in experimental animals (Tsagarakis et al., 1989) and in humans (Rittmaster et al., 1985; Allolio et al., 1986). Blunted ACTH and cortisol responses to CRH (1 $\mu g\,kg^{-1}$) injection have been observed following subcutaneous morphine (0.14 $mg\,kg^{-1}$) pretreatment in subjects who were not drug dependent (Rittmaster et al., 1985). Other ACTH-releasing factors, such as vasopressin and oxytocin, may also be involved in drug-induced disruption of pituitary ACTH secretion, and these neurohormones may be altered by chronic opioid and cocaine administration (Sarnyai and Kovacs, 1994).

34.2.4 Clinical Studies of the HPA Axis and Cocaine's Behavioral Effects

The temporal concordance between peak plasma cocaine levels, peak levels of ACTH, and reports of subjective high have prompted speculation that perturbation of the HPA axis may contribute to cocaine's abuse-related effects (Mendelson et al., 1989a, 1992b; Goeders and Guerin, 1995; Goeders, 1997, 2002a). In a study designed to further examine the covariance between cocaine's subjective effects and activation of the HPA axis, the time course of cocaine's effects on anterior pituitary and adrenal hormones were examined in cocaine abusers using a rapid-sampling procedure (Mendelson et al., 2002). Blood samples for hormone and plasma cocaine analyses were collected at 2-min intervals for the first 20 min after IV administration of cocaine (0.4 $mg\,kg^{-1}$) or placebo over 1 min. Rapid increases in plasma cocaine levels after IV cocaine administration have been measured in several studies, and peak cocaine concentrations have been detected within 15 s in arterial samples and within 4 min in venous samples (Evans et al., 1996). The temporal relationship between cocaine-induced euphoria (high) and plasma levels of epinephrine, ACTH, cortisol, and LH is shown in **Figure 3** (Mendelson et al., 2002). Feeling high was reported between 1 and 20 min following completion of the cocaine injection, then decreasing rapidly to baseline. Epinephrine and ACTH levels increased within 4 min and paralleled increases in cocaine-induced euphoria. LH and cortisol levels increased more slowly and remained elevated for longer periods of time. Cocaine-induced tachycardia was accompanied by a significant increase in plasma epinephrine and activation of the HPA axis. This pattern is similar to the physiologic correlates of stress. It is interesting that the cocaine-induced high and concurrent neuroendocrine and physiologic activation appear to be somewhat analogous to the high associated with self-imposed risk behaviors. For example, potentially dangerous activities such as parachute jumping, mountain-bike jumping, and roller-coaster riding are reported to induce intense feelings of pleasure, excitement, and brief euphoria. The extent to which the induction of a stress-like response may be important for the reinforcing effects of cocaine self-administration is unclear (Mendelson et al., 2002).

Depression is often associated with cocaine dependence (Schmitz et al., 2000), but there has been relatively little experimental attention given to the effects of cocaine in depressed cocaine abusers. Severity of depression is often correlated with high cortisol levels and diminished ACTH responsivity to a CRH challenge (Rubin et al., 1987; Nemeroff, 1996). When cocaine-dependent men with mild depressive symptoms were given a challenge dose of 0.6 $mg\,kg^{-1}$ IV cocaine, ACTH increased by 261% and cortisol increased by 73% above baseline levels (Elman et al., 1999). Moreover, there was a positive correlation between cocaine-induced elevations in ACTH and cortisol and scores on the Hamilton rating scale for depression. Interestingly, the reported frequency of cocaine use during the past 30 days was also correlated with total scores on the Hamilton rating scale for depression (Elman et al., 1999). The generality of these findings to cocaine-dependent persons who meet diagnostic criteria for major depression remains to be determined. However, the hypothesis that the neurobiologic mediators of stress responses may also mediate the abuse-related effects of cocaine has implications for possible approaches to treatment. Taken together, these findings suggest that attenuation of cocaine-induced activation of the HPA axis may be beneficial for the treatment of cocaine abuse and dependence.

At present, there is no uniformly effective medication for treatment of cocaine abuse and dependence (Mendelson and Mello, 1996, 2008; Vocci et al., 2005; Vocci and Ling, 2005). Buprenorphine is an opioid-mixed agonist–antagonist that may be useful for the treatment of opioid abuse and dual dependence on cocaine and opioid drugs (see Mello et al. (1993a) and Mello and Mendelson (1995) for review). Interestingly, chronic treatment with buprenorphine (4 $mg\,day^{-1}$, s.l.) for at least 10 days reduced cocaine-related reports of euphoria and stimulation of ACTH in six cocaine- and opioid-dependent men. Basal ACTH levels were not significantly different before and

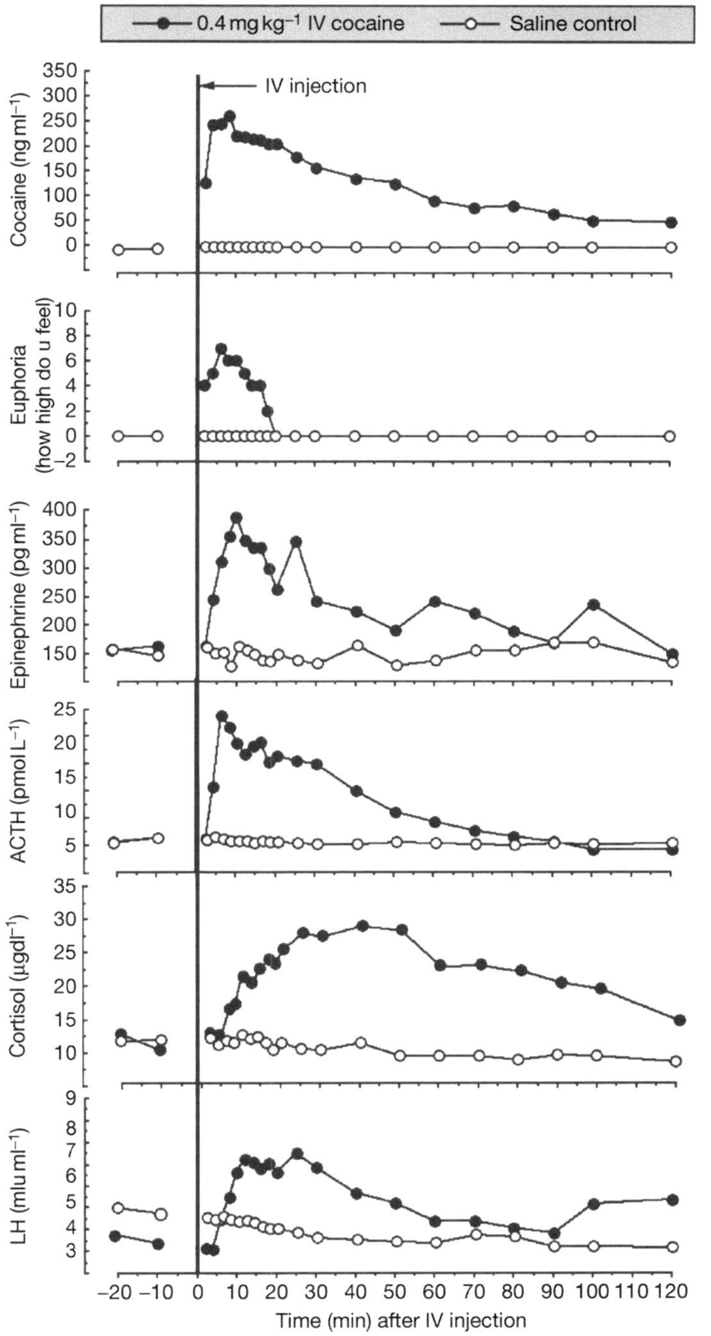

Figure 3 The effects of cocaine on euphoria, epinephrine, ACTH, cortisol, and LH in a male cocaine abuser. T me (min) after IV injection of cocaine is shown on the abscissa. The vertical line indicates when cocaine or placebo was administered. Plasma cocaine levels (ng ml^{-1}) before and after administration of 0.4 mg kg^{-1} cocaine (closed circles) or placebo (open circles) are shown in row 1. Euphoria reports before (BL) and after administration of cocaine or placebo are shown in row 2. Epinephrine levels (pg/ml) before (BL) and after administration of cocaine or placebo are shown in row 3. ACTH levels (pmol L^{-1}) before (BL) and after administration of cocaine or placebo are shown in row 4. Cortisol levels (μg dl^{-1}) before (BL) and after administration of cocaine or placebo are shown in row 5. LH levels before (BL) and after administration of cocaine or placebo are shown in row 6. This man is one of a group of individuals reported in Mendelson JH, Mello NK, Sholar MB, Siegel AJ, Mutschler N, Halpern J (2002) Temporal concordance of cocaine effects on mood states and neuroendocrine hormones. *Psychoneuroendocrinology* 27: 71–82.

after buprenorphine treatment and plasma cocaine levels after 30 mg IV cocaine were equivalent before and during buprenorphine treatment (Mendelson et al., 1992c). Outpatient clinical studies have found that buprenorphine is effective in reducing concurrent opioid and cocaine dependence in polydrug abusers (Gastfriend et al., 1993; Schottenfeld et al., 1993; Montoya et al., 2004). The observed reduction in the ACTH response to cocaine after chronic buprenorphine treatment is consistent with preclinical reports that opioids attenuate CRH-stimulated ACTH release (Rittmaster et al., 1985; Allolio et al., 1986). These data are also consistent with the notion that cocaine's rapid reinforcing effects in humans may be mediated, in part, by CRH activation of ACTH secretion (Mendelson et al., 1992a).

These clinical data are consistent with preclinical studies in rodents that also suggest that the reinforcing properties of cocaine, as well as amphetamine, may be influenced by HPA-axis activity (see Goeders and Guerin (1995), Piazza and LeMoal (1996), Goeders (1997, 2002a,b), Marinelli and Piazza (2002), Mello and Mendelson (2002), and Mendelson and Mello (2008) for review). At present, the contribution of HPA-axis activation to the abuse-related effects of cocaine is unclear, and may be both species and procedure dependent. However, there is increasing evidence that CRH-1 antagonists may modulate some aspects of drug withdrawal (Sinha, 2001; Koob et al., 2004; Stinus et al., 2005; Sinha et al., 2006) and are effective in treating anxiety and depression in humans (Holsboer, 2000, 2003; Grammatopoulos and Chrousos, 2002; Kehne and De Lombaert, 2002; Ising and Holsboer, 2007).

34.2.4.1 CRH antagonists: Development and behavioral implications

An interest in developing CRH antagonists has paralleled an increase in knowledge about the involvement of the HPA axis in a number of clinical disorders including depression, cardiovascular disease, and immune-system dysfunction. This vast and complex literature has been the topic of several comprehensive reviews (Gold et al., 1988; Dunn and Berridge, 1990; Chrousos and Gold, 1992; Heit et al., 1997; McEwen, 1998; Nemeroff, 1998; Papanicolaou, 1998; Bornstein and Chrousos, 1999), and it is beyond the scope of this chapter to attempt a summary here. However, some issues related to the development of CRH antagonists and their potential usefulness in the treatment of cocaine abuse and dependence are summarized below.

Two CRH receptors, CRH-1 and CRH-2, have been identified (Chalmers et al., 1996; Liaw et al., 1997; Martinez et al., 1998). CRH-1 receptors are widely distributed in the brain and predominate in the pituitary, whereas CRH-2 receptors are also found in the gastrointestinal tract and the heart (Chalmers et al., 1996; Martinez et al., 1998). Both peptidic compounds, such as D-Phe-CRH12–41 and astressin (Gulyas et al., 1995; Rivier et al., 1998), and nonpeptidic compounds, such as CP-154,526 and antalarmin, (Webster et al., 1996; Chen and Pollack, 1997; Arvanitis et al., 1999) are available for study. Antagonists selective for CRH-1 receptors may be most useful for influencing the abuse-related effects of cocaine, because CRH-1 receptors are thought to mediate CRH-induced stimulation of ACTH release (Chalmers et al., 1996).

CRH-1 receptor antagonists have been reported to produce both anxiolytic effects and alterations in drug self-administration in rodents (Schultz et al., 1996) and reductions in social-stress-induced fear and anxiety in rhesus monkeys (Habib et al., 2000; Ayala et al., 2004). The CRH-1 antagonist, antalarmin (10, 20, and 40 mg kg^{-1}) reduced fear responses related to social aggression in male rhesus monkeys, and these effects were not related to sedation (Habib et al., 2000). These data suggested that antalarmin had central effects because only modest decreases in ACTH and cortisol were measured in comparison to placebo–antalarmin conditions (Habib et al., 2000). Preclinical studies have shown that the CRH-1 antagonist, CP-154,526, binds with high affinity to CRH receptors and blocks adenylate cyclase activity stimulated by CRH in rat brain membranes (Lundkvist et al., 1996; Schultz et al., 1996). The majority of preclinical studies conducted in rodents are consistent with the notion that antagonism of stress-induced CRH activity may diminish anxiety-related behaviors in rodents. In behavioral studies, CP-154,526 reduced fear-potentiated startle, and reversed the escape deficit in rats previously exposed to inescapable footshock in a learned-helplessness paradigm (Schultz et al., 1996; Mansbach et al., 1997). Anxiolytic-like activity was also inferred from performance on an elevated plus-maze test by rats treated with CP-154,526 (1 mg kg^{-1}, IP; Lundkvist et al., 1996). However, there were some inconsistencies in the findings across laboratories. CP-154,526 (0.6–20 mg kg^{-1}, IP) did not affect performance on an elevated plus-maze or in conflict tests in rats (Griebel et al., 1998). However, in mice, CP-154,526 (5–20 mg kg^{-1}, IP) was more effective in reducing anxiety-related behaviors than the 5-HT1A receptor partial agonist buspirone (Griebel et al., 1998).

In a paradigm of stress-induced reinstatement of drug self-administration, footshock usually reinstates drug self-administration after extinction (defined as responding with no drug-reinforcer delivery; Erb et al., 1996; Ahmed and Koob, 1997; see Shaham et al. (2000) for review). However, when rats were pretreated with CP-154,526 (15 and 30 mg kg^{-1}, s.c.), footshock did not reinstate cocaine or heroin self-administration (Shaham et al., 1998). The failure of footshock to reinitiate drug-maintained responding could not be attributed to suppression of operant behavior by CP-154,526, because the same doses of this CRH antagonist did not change responding for a sucrose solution (Shaham et al., 1998). Similar findings were reported when the peptidic CRH antagonist, D-Phe-CRH12–41, was administered to rats (Erb et al., 1998). Footshock did not reinstate cocaine self-administration by normal rats after CRH-antagonist administration (Erb et al., 1998). In addition, footshock alone did not reinstate cocaine self-administration in adrenalectomized rats, as it did in intact rats or adrenalectomized rats with corticosterone replacement (Erb et al., 1998). One implication of these findings is that CRH antagonist treatment may also reduce relapse to cocaine and heroin self-administration (Shaham et al., 1998).

Consistent with findings in the relapse studies, a CRH-1 antagonist, CP-154,526, reduced cocaine self-administration with minimal effects on food-maintained responding (Goeders and Guerin, 2000). Rats were trained to self-administer food and IV cocaine in alternating 15-min food and drug components during 2-h sessions. Saline was substituted for cocaine in two sessions each week until saline-maintained responding decreased to low levels. After cocaine- and saline-maintained behaviors were stable, CP-154,526 (10–40 mg kg^{-1}, IP) or vehicle control was administered 30 min before the session. At doses of 20–40 mg kg^{-1}, IP, CP-154,526 significantly decreased cocaine-maintained responding to below saline-extinction levels. Subsequently, separate groups of rats were trained to self-administer 0.125, 0.25, or 0.5 mg kg^{-1}/inj cocaine and the effects of 20 mg kg^{-1}, IP, CP-154,526 on cocaine- and food-maintained responding were evaluated. This CRH antagonist produced dose-related decreases in cocaine self-administration and shifted the cocaine dose–effect curve downward. CP-154,526 reduced cocaine-maintained responding significantly below vehicle control treatment. Moreover, cocaine-maintained responding after CP-154,526 was lower than saline-maintained responding. These effects of CP-154,526 appeared to be selective for cocaine, because food-maintained responding was not affected. Moreover, the same CRH-1 antagonist blocked CRH stimulation of ACTH in rats (Schultz et al., 1996). When rats were given extended (6-h) access to cocaine, the CRH-1 antagonist antalarmin (25 mg kg^{-1}, IP) significantly reduced cocaine self-administration (Specio et al., 2008). However, CRH-1 was not effective at lower doses and did not decrease cocaine self-administration in rats given only 1 h of access to cocaine each day (Specio et al., 2008). These data were interpreted to suggest that extended access to cocaine may induce hypersensitivity of the CRF system, which in turn contributes to cocaine self-administration (Specio et al., 2008). To date, there have been no clinical studies of the effects of a CRH antagonist on cocaine-induced euphoria or cocaine self-administration.

Taken together, these data are consistent with the possibility that attenuation of cocaine-induced activation of the HPA axis could be a safe and effective approach to cocaine-abuse treatment. Moreover, CRH antagonists may also have potential therapeutic effects within the broader context of psychiatric disorders that are comorbid with cocaine abuse and dependence. Depression is often one component of the clinical profile of cocaine abuse and dependence (Rounsaville et al., 1991; Schmitz et al., 2000; Mendelson and Mello, 2001, 2008). Depression following cocaine withdrawal in cocaine-dependent persons is very similar to symptoms of depression observed in other neuropsychiatric disorders (*Diagnostic and Statistical Manual of Mental Disorders, Fourth Edition*, 292.9 (DSM IV, 292.9)) that are associated with dysregulation of the HPA axis (APA, 1994). For example, CRH hypersecretion, associated with adrenal and pituitary enlargement, is thought to be associated with the pathophysiology of depression (see Nemeroff (1996), Heit et al. (1997), and Holsboer (2003) for review). In a seminal review, Nemeroff suggests that depression can be conceptualized as a "pathological stress response gone awry" (Nemeroff, 1996). The effectiveness of CRH-1 antagonists in reducing depression remains to be determined. However, it has been reported that administration of a single dose of hydrocortisone reduced symptoms of depression more effectively than a CRH agonist, ovine CRH (DeBattista et al., 2000). Clearly, brain corticotropin-releasing factors and the HPA axis play an important role in the regulation of subjective states as well as modulation of responses to stress. Preclinical studies suggest that drug withdrawal also may be associated with activation of the brain CRH systems (Koob et al., 1993; Rodriguez de Fonseca et al., 1997; Richter and Weiss, 1999). Thus, treatment with a CRH antagonist during cocaine withdrawal might normalize CRH activation and ameliorate withdrawal-related anxiety and anhedonia that is thought to contribute to relapse to drug use (Koob and Le Moal, 1997).

34.3 Cocaine's Effects on Gonadotropins and Gonadal Steroid Hormones

34.3.1 Background

The gonadotropin hormones, luteinizing hormone (LH), and follicle-stimulating hormone (FSH), are released from gonadotropes located in the anterior pituitary. Most gonadotropes (60%) contain both LH and FSH, but some small gonadotropes contain only LH or FSH. The release of both LH and FSH is stimulated by a hypothalamic peptide hormone, luteinizing-hormone-releasing hormone (LHRH), also known as gonadotropin-releasing hormone (GnRH). LH and FSH release is pulsatile and covaries with the episodic secretory release of hypothalamic LHRH (Knobil, 1974, 1980; Yen, 1999a,b). LH and FSH regulate, and are regulated by, gonadal steroid hormones. The ovaries are the primary source of the steroid hormones estradiol and progesterone in women. Estradiol is produced by the ovarian follicles, and estradiol production is maximal before the onset of the LH surge. Progesterone is secreted from the preovulatory ovarian follicle and from the corpus luteum after ovulation. Progesterone is also secreted by the adrenal gland in both men and women (Risner et al., 1985; De Geyter et al., 2002). The testes are the primary source of testosterone and estradiol in men. Estradiol is produced by the Sertoli cells and testosterone by the Leydig cells in the testes; however, testosterone and androstenedione may be converted to estradiol under some conditions (Veldhuis, 1999).

34.3.1.1 Changes in gonadotropin and gonadal steroid hormone levels across the menstrual cycle

Changes in basal levels of gonadotropins and gonadal steroid hormones define the functionally distinct phases of the menstrual cycle, the follicular phase, the ovulatory phase, and the luteal phase (Hotchkiss and Knobil, 1994; Yen, 1999b). The patterns of hormonal changes across a typical menstrual cycle are shown schematically in **Figure 4**. The follicular phase of the cycle begins on the first day of menstruation, lasts for about 13 days, and is followed by the periovulatory phase at mid-cycle. During the follicular phase, adequate FSH levels are necessary for normal development and maturation of the ovarian follicles (Goodman and Hodgen, 1983; Ross, 1985). Studies of folliculogenesis in the primate ovarian cycle indicate that the recruitment of the dominant follicle occurs during days 1 to 4 of the menstrual cycle; a single follicle is selected to ovulate during days 5 to 7, and the selected follicle achieves dominance during cycle days 8 to 12 (diZerega and Hodgen, 1981a; Hodgen, 1982; Goodman and Hodgen, 1983; Gougeon, 1996). The dominant follicle selected for ovulation presumably inhibits development of other competing follicles, but the factors that determine selection and dominance of a single ovulatory follicle are unclear (Gougeon, 1996). A preovulatory surge in LH results in maturation of the dominant ovarian follicle and is essential for

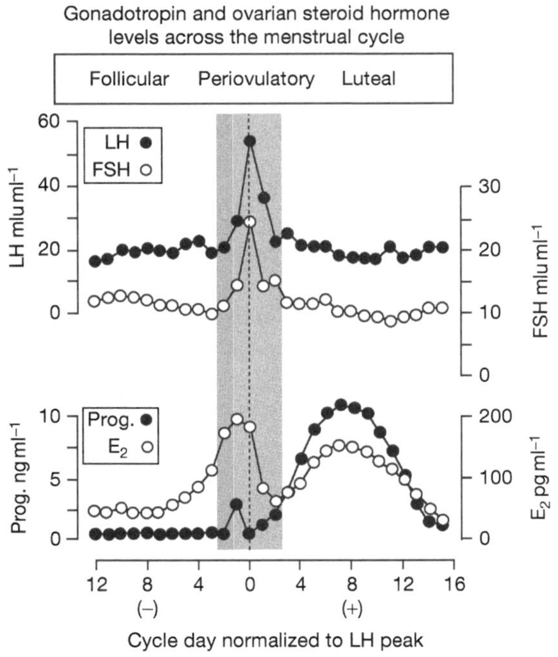

Figure 4 Changes in the gonadotropins (LH and FSH) and the ovarian steroid hormones (estradiol and progesterone) across a typical menstrual cycle in rhesus monkeys and human females. The successive phases of the menstrual cycle are labeled at the top of the diagram. Menstruation defines the beginning of a cycle. Hormone levels are shown as days from the LH peak. This schematic was constructed from data on the human menstrual cycle (Yen, 1991). The rhesus monkey's menstrual cycle is virtually identical to that of humans except that it has not been possible to detect the preovulatory increase in progesterone (Hotchkiss and Knobil, 1994; Knobil, 1980; Knobil and Hotchkiss, 1988).

ovulation to occur. Rupture of the oocyte from the dominant follicle occurs approximately 36 h after the preovulatory LH surge begins. After ovulation, the site of the dominant follicle becomes highly vascularized and forms the corpus luteum. The corpus luteum lasts for approximately 14 days and then spontaneously regresses unless pregnancy occurs. Secretion of progesterone from the corpus luteum defines the luteal phase of the menstrual cycle and indicates whether ovulation occurred. The development of ovarian follicles for the next menstrual cycle is dependent upon the regression of the corpus luteum. The end of the luteal phase, when FSH levels rise and initiate follicle recruitment, is often referred to as the luteal–follicular transition (Yen, 1999b).

34.3.1.2 Interactions between gonadotropins and gonadal steroid hormones

Gonadotropin and ovarian steroid hormones levels are controlled by a complex and changing pattern of reciprocal stimulation and inhibition across the menstrual cycle. Both estradiol and progesterone may have inhibitory or stimulatory effects on gonadotropin release at different phases of the menstrual cycle. Changes in gonadotropin levels also reflect changes in pulsatile release patterns as described below. Abnormally high or low hormone levels can have a variety of functional consequences. For example, during the follicular phase, FSH is one important determinant of follicle development, and low levels of FSH may delay follicle maturation and ovulation or may result in luteal-phase dysfunction after timely ovulation (Wilks et al., 1977; diZerega and Hodgen, 1981b; Goodman and Hodgen, 1983; diZerega and Wilks, 1984). FSH levels also are influenced by estradiol and LH. An increase in estradiol levels during the early follicular phase suppresses FSH, inhibits preovulatory follicular growth, and prolongs the follicular phase (Dierschke et al., 1985, 1987; Hutz et al., 1990). Clinical evidence suggests that high levels of LH during the early follicular phase may also impair normal folliculogenesis and lead to luteal-phase defects (Soules et al., 1987, 1989a; McNeely and Soules, 1988).

During the periovulatory phase, the preovulatory LH and FSH surge is stimulated by increases in estradiol levels that occur over 2–3 days. In women, an abrupt increase in progesterone occurs about 12 h before the onset of the LH surge (Yen, 1999b). After ovulation, LH levels decrease and the luteal phase of the menstrual cycle begins (Soules et al., 1984; Nippoldt et al., 1989). Progesterone levels gradually increase and remain elevated until the postovulatory corpus luteum regresses. The inhibitory feedback actions of estradiol on the gonadotropins presumably occur at the level of the anterior pituitary, because after destruction of the arcuate nucleus in the hypothalamus or transection of the pituitary, estradiol administration still resulted in a decrease in pulsatile LH release and a decrease in LH levels (Knobil, 1980; Ferin et al., 1984). The co-modulatory interactions between ovarian steroid hormones and the gonadotropins also are illustrated by the fact that after ovariectomy or natural menopause, when the inhibitory influence of ovarian steroids is removed, LH and FSH remain at high levels comparable to the periovulatory period. Similarly, in males, the gonadal steroid hormones, testosterone and estrogen, modulate LH release, and disruption of the hypothalamic–pituitary–testicular axis results in elevated LH levels (Veldhuis, 1999). LH stimulates increased secretion of testosterone, and testosterone in turn inhibits LH secretion. Acute administration of testosterone decreases the frequency of LH pulsatile release, whereas acute administration of estradiol decreases the amplitude of LH pulses in men (Veldhuis, 1999).

34.3.1.3 Regulation of pulsatile gonadotropin release patterns

The changing levels of gonadotropins across the menstrual cycle reflect changes in pulsatile release patterns. These patterns are controlled by ovarian steroid hormones and hypothalamic release of LHRH. Current understanding of the neuroendocrine regulation of the menstrual cycle is based on the fundamental discovery that pulsatile gonadotropin release is essential for normal reproductive function (Knobil, 1974, 1980; Knobil and Hotchkiss, 1988). When hypothalamic release of endogenous LHRH was disrupted by lesions of the arcuate nucleus and the median eminence in ovariectomized rhesus monkeys, LH- and FSH-secretory activities were abolished. Pulsatile administration of synthetic LHRH restored the pulsatile release patterns of LH and FSH, whereas continuous administration of LHRH did not (Knobil, 1974, 1980; Knobil and Hotchkiss, 1988). These important findings in rhesus monkeys were rapidly translated into clinical treatment for infertility disorders in women. It was discovered that suppression of pulsatile gonadotropin release was often associated with amenorrhea (a failure to menstruate) and that normal ovulatory function could be restored by the pulsatile infusion of synthetic LHRH (Lyendecker and Wildt, 1983; Hurley et al., 1984;

Crowley et al., 1985; Santoro et al., 1986a,b; Conn and Crowley, 1991).

Recognition of the importance of pulsatile gonadotropin release for normal reproduction (Knobil, 1974, 1980) led to studies of how gonadotropin release patterns changed across the menstrual cycle (Ferin et al., 1984; Veldhuis et al., 1984). LH-pulse frequency increases across the follicular phase of the menstrual cycle from one to two pulses per hour. During the mid-cycle LH surge, there is an increase in LH-pulse amplitude, but no change in the frequency of LH pulsatile release (Adams et al., 1994). The observed increase in LH-pulse amplitude may reflect an increase in pituitary sensitivity to LHRH release (Adams et al., 1994). After ovulation, LH-pulse frequency decreases to approximately one pulse every 4 h during the luteal phase (Ferin, 1984; Soules et al., 1984; VanVugt et al., 1984). Although LH pulsatile release is slower during the luteal phase, the amplitude of LH pulses is almost double of that measured during the early follicular phase in rhesus monkeys (Ferin et al., 1984).

Several lines of evidence indicate that the pulsatile release of LHRH from the hypothalamus is under inhibitory control by endogenous opioid peptides, but the role of norepinephrine and dopamine in primates is unclear (Yen, 1999b). Although LHRH cannot be measured directly in peripheral blood, LH release mirrors LHRH release in preparations where LHRH is measured in cerebrospinal fluid (Xia et al., 1992). Opioid drugs decrease circulating levels of LH (Ferin, 1984; Ferin et al., 1984). Levels of an endogenous opioid, β-endorphin, measured in pituitary stalk blood, increased from the onset of menstruation through the follicular phase and were highest at the luteal phase (Ferin et al., 1984). The administration of an opioid antagonist, such as naloxone or naltrexone, is followed by an increase in LH release during the luteal phase of the menstrual cycle. Opioid-antagonist administration can be used as a provocative test of hypothalamic–pituitary function (Yen et al., 1985; Mendelson et al., 1986). However, opioid antagonists do not change gonadotropin release patterns after menopause (Reid et al., 1983), and these findings are usually interpreted to suggest that gonadal steroids as well as opioid peptides are important in the inhibitory regulation of LHRH release (Yen et al., 1985; Yen, 1999b). LHRH neurons synapse with neurons in the medial basal hypothalamus that contains β-endorphin and dopamine. Axons from cells containing β-endorphin, dopamine, and LHRH, all terminate in the median eminence. It is thought that ovarian steroids influence LHRH release through the estrogen and progesterone receptors on β-endorphin and dopamine neurons, rather than by directly affecting LHRH neurons, but the basis for these co-modulatory interactions remains to be clarified (Yen, 1999b).

In the remainder of this section, the acute effects of cocaine on gonadal steroid hormones and gonadotropin release will be described. Some possible consequences of cocaine's effects on the interactions between gonadal steroid hormones and gonadotropins are discussed. One of the most consistent findings to emerge from both clinical and preclinical studies is that acute cocaine administration is followed by a rapid increase in LH release. This effect was not predictable from the pharmacology of cocaine or from the clinical literature on dopamine–gonadotropin interactions. The functional implications of a cocaine-related increase in LH are unclear, but if repeated episodes of cocaine intoxication are accompanied by sustained elevations in LH, this could compromise folliculogenesis and/or prompt early ovulation. Until recently, there has been relatively little attention given to cocaine's effects on gonadal steroid hormones despite their obvious importance in the control of gonadotropin release throughout the menstrual cycle. Now, it appears that cocaine also stimulates release of estradiol of rhesus females during the follicular phase of the menstrual cycle (Mello et al., 2000, 2004). However, cocaine did not alter progesterone levels during either the follicular or the luteal phase in rhesus females (Mello et al., 2000).

34.3.2 Clinical Studies of Cocaine Effects on Gonadotropin Hormones

34.3.2.1 Acute effects of cocaine on LH in men and women

The finding that cocaine stimulates LH in gonadally intact rhesus monkeys (Mello et al., 1990a,b, 1993b) was confirmed in human male cocaine abusers (Mendelson et al., 1992a) and cocaine-naive men (Heesch et al., 1996). Eighteen men who reported more than 10 years of concurrent cocaine and opioid abuse were studied in a clinical research ward after detoxification (Mendelson et al., 1992b). Acute administration of 30 mg, IV, of cocaine was followed by a significant increase in LH within 5 min when plasma cocaine levels averaged 260 ng ml^{-1}. LH levels were maximal at 15 min, a time course comparable to that observed in rhesus monkeys. There were no changes in LH after placebo–cocaine administration (Mendelson et al., 1992b). These cocaine-related changes in LH did not appear to be a function of cocaine experience because similar findings were reported in cocaine-naive men (Heesch et al., 1996). After administration of intranasal cocaine (2 mg kg^{-1}), LH increased significantly within 30 min and reached peak levels within 60 min. Plasma cocaine levels averaged 102 ng ml^{-1} within 40 min after intranasal cocaine and reached peak levels of 142 ng ml^{-1} at 80 min (Heesch et al., 1996).

The acute effects of cocaine (0.2 and 0.4 mg kg^{-1}, IV) on LH were studied in men and women who met *Diagnostic and Statistical Manual of Mental Disorders, Fourth Edition* (DSM-IV) criteria for cocaine abuse (Mendelson et al., 2001). **Figure 5** shows LH and plasma cocaine levels after 0.2 mg kg^{-1} and 0.4 mg kg^{-1} in men (row 1) and women studied during the follicular and the luteal phases of the menstrual cycle (rows 2 and 3). Menstrual-cycle phase was confirmed by determination of estradiol and progesterone levels. Cocaine was more potent in stimulating LH in men than in women studied under identical conditions. In women, LH levels did not change significantly from baseline after administration of 0.2 mg kg^{-1} cocaine during the follicular or the luteal phase of the menstrual cycle. After administration of 0.4 mg kg^{-1} cocaine, LH increased significantly during both phases of the menstrual cycle. Average LH levels exceeded 80 ng ml^{-1} within 16 min after IV cocaine administration when plasma cocaine levels exceeded 220 ng ml^{-1}. LH increased significantly in men after administration of both a low (0.2 mg kg^{-1}) and a high (0.4 mg kg^{-1}) dose of cocaine. Although peak plasma cocaine levels differed and averaged 86 and 169 ng ml^{-1} after the low and high cocaine dose, the magnitude of the LH increase was equivalent (77 and 72 ng ml^{-1}) in men. The factors that account for these sex differences are unclear (Mendelson et al., 2001).

34.3.3 Clinical Studies of Chronic Cocaine Effects on LH

As noted earlier, it has been difficult to conduct clinical studies in individuals who abuse only cocaine, and multiple-drug use has an undetermined effect on the endocrine variables measured. Despite this caveat, a history of cocaine abuse was not associated with abnormal LH and testosterone pulsatile release patterns in men. The pulsatile release patterns of LH and testosterone were examined in eight normal control men and eight men who met *Diagnostic and Statistical Manual of Mental Disorders, Third Edition, Revised* (DSM-III-R) diagnostic criteria for cocaine abuse (Mendelson et al., 1989b). These men reported 1–7 years of cocaine abuse and use of 1–14 g of cocaine per week. No opioid use was reported, but all subjects used marijuana and alcohol. Although subjects were cocaine-abstinent during the sample collection period, all had used cocaine recently as indicated by qualitative urine analyses. Subjects were admitted to a clinical research ward and endocrine studies were conducted after an overnight fast. Samples for hormone analyses were collected at 10-min intervals over six consecutive hours. LH and testosterone pulsatile release patterns were analyzed with the Cluster Analysis program (Veldhuis and Johnson, 1986). There was no significant difference in LH-pulse frequency or peak duration between cocaine abusers and eight control subjects. To the best of our knowledge, similar studies have not been conducted in women who abuse cocaine.

34.3.3.1 Implications of cocaine's stimulation of LH

The physiologic and/or clinical significance of cocaine-related increases in LH remains to be determined. In women, high levels of LH during the follicular phase can impair folliculogenesis (Soules et al., 1987, 1989a; McNeely and Soules, 1988), and recurrent episodes of cocaine intoxication could have a similar effect. Cumulative increases in LH could also trigger early ovulation. Increases in LH may also interact with some of cocaine's purported effects on sexual arousal. Clinical studies indicate that increases in LH levels were significantly related to the degree of sexual arousal reported by young men during an erotic film in comparison to a neutral control film (La Ferla et al., 1978). Measures of penile tumescence also were correlated with reports of sexual arousal (La Ferla et al., 1978). Clinical descriptions of cocaine's acute effects often include increased sexual feelings and energy, as well as intense euphoria (Siegel, 1982; Van Dyke and Byck, 1983; Smith et al., 1984; Gawin and Ellinwood, 1988). Consequently, it is possible that the cocaine-related increase in LH may covary with sexual arousal. However, it is unlikely that there is a simple relationship between increased LH levels and sexual arousal or sexual behavior, and the possible behavioral significance of these findings is unknown.

In human cocaine abusers, repeated binge cocaine self-administration is the most common use pattern (Ward et al., 1997). Only a single dose of cocaine was administered in our studies (Mello et al., 2000; Mendelson et al., 2001), and the effects of repeated

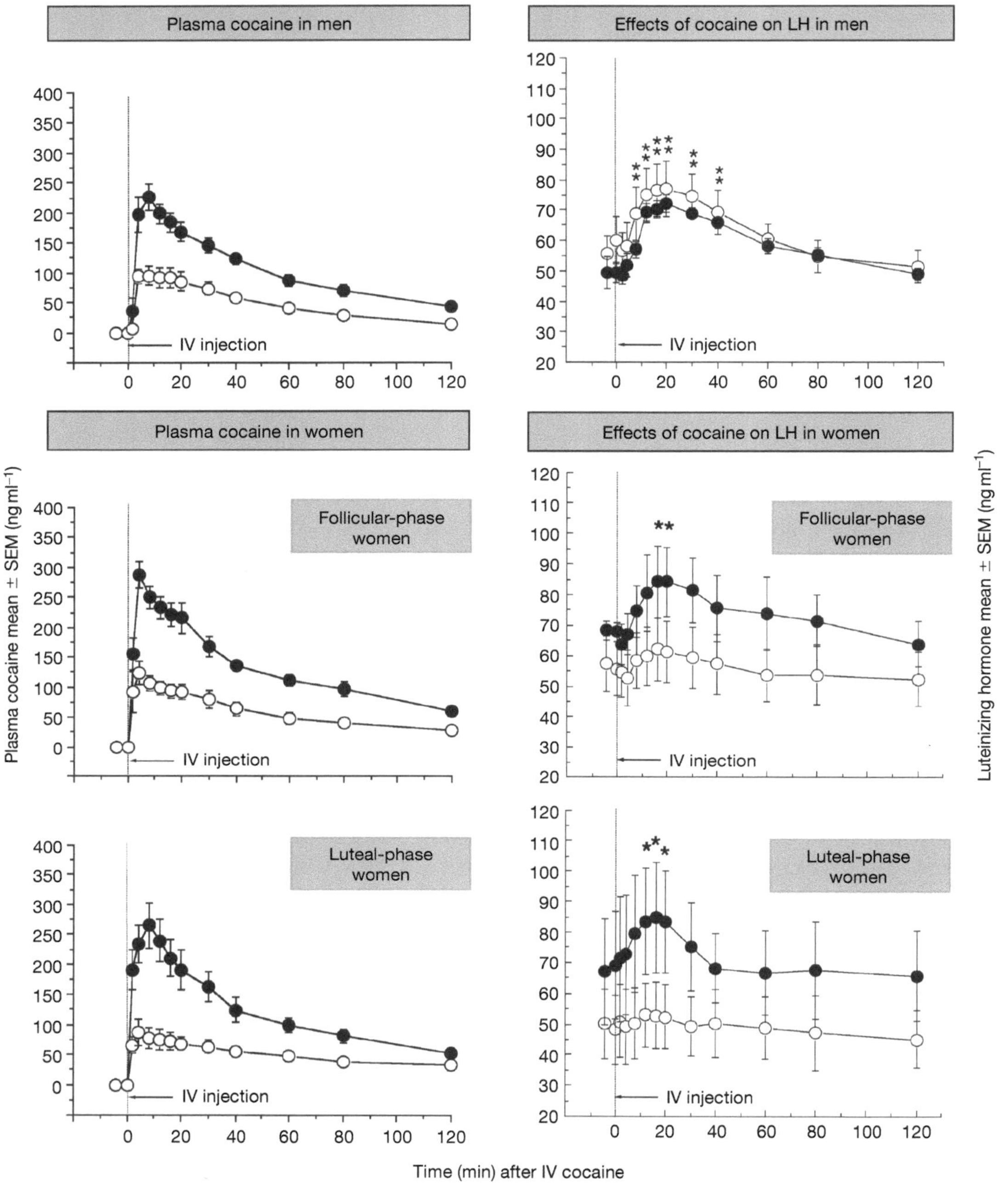

Figure 5 Plasma cocaine and LH levels in men and women after IV cocaine administration. Plasma cocaine levels are shown in the left panel and LH levels are shown in the right panel. Cocaine levels and LH data collected in men are shown in row 1. Cocaine levels and LH data collected in women during the follicular phase of the menstrual cycle are shown in row 2 and cocaine levels and LH data collected during the luteal phase of the menstrual cycle are shown in row 3. Plasma cocaine levels (ng ml^{-1}) are shown on the left ordinates in column 1. LH levels (ng ml^{-1}) are shown on the left ordinates in column 2. The vertical lines indicate when cocaine was administered. Time (min) after cocaine administration is shown on the abscissa. Plasma cocaine and LH levels after 0.2 mg kg^{-1} IV cocaine are shown as open circles. Plasma cocaine and LH levels after 0.4 mg kg^{-1} IV cocaine are shown as closed circles. After administration of 0.2 mg kg^{-1} IV cocaine, each data point is the average (± S.E.M.) of six men and five women. After administration of 0.4 mg kg^{-1} IV cocaine, each data point is the average (± S.E.M.) of six men and six women. Adapted from Mendelson JH, Sholar M, Siegel AJ, Mello NK (2001) The effects of cocaine on luteinizing hormone in women during the follicular and luteal phases of the menstrual cycle and in men. *Journal of Pharmacology and Experimental Therapeutics* 296: 972–979.

cocaine injections on LH and estradiol are unknown. However, if repeated cocaine administration continued to stimulate comparable increases in LH and estradiol, this could contribute to the menstrual-cycle disruptions seen during chronic cocaine exposure in rhesus monkeys (Mello et al., 1997; Chen et al., 1998; Potter et al., 1998, 1999). For example, continuous exposure to estradiol for only 24 h on day 6 of the menstrual cycle resulted in atresia of the dominant ovarian follicle in rhesus monkeys (Dierschke et al., 1985). Repeated 24-h exposures to estradiol at 10-day intervals resulted in recurrent atresia of the dominant ovarian follicle, but normal ovulatory menstrual cycles occurred when estradiol treatments were separated by 14 days (Dierschke et al., 1987). Atresia of the dominant ovarian follicle usually results in an anovulatory cycle. Anovulatory cycles were often observed during chronic cocaine self-administration (Mello et al., 1997) or chronic cocaine administration (Chen et al., 1998; Potter et al., 1998, 1999).

34.4 Interactions between Cocaine, Sex, and Gonadal Steroid Hormones

34.4.1 Background

In addition to cocaine's effects on gonadal steroid hormones, there is increasing evidence that gonadal steroid hormones influence behavioral and psychological responses to cocaine. The importance of estradiol in modulating cocaine's effects was first suggested by early observations that female rats were more resistant to the toxic effects of cocaine than male rats (Downs and Eddy, 1932; Guerrero et al., 1965). Moreover, female rats given estradiol were more resistant to cocaine-induced toxicity than control females (Selye, 1971; Rapp et al., 1979). Cardiovascular toxicity after cocaine administration was significantly greater in males and ovariectomized rats, than in intact female rats (Dickerson et al., 1991; Morishima et al., 1993). These data were consistent with a now-compelling scientific literature on the protective effects of estrogens on cardiovascular function (Farhat et al., 1996; Skafar et al., 1997; Minshall et al., 1998).

Consistent with these cardiovascular effects of estrogen, changes in estrogen and progesterone levels across the rodent estrous cycle appear to modulate cocaine's effects on several behavioral measures. Both estradiol and cocaine increase extracellular dopamine levels measured in microdialysis studies in rodents (Di Chiara, 1995; Hemby et al., 1997; Becker, 1999; Becker et al., 2001; Smith et al., 2006). Dopamine is generally thought to subserve the reinforcing effects of cocaine (Kuhar et al., 1991; Koob, 1992; Woolverton and Johnson, 1992), and the extent to which cocaine-induced increases in estradiol levels may contribute to cocaine-related increases in extracellular dopamine in the nucleus accumbens and ventral striatum is unknown. In the first study to examine cocaine self-administration and phases of the estrous cycle, it was found that high estrogen levels during estrus enhanced the reinforcing saliency of cocaine in female rats (Roberts et al., 1989). Rats were trained to self-administer food, then cocaine on a fixed ratio of 1 (FR I) schedule for one complete estrous cycle (4 days). Then, the reinforcing efficacy of cocaine was evaluated on a progressive ratio schedule in males and females. In this procedure, the response requirement for a single injection of cocaine was gradually increased from one to a maximum of 999 responses until rats stopped responding for 1 h. The ratio at which responding ceased was defined as the break point. Progressive ratio break points were significantly higher during estrus than during other stages of the estrous cycle. Moreover, females reached significantly higher progressive ratio break points (264 responses/injection) than males (48 responses/injection). However, this effect of estrus was only apparent when response requirements were high. When animals could acquire cocaine for a single response per injection, cocaine self-administration did not vary with stages of the estrous cycle (Roberts et al., 1989).

A similar effect of estrus on cocaine self-administration was observed when rats could select the dose of cocaine available by changing the infusion duration (Lynch et al., 2000). Responses on one lever increased the dose of cocaine by increasing infusion duration and responses on the other lever decreased the dose of cocaine by decreasing the infusion duration. Each 5-h cocaine self-administration session began with an infusion duration of 12 s which yielded a cocaine dose of 1.2 mg kg^{-1}. Rats could adjust doses in 3-s increments or decrements over a range of 0–24 s (0–2.4 mg kg^{-1} cocaine). The stage of the estrous cycle was monitored with vaginal swabs twice a day and only data from sessions in which females remained in the same phase for 5 h were analyzed. Under these conditions, females selected significantly higher cocaine doses during estrus than during metestrus/diestrus. Total cocaine intake per session was also higher during estrus than metestrus/diestrus or proestrus, but these differences were not statistically significant. During estrus, over 80% of responses were on the lever that increased the cocaine dose, whereas response patterns were more variable during other phases of the estrous cycle (Lynch and Carroll, 2000).

Although cocaine dose selection was modulated by the stage of the estrous cycle, there were no significant sex differences in cocaine-dose-selection patterns or in the rate of acquisition of cocaine self-administration and stability of cocaine intake (Lynch and Carroll, 2000). These findings were inconsistent with an earlier report from the same laboratory that females acquired cocaine self-administration more rapidly than males (Lynch and Carroll, 1999). In that study, a relatively low dose of cocaine (0.2 mg kg^{-1}) was available on an FR1 in an autoshaping procedure (Lynch and Carroll, 1999). In adolescent rats, females acquired cocaine self-administration more readily than males when high unit doses (0.75 mg kg^{-1}/inj) were available (Lynch, 2008). In addition, females reached higher progressive ratio break points for cocaine, and cocaine self-administration was greatest during estrus when estradiol levels were highest (Lynch, 2008).

Sex differences were detected when the dependent variable was the reinstatement of cocaine self-administration and a priming dose of cocaine was administered after extinction of cocaine-maintained responding. Resumption of drug self-administration after extinction of drug-maintained responding is a commonly used animal model of relapse (de Witt and Stewart, 1981). Exposure to drug-related cues as well as stress has been shown to reinstate drug-seeking behavior (Shaham et al., 2000). In one study, male and female rats were trained to self-administer cocaine (0.2 mg kg^{-1}/inj) on an FR1 schedule (Lynch and Carroll, 1999). After at least five training sessions, cocaine access was restricted to the first 2 h of each 7-h session and saline was available during the remaining 5 h. Saline-priming injections were administered to clear the catheters of cocaine at the beginning of each saline-availability period. Training continued until rats self-administered no more than ten saline infusions over 5 h. Subsequently, priming doses of cocaine (0.32, 1.0, and 3.2 mg kg^{-1}) were administered at the beginning of hour 4 and the catheter was cleared with 0.3 ml saline. Males and females did not differ significantly in baseline cocaine self-administration or in extinction of saline-maintained responding after a saline-priming injection, or after a low priming dose of cocaine (0.32 mg kg^{-1}/inj). However, after higher cocaine priming doses (1.0 and 3.2 mg kg^{-1}), females self-administered significantly more saline injections than males (Lynch and Carroll, 1999). The stage of the estrous cycle was not monitored in that study; therefore, the contribution of the hormonal milieu to the gender differences reported cannot be determined.

If high estrogen levels facilitate cocaine self-administration in gonadally intact female rats (Roberts et al., 1989; Lynch et al., 2000), this suggests that ovariectomy might attenuate cocaine's reinforcing effects, and that estrogen replacement might increase cocaine self-administration by ovariectomized females. However, contrary to this hypothesis, ovariectomy did not change cocaine self-administration dose–effect curves in female rats studied in an own-control design (Caine et al., 2004). Rats were trained to respond for food on an FR5 schedule of reinforcement, then implanted with an IV catheter, and trained to self-administer cocaine (1 mg kg^{-1}/inj) on an FR5 in 3-h sessions. Once cocaine self-administration was stable, dose–effect curves were determined, and cocaine unit doses (0.03–3.0 mg kg^{-1}/inj) were presented in separate sessions in an irregular order. After ovariectomy, 2 weeks of recovery preceded testing to ensure that hormone levels had stabilized following ovariectomy. When rats were re-exposed to cocaine after ovariectomy, the dose–effect curves for cocaine were equivalent to those acquired before ovariectomy (Caine et al., 2004).

Estradiol replacement in ovariectomized rats also has had inconsistent effects across studies. Ovariectomized rats given estradiol met acquisition criteria for cocaine self-administration more readily, and took more cocaine than ovariectomized rats without estradiol replacement (Lynch et al., 2001; Hu et al., 2004; Jackson et al., 2006; Hu and Becker, 2008). In gonadally intact rats, an antiestrogen, tamoxifen, reduced the number of rats that met acquisition criteria for cocaine self-administration from 80 to 30% (Lynch et al., 2001). However, in other studies, estrogen replacement did not enhance cocaine self-administration in ovariectomized rats (Grimm and See, 1997). Ovariectomized rats were trained to self-administer cocaine (0.263 mg/inj) on a progressive ratio schedule in which response requirements increased in an exponential series from 1 to 999. Under these conditions, rats self-administered an average of 10 mg kg^{-1} day^{-1} cocaine. When estrogen (37 μg kg^{-1} E_2 in 100 μg ml^{-1} safflower-oil solution) was administered in a subcutaneous silastic tube, cocaine self-administration decreased significantly below baseline levels. Cocaine self-administration was variable, but usually below baseline levels over 12 days of observation. Ovariectomized animals with estrogen replacement self-administered an average of nine cocaine injections per day, whereas ovariectomized controls self-administered an average of 11.78 cocaine injections per day. Cocaine self-administration did not change in the vehicle control group (Grimm and See, 1997).

There were also no differences in the acquisition of cocaine self-administration (1.5 mg kg^{-1}/inj) between ovariectomized females with and without estradiol replacement (Lynch and Taylor, 2005). Moreover, there were no differences in the shape and position of the cocaine self-administration dose–effect curve (0.032–3.2 mg kg^{-1}/inj) determined in ovariectomized females before and after estradiol replacement (Caine et al., 2004). Similarly, although ovariectomized female rats with estrogen replacement self-administered significantly more cocaine then ovariectomized females without estrogen replacement at cocaine unit doses of 0.3 and 0.4 mg kg^{-1}/inj, there were no differences in responding for a higher unit dose of 0.5 mg kg^{-1}/inj cocaine (Hu et al., 2004). Daily administration of an acute low dose of estradiol (1 and 2 μg) was more effective than a higher dose (5 μg) in facilitating acquisition of cocaine self-administration in ovariectomized rats (Hu and Becker, 2008). Chronic estradiol administration with a subcutaneous pellet (1.5 mg) also did not facilitate acquisition of cocaine self-administration in comparison to vehicle control treatment (Hu and Becker, 2008). However, all ovariectomized females, including vehicle-control treated females, self-administered cocaine and unit doses of 0.4–0.5 mg kg^{-1}/inj were more reinforcing than 0.2, 0.3, and 0.75 mg kg^{-1}/inj under most estradiol-treatment conditions (Hu and Becker, 2008). Taken together, these studies suggest that the presence of estradiol hormones is not essential for stable cocaine self-administration.

34.4.2 Interactions between Cocaine, Sex, and Menstrual-Cycle Phase

These systematic preclinical studies of cocaine's interactions with sex and gonadal steroid hormones have been paralleled by a series of clinical investigations. Evidence that changes in hormonal levels across the menstrual cycle may influence the pharmacokinetics and pharmacodynamics of treatment medications, as well as possible sex differences in drug efficacy, and toxicity led to changes in the Food and Drug Administration policy, and the development of clinical guidelines for inclusion of women in clinical trials (Merkatz et al., 1993). The extent to which sex differences in the biologic effects of drugs may reflect the differences in hormonal milieu between men and women is poorly understood. However, recent clinical studies suggest the possible importance of estradiol and progesterone in modulating cocaine's behavioral and biologic effects.

34.4.2.1 Sex, menstrual-cycle phase, and cocaine pharmacokinetics

Studies of cocaine's effects on subjective-effect reports and neuroimaging endpoints often suggest that sex and menstrual-cycle phase may influence the data obtained. Detection of sex or menstrual-cycle-phase differences in cocaine's acute effects raises the question of whether or not such findings could reflect sex differences in cocaine's pharmacokinetics. Most studies of cocaine pharmacokinetics have been conducted in men (Cone, 1995; Evans et al., 1996). One study compared the pharmacokinetics of IV cocaine (0.2 and 0.4 mg kg^{-1}) in 12 men and 22 women with a history of cocaine abuse (Mendelson et al., 1999b). Menstrual-cycle phase was verified by analyses of estradiol and progesterone levels. Importantly, subjects were matched for BMI (kg m^{-2}) as well as for age. Cocaine was administered over 1 min into the antecubital vein of one arm, and samples for analyses of plasma cocaine were collected from the opposite arm at 2, 4, 8, 12, 16, 20, 30, 40, 60, 80, 120, 180, and 240 min. Pharmacokinetic analysis revealed no differences in IV cocaine disposition between men and women, or between women studied at the mid-follicular and luteal phases of the menstrual cycle (Mendelson et al., 1999b). Cocaine produced dose-dependent increases in plasma cocaine levels, but there were no significant differences in peak plasma cocaine levels (Tmax) between men and women or in women as a function of phase of the menstrual cycle (Mendelson et al., 1999b). There were also no sex or menstrual-cycle-phase differences in cardiovascular or subjective high measures. The time to reach peak plasma cocaine levels (Tmax) did not differ between men and women after 0.2 mg kg^{-1} IV cocaine administration. After 0.4 mg kg^{-1} IV cocaine, follicular-phase women reached maximal cocaine concentrations significantly more rapidly than men, but there were no significant menstrual-cycle differences in Tmax (Mendelson et al., 1999b). The lack of discernible sex differences in cocaine's pharmacokinetics in humans is consistent with findings in rats (Bowman et al., 1999) and rhesus monkeys (Mendelson et al., 1999a; Mello et al., 2004). It is now generally agreed that sex differences in behavior after cocaine cannot be attributed to differences in cocaine metabolism.

34.4.2.2 Sex, menstrual-cycle phase, and neuroimaging studies

Early clinical neuroimaging studies of cocaine abusers revealed that sex may influence the cerebrovascular effects of cocaine (Levin et al., 1994). There are some instances in which women were less vulnerable than men to the adverse cerebrovascular consequences of cocaine (Levin et al., 1994). Cocaine abuse has been associated with a number of cerebrovascular disorders, including ischemic stroke and intracerebral and subarachnoid hemorrhage. Chronic cocaine abuse can lead to perfusion abnormalities in the brain which can be measured with single photon emission computed tomography (SPECT). SPECT analysis showed that cocaine abusers had more focal cerebral perfusion defects than normal controls (Levin et al., 1994). Cerebral perfusion defects were found in the frontal, temporal, and parietal lobes and in the basal ganglia. However, an unanticipated finding was that nine female cocaine abusers had fewer cerebral perfusion defects than nine men who were matched in terms of age. Moreover, these women reported having used more cocaine for a longer time than the men, an average of 15.3 years as opposed to 8.2 years (Levin et al., 1994). It was concluded that these sex differences in cerebral perfusion defects could not be explained by differences in age, race, BMI ($kg\,m^{-2}$), alcohol use, cocaine use, or route of drug administration (Levin et al., 1994). Rather, it was postulated that estrogens and progestins may protect women from cocaine-associated cerebral vasospasm (Levin et al., 1994). Sex differences in regional cerebral blood flow detected by SPECT were also detected in cocaine-dependent men and women (Adinoff et al., 2006). The major finding was that regional cerebral blood flow in the bilateral orbital frontal cortex was lower in men, whereas cerebral blood flow in the medial orbital frontal cortex was lower in women (Adinoff et al., 2006). Testosterone alone, and in combination with cocaine, also may increase vasoconstriction (Yamamoto et al., 2007). An antiandrogen, flutamide (250 mg, PO), reduced peak cocaine levels after $0.4\,mg\,kg^{-1}$ IV cocaine and was associated with a more rapid decrease in heart rate; however, peripheral blood flow and vascular resistance were not measured (Yamamoto et al., 2007). Nitric oxide from vascular endothelial cells is important for normal vascular function and also may contribute to sex differences in cerebral perfusion defects. Cocaine-dependent men had significantly lower vascular nitric oxide end products than cocaine-dependent women or normal controls (Kaufman et al., 2007).

Menstrual-cycle-phase differences in the acute cerebral vasoconstrictive effects of cocaine have been detected with functional magnetic resonance imaging (fMRI) procedures (Kaufman et al., 2001). Cocaine's constrictive effects on the cerebral vasculature were less during the follicular phase than during the luteal phase of the menstrual cycle, and these results were interpreted to reflect the vascular protective effects of estrogen during the follicular phase, when it was not opposed by progesterone. These findings were consistent with evidence that estradiol has cardiovascular protective effects, mediated in part by its direct effects on blood vessels (Farhat et al., 1996; Skafar et al., 1997; Minshall et al., 1998). Moreover, this interpretation is consistent with evidence that, before menopause, women have less atherosclerotic disease and less cerebrovascular disease than men. After menopause, estrogen-replacement therapy reduces the risk for premature cardiac disease and osteoporosis (Jaffe, 1991). Positron emission topography (PET) has been used to examine sex and menstrual effects on striatal dopamine release after acute administration of another psychostimulant, d-amphetamine (Munro et al., 2006). Men released more dopamine than women at either the follicular or the luteal phase of the menstrual cycle, and also reported greater positive subjective effects. Women studied at the follicular and the luteal phase of the menstrual cycle did not differ in dopamine release or subjective-effect ratings (Munro et al., 2006).

34.4.2.3 Sex, menstrual-cycle phase, and cocaine's subjective effects

It has been difficult to detect consistent sex differences in the subjective and cardiovascular effects of cocaine. Often, subjects were not matched for BMI and plasma cocaine levels were not measured; therefore, it was difficult to determine the effective dose of cocaine. Studies of the influence of sex on the acute effects of intranasal cocaine have yielded conflicting findings. In one report, there were no significant sex differences in heart rate, blood pressure, or subjective-effect measures after administration of $2\,mg\,kg^{-1}$ cocaine intranasally to 23 men and 11 women (Kosten et al., 1996). These men and women were of comparable weight (68 vs. 78 kg), but were not matched for BMI, and plasma cocaine levels were not reported (Kosten et al., 1996). In another report, men had higher peak plasma cocaine levels and appeared to have greater subjective responses to intranasal cocaine than women, but there were no sex differences in peak heart rate (Lukas et al., 1996). However, these men and women were not matched for BMI. In women, peak plasma cocaine levels after intranasal administration were greater during the follicular phase than during the luteal phase of the menstrual cycle, but subjective reports of cocaine's effects were equivalent (Lukas et al., 1996). Plasma cocaine levels were measured for only 90 min, and pharmacokinetic analyses were not performed (Lukas et al., 1996). A number of methodological differences (effective dose and route of administration) may have contributed to the different findings observed after intranasal (Lukas et al., 1996) and IV cocaine administration (Mendelson et al., 1999b). One important difference is the greater variability inherent in intranasal cocaine administration where subject-specific variables such as depth of inhalation and vital capacity can affect the resulting levels of cocaine in plasma.

A retrospective analysis of crack-cocaine abusers detected sex and menstrual-cycle-phase differences in subjective responses to smoked cocaine, but there were no differences in peak plasma cocaine levels, cardiovascular responses, or the number of cocaine doses acquired (Sofuoglu et al., 1999). Men and women were studied in an inpatient clinical research ward and worked to solve arithmetic problems for tokens that could be exchanged for cocaine (0.4 $mg\,kg^{-1}$) available at 30-min intervals. Cocaine was administered with a smoking device that required subjects to inhale for 10 s and hold their breath for 15 s. Plasma cocaine levels exceeded 200 $ng\,ml^{-1}$ within 6 min and did not differ as a function of sex or menstrual-cycle phase. The phase of the menstrual cycle was verified by analysis of estradiol and progesterone. There were no sex or menstrual-cycle-phase differences in cardiovascular responsivity to cocaine in terms of number of tokens earned or exchanged for cocaine. Subjective effects were measured on a visual analog scale presented before and at 2.5, 10, and 15 min after the first dose of cocaine, and at 4.5 min before and 2.5 min after each subsequent dose. In general, men rated most subjective effects of cocaine higher than women, but were most similar to women studied during the follicular phase. Women studied during the follicular phase reported feeling higher after cocaine than women studied during the luteal phase (Sofuoglu et al., 1999).

Another retrospective analysis of cocaine smokers also found minimal sex differences in the dependent variables measured (Evans et al., 1999). This study was conducted in a clinical research ward and men and women were asked to choose between cocaine (50 mg base) and nothing. Six choices were available at 14-min intervals. Two choice sessions were conducted on each of 2 days at noon and at 4 p.m. Women were studied at several phases of the menstrual cycle as determined by self-report, but menstrual-cycle phase was not verified by hormone analysis. Men and women were not matched for BMI and cocaine was not administered on an $mg\,kg^{-1}$ basis. Under these conditions, plasma cocaine levels were significantly higher in women than in men, because women weighed less and received effectively higher cocaine doses than the men. There were no significant differences between men and women in the number of cocaine doses selected. Cardiovascular measures and subjective effects measures also did not differ in subgroups of six men and women who were matched for plasma cocaine levels. Taken together, these clinical studies suggest that sex usually does not have a significant influence on the pharmacokinetics or the pharmacodynamics of cocaine after IV or intranasal administration or after crack smoking.

Clinical studies of the subjective effects of cocaine at different phases of the menstrual cycle also have been inconsistent. In some studies, no menstrual-cycle-phase differences were detected in the subjective effects of cocaine (Lukas et al., 1996; Mendelson et al., 1999b; Munro et al., 2006; Collins et al., 2007). In other studies, the subjective effects of both cocaine and d-amphetamine were reported to be greater during the follicular phase of the menstrual cycle, when estradiol and progesterone levels are low, than during luteal phase, when both estradiol and progesterone levels are high (Justice and de Wit, 1999; Sofuoglu et al., 1999; Evans et al., 2002; White et al., 2002). Recent studies from several laboratories suggest that progesterone may have a more consistent effect on subjective effects of cocaine than estradiol (Sofuoglu et al., 1999, 2002; Evans and Foltin, 2006; Evans, 2007). Moreover, administration of progesterone to women during the follicular phase reduced the positive subjective effects of smoked and IV cocaine (Sofuoglu et al., 2002, 2004; Evans and Foltin, 2006).

In many studies of the effects of menstrual-cycle phase on cocaine's effects, it has been difficult to determine the relative influence of estradiol and progesterone except by inference. Any difference in subjective responses to cocaine or to d-amphetamine as a function of menstrual-cycle phase is usually interpreted as reflecting the fact that progesterone acts as an antagonist of estradiol under several conditions (Dierschke et al., 1973; Wildt et al., 1981; VanVugt et al., 1992; Clark and Mani, 1994). Relatively fewer studies have examined the effects of direct gonadal hormone administration, and some illustrative studies of the effects of estradiol and progesterone administration are summarized below.

One clinical study evaluated the effects of estradiol administration on the subjective effects of d-amphetamine in follicular-phase women (Justice and de Wit, 2000b). Two groups of women were given d-amphetamine (10 mg, PO) or placebo tablets during treatment with either a transdermal estradiol patch or a placebo patch. Estradiol treatment increased plasma estradiol to levels that were approximately 10 times higher than those during the normal early follicular phase. Ratings of pleasant stimulation were higher after d-amphetamine during estradiol treatment, in comparison to placebo treatment, but other positive subjective-effect ratings (euphoria, like drug, and feel high) were not significantly increased. Interestingly, estradiol alone also increased ratings of pleasant stimulation in women given placebo amphetamine (Justice and de Wit, 2000b). These data confirm and extend findings from a study designed to examine the influence of estradiol on subjective responses to d-amphetamine, independent of the effects of progesterone (Justice and de Wit, 2000a). Women were studied during the early and the late follicular phase of the menstrual cycle in an own-control design. During the early follicular phase, estradiol levels averaged 57 pg ml^{-1}, and during the late follicular phase, estradiol levels averaged 175 pg ml^{-1}. Progesterone levels were low at both the early and late follicular phase of the menstrual cycle and averaged 1.3 ng ml^{-1}. Although d-amphetamine (15 mg, PO) significantly increased heart rate and positive subjective effects in comparison to placebo, differences in estradiol levels as a function of menstrual-cycle phase did not alter the physiologic or subjective effects of d-amphetamine (Justice and de Wit, 2000a).

Studies of the effects of oral progesterone on subjective reactions to smoked cocaine have consistently found that progesterone attenuates ratings of positive responses to cocaine (Sofuoglu et al., 1999, 2002; Evans and Foltin, 2006). In one carefully controlled study, women with a current history of cocaine abuse were given up to six doses of smoked cocaine (6, 12, or 25 mg) during their normal follicular and luteal phases, and during a subsequent follicular phase after progesterone administration (Evans and Foltin, 2006). The dose of progesterone was selected to produce plasma levels comparable to each woman's natural luteal phase (Evans and Foltin, 2006). Under these conditions, administration of progesterone attenuated positive subjective responses to cocaine in comparison to the normal follicular phase when progesterone levels were low (Evans and Foltin, 2006). Similar effects of progesterone were reported in women after a single dose of smoked cocaine (0.4 mg kg^{-1}) and IV cocaine (0.3 mg kg^{-1}) (Sofuoglu et al., 2002, 2004).

There may be sex differences in the interactions between progesterone and cocaine. Interestingly, progesterone did not significantly attenuate subjective responses to cocaine in men (Sofuoglu et al., 2004; Evans and Foltin, 2006) and did not significantly reduce cocaine-positive urines over a 10-week placebo-controlled trial in methadone-stabilized opioid-dependent men who abused cocaine (Sofuoglu et al., 2007). In women, both cues and stress-induced cocaine craving and anxiety were lower during the luteal phase when progesterone levels were high, than during the follicular phase (Sinha et al., 2007). These findings may have implications for treatment because cocaine craving associated with stress and concomitant increases in HPA hormones are thought to be critical determinants of relapse (Sinha, 2001; Sinha et al., 2003, 2006). Thus far, only two clinical studies have examined the effects of progesterone on cocaine self-administration. In one study, there was no effect of progesterone on the number of cocaine injections self-administered (Sofuoglu et al., 2004). In an ongoing study, progesterone produced a decrease in the self-administration of smoked cocaine in women (Evans, 2007).

Parallel studies of the interactions between cocaine, sex, and menstrual-cycle phase have been conducted with nonhuman primates and generally the findings confirm and extend the data obtained in humans. Neuroendocrine control of the menstrual cycle of macaque monkeys is very similar to that of women, and the rhesus monkey has long been a model of choice in reproductive biology (Knobil, 1980; Knobil and Hotchkiss, 1988; Hotchkiss and Knobil, 1994). Accordingly, this model has been used to explore the influence of estradiol and progesterone on cocaine's abuse-related effects. The effects of estradiol administration on cocaine's reinforcing and discriminative stimulus effects were studied in female rhesus monkeys (Mello et al., 2008). A wide range of $E_2ß$ doses and cocaine doses were studied in the same females across conditions. To mimic the rapid onset, nongenomic actions of estradiol (Wong et al., 1996; Moore and Evans, 1999; Falkenstein et al., 2000; Vasudevan and Pfaff, 2007), $E_2ß$ was prepared in a cyclodextrin vehicle. Estradiol had variable effects on the cocaine self-administration dose–effect curves in female rhesus monkeys. Although low doses of estradiol shifted the peak of the cocaine dose–effect curve to the left in some monkeys, these effects were not estradiol dose-dependent. The highest dose of estradiol studied did not alter the cocaine dose–effect curve appreciably from control conditions (Mello et al., 2008). Similarly, cynomolgus female monkeys did not alter patterns of cocaine self-administration as a function of spontaneous fluctuations in estradiol and progesterone levels across the menstrual cycle when reinforcing unit doses of cocaine were available (0.01 and 0.032 mg kg^{-1}/inj) (Mello et al., 2007). Progressive ratio break points were stable across 44 menstrual cycles in which ovulation was verified by a mid-luteal-phase elevation of progesterone (Mello et al., 2007). However, when a low dose of cocaine (0.0032 mg kg^{-1}/inj) that did not maintain reliable or robust cocaine self-administration was available, progressive ratio break points were significantly higher during the early- and mid-follicular phase than during the late luteal phase of ten ovulatory menstrual cycles (Mello et al., 2007). These findings in rhesus females are consistent with clinical studies in which estradiol failed to enhance responses to d-amphetamine (Justice and de Wit, 2000a,b). When cocaine unit doses that maintained consistent self-administration were available in rodents, the effects of estradiol often were not significant (Grimm and See, 1997; Caine et al., 2004; Hu et al., 2004; Lynch and Taylor, 2005). Taken together, these findings suggest that the effects of estradiol on the abuse-related effects of cocaine are subtle, at best.

In contrast to estradiol, progesterone consistently reduces cocaine self-administration in rhesus monkeys (Mello et al., under review) and in rodents (Jackson et al., 2006; Anker et al., 2007; Larson et al., 2007; Feltenstein et al., 2009). Allopregnanolone, a metabolite of progesterone, also decreased cocaine self-administration in a reinstatement paradigm (Anker et al., 2009). Similarly, when the effects of progesterone on cocaine self-administration were studied in rhesus monkeys, there was a progesterone dose-dependent downward and rightward shift in the cocaine dose–effect curves (Mello et al., under review). These data are consistent with clinical reports that progesterone may decrease ratings of positive subjective effects of cocaine in women as described earlier in this chapter (Sofuoglu et al., 2002, 2004; Evans and Foltin, 2006; Evans, 2007). However, progesterone pretreatment did not significantly alter the cocaine discrimination dose–effect curve (Mello et al., under review). Further studies will be necessary to clarify the conditions under which gonadal steroids modulate the abuse-related effects of cocaine.

34.5 Effects of Cocaine on Reproductive Function

34.5.1 Background

We have seen that acute cocaine administration changes anterior pituitary and gonadal hormone levels (Sections 34.2 and 34.3), but with the exception of prolactin, it has been difficult to detect changes in basal hormone levels after chronic cocaine abuse or short-term cocaine exposure in animal models. We infer that persistent changes in neuroendocrine function do occur on the basis of clinical and preclinical evidence of cocaine-related disruptions in reproductive function (Siegel, 1982; Cocores et al., 1986; Smith and Smith, 1990; Teoh et al., 1994a; Mello et al., 1997; Mello, 1998; Mello and Mendelson, 2002). However, it is not clear how chronic cocaine abuse interacts with gonadotropin and gonadal steroid hormones to disrupt the menstrual cycle in women and compromise reproductive function in men. Anterior pituitary and gonadal steroid hormones have both positive- and negative-feedback effects on each other and disruption of these feedback relationships affects the functional integration and regulation of the neuroendocrine system. It is apparent that integrative neurobiologic studies in humans and in whole animal models are necessary to analyze these complex interrelationships between cocaine and the HPG axis and the HPA axis.

The hormonal changes that define the phases of the menstrual cycle are one of the most fundamental biological rhythms. The patterns of gonadotropins and ovarian steroid hormones during the normal menstrual cycle were shown earlier in **Figure 4**. In women and in higher primates, a menstrual cycle occurs approximately every 28 days from menarche until menopause. The onset of menstruation defines the beginning of a cycle, the follicular phase, and heralds the development of the ovarian follicle which culminates in ovulation at mid-cycle. Subsequently, in a nonfertile cycle, the ovarian follicle becomes the corpus luteum, and there is a concomitant increase in progesterone during the luteal phase. This postovulatory rise in progesterone is essential to maintain the fertilized ovum when pregnancy occurs. In the nonfertile cycle, the demise of the corpus luteum is followed by menstruation and the beginning of the next menstrual cycle.

The adverse effects of cocaine on reproductive function include disorders of menstrual-cycle duration, which in turn may reflect impairments in folliculogenesis, ovulation, and luteal-phase adequacy in otherwise normal women (Teoh et al., 1992, 1994a;

Mello, 1998; Mello and Mendelson, 2002). Such impairments may result in a series of clinical syndromes that include amenorrhea (the complete cessation of menses for periods of months or years), anovulation (a failure to ovulate), and luteal-phase dysfunction (defined either as a short luteal phase of 8 days or less from ovulation to menses or an inadequate luteal phase in which progesterone levels are abnormally low, but the interval from ovulation to menstruation is of normal length). Both anovulation and luteal-phase dysfunction may occur in women who continue to menstruate. Cocaine abuse also may result in disorders of prolactin regulation which are expressed clinically as abnormally high prolactin levels or hyperprolactinemia. This is sometimes associated with abnormal secretion of breast milk, a condition called galactorrhea (Cocores et al., 1986). Hyperprolactinemia may be a concomitant of amenorrhea; however, both amenorrhea and hyperprolactinemia can occur independently (Sauder et al., 1984). Cocaine abuse also may increase the risk for spontaneous abortion once pregnancy occurs (Cregler and Mark, 1986). Cocaine's effects on prolactin were described in the previous version of this chapter (Mello and Mendelson, 2002) and are not included here due to space limitations.

Understanding the basis for the most commonly observed disorders (amenorrhea, luteal-phase dysfunction, and anovulation) is complicated by the fact that each disorder may result from events that occurred earlier in the menstrual cycle. For example, although FSH is only one determinant of normal folliculogenesis, adequate FSH levels during the late luteal and the follicular phases are necessary for normal follicle development and maturation (diZerega and Hodgen, 1981a; Goodman and Hodgen, 1983). Suppression of FSH may delay follicle maturation and subsequent ovulation, or result in luteal-phase dysfunction after timely ovulation. Cocaine stimulates LH release and abnormally high levels of LH and/or estradiol during the follicular phase may suppress FSH and result in anovulation and/or luteal-phase dysfunction (Zeleznik, 1981; Dierschke et al., 1985). If estrogen levels are high during the early luteal phase, this may shorten the menstrual cycle by 5 or 6 days, resulting in a short luteal phase (Hutchison et al., 1987). Acute cocaine administration suppresses prolactin levels and abnormally low or high prolactin levels may also be associated with luteal-phase dysfunction (McNeely and Soules, 1988; see Mello and Mendelson (2002) for review). The continuing controversies and unresolved issues concerning the prevalence, differential diagnosis, and pathogenesis of luteal-phase dysfunction have been critically examined elsewhere (McNeely and Soules, 1988; Stouffer, 1990). A number of medical disorders and malnutrition or fasting, as well as strenuous exercise, may also result in amenorrhea (Frisch, 1982; Bullen et al., 1985; Warren, 1992).

In addition to cocaine's direct and indirect effects on the gonadotropins and prolactin, cocaine abuse also may disrupt reproductive function by stimulation of the HPA axis. The considerable evidence that cocaine stimulates ACTH and cortisol/corticosterone and, by inference, CRH, was described earlier in Section 34.2 of this chapter. Although the contribution of cocaine-related increases in CRH to menstrual-cycle abnormalities is unclear, it is well established that administration of synthetic CRH inhibited pulsatile release of LH and FSH in ovariectomized rhesus females (Olster and Ferin, 1987), whereas ACTH infusions did not (Xiao and Ferin, 1988). Synthetic-CRH administration also suppressed endogenous LHRH levels measured in rat portal blood (Petraglia et al., 1987). Since synthetic CRH also reduced LH and FSH levels in adrenalectomized ovariectomized rhesus females (Xiao et al., 1989), this appears to be a central effect mediated through the hypothalamic/pituitary axis rather than through adrenal activation (Xiao and Ferin, 1988). Subsequent studies have confirmed that administration of CRH (200 μg, IV) decreased LH levels as well as electrophysiological activity in the mediobasal hypothalamus in rhesus monkeys (Williams et al., 1990). The inhibitory effects of CRH on electrophysiological volley frequency (but not duration) were prevented by a continuous infusion of naloxone ($8.0\ \mathrm{ml\ h^{-1}}$) and these data were interpreted to suggest that endogenous opioids may mediate the inhibitory effects of CRH (Williams et al., 1990).

Clinical evaluation of cocaine's effects on reproductive function has been complicated by the fact that many cocaine abusers also use a number of other drugs including opioids, alcohol, and marijuana (Teoh et al., 1992, 1994a; Mello, 1998; NIDA, 2000; Mendelson and Mello, 2008). Moreover, cocaine, opioids, and alcohol appear to disrupt the menstrual cycle in very similar ways, although perhaps by different mechanisms (Smith and Smith, 1990; Mello et al., 1992; Teoh et al., 1994a; Mello, 1998). Consequently, it is impossible to attribute disorders of the menstrual cycle to cocaine alone in clinical studies. Fortunately, the neuroendocrine control of the menstrual cycle is very similar in female rhesus monkeys and in women, and as a result, the rhesus monkey is a model of choice in reproductive biology (Knobil, 1974, 1980; Goodman and Hodgen, 1983). Studies in the rhesus monkey model have clarified the basis for some clinical disorders of endocrine function and have led to new approaches to the treatment of infertility. For example, after the discovery of the importance of pulsatile gonadotropin release in neuroendocrine control of the menstrual cycle in rhesus monkeys (Knobil, 1974, 1980), it was found that many infertility disorders in women are associated with infrequent LH pulses of low amplitude throughout the menstrual cycle or no LH pulses at all (Crowley et al., 1985; Santoro et al., 1986b). These abnormal LH pulsatile release patterns are associated with amenorrhea, the failure to menstruate. In ovariectomized rhesus monkeys, with lesions of the arcuate nuclei, as well as in women with infertility disorders, administration of synthetic hypothalamic LHRH restored normal LH release patterns (Knobil, 1974, 1980; Crowley et al., 1985; Santoro et al., 1986b; Martin et al., 1993; Filicori et al., 1994). These findings suggest the possibility that cocaine, as well as other abused drugs, may disrupt the pulsatile release of gonadotropin hormones to result in amenorrhea, and preclinical data consistent with this hypothesis are described below. However, chronic cocaine abuse did not appear to alter the pulsatile release patterns of LH in men (Mendelson et al., 1989b).

34.5.2 Studies of the Effects of Chronic Cocaine Administration on Reproductive Function

The primate model of cocaine and other drug self-administration is especially valuable for studying the effects of chronic drug use on the neuroendocrine system. Rhesus monkeys self-administer most drugs that are abused by humans, and the patterns of drug self-administration are often similar (Mello, 1979; Griffiths et al., 1980; Brady and Lukas, 1984; Mello and Negus, 1996). Thus, the neuroendocrine effects of a single drug, such as cocaine, can be studied in rhesus monkeys without the influence of polydrug abuse, malnutrition, and concurrent medical disorders that may complicate interpretation of clinical studies. One important advantage of drug self-administration procedures, in comparison to investigator-determined drug administration, is that each monkey can control the frequency and amount of cocaine injected. Previous studies have shown that investigator-administered cocaine resulted in higher rates of lethality and lower levels of extracellular dopamine than self-administration of the same amounts of cocaine in rats (Dworkin et al., 1995; Hemby et al., 1997). Similar findings have also been reported in rhesus monkeys (Johanson and Schuster, 1981).

We examined the effects of two or more years of chronic cocaine self-administration on menstrual-cycle duration and anterior pituitary and gonadal hormone levels in adult rhesus females (*Macaca mulatta*; Mello et al., 1997). Eight monkeys maintained on cocaine and food self-administration were compared with six control females that were occasionally exposed to an acute dose of cocaine (0.4 or 0.8 mg kg^{-1}, IV) once every 2 or 3 months for 2–3 years. All drug-naive rhesus females were adapted to the laboratory for several months until stable ovulatory menstrual cycles occurred. Then monkeys in the cocaine self-administration group were first trained to self-administer food pellets, then implanted with IV catheters under aseptic conditions, and trained to self-administer IV cocaine on a simple operant task. Monkeys were given access to cocaine (0.1 mg kg^{-1}/inj) in four sessions each day and were limited to 20 injections per session to minimize any possible adverse drug effects. Under these conditions, monkeys could self-administer up to 8 mg kg^{-1} day^{-1} of cocaine. Food self-administration sessions preceded drug self-administration sessions so that cocaine intoxication would not compromise food intake. Nutritionally fortified banana-flavored pellets were supplemented with fresh fruits and vegetables, multiple vitamins, and Purina monkey chow. Water was continuously available, and a 12-h light–dark cycle was in effect. Monkeys remained healthy and active, and food intake and body weight were normal under these conditions of limited access to cocaine self-administration. Blood samples for analysis of LH and progesterone were collected every other day starting at day 8 of each menstrual cycle. The onset and duration of menstrual bleeding was monitored with daily vaginal swabs. Each menstrual cycle was classified as normal, anovulatory, with luteal phase defects or amenorrheic according to the following criteria:

- Normal ovulatory menstrual cycles were cycles of normal duration relative to the precocaine baseline cycles in which there was evidence of a mid-cycle periovulatory surge in LH followed by an elevation in progesterone to levels of 8.5 ng ml^{-1} or higher (Filicori et al., 1984).
- Anovulation or failure to ovulate was inferred from low levels of progesterone during the mid-luteal phase (<5 ng ml^{-1}) and absence of a mid-cycle LH surge.

- Two types of luteal-phase defects have been described in the clinical and nonhuman primate literature (diZerega and Hodgen, 1981c; Goodman and Hodgen, 1983; diZerega and Wilks, 1984; Stouffer, 1990).
- An inadequate luteal phase was defined by abnormally low progesterone levels during the mid-luteal phase, but a mid-cycle LH surge was detected and the menstrual cycle was of normal length.
- A short luteal phase was inferred from an abnormally short cycle accompanied by low progesterone levels.
- Amenorrheic cycles were defined as 60 or more days without menstruation.

The effects of chronic cocaine self-administration and withdrawal on menstrual-cycle duration were studied for over 200 menstrual cycles and the control group was studied for over 150 menstrual cycles (Mello et al., 1997). Both cocaine self-administration and cocaine withdrawal were associated with severe disruptions of the menstrual cycle including annovulation, luteal-phase dysfunction, and amenorrhea of 61–190 days duration. Because menstrual-cycle disruption often persisted for several months during cocaine withdrawal, this suggests that chronic cocaine self-administration may induce long-lasting dysregulation of the HPG axis. The average dose of cocaine self-administered did not reliably predict anovulation or changes in duration in the current or the subsequent menstrual cycle. Individual patterns of menstrual-cycle disruptions during chronic cocaine self-administration and withdrawal were quite variable.

In the control group, 94% of the menstrual cycles were of normal duration, whereas in the cocaine self-administration group, approximately one half of all menstrual cycles were of abnormal duration. Abnormally short menstrual cycles, consistent with luteal-phase defects, accounted for half of these aberrant menstrual cycles. Abnormally long menstrual cycles were longer than each monkey's precocaine baseline cycles by one or more standard deviations from the mean. Nineteen cycles were amenorrheic (61–190 days of no menses) in the cocaine self-administration group. There were no amenorrheic cycles in the control group (Mello et al., 1997).

Frequency of ovulation. Sixty-eight percent of the 217 cocaine self-administration cycles and 66% of the 82 cocaine-withdrawal cycles could be classified as ovulatory or anovulatory with certainty on the basis of mid-luteal phase elevations in progesterone. During cocaine self-administration, the frequency of ovulation was significantly higher in menstrual cycles of normal length than in short or long cycles. In menstrual cycles of normal duration, 78% had progesterone elevations consistent with ovulation. In the cocaine-exposed monkeys, peak progesterone levels during menstrual cycles classified as ovulatory and ranged from 11.6 to 16 ng ml^{-1}. In contrast, only 45% of the short cycles and 66% of the long cycles were ovulatory. The frequency of anovulatory cycles was similar during first and second cocaine exposure (26% and 32%). During cycles classified as anovulatory, peak progesterone levels ranged from the lower limit of assay sensitivity to 4.5 ng ml^{-1}. During cocaine withdrawal, the frequency of ovulation did not differ significantly in normal, short, and long cycles. Eighty-seven percent of the menstrual cycles of normal duration and 64% and 78% of the short and long menstrual cycles were ovulatory (Mello et al., 1997).

Amenorrhea during cocaine self-administration and withdrawal. Amenorrheic cycles occurred with approximately equal frequency during cocaine self-administration and withdrawal. Sixteen of the 19 amenorrheic cycles appeared to be anovulatory and peak progesterone values averaged 2.31 ($\pm$0.37) ng ml^{-1} (range 0.69–4.24 ngml^{-1}). For three amenorrheic cycles that occurred during cocaine self-administration, at 5–8 days before the end of the amenorrheic cycle, peak progesterone levels of 14, 14.7, and 15.8 ng ml^{-1} were detected, suggesting the occurrence of ovulation. Because the fall breeding period is sometimes preceded by anovulatory/amenorrheic cycles during the summer in the Northern Hemisphere, we also examined the distribution of amenorrheic cycles across months of the year. The initiation of amenorrheic cycles was distributed across the year.

These data suggest that chronic cocaine exposure disrupts menstrual-cycle regularity in otherwise healthy monkeys studied under controlled laboratory conditions. The menstrual-cycle disruptions observed are consistent with clinical reports in cocaine abusers who are also polydrug abusers (Teoh et al., 1992). It is unlikely that uncontrolled factors other than cocaine self-administration accounted for the menstrual-cycle abnormalities observed in female rhesus monkeys. Normal ovulatory menstrual cycles were observed during the 6-month adaptation to the laboratory. Seasonal factors probably did not contribute because monkeys began cocaine self-administration and developed menstrual-cycle abnormalities at different times of the year and the control group did not develop abnormal menstrual cycles. Monkeys were not food-deprived or malnourished and any anorectic effects of cocaine were transient. Finally, in this cocaine self-administration paradigm, each monkey controlled the frequency of cocaine injections and the total daily dose of cocaine up to a maximum of 2 mg kg^{-1}/session or 8 mg $kg^{-1}day^{-1}$.

Experimenter-determined cocaine administration in nonhuman primates. Cocaine-related disruptions of the menstrual cycle were also reported in a series of studies in which cocaine or saline was administered intravenously to rhesus females during the follicular phase of one menstrual cycle (days 2–14; Chen et al., 1998; Potter et al., 1998, 1999). Only monkeys with normal-length menstrual cycles (26–30 days) for three consecutive months were selected as subjects. All monkeys were implanted with external jugular catheters, then assigned to saline ($N = 6$) or cocaine ($N = 7$) conditions. In the first study, a high dose of cocaine (4 mg kg^{-1}) was administered as an IV bolus once a day between 8 and 9 a.m., and a single blood sample was collected daily for analysis of estradiol, progesterone, LH, and FSH (Potter et al., 1998). All control monkeys had preovulatory LH peaks and ovulation was confirmed by laparoscopy 2 days after a mid-cycle peak in estradiol. Follicular- and luteal-phase lengths were normal except in one control monkey that ovulated on day 22 of the menstrual cycle. In contrast, six of the seven cocaine-treated monkeys had anovulatory cycles of abnormal duration. Two monkeys had short cycles (14 and 17 days) and four monkeys had long cycles of 54–70 days duration. Body weight and caloric intake did not differ between controls and cocaine-treated monkeys. These findings indicated that daily cocaine administration during the follicular phase disrupted folliculogenesis and normal menstrual cyclicity (Potter et al., 1998).

Subsequently, the effects of lower daily doses of cocaine and saline control infusions on menstrual-cycle function were evaluated under similar conditions (Potter et al., 1999). Monkeys were treated with daily injections of 1 mg kg^{-1} once a day, IV cocaine ($N = 7$); 2 mg kg^{-1}, IV cocaine ($N = 7$); or saline ($N = 7$) during the follicular phase of one menstrual cycle (days 2–14; Potter et al., 1999). All seven control monkeys had normal menstrual cycles with preovulatory LH peaks and laparoscopically confirmed ovulation. In the 1 mg kg^{-1} cocaine dose group, two monkeys ovulated and the other five monkeys did not. Of the anovulatory monkeys, two monkeys had abnormally short menstrual cycles (15 and 16 days), then developed amenorrheic cycles of 86 and 91 days, respectively. Three other monkeys had abnormally long menstrual cycles after cocaine treatment (49, 59, and 91 days), but subsequent cycles were disrupted in only one of these monkeys. In the 2 mg kg^{-1} cocaine dose group, three monkeys had ovulatory menstrual cycles and four monkeys did not.

Analysis of pulsatile gonadotropin release patterns throughout the menstrual cycle. Although daily cocaine self-administration (Mello et al., 1997) or single daily cocaine doses (1, 2, or 4 mg kg^{-1}, IV) during the follicular phase of the menstrual cycle (Potter et al., 1998, 1999) can produce anovulation, luteal-phase abnormalities, and amenorrhea, the mechanisms by which cocaine and other abused drugs disrupt the menstrual cycle are poorly understood. Moreover, amenorrheic cycles may persist for several months after cocaine administration has been discontinued. Menstrual-cycle disorders are often characterized by abnormal gonadotropin secretory patterns in women (Crowley et al., 1985; Santoro et al., 1986b). Low rates of LH release and pulsatile patterns have been reported in amenorrheic women. However, administration of synthetic LHRH at a normal physiologic frequency may restore normal menstrual-cycle function in women (Hammond et al., 1979; Crowley et al., 1985; Santoro et al., 1986b). The importance of the frequency of pulsatile infusion of synthetic LHRH for restoring normal patterns of gonadotropin secretory activity was first demonstrated in ovariectomized rhesus monkeys with lesions of the arcuate nucleus and the median eminence (Knobil, 1974, 1980).

The effects of daily cocaine administration on pulsatile release of LH and FSH during the follicular phase of the menstrual cycle were studied in eight female rhesus monkeys treated with cocaine (4 mg kg^{-1}, IV) and eight females treated with saline once each day (Chen et al., 1998). Blood samples (2 ml) for hormone analysis were collected every 15 min for 8 h during the early follicular phase (cycle days 1–5), the mid-follicular phase (cycle days 6–10), and the late follicular phase (cycle days 11–15). The Cluster Analysis program (Veldhuis and Johnson, 1986) was used to identify LH and FSH pulses and to quantify pulse frequency and amplitude. Cocaine exposure resulted in anovulation in seven of eight monkeys, whereas seven of eight control monkeys had normal ovulatory menstrual cycles. However, LH- and FSH-pulse frequency did not change significantly across the early, mid-, and late follicular phases of the menstrual cycle in either group, and were equivalent in the cocaine- and saline-treated groups. LH-pulse amplitude also did not change across successive phases of the menstrual cycle in the anovulatory cocaine-treated monkeys, whereas a normal increase in LH-pulse amplitude occurred during the late follicular phase prior to ovulation in the control monkeys. Estradiol levels also increased significantly during the late follicular phase in the control group but not in the anovulatory cocaine-treated group (Chen et al., 1998). These data confirm and extend findings from the effects of daily doses of 1, 2, and 4 mg kg^{-1} IV cocaine administered during the follicular phase of the menstrual cycle under comparable conditions (Potter et al., 1998, 1999). An unvarying pattern of gonadotropin-pulse frequency and amplitude observed is characteristic of anovulatory menstrual cycles.

Disruptions of the estrous cycle were also observed in rodents when cocaine was self-administered (Roberts et al., 1989) and when it was administered by the investigator (King et al., 1990, 1993). Ovulation, as inferred from oocyte retrieval after sacrifice, was also significantly reduced (King et al., 1990, 1993). Details of these studies in rodents were described in our earlier chapters in this series (Mello and Mendelson, 2002, in press).

34.6 Conclusions

Cocaine interacts with many neuromodulatory systems in the brain and has both direct and indirect effects on anterior pituitary, gonadal, and adrenal hormones. Acute cocaine administration stimulates the release of ACTH, LH, and FSH from the anterior pituitary in several species under a variety of experimental conditions. Cocaine is a monoamine reuptake inhibitor and ACTH release is controlled by CRH, which in turn is regulated by serotonin, dopamine, and norepinephrine. However, gonadotropin stimulation was not predictable from the known pharmacology of cocaine because exogenous dopamine administration suppresses LH in clinical studies, and the basis for cocaine's stimulation of LH is poorly understood. The importance of estradiol and progesterone in modulating cocaine's stimulatory effects can be inferred from observations that in ovariectomized rhesus monkeys, cocaine does not stimulate LH or ACTH release. The possible contribution of gonadal steroid hormones, rather than pituitary dysfunction, to these effects is suggested by the fact that synthetic LHRH and CRH each of which stimulated significant increases in LH and ACTH in the same ovariectomized monkeys. Recent studies indicate that cocaine stimulates estradiol, but not progesterone, in female rhesus monkeys. In rats, cocaine stimulates estradiol and progesterone in males and females. Increases in progesterone in males appear to be of adrenal origin because after adrenalectomy, cocaine did not stimulate progesterone. Species differences as well as methodological differences probably contribute to these discrepant findings between rats and rhesus monkeys.

There is increasing evidence that cocaine's acute hormonal effects influence behavior. Both clinical and preclinical studies are consistent with the hypothesis that cocaine's abuse-related effects are mediated, in part, by stimulation of ACTH and cortisol/corticosterone. Antagonism of CRH-stimulated ACTH with a CRH antagonist significantly reduced cocaine self-administration by rats, but not by nonhuman primates. The gonadal steroid hormones also appear to contribute to cocaine's reinforcing and locomotor effects in rodents. Cocaine self-administration and cocaine-induced locomotion are maximal at the estrus phase of the estrous cycle. But when cocaine doses that maintain robust self-administration are studied, it has been difficult to detect reliable effects of estrous or menstrual-cycle phase or gonadectomy. Both clinical and preclinical studies agree that progesterone reduces cocaine self-administration. Consistent sex differences in responsivity to cocaine on several endpoint measures appear to reflect the contribution of gonadal steroid hormones. Clinical and preclinical studies together suggest that estradiol increases the relative resistance of women to cocaine's toxic cerebrovascular and cardiovascular effects. Much remains to be learned about the interactions between cocaine's endocrine and behavioral effects.

It has been difficult to detect the effects of chronic cocaine exposure on single hormones. With few exceptions, basal levels of specific hormones after chronic cocaine treatment do not differ from control conditions. Furthermore, administration of a challenge dose of cocaine, or other stimulating agent, seldom reveals a significant change in the response of the target hormone after chronic cocaine exposure, although there are some exceptions to this general finding. One interpretation of these observations is that the effects of cocaine are transient and limited to the duration of action of each dose of cocaine. Alternatively, it is possible that repeated hormonal stimulation or suppression by cocaine eventually may disrupt the regulatory feedback system.

There has been relatively little experimental attention given to the consequences of chronic cocaine exposure on reproductive function. Measures of menstrual-cycle adequacy are one index of the integrative actions of anterior pituitary and gonadal hormones over time. Clinical evidence of cocaine's disruptive effects on reproductive function is often complicated by the fact that many human cocaine abusers are polydrug abusers. Since abuse of opiates and alcohol also may be associated with amenorrhea, anovulation, and luteal-phase dysfunction, it is difficult to attribute these menstrual-cycle disorders to cocaine alone. Animal models offer many advantages for the study of cocaine's effects on reproductive function, and neuroendocrine control of the menstrual cycle in rhesus monkeys and in women is very similar. Only a few studies have examined cocaine's effects on the menstrual or estrous cycles, but chronic cocaine exposure consistently results in abnormal cycles in healthy rats and monkeys. In rhesus monkeys, the menstrual-cycle abnormalities observed are similar to clinical reports of amenorrhea, anovulation, and luteal-phase dysfunction. Multidisciplinary integrative studies in animal models should have heuristic value for neurobiology as well as clinical relevance for understanding the functional consequences of chronic cocaine abuse.

It is increasingly evident that the consequences of cocaine exposure on anterior pituitary, gonadal, and adrenal hormones, and the implications for cocaine addiction are important areas for investigation. The interaction between cocaine and the neuroendocrine system and behavior offers many exciting opportunities for research. Recent advances in endocrinology, endocrine pharmacology, neurobiology, and behavioral science should facilitate progress in understanding the ways in which cocaine affects the regulation of neuroendocrine hormones and how the hormonal milieu influences cocaine's behavioral effects.

Acknowledgments

Preparation of this review was supported in part by grants K05 DA 00101 and K05 DA 00064, R01-DA14670 and P01-DA14528 from the National Institute on Drug Abuse, NIH. The authors are grateful to Bruce Stephen and Inge Knudson for preparing the figures.

References

Adams JM, Taylor AE, Schoenfeld DA, Crowley WF, Jr., and Hall JE (1994) The midcycle gonadotropin surge in normal women occurs in the face of an unchanging gonadotropin-releasing hormone pulse frequency. *Journal of Clinical Endocrinology and Metabolism* 79: 858–864.

Adinoff B, Williams MJ, Best SE, Harris TS, Chandler P, and Devous MD, Sr. (2006) Sex differences in medial and lateral orbitofrontal cortex hypoperfusion in cocaine-dependent men and women. *Gender Medicine* 3: 206–222.

Ahmed SH and Koob GF (1997) Cocaine-but not food-seeking behavior is reinstated by stress after extinction. *Psychopharmacology* 132: 289–295.

Allolio B, Deuss U, Kaulen D, Leonhardt U, Kallabis D, Hamel E, and Winkelmann W (1986) FK 33-824, a met-enkephalin analog, blocks corticotropin-releasing hormone-induced adrenocorticotropin secretion in normal subjects but not in patients with Cushing's disease. *Journal of Clinical Endocrinology and Metabolism* 63: 1427–1431.

Anker JJ, Holtz NA, Zlebnik N, and Carroll ME (2009) Effects of allopregnanolone on the reinstatement of cocaine-seeking behavior in male and female rats. *Psychopharmacology* 203: 63–72.

Anker JJ, Larson EB, Gliddon LA, and Carroll ME (2007) Effects of progesterone on the reinstatement of cocaine-seeking behavior in female rats. *Experimental and Clinical Psychopharmacology* 15: 472–480.

APA (1994) *Diagnostic and Statistical Manual of Mental Disorders*, 4th edn. Washington, DC: American Psychiatric Association.

Arvanitis AG, Gilligan PJ, Chorvat RJ, et al. (1999) Non-peptide corticotropin-releasing hormone antagonists: Syntheses and structure–activity relationships of 2-anilinopyrimidines and-triazines. *Journal of Medicinal Chemistry* 42: 805–818.

Ayala AR, Pushkas J, Higley JD, et al. (2004) Behavioral, adrenal, and sympathetic responses to long-term administration of an oral corticotropin-releasing hormone receptor antagonist in a primate stress paradigm. *Journal of Clinical Endocrinology and Metabolism* 89: 5729–5737.

Baumann MH, Becketts KM, and Rothman RB (1995a) Evidence for alterations in presynaptic serotonergic function during withdrawal from chronic cocaine in rats. *European Journal of Pharmacology* 282: 87–93.

Baumann MH, Gendron TM, Becketts KM, Henningfield JE, Gorelick DA, and Rothman RB (1995) Effects of intravenous cocaine on plasma cortisol and prolactin in human cocaine abusers. *Biological Psychiatry* 38: 751–755.

Becker JB (1999) Gender differences in dopaminergic function in striatum and nucleus accumbens. *Pharmacology, Biochemistry, and Behavior* 64: 803–812.

Becker JB, Berkley KJ, Geary N, Hampson E, Herman JP, and Young EA (eds.) (2008) *Sex Differences in the Brain from Genes to Behavior*. New York: Oxford University Press.

Becker JB, Molenda H, and Hummer DL (2001) Gender differences in the behavioral responses to cocaine and amphetamine. Implications for mechanisms mediating gender differences in drug abuse. *Annals of the New York Academy of Sciences* 937: 172–187.

Besedovsky HO and Del Rey A (1996) Immune–neuro-endocrine interactions: Facts and hypotheses. *Endocrine Review* 17: 64–102.

Bornstein SR and Chrousos GP (1999) Adrenocorticotropin (ACTH)- and non-ACTH-mediated regulation of the adrenal cortex: Neural and immune inputs. *Journal of Clinical Endocrinology and Metabolism* 84: 1729–1736.

Borowsky B and Kuhn CM (1991a) Monoamine mediation of cocaine-induced hypothalamo-pituitary-adrenal activation. *Journal of Pharmacology and Experimental Therapeutics* 256: 204–210.

Borowsky B and Kuhn CM (1991b) Chronic cocaine administration sensitizes behavioral but not neuroendocrine responses. *Brain Research* 543: 301–306.

Bowman BP, Vaughan SR, Walkler QD, et al. (1999) Effects of sex and gonadectomy of cocaine metabolism in the rat. *Journal of Pharmacology and Experimental Therapeutics* 290: 1316–1323.

Brady JV and Lukas SE (eds.) (1984) *Testing Drugs for Physical Dependence Potential and Abuse Liability*. Washington, DC: US Government Printing Office.

Bullen BA, Skinnar GS, Beitins IZ, von Mering G, Turnbull BA, and McArthur JW (1985) Induction of menstrual disorders by strenuous exercise in untrained women. *New England Journal of Medicine* 2: 1349.

Cahill L (2006) Why sex matters for neuroscience. *Nature Reviews. Neuroscience* 7: 477–484.

Caine SB, Bowen CA, Yu G, Zuzga D, Negus SS, and Mello NK (2004) Effect of gonadectomy and gonadal hormone replacement on cocaine self-adminsitration in female and male rats. *Neuropsychopharmacology* 29: 929–942.

Carnes M, Lent SJ, Goodman B, Mueller C, Saydoff J, and Erisman S (1990) Effects of immunoneutralization of corticotropin-releasing hormone on ultradian rhythms of plasma adrenocorticotropin. *Endocrinology* 126: 1904–1913.

Chaisson RE, Bacchetti P, Osmond D, Brodie B, Sande MA, and Moss AR (1989) Cocaine use and HIV infection in intravenous drug users in San Francisco. *Journal of the American Medical Association* 261: 561–565.

Chalmers DT, Lovenberg TW, Grigoriadis DE, Behan DP, and De Souza EB (1996) Corticotropin-releasing factor receptors: From molecular biology to drug design. *Trends in Pharmacological Sciences* 17: 166–172.

Chen C and Pollack GM (1997) Blood–brain disposition and antinociceptive effects of [D-penicillamine2,5]enkephalin in the mouse. *Journal of Pharmacology and Experimental Therapeutics* 283: 1151–1159.

Chen EC, Samuels MH, Luther MF, King TS, Eddy CA, Siler-Khodr TM, and Schenken RS (1998) Cocaine impairs follicular phase pulsatile gonadotropin secretion in rhesus monkeys. *Journal of the Society for Gynecologic Investigation* 5: 311–316.

Chrousos GP, Gold PW (1992) The concepts of stress and stress system disorders: Overview of physical and behavioral homeostatis. *Journal of the American Medical Association* 267: 1244–1252.

Chrousos GP and Gold PW (1998) A healthy body in a healthy mind – and vice versa – the damaging power of "uncontrollable" stress. *Journal of Clinical Endocrinology and Metabolism* 83: 1842–1845.

Clark UH and Mani SK (1994) Actions of ovarian steroid hormones. In: Knobil E and Neill JD (eds.) *The Physiology of Reproduction*, 2nd edn., pp. 1011–1059. New York: Raven Press.

Cocores JA, Dackis CA, and Gold MS (1986) Sexual dysfunction secondary to cocaine abuse in two patients. *Journal of Clinical Psychiatry* 47: 384–387.

Collins SL, Evans SM, Foltin RW, and Haney M (2007) Intranasal cocaine in humans: Effects of sex and menstrual cycle. *Pharmacology, Biochemistry, and Behavior* 86: 117–124.

Cone EJ (1995) Pharmacokinetics and pharmacodynamics of cocaine. *Journal of Analytical Toxicology* 19: 459–478.

Conn PM and Crowley WFJ (1991) Gonadotropin-releasing hormone and its analogues. *New England Journal of Medicine* 324: 93–103.

Cregler LL and Mark H (1986) Medical complications of cocaine abuse. *New England Journal of Medicine* 315: 1495–1500.

Crowley WF, Jr., Filicori M, Spratt DI, and Santoro NF (1985) The physiology of gonadotropin-releasing hormone (GnRH) secretion in men and women. *Recent Progress in Hormone Research* 41: 473–531.

DAWN (2007) *Drug Abuse Warning Network, 2005: National Estimates of Drug-Related Emergency Department Visits*. Substance Abuse and Mental Health Services Administration Office of Applied Studies, National Clearinghouse for Alcohol and Drug Information (NCADI), Rockville, MD.

DeBattista C, Posener JA, Kalehzan BM, and Schatzberg AF (2000) Acute antidepressant effects of intravenous hydrocortisone and CRH in depressed patients: A double-blind, placebo-controlled study. *American Journal of Psychiatry* 157: 1334–1337.

Di Chiara G (1995) The role of dopamine in drug abuse viewed from the perspective of its role in motivation. *Drug and Alcohol Dependence* 38: 95–137.

De Geyter C, De Geyter M, Huber PR, Nieschlag E, and Holzgreve W (2002) Progesterone serum levels during the follicular phase of the menstrual cycle originate from the crosstalk between the ovaries and the adrenal cortex. *Human Reproduction* 17: 933–939.

Dickerson GA, Waters IW, Wilson MC, and Davis WM (1991) Effect of sex steroids on cocaine lethality in male and female mice. *General Pharmacology* 22: 553–557.

Dierschke DJ, Hutz RJ, and Wolf RC (1985) Induced follicular atresia in rhesus monkeys: Strength–duration relationships of the estrogen stimulus. *Endocrinology* 117: 1397–1403.

Dierschke DJ, Hutz RJ, and Wolf RC (1987) Atretogenic action of estrogen in rhesus monkeys: Effects of repeated treatment. *American Journal of Primatology* 12: 251–261.

Dierschke DJ, Yamaji T, Karsch FJ, Weick RF, Weiss G, and Knobil E (1973) Blockade by progesterone of estrogen-induced LH and FSH release in the rhesus monkey. *Endocrinology* 92: 1496–1501.

de Witt H and Stewart J (1981) Reinstatement of cocaine-reinforced responding in the rat. *Psychopharmacology* 75: 134–143.

diZerega GS and Hodgen GD (1981a) Folliculogenesis in the primate ovarian cycle. *Endocrine Reviews* 2: 27–49.

diZerega G S and Hodgen GD (1981b) Follicular phase treatment of luteal phase dysfunction. *Fertility and Sterility* 35: 428–432.

diZerega GS and Hodgen GD (1981c) Luteal phase dysfunction infertility: A sequel to aberrant folliculogenesis. *Fertility and Sterility* 35: 489–499.

diZerega GS and Wilks JW (1984) Inhibition of the primate ovarian cycle by a porcine follicular fluid protein(s). *Fertility and Sterility* 41: 1094–1100.

Downs AW and Eddy NB (1932) The effect of repeated doses of cocaine on the rat. *Journal of Pharmacology and Experimental Therapeutics* 46: 199–200.

Dunn AJ and Berridge CW (1990) Physiological and behavioral responses to corticotropin-releasing factor administraton: Is CRF a mediator of anxiety or stress responses? *Brain Research Reviews* 15: 71–100.

Dworkin SJ, Mirkis S, and Smith JE (1995) Response-dependent versus response-independent presentation of cocaine: Differences in the lethal effects of the drug. *Psychopharmacology* 117: 262–266.

Elman I, Breiter HC, Gollub RL, et al. (1999) Depressive symptomatology and cocaine-induced pituitary–adrenal axis activation in individuals with cocaine dependence. *Drug and Alcohol Dependence* 56: 39–45.

Erb S, Shaham Y, and Stewart J (1996) Stress reinstates cocaine-seeking behavior after prolonged extinction and a drug-free period. *Psychopharmacology* 128: 408–412.

Erb S, Shaham Y, and Stewart J (1998) The role of corticotropin-releasing factor and corticosterone in stress- and cocaine-induced relapse to cocaine seeking in rats. *Journal of Neuroscience* 18: 5529–5536.

Evans SM (2007) The role of estradiol and progesterone in modulating the subjective effects of stimulants in humans. *Experimental and Clinical Psychopharmacology* 15: 418–426.

Evans SM, Cone EJ, and Henningfield JE (1996) Arterial and venous cocaine plasma concentrations in humans:

Relationship to route of administration, cardiovascular effects and subjective effects. *Journal of Pharmacology and Experimental Therapeutics* 279: 1345–1356.

Evans SM and Foltin RW (2006) Exogenous progesterone attenuates the subjective effects of smoked cocaine in women, but not in men. *Neuropsychopharmacology* 31: 659–674.

Evans SM, Haney M, Fischman MW, and Foltin RW (1999) Limited sex differences in response to "binge" smoked cocaine use in humans. *Neuropsychopharmacology* 21: 445–454.

Evans SM, Haney M, and Foltin RW (2002) The effects of smoked cocaine during the follicular and luteal phases of the menstrual cycle in women. *Psychopharmacology* 159: 397–406.

Falkenstein E, Tillmann H-C, Christ M, Feuring M, and Wehling M (2000) Multiple actions of steroid hormones-A focus on rapid, nongenomic effects. *Pharmcological Reviews* 52: 513–555.

Farhat MY, Lavigne MC, and Ramwell PW (1996) The vascular protective effects of estrogen. *FASEB Journal* 10: 615–624.

Feltenstein MW, Byrd EA, Henderson AR, and See RE (2009) Attenuation of cocaine-seeking by progesterone treatment in female rats. *Psychoneuroendocrinology* 34: 343–352.

Ferin M (1984) Endogenous opioid peptides and the menstrual cycle. *Trends in Neurosciences* 7: 194–196.

Ferin M, Van Vugt D, and Wardlaw S (1984) The hypothalamic control of the menstrual cycle and the role of endogenous opioid peptides. *Recent Progress in Hormone Research* 40: 441–485

Filicori M, Butler JP, and Crowley WM, Jr. (1984) Neuroendocrine regulation of the corpus luteum in the human. *Journal Clinical Investigation* 73: 1638–1647.

Filicori M, Flamigni C, Dellai P, et al. (1994) Treatment of anovulation with pulsatile gonadotropin-releasing hormone: Prognostic factors and clinical results in 600 cycles. *Journal of Clinical Endocrinology and Metabolism* 79: 1215–1220.

Frisch RE (1982) Fatness, puberty, menstrual periodicity and fertility. In: Vaitukaitis JL (ed.) *Clinical Reproductive Neuroendocrinology,* pp. l05–135. New York: Elsevier Biomedical.

Gambacciani M, Liu JH, Swartz WH, Tueros VS, Basmussen DD, and Yen SSC (1987) Intrinsic pulsatility of ACTH release from the human pituitary *in vitro*. *Clinical Endocrinology* 26: 557–563.

Gastfriend DR, Mendelson JH, Mello NK, Teoh SK, and Reif S (1993) Buprenorphine pharmacotherapy for concurrent heroin and cocaine dependence. *American Journal of Addictions* 2: 269–278.

Gawin FH and Ellinwood EH (1988) Cocaine and other stimulants: Actions, abuse and treatment. *New England Journal of Medicine* 318: 1173–1182.

Goeders NE (1997) A neuroendocrine role in cocaine reinforcement. *Psychoneuroendocrinology* 22: 237–259.

Goeders NE (2002a) Stress and cocaine addiction. *Journal of Pharmacology and Experimental Therapeutics* 301: 785–789.

Goeders NE (2002b) The HPA axis and cocaine reinforcement. *Psychoneuroendocrinology*. 27: 13–33.

Goeders NE and Guerin GF (1995) The HPA axis and cocaine self-administration. In: Harris LS (ed.) *College on Problems of Drug Dependence, 1994, NIDA Research Monograph 153,* pp. 462. Washington, DC: US Government Printing Office.

Goeders NE and Guerin GF (2000) Effects of the CRH receptor antagonist CP-154,526 on intravenous cocaine self-administration in rats. *Neuropsychopharmacology* 23: 577–586.

Gold PW, Goodwin FK, and Chrousos GP (1988) Clinical and biomedical manifestations of depression: Relationship to the neurobiology of stress (Part II). *New England Journal of Medicine* 319: 413–420.

Goodman AL and Hodgen GD (1983) The ovarian triad of the primate menstrual cycle. *Recent Progress in Hormone Research* 39: 1–67.

Gougeon A (1996) Regulation of ovarian follicular development in primates: Facts and hypothesis. *Endocrine Reviews* 17: 121–155.

Grammatopoulos DK and Chrousos GP (2002) Functional characteristics of CRH receptors and potential clinical applications of CRH-receptor antagonists. *Trends in Endocrinology and Metabolism* 13: 436–444.

Griebel G, Perrault G, and Sanger DJ (1998) Characterization of the behavioral profile of the non-peptide CRF receptor antagonist CP-154,526 in anxiety models in rodents. Comparison with diazepam and buspirone. *Psychopharmacology (Berl)* 138: 55–66.

Griffiths R, Bigelow G, and Henningfield J (1980) Similarities in animal and human drug taking behavior. In: Mello NK (ed.) *Advances in Substance Abuse, Behavioral and Biological Research,* pp. 1–90. Greenwich, CT: JAI Press.

Grimm JW and See RE (1997) Cocaine self-administration in ovariectomized rats is predicted by response to novelty, attenuated by 17-ß estradiol, and associated with abnormal vaginal cytology. *Physiology and Behavior* 61: 755–761.

Guerrero S, Gallardo A, and Munoz C (1965) Gonadal influences upon the sex differences in the resistance to toxic effects of cocaine in rats. *Archivos de Biologia y Medicina Experimentales* 2: 51–54.

Gulyas J, Rivier C, Perrin M, et al. (1995) Potent, structurally constrained agonists and competitive antagonists of corticotropin-releasing factor. *Proceedings of the National Academy of Sciences of the United States of America* 92: 10575–10579.

Habib K E, Weld K P, Rice K C, et al. (2000) Oral administration of a corticotropin-releasing hormone receptor antagonist significantly attenuates behavioral, neuroendocrine, and autonomic responses to stress in primates. *Proceedings of the National Academy of Sciences of the United States of America* 97: 6079–6084.

Halpern JH, Sholar MB, Glowacki J, Mello NK, Mendelson JH, and Siegel AJ (2003) Diminished interleukin-6 response to pro-inflammatory challenge in men and women after i.v. cocaine administration. *Journal of Clinical Endocrinology and Metabolism* 88: 1188–1193.

Hammond C, Wiebe R, Haney A, and Yancy S (1979) Ovulation induction with luteinizing hormone-releasing hormone in amenorrheic infertile women. *American Journal of Obstetrics and Gynecology* 135: 924.

Heesch CM, Negus BH, Bost JE, Keffer JH, Snyder RW, and Eichhorn EJ (1996) Effects of cocaine on anterior pituitary and gonadal hormones. *Journal of Pharmacology and Experimental Therapeutics* 278: 1195–1200.

Heesch CM, Negus BH, Keffer JE, Snyder RW, Risser R, and Eichhorn EJ (1995) Effects of cocaine on cortisol secretion in humans. *American Journal of the Medical Sciences* 310: 61–64.

Heinrichs SC, Menzaghi F, Merlo Pich PE, Britton KT, and Koch GF (1995) The role of CRF in behavioral aspects of stress. *Annals of the New York Academy of Sciences* 771: 92–104.

Heit S, Ownes MJ, Plotsky P, and Nemeroff CB (1997) Corticotropin-releasing factor, stress, and depression. *Neuroscientist* 3: 186–194.

Hemby SE, Co C, Koves TR, Smith JE, and Dworkin SI (1997) Differences in extracelluar dopamine concentrations in the nucleus accumbens during response-dependent and response-independent cocaine administration in the rat. *Psychopharmacology* 133: 7–16.

Hodgen GD (1982) The dominant ovarian follicle. *Fertility and Sterility* 38: 281–300.

Holsboer F (2000) The corticosteroid receptor hypothesis of depression. *Neuropsychopharmacology* 23: 477–501.
Holsboer F (2003) Corticotropin-releasing hormone modulators and depression. *Current Opinion in Investigational Drugs* 4: 46–50.
Hotchkiss J and Knobil E (1994) The menstrual cycle and its neuroendocrine control. In: Knobil E and Neill JD (eds.) *The Physiology of Reproduction*, 2nd edn., pp. 711–749. New York: Raven Press.
Hu M and Becker JB (2008) Acquisition of cocaine self-administration in ovariectomized female rats: Effect of estradiol dose or chronic estradiol administration. *Drug and Alcohol Dependence* 94: 56–62.
Hu M, Crombag HS, Robinson TE, and Becker JB (2004) Biological basis of sex differences in the propensity to self-administer cocaine. *Neuropsychopharmacology* 29: 81–85.
Hurley DM, Brian R, Outch K, et al. (1984) Induction of ovulation and fertility in amenorrheic women by pulsatile low-dose gonadotropin-releasing hormone. *New England Journal of Medicine* 310: 1069.
Hutchison JS, Kubik CJ, Nelson PB, and Zeleznik AJ (1987) Estrogen induces premature luteal regression in rhesus monkeys during spontaneous menstrual cycles, but not in cycles driven by exogenous gonadotropin-releasing hormone. *Endocrinology* 121: 466–474.
Hutz RJ, Dierschke DJ, and Wolf RC (1990) Role of estradiol in regulating ovarian follicular atresia in rhesus monkeys: A review. *Journal of Medical Primatology* 19: 553–571.
IOM (2001) Exploring the biological contributions to human health: Does sex matter? *Journal of Womens Health and Gender-Based Medicine* 10: 433–439.
Iranmanesh A, Lizarralde G, Short E, and Veldhuis JD (1990) Intensive venous sampling paradigms disclose high frequency adrenocorticotropin release episodes in normal men. *Journal of Clinical Endocrinology and Metabolism* 71: 1276–1283.
Ising M and Holsboer F (2007) CRH-sub-1 receptor antagonists for the treatment of depression and anxiety. *Experimental and Clinical Psychopharmacology* 15: 519–528.
Jackson LR, Robinson TE, and Becker JB (2006) Sex differences and hormonal influences on acquisition of cocaine self-administration in rats. *Neuropsychopharmacology* 31: 129–138.
Jaffe RB (1991) The menopause and perimenopausal period. In: Yen SSC and Jaffe RB (eds.) *Reproductive Endocrinology, Physiology, Pathophysiology and Clinical Management*, pp. 389–408. Philadelphia, PA: Saunders.
Jain R, Zwickler D, and Hollander CS, et al. (1991) Corticotropin-releasing factor modulates the immune response to stress in the rat. *Endocrinology* 128: 1329–1336.
Johanson CE and Schuster CR (1981) Animal models of drug self-administration. In: Mello NK (ed.) *Advances in Substance Abuse: Behavioral and Biological Research*, pp. 219–297. Greenwich, CT: JAI Press.
Justice AJ and de Wit H (2000a) Acute effects of *d*-amphetamine during the early and late follicular phahses of the menstrual cycle in women. *Pharmacology, Biochemistry, and Behavior* 66: 509–515.
Justice AJ and de Wit H (2000b) Acute effects of estradiol pretreatment on the response to *d*-amphetamine in women. *Neuroendocrinology* 71: 51–59.
Justice AJH and de Wit H (1999) Acute effects of d-amphetamine during the follicular and luteal phases of the menstrual cycle in women. *Psychopharmacology* 145: 67–75.
Kaufman M J, Levin J M, Maas L C, et al. (2001) Cocaine-induced cerebral vasoconstriction differs as a function of sex and menstrual cycle phase. *Biological Psychiatry* 49: 774–781.
Kaufman MJ, Streeter CC, Barros L, et al. (2007) Reduced plasma nitric oxide end products in cocaine-dependent men. *Journal of Addiction Medicine* 1: 96–103.
Kehne J and De Lombaert S (2002) Non-peptidic CRF1 receptor antagonists for the treatment of anxiety, depression and stress disorders. *Current Drug Targets. CNS Neurological Disorders*1: 467–493.
King TS, Canez MS, Gaskill S, Javors MA, and Schenken RS (1993) Chronic cocaine disruption of estrous cyclicity in the rat: Dose-dependent effects. *Journal of Pharmacology and Experimental Therapeutics* 264: 29–34.
King TS, Schenken RS, Kang IS, Javors MA, and Riehl RM (1990) Cocaine disrupts estrous cyclicity and alters the reproductive neuroendocrine axis in the rat. *Neuroendocrinology* 51: 15–22.
Knobil E (1974) On the control of gonadotropin secretion in the rhesus monkey. *Recent Progress in Hormone Research* 30: 1–46.
Knobil E (1980) The neuroendocrine control of the menstrual cycle. *Recent Progress in Hormone Research* 36: 53–88.
Knobil E and Hotchkiss J (1988) The menstrual cycle and its neuroendocrine control. In: Knobil E, Neill J, Ewing LL, Greenwald GS, Markert CL, and Pfaff DW (eds.) *The Physiology of Reproduction*, vol. 2, pp. 1971–1994. New York: Raven Press
Koob GF (1992) Drugs of abuse: Anatomy, pharmacology and function of reward pathways. *Trends in Pharmacological Sciences* 13: 177–184.
Koob GF, Ahmed SH, Boutrel B, et al. (2004) Neurobiological mechanisms in the transition from drug use to drug dependence. *Neuroscience and Biobehavioral Reviews* 27: 739–749.
Koob GF, Heinrichs SC, Merlo-Pich PE, Menzaghi F, Bladwin H, Miczek K, and Britton KT (1993) The role of corticotropin-releasing factor in behavioral responses to stress. *Ciba Foundation Symposium* 172: 277–289.
Koob GF and Le Moal M (1997) Drug abuse: Hedonic homeostatic dysregulation. *Science* 278: 52–58.
Koob GF and Le Moal M (eds.) (2006) *Neurobiology of Addiction*. London: Academic Press.
Kosten TR, Kosten TA, McDougle CJ, et al. (1996) Gender differences in response to intranasal cocaine administration to humans. *Biological Psychiatry* 39: 147–148.
Kuhar MJ, Ritz MC, and Boja JW (1991) The dopamine hypothesis of the reinforcing properties of cocaine. *TINS* 14: 299–302.
La Ferla J, Anderson DL, and Schalch DS (1978) Psychoendocrine response to sexual arousal in human males. *Psychosomatic Medicine* 40: 166–172.
Larson EB, Anker JJ, Gliddon LA, Fons KS, and Carroll ME (2007) Effects of estrogen and progesterone on the escalation of cocaine self-administration in female rats during extended access. *Experimental and Clinical Psychopharmacology* 15: 461-471.
Laviola G, Wood R, Kuhn C, Francis R, and Spear L (1995) Cocaine sensitization in periadolescent and adult rats. *Journal of Pharmacology and Experimental Therapeutics* 275: 345–357.
Levin JM, Holman BL, Mendelson JH, Teoh SK, Garada B, Johnson KA, and Springer S (1994) Gender differences in cerebral perfusion in cocaine abuse: Technetium-99m-HMPAO SPECT study of drug-abusing women. *Journal of Nuclear Medicine* 35: 1902–1909.
Levy AD, Baumann MH, and Van de Kar LD (1994) Monoaminergic regulation of neuroendocrine function and its modification by cocaine. *Frontiers in Neuroendocrinology* 15: 85–166.

Levy AD, Li Q, Kerr JE, et al. (1991) Cocaine-induced elevation of plasma adrenocorticotropin hormone and corticosterone is mediated by serotonergic neurons. *Journal of Pharmacology and Experimental Therapeutics* 259: 495–500.
Levy AD, Rittenhouse PA, Li Q, et al. (1992) Repeated injections of cocaine inhibit the serotonergic regulation of prolactin and renin secretion in rats. *Brain Research* 580: 6–11.
Liaw CW, Grigoriadis DE, Lorang MT, De Souza EB, and Maki RA (1997) Localization of agonist- and antagonist-binding domains of human corticotropin-releasing factor receptors. *Molecular Endrocrinology* 13: 2048–2053.
Lukas SE, Sholar M, Lundahl LH, et al. (1996) Sex differences in plasma cocaine levels and subjective effects after acute cocaine administration in human volunteers. *Psychopharmacology* 125: 346–354.
Lundkvist J, Chai Z, Teheranian R, et al. (1996) A non peptidic corticotropin releasing factor receptor antagonist attenuates fever and exhibits anxiolytic-like activity. *European Journal of Pharmacology* 309: 195–200.
Lyendecker G and Wildt L (1983) Control of gonadotropin secretion in women. In: Norman RL (ed.) *Neuroendocrine Aspects of Reproduction,* pp. 295–323. New York: Academic Press.
Lynch WJ (2008) Acquisition and maintenance of cocaine self-administration in adolescent rats: Effects of sex and gonadal hormones. *Psychopharmacology* 197: 237–246.
Lynch WJ, Arizzi MN, and Carroll ME (2000) Effects of sex and the estrous cycle on regulation of intravenously self-administered cocaine in rats. *Psychopharmacology* 152: 132–139.
Lynch WJ and Carroll ME (1999) Sex differences in the acquisition of intravenously self-administered cocaine and heroin in rats. *Psychopharmacology* 144: 77–82.
Lynch WJ and Carroll ME (2000) Reinstatement of cocaine self-administration in rats: Sex differences. *Psychopharmacology* 148: 196–200.
Lynch WJ, Roth ME, Mickelberg JL, and Carroll ME (2001) Role of estrogen in the acquisition of intravenously self-administered cocaine in female rats. *Pharmacology, Biochemistry, and Behavior* 68: 641–646.
Lynch WJ and Taylor JR (2005) Decreased motivation following cocaine self-administration under extended access conditions: Effects of sex and ovarian hormones. *Neuropsychopharmacology* 30: 927–935.
Mansbach RS, Brooks EN, and Chen YL (1997) Antidepressant-like effects of CP-154,526, a selective CRF1 receptor antagonist. *European Journal of Pharmacology* 323: 21–26.
Marinelli M and Piazza PV (2002) Interaction between glucocorticoid hormones, stress and psychostimulant drugs. *European Journal of Neuroscience* 16: 387–394.
Markham PD, Salahuddin SZ, Veren K, Orndorff S, and Gallo RC (1986) Hydrocortisone and some other hormones enhance the expression HTLV-III. *International Journal of Cancer* 37: 67–72.
Martin KA, Hall JE, Adams JM, and Crowley WF, Jr. (1993) Comparison of exogenous gonadotropins and pulsatile gonadotropin-releasing hormone for induction of ovulation in hypogonadotropic amenorrhea. *Journal of Clinical Endocrinology and Metabolism* 77: 125–129.
Martinez V, Barquist E, Rivier J, and Tache Y (1998) Central CRF inhibits gastric emptying of a nutrient solid meal in rats: The role of CRF2 receptors. *American Journal of Physiology* 274: 965–970.
McEwen BS (1998) Protective and damaging effects of stress mediators. *New England Journal of Medicine* 338: 171–179.
McNeely MJ and Soules MR (1988) Diagnosis of luteal phase deficiency: A critical review. *Fertility and Sterility* 50: 1–15.
Mello NK (1979) Animal models of alcoholism: Progress and prospects. In: Davidson RS (ed.) *Modification of Pathological Behavior, Experimental Analysis of Etiology and Therapy*, pp. 273–333. New York: Gardner Press.
Mello NK (1998) Cocaine abuse and reproductive function in women. In: Wetherington CL and Roman AB (eds.) *Drug Addiction Research and the Health of Women. National Institute on Drug Abuse,* NIH Publ. No. 98-4290, pp. 131–149. Washington, DC: US Government Printing Office.
Mello NK, Knudson I, Kelly M, and Mendelson JH (under review) Effects of progesterone and testosterone on cocaine self-administration and cocaine discrimination by female rhesus monkeys.
Mello NK, Knudson IM, and Mendelson JH (2007) Sex and menstrual cycle effects on progressive ratio measures of cocaine self-administration in cynomolgus monkeys. *Neuropsychopharmacology* 32: 1956–1966.
Mello NK and Mendelson JH (1995) Buprenorphine treatment of cocaine and heroin abuse. In: Cowan A and Lewis JW (eds.) *Buprenorphine: Combatting Drug Abuse with a Unique Opioid*, pp. 241–287. New York: Wiley-Liss.
Mello NK and Mendelson JH (1997) Cocaine's effects on neuroendocrine systems: Clinical and preclinical studies. *Special Issue: Current Issues in Psychostimulant Abuse Research. Pharmacology, Biochemistry, and Behavior* 57: 571–599.
Mello NK and Mendelson JH (2002) Cocaine, hormones and behavior: Clinical and preclinical studies. In: Pfaff DW, Arnold AP, Etgen AM, Fahrbach SE, and Rubin RT (eds.) *Hormones, Brain and Behavior,* pp. 665–745. New York: Academic Press.
Mello NK and Mendelson JH (2009) Cocaine, hormones and behavior: Clinical and precliinical studies. In: Pfaff DW, Arnold AP, Etgen AM, Fahrbach SE, and Rubin RT (eds.) *Hormones, Brain and Behavior,* 2nd edn., San Diego: Academic Press.
Mello NK, Mendelson JH, Drieze J, and Kelly M (1990a) Acute effects of cocaine on prolactin and gonadotropins in female rhesus monkey during the follicular phase of the menstrual cycle. *Journal of Pharmacology and Experimental Therapeutics* 254: 815–823.
Mello NK, Mendelson JH, Drieze J, and Kelly M (1990b) Cocaine effects on luteinizing hormone-releasing hormone-stimulated anterior pituitary hormones in female rhesus monkey. *Journal of Clinical Endocrinology and Metabolism* 71: 1434–1441.
Mello NK, Mendelson JH, Kelly M, and Bowen CA (2000) The effects of cocaine on basal and human chorionic gonadotropin-stimulated ovarian steroid hormones in female rhesus monkeys. *Journal of Pharmacology and Experimental Therapeutics* 294: 1137–1145.
Mello NK, Mendelson JH, Kelly M, Diaz-Migoyo N, and Sholar JW (1997) The effects of chronic cocaine self-administration on the menstrual cycle in rhesus monkeys. *Journal of Pharmacology and Experimental Therapeutics* 281: 70–83.
Mello NK, Mendelson JH, Lukas SE, Gastfriend D, Teoh SK, and Holman BL (1993a) Buprenorphine treatment of opiate and cocaine abuse: Clinical and preclinical studies. *Harvard Reviews. Psychiatry* 1: 168–183.
Mello NK, Mendelson JH, Negus SS, Kelly M, Knudson I, and Roth ME (2004) The effects of cocaine on gonadal steroid hormones and LH in male and female rhesus monkeys. *Neuropsychopharmacology* 29: 2024–2034.
Mello NK, Mendelson JH, and Teoh SK (1992) Alcohol and neuroendocrine function in women of reproductive age. In: Mendelson JH and Mello NK (eds.) *Medical Diagnosis and Treatment of Alcoholism,* 1st edn., pp. 575–621. New York: McGraw-Hill.
Mello NK and Negus SS (1996) Preclinical evaluation of pharmacotherapies for treatment of cocaine and opiate

abuse using drug self-administration procedures. *Neuropsychopharmacology* 14: 375–424.

Mello NK, Negus SS, Knudson IM, Kelly M, and Mendelson JH (2008) Effects of estradiol on cocaine self-administration and cocaine discrimination by female rhesus monkeys. *Neuropsychopharmacology* 33: 783–795.

Mello NK, Sarnyai Z, Mendelson JH, Drieze JM, and Kelly M (1993b) Acute effects of cocaine on anterior pituitary hormones in male and female rhesus monkeys. *Journal of Pharmacology and Experimental Therapeutics* 266: 804–811.

Mendelson JH and Mello NK (1996) Drug therapy: Management of cocaine abuse and dependence. *New England Journal Medicine* 334: 965–972.

Mendelson JH and Mello NK (1998) Cocaine and other commonly abused drugs. In: Fauci AS, Braunwald E, Isselbacher KJ, et al. (eds.) *Harrison's Principles of Internal Medicine,* pp. 2512–2516. New York: McGraw-Hill.

Mendelson JH and Mello NK (2001) Cocaine and other commonly abused drugs. In: Braunwald E, Fauci AS, Kasper DL, Hauser SL, Longo DL, and Jameson JL (eds.) *Harrison's Principles of Internal Medicine*, 15th edn., pp. 2570–2574. New York: McGraw-Hill.

Mendelson JH and Mello NK (2008) Cocaine and other commonly abused drugs. In: Fauci AS, Kasper DL, Braunwald E, Hauser SL, Longo DL, Jameson JL, and Loscaizo J (eds.) *Harrison's Principles of Internal Medicine*, 17th edn., pp. 2733–2736. New York: McGraw-Hill.

Mendelson JH, Mello NK, Cristofaro P, Skupny A, and Ellingboe J (1986) Use of naltrexone as a provocative test for hypothalamic–pituitary hormone function. *Pharmacology, Biochemistry, and Behavior* 24: 309–313.

Mendelson JH, Mello NK, Lukas SE, Woods BT, and Teoh SK (1989a) Promising new biological and behavioral correlates of the reinforcing properties of drugs. In: Fischman MW and Mello NK (eds.) *Testing for Abuse Liability of Drugs in Humans*, DHHS Publ. No. (ADM) 89-1613, pp. 307–340. Rockville, MD: US Government Printing Office.

Mendelson JH, Mello NK, and Negus SS (1999a) Effects of luteinizing hormone releasing hormone on plasma cocaine levels in rhesus monkeys. *Journal of Pharmacology and Experimental Therapeutics* 289: 791–799.

Mendelson JH, Mello NK, Sholar MB, et al. (1999b) Cocaine pharmacokinetics in men and in women during the follicular and luteal phase of the menstrual cycle. *Neuropsychopharmacology* 21: 294–303.

Mendelson JH, Mello NK, Sholar MB, Siegel AJ, Mutschler N, and Halpern J (2002) Temporal concordance of cocaine effects on mood states and neuroendocrine hormones. *Psychoneuroendocrinology* 27: 71–82.

Mendelson JH, Mello NK, Teoh SK, Ellingboe J, and Cochin J (1989b) Cocaine effects on pulsatile secretion of anterior pituitary, gonadal, and adrenal hormones. *Journal of Clinical Endocrinology and Metabolism* 69: 1256–1260.

Mendelson JH, Mello NK, Teoh SK, et al. (1992a) Human studies on the biological basis of reinforcement: A neuroendocrine perspective. In: O'Brien CP and Jaffe JH (eds.) *Addictive States*, pp. 131–155. New York: Raven Press.

Mendelson JH, Sholar M, Mello NK, Teoh SK, and Sholar JW (1998) Cocaine tolerance: Behavioral, cardiovascular, and neuroendocrine function in men. *Neuropsychopharmacology* 18: 263–271.

Mendelson JH, Sholar M, Siegel AJ, and Mello NK (2001) Effects of cocaine on luteinizing hormone in women during the follicular and luteal phases of the menstrual cycle and in men. *Journal of Pharmacology and Experimental Therapeutics* 296: 972–979.

Mendelson JH, Teoh SK, Mello NK, and Ellingboe J (1992c) Buprenorphine attenuates the effects of cocaine on adrenocorticotrophin (ACTH) secretion and mood states in man. *Neuropsychopharmacology* I7: 157–162.

Mendelson JH, Teoh SK, Mello NK, Ellingboe J, and Rhoades E (1992b) Acute effects of cocaine on plasma ACTH, luteinizing hormone and prolactin levels in cocaine-dependent men. *Journal of Pharmacology and Experimental Therapeutics*. 263: 505–509.

Merkatz RB, Temple R, Sobel S, Feiden K, and Kessler DA (1993) Women in clinical trials of new drugs: A change in food and drug administration policy. *New. England Journal of Medicine* 329: 292–296.

Minshall RD, Stanczyk FZ, Miyagawa K, Uchida B, Axthelm M, Novy M, and Hermsmeyer K (1998) Ovarian steroid protection against coronary artery hyperreactivity in rhesus monkeys. *Journal of Clinical Endocrinology and Metabolism* 83: 649–659.

Montoya ID, Gorelick DA, Preston KL, et al. (2004) Randomized trial of buprenorphine for treatment of concurrent opiate and cocaine dependence. *Clinical Pharmacology and Therapeutics* 75: 34–48.

Moore FL and Evans SJ (1999) Steroid hormones use non-genomic mechanisms to control brain functions and behaviors: A review of evidence. *Brain, Behavior and Evolution* 54: 41–50.

Morishima HO, Abe Y, Matsuo M, Akiba K, Masaoka T, and Cooper TB (1993) Gender-related differences in cocaine toxicity in the rat. *Journal of Laboratory and Clinical Medicine* 122: 157–163.

Munro CA, McCaul ME, Wong DF, et al. (2006) Sex differences in striatal dopamine release in healthy adults. *Biological Psychiatry* 59: 966–974.

Nemeroff CB (1996) The corticotropin-releasing factor (CRF) hypothesis of depression: New findings and new directions. *Medicine and Psychiatry* 1: 336–342.

Nemeroff CB (1998) Psychopharmacology of affective disorders in the 21st century. *Biological Psychiatry* 44: 517–525.

NIDA (1991) *The National Commission on Acquired Immune Deficiency Syndrome. Fifth Interim Report to the President and the Congress: The Twin Epidemics of Substance use and HIV.* Washington, DC: National Commission on AIDS.

NIDA (2000) *Epidemiologic Trends in Drug Abuse*, NIH Publ. No. 01-4916, pp. 187.98. Rockville, MD: National Institute on Drug Abuse.

Nippoldt TB, Reame NE, Kelch RP, and Marshall JC (1989) The roles of estradiol and progesterone in decreasing luteinizing hormone pulse frequency in the luteal phase of the menstrual cycle. *Journal of Clinical Endocrinology and Metabolism* 69: 67–76.

Olster DH and Ferin N (1987) Corticotropin-releasing hormone inhibits gonadotropin secretion in the ovariectomized rhesus monkey. *Journal of Clinical Endocrinology and Metabolism* 65: 262–267.

Papanicolaou D (1998) The pathophysiologic roles of interleukin-6 in human disease. NIH Conference. *Annals of Internal Medicine* 128: 127–137.

Pequegnat W, Garrick N, and Stover E (1992) Neuroscience findings in AIDS: A review by the National Institute of Mental Health. *Progress in Neuro-psychopharmacology and Biological Psychiatry* 16: 145–170.

Petraglia F, Sutton S, Vale W, and Plotsky P (1987) Corticotropin-releasing factor decreases plasma LH levels in female rats by inhibiting gonadotropin-releasing hormone release into hypophysial-portal circulation. *Endocrinology* 120: 1083–1088.

Piazza PV and LeMoal M (1996) Pathophysiological basis of vulnerability to drug abuse: Role of an interaction between stress, glucocorticoids, and dopaminergic neurons. *Annual Reviewof Pharmacology and Toxicology* 36: 359–378.

Potter DA, Luther MF, Eddy CA, Siler-Khodr TM, King TS, and Schenken RS (1999) Low-dose follicular-phase cocaine administration disrupts menstrual and ovarian cyclicity in rhesus monkeys. *Journal of the Society for Gynecologic Investigation* 6: 88–94.

Potter DA, Moreno A, Luther MF, Eddy CA, Siler-Khodr TM, King TS, and Schenken RS (1998) Effects of follicular-phase cocaine administration on menstrual and ovarian cyclicity in rhesus monkeys. *American Journal of Obstetrics and Gynecology* 178: 118–125.

Rapp U, Kourounakis P, and Selye H (1979) Effect of steroids and diethylstilbestrol on cocaine toxicity, plasma concentrations and urinary excretion. *Drug Research* 29: 48–50.

Rawson RA, Obert JL, McCann MJ, Castro FG, and Ling W (1991) Cocaine abuse treatment: A review of current strategies. *Journal of Substance Abuse* 3: 457–491.

Reichlin S (1993) Neuroendocrine–endocrine–immune interactions. *New England Journal of Medicine* 329: 1246–1253.

Reid RL, Quigley ME, and Yen SSC (1983) The disappearance of opioidergic regulation of gonadotropin secretion in post-menopausal women. *Journal of Clinical Endocrinology and Metabolism* 57: 1107–1110.

Richter RM and Weiss F (1999) *In vivo* CRF release in rat amygdala is increased during cocaine withdrawal in self-administering rats. *Synapse* 32: 254–261.

Risner ME, Goldberg SR, Prada JA, and Cone EJ (1985) Effects of nicotine, cocaine and some of their metabolites on schedule-controlled responding by beagle dogs and squirrel monkeys. *Journal of Pharmacology and Experimental Therapeutics* 234: 113–119.

Rittmaster RS, Cutler GB, Sobel DO, Goldstein DS, Koppelman MCS, Loriaux DL, and Chrousos GP (1985) Morphine inhibits the pituitary–adrenal response to ovine corticotropin-releasing hormone in normal subjects. *Journal of Clinical Endocrinology and Metabolism* 60: 891–895.

Rivier J, Gulyas J, Corrigan A, et al. (1998) Astressin analogues (corticotropin-releasing factor antagonists) with extended duration of action in the rat. *Journal of Medicinal Chemistry* 41: 5012–5019.

Rivier C and Vale W (1987) Cocaine stimulates adrenocorticotropin (ACTH) secretion through a corticotropin-releasing factor (CRF)-mediated mechanism. *Brain Research* 422: 403–406.

Roberts DCS, Loh EA, and Vickers G (1989) Self-administration of cocaine on a progressive ratio schedule in rats: Dose–response relationship and effect of haloperidol pretreatment. *Psychopharmacology* 97: 535–538.

Rodriguez de Fonseca F, Carrera MRA, Navarro M, Koob GF, and Weiss F (1997) Activation of corticotropin-releasing factor in the limbic system during cannabinoid withdrawal. *Science* 276: 2050–2054.

Ross GT (1985) Disorders of the ovary and female reproductive tract. In: Wilson JD and Foster DW (eds.) *Williams Textbook of Endocrinology*, 7th edn., pp. 206–258. Philadelphia, PA: Saunders

Rounsaville B, Anton S, Carroll K, Prusoff B, and Gawin F (1991) Psychiatric diagnoses of treatment-seeking cocaine abusers. *Archives of General Psychiatry* 18: 48–61.

Rubin RT, Poland RE, Lesser IM, Martin DJ, Blodgett AL, and Winston RA (1987) Neuroendocrine aspects of primary endogenous depression. III. Cortisol secretion in relation to diagnosis and symptom patterns. *Psychological Medicine* 17: 609–619.

SAMHSA (2006) *Substance Abuse and Mental Health Services Administration. Results from the 2006 National Survey on Drug Use and Health: National Findings*. Washington, DC: SAMHSA

Santoro N, Filicori M, and Crowley J (1986b) Hypogonadotropic disorders in men and women: Diagnosis and therapy with pulsatile gonadotropin-releasing hormone. *Endocrine Reviews* 7: 11–23.

Santoro N, Wierman ME, and Filicori M (1986a) Intravenous administration of pulsatile gonadotropin-releasing hormone in hypothalamic amenorrhea: Effects of dosage. *Journal of Clinical Endocrinology and Metabolism* 62: 109–116.

Sarnyai Z, Biro E, Penke B, and Telegdy G (1992a) The cocaine-induced elevation of plasma corticosterone is mediated by endogenous corticotropin-releasing factor (CRF) in rats. *Brain Research* 589: 154–156.

Sarnyai Z and Kovacs GL (1994) Role of oxytocin in the neuroadaptation to drugs of abuse. *Psychoneuroendocrinology* 19: 85–117.

Sarnyai Z, Mello NK, Mendelson JH, Eros-Sarnyai M, and Mercer G (1996) Effects of cocaine on pulsatile activity of the hypothalamic–pituitary–adrenal axis in male rhesus monkeys: Neuroendocrine and behavioral correlates. *Journal of Pharmacology and Experimental Therapeutics* 277: 225–234.

Sauder S, Frager M, Case G, Kelch R, and Marshall J (1984) Abnormal patterns of pulsatile luteinizing hormone secretion in women with hyperprolactinemia and amenorrhea: Responses to bromocriptine. *Journal of Clinical Endocrinology and Metabolism* 59: 941–948.

Schmitz JM, Stotts AL, Averill PM, Rothfleisch JM, Bailey SE, Sayre SL, and Grabowski J (2000) Cocaine dependence with and without comorbid depression: A comparison of patient characteristics. *Drug and Alcohol Dependence* 60: 189–198.

Schoenbaum EE, Hartel D, Selwyn PA, et al. (1989) Risk factors for human immunodeficiency virus infection in intravenous drug users. *New England Journal of Medicine* 321: 874–879.

Schottenfeld RS, Pakes J, Ziedonis D, and Kosten TR (1993) Buprenorphine: Dose-related effects on cocaine and opioid use in cocaine-abusing opioid-dependent humans. *Biological Psychiatry* 3: 66–74.

Schultz DW, Mansbach RS, Sprouse J, et al. (1996) CP-154,526: A potent and selective nonpeptide antagonist of corticotropin releasing factor receptors. *Proceedings of the National Academy of Sciences of the United States of America* 93: 10477–10482.

Selye H (1971) Protection by estradiol against cocaine, coniine, ethylmorphine, LSD, and strychnine. *Hormones and Behavior* 2: 337–341.

Shaham Y, Erb S, Leung S, Buczek Y, and Stewart J (1998) CP-154,526: A selective, non-peptide antagonist of the corticotropin-releasing factor1 receptor attenuates stress-induced relapse to drug seeking in cocaine- and heroin-trained rats. *Psychopharmacology (Berl)* 137: 184–190.

Shaham Y, Erb S, and Stewart J (2000) Stress-induced relapse to heroin and cocaine seeking in rats: A review. *Brain Research Reviews* 33: 13–33.

Sholar MB, Mendelson JH, Mello NK, et al. (1998) Concurrent pharmacokinetic analysis of plasma cocaine and adrenocorticotropic hormone (ACTH) in men. *Journal of Clinical Endocrinology and Metabolism* 83: 966–968.

Siegel RK (1982) Cocaine and sexual dysfunction: The curse of mama coca. *Journal of Psychoactive Drugs* 14: 71.

Sinha R (2001) How does stress increase risk of drug abuse and relapse? *Psychopharmacology (Berl)* 158: 343–359.

Sinha R, Fox HC, Hong KI, Sofuoglu M, Morgan PT, and Bergquist KL (2007) Sex steroid hormones, stress response and drug craving in cocaine dependent women: Implications for relapse susceptibility. *Experimental and Clinical Psychopharmacology* 15: 445–452.

Sinha R, Garcia M, Paliwal P, Kreek MJ, and Rounsaville BJ (2006) Stress-induced cocaine craving and hypothalamic–pituitary–adrenal responses are predictive of cocaine relapse outcomes. *Archives of General Psychiatry* 63: 324–331.

Sinha R, Talih M, Malison R, Anderson GA, Cooney N, and Kreek MJ (2003) Hypothalamic–pituitary–adrenal axis and sympatho-adreno-medullary responses during stress-induced and drug cue-induced cocaine craving states. *Psychopharmacology* 170: 62–72.

Skafar DF, Xu R, Morales J, Ram J, and Sowers JR (1997) Female sex hormones and cardiovascular disease in women. *Journal of Clinical Endocrinology and Metabolism* 82: 3913–3918.

Smith CG and Smith MT (1990) Substance abuse and reproduction. *Seminar in Reproductive Endocrinology* 8: 55–64.

Smith DE, Wesson DR, and Apter-Marsh M (1984) Cocaine- and alcohol-induced sexual dysfunction in patients with addictive disease. *Journal of Psychoactive Drugs* 16: 359–361.

Smith JE, Co C, Coller MD, Hemby SE, and Martin TJ (2006) Self-administered heroin and cocaine combinations in the rat: Additive reinforcing effects–supra–additive effects on nucleus accumbens extracellular dopamine. *Neuropsychopharmacology* 31: 139–150.

Sofuoglu M, Babb DA, and Hatsukami DK (2002) Effects of progesterone treatment on smoked cocaine response in women. *Pharmacology, Biochemistry, and Behavior* 72: 431–435.

Sofuoglu M, Dudish-Poulsen S, Nelson D, Pentel PR, and Hatsukami DK (1999) Sex and menstrual cycle differences in the subjective effects from smoked cocaine in humans. *Experimental and Clinical Psychopharmacology* 7: 274–283.

Sofuoglu M, Mitchell E, and Kosten TR (2004) Effects of progesterone treatment on cocaine responses in male and female cocaine users. *Pharmacology, Biochemistry, and Behavior* 78: 699–705.

Sofuoglu M, Poling J, Gonzalez G, Gonsai K, Oliveto A, and Kosten TR (2007) Progesterone effects on cocaine use in cocaine users maintained on methadone: A randomized, double-blind pilor trial. *Experimental and Clinical Psychopharmacology* 15: 453–460.

Soudeyns H, Geleziunas R, Shyamala G, Hiscott J, and Weinberg MA (1993) Identification of a novel glucocorticoid response element within the genome of the human immunodeficiency virus type 1. *Virology* 194: 758–768.

Soules MR, Clifton DK, Bremner WJ, and Steiner RA (1987) Corpus luteum insufficiency induced by a rapid gonadotropin-releasing hormone-induced gonadotropin secretion pattern in the follicular phase. *Journal of Clinical Endocrinology and Metabolism* 65: 457–464.

Soules MR, McLachlan RI, Ek M, Dahl KD, Cohen NL, and Bremner WJ (1989a) Luteal phase deficiency: Characterization of reproductive hormones over the menstrual cycle. *Journal of Clinical Endocrinology and Metabolism* 69: 804–812.

Soules MR, Steiner RA, Clifton DK, Cohen NL, Aksel S, and Bremner WJ (1984) Progesterone modulation of pulsatile luteinizing hormone secretion in normal women. *Journal of Clinical Endocrinology and Metabolism* 58: 378.

Specio SE, Wee S, O'Dell LE, Boutrel B, Zorrilla EP, and Koob GF (2008) CRF1 receptor antagonists attenuate escalated cocaine self-administration in rats. *Psychopharmacology* 196: 473–482.

Steel E and Haverkos HW (1992) Epidemiologic studies on HIV/AIDS and drug abuse. *American Journal of Drug and Alcohol Abuse* 18: 167–175.

Stinus L, Cador M, Zorrilla EP, and Koob GF (2005) Buprenorphine and a CRF_1 antagonist block the acquisition of opiate withdrawal-induced conditioned place aversion in rats. *Neuropsychopharmacology* 30: 90–98.

Stouffer RL (1990) Corpus luteum function and dysfunction. *Clinical Obstetrics and Gynecology* 33: 668–689.

Teoh SK, Lex BW, Mendelson JH, Mello NK, and Cochin J (1992) Hyperprolactinemia and macrocytosis in women with alcohol and polysubstance dependence. *Journal of Studies on Alcohol* 53: 176–182.

Teoh SK, Mello NK, and Mendelson JH (1994a) Effects of drugs of abuse on reproductive function in women and pregnancy. In: Watson R (ed.) *Addictive Behaviors in Women*, pp. 437–473. Totowa, NJ: Humana Press.

Teoh SK, Sarnyai Z, Mendelson JH, et al. (1994c) Cocaine effects on pulsatile secretion of ACTH in men. *Journal of Pharmacology and Experimental Therapeutics* 270: 1134–1138.

Torres G and Rivier C (1992c) Cocaine-induced ACTH secretion: Dependence of plasma levels of the drug and mode of exposure. *Brain Research Bulletin* 29: 51–56.

Tsagarakis S, Navara P, Rees LH, Besser M, and Grossman A (1989) Morphine directly modulates the release of stimulated corticotrophin-releasing factor-41 from rat hypothalamus *in vitro*. *Endocrinology* 124: 2330–2335.

Tutton CS and Crayton JW (1993) Current pharmacotherapies for cocaine abuse: A review. *Journal of Addictive Diseases* 12: 109–127.

Van de Kar L, Bonadonna A, Rittenhouse P, et al. (1992) Prior chronic exposure to cocaine inhibits the serontonergic stimulation of ACTH and secretion of corticosterone. *Neuropharmacology* 31: 169–175.

Van Dyke C and Byck R (1983) Cocaine use in man. In: Mello NK (ed.) *Advances in Substance Abuse, Behavioral and Biological Research*, pp. 1–24. Greenwich, CT: JAI Press.

VanVugt DA, Heisler LE, and Reid RL (1992) Progesterone inhibits the estrogen-induced gonadotropin surge in the rhesus monkey independent of endogenous opiates. *Journal of Clinical Endocrinology and Metabolism* 74: 1312–1319.

VanVugt DA, Lam NY, and Ferin M (1984) Reduced frequency of pulsatile luteinizing hormone secretion in the luteal phase of the rhesus monkey. Involvement of endogenous opiates. *Endocrinology* 115: 1095–1101.

Vasudevan N and Pfaff DW (2007) Membrane-initiated actions of estrogens in neuroendocrinology: Emerging principles. *Endocrine Reviews* 28: 1–19.

Veldhuis JD (1999) Male hypothalamic–pituitary–gonadal axis. In: Yen SSC, Jaffe RB, and Barbieri RL (eds.) *Reproductive Endocrinology*, 4th edn., pp. 622–631. Philadelphia, PA: Saunders.

Veldhuis JD, Beltins IZ, Johnson MD, Serabian MA, and Dufau MB (1984) Biologically active luteinizing hormone is secreted in episodic pulsations that vary in relation to stage of the menstrual cycle. *Journal of Clinical Endocrinology and Metabolism* 58: 1050–1058.

Veldhuis JD and Johnson ML (1986) Cluster analysis: A simple, versatile and robust algorithm for endocrine pulse detection. *American Journal of Physiology* 250: E486–E493.

Vocci FJ, Acri J, and Elkashef A (2005) Medication development for addictive disorders: The state of the science. *American Journal of Psychiatry* 162: 1432–1440.

Vocci FJ and Ling W (2005) Medications development: Successes and challenges. *Pharmacology and Therapeutics* 108: 94–108.

Ward AS, Haney M, Fischman MW, and Foltin R (1997) Binge cocaine self-administration in humans: Intravenous cocaine. *Psychopharmacology* 132: 375–381.

Warren MP (1992) Amenorrhea in endurance runners (Clinical Review 40). *Journal of Clinical Endocrinology and Metabolism* 75: 1393–1397.

Webster EL, Lewis DB, Torpy DJ, Zachman EK, Rice KC, and Chrousos GP (1996) *In vivo* and *in vitro* characterization of antalarmin, a nonpeptide corticotropin-releasing hormone (CRH) receptor antagonist: Suppression of pituitary ACTH

release and peripheral inflammation. *Endocrinology* 137: 5747–5750.
Wetherington CL (2007) Sex-gender differences in drug abuse: A shift in the burden of proof? *Experimental and Clinical Psychopharmacology* 15: 411–417.
White TL, Justice AJ, and de Wit H (2002) Differential subjective effects of d-amphetamine by gender, hormone levels and menstrual cycle pahse. *Pharmacology, Biochemistry, and Behavior* 73: 729–741.
Wildt L, Hutchison JS, Marshall G, Pohl CR, and Knobil E (1981) On the site of action of progesterone in the blockade of the estradiol-induced gonadotropin discharge in the rhesus monkey. *Endocrinology* 109: 1293–1294.
Wilkins JN, Gorelick DA, Nademanee K, Taylor A, and Herzberg DS (1992) Hypothalamic–pituitary function during alcohol exposure and withdrawal and cocaine exposure. In: Galanter M (ed.) *Recent Development in Alcoholism: Alcohol and Cocaine: Similarities and Differences,* pp. 57–71. New York: Plenum Press.
Wilks JW, Hodgen GD, and Ross GT (1977) Anovulatory menstrual cycles in rhesus monkeys: The significance of serum, follicle stimulating hormone/luteinizing hormone ratios. *Fertility and Sterility* 28: 1094–1101.
Williams C, Nishihara M, Thalabard J-C, Grosser P, Hotchkiss J, and Knobil E (1990) Corticotropin-releasing factor and gonadotropin hormone pulse generator activity in the rhesus monkey. *Neuroendocrinology* 52: 133–137.
Wong M, Thompson TL, and Moss RL (1996) Nongenomic actions of estrogen in the brain: Physiological significance and cellular mechanisms. *Critical Reviews in Neurobiology* 10: 189–203.
Woolverton WL and Johnson KM (1992) Neurobiology of cocaine abuse. *Trends in Pharmacological Sciences* 13: 193–200.
Xia L, Van Vugt D, Alston E, Luckhaus J, and Ferin M (1992) A surge of gonadotropin-releasing hormone accompanies the estradiol-induced gonadotropin surge in the rhesus monkey. *Endocrinology* 131: 2812–2820.
Xiao E and Ferin M (1988) The inhibitory action of corticotropin-releasing hormone on gonadotropin secretion in the ovariectomized rhesus monkey is not mediated by adrenocorticotropic hormone. *Biology of Reproduction* 38: 763–767.
Xiao E, Luckhaus J, Niemann W, and Ferin M (1989) Acute inhibition of gonadotropin secretion by corticotropin-releasing hormone in the primate: Are the adrenal glands involved? *Endocrinology* 124: 1632–1637.
Yamamoto RT, Teter CJ, Barros TL, et al. (2007) Antiandrogen pretreatment alters cocaine pharmacokinetics in men. *Journal Addiction Medicine* 1: 198–204.
Yen SSC (1999a) Neuroendocrinology of reproduction. In: Yen SSC, Jaffe RB, and Barbieri RL (eds.) *Reproductive Endocrinology*, 4th edn., pp. 30–80. Philadelphia, PA: Saunders.
Yen SSC (1999b) The human menstrual cycle: Neuroendocrine regulation. In: Yen SSC, Jaffe RB, and Barbieri RL (eds.) *Reproductive Endocrinology*, 4th edn., pp 191–217. Philadelphia, PA: Saunders.
Yen SSC, Quigley ME, Reid RL, Ropert JF, and Cetel NS (1985) Neuroendocrinology of opioid peptides and their role in the control of gonadotropin and prolactin secretion. *American Journal of Obstetrics and Gynecology* 152 485–493.
Zeleznik AJ (1981) Premature elevation of systemic estradiol reduces serum levels of FSH and lengthens the follicular phase of the menstrual cycle in rhesus monkeys. *Endocrinology* 109: 352–355.

Further Reading

Acri J (2008) Introduction to a symposium on recent advances in the development of medications for drug abuse treatment in honor of Jack H. Mendelson, MD. *Experimental and Clinical Psychopharmacology* 16: 443–445.
Jaffe JH (2008) Obituary: Jack Harold Mendelson 1929–2007. *Addiction* 103: 509–510.
Mello NK (2008) Obituary Jack Harold Mendelson 1929–2007. *Neuropsychopharmacology* 33: 461–462.
Schuckit MA (2008) Jack H. Mendelson: An appreciation 1929–2007. *Journal of Studies on Alcohol* 69: 181–182.

35 Short-Acting Opiates vs. Long-Acting Opioids

M J Kreek, L Borg, Y Zhou, and I Kravets, The Rockefeller University, New York, NY, USA

Chapter Outline

35.1 Laboratory Research Update and Overview

35.1.1 Hypothalamic–Pituitary–Adrenal Axis

Opioids are important in the control of the secretion of hypothalamic–pituitary–adrenal (HPA)-axis stress hormones in rodents as well as in humans. In the rat, acute morphine administration (1 or 2 days) results in increased adrenocorticotropin hormone (ACTH) and corticosterone secretion, while animals treated on a chronic basis (more than 5 days) with morphine show attenuation of the morphine-induced ACTH and corticosterone response, with the development of tolerance by 5 days of morphine treatment (Ignar and Kuhn, 1990; Zhou et al., 2006). In man, in sharp contrast, both acute and chronic morphine administration inhibits the HPA axis (see more details in Section 35.2). For instance, basal levels of ACTH and cortisol are significantly disrupted in active heroin addicts with suppression of ACTH and cortisol and abnormal diurnal rhythms (Kreek, 1973a, 1978; Cushman and Kreek, 1974).

The effects of morphine on HPA-axis activity also depend on the presence or absence of stress. For example, we examined the effect of acute intermittent morphine on pituitary–adrenal function in the rat under mild stress condition of the water restriction (Zhou et al., 1999). Either morphine or water restriction alone increased ACTH secretion as an independent stimulus. When the animals received both morphine- and water-restriction stress, as two combined stimuli, however, there was no increase in plasma ACTH levels. Thus, morphine effectively blunts the classical HPA-axis activity caused by this water-restriction stressor, probably indicating that opioids play a counter-regulatory role on the stress response by inhibiting the stress response cascade.

Methadone (a selective, μ-opioid receptor (MOP-r) agonist, long acting in humans) is widely used in the treatment of short acting opiate (primarily heroin) addiction as well as for the management of chronic pain (see more details in Section 35.2). Methadone has a short half-life in rodents, about 60min in mice, and 90min in rats (Burstein et al., 1980; Kreek, 1983). In several of our animal studies (Zhou et al., 1996), therefore, methadone is delivered through osmotic pumps to mimic steady-state methadone maintenance in humans (Kreek, 1973b). A very early study from our group found that chronic (36days) methadone administration did not alter concentrations of immunoreactive β-endorphin in the rat amygdala and hypothalamus (Ragavan et al., 1983). The steady-state administration of methadone (10mg $kg^{-1}day^{-1}$) by osmotic pumps achieved a mean plasma level of 123 $ngml^{-1}$ with a range from 100 to 150 $ngml^{-1}$ (Zhou et al., 1996), which is comparable to levels achieved and sustained at 24h after last administration during chronic methadone maintenance treatment at lower doses in humans on daily oral doses of 60–120 $mgday^{-1}$, that yield a range of plasma levels of 74–732 $ngml^{-1}$ (Borg et al., 1995). In this rat model, we did not find any effect of steady-state methadone treatment on the mRNA levels of hypothalamic corticotropin-releasing factor (CRF) and proopiomelanocortin (POMC), anterior pituitary CRF-R1 and POMC, or circulating corticosterone levels (Zhou et al., 1996). This demonstrates that steady-state occupancy of μ-opioid receptors with methadone does not have any significant effect on the rat CRF/CRF-R1 or POMC system. Our results support the hypothesis (Kreek, 1973a; Kreek et al., 1983b) that there is no disruption of the HPA-axis activity during steady-state administration of the exogenous opioid methadone.

35.1.2 Steady-State Methadone by Osmotic Pumps Decreases Cocaine-Seeking Behavior in Animal Models

Methadone maintenance at appropriate doses may also be effective in reducing cocaine abuse in heroin-dependent individuals. There is a growing body of evidence, in fact, that at appropriate doses, methadone maintenance can decrease cocaine use in heroin–cocaine co-users. Borg et al. (1995) found that 69% of patients stopped regular cocaine abuse when maintained on an average methadone dose of 67.1 mg day^{-1} (± 2.1). Similarly, in a clinical trial by Strain et al. (1994), it was found that rates of cocaine use were lower among patients maintained on 50 $mgday^{-1}$ (52.6%) compared to those maintained on 20mg day^{-1} (62.4%). Stine et al. (1991) developed a contingency treatment program in which the patient's methadone dose increased by 5mg in response to each cocaine-positive urine screen, up to a maximum dose of 120 $mgday^{-1}$. Under these conditions, when the methadone dose achieved an average of 115 $mgday^{-1}$, cocaine-positive urine samples decreased by 89.2%. Schottenfeld et al. (2005) reported that methadone maintenance (65–85 $mgday^{-1}$), without formal contingency management, promoted retention in treatment, significantly reduced both opioid and cocaine use, and enhanced length of abstinence from both drugs. A recent study by Peles et al. (2006) reported that 68.8% of patients stopped cocaine use after 1 year of methadone maintenance treatment when mean stabilized doses of methadone reached 176.1 $mgday^{-1}$ (± 42.1).

Higher doses of methadone may be required in patients who display severe cocaine abuse, because cocaine exposure can increase MOP-rs in specific brain regions. In our early and recent animal studies, we have found that there is significant increase in MOP-r density after chronic (14days) binge-pattern cocaine administration, which is found in both the mesocorticolimbic and nigrostriatal dopaminergic systems, including the nucleus accumbens (NAc), amygdala and anterior cingulate, and caudate-putamen (Unterwald et al., 1992, 1994, 2001). After acute cocaine exposure (1 or 5days), an increase in MOP-r mRNA levels has been found in the same, specific brain regions (Azaryan et al., 1996; Yuferov et al., 1999). Using positron emission tomography (PET) technology, upregulation of MOP-r binding has been observed in cocaine-dependent individuals, and it has been associated with cocaine craving (Gorelick et al., 2005; Zubieta et al., 1996). We have shown that there is a relative endorphin deficiency in cocaine addicts and also in chronically cocaine-abusing, methadone-maintained former heroin addicts, just as we showed years ago that there is a persistent endorphin deficiency in medication-free, illicit-heroin-free, former heroin addicts (Borg et al., 1995; Kreek, 1996b,c; Kreek et al., 1984a; Schluger et al., 1998a, 2001).

In collaboration with Leri, we investigated the effect of high-dose, steady-state methadone on cocaine-conditioned place preference (CPP) and cocaine intravenous (IV) self-administration, the rodent models for cocaine-taking and -seeking behaviors. We found that (1) rats implanted with osmotic pumps delivering steady-state methadone prior to cocaine conditioning did not express cocaine CPP; (2) rats with steady-state methadone after cocaine conditioning displayed neither spontaneous nor cocaine-precipitated CPP; and (3) steady-state methadone did not alter the IV self-administration (continuous schedule of reinforcement) of various doses of cocaine. Further, MOP-r mRNA levels in the NAc core were significantly elevated in rats after exposure to cocaine conditioning. However, the upregulation of MOP-r mRNA levels was reduced by steady-state methadone in a dose-dependent manner. In conclusion, our results suggest that high-dose, steady-state methadone does not alter the direct reinforcing effect of cocaine but blocks spontaneous and cocaine-precipitated cocaine-seeking, possibly by preventing MOP-r alterations in the NAc core induced by cocaine conditioning (Leri et al., 2006). Our results in animal studies parallel and support the clinical findings in former heroin- and cocaine-co-dependent individuals maintained on high-dose methadone who consume less cocaine. Notably, gas liquid chromatography revealed that steady-state methadone at $30\,mg\,kg^{-1}day^{-1}$ in rats yielded mean plasma methadone levels (450–490 $ng\,ml^{-1}$) within the range found to be effective in reducing cocaine abuse in opiate-dependent individuals (Peles et al., 2006).

35.1.3 Involvement of μ-Opioid Receptor, Orexin, and Preprodynorphin Gene Expression in the Lateral Hypothalamus in Animal Models of Opioid Dependence

Different studies have examined the effect of chronic opioid agonists or antagonists with their withdrawal on MOP-r mRNA levels in different brain regions. The results obtained are conflicting; a decrease, an increase, or no changes have been reported, and these apparently contradictory results may depend on differences in the dose and route of the opioid agonist administered, exposure time, and the brain regions examined. In many previous studies, morphine or opioid agonists were chronically administered by mini-pumps or pellets. This differs from heroin addicts who use an intermittent pattern of self-administration, in which they could experience both rewarding effects and chronic stress induced by repeated heroin injection and withdrawal (Kreek and Koob, 1998). Therefore, an administration paradigm of chronic intermittent escalating-dose morphine has been developed in our laboratory in order to mimic the multiple and escalating doses that human heroin abusers seek daily to achieve rewarding effects and suffer the symptoms of withdrawal from between-dose intervals (see Section 35.2).

As reviewed in Kreek and Koob (1998) and Koob and Kreek (2007), endogenous opioidergic systems, especially POMC-derived β-endorphin, exert inhibitory effects on the HPA axis in both humans and rodents. Furthermore, the lateral portion of the hypothalamus is an important brain region for reward and other motivated behaviors. However, the question of whether morphine withdrawal influences MOP-r gene expression in the lateral hypothalamus (LH) has not yet been studied. We conducted this particular study to determine the effects of acute single-dose morphine administration, 10-day chronic intermittent escalating-dose morphine administration, or its 12-h spontaneous withdrawal on MOP-r mRNA levels (Zhou et al., 2006). We specifically selected several brain regions considered to play an important role in the reinforcing or motivational effects of drugs of abuse, such as the NAc core, amygdala, LH, and caudate-putamen (Kreek and Koob, 1998). We found that either acute or chronic morphine did not modify MOP-r mRNA levels in the brain regions listed above. In contrast, acute spontaneous morphine withdrawal led to an increase in MOP-r mRNA levels in the NAc core, LH, and caudate-putamen, but not in the amygdala. Our data clearly showed that morphine withdrawal increases MOP-r mRNA levels in a region-specific manner. Our finding of increased MOP-r gene expression by morphine withdrawal suggests that endogenous opioid agonists and exogenous opiates also have an inhibitory effect on MOP-r gene expression (Zhou et al., 2006).

Our laboratory has recently investigated the modulation of orexin gene expression in animal models of drug addiction. Most of the LH orexin neurons express the preprodynorphin (ppDyn) gene, and around half of LH orexin neurons also express MOP-r. Therefore, we examined levels of the ppDyn and orexin mRNAs in the LH. During the aversive state of acute withdrawal from chronic escalating-dose morphine, we found increased orexin mRNA levels in rat LH, supporting our hypothesis that enhanced LH orexin neuronal activity, resulting from its increased gene expression, contributes to negative affective states in opiate withdrawal. Under this withdrawal-related stress condition, there was no change in LH ppDyn mRNA levels (Zhou et al., 2006).

35.1.4 Involvement of Arginine Vasopressin and V1b Receptor in Drug Withdrawal and Heroin Seeking Precipitated by Stress and by Heroin

A growing body of evidence suggests that vasopressinergic neuronal activity in the amygdala and hypothalamus represents an important element in the neurobiology of stress-related behaviors. Our laboratory has recently begun to explore the role of arginine vasopressin (AVP) in drug addiction, by examining AVP gene expression in the rat amygdala and hypothalamus after chronic intermittent escalating-dose heroin or during early and late spontaneous withdrawal. We found that amygdalar AVP mRNA levels were increased during early heroin withdrawal only. Of note, this heroin-withdrawal-induced AVP mRNA increase was replicated in a separate study after early spontaneous morphine withdrawal (Zhou et al., 2008).

In collaboration with Leri of the University of Guelph in Canada, we found that footshock stress increased AVP mRNA level in the amygdala of rats withdrawn from heroin self-administration, but not in heroin-naive rats, suggesting that AVP and its receptors may be involved in stress-induced reinstatement of heroin-seeking behavior (an animal model for studying drug relapse in humans). We immediately went on to investigate whether the blockade of central AVP receptors (V1a or V1b receptor) would alter heroin seeking during tests of reinstatement induced by footshock stress and by heroin primes, and HPA-axis hormonal responses to footshock. The selective V1b receptor antagonist SSR149415 (but not V1a antagonist) dose-dependently attenuated footshock-induced reinstatement, blocked heroin-induced reinstatement, and blunted the HPA-axis activation by footshock. In conclusion, these data in rats suggest that the stress-responsive AVP/V1b receptor system (including the amygdala) may be critical components of the neural circuitry underlying the aversive emotional consequences of drug withdrawal and the effect of negative emotional states on drug-seeking behavior. To our knowledge, V1b antagonist is the only systemically available compound identified in reinstatement studies to reduce stress- and drug-induced reinstatement of heroin seeking. It may be worthwhile to explore the usefulness of systemically effective V1b receptor antagonists for the management of withdrawal and for the prevention of drug relapse (Zhou et al., 2008).

35.2 Clinical Research Update and Overview

The important possible relationships between endocrine function, substance abuse, and the addiction syndromes in general, including the early hypothesis that abnormalities in stress responsivity might contribute to the pathophysiology of these disorders, were first formally recognized in 1964 (Dole et al., 1966b); research reports focused on the impact of one of the most devastating illicit drugs of abuse, heroin, on various specific components of endocrine function during cycles of addiction, as well as on the very early reports on normalization of the perturbations caused by this short-acting opiate on endocrine function during long-term pharmacotherapy with the long-acting opioid agonist, methadone, first appearing in the early 1970s (Kreek, 1972, 1973a, 1978; Cushman and Kreek, 1974; Kreek et al., 2004b) Clinical research focused on the developing of a pharmacotherapeutic intervention for heroin addiction was conceptualized by Dole et al. (1966a) in late 1963 and begun in early 1964 at the Rockefeller University. Because of the lack of efficacy for most persons of even the best abstinence-based programs, with less than 30% of long-term heroin addicts sustaining abstinence for 1 year or more after management in any type of humane or inhumane residential or outpatient program, or following discharge from short- or long-term incarceration, it was recognized by Professor Vincent P. Dole that it would probably be essential to develop a different therapeutic approach to address the problem of heroin addiction. Therefore, in late 1963, he recruited, and in early 1964 was joined by, the late Dr. Marie Nyswander, a psychiatrist who had worked in the management of heroin addiction in New York City, and also one of the authors of this chapter (M.J Kreek), a physician-scientist, with training in neuroendocrinology as well as related topics within clinical research, internal medicine, and laboratory-based science (Dole et al., 1966a).

Part of the initial research conducted at the Rockefeller University Hospital in 1964, in addition to the clinical studies, focused on developing a novel pharmacotherapeutic, chronic treatment approach. This research led to the early identification that the long-acting opioid, methadone, administered orally on a daily basis, in addition to preventing the signs and symptoms of opiate withdrawal, could be very effective in reducing drug craving and thus altering the behavior of drug seeking and compulsive drug self-administration, one member of the team (M.J Kreek) began to conduct prospective and also special studies focused on determining the effects of heroin on a variety of physiological systems and on determining the, possibly, very different effects of the long-acting opioid, methadone, on these same functions (Kreek, 1973a,b; Kreek et al., 2004b). These studies were conducted for the purposes of both establishing the safety of methadone as used in the proposed chronic treatment, and also to gain insights into the possible mechanisms underlying the development and persistence of, and relapse to, opiate addiction (Kreek, 1996b, 1972, 1973a, 1975a, 1978, 1991, 1992a, 1996a,c; Cushman and Kreek, 1974; Dole et al., 1966a; Stimmel and Kreek, 1975a,b; Rettig and Yarmolinsky, 1995; National, 1998).

These early studies revealed that the on–off effects of heroin, a short-acting opiate in humans, profoundly perturb many different specific aspects of normal physiology. In contrast, these studies also revealed that the steady-state effects of methadone (that was later demonstrated by pharmacokinetic studies to be a very long-acting opioid in humans) allowed normalization of many of those aspects of physiology which had been disrupted during chronic heroin use (see **Figure 1**). In fact, these early studies served as a roadmap for identifying the role of specific components of the endogenous opioid system unequivocally defined in 1975 and onward, acting at the three types of opiate receptors, μ, δ, and κ, in normal physiology, as well as in a variety of pathological states, including heroin addiction (see **Table 1**) (Kreek, 1978, 1992a, 1996a,b,c). Although that original research team hypothesized the

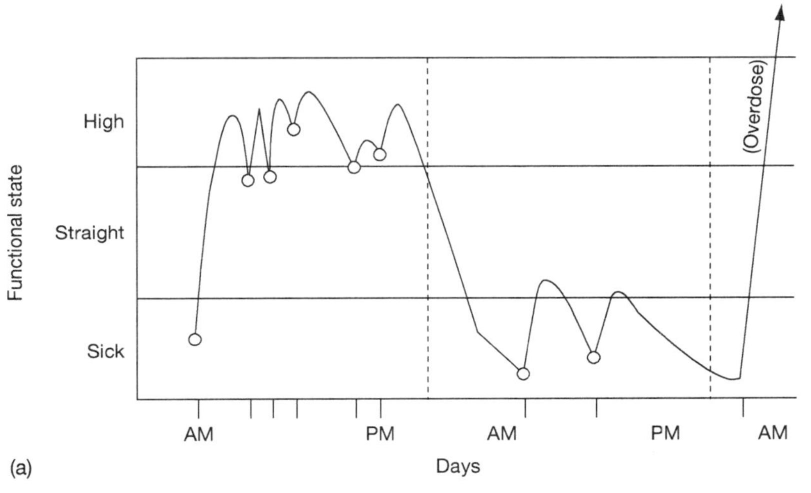

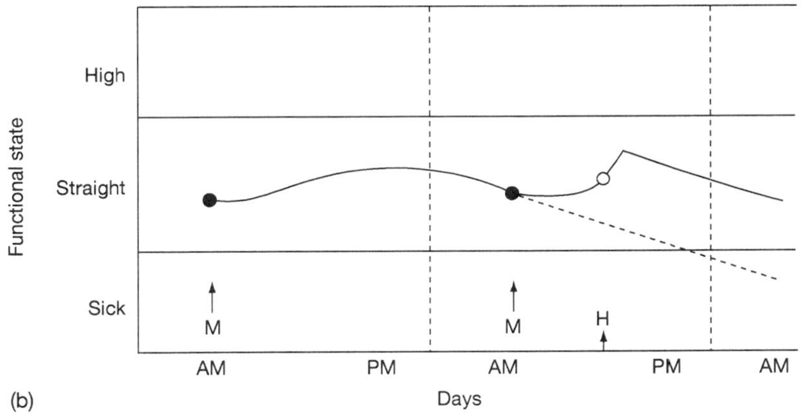

Figure 1 (a) Diagrammatic summary of the functional state of a typical heroin addict. Open circles indicate the repetitive self-administration of heroin. Note that the minority of time is spent in the straight state. (b) Stabilization of a patient in a state of normal functioning during methadone maintenance treatment. Single, daily, oral administration of methadone prevents symptoms of withdrawal (sick) or euphoria (high), even after superimposed heroin administration. From Dole VP, Nyswander ME, and Kreek MJ (1966a) Narcotic blockade. *Archives of Internal Medicine* 118: 304–309, with permission of the Archives of Internal Medicine; and Dole VP, Nyswander ME, and Kreek MJ (1966b) Narcotic blockade: A medical technique for stopping heroin use by addicts. *Transactions of the Association of American Physicians* 79: 122–136.

Table 1 Role of μ-opioid receptor and related endorphin systems in normal human physiological functions

Neuroendocrine functions
Stress-responsive systems, including hypothalamic–pituitary–adrenal axis
Reproductive function, including hypothalamic–pituitary–gonadal axis
Endogenous response to pain
Immunological function
Gastrointestinal function
Cardiovascular function
Pulmonary function
Thermal regulation
? Mood, affect; cognition; memory

All disrupted by chronic abuse of or addiction to the short-acting opiate, heroin. Reproduced from Kreek MJ, Borg L, Zhou Y, and Schluger J (2002) Relationships between endocrine functions and substance abuse syndromes: Heroin and related short-acting opiates in addiction contrasted with methadone and other long-acting opioid agonists used in pharmacotherapy of addiction, In: Pfaff DW, Arnold AP, Fahrbach SE, Etgen AE, and Rubin RT (eds.). *Hormones, Brain and Behavior*, pp. 781–830. San Diego, CA: Academic Press, with permission from Elsevier.

existence of specific opioid receptors, Dole with Ingoglia, and also Goldstein and colleagues, subsequently reported their attempts to elucidate the existence of these receptors, the first conclusive demonstration of the existence of specific opioid receptors, by using the research approach of Dole (i.e., stereoselective binding of opioid ligands by specific receptors), was not accomplished until 1973 (Dole et al., 1966a; Dole, 1970; Ingoglia and Dole, 1970; Goldstein et al., 1971; Pert and Snyder, 1973; Simon et al., 1973; Terenius, 1973). The first type of endogenous opioids were defined by Kosterlitz's group in 1975 (enkephalins), with subsequent definition of β-endorphin in 1976 and also dynorphin peptides in 1979 by various laboratories (Kosterlitz and Hughes, 1975; Hughes et al., 1975; Bradbury et al., 1976; Cox et al., 1976; Li and Chung, 1976; Goldstein et al., 1979).

At the time of the initiation of the treatment research work at the Rockefeller University Hospital, in 1964, the original research team hypothesized that heroin addiction is a metabolic disease – a disease of the brain – with peripheral physiological alterations, along with profound behavioral changes (Kreek, 1992a, 1996c; Dole et al., 1966a). This early concept was soon expanded to pose the specific hypothesis that an atypical responsivity to stress and stressors may contribute to the acquisition and persistence of, and relapse to, use of the illicit drugs, such as the short-acting opiate heroin, and also illicit or excessive use of other drugs and alcohol (Kreek, 1996b,c, 1973a,b 1978, 1987, 1992a; Kreek et al., 2004b; Kreek and Hartman, 1982). It was also hypothesized that this atypical responsivity to stress and stressors might be due to a combined genetic and environmental basis, along with the additionally hypothesized short-acting drug-induced changes, on relevant components of the endocrine stress response systems. Therefore, even the earliest prospective and special studies (started in 1964 and conducted until 1973) included assessments of the HPA-axis component of the stress-responsive systems (Kreek, 1973a,b; Kreek et al., 2004b). Indeed, these very early studies showed profound perturbations of the HPA axis during cycles of heroin addiction and normalization during methadone maintenance treatment (Kreek, 1973a,b, 1978; Cushman and Kreek, 1974; Kreek et al., 2004b; Stimmel and Kreek, 1975a,b; Kreek and Hartman, 1982; Renault et al., 1972).

Additional early physiological studies identified that exogenous short-acting opiates themselves may profoundly alter the hypothalamic–pituitary–gonadal (HPG) axis (Kreek, 1973a,b, 1978; Schluger et al., 1998a; Martin et al., 1973; Santen, 1973; Santen et al., 1975). These findings were made both in prospective clinical research, with findings of abnormal function in high percentages of males, as well as females, during cycles of heroin addiction and during the first 3–6 months of ascending-dose methadone treatment, but with normalization of the aspect of reproductive biological function during long-term (12 months or more) treatment with the long-acting opioid methadone (Kreek, 1973a,b, 1978; Stimmel and Kreek, 1975a; Martin et al., 1973; Santen, 1973; Santen et al., 1975; Cicero et al., 1975; Mendelson and Mello, 1975).

Other studies were conducted on related components of neuroendocrine function, including release, as reflected by peripheral serum levels of prolactin, a hormone now known to be involved not only in lactation, but also in many other aspects of physiology (including immune function), as well as studies on levels of growth hormone and thyroid function (Kreek, 1973a,b, 1978; Kreek et al., 2004b). In each one of these domains, modest-to-significant abnormalities were found during cycles of heroin addiction which were found to be reversed toward normal status during moderate-term (e.g. 2–3 months for the HPA axis) or long-term (e.g., 12–18 months for the HPG axis) treatment with moderate-to-high doses (80–120 mg day^{-1} oral administration) of methadone (Kreek, 1973a,b, 1978; Kreek et al., 2004b).

Extremely provocative were the findings that, although heroin and other μ-opioid receptors directed short-acting opiates predictably in humans, as well as in animal models, and they cause elevations in serum prolactin levels through their effects in lowering dopaminergic tone in the tuberoinfundibular dopaminergic region, it was found that adaptation develops to this phenomenon during long-term steady-dose methadone treatment (Kreek and Hartman, 1982). However, peak plasma levels of methadone (2–4 h after oral dosing) were found to be accompanied by peak serum prolactin levels (but which remain within normal limits), with gradual rise in serum prolactin levels observed to begin around 2 h after oral dosing with methadone, and with peak levels achieved soon thereafter, despite other causes of pulsatile changes in prolactin levels throughout the day (Kreek, 1978; Kreek and Hartman, 1982). Thus, these findings suggested that even during long-term treatment, methadone may act, in part, by lowering dopaminergic tone, although direct documentation of such alterations in dopamine (DA) levels can be observed only indirectly in humans and only for the tuberoinfundibular dopaminergic region. Studies in healthy human volunteers have shown that the natural κ-opioid receptor-selective peptide, dynorphin A, administered IV, similarly effects elevations in serum levels of prolactin (see **Figure 2**) (Kreek et al., 1999). Therefore, apparently both μ-, δ-, and also κ-opioid receptor activation in the tuberoinfundibular dopaminergic system cause a lowering of dopaminergic tone and thus an increase in release and peripheral levels of prolactin (see **Figure 2**) (Kreek et al., 1999).

All these findings suggested that heroin acts in its short-acting pharmacokinetic and pharmacodynamic manner to intermittently result in relatively high levels of occupancy of μ-opioid receptors, which have been defined to be the major binding site of heroin and its major metabolite morphine, followed by rapid offset from those receptor sites, with, undoubtedly, disruption of all signal transduction mechanisms that ensue. In contrast, methadone, with its long-acting pharmacokinetic and pharmacodynamic properties (see below), provides steady-state perfusion of the μ-opioid receptors, the opioid receptor type to which methadone has been found to bind, essentially exclusively, and it acts as a full agonist, with slow onset and slow offset at those receptor sites, with sustained plasma levels, as well as presumably brain μ-opioid receptor occupancy, for over 24 h following oral dosing (see **Figure 1**, and **Tables 1** and **2**) (Kreek, 1973a,b,c, 1992a, 1996a,b,c; Kreek et al., 1979).

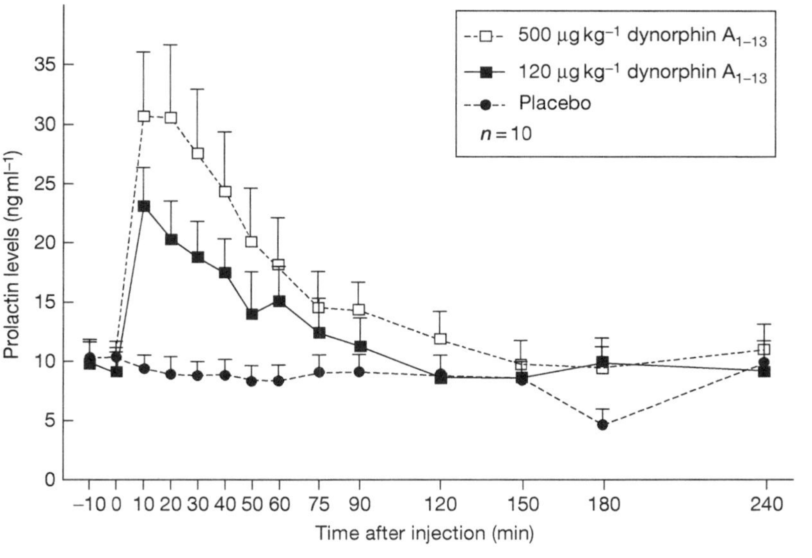

Figure 2 Serum prolactin levels observed in ten normal healthy volunteers without histories of drug or alcohol dependence, after the IV administration of two doses of dynorphin A_{1-13} peptide, a shortened form of the naturally occurring κ-opioid ligand dynorphin A_{1-17}, compared with saline placebo. Note the significant increases in prolactin and dose–response effect. Reproduced from Kreek MJ, Schluger J, Borg L, Gunduz M, and Ho A (1999) Dynorphin A1–13 causes elevation of serum levels of prolactin through an opioid receptor mechanism in humans: Gender differences and implications for modulation of dopaminergic tone in the treatment of addictions. *Journal of Pharmacology and Experimental Therapeutics* 288: 260–269, with permission from American Society for Pharmacology and Experimental Therapeutics.

Table 2 Neuroendocrine effects of opiates, cocaine, and alcohol in human

Agent	*Effect*
Opiates, acute	Suppression of HPA axis
Short-acting opiates (e.g., heroin addiction), chronic	
Opiates, withdrawal	Activation of HPA axis
Opioid antagonists	
Cocaine	
Alcohol	
Long-acting opiates (e.g., methadone maintenance treatment), chronic	Normalization of HPA axis

HPA, hypothalamic–pituitary–adrenal axis (involved in stress response). Reproduced from Kreek MJ, Borg L, Zhou Y, and Schluger J (2002) Relationships between endocrine functions and substance abuse syndromes: Heroin and related short-acting opiates in addiction contrasted with methadone and other long-acting opioid agonists used in pharmacotheraphy of addiction. In: Pfaff DW, Arnold AP, Fahrbach SE, Etgen AE, and Rubin RT (eds.). *Hormones, Brain and Behavior*, pp. 781–830. San Diego, CA: Academic Press, with permission from Elsevier.

35.2.1 Clinical Studies of Pharmacokinetics of Heroin and Morphine as Contrasted with Methadone

For an understanding of the multiple studies in humans, as well as animal models, to be discussed herein, it is essential to understand the profound differences in the pharmacokinetics of the opioid compounds under consideration in humans, as well as the pharmacokinetic and pharmacodynamic differences of each in humans-versus-rodent models. Pharmacokinetics will, therefore, be briefly discussed. It was possible by the early 1970s to develop sensitive and specific analytical methods for measuring plasma, as well as urine, levels of these opioid compounds (Kreek, 1973c, 1979b; Inturrisi and Verebely, 1972a,b; Inturrisi et al., 1984). First it was found that a single dose of methadone in an opioid-naive human volunteer would have approximately a 6-h pharmacokinetic half-life, only slightly longer than the effects of a single dose of morphine when given to a naive subject (and length of morphine half-life that were to be documented later) (Inturrisi and Verebely, 1972a). These findings paralleled several much-earlier pharmacodynamic studies where it had been shown that a single dose of methadone used for the management of pain in an opiate-naive person would have a duration of action very similar to morphine, that is, 4–6h. However, when subsequent studies were conducted to determine methadone disposition in persons chronically maintained on methadone, it became apparent that the lack of sensitivity of the earlier methods had prevented appropriate elucidation of its terminal half-life. The half-life of methadone was found to be approximately 24h when the racemic dl-(SR) form was administered, the form used initially (and in most places in the world exclusively), for the management of addiction or pain (Kreek, 1973c, 1979b; Inturrisi and Verebely, 1972b).

Early studies showed that the half-life of any single dose of methadone is around 24h (Hachey et al., 1977). In further studies, for which the two enantiomers of methadone were separately labeled using stable isotope technology, yielding 1-(R)-pentadeuteromethadone and d-(S)-trideuteromethadone, along with dl-(SR)-octadeuteromethadone, it was possible to determine further the pharmacokinetics of each of the two enantiomers of methadone. In these studies, it was found that the biologically active 1-(R) enantiomer had a much longer half-life than the inactive d-(S) enantiomer, with a half-life of around 36h for the active enantiomer and a half-life of around 16h for the inactive enantiomer (Kreek et al., 1979; Nakamura et al., 1982). After other appropriately sensitive and specific techniques were developed, studies were conducted to determine the pharmacokinetics of heroin and its major metabolite, morphine, in humans (Inturrisi et al., 1984). It was found that heroin has a half-life of around 3min in humans. Its first and biologically active metabolite, 6-acetylmorphine, has a half-life of approximately 30min in humans and the major metabolite, morphine, had a half-life of about 3–4h in humans (Inturrisi et al., 1984).

In the studies of methadone disposition, it was found that after oral administration, methadone has a very slow onset of action, with a peak approximately 2–4h after oral dose administration (Kreek, 1973c). Even following IV administration of methadone, the plasma disappearance curves very rapidly converge with the curves seen after oral administration (Nakamura et al., 1982). This is probably due to the very rapid hepatic extraction, which has been studied directly in a rabbit liver perfusion model, and indirectly modeled in human pharmacokinetic studies (Kreek, 1978, 1979a,b; Hachey et al., 1977; Nakamura et al., 1982; Kreek et al., 1980). In these studies, it has been shown that the liver avidly takes up methadone, but then slowly releases it in unchanged form. This possible role of the liver as a storage site, that is, with uptake, binding, and slow release of unchanged methadone and not simply for the more usually observed first-pass phenomenon of uptake, biotransformation, and extraction, was able to be addressed in human studies in persons with a bile fistula, where the hepatobiliary secretion of unchanged methadone was directly documented (Kreek et al., 1980). Other studies have shown that methadone is avidly bound to most or all types of plasma proteins in humans, whereas it is much more modestly bound, and primarily to albumin fractions, only in other species (Pond et al., 1985).

In sharp contrast to methadone, after IV administration, heroin has a very rapid peak and then is extremely rapidly biotransformed to its successive metabolites 6-acetylmorphine and then morphine. Both of these compounds have relatively short half-lives, and therefore, one sees a rapid fall in levels of all biologically active opioids. In general, it has been found that the rewarding or reinforcing effects of drugs of abuse are related to the rate of rise in blood which is presumed to be correlated to the rate of rise at specific sites of action in the brain (Kreek, 2000). Thus, heroin, with its rapid-rate rise of blood levels and onset of effects, would be expected to and does have very great reinforcing or rewarding effects in humans; whereas methadone, with its slow rate of rise and extremely slow onset of action has essentially no rewarding or reinforcing, or any other desired effects, other than the well-documented prevention of signs and symptoms of opiate withdrawal, along with the subsequently determined reduction or elimination of craving, and also facilitation of normalization of physiological functions which have become persistently disrupted by the on–off effects of heroin (Kreek, 1973a, 1992a, 2000). Similarly, the rates of fall of drugs of abuse from blood, and presumably from the specific opioid receptor sites of action in the brain, have been related to the onset of opiate withdrawal signs and symptoms which are of especial importance in opiate addiction (Kreek and Koob, 1998; Stimmel and Kreek, 1975a,b). The rapid fall of levels of heroin, 6-acetylmorphine, and morphine are related to the onset of withdrawal symptoms, whereas during steady-dose methadone treatment, methadone levels are sustained after a modest decline over 2–4h after the peak, and there is a sustained plasma level at the remainder of the 24-h dosing period (Kreek, 1973c, 1976; Kling et al., 2000). Thus, signs and symptoms of opioid abstinence or withdrawal are prevented throughout the 24-h dosing interval. These profound differences in pharmacokinetics in humans mean that heroin and its metabolites are bombarding MOP-rs in an on–off fashion, whereas methadone allows a steady-state perfusion of those MOP-rs over a 24-h dosing interval. More recent studies have shown that although the on–off effects of heroin or morphine significantly alter not only physiological function, but also molecular events, including induction of changes in gene expression, neurochemical alterations, and related abnormal behaviors, methadone delivered in a steady-state mode does not cause these changes at any level in appropriate animal models (Zhou et al., 1999, 1996; Kreek, 1973b).

Similarly, the other two compounds that have been shown to be like methadone – very effective in the long-term pharmacotherapy of opiate addiction and have long-acting properties in humans – both have either long pharmacokinetic or pharmacodynamic properties in humans. 1-α-Acetylmethadol (LAAM), also a MOP-r agent, has a half-life of approximately 48h and its two major N-demethylated metabolites, norLAAM and dinorLAAM, have even longer half-lives of 48–96h. Thus, in human pharmacotherapeutics, LAAM can be administered every 2 or even every 3days. LAAM offered to methadone-maintained patients who continued to abuse heroin did not lead to heroin abstinence (Borg et al., 2002). The MOP-r partial agonist, buprenorphine, which has also been found to be effective for the management of opiate addiction (and especially when its abuse liability is minimized by addition of the opiate antagonist naloxone to the sublingual preparation used in therapeutics of addiction), has a relatively short pharmacokinetic profile in humans but has long-acting properties due to prolonged μ-opioid receptor occupancy (Kreek, 1996b,c, 1991; Kreek et al., 1983a; Johnson et al., 2000).

35.2.2 Clinical Studies of HPA Axis

The HPA-axis function is one component of stress-responsive function. It is a component that can be measured readily in humans, since some of the products of this system, including ACTH and β-endorphin from the anterior pituitary, as well as cortisol from the adrenal cortex, circulate peripherally and can thus be sampled. (The role of each component of this axis in a variety of critical functions related to survival is well known and is discussed elsewhere in this volume.)

The HPA axis is a major component of the stress-responsive system in humans as well as in all mammalian species. A variety of rigorously validated neuroendocrine tests have been developed for assessment of integrity of each of the three components: the hypothalamic production of CRF, the anterior pituitary production of POMC-derived peptides, including ACTH and β-endorphin, and the adrenal component, which in humans is characterized by the release of the critically important glucocorticoid, cortisol. The first studies which addressed the effects of opiates in the setting of heroin addiction on the HPA axis were reported in the early 1970s (Kreek, 1973a,b; Cushman and Kreek, 1974; Kreek et al., 2004b). All of these initial studies were a part of the work focused on developing a pharmacotherapy, using a long-acting opioid, methadone, for the chronic management of heroin addiction. They also addressed the hypothesis that an atypical responsivity to stress or stressors may contribute to the persistence of, and relapse to, addiction after achieving the abstinent state and that such an atypical responsivity may play a role in the initial development of opiate addiction (Kreek, 1973a,b, 1978, 1992a, 1996b,c, 2000; Kreek and Koob, 1998; Kreek et al., 2004b). However, the first studies of the effects of heroin on adrenal function were conducted at the US Public Health Service Hospital in Lexington, Kentucky (Eisenman et al., 1958, 1961, 1969).

The first prospective and special studies to determine the effects of heroin in addicts versus methadone in patients receiving maintenance treatment confirmed lowered levels of urinary glucocorticoids in heroin addicts coming into treatment and in patients during early methadone maintenance treatment. However, following stabilization on a steady, moderate- to high-dose (80–120 $mg\,day^{-1}$) of methadone in treatment, urinary levels of glucocorticoids returned to normal. These early studies also determined that plasma levels of the major adrenal glucocorticoid in humans, cortisol, were modestly reduced in some heroin addicts, often with disrupted circadian rhythm of levels; following stabilization on methadone treatment, plasma levels of glucocorticoids came within normal limits (Kreek, 1973a,b; Cushman and Kreek, 1974; Kreek et al., 2004b; Stimmel and Kreek, 1975a,b). In those very early studies, plasma levels of cortisol and urinary excretion of 17-ketosteroids and 17-hydroxycorticosteroids were all within normal limits in stabilized methadone-maintained patients. In a further study of former heroin addicts treated in an abstinence-based residential community, urinary levels of 17-ketosteroids and 17-hydroxycorticosteroids, as well as plasma cortisol levels, were also found to be within normal limits (Cushman and Kreek, 1974).

To determine whether the hypothalamic–pituitary part of the HPA axis would suppress normally, dexamethasone suppression tests were also performed in these early studies; all stabilized methadone-maintained patients were found to be within normal limits when the standard 1- or 2-mg test dose of dexamethasone was used (Kreek, 1973a,b). ACTH stimulation tests were also performed to determine the integrity of the adrenal gland; response was within normal limits in all methadone-maintained subjects studied (Kreek, 1973a,b).

Metyrapone tests were also conducted as part of the 1967–73 research (Kreek, 1973a,b). The metyrapone test was already a standardized test of hypothalamic–pituitary reserve. Abnormal test results were found in patients during the first 2 months of methadone treatment when methadone gradually increased to full-treatment doses which block the euphorogenic and other effects of any superimposed heroin. However, normal metyrapone test results were found in patients who had been stabilized for more than 3 months in steady, moderate- to high-dose methadone treatment (Kreek, 1973a,b; Kreek et al., 2004b). Some of these very early studies included chronic heroin addicts first entering the pharmacotherapy research and staying in the inpatient unit of the Rockefeller Hospital during induction into methadone maintenance. These subjects were studied using both baseline assessments and metyrapone tests at the time of admission and with repeated measures during the first 3 months of treatment. Oral metyrapone tests showed normalization after 2–3 months of treatment with steady, moderate- to high-dose methadone treatment in all subjects (Cushman and Kreek, 1974).

In a related early research, Renault et al. (1972) at the University of Chicago found that although the stress response was normal in methadone-maintained patients 1 h after methadone treatment, 21 h after the last dose of methadone the patients became hyperresponsive to cold-induced stress, possibly due to relative opioid abstinence.

Hellman et al. (1975) found low-normal or subnormal production rates of cortisol during short-term methadone treatment and at the end of a cycle of heroin addiction during the beginning of methadone, as used in the detoxification protocol. Also, a sharp rise in the glucocorticoids during early abstinence was observed with return toward normal during longer-term abstinence (Hellman et al., 1975), a finding similar to other studies (Stimmel and Kreek, 1975a,b). Mendelson, Meyer, and colleagues found no changes in plasma cortisol levels either during the period of heroin administration or during methadone withdrawal. However, with the study design used, any elevations of plasma cortisol, which have subsequently been documented by many groups during opiate withdrawal, could have been offset by either the heroin self-administration or the methadone as used in detoxification (Mendelson and Mello, 1975).

From the mid-1970s to the mid-1980s, additional studies were conducted, both by the Rockefeller University research team as part of further development of maintenance treatment and by other research groups. In one research extension of the earlier studies, it was found that although plasma levels are within normal range during chronic long-term methadone maintenance, there is a brisk, though normal, circadian rhythm, that is, the levels fall promptly from the morning to the evening following oral methadone dose administration (Kreek and Hartman, 1982). After radioimmunoassay techniques had been developed to allow direct measurement of plasma levels of ACTH, it was found that in stabilized methadone-maintenance patients, levels of ACTH are within normal limits (Kreek et al., 1981).

By the early 1980s, it had been established that β-endorphin, a major endogenous opioid and the longest (31-amino-acid residues) of this class of peptides, is derived from β-lipotropin, and that β-lipotropin and ACTH are processed from a common 31K precursor, POMC. It had also been shown by several groups in animal models, as well as in humans, that ACTH and β-lipotropin, along with β-endorphin processed from β-lipotropin, are released in parallel with each other. This parallelism was observed both in the setting of stress induced with one of several types of stressors, and under normal conditions, including during the normal circadian rhythm of level fluctuations, with highest levels in the morning and lowest levels in the evening. Frantz and Wardlaw developed the first rigorously characterized and validated radioimmunoassay for measuring β-endorphin. This technique allowed researchers to determine that methadone-maintained patients had β-endorphin levels that were within normal range during the 24-h period after the last oral dose of methadone. In addition, the studies showed that methadone-maintained patients, similarly to normal healthy volunteers, had a circadian rhythm of β-endorphin levels, with reduction of plasma β-endorphin levels as the day progressed. However, these initial studies again showed that the circadian rhythm was more brisk in the methadone-maintained patients than in the healthy subjects, suggesting that the exogenous opioid, methadone, might be playing a role in reducing plasma levels of β-endorphin more rapidly than in healthy volunteers with only endogenous opioids performing this modest, tonic inhibitory function (a mechanism that was only later to be elucidated) (Kreek and Hartman, 1982; Kreek et al., 1981). However, it was concluded that methadone, used on a long-term basis, had no apparent effects on peripheral levels of β-endorphin. In collaboration with Wardlaw and Frantz, plasma levels of ACTH were also determined using the newly developed radioimmunoassay techniques. All ACTH levels in methadone-maintained patients were similarly found to be within normal limits.

Concomitantly measured cortisol levels again documented that all cortisol levels were within normal limits and with a normal but brisk diurnal variation (Borg et al., 1995; Kreek and Hartman, 1982; Kreek et al., 1981). A second set of studies was conducted by the same collaborative research team, showing that under basal conditions, plasma levels of β-endorphin, ACTH, and cortisol were all within normal limits and observed a normal circadian rhythm of levels in all methadone-maintained former heroin addicts (Borg et al., 1995). In 1973, we proposed that the HPA axis, whether normal or abnormal on a genetic or environmental basis prior to exposure to heroin, might be adversely affected during cycles of heroin addiction, probably due to the early much-higher peak levels of heroin and its major metabolite, morphine, yet with the much-shorter duration of action of morphine, as compared with methadone. This hypothesis was based on clinical observations in a controlled setting, since no analytical technology was available. In contrast, methadone with its (then) apparent long duration of action (later documented by pharmacokinetic studies) allowed normalization of the HPA axis, including responses to a chemically induced stress of metyrapone challenge (Kreek, 1973a,b).

In related early studies, Cushman and colleagues similarly found both normal urinary secretion of glucocorticoids and a normal rise in the adrenal cortical precursor of glucocorticoids, following metyrapone administration in former heroin addicts stabilized on methadone-maintenance treatment (Cushman, 1970). In other studies, they found that baseline plasma 17-hydroxycorticosteroid levels were within normal limits and the levels rose normally after the stress of insulin-induced hypoglycemia (Cushman, 1970). However, they found that abnormal circadian rhythm persisted in some after a few months of methadone treatment. In related studies, some heroin addicts, but not others, had a subnormal rise in plasma 17-hydroxycorticosteroids after the insulin-induced hypoglycemic stress (Cushman, 1970).

The initial studies (discussed earlier) done in collaboration with the group of Frantz and Wardlaw showed that both the actual levels and, also importantly, the circadian rhythm of levels of ACTH, β-endorphin, and cortisol were normal in stabilized long-term methadone-maintained patients. Therefore, in further studies, single-dose metyrapone tests were conducted. A normal response to metyrapone was found in persons well stabilized on methadone-maintenance treatment, with no ongoing abuse of any other drug or alcohol, irrespective of dose over a wide range of doses, from low- to moderate-dose methadone treatment (Kreek et al., 1984a). In those treated with high-dose methadone, normal responses were found as well. However, failure to respond to metyrapone was found in two subjects who were not in steady state and whose doses of methadone were being increased, a similar finding to the early studies of persons undergoing induction into full-dose methadone treatment. Very provocatively, an abnormal response of the opposite type, that is, an apparent hyperresponsivity to the metyrapone-induced stress was observed in medication-free, drug-free, former methadone-maintained patients, who had not received methadone treatment for several months and had used no further illicit heroin for those several months (Kreek et al., 1984a). At that time, in 1983–84, there had been no precedent of hyperresponsivity found in response to metyrapone testing. This led to further studies of this type, although finding persons who met the very strict, but essential, criteria of being both methadone maintenance medication-free and also illicit-opiate-free in addition to being free from excessive use of alcohol, or any other psychotropic drug abuse, or psychotropic prescription drug abuse, continues to be extremely difficult because of the very high rate of relapse (or use or misuse of other psychotropic drugs) by such persons. However, to date, a hyperresponsivity to this chemically induced metyrapone stress has repeatedly been found.

Related studies were conducted to determine what levels of β-endorphin would be found in cerebrospinal fluid (CSF) during steady-dose methadone treatment. To perform these studies, advantage was taken of the fact that for minor surgical procedures, a small amount of CSF is removed for the infusion of anesthesia agents. In these studies, the CSF levels of β-endorphin were significantly higher in the methadone-maintained patients than in the control subjects, whereas plasma levels of β-endorphin were not different between the two groups (Kosten et al., 1987). These findings suggested that early opioid abstinence or withdrawal, as evidenced by enhanced levels of either brain or pituitary source β-endorphin, might be found near the end of a dosing interval, at a time when plasma β-endorphin levels in methadone-maintained patients were not yet significantly different from control subject plasma β-endorphin levels. This was the first suggestion that activation of the endogenous opioid system, possibly in part through the activation of the anterior pituitary component of the HPA stress–response axis, and in part through other brain region components contributing to stress responsivity, might occur as a prelude to the onset of any other objective signs or any symptoms of opiate withdrawal in persons in stabilized methadone treatment, when peripheral levels of β-endorphin and ACTH were within normal limits (Kosten et al., 1987). Further studies (see below) have indeed suggested that activation of at least the HPA-axis component of stress responsivity may occur prior to the onset of any other objective signs or symptoms of opiate withdrawal in opioid-dependent persons (Culpepper-Morgan et al., 1992; Culpepper-Morgan and Kreek, 1997).

All of these studies conducted in the late 1960s, 1970s, and early 1980s showed that, whereas there may be significant disruption of function of the HPA-axis component of the stress-responsive axis during cycles of heroin addiction, with suppression of levels of hormones, and also blunting of the circadian rhythm of hormones of this axis during cycles of heroin addiction (conversely) activation of this axis may occur during acute spontaneous or opioid-antagonist-precipitated opiate withdrawal, with the early interpretation (which now has been revisited and changed), that this activation reflected a response to the actual stress of opiate withdrawal (Stimmel and Kreek, 1975a,b). Also, all these early studies found that during steady-dose, long-term (more than 3 months) methadone-maintenance treatment, normalization of both levels and circadian rhythm of levels, as well as response to the chemically induced stress and test of hypothalamic–pituitary reserve, that is, the metyrapone test, as well as response to the standard (1 or 2 mg) dexamethasone suppression test, were all within normal limits. In contrast, modestly suppressed hormone levels, flattened circadian rhythm of levels, and/or suppressed response showing a lower-than-normal hypothalamic–pituitary reserve in response to the metyrapone test was found during cycles of heroin addiction, whereas a hyperresponsivity to this chemically induced provoked stress was found in states of medication-free, opioid-medication-free, illicit and opiate-free, and other-medication-and-alcohol-abuse-free status in former long-term heroin addicts.

Studies published between 1975 and 1985 showed that the plant-derived opiate morphine, synthetic heterocyclic compound opiates, and analogs of natural endogenous opioid peptides all suppress the HPA axis in humans (Morley et al., 1980). Tolis et al. (1975) showed that although morphine blocked ACTH response to surgical stress, a single 10-mg dose of morphine in volunteer subjects did not suppress basal serum levels of cortisol (Tolis et al., 1975). Rolandi et al. (1983) reported that buprenorphine (MOP-r partial agonist) significantly lowered serum levels of cortisol for 6h after that single dose. The group of Zis and Carroll found that cortisol levels were significantly reduced by the infusion of 5mg of morphine over 1min, a protocol similar to that of Tolis (Zis et al., 1984). However, their subjects were healthy volunteers, not preparing to undergo any surgical procedure. The study subjects of Tolis, in retrospect, were undoubtedly under modest stress, and thus the low dose of morphine failed to suppress cortisol which had been observed in earlier studies, as well as in a more recent study from the group of Carroll (Tolis et al., 1975). Rittmaster and colleagues found that morphine blunted ACTH response to CRF for the first 60min and cortisol response for 90min (Rittmaster et al., 1985). These studies, along with related animal studies, suggested that the effect was, probably, due to the suppression of CRF at the hypothalamic level, as well as due to a possible direct action at the anterior pituitary.

Delitala, Besser, and colleagues found that 10mg of morphine, 10mg of methadone, or 0.25 mg of the Met-enkephalin analog (0)-01-enkephalin (DAMME) all lowered serum cortisol levels, as did the mixed agonist–antagonist pentazocine, at a dose of 30mg acting at both μ and κ receptors, and the mixed agonist–antagonist nalorphine administered at 10mg (Delitala et al., 1983a). They also found that a very low dose, 4mg, of the opioid antagonist naloxone did not alter cortisol levels. These studies extended earlier studies by Delitala et al. (1981) in which they had similarly found that opioid agonists, specifically an analog of a natural peptide (the synthetic peptide compound (DAMME)), altered cortisol levels. In the 1983 report, they correctly identified morphine as a μ-agonist; however, they incorrectly state that methadone is both μ and σ agonist, which is now known definitely not to be the case (Delitala et al., 1983a). They ascribed the agonist action of pentazocine and nalorphine to κ-receptor action and the antagonism to MOP-rs. The latter is clear; however, there seems to be some μ-agonism as well as κ-agonism for these compounds. In 1986, they also found that the Met-enkephalin analog DAMME attenuated pituitary–adrenal activation by CRF (Grossman et al., 1986). In 1987, Allolio found that morphine administered to seven naive subjects in a 30-mg slow-release preparation, or a placebo, prior to administration of CRF (0.1 mg), resulted in suppression of basal cortisol levels and also a blunted response to CRF administration (Allolio et al., 1987). However, after a 4-mg dose of naloxone, there was no change in basal levels and no change in response to CRF. Further studies showed that DAMME attenuated the activation by an α-adrenergic agent which stimulated the HPA axis (Delitala et al., 1991).

The group of Gold at the National Institute of Mental Health (NIH) in 1980, built on the earlier findings and also on the then-recent findings that in some types (yet to be fully defined) of major depressive illness, feedback inhibition-independent activation of the HPA axis is found, as evidenced by failure to suppress with standard-dose (1 or 2 mg delivered the evening before) dexamethasone suppression (Gold et al., 1980a). They found that a 5-mg dose of methadone in opioid-naive depressed patients caused a significant reduction of cortisol secretion, as would have been anticipated. This showed that in depressed patients this μ-opioid agonist, methadone, suppressed cortisol production through feedback inhibition of CRF and/or ACTH, whereas the glucocorticoid dexamethasone was unable to do so (Gold et al., 1980b). These studies thus extended earlier findings in which it had been clearly shown that short-acting opiates, such as morphine, suppress hormones of the HPA axis in opioid-naive healthy volunteers, as well as in heroin addicts, and gave further evidence that the hypothalamic–pituitary axis is under tonic or feedback inhibition by the endogenous opioid system, as well as by glucocorticoids.

Studies conducted by the group of Facchinetti, with reports starting in 1984, confirmed earlier findings from the Rockefeller group. They found an impaired circadian rhythm of β-lipotropin, β-endorphin, and ACTH in heroin addicts (Facchinetti et al., 1984). In further studies of the HPA axis in heroin addicts, this group reported significantly reduced cortisol levels in the heroin-addicted group; in those still taking heroin at the time of admission, cortisol levels were found to be significantly lower than in controls in the evening. In a parallel, but separate, study, significantly reduced plasma cortisol levels were found within 60min of morphine (0.1 mg kg^{-1}) administration to drug-naive healthy volunteers, leading authors to a conclusion that chronic opiate abuse leads to hypoadrenalism, possibly due to direct morphine effects at the hypothalamic level (Facchinetti et al., 1985a).

In other studies, the group of Facchinetti showed impaired adrenergic-induced POMC-related peptide release in heroin addicts (Facchinetti et al., 1985b).

Another group, that of Gil-Ad in Israel, studied the effects of clonidine on plasma levels of β-endorphin and cortisol in opiate-addicted subjects and found, in contrast to Facchinetti, that basal (before clonidine administration) morning β-endorphin levels were lower in the addicted group than in the control group. Following clonidine administration, there was an increase to normal values in the heroin-addicted subjects. In control subjects, there was no change in β-endorphin levels. In this study, they also found that basal levels of cortisol, in contrast to the basal levels of β-endorphin, were higher in the addicted subjects than in the controls (although not discussed, this was probably due to the fact that the opiate-dependent persons were in modest early opiate withdrawal) (Gil-Ad et al., 1985). In the control subjects, again in contrast with the findings of Facchinetti, who found only modest and transient (16min) reduction in cortisol levels in controls, they found that clonidine caused a significant (50%) decrease in plasma cortisol levels. In the heroin-addicted subjects, Gil-Ad et al. (1985) found a modest but not significant decrease in cortisol levels, thus a lesser response in heroin addiction, rather than a greater cortisol-lowering response in this opiate-dependent group, as found by Facchinetti et al. (1985b). These authors concluded that in the heroin-addicted subjects, there was impaired activity of the endogenous opioid peptides, leading to alteration of β-endorphin and cortisol secretion, a hypothesis proposed by the laboratory of Kreek (1973a,b), Kreek et al. (2004b), Kreek and Hartman (1982), and Gil-Ad et al. (1985).

In a study from the United States, Dackis et al. (1984) reported the results of dexamethasone suppression tests in heroin addicts who also had concomitant major depression. They found that of the 15 patients with major depression, two patients had an abnormal dexamethasone suppression test. Also, two of the 27 addicts without major depression had an abnormal dexamethasone test. This latter result is surprising, since both the early studies of Kreek and colleagues, and the much more recent studies from her group, have shown that dexamethasone suppression is achieved in all heroin addicts and methadone-maintained former heroin addicts without depression and is also achieved in those with modest depressive symptoms but not meeting clinical criteria for diagnosis of major depression (Kreek, 1973a,b; Kreek et al., 2004b; Kreek and Hartman, 1982; Aouizerate et al., 2006).

Vescovi et al. (1989) evaluated the impact of stressors of a variety of types on the HPA-axis response. In the heroin addicts, as compared with the control subjects, they found a decrease of basal levels of β-endorphin before a modest stressor (sauna-induced hyperthermia), absence of the normal increase of β-endorphin and ACTH after the sauna, and a lower increase of systolic blood pressure at that time. They concluded that there was an impairment of the adaptive response to stress in heroin addicts, even after a drug-free period of 14days. These studies are provocative in the context of the 1972–73 studies from the laboratory of Kreek showing reduced responsivity to stressors in active heroin addicts and the 1984 studies from the laboratory of Kreek showing hyperresponsivity to stressors in the protracted opioid-abstinent state several weeks or months following last use of opiates or methadone treatment (Kreek, 1973a,b; Cushman and Kreek, 1974; Ignar and Kuhn, 1990; Bart et al., 2003; Kreek and Hartman, 1982). Vescovi et al. (1990) also conducted metyrapone testing of heroin addicts who had been on methadone-maintenance treatment for 6months, compared with healthy controls. They found that basal levels of ACTH and cortisol were significantly decreased in patients maintained on methadone for approximately 6months, as compared with the healthy volunteers. They also found a blunted response with respect to plasma levels of ACTH and β-endorphin following metyrapone administration, compared with the healthy volunteers (Vescovi et al., 1990). These findings were similar to the earlier findings from the Kreek laboratory in persons during ascending-dose methadone treatment, and in treatment for less than 3months; however, they were different from the earlier results from the Kreek group in two separate sets of studies in subjects who were maintained on steady-dose moderate- to high-dose methadone treatment for more than 3months (Kreek, 1973a,b; Cushman and Kreek, 1974; Kreek et al., 2004b; Kreek and Hartman, 1982; Vescovi et al., 1990).

The group of Cami in Barcelona, Spain, studied the impact of the α-adrenergic agonist, clonidine, on the circadian rhythm and release of cortisol in heroin addicts undergoing detoxification. They conducted this study because of the observations in all studies that methadone was more effective than clonidine in detoxification, with respect to decreasing signs and symptoms of withdrawal as well as related unpleasant feelings. In this study, urine monitoring was conducted 3 times a week; two different α2-adrenergic compounds were used in the management of detoxification, clonidine and guanfacine, along with a parallel group in which methadone was used. Salivary levels of cortisol were measured. In the 28 active heroin addicts entering the study prior to detoxification, dramatic individual differences in salivary cortisol were noted (Cami et al., 1992). A correlation analysis determined that the larger the heroin consumption, the lower the salivary cortisol levels, and also the longer the time that had elapsed after self-administration, smaller reductions in cortisol levels were determined (Cami et al., 1992). This study found again an activation of the HPA axis in the opiate-abstinent state in persons not receiving any further treatment. However, methadone administered on a very short-term basis during the detoxification protocol lowered the levels of cortisol, whereas clonidine and guanfacine did not. The authors interpreted these data to show that the persistent signs and symptoms of withdrawal, as well as the general feelings of discomfort during α_2-adrenergic agonist treatment, might be related to the activation of the stress-responsive HPA axis. A subsequent additional study found that following cessation of clonidine or methadone, salivary cortisol levels were elevated and remained elevated up to 16 days after the last dose of either medication (Cami et al., 1992; Folli et al., 1992).

In further studies, Vescovi and colleagues found that healthy normal control volunteers had significant and consistent increases in both plasma ACTH and β-endorphin in the setting of performance testing requiring sustained attention. However, heroin addicts, despite preservation of full and normal performance capabilities, had a blunted response of ACTH to this mild stressor, along with a paradoxical decrease in β-endorphin levels, reflecting possibly the attenuation by the opioid of the normal stress response (Mutti et al., 1992).

A major observation was made first in 1979, with multiple studies carried out over the next 20 years, which revealed the fact that the endogenous opioid system plays a major normal role in regulating the HPA axis in healthy subjects. This also led to further studies in persons with defined addictive diseases. The first observation was made by the group of Volavka who found that the MOP-r-selective antagonist, naloxone, would result in an abrupt elevation in plasma levels of both ACTH and cortisol (Volavka et al., 1979). In further studies, Volavka et al. (1980) showed that 20 mg of the MOP-r-selective antagonist naloxone administered to 13 normal volunteers in a placebo-controlled, double-blinded study again resulted in significant increases in ACTH and cortisol levels as well as luteinizing hormone levels. The subjective responses were not different from the responses to placebo. In studies from a different laboratory, 10 mg of naloxone given IV to eight normal healthy volunteers also showed that naloxone produced a significant rise in cortisol levels (Morley et al., 1980). In this study, there were no effects of naloxone either on memory performance on a memory test or on the Profile of Moods States Assessment (Morley et al., 1980). Another group found that 8 mg of naloxone elicited significant increases in plasma levels of cortisol, but did not change levels of β-endorphin or ACTH. However, in further studies in two subjects, they found that 20 mg of naloxone administered IV would cause an elevation of β-endorphin levels (Naber et al., 1981). In other studies which addressed the possible utility of administering naloxone orally (since this MOP-r-selective antagonist has extremely limited systemic bioavailability after oral administration) to reverse opiate-induced gastrointestinal dysmotility disorders, and in which doses of naloxone up to 29 mg were administered orally, no impact on plasma levels of ACTH and cortisol were found. However, when the same patients received 29 mg of naloxone IV, elevations in levels of the HPA-axis stress-responsive hormones were found (Kreek et al., 1984b; Kreek and Culpepper, 1991). In further studies conducted to determine whether the opiate antagonist naloxone would have an effect if it were administered in the steady-state mode rather than by acute IV bolus, 13 healthy volunteer subjects received either naloxone in a bolus of 29.2 mg per 24 h, or in divided doses over a 24-h infusion basis. No changes in ACTH or cortisol were found following the steady-state infusion, whereas elevations in levels of the hormones were found following bolus administration (Kreek et al., 1984b). The doses of naloxone used ranged from 1.3 up to 6.0 mg kg^{-1} body weight. In this study, it was only after administration of very high doses of naloxone that elevations in serum cortisol levels were found (Cohen et al., 1983). In studies from yet another group, it was found that naloxone, 6 mg administered IV, resulted in an increase in plasma levels of ACTH, β-endorphin, and cortisol. Also, the administration of 10 mg of naloxone significantly enhanced the release of ACTH, β-endorphin, and cortisol after human CRF administration (Allolio et al., 1987).

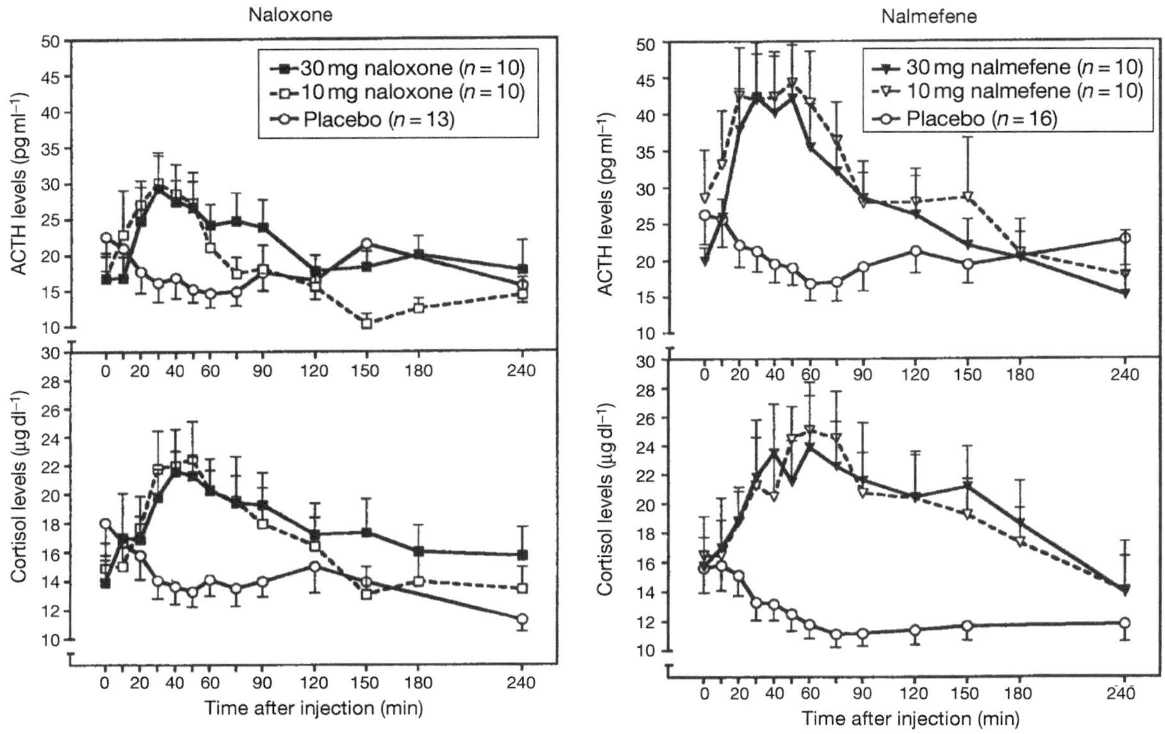

Figure 3 Plasma ACTH (upper panels) and cortisol (lower panels) after IV administration of two doses of the opioid antagonists naloxone (left panels) and nalmefene (right panels), as well as saline placebo, to healthy, normal volunteers without histories of drug or alcohol dependence. Both antagonists are primarily μ-opioid selective; however, nalmefene is more potent at the μ-opioid receptor and has a higher affinity for κ- and δ-opioid receptors, as well as a longer half-life. Note that a ceiling of HPA-axis activation is reached at the 10- and 30-mg doses of either antagonist; however, nalmefene causes significantly greater activation than naloxone at equivalent doses, indicating that κ- as well as δ-opioids may play important roles in the regulation of HPA-axis function. From Schluger J, Ho A, Borg L, et al. (1998b) Nalmefene causes greater hypothalamic–pituitary–adrenal axis activation than naloxone in normal volunteers: Implications for the treatment of alcoholism. *Alcoholism: Clinical and Experimental Research* 22: 1430–1436.

Studies of healthy volunteers demonstrated that not only the MOP-r, but also the κ-opioid receptor are involved in the normal tonic inhibition of the HPA axis (see **Figure 3**; Schluger et al., 1998b). In studies conducted in healthy volunteers, two doses of naloxone, 10 and 30 mg, were administered IV, both of which gave maximal plasma levels of ACTH and cortisol (Schluger et al., 1998b). In parallel studies, two doses of a different opioid antagonist – one which has been shown to have not only MOP-r, but also κ-opioid receptor selectivity of substantial amounts when studied using human membranes or in living humans – nalmefene, was used. Nalmefene resulted in significantly higher activation of the HPA axis, suggesting that the hypothalamic–pituitary part of the HPA stress-responsive axis is under tonic inhibition by not only activation of the MOP-r, presumably by β-endorphin and Met-enkephalin-Arg-Phe, but also under tonic inhibition by the κ-opioid receptor, for which dynorphin A and dynorphin B are the two major natural ligands (see **Figure 3**; Schluger et al., 1998b).

As stated earlier, since the late 1960s it has been hypothesized that an atypical responsivity to stress and stressors contribute to the acquisition of specific addictive diseases, including heroin addiction, and it had also been shown that chronic use of the short-acting opiate heroin itself could profoundly disrupt this HPA axis and in a direction which would serve to reduce activity of this axis by suppressing release of hormones of each component of the axis. In very provocative studies reported in 1990, it was suggested that activation of this axis may, in fact, precede onset of the signs and symptoms of opiate withdrawal in opiate-dependent individuals (Kennedy et al., 1990). In these studies, well-stabilized methadone-maintained patients, with no ongoing polydrug or alcohol abuse, were entered into metyrapone studies. Approximately 30 min after the metyrapone dose, at the time when cortisol levels are significantly reduced below baseline, but when ACTH and β-endorphin levels are beginning to rise rapidly, most of the stabilized patients complained of mild-to-moderate or severe signs and symptoms of opiate withdrawal. However, these symptoms abated after about 2 h, much less than the time course of action of metyrapone, and less than the times of elevations of plasma levels of ACTH and β-endorphin, but after the rapid escalation of levels of ACTH and β-endorphin is achieved. Further studies confirmed that metyrapone itself is not an opioid antagonist and that plasma levels of methadone were not changed by metyrapone. Thus, these studies suggested that activation of the hypothalamic–pituitary part of the stress-responsive HPA axis served as a cue in the former heroin addicts, who had experienced opiate withdrawal several times every day during cycles of heroin addiction. It had been well established that activation with increased levels of the hypothalamic–pituitary and adrenal hormones are observed in opiate withdrawal. Up to the time of this report of Kennedy et al. (1990), it had been assumed that activation of the HPA axis during opiate withdrawal was due to the stress of the opiate withdrawal itself, with very adverse negative signs and symptoms that characterize the abstinence syndrome. An alternative hypothesis was then articulated, that the activation of the HPA axis plays an actual active role in the development of the abstinence syndrome. Activation of the HPA axis in opioid-dependent persons with abrupt discontinuation of opiate administration is directly due to the removal of the exogenous opioids from their action of providing artificial tonic inhibition of the HPA axis, an inhibition similar to that normally played by the endogenous opioid system. The finding of transient onset and prompt remission of opiate withdrawal symptoms during the time of rapidly elevating levels of ACTH and β-endorphin in stabilized methadone-maintained patients gave support to this new hypothesis, since, in this case, the rapid rise of the stress-responsive hormones was identical to the rise occurring in the setting of abrupt withdrawal of exogenous opiates.

Studies of oral opioid antagonists as a treatment of opiate-induced constipation both in chronic-pain patients and patients in chronic methadone treatment for heroin addiction confirmed the findings discussed above by demonstrating that elevations in ACTH and cortisol actually preceded the onset of any signs and symptoms of opiate withdrawal when oral naloxone was given in increasing doses (Culpepper-Morgan et al., 1992). Further studies, in which ascending-dose titration of oral naloxone was administered to opioid-dependent persons, confirmed that not only did the activation of the HPA axis precede the signs and symptoms of opiate withdrawal, but it could, in fact, occur when the opioid antagonist naloxone's plasma and, thus, brain levels were insufficient to produce any signs and symptoms of withdrawal (Culpepper-Morgan and Kreek, 1997). Studies in heroin addicts undergoing withdrawal, stabilized on very low doses of morphine, which was used for detoxification, and given extremely low doses of naloxone, showed activation of the HPA axis when the first very modest signs or symptoms of opiate withdrawal occurred, and long before the signs and symptoms became noxious (Rosen et al., 1995). During the last phases of detoxification management, at a time when hypersensitivity to opioid antagonists would be expected to be found, activation of the HPA axis occured before, or simultaneously with, but not following the onset of signs and symptoms of opioid withdrawal (Rosen et al., 1995).

Studies conducted in illicit-drug-free former heroin addicts treated with the opioid antagonist naltrexone for at least 15 weeks demonstrated that activation of the HPA axis continued to occur even during chronic naltrexone treatment (Kosten et al., 1986a,b). Also, naltrexone had been found by several groups to be effective in the management of 30–50% of unselected alcoholics, whereas it has been found to be effective in less than 15% of unselected heroin addicts (Kreek, 1996d). A very modest dose of 25 mg of naltrexone administered orally, as well as higher doses of 50 mg (the usual treatment dose for alcoholism) and 100 mg (the usual treatment dose for heroin addiction) all activated the HPA axis in abstinent alcoholics (Farren et al., 1999). Other studies demonstrated activation of the HPA axis by naloxone and depression of that axis by heroin; the levels of the HPA-axis hormones normalized in patients maintained on methadone (Kosten et al., 1992). The same studies suggested that, as with methadone, a time longer than 1 month may be required for normalization of the HPA axis during treatment with the partial agonist, buprenorphine (Kosten et al., 1992).

Metyrapone studies in patients with cocaine dependency only and in cocaine-abusing methadone-maintained former heroin addicts showed that in the stabilized methadone-maintained patients, response to metyrapone was normal (**Figures 4** and **5**) (Schluger et al., 1998a, 2001). During cycles of opiate addiction, there appears to be a relative endogenous opioid deficiency, as reflected by the failure of the endogenous opioid system to counter-regulate the HPA axis when

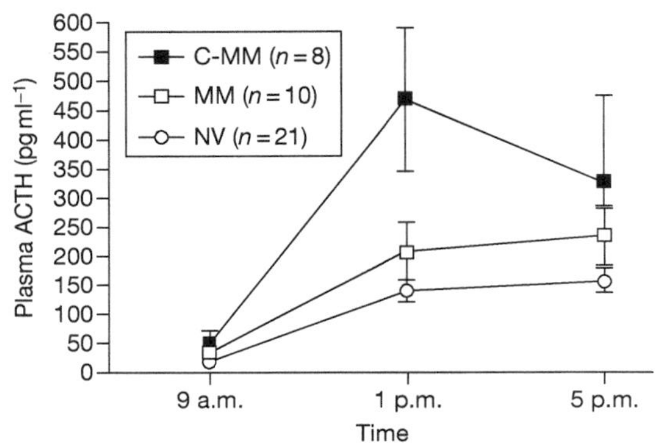

Figure 4 Mean plasma ACTH levels (±SEM) at 9 a.m. (just before metyrapone administration), 1, and 5 p.m. in normal volunteers (NV), illicit-drug-free methadone-maintained patients (MM) and methadone-maintained patients abusing cocaine (C-MM). Reproduced from Schluger JH, Borg L, Ho A, and Kreek MJ (2001) Altered HPA axis responsivity to metyrapone testing in methadone maintained former heroin addicts with ongoing cocaine addiction. *Neuropsychopharmacology* 24: 568–575, with permission from Nature Publishing Group.

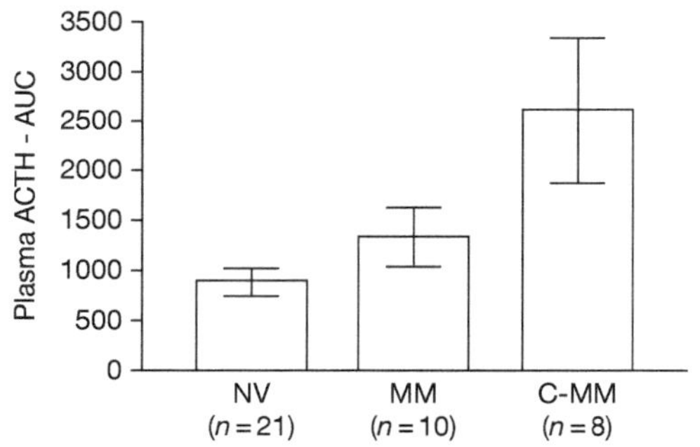

Figure 5 AUC for plasma ACTH (±SEM) from 9 a.m. to 5 p.m. in normal volunteers (NV), illicit-drug-free methadone-maintained patients (MM) and methadone-maintained patients abusing cocaine (C-MM). Reproduced from Schluger JH, Borg L, Ho A, and Kreek MJ (2001) Altered HPA axis responsivity to metyrapone testing in methadone-maintained former heroin addicts with ongoing cocaine addiction. *Neuropsychopharmacology* 24: 568–575, with permission from Nature Publishing Group.

the normal negative feedback control by the glucocorticoid cortisol is abruptly interrupted because the synthesis of cortisol in the adrenal cortex is blocked by metyrapone (Schluger et al., 2001).

CRF testing in methadone-maintained former heroin addicts with no ongoing drug or alcohol dependence revealed that, in comparison to normal volunteers, the former addicts had a significantly greater increase in plasma ACTH levels in response to IV administration of high (2.0 $\mu g\ kg^{-1}$), but not low (0.5 $\mu g\ kg^{-1}$), doses of CRF (Schluger et al., 2003).

Dexamethasone suppression tests in methadone-maintained patients, with and without ongoing cocaine dependence, showed significantly lower plasma ACTH and/or cortisol levels than normal volunteers at 17h after dexamethasone dose; there was no difference in ACTH and cortisol levels between methadone-maintained patients free of illicit-drug abuse and methadone-maintained patients abusing cocaine (Aouizerate et al., 2006).

The physiological function of the HPA axis must be considered in the process of development of new forms of therapy for opioid dependence. For example, a design of a sustained-release formulation of methadone should provide for an even more modest rise every 24h in the morning to counteract the circadian rhythm of cortisol levels (highest in the morning and lower in the evening) with changes in plasma ACTH levels (Kreek et al., 2002).

35.2.3 Tuberoinfundibular Dopaminergic/ Prolactin System Interactions

Potential alterations in the physiology of prolactin secretion associated with opiate addiction have been of interest from two main perspectives. Clinically, amenorrhea–galactorrhea has been observed in heroin addicts. This observation was one of the earliest suggestions of endocrine disturbance associated with opiate addictions (Pelosi et al., 1974). Long-term treatment in methadone maintenance, however, has been associated with normalization of menses. Galactorrhea has not been reported in methadone-maintenance treatment (Cushman and Kreek, 1974). Efforts to elucidate the neurobiology of addiction and its treatment led to focusing intense interest on dopaminergic systems. Prolactin release is under the tonic inhibition of tuberoinfundibular DA, and alterations in prolactin levels have been used as a potential window into central dopaminergic tone. Further, prolactin secretion is known to be a marker of stress, also, in part, mediated by decreased tuberoinfundibular dopaminergic tone (Freeman et al., 2000), and alterations in stress responsivity have been hypothesized to be related to vulnerability to addictive diseases (Kreek, 1992b). In humans, opioid agonists administered acutely increase prolactin levels, presumably through an MOP-r-mediated inhibition of tuberoinfundibular DA (Rolandi et al., 1983; Delitala et al., 1983a,b; Mendelson et al., 1989). κ-Opioid agonists administered acutely have been demonstrated to increase prolactin levels as well (see **Figure 2**; Kreek et al., 1999). An intriguing observation was an induction of increased serum prolactin levels in normal human volunteers by nalmefene. Although the latter was originally known as a μ- and κ-opioid-selective antagonist, we have shown that nalmefene's effect in raising serum prolactin levels is, probably, due to partial κ-opioid-agonist activity. Bidlack and colleagues, in collaboration with our laboratory, have shown that nalmefene is not only a pure μ-opioid antagonist, but also a partial κ-opioid agonist, by conducting studies in appropriate cell constructs. Therefore, the κ-opioid activity leads to lowering of tuberoinfundibular DA tone, resulting in increased prolactin (Bart et al., 2005).

In a 1978 study, Kreek demonstrated that daily methadone administration lowers tuberoinfundibular DA tone, which may be related to its mechanism of action and its continuing effectiveness (Kreek, 1978). Compared with normal volunteers, stable non-drug-abusing methadone-maintained patients demonstrate an attenuated prolactin response to IV administration of dynorphin A_{1-13}, an endogenous opioid peptide binding to κ-opioid receptors (Bart et al., 2003). Although IV administration of $dynorphin_{1-13}$ results in dose-dependent increase in prolactin just as in normal healthy volunteers, the actual level reached at each time point and each dose was lower (Bart et al., 2003).

In 1980, M.S. Gold et al. reported on prolactin levels in a group of heroin addicts maintained on methadone, during methadone discontinuation. Twenty-one male opiate addicts, who had been maintained on 15–75 mg of methadone for 6 months to 11 years, were first studied 36 h after methadone discontinuation, and then after the administration of clonidine for the amelioration of withdrawal symptoms. Prolactin levels were compared to those obtained 4 weeks later, after clonidine discontinuation, as an opiate-free baseline. Prior to the administration of clonidine, prolactin levels were found to be lower in comparison with levels obtained 4 weeks later in the drug-free state. Clonidine administration had no effect on subsequent prolactin levels. The decreased prolactin levels 36 h after methadone administration, in comparison to the drug-free state, were hypothesized to be indicative of dopaminergic hyperactivity during opiate withdrawal. Further, the absence of an effect of clonidine on prolactin levels, despite the observation of a decrease in objective signs and symptoms of withdrawal attributed to clonidine, was hypothesized to indicate that dopaminergic hyperactivity during withdrawal might not be directly related to the signs and symptoms of withdrawal (Gold et al., 1980b).

Spagnolli et al. (1987) found significantly elevated prolactin levels 3–24 h after the last administration of heroin in females who had been addicted to heroin for a minimum of 2 years before entering a short-term detoxification protocol with methadone (as compared to healthy volunteers). However, following thyrotropin-releasing hormone (TRH) administration on the first day of the methadone taper, the increase in prolactin levels was significantly lesser than in normal volunteers (Spagnolli et al., 1987).

Vescovi et al. (1990) reported increased (as compared to healthy controls) baseline prolactin levels in males who had been addicted to heroin for 12–48 months and were studied after 14–24 days of abstinence during treatment in a therapeutic community. The prolactin response to hyperthermia (induced by 30 min in a sauna at 90 °C) was blunted (Vescovi et al., 1990). Mutti et al. (1992) also studied prolactin levels during abstinence from heroin self-administration. The group with the shortest period of abstinence had elevated baseline prolactin levels in comparison with normal volunteers, whereas levels in the two groups with longer periods of abstinence were comparable to normal volunteers.

Little is known about prolactin regulation during treatment with the μ-opioid partial agonist, buprenorphine, an opioid with mixed agonist–antagonist properties approved in the United States for the treatment of heroin addiction. Mendelson and Mello (1984) reported on a study of three adult male heroin addicts who had histories of an average of 10 years of heroin dependence and had failed conventional therapies, who were then treated with buprenorphine. After an initial 5-day drug-free period, subjects received subcutaneous doses of buprenorphine, starting at 0.5 $mg\,day^{-1}$, which were increased gradually over 14 days to a dose of 8 mg. Following this induction, subjects received 8 mg of buprenorphine daily for 10 days, during which time they had access to heroin, which they could administer in doses up to 21 $mg\,day^{-1}$ for the first 5 days, then up to 40.5 mg for the latter 5 days. Data were reported for the three subjects who only administered heroin on a single occasion during the 10-day maintenance period. Prolactin levels during the initial 5-day drug-free period were reported to be normal. During the induction period, prolactin levels were increased compared to baseline and remained elevated during the maintenance period in two of the subjects. Prolactin levels were only increased in comparison to each subject's baseline, and the magnitude of the increase was small; overall levels remained within normal limits (Mendelson and Mello, 1984).

In summary, clinically significant hyperprolactinemia associated with heroin addiction, yet appears to resolve during treatment in long-term methadone maintenance. Acute administration of either μ- or κ-opioid agonists causes prolactin release through inhibition of tuberoinfundibular DA. Abstinence from chronic administration of short-acting opioids appears to be associated, at least in the short term, with increased basal prolactin levels and perhaps with a blunted prolactin response to stimulation. Increases in prolactin levels during long-term methadone maintenance have been reported in some studies; however, this finding may be related to a shift in the pattern of secretion with entrainment to the timing of daily methadone administration, and average prolactin secretion throughout the day is normalized during long-term methadone-maintenance treatment.

35.2.4 Hypothalamic–Pituitary–Gonadal Axis

Men addicted to heroin often complain of sexual dysfunction, which usually resolves with abstinence from heroin or after at least 1 year of methadone-maintenance treatment. LH levels are usually normal (Cushman, 1972), while testosterone is typically either low-normal or low on heroin. Well-controlled studies demonstrate normal measurements of testosterone on methadone maintenance (Mendelson and Mello, 1975). The role of heavy alcohol use and/or ongoing heroin or other illicit-drug use, which may not be documented by history and/or toxicology, may be extremely important in patients reporting persistent sexual dysfunction.

Women addicted to heroin may also complain of sexual dysfunction as well as abnormal menses. These problems may resolve with long-term (more than 1 year) methadone therapy (Santen et al., 1975).

35.2.5 Growth Hormone and Opioid Addiction

Clinically significant abnormalities in the regulation of growth hormone secretion have not been observed in association with opiate addiction and its treatment (Cushman, 1973). Rather, interest in growth hormone regulation has been related to its utility as a marker of hypothalamic–pituitary function, as well as insulin-mediated glucose homeostasis. Growth hormone secretion is stimulated by acute administration of MOP-r agonists (Delitala et al., 1983a).

35.2.6 Thyroid Function and Opioid Addiction

Heroin addiction and pharmacotherapy have not been associated with clinically significant alterations in hypothalamic–pituitary–thyroid-axis function in humans. While there have been no reports of clinically evident thyroid dysfunction due to heroin addiction or methadone therapy in drug-free, medication-free state, some abnormalities in thyroid function have been reported in association with different stages of the addiction and its treatment; some of these abnormalities included increased total T_3, increased total T_4, and increased thyroxine-binding globulin (TBG) (which can explain an increase in total T_3 and T_4) (Cushman and Kreek, 1974; Chan et al., 1979).

35.2.7 μ-Opioid Receptor Binding in Healthy Normal and Methadone-Maintained Volunteers

It was hypothesized that a substantial percentage of MOP-rs must remain unoccupied during chronic, steady moderate- to high-dose (now 60–150 mg day^{-1}) methadone-maintenance treatment and thus are available for performance of their normal physiological functions, since normalization of so many functions is now known to be under control of the μ-opioid system, and is disrupted by chronic on–off bombardment by the short-acting opiate heroin, and which have also been found to normalize during methadone-maintenance treatment (Kreek, 1992a, 1996a,b,c, 1997). A study using PET and a selective, radiolabeled opioid receptor-directed antagonist, [18F]cyclofoxy, has indeed shown that this is the case (Kling et al., 2000). This work was also the first study in which the mapping of a large number of different brain regions for primarily μ-opioid receptor binding was accomplished in healthy young and middle-aged humans (Kling et al., 2000). Building on a limited number of studies using techniques of quantitative autoradiography in postmortem human brains, as well as closely related studies in animal models, it was found that the areas of the living human brain with the highest density of opioid receptors (primarily μ- and also some κ-type, reflecting the binding-selectivity of the PET ligand used) are, in descending order, the thalamus (a center for control of responses to pain); the amygdala, caudate, insula, anterior cingulate, and putamen (all areas of the mesolimbic–mesocortical and nigrostriatal dopaminergic projections, documented to be involved in specific aspects of drug self-administration acquisition and addiction in animal models, as well as in some affective components of responses to pain); followed by the middle temporal cortex, middle frontal cortex, and parietal cortex; and with lesser amounts of specific opioid receptor binding in the cerebellum, inferior temporal cortex, and hippocampus (see **Figure 6**; Kling et al., 2000).

As hypothesized, during steady-dose long-term methadone treatment, there was a modest 19–32% reduction of specific PET F-labeled ligand opioid antagonist binding (Kling et al., 2000). Also, yet again, it was found that steady-state plasma levels of methadone pertain during the interval of 22–23.5 h after the last oral dose of methadone in these long-term maintained patients, which was the time of the 90-min PET scans (see **Figures 6** and 7) (Kling et al., 2000). Of interest is that four of the six regions of the brain which were found to have the highest density of specific opioid receptors – the amygdala, the caudate, anterior cingulate, and putamen – have all been implicated in animal studies as major regions for the neurobiological effects, including the so-called rewarding effects, of opiates, cocaine, and several other specific drugs of abuse. In many of these regions, due to the neuroplasticity of the brain, significant changes in expression of many genes, including several of the endogenous opioid systems and neurochemical events, have been found (Kreek, 1996a,b,c, 1997, 2000; Kreek and Koob, 1998; Kling et al., 2000). Also, specific locomotor activity and other behaviors related to the reinforcing effects of drugs of abuse, as well as their sensitization effects, have been linked with these regions. Some of these regions have been identified in other human PET studies, using blood flow or glucose metabolism, as an overall indicator of activation or reduction in brain function following administration of drugs of abuse and also in functional magnetic resonance imaging (fMRI) studies, in which changes in levels of brain activation may be sensitively monitored, as regions which similarly are altered by administration of such drugs. Thus, in this PET study it was found that over 65% of primarily MOP-rs apparently remain unoccupied by methadone during chronic pharmacotherapy and are thus available to resume their normal physiological roles as essential components of the endogenous opioid modulatory system (see **Table 1** and **Figure 6**) (Kling et al, 2000).

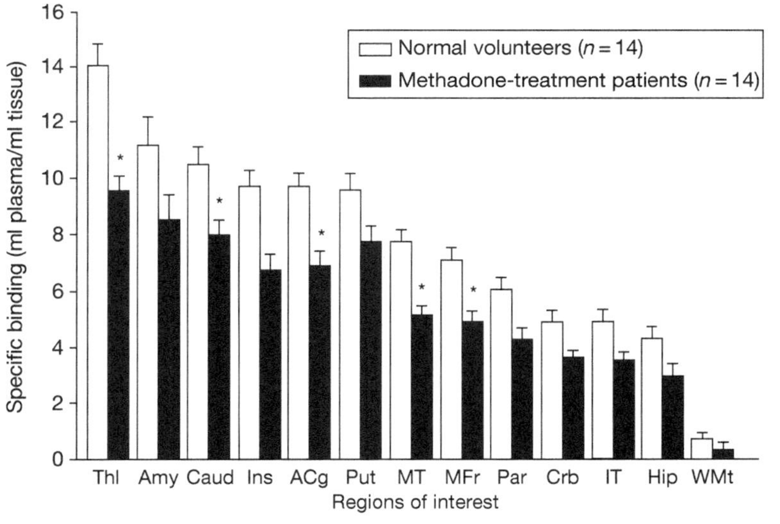

Figure 6 Specific binding of [18F]cyclofoxy, a μ/κ ligand, (mean ± SEM) in 13 brain regions of normal volunteers (NV, open bars) and long-term stabilized methadone-maintained former heroin addicts (MM, shaded bars). Regions of interest are placed in descending order of level of specific binding in normal volunteers. Asterisks indicate regions with significant differences in binding between NV and MM. ACg, anterior cingulate gyrus; Amy, amygdala; Caud, caudate; Crb, cerebellum; Hip, hippocampus; Ins, insula; IT, inferior temporal cortex; MFr, middle frontal cortex; MT, middle temporal cortex; Par, parietal cortex; Put, putamen; Thl, thalamus; WMt, white matter. Reproduced from Kling MA, Carson RE, Borg L, et al. (2000) Opioid receptor imaging with PET and 18F cyclofoxy in long-term methadone treated former heroin addicts. *Journal of Pharmacology and Experimental Therapeutics* 295: 1070–1076, with permission from American Society for Pharmacology and Experimental Therapeutics.

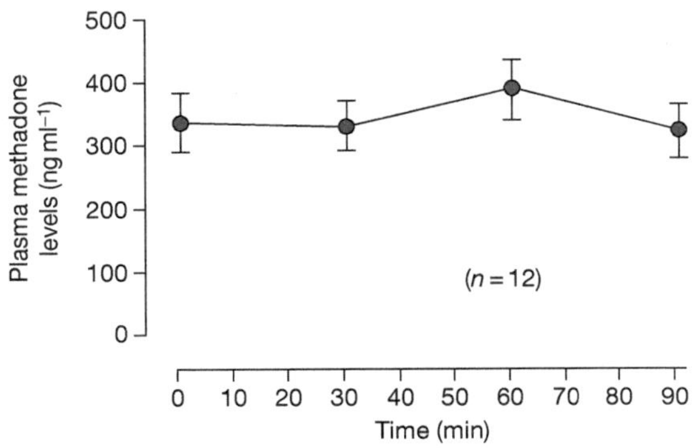

Figure 7 Plasma methadone levels measured during a 90-min PET scan session, in long-term stabilized, methadone-maintained former heroin addicts, approximately 22 h after the last methadone dosage. Reproduced from Kling MA, Carson RE, Borg L, et al. (2000) Opioid receptor imaging with PET and 18F cyclofoxy in long-term methadone-treated former heroin addicts. *Journal of Pharmacology and Experimental Therapeutics* 295: 1070–1076, with permission from American Society for Pharmacology and Experimental Therapeutics.

35.2.8 Human Molecular Genetics of Heroin Addiction of the Endogenous Opioid Systems and Polymorphisms of Genes

Opioid receptors were clearly elucidated in 1973, followed by studies with increasingly selective synthetic chemical ligands, which delineated the existence of probably three separate types of opioid receptors. Despite innumerable attempts by many investigators, there was no successful cloning of the endogenous opioid receptors until late 1992, when two groups, working simultaneously but independently, first cloned the δ-opioid receptor from a hybrid rat–mouse cell line (Evans et al., 1992; Kieffer et al., 1992). This work was subsequently followed by the successful cloning of the other two opioid receptor types, the μ- and the κ-opioid receptors in rodent models (Chen et al., 1993; Wang et al., 1994a; Li et al., 1993; Minami et al., 1993; Yasuda et al., 1993). Subsequently, each type of opioid receptor was also cloned in humans (Bare et al., 1994; Mestek et al., 1994; Wang et al., 1994b; Knapp et al., 1994; Zhu et al., 1995). With the cloning of each of these receptors, a variety of new, molecular neurobiological studies could be conducted, with special attention on alterations of gene expression that may occur during various perturbations, such as those caused by short-acting opiates (Wang et al., 1999). Also, the cloning of the human μ-opioid receptor allowed studies to be initiated to determine the presence of any polymorphisms, including single nucleotide polymorphisms (SNPs), in the opioid system, and to determine if there are any associations of this gene's polymorphisms with specific addictive diseases (see **Figure 8**) (Kreek, 2000; Bergen et al., 1997; Berrettini et al., 1997; Bond et al., 1998; LaForge et al., 2000a,b,c).

A variety of studies have now shown that not only opiate addiction, but also alcoholism and cocaine addiction, involve perturbations of the endogenous opioid system (Kreek, 1997, 2000). Other studies have shown that addictions in general have a 25–45% relative risk contributed by genetic factors and, furthermore, that the contribution of genetic factors to heroin addiction are, in fact, higher than for the other addictions (Tsuang et al., 1998). It has been hypothesized that some of the individual variability and susceptibility, or vulnerability, to the development and persistence of relapse to opiate addiction may be due, in part, to polymorphisms in the μ-opioid receptor, which has been the basis of several of these studies (Kreek et al., 2004a, 2005). The rationale behind this hypothesis lies in clinical studies that demonstrated that the endogenous opioid system tonically inhibits the HPA axis, the major component of the stress-response system, via μ-opioid receptors, and that atypical responsivity to stress and stressors contributes to perpetuation and relapse of specific addictions (Kreek, 1992a, 2001, 2006, 2007; Kreek et al., 2004a, 2005; Kreek and LaForge, 2007). Genetic studies have been conducted of individuals with A118G polymorphism (overall ethnic cultural allelic frequency of 10.5%), which results in a change from asparagine to aspartic acid in position 40 in the extracellular N-terminus of μ-opioid receptor across ethnic and cultural groups, the site of binding of endogenous and exogenous opioids and opiates. As we hypothesized, the presence of one of two copies of this SNP alters the stress responsivity in humans. We have shown that the presence of one of two copies of this SNP leads to significantly elevated basal levels of serum cortisol in normal healthy volunteers (Kreek and LaForge, 2007; Bart et al., 2006) as well as to increased cortisol response following administration of the opioid antagonist naloxone or naltrexone (Wand et al., 2002; Chong et al., 2006). These findings suggest that altered HPA-axis stress responsiveness may predispose the affected individuals to addictions. A significant association was demonstrated between A118G polymorphism and heroin addiction in a Caucasian population in central Sweden (Bart et al., 2004). The A118G SNP is associated not only with opiate addiction, but also

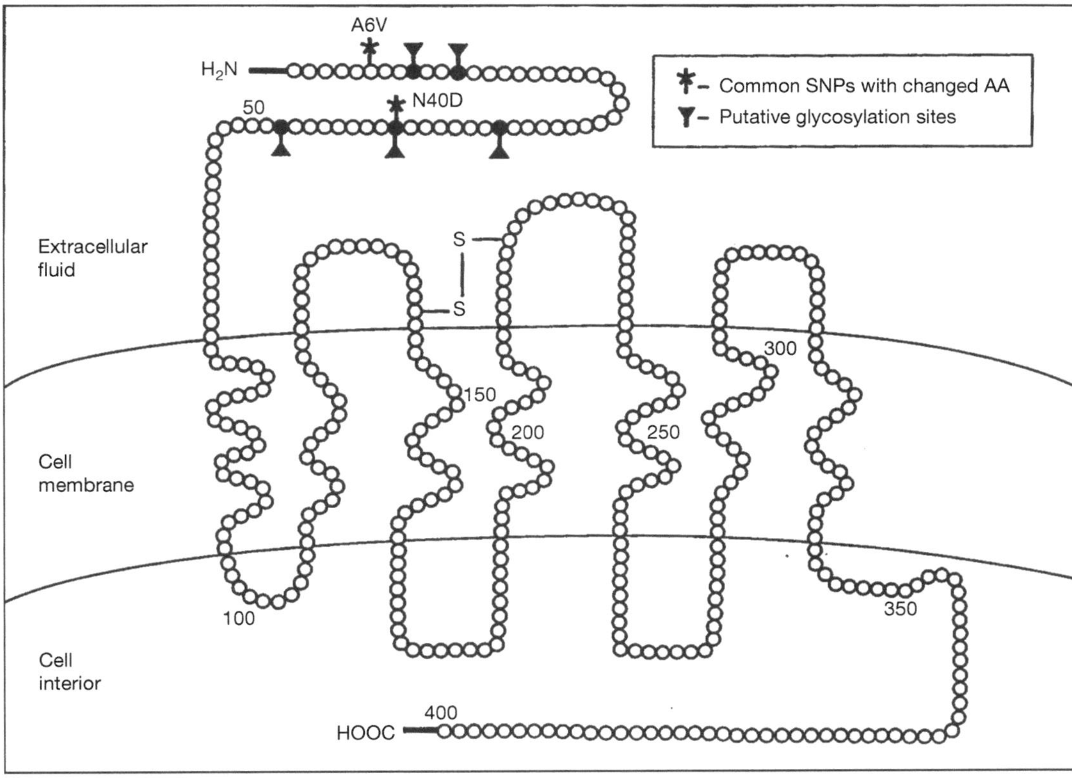

Figure 8 Diagram showing the amino acid structure of the human μ-opioid receptor, a seven-transmembrane G-protein-coupled receptor and also showing the two most frequent allelic variants, both of which are single nucleotide polymorphisms (SNPs), resulting in changed amino acids. Note that one of the SNPs, N40D, is located at a putative glycosylation site. Modified from LaForge KS, Yuferov V, and Kreek MJ (2000c) Opioid receptor and peptide gene polymorphisms: Potential implications for addictions. *European Journal of Pharmacology* 410: 249–268.

with alcohol dependence, as was shown in the central Sweden population (Bart et al., 2005).

Further molecular studies showed that stable, but not transient, expression of the A118G variant and prototype μ-opioid receptor in cell cultures was characterized by different levels of cell-surface-binding capacity, forskolin-induced cAMP accumulation, and agonist-induced accumulation of cAMP for several agonists (Kroslak et al., 2007). It was also recognized that individual differences in response to endogenous opioids might occur, or physiogenetics, where individual responses to pharmacotherapies might be mediated, in part, by variant forms of the μ-opioid receptor (Kreek, 2000, 2008; LaForge et al., 2000a,b,c). The terms pharmacogenetics and pharmacogenomics have been used for many years to refer to the latter phenomena, long before human molecular genetics had advanced to the current state of the art, since genetic epidemiology indicated such differences in responses to medications, along with specific challenges with test compounds and other approaches that could be used to define phenotypic differences in drug metabolism, disposition, and dynamics, without necessarily knowing the molecular genetic differences (Tyndale et al., 1984; Zawertailo et al., 1998; Romach et al., 2000). For altered responses to endogenous peptides, neurotransmitters, or other compounds on a genetic basis, we have coined the terms physiogenetics and physiogenomics (Kreek, 2000; LaForge et al., 2000c).

With respect to the opioid system, it is possible to conduct molecular genetics studies in parallel with the phenotypic characterization. The μ-opioid receptor is the first site of action of the drugs of abuse, heroin, the major metabolite, morphine, and also the major drug for the treatment of opiate addiction, methadone, as well as other treatment agents acting at the tt receptor, such as LAAM and, in part, buprenorphine. Therefore, the μ-opioid receptor was the logical site for focusing initial studies for polymorphisms or variants

(Kreek, 2000; Bond et al., 1998; LaForge et al., 2000a,b,c). To date, 15 SNPs in the coding region of the μ-opioid receptor have been identified and reported (LaForge et al., 2000a,c). Two of these occur with very high allelic frequencies; in one study an allelic frequency of 10.5% was found for the A118G SNP and 6.6% was the allelic frequency found for the C17T variant (Bond et al., 1998). Both of these polymorphisms have been found to have different allelic frequencies in different populations (see **Figure 8**; Bergen et al., 1997; Berrettini et al., 1997; Bond et al., 1998; LaForge et al., 2000a). Molecular and cell biological studies of the receptor peptide resulting from the A118G allelic variant have shown that this variant apparently binds the longest (31 residues) of the endogenous opioids, β-endorphin, which also has the longest half-life in humans (≈30 min), 3 times more avidly than does the prototypic /t-receptor sequence (Bond et al., 1998). Using further molecular–cellular constructs, it has been shown that the G-protein-coupled potassium inwardly rectifying channel (GIRK) activation is threefold greater after binding of β-endorphin to the receptors resulting from the Al18G variant than to the prototype receptors (Bond et al., 1998). Further studies of a molecular and cellular type, as well as potential studies in animal models and, of special importance, future studies in humans who are heterozygous or homozygous for one or more of these polymorphisms of the μ receptor, will be needed to determine whether there are indeed alterations of clinical significance in binding of and activation by β-endorphin acting at either (or both) of these very common variants, or by other endogenous opioids acting at other variants (for which the term physiogenetics was coined). In addition, further studies will be needed to determine whether there are any indications of pharmacogenetic differences, based on allelic variants of either the μ-opioid receptor or possibly other genes, such as those for enzymes involved in the metabolism and biotransformation of exogenous opioids. To date, variants of genes of specific components of the P450-related system enzymes have been shown to lead to altered metabolism of some other specific opiates, such as codeine and dextromethorphan, but not involving any of the major drugs of abuse such as heroin or morphine, nor significantly altering the opioid agonists used in the pharmacotherapy of addiction, including primarily methadone, and also LAAM and buprenorphine (Tyndale et al., 1984; Zawertailo et al., 1998; Romach et al., 2000).

Receptors other than μ-opioid ones also harbor SNPs associated with opiate addiction. For example, the 36G>T polymorphism in the human κ-opioid receptor gene (*OPRK1*) has such an association. (Yuferov et al., 2004) On the other hand, another study showed a pointwise significant association of haplotype pairs containing allele G at position 1180 in the serotonin 1B receptor gene (*HTR1B*) with a protective effect from heroin addiction in Caucasians (Proudnikov et al., 2005).

Acknowledgments

Support for this chapter has been provided by grants from the National Institutes of Health (NIH): National Institute of Drug Abuse (NIDA) Grant numbers P60-DA05130 and K05-DA00049; the New York State Office of Alcoholism and Substance Abuse Services, Grant Number C003189 (Kreek); and the National Institutes of Health and NIH UL1RR024143 and from the National Center for Research Resources (NCRR) (Coller). The authors also thank Susan Russo and Jack Varon for assistance in preparation of the manuscript.

References

Allolio B, Schulte HM, Deuss U, Kallabis D, Hamel E, and Winkelman W (1987) Effect of oral morphine and naloxone on pituitary–adrenal response in man induced by human corticotropin-releasing hormone. *Acta Endocrinologica (Copenhagen)* 114: 509–514.

Aouizerate B, Ho A, Schluger JH, et al. (2006) Glucocorticoid negative feedback in methadone maintained former heroin addicts with ongoing cocaine dependence: Dose-response to dexamethasone suppression. *Addiction Biology* 11: 84–96.

Azaryan AV, Coughlin LJ, Buzas B, Clock BJ, and Cox BM (1996) Effect of chronic cocaine treatment on mu- and delta-opioid receptor mRNA levels in dopaminergically innervated brain regions. *Journal of Neurochemistry* 66: 443–448.

Bare LA, Mansson E, and Yang D (1994) Expression of two variants of the human mu-opioid receptor mRNA in SK-N-SH cells and human brain. *FEBS Letters* 354: 213–216.

Bart G, Borg L, Schluger JH, Green M, Ho A, and Kreek MJ (2003) Suppressed prolactin response to dynorphin A(1–13) in methadone maintained versus control subjects. *Journal of Pharmacology and Experimental Therapeutics* 306: 581–587.

Bart G, Heilig M, LaForge KS, Pollak L, Leal SM, Ott J, and Kreek MJ (2004) Substantial attributable risk related to a functional mu-opioid receptor gene polymorphism in association with heroin addiction in central Sweden. *Molecular Psychiatry* 9: 547–549.

Bart G, Kreek MJ, Ott J, LaForge KS, Proudnikov D, Pollak L, and Heilig M (2005) Increased attributable risk related to a functional mu-opioid receptor gene polymorphism in association with alcohol dependence in central Sweden. *Neuropsychopharmacology* 30: 417–422.

Bart G, LaForge KS, Borg L, Lilly C, Ho A, and Kreek MJ (2006) Altered levels of basal cortisol in healthy subjects with a 118G allele in exon 1 of the mu-opioid receptor gene. *Neuropsychopharmacology* 31: 2313–2317.

Bergen AW, Kokoszka J, Peterson R, Long JC, Virkkunen M, Linnoila M, and Goldman D (1997) Mu-opioid receptor gene variants: Lack of association with alcohol dependence. *Molecular Psychiatry* 2: 490–494.

Berrettini WH, Hoehe M, Ferraro TN, DeMaria PA, and Gottheil E (1997) Human mu-opioid receptor gene polymorphisms and vulnerability to substance abuse. *Addiction Biology* 2: 303–308.

Bond C, LaForge KS, Tian M, et al. (1998) Single nucleotide polymorphism in the human mu-opioid receptor gene alters beta-endorphin binding and activity: Possible implications for opiate addiction. *Proceedings of the National Academy of Sciences of the United States of America* 95: 9608–9613.

Borg L, Ho A, Peters JE, and Kreek MJ (1995) Availability of reliable serum methadone determination for management of symptomatic patients. *Journal of Addictive Diseases* 14: 83–91.

Borg L, Ho A, Wells A, Joseph H, Appel P, Moody D, and Kreek MJ (2002) The use of levo-alpha-acetylmethadol (LAAM) in methadone patients who have not achieved heroin abstinence. *Journal of Addictive Diseases* 21: 13–22.

Bradbury AF, Smyth DG, Snell CR, Birdsall NJM, and Hulme EC (1976) C fragment of lipotropin has a high affinity for brain opiate receptors. *Nature* 260: 793–795.

Burstein Y, Grady RW, Kreek MJ, Rausen AR, and Peterson CM (1980) Thrombocytosis in the offspring of female mice receiving dl-methadone. *Proceedings of the Society for Experimental Biology and Medicine* 164: 275–279.

Cami J, Gilabert M, San L, and de la Torre R (1992) Hypercortisolism after opioid discontinuation in rapid detoxification of heroin addicts. *British Journal of Addiction* 87: 1145–1151.

Chan V, Wang C, and Yeung RT (1979) Effects of heroin addiction on thyrotrophin, thyroid hormones and prolactin secretion in men. *Clinical Endocrinology (Oxf.)* 10: 557–565.

Chen Y, Mestek A, Liu J, Hurley JA, and Yu L (1993) Molecular cloning and functional expression of a mu-opioid receptor from rat brain. *Molecular Pharmacology* 44: 8–12.

Chong RY, Oswald L, Yang X, Uhart M, Lin PI, and Wand GS (2006) The micro-opioid receptor polymorphism A118G predicts cortisol responses to naloxone and stress. *Neuropsychopharmacology* 31: 204–211.

Cicero TJ, Bell RD, Wiest WG, Allison JH, Polakoski K, and Robins E (1975) Function of the male sex organs in heroin and methadone users. *New England Journal of Medicine* 292: 882–887.

Cohen MR, Cohen RM, Pickar D, Weingartner H, and Murphy DL (1983) High-dose naloxone infusions in normals. *Archives of General Psychiatry* 40: 613–619.

Cox BM, Goldstein A, and Hi CH (1976) Opioid activity of a peptide, betalipotropin-(61-91), derived from beta-lipotropin. *Proceedings of the National Academy of Sciences of the United States of America* 73: 1821–1823.

Culpepper-Morgan JA, Inturrisi CE, Portenoy RK, Foley K, Houde RW, Marsh F, and Kreek MJ (1992) Treatment of opioid induced constipation with oral naloxone: A pilot study. *Clinical Pharmacology and Therapeutics* 23: 90–95.

Culpepper-Morgan JA and Kreek MJ (1997) HPA axis hypersensitivity to naloxone in opioid dependence: A case of naloxone induced withdrawal. *Metabolism – Clinical and Experimental* 46: 130–134.

Cushman P (1970) Hypothalamic–pituitary–adrenal axis in methadone-treated heroin addicts. *Journal of Clinical Endocrinology and Metabolism* 30: 24–29.

Cushman P (1972) Sexual behavior in heroin addiction and methadone maintenance. *New York State Journal of Medicine* 72: 1261–1265.

Cushman P (1973) Plasma testosterone in narcotic addiction. *American Journal of Medicine* 55: 452–458.

Cushman P and Kreek MJ (1974) Some endocrinologic observations in narcotics addicts. In: Zimmerman E and George R (eds.) *Narcotics and the Hypothalamus*, pp. 161–173. New York: Raven Press.

Dackis CA, Pottash AL, Gold MS, and Annitto W (1984) The dexamethasone suppression test for major depression among opiate addicts. *America Journal of Psychiatry* 141: 810–811.

Delitala G, Grossman A, and Besser M (1981) Changes in pituitary hormone levels induced by met-enkephalin in man – the role of dopamine. *Life Sciences* 29: 1537–1544.

Delitala G, Grossman A, and Besser M (1983a) Differential effects of opiate peptides and alkaloids on anterior pituitary hormone secretion. *Neuroendocrinology* 37: 275–279.

Delitala G, Grossman A, and Besser M (1983b) The participation of hypothalamic dopamine in morphine-induced prolactin release in man. *Clinical Endocrinology (Oxford)* 19: 437–444.

Delitala G, Palermo M, Tomasi P, Besser M, and Grossman A (1991) Adrenergic stimulation of the human pituitary–adrenal axis is attenuated by an analog of met-enkephalin. *Neuroendocrinology* 53: 41–46.

Dole VP (1970) Biochemistry of addiction. *Annual Review of Biochemistry* 39: 820–840.

Dole VP, Nyswander ME, and Kreek MJ (1966a) Narcotic blockade. *Archives of Internal Medicine* 118: 304–309.

Dole VP, Nyswander ME, and Kreek MJ (1966b) Narcotic blockade: A medical technique for stopping heroin use by addicts. *Transactions of the Association of American Physicians* 79: 122–136.

Eisenman AJ, Fraser HF, and Brooks JW (1961) Urinary excretion and plasma levels of 17-hydroxycorticosteroids during a cycle of addiction to morphine. *Journal of Pharmacology and Experimental Therapeutics* 132: 226–231.

Eisenman AJ, Fraser HF, Sloan J, and Isbell H (1958) Urinary 17-ketosteroid excretion during cycle of addiction to morphine. *Journal of Pharmacology and Experimental Therapeutics* 124: 305–311.

Eisenman AJ, Sloan JW, Martin WR, Jasinski DR, and Brooks JW (1969) Catecholamine and 17-hydroxycorticosteroid excretion during a cycle of morphine dependence in man. *Journal of Psychiatric Research* 7: 19–28.

Evans CJ, Keith DE, Jr., Morrison H, Magendzo K, and Edwards RH (1992) Cloning of a delta opioid receptor by functional expression. *Science* 258: 1952–1955.

Facchinetti F, Grasso A, Petraglia F, Parrini D, Volpe A, and Genazzini AR (1984) Impaired circadian rhythmicity of beta-lipotrophin, beta-endorphin and ACTH in heroin addicts. *Acta Endocrinologica (Copenhagen)* 105: 149–155.

Facchinetti F, Volpe A, Farci G, et al. (1985a) Hypothalamus–pituitary–adrenal axis of heroin addicts. *Drug and Alcohol Dependence* 15: 361–366.

Facchinetti F, Volpe A, Nappi G, Petraglia F, and Genazzini AR (1985b) Impairment of adrenergic-induced proopiomelanocortin-related peptide release in heroin addicts. *Acta Endocrinologica (Copenhagen)* 108: 1–5.

Farren CK, O'Malley S, Grebski G, Maniar S, Porter M, and Kreek MJ (1999) Variable dose naltrexone induced HPA stimulation in abstinent alcoholics: A preliminary study. *Alcoholism: Clinical and Experimental Research* 23: 502–508.

Folli D, Mutti A, Van der Venne MT, et al. (1992) Neuroendocrine response to psychological performance testing. *Psychoneuroendocrinology* 17: 467–474.

Freeman ME, Kanyicska B, Lerant A, and Nagy G (2000) Prolactin: Structure, function, and regulation of secretion. *Physiological Review* 80: 1523–1631.

Gil-Ad I, Bar-Yoseph J, Smadja Y, Zohar M, and Laron Z (1985) Effect of clonidine on plasma beta-endorphin, cortisol and growth hormone secretion in opiate-addicted subjects. *Israel Journal of Medical Sciences* 21: 601–604.
Gold MS, Pottash AL, Finn LB, Kleber HD, and Extein I (1980b) Serum prolactin and opiate withdrawal. *Psychiatry Research* 2: 205–210.
Gold PW, Extein I, Pickar D, Rebar R, Ross R, and Goodwin FK (1980a) Suppression of plasma cortisol in depressed patients by acute intravenous methadone infusion. *American Journal of Psychiatry* 137: 862–863.
Goldstein A, Lowney LI, and Pal BK (1971) Stereospecific and nonspecific interactions of the morphine congener levorphanol in subcellular fractions of mouse brain. *Proceedings of the National Academy of Sciences of the United States of America* 68: 1742–1747.
Goldstein A, Tachibana S, Lowney LI, Hunkapiller M, and Hood L (1979) Dynorphin-(1-13), an extraordinarily potent opioid peptide. *Proceedings of the National Academy of Sciences of the United States of America* 76: 6666–6670.
Gorelick DA, Kim YK, Bencherif B, et al. (2005) Imaging brain mu-opioid receptors in abstinent cocaine users: Time course and relation to cocaine craving. *Biological Psychiatry* 57: 1573–1582.
Grossman A, Delitala G, Mannelli M, Al-Damluji S, Coy DH, and Besser GM (1986) An analogue of met-enkephalin attenuates the pituitary–adrenal response to ovine corticotrophin releasing factor. *Clinical Endocrinology (Oxford)* 25: 421–426.
Hachey DL, Kreek MJ, and Mattson DH (1977) Quantitative analysis of methadone in biological fluids using deuterium-labeled methadone and GLC-chemical-ionization mass spectrometry. *Journal of Pharmaceutical Sciences* 66: 1579–1582.
Hellman L, Fukushima DK, Roffwarg H, and Fishman J (1975) Changes in estradiol and cortisol production rates in men under the influence of narcotics. *Journal of Pharmacology and Experimental Therapeutics* 41: 1014–1019.
Hughes J, Smith TW, Kosterlitz HW, Fothergill LA, Morgan BA, and Morris HR (1975) Identification of two related pentapeptides from the brain with potent opiate agonist activity. *Nature (London)* 258: 577–579.
Ignar DM and Kuhn CM (1990) Effects of specific mu and kappa opiate tolerance and abstinence on hypothalamo–pituitary–adrenal axis secretion in the rat. *Journal of Pharmacology and Experimental Therapeutics* 225: 1287–1295.
Ingoglia NA and Dole VP (1970) Localization of *d*- and *l*-methadone after intraventricular injection into rat brain. *Journal of Pharmacology and Experimental Therapeutics* 175: 84–87.
Inturrisi CE, Max MB, Foley KM, Schultz M, Shin SU, and Houde RW (1984) The pharmacokinetics of heroin in patients with chronic pain. *New England Journal of Medicine* 310: 1213–1217.
Inturrisi CE and Verebely K (1972a) A gas–liquid chromatographic method for the quantitative determination of methadone in human plasma and urine. *Journal of Chromatography* 65: 361–369.
Inturrisi CE and Verebely K (1972b) The levels of methadone in the plasma in methadone maintenance. *Clinical Pharmacology and Therapeutics* 13: 633.
Johnson RE, Chutuape MA, Strain EC, Walsh SL, Stitzer ML, and Bigelow GE (2000) A comparison of levomethadyl acetate, buprenorphine, and methadone for opioid dependence. *New England Journal of Medicine* 343: 1290–1297.
Kennedy JA, Hartman N, Sbriglio R, Khuri E, and Kreek MJ (1990) Metyrapone-induced withdrawal symptoms. *British Journal of Addiction* 85: 1133–1140.
Kieffer BL, Befort K, Gaveriaux-Ruff C, and Hirth CG (1992) The delta-opioid receptor: Isolation of a cDNA by expression cloning and pharmacological characterization. *Proceedings of the National Academy of Sciences of the United States of America* 89: 12048–12052.
Kling MA, Carson RE, Borg L, et al. (2000) Opioid receptor imaging with PET and 18F cyclofoxy in long-term methadone treated former heroin addicts. *Journal of Pharmacology and Experimental Therapeutics* 295: 1070–1076.
Knapp RJ, Malatynska E, Fang L, et al. (1994) Identification of a human delta opioid receptor: Cloning and expression. *Life Sciences* 54: PL463–PL469.
Koob G and Kreek MJ (2007) Stress, dysregulation of drug reward pathways, and the transition to drug dependence. *American Journal of Psychiatry* 164: 1149–1159.
Kosten TR, Kreek MJ, Swift C, Carney MK, and Ferdinands L (1987) Beta-endorphin levels in CSF during methadone maintenance. *Life Sciences* 41: 1071–1076.
Kosten TR, Kreek MJ, Raghunath J, and Kleber HD (1986a) A preliminary study of beta-endorphin during chronic naltrexone maintenance treatment in ex-opiate addicts. *Life Sciences* 39: 55–59.
Kosten TR, Kreek MJ, Raghunath J, and Kleber HD (1986b) Cortisol levels during chronic naltrexone maintenance treatment in exopiate addicts. *Biological Psychology* 21: 217–220.
Kosten TR, Morgan C, and Kreek MJ (1992) Beta-endorphin levels during heroin, methadone, buprenorphine and naloxone challenges: Preliminary findings. *Biological Psychology* 32: 523–528.
Kosterlitz HW and Hughes J (1975) Some thoughts on the significance of enkephalin, the endogenous ligand. *Life Sciences* 17: 91–96.
Kreek MJ (1972) Medical safety, side effects and toxicity of methadone. *Proceedings of the 4th National Conference on Methadone Treatment,* pp. 171–174. National Association for the Prevention of Addiction to Narcotics (NAPAN)-NIMH.
Kreek MJ (1973a) Medical safety and side effects of methadone in tolerant individuals. *Journal of the American Medical Association* 223: 665–668.
Kreek MJ (1973b) Physiological implications of methadone treatment. *Proceedings of the 5th National Conference on Methadone Treatment,* pp. 824–836. National Association for the Prevention of Addiction to Narcotics (NAPAN)-NIMH.
Kreek MJ (1973c) Plasma and urine levels of methadone. *New York State Journal of Medicine* 73: 2773–2777.
Kreek MJ (1978) Medical complications in methadone patients. *Annals of the New York Academy of Sciences* 311: 110–134.
Kreek MJ (1979a) Methadone disposition during the perinatal period in humans. *Pharmacology, Biochemistry, and Behavior* 2(supplement): 1–7.
Kreek MJ (1979b) Methadone in treatment: Physiological and pharmacological issues. In: Dupont RL, Goldstein AJ, and O'Connell J (eds.) *Handbook on Drug Abuse*, pp. 57–86. Washington, DC: National Institute of Drug Abuse-ADAMHA-DEW-ODAP Executive Office of the President.
Kreek MJ (1983) Discussion on clinical perinatal and developmental effects of methadone. In: Cooper JR, Altman F, Brown BS, and Czechowicz D (eds.) *Research in the Treatment of Narcotic Addiction: State of the Art, NIDA Monographs.* vol. pp. 83–1281. pp. 444–453. Rockville, MD: National Institute on Drug Abuse.
Kreek MJ (1987) Multiple drug abuse patterns and medical consequences. In: Meltzer HY (ed.) *Psychopharmacology: The Third Generation of Progress*, pp. 1597–1604. New York: Raven Press.
Kreek MJ (1991) Using methadone effectively: Achieving goals by application of laboratory, clinical, and evaluation research

and by development of innovative programs. *NIDA Research Monograph* 106: 245–266.
Kreek MJ (1992a) Rationale for maintenance pharmacotherapy of opiate dependence. In: O'Brien CP and Jaffe JH (eds.) *Addictive States*, pp. 205–230. New York: Raven Press.
Kreek MJ (1992b) The addict as a patient. In: Lowinson JH, Ruiz P, Millman RB, and Langrod JG (eds.) *Substance Abuse: A Comprehensive Textbook*, pp. 997–1009. Baltimore, MD: Williams & Wilkins.
Kreek MJ (1996a) Long-term pharmacotherapy for opiate (primarily heroin) addiction: Opiate agonists. In: Schuster CR and Kuhar MJ (eds.) *Pharmacological Aspects of Drug Dependence: Toward an Integrated Neurobehavioral Approach*, pp. 487–562. Berlin: Springer.
Kreek MJ (1996b) Opiates, opioids and addiction. *Molecular Psychiatry* 1: 232–254.
Kreek MJ (1996c) Opioid receptors: Some perspectives from early studies of their role in normal physiology, stress responsivity and in specific addictive diseases. *Journal of Neurochemical Research* 21: 1469–1488.
Kreek MJ (1996d) Long-term pharmacotherapy for opiate (primarily heroin) addiction: Opiate antagonists and partial agonists. In: Schuster CR and Kuhar MJ (eds.) *Pharmacological Aspects of Drug Dependence: Toward an Integrated Neurobehavioral Approach*, pp. 563–592. Berlin: Springer.
Kreek MJ (1997) Opiate and cocaine addictions: Challenge for pharmacotherapies. *Pharmacology, Biochemistry, and Behavior* 57: 551–569.
Kreek MJ (2000) Opiates, opioids, SNP's and the addictions: Nathan B. Eddy Memorial Award for lifetime excellence in drug abuse research lecture. *NIDA Research Monograph* 180: 3–22.
Kreek MJ (2001) Drug addictions. Molecular and cellular endpoints. *Annals of the New York Academy of Sciences* 937: 27–49.
Kreek MJ (2005) Impact of bidirectional translational research on treatment of addiction. *Clinical Neuroscience Research* 5: 123–139.
Kreek MJ (2006) Endorphins, gene polymorphisms, stress responsivity and specific addictions: Selected topics. In: Madras BK, Colvis CM, Pollock JD, Rutter JL, Shurtleff D, and von Zastrow M (eds.) *Cell Biology of Addiction*, pp. 63–92. Cold Spring Harbor, NY: Cold Spring Harbor Laboratory Press.
Kreek MJ (2007) Opioids, dopamine, stress and the addictions. *Dialogues in Clinical Neurosciences* 9: 363–378.
Kreek MJ (2008) Role of a functional human gene polymorphism in stress responsivity and addictions. *Clinical Pharmacology and Therapeutics* 83: 615–618.
Kreek MJ, Bart G, Lilly C, LaForge KS, and Nielsen DA (2005) Pharmacogenetics and human molecular genetics of opiate and cocaine addictions and their treatments. *Pharmacological Reviews* 57: 1–26.
Kreek MJ and Culpepper-Morgan J (1991) Neuroendocrine (HPA) and gastrointestinal effects of opiate antagonists: Possible therapeutic application. In: Harris LS (ed.) *Problems of Drug Dependence, 1990: Proceedings of the 52nd Annual Scientific Meeting of the Committee on Problems of Drug Dependence. National Institute of Drug Abuse Research Monograph Series,* vol. 105, pp. 168–174. DHHS Pub. No. (ADM) 91-1753. Washington, DC: Supt. of Docs., US Government Printing Office.
Kreek MJ, Gutjahr CL, Garfield JW, Bowen DV, and Field FH (1976) Drug interactions with methadone. *Annals of the New York Academy of Sciences* 281: 350–371.
Kreek MJ, Hachey DL, and Klein PD (1979) Stereoselective disposition of methadone in man. *Life Sciences* 24: 925–932.
Kreek MJ and Hartman N (1982) Chronic use of opioids and antipsychotic drugs: Side effects, effects on endogenous opioids and toxicity. *Annals of the New York Academy of Sciences* 398: 151–172.
Kreek MJ, Kalisman M, Irwin M, Jaffery NF, and Scheflan M (1980) Biliary secretion of methadone and methadone metabolites in man. *Research Communications in Chemical Pathology and Pharmacology* 29: 67–78.
Kreek MJ and Koob GF (1998) Drug dependence: Stress and dysregulation of brain reward pathways. *Drug and Alcohol Dependence* 51: 23–47.
Kreek MJ and LaForge KS (2007) Stress responsivity, addiction, and a functional variant of the human mu-opioid receptor gene. *Molecular Interventions* 7: 7478.
Kreek MJ, LaForge KS, and Butelman E (2002) Pharmacotherapy of addictions. *Nature Reviews Drug Discovery* 1: 710–726.
Kreek MJ, Nielsen DA, and LaForge KS (2004a) Genes associated with addiction. Alcoholism, opiate and cocaine addiction. *Neuromolecular Medicine* 5: 85–108.
Kreek MJM, Raghunath J, Plevy S, Hamer D, Schneider B, and Hartman N (1984a) ACTH, cortisol and beta-endorphin response to metyrapone testing during chronic methadone maintenance treatment in humans. *Neuropeptides* 5: 277–281.
Kreek MJ, Schaefer RA, Hahn EF, and Fishman J (1983a) Naloxone, a specific opioid antagonist, reverses chronic idiopathic constipation. *Lancet* 5: 261–262.
Kreek MJ, Schluger J, Borg L, Gunduz M, and Ho A (1999) Dynorphin A1-13 causes elevation of serum levels of prolactin through an opioid receptor mechanism in humans: Gender differences and implications for modulation of dopaminergic tone in the treatment of addictions. *Journal of Pharmacology and Experimental Therapeutics* 288: 260–269.
Kreek MJ, Schlussman SD, Bart G, LaForge KS, and Butelman ER (2004b) Evolving perspectives on neurobiological research on the addictions: Celebration of the 30th anniversary of NIDA. *Neuropharmacology* 47: 324–344.
Kreek MJ, Schneider BS, Raghunath J, and Plevy S (1984b) Prolonged (24 hour) infusion of the opioid antagonist naloxone does not significantly alter plasma levels of cortisol and ACTH in humans. *Internatonal Congress Series: Excerpta Medica* 652: 845.
Kreek MJ, Wardlaw SL, Friedman J, Schneider B, and Frantz AG (1981) Effects of chronic exogenous opioid administration on levels of one endogenous opioid (beta-endorphin) in man. In: Simon E and Takagi H (eds.) *Advances in Endogenous and Exogenous Opioids*, pp. 364–366. Tokyo: Kodansha Ltd.
Kreek MJ, Wardlaw SL, and Hartman N (1983b) Circadian rhythms and levels of beta-endorphin, ACTH, and cortisol during chronic methadone maintenance treatment in humans. *Life Sciences* 33: 409–413.
Kroslak T, LaForge KS, Gianotti RJ, Ho A, Nielsen DA, and Kreek MJ (2007) The single nucleotide polymorphism A118G alters functional properties of the human mu-opioid receptor. *Journal of Neurochemistry* 103: 77–87.
LaForge KS, Kreek MJ, Uhl GR, et al. (2000a) Symposium XIII: Allelic polymorphisms of human opioid receptors: Functional studies: Genetic contributions to protection from, or vulnerability to, addictive diseases. *NIDA Research Monographs* 180: 47–50.
LaForge KS, Shick V, Spangler R, et al. (2000b) Detection of single nucleotide polymorphisms of the human mu opioid receptor gene by hybridization or single nucleotide extension on custom oligonucleotide gelpad microchips: Potential in studies of addiction. *Neuropsychiatric Genetics* 96: 604–615.

LaForge KS, Yuferov V, and Kreek MJ (2000c) Opioid receptor and peptide gene polymorphisms: Potential implications for addictions. *European Journal of Pharmacology* 410: 249–268.
Leri F, Zhou Y, Goddard B, Cummins E, and Kreek MJ (2006) Effects of high dose methadone maintenance on cocaine place conditioning, cocaine self-administration, and mu-opioid receptor mRNA expression in the rat brain. *Neuropsychopharmacology* 31: 1462–1474.
Li CH and Chung D (1976) Isolation and structure of an untriakontapeptide with opiate activity from camel pituitary glands. *Proceedings of the National Academy of Sciences of the United States of America* 73: 1145–1148.
Li S, Zhu J, Chen C, Chen YW, Deriel JK, Ashby B, and Liu-Chen LY (1993) Molecular cloning and expression of a rat kappa-opioid receptor. *Biochemical Journal* 295: 629–633.
Martin WR, Jasinski DR, Haertzen CA, Kay DC, Jones BE, Mansky PA, and Carpenter RW (1973) Methadone: A reevaluation. *Archives of General Psychiatry* 28: 286–295.
Mendelson JH and Mello NK (1975) Plasma testosterone levels during chronic heroin use and protracted abstinence. A study of Hong Kong addicts. *Clinical Pharmacology and Therapeutics* 17: 529–533.
Mendelson JH and Mello NK (1984) Hormones and psycho-sexual development in young men following chronic heroin use. *Neurobehavioral Toxicology and Teratology* 4: 441–445.
Mendelson JH, Mello NK, Teoh SK, Lloyd-Jones JG, and Clifford JM (1989) Naloxone suppresses buprenorphine stimulation of plasma prolactin. *Journal of Clinical Psychopharmacology* 9: 105–109.
Mestek A, Hurley JH, Bye LS, et al. (1994) The human mu-opioid receptor: Modulation of functional desensitization by calcium/calmodulin-dependent protein kinase and protein kinase C. *Journal of Neuroscience* 15: 2396–2406.
Minami M, Toya T, Katao Y, et al. (1993) Cloning and expression of a cDNA for the rat kappa-opioid receptor. *FEBS Letters* 329: 291–295.
Morley J, Baranetsky NG, Wingert TD, et al. (1980) Endocrine effects of naloxone-induced opiate receptor blockade. *Journal of Clinical Endocrinology and Metabolism* 50: 251–257.
Mutti A, Folli D, Van der Venne MT, Berlin A, et al. (1992) Long-lasting impairment of neuroendocrine response to psychological stress in heroin addicts. *Neurotoxicology* 13: 255–260.
Naber D, Pickar D, Davis GC, et al. (1981) Naloxone effects on beta-endorphin, cortisol, prolactin, growth hormone, HVA and MHPG in plasma of normal volunteers. *Psychopharmacology (Berl.)* 74: 125–128.
Nakamura K, Hachey DL, Kreek MJ, Irving CS, and Klein PD (1982) Quantitation of methadone enantiomers in humans using stable isotope-labeled zH3,ZH5,2H8 methadone. *Journal of Pharmaceutical Sciences* 71: 39–43.
National Conference Development Panel on Effective Medical Treatment of Opiate Addiction (1998) Effective medical treatment of opiate addiction. *Journal of the American Medical Association* 280: 1936–1943.
Peles E, Kreek MJ, Kellogg S, and Adelson M (2006) High methadone dose significantly reduces cocaine abuse in Methadone Maintenance Treatment (MMT) patients. *Journal of Addictive Diseases* 25: 43–50.
Pelosi MA, Sama JC, Caterini H, and Kaminetzky HA (1974) Galactorrhea–amenorrhea syndrome associated with heroin addiction. *American Journal of Obstetrics and Gynecology* 118: 966–970.
Pert CB and Snyder SH (1973) Opiate receptor: Demonstration in nervous tissue. *Science* 179: 1011–1014.
Pond SM, Kreek MJ, Tong TG, Raghunath J, and Benowitz NL (1985) Altered methadone pharmacokinetics in methadone-maintained pregnant women. *Journal of Pharmacology and Experimental Therapeutics* 233: 1–6.
Proudnikov D, LaForge KS, Hofflich H, et al. (2005) Association analysis of polymorphisms in serotonin 1B receptor (*HTR1B*) gene with heroine addiction: A comparison of molecular and statistically estimated haplotypes. *Pharmacogenetics and Genomics* 16: 25–36.
Ragavan VV, Wardlaw SL, Kreek MJ, and Frantz AG (1983) Effects of chronic naltrexone and methadone administration on brain immunoreactivity betaendorphin in the rat. *Neuroendocrinology* 37: 266–268.
Renault PF, Schuster CR, Heinrich RL, and Van der Kolk B (1972) Altered plasma cortisol response in patients on methadone maintenance. *Clinical Pharmacology and Therapeutics* 13: 269–273.
Rettig RA and Yarmolinsky A (1995) *Federal Regulation of Methadone Treatment*. Washington, DC: National Academy of Sciences, National Academy Press.
Rittmaster RS, Cutler GB, Jr., Sobel DO, Goldstein DS, Koppelman MC, Loriaux DL, and Chrousos GP (1985) Morphine inhibits the pituitary–adrenal response to ovine corticotropin-releasing hormone in normal subjects. *Journal of Clinical Endocrinology and Metabolism* 60: 891–895.
Rolandi E, Marabini A, Franceschini R, Messina V, Bongera P, and Barreca T (1983) Changes in pituitary secretion induced by an agonist-antagonist opioid drug, buprenorphine. *Acta Endocrinologica (Copenhagen)* 104: 257–260.
Romach MK, Otton SV, Somer G, Tyndale RF, and Sellers EM (2000) Cytochrome P450 2D6 and treatment of codeine dependence. *Journal of Clinical Psychopharmacology* 20: 43–45.
Rosen MI, McMahon TJ, Margolin A, et al. (1995) Reliability of sequential naloxone challenge tests. *American Journal of Drug and Alcohol Abuse* 21: 453–467.
Santen RJ (1973) How narcotics addiction affects reproductive function in women. *Contemporary Obstetrics and Gynaecology* 3: 93–96.
Santen RJ, Sofsky J, Bilic N, and Lippert R (1975) Mechanism of action of narcotics in the production of menstrual dysfunction in women. *Fertility and Sterility* 26: 538–548.
Schluger J, Bodner G, Gunduz M, Ho A, and Kreek MJ (1998a) Abnormal metyrapone tests during cocaine abstinence. *NIDA Research Monograph* 178: 105.
Schluger J, Ho A, Borg L, et al. (1998b) Nalmefene causes greater hypothalamic–pituitary–adrenal axis activation than naloxone in normal volunteers: Implications for the treatment of alcoholism. *Alcoholism: Clinical and Experimental Research* 22: 1430–1436.
Schluger JH, Bart G, Green M, Ho A, and Kreek MJ (2003) Corticotropin-releasing factor testing reveals a dose-dependent difference in methadone maintained vs. control subjects. *Neuropsychopharmacology* 28: 985–994.
Schluger JH, Borg L, Ho A, and Kreek MJ (2001) Altered HPA axis responsivity to metyrapone testing in methadone maintained former heroin addicts with ongoing cocaine addiction. *Neuropsychopharmacology* 24: 568–575.
Schottenfeld RS, Chawarski MC, Pakes JR, Pantalon MV, Carroll KM, and Kosten TR (2005) Methadone versus buprenorphine with contingency management or performance feedback for cocaine and opioid dependence. *American Journal of Psychiatry* 162: 340–349.
Simon EJ, Hiller JM, and Edelman I (1973) Stereospecific binding of the potent narcotic analgesic 3H Etorphine to rat-brain homogenate. *Proceedings of the National Academy of Sciences of the United States of America* 70: 1947–1949.

Spagnolli W, DeVenuto G, Mattarei M, DalRi P, and Miori R (1987) Prolactin and thyrotropin pituitary response to thyrotropin releasing hormone in young female heroin addicts. *Drug and Alcohol Dependence* 20: 247–254.

Stimmel B and Kreek MJ (1975a) Pharmacologic actions of heroin. In: Stimmel B (ed.) *Heroin Dependency: Medical, Economic and Social Aspects*, pp. 71–87. New York: Stratton Intercontinental Medical Book Corporation.

Stimmel B and Kreek MJ (1975b) Dependence, tolerance and withdrawal. In: Stimmel B (ed.) *Heroin Dependency: Medical, Economic and Social Aspects*, pp. 88–97. New York: Stratton Intercontinental Medical Book Corporation.

Stine SM, Burns B, and Kosten T (1991) Methadone dose for cocaine abuse. *American Journal of Psychiatry* 148: 1268–1273.

Strain EC, Stitzer ML, Liebson IA, and Bigelow GE (1994) Buprenorphine versus methadone in the treatment of opioid-dependent cocaine users. *Psychopharmacology* 116: 401–406.

Terenius L (1973) Stereospecific interaction between narcotic analgesics and a synaptic plasma membrane fraction of rat cerebral cortex. *Acta Pharmacologica et Toxicologica* 32: 317–320.

Tolis G, Hickey J, and Guyda H (1975) Effects of morphine on serum growth hormone, cortisol, prolactin and thyroid stimulating hormone in man. *Journal of Clinical Endocrinology and Metabolism* 41: 797–800.

Tsuang MT, Lyons MJ, Meyer JM, et al. (1998) Co-occurrence of abuse of different drugs in men. *Archives of General Psychiatry* 55: 967–972.

Tyndale RF, Droll KP, and Sellers EM (1984) Genetically deficient CYP2D6 metabolism proves protection against oral opiate dependence. *Pharmacogenetics* 7: 375–379.

Unterwald EM, Horne-King J, and Kreek MJ (1992) Chronic cocaine alters brain mu-opioid receptors. *Brain Research* 584: 314–318.

Unterwald EM, Kreek MJ, and Cuntapay M (2001) The frequency of cocaine administration impacts cocaine-induced receptor alterations. *Brain Research* 900: 103–109.

Unterwald EM, Rubenfeld JM, and Kreek MJ (1994) Repeated cocaine administration upregulates mu and kappa, but not delta, opioid receptors. *NeuroReport* 5: 1613–1616.

Vescovi PP, Gerra G, Maninetti L, et al. (1990) Metyrapone effects on beta-endorphin, ACTH and cortisol levels after chronic opiate receptor stimulation in man. *Neuropeptides* 15: 129–132.

Vescovi PP, Pedrazzoni M, Gerra G, Pioli G, Maninetti L, Michelinni M, and Passeri M (1989) Impaired ACTH and beta-endorphin response to sauna-induced hyperthermia in heroin addicts. *Acta Endocrinologica (Copenhagen)* 121: 484–488.

Volavka J, Bauman J, Pevnick J, Reker D, James B, and Cho D (1980) Short-term hormonal effects of naloxone in man. *Psychoneuroendocrinology* 5: 225–234.

Volavka J, Cho D, Mallya A, and Bauman J (1979) Naloxone increases ACTH and cortisol in man. *New England Journal of Medicine* 300: 1056–1057.

Wand GS, McCaul M, Yang X, Reynolds J, Gotjen D, Lee S, and Ali A (2002) The mu-opioid receptor gene polymorphism (A118G) alters HPA axis activation induced by opioid receptor blockade. *Neuropsychopharmacology* 26: 106–114.

Wang JB, Imai Y, Eppler CM, Gregor P, Spivak CE, and Uhl GR (1994a) Mu-opiate receptor: cDNA cloning and expression. *Proceedings of the National Academy of Sciences of the United States of America* 90: 10230–10234.

Wang JB, Johnson PS, Persico AM, Hawkins AL, Griffin CA, and Uhl GR (1994b) Human mu opiate receptor: cDNA and genomic clones, pharmacologic characterization and chromosomal assignment. *FEBS Letters* 338: 217–222.

Wang XM, Zhou Y, Spangler R, Ho A, Han JS, and Kreek MJ (1999) Acute intermittent morphine increases preprodynorphin and kappa opioid receptor mRNA levels in the rat brain. *Molecular Brain Research* 66: 184–187.

Yasuda K, Raynor K, Kong H, Breder CD, Takeda J, Reisine T, and Bell GI (1993) Cloning and functional comparison of kappa and delta opioid receptors from mouse brain. *Proceedings of the National Academy of Sciences of the United States of America* 90: 6736–6740.

Yuferov V, Fussell D, LaForge KS, et al. (2004) Redefinition of the human kappa opioid receptor gene (*OPRK1*) structure and association of haplotypes with opiate addiction. *Pharmacogenetics* 14: 793–804.

Yuferov V, Zhou Y, Spangler R, Maggos CE, Ho A, and Kreek MJ (1999) Acute binge cocaine increases mu-opioid receptor mRNA levels in areas of the rat mesolimbic mesocortical dopamine system. *Brain Research Bulletin* 48: 109–112.

Zawertailo LA, Kaplan HL, Busto UE, Tyndale RF, and Seller EM (1998) Psychotropic effects of dextromethorphan are altered by the CYP2D6 polymorphism: A pilot study. *Journal of Clinical Psychopharmacology* 18: 332–337

Zhou Y, Bendor J, Hofmann L, Randesi M, Ho A, and Kreek MJ (2006) Mu opioid receptor and orexin/hypocretin mRNA levels in the lateral hypothalamus and striatum are enhanced by morphine withdrawal. *Journal of Endocrinology* 191: 137–145.

Zhou Y, Leri F, Cummins E, Hoeschele M, and Kreek MJ (2008) Involvement of arginine vasopressin and V1b receptor in heroin withdrawal and heroin seeking precipitated by stress and by heroin. *Neuropsychopharmacology* 33: 226–236.

Zhou Y, Spangler R, LaForge KS, Maggos CE, Ho A, and Kreek MJ (1996) Steady-state methadone in rats does not change mRNA levels of corticotropin-releasing factor, its pituitary receptor or proopiomelancortin. *European Journal of Pharmacology* 315: 31–35.

Zhou Y, Spangler R, Maggos CE, Wang XM, Han JS, Ho A, and Kreek MJ (1999) Hypothalamic–pituitary–adrenal activity and pro-opiomelanocortin mRNA levels in the hypothalamus and pituitary of the rat are differentially modulated by acute intermittent morphine with or without water restriction stress. *Journal of Endocrinology* 163: 261–267.

Zhu J, Chen C, Xue JC, Kunapuli S, DeRiel JK, and Liu-Chen LY (1995) Cloning of a human kappa-opioid receptor from the brain. *Life Sciences* 56: PL201–PL207.

Zis AP, Haskett RF, Albala AA, and Carroll BJ (1984) Morphine inhibits cortisol and stimulates prolactin secretion in man. *Psychoneuroendocrinology* 9: 423–427.

Zubieta JK, Gorelick DA, Stauffer R, Ravert HT, Dannals RF, and Frost JJ (1996) Increased mu opioid receptor binding detected by PET in cocaine-dependent men is associated with cocaine craving. *Nature Medicine* 2: 1225–1229.

36 Pain: Sex/Gender Differences

A Z Murphy, Georgia State University, Atlanta, GA, USA
K J Berkley, Florida State University, Tallahassee, FL, USA
A Holdcroft, Imperial College London, London, UK

Chapter Outline

36.1 Overview

The influence of sex has been neglected in basic and clinical studies on pain, with the vast majority of research conducted exclusively in males. However, it is becoming increasingly clear that males and females differ in both the anatomical and physiological composition of the central and peripheral nervous systems. These differences influence not only our response to noxious stimuli, but also the ability of pharmacological agents to modify this response. Recently, the Sex, Gender and Pain Special Interest Group of the International Association for the Study of Pain (IASP) issued a consensus paper highlighting the need for inclusion of both males and females in basic and clinical studies on pain (Greenspan et al., 2007). This multidisciplinary consensus was triggered by the need for application of basic science to clinical problems to continue in advancing our understanding of how one's biological sex influences potential pain mechanisms and therapeutic strategies. A large component of this accumulating understanding is recognition of the enormous importance of individual differences in pain and how they develop across the life span (LeResche, 2001). Thus, significant effort is underway among pain researchers and clinicians to understand how mechanisms of pain might differ between the young and the aged, females and males, those living in different cultures, those suffering from visceral versus somatic pain, those suffering from acute, episodic, or chronic pain, those with different types of pathophysiology (e.g., nociceptive, inflammatory, and neuropathic), and variations associated with pain in different settings (e.g., workplace, clinic, and home).

This chapter focuses on one of these large categories: sex and gender. At the outset, it is important to note two things. First, the terms sex and gender are not interchangeable and are inconsistently if at all defined, creating problems in the field (Wizemann and Pardue, 2001). In the biological continuum of levels – that is, genetic, molecular, hormonal, physiological, psychosocial, and sociocultural – sex difference is meant to signify genetic-to-physiological differences, whereas gender difference is usually meant to signify psychosocial–sociocultural differences. Here, we define gender as the female or male sex with which an individual identifies. Using this common definition, it is evident that most studies, especially those on nonhuman animals, are assessing sex differences. Second, due to the subject matter of this work, substantive attention is paid in the following to the influence of hormonal factors. Importantly, however, hormonal differences between the sexes constitute only one among the many potential interacting factors that can contribute to sex/gender differences in pain mechanisms and responses to therapy as these differences evolve across the life span of an individual.

36.2 Pain: A Summary

36.2.1 What Is Pain?

This question is not easy to answer, as was debated by a multidisciplinary group of pain experts, who settled upon a long-winded definition: "Pain is an unpleasant sensory and emotional experience associated with actual or potential tissue damage, or described in terms of such damage" (Merskey and Bogduk, 1994). The reason for all the qualifiers in this definition was, and still is, to ensure that we maintain a distinction between a stimulus and a perception. In other words, a noxious or injurious or potentially injurious stimulus is not necessarily a painful one, and an innocuous stimulus can be painful. The assumption of damage is now frequently challenged because nociceptive effects can be demonstrated without tissue damage and there is often discordance between physical findings and a patient's verbal report of pain sensations (Luyten and van Houdenhove, 2005). In humans, affective responses differ between males and females (Sela et al., 2002) and this may change pain language or visual constructs in communication leading to diagnostic and therapeutic variations between male and female physicians and patients. The definition also emphasizes the motivating aspects of the experience of pain, which implies that the organism must do something about the situation, and Wall (1999) has called that pain's action plan.

36.2.2 How Is Pain Classified?

Pain has been classified along the dimensions of time, location, and etiology (i.e., type of pathology). One aspect of time is duration, whether the pain is recent (or acute), episodic persistent, or chronic (defined arbitrarily as being experienced for more than 6 months; Merskey and Bogduk, 1994). Another aspect of time is chronobiology; that is, whether the pain occurs, or is greater, at a certain time of day, week, month (menstrual), season, or portion of the life span (i.e., prenatal, neonatal, child, adolescent, adult, elderly, and very elderly). Pain is also classified as to its presumed general bodily location, that is, whether it is visceral, musculoskeletal, or superficial. Another aspect of location includes specific anatomy, such as arm, leg, shoulder, chest, heart, bladder, etc. Finally, pain can be classified as to the presumed source of the pathophysiology that produces it (e.g., traumatic, inflammatory, degenerative, and ischemic) and whether or not this source is within the nervous system. Nociceptive refers to pathophysiology of peripheral non-neural tissue, while neuropathic refers to pathophysiology of some portion of the nervous system, either peripheral nerves (e.g., diabetic neuropathy) or the central nervous system (e.g., stroke). This need for such a classification system has been recognized for some time (Bonica, 1979), and revised for chronic pain (Merskey and Bogduk, 1994). However, different strategies are continually being proposed (Jänig and Baron, 2006; Woolf and Ma, 2007; Boulton, 2007).

36.2.3 How Is Pain Measured?

Many tests have been used to measure pain in nonhuman animals and humans. Sex/gender differences between pain reports may be explained by both psychosocial factors and first-order, biological differences. For example, in experimental pain where pain tolerance or thresholds are measured, gender role expectation was a predictor for scores (Wise et al., 2002; Robinson and Wise, 2004); and the interpretation of maximal endpoints of numerical pain scales, such as the visual analog scale, exhibits gender variations (Robinson et al., 2004). Recent research has suggested improvements based on factors, such as pain history, that a person uses to conceptualize the pain scale anchors (Dannecker et al., 2007).

Some of these measures are listed in **Tables 1a–1c** and have recently been reviewed with respect to their value in determining sex/gender differences (Greenspan et al., 2007). The plethora of strategies contributes to seeming inconsistencies in the literature because a particular pain measurement method can have a significant impact on conclusions from both experimental and clinical studies. For example, if, in two experimental studies, one study measured differences in pain thresholds between different groups, while the other study, using the same subjects and stimuli, measured differences in pain tolerance, the conclusions with respect to differences in pain sensation between the two studies could easily be different. Similarly, conclusions derived from two clinical studies assessing the pain-relieving effect of some drug could easily differ by sex/gender if the outcome measure in one study was a comparison of how the new drug reduced usage of another pain-relieving drug (e.g., morphine) but the outcome measure in the other study was a direct assessment of reduction in pain intensity scores. Another confounding factor in many clinical studies is that baseline pain intensity may differ between men and women.

Table 1a Examples of tests to measure pain in animals

Animal tests – Experimental test type [measure] (stimulus)

Simple reflexes
- Tail-flick reflex [latency and electrophysiology] (heat or pressure to tail)
- Limb-withdrawal reflex [latency] (mechanical, thermal, or electrical stimulation of skin)
- Jaw-opening reflex [electromyography of jaw muscles] (electrical stimulation of tooth)

Organized unlearned behaviors
- Hot-plate test [paw licking] (heat to paws)
- Guarding of injured structure [scale] (thermal, mechanical, chemical, electrical stimulation of skin or muscles, and nerve injury)
- Spontaneous pain behaviors [frequency] (various types of models of human conditions; e.g., kidney stone, inflammation, and nerve injury)
- Passive avoidance behavior [probability] (visceral stimulation)

Organized learned behaviors
- Escape response-trained [probability] (various types of stimuli to skin, muscles, and viscera)
- Reaction time [latency] (various stimuli to skin and viscera)
- Conflict paradigm [probability of choice-reward vs. aversive stimulus] (various stimuli to skin)

Table 1b Human tests – experimental and clinical (using verbal reporting)

Strategies
- Method of limits
- Method of adjustment
- Method of constant stimuli
- Cross-modality matching
- Rating intensity
- Forced choice

Scales
- Visual-analog scale (VAS)
- Verbal rating scale (VRS)
- Sensory-decision theory
- Magnitude estimation
- Rank order (e.g., McGill Pain Questionnaire (MPQ))

Measures
- Threshold
- Tolerance
- Sensory intensity
- Sensory unpleasantness
- Verbal descriptors (e.g., MPQ)

Table 1c Human tests – nonverbal reporting

Adults
- Physiological measures (e.g., heart rate, sweating, respiration, blood pressure, insulin, adrenalin, and hypothalamic–pituitary–adrenal hormones)
- Reflexes
- Microneurography
- Electroencephalography
- Magnetic methods (e.g., somatosensory-evoked magnetic fields)
- Brain imaging (e.g., positron emission tomography and functional magnetic resonance imaging)

Young age
- Physiological measures (heart rate, sweating, respiration, blood pressure, etc.)
- Children's pain checklist (combination of vocal, eating, sleeping, social, facial expression, activity, physiological – shivering, pallor, sweating, gasping, and breath holding)
- Grimacing patterns
- VAS faces scale

VAS, visual analog scale.

Although it is beyond the scope of this chapter, it should at least be noted here that some aspects of pain measurement involve the translation process between basic science and the clinic (Holdcroft, 2006). This issue of how information gained from basic research on either nonhuman animals or healthy humans can be applied clinically to women and men in pain is a complex one and includes the controversial problem of what constitutes an appropriate model for a given disease.

36.2.4 What Are the Mechanisms of Pain?

There are two distinct lines of thought on how the nervous system produces pain: the fascicular view and the ensemble view. The more traditional fascicular view was inspired by the gate control theory, which stated that information arriving at the spinal cord from peripheral somatosensory receptors could be modulated and modified by converging inputs from other peripheral afferents, from other spinal segments, and/or from fibers descending from the brain. This view specifies that there exists a complex pathway from the spinal cord through various regions in the brainstem to cortex, whose neurons process nociceptive stimuli and whose function is the creation of pain. In addition, there are descending influences from several regions in the brain that act to either inhibit or facilitate the activity of neurons in the spinal cord, thereby modulating nociception at its source.

In contrast to the fascicular view, the ensemble view incorporates concepts of ensemble, distributed, and dynamic information processing now common in conceptualizing other neural systems (e.g., Deadwyler and Hampson, 1997). This view specifies that all bodily perceptions, including pain, arise not out of the activity in a single pathway of information flow, but rather out of a cooperatively controlled balance of the information flowing through multiple systems (Berkley and Hubscher, 1995; Melzack, 1989; Price, 2000). In other words, pain (and other bodily) perceptions occur as a consequence of distributed ensembles of central neuronal activity that result from a balance of cooperatively controlled information about bodily stimuli (in the skin, muscles, viscera etc.) flowing through all three entry ports (i.e., spinal cord, dorsal column nuclei, and solitary nucleus) into an active and dynamic brain. This shifts the creation of pain as being determined less by the nature of the incoming stimulus than by the nature of the brain activity into which incoming stimulus information finds itself arriving.

36.2.5 How Is Pain Managed?

There are three broad categories of pain therapies – drugs, physical interventions, and situational adjustments – some of which are shown in **Table 2**. Despite this vast array, pain remains a huge clinical problem, particularly for those who suffer pain long term. Part of the reason may lie in how therapies are used. If single therapies are applied serially, for example, if acetaminophen does not help, try a cyclo-oxygenase inhibitor; if neither helps, try counselling; the effect is unlikely to be as effective as using them in parallel because pain has multiple mechanisms and using a combination of therapies, for example, massage plus morphine plus cognitive therapy, is more logical (Holdcroft and Berkley, 2006).

36.3 Sex Differences in Pain

36.3.1 Pain, Epidemiology, and Sex/Gender Differences

Epidemiological studies consistently show that women report more frequent pain, more severe levels of pain, pain in more areas of the body, and pain of longer duration than men (von Korff et al., 1988; Unruh, 1996; Berkley, 1997a; Bingefors and Isacson, 2004). In addition, as shown in **Table 3**, there are many more painful diseases that have a documented female prevalence, particularly those of the head and neck, of musculoskeletal or visceral origin, and of autoimmune etiology. Some authors have attributed

Table 2 Pain management modalities for multidisciplinary clinics and hospices

Drugs	*Physical interventions*	*Situational approaches*
Primary analgesics	*Simple*	*Patient generated (lifestyle)*
Acetaminophen (paracetamol)	Exercise	Aromatherapy
NSAIDs	Heat/cold	Art, music, poetry, performing arts, etc.
Opioids	Massage	Diet (nutrition)
	Relaxation	Education
Other analgesics	Vibration	Humor
α_2-Adrenergic agonists		Meditation
β-Adrenergic antagonists	*Minimally invasive*	Pets
Antidepressants	Acupuncture	Religion
Anticonvulsants	Manipulation	Sports, gardening, hobbies, etc.
Antiarrhythmics	Physical therapy	
Channel blockers (e.g., Na, Ca)	TENS	*Clinician*
Cannabinoids	Traction	Attitude
Capsaicin	Ultrasound	Clinical setting and arrangement
Corticosteroids		Education
COX-2 inhibitors	*Invasive*	
$GABA_B$ agonists	Brain lesions	*Interactive/structured*
Serotonin (5HT) agonists	Brain stimulation	Advocacy groups
Steroid hormones	Commissural myelotomy	Behavioral therapy
	Cordotomy	Biofeedback
Symptom management	Dorsal column stimulation	Cognitive therapy
Antihistamines	DREZ lesions	Family counseling
Laxatives	Limited myelotomy	Group therapy
Neuroleptics	Local anesthetic infiltration	Hypnosis
Phenothiazines	Local ganglion blocks	Job counseling
	Nerve blocks	Networking
	Neurectomy	Psychotherapy
	Punctate midline myelotomy	Self-help groups
	Radiation therapy	Support groups
	Rhizotomy	
	Sympathectomy	

NSAIDs, nonsteroidal anti-inflammatory drugs; COX, cyclo-oxygenase inhibitors; GABA, gamma aminobutyric acid; HT, hydroxytryptamine; TENS, transcutaneous electrical nerve stimulation; DREZ, dorsal root entry zone.

this female vulnerablity to an increased willingness for women to report pain and seek healthcare. While seeking healthcare may be a factor, it is only one of many. Much of the female prevalence, for example, can be accounted for by fecundity or gynecological problems. In addition, overall prevalence patterns in both sexes for many types of pain (such as those of temporomandibular disorder, fibromyalgia, migraine, chest, abdomen, and joint) change across the life span (LeResche, 1999, 2001; Cairns, 2007). Thus, the sex prevalence of some painful disorders diminishes or reverses with age. Further complicating the issue is that in some disorders, such as irritable bowel syndrome (IBS), migraine headaches, rheumatoid arthritis, and coronary heart disease, the clinical signs differ between the sexes and vary with reproductive status (Giamberardino et al., 1997, 2000; Jawaheer et al., 2006; Cucurachi et al., 2006; Mayer et al., 2004; Canto et al., 2007; Naliboff, 2007). In other contexts, for example, life-threatening disease, sex differences in pain reports may disappear (Turk and Okifuji, 1999; Robinson et al., 2000). Thus, while, overall, women are more vulnerable than men to acute, intermittent, and chronic pain, an individual's pain report may not reflect the more general population findings.

36.3.2 Pain, Nociception, and Sex/Gender Differences

Remarkably, there is still no consensus as to whether males and females differ in their sensitivity to noxious stimuli. In basic science studies of healthy males and females, somatic pain thresholds and tolerance across various stimulus modalities, including heat, pressure, and chemical irritants, are generally lower in women than men (Chesterton et al., 2003; Craft, 2007; Frot et al., 2004; Sarlani et al., 2003). However,

Table 3 Painful syndromes demonstrating sex prevalence

Female prevalence	*Male prevalence*
Head and neck	
Atypical odontalgia	Cluster headache
Burning tongue	Migraine without aura
Carotidynia	Paratrigeminal syndrome
Cervicogenic headache	Post-traumatic headache
Chronic paroxysmal hemicrania	
Chronic tension headache	
Migraine headache with aura	
Occipital neuralgia	
Postdural puncture headache	
Temporal arteritis	
Temporomandibular disorder	
Tic douloureux	
Limbs	
Carpal tunnel syndrome	Brachial plexus neuropathy
Chilblains	Hemophilic arthropathy
Chronic venous insufficiency	Thromboangiitis obliterans
Diabetic neuropathy	
Peroneal muscular atrophy	
Piriformis syndrome	
Raynaud's disease	
Reflex sympathetic dystrophy	
Internal organs	
Chronic constipation	Duodenal ulcer
Esophagitis	Pancoast tumor[a]
Gallbladder disease	Pancreatic disease[a]
Irritable bowel syndrome	
Interstitial cystitis	
Proctalgia fugax	
General	
Acute intermittent porphyria	Postherpetic neuralgia
Diabetic neuropathy	
Fibromyalgia syndrome	
Lupus erythematosis	
Multiple sclerosis	
Rheumatoid arthritis	

[a]Frequency changes with disease pattern alterations following population changes in lifestyle factors such as alcohol/smoking.

reports to the contrary are also prevalent, and it is becoming increasingly clear that psychological and sociocultural factors act interdependently with physiological, hormonal, and genetic differences to influence pain responding (Myers et al., 2003). In basic science studies utilizing rodents, similar inconsistencies in the literature are observed. In our own studies using Sprague-Dawley rats, we observe no sex differences in withdrawal latencies in response to a noxious thermal or mechanical stimulus, or in the hyperalgesia produced by intraplantar administration of complete Freund's adjuvant (Loyd and Murphy, 2006; Wang et al., 2006). Interestingly, while no sex differences are observed in terms of withdrawal latencies, we do indeed see profound sex differences in how information regarding a nociceptive stimulus is relayed centrally. Sex differences in the central relay of nociceptive information have been observed for both somatic (Loyd and Murphy, 2006; Loyd et al., 2007) and visceral (Suckow et al., 2005) models of nociception, and are discussed in more detail below.

36.3.3 Pain Therapies and Sex/Gender Differences

An important aspect of sex differences in the prevalence patterns of painful conditions is that the sex differences are accompanied by differences in how women and men alleviate their pains through drugs, physical activities, and/or behavioral changes. Population studies have identified that women use more medications (including hormones) and seek more medical consultation than men (Isacson and Bingefors, 2002; Roe et al., 2002). Outcome studies for interventions may be few and lack clinical trial methodological rigor but compensate for this by the size of the population and statistical control of confounding variables. In musculoskeletal pain management standardized across a cohort of 1827 patients (63% men), after a year women had more depression than men and a poorer socioeconomic outcome (McGreary et al., 2003).

Thus, not only are women more likely than men to combine different types of therapies, but also women appear to be able to derive more benefit from this self-motivated poly-therapeutic strategy (France et al., 2004; Myers et al., 2003). Ironically, therefore, the multiple physical, psychosocial, and sociocultural factors that might contribute to more pain in women than men may at the same time contribute to their engagement of a wider array of strategies for coping with pain more effectively (Unruh et al., 1999; Berkley, 1997b; Keefe et al., 2004).

36.4 Pain Mechanisms and Sex/Gender Differences

It seems self-evident that many factors likely contribute to sex differences in pain summarized above, and, importantly, that they do not act independently. As diagramed in **Figure 1**, some of these interacting factors include differences in genetics, organ physiology, body structure, neuroactive agents, sex steroid hormones, central nervous system

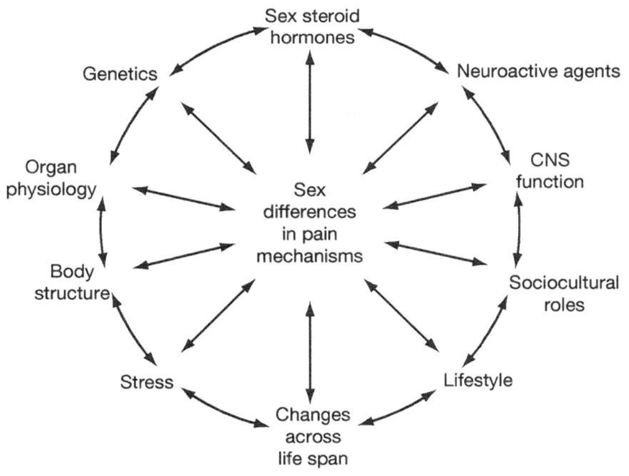

Figure 1 Some of the factors that contribute to the development of sex differences in pain. Note that these factors exert their influence independently as well as by their interactions with each other. CNS, central nervous system. Reproduced from Berkley KJ, Hoffman G, Murphy AZ, and Holdcroft A (2002) Pain: Sex/gender differences. In: Pfaff DW, Arnold AP, Fahrbach SE, Etgen AM, and Rubin RT (eds.) *Hormones, Brain and Behavior*. Vol. 5, pp. 409–442. San Diego, CA: Academic Press, with permission from Elsevier.

function, stress, lifestyle, sociocultural roles, and changes across the life span.

36.4.1 Genetics

John Liebeskind presented to the National Academy of Sciences the challenge that "the study of the genetic differences in pain-related traits has been largely neglected" (Mogil et al., 1996). This neglect is being rapidly remedied, as evidenced by several recent reviews of this issue (Diatchenko et al., 2007; Dib-Hajj et al., 2007; Stamer and Stuber, 2007). There are at least four means by which pain-related traits could manifest themselves in sex-specific ways. First are sex-linked genetic diseases that underlie some of the pain syndromes listed in **Table 3**, such as hemophilia, porphyria, and the X-linked recessive type of perineal muscular atrophy (Charcot–Marie–Tooth). Second are sex-related genetic variations of metabolizing enzyme systems that would contribute to differences in both pain mechanisms and response to therapies. An example includes the cytochrome P450 (CYP) enzyme families, the major enzyme system involved in the metabolism of numerous drugs, in which genetic variations in the CYP2 family evidence themselves differently in females and males (Ciccone and Holdcroft, 1999; Sioud and Melien, 2007). Third are sex-linked genetic differences in both nociception and responses to exogenous and endogenous analgesics manipulations discovered by several investigators in mice and rats. For example, using quantitative trait locus (QTL) mapping techniques, Mogil et al. (1997) localized a male-specific QTL on chromosome 4 that appears to account for variability between two strains of mice in delta-opioid-associated nociception as assessed by the hot-plate assay method. This same group also localized, on chromosome 8, a female-specific QTL that accounts for variability in stress-induced analgesia (SIA; Mogil et al., 1997). They concluded that it may be part of the basis for a female-specific SIA mechanism in rats that is ontogenetically organized, nonopioid, varies with reproductive status, and is estrogen dependent. Fourth are steroidally mediated transcriptional changes in neurotransmitter expression and release (Herbison, 1997; Herbison and Fenelon, 1995; McCarthy et al., 1995; McQueen et al., 1999). For example, the expression of preproenkephalin (Lauber et al., 1990; Priest et al., 1995), gamma-aminobutyric acid (GABA) (Blurton-Jones and Tuszynski, 2006; McCarthy et al., 1995; Nakamura et al., 2004), cholecystokinin (Amandusson et al., 1999; Frankfurt et al., 1986; Sinchak et al., 2000), and μ-opioid receptor (Loyd et al., 2007) is all influenced by changes in gonadal steroids, and in many instances, steroid influence is exerted at multiple sites along a functional pathway (Popper et al., 1995) (**Table 4**).

36.4.2 Body Physiology and Structure

36.4.2.1 Physiology: General

Relative to men, women on average have a higher percentage of body fat, smaller muscle mass, lower blood pressure, and fluctuations associated with hormone status in gastrointestinal transit time, gastric acid secretion, urinary creatinine clearance, metabolism, and thermoregulation. Their pelvic organs also differ significantly. These differences have huge implications not only for sex prevalence differences in various pain conditions (e.g., musculoskeletal), but also, as discussed below, for the pharmacodynamics and pharmacokinetics of analgesics, anesthetics, and adjuvants used for pain management (Beierle et al., 1999; Ciccone and Holdcroft, 1999; Schwartz, 2003).

36.4.2.2 Physiology: Cardiovascular system as an example

How physiological factors are influenced by hormones is also important. Let us consider blood pressure control because it is closely related to the autonomic nervous and adrenocortical systems that respond to stress. Blood pressure is inversely related

Table 4 Major neuroactive chemicals and receptor sites that are influenced by reproductive status or hormone manipulation in females or show sex differences in brain or spinal cord distribution and have drugs either clinically available or are in development that act in association with them

Neurochemical/receptor	*Clinical drug (potential drug)*
Adrenocorticotropic hormone (ACTH)	Cortisol
Alpha-melanocyte-stimulating hormone	(Melanocortin receptor antagonist)
Angiotensin	Clonidine
Bradykinin	(Bradykinin antagonists)
Cannabinoids	Cannabinoids
Cholecystokinin (CCK)	(CCK antagonist)
Corticotropin-releasing hormone	Cortisol
Dopamine	(Dopamine agonist)
Enkephalins	Opioids
GABA	Anticonvulsants, Baclofen
Glutamate	Ketamine (NMDA receptor antagonists)
Growth hormone	(combined with calcitonin)
N-methyl-D-aspartate (NMDA)	Ketamine (NMDA glycine antagonists)
Nerve growth factor	(NGF antagonists)
Nitric oxide	Nitric oxide – NSAIDs
Pre-proenkephalin	Opioids
Serotonin (5-hydroxytryptamine [HT])	5-HT antagonists, antidepressants
Somatostatin	Somatostatin analogs

NSAIDs, nonsteroidal anti-inflammatory drugs.

to pain sensitivity in hypertensive patients with coronary artery disease (Falcone et al., 1997). However, as is common in volunteer studies, males have been studied more than females, and healthy, stressful, or pathological states may elicit different results for men and women. Thus, Fillingim and Maixner (1996) showed that blood pressure is inversely related to pain sensitivity in normotensive subjects in males but not females, and that the higher pain ratings of thermal and ischemic stimuli that had been made by the women in their study were completely accounted for by the women's lower blood pressure. On the other hand, blood pressure was correlated with the unpleasantness but not the intensity of painful stimuli in women (Fillingim et al., 1998). In other subject groups, hypoalgesia has been reported for young men but not for young women at risk of hypertension (al'Absi et al., 1999). These patterns of association between pain perception and blood pressure have led to further volunteer studies measuring cortisol concentrations during experimental pain with and without additional stress. The blood pressure changes in response to stress mediated the stress-induced reduction in pain perception (al'Absi and Petersen, 2003). The proposed mechanisms for these effects include baroreceptor stimulation, cardiovascular reactivity, and endogenous opioid effects. To test the latter, an appropriately designed study with an opioid antagonist demonstrated that hypoalgesia may identify men at risk of hypertension but not women (al'Absi et al., 2006). Further work is planned in rodents and humans to determine if estradiol contributes to the within-female variations (e.g., Butkevich et al., 2000; Saleh and Connell, 2007).

36.4.3 Pelvic Organs

Reproductive organs differ between the sexes, thereby affecting the relative arrangements of other pelvic organs. For example, female urethras are shorter and less tortuous than those of males. These differences create sex-specific pains of pelvic origin (Wesselmann and Burnett, 1999). In addition, as argued by Berkley (1997a), sex differences in the reproductive tract create, in females, a greater vulnerability to both local and remote central sensitization. This situation could partly explain the greater vulnerability in women for multiple referred pains, particularly in muscles and head/neck shown in **Table 3**, and may be a factor associated with the coexistence of several chronic pain syndromes, such as IBS interstitial cystitis, and fibromyalgia (Berkley, 2000; Naliboff, 2007). It might also be part of the underlying basis for sex differences in clinical signs of some diseases, for example, the more widespread neck and shoulder pains in women with coronary heart disease (Patel et al., 2004). Support for this argument lies in considering a number of mechanisms:

1. Afferent innervation of internal pelvic organs is extensive and mainly by C-fibers, many of which are activated specifically by noxious stimuli.
2. Dorsal root axons of C-fibers diverge as they enter the spinal cord, giving rise to long-ranging axonal branches that synapse with dorsal horn neurons along many remote segments above and below their level of entry, including a large component at upper cervical levels.
3. C-fibers are activated and sensitized by trauma, injury, or disease. This peripheral sensitization can persist long after the organ damage abates, thereby helping to maintain central sensitization

of dorsal horn neurons and consequent referred hyperalgesia.

4. A greater proportion of the reproductive tract is internal in females than in males (e.g., vaginal canal and cervix vs. penis).
5. Internal reproductive organs in female are more frequently subject to trauma, injury, and disease than those in men (e.g., vagina – tampons, intercourse, examination, parturition; uterus – periodic strong contractions, pregnancy, and parturition).
6. Uterine disorders give rise to widespread hyperalgesia much greater in muscles than in skin (Giamberardino et al., 1997).
7. Clinical co-morbidities and pain history (Arslanian-Engoren et al., 2006).

36.4.4 Brain Function

While sex differences in pain sensitivity are far from consistent, observations from both animals and human studies suggest that the central and peripheral mechanisms underlying nociceptive processing differ between males and females. Although previous studies on brain function and nociception traditionally included males only, that situation is changing, and it is becoming increasingly clear that there are profound sex differences in both the anatomical and physiological organization of the brain and spinal cord that may contribute to the sex-based differences in pain and analgesia. For example, as discussed below, the midbrain periaqueductal gray (PAG), and its descending projections to the rostral ventromedial medulla (RVM) and spinal cord, has been shown to have a distinct organization in males and females, both in terms of the anatomy as well as in the functional activation by pain (Loyd and Murphy, 2006; Loyd et al., 2007). Similar sex differences have been reported for the spinoparabrachial pathway, which may contribute to the sexually dimorphic incidence rate of various visceral pain syndromes, including IBS (Suckow et al., 2005). To date, it is not known if these sex differences are due to the organizational or activational influences of gonadal steroids; however, both are likely to contribute.

Brain imaging studies using positron emission tomography (PET) or functional magnetic resonance imaging (fMRI) to study changes in brain activity following application of a noxious stimulus have also failed to consistently report sex differences. Using PET, Berman et al. (2002) reported significantly greater activation in men in comparison to women within the somatosensory cortex, thalamus, insula, and midcingulate cortices in response to equi-painful laser stimuli. Other PET studies have reported sex differences in brain activation patterns in response to colorectal distension using normal and chronic visceral pain patients (Berman et al., 2000; Naliboff et al., 2003, 2006). Silverman et al. (1997) used PET to compare brain activation patterns produced by distention of the distal intestine in men and women suffering from IBS with normal subjects. There were no sex differences in the activation pattern of normal men and women. This activation pattern differed from that of males and females with IBS in whom there was an additional activation of left prefrontal cortex and an absence of perigenual anterior cingulate cortex activation. Importantly, however, among the IBS patients, regional activations were much stronger for the males, and significant activation of the insula was observed bilaterally only in males (Berman et al., 2000). Lastly, Becerra et al. (1999) used fMRI to compare brain activation patterns induced by noxious thermal stimulation in women in two stages of their menstrual cycle (mid-follicular stage – low progesterone:estradiol ratio and high testosterone: mid-luteal stage – high progesterone:estradiol ratio and low testosterone) and in men. Their results showed similar patterns of activation of multiple regions in women in their mid-follicular stage and in men but significant reduction of activity in the anterior cingulate, insula, and frontal lobes in women during their mid-luteal phase, despite identical pain ratings in the two phases (see Mayer et al. (2008) for review).

36.5 The Influence of Sex Steroid Hormones on Pain and Nociception

It has become increasingly clear that the steroid hormones produced by the gonads contribute to the variability in pain responsiveness seen in men and women. However, attempts to understand exactly how gonadal steroids influence pain perception have produced a literature filled with inconsistencies (Craft, 2007; Fillingim and Ness, 2000). For example, studies examining changes in pain sensitivity across the menstrual cycle have reported increased thresholds to pain during the luteal phase, when progesterone levels are high relative to estrogen. However, increased pain thresholds have also been reported during the immediate premenstrual phase, when progesterone levels are decreasing, as well as at the time of ovulation when estrogen is high and progesterone is low.

Controlled studies using animal models have proved no less illuminating, with studies reporting increases, decreases, and no change in pain sensitivity and responsiveness to noxious stimuli in the presence or absence of estradiol (see Craft (2007) for review). Studies examining estrous cycle-induced changes in nociceptive thresholds have also failed to identify if a specific stage is associated with a specific directional change in sensitivity.

A major contributing factor to the discrepancies in the literature regarding the influence of gonadal steroids on pain sensitivity is the use of a variety of paradigms for steroid replacement, including daily injections of estradiol that produce supraphysiological plasma levels (Ji et al., 2003) and induce significant increases in corticosterone release (Priest et al., 1995). Alternatively, estradiol capsules or pellets are implanted; these maintain estradiol levels at an unnatural steady state. In addition, naturally cycling females will never experience estrogens in the absence of progesterone, yet most studies fail to co-administer the hormone. The consensus report by the IASP (Greenspan et al., 2007) has provided guidelines on how to examine the influence of gonadal steroids on pain and nociceptive sensitivity; however, it is becoming increasingly clear that more studies examining the influence of estradiol and/or progesterone replacement on the withdrawal latencies of ovariectomized rats to acute nociceptive stimuli are not what is needed. Rather, studies examining the potential mechanisms whereby gonadal steroids may influence the peripheral and central relay of nociceptive information are needed in order to advance the field.

36.5.1 Potential Mechanisms: The Descending Pain Modulatory Circuit

Estrogens produce a physiological response by binding to either estrogen receptor-alpha (ERα) or -beta (ERβ). These receptors are localized in several regions that participate in the transduction and central relay of nociceptive stimuli, including the dorsal and trigeminal root ganglia (Bereiter et al., 2005; Luo et al., 2008), the dorsal horn of the spinal cord (Evrard and Balthazart, 2003; Tang et al., 2007; Vanderhorst et al., 2005), and the midbrain PAG (Murphy and Hoffman, 1998, 1999), and therefore are poised to potentially modulate nociceptive signaling. Recently, PAG neurons that project to the RVM and are considered part of the endogenous pain modulatory circuit have been shown to contain ERα, thereby providing a direct mechanism for steroid modulation of nociceptive sensitivity (Loyd and Murphy, 2008). Gonadal steroids have also been shown to modulate the expression of preproenkephalin, enkephalin, dopamine, serotonin, galanin, *N*-methyl-D-aspartate (NMDA), GABA, glutamate, cholecystokinin, substance P, as well as a variety of growth factors, including BDNF (Bake et al., 2008; Eckersell et al., 1998; Hammer et al., 1994; Herbison, 1997; McCarthy et al., 1995; Nakamura et al., 2004; Priest et al., 1995; Sinchak et al., 2000). Clearly, changes in any of these substances would have an impact on pain sensitivity.

As stated above, the midbrain PAG is a crucial region that is intimately involved in processing nociceptive information. Indeed, it was reported over 30 years ago that electrical stimulation of the PAG allowed for surgery to take place in the absence of anesthesia (Reynolds, 1969). The PAG is densely populated with both estrogen and androgen receptors (ARs), and these receptors are localized preferentially among PAG neurons that project to the RVM. Indeed, in both male and female rats, approximately 25–50% of AR-IR neurons and 20–50% of ERα-IR neurons were co-localized with PAG-RVM neurons (Loyd and Murphy, 2008). PAG-RVM neurons also express the endogenous opiate enkephalin (Fukunaga et al., 1998; Iadarola et al., 1989; Manzanares et al., 1998), and enkephalin expression is directly related to estrogens, such that enkephalin levels are high when estradiol serum levels are high (Sinchak et al., 2000).

The PAG enkephalinergic circuit is a critical component of the endogenous descending analgesia system (Basbaum and Fields, 1978, 1984). Noxious stimulation induces ENK release within the PAG (Williams et al., 1995), which inhibits tonically active GABAergic interneurons (Osborne et al., 1996; Renno et al., 1992; Vaughan and Christie, 1997; Vaughan et al., 1997). Disinhibition of local GABAergic neurons within the PAG ultimately results in the activation of the PAG-RVM descending pain inhibitory system, resulting in a decreased sensitivity to nociceptive stimulation. Estradiol-induced increases in dorsal horn PPE mRNA levels have also been reported which would also be predicted to have a direct influence (i.e., increase) on nociceptive thresholds.

The PAG also contains a large number of GABA immunoreactive neurons (Commons et al., 2000; Kalyuzhny and Wessendorf, 1997; Reichling and Basbaum, 1990), and estradiol administration increases the proportion of PAG cells expressing mRNA for the GABA rate-limiting enzyme glutamic acid decarboxylase (Castaneyra-Perdomo et al., 1992), both the GAD65 and GAD67 subunits (McCarthy et al., 1995). Administration of estradiol also increases GABA-A receptor binding in the PAG and spinal cord of the rat (McCarthy et al., 1991). Together, these studies suggest that estradiol produces an overall increase in central GABA tone within the PAG, which would obviously have a direct influence on nociceptive sensitivity. Unfortunately, the potential influence of testosterone or AR activation on PAG enkephalinergic or GABAergic circuits in the male rat remains to be elucidated.

36.6 Stress and Pain

Pain and stress are intimately linked, such that pain is viewed by most as a potent stressor, and, conversely, stress highly influences pain (Watkins and Maier, 2000). Of relevance here is that there are enormous sex differences at every biological level (i.e., genetic through sociocultural) both in the situations that give rise to stress and in the effects of stress. Also pertinent is that major life events, which are stressful, are accompanied by changes in sex steroid hormones (puberty, pregnancy, parturition, menopause, and gonadal aging). Studies in this field relate to nervous system mechanisms and include those on:

1. The hypothalamic–pituitary–adrenal (HPA) and gonadal axes (Kajantie and Phillips, 2006). These systems are most well known for integration of sexual and reproductive functions. In rats, pain behavior is differentially affected by stress in males and female (Aloisi et al., 1996; Aloisi, 1997) while high vasopressin concentrations coupled with low-pressure visceral stimulation during proestrus demonstrate sex-hormone-dependent integration of the neuroendocrine response (Holdcroft et al., 2000). In humans, using opioid blockade, although the HPA-axis hormone response was similar, pain perception demonstrated sex differences (al'Absi et al., 2004).
2. Exercise-induced cardiovascular, respiratory, and pain responses, studied mainly in the context of angina, that differ in males and females such that symptoms of heat or burning pain were characteristic of ischemic pain in women as well as more frequent pain and other sensations (D'Antono et al., 2006a,b).
3. Mechanisms of SIA that exhibit sex differences in opioid/nonopioid involvement (Romero et al., 1988; Sternberg and Liebeskind, 1995; al'Absi et al., 2004) and descending modulation of the functions of different opioid receptors (Tershner et al., 2000).

36.7 Life Span Events, Lifestyle, and Sociocultural Roles

As previously mentioned, individual differences in pain are clearly influenced across the life span by (Myers et al., 2003; LeResche, 2001):

- major events (both individually unique and universally associated with reproductive status and aging);
- personal characteristics that can be loosely termed lifestyle; and
- changing sociocultural roles.

Despite their obvious importance, these factors have not been clearly integrated into our conceptualizations of pain mechanisms and individualized treatment strategies. However, as society moves away from disease models of illness toward holistic ones, this situation is changing. Although it remains difficult to assess in structured experiments how the factors might participate in generating sex differences, there are some helpful constructive directions. One perspective is to consider how the changing characteristics during each stage of human development (i.e., fetus, childhood, puberty/adolescence, fertile adulthood, menopause, and senescence) accumulate their potential influences on sex differences across the life span (Wizemann and Pardue, 2001).

36.7.1 Fetus, Childhood, and Puberty

Major sex differences in future bodily structure, physiology, and brain function (including some aspects of nociceptive sensitivity and stress modulation) are established during fetal life, gradually evidencing themselves as childhood progresses (Kajantie and Phillips, 2006). These factors together with parenting, family lifestyle, and schooling, all influenced by sociocultural sex roles, operate uniquely on each child, gradually producing overall sex-specific patterns of reported pain sensitivity and other behavior (Berde and Masek, 1999). During puberty and adolescence, dramatic alterations in hormonal status (Grumbach and Styne, 1998) exert their effects within the nervous system and other bodily organs, rapidly producing enormous sex differences in body structure, physiology, and behavior that add to the childhood lifestyle influences to bring about lifelong sex-specific traits. It is at this time that some pain-relevant, sex-different lifestyle patterns begin to be established, all influenced by the cultural milieu, such as tendency toward risk-taking behaviors (smoking, dangerous activities, violent behaviors, etc.), occupational goals, social roles, and attitudes toward injury and disease both in oneself and others. Sex differences in some painful disorders also emerge at this time (e.g., cluster headaches in males, migraine in females) and these can be replicated in a subgroup of sex-change patients (Aloisi et al., 2007).

36.7.2 Fertile Adulthood

It is during the long period of fertile adulthood that an individual's occupation, social roles, and lifestyle, while slowly changing, become entrenched. Although changes in societal attitudes are rapidly occurring in many cultures, it remains the case to varying degrees in most of them that women still predominate as caregivers and organizers with wide-ranging obligations and duties spanning family and workplace realms, while men still predominate in aggressive and focused, physically demanding occupational and leisure activities with a relatively narrower range than women of social obligations and duties. Such differences have enormous implications for sex differences in the following: vulnerability to different types of injury (work- or sport-related for men; assault/rape for women); greater female duty, obligations, and willingness to seek healthcare; higher female sensitivity toward the recognition of conditions as being painful (i.e., those that demand tending) both in others and in themselves; greater female freedom to access healthcare; and, finally, sex-specific attitudes of healthcare givers toward the significance of pain in their patients.

Some of the lifestyle differences evidence themselves specifically in sex prevalence of some diseases (**Table 3**). One example relates to injury-induced conditions, such as brachial plexus neuropathy, more common in men (e.g., motorbike crashes), and pelvic pains, more common in women (e.g., physical assault; Poleshuck et al., 2005). A second includes smoking- and alcohol-associated disorders, more common in men, such as thromboangiitis obliterans (Buerger's disease), which, interestingly, is increasing in women because more are smoking. A third includes gallbladder disorders, more common in women, in which sex differences in lifestyle are but part of a complex multifactorial etiology that can be traced to a combination of nutritional, metabolic pathways and hormone effects.

Sex differences in lifestyle are also one of the big components underlying the overall female vulnerability to pain with its companion strategy of marshalling multiple approaches to deal with it. These sex differences may be exaggerated by the repetitive pain cycles of dysmenorrhea, as well as by pregnancy and parturition. Dysmenorrhea is suffered by up to 80% of Caucasian women (depending on age (Weissman et al., 2004)) and may induce a constant and generalized muscle hyperalgesia (Giamberardino et al., 1997). During pregnancy and premenstrually, fluid retention can increase tissue pressure around nerves, thereby precipitating or exacerbating painful compression neuropathies such as carpel tunnel syndrome and lateral femoral cutaneous nerve pain (Zager et al., 1998). The trauma associated with parturition clearly generates its own extreme pain (Melzack, 1993), and can increase the severity of postpartum pain after subsequent deliveries (Murray and Holdcroft, 1989) as well as sensitivity to noxious stimulation in a manner similar to that in men resulting from comparable experiences, such as severe injury sustained during war, sport, a phenomenon that is likely associated with long-term changes in peripheral nerves and central neurons that is induced by injury in animals (Devor, 2006) and related to the mechanisms described in Section 36.4.2.

36.7.3 Gonadal Aging and Senescence

Following the fertile years, there is at first, for women, a 5–10-year period of menopausal alterations in hormone patterns terminating in their sharp decline, while gonadal aging in men is more inconsistent. Accompanying these changes are additional changes in general metabolism, physiology, and structure. Lifestyle changes also occur relatively rapidly during this period as children leave home and both occupational duties and leisure activities are altered. These changes together give rise to an increasing disease burden for both sexes, alterations of drug metabolism (which, although most dramatic for women, also occur in men), and modified attitudes toward healthcare and access to it. Although the net result is a decrease in many sex differences in pain, especially as the burden of chronic disease becomes significant in both sexes (Warren and Fried, 2001), women remain more vulnerable to disability than men, while their longevity is increased (Leveille et al., 2000, 2005).

Very few basic science studies have examined the impact of advanced age on nociceptive sensitivity (Raut and Ratka, 2007; Smith and French, 2002; Smith and Gray, 2001). Unfortunately, the few studies that have examined the impact of age on nociceptive thresholds were all conducted exclusively in male rats. Our recent studies indicate that while age does not impact significantly on baseline withdrawal responses to noxious mechanical and thermal stimulation, it does impact on the hyperalgesia produced by intraplantar administration of complete Freud's adjuvant. In particular, aged males and females display significantly higher levels of thermal hyperalgesia and decreased morphine-induced anti-hyperalgesia than adult control animals (Hanberry and Murphy, 2008). Clearly, additional studies on senescence, the central processing of nociceptive information, and pain are needed.

36.8 Clinical Implications

It is becoming increasingly accepted that diagnostic and treatment strategies are best focused on an individual, regardless of sex. However, as information accumulates particularly from large clinical studies, and is codified, sex differences in pain reports and experiences as well as the influence of the patient's hormonal history and current milieu begin to have increasingly potent and useful implications for diagnosis, clinical management of pain, and, for healthcare providers, an economic impact.

As shown in **Table 2**, therapies currently available to alleviate pain can be divided into three main categories: drugs, physical interventions, and situational adjustments. Sex differences in mechanisms of action, efficacy, and/or usage strategy apply to all of these categories.

36.8.1 The Diagnostic Process

During medical consultations, the diagnostic process begins by questioning for symptoms, observing signs, and recording drug usage. It is the first stage in the identification and application of sex differences to pain medicine.

Structuring a medical history and investigation in a manner that encourages a partnership to be developed between the clinician and patient and focusing questions so they are relevant to the sex of the patient is valuable for gathering information that might otherwise not have been obtained because of the patient's and doctor's separate contextual frameworks. A pain history that takes such holistic issues into account has the following advantages:

- Because knowledge of general hormonal influences on pain is rapidly accumulating, gathering detailed knowledge (including hormonal assay where relevant) of a woman patient's menstrual, menopausal, and pregnancy history, and for both men and women a history of the use of supplementary hormones (as relevant to disease treatments, hormone replacement and leisure activities, i.e., body building) may be of relevance for diagnosis and treatment design.
- Signs and symptoms that are distant from a disease source (and possibly considered irrelevant from the focus of the presenting illness) are elicited. This broader description of pain/discomfort may generate a more complete picture of pain, particularly in women where visceral symptoms present in referral patterns that are different from those in men, such as coronary heart disease (Section 36.6).
- Knowing the time characteristics of the pain-related symptoms, either from the patient's memory, but preferably gathered from a diary, may identify chronobiological patterns in pain and result in appropriately timed diagnostic measures and therapies.
- Gathering more complete knowledge of a family history of painful disease/s and how family members deal/t with it may alert clinicians to underlying etiologies such as genetic, psychosocial, or cultural factors that can then be targeted in the design of therapy.
- More information about major events in the patient's life is also valuable. For example, stress associated with some of these events may manifest as disease states, affecting pain symptoms, and could also influence treatment efficacy (Section 36.5).
- Knowledge of past or present history of physical abuse allows insight into understanding sources contributing to chronic pain.
- Having more information about lifestyle characteristics, such as risk-taking behaviors, smoking, occupational goals, leisure activities, attitudes toward healthcare, is important. If those behaviors that have the potential for exacerbating or reducing pain sensations are identified, they can then be more appropriately managed, for example, by education directed to increase men's use of the healthcare system to better advantage.

36.8.2 Pharmaceutical Therapies

Hormones that control growth and sexual development also influence many of the body composition-dependent effects and functional actions of drugs. Reviews by Ciccone and Holdcroft (1999), Jensvold et al. (1996), Beierle et al. (1999), and the Institute of Medicine of the US National Academy of Sciences (Wizemann and Pardue, 2001) have highlighted these influences on drug pharmacokinetics and pharmacodynamics. The most important pharmacokinetic differences are caused by sex differences in oral bioavailability. For example, changes in gastric function in women during the menstrual cycle lead to decreased drug absorption during the periovulatory stage (mid-cycle) of drugs such as aspirin and alcohol.

Other effects on local anesthetic and opioid drugs include: (1) alterations in protein binding that increase free lidocaine when sex steroid levels are increased, and sex-specific expression of metabolic enzyme systems for lidocaine such as the cytochrome CYP3A family (Hunt et al., 1992); and (2) sex differences in alfentanil clearance with age such that in females and males of similar weight, total median alfentanil clearance is highest in women less than 50 years of age, decreasing thereafter to rates similar to those for men at any age (Lemmens et al., 1990).

36.8.2.1 Adverse drug events

The prevalence of adverse drug effects is generally twice as much for women as men (Cepeda et al., 2003; Richardson and Holdcroft, 2007). Postmarketing pharmacovigilance reporting confirms this difference for analgesics, for example, women have more adverse effects to acute administration of morphine than men (Bijur et al., 2008; Hill and Zacny, 2000). Factors responsible for these differences include body composition where large differences in weight may be overlooked, potentially leading to overdose and increased frequency of use. Women overall use more medications than men as well as regular supplementary hormones, such as oral contraceptives (Jensvold et al., 1996), but, even where frequency of use has been measured, it does not wholly explain the sex differences in adverse drug events. Multiple drug use may be more common, for example, during the menstrual cycle when women may treat both anemia and dysmenorrhea and at other times urinary tract infections are more common than in men and require antibiotics and analgesics. Because drug interactions are relatively common, it is of obvious importance to routinely collect information about a patient's use of over-the-counter medications as well as prescription drugs and supplemental sex hormones. Drug choice and dosage can then be individualized.

Specific examples are as follows: (1) A series of studies on anesthetics, such as bupivacaine and lidocaine, have shown that their pharmacokinetics are influenced by the hormones of pregnancy, both progesterone and estradiol, leading to greater adverse effects (e.g., Wulf et al., 1991; Moller et al., 1992, 1999). Such results have led to dosage adjustments. (2) The slope of the ventilatory response curve to carbon dioxide is reduced by morphine in young women, compared to men (Dahan et al., 1998; Sarton et al., 2000). Thus, the danger associated with a higher magnitude of central nervous system depression with opioids in normal humans is potentially higher in women than men.

36.8.2.2 Drug development

It has been to the detriment of women that they have, in the past, been excluded from pharmaceutical drug development studies. This exclusion has prevented observations on drug toxicity, side effects, and differences in pharmacokinetics and pharmacodynamics between men and women. In the US, but not Europe, clinical trials are now beginning to be routinely carried out in both sexes, and the development of tests of hormonal status is improving the situation but the pace is slow (Geller et al., 2006).

36.8.2.3 Drug selection

In postmarketing studies examining postoperative pain in young women and men following dental extraction, Gear et al. (1999, 2003) found that the analgesic effectiveness of opioids such as nalbuphine, buprenorphine, and pentazocine was significantly better in young nulliparous females than in young males and was dose-related. Such differences may encourage their preferential use in young women, and these data have prompted this research group to suggest that some opioids might be more effective in women than men in a postsurgical context (Miaskowski et al., 2000). However, the drugs used in that study were mixed agonists/antagonists, and are not generally used for the alleviation of pain associated with third molar extraction.

36.8.2.4 Sex differences in short- and longer-term effects of opioids

As a result of a literature review of postoperative patient controlled analgesia with opioids that reported that men had a higher opioid consumption than women even after controlling for differences in body size (Miaskowski et al., 2000), prospective studies were initiated to control for differences in reported pain intensity. Confirmation of these results appears to relate to the postoperative period after initial recovery from surgery (Chia et al., 2002). However, as most of the negative side effects associated with morphine consumption are exacerbated in women, these studies must be interpreted with caution. When levels of pain relief are controlled for, males experience a significantly greater alleviation of pain by morphine than females (Cepeda and Carr, 2003; Aubrun et al., 2005). The mechanisms for such variations are not known but may relate to differences in opioid pharmacology (Sarton et al., 2000; Murthy et al., 2002), stress responses (see HPA axis), drug adverse effects, and in women recovering faster after anesthesia than men (Gan et al., 1999; Buchanan et al., 2006).

36.8.2.5 Physical interventions

Little is known about sex differences in the effectiveness of some of the simpler forms of nontherapeutic interventions, but, for some effective forms such as relaxation, heat or cold application, massage, and vibration, women are generally more willing to use them (Robinson et al., 2000), suggesting that men might benefit from further encouragement and education on their value. In addition, psychophysical studies in healthy women show that their sensitivity to some forms of somatic stimuli (pressure and thermal) can change with menstrual status, which could affect efficacy when heat, cold, vibration, or massage are used for pain alleviation by women. Regarding exercise and physical therapy, sports medicine studies report greater risks of injury in women than men (Bell et al., 2000). In addition, the differences in exercise-induced cardiovascular responses in males and females may affect pain either by way of the relationship between blood pressure and pain discussed earlier in Section 36.4.2.2 (Fillingim and Maixner, 1996).

Following specific surgical interventions, hormonal status may affect outcome from breast surgery through TRPV-1 receptor activation (Gopinath et al., 2005), and gender differences in management and outcomes are reported after cardiac surgery for chest pain with women suffering worse morbidity and mortality than men (Eastwood and Doering, 2005). In general, after surgery in the immediate postoperative period, a prospective study of 4173 patients (45% female) showed that women evidence more minor postoperative pains of sore throat, headache, and backache than men (Myles et al., 1997). This difference may relate to drugs used for women during anesthesia (such as suxamethonium, known to produce more muscle pain in women than men), or to the choice of postoperative analgesics, or to reporting bias, or some combination of these factors.

36.8.2.5 Situational manipulations

When pain is chronic, situational changes in the patient's environment and social interactions may not eliminate the pain, but there is much evidence that they may reduce it by changing expectations, enhancing productive activity, and improving quality of life and psychophysiological responses. One example is in the use of music therapy to reduce pain after surgery (Voss et al., 2004), yet since women and men have different psychophysiological responses to music, and pain perception in volunteers demonstrates gender differences in response to music arousal, gender issues still require investigation (Nater et al., 2006; Kenntner-Mabiala et al., 2007). Another example is aromatherapy, since citrus lemon essential oil delays neurochemical release in the hippocampus of female rats in response to painful stimuli (Ceccarelli et al., 2002).

Several studies indicate that not only women in pain are more likely than men to make use of multiple therapies, but they may also derive more benefit as well. Part of the basis for this difference may lie in women's greater verbal abilities and in their engagement of more diffuse brain structures for language (Section 36.4.4) as well as in their lifestyles and occupations. But whatever the cause(s), it seems clear that, for the clinician, it is important to consider that male more than female patients might need more education and encouragement on the potential value of considering situational adjustments, including not only changes in their personal lifestyle (music, diet, meditation, etc.), but also in their use of health services. Of obvious relevance here is recognition by the clinician that her/his attitudes also have an important role in diagnosis, motivation, and therapy selection (Sierpina et al., 2007). These attitudes, warranting continual reassessment, include the clinician's general beliefs about the differences between women and men in the validity of their pain reports or responses to treatment, and about the value of nonpharmacological and nonphysical intervention.

36.8.2.6 Advantages of varying and combining therapies

When a person develops pain and seeks medical advice, attempts to alleviate it initially often involve the use of single drug therapies. This strategy may not even be sufficiently effective for acute pain conditions, and it regularly fails when pain continues more long term.

From the multimodal dimensions of peripheral and central nervous system mechanisms that underlie nociception, the most effective approach would be a strategy that involves a flexible combination of drugs, physical interventions, and situational manipulations. For chronic pain, such an approach often may seem possible only through so-called multidisciplinary pain clinics, but the principle can certainly be applied in primary care settings. It has been applied to patients with long-term pain, such as those with irritable bowel disease (Naliboff et al., 2000), or fibromyalgia (Bradley and Alarcón, 2000), or conditions without an obvious diagnosis (Loeser and Turk, 2001). For other pains, multiple drugs may be prescribed each with different mechanisms of action. For example, for pain related to cancer, nonsteroidal anti-inflammatory analgesic drugs and other drugs are often combined with opioids so that opioid dosage can be reduced (Jacox et al., 1994). The list of drugs is shown in **Table 2** and includes steroid hormones.

One of the best examples of the advantages to adopting a poly-therapeutic concept of pain management to more than just long-term pain conditions is labor pain. The acute pain of labor is of moderate-to-severe intensity in 60% of primiparous women (Melzack, 1993). In expectation of pain, preemptive strategies can be accessed and practised. Education, behavioral modifications, and preparation for childbirth often occupy significant periods of time before birth. Coping strategies and postural adjustments are recruited. During labor a combination of these approaches, including drugs, is chosen. Despite wide variations of age, socioeconomic status, previous dysmenorrhea, body size, time of day, and parity, such combined therapies may be successful during parturition and reduce the demand for analgesics. In this context, the partner also has a role by reinforcing learned behavior and participating in the interventions. This preparation and planned management of pain can enable both partners to develop strategies for pain control in the future based on learned experiences.

36.8.3 Hormones, Pain and the Clinic: Two Examples

36.8.3.1 Diabetes

One of the main hormone-deficiency diseases frequently complicated by chronic painful neuropathy is diabetes. Diabetic neuropathy is clinically more common in females and hypertensive patients (Maser et al., 1990). In diabetic patients less than 60 years old, females report more gastrointestinal symptoms (e.g., pain-related heartburn) than their male equivalents (Spangeus et al., 1999). The etiology of the neuropathic disorders in diabetes can be traced to nerve growth factor (NGF), which is required for survival of sympathetic neurons. NGF activity is influenced by sex hormones:

- Estrogens regulate NGF receptor mRNA in sensory neurons (Sohrabji et al., 1994).
- Testosterone and aldosterone reduce levels of NGF mRNA in fibroblasts (Siminoski et al., 1987).

The clinical relevance of neurotrophins and their receptors, their potential regulation by sex hormones, and the therapeutic potentials for pain that accrue make this an emerging area for neurohumoral research (Allen and Dawbarn, 2006).

36.8.3.2 Coronary artery disease

Sex differences in coronary artery disease pervade all aspects of medical care from prevention (using drugs or lifestyle changes) through investigation to chronic illness. There are gender and age differences in prevalence, prognosis, and manifestations of cardiovascular disease (Eastwood and Doering, 2005). Coronary heart disease is the major cause of death in women, and the postmenopausal state increases the risk for cardiovascular events. Sex-specific behavioral strategies to cope with cardiac disease are slowly being developed. For example, using a structured interview in 317 patients after stenting of coronary vessels, Ladwig et al. (2000) found that a negative health perception predominated in women because of frequent chest pain and sleeping disorders, suggesting that such postoperative problems need more aggressive attention in women than men.

When a patient presents with chest pain, the assessment and demand for invasive testing for coronary heart disease often depends on the characteristics of the pain such as whether or not it is in the chest and, if it is in the chest, its characteristics. In acute coronary syndrome, women are less likely than men to complain of chest pain, although there is a lack of standardization of reporting of chest-pain symptoms (Canto et al., 2007). One approach has been to perform subgroup analysis on symptoms related to hormonal status. In premenopausal women, chest pressure and chest pain in locations other than substernal were more frequent than in postmenopausal women or those taking hormone replacement therapy (Méthot et al., 2004). Another has been to investigate symptom severity. Men interpret pain symptom severity with myocardial infarction but not women (Fukuoka et al., 2007). Thus, behavioral patterns in response to chest pain may affect how men and women seek medical care.

Researchers have also recognized sex differences in the co-morbidity associated with coronary artery disease. For example, hormonal disorders of the metabolic and cardiovascular systems, such as diabetes and hypertension, have been identified. More women than men have diabetes and coronary artery disease and there is a similar association for hypertension in the elderly population (Pilote et al., 2007; McBride et al., 2005). The recognition of sex differences in symptoms, morbidity, and outcome is important because mechanisms for these effects are still not clear, inadequate diagnoses and suboptimal management result in poor outcomes for women, and many clinical trials do not include sufficient women to allow sex-difference analysis (Greenspan et al., 2007).

Thus, in the past, when a person presented with pain from suspected coronary heart disease, it might not have been considered relevant to ask about pain distribution. The patient might have considered the pains to be due to arthritis or of gastric origin and irrelevant to a cardiac consultation. However, by including as diagnostic a detailed history of pain complaints and their distribution, as well as for the postmenopausal woman, whether or not she is diabetic and/or hypertensive can be lifesaving.

36.9 Conclusion

Given the fact that the book in which this chapter is published is focused on how hormones affect behavior, it should be clear from the discussion above that in dealing with complex clinical issues such as pain, or more specifically sex/gender differences in pain, the hormonal milieu, while important, remains only one of many interacting components that contribute to the dynamic and ever-changing pain experience and behavioral response, as well as its mechanisms and treatment, as individuals progress through life.

References

al'Absi M, Buchanan TW, Marrero A, and Lovallo WR (1999) Sex differences in pain perception and cardiovascular responses in persons with parental history for hypertension. *Pain* 83: 331–338.

al'Absi M, France C, Harju A, France J, and Wittmers L (2006) Adrenocortical and nociceptive responses to opioid blockade in hypertension-prone men and women. *Psychosomatic Medicine* 8: 292–298.

al'Absi M and Petersen KL (2003) Blood pressure but not cortisol mediates stress effects on subsequent pain perception in healthy men and women. *Pain* 106: 285–295.

al'Absi M, Wittmers LE, Ellestad D, Nordehn G, Kim SW, Kirschbaum C, and Grant JE (2004) Sex differences in pain and hypothalamic–pituitary–adrenocortical responses to opioid blockade. *Psychosomatic Medicine* 66: 198–206.

Allen SJ and Dawbarn D (2006) Clinical relevance of the neurotrophins and their receptors. *Clinical Science (London)* 110: 175–191.

Aloisi AM (1997) Sex differences in pain-induced effects on the septo-hippocampal system. *Brain Research Reviews* 25: 397–406.

Aloisi AM, Albonetti ME, and Carli G (1996) Formalin-induced changes in adrenocorticotropic hormone and corticosterone plasma levels and hippocampal choline acetyltransferase activity in male and female rats. *Neuroscience* 74: 1019–1024.

Aloisi AM, Bachiocco V, Costantino A, Stefani R, Ceccarelli I, Bertaccini A, and Meriggiola MC (2007) Cross-sex hormone administration changes pain in transsexual women and men. *Pain* 132(supplement 1): S60–S67.

Amandusson A, Hallbeck M, Hallbeck AL, Hermanson O, and Blomqvist A (1999) Estrogen-induced alterations of spinal cord enkephalin gene expression. *Pain* 83: 243–248.

Arslanian-Engoren C, Patel A, Fang J, Armstrong D, Kline-Rogers E, Duvernoy S, and Eagle KA (2006) Symptoms of men and women presenting with acute coronary syndromes. *American Journal of Cardiology* 98: 1177–1181.

Aubrun F, Salvi N, Coriat P, and Riou B (2005) Sex- and age-related in morphine requirements for postoperative pain relief. *Anesthesiology* 103: 156–160.

Bake S, Ma L, and Sohrabji F (2008) Estrogen receptor-alpha overexpression suppresses 17beta-estradiol-mediated vascular endothelial growth factor expression and activation of survival kinases. *Endocrinology* 149: 3881–3889.

Basbaum AI and Fields HL (1978) Endogenous pain control mechanisms: Review and hypothesis. *Annals of Neurology* 4: 451–462.

Basbaum AI and Fields HL (1984) Endogenous pain control systems: Brainstem spinal pathways and endorphin circuitry. *Annual Review of Neuroscience* 7: 309–338.

Becerra LR, Breiter HC, Stojanovic M, et al. (1999) Human brain activation under controlled thermal stimulation and habituation to noxious heat: An fMRI study. *Magnetic Resonance Medicine* 41: 1044–1057.

Beierle I, Meibohm B, and Derendorf H (1999) Gender differences in pharmacokinetics and pharmacodynamics. *International Journal of Clinical Pharmacology and Therapeutics* 37: 529–547.

Bell NS, Mangione TW, Hemenway D, Amoroso PJ, and Jones BH (2000) High injury rates among female army trainees: A function of gender? *American Journal of Preventive Medicine* 18: 141–146.

Berde CB and Masek B (1999) Pain in children. In: Wall PD and Melzack R (eds.) *Textbook of Pain,* 4th edn., pp. 1463–1477. Edinburgh: Churchill Livingstone.

Bereiter DA, Cioffi JL, and Bereiter DF (2005) Oestrogen receptor-immunoreactive neurons in the trigeminal sensory system of male and cycling female rats. *Archives of Oral Biology* 50: 971–979.

Berkley KJ (1997a) Sex differences in pain. *Behavioural Brain Science* 20: 371–380.

Berkley KJ (1997b) Female vulnerability to pain and the strength to deal with it. *Behavioural Brain Science* 20: 473–479.

Berkley KJ (2000) Female pain versus male pain? In: Fillingim RB (ed.) *Sex, Gender, and Pain,* vol. 17, pp. 373–381. Seattle, WA: IASP Press.

Berkley KJ, Hoffman G, Murphy AZ, and Holdcroft A (2002) Pain: Sex/gender differences. In: Pfaff DW, Arnold AP, Fahrbach SE, Etgen AM, and Rubin RT (eds.) *Hormones, Brain and Behavior*. Vol. 5, pp. 409–442. San Diego, CA: Academic Press.

Berkley KJ and Hubscher CH (1995) Are there separate central nervous system pathways for touch and pain? *Nature Medicine* 8: 766–773.

Berman SM, Chang L, Suyenobu B, et al. (2002) Condition-specific deactivation of brain regions by 5-HT3 receptor antagonist alosetron. *Gastroenterology* 123: 969–977.

Berman S, Munakata J, Naliboff BD, et al. (2000) Gender differences in regional brain response to visceral pressure in IBS patients. *European Journal of Pain* 4: 157–172.

Bijur PE, Esses D, Birnbaum A, Chang AK, Schechter C, and Gallagher EJ (2008) Response to morphine in male and female patients: Analgesia and adverse events. *Clinical Journal of Pain* 24: 192–198.

Bingefors K and Isacson D (2004) Epidemiology, co-morbidity, and impact on health-related quality of life of self-reported headache and musculoskeletal pain – a gender perspective. *European Journal of Pain* 8: 435–450.

Blurton-Jones M and Tuszynski MH (2006) Estradiol-induced modulation of estrogen receptor-beta and GABA within the adult neocortex: A potential transsynaptic mechanism for estrogen modulation of BDNF. *Journal of Comparative Neurology* 499: 603–612.

Bonica J (1979) The need of a taxonomy. *Pain* 6: 247–252.

Boulton AJ (2007) Diabetic neuropathy: Classification, measurement and treatment. *Current Opinion in Obstetrics and Gynecology* 14: 141–145.

Bradley LA and Alarcón GS (2000) Sex-related influences in fibromyalgia. In: Fillingim RB (ed.) *Sex, Gender, and Pain,* vol. 17, pp. 281–307. Seattle, WA: IASP Press.

Buchanan FF, Myles PS, Leslie K, Forbes A, and Cicuttini F (2006) Gender and recovery after general anaesthesia combined with neuromuscular blocking drugs. *Anesthesia and Analgesia* 102: 291–297.

Butkevich A, Abraham C, and Phillips RA (2000) Hormone replacement therapy and 24-hour blood pressure profile of postmenopausal women. *American Journal of Hypertension* 13: 1039–1041.

Cairns BE (2007) The influence of gender and sex steroids on craniofacial nociception. *Headache* 47: 319–324.

Canto JG, Goldberg RJ, Hand MM, Bonow RO, Sopko G, Pepine CJ, and Long T (2007) Symptom presentation of women with acute coronary syndromes: Myth vs. reality. *Archives of Internal Medicine* 167: 2405–2413.

Castaneyra-Perdomo A, Perez-Delgado M, Montagnese C, and Coen C (1992) Brainstem projections to the medial preoptic region containing the luteinizing hormone-releasing hormone perikarya in the rat. An immunohistochemical and retrograde transport study. *Neuroscience Letters* 139: 135–139.

Ceccarelli I, Masi F, Fiorenzani P, and Aloisi AM (2002) Sex differences in the citrus lemon essential oil-induced increase of hippocampal acetylcholine release in rats exposed to a persistent painful stimulation. *Neuroscience Letters* 330: 25–28.

Cepeda MS and Carr DB (2003) Women experience more pain and require more morphine than men to achieve a similar degree of analgesia. *Anesthesia and Analgesia* 97: 1464–1468.

Cepeda MS, Farrar JT, Baumgarten M, Boston R, Carr DB, and Strom BL (2003) Side effects of opioids during short-term administration: Effect of age, gender and race. *Clinical Pharmacology and Therapeutics* 74: 102–112.

Chesterton LS, Barlas P, Foster NE, Baxter GD, and Wright CC (2003) Gender differences in pressure pain threshold in healthy humans. *Pain* 101: 259–266.

Chia YY, Chow LH, Hung CC, Liu K, Ger LP, and Wang PN (2002) Gender and pain upon movement are associated with the requirements for postoperative patient-controlled iv analgesia: A prospective survey of 2,298 Chinese patients. *Canadian Journal of Anaesthesia* 49: 249–255.

Ciccone G and Holdcroft A (1999) Drugs and sex differences – a review. *British Journal of Anaesthesia* 82: 255–265.

Commons KG, Aicher SA, Kow LM, and Pfaff DW (2000) Presynaptic and postsynaptic relations of mu-opioid receptors to gamma-aminobutyric acid-immunoreactive and medullary-projecting periaqueductal gray neurons. *Journal of Comparative Neurology* 419: 532–542.

Craft RM (2007) Modulation of pain by estrogens. *Pain* 132 (supplement 1): S3–S12.

Cucurachi L, Devetak M, Torelli P, Lambru G, and Manzoni GG (2006) Gender ratio of migraine without aura: Observations over time. *Neurological Sciences* 27: 47–50.

D'Antono B, Dupuis G, Fortin C, Arsenault A, and Burelle D (2006a) Detection of exercise-induced myocardial ischemia from symptomatology experienced during testing in men and women. *Canadian Journal of Cardiology* 22: 411–417.

D'Antono B, Dupuis G, Fortin C, Arsenault A, and Burelle D (2006b) Angina symptoms in men and women with stable coronary artery disease and evidence of exercise-induced myocardial perfusion defects. *American Heart Journal* 151: 813–819.

Dahan A, Sarton E, Teppema L, and Olievier C (1998) Sex-related differences in the influence of morphine on ventilatory control in humans. *Anesthesiology* 88: 903–913.

Dannecker EA, George SZ, and Robinson ME (2007) Influence and stability of pain scale anchors for an investigation of cold pressor pain tolerance. *Journal of Pain* 8: 476–482.

Deadwyler SA and Hampson RE (1997) The significance of neural ensumble codes during behavior and cognition. *Annual Review of Neuroscience* 20: 217–244.

Devor M (2006) Response of nerves to injury in relation to neuropathic pain. In: McMahon SB and Koltzenburg M (eds.) *Wall and Melzack's Textbook of Pain,* 5th edn., pp. 905–927. Philadelphia, PA: Elsevier Churchill Livingstone.

Diatchenko L, Nackley AG, Tchivileva IE, Shabalina SA, and Maixner W (2007) Genetic architecture of human pain perception. *Trends in Genetics* 23: 605–613.

Dib-Hajj SD, Cummins TR, Black JA, and Waxman SG (2007) From genes to pain: Na v 1.7 and human pain disorders. *Trends in Genetics* 30: 555–563.

Eastwood J-A and Doering LV (2005) Gender differences in coronary artery disease. *Journal of Cardiovascular Nursing* 20: 340–351.

Eckersell CB, Popper P, and Micevych PE (1998) Estrogen-induced alteration of mu-opioid receptor immunoreactivity in the medial preoptic nucleus and medial amygdala. *Journal of Neuroscience* 183: 3967–3976.

Evrard HC and Balthazart J (2003) Aromatase (estrogen synthase) activity in the dorsal horn of the spinal cord: Functional implications. *Annals of the New York Academy of Sciences* 1007: 263–271.

Falcone C, Auguadro C, Sconocchia R, and Angoli L (1997) Susceptibility to pain in hypertensive and normotensive patients with coronary artery disease: Response to dental pulp stimulation. *Hypertension* 30: 1279–1283.

Fillingim RB and Maixner W (1996) The influence of resting blood pressure and gender on pain responses. *Psychosomatic Medicine* 58: 326–332.

Fillingim RB, Maixner W, Bunting S, and Silva S (1998) Resting blood pressure and thermal pain responses among females: Effects on pain unpleasantness but not pain intensity. *International Journal of Psychophysiology* 30: 313–318.

Fillingim RB and Ness TJ (2000) Sex-related hormonal influences on pain and analgesic responses. *Neuroscience and Biobehavioral Reviews* 4: 485–501.

France CR, Keefe FJ, Emery CF, et al. (2004) Laboratory pain perception and clinical pain in post-menopausal women and age-matched men with osteoarthritis: Relation to pain coping and hormonal status. *Pain* 112: 274–281.

Frankfurt M, Siegel RA, Sim I, and Wuttke W (1986) Estrous cycle variations in cholecystokinin and substance P concentration in discrete areas of the rat brain. *Neuroendocrinology* 42: 226–231.

Frot M, Feine JS, and Bushnell MC (2004) Sex differences in pain perception and anxiety. A psychophysical study with topical capsaicin. *Pain* 108: 230–236.

Fukunaga Y, Inoue N, Miyamoto M, Kishioka S, and Yamamoto H (1998) Effects of peptidase inhibitors, [D-Ala2, Met5]-enkephalinamide and antiserum to methionine-enkephalin microinjected into the caudal periaqueductal gray on morphine withdrawal in rats. *Japanese Journal of Pharmacology* 78: 455–461.

Fukuoka Y, Dracup K, Moser DK, McKinley S, Ball C, Tamasaki K, and Kim CJ (2007) Is severity of chest pain a cue for women and men to recognize acute myocardial infarction symptoms as cardiac in origin? *Progress in Cardiovascular Nursing* 22: 132–137.

Gan TJ, Glass PS, Sigl J, Sebel P, Payne F, Rosow C, and Embree P (1999) Women emerge from general anaesthesia with propofol/alfentanil/nitrous oxide faster than men. *Anesthesiology* 90: 1283–1287.

Gear RW, Gordon NC, Miaskowki C, Paul SM, Heller PH, and Levine JD (2003) Sexual diamorphism in very low dose nalbuphine postoperative analgesia. *Neuroscience Letters* 339: 1–4.

Gear RW, Miaskowski C, Gordon NC, Paul SM, Heller PH, and Levine JD (1999) The kappa opioid nalbuphine produces gender- and dose-dependent analgesia and antianalgesia in patients with postoperative pain. *Pain* 83: 339–345.

Geller SE, Adams MG, and Carnes M (2006) Adherence to federal guidelines for reporting of sex and race/ethnicity in clinical trials. *Journal of Womens Health* 15: 1123–1131.

Giamberardino MA (2000) Sex-related and hormonal modulation of visceral pain. In: Fillingim R (ed.) *Sex, Gender, and Pain*, pp. 135–164. Seattle, WA: IASP Press.

Giamberardino M, Berkley K, Iezzi S, deBigontina P, and Vecchiet L (1997) Pain threshold variations in somatic wall tissues as a function of menstrual cycle, segmental site and tissue depth in non-dysmenorrheic women, dysmenorrheic women and men. *Pain* 71: 187–197.

Gopinath P, Wan E, Holdcroft A, et al. (2005) Increased capsaicin receptor TRPV1 in skin nerve fibres and related vanilloid receptors TRPV3 and TRPV4 in keratinocytes in human breast pain. *BMC Womens Health* 5: 2.

Greenspan JD, Craft RM, LeResche L, et al. (2007) Studying sex and gender differences in pain and analgesia: A consensus report. *Pain* 132: S26–S45.

Grumbach MM and Styne DM (1998) Puberty: Ontogeny, neuroendocrinology, physiology, and disorders. In: Wilson JD, Foster DW, Kronenberg HM, and Larsen PR (eds.) *Williams Textbook of Endocrinology,* 9th edn., pp. 1509–1625. Philadelphia, PA: WB Saunders Co.

Hammer RP, Jr, Zhou L, and Cheung S (1994) Gonadal steroid hormones and hypothalamic opioid circuitry. *Hormones and Behavior* 28: 431–437.

Hanberry R and Murphy AZ (2008) Sexually dimorphic effects of aging on morphine analgesia and associated changes in μ-opioid receptor expression. *Society for Neuroscience* Abstract.

Herbison A (1997) Estrogen regulation of GABA transmission in rat preoptic area. *Brain Research Bulletin* 44: 321–326.

Herbison A and Fenelon V (1995) Estrogen regulation of GABA receptor subunit mRNA expression in preoptic area and bed nucleus of the stria terminalis of female rat brain. *Journal of Neuroscience* 15: 2328–2337.

Hill JL and Zacny JP (2000) Comparing the subjective, psychomotor, and physiological effects of intravenous hydromorphone and morphine in healthy volunteers. *Psychopharmacology (Berl.)* 152: 31–39.

Holdcroft A (2006) Gender differences in clinical pain management. In: Mao J (ed.) *Translational Pain Research, Comparing Preclinical Studies and Clinical Pain Management. Lost in Translation?,* vol. 2, ch. 15, pp. 79–97. New York: Nova Biomedical Books.

Holdcroft A and Berkley KJ (2006) Sex and gender differences in pain and its relief. In: McMahon S and Koltzenburg M (eds.) *Wall and Melzack's Textbook of Pain,* 5th edn., ch. 75, pp. 1181–1197. Edinburgh: Elsevier Churchill Livingstone.

Holdcroft A, Sapsed-Bryne S, Ma D, Hammal D, and Forsling ML (2000) Sex and oestrous cycle differences in visceromotor responses and vasopressin release in response to colonic distension in male and female rats anaesthetised with halothane. *British Journal of Anaesthia* 85: 907–910.

Hunt CM, Westerkam WR, and Stave GM (1992) Effect of age and gender on the activity of human hepatic CYP3A. *Biochemical Pharmacology* 44: 275–283.

Iadarola MJ, Naranjo JR, Duchemin AM, and Quach TT (1989) Expression of cholecystokinin and enkephalin mRNA in discrete brain regions. *Peptides* 10: 687–692.

Isacson D and Bingefors K (2002) Epidemiology of analgesic use: A gender perspective. *European Journal of Anaesthesiology, Supplement* 26: 5–15.

Jacox A, Carr DB, and Payne R (1994) Management of Cancer Pain. *Clinical Practice Guideline No.9.* AHCPR Publication No. 94-0592. Rockville, MD: Agency for Health Care Policy and Research, US Department of Health and Human Services, Public Health Service.

Jänig W and Baron R (2006) Is CRPSI a neuropathic pain syndrome? *Pain* 120: 227–229.

Jawaheer D, Lum RF, Gregersen PK, and Criswell LA (2006) Influence of male sex on diease phenotype on familial rheumatoid arthritis. *Arthritis and Rheumatism* 54: 3087–3094.

Jensvold MF, Halbreich U, and Hamilton JA (1996) *Psychopharmacology and Women. Sex Gender and Hormones*. Washington, DC: American Psychiatric Press.

Ji Y, Murphy AZ, and Traub RJ (2003) Estrogen modulates the visceromotor reflex and responses of spinal dorsal horn neurons to colorectal stimulation in the rat. *Journal of Neuroscience* 23: 3908–3915.

Kajantie E and Phillips DI (2006) The effects of sex and hormonal status on the physiological response to acute psychosocial stress. *Psychoneuroendocrinology* 31: 151–178.

Kalyuzhny AE and Wessendorf MW (1997) CNS GABA neurons express the mu-opioid receptor: Immunocytochemical studies. *Neuroreport* 8: 3367–3372.

Keefe FJ, Affleck G, France CR, et al. (2004) Gender differences in pain, coping, and mood in individuals having osteoarthritic knee pain: A within-day analysis. *Pain* 110: 571–577.

Kenntner-Mabiala R, Gorges S, Alpers GW, Lehmann AC, and Pauli P (2007) Musically induced arousal affects pain perception in females but not in males: A psychophysiological examination. *Biological Psychology* 75: 19–23.

Ladwig KH, Mulberger N, Walter H, et al. (2000) Gender differences in emotional disability and negative health perceptions in cardiac patients 6 months after stent implantation. *Journal of Psychosomal Research* 48: 501–508.

Lauber AH, Romano GJ, Mobbs CV, Howells RD, and Pfaff DW (1990) Estradiol induction of proenkephalin messenger RNA in hypothalamus: Dose–response and relation to reproductive behavior in the female rat. *Brain Research – Molecular Brain Research* 8: 47–54.

Lemmens HJM, Burm AGL, Hennis PJ, Gladines MPPR, and Bovill JG (1990) Influence of age on the pharmacokinetics of alfentanil: Gender dependence. *Clinical Pharmacokinetics* 19: 416–422.

LeResche L (1999) Gender considerations in the epidemiology of chronic pain. In: Crombie IK, Croft PR, Linton SJ, LeResche L, and von Korff M (eds.) *Epidemiology of Chronic Pain*, pp. 43–52. Seattle, WA: IASP Press.

LeResche L (2001) Gender, cultural, and environmental aspects of pain. In: Loeser JD (ed.) *Bonica's Management of Pain,* 3rd edn., pp. 191–195. Philadelphia, PA: Lippincott Williams and Wilkins.

Leveille SG, Penninx BW, Melzer D, Ismirlian G, and Guralnik JM (2000) Sex differences in the prevalence of mobility disability in old age: The dynamics of incidence, recovery and mortality. *Journal of Gerontology B* 55: S41–S50.

Leveille SG, Zhang Y, McMullen W, Kelly-Hayes M, and Felson DT (2005) Sex differences in musculoskeletal pain in older adults. *Pain* 116: 332–338.

Loeser JD and Turk DC (2001) Multidisciplinary pain management. In: Loeser JD (ed.) *Bonica's Management of Pain,* 3rd edn., pp. 2069–2079. Philadelphia, PA: Lippincott Williams and Wilkins.

Loyd D and Murphy AZ (2006) Sex differences in the anatomical and functional organization of the periaqueductal gray-rostral ventromedial medullary pathway in the rat: A potential circuit mediateing the sexually dimorphic actions of morphine. *Journal of Comparative Neurology* 496: 723–738.

Loyd DR, Morgan MM, and Murphy AZ (2007) Morphine preferentially activates the periaqueductal gray-rostral ventromedial medullary pathway in the male rat: A potential mechanism for sex differences in antinociception. *Neuroscience* 147: 456–468.

Loyd DR and Murphy AZ (2008) Androgen and estrogen (α) receptor localization on periaqueductal gray neurons projecting to the rostral ventromedial medulla in the male and female rat. *Journal of Chemical Neuroanatomy* 36(3–4): 216–226.

Luo H, Liu J, Kang D, and Cui S (2008) Ontogeny of estrogen receptor alpha, estrogen receptor beta and androgen receptor, and their co-localization with Islet-1 in the dorsal root ganglia of sheep fetuses during gestation. *Histochemistry and Cell Biology* 129: 525–533.

Luyten P and van Houdenhove B (2005) Pain: In search of a common language. *Pain* 116: 170–171.

Manzanares J, Corchero J, Romero J, Fernandez-Ruiz JJ, Ramos JA, and Fuentes JA (1998) Chronic administration of cannabinoids regulates proenkephalin mRNA levels in selected regions of the rat brain. *Brain Research – Molecular Brain Research* 55: 126–132.

Maser RE, Pfeifer MA, Dorman JS, Kuller LH, Becker DJ, and Orchard TJ (1990) Diabetic autonomic neuropathy and cardiovascular risk. Pittsburgh Epidemiology of Diabetes Complications study III. *Archives of Internal Medicine* 150: 1218–1222.

Mayer EA, Berman S, Chang L, and Naliboff BD (2004) Sex based differences in gastrointestinal pain. *European Journal of Pain* 8: 451–453.

Mayer EA, Bradesi S, Chang L, Spiegel BM, Bueller JA, and Naliboff BD (2008) Functional GI disorders: From animal models to drug development. *Gut* 57: 384–404.

McBride SM, Flynn FW, and Ren J (2005) Cardiovascular alteration and treatment of hypertension: Do men and women differ? *Endocrine* 28: 199–207.

McCarthy M, Coirini H, Schumacher M, Pfaff D, McEwen B, and Schwartz-Giblin S (1991) Ovarian steroid modulation of [3H] muscimol binding in the spinal cord of the rat. *Brain Research* 556: 321–323.

McCarthy M, Kaufman L, Brooks P, Pfaff D, and Swartz-Giblin S (1995) Estrogen modulation of mRNA levels for the two forms of glutamic acid decarboxylase (GAD) in female rat brain. *Journal of Comparative Neurology* 360: 685–697.

McGreary DD, Mayer TG, Gatchel RJ, Anagnostis C, and Proctor TJ (2003) Gender-related differences in treatment outcomes for patients with musculoskeletal disorders. *Spine Journal* 3: 197–203.

McQueen JK, Wilson H, Sumner BEH, and Fink G (1999) Serotonin transporter (SERT) mRNA and binding site densities in male rat brain affected by sex steroids. *Brain Research – Molecular Brain Research* 63: 241–247.

Melzack R (1993) Labor pain as a model of acute pain. *Pain* 53: 117–120.

Melzack R (1989) Labat lecture. Phantom limbs. *Regional Anesthesia* 14: 208–211.

Merskey H and Bogduk N (1994) *Classification of Chronic Pain: Descriptions of Chronic Pain Syndromes and Definitions of Pain Terms*, 2nd edn. Seattle, WA: IASP Press.

Méthot J, Hamelin BA, Bogaty P, Arsenault M, Plante S, and Poirer P (2004) Does hormonal status influence the clinical presentation of acute coronary syndromes in women? *Journal of Womens Health* 13: 695–702.

Miaskowski C, Gear RW, and Levine JD (2000) Sex-related differences in analgesic responses. In: Fillingim RB (ed.) *Sex, Gender, and Pain,* vol. 17, pp. 209–230. Seattle, WA: IASP Press.

Mogil JS, Richards SP, O'Toole LA, Helms ML, Mitchell SR, Kest B, and Belknap JK (1997) Identification of a sex-specific quantitative trait locus mediating nonopioid stress-induced analgesia in female mice. *Journal of Neuroscience* 17: 7995–8002.

Mogil JS, Sternberg WF, Marek P, Sadowski B, Belknap JK, and Liebeskind JC (1996) The genetics of pain and pain inhibition. *Proceedings of the National Academy of Sciences of the United States of America* 93: 3048–3055.

Moller RA, Datta S, Fox J, Johnson M, and Covino BG (1992) Effects of progesterone on the cardiac electrophysiologic action of bupivacaine and lidocaine. *Anesthesiology* 76: 604–608.

Moller RA, Datta S, and Strichartz GR (1999) Beta-estradiol acutely potentiates the depression of cardiac excitability by lidocaine and bupivacaine. *Journal of Cardiovascular Pharmacology* 34: 718–727.

Murphy AZ and Hoffman GE (1999) Distribution of androgen and estrogen receptor containing neurons in the for the initiation of male rat periaqueductal gray. *Hormones and Behavior* 36: 98–108.

Murphy AZ and Hoffman GE (2001) Distribution of gonadal steroid receptor-containing neurons in the preoptic–periaqueductal gray-brainstem pathway: A potential circuit for the initiation of male sexual behavior. *Journal of Comparative Neurology* 438: 191–212.

Murray A and Holdcroft A (1989) Incidence and intensity of postpartum lower abdominal pain. *British Medical Journal* 298: 1619.

Murthy BR, Pollack GM, and Brouwer KL (2002) Contribution of morphine-6-glucuronide to antinociception following intravenous administration of morphine to healthy volunteers. *Journal of Clinical Pharmacology* 42: 569–576.

Myers CD, Riley JL, III, and Robinson ME (2003) Psychosocial contributions to sex-correlated differences in pain. *Clinical Journal of Pain* 19: 225–232.

Myles PS, Hunt JO, and Moloney JT (1997) Postoperative 'minor' complications: Comparison between men and women. *Anaesthesia* 52: 300–306.

Nakamura NH, Rosell DR, Akama KT, and McEwen BS (2004) Estrogen and ovariectomy regulate mRNA and protein of glutamic acid decarboxylases and cation-chloride cotransporters in the adult rat hippocampus. *Neuroendocrinology* 80: 308–323.

Naliboff BD (2007) Towards a nondualistic approach to multisystem illness. *American Journal of Gastroenterology* 102: 2777–2780.

Naliboff BD, Berman S, Chang L, et al. (2003) Sex-related differences in IBS patients: Central processing of visceral stimuli. *Gastroenterology* 124: 1738–1747.

Naliboff BD, Berman S, Suyenobu B, et al. (2006) Longitudinal change in perceptual and brain activation response to visceral stimuli in irritable bowel syndrome patients. *Gastroenterology* 131: 352–365.

Naliboff BD, Heitkemper MM, Chang L, and Mayer EA (2000) Sex and gender in irritable bowel syndrome. In: Fillingim RB (ed.) *Sex, Gender, and Pain,* vol. 17, pp. 327–353. Seattle, WA: IASP Press.

Nater UM, Abbruzzese E, Krebs M, and Ehlert U (2006) Sex differences in emotional and psychophysiological responses to musical stimuli. *International Journal of Psychophysiology* 62: 300–308.

Osborne PB, Vaughan CW, Wilson HI, and Christie MJ (1996) Opioid inhibition of rat periaqueductal grey neurones with identified projections to the rostral ventromedial medulla *in vitro*. *Journal of Physiology (London)* 490: 383–389.

Patel H, Rosengren A, and Ekman I (2004) Symptoms in acute coronary syndromes: Does sex make a difference? *American Heart Journal* 148: 27–33.

Pilote L, Dasgupta K, Guru V, et al. (2007) A comprehensive view of sex-specific issues related to cardiovascular disease. *Canadian Medical Association Journal* 176: S1–S44.

Poleshuck EL, Dworkin RH, Howard FM, Foster DC, Shields CG, Giles DE, and Tu X (2005) Contributions of physical and sexual abuse to women's experiences with chronic pelvic pain. *Journal of Reproductive Medicine* 50: 91–100.

Popper P, Priest CA, and Micevych PE (1995) Effects of sex steroids on the cholecystokinin circuit modulating reproductive behavior. In: Micevych PE and Hammer RP (eds.) *Neurobiological Effects of Sex Steroid Hormones,* pp. 160–183. Cambridge: Cambridge University Press.

Price D (2000) Sex and gender in irritable bowel syndrome. In: Fillingim RB (ed.) *Sex, Gender, and Pain,* vol. 17, pp. 327–353. Seattle, WA: IASP Press.

Priest CA, Eckersell CB, and Micevych PE (1995) Estrogen regulates preproenkephalin-A mRNA levels in the rat ventromedial nucleus: Temporal and cellular aspects. *Brain Research – Molecular Brain Research* 28: 251–262.

Raut A and Ratka A (2007) Oxidative damage and sensitivity to nociceptive stimulus and opioids in aging rats. *Neurobiology of Aging* doi:10.1016/j.neurobiolaging.2007.09.010.

Reichling DB and Basbaum AI (1990) Contribution of brainstem GABAergic circuitry to descending antinociceptive controls: I. GABA-immunoreactive projection neurons in the periaqueductal gray and nucleus raphe magnus. *Journal of Comparative Neurology* 302: 370–377.

Renno WM, Mullett MA, and Beitz AJ (1992) Systemic morphine reduces GABA release in the lateral but not medial portion of the midbrain periaqueductal gray of the rat. *Brain Research* 594: 221–232.

Reynolds DV (1969) Surgery in the rat during electrical analgesia induced by focal brain stimulation. *Science* 164: 444–445.

Richardson J and Holdcroft A (2007) Results of forty years Yellow Card reporting for commonly used peri-operative analgesic drugs. *Pharmacoepidemiology and Drug Safety* 16: 687–694.

Robinson ME, George SZ, Dannecker EA, Jump RL, Hirsh AT, Gagnon CM, and Brown JL (2004) Sex differences in pain anchors revisited: Further investigation of most intense and common pain events. *European Journal of Pain* 8: 299–305.

Robinson ME, Riley JL, III, and Myers CD (2000) Psychosocial contributions to sex-related differences in pain responses. In: Fillingim RB (ed.) *Sex, Gender, and Pain*, pp. 41–70. Seattle, WA: IASP Press.

Robinson ME and Wise EA (2004) Prior pain experience: Influence on the observation of experimental pain in men and women. *Journal of Pain* 5: 264–269.

Roe CM, McNamara AM, and Motheral BR (2002) Gender- and age-related prescription drug use patterns. *Annals of Pharmacotherapy* 36: 30–39.

Romero MT, Cooper ML, Komisaruk BR, and Bodnar RJ (1988) Gender-specific and gonadectomy-specific effects upon swim analgesia: Role of steroid replacement therapy. *Physiology and Behavior* 44: 257–265.

Saleh TM and Connell BJ (2007) Role of oestrogen in the central regulation of autonomic function. *Clinical and Experimental Pharmacology and Physiology* 34: 827–832.

Sarlani E, Farooq N, and Greenspan JD (2003) Gender and laterality differences in thermosensation throughout the perceptible range. *Pain* 106: 9–18.

Sarton E, Olofsen E, Romberg R, et al. (2000) Sex differences in morphine analgesia: An experimental study in healthy volunteers. *Anesthesiology* 93: 1245–1254.

Schwartz JB (2003) The influence of sex on pharmacokinetics. *Clinical Pharmacokinetics* 42: 107–121.

Sela RA, Bruera E, Conner-spady B, Cumming C, and Walker C (2002) Sensory and affective dimensions of advanced cancer pain. *Psychooncology* 11: 23–34.

Sierpina V, Levine R, Astin J, and Tan A (2007) Use of mind-body therapies in psychiatry and family medicine faculty and residents: Attitudes, barriers, and gender differences. *Explore (NY)* 3: 129–135.

Silverman DH, Munakata JA, Ennes H, Mandelkern MA, Hoh CK, and Mayer EA (1997) Regional cerebral activity in normal and pathological perception of visceral pain. *Gastroenterology* 11: 264–272.

Siminoski K, Murphy RA, Rennert P, and Heinrich G (1987) Cortisone, testosterone, and aldosterone reduce levels of nerve growth factor messenger ribonucleic acid in L-929 fibroblasts. *Endocrinology* 121: 1432–1437.

Sinchak K, Eckersell C, Quezada V, Norell A, and Micevych P (2000) Preproenkephalin mRNA levels are regulated by acute stress and estrogen stimulation. *Physiology and Behavior* 69: 425–432.

Sioud M and Melien O (2007) Treatment options and individualized medicine. *Methods in Molecular Biology* 361: 327–340.

Smith MA and French AM (2002) Age-related differences in sensitivity to the antinociceptive effects of kappa opioids in adult male rats. *Psychopharmacology (Berl.)* 162: 255–264.

Smith MA and Gray JD (2001) Age-related differences in sensitivity to the antinociceptive effects of opioids in male rats. Influence of nociceptive intensity and intrinsic efficacy at the mu receptor. *Psychopharmacology (Berl.)* 156: 445–453.

Sohrabji F, Miranda RC, and Toran-Allerand CD (1994) Estrogen differentially regulates estrogen and NGF receptor mRNAs in adult sensory neurones. *Journal of Neuroscience* 14: 459–471.

Spangeus A, El Salhy M, Suhr O, Eriksson J, and Lithner F (1999) Prevalence of gastrointestinal symptoms in young and

middle-aged diabetic patients. *Scandinavian Journal of Gastroendocrinology* 34: 1196–2002.
Stamer UM and Stuber F (2007) Genetic factors in pain and its treatment. *Current Opinion in Anaesthesiology* 20: 478–484.
Sternberg WF and Liebeskind JC (1995) The analgesic response to stress: Genetic and gender considerations. *European Journal of Anaesthesiology* 10: 14–17.
Suckow SK, Traub RJ, and Murphy AZ (2005) Sex differences in the activation of the spinoparabrachial circuit and morphine analgesia in a model of visceral pain. *Society for Neuroscience*, Abstract.
Tang B, Ji Y, and Traub RJ (2008) Estrogen alters spinal NMDA receptor activity via a PKA signaling pathway in a visceral, pain model in the rat. *Pain* 137(3): 540–549.
Tershner SA, Mitchell JM, and Fields HL (2000) Brainstem pain modulating circuitry is sexually dimorphic with respect to mu and kappa opioid receptor function. *Pain* 85: 153–159.
Turk DC and Okifuji A (1999) Does sex make a difference in the prescription of treatments and the adaptation to chronic pain by cancer and non-cancer patients? *Pain* 82: 139–148.
Unruh AM (1996) Gender variations in clinical pain experience. *Pain* 65: 123–167.
Unruh AM, Ritchie J, and Merskey H (1999) Does gender affect appraisal of pain and pain coping strategies? *Clinical Journal of Pain* 15: 31–40.
Vanderhorst VG, Gustafsson JA, and Ulfhake B (2005) Estrogen receptor-alpha and-beta immunoreactive neurons in the brainstem and spinal cord of male and female mice: Relationships to monoaminergic, cholinergic, and spinal projection systems. *Journal of Comparative Neurology* 488: 152–179.
Vaughan CW and Christie MJ (1997) Presynaptic inhibitory action of opioids on synaptic transmission in the rat periaqueductal grey *in vitro*. *Journal of Physiology* 498: 463–472.
Vaughan CW, Ingram SL, Connor MA, and Christie MJ (1997) How opioids inhibit GABA-mediated neurotransmission. *Nature* 390: 611–614.
von Korff M, Dworkin SF, LeResche L, and Kruger A (1988) An epidemiologic comparison of pain complaints. *Pain* 32: 173–183.
Voss JA, Good M, Yates B, Baun MM, Thompson A, and Hertzog M (2004) Sedative music reduces anxiety and pain during chair rest after open-heart surgery. *Pain* 112: 197–203.
Wall PD (1999) *Pain: The Science of Suffering*. London: Weidenfeld and Nicolson.
Warren MP and Fried JL (2001) Temporomandibular disorders and hormones in women. *Cells Tissues Organs* 169: 187–192.
Watkins LR and Maier SF (2000) The pain of being sick: Implications of immune-to-brain communication for understanding pain. *Annual Review of Psychology* 51: 29–57.
Weissman AM, Hartz AJ, Hansen MD, and Johnson SR (2004) The natural history of primary dysmenorrhoea: A longitudinal study. *British Journal of Obstetrics and Gynaecology* 111: 345–352.
Wesselmann U and Burnett AL (1999) Genitourinary pain. In: Wall PD and Melzack E (eds.) *Textbook of Pain,* 4th edn., pp. 689–709. Edinburgh: Churchill Livingstone.
Williams FG, Mullet MA, and Beitz AJ (1995) Basal release of Met-enkephalin and neurotensin in the ventrolateral periaqueductal gray matter of the rat: A microdialysis study of antinociceptive circuits. *Brain Research* 690: 207–216.
Wise EA, Price DD, Myers CD, Heft MW, and Robinson ME (2002) Gender role expectations of pain: Relationship to experimental pain perception. *Pain* 96: 335–342.
Wizemann TM and Pardue M-L (2001) *Exploring the Biological Contributions to Human Health: Does Sex Matter?* Washington, DC: Institute of Medicine, National Academy Press.
Woolf CJ and Ma Q (2007) Nociceptors – noxious stimulus detectors. *Neuron* 55: 353–364.
Wulf H, Munstedt P, and Maier C (1991) Plasma protein binding of bupivacaine in pregnant women at term. *Acta Anaesthesiologica Scandinavica* 35: 129–133.
Zager EL, Pfeifer SM, Brown MJ, Torosian MH, and Hackney DB (1998) Catamenial mononeuropathy and radiculopathy: A treatable neuropathic disorder. *Journal of Neurosurgery* 88: 827–830.

Further Reading

Baker L and Ratka A (2002) Sex-specific differences in levels of morphine, morphine-3-glucuronide, and morphine antinociception in rats. *Pain* 95: 65–74.
Becker JB, Arnold AP, Berkley KJ, et al. (2005) Strategies and methods for research on sex differences in brain and behaviour. *Endocrinology* 146: 1650–1673.
Cicero TJ, Nock B, and Meyer ER (1996) Gender-related differences in the antinociceptive properties of morphine. *Journal of Pharmacology and Experimental Therapeutics* 279: 767–773.
Craft RM, Mogil JS, and Aloisi M-A (2004) Sex differences in pain and analgesia: The role of gonadal hormones. *European Journal of Pain* 8: 397–411.
Milne RW, Nation R, and Somogyi AA (1996) The disposition of morphine and its 3- and 6-glucuronide metabolites in humans and animals, and the importance of the metabolites to the pharmacological effects of morphine. *Drug Metabolism Reviews* 28: 345–472.
Mogil JS and Chanda ML (2005) The case for the inclusion of female subjects in basic science studies of pain. *Pain* 117: 1–5.
Mogil JS, Ritchie J, Smith SB, et al. (2005) Melanocortin-1 receptor gene variants affect pain and mu-opioid analgesia in mice and humans. *Journal of Medical Genetics* 42: 583–587.
Mogil JS, Wilson SG, Chesler EJ, et al. (2003) The melanocortin-1 receptor gene mediates female-specific mechanisms of analgesia in mice and humans. *Proceedings of the National Academy of Sciences of the United States of America* 100: 4867–4872.
Sherman JJ and LeResche L (2006) Does experimental pain response vary across the menstrual cycle? A methodological review. *AJP – Regulatory, Integrative and Comparative Physiology* 291: R245–R256.
Shinal RM and Fillingim RB (2007) Overview of orofacial pain: Epidemiology and gender differences in orofacial pain. *Dental Clinics of North America* 51: 1–18.
Veldhuijzen DS, Greenspan JD, Kim JH, Coghill RC, Treede RD, Ohara S, and Lenz FA (2007) Imaging central pain syndromes. *Current Pain and Headache Reports* 11: 183–189.
Wang X, Traub RJ, and Murphy AZ (2006) Systemic morphine produces a greater degree of analgesia in male versus female rats in a model of persistent pain. *AJP – Regulatory, Integrative and Comparative Physiology* 291: R300–R306.

37 Traumatic Brain Injury

B E Masel and R Temple, Transitional Learning Center at Galveston, Galveston, TX, USA

Chapter Outline

37.1 Incidence

Traumatic brain injury (TBI) is a leading cause of death and disability in the United States. Each year, over 1 million individuals who sustain a TBI are treated and released from hospital emergency departments. Of those, 230000 are hospitalized and survive; 80000 are disabled; 50000 die (Centers for Disease Control, 2007a). The incidence of those who do not seek medical attention is unknown, but most likely quite significant.

Structural abnormalities in the pituitary, pituitary stalk, and/or hypothalamus have been found in approximately two-thirds of patients who come to autopsy after dying from a severe TBI (Gupta, 2000). Therefore, any of the hormones produced from the hypothalamus, pituitary or regulated by the pituitary axis can be impacted by a TBI. The clinical manifestations of hormone deficiencies can be subtle or quite obvious. The consequences of these deficiencies may be minor, or they may be fatal – and they can be masked by what has been previously attributed to the intrinsic signs and symptoms of the TBI itself. The diagnosis and treatment of post-traumatic hypopituitarism (PTH) may therefore play a significant role in the recovery from a brain injury.

Morris Simmonds, a pathologist at the University of Hamburg, published the first clinical report of hypopituitarism in 1914 (Simmonds, 1914). He described septic pituitary necrosis in a previously healthy woman who developed severe postpartum (puerperal) sepsis. She developed menopause, muscle weakness, dizziness, anemia, and premature aging before lapsing into coma and dying. Cryan (1981) described the first case of post-traumatic anterior lobe hypopituitarism.

Escamilla and Lisser (1942) published a clinical review on hypopituitarism, then known as Simmonds' disease. The major features were loss of sexual function, marked weight loss, and a low basal metabolic rate. Of the 595 cases cited, only four were due to head injury. As most individuals with a moderate to severe head injury probably died in that era, and given that autopsy reports were not readily available, the number of individuals with PTH was probably understated. Edwards and Clark (1986) published a review of the 47 cases of PTH in the literature and added six new cases. Although skull fractures and prolonged unconsciousness were common, eight cases had either a brief or no loss of consciousness. The symptoms of hypogonadism were most notable with amenorrhea, impotence, loss of libido, and secondary sex characteristics. Of the 20 cases that had specific endocrine testing, 85% were hypothyroid, 40% had a history of diabetes insipidus (DI), all but one had an abnormal cortisol response and none had a normal growth hormone (GH) response to the insulin tolerance test (ITT).

Benvenga et al. (2000) published a large review of PTH in 2000, citing 299 additional cases in the literature in addition to 15 new cases of their own. Of those 15 new cases, 11 of 12 who had imaging studies had an abnormal magnetic resonance imaging (MRI) or computed tomography (CT). Interestingly, only three had a history of loss of consciousness. Eight cases were diagnosed more than 10years post-injury, and in several cases, the patients did not remember the history of a TBI. The authors felt that many cases of previously diagnosed idiopathic hypopituitarism might really be due to PTH.

37.2 Anatomy and Physiology of the Pituitary and Hypothalmus

Lying beneath the brain in the middle cranial fossa, the adult pituitary is a pea-sized structure weighing approximately 600mg. It sits within a boney cave called the sella turcica (Turks saddle) due to its shape, and is connected to the hypothalamus by the thin pituitary stalk. The hypothalamus is supplied by the superior hypophyseal arteries which branch from the internal carotid. The pituitary's blood supply is from the short and long hypophyseal vessels which form the hypothalamic–portal circulation. The long hypophyseal portal veins arise above the diaphragma sella from the median eminence and the stalk. They course downward along the anterior portion of the stalk to supply the anterior lobe of the pituitary with 70–90% of its blood supply, predominantly to the lateral wings. The short portal veins arise below the diaphragma sella, and supply predominantly the medial portion of the anterior pituitary gland. Although the process may be slow and incomplete, severed portal vessels are capable of regeneration, and therefore may allow some resumption of anterior pituitary function (Daniel et al., 1959; **Figure 1**).

The pituitary gland is actually two closely associated functionally and anatomically distinct endocrine organs. The posterior lobe is really an outgrowth from the floor of the hypothalamus. It is innervated directly by hypothalamic neurons which produce hormones that travel via the pituitary stalk to the posterior lobe. Its blood supply is from the short hypophyseal vessels. The posterior lobe is responsible for the storage and release of oxytocin and vasopressin into the surrounding capillary circulation. Vasopressin causes constriction of the arterioles with a resultant rise in blood pressure, and also controls water secretion by the kidneys. Oxytocin is responsible for contraction of the smooth muscle of the uterus and for lactation. As there is some release at the hypothalamus and the stalk itself, vasopressin and oxytocin are very sensitive to neuronal damage at the level of the hypothalamus or stalk, as well as the pituitary gland itself.

The anterior lobe is more glandular than neuronal in appearance, and is the structurally largest part of the pituitary. Neural cells within the hypothalamus synthesize specific releasing and inhibiting hormones which are then secreted directly into portal vessels within the pituitary stalk. These vessels then carry these hormones to the secretory cells within the anterior lobe. Adrenocorticotropic hormone (ACTH) is produced by the corticotrophs which constitute approximately 20% of the anterior pituitary, and are located mainly in the central median pituitary wedge. GH is produced by the somatotrophs which constitute approximately 40% of the pituitary cells, and are located predominantly in the lateral wings of

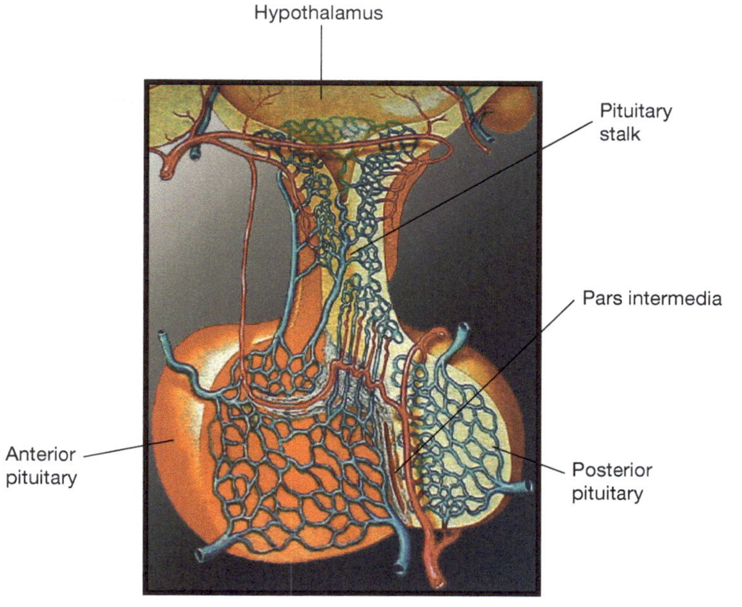

Figure 1 Pituitary and pituitary stalk.

the anterior lobe of the pituitary. TSH is produced by the thyrotrophs which are located in the anterior medial region and constitute 5% of the anterior pituitary. Follicle-stimulating hormone (FSH) and lutenizing harmone (LH) are secreted by the gonadotrophs and constitute 10–15% of the anterior pituitary. Like the somatotrophs, the gonadotrophs are also situated in the lateral wings of the anterior lobe. Prolactin (PRL) is produced by the lactotrophs which constitute 15–25% of the anterior lobe. The lactotrophs are clustered mainly in the median wedge and the lateral wings but are also scattered throughout the anterior lobe (Larsen et al., 2007).

The vascular supply to the pituitary and the location of the secretory cells within the pituitary play a key role in PTH. Only a thin layer of surface cells of the anterior lobe receives arterial blood. The hypophyseal portal system of veins provides the overwhelming majority of the blood supply to the anterior lobe. As the long portal veins arise within the subarachnoid space and then travel through the diaphragma sella, these blood vessels are extremely vulnerable to direct mechanical trauma, direct brain and pituitary swelling, intracranial hypertension, and low cerebral blood flow. As the short portal veins arise below the diaphragma sella from capillaries in the lower part of the stalk, they are less vulnerable to trauma. When the pituitary stalk (and subsequently the long portal vessels) is cut surgically, approximately 90% of the anterior lobe becomes infarcted (Edwards and Clark, 1986).

Not only is the pituitary's vascular supply tenuous following trauma, but the location of the pituitary encased within the sella turcica beneath the diaphragma sella renders the infundibulum and stalk vulnerable to shearing. Also, swelling of the gland following trauma is limited by its bony encasement, therefore causing compression of the pituitary as well as compression of the long portal vessels between the stalk and the free edge of the diaphragma sella. The fragile pituitary vessels are also susceptible to pituitary stalk transection or rupture, as well as hypotension.

A fundamental feature of the endocrine system is its mechanism for maintaining a tight range of hormonal levels through a process of negative and positive feedback control. As demonstrated in **Figure 2**, the hypothalamus produces and secretes releasing factors which stimulate the trophic cells of the anterior lobe to release a particular hormone. That hormone then acts on a target gland to release its hormone. If that target hormone level is low, positive feedback to the hypothalamus increases the releasing factors. If that level is too high, negative feedback to the hypothalamus decreases the output of releasing factors.

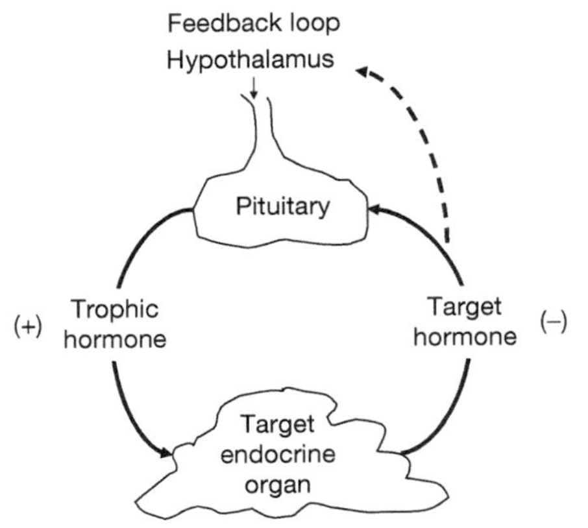

Figure 2 Hypothalamic–pituitary–target feedback loop.

37.3 Prevalence Studies

37.3.1 Acute TBI

Data on the presence of anterior pituitary dysfunction acutely after a TBI are limited. Agha et al. (2004a) studied 50 consecutive patients in the NICU following moderate to severe TBIs. Gonadotropin deficiencies were the most common finding (80%). A low GH response was seen in 18%. A low peak cortisol was seen in 16%, and one patient had thyroid-stimulating hormone (TSH) deficiency. Tanriverdi et al. (2006) studied 104 patients within 24h of their TBI. Eighty-four patients survived their injury. Gonadotropin deficiency was the most common finding (40%). GH deficiency was found in 20% and a low cortisol response was present in 8.8%. There was no relationship of basal hormonal levels to mortality.

37.3.2 Chronic TBI

Kelly et al. (2000) published on the prevalence of PTH in 24 patients with a chronic TBI, including two patients who had sustained a nontraumatic subarachnoid hemorrhage. All had complicated mild, moderate, or severe TBIs. Eight patients (36.4%) had a subnormal response in at least one hormonal axis, including five patients (22.7%) who were deficient in two axes. A history of a hypotensive episode or hypoxic insult was obtained in seven of the eight patients with PTH. All patients with PTH had an initial Glasgow Coma Scale (GCS) score of ten or less and had diffuse swelling on the initial CT scan. They concluded that PTH was more likely to be seen in patients with moderate to severe TBIs.

Lieberman et al. (2001) evaluated 70 adults with moderate to severe chronic TBIs recruited from a residential postacute brain injury program. Of the 38 patients who had initial GCS scores in the records, 32 were eight or less, indicative of a severe injury. Severe GH deficiency (GHD) was seen in 14.6%. The mean insulin-like growth factor-1 (IGF-1) level was below normal in 12.2% of patients without GHD and 57.1% of patients with GHD. A low serum thyroxine (T4) was present in 11%, and 10% had normal T4 levels with low TSH levels. Overall, 87% fell below the mid-normal value for both TSH and T4. Basal morning cortisol levels were found to be below normal 45.7%. Five patients (7.1%) did not respond adequately to ACTH stimulation. None of the patients studied had unexplained deficiencies of the pituitary–gonadal axis. Overall, 36 patients (51.4%) had a single abnormal axis and 12 patients (17.1%) had two abnormalities. No patients had evidence of syndrome of inappropriate antidiuretic hormone (SIADH or DI). As opposed to the Kelly study (Kelly et al., 2000), there was no relationship between hypopituitarism and GCS score.

Aimaretti et al. (2004) published a multicenter study on hypopituitarism in the acute phase (3months post-TBI) and then reevaluated those patients 1-year post-injury (Aimaretti et al., 2005). Fifteen had a severe TBI; 22 had a moderate TBI; and 33 had a mild TBI. At 3 months, 32.8% were found to have PTH. Panhypopituitarism (deficiencies in all axes) were found in 5.7%. DI was present in 4.2%. Of those deficient in at least one axis, 22.8% were defined as having severe GHD and 15.7% were classified as partial GHD. Only 12.8% of those studied had IGF-1 levels below 25% of the age-related norm.

At 12months retesting, some degree of hypopituitarism was present in 22.7%. All subjects with panhypopituitarism at 3months had the same diagnosis upon retesting at the 12month follow-up. Seventy-five percent of individuals with single or multiple axis abnormalities at 3months were normal upon retesting at 12months. Conversely, 2 of the 36 subjects who were normal at 3months developed single-axis hypopituitarism at the 1-year follow-up, and 2 of 15 who had single-axis deficiencies at 3months developed multiple deficiencies by 1year.

The results of this important study showed that after a TBI, early panhypopituitarism does not resolve; however, single and multiple axis pituitary dysfunction may resolve over time. Conversely, normal pituitary function early after a TBI may become impaired by 12 months, perhaps a reflection of inadequate pituitary reserve.

Bondanelli et al. (2004) studied 50 individuals 1–5 years following TBI. Twenty-nine patients were more than 4 years from their TBI. PTH was found in 57.1% of those studied. There was no relationship of PTH to injury severity, time from injury or type of injury.

As indicated in **Table 1**, several other studies have confirmed the findings in these initial reports (Kelly et al., 2000; Lieberman et al., 2001; Agha et al., 2004a; Aimaretti et al., 2005; Bondanelli et al., 2004; Popovic et al., 2004; Tanriverdi et al., 2006). Although there were differences in patient selection and diagnostic methodology, there is broad agreement that pituitary screening is indicated after a TBI, as PTH is common in adults.

37.4 Pediatric TBI

In the United States the annual incidence of TBI in children less than 14 years of age is approximately 475 000. There are 37 000 hospitalizations and over 2500 deaths. Over 30 000 children are disabled due to a TBI on an annual basis. It is also estimated that the annual incidence of mild TBI (mTBI) in children is over 900 000 (Centers for Disease Control, 2007b). Children are obviously also at risk for PTH. Acerini et al. (2006) published a reviewed 20 cases of pediatric PTH. All except one had multiple deficiencies. Growth failure and delayed or arrested puberty were the most common presenting symptoms. Delay of diagnosis was extreme in many cases, as PTH was not considered as a possible complication of a TBI.

Einaudi et al. (2006) evaluated pediatric PTH in retrospective and prospective studies. The prospective group consisted of 30 patients screened acutely, and then at 6 and 12 months. The retrospective group consisted of 22 patients. The overall incidence of PTH was in 10.4% in both studies. Their hypothesis for the lower incidence of PTH than what has been found in the adult population was due to differences in neuronal plasticity which therefore might allow adolescents and children to recover without endocrine dysfunction.

Niederland et al. (2007) also reviewed pediatric TBI, studying 26 children 30.6±8.3 months post-injury, against 21 age-matched controls. All had abnormal CT scans, although seven children had no history of loss of consciousness. As a group, the children with TBIs were slightly shorter when compared to age-matched controls, but not to statistical significance. Although counter-intuitive, there was no correlation between the severity of the head injury and the incidence of hypopituitarism (Niederland et al., 2007).

Aimaretti et al. (2005) reviewed PTH in 23 adolescents and young adults (16–25 years) 3 and 12 months post-injury. The patients were classified as having mild TBI (10), moderate (6), and severe (7) based on their initial GCS scores. At 3 months, PTH was found in 34.6%. All patients with panhypopituitarism or multiple deficits at 3 months had the same findings at 12 months. Notably, it was also found that isolated pituitary deficits may improve over time (21.7% at 3 months vs. 17.4% at 12 months). They concluded that neuroendocrine testing should be performed at 3 and 12 months in this age group. Isolated deficits may improve over time; however, multiple deficits at 3 months may be irreversible.

37.5 Imaging Following TBI

There are limited studies on imaging of the pituitary and hypothalamus following TBI. Normal pituitary CT scans were present in 10 of 11 cases with PTH in a study of women with transient amenorrhea after a TBI (Cytowic et al., 1986). Benvenga and colleagues found abnormal pituitary/hypothalamic imaging (CT/MRI) in over 90% of 76 patients reviewed (Benvenga et al., 2000), and in 10 of 13 patients in a later study (Benvenga et al., 2004). This contrasts with Bondanelli et al. (2004) who found abnormal pituitary–hypothalamic MRIs in only 2 of 27 cases of PTH. Hypoxic insult as well as axonal shearing might result in injuries undetectable by CT or MRI, especially during the period immediately following trauma. Based on limited studies, it appears that normal routine CT or MRI imaging does not rule out PTH.

37.6 Pituitary Hormones

37.6.1 Prolactin

PRL suppresses sexual drive, decreases reproductive function, and induces and maintains lactation. Its central control mechanism is inhibitory, blocking the release of gonadotropin-releasing hormone (GnRH) from the hypothalamus. In the ovary, the result is ultimately anovulation, and in the males, the result is low testosterone and decreased libido. Measurement of PRL is by a basal fasting morning level. Trauma, chronic renal failure, exercise, and

Table 1 Anterior pituitary dysfunction (%)

Authors	*GCS scores*	*Injury to testing times (months)*	*Total*	*GHD (severe)*	*ACTHD*	*GnTD*	*TSHD*	*↑PRL*	*Association with TBI severity*
Kelly et al. $n = 22$	3–15	Median 26	36.4	18.2	4.5	22.7	4.5	0	Diffuse brain swelling
Lieberman et al. $n = 70$	NA	Median 13	68.5	14.6	45.7	1.4	21.7	10	None
Agha et al. $n = 102$	3–13	Median 17	28	10.7	12.7	11.8	1	11.8	None
Aimaretti et al. $n = 100$	3–15	3	35	37	8	17	5	10	None
Bondanelli et al. $n = 50$	3–15	Range 12–64	54	28	0	14	10	8	Lower GCS scores
Popovic et al. $n = 67$	9–13	Median 44	34	15	7	9	4	3.5	None
Leal-Cerro et al. $n = 170$ (99 had biochemical testing)	<8	>12	24.7	5.8	6.4	17	5.8	NA	None
Tanriverdi et al. $n = 52$	3–15	<12	50.9	37.7	19.2	7.7	5.8	3.8	None

ACTHD, adrenocorticotropin deficiency; GnTD; gonadotropin deficiency; GCS, Glasgow Coma Scale; GHD, growth hormone deficiency; TBI, traumatic brain injury; TSHD, thyroid-stimulating hormone deficiency; PRL, prolactin.

Adapted from Agha A and Thompson E (2006) Anterior pituitary dysfunction following traumatic brain injury (TBI). *Clinical Endocrinology* 64: 481–488.

seizures will increase PRL levels. PRL levels are also increased by medications, including oral contraceptives, antihypertensives, neuroleptics, dopamine agonists, and serotonin re-uptake inhibitors (Kasper et al., 2004). As PRL levels are sensitive to so many different disease states, some endocrinologists consider an elevated PRL level as a nonspecific indicator for the presence of disease.

37.6.2 Thyroid Hormone

The thyroid gland produces two related hormones: T4 which then breaks down to triiodothyronine (T3). Both hormones are highly protein bound. As only the unbound portion is active, routine thyroid screening will usually involve a free T4 and TSH level. Replacement dosing with T4 is adjusted by measurement of the TSH level. In the presence of normal pituitary function, a high TSH indicates the need to increase the replacement thyroid dose.

The common signs of hypothyroidism include: weight gain, coarse hair and skin, puffy face, hypothermia, intellectual impairment, and hoarse voice. Common symptoms of hypothyroidism include: hair loss, weakness, dyspnea, dry skin, and feeling cold.

37.6.3 Thyroid Hormone and Cognition

Although the literature on the cognitive and psychiatric effects of abnormal thyroid function following brain injury is sparse, much has been written about the effects of thyroid dysfunction and replacement therapy in general. Similar to the findings with GH, the cognitive effects of thyroid dysfunction appear to mimic deficits seen following brain injury. Hypothyroid patients typically demonstrate deficits in speed of information processing, executive functioning, and aspects of memory (Denicoff et al., 1990). In a comprehensive review of the literature, Davis and Tremont (2007) concluded that "overall, hypothyroidism appears to be associated with neuropsychological deficits in attention, some aspects of executive functioning, and verbal and visual memory in both older and younger adults and across a spectrum of thyroid disease severity." Thyroid disease has also been implicated in depressive symptomatology and described as a modifiable risk factor for depression (Haggerty et al., 1993). This relationship appears to be true for clinical as well as subclinical hypothyroidism (Davis et al., 2003).

The extent to which thyroid replacement improves cognitive and psychiatric symptoms is somewhat unclear, and may be dependent upon the severity of hypothyroidism prior to treatment (e.g., subclinical vs. primary hypothyroidism). Davis and Tremont (2007) reviewed several treatment studies of subclinical hypothyroidism, typically with a majority of female patients. Results were generally indicative of improvement in symptoms, at times to the point of performing equivalently to healthy controls. However, recent studies of overt hypothyroidism suggest that recovery following treatment may be incomplete. For example, Wekking et al. (2005) assessed cognitive functioning and psychological well-being in 141 patients with primary hypothyroidism who had been treated with T4 for at least 6 months. Patients demonstrated poor performance in a number of cognitive domains, particularly on tests of verbal memory and complex attention, compared to normative reference data. Levels of psychological well-being were also below expected levels based on normative data. Similarly, Samuels et al. (2007) found that primary hypothyroidism patients treated with T4 reported more symptoms and poorer quality of life compared to euthyroid controls, and demonstrated poorer performance on measures of working memory and motor learning.

Evidence from the animal literature provides evidence for a possible mechanism linking thyroid function to cognition as well as for the positive effects of thyroid hormone replacement on cognition following brain injury. For example, there is some suggestion that thyroid hormone regulates neurogenesis in the rat hippocampus, providing a logical role for thyroid hormone in learning and memory (Desouza et al., 2005). Smith et al. (2001) administered either T4 or a placebo to rats prior to learning a water maze. T4 was administered either subchronically (every day for 4 days) or chronically (every third day for 28 days). A cognitive deficit was then induced in half of the rate with scopolamine. Both treatment regimens were found to facilitate the initial learning of the spatial task as well as to mitigate the deleterious effects of scopolamine.

37.6.4 Steroids

The adrenal cortex produces three major classes of steroids in response to ACTH stimulation: mineralocorticoids, adrenal adrogens, and glucocorticoids. Adrenal androgens are responsible for secondary sexual characteristics in females. Glucocorticoids help modulate metabolism and immune responses. The mineralocorticoids modulate blood pressure, vascular volume, and electrolytes, and are critical in the stress response.

In the pituitary, the corticotrophs synthesize and secrete ACTH under the feedback control of hypothalamic corticotropin-releasing hormone (CRH). Other factors – including stress, hypoglycemia, sleep–wake cycle, exercise, and free cortisol levels – are also involved in ACTH regulation. Secreted in a pulsatile fashion, ACTH tends to follow a circadian rhythm. The levels are lowest several hours after normal nocturnal sleep and highest around the time of normal morning awakening. The half-life of circulating ACTH is less than 10min, and its actions are within minutes of its release (Kasper et al., 2004). Free urinary cortisol and basal plasma cortisol are often used for diagnostic testing of the adrenal axis; however, levels are frequently in the low normal range even with proven disease (Larsen et al., 2007). The ACTH stimulation test with a low dose of ACTH (cosyntropin) is more definitive.

As cortisol levels will normally increase in response to hypoglycemia, the ITT can be used for diagnosing secondary hypoadrenalism, as well as GHD. Common clinical signs and symptoms of cortisol deficiencies include fatigue, weakness, anorexia, nausea, and hypotension.

There is some evidence to suggest that basal cortisol levels following brain injury may be predictive of fatigue severity. Bushnik et al. (2007) examined the relationship between neuroendocrine function and fatigue in 64 individuals one or more years post-TBI. A trend was found between lower basal cortisol and greater fatigue severity on both the Fatigue Severity Scale and the Global Fatigue Index.

Given its role as a stress hormone, a relationship between cortisol levels and psychiatric symptoms, particularly symptoms of anxiety, seems tenable. In fact, there is evidence in the literature for a complex relationship between post-injury cortisol levels, injury severity, and the development of anxiety. Tanriverdi et al. (2007) measured pituitary functions within 24h of trauma in a sample of 104 patients with TBI. They found a positive relationship between cortisol levels and GCS. In other words, less severe injuries were associated with higher cortisol levels. Flesher et al. (2001) examined the relationship between serum cortisol levels, amnesia, and the development of post-traumatic stress disorder (PTSD) symptoms at 1month in 70 motor vehicle accident victims. Amnestic patients (i.e., those who could not identify events shortly before, during, or shortly after the crash) demonstrated lower norepinephrine/cortisol ratios than nonamnestics, were less likely to meet criteria for PTSD, and displayed fewer symptoms of PTSD than those who were not amnesic. Taken together, the results of these two studies suggest that factors associated with injury severity are antagonistic to an elevated cortisol response. Perhaps cortisol response is related to stressful recollections of the incident that caused the injury. Factors such as loss of consciousness and post-traumatic amnesia may prevent awareness of the injury and its potential tragic consequences, thus precluding an associated stress response.

37.6.5 Gonadotropins

The gonadotrope cells make up approximately 10% of the anterior pituitary, producing two hormones: FSH and LH. Their synthesis and release is dynamically regulated by GnRH from the hypothalamus. FSH and LH initiate and regulate germ cell development in concert with peripheral hormones, and regulate gonadal steroid–hormone biosynthesis (Larsen et al., 2007). Feedback to both the hypothalamus and the pituitary is provided by testosterone and estrogen. Testosterone deficiency in adult males will result in decreased fertility and libido, decreased muscle mass and muscular weakness, as well as loss of secondary sexual characteristics. Women will also exhibit decreased libido and fertility, as well as amenorrhea and decreased bone density.

Central hypogonadism is associated with low or inappropriately normal serum gonadotropin levels despite low estradiol levels in females and low testosterone in males. Assessment of the axis is achieved by measurement of the total serum testosterone, FSH, and LH. In females, a measurement of 17β-estradiol-2 may be obtained. In the appropriate clinical setting, a history of normal menstrual cycles may also be considered adequate.

In the late 1980s, Stein (2001) observed that female rats tended to recover better than males from a TBI. This led to a series of studies that found the female rats recovered best during the estrous phase of their cycle when the level of progesterone was highest (Roof et al., 1993). Later studies on progesterone-treated rats showed that decreased cerebral edema (Wright et al., 2001) and decreased axonal injury (O'Connor et al., 2007) were inversely proportional to the progesterone level. Progesterone-treated rats after a TBI also functioned better in the Morris water maze than untreated rats, suggesting that there might be improved recovery with this treatment (Roof et al., 1994). It has been found that progesterone treatment initiated as long as 24h after the injury might still provide neuroprotection by reducing glutamate toxicity and upregulating gamma-aminobutyric acid (GABA; Roof et al., 1996). A pilot study in 100 acute TBIs showed progesterone to be safe, and produced a marginally significant improvement in 30-day survival (Wright et al., 2007).

Testosterone levels may influence rehabilitation outcomes in males with brain injury. Young et al. (2007) examined the relationship between serum testosterone, length of stay, and functional independence in 54 male TBI victims consecutively admitted to an inpatient rehabilitation setting. Low serum testosterone at the time of admission was associated with longer lengths of stay, lower functional independence ratings, and less improvement in functional independence over time. Individuals with low serum testosterone on admission stayed 26 days longer, on average, than those with normal levels.

There are no large studies on the results of sex hormone replacement after TBI, and therefore any changes must be inferred from other disease states. Presently, the issue of cognitive changes with sex hormone supplementation or replacement remains unresolved (Cherrier et al., 2005).

37.6.6 Growth Hormone

Located in the lateral wings of the pituitary, the GH-secreting cells make up approximately 50% of the total anterior pituitary cell population. GH is secreted in a pulsatile fashion, predominantly during the first episode of slow-wave sleep. GH secretion declines throughout adulthood, paralleling the gradual age-related decline in slow-wave sleep (Kasper et al., 2004). GH levels are increased with fasting, hypoglycemia, exercise, and physical stress. GH secretion is suppressed by obesity, hyperglycemia, hypothyroidism, and an elevated IGF-1.

Although GH has direct effects in target tissues, many of its metabolic and physiologic effects are mediated through the stimulation of the synthesis and release of IGF-1. Produced predominantly in the liver, IGF-1 levels are also profoundly affected by age.

Levels increase during puberty, peak at adolescence, and then gradually decline through middle age.

37.6.7 Diagnosis and Treatment

Due to the pulsatile nature of GH secretion and the short half-life (16 min), a random GH assay to determine GH status is inadequate. Random GH levels are undetectable in 50% of the samples taken from healthy adults and are undetectable in most obese and elderly subjects (Larsen et al., 2007). The diagnosis is established with the appropriate clinical history and a GH peak concentration of less than $5\,\mu m l^{-1}$ after adequately produced hypoglycemia ($<40\,mg\ dl^{-1}$). The rapid and marked hypoglycemia produced by ITT can be epileptogenic. As the TBI population has a greater incidence of seizures, there is reluctance on the part of many, if not most, endocrinologists to use the ITT. The ITT is the gold standard; however, provocative testing with arginine+GHRH (normal peak $>16.5\,\mu m l^{-1}$), arginine+GH-releasing peptide-6 (normal peak $>20\,\mu m l^{-1}$), and glucagon (normal peak $>5\,\mu m l^{-1}$) have also been used (Ghigo et al., 2005). An age-adjusted IGF-1 level can be used as merely a rough measure of GH status, as it only provides an integrated average measure of GH secretion. Moreover, in adults, a normal IGF-1 level does not necessarily exclude the diagnosis of GHD. As a rough rule of thumb, patients with an IGF-1 level more than 2 standard deviations below the norm for their age should be considered abnormal. Regardless of the IGF-1 level, provocative testing is also recommended when another pituitary deficit is identified, due to the likelihood of GHD as well.

37.6.8 Treatment of GHD

GH is species specific, and therefore, GH from nonprimates cannot be used in humans. It is now made by recombinant DNA technology. As it is a protein, it cannot be given in a pill or liquid form, and must be injected subcutaneously. Although the circulating half-life is 20 min, the biological half-life is 9–17 h, and GH can be given once a day. Children require 20–40 $\mu g\ kg^{-1}$ daily, targeting a normal IGF-1 level for age and sex. Adults are frequently started on 150–300 μg and then increased after 2–3 months if the IGF-1 level does not normalize when adjusted for age and sex. GH deficiency should always be treated. The dose is reduced if the patient has any major side effects such as myalgias, hypoesthesias, arthralgias, peripheral edema, or paresthesias (Parker and Schwimmer, 2001).

37.6.9 Metabolic Effects of GHD

Isolated GH deficiency in children is characterized by short stature, increased fat, and a tendency toward hypoglycemia. Adult GHD can produce many metabolic disturbances which may compromise the health and quality of life of the patients, as well as increase their risk for cardiovascular and cerebrovascular disease. Adult GHD may result in decreased bone mass and lean body mass, as well as increased visceral and subcutaneous fat (Bengtsson et al., 1993). As GH is anabolic, it causes an increase in muscle mass and fat mobilization with resultant decreased fat deposition. GHD is also associated with decreased exercise capacity (Colao et al., 2004). GHD has been linked to a higher risk of bone fractures and increased lipid levels (Colao et al., 1999). Adult GHD is also associated with decreased life expectancy (Rosén and Bengtsson, 1990).

37.6.10 Metabolic Effects of GH Replacement

Replacement of GH in deficient adults will result in decreased lipids and body fat (Colao et al., 2004). GH replacement in adults with nontraumatic GHD has been shown to improve exercise capacity, increase lean body mass, and decrease fat mass without altering carbohydrate tolerance (Whitehead et al., 1992; Gibney et al., 1999; Bengtsson et al., 1999). Patients with hypopituitarism have a 1.5- to 6.7-fold increase in mortality from vascular disease (Rosén and Bengtsson, 1990). Using media thickness as a marker for vascular disease, Pfeifer et al. (1999) evaluated intimal thickness in 11 males with nontraumatic GHD compared to matched controls, and followed them over 18 months of treatment. At baseline, the GHD subjects had significantly greater intimal thickness when compared to controls, and with treatment, these early changes of atherosclerotic were reversed. These findings suggest that treatment of nontraumatic GHD may reduce vascular morbidity and mortality (**Table 2**).

Table 2 Signs and symptoms of growth hormone deficiency

Signs
Abnormal lipid profile
Increased cholesterol
Increased LOL, VLDC, triglycerides
Decreased HDL
Decreased bone density
Reduced strength
Altered body composition
Decreased lean body mass
Increased truncal fat
Symptoms
Fatigue
Impaired psychological function
Poor memory
Poor concentration
Depression
Anxiety
Reduced exercise performance
Increased abdominal fat

37.6.11 Cognitive Impact of Post-Traumatic GHD

Greater cognitive dysfunction has been reported in traumatic brain injury patients with GH deficiency compared to those with normal GH levels. However, an important question is whether this observation reflects specific effects of GH deficiency or simply a reflection of injury severity. León-Carrión et al. (2007) examined cognitive and emotional functioning in 22 patients with severe traumatic brain injury: 11 with isolated GH deficiency and 11 without pituitary deficiencies. The GH-deficient group demonstrated greater deficits in simple attention, more intrusions and repetitions on a memory task, increased reaction time, and greater emotional disruption. The results were interpreted as supporting the notion that some deficits following TBI may be the direct result of GH deficiency, rather than being attributable more generally to the brain injury *per se*. These results must be interpreted with caution, as there was no indication that injury severity was similar across groups.

Similarly, Kelly et al. (2006) evaluated neurobehavioral and quality-of-life issues in 44 patients with TBI 6–9 months post-injury. Compared to individuals with normal pituitary function, those with deficits within the GH axis had higher rates of at least one marker of depression, as well as reduced quality of life in the domains of physical health, general health, emotional health, pain, energy, and fatigue. The authors noted a weak trend toward the GH-deficient/insufficient group having a more severe injury as seen on CT scans.

Popovic et al. (2004) evaluated the relationship of GHD to cognitive disabilities and mental distress in 67 survivors of a moderate to severe TBI. They found a significant correlation with the peak GH levels to short-term and long-term memory deficits, paranoid ideation and somatization. They also found a correlation between lower IGF-1 levels and impaired visual memory.

There are presently only anecdotal reports relative to GH replacement in PTH. Although PTH may be included in the study cohorts, published reports on GH replacement are predominantly from other disease entities such as radiation therapy, tumors, and vascular insults. Elucidation of the independent effects of GH deficiency on cognition may also be found in GH-deficient individuals with other etiologies. In fact, GH deficiency has been demonstrated to interfere with cognitive functioning in individuals with nontraumatic conditions. In a meta-analytic study, Falleti et al. (2006) reviewed five cross-sectional studies investigating GH deficiency. Patients studied included those presenting with either isolated or multiple pituitary deficiencies. Etiologies of the former group included pituitary tumors treated either surgically or with medication. Compared to matched controls, analysis of effect sizes revealed moderate to large impairments in attention, memory, and executive functioning. Clearly, this pattern of deficits is very similar to the expected pattern seen in individuals who have sustained a TBI, further complicating attempts to parse out the independent effects of brain injury and GH deficiency on cognition.

37.6.12 Cognitive Impact of GH Replacement

The positive effects of GH and GHRH administration on cognitive functioning have been examined in adults with and without GH deficiency. Vitiello et al. (2006) administered GHRH to 89 healthy older adults in a prospective randomized design. Following 6 months of treatment, significant improvement was noted in nonverbal intellectual functioning, psychomotor speed, and working memory. These findings were independent of gender, estrogen status, or baseline cognitive abilities.

Burman et al. (1995) found improvement with GH replacement in quality-of-life measures, including energy and emotion in a study of patients with nontraumatic GHD. Gibney et al. (1999) also evaluated psychological well-being over a 10-year span in 21 adults with nontraumatic GHD. They found improvement compared to placebo in the overall score, energy levels, and emotional reaction. Arwert et al. (2005) also followed 23 males with treated nontraumatic GHD over a 10-year period. They found improvement in anxiety, mood, as well as short- and long-term memory. However, the results of a meta-analysis of 15 studies on GH replacement and patient reported outcomes suggested that the impact of GH treatment on cognition was inconclusive (Arwert et al., 2005). Oertel et al. (2004) found that GH treatment resulted in a significant improvement in attentional performance but without change in nonverbal intelligence and long-term verbal memory. Utilizing functional MRI (fMRI), Arwert et al. (2006) studied the effects of GH replacement in 13 childhood onset nontraumatic GHD adults. After 6 months of treatment, they found improved working-memory and long-term memory. During working memory tasks, the fMRI showed activation in the parietal, prefrontal, occipital, motor, and anterior cingulate cortices and right thalamus. Following GH treatment, decreased activation was seen in the ventrolateral prefrontal cortex, suggesting decreased effort and more efficient recruitment of the involved neural system.

Falleti et al. (2006) reviewed eight cross-sectional and prospective studies of GH replacement in adults. Cross-sectional results revealed that patients treated for GH deficiency performed significantly worse than controls on measures of attention and memory, and slightly worse on executive function tests. But it is perhaps the prospective studies that are most enlightening. Moderate to large improvements were noted in attentional functioning at both 3–6 months and 9–12 months of treatment. Similar results were observed for memory functioning, with some improvement noted following 1 month of treatment. GH replacement also improved spatial functioning, with moderate effects noted at 1 and 6 months. Large effects were noted in executive functioning at 3–6 months, though only a small improvement was noted after 9–12 months of treatment. The authors accounted for practice effects as a potential confound to these results and reported that the improvements noted at 3–6 months were greater than would be expected from practice effects alone. The results suggest that cognitive improvement takes place for at least a year with continued GH therapy.

The findings from studies of the effects GH replacement has on cognitive functioning provide promise as a treatment for cognitive impairment following brain injury. The results from nontraumatically injured patients suggest that some of the deficits observed following TBI may in fact be the direct result of GH deficiency. Further, the GH replacement literature suggests that at least some of these deficits may be amenable to treatment. Prospective randomized studies of GH replacement with TBI patients are needed to solidify this hypothesis.

37.7 Posterior Pituitary Dysfunction Following TBI

Formed by axons in the paraventricular and supraoptic nuclei of the hypothalamus, the posterior pituitary gland (the neurohypophysis) stores and releases two hormones produced by the hypothalamus. Arginine vasopressin (AVP), also known as antidiurectic hormone (ADH), reduces water loss by concentrating the urine at the level of the renal tubules. Oxytocin, the other hormone produced by the neurohypophysis, stimulates uterine contraction and is responsible for lactation.

37.7.1 Arginine Vasopressin

Synthesized in the hypothalamus, AVP is then packaged in neurosecretory vesicles and stored in the posterior pituitary until released into the peripheral blood in response to osmotic changes. Plasma AVP will rise and cause an antidiuresis in response to an elevated plasma osmolarity. Nausea, smoking, glucocorticoid deficiency, and acute hypoglycemia will also stimulate AVP secretion (Kasper et al., 2004). Plasma AVP decreases in response to a decreased plasma osmolarity, resulting in a diuresis.

37.7.2 Diabetes Insipidus

There are two types of DI. Central DI is due to a malfunction or injury to the neurohypophysis. The subsequent decreased secretion of AVP usually results in abnormally large volumes of dilute urine. Deficiencies in the antidiuretic actions of AVP (nephrogenic DI) can be genetic, acquired, or caused by various drugs. DI of either type is usually associated with plasma hyperosmolarity and hypernatremia.

Uncomplicated pituitary DI can be treated, but not cured, with desmopressin (DDAVP), a synthetic analog of AVP. DDAVP can be administered IV, subcutaneously, by pill or by nasal spray. Nephrogenic DI does not respond to DDAVP, but may improve with amiloride or thiazide diuretics in conjunction with a low sodium diet (Kasper et al., 2004).

37.7.3 Syndrome of Inappropriate Antidiuretic Syndrome

Plasma hypo-osmolarity and hyponatremia with the production decreased volumes of hyperosmolar urine suggests a diagnosis of SIADH, as AVP levels are inappropriately elevated when they should be suppressed. SIADH may be caused by infections, neoplasms, trauma, drugs, and ischemia. Based on history, physical, and laboratory findings, the diagnosis is one of exclusion. The treatment for acute SIADH is to reduce the fluid intake to less than urine output and insensible loss. The treatment for chronic SIADH is with either fludrocortisone or demeclocycline (Agha et al., 2004b).

37.7.4 Incidence of Posterior Pituitary Dysfunction

Chronic posterior pituitary function after a TBI has not been extensively studied. Agha et al. (2004b) studied 102 individuals for posterior pituitary dysfunction at a median of 17 months (range 6–36 months) post-injury. Twenty-two patients (21.6%) developed acute DI of whom seven patients (6.9%) developed chronic DI. The group with the acute DI and ultimately permanent DI had cerebral edema and lower GCS scores.

Of the study subjects, 13 patients (12.7%) had acute SIADH, but there was no relationship to cerebral edema or GCS score. Only two patients developed permanent DI. Other studies have yielded conflicting data on the incidence of SIADH after a TBI. Figures have ranged from 2.3% to 36.6% (Becker and Daniel, 1973; Born et al., 1985; Dóczi et al., 1982; Twijnstra and Minderhoud, 1980; Vingerhoets and de Tribolet, 1988).

37.8 Treatment

37.8.1 When to Screen

In 2004, consensus guidelines on screening for PTH were published by a group of endocrinologists and rehabilitation specialists (Ghigo et al., 2005). A study by Aimaretti et al. (2005) published shortly thereafter confirmed the consensus guidelines. Hormonal testing is recommended during hospitalization when clinically indicated. If studies are negative, follow-up evaluation is recommended at 3 months, and again at 12 months. Patients with adrenal insufficiency, DI, or other symptoms of hypopituitarism should undergo testing of the entire pituitary axis without waiting for 3 months. For those more than 12 months post-TBI, a baseline hormonal workup is recommended at first visit.

37.8.2 How to Screen

Table 3 shows routine basal hormonal screening recommendations for PTH. As noted previously, GHD can be present despite a normal IGF-1 level. When other pituitary deficits are present or the index of suspicion for GHD is high, provocative testing should be conducted. Although baseline screenings can be conducted and evaluated by nonendocrinologist, patients should be referred to an endocrinologist for provocative testing.

37.8.3 When to Treat

Previously noted studies have shown that some deficits of the hypothalamic–pituitary axis may be transient or, conversely, may develop as long as a year following a TBI. All patients with multiple pituitary deficits or panhypopituitarism should undergo immediate replacement of all pituitary deficiencies, except for GH, as replacement of other pituitary deficits may restore a normal GH response to provocative testing. Therefore, it is recommended that appropriate replacement of other deficits be provided first to avoid unnecessary GH therapy in patients with transient GHD that simply reflects other pituitary impairments. As with GHD, gonadal deficits may be transient, and merely a reflection of a stress-induced impairment. Secondary hypogonadism does not represent a clinical emergency. Patients with isolated gonadal deficits should be retested before hormonal replacement is initiated. Due to the anabolic actions of testosterone, however, replacement therapy might be advantageous in males. In women, with secondary amenorrhea, it may be prudent to monitor menses over time to forestall hormone therapy. These recommendations relative to moderate-severe TBI are represented in an algorithm in **Figure 3**.

Table 3 Routine basal hormonal screening tests for post-traumatic hypopituitarism

Basal hormone test	*Test time (h)*
Serum cortisol (morning)	0900
fT3[a], free T_4, thyroid-stimulating hormone (TSH)	0900
IGF-I	0900
Follicle-stimulating hormone (FSH), luteinzing hormone (LH), testosterone (in men), or 17βE2 (in women)	0900
Prolactin (PRL)	0900
Urinary free cortisol (UFC)	24
Patients with polyuria: diuresis, urine density, Na^{2+}, and plasma osmolality	

[a]May be omitted per physician discretion.
Reproduced from Ghigo E, Masel B, Aimaretti G, et al. (2005) Consensus guidelines on screening for hypopituitarism following traumatic brain injury. *Brain Injury* 19: 711–724, with permission of Taylor and Francis, Ltd.

37.9 Symptoms of a TBI and PTH

The symptoms of a TBI can be markedly varied and difficult to document objectively. Nolin et al. (2006) grouped the symptoms into three categories:

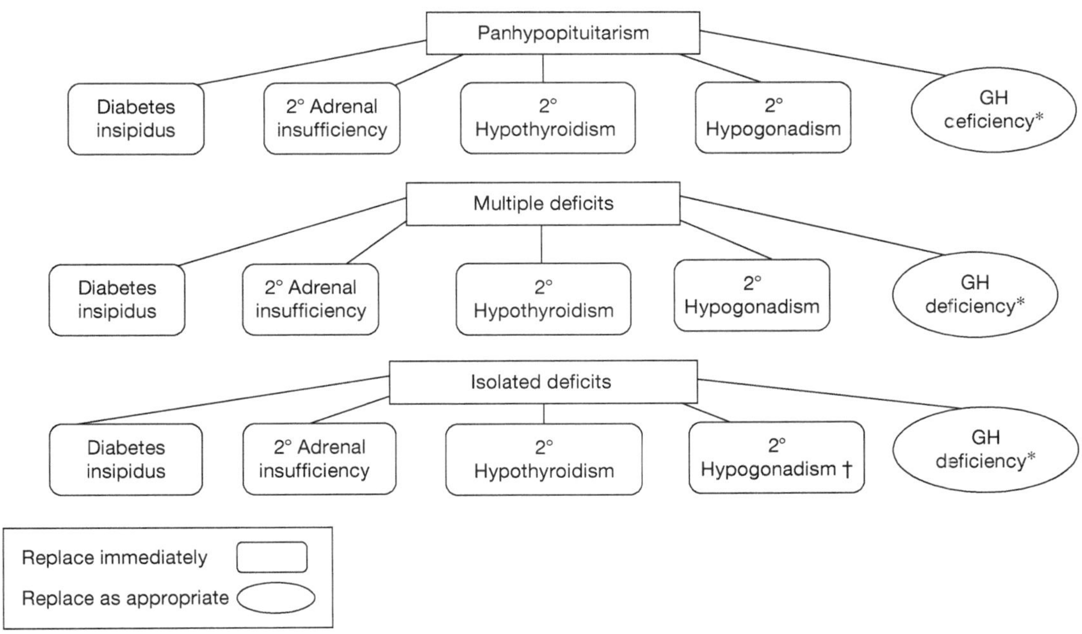

Figure 3 Recommended therapeutic options for patients less than 1-year post-injury based on type of deficit. Note: The indications used in the figure do not rule out any hormone replacement therapy (HRT) when definitely indicated. Reproduced from Ghigo E, Masel B, Aimaretti G, et al. (2005) Consensus guidelines on screening for hypopituitarism following traumatic brain injury. *Brain Injury* 19: 711–724, with permission of Taylor & Francis, Ltd.

(1) affective symptoms which included anger, irritability, anxiety, depression, and altered social functioning; (2) cognitive symptoms involving concentration and attention, slowness of processing speed, loss of memory, fatigability, and altered problem-solving skills; and (3) physical symptoms such as headaches, sleep problems, dizziness, nausea, blurred vision, and light and noise intolerance. In a study of 2668 Scottish survivors of a TBI over a year post-injury, 58% continued to complain of physical problems, 43% complained of cognitive issues, and 47% complained of mood issues (Thornhill et al., 2000). Although cognitive rehabilitation with behavior modification, social learning, modeling, and psychotherapeutic treatment may help some patients with problem solving and adaptation, treatment unfortunately remains empirical and unrewarding. A recent literature review on mTBI stated: "specific to drug interventions, this review has failed to produce solid evidence that any specific drug treatment is effective for one or more symptoms of mTBI" (Comper et al., 2005).

It is clear that the effects of untreated PTH can be both psychological and physical, as can the signs and symptoms of TBI. Many of the signs and symptoms previously attributed to the generalized effects of a TBI could be due to PTH. Although there are presently no large studies on the incidence of PTH in TBI, previously identified screening studies (Benvenga et al., 2000; Aimaretti et al., 2004, 2005; Bondanelli et al., 2004; Tanriverdi et al., 2006; Niederland et al., 2007) have indicated that the number is probably not insignificant. This suggests that those treating an individual with chronic unrelenting symptoms referable to a TBI should have an elevated index of suspicion for the presence of PTH. Instead of merely treating the symptoms, the healthcare provider has the opportunity to treat the underlying root cause of the problem, and potentially produce a better outcome.

References

Acerini CL, Tasker RC, Bellone S, Bona G, Thompson CJ, and Savage MO (2006) Hypopituitarism in childhood and adolescence following traumatic brain injury: The case for prospective endocrine investigation. *European Journal of Endocrinology* 155(5): 663–669.

Agha A, Rogers B, Mylotte D, Taleb F, Tormey W, Phillips J, and Thompson CJ (2004a) Neuroendocrine dysfunction in the acute phase of traumatic brain injury. *Clinical Endocrinology (Oxf)* 60(5): 584–591.

Agha A and Thompson E (2006) Anterior pituitary dysfunction following traumatic brain injury (TBI). *Clinical Endocrinology* 64: 481–488.

Agha A, Thornton E, O'Kelly P, Tormey W, Phillips J, and Thompson CJ (2004b) Posterior pituitary dysfunction after traumatic brain injury. *Journal of Clinical Endocrinology and Metabolism* 89(12): 5987–5992.

Aimaretti G, Ambrosio MR, Di Somma C, et al. (2004) Traumatic brain injury and subarachnoid haemorrhage are conditions at high risk for hypopituitarism: Screening study at 3 months after the brain injury. *Clinical Endocrinology (Oxf)* 61(3): 230–236.

Aimaretti G, Ambrosio MR, Di Somma C, et al. (2005) Residual pituitary function after brain injury-induced hypopituitarism: A prospective 12-month study. *Journal of Clinical Endocrinology and Metabolism* 90(11): 6085–6092.

Arwert LI, Deijen JB, Witlox J, and Drent ML (2005) The influence of growth hormone (GH) substitution on patient-reported outcomes and cognitive functions in GH-deficient patients: A meta-analysis. *Growth Hormone and IGF Research* 15(1): 47–54.

Arwert LI, Veltman DJ, Deijen JB, van Dam PS, and Drent ML (2006) Effects of growth hormone substitution therapy on cognitive functioning in growth hormone deficient patients: A functional MRI study. *Neuroendocrinology* 83(1): 12–19.

Becker RM and Daniel RK (1973) Increased antidiuretic hormone production after trauma to the craniofacial complex. *Journal of Trauma* 13: 112–115.

Bengtsson BA, Abs R, Bennmarker H, et al. (1999) The effects of treatment and the individual responsiveness to growth hormone (GH) replacement therapy in 665 GH-deficient adults. KIMS Study Group and the KIMS International Board. *Journal of Clinical Endocrinology and Metabolism* 84(11): 3929–3935.

Bengtsson BA, Edén S, Lönn L, et al. (1993) Treatment of adults with growth hormone (GH) deficiency with recombinant human GH. *Journal of Clinical Endocrinology and Metabolism* 76(2): 309–317.

Benvenga S, Campenní A, Ruggeri RM, and Trimarchi F (2000) Clinical review 113: Hypopituitarism secondary to head trauma. *Journal of Clinical Endocrinology and Metabolism* 85 (4): 1353–1361.

Benvenga S, Vigo T, Ruggeri RM, et al. (2004) Severe head trauma in patients with unexplained central hypothyroidism. *American Journal of Medicine* 116(11): 767–771.

Bondanelli M, De Marinis L, Ambrosio MR, et al. (2004) Occurrence of pituitary dysfunction following traumatic brain injury. *Journal of Neurotrauma* 21(6): 685–696.

Born JD, Hans P, Smitz S, Legros JJ, and Kay S (1985) Syndrome of inappropriate secretion of antidiuretic hormone after severe head injury. *Surgical Neurology* 23(4): 383–387.

Burman P, Broman JE, Hetta J, Wiklund I, Erfurth EM, Hagg E, and Karlsson FA (1995) Quality of life in adults with growth hormone (GH) deficiency: Response to treatment with recombinant human GH in a placebo-controlled 21-month trial. *Journal of Clinical Endocrinology and Metabolism* 80 (12): 3585–3590.

Bushnik T, Englander J, and Katznelson L (2007) Fatigue after TBI: Association with neuroendocrine abnormalities. *Brain Injury* 21(6): 559–566.

Centers for Disease Control (2007a) Facts about TBI. 2007, Centers for Disease Control and Prevention. http://www.cdc.gov (accessed in May 2008).

Centers for Disease Control (2007b) TBI in the United States children (0–14yrs.), Centers for Disease Control and Prevention. http://www.cdc.gov (accessed in May 2008).

Cherrier MM, Matsumoto AM, Amory JK, et al. (2005) The role of aromatization in testosterone supplementation: Effects on cognition in older men. *Neurology* 64(2): 290–296.

Colao A, Di Somma C, Cuocolo A, et al. (2004) The severity of growth hormone deficiency correlates with the severity of cardiac impairment in 100 adult patients with hypopituitarism: An observational, case-control study. *Journal of Clinical Endocrinology and Metabolism* 89(12): 5998–6004.

Colao A, Di Somma C, Pivonello R, et al. (1999) Bone loss is correlated to the severity of growth hormone deficiency in adult patients with hypopituitarism. *Journal of Clinical Endocrinology and Metabolism* 84(6): 1919–1924.

Comper P, Bisschop SM, Carnide N, and Tricco A (2005) A systematic review of treatments for mild traumatic brain injury. *Brain Injury* 19(11): 863–880.

Cyran E (1918) Hypophysenschädigung durch Schädelbasisfraktur. *Deutsche Medizinische Wochenschrift* 44: 1261.

Cytowic RE, Smith A, and Stump DA (1986) Transient amenorrhea after closed head trauma. *New England Journal of Medicine* 314(11): 715.

Daniel PM, Prichard MM, and Treip CS (1959) Traumatic infarction of the anterior lobe of the pituitary gland. *Lancet* 2: 927–931.

Davis JD, Stern RA, and Flashman LA (2003) Cognitive and neuropsychiatric aspects of subclinical hypothyroidism: Significance in the elderly. *Current Psychiatry Reports* 5(5): 384–390.

Davis JD and Tremont G (2007) Neuropsychiatric aspects of hypothyroidism and treatment reversibility. *Minerva Endocrinologica* 32(1): 409–465.

Denicoff KD, Joffe RT, Lakshmanan MC, Robbins J, and Rubinow DR (1990) Neuropsychiatric manifestations of altered thyroid state. *American Journal of Psychiatry* 147(1): 94–99.

Desouza LA, Ladiwala U, Daniel SM, Agashe S, Vaidya RA, and Vaidya VA (2005) Thyroid hormone regulates hippocampal neurogenesis in the adult rat brain. *Molecular and Cellular Neuroscience* 29(3): 414–426.

Dóczi T, Tarjányi J, Huszka E, and Kiss J (1982) Syndrome of inappropriate secretion of antidiuretic hormone (SIADH) after head injury. *Neurosurgery* 10(6 pt 1): 685–688.

Edwards OM and Clark JD (1986) Post-traumatic hypopituitarism. Six cases and a review of the literature. *Medicine* 65(5): 281–290.

Einaudi S, Matarazzo P, Peretta P, et al. (2006) Hypothalamo-hypophysial dysfunction after traumatic brain injury in children and adolescents: A preliminary retrospective and prospective study. *Journal of Pediatric Endocrinology and Metabolism* 19(5): 691–703.

Escamilla RF and Lisser H (1942) Simmonds' disease: A clinical study with review of the literature. *Journal of Clinical Endocrinology and Metabolism* 2: 65–96.

Falleti MG, Maruff P, Burman P, and Harris A (2006) The effects of growth hormone (GH) deficiency and GH replacement on cognitive performance in adults: A meta-analysis of the current literature. *Psychoneuroendocrinology* 31(6): 681–691.

Flesher MR, Delahanty DL, Raimonde AJ, and Spoonster E (2001) Amnesia, neuroendocrine levels and PTSD in motor vehicle accident victims. *Brain Injury* 15(10): 879–889.

Ghigo E, Masel B, Aimaretti G, et al. (2005) Consensus guidelines on screening for hypopituitarism following traumatic brain injury. *Brain Injury* 19(9): 711–724.

Gibney J, Wallace JD, Spinks T, et al. (1999) The effects of 10 years of recombinant human growth hormone (GH) in adult GH-deficient patients. *Journal of Clinical Endocrinology and Metabolism* 84(8): 2596–2602.

Gupta N (2000) Classification and features of closed head injury. *NeuroReview* [online University of Chicago educational resource]. http://www.ucch.org/sections/neurosurg/NeuroReview/08-CNSTrauma/HeadInjuryTypes.html (accessed on 14 January 2004).

Haggerty JJJ, Stern RA, Mason GA, Beckwith J, Morey CE, and Prange AJJ (1993) Subclinical hypothyroidism: A modifiable risk factor for depression? *American Journal of Psychiatry* 150(3): 508–510.

Kasper DL, Wiener CM, Braunwald E, et al. (2004) *Harrison's Principles of Internal Medicine.* New York: McGraw-Hill.

Kelly DF, Gonzalo IT, Cohan P, Berman N, Swerdloff R, and Wang C (2000) Hypopituitarism following traumatic brain injury and aneurysmal subarachnoid hemorrhage: A preliminary report. *Journal of Neurosurgery* 93(5): 743–752.

Kelly DF, McArthur DL, Levin H, et al. (2006) Neurobehavioral and quality of life changes associated with growth hormone insufficiency after complicated mild, moderate, or severe traumatic brain injury. *Journal of Neurotrauma* 23(6): 928–942.

Larsen PR, Kronenberg HM, Melmed S, and Polonsky KS (2007) *Williams Textbook of Endocrinology.* New York: Saunders.

León-Carrión J, Leal-Cerro A, Cabezas FM, et al. (2007) Cognitive deterioration due to GH deficiency in patients with traumatic brain injury: A preliminary report. *Brain Injury* 21(8): 871–875.

Lieberman SA, Oberoi AL, Gilkison CR, Masel BE, and Urban RJ (2001) Prevalence of neuroendocrine dysfunction in patients recovering from traumatic brain injury. *Journal of Clinical Endocrinology and Metabolism* 86(6): 2752–2756.

Niederland T, Makovi H, Gál V, Andréka B, Abrahám CS, and Kovács J (2007) Abnormalities of pituitary function after traumatic brain injury in children. *Journal of Neurotrauma* 24 (1): 119–127.

Nolin P, Villemure R, and Heroux L (2006) Determining long-term symptoms following mild traumatic brain injury: Method of interview affects self-report. *Brain Injury* 20(11): 1147–1154.

O'Connor CA, Cernak I, Johnson F, and Vink R (2007) Effects of progesterone on neurologic and morphologic outcome following diffuse traumatic brain injury in rats. *Experimental Neurology* 205(1): 145–153.

Oertel H, Schneider HJ, Stalla GK, Holsboer F, and Zihl J (2004) The effect of growth hormone substitution on cognitive performance in adult patients with hypopituitarism. *Psychoneuroendocrinology* 29(7): 839–850.

Parker KL and Schwimmer BP (2001) In: Hardman JG and Limbird LE (eds.) *Goodman and Gilman's: The Pharmacological Basis of Therapeutics,* 10th edn., pp. 1546–1547. New York: McGraw-Hill.

Pfeifer M, Verhovec R, Zizek B, Prezelj J, Poredos P, and Clayton RN (1999) Growth hormone (GH) treatment reverses early atherosclerotic changes in GH-deficient adults. *Journal of Clinical Endocrinology and Metabolism* 84(2): 453–457.

Popovic V, Pekic S, Pavlovic D, et al. (2004) Hypopituitarism as a consequence of traumatic brain injury (TBI) and its possible relation with cognitive disabilities and mental distress. *Journal of Endocrinological Investigation* 27(11): 1048–1054.

Roof RL, Duvdevani R, Braswell L, and Stein DG (1994) Progesterone facilitates cognitive recovery and reduces secondary neuronal loss caused by cortical contusion injury in male rats. *Experimental Neurology* 129(1): 64–69.

Roof RL, Duvdevani R, Heyburn JW, and Stein DG (1996) Progesterone rapidly decreases brain edema: Treatment delayed up to 24 hours is still effective. *Experimental Neurology* 138(2): 246–251.

Roof RL, Duvdevani R, and Stein DG (1993) Gender influences outcome of brain injury: Progesterone plays a protective role. *Brain Research* 607(1–2): 333–336.

Rosén T and Bengtsson BA (1990) Premature mortality due to cardiovascular disease in hypopituitarism. *Lancet* 336(8710): 285–288.

Samuels MH, Schuff KG, Carlson NE, Carello P, and Janowsky JS (2007) Health status, psychological symptoms,

mood, and cognition in L-thyroxine-treated hypothyroid subjects. *Thyroid* 17(3): 249–258.
Simmonds M (1914) Ueber Hypophysisschwund mit tödlichem Ausgang. *Deutsche Medizinische Wochenschrift* 40: 22–23.
Smith JW, Evans AT, Costall B, and Smythe JW (2001) Thyroid hormones, brain function and cognition: A brief review. *Neuroscience and Biobehavioral Reviews* 26(1): 45–60.
Stein DG (2001) Brain damage, sex hormones and recovery: A new role for progesterone and estrogen? *Trends in Neurosciences* 24(7): 386–391.
Tanriverdi F, Senyurek H, Unluhizarci K, Selcuklu A, Casanueva FF, and Kelestimur F (2006) High risk of hypopituitarism after traumatic brain injury: A prospective investigation of anterior pituitary function in the acute phase and 12 months after trauma. *Journal of Clinical Endocrinology and Metabolism* 91(6): 2105–2111.
Tanriverdi F, Ulutabanca H, Unluhizarci K, Selcuklu A, Casanueva FF, and Kelestimur F (2007) Pituitary functions in the acute phase of traumatic brain injury: Are they related to severity of the injury or mortality? *Brain Injury* 21(4): 433–439.
Thornhill S, Teasdale GM, Murray GD, McEwen J, Roy CW, and Penny KI (2000) Disability in young people and adults one year after head injury: Prospective cohort study. *British Medical Journal* 320(7250): 1631–1635.
Twijnstra A and Minderhoud JM (1980) Inappropriate secretion of antidiuretic hormone in patients with head injuries. *Clinical Neurology and Neurosurgery* 82(4): 263–268.
Vingerhoets F and de Tribolet N (1988) Hyponatremia hypo-osmolarity in neurosurgical patients. 'Appropriate secretion of ADH' and 'cerebral salt wasting syndrome'. *Acta Neurochirurgica (Wien)* 91(1–2): 50–54.
Vitiello MV, Moe KE, Merriam GR, Mazzoni G, Buchner DH, and Schwartz RS (2006) Growth hormone releasing hormone improves the cognition of healthy older adults. *Neurobiology of Aging* 27(2): 318–323.
Wekking EM, Appelhof BC, Fliers E, Schene AH, Huyser J, Tijssen JG, and Wiersinga WM (2005) Cognitive functioning and well-being in euthyroid patients on thyroxine replacement therapy for primary hypothyroidism. *European Journal of Endocrinology* 153(6): 747–753.
Whitehead HM, Boreham C, McIlrath EM, Sheridan B, Kennedy L, Atkinson AB, and Hadden DR (1992) Growth hormone treatment of adults with growth hormone deficiency: Results of a 13-month placebo controlled cross-over study. *Clinical Endocrinology (Oxf)* 36(1): 45–52.
Wright DW, Bauer ME, Hoffman SW, and Stein DG (2001) Serum progesterone levels correlate with decreased cerebral edema after traumatic brain injury in male rats. *Journal of Neurotrauma* 18(9): 901–909.
Wright DW, Kellermann AL, Hertzberg VS, et al. (2007) ProTECT: A randomized clinical trial of progesterone for acute traumatic brain injury. *Annals of Emergency Medicine* 49(4): 391–402.
Young TP, Hoaglin HM, and Burke DT (2007) The role of serum testosterone and TBI in the in-patient rehabilitation setting. *Brain Injury* 21(6): 645–649.

38 Human Immunodeficiency Virus and AIDS

Y Miyasaki, M B Goetz, and T F Newton, VA Greater Los Angeles Healthcare System, Los Angeles, CA, USA

Published by Elsevier Inc.

Chapter Outline

Glossary

highly active antiretroviral therapy (HAART) Combination antiretroviral therapy involving the use of combinations of three or more antiretroviral agents from two or more classes of active agents. Patients generally receive combinations of two nucleoside reverse transcriptase inhibitors plus a non-nucleoside reverse transcriptase inhibitor or a protease inhibitor.

HIV-associated dementia (HAD) Cognitive disorder caused by infection of the brain by HIV.

immune reconstitution inflammatory syndrome (IRIS) An acute inflammatory response to subclinical infection that accompanies immunological recovery following the initiation of combination antiretroviral therapy.

viral load The number of HIV RNA copies/ml of plasma.

38.1 Human Immunodeficiency Virus Natural History

The course of HIV infection has been greatly improved by the use of highly active antiretroviral therapy (HAART), which became widely available in the mid-1990s (Palella et al., 1998). The clinical effectiveness of such therapy is mediated by treatment-induced reduction of HIV viral replication as assessed by measurements of the number of copies of HIV RNA in the blood (the plasma viral load). Successful suppression of HIV replication is determined by the intrinsic potency of the prescribed regimen, patients' adherence to treatment, and the resistance of patients' HIV strain to antiretroviral agents (DHHS Panel on Antiretroviral Guidelines for Adults and Adolescents, 2006).

In contrast, chronically untreated HIV infection results in an inexorably progressive disease manifest by complications that are primarily related to defects of cell-mediated immunity due to the depletion of $CD4^+$ lymphocytes (Simon et al., 2006). The vast majority of complications related to opportunistic infections and opportunistic malignancies occur after the $CD4^+$ cell count has decreased to fewer than 200 cells/μl, which satisfies the definition of acquired immunodeficiency syndrome (AIDS, 1992). HIV infection of the brain occurs during the weeks following acquisition of HIV infection, that is, during acute seroconversion reactions (Kassutto and Rosenberg, 2004). Complications related directly to HIV infection of the brain may occur immediately following infection or be delayed.

38.1.1 Clinically Latent Period

Following the resolution of the acute seroconversion syndrome (described in detail in Section 38.2.1), most HIV-infected patients enter a prolonged period during which clinical symptoms are minimal despite ongoing viral replication (Simon et al., 2006). In the absence of effective intervention with antiretroviral therapy, during the 10-year period following infection, approximately 60% of patients will develop an HIV-related opportunistic infection or malignancy (Rutherford et al., 1990) with the rate of progression being strongly dependent on the number of HIV-1 RNA copies/ml of plasma (O'Brien et al., 1996; Mellors et al., 1997). Although there is substantial interpatient variability associated with differences in the number of HIV RNA copies/ml of plasma, patients' age, behaviors, and co-morbidities, on average the rate of $CD4^+$ decline is approximately 20–80 cells/μl/year (CASCADE collaboration, 2003; Rodriguez et al., 2006).

38.1.2 $CD4^+$ Cell Count versus Clinical Complications of HIV Infection

Although HIV-mediated neurocognitive decline and Kaposi's sarcoma may occur in persons with relatively normal $CD4^+$ cell counts, the opportunistic complications of greatest neurological importance, for example, those due to *Toxoplasma gondii*, progressive multifocal leukoencephalopathy (PML), cytomegalovirus (CMV), *Mycobacterium avium* complex (MAC), *Cryptococcus neoformans*, extrapulmonary tuberculosis, and primary brain lymphomas are unusual until the $CD4^+$ cell count is below 100–200 cells/μl (1995; Mocroft et al., 1998). Persons with higher $CD4^+$ cell counts may also develop infectious complications due to the immune reconstitution inflammatory syndrome (IRIS), which is an acute inflammatory response to subclinical infection that accompanies immunological recovery following the initiation of combination antiretroviral therapy (French et al., 2000; Shelburne et al., 2002; Podlekareva et al., 2006). Severe or fatal outcomes of IRIS have been reported when neurologic structures are involved (Safdar et al., 2002; Lortholary et al., 2005), as may occur in association with advanced HIV (Miller et al., 2004), intracranial tuberculomas (Afghani and Lieberman, 1994; Crump et al., 1998; Nicolls et al., 2005), cryptococcal meningitis (Skiest et al., 2005; Venkataramana et al., 2006), and PML (Safdar et al., 2002; Venkataramana et al., 2006).

38.2 Primary Neuropsychiatric Disorders Related to HIV Infection *per se*

38.2.1 Neuropsychiatric Syndromes during Acute HIV Seroconversion Reactions

Approximately 40–90% of patients develop a clinically apparent acute illness within 10–14days of acquiring HIV infection (Kassutto and Rosenberg, 2004). The most common manifestations of primary HIV infection are fever, malaise, myalgia, and rash (Daar et al., 2001; Zetola and Pilcher, 2007). These findings are often accompanied by headache and meningeal signs (Schacker et al., 1996; Daar et al., 2001) as well as by occasional cranial nerve palsies (especially involving cranial nerve VII), radiculopathy, encephalopathy and the Guillain–Barré syndrome (Kassutto and Rosenberg, 2004). In one series, a quarter of patients with acute HIV seroconversion reactions developed symptomatology suggestive of aseptic meningitis (Schacker et al., 1996). Fatal encephalopathy and brain necrosis may occur from acute HIV infection (Silver et al., 1997; Meersseman et al., 2005).

During symptomatic seroconversion reactions the number of HIV RNA copies/ml of plasma often exceeds 1million copies/ml and is almost invariably greater than 50000 copies/ml (Kahn and Walker, 1998). There is strong evidence that it is during this period that HIV enters the brain and infects perivascular macrophages and microglia (Gonzalez-Scarano and Martin, 2005; Spudich et al., 2007). Many of the manifestations of HIV seroconversion are related to the state of immune activation induced by HIV infection and the proliferation of activated $CCR5^+$ $CD4^+$ cells (Zetola and Pilcher, 2007). The appearance of specific anti-HIV cytotoxic $CD8^+$ T lymphocytes 2–4weeks after infection is accompanied by a precipitous drop in the viral load, an increase in the $CD4^+$ cell count from the dip observed during the acute seroconversion reaction, and resolution of acute clinical findings.

38.2.2 Neurocognitive Impairment Associated with HIV Infection

Chronic HIV infection of the central nervous system (CNS) is associated with a variety of neurocognitive disorders. These include HIV-associated dementia (HAD) and minor cognitive motor disorder (MCMD), a milder form of neuropsychological dysfunction. Important risk factors for HAD include increasing age (Janssen et al., 1992), low CD4 count, and high viral load (Childs et al., 1999). Of note is that in a cohort study of 329 AIDS patients, HAD was an independent predictor of time to death (Sevigny et al., 2007).

Early in the HIV epidemic, and prior to the development of effective antiretroviral therapies, the prevalence of HAD was thought to approach 20%. Since the development of HAART the prevalence of HAD in North America and Europe has fallen to less than 5% (Ances and Ellis, 2007). Despite the introduction of HAART, however, patients often fail to recover full cognitive function and thus, the proportions of patients with milder degrees of HIV-related neurocognitive disorders have increased (Ellis et al., 2007).

HAD is largely due to direct invasion of the CNS by HIV. HIV surface glycoprotein gp120 mediates the viral attachment to the CD4 receptor and to either the CXCR4 or CCR5 co-receptor (McArthur et al., 2005). The CCR5 co-receptor facilitates HIV infection of macrophages (Stevenson, 2003), which subsequently transport HIV into the brain (Ellis et al., 2007). Thus, CCR5-tropic HIV quasispecies account for the majority of strains in the CNS (Ellis et al., 2007). Brain macrophages, astrocytes, and microglia are the key cell types that are infected by HIV. These cells mediate the neuroinflammation and neurodegeneration seen in patients with HAD (Gonzalez-Scarano and Martin, 2005; McArthur et al., 2005). Infected perivascular macrophages produce viral proteins, including gp120, transcriptional transactivator (Tat), or viral protein R (Vpr), which are toxic to neurons and/or astrocytes. The infected cells also produce proinflammatory, neurotoxic cytokines such as tumor necrosis factor-alpha (TNF-α), quinolinic and arachidonic acid, platelet-activating factor, and nitric oxide (Gonzalez-Scarano and Martin, 2005; McArthur et al., 2005).

Neurocognitive impairment in HIV patients can be further exacerbated by concomitant substance abuse such as alcohol (Pfefferbaum et al., 2007), methamphetamine (Langford et al., 2003), and cocaine (Tyor and Middaugh, 1999).

38.2.2.1 Clinical manifestations of HAD

The clinical features of HAD can be subtle, and include mild problems with short-term memory, cognitive slowing, poor attention resulting in complaints, such as difficulty reading, and apathy. Motor dysfunctions can include difficulty walking with stumbling, incoordination, and impaired dexterity. Advanced dementia can lead to more global cognitive dysfunction (Baldewicz et al., 2004; McArthur et al., 2005).

38.2.2.2 Diagnostic strategies and therapeutic considerations

Sensitive neuropsychological testing, with particular attention to HIV-related neurologic deficits, is the gold standard for detection of HAD (Ellis et al., 2007). Unless neuroimaging studies or clinical evaluation suggest alternative diagnoses, examination of the cerebrospinal fluid (CSF) is generally not required in HIV-infected patients who manifest typical slowly progressive signs of HAD. No definitive CSF criteria for HAD have been established, and the significance of CSF concentrations of HIV RNA or of antiretroviral agents is uncertain (McArthur et al., 2005). In untreated patients, the severity of neuropsychiatric disease correlates with the concentration of HIV RNA in the CSF (McArthur et al., 2005). Improvements in neurologic status are seen as the CSF concentration of HIV RNA decreases (McArthur et al., 2005). While some recent studies have demonstrated no association between HIV dementia and the CSF viral load in patients treated with HAART (Sevigny et al., 2004; McArthur et al., 2004), Letendre et al. found that CSF HIV RNA suppression, baseline antiretroviral history, and their interaction to be independent predictors of reduction in global dementia scale in HIV-infected patients (Letendre et al., 2004). The possible relationship between the CSF viral load and HAD suggests that antiretroviral agents with good brain penetration may be beneficial in the treatment of HAD. However, as it remains uncertain as to how well CSF concentrations of antiretroviral agents correlate with those in the brain parenchyma, and as direct measurement of brain parenchymal concentrations is not feasible, the utility of using combinations of antiretroviral agents with better brain penetration to prevent or to treat HAD remains uncertain and is the focus of several ongoing cohort studies (McArthur et al., 2005).

Aside from HAART, novel approaches to prevent or to treat HIV-related neurodegeneration or neuroinflammation have been evaluated. For example, in a murine model valproic acid has been shown to protect neurons against neurotoxicity from HIV-1 infected macrophages (Dou et al., 2003). In an experimental simian immunodeficiency virus model of HIV CNS disease, minocycline, which has anti-inflammatory as well as antimicrobial activity, reduced the severity of encephalitis, suppressed viral load in the brain, and reduced CNS expression of neuroinflammatory markers (Zink et al., 2005). In a randomized, double-blind, placebo-controlled, multicenter trial, administration of memantine, a noncompetitive antagonist of the *N*-methyl-D-aspartate receptor that prevents Tat- and gp120-induced intracellular calcium increases and glutamine toxicity (Ellis et al., 2007), improved frontal white matter and parietal cortex metabolism as assessed by magnetic resonance spectroscopy, although no significant differences in cognitive performance were seen in patients with HAD (Schifitto et al., 2007). Other experimental approaches for the treatment and prevention of HAD are listed in **Table 1**.

Table 1 A partial list of treatment options that are under study for HAD

Lithium: used for treatment of bipolar disorder, possible neuroprotectant
Nimodipine: calcium channel blocker, possible neuroprotectant
Selegiline: monoamine-oxidase inhibitor, possible neuroprotectant
Peptide T (D-Ala-1-peptide-T-amide): possible HIV entry inhibitor by targeting selectively CCR5
CPI-1189: antioxidant, TNF-α antagonist
OPC 14117: antioxidant, possible neuroprotectant
Lexipafant: platelet-activating factor antagonist

38.3 Secondary Neuropsychiatric Processes Related to HIV Infection

Apart from HIV-related neuropathology, a number of opportunistic infections and malignancies frequently involve the brain in HIV-infected patients (**Table 2**). CNS complications were observed in 63% of 390 autopsies in AIDS patients between 1982 and 1998 (i.e., prior to the widespread use of HAART) (Masliah et al., 2000). Among patients with CNS abnormalities, aside from HIV-related neuropathology (seen in 28.3% of cases) evidence of infection by CMV was common (18%), followed by fungal infection (4.9%), PML (3.4%), *toxoplasmosis* encephalitis

Table 2 A list of etiologies of common neurological diseases in HIV-infected patients

Primary processes		
Neurocognitive disorders		
Aseptic meningitis and encephalitis		
Secondary processes		
Opportunistic infections and co-infection		
	Focal Infections	
		Toxoplasmosis
		Brain abscess
		Neurocystercosis
		Nocardia asteroides infection
		Aspergillus infection
	Encephalitis	
		Cytomegalovirus
		Herpes simplex virus
		Herpes zoster
	Infiltrative processes	
		Progressive multifocal leukoencephalopathy (PML)
	Meningitis	
		Cryptococcus neoformans
		Mycobacterium tuberculosis meningitis
		Histoplasma capsulatum
	Other	
		Neurosyphilis
		Immune reconstitution inflammatory syndrome
Primary CNS lymphoma		
Other HIV-associated conditions		
	Adverse effects of medications that directly impact neurocognition	
	Substance abuse and withdrawal	

(2.5%), and infection by MAC (1.6%). Other important secondary neurological complications in HIV-infected patients include primary CNS lymphoma. As previously noted, the incidence of these complications is reduced by the administration of effective antiretroviral therapy.

38.3.1 Adverse Neuropsychiatric Side Effects of Medications Used in the Treatment of HIV-Infected Individuals

Among the medications used to decrease the replication of HIV, efavirenz, which is a non-nucleoside reverse transcriptase inhibitor (NNRTI), is most commonly associated with neuropsychiatric symptoms. Although anxiety and vivid dreams are the most common CNS side effects of efavirenz, psychosis may also occur. Efavirenz is primarily metabolized by cytochrome P450 2B6 (CYP2B6). A double-blinded, placebo-controlled, randomized study found the CYP2B6 allelic variant (G→T at position 516), which is more common in African-Americans than in European-Americans, to be associated with greater efavirenz plasma exposure and with increased CNS side effects (Haas et al., 2004).

As mentioned, the inflammatory consequences of HIV infection may be responsible for many of the observed neuropsychiatric complications. Thus, it is relevant that HIV-infected patients are often treated with agents that modulate the inflammatory response, particularly interferon-alpha (IFN-α) or interleukin-2 (IL-2). IFN-α, which is used in the treatment of hepatitis C (HCV) (Dienstag and McHutchison, 2006), commonly produces depressive symptoms (Capuron et al., 2002). Monoamine dysfunction, including decreased baseline serotonin levels, is suggested to be partly responsible for IFN-α-induced depression (Asnis and De La Garza, 2006). The antidepressant paroxetine has been shown to reduce the incidence of symptoms (Musselman et al., 2001; Asnis and De La Garza, 2005). IL-2, which is under investigation for its ability to increase $CD4^+$ cells in HIV-infected patients (Mitsuyasu, 2001), has been associated with the development of confusion and delirium followed by coma, ataxia, hemiparesis, seizure, and cortical syndromes accompanied by imaging studies showing multiple cerebral lesions (Karp et al., 1996). The etiology of this syndrome is unclear.

38.4 Specific Endocrinological Complications

38.4.1 Adrenocortical Dysfunction

Although mineralocorticoid deficiency is also found among HIV patients, little is known regarding the neurobehavioral complications of mineralocorticoid insufficiency or excess.

38.4.1.1 Adrenal insufficiency (Addison's disease)

The adrenal gland is the most common endocrine organ involved, at autopsy, in HIV patients (Findling et al., 1994). Clinically significant adrenal insufficiency, most of which is due to primary adrenal failure, affects approximately 5–10% of HIV-infected patients (Danoff, 1996). In a study of 39 AIDS patients by Dobs et al., only 3 (8%) had evidence of impaired adrenal cortisol reserve, that is, below the normal plasma cortisol response to ACTH administration (Dobs et al., 1988). Importantly, many patients with biochemical abnormalities consistent with primary adrenal insufficiency remain clinically asymptomatic. In one study, although 9 of 98 HIV-infected patients had decreased baseline or cortisol responses to stimulation, none had clinical adrenal insufficiency (Raffi et al., 1991).

Since clinical adrenal insufficiency does not occur until more than 80–90% of the gland is affected (Grinspoon and Bilezikian, 1992; Mayo et al., 2002), histopathological involvement of the adrenal gland is more prevalent than clinically overt adrenal insufficiency (Mayo et al., 2002). Thus, pathological evidence of direct invasion of the adrenal gland by opportunistic infections or by malignancy occurs in up to 80% of untreated HIV-infected patients coming to autopsy (Bricaire et al., 1987; Duch et al., 1998; Welch et al., 1984). Overall, CMV is the most common pathogen to involve the adrenal gland (Bricaire et al., 1987; Guarda et al., 1984; Shibata and Klatt 1989; Welch et al., 1984). Other HIV-related processes reported to invade the adrenal glands include metastatic Kaposi's sarcoma, infection by *Mycobacterium tuberculosis* (MTB), *C. neoformans*, *T. gondii*, MAC, *Pneumocystis jiroveci* (previously termed *Pneumocystis carinii*), herpes simplex virus, and *Histoplasma capsulatum* (Glasgow et al., 1985; Bricaire et al., 1987; Amin et al., 1990; Duch et al., 1998).

Although uncommon, secondary or tertiary adrenal insufficiency due to functional derangement of the hypothalamic–pituitary–adrenal (HPA) axis has been described. In one study, 14 of 22 patients with a clinical AIDS diagnosis had secondary adrenal insufficiency as defined by the presence of normal ACTH stimulation tests in the setting of an abnormal response to corticotropin-releasing hormone (CRH) (Lortholary et al., 1996). In this study the mean basal serum cortisol level showed an inverse relationship with the $CD4^+$ cell count.

Notwithstanding the observation that most neurologic manifestations of HIV infection do not affect neuroendocrine tissues (Hofbauer and Heufelder, 1996), in one pre-HAART autopsy series of HIV-infected patients, focal to widespread necrosis and/or fibrosis of the anterior pituitary gland was seen in 10 of 88 cases (Ferreiro and Vinters, 1988). In this and other series, CMV infection was the most common process (Ferreiro and Vinters, 1988; Sullivan et al., 1992; Merenich, 1994). Other HIV-related diseases found to involve the pituitary gland include infections by *C. neoformans* (Ferreiro and Vinters, 1988), *P. jiroveci* (Northfelt et al., 1990), MTB (Eyer-Silva et al., 1994), and *T. gondii* (Milligan et al., 1984).

38.4.1.2 Adrenal excess and Cushing's syndrome

Basal and ACTH-stimulated cortisol levels in HIV-infected and HIV-uninfected patients have been compared in a number of studies. Basal serum cortisol levels have often been found to be higher among HIV-infected patients in many studies, but this result is not consistent. For example, Villette et al. found that basal cortisol levels were significantly higher in both asymptomatic HIV-infected patients and those with AIDS than in HIV-uninfected control subjects (Villette et al., 1990). Similarly, Membreno et al. (1987) found that even though ACTH-stimulated cortisol levels were significantly lower in patients with symptomatic HIV infection or AIDS than in normal subjects, basal cortisol levels in hospitalized AIDS patients were significantly higher than in normal subjects. In contrast, Merenich et al. (1990) found that among 40 HIV-infected patients without historical or clinical evidence of endocrine dysfunction, the mean basal cortisol levels and ACTH-stimulated cortisol levels were in the normal range but nonetheless significantly lower than found in HIV-uninfected subjects. Similarly, in a longitudinal evaluation of 25 HIV-infected patients, 7 of whom had prior AIDS-defining conditions, Findling et al. (1994) found that neither the mean basal nor ACTH-stimulated cortisol levels of HIV-infected patients were different from those in normal subjects and that there were no differences between non-AIDS and AIDS patients.

In addition, HIV may directly affect cortisol physiology or do so through the activity of proinflammatory cytokines such as TNF-α, IL-1, and IL-6. These cytokines stimulate the HPA axis (Chrousos, 1995) and are increased in HIV-infected patients (Kedzierska and Crowe, 2001). Infusion of the HIV envelope protein gp120 into the brain of rats induces IL-1 production in the brain (Sundar et al., 1991). As IL-1 may stimulate the release of both CRH from the hypothalamus (Sapolsky et al., 1987) and ACTH by pituitary cells (Woloski et al., 1985), these results have been interpreted as suggesting that increased IL-1 might increase the activity of the pituitary–adrenal axis in HIV-infected subjects (Sundar et al., 1991). In a study by Mastorakos et al. (1993), IL-6 was shown to increase ACTH and cortisol levels in six HIV-uninfected cancer patients. Furthermore, the HIV-1 protein Vpr has glucocorticoid receptor coactivator activity, thereby potentially increasing the sensitivity of glucocorticoid target tissues to cortisol (Mirani et al., 2002).

Variable results have been found in studies measuring plasma ACTH levels. Among 13 male HIV-infected patients, Villette et al. (1990) found that when compared with normal subjects, plasma ACTH levels were significantly lower in HIV-infected patients regardless of the presence of a prior AIDS diagnosis. In the above-mentioned longitudinal study by Findling et al. (1994), in eight subjects plasma ACTH levels rose to exceed the normal range during the 2-year period of observation.

Nonpituitary factors also contribute to hypercortisolemia in HIV-infected patients (Villette et al., 1990). In a study of 58 asymptomatic HIV-infected men, Laudat et al. found a significant inverse relationship between $CD4^+$ cell counts and cortisol/dehydroepiandrosterone ratios suggesting a shift from androgen to glucocorticoid production as the disease progresses. In a study of nine HIV-infected patients with hypercortisolemia despite demonstrating signs and symptoms of Addison's disease, Norbiato et al. (1992) found evidence of peripheral cortisol resistance as documented by the lack of cortisol suppression by the administration of dexamethasone, higher glucocorticoid receptor density, and decreased glucocorticoid affinity for the substrate.

38.4.1.3 Common iatrogenic causes of adrenal disease in HIV-infected patients

Several medications that are commonly used by HIV-infected patients disturb corticosteroid metabolism. Although no longer widely used, the azole antifungal medication ketoconazole inhibits steroidogenesis and may precipitate adrenal insufficiency (Pont et al., 1982). In contrast, while other azoles, that is, fluconazole and itraconazole, do not have an effect on steroidogenesis (Phillips et al., 1987; Magill et al., 2004), by inhibiting the activity of cytochrome P450 isoenzyme 3A4 (CYP3A4), itraconazole leads to a state of hypercortisolemia in patients receiving supplemental corticosteroid with budesonide, dexamethasone, or methylprednisolone (Varis et al., 1999). In addition, case reports have described the development of adrenal insufficiency in critically ill patients who receive fluconazole (Albert et al., 2001; Santhana Krishnan and Cobbs, 2006).

Due to direct pharmacological effects and drug–drug interactions, HIV-infected individuals have an increased risk for iatrogenic adrenal excess. Although the antitubercular medication rifampin does not directly cause adrenal disease, it induces steroid metabolism and thereby may increase the needed dose of steroid replacement therapy in persons with preexisting adrenal insufficiency (Mayo et al., 2002). Cushing's syndrome has been reported in HIV-infected patients receiving megestrol acetate, a synthetic progesterone agent with intrinsic glucocorticoid activity that is often used as an appetite stimulant in AIDS patients with anorexia, cachexia, or unexplained significant weight loss (Steer et al., 1995; Mann et al., 1997). Cushing's syndrome also occurs in HIV-infected patients receiving the commonly used HIV-protease inhibitor, ritonavir, which is a potent inhibitor of the CYP3A4 isoenzyme, in combination with the inhaled/intranasal fluticasone (Samaras et al., 2005; Johnson et al., 2006; Arrington-Sanders et al., 2006; St Germain et al., 2007). Clinically apparent adrenal insufficiency may follow discontinuation of prolonged megestrol use (Leinung et al., 1995) or of fluticasone use (Samaras et al., 2005; Johnson et al., 2006; Arrington-Sanders et al., 2006; St Germain et al., 2007).

38.4.1.4 Clinical manifestations of adrenal insufficiency and excess in HIV-infected patients

In acute adrenal insufficiency, patients often present in cardiovascular shock. In contrast, with chronic adrenal insufficiency patients often have vague symptoms such as fatigue, weakness, anorexia, nausea, vomiting, and abdominal pain. In addition, a wide variety of neuropsychiatric disorders, including depression, apathy, sleep disturbances, cognitive impairment, and delirium, have been associated with both adrenal insufficiency and excess. The psychiatric differential diagnosis of HIV-infected patients with these nonspecific symptoms can be difficult. In general, the suspicion of medical, rather than psychiatric, causes increases with the severity of the underlying HIV disease. In patients with advanced disease, a medical workup should be undertaken prior to initiating psychiatric treatment.

As is commonly observed in patients receiving corticosteroid treatment, there is marked variability in psychiatric changes due to corticosteroid excess. Some patients develop hypomania or mania during treatment with apparently modest corticosteroid doses, whereas others develop only mild irritability or dysphoria. There are no known predictors of response, though one would expect that those with histories of mania or depression may be more susceptible to steroid-induced changes.

38.4.1.5 Diagnostic strategies and therapeutic considerations

The diagnostic strategies and therapeutic considerations for adrenal dysfunction in HIV-infected individuals do not differ substantially from those for HIV-uninfected patients (Cooper et al., 2003). Hypercortisolemia without clinical signs of Cushing's syndrome is common among HIV-infected patients and does not necessarily warrant a complete workup of adrenal dysfunction (Mayo et al., 2002).

38.4.2 Gonadal Dysfunction

38.4.2.1 Hypogonadism

Hypogonadism is common among HIV-infected men (Arver et al., 1999), as it affects 6–21% of HIV-infected patients in the post-HAART era (Laudat et al., 1995; Rietschel et al., 2000; Dube et al., 2007) and up to 50% of patients with AIDS (Dobs et al., 1988). In HIV-infected men, low testosterone levels are associated with low $CD4^{+}$ counts (Laudat et al., 1995; Dube et al., 2007), weight loss (Dobs et al., 1988; Coodley et al., 1994), and increased age (Dube et al., 2007).

Although the pathophysiology of hypogonadism in HIV-infected patients appears to involve all levels of the hypothalamic–pituitary–gonadal axis (Bhasin et al., 2001), a variety of studies suggest that dysfunction of the hypothalamic–pituitary axis is the primary cause of hypogonadism. In particular, a study of 148 HIV-infected patients also showed a higher prevalence of biochemically defined hypogonadotropic hypogonadism (81%) than hypergonadotropic hypogonadism, that is, primary testicular failure (19%) (Arver et al., 1999). Similar results were found in an analysis of participants in the Swiss HIV Cohort Study (Wunder et al., 2007). A more detailed study of 24 hypogonadal HIV-infected men found that 18 (75%) had inappropriately normal or low serum gonadotropin levels; seven of eight hypogonadal men (88%) had a normal gonadotropic response to gonadotropin-releasing hormone (GnRH) challenge (Dobs et al., 1988).

In addition, hypogonadal HIV patients may have primary testicular failure caused by a variety of etiologies. In 57 autopsy cases of HIV-infected patients, opportunistic infections were found in 18 patients, including CMV (eight cases), MAC (six cases), and *T. gondii* (five cases). The mean sperm score in the testes of AIDS patients was significantly lower than that of the HIV-uninfected control subjects. This result was consistent with the higher incidence of nonspecific interstitial inflammation and interstitial fibrosis in the testes of the AIDS patients compared with those of the control group (De Paepe and Waxman, 1989). Sex hormone-binding globulin levels were found to be normal in both male asymptomatic HIV-infected patients and AIDS patients (Dobs et al., 1988; Merenich et al., 1990).

Furthermore, hypogonadism among men has also been shown to be due to shunting from androgen production to cortisol secretion by the adrenal gland (Laudat et al., 1995). The same mechanism has been shown among HIV-infected female patients with wasting (Grinspoon et al., 2001). Finally, production of cytokines as well as systemic illness itself may impair the secretion of GnRH or gonadotropins (Grinspoon and Bilezikian, 1992).

Although hypogonadism among HIV-infected patients has been primarily studied in men, significant androgen deficiency also occurs in women (Sinha-Hikim et al., 1998) and has been observed in as many as 66% of HIV-infected women with HIV-related wasting (Grinspoon et al., 1997). In a study that made use of highly sensitive assays, greater than 90% of HIV-infected women had total and free testosterone levels below the median for HIV-uninfected women (Sinha-Hikim et al., 1998). This study also showed that serum-total and free-testosterone concentrations correlate inversely with plasma HIV RNA concentration. In a study of HIV-infected and HIV-uninfected women under the age of 55, 120 of 1139 HIV-infected women (10.5%) versus 16 of 292 (5.5%) HIV-uninfected women reported no menstruation for a 12-month period (Cejtin et al., 2006). Of the HIV infected women, 47% had FSH levels consistent with menopause whereas 69% of the HIV-uninfected women had such values. HIV infection status was significantly associated with amenorrhea due to causes other than menopause after controlling for patients' age. Among HIV-infected women, there was no significant association between the prevalence of menopause and the $CD4^{+}$ cell count, viral load, prior AIDS event, wasting, or use of antiretroviral therapy.

38.4.2.2 Common iatrogenic causes of hypogonadism in HIV-infected patients

High doses of the azole antifungal medication, ketoconazole (Sonino, 1987), directly inhibit steroidogenesis, thereby reducing testosterone synthesis. Similarly, megestrol acetate, a synthetic, orally active progestational agent that is used as an appetite stimulant for treatment of HIV-related wasting, suppresses testosterone production and induces hypogonadism (Geller, 1985; Wagner et al., 1995). Finally, licit or illicit use of opioids may activate an endogenous opioid-peptide pathway that inhibits

GnRH and LH release, thereby reducing testosterone production (Wahlstrom and Dobs, 2000; Cooper et al., 2003).

Because hypogonadism may be a factor contributing to wasting in HIV infection, a wide variety of licit and illicit anabolic steroids are used and abused by HIV-infected patients to maintain or increase lean muscle mass (Johns et al., 2005). Patients typically are prescribed low replacement doses of testosterone. A fairly substantial number of patients also abuse steroids to enhance their appearance or to improve their athletic performance. In a placebo-controlled study, illicit anabolic steroid use was associated with significant increases in hostility, paranoia, and guilt (Pagonis et al., 2006). Other reports identify aggression, mania, psychosis, and rarely, suicide as manifestations of chronic illicit steroid abuse (Trenton and Currier, 2005). Withdrawal from androgen overdose is associated with hypogonadotropic hypogonadism (Alen et al., 1987).

38.4.2.3 Clinical manifestations of hypogonadism in HIV-infected patients

Hypogonadism contributes to sexual dysfunction, decreased body mass, depression, and reduced bone density in HIV-infected individuals. Increased depression scores in association with hypogonadism in HIV-infected men with wasting have been found to be independent of weight, virologic status, and other disease factors (Grinspoon et al., 2000).

However, while 33% of a sample of 668 HIV-infected patients reported erectile dysfunction and 24% reported reductions in sexual desire (Asboe et al., 2007), for many such patients, a medical etiology cannot be found (Liu et al., 2006; Dube et al., 2007). Furthermore, the correlation between sex hormones and sexual dysfunction is imprecise, and many patients with low testosterone levels do not report sexual problems and conversely many reporting sexual dysfunction have normal testosterone levels (Asboe et al., 2007). Factors other than low testosterone levels that contribute to sexual dysfunction in HIV-infected patients include knowledge of having HIV, altered perceptions of sex, and nonspecific factors associated with having a chronic illness. For example, Liu et al. (2006) reported that general quality-of-life ratings declined following seroconversion. Although not assessed, it is likely that sexual quality of life followed a similar pattern.

In addition, depression may be a consequence of hypotestosteronism. This was demonstrated in studies of the effects of chronic treatment with leuprolide, an LH-releasing hormone agonist, which acts as an inhibitor of gonadotropin secretion when given continuously in therapeutic doses. Although experimentally induced hypotestosteronism has minimal effects in most subjects, depressive symptoms develop in about 10% (Schmidt et al., 2004). Consistent with much earlier animal studies (Grunt and Young, 1952; Moore et al., 1978), higher pretreatment sexual activity predicted greater reductions in sexual activity when testosterone activity was artificially reduced.

38.4.2.4 Diagnostic strategies and therapeutic considerations

Many symptoms of hypogonadism are nonspecific and the presence of these symptoms and concomitant low or borderline serum testosterone levels in HIV-infected patients do not mean hypogonadism being solely responsible for these symptoms. In addition, the range of normal serum testosterone levels in men is quite wide, the cut-off for a low total testosterone level for HIV-infected men is controversial, and the threshold testosterone level below which symptoms of androgen deficiency and adverse health outcomes occur may be age dependent (Crum et al., 2005; Bhasin et al., 2006).

When testing is done, an early-morning specimen is preferable due to diurnal variation of testosterone levels (Grinspoon, 2005a). Patients with low testosterone levels should undergo further testing to determine if the hypogonadism is primary (i.e., testicular dysfunction) or secondary (i.e., hypothalamic or pituitary failure).

Administration of testosterone resulted in a significant improvement in depression in a randomized, double-blind, placebo-controlled study among HIV-infected men with wasting (Grinspoon et al., 2000). Similarly, administration of testosterone preparations to HIV-infected men with hypogonadal symptoms such as diminished libido, depressed mood, and low energy can significantly improve quality of life, libido, energy, mood, and depression scores (Mylonakis et al., 2001). This was shown in a double-blind, placebo-controlled 6-week trial with biweekly testosterone injections, followed by 12 weeks of open-label maintenance treatment among 74 symptomatic HIV-infected men (Rabkin et al., 2000). Testosterone treatment is well-tolerated in the short-term, but long-term risks such as its potential impact on prostate cancer, which is an androgen-dependent malignancy, as well as its potential to facilitate the progression of atherosclerotic heart disease (Bhasin et al., 2001) are not well-defined in HIV-infected patients.

Treatment of sexual dysfunction with testosterone replacement remains empirical in patients with subnormal and low normal testosterone levels. When patients with low testosterone levels and sexual dysfunction receive testosterone supplementation, sexual functioning improves to a variable degree (Bhasin et al., 2006). This is complicated by strong placebo effects, and by other effects of sex steroids (e.g., increases in muscle mass) (Bhasin et al., 2006). Finally, the benefit of HAART for hypogonadism is debatable (Collazos, 2007). In a study by Collazos et al. (2002) patients on protease inhibitor-based HAART experienced an increase in testosterone levels compared to their own pretreatment levels. In a study of 213 antiretroviral-naive men, after 64 weeks of HAART, testosterone levels increased on average, with participants receiving zidovudine and lamivudine showing greater increases than participants receiving stavudine and didanosine; all participants also received either efavirenz or nelfinavir (Dube et al., 2007). In another prospective study of 139 antiretroviral-naive men who started on zidovudine and lamivudine-based HAART, free testosterone levels did not change significantly after 2 years of HAART (Wunder et al., 2007).

An overall review of the diagnostic and therapeutic approaches to androgen deficiency in men, including those infected by HIV, has been recently published (Bhasin et al., 2006).

38.4.3 Thyroid Hormone Abnormalities

38.4.3.1 HIV-related hypothyroidism

Hypothyroidism may be manifest as overt hypothyroidism, that is, high thyroid-stimulating hormone (TSH) and low free T_4 (FT_4) levels, subclinical hypothyroidism, that is, a high TSH level and a normal FT_4 level or isolated low FT_4 levels with a normal TSH value. In a cross-sectional study of 350 French HIV-infected patients, approximately 16% had evidence of hypothyroidism of some sort; among these patients 2.6% had overt hypothyroidism, 6.6% had subclinical hypothyroidism, and 6.8% had a low FT_4 level (Beltran et al., 2003).

Another French cross-sectional study of 212 HIV-infected patients similarly found that 12.3% of patients had abnormal results of thyroid function tests, including 1.9% who had overt hypothyroidism and 8.5% who had subclinical hypothyroidism (Grappin et al., 2000). Other studies have found isolated decreases in the serum levels of free T_4 or T_3 in 1.3–6.8% of HIV-infected patients (Beltran et al., 2003; Madeddu et al., 2006; Hoffmann and Brown, 2007). Patients with such isolated, low FT_4 levels do not have a higher incidence of hypothyroidal symptoms (Hoffmann and Brown, 2007). Finally, the euthyroid sick syndrome, which is a physiological response to illness rather than a manifestation of abnormal thyroid functions (Hoffmann and Brown, 2007), wherein T_4 is converted into rT_3 (inactive form) instead of T_3 (active form) is common among severely ill HIV-infected patients – being found in as many as 16% of patients with AIDS (Raffi et al., 1991). Although the TSH response to thyrotropin-releasing hormone (TRH) is generally normal (Merenich, 1994), central hypothyroidism due to panhypopituitarism (Milligan et al., 1984) or hypothalamic dysfunctions (Sullivan et al., 1992) does occur.

Several risk factors are associated with abnormalities in thyroid hormones. Two studies have found stavudine, a nucleoside reverse transcriptase inhibitor (NRTI) that has been widely employed in HIV-infected patients, to be associated with hypothyroidism (Grappin et al., 2000; Beltran et al., 2003). The mechanism by which stavudine may cause hypothyroidism is ill-defined (Hoffmann and Brown, 2007). In addition, hypothyroidism may emerge in the later stages of HIV infection. Thus, one study found a low $CD4^+$ cell count (i.e., <200 cells/μl) to be a significant risk factor for hypothyroidism (Beltran et al., 2003) and, in a longitudinal study of HIV-infected patients on HAART, TSH levels were negatively associated with CD4 nadir and positively with the duration of HAART (Madeddu et al., 2006). However, in an Iranian case-control study $CD4^+$ cell depletion was not found to be a risk factor for hypothyroidism (Afhami et al., 2007).

HIV-mediated increases in the levels of circulating proinflammatory cytokines such as TNF-α, IL-1, and IL-6 (Kedzierska and Crowe, 2001) may also affect thyroid hormone homeostasis (Madeddu et al., 2006). In an *in vivo* study of six healthy males, infusion of recombinant human TNF produced the changes in levels of circulating thyroid hormones and TSH characteristic of the sick euthyroid syndrome (van der et al., 1990).

As several large series show that nearly 20% of HIV-infected persons are co-infected by HCV (Amin et al., 2004; Goulet et al., 2005), it is relevant that thyroid autoantibodies may occur more frequently among HCV co-infected patients than in HCV-uninfected patients. The presence of thyroid autoantibodies has been associated with both

Hashimoto's thyroiditis and Graves' disease in some (Huang et al., 1999; Ganne-Carrie et al., 2000), but not all (Afhami et al., 2007), studies.

Primary thyroidal failure due to destruction of the thyroid gland by opportunistic infections or HIV-related malignancy rarely occurs in advanced HIV infection (Hofbauer and Heufelder, 1996). Autopsy studies have demonstrated infiltration of the thyroid gland by CMV, MAC, *C. neoformans*, *P. jiroveci*, and Kaposi's sarcoma. The consequences of such infections, which have become unusual in persons receiving HAART, are highly variable. Most of these patients did not have clinically significant thyroid abnormalities antemortem (Welch et al., 1984). However, infection may lead to sufficient necrosis of the thyroid gland to result in the development of clinically significant hypothyroidism (Battan et al., 1991).

38.4.3.2 *HIV-related hyperthyroidism*

Graves' disease is the most common cause of hyperthyroidism among both HIV-infected and HIV-uninfected patients (Pearce et al., 2003). While Graves' disease was reported in HIV-infected individuals prior to the availability of HAART, this entity is a prominent complication of IRIS, occurring approximately 17–20 months after commencement of HAART (Gilquin et al., 1998; Jubault et al., 2000; Chen et al., 2005). In a study of 1289 male and 234 female HIV-infected patients, prevalence of *de novo* Graves' disease in the setting of immune reconstitution was 3% for males and 0.2% for females (Chen et al., 2005). Autoimmune thyroid disease may occur as naive $CD4^+$ cell subsets, especially autoreactive clones and regulatory T-cell subsets, and increase during restoration of the immune system by HAART (Chen et al., 2005). Finally, direct infection of the thyroid occasionally causes sufficiently robust inflammation and necrosis to lead to clinically significant thyroid-hormone release (Drucker et al., 1990; Hoffmann et al., 2007).

38.4.3.3 *Common iatrogenic causes of thyroid disease in HIV-infected patients*

Among the other medications often administered to HIV-infected patients, rifampin decreases peripheral thyroid hormone levels by inducing hepatic microsomal enzymes without inducing clinically apparent hypothyroidism or increases in serum TSH levels (Hofbauer and Heufelder, 1996). However, a higher dose of thyroxin-replacement therapy may be required in hypothyroidal patients who are receiving concomitant therapy with rifampin (Sellmeyer and Grunfeld, 1996). IFN-α use in the treatment of HCV can result in hypothyroidism (Primo et al., 1993) or hyperthyroidism (Schultz et al., 1989). IFN-α-associated hyperthyroidism is thought to occur due to IFN-α-induced thyroid autoimmunity. Similarly, IL-2 therapy may lead to Graves' disease by expanding the pool of naive $CD4^+$ cells (Jimenez et al., 2004). Heroin and methadone, drugs commonly used by HIV-infected patients, may increase thyroid hormone-binding proteins, thereby increasing total T_3 and total T_4 levels (Danoff, 1996). Finally, there have been individual case reports wherein the use of protease inhibitors, specifically ritonavir, indinavir, and the combination of lopinavir/ritonavir resulted in the need for increased doses of levothyroxine supplementation, possibly due to induction of levothyroxine glucuronidation (Hoffmann and Brown, 2007).

38.4.3.4 *Clinical manifestations of hypothyroidism in HIV-infected patients*

Patients with hypothyroidism may present with gradual onset of fatigue, depression, weakness, dry skin, alopecia, cold intolerance, bradycardia, and delayed deep tendon reflex. Subclinical hypothyroidism is characterized by a mildly elevated TSH level with a normal FT_4 level and either no or mild nonspecific symptoms (Hoffmann and Brown, 2007). As in other patient populations, HIV-infected patients with hyperthyroidism may present with irritability, heat intolerance, moist skin, weight loss despite increased appetite, tremor, and hyperreflexia. Subclinical hyperthyroidism may precede clinically overt hyperthyroidism (Hoffmann and Brown, 2007).

38.4.3.5 *Diagnostic strategies and therapeutic considerations*

The diagnostic strategies and therapeutic considerations for HIV-infected individuals do not differ substantially from those for HIV-uninfected patients (Hoffmann and Brown, 2007).

38.4.4 Morphologic and Metabolic Abnormalities in HIV-Infected Patients

HIV infection is associated with heterogeneous changes in body composition. Prior to the availability of HAART, wasting, which is defined as a $\geq$10% involuntary weight loss, was a common manifestation of the advanced stages of HIV disease (Moyle et al., 2004).

After HAART became commonly available (at least in developed countries), lipodystrophy (LD), which unlike the wasting syndrome is not associated with protein-energy malnutrition (Fisher, 2001), has become the prominent cause of changes in body morphology among HIV-infected patients. Importantly, LD is an umbrella term for multiple morphologic changes of body fat. Lipoatrophy refers to loss of subcutaneous fat in the face, upper and lower extremities, abdomen, and/or buttocks. In contrast, lipohypertrophy is fat accumulation in the anterior or posterior neck (i.e., dorso-cervical fat pads), abdomen, and/or breasts. Patients may have lipoatrophy and lipohypertrophy at the same time.

Patients with HIV-related LD often develop concomitant dyslipidemia, insulin resistance, and visceral fat accumulation, thereby satisfying criteria for diagnosis of the metabolic syndrome (Garg, 2000; Carr, 2003; Grinspoon, 2005b), that is, the presence of 3 or more of a waist circumference of >102 cm in men or >88 cm in women; a fasting triglyceride level of >150 mg dl^{-1}; an HDL cholesterol level of <40 mg dl^{-1} in men or <50 mg dl^{-1} in women; blood pressure of >130/85 mmHg; and a fasting glucose level of >110 mg dl^{-1} (Hadigan et al., 2003). In a 5-year cohort analysis of 221 HIV-infected patients, the cumulative incidence of hyperglycemia, hypercholesterolemia, hypertriglyceridemia, and LD was 5%, 24%, 19%, and 13%, respectively (Tsiodras et al., 2000). Dyslipidemia associated with increased cardiovascular disease risk occurs in about 70% of HIV-infected patients; diabetes mellitus occurs in about 8–10%; and a further 15% have impaired glucose tolerance (Carr, 2003).

Pathogenic mechanisms for LD are likely the result of complex interactions between host, disease, and drug factors (Lichtenstein, 2005). LD is observed in both patients who received protease inhibitors (Carr et al., 1998) and protease inhibitor-naive patients treated with NRTIs (Galli et al., 2002). Major risk factors for lipoatrophy include use of thymidine analog NRTIs, most commonly stavudine, duration of therapy, white race, and older age (Lichtenstein, 2005; Wohl et al., 2006). In particular, mitochondrial DNA toxicity associated with nucleoside reverse transcriptase inhibitors has been strongly suggested as an etiology for lipoatrophy (Brinkman et al., 1999; Kakuda et al., 1999; Shikuma et al., 2001; Walker et al., 2002). In addition, although causal relationships remain to be determined, expression of TNF-α in adipocytes was higher in patients with lipoatrophy associated with use of protease inhibitors and NRTIs than that in adipocytes from HIV-uninfected patients (Bastard et al., 2002). Major risk factors for lipohypertrophy include use of protease inhibitor, duration of therapy, and older age (Lichtenstein, 2005; Wohl et al., 2006). Finally, impaired growth hormone (GH) secretion has been shown to correlate with LD both among adults (Rietschel et al., 2001; Koutkia et al., 2006) and among pubertal children (Rietschel et al., 2001; Vigano et al., 2003).

Many of the clinical findings in patients with LD resemble those observed in patients with Cushing's syndrome (Bhasin et al., 2001) such as insulin resistance, hypertension, central adiposity, and the development of dorso-cervical fat pads. However, patients with LD rarely have the more specific laboratory abnormalities observed in Cushing's syndrome (Mayo et al., 2002). In particular, Lo et al. studied eight HIV-infected patients who were on stable HAART regimens and had developed dorso-cervical fat pads. Twenty-four-hour urinary free cortisol excretion was normal in seven patients and slightly elevated in one patient. All eight patients had normal suppression of cortisol levels after 1 mg dexamethasone administration (Lo et al., 1998). In addition, Yanovski et al. compared functions of the HPA axis in 12 HIV-infected patients with protease inhibitor-associated LD, 28 patients with Cushing's syndrome, and 43 healthy HIV-uninfected patients. Patients with LD had normal diurnal cortisol secretion, normal cortisol secretory dynamics in response to ovine CRH infusion, normal cortisol-binding globulin levels, and normal glucocorticoid receptor number and affinity (Yanovski et al., 1999). Based on these and other studies, dysregulation of HPA axis does not appear to be the cause of LD.

38.4.4.1 Neuropsychiatric impact of LD in HIV-infected patients

LD itself is not known to cause a direct effect on the brain of HIV-infected patients. However, psychosocial implications of LD are of great concern for patients because LD affects their body image. Surveys and interviews on HIV-infected patients demonstrate that LD has a substantial deleterious impact on quality of life, sexual life, and depression (Power et al., 2003), less physical well-being and less confidence in relationships (Dukers et al., 2001). Compared to HIV-infected men who have sex with men (MSM) who denied LD, HIV-infected MSM with self-reported LD were found to have poor body image as measured by Body Image Quality of Life Scale and Situational Inventory of Body

Image Dysphoria Short-Term Score (Huang et al., 2006). In addition, Oette et al. (2002) found that HIV-infected patients with LD were about twice as likely to feel recognizable as being HIV infected by their physical appearance. Finally, approximately two-thirds of HIV-infected patients with LD stated that they would choose loss of 1 year of life rather than developing LD (Lenert et al., 2002). Consequently, it is not surprising that LD has been associated with nonadherence to therapy (Duran et al., 2001; Ammassari et al., 2002).

38.4.4.2 Diagnostic strategies and therapeutic considerations

The diagnostic criteria for LD are imprecise (Steel et al., 2006). Some clinical studies have classified patients on the basis of self-reported changes in body fat while others have relied upon subjective judgments made by physicians or ill-defined findings on physical examination. Anthropometric measurements, such as skin folds, waist circumference, or waist-to-hip ratio, are useful but are operator dependent. Imaging modalities, including ultrasound, computed tomography scan, magnetic resonance imaging scan, and dual-energy X-ray absorptiometry, give objective data but have not been observed to provide a clinical advantage over self-report and physical examination assessments (Wohl et al., 2006).

A number of treatment options are available for LD although their long-term effects are still unknown. For lipoatrophy, treatment options include surgical implants – temporary injectable implants including collagen and hyaluronic acid (Engelhard, 2006). Interventions that have been studied for metabolic syndrome and lipohypertrophy include pharmacologic interventions such as lipid-lowering agents, sulfonylurea, metformin, thiazolidenediones (Grinspoon, 2003), testosterone (Bhasin et al., 2007), GH and growth-hormone-releasing hormone (GHRH) as described below; and surgical removal (Gervasoni et al., 2002). In randomized, double-blind, placebo-controlled trials enrolling HIV-patients with LD, total and regional body composition improved in response to GHRH (Koutkia et al., 2004; Falutz et al., 2005). Similarly, in a prospective, open-label trial, 24 weeks of supraphysiologic doses of recombinant human GH (rhGH) reduced excess visceral adipose tissue in HIV-infected patients with LD (Engelson et al., 2002). However, adverse effects such as myalgias, paresthesias, and fluid retention were common, and improvements in body composition largely reversed within 12 weeks of discontinuing rhGH therapy (Engelson et al., 2002). The results of quality-of-life measures during GH or GHRH treatments showed inconsistent benefits, probably as a result of both negative consequences (i.e., adverse effects of therapy) and positive consequences (i.e., improved body shape) (Engelson et al., 2002; Falutz et al., 2005).

Numerous studies have investigated which modification of HAART regimen ameliorates LD and quality-of-life indices. Substitution of other agents for a protease inhibitor has not been associated with reversal or improvement in fat redistribution (Drechsler and Powderly, 2002). Numerous open-label, randomized, or switch studies were performed in the following designs: switching zidovudine or stavudine to abacavir (Carr et al., 2002; Martin et al., 2004); zidovudine or stavudine to abacavir or tenofovir (Moyle et al., 2006); stavudine to abacavir or zidovudine (McComsey et al., 2004); and stavudine and/or a protease inhibitor to combination therapy with zidovudine, lamivudine, and abacavir (John et al., 2003). These trials have shown that such replacements result in a gradual gain in limb fat but without consistent decreases in previously accumulated visceral fat or cholesterol profile. The improvement in lipoatrophy by substitution of stavudine with abacavir or zidovudine may be due to the lesser mitochondrial toxicity and fat apoptosis induced by the latter agents (Leonard and McComsey, 2005; McComsey et al., 2005).

References

Afghani B and Lieberman JM (1994) Paradoxical enlargement or development of intracranial tuberculomas during therapy: Case report and review. *Clinical Infectious Diseases* 19: 1092–1099.

Afhami S, Haghpanah V, Heshmat R, et al. (2007) Assessment of the factors involving in the development of hypothyroidism in HIV-infected patients: A case-control study. *Infection* 35: 334–338.

Albert SG, DeLeon MJ, and Silverberg AB (2001) Possible association between high-dose fluconazole and adrenal insufficiency in critically ill patients. *Critical Care Medicine* 29: 668–670.

Alen M, Rahkila P, Reinila M, and Vihko R (1987) Androgenic-anabolic steroid effects on serum thyroid, pituitary and steroid hormones in athletes. *American Journal of Sports Medicine* 15: 357–361.

Amin J, Kaye M, Skidmore S, Pillay D, Cooper DA, and Dore GJ (2004) HIV and hepatitis C coinfection within the CAESAR study. *HIV Medicine* 5: 174–179.

Amin MB, Abrash MP, Mezger E, and Sekerak GF (1990) Systemic dissemination of *Pneumocystis carinii* in a patient with acquired immunodeficiency syndrome. *Henry Ford Hospital Medical Journal* 38: 68–71.

Ammassari A, Antinori A, Cozzi-Lepri A, et al. (2002) Relationship between HAART adherence and adipose

tissue alterations. *Journal of Acquired Immune Deficiency Syndromes* 31(supplement 3): S140–S144.
Ances BM and Ellis RJ (2007) Dementia and neurocognitive disorders due to HIV-1 infection. *Seminars in Neurology* 27: 86–92.
Arrington-Sanders R, Hutton N, and Siberry GK (2006) Ritonavir–fluticasone interaction causing Cushing syndrome in HIV-infected children and adolescents. *Pediatric Infectious Disease* 25: 1044–1048.
Arver S, Sinha-Hikim I, Beall G, Guerrero M, Shen R, and Bhasin S (1999) Serum dihydrotestosterone and testosterone concentrations in human immunodeficiency virus-infected men with and without weight loss. *Journal of Andrology* 20: 611–618.
Asboe D, Catalan J, Mandalia S, et al. (2007) Sexual dysfunction in HIV-positive men is multi-factorial: A study of prevalence and associated factors. *AIDS Care* 19: 955–965.
Asnis GM and De La Garza (2005) Interferon-induced depression: Strategies in treatment. *Progress in Neuropsychopharmacology and Biological Psychiatry* 29: 808–818.
Asnis GM and De La Garza (2006) Interferon-induced depression in chronic hepatitis C: A review of its prevalence, risk factors, biology, and treatment approaches. *Journal of Clinical Gastroenterology* 40: 322–335.
Baldewicz TT, Leserman J, Silva SG, et al. (2004) Changes in neuropsychological functioning with progression of HIV-1 infection: Results of an 8-year longitudinal investigation. *AIDS and Behavior* 8: 345–355.
Bastard JP, Caron M, Vidal H, et al. (2002) Association between altered expression of adipogenic factor SREBP1 in lipoatrophic adipose tissue from HIV-1-infected patients and abnormal adipocyte differentiation and insulin resistance. *Lancet* 359: 1026–1031.
Battan R, Mariuz P, Raviglione MC, Sabatini MT, Mullen MP, and Poretsky L (1991) *Pneumocystis carinii* infection of the thyroid in a hypothyroid patient with AIDS: Diagnosis by fine needle aspiration biopsy. *Journal of Clinical Endocrinology and Metabolism* 72: 724–726.
Beltran S, Lescure FX, Desailloud R, et al. (2003) Increased prevalence of hypothyroidism among human immunodeficiency virus-infected patients: A need for screening. *Clinical Infectious Diseases* 37: 579–583.
Bhasin S, Cunningham GR, Hayes FJ, Matsumoto AM, Snyder PJ, Swerdloff RS, and Montori VM (2006) Testosterone therapy in adult men with androgen deficiency syndromes: An endocrine society clinical practice guideline. *Journal of Clinical Endocrinology and Metabolism* 91: 1995–2010.
Bhasin S, Parker RA, Sattler F, et al. (2007) Effects of testosterone supplementation on whole body and regional fat mass and distribution in human immunodeficiency virus-infected men with abdominal obesity. *Journal of Clinical Endocrinology and Metabolism* 92: 1049–1057.
Bhasin S, Singh AB, and Javanbakht M (2001) Neuroendocrine abnormalities associated with HIV infection. *Endocrinology Metabolism Clinics of North America* 30: 749–764, viii.
Bricaire F, Marche C, Zoubi D, Perronne C, Matheron S, Rouveix E, and Vittecoq D (1987) Adrenal lesions in AIDS: Anatomopathological study. *Annales de Medecine Interne (Paris)* 138: 607–609.
Brinkman K, Smeitink JA, Romijn JA, and Reiss P (1999) Mitochondrial toxicity induced by nucleoside-analogue reverse-transcriptase inhibitors is a key factor in the pathogenesis of antiretroviral-therapy-related lipodystrophy. *Lancet* 354: 1112–1115.
Capuron L, Gumnick JF, Musselman DL, Lawson DH, Reemsnyder A, Nemeroff CB, and Miller AH (2002) Neurobehavioral effects of interferon-alpha in cancer patients: Phenomenology and paroxetine responsiveness of symptom dimensions. *Neuropsychopharmacology* 26: 643–652.
Carr A (2003) HIV lipodystrophy: Risk factors, pathogenesis, diagnosis and management. *AIDS* 17(supplement 1): S141–S148.
Carr A, Samaras K, Burton S, Law M, Freund J, Chisholm DJ, and Cooper DA (1998) A syndrome of peripheral lipodystrophy, hyperlipidaemia and insulin resistance in patients receiving HIV protease inhibitors. *AIDS* 12: F51–F58.
Carr A, Workman C, Smith DE, et al. (2002) Abacavir substitution for nucleoside analogs in patients with HIV lipoatrophy: A randomized trial. *Journal of the American Medical Association* 288: 207–215.
CASCADE collaboration (2003) Differences in CD4 cell counts at seroconversion and decline among 5739 HIV-1-infected individuals with well-estimated dates of seroconversion. *Journal of Acquired Immune Deficiency Syndromes* 34 (supplement 1): 76–83.
Cejtin HE, Kalinowski A, Bacchetti P, et al. (2006) Effects of human immunodeficiency virus on protracted amenorrhea and ovarian dysfunction. *Obstetrics and Gynecology* 108: 1423–1431.
Chen F, Day SL, Metcalfe RA, et al. (2005) Characteristics of autoimmune thyroid disease occurring as a late complication of immune reconstitution in patients with advanced human immunodeficiency virus (HIV) disease. *Medicine (Baltimore)* 84: 98–106.
Childs EA, Lyles RH, Selnes OA, et al. (1999) Plasma viral load and CD4 lymphocytes predict HIV-associated dementia and sensory neuropathy. *Neurology* 52: 607–613.
Chrousos GP (1995) The hypothalamic–pituitary–adrenal axis and immune-mediated inflammation. *New England Journal of Medicine* 332: 1351–1362.
Collazos J (2007) The effects of antiretroviral therapy on testosterone levels and fat-free mass. *Clinical Infectious Diseases* 45: 1525–1527.
Collazos J, Martinez E, Mayo J, and Ibarra S (2002) Sexual hormones in HIV-infected patients: The influence of antiretroviral therapy. *AIDS* 16: 934–937.
Coodley GO, Loveless MO, Nelson HD, and Coodley MK (1994) Endocrine function in the HIV wasting syndrome. *Journal of Acquired Immune Deficiency Syndromes* 7: 46–51.
Cooper OB, Brown TT, and Dobs AS (2003) Opiate drug use: A potential contributor to the endocrine and metabolic complications in human immunodeficiency virus disease. *Clinical Infectious Diseases* 37(supplement 2): S132–S136.
Crum NF, Furtek KJ, Olson PE, Amling CL, and Wallace MR (2005) A review of hypogonadism and erectile dysfunction among HIV-infected men during the pre- and post-HAART eras: Diagnosis, pathogenesis, and management. *AIDS Patient Care and STDs* 19: 655–671.
Crump JA, Tyrer MJ, Lloyd-Owen SJ, Han LY, Lipman MC, and Johnson MA (1998) Military tuberculosis with paradoxical expansion of intracranial tuberculomas complicating human immunodeficiency virus infection in a patient receiving highly active antiretroviral therapy. *Clinical Infectious Diseases* 26: 1008–1009.
Daar ES, Little S, Pitt J, et al. (2001) Diagnosis of primary HIV-1 infection. Los Angeles County Primary HIV Infection Recruitment Network. *Annals of Internal Medicine* 134: 25–29.
Danoff A (1996) Endocrinologic complications of HIV infection. *Medical Clinics of North America* 80: 1453–1469.
DePaepe ME and Waxman M (1989) Testicular atrophy in AIDS: A study of 57 autopsy cases. *Human Pathology* 20: 210–214.

DHHS Panel on Antiretroviral Guidelines for Adults Adolescents (2006) Guidelines for the use of antiretroviral agents in HIV-infected adults and adolescents. http://aidsinfo.nih.gov/ContentFiles/AdultandAdolescentGL.pdf (accessed March 2008).
Dienstag JL and McHutchison JG (2006) American Gastroenterological Association medical position statement on the management of hepatitis C. *Gastroenterology* 130: 225–230.
Dobs AS, Dempsey MA, Ladenson PW, and Polk BF (1988) Endocrine disorders in men infected with human immunodeficiency virus. *American Journal of Medicine* 84: 611–616.
Dou H, Birusingh K, Faraci J, et al. (2003) Neuroprotective activities of sodium valproate in a murine model of human immunodeficiency virus-1 encephalitis. *Journal of Neuroscience* 23: 9162–9170.
Drechsler H and Powderly WG (2002) Switching effective antiretroviral therapy: A review. *Clinical Infectious Diseases* 35: 1219–1230.
Drucker DJ, Bailey D, and Rotstein L (1990) Thyroiditis as the presenting manifestation of disseminated extrapulmonary *Pneumocystis carinii* infection. *Journal of Clinical Endocrinology and Metabolism* 71: 1663–1665.
Dube MP, Parker RA, Mulligan K, Tebas P, Robbins GK, Roubenoff R, and Grinspoon SK (2007) Effects of potent antiretroviral therapy on free testosterone levels and fat-free mass in men in a prospective, randomized trial: A5005s, a substudy of AIDS Clinical Trials Group study 384. *Clinical Infectious Diseases* 45: 120–126.
Duch FM, Repele CA, Spadaro F, dos Reis MA, Rodrigues DB, Ferraz ML, and Teixeira VP (1998) Adrenal gland morphological alterations in the acquired immunodeficiency syndrome. *Revista da Sociedade Brasileira de Medicina Tropical* 31: 257–261.
Dukers NH, Stolte IG, Albrecht N, Coutinho RA, and de Wit JB (2001) The impact of experiencing lipodystrophy on the sexual behaviour and well-being among HIV-infected homosexual men. *AIDS* 15: 812–813.
Duran S, Saves M, Spire B, et al. (2001) Failure to maintain long-term adherence to highly active antiretroviral therapy: The role of lipodystrophy. *AIDS* 15: 2441–2444.
Ellis R, Langford D, and Masliah E (2007) HIV and antiretroviral therapy in the brain: Neuronal injury and repair. *Nature Reviews Neuroscience* 8: 3344.
Engelhard P (2006) Correction options for lipoatrophy in HIV-infected patients. *AIDS Patient Care and STDs* 20: 151–160.
Engelson ES, Glesby MJ, Mendez D, Albu JB, Wang J, Heymsfield SB, and Kotler DP (2002) Effect of recombinant human growth hormone in the treatment of visceral fat accumulation in HIV infection. *Journal of Acquired Immune Deficiency Syndromes* 30: 379–391.
Eyer-Silva WA, Morais de Sa CA, de Oliveira CA, Pinto JF, Vaz LF, and Gameiro CA (1994) Hypophyseal tuberculosis in a patient with AIDS. *Clinical Infectious Diseases* 19: 550–551.
Falutz J, Allas S, Kotler D, et al. (2005) A placebo-controlled, dose-ranging study of a growth hormone releasing factor in HIV-infected patients with abdominal fat accumulation. *AIDS* 19: 1279–1287.
Ferreiro J and Vinters HV (1988) Pathology of the pituitary gland in patients with the acquired immune deficiency syndrome (AIDS). *Pathology* 20: 211–215.
Findling JW, Buggy BP, Gilson IH, Brummitt CF, Bernstein BM, and Raff H (1994) Longitudinal evaluation of adrenocortical function in patients infected with the human immunodeficiency virus. *Journal of Clinical Endocrinology and Metabolism* 79: 1091–1096.
Fisher K (2001) Wasting and lipodystrophy in patients infected with HIV: A practical approach in clinical practice. *AIDS Read* 11: 132–140, 147.
French MA, Lenzo N, John M, et al. (2000) Immune restoration disease after the treatment of immunodeficient HIV-infected patients with highly active antiretroviral therapy. *HIV Medicine* 1: 107–115.
Galli M, Ridolfo AL, Adorni F, et al. (2002) Body habitus changes and metabolic alterations in protease inhibitor-naive HIV-1-infected patients treated with two nucleoside reverse transcriptase inhibitors. *Journal of Acquired Immune Deficiency Syndromes* 29: 21–31.
Ganne-Carrie N, Medini A, Coderc E, et al. (2000) Latent autoimmune thyroiditis in untreated patients with HCV chronic hepatitis: A case-control study. *Journal of Autoimmunity* 14: 189–193.
Garg A (2000) Lipodystrophies. *American Journal of Medicine* 108: 143–152.
Geller J (1985) Rationale for blockade of adrenal as well as testicular androgens in the treatment of advanced prostate cancer. *Seminars in Oncology* 12: 28–35.
Gervasoni C, Ridolfo AL, Rovati L, Vaccarezza M, Carsana L, and Galli M (2002) Maintenance of breast size reduction after mastoplasty and switch to a protease inhibitor-sparing regimen in an HIV-positive woman with highly active antiretroviral therapy-associated massive breast enlargement. *AIDS Patient Care and STDs* 16: 307–311.
Gilquin J, Viard JP, Jubault V, Sert C, and Kazatchkine MD (1998) Delayed occurrence of Graves' disease after immune restoration with HAART. Highly active antiretroviral therapy. *Lancet* 352: 1907–1908.
Glasgow BJ, Steinsapir KD, Anders K, and Layfield LJ (1985) Adrenal pathology in the acquired immune deficiency syndrome. *American Journal of Clinical Pathology* 84: 594–597.
Gonzalez-Scarano F and Martin-Garcia J (2005) The neuropathogenesis of AIDS. *Nature Reviews Immunology* 5: 69–81.
Goulet JL, Fultz SL, McGinnis KA, and Justice AC (2005) Relative prevalence of comorbidities and treatment contraindications in HIV-mono-infected and HIV/HCV-co-infected veterans. *AIDS* 19(supplement 3): S99–S105.
Grappin M, Piroth L, Verges B, et al. (2000) Increased prevalence of subclinical hypothyroidism in HIV patients treated with highly active antiretroviral therapy. *AIDS* 14: 1070–1072.
Grinspoon S (2003) Mechanisms and strategies for insulin resistance in acquired immune deficiency syndrome. *Clinical Infectious Diseases* 37(supplement 2): S85–S90.
Grinspoon S (2005a) Androgen deficiency and HIV infection. *Clinical Infectious Diseases* 41: 1804–1805.
Grinspoon S, Corcoran C, Miller K, et al. (1997) Body composition and endocrine function in women with acquired immunodeficiency syndrome wasting. *Journal of Clinical Endocrinology and Metabolism* 82: 1332–1337.
Grinspoon S, Corcoran C, Stanley T, Baaj A, Basgoz N, and Klibanski A (2000) Effects of hypogonadism and testosterone administration on depression indices in HIV-infected men. *Journal of Clinical Endocrinology and Metabolism* 85: 60–65.
Grinspoon S, Corcoran C, Stanley T, Rabe J, and Wilkie S (2001) Mechanisms of androgen deficiency in human immunodeficiency virus-infected women with the wasting syndrome. *Journal of Clinical Endocrinology and Metabolism* 86: 4120–4126.

Grinspoon SK (2005b) Metabolic syndrome and cardiovascular disease in patients with human immunodeficiency virus. *American Journal of Medicine* 118(supplement 2): 23S–28S.
Grinspoon SK and Bilezikian JP (1992) HIV disease and the endocrine system. *New England Journal of Medicine* 327: 1360–1365.
Grunt JA and Young WC (1952) Differential reactivity of individuals and the response of the male guinea pig to testosterone propionate. *Endocrinology* 51: 237–248.
Guarda LA, Luna MA, Smith JL, Jr., Mansell PW, Gyorkey F, and Roca AN (1984) Acquired immune deficiency syndrome: Postmortem findings. *American Journal of Clinical Pathology* 81: 549–557.
Haas DW, Ribaudo HJ, Kim RB, et al. (2004) Pharmacogenetics of efavirenz and central nervous system side effects: An Adult AIDS Clinical Trials Group study. *AIDS* 18: 2391–2400.
Hadigan C, Meigs JB, Wilson PW, et al. (2003) Prediction of coronary heart disease risk in HIV-infected patients with fat redistribution. *Clinical Infectious Diseases* 36: 909–916.
Hofbauer LC and Heufelder AE (1996) Endocrine implications of human immunodeficiency virus infection. *Medicine (Baltimore)* 75: 262–278.
Hoffmann CJ and Brown TT (2007) Thyroid function abnormalities in HIV-infected patients. *Clinical Infectious Diseases* 45: 488–494.
Huang JS, Lee D, Becerra K, Santos R, Barber E, and Mathews WC (2006) Body image in men with HIV. *AIDS Patient Care and STDs* 20: 668–677.
Huang MJ, Tsai SL, Huang BY, Sheen IS, Yeh CT, and Liaw YF (1999) Prevalence and significance of thyroid autoantibodies in patients with chronic hepatitis C virus infection: A prospective controlled study. *Clinical Endocrinology (Oxf)* 50: 503–509.
Janssen RS, Nwanyanwu OC, Selik RM, and Stehr-Green JK (1992) Epidemiology of human immunodeficiency virus encephalopathy in the United States. *Neurology* 42: 1472–1476.
Jimenez C, Moran SA, Sereti I, et al. (2004) Graves' disease after interleukin-2 therapy in a patient with human immunodeficiency virus infection. *Thyroid* 14: 1097–1102.
John M, McKinnon EJ, James IR, et al. (2003) Randomized, controlled, 48-week study of switching stavudine and/or protease inhibitors to combivir/abacavir to prevent or reverse lipoatrophy in HIV-infected patients. *Journal of Acquired Immune Deficiency Syndromes* 33: 29–33.
Johns K, Beddall MJ, and Corrin RC (2005) Anabolic steroids for the treatment of weight loss in HIV-infected individuals. *Cochrane Database of Systematic Reviews* 19: CD005483.
Johnson SR, Marion AA, Vrchoticky T, Emmanuel PJ, and Lujan-Zilbermann J (2006) Cushing syndrome with secondary adrenal insufficiency from concomitant therapy with ritonavir and fluticasone. *Journal of Pediatrics* 148: 386–388.
Jubault V, Penfornis A, Schillo F, et al. (2000) Sequential occurrence of thyroid autoantibodies and Graves' disease after immune restoration in severely immunocompromised human immunodeficiency virus-1-infected patients. *Journal of Clinical Endocrinology and Metabolism* 85: 4254–4257.
Kahn JO and Walker BD (1998) Acute human immunodeficiency virus type 1 infection. *New England Journal of Medicine* 339: 33–39.
Kakuda TN, Brundage RC, Anderson PL, and Fletcher CV (1999) Nucleoside reverse transcriptase inhibitor-induced mitochondrial toxicity as an etiology for lipodystrophy. *AIDS* 13: 2311–2312.
Karp BI, Yang JC, Khorsand M, Wood R, and Merigan TC (1996) Multiple cerebral lesions complicating therapy with interleukin-2. *Neurology* 47: 417–424.
Kassutto S and Rosenberg ES (2004) Primary HIV type 1 infection. *Clinical Infectious Diseases* 38: 1447–1453.
Kedzierska K and Crowe SM (2001) Cytokines and HIV-1: Interactions and clinical implications. *Antiviral Chemistry and Chemotherapy* 12: 133–150.
Koutkia P, Canavan B, Breu J, Torriani M, Kissko J, and Grinspoon S (2004) Growth hormone-releasing hormone in HIV-infected men with lipodystrophy: A randomized controlled trial. *Journal of the American Medical Association* 292: 210–218.
Koutkia P, Eaton K, You SM, Breu J, and Grinspoon S (2006) Growth hormone secretion among HIV infected patients: Effects of gender, race and fat distribution. *AIDS* 20: 855–862.
Langford D, Adame A, Grigorian A, et al. (2003) Patterns of selective neuronal damage in methamphetamine-user AIDS patients. *Journal of Acquired Immune Deficiency Syndromes* 34: 467–474.
Laudat A, Blum L, Guechot J, Picard O, Cabane J, Imbert JC, and Giboudeau J (1995) Changes in systemic gonadal and adrenal steroids in asymptomatic human immunodeficiency virus-infected men: Relationship with the CD4 cell counts. *European Journal of Endocrinology* 133: 418–424.
Leinung MC, Liporace R, and Miller CH (1995) Induction of adrenal suppression by megestrol acetate in patients with AIDS. *Annals of Internal Medicine* 122: 843–845.
Lenert LA, Feddersen M, Sturley A, and Lee D (2002) Adverse effects of medications and trade-offs between length of life and quality of life in human immunodeficiency virus infection. *American Journal of Medicine* 113: 229–232.
Leonard EG and McComsey GA (2005) Antiretroviral therapy in HIV-infected children: The metabolic cost of improved survival. *Infectious Disease Clinics of North America* 19: 713–729.
Letendre SL, McCutchan JA, Childers ME, et al. (2004) Enhancing antiretroviral therapy for human immunodeficiency virus cognitive disorders. *Annals of Neurology* 56: 416–423.
Lichtenstein KA (2005) Redefining lipodystrophy syndrome: Risks and impact on clinical decision making. *Journal of Acquired Immune Deficiency Syndromes* 39: 395–400.
Liu C, Ostrow D, Detels R, Hu Z, Johnson L, Kingsley L, and Jacobson LP (2006) Impacts of HIV infection and HAART use on quality of life. *Quality of Life Research* 15: 941–949.
Lo JC, Mulligan K, Tai VW, Algren H, and Schambelan M (1998) 'Buffalo hump' in men with HIV-1 infection. *Lancet* 351: 867–870.
Lortholary O, Christeff N, Casassus P, et al. (1996) Hypothalamo-pituitary–adrenal function in human immunodeficiency virus-infected men. *Journal of Clinical Endocrinology and Metabolism* 81: 791–796.
Lortholary O, Fontanet A, Memain N, Martin A, Sitbon K, and Dromer F (2005) Incidence and risk factors of immune reconstitution inflammatory syndrome complicating HIV-associated cryptococcosis in France. *AIDS* 19: 1043–1049.
Madeddu G, Spanu A, Chessa F, et al. (2006) Thyroid function in human immunodeficiency virus patients treated with highly active antiretroviral therapy (HAART): A longitudinal study. *Clinical Endocrinology (Oxf)* 64: 375–383.
Magill SS, Puthanakit T, Swoboda SM, Carson KA, Salvatori R, Lipsett PA, and Hendrix CW (2004) Impact of fluconazole prophylaxis on cortisol levels in critically ill surgical patients. *Antimicrobial Agents and Chemotherapy* 48: 2471–2476.
Mann M, Koller E, Murgo A, Malozowski S, Bacsanyi J, and Leinung M (1997) Glucocorticoidlike activity of megestrol.

A summary of Food and Drug Administration experience and a review of the literature. *Archives of Internal Medicine* 157: 1651–1656.
Martin A, Smith DE, Carr A, et al. (2004) Reversibility of lipoatrophy in HIV-infected patients 2 years after switching from a thymidine analogue to abacavir: The MITOX Extension Study. *AIDS* 18: 1029–1036.
Masliah E, DeTeresa RM, Mallory ME, and Hansen LA (2000) Changes in pathological findings at autopsy in AIDS cases for the last 15 years. *AIDS* 14: 69–74.
Mastorakos G, Chrousos GP, and Weber JS (1993) Recombinant interleukin-6 activates the hypothalamic–pituitary–adrenal axis in humans. *Journal of Clinical Endocrinology and Metabolism* 77: 1690–1694.
Mayo J, Collazos J, Martinez E, and Ibarra S (2002) Adrenal function in the human immunodeficiency virus-infected patient. *Archives of Internal Medicine* 162: 1095–1098.
McArthur JC, Brew BJ, and Nath A (2005) Neurological complications of HIV infection. *Lancet Neurology* 4: 543–555.
McArthur JC, McDermott MP, McClernon D, et al. (2004) Attenuated central nervous system infection in advanced HIV/AIDS with combination antiretroviral therapy. *Archives of Neurology* 61: 1687–1696.
McComsey GA, Paulsen DM, Lonergan JT, et al. (2005) Improvements in lipoatrophy, mitochondrial DNA levels and fat apoptosis after replacing stavudine with abacavir or zidovudine. *AIDS* 19: 15–23.
McComsey GA, Ward DJ, Hessenthaler SM, et al. (2004) Improvement in lipoatrophy associated with highly active antiretroviral therapy in human immunodeficiency virus-infected patients switched from stavudine to abacavir or zidovudine: The results of the TARHEEL study. *Clinical Infectious Diseases* 38: 263–270.
Meersseman W, Van LK, Lagrou K, et al. (2005) Fatal brain necrosis in primary HIV infection. *Lancet* 366: 866.
Mellors JW, Muñoz A, Giorgi JV, et al. (1997) Plasma viral load and $CD4^+$ lymphocytes as prognostic markers of HIV-1 infection. *Annals of Internal Medicine* 126: 946–954.
Membreno L, Irony I, Dere W, Klein R, Biglieri EG, and Cobb E (1987) Adrenocortical function in acquired immunodeficiency syndrome. *Journal of Clinical Endocrinology and Metabolism* 65: 482–487.
Merenich JA (1994) Hypothalamic and pituitary function in AIDS. *Bailliere's Clinical Endocrinology and Metabolism* 8: 757–767.
Merenich JA, McDermott MT, Asp AA, Harrison SM, and Kidd GS (1990) Evidence of endocrine involvement early in the course of human immunodeficiency virus infection. *Journal of Clinical Endocrinology and Metabolism* 70: 566–571.
Miller RF, Isaacson PG, Hall-Craggs M, Lucas S, Gray F, Scaravilli F, and An SF (2004) Cerebral CD8+ lymphocytosis in HIV-1 infected patients with immune restoration induced by HAART. *Acta Neuropathology* 108: 17–23.
Milligan SA, Katz MS, Craven PC, Strandberg DA, Russell IJ, and Becker RA (1984) Toxoplasmosis presenting as panhypopituitarism in a patient with the acquired immune deficiency syndrome. *American Journal of Medicine* 77: 760–764.
Mirani M, Elenkov I, Volpi S, Hiroi N, Chrousos GP, and Kino T (2002) HIV-1 protein Vpr suppresses IL-12 production from human monocytes by enhancing glucocorticoid action: Potential implications of Vpr coactivator activity for the innate and cellular immunity deficits observed in HIV-1 infection. *Journal of Immunology* 169: 6361–6368.
Mitsuyasu RT (2001) The potential role of interleukin-2 in HIV. *AIDS* 15(supplement 2): S22–S27.
Mocroft A, Youle M, Phillips AN, Halai R, Easterbrook P, Johnson MA, and Gazzard B (1998) The incidence of AIDS-defining illnesses in 4883 patients with human immunodeficiency virus infection. Royal Free/Chelsea and Westminster Hospitals Collaborative Group. *Archives of Internal Medicine* 158: 491–497.
Moore FL, Specker JL, and Swanson L (1978) Effects of testosterone and methallibure on courtship and plasma androgen concentrations of male newts, *Taricha granulosa*. *General and Comparative Endocrinology 34: 259–264.*
Moyle GJ, Daar ES, Gertner JM, Kotler DP, Melchior JC, O'brien F, and Svanberg E (2004) Growth hormone improves lean body mass, physical performance, and quality of life in subjects with HIV-associated weight loss or wasting on highly active antiretroviral therapy. *Journal of Acquired Immune Deficiency Syndromes* 35: 367–375.
Moyle GJ, Sabin CA, Cartledge J, et al. (2006) A randomized comparative trial of tenofovir DF or abacavir as replacement for a thymidine analogue in persons with lipoatrophy. *AIDS* 20: 2043–2050.
Musselman DL, Lawson DH, Gumnick JF, et al. (2001) Paroxetine for the prevention of depression induced by high-dose interferon alfa. *New England Journal of Medicine* 344: 961–966.
Mylonakis E, Koutkia P, and Grinspoon S (2001) Diagnosis and treatment of androgen deficiency in human immunodeficiency virus-infected men and women. *Clinical Infectious Diseases* 33: 857–864.
Nicolls DJ, King M, Holland D, Bala J, and del RC (2005) Intracranial tuberculomas developing while on therapy for pulmonary tuberculosis. *Lancet Infectious Diseases* 5: 795–801.
Norbiato G, Bevilacqua M, Vago T, et al. (1992) Cortisol resistance in acquired immunodeficiency syndrome. *Journal of Clinical Endocrinology and Metabolism* 74: 608–613.
Northfelt DW, Clement MJ, and Safrin S (1990) Extrapulmonary pneumocystosis: Clinical features in human immunodeficiency virus infection. *Medicine (Baltimore)* 69: 392–398.
O'Brien TR, Blattner WA, Waters D, et al. (1996) Serum HIV-1 RNA levels and time to development of AIDS in the multicenter hemophilia cohort study. *Journal of the American Medical Association* 276: 105–110.
Oette M, Juretzko P, Kroidl A, Sagir A, Wettstein M, Siegrist J, and Haussinger D (2002) Lipodystrophy syndrome and self-assessment of well-being and physical appearance in HIV-positive patients. *AIDS Patient Care and STDs* 16: 413–417.
Pagonis TA, Angelopoulos NV, Koukoulis GN, and Hadjichristodoulou CS (2006) Psychiatric side effects induced by supraphysiological doses of combinations of anabolic steroids correlate to the severity of abuse. *European Psychiatry* 21: 551–562.
Palella FJ, Delaney KM, Moorman AC, Loveless MO, Fuhrer J, and Satten GA (1998) Declining morbidity and mortality among patients with advance human immunodeficiency virus infection. *New England Journal of Medicine* 338: 853–860.
Pearce EN, Farwell AP, and Braverman LE (2003) Thyroiditis. *New England Journal of Medicine* 348: 2646–2655.
Pfefferbaum A, Rosenbloom MJ, Adalsteinsson E, and Sullivan EV (2007) Diffusion tensor imaging with quantitative fibre tracking in HIV infection and alcoholism comorbidity: Synergistic white matter damage. *Brain* 130: 48–64.
Phillips P, Graybill JR, Fetchick R, and Dunn JF (1987) Adrenal response to corticotropin during therapy with itraconazole. *Antimicrobial Agents and Chemotherapy* 31: 647–649.

Podlekareva D, Mocroft A, Dragsted UB, et al. (2006) Factors associated with the development of opportunistic infections in HIV-1-infected adults with high CD4+ cell counts: A EuroSIDA study. *Journal of Infectious Diseases* 194: 633–641.
Pont A, Williams PL, Loose DS, Feldman D, Reitz RE, Bochra C, and Stevens DA (1982) Ketoconazole blocks adrenal steroid synthesis. *Annals of Internal Medicine* 97: 370–372.
Power R, Tate HL, McGill SM, and Taylor C (2003) A qualitative study of the psychosocial implications of lipodystrophy syndrome on HIV positive individuals. *Sexually Transmitted Infections* 79: 137–141.
Primo J, Hinojosu J, Moles JR, Fernandez J, Martinez C, Miralles A, and Otte A (1993) Development of thyroid dysfunction after alpha-interferon treatment of chronic hepatitis C. *American Journal of Gastroenterology* 88: 1976–1977.
Rabkin JG, Wagner GJ, and Rabkin R (2000) A double-blind, placebo-controlled trial of testosterone therapy for HIV-positive men with hypogonadal symptoms. *Archives of General Psychiatry* 57: 141–147.
Raffi F, Brisseau JM, Planchon B, Remi JP, Barrier JH, and Grolleau JY (1991) Endocrine function in 98 HIV-infected patients: A prospective study. *AIDS* 5: 729–733.
Rietschel P, Corcoran C, Stanley T, Basgoz N, Klibanski A, and Grinspoon S (2000) Prevalence of hypogonadism among men with weight loss related to human immunodeficiency virus infection who were receiving highly active antiretroviral therapy. *Clinical Infectious Diseases* 31: 1240–1244.
Rietschel P, Hadigan C, Corcoran C, Stanley T, Neubauer G, Gertner J, and Grinspoon S (2001) Assessment of growth hormone dynamics in human immunodeficiency virus-related lipodystrophy. *Journal of Clinical Endocrinology and Metabolism* 86: 504–510.
Rodriguez B, Sethi AK, Cheruvu VK, et al. (2006) Predictive value of plasma HIV RNA level on rate of CD4 T-cell decline in untreated HIV infection. *Journal of the American Medical Association* 296: 1498–1506.
Rutherford GW, Lifson AR, Hessol NA, et al. (1990) Course of HIV-I infection in a cohort of homosexual and bisexual men: An 11 year follow up study. *British Medical Journal* 301: 1183–1188.
Safdar A, Rubocki RJ, Horvath JA, Narayan KK, and Waldron RL (2002) Fatal immune restoration disease in human immunodeficiency virus type 1-infected patients with progressive multifocal leukoencephalopathy: Impact of antiretroviral therapy-associated immune reconstitution. *Clinical Infectious Diseases* 35: 1250–1257.
Samaras K, Pett S, Gowers A, McMurchie M, and Cooper DA (2005) Iatrogenic Cushing's syndrome with osteoporosis and secondary adrenal failure in human immunodeficiency virus-infected patients receiving inhaled corticosteroids and ritonavir-boosted protease inhibitors: Six cases. *Journal of Clinical Endocrinology and Metabolism* 90: 4394–4398.
Santhana Krishnan SG and Cobbs RK (2006) Reversible acute adrenal insufficiency caused by fluconazole in a critically ill patient. *Postgraduate Medical Journal* 82: e23.
Sapolsky R, Rivier C, Yamamoto G, Plotsky P, and Vale W (1987) Interleukin-1 stimulates the secretion of hypothalamic corticotropin-releasing factor. *Science* 238: 522–524.
Schacker T, Collier AC, Hughes J, Shea T, and Corey L (1996) Clinical and epidemiologic features of primary HIV infection. *Annals of Internal Medicine* 125: 257–264.
Schifitto G, Navia BA, Yiannoutsos CT, et al. (2007) Memantine and HIV-associated cognitive impairment: A neuropsychological and proton magnetic resonance spectroscopy study. *AIDS* 21: 1877–1886.
Schmidt PJ, Berlin KL, Danaceau MA, Neeren A, Haq NA, Roca CA, and Rubinow DR (2004) The effects of pharmacologically induced hypogonadism on mood in healthy men. *Archives of General Psychiatry* 61: 997–1004.
Schultz M, Muller R, von zur MA, and Brabant G (1989) Induction of hyperthyroidism by interferon-alpha-2b. *Lancet* 1: 1452.
Sellmeyer DE and Grunfeld C (1996) Endocrine and metabolic disturbances in human immunodeficiency virus infection and the acquired immune deficiency syndrome. *Endocrine Reviews* 17: 518–532.
Sevigny JJ, Albert SM, McDermott MP, et al. (2004) Evaluation of HIV RNA and markers of immune activation as predictors of HIV-associated dementia. *Neurology* 63: 2084–2090.
Sevigny JJ, Albert SM, McDermott MP, et al. (2007) An evaluation of neurocognitive status and markers of immune activation as predictors of time to death in advanced HIV infection. *Archives of Neurology* 64: 97–102.
Shelburne SA, III, Hamill RJ, Rodriguez-Barradas MC, et al. (2002) Immune reconstitution inflammatory syndrome: Emergence of a unique syndrome during highly active antiretroviral therapy. *Medicine (Baltimore)* 81: 213–227.
Shibata D and Klatt EC (1989) Analysis of human immunodeficiency virus and cytomegalovirus infection by polymerase chain reaction in the acquired immunodeficiency syndrome. An autopsy study. *Archives of Pathology and Laboratory Medicine* 113: 1239–1244.
Shikuma CM, Hu N, Milne C, Yost F, Waslien C, Shimizu S, and Shiramizu B (2001) Mitochondrial DNA decrease in subcutaneous adipose tissue of HIV-infected individuals with peripheral lipoatrophy. *AIDS* 15: 1801–1809.
Silver B, McAvoy K, Mikesell S, and Smith TW (1997) Fulminating encephalopathy with perivenular demyelination and vacuolar myelopathy as the initial presentation of human immunodeficiency virus infection. *Archives of Neurology* 54: 647–650.
Simon V, Ho DD, and Abdool KQ (2006) HIV/AIDS epidemiology, pathogenesis, prevention, and treatment. *Lancet* 368: 489–504.
Sinha-Hikim I, Arver S, Beall G, et al. (1998) The use of a sensitive equilibrium dialysis method for the measurement of free testosterone levels in healthy, cycling women and in human immunodeficiency virus-infected women. *Journal of Clinical Endocrinology and Metabolism* 83: 1312–1318.
Skiest DJ, Hester LJ, and Hardy RD (2005) Cryptococcal immune reconstitution inflammatory syndrome: Report of four cases in three patients and review of the literature. *Journal of Infection* 51: e289–e297.
Sonino N (1987) The use of ketoconazole as an inhibitor of steroid production. *New England Journal of Medicine* 317: 812–818.
Spudich S, Gisslen M, Hagberg L, et al. (2007) Predictors of cerebrospinal fluid viral burden during primary HIV-1 infection. *Conference on Retroviruses and Opportunistic Infections* 14, Abstract 114.
St Germain RM, Yigit S, Wells L, Girotto JE, and Salazar JC (2007) Cushing syndrome and severe adrenal suppression caused by fluticasone and protease inhibitor combination in an HIV-infected adolescent. *AIDS Patient Care and STDs* 21: 373–377.
Steel JL, Landsittel D, Calhoun B, Wieand S, and Kingsley LA (2006) Effects of lipodystrophy on quality of life and depression in HIV-infected men on HAART. *AIDS Patient Care and STDs* 20: 565–575.
Steer KA, Kurtz AB, and Honour JW (1995) Megestrol-induced Cushing's syndrome. *Clinical Endocrinology (Oxf)* 42: 91–93.
Stevenson M (2003) HIV-1 pathogenesis. *Nature Medicine* 9: 853–860.
Sullivan WM, Kelley GG, O'Connor PG, Dickey PS, Kim JH, Robbins R, and Shulman GI (1992) Hypopituitarism associated with a hypothalamic CMV infection in a

patient with AIDS. *American Journal of Medicine* 92: 221–223.
Sundar SK, Cierpial MA, Kamaraju LS, et al. (1991) Human immunodeficiency virus glycoprotein (gp120) infused into rat brain induces interleukin 1 to elevate pituitary–adrenal activity and decrease peripheral cellular immune responses. *Proceedings of the National Academy of Sciences of the United States of America* 88: 11246–11250.
Trenton AJ and Currier GW (2005) Behavioural manifestations of anabolic steroid use. *CNS Drugs* 19: 571–595.
Tsiodras S, Mantzoros C, Hammer S, and Samore M (2000) Effects of protease inhibitors on hyperglycemia, hyperlipidemia, and lipodystrophy: A 5-year cohort study. *Archives of Internal Medicine* 160: 2050–2056.
Tyor WR and Middaugh LD (1999) Do alcohol and cocaine abuse alter the course of HIV-associated dementia complex? *Journal of Leukocyte Biology* 65: 475–481.
van der PT, Romijn JA, Wiersinga WM, and Sauerwein HP (1990) Tumor necrosis factor: A putative mediator of the sick euthyroid syndrome in man. *Journal of Clinical Endocrinology and Metabolism* 71: 1567–1572.
Varis T, Kivisto KT, Backman JT, and Neuvonen PJ (1999) Itraconazole decreases the clearance and enhances the effects of intravenously administered methylprednisolone in healthy volunteers. *Pharmacology and Toxicology* 85: 29–32.
Varis et al. (2007) Sporanox (Intraconazole) Injection. U.S. Patent 4,727,064. Janssen Pharmaceutica Products, L.P.
Venkataramana A, Pardo CA, McArthur JC, et al. (2006) Immune reconstitution inflammatory syndrome in the CNS of HIV-infected patients. *Neurology* 67: 383–388.
Vigano A, Mora S, Brambilla P, Schneider L, Merlo M, Monti LD, and Manzoni P (2003) Impaired growth hormone secretion correlates with visceral adiposity in highly active antiretroviral treated HIV-infected adolescents. *AIDS* 17: 1435–1441.
Villette JM, Bourin P, Doinel C, et al. (1990) Circadian variations in plasma levels of hypophyseal, adrenocortical and testicular hormones in men infected with human immunodeficiency virus. *Journal of Clinical Endocrinology and Metabolism* 70: 572–577.
Wagner G, Rabkin JG, and Rabkin R (1995) Illness stage, concurrent medications, and other correlates of low testosterone in men with HIV illness. *Journal of Acquired Immune Deficiency Syndromes and Human Retrovirology* 8: 204–207.
Wahlstrom JT and Dobs AS (2000) Acute and long-term effects of AIDS and injection drug use on gonadal function. *Journal of Acquired Immune Deficiency Syndromes* 25 (supplement 1): S27–S36.
Walker UA, Bickel M, LutkeVolksbeck SI, et al. (2002) Evidence of nucleoside analogue reverse transcriptase inhibitor – associated genetic and structural defects of mitochondria in adipose tissue of HIV-infected patients. *Journal of Acquired Immune Deficiency Syndromes* 29: 117–121.
Welch K, Finkbeiner W, Alpers CE, Blumenfeld W, Davis RL, Smuckler EA, and Beckstead JH (1984) Autopsy findings in the acquired immune deficiency syndrome. *Journal of the American Medical Association* 252: 1152–1159.
Wohl DA, McComsey G, Tebas P, et al. (2006) Current concepts in the diagnosis and management of metabolic complications of HIV infection and its therapy. *Clinical Infectious Diseases* 43: 645–653.
Woloski BM, Smith EM, Meyer WJ, III, Fuller GM, and Blalock JE (1985) Corticotropin-releasing activity of monokines. *Science* 230: 1035–1037.
Wunder DM, Bersinger NA, Fux CA, et al. (2007) Hypogonadism in HIV-1-infected men is common and does not resolve during antiretroviral therapy. *Antiviral Therapy* 12: 261–265.
Yanovski JA, Miller KD, Kino T, Friedman TC, Chrousos GP, Tsigos C, and Falloon J (1999) Endocrine and metabolic evaluation of human immunodeficiency virus-infected patients with evidence of protease inhibitor-associated lipodystrophy. *Journal of Clinical Endocrinology and Metabolism* 84: 1925–1931.
Zetola NM and Pilcher CD (2007) Diagnosis and management of acute HIV infection. *Infectious Disease Clinics of North America* 21: 19–48, vii.
Zink MC, Uhrlaub J, DeWitt J, et al. (2005) Neuroprotective and anti-human immunodeficiency virus activity of minocycline. *Journal of the American Medical Association* 293: 2003–2011.

Further Reading

Anonymous (1992) 1993 Revised classification system for HIV infection and expanded surveillance case definition for AIDS among adolescents and adults. *MMWR Recommendations and Reports* 41: 1–19.
Anonymous (1995) Multicenter AIDS Cohort Study (MACS) Public Dataset: Release PO4.
Anonymous (2003) Differences in CD4 cell counts at seroconversion and decline among 5739 HIV-1-infected individuals with well-estimated dates of seroconversion. *Journal of Acquired Immune Deficiency* 34: 76–83.

Subject Index

Notes

Abbreviations used in subentries are defined in the main index.

Cross-reference terms in italics are general cross-references, or refer to subentry terms within the main entry (the main entry is not repeated to save space). Readers are also advised to refer to the text of each article for additional cross-references - not all of these cross-references have been included in the index cross-references.

The index is arranged in set-out style with a maximum of three levels of heading. Major discussion of a subject is indicated by bold page numbers. Page numbers suffixed by T and F refer to Tables and Figures respectively. vs. indicates a comparison.

Species names, in order to facilitate the reader, species have been listed where possible under both their scientific and common names however if there is significant text pertaining to that particular species, readers will be directed to the most commonly used name (for example African clawed frog *see Xenopus laevis*).

This index is in letter-by-letter order, whereby hyphens and spaces within index headings are ignored in the alphabetization. Prefixes and terms in parentheses are excluded from the initial alphabetization.

Where index subentries and sub-subentries pertaining to a subject have the same page number, they have been listed to indicate the comprehensiveness of the text.

A

B

C

D

E

F

G

H

J

K

L

M

N

O

P

Q

R

S

T

U

V

W

X

Y

Z

FSC
www.fsc.org
MIX
Papier | Fördert
gute Waldnutzung
FSC® C083411

Kolibri 360 GmbH
Ferdinand-Jühlke-Straße 7
99095 Erfurt, Deutschland
produktsicherheit@kolibri360.de